PLASMA PHYSICS

Related Titles from AIP Conference Proceedings

651   Dense Z-Pinches: 5th International Conference on Dense Z-Pinches
Edited by Jack Davis, Christopher Deeney, and Nino R. Pereira, December 2002, 0-7354-0108-X

649   Dusty Plasmas in the New Millennium: Third International Conference on the Physics of Dusty Plasmas
Edited by R. Bharuthram, M. A. Hellberg, P. K. Shukla, and F. Verheest, December 2002, 0-7354-0106-3

635   Atomic Processes in Plasmas: 13th APS Topical Conference on Atomic Processes in Plasmas
Edited by David R. Schultz, Fred W. Meyer, and Fay Ownby, October 2002, 0-7354-0090-3

611   Superstrong Fields in Plasmas: Second International Conference on Superstrong Fields in Plasmas
Edited by Maurizio Lontano, Gérard Mourou, Orazio Svelto, and Toshiki Tajima, April 2002, 0-7354-0057-1

606   Non-Neutral Plasma Physics IV: Workshop on Non-Neutral Plasmas
Edited by François Anderegg, Lutz Schweikhard, and Fred Driscoll, February 2002, 0-7354-0050-4

595   Radio Frequency Power in Plasmas: 14th Topical Conference
Edited by Tak Kuen Mau and John deGrassie, November 2001, 0-7354-0038-5

563   Plasma Physics: IX Latin American Workshop
Edited by Hernán Chuaqui and Mario Favre, May 2001, 1-56396-999-8

537   Waves in Dusty, Solar, and Space Plasmas
Edited by F. Verheest, M. Goossens, M. A. Hellberg, and R. Bharuthram, October 2000, 1-56396-962-9

To learn more about these titles, or the AIP Conference Proceedings Series, please visit the webpage **http://proceedings.aip.org/proceedings**

PLASMA PHYSICS

11th International Congress on
Plasma Physics: ICPP2002

Sydney, Australia   15-19 July 2002

ICPP2002

EDITORS
Ian S. Falconer
The University of Sydney
Sydney, Australia

Robert L. Dewar
The Australian National University
Canberra, Australia

Joe Khachan
The University of Sydney
Sydney, Australia

SPONSORING ORGANIZATIONS
International Union of Pure and Applied Physics (IUPAP)
Australian Institute of Physics
New South Wales Department of State and
  Regional Development

Melville, New York, 2003
AIP CONFERENCE PROCEEDINGS ■ VOLUME 669

Editors:

Ian S. Falconer
School of Physics A28
University of Sydney
Sydney, NSW 2006
AUSTRALIA

E-mail: i.falconer@physics.usyd.edu.au

Robert L. Dewar
Department of Theoretical Physics
Research School of Physics, Science and Engineering
Australian National University
Canberra, ACT 0200
AUSTRALIA

E-mail: robert.dewar@anu.edu.au

Joe Khachan
School of Physics A28
University of Sydney
Sydney, NSW 2006
AUSTRALIA

E-mail: j.khachan@physics.usyd.edu.au

Authorization to photocopy items for internal or personal use, beyond the free copying permitted under the 1978 U.S. Copyright Law (see statement below), is granted by the American Institute of Physics for users registered with the Copyright Clearance Center (CCC) Transactional Reporting Service, provided that the base fee of $20.00 per copy is paid directly to CCC, 222 Rosewood Drive, Danvers, MA 01923. For those organizations that have been granted a photocopy license by CCC, a separate system of payment has been arranged. The fee code for users of the Transactional Reporting Service is: 0-7354-0133-0/03/$20.00.

© 2003 American Institute of Physics

Individual readers of this volume and nonprofit libraries, acting for them, are permitted to make fair use of the material in it, such as copying an article for use in teaching or research. Permission is granted to quote from this volume in scientific work with the customary acknowledgment of the source. To reprint a figure, table, or other excerpt requires the consent of one of the original authors and notification to AIP. Republication or systematic or multiple reproduction of any material in this volume is permitted only under license from AIP. Address inquiries to Office of Rights and Permissions, Suite 1NO1, 2 Huntington Quadrangle, Melville, NY 11747-4502; phone: 516-576-2268; fax: 516-576-2450; e-mail: rights@aip.org.

L.C. Catalog Card No. 2003105613
ISBN 0-7354-0133-0
ISSN 0094-243X

CD-ROM available: ISBN 0-7354-0134-9

Printed in the United States of America

CONTENTS

Preface from Congress Chair ... xvii
Editorial Preface ... xix
Sponsors ... xx
Local Organizing Committee and ICPP History ... xxi
International Advisory Committee ... xxii
Program Committees ... xxiii

EXPERIMENTS ON LABORATORY DEVICES

1. Low-Temperature Plasmas

Observation of Transition and Bifurcation in a Magnetized Plasma Column with Electron Flow Channel ... 3
    R. Hatakeyama, M. Sato, W. Oohara, and T. Takado

Ion-Acoustic Shock Waves Formed with Landau Damping ... 7
    Y. Nakamura and V. N. Tsytovich

Effect of Light Ions on Ion-Ion Instability ... 10
    Y. Saitou and Y. Nakamura

Electrostatic Shocks Formed by Ion-Beam Velocity Modulation in a Q-Machine Plasma ... 14
    T. Gohda, S. Ishiguro, S. Iizuka, and N. Sato

Fluctuation of Electron Temperature due to Global Alfvén Eigenmode ... 18
    Y. Amagishi, T. Hishida, M. Kobayashi, and A. Tsushima

The Best Arc Heater Regime for Minimum Copper Cathode Erosion ... 22
    A. Marotta, L. I. Sharakhovsky, and A. M. Essiptchouk

Investigation of the Influence of Different Boundary Conditions on Helicon Discharges ... 26
    C. M. Franck, T. Klinger, and O. Grulke

Production of Helicon Wave Plasmas in Heliotron Magnetic Field ... 30
    S. Morimoto, K. Matsuda, T. Mizuuchi, H. Okada, F. Sano, and T. Obiki

Investigation of Megaampere Discharge in Superdense Gas Media in Order to Obtain a Plasma Source for Thermonuclear Research ... 34
    P. G. Rutberg, A. A. Bogomaz, A. V. Budin, V. A. Kolikov, and A. F. Savvateev

High-Density Hydrogen Plasma Production in a Simple Torus Using Helicon Waves ... 40
    Y. Sakawa and T. Shoji

Generation of High-Density, Uniform Plasmas by Low-Frequency RF Currents: Key Concepts, Experiments, and Applications ... 44
    S. Xu, Z. L. Tsakadze, E. L. Tsakadze, and K. Ostrikov

Wave Propagation and Plasma Production in Planar Microwave Discharges Using a Slot Antenna of Rotating Mode ... 48
    Y. Yasaka, N. Ishii, T. Yamamoto, M. Ando, and M. Takahashi

Investigation on the Characteristics of Pellet Ablation in a Toroidal Plasma ... 52
    K. N. Sato, H. Sakakita, and H. Fujita

Formation of Carbon Oxides in CH_4/O_2 Plasmas Produced by Inductively Coupled RF Discharges at Low Pressure ... 56
    I. Möller and H. Soltwisch

Application of Langmuir Probe for Study of Recombination of D_3^+ Ions with Electrons in He-Ar-D_2 Stationary and Flowing Afterglow Plasma ... 60
    M. Tichy, V. Poterya, R. Plasil, A. Pysanenko, P. Kudrna, O. Novotny, P. Zakouril, and J. Glosik

Comparison of Emissive and Plugged Probes DC Plasma Potential Measurements in a Magnetised Plasma ... 64
    F. Greiner, D. Block, A. Piel, S. Ratynskaja, G. Helblom, and K. Rypdal

Development of Frequency Tunable, Submillimeter Wave Gyrotron FU Series for Plasma Diagnostics ... 68
    T. Idehara, S. Mitsudo, R. Pavlichenko, I. Ogawa, and K. Kawahata

Diagnostics and Simulation of Low-Frequency Inductively Coupled Plasmas ... 71
    K. N. Ostrikov, I. B. Denysenko, E. L. Tsakadze, S. Xu, N. A. Azarenkov, and R. G. Storer

Plasma Characteristics between Biased Facing Electrodes 75
  Y. Saitou and A. Tsushima
**Experimental Study of Axial Plasma Parameter Variations in the
Cylindrical Magnetron Discharge** 79
  P. Kudrna, M. Holik, O. Bilyk, I. A. Porokhova, Y. B. Golubovskii, M. Tichy, and J. F. Behnke
**Experimental Results of Growth and Breakdown Mechanisms in the Final Stage of a
Pseudospark Discharge** 83
  M. Zambra, J. Moreno, L. Soto, and L. Rosales
Coulomb Clusters in Axial Magnetic Field 87
  F. M. H. Cheung, A. A. Samarian, and B. W. James
Dust Grains as a Diagnostic Tool for RF-Discharge Plasma 90
  B. W. James, A. A. Samarian, and W. Tsang
Observation of Naturally-Occurring Waves in a Strongly Coupled Plasma 93
  S. Nunomura, J. Goree, S. Hu, X. Wang, A. Bhattacharjee, and K. Avinash
Confining Instabilities in Complex Plasma 97
  A. A. Samarian, S. V. Vladimirov, and B. W. James
Influence of the Ion Beam on the Behavior of Dust Particles Trapped in an Ion Sheath 101
  T. Murakami, M. Sawai, T. Misawa, N. Ohno, and S. Takamura
Spatio-Temporal Dynamics of Driven Drift Waves 105
  D. Block and A. Piel
Transport Studies of Driven Drift Waves 109
  D. Block and A. Piel
Sound Wave Propagation in Gases at Low Pressure 113
  T. Tonogaki, N. Ohno, S. Takamura, and A. Tsushima
**Wave Properties in a Weakly-Magnetized Plasma Produced by Rotating Radio-Frequency
Electromagnetic Fields** 117
  G. Sato, W. Oohara, and R. Hatakeyama
Observation of Ion Acoustic Waves Excited by Drift Waves in a Weakly Magnetized Plasma 121
  I. Tsukabayashi, S. Sato, and Y. Nakamura
Electron Acceleration by an Oscillating Potential Wall in the Presence of Collisions 125
  A. Tsushima and O. Ishihara
Plasma Science and Environmental Remediation in the Arctic Region 129
  A. Y. Wong
Experimental Evidence of Nonlinear Spectral Power Transfer in Zonal Flow Generation 133
  H. Xia, M. G. Shats, and W. Solomon
Excitation of an Axisymmetric Shear Alfvén Wave by a Rogowski-Type Antenna 137
  T. Yagai, R. Kumagai, Y. Hosokawa, K. Hattori, A. Ando, and M. Inutake
**Nonlinear Evolution of Unstable Waves in an Electron-Beam Plasma: Relation to Beam
Energy Distribution** 141
  K. Yamagiwa, N. Ikeya, T. Takeda, and T. Takai
Measurement of Carrier Properties in InSb using Faraday Rotation 145
  I. Ogawa, K. Yamada, T. Idehara, A. Tsushima, and S. Yamaguchi

2. Magnetic Confinement

Flow of Magnetized Plasma in a Linear Device 149
  G. Fussmann, W. Bohmeyer, Z. Kiss'ovski, and B. Koch
Penetration of Electrons Inside Helical Field Configuration via Stochastic Magnetic Layer 154
  H. Himura, H. Wakabayashi, M. Fukao, Z. Yoshida, M. Isobe, K. Matsuoka, and H. Yamada
Overview and Results from the H-1 National Facility 158
  B. D. Blackwell, J. H. Harris, J. Howard, M. G. Shats, C. Charles, S. M. Collis, H. J. Gardner, F. J. Glass,
  X. Hua, C. A. Michael, D. G. Pretty, H. Punzmann, W. M. Solomon, and G. G. Borg
Transport Properties in the TJ-II Flexible Heliac 162
  F. Castejón, E. Ascasíbar, C. Alejaldre, J. Alonso, L. Almoguera, A. Baciero, R. Balbín, E. Blanco,
  M. Blaumoser, J. Botija, B. Brañas, A. Cappa, R. Carrasco, J. R. Cepero, A. A. Chmyga, J. Doncel,
  N. B. Dreval, S. Eguilior, L. Eliseev, T. Estrada, O. Fedyanin, A. Fernández, C. Fuentes, A. García,
  I. García-Cortés, B. Gonçalves, J. Guasp, J. Herranz, A. Hidalgo, C. Hidalgo, J. A. Jiménez, I. Kirpitchev,
  S. M. Khrebtov, A. D. Komarov, A. S. Kozachok, L. Krupnik, F. Lapayese, K. Likin, M. Liniers,

D. López-Bruna, A. López-Fraguas, J. López-Rázola, A. López-Sánchez, E. de la Luna, A. Malaquias, R. Martín, M. Medrano, A. V. Melnikov, P. Méndez, K. J. McCarthy, F. Medina, B. van Milligen, I. S. Nedzelskiy, M. Ochando, L. Pacios, I. Pastor, M. A. Pedrosa, A. de la Peña, A. Petrov, A. Portas, J. Romero, L. Rodríguez-Rodrigo, A. Salas, E. Sánchez, J. Sánchez, K. Sarksian, S. Schchepetov, N. Skvortsova, F. Tabarés, D. Tafalla, V. Tribaldos, C. F. A. Varandas, J. Vega, and B. Zurro

Effects of the New Island Divertor on the Plasma Performance in the W7-AS Stellarator 166
P. Grigull, K. McCormick, J. Baldzuhn, R. Burhenn, R. Brakel, H. Ehmler, Y. Feng, F. Gadelmeier, L. Giannone, D. Hartmann, D. Hildebrandt, M. Hirsch, R. Jaenicke, J. Kisslinger, T. Klinger, J. Knauer, R. König, D. Naujoks, H. Niedermeyer, E. Pasch, N. Ramasubramanian, F. Sardei, F. Wagner, A. Weller, and the W7-AS Team

Excitation of Alfvén Waves in Mirror Plasmas with a Strong Temperature Anisotropy 170
M. Ichimura, H. Higaki, M. Nakamura, S. Saosaki, H. Kano, S. Kakimoto, K. Horinouchi, Y. Yamaguchi, H. Hojo, Y. Nakashima, T. Watanabe, and K. Yatsu

First HIBP Measurement of Plasma Potential During the H-Mode Transition on the TUMAN-3M Tokamak 175
L. G. Askinazi, A. A. Chmyga, N. B. Dreval, V. E. Golant, S. M. Khrebtov, A. S. Komarov, V. A. Kornev, L. I. Krupnik, S. V. Lebedev, G. Van Oost, E. A. Shevkin, M. Tendler, A. S. Tukachinsky, and N. A. Zhubr

Design of the Collective Thomson Scattering Diagnostics for Large Helical Device Using a Quasi-Optical Frequency Tunable Gyrotron as a Radiation Source 179
R. Pavlichenko, T. Idehara, I. Ogawa, K. Kawahata, and M. Sato

Numerical Simulation in Dense Plasma Diagnosis with an Amplitude-Division Soft-X-Ray Laser Interferometer 183
W. Zheng and G. Zhang

ECH Power Deposition Study in the Collisionless Plasma of LHD 187
S. Kubo, H. Idei, T. Shimozuma, Y. Yoshimura, T. Notake, K. Ohkubo, S. Inagaki, Y. Nagayama, K. Narihara, I. Yamada, S. Muto, S. Morita, and the LHD Experimental Group

Transport Due to Intermittent Events and Plasma Flow Shear in Magnetized Plasmas 191
V. Antoni, G. Regnoli, M. Spolaore, G. Serianni, N. Vianello, R. Cavazzana, E. Spada, and E. Martines

Recent Results of Alfvén Wave Studies in TCABR 195
E. A. Lerche, L. F. Ruchko, E. M. Ozono, R. M. O. Galvão, A. G. Elfimov, A. M. M. Fonseca, and R. P. da Silva

Diagnosing Electron Heat Transport in ASDEX Upgrade with ECH Power Modulation 199
F. Ryter, W. Suttrop, H.-U. Fahrbach, K. K. Kirov, F. Leuterer, A. G. Peeters, G. Pereverzev, G. Tardini, and the ASDEX Upgrade Team

MHD Activities in High-β Plasmas of LHD 203
S. Sakakibara, H. Yamada, K. Y. Watanabe, Y. Narushima, K. Toi, S. Ohdachi, S. Yamamoto, K. Narihara, K. Tanaka, K. Ida, and the LHD Experimental Group

Measurements of the Fluctuation-Induced Flux with Emissive Probes in the CASTOR Tokamak 207
R. Schrittwieser, J. Adámek, P. Balan, I. Ďuran, M. Hron, C. Ioniţă, E. Martines, J. Stöckel, M. Tichý, and G. Van Oost

Applicability of Electron Emissive Probes for Plasma Potential and Electric Field Measurements in Magnetized Plasmas 211
R. Schrittwieser, J. Adámek, P. Balan, J. A. Cabral, H. Fernandes, H. F. C. Figueiredo, C. Hidalgo, M. Hron, C. Ioniţă, E. Martines, M. A. Pedrosa, J. Stöckel, M. Tichý, G. Van Oost, and C. Varandas

Investigation of the Effects of the Radial Electric Field by Electrode Biasing in a Toroidal Plasma 215
Z. Wang, C. Wang, G. Pan, Y. Wen, C. Yu, S. Wan, W. Lui, R. Lu, J. Wang, and H. Gao

Plasma Velocity Profile During the Pulsed Poloidal Current Drive in the MST RFP Plasma 219
H. Sakakita, D. Craig, J. K. Anderson, T. M. Biewer, S. D. Terry, B. E. Chapman, and D. J. Den-Hartog

ELM Dynamics in TCV H-Modes 223
A. W. Degeling, Y. R. Martin, J. B. Lister, X. Llobet, and P. E. Bak

Stochastic Transition of a Turbulent Plasma 228
M. Kawasaki, S.-I. Itoh, M. Yagi, and K. Itoh

Models for H-Mode Pedestal Temperature and Predictions for Future Tokamak Designs 232
T. Onjun, A. H. Kritz, G. Bateman, and V. Parail

Density Transition Phenomena in Magnetized Plasma by Voltage Biasing 236
S. Shinohara, S. Matsuyama, and K. Sugimori

Fluctuation Characteristics in Detached Recombining Plasmas 240
    N. Ohno, N. Tanaka, S. Takamura, and V. Budaev

Dipole Map for Divertor Tokamaks 244
    H. Ali, A. Punjabi, and A. Boozer

Improvement of Carbon Properties used as PFCs in Tokamaks by Nitrogen Irradiation 248
    G. G. Ross and D. Bourgoin

3. Inertial Confinement and Beams

Numerical Simulation of Non-Spherical Implosion Related to Fast Ignition 253
    H. Nagatomo, N. Ohnishi, K. Mima, K. Nishihara, S. Yamada, K. Sawada, and H. Takabe

Fast Ignition Research at The Institute of Laser Engineering, Osaka University 257
    K. A. Tanaka, R. Kodama, Y. Kitagawa, H. Fujita, T. Jitsuno, K. Mima, N. Miyanaga, T. Norimatsu,
    Y. Sentoku, K. Shigemori, A. Sunahara, T. Miyakoshi, F. Otani, T. Sato, M. Tanpo, Y. Tohyama,
    T. Yamanaka, K. Krushelnick, P. A. Norreys, and M. Zepf

Nonlinear State of Sausage-like Instability of Electron Current Channels in Fast Ignition Concept of Inertial Fusion 261
    N. Jain, A. Das, P. Kaw, and S. Sengupta

The Features of Craters Formation on the Target under the Action of Powerful Laser Pulse 265
    A. A. Rupasov, E. A. Bolkhovitinov, I. Y. Doskach, A. A. Erokhin, S. I. Fedotov, L. P. Feoktistov,
    S. Y. Gus'kov, B. V. Kruglov, M. V. Osipov, V. N. Puzirev, V. B. Rozanov, A. S. Shikanov,
    V. B. Studenov, B. L. Vasin, and O. F. Yakushev

Hydrodynamic Instability Experiments on the HIPER Laser 269
    K. Shigemori, H. Azechi, S. Fujioka, Y. Kanai, N. Miyanaga, M. Murakami, T. Muranaka, H. Nagatomo,
    M. Nakai, M. Nishikino, K. Nishihara, H. Nishimura, T. Sakaiya, H. Shiraga, A. Sunahara, H. Takabe,
    T. Takayama, Y. Tamari, M. Tanaka, and T. Yamanaka

Potential Formation in Front of an Electron Emitting Electrode in a Two-Electron Temperature Plasma 273
    T. Gyergyek, M. Cercek, and D. Erzen

Portable Neutron Probe for Soil Humidity Measurements 277
    J. Pouzo, M. Milanese, and R. Moroso

Effect of Electrode Material on a Vacuum Pinch 281
    A. Robledo-Martinez and F. P. Espino

Impedance Coupling in a 32 kj Wire-Array Z-Pinch 285
    A. Robledo-Martinez, J. Nieto, and F. P. Espino

Acceleration Dynamics of Laser-Driven MeV-Ion Jets 289
    M. Hegelich, M. Allen, P. Audebert, A. Blazevic, T. Cowan, J. Fuchs, J. C. Gauthier, M. Geissel,
    W. Guenther, D. Habs, S. Karsch, A. Kemp, G. Pretzler, M. Roth, and K. J. Witte

4. Plasma Applications

Investigation of the Discharge Characteristics of the T6 Hollow Cathode Operating on Several Inert Gases and a Kr/Xe Mixture 294
    I. F. M. Ahmed Rudwan and S. B. Gabriel

An Ion Heating Experiment in a Supersonic Plasma Flow 298
    A. Ando, R. Kumagai, S. Fujimura, T. Yagai, K. Hattori, and M. Inutake

Plasma Fluctuations in an Applied Field MPD Thruster 302
    V. Antoni, M. Bagatin, G. Serianni, N. Vianello, M. Zuin, F. Paganucci, P. Rossetti, and M. Andrenucci

Magnetic-Laval-Nozzle Effect on a Magneto-Plasma-Dynamic Arcjet 306
    M. Inutake, H. Miyazaki, K. Yoshino, H. Tobari, K. Hattori, and A. Ando

Properties of Heat Flow in the JxB Gas Arc Discharge for the Production of Fullerenes 310
    N. Matsumoto and T. Mieno

Evaluation of Electromagnetic Forces in an Axially-Magnetized MPD Arcjet Plasma 314
    H. Tobari, K. Yoshino, K. Hattori, A. Ando, and M. Inutake

Investigations of a Pulsed Cathodic Vacuum Arc 319
    T. W. H. Oates, J. Pigott, P. Denniss, D. R. Mckenzie, and M. M. M. Bilek

Instability of a Vacuum Arc Centrifuge 323
    M. J. Hole, R. S. Dallaqua, S. W. Simpson, and E. Del Bosco
Nano-Microelectronics by Plasma and Electron Beam Techniques 327
    M. Ghoranneviss, G. Benstetter, H. Hora, R. Höpfl, M. R. Hantehzadeh, M. Mardanian, and A. H. Sari
Surface Modification of Blood Contacting Biomedical Implants by Plasma Processes 330
    N. Huang, P. Yang, Y. X. Leng, J. Y. Chen, J. Wang, H. Sun, G. J. Wan, P. K. Chu, and Y. Leng
Effect of the High-Temperature Pulse Deuterium Plasma on the Surface Structure of Vanadium and its Physical-Mechanical Characteristics 335
    I. V. Borovitskaya, A. I. Dedurin, L. I. Ivanov, O. N. Krokhin, V. Y. Nikulin, A. A. Tikhomirov, and A. S. Fedotov
Material Processing by Plasma Shock Wave Generated in an Inverse Z-Pinch "Plasma Expander" 339
    E. A. Aramaki, M. B. de Moraes, and M. Machida
Argon Ion Induced Changes on Cadmium Iodide Thin Films Using Dense Plasma Focus Device 343
    R. S. Rawat, P. Lee, S. Lee, P. Arun, and A. G. Videshwar
The Effect of Carbon Dioxide and Nitrogen Ion Implantation of AISI 52100 Steel 347
    A. H. Sari, M. Ghoranneviss, M. Mardanian, M. R. Hantehzadeh, and H. Hora
The Evolution of Cone-Like Formations on the Cathode of Abnormal Glow Discharge of Ar 350
    V. V. Bobkov, S. S. Alimov, V. V. Andreiev, Y. V. Slyusarenko, and R. I. Starovoitov
Amorphous Hydrogenated Carbon Films Deposited by PECVD: Nitrogen Incorporation during Film Growth and by Plasma Surface Processing 354
    F. L. Freire Jr.
Fore-Vacuum Plasma Electron Gun of Ribbon Beam 358
    V. Burdovitsin, Y. Burachevsky, E. Oks, and M. Fedorov
Hydrogenated Amorphous Silicon by d.c. Plasma Glow Discharge of Argon Diluted Silane 361
    S. A. Rahman, A. Azis, and C. K. Lim
Deposition of Carbon Thin Film at Room Temperature Using Dense Plasma Focus Device 365
    R. S. Rawat, G. Macharaga, P. Lee, S. Xu, and S. Lee
Charged Particles in the Reactor Chamber of a Remote Plasma Enhanced CVD System 369
    L. Sirghi, G. Popa, and Y. Hatanaka
Rotating Cylindrical Magnetrons and Accelerators with Anode Layer for Large-Area Film Deposition Technologies 373
    S. P. Bugaev, A. N. Zakharov, K. V. Oskomov, N. S. Sochugov, and A. A. Solovjev
Development and Application of Vacuum Arc Ion Source at HCEI 377
    A. S. Bugaev, V. I. Gushenets, A. G. Nikolaev, E. M. Oks, K. P. Savkin, and G. Y. Yushkov
Bulk Plasma Production Using Gaseous Discharge System with External Electron Injection 380
    M. V. Shandrikov, A. V. Vizir, E. M. Oks, and G. Y. Yushkov
Decomposition of Benzene by a Low Pressure Glow Discharge 383
    K. Satoh, T. Sawada, T. Naitoh, H. Itoh, M. Shimozuma, and H. Tagashira
Coupling Power and Information to a Plasma Antenna 388
    M. Hargreave, J. P. Rayner, A. D. Cheetham, G. N. French, and A. P. Whichello
Physical Characteristics of a Plasma Antenna 392
    J. P. Rayner, A. P. Whichello, and A. D. Cheetham
Plasma Antenna Radiation Patterns 396
    A. P. Whichello, J. P. Rayner, and A. D. Cheetham

5. Experimental Simulation of Phenomena Observed in Natural Plasmas

Study on Laser Induced Plasma Produced in Liquid 400
    N. Tsuda and J. Yamada
Ion and Electron Whistler Wave Dispersion Experiments 404
    C. M. Franck, T. Klinger, and O. Grulke
Experiments and Observations on Intense Alfvén Waves in the Laboratory and in Space 408
    W. Gekelman, M. VanZeeland, S. Vincena, and P. Pribyl
Excitation and Propagation of Alfvén Waves in a Helicon Discharge 412
    O. Grulke, C. M. Franck, and T. Klinger

Control of Parallel and Perpendicular Flow Shears and Related Low-Frequency Plasma Instabilities 416
    T. Kaneko, E. Tada, H. Tsunoyama, and R. Hatakeyama

Periodical Motion of Ionized Front in a Closed Divertor Simulator 420
    A. Matsubara, T. Sugimoto, T. Shibuya, K. Kawamura, S. Sudo, and K. Sato

Particle Acceleration Due to Electric Field Bursts Close to the Lower Hybrid Frequency in a High-Voltage Linear Plasma Discharge 424
    Y. Takeda and H. Inuzuka

Effect of Neutrals on Alfvén Wave Propagation 427
    C. Watts and J. Hanna

Ion Confinement Due to Radial Electric Field in a Magnetized Plasma 431
    Y. Kawamoto, S. Yasuda, Masashi Kondo, Masuo Kondo, and K. Saeki

Laboratory and Computer Simulations of Non-MHD Flute Instability Structuring the Plasma Clouds During Their Artificial Releases at Near-Earth Space 435
    Y. P. Zakharov, H. Nakashima, V. M. Antonov, E. L. Boyarintsev, A. V. Melekhov, V. G. Posukh, I. F. Shaikhislamov, D. Mourenas, and F. Simonet

PLASMA THEORY AND SIMULATION

1. Kinetic Theory and Transport Modeling

Non-Local Kinetic Fluctuations in Plasma 443
    V. V. Belyi

A Space- and Time- Nonlocal Kinetic Equation for a Polarizable Plasma 445
    V. V. Belyi, Y. A. Kukharenko, and J. Wallenborn

Electroweak Interactions in Dense Plasmas 449
    L. O. Silva, R. Bingham, and W. B. Mori

Simulation of Hysteresis in Glow Discharge 453
    N. Mizuno, U. Tomioka, T. Hayashi, and T. Kawabe

Theory of RF-Induced Current Profile in the Presence of Fluctuations 455
    M. Taguchi

Stabilization of Burn Conditions in a Two-Temperature Fusion Reactor: Preliminary Results ... 459
    J. E. Vitela and J. J. Martinell

Direct Kinetic Simulations of Ion Temperature Gradient Driven Turbulence 463
    T.-H. Watanabe and H. Sugama

Radio Frequency Dielectric Characteristics of a Collisionless Laboratory Dipole Plasma 467
    N. I. Grishanov, A. F. D. Loula, C. A. de Azevedo, and J. P. Neto

Wave Dissipation by Electron Landau Damping in Axisymmetric D-Shaped Tokamaks 471
    N. I. Grishanov, A. F. D. Loula, C. A. de Azevedo, and J. P. Neto

Some Remarks on the Theory of the Plasma-Wall Transition (PWT) Layer 475
    D. D. Tskhakaya and S. Kuhn

New Physics of the Positive Column 479
    G. G. Lister

Kinetic Modeling of Axially Non-Uniform Cylindrical Magnetron Discharge 482
    I. A. Porokhova, Y. B. Golubovskii, P. Kudrna, M. Tichy, and J. F. Behnke

PIC-MCC Modeling of the Cylindrical Magnetron Discharge 486
    P. Kudrna, M. Holík, I. A. Porokhova, Y. B. Golubovskii, and M. Tichy

Model Kinetic Description in Plasma 490
    V. V. Belyi

Evaluation of the Bootstrap Current in Stellarators 492
    W. Kernbichler, S. V. Kasilov, V. V. Nemov, G. Leitold, and M. F. Heyn

Influence of Nongyrotropy in the Electron Beam-Plasma Interaction 496
    M. A. E. de Moraes, Y. Omura, M. Virgínia Alves

PIC Simulation of Collisionless Negative Ion Plasma Expansion into a Vacuum 500
    M. Cercek, T. Gyergyek, and V. Ignatescu

Calculation of Self-Consistent Radial Electric Field in Presence of Convective Electron Transport in a Stellarator 504
    W. Kernbichler, S. V. Kasilov, and M. F. Heyn

Particle-In-Cell Simulations of Laser-Produced Plasma Experiments to Study Thrust Conversion Processes in a Laser Fusion Rocket 508
    H. Nakashima, K. V. Vchivkov, T. Esaki, Y. P. Zakharov, T. Kawano, and T. Muranaka

Integration of ALE Hydro and Collective PIC Codes for Fast Ignition Simulations 512
    H. Sakagami and K. Mima

A Kinetic Model of Solar Wind 516
    Y. M. Vasenin, N. R. Minkova, and A. Shamin

Evolution of Electron-Acoustic Wave in Auroral Region 520
    P. H. Sakanaka and R. da Trindade Faria Jr.

Effect of Electron Heat Transport on Spatial Distribution of Surface Wave Plasma Parameters 524
    I. B. Denysenko, K. N. Ostrikov, N. A. Azarenkov, M. Y. Yu, and S. Xu

2. Statistical Mechanics, Atomic, and Molecular Processes

Dust Growth and Gravitation-like Instabilities in Astrophysical Plasmas 528
    R. Bingham and V. N. Tsytovich

Thermal Instability of an Optically Thin Dusty Plasma 532
    M. P. Bora

Dynamics of Dust Particulates in Magnetic Field 536
    O. Ishihara, T. Kamimura, K. Hirose, and N. Sato

Algebraic Improvement for the Numerical Treatment of the Ion Acoustic Modes in Ion Beam Dusty-Plasma Considering Dust Charging Effects 540
    J. Puerta, J. Silva, and C. Cereceda

Theory of Dust Ion-Acoustic Solitary and Shock Waves 543
    P. K. Shukla and A. A. Mamun

Finite Amplitude Double-Layer Solutions in Two-Component-Charged Dusty Plasmas with Trace Electrons and Ions 546
    I. Spassovska, P. H. Sakanaka, and P. K. Shukla

Charge Variations in Planar RF Discharge 550
    O. S. Vaulina and A. A. Samarian

3. Macroscopic Equilibrium and Stability

Control of Equilibrium Structure of a Toroidal Non-Neutral Plasma in Proto-RT 553
    H. Saitoh, Z. Yoshida, H. Himura, C. Nakashima, J. Morikawa, and M. Fukao

Simply Connected High-β Magnetic Configurations 557
    F. Rogier, G. Bracco, A. Mancuso, P. Micozzi, and F. Alladio

Constructing Integrable Full-Pressure Full-Current Free-Boundary Stellarator Magnetohydrodynamic Equilibria 561
    S. R. Hudson, D. A. Monticello, A. H. Reiman, D. J. Strickler, and S. P. Hirshman

Transport and Stability Analysis of Low q_a Discharges 565
    S. Lahiri and S. Mukhopadhyay

Magnetic Island Growth—A Comparison of Local and Global Effects 569
    S. S. Lloyd and H. J. Gardner

Calculation of Eddy Currents in the ETE Spherical Torus 573
    G. O. Ludwig

A Broken Degeneracy in the Resistive MHD Spectrum 577
    B. F. McMillan, R. L. Dewar, and R. G. Storer

On the Nature of Visco-Resistive MHD Steady States 581
    L. P. J. Kamp and D. C. Montgomery

Effect of Equilibrium Flow on Plasma Parameters 585
    S. Mukhopadhyay, S. Lahiri, P. H. Sakanaka, and B. Dasgupta

Neoclassical MHD Equilibria in Low Aspect Ratio Reversed Field Pinch 589
Y. Nagamine, Y. Yoshioka, M. Harada, Y. Osanai, Y. Kondoh, M. Taguchi, E. Uchimoto, K. H. Saito, K. N. Saito, and S. Shiina

An Equilibrium Equation of a Magnetized Rotating Plasma 593
K. Saeki, A. Tsushima, and H. Sanuki

High-Mode Configuration Solutions from Generalized Beltrami Equation 597
A. M. A. Taveira, P. H. Sakanaka, C. E. Scussiatto, and B. Dasgupta

Filamentary Magnetohydrodynamic Simulation Using MDGRAPE-2 601
Y. Yatsuyanagi, T. Ebisuzaki, T. Hatori, and T. Kato

Advances in Sun-Earth Connection Modeling 605
S. B. Ganguli and V. V. Gavrishchaka

4. Waves and Nonlinear Dynamics

Dynamics of Vortex Type Wave Structures in Plasmas and Fluids 609
V. Y. Belashov and R. M. Singatulin

Electrostatic Turbulence and Transport in Edge Plasmas: Bursts and Zonal Flows, Stochastic Field Lines, and Transport Barriers 613
P. Beyer, S. Benkadda, X. Garbet, P. Ghendrih, and Y. Sarazin

Turbulent Evolution of a Plasma Described Through Classical Mechanics Only 617
D. F. Escande and Y. Elskens

A Theory of Longitudinal Plasma Waves with the Motion of Ions Taken into Account 622
G. N. Kichigin

Electromagnetic Transport Components and Sheared Flows in Plasma Edge Turbulence 626
V. Naulin

Self-Organization and Coupling of Waves in a Plasma 630
L. Nikolić, S. Ishiguro, and T. Sato

Modeling of Small Dense-Plasma Helicon Source 634
S. Shinohara and K. P. Shamrai

Gyrokinetic Theory and Simulations of Alfvénic Instabilities in Dipole Plasmas 638
S. Dettrick and L. Chen

Analogies of Rapidly Rotating Tokamaks and Accretion Disks 642
J. P. Goedbloed, A. J. C. Belien, and B. van der Holst

Kinetic Effects on the Parametric Decay of Circularly Polarized Electromagnetic Waves in a Relativistic Pair Plasma 646
V. Muñoz and L. Gomberoff

Generation of Zonal Flows by Interchange Mode Turbulences 650
Z. N. Andrushchenko, V. P. Pavlenko, K. Schoepf, and S. Kuhn

A Model of Self-Consistent L-H Transition Based on Finite-β Drift Waves 654
J. J. Martinell and P. N. Guzdar

Dynamics of Transport Barriers and ELM-like Behaviour in Electrostatic Turbulence 658
V. Naulin, J. Nycander, and J. J. Rasmussen

Modelling the Formation of Large Scale Zonal Flows in Drift Wave Turbulence in a Rotating Fluid Experiment 662
V. Naulin, J. J. Rasmussen, B. Stenum, L. J. A. van Bokhoven, and J. van de Konijnenberg

Destabilizing Effect of Shear Flow—Beyond Kelvin-Helmholtz Instability 666
T. Tatsuno, M. Hirota, A. Ito, S. Kondoh, Z. Yoshida, and S. M. Mahajan

Dynamics and Stability of Multidimensional Solitons in a Plasma 670
V. Y. Belashov and S. V. Vladimirov

Laser Envelope Solitons in Plasmas 675
S. Poornakala, A. Das, A. Sen, P. K. Kaw, Z. M. Sheng, Y. Sentoku, K. Mima, and K. Nishikawa

Ion-Acoustic Cnoidal Waves in a Plasma with Negative Ions 679
L. L. Yadav

Turbulence-Double-Layer Synergetic Auroral Electron Acceleration 684
A. Souza de Assis

Sub-Grid-Scale Parameterisations for Large-Scale Eddy Simulations 687
J. S. Frederiksen

Understanding the Simple Magnetized Torus 691
  J.-V. Paulsen, O. E. Garcia, and K. Rypdal
Impact of Large Scale Flows on Turbulent Transport 696
  Y. Sarazin, P. Ghendrih, S. Benkadda, P. Beyer, G. Falchetto, C. Figarella, X. Garbet, and V. Grandgirard
Electron Beam Driven Cyclotron Maser Radiation 700
  R. Bingham, R. A. Cairns, and B. J. Kellett
A Detailed Model of the X-Ray Emission from Comets 704
  B. J. Kellett, R. Bingham, C. M. Lisse, M. Torney, H. P. Summers, and V. D. Shapiro
Radiation Resonance Emission from Steep Overcritical Plasma Profiles Illuminated by Femtosecond Laser Pulses 708
  R. Ondarza-Rovira and T. J. M. Boyd
Bifurcation and Metamorphosis of Plasma Turbulence—Shear Flow Dynamics: The Path to the Top of the Hill 711
  R. Ball, R. L. Dewar, and H. Sugama
Onset of Alfvén Turbulence via Boundary Crisis 715
  A. C.-L. Chian, F. A. Borotto, and E. L. Rempel
Nonlinear Dynamical Analysis of Two Current-Driven Low-Frequency Instabilities in a Magnetised Plasma Column 719
  D. Dimitriu, V. Ignatescu, C. Ionita, E. Lozneanu, M. Sanduloviciu, and R. Schrittwieser
High-Dimensional Interior Crisis in Plasmas 723
  E. L. Rempel, A. C.-L. Chian, E. E. Macau, R. R. Rosa, and F. Christiansen
Reconnection Bifurcation in Tokamaks 727
  M. Roberto, E. C. Silva, and I. L. Caldas
Nonlinear Shear Flow Structures in Magnetic Curvature Driven Rayleigh Taylor Instability 731
  A. Das, A. Sen, and P. K. Kaw
Two-Dimensional Particle Simulation of Electrostatic Solitary Waves with an Open Boundary Condition 735
  T. Umeda, Y. Omura, H. Matsumoto, and H. Usui

5. Flowing and Rapidly Changing Plasmas

Laser Giant Ion Source and the Prepulse Effects for Picosecond Interaction for High Gain Laser Fusion 739
  H. Hora, J. Badziak, F. P. Boody, R. Höpfl, K. Jungwirth. B. Králikova, J. Krása, L. Láska, P. Parys, V. Perina, M. Pfeifer, K. Rohlena, J. Skála, J. Ullschmied, J. Wolowski, and E. Woryna
Laser ICF with Single Event Solution 744
  H. Hora, P. Toups, P. Evans, F. Osman, R. Castillo, K. Mima, M. Murakami, S. Nakai, K. Nishihara, C. Yamanaka, T. Yamanaka, and G. H. Miley
Dynamics of a Supersonic Plume Moving Along a Magnetized Plasma 749
  G. J. Morales, F. S. Tsung, and J. N. Leboeuf
MHD Simulation of Reflection Dynamics of Field-Reversed Configuration Plasma 753
  T. Kanki, S. Okada, and S. Goto
Electron Heating in Thomson Scattering Measurements of Plasma Temperature: Are Thermal Plasmas Thermal? 757
  A. B. Murphy, J. Aubreton, and M. F. Elchinger
Magnetic Field Generation and Electrostatic Shock Wave Formation Driven by Counter-Streaming Pair Plasmas 762
  T. Haruki and J. I. Sakai
Extreme Plasmas near Pulsars and Strange Stars 766
  V. V. Usov
Driven Reconnection Controlled by Particle Dynamics in a Collisionless Open System 769
  R. Horiuchi, W. B. Pei, and T. Sato
Magnetic Reconnection: MHD and Beyond 773
  G. Lapenta and J. U. Brackbill
Alfvén Resonances, Forced Magnetic Reconnection, and Model of Solar Flares 777
  C. Uberoi

Dynamo Activity in Imposed DC Magnetic Fields 781
    D. C. Montgomery, W. H. Matthaeus, L. J. Milano, and P. Dmitruk

6. Particle Acceleration and Transport

Explosion of Plasma Foils in the Petawatt Regime: Generation of MeV Particle Beams 784
    M. Eloy, A. Guerreiro, J. T. Mendonça, and R. Bingham

Theory of Anomalous Bohm Diffusion and the Related Concept of Photo-Field Fusion Phenomena 788
    V. S. Belyaev and V. N. Mikhaylov

High Energy Electrons Formation in Laser-Produced Plasma: Theory and Experiment 792
    V. S. Belyaev and V. N. Mikhaylov

Millimeter Wave Generation by a Relativistic Electron Beam in a Plasma Filled Sheath Helix Loaded Waveguide 796
    N. K. Jaiman and V. K. Tripathi

Non-Thermal Particle Populations in Space Plasmas 800
    M. P. Leubner

Collimation of Hot Electrons by Spontaneous Magnetic Fields in the Interaction of a Short-Pulse and High-Intensity Laser with a Relativistic Plasma 804
    H. Liu and X.-T. He

Cross Field Diffusion of Cosmic Rays: Dependence on 2-D Field Turbulence Models 808
    F. Otsuka and T. Hada

Resonant Solar Acceleration and its Astrophysical Implications 812
    I. Roth

Cross-Field Plasma Acceleration and Potential Formation Induced by Electromagnetic Waves in a Relativistic Magnetized Plasma 816
    R. Sugaya

Nonlinear Development of Current-Driven Instabilities and Energy Transport to Heavy Ions 820
    M. Toida and H. Okumura

Stochastic Motion of Relativistic Particles in the Field of a Wide Wave Packet 824
    E. Nagornykh and A. Tel'nikhin

OBSERVATIONS OF NATURAL PLASMAS

1. Planetary Atmospheres, Ionospheres and Auroral Regions

On Property of Scattering Structure in Random Continua 831
    J.-S. Guo, M.-L. Zhang, J.-K. Shi, S.-P. Shang, X.-G. Luo, and H. Zheng

2. Planetary Magnetospheres and Space Weather

Ionospheric Response to Flux Transfer Events at the Earth's Magnetopause 835
    F. Pitout and P. L. Blelly

3. Solar Wind and Outer Heliosphere

Phase Coherence of Large Amplitude MHD Waves in the Earth's Foreshock: Geotail Observations 840
    T. Hada, E. Yamamoto, and D. Koga

4. Solar and Coronal Physics

Relationship between Horizontal Flow Velocity and Cell Size for Supergranulation using SOHO Dopplergrams 844
    U. Paniveni, V. Krishan, J. Singh, and R. Srinkanth

Author Index 847

Preface from Congress Chair

The biennial International Congresses on Plasma Physics (ICPPs) are the most international of the full-spectrum plasma physics conferences, having no permanent tie to any one regional physical society, but rather making different regional links as they move around the world. The 2002 Congress, held in the southern Asia-Pacific region, incorporated the 6th Asia Pacific Plasma Theory Conference, the 24th Australian Institute for Nuclear Science and Engineering (AINSE) Plasma Science and Technology Conference, and the 6th Japan-Australia Plasma Theory and Computation Workshop.

In an age of specialization, ICPP 2002 was a general conference in plasma physics. A theme of the conference was the unity of plasma physics, covering all aspects, from low-temperature plasmas to fusion plasmas, from industrial applications to astrophysics, and all three approaches to investigating physical phenomena: experiment, observation and theory.

The Congress was organized on behalf of the ICPP International Advisory Committee (IAC) by the Local Organizing Committee (LOC). The International Program Committee, co-chaired by Robert Dewar of the ANU and Michael Tendler of the Royal Institute of Technology, Stockholm, met in Sydney in February 2002 to select the program of 49 invited Plenary and Review talks for the morning program. These were selected from nominations submitted by the IAC, the LOC, the Program Committee itself, and self-nominations in abstracts submitted. An innovation at ICPP 2002 was the interesting program of afternoon invited topical talks assembled by four autonomous specialist subcommittees chaired by Iver Cairns and Marcela Bilek of the University of Sydney, Sadrahuddin Benkadda of University of Provence and Michael Tendler, Stockholm. The themes of the sessions were Space and Astrophysical Plasmas, Plasma Applications, Transport and Complexity in Magnetically Confined Plasmas, and Fusion Plasmas.

The conference, attended by 292 delegates, was opened by Professor Lawrence Cram of the Australian Research Council.

It is also a tradition of the International Congresses on Plasma Physics to hold a forum on the last day, to reflect on where we are now and where we are going in plasma physics — what are the most important scientific, technological and political challenges over the two years to the next ICPP, and beyond? The forum in Sydney was chaired by Dr Jean Jacquinot, Head of the Research Unit of the Association Euratom-CEA, Controlled Fusion Research Department, Cadarache, France. Panelists selected broadly to represent the four strands of the afternoon sessions gave short presentations, followed by lively discussions involving the audience, including students.

The main sponsors were the International Union of Pure and Applied Physics (IUPAP), USD 14,000, and the New South Wales Department of State and Regional Development, AUD 10,000. The IUPAP grant was predominantly used to encourage participation by delegates from the most underdeveloped countries (those classified Low-Income or Lower-Middle-Income by the World Bank). The Manly Sunrise Rotary Club hosted a reception for delegates from developing countries on the Sunday prior to the opening of the Congress.

The Congress was underwritten by the Australian Institute of Physics with a $15,000 loan, supplemented by a risk/profit sharing agreement with the Plasma Research Laboratory and Department of Theoretical Physics at the Research School of Physical Sciences and Engineering, ANU, and Professor Don Melrose's consulting account at the School of Physics, University of Sydney.

ICMS Australasia were appointed as Professional Conference Organizers, the contract being signed by the President of the Australian Institute of Physics, A/Prof John O'Connor.

The venue, the Manly Pacific Parkroyal Hotel, overlooking the surf beach at suburban Manly, proved to be ideal for a conference of the size of ICPP 2002 and many delegates expressed their satisfaction with the organization of the Congress. Social and cultural highlights were the conference dinner at the Manly Pacific (with an after-dinner talk by Dr Karl Kruszelicki) an organ recital by well-known UK plasma physicist Professor Malcolm Haines in the Great Hall of the University of Sydney, and a public lecture by Academician Vladimir Fortov, Russia, reporting on fascinating micro-gravity dusty plasma experiments in the MIR space station and the International Space Station.

At a meeting of the IAC during the Congress, the host country for ICPP 2004 was chosen to be France.

Robert Dewar
Australian National University
Congress Chair

Editorial Preface

This Proceedings of short papers (approximately four pages) is based mainly, but not exclusively, on the afternoon invited Topical Talks and the Poster presentations. In a few cases papers have been accepted from intending delegates who were not able to attend the meeting for one reason or another. The papers have been rigorously selected and edited, but not formally refereed. IOP Publishing Limited has agreed to publish a cluster of papers from the morning Plenary and Review talks in an issue of Plasma Physics and Controlled Fusion as regular refereed papers.

The editors were helped in preliminary screening by a number of academics in particular, Rowena Ball, Boyd Blackwell, and Neil Cramer. The editing process was greatly helped by the efforts of Departmental Administrators, Helen Hawes, Renée Vercoe and Wendy Quinn.

Sequencing of the papers has been done by using one of the three codes developed for Experiment, Observation or Theory, depending on which of the three was most heavily weighted (see later for an explanation of this three-dimensional topic code system). The papers divided mainly into Experiment or Theory and Computation, though there were a few papers (listed at the end) that were primarily observational.

In order to make maximal use of the topic code data, a spreadsheet, PaperLocator.xls, containing the data for each paper and a hyperlink to it, has been provided on the CD. This allows the reader to create a customized Table of Contents by sorting on the "distance" of each paper from an arbitrarily specified point in three-dimensional topic space, and providing weights to indicate the relative importance of Experiment, Observation, and Theory to the reader.

Ian F. Falconer
Robert L. Dewar
Joe Khachan

SPONSORS

ICPP2002 thanks the following organizations for their financial assistance

International Union of Pure and Applied Physics

New South Wales Department of State and Regional Development

ICPP2002 also thanks the following organizations for their support

 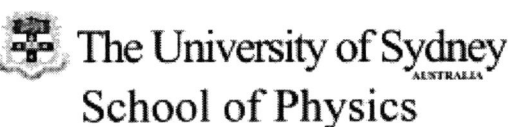

Local Organizing Committee

Prof R L Dewar (**Chair**), The Australian National University, robert.dewar@anu.edu.au
Dr I S Falconer (**Secretary**), University of Sydney, i.falconer@physics.usyd.edu.au
A/Prof R C Cross (**Treasurer**), University of Sydney, cross@physics.usyd.edu.au
A/Prof A D Cheetham (**Web Master**), University of Canberra, andrew.cheetham@canberra.edu.au

Prof M M M Bilek, University of Sydney
Prof R W Boswell, The Australian National University
Dr I H Cairns, University of Sydney
Dr R L Carman, Macquarie University
Dr R Castillo, UWS Macarthur
Prof K D Cole, Latrobe University
Dr G Collins, Australian Nuclear Science and Technology Organisation (ANSTO)
Dr N F Cramer, University of Sydney
Prof B J Fraser, University of Newcastle
Prof J H Harris, The Australian National University
Prof H Hora, University of New South Wales
Dr J M Howard, The Australian National University
A/Prof B W James, University of Sydney
Dr J J Lowke, Commonwealth Scientific and Industrial Research Organisation (CSIRO)
Prof P A Robinson, University of Sydney
A/Prof S W Simpson, University of Sydney
Prof R G Storer, Flinders University

ICPP HISTORY

The first *International Conference on Plasma Physics* was held in 1980 in Nagoya, when the fourth *Kiev International Conference on Plasma Theory* (this series having been held in Kiev in 1971 and 1974 and Trieste in 1977) was combined with the fourth *International Congress on Waves and Instabilities in Plasmas* (this series having been held in Innsbruck in 1973 and 1975 and Palaiseau in 1977).
Subsequent International Conferences/Congresses on Plasma Physics were held in:

Göteborg (1982)
Lausanne (1984)
Kiev (1987)
New Delhi (1989)
Innsbruck (1992)
Foz do Iguaçu (1994)
Nagoya (1996)
Prague (1998)
Quebec City (2000)

International Advisory Committee

F. Gratton	ARGENTINA
R. L. Dewar	AUSTRALIA
S. Kuhn	AUSTRIA
F. G. Verheest	BELGIUM
P. H. Sakanaka	BRAZIL
R. Décoste	CANADA
A. Hirose	CANADA
T. Johnston	CANADA
X.-T. He	CHINA
P. Sunka	CZECH REPUBLIC
P. K. Michelsen	DENMARK
M. Chatelier	FRANCE
C. Deutsch	FRANCE
D. Gresillon	FRANCE
G. C. Laval	FRANCE
P. K. Shukla	GERMANY
H. K. A. Soltwisch	GERMANY
A. Sen	INDIA
M. P. Srivastava	INDIA
C. Uberoi	INDIA
R. Pozzoli	ITALY
T. Yamanaka	JAPAN
A. Iiyoshi	JAPAN
O. Motojima	JAPAN
T. Sato	JAPAN
K. Sato	JAPAN
N. Sato	JAPAN
M. Wakatani	JAPAN
J. K. Lee	S. KOREA
J. E. Herrera	MEXICO
M. Sadowski	POLAND
V. E. Fortov	RUSSIA
V. E. Golant	RUSSIA
A. G. Litvak	RUSSIA
S. Lee	SINGAPORE
M. Hellberg	SOUTH AFRICA
C. Alejaldre	SPAIN
M. Tendler	SWEDEN
F. W. Sluijter	The NETHERLANDS
A. G. Sitenko	UKRAINE
R. Dendy	UNITED KINGDOM
M. G. Haines	UNITED KINGDOM
D. C. Robinson	UNITED KINGDOM
J. F. Drake	USA
N. Fisch	USA
R. D. Hazeltine	USA
G. J. Morales	USA
T. M. O'Neil	USA
D. D. Ryutov	USA
M. Yamada	USA

Program Committees

MAIN PROGRAM COMMITTEE

Chair: Dewar, Robert L (ANU, Canberra, Australia, robert.dewar@anu.edu.au)

Co-Chair: Tendler, Michael (Royal Inst. of Tech., Sweden, tendler@fusion.kth.se)
Campbell, David J (EFDA, MPIPP Garching, Germany, djc@ipp.mpg.de)
Chang, HY (KAIST Korea, hychang@mail.kaist.ac.kr)
Cheng, CZ (PPPL, Princeton, USA, fcheng@pppl.gov)
Diamond, Patrick H (UCSD, La Jolla, USA, sconover@physics.ucsd.edu)
Fraser, Brian J (Univ. of Newcastle, Australia, phbjf@cc.newcastle.edu.au)
Fussmann, Gerd (Humboldt Univ., Berlin, Germany, fussmann@ipp.mpg.de)
Gratton, Fausto T (INFIP, Buenos Aires, Argentina, faustogratton@infip.org)
He, Xian-Tu (IAPCM, Beijing, China, xthe@mail.iapcm.ac.cn)
Ongena, Jef (JET, Abingdon, UK, j.ongena@fz-juelich.de)
Redi, Martha H (PPPL, Princeton, USA, redi@pppl.gov)
Sato, Noriyoshi (Tohoku Univ., Japan, nsato@ecei.tohoku.ac.jp)
Shukla, Padma K (Ruhr-Univ., Bochum Germany, ps@tp4.ruhr-uni-bochum.de)
Speller, Carlos V (Federal University of Santa Catarina, Brazil, speller@materiais.ufsc.br)
Yamanaka, Tatsuhiko (ILE, Osaka, Japan, tyama@ile.osaka-u.ac.jp)

SPECIAL TOPIC SUBCOMMITTEES:

Plasma Applications

Chair: Marcela Bilek, University of Sydney, Australia
Rod Boswell: Australian National University, Australia
Ian Brown: Lawrence Berkeley National Lab., USA
Hong-Young Chang, Korea Advanced Institute of Science and Technology, Korea
Nan Huang: Southwest Jiaotong University, China
Marek J. Sadowski: Andrzej Soltan Inst. for Nuclear Studies, Poland
Carlos V. Speller: Federal University of Santa Catarina, Brazil
Thiraphat Vilaithong: Chiang Mai University, Thailand

Space and Astrophysical Plasmas

Chair: Iver Cairns, University of Sydney, Australia
Estelle Asseo: Ecole Polytechnique, France
Geoff Bicknell: Australian National University, Australia
Attilio Ferrari: Universita di Torini, Italy
Brian Fraser: University of Newcastle, Australia

Alice Harding: NASA/Goddard Space Flight Centre, USA
Janet Kozyra: University Michigan, USA
Hiroshi Matsumoto: Kyoto University, Japan
Bruce Remington: Lawrence Livermore National Laboratory, USA
Steve Spangler: University of Iowa, USA
Lev Zelenyi: Space Research Institute, Russian Federation

Transport and Complexity in Magnetized Plasmas

Chair: Sadruddin Benkadda, Universite de Provence, France
Robert L. Dewar: Australian National University, Australia
Patrick H. Diamond: University of California, San Diego, USA
Sanae-I. Itoh: Kyushu University, Japan
Abhijit Sen: Institute For Plasma Research, India
George Zaslavsky: New York University, USA

Fusion Plasmas

Chair: Michael Tendler, Royal Inst. of Tech. Stockholm, Sweden
Guido Van Oost, Ghent University, Belgium
Martha H Redi PPPL, Princeton, USA
Vladimir Rozhansky: St Petersburg Technical Univ, Russian Federation
Rainer Salomaa: Helsinki University of Technology, Finland
Michael Shats: Australian National University. Australia

EXPERIMENTS ON LABORATORY DEVICES

1. Low-Temperature Plasmas

2. Magnetic Confinement

3. Inertial Confinement and Beams

4. Plasma Applications

5. Experimental Simulation of Phenomena Observed in Natural Plasmas

Observation of Transition and Bifurcation in a Magnetized Plasma Column with Electron Flow Channel

R. Hatakeyama*, M. Sato*, W. Oohara* and T. Takado*

Department of Electronic Engineering, Tohoku University, Sendai 980-8579, Japan

Abstract. When electrons are emitted in the form of current channel along magnetic-field lines into a fully-ionized collisionless plasma column, the ion-density and low-frequency fluctuations are experimentally observed to increase greatly and grow in the channel, respectively, both of which appear to have a causal relation to the inward cross-field transport of ions. A transition from widely-flattened to sharply-peaked density and potential profiles is caused with an increase in the electron-emission quantity. The ion-density increase in the channel saturates as the radial density profile steepens.

INTRODUCTION

Transition and bifurcation phenomena in plasmas have attracted special interest from the viewpoints of basic nonlinear physics [1] and fusion-oriented physics. The transition from regular to stochastic behavior, in the former, is one of the basic subjects concerning stochastic motion in nonlinear oscillator systems. The discovery of the H mode (high confinement) in a divertor tokamak [2], in the later, was a clue for investigating the transition between the L mode (low confinement) and H mode. The great variety of plasma situations in which the transition to improved confinement is observed indicates that there may be some universal mechanism responsible for the formation of a transport barrier and the decrease in the particle outflow. From this viewpoint a unified bifurcation model to explain the L−H transitions is expected to be developed, which includes effects of fluctuations, radial electric field, its gradient, poroidal and toroidal rotation velocities (flows).

Transition and bifurcation phenomena in plasmas lie behind many kinds of experimental situations and configurations. Our attention is focused on basic features of transition and bifurcation phenomena in a fully-ionized collisionless plasma column, which are caused when electrons are excessively supplied along magnetic-field lines to the background plasma from its narrow open end. It is the key in this study to investigate the relations among field-aligned electron-flow channel, low-frequency instabilities, cross-field ion transport, transition, and bifurcation [3]. In the present experiment an electron-emitting small electrode is installed in order to form the electron-flow channel in the plasma instead of a non-emitting small cold electrode, which was previously used for the excitation of potential-driven ion cyclotron oscillations by biasing positively to the small cold electrode [4-6].

EXPERIMENTAL ARRANGEMENT

The experiment is performed in a single-ended Q machine with a vacuum chamber of 15.7 cm in diameter and 200 cm in length, as shown schematically in Fig. 1. A plasma consisting of electrons and potassium positive ions K^+ is produced by thermal emission of electrons and contact ionization of potassium atoms on a hot tungsten plate of 5.2 cm in diameter. The plasma density is about 1×10^9 cm^{-3} and the temperatures of electrons T_e and positive ions T_+ are $T_e \simeq 0.2$ eV $> T_+$. The background gas pressure is 2×10^{-4} Pa. Thus, the plasma is considered to be fully ionized and collisionless. The plasma column is confined radially by a uniform magnetic field ($B = 0.2$ T), being terminated by an endplate situated at 110 cm from the hot plate. Ion and electron Larmor radii are $\rho_+ \simeq 0.2$ cm and $\rho_e \simeq 7 \times 10^{-3}$ cm in $B = 0.2$ T, respectively. The endplate consists of a 10-cm-diam floating plate and a 1-cm-diam cathode with BaO coated. The biasing voltage and current of the cathode are denoted by V_k and I_k, respectively. The electron-flow channel of 1-cm-diam is formed in the background plasma of comparatively large diameter ($\simeq 5.2$ cm). The emitted-electron density is controlled by electric power supplied to the cathode heater. Plasma parameters including a plasma potential are measured by Langmuir probes.

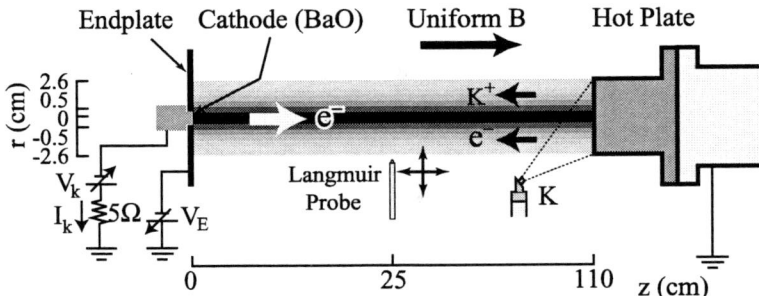

FIGURE 1. Schematic diagram of experimental setup. Electrons are injected from the cathode into a collisionless magnetized plasma.

EXPERIMENTAL RESULTS

Figure 2 (I) shows the cathode current I_k (a), saturation currents of ions I_+ (b) and electrons I_e (c), and a floating potential ϕ_f (d) depending on the cathode biasing voltage V_k, which are measured by a Langmuir probe at $r = 0$ cm and $z = 25$ cm. When the cathode is a cold plate without electron injection, the cathode current I_k for the negative-bias of V_k is small compared with the plasma electron flux given around $V_k = 0$ V, which is absorbed into the cathode, and the characteristic curve $V_k - I_k$ is smoothly traced. As electrons injected from the cathode into a stationary collisionless plasma gradually increase, the emitted electron flux becomes larger than the absorbed flux, and a sudden jump appears in the $V_k - I_k$ curve at a certain value of V_k (~ -4.2 V). When the cathode is forced to bias negatively, the increment of the electron supply to the plasma is allowed even after the saturation of the ion-density increase. However, the further increase in the electron supply leads to an electron-rich condition inside the channel, and the suppression mechanism of electron-beam injection such as a potential dip is formed in front of the cathode to keep charge neutrality of the plasma.

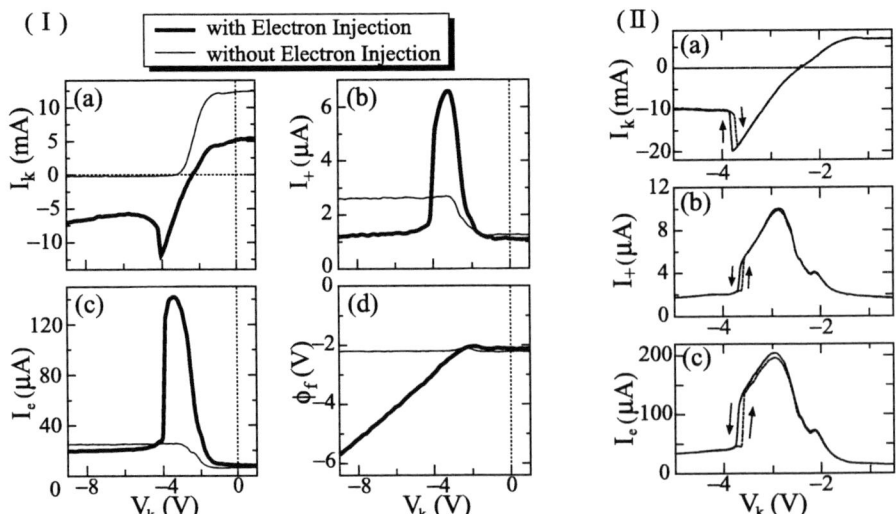

FIGURE 2. (I) Cathode current I_k (a), probe saturation-currents of ions I_+ (b) and electrons I_e (c), and floating potential ϕ_f (d) versus cathode voltage V_k. A comparison between with and without electron flow is made. (II) Hysteresis curves of I_k, I_+, and I_e versus V_k.

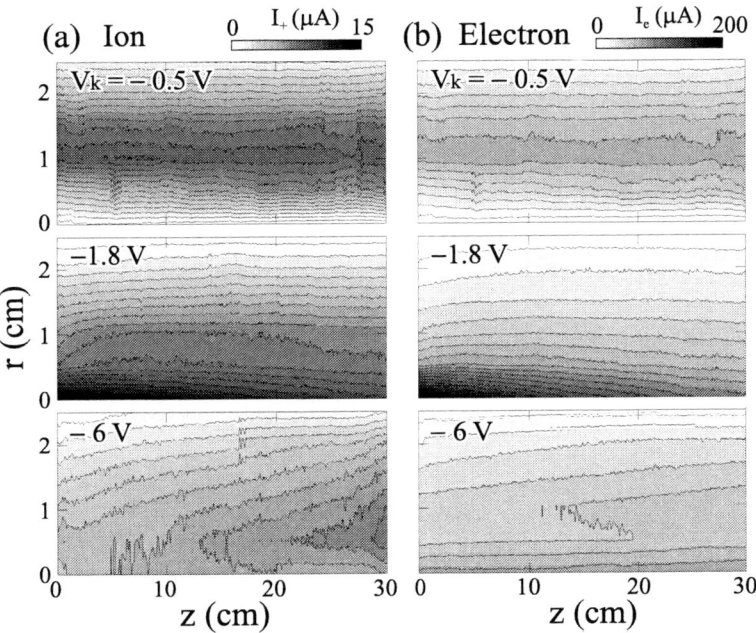

FIGURE 3. Two-dimensional $(r-z)$ spatial distributions of saturation currents of ions (a) and electrons (b) for typical V_k.

Since the ion-saturation current I_+ is proportional to the ion density, we refer to I_+ as the ion density hereafter. When V_k is decreased and the electron flux injected gradually increases, the densities of ions and electrons in the electron channel increases remarkably in spite of the constant supply of the ion flux from the hot plate. The further decrease of V_k results in the rapid decrease of the electron and ion densities. The characteristic curve $V_k - I_k$ are not so sensitive to the magnetic-field strength in the range of $B = 0.15 - 0.3$ T under the configuration where both electrons and ions are magnetized. When we sweep V_k from a positive value to a negative value and contrariwise from a negative value to a positive value, on the other hand, the hysteresis curves are measured in I_k, I_+, and I_e, as shown in Fig. 2 (II) (a), (b), and (c), respectively. The fine structure of the transitional curves indicating the hysteresis and bifurcation susceptibly depends on the geometric configuration among the hot plate, the endplate, and the cathode. As is often referred, such a hysteresis generation is closely connected to the bifurcation nature.

Figure 3 shows two-dimensional $(r-z)$ spatial distributions of I_+ (a) and I_e (b) for typical values of $V_k = -0.5$, -1.8, -6 V (far ahead of, just ahead of, and after the bifurcation). The steep gradient around the radial boundary of the channel is generated along the plasma column in the stage ahead of the bifurcation. The ion density inside the channel increases towards the cathode but that outside the channel decreases towards the cathode along the ion flow in the stage ahead of the bifurcation. This result implies that ions inwardly diffuse from outside to inside of the channel across the magnetic-field lines in the vicinity of the cathode.

The floating potential inside the channel rises for a moment when V_k enters upon a stage of the ion-density increase, and after then falls, as shown in Fig. 2 (d). At the channel center, a peak of electron energy distribution function parallel to the magnetic field yielding the plasma potential at first shifts to the positive-potential region before the bifurcation, then backs away just before the bifurcation and finally falls into the negative-potential region after the bifurcation. When V_k is gradually decreased, V_k approaches the plasma potential inside the channel ϕ_{si} in the stage of the ion-density increase ahead of the bifurcation. However, ϕ_{si} becomes negatively in proportion to V_k after the bifurcation. The plasma potential outside the channel ϕ_{so} is almost constant and the endplate potential is fixed at $V_E = -3.2$ V for any V_k. The potential difference between ϕ_{si} ($r = 0$ cm) and ϕ_{so} ($r = 1$ cm) as a function of V_k is shown in Fig. 4 (a) [$V_k - I_k$ curve is also shown (b)]. $\phi_{si} - \phi_{so} = 0$ indicates the bifurcation point. The ion flux supplied from the hot plate is constantly kept, flows toward the endplate and is never reflected under a normal single-ended Q-machine. Therefore the results of the ion-density increment and the radial potential difference, shown in Fig. 4 (a), give the clear indication of the existence of inward ion-transport across the magnetic-field lines.

It is generally expected that an electron emission and injection into a background plasma is accompanied by the excitation of plasma fluctuations. Low-frequency (≤ 10 kHz) spectra inside ($r = 0$ cm) and outside ($r = 1$ cm) the

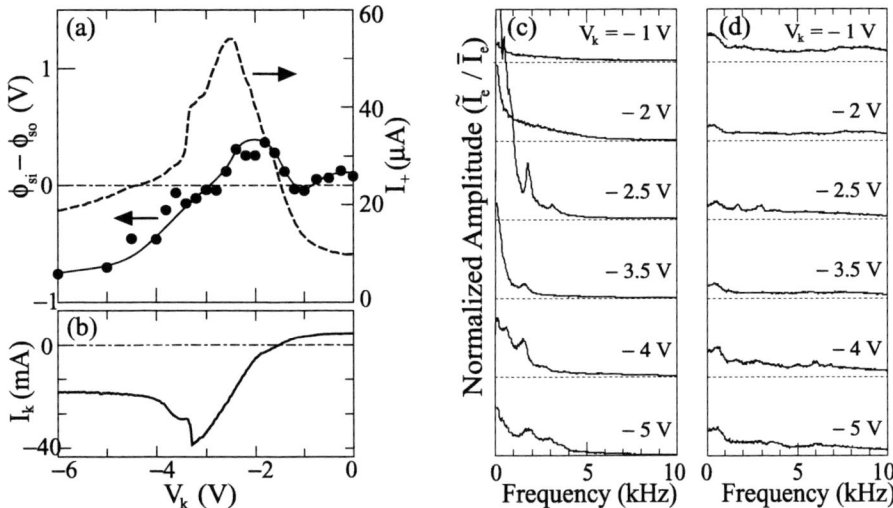

FIGURE 4. (a) Plasma potential difference between ϕ_{si} ($r = 0$ cm) and ϕ_{so} ($r = 1$ cm), and (b) cathode current I_k depending on cathode voltage V_k. Low-frequency spectra of density fluctuations for various V_k (c) inside the channel ($r = 0$ cm) and (d) outside ($r = 1$ cm).

channel for various V_k are shown in Fig. 4 (c) and (d), respectively. No appreciable instability and spectrum change is observed for any value of V_k outside the channel. On the other hand, a very strong instability with frequencies below 5 kHz is generated inside the channel in the stage of the ion-density increase ahead of the bifurcation. The fluctuation amplitude appears to be maximum around the edge of the channel ($r = 0.5$ cm) in the range of $\phi_{si} - \phi_{so} > 0$, where the radial density and potential gradients are very large, and the ion transport occurs even though $\phi_{si} - \phi_{so} > 0$. Its frequency range seems to correspond to a drift-wave frequency or a $\mathbf{E} \times \mathbf{B}$ drift frequency ($\mathbf{E}$: radial electric field). This low-frequency instability is expected to be directly concerned with the ion transport.

CONCLUSIONS

Transition and bifurcation phenomena have been investigated experimentally in the system where an electron flow is externally supplied by an electron emitter in the form of channel from a magnetic-field-aligned open end of a fully-ionized collisionless cylindrical-plasma. The ion density inside the electron-flow channel greatly increases and low-frequency fluctuations such as the Kelvin-Helmholtz and drift-wave instabilities grow with an increase in the electron-emission quantity. This ion-density increase appears to be attributed to the inward cross-field transport of ions, causing a transition in the radial plasma structure from widely-flattened to sharply-peaked density and potential profiles. The enhanced steepening of radial density profile is accompanied by the saturation of ion-density increase in the channel. The further increase in the electron-flow flux results in the formation of current-suppression potential dip in front of the cathode, triggering the bifurcation into a new plasma state with the degraded ion density and current-limited electron flux in the channel due to a charge-neutrality requirement of the plasma.

REFERENCES

1. A. J. Lichtenberg and M. A. Lieberman, *Regular and Stochastic Motion* Springer-Verlag, New York, 1983.
2. F. Wagner, G. Becker, K. Behringer, D. Cambell, A. Eberhagen, W. Engelhardt, et al, *Phys. Rev. Lett.* **49**, 1408 (1982).
3. R. Hatakeyama, M. Sato, W. Oohara, and T. Takado, *Bifurcation Phenomena in Plasmas*, eds S. -I. Itoh and Y. Kawai, Kyushu University, Japan, 2002 p. 128.
4. R. Hatakeyama, N. Sato, H. Sugai and Y. Hatta, *Plasma Phys.* **22**, 25 (1980).
5. N. Sato and R. Hatakeyama, *J. Phys. Soc. Jpn.* **54**, 1661 (1985).
6. R. Hatakeyama. F. Muto and N. Sato, *Jpn. J. Appl. Phys* **24**, L285 (1985).

Ion-Acoustic Shock Waves Formed with Landau Damping

Y. Nakamura[*] and V. N. Tsytovich[†]

[*]*The Institute of Space and Astronautical Science, Kanagawa 229-8510 JAPAN (e-mail:n-yoshi@pub.isas.ac.jp)*
[†]*General Physics Institute, Russian Academy of Science Moscow, 117942 Moscow, RUSSIA*
(e-mail:tsytov@td.lpi.ac.ru)

Abstract. Ion-acoustic shock waves have been investigated experimentally in a double-plasma device. An initial step signal with a ramp shape steepens to form oscillations at the leading part due to dispersion. The wave becomes an oscillatory shock wave when hydrogen gas is mixed in the Ar plasma to increase the Landau damping and when the density n_{H_2} of H_2^+ ions is smaller than a critical value. When n_{H_2} is larger than the critical value, a monotonic shock is formed.

INTRODUCTION

Increasing the ratio T_e/T_i where T_e and T_i are temperatures of electrons and ions, respectively, by cooling ions in a Q-machine plasma so as to reduce Landau damping, Anderson et al. [1] have observed steepening of a ramp signal. The steepend ramp signal became a monotonic structure. Ion-acoustic shock-like waves were observed in a novel device called a double-plasma device [2]. However, we suppose the waves are not shock waves since there existed no dissipation in this case. The waves might be a train of solitons. Recently steepenings of leading edges of ramp signals were observed to propagate in θ-machine plasmas with negative ions [3]. They state that the steepening is due to the absence of Landau damping. The reason why Landau damping is reduced is that the phase velocity of the ion-acoustic wave increases when negative ions are introduced into the plasma. The increases of the phase velocity reduces the Landau damping so that the nonlinear steepening was observed at the leading part of a density pulse, which was considered to be a shock wave [3]. It must be mentioned here that the nonlinear steepening at the leading part is a necessary but not sufficient condition for the formation of shock waves. For the steepened ramp signal to become a shock wave, a dissipation process is needed.

The purpose of the present report is to describe the experimental observation of ion-acoustic shock waves in a double-plasma device. The Landau damping of the wave is enhanced by introducing light ions into the plasma [4].

EXPERIMENTAL PROCEDURE

The experiment was performed in a multi-dipole double plasma device of 90 cm in length and 50 cm in diameter [5]. The system was separated into a source and target section by a mesh grid of 81% transparency, which was kept at a floating potential. The chamber was evacuated down to 6×10^{-5} Pa with a turbomoleculor pump backed by a rotary pump. Then Ar gas was fed into the system at a pressure $P_{Ar} \sim (2-4) \times 10^{-2}$ Pa under continuous pumping. Hydrogen gas was also independently introduced into the chamber through a needle valve, whose pressure P_{H_2} was $(0-7) \times 10^{-3}$ Pa. The source and target plasma were produced by dc discharges between tungsten filaments of 0.1 mm diameter and magnetic cages. The magnetic cage of the target section was grounded. The discharge voltage and carrent were 60 V and $(50-60)$ mA, respectively. The electron density $n_e \approx 5 \times 10^8$ cm^{-3}.

An ion-acoustic wave was excited with a continuous sinusoidal signal for a dispersion measurement or with a positive ramp voltage both of which were applied to the source anode. Signals were detected by an axially movable Langmuir probe of 6 mm in diam which was biased at +4V with respect to the plasma potential in order to detect perturbations in the electron saturation current. The current was then converted to a voltage by a 100 Ω resistance and the resultant signal was fed to a balanced mixer for an interferometer or to a digital oscilloscope for observations of shock waves.

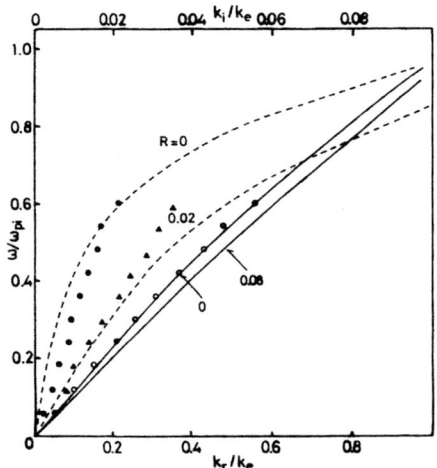

FIGURE 1. Dispersion relations. Solid and dotted curves are calculated real and imaginary parts of K, respectively when $T = 0.1$. Closed circles are experimental k_i/k_e when $P_{H_2} = 0$. Closed triangles are experimental k_i/k_e when $P_{H_2} = 2.0 \times 10^{-3}$ Pa. Open circles are experimental k_r/k_e when $P_{H_2} = 0$. $P_{Ar} = 3.2 \times 10^{-2}$ Pa.

RESULTS AND DISCUSSIONS

The kinetic dispersion relation of the two-ion plasma is given by the following relation,

$$2 + 2K^2 - T(1-R)Z'\left(\sqrt{\frac{T}{2}}\frac{\Omega}{K}\right) - RTZ'\left(\sqrt{\frac{TM}{2}}\frac{\Omega}{K}\right) = 0, \tag{1}$$

where $K = k(\kappa T_e/4\pi n_e e^2) = k/k_e$, $T = T_i/T_e$, $R = n_l/n_e$, $\Omega = \omega/(4\pi n_e e^2/M_h)^{1/2} = \omega/\omega_{pi}$, $M = M_l/M_h$, n_e and n_l are densities of electrons and light ions, respectively, and M_l and M_h are masses of light and heavy ions, respectively. To obtain Eq. (1), we have assumed that the phase velocity is much faster than the electron thermal velocity and $n_e = n_l + n_h$ where n_h is the density of heavy ions. Eq. (1) is solved numerically asssuming a real Ω and a complex K.

Examples of numerical results are shown in Fig. 1 when $T = 0.1$. When $R = 0$, the Landau damping is due to A^+ ions. When H_2^+ ions are added as a contaminant, the damping increases. The phase velocity decreases a little with addition of H_2^+. Real and imaginary parts of k measured with interferometer technique are also plotted in Fig. 1. The damping when $R = 0$ is due to not only Ar^+ ions but also impurity ions produced by ionization of remaining gas of

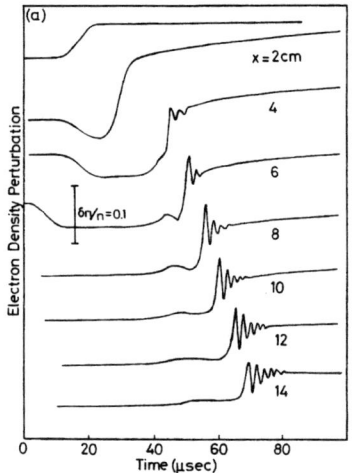
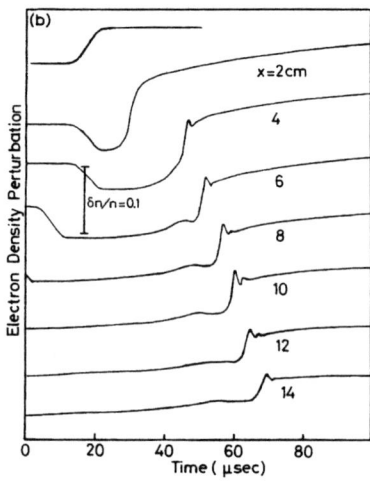

FIGURE 2. Observed signals at different probe positions x from the grid. From $x = 6$ cm, the succesive signals ($6 - 14$ cm) are shifted by $10\,\mu$s to the left. (a) $P_{Ar} = 3.2 \times 10^{-2}$ Pa. $P_{H_2} = 0$. (b) $P_{Ar} = 3.2 \times 10^{-3}$ Pa. $P_{H_2} = 1.5 \times 10^{-3}$ Pa.

the finite base pressure. It is evident from Fig. 1 that the Landau damping is enhanced by H_2^+ ions. Measured damping rates (k_i) are considered to be proportional to k_r.

A nonlinear ion-acoustic wave was excited with a positive ramp voltage. Observed wave signals at different positions are shown in Fig. 2(a) when $R = 0$. As soon as the ramp signal propagates in the target plasma, the leading part whose width is much wider than $1/k_e$ steepens ($x = 2$ cm) owing to the positive nonlinearity, which makes the effect of dispersion larger. Then an oscillatory structure is formed due to the dispersion ($x = 6 - 10$ cm). This observation is similar to the first observation of ion-acoustic shock waves [2]. The oscillation at the leading part developes and does not retain a steady shape. It is similar to results of numerical integration of the Korteweg-de Vries (K–dV) equation with an initial condition of a ramp signal [6]. The hump seen in front of the first peak is a burst of ions. The burst is found to be excited by the external signal in the transient sheath in front of the grid.

Modifications of the propagation characteristics of the nonlinear wave is observed by introducing H_2 gas into the chamber [Fig. 2(b)]. Near the grid ($x = 2$ cm), the signals resemble to the case of no H_2^+ ion condition. The leading part steepens to form an oscillation ($x = 4 - 8$ cm). The form of the oscillaton is keeping the same shape while the wave propagates farther ($x = 10 - 14$ cm). It is different from Fig. 2(a). When P_{H_2} is increased, the oscillation at the leading part becomes less evident and a monotonic shock is observed at a critical $\tilde{P}_{H_2}$.

Observed phenomena, i.e., change of shock waves from oscillatory to monotonic, are similar to those observed when dust particles are mixed into a plasma [5]. For the dusty plasma, dissipation is due to collision of ions with dust particles [5].

The effect of light ions on propagations of ion-acoustic solitons were investigated experimentally by Kozima *et al.* [7]. They have numerically integrated a modified K–dV equation which has a term of Landau damping. The equation was derived by Ott and Sudan [8]. Kozima *et al.* [7] did not measure the damping of linear waves. Therefore, it is uncertain that experimental damping of the ion-acoustic wave in their plasma agrees with that of the modified K–dV equation.

SUMMARY

An initial ramp signal steepens at the leading part at the initial stage of propagation. The steepening developes to an oscillation due to dispersion. When H_2^+ ions are introduced in the plasma, the wave becomes an oscillatory shock wave.

For our study in the future, we will have to compare quantitatively experimental results with results of the modified K–dV equation or with another theory.

REFERENCES

1. Andersen, H. K. *et al.*, *Phys. Rev. Lett.*, **19**, 149 (1967).
2. Taylor, R. J. *et al.*, *Phys. Rev. Lett.*, **24**, 206 (1970).
3. Takeuch, T. *et al.*, *Phys. Rev. Lett.*, **80**, 77 (1988). Luo, Q-Z. *et al.*, *Phys. Plasmas*, **5**, 2868 (1988).
4. Nakamura, M. *et al.*, *Phys. Fluids*, **18**,651 (1975).
5. Nakamura, Y., *Phys. Plasmas* , **9**, 440 (2002).
6. Nakamura, Y., *Nonlinear and Environmental Electromagnetics*, edited by H. Kikuchi (Elsevier, Amsterdam, 1985), p. 139.
7. Kozima, H. *et al.*, *J. Phys. Soc. Japan*, **58**, 504 (1989).
8. Ott, E. and Sudan, R. N., *Phys. Fluids*, **12**, 2388 (1969)

Effect of Light Ions on Ion-Ion Instability

Y. Saitou* and Y. Nakamura[†]

*Utsunomiya University, Tochigi 321-8585, Japan
[†]The Institute of Space and Astronautical Science, Kanagawa, Japan 229-8510

Abstract. Effects of light ions on an ion-ion instability are investigated in laboratory experiments. Spatial growth rates of the instability are measured by an interferometer method. The results show that the growth rate of the instability is affected by light ions and is reduced by increasing their amount. The experimental results are compared with the ones obtained by numerical calculations of the dispersion relation with the plasma dispersion function Z.

INTRODUCTION

An ion-ion instability is one of instabilities excited in an ion beam-plasma system. The system consists of three components, that is, electrons, ions, and ion beams. Basic characters of the ion-ion instability have been investigated in the three components plasmas [1] but not in plasmas with additional component such as different kind of ions which have different mass.

In this study, we will excite the ion-ion instability in an argon-hydrogen mixed gas plasma and investigate an effect of the light ions (hydrogen ions) on the instability. Especially, we will concentrate on an effect of the light ions on the spatial growth rate of the instability. Experimental results will be compared with numerical ones. On the other hand, we have reported interactions of dust grains with the ion-ion instability in our previous work [2]. The negatively charged dust grains also reduce the spatial growth rate of the instability due to collisions between the dust and plasma particles. We will compare the present results to the previous ones and will give a picture of mechanisms brought by the light ions.

EXPERIMENTAL SETUP

Experiments are performed in a homogeneous multi-dipole double-plasma device [1, 3]. Schematic drawing of the device is shown in Fig. 1. The device is 90 cm in length and 50 cm in diameter, and separated into a source and a target chamber by a floating mesh grid. An ion beam is injected from the source to the target by applying a potential difference between the two sections. To observe the ion-ion instability and to measure plasma parameters and noise exited in the target chamber, a Langmuir probe which is movable along the chamber axis – x-axis –, is inserted in it.

Ar gas is fed into the chambers at a pressure of $(2-4) \times 10^{-4}$ Torr. Plasmas in the both sections are produced independently by discharges between tungsten filaments and magnetic cages. H_2 gas is additionally fed into it at a pressure of $(0-5) \times 10^{-5}$ Torr. As a result, the plasma produced in this system consists of electrons, Ar and H_2 ions, and Ar and H_2 ion beams. It is known that hydrogen is ionized as H_2^+ but not as H^+ in plasmas produced by lower discharge voltage like in this device. The H_2 ion beam component is ignorable because the density ratio n_b/n_i in conventional double plasmas is less than 10 % and the density ratio n_{H_2}/n_{Ar} is less than a few percent, where n_I, n_B, n_{H_2}, and n_{Ar} are the densities of ions, ion beams, and H_2 and Ar ions, respectively. The H_2 ion density is estimated by taking into account production based on each gas pressure and loss based on the difference in mobility of both kinds of ions. Typical plasma parameters are as follows: the electron temperature $T_e \simeq 0.9$ eV, and the electron density $n_e \sim 3 \times 10^8$ cm^{-3}. The density ratio n_{H_2}/n_{Ar} is ranged from 0 to approximately 5 %.

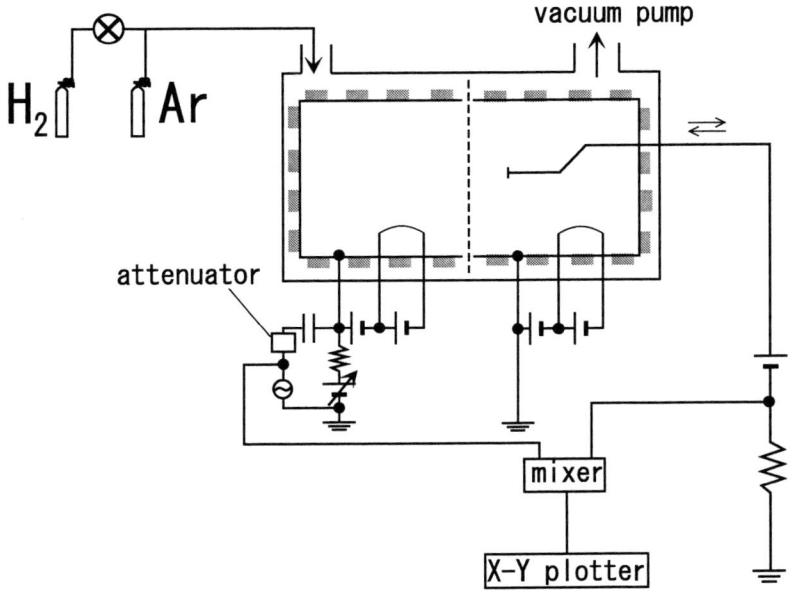

FIGURE 1. Schematic drawing of the experimental device with a circuit for the interferometer method. The left-hand (right-hand) side section of the chamber is the source (target) plasma.

EXPERIMENTAL RESULTS

When the ion beam whose normalized velocity u_B/C_s, C_s is the ion acoustic velocity, is approximately from 1 to 2 is injected to the target plasma, the ion-ion instability is excited. The frequency which gives the maximum amplitude at $x \simeq 3-4$ (cm) is around 300 kHz. Figure 2 are typical examples of spectra and waveforms of excited ion-ion instability. By feeding H_2 gas with keeping the total gas pressure p_{total} constant, it is found from the left-hand side figures of Fig. 2 that power of the noise corresponding to the instability decreases with increasing p_{H_2}. In fact, as shown in right-hand side figures of Fig. 2, the spatial growth rate of the instability becomes smaller for larger p_{H_2}. The spatial growth rate of the instability, k_i, is obtained from the waveform. The results are shown as closed and open circles in Fig. 3.

DISCUSSION

In order to investigate why the growth rate of the instability is suppressed by mixing the light ions, we will compare the growth rates obtained by the experiments and the numerical calculations. The dispersion relation of the ion-ion instability for the mixed gas is given by

$$1 = \frac{\omega_{pe}^2}{k^2 v_{th,e}^2} Z'\left(\frac{\omega}{kv_{th,e}}\right) + \frac{\omega_{pi,H}^2}{k^2 v_{th,H}^2} Z'\left(\frac{\omega}{kv_{th,H}}\right) + \frac{\omega_{pi,L}^2}{k^2 v_{th,L}^2} Z'\left(\frac{\omega}{kv_{th,L}}\right)$$

$$+ \frac{\omega_{pB}^2}{k^2 v_{th,B}^2} Z'\left(\frac{\omega - ku_B}{kv_{th,B}}\right) + \frac{\omega_{pb}^2}{k^2 v_{th,b}^2} Z'\left(\frac{\omega - ku_b}{kv_{th,b}}\right), \quad (1)$$

where ω_{pj} and $v_{th,j}$ are the angular plasma frequency and the thermal velocity of the j-th component, u_j the ion beam velocity (subscriptions H, L, B, b correspond to Ar and H_2 ions, and Ar and H_2 ion beams), and Z is the plasma dispersion function [4]. The last term of the right-hand side of eq. (1) is negligibly small. When we substitute a real ω into eq. (1), we can obtain the spatial growth rate k_i. The results are shown in Fig. 3 with a solid line for $\omega/\omega_{pi,Ar} = 0.6$, and $T_e/T_i = 25$. Both spatial growth rates obtained from the experimental and numerical results reasonably agree and decrease with increasing light ion density. The disagreement for lower H_2 ion density may be thought to be an effect

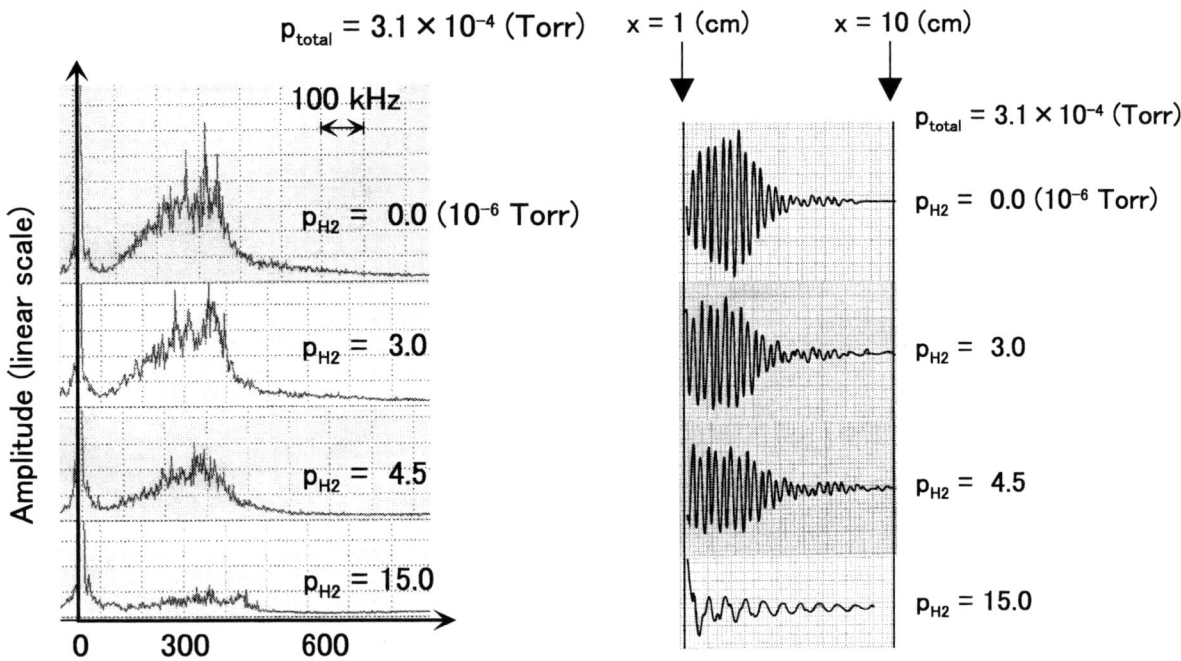

FIGURE 2. Typical examples of spectra (left), and the waveforms whose frequency is 350 kHz (right) of the ion-ion instability as functions of the H$_2$ gas pressure.

of impurities due to non-zero base pressure of the chamber, 1.6×10^{-6} Torr. The fed H$_2$ gas pressure is less than approximately 2×10^{-5} Torr. Under higher pressure (approximately 10 times higher than the base pressure), the experimentally obtained k_i coincides with numerically obtained one. On the other hand, under lower pressure which is comparable to the base pressure, the displacement of the experimental result from the numerical one is large. The charge exchange mean free path is some 10 cm in the present experimental condition. It is clear that the reduction in the spatial growth rate is not due to the collisional damping.

To explain the reduction of the instability, the dispersion relation for this system is considered. The dispersion relation (1) can be approximated as follows:

$$k_i \simeq \frac{\omega_i}{v_g}$$
$$\simeq \frac{kC_s}{v_g}\sqrt{\frac{\pi}{8}}\left[-\left(\frac{T_e}{T_{i,H}}\right)^{3/2}\exp\left(-\frac{T_e}{2T_{i,H}}\right) - \frac{n_{i,L}}{n_{i,H}}\sqrt{\frac{m_{i,L}}{m_{i,H}}}\left(\frac{T_e}{T_{i,L}}\right)^{3/2}\exp\left(-\frac{m_{i,L}}{m_{i,H}}\frac{T_e}{2T_{i,L}}\right)\right.$$
$$\left. + \frac{n_B}{n_{i,H}}\left(1-\frac{u_B}{C_s}\right)\left(\frac{T_e}{T_B}\right)^{3/2}\exp\left\{-\frac{T_e}{2T_B}\left(1-\frac{u_B}{C_s}\right)^2\right\}\right] \quad (2)$$

under assumptions of $v_{th,i} \ll \omega/k_r \ll v_{th,e}$, $|(\omega_r - ku_B)/k_r v_{th,B}| \gg 1$, $\omega_r \simeq k_r C_s$, and $\omega_i \simeq k_i v_g$, where v_g is a group velocity. Furthermore, by assuming $T_{i,L} = T_{i,H} = T_i$, eq. (2) is rewritten as follows:

$$k_i \simeq \frac{kC_s}{v_g}\sqrt{\frac{\pi}{8}}\left[-A\left(\frac{T_e}{T_i}\right)^{3/2}\exp\left(-\frac{T_e}{2T_i}\right) + \frac{n_B}{n_{i,H}}\left(1-\sqrt{\frac{2E_B}{k_B T_e}}\right)\left(\frac{T_e}{T_B}\right)^{3/2}\exp\left\{-\frac{T_e}{2T_B}\left(1-\sqrt{\frac{2E_B}{k_B T_e}}\right)^2\right\}\right] \quad (3)$$

$$A = 1 + \frac{n_{i,L}}{n_{i,H}}\sqrt{\frac{m_{i,L}}{m_{i,H}}}\exp\left[\left(1-\frac{m_{i,L}}{m_{i,H}}\right)\frac{T_e}{2T_i}\right],$$

where E_B is the ion beam energy. The first term of the right-hand side of this equation gives the Landau damping. The factor A is interpreted as representing an influence of the mixed gas of Ar and H$_2$. The Landau damping is enhanced

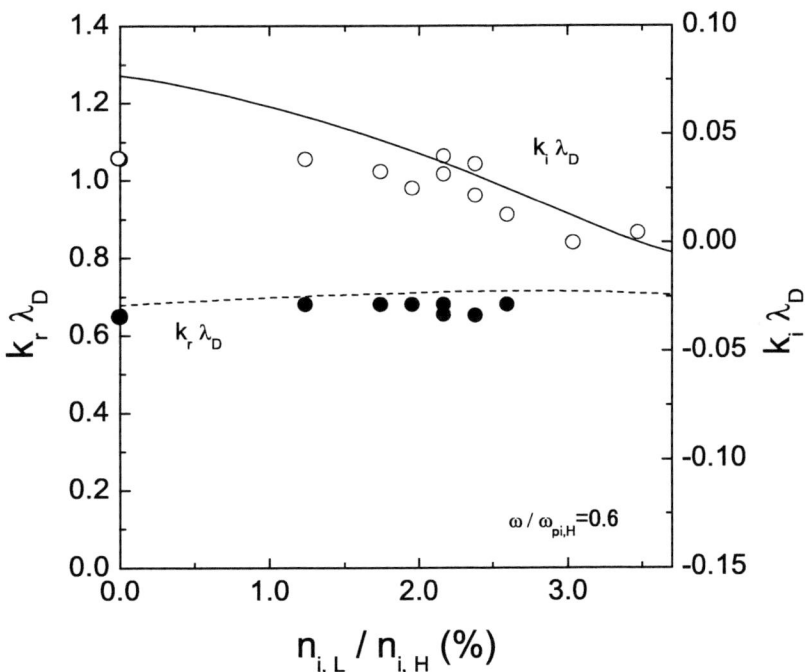

FIGURE 3. Wave number and Spatial growth rate of the ion-ion instability as a function of the density ratio of n_{H_2}/n_{Ar} for a constant $\omega/\omega_{pi,Ar} = 0.6$. Closed circles and a broken line denote experimentally and numerically obtained k_r, and open circles and a solid line denote experimentally and numerically obtained k_i. In numerical calculations, it is assumed that $T_{H_2} = T_{Ar} = T_i$, $T_i/T_e = 1/25$, $T_B/T_e = 1/250$, $n_B/n_{Ar} = 1/100$, and $u_B/C_s = 1$.

by increasing amount of resonant particles by feeding the light H_2 ions to the Ar plasma. Due to this, the intensity of the instability is reduced.

As described above, it is considered that the spatial growth rate of ion-ion instability is reduced by the Landau damping due to the increase of resonant particles. Mixing the light ions is equivalent to increase in the number of the resonant particles. On the other hand, the decrease in the growth rate of the instability in dusty plasmas is considered to be due to the influence of the space potential change and/or the collision between the dust grains and the plasma particles [2]. On both cases, in the mixed gas plasmas and the dusty plasmas, the spatial growth rate seems to decrease due to different mechanisms although it is required to examine in detail in experiments in order to specify the mechanism for each case.

SUMMARY

The ion-ion instability was excited in the Ar-H_2 mixed gas plasma and the spatial growth rate of the instability was measured using the interferometer method. When H_2 ion density increases, the growth rate is reduced. The reduction is due to the enhancement of the Landau damping by feeding H_2 gas.

REFERENCES

1. e.g.: Saitou, Y., Nakamura, Y., et al., *Plasma Phys. Control. Fusion*, **35**, 1755 (1993).
2. Saitou, Y., and Nakamura, Y., "Observation of Ion-Ion Instability in Dusty Plasmas," in *3rd International Congress on the Dusty Plasma Physics (to be published)*, AIP Conference Proceedings, American Institute of Physics, 2002.
3. Taylor, R. J., MacKenzie, K. R., et al., *Rev. Sci. Instrum.*, **43**, 1675 (1972).
4. Fried, B. D., and Conte, S. D., *The Plasma Dispersion Function*, Academic Pess, New York, 1961, pp. 1–8.

Electrostatic Shocks Formed by Ion-Beam Velocity Modulation in a Q-Machine Plasma

Takuma Gohda, Seiji Ishiguro*, Satoru Iizuka, and Noriyoshi Sato**

Graduate School of Engineering, Tohoku University, Sendai 980-8579, Japan
* National Institute for Fusion Science, Toki 509-5292, Japan
** Professor Emeritus, Tohoku University

Abstract. Nonlinear spatial evolutions of velocity-modulated ion beams along a magnetized plasma column are investigated by computer simulation for a Q machine where no electrostatic shock formation is observed for density-modulated ion perturbations. In case of the velocity modulation, the perturbations grow spatially with subsequent saturation even if the system is stable. This is due to the ion bunching along the plasma column. With an increase in the velocity modulation, an electrostatic shock formation is observed to evolve at the propagating front along the plasma column. The observation is compared with the experimental results with the bunching process under the Landau damping. The phenomena are explained by taking into account of the nonlinear collective effect for the bunching of velocity-modulated large-amplitude ion perturbations.

INTRODUCTION

Various kinds of nonlinear waves such as solitons and shocks have been investigated in discharge plasmas. Since the electron temperature surpasses the ion temperature in low-pressure discharge plasmas, the effect of Landau damping is small, so the waves are fully developed according to the electron and ion dynamics. Under these conditions ion acoustic shocks were excited in double-plasma (DP) devices1 by modulating density of beam ions injected from driver to target plasmas.2 The behaviors of velocity-modulated ion beam are of current interest in plasmas, particularly in connection with plasma instability and heating. Spatial evolution of density perturbations produced by velocity modulation in an ion beam-plasma system have been demonstrated by Sato et al., using a double-ended Q-machine plasma.3-5 In case of small amplitude modulation, the phenomenon are explained by the linear wave theory with fast and slow beam modes. However, the behaviors of ion perturbations in case of large amplitude modulation have not been investigated. In order to excite the shocks in the Q-machine plasma, several attempts are examined to increase the temperature ratio between electrons and ions. 6-12

In this paper, we investigate the characteristics of nonlinear evolution of density perturbations excited by a velocity-modulated ion-beam by employing a particle simulation for a Q-machine plasma without negative ions. For a comparison we also employ a density-modulation simulation for the nonlinear wave excitation under the condition of the Q-machine plasma. The simulation results are compared with the experimental results[13] using the velocity modulation.

EXPERIMENTS

The experiment is carried out in a double-ended Q-machine with a vacuum chamber of 20.8 cm in diameter and 167 cm long. Potassium ion plasmas produced by contact ionization at 52-mm-diameter hot tungsten plates (HP) of 2300K, placed at both ends of chamber under electron-rich condition, are confined by axial magnetic field of 2 kG. The machine is operated as a DP device. One of them, the ``driver plasma'', is surrounded by a small metal cylinder connected electrically to the hot plate HP_D. The other hot plate HP_T in the "target plasma" is grounded. Electrons of the two plasmas are separated from each other by a negatively biased grid of 100 mesh/inch. By applying positive bias V_b to HP_D, ions in the driver plasma flow into the target plasma as a beam. The ramp modulation bias V_m is -5~15V and the rise time τ is 5~20 μs. When $V_m > 0$, compressional pulses with positive

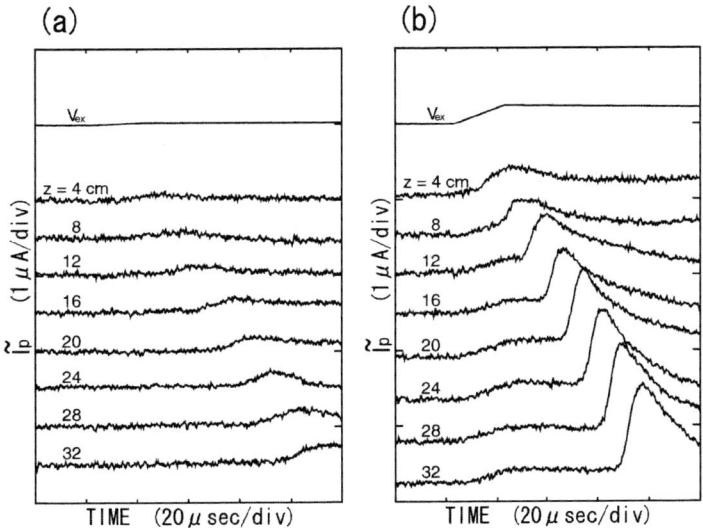

FIGURE1. Ion-beam current perturbations measured at different axial positions Z when (a) $v_m = 0.5$ V and (b) $v_m = 5$ V.

density slope are exited. On the other hand, when $V_m < 0$, rerafactive pulses with negative density slope are generated. Plasma densities of the driver and target plasmas are $5 \times 10^8 \sim 1 \times 10^9$ cm^{-3} and the electron temperatures T_e are about 0.2 eV which is comparable to positive ion temperature T_p. Under our condition, collision mean free paths of charged particles are longer than the plasma column length. The evolutions of perturbations are measured by axially movable mesh probe (6-mm–diameter, 200 mesh/inch) which is biased negatively to detect ion current.

Figure 1(a) shows the evolutions of ion density perturbations when the beam velocity is positively modulated by applying ramp voltage $V_{ex} = V_b + V_m$ to the HP_D. Here, τ is 20 μs, initial beam energy V_b is 4V, and V_m is 0.5 V. Typical bunching position is defined by $L = v_1 v_2 \tau /(v_2 - v_1)$, where v_1 and v_2 are the velocities of ions before and after the modulation, respectively. When $V_m = 0.5$ V and $V_b = 4$ V, $L \sim 23$ cm, which is about the middle of the experimental region between the grid and HP_T. The perturbations excited grow gradually with propagating along the beam. Since the modulation amplitude of beam energy is small, the evolution of perturbation is interpreted by the linear wave theory as in Ref. 5. On the other hand, when the modulation amplitude of beam energy is $V_m = 5$ V, a large amplitude density pulse is generated around a bunching position of $L \sim 12$ cm, accompanied by a steepening of the front slope as shown in Fig. 1(b). Finally, a shock-like structure with positive density slope (positive shock) is formed and propagates with an almost constant velocity comparable to the average speed of ion beam, $v_0 = (v_1 + v_2)/2$.

SIMULATIONS AND DISCUSSIONS

In order to understand the phenomena in the experiments, we have carried out particle simulations. In our simulation the plasma system is divided into two parts as in a double-ended Q-machine plasma.[13] One is the positive ion beam generator regarded as a driver plasma and the other is the beam propagation region regarded as a target plasma. The potential of the driver plasma can be controlled and is varied to modulate the ion beam velocity from v_b to $v_b + v_m$ within a time τ. Therefore, the modulation ratio of beam velocity is given by v_m/v_b. In this simulation v_m is normalized by electron thermal velocity v_{Te}, i.e. v_m/v_{Te} which also has a relation with ion thermal velocity through $v_m/v_{Tp} = ((T_e/T_i)(m_i/m_e))^{1/2}(v_m/v_{Te})$. Here, mass ratio of ions to electrons is $m_i/m_e = 400$. The target plasma potential is fixed at ground potential. The length of the boundary region between diver and target plasmas is $200\lambda_D$. Here, λ_D is the Debye length. The propagation of perturbation is observed in a moving frame with the beam velocity of v_b. Therefore, the beam ions in the target plasma are plotted like a stationary in the coordinates of v-x phase space. Since the electron thermal velocity is much faster than the ion beam velocity the electrons are regarded as stationary electrons. The space and time are counted by x/λ_D and $\omega_{pe}t$, respectively, and the origin of space coordinate is the edge of the target plasma.

The temperature ratio T_e/T_i of the Q-machine plasma is considered to be larger than unity, because the ions ionized by contact surface ionization on the hot plate of the Q-machine are accelerated by a sheath potential formed in front of the hot plate. Initial velocity difference Δv of the ions is reduced by accelerating beam ions with the same energy. Therefore, the spread of the velocity distribution function of beam ions, i.e., ion beam temperature T_i, is decreased with an increase in the beam velocity. Here, we assume that effective temperature ratio is $T_e/T_i = 3$.

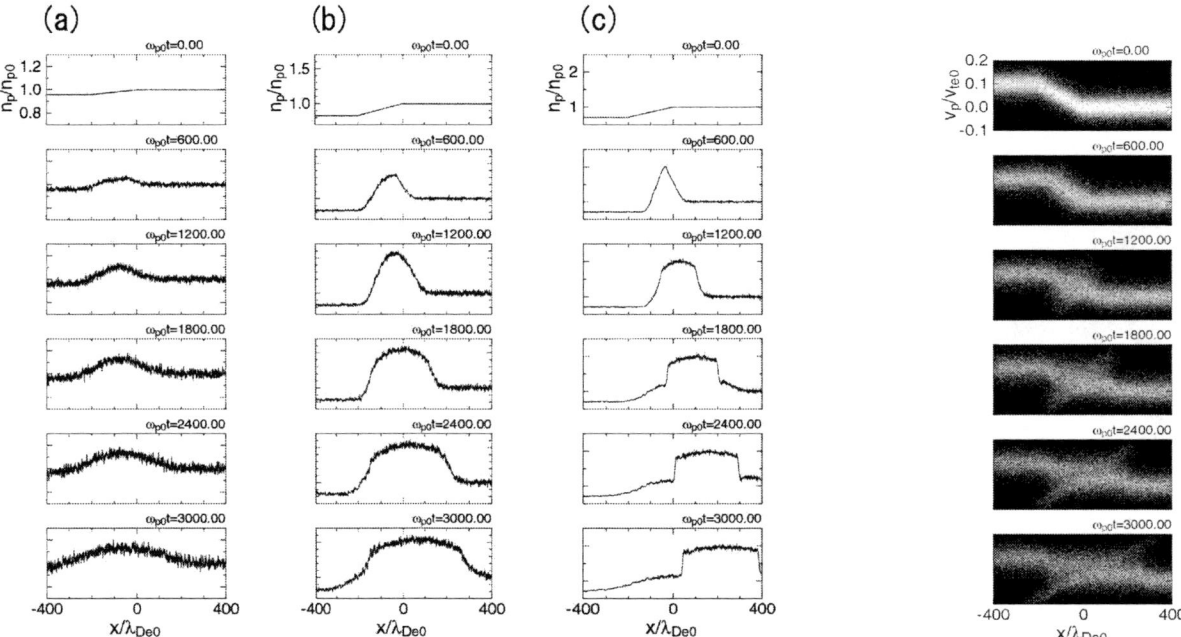

FIGURE 2. Evolutions of ion density perturbations at different velocity modulation ratios (a) $v_m/v_{Te} = 0.02$, (b) $v_m/v_{Te} = 0.1$, and (c) $v_m/v_{Te} = 0.2$. $T_e/T_i = 3$.

FIGURE 3. Ion reflection in phase spaces when $T_e/T_i = 3$ and $v_m/v_{Te} = 0.1$, corresponding to Fig. 2(b).

Typical simulation results are shown in Fig. 2 with velocity modulation ratio as a parameter. It is found that the velocity modulation ratio is an important factor for the shock excitation. When $v_m/v_{Te} = 0.02$, no remarkable growth and formation of the density jump are observed as shown in Fig. 2(a). Initial growth of the density perturbation driven by the ion bunching due to the velocity modulation is heavily damped. When $v_m/v_{Te} = 0.1$, however, the density perturbation grows and the front of the perturbation starts to move with almost constant velocity, keeping the front shape almost constant as shown in Fig. 2(b). In case of $v_m/v_{Te} = 0.3$, a more clear density jump is formed as shown in Fig. 2(c). Figure 3 shows the evolutions in the phase spaces for the case of $v_m/v_{Te} = 0.1$ and $T_e/T_i = 3$. The velocity-modulated beam ions are first accumulated in a narrower region. Then, for $\omega_{pe}t > 1200$ some parts of ions moving ahead of the jump are reflected at the front of the jump due to the evolution of positive potential growth as usually seen in the shock structures. Therefore, the ion bunching is very important to create a shock-like structure in the presence of the strong dissipation. When the velocity modulation ratio v_m/v_{Te} is reduced, the growth of perturbation is limited and suppressed by the heavy Landau damping.

We also try to excite the density jump by using the density modulation. In the simulation the ion density in the driver plasma is set 3 times as much as that in the target plasma. Therefore, the density jump appears initially in the boundary region. Here, $T_e/T_i = 3$ as in the case of the velocity modulation. In this case the beam ions in the driver plasma start to diffuse toward the target plasma with almost ion thermal velocity. Therefore, ion density compression toward the downstream region is rather weak compared to the velocity modulation. The initial density jump between the driver and target plasmas is gradually decaying during the propagation as shown in Fig. 4 and no clear stable structure with a constant slope is observed, which is in contrast to the velocity modulation shown in Fig. 2(b). Therefore, the density modulation method is not effective for the shock formation in the plasma with heavy Landau damping as in the Q-machine plasma.

CONCLUSIONS

We have investigated the evolution of electrostatic shocks by using particle simulations. We find that the excitation method using velocity modulation is quite effective compared to the density modulation in the plasmas with heavy Landau damping as in the Q-machine plasma. This is due to the ion bunching along the plasma column. With an increase in the velocity modulation, density perturbation grows markedly to drive electrostatic shocks propagating along the plasma column. The shocks observed in the experiments are generated by the nonlinear collective effect for the bunching of velocity-modulated large ion perturbations.

The authors thank H. Ishida for his technical support.

REFERENCES

1. R. J. Taylor, *et al.*, Rev. Sci. Instrum., **43**, 1675 (1972).
2. R. J. Taylor, D. R. Baker and H. Ikezi, Phys. Rev. Lett. **24**, 206 (1970).
3. H. Ikezi, *et al.*, Phys. Fluids **16**, 2167 (1973)
4. T. Honzawa and T. Nagasawa, Phys. Plasmas **4**, 3954 (1997).
5. N. Sato, H. Sugai, and R. Hatakeyama, Phys. Rev. Lett. **34**, 931 (1975).
6. A. Y. Wong, *et al.*, Phys. Rev. **133**, 436 (1964).
7. V. Vanek and T. C. Marshall, Plasma Phys. **14**, 925 (1972).
8. H. K. Andersen, *et al.*, Phys. Rev. Lett., **19**, 149 (1967).
9. D. P. Sheehan and N. Rynn, Rev. Sci. Instrum., **59**, 1369 (1988).
10. Y. Nakamura, , *et al.* Phys. Rev. Lett. **83**, 1602 (1999).
11. B. Song, N. D'Angelo, and R.L.Merlino, Phys. Fluids **B3**, 284 (1991).
12. T. Takeuchi, S. Iizuka, and N. Sato, Phys. Rev. Lett. **80**, 77 (1998).
13. T. Gohda, H. Nagaoka, S. Iizuka and N. Sato J. Plasma Fusion Res. SERIES, **4**, 570 (2001).

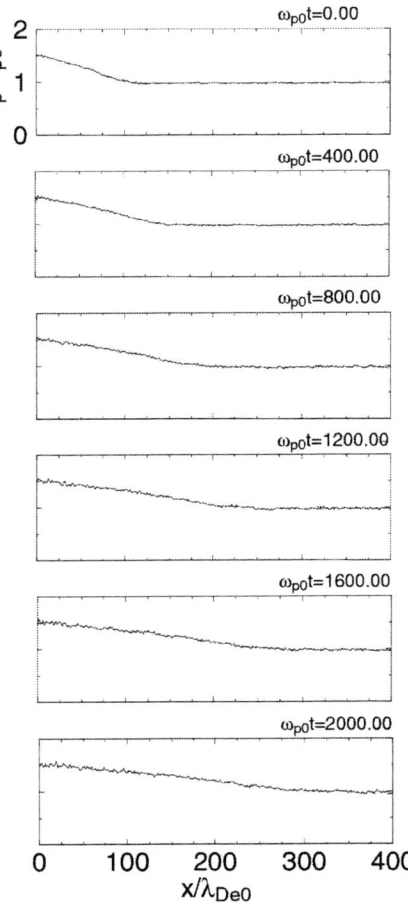

FIGURE 4. Evolution of ion density perturbation produced by using density modulation for $T_e/T_i = 3$.

Fluctuation of Electron Temperature due to Global Alfvén Eigenmode

Yoshimitsu AMAGISHI, Takanori HISHIDA, Motoshi KOBAYASHI and Akira TSUSHIMA[1]

Department of Physics, Faculty of Science, Shizuoka University, 836 Ohya, Shizuoka 422-8529, Japan
[1]*Department of Physics, Faculty of Engineering, Yokohama National University, Yokohama 240-8501, Japan*

Abstract. Two kinds of magnetohydrodynamic (MHD) modes were observed in a cylindrical magnetized plasma produced by a magneto-plasma-dynamic (MPD) arcjet plasma source. Each mode belongs to the "Compressional Alfvén Wave (CAW)" with the azimuthal mode number of m = 1 and the "Global Alfvén Eigenmode (GAE)" with m = 0, respectively. We found that the latter was responsible for the fluctuation of electron temperature.

INTRODUCTION

The numerical calculation for the MHD instabilities which occur in a current-carrying plasma and develop to the MHD wave modes was first accomplished by Appert et al.[1] They treated the problem in a cylindrical coordinate system (r, θ, z) and with the condition that the radial distributions of plasma and current density are parabolic. According to their calculation for |m| = 1 mode, the kink instabilities ($\omega^2 < 0$) develop to the MHD surface wave modes in low-k_z region. Here, we express the waves or temporarily developing perturbations as $\exp[i\{k_z z + m\theta - \omega t\}]$.

In this paper, we report the observation of the fluctuation of electron temperature accompanied with the MHD wave of m = 0 excited during the plasma production process by the MPD arcjet.

EXPERIMENTAL SETUP

The experiment was performed in the "TPH" device (Fig.1(a)) in Shizuoka University. It mounts the "MPD arcjet" plasma source at the end of it and produces a high-density magnetized plasma, cylindrical in shape (15.0 cm in the diameter, 3 m long), singly-ionized helium plasma. The maximum values of electron temperature and plasma density are $T_e \sim 5$ eV and $n \sim 5\times10^{21}$ m^{-3} on the axis, respectively. Axial magnetic field is $B_z = 0.3$ T (the ion cycrotron frequency ω_{ci} is 7.2×10^6 rad/sec).

The schematic of MPD arcjet is also shown in Fig.1(a). It consists of a rod-shaped central cathode of 1.0 cm Φ and a surrounding doughnut-shaped anode of 4.0 cm inner diameter and a gas puff system. The conceptual figure of plasma production is shown in Fig.1(b). Plasma is produced by the arc discharge between electrodes. The "total" discharge current $I_T = (-4.0 \sim -4.3)\times10^3$ A concentrates into the cathode. On the other hand, the current flows not only from the anode but also from the stainless chamber wall, because they're both connected to the earth. As a result, in the region near the MPD arcjet (we will call it "Region I", and call the "Region II" for z > 80 cm, where the current from the anode is negligible.), the integration of the axial plasma current density over the plasma cross section is not zero, i.e. the "net current (I_w)" exists in Region I. The ratio of I_w to I_T is about 10%. We consider that the m = 0 mode should be related to the total current as mentioned in Ref. 2.

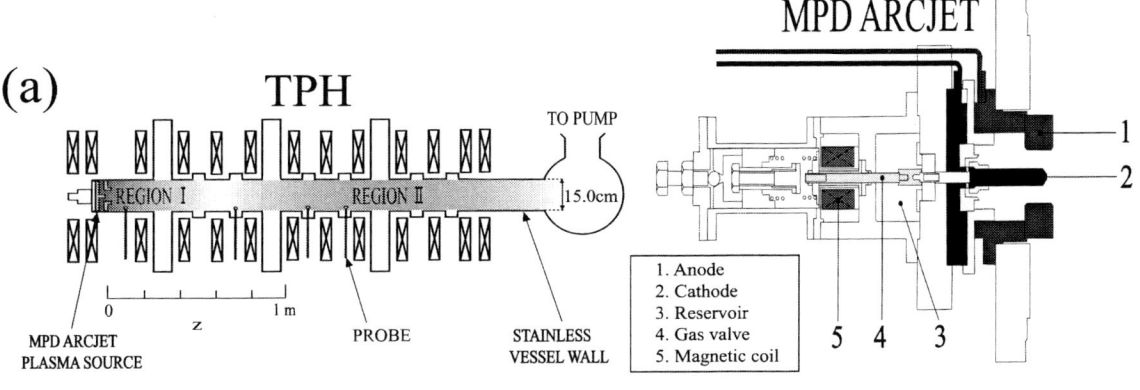

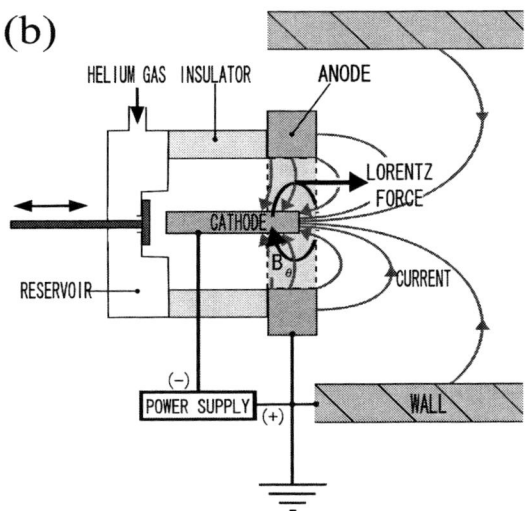

FIGURE 1. (a) Schematic drawing of the experimental apparatus TPH and MPD arcjet.
(b) Conceptual of the arc discharge for the plasma production.

We measured the fluctuating components of azimuthal magnetic field B_θ, electron temperature T_e and ion saturation current I_{is} in Region II. The magnetic field B_θ was measured by using a small magnetic probe. The fluctuating parameters T_e and I_{is} were measured by using a triple probe and a double probe, respectively.

EXPERIMENTAL RESULTS

Fluctuating Magnetic Fields (ω_2)

Experimental results of azimuthal magnetic field fluctuation ($\delta B\theta$) in Region II ($z = 1.1 \sim 1.3$ m from the MPD) are shown in Fig.2(a) ~ Fig.2(c) and Fig.3(a). Figure 2(a) shows the spectrum of $\delta B\theta$ at the radial position of $r = 2.0$ cm. From this figure, it is obviously seen that the perturbation contains two distinct modes of higher and lower frequency bands (we will call these frequency band "ω_1"(higher) and "ω_2"(lower), respectively). In this paper we focus on the lower frequency ω_2, because the m = 1 mode is not related to the fluctuation of the electron temperature. Figure 2(b) shows dispersion relations of δB_θ in the axial direction z. In this figure, a theoretical dispersion relation of the shear Alfvén wave (SAW) for the condition of the maximum plasma density on the axis is also drawn with a solid line. The dispersion relation of ω_2 mode locates slightly below that of SAW, from which, together with a global field distribution in the radial direction as shown in Ref.2, we may conclude the ω_2 mode is the "Global Alfvén Eigenmode (GAE)" because the SAW can no longer propagate so slowly; the solid curve in the figure indicates the "slowest phase velocity" for SAW in an inhomogeneous density plasma.

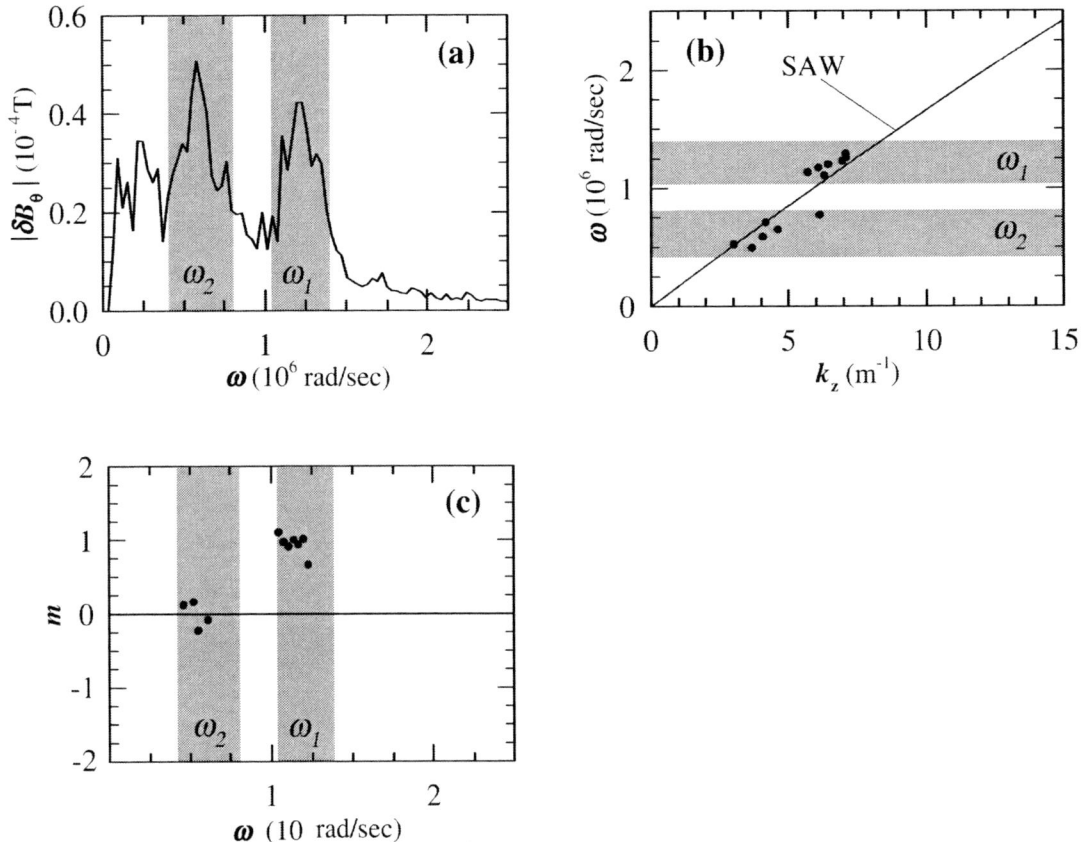

Figure 2. Magnetic fluctuation measured at r = 2.0 cm in Region II.
(a) Spectrum of azimuthal magnetic field $\delta B\theta$, (b) dispersion relation and (c) azimuthal mode number.

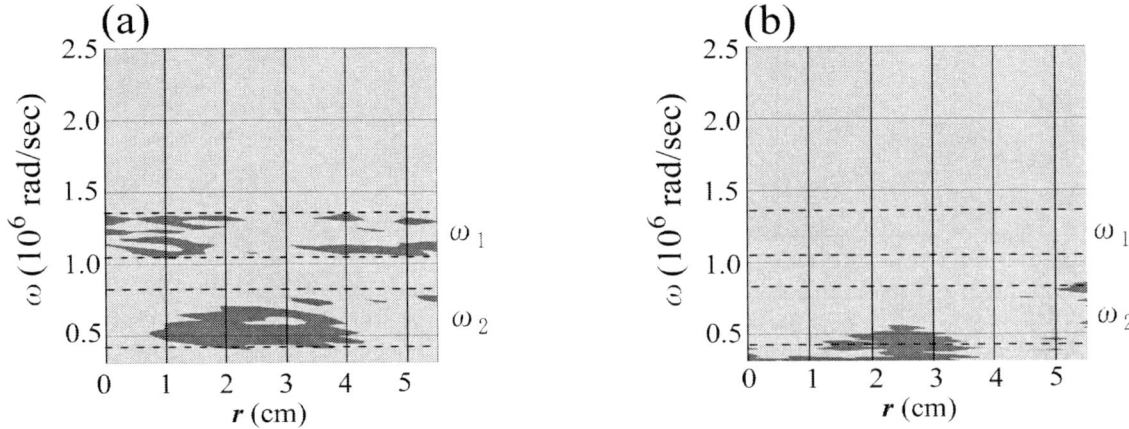

Figure 3. Radial distribution of the spectrum measured in Region II.
(a) The magnetic field $\delta B\theta$, (b) The electron temperature δTe. More darkly drawn area indicates stronger perturbation.

Experimental Results from the Electrostatic Probes

The spectrums of measured electron temperature T_e and ion saturation current I_{is} are shown in Fig.4(a) and Fig.4(b), respectively. In each figure, the spectrum has oscillating component only in the frequency region of ω_2. Obviously the m = 0 GAE is related to the modulation of electron temperature. Comparing Fig.3(a) with Fig.3(b), radial distributions of spectrum of both δB_θ and δT_e have their peak at the radial position r = 2.0 ~ 3.0 cm.

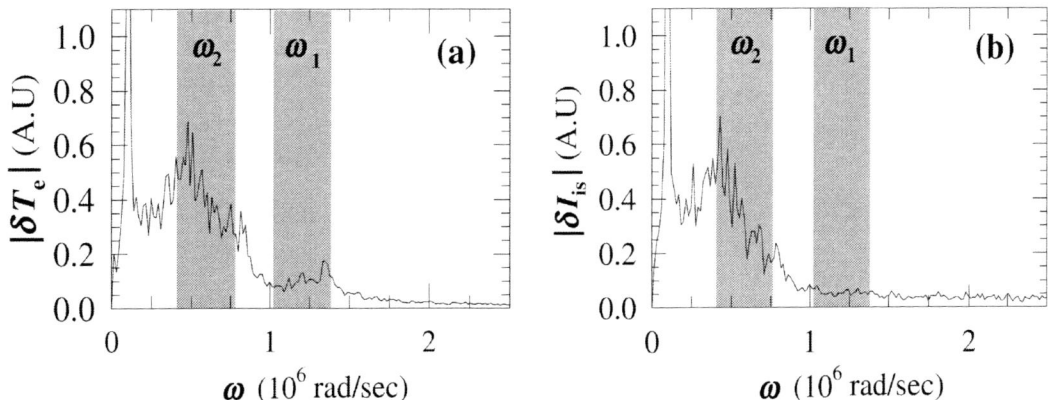

FIGURE 4. Electrostatic probe data measured at r = 3.0 cm in Region II.
Spectrum of fluctuating electron temperature T_e measured by using a triple probe.
Spectrum of ion saturation current I_{is} measured by using a double probe.

CONCLUSION

We have observed two kinds of MHD modes excited in the plasma production process by the MPD arcjet, and the m = 0 GAE mode is obviously related to the modulation of electron temperature. So, it will be expected to make application of this mode for the current drive or plasma heating in nuclear fusion devices. The mechanism of the generation of m = 0 instability is not clear so far but undoubtedly depends on the intensity of total current density.

REFERENCES

1. K. Appert, J. Vaclavik and L. Villard: Phys. Fluids **27**, 1984, 432.
2. Y. Amagishi, T. Hihsida, T. Ino, M. Kobayashi and A. Tsushima: J. Phys. Soc. Jpn **71**, 2002 (in press).

The Best Arc Heater Regime for Minimum Copper Cathode Erosion

A. Marotta[1], L. I. Sharakhovsky[2], and A. M. Essiptchouk[1,2]

[1] *Instituto de Física "Gleb Wataghin", Universidade Estadual de Campinas, Unicamp 13083-970, Campinas, São Paulo, Brazil*
[2] *Permanent address: The Luikov Heat & Mass Transfer Institute, P. Brovki Street, 15, 220072, Minsk, Rep. Belarus*

Abstract. On the basis of the experimental investigations and simple theoretical model, which regards electrode erosion as ablation of the electrode material under the action of intensive heat fluxes in the arc spot, we give estimations of the range of operating regimes of copper cathode with the minimum erosion. We show, that application of a magnetic field for displacing of arc over the electrodes surface has a number of specific limitations from point of view of electrode erosion. These limitations are related with a variation in the energy parameters of the arc spot and heat transfer between electric arc and the electrodes under the variation of magnetic field. As a result, an optimal magnetic field exists, which is a function of an electrode temperature θ_0 and derivative $d\theta/ds$. Here $\theta_0 = T/T_f$; T_f - melting point of electrode material; $s = v/I^{0.5}$; v - arc velocity; I - current. We give here relationships and diagrams, which make it possible to evaluate the range of the operation of copper cathode with the minimum erosion.

INTRODUCTION

The insufficient reliability of electric arc heaters (EAHs) due to the uncertainty in the lifetime of its cold copper electrodes remains main obstruction for the wider application of plasma technology in the industry. Due to the extra high heat-flux density in the arc spots on such electrodes (about 1 MWcm^{-2}), they can operate only with the rapid displacement of spots along electrode surface. The most easily high velocities of displacement of arc (up to hundred of ms^{-1}) are obtained with the aid of the magnetic field.

For comparison, with the displacement of arc by vortex gas flow [1,2], arc velocity does not exceed few ten meters per second. This leads to the different erosion behaviour as function of arc velocity for case with the gas-dynamic or magnetic displacement of arc. The most complex character has magnetic displacement, since it substantially affects on the arc spots parameters (see [3-5]) and the characteristic features of heat transfer between plasma and the electrode surface [6,7]. Earlier in [5] a method and some preliminary analysis of the arc velocity effect on the erosion of the cold electrodes of EAH were reported. The dependences of the current density j and thermal volt-equivalent U of arc spot on the magnetic field B were not taken into account. Furthermore, since in [5] the arc velocity and magnetic field interrelation was not considered also, the authors cannot advise the optimum magnetic field for minimal erosion of electrodes. Due to complex relationship between the arc velocity and EAH operation parameters, the magnetic field but not arc velocity is more convenient for practical control of optimum regimes. This paper is devoted to a more elaborate analysis of optimum operational regimes, taking into account the effect of magnetic field on the arc velocity, arc spot thermal volt-equivalent and current density.

BACKGROUND OF EROSION MODEL

In accordance with thermal model [3] the cold electrode erosion g (in kgC^{-1}) is given by the expression

$$g = g_0 + \frac{UW}{h_{ef}}, \qquad (1)$$

where g_0 (called micro-erosion) is introduced to take into account the experimental fact that a certain minimum value of erosion g_0 is observed under optimal conditions. In equation (1), h_{ef} is the erosion enthalpy and dimensionless erosion energy W is a function of the dimensionless parameter f

$$f = \frac{\pi^{1.5} s \lambda^2 (T_f - T)^2}{8aj^{1.5}U^2}. \qquad (2)$$

Here λ, a and T_f are the thermal conductivity, diffusivity and fusion temperature of the electrode material, respectively, T is the electrode temperature, and $s = vI^{0.5}$ is the normalized velocity. From the erosion model [3] follows that for $f > 1$ the micro-erosion regime is attained and when $f < 1$ – the macro-erosion one ($g > g_0$). From (2) we can write

$$f = As(1-\theta)^2, \qquad (3)$$

where $\theta = T/T_f$ is the dimensionless temperature of the electrode. The parameter $A = \pi^{1.5}\lambda^2 T_f^2 / (8aU^2 j^{1.5})$ take into account the obtained experimentally [6,7] dependences of U and j on the magnetic field strength B:

$$U = 6.5 + 4.9B$$
$$j = \left(2.42 - 1.59 \exp\left[-\frac{B}{0.17}\right]\right) \times 10^9, \qquad (4)$$

where B is in Tesla, j is in Am^{-2} and U is in Volts. The velocity of the arc displacement in EAHs under the action of magnetic field was investigated in [8] and there was proposed the following relationship for the normalized velocity

$$s = 54.8 B^{0.6} \rho_0^{-8/9} \varphi^{-1/3}. \qquad (5)$$

Here ρ_0 is the gas density and $\varphi = 1/(1+w) + w$, where w is the mass-averaged axial gas velocity.

The condition for minimum erosion can be expressed as

$$f = As\left(1 - \theta_0 - s\frac{d\theta}{ds}\right)^2 \geq 1, \qquad (6)$$

where it was assumed, as in [5], a linear increase of the electrode temperature with the arc velocity

$$\theta = \theta_0 + s\frac{d\theta}{ds}. \qquad (7)$$

Here $\theta_0 = T_{00}/T_f$ is some initial "virtual" temperature of the electrode (which corresponds to $d\theta/ds=0$) and characterizes the perfection of the electrode cooling and its relative dimensions.

RESULTS AND DISCUSSIONS

For the subsequent approximate analysis of optimal electrode regimes with minimal erosion we will assume a copper as electrode material ($\lambda=377$ $Wm^{-1}K^{-1}$, $T_f = 1356$ K, $a=10^{-4}$ m^2s^{-1}) and air ($\rho_0 = 1.29$ kgm^{-3}) as plasma-forming gas. To compare new results with ones obtained in [5] and make our erosion analysis more universal, suitable both for the cases of magnetic and gasdynamic arc rotation, we will apply here parameter $d\theta/ds$, but not $d\theta/dB$. Furthermore, we will assume $d\theta/ds$ = const, quite the same as in [5]. In our analysis, in accord with relation $d\theta/dB=(d\theta/ds)(ds/dB)$, it means $d\theta/dB$ is variable. So, using (5), we obtain:

$$\frac{d\theta}{dB} = 26.22 B^{-0.4} \varphi^{-1/3} \frac{d\theta}{ds} \qquad (8)$$

It follows from (8), that if $d\theta/ds$=const, $d\theta/dB$ is a function of B and φ.

For copper electrode in air we obtain condition for optimal regime with minimal erosion:

$$f = 43.7 A B^{0.6} \varphi^{-1/3} \left(1 - \theta_0 - 43.7 B^{0.6} \varphi^{-1/3} \frac{d\theta}{ds}\right)^2 \geq 1 \qquad (9)$$

Thus, from (9) it follows that the parameter f depends on magnetic field B, axial velocity of the plasma-forming gas w, electrode temperature θ_0 and $d\theta/ds$. While during experiments we can perform control of B and w, it is difficult to manipulate the last two parameters. Since, the parameters θ_0 and $(d\theta/ds)$ depend on the design features of the particular installation (dimensions and internal gas dynamics of electrodes, configuration of magnetic field and others).

Initially we carry out analysis assuming as known θ_0 and $d\theta/ds$. Figure 1a shows the curves, which correspond to the condition $f = 1$, in the coordinates of axial velocity of plasma-forming gas w and magnetic field B. In the

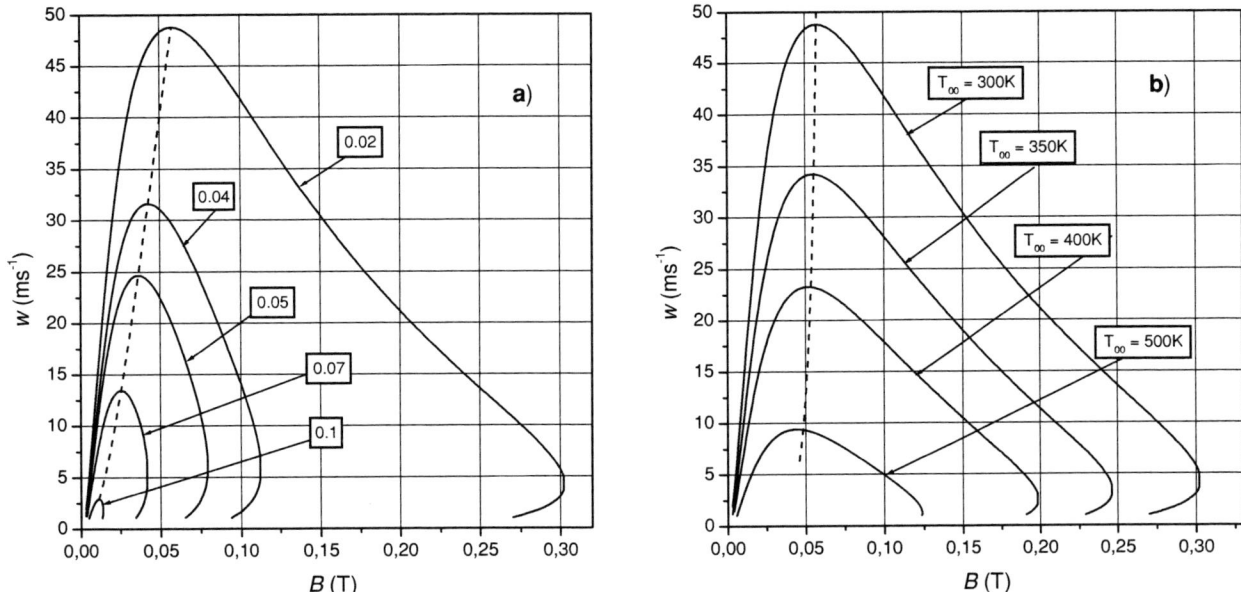

FIGURE 1. The theoretical prediction for the minimum copper-cathode erosion. Solid lines – $f = 1$ dashed lines – $f =$ max. a) $\theta_0 = 0.2212$, the numbers - magnitudes of derivative $d\theta/ds$; b) $d\theta/ds = 0.02$.

calculations we used the constant value of $\theta_0=0.2212$ that corresponding to virtual temperature $T_{00}=300$ K. Numbers near the curves show the values of the derivative $d\theta/ds$. Area inside of the solid curves corresponds to the operational regime of the electrode with minimal erosion (that is micro-erosion mode) when the condition $f > 1$ is satisfied. As it follows from Figure 1a, the domain with the minimum erosion increases with the decrease of $d\theta/ds$.

For certain chosen axial velocity of the plasma-forming gas, the erosion character passes different stages with a variation of the magnetic field. In the region of low magnetic fields an increase in B leads to the transition from the macro-erosion mode (region of $f < 1$) to the micro-erosion one ($f > 1$). The minimum value of erosion is attained at the optimal magnetic field B_{opt} when parameter f attains maximal value (dashed line in Figure 1a). Further increase in B leads to decrease in value of f and micro-erosion increases up to transformation into macro-erosion mode ($f < 1$). In the case of poor design of EAH (if $d\theta/ds$ is high), the micro-erosion mode cannot be reached by any magnetic field.

Figure 1b shows the regions of optimal regimes for different values of virtual temperature θ_0. To carry out calculations, the derivative $d\theta/ds$ was chosen constant $d\theta/ds=0.02$. As it was expected, with an increase in the value of θ_0 the area of optimal mode (bounded by the curves) decreases. This is caused by the fact that the parameter f (see eqs. (3), (6)) decreases with an increase in θ_0. The magnetic field B effect on the parameter f for different values of the virtual temperature θ_0 is shown in Figure 2. In this case the calculations was carried out for $d\theta/ds = 0.02$ and two values of the axial gas velocity: $w = 10$ ms^{-1} and $w = 1.0$ ms^{-1}. The optimal magnetic field corresponds to the maximum value of the parameter f and has poor dependence on the value of θ_0. It is obvious that B_{opt} slightly decreases with an increase in θ_0.

In the Figure 3 we show curves $f = 1$ for different values $d\theta/ds$ in the coordinates of $\theta_0 - B$ (virtual temperature – magnetic field). For the comparison we used two axial gas velocities: $w = 10$ ms^{-1} (solid lines) and $w = 1.0$ ms^{-1} (dotted lines). The area limited by curves corresponds to the optimal operational regimes of electrode with minimal erosion. Line for maximum reliable operation outermost from macroerosion (maximum value of parameter f) is shown by dashed lines. It is obvious that B_{opt} (dashed lines) is displaced into lower magnetic fields (the displacement shown by arrows) with decrease in the axial gas velocity.

CONCLUSION

A magnetic field is widely applied for the rotation of electric arc in EAHs for decrease in electrode erosion. In this report we show, on the basis of experimental data and theoretical analysis, that positive effect of magnetic field is limited for the number of reasons. They are: the first, a magnetic field causes an increase in important energy parameters of the arc spot - thermal volt-equivalent U and current density j in the arc spot. The second, fast rotation

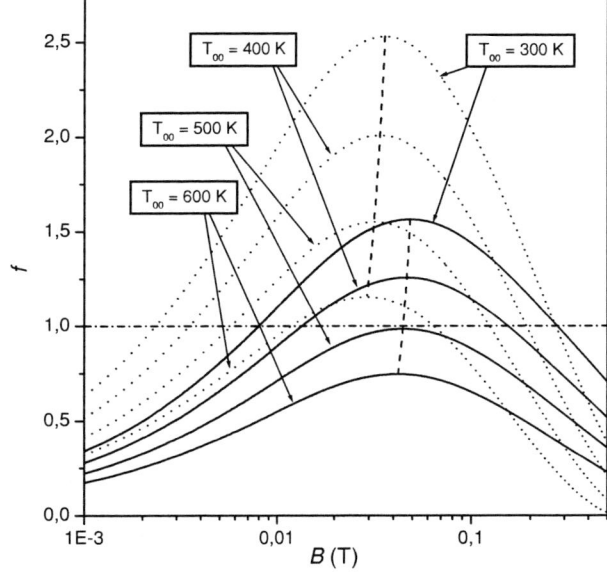

FIGURE 2. Parameter f as function of the magnetic induction B for different values of virtual temperature $\theta_0 = T_{00}/T_f$ and $d\theta/ds = 0.02$. Axial gas velocity $w = 1.0$ ms^{-1} (dotted lines) and $w = 10$ ms^{-1} (solid lines). Optimal B_{opt} (dashed lines) correspond to maximal value of parameter f.

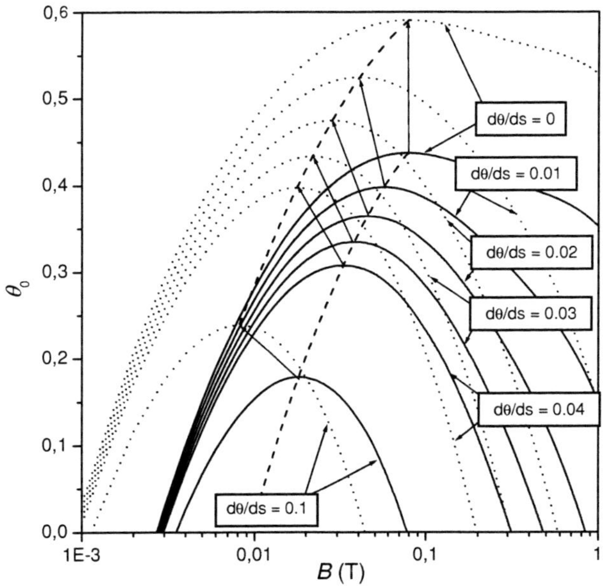

FIGURE 3. Optimal regimes for copper cathode for different magnitudes of $d\theta/ds$. Solid lines – $w = 10$ ms^{-1}, dotted lines $w = 1$ ms^{-1}. Optimal B_{opt} (dashed lines) correspond to maximal value of parameter f.

of arc in magnetic field causes a supplementary increase of temperature of electrode due to forced turbulent raising the rate of convective heat transfer between electric arc and electrode surface. The third, at high velocities of the arc rotation the residual heating of electrode from the preceding passages of arc becomes substantial. These factors leads to the positive value of derivative $d\theta/ds$ during magnetic rotation of arc. That causes (together with increase in U and j) above mentioned limitations in permissible magnetic field and electrode temperature from point of view of minimum erosion.

ACKNOWLEDGMENTS

We acknowledge the financial support of FAPESP, CNPq and FINEP of Brazil.

REFERENCES

1. R.N. Szente, R.J. Munz, and M.G. Drouet, "Electrode erosion in plasma torches with gas vortex driven arcs". In U. Ehlemann, H.G. Lergon, and K. Wiesemann, editors, *Proc. of 10th Int. Symp. on Plasma Chem.*, pp. 1.3-14 p.1 - 1.3-14 p.7, Bochum, Germany, 1991. Int. Union of Pure and Appl. Chem.
2. X. Zhang, Y. Fan, B. Dai, J. Xiao, J. Yang, and X. Cui, "Measurement of rotational velocity of arc root in hollow electrode plasma generator" in C.K. Wu, editor, *Proc. of 13th Intern. Symp. on Plasma Chemistry, vol.1*, pages 196-200, Beijing, China, August 18-22 1997. IUPAC, Peking University Press.
3. A. Marotta and L.I. Sharakhovsky, *J. of Phys. D: Appl. Phys.* 29, 2395-2403 (1996).
4. L.I. Sharakhovsky, A. Marotta, and V.N. Borisyuk, *J. of Phys. D: Appl. Phys.* 30, 2018-2025 (1997).
5. L.I. Sharakhovsky, A. Marotta, and V.N. Borisyuk, *J. of Phys. D: Appl. Phys.* 30, 2421-2430 (1997).
6. A. M. Esipchuk, A. Marotta, and L. I. Sharakhovsky, Journal of Engineering Physics and Thermophysics 73, 1205-1213, November 2000.
7. A. M. Esipchuk, A. Marotta, and L. I. Sharakhovsky, Journal of Engineering Physics and Thermophysics, 74, 813-824, May - June 2001.
8. A.M. Essiptchouk, L.I. Sharakhovsky, and A. Marotta, *J. of Phys. D: Appl. Phys* 33, 2591-2597 (2000).

Investigation of the Influence of Different Boundary Conditions on Helicon Discharges

Christian M. Franck*[†], Thomas Klinger*[†] and Olaf Grulke*[†]

Max-Planck-Institut für Plasmaphysik Greifswald, EURATOM Association, Germany
[†]*Ernst-Moritz-Arndt Universität Greifswald, Germany*

Abstract. In this paper we present investigations on the influence of boundary conditions on helicon discharges. This is done using two approaches: Firstly, the influence of the plasma size is studied by excitation of whistler waves with different wavelengths (frequency range $100 - 1000\,\mathrm{MHz}$). For wavelengths much smaller than the plasma dimensions (4.5 m in length and up to 40 cm in diameter), the unbounded whistler dispersion relation turns out to be appropriate. For increasing wavelengths, the measured dispersion deviates more and more from the predictions of unbounded plasma theory and the helicon wave dispersion relation must be used. Secondly, the boundary condition between the antenna and the plasma is investigated. Standard right-helical antennae are used to operate an rf helicon plasma at a typical frequency of 13.56 MHz. In the linear magnetised plasma experiment VINETA, two different setups for the RF plasma source are available: a standard helicon setup with a glass cylinder attached to one end of the chamber, and a helicon antenna inserted into the stainless steel chamber. The spatial distribution of magnetic fluctuations as well as the plasma parameters are measured with high resolution in a plane perpendicular to the magnetic field as well as in a parallel plane. The influence of the different electric boundaries in the two setups is studied extensively under different operational conditions. This is done by studying the transitions between the capacitive, the inductive and the helicon mode operation.

EXPERIMENTAL SETUP

The experiments were conducted in the VINETA plasma experiment. Figure 1 shows a diagram of the experimental setup. The vacuum vessel of 4.5 m length and 40 cm diameter is immersed in a linear magnetic field of strength up to $B_0 = 100\,\mathrm{mT}$. There are two operational modes of plasma production. Firstly, a plasma is produced by an external helical antenna placed around a vacuum glass extension with 10 cm diameter driven at a frequency of $f = 13.56\,\mathrm{MHz}$ with powers of $P = 300\ldots 2500\,\mathrm{W}$. Depending on the applied rf power, three qualitatively different discharge modes can be established, capacitive, inductive, and helicon wave sustained modes. Plasma densities vary between $n_e = 10^{16}\ldots 10^{19}\,\mathrm{m^{-3}}$. Secondly, an antenna is directly inserted centrally into the chamber. The rf power is coupled to the antenna with vacuum feedthroughs and the antenna legs are coaxially shielded with grounded tubes to ensure plasma production at the antenna only. With powers of up to $P \leq 2500\,\mathrm{W}$ densities of only up to $n_e \leq 8 \times 10^{16}\,\mathrm{m^{-3}}$ can be achieved. The VINETA device is explained in more detail elsewhere [1].

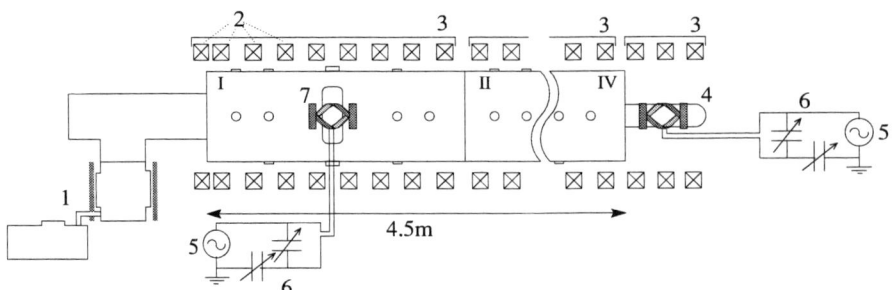

FIGURE 1. Schematic plot of the VINETA experiment. Vacuum chambers I-IV, vacuum pumps (1), field coils (2) with their dc power supplies (3), glass vacuum extension with helical antenna (4), rf power supply (5), and matching unit (6). In addition, an internal helicon antenna (7) can be used to produce a plasma.

EXPERIMENTAL RESULTS AND DISCUSSION

boundary condition - plasma dimensions:. Compared to satellite or rocket experiments, laboratory whistler wave experiments have the advantage of measuring with high spatial resolution and of studying systematically the influence of basic plasma parameters. Unfortunately, all laboratory experiments are necessarily always bounded. Up to now, this boundary condition in whistler wave propagation has mostly been circumvented by going to large sized experiments and using small wavelength waves. A systematic study to investigate the influence of the plasma dimension on the wave dispersion has been carried out for the first part of this paper. Waves are excited from an additional exciter loop antenna at frequencies $f_{ex} = 100\ldots1000\,\text{MHz}$ with power levels below $P_{ex} \leq 1\,\text{W}$. The wave propagation is measured with three axially movable magnetic fluctuation probes at numerous different distances to the exciter antenna. The probes are carefully compensated for electrostatic pickup and calibrated in a magnetic test field of Helmholtz arrangement [2]. From a linear best fit to the phase shift of the measured signal along the axis away from the exciter loop antenna, the parallel wavelength of the propagating wave is determined. To obtain a full dispersion relation, the wavelengths of

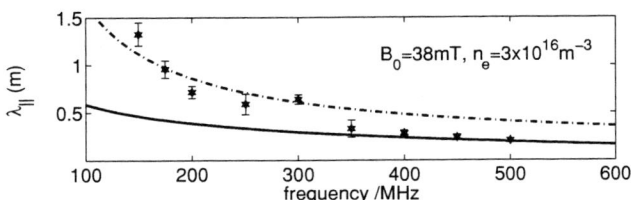

FIGURE 2. Dispersion wave diagram for one set of plasma parameters. Shown are the measured points (markers) and the theoretical plots from unbounded plasma whistler wave dispersion (solid line) and the dispersion of an electromagnetic wave in a plasma filled waveguide (dashed line).

waves excited with different frequencies are measured. Figure 2 shows the whistler wave dispersion measurements for a plasma with density of $n_e = 3 \times 10^{16}\,\text{m}^{-3}$ and a homogeneous magnetic field of $B_0 = 38\,\text{mT}$ strength. The measurements show a clear increase in wavelength for decreasing excitation frequency. Plotted with a solid line is the theoretical curve for unbounded whistler waves. Note that there is no fit parameter left as the plasma density and the magnetic field are determined independently with Langmuir probes and Hall sensors, respectively. Only the high frequency measurements, respectively the small wavelength measurements are satisfyingly described with unbounded plasma dispersion theory. For larger wavelengths, the plasma dimensions have to be taken into account. Plotted with a dashed line is the theoretical expression for a wave propagating in a plasma filled waveguide of radius $R_c = 0.2\,\text{m}$ [3]. It turns out that the measurements at lower frequency are sufficiently well described by this approach. Unfortunately, the exact shape of the dispersion measurements cannot be recovered over the entire frequency range. This is due to the fact that the theoretical curve is only a simple analytic formula approximated for simplifying assumptions that are not completely met in our experimental situation. A more detailed in survey of these experiments is given in [1, 4]

boundary condition - operational regime:. The standard helicon source arrangement (helical $m = +1$-antenna around a dielectric barrier) can be operated in qualitatively different regimes. At low rf powers, the discharge mechanism is a capacitive coupling of the rf power into the plasma at the antenna edges (capacitive mode). For increasing rf powers, the plasma density is high enough to fit a skin depth into the antenna dimensions. The plasma is basically produced from the inductive currents generated by the antenna current (inductive mode). A high density mode can be reached at even higher rf powers via wave heating (helicon wave sustained mode). Although the efficient coupling of the wave into the plasma is still unexplained, helicon sources are widely used in research institutes and in industrial laboratories. The transition between these three modes is not a gradual one but sharp density jumps occur. Figure 3 shows two dimensional plots of the plasma densities measured in a plane perpendicular to the magnetic field in all three operational modes. As expected from the above mentioned discharge mechanism, the density in the capacitive mode is largest close to the antenna edge and almost homogeneous in the inductive case, where the plasma density is about ten times higher (note the different axis scaling). The profile in the helicon wave sustained modes is centrally peaked with an even higher amplitude.

The propagation of waves excited by the antenna at the driver frequency of $f_{rf} = 13.56\,\text{MHz}$ is measured for all three operational modes by magnetic fluctuation probes [2] in a plane parallel to the magnetic field. The evolution of the relative wave amplitude along the axis parallel to the magnetic field is shown in figure 4 (solid lines). The amplitude modulation is clearly seen. Plotted with dotted lines are positions where the phase shift of the wave is an integer multiple of 2π compared to the antenna current signal. A parallel wavelength of $\lambda_\parallel = 1050\,\text{mm}$ is determined in the

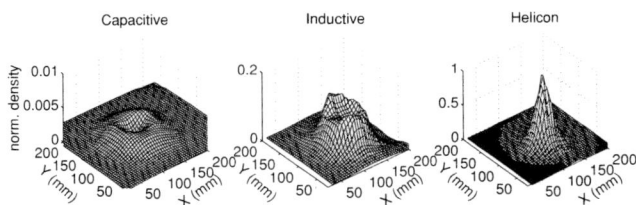

FIGURE 3. Two dimensional plots of the relative plasma densities measured in a plane perpendicular to the magnetic field for the three different operational discharge modes.

case of the capacitive mode and of $\lambda_\| = (155 \pm 16)$ mm in the helicon mode. Both parallel wavelengths agree well with estimates obtained from simple helicon wave dispersion theory. Nevertheless, the wave is not completely described by the parallel wavelength only. Therefore, profiles of the magnetic fluctuation in the different operational modes are measured in a plane perpendicular to the magnetic field. Fig. 5 shows on the left-hand-side the grey-scaled amplitudes of the magnetic fluctuations caused by the wave at the frequency $f_{rf} = 13.56$ MHz. Both the field component parallel ($\dot{B}_z$) and perpendicular ($\dot{B}_y$) to the the magnetic field are shown. Due to the centrally focused plasma production in the helicon mode, the region of interest is smaller in this case. Clearly seen is the double humped structure of $\dot{B}_y$ in the capacitive mode. In fact, this corresponds to a minimum and a maximum as deduced from the phase profile (not shown here). The $\dot{B}_z$ profile in turn shows a single axissymmetric peak. The situation is reversed for both profiles in the helicon mode. For a better visual inspection, the measured amplitude (markers) is plotted against radial position on the right-hand side of Fig. 5 for a horizontal section through the centre of the poloidal plots. The measurements agree well with the theoretically predicted radial mode-structure (solid lines) from helicon wave dispersion theory [5]. Plotted together with the measurements in the capacitive mode is the radial mode-structure for an $m = 0$-mode (plasma radius $R = 0.1$ m). For the plots in the helicon mode, the mode-structure of an $m = 1$-mode with plasma radius $R = 0.04$ m fits the measurements best. On the first glimpse, it is unexpected that an $m = 1$-antenna excites an $m = 0$-wave in the capacitive mode operation. But taking into account the long parallel wavelength it is reasonable to assume that the wave is insensitive to the smaller structure of the antenna (length $l = 300$ mm). The helical structure of the antenna is in that case not of importance and finally acts like a single loop antenna with azimuthal symmetry and excites an $m = 0$-wave.

boundary condition - plasma / antenna coupling:. A further influence of the boundary condition on the discharge deals with the antenna-plasma coupling. Therefore, a second helicon antenna is placed centrally into the vacuum vessel. Whereas in the standard helicon source setup, the antenna is dielectrically insulated from the plasma, the antenna has here a direct contact to the plasma. This setup is much easier to realise and therefore sometimes used for plasma production [6, 7]. To identify the wave and its mode, again the plasma parameters and the magnetic fluctuations

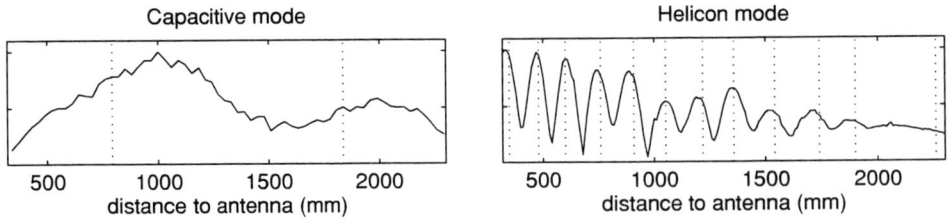

FIGURE 4. Axial evolution of the magnetic fluctuation amplitude parallel to the magnetic field of the wave launched by the rf discharge antenna in two different modes of operation.

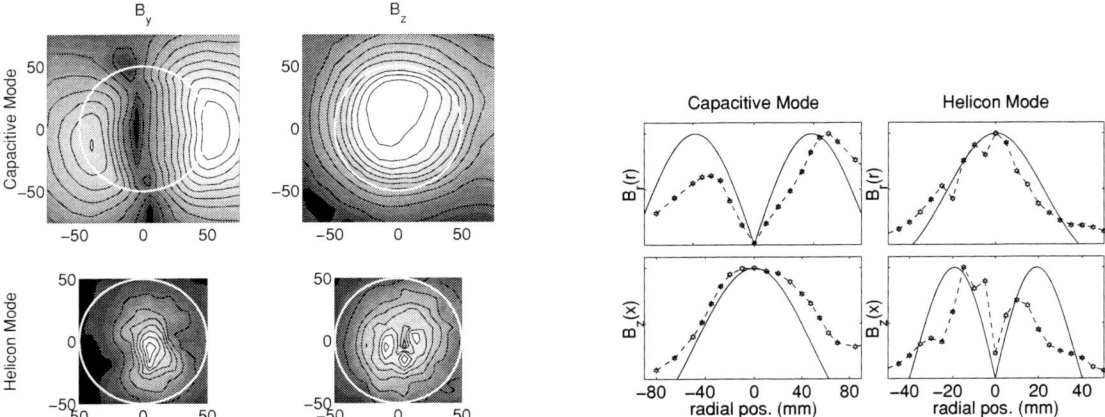

FIGURE 5. Plots of the magnetic fluctuation amplitude in directions parallel and perpendicular to the magnetic field of the wave launched by the rf discharge antenna in two different modes of operation. Shown are two-dimensional measurements of the mode structure (left) and the radial cuts horizontally through the centre (right). Together with the measurements (markers) are shown a theoretical plots (solid lines) for an $m = 0$-mode in the capacitive case and a theoretical $m = 1$-mode for the helicon case.

are measured in a plane perpendicular to the magnetic field. The density profile exhibits the same behaviour with maximum density close to the antenna edges as was found in the capacitive mode of the external antenna setup (not shown here). The measured mode structure (shown in Fig. 6) indicates again an $m = 0$-wave, like in the capacitive operational mode of the external antenna setup. Unfortunately, the mode structure is not as clearly measurable as there is now a direct current between the conducting internal antenna and the grounded vacuum vessel that spoils the fluctuation measurements. All plasma parameter and mode structure measurements confirm that with an internal antenna, the helicon wave sustained plasma mode is not established, but rather a capacitive discharge mode.

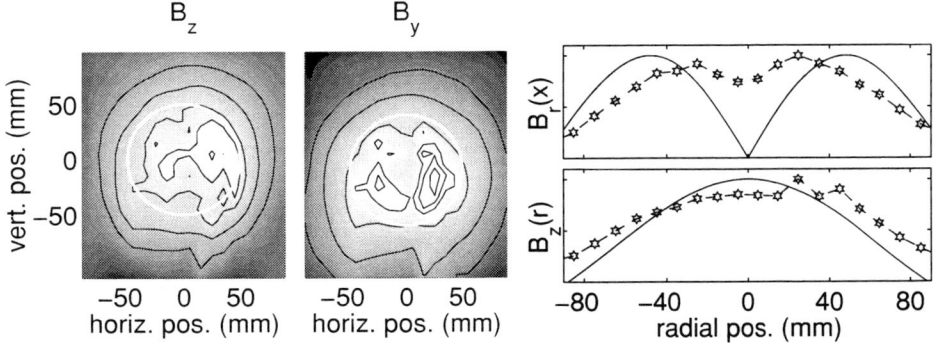

FIGURE 6. Plots of the mode structure of the waves launched by the internal antenna in the plane perpendicular to the magnetic field. Shown are the two dimensional plots (left) and radial cuts horizontally through the centre together (right) with the measurements (markers) and the theoretical plots for a ($m = 0$)-mode wave (solid lines).

REFERENCES

1. Franck, C., Grulke, O., and Klinger, T., *Phys. of Plasmas*, **9** (2002).
2. Franck, C., Grulke, O., and Klinger, T., *submitted to Rev. Sci. Instrum.* (2002).
3. Uhm, H., Hguyen, K., Schneider, R., and Smith, J., *J. Appl. Phys.*, **64**, 1108 (1988).
4. Franck, C., Grulke, O., and Klinger, T., *Proceedings of ICPP 2002*, p. #525 (2002).
5. Chen, F., *Plasma Phys. Contr. Fusion*, **33**, 339 (1991).
6. Lechte, C., Stöber, J., and Stroth, U., *Phys. of Plasmas*, **9**, 2839 (2002).
7. Grulke, O., Greiner, F., Klinger, T., and Piel, A., *Plasma Phys. Contr. Fusion*, **43**, 525 (2001).

Production of Helicon Wave Plasmas in Heliotron Magnetic Field

Shigeyuki Morimoto*, Kazuaki Matsuda*, Tohru Mizuuchi[†],
Hiroyuki Okada[†], Fumimichi Sano[†], and Tokuhiro Obiki[†]

Kanazawa Institute of Technology, Nonoichi, 921-8501 Ishikawa, Japan
[†]*Institute of Advanced Energy, Kyoto University, Uji, 611-0011, Japan*

Abstract. RF (f = 8 MHz, $P_{rf} \leq 8$ kW) argon plasmas have been produced in Heliotron DR (R = 0.9 m, $a_p \sim 0.05$ m, B ≤ 0.17 T). Obtained plasma parameters are $n_e = (1-5) \times 10^{17}$ m^{-3} and $T_e = (8-14)$ eV. The dependencies of the plasma parameters on argon gas pressure, RF power, total and helical magnetic field intensities have been investigated.

INTRODUCTION

High-density ($n_e \geq 10^{19}$ m^{-3}) helicon wave plasmas have been produced in numerous magnetic devices with open-field geometries [1-5]. In external conductor systems such as heliotrons, closed magnetic surfaces exist and there are no particle losses along the field lines. Therefore, efficient production of helicon wave plasmas is expected. A few experiments have been done in closed-field systems [6-8]. Heliotron DR is a conventional heliotron/torsatron device with a planer magnetic axis. It has also toroidal coils and we can examine plasma performance in various magnetic configurations including the simple torus. Main subjects of the present experiment are to clarify the dependencies of the RF plasma parameters on the gas pressure, magnetic field intensity and the existence of the closed magnetic surfaces and/or rotational transform.

EXPERIMENTAL SETUP

Figure 1 shows schematically the experimental arrangement. The experiment has been done in Heliotron DR which is a conventional heliotron/torsatron device with a continuously wound helical winding (poloidal pitch number l = 2, toroidal pitch number m = 15). The vacuum vessel is made of Inconel-625 and its major and minor radii are 0.9 m and 0.11 m, respectively. This device is also equipped with the 30 toroidal coils. Thus, experiments at various magnetic configurations (including the simple torus) are possible. Both helical and toroidal coils are energized by the capacitor banks. The maximum magnetic fields (on the minor axis) produced with the helical and toroidal coils are B_h = 0.18 T and B_t = 0.06 T, respectively. The resultant 90% flat top time of the magnetic field is ~ 13 ms. We define a parameter α^* as the ratio B_t/B_h. When α^* is increased, the rotational transform is decreased.

A one-turn loop antenna was installed inside the vacuum chamber as shown in Fig.2. It has a race track cross section and contacts with an elliptical magnetic surface with average minor radius of 5.3 cm. Thus, the aspect ratio of plasma is rather high (~17). The antenna is made of a copper strip (3 mm thick and 30 mm wide) without insulation tubing. An RF oscillator (frequency f = 6-20 MHz, maximum output power P_{rf} = 50 kW, maximum pulse length T_{pulse} = 10 ms) is used for plasma production. However, in the present experiment, it is operated at f = 8 MHz, $P_{rf} \leq 8$ kW, $T_{pulse} \leq 3$ ms. The RF power is supplied to the antenna through a matching box and a directional coupler. The matching circuit is composed from the variable vacuum capacitors with C_p = 500-720 pF and C_s = 0-220 pF. The injected and reflected powers (P_{inj}, P_{ref}) were estimated using the directional coupler. The working gas is argon and is fed continuously at the filling pressure p = 0.04-2 mTorr. Two single Langmuir probes are used to measure electron temperature and density of the RF plasma. They are installed at the toroidal positions 24° (probe #1) and 120° (probe #2) from the antenna. Magnetic probes to measure RF wave fields are not installed yet.

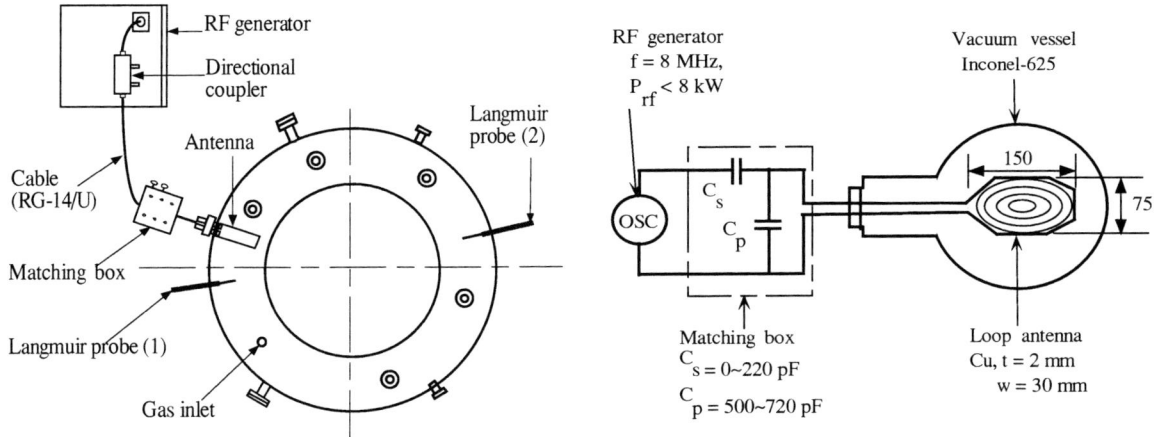

FIGURE 1. Experimental arrangement.

FIGURE 2. Antenna and matching circuit.

EXPERIMENTAL RESULTS

Typical Discharge and Dependence on Filling Gas Pressure

Figure 3 shows time evolutions of injected (P_{inj}) and reflected powers (P_{ref}) and the probe (#2) current (here, electron saturation current is measured to reduce the effect of RF noise) for a typical discharge at p = 0.2 mTorr, B = 0.09 T, and V_{RF} (anode voltage of the oscillator tube) = 6 kV. During the discharge, the ratio of P_{ref} / P_{inj} is almost constant and 0.10-0.15. The probe current saturates at t ~ 1 ms after the onset of RF pulse. In Fig.4, the probe current densities measured with the probes #1 and #2 are plotted against the argon gas pressure. They are nearly close to each other at low gas pressures (p ≤ 0.4 mTorr). However, as the gas pressure is increased, the plasma becomes toroidally non-uniform and at p ~ 1 mTorr, the probe signals show jumps indicating a change of discharge mode. This specific pressure decreases at lower magnetic field and ~ 0.3 mTorr at B = 0.03 T.

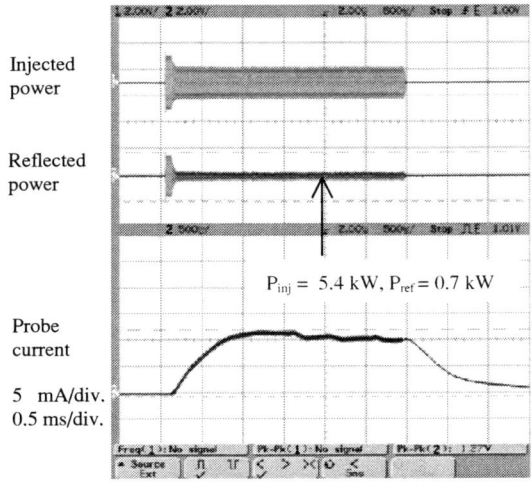

FIGURE 3. A typical discharge ($\alpha^* = 0.2$, B = 0.09 T, p = 0.2 mTorr, V_{RF} = 6 kV).

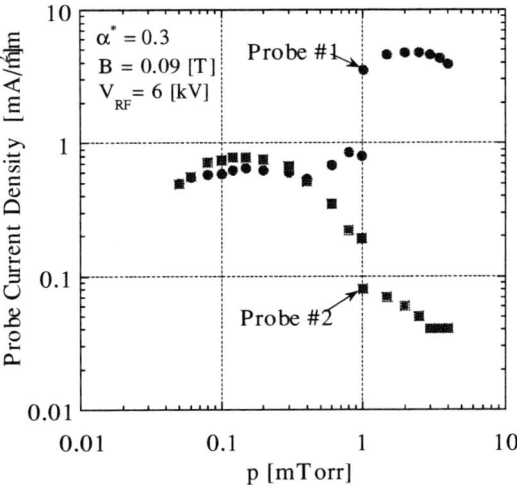

FIGURE 4. Probe currents vs. argon gas pressure.

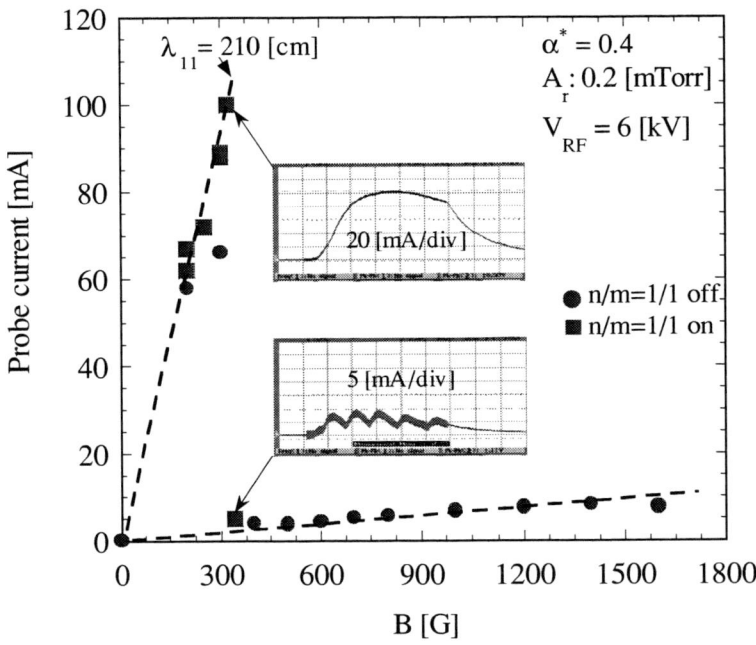

FIGURE 5. Probe (#2) current vs. magnetic field intensity.

Dependence on Magnetic Field Intensity

Figure 5 shows the probe current (#2) as a function of the total magnetic field intensity $B = B_t + B_h$ at a relatively low gas pressure of p = 0.2 mTorr. In this figure, the square data points correspond to discharges under a small auxiliary n/m (m and n are poloidal and toroidal pitch numbers, respectively) = 1/1 helical coil current of 11 A. This current minimizes the size of n/m = 1/1 island produced by the earth field [9].

When B is decreased from B = 0.16 T, the probe current decreases almost in proportion to B. However, at B ~ 0.03 T, it shows oscillations (Fig.5, inset) and finally jumps up to a higher (more than one order) and stable current level which suggests a change of discharge mode. At $B \leq 0.03$ T, the current decreases again almost in proportion to B. If we assume that dispersion relation for m = 0 helicon wave is applicable to these plasmas and use the electron density (shown below), the wavelength along with the magnetic field line λ_{11} becomes ~ 2.1 m at B = 0.03 T.

Next, we determined the electron temperature T_e and density n_e from single probe characteristics in these low- and high-density plasmas. The #2 probe was used and its tip was located on the minor axis. The obtained results are as follows.

For low-density plasma (B = 0.09 T, $\alpha^* = 0.2$), n_e ~ 1.4×10^{17} m^{-3} and Te ~ 14 eV.

For high-density plasma (B = 0.03 T, $\alpha^* = 0.4$), n_e ~ 4.5×10^{17} m^{-3} and Te ~ 8 eV.

Radial distribution of the probe current was measured. Figure 6 shows a horizontal distribution of the #2 probe current (electron saturation current) for the low-density plasma. The peak is shifted inward by 2-3 cm from the magnetic axis (R = 90 cm) but the reason is not understood.

Dependence on Helical Magnetic Field Intensity

Effect of the helical field on the plasma performance was examined. Figure 7 shows the ion saturation current into the #2 probe as a function of the helical magnetic field intensity B_h. Here, the total magnetic field intensity is kept at B = 0.08 T. In this series of experiment, the probe current is not observed at $B_h \leq 0.03$ T. It suggests that the plasmas don't circulate around the torus. This figure also shows that the probe current tends to decrease with increase of the helical field at $\alpha^* \leq 0.3$.

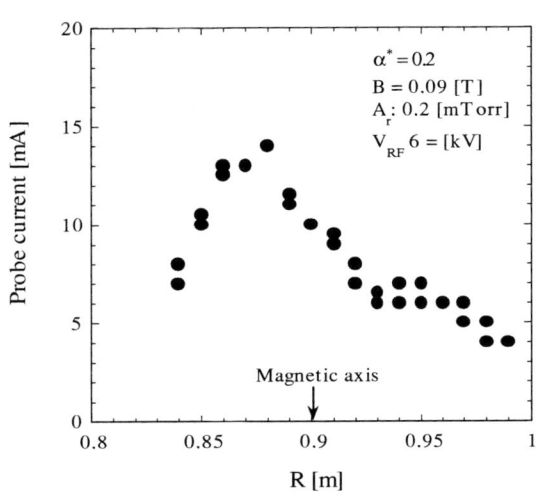
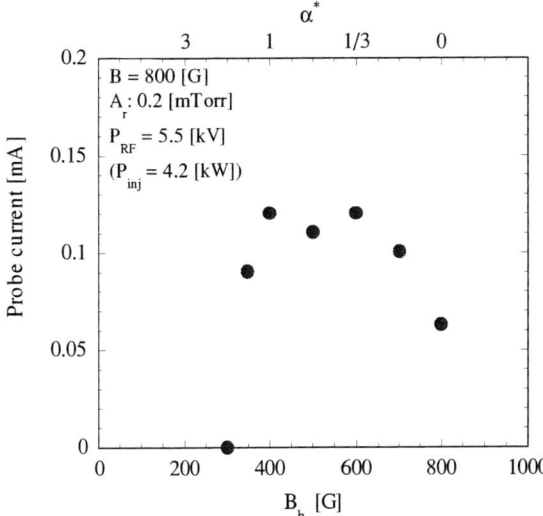

FIGURE 6. Radial distribution of electron saturation current into #2 probe.

FIGURE 7. Ion saturation current into #2 probe vs. helical field intensity.

SUMMARY

RF (f = 8 MHz, $P_{rf} \le 8$ kW) argon plasmas have been produced in Heliotron DR (R = 0.9 m, $a_p \sim 0.05$ m, B $\le$ 0.17 T) intending to heat the plasmas by helicon waves. The obtained plasma parameters are $n_e = (1\text{-}5) \times 10^{17}$ m^{-3} and $T_e = (8\text{-}14)$ eV. The plasmas tend to be non-uniform in toroidal direction at higher filling gas pressure and lower magnetic field intensities. During the B-scan experiment, a discharge-mode change has been observed along with the probe current jump. It has been also shown that the toroidal uniformity of the plasma depends not only on gas pressure and magnetic field intensity but also on the helical field intensity. Measurement of RF fields is not carried out and is left as a main subject in the future experiment.

ACKNOWLEDGMENTS

This work was supported in part by a Grant-in-Aid for Scientific Research from the Ministry of Education, Science, Sports, and Culture of Japan.

REFERENCES

1. Boswell, R.W., Plasma Phys. Control. Fusion **26**, 1147 (1984).
2. Chen, F. F., Laser and Particle Beams **7**, 551 (1989).
3. Komori, A., Shoji, T., Miyamoto, K., Kawai, J., and Kawai, Y., Phys. Fluids **B3**, 893-898 (1991).
4. Shoji, T., Sakawa, Y., Nakazawa, S., Kadota, K., and Sato, T., Plasma Sources Sci. Tech. 2, 5-10 (1993).
5. Shinohara, S., Miyauchi, Y., and Kawai, Y., Plasma Phys. Control. Fusion **37**, 1015-1030 (1995).
6. Loewenhardt, P. K., Blackwell, B. D., Boswell, R.W., Conway, G. D., and Hamberger, S. M., Phys. Rev. Letters **67**, 2792-2794 (1991).
7. Shats, M. G., Rudakov, D.L., Blackwell, B. D., Borg, G. G., Dewar, R. L., Hamberger, S. M., Howard, J., and Sharp, L. E., Phys. Rev. Letters **77**, 4190-4193 (1996).
8. Stroth, U., Ascasibar, E., Krause, N., Lechte, C., Niedner, S., "Turbulence and Wave Experiments with the Low-Temperature Plasma in the Torsatron TJ-K," in *The 13th International Stellarator Workshop – 2002*, Paper No. OIII.
9. Morimoto, S., Matsushita, K., Niwa, S., Nayuki, M., Minamigawa, T., Yamashita, F., Masaki, M., and Obiki, T., Journal of Plasma and Fusion Research Series **1**, 195-198 (1997).

Investigation of Megaampere Discharge in Superdense Gas Media in Order to Obtain a Forplasma Source for Thermonuclear Research

Ph. G. Rutberg, A.A. Bogomaz, A.V. Budin, V.A. Kolikov, A.F. Savvateev

Institute of Problems of Electrophysics of Russian Academy of Sciences (IPE RAS),
Dvortsovaya nab. 18, St.-Petersburg, 191186, Russia.

Abstract. The investigation results, represented in the paper, have been directed to production of dense hydrogen plasma with concentration of charged particle up to 10^{21} cm^{-3} and temperature up to 50 eV. This kind of plasma can be used as forplasma source for thermonuclear researches. It is also a source of ultra-violet and mild x-ray with a high-energy yield. A combined two-stage unit was developed and created to carry out investigations. The unit allows achieving the initial concentration of particle in the discharge chamber up to 5×10^{22} cm^{-3} the expense of preliminary adiabatic compression. Plasma was generated at strong current pulse discharge in super dense hydrogen. The experiments were conducted at currents of 100-500 Ka, energy of storage of 100-500 kJ and discharge time of 100-500 µs. The stable discharge is obtained at the mentioned above conditions. The estimations show that near the current maximum the concentration of charged particle in the discharge column is about 10^{20} cm^{-3} and temperature reaches 15 eV. The experiments demonstrate that with increase of the initial concentration of particle from 0.5 to 3×10^{22} cm^{-3} the field strength in the discharge column rises from 0.7 to 1.6 kV/cm. Higher temperature in the discharge column in comparing with the discharges at lower initial density, to our mind, is connected with locking of radiation from the central area of the discharge. It appears due to the slow quantum diffusion through the transition regions of the discharge. Slow velocity of quantum diffusion is stipulated by small length of Rosseland path of quanta in the transition region. According to estimation this quantity is rather less than characteristic size of the discharge column. This process causes the overheating of central region of the discharge. Further temperature increase can be reached at the expense of increase of the initial density of hydrogen.

INTRODUCTION

The experiments for the studying of a high-current discharge in dense gas are been conducting now at the IPE RAS. The experimental facility, which allows achieving the initial particle density of hydrogen up to 5×10^{22} cm^{-3} at pressure of 350 MPa and temperature about 700 K by means of previous piston compression, was developed and manufactured. The first experiments [1] had revealed the peculiarities of the electric discharge at the specified conditions. It was shown that the initial particle density had the strong influence on the basic discharge parameters, such as strength of field and magnitude of near-electrode voltage drops. The strength of field E is about 700 V/cm at the initial particle density of $1.0-1.5\times10^{22}$ cm^{-3}, that is much more than strength of field at the initial particle density $\approx 10^{20}$ cm^{-3} (E=180-280 V/cm [2, 3]). The magnitude of near-electrode voltage drops decreases as particle density n increases. At $n\approx 10^{20}$ cm^{-3} the overall magnitude of voltage drops was about 1 kV [4, 5], but in conducted experiments at the $n\approx 2\times 10^{22}$ cm^{-3} – 700 V.

The following experiments, which results are presented in the paper, were performed for the estimation of such electric arc parameters as its temperature, conductivity, particle density in the discharge channel and its geometry. Results of the experiment permit to define the dependence of strength of field on initial particle density. It was revealed that there is the strong increasing of strength of field as n increases at the initial particle density higher than 10^{22} cm^{-3}.

The design of improved facility allows carrying out the spectroscopic measurements.

EXPERIMENTAL FACILITY DESIGN

At the IPE RAS the series of pulse plasma generators (PPG) had been developed and manufactured during the past twenty years. They operated on the different kinds of gases at the initial pressure 1-40 MPa. The results of numerical experiments had shown that initial gas density has the strong influence on discharge parameters. The most important appearance for the practical application of PPG is the increasing of energy transfer efficiency from the arc to gas as gas density increasing. It was the reason for the development of experimental facility to study the electric discharge at the high initial gas density. It is the two-stage facility, which joints the stage of quick adiabatic compression of gas and the discharge chamber, where the powerful electric discharge of capacitive storage performs. Design of the first facility did not permit to carry out the direct measurements of temperature of the discharge channel surface. That is why the new facility was developed and manufactured. The design of this facility is presented in Fig. 1.

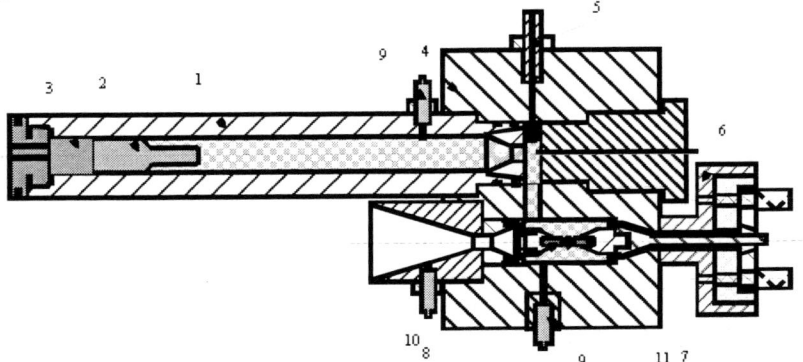

FIGURE 1. Design of the experimental facility.

The two-stage diagnostic bench consists of the compression channel 1, discharge chamber 4 and current collector 6. The piston 2 and powder charge 3 are placed inside the compression channel. The discharge chamber is equipped with the gas input 5, the cathode 7 and the grounded anode 8. To measure the gas parameters in the discharge chamber it is equipped with the pressure transducers 9 and optical windows 11. Before the launch the discharge chamber and exhausting nozzle are separated with the diaphragm 10.

The discharge chamber volume is $V_D=70 cm^3$ and compression channel volume is $V_C=640 cm^3$ so it permits to obtain the maximum gas compression degree $K_c=(V_C+V_D)/V_D=10$. The initial gas pressure been varied from 6 to 20 MPa. The capacitive storage was triggered by the ignition system when the piston achieved its final position and closed the contact.

Before the launch the discharge chamber and the compression channel are pumped up to the pressure of 6-20 Mpa. In the first stage the adiabatic compression of gas by the piston and powder charge is performed. By the action of compression, the gas pressure in the discharge chamber achieves 250-350 MPa at the particle density of $(1.0-3.3) \times 10^{22}$ cm-3. Intermediate unit design provides the piston fixing in the final position to prevent its back moving. A fuse wire is installed between the cathode and anode. Transducer signal, generated after the diaphragm rupture (at the maximum of achieved pressure), can trigger the ignition system of capacitive storage too (it is used as a reserve triggering system). The design of the electrode system provides the constant interelectrode gap during the discharge and allows varying its shape and dimensions in wide range. Design of this facility permits to vary the initial experimental conditions in the following range:

- initial gas pressure 6-25 MPa;
- maximum compression degree 5-10;
- pulse pressure 200-600 MPa;
- average gas temperature up to 700-3000 K.

EXPERIMENT RESULTS

The most important task of conducted experiments was the definition of quantitative influence of initial particle density on the basic discharge parameters. It was necessary to estimate the diameter of the discharge channel, its temperature and conductivity on the base of obtained experimental data. The typical curves of discharge current and arc voltage are presented in Fig. 2, 3.

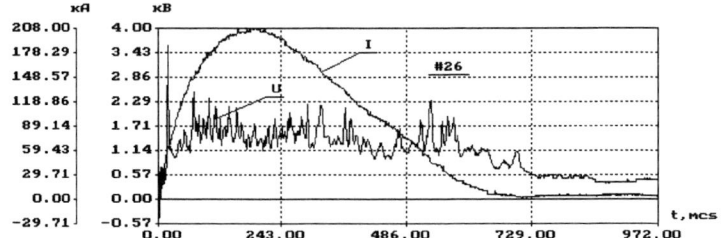

FIGURE 2. Discharge current and arc voltage, $l=9$ mm.
initial particle density - $n=1.04\times10^{22}$ cm^{-3};

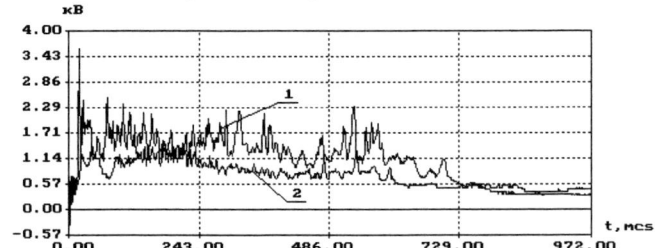

FIGURE 3. Arc voltage at the different initial particle density, $l=9$ mm.
1 - $n=1.04\times10^{22}$ cm^{-3}; 2 - $n=0.47\times10^{22}$ cm^{-3}.

The absence of strong changing of electric parameters and sharp voltage jumps can testify about the stable arc burning at high n_h. The essential peculiarity of the electric discharge at the considered parameters is very weak changing of arc voltage during the discharge process (after the arc stabilization). The discharge current is changing in several times in this time interval (Fig. 2). This phenomenon testifies that the arc voltage is depending weakly on the discharge current at the high initial particle density. The magnitude of the arc voltage is defined, mainly, by the initial particle density (Fig. 3).

The interelectrode gap l was varying during the experiments from 2 to 18 mm. The dependencies of the arc voltage on the interelectrode gap are presented in Fig. 4.

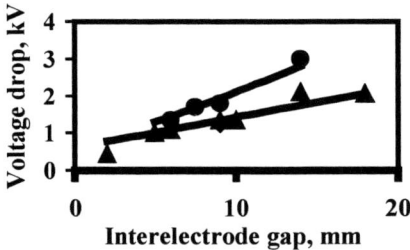

FIGURE 4. Dependence of arc voltage drop at the interelectrode gap at various initial particle density:
▲ - $n \approx (0.9\text{-}1.2)\times10^{22}$ cm^{-3}; ● - $n \approx (2.5\text{-}3.0)\times10^{22}$ cm^{-3}; ◆ - $n \approx 0,5\times10^{22}$ cm^{-3}.

It was observed that in the considered range of experimental parameters the dependence of the arc voltage drop on the interelectrode gap is close to linear. The analysis of obtained data has shown that the initial particle density has the strong influence on the strength of field and the weaker one on the magnitude of overall near-electrode voltage drops. At the initial particle density $n\approx1.0\times10^{22}$ cm^{-3} the average magnitude of strength of field is 700 V/cm and the overall near-electrode voltage drops is about 700 V. At the initial particle density $n\approx2.5\times10^{22}$ cm^{-3} the average magnitude of strength of field is 1300 V/cm and the overall near-electrode voltage drops is about 650 V. Results of present experiments and data obtained earlier show that increasing in initial particle density causes monotonic increasing in strength of field. It was revealed that in range of initial particle density $n>10^{22}$ cm^{-3} this dependence becomes more sharp (Fig.5). In all conducted experiments the size of arc binding spot on the cathode surface does not exceed 4 mm (cathode diameter is 8 mm). This data permit to limit the arc diameter in the following calculations and define the current density.

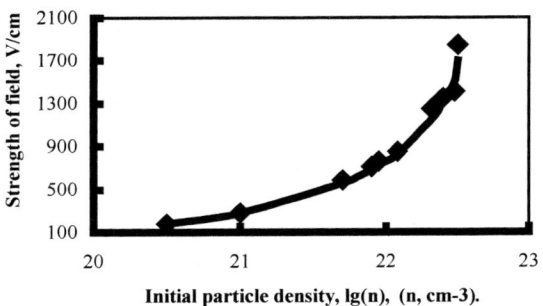

FIGURE 5. Strength of field vs. initial particle density.

ANALYSIS AND DISCUSSION

Some assumptions about the arc parameters were made on base of obtained experimental data. First is high temperature of arc. It has been known that at the temperature of the arc channel more than 10 eV the energy is removed from the arc mainly by the radiation. The energy transferred by the electron heat conductivity $W_e \sim T^{3/4}$, and the energy transferred by the radiation $W_r \sim T^4$. Moreover, high efficiency of energy transfer from the arc to a gas (usually more than 90 %), when arc is motionless, can be explained only by the high gas absorption ability. It is possible if energy of photons, radiated by the arc plasma, exceeds the ionization potential for hydrogen – 13.6 eV. It means that temperature of radiating plasma (taking into account Maxwell distribution of energy of photons) is no less than 5 eV. In this case, lets assume that diameter of arc channel does not exceed the size of arc binding spot on the cathode surface, it is possible to calculate the minimum arc temperature.

Estimation of arc channel parameters was performed on the base of obtained experimental data: discharge current I, strength of field E and pressure P in the moment of current maximum. Temperature of arc surface T_s and radius of the arc channel r_0 were estimated preliminary from the system of equations:

$$IE = A\sigma_s T_s^4 2\pi r_0 \qquad (1)$$

$$\frac{E}{I} = \frac{1}{\pi r_0^2 \sigma(T)} \qquad (2)$$

where, σ_s – Stephan-Boltzmann constant, σ - conductivity.

The first equation describes the energy removal from the arc surface by means of radiation in band of hydrogen transparency – $\int_0^{h\nu/kT} \frac{x^3}{e^x - 1} dx$, where $h\nu$=13.6 eV – potential of ionization of hydrogen. The coefficient A can be calculated as: $A = \int_0^{h\nu/kT} \frac{x^3}{e^x - 1} dx \bigg/ \int_0^{\infty} \frac{x^3}{e^x - 1} dx$

For temperature of 7-10 eV $A \approx 0.1$. The conductivity $\sigma(T)$ is proportional to $T^{1.5}/\ln\Lambda$, and in the first approximation can be expressed by Spitzer formula for the full-ionized gas. Coulomb logarithm for the considered conditions is assumed equal to 3.5-5.

Solution of its system gives the first rough approximation $T_s \sim 10$ eV and $r_0 \sim 0.15$ cm (for experiment conditions $I_{max}=10^5$ A and E=1000 V/cm). Taking into account these preliminary estimations it is expedient to use the channel model of the discharge for the following calculations. Due to the high temperature of discharge channel it is transparent for the own radiation, because the length of run of quanta of $3kT$ energy (here T=1.1×10^5 K) is more than the channel diameter. Inside the transition layer the energy transfer is performed with the beam heat conductivity (diffusion of photons) and thickness of this layer on some orders it is less than the channel diameter. The small length of Roseland run defines small dimension of this layer. For the temperature of transition layer ~ 5 eV and particle density about 10^{21} cm^{-3} the magnitude of Roseland run ~ 10^{-3} cm for the quantum energy $4kT$=20 eV [6]. The run length of quantum increasing as temperature decreasing and plasma becomes semi-transparent. The estimations in which were used the data of calculation for the conductivity of hydrogen plasma [7] show that semi-transparent zone does not have a noticeable contribution into overall discharge current. That is why the discharge current passes through the central zone of T_0 temperature and of r_0 diameter.

It is possible to define the arc channel parameters more correct if use the following system of equations.

$$Q_{rec}\pi r_0 = 5.35 \times 10^{-22} z^4 n_e n_i T_o^{-0.5} \pi r_0^2 = IE \quad (3)$$

$$\frac{E}{I} = \frac{1}{\pi r_0^2 \sigma(T_0, n_0)} \quad (4)$$

$$(1+\alpha)n_0 k T_0 = P_g + P_m, \text{ where } P_m = 1.6 \times 10^{-4} \frac{I^2}{r_0^2} \quad (5)$$

r_0 (cm), T_o (K), and n_0 (cm^{-3}) – radius of central zone, its temperature and particle density inside zone; n_e and n_i – electron and ion concentration; z – charge of ion; $\sigma(T_0, n_0)$ – conductivity [7]; k – Boltzmann constant; P_g – gas pressure in discharge chamber; P_m – magnetic pressure into discharge channel.

The first equation in this system (Eqn. 3) describes the removal of inputted electric energy by means of recombination radiation through the nontransparent transition layer. Ranges of temperature and particle density inside the discharge channel, estimated from the first system of equations, are corresponding to the almost full ($\alpha \sim 0.9$) ionization. When $n_e = n_i = n_0$ we can use Eqn. 5 for the pressure description in the discharge channel because the influence of plasma imperfect on magnitude of pressure is less than on the magnitude of conductivity (in comparison with Spitzer formula). It can be supposed that expansion of arc channel is negligible (size of spot of arc binding on the cathode surface does not exceed 4 mm) and the pressure inside the discharge channel is equal to sum of gas static pressure and magnetic pressure.

Calculation for some experiment conditions was performed by the following way. The system of equations 3 and 5 was solved relatively T_0 and r_0 at the several fixed values of n_0. Sample of calculation in Tab.1.

TABLE 1.

n_0, cm^{-3}	r_0, cm	$T_0 \times 10^4$, K	σ_{exp}, Eqn. 6 ($\Omega \cdot$cm)$^{-1}$	σ_{calc}, [7] ($\Omega \cdot$cm)$^{-1}$
10^{20}	0.145	13	1620	2200
1.1×10^{20}	0.13	12.4	1880	2000
1.4×10^{20}	0.1	11	3184	1150

After that, the magnitude of conductivity calculated on the base of experimental data σ_{exp} according to Eqn. 4 (column 4 of Tab. 1) was compared with the calculated magnitude of conductivity σ_{calc} [7] (column 5 of Tab. 1). Imperfect of plasma caused by the high particle density is taken into account at the calculation of value σ_{calc}. It is easy to see that coincidence of these magnitudes corresponds to the following arc parameters: $r_0 \approx 0.13$ cm, $T_0 \approx 124000$ K and $n_0 \approx 1.1 \times 10^{20}$ cm^{-3}. The same estimations were conducted for the several performed experiments (when the interelectrode gap and initial particle density were varying). Results of these calculations show that in any case the diameter of arc channel does not exceed of 4 mm and temperature of arc channel is ranged from 10 to 15 eV.

Results of these calculations show that the diameter of arc channel reduces as particle density increasing. The main peculiarity of the discharge at the considered conditions is higher temperature of arc channel and its smaller diameter in comparison with the arc at the lower initial particle density [4, 5].

In performed estimations we assumed that power input into the arc is equal to the output power. However, if this condition does not executed due to the locking radiation caused by the long time of photons diffusion in the transition layer, then the channel temperature can be significantly higher in comparison with performed estimations. It is known [8], that such occurrence is quite possible at the high initial particle density.

CONCLUSION

The installation, which was developed and manufactured at the IPE RAS is the reliable scientific tool for the exploration of powerful electric discharge in superdense gas.

The results of present experiments and calculations allow making the next suggestions about the high-current discharge at the high initial particle density:

arc channel diameter is small and in performed experiments did not exceeded 3-4 mm;

the temperature of arc channel increases as initial particle density increases and is achieving of 15 eV;

discharge current is concentrated in the central hot zone of arc unlike experiments at the lower initial particle density (channel model of the discharge).

These features give a good prospect for further researches. Increasing in initial particle density and discharge current causes reducing of the arc diameter. It is quite possible to achieve by this way the overheating of the arc channel due to the locking of radiation. We plan to carry out the series of experiment at the discharge current up to 0.5 MA and at the initial particle density up to 4×10^{22} cm^{-3}.

REFERENCES

[1] A.F. Savvateev, A.V. Budin, V.A. Kolikov, Ph.G. Rutberg, "Features of electric discharge in gas of high density," Proceedings of Conference "Pulsed Power Plasma Science – 2001," Las Vegas, Nevada, USA, June 17-22, 2001.

[2] Ph.G. Rutberg, A.A. Bogomaz, V.A. Kolikov, "Powerful pulse generator of dense plasma with high concentration of metal vapor," in Proceeding of Hypervelocity Impact Symposium, Santa Fe, MM, USA, October 17-19, 1994.

[3] V.P.Ignatko, G.M.Chernyavsky, Materials of the 1st Ull Union Workshop on Dynamics of Strong Current Arc Discharge in Magnetic Field, 1990, p.88-110.

[4] Ph.G. Rutberg, A.A. Bogomaz, A.V. Budin, V.A. Kolikov, A.G. Kuprin, A.A. Pozubenkov, "Estimation of some parameters of the discharge chamber of powerful electric discharge launchers," 27th AIAA Plasmadynamics and Laser Conference, AIAA-96-2328, New Orlean, L.A., USA, June 18-20, 1996.

[5] Ph.G. Rutberg, A.A. Bogomaz, A.V. Budin, V.A. Kolikov, A.G. Kuprin, A.A. Pozubenkov, "Experimental study of hydrogen heating in the discharge chamber of powerful electric discharge launcher," Journal of Propulsions and Power, vol.13, N5, 1996

[6] Y.B. Zeldovich, Y.P. Ryzer, "Physics of shock waves and high-temperature hydrodynamic phenomenon," "Science", Moscow, 1966.

[7] O.V. Korishev, D.O. Nogotkov, Y.Y. Protasov, V.D. Teleh, "Thermodynamic, optic and transfer properties of working substances of plasma and photon energy installations," MGTU named N.E. Bauman, Moscow, 1999.

[8] E.S. Borovik, V.I. Petrenko, R.V. Mitin, V.P. Kancedal and Y.R. Knyazev, *"Exploration for the pulse arc in argon and helium at the superhigh pressure,"* (in Russian), Journal of Technical Physics, vol. 39, pp. 1416-1424, 1969.

High-Density Hydrogen Plasma Production in a Simple Torus using Helicon Waves

Youichi Sakawa* and Tatsuo Shoji*

*Department of Energy Engineering and Science, Nagoya University, Nagoya 464-8603, Japan.

Abstract. High-density hydrogen plasma production in a simple torus using helicon waves has been investigated. The measured plasma density versus external magnetic field peaks at a condition close to the lower-hybrid resonance. The measured dispersion relation of helicon waves in the high-density region shows deviation from that of the $m = +1$ mode derived using the uniform plasma assumption and agrees relatively well with that using the parabolic density profile. The difference in the dispersion relation of the $m = +1$ mode between the uniform and non-uniform plasma models is explained by adding the vacuum region in the calculation of the boundary condition.

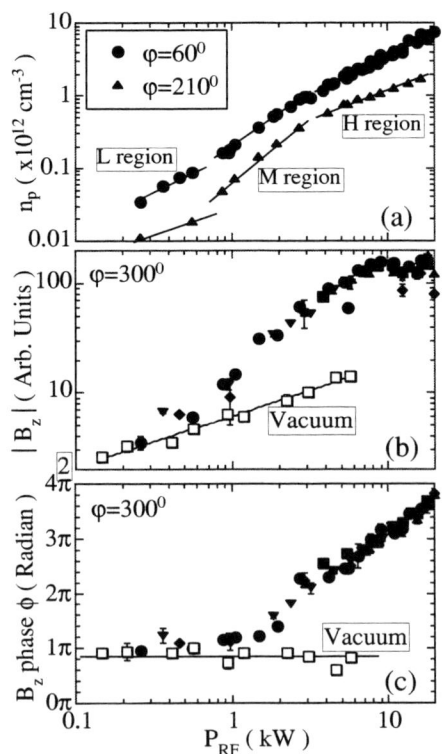

Plasma production based on inductively coupled rf excitation of a fast wave under a static magnetic field B_0 in the angular frequency ω range of $\omega_{ci} \ll \omega \ll \omega_{ce}$ (ω_{ci} and ω_{ci} are the ion and the electron cyclotron frequencies, respectively) is known as a helicon-wave discharge [1, 2, 3].

Because of its ability to produce high-density ($n_p \geq 10^{13}$ cm^{-3}) plasmas at relatively low gas pressure of a few mTorr, the helicon-wave discharge is getting a broad interest in applications [4]. On the other hand, torus plasma production using rf is important in applications such as pre-ionization and discharge-cleaning in Tokamak, and plasma production in helical devices. Helicon waves were applied to toroidal heliac [5]. Whereas helicon waves are also tried to launch in simple magnetized torus systems [6], because of the limitation in the rf power P_{rf}, no clear evidence for the helicon-wave discharge is observed. Investigation of high-density H$_2$ plasma production is especially important in toroidal helicon-discharge because of its relevance to fusion research.

In this paper, we document the first experimental investigation of the high-density ($n_p \simeq 8 \times 10^{12}$ cm^{-3}) hydrogen plasma production in a simple torus by the $m = +1$ azimuthal mode of helicon waves. The dispersion relation of the $m = +1$ mode calculated for the parabolic density profile shows deviation from $k_z \propto n_p$ dependence (k_z is the wave number parallel to B_0, which is in the toroidal or axial direction), and agrees relatively well with the measured one. It is observed in many experiments that the measured dispersion relation for the $m = +1$ mode satisfies $k_z \propto n_p^{0.5}$ dependence and behave as if $k_\perp = 0$ ($k_\perp$ is the wavenumber perpendicular to B_0) even though the predicted $k_\perp$ from the first zero of the first-order Bessel function, i.e., $k_\perp = 3.83/a$ (a is the radius of the plasma), is larger than k_z [7]. However, no clear explanation is given so far. The numerical results shown here indicate that if n_p is non-uniform, $k_z \propto n_p$ dependence is not satisfied for the $m = +1$ mode. The reason for the deviation from the $k_z \propto n_p$ dependence is explained by adding the vacuum region in the calculation of the boundary condition.

FIGURE 1. P_{rf} dependence of (a) n_p measured at $\varphi = 60$ and 210 degrees, (b) amplitude of B_z ($|B_z|$), and (c) phase difference between I_{rf} and B_z (ϕ) measured at $\varphi = 300$ degrees (closed symbols) at $B_0 = 120$ G. Different closed symbols in (b) and (c) represent different series of the measurements. $|B_z|$ and ϕ in vacuum (open squares) are also shown in (b) and (c), respectively.

A pulse modulated rf (power $P_{rf} \leq 20$ kW, frequency $f = 9$ MHz, pulse length = 5 ms, and repetition time = 2 s) is applied to an antenna surrounding a Pyrex discharge tube (major radius R = 25 cm and minor radius a = 4 cm) through a matching circuit. The antenna used in this experiment is Nagoya type III antenna [8] with the antenna length $L_A = 16$ cm. A static magnetic field of $B_0 \leq 950$ G at radially center ($r = 0$) of the discharge tube is generated via 12 magnetic coils. H_2 gas is used at the pressure P = 3.7 mTorr (the base pressure is $P \simeq 10^{-4}$ Torr). The zero-to-peak rf current applied to the antenna (I_{rf}) is measured by a small one-turn loop located at the matching box. The radial profile of n_p is measured by double probes located at toroidal position of $\varphi = 60$ and 210 degrees. Here, $\varphi = 0$ is defined as the position of the antenna center and increases in the clock-wise direction. The radial and axial profiles of the oscillating magnetic field are measured by one-turn magnetic probes located at $\varphi = 300$ degrees and just outside the discharge tube [9] (located at $R_{probe} = 29.5$ cm from the axis of the torus), respectively. We define the distance from the antenna in toroidal direction measured at $R_{probe} = 29.5$ cm as z. The radially movable magnetic probe is calibrated using Helmholtz coil.

The measured n_p peaked at an optimum value of B_0 (B_{0-MAX}) = 200 G, which is close to the calculated B_{LH-HD} = 140 G that satisfies the high-density limit of the lower-hybrid resonance condition, $\omega = \omega_{LH-HD} = \sqrt{\omega_{ci}\omega_{ce}}$ [10].

Figure 1(a) shows P_{rf} dependence of n_p. It is clear that the discharge is separated into three regions. We define the regions from the lower P_{rf} as the low-density (L), the medium-density (M), and the high-density (H) regions, respectively. The slope of the variation of n_p is the largest in the M region, and n_p tends to saturate in the H region.

In our previous experiments with a linear machine in At, in which half-turn helical and one-turn loop antennas are used [10, 11], we observed two discharge modes; the LD mode, in which the discharge is sustained by antenna induction field, and the HD mode, in which plasmas are produced by helicon waves. As will be shown later, the L region is equivalent to the LD mode, and the M and H regions correspond to the HD mode.

Amplitude of the axial (toroidal) component of the oscillating magnetic field B_z ($|B_z|$) and phase difference between I_{rf} and B_z (ϕ) are shown in Figs. 1(b) and 1(c), respectively. In the L region, ϕ is constant at $\phi \simeq \pi$. Whereas in the M region, ϕ starts to increase with P_{rf}. Note that when the transition from the L to M regions occurs, not only the slope of n_p but also that of $|B_z|$ increases. $|B_z|$ and ϕ in vacuum (open circles) are also shown in Figs. 1(b) and 1(c), respectively. The slope of $|B_z|$ and ϕ in vacuum ($\phi \simeq \pi$) is nearly the same as those in the L region.

When helicon waves that are traveling from the antenna are excited, ϕ represents $k_z z$, where z is the distance from the antenna to the magnetic probe. Therefore, ϕ increases with P_{rf} because k_z increases by increasing P_{rf} or n_p. $|B_z|$ increases when helicon waves are excited because the axial damping of helicon wave is smaller than that of the antenna field [12]. Therefore, it is postulated that whereas no waves are excited in the L region, helicon wave are excited in the M and H regions. In the H region, not only n_p but also $|B_z|$ tend to saturate. P_{rf} dependence of $|B_z|$, which is normalized by I_{rf} ($|B_z|/I_{rf}$) and ϕ, versus the distance from the antenna in the toroidal direction (z) measured just outside the discharge tube are shown in Figs. 2(a) and 2(b), respectively. Since the direction of the azimuthal antenna current at the end of the antenna is 180 degrees out of phase from one to the other, $\phi \simeq 0$ and π at the two ends of the antenna. ϕ is nearly constant in the L region (open squares), and the axial variation is quite similar to that in vacuum (closed squares). Therefore, it is confirmed that in the L region helicon wave is not excited. In the M and H regions, $|B_z|$ decreases and ϕ increases with the distance from the both end of the antenna. Helicon waves are excited in the both sides of the antenna and traveling from the antenna to both the positive and negative z directions.

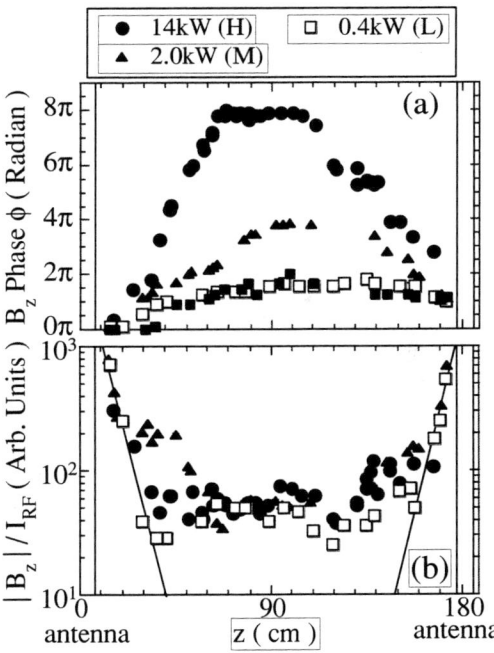

FIGURE 2. P_{rf} dependence of (a) ϕ and (b) $|B_z|/I_{rf}$ versus distance from the antenna in the toroidal direction (z) measured just outside the discharge tube ($z = 0 - 185.4$ cm) at $B_0 = 120$ G. ϕ (closed squares) and $|B_z|/I_{rf}$ (thick solid lines) in vacuum are also shown in (a) and (b), respectively. Vertical lines represent the location of the two ends of Nagoya type III antenna, $z = 8$ and 177.4 cm.

In the H region, n_p peaks at $r \simeq -1.0$ cm both at $\varphi = 60$ and 210 degrees. In the M regions, n_p profile at $\varphi = 60$ degrees is broader than that in the H region, and the peak position of n_p at $\varphi = 210$ degrees shifts to $r \simeq -2.5$ cm. In the L region, n_p profiles were broader than those in the M region, and n_p peaked at $r \simeq 0.5$ cm and $\simeq -2.5$ cm at $\varphi = 60$ and 210 degrees, respectively. $|B_r|$ peaks at plasma center ($r \simeq -1.0$ cm) both for $P_{rf} = 6$ and 13 kW, which are in the H region. Whereas $|B_r|$ monotonically decreases from $r \simeq -1.0$ to 4.0 cm for $P_{rf} = 6$ kW, it shows minimum value at $r \simeq 1.5$ cm and a second peak at $r \simeq 3.0$ cm for $P_{rf} = 13$ kW. We have confirmed that phase shift of $\simeq 180$ degrees occurs at $r \simeq 1.5$ cm for $P_{rf} = 13$ kW. Therefore, the first-radial (second-radial) mode of the $m = +1$ helicon wave is excited at $P_{rf} = 6$ kW (13 kW).

The dispersion relation, k_z versus n_p, reduced from the measured P_{rf} variation of ϕ and n_p is shown in Fig. 3(a). k_z is derived from $k_z = (\phi - \phi_0)/L$, where L (= 14 cm) is the distance from the antenna center ($\varphi = 0$) to the probe ($\varphi = 300$ degrees) and ϕ_0 (= 1.1π Radian) is the phase when no helicon waves are excited, i.e., ϕ in the L region. In the M region, k_z is proportional to n_p. While in the H region, k_z fits relatively well to $n_p^{0.4}$ dependence.

In the angular frequency ω range of $\omega_{ci} \ll \omega \ll \omega_{ce}$, the dispersion relation of the helicon wave in a uniform plasma is approximated as [2, 3] $k_z\sqrt{k_\perp^2 + k_z^2} \simeq \omega\omega_{pe}^2/(\omega_{ce}c^2)$, and the boundary condition is expressed as $m\sqrt{k_\perp^2 + k_z^2}J_m(k_\perp a) + k_z a J_m'(k_\perp a) = 0$, where J_m is the Bessel function of the first kind and $\prime = \partial/\partial r$. The azimuthal mode dependence of k_z versus n_p calculated from the uniform-plasma dispersion relation is shown in Fig. 3(a). Disagreement between the measured and the calculated dispersion relations are clearly seen.

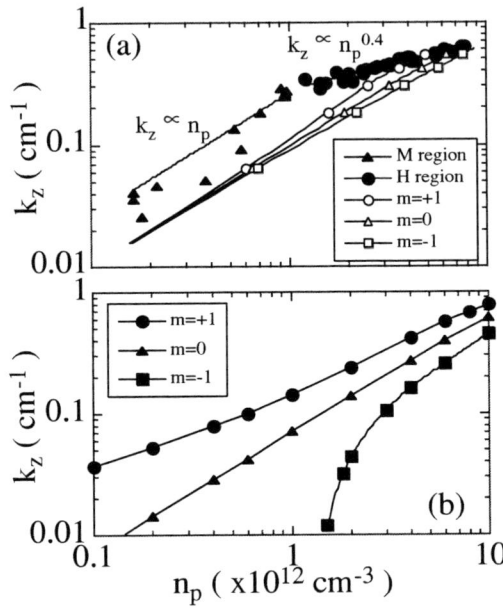

FIGURE 3. (a) The measured dispersion relation k_z versus n_p for the M (closed triangles) and H (closed circles) regions at $B_0 = 120$ G: Azimuthal mode dependence of the calculated k_z versus n_p for the first-radial mode for (a) a uniform and (b) parabolic density profiles.

The calculated threshold k_z (k_z^{th}) just before the transition from M to H regions is close to that determined by the antenna length $L_A = 16$ cm, i.e., $k_z = \pi/L_A = 0.20$ cm^{-1}. Nagoya type III antenna contains two half-turn loops at the end of the antenna, and rf current flows in the opposite direction. Therefore, the best antenna-coupling occurs for $k_z = \pi/L_A$ that corresponds to the axial wavelength $= 2L_A$.

In order to derive the dispersion relation in non-uniform n_p profiles, we have solved the wave equation for B_z

$$B_z'' + (\frac{1}{r} - \frac{2\alpha\alpha'}{\alpha^2 - k_z^2})B_z' + [\alpha^2 - k_z^2 - (\frac{m}{r})^2 - \frac{m\alpha'(\alpha^2 + k_z^2)}{rk_z(\alpha^2 - k_z^2)}]B_z = 0, \qquad (1)$$

satisfying the boundary condition $B_r(r = a) = 0$, using shooting method [13, 14]. Here, $\alpha = \alpha(r) = \omega\omega_{pe}^2(r)/(k_z\omega_{ce}c^2)$, $a = 4$ cm is the radius of the chamber wall, and $B_r = i(\frac{\alpha}{\alpha^2 - k_z^2}\frac{m}{r}B_z + \frac{k_z}{\alpha^2 - k_z^2}B_z')$. The dispersion relation k_z versus n_p is obtained by seeking k_z that satisfies both the wave equation and $B_r(a) = 0$. We have confirmed that when the uniform n_p is used, the dispersion relation obtained using shooting method and the uniform-plasma dispersion relation agree each other quite well. Figure 3(b) shows the azimuthal mode dependence of k_z versus n_p using parabolic $n_p(r) = n_p[1 - (\frac{r}{a})^2]$ density profile. For the $m = 0$ mode, k_z of parabolic profile is by a factor of 1.4 smaller than that of the uniform case, and $k_z \propto n_p$ for both parabolic and uniform profiles is obtained. For the $m = -1$ mode, k_z shows stronger n_p dependence, especially at the lower n_p, compared with the $m = 0$ mode. This is because the $m = -1$ mode represents cutoff ($k_z = 0$) below a critical density. This cutoff occurs only for the negative values of m mode when non-uniform n_p profiles are considered [14]. For the $m = +1$ mode, on the other hand, k_z shows weaker dependence on n_p.

In order to clarify the reason for the deviation of k_z from $k_z \propto n_p$ dependence for the $m = +1$ mode using parabolic profile, we have considered the the vacuum region between the plasma and the wall. As described by Shoji et al. [3], we consider a case where a cold uniform plasma of radius a is placed in a conducting cylinder of radius b. As a result, we obtain the dispersion relation for uniform plasma with the boundary condition

$$k_\perp a \frac{I'_m(k_\perp a)K'_m(k_\perp b) - K'_m(k_\perp a)I'_m(k_\perp b)}{I_m(k_\perp a)K'_m(k_\perp b) - K_m(k_\perp a)I'_m(k_\perp b)} + m\sqrt{k_\perp^2 + k_z^2} + k_z a \frac{J'_m(k_\perp a)}{J_m(k_\perp a)} = 0, \qquad (2)$$

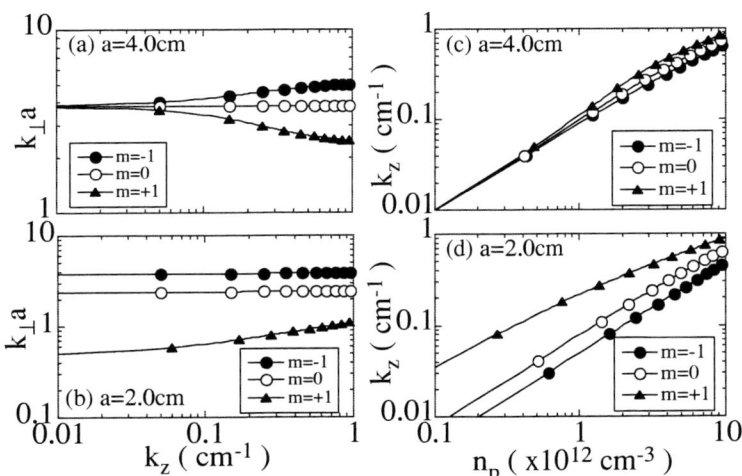

where, I_m and K_m are the modified Bessel functions of the first and second kind, respectively. When $a = b$, the boundary condition Eq. (2) and that for the uniform plasmas are equivalent. The first and the third terms in the left-hand side of Eq. (2) are identical for the $m = -1$ and $+1$ modes. Figures 4(a) and 4(c) show the azimuthal mode dependence of $k_\perp a$ versus k_z calculated using Eq. (2) by varying a. Here, b is fixed at 4.0 cm. When $k_z = 1.0$ cm^{-1}, by decreasing a from 4.0 to 2.0 cm, $k_\perp a$ decreases from 5.0 to 3.83 (3.8 to 2.48) for the $m = -1$ ($m = 0$) mode. The values of 3.83 and 2.4 are the first zero of J_1 and J_0, respectively. For the $m = +1$ mode, on the other hand, $k_\perp a$ decreases below 2.4. This small $k_\perp a$ solution appears only for the $m = +1$ mode when vacuum region is considered. Figures 4(c) and 4(d) show the azimuthal mode dependence of k_z versus n_p calculated from the uniform plasma dispersion relation using the results shown in Figs. 4(a) and 4(b). We see that $k_z \propto n_p$ dependence is nearly satisfied for all the a values both for the $m = -1$ and 0 modes. Note that no cutoff appears in the $m = -1$ mode since n_p in the plasma region is uniform. However, for the $m = +1$ mode, deviation from $k_z \propto n_p$ dependence occurs at small values of a, and k_z shows more like $k_z \propto n_p^{0.5}$ dependence. The dispersion relation obtained for the $m = +1$ mode with $a = 2.0$ cm is quite similar to the measured one.

FIGURE 4. Azimuthal mode dependence of (a), (b) $k_\perp a$ versus k_z and (c), (d) k_z versus n_p, calculated from Eq. (2) for the first-radial mode by varying a. b is fixed at 4.0 cm.

REFERENCES

1. R. W. Boswell, Plasma Phys. Controlled Fusion **26**, 1147 (1984).
2. F. F. Chen, Plasma Phys. Controlled Fusion **33**, 339 (1991).
3. T. Shoji et al., Plasma Sources Sci. Technol. **2**, 5 (1993).
4. J. A. Perry, D. Vender, and R. W. Boswell, J. Vac. Sci. Technol. B **9**, 310 (1991); S. Miyake et al., Surface and Coatings Technology. **116-119**, 11 (1999).
5. P. K. Loewenhardt et al., Phys. Rev. Lett. **67**, 2792 (1991).
6. A. Mukherjee and D. Bora, in *Proceedings of the 1994 International Conference on Plasma Physics*, Brazil 1994, p. 49; O. Grulke et al., Plasma Phys. Controlled Fusion **43**, 525 (2001).
7. A. W. Degeling et al., Phys. Plasmas **3**, 2788 (1996); D. A. Schneider, G. G. Borg, and I. V. Kamenski, Phys. Plasmas **6**, 703 (1999); Y. Yasaka and Y. Hara, Jpn. J. Appl. Phys. **33**, Part 1, 5950 (1994).
8. T. Watari et al., Phys. Fluids 21, 2076 (1978).
9. R. T. S. Chen et al., Plasma Sources Sci. Technol., **4**, 337 (1995).
10. Y. Sakawa, T. Takino, and T. Shoji, Phys. Plasmas **6**, 4759 (1999).
11. Y. Sakawa, N. Koshikawa, and T. Shoji, Appl. Phys. Lett. **69**, 1695 (1996).
12. Y. Sakawa, N. Koshikawa, and T. Shoji, Plasma Sources Sci. Technol. **6**, 96 (1997).
13. F. F. Chen, M. J. Hsieh, and M. Light, Plasma Sources Sci. Technol. **3**, 49 (1994).
14. M. Krämer, Phys. Plasmas **6**, 1052 (1999).

Generation of High-Density, Uniform Plasmas by Low-Frequency RF Currents: Key Concepts, Experiments and Applications

S. Xu[1], Z. L. Tsakadze[1], E. L. Tsakadze[1,2], and K. Ostrikov[1,3]

[1] *Plasma Sources and Applications Center, NIE, Nanyang Technological University, 637616 Singapore*
[2] *Optics and Fluid Dynamics Department, RISOE National Laboratory, P.O. Box 49, DK-4000 Roskilde, Denmark*
[3] *SOCPES, Flinders University, GPO Box 2100, Adelaide 5001, Australia*

Abstract. The results on generation and technological applications of high-density and highly uniform low-frequency (~460-500 kHz) inductively coupled plasmas (LF ICP) are reviewed. The ICP source is capable of sustaining stable discharges in electrostatic (E) or electromagnetic (H) regimes. In the H operation mode, plasma densities as high as 8×10^{12} cm^{-3} are achievable with moderate (~1 kW) Rf powers at low and intermediate pressures (10-100 mTorr). The ICP also features high degree of uniformity over large processing volumes and surfaces. It is shown that the plasma source has a great potential for a number of industrial applications. As an example, the efficiency of the LF ICP for synthesis of silicon nitride (Si_3N_4) on Si(111) substrate in low-pressure discharges of Ar+N$_2$ gas mixtures, is demonstrated. The main mechanism involved in the process is the extensive nitrogen ion implantation combined with plasma enhanced thermal diffusion. The crystalline structure, bonding states and chemical composition of the silicon nitride film are studied.

INTRODUCTION

There has been a continuously increasing recent interest of microelectronic, automotive, optical, and other industries in price-efficient, stable, easy-to-handle sources of low and intermediate pressure plasma sources satisfying a number of basic requirements, such as high density and uniformity of distribution of reactive species, high product yield, low product damage, stability, reproducibility, and several others. Sources of inductively coupled plasmas with external flat spiral (commonly referred as a "pancake") inductive coil do satisfy most of the above requirements and have recently been adopted as reference plasma reactors for a number of applications. Operation of ICP sources at lower (~500 kHz) frequency has proved instrumental in overcoming many undesirable features of conventional 13.56 MHz devices, such as strong electrostatic coupling through a dielectric window, excessive window sputtering, standing wave effects, limitations in device up-scaling, and some others [1-3]. In this work, we emphasize the key features of low-frequency inductively coupled plasma sources of Nanyang Technological University [2,3] and Flinders University [1], and report on recent progress in applications of LF ICPs made at NTU. To demonstrate the unique feature of the LF ICP source, low temperature synthesis of crystalline silicon nitride by means of extensive nitrogen ion implantation is carried out. Silicon nitride has wide applications in microelectronics device fabrication as passivation layers, masks, and insulators in metal-insulator-semiconductor devices [4,5]. In the synthesis of silicon nitride, most widely used processes employ mixtures of SiH_4 and NH_3 or N_2 as a source of nitrogen. In this work, by employing the LF ICP nitriding technique, the use of hazardous silane gases is completely eliminated.

EXPERIMENTAL DETAILS

Experimental measurements have been carried out in a low-frequency (~460-500 KHz) inductively coupled plasma source described in detail elsewhere [1-3]. The plasma is generated in a cylindrical, stainless steel vacuum chamber with inner radius of R= 16 cm and length L = 20 cm. A schematic diagram of the discharge chamber is given in Fig. 1. The chamber is chilled by a continuous water flow in between the inner and outer walls of the double-walled chamber. The top plate of the chamber is a quartz dielectric disk, 35 cm in diameter and 1.2 cm thick. A 450 l/s turbo-molecular pump backed by a two-stage rotary pump is used to evacuate the plasma chamber. The inflow rate and pressure of the working gas are regulated by a combination of a gate valve and MKS mass-flow controllers. The pressure is measured by an MKS Baratron capacitance manometer. The operating pressure of argon gas feedstock is typically maintained in the range $P = 10 - 100$ mTorr. The global plasma parameters, such as the electron/ion number densities, plasma potential and effective electron temperature have been estimated by means of a time-resolved RF-compensated single Langmuir probe technique. A number of holes in four rectangular side ports and in the bottom plate of the chamber allow one to insert the Langmuir probe into the plasma and move it in radial and axial directions.

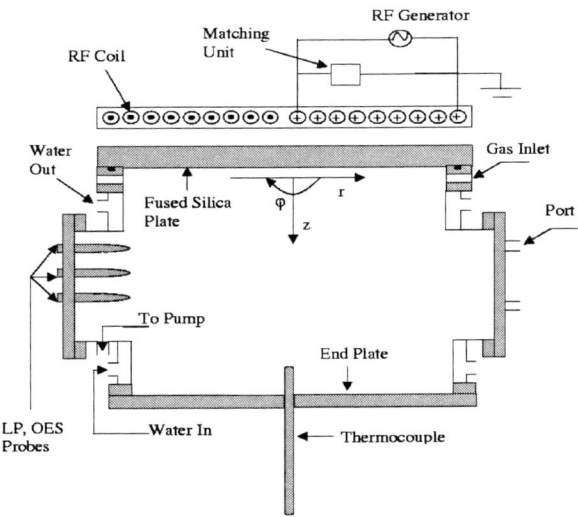

FIGURE 1. Schematic diagram of the discharge chamber

PLASMA PARAMETERS

The main plasma parameters have been investigated using the detailed Langmuir probe scans inside the chamber. In such a way, the spatial profiles of the electron number density n_e, electron temperature T_e, and plasma potential V_p have been measured for various RF power inputs and operating gas pressures. Radial scans were taken at 15 radial positions sweeping the probe along the radius at z=10 cm position (central cross-section), and in 7 axial positions using 7 available portholes in the side port. Fig.2 shows the radial and axial profiles of n_e, T_e, and V_p for the power deposited to the plasma of ~1.5 kW in a 22 mTorr argon discharge in the electromagnetic mode. The profiles suggest that a high-density, highly uniform over large areas and volumes, plasmas with quite low plasma potential and electron temperature are efficiently generated. It is worth noting that the plasma source simultaneously embodies the sources of capacitively (in E-mode) and inductively (in H-mode) coupled plasmas, which enhances the flexibility of the source in a number of materials processing and synthesis applications. Furthermore, in the H-mode, the source features low sheath potentials near the substrate, independent control of the plasma density and energy of impinging ions, remarkable power transfer efficiency, low skin-effect caused circuit loss, absence of any external magnetic field confinement, and some others. For further details of the source operation and plasma diagnostics, the reader can be referred to [3]. Below, we discuss a new application of the LF ICP for synthesis of silicon nitride, a material of major interest for applications in microelectronics device fabrication and as a gate dielectric in thin-film transistors for flat panel displays.

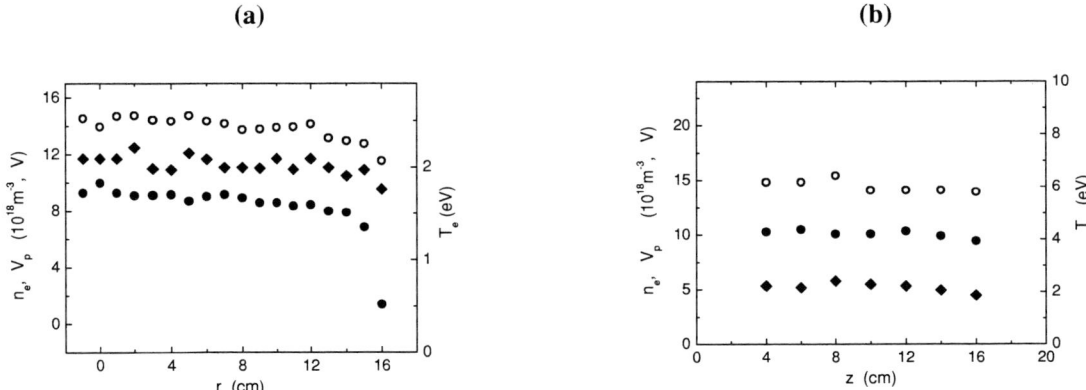

FIGURE 2. Radial (a) and axial profiles of the electron number density (solid circles), temperature (solid squares), and plasma potential (open circles) in a 22 mTorr, H-mode discharge in argon at Rf power input of ~1.5 kW.

APPLICATION OF LF ICP FOR SYNTHESIS OF SILICON NITRIDE

The silicon nitride layer is formed on lightly doped n-type single crystal Si (111) wafers. The substrates are cleaned by the RCA method. Based on intensive preliminary experimentation, the synthesis condition have been set as $N_2:H_2:Ar=60:40:15$ sccm, total working gas pressure: 60 mTorr, substarte temperature: 450 deg C. The DC substrate bias and processing time were varied. Hydrogen promotes bonding in the silicon nitride films and it also plays a key role in determining the film properties such as defect density, film structure and electrical/optical properties. The films were characterized with various advanced analytical tools, including x-ray photoelectron spectroscopy (XPS), infrared absorption spectroscopy (FTIR), and glancing angle x-ray diffractometer (XRD).

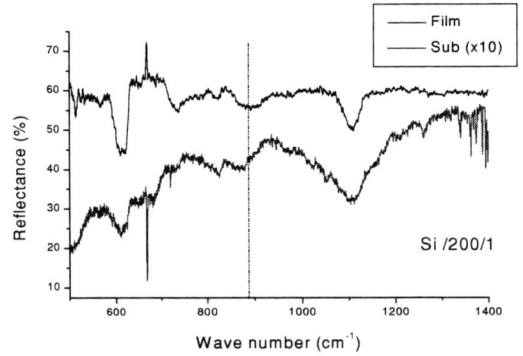

FIGURE 3. FTIR spectra of the silicon nitride film showing formation of Si-N bond.

The chemical states of the film have been assessed using a PERKIN ELMER FT-IR 1725X spectrometer. Recent works indicated that the Si-N bands yield a peak around 880 cm^{-1}. Fig. 3 shows the absorption spectra of a processed sample together with a silicon substrate. The sample has been processed for 2 hours at –300 volts bias. For the processed sample, an absorption band in the range of 850 - 890 cm^{-1} is clearly seen, which verifies the formation of the Si-N stretching vibration mode. This band is not found in unprocessed silicon substrates.

The chemical composition and binding states of the deposited thin films have been studied by VG ESCALAB 220i-XL spectrometer (XPS) and Mg Kα (1253.6eV) x-ray source. To remove the adventitious contaminants, the surface of the sample has been sputter-cleaned using 2 keV argon ions for 2 minutes prior to the XPS analysis. A survey scan shows that the film composes primarily Si and N with a small amount of O. Fig. 4 shows the narrow scan XPS spectra of Si 2p (a) and N 1s (b) states of an Ar-ion-sputter-cleaned sample. The two peaks displayed in Fig.4a confirm the presence of chemical states of Si with BEs of 101.5 eV and 103.4 eV, respectively. The peak with lower binding energy corresponds to the nitride state This peak can thus be related to Si-N bonds. The chemical state at 103.2 eV is due to characteristic Si-O bonds. Similarly, the single peak in N 1s having a binding energy 399.2 eV can be attributed to sp^{3}Si-N bonding.

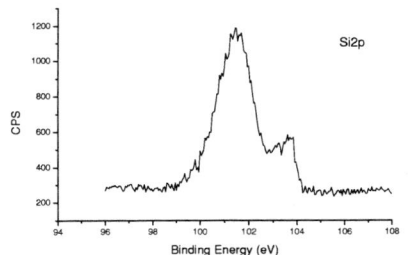 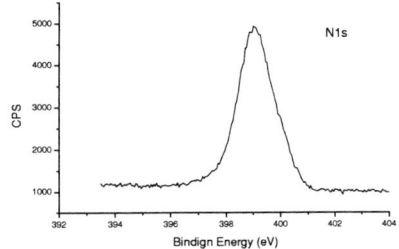

FIGURE 4. XPS narrow scan spectra of (a) Si 2p and (b) N1s.

The crystalline structure of the composite has been analyzed with SIEMENS D5005 X-Ray diffractometer (XRD) in the glancing angle scanning mode with an incident x-ray wavelength of 1.540Å (Cu Kα line). Fig.5 shows the XRD patterns of the processed sample. It is clear that hexagonal β-Si_3N_4 crystallite are present in the nitrided layer. The lines corresponding to Si_3N_4 (201), (103) and (321) are labeled in the spectrum. We note that we have not recorded any lines corresponding to the substrate silicon (111).

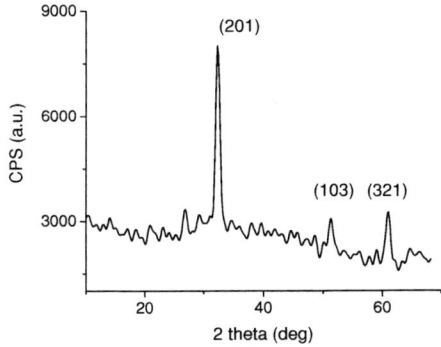

FIG. 5 Glancing angle (0.5 deg) XRD spectrum showing the crystalline structure of the silicon nitride film.

CONCLUSION

Generation and technological applications of high-density and highly uniform low-frequency (~460-500 kHz) inductively coupled plasmas (LF ICP) have been discussed. The ICP source has shown to be capable of sustaining stable discharges in both E and H regimes. In the high density, H mode discharge regime, crystalline silicon nitride films have been successfully grown on single crystal Si (111) substrates, without using any silicon containing precursors like silane. Structural and bonding state investigation confirm that the films are in the form of β-Si_3N_4.

ACKNOWLEDGMENTS

This work was supported in part by the A*STAR, Singapore (Project No. 012 101 00247).

REFERENCES

[1] I. M. El-Fayoumi and I. R. Jones, *Plasma Sources Sci. Technol.* **7**, 162 (1998).
[2] S. Xu, K. N. Ostrikov, W. Luo, and S. Lee, *Journ. Vac.Sci. Technol.* **A 18**, 2185 (2000).
[3] S. Xu, K. N. Ostrikov, Y. Li, E. L. Tsakadze, and I. R. Jones, *Phys. Plasmas* **8**, 2549 (2001).
[4] S. S. He, M. J. Williams, D. J. Stephens, and G. Lucovsky, *J. Non-Cryst. Solids* **164–166**, 731 (1993).
[5] G. Lucovsky and J. C. Phillips, *J. Non-Cryst. Solids* **227**, 1221 (1998).

Wave Propagation and Plasma Production in Planar Microwave Discharges Using a Slot Antenna of Rotating Mode

Yasuyoshi Yasaka, Nobuo Ishii[*], Tetsuya Yamamoto[+],
Makoto Ando[**], and Masaharu Takahashi[++]

Department of Electronic Science and Engineering, Kyoto University, Kyoto 606-8501, Japan
[*]*Kansai Technology Development Center, Tokyo Electron Ltd., Amagasaki 660-0891, Japan*
[+]*Department of Electrical and Electronic Eng., Yamagata University, Yamagata 992-8501, Japan*
[**]*Department of Electrical and Electronic Eng., Tokyo Inst. Technol., Tokyo152-8552, Japan*
[++]*Department of Electrical and Electronic Eng., Tokyo Univ. A&T., Koganei 184-8588, Japan*

Abstract. Planar microwave discharges using a multi-slotted planar antenna are investigated. A three-dimensional finite difference time domain code is developed for the calculation of wave propagation in nonuniform plasmas. The enhancement of the microwave fields near the plasma resonance is revealed in accordance with the theory of the resonant absorption. The global wave electric field is obtained for various excitation modes of the antenna. By operating the antenna with transverse electric mode, where the field is rotating in the azimuthal direction, and by minimizing the reflected power from the antenna, highly uniform overdense plasmas can be produced in a wide range of gas pressures and microwave powers.

INTRODUCTION

Planar microwave discharges are widely investigated as a large diameter plasma source for material processing. Microwave fields are radiated from a variety of slot antennas located on a vacuum window made of dielectric materials, and are propagated and absorbed in a plasma of a small aspect ratio. In so-called surface-wave excited plasmas, where the wave is propagating between the slot plate of a few apertures and the plasma surface forming standing wave eigenmodes in the area bounded by radial walls, the nature of discrete series of eigenmodes gives rise to abrupt changes in plasma density when changing the incident power.

On the other hand, if the microwave field is radiated uniformly from the antenna of multiple slots and is strongly damped in the plasma within a short distance, the plasma is free from eigemodes allowing uniform and stable plasma production. As for the strong damping, we can employ resonant absorption,[1] in which the field incident obliquely to the direction of the plasma density gradient self-excites plasma oscillations at the plasma resonance and is strongly damped. We have presented[2] that, by using this scheme, uniform overdense plasmas of 30 cm in diameter can be produced in milliTorr ranges using microwave discharges of 1--2.5 kW at a frequency of 2.45 GHz. We used a transverse electromagnetic (TEM) excitation or transverse electric (TE) excitation with azimuthally rotating or non-rotating fields for the multi-slotted antenna, and observed differences in radial density profiles.

In this paper, we develop a three-dimensional (3-D) finite difference time domain (FDTD) code that calculates wave propagation in nonuniform plasmas. The field excitation used in the experiment is modeled by a cylindrical waveguide and a metal plate with concentric annular slots with a given transverse electric fields on the axis of the waveguide rotating or non-rotating in time. The enhancement of the microwave fields near the plasma resonance is revealed in accordance with the theory of the resonant absorption. The global wave electric field is obtained for various excitation modes.

In the experiment, it is observed that the light emission from Argon plasmas is very uniform in radial and azimuthal directions indicating that there is little eigenmode formation. We can produce uniform overdense plasmas over 30 cm in diameter in a wide range of gas pressures and microwave powers. The wave field in the plasma is measured by a movable probe antenna and found to propagate in the azimuthal direction. A fraction of standing wave component is observed depending on the matching conditions of the antenna and the plasma. By improving the matching, small local peaks in the light emission intensities are eliminated yielding a formation of highly uniform plasmas.

CALCULATION OF THE GLOBAL WAVE FIELD

We perform the calculation of the wave field in nonuniform plasma located in the system shown in Fig. 1 using a standard FDTD scheme in 3-D inhomogeneity. The plasma chamber is 50 cm in diameter and has a quartz glass window of 30-mm thick on the top. The multi-slotted antenna is approximated by a conducting plate with several annular slits placed 10-mm above the glass window. The distance between the plate and the top wall is 17 mm. A short waveguide is attached on the top, and the source field of TEM mode or TE mode of azimuthally rotating or non-rotating at a frequency of $\omega = 2.45$ GHz is applied on the z-axis of the waveguide. It is assumed that the relative dielectric constant of the glass is 4, and the plasma density is uniform in the radial direction but nonuniform in axial direction with a given profile.

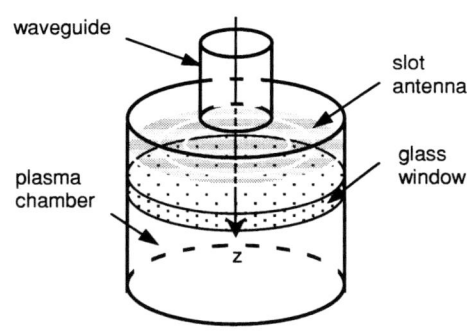

FIGURE 1. Model of calculations. The bottom surface of the glass window is at $z = 0$. The source field is given in the waveguide.

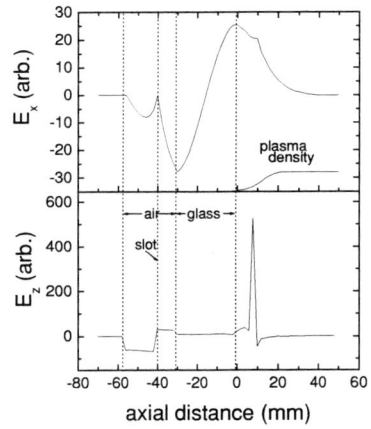

FIGURE 2. Axial distribution of E_x and E_z at the radius of 20 mm for TEM mode excitation.

Maxwell equations and equation of motion of electrons are solved iteratively in time domain in Cartesian coordinate with boundary conditions that the tangential components of the electric field E are zero on the wall. Figure 2 shows the axial distribution of the calculated field components E_x and E_z for the density of 7.0×10^{11} cm^{-3} with a profile given in the figure and TEM excitation. It is noted that E_z has a very large amplitude at $z \sim 8$ mm where ω equals the plasma frequency ω_p. The amplitude of E_x decays exponentially beyond the plasma resonance.

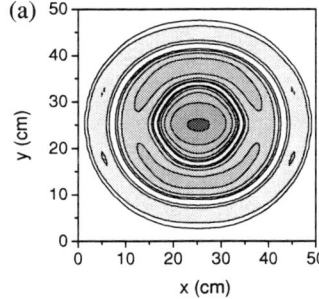 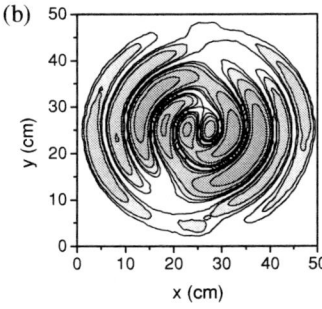 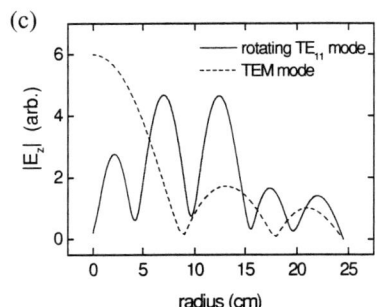

FIGURE 3. Distribution of E_z in a plane perpendicular to the z-axis for (a) TEM and (b) TE$_{11}$ mode excitation. (c) Radial distribution of $|E_z|$ for the TEM (dotted curve) and the rotating TE$_{11}$ (solid curve) mode excitations.

The resonant absorption is thus confirmed in the present 3-D calculation extended over the previous calculation.[3]

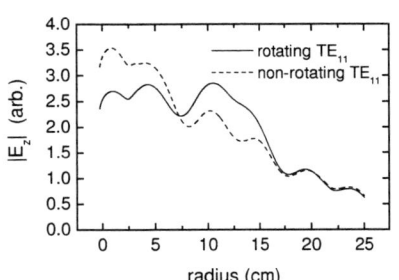

FIGURE 4. Radial distribution of $|E_z|$ for the rotating (solid curve) and non-rotating (dotted curve) TE_{11} mode excitations.

The field distribution on a plane perpendicular to the z-axis is calculated for various excitation modes for the same condition as in Fig. 2. Figures 3 (a) and (b) are the profile of $|E_z|$ at z = 8 mm for the TEM and the rotating TE_{11} mode, respectively. Both profiles exhibit radial eigenmodes determined by the boundary conditions in the slot antenna. The azimuthal mode number is 0 for (a) and 1 for (b). The profile of (b) shows a spiral pattern that is rotating in the azimuthal direction. The radial distributions of $|E_z|$ for the two excitation modes are compared in Fig. 3 (c), where the TEM excitation gives a profile peaked on axis, while the rotating TE_{11} excitation produces the off-axis peak.

We also compare the radial profiles of $|E_z|$ for the rotating and non-rotating TE_{11} modes as shown in Fig. 4, where the value of $|E_z|$ at each radius is time-averaged over the period of the microwave. It is found that $|E_z|$ for the non-rotating mode is more peaked than for the rotating mode. Furthermore, the non-rotating mode has nonuniform distribution in the azimuthal direction since the mode is standing in this direction.

EXPERIMENTAL OBSERVATION

The experiment is performed by using a device shown in Fig. 5, the size of which is the same as the model chamber in Fig. 1. We use a coaxial feeder for the TEM mode excitation, and a waveguide with a polarizer for the TE_{11} mode excitation. For the latter mode, a quadruple three stub tuner is inserted between the polarizer and the antenna to reduce the power reflection from the antenna, which would otherwise produce oppositely polarized components. Plasma parameters are measured by a fast-scanning Langmuir probe and the microwave fields by several types of small antennas connected to a spectrum analyzer or a double balanced mixer. Light emission from the plasma is monitored using a CCD camera.

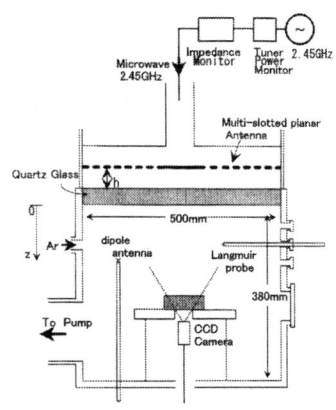

FIGURE 5. Experimental setup.

Resonant Enhancement of the Field

Typical plasma densities and electron temperatures are n_e = 4-7 $\times 10^{11} cm^{-3}$ and T_e = 1.5-2 eV at a location z = 55 mm from the bottom surface of the glass window for an incident microwave power P_μ = 1.5-2 kW with the TEM or TE_{11} excitation, Ar gas pressure p = 30 mTorr, and a flow rate of 60-80 sccm. The microwave field strength in the plasma near the glass window is measured by using the antenna movable in the z-direction. The amplitude has a sharp peak at around z = 10 mm and decays out for z > 30 mm. The position of the peak shifts with the bulk plasma density so that the peaking takes place at a fixed value of the density, $\omega_p = \omega$. This observation agrees well with the calculation result shown in Fig. 2.

Comparison between TEM and TE_{11} Mode Excitations

Figure 6 compares the radial profiles of the ion saturation current I_{is} for the TEM and TE_{11} excitation modes for P_μ = 2.5 kW and p = 30 mTorr. It is apparent that I_{is} for the TE_{11} excitation does not decrease for outer radii forming more uniform distribution than for the TEM excitation. The value of I_{is} for the TE_{11} excitation is, as a whole, larger than for the TEM excitation. This experimental result can be compared with the calculation given in Fig. 3. The value of $|E_z|$ for the TEM mode peaks on the axis and decreases steeply toward the outer radii, while for the TE_{11} mode, the field has maximum in the off-axis region. If we assume that the rate of plasma production is roughly proportional to E_z^2, we see a reasonable correspondence between the calculation in Fig. 3 (c) and the experimental profiles in Fig. 6, where the radial profile of I_{is} for the TEM mode has a hill shape, while that for the TE$_{11}$ mode has a hollow shape.

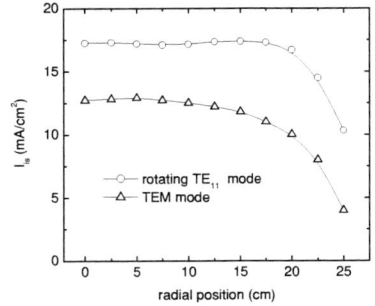

FIGURE 6. Radial distribution of I_{is} for the TEM (triangles) and the rotating TE$_{11}$ (circles) mode excitations.

FIGURE 7. Light emission from the plasma for the TE11 mode with (a) 10-20% reflected power and (b) < 5% reflection. The dotted circle indicates the radius of 15 cm.

In the rotating TE$_{11}$ mode excitation, it is sometimes observed that the light emission from the plasma has two bright spots near the center axis as shown in Fig. 7 (a). We measured the azimuthal phase difference of the microwave field in the plasma with respect to a reference signal of the power source with moving the dipole antenna in Fig. 5 along the azimuth. The plot of the phase difference versus the azimuthal angle exhibits a linear curve but with a sinusoidal modulation, indicating that the field is not fully traveling in the azimuthal direction. It is estimated that 10-20% of the incident power is reflected from the antenna and constitutes standing wave components in the slot antenna. In this case, the radial distribution of $|E_z|$, predicted by the dotted curve in Fig. 4, has a weakly peaked profile near the axis and may give rise to a power absorption that leads to form a plasma as shown in Fig. 7 (a). We then made a careful adjustment of the stub tuner so that the phase difference along the azimuth is a linear function of the angle within 5% modulation. It is expected that the oppositely rotating component of the microwave is almost eliminated and the resultant purely rotating field would produce the distribution of $|E_z|$ as shown by the solid curve in Fig. 4. We confirmed this by the uniform pattern of the light emission from the plasma shown in Fig. 7 (b).

When the stub tuner is in the best position, a slight change of the distance between the slot plate of the antenna and the glass window changes the fraction of the reflected power. By utilizing this effect, we can adjust the radial distribution of I_{is} between peaked to hollow profiles to some extent. This ability of tailoring the profile of uniform plasmas with large diameter is very useful for a plasma processing device that requires very uniform etching or deposition.

SUMMARY

A 3-D FDTD code is developed for the calculation of wave propagation in nonuniform plasmas. The enhancement of the microwave fields near the plasma resonance is revealed in accordance with the theory of the resonant absorption. This phenomenon is also observed in the experiment. Measured radial profile of the plasma depends on the excitation mode of the antenna, which is well explained by the calculated global wave electric fields in the system. The wave field in the plasma is found to be traveling in the azimuthal direction. A fraction of standing wave component is observed depending on the matching conditions. By improving the matching, small local peaks in the light emission intensities are eliminated yielding a formation of highly uniform plasmas.

REFERENCES

1. Ginzburg, V., *The Propagation of Electromagnetic Waves in Plasmas*, Pergamon Press., Oxford, 1964.
2. Yasaka, Y., Nozaki, D., Koga, K., Ando, M., Yamamoto, T., Goto, N., Ishii, N., and Morimoto, T., *Plasma Sources Sci. Technol.* **8**, 530 (1999).
3. Yasaka, Y. and Hojo, H., *Phys. Plasmas* **7** 1601 (2000).

Investigation on the Characteristics of Pellet Ablation in a Toroidal Plasma

K. N. Sato[1], H. Sakakita[2] and H. Fujita[3]

[1]*Research Institute for Applied Mechanics, Kyushu University, Kasuga, Fukuoka 816-8580, JAPAN*
[2]*Energy Electronics Institute, National Institute of A. I. S. T., Tsukuba, Ibaraki 305-8568 JAPAN*
[3]*Faculty of Science and Engineering, Saga University, Saga 840-8502, JAPAN*

Abstract. Characteristics of a cloud ablated from an ice pellet has been investigated in detail in the JIPP T-IIU tokamak plasma by utilizing a new scheme of pellet injection system, "the injection-angle controllable system". A long "helical tail" of ablation light has been observed using CCD cameras and a high speed framing photograph in the case of on-axis and off-axis injection with the injection angle smaller than a certain value. The direction of the helical tail is found to be independent to that of the total magnetic field lines of the torus. From the experiments with the combination of two toroildal filed directions and two plasma current directions, it is considered that the tail seems to rotate, in most cases, to the electron diamagnetic direction poloidally, and to the opposite to the plasma current direction toroidally. Consideration on various cross sections including charge exchange, ionization and elastic collisions leads us to the conclusion that the tail-shaped phenomena may come from the situation of charge exchange equilibrium of hydrogen ions and neutrals at extremely high density regime in the cloud. The relation of ablation behavior with plasma potential and rotation has also been studied. Potential measurements of pellet-injected plasmas using heavy ion beam probe (HIBP) method were carried out for the first time. In the case of an injection angle to be anti-parallel to the electron diamagnetic direction in the poloidal plane, the result shows that the direction of potential change is negative, and consequently the potential after the injection should be negative because it has been measured to be negative in usual ohmic plasmas without pellet injection. Thus, the direction of the "tail" structure seems to be consistent to that of the plasma potential measured, if it is considered that tail structure may be caused by the effect of the plasma potential and the rotation.

INTRODUCTION

In these years, ice pellet injection has become one of the most important issues in the toroidal plasma research not only for particle supplying, but also for profile control and confinement improvement [1]. This system has also been used diagnostic tools for particle transport, thermal transport [2] and safety factor studies. However, the other important aspect of ice pellet injection is as a useful tool for the study of fundamental properties of toroidal plasmas. Several interesting phenomena have been reported about pellet injected plasmas, for example, ablation and striation [3], snake oscillations [4], and rapid precooling of plasma core [2].

Moreover, in order to carry out various basic investigation of toroidal plasmas by pellet injection, the degree of freedom concerning the pellet injection is considered to be a very important and essential factor. As the example of the degree of freedom, pellet size, pellet injected position, high repetition rate and pellet speed are represented. A new technique of injection system, "the injection-angle controllable system", has been developed and installed to the JIPP T-IIU tokamak. By using this system, one can control the injection-angle easily and successfully during an interval of two plasma shots in the course of an experiment, not moving the whole injector. Thus, we can carry out and various basic experiments by changing the pellet deposition profile actively and drastically.

In the present paper, details of the pellet ablation properties with various injection angles have been studied by multi-dimensional observations with two CCD cameras and a high speed framing photograph, and an interesting phenomenon has been found concerning the flow characteristics of ablation cloud. In the case of an injection angle (θ) larger than a certain value ($\theta \geq 4°$), a pellet penetrates straightly through the plasma with a trace of straight ablation cloud, which has been expected from usual theoretical consideration. On the other hand, in the cases of on-axis and off-axis injections with the angle smaller than the certain value ($\theta \leq 4°$), a long helical shape ("tail") of ablation light has been observed. In the present study, the interaction mechanism among high density particles of ablation cloud in the complex electromagnetic field will be described.

EXPERIMENTAL SETUPS

JIPP T-II U is a small sized tokamak machine with plasma major (R) and minor (a) radii of 0.93 and 0.23 m [5]. The maximum plasma current (I_P) is 320 kA, toroidal magnetic field (B_T) 3 T, and the longest duration time 0.5 s. Hydrogen gas is used in these experimental series. Present experimental conditions are as follows; the plasma current is ~100 - 250 kA, toroidal magnetic field 2.5 - 3 T, the typical line average density (n_{e0}) ~1 - 4 × 10^{19} m^{-3}, and the typical core electron temperature (T_{e0}) ~1.0 - 1.2 keV. Arrangements of pellet injector and various diagnostics used in the experiments are shown in Fig. 1.

In the present experiment, an ice pellet injector of the helium gas gun type has been used. Ice pellet injection system is composed of a liquid helium supply line, a hydrogen gas supply line, helium gas supply line for pellet acceleration, an ice pellet making part, a pellet acceleration part, the differential pumping system and an electric controller system. Ice pellet size is 1.4 mm in length and 1.4 mm in diameter (or 1.0 mm in length and 1.0 mm in diameter). The number of hydrogen atoms is estimated to be about 1 - 4 × 10^{19} particles. Pellet velocity is almost 600 - 700 m/s in the present experimental series. Photograph of an ice pellet (1.4 × 1.4 mm) taken by the high speed framing camera is shown in Fig. 2.

The "injection-angle controllable system" enable us to vary the injection angle poloidally from -6 to 6 degrees by changing the angle of the last stage guide tube using a micrometer, whose injection angle is limited by the port structure of the device. This situation makes possible for pellets to aim at about from z = -2a/3 to 2a/3 of the plasma, here the symbol 'z'-axis and 'a' designate the vertical position and the plasma minor radius, respectively. When we change the graduation of the micrometer, 1mm, the injection angle and the injected position vary 0.7 degree and 1.7 cm, respectively. On an equatorial plane, pellet injection angle has an inclination of 7 degrees toroidally from the direction of major radius.

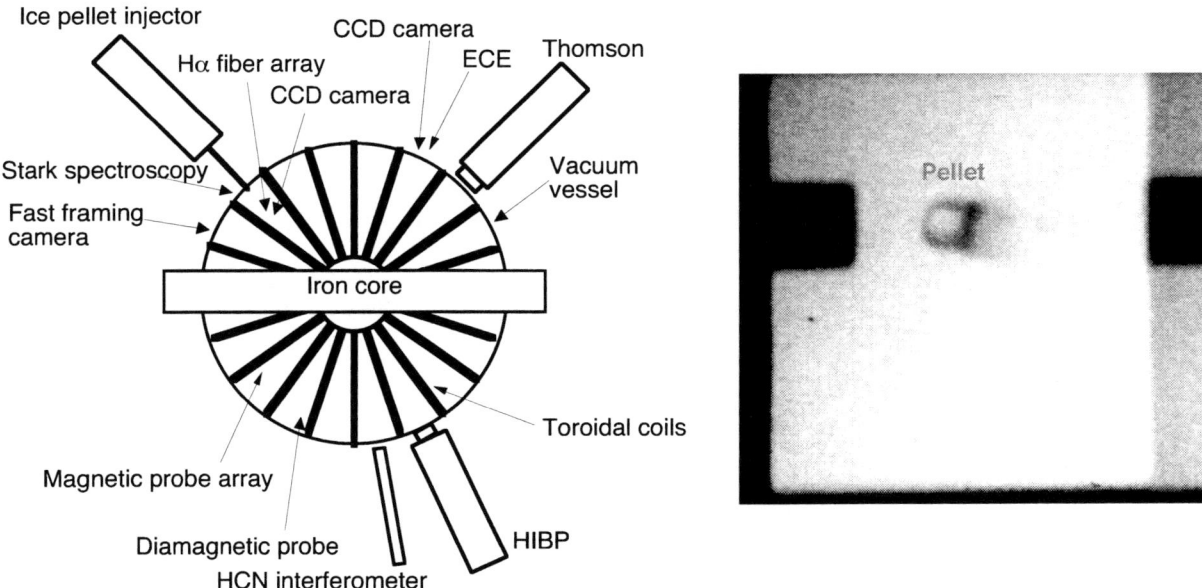

FIGURE 1. Schematic drawing of diagnostics arrangements.

FIGURE 2. High speed framing photograph of an ice pellet in flight.

RESULTS

Figures 3 (a) and (b) show typical example of ablation cloud when an ice pellet is injected at the plasma vertical position of about Z = +12 cm. Here, the direction of toroidal magnetic field is counter-clockwise (CCW), and that of plasma current is clockwise (CW), and Figs. 3(a) and (b) represent photographs of the tangential view and of the top view taken by the CCD camera, respectively. A circle line on the Fig. 3(a) indicates the plasma region around the pellet injected port section. From these pictures we can recognize that the ablation cloud has a trace being straight along the pellet path, as is expected from usual theoretical consideration. The white color region in the left-hand side of the figure is due to the halation of the CCD camera by the vacuum chamber wall.

Figures 3(c) and (d) show typical example of ablation cloud when an ice pellet is injected at the plasma position of about Z ~ +7 cm. Here, the direction of toroidal magnetic field is CCW, and that of plasma current is CW. From Fig. 3(c), it is found that the ablation cloud has a downward curved trace not along the pellet path. The tip of the emission region crosses over the plasma center. From Fig. 3(d), a pellet finishes ablating before arriving at the plasma center. It is found that the ablation cloud performs an asymmetric behavior toroidally, and spreads over in the inverse to the plasma current direction (the upside direction in the figure). From both Figs. 3(c) and (d), we can easily understand that the direction of this helical "tail" is independent to that of the total magnetic field lines of the torus. It is found that the ablation cloud poloidally rotates to the electron diamagnetic direction, and toroidally to the opposite to the plasma current direction as to almost all conditions of injection angles.

In order to understand details of an ice pellet ablation structure, a spectroscopy system has been developed to obtain the local parameters within the ablation cloud. Through the analysis using a multi-Lorenzian fitting method, it has been found that the typical cloud density is in the range of 10^{16} - 10^{17} cm^{-3}, and the typical temperature is in the range of 1 - 4 eV [6]. Potential measurements of pellet injected plasmas using HIBP method has been carried out.

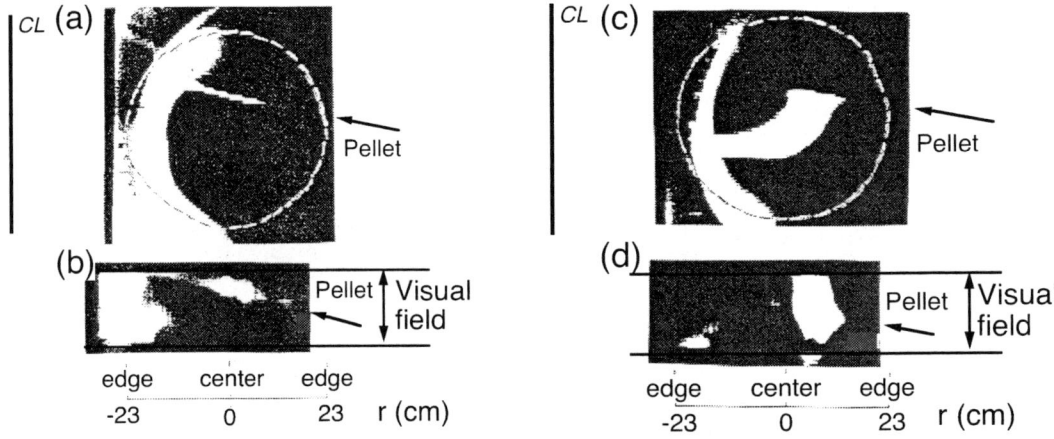

FIGURE 3. Photographs using two-directional CCD cameras. Pellet injection to the plasma vertical position of ~+12 cm, (a) tangential view, (b) top view. Pellet injection to the plasma vertical position of ~+7 cm, (c) tangential view, (d) top view.

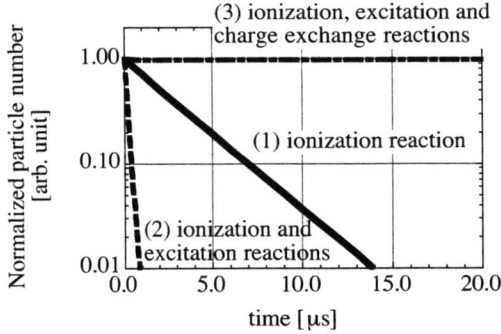

FIGURE 4. Calculated neutral particles as a function of time with various cases. (1) Only ionization reaction is included, (2) only ionization and excitation reactions are included, and (3) ionization, excitation and charge exchange reactions are included.

In the case of an injection angle to be anti-parallel to the electron diamagnetic direction in poloidal plane, the result shows that the direction of potential change is negative, and consequently the potential after the injection should be negative because it has been measured to be negative in usual ohmic plasmas without pellet injection [6, 7]. Thus, the direction of the "tail" structure in this case seems to be consistent to that of the plasma rotation which depends on $E_r \times B$ flow [7].

The characteristic times for various processes such as charge exchange (τ_{cx}), elastic collision, excitation (τ_{ex}), ionization (τ_{ion}) and recombination are analyzed. Simultaneous differential equations (1 - 4) for ground level hydrogen neutral ($N_{n=1}$), hydrogen ion (N_+), excited hydrogen neutral of n= 2 ($N_{n=2}$) and excited hydrogen neutral of n= 3 ($N_{n=3}$) are solved. Here, electron temperature is assumed to be constant. As shown in Fig.4, neutral particle reduction is much faster in the case where only the ionization reaction is included (case (1) in Fig. 4), or in the case where only the ionization and excitation reactions is included (case (2) in Fig. 4). However, if the charge exchange reaction is included in the equations, neutral particles do not quickly decrease (case (3) in Fig. 4). Therefore, "tail-shaped" phenomena may come from continuation/maintenance of the charge exchange equilibrium of hydrogen ions and neutrals at extremely high density regime in the ablation cloud during a certain period.

$$\frac{d[N_{n=1}(t)]}{dt} = -\frac{N_{n=1}(t)}{\tau_{ion(n=1\to+)}} - \frac{N_{n=1}(t)}{\tau_{ex(n=1\to 2)}} - \frac{N_{n=1}(t)}{\tau_{cx}} + \frac{N_+(t)}{\tau_{cx}} \quad (1)$$

$$\frac{d[N_+(t)]}{dt} = +\frac{N_{n=1}(t)}{\tau_{ion(n=1\to+)}} + \frac{N_{n=3}(t)}{\tau_{ion(n=3\to+)}} + \frac{N_{n=1}(t)}{\tau_{cx}} - \frac{N_+(t)}{\tau_{cx}} \quad (2)$$

$$\frac{d[N_{n=2}(t)]}{dt} = -\frac{N_{n=2}(t)}{\tau_{ex(n=2\to 3)}} + \frac{N_{n=1}(t)}{\tau_{ex(n=1\to 2)}} + \frac{N_{n=3}(t)}{\tau_{spontaneous(n=3\to 2)}} \quad (3)$$

$$\frac{d[N_{n=3}(t)]}{dt} = -\frac{N_{n=3}(t)}{\tau_{ion(n=3\to+)}} - \frac{N_{n=3}(t)}{\tau_{spontaneous(n=3\to 2)}} + \frac{N_{n=2}(t)}{\tau_{ex(n=2\to 3)}} \quad (4)$$

SUMMARY

A new technique for an ice pellet injection system with the controllability of injection angle has been developed and installed with the JIPP T-IIU tokamak in order to vary deposition profile of ice pellets within a plasma easily. An interesting phenomenon has been obtained concerning the flow characteristics of ablation cloud using this "injection-angle controllable system" and multi-dimensional observations with two CCD cameras. In the case of an injection angle (θ) larger than a certain value ($\theta \geq 4°$), a pellet penetrates straightly through the plasma with a trace of straight ablation cloud, which has been expected from usual theoretical consideration. On the other hand, in the cases of on-axis and off-axis injections with the angle smaller than the certain value ($\theta \leq 4°$), a long helical shape ("tail") of ablation light has been observed. The direction of this helical "tail" is independent to that of the total magnetic field lines of the torus. The "tail" poloidally rotates to the electron diamagnetic direction, and toroidally to the opposite to the plasma current direction as to almost all conditions of injection angles. From the experimental results and consideration, the "tail" structure may be caused both by the situation of charge exchange equilibrium of hydrogen ions and neutrals at extremely high density regime in the cloud, and by the effect of the plasma potential and the rotation. The present study strongly suggests that the importance of charge exchange reaction and the plasma potential ($E_r \times B$ flow) should be considered and taken into account in the pellet ablation model.

REFERENCES

1. Greenwald, M., et al., Phys. Rev. Lett. **53**, 352-355 (1984).
2. Sakamoto, M., et al., Plasma Phys. and Control. Nucl. Fusion **33**, 583-594 (1991).
3. TFR Group, Nuclear Fusion 27, 1975-1999 (1987).
4. Sato, K.N., et al., Proc. 18th EPS Conf. 15C, Part I, 333-336 (1991).
5. Toi, K., et al., Phys. Rev. Lett. **64**, 1895-1898 (1990).
6. Sakakita, H., et al., Fusion Engineering and Design 34-35, 329-332 (1997).
7. Sakakita, H., et al., Proc. ICPP Conf. 1, 710-713 (1996).

Formation Of Carbon Oxides In CH_4/O_2 Plasmas Produced By Inductively Coupled RF Discharges At Low Pressure

Ivonne Möller and Henning Soltwisch

Institut für Experimentalphysik, AG Laser- und Plasmaphysik,
Ruhr-Universität Bochum, 44780 Bochum, Germany

Abstract. The formation of CO and CO_2 has been studied in inductively coupled rf (13.56 MHz) discharges with varied mixtures of CH_4 and O_2 as feed gases at a total pressure of 10 Pa, flow rates of <10 sccm, and input powers of <500 W. The primary diagnostic tool has been TDLAS (tunable diode laser absorption spectroscopy) to measure absolute concentrations of molecular species as well as their kinetic and rovibrational temperatures. Of particular interest is the sudden transition between different modes of power coupling (capacitive and inductive mode, resp.) and the related changes of the plasma composition. We have found that the power threshold for this transition exhibits a clear hysteresis and depends on the oxygen content. Comparing the ratio of the CO- and CO_2-concentrations in capacitive mode with corresponding data from a parallel-plate discharge, clear differences have been observed. The findings can partly be explained on the basis of plasma-chemical reaction chains using tabulated cross-sections in combination with estimations of the electron energy distribution function. Some observations (as, e.g. the presence of CO in inductively coupled plasmas that are fed by pure oxygen) cannot be understood from volume reactions only but point to an important role of surface processes, which depend on the materials of the discharge chamber and on its history and cleaning method.

INTRODUCTION

Radio-frequency (rf) discharges in reactive molecular gases play an important role in many technological applications. For example, low-pressure hydrocarbon plasmas are commonly used to deposit thin films of hydrogenated amorphous carbon (a-C:H) or polycrystalline diamond (see, e.g. [1]). However, owing to the complexity of the chemical reactions and the absence of thermodynamic equilibrium, the understanding of the discharge behavior is still far from being complete. The development of realistic plasma models depends heavily on further experimental investigations of the plasma chemistry and kinetics (in particular on detailed measurements of transient and stable reaction products) under well-defined operational conditions.

For this reason we have studied the formation of carbon oxides in CH_4/O_2 plasmas in a standard Gaseous Electronic Conference (GEC) reactor. The carbon oxides were chosen because their production from the feed gases involves a large number of intermediate species and chemical reactions, which makes them good candidates for testing the overall reliability of computer modeling. Moreover, the molecular spectra and individual line strengths of carbon oxides are sufficiently well known to warrant precise measurements of their concentration by means of tunable diode laser absorption spectroscopy (TDLAS).

Using a planar coil for inductive coupling of rf power to the plasma, we observe sharp transitions between two different operational regimes at certain power levels. The physics of these so-called *E*- to *H*-mode transitions has been investigated for long in discharges of simple gas composition by various groups (see e.g. [2]) and has been attributed to a change-over of the electron heating by longitudinal and azimuthal rf electric field components within the plasma. In our discharges of rather complex gas composition we find a clear dependence of the power threshold on the mixing ratio of the feed gases. In addition, the mode transition exhibits a hysteresis whose width is also a function of the mixing ratio. The CO/CO_2 balance varies from one regime to the other, indicating different chemical reaction chains that are presumably altered by changes of the electron energy distribution function (EEDF).

Although the inductively coupled GEC reactor in *E*-mode resembles a capacitively coupled parallel-plate discharge, we have found surprisingly large differences in the dissociation of the feed gases and the production of carbon oxides, which may be due to a quartz window separating the antenna coil from the plasma chamber.

EXPERIMENTAL SET-UP AND METHODS

Our plasma source is a standard GEC reference cell [3]. The stainless-steel chamber is pumped by a turbo-molecular pump to obtain a background pressure below 10^{-4} Pa. The feed gases CH_4 and O_2 are supplied by a mass flow control system at varied mixing ratios. The gas inlet is ring-shaped in order to achieve a more uniform gas flow into the chamber at a rate between 2 and 10 sccm. The operation pressure is regulated by means of a throttle valve in the exhaust and kept constant at 10 Pa for the experiments reported here. Rf power of up to 500 W is provided by a stabilized 13.56 MHz generator and coupled (via a matching network) to a standard five-turn planar coil, which is placed outside the vacuum vessel behind a quartz window (thickness 9 mm) and a Faraday shield. The latter consists of 62 star-shaped radial spikes at ground potential. With this shield in place, the discharge has to be ignited by a short high-voltage pulse onto a pin electrode protruding into the vessel.

For measuring the concentrations of carbon oxides formed in the discharge we employ two diode laser absorption spectrometers at wavelengths around 4.9 µm and 15.3 µm for CO and CO_2, respectively. A detailed description of the optical system may be found in [4]. In order to increase the sensitivity by lengthening the absorption path we have installed a Herriott-type multiple-reflection cell [5] to pass the laser beams several ten times through the plasma. In addition, we apply the technique of derivative spectroscopy where the injection current (and hence the wavelength) of the diode laser is modulated sinusoidally and the resulting modulation of the absorption signal is detected by a lock-in amplifier. In this way the first derivative of the absorption line profile (rather than the profile itself) is measured with a strongly reduced noise level (for details see, e.g. [6]).

Provided that the population densities of the molecular energy levels remain unchanged by the laser beam, the transmitted intensity as a function of wavenumber $\tilde{v} = \lambda^{-1}$ is given by the Beer-Lambert law

$$I(\tilde{v}) = I_0(\tilde{v}) \cdot \exp\left(-\int_0^L S(T_{rv}) g(\tilde{v}, T_t) n(x) dx\right) \quad (1)$$

where n is the sought-after number density of the absorbing species, S is the line strength (depending on the rovibrational temperature T_{rv}), and g is the line shape function (depending on the translational temperature T_t due to Doppler broadening under our operational conditions). In the non-equilibrium plasma environment a distinction between T_{rv} and T_t has to be made since selective pumping of energy levels by electron impact cannot be excluded. While the translational temperature is readily determined from the width of the absorption profile, the rovibrational temperature has to be obtained with the help of a Boltzmann plot [7] which gives the population distribution of the lower energy levels of the probed transitions. For all conditions in our experiments it was found that T_{rv} and T_t rise only little above room temperature. To solve equation (1) for the average particle density along the laser beam path, the corresponding line strength S was taken from the HITRAN database [8].

EXPERIMENTAL RESULTS AND DISCUSSION

A characteristic feature of inductively coupled rf discharges is the existence of two distinct operational regimes. The change-over occurs at a certain threshold value of the rf power and manifests itself in a sudden alteration of the plasma luminosity due to a drastic change of the electron density by one or two orders of magnitude. In our

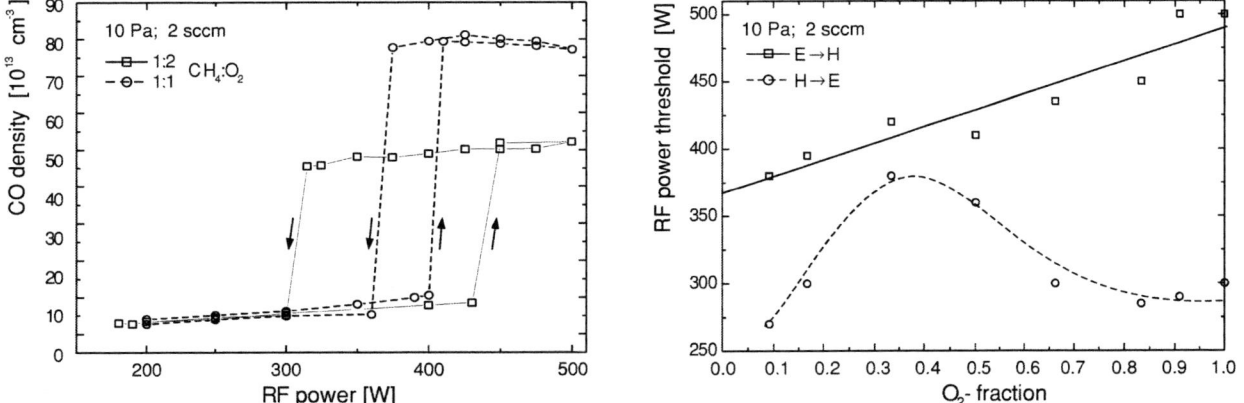

FIGURE 1. $E - H$ mode transition: sharp change of the CO particle density as a function of rf power (left), and dependence of the power thresholds on the oxygen admixture to the filling gas (right).

experiments we observe not only a related change of the plasma composition but, conversely, also a pronounced influence of the plasma composition on the threshold levels. As shown in Figure 1, the E- to H-mode transition takes place at higher rf power than the H- to E-mode back-transition. The absolute values of the thresholds (P_{EH} and P_{HE}, respectively) depend clearly on the amount of oxygen in the filling gas. While P_{EH} increases linearly with the fraction of O_2, P_{HE} has a maximum for a mixing ratio CH_4/O_2 of about 2:1.

In Figure 2 we show measured concentrations of CO and CO_2 as a function of input power in E- and H-mode for various gas mixtures at a flow rate of 2 sccm and a fixed pressure of 10 Pa (note that the densities represent line-averaged data along the laser beam, which are probably somewhat lower than the local values within the chemically active plasma bulk). Generally, we find significantly higher concentrations of CO than of CO_2. In E-mode the dependence on P_{rf} is linear with a positive slope that changes with the feed gas composition. Following the predictions of "global" plasma models (as, e.g. [9]), the electron density n_e scales in proportion to P_{rf} whereas the temperature remains more or less constant. So, the plasma behavior in E-mode reflects rather stable chemical reaction chains with production and loss rates mainly determined by n_e.

In H-mode the observed dependencies are much more involved, indicating significant shifts of the CO/CO_2 particle balance with varying input power. Qualitatively, this may be understood by alterations of the chemical reaction chains brought about by changes of the electron energy distribution function (EEDF). We have tried to test this hypothesis by Langmuir-probe measurements, but technical problems in our reactive plasma environment have hindered us so far from reaching a conclusion.

Drastic changes of the carbon oxide concentrations can be achieved in E- and H-mode by changing the flow rate of the filling gas (see Figure 3). Lowering the rate from 10 to 2 sccm can cause an increase of the particle densities by a factor of up to 6. We attribute this effect to a longer residence time of the species in the active plasma volume. In that case the maximum production of CO and CO_2 occurs in H-mode at mixing ratios CH_4:O_2 close to 2:1 and 1:1, respectively, as one might expect on stoichiometric grounds. In E-mode and at higher flow rate the optimum admixture of O_2 does not follow this expectation, as can be seen in Figure 3. Using pure oxygen as a filling gas, carbon oxides are still present in the discharge, which we explain by an influx of carbon atoms from deposits on the grounded electrode, the quartz window and the walls of the vacuum chamber. So, to a certain extent our measurements are affected by surface processes which depend on the feed gases and wall materials but also on the history and cleaning methods of the discharge vessel.

As has been mentioned before, we find under all conditions in our experiments higher densities of CO than of CO_2. This result is in contrast with earlier investigations in a capacitively coupled GEC cell (parallel-plate reactor) under similar operational conditions. Here the balance was clearly shifted towards the formation of CO_2, and the relative number densities of both carbon oxides with respects to the total density of particles within the discharge were higher by more than one order of magnitude (further details may be found in [4]). The more efficient production of CO_2 and CO in the parallel-plate reactor goes along with a much higher dissociation of the feed gases as was confirmed by measurements of the CH_4 concentration. For E-mode operation of the inductively coupled GEC cell it appears that the Faraday shield and the dielectric quartz window take a strong influence on the discharge.

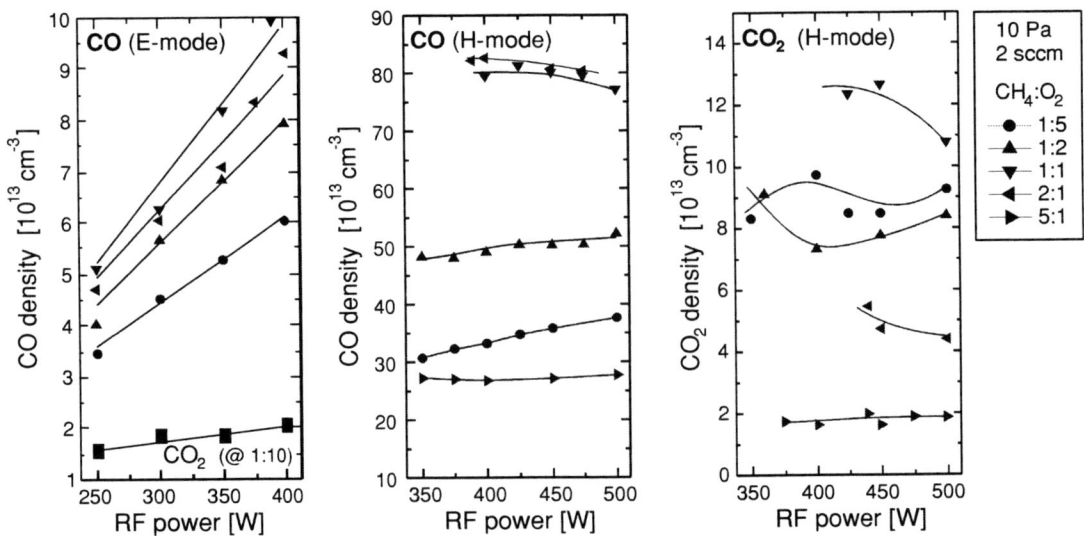

FIGURE 2. Measured concentrations of CO and CO_2 as a function of input power in E- and H-mode.

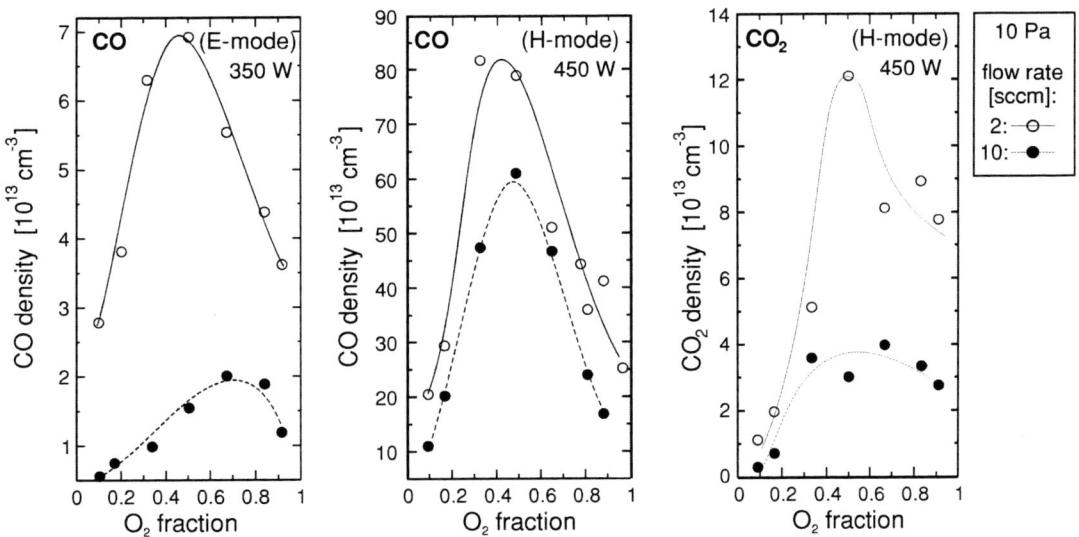

FIGURE 3. Measured concentrations of CO and CO_2 as a function of the feed gas composition in *E*- and *H*-mode.

ACKNOWLEDGMENTS

The support of this work by the Sonderforschungsbereich 191 of the Deutsche Forschungsgemeinschaft and by the German Ministry for Research and Technology under grant 13N8049 is gratefully acknowledged.

REFERENCES

1. Jacob, W., *Thin Solid Films* **326**, 1 (1998).
2. Kortshagen, U., Gibson, N. D., and Lawler, J. E., *J. Phys. D: Appl. Phys.* **29**, 1224 (1996).
3. Miller, P. A., Hebner, G. A., Greenberg, K. E., et al., *J. Res. Nat. Inst. Stand. Technol.* **100**, 427 (1995).
4. Busch, C., Möller, I., and Soltwisch, H., *Plasma Sources Sci. Technol.* **10**, 250 (2001).
5. Herriott, D., Kogelnik, H., and Kompfer, R., *Appl. Opt.* **3**, 523 (1964).
6. Webster, C.R., Menzies, R. T., and Hinkley, E. D., "Infrared Laser Absorption: Theory and Applications", in *Laser Remote Chemical Analysis,* edited by R. M. Measures, Wiley & Sons, New York, 1987, vol. 94, ch. 3, p. 163.
7. Wiese, W. L., *Spectrochim. Acta* B **46**, 831 (1991).
8. Rothman, L. S., Rinsland, C. P., Goldman, A., et al., *J. Quant. Spectrosc. Radiat. Transfer* **60**, 665 (1998).
9. Lieberman, M. A., and Lichtenberg, A. J., *Principles of Plasma Discharges and Materials Processing*, Wiley & Sons, New York, 1994.

Application of Langmuir Probe for Study of Recombination of D_3^+ Ions with Electrons in He-Ar-D_2 Stationary and Flowing Afterglow Plasma

M. Tichy, V. Poterya, R. Plasil, A. Pysanenko, P. Kudrna, O. Novotny, P. Zakouril, J. Glosik

Charles University in Prague, Faculty of Mathematics and Physics, Department of Electronics and Vacuum Physics, V Holesovickach 2, Prague 8, Czech Republic

Abstract. We report measurements of the rate coefficient for recombination of D_3^+ and D_5^+ with electrons in He-Ar-D_2 plasma. Two afterglow experiments, flowing afterglow and stationary afterglow were used to cover large extent of pressures of He buffer gas (2-10 Torr) and large extent of partial number densities of D_2 (5×10^{10}-3×10^{15}cm^{-3}). Langmuir probes and mass spectrometers were used to monitor decay of the plasma during the afterglow. The observed rate coefficient is dependent on the deuterium number density indicating that third-body-assisted recombination is efficient and significantly contributes to recombination when sufficient number density of deuterium is present.

At low D_2 number densities the ions D_3^+ dominate the ion composition and electron density decay is controlled by recombination of D_3^+ with recombination rate coefficient $\alpha(D_3^+)$. At higher D_2 number densities and lower temperatures D_5^+ are formed and electron density decay is controlled by recombination of D_5^+ ions with recombination rate coefficient $\alpha(D_5^+)$. The overall effective recombination rate coefficient α_{eff} as a function of D_2 number density was measured and from this dependence the rates $\alpha(D_3^+)$ and $\alpha(D_5^+)$ at several temperatures were determined. Obtained pressure dependencies are in good agreement with thermodynamic data. When the deuterium number density is decreased down to 5×10^{10}cm^{-3}, the rate coefficient also decreases to $\alpha_{eff} \sim 4\times10^{-9}$ cm^{3}s^{-1}. These data indicate that the binary dissociative recombination of D_3^+ is very slow with $\alpha_{DR} < 4\times10^{-9}$ cm^{3}s^{-1}. The observation of an additional de-ionization process proceeding via formation of D_5^+ and its recombination is also reported.

INTRODUCTION

The three-atomic ions H_3^+, H_2D^+, HD_2^+ and D_3^+ have been the subject of a number of studies. The H_3^+ ion and its deuterated analogue D_3^+ play important roles in the kinetics of media of astrophysical interest (interstellar molecular clouds [1,2], and planetary atmospheres [3]) and also in laboratory produced plasmas (glow, rf and microwave discharges, fusion plasmas at walls [4] etc.). Recent discovery of interstellar H_3^+ in molecular clouds has confirmed the presence of H_3^+ and opened a problem of observed high column density, which critically depends on rate of recombination of H_3^+ with electrons [1,2,5].

Among the recent measurements on D_3^+ recombination we wish to mention explicitly Flowing Afterglow (FALP) studies of recombination of D_3^+. Smith and Spanel obtained in FALP studies $\alpha \sim 2\times10^{-8}$ cm^{3}s^{-1} [6]. In 1995 Gougousi et al. [7] obtained in the FALP study $\alpha \sim (0.7-1.3)\times10^{-7}$ cm^{3}s^{-1} dependent on deuterium partial pressure. Laube et al. [8] obtained in FALP-MS with consideration of ionic composition 0.7×10^{-7} cm^3 s^{-1}. From recent studies in CRYRING the rate coefficient 2.7×10^{-8} cm^{3}s^{-1} was deduced [9]. All experimental recombination rate coefficients obtained for both H_3^+ and D_3^+ are significantly larger than the values predicted by theory, see e.g. [10].

To address discrepancies between different experiments and between experiments and theory we recently designed and built new afterglow experiment - Advanced Integrated Stationary Afterglow (AISA) - utilizing advantages of both the stationary and the flowing afterglow techniques [11]. In our recent studies of recombination of H_3^+ and D_3^+ ions with electrons we observed in the He-Ar-H_2 and He-Ar-D_2 afterglow dependence of "apparent" recombination rate, α_{eff}, on number density of H_2, D_2. We concluded that the observed recombination is a three-body process proceeding via formation of long-lived neutrals H_3^* or D_3^*, which are stabilized against reverse auto-ionization by collisions with neutral molecules of H_2 or D_2.

In the present study we extended the range of deuterium number densities up to $10^{16} cm^{-3}$ and the range of He pressures up to 10 Torr. To examine the eventual role of D_5^+ ions the experiments were carried out at several neutral gas temperatures to enhance formation of these ions. The experiments were carried out with our High Pressure Flowing Afterglow (HPFA) system.

EXPERIMENT

The Flowing Afterglow with Langmuir Probe (FALP) technique was primarily developed to study rate coefficients of ion-molecule reactions, electron attachment reactions and electron ion and ion-ion recombinations. The apparatus is based on simple principle: the buffer gas flows down the flow tube and carries decaying plasma created upstream in the plasma source. At different positions along the flow tube there is plasma in different degree of decay. We have built a high pressure version of this apparatus - High Pressure Flowing Afterglow – HPFA. In the present UHV version of HPFA the differential pumping of mass spectrometer chamber is used to enable the increase of the operational buffer gas pressure. The HPFA has been described in detail previously [12,13], hence only a short description will be given here. The schematic diagram in Figure 1 outlines the essential parts of the used HPFA. The Langmuir probe (W wire 14 μm in diameter and 7 mm long) is applied to determine the electron number density along the flow tube. Region of the saturated electron current of probe characteristics is used to calculate the electron number density. The validation of probe method at the used pressures was made by the measurements of the known rate of recombination of O_2^+ with electrons at the pressure 9 Torr. In the validation it was assumed that $\alpha(O_2^+,300K) = 2 \times 10^{-7} cm^3 s^{-1}$ (for details see our previous study [13] and previous studies of this rate coefficient [14,15]).

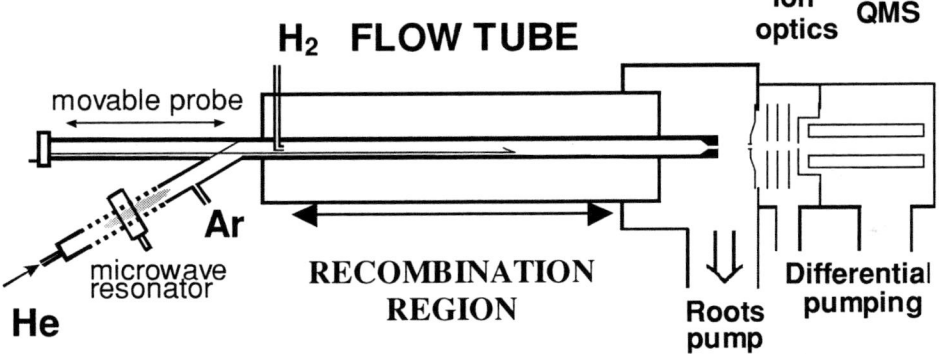

FIGURE 1. Schematics of High Pressure Flowing Afterglow experiment. QMS indicates the downstream mass spectrometer with detector.

Formation of $D_3^+/D_5^+/e^-$ plasma: The pure He flows through the microwave cavity and plasma containing He^+, He^m and electrons is formed. Downstream of the discharge region (at position corresponding to approximately 1ms of the decay time) Ar gas is added to the flow and the plasma is converted to the Ar^+ dominated plasma. Further downstream (approximately 2ms of the decay time) deuterium is added to the already thermally-relaxed Ar^+/e^- plasma and $D_3^+/D_5^+/e^-$ plasma is formed in a sequence of ion-molecule reactions. The kinetics of this formation is well understood and the rate coefficients are known [16]. Briefly, deuterium reacts with Ar^+ ions to form ArD^+ and D_2^+. Both ions react further with D_2 and D_3^+ ions are formed.

In present experiments higher pressure and lower temperature of the buffer gas is essential to enhance formation of D_5^+. We assumed that the plasma is fully thermalized; this assumption has been confirmed by the probe measurements of the electron temperature in the decaying plasma. The axially movable Langmuir probe enables the measurement of electron number density, n_e, and electron temperature at various points along the axis of the flow tube. The Langmuir probe is therefore an essential tool for determination of the recombination rate coefficients.

DATA ANALYSIS

If D_3^+ ions are the dominant ions in decaying plasma and the recombination and diffusion are the only loss processes then the determination of the rate coefficient (α_3) is straightforward. The balance equation for electrons is:

$$w_p \frac{dn_e(z)}{dz} = \frac{dn_e(t)}{dt} = -\alpha_3 \cdot [D_3^+] \cdot n_e - v_D \cdot n_e = -\alpha_3 \cdot n_e^2 - v_D \cdot n_e. \tag{1}$$

In this equation z, w_p and t denote the position within the flow tube, the plasma velocity and the decay time, respectively. These parameters are coupled by equation: $z = w_p t$. $[D_3^+]$ indicates number density of D_3^+ ions, and α_3 is the recombination rate coefficient. Last equality is given by quasineutrality of the plasma, $n_e = [D_3^+]$. The parameter v_D is coupled with characteristic diffusion time by relation: $v_D = 1/\tau_D$. The value of τ_D is given by the buffer gas pressure and by the flow tube radius. The solution of Eq. (1) that accounts also of diffusion hence reads:

$$\frac{1}{n_e} = \frac{1}{n_{e0}} e^{v_D \tau} + \alpha \frac{e^{v_D \tau} - 1}{v_D}. \tag{2}$$

This formula was used in the present study to fit the measured decay curves.

In HPFA at a relatively low temperature and at higher pressure of He and D_2 the formation of cluster ions D_5^+ in three-body association reaction and their simultaneous destruction by a collision induced dissociation (CID) in collisions with He atoms [17] have to be considered:

$$D_3^+ + D_2 + He \underset{k_{-3}}{\overset{k_3}{\leftrightarrow}} D_5^+ + He. \tag{3}$$

Here k_3 is the rate coefficient of the three-body association (forward process) and k_{-3} is the binary rate coefficient of the CID (reverse process). At the same time both ions recombine with electrons:

$$D_3^+ + e \xrightarrow{\alpha_3} products, \tag{4}$$

$$D_5^+ + e \xrightarrow{\alpha_5} products, \tag{5}$$

where α_3 and α_5 are the respective recombination rate coefficients. The rate coefficient k_3 is known, the rate coefficient k_{-3} of the reverse CID can be calculated from thermodynamic data by application of van't Hoff equation (see discussion below). When considering processes (3), (4) and (5), the balance equations for electrons converts to:

$$\frac{dn_e}{dt} = -\alpha_3 \cdot [D_3^+] \cdot n_e - \alpha_5 \cdot [D_5^+] \cdot n_e - v_D \cdot ([D_3^+] + [D_5^+]) \tag{6}$$

For a quasineutral plasma in an equilibrium we can write: $n_e = [D_3^+] + [D_5^+]$ and $R = [D_5^+]/[D_3^+] = k_3[D_2]/k_{-3} = K_C[D_2]$, where K_C is the equilibrium constant, given by an entropy and an enthalpy change in the reaction [18] (van't Hoff equation). Using ratio R we can rewrite Eq. (6) to the form:

$$\frac{dn_e}{dt} = -(\alpha_3 + R \cdot \alpha_5) \frac{n_e^2}{1+R} - v_D \cdot n_e = \alpha_{eff} n_e^2 - v_D \cdot n_e, \tag{7}$$

where we introduced the effective recombination rate coefficient:

$$\alpha_{eff} = (\alpha_3 + R \cdot \alpha_5) \frac{1}{1+R} = (\alpha_3 + \alpha_5 K_C [D_2]) \frac{1}{1 + K_C [D_2]}. \tag{8}$$

If the ratio R is constant during the decay of the plasma along the flow tube then α_{eff} is constant and decay typical for plasma governed by recombination can be observed. Such situation occurs e.g. if the ions are in equilibrium described by the reaction scheme (3) during the decay of the plasma. The equation (8) describes the influence of a deuterium partial pressure on the overall (apparent) rate of the recombination in $D_3^+/D_5^+/e^-$ plasma. From equation (8) follows: for a very small $[D_2]$ $\alpha_{eff} \approx \alpha_3$, and for very high $[D_2]$ $\alpha_{eff} \approx \alpha_5$. In the intermediate region of pressures the value α_{eff} is given by relation $\alpha_{eff} = \alpha_3 + \alpha_5 K_C [D_2]$.

RESULTS AND DISCUSSION

We have measured decay of the plasma in He/Ar/D_2 mixture at several temperatures ranging from 170 K up to 260 K. The deuterium number density was changed systematically from $[D_2] = 1 \times 10^{13}$ cm^{-3} up to $[D_2] = 1 \times 10^{15}$ cm^{-3}. This large variation makes it possible to cover plasmas dominated: by the recombination of D_3^+ ions, by the recombination in the mixture of D_3^+ and D_5^+ ions and also by the recombination of D_5^+ ions only. The effective recombination rate coefficients obtained at four different temperatures are plotted in Figure 2 together with previous data obtained in AISA experiment. In agreement with formula (8), at $[D_2] < 5 \times 10^{13}$ cm^{-3} the term with α_5 can be neglected and $\alpha_{eff}(T) \approx \alpha_3(T)$. In opposite at $[D_2] > 1 \times 10^{15}$ cm^{-3} the term with α_3 can be neglected and $\alpha_{eff}(T) \approx \alpha_5(T)$. By fitting the data over large range of values of $[D_2]$ the equilibrium constant K_C can be calculated. The obtained K_C

is in agreement with thermodynamic data obtained in high-pressure mass spectrometer. There is, however, not enough experimental data for final conclusion on temperature dependencies of $\alpha_3(T)$ and $\alpha_5(T)$ yet.

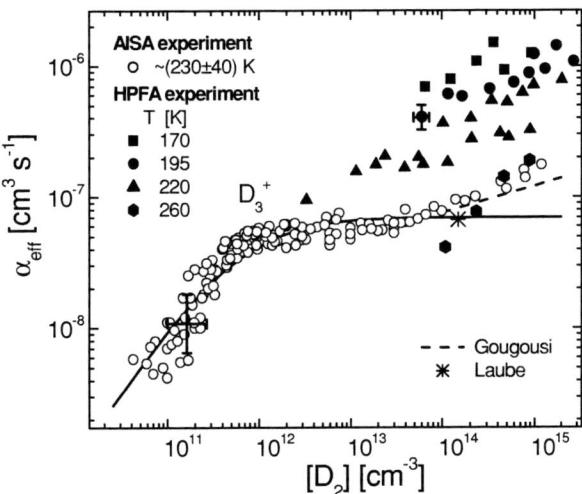

FIGURE 2. The effective recombination rate coefficient α_{eff} measured over very broad range of deuterium number densities on AISA (open symbols) and on HPFA (full symbols). The AISA measurements were carried out at the temperature (230±40) K.

ACKNOWLEDGMENTS

The work was partially financially supported by the Grant Agency of Czech Republic, Grant No. 202/00/1689, 202/01/D095, 202/99/D061, 205/02/0610, 202/02/0948, by the Grant Agency of Charles University, Grant No. 171/2000 B FYZ MFF and 146/2000 B FYZ MFF, and by the Ministry of Education, Youth and Sports, Research plan MSM 113200002. The experiments were carried out with financial support from EC's Research Training Network under contract HPRN-CT-2000-0142, ETR and with support from EURATOM.

REFERENCES

1. Geballe, T. R., *Phil. Trans. R. Soc. London*, **A 358**, 2503 (2000).
2. Millar, T. J., Roberts, H., Markwick, A. J., Charnley, S. B., *Phil. Trans. R. Soc. London*, **A 358**, 2535 (2000).
3. Miller, S., Achilleos, N., Bellester, G. E., Geballe, T. R., Joseph, R. D., Prange, R., Rego, D., Stallard, T., Tennyson, J., Trafton, L. M., Waite, J. H., *Phil. Trans. R. Soc. London*, **A 358**, 2485 (2000).
4. Janev, R.K., in Proceedings of the *Conference on Dissociative recombination, Theory, Experiment and Applications IV-1999*, edited by M. Larsson, J.B.A. Mitchell, I. F. Schneider, World Scientific, Singapore, 40 (1999), p. 40.
5. Black, J. H., *Phil. Trans. R. Soc. London*, **A 358**, 2515 (2000).
6. Smith, D., and Španěl, P., *Chem. Phys. Letters*, **211**, 454 (1993).
7. Gougousi, T., Johnsen, R., Golde, M. F., *Int. J. Mass Spectr. Ion Proc.*, **149/150**, 131 (1995).
8. Laube, S., Le Padelleck, A., Sidko, O., Rebrion-Rowe, C., Mitchell J. B. A., and Rowe, B. R., *J. Phys. B: At. Mol. Opt. Phys.*, **31**, 2111 (1998).
9. Le Padellec, A., Larsson, M., Danared, H., Larson, A., Peterson, J. R., Rosen, S., Semeniak, J., Stromholm, C., *Phys. Scripta*, **57**, 215 (1998).
10. Orel, A. E., Schneider, I.F., Suzor-Weiner, A., *Phil. Trans. R. Soc. Lond.* **A 358**, 2445 (2000).
11. Glosik, J., Plasil, R., Poterya, V., Kudrna, P., Tichy, M., *Chem. Phys. Letters*, **331**, 209 (2000).
12. Glosik, J., Zakouril, P., Hanzal, V., Skalsky, V., *Int. J. Mass Spectr. Ion Proc.*, **149/150**, 187 (1995).
13. Glosik, J., Bano, G., Plasil, G., Luca, A., Zakouril, P., *Int. J. Mass Spectr. Ion Proc.*, **189**, 103-113 (1999).
14. Spanel, P., Dittrichova, L., Smith, D., *Int. J. Mass Spectr. Ion Proc.*, **129**, 183 (1993).
15. Johnsen, R., *Int. J. Mass Spectr. Ion Proc.*, **81**, 67 (1987).
16. Plasil, R., Glosik, J., Poterya, V., Kudrna, P., Rusz, J., Tichy, M., Pysanenko, A., *Int. J. Mass Spectr. Ion Proc.*, **12166**, 1-26 (2002).
17. Glosik, J., Skalsky, V., Praxmarer, C., Smith, D., Freysinger, W., and Lindinger, W., *J. Chem. Phys.*, **101**, 3792 (1994).
18. Atkins, P.W., *Physical Chemistry*, Oxford University Press, Oxford, 1988.

Comparison of Emissive and Plugged Probes DC Plasma Potential Measurements in a Magnetised Plasma

F. Greiner[*], D. Block[*], A. Piel[*], S. Ratynskaja[†], G. Helblom[†] and K. Rypdal[†]

[*]*Institut für Experimentelle und Angewandte Physik, Christian-Albrechts-Universität, D-24098 Kiel, Germany*
[†]*Department of Physics, University of Tromsø, N-9037 Tromsø, Norway*

Abstract. Measurements of DC plasma potential profile with emissive and plugged probe were carried out for different experimental conditions in the Simple Magnetised Torus Devices BLAAMANN and TEDDI. The comparison of the data shows, that after a proper check of the probe alignment, the plugged probe can be an durable and easy to use alternative.

INTRODUCTION

The measurement of the plasma potential with Langmuir probes is always a delicate task. If the whole current voltage characteristic is measured, the potential of the turning point of the characteristic can be taken as an estimate of the plasma potential. The measurement of the whole characteristic however is time consuming and the the plasma might be disturbed. Therefore, one is interested in a direct measurement of the plasma potential. The floating potential, i.e. the potential of the current-less probe, is only a rough estimate of the plasma potential, because it differs by some kT_e from the true plasma potential. In addition non-thermal species of charged particles can have a strong influence on the floating potential. A standard method to measure the plasma potential without need of measuring the whole characteristics is the use of an emissive probe [1]. The probe tip is heated to emit electrons. A sharp zero crossing is observed in the characteristic, which is a good measure of the plasma potential. This technique gives a robust estimate of the plasma potential even in cases where the characteristic is strongly influenced by electron or ion beams. An newer method to measure plasma potential in a magnetised plasma is the use of a Langmuir probe which is aligned along the magnetic field and has isolated end plugs (plugged probe) [2]. The electron current to the probe is strongly decreased, i.e. the characteristic becomes nearly symmetric and is mainly influenced by the ion temperature. If the ions are cold, this leads to a floating potential, which is very close to plasma potential. A critical point of the plugged probe is the shielding of the electrons which depends on the exact alignment of the probe and on collisions, which refill the flux-tube around the probe.

COMPARISON OF EMISSIVE AND PLUGGED PROBE

All experimental data presented in this paper is collected in the simple magnetised torus (SMT) experiments TEDDI in Kiel [3] and BLAAMANN in Tromsø [4]. The plasma is produced with a hot cathode filament. This filament emits thermionic electrons and is negatively biased (-100 to -140 Volts) with respect to the chamber walls. Ions are produced by ionisation of the neutral gas by the fast electrons. The operating gas is argon or helium at low pressure (0.01···0.3Pa) and the plasma is strongly magnetised (0.15T). The measurement of plasma potential is complicated in such a plasma, because: (i) A fast electron component is present and (ii) the plasma is magnetised, i.e. even small additional loss surfaces or particle sources (e.g. an emissive probe) inserted into the plasma can change the plasma state on a magnetic flux tubes. The presence of fast electrons corrupts the floating potential and makes it a poor measure of the plasma potential. Therefore it is necessary to use more sophisticated technique for plasma potential measurements.

Emissive probes are a widely used probe technique to get a direct measurement of the plasma potential [1]. Figure 1 shows how the I(U)-characteristic of an emissive probe changes from a typical Langmuir form ($I_{i,sat}$ much smaller than

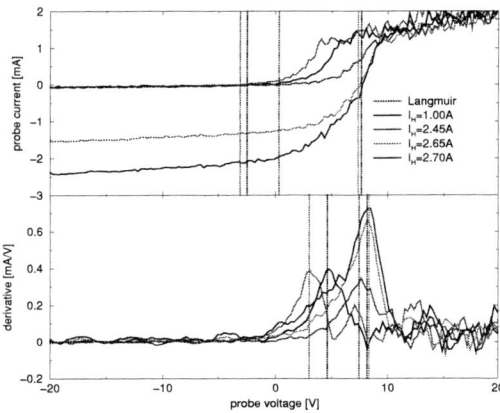

FIGURE 1. Development of the characteristic of an emissive probe, when heating current is increased. TOP: The ion saturation current increases if the emissive probe is heated. The increase of the emission current shifts the floating potential (vertical lines) to the right. Bottom: The maximum of the first derivative (vertical lines) of the characteristic give an estimate for the plasma potential. The difference between plasma potential and floating potential vanishs if the electron emission is high enough.

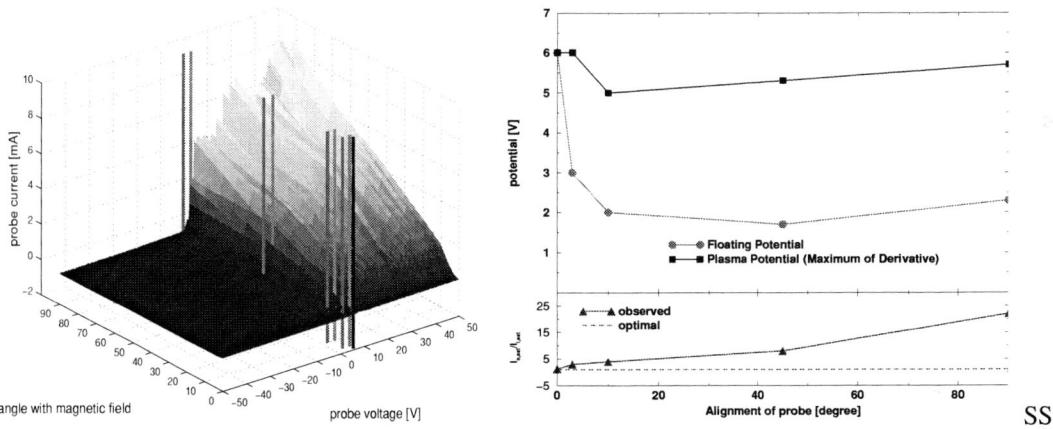

FIGURE 2. Left: Development of the characteristic of a plugged probe, when alignment is changed from parallel to magnetic field (0°) to perpendicular to the magnetic field (90°). Vertical lines give position of floating and plasma potential. Top-Right: Evolution of floating potential and plasma potential when alignment to magnetic field is changed from parallel to perpendicular. Bottom-Right: $I_{e,sat}(+50V)/I_{i,sat}(-50V)$ over the angle between probe and electric field. The dashed line gives the optimal value $I_{e,sat}(+50V)/I_{i,sat}(-50V) = 1$, the straight line with triangle gives the value observed for the specific angle.

$I_{e,sat}$) to an symmetric ($I_{i,sat} = I_{e,sat}$) characteristic. Taking the maximum of the first derivative as a estimate for plasma potential, Figure 1 shows that for higher heating current (i.e. electron emission) the floating potential is shifted towards the plasma potential. Because electron emission increases exponentially above a certain threshold, the emission of the probe has to be chosen carefully. One the one hand, a high emission current shortens probe lifetime, on the other hand an undersized emission current does not shift floating potential to the plasma potential. Therefore, as a rule-of-thumb the emission current is adjusted to get a symmetric characteristic. As seen from Figure 1, the plasma potential seems to change, when the heating current is increased. This observation is due to the effect of the additional voltage drop of the heating power-supply. If this is not compensated for (as recommended in Ref. [1]) a substantial measurement error is produced. In daily work the emissive probe is a somewhat touchy tool, because its mechanical production is complicated and the lifetime of the probe is limited.

In a magnetised plasma the so called plugged probe can be an alternative for plasma potential measurements. The probe consists of a tungsten wire with insulated end plugs. If the probe is aligned parallel to the magnetic field, the plugs inhibit the electrons from streaming to the probe whereas the ions still hit the conducting probe surface due to their much larger gyro-radius. Thus the electron current to the probe is significantly reduced. For an ordinary Langmuir probe the voltage difference between floating potential and plasma potential mainly depends on the electron

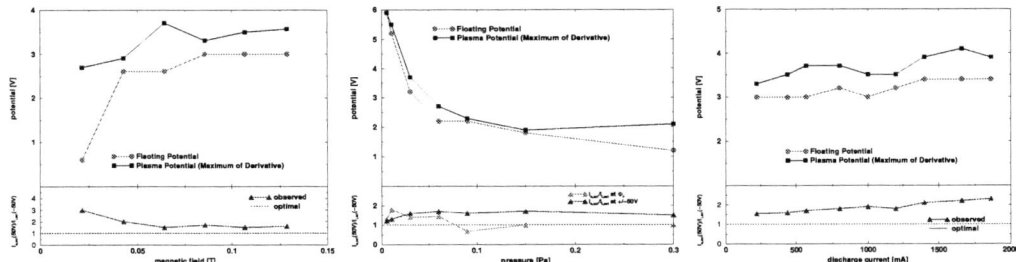

FIGURE 3. Validity of plugged probe plasma potential estimate, when magnetic field is increased (left); when pressure is increased (middle); when the discharge current is increased (right). On bottom of each figure the deviation of $I_{e,sat}(+50V)/I_{i,sat}(-50V)$ from the optimal value (straight line) is shown.

temperature. However when the electrons do not reach the probe this difference collapses because its dominated by the ion temperature now, i.e. for cold ions the floating potential becomes a good estimate for plasma potential. This is valid even if beam electrons are present in the plasma. Figure 2 shows the evolution of the probe characteristic of a plugged probe when rotated with respect to the magnetic field. If the probe tip is perpendicular to the magnetic field the characteristic of the probe is the usual Langmuir characteristic. Floating potential and plasma potential are clearly separated. When the probe is aligned to the magnetic field floating potential and plasma potential give identical values. The allowed error of the angle is about 1-2 ^{o} and depends on the diameter of the ceramic end plugs. However, if the plugs are too large, high electric fields can be established in the electron-free space between the plugs and parasitary discharges obscure the potential measurements. Again, as a rule of thumb, the floating potential is a good estimate for the plasma potential, if $I_{e,sat}/I_{i,sat} \approx 1$. This is seen from Figure 2. If the alignment becomes bad, $I_{e,sa}/I_{i,sat}$ leaves the optimal value and increases rapidly. A cross check with emissive probes shows that for the plugged probe, the maximum of the first derivative gives an identical estimate of plasma potential if $I_{e,sat}/I_{i,sat} \approx 1$ is fulfilled. This means, that for a valuable check of the plasma-potential measurement of a plugged probe, a comparison of its floating-potential with the plasma-potential estimated from the first derivative of the characteristic is sufficient.

Figure 3 shows the dependency of plasma-potential measurements of the plugged probe, on parameter variations of the plasma device. For the given geometry of the plugged probe (diameter of plugs, angle between probe tip and probe holder) it was not possible to make $I_{e,sat}(+50V)/I_{i,sat}(-50V)$ smaller than 1.5. The variation of the magnetic field Figure 3 (left) shows that the plugged probe leaves its regime of operation when the electron gyro radius becomes so large, that the electrons can reach the probe surface again (at $B < 0.05T$). This is seen from both indicators maximum of first derivative shows large deviations from floating potential and $I_{e,sat}/I_{i,sat}$ becomes larger than 2. Figure 3 (middle) shows the dependency of the plugged probe estimate from neutral gas pressure. Over the whole investigated region (0.005 to 0.3 Pa) the plugged probe gives good plasma potential measurements. This is also valid for variation of the discharge current as given by Figure 3 (right). The estimate has a constant offset from plasma potential. This is probably a result of the residual misalignment of the probe. However, compared to the offset of floating potential and plasma potential of a conventional probe of 4 Volt (see figure 2 for 90 ^{o}), the deviation is small.

The different methods to measure plasma potential are compared to each other in Figure 4 for a 2d cross section of a helium plasma in the BLAAMANN device. The relevant plots for the plasma potential as given from the first derivative of the characteristic (4,top-middle), the floating potential of the plugged probe (Figure 4, bottom-middle), and the floating potential of the emissive probe (Figure 4,bottom-right)) give the same absolute values and reveal the same spatial structure of the potential. In contrast to the floating potential of the Langmuir probe (Figure 4,top-left and bottom-left) the profiles are much flatter. The differences in spatial structure between emissive and plugged probe (Figure 4, top-right) are due to the much smaller probe tip of the plugged probe in contrast to the wire loop of the emissive probe. This results in a much sharper visualisation of regions of large potential gradients, whereas the emissive probe tends to smear out these subtle spatial structures.

SUMMARY AND CONCLUSION

In most plasmas the floating potential of al Langmuir probe is a inadequately estimate for the plasma potential. There, the emissive probe is a well established technique for plasma potential measurements. However, in magnetised plasmas

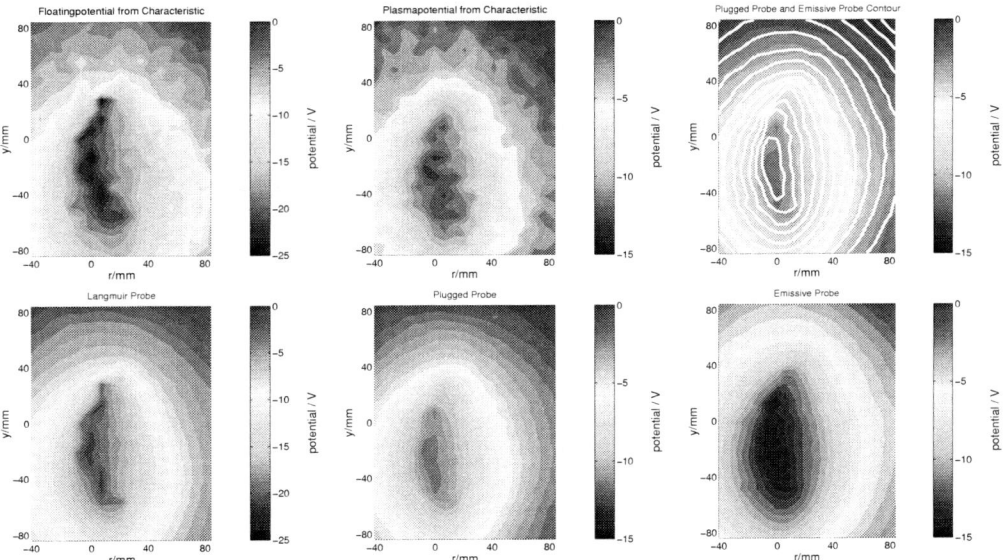

FIGURE 4. Comparison of 2d potential measurements. Top-Left: Floating Potential of the characteristic of a usual Langmuir probe. Top-Middle: Plasma Potential from maximum of first derivative of the characteristic of a usual Langmuir probe. Top-Right: Floating potential of a plugged probe with floating potential contours of the emissive probe. Bottom-Left: Floating potential for a Langmuir probe (single measurement). Bottom-Middle: Floating potential of plugged probe (single measurement). Bottom-Right: Floating potential of emissive probe (single measurement)

the plasma in a flux tube can be strongly influenced by the electron emission of the probe. In addition, the fabrication of emissive probes is complicated and it has a relatively short lifetime. In contrast, the plugged probe is easy to build and more durable. If the plasma is magnetised and the ions are cold, the plugged probe seems to be an alternative. If well aligned, the floating potential of a plugged probe gives an good estimate of plasma potential for a wide range of magnetic field and neutral gas pressures. Its has been shown, that the quotient of $I_{e,sat}$ and $I_{i,sat}$ can used to prove the alignment of the probe. If $I_{e,sa}/I_{i,sat} \approx 1$, the errors of the plasma potential measurements are very small. The maximum of the first derivative of the characteristic gives even for the plugged probe a valid measure of plasma potential and can be used for cross checks. The spatial resolution of plugged probes is much higher than the resolution of emissive probes due to their typically smaller size. The plugged probe is also an invasive technique, however, the disturbance of the plasma is minimised. Modern stepping motor and servo motor technique can be used to automatically align the plugged probe in devices with a magnetic field that changes its spatial angle to the probe tip. The sensitivity of the plugged probe to the alignment with the magnetic field can be used for time resolved measurements of the direction of magnetic field.

ACKNOWLEDGEMENTS

This work was supported by Deutsche Forschungsgemeinschaft under Contract No. Pi 185/22–1.

REFERENCES

1. N. Hershkowitz and M. H. Cho, *J. Vac. Sci. Tech.*, **A 6**, (1998).
2. V.I. Demidov *et al.*, *Rev. Sci. Instrum*, **70**, (1999)
3. O. Grulke *et al.*, *Plasma Phys. Controlled Fusion*, **43**, no. 4, (2001)
4. K. Rypdal *et al.*, *Plasma Phys. Controlled Fusion*, **36**, no. 7, 1994

Development of Frequency Tunable, Submillimeter Wave Gyrotron FU Series for Plasma Diagnostics

T. Idehara, S. Mitsudo, R. Pavlichenko, I. Ogawa[1] and K. Kawahata[2]

Research Center for Development of Far-Infrared Region, Fukui University
Fukui 910-8507, Japan
[1]*Cryogenic Laboratory, Faculty of Engineering, Fukui University*
Fukui 910-8507, Japan
[2]*National Institute for Fusion Science*
322-6, Oroshi-cho, Toki 509-5292, Japan

Abstract. High frequency, frequency tunable, medium power gyrotrons (Gyrotron FU series) are being developed in Fukui University as millimeter to submillimeter wave sources. The gyrotron series has achieved frequency tunability in a wide range from 38 to 889 GHz and medium output power from 0.1 kW to several kW. For the application to plasma diagnostics, modulations and stabilizations of both amplitude and frequency of their outputs have been achieved. The present status of the Gyrotron FU series is described.

INTRODUCTION

The submillimeter wave gyrotron FU series have achieved frequency tunability from 38 GHz to 889 GHz,[1] high harmonic operation up to fourth, modulation of amplitude and frequency[2], high stabilization in amplitude and frequency and high purity mode operations. Its maximum frequency of 889 GHz is the current record for high frequency operation of a gyrotron. The corresponding wavelength is 337 _m. The gyrotrons are also applied as a radiation source to plasma scattering measurement. In this paper, the present status of Gyrotron FU Series is summarized,

THE MAIN RESULTS OF GYROTRON FU SERIES

Gyrotron FU series includes 8 gyrotrons. Each gyrotron consists of a sealed-off gyrotron tube and a superconducting magnet, except Gyrotrons FU III and FU V, which have demountable tubes. Table 1 summarizes the main results of the gyrotrons included in Gyrotron FU series.

The design of each gyrotron was carried out by computer simulations. We are using narrow cavities to get a good mode separation and then to operate the gyrotrons in many single modes on fundamentals, second even third harmonics. Such a situation is important for our high frequency, harmonic gyrotrons.

The electron cyclotron frequency is $f_c = eB_0/(2\pi\gamma m_0)$, where e is the electric charge unit and m_0 the rest mass of the electron, B_0 the static magnetic field and $\gamma = (1-v^2/c^2)^{-1/2}$ the relativistic factor. When the energy of an electron is changed, γ which is related to the beam voltage V by $(\gamma-1)m_0c^2 = eV$ changes and, as a consequence, the electron cyclotron frequency f_c changes. This means that modulation of a cathode voltage enables a fast frequency modulation of gyrotron output. Gyrotron FU IV has achieved the frequency modulation within the limit of resonance frequency width of a cavity mode. The frequency modulation amplitude Δf versus the body potential modulation amplitude ΔV_b is plotted in Fig. 1 for several values of the modulation frequency f_m. There is an almost linear dependence between Δf and ΔV_b for all values of f_m. The efficiency of frequency modulation is $\Delta f/\Delta V_b$=0.247 MHz/V. The simulation results are also indicated in the figure. The estimated efficiency is distributed close to the experimentally obtained values.

TABLE 1. The results of Gyrotron FU series

Name of gyrotron	Frequency range	Items which each gyrotron has achieved
Gyrotron FU I	38-220 GHz	High frequency operation at 100GHz and 9 kW output power
Gyrotron FU E	90-300 GHz	Radiation source for the first experiment on ESR
Gyrotron FU IA	38-215 GHz	Radiation source for plasma scattering measurement of WT-3
Gyrotron FU II	70-402 GHz	Studies on mode competition and mode cooperation, Radiation source for plasma scattering measurement of CHS
Gyrotron FU III	100-636 GHz	3rd harmonic operation in single modes, Amplitude modulation, Frequency step switching
Gyrotron FU IV	160-847 GHz	Frequency modulation, cw operation for high stability of amplitude and frequency
Gyrotron FU IVA	160-889 GHz	Higher frequency operations by 3rd harmonics, High purity mode operation, Radiation source for ESR experiment
Gyrotron FU V	186-222 GHz	cw operation for long time using a He free magnet, High stabilizations of frequency and amplitude, High purity mode operation

HIGH FREQUENCY STABILIZATION BY PHASE LOCK CONTROL

The success of the frequency modulation suggests the possibility of high stabilization of the output frequency by the phase lock control of beam electron energy. We tried an experimental work on this subject. Fig. 2 shows a block diagram of the experimental setup.

FU IV operates in cw mode using stabilized high voltage power supplies for gun anode and cathode. The output frequency f_0 and power P_0 are 301.128 GHz and around 20 W, respectively. The IF signal with a frequency of around 40 MHz is fed on the phase lock loop circuit (PLL) and compared with the signal from a reference oscillator (a crystal oscillator) with a frequency of 40 MHz. The output voltage from PLL is amplified and then applied to the gyrotron body. A gyrotron cavity is included in the body. Therefore, the energy of beam electron injected into the cavity is modulated by the feedback control voltage, which is generated in PLL. The modulation sensitivity is adjusted at 1.6×10^4 Hz/V, by controlling the amplification factor. Fig. 3 shows typically observed frequency spectrum with phase lock control. During the phase lock control, a half value width of the frequency spectrum Δf and frequency fluctuation width δf are both smaller than 0.5 kHz and 0.1 kHz, respectively. This means that $\Delta f/f$ is smaller than 1.5×10^{-9} and $\delta f/f$ than 3×10^{-10}. A phase lock control of the beam electron energy makes the gyrotron operation very stable. This is the first experimental result on phase lock stabilization of frequency in a gyrotron.

High stabilization of output frequency enables us to use the gyrotron as a radiation source of spectroscopy in many fields. In addition, Gyrotron FU series is a frequency step-tunable radiation source. The frequency can be

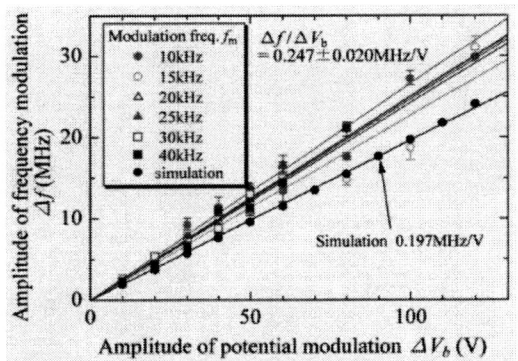

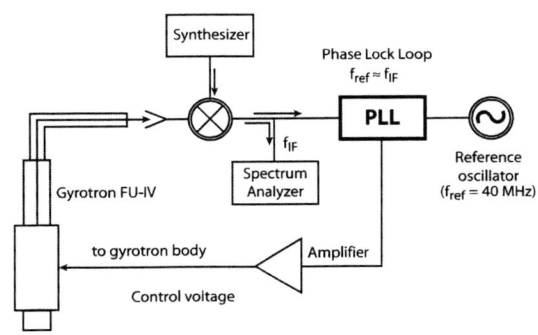

changed in the wide range from 38 GHz to 889 GHz. The output power is several ten watts, when it operates in cw mode. Therefore, a high power, high stable submillimeter wave source was just realized.

FIGURE. 1 The frequency modulation amplitude Δf versus the body potential modulation amplitude ΔV_b

FIGURE. 2 A block diagram of the phase lock control System. Modulation sensitivity 16 kHz/V.

APPLICATION TO PLASMA SCATTERING MEASUREMENT

As it is shown in Table 1, Gyrotron FU II is employed as a radiation source for plasma scattering measurements on the Compact Helical System (CHS) of the National Institute for Fusion Science (NIFS). The gyrotron output is transmitted by conventional over-sized circular waveguides and converted to a two-dimensionally focused, linearly polarized, quasi Gaussian beam, before it is injected into plasma. Scattered wave from plasma is received by horn antennae installed in the plasma vessel. It is converted into low frequency signal by a homodyne detection system.

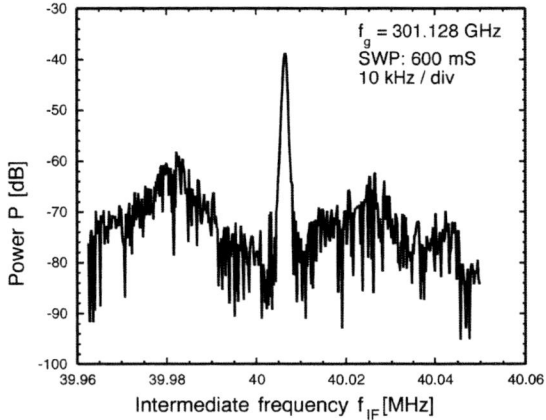 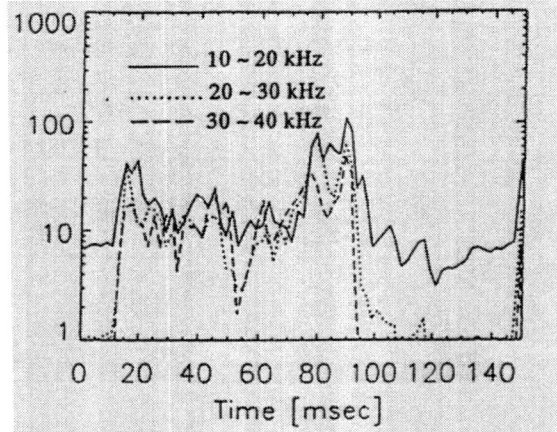

FIGURE. 3 Experimental and simulation results for frequency modulation amplitude $\_f$ versus amplitude of body potential modulation $\_V_b$

FIGURE. 4 Time evolution of scattered wave power for respective frequency interval

Fig. 4 shows a scattering measurement result for ICRF heated plasma. The plasma is fired at the time of 10 msec by an ECH pulse and an ICRF heating pulse is applied in the time interval from 40 to 90 msec. In Fig. 4, each curve shows time evolution of scattered wave power for respective frequency interval. The scattering angle of 8.8 degree corresponds to wave number of 11.4 cm^{-1}. The increase in scattered wave power is observed during the ICRF heating. The results of the reflection measurement are in reasonable agreement with the scattering measurement and support the availability of both measurements.

Observed scattered signals suggest that some instability connected with the drift wave occurs during ICRF heating. This phenomenon will be dangerous for a good plasma confinement. We will continue the measurement under various plasma parameters, to study the effect of this instability on plasma confinement.

SUMMARY

Gyrotron FU series in Fukui University has achieved frequency tenability in the wide range from 38 GHz to 889 GHz, fast modulation and high stabilization of its output frequency. The Gyrotron FU II has already been applied to plasma scattering measurement for CHS in NIFS. A drift wave excited during ICRF heating can be observed by the measurement.

The work was partially supported by the Grant in Aid from the Japan Society for Promotion of Science.

REFERENCES

1. T. Idehara, I. Ogawa, S. Mitsudo, IEEE trans. Plasma Sci. **27**, 340-354 (1999).
2. T. Idehara, M. Pereyaslavets, N. Nishida, Phys. Rev. Lett. **81**, 1973-1976 (1998).

Diagnostics and Simulation of Low-Frequency Inductively Coupled Plasmas

K. N. Ostrikov[1,2], I. B. Denysenko[3], E. L. Tsakadze[1,4], S. Xu[1], N. A. Azarenkov[3], and R. G. Storer[2]

[1]*Plasma Sources and Applications Center, NIE, Nanyang Technological University, 1 Nanyang Walk, 637616 Singapore;*
[2]*SOCPES, Flinders University, GPO Box 2100, Adelaide 5001, Australia;*
[3]*Faculty of Physics & Technology, Kharkiv National University, 4 Svobody sq., Kharkiv 61077, Ukraine;*
[4]*Optics and Fluid Dynamics Department, Risoe National Laboratory, P.O. Box 49, DK-4000, Roskilde, Denmark*

Abstract. The results on the diagnostics and numerical modeling of low-frequency (~460 KHz) inductively coupled plasmas generated in a cylindrical metal chamber by an external flat spiral coil are presented. Experimental data on the electron number densities and temperatures, and optical emission intensities of the abundant plasma species in low/intermediate pressure argon discharges are included. The spatial profiles of the plasma density, electron temperature, and excited argon species are computed, for different RF powers and working gas pressures, using the 2D fluid approach. The model allows one to achieve a reasonable agreement between the computed and experimental data. The effect of the neutral gas temperature on the plasma parameters is also investigated. It is shown that neutral gas heating at higher (> 1 kW) RF powers is among the key factors that control the electron number density and temperature. The dependence of the average RF power loss, per electron-ion pair created, on the working gas pressure shows that the electron heat flux to the walls appears to be a critical factor in the total power loss in the discharge.

INTRODUCTION

Experimental measurements have been carried out in the low-frequency (~460 KHz) inductively coupled plasma source described in detail elsewhere [1,2]. The plasma is generated in a cylindrical, stainless steel vacuum chamber with inner diameter 2R= 32 cm and length L = 20 cm (Schematic diagram of the discharge chamber see in [1]). The chamber is cooled by a continuous water flow in between the inner and outer walls of the chamber. The top plate of the chamber is a fused silica disk, 35 cm in diameter and 1.2 cm thick. A 450 l/s turbo-molecular pump backed by a two-stage rotary pump is used to evacuate the plasma chamber. The inflow rate and pressure of the working gas are regulated by a combination of a gate valve and MKS mass-flow controllers. The pressure is measured by an MKS Baratron capacitance manometer. The operating pressure of argon gas feedstock is typically maintained in the range $P = 20 - 200$ mTorr. The global plasma parameters, such as the electron/ion number densities, plasma potential and effective electron temperature have been estimated by means of a time-resolved RF-compensated single Langmuir probe technique. A number of holes in four rectangular side ports and in the bottom plate of the chamber allow one to insert the Langmuir probe into the plasma and move it in radial and axial directions. The optical emission from the ICP discharge has been collected using a light receiver mounted on different port-holes and transmitted via the optical fiber to a SpectroPro-750 spectrometer (Acton Research Corporation) with the resolution of 0.023nm. The optical emission spectra (OES) of excited neutral and/or ionized argon atoms have been investigated in the wavelength range 350 – 850 nm. Further details of the Langmuir probe and optical emission intensity measurements can be found elsewhere [1,2].

MODEL DESCRIPTION

Theoretically, the plasma chamber is modeled by considering a metal cylinder of the inner radius R and length L, with a dielectric disk of width d and permittivity ε_d atop. The components of the electromagnetic field are calculated assuming that the chamber is uniformly filled by the plasma with the electron/ion number density equal to the spatially averaged plasma density n. Since the column and inductive coils are fairly uniform in azimuthal direction, the problem is two-dimensional and the plasma parameters depend on r and z only. It is assumed that the ion temperature T_i is equal to the temperature of the working gas T_g. The latter was a variable parameter in the computations. The plasma is treated within the ambipolar model that assumes the plasma quasineutrality and the equality of electron and ion fluxes. Accordingly, the particle balance equation for the electrons or ions is $\partial_t n + \vec{\nabla} \cdot (n\vec{v}) = n\nu_i$, where ν_i is ionization rate, $\vec{v} \approx -\nabla(nT_e)/nm_i\nu_{in}$, $\nu_{in} \approx \sqrt{v^2 + v_{Ti}^2}/\lambda$ is the ion-neutral collision frequency, T_e is electron temperature, m_i, v_{Ti} and λ are the ion mass, the average ion thermal velocity and ion mean free path, respectively. The RF power balance in the discharge is described by [4]: $(3/2)n\partial_t T_e + \vec{\nabla} \cdot q_e \approx -nI_e + S_{ext}$, where I_e is the collision integral for the electrons, and $q_e \approx -(5/2-\Delta)nT_e/(m_e\nu_{en})\vec{\nabla}T_e$ is the heat flux density with $\Delta = (T_e/\nu_{en})\partial\nu_{en}/\partial T_e$, ν_{en} is electron-neutral collision frequency that is function of T_e. The term $S_{ext} \approx n\nu_{en}m_e u_{osc}^2$ denotes Joule heating of the electrons by the RF field. $u_{osc} \approx \left|eE_\phi^p\right|/[2m_e^2(\omega^2 + \nu_{en}^2)]^{1/2}$, Here E_ϕ^p is azimuthal component of electric field that has been calculated from Maxwell equations. Finally, the boundary conditions for the integrating of the particle and power balance equations have to be specified. First, because of symmetry, the radial gradients of the electron temperature and density are equal to zero at the chamber axis (r=0). It is assumed that at the column edge $r = R$ ($z=z_s$, where $z_s = 0$ or L) the radial (axial) component of the fluid velocity is equal to Bohm velocity. The computation has been carried out for the two different boundary conditions for the electron heat flux: i) $q_{er}(R,z) = T_e(R,z)(2+\ln\sqrt{m_i/m_e})n(R,z)\sqrt{T_e(R,z)/m_i}$, and $q_{ez}(r,z_s) = T_e(r,z_s)(2+\ln\sqrt{m_i/m_e})n(r,z_s) \times \sqrt{T_e(r,z_s)/m_i}$, where q_{er} and q_{ez} are the radial and axial components of the heat flux density, respectively, and ii) The electron temperature gradient vanishes at the boundary ($\vec{\nabla}T_e = 0$).

NUMERICAL AND EXPERIMENTAL RESULTS

The calculated 2D profiles of the electron number density and temperature are shown in Fig.1. The computed electron number densities and temperatures have been also compared with the ones measured by the Langmuir probe, with the tip positioned at z=5.6 cm and r=4.0 cm. Figs.1c,d display the computed and measured values of n_e and T_e as functions of the input RF power P_{in}.

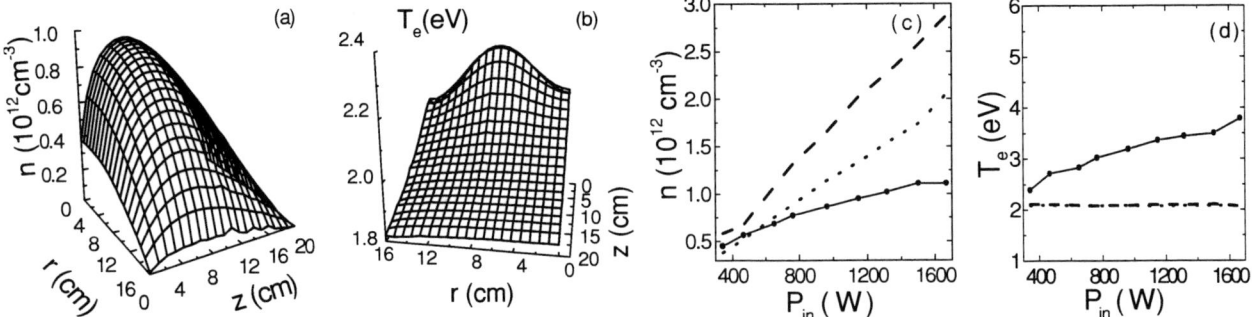

FIGURE 1. The computed electron density (a) and temperature (b) profiles in the chamber for the boundary condition for the heat flux i) at P_{in}=612.4 W. The computed and measured electron density (c) and temperature (d) in the chamber at z=5.6 cm and r=4.0 cm. Dotted and dashed curves are calculated for the boundary condition i) and ii), respectively. It the all cases P=28.5 mTorr, and T_g = 543K.

From Fig.1c one can see that the plasma density obtained using boundary conditions i) is closer to the experimentally measured one. Indeed, for non-vanishing heat flux boundary conditions i), in the RF power range of P_{in}< 800 W the computed plasma density almost fully recovers the experimental data (Fig. 1c). As the power increases (P_{in}> 800 W), the calculated plasma density becomes larger than the one obtained in experiment. The discrepancy can be attributed to the observed increase of T_e in the subsequent power range (Fig.1d). The electron number density calculated in case of the vanishing heat flux onto the chamber walls ($\nabla T_e = 0$) appears to be 20-30% larger when the one obtained assuming $\nabla T_e \neq 0$. To elucidate the role of the electron heat flux in the plasma column we have separated the average power loss per electron θ into the collisional θ_c and heat flux θ_f components, so that $\theta = \theta_c + \theta_f$. Here $\theta_c = (\pi L R^2 \bar{n})^{-1} \int_0^R \int_0^L n I_e 2\pi r dr dz$ and $\theta_f = (\pi L R^2 \bar{n})^{-1} \int_S q_{es} dS$ where S is the total area of the RF discharge, and q_{es} is the component of electron flux density perpendicular to the surface. The dependence of θ_c and θ_f on the working gas pressure are displayed in Fig.2a. The relative contribution of the heat transport component in the total power loss per electron $\theta_f / \theta \times 100\%$ is shown in Fig.2b.

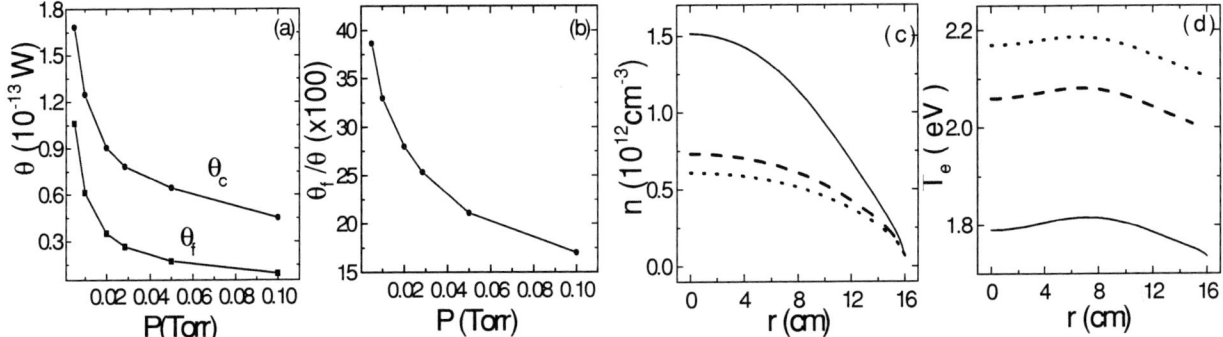

FIGURE 2. Power losses into collisions θ_c and due to electron heat flux θ_f (a), and the percentage of power lost through the electron heat flux θ_f / θ (b) for Tg =543 K. Radial profiles (z=6cm) of the electron density (c) and temperature (d) for different temperatures of neutral gas Tg. Solid, dashed, and dotted curves correspond to Tg =300, 543, and 700K, respectively. The other parameters are the same as in Fig.1a.

One can see that the heat flux contribution to the total power loss varies from 40% at low pressures ($P \sim 10$ mTorr) to 17% at 100 mTorr. It should be noted that both θ_c and θ_f decline with pressure, as does the contribution of the heat flux to the total power loss per electron. The results of Figs.2a,b clearly confirm the importance of the power loss through the heat flux to the chamber walls, which also strongly affects the value of the plasma density (Fig.1c). We remark that the discharge in question is powered with relatively high RF powers. Therefore, in the plasma regions remote from the water-chilled chamber walls, areas of elevated neutral gas temperature may appear. To study the effect of the neutral gas temperature on the plasma properties, T_g was varied in computations. The radial electron density and temperature profiles calculated at z=6 cm, P =28.5 mTorr, P_{in} =612.4 W for the gas temperatures T_g =300, 543, and 700 K relevant to our previous plasma processing and materials synthesis experiments [1,2] are presented in Figs.2c and 2d. One can see from Figs.2c,d that an increase of the neutral gas temperature exerts an almost similar effect on the plasma parameters as a decrease of the working gas density. This can best be understood by noting the apparent link $P = n_N T_g$ (where n_N is neutral gas density), which means that the fixed pressure conditions require that the neutral gas density, which enters the expressions for the most of the reaction rate coefficients, has to diminish when T_g rises. We have examined the spatial distribution of the excited argon atoms in the $3p^5 5p$ configuration. The profile of the excited atom density $n^*(r,z)$ has be calculated from the following equation: $n^*(r,z) \sim n(r,z) v^*(r,z)$, where $v^*(r,z)$ is the excitation rate. Figs.3a,b present the comparison of the radial and axial OEI profiles obtained experimentally and numerically.

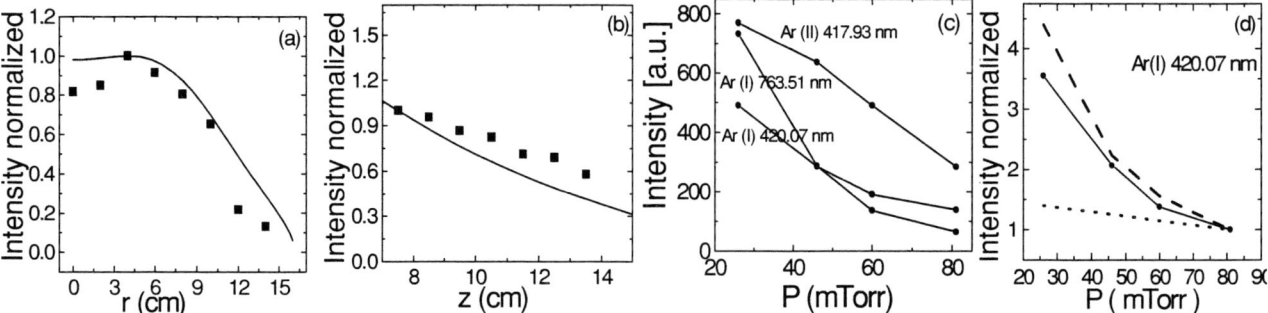

FIGURE 3. The measured (dots) and computed (solid curve) radial (a) at P=29.3 mTorr, P_{in}=536W and axial (b) at P=40 mTorr, P_{in}=960W profiles of the optical emission intensity of 420.07 nm atomic argon line. In both cases T_g=543 K. OEI of selected argon lines (c) *versus* pressure for P_{in}=630W. The position of the optical probe is z = 9.6 cm. Non-dimensional optical emission intensity (d) of Ar(I)420.07 nm line (normalized on its value at P=81 mTorr). Other parameters are the same as in Fig.3c. The solid, dashed, and dotted curves correspond to the experiment, calculations with $v^*(n_N) = const$ and $v^* \sim n_N$, respectively.

One can notice a remarkable agreement of the calculated emission intensities with the experimental data. However, a minor discrepancy can be seen in the radial profile in the vicinity of the discharge center. Fig.3c displays the experimental data on variation of the optical emission spectra with the operating pressure. The experiment has been carried out in the pressure range of 26 - 80 mTorr (argon gas flow rates were varied from 4 to 84 sccm). In the pressure range concerned it is seen that there is a consistent general trend of diminishing of the OEI with P. For computation of the dependence OEI(P), the 420.07 nm line of neutral argon has been selected. This dependence has been computed for the same axial position (z=9.6 cm) as in Fig.3c and is shown in Fig.3d. In calculations, two different assumptions have been made. The first one assumes that $v^*(r,z) \sim n_N$ and corresponds to the dotted curve in Fig.3d. The dashed curve is obtained assuming that $v^*(r,z)$ does not depend on the density of neutrals. In the first case, the agreement between the theory and experiment is satisfactory for the gas pressures exceeding 60 mTorr. The latter assumption provides much better consistency of the computational and experimental data within the entire range of operating pressures.

ACKNOWLEDGMENTS

This work was supported in part by the Agency for Science, Technology, and Research of Singapore (Project No. 012 101 00247), The Flinders Institute for Research in Science and Technology, and the Science and Technology Center in Ukraine (Project No. 1112).

REFERENCES

1. S. Xu, K. N. Ostrikov, W. Luo, and S. Lee, Journ. Vac.Sci. Technol. **A 18**, 2185 (2000).
2. K. N. Ostrikov, S. Xu, and A. B. M. Shafiul Azam, Journ.Vac. Sci. Technol. **A 20**, p.251 (2002).
3. M. A. Lieberman and A. J. Lichtenberg, *Principles of Plasma Discharges and Materials Processing*, Wiley, New York, 1994.
4. V. E. Golant, A. P. Zhilinskii, and I. E. Sakharov, *Fundamentals of Plasma Physics*, Wiley, New York, 1980.

Plasma Characteristics between Biased Facing Electrodes

Y. Saitou* and A. Tsushima[†]

Faculty of Engineering, Utsunomiya University, Tochigi 321-8585, Japan
[†]*Faculty of Engineering, Yokohama National University, Kanagawa 240-8501, Japan*

Abstract. A new diagnostic method, which is named facing double probe (FDP) method, for measuring plasma flow in a magnetized plasma has been developed. The method uses a pair of facing electrodes located along the magnetic field line and connected to each other by a power supply. Its shape is similar to a conventional double probe but the surface of each electrode is open only its facing side. The characteristic properties of the plasma between the facing electrodes with different potentials are investigated using one-dimensional particle-in-cell simulation in order to obtain better understanding of the FDP method.

INTRODUCTION

Measurement of plasma flow is one of the important approaches to understand basic physics related to plasma transport phenomena *etc*. There are various kinds of methods for measuring plasma flow, and each method has both advantages and disadvantages. We have recently developed a diagnostic method named facing double probe (FDP) method for measuring plasma flow in a magnetized plasma [1]. The probe consists of a pair of facing electrodes and a power supply as shown in Fig. 1. Although the surface of each electrode is open only its facing side, its shape is similar to a conventional double probe. According to the analysis based on a fluid model, a potential difference between the electrodes is proportional to a Mach number of the plasma flow outside the probe, that is, $\Delta V \propto M_\infty$ if the electrodes are floating independently and if the Mach number is less than a certain value. Besides, behavior of plasmas between the electrodes with different potentials is one of basic and interesting subjects. In many cases, as for the studies of sheath and presheath problems, a plasma is bounded at only one end and the other end is extended far away. Some cases treat a plasma bounded at the both ends, but boundary walls are assumed to have a same potential as far as authors know. In the model of FDP, however, we have to consider boundary walls with different potentials.

In this report, we investigate characteristic properties of the plasma between the facing electrodes with different potentials using one-dimensional particle-in-cell (PIC) simulation [2] in order to obtain better understanding both of the FDP method and of the behavior of plasmas bounded by walls with different potentials. Especially, we concentrate on how velocity distributions at the center and at both the boundaries depend on the external flow and the potential difference. Dependence of temperature at the central position on the flow velocity and on the potential difference is investigated as well.

SIMULATION MODEL

A plasma in an inner region between the facing electrodes is simulated using a one-dimensional PIC with generation and loss of particles [3, 4]. Ions with flow velocity, v_∞, and electrons move along a magnetic field between the electrodes. The electrons and ions are generated due to diffusion from the external region into the inner region, and they are lost at the electrodes due to recombination. By balance of the generation and the loss, the particle number at the central cell settles down at a constant after sufficient long time calculation. A distance from an electrode to another is approximately 200 times Debye length ($L \simeq 200\lambda_D$). The ion and the electron temperatures in the external region are assumed to be $T_{i,\infty} = 0.1 T_{e,\infty}$ and the mass ratio adapted in the simulation is $m_i/m_e = 200$ (m_i: the ion mass, and m_e: the electron mass). The potential of the electrode at the right-hand side is set to be V_R against that of the electrode at the left-hand side, which is always grounded ($V_L = 0$). By changing the value of the normalized potential $eV_R/k_B T_{e,\infty}$ from -5 to $+5$, the ion and the electron velocity distributions at several positions are investigated.

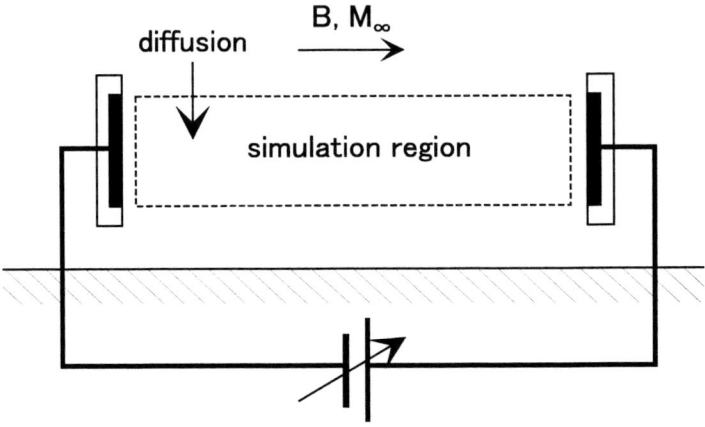

FIGURE 1. Schematic drawing of the facing double probe (FDP). Electrodes are connected to each other by a power supply and are placed along the magnetic field. The simulation region is also shown in this figure.

RESULTS AND DISCUSSION

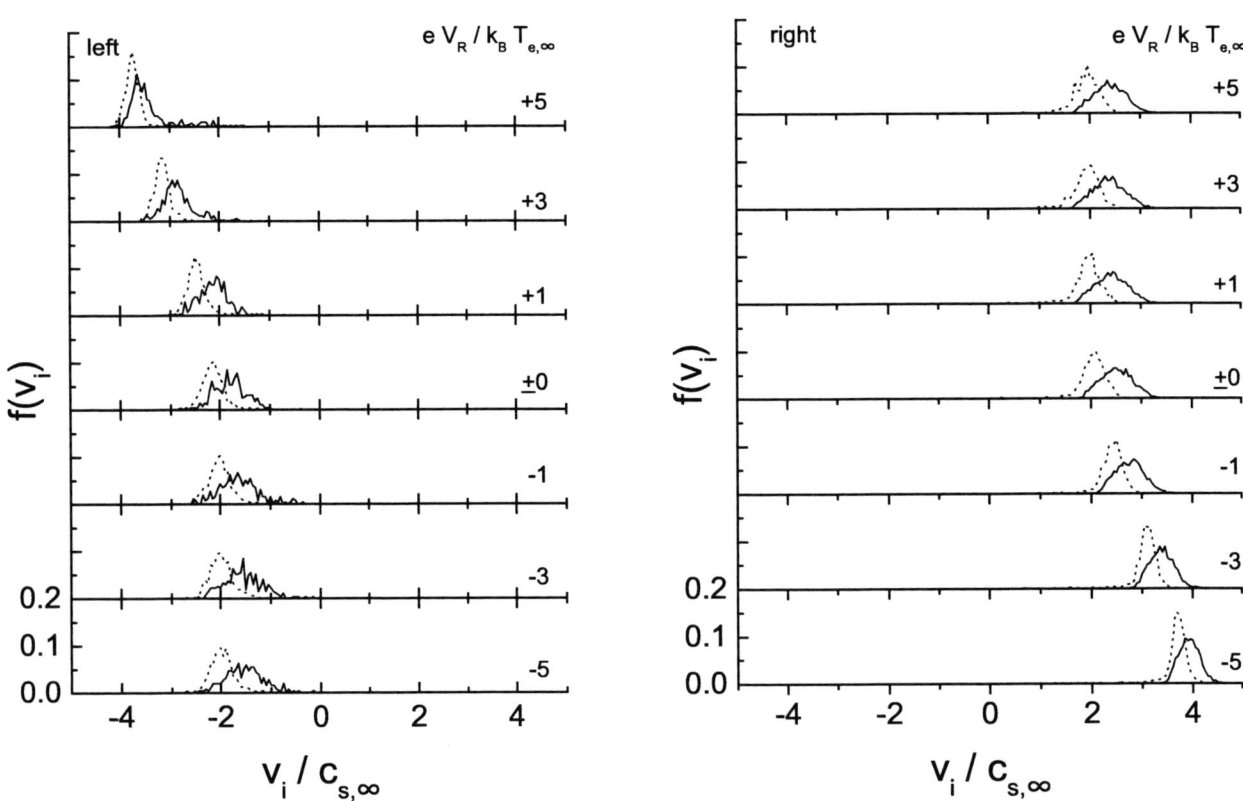

FIGURE 2. Ion velocity distributions just in front of the left-hand side and the right-hand side electrodes. Dotted and solid curves correspond to the cases of $v_{i,\infty}/c_{s,\infty} = 0.0$ and 1.8, where $v_{i,\infty}$ is the flow velocity of the external plasma. The potential of the right-hand side electrode is ranged $-5 \leq eV_R/k_B T_{e,\infty} \leq 5$ against that of the left-hand side electrode, $V_L = 0$.

At first, we investigate the ion velocity distribution at positions just in front of the left-hand and the right-hand side electrodes by biasing the right electrode, $eV_R/k_B T_{e,\infty}$, from -5 to $+5$ for $v_{i,\infty}/c_{s,\infty} = 0.0$ and 1.8, as shown in

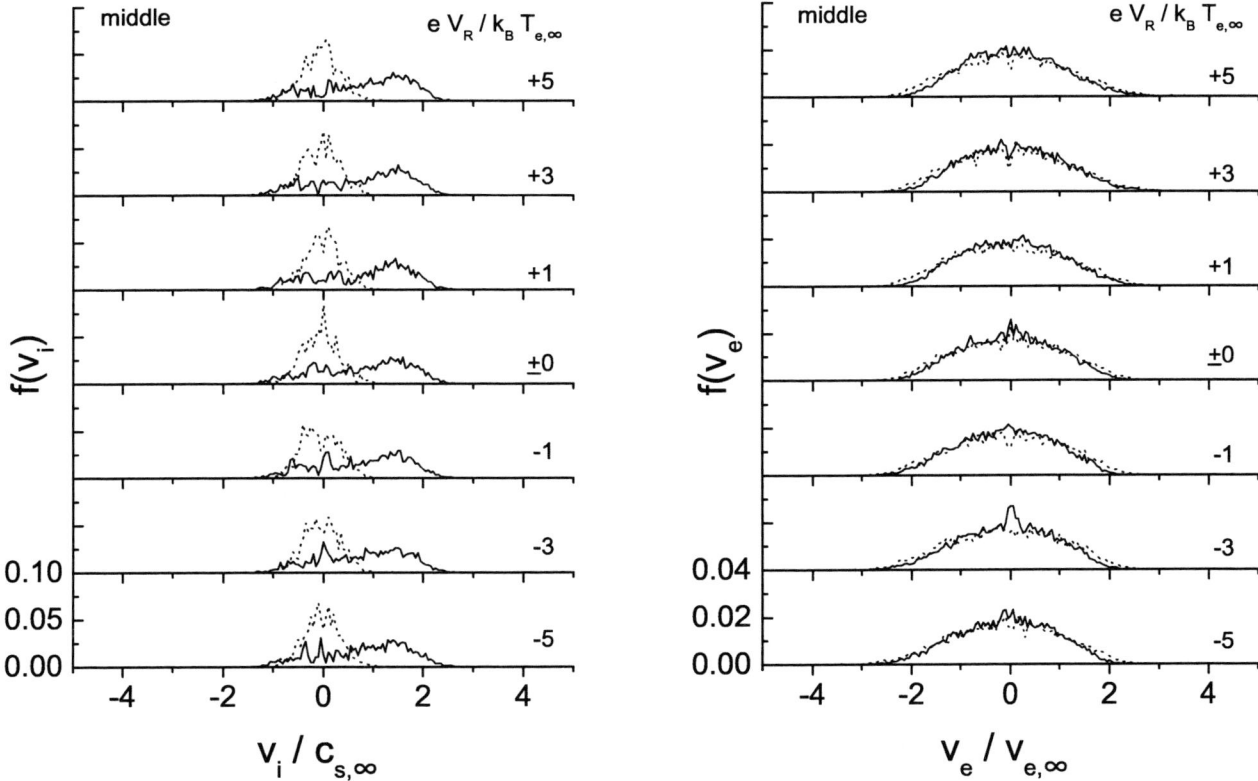

FIGURE 3. Velocity distributions of the ions and the electrons at the central part of FDP, where v_i is normalized by $c_{s,\infty} = \sqrt{k_B T_{e,\infty}/m_i}$ and v_e by $v_{e,\infty} = \sqrt{k_B T_{e,\infty}/m_e}$. Dotted and solid curves correspond to the cases of $v_{i,\infty}/c_{s,\infty} = 0.0$ and 1.8. The potential of the right-hand side electrode is ranged $-5 \leq eV_R/k_B T_{e,\infty} \leq 5$ against that of the left-hand side electrode, $V_L = 0$.

Fig. 2, where $c_{s,\infty} = \sqrt{T_{e,\infty}/m_i}$. In all cases, the averaged ion velocity much exceeds the ion acoustic velocity. The difference in each distribution, which appears as a shift of the distribution function, is due to a difference in an extent of acceleration by the sheath potential since the averaged ion velocity at the sheath edge has to be equal to the ion acoustic velocity according to the Bohm criterion [5]. The total shift of the ion distribution in each case corresponds to a depth of the sheath potential. So, it is found that the space potential in the inner region, measured from the left-hand side electrode($V_L = 0$), linearly rises with V_R when $V_R > 0$ because the total shift of the ion velocity distribution at the left increases with V_R while that at the right does not change when $V_R > 0$. On the other hand, it is found that the space potential in the inner region does not change when $V_R < 0$, because the total shift of the ion distribution at the left does not change when $V_R < 0$. If we see the above finding for $v_{i,\infty}/c_{s,\infty} = 0.0$ in detail, the total shift of the distribution at the left for $V_R > 0$ is similar to that at the right for $V_R < 0$. It means that the sheath formation at the left is the same with that at the right when $V_R = 0$. However, the velocity distribution for $v_{i,\infty}/c_{s,\infty} = 1.8$ shifts to the positive direction due to v_i in comparison with the case for $v_{i,\infty}/c_{s,\infty} = 0$. In Fig. 2, solid curves, which represent the velocity distributions for $v_{i,\infty}/c_{s,\infty} = 1.8$, are displaced to rightward from the broken curves, which represent those for $v_{i,\infty}/c_{s,\infty} = 0$. At the same time, it is found that the velocity distributions for $v_{i,\infty}/c_{s,\infty} = 1.8$ become a little broader than those for $v_{i,\infty}/c_{s,\infty} = 0$.

Now, as shown in Fig. 3, let us see the ion and the electron velocity distributions at the central part of the inner region. We note that the central part is, so-called, a presheath, which is a transition region where the ions are accelerated to satisfy the Bohm criterion at the sheath edge. The electron velocity distribution is deformed by applying neither V_R nor $v_{i,\infty}$. It means that the presheath region is shielded by the sheath and the effect of the biased potential V_R does not penetrate into this bulk of plasma region. That is a reason for the plasma potential rises linearly when $V_R > 0$ and does not respond when $V_R < 0$. In other words, the potential difference is localized in the left sheath when $V_R > 0$ and in the right sheath when $V_R < 0$. However, if we see the electron velocity distribution in detail, a tail of the distribution is deformed when $V_R \neq 0$: a population of the positive (negative) tail is magnified and a population of the negative (positive) tail is reduced with increasing the absolute value of V_R when $V_R > 0$ (< 0). This agrees with the fact that the potential difference at the left (right) sheath increases when $V_R > 0$ (< 0), where most of the electrons are reflected. Therefore, when $V_R \neq 0$ or the electrodes have different potentials by biasing one of the electrodes, a current flows due to the deformation of the tail of the electron velocity distribution function. This is the same for both cases of $v_{i,\infty}/c_{s,\infty} = 0$ and 1.8, although the electron temperature for $v_{i,\infty}/c_{s,\infty} = 1.8$ is a little smaller than that for $v_{i,\infty}/c_{s,\infty} = 0$. As for the ion velocity distribution at the central part, it is not deformed by V_R. However, the ion velocity distributions for $v_{i,\infty}/c_{s,\infty} = 1.8$ is deformed remarkably to the direction of $v_{i,\infty}$ if compared with those for $v_{i,\infty}/c_{s,\infty} = 0$. Simultaneously, the ion temperature, which is determined from the standard deviation of the ion velocity distribution, is much larger for $v_{i,\infty}/c_{s,\infty} = 1.8$ than that for $v_{i,\infty}/c_{s,\infty} = 0$.

SUMMARY

The behavior of both the ions and the electrons in the space between the two electrodes with different potentials, placed along the magnetic field in a flowing plasma, is investigated using the one-dimensional particle-in-cell simulation code. From the shift of the ion distribution function at the sheath regions, we consider the response of the sheath potential by biasing one of the electrodes whose potential is V_R and find that the plasma potential is determined by the higher potential of the electrodes. We also find that the difference in the ion velocity distributions at the left and the right sheaths is introduced by the plasma flow whose velocity is $v_{i,\infty}$. The fact that the tail of the electron velocity distribution at the central part is influenced by the biased potential V_R is related to this response of the sheath potential. That is why the current arises by the biased potential V_R. On the other hand, the ion velocity distribution at the central part is not deformed by the biased potential V_R but is deformed remarkably by the ion flow when $v_{i,\infty} \neq 0$.

From the viewpoint of the FDP, the ion flow of the external region changes the ion velocity distribution at the central part but not the electron one. The biased potential between the electrodes changes the tail of the electron velocity distribution but not the ion one. In terms of the current streaming between the electrodes, the biased potential causes the electron current, and the ion flow in the external region causes the ion current, with hardly deforming the ion velocity distributions.

REFERENCES

1. Saitou, Y., and Tsushima, A., *Jpn. J. Appl. Phys.*, **40**, L1387–L1389 (2001).
2. Birdsall, C. K., and Langdon, A. B., *Plasma Physics via Computer Simulation*, Adam Hilger, New York, 1991, pp. 1–301.
3. Saitou, Y., and Tsushima, A., *J. Phys. Soc. Jpn.*, **70**, 3201–3204 (2001).
4. Tsushima, A., and Kabaya, S., *J. Phys. Soc. Jpn.*, **67**, 2315–2321 (1998).
5. Chen, F. F., *Introduction to Plasma Physics and Controlled Fusion, 2nd edition, Volume 1: Plasma Physics*, Plenum, New York, 1985, pp. 292–294.

Experimental Study of Axial Plasma Parameter Variations in the Cylindrical Magnetron Discharge

P. Kudrna[1,2], M. Holik[1], O. Bilyk[1], I.A. Porokhova[1,3], Yu.B. Golubovskii[3], M. Tichy[1,2], J.F. Behnke[1]

[1]*University of Greifswald, Institute of Physics, Domstrasse 10a, 17487 Greifswald, FRG*
[2]*Charles University in Prague, Faculty of Mathematics and Physics, Department of Electronics and Vacuum Physics, V Holesovickach 2, 180 00 Praha 8, Czech Republic*
[3]*St. Petersburg State University, Department of Optics, Ulianovskaia 1, St.Petersburg, 195904, Russia*

Abstract. In the cylindrical magnetron the electric field is applied in radial direction and the magnetic field in axial direction. In this paper we present a study of the variations of plasma parameters in both the axial as well as in radial directions in the novel construction of cylindrical magnetron developed in the University of Greifswald, FRG. Six evenly distributed coils create the axial magnetic field. The homogeneity of the magnetic field ±0.2 % has been achieved over the whole discharge vessel length 300 mm (vessel diameter 58 mm). The system is equipped with three cylindrical Langmuir probes movable in radial direction, placed in ports located in between each couple of coils in distance 60 mm from each other. In order to measure the axial variations of the discharge current, one half of the cathode length is segmented into 14 segments, i.e. one segment has a length of about 10 mm. This enables the measurement of the axial variations of the discharge current. We present measurements of the axial distribution of the discharge current at different magnetic fields. We also demonstrate measurements of the axial and radial variations of the plasma density.

INTRODUCTION

The cylindrical magnetron system is being commercially exploited [1] as well as intensively studied, see e.g. [2,3]. During past several years some progress has been achieved in description of the dc discharge in cylindrical magnetron at the University of Greifswald, FRG, see e.g. [4,5,6,7]. In these studies the extensive experimental investigation of the dc discharge in cylindrical magnetron system (110 mm in length and 60 mm in diameter) by Langmuir probe as well as kinetic, PIC and fluid modeling has been performed [8]. The comparison of the measured and calculated results revealed satisfactory agreement in most of the investigated plasma parameters, namely the plasma density, the electron mean-energy, and the tail of the electron energy distribution function. Nevertheless the discrepancy between the experimentally measured and theoretically modeled radial course of the electric field (plasma potential) lead the investigators to the suggestion that the cause of the disagreement consists in the end effects which were included neither in the fluid nor the PIC-MCC 1-D models [9]. In order to get closer to the ideal supposition of the infinitely long system a novel system of the cylindrical magnetron of 300 mm in length and 40 mm in diameter has been designed in the University of Greifswald, FRG [10]. The experimental investigation of this new system showed, however, that there are also significant axial changes of the plasma parameters (former system did not allow the investigation of plasma parameters in axial direction). Moreover, the dc discharge in the mentioned both versions of the cylindrical magnetron, which we have investigated, demonstrates instabilities, the amplitude of which depends on the magnitude of the magnetic field with the onset being at about 15 mT. Instabilities of different sorts and physical mechanisms have already been observed in magnetically confined dc discharges, see e.g. [11,12]. In view of the new experimental facts the novel experimental system in Greifswald has been further modified to enable investigation also of the longitudinal variations of the plasma parameters. This was performed by constructing the segmented cathode, which enabled the measurement of the longitudinal variations of the discharge current with spatial resolution of about 10 mm. In addition three Langmuir probes were placed at different distances along the discharge tube length (in between the magnetic field coils), which enabled simultaneous measurement of the radial variations of plasma parameters at three axial positions within the discharge vessel.

EXPERIMENTAL

The new system of cylindrical magnetron is schematically depicted in Figure 1. The system has been in detail described in [13], so that only its basic data will be given here. The system is made of stainless steel. The length of cylindrical discharge vessel is 300 mm and its diameter is 58 mm. The diameter of cathode is 18 mm, and it is placed co-axially with the discharge vessel that serves as anode. Half of the cathode length is segmented into 14 segments of the length approximately 10 mm each. The other half of the cathode is not segmented. The cathode is water-cooled in order to prevent its overheating due to the large power density on its surface. The strength of the

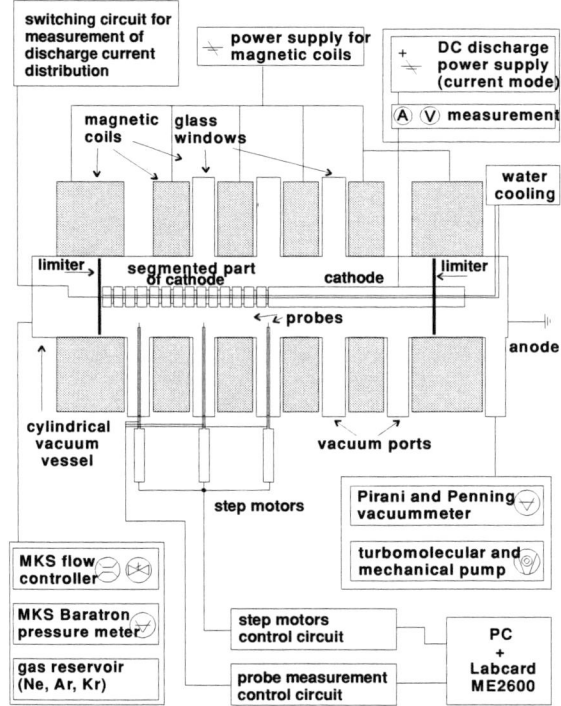

FIGURE 1. Schematic diagram of the cylindrical magnetron system.

axial magnetic field can be changed from 0 to 40 mT. The measured homogeneity of the magnetic field is better than ±0.2 % along all the vessel length. The system is equipped with 5 vacuum ports placed 6 cm apart from each other. In present experiments three of them have been used for Langmuir probes. The installed Langmuir probes have cylindrical tips (tungsten, 47 μm in diameter, 2.5 mm in length) and are radially movable by means of stepping motors. Their construction is insensitive to the sputtering of the cathode material, which occurs intensively at higher discharge currents (approximately above 100 mA). In all measurements the probe wires were oriented perpendicular to the magnetic field lines. The probe characteristics were measured by a data acquisition system that consisted of a lab-card ME2600, and a circuit for generation of the probe bias and for probe current-to-voltage conversion.

RESULTS OF MEASUREMENTS

Sample of the longitudinal distribution of the discharge current as measured by using the segmented cathode in argon is depicted in Figure 2. The measurements were taken at 15 combinations pressure-magnetic field in the range p=(1.5÷4) Pa, B=(15÷30) mT. We selected results at magnetic field 20 mT and pressure 4 Pa. These results are typical for the axial distribution of the discharge current in our magnetron system. The parameter of the dependencies is the total discharge current measured by the external ammeter. Generally, at lower pressures the effect of the vessel end protrudes at lower discharge currents deeper to the discharge volume than in case of higher pressures. These measurements clearly showed that the diaphragms (shields, limiters) at the vessel ends significantly affect the discharge into the depth of several cm. The kinetic model of the influence of the vessel ends on the plasma parameters within the discharge vessel is given in another contribution to this conference [14].

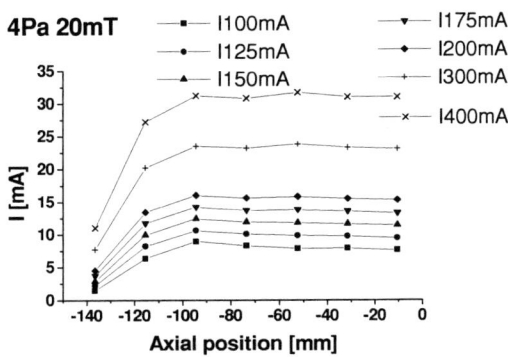

FIGURE 2. The longitudinal distribution of the discharge current as measured by using the segmented cathode in argon at p=1.5 Pa and B=20 mT. Parameter is the total discharge current.

We performed also probe measurements of the radial variations of electron density at three different positions along the discharge vessel simultaneously. The measurements were taken at different pressures and magnetic fields in the similar range as above. The selected results at argon pressure p=3 Pa and B=30 mT are depicted in Fig. 3.

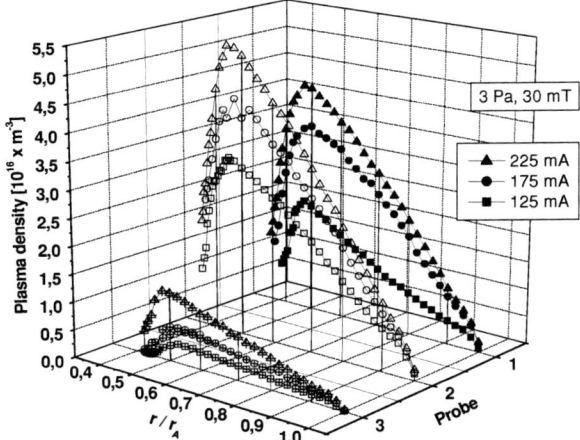

FIGURE 3. The radial variations of electron density at three different positions along the discharge vessel in argon at p=3 Pa and B=30 mT. Parameter is the total discharge current. Probe positions - see text.

Parameter of the graphs is the total discharge current. The probe 1 is the probe located in the middle of the discharge vessel length, the probe 2 is positioned 60 mm apart from probe 1 and the probe 3 is placed 120 mm apart from probe 1. The positions of the probes cover then one symmetric half of the vessel length; the outermost probe being placed at approximately 3 cm distance from the diaphragm, which limits the length of the discharge vessel. The volt-ampere characteristics were recorded simultaneously at all three probes in order to exclude possible variations of the plasma parameters with time. The data acquisition for a set of three characteristics took less than 30 seconds; the whole measurement of three radial dependencies took approximately quarter of an hour. From these data it is evident that there is a distinct minimum of electron density in the middle of the discharge vessel at all three investigated discharge currents. Such minimum has been observed at certain experimental conditions also on the axial discharge current distribution.

The cylindrical magnetron dc discharge is also affected by instabilities, which arise in decent amplitudes at the magnetic fields higher that approx. 15 mT. The character of these instabilities is not yet known, some hints may be found in [11,12]. We attempted to characterise the fluctuations of the floating potential of the Langmuir probe in three rare gases, namely in Ne, Ar and Kr. Measurements were carried out by means of the sampling oscilloscope at the pressure 4 Pa and with a magnetic field ranging from 15 to 40 mT. The data from the oscilloscope (400 000 data points for each measurement) were downloaded to a computer and with help of FFT the spectra were calculated. At low magnetic field (<20 mT) the spectra of oscillations were found flat with amplitudes below 1 mV. However, at certain discharge conditions and magnetic fields above 20 mT two distinct peaks appeared on the spectrum. The frequency of these peaks may be related (within experimental error) to gyrofrequencies of atomic and molecular working gas ions in all three cases of the used working gases Ne, Ar and Kr. The interesting continuation of these investigations may be the pursuit of the two-point spectral analysis as it was done in planar dc magnetron in [11].

DISCUSSION

It is possible to qualitatively explain the axial distribution of the electron density at the ends of the discharge vessel by the influence of the diaphragms (shields, limiters), which axially terminate the discharge vessel. The diaphragms are connected to the cathode, and hence they are biased at the cathode potential. There is a sheath in the vicinity of the shields, which screens the plasma inside the vessel from the highly negative potential of the shields. The thickness of the sheath around a charged body immersed in a plasma depends on the Debye length, and on the difference between the potential of the body and of the plasma at the place where the body is located. Sheath thickness can be calculated from a theory, see e.g. [15], or numerically, see e.g. [16]. Typically, at moderate potentials across the sheath, the influence of the charged body spans into plasma to the depth approximately twenty times the Debye length [16]. As the diaphragms are at the cathode potential, the electron density in their vicinity (the same as at the cathode) is very small. For the estimation of the depth of penetration of the influence of the shields into the plasma we hence take the approximate electron density 10^{14} m^{-3}. At such electron density and $T_e=1$eV the Debye length amounts to approximately 0.95 mm, and the charged body will then influence the plasma to the depth approximately 20 mm. By assuming the electron density in the region of the shields 10^{13} m^{-3} we would increase this figure approximately by factor three. This is, however, only very rough estimate, but together with the longitudinal diffusion and recombination at the diaphragm surface it could, qualitatively, account for the decrease of the plasma density towards both ends of the vessel depicted in figures 2 and 3.

ACKNOWLEDGMENTS

The work in Greifswald was financially supported by the Deutsche Forschungs-Gemeinschaft (DFG) in frame of the project SFB 198 Greifswald "Kinetik partiell ionisierter Plasmen" and by DAAD trilateral project of the University in Greifswald, Paris-Sud and St. Petersburg. The work in Prague was partially financially supported by the Grant Agency of Czech Republic, Grant No. 202/00/1689, 202/00/1217, 202/01/D095, by the Grant Agency of Charles University, Grant No. 171/2000/B FYZ /MFF, by project COST action 527.70 and by the Ministry of Education, Youth and Sports, Research plan MSM 113200002.

REFERENCES

1. Zhang, Qi-Chu, Zhao, K., Zhang, B.-C., Wang, L.-F., Shen, Z.-L., Lu, D.-Q., Xie, D.-L., Zhou, Z.-J., Li, B.-F., *J. Vac. Sci. Technol.*, **A16**, 628-632 (1998).
2. van der Straaten, T. A., Cramer, N. F., *Physics of Plasmas*, **7**, 391-402 (2000).
3. Gracin, D., Denkelmann, R., Maurmann, S., Andreic, Z., *Contrib. Plasma Phys.*, **40**, 120-125 (2000).
4. Behnke, J. F., Passoth, E., Csambal, C., Tichy, M., Kudrna, P., Trunec, D., Brablec, A., *Czech. J. Phys. B*, **49**, 483-498 (1999).
5. Passoth, E., Kudrna, P., Csambal, C., Behnke, J. F., Tichy, M., Helbig, V., *J. Phys. D: Appl. Phys.*, **30**, 1763-1777 (1997).
6. Passoth, E., Behnke, J. F., Csambal, C., Tichy, M., Kudrna, P., Golubovskii, Yu. B., Porokhova, I. A., *J. Phys. D: Appl. Phys.*, **32**, 2655-2665 (1999).
7. Golubovskii, Yu. B., Porokhova, I. A., Csambal, C., Behnke, J., Behnke, J. F., in *Proc. 14th International Symposium on Plasma Chemistry*, edited by M. Hrabovsky, M. Konrad, V. Kopecky, Praha (Czech Republic), August, 2-6, 1999, ISBN 80-902724-0-1, Vol. II (ISBN 80-902724-2-8), pp. 669-674.
8. Porokhova, I. A., Golubovskii, Yu. B., Bretagne, J., Tichy, M., Behnke, J. F., *Physical Review E*, **63**, (2001) art. No 056408 (9 pages), May issue.
9. Behnke, J. F., Csambal, C., Rusz, J., Kudrna, P., Tichy, M., *Czech. J. Phys.*, **50/S3**, 427-432 (2000).
10. Behnke, J. F., Rusz, J., Kudrna, P., Tichy, M., in *Proc. ICPP-2000*, edited by R. Decoste et al., Quebec (Canada), Quebec University 2001, Vol. III, pp. 944-947.
11. Gyergyek, T., Cercek, M., Jelic, N., Stanojevic, M., *Contrib. Plasma Phys.*, **33**, 53-72 (1993).
12. Martines, E., Cavazzana, R., Serianni, G., Spolaore, M., Tramontin, L., Zuin, M., Antoni, V., *Physics of Plasmas*, **8**, 3042-3050 (2001).
13. Holik, M., Kudrna, P., Bilyk, O., Rusz, J., Tichy, M., Behnke, J. F., Porokhova, I. A., Golubovskii, Yu. B., *Czech. J. Phys.*, **52**, Suppl. D, pp. D673-D680 (2002).
14. Porokhova, I. A., Golubovskii, Yu. B., Kudrna, P., Tichy, M., Behnke, J. F., "Kinetic Modelling of Axially Non-Uniform Cylindrical Magnetron Discharge" in *Proc. International Congress on Plasma Physics 2002 (this volume)*, accepted.
15. Bettinger, R. T., Walker, E. M., *Phys Fluids*, **8**, 748-751 (1965).
16. Rohmann, J., Klagge, S., *Contrib. Plasma Phys.*, **33**, 111-123 (1993).

Experimental Results of Growth and Breakdown Mechanisms in the Final Stage of a Pseudospark Discharge

M. Zambra[1], J. Moreno[1], L. Soto[1], and L. Rosales[2]

[1]*Research and Development Department, Chilean Nuclear Energy Commission, Amunátegui 95, Casilla 188-D, Santiago, Chile*
[2]*Physics Faculty, Santiago de Chile University, Santiago, Chile*

Abstract. The Pseudospark Transient Hollow Cathode Discharge (THCD) is a high-voltage low-pressure discharge, which is characterized by an axial hollow cathode electrode. This work presents optical and electric experimental diagnostics in the final stage of THCD in Hydrogen gas at working pressure between 100 *mTorr* and 250 *mTorr*. The characteristic signals of the electron and ion beam emission, growing plasma diagnosed by means of the H_α light emission at 656 *nm* from a point behind the cathode aperture, and the arrival to the cathode of the moving virtual anode are temporal and spatial correlated with the ICCD image of the growing plasma in the Hollow Cathode Region (HCR), and compared with our previous results. A performed statistical method lead us to suggest a sequence of ionization growth events which is analyzed to understand the mechanisms involved in the final stages just before breakdown.

INTRODUCTION

The fundamentals of the Transient Hollow Cathode Discharge (THCD) were described for the first time by Frank and Christiansen [1]. In the time, two regions become center of interest: the interelectrode region (A-K gap) and the space behind the hollow cathode called the Hollow Cathode Region (HCR), the axial hollow in the cathode electrode communicates both. It was found that electric breakdown formation at low pressure can be separated in three distinguishable regimes of ionization growth [2]: (i) initial ionization growth with plasma formation close to the anode which creates the virtual anode potential; (ii) extension of the mentioned ionization growth in the A-K gap, leading to a displacement of the anode potential from the anode toward the cathode; and (iii) ionization growth inside the HCR under enhanced field due to the close proximity of the anode potential. Our recent experimental results showed in this paper give us further knowledge about the ionization growth and breakdown formation mechanisms in the third regime described above.

The third regime of ionization growth in a THCD is characterized by five sequential events [3] which are associated with the breakdown formation mechanisms. These events are: (a) arrival of the virtual anode at the proximity of the cathode; (b) onset of plasma formation inside the HCR, (c) penetration of the anode potential in the HCR through the cathode aperture, (d) onset of a fast electron density growth inside the HCR, final electric breakdown in the A-K gap. Although the physical mechanisms of the pseudospark are not yet fully understood, a statistical analyze lead us to suggest a time sequence of the events in the last phase prior to breakdown [3,4]. Further experimental results presented in this paper enrich the description mentioned above. Images of the HCR, before and after electrical breakdown, are captured by mean of an 2D-ICCD camera and correlated in time with the mentioned events. Also, the ion beam emitted from the HCR is detected using a simple faraday cup placed axially and close to the cathode hole in the HCR.

Characteristic delay times from the application of the voltage to electric breakdown can range from a few microseconds to a few hundreds of nanosecond. The experiments have been performed in Hydrogen at pressure between 100 *mTorr* and 250 *mTorr*, with a cathode aperture of 5-*mm*.

EXPERIMENTAL SETUP

A scheme of the experimental setup is shown in Fig. 1a. The discharge apparatus uses a pulsed charged capacitor scheme to produce up to a 30 *kV* step across the electrodes with a rise time of less than 50 *ns*. Further details of the experimental setup have been published elsewhere [2]. The diagnostics used to measure the relevant events associated with this study have been described in detail in reference [2]. The presence of the virtual anode inside the main gap and close to the hollow cathode has been diagnosed using a capacitive probe (A_6). Electron beams were monitored using a beam-target optical fiber scintillator-photomultiplier combination, sensitive to electrons with energies above 5 *keV* (SC). Ion beams were detected using a simple faraday cup (FC) consisting in a copper disc which diameter was equal to 12.2-*mm* and placed 10-*mm* behind the cathode electrode in the axial position inside the HCR. The faraday cup bias voltage was –206 *V*. The plasma formation (H_α) inside the HCR, and close to the cathode aperture, is monitored by collecting the H_α light from a small region on the axis of the discharge using an optical fiber array. The electronic density (n_e) inside the HCR is obtained from the Stark broadening of the H_β line (486.13 *nm*) measured by means a spectrometer. Typical signals obtained with the different electric diagnostics are shown in Fig. 1b, in which t=0 corresponds to the instant of electric breakdown. Images of the emitted light inside the HCR were detected using a 2D-ICCD camera.

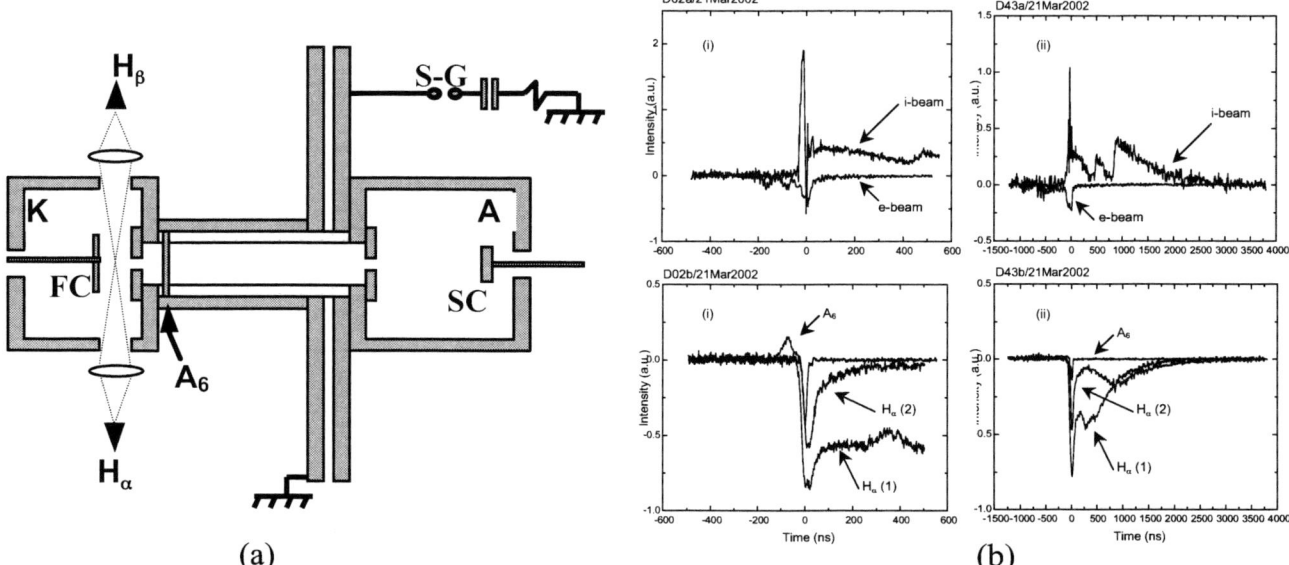

FIGURE 1. (a) Scheme of the experimental device: anode (A), cathode (K), capacitive probe (A_6), electron beam probe (SC), ion beam probe (FC), emission probes H_α and H_β. (b) Typical diagnostic signals at (i) 250 *mTorr* and (ii) 100 *mTorr*, obtained with a 5-*mm* cathode aperture. From the top to bottom; e-b, high-energy electron beam signal; i-b, ion beam signal; A_6, temporal evolution of electrical signal, arrival of the virtual anode close to the cathode in the main gap; and H_α, the plasma formation inside the HCR: $H_\alpha(1)$ and $H_\alpha(2)$ are focalized 3-*mm* and 9-*mm* from the edge of the cathode electrode respectively. A negative time value represents the time before breakdown (t = 0).

EXPERIMENTAL RESULTS

A single characteristic time has been chosen to represent the occurrence of each event associated to the different part of ionization growth in breakdown formation [2]: the maximum value for the virtual anode potential signals A_6, the maximum negative intensity value for the electron beam signal e-b, the onset and subsequent evolution of the i-b signal, and the onset and subsequent increase of H_α signal. Fig. 2a shows the characteristic mean times, over more than 300 consecutive shots, associated with these four characteristic ionization events, in the 100 to 250 *mTorr* pressure range, with 5-*mm* cathode aperture. In Fig. 2a, all characteristic mean times correspond to the elapsed time between the occurrence of the particular event and the electric breakdown in the A-K gap. Also the respective characteristic mean times are in agreement with the characteristic times obtained from an improved statistical analysis, using the von Laue formalism [3,4]. So Fig. 2a represents a global sequence of events, which starts with the arrival of the virtual anode close to the cathode, follows with the detected onset ion beam signal, continues with the onset of plasma formation inside the HCR, and finally the occurrence of the high-energy electron beam.

Fig. 2b shows the temporal evolution of the electron density inside the HCR, close to the cathode aperture at different pressures. The time equal to zero represents the instant of electric breakdown. The electron density is obtained from the Stark broadening of the H_β line. In this case, the calculation of half width of the H_β line have been used to infer the electron density from the measured spectra.

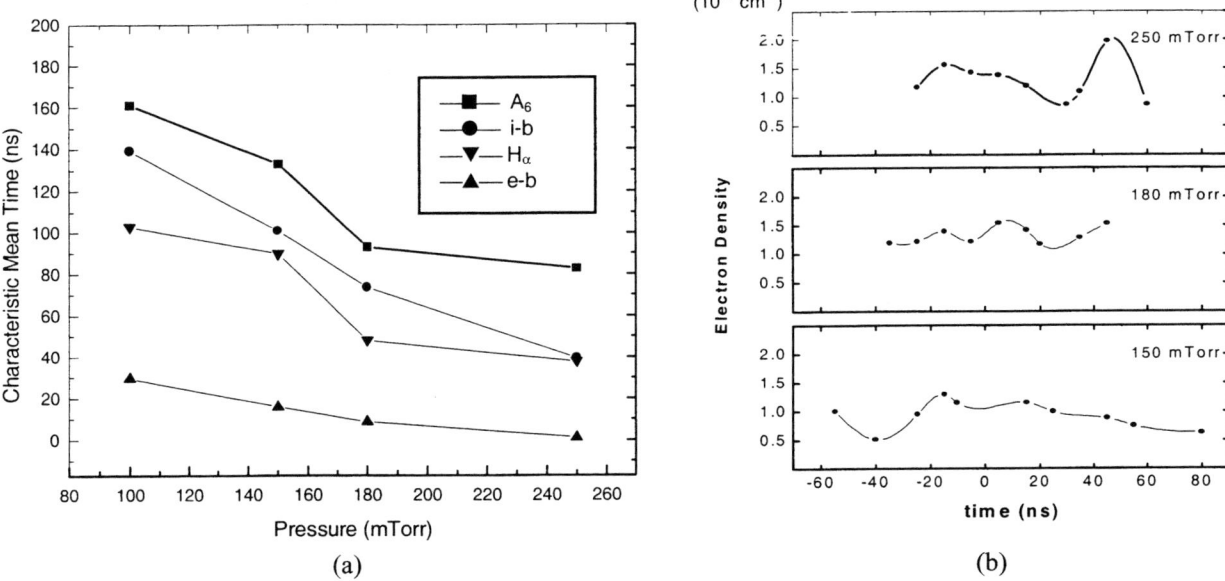

FIGURE 2. (a) Characteristic mean times associated with the different events, as a function of pressure at 5 mm cathode aperture: (1) arrival of the virtual anode (A_6) close to the cathode; (2) onset of the ion beam activity; (3) onset of plasma formation (H_α) inside the HCR; and (4) peak in final period of high energy electron beam emission. In this case, positive value indicates time before breakdown. (b) Time evolution of the electron density inside the HCR, relative to electric breakdown at t = 0, for different pressures at 5-*mm* cathode aperture. In this case, negative value indicates time before breakdown.

DISCUSSION

In a previous work [2,3] it was suggested that the third regime, which corresponds to the condition when the hollow cathode is less effective, the sequence of events in the prebreakdown phase indicates that more then one characteristic time should be associated with any of the individual processes involved. Both the growth of an ion space charge and the penetration of the anode potential inside the HCR will produce the initial conditions for to start the last phase prior to breakdown. Contraction of the sheath over the HCR walls leads to a reduction in electron multiplication, which combined with an expansion of the HCR plasma can explain the observed decrease in electron density before electric breakdown in the main gap. The expansion of the HCR plasma was observed in 2D-ICCD camera measurements (Fig. 3). For a fixed external potential, the magnitude of the ion space charge field that shields the electric field entering through the cathode hole does not depend on the initial density of neutrals, which is determined by the operating pressure. This explains the observation of a maximum electron density, which does not vary significantly over the pressure range investigated (Fig. 2b). Moreover, for a given applied voltage, there is a minimum pressure below which, although the virtual anode forms and moves to the proximity of the cathode, electric breakdown is not achieved [3]. This limiting pressure might be defined by an insufficient electron density inside the HCR, as to provide the electron source for the injection of the final electron beam, prior to electric breakdown. In this situation, despite the fact that the anode potential has moved close to the cathode electrode, breakdown does not occur. This might correspond to a situation in which plasma begins to grow inside the HCR and the anode potential penetrates into this region, but local ionization does not reach a situation that allows the necessary electron beams to be extracted to assist breakdown in the A-K gap. Upon emission of the electron beam, and space charge is left behind, with a reduction in plasma density. This situation favors plasma formation again (see

Fig. 3), and the above situation repeats, but with a reduction in the maximum plasma density that can achieved, due to reduction in the local potential due to the combination of ion space charge and geometric field. Which each oscillation, the actual magnitude of the electron beams decreases, and after a few cycles becomes insufficient to induce final ionization and subsequent breakdown in the A-K gap.

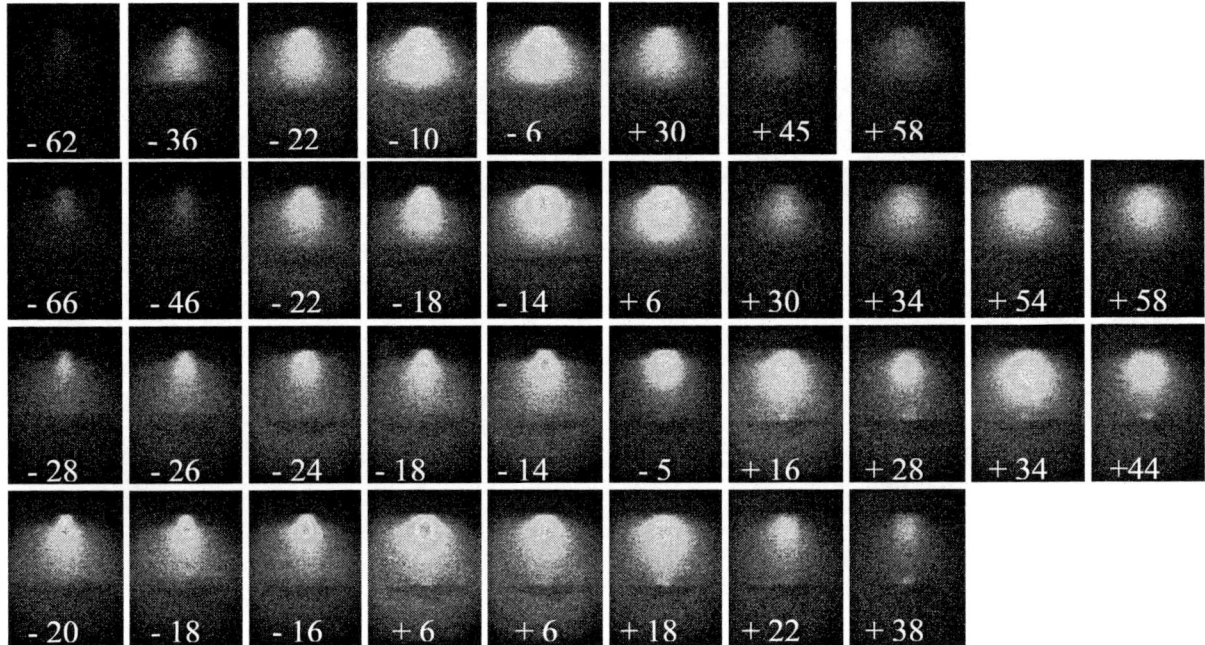

FIGURE 3. Images obtained by mean of a 2D-ICCD camera for different pressures. From the first to the fourth row corresponds to 100 *mTorr*, 150 *mTorr*, 180 *mTorr* and 250 *mTorr* respectively. Values in each figure represent the time, in nanoseconds, before (negative values) and after (positive values) breakdown.

The previous explanation could be supported by the observed fluctuation in our actual measures of ion beams and H_α signals (Fig. 1b). Also, the images of the growing plasma inside the HCR (Fig. 4) show this singular behaviour. In fact, at 180 and 150 *mTorr*, the plasma growing in the HCR shows a second attempt of growth which is consistent with the fact that at lower pressure the hollow cathode effect is less effective. More data is needed at 100 *mTorr* for to consolidate this explanation.

ACKNOWLEDGMENTS

The authors wish to thank Patricio Silva, from Comisión Chilena de Energía Nuclear, for fruitful discussions. FONDECYT under Grant 1000183, and a *Cátedra Presidencial en Ciencias*, awarded to L. Soto supported this work.

REFERENCES

[1] K. Frank and J. Christiansen, IEEE Trans. Plasma Sci., **17**, 748 (1989)
[2] M. Zambra, M. Favre, J. Moreno, H. Chuaqui, E. Wyndham, and P. Choi, IEEE Trans. Plasma Sci., **27**, 746 (1999)
[3] J. Moreno, M. Zambra, and M. Favre, IEEE Trans. Plasma Sci., **30**, 417 (2002)
[4] P. Choi, h. Chuaqui, M. Favre, and V. Colas, IEEE Trans. Plasma Sci., **23**, 221 (1995)

Coulomb Clusters in Axial Magnetic Field

F.M.H. Cheung, A.A. Samarian, and B.W. James

School of Physics, University of Sydney, NSW 2006, Australia

Abstract. The rotation of Coulomb clusters with different numbers of micron-sized particles is observed in an inductively coupled dusty plasma in the presence of an axial magnetic field. The rotation is found to be dependent on the particle number and configuration. Clusters with smaller numbers of particles require a higher magnetic field strength in order to initiate the rotation, the threshold magnetic field at which the cluster begins to rotate being proportional to the square of the number of particles in the cluster. The angular velocity of clusters increases and the radius of the clusters decreases as the magnetic field strength is increased. The relation between angular velocity and magnetic field is dependent on the number of particles in the cluster.

INTRODUCTION

When micron-sized dust particles are released into the plasma, the dust particles undergo frequent collisions with the highly mobile electrons rather than the slow and heavy ions within the plasma. As a result, these dust particles accumulate thousands of electrons on their surface and become negatively charged. It is possible to levitate these dust particles in the plasma to form a 2-dimensional or a 3-dimensional lattice with an ordered structure similar to a crystal. These crystals of many (more then a thousand) particles are commonly known as dust Coulomb crystals. These systems have been shown to be an ideal model for studying strongly coupled systems, because of their unique nature, relative ease of production and simple optical imaging required. A dust Coulomb crystal with one to several particles is a dust Coulomb cluster.

The Coulomb clusters can be considered as a system of a small number of particles confined by external electric field. Such systems have been a topic of theoretical and experimental studies with respect to particle ordering, phase transition, energy spectra etc [1-4]. But up until now, no experiments had been reported on properties of such clusters in magnetic field, while some papers were concerned with the investigation of the behaviour of large plasma crystal in axial magnetic field [5-9].

EXPERIMENTAL RESULTS

Here we report the dynamical behaviour of Coulomb clusters with N number of particles, N equal from 2 to 12 in inductively coupled magnetised plasma. To our knowledge this is the first time such small crystal systems are experimentally observed to exhibit such rotational behaviour.

The experiment is conducted in a radio-frequency (rf) discharge with a printed-circuit board (PCB) electrode system. Fig. 1 shows the interior of the experimental apparatus. The dust clusters formed above the confining electrode are illuminated by a fully height adjustable He-Ne laser for observation. The motion of the dust crystals was observed under the microscope and on the televisions from the video images generated from the cameras. The images of the rotational motion were then recorded on videotapes at a frame rate of 50 fps and a shutter speed of 0.008 seconds. The particles are then tracked with a software program that outputs the x and the y-coordinates of their trajectories as a function of time for analysis.

Coulomb clusters of different number of particles in a plane were formed in the experiment. When the magnetic field was off, the clusters exhibited small random fluctuation but always remained around their equilibrium position. However when the magnetic field was switched on, the small clusters were observed to be going under rotational

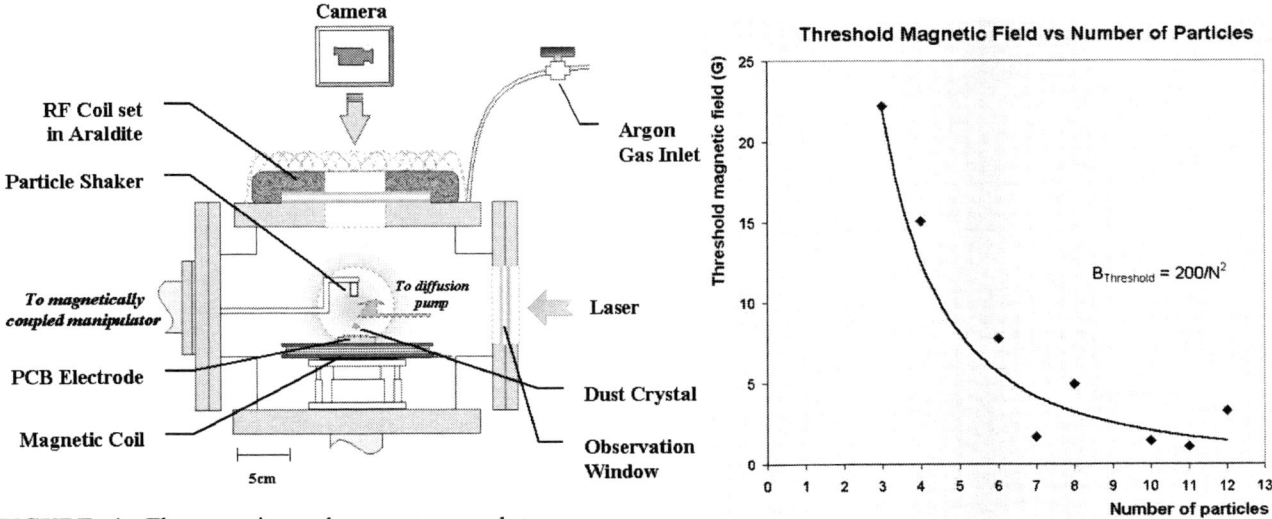

FIGURE 1. The experimental apparatus used to produce the coulomb dust clusters.

FIGURE 2. Threshold magnetic field versus number of particles.

motion. The direction of the rotation was in the left-handed direction with respect to the magnetic field. And a change in the direction of the magnetic field will cause the rotation to go counterclockwise.

Fig. 2 shows how the clusters with smaller number of particles like planar-3 and planar-4 requires a higher magnetic field strength in order to initiate the rotation. And planar-2 exhibits the highest resistance to the change in magnetic field with only momentary pauses at higher field strength. The threshold magnetic field strength at which the particles will start to rotate decreases as the number of particle increases. It should be mentioned that, in contrast with experiments in DC discharge and capacitive rf-discharge, we need only relatively small magnetic field (~ few Gauss) to rotate Coulomb clusters. It has been found that threshold value is inversely proportional to the square of the number of particle in the cluster. This means that the rotation is generated in the presence of the strong Coulomb coupling among the particles.

When the magnetic field was increased, two events occurred. Firstly, in general, there was an increase in the angular velocity of the cluster as shown on Figure 3. In particular, the angular velocity of the single-ring clusters (planar-2 to planar-8) seems to increase linearly as magnetic field strength increases. And for the double-ring clusters (planar-10 to planar-12), the angular velocity of the cluster increases very quickly and then saturates even when magnetic strength increases. This phenomenenon is similar to the saturation under the kG magnetic fields reported in [8]. In [10] the possible explanation of saturation was proposed, below we suggest a new explanation of this fact.

Secondly, there was a decrease in the radius of the cluster. The total angular momentum L, which is summed over all particles in a particular crystal structure is independent on the magnetic field. This is an indication that the increase in angular velocity, accompanied by a contraction of the cluster is a result of the conservation of angular momentum. The angular momentum per particle is obviously independent to the change in magnetic field. But it is also independent on the number of particles there are in the cluster going under rotation.

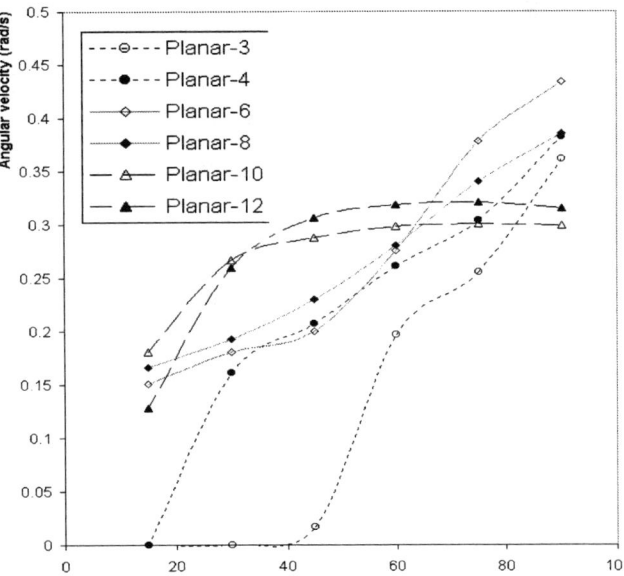

FIGURE 3. The variation of angular velocity of the cluster with the magnetic field strength.

Now the magnitude of the driving force for the cluster rotation can be easily calculated from the neutral friction force in the azimuthal direction. The obtained value is about 10-16N. On the other hand, there are many models in calculating the value of the ion drag force. But the upper limit of the ion drag force from all these models are less than 10-17N (under the assumption that the gas is at rest). It is easy to see that in comparison with the friction force, the ion drag force is about an order less. Thus the explanation using azimuthal ion drag to attribute to the rotation of plasma crystals is close but not satisfactory.

Another item that we should take into account is the divergence of the axial magnetic field. Since the sheath electric field is at least ten times the magnitude to the radial electric field, only a small tilt in the angle at which the magnetic field bends away from the vertical axis will affect the ion drag drift velocity.

From the experiment, dependence of radial electric field on magnetic field was observed (see Figure 4). Obtained data show that radial electric field is linearly proportional to the magnetic field. Since electric field is modified by the magnetic field, it must be taken into account in the analysis of the driving force of cluster rotation.

The Larmor radii for electrons and ions are 6×10^{-5}m and 2×10^{-2}m respectively. So the electrons are highly magnetized compared to the ions. And because the ratio of electron gyrofrequency to electron-neutral collisional frequency is about 1.5 (for ions, this ratio <0.01), the electrons will tend to be localized at the center of the system. As a result, the non-uniform distribution of the charge density will lead to a change in the electric potential profile.

In fact, this phenomenon can explain the saturation in angular velocity for double-ring clusters mentioned earlier. As the magnetic field increases, the radial electric field at the center of the cluster will change into the opposite direction. Consequently, the particles on the inner ring of the cluster will attempt to rotate in the opposite direction. However, due to the strong interparticle force, the cluster remains as a rigid body. And so the net torque on the whole cluster will decrease. Thus saturation of double-ring cluster rotation occurs.

It was demonstrated from the experiment that the rotation of small dust coulomb clusters is possible with the application of an axial magnetic field. It is easier to initiate the rotation of the clusters with larger number of particles than smaller number of particles at very low magnetic field strength. The angular velocity of the clusters increases while the radius of the clusters decreases as the magnetic field strength increases.

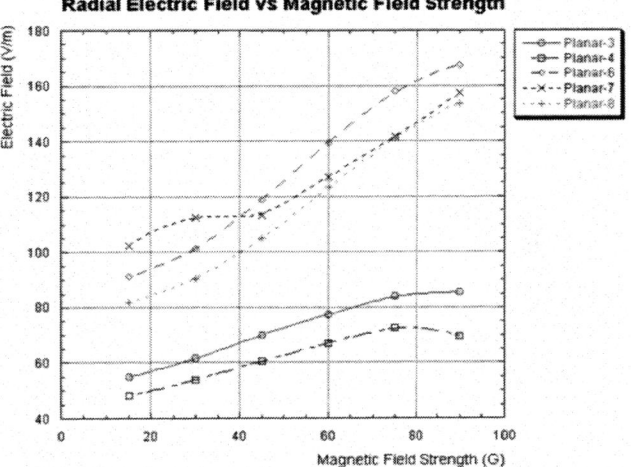

FIGURE 4. Radial electric field dependence on axial magnetic field.

ACKNOWLEDGMENTS

This work was supported by Australian Research Council, the Science Foundation for Physics within the University of Sydney. A.A. Samarian was supported by University of Sydney U2000 Fellowship.

REFERENCES

1. V.M. Bedanov and F.Peeters, Phys. Rev. B **49**, 2667 (1994)
2. V.A. Schwegeirt and F.Peeters, Phys. Rev. B **61**, 7700 (1995)
3. W.T. Juan, J.W. Hsu, Z.H. Huang, Y.J. Lai and Lin I, Phys. Rev.E **58**, 6947 (1998)
4. M. Klindworth, A.Melzer, A. Piel and V.A. Schwegeirt, Phys. Rev. B **61**, 8404 (2000)
5. G. Uchida, R. Ozaki, S. Iizuka and N. Sato, ICPP & 25th EPS, Praha,. ECA **22C** 2557 (1998)
6. N. Sato, G. Uchida, T. Kamimura, and S. Lizuka, p.239 in *Physics of Dusty Plasmas*, AIP Institute of Physics, NewYork, (1998)
7. U. Konopka, D. Samsonov, A.V. Ivlev, J. Goree and V. Steinberg, Phys. Rev. E **61**, 1890 (2000)
8. Sato N, Uchida G, Kaneko T, Shimizu S, Iizuka S. Physics of Plasmas, 8, 1786 (2001)
9. Shimizu S, Uchida G, Kaneko T, Iizuka S, Sato N, *25th ICPIG,* Nagoya Univ. 3, 39, Nagoya, Japan. (2001).
10. Kaw PK, Nishikawa K, Sato N., Physics of Plasmas, 9,.387 (2002).

Dust grains as a Diagnostic Tool for RF-Discharge Plasma

B.W. James, A.A. Samarian and W. Tsang

School of Physics, University of Sydney, NSW 2006, Australia

Abstract. Dust particles can be a useful plasma diagnostic tool. Sufficiently dust particles can be used as test grains for the visualization of the plasma potential distribution; larger dust particles can be used as specific probes to determine electron temperature. Here we report on diagnostic measurements carried out in a capacitively coupled planar rf discharge. The location of the sheath edge has been determined using test dust grains. Our diagnostic technique is based on measuring the equilibrium position of fine grains levitated above the powered electrode in an rf-discharge. Estimates show that for grains with radii less than 500 mm the grain equilibrium position and sheath edge location differ by less than 5 percent, and this difference continues to decrease with decreasing investigate grain radius. We use this technique to diagnose the sheath in an argon plasma which was generated at pressures in the range 20–100 mTorr by applying a 15 MHz signal to the power electrode. The shape of the potential well above the confining electrode was also visualized using even smaller dust grains that were generated in the discharge. The well shape was found to depend strongly on the confining potential.

INTRODUCTION

The possibility of using the dust grains as a diagnostic tool based on the fact that the grains in a discharge plasma acquire electric charge by collecting electrons and ions from the surrounding plasma. The charge on a grain can be extremely high (say 10^3 - 10^4 electron charges for a micron-sized particles) and depends on the ion and electron fluxes to the particle surface. The equilibrium position of such test grains is determined mainly by gravity and electrostatic forces in the sheath region. This means that the equilibrium position and motion of the grains is strongly dependent on plasma and sheath conditions. Analysis of grain behaviour can provide information on the spatial profile of electric field, ion density and velocity, and the sheath edge location.

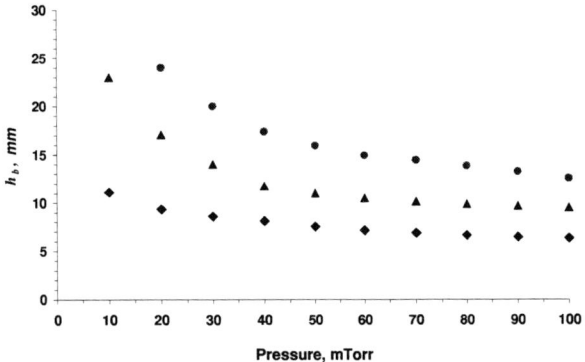

FIGURE 1. Position of sheath edge, h_b, as a function of pressure for different rf-input powers: squares – 100W. triangles – 60W. circles – 35W.

The location of the sheath edge in a planar rf-discharge has been determined using test dust grains. The diagnostic technique is based on measurement of the equilibrium position of fine dust grains levitated above the powered electrode in an rf-discharge. Estimates show that for grains with radii less than 500 nm the grain equilibrium position and sheath edge location differ by less than 5 percent, and this difference continues to decrease with grain radius. We have used this technique to diagnose the sheath in an argon plasma which was generated at pressures in the range 20 – 100 mTorr by applying a 15 MHz signal to the power electrode [1,2].

The test grains are generated in the discharge by electrode sputtering under high power (up to 200W) and high pressure (up to 1 Torr) conditions. The dust grains are illuminated using a Helium-Neon laser, which enters the discharge chamber through a 40-mm diameter window. Windows mounted on a side port and on the top of the chamber provide a view of the light scattered at different angles by the suspended dust particles. The size of the growing grains was estimated from analysis of the scattered light using techniques proposed in [3]. In our experiments grains grown to 300 – 500 nm. To obtain a vertical cross-section of the dust grain layer and provide a sheath dimension measurement, the laser beam was expanded in the vertical directions into a sheet of light by a system of cylindrical lenses. Images of the illuminated dust layer are obtained using a charged-coupled device (CCD) camera with a micro lens. The video signals are stored on a videotape recorder or are transferred to a computer. The resulting images allow direct measurement of test grain equilibrium position and therefore a determination of the sheath edge location. The variation of sheath thickness with pressure for different rf-input powers is shown in Figure 1.

The radial potential profile was measured using a transient motion technique (TMT) in which a dust particle levitated in the sheath was displaced from its equilibrium position by applying a negative voltage to a two pin electrode. When the voltage is switched off the particle returns to its original position due to the action of the radial electrostatic force $F_{el}=Z_d E_r$. By tracking the particle motion it is possible measure instantaneous velocity and acceleration. Then, using the equation of motion the potential difference between two points can be found as follows:

$$\Delta\varphi = \frac{ma\Delta s + F(v_1)\Delta s_1 + F(v_2)\Delta s_2}{Z_d}$$

Where $F = 4/3 \delta m n_n v_{T_n} \pi a^2 v$ is the neutral drag force. The potential variation obtained in this way was normalised to the plasma potential at $r = 0$, as measured by the Langmuir probe.

The value of Z_d, the particle charge was obtained by the vertical resonance method [4]. As the particle motion is in the horizontal plane we can assume that the charge is constant. The resulting profile is shown on the Figure 2. The profile obtained using a rf-compensated single Langmuir probe is also shown. The two results show good agreement.

The radial electric field, calculated from the potential data, is shown in Figure 3. The high degree of scatter among the points is due to the use of simple numerical differentiation. The result, along with linear fits to the data, is shown, however, to indicate the possibilities of the technique. By fitting functions to the potential variation it would be possible to obtain more reliable estimates of radial electric field. The TMT itself could also be improved by using

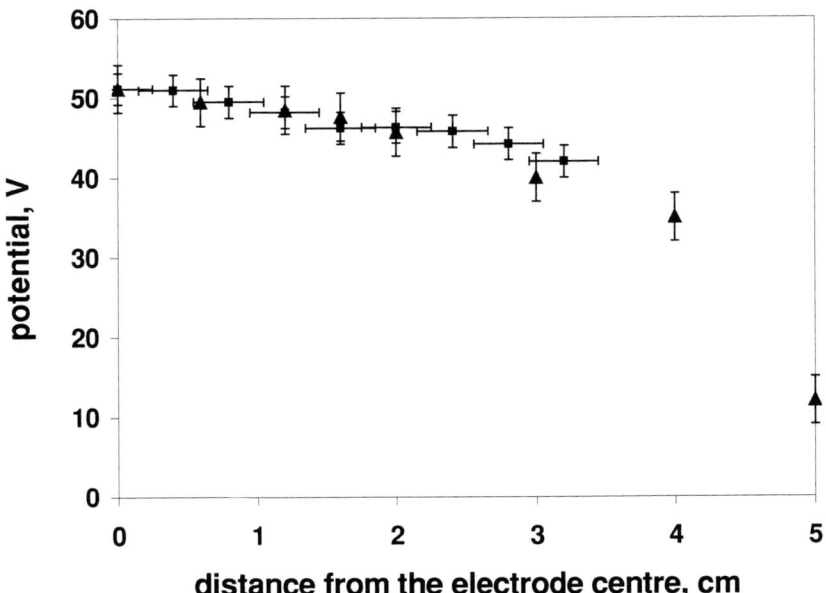

FIGURE 2. Radial potential profile: - Langmuir probe technique, Δ-transient motion technique

a laser, instead of a pin electrode, to displace the particle, causing minimal disturbance to the plasma.

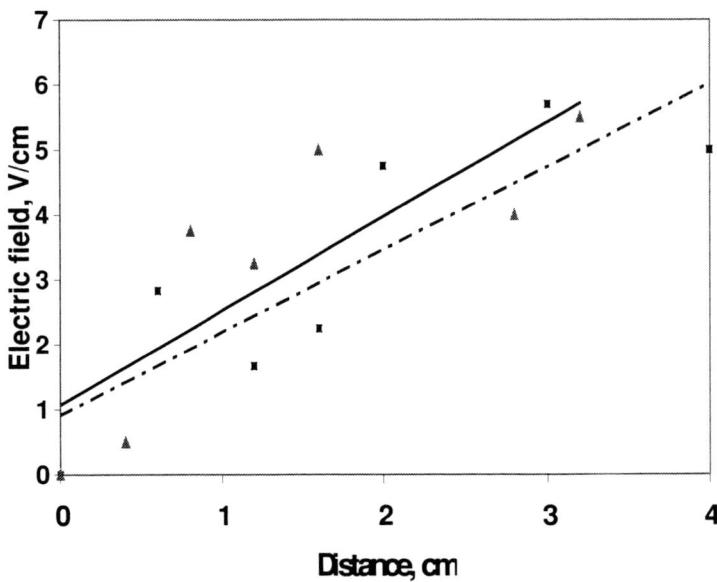

FIGURE 3. Electric field dependence versus distance from the electrode centre. - Langmuir probe technique, Δ-transient motion technique. The solid lane is trend for TMT data; the dashed line is mean square fit for probe data.

Even smaller particles than those used above find no equilibrium position in the sheath. Instead, they fill the plasma volume leaving, the sheath as a dust void. The shape of the potential well above the confining electrode in a radio-frequency (rf) discharge with a printed-circuit board electrode system [5] was visualised using fine dust grains that were generated in the discharge. The well shape was found to depend strongly on the confining potential.

ACKNOWLEDGMENTS

This work was supported by Australian Research Council and the Science Foundation for Physics within the University of Sydney. ASS was supported by a University of Sydney U2000 Fellowship.

REFERENCES

1. .Samarian, A., James, B., Vladimirov, S., and Cramer, N., *Phys. Rev.* **64**, 025402(R) (2001).
2. Samarian, A.A. and James, B.W., *Phys. Lett.* **A287**, 125 (2001).
3. Nefedov, A.P, Vaulina O.S., Petrov O.F., *Appl. Opt.* **36**, 1357 (1997).
4. Melzer A., T. Trottenberg T., A. Piel A., *Phys. Lett.* **A191**, 301 (1994).
5. Cheung F, A. Samarian, B. James, *Physica Scripta,* **T98**, 143 (2002).

Observation of Naturally-Occurring Waves in a Strongly Coupled Plasma

S. Nunomura[*], J. Goree, S. Hu[†], X. Wang[‡], A. Bhattacharjee, and K. Avinash[§]

Department of Physics and Astronomy, The University of Iowa, Iowa City, Iowa 52242

Abstract. The Fourier spectra of longitudinal and transverse waves corresponding to random particle motion were measured in a two-dimensional strongly coupled plasma. As a model of a strongly coupled plasma, a plasma crystal, i.e., lattice composed of negatively charged microspheres immersed in a plasma, is used. The phonons were found to obey a dispersion relation that assumes a Yukawa inter-particle potential. The crystal was in a non-thermal equilibrium, nevertheless phonon energies were almost equally distributed with respect to wavenumber over the entire first Brillouin zone.

INTRODUCTION

In a lattice at a finite temperature, particles move about in configuration space with random velocities and energies that have a Maxwellian distribution. The particle velocities, when Fourier transformed, correspond to a spectrum of waves, or phonons, with a wide range of frequency ω and wavenumber **k**.

The energy of phonons corresponding to thermal motion is not distributed broadly over all **k** - ω space; instead, it is concentrated at values permitted by a dispersion relation. In general, dispersion relations describe the preferred frequencies for normal modes, and they are valid for low amplitude, i.e., linear waves. The manner in which the energy is distributed among the wavenumbers depends on the lattice's energetics; a lattice in thermal equilibrium will exhibit equipartition, so that each mode has the same energy, when averaged over time [1].

Here we report observations of phonons that we did not excite intentionally; they were present naturally, due to random particle motion. Our system was not in thermal equilibrium, but it nevertheless exhibited several characteristics typical of thermal equilibria. These include a Maxwellian distribution of velocities as measured in configuration space, and a phonon spectrum with wave energy distributed almost equally with respect to wavenumber.

Our system is a two-dimensional lattice formed by levitating polymer microspheres in a glow discharge plasma. The particles are immersed in a background of electrons and ions, which cause the particles to become negatively charged and interact through a screened Coulomb repulsion [2]. They are also immersed in a rarefied neutral gas, which induces a thermal Brownian motion and damps any organized particle motion. This kind of particle suspension is termed a dusty plasma [3]. Particles are levitated in the vertical direction, and trapped in the horizontal direction, by an electric field in the sheath region above a horizontal electrode. Our suspension was two dimensional because it had only enough particles to form a single layer, and the confining forces allowed measurable particle motion only in the horizontal plane. When their kinetic energy is limited by neutral gas damping, the particles tend to arrange themselves in an ordered structure [4-7], which is called a "plasma crystal" in analogy to a colloidal crystal. In our experiment, the particles formed a triangular lattice with hexagonal symmetry. The particle motion was observed directly by video microscopy, allowing us to determine the velocities of all particles in a sample area viewed by the camera.

---

[*] Present address: Max-Plank-Institut für Extraterrestrische Physik, D-85740 Garching, Germany
[†] Present address: Department of Physics and Astronomy, University of California at Irvine, CA 92697
[‡] Permanent address: Department of Physics, Dalian University of Technology, Dalian, China 116024
[§] Permanent address: Institute for Plasma Research, Bhat, Gandhinagar 282428 India

Our lattice was a driven system. We measured the velocity distribution function, and found that it was Maxwellian, but with a temperature higher than the ambient neutral gas. The extra heating is presumably due to a fluctuating Coulomb force, but its exact mechanism is unknown. What is known with greater confidence is that this energy input is balanced by an energy loss due to neutral gas drag. Because the particles in the lattice simultaneously gain energy and lose energy by these external mechanisms, they form a driven system, which is a type of non-thermal equilibrium.

The experimental apparatus is sketched in Fig. 1. An argon plasma was generated by rf glow discharge at a gas pressure of 18.6 mtorr, using at 13.56 MHz and 62 V peak-to-peak. The rf power was applied between a grounded vacuum chamber, which is not shown in Fig. 1, and a horizontal electrode. In the glow that formed above the

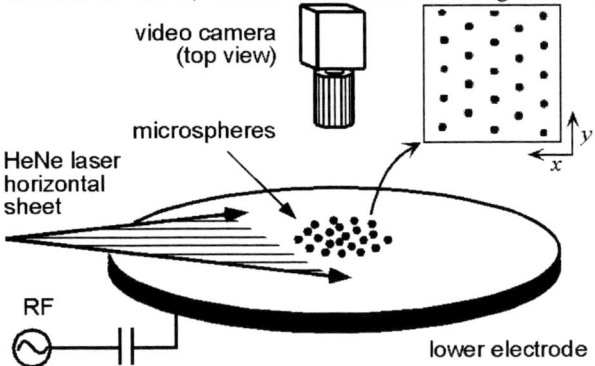
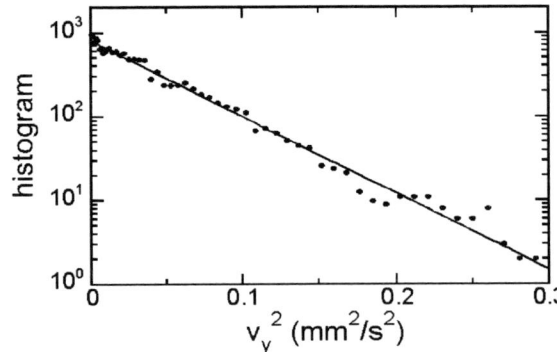

FIGURE 1. Sketch of experimental setup. The inset is an image of part of the 2D lattice.

FIGURE 2. Distribution function of particle velocities, as measured in configuration space. The line is a Maxwellian fit with $T = 0.046$ eV, which is hotter than the room temperature gas.

horizontal electrode, the plasma had a density $\approx 3.5 \times 10^8$ cm^{-3} and an electron temperature ≈ 1 eV, as measured with a compensated Langmuir probe inserted into the main plasma region above the sheath.

The particles that were dispersed into the plasma were melamine-formaldehyde microspheres with a 4.04 μm radius, a size dispersion with a standard deviation of ± 2.2%, and a mass density 1.51 g/cm^3, corresponding to a particle mass $m = 4.17 \times 10^{-13}$ kg/m^3. Approximately 5500 particles were introduced into the plasma, where they arranged themselves in a monolayer triangular crystal, approximately 7 cm in diameter. The inter-particle spacing was $a = 0.90$ mm, as determined from the first peak of the pair correlation function, $g(\mathbf{r})$. Particles were illuminated with a low-power horizontal laser sheet, and they were imaged by a video camera with a field-of-view of 24 × 18 mm, including approximately 600 particles.

To analyze the particle motion, we digitized video images and identified each particle's x-y coordinates with sub-pixel resolution. This was repeated for 128 frames at 15 frames per second. The particle velocity was calculated from the difference of particle positions in consecutive frames.

The particles had a velocity distribution that was nearly Maxwellian. This is shown in Fig. 2, which is a histogram of the squared particle velocity v_x^2, for only the x-component of velocity. It exhibits an exponential decay over a wide range of v_x^2, which is fitted by a temperature $T = 0.046$ eV. We also calculated the kinetic energy from the mean square velocity, yielding $\langle mv_x^2 \rangle = 0.038$ eV. (The discrepancy in these two values reflects the uncertainty in the fit of Fig. 2.) For comparison, if our lattice were in thermal equilibrium with the gas, the particles would undergo Brownian motion with a temperature equal to the room temperature of the gas, 0.026 eV. This higher temperature indicates that our lattice was a driven system.

Using a Fourier transform method, the phonon spectrum is computed. From the components of the particle velocity $\mathbf{v}(\mathbf{r},t)$, we calculated the wave amplitude $V_{\mathbf{k},\omega} = 2/(TL)\int\int v(\mathbf{r},t)\exp[-i(\mathbf{k}\cdot\mathbf{r} - \omega t)]d\mathbf{r}dt$, where L and T are the length and period over which a particle's motion is sumffled. This method is repeated for two kinds of polarization selections. First, we selected the angle θ of the wave propagation direction $\mathbf{k}$ with respect to a primitive translation lattice vector $\mathbf{a}$. Second, we selected either the longitudinal or transverse mode by using only the component of $\mathbf{v}(\mathbf{r},t)$ that is parallel or perpendicular to the wave propagation direction $\mathbf{k}$, respectively.

Our results for the natural phonon spectrum are shown in Fig. 3(a) - (d) [8]. These are maps of the wave energy density in various subspaces of the full $\mathbf{k}$ - ω space. Results are shown separately for two different propagation directions, $\theta = 0°$ and $90°$. The data cover the entire first Brillouin zone, i.e., $-2 < ka/\pi < 2$ for $\theta = 0°$, and $-2/\sqrt{3} < ka/\pi < 2/\sqrt{3}$ for $\theta = 90°$.

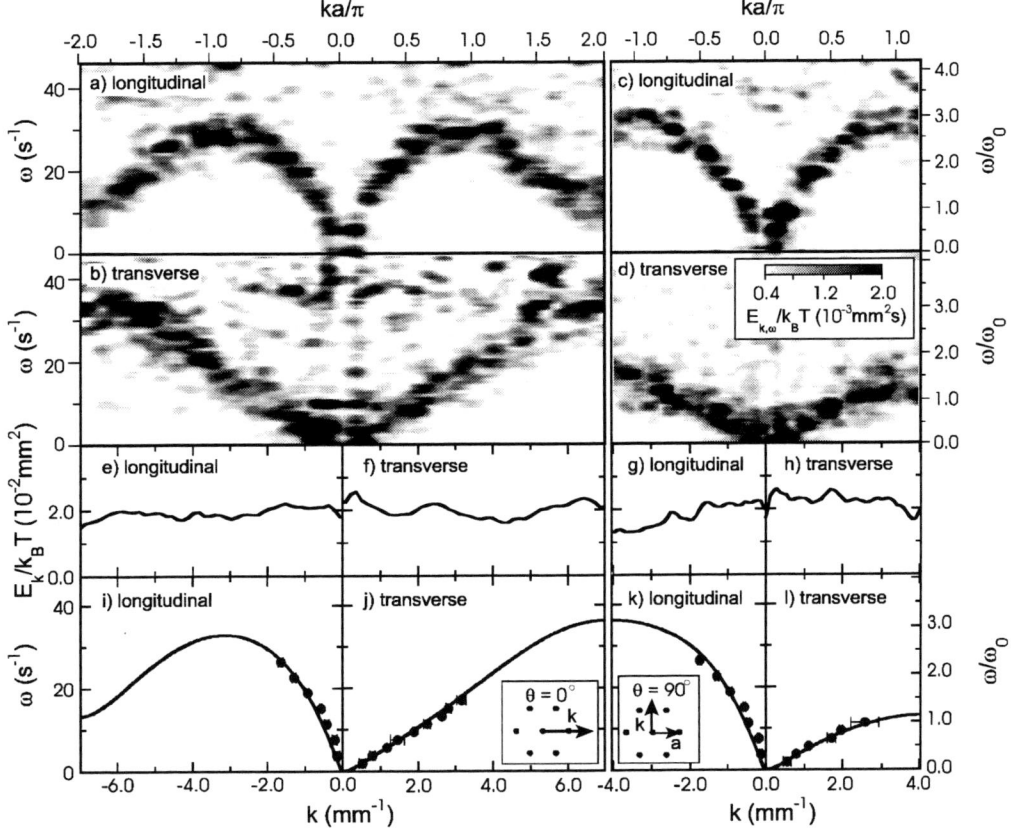

FIGURE 3. Phonons in the first Brillouin zone. (a) - (d) Spectra of wave energy density in **k** - ω space, in the absence of any intentional wave excitation. Darker grays correspond to higher wave energy. Energy is concentrated along a curve corresponding to a dispersion relation. (e) – (h) Spectra integrated over ω, showing that wave energy is distributed nearly uniformly with respect to wavenumber. (i) - (l) Theoretical dispersion relation (curves) shown fitted to the experimental dispersion relation for waves excited intentionally using a laser (circles). The angle θ between **k** and **a** is: $0°$ in the left panels, $90°$ in the right panels.

In Fig. 3, darker grays correspond to higher wave energy density. The grayness appears spotty due to statistics arising from the finite length of data's time series. The energy density was computed for each frequency and wavenumber as $m V_{k,\omega}^2 / 2 k \delta\theta \delta k \delta\omega k_B T$. Here, $\delta\theta$, δk and $\delta\omega$ quantify the resolution by which we are able to distinguish waves with different values of θ, k and ω, respectively. In our experiment, the product $k\delta\theta \delta k \delta\omega$ was 0.0729 $mm^{-2}s^{-1}$ for all the **k** - ω space shown in Fig 3.

Examining Figs. 3 (a)-(d), we note that the wave energy is concentrated on distinct curves in the various **k** - ω subspaces; we identify these curves as the dispersion relations. The dispersion relations depend on the mode's propagation angle and on its polarization, longitudinal or transverse.

For comparison, we also measured the dispersion relations of the longitudinal and transverse modes using the laser-excitation method of Ref. [9], and this yielded the data points plotted in Fig. 3 (i) - (l). We find that these data for externally-excited sinusoidal waves agree well with the spectra in Fig. 3 (a) - (d) for the natural phonons in the absence of any external stimulation. For the longitudinal wave in Fig. 3 (a), it includes maxima in the frequency at $|ka/\pi| \approx 1$. For $|ka/\pi| > 1$ in Fig. 3 (a), we note a downward slope, $(\omega/k) (d\omega/dk) < 0$, indicating a backward wave, with phase and group velocities that point in opposite directions. In contrast, the transverse wave in Fig. 3 (b) is essentially dispersionless, i.e., $\omega \propto k$, over nearly the entire first Brillouin zone.

The waves due to random motion in Fig. 3 (a)-(d) were detected for all allowed wavenumbers. In contrast, the laser was unable to excite the largest wavenumbers in Fig. 3 (i)-(l), which we attribute partly to the laser beam's finite size. The wavenumbers that the laser can excite efficiently are those that are comparable to the reciprocal of the laser beam's width. Laser-excited waves also experience heavy damping at frequencies where the group velocity is zero, $d\omega/dk=0$ [10].

Our spectra agree well with the theoretical dispersion relations shown as solid curves in Fig. 3 (i) – (l). We use the theory derived by Wang et al. [10] from the equation of motion for a 2D triangular lattice with a Yukawa potential. This agreement indicates that the Yukawa potential is a good approximation for the particle interaction in our 2D crystal. We used two adjustable parameters, $\kappa \equiv a/\lambda_D$ and particle charge Q, to fit theory to experiment in Figure 3, yielding $\kappa = 1.2$, $Q/e = 13\,000$. The damping rate was assumed to be 3.7 s^{-1}, according to the particle radius and the gas pressure. These results, together with our measurements of a and T, yield the Coulomb coupling parameter $\Gamma \equiv Q^2/4\pi\varepsilon_0 a k_B T = 7500$, and $\omega_0 \equiv (Q^2/4\pi\varepsilon_0 m a^3)^{1/2} = 11.6$ s^{-1}. Finally, we note that the theoretical dispersion relation is valid for linear or small amplitude waves, and is able to show agreement with our spectra in Figure 3 (a) – (d) because our phonons have a small amplitude.

We now turn our attention to quantitative measurements of the wave energy in the spectra of Fig. 3 (a) – (d). The results presented below reveal that our driven system seems to have properties of a system in thermal equilibrium.

We find empirically that the wave energy was distributed equally over all values of **k** in the first Brillouin zone. This conclusion is drawn from Fig. 3 (e) – (h), which are **k**-spectra of the phonons, computed by integrating the spectra in Fig. 3(a) - (d) over frequency to yield a wave energy E_k as a function of **k**. Because all reciprocal lattice vectors are placed at regular intervals in **k**-space, this result indicates that the thermal energy of the plasma crystal is distributed nearly equally into each possible mode. Such a result would be expected if the lattice were in thermal equilibrium, which would obey equipartition, although our lattice is a driven system.

We can compare the total energy E in our wave spectrum with the kinetic energy K as measured in configuration space. For 2-D systems, $E = N \sum_{polarization}^{2} \int\int E_{\mathbf{k},\omega} d\omega d\mathbf{k}$, and $K = Nk_B T$ using the temperature determined from particle velocities in configuration space. Here, N is the total number of particles, and "polarization" refers to the two modes, longitudinal and transverse. The area of integration in 2-D **k** space is the first Brillouin zone, and the range of integration on the ω axis is zero to infinity. For our plasma crystal, we use our conclusion from Fig. 3 (e) – (h), that E_k is constant over the entire area of the first Brillouin zone in a 2D reciprocal lattice. Repeating the experiment at various conditions, with a shielding parameter κ in the range 0.78 – 1.32 and gas pressure in the range 18 – 33 mtorr, we found that E/K was in the range 1.5 – 3.1. (For the experimental conditions corresponding to Fig. 2 and 3, $E/K \approx 2$.) For comparison, we note that for systems in thermal equilibrium, which obey the Virial theorem, $P = K$ and $E/K = 2$.

To summarize, we measured a phonon spectrum due to random particle motion. It obeys the dispersion relations for longitudinal and transverse waves over the entire first Brillouin zone. We have shown three results for our driven system that seem to match what is expected for a thermal equilibrium. First, the particle velocities as measured in configuration space have a Maxwellian velocity distribution. Second, the phonons, as measured in Fourier space, have energies that are almost equally distributed with respect to wavenumber. Third, the wave energy measured in Fourier space is approximately twice as large as the particle kinetic energy measured in configuration space.

We thank V. Nosenko and F. Skiff for useful discussions and L. Boufendi for TEM measurements of our particle size. This work was supported by NASA, NSF and DOE. S. N. acknowledges financial support from the Japan Society of the Promotion of Science.

REFERENCES

1. C. Kittel and H. Kroemer, Thermal Physics, 2nd ed. (W. H. Freeman, San Francisco, 1980), Chap. 4.
2. U. Konopka, G. E. Morfill, and L. Ratke, Phys. Rev. Lett. **84**, 891 (2000).
3. H. Ikezi, Phys. Fluids **29**, 1765 (1986).
4. J. H. Chu and Lin I, Phys. Rev. Lett. **72**, 4009 (1994).
5. H. Thomas et al., Phys. Rev. Lett. **73**, 652 (1994).
6. Y. Hayashi and K. Tachibana, Jpn. J. Appl. Phys. **33**, L804 (1994).
7. A. Melzer, T. Trottenberg, and A. Piel, Phys. Lett. A **191**, 301 (1994).
8. S. Nunomura et al., Phys. Rev. Lett. **89**, 35001 (2002).
9. S. Nunomura et al., Phys. Rev. E **65**, 66402 (2002).
10. X. Wang, A. Bhattacharjee, and S. Hu, Phys. Rev. Lett. **86**, 2569 (2001).

Confining Instabilities in Complex Plasma

A.A.Samarian*, S.V.Vladimirov* and B.W.James*

*School of Physics, University of Sydney, New South Wales 2006, Australia

Abstract. It is shown that the stability of the vertical and horizontal confinement of colloidal "dust" particles levitating in a complex plasma appears as a non-trivial interplay of the external confining forces as well as the interparticle interactions and collective processes such as the plasma wake.

In the laboratory experiments, the micrometer sized highly charged dust grains levitate in the sheath region under the balance between the gravitational and electrostatic forces acting in the vertical direction as well as the externally imposed confining potential applied in the horizontal plane [1]. The vertical confinement involving the gravity force and the electrostatic force acting on the dust particles with variable charges is a complex process exhibiting oscillations, disruptions and instabilities [2-6]. A characteristic feature of the particle confinement is also the strong influence of plasma collective processes such as the plasma wake [7, 8].

Consider two colloidal particles of mass M and charges Q, separated by the distance x_d horizontally (i.e., aligned along the x-axis), see Fig. 1a or z_d vertically (aligned along the z-axis), see Fig. 1b. In the simplest approximation, the

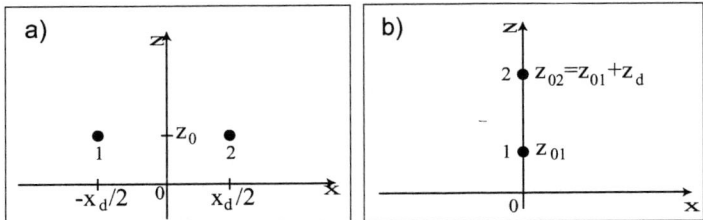

FIGURE 1. Sketch of the particle configurations.

particles interact via the screened Coulomb (Debye) potential $\phi_D = Q^2 \exp(-|\mathbf{r}|/\lambda_D)/|\mathbf{r}|$ where λ_D is the plasma Debye length. For particles levitating in the plasma sheath, the interaction potential in the vertical direction is asymmetric because of the ions flowing towards the negatively charged electrode. However, it is also instructive to consider the case with Debye only interaction even in the vertical direction.

We assume that the external confining force acting in the x-direction can be written as $F^{ext} = -\gamma_x(x - x_0)$, where $\gamma_x \sim QdE_x^{ext}/dx$ is a constant and obtain the balance of the external confining and Debye repulsion forces

$$\frac{2Q^2}{x_d^2}\left(1 + \frac{x_d}{\lambda_D}\right)\exp\left(-\frac{x_d}{\lambda_D}\right) = \gamma_x x_d. \qquad (1)$$

The balance of forces in the vertical direction includes also the gravitational force $F_g = Mg$ and the sheath electrostatic force $F_{el} = QE_z^{ext}(z)$. In equilibrium, we assume that the sheath electric field near the equilibrium position can be linearly approximated $F_{el} - Mg = -\gamma_z(z - z_0)$, where $\gamma_z \sim QdE_z^{ext}/dz$ is a constant. For the vertically aligned particles (Fig. 1b) the lower and upper equilibrium positions are z_{01} and $z_{02} = z_{01} + z_d$, respectively. In this case, the equilibrium balance of the forces in the vertical direction can be written as $F_{el,1(2)}(z_{01(2)}) - M_{1(2)}g + F_{1(2)}^{D,W}(z_{02} - z_{01}) = 0$, where $F_{1,2}^{D,W}$ are the forces of the interaction between the particles due to their interaction Debye and/or asymmetric (wake) potentials Φ_D and/or Φ_W, respectively: $F_1^D(z_{02} - z_{01}) = Qd\Phi_D(|z|)/d|z||_{|z|=z_d}$, and $F_2^{D,W}(z_{02} - z_{01}) = -Qd\Phi_{D,W}/(|z|)d|z||_{|z|=z_d}$. In the case of Debye only interaction between the particles, we obtain equation similar to (1), with the obvious change of x to z. In the case of the asymmetric wake potential, the equilibrium condition for the levitation of two identical particles gives us

$$\frac{Q^2}{z_d^2}\left(1+\frac{z_d}{\lambda_D}\right)\exp\left(-\frac{z_d}{\lambda_D}\right) - \gamma_z^W(z_d - z_W) = \gamma_z z_d, \qquad (2)$$

where z_W is the distance between the minimum of the asymmetric attracting potential characterized by γ_z and the upper particle.

Now, consider horizontal perturbations of two horizontally aligned particles, Fig. 1a. By including the phenomenological damping β and linearly expanding the interaction forces, we obtain two oscillation modes with the frequency

$$\omega_{xx,1} = -\frac{i\beta}{2} + \left(-\frac{\beta^2}{4} + \frac{\gamma_x}{M}\right)^{1/2}, \qquad (3)$$

for the two particle oscillating in phase with equal amplitudes $A_1 = A_2$, and

$$\omega_{xx,2} = -\frac{i\beta}{2} + \left[-\frac{\beta^2}{4} + \frac{\gamma_x}{M}\left(3 + \frac{x_d^2/\lambda_D^2}{1+x_d/\lambda_D}\right)\right]^{1/2} \qquad (4)$$

for the particles oscillating counter phase ($A_1 = -A_2$). Both modes are always stable.

The next case involves vertical oscillations of two horizontally aligned particles, Fig. 1a. We obtain that the two oscillation modes have the frequency similar to (3), with the change of x to z, for the two particle oscillating in phase ($A_1 = A_2$), and

$$\omega_{xz,2} = -\frac{i\beta}{2} + \left(-\frac{\beta^2}{4} + \frac{\gamma_z}{M} - \frac{\gamma_x}{M}\right)^{1/2} \qquad (5)$$

for the two particles oscillating counter phase ($A_1 = -A_2$). While the first mode is always stable, the counter phase mode *can now be unstable*, depending on the ratio γ_x/γ_z. This instability arises because of the *confining* potential in the direction *perpendicular* to the direction of particle oscillations.

By introducing small vertical perturbations δz_i of the vertically aligned particles, we obtain for the case of Debye only interactions equations analogous to the first case of horizontal vibrations of horizontally aligned particles. There are two oscillations modes; the first one has the frequency similar to (3) for the two particle oscillating in phase with equal amplitudes $A_{1,2}$, and the second mode's frequency is similar to (4), both with the obvious change of x to z. Taking into account the asymmetry of the interaction potential, we obtain that the first oscillation mode, for the particles moving in phase with equal amplitudes $A_1 = A_2$, is unchanged while the second frequency is now given by

$$\omega_{zz,2}^W = -\frac{i\beta}{2} + \left[-\frac{\beta^2}{4} + \left(\frac{\gamma_z}{M} + \frac{\gamma_z^W}{M}\left(1 - \frac{z_W}{z_d}\right)\right)\left(3 + \frac{z_d^2/\lambda_D^2}{1+z_d/\lambda_D}\right) + \frac{\gamma_z^W}{M}\frac{z_W}{z_d}\right]^{1/2} \qquad (6)$$

for the counter phase oscillations; their amplitudes are not equal in magnitude: $A_1 = -\left(2 + \frac{z_d^2/\lambda_D^2}{1+z_d/\lambda_D}\right)\left(1 - \frac{z_W}{z_d} + \frac{\gamma_z}{\gamma_z^W}\right)A_2$.

Both modes are always stable.

Now, consider horizontal oscillations of two vertically aligned particles. When the particle interaction is symmetric of Debye type, we obtain two modes of oscillations, the first one corresponds to to the particles oscillate in phase (with equal amplitudes), and its frequency is equal to (3). The second one is similar to (5), with the frequency

$$\omega_{zx,2} = -\frac{i\beta}{2} + \left(-\frac{\beta^2}{4} + \frac{\gamma_x}{M} - \frac{\gamma_z}{M}\right)^{1/2}, \qquad (7)$$

and $A_1 = -A_2$. We see that the counter phase mode can be unstable, the condition for this instability is somewhat opposite to the condition of the instability of the mode of vertical vibrations of two horizontally arranged particles (5). If to take into account the plasma wake, the equation of horizontal motion of the upper particle in this case is the same as for the symmetric Debye only interaction, while the lower particle is oscillating in the wake potential characterized by γ_x^W which is its horizontal strength in the parabolic approximation. For our purposes here it is sufficient to assume that γ_x^W is a positive constant of order (or slightly more) than γ_z^W, see, e.g., numerical simulations [8]. The frequency of the first mode coincides with (3) while the frequency of the second mode is given by

$$\omega_{zx,2} = -\frac{i\beta}{2} + \left[-\frac{\beta^2}{4} + \frac{\gamma_x}{M} + \frac{\gamma_x^W}{M} - \frac{\gamma_z}{M} - \frac{\gamma_z^W}{M}\left(1 - \frac{z_W}{z_d}\right)\right]^{1/2}. \qquad (8)$$

Now, we see that the wake potential can *stabilize* possible horizontal instability of two vertically algned particles; note that for the supersonic wake potential this stabilization occures only within the Mach cone. The amplitudes of the second mode of oscillations are related by $A_1 = \frac{\gamma_x^W A_2}{\gamma_z + \gamma_z^W(1 - z_W/z_d)}$. Thus for the asymmetric interaction potential, the second mode of oscillations does not correspond to the counter phase motions: the vibrations of particles are *in phase* now, with unequal amplitudes.

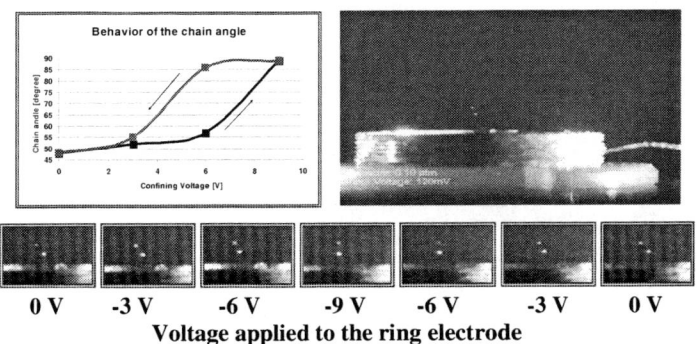

FIGURE 2. Experiment: levitation of two particles.

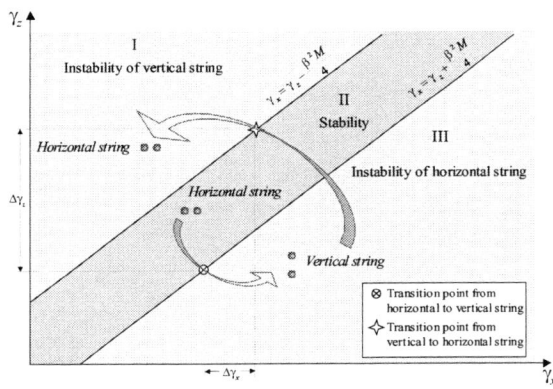

FIGURE 3. Stability diagram of the particle arrangements.

The proposed mechanism can be related to experimentally observed phenomena, for example, for the two-particle system in planar rf-discharge, involving horizontal oscillations of two particles aligned in the vertical string and hysteretic phenomena in the disruptions of the horizontal and vertical arrangements, see Fig.2. The stability diagram for the two-particle system, Fig.3, reveals two extreme regions: one is the region (I) where $\gamma_z > \gamma_x + M\beta^2/4$, corresponding to the vertical string unstable with respect to the horizontal motions of the particles, another is the region (III) where $\gamma_x > \gamma_z + M\beta^2/4$ corresponding to the horizontal string unstable with respect to the vertical motions of the particles, as well as the central region (II) where both structures can be stable.

ACKNOWLEDGMENTS

This work was supported by the Australian Research Council.

REFERENCES

1. Thomas, H., and Morfill, G., *Nature (London)* **379**, 806 (1996).
2. Vladimirov, S.V., et al, *Phys. Rev.* **56**, R74 (1997); ibid **60**, 7369 (1999); ibid **62**, 2754 (2000).
3. Nunomura, S., et al, *Phys. Rev. Lett.* **83**, 1970 (1999).
4. Melzer, A.., et al, *Phys. Rev. Lett.* **83**, 3194 (1999).

5. Steinberg, V., *et al*, *Phys. Rev. Lett.* **86**, 4540 (2001).
6. Samarian, A.A., *et al*, *Phys. Rev. E* **64**, 025402(R) (2001).
7. Vladimirov, S.V., *et al*, *Phys. Plasmas* **3**, 444 (1996).
8. Maiorov, S.A., *et al*, *Phys. Rev. E* **63**, 017401 (2001)

Influence of the Ion Beam on the Behavior of Dust Particles Trapped in an Ion Sheath

T. Murakami*, M. Sawai*, T. Misawa*, N. Ohno* and S. Takamura*

Department of Energy Engineering and Science, Graduate School of Engineering, Nagoya University, Nagoya 464-8603, Japan

Abstract. Influence of the ion beam on the levitation position of a dust particle trapped in an ion sheath has been investigated in a simple calculation and experiment by using a double plasma device. In the numerical analysis, when the ion beam density is relatively small, the levitated height of the dust particle is found to decrease with an increase in the ion beam density until the ion drag force becomes as large as the gravity force, because the existence of the ion beam component in the ion sheath modifies the potential profile and the charge of a dust particle. The experimental observation in the double plasma device agrees with this calculation results.

INTRODUCTION

Recently, the interaction between an ion flow and negatively charged dust particles in plasmas becomes one of the most important issues in the dust plasma physics. There are several interesting works mainly related to the wake potential induced by this interaction[1, 2]. The wake potential is thought to play an important role in the structural formation of Coulomb lattice. In order to get a much deeper understanding of the phenomena on the interaction between the ion flow and the dust particles, it is necessary to develop an experimental method to control the ion flow velocity and density externally. Double plasma device can make an ion beam-plasma system, in which the ion beam energy and density can be controlled. Figure 1(a) shows the schematics of the double plasma device "Kagerou" to arrange the driver and target chambers vertically. The dust particles are levitated in the ion sheath formed in front of the negatively biased meshed electrode in the target plasma region. The ion beam is injected from the driver to target plasma region. One of the most convenient way to observe the interaction between the ion beam and the dust particles is to measure the change of the levitation height of the dust particle from the meshed electrode depending on the ion beam density and/or energy[3]. However, the ion beam passing through the ion sheath could modify the potential profile there and the charge of the dust particles, which also leads to a change of the levitation height. Then, a self consistent calculation is required to estimate the change of the levitation height of the dust particles in the presence of the ion beam. In this paper, we have carried out the calculation based on a one dimensional sheath model by taking account of the ion beam components and have compared the numerical results with the experimental observation in the double plasma device.

NUMERICAL ANALYSIS

The levitated position of a dust particle is determined by a balance between gravitational force, electrostatic force due to electric field in the ion sheath, and the ion drag force due to the upward ion beam. Figure 1(b) show the ion drag force as a function of the ion beam energy. The ion drag force acting upon a charged particle has two parts, the collection force induced by the direct collision between the dust particle and ions, and the orbit force due to the deflection of the ion orbits. In our experimental condition (the ion beam energy E_b >10 eV), the collection force is found to be a dominating part in the ion drag force.

In order to estimate the electrostatic force on a dust particle levitated in an ion sheath, we need to calculate the potential profile in the ion sheath by taking account of the influence of the ion beam component. Figure 2 shows a simple numerical model. Bulk electrons are assumed to obey Boltzmann distribution. Ion beam is injected through the meshed electrode with density n_{ib0} and upward velocity v_{ib0}. In the bulk plasma, charge neutrality is assumed to be

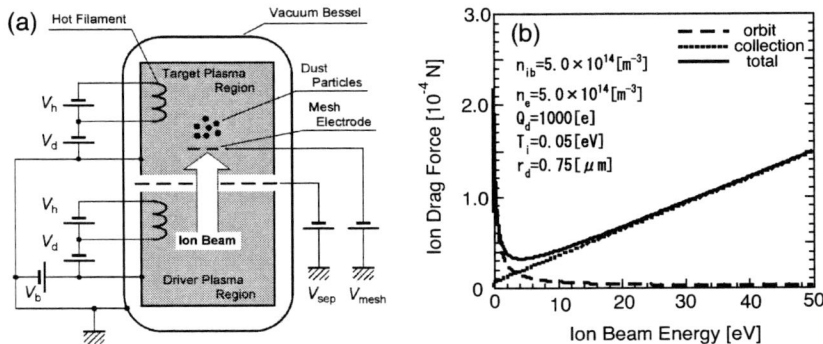

FIGURE 1. (a) Schematics of the experimental device. (b) dependence of ion drag force on the ion beam energy.

$n_e = n_i + n_{ib}$. Poisson equation is given by:

$$\frac{d^2 V(z)}{dz^2} = \frac{e}{\varepsilon_0}\{n_e(z) - n_i(z) - n_{ib}(z)\}, \quad (1)$$

where

$$n_i(z) = n_{si}\left\{\frac{V_0}{V(z)}\right\}^{\frac{1}{2}}, \quad n_{ib}(z) = n_{ib0}\left\{1 - \frac{e(V_0 - V_{mesh})}{\frac{1}{2}m_i v_{ib0}^2}\right\}^{-\frac{1}{2}}, \quad n_e(z) = n_{se}\exp\left\{\frac{V(z) - V_0}{k_B T_e}\right\}. \quad (2)$$

This equation is numerically solved to obtain the potential profile in the ion sheath.

The charge Q on a dust particle is given by the product of the capacitance C_d and the floating voltage V_f. C_d is generally given by $4\pi\varepsilon_0 r_d$ for a spherical dust with the radius of r_d. It should be noted that the V_f is defined as the potential difference between the surface of the dust particle and the space around it. The V_f is determined by the balance between the electron and the ion currents into the dust surface. If $V(z)$ is assumed to be space potential distribution in the ion sheath, the electron current into a spherical dust at a position z is described by:

$$I_e(z) = -4\pi r_d^2 e \frac{n_{se}}{4}\left(\frac{8k_B T_e}{\pi m_e}\right)^{\frac{1}{2}}\exp\left\{\frac{e(V(z) + V_f(z) - V_0)}{k_B T_e}\right\}. \quad (3)$$

On the other hand, the bulk ion current into the dust particle is given by:

$$I_i(z) = \pi r_d^2 e n_{is} v_{is}\left(1 - \frac{eV_f(z)}{\frac{1}{2}k_B T_{it} + eV(z)}\right), \quad (4)$$

and the beam ion current is:

$$I_{ib}(z) = \pi r_d^2 e n_{is} v_{is}\left(1 - \frac{eV_f(z)}{\frac{1}{2}k_B T_{id} + \frac{1}{2}m_i v_{ib0}^2 + e(V_0 - V_{mesh})}\right), \quad (5)$$

where T_{it} and T_{id} are ion temperature in the target and driver regions, and the change of ion collection area due to the modification of ion orbit is taken into account. The floating voltage $V_f(z)$ is obtained by $I_e + I_i + I_{ib} = 0$, which is numerically solved with Newton method. Finally, the force balance equation is given by:

$$\frac{4}{3}\pi r_d^3 \rho_d g = 4\pi\varepsilon_0 r_d V_f(z) E(z) + f_i(z), \quad (6)$$

where ρ_d is the mass density of the dust particle and f_i represents the ion drag force. Solving this equation gives the levitation position of a dust particle in the ion sheath.

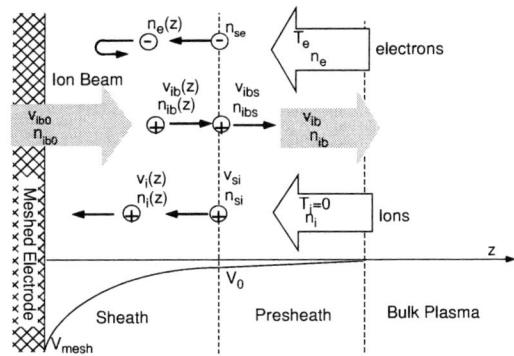

FIGURE 2. Numerical Model.

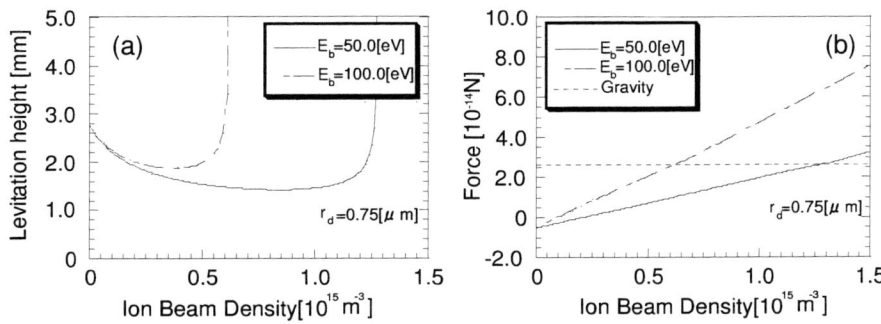

FIGURE 3. Ion beam density dependence of (a): levitation height from the negatively biased meshed electrode (V_{mesh}) and (b): ion drag force for three different beam energy E_b.

Figure 3 shows the levitation height from the meshed electrode and the ion drag force as a function of the ion beam density for three different ion energies. First, the levitation height is found to decrease even when the ion drag force is increasing with an increase in the ion beam density. When the ion drag force becomes as large as the gravity force, the levitation height rapidly increases with the ion beam density as shown in Fig. 3(a).

There are two reasons for reduction of the levitation height. Figure 4 (a) shows a change of the potential distribution with and without the ion beam component. It should be noted that the injection of the ion beam automatically leads to an increase of the bulk electron density n_e because of the boundary condition: $n_e = n_i + n_{ib}$ and n_i ($= n_{eo}$) is assumed to be constant not depending on the ion beam density. The increase of n_e results in a narrower sheath thickness(see dotted line in Fig. 4(a)). In addition, the beam ion component in the ion sheath also modifies the potential profile to reduce the sheath thickness (dashed line in Fig. 4(a)) even at almost same n_e. As shown in Fig. 4(b), the charge of the dust particle becomes less negative with the ion beam injection. Due to these effects, the levitation height of the dust particle decreases with the ion beam density.

EXPERIMENTAL RESULTS

The experiment was carried out in a large unmagnetized device "KAGEROU", 0.4m in diameter and 1.1m in height consisted of multi-dipole magnets shown in Fig. 1(a). Ar plasmas were generated by a dc discharge at a pressure of less than 4mtorr. Typical plasma parameters measured with a Langmuir probe are; electron temperature $T_e \sim 0.5$ eV and plasma density $n_e \sim 5 \times 10^{13} - 5 \times 10^{14}$ m^{-3}. Plasma potential is almost zero. The diameter of the employed dust particle is $0.75 \mu m$ and its mass density is 1.2 g/cc. Dust particles can be levitated above a negatively biased mesh electrode and confined horizontally by a negatively biased ring electrode. Both diameters of the mesh and ring electrode are 80mm. The dust particles were illuminated by a vertical sheet of He-Ne laser light and detected by an intensified charge coupled device (ICCD) camera.

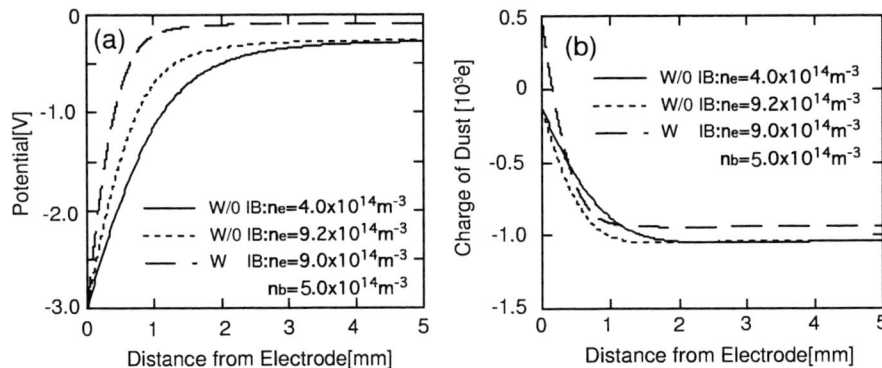

FIGURE 4. (a): Potential profile and charge of a dust particle plotted as a function of distance from the meshed electrode with and without ion beam component, where T_e=0.5 eV, V_{mesh}= -3.0 V and E_b = 38.0 eV.

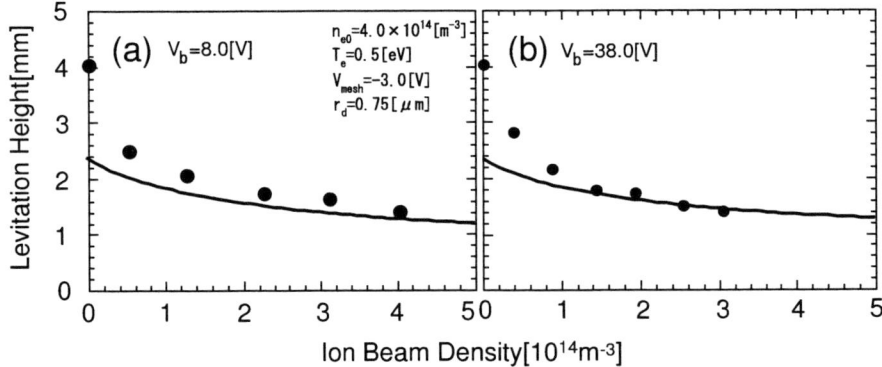

FIGURE 5. Experimental results on the levitation height of the dust particles as a function of the ion beam density at (a):V_b =8.0 V and (b): 38.0 V.

Ion beam is injected from the driver plasma region to target one by applying the DC voltage V_b. The ion beam energy E_b can be controlled by changing V_b. Dust particles are trapped in the ion sheath above the negatively biased mesh electrode. Figure 5 show the typical experimental results. Levitating height of the dust particles is found to decrease with an increase in the ion beam density. The experimental result of levitating position of dust particles agrees almost with the calculated results in Fig. 3. In order to levitate the dust particles by the ion drag force, much higher ion beam density or energy are required. However, such ion beam density and energy have not been achieved at this moment because of the problem of breakdown between the driver and target chambers.

CONCLUSIONS

We have investigated the dependence of the levitation height of dust particles on the ion beam density in a simple calculation and experiment. At relatively small ion beam density, the levitated height of the dust particle decrease with the ion beam density both in numerical analysis and experiment. This can be explained by the change of the ion sheath potential profile and the charge of a dust particle due to the ion beam injection.

REFERENCES

1. V. A. Schweigert *etal*., Phys. Rev. E, **54**, 4155(1996).
2. A. Melzert *etal*., Phys. Rev. Lett., **83**, 3194(1996).
3. T. Misawa *etal*., IEE of Japan, **A12**, 180(2000).

Spatio-temporal Dynamics of Driven Drift Waves

D. Block and A. Piel

Institut für Experimentelle und Angewandte Physik, Christian-Albrechts-Universität, D-24098 Kiel, Germany

Abstract. The response of drift modes to spatio-temporal driver signals is reported. The phenomenon of periodic pulling, which is typical for driven, nonlinear systems is investigated in detail. The emphasis of the investigation lies on the spatial aspect of the observed nonlinear dynamics. Beyond the well known temporal behavior a novel, spatial component is clearly identified using a probe array to obtain spatio-temporal information on the plasma fluctuations.

INTRODUCTION

It is well known that periodically driven oscillatory systems exhibit characteristic features like complete and incomplete synchronization. These phenomena have been observed in a number of different plasma experiments [1, 2, 3, 4, 5] and computer simulations [6, 7]. In this context complete synchronization means frequency entrainment, i.e. the oscillation frequency is shifted towards the frequency of the external driver. During incomplete synchronization the coupling of the external driver is too weak to establish full synchronization. Instead the frequency is periodically pulled towards the exciter frequency. The concept of incomplete synchronization has been used for various technical applications [8]. In the past, most fundamental investigations concentrated on temporal aspects of synchronization, now the interest is shifting towards spatio-temporal behavior, including wave phenomena. At present the understanding is most advanced for metastable–guided ionization waves in neon gas discharges [9]. It has been found that spatial and temporal features of incomplete synchronization play an important role for transitions between competing eigenmodes. A direct observation of the spatial component has been realized only in one case [10]. A different approach to synchronize wave-like systems has been developed for drift waves recently [11]. It has been shown that spatiotemporal driver signals can synchronize drift modes even in the case of weak turbulence [12]. Further, complete and incomplete synchronization have been studied in detail [13].

This paper deals with spatio-temporal measurements of driven drift waves to resolve the full spatio-temporal evolution during incomplete synchronization. Therefore, we first give a brief overview of the experimental setup and the applied diagnostics. Next we review the temporal evolution for use as reference in the subsequent investigation of the spatial evolution.

DIAGNOSTICS AND EXPERIMENTAL SETUP

The experiments were conducted in the linear triple plasma device KIWI (Kiel Instrument for Wave Investigations). Two plasma source chambers are connected by a homogeneously magnetized midsection where drift waves are observed. The plasma is produced in argon gas by a thermionic hot-cathode discharge in steady-state operation with grounded anode. The plasma source chambers are separated from the midsection by stainless-steel mesh grids. For the present study, only one source chamber is operated and the separation grid at the active chamber is positively biased with respect to ground, while the other grid at the far end is grounded. For this configuration, the grid bias acts as control parameter for destabilizing drift waves [14]. The confinement parameter β is low ($\beta \approx 10^{-6}$) and the plasma is weakly ionized ($n_e/n_0 \leq 0.1\%$). Further information on the plasma equilibrium and the unperturbed drift wave dynamics can be found in [13]. For the present investigations the KIWI is operated at $U_g = 6\,\text{V}$ grid voltage, where monochromatic $m_{dw} = 2$ drift modes are observed, with m_{dw} being the azimuthal mode number. Throughout the paper we use a cylindrical coordinate system (r, z) with the z-axis along the magnetic field and $r = 0$ at the center of the plasma column.

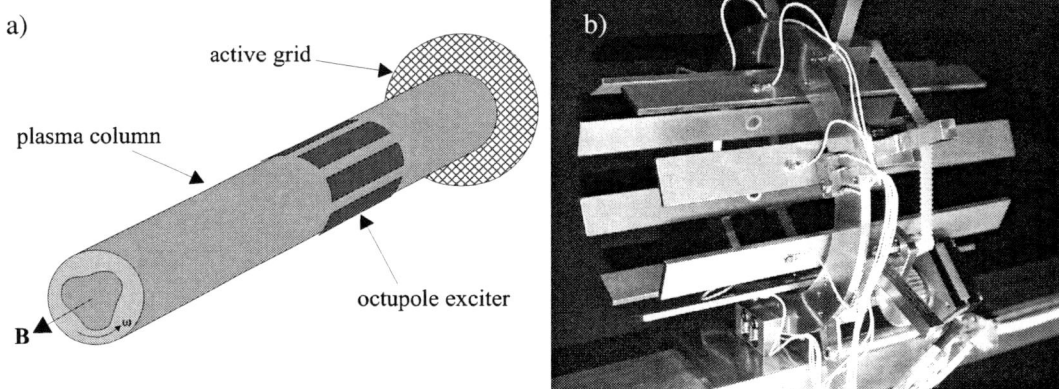

FIGURE 1. (a) Schematic drawing including the plasma column and (b) a photo of the octupole exciter arrangement of electrodes. The radial position of the electrodes can automatically be adjusted to achieve optimized coupling to the plasma. During the measurements the position is kept fixed and all related electronics are switched off.

Synchronization of drift waves is achieved by means of an octupole arrangement of electrodes (Fig. 1). Each electrode is 20 cm long and 2.5 cm wide. The radial position of the electrodes is carefully chosen to provide a reasonable contact to the plasma column without acting as a limiter. Sinusoidal signals of typically 5 volts amplitude are applied to each electrode. The phase shift between the neighboring electrodes is adjusted to a periodic $m_{ex} = 2$ wave pattern in azimuthal direction. With a proper choice of sign of phase shift and exciter frequency an almost identical propagation of the $m_{ex} = 2$ wave pattern with the drift wave can be realized. For the presented investigations the frequency is chosen $\Delta f = 250$ Hz lower than the unperturbed drift wave frequency $f_{dw} \approx 16$ kHz. For further technical details we refer to [13, 11].

The central diagnostic tool for the investigations is a probe array built of 15 Langmuir probes which are equally distributed over half the circumference at radial position $r = 23$ mm. The probes are all operated in the ion saturation current regime to measure density fluctuations.

SPATIO-TEMPORAL DYNAMICS

The temporal interaction of drift wave and exciter were already reported in [13]. Therefore, only the relevant results for the current investigation will be briefly summarized. It was found that driven drift waves show typical temporal characteristic of driven nonlinear systems, i.e. complete and incomplete synchronization. The latter is also known as periodic pulling, which is originating from the periodical frequency entrainment observed in driven, nonlinear, oscillatory systems [15].

In Fig. 2 a data set from the probe array is shown for different frames of reference. Usually spatio-temporal data sets are plotted as contour plots with time versus azimuthal position. The frame of reference is the laboratory frame. For a detailed investigation of the slow process of periodic pulling the laboratory frame is not well suited. The fast $E \times B$ rotation of the whole plasma column and the additional diamagnetic drift velocity of the waves would dominate the plot and cover the slow processes. Therefore, the exciter field has been recorded simultaneously to transform the recorded density fluctuations into the frame which is co-rotating with the *exciter wave*. The result is depicted in Fig. 2 a. The positive density perturbations are white and the negative are black. In the co-rotating frame the position of maximum exciter signal is at zero. The exciter wave length is 7 cm, i.e. the azimuthal axis covers a full exciter wave length. For synchronization one expects the drift wave to propagate with the same velocity as the exciter signal, i.e. it should rest within the co-rotating frame with a fixed phase shift with respect to the exciter wave. For the presented case of incomplete synchronization the drift waves propagates relative to the exciter wave. That the propagation velocity is not constant as well is directly seen from the bending of the black and white areas (indicating the maximum and minimum of the drift wave). An important feature are the different time scales for fast and slow propagation. For $2 \text{ s} \leq t \leq 5 \text{ s}$ the position of the drift wave is almost constant corresponding to a slow propagation.

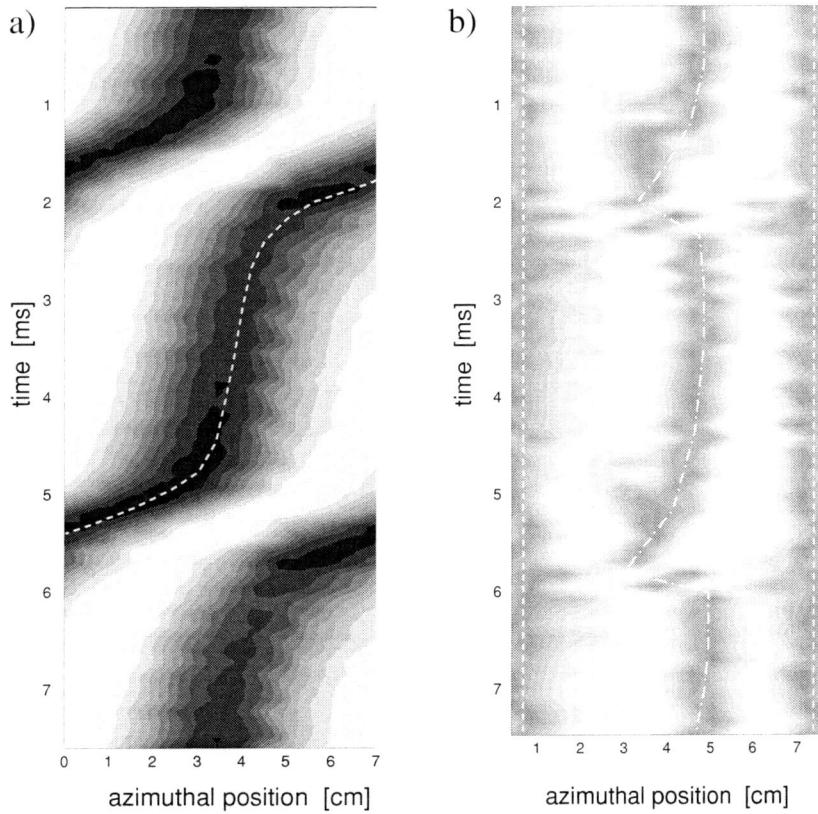

FIGURE 2. (a) Contour plot of a $m = 2$ drift mode in a frame which is co-rotating with the $m_{ex} = 2$ *exciter signal*. Positive (negative) density perturbations are marked white (black). The exciter signal maximum is located at the origin of the azimuthal position axis. The wavelength of the exciter signal is $\lambda_{ex} \approx 7$ cm. The dashed line marks the temporal evolution of the position of a drift wave minimum relative to the exciter signal maximum. The contour plot (b) shows the same data set but transformed into the frame which is co-rotating with the *drift wave*. Additionally only the absolute wave amplitude is plotted. Large (zero) amplitudes are plotted white (black). The dashed and dashed-dotted lines mark the temporal evolution of the positions of zero wave amplitude. The drift wave wavelength is $\lambda_{dw} \approx 7$ cm.

The following period of fast propagation is much shorter. A quantitative analysis [13] reveals that these measurements clearly identify the incomplete synchronization of drift waves as a nonlinear modulation of the drift wave propagation velocity. The periodicity of the process of incomplete synchronization is directly given by the frequency mismatch of the unperturbed drift wave and the exciter wave. An important point is that these measurements relate the temporal effect of incomplete synchronization with the spatial propagation of the drift wave. Therefore, it is rather suprising that the temporal dynamics of such a spatiotemporal system are still well described by a model of a driven van der Pol oscillator [15].

From these investigations, however, it is not clear, whether the spatial structure is modulated due to the incomplete synchronization process as well. The answer can be given by a transformation of the spatio-temporal data set into the frame which is co-rotating with the *drift wave*. The result of this operation is depicted in Fig. 2 b where the absolute wave amplitude is given as a function of time and azimuthal position in the co-rotating frame of the drift wave. Zero amplitudes are marked black, while maximum amplitudes are white. At first it can be seen that the spatial extent of a full wave length (covering two extrema) is constant ($\lambda_{dw} \approx 7$ cm). A temporal wavelength modulation is not observed. However, the spatial extent of minimum and maximum itself is obviously not constant. Although the total wavelength is not modulated the spatial extent of driftwave minima λ_- and maxima λ_+ are. The shift is about 1cm, e.g. 15% of λ_{dw}. The temporal evolution of this shift is clearly related with the periodic pulling process (Fig. 2 a). The periods of strong changes in drift wave propagation correspond to strong changes in λ_- and λ_+. During the slow propagation periods the ratio of λ_- and λ_+ is nearly unchanged. The reason why the modulation of λ_- and λ_+ is in phase opposition are periodic boundary conditions in azimuthal direction. For a $m = 2$ mode the wave length has to be half the circumference of the plasma column to achieve a stable wave pattern. Therefore, any modulation of λ_{dw} would have to be compensated by an opposite modulation within the next wave front. Due to symmetry arguments this is not likely. To come back to the initial question, the measurements clearly show that despite of temporal characteristics a pure spatial modulation of the drift wave extrema is related with incomplete synchronization of drift waves.

SUMMARY AND CONCLUSIONS

We have studied the spatio-temporal evolution of driven drift waves by using probe arrays. For this purpose spatio-temporal wave-like perturbation is coupled into the plasma by an octupole arrangement of electrodes. For resonant *spatial* and nearly resonant *temporal* exciter waves incomplete synchronization is observed. It is shown that the nonlinear evolution of the driven system consists of two components, temporal and spatial. The novel and significant spatial component is clearly observed by spatio-temporal recordings. The modulation is found to be caused by the incomplete synchronization as the nonlinear evolution of temporal and spatial modulation coincide. Further, the periodic boundary conditions in azimuthal direction seems to play an important role. They are supposed to inhibit a modulation of λ_{dw}.

A more general importance arises from these observations when they are related to actual experiments on driven ionization waves [10, 16]. There, a spatial modulation has been observed as well. As ionization waves and drift waves are based on completely different mechanisms it is concluded that a spatial component is an inherent feature of driven wave-like systems.

ACKNOWLEDGMENTS

This work was supported by Deutsche Forschungsgemeinschaft under Contract No. Pi 185/14–3. Technical assistance of V. Rohwer is gratefully acknowledged.

REFERENCES

1. Klostermann, H., Rohde, A., and Piel, A., *Phys. Plasmas*, **2**, 2406 (1997).
2. Klinger, T., Greiner, F., Rohde, A., Piel, A., and Koepke, M. E., *Phys. Rev. E*, **52**, 4316 (1995).
3. Amemiya, H., *Plasma Phys.*, **25**, 735 (1983).
4. Abrams, R. H., Yadlowsky, E. J., and Lashinsky, H., *Phys. Rev. Lett.*, **22**, 275 (1969).
5. Koepke, M. E., Alport, M. J., Sheridan, T. E., Amatucci, W. E., and J. J. Carroll III, *Geophys. Res. Lett.*, **21**, 1011 (1994).
6. Greiner, F., Klinger, T., and Piel, A., *Phys. Plasmas*, **2**, 1810 (1995).
7. Rohde, A., Klostermann, H., and Piel, A., *Phys. Plasmas*, **4**, 3933 (1997).
8. Kurokawa, K., *Proc. IEEE*, **61**, 1386 (1973).
9. Koepke, M. E., Klinger, T., Seddighi, F., and Piel, A., *Plasma Phys.*, **3**, 4421 (1996).
10. Koepke, M. E., Dinklage, A., Klinger, T., and Wilke, C., *Phys. Plasmas*, **8**, 1432 (2001).
11. Block, D., Schröder, C., Klinger, T., and Piel, A., *Contrib. Plasma Phys.*, **41**, 455 (2001).
12. Schröder, C., Klinger, T., Bonhomme, G., Block, D., and Piel, A., *Phys. Rev. Lett.*, **86**, 5711 (2001).
13. Block, D., Piel, A., Schröder, C., and Klinger, T., *Phys. Rev. E*, **63** (2001).
14. Klinger, T., Latten, A., Piel, A., Bonhomme, G., Pierre, T., and de Wit, T. D., *Phys. Rev. Lett.*, **79**, 3913 (1997).
15. van der Pol, B., *Phil. Mag.*, **3**, 65 (1927).
16. Weltmann, K.-D., Koepke, M. E., and Selcher, C. A., *Phys. Rev. E*, **62**, 2773 (2000).

Transport Studies of Driven Drift Waves

D. Block and A. Piel

Institut für Experimentelle und Angewandte Physik, Christian-Albrechts-Universität, D-24098 Kiel, Germany

Abstract. Recent results show that drift waves can be synchronized by spatio-temporal driver signals [1, 2]. They show that a spatio-temporal driver signal allows to drive drift modes as soon as the driver signal is spatio-temporally resonant with the drift mode. This paper reports about experimental investigations concerning the related changes in fluctuation induced transport.

INTRODUCTION

During the last 50 years, considerable effort has been spent on the investigation of anomalous transport in magnetically confined plasmas, on its origin and on possibilities to reduce it. The observed transport is usually not satisfactorily described by classical or neoclassical transport theory. Therefore, additional transport processes driven by universal instabilities are made responsible for the observed anomalous transport [3, 4, 5]. Among the most prominent instabilities are drift waves, which propagate mostly azimuthally (in cylindrical plasmas) and have a radial fluctuation maximum close to the position of maximum density gradient. Research on drift waves has been linked with the research on anomalous transport from the very beginning [6]. Parallel to the investigation of drift wave dynamics and the related transport, there have been attempts to control transport. Already at the beginning of the 70's first investigations with feedback techniques were started [7, 8]. Till today this technique has been refined and tested in fusion experiments [9, 10], but the observed reduction of fluctuations is localized and the efficiency is not satisfying. First experiments in Kiel and Nancy showed that drift wave dynamics can be influenced with spatio-temporal driver signals [1, 2]. SCHRÖDER et al. demonstrated that this control method is capable to synchronize selected drift modes in a weakly turbulent plasma state. As any change of drift wave dynamics can cause severe changes of the related anomalous transport, this paper reports on the influence of spatio-temporal driver signals on anomalous transport caused by monochromatic drift waves.

EXPERIMENT AND DIAGNOSTICS

The experiments were conducted in the linear triple plasma device KIWI (Kiel Instrument for Wave Investigations). Two plasma source chambers are connected by a homogeneously magnetized midsection where drift waves are observed. The plasma is produced in argon gas by a thermionic hot-cathode discharge in steady-state operation with grounded anode. The plasma source chambers are separated from the midsection by stainless-steel mesh grids. Further information on the plasma equilibrium and the unperturbed drift wave dynamics can be found in [1]. Table 1 gives an overview of the most important plasma parameters. For the present investigations only one source chamber is operated. The grid voltage is set to $U_g = 6$ V, where monochromatic $m_{dw} = 2$ drift modes are observed, with m_{dw} being the azimuthal mode number.

Synchronization of drift waves is achieved by means of an octupole arrangement of electrodes (see [11] for details). Sinusoidal signals of typically 5 volts amplitude are applied to each electrode. The phase shift between the neighboring electrodes is adjusted to a periodic $m_{ex} = 2$ wave pattern in azimuthal direction. With a proper choice of sign of phase shift and exciter frequency f_{ex} an almost identical propagation of the $m_{ex} = 2$ wave pattern with the drift wave can be realized. For the present investigations the exciter frequency is varied around the frequency f_0, the frequency of the unperturbed drift wave.

To measure the fluctuation induced transport with high resolution, an electrostatic triple probe construction is used in combination with a transient recorder (1 MHz, 256 kWords per channel). The triple probe consists of three

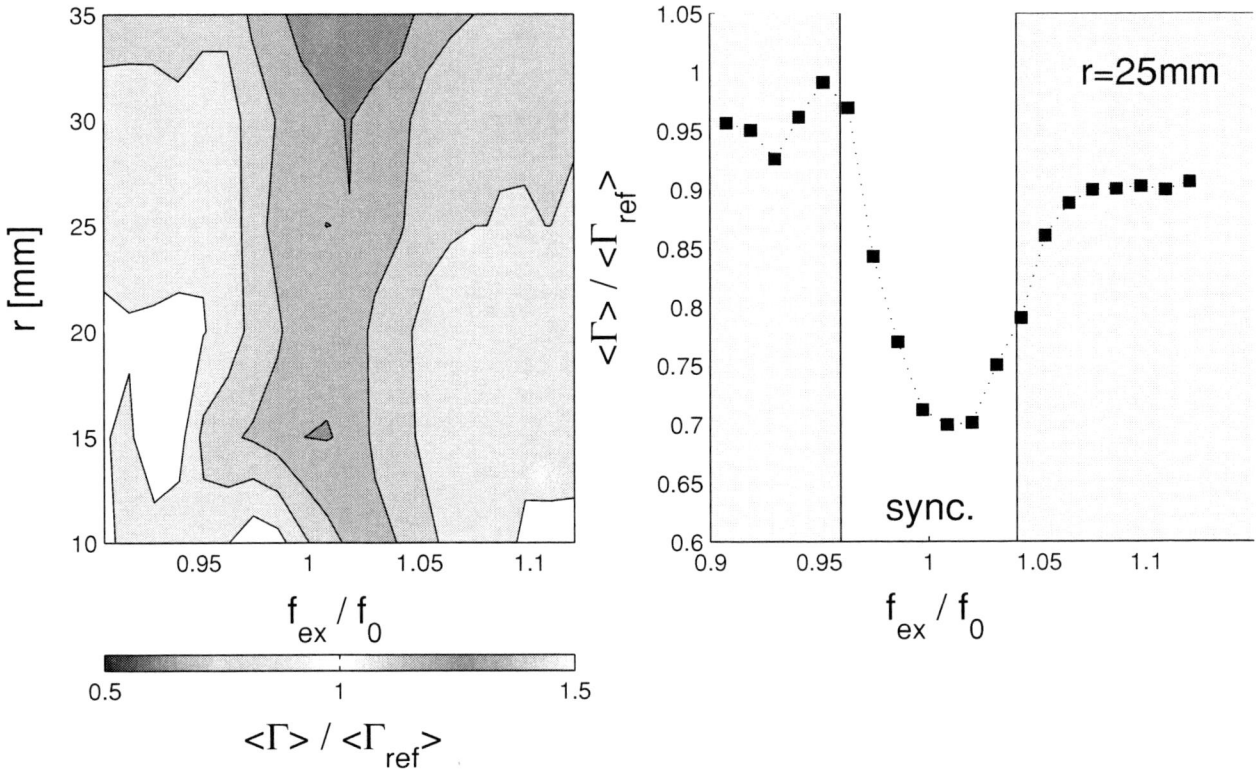

FIGURE 1. The contour plot shows the average transport for a co-rotating $m_{ex} = 2$ exciter signal applied to a $m = 2$ mode as function of radius and exciter frequency. The transport is normalized to the average transport without exciter signal. f_0 denotes the drift wave frequency without applied exciter signal. The right plot shows the normalized average transport as function of exciter frequency at $r = 25$ mm. The region, where complete synchronization is observed, is marked.

Langmuir probes which are aligned in azimuthal direction. Their radial position with respect to the center of the plasma column can be varied. The distance between two neighboring probes is about 4 mm. The two outer probes are used to record floating potential fluctuations $\tilde{\phi}_f$ while the central probe is negatively biased to record ion saturation current fluctuations $\tilde{n}$. The floating potential measurements are used to calculate electric field fluctuations. Using $\tilde{\Gamma} = \tilde{n}\tilde{E}_\theta = \tilde{n}\nabla\tilde{\phi}_f$ the average fluctuation induced transport in azimuthal direction $\langle \Gamma \rangle$ is obtained. Subsequently, Γ is used for the transport observed with applied exciter signals and Γ_0 for the transport in the unperturbed system.

TABLE 1. Plasma and discharge parameters of the triple plasma device KIWI. $i-n$ stands for ion-neutral collisions, $e-n$ for electron-neutral collisions. Values with a radial dependence are taken at plasma center. L_n is taken at the position of maximum gradient

column length	$l = 2$ m
magnetic field	$B = 0.07$ T
argon gas pressure	$P = 8 \cdot 10^{-2}$ Pa
$e-n$ mean-free path (elastic collisions)	$\lambda_{en} = 1$ m
$i-n$ mean-free path (charge exchange collisions)	$\lambda_{in} = 0.3$ m
electron density	$n_e = 1\ldots 2 \cdot 10^{16}$ m^{-3}
electron temperature	$T_e = 1.2$ eV
ion temperature	$T_i \approx 0.03$ eV
density gradient length	$L_n = 0.02$ m

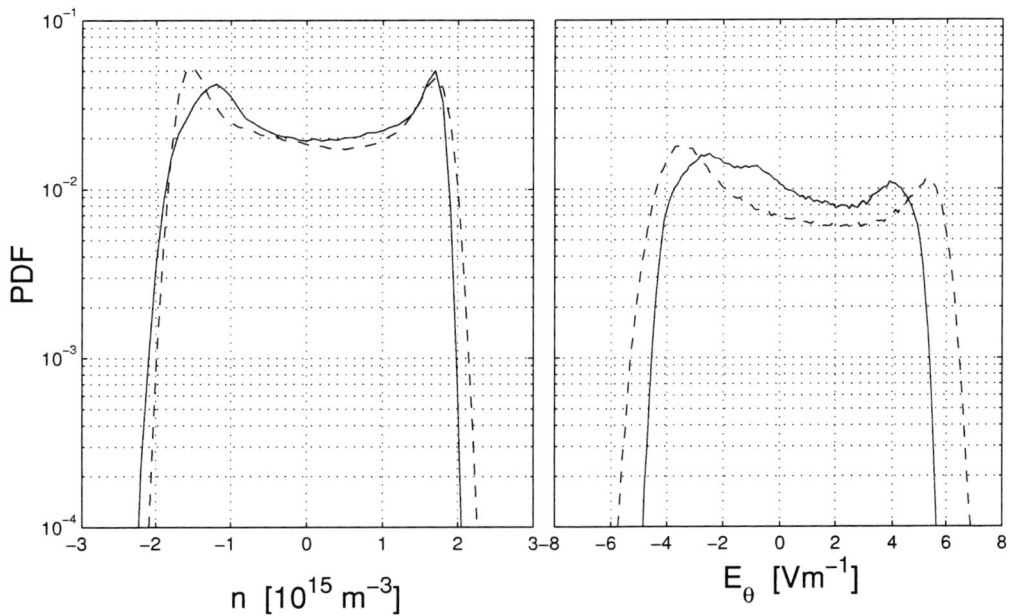

FIGURE 2. PDFs of density and azimuthal electric field for co-rotating $m_{ex} = 2$ exciter signal applied to a $m = 2$ drift mode (solid) and undriven system (dashed). The time series are recorded at $r = 25$ mm.

TRANSPORT MEASUREMENTS

To investigate the influence of the exciter signal on transport, the transport is recorded at different radial positions for $m_{ex} = 2$ exciter signals with frequencies from 16 to 20 kHz. The frequency step width is $\Delta f = 200$ Hz. For each radial position a data set without applied exciter signal is recorded as well, to take small plasma parameter drifts into account. The results are plotted in Fig. 1. The contour plot shows the average transport of the driven system, which has been normalized to the average transport of the undriven system. The frequency scale has been normalized to the drift wave frequency f_0 without exciter signal. The second plot in Fig. 1 shows the normalized average transport at radial position $r = 25$ mm. The frequency domain of complete synchronization is marked. It is seen that the average transport is reduced significantly for almost matching frequencies at all radial positions. The transport reduction is about 30 %. The errors are estimated to be ± 5 %. An enhancement of transport is not observed.

Due to its definition, a change of transport must be related to changes of density and potential fluctuations. Basically three effects can cause a change in transport:

1. A change of density and/or potential **fluctuation amplitude**.
2. A change of **phase shift** between density and potential fluctuations.
3. A **decorrelation** of density and potential fluctuations.

To check to what extent amplitude changes are responsible for the observed transport reduction, Fig. 2 depicts the PDFs for density and azimuthal electric field fluctuations recorded at $r = 25$ mm for matching frequencies of driver and drift wave. The plot on the left hand side shows the PDF of the density fluctuations for the driven (solid line) and the undriven system (dashed line). Clearly only small deviations are observed between the two states. These slight changes of fluctuation amplitude do not change the average amplitude significantly, because the average amplitude of the density fluctuations has not changed. Therefore, the observed transport reduction can not be caused by the density fluctuations, if an unchanged phase relation of density and electric field is assumed . For the azimuthal electric field the PDF is plotted on the right hand side. Here, the changes of the fluctuation amplitudes are significant. The average amplitude of the electric field fluctuations is reduced by about 25 %. The similar shape of the PDF for driven and undriven system indicates that this reduction is due to an overall reduction of azimuthal electric field fluctuations. With the assumption of unchanged phase between density and potential the observed amplitude reduction would lead to a transport reduction of 25 %. This is of the order of the observed overall transport reduction.

However, transport is very sensitive to phase changes of density and electric field fluctuations for certain phase values. A claculation of the crossphase of density and electric field reveals, that a significant change of phase shift is not observed. The phase shift of density and electric field fluctuations is found to be $\alpha_{\tilde{n}\tilde{E}} \approx \pi/6$. However, the uncertainty of this phase calculation is estimated to $\Delta\alpha_{\tilde{n}\tilde{E}} \approx \pi/18$. This means such an accuracy in phase measurement results in an absolute accuracy for the average transport of $\Delta\Gamma = 15\%$. This sensitivity shows that for the presented investigations an influence on transport due to phase changes cannot be generally excluded. To estimate the influence of decorrelation of density and potential fluctuations the cross-correlation is calculated. This analysis reveals no significant decorrelation of density and electric field for the driven and undriven system.

Therefore, the observed changes in transport are mainly due to the modification of the amplitude of the electric field fluctuations. The other possibile mechanisms may contribute to the change of transport but are not dominant, as the reduced electric field fluctuation are capeable to explain the observed average transport reduction on their own.

SUMMARY AND CONCLUSIONS

To summerize, we have shown by radially resolved and systematic scan of driven drift waves that spatio-temporal driver signals do have an influence on transport. The investigated transport characteristics of driven monochromatic drift waves show transport reductions of up to 30 % for synchronization. The maximum transport reduction is observed for matching frequencies of drift wave and exciter signal.

The careful investigation of the density and potential fluctuations revealed that the electric field fluctuations are significantly reduced by the exciter signal. The resulting transport reduction is constent with the observations.

The reduction of electric field fluctuations by the external driver is not that surprising, as the exciter signal itself is electrostatic. Further investigations show that the azimuthal electric field plays a significant role for the synchronization process [12].

ACKNOWLEDGMENTS

This work was supported by Deutsche Forschungsgemeinschaft under Contract No. Pi 185/14–3. Technical assistance of V. Rohwer is gratefully acknowledged.

REFERENCES

1. Block, D., Piel, A., Schröder, C., and Klinger, T., *Phys. Rev. E*, **63** (2001).
2. Schröder, C., Klinger, T., Block, D., A. Piel, G. B., and Naulin, V., *Phys. Rev. Lett.*, **86**, 5711–5714 (2001).
3. Wootton, A. J., Carreras, B. A., Matsumoto, H., McGuire, K., Peebles, W. A., Ritz, C. P., Terry, P. W., and Zweben, S. J., *Phys. Fluids B*, **2**, 2879–2903 (1990).
4. Rudyi, A., Untersuchung transportrelevanter Fluktuationen in der Randschicht von ASDEX, Tech. Rep. III/160, Max-Planck-Institut für Plasmaphysik, Garching (1990).
5. Wagner, F., and Stroth, U., *Plasma Phys. Controlled Fusion*, **35**, 1321–1371 (1993).
6. Hendel, H. W., Chu, T. K., and Politzer, P. A., *Phys. Fluids*, **11**, 2426–2439 (1968).
7. Thomassen, K. I., *Nucl. Fusion*, **11**, 175–186 (1971).
8. Arsenin, V. V., and Chuyanov, V. A., *Sov. Phys.-Usp.*, **20**, 736–762 (1977).
9. B. Richards, T. U., Wootton, A. J., Carreras, B. A., Bengtson, R. D., Hurwitz, P., Li, G. X., Sen, A. K., , and Uglum, J., *Phys. Plasmas*, **1**, 1606–1611 (1994).
10. Sen, A. K., *Phys. Plasmas*, **7**, 1759–1766 (2000).
11. Block, D., and Piel, A. (2002).
12. Block, D., *Synchronization of drift waves and its effect on fluctuation induced transport*, Ph.D. thesis, Christian-Albrechts-Universität Kiel (2001).

Sound Wave Propagation in Gases at Low Pressure

T. Tonogaki*, N. Ohno*, S. Takamura* and A. Tsushima[†]

*Department of Energy Engineering and Science, Graduate School of Engineering, Nagoya University, Nagoya 464-8603, Japan
[†]Faculty of Engineering, Yokohama National University, Yokohama 240-8501, Japan

Abstract. We have investigated sound wave propagation in helium and argon gases at low gas pressure. At pressure around 50 mtorr for argon and 30 mtorr for helium and sound source frequency of 40 kHz, the phase velocity of sound wave increases with a decrease in pressure, which agrees well with the theoretical prediction based on Navier-Stokes equation. At a very low pressure (5 mtorr) of helium gas, the wave propagation shows pseudo wave property that the phase velocity is increasing as distance from sound source increases.

INTRODUCTION

In recent experiments associated with fusion-related edge plasmas, high density Rydberg gases can be generated by volumetric plasma recombination, which occurs by interaction between high density plasma ($> 10^{19} m^{-3}$) and neutral gases at pressure of a few ten mtorr[1]. As increasing neutral gas pressure from 3 mtorr to 15 mtorr shown in Fig. 1, high density helium plasmas, which transport along magnetic field, are cooled down to be less than 1 eV due to elastic, inelastic and charge exchange collisions with neutral gases, resulting in strong volumetric plasma recombination including three body and radiative recombination. It is seen in Fig. 1(b) that the plasma is extinguished in front of the target plate, which is so-called "detached plasma". Singly ionized helium ions are recombined with electrons, which are captured at highly excited states, to be Rydberg atoms. Highly excited helium atoms with a principal quantum number n above 20 has been observed with light emission from the highly excited states as shown in Fig. 1(c). Such high density Rydberg gas is thought to be quite interesting and new medium because the Rydberg atoms have large atomic radius, which is proportional to n^2 and they strongly interact each other by dipole-dipole interaction. One of the most convenient way to study the collective phenomena in the Rydberg gas is to measure sound wave propagation. We plan to investigate the sound wave propagation in the Rydberg gas. However, the high density Rydberg gas is generated at very low gas pressure around 10 mtorr in our experiment, then first it is necessary to know the propagation property of the sound wave in such low gas pressure because the propagation property should be changed depending on neutral gas pressure[2,3]. In this report, we will presents sound wave propagation in helium and argon gases at low gas pressure.

EXPERIMENTAL SETUP

Figure 2 shows schematic sketch of experimental setup. Vacuum vessel is made of stainless steal with a diameter of 40 cm and a hight of 35 cm, pumped out by diffusion and rotary pumps. An electrostatic transducer with a response frequency of 40 kHz is employed as a sound source. A 1/8 inch microphone (Breul and Kjaer Model 4138) is suspended above the transducer. The distance d between microphone and transducer can be varied by using micrometer. Detected signal is amplified by pre and main amplifiers up to 50 dB and stored in digital oscilloscope with 14 bit resolution. Typical waveforms measured in helium gas at a pressure of 50 mtorr are shown in the right hand side of Fig. 2. In order to improve signal to noise ratio(S/N), averaging of detected signals is employed. At very low gas pressure around 5 mtorr, we need to average more than 70000 signals to obtain sufficient S/N.

Sound wave propagation can be characterized by the two quantity, d/λ_{mfp} and f_c/f, where λ_{mfp} is mean free path of neutral particle and f_c is inter-particle collision frequency. As shown in Fig. 3, the parameter space of d/λ_{mfp}

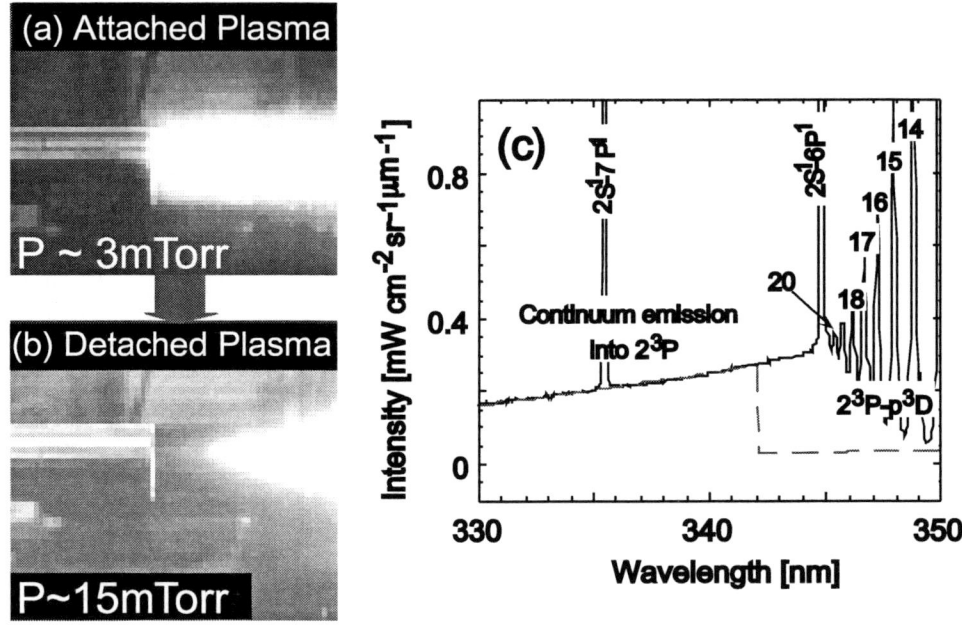

FIGURE 1. Production of Rydberg gas by interaction between high density helium plasma and neutral gas. High density helium plasma are recombined in front of the target plate as increasing neutral gas pressure from (a): 3 mtorr (attached plasma) to (b): 15 mtorr(detached plasma). (c) : Spectrum of optical emission from the detached plasma.

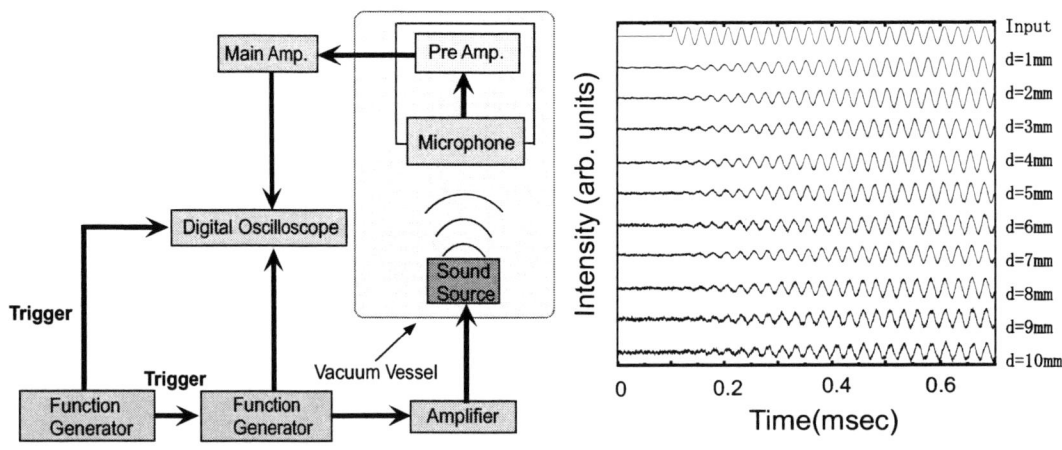

FIGURE 2. Schematic of experimental setup and detected typical waveforms in helium gas at a pressure of 50 mtorr.

and f_c/f is divided to three regions A, B, and C. In the region A where $d/\lambda_{mfp} > 1$ and $f_c/f > 10$, sound wave can propagate with constant phase velocity. When $f_c/f < 10$, the phase velocity depends on f_c/f. This region B is frequency relaxation region. At $d/\lambda_{mfp} < 1$ (region C), neutral particles, which are oscillated by the transducer, can reach the microphone directly. This propagation is ballistic mode, which is sometimes called "pseudo wave". Hatched region in Fig. 3 shows our experimental parameters. As decreasing neutral pressure, the experimental parameters d/λ_{mfp} and f_c/f in Fig. 3 move from region A to B (C). Especially, we focus on sound wave propagation in the boundary between region A and B, and pseudo wave propagation in region C.

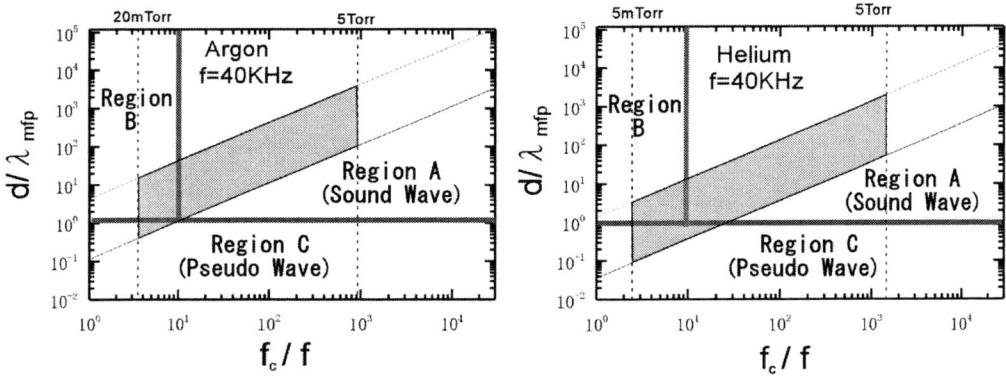

FIGURE 3. Classification of sound wave propagation. Hatched areas indicate the our experimental parameters for argon and helium gases.

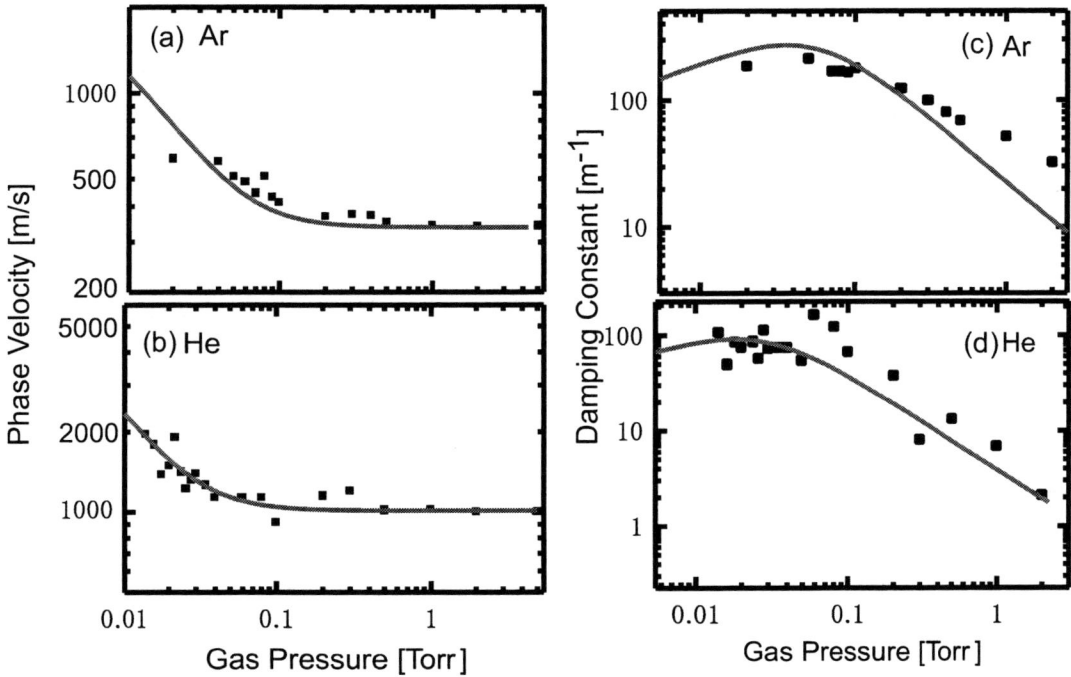

FIGURE 4. Dependence of phase velocity and damping constant on gas pressure. Solid curves are the theoretical prediction.

EXPERIMENTAL RESULTS

Figure 4 show the typical experimental results, corresponding to the boundary between region A and B in Fig. 3. The phase velocity is constant at pressure above 100 mtorr both in argon and helium gases. On the other hand, below 100 mtorr, the phase phase velocity is found to increase with a decrease in pressure. The boundary ($f_c/f = 10$) is corresponding to 50 mtorr for argon gas and 30 mtorr for helium gas, respectively. Then, it is found that the phase velocity start to increase near this boundary. On the other hand, damping constant is increasing with a decrease in gas pressure and saturated near the boundary ($f_c/f = 10$). Solid curves are theoretical calculated results based on Navier-Stokes equation by taking account of viscosity and thermal conductivity[4]. Theoretical predictions are in good agreement with the experimental results. This means that viscosity and thermal conductivity plays a role in relatively low gas pressure near the boundary between region A and B.

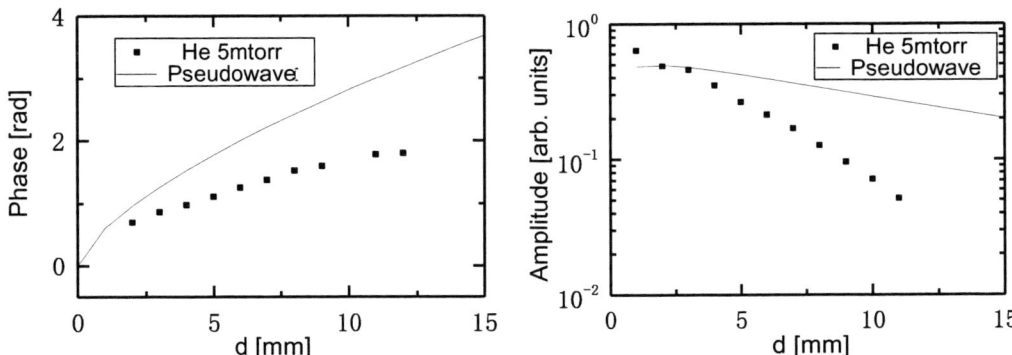

FIGURE 5. Observation of the pseudo wave in helium gas at a pressure of 5 mtorr. (a): phase variation and (b): amplitude as a function of the distance d between transducer and microphone.

Figure 5 shows phase variation and amplitude as a function of the distance d in helium gas at a pressure of 5 mtorr. The mean free path of inter-particle collision is larger than 15 mm. Then, helium atoms, vibrated at surface of the transducer, can reach the microphone without collision. It is found that the phase delay becomes smaller with an increase in d, which means the phase velocity increases in d. Theoretical analysis with simple model gives phase θ of the pseudo wave gives as follows;

$$\theta = \frac{3\sqrt{3}}{2}\left(\frac{s}{2}\right)^{\frac{2}{3}}, \qquad (1)$$

where $s = \omega d/\sqrt{2}v_t$ and $v_t = \sqrt{kT/m}$. Solid curve is calculated by using eq. (1). Experimental results qualitatively agree with the theoretical predictions. Amplitude variation in d also shows the same tendency predicted by the theoretical analysis. It can be concluded the pseudo wave was observed although there is some qualitative discrepancy between experimental results and theoretical predictions.

CONCLUSIONS

We have investigated propagation property of sound wave in argon and helium gases with low gas pressure. At relatively small pressure around 50 mtorr for argon and 30 mtorr for helium, the phase velocity increases with a decrease in pressure, which agrees well with the theoretical prediction based on Navier-Stokes equation. At a very low pressure (5 mtorr) of helium gas, the phase velocity increase in d. This matches propagation property of pseudo wave.

In near future, we will investigate sound wave propagation in the Rydberg gas based on this work.

REFERENCES

1. N. Ohno *et al*., Nucl. Fusion, **41**, 1055(2001).
2. G. Maidanik *et al*., Phys. Fluids, **8**, 266(1965).
3. G. Maidanik *et al*.,Phys. Fluids, **8**, 269(1965).
4. M. Greenspan *et al*., Acoust. Physic, **2**, 1(1965).

Wave Properties in a Weakly-Magnetized Plasma Produced by Rotating Radio-Frequency Electromagnetic Fields

G. Sato*, W. Oohara* and R. Hatakeyama*

Department of Electronic Engineering, Tohoku University, Sendai 980-8579, Japan

Abstract. Ar plasmas are produced by using spatially- and temporally-rotating radio-frequency (f_{RF}) electromagnetic fields in very weak uniform magnetic fields. The plasma density is observed to increase greatly compared with ICP mode when the RF fields rotate in the electron-diamagnetic direction and $f_{ce}/f_{RF} \sim 4$ is satisfied (f_{ce}: electron cyclotron frequency). Then the Trivelpiece-Gould wave with large amplitude propagates and damps toward the downstream. The existence of the damped wave appears to contribute to the efficient plasma production.

INTRODUCTION

In recent years there has been a diversification tendency in plasma processings used in not only materials processing but also biomedical application and treatment of hazardous waste. The development of efficient plasma sources (high density and low-electric-power consumption) is important in such all fields. So far we have investigated theoretically and experimentally the control method of time-averaged radial plasma-profiles using azimuthally-traveling (rotating) radio-frequency (RF) electromagnetic fields in a stationary magnetic field from the viewpoint of basic plasma physics [1,2]. This control method may also be applicable to the development of efficient plasma production as results of plasma-profile control and particle-loss reduction. Here we present a plasma dynamics in the case that the production and control of the plasma are simultaneously performed in a very weak magnetic field (< 50 G) under the influence of the spatially- and temporally-rotating RF fields in the electron-cyclotron range of frequencies.

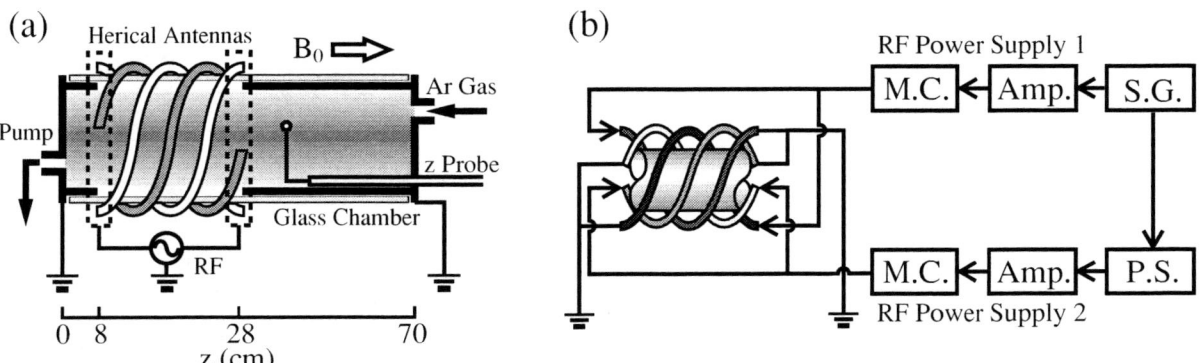

FIGURE 1. Schematic diagrams of (a) experimental setup and (b) RF power-supply circuit.

EXPERIMENTAL APPARATUS

Figure 1 shows schematic diagrams of (a) experimental setup and (b) RF power-supply circuit. The experiment is performed in a pyrex tube (70 cm in length and 10 cm in diameter) filled with argon gas (0.5 mTorr). A uniform magnetic field is applied $|B_0| < 200$ G along z axis by solenoidal coils, and $B_0 < 0$ denotes that the direction of magnetic field lines is along $-z$ axis. Helical antennas, as shown in Fig. 1 (a), consist of four elements, and each element is a one-turn antenna with 20-cm-length and set on the outer surface of the tube. Concerning the RF power supply 1 shown in Fig. 1 (b), an RF signal generated in a signal generator (S.G.) is amplified and supplied to the antennas through an impedance matching circuit (M.C.). The RF power is divided into two branches and transmitted to the opposite terminals of the two helical antennas spaced azimuthally 180°. In the case of RF power supply 2, an RF signal phased temporally by a phase shifter (P.S.) is amplified and supplied to the others, same as the RF power supply 1. When the phase difference between one antenna and the next antenna is 90° (−90°), the RF fields temporally rotating in the right-hand (left-hand) direction with respect to $+z$ axis are generated. As a result, the spatially- and temporally-rotating RF fields with dominant mode number $|m| = 1$ are generated by the four helical antennas. The total RF power is 400 W and the frequency is $f_{RF} (= \omega_{RF}/2\pi) = 7 - 28$ MHz. Plasma parameters are measured by z probes, the point part of which functions as a Langmuir probe or a magnetic probe.

PLASMA PRODUCTION DEPENDING ON B_0

Dependencies of the electron density on B_0 for f_{RF}=14 MHz are shown in Fig. 2 (a) in the vicinity of the antenna region ($z = 28$ cm) and the center of the downstream region ($z = 43$ cm). Closed and open circles indicate the densities in the right-hand- and the left-hand-rotating RF electromagnetic fields, respectively. For the case of the right-hand-rotating RF fields, the electron density for $B_0 \sim +20$ G increases greatly compared with that for $B_0 = 0$ G (ICP mode), where the rotating direction of RF fields corresponds to the electron-diamagnetic direction. The peak appearance of the electron density become more clear in the further downstream. For the case of the left-hand-rotating RF fields, the density peak, on the other hand, is observed for $B_0 \sim \pm 20$ G because the launched axial-direction of the RF

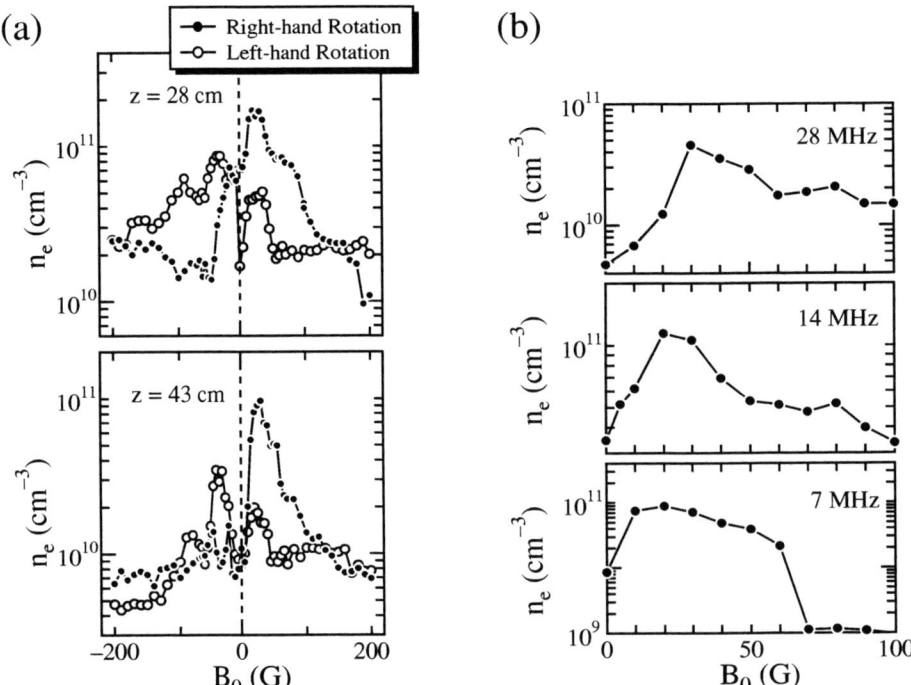

FIGURE 2. Electron density dependences on magnetic field B_0 measured at $z = 28, 43$ cm for $f_{RF} = 14$ MHz (a) and at $z = 28$ cm for $f_{RF} = 7, 14, 28$ MHz (b). The increment of plasma density occurs around $|B_0| = 20$ G in the case of $f_{RF} = 14$ MHz.

fields is decided by the rotating direction. In our case, the RF fields are launched in the $+(-)$ z direction if the rotation direction is right-hand (left-hand) one. The RF fields, which are launched in $-z$ direction and reflected by the endplate, propagate toward the downstream region and give rise to the small increase of the density when the RF fields rotate in the left-hand direction. Figure 2 (b) shows electron density dependences on B_0 (>0) in the right-hand-rotating RF fields for various RF frequencies. For each RF frequency, the density increases as a result of applying a weak magnetic field. Here the magnetic field corresponding to the electron cyclotron frequency f_{ce} ($=eB_0/2\pi m_e$) is $B_0 = 5$ G in the case of $f_{RF} = 14$ MHz. Considering that the density peaks are observed at $B_0 \simeq 10, 20, 40$ G for $f_{RF} = 7, 14, 28$ MHz, respectively, the condition of $f_{ce}/f_{RF} \simeq 4$ is always satisfied in this phenomenon.

SPATIAL STRUCTURE OF TRAVELING WAVE

The electron density is observed to increase in the downstream region, as shown in Fig. 2 (a). Axial magnetic fluctuations $\tilde{B}_z$ are measured by magnetic probes in the case of right-hand-rotating RF fields for $f_{RF} = 14$ MHz. In the spectrum analysis in the vicinity of the antennas, the rotating RF fields contain higher harmonics ($f = nf_{RF}$), but no harmonics exists only in the case that the density increases greatly ($20 \leq B_0 \leq 50$ G). The RF fields change to a traveling wave satisfying a dispersion relation in the plasma and propagate toward the downstream in the range of $20 \leq B_0 \leq 50$ G. Figure 3 (a) shows a typical two-dimensional (r-z) distribution of $\tilde{B}_z$ under the condition that the density remarkably increases ($B_0 = +25$ G). It is found that the large amplitude of $\tilde{B}_z$ with the opposite phase alternately appears along B_0 in the periphery region of the plasma column. Such a spatial structure of $\tilde{B}_z$ indicates the formation of helicity along B_0, giving rise to a transverse electric field. Since $\tilde{B}_z$ propagates toward the downstream with the spatial-distribution pattern kept orderly, the transverse E-field is observed to be circularly polarized at a fixed z point.

Figure 3 (b) gives axial profiles of the amplitude of $\tilde{B}_z$ for various magnetic fields at $r = 3$ cm. A clear amplitude oscillation is observed in the downstream region in the range of $B_0 = 20 - 50$ G, and the wave amplitude is large near the antennas, decaying downward. A dependence of the axial wave number k on B_0 is obtained from Fig. 3 (b) and the dispersion relation is shown in Fig. 4, which is compared with the theoretical one. The theoretical dispersion relation of the electromagnetic wave in a plasma column with a conducting boundary has been derived in the study of helicon discharge[3,4]:

$$\sqrt{k_\perp^2 + k_\parallel^2} = \frac{k_\parallel \omega_{ce}}{2\omega_{RF}} \left[1 \pm \sqrt{1 - \frac{4\omega_{RF}^2 \omega_{pe}^2}{k_\parallel^2 \omega_{ce}^2 c^2}} \right] \quad (1)$$

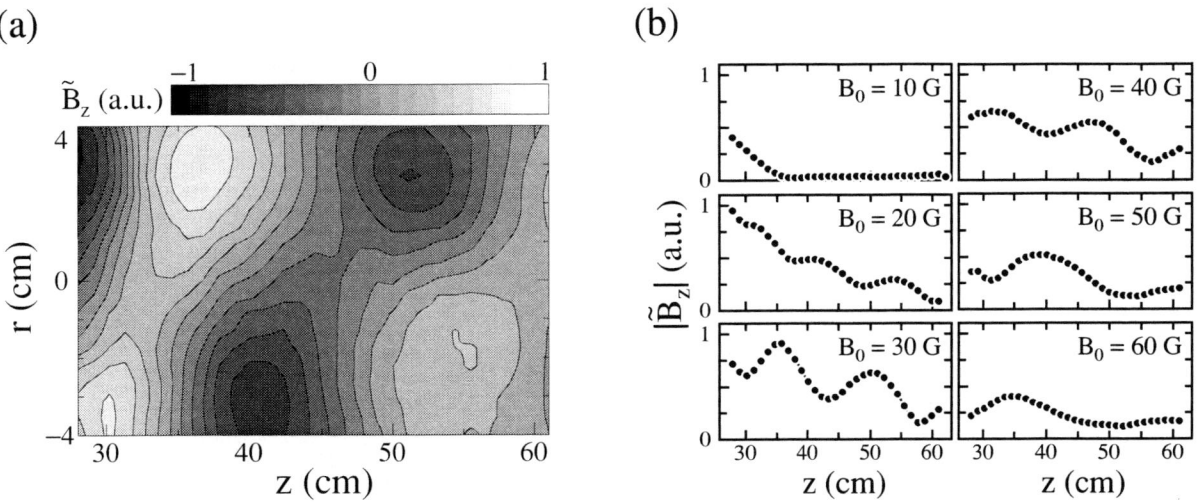

FIGURE 3. (a) Typical two-dimensional spatial distribution of $\tilde{B}_z$ for $B_0 = +25$ G and (b) axial profiles of the $\tilde{B}_z$ amplitude for various magnetic fields.

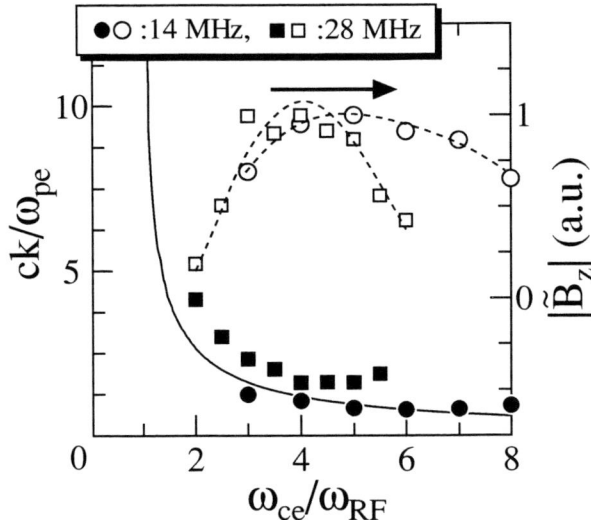

FIGURE 4. Dispersion relation of $\tilde{B}_z$: calculated (line) and measured (closed marks) for 14 MHz and 28 MHz. Open circles denote wave amplitude.

Here the plus and minus signs denote the Trivelpiece-Gould (TG) wave dispersion relation and helicon wave dispersion relation, respectively. The radial profile of $\tilde{B}_z$ almost fits to the first Bessel function J_1, yielding $k_\perp$ (transverse wave number) = 0.85, which is calculated from $k_\perp = J_1/a$ under the assumption of $\tilde{B}_z = 0$ at the wall of discharge tube. When we substitute the experimental parameters into Eq. (1), only one solution corresponding to the TG wave is derived under our condition of low magnetic field and density. In Fig. 4, the closed marks and solid curve indicate the measured dispersion relation and theoretical curve of the TG wave, respectively. The open marks show the maximum amplitude of $\tilde{B}_z$ in the downstream. The measured dispersion relation is similar to the theoretical one, and the wave-amplitude dependence on B_0 also coincides with the electron density property as shown in Fig. 2(b). Therefore, it can be said that the wave measured in the downstream is the TG wave and is considered to stimulate the plasma production.

CONCLUSION

We have investigated the properties of plasma production and associated wave structure in the case where the spatially- and temporally-rotating radio-frequency (f_{RF}) electromagnetic fields are generated in a uniform magnetic field B_0. When the RF fields rotate in the electron-diamagnetic direction, the plasma density increases tenfold around the magnetic field satisfying the condition of $f_{ce}/f_{RF} \sim 4$. An electromagnetic wave with the maximum amplitude in the periphery of the plasma column propagates and damps toward the downstream. The wave is characterized by the Trivelpiece-Gould mode in making a comparison between the calculated dispersion relation and the measured one. Since the dependence of the wave amplitude on B_0 is similar to that of the electron density, the TG wave appears to promote the plasma production.

REFERENCES

1. R. Hatakeyama, N. Y. Sato, and N. Sato, *Phys. Rev. E* **52** 6664 (1995).
2. R. Hatakeyama, N. Hershkowits, R. Majeski, Y. J. Wen, D. B. Brouchous, P. Proberts, R. A. Breun, D. Roberts, M. Vukovic, and T. Tanaka, *Phys. Plasmas* **4** 2947 (1997).
3. F. F. Chen and D. Arnush, *Phys. Plasmas* **4** 3411 (1997).
4. T. Lho, N. Hershkowitz, J. Miller, W. Steer, and G. H. Kim, *Phys. Plasmas* **5** 3135 (1998).

Observation of Ion Acoustic Waves Excited by Drift Waves in a Weakly Magnetized Plasma

Isao Tsukabayashi[1], Sugiya Sato[1], and Yoshiharu Nakamura[2]

[1]*Department of Engineering, Nippon Institute of Technology, JAPAN*
[2]*Institute of Space and Astronautical Science, JAPAN*

Abstract. Spontaneous fluctuations excited by drift waves are investigated experimentally in magnetic multi-pole plasma. The magnetic multi-pole has been widely used in DP devices and so on. It was observed that the high level of density fluctuations was generated by the drift instability near a magnetic multi-pole or a dipole magnet. The waves propagate to the middle plasma region forming the envelope train waves.

INTRODUCTION

The magnetic multi-pole devices have been widely used in plasma experiments such as double plasma device and negative ion source chambers. Such devices, using multi-cusp magnet confinement at the chamber walls, effectively produce quiescent steady plasma without magnetic field except for the vicinity of the multi-pole. There has been great deal of interest in studying the electrostatic solitary waves by means of double plasma device. And solitary waves in the magnetic field are subject to understand the phenomenon in the auroral or magnetospherical plasma.

When the bi-polar sinusoidal pulse of electric potential is applied to the grid immersed in the multi-cusp-confined plasma, the positive pulse propagates as supersonic compressive ion acoustic solitons. On the other hand the negative pulse becomes rarefactive subsonic waves due to the nonlinear effect of finite amplitude and appears as the trailing slow envelope [1]. But the subsonic waves had not noticed because rarefactive wave has no soliton solution theoretically and it is difficult to look for experimentally in the field free plasmas.

We found that if the weak parallel weak magnetic field is applied the subsonic envelope clearly appears. Characteristics feature of the subsonic envelope are summarized as follows; the envelope is generated behind a rarefactive perturbation, the group velocity of the envelope is slower than the ion acoustic velocity, the carrier frequency of the envelope is about the harmonics of initial perturbation and the amplitude of envelope increases with the increase of the magnetic field [2]. With it, with parallel magnetic filed, spontaneous oscillations are obliquely propagate from wall (multi-pole) region to the middle plasma with ion acoustic speed. There are significant density gradient around the multi-pole magnet, so the diamagnetic drift current might be driven such spontaneous fluctuations and waves [3]. Also such the spontaneous unstable waves are found at surround of a dipole magnet in weakly magnetized plasmas [4]. It is consider that the spontaneous fluctuations generated in front of multi-pole magnet can be propagate and take some effect on such the envelope trails.

EXPERIMENTAL SETUP AND RESULT

The experiment was performed in magnetic multi-pole plasma with 1.2m length and 0.6m diameter. The inside chamber wall is covered with permanent magnets of the surface magnetic field 1kGauss. The plasma is produced by a hot cathode supported dc discharge in argon gas at $1-2 \times 10^{-4}$ Torr. The typical plasma parameters are: plasma density $n_e = 3-5 \times 10^8 \text{cm}^{-3}$, electron temperature $T_e = 1-1.5$ eV and electron temperature to ion temperature ratio $T_e/T_i > 10$.

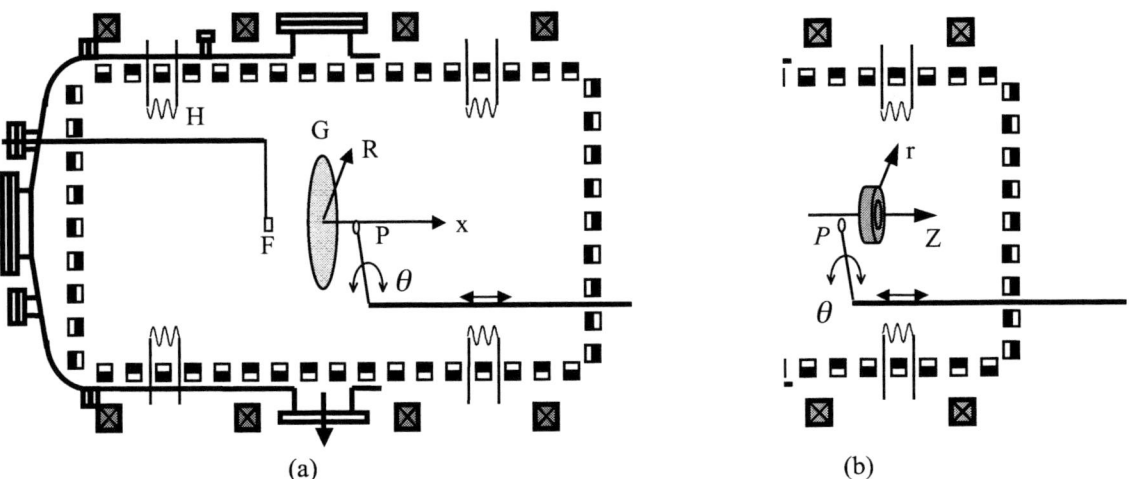

FIGURE 1. (a) Schematic of the multi-pole magnet device. G: an excitation grid (30cm diam.). H: hot filament (cathode). P: electrostatic probe. F: Faraday cup. 4 coils are placed outside of chamber to apply weak parallel magnetic field. The probe can move axial direction (x) and azimuthal direction (θ). By changing place of probe or combination of these two direction mobility allows to probe radial direction (r). (b) Schematic of survey of surround of a ringed magnet.

Subsonic Envelope and Spontaneous Wave Propagation with Weak Magnetic Field

The schematic is shown in figure 1(a). A weak axial magnetic field (2~5Gauss) is applied by external 4 coils. When one period of sinusoidal negative potential pulse (V_{pp}=3-5V, f=100-120kHz$\approx f_{pi}/5$) is applied, the ion acoustic perturbations are detected by the electrostatic probe as perturbation in electron saturation current. The detected signals at x=20cm from the grid are shown in figure 2 as a parameter of parallel magnetic field.

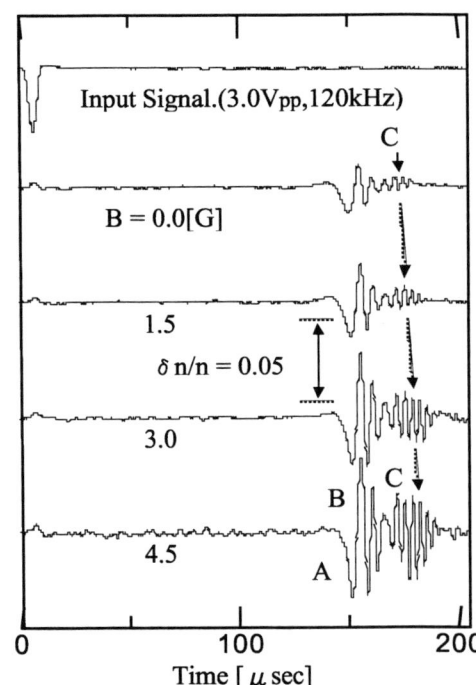

FIGURE 2. Variation of the ion acoustic perturbations in a parallel magnetic field

A leading rarefactive perturbation (A) is excited by the initially applied negative pulse, and large amplitude oscillating waves (B) are followed by the first rarefactive perturbation. Both waves (A) and (B) are propagated to coalesce with the ion acoustic velocity. Behind the oscillating waves, a small trailing perturbation (C) which is also observed in upper trace (B=0) of figure 2 increases the amplitude and forms a envelope in parallel weak magnetic field (B=3-5Gauss). The group velocity of the envelope is slower than the ion acoustic velocity. The frequency of envelope is composed of the second harmonic of the precedent oscillating waves.

It is found that when the external magnetic field is applied, spontaneous waves similar to these envelope characteristics propagate obliquely from in front of multi-pole to middle plasma region. The interferometory shows its standing wave like spatial profile.

Spontaneous Fluctuation in front of Multi-pole and around Ringed Magnet

To investigate the origin of spontaneous waves, fluctuation level is measured in front of multi-pole magnets by means of ac power meter (10Hz-1MHz) with probe signal, and also density profile is measured. The fluctuation of 50-300kHz is spontaneously excited in the place left 3-6cm from multi-pole magnets. It is found that the fluctuation level is high ($\delta n/n = 0.1 - 0.3$) at where density gradient is maximum value and the fluctuation propagates to middle plasma with ion acoustic speed. From these information, the magnetic field and density gradient seem to be necessary for the excitation of such fluctuations.

The ring magnet was put like figure 1(b) in the plasma in order to clarify this point. The magnetic outer diameter is 10cm, 4cm inside diameter, 2.4cm thickness, surface magnetic field strength 1kG. The measured profile of density, magnetic field strength, oscillation strength around the magnet are shown in figure 3 (a), (b) and (c) respectively. The field strength profile is calculated result from the measured three-dimensional component and the contribution of the multi-pole field is not contained. In figure 3, the boundary of the r direction is correspondent to the position of the multi-pole magnets. Figure 3(c) shows that there are three regions with the strong fluctuation. Region A is the spontaneous fluctuation by the multi-pole field in the DP equipment circumference as mentioned earlier. Region B exists in the form that surrounds the ring magnet and also observed by Yagra et. al. [4]. Region C was excited at central axis part of the ringed magnet newly observed.

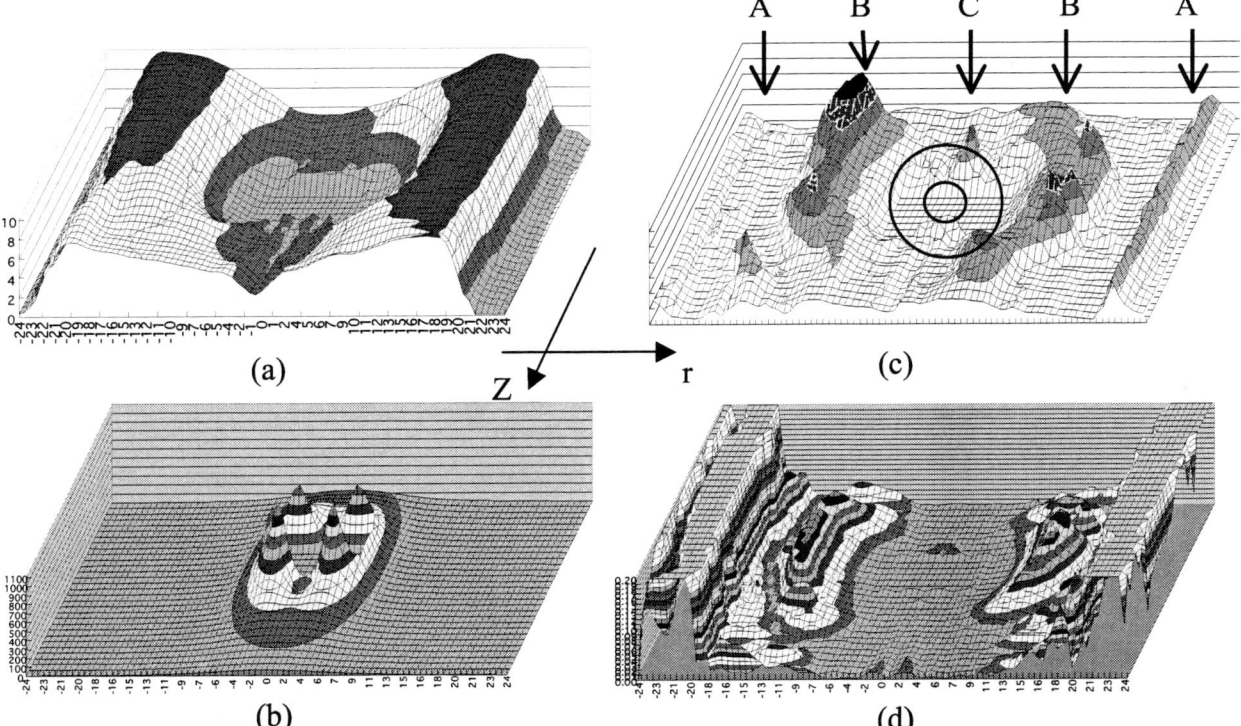

FIGURE 3. Measured profile around a ringed magnet. (a) density, (b) magnetic field strength, (c) fluctuation level, (d) calculated $\vec{B} \times \nabla n / B^2$. A ringed magnet is at center as shown in (c).

Because these spontaneous fluctuations are related to magnetic field and density gradient, the diamagnetic current may drive it. The diamagnetic current is written as

$$J_D = (KT_i + KT_e) \frac{\vec{B} \times \nabla n}{B^2}.$$

In our condition, temperatures are almost a constant. Figure 3(c) shows the profile of $\vec{B} \times \nabla n / B^2$ calculated from measured density and magnetic field components. Region B and C have been clearly reproduced, though the value can not be trusted near the wall because there is no data of the magnetic field.

The result of measuring propagation characteristics as a wave in region B by the interferometry is shown in figure 4. The interferometry is taken by filtered probe signals. From this, the wavelength is about 2cm at 100kHz of frequency, means the phase velocity is 2×10^5cm/s. So this spontaneous oscillation propagates in azimuthal direction with ion acoustic speed, or ion acoustic waves is radiated.

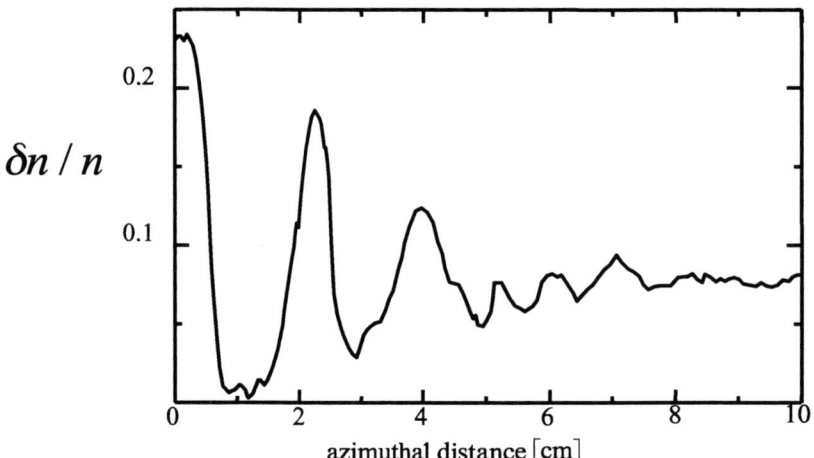

FIGURE 4. Interferometry of region B fluctuation in azimuthal direction, measured at r=11cm and f=100kHz.

CONCLUSION

The spontaneous excited fluctuation in front of multi-pole magnets and around a ringed magnet are investigated experimentally in magnetic multi-pole plasma. These fluctuations have maximum amplitude in the place that magnetic field exists and that density gradient is maximum. The waves can propagate from these region that fluctuations are excited. Especially, it was observed to propagate in the circumference of the ringed magnet in the azimuthal direction. The profile of $\vec{B} \times \nabla n / B^2$ calculated from precise measurement result of magnetic field and density reproduced clearly the regions where fluctuations are excited. Though the diamagnetism current seems to have caused these vibrations, the measurements which are more precise than the future are necessary.

REFERENCES

1. Okutsu, E., and Nakamura, Y., Plasma Phys. **21**, 1053 (1979).
2. Tsukabayashi, I. Sato, S., and Nakamura, Y., Proc. of XXIII ICPIG (Toulouse), vol. 1, 242 (1997).
3. Gauthereau, C., and Mathieussent, G., Phys. Lett. A **121**, 342 (1987).
4. Yagura, S., and Fujita, H., Phys. Fluids B **1**, 72 (1989)

Electron Acceleration by an Oscillating Potential Wall in the Presence of Collisions

A. Tsushima* and O. Ishihara*

Faculty of Engineering, Yokohama National University, Yokohama 240-8501, Japan

Abstract. Electrons were injected along magnetic field from an electron source and trapped between an oscillating potential wall and a fixed electrostatic potential wall. Some electrons in such a trap may be accelerated enough to go over the fixed potential wall in the course of bouncing motion. We measured a current of electrons escaped over the fixed potential, height of which was kept lower than that of the oscillating potential, by varying the frequency of the oscillating potential and the height of the fixed potential wall in the presence of collisions between electrons and neutrals.

INTRODUCTION

Cosmic ray particles with extremely high energy range between 10^8 eV and 10^{21} eV have been observed [1] and a power law distribution in the medium energy range has been explained by a diffusive shock acceleration model at supernova. The model is based on Fermi acceleration, the original of which was the acceleration by particle scattering against moving magnetic inhomogeneities [2].

In laboratory plasmas, Fermi acceleration has been investigated in relation with wave-particle interactions or interactions between electric field and electrons, while a standing wave has been used in some experiments [3]. Recently, the experiment of electron heating has been performed in an electron trap, where oscillating electric fields interacted with bouncing electrons [4]. Similar work was reported in the past in an after glow plasma as a model experiment of the transit time heating based on Fermi acceleration [5]. In these experiments, electrons were thermalized due to frequent electron collisions.

In our experiment, electrons were injected along magnetic field from an electron source and trapped between an oscillating potential wall and a fixed electrostatic potential wall, which were apart by a distance L. The collision of electrons was controlled by introducing helium gas and increased gradually from an almost collisionless condition, that is, the ratio of the mean-free-path, λ, to the length of electron bounce motion changes from 1000 to 1; or $1 < \lambda/(2L) < 1000$. Moreover, accelerated electrons were measured by detecting a small amount of electrons escaped over the fixed potential wall in a steady condition, while electrons were continuously supplied by the electron source placed in the trap. Some preliminary experimental results with numerical studies based on standard mapping were reported earlier [6].

EXPERIMENTAL SETUP

Electrons were injected in a trap consisting of cylindrical electrodes with an inner diameter of 20 mm along magnetic field with an intensity of $B = 0.045$ T by an electron source placed at the center between an oscillating and a fixed electrostatic potential wall in a vacuum vessel (base pressure: $p \simeq 1 \times 10^{-4}$ Pa). The sketch of the experimental setup is shown in Fig. 1, where the distribution of electrostatic potential, ϕ_0, on the axis is also shown for $V_A = -150$ V, $V_B = -50$ V and all other electrodes grounded. The bias potential for electron injection, V_{in}, can be varied between -10 V and -30 V. Then, the injected electrons move between two electrostatic potential walls and form an electron cloud of a density of $\sim 2 \times 10^{14}$ m^{-3}, a diameter of ~ 2 mm and a length of ~ 300 mm. If an oscillating potential with a frequency f_A and an amplitude of $\tilde{V}_A$ is superimposed to the electrode A, it is expected that some electrons in the trap are accelerated or decelerated when the electrons are reflected from the left oscillating potential wall. Such electrons may gain sufficiently high energy if they are successively accelerated to escape from the trap over the right fixed potential wall. The escaped electrons were detected as current, I_{out}, by a collector placed farther right of the fixed potential wall. Thus, the height of the fixed potential wall was set lower than that of the oscillating potential wall, that is, $|V_A| > |V_B|$, in order to determine the lowest energy of the escaped electrons, E_{out}, by V_B, that is, $E_{\text{out}} \simeq e(|V_B| - |V_{\text{in}}|)$ since the space charge of the electron cloud is almost V_{in}, where e is the electronic charge. The magnitude of the current of the escaped electrons or I_{out} indicates a rate that electrons are accelerated to energy high enough to go over the fixed potential wall. In the experiment, helium gas was introduced into the vacuum vessel so that the number of the interaction of a bouncing electron with the oscillating potential wall was varied via the electron mean-free-path: $\lambda \simeq 0.13/p$ (λ in m and p in Pa). Throughout the experiment in this paper, we set $V_A = -150$ V, $\tilde{V}_A = 6$ V and $V_{\text{in}} = -10$ V; and change V_B, f_A and p.

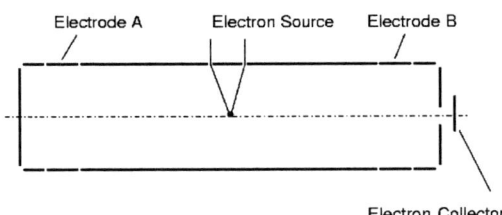

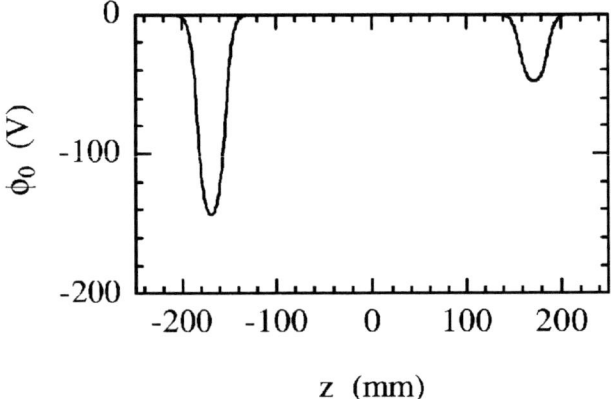

FIGURE 1. Sketch of electron trap and distribution of electrostatic potential on the axis of the trap for $V_A = -150$ V and $V_B = -50$ V.

RESULTS

At the base pressure ($p \simeq 1 \times 10^{-4}$ Pa), a relation between the current of escaped electrons, I_{out}, and the frequency of of the oscillating potential, f_A, for $V_B = -30, -50$ and -70 V was measured and the results are shown in Fig.2(a). It is clearly shown that I_{out} is enhanced at $f_A \simeq 40$ and 86 MHz for $V_B = -30$ V, but at only $f_A \simeq 86$ MHz for $V_B = -70$ V. Then, a relation between I_{out} and V_B at $f_A \simeq 40$ and 86 MHz was measured and the results are shown in Fig.2(b), where the decrease of I_{out} with $|V_B|$ at $f_A \simeq 86$ MHz is smaller than that of I_{out} with $|V_B|$ at $f_A \simeq 40$ MHz. Our previous studies indicate that the enhancement of I_{out} at some frequencies is closely related with the presence of a separatrix around a large island structure in the phase space [6].

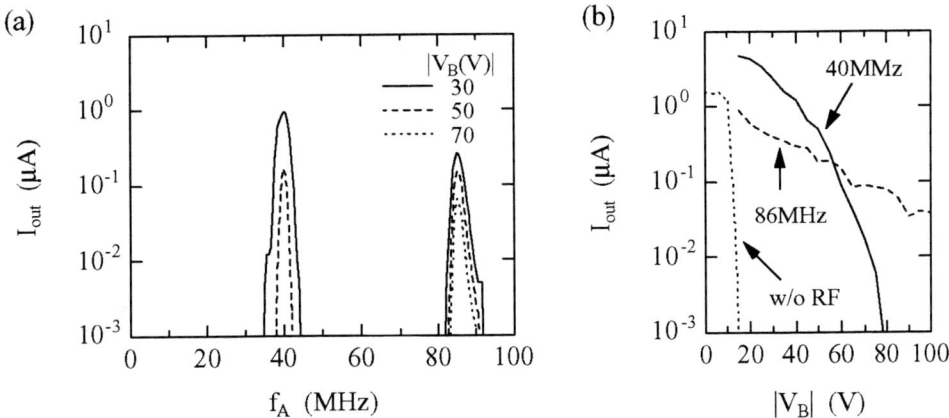

FIGURE 2. (a) I_{out} versus f_A for $V_B = -30, -50$ and -70 V and (b) I_{out} versus V_B at $f_A \simeq 40$ and 86 MHz when $p \simeq 1 \times 10^{-4}$ Pa (base pressure).

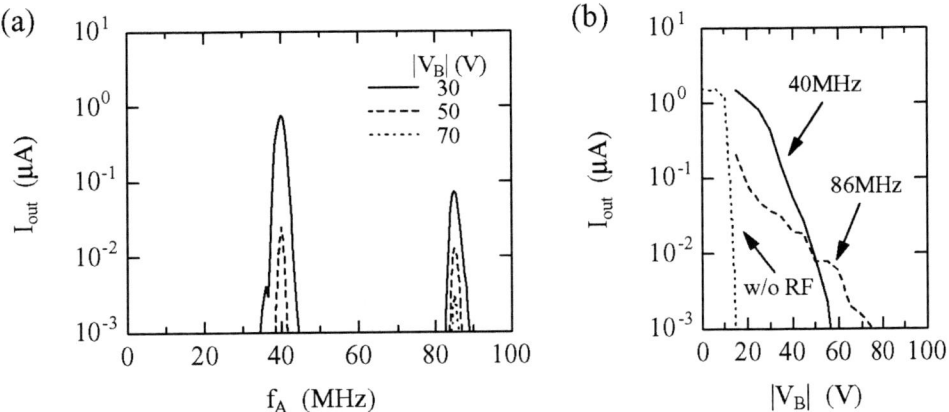

FIGURE 3. (a) I_{out} versus f_A for $V_B = -30, -50$ and -70 V and (b) I_{out} versus V_B at $f_A \simeq 40$ and 86 MHz when helium gas is filled at $p \simeq 5 \times 10^{-3}$ Pa.

When helium gas was filled at $p \simeq 5 \times 10^{-3}$ Pa, the relation between I_{out} and f_A for $V_B = -30, -50$ and -70 V was measured and the results are shown in Fig.3(a), where the enhancement of I_{out} at $f_A \simeq 40$ and 86 MHz is also seen. So, the relation between I_{out} and V_B at $f_A \simeq 40$ and 86 MHz was measured and the results are shown in Fig.3(b). The situation that the decrease of I_{out} with $|V_B|$ at $f_A \simeq 86$ MHz is smaller than that of I_{out} with $|V_B|$ at $f_A \simeq 40$ MHz remains the same as before. However, when helium gas is filled, the magnitude of I_{out} becomes smaller for larger $|V_B|$ if Fig.3(a) is compared with Fig.2(a) and the decrease of I_{out} with $|V_B|$ becomes steeper if Fig.3(b) is compared with Fig.2(b). It indicates that accelerated electrons are lost due to collisions with helium gas.

Since the current resulting from escaped electrons may be expressed as

$$I_{out} \sim \exp(-2LN_{out}/\lambda) , \qquad (1)$$

where N_{out} is the number of the reflection of an electron at the oscillating potential to achieve the escaping energy, E_{out}, the relation of I_{out} with filled helium gas pressure, p, was measured to estimate N_{out}. Figures

4(a) and (b) show the relation of I_{out} with p at $f_A \simeq 40$ and 86 MHz, respectively, for $V_B = -30$, -40, -50 and -60 V, where closed and open circles are depicted by the measurement and each solid curve is depicted from Eq. (1) by choosing an appropriate value of N_{out}. Then, the evaluated values of N_{out} can be represented as a function of E_{out}, as shown in Fig. 4(c). The closed circles may be approximated by $N_{\text{out}} \simeq 1.5 \times 10^{-3}(E_{\text{out}})^3$ and the open circles by $N_{\text{out}} \simeq 2.5 \times 10^{-2}(E_{\text{out}})^2$, where the unit of E_{out} is eV. These relations are also represented as solid lines in Fig. 4(c).

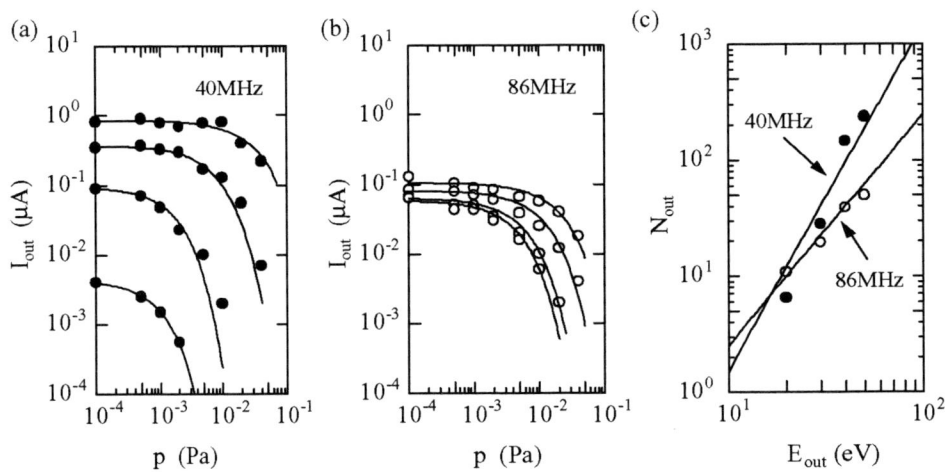

FIGURE 4. (a) I_{out} versus p at $f_A \simeq 40$ MHz for $V_B = -30$, -40, -50 and -60 V (from top to bottom), (b) I_{out} versus p at $f_A \simeq 86$ MHz for $V_B = -30$, -40, -50 and -60 V (from top to bottom), and (c) N_{out} versus E_{out} at $f_A \simeq 40$ and 86 MHz.

CONCLUSION

The enhancement of the current of escaped electrons, I_{out}, was experimentally observed at $f_A \simeq 40$ and 86 MHz, although these frequencies are much higher than a bounce frequency of an electron: $f_A \gg f_b = v_e/(2L) \sim 3$ MHz (v_e is a electron velocity and f_b is evaluated for an electron energy of 10 eV). Then, the dependence of I_{out} with p was measured for various values of V_B. The results show that $N_{\text{out}} \sim (E_{\text{out}})^3$ at $f_A \simeq 40$ MHz and $N_{\text{out}} \sim (E_{\text{out}})^2$ at $f_A \simeq 86$ MHz. Here, if we assume that the average energy increase at each reflection is $(\Delta v)_{ave} \simeq [\delta/(1-\alpha)](E_{\text{out}})^\alpha$ with constants α and δ, we have $N_{\text{out}} \simeq (1/\delta)(E_{\text{out}})^{1-\alpha}$. It indicates $\alpha \simeq -2$ at $f_A \simeq 40$ MHz and $\alpha \simeq -1$ at $f_A \simeq 86$ MHz. Such an observed energy dependence is quite different from the expected value of $\alpha = 1/2$ from a model of elastic collisions from a coming wall. Further study is needed to understand these experimental results of a stochastic nature.

REFERENCES

1. Cronin, J. W., Rev. Mod. Phys., **71**, S165-172 (1999).
2. Fermi, E., Phys. Rev., **75**, 1169-1174 (1949).
3. Doveil, F., Phys. Rev. Lett., **46**, 532-534 (1981).
4. Cluggish, B. P., Danielson, J. R., and Driscoll, C. F., Phy. Rev. Lett., **81**, 353-356 (1998).
5. Sugai, H., Ido, K., and Takeda, S., J. Phys. Soc. Jpn., **46**, 228-234 (1979).
6. Tsushima, A., and Ishihara, O., Proceedings of 25th International Conference on Phenomena in Inonized Gases, Nagoya, 2001, Vol.1, pp. 347-348; Tsushima, A., Ishihara, O., Takahashi, H., and Miyahara, K., J. Plasma Fusion Res., **S4**, 591-594 (2001).

Plasma Science and Environmental Remediation in the Arctic Region

Alfred Y. Wong

UCLA HIPAS Observatory, Fairbanks, Alaska, &
Dept of Physics and Astronomy, UCLA, Los Angeles, CA 90024, USA

Abstract. A concept to produce negative charges in the stratosphere to disable chlorine radicals in its catalytic destruction of ozone is presented. An outdoor experiment in the Arctic region is described which uses a 200J laser and a 2.7 m rotating mercury mirror to ionize atmospheric dusts to produce the required charges. This large dish will also function as a sensitive detector of the ozone profile using the DIAL method. The fast resolution of the ozone profile will enable us to better measure the effectiveness of our remediation concept. A LIDAR system assembled by Wuerker [1] makes this possible. A second concept is presented based on laboratory experiments that demonstrate the selective acceleration of CO_2- at its ion cyclotron frequency in a divergent magnetic field, simulating the process of extracting CO_2 as a minority species in the Arctic region. This experiment supports the concept [2] of using ion cyclotron waves to selectively accelerate unwanted species in the upper atmosphere. The natural free energy in the form of electron currents along the earth's magnetic field line can be channeled into supporting the excitation of these ion cyclotron waves.

BACKGROUND

Ozone holes [3], occurring in the early spring in the stratosphere of Antarctica and the North Pole, are caused by free chlorine atoms, which find their origin in CFC's such as Freon (CFC-11). The Antarctic ozone hole is a region of extreme ozone loss that has been appearing annually since the 1970s. Recent observations show that ozone holes are now being observed in the northern hemisphere as well and their durations are even longer than those in the southern hemisphere.

According to a NASA press release [4] on May 2002, it is reported that in some parts of the Arctic stratosphere, located from about 10 miles to 30 miles above Earth, ozone concentrations declined as much as 60 percent from November 1999 through March 2000. The significant decline over the Arctic region was due to an increase in the area and longevity of polar stratospheric clouds (PSCs), according to a group of researchers who participated in a large, international atmospheric science campaign by NASA's SAGE III Ozone Loss and Validation Experiment, or SOLVE.

As the stratosphere cools to very cold temperatures over the Arctic during the winter, polar stratospheric clouds (PSCs) form. Most chlorine compounds pumped into Earth's atmosphere in recent decades by human activity initially were tied up as chlorine nitrate or hydrochloric acid, both of which are non-reactive. But if there is a surface area to attach to like the polar stratospheric cloud ice crystals, the chlorine compounds change into ozone-gobbling chlorine radicals in late winter and early spring after reacting with sunlight. PSCs convert inorganic chlorine from reservoir species (HCl and ClONO2) to free radical form as in the Antarctic.

A northern ozone hole could be significant since more people live in Arctic regions than near the South Pole. Although seasonal ozone loss is more severe in the Antarctic, the ozone loss in the Arctic presents potentially more serious health problems to human beings. Ozone-depleted air from the Arctic drifts south toward North America, Europe and Russia each spring, increasing the amounts of ultraviolet light reaching Earth's surface in the highly populated mid-latitudes and potentially causing increases in several types of cancer. The earth's magnetic field line in the Arctic region is nearly vertical, making it easier to propagate low-frequency plasma waves upward into the ionosphere. There are a number of free-energy sources in the auroral ionosphere. On account of these unique characteristics, we are targeting the Arctic region to demonstrate the feasibility of environmental remediation.

LABORATORY EXPERIMENTS

In our plasma laboratory we have performed experiments [5] to confirm that highly energetic electrons and short-wavelength UV radiation dissociate ozone molecules while low-energy electrons disable halogen atoms by attaching to them, forming negative ions. In order to conduct controlled studies of catalytic destructive processes and mitigation measures in the upper atmosphere, we have built a laboratory chamber of sufficient dimensions in which ozone and reactive species have sufficient lifetimes to react repetitively. The neutral pressures and UV spectrum were chosen to simulate the stratospheric conditions and at the same time satisfy detectability. The catalytic destruction by Cl dissociated from CFC is verified experimentally. When negative charges are injected into the medium and Cl⁻ ions are formed, the recovery of the ozone proceeds as the chlorine radical is disabled and collected. These negative chlorine ions do not destroy ozone catalytically. Based on these laboratory results, we are prepared to carry out this experiment on a much larger scale in the Arctic Region where our goal would be to use lasers to generate enough low-energy electrons in the atmosphere to disable the chlorine radicals and prevent further ozone destruction.

ARCTIC ATMOSPHERE

At the HIPAS Observatory in Alaska (64.9° N latitude and 146.8° W), we plan to use a large optical collector and a powerful laser to detect the ozone profile in fast time scales as well as to generate negative charges in the stratosphere and troposphere. This collector is a rotating Liquid Mirror with a parabolic mercury- reflecting surface with a diameter of 2.7 m. Its collecting area is twenty times larger than conventional collectors and when used in DIAL (Differential Absorption Lidar) experiments makes it possible to record the ozone profile in one minute compared to upwards of 20 minutes as required by other lidars with smaller collectors.

More rapid measurements of the ozone profile allow us to distinguish other effects that could be causing the destruction of ozone from the destruction caused by Cl. The ozone layer extends from the stratosphere to the mesosphere. In the mesosphere both the density of charges increases and the velocity distribution of charges broadens due to higher temperature and energy. As it turns out, the aurorae, which are caused by the solar wind-magnetosphere interaction, can generate highly energetic electrons at the mesospheric level that would dissociate ozone molecules. It is estimated that aurorae, could produce enough UV radiation to photo-dissociate on the order of 10^{33} ozone molecules per year (a decline of ~ 100 DU over an area as large a 10^{15} cm²) in the mesosphere. Monitoring ozone systematically in the Arctic is important due to the fact that both the electrojet and aurora penetrate to the top of the ozone layer at 80 km altitude. Thus fast-time resolution is needed in order to ascertain cause and effect in the study of the variability of the ozone layer. The higher sensitivity of the HIPAS Lidar allows us to probe the stratospheric and mesospheric ozone distribution and its response to auroral precipitations and other environmental factors.

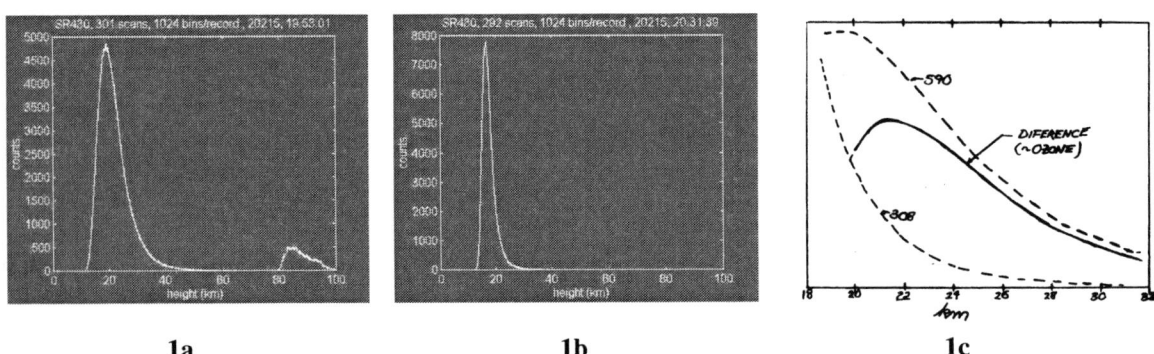

FIGURE 1. Lidar recordings of ozone profile at 590 nm (Fig. 1a) and 308 nm (Fig. 1b). Subtraction of the first and second curves (Fig. 1c) shows the ozone layer with an expected peak at 21.5 km altitude (R. Wuerker, private communications).

Besides acting as a lidar collector, the Liquid Mirror Telescope (LMT) at HIPAS is large enough to focus a laser beam at distances of 100 km (the beginning of the ionosphere) to a small enough spot to possibly initiate plasma production at 100 km distances. The LMT can focus a several hundred Joule - nanosecond duration laser pulse to 100 km altitudes, for the purpose of creating multi kilometer long plasma columns in the sky for direct electrojet modification experiments [1]. The new laser (Fig. 2) will employ a 15 cm aperture Nova isolator, a single 15 cm aperture amplifier, a 15-9.2 cm beam expander, and a 9.2 cm aperture amplifier all in a double pass design, which includes a stimulated Brillouin scattering (SBS) cell which will use the phase conjugation properties of the SBS process to keep the output beam uniform in phase. This study conducted for the 100Km height led to our view that the production of negative charges by ionization of dust particles is possible at 30 Km.

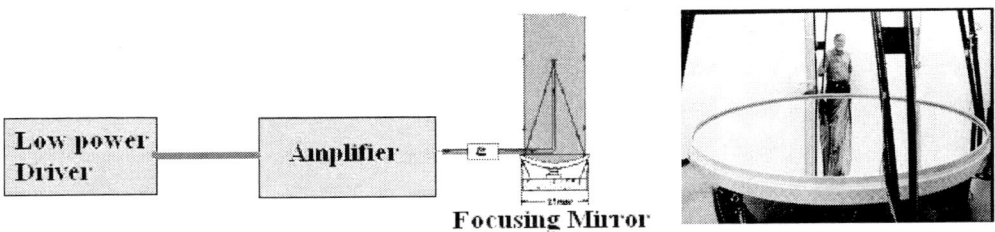

FIGURE 2. Laser Beam (200 J laser, ns pulse width 2×10^{11} watts) directed by large mirror upward to the atmosphere 30-100 km above to ionize dust particles with intensity greater than 200 MW/cm^2. The 2.7 m diameter rotating mercury mirror at HIPAS with its builder R. Wuerker is shown on the right.

In the effort to remediate the ozone depletion in the stratosphere (20-30 Km) the high-powered laser at HIPAS would essentially ionize the dust to generate the necessary negative charges to attach them to chlorine atoms. This will be the first attempt ever to focus a laser beam to high altitudes with a large telescope, with an LMT that cannot be damaged by high-energy laser pulse. Our goal is to use the high-powered laser at HIPAS to produce negative charges in the stratosphere where PSCs are formed and Cl radicals are destroying ozone molecules. In addition, negative charges can be generated in the troposphere to intercept and dissociate CFC's molecules on their way up to the stratosphere. Electrons, produced in the troposphere, readily attach to oxygen and other molecules to form negative ions.

In order to suppress the destruction of ozone molecules by energetic electrons from the auroral region, the HIPAS facility has a powerful HF array that irradiates electromagnetic waves to generate large-amplitude electric fields at an altitude where the radiating frequency matches the plasma resonant frequency. The strong interaction between this resonant layer and the ambient electrons has produced strong optical emission according to the observation by Wuerker and Sentman. This layer of strong oscillating electric fields can be used to scatter energetic electrons originating from the aurorae, which would have otherwise destroyed ozone molecules at the lower height.

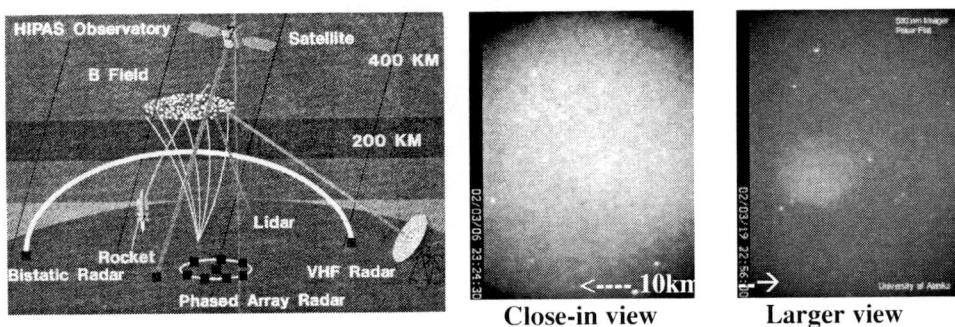

FIGURE 3. Overview of the HIPAS facility site. Optical emission data, on the right, collected at HIPAS showing electron excitation of molecules at the resonant layer (Wuerker and Sentman, March, 2002).

SELECTIVE ACCELERATION OF CO_2^- IONS IN A MAGNETIC MIRROR

ELF experiments performed in Alaska have shown that the EM waves at the cyclotron frequency of negative CO_2 ions can be excited at a sufficiently high level (1-10 pT) to allow such ions to be accelerated to escape earth's gravity. Our laboratory device is shown in Fig. 4a, which depicts an excitation coil and a magnetic mirror of 6%. The resonance of CO_2^- is shown in Fig 4b at different magnetic field of the central region of uniform field. These experiments demonstrate that CO_2 forms negative ions readily in a plasma and can be preferentially excited as a minority species by waves at its cyclotron frequency to energy 100 times its initial energy. This perpendicular energy can be converted to axial energy by a divergent magnetic field. The laboratory experiments demonstrate (Fig. 4c) that when there is a divergent magnetic field there is a strong outflow of CO_2^- ions and when a magnetic mirror is imposed, this outflow stops because CO_2^- ions with high transverse energies are trapped. Likewise the divergent earth's magnetic field in the polar region converts the perpendicular ion motion into upward outflow along the magnetic field (Fig. 5). The free energy source of currents flowing along the earth's magnetic field might make this method economically feasible [2].

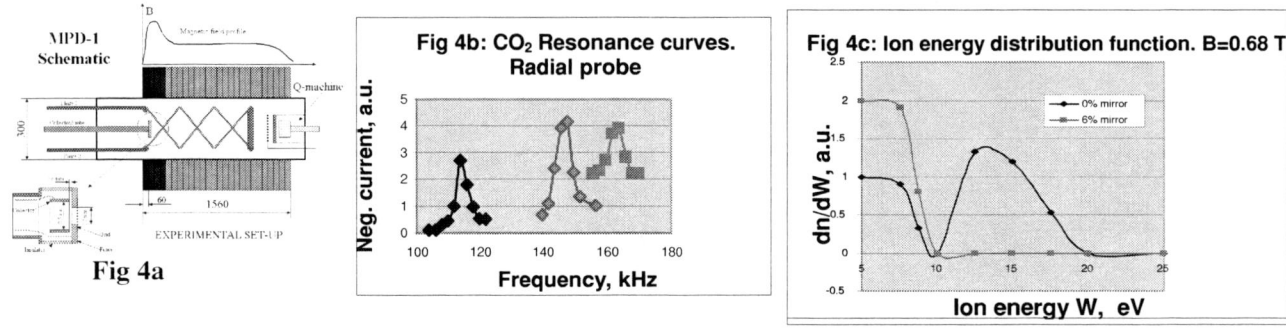

FIGURE 4. Laboratory experiments on the excitation of CO_2^- ions

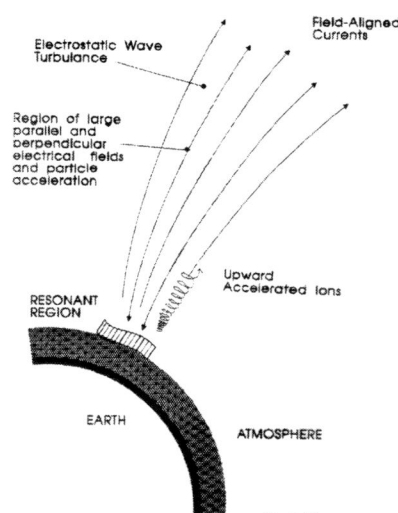

FIGURE 5. Schematic showing ions accelerated upward along the divergent magnetic field lines.

ACKNOWLEDGEMENTS

I wish to thank Dr. R. Wuerker, G. Rosenthal, G. Paskalov, D. Karfidov and HIPAS staff.

REFERENCES

1. R.F. Wuerker, Arctic Lidar at University of California, Los Angeles' HIPAS Observatory, Proceeding of the SPIE. Vol 4485, paper 52, SPIE 46th Annual Meeting, San Diego, CA 2001.
2. A.Y. Wong et al., Ionospheric Modification and Environmental Research in the Auroral Region in Plasma Science and the Environment, Chapter 3, pp. 41-75, 1997, AIP Press, edited by: W. Manheimer, L. Sugiyama, T. Stix;
   A.Y. Wong, Nonlinear Interactions of Electromagnetic Waves with the Auroral Ionosphere, Radio Frequency Power in Plasmas, edited by S. Bernabei and F. Paoletti, AIP CP485 p 18-34, 1999.
3. Toon, Owen B. and Richard P. Turco, 1991, Polar Stratospheric Clouds and Ozone Depletion, Scientific American 264(6): 68.
4. NASA press release "Arctic Ozone Depletion" May 18, 2002: http://svs.gsfc.nasa.gov/stories/arctic/solve2_press.html.
5. A.Y. Wong et al., Observation of Charge-Induced Recovery of Ozone Concentration after Catalytic Destruction by Chlorofluorocarbons, Physical Review Letters 72, 19 (1994).

Experimental Evidence of Nonlinear Spectral Power Transfer In Zonal Flow Generation

H. Xia, M.G. Shats and W.M. Solomon

RSPhysSE, The Australian National University, Canberra, ACT 0200, Australia

Abstract. Strong fluctuations observed in the low-temperature plasma in the H-1NF heliac are analyzed using the power transfer function in order to understand energy transfer mechanisms between short and long wavelengths, including zonal flows.

INTRODUCTION

Zonal flows (ZF) are turbulence-generated poloidally symmetric low-frequency potential structures, which play an important role in the self-regulation of turbulence[1]. Experimental evidence of ZF like structures generated by strong fluctuations has been found in the H-1 Heliac[2]. Here we present results on the mechanism by which energy transfers from the background fluctuations to zonal flow structures.

Since ZFs are an azimuthally symmetric structures that can not tap free energy from the radial gradients[1], they are excited only via nonlinear processes, hence estimating the non-linear power transfer function gives important information about the generation of zonal flow.

From calculations of the bispectrum and bicoherence (or summed bicoherence), the degree of nonlinear interactions between different fluctuations can be estimated. It has been found from the experimental measurements of the poloidal and the radial wave numbers of the plasma potential fluctuations, that the enrichment of the fluctuation spectrum coincides with an increase in the non-linear mode coupling and results in the formation of the zonal flow like structure[2]. However the bispectrum estimations do not provide the information on the direction of the energy flow, which is important in confirming the generation of zonal flow. It is therefore essential to estimate the energy transfer function from the measured fluctuation components.

In this work, the power transfer function is estimated using the measured fluctuations. Computations of the power transfer function are discussed and the results show that the energy is transferred nonlinearly from the broadband high wave number fluctuations to the low wave number modes, including zonal flows.

POWER TRANSFER FUNCTION ESTIMATION

To understand the energy flow within the spectra, we estimate the power transfer function, which quantifies the strength of three-wave coupling and the direction of energy cascade[3].

A simple fluctuation model of a single input and single output can be modeled in the spatial or temporal frequency domain by linear and quadratic coupling:

$$Y_k = L_k X_k + \frac{1}{2} \sum_{\substack{k_1,k_2 \\ k=k_1+k_2}} Q_k(k_1,k_2) X_{k_1} X_{k_2} \qquad (1)$$

Here, Y_k and X_k are the Fourier transforms of the measurable output and input signal, L_k and $Q_k^{k_1,k_2}$ are the linear and non-linear coupling coefficient.

The wave kinetic equation can be written as:

$$\frac{\partial P_k}{\partial t} \approx 2\gamma_k P_k + \sum_{\substack{k_1,k_2 \\ k=k_1+k_2}} T_k(k_1,k_2) \tag{2}$$

where $T_k(k_1,k_2)$ is the non-linear power transfer function, which quantifies the energy exchanged between the different waves in the spectrum due to three-wave interactions. γ_k is the linear growth rate of the modes. The time change of the spectral power of a wavenumber k, $\frac{fP_k}{ft}$, is due to the growth or damping rate and the sum over all components of the power transfer function, $T_k(k_1,k_2)$ at the wavenumber k.

First we estimate the linear and non-linear coupling coefficient L_k and $Q_k^{k_1,k_2}$ in equation (1) from the experimental fluctuation data. Both the linear growth rate and power transfer function can be estimated through the coupling coefficients.

To test the accuracy of the estimation, first we generate some model data with known linear and nonlinear coupling coefficients[3]. Then we get the photon noise signal from the photomultiplier and apply the coupling coefficients to it. The output is then used as the input again to make sure that the resulting signal is non-Gaussian.

Fig 1 (a) shows the real part of the linear transfer function with the defined and estimated values shown in dash and solid line respectively. Fig 1 (b) similarly shows the imaginary part of the linear transfer function. The result in fig. 1 shows that the estimation can recover the linear transfer function very well. Note that the results are obtained using 400 realizations with the data length of each segment of 80 points.

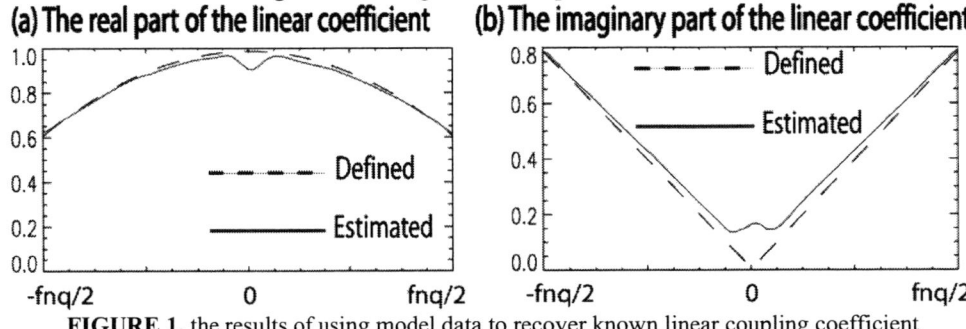

FIGURE 1. the results of using model data to recover known linear coupling coefficient

Following Ritz et al[3], we confirm that averaging large number of realizations is essential for the statistical error of the higher-order cumulant estimation to be small with respect to the signal of interest.

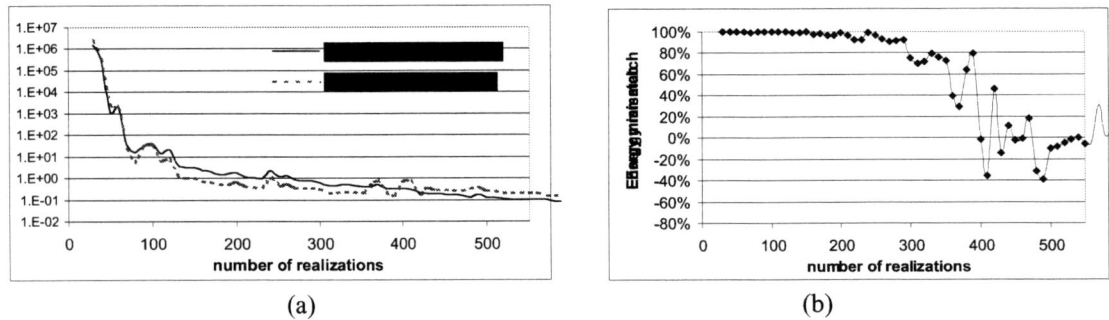

FIGURE 2. Convergence of the power transfer estimation as a function of the number of realizations

Fig.2 (a) shows the relation between the number of realizations and (1): the total positive value of the power transfer function (solid line) and (2): the maximum value of the estimated linear growth rate (dash line). Stable solution are achieved with realization numbers of more than 500.

It is important to check the accuracy of the results using some physically meaningful criterion since it is difficult to estimate the error involved in the estimation of the power transfer function from experiment data[3].

The energy stored in the potential fluctuations φ_k of plasma can be written as[4]:

$$W_k = (1+k_\perp^2)\langle \varphi_k \varphi_k^* \rangle \tag{3}$$

Here we will use the energy conservation law as the 'test of goodness' of our estimation. We define the error in the energy conservation as the percentage of energy mismatch part between the total energy transfer out of the spectrum and into the spectrum. In fig 2 (b) we show the result obtained from the experiment using the same parameters as that in fig 2 (a). At low numbers of realization, the result is totally meaningless since the calculation is not convergent (fig 2 (a)). After the number of realizations reaches 500, the energy mismatch decreases to less than 10%.

EXPERIMENTAL RESULTS

H-1 heliac is a helical axis stellarator with major radius of 1.0 m and minor radius of less than 0.2 m. The machine was operated at low magnetic fields (< 0.2 T) with current free plasma produced by the pulsed radio-frequency power of less than 100 kW at 7 MHz. The electron temperature is low enough (Te = 5~40 eV) that a number of electric probes can be inserted into the plasma. We use several combinations of triple probes. The radial and poloidal electrical fields are measured by radially and poloidally separated probes.

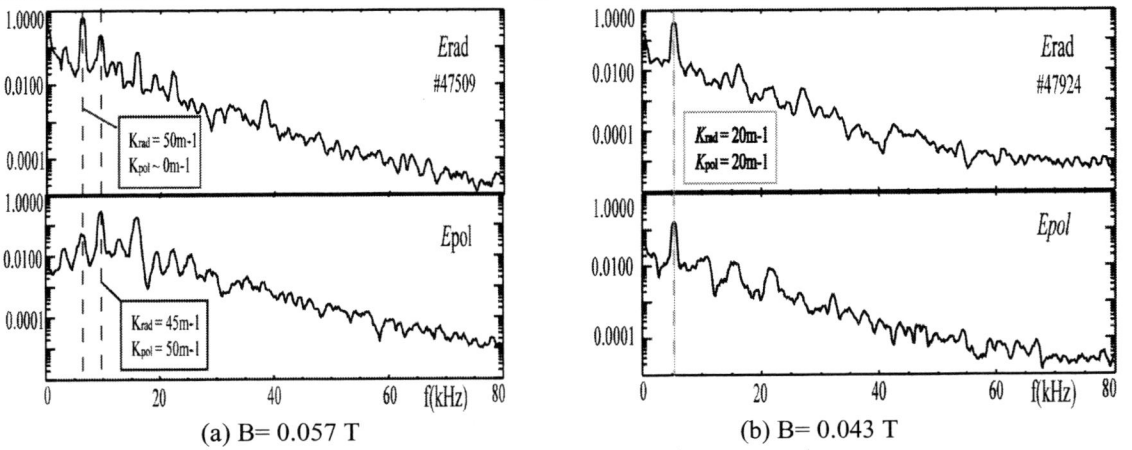

(a) B= 0.057 T (b) B= 0.043 T

FIGURE 3.: Power spectrum of the *Erad* and *Epol* with the wave number measurements

Fig 3 shows the power spectra of the radial and poloidal electrical field with the measured wave numbers of different modes at the magnetic field of 0.057 T and 0.043 T, respectively. In Fig.3 (a), at the frequency of about 6.4 kHz, the fluctuation has the radial wavenumber of 50 m^{-1} and the poloidal wavenumber close to 0, which satisfy the zonal flow property of $krad \gg kpol \sim 0$. In fig.3 (b), only coherent modes are observed with no signatures of the ZF structure.

For the broadband components of the fluctuations we observe a linear poloidal dispersion relation. All the calculations are performed in the frequency domain instead of the wavenumber domain, since $k=\omega/v_p$.

In the estimation we use the floating potential measured from one of the quadruple probes. The data from the time interval of 13 ms to 73 ms, digitised at 1 MHz sample frequency is re-sampled to 300 kHz. The whole data range is divided into 550 segments with 50% overlapping of each segment. Each segment has 65 data points.

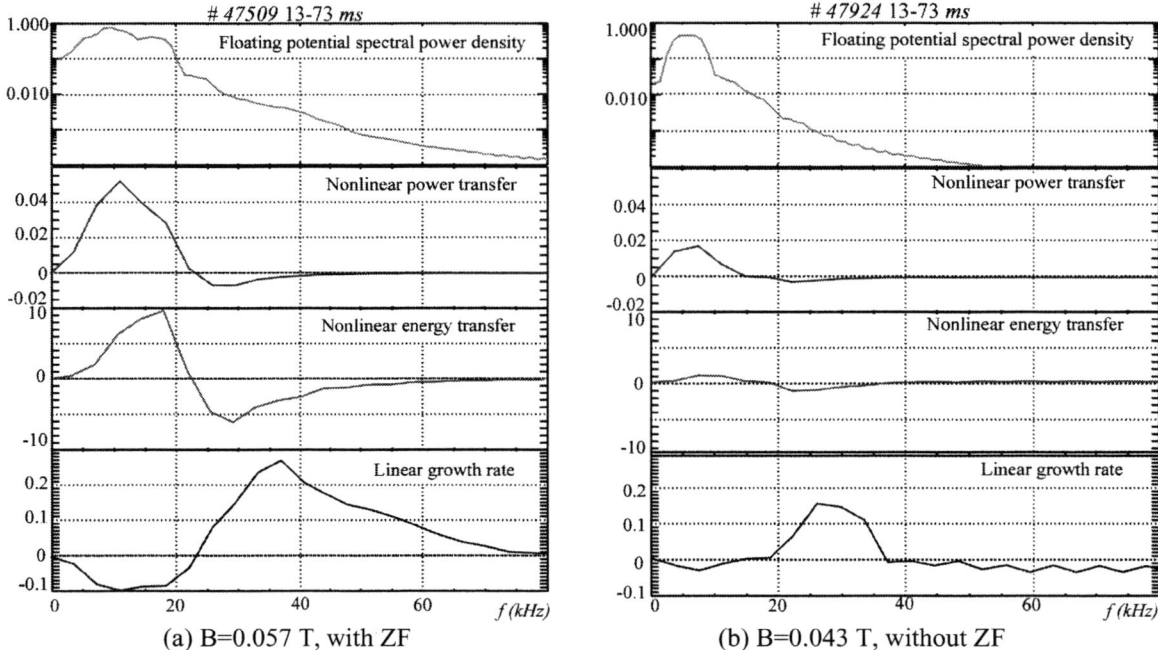

FIGURE 4. Power transfer function and linear growth rate estimation for two different plasma discharges

Fig.4 shows the result of the power transfer function and linear growth rate estimation for two different discharges shown in Fig.3. Fig.4 (a) shows the discharge at B= 0.057 T, with the zonal flow structure in the fluctuations, while fig.4 (b) is the discharge at B=0.043 T, without the ZF like structure. In the figures, we show the power spectrum of the floating potential, the estimated power transfer function, the estimated energy transfer function from equation (3) and the estimated linear growth rate.

From Fig.4, it is clear that:

1). In both situations, there is nonlinear energy flow from high frequency (higher than 25 kHz) to low frequency (up to 20 kHz), and there is linear energy growth in the high frequency range with a very small damping in the low frequency range.

2). There is much higher non-linear energy transfer from high frequency to low frequency in the case of B=0.057 T associated with the presence of ZF. Since ZFs are low frequency structures (see fig.3) and located in the frequency range which receive energy nonlinearly, we may hypothesise that this confirms the generation of ZFs through nonlinear wave interaction. However, the frequency resolution in our estimate is not high enough to precisely verify that the energy flow is into the ZF structure.

SUMMARY

This work is the first that gives the experimental evidence of the nonlinear power transfer into the ZF structure. Further works need to be done to improve the frequency resolution. In the next step, we will use the modified Ritz model[5], to improve our estimation.

REFERENCE

[1] P. H. Diamond et. al, in *17th IAEA Fusion Energy Conference*, Yokohama, Japan, 1998), Vol. IAEA-CN-69/TH3/1.
[2] M. G. Shats and W. M. Solomon, Physical Review Letters **88**, 045001 (2002).
[3] C. P. Ritz, E. J. Powers, and R. D. Bengtson, Physics of Fluids B-Plasma Physics **1**, 153 (1989).
[4] A. Hasegawa and C. G. Maclennan, Physics of Fluids **22**, 2122 (1979).
[5] J. S. Kim, R. D. Durst, R. J. Fonck, et al., Physics of Plasmas **3**, 3998 (1996).

Excitation of an Axisymmetric Shear Alfvén Wave by a Rogowski-Type Antenna

Tsuyoshi Yagai, Ryosuke Kumagai, Yohei Hosokawa,
Kunihiko Hattori, Akira Ando and Masaaki Inutake

Department of Electrical Engineering, Tohoku University
Aoba05, Sendai, 980-8579, Japan

Abstract. Characteristics of wave excitation in a high speed plasma flow are investigated experimentally. Axisymmetric, azimuthal mode number m = 0 shear Alfvén waves (SAW) are excited by a Rogowski-type antenna in a high density (more than 10^{13} cm^{-3}), high speed, magnetized helium plasma flow with ion Mach number ($M_i \sim 1$). The Rogowski-type antenna with Faraday-shield for the SAW excitation consists of a solenoid coil, so the magnetic field is not generated around the antenna in vacuum. Dispersion relations of excited SAW are in good agreement with the theoretical curve which is obtained by taking into account the effect of a Doppler shift due to a high speed plasma flow. Ratio $\delta B_y / B_0$ of the excited magnetic perturbation δB_y to the uniform magnetic field B_0 is as large as 1%. Spatial profiles and damping rate of the SAW are also obtained experimentally.

INTRODUCTION

Alfvén waves are ubiquitous and play a crucial role in MHD phenomena in space plasmas. Since Alfvén has proposed the existense of a hydromagnetic wave to explain moving phenomena in the sunspot zone[1], Reserchers in both plasma physics and geophysics have been interested in these waves. In space, it is difficult to obtain a spatial profile of Alfvén waves by using the spacecraft data because the waves are slowly varying temporally and have very long wavelength in a thin density plasma in space. In laboratory plasmas, several excitation methods were proposed to excite shear Alfvén waves in suitably dense and magnetized plasmas, with an azimuthal mode number m = −1 (non−axisymmetric) excited inductively by a Nagoya-type-III antenna[2,3], and a helical antenna[4]. As for an axisymmetric m = 0 shear Alfvén wave, many excitation methods have been reported such as a Stix coil[5], a coaxial electrode[6], and a pair of field-aligned grid electrodes[7] immersed in a plasma. Recently it stimulate much interest of many reserchers that large amplitude excitation (the ratio of wave field δB_{wave} to uniform magnetic field B_0, $\delta B_{wave}/B_0 \approx 10^{-3}$) of shear Alfvén waves with a density depression $\delta n / n \approx 15\%$ is observed by using a helical antenna[8].

Through a series of these experiments, electrodes immersed in a plasma were used for the excitation of a shear Alfvén wave. This method is not always adequate for a large amplitude excitation because the current which drives δB_{wave} is restricted by a limited ion saturation current.

We have proposed a unique antenna, a Rogowski-type antenna for an inductive excitation of a shear Alfvén wave without any plasma-electrode contact in a suitably dense, magnetized streaming plasma flow. The purpose of this paper is to report excitation and propagation properties of a shear Alfvén wave excited in a fast-flowing plasma.

PROPERTIES OF SHEAR ALFVÉN WAVE IN A FLOWING PLASMA

At frequencies lower the ion cyclotron frequency ω_{ci}, $\omega < \omega_{ci} \ll \omega_{ce}$, ω_{pe}, the dispersion relation of the Alfvén wave propagating along the external magnetic field B_0 is

$$\omega^2 = k_{//}^2 V_A^2 \left(1 \pm \frac{\omega}{\omega_{ci}}\right) \tag{1}$$

where $k_{//}$ is the field aligned component of the wave number k, and V_A is Alfvén velocity. The notations +, − in eq. (1) mean fast (compressional) wave and slow (shear) wave, respectively. In this paper, we are concerned only with the shear Alfvén wave.

In the HITOP device, a high density plasma flows with an ion Mach number M_i of nearly unity in the uniform B_0. The dispersion relation should be modified by taking into account an effect of the Doppler shift $\omega \rightarrow \omega - k_{//}U$, where U is a plasma flow velocity. Then the dispersion relation is modified as follows:

$$(\omega - k_{//}U)^2 = k_{//}^2 V_A^2 \left(1 - \frac{(\omega - k_{//}U)}{\omega_{ci}}\right). \tag{2}$$

Now, we introduce the following normalized parameters: of $\tilde{\omega}/\omega_{ci} = \Omega_i$, $(k_{//}c)/\omega_{pi} = K$, $U/V_A = M_A$ (Alfvén Mach number). By using the relation that $\omega_{ci}/\omega_{pi} = c/V_A$, where c is the light velocity, eq. (2) becomes as follows:

$$(\Omega_i - KM_A)^2 = K^2(1 - \Omega_i + KM_A). \tag{3}$$

Solving the eq. (3), we obtain

$$\Omega_i = \frac{K}{2}\left(2M_A - K + \sqrt{K^2 + 4}\right) \tag{4}$$

Eq. (4) is the dispersion relation of a shear Alfvén wave in a streaming plasma with Alfvén Mach number MA.

EXPERIMENTAL ARRANGEMENT

Experiments of the excitation of a shear Alfvén wave (SAW) are performed in the HITOP (HIgh density TOhoku Plasma) device of Tohoku University. The device is a cylindrical stainless steel chamber with 0.8m in diameter and 3.3m in length, surrounded in 17 external magnetic coils which can form various types of magnetic field configurations by adjusting the coil current. In the present experiments, an uniform magnetic field B_0 is formed externally.

A high density ($< 2 \times 10^{14}$ cm^{-3}), quasi-steady (1ms in duration), high-speed (ion Mach number $M_i \sim 1$) plasma flow is produced by a magneto-plasma-dynamic arcjet (MPDA). It is installed at one end of the vacuum chamber and has a coaxial structure with a center tungsten-rod cathode and an annular molybdenum anode. In the present experiment, typical electron temperature in the downstream region is 2eV which is almost equal to ion temperature, and the ratio of ion Larmor radius to plasma column radius (ρ_i / R_p) is less than 0.33 for the B_0 strength of 115G to 700G used in these experiment.

A Rogowski-type antenna which excites an axisymmetric, an azimuthal mode number m = 0 SAW is installed at 139 cm downstream of the MPDA. The antenna consists of a Rogowski coil (major radius R = 30 mm , minor radius a = 4.5 mm) and a stainless-steel Faraday shield as shown in Fig. 1(a). An auxiliary coordinate Δz is defined in the z-direction, which is a relative distance measured from the antenna. The notations of $-\Delta z$ and $+\Delta z$ stand for the positions upstream (toward the MPDA) and downstream of the antenna, respectively. It is a unique feature of the antenna that any vacuum magnetic field generated inside the Rogowski coil dose not leak out at all, so the antenna-near-field does not exist in a plasma.

Figure 1(b) shows the mechanism of δB_θ excitation. δB_θ (δB_y on the x–z plane) is formed in response to the changing magnetic field dB_{in}/dt inside the antenna coil as shown in Fig. 1(b). The electric field **E** is induced around the antenna according to Faraday's law, rot **E** = −d**B**/dt. On the x–z plane, the electric field drives an axial current J_z in the plasma according to $J_z = \sigma E_z$, where σ is the conductivity. Then, according to Ampere's law, rot **B** = μ_0 **J**, azimuthal magnetic field perturbation δB_θ is induced in a plasma flow. In this excitation method, it is confirmed that no magnetic perturbation is observed without a plasma.

The magnetic field perturbations are measured by three orthogonally-arranged magnetic probe tips, which consists of multi-layer tip inductors. Precise measurements of a spatial profile of dB/dt is difficult due to the lack of reproducibility of the plasma and so we introduced a multi-channel dB/dt probe in the radial direction, 3-D vector components of dB/dt and a spatial profile of SAW are measured simultaneously without suffering from the shot-to-shot plasma reproducibility.

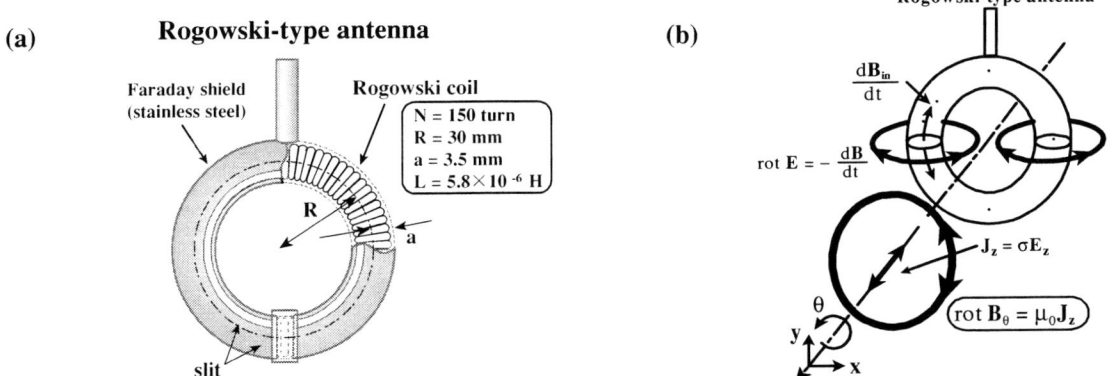

FIGURE 1. (a) Schematic of a Rogowski-type antenna. The generated vacuum magnetic field perturbations only exist inside the antenna coil. (b) Excitation mechanism of SAW. Induced-electric field in the plasma drives axial (poloidal) current $\mathbf{J}_z$, then $\mathbf{J}_z$ drives magnetic field oscillations $\delta\mathbf{B}_\theta$ ($\delta\mathbf{B}_y$ on the x–z plane).

EXPERIMENTAL RESULTS

Figure 2 shows typical time evolutions of the discharge current I_d, ion saturation current J_{is} (electron density n_e) measured by a Langmuir probe, antenna current I_{ex} and magnetic field perturbation of y-component δB_y orthogonal to uniform magnetic field $\mathbf{B}_0$ measured by a multi-channel magnetic probe. A highly-ionized plasma flow is produced quasi-steadily (~ 1ms) and the antenna current for excitation of δB_y is provided at 0.6ms – 0.9ms during the quasi-steady period of the arc discharge.

Contour plot of a time-varying radial profile of δB_y for one-cycle is shown in Fig. 3. Phase of the excited δB_y is inverted at x = 0 cm, and so an axisymmetric, m=0 SAW is excited successfully. The radial position of the δB_y peak is ~ 3 cm which corresponds to the mean radius of the antenna. The peak value is about 5 G, which corresponds to $\delta B_y / B_0 \approx 10^{-2}$. Wave amplitude is by one or two-order larger than those observed in other experiments, where low-amplitude ($\delta B_y / B_0 \approx 10^{-4}$) wave has been launched by modulating a electron-skin-depth-size current channel and a high-power ($\delta B_y / B_0 \approx 10^{-3}$) wave accompanied by nonlinear phenomena such as a density depression by a helical antenna[8]. The wave amplitude of SAW in the present experiment is rather large, but they produces no significant density perturbation.

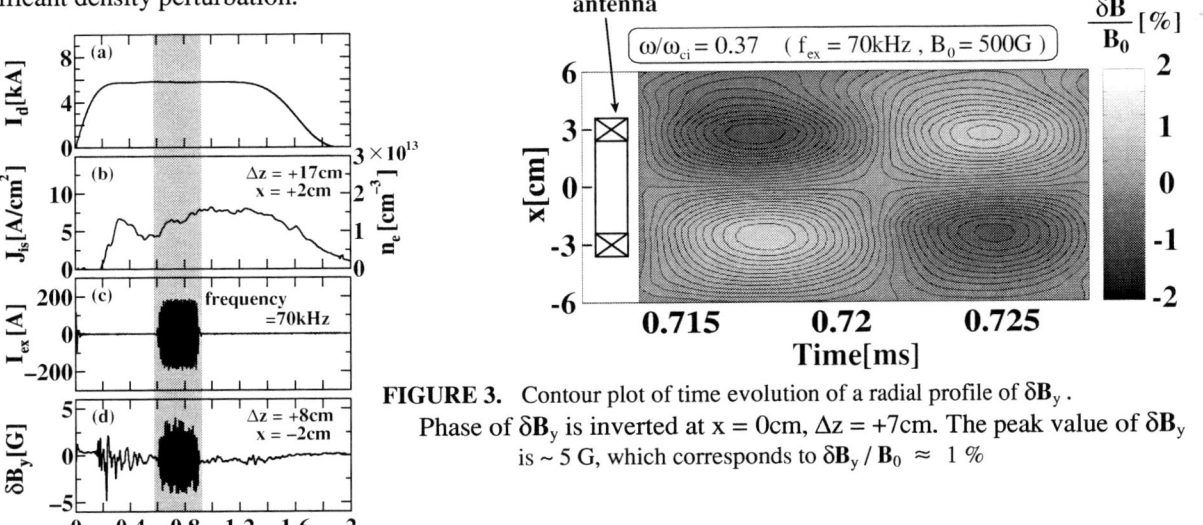

FIGURE 3. Contour plot of time evolution of a radial profile of δB_y. Phase of δB_y is inverted at x = 0cm, Δz = +7cm. The peak value of δB_y is ~ 5 G, which corresponds to $\delta B_y / B_0 \approx 1\%$

FIGURE 2. Time evolutions of (a) discharge current I_d, (b) ion saturation current J_{is} (electron density n_e), (c) Rogowski-type antenna current I_{ex}, and (d) magnetic oscillation δB_y. The antenna energized from 0.6ms to 0.9ms.

As discussed above, dispersion relation of SAW in a streaming plasma is modified by an effect of a Doppler shift. Fig. 4(a) shows measured dispersion relations of SAW in the downstream of the antenna. The dispersion curves of Eq. (4) and asymptotic lines are also shown in Fig. 4(a). In a streaming plasma, ion cyclotron resonance depends on the product of normalized wave number $K = k_{//} c / \omega_{pi}$ and Alfvén Mach number M_A. The excited δB_y by a Rogowski-type antenna is confirmed as SAW. Figure 4(b) shows $\delta B_y / B_0$ as a function of Δz, which indicates the damping length L_d of SAW is nearly 0.4 m. We can estimate the resistive damping length of Alfvén waves L_r by deriving the dispersion relation from the single-fluid equations and Maxwell's equations. We obtain $L_r = 2 \mu_0 V_A / (\eta k^2)$, where η is resistivity of the plasma and resulting in $L_r \approx 13.5$ m by using plasma parameters. Then, the ratio is $L_d / L_r \approx 0.03$. It seems that other wave damping mechanisms should be taken into account.

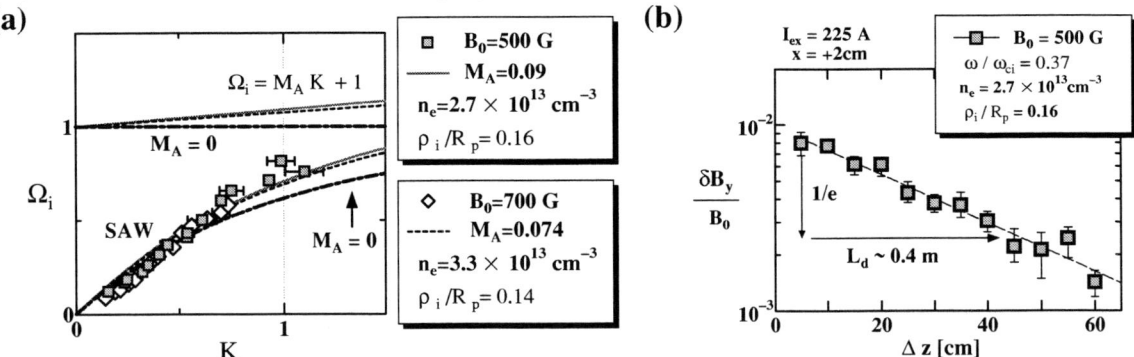

FIGURE 4. (a) SAW dispersion relations for different B_0 cases compare. Ω_i is a normalized angular frequency ω/ω_{ci} and K is a normalized wave number parallel to B_0 $(k_{//}c) / \omega_{pi}$. (b) Ratio $\delta B_y / B_0$ as a function of Δz ($B_0 = 500$G). The wave amplitude decreases exponentialy.

CONCLUSIONS

Axisymmetric m=0 shear Alfvén waves with various frequencies are excited inductively by a Rogowski-type antenna. The contour plot of a time-varying radial profile of the SAW shows that the peak value of δB_y is located at $x = \pm 3$ cm, which corresponds to the mean radius of the antenna and the ratio of δB_y to B_0 is $\delta B_y / B_0 \approx 10^{-2}$. The amplitude seems to be large enough to produce some nonlinear phenomena such as a density depression observed in UCLA[8], but there is no significant indication. The measured dispersion relations are in good agreeement with the theoretical curves predicted in a plasma flow. The observed wave damping length is much shorter than the resistive damping length.

ACKNOWLEDGMENTS

This work was supported in part by a Grant-in-Aid for Scientific Research from Japan Society for the Promotion of Science, Science and Culture of Japan. Part of this work was carried out under the Cooperative Research Project Program of the Research Institute of Electrical Communication, Tohoku University.

REFERENCES

1. Alfvén, H., Nature **150**, 405-406 (1942).
2. Watari, T., Hatori, T., Kumazawa, R.,*et al.*, Phys. Fluids **21**, 2076 (1978).
3. Ohsawa, Y., Inutake, M., Tajima, T., Hatori, T. and Kamimura, T., Phys. Rev. Lett. **43**, 1246-1249 (1979).
4. Amagishi, Y., Inutake, M., Akitsu, T. and Tsushima, A., Jap. J. of Appl. Phys. **20**, 2171-2179 (1981).
5. Tsushima, A., Amagishi, Y. and Inutake, M., Phys. Letters **88A**, 457-460 (1982).
6. Jephcott, F. D., Stocker, M. P., J. Fluid Mech. **13**, 587-596 (1962).
7. Leneman, D., Gekelman, W., and Maggs, J., Phys. Plasmas **7**, 3934-3946 (2000).
8. Gekelman, W., Vincena, S., Palmer, N., Pribyl, P., Leneman, D., C. Mitchell and J. Maggs, Plasma Phys. Control. Fusion **42**, B15-B26 (2000).

Nonlinear Evolution of Unstable Waves in an Electron-Beam Plasma: Relation to Beam Energy Distribution

K. Yamagiwa, N. Ikeya, T. Takeda and T. Takai

Department of Physics, Faculty of Science, Shizuoka University,
836 Ohya, Shizuoka 422-8529, Japan
e-mail: spkyama@ipc.shizuoka.ac.jp

Abstract. Experimental studies were performed to investigate one-dimensional evolution and formation of nonlinear structure of linearly unstable waves in an electron-beam plasma. The beam energy distribution function was also measured in connection with the formation of the nonlinear structure.

INTRODUCTION

There has been a growing interest [1-5] in nonlinear phenomena of unstable electron-beam modes. Their dispersion properties are quite different from Langmuir waves. An electron-beam plasma is linearly unstable against electrostatic perturbations with frequencies lower than the critical frequency ($\approx$ plasma frequency). Yajima and Tanaka [4] predicted the existence of soliton modes in the beam plasma system. The present authors have experimentally studied nonlinear wave phenomena of unstable beam modes and found that the nonlinear structure is determined by the linear growth rate [5].

In this paper we will present spatial nonlinear evolution of beam waves and the energy distribution function of beam electrons in connection with the nonlinear structure.

PLASMA DEVICE AND EXPERIMENTAL SETUP

A target plasma is produced in argon gas of 1.4×10^{-5} Torr by a dc discharge in a so called magnetic multi-pole plasma device [5] shown in Fig. 1. An additional magnetic field (= 100G) is externally applied parallel to the axis of the plasma chamber to observe the one-dimensional behavior of the beam modes.

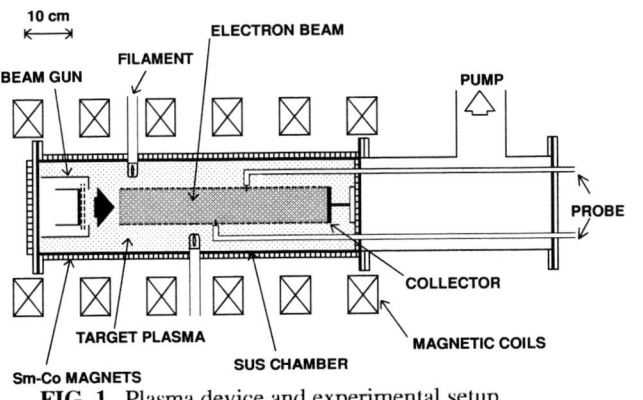

FIG. 1. Plasma device and experimental setup.

TABLE 1. Experimental parameters

Plasma density	$n_0 = 2.3 \times 10^{14}$ m^{-3}
Electron temperature	$T_e = 1.3$ eV
Plasma frequency	$f_{pe} = \omega_{pe}/2\pi = 135$ MHz
Debye length	$\lambda_D = 0.5$ mm
Beam density	$n_b/n_0 = 0.04 \sim 0.2$ %
Beam velocity	$v_b/v_T = 6 \sim 7$ (v_T thermal velocity)
Wave number	$k\lambda_D = k/k_D < 0.2$
Beam duration	$t = 4.5 \times 10^{-6}$ s ($\omega_{pi} t = 14$)
Beam diameter	$2r_b = 45$ mm ($kr_b \approx 8$)

We inject a pulse electron beam with the time width of 4.5×10^{-6}s into a target plasma along the external magnetic field. The beam current I_b passing through the plasma is measured by a collector located at the opposite end from the beam gun. Fluctuation signals $\tilde{n}(t)$ are picked up with two plane probes (Mo disks of 3mm in diameter) located at different positions, z_1 and z_2, measured from the beam gun and then two sets of real-time data are captured by a fast digitizing oscilloscope (1GSa/s, 2ch, 5kW/ch). Typical experimental parameters are summarized in Table 1. In order to observe spatial evolution of test waves we repeatedly inject pulse beams into the target plasma and simultaneously apply small rf-perturbations with the time width of 4.5×10^{-6}s to the control grid of the beam gun as shown in Fig. 2.

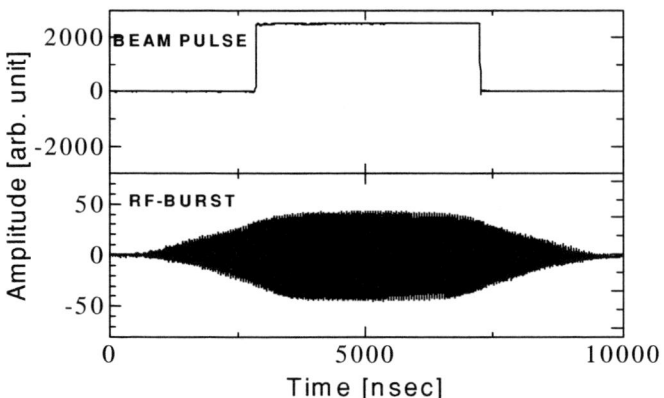

FIG. 2. Time sequence of beam pulse and rf-burst.

The spatial development of the test waves along the beam stream is measured by interferometer with a boxcar integrator as a low pass filter with a gate time of 4×10^{-6}s. Typical wave patterns observed in the case of beam density $n_b/n_0=0.15\%$ and beam velocity $v_b/v_T=6.3$ are shown in Fig.3. The axial position z is the distance measured from the beam gun.

EXPERIMENTAL RESULTS AND DISCUSSION

Spatial development and its nonlinear structure of test waves

In the initial stage of propagation (z<180mm), excited test waves are linearly unstable, that is, amplitudes increase exponentially along the beam stream. But they are stabilized in the nonlinear stage and reduced in the downstream region. Finally symmetric nonlinear structures are formed. The growth rate k_i increases with the wave frequency, but decreasing around the plasma frequency (=135MHz). Above the plasma frequency, amplitudes abruptly decrease. Such a nonlinear structure is not observed in the case of the weak beam, $n_b/n_0<0.05\%$.

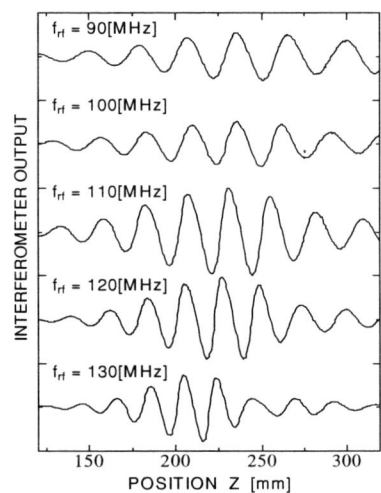

FIG.3. Spatial development and formation of nonlinear structure of linearly unstable beam waves

The dispersion data extracted from the linear phase are plotted with closed circles and triangles in Fig. 4, where closed circles and triangles are real and imaginary wave numbers, respectively. Solid curves indicate the dispersion relation for linear waves in an electron-beam plasma given by [6]

$$1 = \frac{\omega_{pe}^2}{(\omega^2 - k^2 v_T^2)} + \frac{\omega_{pb}^2}{(\omega - kv_b)^2} \quad (1)$$

Here v_T and v_b are thermal velocity of plasma electrons and beam velocity, and ω_{pe} and ω_{pb} are the plasma frequencies for plasma and beam electrons, respectively. The equation (1) has a pair of complex conjugate roots, $k_r \pm ik_i$ for a real frequency ω less than the critical frequency ($\approx \omega_{pe}$). One of these roots corresponds to unstable modes. Experimental sets of data agree well with unstable beam modes calculated from eq. (1).

As the frequency increases from 90MHz to 130MHz, the growth rate k_i increases and the amplitude becomes large. On the other hand, the width of the wave envelope Δz becomes narrow. As a result the width inversely varies as the amplitude. Figures 5(a) and 5(b) show that how the amplitude and the inverse of width change with the linear growth rate k_i. They suggest that the envelope of wave patterns $A(z)$ is empirically described by

$$k_i \operatorname{sech}[k_i(z - z_0)] \tag{2}$$

The z_0 is the position of the maximum amplitude. Equation (2) shows that the localized structure of unstable beam waves is characterized by the initial growth rate k_i, which is proportional to the cubic root of the beam density ($\propto (n_b/n_0)^{1/3}$) [6].

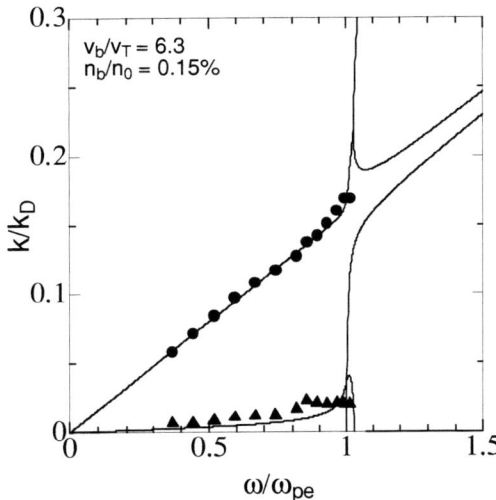

FIG. 4. Dispersion relation. Closed circles and triangles are the real k and imaginary k_i observed in the linear phase, respectively. The k_D means Debye wave number.

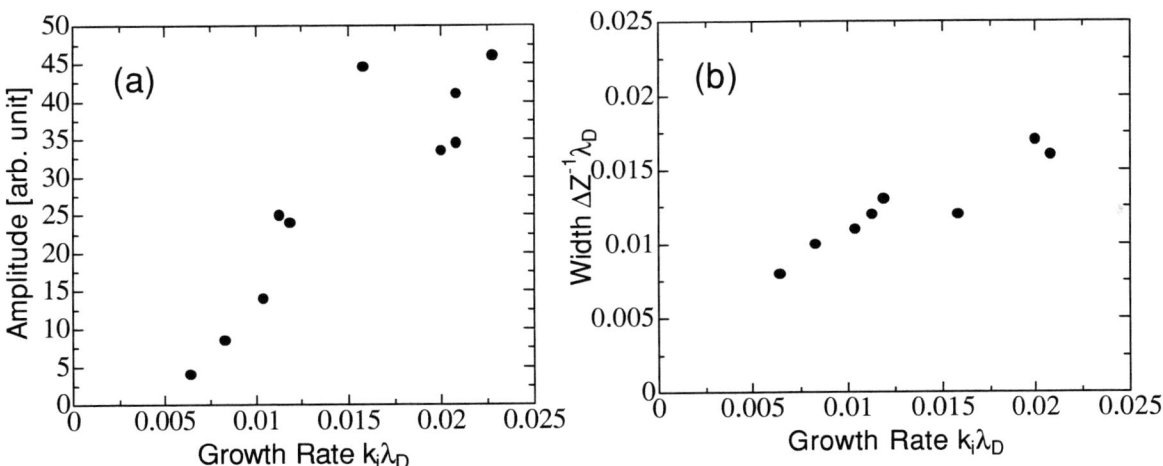

Fig. 5. Structure of nonlinear waves shown in Fig. 3. (a) Amplitude as a function of growth rate k_i/k_D. (b) Inverse of the structure width $\Delta z^{-1}\lambda_D$ as a function of growth rate k_i/k_D.

Energy distribution of beam electron in connection with the nonlinear structure

In one dimensional flow, the first derivative of the Langmuir probe current I_p with respect to the probe voltage V_p, dI_p/dV_p, is proportional to the energy distribution function $f(v)$, where $v=[2e(V_s-V_p)/m]^{1/2}$, V_s the space potential and V_p the probe potential. Figures 6(a) and (6b) show how the energy distribution functions vary along the axial position z. Without test waves, in Fig. 6(a), the energy distributions are not much changed and slightly deformed around the mid position, $z \approx 230$mm, where weakly self-exited beam instabilities are observed.

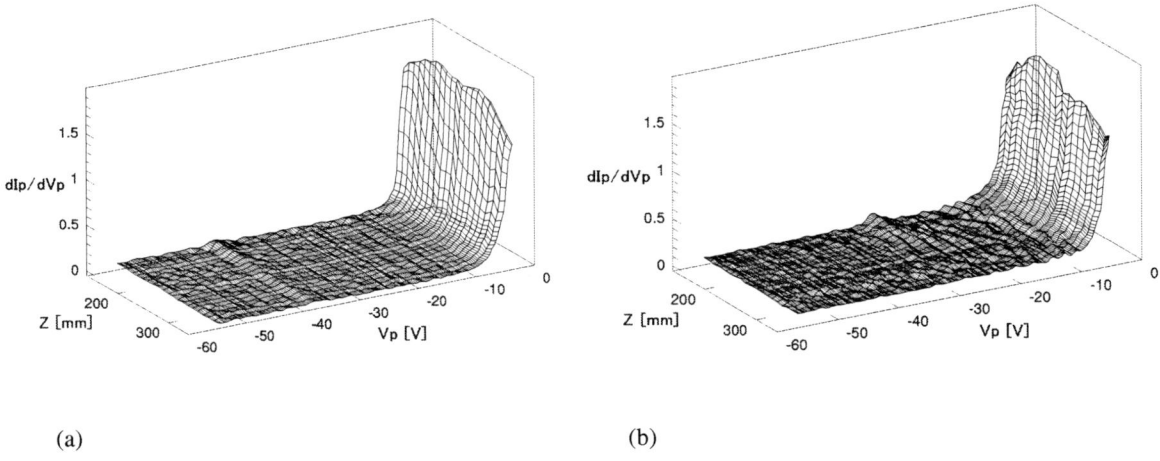

FIG. 6.. Changes of energy distribution function ($\propto dI_p/dV_p$) along the beam path z. (a) Without test waves, v_b/v_T =7 and n_b/n_0=0.1%. (b) Test waves (f_{rf}=130MHz) are generated, v_b/v_T=6.3 and n_b/n_0=0.15%.

Figure 6(b) shows that when the test waves are generated, beam electrons are scattered by intense waves and lose their energy. As a result their energy distributions are seriously deformed and wake-like structures are observed after forming nonlinear wave structures. The trapping effects on beam electrons due to nonlinear waves may be responsible for the wake structures. These will be discussed elsewhere.

CONCLUSION

The present experiment showed that soliton-like excitations exist in an electron-beam plasma, which is a weakly dispersive medium. The localized nonlinear structures are characterized by linear growth rate k_i. These are completely different from Zakharov's Langumuir solitons [7], which arise in a strong dispersive medium.

REFERENCES

[1] M. V. Goldman, Rev. Mod. Phys. **56**, 709 (1984).
[2] A. Y. Wong and P. Y. Cheung, Phys. Rev. Lett. **52**, 1222 (1984).
   Cheung P. Y. and Wong A. Y., Phys. Fluids **28**, 1538 (1985).
[3] T. Intrator, C. Chan, N. Hershkowitz and D. Diebold, Phys. Rev. Lett. **53**, 1233 (1984).
[4] N. Yajima and M. Tanaka, Prog. Theor. Phys. Suppl. No. **94**, 138 (1988).
[5] K. Yamagiwa, T. Itoh and T. Nakayama, Invited paper of ICPIG (Toulouse, 1997),
   Journal de Physique IV (France) **7**, C4-413 (1997).
[6] R. J. Briggs, *"Electron-Stream Interaction with Plasmas"* in Research Monograph No. **29**,
   (M.I. T. Press, Cambridge, 1964).
[7] V. E. Zakharov, Sov. Phys. JETP **35**, 908 (1972).

Measurement of Carrier Properties in InSb using Faraday Rotation

I. Ogawa[1], K. Yamada[1], T. Idehara[2], A. Tsushima[3] and S. Yamaguchi[4]

[1] *Faculty of Engineering, Fukui University, Fukui 910-8507, Japan*
[2] *Research Center for Development of Far-Infrared Region, Fukui 910-8507, Japan*
[3] *Faculty of Engineering, Yokohama National University, Yokohama 240-8501, Japan*
[4] *College of Engineering, Chubu University, Kasugai 487-8501, Japan*

Abstract. Carrier properties of n-type InSb doped with tellurium at liquid nitrogen temperature have been estimated by means of transmission of a millimeter wave (119.3 GHz) propagating along a magnetic field up to 4 T. The information of carrier properties is obtained from the Faraday rotation arising from the difference between phase velocities of two circularly polarized components. The advantage of this method lies in the fact that it allows both the electron density and effective mass to be measured simultaneously.

INTRODUCTION

Measurement of carrier mass and density of a semiconductor is one of the important issues for thermoelectric energy conversion in a magnetic field because a semiconductor with heavier electrons is expected to have a larger Nernst effect. In order to measure the carrier mass by means of optical properties of microwave, studies of cyclotron absorption of a cavity with a sample have been made [1-2].

This paper presents the measurement of the carrier mass and density by observing the rotation of the plane of polarization (Faraday rotation) in n-type InSb doped by tellurium (T_e) with millimeter wave (119.3 GHz).

The millimeter wavelength corresponds to the range where the plasma frequency ω_p and the cyclotron frequency ω_c are much larger than the probe beam frequency ω. In the absence of the magnetic field, the sample of InSb is opaque to electromagnetic waves of frequency ω because $\omega \ll \omega_p$. When the magnetic field is applied to the sample whose carrier collision frequency is much lower than the cyclotron frequency ω_c, the incident millimeter wave propagates the sample. Transmission of one circular polarization (Helicon wave) begins and finally, the opposite circular polarization propagates and a Faraday rotation effect arising from the difference of phase velocities appears. This phenomenon provides a method of measuring various parameters of the medium.

EXPERIMENTAL SETUP

A thin circular sample is set in the magnetic field up to 4 T produced by superconducting magnet (Fig. 1). A millimeter wave output (119.3 GHz, rectangular waveguide TE_{10} mode) from a gunn oscillator is converted into circular waveguide TE_{11} mode and incident on the sample through an uptaper, oversized circular waveguides and Teflon window. The initially linearly polarized wave incident on the sample will be rotated due to the difference in velocity of its two circularly polarized components. The power passing through the sample is reversely converted into rectangular waveguide TE_{10} mode and detected by a crystal detector. A down taper is connected with an oversized circular waveguide using a rotary flange. The angle of Faraday rotation is obtained from data of the transmitted power versus magnetic field with the polarization set at various orientations by means of the rotary flange.

In order to decrease the transmission loss of millimeter wave, we use waveguides made of cupper or brass except in the vicinity of the sample. The sample is hold in oversized circular waveguides (internal radius of 10 mm) made of stainless steel and keeps in high thermal contact with a liquid nitrogen container (Fig. 2). The inside of the stainless steel waveguide is kept at a vacuum.

For such arrangement, we can easily cool the sample around 77 K by flowing liquid nitrogen into the liquid nitrogen container. The cooling of the sample is effective for decreasing carrier collision frequency and realizing the condition that carrier collision frequency is much lower than the cyclotron frequency ω_c.

The magnetic field intensity produced by the superconducting magnet is calibrated by means of electron spin resonance (ESR) of DPPH powder with g=2.0036 for a millimeter wave output of 88.4 GHz. An absorption is observed at coil current of 44.295 A in the case of increasing the magnetic field. While the absorption appears at slightly higher current (44.318 A) when the magnetic field is decreased. As it can be seen from $\hbar\omega=\mu_B gB$, these absorptions correspond to the magnetic field of 3.16 T. Calibration factor of 0.0712 T/A is estimated by using their average (44.3 A).

A data acquisition system is used to monitor coil current and transmitted millimeter wave power. After the signals are introduced into different isolation amplifiers, they are recorded into a RAM disk by a 16 bit plug-in A/D converter.

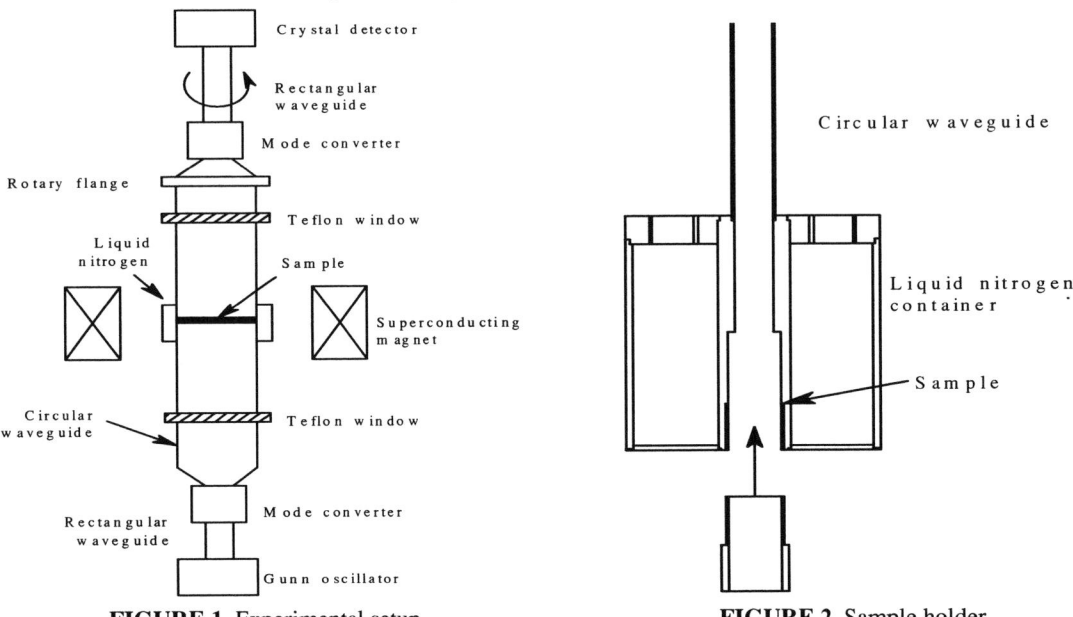

FIGURE 1. Experimental setup. **FIGURE 2.** Sample holder

EXPERIMENTAL RESULTS AND ANALYSIS

Figure 3 shows the transmitted power of the millimeter wave through a sample of InSb doped by tellurium with a thickness d = 0.955 mm at a temperature T=77 K as a function of the magnetic field B. We have measured the transmitted powers versus the magnetic field for every 15 degrees of rotary flange angle. The angle of Faraday rotation is obtained from these data as a function of magnetic field (Fig. 4). Faraday rotation θ is given by

$$\theta = \frac{k_L - k_R}{2} d, \qquad (1)$$

where k_L and k_R are wavenumbers of left- and right-hand circularly polarized waves, d a thickness of the sample. The wavenumbers k_L and k_R are given by using the dispersion relations of right- and left-hand circularly polarized waves, respectively. Since there is one type of carrier, i.e., electrons, in the sample, they are given by [3]

$$\frac{c^2 k_R^2}{\omega^2} = \varepsilon_r \left\{ 1 - \frac{\omega_p^2}{\omega(\omega - \omega_c)} \right\}, \qquad (2)$$

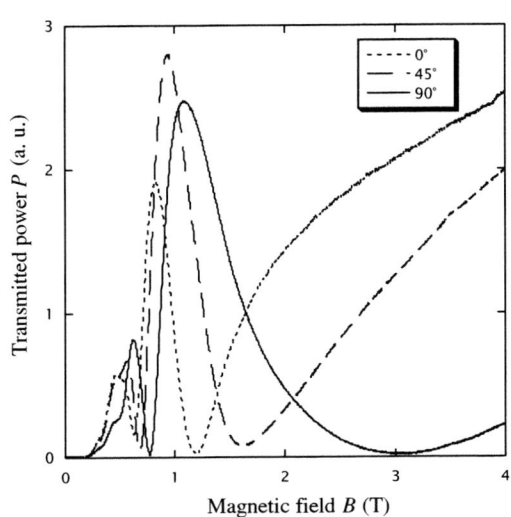

FIGURE 3. Transmitted power P at the given angle of a rotary flange as a function of magnetic field B.

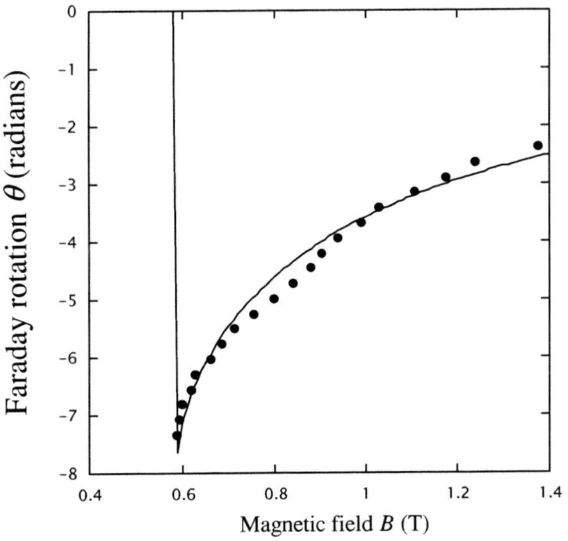

FIGURE 4. Angle of Faraday rotation as a function of magnetic field B.

$$\frac{c^2 k_L^2}{\omega^2} = \varepsilon_r \left\{ 1 - \frac{\omega_p^2}{\omega(\omega + \omega_c)} \right\}, \qquad (3)$$

where ε_r is the relative dielectric constant of the sample, $\omega_c = eB/m^*$ and $\omega_p = (ne^2/\varepsilon_0 \varepsilon_r m^*)^{1/2}$. e is the electron charge, m^* the effective mass of the carrier, n the carrier condensation, ε_0 the absolute dielectric constant in vacuum. Therefore, we can obtain three unknowns, say, m^*, n and ε_r. The results are $m^* \sim 0.02$ in the units of the electron mass m_0, $n \sim 5 \times 10^{20}$ m^{-3} and $\varepsilon_r \sim 17.9$.

As it can be seen from Fig. 4, Faraday rotation effect is observed for the magnetic field exceeding the value of 0.57 T.

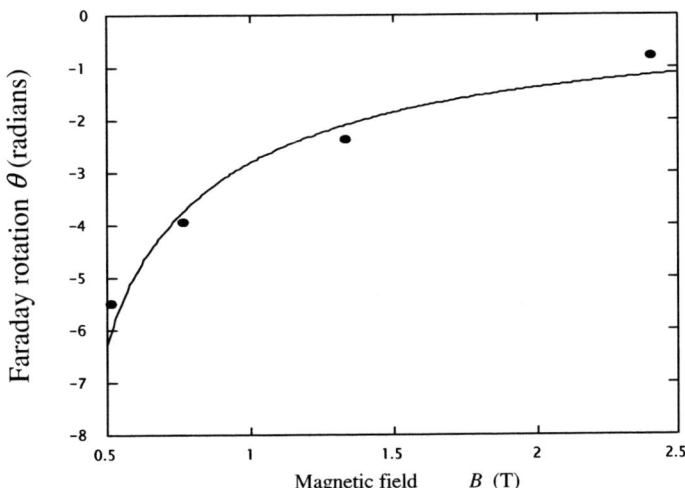

FIGURE 5. Angle of Faraday rotation as a function of magnetic field B.

At the magnetic field intensity less than the value, left-hand circularly polarized waves can't propagate because of cut off.

We also measured the dependence of the transmitted power of the millimeter wave on magnetic field in case of InSb doped by a small amount of tellurium with $d = 0.714$ mm at $T=77$ K. The angle of Faraday rotation is obtained from these data as a function of magnetic field (Fig. 5). As a result, we have $m^* \sim 0.02$ in units of the electron mass m_0, $n \sim 4 \times 10^{20}$ m^{-3} and $\varepsilon_r \sim 17.9$.

CONCLUSION

We measured the properties of the two samples of InSb doped by the different amount of tellurium by the transmission of the millimeter wave along a magnetic field. The results of the one sample were $n \sim 5 \times 10^{20}$ m^{-3}, $m^* \sim 0.02$ in units of m_0 and $\varepsilon_r \sim 17.9$. Those of the other sample were $m^* \sim 0.02$ in units of the electron mass m_0, $n \sim 4 \times 10^{20}$ m^{-3} and $\varepsilon_r \sim 17.9$. The advantage of this optical method is in the fact that both the electron density and effective mass can be measured simultaneously.

REFERENCES

[1] G. Dresselhaus, A. F. Kip, and C. Kittel: Phys. Rev., **98**, 368 (1955).
[2] J. K. Galt, W. A. Yager, F. R. Merritt, B. B. Cetlin, and A. D. Brailsford: Phys. Rev., **114**, 1396 (1959).
[3] H. Ibach and, *Solid-State Physics* (Springer-Verlag, Berlin, 1991), p.247.

Flow of Magnetized Plasma in a Linear Device

G. Fussmann, W. Bohmeyer, Zh. Kiss'ovski[*], B. Koch

Max-Planck Institut für Plasmaphysik, Bereich Plasmadiagnostik, Mohrenstrasse 41, 10117 Berlin, Germany

Abstract. Streaming of the magnetized plasma produced in the stationary plasma generator PSI-2 has been investigated by means of a Mach probe. We noticed that the streaming can be strongly influenced by changing the magnetic configuration in this region. The Mach numbers measured, however, are in the range of 0.05-0.2 only, i.e. far below supersonic flow. There are some analogies to the flow of a gas in a Laval nozzle. However, in the magnetized plasma, the Bohm criterion at the neutralizer plate should impose a special boundary condition and force a transition to supersonic flow at the region of maximum magnetic field strength. The experimental results (M << 1) appear therefore to be in disagreement with Bohm's criterion, but further measurements – immediately in front of the neutralize plate – are required to exclude the influence of ionization in the considered region.

INTRODUCTION

Plasma flow along the magnetic field lines is an essential issue for the proper functioning of divertors used in tokamaks and stellarators in particular with respect to impurity exhaust. Reliable determination of the streaming velocities is therefore a highly relevant matter.

An important question in this context is whether, and if so, in which regions, the flow is supersonic. Supersonic flow is postulated to be attained at any target by the Bohm-criterion which enters the simulations as a boundary condition. In this paper radial profiles of the flow velocity in an argon discharge and the influence of a corrugated magnetic field on the streaming are investigated. The results are discussed by referring to 1D equations of continuity and momentum.

Plasma rotation and flow are most frequently determined by passive spectroscopy, measuring the Doppler shift of ions emission lines [1]. Here instead the Mach probe technique [2-4] has been applied. This method is confined to relatively low temperatures but has the benefit of high spatial resolution.

EXPERIMENTAL SET-UP AND DIAGNOSTIC METHODS

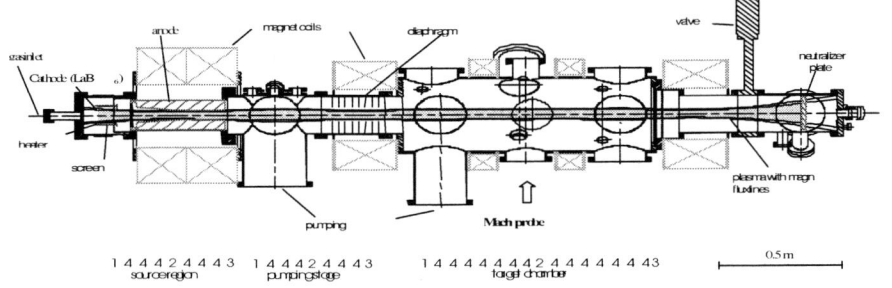

FIGURE 1. Plasma generator PSI-2.

The plasma in the PSI-2 device is generated in the source region between a hollow cylindrical cathode made from LaB_6 and a hollow anode as shown in Fig. 1. The discharge region is followed by a drift region extending over a length of 2.3 m which is finally terminated by a neutralizer plate at floating potential.

* Permanent address: Faculty of Physics, Sofia University, BG-1164 Sofia, Bulgaria

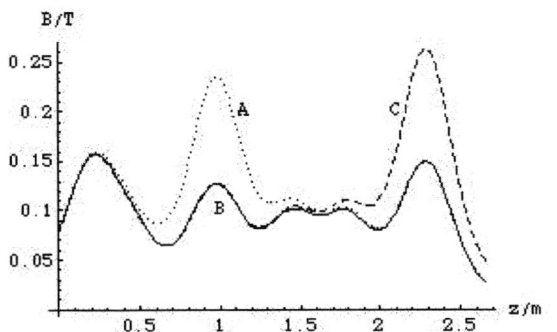

FIGURE 2. Axial magnetic field strength for the three configurations A (dotted), B (solid), and C (dashed).

The discharge region is operated in the low-pressure regime (0.3 - 3 Pa in the cathode-anode region) at voltage and current ranges of U = 15-60 V, I = 50-1000 A. The neutral gas pressure is reduced in the pumping region and further in the target chamber where typical values are in the range of 0.01 – 1 Pa. Measurements of the plasma parameters, as well as flow and rotation velocities were carried out with a Mach probe along a central chord in the target chamber at the position indicated in Fig. 1

The PSI-2 chamber is surrounded by six coils providing an axial magnetic field of B = 0.1 – 0.25 T. Due to the arrangement and the small number of coils there are noticeable inhomogeneities leading to more or less pronounced magnetic mirror configurations in the target chamber which are important with respect to plasma flow.

The magnetic field can be calculated from the magnetic flux surfaces $\psi(r, z)$ = const. Close to the axis these are given by $\psi = B(z) r^2/2$ where $B(z)$, the axial field produced by the set of coaxial current loops, is obtained from

$$B(z) = \frac{\mu_0}{2} \sum_{k=1}^{6} \frac{I_k r_k^2}{[r_k^2 + (z - z_k)^2]^{3/2}}. \quad (1)$$

In order to investigate the influence of axial magnetic variations on flow the three different profiles (A, B, C) plotted in Fig. 2 have been realized.

The position z = 0 coincides with the emitting annulus of the cathode while the neutralizer plate is at z = 2.65 m. In the case of profile of type A the magnetic field has been enhanced around z = 1 m by increasing the current in the third coil from the left. The standard case B shows the most homogeneous situation achievable, while in case C the magnetic field toward the end of the target chamber has been enhanced considerably by rising the current in the last coil. At the position of the

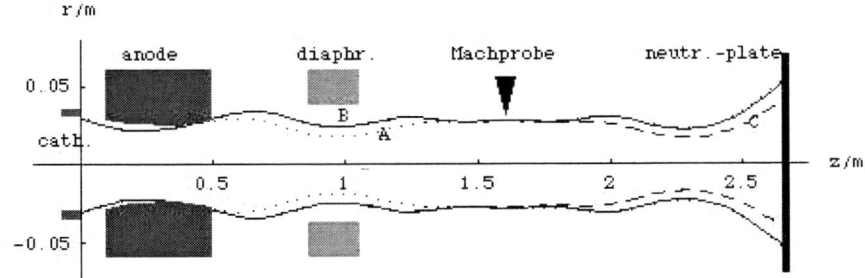

FIGURE 3. Magnetic flux contours for the cases A (dotted), B (solid), and C (dashed).

Mach probe (z = 1.6 m), however, the field variations are very weak for all three cases. In Fig. 3 we show the flux contours ψ = const. for the three configurations. These are obtained by choosing the particular contour touching the anode exit.

At the position of the Mach probe the nominal plasma radius determined this way is obtained to r_p = 2.9 cm. This fits quite well with optical measurements: scanning the Ar II intensity profile over the full diameter at this position a sharp drop is observed for 2r = 6.5 cm, i.e. an effective plasma radius of 3.25 cm.

For measurements of flow and rotation velocities a Mach probe as described by Peterson et al. [4] and depicted in Fig. 4 with two collecting plates has been used. The circular plates (r_p = 2.25 mm) of identical areas are made from tungsten and are flush mounted in a ceramic body separated by an insulator (1.5 mm thick). These disks are positioned perpendicular to the magnetic field lines. Both plates behave as single Langmuir probes on the ion saturation current part of the current-voltage probe characteristic. The ratio of the upstream and downstream ion saturation currents of the two plates is related

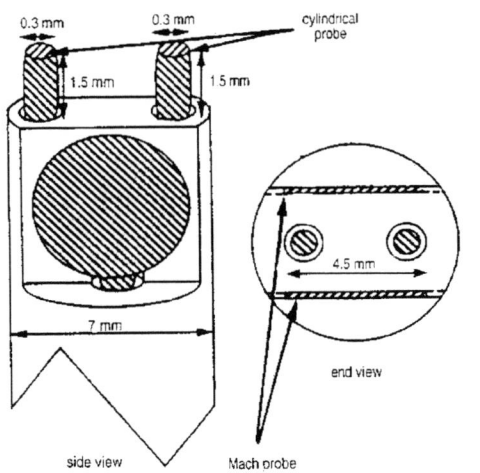

FIGURE 4. The measuring head: Mach probe and double probe.

to the plasma flow velocities in front of the collecting regions of the probe. As can be seen in Fig. 4, the measuring head of the Mach probe is equipped in addition with a double Langmuir probe to allow independent determination of the plasma density and electron temperature. Electron temperature profiles are taken by using the probe tips as single probes.

Following the discussion given in [4], the ion saturation current density collected in the upstream (+) or downstream (-) directions along to the magnetic field lines are given by $j_{\pm} = e\, n_{\infty}\, (T_e/m_i)^{1/2} \exp[m_i(v_i \pm u)^2/2T_e - 1/2]$. Here e is the charge of the ions, n_{∞} the plasma density outside of the collection regions. T_i, T_e are the temperatures of ions and electrons (in eV), m_i the ion mass, $v_i = (2T_i/m_i)^{1/2}$ the thermal ion velocity, and u the plasma flow velocity. The ratio of these current densities

$$\frac{j_+}{j_-} = \exp\left(2\frac{v_i\, u}{T_e/m_i}\right) = \exp\left(2M\sqrt{2\frac{T_i}{T_e}\left(1+\frac{T_i}{T_e}\right)}\right) \quad (2)$$

is thus related to the Mach number $M = u/c_s$ being defined as the quotient of flow and ion sound velocity $c_s = [(T_e+T_i)/m_i]^{1/2}$.

EXPERIMENTAL RESULTS

In order to evaluate the profiles of flow velocity from the ratio of the ion saturation currents the radial profiles of electron and ion temperatures are required. Typically, hollow profiles are found for T_e and the plasma density due to the hollow shape of the cathode and the anode [5]. The measured radial profiles of T_e and n_e are obtained at a very low neutral pressure of 9 mPa in the target chamber for the three magnetic configurations discussed before are shown in Fig. 5 and Fig. 6. The highest values of T_e are obtained for the normal case B, whereas the magnetic peak occurring upstream in case A is seen to lower the peak temperature by about a factor of two. A similar tendency is noticed in case of the ion temperatures also included in the figures.

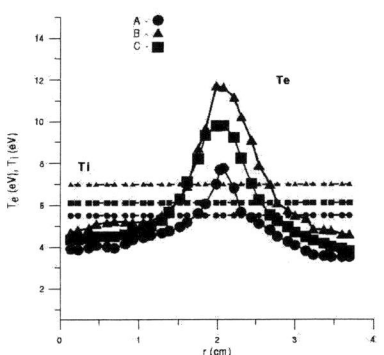

FIGURE 5. Radial profiles of T_e and T_i for the three magnetic configurations (A,B,C). Spectroscopic measurements of T_i are approximated by horizontal lines.

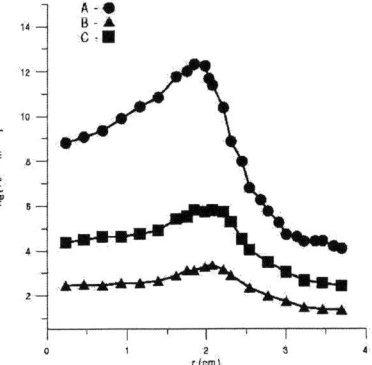

FIGURE 6. Radial profiles of plasma density for the magnetic configurations A,B and C.

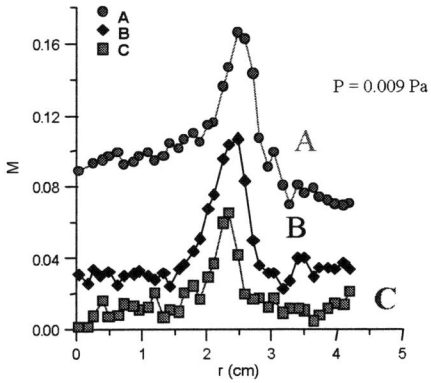

FIGURE 7. Mach number vs. radius.

The profiles of the plasma density shown in Fig. 6 are also affected by the axial magnetic field profile. Lowest densities are obtained in case of configuration B, while about four times higher values are found in case A. In the case C the magnetic field is enhanced by a factor of 1.76 at the exit of the target chamber. As a consequence the density is seen to be approximately doubled compared to the normal case B.

The plasma produced in the discharge region is forced to stream along the magnetic field lines due to a small axial pressure gradient. The radial profiles of the Mach numbers are shown in Fig. 7. They are seen again to be hollow in the target chamber for all three magnetic configurations (A, B, C). The highest flow velocities are obtained in case of configuration A ($M \leq 0.16$, $u \leq 800$ m/s) whereas in case C the smallest values ($M \leq 0.05$, $u \leq 250$ m/s) are found.

Since the variation of temperature is much reduced compared to that of density in the three cases (A, B, C), we infer that case A is characterized by the highest plasma pressure at the position of measurement, while in case B it is the lowest. We see that at the chosen plasma cross-section the flow is always subsonic with typical values around M = 0.1.

THEORY OF PLASMA FLOW

In what follows we use the 1D equations for continuity and momentum but replace the energy equation by an adiabatic pressure law $p = p_e + p_i \sim n^\gamma$. In contrast to the usual treatment, however, we will take into account the curvature of the field lines to which the magnetized plasma is tied. The flow velocity is then given by $\mathbf{u} = u\, \mathbf{B}/B$. Neglecting ionization and viscosity we get from the continuity and momentum equations the relations

$$\frac{nu}{B} = const.; \quad m_i nu \frac{du}{ds} + \frac{dp}{ds} = 0. \tag{3}$$

By means of the adiabatic law the second equation can be solved and the quantity being conserved under such streaming conditions is known as the *total specific enthalpy*. In particular the streaming of a noble gas ($\gamma = 5/3$) through a Laval nozzle is, as far as the flux relations are concerned, quite similar to the streaming of a plasma through the neck of a magnetic bottle. A pronounced peak of B may thus be denoted as "magnetic nozzle". Most important is the expression for the particle flux density $\Gamma = n\, u$ which is obtained as a function of the Mach number M. With respect to a reference position "a" (anode or position of maximum B) the result reads

$$\frac{\Gamma}{\Gamma_a} = \frac{M}{M_a} \left(\frac{1 + \frac{\gamma-1}{2} M^2}{1 + \frac{\gamma-1}{2} M_a^2} \right)^{-\frac{\gamma+1}{2(\gamma-1)}}. \tag{4}$$

Its main feature is that for all values of $\gamma \geq 1$ it attains a maximum at M = 1. This means that flux enhancement is achieved by increasing the Mach number if the plasma is subsonic (M < 1), but, conversely, a reduction of the Mach number is to postulated if it is supersonic (M > 1). Because of this property the plasma will attain sound velocity (M = 1) at the position of maximum flow, i.e. where $B = B_{max}$. Quantitatively this behavior is obtained by combining Eq. (4) with Eq. (3) yielding

$$\frac{B}{B_{max}} = M \left(\frac{2 + (\gamma-1)M^2}{\gamma+1} \right)^{-\frac{\gamma+1}{2(\gamma-1)}}, \tag{5}$$

which is actually an equation that specifies the inverse function B = B(M). Figure 8 shows the plot of this function for three values of the adiabatic coefficient γ. As can be seen there are two branches, a subsonic branch (M < 1), where all three curves practically coincide, and a supersonic branch (M > 1), where M is a monotonically rising with γ.

An important consequence of Eq.(5) is obtained when we apply Bohm's law to define the boundary condition at the neutralizer plate. This law states that any floating wall in contact with the plasma is charged in such a manner that the ions will enter the electrostatic sheath region with $M \geq 1$ (see for instance ref. [6]). Looking at Fig. 8 we realize that this means $M_{plate} = 1$ if the maximum of the magnetic field, within the whole plasma region, is attained at the neutralizer plate. In any other case, however, when the maximum field strength is reached left from the plate at $z = z_{max}$, the flow in the region $z_{max} < z \leq z_{plate}$ is necessarily supersonic, and at the plate proper Mach numbers much in excess of unity can occur. The region $z < z_{max}$, on the other hand, is subsonic and described by the lower branch of the curve shown in Fig. 8.

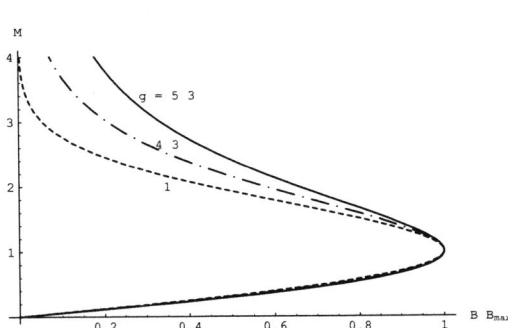

FIGURE 8. The Mach number as a function of the magnetic field strength for three different values of the adiabatic coefficient according to Eq. (5) $\gamma = 1$ corresponds to isothermal flow conditions, $\gamma = 5/3$ describes the case where both ions and electrons are adiabatic, and $\gamma = (1+5/3)/2 = 4/3$ may be chosen when electrons behave isothermal but ions adiabatic.

Since we know the magnetic field strength B as a function of z we are now in the position to calculate axial profiles of the Mach number and all other quantities of interest normalized to their values at z_{max}. Examples for such calculations for the configurations A and C are shown in Fig. 9 and Fig. 10 assuming $\gamma = 4/3$. As can be seen from the figures the supersonic region with M > 2 extends over the major part of the target chamber in case A (see Fig. 2) whereas in case C the Mach number is around 0.2 over the larger part of this region, and M > 1 is reached only behind the last coil, i.e. relatively close to the neutralizer plate.

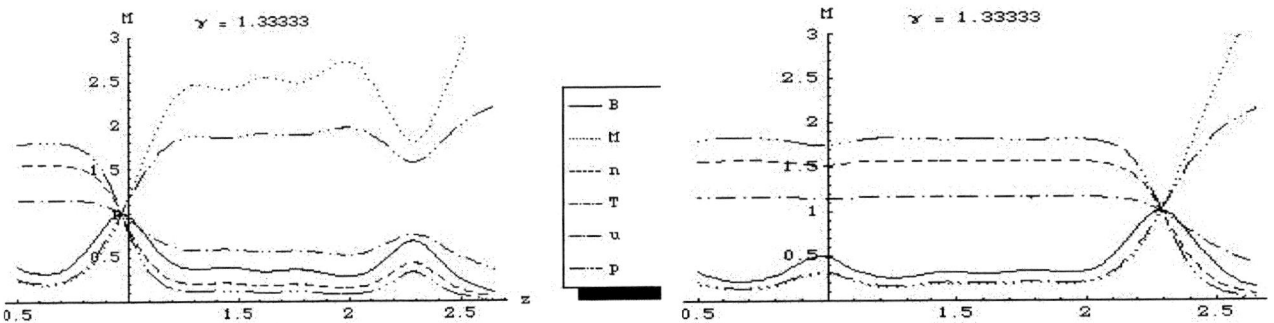

FIGURE 9. Axial profiles calculated for case A. (Mach number dotted).

FIGURE 10. The same as in Fig. 9 for case C.

SUMMARY

Magnetic field rippling can be of essential influence with respect to plasma flow. In regions of enhanced magnetic field strength the plasma is accelerated and, akin to normal gases in a Laval nozzle, a transition into supersonic flow may be enforced. In a magnetized plasma such a transition should always take place because of Bohm´s criterion. In qualitative accordance with these ideas the Mach number is seen in Fig. 7 to rise substantially when the leading magnetic nozzle (position of maximum B) is shifted from the neutralizer plate (case C) to a position upstream (left) from observation (case A). The comparison between experiment and theory is, however, not satisfying since no transition into the supersonic branch was observed. In fact, Mach numbers as large as $M \approx 2.5$ were expected (Fig. 9) but only $M \approx 0.15$ were measured.

REFERENCES

[1] Field A. R., Fussmann G., Hofmann J. V. and ASDEX team, Nuclear Fusion **32** (1992), 1191
[2] Hudis M. and Lidsky L. M., J. Appl. Phys. **41** (1970), 5011
[3] Hutchinson I. H., Phys. Rev. **A37** (1988), 4358
[4] Peterson B. J. et. al., Rev. Sci. Instrum. **65** (1994), 2599
[5] Fussmann G., Meyer H., Pasch E., Contr. Plasma Phys. **36** (1996), 501
[6] Chen F. F., Introduction to Plasma Physics and Controlled Fusion, (Plenum Press, New York, 1988)

Penetration of Electrons inside Helical Field Configuration via Stochastic Magnetic Layer

H. Himura*, H. Wakabayashi*, M. Fukao*, Z. Yoshida*, M. Isobe[†], K. Matsuoka[†] and H. Yamada[†]

The University of Tokyo, Graduate School of Frontier Sciences, Hongo, Tokyo 113-0033, Japan
[†]*National Institute for Fusion Science, Toki, Gifu 509-5292, Japan*

Abstract. An experimental study on motions of electrons in a stochastic magnetic field has been performed on the Compact Helical System (CHS) device [1]. The purpose of this research is to experimentally investigate how strongly magnetized electrons move in the stochastic region where the field line itself is chaotic. As an experiment, electrons with 1.2 keV directed energy are launched from the stochastic magnetic layer. Remarkably, the electrons penetrate the stochastic layer and moreover, reach near the central core of the helical configuration. No theory on the loss cone orbit can be applied to the experimental data. The result may reflect some chaotic or collective effect which allows the electrons to move completely in a non-ordered way in the stochastic magnetic region.

INTRODUCTION

Recently, an equilibrium theory of two-fluid plasmas is derived [2]. One of the interesting forecasts is that the extremely high-β equilibrium exists with fast perpendicular flow of ions. In fact, sample equilibria of the hydrodynamic (relaxed) states resemble field-reversed configurations and H-mode tokamaks. Thus, studies on flowing plasmas have attracted considerable interest in experimental plasma physics. In order to drive a fast flow by means of an electric field **E** inside toroidal plasmas, a technique to generate **E** is called for. The simplest way among them is to use a pair of electrodes in which a high voltage is externally applied between the electrodes to produce **E**. This method has worked well to produce a flow in boundary layer of plasmas which actually seems to trigger a transition of confinement modes of small tokamak plasma from L- to H-mode [3]. However, in the method, at least one of the electrodes must be inserted inside the plasmas, which obviously enhances the presence of impurities. Thus, another way of generating **E** in plasmas has been required.

An innovative idea of producing **E** is proposed. The method applies the toroidal non-neutral plasmas in which strong **E** exists due to the completely non-neutral condition of charge. In a sense, the concept is based totally on the electrostatic confinement of magnetized plasmas in a deep potential cavity. And, the background cloud of magnetized electrons is a substitute to form the negative well, which inherently solves the impurity problem. To obtain such a non-neutral condition in the boundary layer, we have proposed to inject excess electrons there. If the excess electrons are successfully trapped just inside the separatrix surface of the toroidal plasmas. a strong **E** would appear inside the plasmas, causing a fast $E \times B$ flow.

One of key issues of the above method is how to inject electrons across the separatrix surface. There is a long history of experiments on cross-field injection of charged particles or plasmas. For the case of electrons, they usually cannot move across the separatrix. Thus, some method needs to be applied to have electrons drifted. Several ideas can be considered. And, one of promising methods is to cause electrons to move chaotically around the boundary layer. This is because in this case even the magnetic moment (μ) of electron is not conserved, which operates electrons in a non-ordered way. Thus, a part of electrons are expected to travel across the separatrix surface. To explore this innovative way, an experimental project [4] has been performed on the Proto-RT (Prototype Ring Trap) device. In the machine, an inhomogeneous AC electric field is applied to electrons around the separatrix in order to violate the conservation of μ there, because Proto-RT is an axisymmetric device. On the other hand, a helical magnetic configuration is asymmetry. This means that no (global) adiabatic invariant is inherently existed. Thus, this property suggests the possibility of the penetration of electrons across the separatrix even in an electrostatic operation. And, as a result, a strong radial electric field due to the injected electrons may be produced in the boundary layer of the helical configuration. In order to test that experimentally, we have initiated experiments on the Compact Helical System (CHS) device [1]. At the first series of experiments, electrons are injected into the CHS vacuum field. As expected, the initial result shows broad profiles of electron flux formed inside the separatrix surface, despite they are fired from the stochastic region outside the separatrix. This means that the electrons penetrates inside the separatrix via stochastic magnetic layer. The corresponding electric field formed by the electrons is $4-7$ kV/m in the boundary layer just inside the separatrix. The strength of E_z is already beyond the critical value that is required to maintain $E \times B$ toroidal flow ($\sim 10^{3-4}$ m/s for ~ 1 T of magnetic field strength, although E_r is formed in the vacuum field, not in helical plasmas. In the following, we will present the initial data taken from the first experiment.

APPARATUS

CHS is a medium size device whose major and averaged minor radii are 1.0 and 0.2 m, respectively. A detail explanation of the device can be found in Ref. 1. On the CHS experiments, there are two parameters which are essential to form the helical configuration. One of these is the radial position of magnetic axis (R_{ax}). For the conventional experiments focusing on the study of hot plasmas, R_{ax} is usually set around 92.1 cm. However, for this setting, the inside edge of separatrix surface intersects the vacuum chamber that works as a limiter. Thus, at the presented experiment, we have shifted R_{ax} outwardly in radial direction and fixed at 101.6 cm in order to completely isolate the separatrix surface from the chamber. The other is the magnetic field strength of helical configuration. The field strength at R_{ax} is usually more than ~ 0.9 T for hot plasma experiments in pulsed operation. The presented experiments, on the other hand, have been performed under DC operation to both create a well-controlled helical field and avoid experimental difficulties. Typical field strength for the presented experiments at R_{ax} is ~ 0.05 T for $R_{ax} = 101.6$ cm where the gyroradius of electron re is calculated to be ~ 0.23 cm for the case of $v_\perp \sim 2 \times 10^7$ m/s ($E_\perp \sim 1.2$ kV). Electrons are launched from a typical diode-type electron gun with LaB_6 cathode which can be operated in relatively strong field. It also achieves higher current densities with lower temperature than pure tungsten because of lower work function. The LaB_6 cathode employed is quadrate (1.5 cm each). And, the emitter has also a quadrille shape with tungsten wires. The beam current (j_b) of the electrons is variable. However, for the present research we have kept j_b at a constant value of ~ 10 mA for $V_{acc} = 1.2$ kV where V_{acc} is the potential difference between the LaB_6 cathode and the (grounded) emitter.

INITIAL RESULT AND DISCUSSION

At the first series of the experiment, electrons are injected into the CHS vacuum helical field (no plasmas) via stochastic field region. The electrons are launched from the 1.7 cm outside of the separatrix surface. Since they form pure electrons plasmas if being successfully confined in the helical field, a high impedance emissive probe is installed to measure the value of space potential ϕ_p due to the presence of electrons at the measurement point.

Figure 1 shows three profiles of ϕ_p for different cases of r_{gun}. Substantial values of ϕ_p is observed inside the separatrix of the helical field. For the case of $r_{gun} = 25.5$ cm (+1.7 cm outside the separatrix), ϕ_p distributes broadly astride the separatrix. In fact, ϕ_p is finite (up to 300 V) even outside the separatrix which is called "stochastic region" in helical plasmas. This is probably due to the position of the electron gun. A large amount of electrons stream continuously in the stochastic region. Since the number of the electrons decreases in accordance with the transport coefficient defined in the region, a part of them could still remain in the measurements. Here, we should mention about the possibility of connection lengths in the stochastic region. Since the lengths of field lines in the region is significantly long, one might think that the result may reflect good confinement of electrons there. This does not seem to hold in this case. If the confinement is so good, values of ϕ_p for other cases where the gun is inside the separatrix should be finite. This is because even for the cases electrons, flowing out from the inside, could be trapped again in the stochastic region, which results in some finite value of ϕ_p. However, as recognized in Fig. 1, the data for the (two) cases are not finite but ~ 0 V outside the separatrix. Inside the separatrix, the value of ϕ_p gradually increases up to ~ 700 V at $z \sim 15$ cm, about 9 cm inside from the separatrix. For $z < 15$ cm, ϕ_p does not keep its value but decreases down to ~ 500 V near the magnetic axis. This tendency is basically same for the cases of $r_{gun} < 23.8$ cm with several differences. For the $r_{gun} = 22.5$ cm case, the maximum value of ϕ_p is about 1100 V, slightly less than the acceleration voltage (1175 V) of the electrons. This value also decreases at $z < 15$ cm down to 700 V, thus the drop of ϕ_p (400

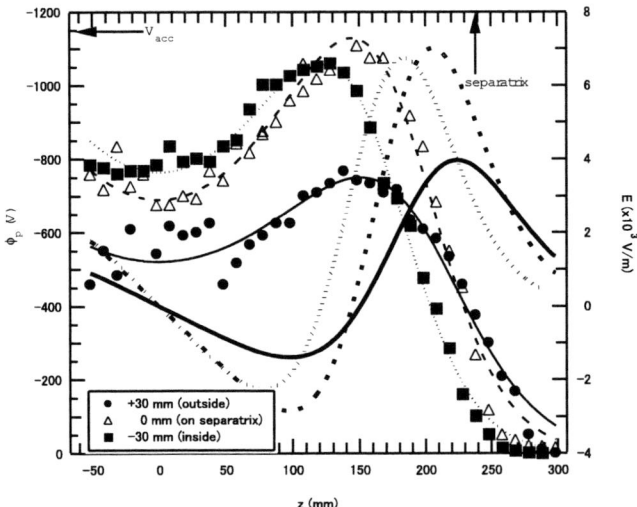

FIGURE 1. Profiles of space potential and the corresponding electric field formed by the injected electrons in the helical field.

V) is bigger than the case of $r_{gun} = 25.5$ cm (200 V). For the case of $r_{gun} = 19.5$ cm two distinctions can clearly be recognized, although the shape of the ϕ_p profile itself is almost same as that of $r_{gun} = 22.5$ cm. One of those is that the profile shifts about 3 cm toward the center, possibly due to the gun position. The other is the drop of ϕ_p at $r < 12$ cm. It is about 250 V which is smaller than the case of $r_{gun} = 22.5$ cm (400 V).

Since the profile of ϕ_p is measured, the corresponding E_z can be calculated by $E_z = -\nabla \phi_p$. Solid lines in Fig. 1 indicate the profiles of E_z which are obtained from the fitting curves (dashed lines) on the measured ϕ_p profiles. Considerable E_z is established in the boundary layer (just inside the separatrix). For the $r_{gun} = 25.5$ cm case, the value of E_z takes its maximum ~ 4 kV/m at $z = 22.5$ cm, about 3 cm inside from the electron gun. This value increases up to $E_{max} \sim 7$ kV/m for other cases because of steep gradient of ϕ_p. And, the position of E_{max} also shifts ~ 3 cm towards the center for each case. One notes that in the region of $r < \sim 15$ cm, E_z is flipped for all cases. We will discuss about this result later.

The above results of ϕ_p strongly suggest that the injected electrons penetrate deeply inside the separatrix. To verify it, we measured particle flux Γ by the same emissive probe but grounded to the chamber with a 10 kilo ohm resistance. Data in Fig. 2 show time histories of Γ obtained at different coordinates between $z = 1$ and -18 cm, where the separatrix surface is at $z = 0$ cm. The electrons are fired at $t = 0.5$ ms from the electron gun which is located 1.7 cm outside the separatrix. Despite the gun is located outside the separatrix, substantial Γ is recognized inside the separatrix. The value of Γ at $z = +1$ cm is $\sim 8 \times 10^{16}$ /m²s which is not so large as compared with those measured inside the separatrix, as recognized in Fig. 2. Inside the separatrix, Γ is in the range between 10^{17} and 10^{18} /m²s. This is probably attributed that the field lines are opened outside the separatrix: no magnetic surface at $z > 0$ cm, as mentioned. Thus, only a part of electrons just launched from the gun may be detected at $r = 1$ cm. In contrast, the measured Γ significantly enlarges inside the separatrix ($z < 0$ cm). Around $z \sim -(1-3)$ cm, the value of Γ increases up to $\sim 5 \times 10^{19}$ /m²s, which is 50 times as large as the data at $z = 1$ cm. This suggests the confinement of the injected electrons inside closed magnetic surfaces. Then, the value of Γ gradually decreases at $z \sim -5$ cm, which is about 6×10^{18} /m²s. It, however, slightly increases again at $r \sim -7$ cm up to 1.5×10^{18} /m²s and then, decreases down to $\sim 1 \times 10^{17}$ /m²s around the magnetic axis of the CHS helical field. One notes that Γ is not zero but still finite which is almost comparable with the value measured outside the separatrix. Regarding the penetration length δ of electrons, data of the measured Γ indicates that δ is at least ~ 8 cm. This value is about 40 times greater than ρ_e.

Since $\rho_e \sim 0.5$ cm around the separatrix, this result cannot be explained by a finite Larmor radius effect. On the other hand, the rise-up time of Γ inside the separatrix is about ~ 300 μs which is comparable with a single collision time of electron against background neutral particles τ_{en} (~ 330 μs) at $P_0 \sim 1 \times 10^{-7}$ Torr. Therefore, the diffusion process due to the classical collisions is insufficient to explain the penetration of injected electrons. However, the classical process is insufficient to explain such a deep length of penetration of electrons too long to be accomplished in such a single collision time. Thus, some collision-free effect should occur to bring about the result.

Another possible mechanism is some electron orbit which circulates across the separatrix. The loss cone orbits [5] is

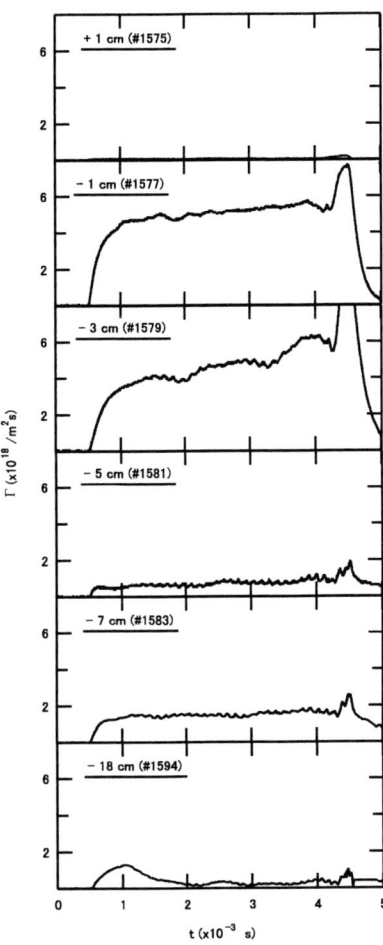

FIGURE 2. Time histories of electron flux. These data show that the electrons launched from the outside of the separatrix traverse across there and furthermore, penetrate deeply inside the separatrix.

a typical example for it. Actually, the measured profiles of ϕ_p have a dependence on the angle between the gun and field lines so that some orbital motion or related phenomenon may play a role for the result. However, the calculation of a single particle orbit for this experimental setting does not show any orbits which circulate across the separatrix. Thus, further studies are required to explain this experimental result.

ACKNOWLEDGMENTS

This work is performed with the support and under the auspices of the NIFS LHD Project Research Collaboration.

REFERENCES

1. Matsuoka, K. et al., Plasma Physics and Controlled Nuclear Fusion Research 1988, (IAEA, Vienna, 1989), Vol. 2, p. 441; Matsuoka, K. et al., Fusion Technol. **17**, 86 (1990).
2. Mahajan, S. and Yoshida, Z., Phys. Rev. Lett. **81**, 4863 (1998).
3. Taylor, R. , Phys. Rev. Lett. **63**, 2365 (1989).
4. Himura, H. et al., Phys. Plasmas **8**, 4651 (2001).
5. Motojima, O. et al., Nucl. Fusion **40**, 833 (2000)

Overview and Results from the H-1 National Facility

B.D. Blackwell, J.H. Harris, J. Howard, M.G. Shats, C. Charles, S.M. Collis,
H.J. Gardner, F.J. Glass, X. Hua, C.A. Michael, D.G. Pretty, H. Punzmann,
W.M. Solomon and G.G. Borg

*Plasma Research Laboratory, Research School of Physical Sciences and Engineering,
Institute of Advanced Studies, Australian National University, ACT 0200 AUSTRALIA.*

Abstract. The H-1 heliac has been substantially upgraded under the Australian Government's Major National Research Facility Program. Enhancements include precision magnet power supplies, a 250kW RF source, and a 200kW 28GHz ECH system in collaboration with NIFS and Kyoto University. The power supply allows operation to fields ~1 T with computer-controlled configurations, and ripple < 0.01% to eliminate induced currents. Up to 200kW of RF power at 7 MHz is used at present to produce plasmas using helicon waves, and to produce target plasma for second harmonic ECH at ~100kW. Local and remote data access methods include MDSPlus, a Java control interface and an advanced SQL-based electronic log.

At low fields in argon, tomographic interferometry and probe studies show low-mode-number coherent oscillations in electron density and electron and ion temperatures. These are suppressed when the plasma enters a higher confinement mode. Multiple diagnostic studies and comparison experiments with a linear helicon device suggest that the near fields of the RF antennas may be responsible for relatively high ion temperatures (>20eV) in this mode. In many cases, probe and spectroscopic results indicate that in these plasmas, the mass flow velocities are less than the E×B velocity. Correlation studies of probe and spectroscopic data yield important information about flow structures in these plasmas.

At 0.5 tesla, RF (20 ~150kW, $\omega \sim \omega_{cH}$) produces plasma in H:He and H:D mixtures at densities up to $<n_e> \sim 2\times10^{18}m^{-3}$, with temperatures initially limited to < 50eV by low-Z impurities. ECH ($\omega = 2\omega_{ce}$) produces considerably higher temperatures and centrally-peaked density profiles. Magnetic configuration scans show a strong, detailed dependence of plasma density on rotational transform. Magnetic fluctuations are stronger in RF-produced plasma, and spectra depend on magnetic configuration.

INTRODUCTION

The H-1 heliac is a current-free stellarator with a helical magnetic axis which twists around the machine axis (radius 1m), a circular ring conductor, three times in one toroidal rotation. It is a "flexible" heliac[1] composed almost entirely of circular coils with the exception of the control winding, a helical conductor. This takes the form of a toroidal helix, which also wraps around the ring conductor, in phase with the magnetic axis of the plasma, but with a smaller swing radius (95 mm c.f. ~230mm). Control of this current produces a range of rotational transform ι from 1 to 1.5 at full specification (B_0 =1T, average minor radius $<r>$ > 0.15-0.2 m, and 0.7 to 2.2 at relaxed specification ($B_0 \sim 0.5$ T, $<r>$ > 0.1 m). The magnetic well, $V' \equiv 1/N \int dl/B$, also varies, but when currents in the two vertical field sets are controlled as well, this flexibility is enhanced to allow almost independent control of two the three parameters: ι, magnetic well and shear in rotational transform. Magnetic well can be controlled between −2% to +6%, and shear can be made positive (stellarator-type) negative (tokamak-like), or reduced to a small value (<0.1). The magnetic configuration was optimized to obtain a low aspect ratio (5-7), and as a secondary goal, to minimize toroidal ripple. A further criterion was that magnetic surfaces of high quality be obtained without energizing the helical conductor ("Standard" configuration). The 36 toroidal coils (TFCs) inner vertical field coil pair (IVF), ring and helical conductor are all contained in a 33m³ stainless-steel vacuum vessel, sharing the vacuum with the plasma (~1m³). This creates some engineering challenges: the coils are all encased in stainless steel with a carefully finished surface, and a 90C baking system and 100,000 l/s "PolyCold" cryo-pump are used to control water vapour.

FIGURE 1. Representation of the H-1 heliac: The plasma twists around the ring conductor 3 times, as does the helical conductor between it and the ring. Outer vertical field coils are in the background.

The coils are powered by precision regulated DC supplies of 800V(main) and 100V(control) at 1-14kA. Ripple is kept well below 1A to allowing precise control of configuration, prevent "shimmer" in the magnetic surfaces and to avoid induction of plasma currents.

Plasma Production and Heating

H-1 can be run with a steady magnetic field up to 0.15 T, or a pulsed field to 1 T (0.6 T to-date). At low fields, plasma is produced in argon, helium, hydrogen and neon by a bare "picture-frame" antenna, designed to have the characteristics of a helicon antenna[2] as much as possible, without interfering with the nearby TFC coils. This helicon mode of operation is non-resonant, and provides plasma over a wide range of magnetic fields. Maximum density occurs around 0.05 – 0.1T, falling above this. Depending on rf power, plasma formation ceases at 0.15 (20kW) -0.3T(200kW). Confinement improves markedly at lower fields when a transition occurs from a mode with large coherent drift-type modes and poor confinement to a (usually, but not always) more quiescent higher confinement mode.

At still higher fields, a resonant mode occurs at the ion cyclotron frequency (7MHz@0.45T for $^1H_1$). Density returns to the maximum value at low field, but depending on the concentration of the resonant species the range of magnetic field is quite limited (see fig. 2).

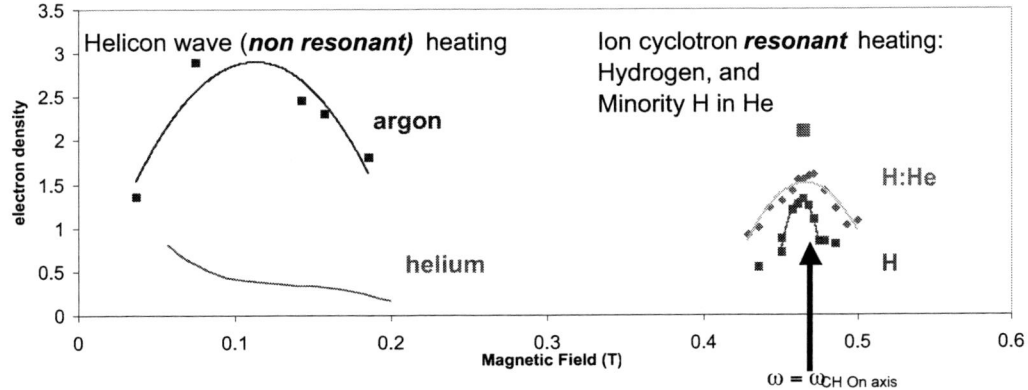

FIGURE 2. RF heating modes in H-1: non-resonant (helicon) at low field, ion-cyclotron resonant at 0.5Tesla.

Impurities are not noticeable in argon plasma, because of the lower electron temperatures, but can readily be observed in H:He or H:D plasma for longer discharges (>50ms) at higher densities (> $10^{18}m^{-3}$). A reduction in low-Z impurities and an even greater reduction in heavy impurities is seen near $\iota=1$, when density drops, but temperature increases to about 50eV.

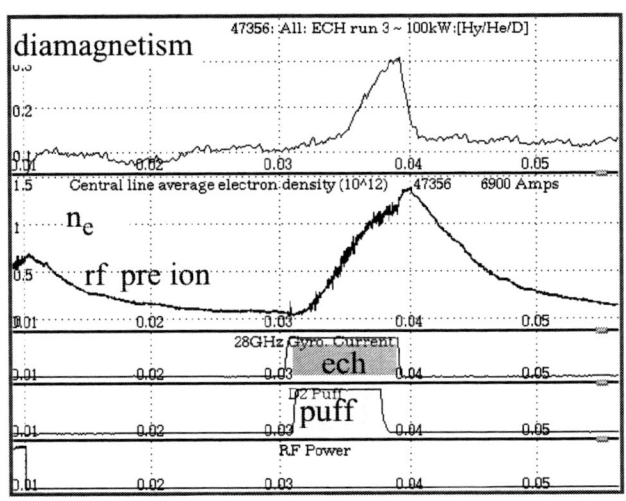

FIGURE 3. EC heating in H-1, B~0.5T, H:D 1:1, >100kW ECH.

Electron cyclotron heating: Close to full power from a 200kW 28GHz gyrotron ($2\omega_{CE}$ at 0.5T) was applied through a simple open-ended helical waveguide launcher. With a 10ms pulse, and rf preionization of $\sim 1\times 10^{17} m^{-3}$, a diamagnetic temperature of 100-200eV was observed provided gas feed was carefully controlled ($p < 2\times 10^{-6}$ Torr). A highly localised ionization rate was observed in the emission from argon doping, and at higher gas fills, a peak density in excess of 3×10^{18} was obtained, with a lower temperature. Impurity levels, estimated by comparative spectral line intensities, were lower than in the rf discharges.

Configuration studies

In RF-produced plasma at 0.5T, electron density depends in detail on magnetic configuration as shown in Figure 4. There are broad regions of low or zero density when central ι is near $\iota_0 \sim 5/4$ and $4/3$, and other narrower drops in density less clearly identifiable with rational iota (e.g. k_h=0.58, Fig. 4). A careful comparison with ι profiles suggests that the presence of a lower order rational at the edge ($\iota_a = 7/5$) in the region $\kappa_h \sim 0.7$ (associated above with $\iota_0 \sim 4/3$) may instead be the dominant factor. It is possible that in addition to indicating particle confinement times, this phenomenon may be sensitive to plasma generation efficiency. ECH also produced no density at k_h = 0.75, but as under present conditions, ECH is dependent on rf pre-ionization, this measurement will be repeated with a independent pre-ionization source.

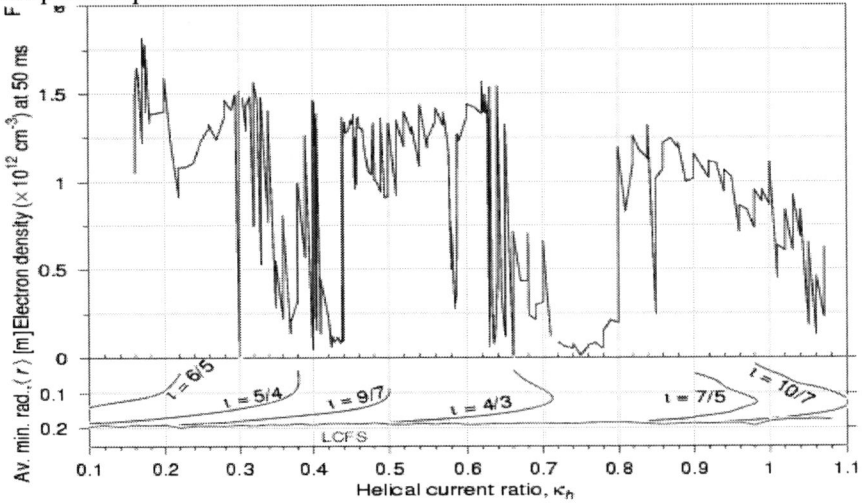

FIGURE 4. Dependence of electron density on helical conductor current ratio k_h for H:He plasma at 0.5T. The rotational transform is shown underneath as a function of average minor radius $\langle r \rangle$

Magnetic fluctuations are seen at a level $\Delta B/B \sim 2\times 10^{-5}$ at frequencies in the range of at least 20-120kHz. These consist of broadband and coherent components, with (low 1-8) poloidal mode numbers often related to the denominator of lower order rational ɩ. A larger array of Mirnov coils (20) has been installed to remove ambiguities. Changes in fluctuation spectra may often be related to changes in the plasma density.

Innovative Diagnostics

Langmuir probe arrays are useful in low field, low temperature plasma. Probe studies of the dynamics of confinement transitions provide data not otherwise obtainable[3]. In argon at ~0.1T a combination of triple probes and a radial Mach probe allow the electron and ion components of the fluctuation induced flux to be separated. It has been shown that these are unequal under conditions with large drift-type fluctuations and poor confinement, violating ambipolarity[4]. The implied radial currents are consistent with electric field changes that occur when the plasma makes a transition to a higher confinement mode. Similar studies of flows and cross-spectra have led to the identification of self-regulation by zonal flows[5] in H-1.

"Correlation Spectroscopy" has been demonstrated to provide an estimate of plasma poloidal velocity based on the velocity at which fluctuations move across the field of view of an array of spectrally (visible) selective detectors. This data is in agreement with data from poloidal Mach probes. Once proven in low temperature plasma, this technique will be applied to helium and impurity emission from high temperature plasma.

A range of applications of high throughput, high resolution, imaging coherence Doppler spectroscopy (MOSS[6]) have been demonstrated for application to high temperature plasma. The ratio of hydrogen isotope densities in icrf heating at 0.5T has been estimated from ratios of the D and H Balmer β line intensities. The results show the evolution in space and time – both isotopes change in time from central to more edge-peaked emission, and the hydrogen shows a stronger source effect at the plasma edge.

A study at low fields (0.05-0.1T) with different coherence lengths revealed some anisotropy and non-Maxwellian nature in argon ion distribution functions. Anisotropy decreased with increased ion-ion collisionality (lower Ti). There are suggestions of edge-cooling by Ar^+-Ar^0 charge exchange.

CONCLUSION

We have demonstrated the flexible heliac properties of the H-1NF by a detailed study of particle confinement and magnetic fluctuations as a function of rotational transform, yielding a rich, detailed spectrum of behavior. The combination of multiple probes, probe arrays and flexibility of plasma parameters provides a powerful tool for investigation of basic phenomena of toroidal plasma physics: turbulence, (zonal) flows, and radial force balance.

Future studies include higher power RF and ECH, directional gas-puffing, modulated gas puffing for transport analysis and a soft-Xray camera. Both ECH and RF operation shows promise at higher magnetic fields, and progression to regular high temperature operation will see MOSS, line-ratio and correlation spectroscopy alternatives to probe measurements, and a LIF system for direct measurement of radial electric field.

ACKNOWLEDGMENTS

The authors gratefully acknowledge support of the H-1 technical team, G.C Davies, J. Wach, R.J. Kimlin, R. Davies and C. Costa, and the financial support of the Australian Government through the Institute of Advanced Studies ANU, DISR, and the Australian Institute for Nuclear Science and Engineering (AINSE).

REFERENCES

1. Harris, J.H., Cantrell, J.L. Hender, T.C., Carreras, B.A. and Morris, R.N. *Nucl. Fusion* **25**, 623 (1985).
2. Boswell, R.W. *Plasma Physics and Controlled Fusion.* **26**, 1147 (1984).
3. Rudakov, D.L., Shats, M.G., Harris, J.H. and Blackwell, B.D., *Plasma Physics and Controlled Fusion.* **43**, 559 (2001).
4. Solomon, W. M. and Shats, M.G. *Phys. Rev. Lett.* **87** (2001):195003
5. Shats, M. G., Solomon W. M. *Phys. Rev. Lett.* **88** (2002):045001
6. Howard, J., Michael, C., Glass, F. and Cheetham, A. *Rev. Sci. Instr.* **72**, 888-897 (2001)

Transport Properties in the TJ-II Flexible Heliac

F. Castejón, E. Ascasíbar, C. Alejaldre, J. Alonso, L. Almoguera, A. Baciero, R. Balbín, E. Blanco, M. Blaumoser, J. Botija, B. Brañas, A. Cappa, R. Carrasco, J. R. Cepero, A.A. Chmyga[1], J. Doncel, N.B Dreval[1], S. Eguilior, L. Eliseev[2], T. Estrada, O. Fedyanin[4], A. Fernández, C. Fuentes, A. García, I. García-Cortés, B.Gonçalves[3], J. Guasp, J. Herranz, A. Hidalgo, C. Hidalgo, J. A. Jiménez, I. Kirpitchev, S.M. Khrebtov[1], A.D. Komarov[1], A.S. Kozachok[1], L. Krupnik[1], F. Lapayese, K. Likin, M. Liniers, D. López-Bruna, A. López-Fraguas, J. López-Rázola, A López-Sánchez, E. de la Luna, A. Malaquias[3], R. Martín, M. Medrano, A.V.Melnikov[2], P. Méndez, K.J. McCarthy, F. Medina, B. van Milligen, I. S. Nedzelskiy[3], M. Ochando, L. Pacios, I. Pastor, M.A. Pedrosa, A. de la Peña, A. Petrov[4], A. Portas, J. Romero, L. Rodríguez-Rodrigo, A. Salas, E. Sánchez, J. Sánchez, K. Sarksian[4], S. Schchepetov[4], N. Skvortsova[4], F. Tabarés, D. Tafalla, V. Tribaldos, C.F.A. Varandas[3], J. Vega and B. Zurro

Laboratorio Nacional de Fusion. Asoc. Euratom/Ciemat para Fusión .Av. Complutense, 22. 28040, Madrid (Spain)
(1) Institute of Plasma Physics, NSC KIPT, 310108 Kharkov, Ukraine
(2) Institute of Nuclear Fusion, RNC Kurchatov Institute, Moscow, Russia
(3) Associação EURATOM/IST, Centro de Fusão Nuclear, 1049-001 Lisboa, Portugal
(4) General Physics Institute, Russian Academy of Sciences, Moscow, Russia

Abstract. TJ-II flexibility is exploited to perform an investigation on the transport properties of this device. Rotational transform can be varied in a wide range, which allows one to introduce low order rationals and to study their effect on transport. On the other hand, confinement properties can be studied at very different rotational transform values and for different values of magnetic shear: Experiments on influence of the magnetic shear on confinement are reported. Plasma potential profiles have been recently measured in some configurations up to the plasma core with the Heavy Ion Beam Probe (HIBP) diagnostic and the electric field values measured in low-density plasmas are consistent with neoclassical calculations near the plasma core. Plasma edge turbulent transport has been studied in configurations that are marginally stable due to decreased magnetic well. Results show a dynamical coupling between gradients and turbulent transport.

INTRODUCTION

TJ-II is a medium size (R=1.5 m, a<0.22 m, l=1, M=4) helical axis stellarator[1], which main characteristic is its flexibility. A wide range of rotational transform, ι, values can be achieved ($0.9 \leq \iota(0)/2\pi \leq 2.2$), that permits to study the dependence of confinement on ι and to study the effect of low order rational values on transport. Moreover, the rotational transform profile can be varied inducing moderate OH current in the plasma, that has the effect that magnetic shear can be varied from the shearless vacuum value to positive and negative ones. This capability has been used to explore the effect of magnetic shear on confinement.

The magnetic well, which is the main stabilizing mechanism in TJ-II, can also be varied keeping almost constant the rotational transform profile. The magnetic well can vary form values of 6% at the edge to be almost suppressed. This flexibility is used to destabilize the plasma and study the transport in these conditions.

TJ-II is now equipped with a Heavy Ion Beam Probe (HIBP) diagnostic that can measure plasma potential profiles in all the TJ-II configurations. It is admitted that electric field plays a key role on transport and confinement, therefore, such a diagnostic together with the commented flexibility mechanisms allows one to perform clarifying experiments on TJ-II stellarator plasmas. This paper is organized as follows: section 2 is devoted to the effect of low order rationals on electric field and transport. The influence of magnetic shear on confinement is shown in Sect. 3 and the results of magnetic well scan are shown in Sect. 4. The conclusions are drawn in every section.

ELECTRIC FIELD, LOW ORDER RATIONALS AND TRANSPORT

HIBP Results. Comparison with Neoclassical Predictions

Two detectors for the secondary ions are used simultaneously in the HIBP system of TJ-II: a 30° Proca-Green electrostatic energy analyzer and a multiple cell array detector (MCAD)[2]. During operation with electrostatic energy analyzer the sample (ionization) volume position is controlled by changing the entrance angle of the primary beam to the plasma, using electrostatic sweep plates. The operation with two detectors allows enlarging the number of the sample volumes inside the plasma to obtain plasma profiles and their fluctuations. A series of plasma potential radial profiles is shown in Figure 1 (left panel) for several plasma densities. It has been found that bulk plasma potential decreases as plasma density increases. The positive values of radial electric field measured in low-density plasmas agree qualitatively with neoclassical estimations[3] in the plasma center, as shown in Figure 1 (right panel). The level of fluctuations in the plasma potential profile increases towards the plasma boundary region. However, in some cases, localized turbulent bursts showing a quasi-coherent mode (30-40 kHz) have been observed in the plasma core region. This turbulence might be related to existence of rational surfaces as suggested by the correlation that has been found between the HIBP and the Mirnov coils signals. Morevoer, the root mean square (r.m.s.) value of fluctuations in the secondary ion current increases in the vicinity of the radial location where the maximum correlation Φ_{HIBP}-V_{Mirnov} is found.

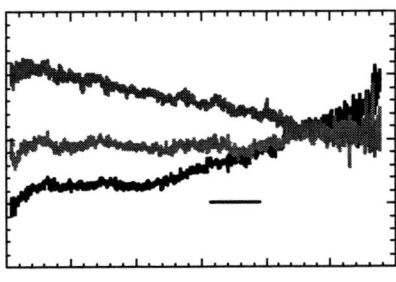
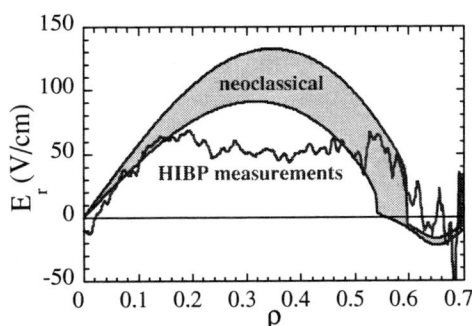

FIGURE 1. Left panel: Measured plasma potential profiles for several plasma densities. Right panel: Measured electric field for the low density case, compared with neoclassical estimations.

Influence of Low Order Rationals on Heat Transport in TJ-II Central Plasma

Electron Cyclotron Resonance Heating, ECRH, in low-density plasmas gives rise to a set of characteristic features commonly observed in several stellarators [4,5,6]. Perpendicular ECRH produces an enhanced outward flux of the ripple trapped electrons that are pushed into the loss cone and, as a consequence, a positive radial electric field builds up to reduce the outward electron flux and maintain ambipolarity. This radial electric field may be perturbed by the existence of rational surfaces in the plasma region [6,7] and, because of this fact, the plasma transport properties will change. The steep pressure gradient observed in the Enhanced Confinement Regime (EHC) regime previously observed in TJ-II can be even increased by positioning a low order resonance near the plasma center. This point has been investigated by inducing a current using the OH transformer. In this way, the rational surface 3/2, that is absent in the vacuum configuration, has been positioned overlapping with the power deposition area. This hypothesis has been confirmed by VMEC calculations. Figure 2 (left panel) shows temporal traces of ECE channels, line density and plasma current. It is observed that for a given value of plasma current (about -1.5 kA) an improvement of heat confinement occurs. This transition to enhanced heat confinement can be observed in the clear rise of the central ECE channels. Figure 2 (right panel) shows the vacuum ι profile and the electron temperature profiles before and after the transition. It can be seen a clear increase of the temperature gradient in the core.

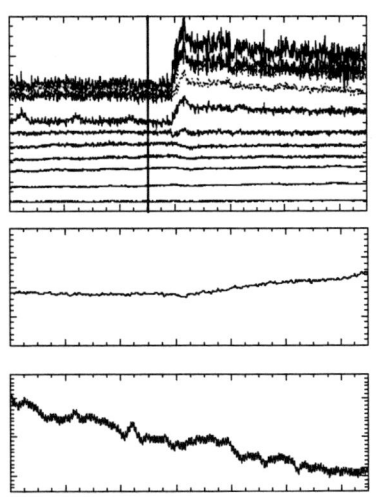 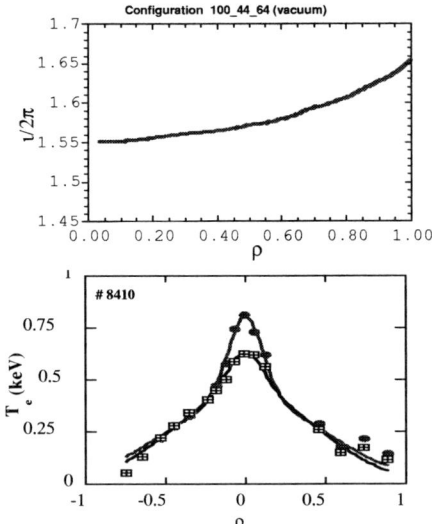

FIGURE 2. Left panel: From top to bottom: time evolution of ECE signals, showing a clear transition, line density, and plasma current. Right panel: From top to bottom: Vacuum rotational transform profile for this TJ-II configuration and electron temperature profile before (squares) and after (circles) the transition.

INFLUENCE OF MAGNETIC SHEAR ON CONFINEMENT

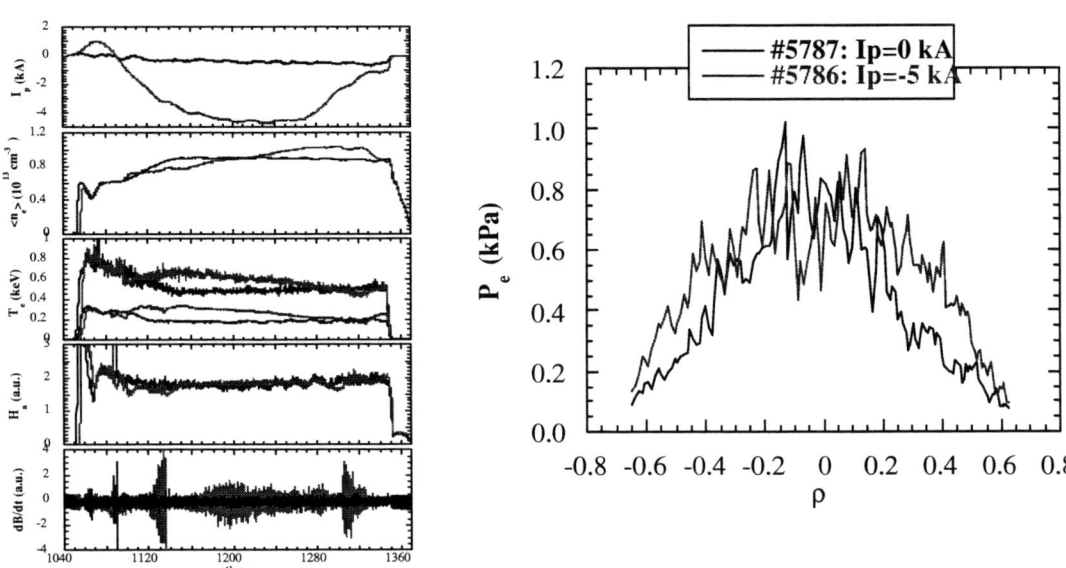

FIGURE 3. Left panel: From top to bottom: time evolution of plasma current, line density two ECE channels, Hα signal and Mirnov coil signals, showing an improvement of confinement in the discharge with negative OH induced current. Right panel: Pressure profiles for the same both discharges as before, showing a clear improvement of confinement in the discharge with negative current.

It has been observed in W7-AS stellarator[8], that a moderate shear imposed by induced toroidal current produce a confinement improvement, independent of the sign of the magnetic shear. In order to investigate this issue in TJ-II a series of experiments has been performed in which profiles are swept with the aid of the transformer at lower rate than the plasma particle (~10 ms) and energy confinement time (~3 ms)[9]. Figure 3 (left panel) shows temporal traces of two discharges with the same magnetic configuration $\iota/2\pi(a) = 1.68$. One is a reference discharge, with no additional OH current. In the second one, a negative plasma current up to about -5 kA is induced. The discharge with OH-induced plasma current shows a clear increase in electron temperature (as can be seen in the ECE channels). The symmetric features observed in the magnetic fluctuations temporal trace of the OH-discharge can be interpreted as the two intersections (first "downwards" and then "upwards") of the rotational transform profile with the 3/2 resonant surface, as the magnetic configuration is swept by the plasma current. A remarkable broadening of the pressure profile can be observed in the case of negative plasma current, as can be seen in Figure 3 (right), where pressure profiles measured at the plasma flat-top (1200 ms) are shown in for both discharges. A positive current has been also induced and the results indicate that there is not comparable confinement improvement in this case, showing that anomalous transport is reduced only by negative shear in TJ-II in this range of plasma pressure.

MAGNETIC WELL SCAN

Three magnetic configurations with very similar ι profiles and different magnetic well values (2.4, 0.6 and 0.2%) have been investigated. Results show that the level of edge fluctuations and the degree of intermittency show a significant increase when magnetic well is reduced in TJ-II[10]. Results also prove that the turbulent E×B particle flux decreases as the well is increased. A significant fraction of the total E×B turbulent flux can be assigned to large and sporadic transport bursts whose amplitude increase as well depth is decreased. This bursty behavior of turbulent transport is strongly coupled with fluctuations in density gradients: As the density gradient increases above the most probable value, the E×B turbulent driven transport increases until the system relaxes back to the initial marginally stable situation, as shown in figure 4.

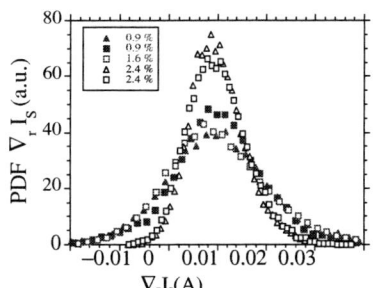 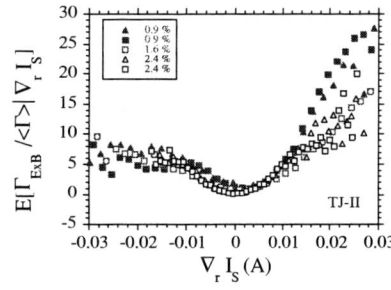

FIGURE 4. Left panel: Probability distribution function of the occurrence of a gradient. Right panel: the most probable flux associated to a given gradient.

REFERENCES

[1] C. Alejaldre et al. 1999 Plasma Phys. Control. Fusion **41** A539-A548
[2] I. S. Bondarenko et al., Chechoslovak Journal of Physics **50** (2000)1397
[3] V. Tribaldos et al., Physics of Plasmas **8** (2001) 1229
[4] A. Fujisawa et al. Phys. Rev. Lett. **81** (1998) 2256-2259
[5] H. Maassberg, et al. Phys. Plasmas **7** (2000)295-311
[6] F. Castejón et al. Nucl. Fusion **42** (2002) 271
[7] U. Stroth et al., Phys. Rev. Lett. **86** (2001) 5910
[8] R. Brakel et al. Plasma Phys. Control. Fusion **39** (1997) B273-B286
[9] J. A. Romero et al., submitted to Nuclear Fusion
[10] J. Castellano et al.,Phys Plasmas **9** (2002) 713

Effects of the New Island Divertor on the Plasma Performance in the W7-AS Stellarator

P. Grigull, K. McCormick, , J. Baldzuhn, R. Burhenn, R. Brakel, H. Ehmler,
Y. Feng, F. Gadelmeier, L. Giannone, D. Hartmann, D. Hildebrandt, M. Hirsch,
R. Jaenicke, J. Kisslinger, T. Klinger, J. Knauer, R. König, D. Naujoks,
H. Niedermeyer, E. Pasch, N. Ramasubramanian, F. Sardei, F. Wagner, A. Weller,
and the W7-AS Team

Max-Planck-Institut für Plasmaphysik, EURATOM Ass., D-85748 Garching, Germany

Abstract. The island divertor in the W7-AS stellarator enables access to a new NBI-heated, high density operating regime with promising confinement properties. This regime – the High Density H-Mode – displays no evident mode activity, is extant above a threshold density and characterized by flat density profiles, high energy- and low impurity-confinement times and edge localized radiation. Impurity accumulation, normally associated with ELM-free H-modes, is avoided. Quasi steady-state discharges with $\bar{n}_e$ up to $4 \ 10^{20}$ m^{-3}, edge radiation levels up to 90%, and partial plasma detachment at the divertor targets can be simultaneously realized.

INTRODUCTION

The stellarator line of fusion research unquestionably has to incorporate divertors with plasma exhaust capabilities similar to those of tokamak poloidal field divertors. However, in contrast to the extensive experience gained with tokamak divertors over the last decade, related research on stellarators is at the very beginning. An option is the so-called island divertor, where flux diversion is realized through inherent or externally induced magnetic islands at the edge. This paper reports on advances associated with the first-ever realization of an island divertor on the W7-AS stellarator. The arrangement is similar in major aspects to that foreseen for the W7-X stellarator ($R = 5.5$ m, $a_{eff} \approx 0.5$ m) now under construction [1] and serves as a test bed. W7-AS ($R = 2$ m, $a \leq 0.16$ m, $B \leq 2.5$ T) is a modular, low-shear stellarator with five magnetic field periods. Per period, the plasma shape varies from a standing ellipse to a triangle and back again. Depending on the adjustable rotational transform, the plasma is bounded either by smooth flux surfaces or by a separatrix formed from naturally occurring magnetic islands at values of the edge rotational transform $t_a = 5/m$ ($m=8, 9,10...$). The divertor is optimized for $t_a = 5/9$, but works as well for $t_a = 5/8$ or $5/10$. It consists of five (top-bottom) module pairs arranged at the elliptical cross sections, Fig.1a. Each module is composed of a 3D shaped, inertially cooled target (CFC) intersecting the islands, and of baffles (isotropic graphite). The sub-divertor chambers are equipped with titanium evaporators for gettering of neutrals (not yet activated). For further details see Refs. [2, 3]. The new divertor has generally improved the impurity and recycling behaviour in W7-AS (reduced core radiation, improved density control) and enables, in particular, access to an exciting new confinement regime with NBI heating at very high density [2, 4, 5].

HIGH DENSITY H-MODE

Previously on W7-AS under limiter conditions, it was impossible to produce high-power, high density, quasi-stationary NBI discharges. Since both particle and impurity confinement times usually increased with density, the discharges tended to evince impurity accumulation, lack of density control and subsequent radiation collapse. This behaviour has been dramatically changed after installation of the new divertor, especially at high density. Below a certain threshold line-averaged density $\bar{n}_e^{thr}$, density control is in fact improved compared to limiter operation, but impurities accumulate, and stored energies typically develop after an initial maximum to normal levels (normal confinement, NC), Fig.1b. The edge dynamics typically develops from an initial ELMy over a short quiescent phase towards a 'grassy' state as seen in the respective H_a trace. Above $\bar{n}_e^{thr}$, impurity accumulation vanishes, and high-power, high-energy discharges can be quasi steadily maintained over many confinement times, including also discharges with partial detachment from the divertor targets. At present, the pulse length is only technically limited to about 1.2 s by the NBI performance. Since this regime shows certain similarities to the quiescent H-mode (H*, see below) previously observed in W7-AS, it is termed High Density H-mode (HDH).

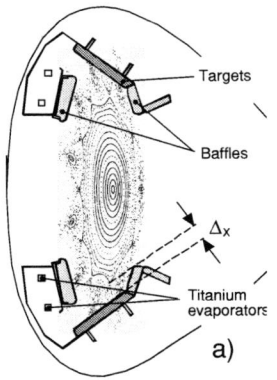
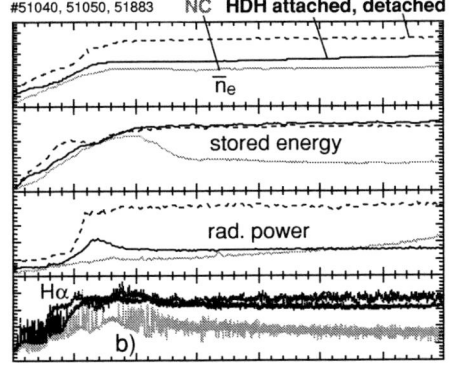

FIGURE 1.
a) Elliptical plasma cross section with island divertor modules. The targets are 3D shaped. The toroidal lengths of each module is 0.7 m. Δ_x indicates the minimum distance between x-points and targets.
b) Examples for NBI (H°→H⁺) heated normal confinement (NC) and high density H-mode (HDH) discharges. Standard divertor configuration (SDC, ι_a = 5/9, $a \approx$ 12 cm, Δ_x = 3.8 cm), absorbed NBI power P_{abs} = 1.4 MW.

Figure 2 shows energy confinement times τ_E, impurity confinement times τ_{imp} (from laser blow-off injection of Al), and radiated power fractions $f_r = P_r/P_{abs}$ (with P_r and P_{abs} being the radiated power and the absorbed NBI power, respectively) obtained from quasi steady-state discharges at three different NBI powers. At the NC → HDH transition, τ_E jumps to values exceeding the International Stellarator Scaling ISS95 [6] by a factor of about two, whereas τ_{imp} drops to values close to τ_E. The threshold density $\bar{n}_e^{thr}$ increases with heating power. Radiated power fractions are low to moderate, but reach up to about 90% in detached discharges.

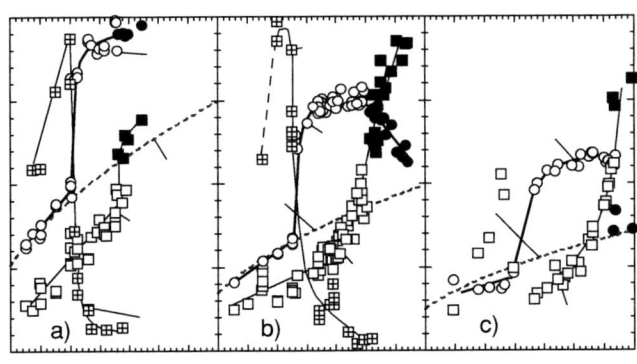

FIGURE 2. Energy confinement times τ_E compared with the ISS95 scaling, impurity confinement times τ_{Al} (from laser blow-off injection of aluminum), and radiated power fractions f_r (from bolometry) as functions of the line-averaged density $\bar{n}_e$. SDC, NBI (H°→H⁺).
a) P_{abs} = 0.7 MW, b) P_{abs} = 1.4 MW, and c) P_{abs} = 2.45 MW. Solid symbols indicate detached discharges. τ_{Al} values at the highest power have not yet been measured.

$\bar{n}_e$ /10²⁰ m⁻³

Radiative losses from NC discharges are dominated by core radiation from higher ionization states of iron, oxygen and chlorine. Consistent with long impurity confinement times, the radiative losses always increase with time even at constant density as demonstrated by radial profiles of the radiated power density in Fig.3a. In contrast, the radiation in HDH discharges originates mainly from lower ionization states of carbon and oxygen and emanates always from the edge. At stronger detachment, the edge radiation becomes increasingly asymmetric. Concomitant with the NC → HDH transition, the n_e radial profiles typically change from centre peaked to flat with steep gradients at the edge, which indicates that edge refueling exceeds central fueling by the beams. The shape of the T_e profiles is not significantly changed, Figs.3b,c. There is some uncertainty as to the r_{eff} scale in Fig.4 since reliable equilibria for this particular configuration with large islands at the edge are not yet available, but this should not crucially influence the profile shapes. Based on such profiles, an analysis of the spatiotemporal behaviour of highly ionized states of laser-ablated aluminum [7] yield nearly identical impurity diffusion coefficients in the core for both NC and HDH discharges, but a reduction of the convective inward velocity v_{in} by a factor of about five in HDH regimes, thus relating the supression of impurity peaking during HDH to reduced inward pinching. Except for the existence of a threshold density increasing with power, the global features of the HDH-mode show some similarities with the EDA-mode found in the Alcator C-Mod tokamak [8].

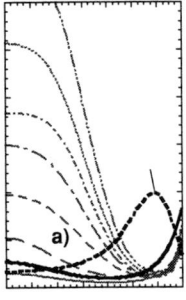 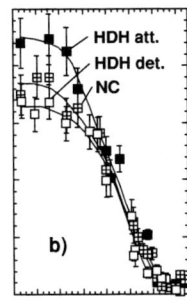 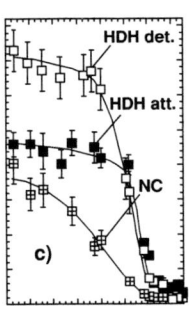

FIGURE 3. SDC, NBI (H°→H⁺) discharges with P_{abs}=1.4 MW.
a) Radiated power profiles from bolometry. Temporal evolution for a NC discharge (black lines) at constant $\bar{n}_e \approx 1.7\ 10^{20}$ m^{-3}, and profiles from quasi steady-state HDH discharges (coloured lines) at $\bar{n}_e = 2.4\ 10^{20}$ m^{-3} (attached) and $3.2\ 10^{20}$ m^{-3} (detached).
b) and **c)** Typical electron temperature T_e and density n_e radial profiles (from Thomson scattering), respectively, for NC and HDH discharges. Note the strong n_e profile widening in the HDH regime.

Similarities and differences between the HDH and H* modes studied so far are elucidated in Fig.4, which displays characteristics of two discharges in a modified field configuration where both regimes can be established in a rather clear fashion. Discharge #56147 runs from an ELMy regime (density ramp) through an H* phase (t > 0.4 s, constant density) into radiation collapse, whereas discharge #56156 is driven by a strong gas puff from H* to the HDH regime (attached) leading to drastically reduced core radiation (Fig.4b) and quasistationarity.

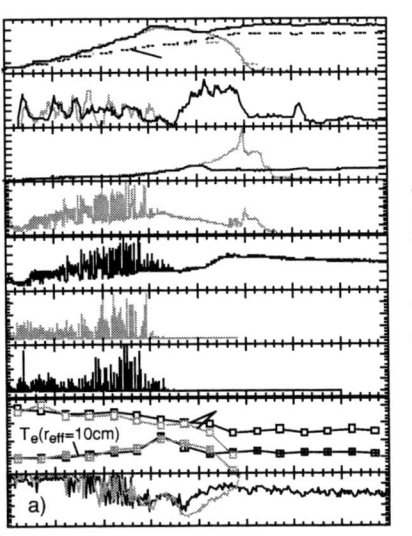 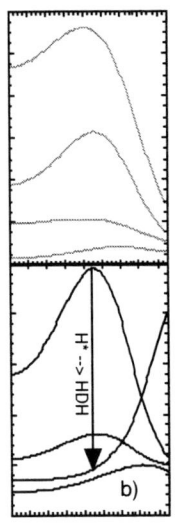

FIGURE 4. Comparison of H*- and HDH-mode features. NBI (H°→H⁺) discharges with P_{abs} = 1.1 MW, t_a = 5/9, $a \approx 14$ cm, Δ_x = 0.3 cm.
a) Parameters versus time. Grey lines: data from a discharge running through an ELMy phase (density ramp) into the H*-mode (density plateau, nearly vanishing edge dynamics and magnetic activity) and subsequent radiation collapse. Black lines: a discharge which runs through the same cycle, but avoids radiation collapse by further increasing the density from the H*-mode level to above the HDH mode threshold density. The radial E-field was obtained from passive BIV spectroscopy at $r_{eff} \approx 11.3$ cm.
b) Profiles of the radiated power indicating impurity accumulation during H* phases and a dramatic reduction of the core radiation after entering the HDH-mode (0.6 s in Fig.b).

In both regimes, H* and HDH, the edge localized dynamics as well as the magnetic activity is reduced to nearly zero. Corresponding Mirnov spectra (≤ 125 kHz) do not show a clear and specific difference and, in particular, an edge localized, coherent mode at about 100 kHz (characteristic for the EDA-mode in Alcator C-Mod) is not observed in HDH regimes. Apart from the absolute values, the H* and HDH radial density profiles are also quite similar: in both cases they are flat over most of the radius. Specific features of the HDH-mode compared to H*-scenarios are: higher threshold densities, lower radial E-fields – at least at the position where it could be measured– and the absence of impurity accumulation. Pedestal temperatures (T_e at r_{eff} = 10 cm in the example shown) seem to be generally lower in HDH than in H* scenarios. Detailed conclusions on the n_e and T_e values in the gradient region are limited by the 2 cm radial resolution of the YAG laser Thomson system. The combination with edge-specific diagnostics from different toroidal locations (edge ruby laser system, Li-beam) suffers of uncertainties due to magnetic mapping.

The energy and particle deposition zones on the divertor targets (from thermography, target-integrated Langmuir probe arrays and H_α camera data, not shown) do not simply reflect the island geometry, but are strongly influenced by *ExB* drift effects [3]. Detachment is always partial meaning that it extends over most of the target area except a small region where the plasma stays attached (downstream $T_e > 10$ eV) even at the highest densities. There are indications that this is due to an inside/outside asymmetry of T_e at the upstream separatrix position, which is finally caused by the specific field geometry at the edge [3]. Nevertheless, compared with fully attached scenarios, the peak heat load even on this particular target region is reduced at partial detachment by factors of up to five. The asymmetry probably prevents shrinking of the hot plasma cross section and supports the stabilization of detachment regimes.

The HDH regime could be established in a variety of configurations at t_a = 5/8, 5/9 and 5/10 with different plasma radii and separations Δ_x between x-points and targets from zero up to about 4 cm (for examples see Ref. [9]). The mode is clearly observed also in deuterium plasmas ($D^o \rightarrow D^+$ injection), but the NC $\rightarrow$ HDH transition is more continuos than in hydrogen plasmas [5]. Due to uncertainties in the deposition profiles, it is not yet clear to what degree this reflects a difference in transport. Stable partial detachment extends the accessible density range towards higher values, but is restricted to configurations with $\Delta_x > 2$ cm.

SUMMARY AND CONCLUSIONS

The island divertor enables access to a new NBI heated, high density operating regime on W7-AS with high energy- and low impurity confinement times (HDH regime). The regime shows some similarity with the quiescent H-mode (H*), but avoids impurity accumulation. It occurs above a certain threshold density, increasing with heating power, and allows full density control – already without Ti gettering – and stable quasi steady-state operation, also under conditions of partial detachment from the divertor targets. Radiation is always edge-localized; radiated power fractions are low to moderate in attached regimes and reach up to about 90% in detachment scenarios. The avoidance of central impurity peaking is related to a reduced inward pinch for impurities. The physics background is, however, not yet clear, but flat density profiles requiring an adequate balance between central fueling by the beams and edge density build up by recycling and gas puffing seem to be at least one of the key ingredients. Nevertheless, similar profiles are found also in the H*-mode, thus indicating that they are a necessary but not a sufficient pre-condition. A further important aspect concerning the universality of the HDH-mode is the role of the specific collisionality range accessible with NBI in W7-AS (strongly collisional impurities, central ν^* of the background plasma $\approx$ 0.1-0.3). An improved understanding is expected from a quantitative analysis of the balance between inwards and outwards driving terms, which is under way, and from future experiments on W7-X. Divertor detachment is always partial in the sense that it extends over most of the target area except a small part where the plasma stays attached, but at reduced power load. This asymmetry is ascribed to an inside/outside temperature difference at the upstream separatrix position which probably helps to stabilize such scenarios. The HDH regime is robust against changes of the magnetic field configuration, whereas stable partial detachment is restricted to typical divertor configurations with sufficiently large separations between x-points and targets (large islands).

REFERENCES

[1] Renner, H. *et al.*, *Nucl. Fusion* **40**, 1083 (2000).
[2] Grigull, P. *et al.*, *Plasma Phys. Control. Fusion* **43**, A175-A193 (2001)
[3] Grigull, P. *et al.*, Proc. 15th Int. Conf. on Plasma Surface Interaction in Fusion Devices, Gifu, Japan, 2002, to be published in *J. Nucl. Mater.*
[4] McCormick, K. *et al.*, *PRL* **89**, 015001 (2002).
[5] McCormick, K. *et al.*, Proc. 15th Int. Conf. on Plasma Surface Interaction in Fusion Devices, Gifu, Japan, 2002, to be published in *J. Nucl. Mater.*
[6] Stroth, U. *et al.*, *Nucl. Fusion* **36**, 1063 (1996).
[7] Burhenn, R. *et al.*, 29th Europ. Conf. on Contr. Fusion and Plasma Physics, Montreux, Switzerland, 2002, *Europhys. Conf. Abstracts* P4.03
[8] Greenwald, M. *et al.*, *Phys. Plasmas* **6**, 1943 (1999); Greenwald, M. *et al.*, *Plasma Phys. Control. Fusion* **42**, A263 (2000); Snipes, J. A. *et al.*, *ibid* **43**, L23 (2001).
[9] Grigull, P. *et al.*, Proc. 12th Int. Toki Conf. on Plasma Physics and Controlled Nuclear Fusion, Toki City, Japan, 2001, accepted for publication in *J. Plasma Fusion Res.*

Excitation of Alfvén Waves in Mirror Plasmas with a Strong Temperature Anisotropy

M. Ichimura, H. Higaki, M. Nakamura, S. Saosaki, H. Kano, S. Kakimoto, K. Horinouchi, Y. Yamaguchi, H. Hojo, Y. Nakashima, T. Watanabe[1], K. Yatsu

Plasma Research Center, University of Tsukuba, Tsukuba, Ibaraki 305-8577, Japan
[1]*National Institute for Fusion Science, Toki, Gifu 509-5292, Japan*

Abstract. In the GAMMA 10 tandem mirror, fAlfvén ion cyclotron modes are spontaneously excited due to the strong temperature anisotropy. The plasma parameters needed for the excitation are explained by the theoretical prediction with the effective mirror ratio. It is suggested that the power of ICRF waves applied externally branches off spontaneously excited Alfvén eigenmodes directly.

INTRODUCTION

In fusion-oriented devices with a toroidal configuration, Alfvén eigenmodes become unstable, for example, a toroidicity-induced Alfvén eigenmode (TAE) [1] and so on. On the while, in mirror plasmas which have the configuration without magnetic shear, a global Alfvén eigenmode (GAE) [2] is predicted to be unstable. These spontaneously excited eigenmodes will have an influence on the particle transport and confinement. Those modes have frequencies in the range of MHD modes and much less than the ion cyclotron frequencies. In the GAMMA 10 tandem mirror, the ion cyclotron range of frequency (ICRF) heating is used not only for producing an initial plasma but for sustaining MHD stability and heating of central cell ions. An hot-ion mode of operation [3] has been realized and plasmas with a strong temperature anisotropy (which is defined as the temperature ratio of perpendicular to parallel to the magnetic field line) are formed. In the typical discharges, Alfvén ion cyclotron (AIC) modes are spontaneously excited due to the strong temperature anisotropy [4]. The AIC mode is one of the micro-instabilities and is formed as an eigenmode in the axial direction [5]. The conditions for the AIC mode excitation in the mirror plasmas have been predicted as $\beta A^2 = 3.52$, where β is the ratio of the plasma pressure to magnetic field strength and A is the anisotropy, by a theory [6]. In previous paper [4], it has been reported the excitation condition observed experimentally was almost one order of magnitude smaller than that of the theoretical prediction. When an effective mirror ratio is introduced to the theoretical calculation, it becomes clear the experimental condition agrees with the theoretical one.

In GAMMA 10, the effects of the AIC modes to plasma parameters; for electrons [7], for bulk ions [8] and for high energy ions [9], are observed. When the RF2 power is increased, the saturation of the diamagnetism is sometimes observed in the experiments. The parametric mode coupling among RF2, the AIC modes and low frequency Alfvén waves is discussed [10]. In this paper, the parameter dependence of the low frequency fluctuations are studied.

EXPERIMENTAL SETUP

For the initial plasma production (RF1), so-called Nagoya Type-III antennas are used in combination with the hydrogen gas puffing in the central cell. Fast Alfvén waves are excited in the central cell and propagates to the anchor cell [11] The MHD stability of GAMMA 10 is kept by the averaged minimum-B configuration due to the

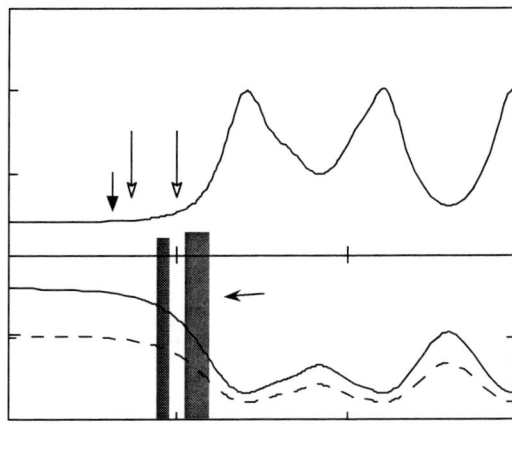

FIGURE 1. (a) Magnetic field profile and locations of ICRF antennas and diagnostics (b) Applied frequency normalized by the local ion cyclotron frequency.

anchor heating. Another ICRF source (RF2) with conventional double half-turn (DHT) antennas is also applied in the central cell for the main plasma heating. The ion temperature becomes several keV in the hot-ion mode, where a relation of $T_i/T_e > (m_i/m_e)^{1/3}$ is hold [3]. The magnetic field profile and the location of antennas and diagnostics are shown in Fig. 1. The applied frequencies of RF1 and RF2 normalized by the local ion cyclotron frequencies are indicated in Fig. 1(b). As predicted in Fig. 1, the fundamental cyclotron resonance layer for RF2 exists near the midplane of the central cell. The radiated power of RF2 antennas is typically 100kW.

The typical plasma parameters are the density of 2×10^{18} m^{-3}, the ion temperature of 5 keV and the temperature anisotropy of more than 10. The temperature anisotropy is estimated from signals of the diamagnetic loop array indicated in the figure. Magnetic fluctuations generated in plasmas are mainly detected by using small pick-up coils with 2 mm in diameter. With a conventional fast Fourier transform (FFT) method, signals of magnetic probes are converted into the frequency spectrum. Magnetic probe arrays both in the axial- and azimuthal- directions are also used.

EXCITATION OF THE AIC MODES

The AIC modes excited in the central cell of GAMMA 10 have several discrete peaks as shown in Fig. 2. A frequency of 6.36MHz in the figure is an applied RF2 frequency. The frequency of the AIC mode is just below the ion cyclotron frequency at the midplane of the central cell. The spatial mode structures of each discrete peak in radial and azimuthal directions have been measured by magnetic probes and confirmed to be the same structure[5]. The AIC modes are excited as eigenmodes in the axial direction. The profiles of the excited modes are consistent

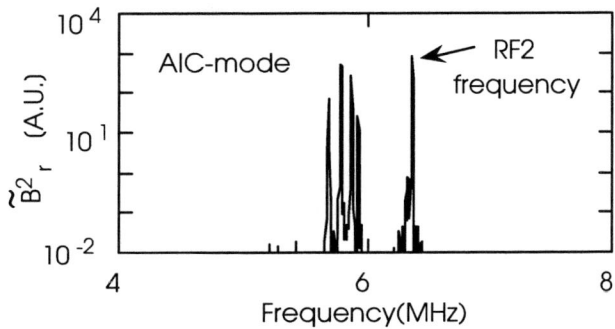

FIGURE 2. Frequency spectrum of the magnetic probe. The frequency of the AIC modes are just below the cyclotron frequency.

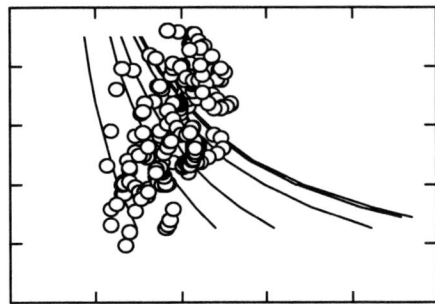 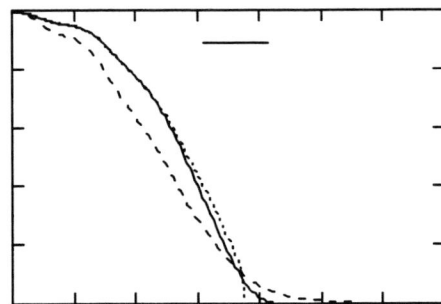

FIGURE 3. (a) Experimental data (open circles) with the AIC modes in $\beta_\perp$-A^2 space and the theoretical boundary (solid lines) between absolute and convective instabilities as a function of the mirror ratio R. (b) Estimated pressure profile from the diamagnetic loop array (solid line), and calculated profiles in cases of R=5 (chain line) and R=1.2 (dotted line).

with the waves in the shear Alfvén branch. The boundary between the absolute and convective instabilities has been evaluated in an uniform and an infinite plasma by the theory[6]. The dispersion relation of the AIC modes is written as Eq. (11) and the ion susceptibility is written as Eq. (13) in Ref.6. In the present conditions, that is, A>>1, the boundary is given as βA^2=3.52. In Fig. 3(a), experimental data are plotted in $\beta_\perp$-A^2 space. Open circles in the figure represent the experimental points with the AIC modes. The value $\beta_\perp$ is evaluated from $n(0)T_\perp(0)$, where $n(0)$ and $T_\perp(0)$ are the density and the perpendicular ion temperature on the axis, respectively. The solid curves drawn in the figure indicate the boundary between the absolute and convective instabilities as a function of the mirror ratio R. The mirror ratio, R, of the central cell is 5. In the case of R=∞, the boundary is the line of βA^2=3.52. As clearly shown in Fig. 3(a), the AIC modes in GAMMA 10 are observed experimentally in much smaller region of βA^2 than the theoretical prediction.

As previously predicted, the pressure profile in the axial direction is evaluated from the diamagnetic loop array. The evaluated profile is significantly localized near the midplane of the central cell as shown in Fig. 3(b)[12]. In GAMMA 10, the hydrogen gas puffing for the plasma sustainment is installed near the mirror throat of the central cell to avoid the charge exchange loss of the high energy ions near the midplane. The localized pressure profile will be due to the result of such the gas injection. The distribution function for ions which is used for the calculation of the dispersion relation holds a parameter of R. As shown in Fig. 3(b), the pressure profile obtained experimentally (solid line) can not be explained by the distribution function with R=5 (chain line). In the case of R=1.2 (a dotted line), the calculated profile agrees with the experimental observation. The boundary in the case of R=1.2 is also plotted in Fig. 3(a). The boundary is quite sensitive to the mirror ratio [13]. The experimental observation of the AIC mode excitation can be explained by introducing the effective mirror ratio to the distribution function for the theoretical calculation

EXCITATION OF LOW FREQUENCY MODES

When the RF2 power for the ion heating is increased, the diamagnetism increases and the anisotropy becomes strong. However, the saturation of the diamagnetism is sometimes observed in the experiments as indicated in Fig. 4. The excitation of the low frequency magnetic fluctuations has been detected in the experiments with high diamagnetism. In Fig. 5(a), the frequency spectrum of the magnetic probe which is used for the low frequency magnetic fluctuations in the range below 2 MHz is indicated. There are three peaks in the spectrum. In the range from 600 to 800 kHz, the fluctuations of which frequencies satisfy the relation of $f_{RF2} - f_{AIC}$ are observed. These modes are excited in the different parameter range from the AIC modes as shown in Fig. 5(b). The excitation of these low frequency modes are related to the saturation of the diamagnetism. The parametric mode coupling between RF2 and the AIC modes is a possible mechanism of the excitation of the low frequency fluctuations. The

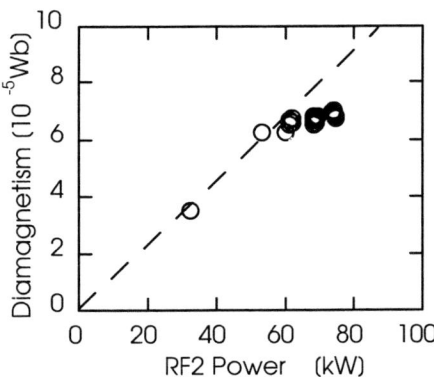

FIGURE 4. The RF2 power dependence of the diamagnetism.

power of RF2 is suggested to branch off the spontaneously excited Alfvén eigenmodes directly. Further studies to identify such a nonlinear mode coupling are needed from a view point of higher beta plasma formations.

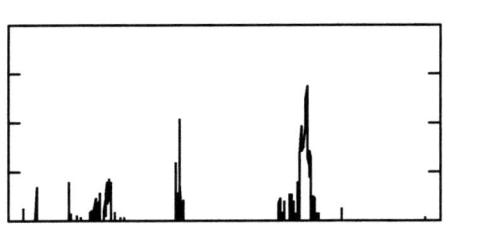 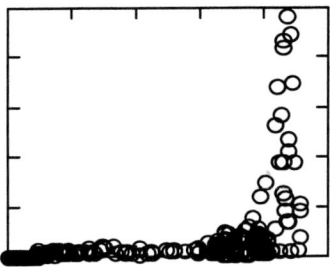

FIGURE 5. (a) Frequency spectrum of the magnetic probe for low frequency fluctuations, (b) Diamagnetism dependence of the amplitude of the fluctuations in the range from 600 to 800 kHz.

SUMMARY

In the central cell of the GAMMA 10 tandem mirror, plasmas with the strong temperature anisotropy are formed. The AIC modes are spontaneously excited due to the anisotropy. It becomes clear that the condition for the AIC mode excitation agrees with the theoretical prediction by introducing the effective mirror ratio into the distribution function.

The low frequency magnetic fluctuations of which frequencies are the difference between the frequencies of the applied ICRF and the AIC modes. It is suggested the power of RF2 branches off the spontaneously excited Alfvén eigenmodes directly due to the nonlinear wave coupling.

REFERENCES

[1] Cheng, C.Z. and Chance, M.S., Phys. Fluids, **29**, 3695 (1986).
[2] Li, Y.M., Mahajan, S.M., and Ross, D.W., Phys. Fluids, **30**, 1466 (1987).
[3] Tamano, T., et al., in *Plasma Phys. and Controlled Nuclear Fusion Research*, Vol.II, IAEA, p.399 (1995).
[4] Ichimura, M., et al., Phys. Rev. Lett. **70**, 2734 (1993).
[5] Katsumata, R., et al., Phys. Plasmas, **3**, 4489 (1996).
[6] Simith, G., Phys. Fluids, 27, 1499 (1984).

[7] Saito, T., et al., Phys. Rev. Lett., **82**, 1169 (1999).
[8] Ishii, K., et al., Phys. Rev. Lett., **83**, 3438 (1999).
[9] Ichimura, M., Nuclear Fusion, **39**, 1707 (1999).
[10] Higaki, H., et al., to be published in Fusion Science and Technology
[11] Ichimura, M., et al., Nuclear Fusion, **28** (1988) 799.
[12] Katsumata, R., et al., Jpn. J. Appl. Phys., **31**, 2249 (1992).
[13] Nakamura, M., et al., J. Plasma Fusion Res., **75**, 1211 (1999).

First HIBP Measurement of Plasma Potential During the H-Mode Transition on the TUMAN-3M Tokamak

L.G.Askinazi[1], A.A.Chmyga[2], N.B.Dreval[2], V.E.Golant[1], S.M.Khrebtov[2], A.S.Komarov[2], V.A.Kornev[1], L.I.Krupnik[2], S.V.Lebedev[1], G.Van Oost[3], E.A.Shevkin[1], M.Tendler[4], A.S.Tukachinsky[1], N.A.Zhubr[1]

[1] *Ioffe Institute, 194021, St.Petersburg, Russia*
[2] *Institute of Plasma Physics, NSC "KIPT", 61108 Kharkov, Ukraine*
[3] *Department of Applied Physics, Ghent University, Ghent, Belgium*
[4] *Alfven Laboratory, Fusion Plasma Physics, Royal Institute of Technology, Stockholm, Sweden*

Abstract. The difficulty of Heavy Ion Beam Probe (HIBP) application on the TUMAN-3M (R=0.53m, a=0.22m, B_T=0.8T, I_p=140kA, T_e=0.5keV, n<4 $10^{19}m^{-3}$) – significant toroidal shift of beam trajectory – is caused by high ratio of poloidal field to toroidal one. Strong UV radiation from the plasma loads the energy analyzer's detector and complicates the problem even more. This paper presents the results of first measurement of plasma potential evolution in the discharges performed in ohmic H-mode using 80 keV K^+ beam and a Proca-Green secondary ion energy analyzer. Spatial region covered by the diagnostic in the experiments discussed was 0<r<0.6a. Spatial scan was performed utilizing the toroidal field decrease due to capacity power supply battery discharge. The change in plasma potential of the order of 100V has been measured during the H-mode formation. The potential in core plasma (r<0.6a) starts to change simultaneously with L-H transition, and than changes during ~6-8ms after the transition. Thus, the potential changes rather slowly in a comparison with L-H transition timescale (~2ms for TUMAN-3M ohmic H-mode). Possible explanation to the slow change in central plasma potential may be a formation of potential well structure at the plasma edge, in which radial electric field changes direction. This kind of structure is beneficial for the edge turbulent transport suppression because of high $|\partial E_r/\partial r|$, but not necessary requires a strong change in central plasma potential to occur immediately. The results from microwave reflectometry support this hypothesis.

THE HIBP IMPLEMENTATION ON THE TUMAN-3M

It was observed experimentally [1,2] and shown theoretically [3,4] that radial electric field is closely related with confinement improvement in the H-mode transition.

The HIBP was used for the tokamak plasma potential measurement for the first time in 1970 on ST tokamak [5]. Since that time, this diagnostic remains the only one capable of direct measurement of plasma potential, whereas other methods, i.e. ion spectroscopy, require additional modeling of plasma behavior.

Being applied to the TUMAN-3M tokamak, the HIBP experimental set up is as follows. Singly charged ion beam (K+), called the primary beam, with energy up to 100keV, is injected into the tokamak plasma. As a result of collisions with plasma electrons, a fraction of primary beam get additional ionization, forming a secondary (K++) beam fan. The potential energy of the injected ion changes at the point of additional ionization by $\Delta e\Phi_{pl}$, here Δe is the change of ion's charge, Φ_{pL} is the plasma potential at the point of ionization. Since the electrostatic potential everywhere outside the plasma is fixed, and magnetic fields don't affect the energy of the ions, the difference of kinetic energies between secondary and primary beams, when measured outside the plasma vessel, is $W_s - W_p = \Delta e\Phi_{pl}$. Thus, the HIBP gives a method of direct measurement of plasma potential. It should be stressed that a unique feature of this method is that it gives a value of Φ_{pl} utilizing solely fundamental principles like energy and momentum conservation, without any additional modelling of plasma behaviour involved. However, the spatial location of sample volume may be obtained only from numeric calculations of ions trajectories, providing that the exact topology of the magnetic field is known.

Relatively low aspect ratio of the TUMAN-3M tokamak (R/a=2.4), in a combination with bad access to the plasma through rather narrow vacuum ports, imposes severe limitations on the HIBP implementation. Low toroidal field of the tokamak requires K$^+$ primary ions energy to be not higher than 100 keV. On the other hand, the plasma current in the ohmic H-mode shot can be as high as 150 kA. Strong poloidal magnetic field (created by the plasma current and by equilibrium control field coils) gives rise to a toroidal projection of the total [v×B] force, resulting in the toroidal displacement of beam trajectories. Moreover, all magnetic fields and the plasma current are not constant in time in TUMAN-3M due to capacity bank type power supplies used. As a result, both the primary and secondary beam trajectories are essentially 3D curves, moving in space during the plasma shot. It was not possible to guide secondary ions into an energy analyzer throughout the tokamak shot. So, the injector and the analyzer were aligned in such a way as to obtain a time window, during which secondary beam reached the analyzer. This time window typically had ~30...40 ms duration, and was centered on the moment of the L-H transition.

The secondary beam energy was measured using a 30º Proca-Green type [6,7] located rather close to the plasma volume, with its entrance slit facing the tokamak plasma. As a result, the UV radiation coming from the tokamak discharge caused surface photo-effect on the metallic elements of the analyzer's construction, giving rise to stray currents to the detector. These stray currents were many times stronger than secondary beam current. To suppress them, the detector unit was equipped with electrostatic shield and additional electrodes negatively biased. This design allowed the suppression of stray currents below the secondary current level [8].

IN SITU CALIBRATION OF THE ENERGY ANALYZER

When calibrating the analyzer *in situ*, the tokamak vacuum vessel has been filled with neutral gas (D or He) with the full-range toroidal and vertical fields applied, but without the plasma discharge and, hence, a plasma current. Primary ions were additionally ionized due to collisions with target gas atoms, and conserved their energy because of lack of the electric field in a neutral gas. As the toroidal displacement was decreased due to the absence of the plasma current, the toroidal deflector plates biased up to 10 keV were activated to direct the secondary beam into the analyzer's entrance slit. When the detector unit of the analyzer receives the secondary beam, its energy is given – it equals exactly to the primary beam energy. This procedure, being repeated several times with the different tilt angles (i.e. different particle entrance angles θ) of the analyzer, allowed the main calibration functions of the analyzer F(θ) ("off-line processing function") and G(θ) ("gain function") to be determined. It was found that experimentally measured values of G(θ) are close to the calculated [9]. Also, the calibration evidences that in practically important angle region 28°≤θ≤32° angular dependence G(θ) is weak, so G=2.31=const was used when extracting plasma potential Φ_{pl} from detector signals measured in plasma shot using a relation:

$$\Phi_{pl} = 2U_aF(\theta)S + (2U_aG(\theta) - U_s) \qquad (1)$$

Here, U_a and U_s are analyzer and source accelerator voltages, S = $(i_U-i_L)/(i_U+i_L)$, i_U and i_L are secondary beam currents to the upper and lower plates of the detector.

PLASMA POTENTIAL MEASUREMENTS AT THE L-H TRANSITION

The HIBP was used for plasma potential evolution measurements during the ohmic H-mode transition on TUMAN-3M. The transition was initiated on t=53ms of a discharge by a short (~3ms) pulse of additional gas puffing. The example of the row HIBP signals in a typical ohmic H-mode discharge, together with some plasma parameters, is shown in Fig.1a&b.

Note that the vertical field coil current ("control current") is not constant in time in the TUMAN-3M. This causes a variation of displacement of the secondary beam spot along the entrance slit of the analyzer, hence geometrical conditions for the beam penetration through the slit are fulfilled during time interval from 35 until 78 ms only. After a data conditioning and processing two physically significant values were obtained, see Fig. 1b. First was plasma

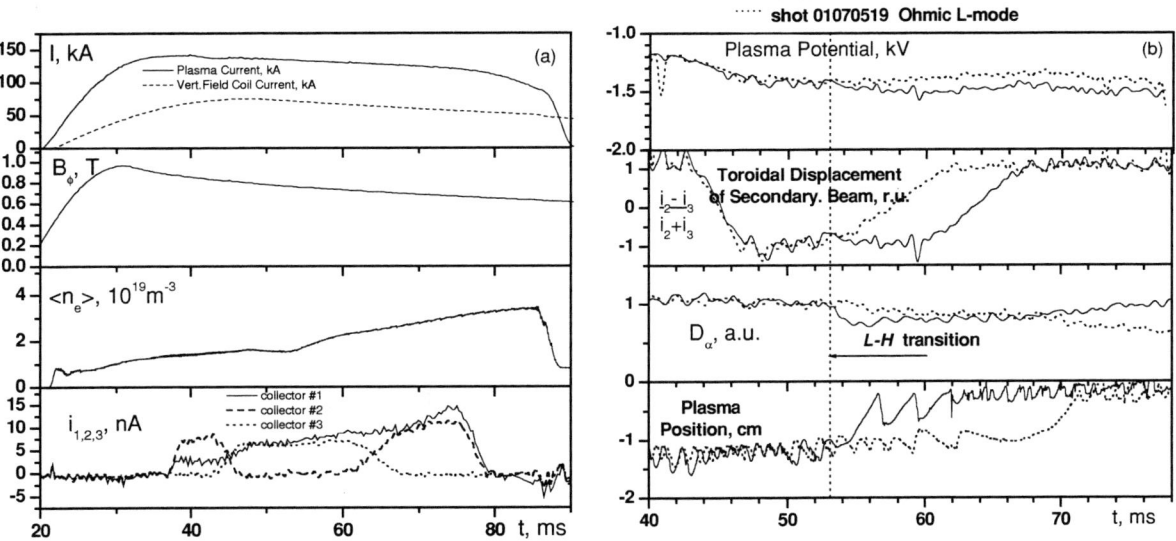

FIGURE 1. (a) Detector Signals in a Typical Ohmic H-mode Shot, (b) Plasma Potential in Shots with and without H-mode Transition

potential evaluated according (1) with $i_U = I_2 + I_3$ and $i_L = i_1$. The potential was found to become ~100-150V more negative after the H-mode transition, as compared to a shot without the transition. This is a manifestation of the negative radial electric field generation at the plasma edge in the H-mode. It is interesting to note, however, that the characteristic timescale of potential evolution is ~7-8 ms, that is very slow when compared to the transition timescale of ~2 ms as seen from D_α trace. The possible way to reconcile this observation with generally approved idea of the H-mode triggering by E_r generation at the edge is discussed below.

Another combination of secondary beam currents to detector elements, namely $\delta = (i_2 - i_3)/(i_2 + i_3)$, is proportional to the toroidal displacement of the beam. It reflects the value of poloidal magnetic field and its spatial distribution along the beam trajectories. As seen from Fig.2b, temporal evolution $\delta(t)$ after the moment of the H-mode transition is quite different in shots with and without the transition. Apparently, this is a result of a small perturbation of plasma equilibrium, which may be concluded from ~1cm perturbation of the plasma position after the transition. Some modification of plasma current profile caused by the transition may add to this effect as well.

As it was mentioned above, the toroidal magnetic field in TUMAN-3M is not constant in time – it decays

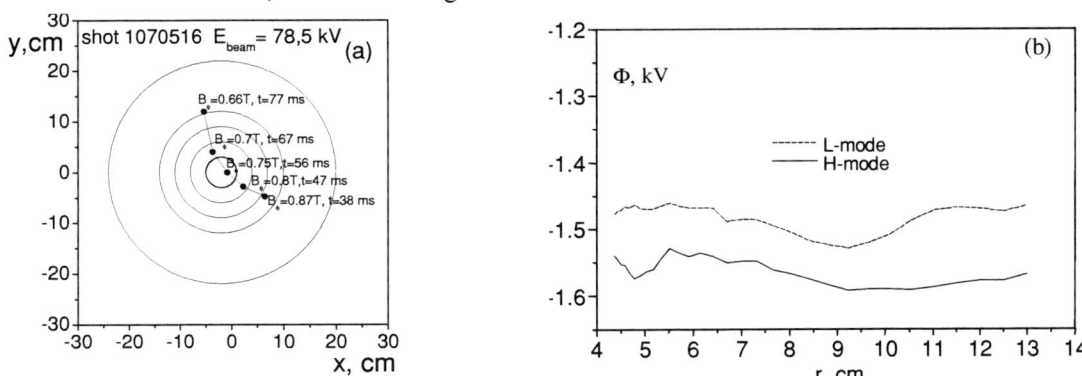

FIGURE 2. (a) Sample Volume Movement Through the Plasma Cross-Section Due to the Toroidal Field Decay, (b) Plasma Potential Profile Obtained Using This Movement.

with a time constant of ~150ms. This results in movement of the sample volume through the plasma cross-section along so-called 'detector line', see Fig. 2a, thus giving the possibility to perform "naturally" the radial scan of the plasma potential. The radial profiles of the plasma potential obtained in this way in two plasma shots are shown in Fig. 2b. Only a part of plasma cross-section with 4cm<r<13cm was accessed by the diagnostic because of geometrical restrictions discussed above. The potential profile in this region was found to be flat in both L and H-modes, yielding the E_r change outside this region – most probably, in peripheral plasma.

DISCUSSION

The core plasma potential in the TUMAN-3M tokamak was found to change very slowly after the H-mode transition. In principle, it may be a result of slow evolution of E_r after the transition. However, this point of view contradicts to the generally approved paradigm of the H-mode triggering by the abrupt change in peripheral E_r. If one take into account an obvious relation between central potential Φ and radial electric field E_r: $\Phi = -\int E_r(r,t)dr + \tilde{\Phi}_0$, here, Φ_0 is reference point potential, and integrating is performed between reference point and sample volume location), one might expect that central plasma potential will response immediately to the fast change in E_r at the plasma periphery. In order to reconcile the surprisingly slow core potential evolution after the H-mode transition in the TUMAN-3M with the model of the H-mode triggering by the E_r generation at the edge, one should admit non-monotonous profile of radial electric field perturbation $\delta E_r(r)$. In this case, two neighboring regions with oppositely directed $\delta E_r(r)$ may appear, forming a kind of a potential well at the plasma edge. Note that this potential well structure is characterized by a high value of $\partial E_r/\partial r$ that is beneficial for the transport barrier formation. The microwave reflectometry measurements during the TUMAN-3M H-mode transition [10] demonstrated that the two regions with counter-directed poloidal rotation velocity appear in peripheral plasma after the transition. This supports the idea of potential well formation at the plasma edge. Accurate measurements of plasma potential at the plasma boundary both using HIBP and Langmuir probes are planned to check the potential evolution at the edge.

SUMMARY

- Negative potential of approx. –1.5kV has been measured in core plasma using HIBP in ohmic L-mode on the TUMAN-3M. Plasma potential profile is flat in spatial region 4cm < r < 13cm accessed by the diagnostic. The potential becomes ~100-150V more negative as a result of ohmic L-H transition, but remains flat in that region.
- The potential in core plasma changes only after the L-H transition, and very slowly. It seems to be a result of potential well structure generation at the edge.

ACKNOWLEDGMENT

This work was jointly supported by RFBR grants 01-02-17922 and 02-02-17597 and INTAS grant N 2056.

REFERENCES

1. Groebner, R.J., Burrell, K.H. and Seraydarian, R.P., *Phys. Rev. Lett* **64**, 3105 (1990).
2. Ida, K, et al., *Phys. Rev. Lett.* **65** 1364 (1990).
3. Biglari, H., Diamond, P.H.and Terry, P.W., *Phys. Fluids* **B 2,** 1 (1990).
4. Rozhansky, V.A. and Tendler M., *Phys. Fluids* **B 4,** 1877 (1992).
5. Jobes, F.C. and Hickok R.L., *Nucl. Fusion* **10**, n.2, 195 (1970).
6. Green T.S. and Proca G.A., *Rev. Sci. Instr.* **41**, n.10, 1409 (1970).
7. Crowley T.P., et al., *IEEE Trans. On Plasma Science* **22**, 291 (1994).
8. Nedzelskiy I.S., Dreval N.B., S.M. Khrebtov et al, *Rev. Sci. Instr.* **72**, n.1 part 2, 575 (2001).
9. Askinazi L., et al., "Results from Heavy Ion Beam Probe Diagnistcics on TUMAN-3M Tokamak" in *Proc.28th EPS Conf. On Contr. Fusion and Plasma Physics-2001*, editted by C.Silva et al., ECA **25A**, EPS 2001, p. 405.
10. Bulanin V.V., et al., *Plasma Physics Reports* **26**, n.10, 813 (2000).

Design of the Collective Thomson Scattering Diagnostics for Large Helical Device using a Quasi-optical Frequency Tunable Gyrotron as a Radiation Source

Rostyslav Pavlichenko*, Toshitaka Idehara*, Isamu Ogawa[†], Kazuo Kawahata** and Motoyasu Sato**

*FIR Center, Fukui University, Bunkyo 3-9-1, Fukui 910-8507, Japan
[†]Faculty of Engineering, Fukui University, Bunkyo 3-9-1, Fukui 910-8507, Japan
**National Institute for Fusion Science, Oroshi-cho 322-6, Toki 509-5292, Japan

Abstract. Development of the collective Thomson scattering (CTS) diagnostic system for LHD is presented. High frequency, tunable (87-97 GHz), medium power (~100 kW) quasi-optical gyrotron will be used as a radiation source. We will show the detailed description of the system as well as initial calculations of the scattered microwave power from the LHD plasma. Using the quasi-optical gyrotron as a radiation source was inspired by the fact that this particular device has the ability of tuning of the operational frequency. Which is made it very attractive for using under the various scenarios for LHD plasmas

INTRODUCTION

Collective Thomson scattering (CTS) has been traditionally used for measuring the electron and ion temperature in laboratory plasmas. The detected signal is the one that originate due to the scattering of the electromagnetic radiation by the Debye cloud of electrons, which effectively surround each ion. These clouds of electrons move with the ions and impart a Doppler shift to the scattered radiation.

Theory shows that the net scattered power is dependent on the amplitude of fluctuations in electron density[1, 2, 3]. The scattered radiation power $P_s(\omega)$ per frequency interval $d\omega$ into solid angle interval $d\Omega$ is:

$$P_s(\omega)d\Omega d\omega = P_i \Psi r_0^2 n_e L_{pl} S(\boldsymbol{k},\omega) d\Omega d\omega, \qquad (1)$$

where P_i is source power, r_0^2 is classical electron radius, Ψ and L_{pl} are geometrical factors. $S(\boldsymbol{k},\omega)$ is the scattering form factor, containing the information about the frequency spectrum at the selected wave vector.

The form of $S(\boldsymbol{k},\omega)$ depends on $\alpha = 1/k\lambda_D$[4] – (the usual scale length to Debye length ratio). For small scale fluctuations ($\alpha \ll 1$) the electrons become uncorrelated. This is the case of incoherent Thomson scattering, for which $S(\boldsymbol{k},\omega)$ is simple Gaussian. In opposite case ($\alpha \gtrsim 1$), collective scattering fluctuations occur. $S(\boldsymbol{k},\omega)$ consist of a low frequency part, called *ion feature* at $\omega_i = k v_{Ti}$, and high frequency part near electron plasma wave frequency.

DESCRIPTION OF THE SYSTEM

Gyrotron and MOU

The proposed system for LHD will utilize the quasi-optical (Q. O.) gyrotron as a power source. The Q. O. gyrotron designed for operation at the fundamental frequency ($\omega = \Omega_{ce}$) in the range of 92 ± 5 GHz and in the range of $2 \times (92 \pm 5)$ GHz at the second harmonic ($\omega = 2\Omega_{ce}$) [1]. In Q.O. gyrotron the resonant structure is a Fabry-Pérot

---

[1] At present time the gyrotron under conditioning at FIR, Fukui University

resonator placed transversely to electron beam and operating in the pure Gaussian TEM$_{00q}$ mode. The advantages of Q.O. concept are the frequency tunability and the geometric separation between the spent electron beam and the microwave output. Above mentioned characteristics offers flexibility for experiments on fusion devices, where it may be advantageous to change the localization of power deposition zone (even for diagnostic usage) without changing the magnetic field and therefore affecting the plasma properties.

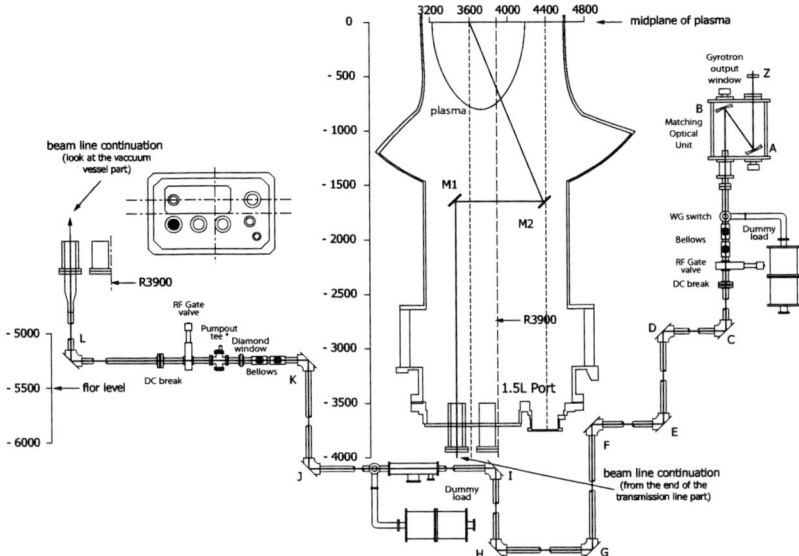

FIGURE 1. Schematic diagram of CTS transmission line

For coupling the output gyrotron radiation to a corrugated circular waveguide matched optical unit (MOU) is used. MOU mirrors (mirrors A and B at the Fig. 1) are designed to reshape beam profile from gyrotron output so that the radiation smoothly couples with HE$_{11}$ waveguide transmission mode.

Transmission line

At present time for the delivery microwave radiation to the plasma we are planned to use one of the present ECH transmission line [2] [5]. CTS transmission line (Fig. 2) is using 31.75 mm evacuated corrugated waveguides. The full path will have 10 miter bends and about 100 m in length. This transmission system was optimized for delivering of 70–90 GHz microwave radiation. The probing microwave beam, according present plans, will be injected through the downside 1.5 L port. The detection part will be placed at corresponding upper side port. The investigation of the possibility of using horisontal port for scattered radiation is under way now.

Detection system

Use of middle power gyrotron require good signal-to-noise ratio. Thus, a heterodyne type detection system will be used (Fig. 3). The scattered radiation from LHD plasma, which consist of the incident carrier and Doppler shifted parts, will be mixed with 75 GHz local oscillator (LO) signal. To suppress high radiation from the incident beam the tunable band-reject filter have to be inserted in front of detector. To obtain frequency spectra at post detection stage the signal will be split into several parts. To evaluate detailed spectrum of the scattered radiation each channel has its own intermediate frequency range that depends on corresponding part of microwave spectrum.

In the real plasma experiment, measurable spectral width is constrained to lie within the bandwidth capability of heterodyne detector and conventional electronics, typically about of several GHz. In the case when both temperatures,

---

[2] Designed at NIFS

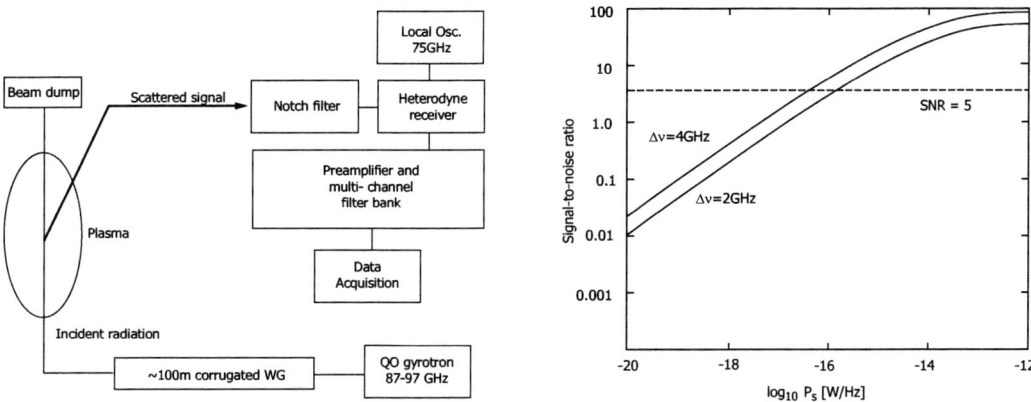

FIGURE 2. Schematic diagram of the heterodyne detection system (left) and its estimated SNR (right)

for electrons and ions, are practically the same ($T_e \simeq T_i$) the spectral width could be given by $kv_i/2\pi = \sqrt{2}\omega_{pi}/2\pi\alpha$. This condition can be easily met by using a sufficiently high α. If heterodyne detector noise dominates, the parameters of frequency analyzing circuit can be derived from the signal-to-noise level at the detector $s = P_s^{\Delta\nu}/NEP$ ($NEP = 8.6 \times 10^{-17} W Hz^{-1/2}$) and the output of frequency analyzer $S = \sqrt{1+\Delta\nu\tau}s/(1+s)$, where $P_s^{\Delta\nu}$ is power of scattered radiation delivered into heterodyne detector per Hz bandwidth, $\Delta\nu$ is the bandwidth of resolution interval and τ is integration time (which equal to gyrotron pulse length). Estimated SNR is shown on the Fig. 2.

DETERMINATION OF THE SCATTERED ANGLES

Using of 90 GHz range radiation allow to achieve larger α values. For the typical LHD plasma parameters ($B_0 = 2.75$ T, $n_e \approx n_i = 5 \times 10^{13} cm^{-3}$, and $T_e \approx T_i = 3$ keV) we conduct the calculation of the possible scattered angles. Those estimation shows that almost any kind of scattering experiment is possible. The scattered angles vary from near very small angles $\sim 0-5°$ (quasi-forward scattering) to almost backward scattering values $\sim 180°$. From the expression of Salpeter parameter:

$$\sin\frac{\theta_{Br}}{2} = \frac{1.08 \times 10^{-4}\lambda_i}{\alpha}\left(\frac{n_e(cm^{-3})}{T_e(eV)}\right)^{1/2} \quad (2)$$

one can see that utilization of lasers (for instance CO_2 laser) is possible for a very small angles only. This fact will be make separation of insident and scattered beams more difficult. A plot of condition $\alpha = 1.5$ for 92 GHz quasi-optical gyrotron and CO_2 laser radiation as function of density and temperature for LHD plasmas (see Fig. 3).

CALCULATION OF EXPECTED SCATTERED SPECTRUM FROM LHD PLASMA

Upon the inspection of the form factor it was found that a wide low part of the spectrum coming from the first term in Eq. ??, and tall narrow part coming from the second term of the same equation. The latter stems from the collective motion of the electrons with the ions, which is what we are concern to measure. Our calculations have been confined near the values of angular frequency ω for which second term makes a substantial contribution to the whole spectrum, i.e. $\omega \lesssim 3kv_i$.

The expected scattered spectrum from LHD neutral heated plasma as a function of a Doppler shifted frequency from the gyrotron centered frequency is shown on the Fig. 5. For spectrum simulation the following plasma parameters are chosen: $B_0 = 2.75$ T, $n_e \approx n_i = 5 \times 10^{13} cm^{-3}$, and $T_e \approx T_i = 3$ keV, $E_{beam} = 150$ keV. We assume that electron and bulk ion velocity distribution functions of the plasma are Maxwellian and (alphas) have an isotropic $1/v^3$ 'slow-down' distribution. For evaluating of the scattered form factor, so-called Salpeter approximation has been used.

The calculation was done for vertically elongated plasma. As a receiver 20 dB gain conical horn antenna was used. The position of antenna was changed according to the values of scattered angle. The results shows (see Fig. 3) that

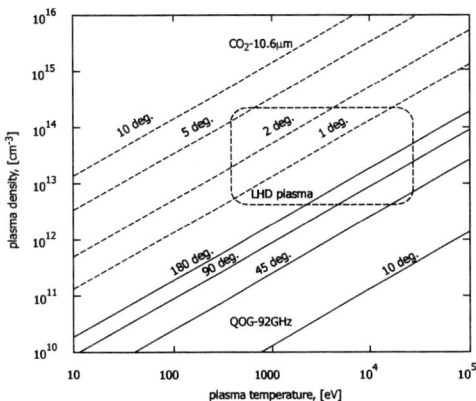
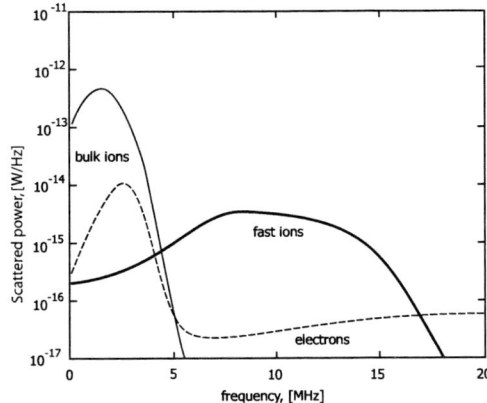

FIGURE 3. Plot of condition $\alpha = 1/k\lambda_D = 1.5$ for 92 GHz quasi-optical gyrotron and CO_2 laser radiation as function of density and temperature for LHD plasmas. Density and temperature for collective type of scattering are located over each line (left) and the expected scattered spectrum from LHD plasma as a function of a Doppler shifted frequency from the gyrotron centered frequency (right)

for the wide frequency range, there is sufficient scattered power, expected from bulk ions, and the contribution from electrons is three orders lower.

SUMMARY

We have performed initial design of the LHD CTS system based on the Q. O. gyrotron as a radiation source. The main advantage of the using gyrotron as a radiation source instead CO_2 laser is as follows. In the case of lasers, the spectrum changes very rapidly with angle, only small-angle scattering can be used. This complicates obtaining high spatial resolution $\triangle r$. In contrast, at gyrotron frequencies, much larger scattering angles are available. ($\triangle r_{las} \simeq 43cm$, $\triangle r_{gyr} \simeq 2.7cm$). The spectra change on slowly with angle, and reasonably large collection solid angles can be used - limited only by coherence requirements for heterodyne detection. The choice of the scattered angle will depend primarily on effects of plasma refraction.

For the significant part of the expected spectrum the fast ions component is dominant. It will be encouraging fact for the successful evaluation of the α-particles temperature directly from the total scattering signal.

Preliminary estimates show that middle-power gyrotron with power of up to hundred watts, and pulse duration about $1-2msec$ is well suited for ion feature evaluation.

REFERENCES

1. J. Sheffield, *Plasma scattering of electromagnetic radiation*, Academic Press, London, (1975)
2. T.P. Hughes and S.R.P. Smith, Rev. Sci. Instrum., **72(6)**, 1988 (1075).
3. H. Bindslev J. Hoekzema, T. Huges and J. Machuzak Phys. Rev. Lett., **83**, 1999 (3206).
4. E.E. Salpeter Physical Reviev, **72**, 1960 (1528).
5. T. Simosuma S. Kubo, M. Sato *et al.*, Fusion Eng. and Design, **53**, 2001 (525).

Numerical Simulation in Dense Plasma Diagnosis with an Amplitude-Division Soft-X-Ray Laser Interferometer

Wudi Zheng and Guoping Zhang

Institute of Applied Physics and Computational Mathematics, China

Abstract. In this paper, we present a design of plasma electron-density measurement using a soft X-ray laser interferometer. Ni-like Ag 13.9nm lasing with ~20ps pulse duration was adopted, plasma was produced by ~85ps laser incidence on CH plane target. In order to measure electron density round critical face, and to obtain large scale, high temperature, dense plasma suited for diagnosis, we simulated the hydrodynamics of plasma under some different driven conditions using JB19 code. The affects of driven laser wavelength, laser power, and flux limit factor etc on plasma situation were analyzed. After investigated the affect of plasma rapid variety on interference fringe blur, the absorption and reflection of XRL by plasma, we gave out the laser driven condition, and plasma scale ~1mm, electron density which can be measured (~$5\times10^{21}cm^{-3}$).

INTRODUCTION

As a probe beam to diagnose the electron density of a dense plasma, the X-ray laser (XRL) has its unique advantages. With the obtaining of saturated output of X-ray lasers, applications of plasma diagnosis using X-ray lasers become possible. The methods of measuring electron density N_e of plasma include deflection method (such as using Moier deflectometer [3]), wavefront splitting method (such as using Lloyd mirror [4]) and amplitude-division method (such as using Mach-Zehnder interferometer [1,2,5]). In this paper, we give a design of plasma electron density measurement using soft XRL skew Mach-Zehnder interferometer under special experimental condition.

Through plasma density diagnosis, we can study many defects in physics modeling such as laser energy deposition, flux-limited heat conduction, hydrodynamics, and non-local thermo-dynamic equilibrium atomic kinetics. Especially, in flux-limited approximation of electronic heat conduction, it is difficulty to give the value of flux limit factor f_e, but it is very important in numerical simulation, so we hope that the hydrodynamics of plasma to be measured should be sensitive to f_e, then it can be found by comparing with experimental result. Additionally, we hope that distribution of N_e near critical face can be measured where many physics phenomena take place. One advantages of Mach-Zehnder interferometer is that large-scale plasma can be measured, so we analyze the plasma scale and the maximum plasma electronic density can be measured in this paper.

In design, Ni-like Ag 13.9nm lasing is adapted as probe beam. The 20ps X-ray laser pulse can be produced by incidence of about 100J, 85ps 1.06μm laser on Ag target. The plasma to be measured is produced by irradiating 10μm CH on Si wafer with a laser beam focused to a spot. Because some reflecting mirrors and two beam splitter with low reflectivity and low transmissivity are used in the Mach-Zehnder interferometer, in order to get high Signal-to-Noise, selecting low Z material should be suitable for its low irradiation. Two driven laser beams each of which consists of 85ps main pulse with a prepulse delayed 4ns irradiate target with 21 degree declining angle. For 1.06μm light (1ω), the energy of each laser beam is less than 100J with 3% prepulse, for 0.53μm light (2ω), less than 50J with 0.09% prepulse because of 50% harmonic efficiency. Numerical simulations are completed by 1D code JB19.

NUMERICAL SIMULATIONS AND RESULTS ANALYSIS

Different frequencies of driven laser

Different driven laser frequencies lead to different plasma hydrodynamics. In the case of driving by $1\omega+1\omega$, $2\omega+2\omega$, $3\omega+3\omega$ and $1\omega+2\omega$, plasma hydrodynamics are simulated, two laser beams get to target at same time. The results illustrated in fig. 1 is 1ns after arrive of main pulse, driven conditions also can be seen in figure, the position of target surface is at r = 125μm, and same to all other figures in this paper. Figure shows that, target will be ablated deeper by driven laser with higher frequency, and less density gradient can be gained near critical face. A shock wave propagates away from the target surface can be seen in fig. 2. Simulation results shows that, in case of high frequency drive, the velocity of shock wave would be much larger, but the electron density at the head of shock wave would be smaller. Electron temperature T_e depends on driven energy mostly, so $T_{e(1\omega+1\omega)} > T_{e(2\omega+2\omega)} \sim T_{e(3\omega+3\omega)}$. Though, the distribution of T_e keeps rather flat in 1mm scale, T_e declines while approaching target surface. After the end of main pulse, T_e begins to descend, but still stays about 500eV at 1ns later. Because of high T_e, ionization of CH becomes easily, ionization degree of large part of plasma is kept much higher than 4.

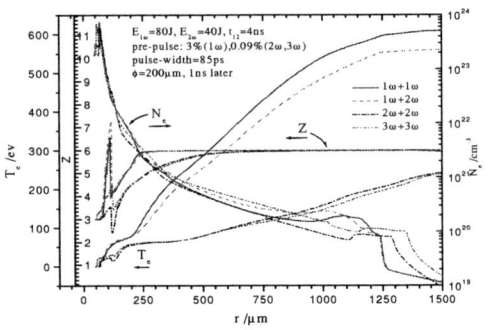

FIGURE 1. The distribution of T_e and N_e in the case of different driven laser frequencies.

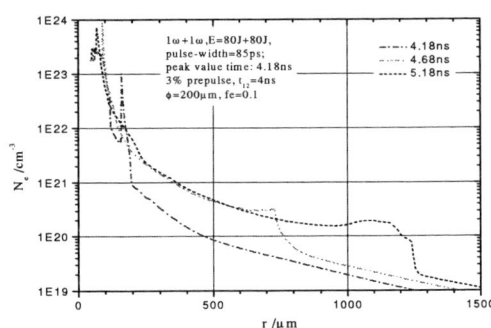

FIGURE 2. The evolution of N_e, in the case of $1\omega+1\omega$, the peak-value time of main pulse is at 4.18ns.

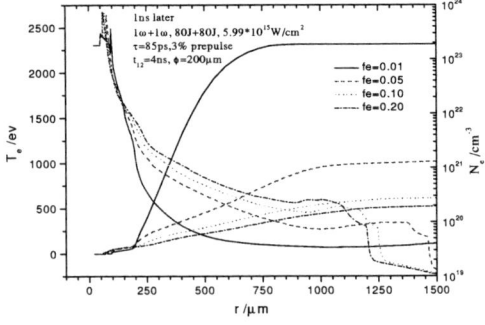

FIGURE 3. The distribution of N_e and T_e in the case of $1\omega+1\omega$ with different f_e.

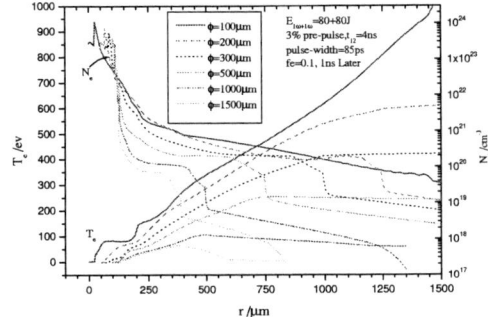

FIGURE 4. The distribution of N_e and T_e in the case of $1\omega+1\omega$ with different focus spot diameters ϕ.

Effect of flux limit factor f_e

As f_e changes, the distributions of T_e and N_e would make much difference, shown in fig. 3. As f_e increases, laser ablation deepness increases, but the gradient of N_e near critical face decreases, as well as average T_e. While driven laser intensity increases, affected by self-generated magnetic field and anomalous transport, the resistance of electronic heat conduction increases, so f_e should be small. While intensity is up to 5×10^{14}W/cm^2, this heat resistance effect becomes notable. In this paper, total intensity of target surface is higher than 10^{15}W/cm^2, f_e must be small.

Different driven laser intensity

The effect of different driven laser intensity on plasma hydrodynamics can be investigated by varying focal spot size. The overwhelming effect of intensity can be seen in fig. 4, intensity increase one order of magnitude, the scale of plasma increases doubly. High intensity will lead to high T_e, and then decrease the absorption of X-ray laser by plasma, but the high X-ray irradiation of plasma will decreases signal-to-noise. The shortcoming of little diameter is that the serious 2D jet of plasma will bring difficulty in later interpretation of experimental results.

Fringe blur by quickly variation of N_e

With arriving of main pulse, the energy absorption of laser occurs around critical face mainly, and then a shock wave forms and propagates away from critical face. The electron density will vary quickly at the position of the head of shock wave, fringe at this position will be blurred. We take interest in electron density distribution near critical face, then the time of diagnosis should be chosen rightly so that shock wave is away from critical face. It is favorable to select Ni-like Ag XRL as probe beam for its short width (20ps). Take the calculating model corresponding to fig. 2 for example, the shift velocity of critical face position is 6~9μm/20ps at the end of main pulse, but 0.5ns later it becomes ~14μm/200ps, 1ns later 3~5μm/200ps. So diagnosis time should be later than 0.5ns. Our simulations also show that higher driven laser intensity leads to quicker shift.

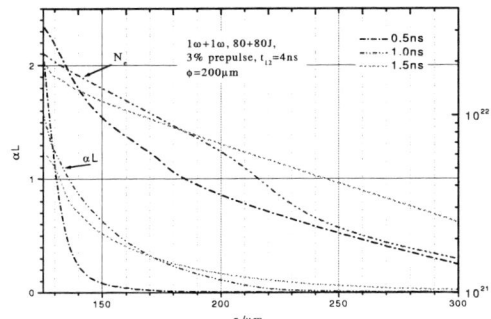 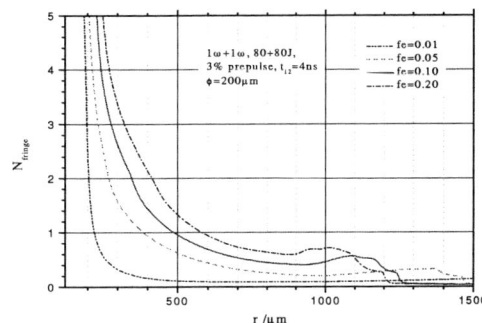

FIGURE 5. The distribution of inverse bremsstrahlung absorption and N_e.

FIGURE 6. The distribution of fringe shifts number with different f_e.

The absorption of XRL by plasma

In order to get high quality interference fringes, the absorption of XRL by plasma should be less. Because CH plasma is highly ionized, as mentioned above, photoelectric absorption can be neglected. In plasma dominated by inverse bremsstrahlung absorption the absorption coefficient α is given approximately by:

$$\alpha \approx 2.44\times 10^{-37}\frac{\langle Z^2\rangle n_e n_i}{\sqrt{kT_e}(h\nu)^3}\left[1-\exp(\frac{-h\nu}{kT_e})\right]\text{cm}^{-1}$$

where the electron temperature kT_e and photon energy $h\nu$ are in eV and electron and ion densities are in cm^{-3}. Assume plasma driven by 85ps laser jets with π/8 solid angle, the absorption of XRL going through plasma parallel to target surface can be calculated. Fig. 5 gives out the absorption in symmetric plane cone. We can see that absorption by plasma at area R > 15μm is small, αL < 1, where R is the distance to the initial target surface and L is plasma transverse scale. At the position of R=15μm electron density is about 1.5×10^{22}cm^{-3}.

The number of fringe shifts and refraction

In an interferometer the number of fringe shifts is given by:

$$N_{fringe} = \frac{\delta\phi}{2\pi} = \frac{1}{\lambda}\int_0^L (1 - n_{ref})dl \approx \frac{n_e}{2n_c}\frac{L}{\lambda},$$

where the integral is along ray trajectories through the plasma, dl is the differential path length and refraction effects are negligible. We calculate some models according to this equation, as shown in fig. 6. While f_e is 0.1, the fringe shifts number is about 20 at R=15μm, that is less than the maximum number measurable ~50 [1], if N_{fringe} of 0.5 can be resolved by interferometer system, then plasma scale measurable should be more than 1mm for 1ω+1ω, ~700μm for 2ω+2ω. As seen in figure, the distribution of N_{fringe} is very sensitive to variation of f_e. In our design, driven laser intensity is chosen too high to get large f_e, if real f_e is litter than 0.05, fringe shifts number of 0.2 must be resolved in experiment to get plasma scale of 1mm.

The refraction effect is a disadvantage factor in experiment, it will bring trouble to interpretation of experiment result too, and it is very difficult to analyze experimental error brought by refraction effect because of complex optic system. According to our calculation, as a conservative estimate, if refraction angle little than 2mrad can be bear, then the electron density of ~5×10^{21}cm^{-3} for 1ω+1ω and ~4×10^{21}cm^{-3} for 2ω+2ω can be measured, correspond position is about 75μm away from target surface.

CONCLUSION

By our numerical simulations, we find that high driven intensity and high energy is needed to get large-scale plasma measurable. Take advantage of smaller focal spot size, the refraction and absorption to XRL by formed plasma can be limited less, and high electron density can be measured, but there is no enough fringe shifts for low density plasma area, and serious 2D effect will be induced, so calculated result by 2D code is needed while interpreting experiment results. Flux limit factor is sensitive to distribution of fringe shifts number, it is good for us to verify and understand flux-limited approximation of electronic heat conduction.

After balance all factor mentioned above, we give out the laser driven condition: plasma to be measured is formed by irradiating CH target by two 1ω laser beams with energy 80+80J, or by a 1ω plus a 2ω laser beam with energy 80+40J, 3% prepulse for 1ω, 0.09% prepulse for 2ω, delay time of 4ns between prepulse and main pulse are proposed, and driving by two 2ω laser beams is not be proposed. Suitable delay no more than 0.5ns between two main pulses is useful in enlarging plasma scale. Diameter of focal spot is about 200μm. The scale of plasma more than 1mm and the maximum electron density of ~5×10^{21}cm^{-3} can be measured.

REFERENCES

1. L.B. Da Silva, T.W. Barbee et al, Phys. Rev. Lett. **74**, 3991-3994 (1995)
2. A.S. Wan, C.A. Back, et al, X-ray Lasers 1996 Conf., Lund, Sweden, 504 (1996)
3. D.Ress, L.B. Dasilva, et al, Rev. Sci. Instrum. **66**, 579-584(1995)
4. C.H. Moreno, M.C. Marconi, et al, Phys. Rev. E **60**, 911-917 (1999)
5. J. Filevich, K. Kanizay, et al, Opt. Lett. **25**, 356-358 (1999)

ECH Power Deposition Study in the Collisionless Plasma of LHD

S. Kubo*, H. Idei*, T. Shimozuma*, Y. Yoshimura*, T. Notake†, K. Ohkubo*, S. Inagaki*, Y. Nagayama*, K. Narihara*, I. Yamada*, S. Muto*, S. Morita* and LHD Experimental Group*

*National Institute for Fusion Science, 322-6 Oroshi cho, Toki 509-5292, Japan
†Dept. of Energy Science, Nagoya Univ., Furo cho, Nagoya 464-8603, Japan

Abstract. Power deposition profile is one of the essential keys in understanding the transport mechanism in high temperature low collisional plasma. Using the power modulation techniques, the power deposition profile and its change in shape and position by the focal position is clearly observed experimentally in Large Helical Device (LHD). These power deposition profiles show fairly in good agreement with the results from ray tracing calculation, as well as discrepancies arising from supra thermal electrons. Such experimentally deduced power deposition profiles are used as the initial inputs of one dimensional time dependent electron transport code for the further investigation of the plasma confinement.

INTRODUCTION

A high central electron temperature of more than 10 keV is achieved by electron cyclotron heating (ECH) in the Large Helical Device (LHD)[1]. This high electron temperature plasma is realized by concentrating the ECH power near the magnetic axis. The dependence of the central electron temperature on the electron density indicates that some nonlinear mechanisms related to the change in the confinement occurs in the collisionless regime in the LHD. The electron temperature profile in such high temperature regime is characterized by a high electron temperature gradient which is often observed in the plasma with the internal transport barrier. In such collisionless regime, plasma confinement properties change drastically in toroidal plasma confinement system, especially in helical system. Local heating by ECH is a key tool not only for heating electrons but also for deforming the distribution function to control the electron flux resulting in generating a radial electric field though ambipolar potential. This potential leads to suppression of neoclassical transport and/or anomalous transport via E×B shear flow.

Obtaining the precise power deposition profile and control it is one of the key factors for the realization of such high temperature and for the detailed study of the physics of such collisionless plasma. The power deposition profile and its change in shape and position by the variation of the focal position are clearly observed during the ECH power modulation experiment.

In the next section, ray tracing calculation applied in LHD is described and the result of ray tracing is compared with the experimentally deduced from power modulation experiment. In far low density regime, the presence of supra-thermal electrons is confirmed and effect of such non-thermal electrons is discussed in conjunction with the power deposition. For more detailed analysis, simple time dependent cylindrical transport code is developed and preliminary results are given in the last section.

RAY TRACING

In order to save the calculation time, data on three dimensional magnetic profile and flux surface on mesh points (65 radial × 65 horizontal square mesh in 17 poloidal plane in helical half pitch). Starting points of each ray is distributed within 1.5 times of waist size so as to represent a equal area on the starting plane. In the antenna system on LHD, injection beam is strongly focused by last second mirror and steered by the last plane mirror. Initial wave-

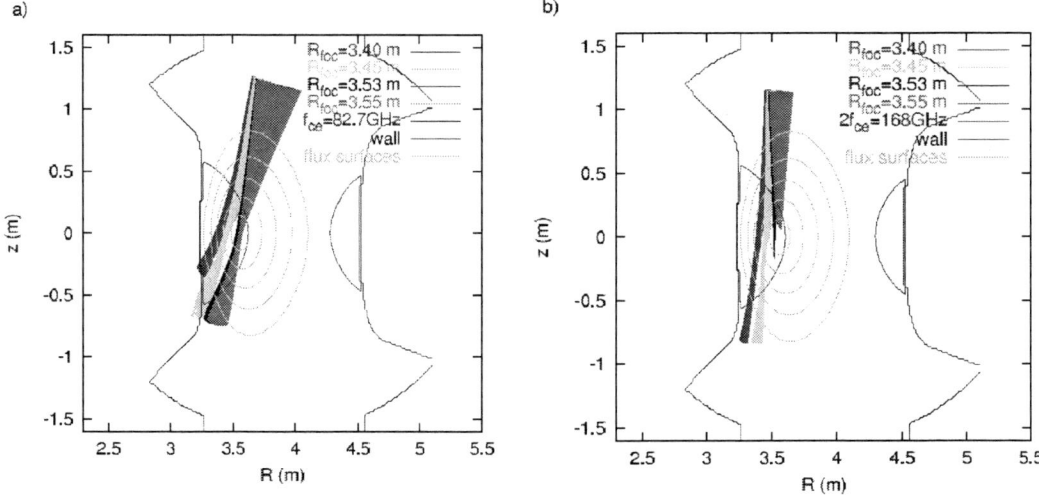

FIGURE 1. Ray tracing results for a) 82.7 GHz beam and b) 168GHz beam. Each beam is represented by bundle of 351 rays, equally distributed within 1.5 times of waist size at the antenna. All rays are projected on the vertically elongated cross section.

vector of each ray is set normal to the equi-phase surface defined by the Gaussian beam parameter of each antenna. Normally, a ray is traced using a cold dispersion relation, and the absorption is estimated from quasi perpendicular, weakly relativistic absorption formula[2]. The total power deposition profile is reduced by summing up the deposition fraction of each ray with the initial weighting function proportional to the power fraction of each representing area. The Gaussian beam parameters and beam position are taken as the same as the designed values of the antenna, which are confirmed by the low power testing [3]. Projected rays on a poloidal cross section from upper port antenna are shown in Fig. 1. Here, bundles of rays are shown for the injection angles corresponding to target focal positions on the mid plane of LHD, R_{foc}=3.40, 3.45, 3.53 and 3.55 m. Normally, the number of rays are selected so as the power deposition profile be smooth and realistic. The bundle contains 351 rays (1 center and 9-radial × 36-azimuthal). Due to the diffraction of the beam, result tend to be narrower than the real one, especially in the low refractive (density) plasma. For the detailed investigation, beam tracing method should be used.

ECH POWER MODULATION EXPERIMENTS

The modulation experiments described here are performed at the averaged electron density of 1.0×10^{19} m^{-3} with flat profile. The electron temperature of parabolic profile and $T_{e0} \approx 2.5$ keV is sustained by NBI. The magnetic field is set at 2.951 Tesla and magnetic axis at 3.50 m. The fundamental and second harmonic resonance layers just cross the magnetic axis under this condition. Both heating powers are injected from the upper port antenna in the vertically elongated poloidal cross section. The 32 channel electron cyclotron emission (ECE) radiometer system is used to diagnose the local temperature response to the modulated ECH power. The viewing chord of the ECE radiometer is on the mid-plane of the horizontally elongated cross section. The direction of the beam injection is labeled by the radial focal position, R_{foc}, where the center of the injected beam crosses the mid-plane. Under the same magnetic field, the radial position and shape of the cross section between resonance layer and injected beam can be varied by changing the R_{foc} as shown in Fig. 1. Both injected powers have elliptical Gaussian beam shape. The waist sizes of each beam are 15 mm and 50 mm in radial and toroidal direction, respectively. The modulated power of both the fundamental and the second harmonic heating is applied on this target plasma. The fundamental heating power of 255 kW at the 82.7 GHz is applied from $t = 1.02$ to 1.42 s with the 100%, 50 Hz square wave power modulation at the latter 300 ms of the pulse. The second harmonic heating power of about 400 kW from two identical upper port antenna at 168 GHz is applied from $t = 1.52$ to 1.92 s with the identical modulation as the fundamental heating. In Fig. 2 are shown the contour plot of the electron temperature. Here, the cases of $R_{foc} = 3.55$ m and 3.4 m are shown in a) and b), respectively. In the

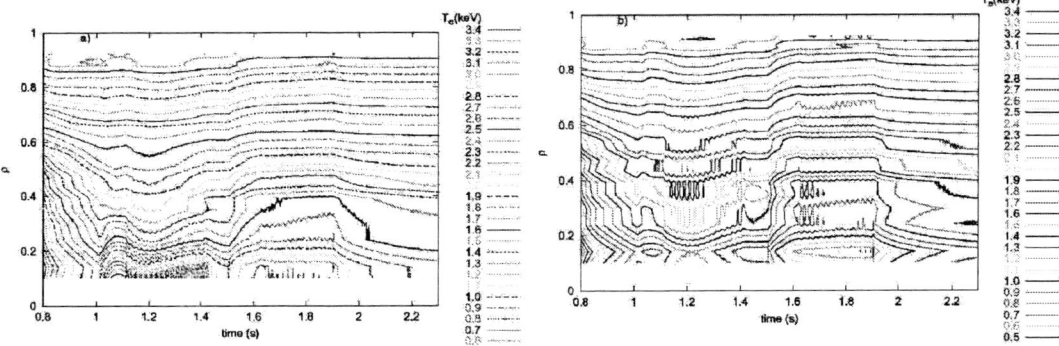

FIGURE 2. Equi-electron temperature contour measured by 32 channel ECE radiometer. The fundamental and second harmonic ECH are applied from t=1.02 to 1.42 and 1.52 to 1.92, respectively. The 50 Hz square wave modulation is applied at the latter 300 ms of each pulse. The focal point is set at a) 3.55 m and b) 3.40 m for both fundamental and second harmonic ECH.

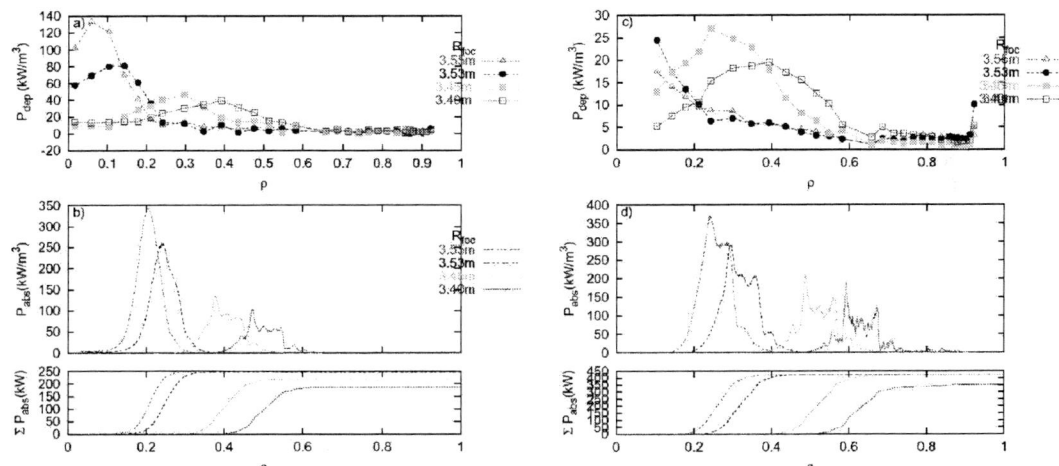

FIGURE 3. a): Power deposition profile for 82.7 GHz estimated from boxcar method for R_{foc} = 3.40, 3.45, 3.53 and 3.55 m. The sample data are restricted within ± 3 ms at each turn on and off timings in order to reduce the diffusion effect. b): Power deposition profile (upper traces) and total absorption fraction within given flux surface (lower traces) estimated from ray tracing calculation for 82.7 GHz. c): Power deposition profile for 168 GHz estimated from boxcar method for R_{foc} = 3.40, 3.45, 3.53 and 3.55 m. The sample data are taken as the same as the case for a). d): Power deposition profile (upper traces) and total absorption fraction within given flux surface (lower traces) estimated from the ray tracing calculation for the 168 GHz.

case of R_{foc} = 3.55 m of the fundamental heating, the central part of the electron temperature shows the fast response to the modulation, while it shows almost no response in the case of R_{foc} = 3.40 m. It is clearly seen that the region where the electron temperature show the fast response is center localized within ρ=0.2. In case of R_{foc} = 3.40 m, the fast response is seen in the region from ρ = 0.2 to 0.6.

Boxcar Analysis

The boxcar technique is used for ±3 ms data points at every turn on and off timings [4, 3] to deduce the first order power deposition profile. The differences of the inclination of each ECE channel are plotted as a function of normalized radius in Fig. 3 a) for the cases of R_{foc} = 3.55, 3.53, 3.45 and 3.40 m. Here, the differences of the inclination (eV/s) are multiplied by the local density (m^{-3}), factors (3/2 and e), and normalized radius to represent the total power deposited at each flux surface.

The discrepancies between calculated and deduced profiles in 168 GHz case may indicate the limitation of the ray tracing calculation using geometrical optics. The calculated power deposition profile is more sensitive to the antenna and resonance configuration for 168 GHz injection. In Fig. 3 are shown the deduced and calculated power deposition profiles for the similar focal point settings in the case of 168 GHz modulation. The central ECE channels are affected by the stray of the gyrotron power and are not available. This shift might be interpreted by the shift of the magnetic axis, since the axis shift due to the finite beta is not included in the ray tracing code. The other possible interpretation might be the presence of the high energy electrons. The calculation of the deposition profile only includes the weakly relativistic effects but not the non thermal component of the plasma. It is noted that In far lower density, intense hard X-ray emission with the effective energy range from 30 to 200 keV is observed. Down shifted non-thermal ECE emission is also observed. The fact that these down shifted non-thermal ECE emissions responds fast to the 168GHz power modulation indicate that behavior of supra-thermal electrons play an critical role in the estimation of the power deposition profile.

TIME DEPENDENT 1-D ELECTRON TRANSPORT CODE

The limit of applying boxcar analysis is that the data points to analyze should be restricted within the time window where the adiabatic condition is satisfied. It is also reported that total modulated power apt to be lower than the experimentally modulated one [5, 6]. In order to clarify this point and more over to deduce electron thermal diffusivity and its dynamic response to the local plasma parameters, a time dependent one dimensional electron transport code is developed. This code solves time dependent diffusion equation

$$\frac{3}{2}n_e(\rho)\frac{\partial T_e(\rho,t)}{\partial t} = -\frac{3}{2}\frac{1}{a^2\rho}\frac{\partial}{\partial \rho}\left(\rho\chi_e\left(\rho,T_e(\rho,t),\frac{\partial T_e(\rho,t)}{\partial \rho}\right)\frac{\partial n_e(\rho)T_e(\rho,t)}{\partial \rho}\right) + \frac{P_0(\rho)+P_{mod}(\rho,t)}{2\pi a^2 R_{axis}}. \quad (1)$$

with the dependence of the electron thermal diffusion coefficient on radius, ρ, local temperature, $T_e(\rho,t)$ and the local temperature gradient, $\frac{\partial T_e(\rho,t)}{\partial \rho}$ included. The thermal diffusivity ranges below 1 m^2/s in both cases near the axis. It is noted that including somewhat negative dependence of the diffusivity on the temperature gradient is necessary to reproduce both on axis and off axis cases. Detailed analysis and comparison with the experimental data are underway. It is noted that including somewhat negative dependence of the diffusivity on the temperature gradient is necessary to reproduce both on axis and off axis cases.

CONCLUSION

The power deposition profile is the essential key n analyzing and understanding the transport mechanisms. The power deposition profiles are deduced by the boxcar method during the power modulation experiment. These profiles are compared with those calculated from the ray tracing, showing good coincidence. The developed one dimensional time dependent transport code is tried to use for the purpose of deducing the diffusion coefficient and its dependence on the local temperature and local temperature gradient.

REFERENCES

1. S. Kubo, *et al.*, *J. Plasma Fusion Res.*, **78**, 99–100 (2002).
2. M. Bornatici, *et al.*, *Nucl. Fusion*, **23**, 1153– (1983).
3. S. Kubo, *et al.*, "RF Experiments in LHD," in *Radio Frequency Power in Plasmas*, edited by S. Bernabei and F. Paoletti, AIP Conference Proceedings 485, American Institute of Physics, New York, 1999, pp. 237–244.
4. M. Iwase, *et al.*, *Japan J. Appl. Phys.*, **37**, 678 –687 (1998).
5. L. Giannone, *et al.*, *Plasma Phys. Control. Fusion*, **38**, 477–488 (1996).
6. U. Stroth, *et al.*, *Plasma Phys. Control. Fusion*, **38**, 611–618,1087 (1996).

Transport Due to Intermittent Events and Plasma Flow Shear in Magnetized Plasmas

V. Antoni[1], G. Regnoli[1,2], M. Spolaore[1], G. Serianni[1], N. Vianello[1,2], R. Cavazzana[1], E. Spada[1], E. Martines[1]

[1] *Consorzio RFX, Associazione Euratom-ENEA sulla Fusione Corso Stati Uniti, 4, 35127 Padova, Italy*
[2] *Department of Electric Energy, University of Padova, via Gradenigo 6/a 35100 Padova Italy*

Abstract. In the Reversed Field Pinch experiment RFX a highly sheared ExB flow is observed in the edge region, with shear value close to the value required for turbulence suppression or reduction. Recent observations have shown that almost 50% of particle flux due to turbulence is due to intermittent events which tend to cluster during magnetic relaxation phase. These events have been associated in RFX to vortex like structures whose rotation direction depends on the local value of the mean ExB flow shear.

INTRODUCTION

ExB flow shear plays an important role in experiments for Thermonuclear Fusion Research where turbulence drives most of the energy and particle transport. Plasma turbulence investigation in Reversed Field Pinch (RFP) configurations has lead to the conclusion that ExB drift velocity affects turbulence properties [1] as well as in other Thermonuclear Fusion devices. A proof of that are biasing experiments of the edge region by plasma guns [2] or electrodes [3] which were successful in reducing anomalous transport via suppression or reduction of the electrostatic and magnetic turbulence in analogy to enhanced confinement regimes observed in tokamaks and stellarators.

The aim of this paper is to contribute to the general understanding of the turbulent transport in RFP's, with special emphasis on the role played by the ExB velocity shear, pointing out the analogies with ordinary fluid dynamics.

EXPERIMENTAL SETUP AND RESULTS

RFX has a minor radius $a = 0.5$ m and a major radius $R = 2$ m. Present data refer to plasmas with density $n \sim 1\text{-}2 \cdot 10^{19}$ m^{-3} and $I < 500$kA in order to minimize the plasma perturbation and the probe damage. An arrays of Langmuir probes have been inserted up to $r/a=0.86$ into the plasma, i.e. in a location close to that where the toroidal field changes sign. The sampling frequency was 1 MHz though the signal bandwidth was limited to ~ 400kHz by the electronic equipment for signal conditioning.

The analysis of the ExB velocity and turbulence in the edge region revealed a double shear structure [4] as shown in fig. 1. In a RFP the magnetic field in the outer region is mainly poloidal so that the drift velocity is in the toroidal direction. In the graph, the radius has been normalized to the radius of the Last Closed Flux Surface.

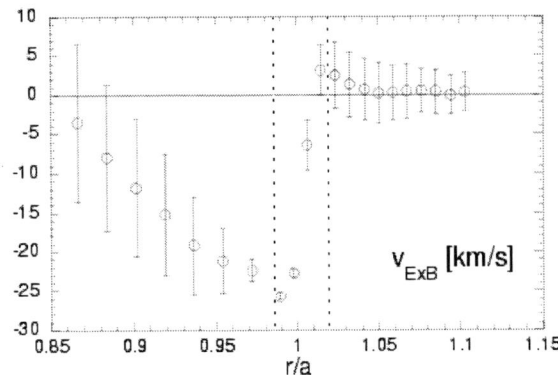
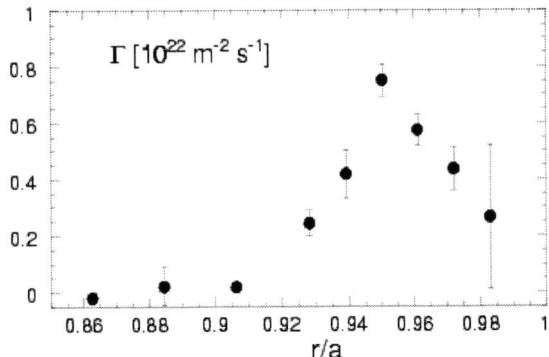

FIGURE 1. Radial profile of the ExB velocity. **FIGURE 2.** Radial profile of electrostatic particle flux.

Locally comparing the H_α emission to the particle flux driven by electrostatic fluctuations has lead to the conclusion that the particle flux is mostly driven by turbulence [1]. In fig. 2 the radial profile of the electrostatic particle flux is shown. Fluctuations contributing to the transport have time scale in the range 5 - 50 µs, while the corresponding toroidal wavelength ranges from 0.15 m to 1 m. These ranges can be compared with characteristic time and lengths; the energy and the particle confinement time are of the order of 1 ms while the pulse length is typically 80 ms, while the major circumference is ~12 m and the Larmor radius for H ions (at the edge) is ~0.5 cm.

According to the turbulence suppression criterion [5], the shearing frequency ω_s (defined as $\omega_s = k_\perp \Delta r_t \, dv_{ExB}/dr$ where dv_{ExB}/dr is the radial derivative of the ExB velocity and Δr_t is the ambient turbulence radial correlation length) must be larger than the ambient turbulence spectrum width $\Delta\omega_t$. Since the velocity shear is ~ $10^6 s^{-1}$ and typical values for $k_\perp \sim 10 m^{-1}$, $\Delta r_t \sim 0.01 m$, $\Delta\omega_t \sim 10^5$ rad/s it has shown that at the location of the maximum flux, velocity shear have values close to those required for turbulence suppression or reduction [1].

An important feature common to other fusion devices, is that time resolved analysis of particle fluxes as well as of density and potential fluctuations reveal the occurrence of bursts. These bursts, that account for less that 20% of the fluctuating power, carry up to 50 % of the total particle transport in RFX [6]. The relevant contribution of these events has motivated a statistical analysis of the fluctuations at the different spatial scales. It has been proved that in the range of frequencies relevant for the particle transport the Probability Distribution Function (PDF) develops even more non gaussian tails at the shorter time scales.

Departure from self-similarity of the PDF is called intermittency. Intermittent events, defined as those events in the non gaussian tail, have been identified at all scales by a wavelet analysis. These events tend to cluster during relaxation processes, i.e. during highly non-linear coupling phase for internal tearing modes [7]. It has been noticed that intermittency tends to become more pronounced moving towards the wall [8]. This effect, which has a remarkable analogy with ordinary fluids, proves that the boundary condition provided by the material wall in fusion devices plays an important role in plasma turbulence features and then in transport mechanisms at the edge. Focusing the analysis on the time scales relevant for transport, the spatial structure of the event has been reconstructed by a frozen turbulence hypothesis from their time behavior [9]. The corresponding ExB velocity drifts have been reconstructed showing features reminiscent of monopolar vortices.

These vortices have been grouped in two classes depending on their rotation direction. It is observed that the prevalent rotation direction depends on the shear of the surrounding mean ExB drift velocity, showing a clear analogy with ordinary fluids.

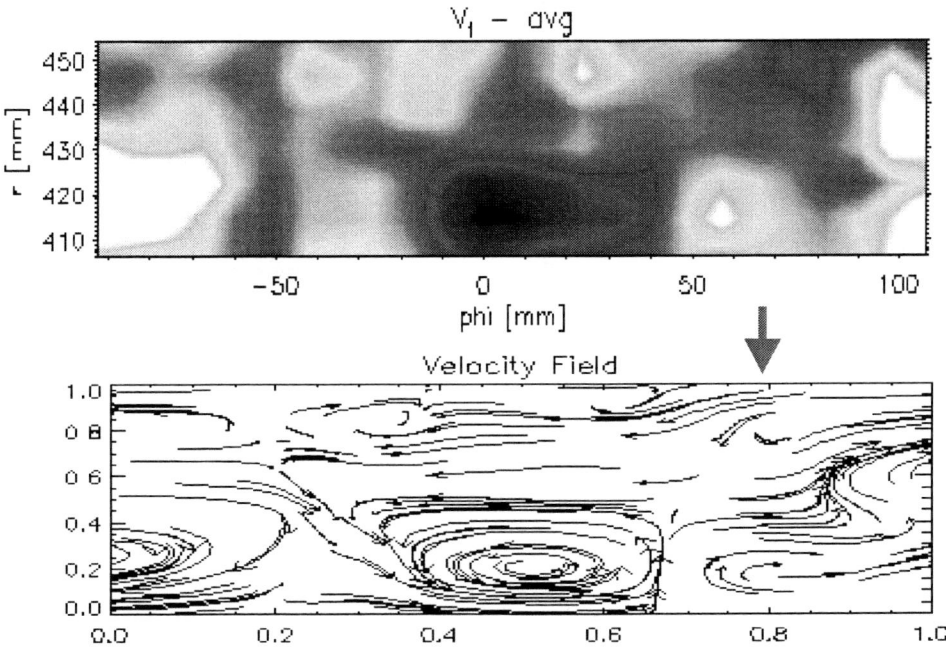

FIGURE 3. Reconstruction of the electrostatic structure corresponding to an intermittent event.

In fig. 3 an example of a ExB velocity structure derived from potential measurements in RFX is shown. From the 2-D reconstruction, the structures with time scale in the range 5 - 50 μs have radial and toroidal extension of the order of 5 and 50 cm approximately. Vortices have been found to be grouped in two classes depending on their vorticity sign. In fig.4 is shown the fraction of vortices belonging to the two groups. The two population equate only at the radial position where the shear vanishes. In the other locations the effect of the velocity shear appears to influence the preferred vorticity. The vorticity of the major part of the detected structures depends on the relative motion of the surrounding medium. This feature has been related to the average lifetime of the structures and to the fragmentation of vortices [9]. It has been suggested that vortices are generated with a random distribution in the plasma and the local shear of the mean flow selects those with vorticity consistent with the vorticity of the mean flow [9].

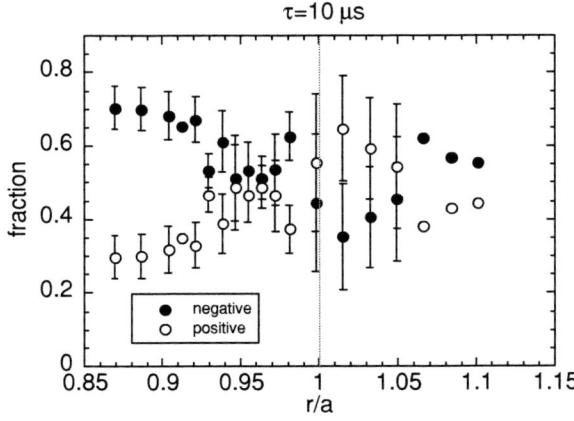

FIGURE 4. Fraction of intermittent events at different radial locations.

CONCLUSION

Experimental results in RFX indicate that ExB flow influences both background and bursts of electrostatic turbulence at the edge, affecting the electrostatic particle flux and the dynamics of monopolar vortices. The emergence of coherent structures has been already observed in non fusion plasmas [10] and their interplay with the mean flow shear investigated [11]. It is worth to note that these results add new similarities between plasmas and ordinary fluids.

As a final remark, the observation of the intermittent nature of electrostatic transport opens new frontiers in the investigation of new techniques based on local and temporal control/suppression of the single electrostatic structures are worth to be investigated as advanced schemes for turbulence control.

REFERENCES

1. V. Antoni,et al.Phys. Rev. Lett. **80**, 4185 (1998).
2. D. Craig, et al., Phys. Rev. Lett. **79** 1865 (1997).
3. V. Antoni, et al., Plasma Phys. Control. Fusion **42** (2000) 83
4. V. Antoni, D. Desideri, E. Martines, G. Serianni, L.Tramontin, Rev. Lett. **79**, 4814 (1997).
5. H. Biglari, P. H. Diamond, and P. W. Terry, Phys. Fluids B **2**, 1 (1990).
6. V. Antoni,et al. Phys Rev Lett , **87**, (2001) 045001 1-4
7. V. Antoni, et al., Europhys. Lett. **54** 51(2001)
8. V. Carbone et al. 2000 Phys.Rev E **62** R49
9. M. Spolaore, et al. submitted to Physics of Plasmas
10. O. Grulke, T. Klinger, A. Piel, Phys. Plasmas , **6** (1999) 788
11. P. S. Marcus, T. Kundu, Phys of Plasmas {\bf 7} (2000) 1630
12. W. Klinger, Handbook of chaos controlH.G. Shuster ed.Wiley VCH (1999)

Recent Results of Alfvén Wave Studies in TCABR

E.A. Lerche, L.F. Ruchko, E.M. Ozono, R.M.O. Galvão, A.G. Elfimov, A.M.M. Fonseca and R.P. da Silva

Laboratório de Física de Plasmas, Instituto de Física da USP, CEP 05315-970, São Paulo, Brasil

Abstract. The results on comparative studies of Alfvén wave plasma heating by two different antenna types in TCABR are presented. Emphasis is placed on the excited wave spectra and parasitic coupling with the edge plasma. The antenna modules have two groups of RF current-carrying straps separated by a toroidal angle of approximately 22°. In type I antenna, each group consists of two circular loops that are cut in two half-turn windings. The feeders of each loop pair are rotated 90° in the poloidal direction with respect to each other, to decrease the mutual coupling between them and make it possible to excite single helicity plasma modes (M=+1 or M=−1). In type II antenna, each group consists of two poloidal straps located at the low-magnetic-field side of the vacuum chamber. The poloidal extension of each strap is around 90° and the angle between straps is also of the same value. In both antenna types, the straps have side protectors of boron nitride. Initial experiments indicate that the parasitic interaction with the edge plasma is quite different for the two antennae. Also the first type has larger self-inductance, making it more difficult to deliver high currents to the antenna without increasing the dynamic polarization voltage up to breakdown limits. Results on the excited spectrum and floating potential at the plasma edge are presented.

INTRODUCTION

Worldwide tokamak research in the last two decades has established that auxiliary plasma heating and current drive are vital to attain thermonuclear ignition, advanced regimes of energy confinement and steady-state operation of fusion reactors. Among many different techniques investigated, the Alfvén wave (AW) heating scheme is particularly attractive because of its characteristic radially localized wave absorption, allowing local plasma heating and the possibility of generating controlled shear flows and internal transport barriers. The last results on high power AW experiments in tokamaks [1,2] were quite successful. Poloidal antennae are used to excite fast waves (FW) at the plasma boundary, which propagate towards plasma center and mode convert to kinetic Alfvén waves (KAW) that transfer energy to the plasma by Landau damping. However, some important questions about the Alfvén wave interaction with plasma, especially in high power regimes, remain unanswered.

With the objective of further advancing this promising branch of plasma physics research, experimental and theoretical investigation of Alfvén waves has been one of the primary research programs in the TCABR tokamak (R_0 = 0.61m, a = 0.18m, B_ϕ = 1.1T, $I_P \leq$ 100kA, $n_0 \leq 4 \times 10^{19} m^{-3}$) [3-8]. One first antenna module with four poloidal loops was installed in the tokamak and both AW plasma heating and current drive in intermediate power regime (P_{RF} < 100kW) have been achieved under different plasma conditions [9,10]. Nevertheless, the power input was limited due to high antenna dynamic polarization and strong interaction with the plasma periphery, in spite of the introduction of lateral protectors of boron nitride (BN). To improve the efficiency of high power heating experiments, the original antenna system of the TCABR was modified. The modifications were directed to reduce the antenna self induction, ensure lateral protection and alter the feeding configuration. The new antenna module has already been installed in the tokamak and the first experiments are being carried out.

ANTENNA STRUCTURE

The Alfvén wave excitation system (AWES) designed for the TCABR tokamak consists of four equally spaced antenna modules, with four poloidal straps each, fed by a four-phase RF oscillator which is able to supply up to

1MW RF power to the antennae in the f = 2 − 8MHz frequency range. Its distinctive feature is the possibility to excite very pure wave spectra, as single helicity traveling modes with definite poloidal wave number M = +1 or M = −1, and toroidal wave numbers in the range N = ±1 to N= ±8. The excited mode spectrum is controlled through the phasing of the antenna elements. Two complete antenna modules with different poloidal structures have been tested in the tokamak. The toroidal distribution of the antenna elements is the same for both types, consisting of two pairs of poloidal straps separated by a toroidal angle of approximately 22°. The poloidal configurations of the original (a) and modified (b) antenna types are indicated in Fig. 1.

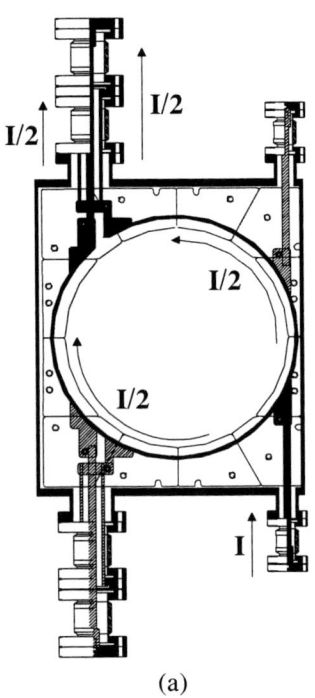
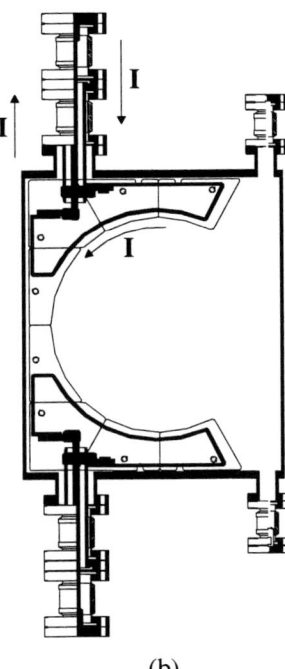

(a) (b)

FIGURE 1. Poloidal cross-section of the TCABR vacuum vessel, showing one loop pair of the original (a) and the modified (b) Alfvén wave antennae.

In type I antenna, each strap is a full loop that is cut in two half-turn windings fed in parallel. The loops of each pair are rotated 90° in the poloidal direction with respect to each other, reducing the mutual coupling between them and allowing the excitation of single helicity traveling modes (M = +1 or M = −1). In type II antenna, the poloidal extension of the windings has been reduced to approximately 90° and their width has practically doubled, decreasing their self-inductance about four times compared to the original antenna. The straps of each pair are disposed symmetrically with respect to the equator plane, in the upper and lower external corners of the vacuum vessel, covering only the low field side of the poloidal cross-section. For the two antenna types, only the poloidal components are rather well defined, in particular for type I, while several toroidal harmonics are simultaneously generated. A more selective toroidal mode excitation will be achieved with the installation of other antenna modules in different toroidal positions.

EXPERIMENTAL RESULTS

Besides the usual tokamak diagnostics, a set of RF magnetic probes and a triple electrostatic probe, both located in the shadow of the limiter, were the principal diagnostics used in these experiments. The RF power input was monitored with a specially developed electronic circuit that computes the power coupled to the plasma multiplying the currents and voltages in the antennae circuits, taking into account the ohmic losses throughout the system. An electron-cyclotron radiometer (ECE), which provides the temporal evolution of the plasma temperature profile, has been installed, and will be an essential tool in the subsequent Alfvén wave experiments.

In Fig. 2, we show the blown-up time traces of various diagnostic signals in two different plasma discharges, where Alfvén wave experiments using the original (a) and modified (b) antenna types were performed. In these experiments, only two of four antenna straps, fed with opposite phases $(0,\pi)$, were used, thus privileging the excitation of standing waves. In both cases, an RF power input of $P_{RF} \approx 80kW$ at $f = 4MHz$ was supplied to the antennae during ~10ms and the plasma density was maintained constant around $\bar{n} = (1.20 \pm 0.05) \times 10^{19} \, m^{-3}$ using a high-speed programmable gas puffing system.

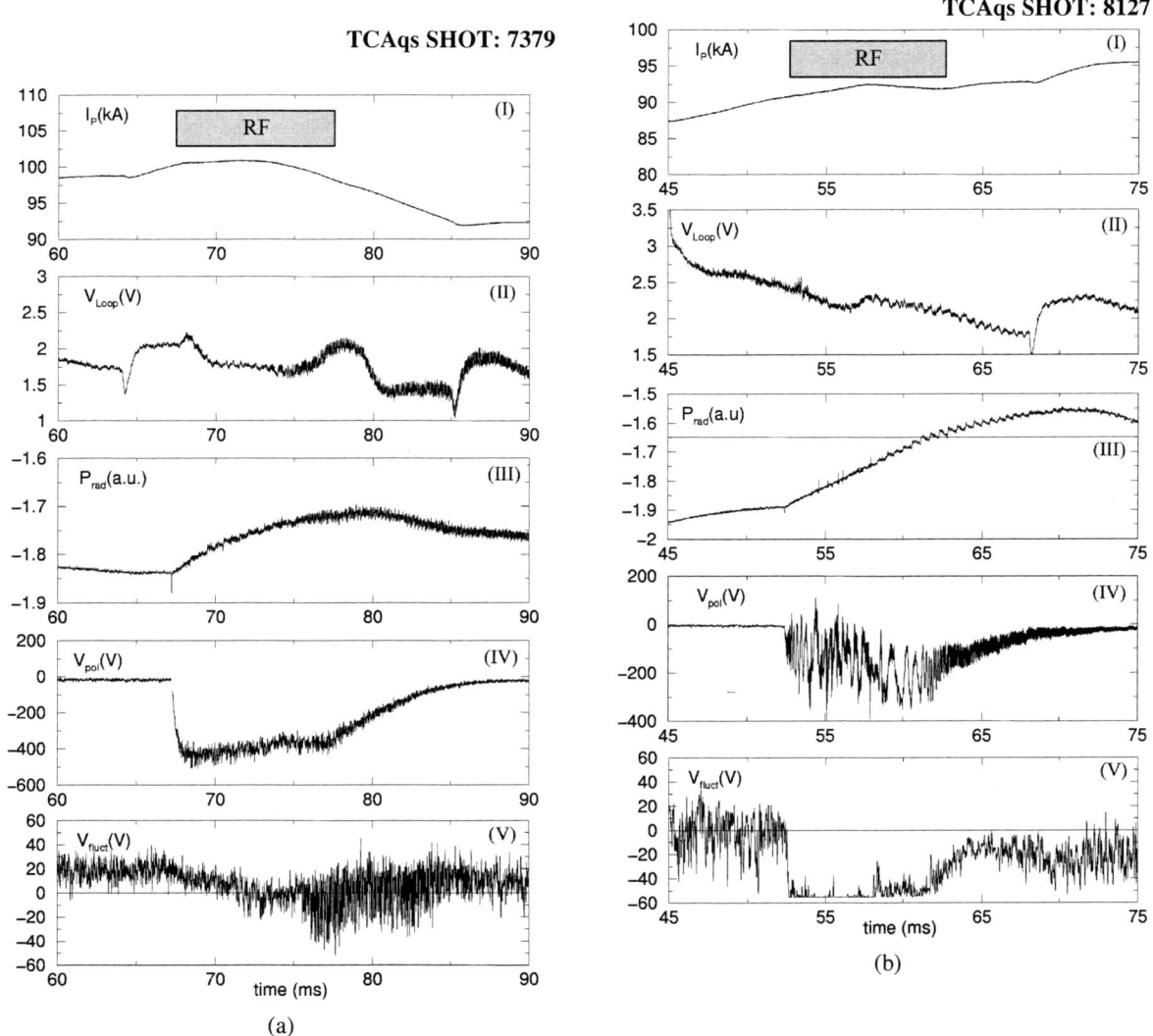

FIGURE 2. Two TCABR plasma discharges with Alfvén wave heating using the original (a) and the new (b) antenna types: (I) plasma current, (II) loop voltage, (III) radiated power, (IV) antenna dynamic polarization voltage, (V) fluctuating potential at scrape-off layer (SOL).

A first indication of plasma heating (or non-inductive current drive) can be extracted from the loop voltage and current traces of the first shot (#7379). As can be seen in signal (II-a), the loop voltage drops significantly during the RF pulse, while the plasma current (I-a) remains rather constant, evidencing an increase in the electron temperature of the plasma. At the end of the RF pulse, the input power starts to decrease and the loop voltage returns to its original value. The dynamic polarization voltage of the antenna (IV) grows rapidly with the RF pulse in both cases, causing an increase of the impurity flow into the plasma, as can be seen by the raise in the radiated power detected by the bolometer (III). Note that the dynamic polarization of the antenna is reduced about two times with the new antenna type (IV-b), demonstrating that the changes in geometry and ensured lateral protection were efficient. In contrast, the perturbation of the fluctuating potential at the plasma periphery has increased (V-b). This can either be a consequence of the broader poloidal spectrum excited by the second antenna type or the result of a strong modification of the plasma equilibrium profiles in this experiment. As a matter of fact, particularly in shot #8127, we observe the manifestation of saw-tooth oscillations in the loop voltage (II-b), radiated power (III-b) and soft X-ray emission (not shown) signals, a few milliseconds after the beginning of the RF pulse. This is a clear indication of core plasma heating and change in the central plasma equilibrium profiles, leading to the existence of the rational surface q = 1 and thus to rotating magnetic islands near the plasma center. Magnetic probe measurements and theoretical simulations of the excited spectrum identified the mode M=−1, N=−2 (q<0) as mainly responsible for this effect, having its mode conversion surface located near x = r/a ≈ 0.2 for this value of plasma density. Recent experiments performed with the ECE radiometer under the same plasma conditions, confirmed a very localized power deposition at x = (0.17±0.03), with a relative increase in the plasma temperature of $\Delta T/T \approx 25\%$ for 80kW RF power input. An increase in the temperature near the plasma periphery has also been observed, as expected in experiments with multi-harmonic Alfvén wave excitation and toroidal mode coupling effects.

CONCLUSION

The results of Alfvén wave experiments performed with two different antenna types in the TCABR tokamak were discussed. In these experiments, the antennae were tuned to excite primarily standing waves at f = 4MHz, and main attention was focused in the efficiency of plasma heating, in the change of the equilibrium profiles and in the interactions with the plasma periphery.

With the first antenna, designed to excite very pure poloidal spectra, plasma heating and current drive in reasonable power regimes were achieved. In contrast with other AW experiments [1,2], no significant density change was observed, in spite of the rather high dynamic polarization voltage induced in the antenna, what limited the power input below 100kW. Many RF phasing configurations were studied in different plasma regimes and the best conditions for plasma heating were identified.

A second antenna module with ensured lateral protectors and shorter elements was recently installed in the tokamak. The initial results showed that the modifications were successful with respect to reducing the dynamic polarization voltage of the antenna as twice for the same power input. However, the broader poloidal spectrum excited by the new antenna influences the plasma periphery more strongly, as indicated by an increase in the variation of the fluctuating potentials in the SOL during the RF pulse. A strong evidence of core plasma heating and alteration of the plasma central profiles was also observed, as the manifestation of saw-tooth oscillations in the loop voltage, soft X-ray and bolometer signals for lower densities. Magnetic probes measurements and theoretical investigations indicate that the mode M = −1, N = −2 is responsible for this central power coupling, however discrepancies between the expected values of the coupling impedance and those obtained experimentally suggest that poloidal side-band harmonics may also be playing a significant role in this case. Last results with a recently installed ECE radiometer showed very localized RF power deposition close to the plasma center, in spite of the rather broad wave spectrum excited. They also showed that the impurities generated by the antenna sputtering penetrate only a few centimeters inside the plasma column, in contrast with the global density rise observed in previous experiments.

The next step is to install three more identical antenna arrays in other sectors of the tokamak and perform experiments with more selective toroidal wave spectra, in order to improve the coupling efficiency and reduce the parasitic interactions with the plasma border.

REFERENCES

1. G.G. Borg, J.B. Lister, S. Dalla et al. , *Nucl. Fusion*, **33**, p.841 (1993).
2. R. Majesky, P.H. Probert, T. Tanaka et al., *Fus. Eng. and Design,* **24**, p.159 (1994).
3. L.F. Ruchko, E. Ozono, R.M.O. Galvão et al.., *Fus. Eng. and Design,* **43**, p.15 (1998).
4. L.F. Ruchko, E. Ozono, E.A. Lerche et al., *In Proc. of ICPP2000*, **3**, p.824, Quebec (2000).
5. R.M.O. Galvão, V. Bellintani, R.D. Bengston et al., *Plasma Phys. Cont. Fusion*, **43**, p.299 (2001).
6. L.F. Ruchko, M. Andrade and R.M.O. Galvão, *Nucl. Fusion*, **36**, p.503 (1996);
7. A.G. Elfimov, R.M.O. Galvão, I.C. Nascimento et al., *Plasma Phys. Cont. Fusion*, **40**, p.451(1998);
8. G. Amarante, A. Elfimov, D. Ross et al., *Phys. Plasmas*, **6**, p.2437 (1999);
9. E.A. Lerche, E. Ozono, L.F. Ruchko et al., *AIP Conf. Proc.*, **563**, p.191 (2001).
10. L.F. Ruchko, E.A. Lerche, R.M.O. Galvão et al., *Brazilian Journal of Physics*, **1**, p.57 (2002).

Diagnosing Electron Heat Transport in ASDEX Upgrade with ECH Power Modulation

F. Ryter, W. Suttrop, H.-U. Fahrbach, K. K. Kirov, F. Leuterer, A.G. Peeters, G. Pereverzev, G. Tardini and ASDEX Upgrade Team

Max-Planck-Institut für Plasmaphysik, EURATOM Association, D-85748 Garching, e-mail: ryter@ipp.mpg.de

Abstract. Diagnosing electron heat transport is achieved in ASDEX Upgrade using Electron Cyclotron Heating in experiments combining steady-state and power modulation. The flexible ECH system and the high quality Electron Cyclotron Emission diagnostic allow detailed experimental investigations of the heat transport characteristics.

INTRODUCTION

Experimental set-up and analyses

Electron heat transport has been extensively investigated in the ASDEX Upgrade tokamak (R=1.65 m, a=0.5 m) [1, 2] using the flexible Electron Cyclotron Heating system and the excellent measurement of the electron temperature T_e provided by the Electron Cyclotron Emission heterodyne radiometer. The ECH system is composed of 4 beams whose power deposition location can be varied independently by mirror launchers. The width of the beam is narrow, about 3 cm, and the single-pass absorbed power of 100% by the electrons is essential to make sure that no unknown spurious power is deposited elsewhere in the plasma. In addition the ECH power can be modulated with various wave forms and frequencies. The electron temperature provided by the 60 channel ECE heterodyne radiometer, which generally covers the whole plasma radius, has a spatial resolution of about 1 cm for each channel and a bandwidth of ≈ 30 kHz. The electron temperature is measured in addition by the Thomson scattering diagnostic with 16 radial channels yielding a profile every 16 ms. These two diagnostics agree within $\pm 10\%$. The other required plasma quantities are provided by the usual diagnostics available on a modern tokamak.

The experimental studies of electron heat transport have been performed in ASDEX Upgrade by combinations of the steady-state and temperature modulation methods. The former yields the usual power balance heat conductivity χ_e^{PB}, whereas the latter yields χ_e^{HP}, the so-called transient transport, which characterises the propagation of the heat pulses excited by the power modulation and is given by the expression (see review [3]):

$$\chi_e^{HP} = \chi_e^{PB} + \frac{\partial \chi_e}{\partial \nabla T_e} \nabla T_e \qquad (1)$$

This coefficient is derived experimentally from the modulated T_e, extracted from the data by Fourier transform, according to the usual expressions for slab geometry, also given in [3]. In the present work these are corrected for cylindrical geometry and density gradient as described in [4]. In our experiments the relative amplitude of the T_e modulation is limited to about 10% and the ECH modulation can be considered as an active diagnostic tool.

In general, conventional tokamak plasmas exhibit sawteeth which produce heat pulses propagating from the inversion radius towards the edge. If the ECH modulation is applied inside the inversion radius, a nonlinear interaction occurs between sawteeth and ECH modulation which strongly perturbs the analysis of χ_e^{HP} [5]. This situation was avoided in our experiments by depositing the ECH modulated power outside of the sawtooth inversion radius. Under these conditions sawteeth might cause a strong broad-band noise in some cases but the nonlinear interaction remains small and the quality of the χ_e^{HP} analysis depends only on the signal to noise ratio with which the T_e modulation can be extracted by the Fourier transform. In usual cases, the minimum detectable amplitude of the T_e modulation is in the range of 1 to 5 eV. The signal to noise ratio, generally above 10, can be optimized by an adequate choice of the ECH modulation frequency and amplitude and of the plasma conditions.

Basics on electron heat transport

The experimental studies in ASDEX Upgrade [1, 2] and other tokamaks [6, 7] suggest that electron heat transport is governed by turbulence increasing above a threshold $(\nabla T_e/T_e)_c = 1/L_{T_c}$, named κ in the following. This agrees with transport theory based on Trapped Electron Modes driven turbulence (coupled with the Ion Temperature Gradient) as one candidate [8], and Electron Temperature Gradient modes as second possibility [9]. These two instabilities indeed both have a respective threshold in $1/L_{T_e}$, normalized R/L_{T_e}. As a consequence of this transport property, the temperature profiles react weakly to changes of the heating power intensity and deposition profile: "profile resilience" of "stiffness". In fact, the temperature profiles exhibit very similar values of R/L_{T_e} in several tokamaks [6]. Equivalently, this is reflected in each device by the fact that plotted on a logarithmic scale the T_e profiles have the same shape and are shifted according to the edge or pedestal temperature, which is therefore a key parameter.

Based on these considerations, a simple analytical transport model has been developed and tested on ASDEX Upgrade data [10]. It is based on the following assumption for the heat diffusivity:

$$\chi_e = \chi_0 + \lambda T_e^\alpha (\nabla T_e/T_e - \kappa) H_\kappa \qquad (2)$$

where λ, α and κ are coefficients to be adjusted, H_κ is the Heaviside function equals to zero for $\nabla T_e/T_e < \kappa$ and to unity for $\nabla T_e/T_e \geq \kappa$. We will show below that good results are obtained with $\alpha = 0.5$. In the remaining of this paper, the units are mks except keV instead of eV for the temperatures.

Using Eq. 1, he expression for χ_e^{HP} can be derived explicitly from Eq. 2:

$$\chi_e^{HP} = \chi_0 + \lambda T_e^\alpha (2\nabla T_e/T_e - \kappa) H_\kappa \qquad (3)$$

under the assumption that χ_0 does not depends on ∇T_e. This expression shows the important property that χ_e^{HP} increases in a step largely above χ_0 as soon as $\nabla T_e/T_e$ is larger than κ, whereas χ_e^{PB} increases continuously with $\nabla T_e/T_e - \kappa$, see [10] and also Fig. 1.b. The physics validity of the empirical model is supported by the good results obtained in ASDEX Upgrade using the Weiland model [8] for NBI heated plasmas [11] and in ECH heated plasma dominated by electron transport [12].

EXPERIMENTAL RESULTS

Variation of heat flux at constant edge flux

According to the considerations of Sect. 1.2, it is essential in transport studies to vary the heat flux in the confinement region at constant edge temperature. This was achieved in new experiments at ASDEX Upgrade where we varied the electron heat flux in the confinement region ($0.35 \leq \rho_t \leq 0.7$) by one order of magnitude while keeping the heat flux at the plasma edge ($\rho_t \geq 0.65$) constant, ρ_t being the normalized toroidal flux radius. For this purpose, we deposited the ECH power at $\rho_1 \approx 0.35$ and $\rho_2 \approx 0.65$ with the respective intensities P_{ECH1} and P_{ECH2}. These were varied while keeping $P_{ECH1} + P_{ECH2}$ constant at about 1.3 MW. The discharges were L modes run at low density $\bar{n}_e = 2 \cdot 10^{19} m^{-3}$ to reduce the electron-ion energy transfer and provide good conditions to study the electron heat transport. In addition, modulation of P_{ECH1} or P_{ECH2} allows to compare transient transport (χ_e^{HP}) to power balance (χ_e^{PB}). In both experiment and modelling the Fourier transform yields profiles of amplitude and phase of the modulated T_e data at the frequency of the power modulation, providing a quantitative comparison.

The steady-state temperature profiles are shown in Fig. 1.a for a selection of these discharges. Indeed, under these conditions a clear variation of ∇T_e and $\nabla T_e/T_e$ can be achieved, which is not the case when central heating only is varied because the edge temperature increases with heating power. At the edge the ECE and Thomson scattering do not coincide perfectly. However, note the respective reproducibility of the edge profiles for $\rho_t \geq 0.65$ independently of the ratio P_{ECH1}/P_{ECH2}. The results of power balance and transient transport at $\rho_t \approx 0.5$ are shown in Fig. 1.b. There, the χ values are divided by the $T_e^{1/2}$ dependence of the model to correct for the (moderate) variation of T_e in the region of analysis. A linear fit through the power balance data, neglecting the very small contribution from χ_0, yields

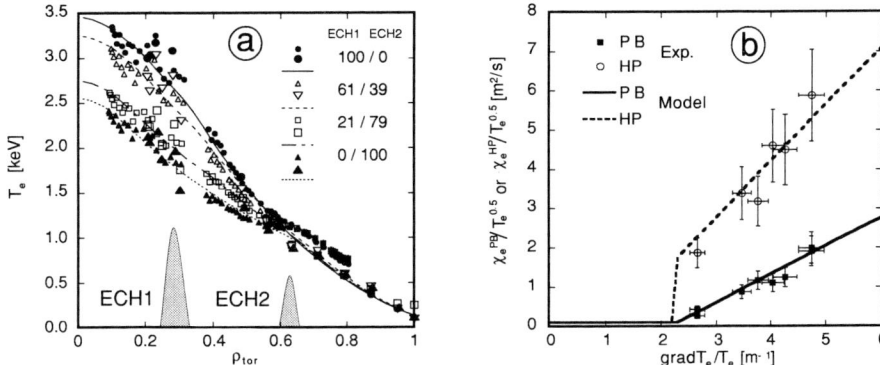

FIGURE 1. a: T_e profiles in the experiments with heat flux variation by ECH in the confinement region with constant edge flux ($P_{ECH1} + P_{ECH2} = 1.3MW$). Small symbols for ECE, large symbols for Thomson scattering, lines for modelling. b: Results from power balance and heat pulse analysis at $\rho_t \approx 0.5$. The lines are given by the model from Eq. 2 with $\lambda = 0.72$ and $\kappa = 2.3$

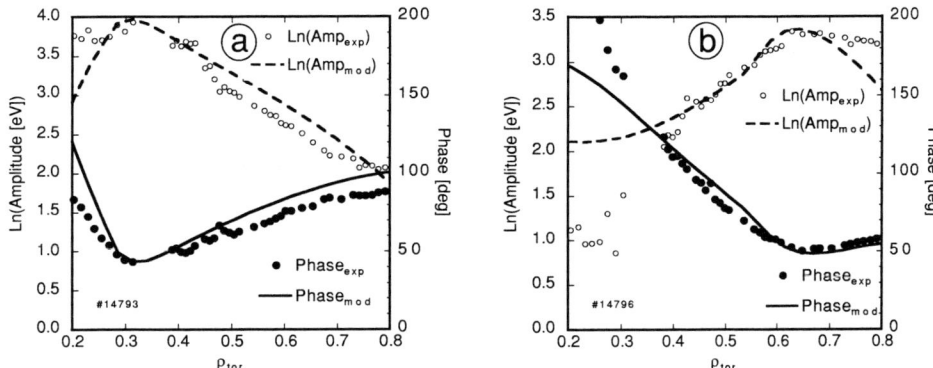

FIGURE 2. Amplitude and phase of the T_e modulation for the two extreme cases of Fig. 1, $P_{ECH1} = 100\%$ (a) or $P_{ECH2} = 100\%$ (b). The points are the data, the lines the empirical model model, both with the same values $\lambda = 0.72$ and $\kappa = 2.3$.

$\lambda = 0.72$ and $\kappa = 2.3$ (solid line in the figure). According to Eq. 3 we can also calculate the corresponding values for transient transport. The result, dashed line in Fig. 1.b, agrees well with the experimental data. These values of λ and κ were then taken for transport simulations with the ASTRA code using Eq. 2. The results, lines in Fig. 1.a, agree very well with the experimental data over the whole radius. The boundary condition is the temperature at $\rho_t = 1.0$. Using, instead of $T_e^{1/2}$, the gyro-Bohm dependence $T_e^{3/2}$ for χ, as in [10], yields poor agreement. However, $T_e^{3/2}q^2$ (q = safety factor) gives satisfactory results and this assumption is still under investigation.

The simulations also include the power modulation made in the experiment. The results of the Fourier transform of the experimental and modelled T_e are illustrated in Fig. 2 by the two extreme cases with central or edge heating. The agreement is quite satisfactory. The intermediate cases give comparably good results. It must be underlined that this agreement for both steady-state and modulation data, which is quite constraining for the model, strongly suggests that the main physics assumptions are correct. It is worthwhile noting that under the present experimental conditions in the pure off-axis heating case the T_e profile inside the ECH2 deposition is just above but close to the threshold κ. Indeed, in such off-axis cases, χ_e^{PB} is very low, but , as expected from the model and also shown by Fig. 1.b χ_e^{HP} stays rather high. We also observed experimentally and in the model that the ratio χ_e^{HP}/χ_e^{PB} goes down to about unity, with very low values of $\approx 0.2 m^2/s$ for both χ_e^{PB} and χ_e^{HP}, when the T_e profile drops below the threshold. It must be emphasized that the rapid variation of χ_e^{HP} for very small changes of χ_e^{PB} and $\nabla T_e/T_e$ is a monitor of the status of the T_e profile: above or below the threshold. Therefore, such conditions deliver a *direct* measurement of the threshold and might allow to discriminate between physics hypotheses on the turbulence involved. In particular, the good results obtained using κ without dependence on radius or plasma parameters suggest that the actual threshold should also be quite insensitive to plasma values. This is the case for the TEM driven turbulence [8], but not for the ETG turbulence for which the threshold depends strongly on T_e/T_i and s/q [9]. As already found for others discharges [12],here also

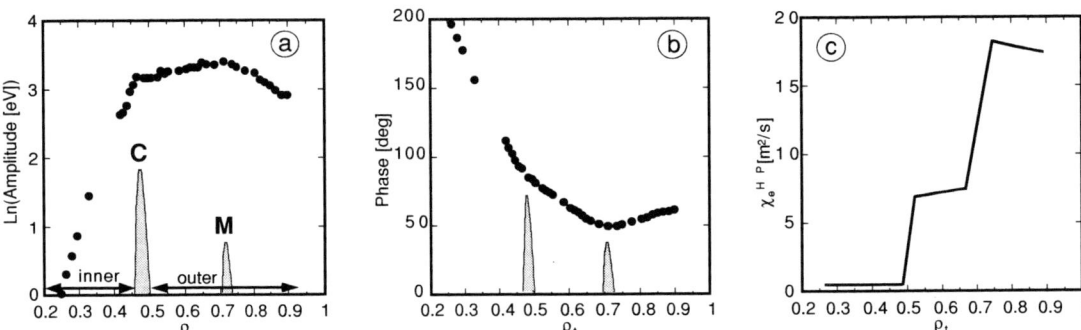

FIGURE 3. Results from heat pulse propagation in the presence of off-axis heating. (a) Natural logarithm of the amplitude of the T_e modulation; (b) Phase of the T_e modulation with respect to the power modulation of (M); (c) Derived values of χ_e^{HP}.

the ITG/TEM Weiland model [8] gives for both the steady-state and modulation data quite good results, which are very similar to those obtained with the empirical model.

Transport with off-axis ECH

According to our understanding of electron heat transport, off-axis ECH heating with deposition at about half radius, separates the plasma into two parts: the inner part inside the ECH deposition with low transport and the outer part with high transport outside of the ECH deposition. In another set of experiments we investigated this behaviour by means of ECH modulation. For this purpose we deposited 800 kW of ECH (C), steady for 2 seconds at $\rho_t \approx 0.5$. In addition we launched heat pulses from the plasma edge by ECH (M) deposited at $\rho_t \approx 0.75$ using an on-off 30 Hz modulation with a peak power of 350 kW. The propagation of the heat pulses is fast from their source up to the position of the beam (C) and slow from that position to the plasma center. This is shown in figure 3 in which the profiles of the amplitude of the T_e perturbation (on logarithmic scale) and the phase are shown. The flat slope of these two quantities in the outer region indicates fast propagation and high transport whereas the steep slope in the inner region reflects very low transport. Figure 3.c gives the quantitative results in the form of χ_e^{HP}. Note the very large difference between the inner and the outer part. The increase of χ_e^{HP} towards the plasma edge is still under investigation. These results, together with the steady-state profiles (not shown here, but see [1]), confirm the physics assumption described above. The empirical model gives good agreement also in this case [10], as well as the Weiland model [12].

Concluding, the combination of ECH power modulation with a high quality ECE diagnostic is a powerful tool to investigate electron heat transport in dedicated experiments. The analyses and the models presented here, which reflect the present understanding in electron heat transport, allow a comprehensive description of the experimental observations *simultaneously* for steady-state and transient transport.

REFERENCES

1. Ryter, F., Leuterer, F., Pereverzev, G., Fahrbach, H.-U., Stober, J. and Suttrop, W., *Phys. Rev. Letters*, **86**, 2325–2328 (2001),
2. Ryter, F., Imbeaux, F., Leuterer, F., Fahrbach, H.-U. and Suttrop, W., *Phys. Rev. Letters*, **86**, 5498–5501 (2001),
3. Lopes Cardozo, N. J., *Plasma Phys. Contr. Fusion*, **37**, 799–853 (1995).
4. Jacchia, A., Mantica, P., De Luca, F., and Gorini, G., *Phys. Fluids*, **B 3**, 3033–3040 (1991).
5. Jacchia, A., De Luca, F., Hogeweij, G. D., Gorini, G., and Konings, J. A., *Nucl. Fus.*, **34**, 1629–1639 (1994).
6. Ryter, F., Angioni, C., Beurskens, M., Cirant, S., Hoang, G. T., Hogeweij, G. M. D., Imbeaux, F., Jacchia, A., Mantica, P., Suttrop, W., and Tardini, G., *Plasma Phys. Control. Fus.*, **43**, A323–A338 (2001),
7. Hoang, G. T., Bourdelle, C., Garbet, X., Giruzzi, G., Aniel, T., et al., *Phys. Rev. Letters*, **87**, 125001–125001 (2001).
8. Nordman, H., Weiland, J., and Jarmen, A., *Nucl. Fus.*, **30**, 983–990 (1990).
9. Jenko, F., Dorland, W., and Hammett, G. W., *Phys. Plasmas*, **8**, 4096–4104 (2001).
10. Imbeaux, F., Ryter, F., and Garbet, X., *Plasma Phys. Control. Fus.*, **43**, 1503–1524 (2001).
11. Tardini, G., Peeters, A. G., Pereverzev, G. V., Ryter, F. and Stober, J., *Nucl. Fus.*, **42**, 258–264 (2002),
12. Tardini, G., Peeters, A. G., Pereverzev, G. V., and Ryter, F., A., *Nucl. Fus., in press*, **40**,(2002).

MHD Activities in High-β Plasmas of LHD

S. Sakakibara*, H. Yamada*, K.Y. Watanabe*, Y. Narushima*, K. Toi*,
S. Ohdachi*, S. Yamamoto¶, K. Narihara*, K. Tanaka*, K. Ida*
and LHD Experimental Group*

National Institute For Fusion Science, Toki 509-5292, Japan
¶ Department of Energy Engineering and Science, Nagoya Univ., Nagoya, Japan

Abstract. MHD activities in high-β plasmas have been investigated in the unfavorable configuration to an ideal interchange instability in the present operational regime of LHD. A volume averaged beta value of over 2 % was achieved in NBI plasmas without disruptive phenomena. The $m/n = 2/1$ mode excited in core region is dominant, and the plasma current to decrease magnetic shear enhances the mode activity. The plasma current exceeding a certain value leads to the disappearance of the mode and improves the plasma confinement by 20 %. Then the beta value increases by 38 %. The moderate plasma current mitigates an affect of MHD activity on the plasma confinement even in such an *unstable* configuration.

INTRODUCTION

An understanding of MHD characteristics, which may lead to a β-limit, is a major subject for realization of an efficient fusion reactor in a magnetic confinement system. In heliotron plasmas, an excitation of pressure-driven mode is key issue for a production of high-β plasma. In particular, a linear theory on ideal interchange mode such as Mercier criterion and low-n analysis is well used for an optimization of a magnetic configuration [1]. The experimental verification of the validity of the prediction is very important for an optimization of a helical fusion reactor.

Large Helical Device (LHD) is the largest superconducting heliotron-type device and a confinement magnetic field is produced by $\ell = 2/m = 10$ continuous helical coils and three pairs of poloidal coils without plasma currents [2]. The magnetic configuration is characterized by a magnetic axis position, R_{ax}, which can be widely set by controlling poloidal coil currents. The β-limit predicted by the linear theory on an ideal interchange mode is higher for the outward-shifted plasma rather than the inward-shifted case because of magnetic well formation, while inward-shifted plasma has an advantage for a neoclassical transport and particle confinement because of good particle orbit. While the shift of the R_{ax} determines the formation of magnetic well/hill, net plasma current directly change the magnetic shear even if they are sufficiently small to activate current driven instabilities. The toroidal currents with direction increasing the rotational transform lead to the decrease in the magnetic shear and the suppression of Shafranov shift, which restrains the formation of magnetic well.

In actual experiments, a configuration with $R_{ax} = 3.6$ m is well used for the production of high-β plasma because an efficiency of plasma confinement is higher than that in the standard configuration with $R_{ax} = 3.75$ m [3]. A volume averaged beta value, $<\beta_{dia}>$, of 3.2 % was obtained in NBI plasmas, where the $<\beta_{dia}>$, measured with a diamagnetic loop, is defined as $2\mu_0<p>/B_{av}^2$ and B_{av} is the averaged toroidal magnetic field in vacuum. While the $m/n = 2/1$ mode, excited in core region, affects the pressure profile in this configuration, violent instabilities which terminate plasmas have not been observed [4]. Recently, the high-β plasma experiments were made in the unfavorable magnetic configuration with $R_{ax} = 3.5$ m on ideal interchange instability in the present operational regime. This configuration is favorable within the scope of the validity of theoretical prediction of linear MHD theory. In this study, a behavior of MHD modes in high-β plasmas of $R_{ax} = 3.5$ m configuration are reported. The affect of net plasma currents on MHD activities and the plasma confinement is also discussed.

EXPERIMENTAL RESULTS

Figure 1 (a) shows the MHD activities in two high-β discharges with different plasma currents. The R_{ax} and operational toroidal field, B_t, are 3.5 m and 0.5 T, respectively. The target plasma is produced by two tangentially aimed neutral beams with total injected power of 6.1 MW and maintained to the end of the discharge. The direction of their injections was changed by reversing B_t. When $B_t > 0$, it means that one co- and two counter beams are injected. When $B_t < 0$, the directions of beams are the opposite of the $B_t > 0$ case. While the $<\beta_{dia}>$ approaches almost the same value of about 2.5 % in both discharges, the generated I_p/B_t is different. In the $B_t = +0.5$ T case, the I_p/B_t is nearly zero because the Ohkawa currents in the counter direction are cancelled out by bootstrap currents flowing in the co- direction [5, 6]. The I_p/B_t in the $B_t = -0.5$ T case, which is generated by co- Ohkawa and co-bootstrap currents, reaches about 70 kA/T at t = 2.15 s. The magnetic fluctuation with $f \leq 50$ kHz was measured in the time range from 0.4 s to 3.0 s, and the coherent modes with $n \leq 4$ and $m \leq 3$ were observed. In the $B_t = +0.5$ T discharges of fig.1 (a), the $m/n = 2/1$ mode with a frequency of about 1 kHz appears at t = 0.76 s and is continuously observed till $<\beta_{dia}>$ starts to decrease at t = 1.85 s. The amplitude of the mode approaches about 10^{-4} and is largest in all of observed modes, although the resonant surface may be located at an innermost region. The $m/n = 2/2, 3/3, 3/4$ and $2/3$ modes, which are excited in the region with $\rho \geq 0.9$, are observed to the end of the discharge, and the amplitudes increase and decrease with $\bar{n}_e$.

In the $B_t = -0.5$ T case, the $m/n = 2/1$ mode abruptly grows at t = 0.8 s and the amplitude is two and a half times as large as the $B_t = +0.5$ T case. This mode suddenly disappears when the I_p/B_t approaches about 38 kA/T. The $<\beta_{dia}>$ is restrained when the $m/n = 2/1$ mode appears, and $<\beta_{dia}>$ increases again when the amplitude of the mode starts to decrease at t = 1.2 s and reaches about 2.4 %. The $\bar{n}_e$ continues to increase from 0.5 to 1.8 s. Although the $m/n = 2/2$ and $3/3$ modes appears with the same amplitude as the $B_t = +0.5$ T case, they disappear together with the $m/n = 2/1$ mode at t = 1.2 s. The $m/n = 3/2$ mode is destabilized just after the disappearance of the $m/n = 2/1$ mode. This mode is not observed in the $B_t = +0.5$ T discharge.

Figure 2 (a) shows the time behavior of $<\beta_{dia}>$, τ_E/τ_{ISS95}, the plasma pressure and the magnetic fluctuation from 0.7 to 1.4 s in fig.1 (b) discharge. Where τ_{ISS95} is energy confinement time estimated by using International Stellarator Scaling 95 (ISS95) [7]. When the $m/n = 2/1$ mode is observed from t = 0.8 to 1.2 s, $<\beta_{dia}>$ saturates around 1.8 % and τ_E/τ_{ISS95} changes from 1.3 to 1.1. At t = 1.28 s, the $<\beta_{dia}>$ and τ_E/τ_{ISS95} dramatically increase with the

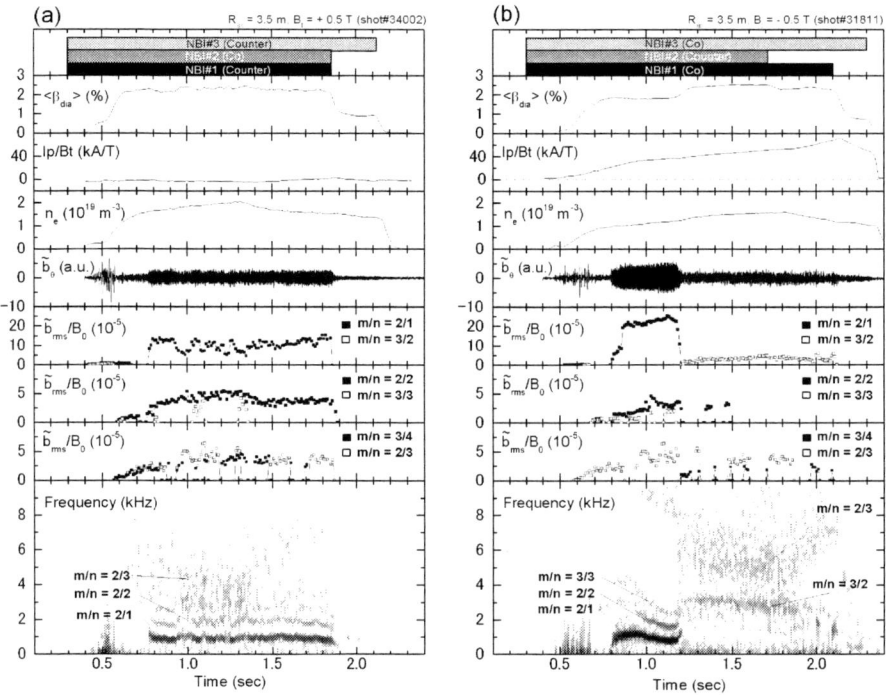

FIGURE 1. MHD activities in high-β discharges with (a) $B_t = +0.5$ T and (b) $B_t = -0.5$ T

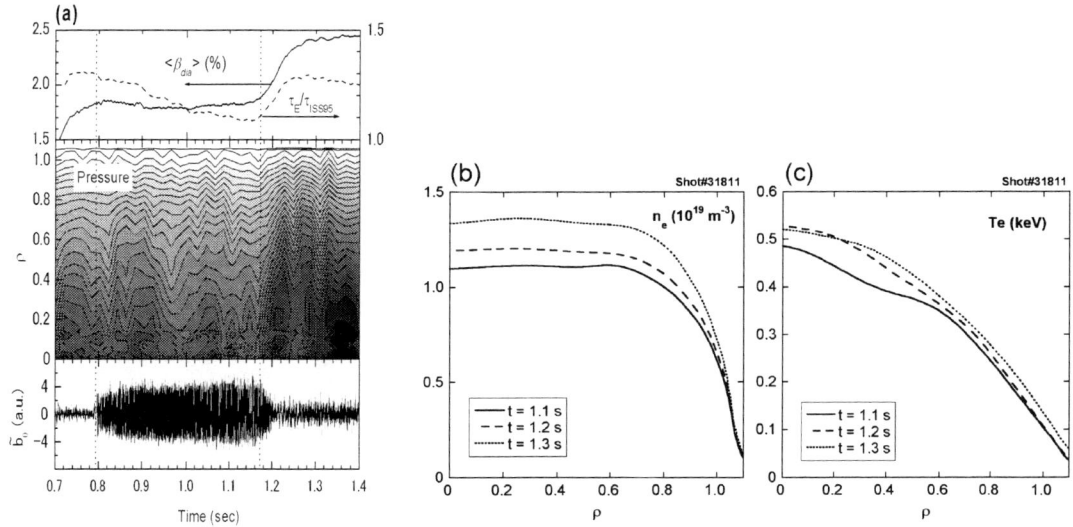

FIGURE 2. (a) Extended view from t = 0.7 to 1.4 s in fig.2 (b) discharge, and (b) electron density and (c) temperature profiles at t =1.1, 1.2 and 1.3 s.

decrease in the amplitude of the mode, and they approach 2.4 % and 1.3 at t = 1.28 s, respectively. The amplitude of the $m/n = 2/1$ mode starts to decrease, the plasma pressure increases in the core region, and the peripheral pressure also increases after the mode disappears (see fig.2 (a)). The profiles of n_e and electron temperature, T_e, at t = 1.1, 1.2 and 1.3 s are shown in fig.2 (b) and (c), respectively. The T_e in the core region with $\rho \leq 0.6$ at t = 1.2 s is higher than the t = 1.1 s, and it corresponds to the decrease in the amplitude of the mode. The T_e profile at t = 1.3 s is almost the same as that at t = 1.2 s. The n_e profile has a flatten one at t = 1.1 s, and it increases with keeping the same form. The ramp-up rate of n_e from 1.2 s to 1.3 s is higher than that from 1.1 to 1.2 s, although the gas puff fueling is constant. The increase with $<\beta_{dia}>$ is caused by the increment of n_e in the peripheral region in addition to the increase in T_e in the core region.

Figure 3 shows the changes of plasma parameters before and after the disappearance of the $m/n = 2/1$ mode as a function of I_p/B_t. The disappearance of the mode was observed in discharges with $I_p/B_t \geq 12$ kA/T. The $<\beta_{dia}>$ is 1.8 ~ 2.0 % before this disappearance, and n_e decreases from 2 to 1×10^{19} m^{-3} when I_p/B_t changes from 12 to 38 kA/T. Therefore the changes of I_p/B_t may correspond to that of Ohkawa currents. The ratio of $<\beta_{dia}>$ before and after the disappearance is less than 1.1 when $I_p/B_t \leq 25$ kA/T. It increases with I_p/B_t in the range of $I_p/B_t \geq 25$ kA/T and approaches 1.4 when $I_p/B_t = 38$ kA/T. The τ_E/τ_{ISS95} before the event is about 1.0 at $I_p/B_t < 25$ kA/T and 1.0 ~ 1.2 at $I_p/B_t \geq 25$ kA/T. One of the reasons for this difference is that the broadening of the deposition profile of neutral

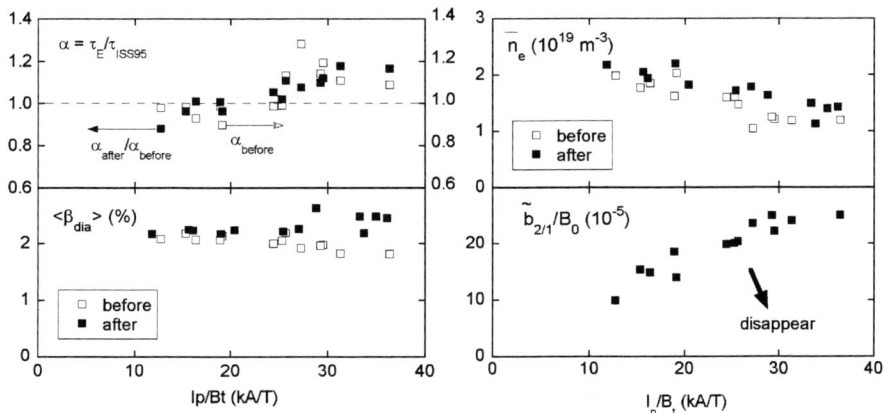

FIGURE 3. Changes of the efficiency of plasma confinement, $<\beta_{dia}>$, ne and amplitude of $m/n = 2/1$ mode before and after a disappearance of the mode as a function of I_p/B_t.

beams. Since the direction of neutral beams are optimized to magnetic configuration with $R_{ax} \geq 3.6$ m, the peak of deposition profile shifts to the outward direction with the increase in n_e. The ratio of the τ_E/τ_{ISS95} before and after the disappearance increases when I_p/B_t exceeds 25 kA/T and reaches 1.2 at $I_p/B_t = 38$ kA/T. The amplitude of the $m/n = 2/1$ mode is about 10^{-4} when I_p/B_t is 12 kA/T and it is almost the same level as that in the currentless plasmas shown in fig.1 (a). The amplitude increases with I_p/B_t.

DISCUSSION

The disappearance of the $m/n = 2/1$ mode may corresponds to exclusion of the resonant surface due to co- I_p. It happens in the wide range of the plasma currents as shown in fig.3. It may be caused by the change of the current profile. If the current profile changes from flatten to peak one as a function of ne, the $\iota = 1/2$ surface can disappear in spite of different net current. The changes of current profile as a function of ne is mainly given by Ohkawa currents rather than bootstrap currents. The analysis for the time development of Ip is required.

The improvement of plasma confinement after the disappearance of MHD activity becomes remarkable when the plasma current exceeds a certain value. A possibility for this phenomena is that the confinement in peripheral region is improved by the stabilization of the $\iota = 1$ resonant modes due to the increase in Ip. In the discharges with low- Ip, the $\iota = 1$ resonant modes are observed just after the m/n = 2/1 mode disappears. Then <βdia> changes by less than 10 %, which is caused by the increase in the Te in the core region. The similar phenomena was observed in high-β experiment in the Rax = 3.6 m configuration 4). The $\iota = 1$ resonant modes disappear with the m/n = 2/1 mode in the discharge with high- Ip as shown in fig.1 (b). Then the pressure in the peripheral region increases after the increase in the core pressure. The stabilization of the $\iota = 1$ resonant modes may lead to improvement of the confinement in the peripheral region, although the mechanism of the stabilization is not clear. The Ip is too small to disappear the $\iota = 1$ resonant surface. Although it is theoretically predicted that the increase in pressure gradient destabilizes the pressure driven mode, the $\iota = 1$ resonant modes disappear in spite of the increase in pressure gradient as shown in fig.2.

In previous experiments in Rax = 3.6 m configuration, the m/n = 2/1 mode has been observed only in the Mercier unstable region 8). The Rax = 3.5 m configuration is more unstable than the Rax = 3.6 m case. Mercier criterion indicates that the ideal m/n = 2/1mode is unstable in the present operational regime. Even in the currentless plasmas, Mercier criterion, DI, at the $\iota = 1/2$ resonant surface approaches about 1.5, which corresponds to about three times as high as the case of Rax = 3.6 m discharge with the same <βdia>. The amplitude of the observed mode in the Rax = 3.5 m case is several times as high as the Rax = 3.6 m case. While the DI at the $\iota = 2/3$ surface is about 0.3 with <βdia> = 2 %, the m/n = 3/2 mode has not been observed as shown in fig.2 (a). When there are Ip in co- direction, which decreases the magnetic shear, the m/n = 3/2 mode appears. These facts may suggest that the threshold of excitation is significantly higher than the Mercier criterion and the saturation of the mode may be related with the absolute value of the criterion.

REFERENCES

1 Ichiguchi K *et al Nucl. Fusion* **33** (1993) 481.
2. Motojima O *et al Phys. Plasmas* **6** (1999) 1843.
3. Yamada H *et al Plasma Phys.Control Fusion* **44** (2002) A245.
4. Sakakibara S *et al Nucl. Fusion* **9** (2001) 1177.
5. Watanabe K.Y *et al Nucl. Fusion* 35 (1995) 335.
6. Sakakibara S *et al Jpn. Plasma Fusion Research SERIES* 3 (2000) 109.
7. Stroth *U et al Nucl. Fusion* **36** (1996) 1063.
8. Sakakibara S *et al Plasma Phys.Control Fusion* **44** (2002) A217.

Measurements of the Fluctuation-Induced Flux with Emissive Probes in the CASTOR Tokamak

R. Schrittwieser,[1] J. Adámek, P. Balan,[1] I. Ďuran, M. Hron, C. Ioniță,[1] E. Martines,[2] J. Stöckel, M. Tichý,[3] G. Van Oost[4]

Academy of Sciences of the Czech Republic, Prague, Czech Republic
[1]*Department of Ion Physics, University of Innsbruck, Innsbruck, Austria,*
[2]*Consorzio RFX, Associazione Euratom-ENEA sulla Fusione, Padova, Italy*
[3]*Charles University, Prague, Czech Republic*
[4]*Department of Applied Physics, Ghent University, Ghent, Belgium*

Abstract. In the edge region of the CASTOR tokamak the radial particle flux, induced by fluctuations of the poloidal electric field and of the density, and the electron temperature have been measured. In contrast to the standard procedure using cold probes, here emissive probes have been used to determine the fluctuations of the plasma potential and of the poloidal electric field. We have used an arrangement consisting of two electron emissive probes and two cold probes, inserted into the edge plasma region. The two emissive probes are arranged in poloidal direction. When they are heated to electron emission, the floating potential of these probes was considered to be a relatively accurate measure of the plasma potential. The cold probes were used to measure the time average electron temperature and the ion saturation current.

INTRODUCTION

At the edge of a magnetically confined hot plasma, especially in the scrape-off layer (SOL), the plasma is turbulent and low frequency fluctuations of plasma density $n_{e,i}$, temperature $T_{e,i}$ and potential Φ_{pl} can lead to an enhanced radial loss of plasma. One of the mechanisms are fluctuations of the poloidal electric field E_θ, which in connection with the magnetic field B_0 lead to a radial drift of particles with a velocity $\tilde{v}_r = \tilde{E}_\theta B_0 / B_0^2$. To understand the actual mechanism of the turbulent transport through the SOL, a reliable and precise knowledge of Φ_{pl} and its fluctuations is vital for a comparison of the experimental observations with theoretical models and numerical simulations.

In order to measure the plasma potential and the electric field as precisely and directly as possible, we have used electron emissive probes, the floating potential of which in principle equals the plasma potential Φ_{pl}, when the electron emission current I_{em} is sufficiently high [1,2]. For a Maxwellian plasma the floating potential of a cold probe is proportional to the plasma potential Φ_{pl} through the relation

$$V_{fl} = \Phi_{pl} - \mu T_e = \Phi_{pl} - T_e \ln\left(\frac{I_{es}}{I_{is}}\right), \quad (1)$$

with $I_{es,is}$ being the electron and ion saturation currents, respectively. For equal temperatures of electrons and ions, i.e., $T_e = T_i$, the factor μ becomes $\mu = 2.04$, if the radius of the probe wire is less or comparable with the ion Larmor radius. Here, we have assumed the electrons to be strongly magnetized so that the effective area for electron collection is just the projection of the probe along the magnetic field lines, whereas the effective area for ion collection is the total area of the probe wire.

If there is an additional electron emission current I_{em} from the probe into the plasma, Eq. (1) becomes

$$V_{fl}^{em} = \Phi_{pl} - \mu_{em} T_e = \Phi_{pl} - T_e \ln\left(\frac{I_{es}}{I_{is} + I_{em}}\right) \quad (2)$$

so that the factor in front of T_e now is: $\mu_{em} = \ln[I_{es}/(I_{is}+I_{em})]$. Therefore the floating potential of such a probe attains the plasma potential for $I_{em} = I_{es} - I_{is}$. In addition, the floating potential of an emissive probe is independent of electron temperature fluctuations and electron drifts. Indeed, also in a realistic experiment, V_{fl}^{em} is observed to increase with the emission current, however, reaching a clear saturation for $I_{em} \cong I_{es}$, i.e., a further increase of I_{em} will not lead to a further growth of V_{fl}^{em}. This saturated value of V_{fl}^{em} is then considered to be a good approximation for Φ_{pl}. This result can, however, be falsified by some effects, among them the formation of a space charge layer around the emissive probe by the emitted electrons [2,3,4].

The main goal of this experiment was (i) to measure the fluctuation-induced flux and (ii) to determine the electron temperature and its fluctuations.

EXPERIMENTAL SET-UP AND CONSIDERATIONS

The CASTOR tokamak has a major radius of 0.40 m and a minor radius of $a = 85$ mm, the latter being determined by a poloidal ring limiter. The typical duration of each shot was 30 ms. The toroidal magnetic field was 1 T, the toroidal plasma current was typically 10 kA. The working gas is hydrogen. The typical line average plasma density was 10^{19} m^{-3}. In the SOL the plasma density was 0.5 - 2×10^{18} m^{-3}, and T_e on the order of 10 eV.

In order to measure the plasma potential and the electric field in this device, we have used a radially movable arrangement consisting of two electron-emissive probes and two cold cylindrical probes located on the same minor radius. Each emissive probe consist of a loop of thoriated tungsten wire of 0.2 mm diameter. Each loop has a total length of about 8 mm which is inserted into a double-bore ceramic tube of Al$_2$O$_3$, forming approximately a half-circle of about 3.5 mm height. Inside the bores, the thoriated tungsten wires are densely covered with a number of thin copper threads [5], which are protruding from the other end of the tube so that there, by means of a feedthrough, they can be connected to an external heating circuit. This has the effect that inside the ceramic tubes, the conductivity of the wires is strongly increased so that only the actual loop will be heated to white glow and electron emission. By heating these probes simultaneously with currents up to 7.4 A, a sufficient electron emission of about 250 mA could be achieved [2]. The heating of the probes was switched on 5 s before each CASTOR shot and has been turned off immediately afterwards. In this way, a long lifetime of the probes was guaranteed. Fig. 1 shows a

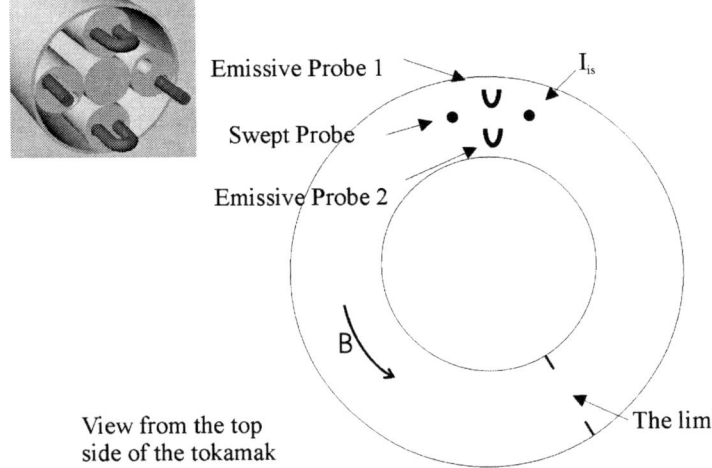

FIGURE 1. Schematic of the probe set-up, which consists of two emissive probes and two cylindrical cold probes. The main figure shows the arrangement of the probes with respect to the torus. The insert shows a detail of the probe head. The poloidal separation between the two emissive probes is 5.4 mm, the two cold probe are 5.6 mm apart from each other. The probe head is introducible into the plasma on a shot-to-shot basis from the top of the torus.

schematic of the experimental arrangement and the probe head used in this experiment.

The two emissive probes are situated along a poloidal meridian so that the poloidal component of the electric field can be determined by heating them simultaneously to sufficient emission by the formula:

$$E_\theta = \frac{(\Phi_{pl,1} - \Phi_{pl,2})}{d}, \qquad (3)$$

where $d = 5.4$ mm is the poloidal spacing between the two emissive probes. Naturally, Eq. (3) is also valid for the poloidal electric field fluctuations $\tilde{E}_\theta$. Measuring at the same time the fluctuations of the plasma density $\tilde{n}_{pl} = \tilde{I}_{i,sat}$ by means of one of the nearby cold probes, the radial fluctuation-induced flux,

$$\tilde{\Gamma}_r = \langle \tilde{n}_{pl} \tilde{v}_r \rangle = \langle \tilde{I}_{i,sat} \tilde{E}_\theta / B_0 \rangle, \qquad (4)$$

can be derived. $I_{i,sat}$ is the ion saturation current to the cold probe, where we have neglected $\tilde{T}_e$.

Another possibility with our probe system is to determine the electron temperature and its fluctuations. This is based on the idea that Eq. (1) is able to deliver $T_e(t)$, provided we can determine V_{fl} and Φ_{pl} independently:

$$T_e = \frac{\Phi_{pl} - V_{fl}}{\mu}. \qquad (5)$$

So, by using the floating potential from one of the cold probes and the plasma potential from one of the emissive probes, we can obtain the electron temperature and its fluctuations with one simultaneous measurement. The two probes have, however, to be sufficiently close to each other in order to make a meaningful measurement.

EXPERIMENTAL RESULTS AND DISCUSSION

Fig. 2 shows radial profiles of the fluctuation-induced flux (calculated by means of Eq. 4), determined in different ways for comparison. Fig. 2a presents the time average values of $\tilde{\Gamma}_r$, Fig. 2b shows the rms (root-mean square) - values of the flux. Although the poloidal electric field E_θ was always determined by the emissive probes, in the case of the solid square symbols the probes were not heated so that they acted as simple Langmuir probes. We remember that in this case E_θ is calculated as the difference between two cold probe floating potentials, and that each of these values still contains also μT_e. If, therefore, there is a difference between the electron temperatures at the two probe positions, this would perturb such a measurement. Likewise, if there is a different electron drift at the two probe positions, also this would give rise to an erroneous result for the poloidal electric field. However, in the case of the star symbols the probes were sufficiently emissive so that the plasma potentials were registered and thus a more realistic value of E_θ was obtained.

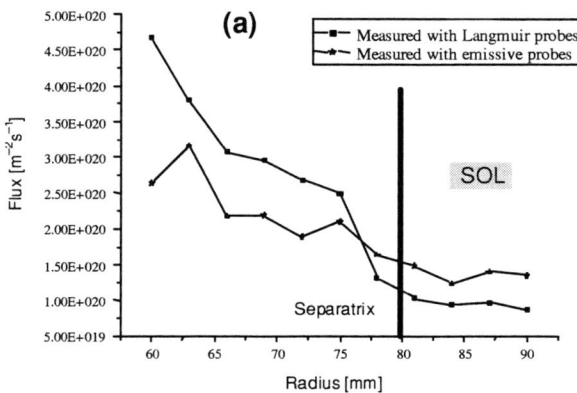
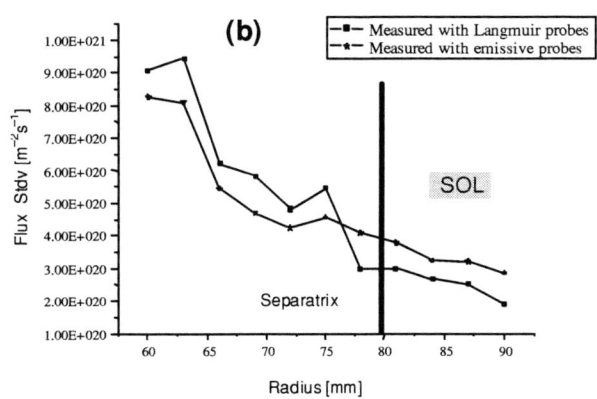

FIGURE 2. Radial profile of the fluctuation-induced flux; (a) average values; (b) rms-values; in both cases the emissive probes have been once not heated ("Langmuir probes") and then heated to electron emission. The vertical solid bars indicate the position of the last closed flux surface, outside of which the scrape-off layer begins.

We observe that in both cases the two lines intersect near the separatrix. Inside the last closed flux surface (LCFS), thus in the core plasma, the measurements with the cold probes ("Langmuir probes") deliver a larger value than outside. In order to explain this result, further analysis is needed since also the phase relation between the plasma potential and the electron temperatures fluctuations in the SOL seem to play a decisive role.

Fig. 3 shows a radial plasma potential profile, taken from the floating potential of one of the emissive probes. Also shown is a radial profile of the floating potential V_{fl}, taken by one of the cold probes. By applying Eq. (5), the difference between the plasma potential and the floating potential of the nearby cylindrical probe gives us an ap-

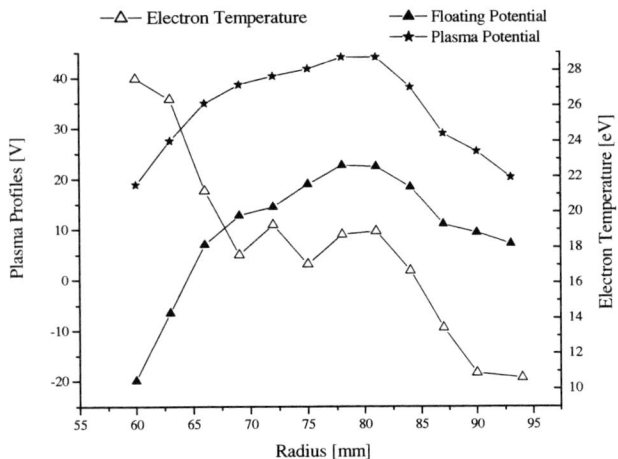

FIGURE 3. Radial profiles of the plasma potential (determined as the floating potential of one of the emissive probes) and of the floating potential profile of one of the cold probes. The electron temperature profile (Eq. 5) was calculated choosing $\mu = 2.04$.

proximation of the electron temperature and its fluctuations [2]. The correlation coefficient between $\tilde{\Phi}$ and $\tilde{V}_{fl}$ is 0.9. This is a proof that we measure with the two probes at essentially at the same spatial position [6].

CONCLUSIONS

We have performed a number of experiments and measurements in the CASTOR tokamak, using electron emissive probes instead of the conventional cold probes. We have observed that our probes can survive the plasma environment in the edge region of such a tokamak as CASTOR and similar devices. The floating potential of a sufficiently emissive probe is a better approximation of the plasma potential than that of a cold probe.

By means of an array of two emissive probes and two cold probes we could determine the fluctuation-induced radial particle flux directly and compare it with the standard technique, which uses three cold Langmuir tips. It is found that the difference is not dramatic. This implies that the electron temperature fluctuations do play some, but not a crucial role if the $\tilde{\Gamma}_r$ is measured by the standard technique, which, therefore, yields a good estimate of the $\tilde{\Gamma}_r$-magnitude. However, the observed differences in $\tilde{\Gamma}_r$ between these two techniques (being either positive or negative by about $\cong 20 - 30\%$, depending on the radial position of the probe head) need further analysis of the experimental data to be interpreted.

ACKNOWLEDGMENTS

This work has been carried out within the Association EURATOM-ÖAW, within the Association EURATOM/ENEA and within the Association IPP.CR under contract. The content of the publication is the sole responsibility of its author(s) and it does not necessarily represent the views of the Commission or its services. The support by the Fonds zur Förderung der wissenschaftlichen Forschung (Austria) under grant No. P-14545 and by the University of Innsbruck and by the Grant Agency of the Czech Academy of Sciences, grant No. 1043101 are also acknowledged.

REFERENCES

[1] R. Schrittwieser, C. Ioniță, P.C. Balan, Jose A. Cabral, F.H. Figueiredo, V. Pohoață, C. Varandas, *Contrib. Plasma Phys.* **41** (2001), 494.
[2] R. Schrittwieser, J. Adámek, P. Balan, M. Hron, C. Ioniță, K. Jakubka, L. Kryška, E. Martines, J. Stöckel, M. Tichý, G. Van Oost, *Plasma Phys. Contr. Fusion* **44** (2002), 567.
[3] K. Reinmüller, *Contrib. Plasma Phys.* **38** (1998), 7.
[4] M.Y. Ye, S. Takamura, *Phys. Plasmas* **7** (2000), 3457.
[5] A. Siebenförcher, R. Schrittwieser, *Rev. Sci. Instrum.* **67** (1996), 849.
[6] J. Adámek, P. Balan, I. Ďuran, M. Hron, C. Ioniță, L. Kryška, E. Martines, R. Schrittwieser, J. Stöckel, M. Tichý, G. Van Oost, 20th *Symp. Plasma Phys. Techn.* (Prague, Czech Republic, 2002), to be published in *Czechoslovak J. Phys.*

Applicability of Electron Emissive Probes for Plasma Potential and Electric Field Measurements in Magnetized Plasmas

R. Schrittwieser,[1] J. Adámek,[2] P. Balan,[1] J.A. Cabral,[3] H. Fernandes,[3] H.F.C. Figueiredo,[3] C. Hidalgo,[4] M. Hron,[2] C. Ioniță,[1] E. Martines,[5] M.A. Pedrosa,[4] J. Stöckel,[2] M. Tichý,[6] G. Van Oost,[7] C. Varandas[3]

[1]*Department of Ion Physics, University of Innsbruck, Innsbruck, Austria;* [2]*Institute of Plasma Physics, Academy of Sciences of the Czech Republic, Prague, Czech Republic;* [3]*Center for Nuclear Fusion, Instituto Superior Técnico, Lisbon, Portugal;* [4]*Particle and Probe Diagnostic Group, CIEMAT, Madrid, Spain;* [5]*Consorzio RFX, Associazione Euratom-ENEA sulla Fusione, Padova, Italy,* [6]*Charles University, Prague, Czech Republic,* [7]*Department of Applied Physics, Ghent University, Ghent, Belgium*

Abstract. In the edge region of magnetised fusion experiments hitherto mainly cold probes were used in order to determine the plasma potential and thereby parameters like electric field turbulence. However, this method causes problems when the electron temperature varies and when the electrons are drifting. We have therefore used electron-emissive probes in (i) the tokamak ISTTOK in Lisbon, Portugal, (ii) the CASTOR tokamak in Prague, Czech Republic, and (iii) the TJ-II stellarator in Madrid, Spain. Our method has the advantage that in principle the electron emission current compensates temperature variations and electron drifts. We discuss the applicability, the advantages and the limits of emissive probes for measurements of the plasma potential and the electric field, taking into account especially the effect of a space charge around the probe wire, formed by the emitted electrons.

INTRODUCTION

The stability of a hot, magnetically confined fusion plasma is essentially determined by the conditions in the edge plasma region, which often is turbulent. This can cause an enhanced radial loss of plasma near the last closed flux surface (LCFS) towards the scrape-off layer (SOL). For a better understanding of turbulence near the LCFS and in the SOL, a more reliable and more direct determination of the plasma potential Φ_{pl}, respectively of the electric field and of their fluctuations is essential.

In magnetised fusion plasmas up to now mainly cold probes have been used to determine Φ_{pl} and/or various electric field components. Here we present a summary of our investigations where electron-emissive probes were used for this purpose. Measurements of Φ_{pl} with cold probes deliver false results when the electron temperature varies and/or when there are electron drifts or electron beams in the plasma. However, emissive probes are in principle not subject to such errors since the emitted electron current compensates electron temperature variations and electron drifts to a certain extent [1,2].

COMPARISON BETWEEN COLD PROBES AND EMISSIVE PROBES

It is usually assumed that Φ_{pl} can be inferred from the floating potential V_{fl} of a cold probe. Indeed, in a purely Maxwellian plasma, V_{fl} and Φ_{pl} are related to each other by $V_{fl} = \Phi_{pl} - \mu T_e$, with $\mu = \ln(I_{es}/I_{is})$. T_e is the electron temperature in energy units, $I_{es,is}$ are the electron and ion saturation currents, respectively. So, when T_e and V_{fl} are determined simultaneously, the plasma potential can in principle be calculated. For equal temperatures of electrons and ions, the factor μ becomes $\mu = 2.04$, where the electrons are assumed to be strongly magnetized so that the effective area for electron collection is just double the projection of the probe along the magnetic field lines, whereas the effective area for ion collection is the total surface area of the probe wire [2]. However, if there are temporal or spatial variations of T_e, the calculation of $\Phi_{pl}(V_{fl})$ can become impossible. In addition, deviations of the electron velocity distribution function (VDF) from a Maxwellian one (e.g. by an electron drift, beam or runaway electrons) can shift the floating potential by a voltage which corresponds to the average kinetic energy of the drifting electrons [1].

These problems can partly be circumvented when we use probes, which actively emit an electron current into the plasma [1,2,3,4,5,6]. An electron emission current I_{em} will flow from the probe to the plasma as long as the probe voltage V_p is smaller than Φ_{pl}, *irrespective* of the plasma electron VDF and of fluctuations of T_e. For $V_p \geq \Phi_{pl}$, the emission current drops and electron collection begins to dominate the probe current. Usually the floating potential V_{fl}^{em} of an emissive probe is considered a reasonably accurate approximation for the plasma potential [5].

In a purely Maxwellian plasma, for $V_p \leq \Phi_{pl}$, the total current I_p to an emissive probe is $I_p = I_{is} + I_{em} - I_{es}[(V_p - \Phi_{pl})/T_e]$. At the floating potential, i.e., for $V_p = V_{fl}^{em}$, I_p equals zero. Then we get:

$$\Delta \equiv \frac{\Phi_{pl} - V_{fl}^{em}}{T_e} = \ln\left(\frac{I_{es}}{I_{is} + I_{em}}\right), \qquad (1)$$

which, for an unheated probe ($I_{em} = 0$) becomes simply $\ln(I_{es}/I_{is}) = \mu \equiv \Delta_0$ (see above). For the case $V_{fl}^{em} = \Phi_{pl}$, we have to put $\Delta = 0$, and then obtain from Eq. (1) the condition $I_{em} = I_{es} - I_{is}$.

Without probe heating, $I_{em} = 0$, and V_{fl} corresponds to that of a cold probe. Thus $\Phi_{pl} = V_{fl} + 2.04T_e$. For increasing probe heating, I_{em} increases and Δ becomes smaller, while the floating potential of the probe grows and approaches Φ_{pl} until $V_{fl}^{em} \cong \Phi_{pl}$ for Iem = Ies − Iis. Whereas Eq. (1) indicates that for stronger heating V_{fl}^{em} becomes larger than Φ_{pl}, in reality the floating potential of a sufficiently emissive probe does not surpass Φ_{pl}, since our simplified treatment is valid only for $V_p \leq \Phi_{pl}$.

The electron emission current I_{em} is given by the Richardson-Dushman equation $I_{em} = A_{em} A^* T_w^2 \exp(-eW_w/k_B T_w)$. Here, A_{em} is the total area of the probe wire, A^* the Richardson constant, and T_w (in K) and W_w the temperature and the work function of the wire material, respectively. In principle, this emission current only depends on T_w but not on V_p, unless space charge effects have to be taken into account (see Sec. 4).

Fig. 1 shows the conditions around an emissive probe at floating potential schematically. In addition to the usual currents $I_{es,is}$ from the plasma to the probe, there is the current I_{em} of emitted electrons from the probe to the plasma, which in the ideal case (Fig. 1a) compensates the electron current from the plasma, thereby rising the floating potential of the probe to the plasma potential.

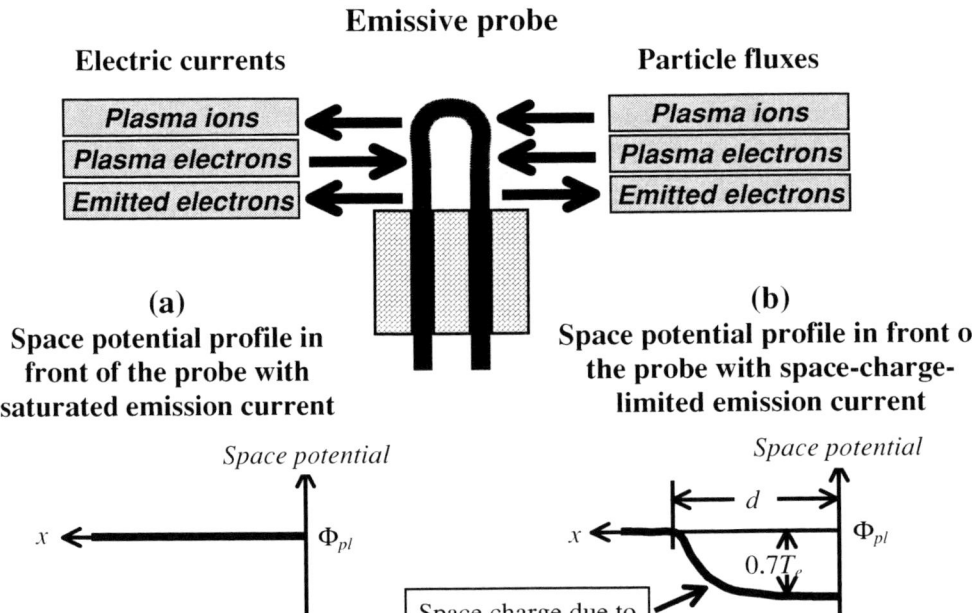

FIGURE 1. Schematic representation of the conditions around an electron emissive probe, where in addition to the particle fluxes from the plasma to the probe, there is a flux of emitted electrons from the probe to the plasma; (a) the conditions around an idealised probe, where the floating potential assumes the value of the plasma potential; (b) the conditions when the emitted electrons form a non-negligible space charge.

EXPERIMENTAL SET-UPS AND EMISSIVE PROBE ARRANGEMENTS

The most important parameters of the three toroidal fusion experiments, in which emissive probes have been used up to now, are listed in Table I. In all cases, hydrogen was the working gas.

	Confinement scheme	Major radius [m]	Minor radius [cm]	Magnetic field [T]	Pulse length [ms]	Toroidal current/RF heating power	Line averaged density [m^{-3}]	Edge plasma density [m^{-3}]	Core electron temperature [eV]	Edge electron temperature [eV]
ISTTOK	Tokamak	0.46	8.5	0.5	40	9 kA	5-10×10^{18}	5-10×10^{16}	80-220	10
CASTOR	Tokamak	0.40	8.5	1.0	30	10 kA	0.5-2.5×10^{19}	0.2-2×10^{18}	150-300	8-25
TJ-II	Flexible heliac stellarator	1.5	ca. 22	1.2	250	600 kW	10^{19} – 10^{20}	2×10^{18}	1000	100

Table I. Typical parameters of the three toroidal devices used.

The design and construction of our electron emissive probes has been described in detail in Refs. [1,2,6]. They are able to deliver an emission current up to 300 mA. Various probe arrangements have been used in the three above mentioned toroidal fusion experiments:

The first measurements have been made in ISTTOK in Lisbon with a set-up of three emissive probes located on the same poloidal meridian, 20 mm apart from each other, but on three different minor radii outside the LCFS. With this arrangement, the plasma potential and its fluctuations could be measured at three radial positions in the edge region simultaneously [1]. Approximate radial potential profiles have been obtained with this arrangement. In addition, the power spectra of these fluctuations were evaluated in the range of 10 Hz < f < 60 kHz and a proportionality with $f^{-\alpha}$ (with α being 1.9 approximately) was found [7].

In the CASTOR tokamak at first a single emissive probe was used for taking radials profiles of Φ_{pl} outside and even inside the LCFS with and without separatrix biasing [2]. In the meantime a more complicated arrangement consisting of a double emissive and a double cold probe has been inserted in the CASTOR. This set-up allows the direct determination of the poloidal electric field and of the plasma density (through the ion saturation current) and was used to measure the fluctuation-induced flux $\widetilde{\Gamma}_r = \langle \widetilde{n}_{pl} \widetilde{v}_r \rangle = \langle \widetilde{I}_{i,sat} \widetilde{E}_\theta / B_0 \rangle$ [8,9]. Moreover, also the electron temperature was directly determined by using the relation $T_e = (\Phi_{pl} - V_{fl})/\mu$, which can be derived from the basic relation between the floating potential of an emissive probe and the plasma potential [10]. We have also performed extensive comparisons of the fluctuations of the floating potentials of the emissive and of the cold probes and of the phase relations in between.

Also in ISTTOK a new probe arrangement has been inserted recently. This is a vertically movable arrangement of three emissive probes, positioned in such a way that the radial and the poloidal component of the electric field in the edge region can be recorded simultaneously. The idea is a direct determination of the Reynolds stress $Re = \langle \widetilde{v}_r \widetilde{v}_\theta \rangle \equiv \langle \widetilde{E}_r \widetilde{E}_\theta \rangle / B_0^2$ [11] and a comparison with measurements with cold probes [12].

In the TJ-II flexible heliac in Madrid, an emissive probe and a cold probe have been mounted on the reciprocal probe carrier, with which the probes were inserted into the edge region of the plasma for about 30 ms. A comparison was made between the floating potentials of both probes and the floating potential fluctuations [13].

In general, we have demonstrated in a number of successful experiments that emissive probes can sustain the conditions in the edge region of smaller toroidal fusion experiments and can deliver useful information on the plasma potential. The probes do not perturb the plasma too much and also no detectable evaporation of wire material was found.

POSSIBLE SPACE CHARGE EFFECTS

Eq. (1) together with the other relations mentioned above is valid only as long as the electron emission current is saturated. However, as pointed out recently [14,15], for strong electron emission a space charge might form around the emissive probe, which would lead to a modification of the floating potential of the probe with respect to the case discussed in Sec. 2. In order to get a rough quantitative idea about this effect, in Eq. (1) we have to complement the Richardson-Dushman law, which yields the current I_{em} depending only on the wire temperature, by the Child-Langmuir law:

$$I_{CL}(V_p) = A_{em} \frac{4\varepsilon_0}{9} \left(\frac{2e}{m_e}\right)^{1/2} \frac{(\Phi_{pl} - V_p)^{3/2}}{d^2}. \tag{2}$$

Now the emission current depends also on the difference between the probe voltage and the plasma potential. Here m_e is the electron mass and d is the thickness of the sheath which forms around the probe wire by the emitted electrons. We assume that this has the order of the Debye length. Fig. 1b shows these conditions schematically. It is obvious that not only an electron-rich sheath is formed but there must also be an ion-rich sheath, so that actually a double layer is created around the probe wires.

By defining an analogous term ΔCL like in Eq. (1), we can derive an equation, which has the solution $\Delta CL \cong 0.7\, Te$ [2]. This means that the difference between V_{fl}^{em} and Φpl cannot become zero anymore, but that the floating potential of an emissive probe remains below the actual value of Φpl by $0.7\, Te$ approximately. Although we have indications that this space charge effect does not play a significant role in our experiments, further investigations have to be performed to clarify this question [2].

ACKNOWLEDGMENT

This work has been carried out within the Association EURATOM-ÖAW, within the Association EURATOM/ENEA and within the Association IPP.CR under contract. The content of the publication is the sole responsibility of its author(s) and it does not necessarily represent the views of the Commission or its services. The support by the Fonds zur Förderung der wissenschaftlichen Forschung (Austria) under grant No. P-14545 and by the Grant Agency of the Czech Academy of Sciences, grant No. 1043101 are also acknowledged.

REFERENCES

1. Schrittwieser, R., Ioniţă, C., Balan, P. C., Cabral, Jose A., Figueiredo, F. H., Pohoaţă, V., Varandas, C., *Contrib. Plasma Phys.* **41**, 494 (2001).
2. Schrittwieser, R., Adámek, J., Balan, P., Hron, M., Ioniţă, C., Jakubka, K., Kryška, L., Martines, E., Stöckel, J., Tichý, M., Van Oost, G., *Plasma Phys. Contr. Fusion* **44**, 567 (2002).
3. R.F. Kemp, J.M. Sellen Jr., *Rev. Sci. Instrum.* **37** (1966), 455; J.R. Smith, N. Hershkowitz, P. Coakley, *Rev. Sci. Instrum.* **50** (1979), 210.
4. Motley, R.W., *J. Appl. Phys.* **43** (1972), 3711; Fujita, H., Yagura, S., *Jpn. J. Appl. Phys.* **22** (1983), 148; Makowski, M.A., Emmert, G.A., *Rev. Sci. Instrum.* **54** (1983), 830.
5. Iizuka, S., Michelsen, P., Rasmussen, J.J., Schrittwieser, R., Hatakeyama, R., Saeki, K., Sato, N., *J. Phys. E: Sci. Instrum.* **14** (1981), 1291.
6. Siebenförcher A., Schrittwieser R., *Rev. Sci. Instrum.* **67** (1996), 849.
7. R. Schrittwieser, J.A. Cabral, P. Balan, H.F.C. Figueiredo, H. Fernandes, C. Ioniţă, C. Varandas, *29th EPS Conference on Plasma Physics and Controlled Fusion*, (Montreux, Switzerland, 2002), in print.
8. Balan, P., Adámek, J., Ďuran, I., Hron, M., Ioniţă, C., Martines E., Schrittwieser, R., Stöckel, J., Tichý, M., Van Oost, G., *Proc. 29th EPS Conf. Plasma Phys. Contr. Fusion*, (Montreux, Switzerland, 2002), in print.
9. Adámek, J., Balan, P., Ďuran, I., Hron, M., Ioniţă, C., Martines, E., Schrittwieser, R., Stöckel, J., Tichý, M., Van Oost, G., *11th Int. Cong. On Plasma Physics* (Sydney, Australia, 2002), Contribution No. 40.
10. Adámek, J., Balan, P., Ďuran, I., Hron, M., Ioniţă, C., Kryška, L., Martines, E., Schrittwieser, R., Stöckel, J., Tichý. M., Van Oost, G., *Proc. 20th Symp. Plasma Phys. Techn.* (Prague, Czech Republic, 2002).
11. Balan, P., Cabral, J.A., Schrittwieser, R., Figueiredo, H.F.C., Fernandes, H., Ioniţă, C., Varandas, C., Adámek, J., Hron, M., Stöckel, J., Martines, E., Tichý, M., Van Oost, G., *14th APS Top. Conf. High Temp. Plasma Diagn.* (Madison, Wisconsin, USA 2002).
12. Hidalgo, C., Silva; C., Pedrosa, M.A., Sanchez, E., Fernandes, H. and Varandas, C.A.F., *Phys. Rev. Lett.*, **11** (1999) p. 2203.
13. Schrittwieser, R., Adámek, J., Balan, P., Cabral, J.A., Fernandes, H., Figueiredo, H.F.C., Hidalgo, C., Hron, M., Ioniţă, C., Jakubka, K., Kryška, L., Martines, E., Pedrosa, M.A., Pohoaţă, V., Stöckel, J., Tichý, M., Van Oost, G., Varandas, C., *Invited Lecture at the XIth Romanian Nat. Conf. Plasma Phys. Applic.* (Constanţa, Romania, 2001).
14. K. Reinmüller, *Contrib. Plasma Phys.* **38** (1998), 7.
15. M.Y. Ye, S. Takamura, *Phys. Plasmas* **7** (2000), 3457.

Investigation of the Effects of the Radial Electric Field by Electrode Biasing in a Toroidal Plasma

Zhijian Wang[1], Cheng Wang[1], Gesheng Pan[1], Yizhi Wen[1], Changxuan Yu[1], Shude.Wan[1], Wandong.Liu[1], Ronghua Lu[1], JunWang[1] and Hui.Gao[2]

[1] *Department of Modern Physics, University of Science and Technology of China, 230027, Hefei,*
[2] *National Synchrotron Radiation Lab, University of Science and Technology of China, 230029, Hefei,*

Abstract. The electrode biasing experiments were carried out on the KT-5C tokamak to investigate the effects of the radial electric field E_r on turbulence in a toroidal plasma. It is observed the radial electric field is mainly contributed by the poloidal flow both in Ohmic and in electrode biasing discharges. The changes of E_r is led by the changes of poloidal flow, and it is the Reynolds stress driving the poloidal flow in Ohmic discharges in the device, but the Reynolds stress is not so important in the electrode biasing discharges. A modestly enhanced $E \times B$ shear layer is formed at the plasma edge by the electrode biasing. In the sheared $E \times B$ layer, reductions in the fluctuation amplitude, and the radial correlation length as well as the turbulent particle flux, are observed, Indicating that the turbulence suppression by de-correlation is due to the modestly enhanced $E \times B$ shear layer.

INTRODUCTION

Great progress has been made in revealing the underlying mechanism responsible for suppressing turbulence and improving confinement [1]. $E \times B$ shear flow is widely considered to be responsible for transport barriers observed in experiments [2-4]. However, there is not yet consensus on the mechanism responsible for generation of the shear flow at present. This statement even applies to externally driven shear flow be electrode biasing, where although there is clearly an external current. The measurement should account for more forces, e.g. the Reynolds stress. Experiments of electrode biasing have been carried out on KT-5C tokamak and some new evidences are observed for verifying the underlying mechanism driving the $E \times B$ flow.

KT-5C tokamak runs at parameters of R=32.5cm, a=8.2cm, B_ϕ=0.4T, and I_P=10kA. 150V positive electron voltage was applied on a biasing electrode placed at r =4.5cm on the top of KT-5C. Electron density n_e, electron temperature T_e, floating potential V_f and their fluctuations ($\tilde{n}_e$, $\tilde{T}_e$ and $\tilde{V}_f$) were measured by triple Langmuir probe. The plasma potential ϕ_P is determined from the measured V_f and T_e by $\phi_P = V_f +2.5\ T_e$. The radial electric field is deduced from the radial derivative of ϕ_P. A retarding field energy analyzer (RFEA) was used to measure the ion temperature T_i. The ion density n_i is approximately the same as n_e. The toroidal flow velocity u_ϕ was obtained using a Mach probe. The value of poloidal flow velocity can not be decided directly, but it is proportional to δC_s, where C_s is the ion acoustic velocity, $\delta=(I_u-I_d)/(I_u+I_d)$ is the ratio of collective ion saturation currents up and down stream along poloidal direction. Shot to shot measurements were conducted to obtain a variety of profiles. The signals are buffered through the high performance opto-coupled isolation amplifiers whatever it is necessary. All the data were sampled at 500kHz by a multi-channel 12 bits digitizer.

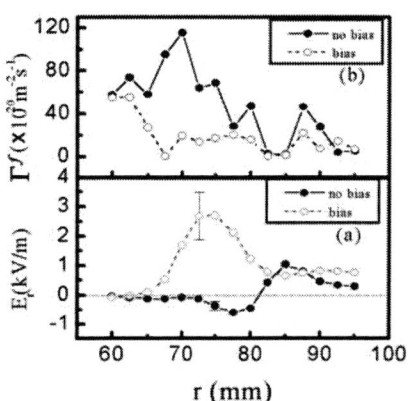

FIGURE 1. Radial profiles of (a) the radial electric field, (b) the turbulent particle flux

The radial electric field E_r is changed from a low negative well to a mild positive hill in the edge region in discharges from Ohmic only to with +150V electrode biasing as shown in Fig. 1(a). Fig. 1(b) indicates the turbulence induced particle flux Γ^f. The Γ^f is obtained by the standard two-point cross-correlation technique. A reduction in Γ^f can be found in the region of 7cm<r<8cm where the E_r shear layer exists. This is in agreement with the well known theoretic predictions of the enhanced $E \times B$ flow can suppress turbulent transport and improving confinement.

It is the ion radial force balance that is of interest to determine the radial electric field. The equilibrium radial force balance for ions is given by

$$E_r = \frac{1}{Z_i e n_i}\frac{\partial P_i}{\partial r} + u_{\phi i} B_\theta - u_{\theta i} B_\phi,$$

where $P_i = n_i T_i$ is the ion pressure, $u_{\phi i}$ the toroidal flow, u_θ the poloidal flow, $Z_i e$ and n_i are the ion electric charge and density, B_θ and B_ϕ are the poloidal and toroidal components of the magnetic filed. All of the three terms are calculated and shown in Fig. 2, where the $-u_\theta B_\phi$ term is derived from the measured E_r, $\nabla P_i / Z_i e n_i$, and $u_\phi B_\theta$. By comparing their values and profiles with E_r, it can be seen that the radial electric field is dominantly upheld by the $-u_\theta B_\phi$ term both in Ohmic and in biasing discharges, although the other two terms can not be neglected for regulating the E_r profile. To firmly establish the causal link, the temporal evolution of these quantities was investigated, which is shown in Fig. 3, in which δC_s represents the poloidal flow velocity. It is found the poloidal flow changes at the first shortly after the rise of electrode current, leading the change of radial electric field 20~30μs; the toroidal flow arises slower than poloidal flow, but still ahead of E_r. This suggests that it is the poloidal flow u_θ driving the radial electric field E_r, not vice versa.

The mode of turbulence-driven flow via the Reynolds stress [5,6] is now been reemphasized. Fig. 4 shows the radial profile of the Reynolds stress calculated by $-\langle \tilde{E}_r \tilde{E}_\theta \rangle / B_\phi^2$ in the time domain. It can be seen that the Reynolds stress is non-zero everywhere and has a radial gradient in the vicinity of the limiter. For the steady-state Ohmic confinement plasma, the poloidal flow profile is governed by the balance of the Reynolds stress gradient and the damping term,

$$-\frac{\partial}{\partial r}\langle \tilde{V}_r \tilde{V}_\theta \rangle - \mu V_\theta = \frac{\partial \langle V_\theta \rangle}{\partial t} = 0.$$

To estimate quantitatively the importance of the Reynolds stress in generating the mean poloidal flow, a comparison of the magnitude and the radial structure between the two terms is required. The above equation indicates that, the damping term μV_θ of the poloidal flow due to magnetic pumping in the plasma edge region can be expressed by the "Stix-like" model [7] as

$$\mu V_\theta = \left(\frac{\sqrt{\pi}}{2} \frac{q v_{th}}{R} \exp(-U_{pm}^2) + \frac{1}{2}\frac{\nu_{ii} q^2}{1+U_{pm}^2} \right)\left(V_\theta - V_\theta^{Neo} \right)$$

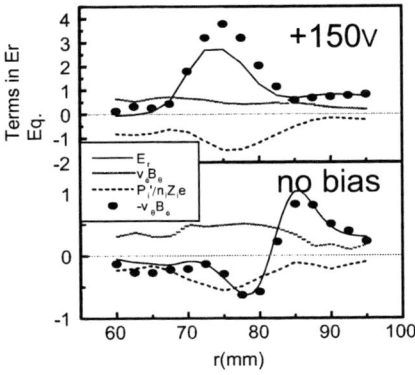

FIGURE 2. The contributions in the E_r equation from the terms of $\nabla_r P_i / Z_i e n_i, u_\phi B_\theta, -u_\theta B_\phi$, together with E_r.

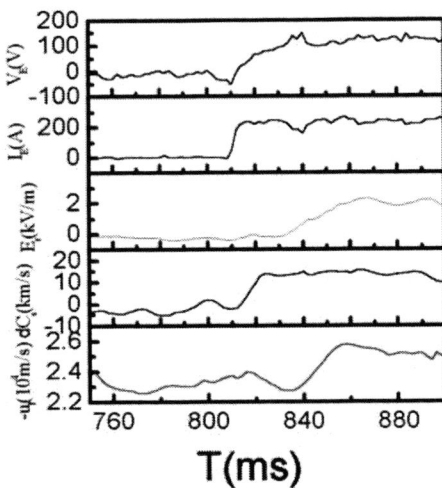

FIGURE 3. Temporal evolution of electrode voltage V_E and current I_E, radial electric field E_r, poloidal ratio δC_s and toroidal Mach number u_{th} at r=7.5cm.

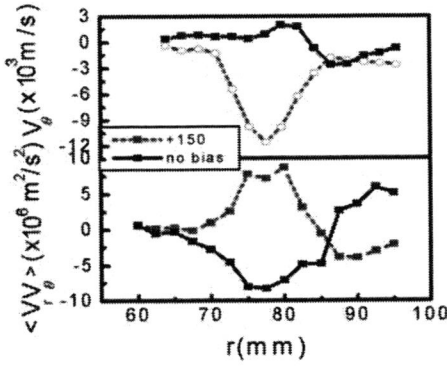

FIGURE 4. Radial profiles of Reynolds stress and poloidal flow (deduced from the ion radial force balance equation.

where q is the safety factor, $v_{th} = (2T_i/m_i)^{1/2}$ the ion thermal velocity, $U_{pm} = -E_r/(B_\theta v_{th})$, ν_{ii} the ion-ion collision frequency, $V_\theta^{Neo} = (-0.5/eB)(dT_i/dr)$ the neoclassical poloidal flow velocity in the plateau regime. Shown in Fig. 5 is a comparison of the profiles of the calculated μV_θ and $-\partial(R_s)/\partial r$ deduced from the measured R_s profile for Ohmic discharges. Fairly good agreement indicates that the sheared poloidal flow in the edge region of the KT-5C tokamak may be generated by the turbulence-induced Reynolds stress. The role of Reynolds stress in biasing discharges is also estimated. The j_r induced by the electrode is about 200A/m^2, and the maximum of j_r caused by Reynold stress is approximately 40A/m^2 calculated, this shows that Reynolds stress does not play a key role in the biasing case.

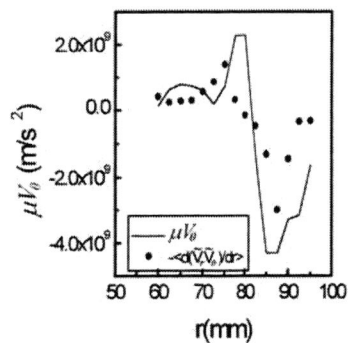

FIGURE 5. Comparison between the damping rate of poloidal flow μV_θ and the radial gradient of Reynolds stress $-\partial(R_s)/\partial r$.

In summary, in discharges with biased electrodes in KT-5C, a mild positive hill of radial electric field is generated by +150V biasing, where a reduction of turbulent transport is observed. The E_r is driven by the poloidal flow, which is drawn from the facts that the change of poloidal flow leads the change of radial electric field 20~30µs and the poloidal flow dominantly maintains the radial electric field. Furthermore, the poloidal flow is driven by Reynolds stress in Ohmic discharges, but Reynolds stress is not so important in biasing discharges.

ACKNOWLEDGMENTS

This work has been supported by the National Natural Science Foundation of China under Grant Nos. 19635020, 19675036, 19775046, the Chinese Academy of Science and the National Doctoral Training Foundation under Grant Nos. 9535806, 9803810.

REFERENCES

1. P.W. Terry, Rev. Mod. Phys. **72**, 109 (2000).
2. H. Biglari, P.H. Diamond, and P.W. Terry, Phys. Fluids B **2**, 1 (1990).
3. E. Wagner, et al., Phys. Rev. Lett. **49**, 1408 (1982)
4. Z. Lin, et al., Science **281**, 1835 (1998)
5. P.H. Diamond and Y. –B. Kim, Phys. Fluids B **3**, 1626 (1991).
6. B.A. Carreras, et al., Phys. Plasmas **1**, 4014 (1994).
7. J. Cornelis, et al., Nucl. Fusion **34**, 171 (1994).

Plasma Velocity Profile During The Pulsed Poloidal Current Drive In The MST RFP Plasma

H. Sakakita[1,2], D. Craig[2], J. K. Anderson[2], T. M. Biewer[2], S. D. Terry[3], B. E. Chapman[2] and D. J. Den-Hartog[2]

[1]*Energy Electronics Institute, National Institute of A. I. S. T., Tsukuba, JAPAN*
[2]*Department of Physics, University of Wisconsin, Madison, USA*
[3]*Department of Electrical Engineering, University of California, Los Angels, USA*

Abstract. We report on the plasma velocity profile measurements during the pulsed poloidal current drive (PPCD) in the Madison Symmetric Torus (MST) reversed-field pinch (RFP). In order to decrease fluctuations due to dynamo activities, PPCD was applied to replace the dynamo electric field. As a result, the magnetic fluctuations have been further suppressed, and considerable increase of energy confinement has been already achieved. In the initial stage of PPCD, accompanying sudden reduction of both magnetic fluctuations and radiation from neutral deuterium atoms, the electron temperature increased rapidly. This improvement may be concerned with a current profile change to more stable region. For this change, we have studied whether plasma velocity profile changes To obtain the plasma toroidal velocity profile, we have measured the Doppler shift of several impurity lines. To make sure of the radial maximum emission location, line intensities for each impurity species have been measured at 10 poloidal chords. The data are inverted using MSTFit to obtain the radial impurity emission profile. As a result, a change of toroidal plasma rotation profile was unclear, since impurity ions shifted to r/a > 0.8 during PPCD.

INTRODUCTION

Reversed-field pinch (RFP) has attractive features as a fusion reactor, since the plasma is confined by weak toroidal magnetic fields. However, RFP plasmas are susceptible to large-amplitude magnetic field fluctuations due to dynamo activities. These fluctuations grow to an amplitude sufficient to cause reconnection and stochastization of the magnetic field lines, thereby degrading energy confinement. In order to decrease these fluctuations, a pulsed poloidal current drive (PPCD) was applied to replace the dynamo electric field in the Madison Symmetric Torus (MST) RFP (major and minor radii, $R/a = 1.5/0.52$ m) [1]. As a result, the magnetic fluctuations with poloidal mode number $m = 1$ and $m = 0$ have been further suppressed, and considerable increase of energy confinement has been achieved [2]. In the initial stage of PPCD, accompanying sudden reduction of both magnetic fluctuations and radiation from neutral deuterium atoms, the electron temperature increased rapidly [2]. Mostly, this sudden reduction is induced together with a small sawtooth crash. This improvement may be concerned with a current profile change to more stable region [3,4]. In tokamaks, H-mode confinement state has been sometimes related with a sawtooth crash [5]. Therefore, for this sudden change, we are attempting to measure spectroscopically the toroidal velocity component of the radial electric field in order to confirm or deny the probe measurements that were previously published [6]. It would be verified a hypothesis that a transport barrier due to a local velocity shear makes electron temperature further increase, and a reduction of the edge resistivity assists the current profile flatten. To obtain the plasma toroidal velocity profile, we have measured the Doppler shift of several impurity lines [7]. To make sure of the radial maximum emission location, line intensities for each impurity species have been measured at 10 poloidal chords. The data are inverted using MSTFit to obtain the radial impurity emission profile [8]. The measurement result of plasma velocity and ion temperature profiles during the initial stage of PPCD is described.

EXPERIMENTAL SETUPS

In order to obtain a toroidal plasma velocity profile, we have measured the Doppler shift of several impurity lines using the ion dynamics spectrometer (IDS) precisely described in ref. 7. (Charge exchange recombination spectroscopy (CHERS) is not available for the toroidal plasma velocity measurement at the present.) This spectrometer is available to measure simultaneously impurity ion temperature and flow velocities with 10 μsec temporal resolution. This device is actually a duo-spectrometer: measurements from toroidally two different chordal views of the plasma can be made simultaneously via two separate quartz input fiber optic bundles coupled to the entrance slit of a spectrometer. We measured flow velocities and ion temperature for CV (227.1 nm), BIV (282.3 nm), OV (278.1 nm), CIII (229.6 nm) and HeII (468.6 nm) lines. Measurements are line-averaged along the toroidal viewing chord which samples the plasma from r/a = 0.3 to 1.0.

To make sure of the radial maximum emission location, line intensities for each impurity species were measured at 10 poloidal viewing chords of r/a = -0.87, -0.58, -0.41, -0.24, -0.09, 0.10, 0.28, 0.45, 0.62 and 0.839. r/a = -0.87 and 0.839 chords are toroidally apart 120 degrees from other 8 chords. But this influence was taken into account in the analysis. Using two fiber bundles of the IDS system, line intensities for two chords were measured at one discharge. We measured CV (227.1 nm), BIV (282.3 nm), OV (278.1 nm) and HeII (468.6 nm) line intensities.

RESULTS

Plasma current was ~210 kA, and PPCD trigger timing was fixed at t = 9.0 ms. The SXR ratio (beryllium filter, 15 μm/7.5 μm) that corresponds to electron temperature increases after a sawtooth crash at t = ~11 ms as shown in Fig. 1(a). Figures 1(b) and (c) show time behaviors of magnetic mode fluctuations and H_α. After triggering of PPCD, single helicity state of toroidal mode number $n = 6$ ($m = 1$) is formed. After the final SXR crash (t = ~11 ms), both magnetic fluctuations and radiation from neutral deuterium atoms decrease. Hereafter, we pay attention to just before and after SXR crash, i.e., t = 10.5 and 13.5 ms.

Electron density profiles measured by a FIR interferometer are shown in Fig. 2. At t = 13.5 ms, electron density gradient becomes steeper at r/a ~ 0.7. Electron temperature gradient also becomes steep at r/a ~ 0.7 as shown in Fig. 7 of Ref. 2.

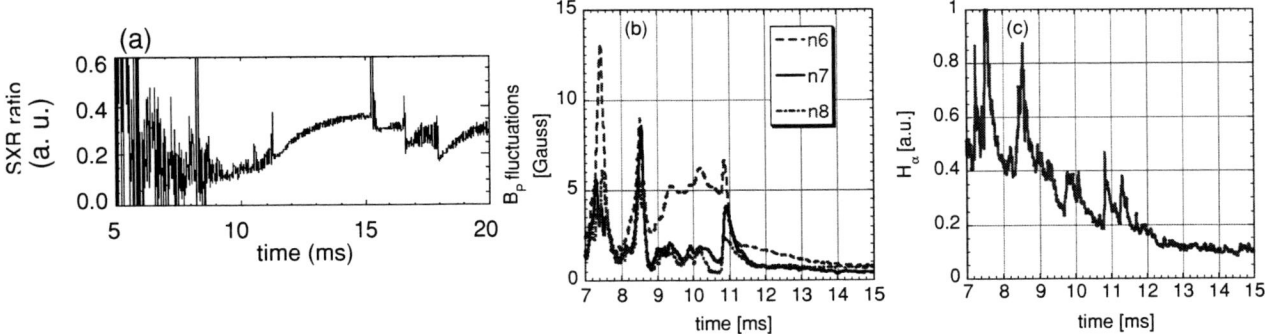

FIGURE 1. Time behaviors of (a) SXR ratio, (b) magnetic mode amplitudes of n = 6, 7 and 8, (c) H_α emission.

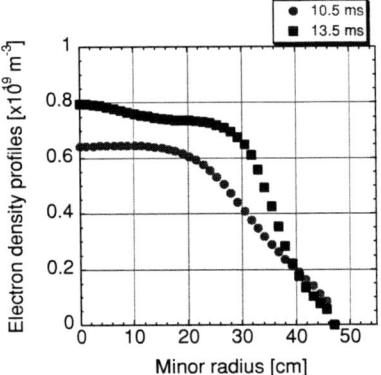

FIGURE 2. Electron density profiles at t = 10.5 ms (solid circle symbol) and 13.5 ms (solid square symbol), respectively.

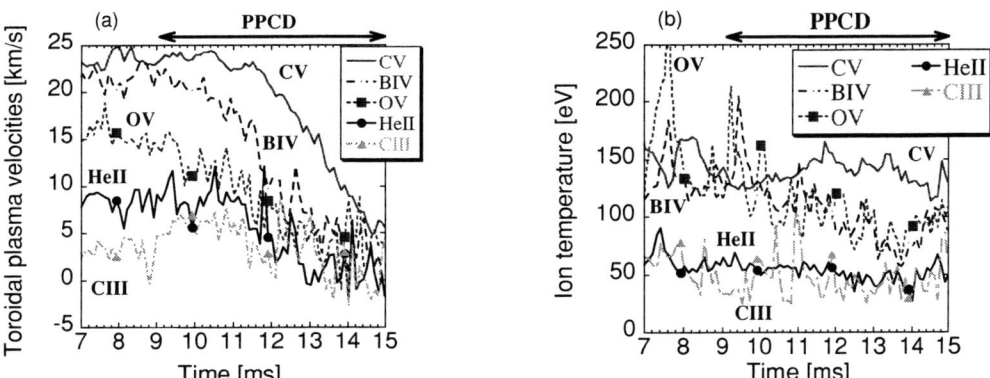

FIGURE 3. Time behaviors of (a) Toroidal plasma velocities, (b) ion temperatures. Solid line (CV), broken line with empty circle (BIV), dashed line with solid square (OV), solid line with solid circle (HeII) and dashed line with solid triangle (CIII).

Figure 3 shows measured toroidal plasma velocities and ion temperatures for CV, BIV, OV, HeII and CIII. 100 µs moving average and ensemble shot average were conducted. Toroidal plasma rotation decreased with PPCD. Ion temperatures of CV, HeII and CIII show almost no change. But ion temperatures of BIV and OV decrease with PPCD.

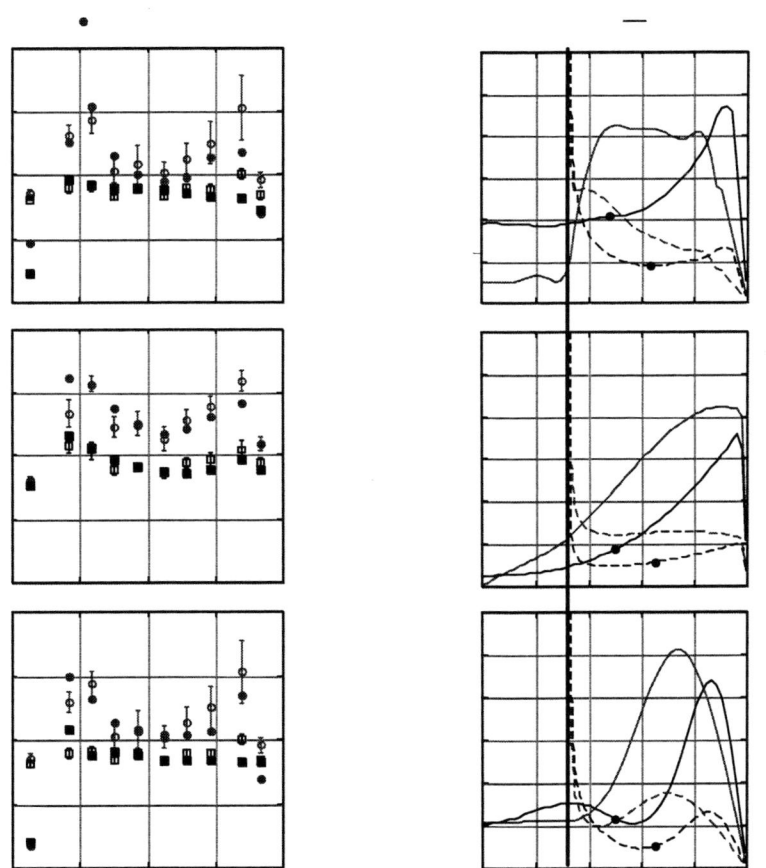

FIGURE 4. Line integrated emission profiles at t = 10.5 and 13.5 ms, (a) CV, (b) BIV and (c) OV. Inverted radial emission profiles and radial emission profiles multiplied by the appropriate geometric sensitivity function, (d) CV, (e) BIV and (f) OV.

Figures 4(a), (b) and (c) show line integrated emission profiles (empty symbols). Solid symbol data means inverse transformed data from Abel inverted emission profile (solid lines) shown in Fig. 4(d), (e) and (f). Inverted data fairly fits to the raw data as shown in Fig. 4(a), (b) and (c). Transformed emission profiles from line integrated emission intensities indicate the outward shift of the peak emission with increased electron temperature from PPCD as shown in Fig. 4(d), (e) and (f). During PPCD, since the equilibrium configuration drastically changes, we reconstructed the equilibrium of the MST RFP using MSTFit [8]. It provides rather good inversion of the line integrated emission. I*W curves (broken lines in Fig. 4(d), (e) and (f)) mean radial emission profiles multiplied by the appropriate toroidal geometric sensitivity function [9]. The outer peak of I*W curve indicates a main existing location of each species. Only for CV data, the raw data was divided by CV data from an another spectrometer, which was measuring the emission from a plasma center, in order to eliminate shot differences. Helium gas was puffed during the shot. But a strong asymmetric emission was measured, especially for the radial direction, since a puffing port was close to the used poloidal viewing chord. The inverted HeII emission looked ugly.

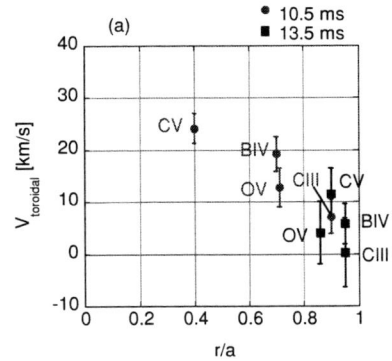
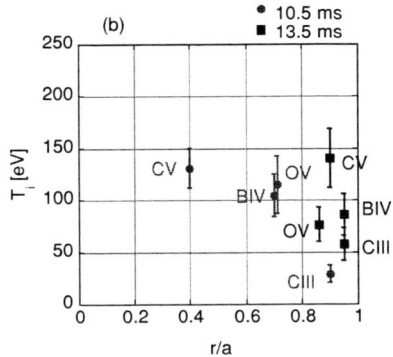

FIGURE 5. (a) Toroidal plasma rotation profiles, (b) ion temperature profiles at t = 10.5 and 13.5 ms. CV, BIV, OV and CIII data are listed. For CIII, $r/a = 0.9$ (t = 10.5 ms) and $r/a = 0.95$ (t = 13.5 ms) are assumed, respectively.

Figure 5(a) shows toroidal plasma velocity profiles. Measured species shifted to $r/a > 0.8$ during PPCD, and broadening of I*W curve corresponds to the error bar of the radial profile resolution. Therefore, the change of plasma rotation profile is unclear due to the insufficient spatial resolution. Figure 5(b) shows ion temperature profiles. It seems that ion temperature increases at the edge region in the case of PPCD. Ion temperature is higher than electron temperature at the edge region, nevertheless the magnetic fluctuations are suppressed. However, the conclusion of an increase of T_i in the edge really depends on the actual impurity location and the CIII assumption.

SUMMARY

In the initial stage of PPCD, accompanying sudden reduction of both magnetic fluctuations and H_α, the electron temperature increased rapidly. This improvement may be concerned with a current profile change to more stable region. Mostly, this sudden reduction is induced together with a small sawtooth crash. For this sudden change, we have studied whether plasma velocity profile changes. However, due to the insufficient spatial resolution, a change of plasma rotation profile is unclear. It seems that ion temperature increases at the edge region in the case of PPCD. However, the conclusion of an increase of T_i in the edge really depends on the actual impurity location and the CIII assumption. The hypothesis described in the introduction was not verified. Improvement of the spatial resolution using CHERS, and to study the correlation with local fluctuations for any frequencies are future problems [10]. Work was supported by USDOE.

REFERENCES

1. Sarff, J. S., et al., Phys. Rev. Letters 72, 3670-3673 (1994).
2. Chapman, B. E., et al., Phys. Plasmas 9, 2061-2068 (2001).
3. Sovinec, C. R. and Prager, S. C., Nuclear Fusion 39, 777-790 (1999).
4. Brower, D. L., et al., Phys. Rev. Letters 88, 185005-1-185005-4 (1994).
5. Kimura, H, and the JT-60 Team, Phys. Plasmas 3, 1943- 1950 (1996).
6. Chapman, B. E., et al., Phys. Plasmas 5, 1848-1854 (1998).
7. Den-Hartog, D. J., et al., Rev. Sci. Instrum. 65, 3238-3242 (1994).
8. J.K. Anderson, Ph.D. thesis, University of Wisconsin, Madison (2001).
9. Den-Hartog, D. J., et al., Phys. Plasmas 6, 1813-1821 (1999).
10. Lanier, N. E., et al., Phys. Plasmas 8, 3402-3410 (2001).

ELM Dynamics in TCV H-modes

A W Degeling, Y R Martin, J B Lister X Llobet and P E Bak[*]

*Centre de Recherches en Physique des Plasmas, Association EURATOM-Confédération Suisse,
Ecole Polytechnique Fédérale de Lausanne, 1015 Lausanne, Switzerland.
JAERI Naka Fusion Research Establishment, Ibaraki-ken, 311-0193, Japan.

Abstract. TCV (Tokamak à Configuration Variable, R = 0.88 m, a < 0.25 m, B_T < 1.54 T) is a highly elongated tokamak, capable of producing limited and diverted plasmas, with the primary aim of investigating the effects of plasma shape and current profile on tokamak physics and performance. L-mode to H-mode transitions are regularly obtained in TCV over a wide range of configurations. Under most conditions, the H-mode is ELM-free and terminates in a high density disruption. The conditions required for a transition to an ELMy H-mode were investigated in detail, and a reliable gateway in parameter space for the transition was identified. Once established, the ELMy H-mode is robust to changes in plasma current, elongation, divertor geometry and plasma density over ranges that are much wider than the size of the gateway in these parameters. There exists marked irregularity in the time interval between consecutive ELMs. Transient signatures in the time-series revealing the existence of an underlying chaotic dynamical system are repeatedly observed in a sizable group of discharges [1]. The properties of these signatures (called unstable periodic orbits, or UPOs) are found to vary systematically with parameters such as the plasma current, density and inner plasma – wall gap. A link has also been established between the dynamics of ELMs and sawteeth in TCV: under certain conditions a clear preference is observed in the phase between ELMs and sawtooth crashes, and the ratio of the ELM frequency (f_{elm}) to sawtooth frequency (f_{st}) is found to prefer simple rational values (e.g. 1/1, 2/1 or 1/2). An attempt to control the ELM dynamics was made by applying a perturbation signal to the radial field coils used for vertical stabilisation. Phase synchronisation was found with the external perturbation, and f_{elm} was found to track limited scans in the driver frequency about the unperturbed value, albeit with intermittent losses in phase lock.

INTRODUCTION

It has been known for many years that under appropriate conditions tokamak plasmas can undergo a sudden increase in the energy confinement time, accompanied by a reduction in the deuterium visible light emission (D_α), which is referred to as the H – mode. H – modes are frequently accompanied by bursts in MHD activity and D_α emission, known as edge localised modes (ELMs), and H-modes of this type are termed ELMy H-modes. ELMs are momentary breakdowns in the edge transport barrier that gives rise to the increased confinement, and result in the pulsed release of up to 10% of the stored plasma energy. The plasma density in an H-mode without ELMs (a so called ELM – free H-mode) is generally non-stationary, and can rise uncontrollably until the plasma disrupts. The periodic degradation in confinement provided by ELMs has a stabilizing effect, and allows stationary H-mode operation. ELMs are also useful for removing impurities that build up in the plasma interior, and will aid the removal of He ash in a working reactor. These beneficial properties of ELMs have resulted in the adoption of the ELMy H-mode as the basic operating scenario for ITER. However, ELMs have one serious drawback: the projected energy released per ELM in ITER presents an unacceptable heat load on plasma facing components. For this reason considerable effort worldwide has gone into investigating the ELM phenomenon, and ways in which the damaging effect of ELMs can be ameliorated.

Ohmic H – modes in TCV

Tokamak à Configuration Variable (TCV) is a highly elongated tokamak with unique plasma shaping capabilities (for example elongation κ ranges from 1 to 2.8, triangularity δ ranges from –0.7 to 1). Up to 4.5 MW of electron cyclotron heating (ECH) is available on TCV, although all the results presented in this paper are taken from ohmic H – modes.

The transition to H – mode is regularly obtained on TCV over a wide range of discharge conditions, however it is found that over most of the operational domain the H – mode is ELM – free, and results in a high density disruption. Systematic investigations lead to the identification of a narrow range in a set of parameter values that reliably produced a transition from L – mode to the ELMy H – mode when the plasma was changed from a limited to a single null diverted configuration [2]. This small portion of the TCV operational domain has been termed the gateway to the ELMy H – mode, and is described in TABLE 1.

TABLE 1: ELMy H – mode gateway parameters

Parameter	Min.	Max.
I_p [kA]	380	420
n_e [10^{19} m^{-3}]	4.5	7.5
δ	0.5	0.6
κ	1.6	1.7

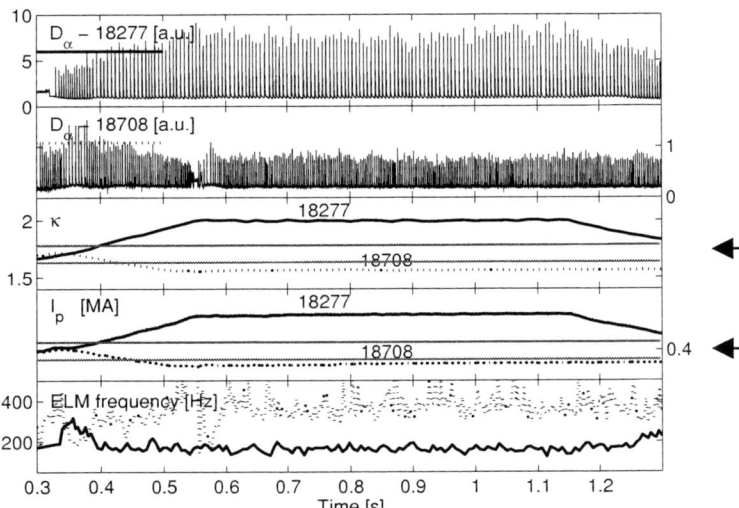

FIGURE 1. The D_α trace showing ELMs, k, I_p and the ELM frequency for two discharges in which I_p and k were ramped above and below the gateway values (the parallel horizontal lines to the left of the arrows).

Once established, it was found that the ELMy H – mode is robust to changes in parameter values over ranges that are much larger than the gateway values. Two examples are shown in Fig. 1, discharge 18277 and 18708, in which both I_p and κ are raised and lowered respectively from their gateway values. Note also that the ELM frequency (f_{elm}) changes by about a factor of two between these discharges, with the low frequency ELMs corresponding to the high κ discharge. This figure also illustrates some important features about ELM dynamics. Firstly, the ELM amplitude is generally found to scale with the interval between ELMs ΔT. This indicates that it may be possible to avoid large amplitude ELMs if the ELM interval is controlled. Secondly, the instantaneous ELM frequency ($1/\Delta T$) is not a constant value, but varies erratically during the stationary part of the discharge. The significance of this feature is the subject of discussion in the next section.

ELM Dynamics

The Current understanding of the ELM dynamical system is as follows: 1) Quiescent phase: During the time interval between ELMs, the edge transport barrier causes a slow increase in MHD energy, which leads to an increase in the edge pressure gradient and the edge current density; 2) Threshold for instability: An MHD stability threshold is exceeded, and the resulting rapidly growing mode gives rise to either ergodicity in the edge magnetic field (e.g. magnetic field island overlap), or turbulence from nonlinear saturation of the mode, or a cascade of unstable modes (sandpile model); 3) Breakdown of the edge transport barrier, which leads to a sudden loss in plasma energy and the observed spike in D_α emission; 4) The drop in stored energy leads to stabilisation of the MHD activity, and the re-establishment of the edge transport barrier.

One might expect this description to reduce to a limit cycle or relaxation oscillator model simply governed by the reduced transport during the H – mode and the threshold condition for instability, which would be expected to give regular periodic behaviour. However, irregular behaviour is observed in the experiment, and it is therefore important to consider how such behaviour can arise in this picture. There are two extreme alternatives: 1) the irregularity is the product of a noisy environment affecting an otherwise regular limit cycle, in which case one might attempt to characterise the low order limit cycle by considering average quantities or filtering; 2) the irregularity is an intrinsic part of a low order system in a chaotic state, hence the irregularity is not ignorable. These two cases may be distinguished experimentally by making use of a feature that is generic to low order chaotic systems, namely, the existence of unstable periodic orbits (UPOs) [3].

Searching for Chaotic Behaviour

An unstable periodic orbit is an observable event that occurs when the state of a chaotic system is close to an unstable equilibrium and is almost periodic for a short number of cycles. The system momentarily behaves in a linear way, giving rise to a characteristic, transient signal in the time-series of any of its dependent variables. There are a number of methods that may be employed to search for UPOs. The method we adopted involved the construction from the D_α time-series of each discharge a sequence of consecutive inter-ELM time intervals ΔT_i, as shown in Fig. 2. When a UPO occurs, a short sequence of consecutive values of ΔT_i (at least 4) oscillate about a constant value ΔT^* (called the fixed point period) with an exponentially growing amplitude. Two examples are shown in the figure. By making a scatter plot of ΔT_i against ΔT_{i+1}, these sequences appear as series of points that alternate about the diagonal and lie on a straight line with a slope (s_+) less than -1.

It must be noted that no single UPO event can be taken as evidence for chaotic behaviour, since the chances of 4 consecutive points resembling this pattern by chance in a noise dominated system is not negligible. However, the parameters ΔT^* and s_+ are properties of the dynamical system if it is chaotic, whereas in a noisy system they will only be subject to the probability distribution of ΔT. Therefore, chaotic and noisy systems are distinguishable by the statistical properties of the UPO signatures found within their time-series [1].

A total of 241 ELMy H-mode discharges spanning a wide range of plasma parameters were searched for these signatures and yielded a total of 536 events. We tested the null hypothesis that these events are chance occurrences by generating 100 randomly re-ordered sequences of ΔT's from the original sequence of each discharge (so called surrogate sets), and searching these for UPO signatures. Note that these surrogates have the same probability distribution of ΔT as the original series from which they were generated. A mean of 393 events with a standard deviation of 16 were found from the surrogate sets of all the discharges. If the irregularity in ELM dynamics was produced by noise then the number of UPOs in the original data should lie within at most 3 standard deviations of the mean of the surrogates, however our result is nearly 9 standard deviations higher. This clearly signifies the existence of chaotic dynamics.

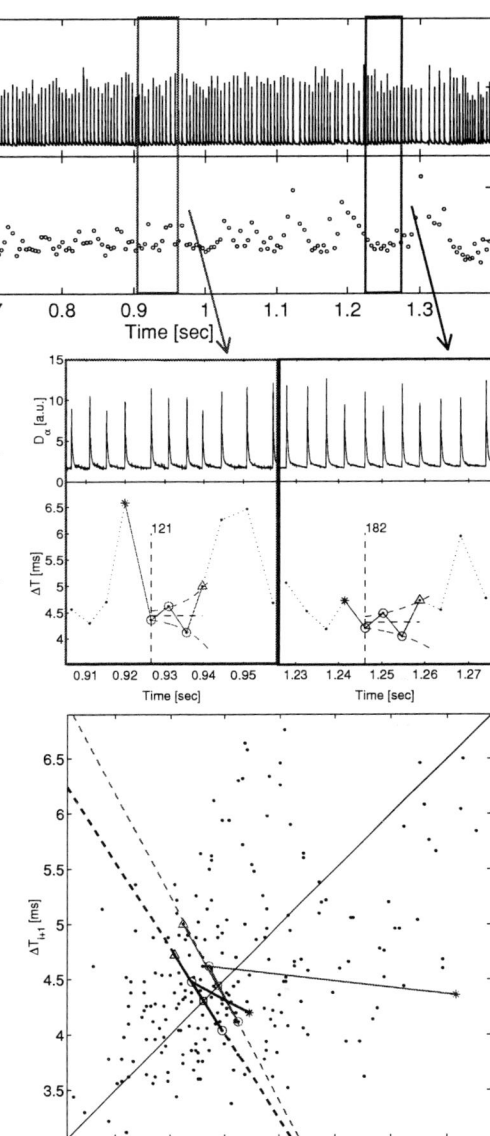

FIGURE 2. Example of UPO detection showing the original D_α time-series, the sequence of ΔTs & the return map.

ELM – Sawtooth Synchronisation

Another interesting phenomenon discovered by examining the ELM time-series is the frequent partial or complete synchronization of ELM and sawtooth cycles [4]. This behaviour is clearly evident in examples of individual discharges, where cases of entrainment of f_{elm} to f_{st}, or its harmonics are easily found, however a more general observation is produced by building a database using the ELM occurrence times from a large number of discharges (100 were used). The time interval with respect to each ELM of the nearest preceding and succeeding sawtooth crash (Δt_{prev} and Δt_{next}) were included in the database, as were amplitudes of the variations in soft X-ray emissivity along a chord sampling the plasma edge (with $\rho > 0.9$) due to each ELM and sawtooth event (A_{elm} and A_{st} respectively). The phase of each ELM in the sawtooth cycle is defined by $\phi_{st \to elm} = \Delta t_{prev}/(\Delta t_{prev} + \Delta t_{next})$. Fig. 3

shows a series of histograms of the phase, in which the data has been segregated into rows using the amplitude of the previous sawtooth A_{st}. Peaks in the histograms indicate a preferred phase that persists even in different discharges, and clearly indicates coupling between the ELMs and the Sawteeth. The appearance of a single peak as a function of $f_{st\rightarrow elm}$ corresponds to synchronisation of the ELMs with the fundamental f_{st}, while multiple peaks suggest a harmonic. This figure shows that as the sawtooth amplitude increases, the peaks in the histogram become more apparent, which suggests that the systems become more strongly coupled, and that the sawtooth acts as a driver signal in the ELM cycle. The top two rows of this figure also show that a transition from a histogram with a double peak to a single peak appears to take place as A_{st} is increased. These cases in which A_{st} was large correspond to a series of shots in which the elongation was scanned. As noted from Fig. 1, the average ELM frequency decreased with elongation, and this enabled strong coupling at the fundamental f_{st} (rather than the harmonic) to occur in these cases. The histogram colours correspond to A_{elm} and show that in general larger amplitude ELMs occur as A_{st} increases. It is also apparent that when two ELMs occur in a sawtooth cycle, the first ELM generally has a higher amplitude than the second.

Active ELM Frequency Modification

The fact that ELMs become synchronized with sawteeth suggests the possibility of actively controlling ELMs with an external driver signal. In this experiment a voltage perturbation was applied to the radial magnetic field coils during a single null ELMy H-mode discharge with stationary conditions. These coils are used for stabilization of the plasma vertical position, and the perturbation signal was added to the control feedback loop. The signal consisted of a series of square pulses of 1 ms duration with a variable delay between the pulses. This produced spike-like pulses in the coil current (up to 2 kA) and resulted in deviations in the vertical plasma position of up to 5 mm. Fig. 4 shows an example in which a constant amplitude pulse train was applied with an increasing driving frequency f_D from 143 to 330 Hz. The four graphs respectively show: the D_α time-series (showing ELM times), the perturbed coil current, the time-series of f_{elm} and f_D, and the phase of the ELMs with respect to the driver signal. Fig. 4 shows an example in which the external perturbation was sufficiently large to cause f_{elm} to track f_D over much of the range of the frequency sweep, albeit with intermittent losses in track. During periods where f_{elm} tracks f_D, the phase of the ELMs are clearly locked in the sawtooth cycle, and when synchronization is lost the phase rapidly passes through 2π radians.

CONCLUSIONS

ELMs exhibit many behaviours consistent with a low dimensional dynamical system. We have demonstrated the

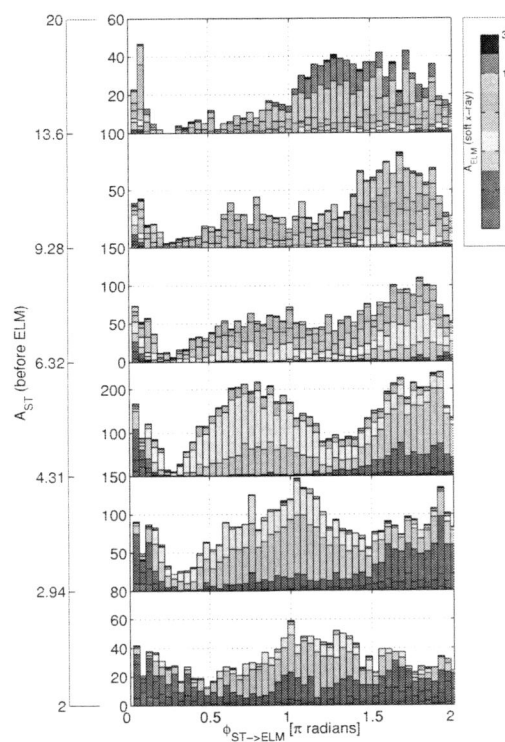

FIGURE 3. Histograms of ELM phase in the sawtooth cycle with rows of increasing sawtooth amplitude.

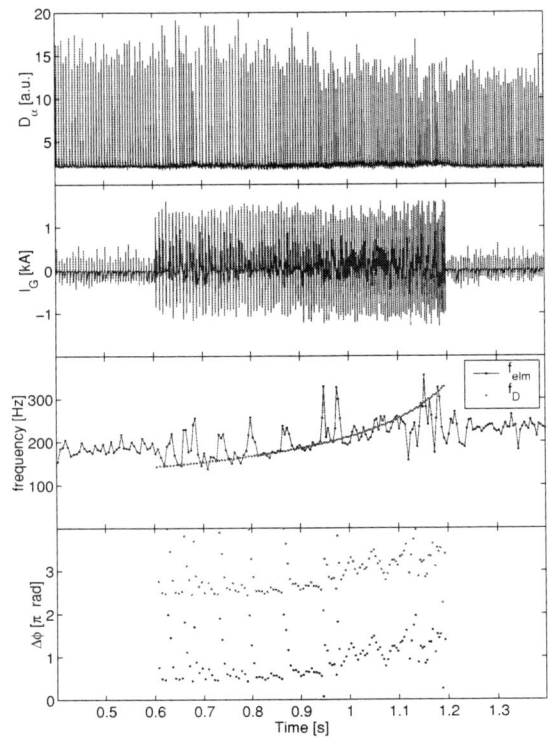

FIGURE 4. Active ELM Perturbation experiment: D_α time-series; Perturbed coil current; f_{elm} (blue) & f_D (red); ELM phase with respect to driver signal

existence of unstable periodic orbits in the ELM time series, which are signatures of chaotic behaviour. The model of ELMs that contains the correct physics must display the correct dynamics. This means that the correct model must be chaotic under the appropriate conditions. Further constraints may be provided by observations of the properties of UPOs. We have shown that ELMs couple with both internal perturbations arriving at the edge from sawtooth oscillations, and also externally applied magnetic perturbations. This will provide further insight to the properties of the underlying dynamical system.

ACKNOWLEDGEMENTS

The authors sincerely thank the TCV team, without whom these results would be unavailable. This work is supported by the Swiss National Science Foundation.

REFERENCES

1. A. Degeling et al, Plasma Phys. Control. Fusion **43** 1671 (2001)
2. Y. R. Martin et al submitted to Nucl. Fus.
3. P. E. Bak et. al, Phys. Rev. Lett., **83**, 1339 (1999)
4. Y. R. Martin, et al, Plasma Phys. Control. Fusion **44** A373 (2002)

Stochastic Transition of a Turbulent Plasma

Mitsuhiro Kawasaki[*], Sanae-I. Itoh[*], Masatoshi Yagi[*] and Kimitaka Itoh[†]

[*]Research Institute for Applied Mechanics, Kyushu University, Kasuga 816-8580, Japan
[†]National Institute for Fusion Science, Toki 509-5292, Japan

Abstract. Transition phenomena between thermal noise state and turbulent state observed in a submarginal turbulent plasma are analyzed with statistical theory. Time-development of turbulent fluctuation is obtained by numerical simulations of Langevin equation which contains hysteresis characteristics. Transition rates between two states are analyzed. Transition from turbulent state to thermal noise state occurs in entire region between subcritical bifurcation point and linear stability boundary.

INTRODUCTION

In high temperature plasmas, the dynamical change often occurs on the short time scale, sometimes triggered by subcritical bifurcation. The feature naturally leads to the concept of transitions. The transition takes place as a statistical process in the presence of statistical noise source induced by strong turbulence fluctuation. As the generic feature the transition occurs with a finite probability when a parameter approaches the critical value.

The nonequilibrium statistical mechanics should be extended for inhomogeneous plasma turbulence [2]. To this end, statistical theory for plasma turbulence has been developed and stochastic equations of motion (the Langevin equations) of turbulent plasma were derived [3]. The framework to calculate the probability density function (PDF), the transition rates, etc. has also been made.

In this paper, which is a part of the published paper [4], we apply the theoretical algorithm to an inhomogeneous plasma with the pressure gradient and the shear of the magnetic field. Micro turbulence is known to be subcritically excited from the thermal noise state [5]. The transition between thermal noise state and turbulent state is studied. We show that the transition occurs stochastically by numerically solving the Langevin equation of the turbulent plasmas. In order to characterize the stochastic nature of the transition, the frequency of occurrence of a transition per unit time (the transition rate) is calculated as a function of the pressure-gradient and the plasma temperature. The results show that the transition from the turbulent state to the thermal noise state occurs in a wide region instead of at a transition point.

THEORETICAL FRAMEWORK

In this section, we briefly review the theoretical framework used in our analysis of turbulent plasmas.

The theory is based on the Langevin equation Eq. (1) derived by renormalization with the direct-interaction approximation of the reduced MHD for the three fields: the electro-static potential, the current and the pressure.

$$\frac{\partial \mathbf{f}}{\partial t} + \hat{L}\mathbf{f} = \mathbf{N}(t), \text{ where } \mathbf{f}(t) \equiv \begin{pmatrix} \phi(t) \\ J(t) \\ p(t) \end{pmatrix}. \tag{1}$$

Since $\mathbf{N}(t)$ is a force which fluctuates randomly in time, the Langevin equation describes the stochastic time-development of the fluctuation of the three fields.

Analysis of the Langevin equation Eq. (1) derives a number of statistical properties of turbulent plasmas: For example, the PDF of the fluctuation level of the electric field, the analytical formulae of the rate of the change of states of plasmas (the transition rates), etc.

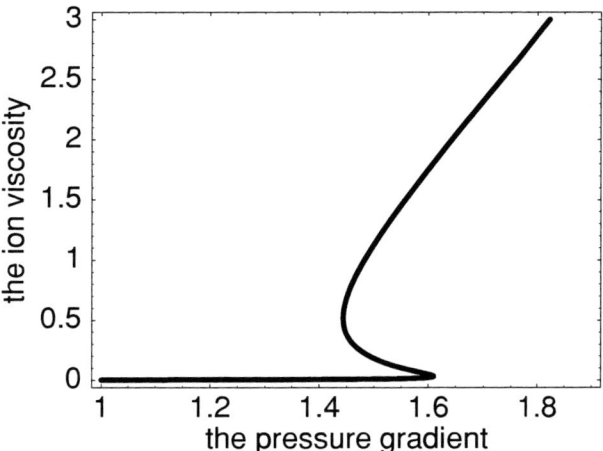

FIGURE 1. The pressure-gradient dependence of the renormalized ion-viscosity. It is clearly seen that the bifurcation between a low viscosity state (the thermal noise state) and a high viscosity state (the turbulent state) occurs.

A MODEL

With the theoretical framework briefly described in the previous section, we analyze a model of inhomogeneous plasmas with the pressure-gradient and the shear of the magnetic field [3]. The model is formulated with the reduced MHD of the three fields of the electro-static potential, the current and the pressure. The shear of the magnetic field is given as $\mathbf{B} = (0, B_0 s x, B_0)$ where $B_0(x) = \text{const} \times (1 + \Omega' x + \cdots)$. The pressure is assumed to change in x-direction.

It has been known that in this system bifurcation due to the subcritical excitation of the current diffusive interchange mode (CDIM) occurs [5] as shown in Fig. (1).

Figure (1) shows the pressure-gradient dependence of the turbulent ion-viscosity which is proportional to the fluctuation level. It is clearly seen that the bifurcation between a low viscosity state and a high viscosity state occurs. Due to the bifurcation, transition between the two states and hysteresis are expected to be observed. We call the low viscosity state "the thermal noise state", since in this state the system fluctuates with thermal noise considered in the model [6]. We call the high viscosity state "the turbulent state", since the fluctuation level is also large in a strong turbulent limit [3]. The ridge point where the turbulent branch ends is denoted "the subcritical bifurcation point". The region between the subcritical bifurcation point and the ridge near the linear stability boundary is called "the bi-stable regime".

From the deterministic point of view, the transition from the thermal noise state to the turbulent state is expected to occur at the ridge point near the linear stability boundary and the transition in the opposite direction is expected to occur at the subcritical bifurcation point.

STOCHASTIC OCCURRENCE OF THE TRANSITION

In order to capture the characteristics of the two states, we concentrate on the time-development of the energy of fluctuation of the electric field, $\varepsilon(t)$. The quantity $\varepsilon(t)$ obeys the coarse-grained Langevin equation Eq. (2) which has been derived in [3].

$$\frac{d}{dt}\varepsilon(t) = -2\Lambda(\varepsilon)\varepsilon(t) + \eta(\varepsilon)R(t). \tag{2}$$

Here, $R(t)$ is assumed to be the Gaussian white noise. For the detailed formulae of $\Lambda(\varepsilon)$ and $\eta(\varepsilon)$, see [6]. The essential point is that the function $\Lambda(\varepsilon)$ takes both a positive and a negative value in the bi-stable regime. So, the fluctuation of the electric field is suppressed when Λ is positive and it is excited when Λ is negative. Consequently, there are two metastable states in the bi-stable regime. In addition, $\eta(\varepsilon)$ is a positive function.

By solving numerically Eq. (2), we obtain samples of a time seriese. When the pressure-gradient takes a value in the bi-stable regime, bursts are observed intermittently as shown in Fig. (2). That is, transition between the thermal noise state and the turbulent state occurs *stochastically*. The bursts corresponds to the turbulent state and the laminar

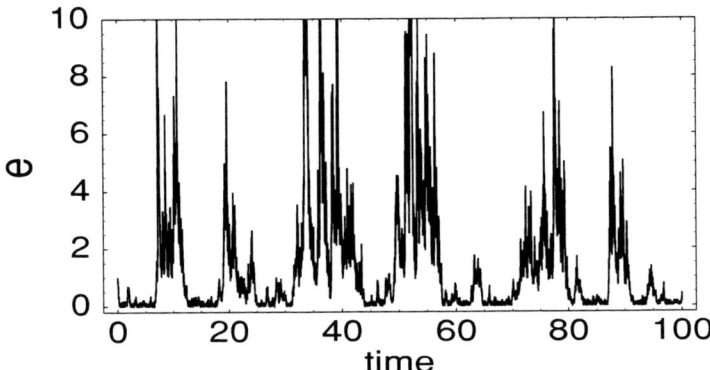

FIGURE 2. A sample of a time-series of $\varepsilon(t)$ (denoted as 'e' in the figure) when the pressure-gradient takes a value in the bi-stable regime. Bursts are observed intermittently. It means the transition between the thermal noise state and the turbulent state occurs stochastically.

corresponds to the thermal noise state. The fact that the residence times at the each states are random leads to the statistical description of the transition with the transition rates described in the next section.

THE TRANSITION RATES

In order to formulate the stochastic transition phenomena in the bi-stable regime, we introduce the transition rates. There are transitions in two opposite direction: the transition from the thermal noise state to the turbulent noise state, which we call "the forward transition, and the transition in the opposite direction is called "the backward transition". There are two transition rates. One is the forward transition rates r_f which is the frequency of occurrence of the forward transition per unit time and the other is the backward transition rate r_b defined similarly as the frequency of occurrence of the backward transition per unit time.

We analyze in which region of the value of the pressure-gradient the transition occurs frequently. The transition rates are calculated with the formulae derived in [7]. Two figures, Fig. (3) and Fig. (4), show the pressure-gradient dependence of the forward transition rate and the backward transition rate in the bi-stable regime respectively.

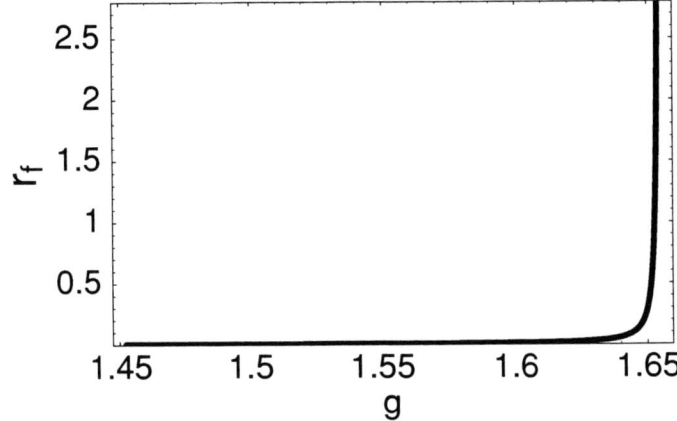

FIGURE 3. The pressure-gradient (g) dependence of the forward transition rate per unit time in the bi-stable regime. The left edge and the right edge of the horizontal axis corresponds to the subcritical bifurcation point and the linear stability boundary. It is seen that the forward transition occurs mainly in the vicinity of the linear stability boundary.

The forward transition triggered by the thermal noise occurs mainly in the vicinity of the linear stability boundary. In contrast, it is clearly seen that the backward transition occurs in the almost entire bi-stable regime. This behavior is due to strong turbulent fluctuation. It is noted that the backward transition, i.e. the transition in a turbulence, occurs in a "region" instead of a "point".

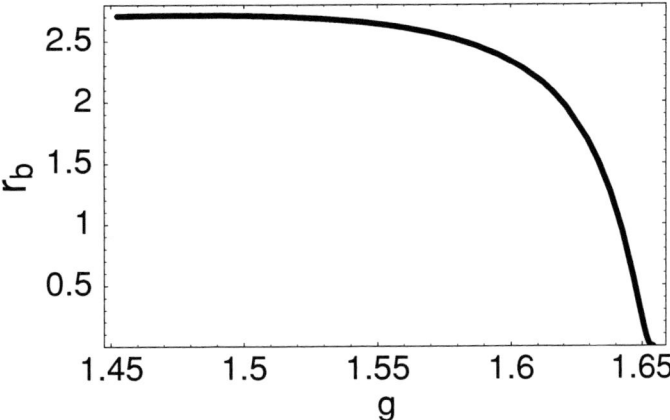

FIGURE 4. The pressure-gradient dependence of the backward transition rate per unit time in the bi-stable regime. It is seen that the backward transition occurs in the almost entire bi-stable regime.

SUMMARY AND DISCUSSION

Summarizing our work, we applied the statistical theory of plasma turbulence to problems of the transition phenomena of submarginal plasma turbulence. By numerically solving the Langevin equation, the typical time-development of fluctuation is obtained. It tells that the transition for the model of inhomogeneous plasma occurs stochastically and suggests how the transition phenomena due to subcritical bifurcation may look in time-serieses obtained in real experiments.

Furthermore, we obtained the pressure-gradient dependence of the transition rates. It is shown that the backward transition occurs with almost equal frequency in the entire bi-stable regime, so the transition occurs in a "region". The concept "transition region" is necessary in the analysis of data obtained by real experiments.

ACKNOWLEDGMENTS

We wish to acknowledge valuable discussions with Atsushi Furuya. We thank Akihide Fujisawa for showing us experimental results which inspired us. The work is partly supported by the Grant-in-Aid for Scientific Research of Ministry of Education, Culture, Sports, Science and Technology of Japan, the collaboration programmes of RIAM of Kyushu University and the collaboration programmes of NIFS.

REFERENCES

1. For a review of theoretical modeling of transport barriers, see, *e.g.*, S.-I. Itoh, K. Itoh and A. Fukuyama, *J. Nucl. Mater.*, **222**, 117 (1995); K. Itoh and S.-I. Itoh, *Plasma Phys. Control. Fusion*, **38**, 1 (1996); M. Wakatani, *ibid.* **40**, 597 (1998); J. W. Connor and H. R. Wilson, *ibid.* **42**, R1 (2000); P. W. Terry, *Rev. Mod. Phys.*, **72**, 109 (2000) ; A. Yoshizawa, S.-I. Itoh, K. Itoh and N. Yokoi, *Plasma Phys. Contr. Fusion*, **43**, R1 (2001).
2. See, *e.g.*, R. Kubo, M. Toda and N. Hashitsume, *Statistical Physics II*, Springer, Berlin, 1985; R. Balescu, *Equilibrium and Nonequilibrium Statistical Mechanics*, John Wiley & Sons, NY, 1975.
3. S.-I. Itoh and K. Itoh, *J. Phys. Soc. Jpn.*, **68**, 1891 (1999); *ibid.* **68**, 2611 (1999); *ibid.* **69**, 408 (2000); *ibid.* **69**, 427 (2000); *ibid.* **69**, 3253 (2000); *Plasma Phys. Control. Fusion*, **43**, 1055 (2001).
4. M. Kawasaki, S.-I. Itoh, M. Yagi and K. Itoh, *J. Phys. Soc. Jpn.* **71**, 1268 (2002).
5. K. Itoh, S.-I. Itoh, M. Yagi and A. Fukuyama, *Plasma Phys. Control. Fusion*, **38**, 2079 (1996).
6. S.-I. Itoh and K. Itoh, *J. Phys. Soc. Jpn.*, **68**, 2611 (1999).
7. S.-I. Itoh and K. Itoh, *J. Phys. Soc. Jpn.*, **69**, 427 (2000).
8. ITER H-mode database working group, *Nucl. Fusion*, **34**, 131 (1994).

Models for H-Mode Pedestal Temperature and Predictions for Future Tokamak Designs

T. Onjun*, A. H. Kritz*, G. Bateman*, and V. Parail†

*Physics Department, Lehigh University, 16 Memorial Drive, Bethlehem, PA 18015, USA
†EURATOM/UKAEA Culham Science Centre, Abingdon OX14 3DB, UK

Abstract. Studies of type 1 ELMy H-mode pedestals are carried out using 0 D and 1.5 D approaches. In the 0 D study, predicted pedestal temperatures are compared with 533 data points. The pedestal temperature models, with pedestal width based on magnetic and flow shear stabilization and neutral penetration, yield an RMSE of 32% and 53.4%, respectively, when compared with data. In the 1.5 D studies, the pedestal models are used together with a core transport model in the integrated predictive transport code JETTO. In simulations of type 1 ELMy H-modes using the JETTO code, when a pedestal width model based on neutral penetration is used, the pedestal width and pressure are significantly under-predicted and the ELM frequency obtained is that expected for type 3 rather than type 1 H-mode plasmas. In contrast, the simulations using a pedestal width based on magnetic and flow shear stabilization yield better agreement for the pressure profiles and the ELM frequency is appropriate for type 1 ELMs. Consequently, the simulations indicate that magnetic and flow shear stabilization plays a more significant role in determining the width of type 1 ELMy H-mode pedestals than does neutral penetration. The pedestal temperature model, using a pedestal width based on magnetic and flow shear stabilization, together with the MMM95 core transport model, is used in the BALDUR code to predict the performance of ITER. At the ITER design point, the simulation yields a pedestal temperature of 2.74 keV and an alpha power of 89.3 MW, corresponding to fusion Q of 11.2.

INTRODUCTION

Studies of type 1 ELMy H-mode pedestals are carried out using 0 D and 1.5 D approaches. In the 0 D approach, a model for the pedestal width and a model for the pedestal pressure gradient that is based on the ideal ballooning first stability boundary limiting the gradient, is used with the experimental pedestal density to predict the pedestal temperature, T_{ped}. The predicted T_{ped} are compared with the corresponding experimental data. For a pedestal width based on magnetic and flow shear stabilization, the comparison with 533 data points yields an RMSE of 32.0%; whereas, for a pedestal width based on neutral penetration length, the comparison yields an RMSE of 53.4%. In the 1.5 D approach, the pedestal models are used together with a mixed Bohm/gyro-Bohm core transport model [1] in the integrated transport modelling code, JETTO. Simulations of four JET type 1 ELMy H-mode discharges, two with low triangularity and two with high triangularity, are carried out and pressure profiles are compared with the experimental profiles. It is found that the simulations using the neutral penetration model significantly under-predict the width of the pedestal and predict an ELM frequency expected for type 3 H-mode plasmas. In contrast, the simulations using magnetic and flow shear stabilization pedestal width model yield better agreement with the experimental pressure profiles and indicate an ELM frequency of type 1 ELMs. Therefore, it appears that magnetic and flow shear stabilization plays a more significant role than neutral penetration in determining the width of type 1 ELMy H-mode pedestals. The pedestal temperature model with the width based on magnetic and flow shear stabilization is then used together with the MMM95 transport code [2] in the BALDUR code to predict the performance of ITER. At the ITER design point, the simulation yields the pedestal temperature of 2.74 keV and the alpha power of 89.3 MW.

0 D APPROACH FOR H-MODE PEDESTAL MODEL

In this study, we assume that the pressure gradient within the pedestal is constant and limited by the high n ballooning instability:

$$\frac{\partial p}{\partial r} \approx \left(\frac{\partial p}{\partial r}\right)_c = -\frac{B^2}{2\mu_0 R q^2}\alpha_c \tag{1}$$

where q is the safety factor and α_c is the normalized critical pressure gradient. The scaling of α_c is assumed to depend on the magnetic shear (s) and on the elongation (κ_{95}) and triangularity (δ_{95}) at the 95% flux surface:

$$\alpha_c = 0.4 s(1 + \kappa_{95}^2(1 + 5\delta_{95}^2)). \tag{2}$$

In the results presented in this paper, s and q are calculated at one pedestal width away from the separatrix [3]. In addition, the effect of the bootstrap current, which reduces the magnetic shear in the steep gradient region, is included in the determination of the magnetic shear.

The pedestal temperature, when the pedestal width is based on magnetic and flow shear stabilization of drift modes [4], that is $\Delta = 2.42\rho s^2$ (where ρ is the ion gyro-radius), is given by

$$T_{\text{ped}}[\text{keV}] = 1.89 \left(\frac{B}{q^2}\right)^2 \left(\frac{M_i}{R^2}\right) \left(\frac{\alpha_c}{n_{\text{ped }19}}\right)^2 s^4 \tag{3}$$

where B is the toroidal magnetic field, M_i is the hydrogenic mass, and $n_{\text{ped }19}$ is the electron density at the top of the pedestal in units of 10^{19} m^{-3}. Note that Eq. (3) is non-linear in T_{ped} because s and q are functions of the pedestal width, which in turn is a function of the pedestal temperature. Also, the bootstrap current that reduces the magnetic shear also depends on the collisionality, which is also a function of the pedestal temperature.

For the other pedestal width model considered in this paper, that is the pedestal width based on the penetration length of neutrals at the edge of plasma, $\Delta = (2.6 \times 10^{27})/n_{\text{ped}}^{3/2}$ [3], the pedestal temperature is:

$$T_{\text{ped}}[\text{keV}] = 10.2 \left(\frac{B}{q}\right)^2 \left(\frac{1}{R}\right) \left(\frac{\alpha_c}{n_{\text{ped }19}^{5/2}}\right) \tag{4}$$

Note that Eq. (4) is also a non-linear equation, as described previously.

Both pedestal width models are used in the prediction of the pedestal temperature for 533 data points of type 1 ELMy H-mode discharges. The data points are obtained from the International Pedestal Database version 3.1 (http://pc-sql-server.ipp.mpg.de/Peddb/). The pedestal temperature model with $\Delta = 2.42\rho s^2$ yields an RMSE of 32.0% while the model with $\Delta = (2.6 \times 10^{27})/n_{\text{ped}}^{3/2}$ yields the RMSE of 53.4 %. The comparison between the pedestal temperatures from both models and experimental data are shown in Fig. 1.

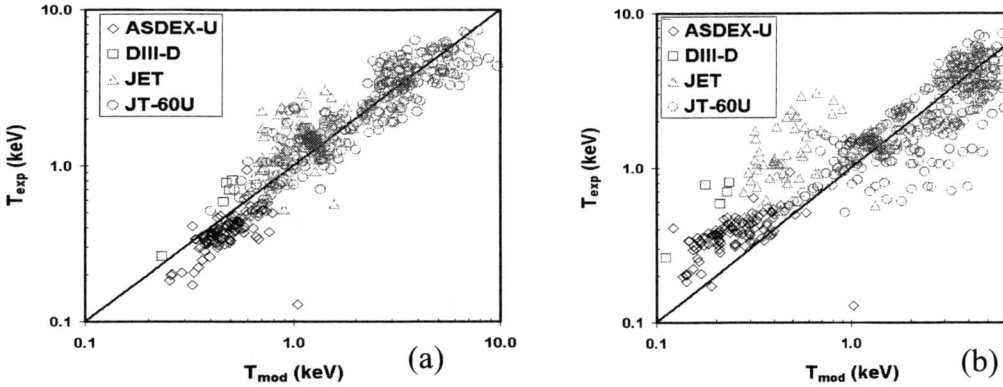

FIGURE 1. Predicted pedestal temperatures using $\Delta = 2.42\rho s^2$ (a) and $\Delta = (2.6 \times 10^{27})/n_{\text{ped}}^{3/2}$ (b) compared with 533 experimental data points from International Pedestal Database. The numerical coefficients in the expressions for Δ are chosen to minimize the RMSE.

1.5 D APPROACH FOR H-MODE PEDESTAL MODEL

Modelling of plasma parameters within edge transport barrier are explicitly included in simulations using the 1.5 D transport modelling JETTO by imposing the boundary conditions at the separatrix. Two main assumptions are made about the edge barrier. First, it is assumed that all anomalous transport within the pedestal is nullified starting from the top of the barrier outward so that all elements of the transport matrix within the barrier are equal to the ion neo-classical thermal conductivity, calculated at the top of the pedestal. The second assumption regards the width of the pedestal. The two pedestal width models, which are used in the 0 D approach, are used in the 1.5 D simulations of the JET discharges described below.

Simulations, have been carried out, using the JETTO code with the pedestal width given either by $\Delta = 2.42\rho s^2$ or by $\Delta = (2.6 \times 10^{27})/n_{\text{ped}}^{3/2}$, for 4 JET discharges. These discharges have approximately the same current, 2.5 MA;

magnetic field, 2.6 to 2.7 T; and elongation, 1.7. However, the triangularity for two of the discharges is low, 0.25 and 0.32, and, for the other two discharges, is high, 0.45 and 0.49. These JET discharges are all type 1 ELMy H-mode discharges. The core transport model, which is used in these simulations, is the mixed Bohm/gyro-Bohm model.

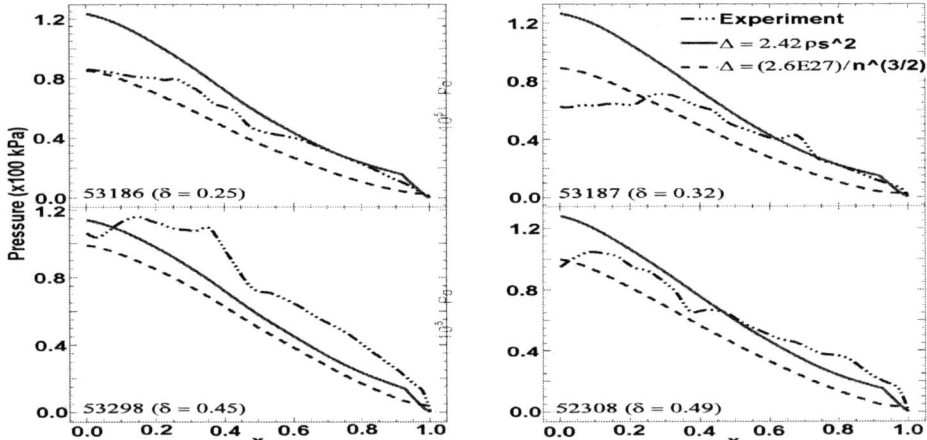

FIGURE 2. Predicted pressure profiles using $\Delta=2.42\rho s^2$ (solid line) and $\Delta=(2.6\times10^{27})/n_{ped}^{3/2}$ (dashed line) compared with the experimental pressure profiles (dot-dashed line) for 4 JET discharges.

The simulations with the pedestal width based on magnetic and flow shear stabilization yields a pedestal width of about 3 cm while the simulations with the width based on neutral penetration results in a much narrower width, less than 1 cm, for all 4 discharges. The narrow pedestal widths that result from using the neutral penetration model yield a significant under-estimation of the pedestal pressure for all 4 discharges, as shown in Fig. 2. Simulations, which use the shear stabilization pedestal model, yield better agreement for the low triangularity discharges than for the high triangularity discharges. This might be due to better access to second stability for high triangularity discharges.

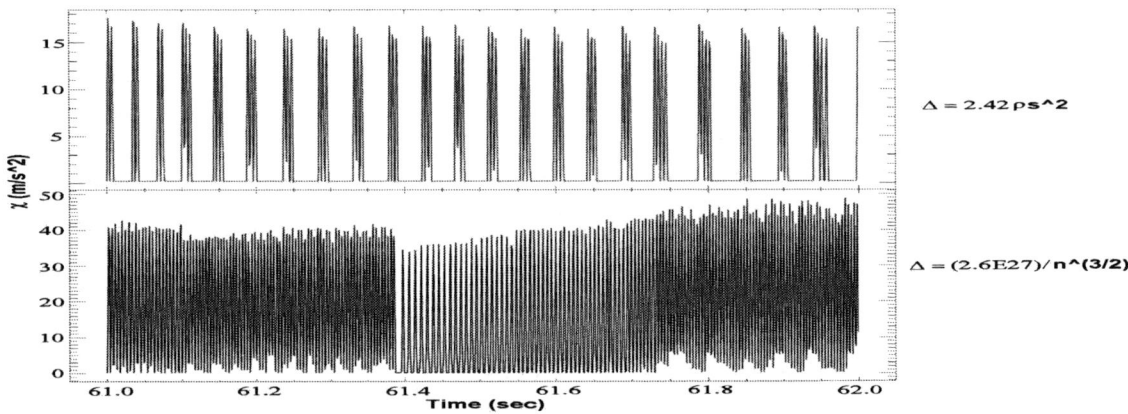

FIGURE 3. Time evolution of the ion thermal diffusivity on the top of the barrier for discharge 53186 from 2 pedestal models using Mixed Bohm/gyro-Bohm core model.

Fig. 3 shows, for the two pedestal models, the time evolution of the ion thermal diffusivity on the top of the barrier for the JET discharge 53186 ($\delta = 0.25$). Note that the ELM frequency, when using the neutral penetration model (bottom panel), is very high and characteristic of type 3 ELMy H-mode plasmas; whereas, the ELM frequency, when using the magnetic and flow shear stabilization (top panel), is characteristic of type 1 ELMy H-mode plasmas. Although the results using the magnetic and flow shear stabilization yield better agreement with experiment, it cannot be concluded that the penetration of neutrals does not play any role in the determination of the width of the pedestal, only that neutral penetration is not the main mechanism, which determines the pedestal width of type 1 ELMy H-mode plasmas.

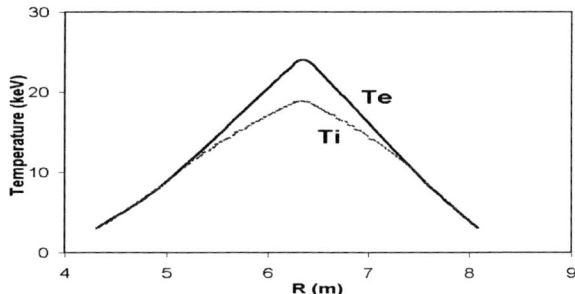

FIGURE 4. Electron and ion temperatures of ITER predicted using the combination of the MMM95 core transport model and the pedestal temperature model.

PREDICTIONS FOR ITER

The simulation for ITER has been carried out using the BALDUR code with the MMM95 core transport model and with the boundary conditions predicted by the pedestal model based on magnetic and flow shear stabilization. The parameters for ITER are $R = 6.2$ m, $a = 2.0$ m, $I_p = 15$ MA, $B = 5.3$ tesla, $\kappa_{95} = 1.7$ and $\delta_{95} = 0.33$. The simulation is carried out with the auxiliary heating power $P_{aux} = 40$ MW and the impurity concentration of 2% Be plus 0.12% Ar plus the accumulation of Helium ash, which remains below 2% in the simulation. At the ITER design point, with the line-average density of 84% of the Greenwald density ($= I_p[\text{MA}]/\pi a[\text{m}]^2 \; 10^{20}$ particles/m^3), the predicted pedestal temperature is 2.74 keV. Fig. 4 shows the electron and ion temperature profiles predicted using the MMM95 core transport model and pedestal temperature based on magnetic and flow shear stabilization. The alpha power, P_α, predicted by the BALDUR code simulations is 89.3 MW, which corresponds to a fusion $Q = 11.2$ ($Q = 5P_\alpha/P_{aux}$).

CONCLUSIONS

In the 0 D approach, two theory-motivated models have been developed for the temperature at the top of the pedestal at the edge of type 1 ELMy H-mode plasmas. The pedestal temperatures predicted from the combination of the pedestal width model (either based on the magnetic and flow shear stabilization or the neutral penetration) together with the pressure gradient, from the ideal ballooning instability, yield RMSE of 32.0% and 53.4%, respectively, when compared with 533 data points. In the 1.5 D approach, these pedestal width models have been used together with the predictive integrated modelling code, JETTO, to simulate the JET triangularity H-mode discharges. It is found that the simulations from the neutral penetration model significantly under-predict the width of the pedestal and result in the ELM frequency of type 3 H-mode plasmas. On the other hand, the simulations from the magnetic shear and flow shear stabilization yield better agreement for the pressure profiles and the ELM frequency of type 1. At the ITER design point, the fusion performance predicted using MMM95 core transport model together with pedestal model based on magnetic and flow shear stabilization yields the alpha power of 89.3 MW, which is corresponding to a fusion $Q = 11.2$.

ACKNOWLEDGMENTS

The authors thank Dr. Gregory Hammett for his help with this research. This work is supported by U.S. DOE contract DE-FG02-92-ER-54141.

REFERENCES

1. Erba, M., Aniel, T., Basiuk, V., et al., *Nuclear Fusion* **38**, 1013 (1998).
2. Bateman, G., Kritz, A.H., Kinsey, J.E., et al., *Phys. Plasmas* **5**, 1793 (1998).
3. Onjun, T., Bateman, G., Kritz, A.H., and Hammett, G., "Models for the pedestal temperature at the edge of H-mode tokamak plasmas," submitted to Physics of Plasmas.
4. Sugihara, M., Igitkhanov Y., Janeschitz G., et al., *Nuclear Fusion* **40**, 1743 (2000).

Density Transition Phenomena in Magnetized Plasma by Voltage Biasing

S. Shinohara, S. Matsuyama and K. Sugimori

Interdisciplinary Graduate School of Engineering Sciences, Kyushu University, Fukuoka, Japan

Abstract. Repeated transition phenomena (flip-flop pattern) with abrupt reductions and jumps of the electron density was observed, by the voltage biasing to an electrode in the cylindrical magnetized plasma. These global, self-excited, dynamical density transitions in bistable system were accompanied by changes of the floating potential profile and the bias current. Furthermore, the staying time probability in one of two states with hysteresis loops, was varied, which showed the importance of sheath region and fine structure pattern.

INTRODUCTION

Nonlinear phenomena such as phase transition, bifurcation, and pattern formation have been providing many interesting physical topics. Control of plasma generation and sustainment is one of the critical issues, and plasma profile is nonlinearly governed by the balance between generation and diffusion mechanisms. In a nuclear fusion field, the structural formation of electric fields and a bifurcation have been major concerns, e.g., a transition from L to H modes in tokamaks [1]. Experiments on this transition and profile change using voltage biasing [2] were also conducted in mirror machines [3-9]. Furthermore, hystereses and mode changes in the various DC and RF discharges were found, varying an input voltage or input power. However, there have been few experiments [5-9] from a basic viewpoint to understand transitions and to control the density and rotation profiles by external electric fields. These transition/transport characteristics are connected with structure formation in a self-organized manner.

Here, we have observed plasma density transitions in bistable system (flip-flop pattern) by voltage biasing to electrodes inserted into the RF (radio frequency) produced, cylindrical magnetized plasma: global characteristics of self-excited density transitions and back ones (oscillatory behaviors) between two states, accompanied by potential changes. Furthermore, we could also control the staying time, i.e., resident time, and staying probability in one of two states, varying the bias voltage, including observations of hystresis loops and fine structure pattern.

EXPERIMENTAL SETUP

Experiments have been carried out in argon plasma at a pressure of $P_0 = 0.1 \sim 10$ mTorr with the continuous RF power and frequency of $P_{RF} = 160 \sim 500$ W and $f_{RF} = 7$ MHz, respectively, in a linear device, 45 cm in diameter and 170 cm in axial length (magnetic field of $B = 500$ G) [10]. Here, mostly, no. 3 electrode (radius is $3.7 \sim 6$ cm) was used from ten concentric, segmented rings used as biased electrodes [6,7,9]. The spatial plasma parameters were measured by scanning the Langmuir probes including the Mach probe. The typical target (before biasing) plasma density n_e was in the range of $4 \times 10^9 \sim 4 \times 10^{10}$ cm^{-3} with the electron temperature $T_e = 3 \sim 6$ eV and estimated ion temperature < 1 eV.

EXPERIMENTAL RESULTS

Figure 1 shows an example of the time evolution of the ion saturation current I_{is}, varying the biased voltage V_3 (voltage at no. 3 electrode). For the case of low V_3 (less than 110 V), no n_e change was observed (state I). With the increase in V_3, an increased number of transitions from higher to lower n_e (from states I to II) were found, in addition to back transitions, i.e., density oscillation (self-excited transitions) between two states (bistable system). Typical transition time (rise and decay times between two states) was 1 ~ 2 ms. Finally, with a further increase in V_3, there appeared a lower n_e state (state II only) without back transitions ($V_3 \geq 140$ V).

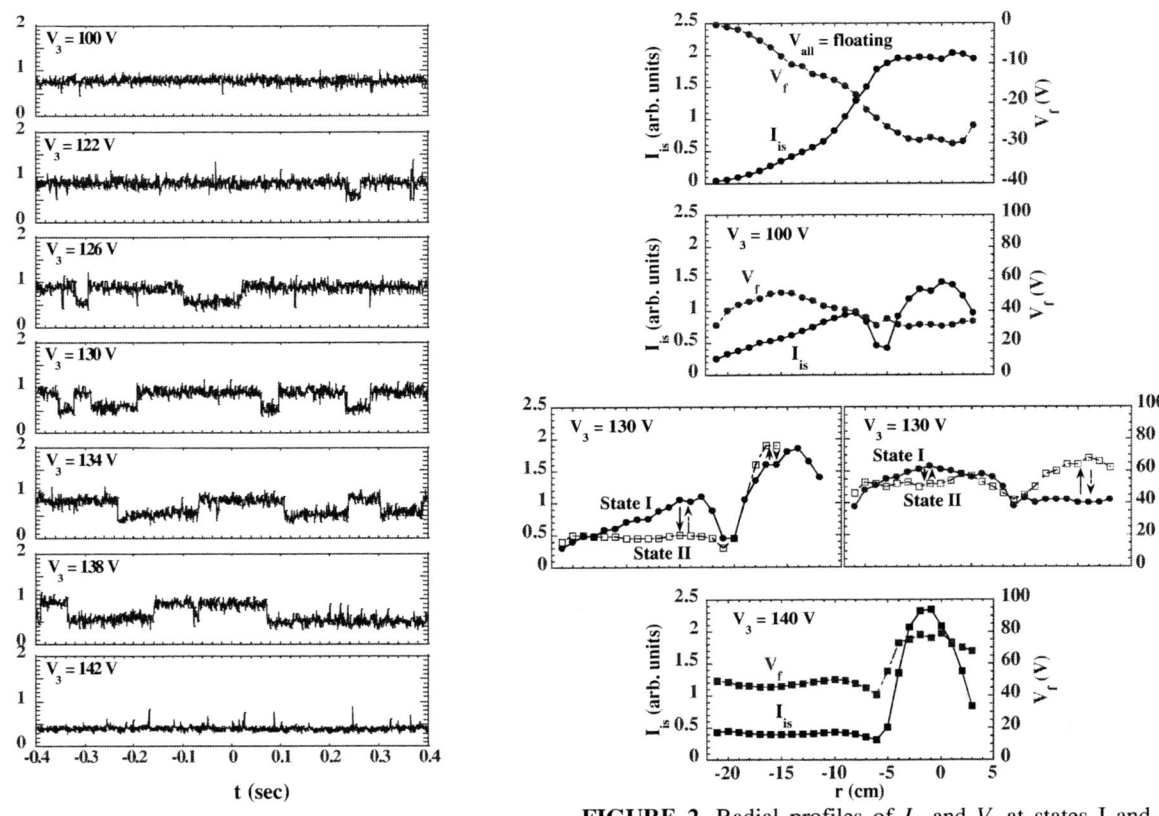

FIGURE 1. Time evolution of ion saturation current I_{is} at $r = -10$ cm, varying bias voltage V_3 ($P_0 = 3.2$ mTorr and $P_{RF} = 160$ W). Here, seven individual traces have different shots.

FIGURE 2. Radial profiles of I_{is} and V_f at states I and II, varying V_3 ($P_0 = 3.2$ mTorr and $P_{RF} = 160$ W).

During a transition, global profile changes were found, as shown in Fig. 2. For the case of the floating biasing to all electrodes, a hollow profile of the floating potential V_f and density peaking were observed. In the outer region of the biased electrode at $V_3 = 130$ V, both n_e and V_f decreased in a transition from states I to II by up to 60 % and 12 V, respectively. Here, a dip of n_e near this electrode was due to the parallel electron current along the magnetic field [5-7,9]. To the contrary, in the inner region, changes of n_e and V_f behaved in opposite ways: increases in n_e and V_f (by up to 30 V) after a transition to the state II. Here, profiles in the state I (II) with $V_3 = 130$ V correspond well with to those with the lower (higher) bias voltage than 130 V where no transitions were observed. However, from changes of n_e and V_f, Boltzmann's relation was not found on the same position as well as along the radial direction. Concerning density fluctuations, odd number modes m along the azimuthal direction dominated in the outer region, and relative fluctuation amplitude from states I to II decreased by up to more than a factor of two.

Figure 3 shows the characteristics of transitions as a function of P_0. Here, average staying time T_{aII}^* in Fig. 3 (b) is defined as the total staying time at the state II divided by frequency (number of events) during observation time, for the case of 50 % staying probability for both states of I and II (the same condition as open triangles in Fig. 3 (a)). From this figure, with the increase in P_0, both mean V_3 to have transitions and region of V_3 to have transitions (vertical distance between a closed circle and a closed box) decreased. On the other hand, with the increase in P_0,

T_{aII}^* decreased from ~ 15 sec at P_0 ~ 0.6 mTorr and had the minimum of ~ 20 ms at P_0 ~ 2.5 mTorr, and then, T_{aII}^* rose slowly to ~ 200 ms at P_0 ~ 5 mTorr, where transitions became negligible. Although this curve of T_{aII}^* has not been explained so far, the electron mean free path (mainly dominated by electron-neutral collisions) is about one half of plasma radius at P_0 ~ 2.5 mTorr, where there seemed to be a boundary between two regions.

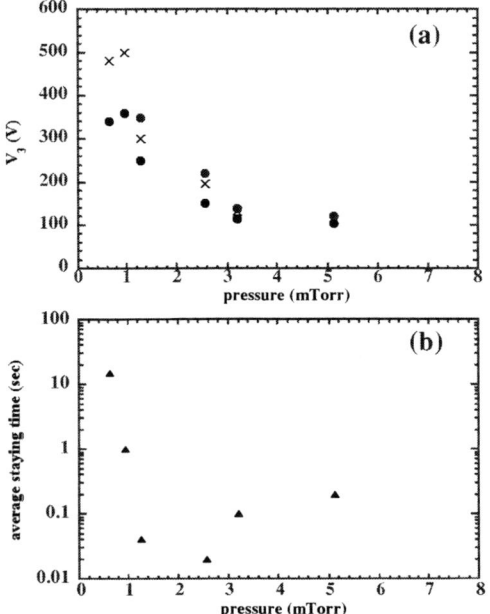

FIGURE 3. (a) V_3 and (b) average staying time at state II as a function of fill pressure measured at $r = -10$ cm ($P_{RF} = 160$ W). Here, closed circles, open triangles, and closed boxes show, respectively, points of the onset from states I to II, points of 50 % staying probability (time) for each state, and points of onset of full transitions to state II.

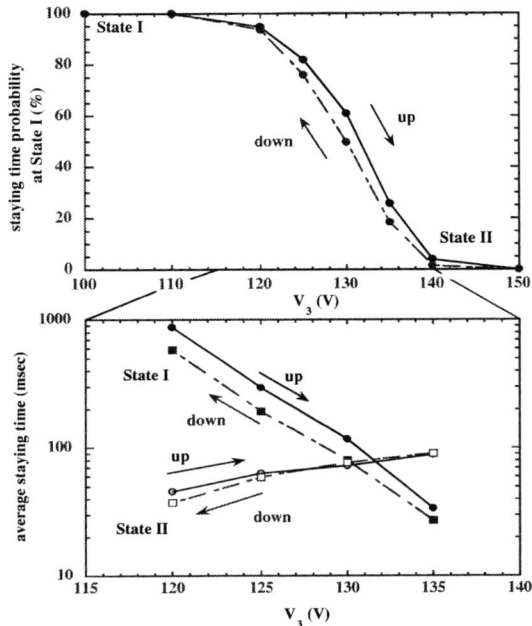

FIGURE 4. Hysteresis curves of staying time probability at state I and average staying time for states I and II, from data of ion saturation current measured at $r = -10$ cm, as a function of bias voltage V_3 ($P_0 = 3.2$ mTorr and $P_{RF} = 160$ W)

Transitions in this bistable system may be classified as a subcritical bifurcation type and be understood based on the effective, asymmetric double well potential structure, e.g., Ginzburg-Landau equation, or Schmitt trigger, including effects of some instabilities (external force) for triggering a transition statistically. This system is also related to the stochastic resonance. However, due to the presence of hysteresis loops described below, this picture must be modified/extended to explain these findings: a search for local but not global parameters may be necessary, and also roles and couplings between the inner and outer region of the electrode must be also investigated. In order to check a hysteresis character, I_{is} was measured by, first, increasing V_3 from the lower to the higher voltages through self-excited transitions between two states. Next, decreasing V_3 in the opposite way have been conducted without a termination of plasma. The timescale to take data was chosen to be much longer than typical staying time. In Fig. 4 (upper figure) staying time probability P_{sI}, defined as the ratio of the time occupied at the state I normalized by total observation time, is shown (the same conditions as Fig. 2). Below V_3 ~ 110 V, plasma had the state I all the time, and P_{sI} decreased with increasing V_3, and finally only the state II existed by high voltage biasing of > 140 V.

Figure 4 (lower figure) shows average staying time T_{aI} (T_{aII}), which was defined as the total staying time at the state I (II) divided by frequency during observation time. While T_{aI} decreased rapidly with V_3 by more than one order of magnitude, T_{aII} increased slowly with V_3. Clear hysteresis loops were found from Fig. 4 (here, hystereses were found not in conventional DC characteristics curves but in transition states (statistical value) which have not been studied so far): P_{sI} and T_{aI} were higher with phases of increasing V_3 (up) than those of decreasing V_3 (down) for the same value of V_3. This hysteresis character was also confirmed by measuring I_{is} (average value) as a function of V_3. Relating to this hysteresis, a histogram expression of I_{is} (PDF: probability density function) was plotted, as in Fig. 5. In the increasing phase of V_3, there was a tendency, especially for the case of the high bias voltage, that probabilities to have the states I and II were higher and lower than those in the decreasing phase of V_3, respectively, which is consistent with Fig. 4. In addition, central values of I_{is} at the states I (larger side of I_{is} value from two main peaks in this figure, with 120 ~ 135 V) and II (lower side with $V_3 = 135$ V) were larger in the increasing phase than that in the decreasing one. These hysteresis characters obtained may come from the fine spatial structure change between increasing and decreasing phases, from other hidden parameters, or from some memory effects, which should be clarified in future studies. Needless to say, sheath effect is also important as mentioned after.

Finally, times of transitions of various parameters were measured in order to find a triggering process. There were neither clear difference of onset of times between Iis and Vf during a transition, nor significant difference of times between measuring radial positions of both Iis and Vf. However, the bias current Ib on the order of 10 mA at the electrode no. 3 increased just before density transitions from states I to II: time of the Ib rise with an overshot positive spike was slightly earlier than that of the Iis drop (Note that changes of V3 and PRF were less than 1 % during a transition). On the back transition, Ib decreased also earlier than the time of the Iis increase. From our measurements, only a change of Ib was earlier than other signals during a transition among measured signals, which indicates the importance of changing parameters in neighboring region (sheath region) of the bias electrode for triggering a transition process. Future work of the clarification of observed transitions and oscillations with hystereses will contribute to the field of high temperature plasma (and many other fields).

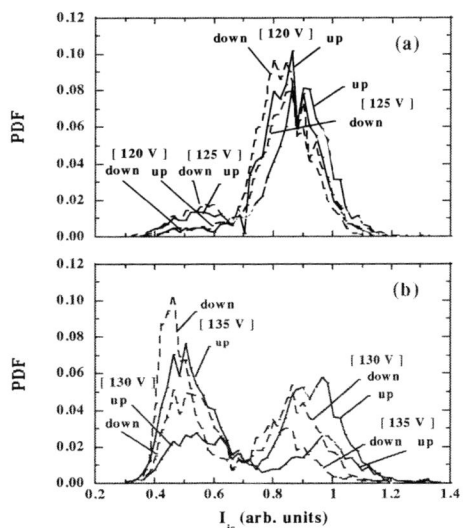

FIGURE. 5. Probability distribution as a function of I_{is} at $r = -10$ cm, for increasing (up) and decreasing (down) phases of V_3: (a) 120 V and 125 V and (b) 130 V and 135 V ($P_0 = 3.2$ mTorr and $P_{RF} = 160$ W).

CONCLUSIONS

Self-excited, repeated transitions between two states (flip-flop pattern) have been observed recently by the voltage biasing to an inserted electrode in the RF produced, magnetized plasma (bistable system). Transitions were accompanied by changes such as the floating potential profile and the bias current. Control of the staying time probability was also tried, and hysteresis characters of this probability, the average staying time, and a probability distribution of ion saturation current were investigated, varying the bias voltage. This indicated the presence of the fine structure pattern, and a change of bias current was earlier than other signals during a transition.

ACKNOWLEDGMENTS

We would like to thank Prof. Y. Kawai for his continuous encouragement.

REFERENCES

1. Wagner, F. *et al.*, Phys. Rev. Lett. 49, 1408-1411 (1982).
2. Taylor, R. J. *et al.*, Phys. Rev. Lett. 63, 2365-2368 (1989).
3. Tsushima, A., Mieno, T., Oertl, M., Hatakeyama, R., and Sato, N., Phys. Rev. Lett. 56, 1815-1818 (1986).
4. Sakai, O., Yasaka, Y., and Itatani, R., Phys. Rev. Lett. 70, 4071-4074 (1993).
5. Shinohara, S., Tsuji, H., Yoshinaka, T., and Kawai, Y., Surf. Coat. Technol. 112, 20-24 (1999).
6. Shinohara, S., Matsuoka, N., and Yoshinaka, T., Jpn. J. Appl. Phys. 38, 4321-4325 (1999).
7. Shinohara, S., Matsuoka, N., and Matsuyama, S., Trans. Fusion Technol. 39, 358-361 (2001).
8. Matsuyama, S., Shinohara, S., and Kaneko, O., Trans. Fusion Technol. 39, 362-365 (2001).
9. Shinohara, S., Matsuoka, N., and Matsuyama, S., Phys. Plasmas 8, 1154-1158 (2001).
10. Shinohara, S., Takechi, S., Kaneda, N., and Kawai, Y., Plasma Phys. Control. Fusion 39, 1479-1486 (1997).

Fluctuation Characteristics in Detached Recombining Plasmas

N. Ohno*, N. Tanaka*, S. Takamura* and V. Budaev[†]

Department of Energy Engineering and Science, Graduate School of Engineering, Nagoya University, Nagoya 464-8603, Japan
[†]*Kruchatov Institute, NFI, Russia*

Abstract. Fluctuation in detached recombining plasmas has been investigated experimentally in the linear divertor plasma simulator, NAGDIS-II. As increasing neutral gas pressure, floating potential fluctuation of the single Langmuir probe and the target plate installed at the end of the NADIS-II device becomes larger and bursty negative spikes are observed in the signal associated with a transition from attached to detached plasmas. The fluctuation property has been analyzed by using Fast Fourier Transform (FFT), probability distribution function (PDF). The PDF of the floating potential fluctuation in the attached plasma condition obeys the Gaussian distribution function, on the other hand, the PDF in detached plasma shows a strong deviation from the Gaussian distribution function, which can be characterized by flatness and skewness.

INTRODUCTION

In recent years, plasma detachment has been one of most critical issues in fusion-related edge plasma physics. Volumetric plasmas recombination, including three body and radiative recombination, plays an essential role for the reduction of the plasma particle and heat flux to the plasma-facing components, which strongly depends on electron temperature and density. In order to understand the particle and energy balance in the detached recombining plasmas (DRP), accurate measurements of electron temperature and density are required. However, it is found in the recent experiments in linear plasma devices and large tokamaks that the Langmuir probe measurements show strong anomaly that the electron temperature measured by the Langmuir probe is much larger than that by the optical measurement[1, 2]. Two reasons are mainly considered: plasma resistivity and fluctuation in in the DRP. In this presentation, characteristics of plasma fluctuation in the DRP will be reported.

EXPERIMENTAL SETUP

Figure 1 shows a schematic view of the NAGDIS-II device[3, 4, 5], corresponding a water-cooled vacuum chamber, 2.5 m in length and 0.18 m in diameter equipped with 21 solenoidal magnetic coils. The magnetic field strength is up to 0.25 T. This device can generate high density plasma with the electron density up to 10^{20} m^{-3} in steady state, relevant to the edge plasma conditions in present fusion experimental devices. For the production of high density plasmas, DC plasma source (based on the TP-D type plasma source) was employed. The plasma source assembly consists of a 108 mm diameter LaB_6 disk cathode, an intermediate hollow SUS electrode, and a hollow anode made of copper. The LaB_6 disk cathode is heated by a carbon heater with a typical heating power of 3 kW. External heating of the cathode facilitates a start up of the discharge and keeps the discharge voltage less than 100V for He plasmas. The cathode is covered with a molybdenum cylinder to improve the efficiency of gas usage and the confinement of heat from the carbon heater. The floating potential is measured by a single Langmuir probe in the downstream near the target plate.

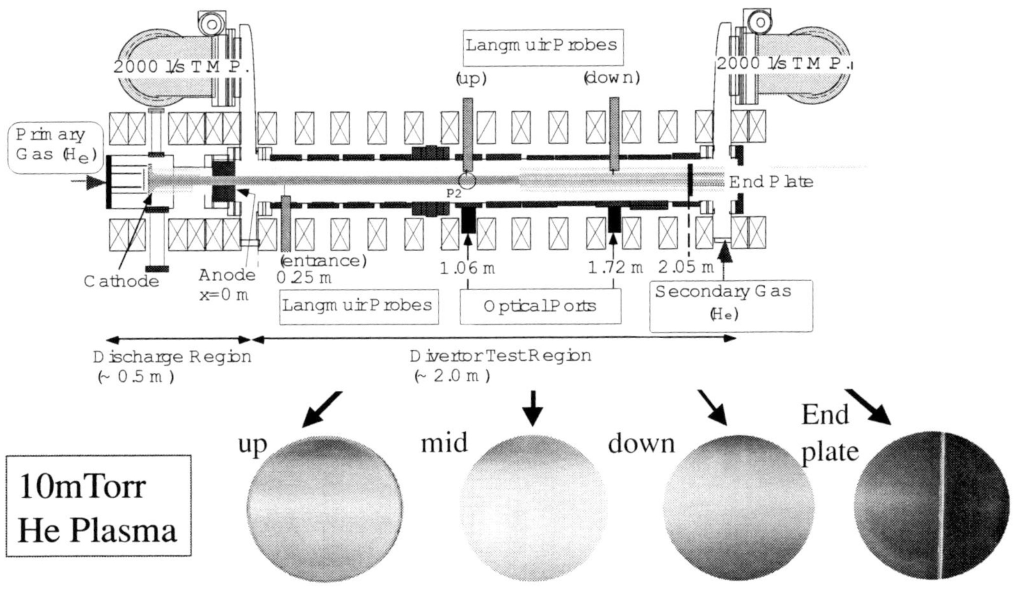

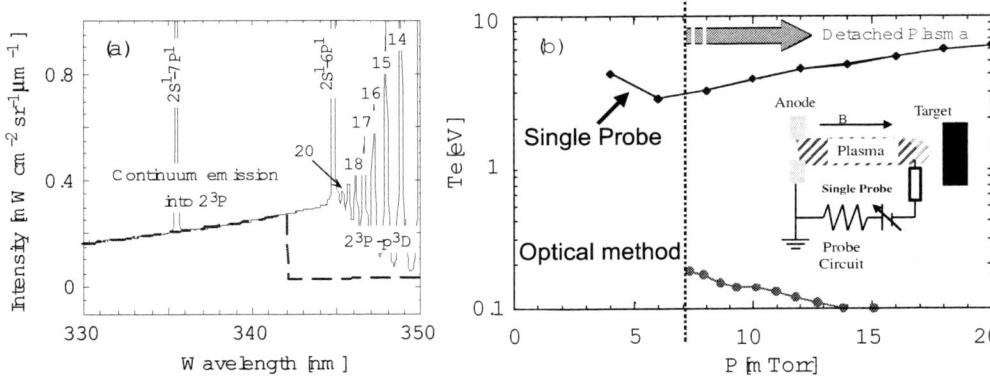

FIGURE 1. Schematics of divertor plasma simulator, NAGDIS-II and typical photograph of detached recombining plasma at a neutral pressure of 10 mtorr.

FIGURE 2. (a): Spectrum of the light emission from DRP and (b):comparison between the electron temperature estimated by the optical method and a single Langmuir probe. The probe circuit is illustrated in the inset.

EXPERIMENTAL RESULTS AND DISCUSSION

Fluctuations in the DRP has been investigated experimentally in NAGDIS-II. As increasing neutral gas pressure, the DRP is formed in front of the target plate as shown in the photograph of Fig. 1. Figure 2(a) shows the light emission from the DRP, indicating radiative and three body recombination. By analyzing the spectrum, electron temperature T_e can be estimated to be less than 0.5eV[6, 7]. Figure 2(b) shows a comparison of T_e obtained by the optical method and a single Langmuir probe measurement. As increasing neutral pressure P, T_e with the optical method is monotonically decreasing. On the other hand, T_e with the single probe is increasing with P and the absolute value is much larger than T_e with the optical method. We consider the potential fluctuation in the DRP has an effect on the modification of the I-V characteristics of the Langmuir probe measurement.

Figure 3 shows the time evolution of the floating potential measured by the single probe located in the downstream. Floating potential fluctuation is found to become larger with P, and bursty negative spikes are often observed in the signal associated with a transition to DRP. The negative spikes have the common property with the rapid drop and slow recovery. The fluctuation property has been analyzed by using fast Fourier transform (FFT), which indicates there are no typical frequency in this fluctuation. We have also carried out the analysis based on the probability distribution

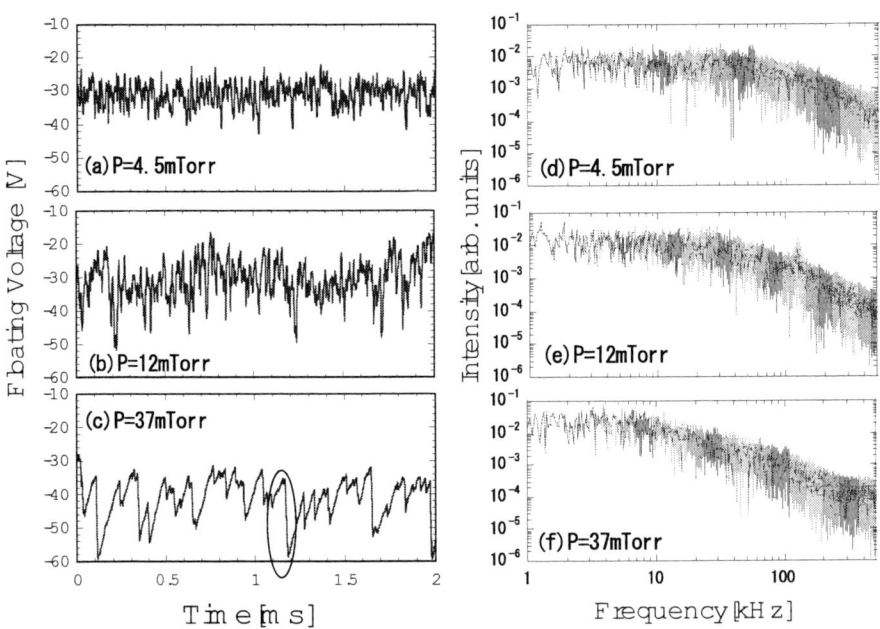

FIGURE 3. Time evolution of the floating potential at a pressure of (a):4.5mtorr, (b) 12mtorr and (c): 37mtorr. (d)-(f) are Fourier spectra, corresponding to (a)-(c), respectively.

function (PDF). The main goal of the probability distribution function is to compare the signal statistics to a Gaussian. In a Kolmogorov-type model (K41), it is assumed that the fluctuation are random and self-similar. The deviation from a Gaussian is due to the existence of coherent events that reduce locally the number of degrees of freedom in space and time. Figure 4 shows the PDF of the floating potential signals at P of 4.5 mtorr and 37 mtorr. It is found that the PDF of the floating potential fluctuation before a transition to DRP almost obeys the Gaussian distribution function. On the other hand, the PDF in the DRP shows a strong deviation from the Gaussian distribution function, which can be characterized by flatness and skewness. The skewness S is determined by: $S = <X^3>/<X^2>^{3/2}$, which describes the asymmetry of the PDF and the flatness $F = <X^4>/<X^2>^2$ measures the tail's weight with respect to the core of the distribution. In the Gaussian distribution function, corresponding to the fully random motion, the flatness and skewness are 3 and 0, respectively. During the transition from attached plasma to fully detached plasma, the strong deviation from the Gaussian distribution was observed, which is schematically illustrated in Fig. 4(e). Based on the PDF in the DRP, the data points during the negative spikes can be removed in the I-V characteristics as shown in Fig. 5(b). The reconstructed I-V characteristics removing the contributions during negative spikes gives a much lower T_e, which almost agrees with T_e with the optical method as shown in Fig. 5(a). Comparison of the I-V characteristics with and without the negative spikes shows the plasma potential is dramatically changed in time, which should have a strong effect on the modification of the I-V characteristics.

CONCLUSIONS

Fluctuation of the floating potential has been investigated in attached and detached recombining plasmas. Bursty negative spikes of the floating potential appears in the detached recombining plasma. The amplitude of floating potential is quite high compared with the electron temperature measured with the spectroscopic method. By removing the influence of the negative spikes form the I-V characteristics, lower electron temperatures can be evaluated. In addition to the fluctuation of the floating potential, positive spikes are observed in the ion saturation current, which could be associated with bursty transport process.

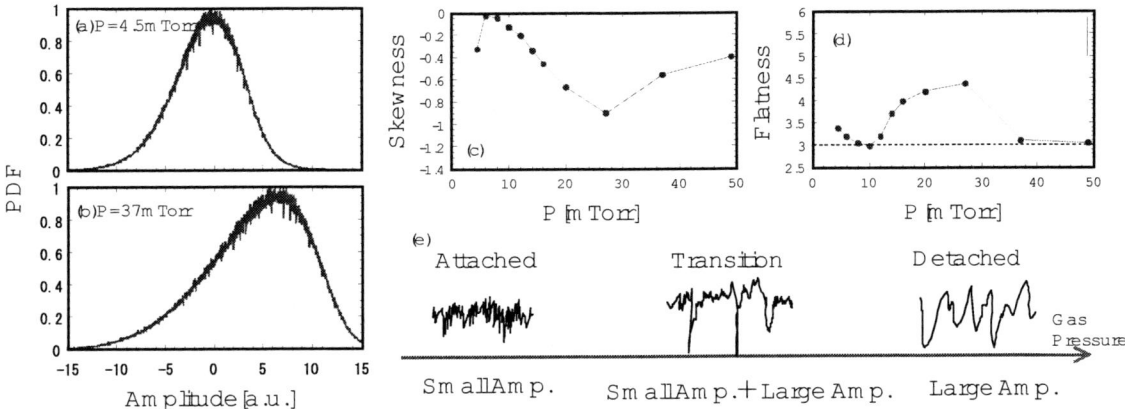

FIGURE 4. Probability distribution function of the floating potential signal at a neutral pressure of (a) :4.5mtorr and (b): 37 mtorr.(c) and (d) are dependence of skewness and flatness on the pressure.

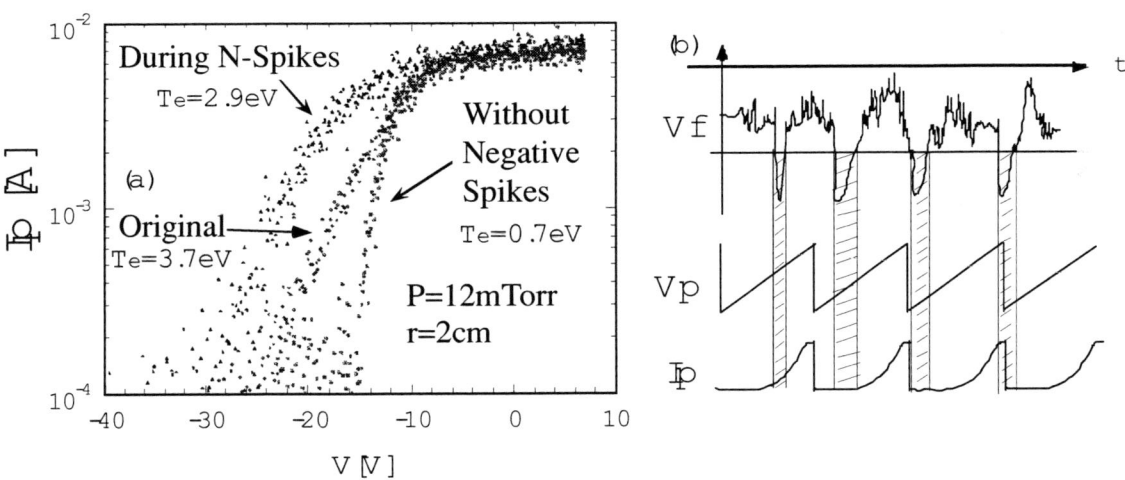

FIGURE 5. (a) Reconstructed I-V characteristics by removing the influence of the negative spikes. The procedure is illustrated in (b).

ACKNOWLEDGMENTS

We wish to acknowledge Dr. Y. Uesugi, Prof. S. Krasheninnikov, Dr. A. Pigarov and Dr. U. Wenzel for very fruitful discussion and Mr. M. Takagi for useful technical supports.

REFERENCES

1. N. Ohno, *et al.*, Contrib. Plasma Phys., **41**, 473(2001).
2. N. Ezumi, *et al.*, Contrib. to Plasma Physics, **38**, 31(1998).
3. N. Ezumi, *et al.*, Proc. of 20th EPS Conf. on Controlled Fusion and Plasma Physics, Bercgtesgardenm, Germany, 1225(1997).
4. S. Narita, *et al.*, Proc. of the 1996 International Conference on Plasma Physics, Nagoya, Japan, 1386(1996).
5. N. Ohno, *et al.*, Nucl. Fusion, **41**, 1055(2001).
6. N. Ohno, *et al.*, Phys. Plasmas, **6**, 2486(1999).
7. D. Nishijima, *et al.*, J. Nucl. Mater., **290-293**, 31(2001).

Dipole Map For Divertor Tokamaks

Halima Ali[1], Alkesh Punjabi[1] and Allen Boozer[2]

[1]Center For Fusion Research & Training, Hampton University, Hampton, VA 23668
[2]Columbia University, New York, NY 10027 and Max Planck Institute For Plasma Research, Garching, Germany

Abstract. Heat flux impinging on the collector plates of divertor tokamaks can be prodigious. Therefore, the problem of spreading the heat flux on plates is a crucial issue for divertor tokamaks such as ITER. Here we use method of maps /1,2/ to investigate this problem. Magnetic field lines in non-axisymmetric divertor tokamaks are a one and a half degree of freedom Hamiltonian system /1-3/. We represent the unperturbed magnetic topology by the Symmetric Simple Map (SSM) /4/ given by

$$y_{n+1} = y_n + 2kx_n - 2k^2 y_n(1-y_n), \quad x_{n+1} = x_n - ky_n(1-y_n) - 2k^2 y_{n+1}(1-y_{n+1}).$$

The effects of a current carrying coil placed externally across from X-point is represented by Dipole Map (DP) /4,5/ given by

$$x_{n+1} = x_n + 2\delta s^3 x_{n+1} \frac{y_n - y_s + s}{[x_{n+1}^2 + (y_n - y_s + s)^2]^2}, \quad y_{n+1} = y_n + \delta s^3 x_{n+1} \frac{(y_n - y_s + s)^2 - x_{n+1}^2}{[x_{n+1}^2 + (y_n - y_s + s)^2]^2}.$$

δ is amplitude of high MN magnetic perturbation, s is the distance of coil from last good surface across from X point, and is the y coordinate of last good surface where it crosses the axis joining X point and O point across from X point. We fix k=0.3 and $s = \frac{1}{2}|y_s|$. We calculate the increase in width of stochastic layer and area of footprint of field lines on divertor plate as δ is increased. We also calculate how connection length, toroidal and poloidal circuits and their fractal structures, the number, location and density of hot spots change with δ. Finally, we make conclusions about how the heat flux can be possibly controlled and reduced by applying external magnetic perturbation in divertor tokamaks.

DISCUSSION

The problem of controlling and predicting the tremendous heat loads ($\sim$ 15-20 MW m-2) striking the divertor plates is of crucial importance in the design of the future large tokamaks. In this paper, the focus will be on the effects of dipole perturbation on the magnetic footprint in a single-null divertor tokamak. For this purpose, we use the Symmetric Simple Map with k = 0.3. In this case, the last good confining surface passes close to x = 0 and y_{lgs}= y = 0.9973, and the safety factor for this surface is q_{lgs} = 30. Thus, if we consider 10 iterations of the SSM to be equivalent to 1 toroidal circuit of tokamak, then SSM with k = 0.3 represents magnetic topology of a single-null divertor tokamak with qedge = 3. We choose this as the unperturbed magnetic topology, and apply the Dipole Map, with y_s = -0.5128 and s=2|y_s|, after every 10 iterations of SSM to calculate effects of high MN dipole perturbation on the stochastic layer, magnetic footprint and heat load on collector plate. As the amplitude δ of dipole perturbation is increased from 0 to 6.8E-6, the width w of stochastic layer increases slowly from $\sim$ 0.003 to $\sim$ 0.001. When δ is further increased to 6.9E-6, the width increases suddenly by a large amount from $\sim$ 0.001 to $\sim$ 0.05, and as δ is increased from 6.9E-6 to 7.3E-6, increase in width of stochastic layer is very fast, reaching up to $\sim$ 0.1. For 0 # δ # 6.8E-6, w % $\delta^{2/5}$, and for 6.9E-6 # δ # 7.3E-6, w % δ^8. As δ increases from 0 to 6.8E-6, the area of magnetic footprint increases by 43 %, fraction of magnetic flux escaping from the stochastic layer decreases by 60 %, number of hot spots on plate decreases from 34 to 9, number of medium hot spots increases from 418 to 496, number of cold spots increases from 184 to 259, the fraction of heat flux going into hot spots decreases by 74 %, fraction of heat flux going into medium hot spots increases by 19 %, and fraction of heat flux going into cold spots increases by 41 %. Note that the rectangular region containing the footprint is divided into 2000 rectangular sub-regions of equal size. Total number of field lines in footprint is 20,000. A subregion is labeled as a cold spot if it contains less than 10 strike points, a medium hot spot if it contains from 11 to 100 points, and hot spot if it contains more than 100 points. Here we have assumed that, to first approximation, plasma particles travel along field lines. These estimates are the worst case scenarios, since particle and heat transport transverse to magnetic field will lead to lesser heat loads. The area of stochastic layer can be roughly approximated by a circular ring of outer radius 1 and inner radius y_{lgs}. Average heat flux density striking the divertor plate, $<H>$, is then

$$<H> \propto \frac{\Phi A_{SL}}{AL_c}$$

where M is the fraction of magnetic flux escaping the stochastic layer, ASL is the area of stochastic layer, A is the area of magnetic footprint, and LC the average connection length). We see that as δ increases from 0 to 6.8E-6, average heat flux density decreases by 16 %. For further increase in δ from 6.9E-6 to 7.3E-6, flux escaping stochastic layer further decreases to 96 % and average heat flux density decreases by 18 %, while other parameters change marginally.

In figures below, we show the width of stochastic layer, area of magnetic footprint, fraction of magnetic flux escaping stochastic layer, number of hot spots as functions of amplitude δ of high MN perturbation. We also show the phase portraits in principal plane and the magnetic footprints.

Number of hot spots, heat flux contained in hot spots and fraction of magnetic flux escaping stochastic layer reduces and the heat load is dispersed more evenly over larger area by applying dipole perturbation in divertor tokamaks.

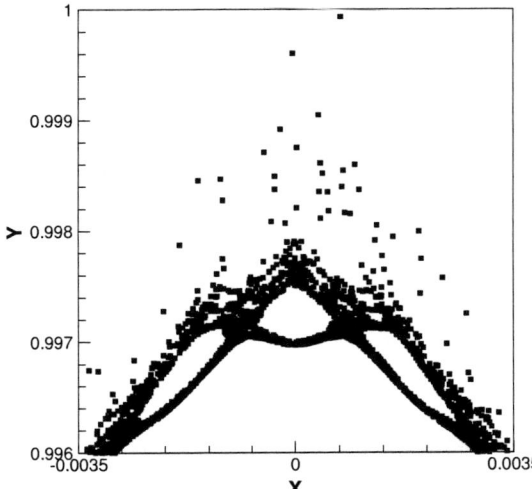

FIGURE 1. Phase portrait of the stochastic layer near the X point for k = 0.3 and * = 0.

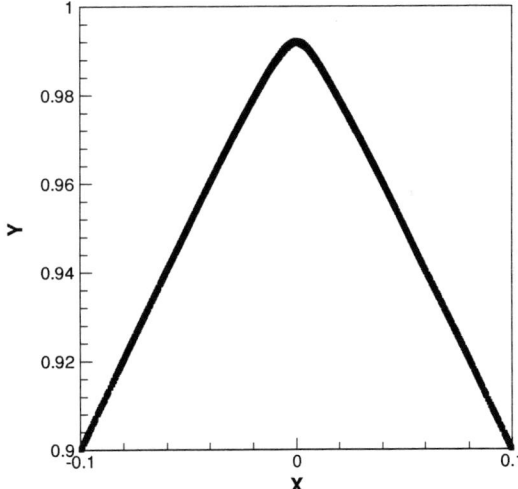

FIGURE 2. Phase portrait of the stochastic layer X point for k = 0.3 and * = 7.3E-6.

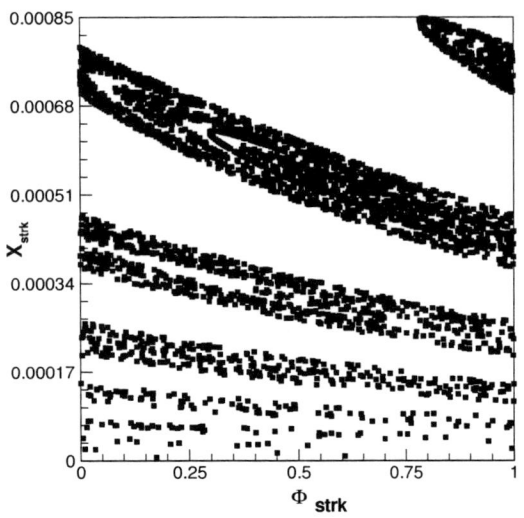

FIGURE 3. Magnetic footprint for k = 0.3 and ∗ = 0.

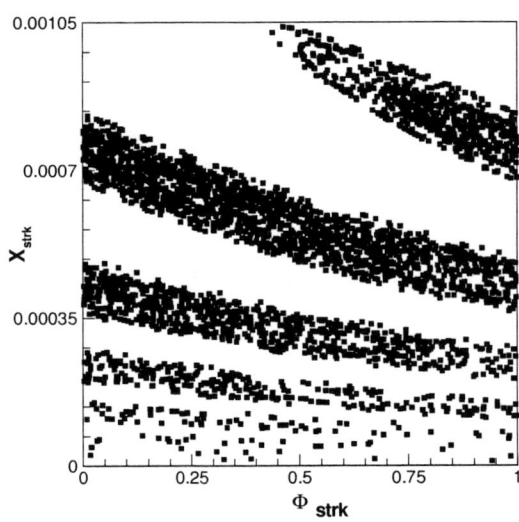

FIGURE 4. Magnetic footprint for k = 0.3 and ∗ = 7.3E-6.

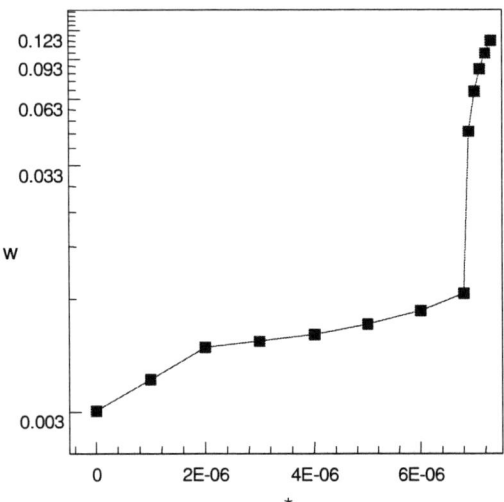

FIGURE 5. Width of stochastic layer as a function of amplitude ∗ of high MN perturbation with k = 0.3.

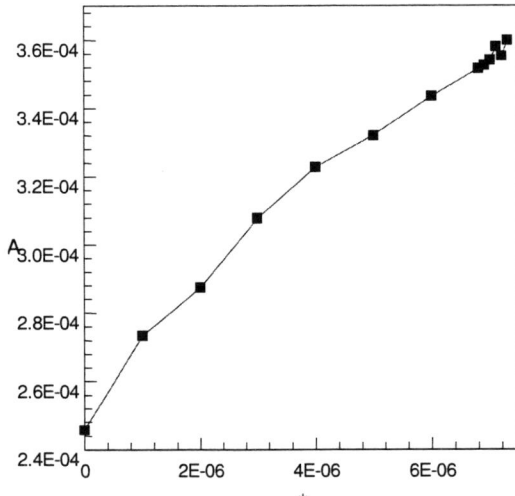

FIGURE 6. Area of magnetic footprint as a function of amplitude ∗ of high MN perturbation with k = 0.3.

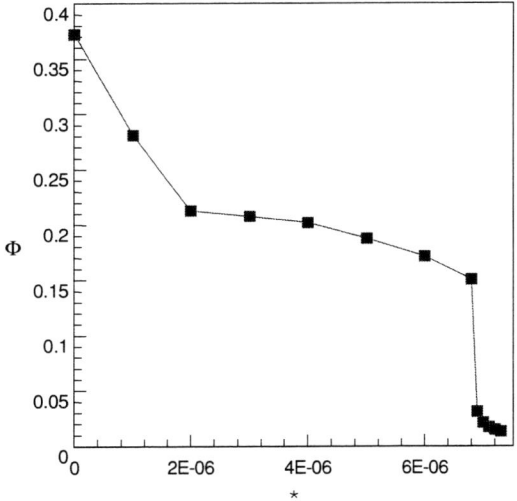

FIGURE 7. Fraction of magnetic flux escaping stochastic layer as a function of *.

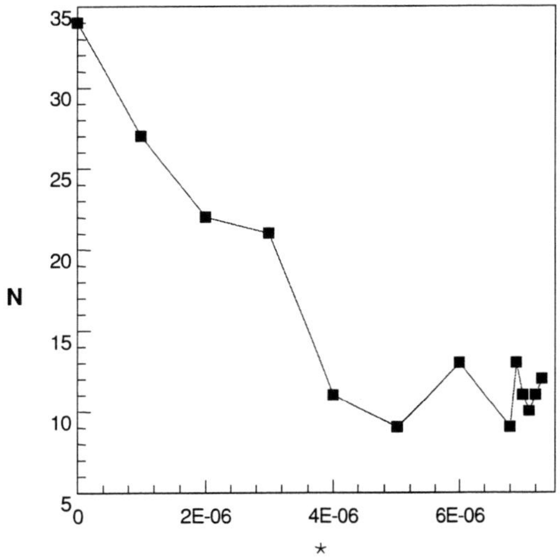

FIGURE 8. Number of hot spots, **N**, on collector plate as a function of *.

ACKNOWLEDGMENTS

This work is supported by US DOE OFES by DE-FG02-ER54624 and DE-FG02-97ER54451. Computations for this work were carried out on Cray computers at NERSC. Authors thank NERSC for this.

REFERENCES

1. Punjabi, A., Verma, A., and Boozer, A., Phys. Rev. Lett., 69, 3322 (1992).
2. Punjabi, A., Verma, A., and Boozer, A., J. Plasma Phys., 52, 91 (1994).
3. Punjabi, A., Verma, A., and Boozer, A., J. Plasma Phys., 56, 569 (1996).
4. Punjabi, A., Ali, H., and Boozer, A., Plasma Phys., 4, 337 (1997).
5. Ali, H., Punjabi, A., and Boozer, A., Bull. Amer. Phys. Soc., 46, 233 (2001).

Improvement of Carbon Properties used as PFCs in Tokamaks by Nitrogen Irradiation

Guy G. Ross and Daniel Bourgoin

INRS–Énergie, matériaux et télécommunications, Varennes, Québec, Canada, J3X 1S2

Abstract. Many properties of carbon make it suitable for plasma facing components (PFC). However, chemical erosion of carbon by hydrogen and oxygen at high temperature causes a dilution of the plasma and an important loss of energy by radiation. Diffusion of hydrogen in carbon as well as its poor retention constitute other major obstacles concerning the use of carbon in PFCs. Inspired by the work done for synthesis of C_3N_4 and its promising properties, we have studied the use of carbon implanted with nitrogen as PFC's. Thermal annealing used in combination with surface analysis tools (ERD ExB, RBS, XPS) were used to characterise in laboratory the influence of N implantation on the retention of hydrogen and the chemical erosion of two kinds of carbon (a-C and CFC). Results from laboratory experiments show that N implantation in carbon slows down the diffusion of hydrogen and causes an improvement of its retention without a significant increase in methane and ethylene emission during thermal annealing for an appropriate implantation dose (N/C~0.2). These results suggest that implantation of N in carbon is a valuable tool to improve the properties of carbon in PFC's.

INTRODUCTION

Carbon has been widely used for PFC's in different tokamaks. However, chemical erosion by hydrogen and oxygen at high temperature, and poor retention constitute the major obstacles to its use (e.g. see ref.[1]). Boron [2], silicon [3] and lithium [4] have been added to carbon, and tungsten [6] and molybdenum [5] have also been used as PFCs, but none of these materials is a perfect solution [7]. Physical sputtering of low-Z (atomic number) materials is large and is responsible for contamination of the plasma. Thermal or mechanical properties of these materials are often inappropriate for the use in a tokamak. Sputtering yield of high-Z materials is low but the few high-Z atoms going into the plasma cause a dilution of the plasma and an important loss of energy by radiation (loss ~ Z^2). The maximum tolerable concentrations in fusion plasma are 10^{-1} and 10^{-3} for C and W, respectively.

The structural and electronic properties of the C_3N_4 hypothetical compound are promising [8]. Its calculated bulk modulus was found to be comparable to that of diamond. The predicted velocity of sound in the C-N compound (1.1 x 10^6 cm/s), suggests a high thermal conductivity. Several techniques have been used for synthesising C_3N_4 [9-10]. Inspired by that work, we propose in this paper the use of carbon implanted with nitrogen as PFC's. N implantation could be made in a tokamak by some conditioning between the discharges. Furthermore, nitrogen is a low-Z element compatible with the constraints that its use as PFC's in a tokamak impose.

EXPERIMENTAL PROCEDURE

Carbon was deposited (~50 nm) on cleaned silicon wafers by sublimation of a carbon filament under vacuum (2×10^{-7} Torr.) Four coupons were implanted under a pressure of 1×10^{-7} Torr at 3 different energies of N_2 ions: 4keV/at, 2 keV/at, 0.5 keV/at in order to obtain a mostly constant N depth over a depth of 13 nm. Four different doses were implanted in such a way that a N/C composition of 15%, 30%, 60% and 150% would have been obtained if a complete absorption occured. Experimental conditions are summarised in Table 1.

TABLE 1. Details of the nitrogen implantation doses.

Name	CN0	CN15	CN30	CN60	CN150
Total dose (10^{16} N/cm^2)	0	1.8	3.5	7.0	17.5
0.5 keV/at (10^{16} N/cm^2)	0	0.24	0.48	0.94	2.4
2.0 keV/at (10^{16} N/cm^2)	0	0.57	1.2	2.3	5.7
4.0 keV/at (10^{16} N/cm^2)	0	0.96	1.9	3.8	9.6

The elemental composition of the deposit was evaluated by means of nuclear microanalysis. A 350 keV accelerator delivered a 1 mm^2 ^{4}He$^+$ beam on our samples. Contents in hydrogen was measured by Elastic Recoil Detection (ERD) [11]. Areal density of the deposited carbon film was measured by Rutherford Back Scattering (RBS) [11], which was also used to evaluate the N/C ratio.

The chemical composition of the deposits (implanted or not) was done by X-ray Photoelectron Spectroscopy (XPS, model ESCALAB 220i-XL) which was equipped with a monochromatic Al K$_{alpha}$ (1486,6 eV) source at a power of 4.05 W/mm^2 (10 kV, 27 mA) and a large area XL. The C1s (285 eV), N1s (400 eV) and O1s (532 eV) signals were measured. Details of the spectra were obtained by means of the Casa software [12].

As deposited and nitrogen implanted samples were thermally annealed under vacuum (2 x 10^{-8} Torr) by means of a cylindrical furnace coupled to a DATAQUAD 100 mass spectrometer. A 30 °C per minute ramp was performed on each sample up to 950 °C. The mass spectrometer monitored masses 2, 12, 14,15,16, 26, 27 and 28 every 5 seconds. Mass 2 monitored H$_2$, masses 15 and 16 monitored essentially methane, masses 14, 26, 27, 28 monitored essentially nitrogen and/or ethylene (and part of their cracking pattern). The precision in temperature for each point recorded by the mass spectrometer is better than 2.5 °C. The background (residual gas pressure without any sample in the vacuum chamber) was measured and subtracted from the recorded signals.

RESULTS AND DISCUSSION

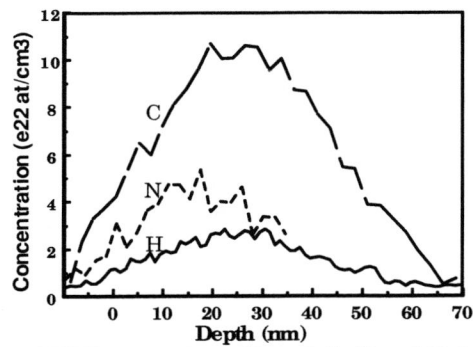

FIGURE 1. Depth profiles of C, N and H in CN60 sample as measured by RBS and ERD.

Measurement on the CN60 sample is shown in Figure 1. The depth distribution of H, N and C contents are presented. No oxygen was found through all the depth of the deposits except at the surface (adsorbed H$_2$O) and at the interface of the deposits with the substrate (SiO$_2$). Table 2 summarises the measurements performed on all the samples. The hydrogen which is present in the sample does not come from residual water vapour in the vacuum chamber or adsorbed on the filament because no oxygen was found in all the depth of the samples. We believe that it comes from the cracking of hydrocarbon molecules already present in the filament.

TABLE 2. Composition of the samples as measured by RBS, ERD and XPS.

		CN0	CN15	CN30	CN60	CN150
Imp. Dose (10^{16} N/cm^2)		0	1.8	3.5	7	17.5
N/C ratio	(before - after annealing)	0 - 0	0.15 - 0.14	0.27 - 0.16	0.41 - 0.18	0.48 - 0.20
H retained (10^{16} H/cm^2)	(before - after annealing)	7.0 - 0.9	6.8 - 1.1	6.6 - 1.5	6.4 - 3.7	3.6 - 0.4
H retention ratio		0.13	0.16	0.23	0.58	0.11
C-C, C-H %		73	60	55	55	55
CH$_3$ %		25	31	33	34	33
C-N			7.6	9.4	9.7	10
C=N			1.4	2.6	1.3	2

A maximum N/C ratio of ~0.48 is obtained even if the samples were implanted at higher doses. This effect has been reported elsewhere [10]. A little H desorption was induced by N implantation as long as the dose is not higher than the optimal N/C (0.41) ratio. However, half of the H contents was desorbed from the sample CN150. RBS and ERD measurements have not shown any erosion of the deposit due to N implantation.

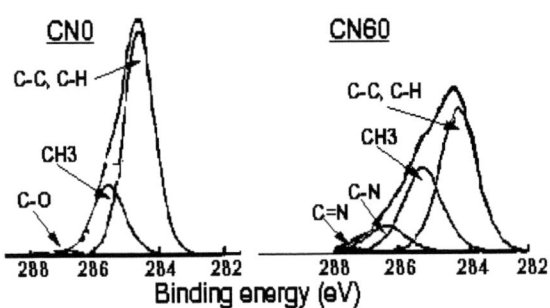

FIGURE 2. XPS C 1s peak, CN0 and CN60 samples.

Figure 2 shows the C1 peaks for the CN0 and CN60 samples as measured by XPS. Similar measurements were performed for all samples. The shape of the C1s peak has been changed significantly by the N implantation. Two peaks located at 286.5 eV and 287.5 eV are larger for the implanted samples than for the as-deposited sample. The 286.5 eV peak is related to C-N like (amines) or C-O chemical bonds [13,14]. In the as-deposited sample a small residual C-O peak is present but it is negligible in the N implanted samples. According to Marton et al. [9], an increase in this peak is an indication of the C_3N_4 phase. The proportion of this peak over the total area is maximum for the CN60 and CN150 samples. The 287.5 eV peak is related to C=N like (imines) chemical bonds [13,14]. This peak is low for all the implanted samples and is maximum for the CN30 sample. Two other peaks are also observed; one located at 285.5 eV, is associated to the presence of CH_3 and the other at 284.5 eV is related to the aromatics (benzene like) bonds [13,14]. The former has increased in the implanted samples while the latter has decreased, suggesting that implantation forces a given quantity of carbon to change their aromatics bonds into different chemical hybridizations. The relative concentration of each peak is included in table 2 for all samples.

Optical microscopy of the surface of the CN150 sample has shown some blisters which could be related to the saturation in N contents and to the formation of N_2 gas bubbles. It is also possible that H can be accumulated in bubbles and blisters prior their desorption. The blisters could also be associated to local deformations which give rise to paths of diffusion followed by the molecules (e.g. H_2 and N_2) desorbed during the N implantation and/or thermal annealing and/or when samples are implanted with N above the saturation limit.

Partial pressures during annealing for the different samples and masses (2, 16 and 28 amu) are shown in Figure 3. A steep increase of the desorption rate of 2 amu molecules at 450 °C is observed. The implantation of N has lowered the H_2 desorption intensity at higher temperatures except for the lowest dose (CN15 sample). A second desorption region at ~750 °C is clearly visible for samples implanted to doses higher than N/C ~ 0.1. Globally, there is an optimal N implantation dose which minimises the H_2 desorption (N/C=~0.4) on the temperature range going from 300 to 950 °C, the initial H contents in the CN150 being the half of that in the other samples (table2). However, for the temperature range going from 300 to 750 °C, the CN30 sample has the lowest H_2 desorption signal.

Table 2 shows the H contents and N/C ratio before and after annealing as measured by ERD and RBS. A large quantity of both H and N has been removed by the annealing. The H retention can be evaluated by calculating the ratio of the final to the initial hydrogen contents, presented in table 2. H retention is higher with an increase of the N implantation dose, the exception being the high dose (CN150) sample for which the H retention is at a lower level. This suggests that with an increase of the N dose, either the bond of H is increased (thus detrapping needs more energy) or the diffusion of H_2 becomes more and more difficult due to larger quantity of defects produced during the implantation. However, when the defects become connected together, they could facilitate the H_2 diffusion to the surface. This could explain why the H retention is so poor for the high dose sample.

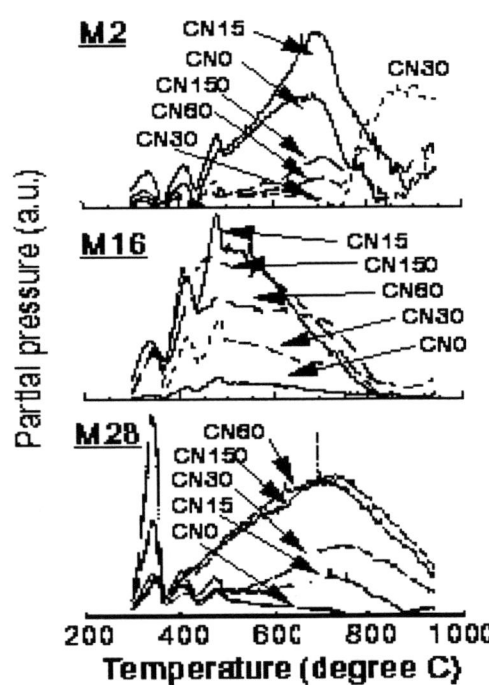

FIGURE 3. TDS measurements. Partial pressure of H_2 (M2), CH_4 (M16) and C_2H_4 or N_2 (M28).

The H_2 desorption as recorded by the TDS spectra is not well related to the final H concentration as measured by ERD. For example, the CN15 sample was found to be the sample with the highest TDS desorption signal (fig.3), but, according to table 2, it seems to have a similar H retention than the CN0 sample. Thermal desorption of atomic H has been suggested by several authors (e.g. see Ref.[15]) as a part of the explanation. Desorption of H at low temperature could empty the H reservoir to a lower level. In order to verify this hypothesis three samples (CN0, CN15 and CN30) were submitted to three different annealing ramps (30 °C/minute) up to 300, 700 and 950 °C. After each ramp, the retained quantity of H was measured by ERD and the results shown in Figure 4. After a ramp to 300 °C, 31%, 55% and 61% of H were retained in the samples CN0, CN15 and CN 30, respectively and after a ramp to 700 °C, the percentages were 12% and 54% for the samples CN0 and CN30, respectively Finally, the retained H percentages in the samples CN0, CN15 and CN30, decreased to 12%, 14% and 22%, respectively for a ramp to 950°C. So, there is evidence that a large quantity of H is desorbed at temperature lower than 300 °C without being detected by TDS. Also, it is clear that N implantation does improve the H retention in the C samples. The former result could be confirmed by the monitoring of the 1 amu particles. Unfortunately, this signal is not reliable in our set-up.

In hydrogenated carbon, emission of methane appears at temperatures around 450 °C. This phenomenon is known as the carbon bloom and has been observed in numerous tokamaks [1-2]. Figure 3 shows that a larger quantity of methane is desorbed from the implanted samples. One possible explanation would be the creation of damages during N implantation which would activate the thermal desorption of methane. As methane should come from the surface, the damage induced by the N implantation could be responsible for the increase of methane desorption. Finally, the CN30 sample has the lowest methane desorption signal among the implanted samples.

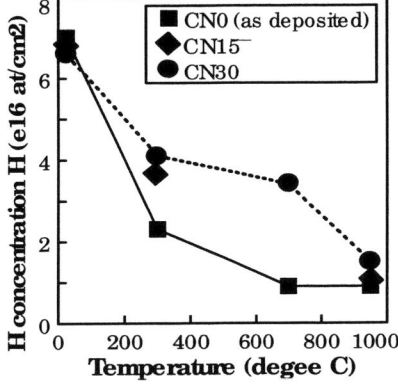

FIGURE 4. H contents in CN0, CN15 and CN30 samples as measured by ERD after annealing to 300, 700 and 950 °C.

RBS measurements have shown that the remaining thickness of deposits after annealing decreases with an increase of the integrated area of methane signal, the high dose (CN150) sample excepted. It suggests that methane desorption is an important process responsible for the erosion of samples submitted to high temperatures. The H concentration seems to play an important role and could explain the rather weak methane desorption for the CN150 sample which has an H concentration of 7% instead of 16%.

The partial pressure of 28 amu molecules shown in figure 3 could be related either to N_2 or to ethylene. The mass 28 spectra can be split in two distinct regions, 1 and 2, lower and higher than the temperature of 375 °C, respectively. The peak in region 1 is low for the CN0, CN60 and CN150 samples but higher for the CN15 and CN30 samples, the CN30 sample being the highest. Analysis of the cracking pattern of this peak revealed that it was mainly ethylene. In region 2, the larger part of the signal comes from N_2 which explains both the low signal for the as deposited sample and the different ratios of N/C before and after as given in Table 2. So, the thermally desorbed nitrogen was probably not well bonded. Finally, figure 3 shows that the desorption of N_2 starts at lower temperature for the samples saturated in nitrogen (CN60 and CN150).

CONCLUSION

After an implantation of to different doses of nitrogen into carbon, a maximum N/C ratio of ~0.48 was measured even if the samples were implanted at higher doses. Significant H desorption induced by implantation was observed only for the sample implanted with a dose of 150%. C-N and C=N bonds were observed in the implanted samples by XPS, suggesting that a given quantity of N is strongly bonded in the samples. The increase of the CH_3 peaks combined with the decrease of the C-C and C-H peaks suggest that N implantation forces a given quantity of C to change their aromatics bonds into different chemical hybridizations.

The hydrogen retention increases with the nitrogen contents of the annealed samples suggesting that the diffusion of H_2 becomes more and more difficult with an increase of the nitrogen concentration. It could be due to a larger quantity of defects produced during the implantation. However, when the defects become connected, like it would be the case for the 150% implantation dose, they could facilitate the H_2, as well as the N_2 diffusion to the surface during implantation and/or thermal annealing.

N_2 desorption induced by thermal annealing occurs at temperature higher than ~400 °C. The released quantity varies according to the ratio of N/C (before annealing) over N/C (after annealing). Similar quantity of nitrogen was desorbed from the samples CN60 and CN150 which is consistent with the relative concentration (N/C) of nitrogen and not with the implanted dose.

The improvement of the hydrogen retention in thermal annealing experiments suggests that nitrogen implantation is a promising tool to improve the properties of carbon used as PFC's. Taking into account, H_2, CH_4 and N_2 desorption, as well as the surface erosion during annealing, the optimal nitrogen dose would be around N/C ~ 30%.

REFERENCES

1. Refke A., Philipps V. and Vietzke E., *J. Nucl. Mater.*, **250**, 13-22 (1997).
2. Winter J., Esser H.G., Könen L., Philipps V., Reimer H., Von Seggern, et al., *J. Nucl. Mater.*, **162-164**, 713-723(1989).
3. Apicella M.L., Bartiromo R., Gabellieri L., Pericoli Ridolfini V., et al., *Plasma Phys. Control. Fusion*, **39**, 1153-1167 (1997).
4. Strachan J.D., Bell M., Janos A., Kaye S., Kilpatrick S., Manos D., et al., *J. Nucl. Mater.*, **196-198**, 28-34 (1992).
5. Rubel M., Larsson D., Philipps V., Kogler U., Unterberg B., Wienhold P., et al., *J. Nucl. Mater.*, **249**, 116-120 (1997).
6. Krieger K., Roth J., Annen A., Jacob W., Pitcher C.S., Schneider W., et al., *J. Nucl. Mater.,* **241-243**, 684-689 (1997).
7. Winter J., *Plasma Phys. Control. Fusion*, **38**, 1503-1542 (1996).
8. Liu A.Y. and Cohen M.C., *Phys. Rev.*, **B41**, 10727 (1990).
9. Marton D., Boyd K.J., Al-Bayati A.H., Todorov S.S. and Rabalais J.W., *Phys. Rev. Letters*, **73**, 118-121 (1994).
10. Withrow S.P., Williams J.M., Prawer S. and Barbara D., *J. Appl. Phys.*, **78 (5)**, 3060-3065 (1995).
11. Ross G.G., Leblanc L., Terreault B., Pageau J.F. and Gollier P.A., *Nucl. Instr. Methods Phys. Res*, **B66**, 17-22 (1992).
12. http://www.casaxps.com
13. Gerenser L.J., *J. Adhesion Sci. Technol.*, **7**, 1019-1040 (1993).
14. Gengenbach T.R., Chatelier R.C., and Griesser H.J., *Surf. Interf. Anal.,* **24**, 611-619 (1996).
15. Franzen P. and Vietzke E., *J. Vac. Sci. Technol.*, **A12**, 820-825(1994).

Numerical Simulation of Non-spherical Implosion Related to Fast Ignition

Hideo Nagatomo*, Naofumi Ohnishi[†], Kunioki Mima*, Katsunobu Nishihara*, Shouichi Yamada*, Keisuke Sawada** and Hideaki Takabe*

*Institute of Laser Engineering, Osaka University, 2-6 Yamada-oka, Suita, Osaka 565-0871 JAPAN
[†]Department of Aeronautics and Space Engineering, Tohoku University,
01 Aramaki-Aza-Aoba, Aoba-ku, Sendai 980-8579, JAPAN
**Department of Aeronautics and Space Engineering, Tohoku University, 01 Aramaki-Aza-Aoba, Aoba-ku, Sendai 980-8579, JAPAN

Abstract. Physics of the inertial fusion is based on a variety of elements such as compressible hydrodynamics, radiation transport, non-ideal equation of state, non-LTE atomic process, and laser plasma interaction. In addition, implosion process is not in stationary state and fluid dynamics, energy transport and instabilities should be solved simultaneously. In order to study such complex physics, an integrated implosion code including all physics important in the implosion process should be developed. Before starting this work, an integrated code based on Hirt's ALE method had been developed. But it needed sophisticated rezoning/remapping algorithm and less dissipative ALE method in hardly distorted mesh. In this work, we have developed 2-D integrated implosion code based on CIP method which was described in ALE formation. In the IFE research, the fast ignition scheme is one of the epoch making new scheme. In the scheme, the formation of the high density core plasma is one of the problem to be solved. In this paper non-spherical implosion for fast ignition is solved using the integrated code.

INTRODUCTION

Physics of the inertial confinement fusion is based on a variety of elements such as compressible hydrodynamics, radiation transport, non-ideal equation of state, non-LTE atomic process, and relativistic laser plasma interaction. In addition, implosion process is not in stationary state and fluid dynamics, energy transport and instabilities should be solved simultaneously.

In order to study such complex laser plasma physics, computational simulation play an important role and an integrated implosion code including all physics important in the implosion process should be developed. The details of physics elements should be studied and the resultant numerical modeling should be installed in the integrated code so that the implosion can be simulated with available computer within realistic CPU time.

In the previous work, the integrated implosion code, ILESTA-2D[1] has been modified to be an implicit arbitrary Eulerian Lagrangian code (ALE) for the robustness and reducing the computational time[2]. However, the difficulty of the rezoning/remapping remained even for the the improved ILESTA-2D when we simulate the complicated problem, such as non-linear Rayleigh-Taylor instability. Therefore, we have developed a new ALE method based on the CIP (Constrained Interpolation Profile) method [3] which has high-order accuracy in space and time. We have developed a new integrated implosion code "PINOCO" with the new ALE CIP method. The feature of this new code is described in the next section.

On the other hand, there is a new approach for inertial fusion energy, which is called the fast ignition scheme[4]. In this scheme, there are two key issues for the plasma physics. One is the controlling the hydrodynamic of imploding target to form a high density core plasma in non-spherical implosion, and the other is heating core plasma efficiency by the short pulse high intense laser. In this paper, the numerical result of the implosion of spherical target with a conical target shows the capability of this new integrated implosion code. And the result suggests that the high density compress can be accomplished with the non-spherical shell with conical target.

NUMERICAL METHOD

In the integrated implosion code, mass, momentum, electron energy, ion energy, equation of states, laser ray-trace, laser absorption, radiation transport, surface tracing and other related equations are solved simultaneously. The hydrodynamic solver is the most important and fundamental algorithm. Because the scale ratio of the expanded plasma to the target shell thickness is extremely large, the implosion must be solved by Lagrangian coordinates to save the computational resources and to capture the large gradient values in the phase space clearly. Therefore, most of the conventional ALE implosion codes are based on Lagrangian method in which the computational grids move along with the material. In general, computational grids are destroyed when applying the Lagrangian method naively. In order to continue the calculation stably, it requires a sophisticated and expensive rezoning/remapping algorithm. In case of complicated simulations, it needs a graphical user interface for rezoning[5].

To avoid such problem, a simple hydrodynamic solver was developed using CIP method. The CIP has some characteristics of Lagrangian method, although the fundamental formulas are done for Eulerian coordinates. To obtain pressure implicitly, we also applied C-CUP(CIP and Combined, Unified Procedure) which is a pressure-based algorithm and rational CIP method. These methods enable us to treat the ablation surface and laser absorption region stably. Because the dynamic range of the implosion is very wide, the calculation cost of the simulation on the Eulerian coordinate might be very expensive. Therefore, we have developed CIP code into the Arbitrary Lagrangian Eulerian code.

This CIP method is also employed to track the interface between the different materials clearly also, which is very useful when multi-material target structures must be considered.

In the energy equations, the Spitzer-Härm type thermal transport model is solved using the implicit 9-point differencing of the diffusion equation with ICCG method. For the radiation transport, diffusion type model is install in the code. But for the limitation of the CPU time, we have ignored the radiation in the calculated result here. For the laser ray-trace, a simple 1-D ray-tracing method is applied.

To show the validation of this code, we have calculated the laser driven Rayleigh-Taylar instability problem, and the reasonable results were obtained[6]

THE IMPLOSION DYNAMICS OF NON-SPHERICAL SHELL WITH CONICAL TARGET

In order to study the dynamic of imploding shell target of fast ignition scheme, an implosion of a non-spherical shell target with a conical target is simulated. For the conventional integrated implosion code, it is very difficult to treat this kind of simulation because of the complex geometry which requires a sophisticated rezoning/remapping method and many times of trial and error during the execution. The newly developed PINOCO enables to simulate the problem without any rezoning/remapping difficulty.

A target shell of polystyrene ($\rho = 1.06$ g/cm^3) which has a uniform thickness of 8μm. The cone with an opening angle of $60°$ is attached to the spherical shell (Fig.1(a)). The similar kind of implosion experiment is performed on GXII laser facility at ILE Osaka University to demonstrate the fast ignitor scheme[7]. To simplify the computational conditions, the conical target is assumed to be made of polystyrene ($\rho = 30$ g/cm^3), although it was gold in the experiment. The target is irradiated by uniform green laser ($\lambda = 0.53$ μm, 1.0×10^{14} W/cm^2 on the initial target surface).

Figure 1. shows the time history of the density contours and the computational mesh shape of the non-spherical implosion simulation at (a) t=0 ns, (b) 1.40 ns, (c) 1.425 ns, and (d) 1.45 ns respectively. Although, the edge of the shell is delayed under the influence of the conical target, the shell target is compressed to more than 200g/cm^3 at the maximum. Because of the effect of the tip of the gold cone and the jet, the hot spot is shifted to the right-hand side of the mass center along the cone surface.

Figure 2. shows the density contours (left) and electron temperature (right) at the maximum compression time. The high density jet is observed which penetrates into the tip of the conical target. The hot spot is shifted to be right-hand side of the mass center.

In the non-spherical case, the mass center and the center of the hot spot is not identical in this condition. Therefore, the compressed shell collides on the axis and create the high density jet flow toward the top of the conical target. Finally, the jet penetrates into the tip of the conical target. This phenomena means that the life time of core plasma get longer than that of the spherical implosion. This will be an advantage to adjust the shot timing of the ignition pulse

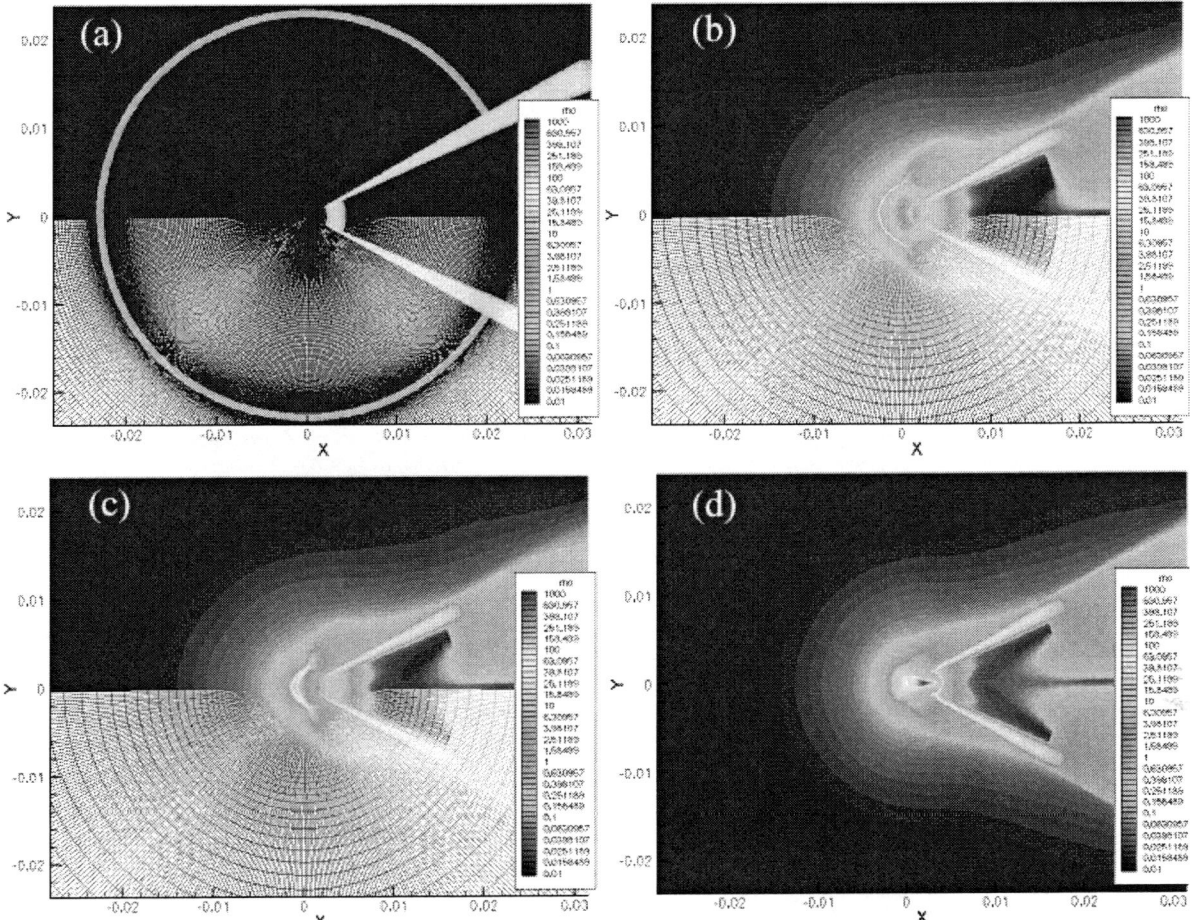

FIGURE 1. Time history of the density contours and the computational mesh shape of the non-spherical implosion simulation. (a) t=0 ns, (b) t=1.40 ns, (c) t=1.425 ns, (d) t=1.45 ns, respectively

laser.

It is difficult to measure the radius-density(ρR) which determines the ignition condition of the inertial confinement fusion because of the shape of core plasma jet. However, this simulation result suggests that the non-spherical shell target can form a high density core plasma.

The position of the hot spot can be determined by the control of the laser irradiation and the shell structures.

In this simulation, the singular point of structual mesh is located at the center of the implosion. We have to consider the effect of the singular point and the geometry carefully.

CONCLUSION

The computational simulation for the fast ignition which has non-spherical shell with conical target is performed using a newly developed integrated implosion code PINOCO. The calculated result suggests that the high density core plasma can be formed in the case of non-spherical implosion with conical target as well as the case of spherical target.

The detail quantitative analysis is necessary in the future works. In that case, very careful treatment of the computational mesh and surface tracing algorithm are required.

In the future, we will design the optimized targets for the fast ignition experiments, as well as the reactor scale targets.

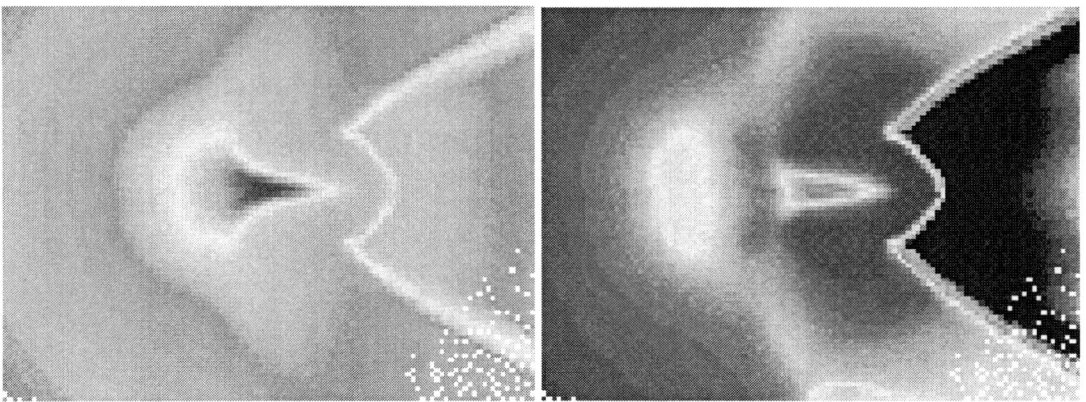

FIGURE 2. Density contours (left) and electron temperature (right) at the maximum compression time. The high density jet penetrates into the tip of the conical target. The hot spot is shifted to be right-hand side of the mass center.

ACKNOWLEDGMENTS

The authors would like to appreciate the techniccal staffs of the super computer room in ILE and the Cyber Media Center, Osaka University.

REFERENCES

1. H. Takabe, *Lecture notes*, ILE Osaka Univ., 1997.
2. H. Nagatomo, *et al.*, *Proc. of IAEA Tech. Comm. Meet. on Drivers and Ignition Facilities for Inertial Fusion*, Elsevier, 1999, pp.187-190.
3. T. Yabe, F. Xiao, and T. Utsumi *J. Comput. Phys.* **169**, 556-593 (2001).
4. M. Tabak, *et. al.*, *Phys. Plasmas* **1**, 1626-1634 (1994).
5. J.A. Harte, *et al.*, *Inertial Confinement Fusion*, LLNL, 1996 **6** 150.
6. H. Nagatomo *et al.*, *Proceedings of 2nd Int. Conf. on Inertial Fusion Sciences and Applications*, Kyoto, 2001.
7. R. Kodama *et al.*, *Nature* **412**, 798-802 (2001).

Fast Ignition Research at the Institute of Laser Engineering, Osaka University

[1,2]K.A. Tanaka, [1]R. Kodama, [1]Y. Kitagawa, [1]H. Fujita, [1]T. Jitsuno, [1]K. Mima,
[1]N. Miyanaga, [1]T. Norimatsu, [1]Y. Sentoku, [1]K. Shigemori, A. Sunahara,
[1]T. Miyakoshi, [1]F. Otani, [1]T. Sato, [1]M. Tanpo, [1]Y. Tohyama, [1]T. Yamanaka,
[3]K. Krushelnick, [4]P.A. Norreys, and [5]M. Zepf

[1]Institute of Laser Engineering, Osaka University,
[2]Graduate School of Engineering, Osaka University, Osaka Japan
[3]Imperial College, [4]Rutherford Appleton Laboratory, and [5]Queens University, Belfast, UK

Abstract. Fast ignition related studies were conducted at the Institute of Laser Engineering, Osaka University. Hot electron production was shown to have enough energy with a good conversion efficiency. Electrons were ejected mainly toward the forward direction. A sophisticated gold cone was inserted to a CD shell to guide a fast heating pulse to a compressed plasma core. The guide cone plays a role to avoid any strongly nonlinear interaction of the heating laser pulse with the surrounding plasma. The results were very interesting to show that the neutron was increased more than 100 times when the PW fast heating laser pulse enforced heating on the core plasma.

INTRODUCTION

Fast ignition is an attractive option to attain the ignition and high gain in inertial fusion energy approach. The attractiveness of the scheme is simply that the lasers for high density compression and enforced heating of the compressed fuel core are different. Thus the laser for compression may not need highly uniform laser irradiation on the fuel shell such as the ones required for the central ignition scheme. Since the ignition energy is not included in the implosion, the energy required for the compression corresponds to the internal energy of the core in a degenerate state. This reduces the energy requirement for lasers for compression almost by a factor of 10. A highly compressed fuel core should be ignited from the outside with an ultra short laser pulse to 5-10 keV via. relativistic hot electrons or via. high energy ions. The density radius product in the core may have to exceed a certain value to stop these highly energetic electrons at least higher than 0.3 g/cm^2.

We have conducted both basic hot electron production and fast ignition model experiments using implosion and fast heating lasers. Hot electrons were measured to have enough energy for fast heating with a good conversion efficiency. In the model experiment neutron number increased by a factor 10 or even higher with our latest results. Hot electron spectra were measured through compressed CD plasmas. At the right timing created hot electron flow appears to be strongly absorbed through the core[1].

HOT ELECTRON EXPERIMENT

Hot electrons were measured by varying the irradiation laser intensity from 10^{17} to 10^{19} W/cm^2 with the use of 100TW laser system. Targets were all plane of Al or CH (plastic). The S/N ratio of the main laser pulse is of the order of 10^{-4} because of the pre-pulses. The dark circles represent Al targets while the open represents for CH. As hot electrons were known to be created via. ponderomotive force of the laser light, the solid line indicates the scaling of the force. The upper dotted line indicates the scaling taken from the betatron acceleration after A. Puhkov[2]. When we have a small amount of plasmas there will be a longer resonant region where electrons were accelerated efficiently by the laser plasma nonlinear interactions such as the betatron. We have not confirm if the main acceleration mechanism for our hot electron generation is really the betatron. Other possible acceleration mechanisms include stimulated Raman forward scattering and stochastic heating. Basically all the data are between the two curves, indicating that hot electrons were created with ponderomotive force and sometime with additional mechanisms such as the betatron when there was a small scale plasma prior to the main pulse. As one can see in Fig. 1, hot electron temperature reaches 1 MeV at above the laser intensity 10^{18} W/cm^2. We studied from another set of experiments that these hot electrons were heavily oriented in the forward direction with a conversion efficiency of 40 %[1]. It was possible to proceed to a fast ignition model experiment with these basic studies of hot electrons.

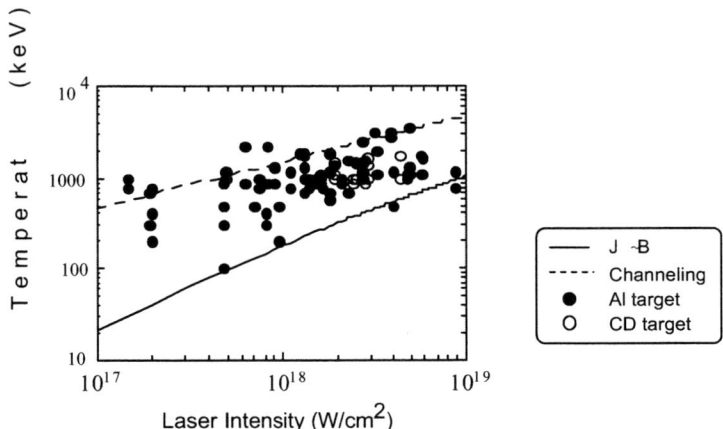

FIGURE 1. Hot electron temperature vs. laser intensity.

MODEL EXPERIMENT FOR FAST IGNITION

A model experiment was conducted of fast ignition. The experiment was conducted with 9 beam, green (532 nm) laser (1.5 kJ total) with 1 ns pulse width for a high density compression and with one beam infrared (1054 nm) laser system (60J total energy) with 0.8 psec pulse width, giving a 80 TW peak power for heating a compressed core. In order to avoid any of the transmission losses via. nonlinear interaction processes such as stimulated Raman scattering, stimulated Brillouin scattering, and modulational instability, we used a CD (deuterated carbon) shell (500 micron dia. 6 micron shell thickness) inserted with an Au cone with 30° opening angle. The head of the Au cone had a Au disk (5 micron thick) with a 50 micron dia. and was set at 50 micron away from the CD shell center. The CD shell compression was measured by an x-ray framing camera with an x-ray backlighting method. Neutron yield was monitored with a time-of-flight method. The imploded core was created about 50 micron away from the tip of the Au cone. There was some x-ray emission on the Au cone where the scattered laser light hit the surface of the cone. The core density was estimated from the x-ray backlit imaged to be $\rho = 60$ g/cm^3. The neutron yield was increased by an order of magnitude with the enforced heating to be 2 x 10^5 (DD neutrons). The estimated temperature rise on the shot was about 100 eV in addition to 300 eV background core temperature. In order to match the neutron yield, 2.6 kJ laser energy was needed for implosion, while 1.2 kJ was used for the implosion with 60 J enforced heating energy. The hot electron spectrum was measured with a permanent magnet (1.2 Kgauss) spectrometer at 20° from the 100TW laser axis in the forward direction.

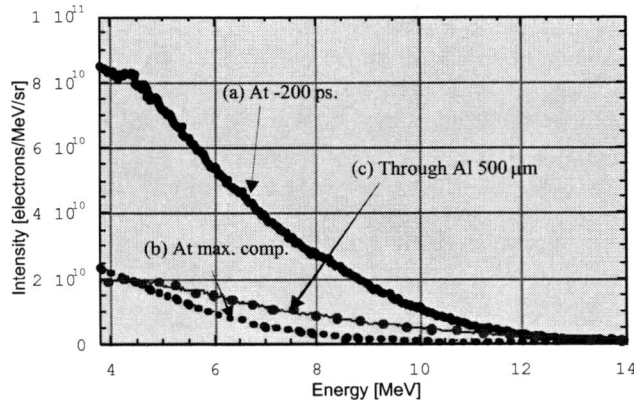

FIGURE 2. Hot electron spectra through compressed cores. At the max compression timing hot electron spectral intensity reduced drastically observed through the core. (a) at – 200 ps, (b) at max. compression, and (c) through Al block (500 μm).

The spectrometer measures up to 20 MeV electrons with imaging plate as a detector. Imaging plates have been calibrated separately using electron beams at a lineac machine of energies from 10 MeV to 100 MeV.

Figure 2 shows the results from the Au-cone implosion experiment. Curve (a) in the graph shows the hot electron spectrum on a shot, when the 100 TW laser pulse was fired 200 psec before the maximum compression. A peak was observed at around 5 MeV and the spectrum is basically comparable to the spectra observed with the other planer target experiment. Curve (b) shows the spectrum from a shot when the ultra-intense laser pulse was fired at the right timing to heat the core. We have observed neutron increase 10 times higher than the shot without fast heating pulse. The spectral intensity was decreased from 7e10 (electrons/MeV/sr) to 1.5e10 at 5 MeV. The core ρR was estimated to be 0.1 g/cm^2 in the experiment. We have shot an aluminum block of 500 μm thickness ($\rho X = 0.135$ g/cm^2) attached onto the Au cone head and observed the hot electron spectrum from the target rear side as a proof shot as shown in the curve (c). The spectral intensity was reduced drastically again on the proof shot. However the high energy spectral component between 5 MeV and 15 MeV did not decrease as much as the curve shown in Fig. 2. From another set of hot electron experiments, we know that the forward lobe of each accelerated hot electron beam has a 30 degrees FWHM divergence angle. Here hot electrons were accelerated via. JxB acceleration and vacuum heating mechanisms. We discuss possibilities of the drastic reduction observed through the compressed core and Al block. Al block of $\rho X = 0.1$ g/cm^2 should transmit electron of energies with more than 1 MeV more than 90 %. When such a transmission curve is drawn in Fig. 2, it should overlap basically with the curve A. If the electrons are absorbed in a classical manner, more reduction in the spectral intensity should be observed toward the low energy end because of the electron mean free path. Another possibility may be due to electron scattering in the materials. If the electron scattering is severe and the angular distribution of the electron is not negligible after some distance in a material then there should be a reduction in the spectral intensity. We have performed an electron scattering experiment using 30 MeV electron pulse with a 5 ns pulse width with a total charge of 0.05 nC using a facility in the Osaka University. The electron pulse with an initial beam diameter 3 mm was injected to an Al plate of 2 mm thick ($\rho X = 0.538$ g/cm^2). After the 450 mm traveling, the electron beam diameter was measured to be 55 mm, resulting in the divergence angle is 6.6° (FWHM). The spatial distribution was measured with an imaging plate. The scattering angle of electrons is proportional as,

$$\Phi \propto Z(\rho X)^{1/2}/p^2 = Z(\rho X)^{1/2}/(\gamma^2-1), \quad (1)$$

where Z is the charge of atom, ρ the density of the material, X the distance of travel, p the momentum of electron. Based upon the measured value of electron scattering angle at 30 MeV, we estimate the scattering angle of electron at 10 MeV through a core with $\rho R=0.1$ g/cm^2 to be 65° and the reduction of the electron spectral intensity is within 10%. The observed large reduction is not explicable with the electron scattering, since electrons with lower energy is subject to more scattering.

3D PIC SIMULATION

3D particle in cell simulation was run to study the electron energy transport. Following are the typical simulation conditions. The target plasma is homogeneous, except for two longitudinal sharp boundaries, and its density is 10 times critical density (n_c). The plasma consists of fully ionized deuterons (Ion mass is 3680 m_e), and the initial electron and ion temperatures are set to 5 keV and zero. The maximum of the normalized vector potential is $a(\equiv eA/m_e c^2)=3.75$, which corresponds to $I_0 = 2 \times 10^{19}$ W/cm^2 approximately, and an amplitude of the oscillating laser magnetic field B_0=400MG. The transverse system size is 2.5λx2.5λ and the longitudinal length is 20λ, here λ is the laser wavelength. The plasma has a length 16λ, and there are 2λ vacuum regions on both sides of the plasma. The total simulation time is 30τ, where τ is one laser oscillation period. When the hot electrons come into the plasma by several μm, a large reduction of the energy flux is observed. Namely the normalized initial flux reduced to 1/e already at the 4 μm depth. The magnetic fields carried energies comparable to the hot electrons at very close to the plasma surface, but little energies after a few μm from the surface. The details of the hot electron behaviors were discussed in a 3D PIC study[3]. Merging longitudinal modulations enhanced the stopping of the electrons. When hot electrons entered into the plasma with a 10 times critical density, they are subject to the Weibel instability and tend to break up into multiple filaments. The return current tends to be scattered with many local B fields, resulting in anomalous resistivity of the main hot electrons. The effective ρX for this energy deposition is $\rho X_{3D} = 2e-5$, which should be compared to a classical value of $\rho X_c = 0.5$ g/cm^2 for the density of plasma. Hot electron energy went mainly to the ambient plasma. In this simulation, the plasma temperature went up to 20 keV heated by the hot electrons after 60 fsec. We apply this heating to the Au cone implosion experiment. The core density was estimated to be 60 g/cm^3, which is 600 times higher than the plasma density used in the simulation. If the energy deposition is same as the one in the simulation, then the temperature becomes correspondingly low to be about 300 eV after 800 fsec. In Ref. [1], we estimated that the ion temperature increase is about 120 eV to explain the ten times neutron increase by the fast heating. This result suggests that the core has been heated with the anomalous hot electron stopping through the core.

MOST RECENT NEWS ON THE ENFORCED HEATING

Our most recent results on the Au cone target was even more striking. With the similar target conditions, implosion laser was increased to 2.4 kJ with 9 beams of green 1.2 nsec pulses. The heating laser pulse energy was 300 J on-target compared to the one above with 60 J. The neutron increased 100 times with the enforced heating shot compared to the one without the heating. The details of the most recent experiment will be reported soon elsewhere.

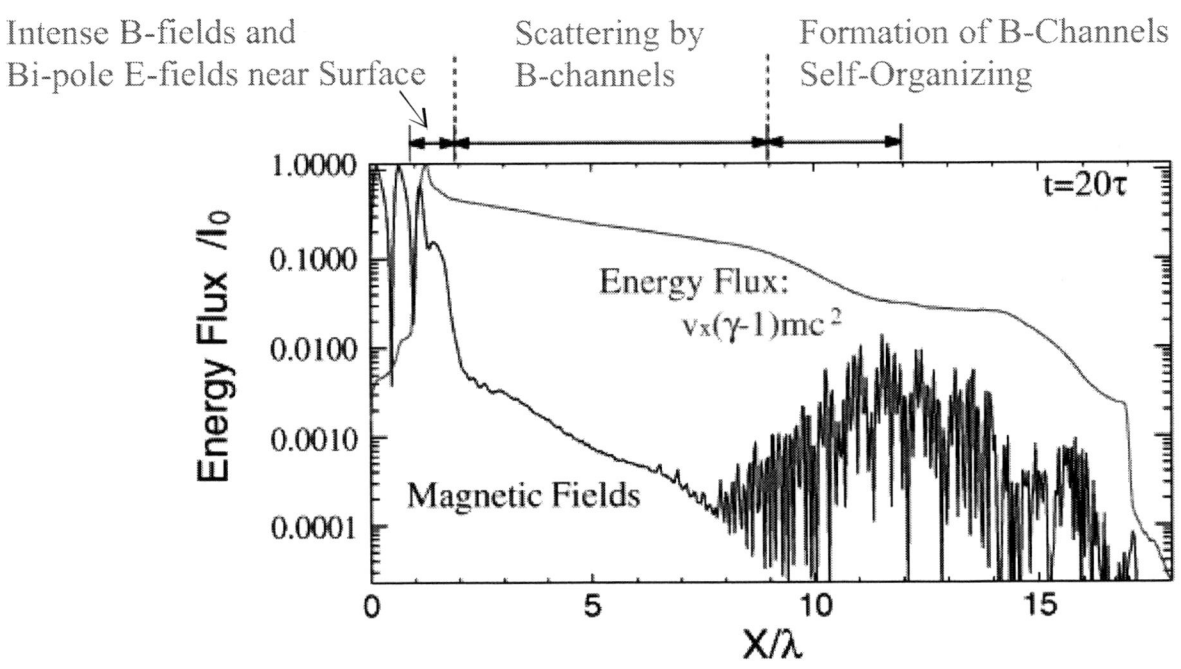

FIGURE 3. Energy flux and magnetic fields versus depth in the plasma. The plots are normalized with the incident laser flux.

ACKNOWLEDGEMENTS

We would like to acknowledge the 100TW laser operation, PW laser operation, GXII laser operation, target fabrication, and MT groups.

REFERENCES

1. Kodama, R et al., Phys. Plasmas **8**, 2268-2272 (2001).
2. Pukhov, A. et al., Phys. Plasmas **7**, 2847-2854 (1999).
3. Sentoku, Y. et al., Phys. Review E, **65**, 046408-1-7 (2002).

Nonlinear State of Sausage-like Instability of Electron Current Channels in Fast Ignition Concept of Inertial Fusion

Neeraj Jain*, Amita Das*, Predhiman Kaw * and Sudip Sengupta*

Institute for Plasma Research, Bhat, Gandhinagar 382428, India

Abstract. This paper deals with a detailed fluid simulation study of linear and nonlinear aspects of the velocity shear modes in electron current channels in a two dimensional geometry. Simulation results clearly show the flattening of flow profile and the development of sausage like structures (kink structures, which are intrinsically three dimensional excitations, are ruled out in the present simulations) which grow linearly and eventually saturate by nonlinear effects. An analytic understanding of the nonlinear saturation mechanism is also provided.

INTRODUCTION

The stability of current channels to EMHD modes is a topic of great interest in several frontier areas of plasma research e.g. fast ignitor concept of laser fusion, fast Z pinches, plasma opening switches, current channels at the center of fast magnetic reconnection region etc[2, 3, 4, 5, 6]. Experiments and multidimensional particle in cell (PIC) simulations of the fast ignitor concept in laser fusion have shown that intense sub-picosecond heating laser pulse incident on a precompressed pellet generates fast electrons on the surface of the pellet which propogate inward. These fast electrons soon organize themselves into current channels. The stability of these current channels to sausage and kink modes on the fast EMHD time scales is thus crucial to our understanding of heating of pellet core in the fast ignitor concept. Whereas some analytical work has been done on the linear physics of such instabilities [1], the nonlinear regime remains relatively unexplored. It is the objective of this paper to provide a two dimensional EMHD fluid simulation of linear and nonlinear aspects of velocity shear driven modes in electron current channels.

Model Equations and Numerical Results

The dynamics of electron current channel can be described appropriately by the EMHD equation given below:

$$\frac{\partial}{\partial t}(\vec{\nabla} \times \vec{P}) = \vec{\nabla} \times \left\{ \vec{v}_e \times (\vec{\nabla} \times \vec{P}) \right\} \tag{1}$$

Effects due to displacement currents are neglected because of low frequencies involved in EMHD physics; namely $\omega < \omega_p^2/\omega_c$. Here time is normalized by ω_c, length by $d_e = c/\omega_p$ and magnetic field by B_{00}, where $\omega_c = eB_{00}/m_e c$. We simplify the simulation by considering a slab model of the cylindrical current channel. In this model, the radial direction is taken as $\hat{x}$, the poloidal direction as $\hat{y}$ and the axial direction as $\hat{z}$. For simulation the initial electron velocity profile has been taken to be of the form $\vec{v}_e = V_0 \tanh\left(\frac{x}{\varepsilon}\right)\hat{z} + \tilde{\vec{v}}_e(x,z)$ The first term represents the equilibrium profile and is similar to the piecewise linear profile for which the analytical dispersion relation was obtained in Das *et al.* [1] earlier. The second term $\tilde{\vec{v}}_e(x,z)$ is a small perturbation which is added initially to hasten the development of instabilities that the system might support.

An integration of $\vec{\nabla} \times \vec{B} = -\vec{v}_e$ yields the initial magnetic field (along $\hat{y}$) profile as $B = -V_0 \varepsilon \log\left\{\cosh\left(\frac{x}{\varepsilon}\right)\right\} + \tilde{B}(x,z)$ The boundary conditions for the evolution of $\vec{B}$ field are chosen to be periodic along z, while at the x boundaries the perturbed field $\tilde{B} = 0$ for all times at all values of z. The analytical studies for some specific shear profiles for the electron current show [1] that such a configuration is unstable to the excitation of a sausage like mode at EMHD time scales. We have obtained the exact growth rates for various k_z and ε, by solving numerically the dispersion relation for the linear velocity profile [1].

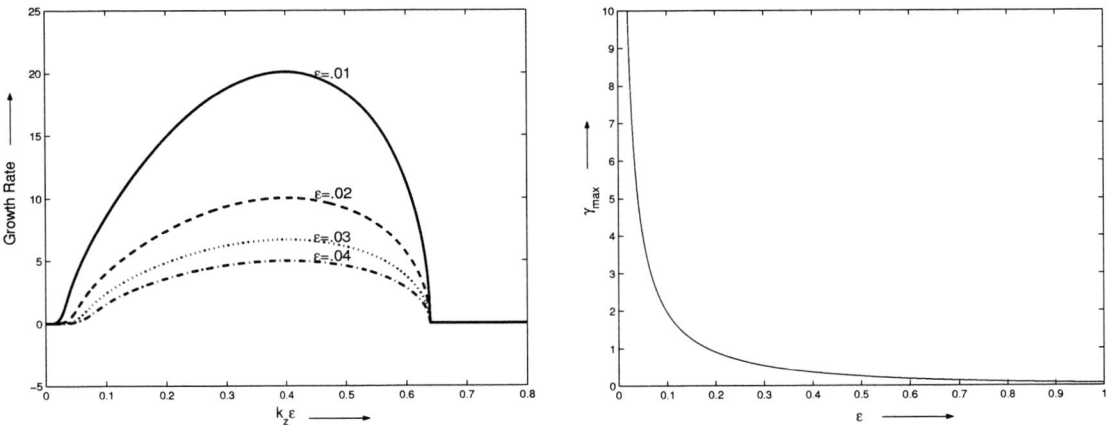

FIGURE 1. Fig(1A):Growth rate vs. $k_z\varepsilon$; Fig(1B):Maximum value of growth rate vs. ε)

It can be seen from Fig.(1A), that growth rate vanishes for $k_z\varepsilon > .639$ and decreases as ε increases. The observed limit of $k_z\varepsilon = 0.639$ can be obtained analytically also. At the point where the growth rate vanishes there is an exchange of stability and $\omega = 0$. Substituting $\omega = 0$ in the dispersion relation [1] we obtain, $\exp(-4k_z\varepsilon) = (2k_z\varepsilon - 1)^2$, which is satisfied when $k_z\varepsilon = 0.639$. The growth rate of the instability reduces drastically as ε becomes comparable or greater than the electron skin depth as can be seen from Fig.(1B). This indicates that electron inertia is critical for driving this instability. Theoretical solutions were earlier obtained for profiles which had discontinuities either in the value of the velocity and/or its derivative. Our numerical simulations have now been able to confirm that the instability persists even for smooth velocity profiles. The simulation studies help us explore the nonlinear regime of this instability for the first time.

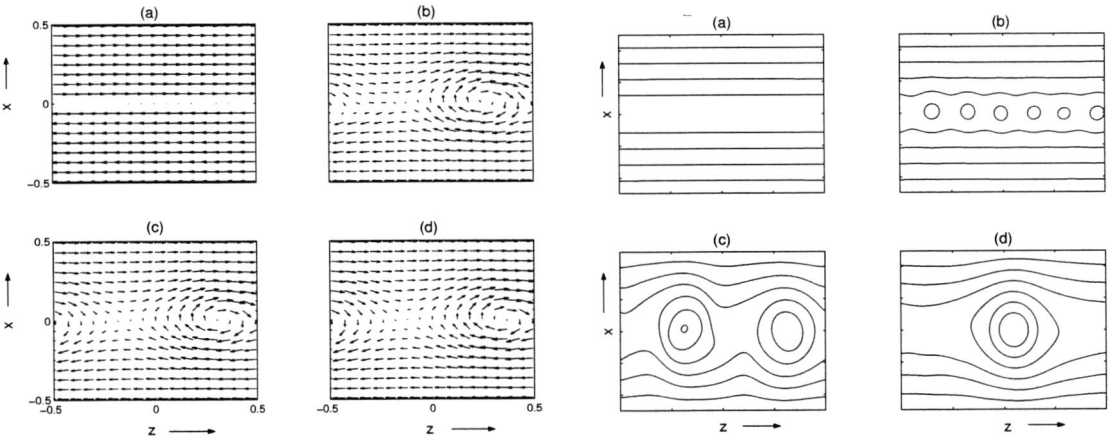

FIGURE 2. Fig.(2A):Convective cell pattern in flow velocity at (a)t=0(b)t=2(c)t=4(d)t=6 ; Fig.(2B):Contours of constant magnetic field at (a)t=0(b)t=5(c)t=10(d)t=40.

The simulations show the development of an instability leading to the formation of convective cell patterns in the flow velocity as shown in the Fig.(2A). The constant contours for the magnetic field can be seen from Fig.(2B). In Fig.(3A) we plot $\ln E(t)$ vs. time, where $E(t) = \int |\tilde{B}|^2 \, dxdz$. The slope of dotted reference line in the figure is twice the analytical growth rate. It can be seen that it compares well with that obtained from simulations. The behavior of individual k_z mode has also been studied by fourier analysing the perturbed field spectrum. There is a good agreement of simulation growth rate with the analytical growth rate obtained from the dispersion relation, as shown in Fig.(3B). However, the simulation points show a finite power accumulation even in those modes which are linearly stable. This undoubtedly occurs due to a nonlinear transfer of power from unstable scales. In Fig.(2B) the coalescing of multiple structures into a single island structure of the size of simulation box length is essentially due to the inverse cascade

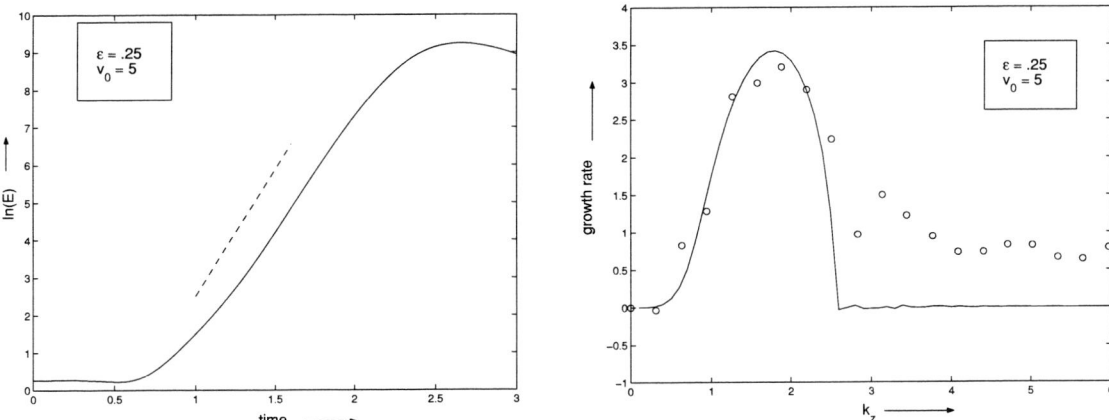

FIGURE 3. Fig.(3A):log of perturbed mag. energy vs time.Dotted line represents the theoretical growth rate. ; Fig.(3B):Comparision of theoretical growth rates with simulation growth rates.Solid line represents theoretical growth rate and circles represent simulation growth rates.

of power towards long scales. Basically, the nonlinear term $\hat{y} \times \vec{\nabla}\psi \cdot \vec{\nabla}\nabla^2\psi$ (here $\hat{y}$ is the direction of symmetry, and ψ represents the electrostatic potential) supports two mean square invariants ($\int\int(\vec{\nabla}\psi)^2 dxdz$ and $\int\int(\nabla^2\psi)^2 dxdz$) in the non dissipative limit. Such a conservation puts some restriction on the process of power transfer, which yields a predominant transfer of power towards long scales [7, 8].

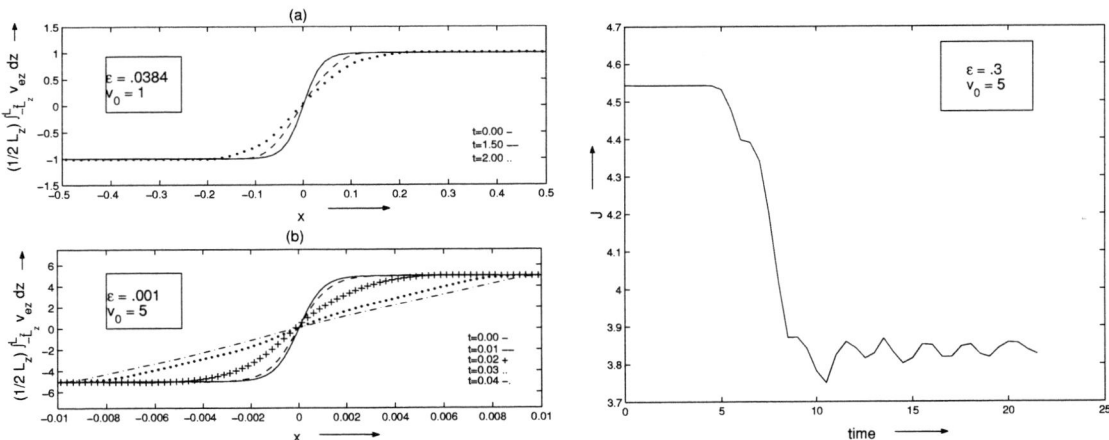

FIGURE 4. Fig.(4A):Flow velocity profile at different times for different values of ε and v_0. ; Fig.(4B); Flow velocity averaged over half x-space vs. time.

The flattening of z-independent electron flow $\bar{V}(x)\hat{z} = \frac{1}{2L_z}\int_{-L_z}^{L_z}\vec{v}_{ez}dz$ due to backreaction of unstable modes on the flow, as shown in Fig.(4A-a), increases the effective shear width (ε_{eff})resulting in the stablization of the profile. In Fig. (4A-b) electron flow profile saturates and becomes straight line ($V'' = 0$) even though $k_{zl}\varepsilon_{eff} < .639$. Thus we found the two mechanism of saturation of instability. One is the disappearance of curvature of flow profile and the other is violation of the condition $k_z\varepsilon < .639$.

QUASILINEAR ANALYSIS

We represent the sheared z independent electron flow velocity by $v_0\hat{z} = v_q\exp(iq_xx) + c.c$ and the corresponding magnetic field by $B_0\hat{y} = B_q\exp(iq_xx) + c.c$. Other scales are represented by $\hat{B}(x,z,t) = \sum_{k_x,k_y}B_k(t)\exp(ik_xx + ik_zz)$.

Now, taking the qth fourier component of Eq.(1) we obtain,

$$(1+q_x^2)\frac{dB_q}{dt} = q_x^2 \tau \sum k_z k_1^2 \left\{ \frac{k_z(k^2-q_x^2)}{(1+k_1^2)} \mid B_k \mid^2 + \frac{k_{1z}(k_1^2-q_x^2)}{(1+k^2)} \mid B_{k1} \mid^2 \right\} B_q \qquad (2)$$

where τ is the correlation time which has been assumed to be independent of k. For simplicity, we assume that typically the spectral strengths are such that we have $\mid B_k \mid^2 \approx \mid B_{k1} \mid^2$. Since $k_{1z} = -k_z$ it is clear from the above expression that the right hand side will have terms which have powers of q_x^4 and higher. This establishes that the z independent flow being $V_q = -iq_x B_q$ suffers a viscous and/or a higher derivative damping. The effective viscosity coefficient can be obtained from Eq.(2) by collecting the coefficient of terms containing powers of q_x^4. This gives

$$\mid \mu_{eff} \mid = \tau \sum \frac{k_z^2}{(1+k^2)^2}(1+2k^2)(k^2+4k_z^2) \mid B_k \mid^2 \qquad (3)$$

DISCUSSION

The flattening of the electron flow profile caused by these shear driven unstable excitations can have crucial implications to several frontier areas of plasma research. The question of paramount importance in the context of the fast ignitor concept is whether the fast electrons are capable of propagating unmolested into the compressed fuel. We observe that the instability in such sheared electron current channels saturates by getting rid of the shear in electron velocity. In Fig.(4B) we show the evolution of the half space average $J = \frac{1}{L_x}\int_0^{L_x} \bar{V} dx$ with time. It is noted that J is significantly reduced but does not go to zero. We ascribe the incompleteness of the stopping to two dimensionality of our simulations. We speculate that the initial development of the velocity profile in $2d$ simulations gives us a reasonable indication of the overall collective stopping power due to the development of EMHD turbulence; this can however best be pinned down only in actual three dimensional EMHD simulations where turbulence would have an opportunity to fully develop.

We now make an estimate of the expected stopping length due to collective effects. >From Fig.(4B), for the case when the typical shear width is a fraction of electron skin depth (i.e. $\varepsilon = 0.3$ in units of c/ω_p) we observe a reduction in J of $\sim \Delta J = 0.7$ in a time duration $\Delta t = 10$. If the deceleration had continued at this rate ($a = 0.7/10 = 0.07$) a complete stopping would have occurred after traversing a length $s = 4.5^2/(2a) = 145$. For the simulation parameters [?] of plasma density $n = 4 \times 10^{21}/cm^3$, we have $\omega_{pe} = 3.6 \times 10^{15}/sec$ and thus the electron skin depth $c/\omega_{pe} = 0.1\mu$. Since length is normalized by skin depth in our simulation, the stopping length estimate obtained above, $s = 145$ corresponds to $\approx 15\mu$. Three dimensional PIC simulations do indeed show collective stopping with an effective friction term [? ?] $\frac{v_{eff}}{\omega_c} \sim \frac{\delta B}{B_0} \sim 10^{-2}$ where δB is the perturbed magnetic field. For typical fast ignitor situations, this gives $v_{eff} \sim 10^{13}$ and hence a stopping length $\sim 15\mu$ which is close to the estimate we gave above based on initial stopping rate in two dimensional simulations.

REFERENCES

1. A. Das and P. Kaw, Phys. Plasmas **8**, 4518 (2001).
2. A. S. Kingsep, K. V. Chukbar and V. V. Yankov, in Reviews of Plasma Physics (Consultant Bureau, New York, 1990) vol 16 and references therein.
3. M. Sarfaty, R. Shpitalnik, R. Arad, A. Weingarten, Ya. E. Krasik, A. Fruchtman and Y. Maron, Phys. Plasmas **2**, 2583 (1995).
4. R. Shpitalnik, A. Weingarten, Ya. Krasik and Y. Maron, Phys. Plasmas **5**, 792 (1998).
5. A. Fruchtman, A. A. Ivanov and A. S. Kingsep, Phys. Plasmas **5**, 1133 (1998).
6. C. E. Seyler, Phys. Fluids **B3**, 2449 (1991).
7. R. Kraichnan, Phys. Fluids **10**, 1417 (1967).
8. G. K. Batchelor, Phys. Fluids, Suppl II **12**, 233 (1969).

The Features of Craters Formation on the Target under the Action of Powerful Laser Pulse

A.A.Rupasov, E.A.Bolkhovitinov, I.Ya.Doskach, A.A.Erokhin, S.I.Fedotov,
L.P.Feoktistov, S.Yu.Gus'kov, B.V.Kruglov, M.V.Osipov, V.N.Puzirev,
V.B.Rozanov, A.S.Shikanov, V.B.Studenov, B.L.Vasin, O.F.Yakushev

Lebedev Physical Institute, RAS, 117924, Leninsky pr., 53, Moscow, Russia

Abstract. The results of experimental and theoretical investigations of craters originating in solid targets of different materials when powerful Nd:glass laser pulse irradiates the surface at flux densities in the range of 10^{10}–10^{14} W/cm^2 are presented. The experimentally observed dependencies of crater depth and ablated mass on laser pulse energy and target material properties are analysed employing the theory of shock wave initiation and propagation under the action of the plasma-producing laser beam. From the comparison of the theoretically deduced and experimentally observed dependencies, a simple formula is derived allowing to determine the pressures in the shock wave and in the plasma corona using the measurements for the crater depth and ablated mass.

INTRODUCTION

Investigations of laser interaction with solid target and crater formation are carried out in this work with the objective to determine the ablation loading efficiency [1], that is the conversion coefficient of absorbed laser energy into shock wave energy. The material destruction due to laser-produced ablation is studied for the condition that the ablation process is purely hydrodynamic. This happens when the hydrodynamic motion is the dominant energy transfer mechanism in the plasma when compared to electron thermal conductivity, fast-electron energy transport, and thermal radiation. These conditions are met at moderate laser intensities $I \leq 10^{14}$ W/cm^2 and sufficiently long laser pulses with duration $\tau \geq 0.1$ ns. If in addition $I\lambda^2 \leq 10^{14}$ W·µm^2/cm^2 condition holds, inverse bremsstrahlung absorption without fast-electron generation is the dominant absorption mechanism. For our experimental conditions, the depth of the solid-matter destruction is due to the decaying shock wave which keeps propagating for quite a long after the completion of the laser pulse. This possibility of post-pulse material destruction was mentioned in [2].

EXPERIMENTAL RESULTS AND DISCUSSION

The experiments were carried out on one-beam laser facility "Kanal" at pulse energies of up to 60 J and a pulse duration of 1 ns. The output laser beam with a diameter of 45 mm, an angular divergence of $2\alpha = 1.3 \cdot 10^{-4}$ rad and an energy contrast of 10^5 has been focused on the target surface by an aspherical lens with a focal length of 48 cm yielding a focal spot diameter of 100 µm and an on-target flux density of up to $5 \cdot 10^{14}$ W/cm^2. In these experiments the target surface was perpendicular to the axis of the laser beam and long-focal optics (D/F ≈ 0.1) was used, so that all the rays irradiated the target at the almost normal angle of incidence. Therefore the bremsstrahlung absorption was a dominant mechanism of the laser energy absorption and according to the measurements the coefficient of absorption was near 0.63. Flat solid targets of different materials (aluminium, copper, titanium, lead, tin, germanium, carbon, polyethylene, etc.) were used. The experimentally obtained dependencies of the crater depth and ablated mass on the laser pulse energy and target material properties were measured. The crater profile was determined using the computerised three-dimensional device Form Talysurf, manufactured by Taylor Hobson, UK. Especially developed software enabled to determine the crater volume and the mass of ablated matter. The accuracy of these measurements was about 10%. Values of the crater depths, L_d, and the ablated matter mass, M_d, are presented in Table 1 for aluminium, copper, and lead targets and different laser pulse energies.

TABLE 1.

Material	E_{Las}, J	Experiment		Theory		M_d^*/M_d^0
		L_d^0, µm	M_d^0, µg	L_d^*, µm	M_d^*, µg	
Al	4.1	89.1	8.26	112.1	8.28	1.00
Al	14.6	126.3	21.6	174.1	24.6	1.14
Al	15.1	164.9	21.5	174.6	25.8	1.19
Al	18.6	103.4	37.3	192.7	32.0	0.86
Cu	12.9	72.6	15.2	105.6	23.9	1.57
Pb	11.8	180.2	123.7	253.5	277.6	2.24

For the analysis and discussion of the results obtained in this work, we use the analytical theory of shock wave propagation in solid matter and ablative destruction in a semi-infinite matter layer irradiated with a laser beam whose intensity corresponds to the hydrodynamic regime [1].

The matter destruction appears in the form of evaporation or melting. We assume that the laser radiation is absorbed in a region within the plasma corona where the electron density is close to the critical density $\rho_{cr} = 1.83 \cdot 10^{-3} \, \mu/z\lambda^2$ g/cm^3, μ and z are the atomic number and the charge of plasma ions respectively, λ is the laser radiation wavelength in µm.

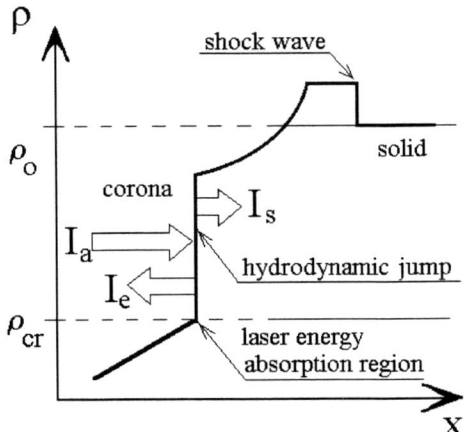

FIGURE 1. Schematic picture of the density profile and energy flows, where **x** is a coordinate along the axis, perpendicular to the target surface.

At the hydrodynamic regime of an ablation processes the region of laser energy absorption is placed near the evaporation boundary. On this reason we use the approximation that the flow of absorbed laser energy comes into the target through the hydrodynamic jump from the side of low density plasma and splits on the jump onto the flow of the evaporated matter latent energy, I_e, and flow of the matter behind the shock wave front propagating in solid part of the target, I_s, (see Fig.1) according [1]:

$$I_e \cong \frac{(3\gamma_c - 1)}{2(\gamma_c - 1)} \cdot \rho_{cr} \cdot c^3, \quad I_s \cong \frac{4}{(\gamma_s + 1)^{1/2}} \cdot \left(\frac{\rho_{cr}}{\rho_0}\right)^{1/2} \cdot \rho_{cr} \cdot c^3, \quad (1)$$

where γ_c and γ_s are the isentropic exponents for evaporated and solid parts of target, c is the isothermal sound velocity at the critical density. Since we take into account that $\rho_c \ll \rho_0$, we obtain the following expression for the ablation loading efficiency:

$$\sigma = \frac{I_s}{I_s + I_e} \cong \frac{4}{\sqrt{\gamma_s + 1}} \cdot \frac{2(\gamma_c - 1)}{3\gamma_c - 1} \sqrt{\frac{\rho_c}{\rho_0}}. \quad (2)$$

The ablation loading efficiency increases with a decrease in the initial density of the material and an increase in the critical plasma density and equals 2.5 – 4% for aluminium, and 2% for lead (assuming $\gamma_s = 5/3$ for evaporated matter and $\gamma_s = 3$ for the non-evaporated one). For plastics, the ablation loading efficiency is 2 – 3 times higher and is estimated to be 6-8 %.

Taking into account that the ablation loading efficiency is small, the calculations of the plasma parameters in the evaporated part of the target can be made with high accuracy in the approximation that all absorbed energy is contained only in the laser plasma corona. The cross-section of the energy flow from the corona into the solid material is calculated on the assumption that the corona is two-dimensional and can be described in terms of a sected cone with the base radius $R_{Las} + \tau \cdot c$, where c is the sound velocity in corona, and that the absorbed energy is homogeneously within the corona.

$$c = \left[\frac{2(\gamma_c - 1)}{3\gamma_c - 1} \cdot \frac{I_a}{\rho_c} \cdot \frac{R_L^2}{\tau}\right]^{1/5} \quad ; \quad p_c = \left[\frac{2(\gamma_c - 1)}{3\gamma_c - 1}\right]^{2/3} \cdot \rho_c^{1/3} \cdot I_a^{2/3}. \qquad (3)$$

We take into account that in the pressure region close to the yield point the shock adiabate of the metals corresponds to the ratio between the elastic and thermal parts of internal energy with the value of approximately 3. Under these conditions, the material destruction stops when the specific internal energy of the matter behind the decaying shock wave has decreased to a value exceeding the sum of the latent heat of melting and the energy for material heating from initially 300K to the melting temperature by a factor of 4. It turned out that in our experiments ($I > 10^{10}$ W/cm^2) the shock wave parameters met this condition for a time long after the end of the laser pulse. Therefore the post-pulse destruction lasts while

$$\sigma \cdot I_a \cdot \tau \cdot S > 4 \cdot E_d \cdot M_d, \qquad (4)$$

where S is the energy flow cross-section area, $M_d = \pi \cdot L_d \cdot (L_d + R_L)^2 \cdot \rho / 3$ is the limit of the destructed material mass, L_d is the maximum destruction depth, $E_d = \varepsilon_d + \int_{300K}^{T_M} c_p(T) dT$, ε_d is the latent heat of melting, and T_M is the melting temperature. From the 2D corona model we approximate S by $\pi \cdot (R_L + 0.45 \cdot c \cdot \tau)^2$. If the melting condition is satisfied the absorbed laser intensity must exceed

$$I_a > (\gamma_s + 1)^{3/2} \cdot \frac{3\gamma_c - 1}{2(\gamma_c - 1)} \cdot \frac{E_d^{3/2}}{\rho_c^{1/2}} \cdot \rho_0^{3/2} \qquad (5)$$

The dependencies of the crater parameters on the laser pulse energy for the aluminium targets obtained from the expressions above have a good agreement with the experimental data on craters depth and volume (Table 1). The part of an energy radiated from the target corona is much bigger for the heavy materials such as copper and lead and it leads to a large decrease of the shock wave energy and therefore to the crater volume decrease in comparison with our model developed without involving into the consideration any radiation losses. That is why the discrepancy between experimental results and our model for the copper targets (about 1.5) turned out essentially less than for the lead ones (about 2.2), as one can see in Table 1.

Since the model presented agrees well with the experiment for light materials, we can estimate some laser plasma parameters from the measurements of the depth and volume of the crater in the solid target. One interesting parameter is the pressure in the plasma corona. Using the approximate formula

$$I_s = \frac{2}{\sqrt{\gamma_s + 1}} \sqrt{\frac{\rho_c}{\rho_0}} \cdot \sqrt{\frac{p_s}{2\rho_0}} p_s = \sigma \cdot I_a = \sqrt{\frac{2}{\gamma_s + 1} \frac{\rho_c}{\rho_0}} \frac{p_s^{3/2}}{\rho_c} \qquad (6)$$

we obtain for the pressure

$$p_s = \left[8(\gamma_s + 1)\right]^{1/3} \left[\frac{E_d \cdot M_d \sqrt{\rho_c \cdot \rho_0}}{\tau \cdot \pi \cdot (R_L + 0.45 \cdot \tau \cdot c)}\right]^{2/3} \quad \text{with} \quad c = \left[\frac{2(\gamma_c - 1)}{3\gamma_c - 1} \cdot \frac{I_a}{\rho_c} \cdot \frac{R_L^2}{\tau}\right]^{1/5} \qquad (7)$$

For aluminium targets, we find pressure values of 2.07, 2.76, 2.80, 2.95 Mbar for pulse energies of 4.1, 14.6, 15.1, and 18.6 J, respectively.

CONCLUSION

The experimentally observed dependencies of a created crater parameters on laser pulse energy and target material properties are analysed on the base of the developed theory of shock wave initiation and propagation under the action of the laser beam on a flat solid target surface. The further development of this theory including the corona radiation and more precise computation for the ionisation degree in corona will allow to determine the laser plasma parameters, such as pressure in corona, by easy methods of the craters measurement.

ACKNOWLEDGMENTS

The work is performed under the Russian Foundation of Basic Research grants № 00-02-16113, № 01-02-17589, № 02-02-16966.

REFERENCES

1. K.S.Gus'kov, S.Yu.Gus'kov. Efficiency of ablation loading and the limiting destruction depth of material irradiated by a high-power laser pulse. *Kvantovaya Elektronica*, **31(4)**, pp.305-310 (2001).
2. S.I.Anisimov, Yu.A.Imas, G.S.Romanov, Yu.A.Khodiko. *The action of high-power radiation on metals*. Nauka, Moscow, 1970.

Hydrodynamic Instability Experiments on the HIPER Laser

K. Shigemori, H. Azechi, S. Fujioka, Y. Kanai, N. Miyanaga, M. Murakami,
T. Muranaka, H. Nagatomo, M. Nakai, M. Nishikino, K. Nishihara, H. Nishimura,
T. Sakaiya, H. Shiraga, A. Sunahara, H. Takabe, T. Takayama, Y. Tamari,
M. Tanaka, and T. Yamanaka

Institute of Laser Engineering, Osaka University
2-6 Yamada-Oka, Suita, Osaka 565-0871, Japan

Abstract. We present recent results on the hydrodynamic instability experiments on the HIPER (High Intensity Plasma Experimental Research) laser facility at Institute of Laser Engineering, Osaka University. We measured the Rayleigh-Taylor growth rate on the HIPER laser. Also measured were all parameters that determine the RT growth rate. We focused on the measurements of the ablation density of laser-irradiated targets, which had not been experimentally measured. The experimental results were compared with calculations with one dimensional simulation coupled with and without Fokker-Planck equation for electron heat transport.

INTRODUCTION

Hydrodynamic instabilities in the inertial confinement fusion (ICF) targets have been of great interest for last two decades because the stability of the imploding ICF targets determines the performance of the ignition and burn conditions. Among the hydrodynamic instabilities of the ICF targets, the Rayleigh-Taylor (RT) instability [1] is the most crucial instability because the RT growth gives the largest growth factor in whole implosion regimes.

The goal of this experimental study is the understanding the Rayleigh-Taylor (RT) growth rate under the standard irradiation condition on the HIPER laser system. From previous RT experiments on the GEKKO green laser system, our understanding is that the RT growth rate is reduced by non-local electron heat transport [2]. However, for the blue laser irradiation on the HIPER laser, the effect of the non-local electron heat transport is expected to be less effective because the temperature at the cutoff density is lower than that for the green laser irradiation (for same laser intensity).

The measurement of the RT means to measure not only the growth rate γ but also all the parameters that determine the RT growth rate. The theoretical RT growth rate is referred to as [3]:

$$\gamma = \sqrt{\frac{kg}{1+kL}} - \beta k \frac{\dot{m}}{\rho_a} \quad [1]$$

where k is the wave number of the perturbation, g is the acceleration, L is the density scale length of the ablation front, $\dot{m}$ is the mass ablation rate per unit surface, ρ_a is the density at the ablation front. In order to verify the RT growth equation, all the parameters in eq. (1) should be experimentally measured. Among these parameters, the ablation density ρ_a was very difficult to measure due to the limitation of spatial resolution of diagnostics system because the density scale length of the ablation front of laser-irradiated targets is typically less than 5 μm. In this experimental research, we focused on the measurements of the ablation density for the understanding of the "ultimate" understanding of the RT growth. The measurement of the ablation density is very important for understanding the effect of the non-local electron heat transport because the ablation density is reduced by the heating with the high-energy electrons. In order to observe the ablation density, we have developed several types of high-resolution diagnostics technique, the penumbral imaging (PI) [4] and Fresnel zone plate (FZP) [5].

The experimental results are compared with the one-dimensional hydrocode ILESTA-1D [6]. The ILESTA code is able to calculate the electron transport both with classical electron heat conduction (Spitzer-Harm: SH) and with non-local electron heat transport (Fokker-Planck: FP), so we compared two results from the ILESTA calculations.

EXPERIMENTS

Experimental Conditions

The HIPER laser facility [7] is one of the irradiation systems of the GEKKO-XII laser facility. Three of the twelve beams from GEKKO are partially coherent light (PCL, λ: 0.53 µm) for the foot pulse. Remaining nine beams are three-directional two-dimensionally smoothed by spectral dispersion (SSD, λ: 0.35 µm) for the main drive pulse. The pulse duration of the foot and the main pulse were 2 ns and 2.5 ns, respectively. Both beams were further smoothed by Kinoform phase plates (KPPs). The focal diameters of the foot and the main pulse were 1000 µm and 600 µm, respectively. Compared to the experiments last year, the 3ω energy is increased by improvement of spatial pattern of SSD at the front end system. Also, the pulse shape, the pumping of the amplifiers and the setting of the KDP crystals are further optimized. Typical 3ω output energy of nine SSD beams was up to 2.5 kJ. However, there is a problem on KPPs about the focusing; the effective energy on the target within the spot diameter was 30 – 40 % of the monitored 3ω energy. Backlight laser beam (2ω) was also employed for x-ray backlighting measurements. For the Rayleigh-Taylor growth rate and the trajectory measurements, the pulse duration and the laser energy were ~ 3 ns and ~ 1 kJ, respectively. For some target-density measurements, the pulse duration and the laser energy were 100 ps and 200 J, respectively.

The material of the laser-irradiated target was polystyrene (CH, $\rho = 1.056$ g/cm^3). The thicknesses of the polystyrene foils were 25 – 40 µm. For the measurements of the RT growth rate, initially-perturbed polystyrene foils were employed. Sinusoidal perturbations were imposed on the laser-irradiated surface. Initial perturbation wavelengths were 12 – 100 µm, and the initial perturbation amplitudes were ~ 1% of the perturbation wavelength.

Measurements of the ablation density

The ablation density of the laser-irradiated targets was measured by two methods: penumbral imaging (PI) and Fresnel phase zone plate (FPZP). These measurements were based on side-on x-ray backlighting showing in Fig.1. The backlighting target material was Ti whose K-line x-ray (~ 4.8 keV) backlit the target from the side of the target. The backlit x-ray intensity is a function of the areal density ($\rho\ell$) and the transmission coefficient of the backlit x-ray. We employed an x-ray CCD camera or an imaging plate as a detector. The pulse width of the backlight laser was 100 ps for "flash" backlighting.

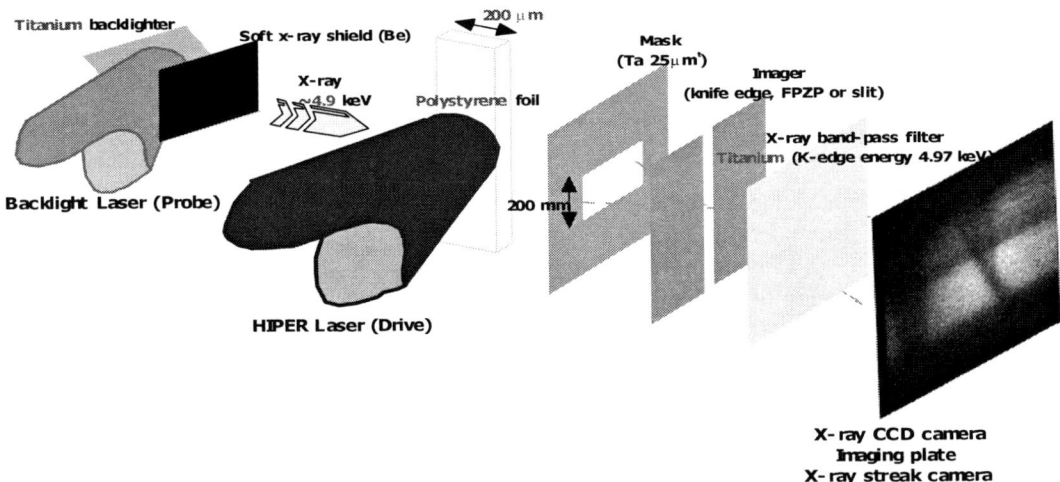

FIGURE 1. Schematic view of experimental arrangement for the measurements of the ablation density

Figure 2(a) shows an example of the FPZP image of backlit polystyrene foil at 1.1 ns after onset of the main pulse. The target was 40-μm thickness polystyrene. The FPZP is optimized for observation of 4.8 keV x-ray. The spatial resolution of the FPZP is 2.2 μm, which was measured with a backlit grid image. Analyzed density profile with the FPZP is shown in Fig. 2(b). Also shown in Fig. 2(b) is calculated density profile from ILESTA-1D simulation with SH (classical) electron heat conduction, indicating that the experiment is in good agreements with the simulation for the SH model.

For the PI, we employed a knife edge coupled with an imaging plate. The measured spatial resolution of the whole diagnostic system was approximately 3 μm. Figure 2(c) shows the analyzed density profiles at 0.6 ns after the onset of the main pulse. Also plotted in Fig. 2(c) is the calculated density profile with ILESTA code with SH electron heat conduction and with FP calculation for electron heat transport (with non-local electron heat transport). The experimental results well agree with the calculated density profile, however, there is little difference between the SH and FP calculations.

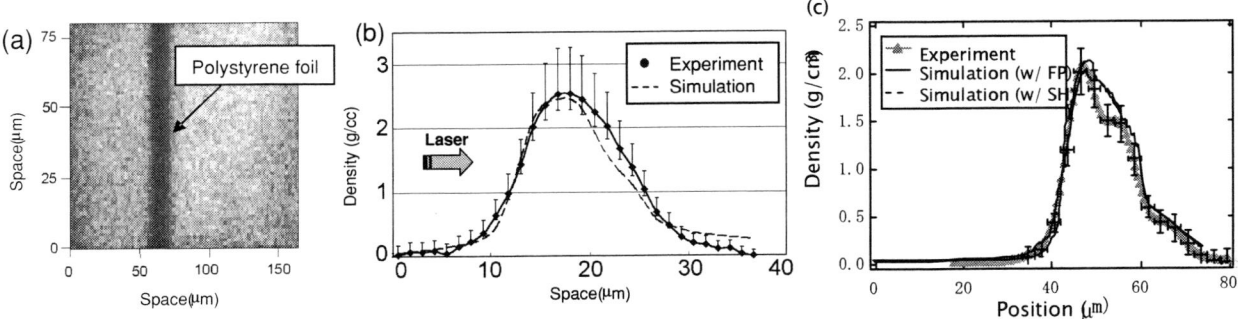

FIGURE 2. (a) An example of FPZP image of backlit CH target. Analyzed density profile (b) from FPZP and (c) from PI.

Measurements of the RT growth rate and other parameters

The RT growth rates γ were measured by face-on x-ray backlighting technique, which had been done in the previous RT experiments. The backlighting target material was zinc (Zn). We obtained growth rates for some short wavelength perturbations with Moiré interferometry technique [8].

The target acceleration g in eq.(1) was measured by side-on x-ray backlighting, which had been carried out previous experiments [2]. The mass ablation rate was measured by face-on backlighting technique using flat CH foil [9]. The measured acceleration and the mass ablation rate are in good agreement with calculation by ILESTA-1D code.

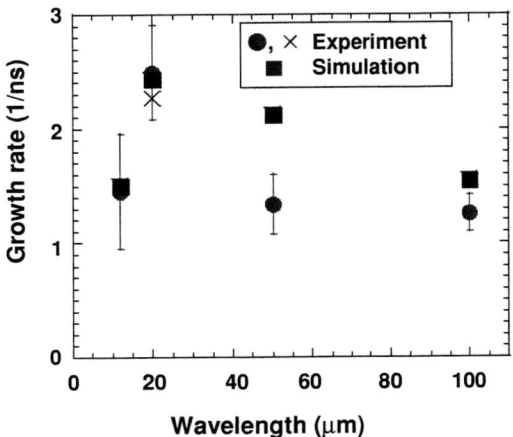

FIGURE 3. The RT growth rates vs. perturbation wavelength. The simulated growth rates are based on the parameters from the SH calculation for electron heat transport.

The RT growth rates vs. perturbation wavelength were shown in Fig. 3. The measured growth rates were compared to the calculated growth rate by Eq. [1] in which the parameters were calculated with the ILESTA-1D.

Although the plotted calculated points in Fig. 1 are with SH calculation, the experimental growth rates well agree with the calculated growth rate. The calculated RT growth rates with FP calculation (not shown here) are almost same as the SH calculation. This fact also means that the non-local electron heat transport is not effective for current HIPER laser irradiation condition.

DISCUSSION

The density profile measured in the experiments by two different instruments shows good agreement with the calculation with 1-D simulation with SH electron heat conduction. This implies that the non-local electron heat transport in not effective for our experimental condition.

We calculated an index of the effect of the non-local electron heat transport, λ_e/L_T, where λ_e is the electron mean free path, and L_T is the temperature scale length at the cutoff point. The parameters were calculated with 1-D ILESTA code. It is shown that the breakdown of the approximation of the classical electron heat conduction happens where $\lambda_e/L_T \sim 0.01$ [10]. In our experimental condition the calculated index λ_e/L_T is very close to ~ 0.01, that suggests the effect of the non-local electron heat transport is unclear.

Our analysis with 1-D simulation also shows that the index increases with increasing the laser intensity. So the effect of the non-local electron heat transport on the RT growth rate is expected at higher laser intensity for blue laser.

CONCLUSION AND OUTLOOK

We have carried out a series of experiments on the HIPER laser system for understanding of the RT instability. We measured all the parameters, especially ablation density of the laser-irradiated target with high-spatial resolution diagnostics. The experimental results indicate that all the results are in good agreements with the 1-D simulation with SH electron heat conduction, indicating that the non-local electron heat transport on the reduction of RT growth is not effective for our experimental condition. Next we are going to carry out an experiment on the reduction of the RT growth with several methods. We also plan to carry out an experiment with planar cryogenic targets, and related developments on diagnostics

REFERENCES

1. Chandrasekhar, S., *Hydrodynamic and Hydromagnetic Stability*, Oxford University Press, London, 1968, Chap. 10.
2. Shigemori, K. et al, *Phys. Rev. Lett.* **78**, 250 (1997); Sakaiya, T. *et al.*, *ibid* **88**, 145003 (2002).
3. Bodner, S., *Phys. Rev. Lett.* **33**, 761 (1974); Takabe, H. *et al.*, *Phys. Fluids* **28**, 3676 (1985); Betti, R. *et al.*, *Phys. Plasmas* **5**, 1446 (1998).
4. Fujioka, S. *et al.*, *Rev. Sci. Instrum.* **73**, 2588 (2002).
5. Tamari, Y. *et al*, *to be submitted to Rev. Sci. Instrum.*
6. Takabe, H. *et al*, *Phys. Plasmas* **31**, 2884 (1988).
7. Miyanaga, N. *et al.*, "The GEKKO XII (High Intensity Plasma Experimental Research) System Relevant to Ignition Targets" in 18th IAEA Fusion Energy Conference, Sorrent, Italy, 4-10 September, IAEA-CN-77/IFP/14 (2000).
8. Matsuoka, M. et al., *Rev. Sci. Instrum.* **70**, 673 (1999).
9. Shigemori, K. *et al.*, *Rev. Sci. Instrum.* **69**, 3942 (1999).
10. Bell, A. *et al.*, *Phys. Rev. Lett.* **46**, 243 (1981).

Potential Formation in Front of an Electron Emitting Electrode in a Two-Electron Temperature Plasma

T. Gyergyek[1,2], M. Cercek[2], D. Erzen[2]

[1]*Faculty of Electrical Engineering, University of Ljubljana, Trzaska 25, 1000 Ljubljana, Slovenia*
[2]*J. Stefan Institute, Jamova 39, 1000 Ljubljana, Slovenia*

Abstract. Plasma potential formation in the pre-sheath region of a floating electron emitting electrode (collector) is studied theoretically in a two-electron-temperature plasma using a static kinetic plasma-sheath model. Dependence of the collector floating potential, the plasma potential in the pre-sheath region, and the critical emission coefficient on the hot electron density and temperature is calculated. It is found that for high hot to cool electron temperature ratio a double layer like solutions exist in a certain range of hot to cool electron densities.

INTRODUCTION

In many plasma devices for material processing plasmas with additional energetic electrons or with a two-temperature electron population are readily produced. Also in fusion machines such electron populations appear due to strong rf fields during ion cyclotron and lower hybrid wave heating and rf current drive. The presence of energetic electrons has a remarkable effect on potential formation in the plasma and consequently on particle losses to the wall.

MODEL AND ASSUMPTIONS

The bounded plasma system is modeled after Schwager and Birdsall [1,2]. The plasma source is located at $x = 0$. The system is bounded at $x = L$ by a floating collector. The plasma that is injected from the source, consists of 4 species of charged particles: singly charged positive ions (index i), cool electrons (index $e1$), hot electrons (index $e2$) and secondary electrons emitted from the collector (index $e3$). It is assumed that the collector absorbs all the ions. We assume that electric field at the source is zero. The potential at the source is set to zero, $\Phi(x=0)=0$. The plasma potential $\Phi_P(x)$ is therefore negative for any x. The floating potential of the collector is $\Phi(x=L)=\Phi_C$. We assume that the plasma is collisionless and that the energy of the particles is a constant of motion. The velocity distribution functions of all 4 particle species are assumed to be Maxwellian with different temperatures and cutoff velocities. For theoretical treatment dimensionless variables are introduced. The potentials and temperatures are normalized to kT_{Se1}, where T_{Se1} is the temperature of cool electrons at the source and k is the Boltzmann constant. The densities are normalized to the density of cool electrons at the source n_{Se1} and velocities are normalized to the electron thermal velocity of cool electrons $v_0 = \sqrt{2kT_{Se1}/m_e}$. The ratio between the electron and ion mass is $\mu = m_e/m_i$. The complete set of nondimensional variables is the following:

$$\mu = \frac{m_e}{m_i}, \quad \tau = \frac{T_{Si}}{T_{Se1}}, \quad \Theta = \frac{T_{Se2}}{T_{Se1}}, \quad \sigma = \frac{T_{Ce3}}{T_{Se1}}, \quad \Psi(x) = \frac{e_0 \Phi(x)}{kT_{Se1}},$$

$$\alpha = \frac{n_{Si}}{n_{Se1}}, \quad \beta = \frac{n_{Se2}}{n_{Se1}}, \quad \varepsilon = \frac{n_{Ce3}}{n_{Se1}}, \quad u = \frac{v}{v_0}, \quad v_0 = \sqrt{\frac{2kT_{Se1}}{m_e}}.$$

Here T_{Si} is ion temperature at the source, T_{Se2} is the hot electron temperature at the source, T_{Ce3} is the secondary electron temperature at the collector, n_{Si} is the ion density at the source, n_{Se2} is the hot electron density at the source and n_{Ce3} is the secondary electron temperature at the collector. With these variables the velocity distribution functions are written as:

$$F_i = \frac{\alpha}{\sqrt{\pi\tau\mu}}\exp\left(-\frac{\Psi}{\tau}\right)\exp\left(-\frac{u^2}{\mu\tau}\right)H(u-u_{mi}), \quad F_{e1} = \frac{1}{\sqrt{\pi}}\exp(\Psi)\exp(-u^2)H(u-u_{me}),$$

$$F_{e2} = \frac{\beta}{\sqrt{\pi\Theta}}\exp\left(\frac{\Psi}{\Theta}\right)\exp\left(-\frac{u^2}{\Theta}\right)H(u-u_{me}), \quad F_{e3} = \frac{\varepsilon}{\sqrt{\pi\sigma}}\exp\left(\frac{\Psi-\Psi_C}{\sigma}\right)\exp\left(-\frac{u^2}{\sigma}\right)H(u_{me}-u).$$

Here H is the Heaviside step function, $u_{mi} = \sqrt{-\mu\Psi}$ is the lowest velocity that an ion can have at the position x if he left the source with zero velocity. Alternatively an electron that has left the collector with zero velocity has at the distance x from the source the velocity $u_{me} = -\sqrt{\Psi-\Psi_C}$. Then we calculate zero (densities) and first moments (fluxes) of the above distribution functions:

$$N_k(\Psi) = \int_0^\infty F_k(u)\,du, \qquad J_k = \int_0^\infty u\,F_k(u)\,du, \tag{1}$$

where k stands for i, $e1$, $e2$, $e3$. We assume that the collector is floating, so the total flux of the charged particles to it must be zero. We also assume, that the flux of secondary electrons from the collector is proportional to the flux of cool and hot electrons to the collector, while the incoming ions don't kick out any secondary electrons from the collector. This last assumption keeps the model solvable. The proportionality constant (emission coefficient) is γ.

$$J_i + J_{e3} = J_{e1} + J_{e2}, \qquad J_{e3} = \gamma(J_{e1} + J_{e2}). \tag{2}$$

From equations (2) α and ε are eliminated. Somewhere between the collector and the source sheaths the plasma potential Ψ_P is characterized by the condition $\nabla^2\Psi_P = N_i - N_{e1} - N_{e2} - N_{e3} = 0$. So setting the net charge density to zero finds the inflection point of the plasma potential. We put $\Psi = \Psi_P$ into the integrals for N_k (eq. 2) and obtain a quasineutrality condition of the form:

$$N_i - N_{e1} - N_{e2} - N_{e3} = 0. \tag{3}$$

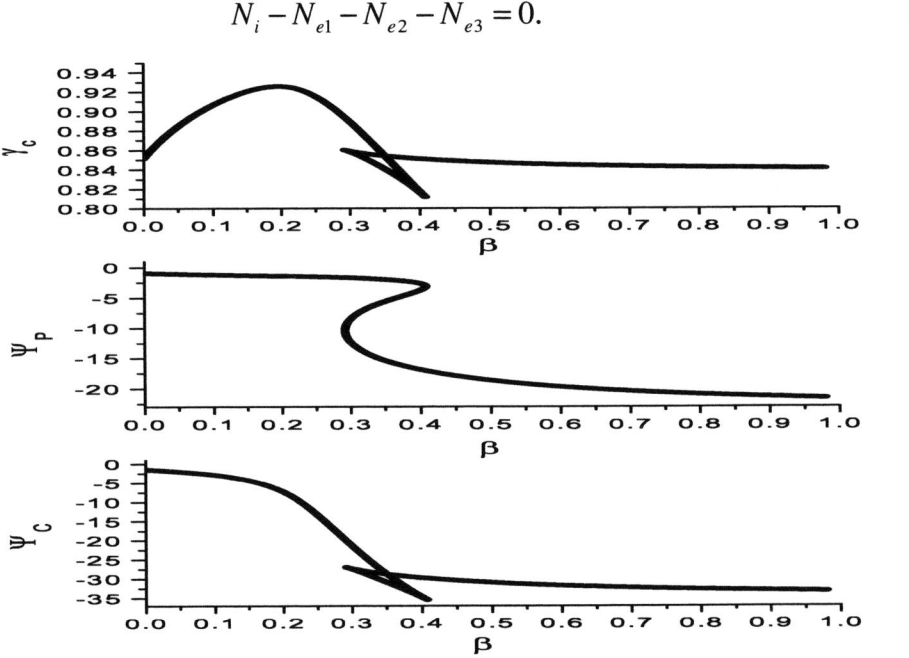

FIGURE 1. Dependence of γ_c, ψ_P and ψ_C on β for $\Theta = 20$. The other parameters are: $\mu = 1/1836$, $\sigma = 0.01$ and $\tau = 0.1$

At the inflection point also the electric field must be zero. The electric field is proportional to $\nabla\Psi$ but this is equivalent to integrating the charge density once in the form:

$$E_k = \int_0^{\Psi_P} N_k(\Psi) d\Psi.$$

In this way a zero field condition at the inflection point is obtained in the following form:

$$E_i - E_{e1} - E_{e2} - E_{e3} = 0. \qquad (4)$$

If the emission coefficient γ increases, eventually the density of secondary electrons in front of the collector becomes so high, that the electric field at the collector becomes zero. The value of γ at which this happens is called the critical emission coefficient γ_c. This fact is used to derive another expression relating Ψ_P and Ψ_C. In order to obtain the electric field at the collector we integrate the density in similar way as before, only the boundaries are changed:

$$P_k = \int_{\Psi_P}^{\Psi_C} N_k(\Psi) d\Psi.$$

The zero field condition at the collector, which is valid only when $\gamma=\gamma_c$ is then written in the form:

$$P_i - P_{e1} - P_{e2} - P_{e3} = 0. \qquad (5)$$

Equations (3), (4) and (5) form a set of 3 equations from which Ψ_P, Ψ_C and γ_c can be calculated if the other parameters (μ, τ, Θ, β and σ) are selected. Equations (3)-(5) are long expressions containing sums of exponential and error functions. For a more detailed derivation and exact form of equations (3)-(5) see [3].

RESULTS AND DISCUSSION

In figure 1 we show the dependence of γ_c, Ψ_P and Ψ_C on β. The other parameters are: $\Theta = 20$, $\mu = 1/1836$, $\sigma = 0.01$ and $\tau = 0.1$. In a certain region of β a triple solution of the system (3)-(5) is obtained. In figure 2 this region is

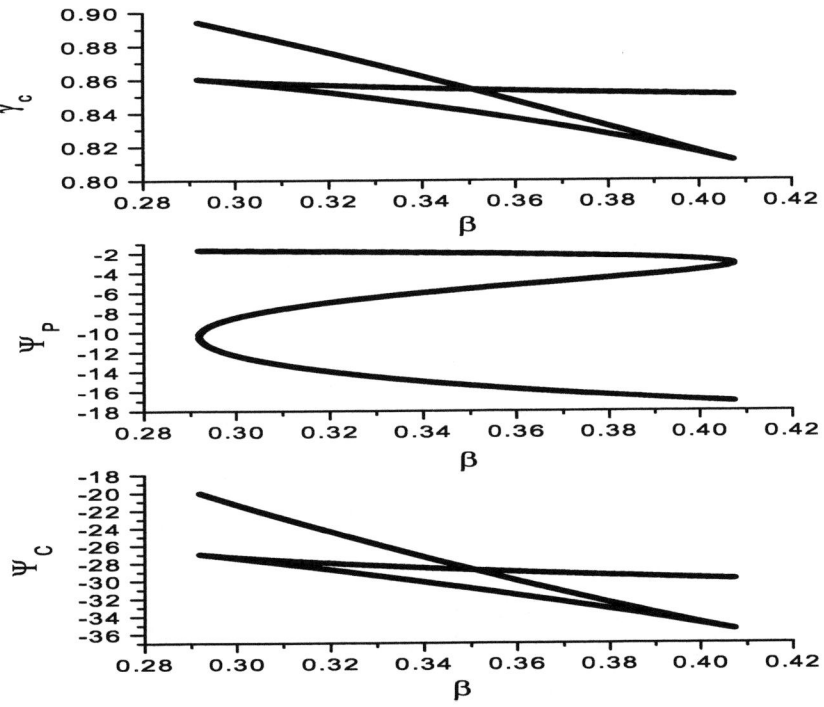

FIGURE 2. Dependence of γ_c, ψ_P and ψ_C on β for $\Theta = 20$. The other parameters are: $\mu = 1/1836$, $\sigma = 0.01$ and $\tau = 0.1$.

shown on an expanded scale. The solutions with high and low values of the plasma potential correspond to 2 different plasmas that can exist between the source and the collector at given parameters. These two plasmas must be separated locally, which means that a double layer like structure is formed between the collector and the source. In our previous work [4] we studied a very similar problem, only without secondary electron emission. The existence of a double layer was verified by PIC simulations. In our future investigations we intend to verify our analytic results with computer simulations and to investigate the effect of secondary emission on the formation of the double layer. The last result that we show is the dependence of the potentials Ψ_P and Ψ_C on the emission coefficient γ. The other parameters are the following: $\mu = 1/1836$, $\sigma = 0.01$, $\tau = 0.1$, $\Theta = 20$ and $\beta = 0.32$. With these values of parameters the system (3)-(5) has a triple solution. The solutions are: $\Psi_P=-13.97$, $\Psi_C=-27.986$, $\gamma_c=0.857$; $\Psi_P=-6.99$, $\Psi_C=-28.72$, $\gamma_c=0.852$ and $\Psi_P=-1.81$, $\Psi_C=-24.41$, $\gamma_c=0.876$. The results shown in figure 3 are obtained by solving equations (3) and (4) for ψ_P and ψ_C while gradually increasing γ from 0 to γ_c which is written on each corresponding small figure. For solutions with the highest and the lowest plasma potential Ψ_P decreases, when γ is increased. For the middle solution Ψ_P increases, when γ is increased. This is an indication that this solution is probably non-physical. Note that Ψ_C increases (becomes less negative) when γ increases. This result was obtained also with fluid models [5].

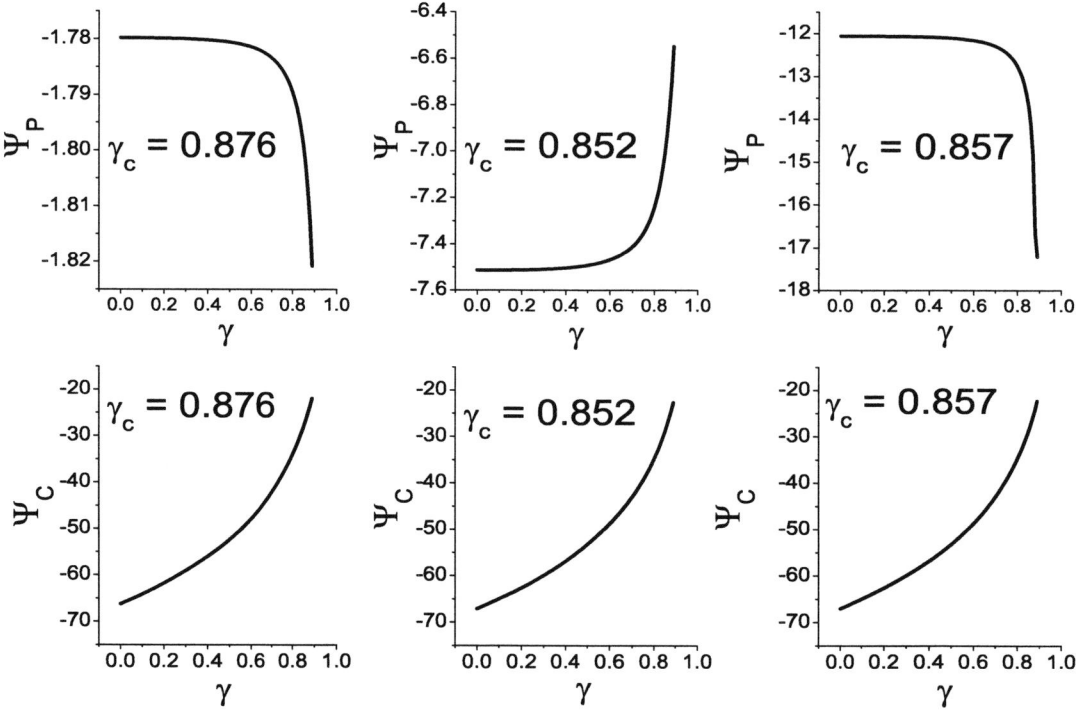

FIGURE 3. Dependence of ψ_P and ψ_C on γ. The other parameters are: $\mu = 1/1836$, $\sigma = 0.01$, $\tau = 0.1$, $\Theta = 20$ and $\beta = 0.32$. See text for additional explanation.

REFERENCES

1. Schwager L. A. and Birdsall C. K., *Phys. Fluids B*, **2**, 1057-1068 (1990)
2. Schwager L. A., *Phys. Fluids B*, **5**, 631-645 (1993).
3. Gyergyek T., Cercek M., *Czechoslovak Journal of Physics*, **52**, 528-543 (2002), Suppl. D.
4. Cercek M., Gyergyek T., Stanojevic M., *Contrib. Plasma Phys.*, **39**, 541-556 (1999)
5. Ye M. J., Takamura S., *Phys. Plasmas* **7**, 3457-3463 (2000).

Portable Neutron Probe for Soil Humidity Measurements

J. Pouzo[1-2], M. Milanese[1-2], R. Moroso[1-2]

[1]*Inst. de Física Arroyo Seco (IFAS), Fac. de Cs. Ex., Univ. Nac. del Centro Prov. Bs. As. (UNCPBA)*
[2] *Member of CONICET*

Abstract. A very small dense plasma focus (142 J) was designed and constructed with the aim to be used as portable pulsed fast neutron source (2.45 MeV in a pulse of 90 ns). The characteristics of this device make it suitable for underground use. The soil humidity can be determined through the detection of thermalized neutrons. In this work we show results of the several prototypes developed in laboratory research to find the best geometric and dynamic parameters for optimize the neutron yield. The diagnostic measurements performed in each discharge include bank current derivative and both time-integrated and time-resolved detection of neutrons emitted by the apparatus. The best neutron yield reached is of the order of 10^7 in 4π sr. flux per pulse. This result introduces a new point in an unexplored range of the neutron yield vs. focus current scaling law.

INTRODUCTION

A plasma focus, called NANOFOCUS is a compact pulsed plasma generator that can give intense pulses of neutrons, X-rays and beams of ions and electrons. Some potential applications are related with nuclear fusion reactions, e. g., studies on reactors, neutron therapies, neutron-graph, and neutron attenuation in soils. This last application has motivated, mainly, this work. In a dense plasma focus (DPF) device [1], the neutron yield Y (in 4π sr. per pulse) scales with the focus current I according with the empirical law $Y \sim I^4$ [2]. This scaling law, for very small DPF, is checked in this work.

The main design characteristics of 125 J NANOFOCUS is presented in this work. The results of a study of the temporally resolved D-D neutron emission of this device, realized with the aim to be applied in soil humidity measurements and some news on the Y scaling law in a very small device, are also informed here.

EXPERIMENTAL SYSTEM AND METHODS

The NANOFOCUS is designed founded in a Mather-type plasma focus [1]. A computational code [2] is used to rise a design optimized respect to the 2.45 MeV neutron yield, product of the Deuterium - Deuterium nuclear fusion. The point of start was to obtain a device with minimal dimensions able to be used in the field, because of one of the objectives mentioned above is the measurement of soil humidity for application in agronomy.

One of the bases of the design is that the device must fulfil the condition that the electromechanical work at the roll-off time must be higher than the specific ionization energy of Deuterium [3]. This energy per unit mass must be high but not too much, because preheating shocks could be present inhibiting a good compression [4]. Summing up, design of NANOFOCUS follows the criteria given in [5]

NANOFOCUS was designed and constructed with the following general characteristics:

We choose the capacitor bank able to be charged by a battery (by instance those used for tractors) for its use in the field. Six condensers connected in parallel compose the bank. The connection of the capacitors was very carefully studied in order to obtain a very low parasitic inductance. Firstly the capacitors were connected following a linear disposition. The parameters in this case were the following: total capacitance: 1.1 µF, parasitic inductance L_o = 74 nH, coaxial electrodes made in brass, anode (central electrode) diameter: 15 mm, anode free length: 22 mm, cathode inner diameter: 42 mm, Pyrex insulator free length: 7 mm. The cathode made in brass constitutes also the vacuum chamber. After the first design, we searched for a more compact one in order to diminish the parasitic inductance in the connection of capacitors, using a "star" disposition (see Fig. 1). We needed in this case to modify also other variables. The working parameters are now: anode diameter: 15 mm, Pyrex cylindrical insulator at the anode level; the parasitic inductance resulted 58.7 nH. We worked with three different anode free lengths: 20, 10 and 15 mm. The spark gap at atmospheric pressure is thought to give a very compact design. A silver activation counter located side-on 0, 5 m far from the NANOFOCUS is used to detect time-integrated neutron pulses. It consists in a 10-cm long Victoreen 1B85 Geiger-Müller counter wrapped in a 0, 5 mm silver foil and insert on a 15 cm-sized paraffin block. The silver activation detector was calibrated with an Am-Be source, that continuously emits 2, 5 MeV neutrons. A scintillator-photomultiplier system located 1-m far from the device detects time-resolved neutron pulses. The plastic scintillator is a NE 102A that have a high response for 2.45 MeV neutrons originated in D-D fusion reactions. It is a solid cylinder whose dimensions are 18 cm of diameter and 5 cm of length. The plastic scintillator has a high efficiency for detecting also hard X-ray pulses (energy over 100 keV). Then, we used a filter for hard X-rays composed by a 3-mm thick lead sheet and a 2-mm thick copper sheath. The scintillator is located at 2 m far from the focus. A Rogowski coil measures the total current flowing through the system. Both time-resolved signals: current derivative and scintillator-photomultiplier are registered simultaneously in a digital Tektronix TDS 3014 oscilloscope; both output cables have the same length. The time precision in the signals coming from scintillator-photomultiplier is about 5 ns (2 ns of rise-time in the scintillator and 3 in the photomultiplier). The Rogowski coil bandwidth is over 200 MHz; then it reproduces signals with rise-time higher than 2 ns. On the other hand, the oscilloscope time resolution is units of ns.

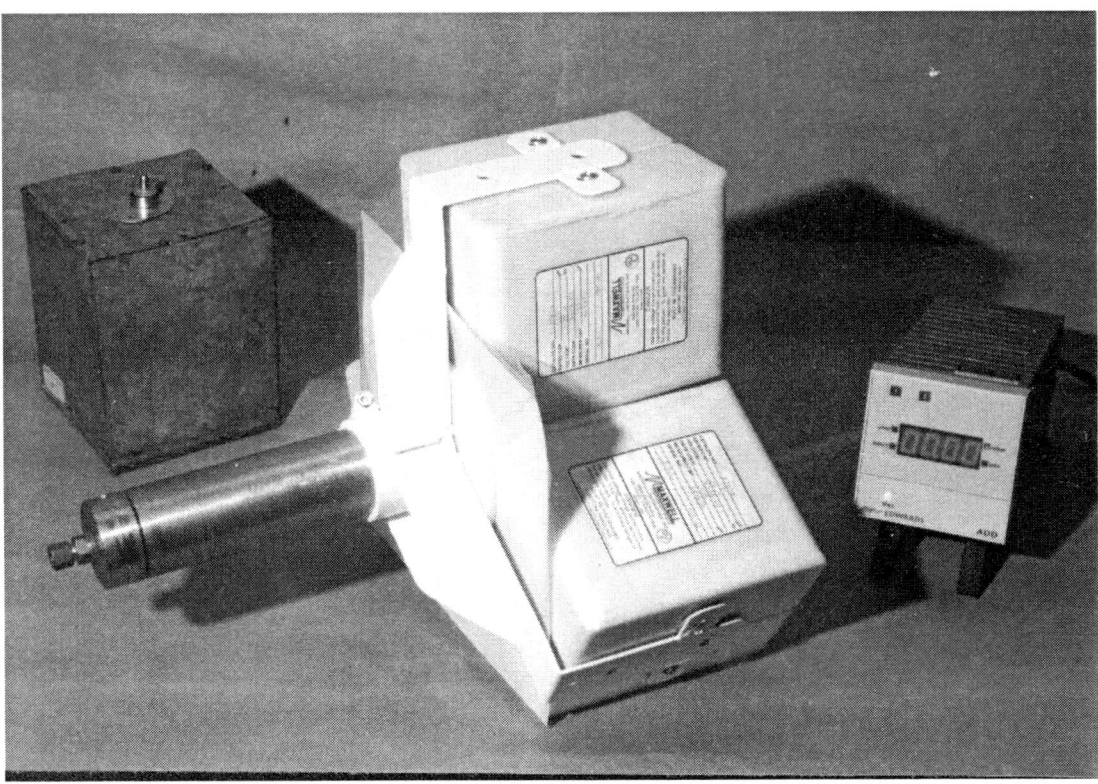

FIGURE. 1. Picture of NANOFOCUS device and neutron integrated detector system. Approximated scale of reduction: 10 to 1.

RESULTS

We will describe now the results obtained in the first device and after those of the second one. For the linear setup of the capacitor series of discharges are made in pure deuterium at a pressure of 1 mb. Previously to the filling of the chamber with deuterium, it is evacuated at a pressure of 10^{-6} mb. The background measured by the silver activation counters would correspond to about 10^6 neutrons emitted in 4π sr. Our successful events are those that exceed at least 7×10^6 neutron yield. In Fig. 2 we show typical signals of X-ray and neutron pulses compared with the dI/dt signal.

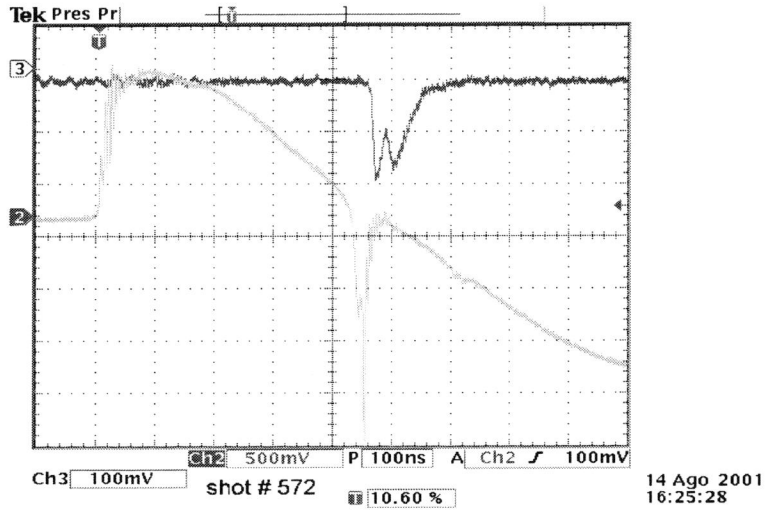

FIGURE 2. Typical OSC signals. Upper beam signals of X-ray (first peak) and neutrons (second one). Lower beam dI/dt signal from the Rogowsky coil.

In Fig. 3 we present the empirical Y vs. I scaling law (from Ref. [1]), with the mean and maximum Y values registered in the NANOFOCUS. These values are two orders of magnitude higher that the predicted by the linear scaling law.

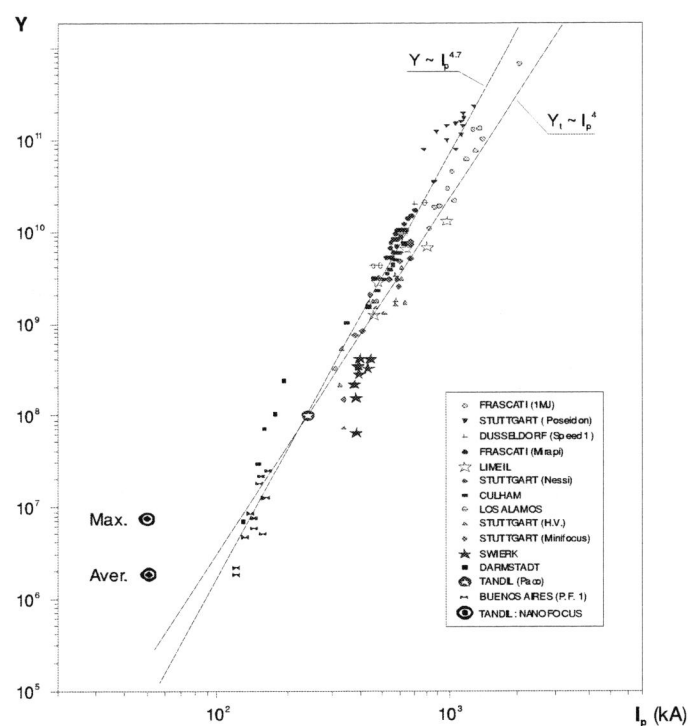

FIGURE 3. Graphic of empirical relation from neutron yield Y and focus current Y (see text).

FINAL REMARKS

The obtained neutron yield Y in the order of 10^6 to 10^7 neutrons per shot in 4π sr, render to NANOFOCUS as a feasible device to be used as neutron probe for measurements of soil humidity, and other applications as neutrongraphy of cement constructions or medical bone studies.

The fact that Y is higher than those predicted by the scaling law, could be interpreted as the influence of non-thermal mechanisms in the D-D fusion reactions, likewise, it is predicted [6] as different scale behaviour for small devices.

ACKNOWLEDGMENTS

This work is supported by Argentine State trough the National Agency of Scientific and Technological Promotion (Agencia Nacional de Promoción Científica y Tecnológica)

REFERENCES

1. J. W. Mather, Phys. of Fluids, **8** (1965) p. 366.
2. M. Milanese, J. Pouzo, "Small Plasma Physics Experiments", World Scientific Publications, London, (1988) p. 66.
3. M. Milanese, R. Moroso, J. Pouzo, IEEE Transactions on Plasma Sc., **21** (1993) p.373.
4. M.Milanese, R. Moroso, J. Pouzo, IEEE Transactions on Plasma Sc., **21** (1993) p.606.
5. M.Milanese, J. Pouzo, Nucl. Fusion **25** (1985) p. 840.
6. J. Pouzo, "Review on a Dense Plasma Focus Research Line and Projections in the Nuclear Fusion Field", Current Trends in International Fusion Research- Proceedings of the Second Symposium – NRC Research Press (1997) pp. 41-57.

Effect of Electrode Material on a Vacuum Pinch

A. Robledo-Martinez[1] and F. P. Espino[2]

[1]*Energy Dept., Universidad Autónoma Metropolitana, Azcapotzalco DF, 02200 Mexico City*
[2]*ESIME-IPN, U. P. López Mateos, Col. Lindavista, 07738 Mexico City*

Abstract. The results obtained in an experimental investigation on the effects of electrode material on a pulsed vacuum arc are reported. For the tests pairs of electrodes made of stainless-steel and brass were employed. These were exchanged to study its effect on the arc dynamics and X-ray emission. It is found that the main effect obtained is in the radiation yield. For the energy employed here (1 kJ) no characteristic emission was observed, the pinching of the arc creates type II micropinches which are bigger in size and emit less characteristic lines than "hot-spots".

INTRODUCTION

During the development of the vacuum arc, the plasma created in the electrode vapors often contracts due to magnetohydrodynamical compression and, if there is enough energy in the magnetic field, radiative collapse takes place. Already one of the pioneering investigation into the properties of the fast vacuum spark demonstrated the existence of midgap dense plasma concentrations in which intense characteristic X-radiation was produced [1]. The radiation was seen to be emitted in coincidence with dips in the di/dt signal and to have energies of the order of several keV. However, a previous investigation into DC vacuum breakdown using time-resolved spectroscopy demonstrated that effectively the initial emission originates in the cathode but that current amplification takes place in anode vapor [2]. There are clearly still some aspects of the evolution of the media in which the vacuum arc propagates that remain unsolved.

The later introduction of both ultra-fast pulsed power and diagnostics helped to understand the short time-scale behavior of the vacuum spark. In one of the standard references on vacuum arcs [3] it is maintained that for pulsed regimes, the initial stages of the vacuum discharge are initiated by explosive emission from the cathode which thus provides material for the initial propagation of the arc. In intermediate stages both electrodes contribute and in the final stage it is the anode the main supplier of conducting medium.

According to results reported in [4,5] the plasma concentrations, "hot spots" or "plasma points", observed in vacuum arcs have sizes ranging from 10 to 100 µm and emit energies that vary inversely to their size. Two types are identified: Type I, smaller in size, called the micropinch, is a stage towards radiative collapse and emits mostly characteristic radiation. Type II is one order of magnitude bigger, radiates less characteristic lines and its radiation is insufficient for a full radiative collapse.

The aim of the present work is to investigate the effect of electrode material on a pulsed arc in a semi uniform electric field configuration. One of the goals is to understand what effect the swapping of electrode material could have on X-ray emission and arc development.

EXPERIMENTAL SET-UP

A 0.1µF-75 kV capacitor was discharged to a pulse-forming line through a low-inductance spark gap. The stored energy was of the order of 1 kJ. The pulse-forming line employed, 3 m long, was of the parallel-plate type and had a 1.2 Ω characteristic impedance. Its single transit-time was 16 ns. A self-trigger transfer switch (two Borda electrodes with a 10 mm gap in ambient air) then transferred the charge stored in the line to the load. This consisted of a 0.7 mm gap, hemispherical tip cathode-rod/anode-plane configuration inside a vacuum chamber. In order to study the effect of electrode material these were made of different alloys: either stainless-steel (SS) or brass. Typically, the vacuum employed was of the order of 10^{-5} mbar and the charging voltage varied from -30 to -50 kV. The time derivative of the current passing through the load (I-dot) was measured through a one-turn Rogowski belt coil. The voltage on the transmission line was measured using a D-dot probe.

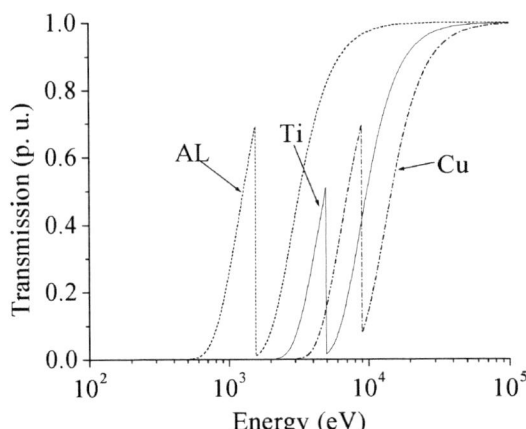

FIGURE 1. X-ray transmission coefficients of the filters employed.

Time resolved detection of X rays was made possible through the use of 3 X-ray PIN diodes fitted with different transmission filters. The diodes employed had a 0-100 keV detecting range. Figure 1 shows the calculated transmission coefficients for the filters employed. They consisted of: Al (thickness 4 µm), Ti (12.5 µm) Cu (10 µm); the last two filters had an additional 4 µm Al filtering to avoid fluorescence. In some of the shots a pinhole camera with a magnification ratio of 0.5 was used to obtain a time-integrated picture of the X-rays emitted. The diagnostics employed also included optical streak and frame photography.

EXPERIMENTAL RESULTS

Figure 2 shows optical streaks obtained in two different shots. Both are axial streaks, i.e. the axis of the slit camera is parallel to the axis of the electrode. Thus, the streak trace is the image of the axis of the gap swept in time. The upper boundary of the dark streak's strip corresponds to the tip of the rod cathode electrode and the bottom to the anode plane. In the majority of the shots light emission began at the cathode and the propagated towards the anode. This was followed by another front appearing at the anode some tens of nanosecond later which propagated towards the cathode. The left-hand frame in figure 2 shows an example of this type of shot. In some other shots light first appeared at the anode plane and propagated towards the anode as shown in the right-hand frame of figure 2. In this frame the slow luminous front from the anode travels towards the cathode at a speed of $1.3 \cdot 10^{-4}$ m/s while the brighter front from the anode propagates at a speed $0.36 \cdot 10^{-4}$ m/s. Exchanging the material of the electrodes did not seem to have any effect on this behavior, it is a feature that seem to be independent of their chemical composition.

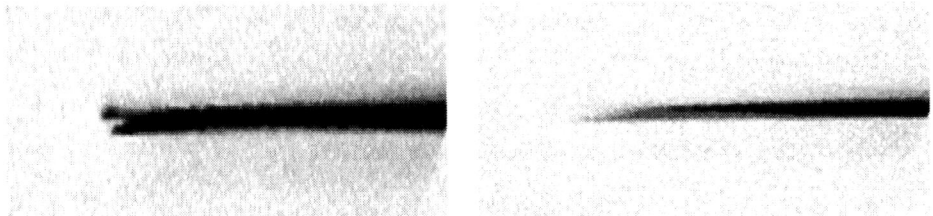

FIGURE 2. Streak photograph (negative) showing light emission beginning from the cathode (left frame) and from the anode (right frame).

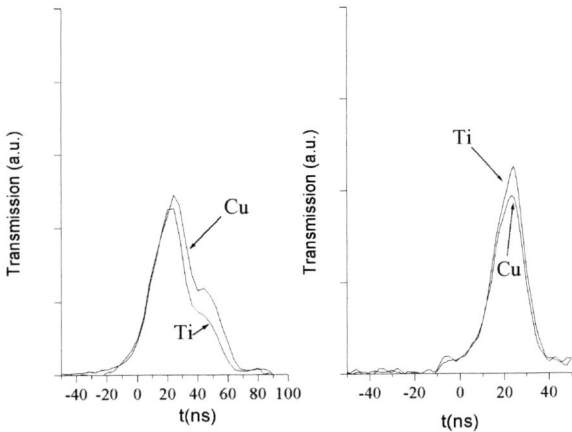

FIGURE 3. Time-resolved detection of X-rays for brass cathode/SS anode (left) and SS cathode/brass anode (right).

The use of filtered PIN diodes allowed the time-resolved measurement of the X-ray emission within the transmission bands of each filter. For every electrode arrangement a series of shots was taken using two different filter-fitted X-ray detectors. The left-hand frame in figure 3 shows the numerical average of a series of measurements obtained under identical conditions for a brass cathode and an SS anode. Notice both signals are very similar in size although the Cu-filtered is slightly bigger than the other. The slight bump to the right of the peak is due to variations of the position in time of the individual peaks. A similar averaging of several shots was carried out for an SS-cathode, brass-anode configuration of electrodes. The results obtained are shown in the right-hand frame in figure 3. Notice this time the Ti trace is slightly bigger than the Cu trace.

The arc lasts several tens of microsecond. In intermediate stages of the arc development a series of light oscillations with a period of about 4-5 ns was observed in the arc column. These match concurrent oscillations in the di/dt signal having the same period. These phenomena has been observed by other authors [6] who report optical oscillations of the arc column with frequencies of hundreds of MHz. They attributed this behavior to an initial, resistive, stage of arc buildup.

DISCUSSION

At the beginning of the discharge a light front first appears at the tip of cathode but this did not move rapidly. Other axial streaks showed front lights appearing at both electrodes, as in figure 2, with typical propagation speeds from the anode to the cathode of about $2 \cdot 10^4$ m/s. This is 2 orders of magnitude below the values reported in [7]. This could be attributed to the lower energy employed here.

The material of the electrode does affect the X-ray yield. Under similar conditions, the SS cathode/brass anode configuration emits more intense (1.1-2 times more) radiation than the opposite configuration. Most of the characteristic X-ray lines of the alloy components employed fall within the energy interval 5-9 keV. Figure 4 shows a calculation of the filter transmission ratio at the $K\alpha_1$ lines of each of the elements that make up the electrodes, from Cr to Zn. The experimentally determined corresponding ratio for the pair of filters Cu-Ti was close to unity and not the 1.4-3.8 calculated. Similarly, the higher experimental ratio of Al/Ti-filtered detection (1.2-1.5) was different from the calculated 2-13. This shows that the plasma created does not contain much characteristic radiation as it does not fit the expected ratios.

The X-ray pinhole photographs did not show much radiation at the anode. This and the fact that the use of additional, thicker, filters for the PIN diodes did not reveal high energy radiation discards the existence of hard X-rays. These would be produced by an e-beam hitting the anode, as in flash X-ray tubes. In fact the pinhole photographs show that the X-ray emission came from an area covering the whole gap with a width of about twice the gap distance.

The above facts indicate that the sources of emission are not the minute, energetic hot-spots of type I but rather the larger micropinches of type II which emit fewer characteristic lines and have more of a continuum in the interval 1-8 keV [4].

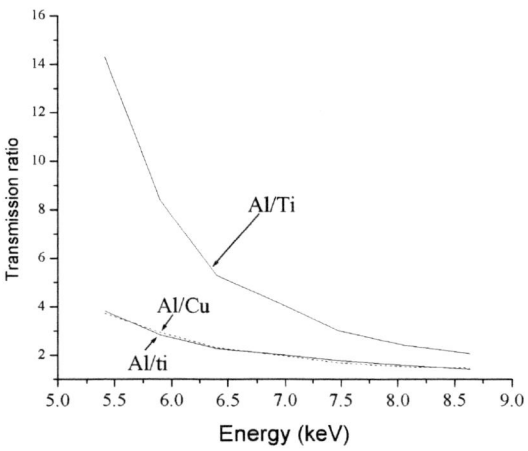

FIGURE 4. Filter transmission ratio at the Kα_1 lines of electrode material.

CONCLUSIONS

The effect of electrode material on the properties of a pulsed vacuum arc was investigated. The results obtained show that whether the cathode or the anode were made of SS or brass did not matter for the initial and intermediate light emission pattern. In most cases a light front was seen traveling from the anode to the cathode followed by another front, some tens of nanosecond later traveling in the opposite direction. For intermediate times (of the order of microseconds) the arc light pulsated with a period of 4-5 ns which matched similar periodic oscillations in the di/dt signal. Not much characteristic X-radiation was observed, the X-ray detected are thought to have their source in type II micropinches which emit few characteristic lines and have more of a continuum in the interval 1-8 keV. The main effect of electrode material was on the X-ray yield; the SS cathode/brass anode configuration's emission intensity was as much as twice that of the reverse configuration.

ACKNOWLEDGMENTS

This work was supported by the Mexican Science Council (CONACyT, grant No. 34376-E). A. Robledo-Martinez is a fellow of Sistema Nacional de Investigadores. The authors would like to thank Dr. J. Herrera and Dr. M. Villagran for lending the trigger generators employed in the experiments.

REFERENCES

1. Cohen, L., Feldman, U., Swartz, M. and Underwood, J. H., J. OSA 58, 843-846 (1968).
2. Davies, D. K. and. Biondi, M. A, , J. Appl. Phys. 48, 4229-4233 (1977).
3. Mesyats, G. A. and Proskurovsky D. I., Pulsed Electrical Discharge in Vacuum, Springer-Verlag, New York, 1989.
4. Koshelev, K. N. and. Pereira, N. R, J. Appl. Phys. 69, R21-R44 (1991).
5. Antsiferov, P. S., Koshelev, K. N., Kramida, A. E. and Panin, A. M., J. Phys. D: Appl. Phys. 22, 1073-1077 (1989).
6. Cross, J. D., IEEE Trans. Elec. Insul. EI-18, 230-233 (1983).
7. Rout, R. K., Auluck, S. K. H., Nagpal, J. S. and Kulkarni, L. V., J. Phys. D: Appl. Phys. 32, 3013-3018 (1999).

Impedance Coupling in a 32 kj Wire-Array Z-pinch

A. Robledo-Martinez[1], J. Nieto[1] and F. P. Espino[2]

[1]*Energy Dept., Universidad Autónoma Metropolitana, Azcapotzalco DF, 02200 Mexico City*
[2]*ESIME-IPN, U. P. López Mateos, Col. Lindavista, 07738 Mexico City*

Abstract. The successful transfer of electrical power to a Z-pinch load is one of the concerns of pulsed power engineering. Some of the power at the load can be reflected back to the generator due to poor impedance matching. We report results obtained in experiments performed with a 32 kJ fast pulser and a wire array load. It is found that the number and spacing of wires determines the amount of current passing through to the load. The wire array's load behavior is compared with short-circuit and air discharges. It is also demonstrated that the X-ray emission depends on the impedance of individual shots. It is found that as the number of wires grows, the chances of pinching the array diminish. For the circuit parameters employed and available energy it is found that a 2-wire configuration gives the optimum coupling.

INTRODUCTION

The Z-pinch is being investigated for its potential to become a powerful source of pulsed X-rays [1]. Ongoing research projects aim to implode arrays of thin conductors through the application of short, powerful current pulses with peak amplitudes of the order of several megampere [2]. One of the problems that researchers face is how much of the electric energy stored in the capacitors is finally converted to radiative power. A good coupling of the generator with the load is essential for maximum radiated power. Poor coupling can lead to failure to pinch and to disruptions of the plasma column [3].

At the onset of any discharge when the plasma begins to grow the dynamic behavior of its column impacts the rest of the circuit. The resistance initially drops at a fast rate matched by a corresponding increment in its inductance [4]. The dimension, chemical composition and number of wires in the array determine the way current flows through the load. Ideally a good load is a low impedance one, as it allows maximum flow of current through it. But if it is physically too big in size it can make implosion of the array more complicated.

The aim of the present work is to investigate the effect of the wire number and its distribution on the initial flow of current through the load in a small plasma machine. This is done through the comparison of the properties of several wire configurations. The impedance of the load is used as a numerical indicator of the quality of current flow through the load. The X-ray yield, the most important indicator of how successfully the load pinches, is linked to the circuit behavior.

EXPERIMENTAL SET-UP

A 32 kJ-stored energy plasma pulser was employed for the experiments. It consists of an 8-stage, 600 kV Marx with a total capacitance of 0.2μF when erect. The Marx feeds a transmission line of 4.8 Ω characteristic impedance. Its single transit-time was 33 ns. A self-trigger transfer switch insulated with SF_6 dumps the charge stored in the line to a second length of transmission line. This has identical parameters to the first one but because of its shorter length its single transit time is 30 ns. This second line segment feeds the load array inside a vacuum chamber. The quarter of wave-short circuit current had a period of 1.2 μs. Typically, the vacuum employed was of the order of 10^{-5} mbar. The charging voltage (per stage) varied from -40 to -50 kV.

The time derivative of the current passing through the load (I-dot) was measured through a one-turn Rogowski belt coil at the ground side of the load. The voltage on the transmissions lines was measured using a D-dot probe. One of them was placed close to the end of the second transmission line, close to the load.

Two brass electrodes with a gap of 3 cm were employed to hold the arrays. The fibers employed were copper wires with a diameter of 25µm. Three configurations were employed: single wire, 2 parallel wires with a 2.4 mm separation and 4 parallel wires in a square disposition, 2.4 mm a side.

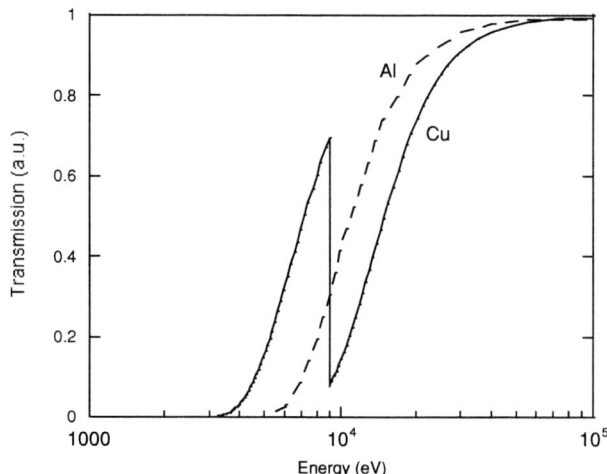

FIGURE 1. Filter transmission ratio.

Time resolved detection of X rays was made possible through the use of 3 Quantrad X-ray PIN diodes fitted with different transmission filters. The diodes have a 0-100 keV detecting range. Figure 1 shows the calculated transmission coefficients for the filters employed. They consisted of: Al (thickness 125 µm), Ti (12.5 µm) Cu (10 µm); the last two filters had an additional 4 µm Al filtering to avoid fluorescence. The diagnostics employed also included optical streak and frame photography (Hamamatsu C2830).

EXPERIMENTAL RESULTS

Figure 2 shows an optical streak taken in a shot on two 25 µm-diameter parallel copper wires separated by a distance of 2.4 mm. The streak is a radial one, i.e. the slit of the camera was perpendicular to the wires and was pointed midgap. The picture shows that the discharge begins concurrently on the 2 wires and pulsates with a period of 4 ns. Later on the two plasmas merge and the merged plasma pinches. It then continues pulsating with the same frequency as before the merge. The observed X-ray pulse was produced soon after the 2 plasmas merge.

FIGURE 2. Radial streak photograph (negative) of 2 parallel wires. Sweep time: 100ns.

During the experiments the voltage was monitored by means of 2 V-dot probes on each section of the transmission line. The second one was placed quite close to the vacuum chamber. The ratio of line voltage to load current provides a good representation of the instantaneous magnitude of power flow to the load. This is conveniently expressed as an impedance. In spite of the fact that it does not fit the classical definition of a passive element in a circuit, it is nevertheless meaningful in terms of gauging load coupling. In fact, the ratio V/I represents more an output parameter of the pulser than a physical characteristic of the load.

In several shots the impedance of the load as a function of time was measured for different array configurations. The results obtained are shown in figure 3. In Fig 3(a) the impedance of a short circuit obtained with an SF_6 pressure of 0.5 kg/m^2 in the transfer switch is shown. The reduced pressure has the effect of switching earlier thus producing a reduced voltage and current on the second section of the transmission line. Fig 3(b) shows the short-circuit current with the normal operating pressure, which is the same for the remaining frames in figure 3. Even though it is not directly related to the subject of this work, for the sake of comparison an air shot was also taken and is shown in figure 3(c). figures 3(d) to 3(f) show the results of the calculated impedance for arrays of one, two and four wires. The vertical scale is the same in all figures. Current starts at t=-5 ns.

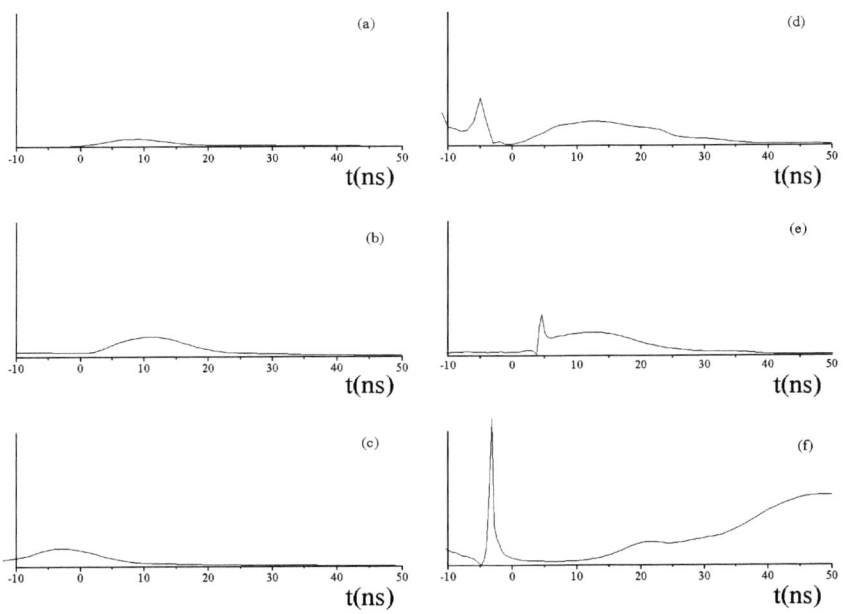

FIGURE 3. Time evolution of load impedance. Current starts at t=-5 ns. Key: a) short-circuit with reduced pressure in transfer switch, b) short-circuit, c) air shot, d) single wire, e) double wire, f) four wires. Applied voltage: 320 kV.

The use of filtered PIN diodes allowed the time-resolved measurement of the X-ray emission within the transmission bands of each filter. For every array configuration a series of shots was taken using two different filter-fitted X-ray detectors. Figure 4 shows the time-resolved detection of the X-ray pulses for a 2-wire bundle. The magnitude of the X-rays emitted by the 4-wire array was rather low. The highest X-ray yield was obtained with a single wire. The 2-wire array gave an slightly lower X-ray yield than the single wire, but not much less.

DISCUSSION

The ideal load, in terms of power transfer is a short circuit as it allows maximum flow of current to the load. Figures 3(a) and 3(b) show that they have a low impedance value and are smooth after current start. The air shot is also a good load: it has a low impedance magnitude and smooth variation. The wires impedances, on the contrary, show a sharp variation when the arc strikes and their magnitudes are larger than short-circuit or air. Even though the 4-wire load might seem initially good its impedance grows very quickly for later times.

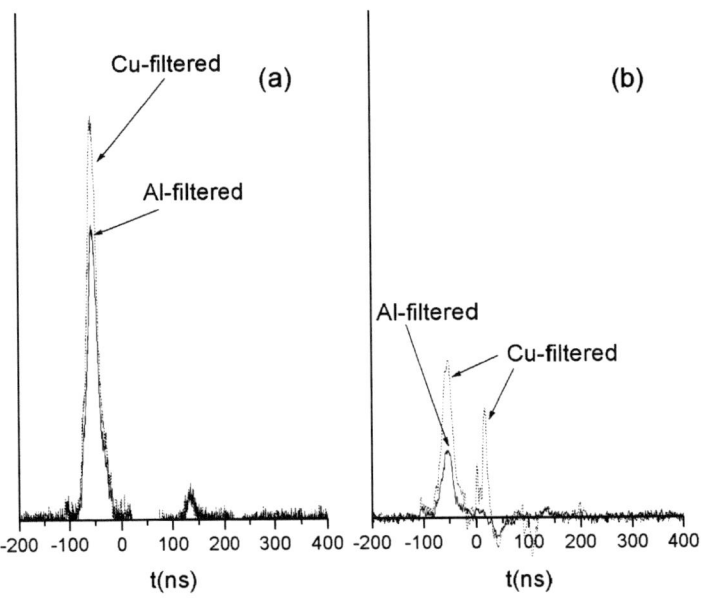

FIGURE 4 X-ray emission for a 2-wire bundle for: a) 50 kV and b) 40 kV charging voltage.

In most shots only one X-ray pulse was detected by both detectors (see figure 4-a). In this case the amplitude of the Cu-filtered diode was always bigger than that of the Al-filtered one. Sometimes double peaks were detected by the Cu-filtered diode while the Al diode would only see one of them. As both filters can detect the characteristic lines of the copper plasma the peak missed by the Al diode must be a pulse of X-rays containing no characteristic lines of copper. This peak is in the energy interval 4-9 keV.

CONCLUSIONS

For the specific experimental conditions employed here, the best load, in terms of impedance matching was the 2 parallel wire. There must be a balance between the desirable electrical characteristics of the load and the X-ray yield. For example, the advantage, in terms of lower impedance, of using a 2-wire array versus a single fiber is offset by the lower X-ray yield of the former.

ACKNOWLEDGMENTS

This work was supported by the Mexican Science Council (CONACyT, grant No. 34376-E). A. Robledo-Martinez is a fellow of Sistema Nacional de Investigadores. The authors would like to thank Messrs. Jaime Alonso, Manuel Negrete, Rafael Puente and Mr Mario García for assistance in the experimental work.

REFERENCES

1. Haines, M.G., Lebedev, S. V., Chittenden, J. P., Beg, F. N., Bland, S. N. and Dangor, A. E., Phys. Plasmas 7, 1672-1680 (2000).
2. Deeney, C. et al, Phys. Rev. E 56, 5945-5958, (1997).
3. Kies, W., Decker, G., Malzig, M., van Calker, C., Westheide, j., Ziethen, G., Bachmann, H., Baumung, K., Bluhm, H., Rusch, D., Ratajczak, W., Stoltz, O. and Bayley, J. M., J. Appl. Phys. 70, 7261-7272 (1991).
4. Persephonis, P., Giannetas, V., Ioannou, A., Parthenios, J. and Georgiades, C., Jpn. J. Appl. Phys 34, 6226-6229.

Acceleration Dynamics of Laser-Driven MeV-Ion Jets

M. Hegelich[1,2], M. Allen[6], P. Audebert[4], A. Blazevic[3], T. Cowan[6], J. Fuchs[4], J.C. Gauthier[4], M. Geissel[3], W. Guenther[5], D. Habs[1], S. Karsch[1,2], A. Kemp[2], G. Pretzler[1,2], M. Roth[3], K.J. Witte[2]

[1] LMU, Muenchen,
[2] MPQ, Garching
[3] GSI, Darmstadt
[4] LULI, Paris
[5] Uni Siegen
[6] GA, San Diego

Abstract. Collimated jets of carbon and fluorine ions up to 5 MeV/nucleon (~100MeV) are observed from the rear surface of thin foils irradiated with laser intensities of up to 5×10^{19} W/cm^2. The normally dominant proton acceleration could be suppressed by removing the hydrocarbon contaminants by resistive heating. This inhibits screening effects and permits effective energy transfer and acceleration of other ion species. The acceleration dynamics and the spatio-temporal distributions of the accelerating E-fields at the rear surface of the target are inferred from the detailed spectra.

INTRODUCTION

Recent advances in high-intensity laser technology have opened new realms of relativistic laser plasma physics, such as the acceleration of charged particles to high energies over very short distances (~ 100µm). For over 25 years, multi-MeV protons and heavier ions have also been generated using large Nd:glass and and CO$_2$ lasers[1,2] by the expansion into vacuum of hot laser-produced coronal plasmas. More recently intense, collimated beams of protons have been accelerated by focusing high-intensity short-pulse lasers on thin foils[3,4]. Wilks et al[5] attribute this production of low-divergence, laser-accelerated proton beams to the so-called target-normal sheath acceleration (TNSA) mechanism[6]. Rear-surface acceleration offers many advantages for producing well-controlled heavy-ion beams, particularly in terms of being able to decouple the complicated and often stochastic laser-plasma interaction from the ion generation and acceleration processes. It differs fundamentally from prior work on energetic ion production from coronal plasmas where the ions are accelerated by the ambipolar electric field present in the quasi-neutral expanding plasma, which is proportional to the plasma density gradient. The rear-surface acceleration dynamics, however, should be largely dominated by the strong transient electrostatic fields during the initial, non-neutral phase of the acceleration. This provides an impulsive acceleration normal to the surface which produces very cold, highly laminar ion beams. Since only a nm-scale layer is accelerated, the source material may be deposited on a target substrate which is separately chosen to optimize the laser-plasma coupling and relativistic electron transport to the rear surface. Finally, the ionic charge state distribution, the laser-to-ion conversion efficiency, and the attainable energy, might be selectable by appropriate target design and choice of laser parameters. In this report, we present the first experimental study of high-quality heavy ion beams accelerated from the *rear surface* of thin foils. We find that the spectra of different ion species provide additional information, not available in the proton signal, about the spatio-temporal evolution of the accelerating fields. The initial dynamics of the electron distribution on a timescale $t \ll \tau_{laser}$, as well as the late expansion and cooling dynamics after the pulse at $t \gg \tau_{laser}$ have to be considered to correctly treat the ionization and acceleration processes and explain the measured spectra.

EXPERIMENTAL SET-UP

The experiments were performed with the 100 TW laser at LULI. The laser pulses (~30 J, ~300 fs, 1.05 μm) were focused at normal incidence on target to an intensity of up to 5×10^{19} W/cm^2 (contrast ~10^{-7} in intensity). Targets were 50-μm thick Al and W foils coated on the rear side with 1-μm carbon or 0.3-μm CaF$_2$, respectively. The accelerated particles were investigated by three complementary diagnostics: (a) a stack of radiochromic film (RCF) located 5 cm behind the target to record the angular distribution of the emitted proton beam; (b) a magnetic proton spectrometer at 13° to the target normal with Kodak DEF X-ray film to measure the proton energy spectra; (c) two Thomson parabola spectrometer (B=0.65 T, E=1.3 MV/m) with CR-39 track detectors at 0° and 6° at a distance of about 1 m (solid angle ~5×10^{-8} sr) to obtain the energy spectra for the different species of heavy ions for which (a) and (b) are almost insensitive due to protective overlayers. CR-39 is sensitive to single ion events but insensitive to electromagnetic radiation and electrons. An ion striking a CR-39 plate destroys the polymer matrix along its path and causes nm-scale damage transformed into cone- or bowl-shaped craters, when the CR-39 is etched in NaOH solution. We analyze each individual track on the detector plate by optical microscopy with custom pattern recognition software[7] to yield position and track size parameters. The absolute energy spectra for each ion species are then obtained from the distribution of pits along the distinct traces.

EXPERIMENTAL RESULTS

Typically, the spectrum of accelerated ions is dominated by protons originating from contaminant hydrocarbon layers, which outrun the heavier C ions and screen the fields behind them. We tried several techniques to remove the hydrocarbons (including radiative and laser heating) and found that resistive heating, as used in Ref. 8, was most effective. We heated Al and W foils up to ~600 K and ~1200 K, respectively. By the partial removal of hydrocarbons, the carbon ion acceleration is strongly enhanced, as shown qualitatively in Figs. 1a and 1c, and quantitatively in the evaluated spectra in Figs. 1b and 1d. The proton spectrometer gave typical yields of ~10^{11} protons with energies of up to 25 MeV, for unheated targets.

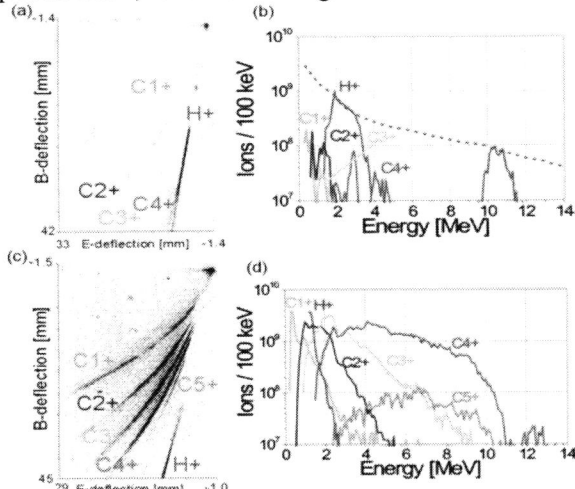

FIGURE 1. (a) Ion traces from an unheated Al/C-target on CR39. (b) Corresponding spectra, the proton signal shows a gap due to detector properties which is optimized for heavier particles. The dotted line illustrates the spectra obtained with the proton spectrometer. (c) Ion traces from heated Al/C-target. The ion signals are strongly enhanced. (d) Corresponding spectra from heated target.

For heated Al-targets, the number of protons is reduced to 1.2×10^{10} with energies of only up to 3 MeV. Removal of the hydrocarbons increased the energy of carbon ions by ~2.5 and their number by two orders of magnitude to 2.1×10^{11} corresponding to a laser-to-ion energy conversion of 0.5%. Acceleration is most efficient for C^{4+} ions, with a spectral cutoff at 1 MeV/nucleon at the high-energy-end. For all spectra, one notes a dependence of the high energy cutoff on the charge state, which rules out recombination as a dominant effect in our experiments. This behavior is fundamentally different from the observed carbon spectra in long-pulse experiments[2], where all charge states had the same cutoff, and the rear-surface ion signal almost vanished for heated targets. The W targets could be heated to higher temperatures than the Al, and the proton spectrometer as well as the CR-39 did not show any protons, while strong fluorine ion tracks are observed originating from the CaF_2 layer at the target rear side (see Fig. 2).

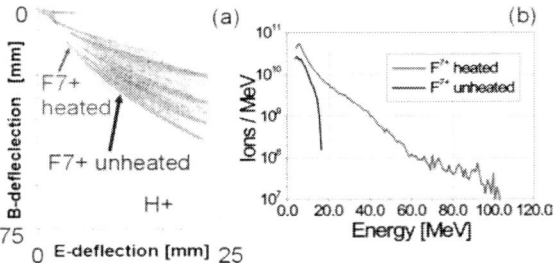

FIGURE 2. (a) Overlayed signals of heated (grey) and not-heated (black) W/CaF_2 targets: The proton signal vanishes for heated targets, the fluorine signals (especially F^{7+}) go up to much higher energies. (b) Corresponding F^{7+}-spectra.

The complete removal of contaminants increased the acceleration of heavier ions again considerably. Quantitative evaluation shows that F^{7+} was accelerated up to 100 MeV, i.e. more than 5 MeV/nucleon at 4% energy conversion. The RCF diagnostic confirmed this finding. We observe a narrow spot in the first layer, which, in the absence of protons, indicates fluorine ions of energies above 4 MeV/nucleon. The analysis of the lower charge state proves to be more difficult in this case due to overlapping F and Ca traces with the same charge-to-mass ratios.

The influence of the presence of different amounts of protons on the ion acceleration mechanism can be seen in Fig. 3. The high energy cutoff of the Helium-like ion charge states of Carbon and Fluorine ions are plotted against the incident laser energy. In the case of unheated targets (black pentagons) where mainly protons are accelerated, the ion signal always cuts of at about 0.5 MeV/nucleon for all laser energies. For Aluminum targets which could be heated resistively to ~600°C and for laser heated targets the number of protons is reduced while the ion signal increases till both signal are of toughly the same magnitude. The ion cutoff energy than doubles to more than 1 MeV/nucleon and a slight increase of the cutoff energy with increasing laser energy can be seen. In the case of the heated Tungsten targets no more protons were present and the highest dominant ion charge state shows the same behavior as the protons in non-heated experiments, i.e. a clear dependence of the high energy cutoff on the laser energy. It is clear that in the presence of protons screening the electric fields other effects are of minor importance to the ion spectra. Also from the measured ion spectra we can instantly see the presence of protons and determine how if the heating process did not work to full extend as was the case with two of tungsten shots, where the heating contacts broke before the laser shot.

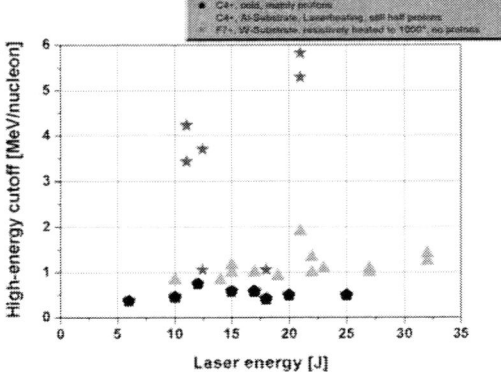

FIGURE 3. Influence of protons on the high-energy cutoff of Helium-like Carbon and Fluorine charge-states: Unheated targets (pentagons, mainly protons): The cutoff energy is constant at 0.5 MeV/nucl. and shows no dependence on the laser energy. Low-heated targets (triangles, reduced proton numbers): Cutoff at ~1MeV/nucl., slight dependence on laser energy. Fully heated targets (stars, no protons) at strong dependency on the incident laser energy can be seen.

Since only the target rear surface is coated with either C or CaF_2 and the cleaning procedure has been shown to be effective, the carbon and fluorine ions found in our experiment originate unambiguously from the *rear surface* of the targets.

IONIZATION PROCESS AND MODELING

Previous work with ns-laser-produced C and CH plasmas demonstrated that the ion spectra can be used to infer the hot electron temperature. As noted previously, the ionization in those experiments was thermal from the direct laser heating, and T_{hot} was nearly constant throughout the expansion. For hot plasmas, only C^{6+} was present, and the other charge states were populated by recombination during the drift phase after the acceleration, leading to very similar energy spectra for each ionization state. In our experiment, Barrier Suppression Ionization (BSI) in the strong transient electric field should dominate, suggesting that shortly after the laser pulse the ion distribution is stratified — i.e. C^+ closest to the unperturbed surface, followed by C^{2+}, C^{3+} and finally C^{4+} extending to the edge of the expanding plasma front. We expect very different energy spectra for the various charge states, but the interaction is highly dynamic and gets increasingly complex as the system cools down after the pulse. By simultaneous target interferometry, we see that the laser pulse is completely absorbed in the preplasma. The target thickness is chosen such that no prepulse-caused shock-breakout will occur until 5 to 10 ns after the main pulse. This rules out ionization by the laser pulse or by a shock front. The influence of recombination and collisional ionization by the hot electron component ($T_{hot} \sim 2$ MeV) on the charge state distribution is investigated using the FLY-code[9] with time dependent ion density, $n_i(t)$, electron density, $n_e(t)$, and electron temperature $T_{hot}(t)$. Starting with solid state density at the target rear surface, $n_i(t)$ in the adjacent half-space evolves through velocity dispersion and angular divergence; $n_e(t)$ is assumed to rise with the laser pulse up to 2.5×10^{19} cm^{-3} and fall again to $<Z>n_i$ ($<Z>$ is the local average ion charge), assuring quasi-neutrality of the plasma cloud. T_{hot} is estimated from the modelled acceleration fields. The total recombination and collisional ionization rates, integrated along the path to the detector are in the range of 1% and 0.1%, respectively. We also considered the influence of heating by possible "warm" return currents with a temperature $T_{ret} \sim 50$ eV[10], which must balance the hot electron flow. This warm electron component is pushed back by the space charge field as the ions are accelerated. Taking the field build up dynamics as calculated for the hot component, the time a C^+ ion will spend in the warm electron component is ~2 fs. Using the above values in an analytical estimate for the collisional ionization with rate coefficients from Ref. 11, we found that the collisional ionization due to a warm component is negligible. We can therefore establish field ionization as the dominant process for our set of parameters, a result recently confirmed by theoretical analysis in Ref. 12. We have developed a 1-D numerical model which self-consistently calculates the electron dynamics behind the target, assuming Boltzmann equilibrium for the electron temperature and a current distribution which follows the laser pulse. The code includes the ionization process (BSI), the acceleration of different ion species as well as screening effects on the potential. In each time step it solves Poisson's equation for the given charge distribution, checks for ionization events, advances the different ion species, and solves again Poisson's equation for the new charge state distribution. When the first laser-heated electrons enter the vacuum behind the target, they create not a static electric field as in standard TNSA but a highly dynamical situation. When more (and hotter) electrons arrive, the field increases until $E_{stat} = E_1$ at the surface (E_1 is the ionization potential of carbon). At this point, BSI sets in and the produced ions and electrons decrease the field at the surface. An equilibrium is created between the growing external field and the surface ionization rate such that the field at the surface is always kept at the value E_1. The created ions are accelerated into the vacuum (while the freed electrons are pushed into the target). The unshielded field in vacuum keeps increasing, ionizing the outermost ions further, while the zone of highest fields will move into vacuum because the fields are shielded by the ions towards the target as can be seen in Fig. 4.

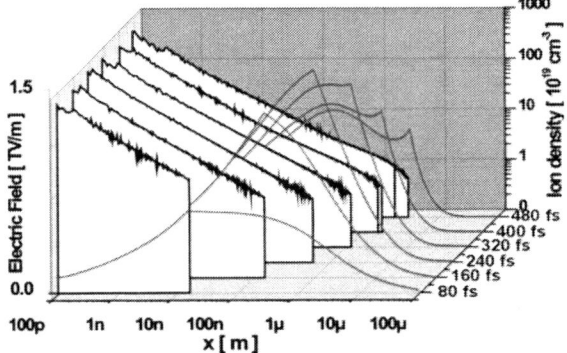

FIGURE 4. Time-space history of the accelerating electric fields (curves) and the ion distribution (white histogram): The region of maximum field moves out into the vacuum.

When the source of hot electrons is on, virtually only the high charge states are created, while the low charge states have to be created after the pulse or in field-fringe-regions requiring 2-D treatment. A model to completely explain the acceleration dynamics is currently under development. Evaluation of the fluorine shot in Fig. 2 shows that E-fields $E_{stat} \sim 2$ TV/m on a time scale of ~ 350 fs are necessary to accelerate F^{7+} ions up to 100 MeV over a scale length of ~ 10 μm. The shot presented in Fig. 2 was virtually without any protons, but the modelled fields can accelerate C^+ up to ~ 25 MeV, as typical with unheated targets.

ACKNOWLEDGMENTS

We thank A. Kemp and H. Ruhl for the fruitful discussions. This work was supported by the EU Programme No. HPRI CT 1999-0052 and in part by grant E 1127 from Région Ile-de-France.

REFERENCES

1. R. Decoste et al., Phys. Rev. Lett. **40**, 34 (1978).
2. C. Joshi et al., Appl. Phys. Lett. **34**, 625 (1979).
3. S. Hatchett et al., Phys. Plasmas **7**, 2076 (2000).
4. E. Clark et al., Phys. Rev. Lett. **84**, 670 (2000).
5. S. Wilks et al., Phys. Plasmas **8**, 542 (2001).
6. M. Roth et al., Rapport LULI 2000, p.13.
7. G. Rusch et al., Nucl. Tracks Radiat. Meas. **19**, 261 (1991).
8. D. Forslund et al. Phys. Rev. Lett. **48**, 1614 (1982).
9. R.W. Lee et al., JQSRT **56**, 535 (1996).
10. L. Gremillet et al., Phys. Rev. Lett. **83**, 5015 (1999).
11. G. Voronov et al. Atom. Dat. Nuc. Data Tables **65**, 1 (1997).
12. V. Tikhonchuk, Phys. Plasmas **9**, 1416 (2002).

Investigation of the Discharge Characteristics of the T6 Hollow Cathode Operating on Several Inert Gases and a Kr/Xe Mixture

I. F. M. Ahmed Rudwan[1] and S.B. Gabriel[1]

[1]*Astronautics Research Group, School of Engineering Sciences, University of Southampton, U.K.*

Abstract. Xenon is currently the propellant of choice for gridded ion thrusters. But in order to make deep space missions feasible, an increase in the Specific Impulse (SI) that these thrusters can achieve is necessary. One method of achieving this is to use a propellant with a lower atomic mass (e.g. argon). However, the feasibility of operating the hollow cathode using these alternative propellants has to be demonstrated. Moreover, interest in decreasing the propellant cost in missions and ground testing (especially life tests) have led to the comprehensive discharge characterisation of several gases that will be presented in this paper. The tests were carried out in diode configuration using a T6 hollow cathode with an enclosed keeper design employing xenon, krypton, argon and a Kr/Xe mix. The discharge initiation tests were undertaken with a view to investigate some of the factors thought to influence the starting potential such as mass flow rate and tip temperature. It was found that, for mass flow rates ranging from 0.2-1.1 mg/s and cathode tip temperatures ranging from 900-1300°C, the breakdown potential was less than 50V for argon, less than 25V for krypton, less than 21V for xenon and less than 35V for the Kr/Xe mix. The discharge initiation results were then compared to those obtained by Fearn et al. [8] with a T5 cathode operating on mercury and with a T6 cathode utilising an open keeper design using xenon propellant [3]. Steady state discharge behaviour was also investigated under a range of operating conditions. Spot to plume mode transitions were observed in argon, krypton and Kr/Xe discharges.

INTRODUCTION

The specific impulse (S.I.) of an ion thruster, for a specific accelerating voltage, is inversely proportional to the square root of the propellant mass, thus for deep space missions, where feasibility rests on increasing the S.I. capability of modern thrusters, an investigation into the use of lighter alternatives to xenon (such as argon and krypton) is necessary. The investigation into alternative propellants also included a krypton/xenon mix in the naturally occurring ratio (12:1, Kr:Xe volumetric ratio) which has the advantage of offering substantial savings in propellant costs for space missions and ground testing (it could offer a 15 fold cost saving when compared to pure Xe and 2-3 fold saving when compared to Kr). The discharge characteristics of these gases were investigated and similar experiments were carried out on Xe for comparison.

This is complemented by an investigation of the discharge initiation phase of ion thruster hollow cathode discharges, which has not been the subject of comprehensive investigation and is not very well understood.

EXPERIMENTAL EQUIPMENT AND PROCEDURE

The Hollow Cathode

In an ion thruster, the hollow cathode is the primary electron production region and has a major influence in determining the efficiency and life expectancy of the thruster. The hollow cathode used in the work reported here is of the same design as that utilised in the T6 Kaufman-type ion thruster [1]. The cathode is essentially a hollow metal tube terminating in a 2mm thick disc with a 1.4mm central orifice. It is manufactured from a single tantalum rod to eliminate the risk of tip-weld related failure. The cathode contains a porous tungsten insert impregnated with a barium-calcium-aluminate mix with the purpose of lowering the work function of the cathode surface. This provides a cylindrical working section 20mm long with an internal diameter of 2mm. To commence operation the propellant gas is fed from the downstream end and a heater is used to raise the cathode body temperature to thermionic emission values, a potential is then applied to the keeper electrode till a discharge is formed, the heater is then switched off, as ion bombardment of the cathode surface supplies the energy input necessary to sustain the electron emission.

Hollow cathodes usually employ a first anode known as the 'keeper' situated a few millimeters from the cathode tip. It serves the dual purpose of initiating the discharge and maintaining it when instabilities are encountered. This wok employed an enclosed-keeper design, which differs from the more common open-keeper configuration in that the keeper completely encloses the inter-electrode space as opposed to having an orificed circular flat plate. The graphite keeper was situated 2.5mm from the cathode tip and had a central orifice of 4mm diameter.

Apparatus

The experiments were carried out in diode configuration. The vacuum system and propellant feed subsystem is essentially the same as the one constructed by Edwards [2]. It achieves a high vacuum, with a background pressure in the 10^{-7}mbar range, which is required for the operation of the hollow cathode to prevent oxygen poisoning of the cathode insert. The anode was made of a stainless steel plate of 100mm diameter situated 40mm from the cathode tip, with a 50mm diameter hole which allows an unobstructed view of the cathode. The cathode tip is observed with an optical pyrometer via a quartz view port and a mirror, facilitating non-intrusive thermometry. The pyrometer was later post-calibrated by an R-type thermocouple inserted into the orifice of the cathode while not in operation; the heater was used to raise the cathode temperature during calibration.

Commercial power supplies were used for the main discharge, the heater and the keeper discharge. These were operated under current controlled mode. Discharge and keeper voltage variation with discharge conditions under steady state investigation as well as the breakdown voltage dependence on cathode conditions were monitored by a Tektronix TDS410A digitizing oscilloscope, with 200MHz bandwidth and a sampling rate of 100MS/s.

Experimental Procedure

The two main independent variables governing hollow cathode discharge initiation were identified to be tip temperature (T) and propellant mass flow rate ($\dot{m}$). Experiments were carried out by varying one variable while keeping the other constant and recording the breakdown voltage (V_b). In this manner the breakdown behaviour of Xe, Kr, Ar and Kr/Xe mix was investigated in the range of 0.2-1.1mg/s flow rate and 900-1300°C tip temperature.

The main independent variables determining the operating characteristics of a hollow cathode are the propellant flow rate ($\dot{m}$) and the discharge current (I_a), with the anode and keeper potential being the dependent variables. The propellant flow rate was investigated at 5A discharge current in the range 0.2(0.147) - 3.61(2.657)mg/s (Aequiv) for Xe, 0.8(0.92) - 4.2(4.83) mg/s (Aequiv) for Kr, 0.83(0.92) - 4.38(4.83)mg/s (Aequiv) for Kr/Xe and 0.8(1.932) - 3.5(8.453)mg/s (Aequiv) for Ar. The flow rate was increased every minute till a gas specific maximum value was reached and then the flow rate was brought down again, repeating the same procedure in reverse. Dependence on discharge current was also investigated employing a similar procedure by varying I_a in steps of 0.5A every minute in the range 0.5 - 5A while keeping $\dot{m}$ constant, for several values of propellant flow rate.

RESULTS AND DISCUSSION

Discharge Initiation Results

Paschen's Law describes almost all gaseous breakdown behaviour. It is well defined for plane parallel electrodes with a uniform pressure profile. A consequence of Paschen's Law is that a unique minimum breakdown voltage exists for each gas at a certain (pd) value known as the Paschen minimum (where p is the gas pressure and d is the distance between the electrodes).

Hollow cathode breakdown differs from the simple, ideal case in several ways: 1)The more complex electrode geometry. 2)The high pressure gradient between the electrodes, with the pressure in the cathode vicinity being approximately two orders of magnitude greater than at keeper. 3)The heated thermionically emitting insert surface. 4)The high flow velocity [3].

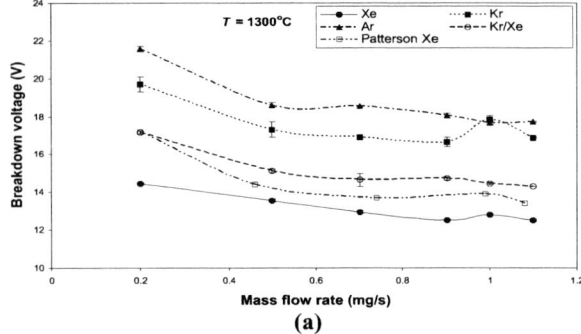 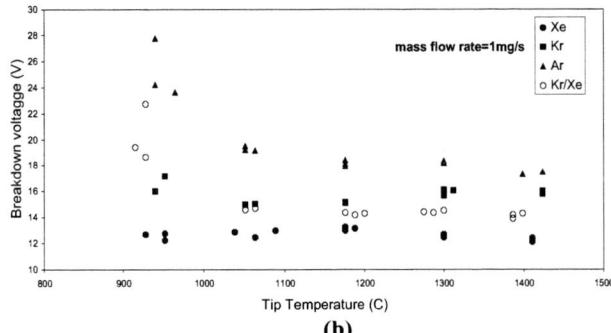

FIGURE 1. (a) V_b as a function of $\dot{m}$ at $T=1300°C$ for Xe, Kr, Ar and Kr/Xe, also included for comparison are the T6 open keeper results [3]. (b) V_b as a function of T at $\dot{m}=1$mg/s for Xe, Kr, Ar and Kr/Xe.

Figure 1(a) compares the breakdown voltage dependence on $\dot{m}$ for Xe, Kr, Ar and Kr/Xe mix. From computer simulations carried out in the University of Southampton's Astronautics Research Group [4] it was seen that the pressure in the cathode vicinity (due to the rapid pressure drop downstream this is the region that determines the (*pd*) integral across the gap) is a linear function of $\dot{m}$. This coupled with the fact that the cathode-keeper separation was kept constant throughout the tests leads to the conclusion that the $\dot{m}$ vs. V_b plots will approximate the Paschen breakdown curves for the above gases.

It can be seen from figure 1(a) &(b) that V_b values are arranged, as would be expected, in the same order as the first ionization potentials of the gases: with Xe lowest followed by the Kr/Xe mix, then Kr and finally Ar having the highest breakdown voltages for a given mass flow rate and tip temperature. Also displayed in figure 1(a) for comparison are the results for xenon breakdown in the open keeper case [3]; they show slightly higher breakdown voltages, this is thought to be due to the higher inter-electrode pressure expected with the enclosed-keeper.

All the gases in figure 1(a) seem to exhibit anomalous breakdown behaviour; exhibiting what seems to be a double minimum instead of the expected one (this is clear in the figure only in the Ar and Kr/Xe case but must also occur at higher $\dot{m}$ for the other gases as V_b is expected to increase at higher $\dot{m}$ due to the decreasing electron mean free path not allowing primary electrons to gain enough energy for ionization). This anomalous behaviour was also exhibited by the open keeper cathode [3]. It is suggested here that an explanation can be given by the 'Penning Effect'.

Penning [5] showed that a rare gas with a very small amount of admixed gas molecules (O~10^{-3}%), satisfying the condition that the energy of the metastable state of the main rare gas exceeds the ionization energy of the admixed molecules, will experience 'Penning ionization' of the admixed gas by the metastable atoms. This leads to a departure from Paschen's law which manifests itself as a Paschen curve with a double minimum (This was observed by Penning in a Ne+10^{-3}%Ar gas), the first minimum corresponds to that due to direct ionization of the rare gas atoms by electrons (similar to the minimum in the pure rare gas), while the second minimum corresponds to the increase in the ionization coefficient due to Penning ionization of the admixed molecules by the metastables.

Barium (with an ionization potential of 5.2eV) satisfies all the above criteria. Its ionization potential is lower than the potential energy of the metastable states of the noble gases (8.28eV and 9.4eV for Xe, 9.86eV and 10.51eV for Kr and 11.49eV and 11.66eV for Ar). Free barium is formed when the cathode insert is activated by heating, some of the barium also evaporates from the cathode surface. Brodie and Jenkins [6] demonstrated and measured its evaporation rate from porous tungsten inserts. The presence of barium is also confirmed by BaII emission lines detected in the T6 discharge in the spectroscopic part of this characterisation work [7]. Thus the presence of barium is thought to be the reason behind departure from Paschen's law in ion thruster hollow cathodes.

V_b however shows weak dependence on T over the range investigated, with virtually no change in the value of V_b for all the gases for tip temperatures above 1000°C (see figure 1(b)).

In general the breakdown behaviour of the T6 thruster was found to be more reproducible than those reported for the T5 [8] (operating on mercury propellant) where discharge initiation fell at random within ranges of the order of hundreds of volts. This improvement is attributed to the increased orifice size in the T6 cathode which is thought to facilitate the penetration of the electric field into the cathode cavity [3].

Steady State Characterisation Results

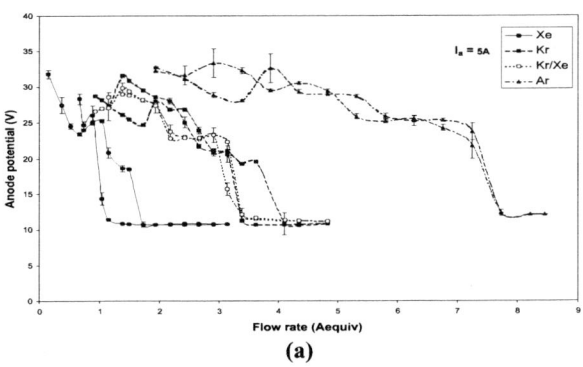 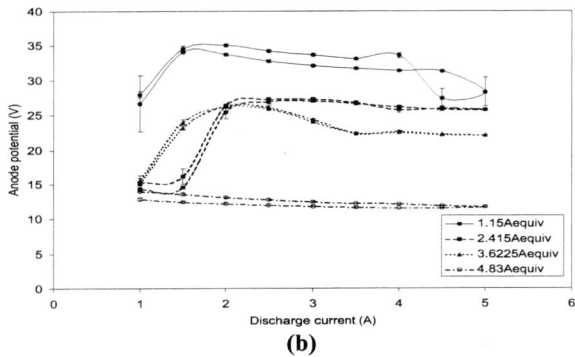

(a) (b)

FIGURE 2. (a) Discharge voltage dependence on $\dot{m}$ at 5A discharge current for Xe, Kr, Kr/Xe and Ar discharges. (b) Discharge voltage dependence on I_a for a Kr/Xe discharge for several flow rates.

Figure 2(a) shows discharge voltage dependence on $\dot{m}$ and demonstrates clear spot to plume transitions for Xe, Kr, Kr/Xe and Ar propellants. Plume to spot transitions were experienced, with increasing $\dot{m}$, at 1.73Aequiv for Xe, 3.37Aequiv for Kr/Xe, 4.11equiv for Kr and at 7.78Aequiv for Ar. Note here that the Kr/Xe mix showed discharge behaviour that is superior to Kr and Ar, the discharge potentials experienced in spot mode are only slightly (<1.5V) greater than those for Xe.

Figure 2(b) shows Kr/Xe discharge dependence on I_a at several $\dot{m}$. In the curve representing $\dot{m}$=4.83Aequiv the discharge is in spot mode, while for $\dot{m}$=1.15, 2.415 and 3.6225Aequiv the discharge is in plume mode. Qualitatively similar results were obtained for V_a dependence on I_a for the other gases at different values of $\dot{m}$. The curves exhibit hysteresis, this can be explained by the discharge not achieving thermal equilibrium within one minute and thus the value of V_a at the next setting is dependent on previous discharge history,.

CONCLUSIONS

This work characterised alternative propellants to Xe for use in Kaufman-type ion thrusters. The results for the Kr/Xe mix were especially encouraging as it can offer significant savings in propellant cost especially in hollow cathode life-tests. Discharge initiation of hollow cathodes was also investigated using inert gas propellants. The xenon results were in agreement with those from Patterson [3]. All the gases exhibited anomalous breakdown behaviour, which is attributed here to Penning ionization of barium atoms by metastable rare gas atoms. The effects of the Penning ionization mechanism will not be limited to the starting phase of hollow cathode discharges and is expected to have an influence on its steady state behaviour as well.

REFERENCES

1. Fearn, D. G., and Patterson, S. W., "The Hollow Cathode-A Versatile Component of Electric Thrusters". ESA Publications, vol.465, pp587-94, 2000
2. Edwards, C. H., Ph.D. thesis, Dept. of Aeronautics & Astronautics, University of Southampton, U.K., 1997.
3. Patterson, S. W., Jugroot, M., and Fearn, D. G., "Discharge Initiation in the T6 Thruster Hollow Cathode". AIAA-2000-3532, 2000.
4. Crawford, F. T. A., and Gabriel, S. B., "Modelling Small Hollow Cathode Discharges for Ion Microthrusters". AIAA-2002-2101, 2002.
5. Penning, F. M., "Anomalous Variations of the Sparking Potential as a Function of $p_o d$". Proc. Royal Ac.Sc., 34, 1305, 1931
6. Brodie, I., Jenkins, R. O., and Trodden, W. G., "Evaporation of Barium from Cathodes Impregnated with Barium-Calcium-Aluminate". J.Electron. & Control., 6, pp. 149-161, 1959.
7. M. Ahmed Rudwan, I. F., Ph.D. Transfer thesis, Astronautics Research Group, University of Southampton, U.K., 2002
8. Fearn, D. G., Cox, A. G., "An Investigation of the Initiation of Hollow Cathode Discharges". RAE Tech. Rept. 76054, 1976

An Ion Heating Experiment in a Supersonic Plasma Flow

<u>Akira Ando</u>, Ryosuke Kumagai, Shinya Fujimura, Tsuyoshi Yagai, Kunihiko Hattori and Masaaki Inutake

*Department of Electrical Engineering Tohoku University,
Aoba05, Aramaki, Aoba, Sendai, 980-8579, JAPAN*

Abstract. Ion heating experiments are performed in a fast flowing plasma with an ion Mach number of nearly unity. RF waves with an ion cyclotron range of frequency was excited by a pair of loop antennas located at a divergent magnetic nozzle. Increase of plasma thermal energy $W_\perp$ measured by a diamagnetic coil is observed when the waves are excited with various azimuthal mode numbers in several magnetic nozzle configurations. It is most effective to heat ions of plasma flow to excite the waves with an azimuthal mode number of $m=\pm1$. The heating efficiency is larger in the magnetic beach configuration than that in the uniform one. A dependence of the ratio $\Delta W_\perp/W_\perp$ on the magnetic field strength shows no clear indication of the ion cyclotron resonance region of thermal ions. It should be caused by the Doppler effect of fast flowing ions.

INTRODUCTION

Recently a plasma flow is found to play an important role in space and fusion plasmas, such as violent activities near the solar surface and jet formation observed in an active galactic nuclei [1]. High-speed flow effect on stability and transport of plasmas is also a current topic in fusion plasmas [2]. Intensive researches to develop a fast flowing plasma with high particle flux and high heat-flux are required for the purpose of basic plasma-physics as well as various industrial and space applications. An electric propulsion system is one of the key elements in future space exploration project and has been developed for various space missions [3]. Development of a high power density plasma thruster with a higher specific impulse and a larger thrust is desired for a manned interplanetary space thruster.

A magneto-plasma-dynamic arcjet (MPDA) is one of the representative devices for the space thruster and is also utilized as a supersonic plasma flow source. The MPDA plasma is accelerated axially by a self-induced electromagnetic force, $F_z=J_r \times B_\theta$, where J_r is a radial discharge current and B_θ is self-induced azimuthal magnetic field. Thrust efficiency of the MPDA is expected to increase by applying an externally-applied magnetic nozzle instead of a solid nozzle because of the addition of j×B rotational and magnetic-nozzle accelerations [4-8]. In order to increase the direct momentum of plasma flow, it is one of the promising methods that the bulk energy of the plasma is added and its azimuthal energy is converted to axial momentum in a magnetic nozzle.

Plasma acceleration in a divergent magnetic nozzle has been successfully demonstrated in the HITOP device, where the transonic plasma flow (M_i is nearly unity) converts into a supersonic one (M_i increases up to 3) through a divergent magnetic nozzle where no j×B acceleration is exerted [9].

Ion heating in a magnetized plasma has been precisely investigated both theoretically and experimentally in many researches. Recently, ion heating in a fast-flowing plasma attracts much attention in view of advanced high beta plasma confinement and the advanced electric propulsion system. In the Variable Specific Impulse Magnetoplasma Rocket (VASIMR) project, the combined system of the ion cyclotron heating and the magnetic nozzle is proposed and tested to modulate a ratio of specific impulse to thrust at constant power [10].

So far, few attempt of direct ion heating to fast-flowing plasmas by radio-frequency waves has been done, where the condition of ion cyclotron resonance is drastically changed by the effect of Doppler shift. In this experiment we have performed an ion heating experiment in a supersonic plasma flow produced in the HITOP device. Axisymmetric (m=0) and non-axisymmetric (m=1 and 2) mode waves near the ion cyclotron frequency are launched

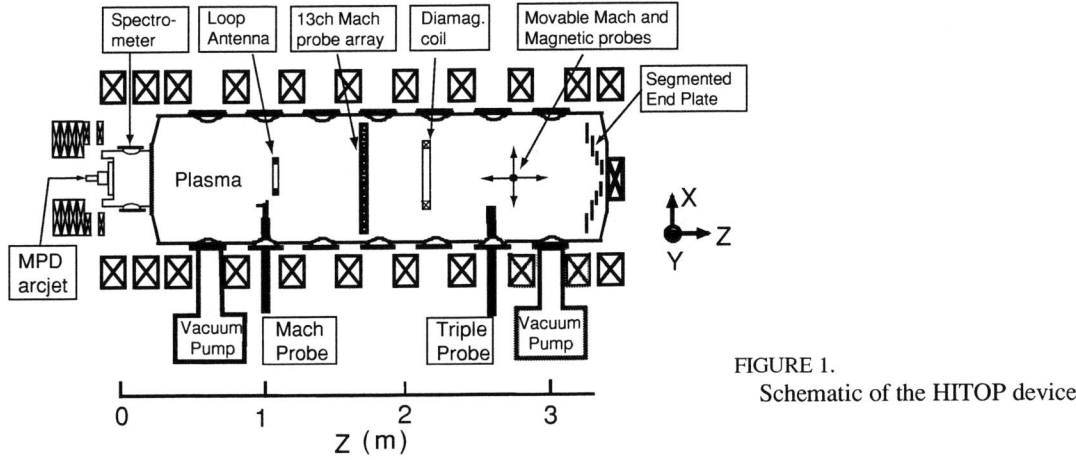

FIGURE 1. Schematic of the HITOP device

by a pair of loop-type antennas set in a diverging magnetic field. It is found that the plasma thermal energy measured by a diamagnetic loop coil increases when the waves are launched as a beach-heating configuration.

In this paper are reported characteristics of a plasma flow produced and results of ion heating experiments. The heating effects are compared in various wave excitation mode numbers and in several magnetic nozzle configurations.

EXPERIMENTAL APPARATUS

A high power, quasi-steady MPDA device is installed in the HITOP device of Tohoku University. The HITOP device consists of a large cylindrical vacuum chamber (diameter D = 0.8 m, length L = 3.3 m) with eleven main and six auxiliary magnetic coils, which can generate a uniform magnetic field up to 0.1 T, as shown in Fig.1. Adjusting the external coil current can form various types of magnetic field configurations.

The MPDA, which is installed at one end port of the HITOP device, has a coaxial structure with a center tungsten rod cathode (10mm in diameter) and an annular molybdenum anode (30mm in diameter). A discharge current I_d up to 10kA is supplied by a pulse-forming network (PFN) system with the quasi-steady duration of 1ms. The current I_d is kept nearly constant during the discharge with a typical voltage of 200V-300V and can be controlled by varying the charging voltage of capacitor banks of the PFN power-supply. It can generate a high density (more than $10^{20} m^{-3}$) and high Mach number (M_i up to 3) plasma flow in an axial magnetic field (B_z up to 1kG)[8,9,11,12].

A pair of loop antennas with 60mm in diameter is used for the wave excitation [13]. Employing the Faraday shield reduces electrostatic coupling between the antenna and plasma. The antenna current is supplied by a pulsed oscillation power-supply, which consists of a condenser and a gap-switch. The antennas are set at Z=1.0m downstream from the MPDA. The azimuthal mode number of the exciting wave can be changed by adjusting the combination of the antennas, as shown in Fig.2. Plasma flow characteristics are measured by several diagnostics installed on the HITOP device. Electron temperature and density profiles are measured by a movable triple probe and a fast-voltage-scanning Langmuir probe. The plasma thermal energy is measured by a diamagnetic loop coil located at Z=2.2m. A plasma flow is characterized by an ion Mach number M_i which is defined as ratio of the plasma flow velocity to ion acoustic velocity. Profiles of M_i and plasma density along and across the field lines are

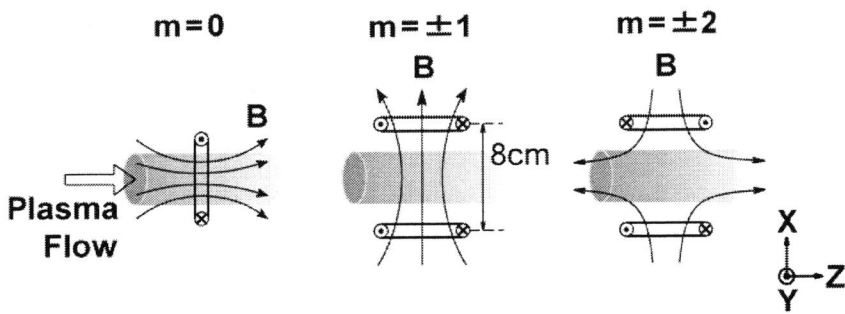

FIGURE 2. Three types of antenna setting to excite the waves with azimuthal mode number m=0,±1, ±2.

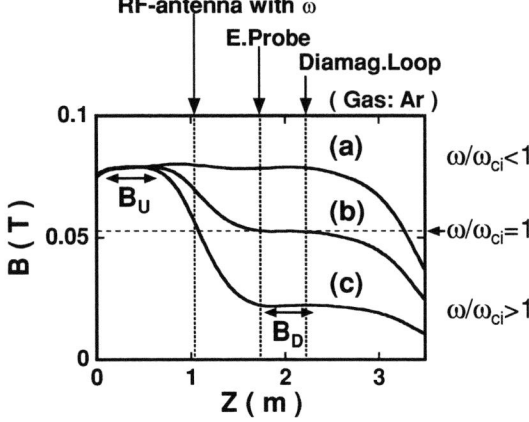

FIGURE 3. Three types of magnetic field configurations. The wave is excited in $\omega/\omega_{ci}<1$ region and propagates downward approaching to the region of (a) $\omega/\omega_{ci}<1$, (b) $\omega/\omega_{ci}=1$, (c) $\omega/\omega_{ci}>1$. The antenna is located at Z=103cm and The diamagnetic coil is located at Z=223cm. Electrostatic probe position is also shown.

measured by a movable Mach probe and an array of 13-channel Mach probes set at 1.7m downstream of the MPDA outlet in the HITOP [14].

EXPERIMENTAL RESULTS

Experiments are performed in various magnetic field configurations as shown in Fig.3. The upstream magnetic field B_U is kept constant and the downstream one B_D is varied to form a magnetic beach configuration. In this experiment an argon gas is used due to the relatively low excitation frequency of about 20kHz. Figure 4 shows typical waveforms of discharge current I_d, the antenna current I_{RF} and observed diamagnetic signal $W_\perp$ under the condition of m=±1 in the wave excitation mode. Though the antenna current damps rapidly because of the lack of power supply capability, the diamagnetic coil signal increases during the excitation.

In the present experiments, the electron and ion temperature are almost equal at about 2eV and the density is $2\times10^{19} m^{-3}$. Plasma flow velocity can be estimated to be about 6km/sec by the Mach probe measurement and time delay of ion saturation current signal. It takes about 0.2ms for the Ar plasmas to flow from the antenna position to the diamagnetic coil position. The $W_\perp$ signal, however, increases without such delay time as shown in Fig. 4 (c), which indicates that the waves propagate and absorbed in the downstream region. The wave phase velocity is about 10^5 m/s, which is measured by the phase shift of two magnetic probe signals located at axially-different positions.

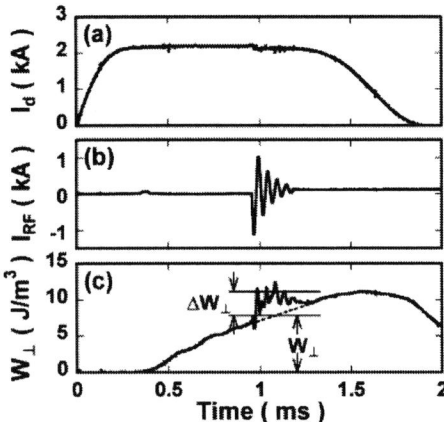

FIGURE 4. Temporal evolutions of (a) discharge current, (b) antenna current, (c) diamagnetic signal. The m=±1 mode waves are excited in the magnetic configuration of case (b) in Fig.3.

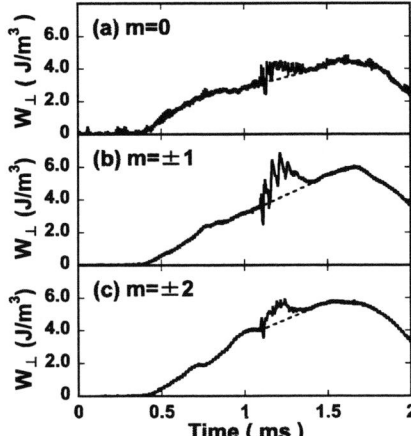

FIGURE 5. Temporal evolutions of diamagnetic signals in the three types of azimuthal mode number, (a) m=0, (b)m=±1, (c)m=±2. The magnetic configuration is the case (b) in Fig.3.

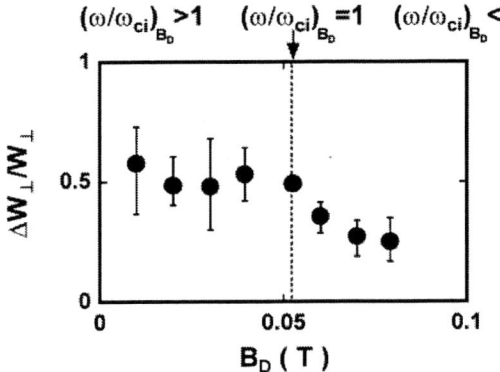

FIGURE 6. The increment ratio of diamagnetic signal as a function of B_D.

We varied an azimuthal mode number of the excited waves by changing the coil arrangement. Figure 5 shows the observed temporal evolutions of diamagnetic signal with three types of azimuthal mode number, m=0,±1, ±2. The increment ratio of the thermal energy ratio $\Delta W_\perp/W_\perp$ is large in the case of m=±1. The dependence of $\Delta W_\perp/W_\perp$ on the magnetic field strength in the downstream region B_D is shown in Fig.6. The heating efficiency is larger in the magnetic beach configuration than that in the uniform one. But, the dependence on the magnetic strength shows no clear indication of the cyclotron resonance region. It should be caused by the Doppler effect of fast flowing ions.

CONCLUSIONS

Ion heating experiments are performed in a fast-flowing plasma, which is produced by an MPDA in the HITOP device. Axisymmetric (m=0) and non-axisymmetric (m=±1and ±2) mode waves near the ion cyclotron frequency are launched by a pair of loop-type antennas set in a diverging magnetic nozzle. Plasma thermal energy $W_\perp$ was measured by a diamagnetic loop coil in the downstream region of the antenna.

We have varied the azimuthal mode number of wave excitation and the magnetic field configuration. It is found that the azimuthal mode number of wave excitation m=±1 is most effective to heat ions of the plasma flow. The heating efficiency is large in the magnetic beach configuration. But, the dependence on magnetic field strength shows no clear indication of the ion cyclotron resonance region of thermal ions. It should be caused by the Doppler effect of fast flowing ions.

ACKNOWLEDGMENTS

This work was supported in part by Grant-in-Aid for Scientific Researches from Japan Society for the Promotion of Science. Part of this work was carried out under the Cooperative Research Project Program of the Research Institute of Electrical Communication, Tohoku University.

REFERENCES

1. Meier D.L., et.al., Science, 291, (2000), pp.84
2. Mahajan S.M. and Yoshida Z., Phys. Rev. Lett., 81, 4863 (1998).
3. Jahn R. G., Physics of Electric Propulsion, McGRAW-HILL, NewYork, 1968, pp. 196-256.
4. Kuriki K. and Inutake M., Phys. Fluids, 5, 92 (1974).
5. Sasoh M. and Arakawa Y., , J. Propulsion and Power, 11, 351 (1995).
6. Tahara H., Kagaya Y., and Yoshikawa, T., J. Propulsion and Power, 13, 651 (1997).
7. Schoenberg K.F., et.al., Phys. Plasmas, 5, 2090 (1998).
8. Inutake M., et.al., Proc. 11th Int. Congress on Plasma Physics, (Sydney, 2002), ICPP-2002, paper No.481.
9. Inutake M., et.al., Proc. 10th Int. Congress on Plasma Physics and 42nd APS Meeting, (Quebec, 2000), ICPP-2000, Vol.1, pp.148
10. F.R.ChangDiaz, *et.al.* Proc. of 36th Joint Propulsion Conf., (Huntsville,2000), **AIAA-2000-3756**,(2000). pp. 1-8.
11. Ando A., et.al., J. Plasma Fusion Res. SERIES, 4, 373 (2001).
12. Tobari H., et.al., Proc. 11th Int. Congress on Plasma Physics, (Sydney, 2002), ICPP-2002, paper No.475.
13. Yagai T., et.al., Proc. 11th Int. Congress on Plasma Physics, (Sydney, 2002), ICPP-2002, paper No.482.
14. Ando A., et.al., Proc. of 25th Int. Conf. on Phenomena in Ionized Gases, (2001,Nagoya) Vol. 2, pp.195.

Plasma Fluctuations in an Applied Field MPD Thruster

V. Antoni[2], M. Bagatin[1,2], G. Serianni[2], N. Vianello[1,2], M. Zuin[1,2], F. Paganucci[3], P. Rossetti[3], M. Andrenucci[3]

[1]*Dipartimento Ingegneria Elettrica, via Gradenigo, 6a, 35131 Padova, (Italy)*
[2]*Consorzio RFX, Associazione Euratom-Enea sulla Fusione, C.so Stati Uniti 4, I-35127 Padova (Italy)*
[3]*Centrospazio, Via A.Gherardesca 5, 56014 Pisa, (Italy)*

Abstract. The mean values of electron temperature and density in the plume of a gas-fed, externally applied field, Magneto-Plasma Dynamic (MPD) thruster, have been measured with a balanced triple probe. The spatial distribution and the temporal behaviour of floating potential fluctuations have been also investigated using multielectrode probes. The addition of an external magnetic field results in an increase of the fluctuation levels of electron temperature, density and floating potential. Mean values in the anode region at the thruster outlet have been related to the Hall parameter to understand the Hall current's contribution to the thrust. Spectral analysis of the floating potential fluctuations shows the importance of the external magnetic field in the dynamic of plasma instabilities.

INTRODUCTION

Magneto-Plasma-Dynamic (MPD) thrusters are electromagnetic plasma accelerators based on the principle of low-pressure arc jets. Due to their high values of exhaust speed and thrust density, they represent a promising propulsion option for high power, primary space missions, ranging from orbit raising to interplanetary transfers of large spacecrafts. Their performance can be furtherly improved by adding an external magnetic field, whose effect is to stabilize the plume and increase the thrust, with a relevant increase of the thrust efficiency. An applied field MPD thruster, called Hybrid Plasma Thruster (HPT), has been recently developed1. On this thruster, it is generally found that the improvement due to the applied field is considerably large only at low current regimes. No significant increase of the thrust is observed when the current goes beyond a critical value, that in some cases corresponds to the condition of full ionization of the injected propellant. For currents higher than the full ionization current a further degradation of performance is normally observed. These phenomena, generally referred as onset phenomena, are macroscopically evidenced by the appearance of large fluctuations of the measured anode-cathode voltage.

THE EQUIPMENT

The HPT[2] is a coaxial device with a central hollow cathode 20 mm in diameter. The anode consists of an aluminum cylinder 200 mm in inner diameter and of eight copper straps, which divide a central chamber from a peripheral ionization chamber. The thruster is powered by a pulse forming network, capable of delivering up to 15 kA for a time duration of 2.5 ms. A constant mass flow rate of 660 mg/s of argon propellant is injected, mostly ($\cong$ 90 %) through the central cathode, while the remaining 10 % is injected in the peripheral ionization chamber. Tests were performed both with and without an external magnetic field, B_{ext}, generated on the thruster axis (z-direction) by a 70-turns coaxial coil. Different power supply conditions have been explored corresponding to plasma current (here defined $I_{discharge}$) values from 4 to 8 kA.

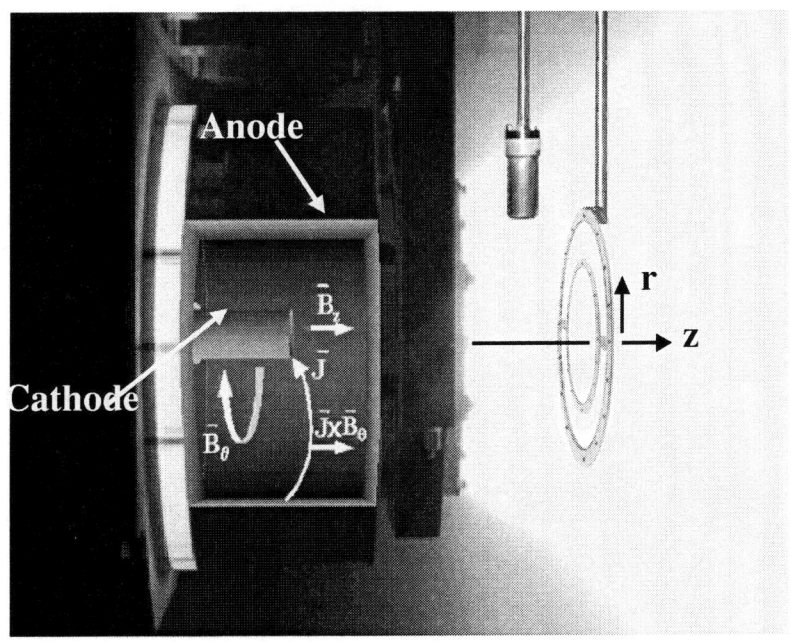

FIGURE 1. Schematic of the experimental arrangement, with both the diagnostic systems inserted in the plume of the thruster.

Mean electron density and temperature have been measured by using a diagnostic system called "rake probe"[3], housing 7 aligned graphite electrodes in a boron nitride case, which was deployed in the plume as illustrated in Fig. 1. The probe was configured as a five-pin balanced triple probe[4], giving the local values of electron density and temperature at the same time. The measurements have been performed with the tip electrode located at r=115 mm off the HPT axis. The space and time correlations of floating potential fluctuations in the plume have been investigated by a proper fluctuation probe instrument[5]. It consists of 32 electrodes, arranged on two concentric rings of respectively 60 and 100 mm radius, placed coaxially on the thruster axis, as shown in Fig. 1. Both the rake and the fluctuation probe systems were placed at a distance of 80 mm from the thruster outlet.

EXPERIMENTAL RESULTS

It was found that the mean thrust values with no external magnetic field result in fairly good agreement with the classical evaluation considering both the interaction of the radial arc current with the self-generated azimuthal field and the so called "electromagnetic pumping" process. The external magnetic field provides an additional contribution to the thrust, which is mainly generated by the interaction of its radial component with the azimuthal electron Hall current. Specific attention has been therefore dedicated to the mean plasma parameters in this crucial region. Fig. 2a) reports the mean electron density and temperature at r=115 mm off the axis. Fig. 2b), where the relative increase of the thrust values $(T_B-T_0)/T_0$ are reported (T_B and T_0 are the thrust values with and without an external magnetic field of 40 mT), clearly shows that the additional thrust provided by the applied field is progressively vanishing at increasing current[2]. The Hall interaction is mainly localized on the anode region at the thruster outlet, where the radial B_{ext} component is more important. The relative importance of the Hall current component to the total thrust may be thus conveniently evaluated in terms of the electron Hall parameter, defined as the ratio of the electron gyro frequency to its collision frequency, computed in the anode region.

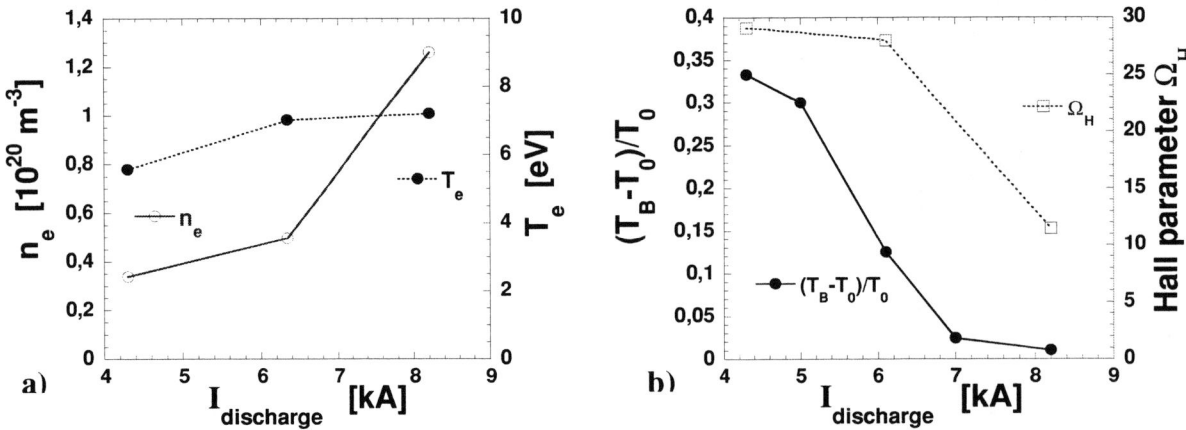

FIGURE 2. a) Electron density and temperature at r=115 mm as a function of current with B_{ext}=40 mT; b) Computed Hall parameter at r=115 mm and relative increase of thrust as a function of current with B_{ext}= 40 mT.

At sufficiently high ionization levels, the electron collision frequency is dominated by electron-ion interaction and the Hall parameter, assuming $n_e \cong n_i$, can be estimated as[6]:

$$\Omega_H = \frac{\omega_B}{\nu_e} \approx 7 \times 10^{21} \frac{BT_e^{3/2}}{n_e} \qquad (1)$$

where T_e is in eV and B is the external magnetic field. The computed Hall parameter in the anode region, plotted in *Fig. 2b)*, drops at higher currents, confirming the expected decreasing role of the Hall current contribution.

The thrust performance degradation at high $I_{discharge}$ was found as being related to the growth of fluctuation amplitudes of global parameters[3]. The analysis of the spectral properties are here focused on the floating potential collected by the annular probe system located at r=60 mm.

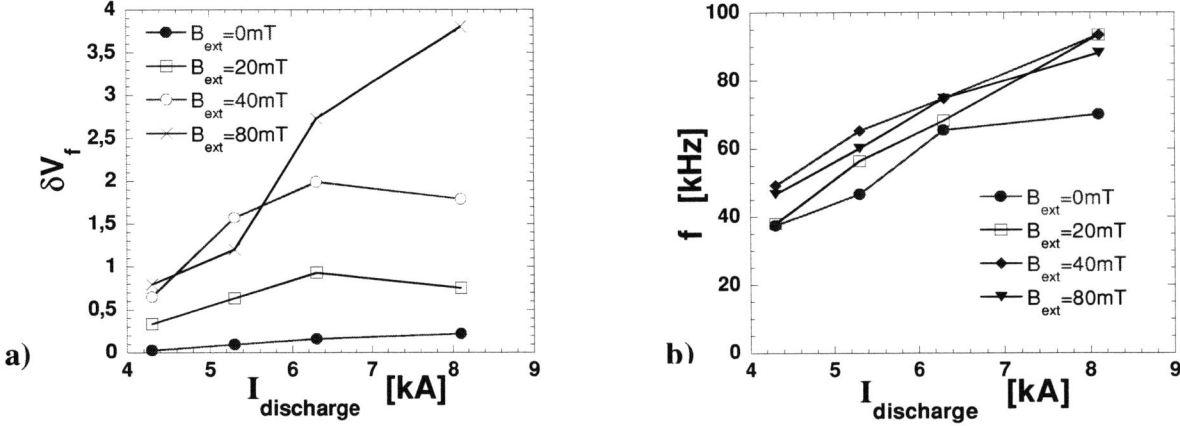

FIGURE 3. Signals rms values *a)* and dominant frequency *b)* dependence on current and applied magnetic field at r=60mm.

The root mean square amplitude of the signals increases with the current, and grows to higher values when the external magnetic field is applied (*Fig. 3a)*). Without external magnetic field the power spectra of the signals show a clear peak whose frequency increases almost linearly with the current from about 40 kHz at $I_{discharge}$ = 4.2 kA up to 70 kHz at 8.2 kA. The application of B_{ext} causes a slight increase of the dominant frequency up to 90 kHz at 8.2 kA, while the dependence on the current doesn't seem strongly affected (*Fig 3b)*). The signals correlation has been studied by Fourier decomposition in the azimuthal direction. The results of the mode dynamics analysis are shown in Fig. 4 in three different conditions. At low $I_{discharge}$ in absence of B_{ext} no phase difference has been measured, then only a global m=0 mode can be seen (*Fig. 4a)*). The application of the external magnetic field causes the development of an azimuthal m=1 mode, that implies a wavelength λ=37.7 cm, at r=6 cm (*Fig. 4b)*). Higher order modes are also evident and they seem to alternate in time with the m=1 mode with a frequency increasing with current, as can be seen by comparing *Fig. 4b)* and *Fig. 4c)*. The frequency peak described above can be assigned to the dominant m=1 mode, while no clear frequency can be associated to higher order modes.

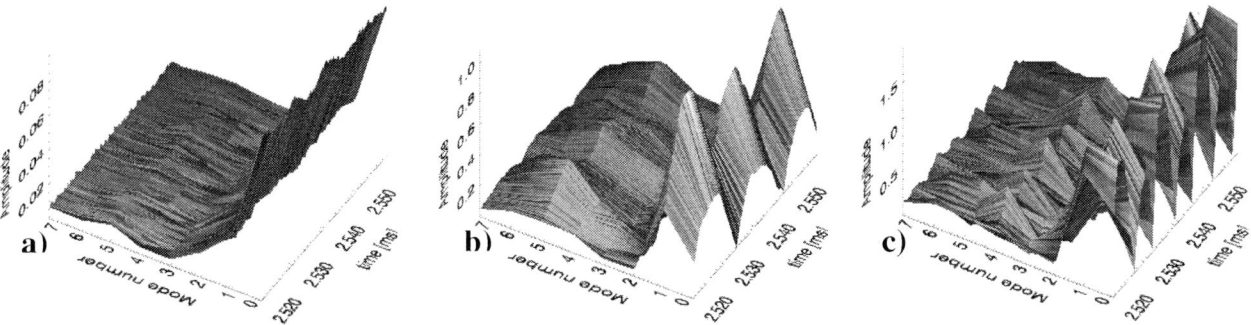

FIGURE 4. Mode amplitude time evolution of floating potential of the probes located at r= 60 mm with: a) B_{ext}=0, $I_{discharge}$=4.3 kA; b) B_{ext}=40 mT, $I_{discharge}$ = 4.3 kA; c) B_{ext}=40 mT, $I_{discharge}$=6.2 kA.

CONCLUSIONS

The mean parameters measurements confirmed that the thrust degradation in HPT at high currents is directly related to a reduced role of the Hall current associated to magnetized electrons in the anode region. The floating potential signals exhibit a pulsing m=1 azimuthal mode which appears either as the power is increased and mainly as a consequence of the application of the external magnetic field. This mode number is associated to a well defined frequency that increases with the current.

REFERENCE

1. V.B.Tikhonov et al., "Investigation on a new type of MPD Thruster", OR21, *27th EPS Conference on Controlled Fusion and Plasma Physics*, 12-16 June 2000, Budapest Hungary
2. F.Paganucci et al., "Performance of an Applied field MPD thrusters", *27th International Electric Propulsion Conference*, October 14-19 2001, Pasadena, CA
3. G.Serianni et al., "Plasma diagnostics on a Applied Magnetic Field MPD thrusters", *27th International Electric Propulsion Conference*, October 14-19 2001, Pasadena, CA
4. H.Y.Tsui et al., *Rev.Sci.Instr.* **63**, 4608 (1992)
5. N. Vianello et al.," Electrostatic fluctuations in a Magneto-Plasma-Dynamics (MPD) Thruster", *Proc. 29th EPS Conf. on Plasma Physics and Controlled Fusion*, 17-21 June 2002, Montreux, Switzerland, paper IDP 4.026
6. R.G. Jahn, *Physics of Electric Propulsion*, McGraw Hill, New York, 1966

Magnetic-Laval-Nozzle Effect on a Magneto-Plasma-Dynamic Arcjet

Masaaki Inutake, Hiroyuki Miyazaki, Kyohei Yoshino,
Hiroyuki Tobari, Kunihiko Hattori and Akira Ando

*Department of Electrical Engineering Tohoku University,
Aoba05, Aramaki, Aoba, Sendai, 980-8579, JAPAN*

Abstract. A magneto-plasma-dynamic arcjet (MPDA) is one of the promising candidates for a manned interplanetary space thruster with a higher specific impulse and larger thrust. An MPDA with an externally-applied magnetic nozzle is investigated to improve the thrust efficiency. From spectroscopic measurements of the MPDA plasma, it is found that with the increase in the discharge current not only the flow velocity but also the ion temperature increase near the muzzle of the MPDA with a uniform axial magnetic field and so the ion acoustic Mach number is limited to a value less than unity. By installing a Laval-type magnetic nozzle near the muzzle of the MPDA, the subsonic flow is successfully accelerated to a supersonic one by converting the ion thermal energy to the flow energy. The results are compared with the prediction by a 1-D isentropic flow model.

INTRODUCTION

Production of a high-beta, supersonic plasma flow is quite important for basic researches on MHD phenomena in space and fusion plasmas. A magneto-plasma-dynamic arcjet (MPDA) is one of the promising devices to produce a supersonic plasma flow and is utilized as an electric propulsion with a higher specific impulse and relatively larger thrust[1]. The MPDA evoked a renewed interest for a manned interplanetary space thruster.

The MPDA plasma is accelerated axially by a self-induced electromagnetic force, $F_z = J_r \times B_\theta$, where J_r is a radial discharge current and B_θ is self-induced azimuthal magnetic field. Thrust efficiency of the MPDA is expected to increase by applying an external applied magnetic nozzle instead of a solid nozzle, though acceleration mechanism becomes very complicated due to the addition of $J_r \times B_z$ rotational, $J_\theta \times B_r$ Hall and magnetic-nozzle accelerations[2-5].

In an externally-applied divergent magnetic nozzle configuration with a longer characteristic length, a supersonic plasma flow with an ion acoustic Mach number up to 3 has been successfully obtained in the far downstream region of the MPDA, where no $J_r \times B_\theta$ acceleration is exerted. In a uniform axial magnetic field configuration, on the other hand, the Mach number is limited at the value near unity[6]. From spectroscopic measurements in the vicinity of the MPDA muzzle with a uniform axial field, it is found that both axial and rotational flow velocities increase linearly with the increase in a discharge current, but, at the same time, ion temperature increases more steeply, resulting in the limitation of the ion Mach number less than 1 in the vicinity of the muzzle[7]. Several attempts such as measurements of spatial profiles of induced magnetic fields by a magnetic probe array have been performed to clarify mechanisms of the ion heating and the Mach number limitation[8].

It is very important to develop an optimum magnetic nozzle configuration in the vicinity of the MPDA muzzle for converting the high ion thermal energy to the axial flow energy and improving thrust efficiency. In order to verify the magnetic nozzle effect in the vicinity of the MPDA, we have attached a small-sized magnetic coil with a shorter characteristic length. A Laval-type magnetic nozzle configuration has been formed in the muzzle region of the MPDA and its effects on the spatial variation of plasma parameters are measured by use of a spectrometer and Mach probes. It is expected that the subsonic flow near the muzzle be converted to a supersonic flow through the magnetic Laval nozzle as in a conventional compressible gas flow.

In this paper are reported characteristics of a plasma flow produced by an MPDA with additional magnetic nozzle configuration. Spatial variations of plasma parameters are discussed in terms of one-dimensional isentropic fluid equations in a compressible gas.

EXPERIMENTAL APPARATUS

A high power, quasi-steady MPDA is installed in the HITOP device of Tohoku University. The HITOP device consists of a large cylindrical vacuum chamber (diameter D = 0.8 m, length L = 3.3 m) with eleven main and six auxiliary magnetic coils, which can generate a uniform magnetic field up to 0.1 T.

The MPDA, which is installed at one end of the HITOP device, has a coaxial structure with a center tungsten rod cathode (10mm diam) and an annular molybdenum anode (30mm diam) as shown in Fig.1. A fast acting gas valve can inject helium gas quasi-steadily for 3ms. Discharge current I_d up to 10kA is supplied by a pulse-forming network (PFN) system with the quasi-steady duration of 1ms. The current can be increased to 15kA with a shorter quasi-steady duration of 0.4ms. The current I_d can be controlled by varying the charging voltage of the PFN power supply.

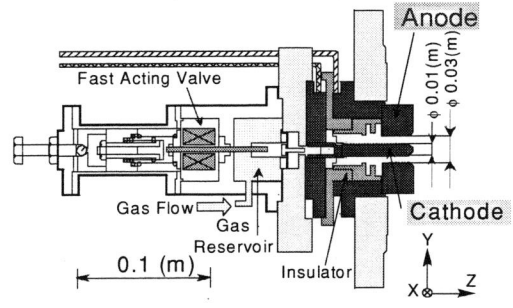

FIGURE 1. Schematic of a quasi-steady pulsed MPD Arcjet.

A small-sized additional coil is installed in the vicinity of the MPDA muzzle. Fig.2 shows the setting of the coil and axial profiles of the generated magnetic field. The coordinate direction of X, Y and Z-axes are shown in the Fig.2. The position of Z=0 is located at the tip of the cathode rod of the MPDA. Two types of the coil (type-A and type-B with the inner diameters of 10cm and 20cm, respectively) are used alternatively.

Plasma flow characteristics are measured by several diagnostics installed on the HITOP device. Electron temperature and density profiles are measured by a movable triple probe and a fast-voltage-scanning Langmuir probe. A plasma flow is characterized by an ion Mach number M_i that is defined as a ratio of the plasma flow velocity to the ion acoustic velocity. The Mach probe has two probe tips facing to the upstream direction and to the perpendicular direction. The ion saturation currents are measured by these tips and indicated as $J_\parallel$ and $J_\perp$, respectively. The ratio of $J_\parallel / J_\perp$ is proportional to M_i. Profiles of M_i and plasma density along and across the field lines are measured by a movable Mach probe and an array of 13-channel Mach probes set at 1.7m downstream of the MPDA muzzle.

Spectroscopic diagnostics are used to measure ion temperature T_i, axial and azimuthal flow velocities U_z and U_θ, respectively, near the muzzle of the MPDA. Emission from the plasma is collected by a quartz lens and is transferred to a spectrometer by a single fiber cable. A spectrum is detected with an image intensifier tube coupled with a CCD camera (ICCD), which is set at the exit plane of a 1m Czerny-Turner spectrometer with a grating of 2400 grooves/mm. HeII line spectra (λ=468.58nm) are obtained in every 0.1msec time interval during a shot with the spectral resolution of 0.02nm. T_i, U_z and U_θ are obtained from Doppler broadening and spectral shift of HeII line.

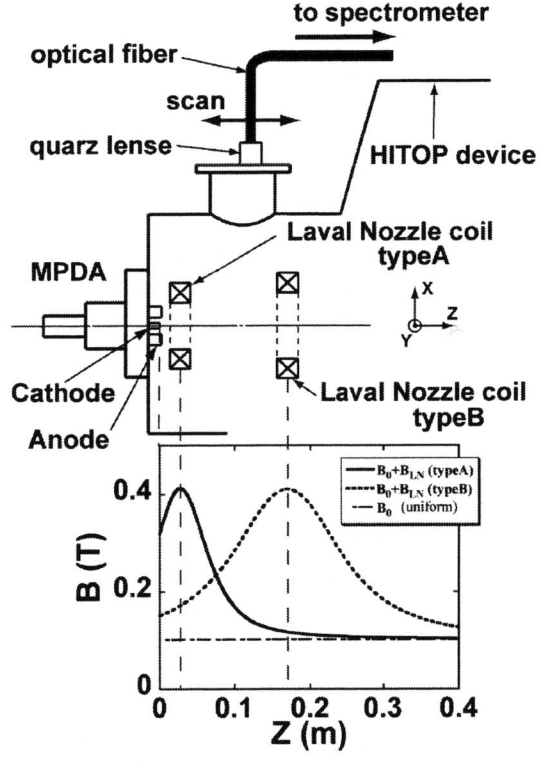

FIGURE 2. Layout of alternatively used two types of Laval-nozzle coils and axial profiles of the magnetic field near the MPDA muzzle.

EXPERIMENTAL RESULTS

Firstly, the type-A coil is installed at Z=3cm downstream of the MPDA. Figure 3(a) shows the axial profile of the magnetic flux density with a coil current I_N of 2.8kA. A pulsed power supply is used for the additional nozzle coil current with the duration of 0.5msec. Time evolutions of the coil current, ion saturation current collected by two tips on a Mach probe and its ratio are shown in Fig.3 (b), (c), (d), respectively.

As shown in Fig.3, the ion Mach number M_i measured at Z=1.73m far downstream of the MPDA is choked with M_i of nearly unity in case of the uniform magnetic field. It is found that M_i increases from 1 to 1.5 by the addition of the type-A coil.

Secondly, the type-B coil is used to measure axial variations of T_i and U in both upstream and downstream regions of the magnetic throat and to clarify mechanism of the increase in the Mach number. Fig. 4 shows axial profiles of effective radius of the magnetic, T_i, U and M_i. The nozzle radius is estimated from the radial profile of light emission intensity. The profile is almost Gaussian and plasma radius is derived from the e-folding length of the profile. This radius agrees well with that calculated under the assumption that the channel area A varies according to the magnetic flux conservation in vacuum that BA=const

The plasma flow velocity U has two components, that is an axial velocity U_z and an azimuthal velocity U_θ. These velocities are measured simultaneously by the spectrometer. Total flow velocity U is derived as

$$U = \sqrt{U_z^2 + U_\theta^2} \quad . \quad (1)$$

The M_i is derived by using T_i and U as,

$$M_i = \frac{U}{C_s} = \frac{U}{\sqrt{(\gamma_i T_i + \gamma_e T_e)/m_i}} \quad . \quad (2)$$

Here γ_i and γ_e are the specific heat ratio of ions and electrons, respectively, and m_i is the mass of helium gas.

It is clearly shown in Fig.4 that the ion temperature decreases and at the same time the flow velocity increases when the plasma passes through the magnetic Laval nozzle.

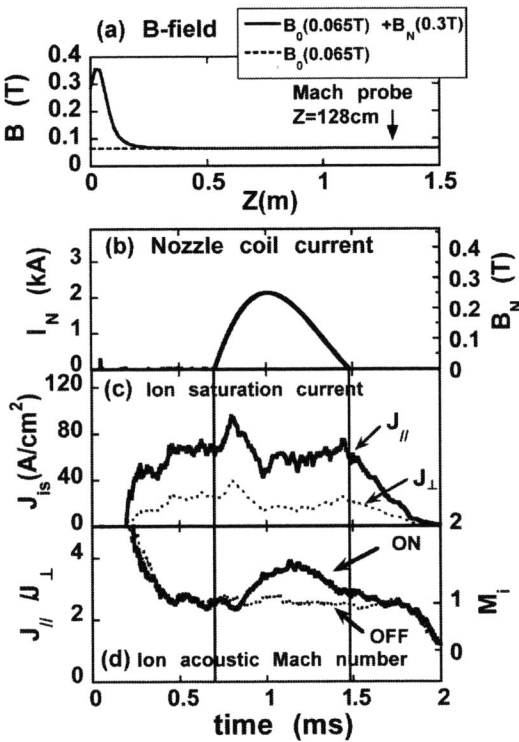

FIGURE 3. Effect of pulsed magnetic Laval nozzle of type-B on the downstream Mach number. (a) axial profiles of magnetic field, (b) pulsed coil current, (c) Mach probe currents, (d) current ratio of Mach probe, i.e., Mach number.

In terms of the one-dimensional isentropic flow model of a compressible gas, the Mach number is related to the variation of the cross-sectional area A of the flow channel,

$$\frac{dM}{M} = \frac{2+(\gamma-1)M^2}{2(M^2-1)} \frac{dA}{A} \quad . \quad (3)$$

Also, the flow velocity U, the temperature T and the mass density ρ vary as follows:

$$\frac{dU}{U} = \frac{1}{(M^2-1)} \frac{dA}{A} \quad , \quad (4)$$

$$\frac{dT}{T} = -\frac{(\gamma-1)M^2}{(M^2-1)} \frac{dA}{A} \quad . \quad (5)$$

$$\frac{d\rho}{\rho} = -\frac{M^2}{(M^2-1)} \frac{dA}{A} \quad . \quad (6)$$

Assuming that the flow-channel area A varies according to the magnetic flux conservation in vacuum BA = const., the axial profiles of U, T and M are calculated from the above Hugoniot equations. Results are shown as solid lines in Fig.4. Here, $\gamma_e = 1$ and $\gamma_i = 1.2$, and $T_e = 5\text{eV}$ are assumed. Dotted lines shown in Fig.4 is the effective nozzle radius calculated by assuming that the flow channel area A varies as the emission profile data measured spectroscopically.

By comparing experimental results with the predicted one, it is found that the plasma flow behaves as a compressible flow and it is accelerated by converting the ion thermal energy to the flow-energy through the Laval-nozzle. This is consistent with the prediction from the one-dimensional isentropic flow model. Considerable azimuthal magnetic field induced by a current extending from the discharge electrodes has been measured in the downstream region of the MPDA muzzle[8]. Effects of the azimuthal field on the nozzle configuration should be taken into consideration for a better prediction.

CONCLUSIONS

A magneto-plasma-dynamic arcjet (MPDA) with an externally-applied magnetic nozzle is investigated to improve the thrust efficiency. A small-sized coil with a shorter characteristic length is attached near the MPDA muzzle to form a Laval-type magnetic nozzle. The subsonic flow near the muzzle is successfully accelerated to a supersonic flow by letting the plasma pass through the magnetic Laval nozzle, as is in a conventional compressible gas flow. The ion Mach number increases from a value less than 1.0 to 1.5. These results are consistent with the prediction from the 1-D isentropic flow model.

ACKNOWLEDGMENTS

This work was supported in part by Grant-in-Aid for Scientific Researches from Japan Society for the Promotion of Science. Part of this work was carried out under the Cooperative Research Project Program of the Research Institute of Electrical Communication, Tohoku University.

REFERENCES

1. Jahn R. G., *Physics of Electric Propulsion*, McGRAW-HILL, NewYork, 1968, pp. 196-256.
2. Sasoh M. and Arakawa Y., J. Propulsion and Power **11**, 351- (1995).
3. Tahara H, Kagaya Y, Yoshikawa T, J. Propulsion and Power **13**, 651 (1997).
4. Schoenberg K.F., *et al.*, Phys. Plasmas **5**, 2090 (1998).
5. Scheuer J.T., *et al.*, IEEE Trans. Plasma Phys. **22**, 1015 (1994).
6. Inutake M., *et al.*, *Proc. 10th Int. Congress on Plasma Physics*, (Quebec, 2000), ICPP-2000, **1**, p.148.
7. Ando A., *et al.*, J. Plasma Fusion Res. SERIES, **4**, 373 (2001).
8. Tobari H., *et al.*, #475 paper in this conference. ICPP 2002 (Sydney).

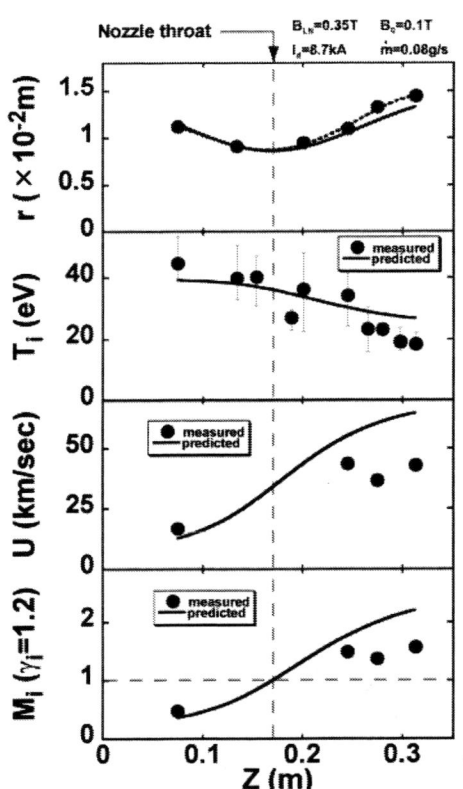

FIGURE 4. Comparison of measured flow characteristics (solid circles) and prediction from 1-D isentropic flow model (solid lines). It is assumed that $M_i=1$ at the throat and the measured T_i near the throat are used. The nozzle radius r calculated from the flux tube in vacuum (solid line) agrees well with the e-folding radius determined from the measured radial profiles of HeII emission intensity (dotted line).

Properties of Heat Flow in the *J*×*B* Gas Arc Discharge for the Production of Fullerenes

Naoki Matsumoto and Tetsu Mieno*

Department of Physics, Shizuoka University, Ooya, Shizuoka-shi, 422-8529, Japan

Abstract. Heat flow of the *J*×*B* gas arc discharge in helium gas is investigated as this discharge has been developed for the efficient production of fullerenes and the control of hot-gas reaction is an important theme. Heat flux around the arc as functions of position, magnetic field and discharge parameters is measured by means of a water-cooled calorimetric probe. As a result, the heat flux from the arc concentrates to the *J*×*B* direction and it is strongly increased by the magnetic field. Also, the heat flux increases by increasing gas pressure and gap distance.

INTRODUCTION

Fullerenes are mainly produced by DC arc discharge using carbon electrodes in helium atmosphere of 10-100 kPa, where heat convection plays an important role in the synthesizing process. By applying steady magnetic field perpendicular to the gas arc current, the arc plasma and the sublimated carbon particles from an anode are jetted out in the *J*×*B* direction. It has been reported that the efficiency of fullerene synthesis increases by the *J*×*B* force [1-3]. However, the mechanism of molecular process in the *J*×*B* arc is not clear. Therefore, it is necessary to clarify the properties of particle motions in the *J*×*B* arc discharge, which would govern the molecular reactions, in order to improve the efficiency of production of fullerenes and carbon nanotubes. It is already reported that discharge voltage and arc flame length increase as an increase of magnetic field. [4]. Here, properties of heat flux in the *J*×*B* gas-arc discharge are investigated. The heat flux per unit area is measured by increase of water temperature in a calorimetric probe, assuming that the heat flux is symmetrical with respect to the vertical arc-flame axis [5-6].

BRIEF THEORY

A simplified model of the *J*×*B* arc jet is as follows. An electron moving perpendicular to magnetic field is forced by the Lorentz force and starts cyclotron motion. If the energy of electron is 1 eV and the magnetic field is 10 mT, the Larmor radius and the cyclotron frequency of the electron are about 0.24 mm and 280 MHz, respectively. But, in a gas atmosphere near 1 atm, the electrons collide with neutral atoms frequently and elastically. When gas pressure p = 30 kPa and gas temperature T_g = 5000 K (which is the typical condition in this experiment), the mean free path and the collision frequency of the electron are about 0.01 mm and 10 GHz, respectively. Therefore, electrons in the *J*×*B* gas arc discharge do not make full Larmor rotation, but continue to be scattered by frequent collisions with neutral atoms, and they diffuse in the *J*×*B* direction. Although ions in the arc move much more slowly, the diffused electrons make electric field and it accelerates ions as the ambipolar diffusion. Neutral atoms are also accelerated to the same direction by their high viscosity. Finally, parts of the electrons, the ions and the neutrals are jetted out to the *J*×*B* direction. By this motion, the discharge voltage increases and additional ionization would supply the plasma particles in the arc region.

---

*Corresponding author. E-mail address: piero@sannet.ne.jp

EXPERIMENTAL SETUP AND METHOD

Schematic of the experimental setup is shown in Fig. 1. The reactor is made of a stainless steel (18 cm in diameter and 20 cm high) with a carbon anode (8 mmϕ) and a carbon cathode (15 mmϕ). A regulated DC power (constant current) is supplied to the electrodes with discharge current I_d= 20- 80 A. The calorimetric probe is made of a copper pipe (inner diameter 1 mm, outer diameter 2 mm) bending back into parallel style, in which water flows at about 2.5 ml/s. The probe can be moved up and down and be rotated horizontally. Temperature increase of the cooling water is measured by small thermocouples. A motor drive is used to adjust the gap distance between the two electrodes automatically. By feeding 0-6 A of DC current to the solenoid coils, 0-3 mT of magnetic field perpendicular to the arc current is applied at the arc region. Here, a cylindrical coordinate is used and the vertical axis is set as z-axis as shown in Fig. 1, which is parallel to the $J \times B$ direction. The directions of J and B are also shown in the figure.

The chamber is evacuated to less than 10 Pa by a rotary pump, after which about 30kPa of helium gas is introduced. The discharge starts by contact ignition. The probe is horizontally rotated into the arc flame at several heights and increase of water temperature is measured. In this experiment, gas pressure p = 13-66 kPa, discharge current I_d = 20 - 60 A, gap distance between the two electrodes d_G = 3-7 mm and magnetic field B = 0-3 mT. By measuring 11 angles of the probe position at one horizontal plane, radial profiles of the heat flux is calculated by assuming that the heat flux is symmetrical with z-axis [5-7]

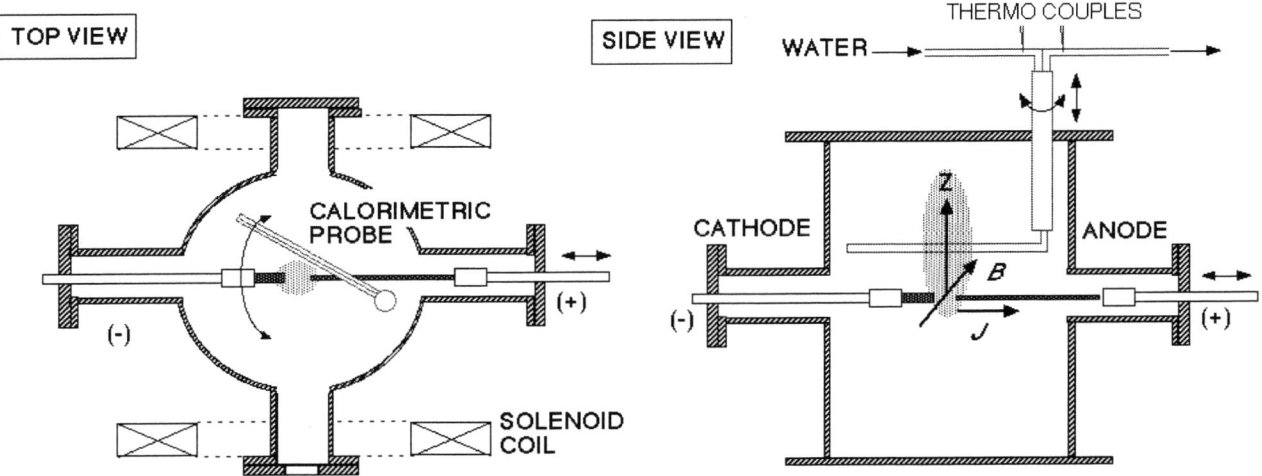

FIGURE 1. Schematic of the experimental setup.

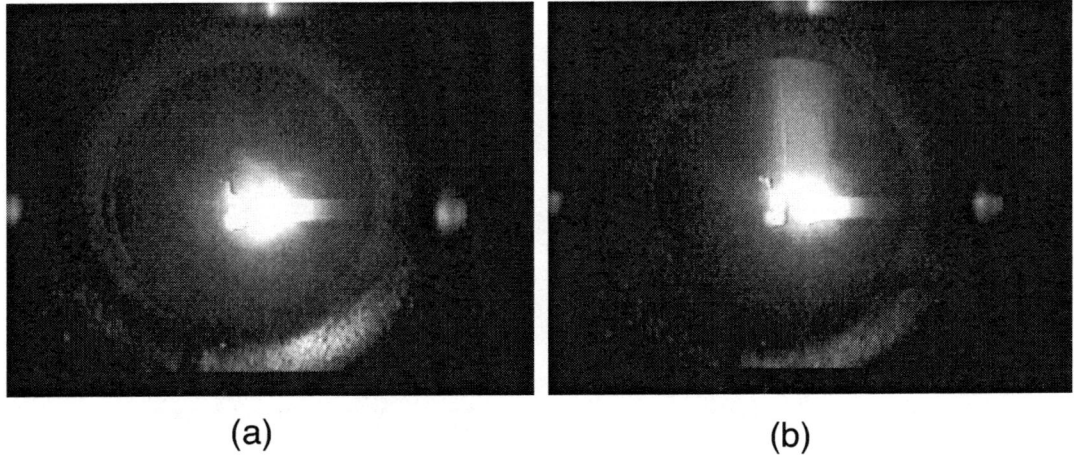

FIGURE 2. Photographs of profiles of the arc for (a) B= 0 and (b) B= 15 mT. p= 40 kPa, I_d= 80 A and d_G= 10 mm.

RESULTS AND DISCUSSION

Photographs of the arc flame from radial direction are shown in Fig. 2. Without the magnetic field (a), the arc flame has nearly spherical form. While with the magnetic field (b), the flame extends to the $J \times B$ direction and the flame length becomes more than 50 mm. It increases as an increase of the magnetic field.

Figure 3 shows radial profiles of the heat flux for 4 magnetic fields at $z = 30$ mm, where $p = 40$ kPa, $I_d = 60$ A, $d_G = 5$ mm and height from the arc center $z = 30$ mm. The heat flux is maximum at $r = 0$ and about 2.5 times larger at this position for $B = 3$ mT. The heat flux increases monotonically with the magnetic field. At $|r| > 25$ mm, the heat flux is almost the same regardless of the magnetic field.

The z direction distribution of the heat flux on the z-axis is shown in Fig. 4. Below $z = 60$ mm, the heat flux monotonically decreases with z. But for $z > 60$ mm, the heat flux is small and almost constant with respect to the magnetic field.

Figures 5 and 6 are heat fluxes as functions of the gas pressure p and gap distance d_G, respectively at $r = 0$ and $z = 40$ mm, where three magnetic fields are used as a parameter, $d_G = 5$ mm and $p = 40$ kPa or $I_d = 60$ A for each case. The heat flux increases monotonically with pressure and gap distance. The increasing rates are stronger when the magnetic field is larger. When $p > 60$ kPa and $d_G > 7$ mm, the arc becomes unstable and the arc spot moves around and the discharge tends to be extinguished for $B > 3$ mT. Radial profiles of the heat fluxes for two discharge currents are measured and shown in Fig. 7. In this case, the heat fluxes for the two conditions are same at $r = 0$, but the cross section of the heat flux by the $J \times B$ force increases by increasing the discharge current.

When the gas pressure is higher, collision frequency between electrons and neutral atoms increases, which would increases the flow rate of the neutral particles and they bring the heat. Totally the heat flow increases with an increase of pressure. When the gap distance is wider, the volume of arc is larger, by which more heat is transported. When the discharge current is increased, cross section of the heat flux by the $J \times B$ force increases. At the higher position as $z > 60$ mm, the plasma is sufficiently quenched and the heat flux by the $J \times B$ force is relaxed to be a broad flux by sufficient collision process. Therefore, the effect of magnetic field at $z > 60$ mm is very small.

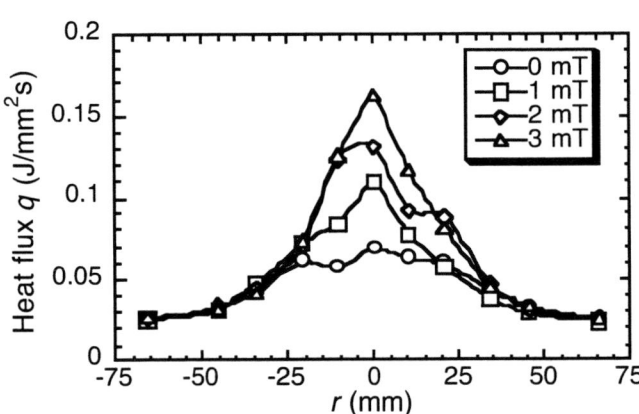

FIGURE 3. Radical profiles of heat flux q for 4 magnetic fields. $p = 40$ kPa, $I_d = 60$ A, $d_G = 5$ mm and height from the arc center $z = 30$ mm.

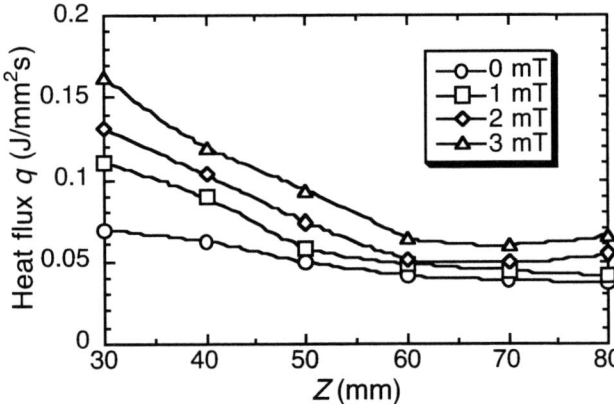

FIGURE 4. Heat flux q versus height from the arc center z for 4 magnetic fields. $p = 40$ kPa, $I_d = 60$ A and $d_G = 5$ mm on the flame-axis.

CONCLUSION

The variation of the heat flux from the DC gas arc as a function of magnetic field is measured. The heat flux to the $J \times B$ direction increases with an increase of the magnetic field near the center of the arc. By increasing gas pressure and gap distance, the heat flux to the $J \times B$ direction increases on z-axis of the arc and the effect of the $J \times B$ arc jet is clear. By increasing the discharge current, the cross-section of the heat flow by the $J \times B$ force increases.

The magnetic field affects the heat flow of the gas arc discharge and would strongly influence on the production of carbon clusters.

ACKNOWLEDGMENTS

We would like to thank Dr. S. Sakiyama of Yamaguchi University for his useful support.

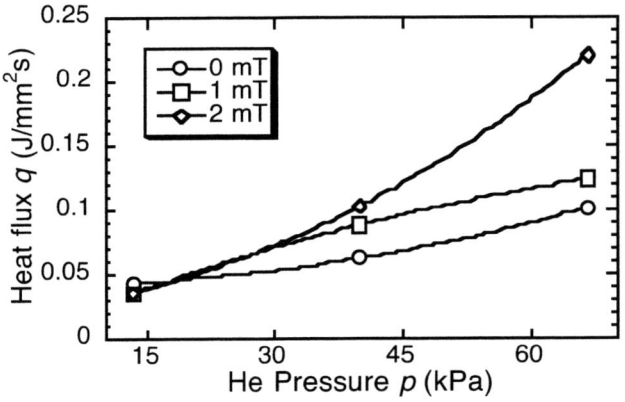

FIGURE 5. Heat flux q versus p for 3 magnetic fields. I_d= 60 A, d_G= 5 mm and z= 40 mm on the flame-axis.

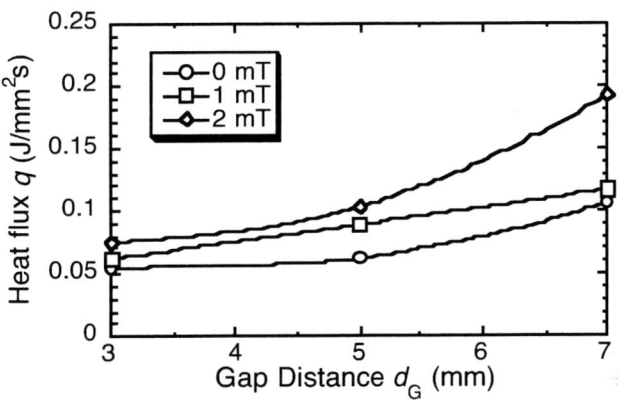

FIGURE 6. Heat flux q versus d_G for 3 magnetic fields. p= 40 kPa, I_d = 60 A and z= 40 mm on the flame-axis.

REFERENCES

1. Mieno, T., Fullerene Sci. Technol. 3, 429-435 (1995).
2. Mieno, T., Sakurai A., and Inoue H., Fullerene Sci. Technol. 4, 913-923 (1996).
3. Aoyama, S. and Mieno, T., Jpn. J. Appl. Phys. 38, L267-L269 (1999).
4. Matsumoto, N. and Mieno, T., Proc. 25th Int. Conf. Phenomena Ionized Gases, Nagoya, Japan, 2001, Vol. 3, pp.343-344.
5. Sakiyama, S. and Fukumasa, O., Jpn. J. Appl. Phys. 38, 4567-4570 (1999).
6. Okada, M., Nishiguchi, K., Tashiro, K., Hayashi, T. and Miyazaki, T., J. Japan Welding Soc. 36, 77-84
(1967) [in Japanese].
7. Matsumoto, N and Mieno, T., (to be published).

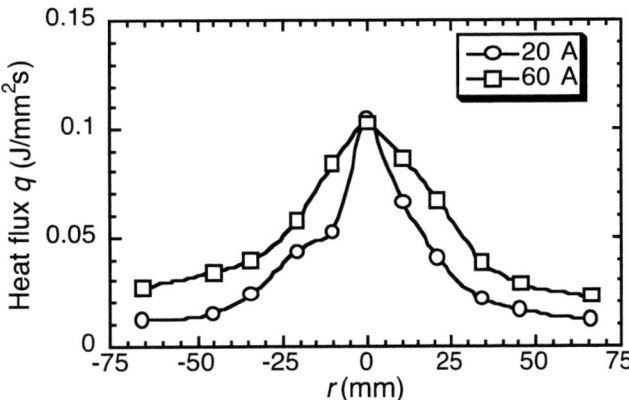

FIGURE 7. Radial profiles of heat flux q for 2 discharge currents. p= 40 kPa, d_G= 5 mm, B= 2 mT and z= 40 mm.

Evaluation of Electromagnetic Forces in an Axially-Magnetized MPD Arcjet Plasma

Hiroyuki Tobari, Kyohei Yoshino, Kunihiko Hattori, Akira Ando
and Masaaki Inutake

Department of Electrical Engineering, Tohoku University, Aoba05, Sendai 980-8579, JAPAN

Abstract. Characteristics of an axially-magnetized plasma flow has been investigated in the vicinity of a magneto-plasma-dynamic arcjet (MPDA) by use of spectroscopy, Mach probes and magnetic probes. Axial and rotational flow velocity and temperature of He ion and atom near the muzzle region of MPDA are measured by Doppler shift and broadening of the HeI (λ = 578.56 nm) and HeII (λ = 468.58 nm) lines. It has been observed that the plasma rotates with a rigid body and that ion temperature increases extraordinarily in a factor of 2-3 at several cm downstream from MPD outlet when a discharge current increases with a lower mass-flow-rate of He gas. Therefore, the ion acoustic Mach number saturates at near unity. To clarify mechanisms of ion heating and electromagnetic acceleration, spatial distribution of induced magnetic fields are measured in the vicinity of MPD outlet by magnetic probes. Spatial structure of magnetic field and plasma current density is clarified experimentally in the muzzle region of MPDA. Among three components of ***j*×*B*** force ***F***, radial component F_r is dominant and axial component F_z is much smaller than F_r because of a generation of a drag force canceling an acceleration force.

INTRODUCTION

Recently, various magneto-hydro-dynamic (MHD) phenomena have been observed in fusion plasma, earth's magnetosphere and sun's surface owing to remarkable development of diagnostic instruments. Dramatic phenomena such as particle acceleration, magnetic reconnection and shock wave formation have been reported. A plasma flow plays an important role in these phenomena. As for a fusion plasma, it has been proposed a high Mach number plasma flow could be confined stably by keeping the force balance between the static plasma pressure and the dynamic pressure of the plasma flow according to Beltrami-Bernoulli's condition.[1] Production of a high-beta and supersonic plasma flow is also useful for the development of a space thruster. Therefore, researches on production and utilization of a fast flowing plasma are important not only in basic researches in space and fusion plasmas but in many applications.

A magneto-plasma-dynamic arcjet (MPDA) is one of plasma sources which can produce a highly-ionized, high-density and supersonic plasma flow and is also expected as one of the promising electric propulsion systems owing to features of a relatively large thrust, high specific impulse and long lifetime which are required for a manned Mars mission [2].

We have investigated characteristics of a high-beta, supersonic plasma flow produced by an MPDA on the HITOP device in Tohoku University. The MPDA plasma is accelerated axially by self-induced $j_r \times B_\theta$ force, where j_r is radial discharge current and B_θ is self-induced azimuthal magnetic field. To enhance the acceleration performance of the MPDA, it has been proposed and examined experimentally to operate it in various types of an externally-applied axial magnetic field [3-6]. In this case, in addition to the axial acceleration, the interaction between the externally-applied axial magnetic field B_z and j_r generates an azimuthal acceleration force, which drives the plasma to rotate azimuthally. Further, an additional axial acceleration is expected by the interaction between azimuthally-induced Hall current j_θ and radial component of the magnetic field B_r under appropriate operating conditions.

Considering the acceleration phenomena mentioned above, it is expected a complicated acceleration force field to be formed. Though an increase in thrust was observed in previous works[3], acceleration mechanisms have not been clarified enough. Therefore, it is quite important to investigate the acceleration force field and flow field of the MPDA plasma in the muzzle region of MPDA not only to clarify the electromagnetic acceleration mechanisms but also to realize a quasi-steady supersonic plasma flow but also for basic researches on complicated MHD phenomena.

In this paper are reported characteristics of a plasma flow in the muzzle region of MPDA. Spatial distribution of plasma parameters such as ion/electron temperatures, axial and rotational flow velocities, induced magnetic field and plasma current density in the region are measured experimentally. The external magnetic field configuration suitable for an efficient acceleration is discussed.

EXPERIMENTAL APPARATUS

Experiments are performed in the HITOP (HIgh density TOhoku Plasma) device of Tohoku University. The HITOP device, consists of a large cylindrical vacuum chamber (diameter $D=0.8$ m, length $L=3.3$ m) with external magnetic coils, which can generate an uniform magnetic field up to 0.1 T. Various types of magnetic field configurations can be formed by adjusting these coil currents. A high-power, quasi-steady MPDA installed at one end of the HITOP device has coaxial structure with a center tungsten rod cathode (10mm in diameter) and an annular molybdenum anode (30mm in inner diameter). A quasi-steady discharge continues for 1ms with a pulse-forming-network (PFN) and a fast acting gas valve puffs helium gas into the discharge region. Discharge current I_d can be controlled by varying the charging voltage of the PFN power supply. Maximum value of I_d is 10kA and a typical discharge voltage is 200 V[7].

A spectroscopic method is adopted for measurements of ion temperature T_i, flow velocity u_z in the muzzle region of the MPDA and an ion acoustic Mach number M_i is calculated with these parameters. T_i and u_z are obtained from Doppler broadening and spectral shift of HeII line ($\lambda=468.58$nm), which is detected with an image intensifier tube coupled with a CCD camera[8].

In the downstream region, u_z is measured by a Mach probe, which consists of two plane current-collecting tips. One of them faces a parallel direction to the plasma flow and the other faces a perpendicular direction to it. M_i is calculated by a ratio of ion saturation current densities collected by the two tips.

Variations of magnetic fields in the plasma flow are measured directly by use of a movable magnetic probe array, which consists of 11 magnetic probes arrayed in the radial direction. Plasma current density can be calculated with the magnetic field data. Each probe has three sets of mutually perpendicular pick up coils wounded to measure the magnetic field components in the r, θ and z directions. Probe signal (B-dot signal) is transferred to differential amplifiers and integrators and digitized with 1M samples per second.

EXPERIMENTAL RESULTS AND DISCUSSION

Spatial contour maps of emission intensity, rotational velocity and particle temperature measured in a uniform external magnetic field $B_0 = 0.1$T is shown in Fig. 1. Emission intensity of HeI line broadens isotropically and that of HeII is constrained in the region of the inner radius of anode. According to detailed measurements, rotational velocity of He ion increases linearly in the core region. These data show that He atom diffuses isotropically and that He ion is trapped by an external magnetic field and flows downstream with a rigid rotation. Ion temperature near the MPDA outlet is around 20eV and increases extraordinarily in factor of 2-3 at 5-15 cm downstream of the MPDA. The ion acoustic Mach number tends to saturate near unity. Several hypotheses are discussed[9], but details have not been clarified enough.

To investigate correlation between the ion heating and electromagnetic force field, magnetic fields in the plasma flow is measured by use of the magnetic probe array and spatial profiles of the plasma current density are derived from the magnetic field data. Axial current density j_z is calculated by Ampere's law in Eq. (1).

$$j_z = \frac{1}{\mu_0}\left\{\frac{1}{r}\frac{\partial}{\partial r}(rB_\theta(r))\right\} \quad (1)$$

Here, the profile of $B_\theta(r)$ is obtained from the measured azimuthal magnetic field ΔB_θ fitted by 7th polynomial function. Azimuthal current density j_θ is calculated from a radial force-balance equation (Eq. (2)) of a plasma column with a rigid body rotation. A density profile is assumed as a Gaussian with the on-axis value of 10^{21}m^{-3}, and electron temperature is estimated as 5eV from a Langumuir probe measurement in downstream region. Subscripts i and e denote ion and electron, respectively.

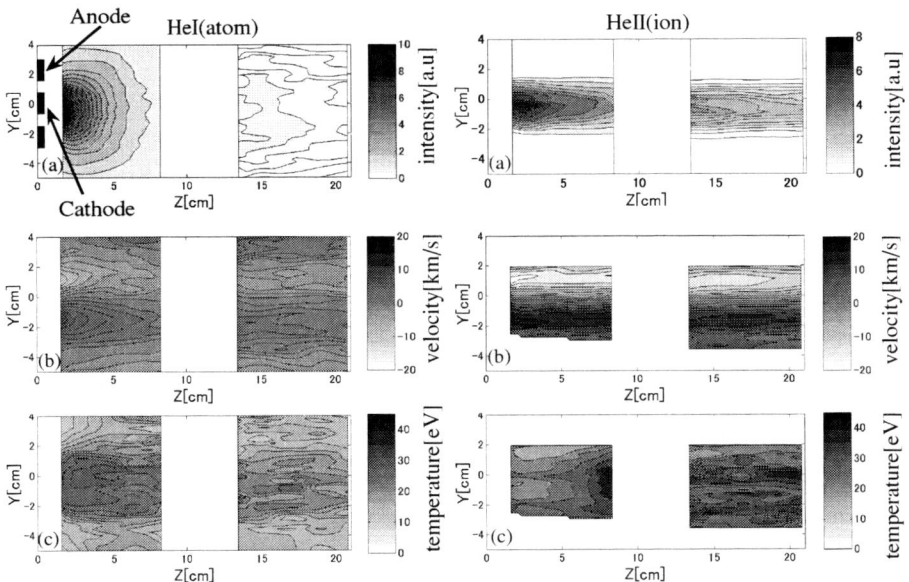

FIGURE 1. Spatial contour maps of (a) emission intensity, (b) rotational velocity and (c) particle temperature of He atom (left) and He ion (right). I_d=7.7kA, B_0=0.1T, dm/dt=0.06g/s.

$$m_i n_i \frac{u_\theta^2}{r} - \frac{\partial(p_i + p_e)}{\partial r} + j_z B_\theta - j_\theta B_z = 0 \qquad (2)$$

Radial current density j_r is calculated from an axial difference of ΔB_θ and radial component of rot $\boldsymbol{B} = \mu_0 \boldsymbol{j}$ (Eq. (3)).

$$\mu_0 j_r = -\frac{\partial B_\theta}{\partial z} \qquad (3)$$

Radial profiles of the magnetic field and plasma current density at Z=9cm are shown in Fig. 2. Diamagnetic signal ΔB_z is in the direction of canceling the external magnetic field B_0 and total axial magnetic field strength decreases to about 1/2. The diamagnetic effect is proportional to a plasma density and it has a maximum near the muzzle of MPDA and weakens gradually with the plasma density decrease in the streamwise direction. The results suggest that the magnetic field lines are expanded by a high-beta plasma in the upstream region and a magnetic flux converges gradually in the downstream, *i.e.* a slightly-converging helical magnetic nozzle should be formed spontaneously. The direction of j_z is reversed in the center and the edge, the j_r is directed inward. This shows the discharge current flows from the anode to the cathode across the magnetic field. The direction of j_θ corresponds to that of the ion diamagnetic drift. Radial profiles of $\boldsymbol{j} \times \boldsymbol{B}$ force (Lorentz's force) are shown in Fig 3. Among three components, the radial force (pinch force) is dominant and the direction of an azimuthal force corresponds to that of the rotational velocity measured by spectroscopy.

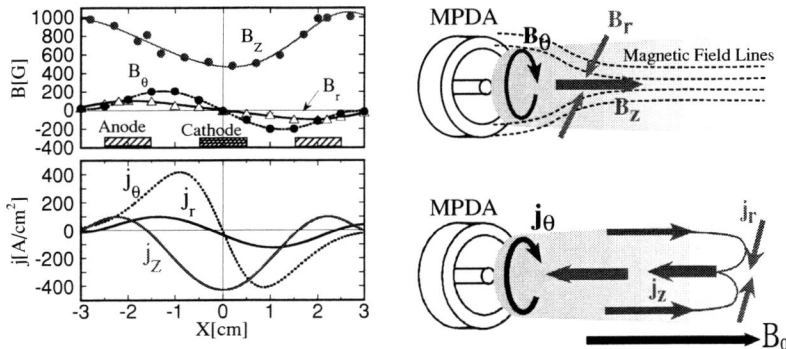

FIGURE 2. Radial profiles of the measured magnetic fields and plasma current densities at Z=9cm from MPDA ; I_d =7.2kA, dm/dt=0.10g/s, B_0=0.1 T(left), and schematic views of the spatial structures of magnetic fields and plasma current densities in MPDA plasma flow (right).

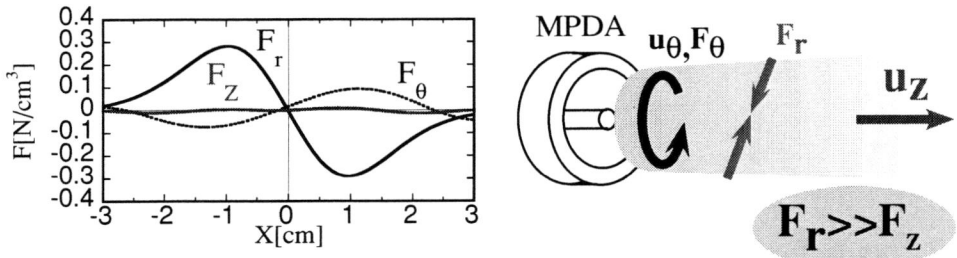

FIGURE 3. Radial profiles of $j \times B$ force in the same conditions with those in Fig. 2 and schematic view of spatial distributions of the electromagnetic forces.

It should be noted that the axial component of $j \times B$ force F_z almost vanishes in the measured region. Here F_z is calculated by the following equation,

$$F_z = j_r B_\theta - j_\theta B_r \qquad (4)$$

The first term is the self-field accelerating force. The second term, however, cancels the first term because of the inward B_r generated by the diamagnetic effect, so it acts as a drag force. It is proposed that a more efficient acceleration force should be generated in an external magnetic field configuration with a divergent magnetic nozzle near the muzzle of MPDA. The externally-applied radial magnetic field is expected to reverse the sign of the second term in Eq. (4) [6].

CONCLUSION

In this work, characteristics of a high density and high Mach number plasma flow produced by a quasi-steady MPDA in the HITOP device. Under the conditions of a uniform magnetic field and a high discharge current with a lower mass-flow-rate, ion temperature increases extraordinarily at several cm downstream of the MPDA muzzle, then ion acoustic Mach number saturates near unity. To investigate the mechanisms of ion heating and electromagnetic acceleration, the spatial structures of the magnetic field and current density in the plasma flow are measured. Due to the weakening diamagnetic effect in the streamwise direction and the induced azimuthal magnetic field, the slightly-converging helical magnetic nozzle is formed spontaneously. In the muzzle region of the MPDA, radial component of $j \times B$ force F_r is dominant, which is balanced with a radial plasma pressure. In a uniform magnetic field, the interaction between the azimuthal current and the radial magnetic field generates the drag force, which prevents from axially accelerating the plasma. The externally-applied diverging radial magnetic field is expected to convert the drag force to the acceleration force.

REFERENCES

1. S.M. Mahajan and Z. Yoshida, Phys. Rev. Lett. **81**, 4863 (1998)
2. R.G. Jahn, *Physics of Electric Propulsion* (McGRAW-HILL, 1968)
3. H. Tahara, Y. Kagaya, T. Yoshikawa, J. Propulsion and Power **13**, 651 (1997)
4. K.F. Schoenberg, *et al.*, Phys. Plasmas **5**, 2090 (1998)
5. J.T. Scheuer, *et al.*, IEEE Trans. Plasma Sci. **22**, 1015 (1994)
6. M.Inutake, *et al.*, presented in this congress (ICPP2002) (paper #481)
7. M.Inutake, *et al.*, *Proc. of 10th Int.Congress on Plasma Physics* (ICPP2000) (Quebec) **1**, 148 (2000)
8. A. Ando *et al.*, J. Plasma and Fusion Res. Series **4**, 373 (2001)
9. H.Tobari, *et al.*, *Proc. of 25th Int.Conf. on Phenomena in Ionized Gases* (Nagoya) **2**, 193 (2001)

Investigations Of A Pulsed Cathodic Vacuum Arc

T.W.H. Oates, J. Pigott, P. Denniss, D.R. Mckenzie, M.M.M. Bilek

Applied and Plasma Physics, University of Sydney, NSW, Australia.

Abstract. Cathodic vacuum arcs are well established as a method for producing thin films for coatings and as a source of metal ions. Research into DC vacuum arcs has been going on for over ten years in the School of Physics at the University of Sydney. Recently a project was undertaken in the school to design and build a pulsed CVA for use in the investigation of plasma sheaths and plasma immersion ion implantation. Pulsed cathodic vacuum arcs generally have a higher current and plasma density and also provide a more stable and reproducible plasma density than their DC counterparts. Additionally it has been shown that if a high repetition frequency can be established the deposition rate of pulsed arcs is equal to or greater than that of DC arcs with a concomitant reduction in the rate of macro-particle formation. We present here results of our investigations into the building of a center-triggered pulsed cathodic vacuum arc. The design of the power supply and trigger mechanism and the geometry of the anode and cathode are examined. Observations of type I and II arc spots using a CCD camera, and cathode spot velocity dependence on arc current will be presented. The role of retrograde motion in a high current pulsed arc is discussed.

MOTIVATION AND DESIGN

The design of our system was based around the requirements for experiments we intend to conduct. These include the production and investigation of nano-laminate multilayer films and investigations of sheath dynamics during plasma immersion ion implantation (PIII) and deposition (PIIID). For the first of these experiments we require a metal plasma source capable of depositing homogenous thin films of the order of a few angstroms thickness. The reproducibility of the deposition process is of critical importance. DC arcs are limited in their ability to reproducibly deposit films of a given thickness below a few nanometers due to periodic fluctuations in the plasma density. Pulsed arcs on the other hand can reproducibly deposit a given amount of material for each arc pulse. By calibrating the amount of material deposited per pulse the thickness of the film can be accurately controlled by simply controlling the number of pulses.

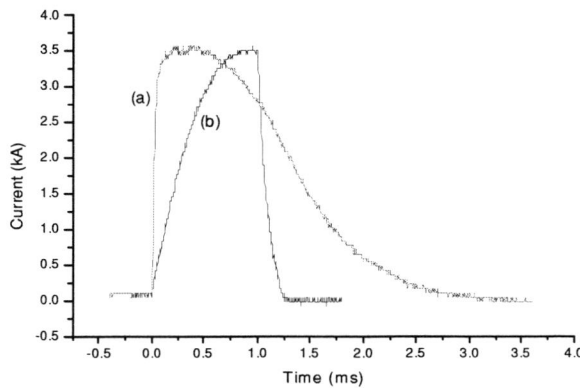

FIGURE 1. Current profiles for the two power supplies used in our prototype; (a) simple 12mF capacitor bank; (b) oscillating LC circuit "crowbarred" after one millisecond.

Measurements of sheath dynamics during PIII require that the plasma does not exhibit large density fluctuations over a timescale of around a few hundred microseconds. Compared to DC arcs the plasma density of pulsed arcs is far more stable for this length of time. High current arcs (currents greater than a few hundred amperes) have also been shown to exhibit a reduction in macroparticle production[1]. This is conjectured to be due to the increase in arc spot velocity, discussed below. A reduction in macroparticles is advantageous for both applications of our plasma source.

We based our prototype on a center-triggered cylindrical cathode design utilised by a number of groups over the years[1,2]. Initially the power supply was a simple 12mF capacitor bank, charged to between 100 and 400V, triggered by a 1ms, 3kV pulse generator, providing the current profile shown in figure 1(a). It was found that such a current profile produced an uneven erosion profile on the cathode surface. Plasma production in a cathodic arc is due to the erosion of the cathode surface by high current density cathode spots. Cathode arc spots are limited in their current carrying capacity to around 100A per spot so a high arc current results in a large number of arc spots. Because the simple capacitor supply produced a fast rising current with a decaying tail the majority of the plasma production was at the beginning of the pulse, close to the trigger pin, resulting in an uneven erosion profile. This is undesirable for the efficient utilization of cathode material.

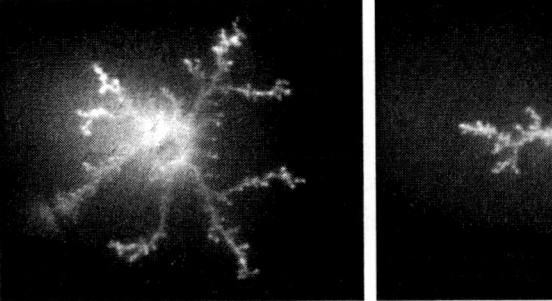

FIGURE 2. CCD images of an aluminium cathode. The image on the left shows the increased arcing near the central trigger for the simple 12mF capacitor bank. The image on the right is the arc trace for the oscillating LC circuit, crowbarred after one millisecond. Arcing is reduced around the central trigger. Image sizes are 5cm x 5cm.

To overcome this problem we designed a second power supply based on an oscillating LC circuit with a resonance period of a few milliseconds. Partway into the cycle the current was interrupted by initiating a "crowbar". This effectively short-circuited the capacitor bank, thus extinguishing the arc current at a specified time. Figure 1(b) shows such a current profile with the crowbar initiated after one millisecond. This current profile reduced the concentration of the erosion in the center of the cathode providing a more even erosion profile and better utilization of the cathode material. Figure 2 consists of two CCD images of arc traces for the two different power supplies, showing the reduction in the arcing and erosion near the center trigger for the oscillating LC power supply.

The main function of the crowbar is to restrict the arcs from running over the edge of the cathode and either arcing on material which would contaminate the plasma, or forming a breakdown to the anode or chamber walls. Multiple arc spots caused by high currents repel one another by magnetic forces. Triggering the arc in the center causes the spots to move outward toward the cathode edge. By monitoring the time taken for the arc spots to reach the edge, and initiating the crowbar at the appropriate time, the erosion can be controlled to utilize as much of the cathode as possible without allowing the arc to run over the edge.

OBSERVATION OF TYPE I AND II ARC SPOTS

Cathode spots have been observed to operate in two distinct modes characterised by fundamental differences in the arc current per spot, the spot velocity and the brightness of the spots. Type I spots are characterised as being of low current, high velocity and low brightness whilst the opposite characteristics distinguish type II spots[3]. Type I spots are associated with contamination of the cathode surface by adsorbed gases and hydrocarbons. These contaminations reduce the work function of the surface and affect the secondary electron emission, thereby altering the arc properties[4]. It is apparent from observations of different spot types that the arc spots will preferentially run on contaminations and subsequently operate in a type I mode until the surface has been cleaned by the arc process itself.

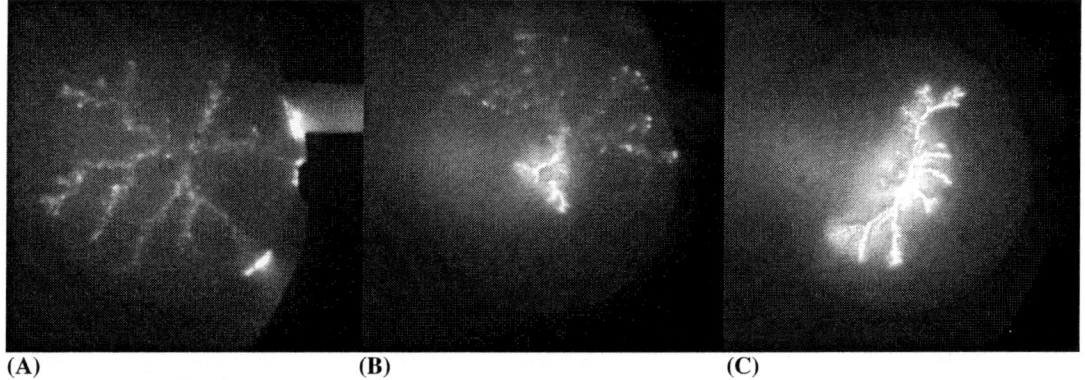

FIGURE 3. CCD images of an aluminium cathode showing transition from type I spot mode to type II mode. Image A is taken 8 pulses after the cathode was inserted into the vacuum system (exposure time 0.2ms). After 49 pulses (B) the arc begins to operate initially in a type II mode before reverting to type I as contaminated regions are reached (exposure time 0.3ms). After 68 pulses the cathode surface has been cleaned of contaminants and the arc is operating purely in type II mode (exposure time 1ms). Image sizes are 5cm x 5cm.

Figure 3 shows the transition between the two modes of operation. The implications of the two modes is that it is imperative for deposition rate reproducibility and plasma purity that the cathode be fully cleaned and operating in the second mode. Residual gas analysis of the gaseous species in the chamber during the initial phase of operation shows an increase in nitrogen, oxygen and hydrocarbons when the arc is operating in a type I mode. These gases would affect the film growth process and plasma properties.

RETROGRADE MOTION

In an external magnetic field, parallel to the cathode surface, movement of cathode spots is observed. Amperian force is expected in the J x B direction. In fact the opposite is observed: spots move in the -J x B direction. This has been termed "retrograde motion"[5]. High current arcs (>100A) contain multiple spots, each producing a magnetic field. Retrograde motion acts to repel the spots away from one another (figure 4).

The velocities of the spots are proportional to the total force exerted upon them by all other spots on the cathode surface. The number of spots depends on the total current supplied by the power supply. A simple 12mF capacitor bank exhibits a fast rising current, quickly reaching a peak and then tailing away (figure 1a). The number of cathode spots follows the same trend as observed in the series of CCD images in figure 5A. The velocities of the spots for this type of power supply also follow a similar trend, initially moving quickly outward before slowing as the current reduces and the retrograde forces are diminished. For the oscillating LC circuit described previously the current profile

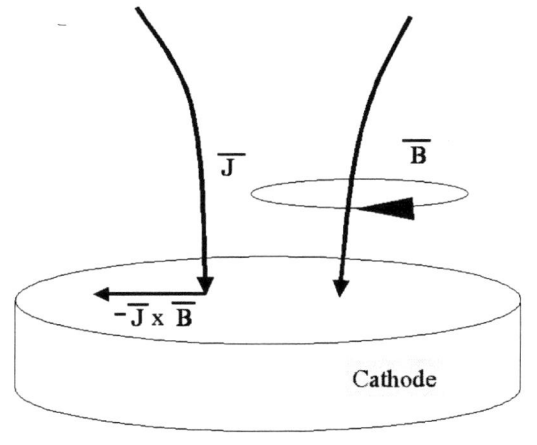

FIGURE 4. Retrograde motion.

slowly rises to a maximum before the crowbar is initiated. The number of arc spots also grows in the same way and the velocity of the spots at the beginning of the pulse is reduced due to a reduction in the repulsive force (figure 5B).

Retrograde repulsion therefore acts to increase the velocity of arc spots when the total arc current is increased. It has been observed that high current arcs exhibit a reduction in macroparticles. Macroparticles are droplets of molten cathode material that are explosively ejected from the cathode surface during the arc process. Because high current arcs produce multiple cathode spots, and therefore greater retrograde repulsion, the arc spots move more quickly on the cathode surface. This reduces the dwell time of the cathode spots in any given area, thus reducing the local heating and reducing the amount of molten cathode material that can be incorporated in the plasma.

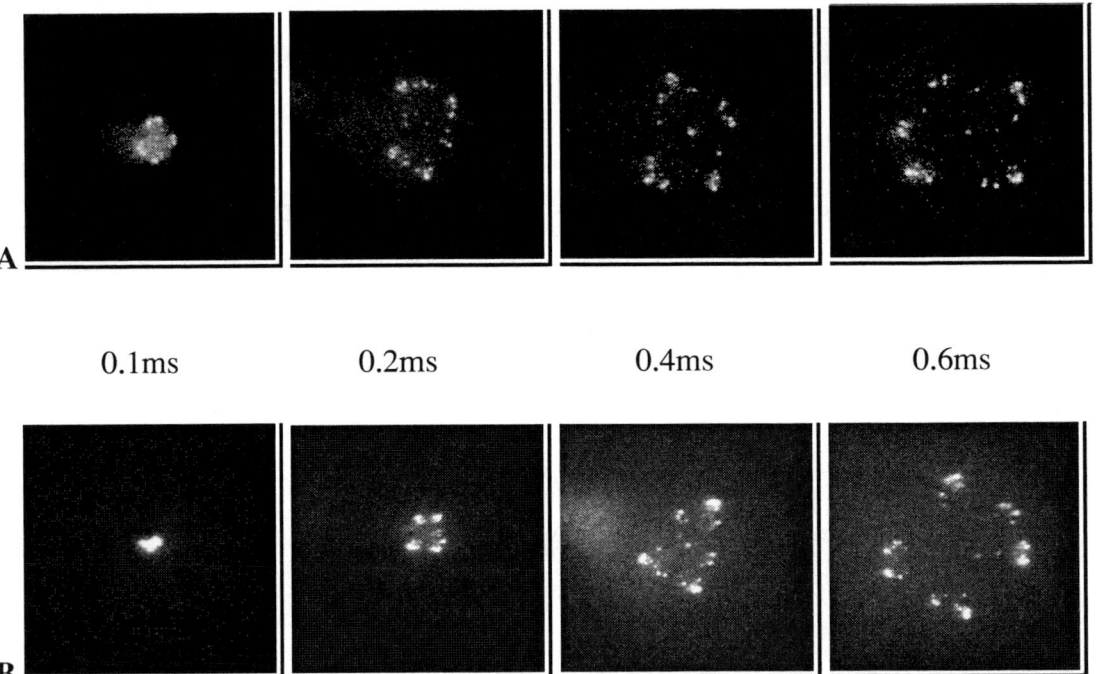

FIGURE 5. CCD images of arc spots during an arc pulse taken 0.1, 0.2, 0.4 and 0.6 ms after arc ignition, from left to right respectively (1µs exposures). The upper series (A) shows the spot locations for the 12mF power supply and the lower series (B) for the oscillating LC circuit. Images are 5cm x 5cm.

CONCLUSIONS

We have presented results of investigations into the design and testing of a new pulsed cathodic vacuum arc. The power supply design is a critical part of the arc, determining the erosion profile and the length of the arc pulses, both are which are important in maximising the utilisation of cathode material. Type I and II spots were observed and the necessity to effectively clean contaminants from the cathode surface was discussed. Repulsive retrograde forces between the multiple cathode spots present in a high current arc were observed. Retrograde forces were used to explain the reduction in macroparticle production in high current arcs.

ACKNOWLEDGEMENTS

The authors gratefully acknowledge financial support from the Australian Research Council.

REFERENCES

1. P. Siemroth, T. Schulke, and T. Witke, Surf. Coat. Technol. **68**, 314 (1994).
2. I. G. Brown, Rev. Sci. Instrum. **65**, 3061 (1994).
3. J. M. Lafferty, Vacuum arcs. Theory and applications. (John Wiley & Sons, New York, 1980).
4. S. Anders and B. Juttner, IEEE Trans. Plasma Sci **19**, 705 (1991).
5. B. Juttner and I. Kleburg, J. Phys. D: Appl. Phys. **33**, 2025 (2000).

Instability of a Vacuum Arc Centrifuge

M. J. Hole, R. S. Dallaqua†, S. W. Simpson* and E. Del Bosco†.

EURATOM/UKAEA Fusion Association, Culham Science Centre, Abingdon, Oxfordshire, OX14 3DB UK
*School of Electrical and Information Engineering, University of Sydney, N.S.W. 2006 Australia
†Laboratório Associado de Plasma (LAP), Instituto Nacional de Pesquisas Espaciais (INPE), CP 515, 12201-970, São José dos Campos, SP, Brazil

Abstract. Ever since conception of the Vacuum Arc Centrifuge (VAC) in 1980, periodic fluctuations in the ion saturation current and floating potential have been observed in Langmuir probe measurements in the rotation region of a VAC. Our theoretical and experimental research suggests that these fluctuations are in fact a pressure-gradient driven drift mode. In this work, we summarise the properties of a theoretical model describing the range of instabilities in the VAC plasma column, present theoretical predictions and compare with detailed experiments conducted on the PCEN centrifuge at the Brazilian National Space Research Institute (INPE). We conclude that the observed instability is a 'universal' instability, driven by the density-gradient, in a plasma with finite conductivity.

INTRODUCTION

In 1981, Krishnan, Geva and Hirshfield[1] presented the first results for a new type of centrifuge, the vacuum arc centrifuge (VAC). The VAC is an axially configured plasma centrifuge, with a uniform plasma column present in a cylindrical vessel. Figure 1 shows a schematic of the PCEN centrifuge at the Brazilian National Space Research Institute (INPE) used for the experimental measurements reported here.

The plasma is created from metal vapour ablated from the cathode of the discharge by the action of vacuum arcs (see Figure 1) with a metal grid forming the anode of the discharge. The discharge is initiated by a high voltage pulse at the cathode. The interaction of the current with the axial magnetic field sets the plasma in rotation in the region to the left of the anode grid in the figure.[2] The rotating metal vapor plasma streams supersonically through the anode mesh, passing through the rotation region and eventually impinging on the far endplate. Table 1 shows typical parameters of the VAC for the experiments with a magnesium cathode described here.

TABLE 1. Typical VAC Plasma Parameters

Parameter	value
cathode	Mg
background gas pressure	1.0×10^{-4} Pa
mean ionization	1.5
axial magnetic field B_z	0.1 T
plasma temperature $T \approx T_i \approx T_e$	5.5 eV
discharge current	1 kA
discharge voltage	5 kV
plasma lifetime	10 ms
on axis ion density in rotation region	5.0×10^{19} m^{-3}
characteristic column radius	15 mm
plasma rotation frequency	150 krad s^{-1}
ion axial streaming velocity	10^4 ms^{-1}

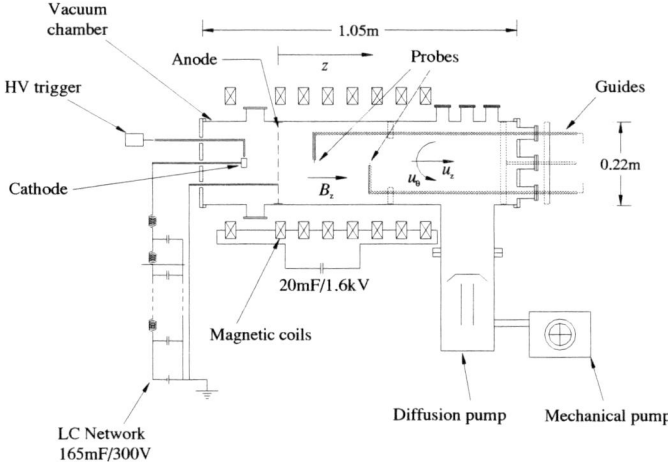

FIGURE 1. The PCEN vacuum arc centrifuge at the Brazilian National Space Research Institute (INPE).

Krishnan, Geva and Hirshfield[1] and other researchers have made Langmuir probe measurements in the rotation region of vacuum arc centrifuges. A key feature of these measurements is the observation of periodic fluctuations in the ion saturation current and floating potential. Figure 2 shows typical oscillations.

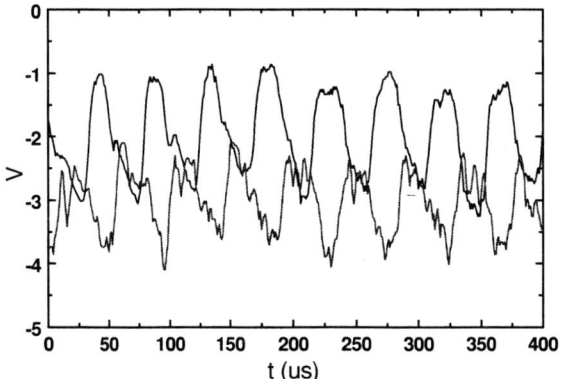

FIGURE 2. Time trace of floating potential: probes 180° apart in azimuth, in rotation region of VAC.

This work presents a summary of experiments and a theoretical model aimed at explaining the observed oscillations. In the following section, the experiments are briefly described, and this is followed by an outline of the theory and comparison with experimental results.

EXPERIMENT

Most of the experimental measurements were made using Langmuir probes: the electron temperature Te was measured in the usual way by sweeping the probe bias voltage, and the ion density n_i was deduced by measuring currents with the probe biased to the ion saturation voltage. To determine phase differences in the oscillation as a function of position in the plasma, a downstream probe at the edge of the plasma column was utilized to establish reference phase information.[3] Figure 3 shows plasma properties as a function of radius across the plasma column.

THEORY AND RESULTS

The model is two-fluid in a cylindrical geometry (r, θ, z), assuming a steady-state plasma which is azimuthally symmetric and has no axial structure in rigid rotation (see figure 3). The ion and electron temperatures are assumed uniform and the steady state ion density distribution has a Gaussian profile. Finite Larmor Radius (FLR) effects are neglected. Small perturbations are considered with unknown radial structure but standing waves in the θ direction and travelling in the z direction. Long axial wavelength solutions with zero or small electron-ion collision frequency are investigated and the radial solutions are Generalised Laguerre polynomials in limiting cases, otherwise a finite difference numerical treatment is used.

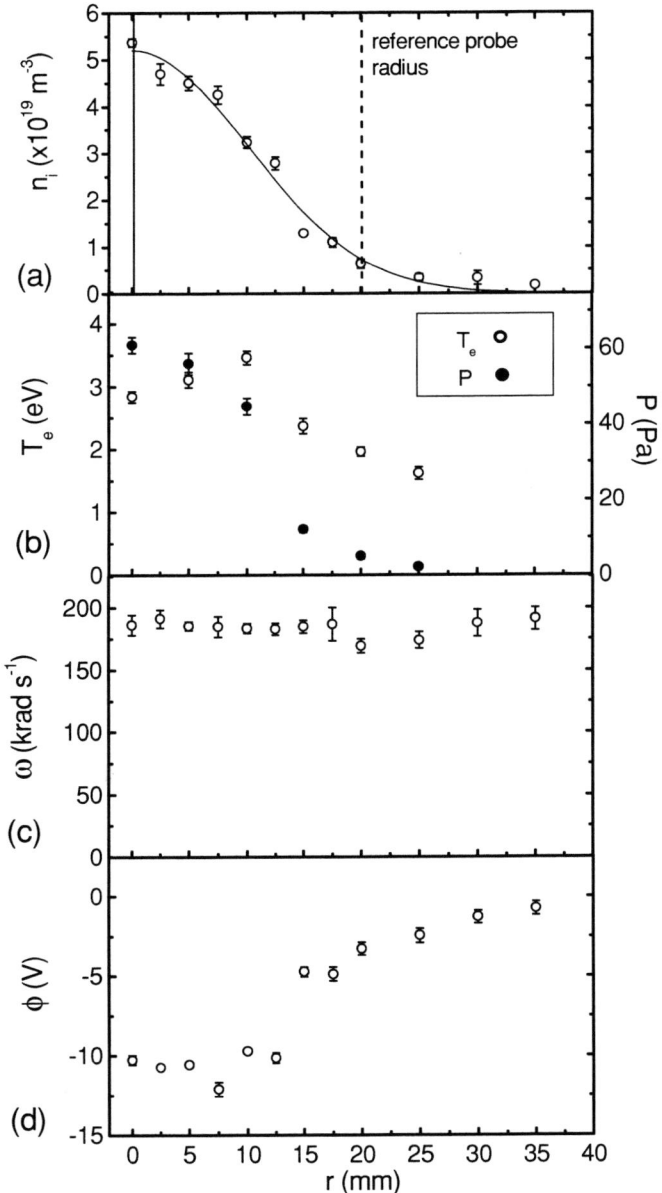

FIGURE 3. Radial profiles for $B_z=0.05T$ and $z=150mm$ (from grid) of (a) steady-state ion density, (b) electron temperature (left axis), and total pressure (right axis), (c) instability frequency and (d) steady-state floating potential.

Figures 4 and 5 compare experimental data with predictions for the density-gradient driven drift wave at maximum growth. The azimuthal mode number m is 1. The wave is close to stationary in the frame of the ion fluid, and propagates downstream in the laboratory frame. The predicted slip of the wave (rotation speed of instability compared to plasma) is 20%, which is comparable to the experimentally measured slip of 17%. Figure 4 shows that the predicted amplitude profiles of the ion density oscillation and the floating potential oscillation are similar to measured profiles, while figure 5 indicates that the predicted radial variation of the phase difference between the ion

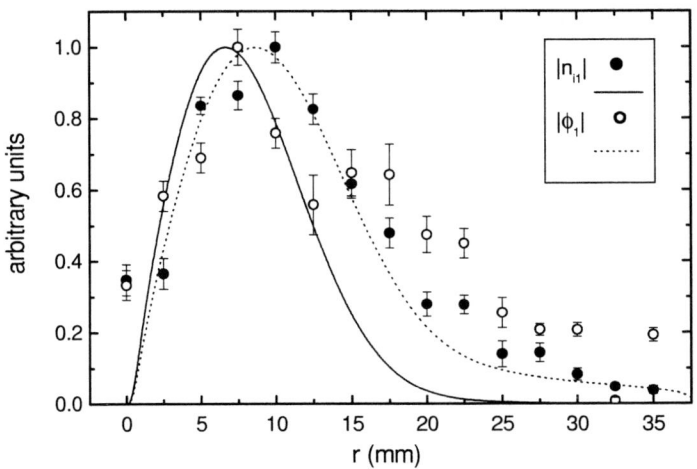

FIGURE 4. Circles are measurements of $|n_{i1}|$ (solid) and $|\phi_1|$ (open). Lines are theory predictions of $|n_{i1}|$ (solid) and $|\phi_1|$ (dotted). Measured and predicted amplitudes have been normalized to peak values.

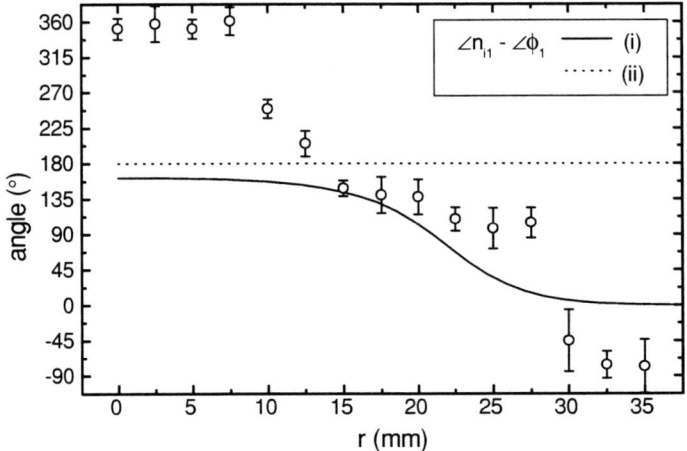

FIGURE 5. Circles are measurements of the phase of n_{i1} with respect to the phase of ϕ_1. Lines are theory predictions of the phase difference for (i) the density gradient driven drift wave (solid) and (ii) the $(m,n)=(1,1)$ centrifugal instability.

density fluctuations and the floating potential fluctuations are also in satisfactory agreement with experiment except for radii less than 10mm. Possible explanations for the discrepancy are either an incomplete model or perturbation of the plasma by the probe. Figure 5 also plots the predicted phase difference for the centrifugal instability, which is not considered a likely candidate to explain the experimental data.

ACKNOWLEDGMENTS

The authors gratefully acknowledge the support of the Department of Education, Training and Youth Affairs, the School of Electrical and Information Engineering, and James King of Irrawang for the provision of Grants-in-aid that enabled one of us (MJH) to work in Brazil. The authors also gratefully acknowledge the financial support of the Research Foundation of the State of São Paulo (FAPESP) towards the PCEN experiment.

REFERENCES

1. Krishnan, M., Geva, M., and Hirshfield, J.L., Phys. Rev. Lett. 46, 36 (1981).
2. Simpson, S.W., Dallaqua, R.S., and Del Bosco, E., J. Phys. D.: Appl. Phys. 29, 1040 (1996).
3. Hole, M.J., Dallaqua, R., Simpson, S.W. and Del Bosco, E., Phys. Rev. E. 65, 046409-1 (2002).

Nano-Microelectronics by Plasma and Electron Beam Techniques

M. Ghoranneviss[1], G. Benstetter[2], H. Hora[2,3], R. Höpfl[2],
M.R. Hantehzadeh[1], Mamoud Mardanian[1], A.H. Sari[1]

[1]*Plasma Physics Research Center, I.A.U. Tehran, Iran*
[2]*Faculty of Electr. Engineering, University of Applied Science, Deggendorf, Germany*
[3]*Dept. Theoretical Physics, University of New South Wales, Sydney 2052, Australia*

Abstract. For production of transistor electronics for lengths below the optical to ultraviolet wavelengths, technologies with plasmas and electron- ion- and neutral beams are studies where a combination with laser driven ion beam sources is involved. The same can be used for production of p-n junctions in silicon, gallium arsenide or conducting polymers diodes and solar cells.

INTRODUCTION

The present technology for miniaturization of microelectronics is limited by the optical wave lengths. A next step is under development to go to the wave length of the far-ultraviolet region. As a next step for a further reduction of the length scale down to about 10 nm is possible by plasma and electron beam techniques using the sub-threshold change of n-silicon into p-silicon by electron beams and its reversion at higher electron beam intensities [1][2]. Similar observations with electron beam irradiation were reported [3]. We developed the application of these methods for producing transistors with these techniques using the electron beams to arrive at about ten times smaller length scales for micro electronics. The problem of producing contacts and wiring in these small dimensions is considered using laser driven ion sources [4] and plasma techniques for ion charge exchange for arriving at highly accurate neutral beam deposition at the necessary high level of accuracy. The laser-plasma driven ion source [4] provides the necessary high ion currents for the deposition.

PROBLEMS

Increasing numbers of smaller and smaller transistor elements based on silicon or similar semiconductor chips are used for production of microelectronics. Smaller and smaller electronic elements are produced where the length dimension reaches the limits of the optical wave length, because masking and etching techniques for doping of the semiconductors and for electrical connections are using optical methods. In order to arrive at smaller dimensions for even higher densities of the switching elements on a chip, application of optical wave lengths form the ultraviolet range are now being developed. With these wave length, an absolute lowest limit of these techniques for the chop production has been reached, if not basically different methods of microelectronics with quantum computation, one-electron-transistors etc. may provide an alternative in the future.

If microelectronics should be produced with the presently available techniques for the range below the limit of the optical wave lengths, the invention is used as described in the following methods using at least partially the application of electron-, ion-, or neutral-beam techniques.

EXPERIMENTS

According to the considered methods, the well known technique of the transition of n- into p-conducting semiconductor crystals (silicon, germanium, gallium arsenide etc.) is used for producing geometry with a smaller dimension than the optical wave length. This transition is well known [1,5,6] and is applied for production of low cost solar cells made from organic semiconductors [7].

Fig. 1 to Fig.3 of Ref. [2] describes at least the partial application of the technique for production of solar cells. A range of the width or diameter a is the clean surface on an n-silicon crystal which otherwise is covered by photoresist material such that an electron beam of a duration and dose selected according [1] is applied to transform the n-silicon up to a given depth into p-silicon. A gold layer is then covering the range. An intense electron beam is applied following the specifications of [1] for re-transformation of the thinner range of n-silicon within b into n-silicon. This is a thermal annealing process or due to plasmon interaction (see Zaikovskaya et al. in Ref. [7]). Then, using photoresist methods, a gold layer is removed from the covered range for the basis and collector contacts are established where the emitter contact is on the silicon base.

The process following this description is a partial application of optical means with photoresist application, though according to the technique, partially electron beam technique was applied in a way as it was never known before. Apart from this production of polar transistors according to this technique, additional steps for ion implantation can be used there the annealing for a thermal incorporation of the implanted additional atoms into the diamond-like semiconductor crystal is performed by electron beam irradiation using the specifications described before [1] for avoiding the difficulties of void generation Since the method using the electron beam cracking of molecular bonds is basically different from the usual method of substituting elements of the 3rd or the 5th group of the periodic system in the molecular structure of the silicon lattice, theoretical studies are concentrating on the process of the electron interaction where the energies of 100 keV or less needed an explanation. The usual threshold for the change of properties in silicon by electron bombardment needed an electron energy of more than 200 keV [1,2]. While it was suggested that the tangling bond generation may have been a purely thermal effect in connection with phonon excitation [1,2,8] it was suggested that the cracking of bonds may have been due to an intense excitation of plasmons [10]. Studies of the phonon and plasmon mechanisms for the generation of tangling bonds are under way.

APPLICATION FOR MICRO ELECTRONICS

In order to arrive at the dimensions below the optical wave length limit, the range of a the masking is being transformed form n- into p-silicon directly by fine focusing of the electron beam without using phtoresist. The problem of the contacting is being solved by patterns of very fine focused neutral beams (network patterns) using an insulator and metallic contacts. It is essential that for the focusing of the neutral beams, ion beams are used for electromagnetic controlling of the ions with subsequent transmitting foils for the ion beam for charge neutralization. The then necessary nearly point-like ion source is being realized by ion beam generation with laser beams [4]. Using relativistic self-focusing, very intense ion beams are produced by the laser where the initial ion beam diameter can have an initial diameter down to one laser wave length.

As described before, ion implantation steps for the chip production can be included. The described techniques can be used also for field effect transistors apart from the polar transistors.

The method has the advantage that the chemical methods with extremely aggressive entities (hydrofluoric acid etc.) are avoided. This is of special importance if a mass production of solar sells is considered. On the other hand the method is based on particle beam producing or thermal reduction of tangling chemical bonds. This may cause some higher sensibility of the electronics against high temperatures than the usual silicon based electronics.

CONCLUSIONS

One possible way to produce micro electronics with the size below the limit by the optical (or UV) wave length is the use of the effect of subthreshold electron beam generated changes of n-silicon into p-silicon and the possible reverting of this change [1]. After confirming this effect by a number of authors and clarifying several aspects of its properties [2,10,11] it seems to be clear that this is a basically different p- to n-changing mechanism than the always used substitution of III or V elements within the (diamond-like-molecular) crystal lattice of the silicon. Instead of the 200 keV electron energy for removing a silicon atom from its lattice place, the 75 keV electron splits only an electronic bound within the lattice such that tangling bonds are created shifting the n-conductivity into the p-conductivity. The stronger heating by the electron beam reverses this tangling bond generation. The details of these processes are being studied now experimentally and theoretically for the application for solar cells including those of organic polymers and for microelectronics. All kinds of processes are being studied, direct collisions, plamon effects or two-step processes [12].

REFERENCES

[1]. H. Hora, Zeitschrift f. Angew. Phys. 14, 9 (1962)
[2] S. Hinckley, H. Hora and J.C. Kelly, Physica Status Solidi (a) 51, 419 (1979)
[3] A.H. Sari, M.R. Hantehzadeh, M. Ghoranneviss, 25th Internat. Conf. Plasma and Ionized Gases, Nagoya, July 2001, paper 17a52
[4] E. Woryna, J. Wolowski, B. Kralikova, J. Krasa, L. Laska, M. Pfeifer, K. Rohlena, J. Skala, V. Perina, F.P. Boody, R. Höpfl and H. Hora, Review of Scientific Instruments 71, 949 (2000); F. Boody, R. Höpfl, H. Hora and J.C. Kelly, Laser and Particle Beams 14, 443 (1996); H. Hora, F. Osman, R. Höpfl, J. Badziak, P. Parys, J. Wolowski, E. Woryna, F. Boody, K. Jungwirth, B. Kralikowa, J.Kraska, L.Laska, M.Pfeifer, K.Rohlena, J.Skala, and J.Ullschmied, Czechoslov. J. Phys.52, D349 (2002); H. Hora, J. Badziak, F.P. Boody, R. Höpfl, K. Jungwirrth, B. Kralikova, J. Kraska, L. Laska, P. Parys, V. Perina, M. Pfeifer, K. Rohlena, J. Skala, J.Ullschmied, J.Wolowski, R. Woryna, Optics Communic. 207, 333 (2002)
[5] H. Hora, US Patent 3,206,336 (United Aircraft Corp.)
[6] S. Hinckley, H. Hora, E.L. Kane, J.C. Kelly, G. Kentwell, P. Lalousis, V.F. Lawrence, R. Mavaddat, M.M. Novak, P.S. Ray, A. Schwartz, and H.A. Ward, Experim. Techn. Phys. 28, 417 (1980).
[7] H. Hora, German Patent 2415399
[8] H. Hora, Appl. Phys. A32, 217, (1983)
[9] H.J. Goldsmid et al. Phys. Stat. Solidi (a)81,K127 (1984)
[10] A. Zhaikovskaya, A.E. Kiv, O.R. Niyazova, and S.V. Starodubtsev, Physica Status Solidi (a)51, 419 (1979)
[11] V.S. Vavilov, A.E. Kiv, and O.R. Niyazove, Phys. Stat. Solidi (a) 32, 11 (1975)
[12] D.E. Hill and K. Lark-Horowitz, Bull. Am. Phys. Soc. 3, 142 (1958)
[13] G. Benstetter, M. Ghoranneviss, M.R. Hantehzadeh, R. Höpfl, H. Hora, A. Sari and H. Savalouni, German Patent Appl. 10206581.0 (15.02.2002)

Surface Modification of Blood Contacting Biomedical Implants by Plasma Processes

N. Huang[1*], P. Yang[1], Y. X. Leng[1], J.Y. Chen[1], J. Wang[1], H.Sun[1], G. J. Wan[1], P. K. Chu[2], Y. Leng[3]

[1]*Institute of Biomaterials Surface Engineering, Dept. of Materials Engineering, Southwest Jiaotong University, Chengdu, 610031 China*
[2]*Dept. of Physics and Materials Science, City University Of Hong Kong, 83 Tat Chee Avenue, Kowloon, Hong Kong, China*
[3]*Department of Mechanical Engineering, Hong Kong University of Science and Technology, Clear Water Bay, Kowloon, Hong Kong, China*

Absract. Surface modification is becoming an increasingly popular method to Improve the surface properties of biomedical materials and implants. Among the techniques plasma processes have been attracting attention because of their high effect and low cost. In this paper some application of plasma processes such as plasma modification, plasma polymerization, films synthesis, and plasma based ion implantation etc to modify blood contacting biomaterials and implants are presented. The authors work on the surface modification of artificial heart valves, stents and catheter etc. are provided. The further development of this discipline is discussed.

INTRODUCTION

Biomedical implants play more and more important role for repairing or replacing human's diseased, damaged tissue or organ. The biocompatibility and durability are key properties of the implants. It was found that a single material is quite difficult to fulfil the both needs. A common approach is to fabricate a biomaterial with adequate bulk properties and followed by surface modification to improve surface properties. Among various surface modification approaches plasma processes play a important role because the effectiveness, lower cost and versatility. In this paper some recent research on plasma surface modification of blood contacting biomaterials and implants as well as the author's work are discussed.

REQUIREMENT OF THE SURFACE CHARACTERISTICS FOR BLOOD CONTACTING IMPLANTS

The ideal surface for the material contacting with blood should not cause coagulation, however up to now the any artificial material is far from the requirement. Fig.1 is the schematic description of the basic consideration of surface modification to modify the surface characteristics and the interaction of the modified material with blood.

The most important controlling factors of the material on the interaction are the surface structure and composition. However, other characteristics, such as surface energy, surface physical state, roughness, etc. could all affect the interaction process. Almost all surface characteristics of a material can be altered by plasma surface modification processes. These techniques can be classified as shown in fig. 3, and are discussed in the following section.

PLASMA PROCESSES

According to the modified surface state, plasma processes applied for blood contacting materials are mainly classified as plasma modification, plasma film synthesis and hybrid process combing with other technique such as biological molecular modification. In the plasma modification processes the matrix surface is directly interact with plasma, no additional coating deposited on the surface. Plasma modification involves excitation of a gas at reduced pressure by radio-frequency (RF) or microwave energy source. As inorganic gas plasma is generated, the ultraviolet produced by plasma and activated gas molecules interact with the organic material surface, activate the surface and produce function groups such as -OH, -OOH groups. the plasma treatments impact only a few molecular layers on the surface of a material. The activated surface provides a favorable condition for further grafting or bimolecular attachment. Plasma grafting is a further treatment of the material in a liquid or gas monomer agent after plasma treating

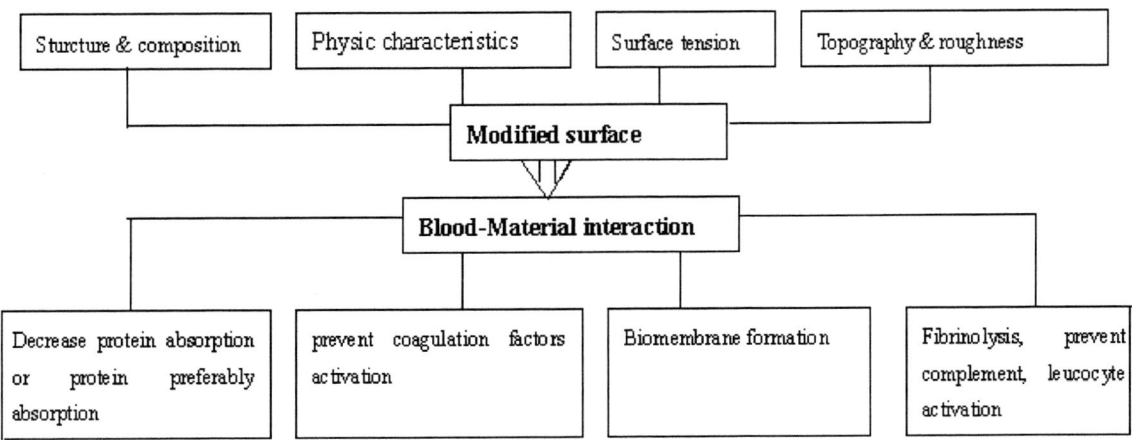

FIGURE 1. Interaction of blood with modified material surface

During plasma immersion ion implantation (PIII) process, high negative pulsed voltage is applied on the sample stage, which is surrounded by plasma, positively charged ions accelerated through plasma sheath are implanted into the target. PIII can provide much thicker modified layer than plasma treatment.

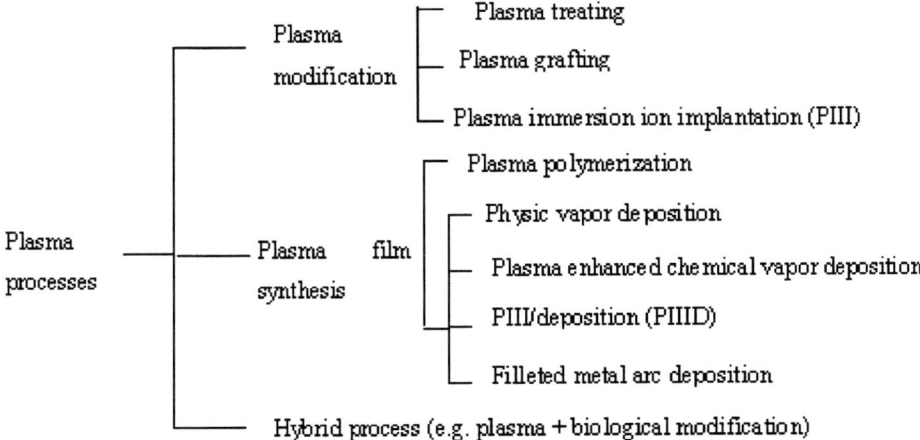

FIGURE 2. Plasma processes applied for blood contacting biomaterials

In the case of plasma films synthesis, a new layer is formed which is totally difference from the matrix. The processes comprise organic films synthesized by plasma polymerization and inorganic hard coatings fabricated by various methods such as physic vapor deposition (PVD), plasma enhanced chemical vapor deposition (PECVD), filleted metal arc deposition(FAD) and plasma immersion ion implantation and deposition (PIIID) etc.

In the plasma polymerization process, plasma generated from organic monomer. The monomer is decomposed as active energy carrying groups, and collided together or onto material surface to form a highly cross-linked thin polymer film.

The hard coatings generally formed by reaction of gas plasma and/or metal plasma with the second source. Many hard films such as TiN film, SiC film, diamond like carbon (DLC) film etc were fabricated with these processes.

MODIFICATION OF BLOOD CONTACTING BIO- MEDICAL IMPLANTS BY PLASMA PROCESSES

Plasma surface modification is very frequently applied on polymer. Polyurethane (PU) is the hopeful material as blood vessel due to its flexibility and higher durability, but it is difficult to bind endothelial cell (EC) on its surface. A PU tube with the diameter of 1.5mm was treated by atmosphere plasma and then EC was bound on the surface of plasma treated and untreated, then a liquid with the shearing stress of 9 Pa was flow over the surface for 90 minutes. It was found that on the plasma treated surface most of EC maintained tightly bound to the matrix, while on untreated PU there was no EC left[1].

ePTFE membranes exhibit the potential for blood filter application. For improving its blood compatibility, molecular weight of 600 (PEG-600) was graft in atmospheric pressure glow discharge plasma. Albumin adsorption on the PEG-modified ePTFE membranes increased with increasing PEG-600 grafted on ePTFE membranes. Fibrinogen adsorption decreased with increasing PEG-600 grafted on ePTFE membranes. On PEG-600 plasma-treated ePTFE adhering platelets evidenced no pseudopodia formation. [2].

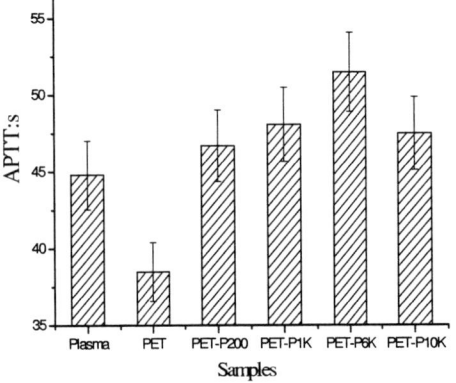

FIGURE 3. APTT of PEG grafted PET

In our recent research [3], Polyethylene glycol (PEG) of different molecular weight from 200 to 10000 was grafted on Polyethylene terephthalate (PET) films by plasma surface grafting. The results shown that the interface free energy between materials and water γ_{sw} of PEG grafted PET decreased from 30.66 mN/m of not treated to the lowest level 6.28 mN/m of PEG grafted PET with molecular weight 6000, activated partial thromboplastin time (APTT) increased from 38.5 seconds to51.5 seconds and platelet adhesion and activation behavior was improved dramatically as the grafted molecular weight of PEG was 6000, which are shown in Fig.3 and Fig. 4. We have done also novel covalent complex grafting PEG-heparin on PU and shown significantly improvement of blood compatibility [4].

The current trends of plasma surface modification of polymer materials presented the hybrid process combing plasma modification with other processes such as seeding of endothelial cell (EC), grafting of poly (ethylene oxide) (PEO) or Polyethylene glycol (PEG), heparin, albumin etc. on the surface. Plasma modification plays an important role to improving the adhesion ability of these biocompatible molecules on the surface. The long term blood compatibility in vivo has been concerned.

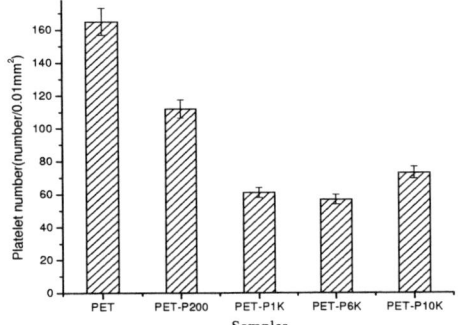

FIG. 4 Platelet adherent on PEG grafted PET

Since last decade many hard coatings such as nitride, oxide, and carbide have been investigated for the candidate of the surface coating of blood contacting implants, some inorganic films such as TiN, SiC, diamond like carbon (DLC), Al2O3 etc. films have been considered as coating materials to modify mechanical heart valve [5-8]. DLC film has also been studied to coat left ventricular assist device [9], vascular stents, and bio-sensor etc. and even be adapted to coating on polymer surface for the artificial heart. The advantages of these hard coating processes are that they can be used with components of complicated shapes, relatively lower cost, and higher hardness and wear resistance. But the blood compatibility of these kinds of films has not reported to be superior to that of LTI-carbon, which has been used as the heart valve material since last two decades.

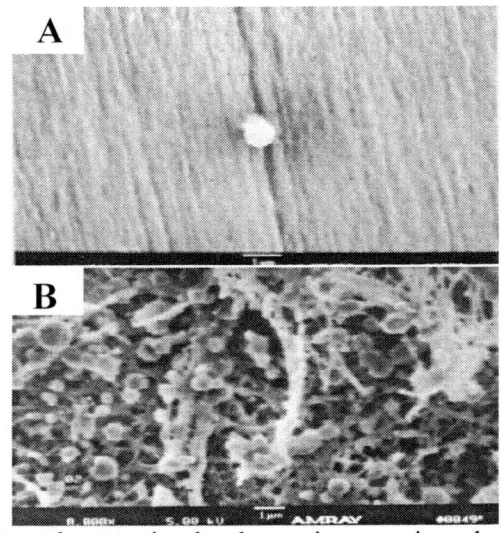

FIG. 5 In vivo results of Ti-O film coated Ti and LTIC implanted into dog's right atrium for 15 weeks, without anticoagulant
A: Ti-O film, no coagulation was formed
B. LTIC, serious thrombus was formed.

Great efforts have been paid by the authors of the present paper to develop titanium oxide material system by PIIID process. It has been proved that crystallization, doping, and non-stoicheometry of Ti-O film can improve the blood compatibility of Ti-O film significantly. Excellent blood compatibility of Ti-O film coated heart valves and stents were obtained in vitro and in vivo. Fig.5 shows the in vivo results of Ti-O film coated metal samples and compared with LTIC. It was found that the anticoagulation properties of Ti-O film were significant superior to that of LTI-carbon. The absorbed protein in single molecular state on Ti-O film surface was observed by atom force microscopy (AFM). It was suggested that the modified surface could prevent the protein from denaturation by decreasing proteins adsorption and prevention of the charges transfer from proteins into the material [10-12].

CONCLUSION

Techniques and examples of plasma surface modification applied for blood contacting implants are described in the paper. The aim of the plasma modification is to obtain a more precisely controlled surface structure for specific functions, such as rejection of proteins absorption, preferably reception of endothelial cell, binding of PEO, heparin etc. blood compatible molecular. Understand the interaction between materials and blood is very important to guide the design of the surface modification. Development and hybrid utilization of the surface modification to obtain better blood contacting biomaterials could be expected. Acknowledgment

ACKNOWLEDGEMENTS

This paper was supported by Key Basic Project of China G1999064705, High Tech project 102-12-09-01 and 2001AA320604

REFERENCES

1. Kawamoto, Y., Nakao, A., Endothelial cells on plasma treated segmented-polyurethane, adhesion strength, antithrombogenisity and cultivation in tubes. J Mater Sci: Mater in Med, 1997, 8:551
2. Zhang, Q., Wang, C., Babukutty, Y., Ohyama, T., Kogoma, M., Kodama, M., Biocompatibility evaluation of ePTFE membrane modified with PEG in atmospheric pressure glow discharg, J Biomed Mater Res 2002 Jun 5; 60(3):502-9
3. ChangJiang, P., Jin, W., Nan, H., Hong, S., Ping, Y., Yongxiang, L., Junying, C., Guojiang, W., Study on Blood Compatibility of Polyethylene terephthalate (PET) Modified by Plasma Surface Grafting, submit to Functional Mater. of China, In Chinese
4. Wang, J., Liu, Q. L., Sun, H., Yang, P., Yang, Z. B., leng, Y., Nan H., Surface modification and antithrombogenicity of medical polyurethane, 5th Asian Symposium on Biomedical Materials (ASBM5), Hong Kong, December 9-12, 2001
5. Dion, I., Roques, X., More, N., Labrousse, L., Ex vivo leucocyte adhesion and protein adsorption on TiN', Biomaterials, 14(9) (1993), 712-719
6. Bolz, A., Schaldach M., Artificial heart valves. improved blood compatibility by PECVD a-SiC coating, Artficial Organs, 1990, 144(4), 260-269
7. Thomson, A., Law, F.G., Rushton, N., Franks, J., Blood compatibility of diamond-like carbon coating',Biomaterials, 1991, 12(1), 37-40
8. Dion, I., Roquey, CH., Baudet, E., Basse, B., More, N., Hemocompatibility of diamond-like carbon coating, Biomed. Mater. Engng., 1993,3,51-55
9. Monties, J.R., Dion, I, Havlik, P., Rouais, F., Trinkl, J., Baquey C., Rotary pump for implantable left ventricular assist device: Biomaterial aspects, Artificial Organs, 1997, 21(7), 730-734

10. Nan. H., Yang, P., Cheng, X., Leng, Y.X.,..Zheng, X.L, Cai, G.J., Zhen, Z.H., Zhang, F., Chen, Y.R., Liu, X.H., Xi, T.F., Blood compatibility of amorphous titanium oxide films synthesized by ion beam enhanced deposition, Biomaterials, 1998, 19, 771-776
11. Yang, P., Huang N., et al, In Vivo Study of Ti-O Thin Films Fabricated by Plasma Immersion Ion Implantation, Surface Coating & technology. In press
12. Chen, J.Y., Huang N., et al, Antithrombogenic Investigation of Surface Energy and Optical Bandgap and Hemocompatibility Mechanism of $Ti(Ta^{+5})O_2$ Thin Films, Biomaterials, In press.

Effect Of The High-Temperature Pulse Deuterium Plasma On The Surface Structure Of Vanadium And Its Physical-Mechanical Characteristics

I.V. Borovitskaya[*], A.I. Dedurin[*], L.I. Ivanov[*], O.N. Krokhin[†], V.Ya. Nikulin[†], A.A. Tikhomirov[†], A.S. Fedotov[¶]

[*] *A.A. Baikov Institute of Metallurgy RAS*
[†] *P.N. Lebedev Physical Institute RAS, Leninsky pr.53, 119991, Moscow, Russia*
[¶] *Institute of Theoretical and Experimental Physics*

Abstract. The effect of the irradiation of the high-temperature pulse deuterium plasma on vanadium is researched. Vanadium is used as one of the base elements for the composition of the low-activated alloys. Maximum energy in the plasma pulse is 3.6-4 kJ, the speed of the plasma flows is 2-$4 \cdot 10^7$ cm/s, the time duration of single pulse is about 100 ns, and number of pulses for each sample is 10. Super-deep penetration of deuterium into vanadium is observed, and as a result, vanadium becomes fragile. Microscopic research of the irradiated surface shows considerable changes in the surface structure of the outer layers, for example the appearance of bands of slips, fractures, a staircase effect of grain-boundaries.

INTRODUCTION

In the process of the operation of the thermonuclear reactor with the plasma magnetic confinement, the pulse action of the deuterium-tritium plasma on the materials of the first wall and the diverter plates of the reactor is possible. This can happen when plasma disrupts due to loss in its stability, for instance when the irregularities of magnetic fields occur. In the inertial thermonuclear fusion, expanded deuterium-tritium plasma effects on the operating ability of chamber materials. The pulse action of the high-temperature plasma on the constructive materials of thermonuclear devices creates the possibility of damage and destruction of these materials. Especially, this can occur if the materials are subjected the periodic elastic deformation, which can occur in cycle reactors. In this work, to study the effects of the high-temperature pulse hydrogen plasma on vanadium which is chosen as one of the basic element for the composition of the low-activated alloys with a fast decay of the induced activity.

THE EXPERIMENTAL SETUP

The high-temperature deuterium plasma stream was formed by the experimental Plasma Focus Installation "Tulip" at the P.N. Lebedev Physical Institute. The total energy of the plasma focus pulse was 3.6 to 4.0 kJ, with a current at 400 kA. The energy range of deuterium ions was between 10 and 200 keV. The speed of the deuterium plasma stream was 2–$4 \cdot 10^7$ cm/s, with the plasma density at $\sim 10^{18}$ cm^{-3}. The time duration of the deuterium plasma pulse did not exceed 100 ns, which is consistent with the experimental values of the period of plasma disruption in the thermonuclear reactor with the magnetic confinement. The investigation of changes in the physical-mechanical characteristics of vanadium was performed within 10 pulses of plasma. The time interval between pulses was 3 minutes. According to the calculations and direct measurement method, the temperature on the rear side of samples did not exceed 600 °C.

The electropolished plane samples of pure vanadium were used in the experiment. The thickness of samples varied from 0.29 to 0.55 mm. The samples were placed at a specified distance apart from the anode of the Plasma Focus installation. Maximum input power on the sample in single pulse was not exceed 10^8 W/cm^2.

RESULTS AND DISCUSSIONS

The experiments show that the central bend of the sample surface after the plasma acts on it depends on their thickness. For example, with the thickness of vanadium sample 0.29 mm and the diameter of the plasma pinch at 11 mm, the bend of the vanadium sample is 0.29 mm. The sample with the thickness 0.55 mm was bent by 0.18 mm. In both cases, the samples were placed 10 mm apart from the anode. Figure 1 shows the surface of the vanadium sample with the thickness of 0.29 mm after it is irradiated. One can see the formation of stretched crests, which form the so-called periodic running waves of deformation. They are especially visible in the peripheral part of the sample. The crests are chaotic and their shapes are changed in the central part of the sample.

Such distribution in the visible surface disturbances show that the intensity of the plasma stream in the Plasma Focus installation is irregular. It is greater in the center [1]. Due to the spread of the periodic waves of deformation, a thickening of the edges of the sample is observed. The vanadium sample with the original thickness of 0.29 mm exposes a thickening of 0.09 mm. This shows that some of the material in the sample is shifted from the center to the periphery under the action of deformation waves.

The physical model of the appearance and the propagation of periodic running waves of deformation in real crystalline structure and their dissipation are discussed in papers [2] and [3]. Deformation of vanadium by the running deformation waves leads to significant changes in the structure of the rear layers: bands of slips appear in grains of poly-crystallized vanadium and grain-boundaries have a stair-case structure (Figure 2a). Also, the deep extended cracks appear (Figure 2b), which are not presented in the non-deformed materials. A multitude of small-size extractions and large separate circular particles are seen on the sample surface also.

Besides, according to the scanner tunnel microscopy, on the surface of grains, directed wave-like structures are formed also. The extraction of spherical particles from these grains is observed also. The size of the some of these particles does not exceed 200 Å (Figure 3). In accordance with a diagram of the state of vanadium-deuterium [4], deuterium with the concentration up to 40 at.% creates interstitial solid solutions with vanadium, in which the δ-phase of vanadium deuteride is present. With the increase of the deuterium concentration in the solid solution, the concentration of the δ-phase increases. Based on this data, one can conclude that the observed extractions belong to hydride formations of vanadium.

It is interesting to evaluate the depth of the diffusive penetration of deuterium into vanadium at typical isothermal conditions at a temperature 900 K° during the one pulse period 100 ns. The diffusion coefficient of deuterium under the chosen temperature is taken from paper [5] and equals $1.5 \cdot 10^{-4}$ cm^2/s. The depth of the deuterium penetration $x = \sqrt{Dt}$ during the period of one pulse will not exceed 0.1 μm. Under the multiple irradiation of the sample (10 pulses with 3 minute interval), because of the dissipation of shock waves that pass through poly-crystallized structure, point defects (vacancies and interstitial atoms) appear and the dislocation structure of the material is changed. This can considerably change the value of the depth of the deuterium penetration into the samples [6, 7]. One can make such conclusion based on the values of the micro-hardness of the irradiation and non-irradiated samples of vanadium (load P=50g). Thus, the micro-hardness of the vanadium samples with a thickness of 0.29 mm under the pulse irradiation at a distance of 10 mm from the anode, on the irradiated and non-irradiated side, equals 219 and 210 kg/mm^2 accordingly; the micro-hardness of the original vanadium sample was 105 kg/mm^2. These changes in the micro-hardness, depending on the original thickness of the samples, correlate with the observed bending of the samples. The bend of a "thick" sample was 1.6 times smaller that that of a "thin" one.

From this one can conclude, shock waves that appear due to the pulse action of deuterium plasma on the surface of vanadium lead to plastic deformations of the samples and stimulate the super-deep penetration of deuterium in comparison with that at the thermal diffusion. As a result of this, vanadium becomes fragile, fractures (cracks) appear on the surface layers, and the hardness is significantly increased.

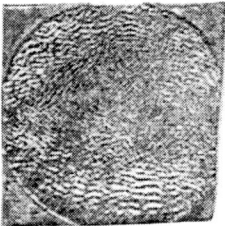

FIGURE 1. The structure of the surface of the vanadium sample exposed to the pulse of deuterium plasma. The thickness of the sample is 0.29 mm. The size of the exposed area is 11 x 7 mm.

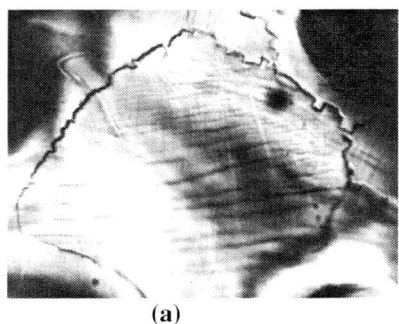   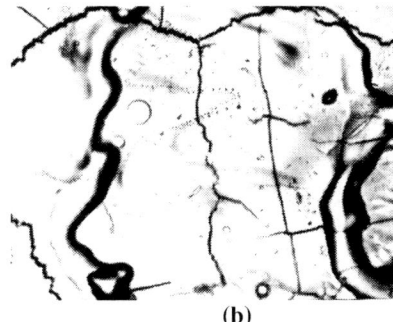
(a)   (b)

FIGURE 2. a -The center of the vanadium sample exposed to the irradiation of plasma: a – the sample with thickness 0.55 mm placed at distance of 10 mm from the anode of the Plasma Focus.   b – the sample with thickness of 0.32 mm placed at a distance of 32 mm. One can see the appearance of cracks.

FIGURE 3. The scanning tunnel microscopy of the center of the vanadium sample with thickness 0.29 mm. The irradiation of the sample was done by pulses of deuterium plasma at a distance of 10 mm from the anode of the Plasma Focus. The size of the area is 1.2 x 1.2 μm.

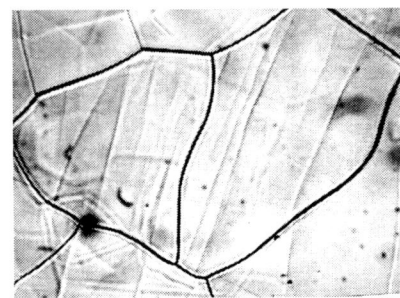

FIGURE 4. The structure of the rear side of the vanadium sample (thickness is 0.32 mm) exposed to 10 pulses of deuterium plasma. The sample was placed 32 mm from the anode. Magnification is 440.

This conclusion is in the agreement with the paper [8] in which they found that the saturation of vanadium with hydrogen at the isothermal conditions with concentration up to 33 at. % causes an increase of hardness up to 240 kg/mm^2. It is worth to point out that the paper [7] experimentally shows that structural defects created by shock waves have irregular volume distribution. This could lead to a significant concentration irregularity in the distribution of deuterium in vanadium. As a result, vanadium deuteride will be distributed irregularly in the investigated samples.

Finally, we would like to note an interesting fact that was observed in our experiment: the shock waves of compression reach the non-irradiated rear vanadium surface and lead to the exposure the change of its structure (Figure 4), i.e. the effect of cumulated etching of the surface is observed.

CONCLUSIONS

The effect of super-deep penetration of deuterium is observed under the pulse action of deuterium plasma with energy levels of up to 4 kJ and the pulse duration of 100 ns. As a result, vanadium becomes considerably more fragile.

Surface morphology of vanadium under the pulse action of deuterium plasma is formed by the periodic running deformation waves propagation from the center of the action to the periphery. This causes the displacement of the material.

The formation of a microwave-oriented structure is observed on the surface of grains of the poly-crystallized vanadium under the pulse action of deuterium plasma.

ACKNOWLEDGMENTS

We are grateful for the support of this work to Ministry of Industry, Science and Technology of Russian Federation (contract # 40.006.1.1.1129) and Center of Integration (project # Б0049).

REFERENCES

1. Gurei, A.E., Krokhin, O.N., Nikulin, V.Ya., Polukhin, S.N., Tikhomirov, A.A., Safronova, T.V, and Volobuev, I.V., "Investigation of cumulative flows in plasma focus" in *Plasma 2001*, Conference Proceedings of Int. Symp., IFPiLM, Warsaw, 2001.
2. Mirzoev F., Shelepin, L., *Journal of Theoretical Physics* **7**, 1-9 (2001).
3. Ivanov, L.I., Litvinova, N.A., Yanushevich, V.A., *Problems of strength* **6**, 99-101 (1978).
4. *Diagrams of state of double metallic systems*, Mashinostroenie, Moscow, v. 2, 1991,1023.
5. Volkl, J., Alefeld, G., *Applied Physics* **28**, 321 (1978).
6. Yanushevich, V.A., *Physics and Chemistry of Material Treatment* **2**, 47-51 (1979).
7. Ivanov, L.I., Litvinova, N.A., Yanushevich, V.A., *Physics and Chemistry of Material Treatment* **2**, 3-6 (1976).
8. Antonova, M.M., *Properties of Hydrides of Metals. Reference book*, Naukova dumka, Kiev, 1975, p.93.

Material Processing By Plasma Shock Wave Generated In An Inverse Z-Pinch "Plasma Expander"

Emilia A. Aramaki*, Mario B. de Moraes, Munemasa Machida

Universidade Estadual de Campinas - UNICAMP, Instituto de Física "Gleb Wataghin" ,Cidade Universitária, Campinas, S.P. Brazil, CEP 13083-970
* Universidade Estadual Paulista "Júlio de Mesquita Filho" , Faculdade de Engenharia, Departamento de Física e Química, Av. Ariberto Pereira da Cunha 333, Pedregulho, 12516-410, Guaratinguetá, SP, Brazil

Abstract. The applicability of plasma shock wave for material processing was investigated using modified inverse Z-pinch device. Shock wave expanding speed and plasma spectral analysis were studied using an internal magnetic probe and spatially collimated light spectroscopy. The material processing capability of the device was shown by many different surface analysis techniques such as AES, IRS, EPM and SEM. The interactions between a plasma shock wave of ~4x10^6 cm/s speed with a Si substrate surface shows some ion implantation capability using a nitrogen plasma and thin film formation using a methane plasma.

INTRODUCTION

Direct current, radiofrequency and microwave plasmas are often used for material processing but other techniques, covering distinct plasma conditions, could be of interest. In this work we investigate the possibility of surface treatment and thin film deposition, using plasma shock waves generated in an inverse Z-pinch machine. These machines have been used for more than 40 years for studies on the generation of plasma shock waves [1,2] with high density plasma column. For our investigation, a Z-pinch machine was modified in order to have higher repetition rate, easiness to set and change sample holder, and still keeping the shock wave characteristics. Such device will be from now on called a Plasma Expander.

For a simple shock wave analysis, we can use the snowplow model [3] with RLC circuit equation from which the radial speed, u_c, of the shock wave can be deduced:

$$u_c = \left(\frac{V_0 \mu_0}{L^2 2\pi^2 \rho_c} \right)^{1/4} \qquad (1)$$

In Eq. (1), V_0 is the maximum charging voltage, L is the circuit inductance, ρ_c is the density at the compression point and μ_0 is the vacuum permitivity. Once the compression speed is determined, the ion temperature can be found using the equation:

$$\frac{1}{2} m_i u_c^2 = k_B T_i \qquad (2)$$

where m_i is the ion mass, T_i is the ion temperature and k_B is the Boltzmann constant.

Typical values obtained from this analysis are: $u_c = (1 - 5) \times 10^6$ cm/s, $T_i = 4\text{-}10$ eV, and $n_e = (1\text{-}4) \times 10^{15}$ cm^{-3}.

EXPERIMENTAL

The construction of a Plasma Expander is easy and of low cost. Figure 1a is an schematic representation of the Plasma Expander used in this work, while Fig. 1b shows, in some detail, the plasma chamber, consisting of two parallel plate electrodes of 30 cm diameter, separated by a 15 cm height cylindrical glass tube. A small central conductor insulated by glass or ceramic tube passes trough a central hole of the lower plate and is connected to upper plate .The power supply and the capacitor (few tenths of µF) is connected to the central conductor through a spark gap switch. The chamber is pumped by a 70 l/s turbomolecular pump to 10^{-5} torr, and filled with N_2, CH_4, He, or Ar.

When the spark switch closes, the current passes through the central copper conductor and returns as a hollow cylindrical arc through the gas to the grounded lower plate electrode. The current in the gas interacts with the self-magnetic field produced by the current in the central conductor producing a Lorentz force directed radially outward as can be seen in Fig. 1b.

The shock wave experimental analysis was carried out using internal magnetic probes and double optical fibers for space-resolved optical measurements.

Figure 2 shows the magnetic signal time variation for several pressures. In the figure, the spikes appearing at each half-cycle characterize the passage of the shock wave at the magnetic probe.

In Fig. 3, the space resolved optical measurements show that up to ~5 cm from the center, the shock wave is still the same, and even at higher pressure of 140 mTorr, not shown here, the shock wave is still formed and possess expanding speed of 1.3×10^6 cm/s.

We notice also that, unlike low-density plasma devices, species such as NII, HeII are easily observed by optical spectroscopy in the Plasma Expander.

For material analysis different techniques were applied. To observe nitrogen plasma interaction with the Si substrate, Auger Electron Spectroscopy (AES) and nanoindentation techniques were used. To characterize the films deposited from the CH_4 plasma in Si substrates, Scanning Electron Microscopy (SEM), IR spectroscopy and Raman Scattering were employed. Perfilometry was also used to check film thickness.

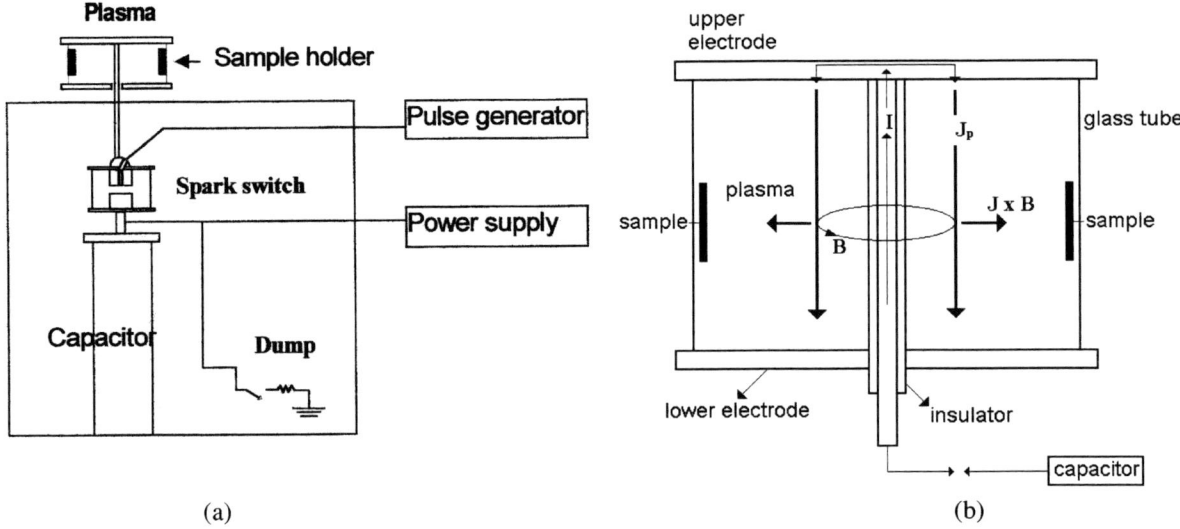

FIGURE 1. (a): Schematic view of Plasma Expander where plasma is produced between two parallel plate conductors. (b): Inside parallel plate conductors, where shown current flow and interaction with self-magnetic field producing always outward **JxB** force.

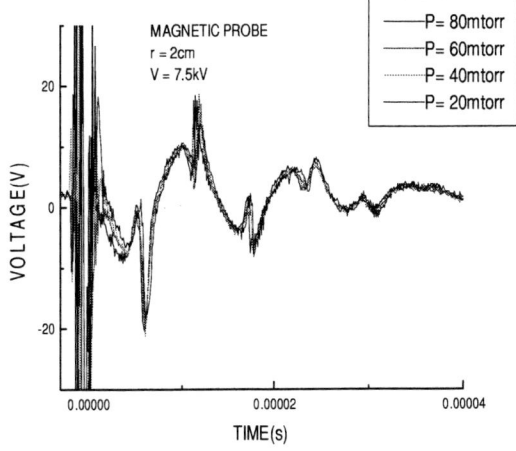 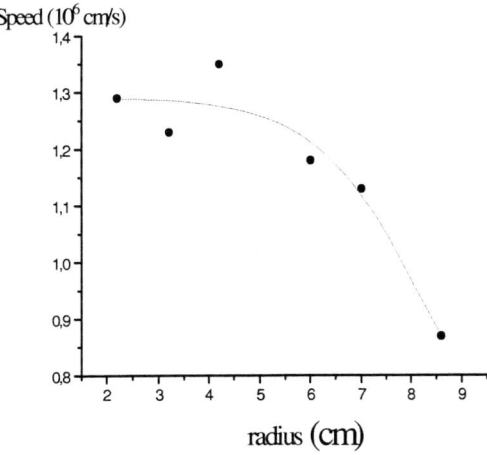

FIGURE 2. Internal magnetic probe signal for four different pressures. Each spike is shock wave passing through the probe.

FIGURE 3. Expanding speed versus radial position (cm) using two position photodiode in N plasma.

In Fig. 4, the AES depth profile of a Si target exposed to 8000 discharges of nitrogen plasma produced by Plasma Expander is shown. A large concentration of O and Fe is present in the first 30 nm of the substrate. For the next 40 nm Si is predominant with about 20 % implantation of other materials. For depths larger than 55 nm the concentrations of O and Fe start to increase, reaching O (40%), Si (40%) and Fe (20%) around 85 nm. Nitrogen is present in a small amount only up to 45 nm, but this can be changed increasing the number of discharges.

Using a Hysitron nanoindenter, the modification in the Si substrate hardness caused by the nitrogen plasma was measured. As can be seen on Fig. 5, the hardness has decreased after the treatment with 8000 discharges. The deepness of the is in the 100 nm range and may be caused by some sputtering effect.

Figure 6 show the SEM image of a Si surface exposed to CH_4 plasma for 8000 discharges. As can be seen, the Si surface becomes covered with new material, forming a thin film of a few microns. The small bright spots have been recognized as Fe deposition.

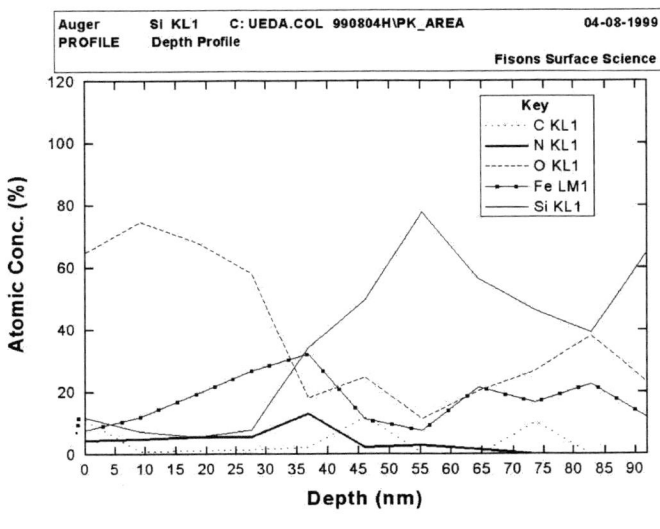 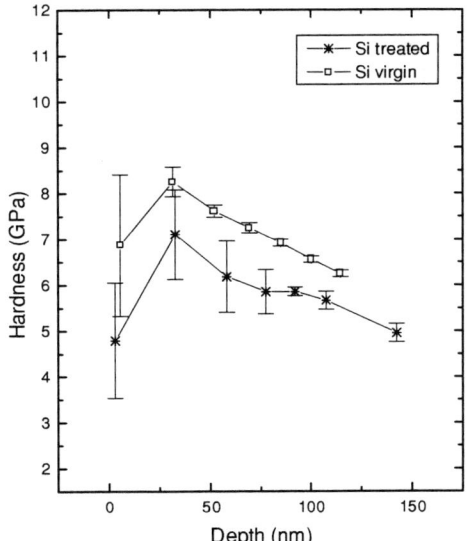

FIGURE 4. Auger deep profile, N plasma, 8000 Discharges and Si substrate.

FIGURE 5. Nanoindenter measurement of surface hardness.

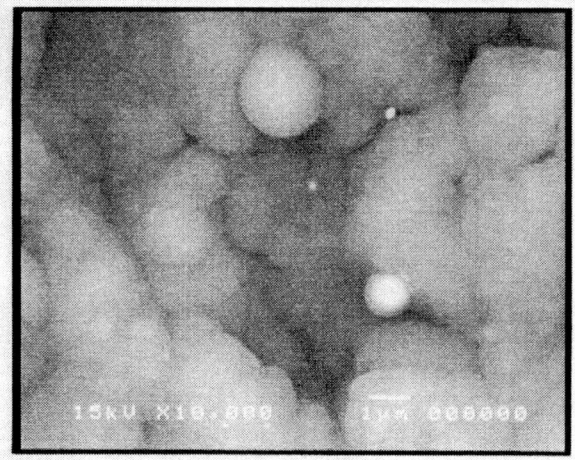
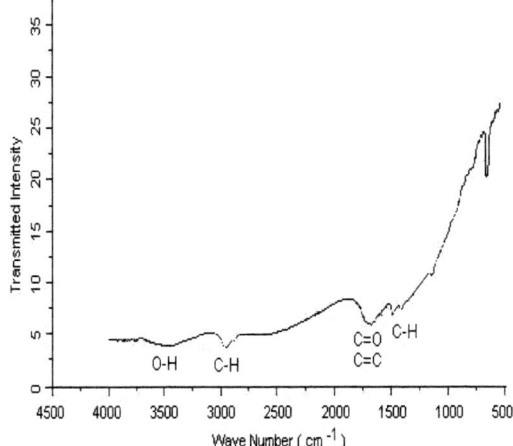

FIGURE 6. SEM image of Si substrate exposed to 8000 discharges of CH_4 plasma.

FIGURE 7. IR spectrum of a film deposited from a CH_4 plasma on a Si substrate. Number of discharges: 8000.

Figure 7 shows the IR spectrum of a film deposited from a CH_4 plasma after 8000 discharges. Formation of an amorphous hydrogenated film is indicated by the absorption bands at 2950 cm^{-1}, due to C-H bonds (stretching modes) and those at around 1650 and 1430 cm^{-1} attributed to C=O, C=C and C-H (bending mode). The O-H band at about 3300 cm^{-1} is probably due to water absorption from ambient air after deposition. Indeed, Raman scattering measurements showed only C=C peak at 1600 cm^{-1} and no DLS.

CONCLUSIONS

The Plasma Expander has proven to be an interesting new option for material treatment. Plasma analysis showed a relatively wide range of plasma parameter operation and production of highly ionized atomic elements not seen in low-density plasma devices. Some ion implantation capability has been presented by Auger analysis of nitrogen-treated Si substrate. Thin film production was tested using a CH_4 plasma on Si substrate and showed formation of amorphous hydrogenated carbon films. Further work is necessary to find Plasma Expander applicabilities to treatment of different materials as well as the effect of changing the electrode material and position, gas pressure, and the number of discharges. Owing to the easiness of device construction and operation, wide free internal space, and high plasma density, this device can bring interesting results in a short time.

ACKNOWLEDGMENTS

This work was supported by FAPESP and FINEP. The authors gratefully acknowledge Dr. H. Reuther of the Institute of Ion Beam Physics and Materials Research, Rossendorf, Dresden, Germany for the AES measurements; Dr. Lucila Cescato of Optics Laboratory., IFGW, UNICAMP, for SEM image; Dr. Elidiane C. Rangel and Dr. Nilson C. da Cruz of DFQ, FEG, UNESP for the nanoindentation measurements; and Dr. Mario Ueda of LAP, INPE for clarifying discussions.

REFERENCES

1. Vlases, G. C., J. Fluid Mechanics 16, 82-96 (1963).
2. Weisbach, M. F., and Ahlstrom, H. G., The Physics of Fluids 15, 1459-1468 (1972).
3. Bittencourt, J. A., Fundamentals of Plasma Physics, Co-Edition FAPESP, São José dos Campos, 1995, pp. 325-341.

Argon Ion Induced Changes On Cadmium Iodide Thin Films Using Dense Plasma Focus Device

R.S. Rawat[1], P. Lee[1], S. Lee[2], P. Arun[3] and A.G. Videshwar[3]

[1]*Natural Sciences, National Institute of Education, Nanyang Technological University, Singapore*
[2]*International Centre for Dense Magnetised Plasmas, Warsaw, Poland*
[3]*Depatment of Physics, University of Delhi, Delhi, India*

Abstract. The pulsed beam of energetic argon ions that is generated in dense plasma focus device is used to irradiate the as-grown cadmium iodide films. The film samples were exposed to energetic argon ions at the different distances from the top of the central electrode. The exposed samples were then analyzed for their structure, surface morphology, optical property and elemental composition using X-ray diffraction (XRD), scanning electron microscopy (SEM), UV-visible spectroscopy and energy dispersive X-ray spectroscopy (EDX). The comparison of the unexposed and energetic ion exposed film samples is discussed in the present paper.

INTRODUCTION

The dense plasma focus [1] is a simple pulsed coaxial plasma accelerator that forms a short but finite two-dimensional non-cylindrical Z-pinch, at the end of the central electrode (anode), to compress the plasma to very high densities ($\approx 10^{25-26}$ m^{-3}) and high temperatures (≈ 1-2 keV). The focus device is a source of highly energetic ions (with energies in the range of ten to hundreds of keV), relativistic electrons and abundant amount of x-rays. The energetic ions from plasma focus device have been used for inducing phase change in materials [2,3] and for deposition of thin films [4,5].

In this paper, we report the application of energetic argon ions from dense plasma focus device for irradiation of cadmium iodide (CdI$_2$) thin film. The cadmium iodide form layered films with anisotropic properties due to its crystal structure. This makes it an interesting material for material scientists. Its photochemical properties make it useful for micro-recording and digital information recording. The films are exposed to energetic ions at different distances from the top of the central electrode and comparison of unexposed (as-grown) and argon ion exposed films is done with the systematic study of their structure, surface morphology, elemental composition and optical property.

EXPERIMENTAL SETUP AND METHODOLOGY

A simple, 3.3 kJ, Mather-type dense plasma focus (DPF) device, designated as United Nation University/International Center for Theoretical Physics Plasma Focus Facility (UNU/ICTP PFF), is used for irradiation of cadmium iodide thin films. The experimental setup along with the focus subsystems is sketched in fig. 1. The film sample was mounted inside the DPF chamber, axial above the anode, with the help of axially movable sample holder. Thin film samples were exposed to DPF shots at distances from 5 to 10 cm from the top of the anode. An aperture assembly, combination of two collinear apertures, was mounted between the film sample and the top of the anode to stop the plasma shock wave as well as to reduce the amount of copper debris (that may be contributed by copper anode) on the sample to be irradiated.

The DPF device was operated at a charging voltage of 14 kV and argon filling gas pressure of 1.5 mbar. It has been noted, that after each fresh loading of film sample at desired distance from the top of the anode, it takes few DPF conditioning shots for strong focusing that is essential for generation of high-fluence energetic ions. A shutter was placed, as shown in fig.1, between the aperture and the film sample to avoid exposing the sample to low-energy low-fluence ions while optimizing the DPF device for strong focusing. The shutter was removed once the strong focusing, as indicated by voltage probe signal, was achieved thereby exposing the sample to energetic ions in the next DPF shot. All the samples were irradiated to only one focus shot.

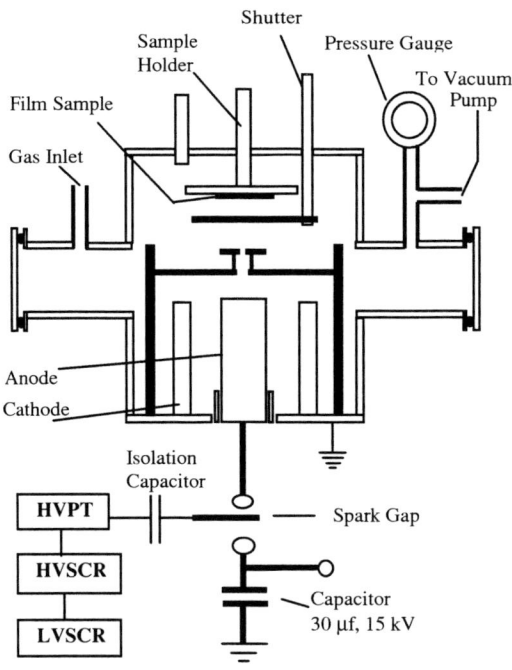

FIGURE 1. Schematic of experimental setup for thin film irradiation.

The CdI_2 thin films were grown on glass substrates at room temperature by thermal evaporation using a molybdenum boat at a pressure better than 10^{-6} Torr. The starting material was analar grade powder, which was palletized for evaporation. The deposited films were translucent as their thickness was more than 100 nm. The crystalline structure of the films was characterized by X-ray Diffraction (XRD) using Siemen's D 5005 X-ray diffractometer. Jeol JSM 5310-V Scanning Electron Microscope (SEM) was used to study surface morphologies of the films. The stoichiometery of the film was established using Energy Dispersive X-ray spectroscopy attached to the SEM. The optical absorption measurements were performed on Varian Cary 50 UV-VIS spectrophotometer.

RESULTS AND DISCUSSION

The X-ray diffraction patterns of the unexposed and exposed CdI_2 thin film samples are shown in fig.2a. The as-grown CdI_2 thin film were polycrystalline with most intense diffraction peak at $2\theta=12.86^0$, which corresponds to (002) plane. Other noticeable reflections are (004), (006), (110), and (008). The absence of (111) reflection and other l-reflections, except (110) peak, indicate the strong c-axis alignment of the as-grown unexposed thin films of CdI_2. The observation of (102) peak indicates the misalignment among the grains. The exposed samples remain polycrystalline with similar diffraction peaks however it has been noticed that the intensity of diffraction peak corresponding to (002) plane initially decreased with the increase in distance of exposure and then started to increase again at higher distances of exposure. The samples exposed at 6.0, 7.0 and 8.0 cm have strong preferred orientation towards (110) and (102) planes as seen by the significant increase in peak intensities for these plane. In fact the intensity of (102) diffraction is most for these distances of exposure. Other l-reflections that have appeared at these distances of exposure are (112) and (114). This shows that exposure of CdI_2 thin film to energetic argon ions has destructive effect on the c-axis alignment of the as-grown film. The grain-size of the unexposed and exposed thin films samples was calculated using Scherrer's formula for the most intense reflection. The grain-size of the unexposed film was calculated to be 50 nm. The grain-size variation of the exposed samples with the distance of exposure is shown in fig.2b. It can be noticed that the grain-size decreases as the c-axis alignment of the exposed samples is tampered by the preferred orientation of the exposed samples to l-reflections.

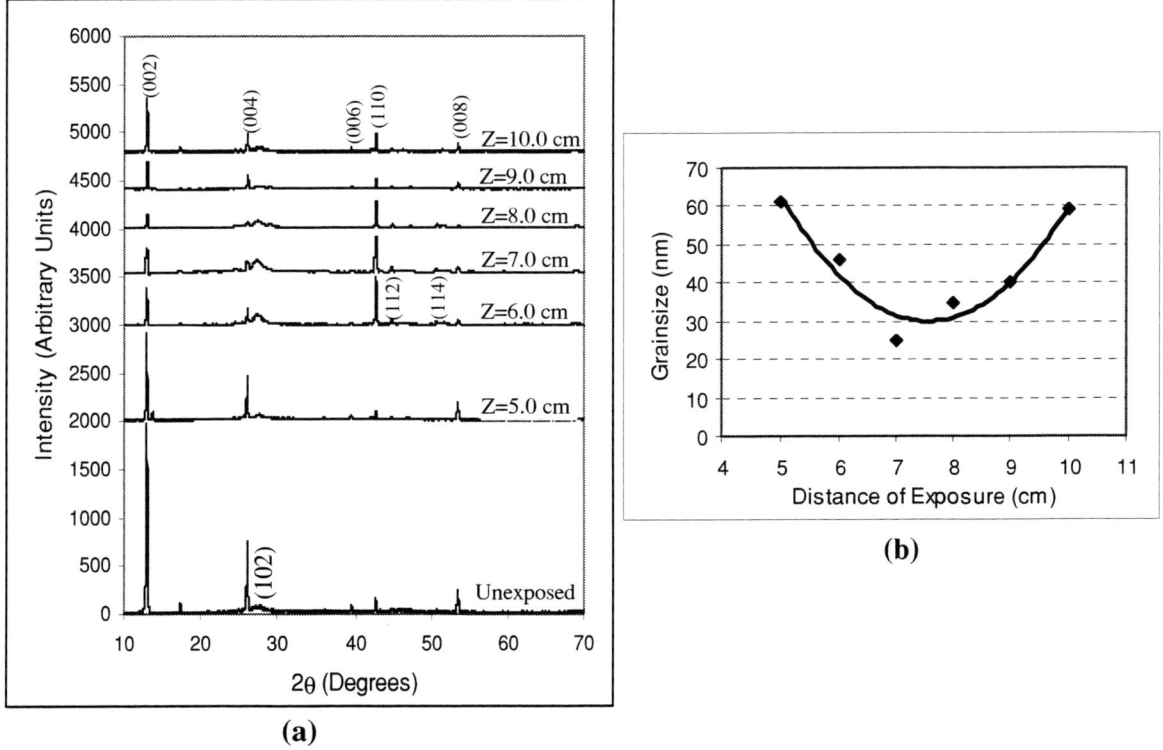

FIGURE 2. (a) XRD pattern of unexposed and exposed film samples. (b) Grain-size variation with the distance of exposure.

The band gap of the unexposed and exposed samples was estimated using the UV/Visible absorption spectra. The typical optical absorption curve as a function of wavelength is shown is fig.3a. The steep rise in the absorbance near the absorption edge hints at a direct type transition. The absorption coefficient α is calculated using $\alpha = (2.303A)/t$ where A is the absorbance and t the film thickness.

The band gap is determined by plotting a graph between $(\alpha h\nu)^{1/n}$ and $h\nu$ and then to look for that value of n which gives best linear graph in the band edge region. The best linear fit for exposed as well unexposed samples was obtained for $n=1/2$, i.e. a direct transition. The variation of band gap (direct) as a function of distance of exposure of the film samples is shown in fig.3b. The energy band gap (direct) of the unexposed sample was estimated to be between 3.25 to 3.3 eV for different unexposed samples.

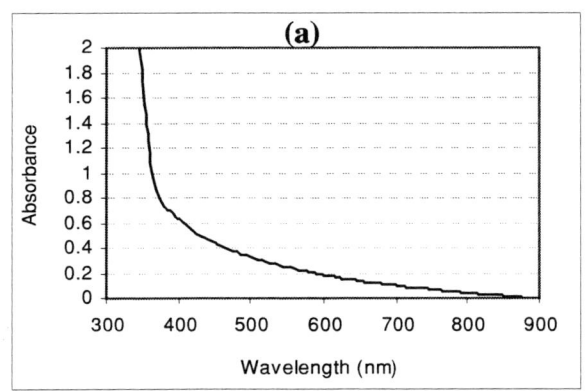
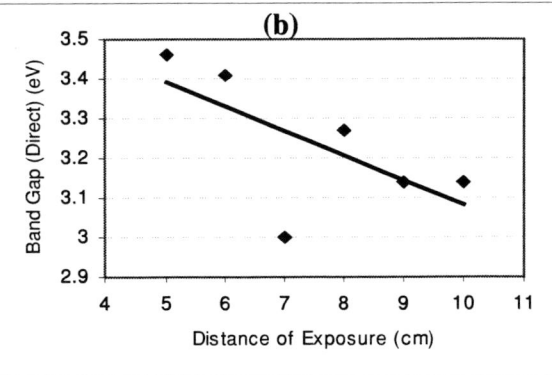

FIGURE 3. (a) Typical optical absorption spectra and (b) estimated direct band gap at different distance of exposure.

The surface morphology of the unexposed and exposed film samples is shown in fig.4. The unexposed film sample has grains, clearly visible, all over the surface. It may be noted that the spherical or flake like structures on the surface are actually the conglomeration of many smaller sized grains that could be seen only at much higher resolution. The whole surface is covered equally with these spherical or cylindrical flake kinds of structures. The exposure of the film to energetic ions of the plasma focus device has resulted in the smoothening of the sample surface to more uniform one as shown in fig.4b. This smoothening has decreased with the increasing distance of

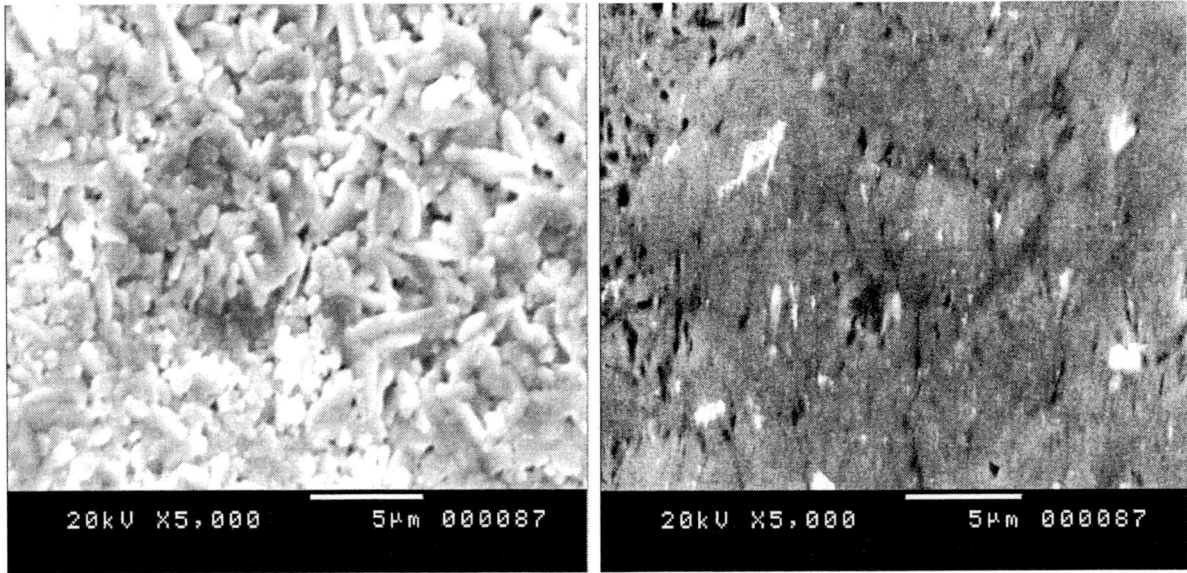

FIGURE 4: SEM pictures of (a) unexposed sample and (b) sample exposed at Z=5.0 cm.

exposure and the surface morphology is similar to the one shown in fig.4a except for the fact that size of grain conglomerate is much bigger. The EDX scanning attached to the SEM machine confirmed that the unexposed as well the exposed samples are stoichiometric.

CONCLUSIONS

The irradiation of thin film of CdI_2 to the energetic argon ion at the different distances from the top of the central electrode has resulted in decrease in c-axis alignment of the as-grown thin films. The grain sizes as well as the energy band gaps of the exposed samples have been affected by ion irradiation. The surface morphology of the exposed samples indicates the smoothening of the film surface due to increase in size of the grain clusters. The stoichiometery of the sample is found to remain unchanged after irradiation to energetic ions.

ACKNOWLEDGMENTS

Authors (RSR and PL) are thankful to National Institute of Education for providing the ARF grant RP17/00/RSR to fund the research project under which this investigation has been performed.

REFERENCES

1. Lee, S., Tou, T.Y., Moo, S.P., Eissa, M.A., Gholap, A.V., Kwek, K.H., Mulyodrono, S., Smith, A.J., Surayadi, Usada, W., and Zakaullah, M., Am. J. Phys. 56, 62-68 (1988).
2. Rawat, R.S., Srivastava, M.P., Tandon, S., and Mansingh, A., Phys. Rev. B 47, 4858-4862 (1993).
3. Rawat, R.S., Arun, P., Videshwar, A.G., Lam, Y.L., Lee, P., Liu, M.H., Lee, S., and Huan, A.C.H., Bull. Mat. Res. 35, 477-486 (2000).
4. Kant, Chayya R., Srivastava, M.P. and Rawat, R.S., Phys. Lett. A 239, 109-113 (1998).
5. Rawat, R.S., Lee, P., White, T., Ying, L., and Lee, S., Surf. Coat. Tech. 138, 159-165 (2001).

The Effect of Carbon Dioxide and Nitrogen ion implantation of AISI 52100 Steel

Amir H. Sari[1], M. Ghoranneviss[1], M. Mardanian[1], M.R. Hantehzadeh[1], H. Hora[2,3]

[1]*Plasma Physics Research Center, Science and Research Branch, I.A.University, Tehran, Iran*
[2]*Faculty of Electr.Engineering, University of Applide Science, Deggendorf, Germany*
[3]*Dept. Theoretical Physics, University of New South Wales, Sydney 2052, Australia*

Abstract. Ion implantation has been used to modify the mechanical properties of a wide range of metals and alloys using plasma techniques for ion sources and plasma surface treatment [1]. In this study AISI 52100 steel disks, containing 1.5 wt% Cr as the major alloying element, were implanted with nitrogen and carbon dioxide ions at the energy of 90 KeV, with dose in the range 1×10^{18} to 1×10^{19} N^+_2 ions cm^{-2}, and 3×10^{18} to 1×10^{19} for co_2^+ ions cm^{-2}. Ion beam current densities and sample temperature, during implantation were 3-6 $\mu A/cm^2$ and 170°C, respectively. Experiments show, hardness of sample, increases 30-49% using N^+_2 ions, and 5-17% using co_2^+ ions. In order to explain the results, formation of beta-CrN and carbide pahses have been carried out using X-ray diffraction technique.

INTRODUCTION

A central goal of physical metallurgy is to improve physical properties through the control of microstructure. Thermal and/or mechanical treatment can have dramatic effects on mechanical properties; the microstructural changes induced by these processes are discernible with standard electron and optical microscope techniques [2].
Ion implantation has been used for producing a modification in the structure of the superficial layers of metals by formation of new crystalline phases, metastable or amorphous, and thus to improve the surface properties; it increases specially hardness and resistance of fatigue, wear and corrosion. Crystalline materials become stronger when the dislocations are made more difficult to move and multiply, and in some complex way the implantation products interact with dislocations. A review is given of the current status of activities as represented by a number of research groups [2].

EXPERIMENTAL DETAILS AND RESULTS

The steels 100 Cr6 (AISI 52100) with nominal composition (in weight percent (wt %1)) : Cr-1.52; C-10 ; Mn-0.35 ; Si-0.25 ; Ni , Cu-0.3 min. were used as disks with a diameter of 38 mm and a thickness of 0.5mm. The samples were ultrasonically cleaned in acetone and methanol. Ion implantation was done with the instrument of Institute of Accelerators Technology (IAEO), using a molecular nitrogen and carbon dioxide ions at the energy of 90 KeV for each ion species. The implantation of nitrogen was performed at the dose of 1×10^{18} and 1×10^{19} cm^{-2} and carbon dioxide ions with dose in the range of 3×10^{18} to 1×10^{19} cm^{-2} respectively . The sample temperature during implantation was 170^0 C, and the target chamber pressure 6×10^{-6} Torr initially and 2×10^{-6} Torr during implantation, and the sample ion beam current densities was between 3 and 6 $\mu A/cm^2$.

The microhardness measurements shows a strong increasing of hardness from 298 HV (unimplanted steel sample) for a load of 50 gf to about 450 HV using nitrogen implantation at dose of 3×10^{18} cm^{-2}, and also the microhardness of martensitic 100 Cr 6 (AISI 52100) increased upon implantation of nitrogen ions at the dose 1×10^{19} cm^{-2} from 298 HV for a load of l0gf to about 390 HV. The ion implantation of carbon doxide at the dose of 3×10^{18} cm^{-2} is caused the increasing of microhardness from 298 HV for a load of l0gf to about 350 HV, also we found a weak increasing of microhardness from 298 HV (unimpanted steel) for a load of 25gf to about 305 HV using carbon dioxide ions at the dose of 1×10^{19} cm^{-2}.

TABLE 1. Results of microhardness measurements performed at AISI 52100 steel substrates which implanted with Nitrogen and carbon dioxide ions, at the energy of 90 KeV.

Ion	Dose [cm^{-2}]	Microhardness [HV]	Relative Microhardness [%]
N^+_2	1×10^{18}	450	49.70
N^+_2	1×10^{19}	390	31.10
Co^+_2	3×10^{18}	350	17.15
Co^+_2	1×10^{19}	305	5.01

A quantitative surface/near - surface analysis technique was performed to establish the compositions and microstructures of four implanted samples and unimplanted one. Figure1. shows the profile of net intensity versus angle of 2θ for unimplanted sample. Lattice parameters are a=2.028 °A, b=1.171 °A (a=2.026 °A , b =1.170 °A as given by JCPDS card), which implies the presence of pure ^{57}Fe in steel. Table 2. shows the strong lines for a bcc crystalline structure (Fe phase) which found in the steel .

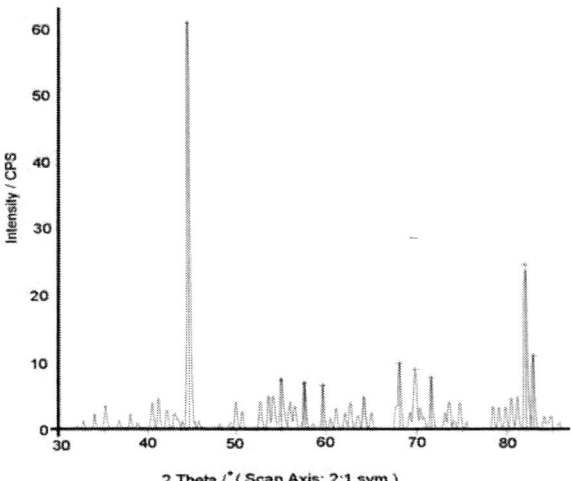

FIGURE 1. The XRD spectrum of unimplanted AISI 52100 steel .

Figure2, supports the interpretation of the microhardness results in the two implanted sample by nitrogen ions with high - dose (1×10^{18} , 1×10^{19} cm^{-2}) there are three important peaks which implies the formation of beta-CrN phase in the AISI 52100 steel . The lattice parameters for implanted steel with dose 1×10^{18} cm^{-2} that determined from figure2, are: a=2.253 °A, b= 2.041° A, c=1.636°A (a = 2.212 °A , b = 2.041°A, c = 1.640 °A as given by JCPDS card). In a similar manner , the lattice parameters for implanted steel with dose of 1×10^{19} cm^{-2} are a = 2.242 °A, b= 2.051 °A, c = 1.675 °A.

TABLE 2. The XRD analysis of unimplanted and Nitrogen - implanted AISI 52100 steel

(hki)	Relative intensity	d-spacing (A^0)	
(110)	x=100	2.0268	
(211)	30	1.1702	**Unimplanted**
(200)	30	1.4332	
(220)	12	1.0134	

(310)	10	0.9064	
(111)	x	2.2120	
(002)	80	2.0419	
(112)	21	1.6405	**Nitrogen-implanted**
(300)	15	1.3829	
(113)	13	1.2629	

In the two implanted samples using carbon dioxide ions at the dose of 3 and 10×10^{18} cm^{-2}, formation of carbide or graphite and also a broad hump in the XRD spectrum which confirm the presence of amorphous carbon was not detected.

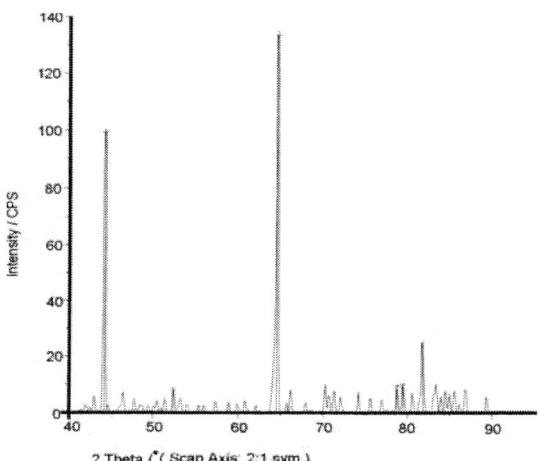

FIGURE 2. The formation of beta-CrN phase in the surface layer using N$^+_2$, at the dose of 1×10^{18} cm^{-2}. These results indicate the increasing of microhardness can be obtained as a consequence of nitrogen and carbon-dioxide ion implantation in AISI 52100 steel, but the nitrogen implantation is not comparable with the carbon dioxide implantation. This conclusion is confirmed by the lattice parameters for implanted steels.

There are too many papers which have been showed that, surface-controlled mechanical properties can be significantly modified through ion implantation [3-7]. However, the microstructual modifications that is responsible for these effects are investigating now, using advance surface microanalysis (e.g.SIMS) including phase transformation, compound formation and carbide/oxide chemical alternations..

REFERENCES

[1] Sari, A.H., Hantehzadeh, M.R, Ghoranneviss, M., "The effect of proton bombardment on electrical and optical properties of GaAs," in *25th Int. Conf. Plasma and Ionized Gases*, Nagoya, Japan, July 2001, paper 17a52.
[2] Herman, H., *Nucl. Inst. And Meth*, **182/183**, 887 –898 (1981).
[3] KUSTAS, F. M.,and MISRA, M.S., WILLIAMSON, D.L., *Nucl, Inst, and Meth B*, **31**, 393-401 (1988).
[4] Singer, I.L., in *Ion implantation and Ion Beam processing of Materials*, Vol. 27 . eds., North - Holland, New York, (1984).
[5] Kobs, K., and Dimigen, H., Ryssel, H., Kluge, A., *Appl. Phys. Lett.* **57**(16),1622, (1981).
[6] Dimigin, H., Kobs, K., Levteneckev, R.,.Ryssel, R., and P.Eichingev, *Mater Sci. Eng.* **69**, 181 (1985)
[7] Follstaedt, D.M., Knapp, J.A., and L.E.P.P. *J. Appl. Phys.* **66**, 2743 (1989).

The Evolution of Cone-Like Formations on The Cathode of Abnormal Glow Discharge of Ar

V.V. Bobkov, S.S. Alimov, V.V. Andreiev, Yu.V. Slyusarenko, R.I. Starovoitov

Kharkov National University, Kurchatov ave. 31, Kharkov 61108, UKRAINE

Abstract. The paper presents results of experimental studies and theoretical description of processes that occur during modification of cathode surface in glow discharge of Ar. The model is proposed for the experimentally observed phenomenon of the cone like microprotrusions growth. We consider the case when a strong electric field stipulated by the cathode layer of abnormal glow discharge was taken into consideration during sputtering. The model takes into account most important physical phenomena that accompany modification of the cathode surface in glow discharge. Within the framework of the model an analytical solution of the equation that describes the dynamics of surface microprofile evolution is obtained. From the analysis of this solution the evolution of cone microbulges on the cathode surface of metal in glow gas discharge is described.

INTRODUCTION

There is a constant interest to study the phenomena of originating and evolution of surface microstructures during sputtering of solid surface by low-energy ion beams. It is supported first besides only academic issues by an opportunity to use properties got by a surface in modern technologies [1, 2]. Among a variety of surface "new born" formations (whiskers [3-5], tubes [6], thin film structures of sheet shape [7] etc.) the cone-like microprotrusions occupy a separate place from the point of view of changing of field-emission properties of a surface. In particular paper [1] considers the procedure of ion-plasma processing of metal and alloy surface that to vary different physical properties of a surface by making on it advanced microrelief consisting of a great number of microprotrusions. Paper [2] discuss the aspects concerned with a change of optical and field-emission properties of cathodes after preliminary "cultivation" of cone-like protrusions on their surface.

RESULTS AND DISCUSSION

Experiment

In our experiments studies of cathode microprotrusions were carried out on the Cu specimen arranged on the Cu cathode of DC magnetron sputtering system (MSS) as shown in Figure 1. Magnetron system allow to increase a flux of sputtering Ar atoms tens times more in comparison to the usual DC glow discharge systems.

Surface morphology at different stages of target sputtering was studied by SEM and also by optical microscope. On the sputtered region of the cathode we found a great amount of cone-like microprotrusions. Such formations were observed in the full range of discharge parameters: I=1…2,5 A; U=400…600 V; p=2…6·10^{-3} Torr. Originating of the microprotrusions starts as early as the first stages of target treatment in plasma of magnetron discharge. In this case the well-known phenomenon of target erosion takes place. It leads to the

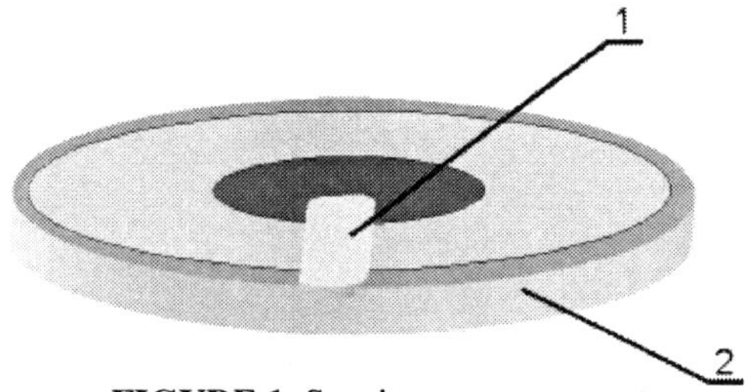

FIGURE 1. Specimen arrangement (1 – specimen; 2 - cathode-target).

CP669, *Plasma Physics: 11th International Congress on Plasma Physics: ICPP 2002*
edited by I. S. Falconer, R. L. Dewar, and J. Khachan

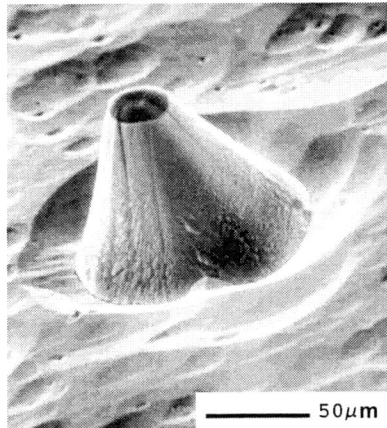
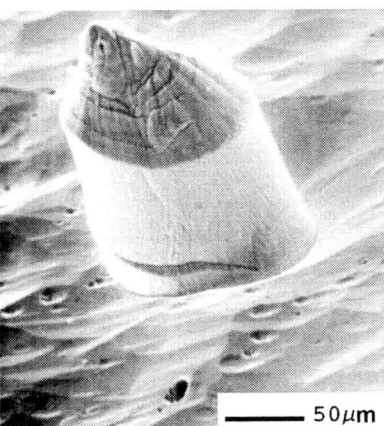

FIGURE 2. Cone like formation grown due to dot contamination on its apex.

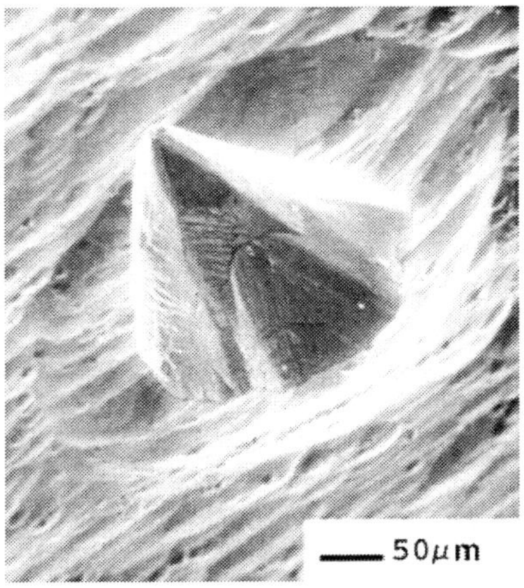

FIGURE 3. Cone like formation grown due to rise of the crystal face.

FIGURE 4. Cone like formation with columnar structure of the base.

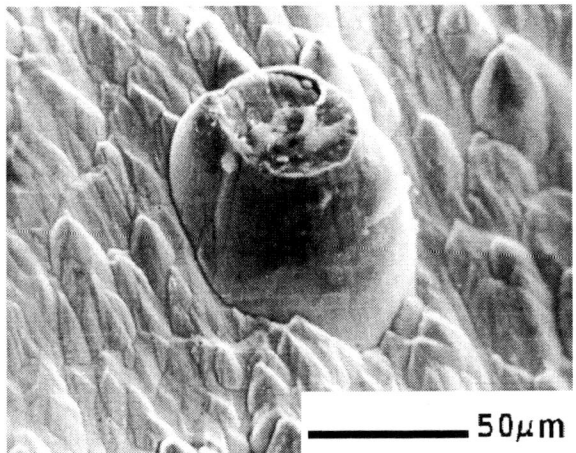
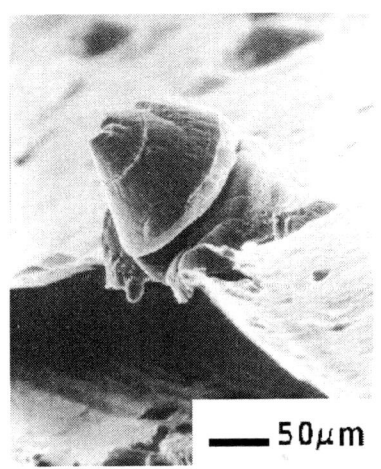

FIGURE 5. Conical formation broken as a result of melting of its apex.

FIGURE 6. Cone like formation on a Cu film substrate.

evolution of the cathode surface topography due to inhomogeneous ion sputtering of regions with different sputtering yields (Y_s). At site of local contamination or crystal face outlet with high Y_s the forming of cone-like microprotrusions may occur (Figures 2, 3 respectively) according to the classical model described in [3-5]. The height of formations increases with cathode surface sputtering under the conditions of considerable mass transfer out of the target surface.

Further evolution of these formations we obtained to go in one of the following ways.

Cone-like formations with height less than 10 microns are sputtered with time. Some part of microprotrusions continues to grow only due to the surface level lowering as a result of ion sputtering. In this case the cone peaks are not overhang the level of initial cathode surface.

Cone-like formations with height more than 10...20 microns further grow due to the grafting of cone apex. In this case the height of cone apex may overhang the level of initial cathode surface and reach the value of several hundreds of microns (Figure 4). This microstructure differs from others also by columnar structure of its base that was not detected on the same samples which being sputtered by ion beam with parameters similar to plasma ions.

We believe that in the last case a strong electric field stipulated by the cathode layer of magnetron discharge defines the mechanism of microprotrusion growth. We suppose that this phenomena run in the following way. Initial microprotrusions having the value of height and apex angle enough to initiate cold emission serves as starting point for these "new born" formations. Mass transfer leading to the growth of the "new born" formations is stipulated by the trapping of particles sputtered from the cathode surface and ionised by the field-emission flow from the apex of microprotrusion. Experiments have shown that the diffusion processes applied by other authors to describe whisker growth do not play a significant role in this case because of low temperature of our target. When the height of the "new born" formation reaches the critical magnitude that depends on discharge conditions the melting of the formation apex may occur owing to the increase of field-emission flow density or micro-breakdown leading to evaporation of the formation apex (Figure 5).

Visual demonstration of above mentioned growing of the formations with the height much more than the level of initial surface has been demonstrated by the following experiment. On the local area of Cu sample the carbon film with the thickness of about 1 mcm first and the Cu film of the same thickness second were deposited. Further the sample was sputtered on the cathode of MSS. After a certain period of time the Cu film together with carbon sublayer was investigated by SEM. The conical "new born" formations have also been detected on its surface. Such formation which have been formed at the border of the film is shown in Figure 6 in comparison with the film thickness of 1 micron.

We observed also growth of cone-like, ridge-like or tubular formations, which are formed perpendicular to the cathode surface, on the metal cathode of usual DC glow discharge in Ar. Shape of "new born" formation depends on the shape of the initial microprotrusion. From the experimental results obtained by SEM we have concluded that microstructures like cones and ridges are formed on the base of the microprotrusions of the same shape. But tubular formations are originated from droplets made by arc spot of the second type.

Theoretical Modelling

We have developed the phenomenological model of cone-like [8] "new born" formation evolution on the cathode surface in abnormal glow discharge. It takes into account most important physical phenomena that accompany modification of cathode surface in glow discharge:
– sputtering of the cathode surface by a flow of positively charged ions or neutral particles of work gas;
– drastic increase of field emission from the apex of microprotrusion during the growth process in comparison with a flat regions of cathode because of increase of drawing electric field intensity near the peaks;
– ionisation of low energy atoms knocked out from a target surface owing to the collisions with electrons emitted from the cathode surface (mainly from bulges) in accordance with previous statement;
– redeposition of said ions of cathode material mainly on the formation apex due to the increase of electric intensity field from the pedestal of formations to their peaks.

In the framework of this model the analytic solution of the equation that represents dynamic of surface profile was obtained. Exact solution of this equation with initial conditions of non-trivial profile in the form of circular cone with a pedestal centre in a point $\rho=0$, height z_0 and pedestal radius z_0/α_0 (value α_0 defines the opening angle of cone) is given by

$$z(\rho;t) = e^{\lambda t} z_0 + \frac{\Lambda}{\lambda}\left\{\left(1+\alpha_0^2 e^{2\lambda t}\right)^{\frac{1}{2}} - e^{\lambda t}\left(1+\alpha_0^2\right)^{\frac{1}{2}}\right\} - \alpha_0 e^{\lambda t}\rho. \qquad (1)$$

Parameter Λ characterizes the rate of flat surface foundering during sputtering. In the case of sputtering of flat polycrystalline target the surface level are known to founder steadily with time. Parameter λ^{-1} defines a typical time of surface level growth because of the effects concerned with relief imperfection.

It's easy to see that type of surface formation originated from initial single cone does not change. The evolution of cone-like bulges should be performed solely with sharpening of their peaks. Cone height z_0 are changed with time as

$$z_0(t) = z_0 + t\Lambda + \frac{\Lambda}{\lambda}\left\{\sqrt{1+\alpha_0^2 e^{2\lambda t}} - e^{\lambda t}\sqrt{1+\alpha_0^2}\right\} \qquad (2)$$

Thus within the framework of the mode defined by the equation (1) and above mentioned initial conditions scenario of the cone-like bulges evolution on a metal surface in glow discharge seems to be following:

Cones with initial parameters α_{0i} and z_{0i} (subscript "i" regulate distribution of cone-shaped microprotrusions on a surface at start) that does not meet the condition

$$z_0 + \frac{\Lambda}{\lambda}\left\{\alpha_0 - \sqrt{1+\alpha_0^2}\right\} > 0 \qquad (3)$$

are sputtered for a limited time τ.

Cones with values of initial parameters α_{0i}, z_{0i} fulfilling the condition (3) grows with the evolution in accordance with the formula (2). The diameters of their pedestals reach certain fixed values. This leads to the decreasing of opening angles, i.e. to the sharpening of their peaks.

However a cones growth could not continue infinitely. It is clear that after a certain period the engines suppressing infinite growth should start a work. The model evidently does not consider this fact. In our opinion the melting of cone peaks due to the local breakdowns in near cathode region may serve as one of such engines. Such breakdowns occur owing to the fact well known from the electrodynamics that intensity of electric field near the cone peak is increased rapidly with decreasing of opening angle and increasing of cone height. Melting of cones may be caused also by the phenomena of explosive cold emission due to the increase of electric field intensity up to a certain critical value close to the peaks of highest and narrowest cones.

CONCLUSION

On a real metal surfaces there exist a lot of microscopic asperities with a very broad range of parameters α_{0i} and z_{0i}. Moreover such a formations may appear during cathode sputtering. We do not consider this fact in the proposed model. It is clear that the growth rate for these cones will be different. Consequently the period of time that necessary for microprotrusions to reach the height and apex angle the value when the cone apex ruined will be also different. Therefore after a certain period of time a quasi-stationary mode should be set in the system. This consists in a certain balance between a number of originating, growing and breaking cones at any time.

One may conclude that in the case of controlled mass transfer on preliminary prepared cathode surface in discharge with parameters defined by our model there is a high probability to obtain microstructures of desired shape.

REFERENCES

1. Begrambekov, L.B., Zakharov, A.M., and Telkovsky, V.G., *Nucl. Instr. and Meth.* **B115**, 456-461 (1995).
2. Begrambekov, L.B., and Zakharov, A.M., "Ion-plasma methods for modifications of optical and field-emission properties of surfaces" in *Plasma Physics and Plasma Technologies-1997*, edited by V.S. Burakov et al., Conference Proceeding, vol.3 Institute of Molecular and Atomic Physics of Byelorussian NAS, Minsk, p. 471-472, 1997 (in Russian).
3. Begrambekov, L.B., *Poverkhnost': Physika, Khimiya, Mekhanika*, **6**, 125-131 (1986) (in Russian).
4. Wehner, G.K., *NASA Report NCR 159549*, Univ. Minnesota, Minneapolis, 1979.
5. Wehner, G.K., *J. Vac. Sci. and Technol.*, **A3**, 1821-1826 (1985).
6. Glushko, V.I., Bobkov, V.V., Ryabchikov, D.L. et al. *Izv. Akad. Nauk. Ser. Phys.*, **58**, 138-142 (1994) (in Russian).
7. Bobkov, V.V., Starovoytov, R.I., *Bulletin of Kharkov University, Ser. Phys. - Cores, particles, fields*, **421**, 201-204 (1998) (in Russian).
8. Alimov, S.S., Bobkov, V.V., Slyusarenko, Yu.V., Starovoytov, R.I., *Bulletin of Kharkov University, Ser. Phys. - Cores, particles, fields*, **481**, No. 2, 71-79 (2000) (in Russian).

Amorphous Hydrogenated Carbon Films Deposited by PECVD: Nitrogen Incorporation during Film Growth and by Plasma Surface Processing

F. L. Freire Jr.

Departamento de Física, Pontifícia Universidade Católica do Rio de Janeiro
Caixa Postal 38071, 22453-970, Rio de Janeiro, RJ, Brazil

Abstract. Amorphous carbon films are being currently used in a wide variety of applications. The properties of the films can be tuned by the deposition technique employed and by the growth conditions. One way to improve these properties is through the incorporation into the amorphous skeleton of several elements, like H, N, F and Si. In this work, we review the effects of nitrogen incorporation into hydrogenated carbon films (a-C:H) deposited by plasma enhanced chemical vapor deposition (PECVD) and by post-deposition nitrogen plasma processing of a-C:H films. Modifications on film microstructure, surface morphology, mechanical and tribological properties are discussed.

INTRODUCTION

Amorphous carbon films (a-C) are presently being used in a wide variety of applications as protective coatings: automotive components, shaving blades, biomedical implants and computer hard disks [1]. The use of hydrogenated carbon films (a-C:H) as the interconnect dielectric in ultra-large scale (ULSI) devices was also proposed [2]. Besides these many applications, a-C films are materials with considerable interest from an intrinsically basic point of view. Their properties are essentially controlled by the ratio between the number of carbon atoms in sp^2 and sp^3 hybridization states. The amorphous atomic network is usually described by sp^2 carbon nanoclusters bonded to each other by sp^3 bonds. The properties of the films, which are closely related to their microstructure, can be tuned by the deposition technique employed and by the growth conditions, with the energy of the impinging species playing the main role in the control of the carbon bonding hybridization [1].

In the last few years there has been a strong interest in the study of nitrogenated amorphous carbon films [3]. One of the main reasons of this research effort was the intent to synthesize the β-C_3N_4 solid, proposed by Liu and Cohen to have mechanical properties comparable to that of diamond [4]. Despite this, no clear experimental evidence of the formation of β-C_3N_4 has been presented until now [3]. An important fraction of this research effort was dedicated to the study of hydrogenated amorphous carbon-nitrogen films (a-C(N):H) deposited by plasma enhanced chemical vapor deposition, PECVD. Nitrogen incorporation in a-C:H films was found to modify the structure and the mechanical properties [5-7] of the films, as well as their electrical and optical properties [8]. It results in a strong decrease of the fraction of carbon atoms in sp^3 hybridization state [6]. In which concerns to the mechanical properties, a strong reduction on the internal stress was observed, with minor changes in the mechanical hardness [5]. Concerning the modification of electrical and optical properties, it was found that nitrogen could electronically dope a-C:H films, with the simultaneous reduction of the electronic defect density [8]. This makes possible the use of a-C(N):H films as a semiconductor material [8]. It is clear that these so many applications are critically dependent on the film surface properties.

One of the ways to modify surface properties in a controlled way is by low-energy ion bombardment that can be achieved by plasma treatment. For example, argon plasma is used to increase the surface roughness of a-C(N):H films thereby creating a more efficient electron emitter [9]. However, there are only a few studies in the surface nitrogen incorporation in a-C:H films by nitrogen plasma processing [10,11]. In this work, we will review some of the effects of the nitrogen incorporation into a-C:H films during film growth and by post-deposition N_2 rf-plasma treatment.

NITROGEN INCORPORATION DURING FILM GROWTH

The incorporation of nitrogen in a-C:H films deposited by PECVD was studied for several precursor gas-mixtures involving different hydrocarbon gases [3,6]. In the case of methane, the use of ammonia or N_2 as a source of nitrogen results in a-C(N):H films with properties governed by the amount of nitrogen incorporated in the amorphous network. In Table 1 we present results obtained from films deposited by PECVD using CH_4-N_2 mixtures with self-bias voltage (V_b) fixed at –350V and total pressure of 8Pa. It is clear that the incorporation of N occurs at the expenses of the carbon content of the films. The film density remains nearly constant at 2.2 ± 0.2 g/cm^3. The observed deposition rate reduction was attributed to the onset of a kind of chemical sputtering process arising from energetic N_2^+ bombardment of the film growing surface [12]. However, there is no justification for the absence of the symmetric situation in carbon-nitrogen film deposition process, i.e., nitrogen atoms being sputtered out by carbon carrying ions. Todorov at al. [13] considered both processes in the simulation of ion beam deposition of carbon-nitrogen films. In addition, they included the possibility of N_2 evaporation during film growth. This situation is likely to occur since N-atom subsurface penetration must favor N-N bond formation with N atoms already in film. The formation of other volatile species, as CN, can also occur. These mechanisms are increasingly more probable for deposition conditions that result in increasing nitrogen contents, reducing the deposition rates. Therefore, the growth kinetics of carbon-nitrogen films may be pictured as a competition between aggregation and erosion resulting from the impinging of different species on the film-growing surface.

TABLE 1. Film composition and deposition rates as functions of the nitrogen partial pressure in CH_4-N_2 atmospheres.

N_2 partial pressure (%)	Composition (at.%)			Deposition rate (nm/s)
	C	N	H	
0.0	86	0	14	12.5
12.5	79	4	16	12.5
25.0	79	8	14	11.0
37.5	76	9	15	6.0
50.0	72	11	17	2.5

In which concerns the structural modifications, the incorporation of nitrogen increases the size or the number of sp^2-carbon clusters, as is clear from Raman and electron energy loss (EELS) results. In fig. 1a, we show the ratio between the intensities of the D- and G-bands, the main features of the Raman spectra obtained from a-C(N):H films. This increase is interpreted as being due to an increase in size or in number of the sp^2-clusters [6]. The increase of the fraction of sp^2-hybridized carbon atoms measured by EELS (fig. 1b) supports this interpretation.

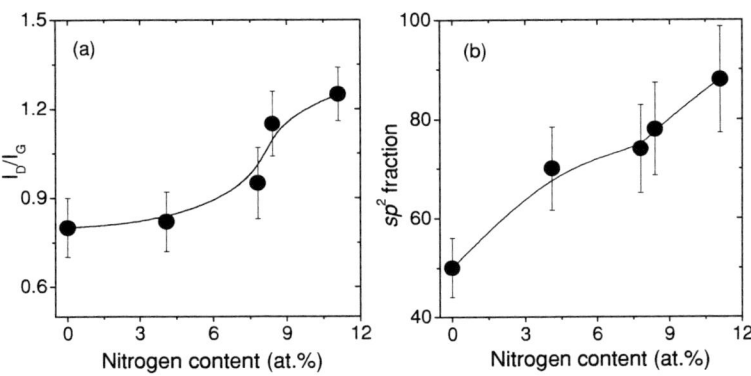

FIGURE 1. (a) I_D/I_G intensity ratio and (b) sp^2 fraction as functions of the incorporated nitrogen. The lines are eye-guides only.

It is well known that the high internal compressive stress observed in a-C:H films may be viewed as a result of the material overconstraining. This mean that the relatively high carbon sp^3 fraction observed in hard a-C:H films causes the mean atomic coordination number to be higher than the ideal value predicted for a fully constrained network [14]. In this scheme, any stress relief process may be strongly coupled to a reduction of the coordination number. In the case of a-C(N):H films, the important decrease in the internal compressive stress upon nitrogen incorporation (see fig. 2a) is conceived as a combination of the chemical composition and hybridization states.

Besides the nitrogen incorporation itself, the observed increase in the fraction of carbon atoms in sp^2-hybridization state is a source of decrease in the mean coordination number. Thus, for 8.4 at.% of nitrogen content in a-C(N):H films, the hardness shows only a slight reduction (fig. 2b) while the internal stress is reduced by a factor of two.

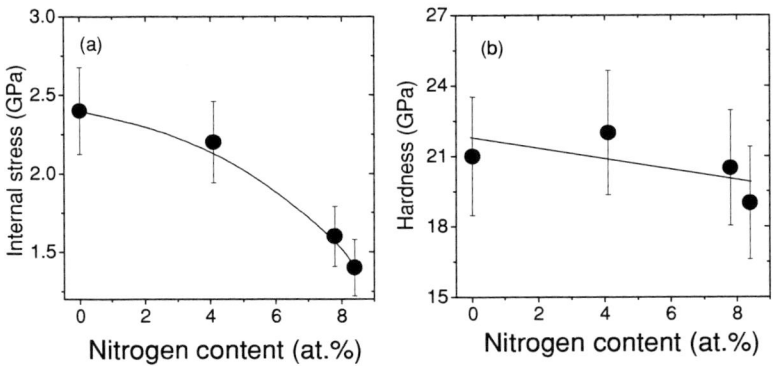

FIGURE 2. (a) Compressive internal stress and (b) Hardness as functions of the incorporated nitrogen. The lines are guide-eyes.

NITROGEN PLASMA SURFACE PROCESSING

The effects of the N_2 rf-plasma treatment of a-C:H films deposited by PECVD (precursor gas: CH_4, P= 8Pa, V_b= -350V) were investigated by a multitechnique approach. The total pressure in the plasma chamber was 3Pa and V_b was in the range between -50 and -500V. At this pressure, ions mean free path is of the order of the plasma sheath. The films were not exposed to air before submitted to plasma processing. No kind of sequential treatment was performed. The results of erosion rate, friction coefficient and amount of nitrogen incorporated as functions of the treatment time indicate that, for these plasma conditions, a steady state regime is achieved within tens of seconds.

In fig. 3a we present the erosion rate as a function of V_b. The increase of the erosion rate with self-bias voltage may be at least partially ascribed to the increased N_2^+ ion current, since higher V_b is achieved by increasing the rf power fed to the substrate electrode (from 4 to 55W, in the range of V_b studied). On the other hand, the self-bias increase also results in the increase of the N_2^+ energy, and this may also affect the erosion process since both chemical and physical sputtering of carbon films by N_2^+ increases with the ion energy [15]. In fig. 3b, we show the depth profiles obtained by medium ion scattering spectrometry (MEIS) from a sample treated 20 minutes at -500V. The total amount of nitrogen at the treated surface is nearly the same (~5×10^{15} atoms/cm^2) and independent of the self-bias. Probably, the increase of the sputtering yield with the ion energy compensates the increase of ion flux due the increase of gas dissociation when higher power was applied to the plasma to achieve higher values of V_b. The chemical bonds were probed by x-ray photoelectron spectroscopy and, as in the case of a-C(N):H films deposited by PECVD, we could identify two chemical environments, one that corresponds to N substitutional in a graphitic-like configuration and the second one that corresponds to N single bonded to C (sp^2) or to C≡N bonds [16].

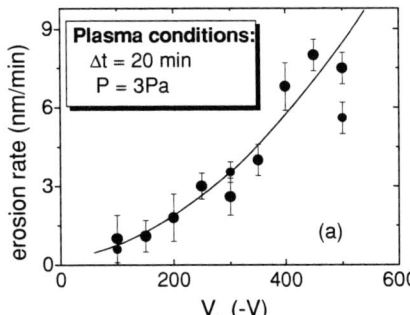 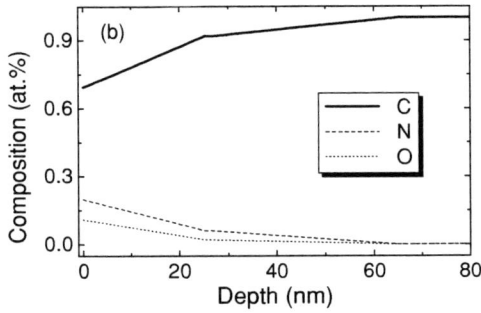

FIGURE 3. (a) Erosion rate as function of the self-bias and (b) depth profiles obtained by MEIS from an a-C:H films treated by N_2 plasma during 20 minutes with V_b= -500V and pressure of 3Pa. The depth profiles are quite similar for all V_b.

The plasma treatment modifies the film surface properties, as is illustrated in fig. 4. A progressive increase of the friction coefficient upon the increase of V_b was measured by atomic force microscopy. Films submitted to plasma treatment with V_b higher than -100V are slightly rougher when compared with as-deposited a-C:H films. However, a factor of three higher roughness values was determined in films treated at V_b = -100V. In our plasma conditions, the energies of the N_2^+ that impinges the film are controlled essentially by the V_b. In these cases, N_2^+ ions break when hit the film surface and each fragment carries half of the incident energy. It is important to note that the displacement energies of graphite and diamond are 25 and 80eV, respectively. Thus, at low N_2^+ energies and depending on the sp^2/sp^3 ratio, subimplantation cannot occur. In this case, surface diffusion tends to generate ordered clusters of high sp^2 content with structures closer to the thermodynamically stable graphite phase. Such clusters that can be nucleated also by post-deposition annealing lead to high surface roughness [17]. However, more experimental investigations are needed in order to understand the friction coefficient behavior of plasma processed films. In fact, no direct correlation between nitrogen content and surface roughness with friction coefficient could be determined.

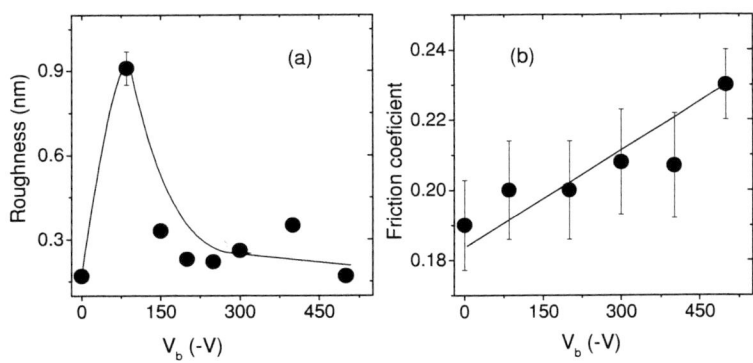

FIGURE 4. (a) RMS surface roughness and (b) friction coefficient as functions of V_b. The lines are to guide the eyes only.

ACKNOWLEDGMENTS

This work is partially supported by the Brazilian agencies: CNPq and FAPERJ.

REFERENCES

1. Robertson, J., Mater. Sci. Eng. R 37, 129 (2002).
2. Grill, A., Diamond Rel. Mater. 10, 234 (2001).
3. Muhl, S. and Mendez, J. M. , Diamond Rel. Mater. 8, 1809 (1999).
4. A. Y. Liu and M. L. Cohen, Science 245, 841 (1989).
5. Franceschini, D. F., Achete, C. A. and Freire Jr., F. L., Appl. Phys. Lett. 60, 3229 (1992).
6. Freire Jr., F. L., Jpn. J. Appl. Phys. 36, 4892 (1997).
7. Freire Jr., F. L., J. Non Cryst. Solids 304, 251 (2002).
8. Silva, S. R. P., Robertson, J., Amaratunga, G. A. J., Raferty, B., Brown, L. M., Schwan, J., Franceschini D. F. and Mariotto, G., J. Appl. Phys. 81, 2626 (1997).
9. Liu, X. W.,Tsai, S. H., Lee, L. H., Yang, M. X., Yang, A. C. M., Lin, I. N. and Shih, H. C., J. Vac. Sci. Technol. B 18, 1840 (2000).
10. Hong, J., Granier, A., Goullet, A. and G. Turban, Diamond Relat. Mater. 9, 573 (2000).
11. Castañeda, S. I., Espinoza, V. A. A., Freire Jr., F. L., Franceschini, D. F. and Jacobsohn, L. G., Nucl. Instr. Meth. B 175-177, 699 (2001).
12. Clay, K. J., Speakman, S. P., Amaratunga, G. A. J., Silva and S. R. P., J. Appl. Phys. 79, 7227 (1996).
13. Todorov, S., Marton, D., Boyd, K. J., Al-Bayati, A. H. and Rabelais, J. W., J. Vac. Sci. Technol. A 12, 3192 (1994).
14. Angus, J. C. and Jansen, F., J. Vac. Sci. Technol A 6, 1778 (1988).
15. P. Hammer, P. and Gissler W., Diamond Relat. Mater. 5, 1152 (1996).
16. Ripalda, J. M., Díaz, N., Román, E., Galán, L., Montero, I., Goldoni, A., Baraldi, A., Lizzit, S., Comelli, G. and Paolucci, G., Phys. Rev. Lett. 85, 2132 (2000).
17. Peng, X. L., Barber, Z. H. and Clyne, T. W., Surf. Coat. Technol. 138, 23 (2001).

Fore-Vacuum Plasma Electron Gun of Ribbon Beam

Viktor Burdovitsin, Yurii Burachevsky, <u>Efim Oks</u> and Michael Fedorov

Tomsk State University of Control Systems and Radio-Electronics,
Lenin Ave, 40, Tomsk, 634050, Russia.

Abstract. Plasma electron gun for ribbon beam generation was designed on the basis of glow discharge with hollow cathode. Electrons were extracted through emission hole in the anode from plasma boundary, stabilized by metal mesh, and accelerated by the voltage applied between the anode and extractor. Electron beam was of 25 cm width, 1 cm thickness. Beam current and energy were of 0.1-1 A and 2-6 keV respectively, at gas pressure of 10 – 60 mTorr. Maximum parameters are defined mostly by the acceleration gap geometry. Current density distribution along the beam width depends on the gas pressure and total beam current. At pressures higher than 30 mTorr local current maximums appear in the electron beam. They look as streams, and their positions are determined by the anode mesh deviation from flatness, but they are always at the edges of the beam. Our experiments show that in the absence of electron emission plasma density distribution in a hollow cathode maintains maximums at edges but their amounts are not more than 5 percents. At the same time, local beam maximums are about two times more. It means there is another reason of non-uniformity. We believe this intensifying is caused by gas ionization in the acceleration gap and back-stream ion flow to discharge plasma. Recharging in plasma, these ions increase plasma density and that, in its turn, leads to stream intensifying and so on. Local plasma density growth is balanced by ion diffusion from this excite zone. Lower pressure, lower ion back flow and lower plasma non-uniformity.

INTRODUCTION

To realize a technology of large-area coating precipitation, the idea of plasma chemical reaction in the "plasma sheet" volume is attractive [1]. This technology requires using an electron source forming a continuous ribbon electron beam in "bad" vacuum, i.e., in the pressure range suitable providing reasonable velocities of chemical reactions. As a rule, this is of $10^{-2} – 1$ Torr. The indicated pressure range excludes possibility to apply hot-cathode sources, making use of electron emission from gas-discharge plasma practically alternative-free. Among the existing plasma sources in the indicated pressure range only the sources based on a high-voltage glowing discharge (HVGD) turn out to be serviceable. However, acceptable currents in the sources of this type are achieved at the accelerating voltages no less than 15 – 20 kV. In plasma chemical setups electron beams with the energy of 1-7 keV are usually used, in particular, for prevention of X-radiation. Our experience in development of sources of a cylindrical electron beam for fore-vacuum pressure range [2] allowed outlining the ways for realization a design allowing to form a beam with ribbon configuration in the given pressure range. This publication presents the results of the design development and investigations of operation modes of the plasma electron gun for fore-vacuum pressure range on the basis of the discharge with the extended hollow cathode.

DESIGN OF THE PLASMA ELECTRON GUN

To the peculiar features of the electron source in the fore-vacuum region one should refer, firstly, gas ionization in the accelerating gap and, as a result, formation of back-stream of the ions getting into plasma and changing its parameters. Secondly, a propagating electron beam forms, as a rule, secondary plasma that becomes a virtual accelerating electrode. The form of the proper accelerating electrode plays no essential part. At last, the third feature is in the increased opportunity of discharge initiation in the accelerated gap that is usually qualified as breakdown. Our investigations [3] have shown that two modes of breakdown are possible: "inter-electrode" and "plasma". The

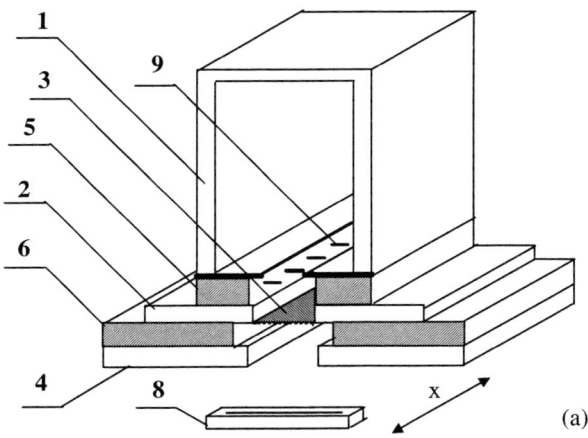
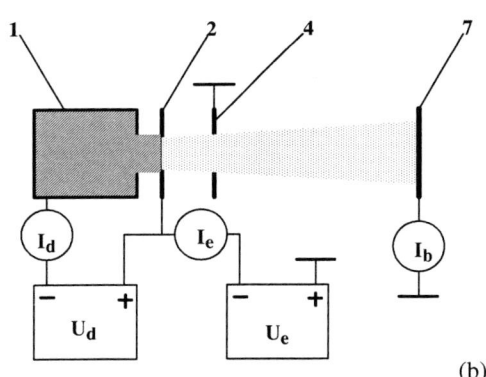

FIGURE 1. Constructive (a) and electrical (b) scheme of electron gun.

first is realized between the electrodes of the accelerating system, the second – between plasma and accelerating electrode. Prevention of the first breakdown type is achieved by screening the peripheral parts of the accelerating system. In order to eliminate the second breakdown type, it is necessary to take measures on plasma boundary stabilization. An electron source based on the glowing discharge with a hollow cathode presented schematically in Fig. 1 satisfies these conditions. The main parts of the source are a discharge chamber and accelerating system. The discharge chamber includes a hollow rectangular-section cathode 1 with the internal dimensions of 300x80x40 mm. The walls of the cavity are cooled with running water. The wall of the cavity turned to the anode 2 has a split with the width of 25 mm and length of 260 mm. In the flat anode 2 the emissive split with the dimension of 250x10 mm covered by a metal grid 3 with the mesh size of 0.5x0,5 mm is made as well. The accelerating system is formed by the anode 2 and flat accelerating electrode 4 with the emissive split of 300x40 mm. Electrodes of the discharging and accelerating systems are separated from each other electrically by polymer insulators 5, 6 that are closed by screens to prevent influence of plasma and particle flows on them. An electron beam is caught by a collector 7. Electric feed circuit is shown in Fig.1 (b). Ampere-meters switched-on into relevant circuit parts are fixing discharge current I_d, emission current I_e, beam current I_b. Required pressure is achieved by gas filling directly into the vacuum chamber as there is no pressure differential between the source and the chamber. Plasma concentration control is made by means of discharge current change. Electron beam is formed owing to the voltage applied between the anode 2 and grounded accelerating electrode 4.

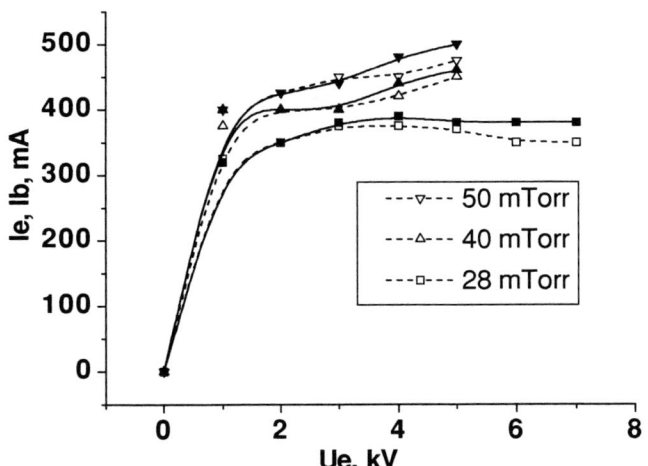

FIGURE 2. Emission I_e (solid) and beam I_b (dash) currents as functions of accelerating voltage U_e (Discharge current I_d =500 мА).

ELECTRON SOURCE CHARACTERISTICS

The described electron source allowed forming the electron beam with the width of 250 mm, thickness of 10 mm with the energy of 2 – 6 keV and current up to 1 A at a gas pressure of 10 – 60 mTorr. Emission efficiency, i.e., ratio of emission current to the discharge one reached from 70 to 100 percents. Typical voltage-current characteristics of the electron source presented in Fig. 2 contain saturation part that can be considered as the proof of the beam existence. At further increase of the accelerating voltage the self-maintained discharge initiation takes place in the accelerating gap that is accompanied by thesharp rise of the current I_e and drop of the voltage U_e to the value of the order of several hundreds volt. We consider this as a breakdown.

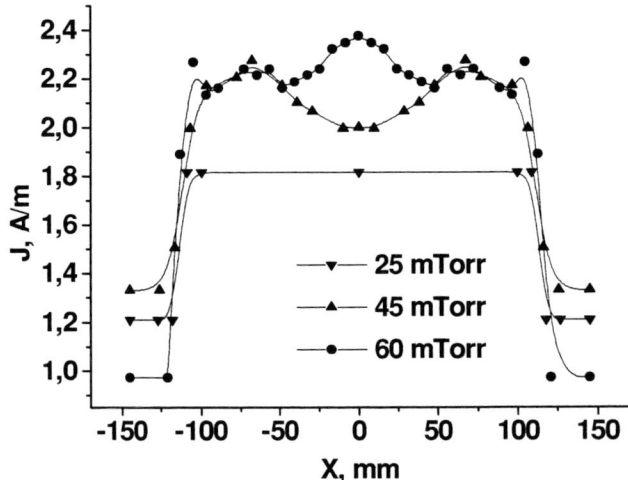

FIGURE 3. Beam current linear density distribution. (U_e = 5 kV, I_d = 500 mA).

An important feature of the ribbon beam is its uniformity along the width. To obtain this characteristic, measurement of the current directed to a displaced probe 8 (Fig. 1a) presenting a 1-mm diameter tungsten wire placed perpendicular to the beam plane at a 2-cm distance from the accelerating electrode was carried out. Characteristic curves of linear current density distribution to the collector are presented in Fig. 3. Coordinate X was measured from the middle of hollow. Beam non-uniformity rises with the pressure increase but it doesn't exceed 10-15 percent. Without electron extraction, plasma density distribution in hollow, measured by probes 9 (Fig. 1a), also has maximums at ends of the hollow cathode, however these maximums are not different from average value of plasma density more then 5 percents. The rise of beam non-uniformity during electron extraction allows us to propose the following explanation of "amplifying" the beam non-uniformity. As it was shown earlier [3], especially for fore-pump pressure range gas ionization in accelerating gap provides intense back-stream ion flow. Penetrated to the plasma these ions provide increase of the plasma density. Because of positive feed back a small initial plasma non-uniformity converts to more stronger one. This results in electron beam non-uniformity. The reason of local beam density maximums is not only because of plasma density distribution, but also due to different thickness of space charge sheath between plasma and mesh. So better beam uniformity can be reached by better stabilization of plasma emission surface.

REFERENCES

1. Manheimer W.M., Fersner R.F., Lampe M.and Meger R.A. Plasma Sources Sci. Technol., **9**, 370-386, (2000).
2. Burdovitsin V.A. and Oks E.M.// Rev.Sci.Instrum.. **70**, 2975-2778, (1999).
3. Burdovitsin V.A., Kuzemchenko M.N. and Oks E.M. Technical Physics, **47**, 926-928, (2002).

Hydrogenated Amorphous Silicon by d.c. Plasma Glow Discharge of Argon Diluted Silane

S.A. Rahman[*], A. Azis[w] and C.K. Lim[*]

[*]Physics Department, University of Malaya,
50603 Kuala Lumpur, Malaysia
[w]University of Technology MARA,
40450 Shah Alam, Selangor, Malaysia.

Abstract. Hydrogenated amorphous silicon (a-Si:H) have been deposited by direct current plasma glow discharge of silane diluted in argon. Films prepared using different argon to silane flow-rate ratios were studied using atomic force spectroscopy (AFM) to investigate the microstructure at surface of the film. Concurrently, the optical and chemical bonding properties were studied from the optical and infrared transmission spectra of the films. The refractive index of the films and the optical energy gap were determined from the optical transmission spectrum of each film while the hydrogen content and microstructure parameters were determined from the integrated intensity of the Si-H wagging and stretching respectively. The effects of argon dilution of silane at fixed rate growth temperature on the observed microstructure, optical properties, hydrogen content and microstructure to the optical properties, hydrogen content and microstructure parameter is discussed.

INTRODUCTION

The formation of microstructures in hydrogenated amorphous silicon(a-Si:H) thin film processed by Plasma Enhanced Chemical Vapour Deposition (PECVD) or Plasma Glow Discharge has been known since early works [1,2]. The deterioration of film quality and the enhancement of surface roughness and porosity of a-Si:H prepared by this technique has always been due to the formation of microstructures in the film structure. The effect of argon or helium dilution of silane in PECVD a-Si:H films has created some interest in this field of research. The growth of high quality homogeneous microstructure or nanostructure a-Si:H films could have promising future applications. It has been shown that material with high structural order at high deposition rates can be obtained using argon-diluted silane due to an enhanced contribution to the growth of Ar^+ and/or Si_x^+ ions [3]. Helium dilution of silane in PECVD a-Si:H have also been shown to produce device quality materials in d.c. PECVD at higher deposition rate than hydrogen diluted silane [4].

In this work, a-Si:H films are fabricated from the discharge of argon-diluted silane at room temperature (25°C) using our in-house direct-current plasma glow system. In this work, the effects of argon dilution of silane on the morphological, H bonding, microstructural and optical properties of the films are reported. This initial work is directed towards obtaining device quality homogeneous microstructural or nanostructural a-Si:H by d.c. plasma glow discharge argon diluted silane at low deposition temperature.

Experimental Details

The a-Si:H films studied in this work were deposited simultaneously on cleaned quartz and crystal silicon substrates at room temperature (25°C) using a home-built direct current (d.c.) plasma glow discharge system. The d.c. potential across the electrodes and the ionization current were maintained at 600 volts and 8mAmp respectively throughout the deposition process. The four different sets of a-Si:H films studied in this work were prepared using different argon to silane flow-rate ratios. The deposition pressures were fixed at the lowest pressure maintained after the gases were admitted into the chamber. The argon and silane flow-rates together with the deposition pressures for the different series of films are as tabulated in table 1.

TABLE 1. The argon and silane flow-rates, argon to silane flow-rate ratio and the pressure maintained during the deposition of the six series of a-Si:H films studied in this work.

SAMPLE	ARGON FLOW-RATE (SCCM)	SILANE FLOW-RATE (SCCM)	ARGON FLOW-RATE: SILANE FLOW-RATE	DEPOSITION PRESSURE (MBAR)
5A/20S	5	20	0.25	0.5
5A/5S	5	5	1	0.3
10A/5S	10	5	2	0.4
15A/5S	15	5	3	0.5

The surface morphology of the a-Si:H films was obtained by operating the atomic force Microscope in tapping mode with a Nanoscope 3000 (Digital Instruments, Inc.) All images were taken with Si_3N_4 tip within a scanning area of $4\mu m^2$. The Fourier Transform infrared (FTIR) spectra were obtained using a Perkin-Elmer FTIR spectroscopy system 2000. The films were scanned within the scanning range of $4000 cm^{-1}$ to $400 cm^{-1}$. The total H%(T) of the film was determined from the integrated intensity of the Si-H wagging band at $650 cm^{-1}$. The microstructure parameter, R was determined from the relation,

$$R = \frac{I_{2090}}{[I_{2090} + I_{2000}]} \quad \ldots (1).$$

where I_{2090} and I_{2000} were the integrated intensities of the $Si-H_2$ stretching band at $2090 cm^{-1}$ and Si-H stretching band at $2000 cm^{-1}$ respectively. The H content bonded as polyhydride or clustered monohydride, H%(P) complex was determined from the product of R and H%(T) and the H content bonded as isolated monohydrides, H%(M) was determined from the difference between H%(T) and H%(P). The optical transmission spectra scanned in the range of 200nm to 2500nm were obtained using a Jasco UV-VIS-NIR 3102-PC double beam spectrophotometer. The refractive index and film thickness were derived from the interference fringes in the high transmission region. The optical energy gap was determined from the extrapolation of Tauc's plot, $(\alpha E)^{1/2}$ versus E where α and E are the absorption coefficient the photon energy respectively.

Results and Discussions

The FTIR transmission spectra showing absorption bands for the Si-H bending and Si-O stretching vibrations for the four samples are presented in Fig. 1(a). Consequently, the variation of the hydrogen content bonded as polyhydride ($Si-H_2/(SiH_2)_n$) or clustered monohydride complexes, $(Si-H)_n$, H content bonded as isolated monohydride (Si-H) and the microstructure parameter represented as H%(P), H%(M) and R respectively as a function of argon to silane flow-rate ratio is shown in Fig. 1(b).

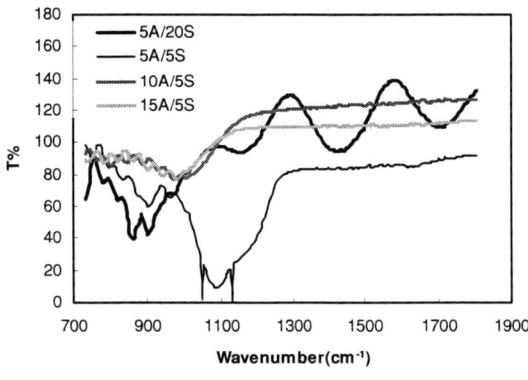
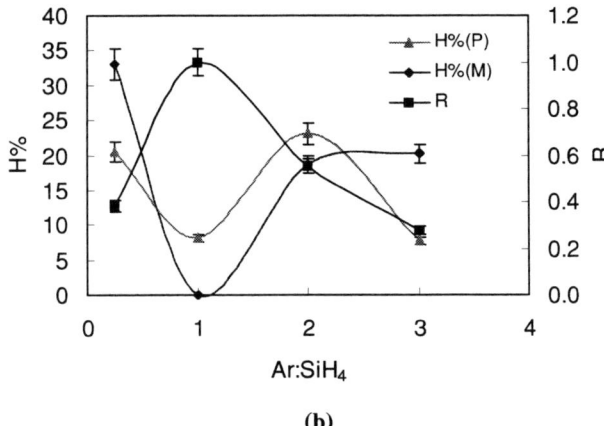

(a) (b)

FIGURE 1. (a) Infrared spectra of a-Si:H films prepared from the d.c. plasma glow discharge of argon diluted silane in the scanning range of $1800 cm^{-1}$ to $600 cm^{-1}$ showing the Si-H bending and Si-O stretching absorption bands. (b) The variation of H content bonded as polyhydride or clustered monohydride complexes, H%(P), H content bonded as isolated monohydride and microstructure parameter, R as a function of argon to silane flow-rate ratio.

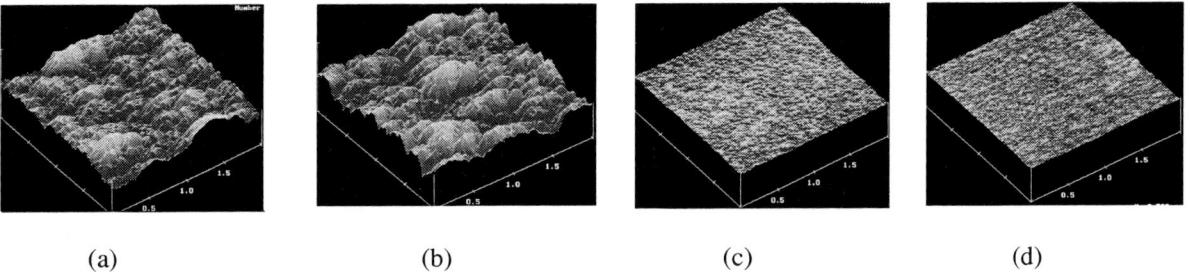

(a)   (b)   (c)   (d)

FIGURE 2. Atomic Force Microscope Images of a-Si:H films prepared argon diluted silane with argon to silane flow-rate ratio of (a) 0.25 (5A/20S), (b) 1 (5A/5S), (c) 2 (10A/5S) and (d) 3 (15A/5S).

A well defined doublet with peaks at 855cm^{-1} and 893cm^{-1} observed for the 5A/20S sample and a shoulder at 855cm^{-1} and a prominent peak at 893cm^{-1} observed for the 5A/5S sample indicate a strong presence of $(Si-H_2)_n$ chains in these films[5]. These $(Si-H_2)_n$ chains usually form columnar microstructures normal to the surface. However, as shown in Fig. 1(b), H bonded as polyhydrides are less dominant in the 5A/20S film as compared to H bonded as isolated monohydrides while the reverse is observed for the 5A/5S film. This is consistent with the surface morphology of these films as shown in the AFM image of this film in Fig. 2(a) and 2(b) where the mountain-like features observed are microvoids formed from $(Si-H_2)_n$ chains and the valleys are the homogeneous amorphous structure where the H atoms are incorporated in the form of isolated Si-H bonds. Comparatively between these two films, the 5A/5S has a rougher surface morphology as shown in figure 3(a) due the more dominant presence of H bonded as polyhydrides compared to H bonded as isolated monohydrides. The absence of the of the $(Si-H_2)_n$ or Si-H_2 bending bands for the a-Si:H films with the two highest argon dilution of silane (10A/5S and 15A/5S) is related to the presence of clustered monohydride, $(Si-H)_n$ groups forming microstructures in the film. In the 10A/5S sample, the clustered $(Si-H)_n$ groups is more dominant than the isolated Si-H bonds, small granular structures uniformly distributed within the amorphous structure are observed from the AFM image of this film as observed in Fig. 2(c). These clusters decrease in size to around 50nm as the argon to silane flow-rate ratio is increased to 2 for sample 10A/5S and is decreased further for sample 15A/5S as shown in Fig 3(a). These nanostructures formed by $(Si-H)_n$ clusters result in a more homogeneous film structure.

The microstructure parameter, R is observed to shown some dependence on the deposition pressure (Refer to Fig. 1(b) and Table 1). Samples produced at lower deposition pressure (samples 5A/5S and 10A/5S) have higher R values suggesting that higher argon ion bombardment on the film surface during deposition results deterioration in structural order in film.

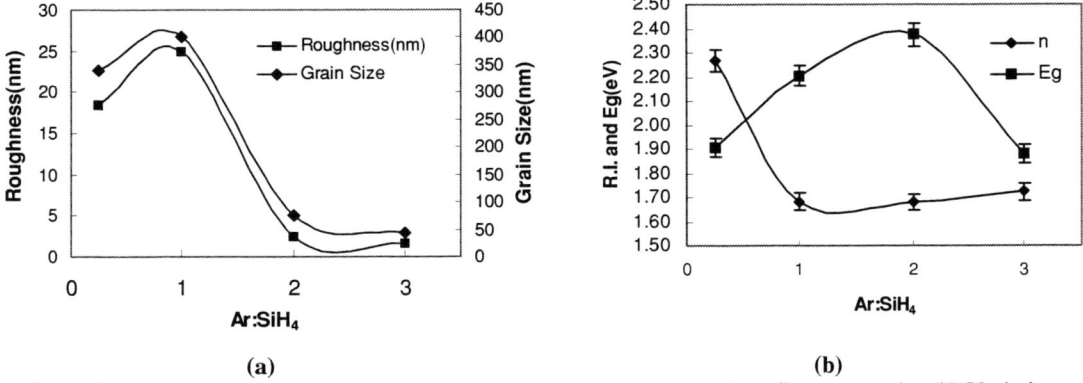

(a)   (b)

FIGURE 3. (a) Variation of surface roughness and grain size with argon to silane flow rate ratio. (b) Variation of refractive index, n, optical energy gap, Eg and deposition rate of the a-Si:H films with argon to silane flow-rate ratio.

The effects of argon dilution on the refractive index and the optical energy gap of the a-Si:H films are presented in Fig. 3(b). Argon dilution of silane results in low refractive index a-Si:H indicating low density films. The refractive index decreases significantly from about 2.28 to 1.65 when the argon to silane flow rate ratio is increased to 1 and increases slightly with further increase in gas ratio. The $(Si-H_2)_n$ chains present in a-Si:H films 5A/20S and 5A/5S result in a more compact structure but the significant decrease in the refractive index for the latter film is due the high oxygen contamination as exhibited by the dominant Si-O stretching band at 1070cm^{-1} (Refer to Fig.1(b)). The source of oxygen contamination in this film could be from the more energetic argon ion bombardment on the film surface resulting in incorporation of oxygen atoms into the film structure as the deposition pressure is the lowest for this film(refer to Table 1). The optical energy gap increases to a maximum when the argon to silane flow-rate ratio is increased to 2 and decreases significantly when the gas ratio is increased to 3 as shown in Fig. 3(b). The optical energy gap is not effected by the H content in the film for a-Si:H films produced from the discharge of argon diluted silane as in a-Si:H films produced from pure silane[6]. It appears that the optical energy gap is larger when the H bonded as polyhydride or clustered monohydride complexes is larger than H bonded as isolated monohydrides (samples 5A/5S and samples 10A/5S) and consequently, for these two samples, the optical energy gap is larger when the size of the microstructures is smaller. Further investigation is needed to determine if this blue shift in the optical energy gap is due to quantum confinement effect.

CONCLUSIONS

An analysis of the effect of argon dilution of silane on the optical, morphological and microstructural properties of the a-Si:H films deposited by d.c. plasma glow discharge was presented using a combination of IR absorption, optical transmission and AFM images of the surface morphology of the film. Lower argon dilution of silane (argon:silane ≤ 1) results in the formation of polyhydride bonds in the film forming columnar structures normal to the film surface due to the formation of $(Si-H_2)_n$ chains. Higher argon dilution films produces a-Si:H with clusters of monohydride bonds. Homogeneous amorphous structure surrounds the microstructures formed by the $(Si-H_2)_n$ chains or clustered $(Si-H)_n$ complexes. The polyhydride bond clusters are larger in size and produce a rougher surface morphology. These microstructures result in the a-Si:H films with low refractive indices. The optical energy gaps of these films are not dependent on the hydrogen content in the film but films with higher proportion of microstructure concentration formed from clustered $(Si-H)n$ complexes have larger energy gaps. However, results in these work indicates that the films produced at high argon dilution are nanostructured, more homogeneous with higher structural order, has low H content and a smoother surface morphology. The decrease from microstructure size to nanostructure size and the blue shift in the optical energy gap for the film produced with argon to silane flow-rate ratio of 2 could be further investigated for interesting photoluminescence properties in future works.

ACKNOWLEDGEMENTS

This work was supported by the Ministry of Science, Technology and Environment, Malaysia under the IRPA research grant (No.09-02-03-0409).

REFERENCES

1. Street,R.A., Knights, J.C. and Biegelson, D.K, Phys. Rev. B. (USA), **18**, no.4 pp.1880-1891 (1978).
2. Fritzche, H., Sol. Energy Mater. (Netherlands), **3**, no. 4, pp. 447-501 (1980).
3. Meiling, H., Bezemer, J., Schropp, R.E.I. and Van Der Weg,W.F., Mater. Res. Soc. Symp.Proc., **467**, pp.459-470 (1997).
4. Middya, A.R., G. Wood, A.R., Lin, G.H. and Carlson, D.E., Mater. Res. Soc. Symp.Proc., **557**, pp. 51-156 (1999).
5. Daouahi, M., Ben Othmani, A., Zellama, K., Zeinert, A., Essamet, M. and Bouchriha, H., Solid State Commun. **120** pp.243-248 (2001).
6. Bruyère, J.C., Deneuville, A. and Mini, A, J. Appl. Phys. **51**(4), pp.2199-2205 (1980).

Deposition Of Carbon Thin Film At Room Temperature Using Dense Plasma Focus Device

R.S. Rawat[1], G. Macharaga[1], P. Lee[1], S. Xu[1], and S. Lee[2]

[1]*Natural Sciences, National Institute of Education, Nanyang Technological University, Singapore*
[2]*International Centre for Dense Magnetised Plasmas, Warsaw, Poland*

Abstract. A 3.3 kJ ICTP UNU pulsed plasma focus device was used to deposit thin films of carbon at room temperature on silicon substrates. For the deposition of carbon thin films, the conventional hollow copper central electrode of plasma focus device was fitted with a solid graphite top. The deposited films were analyzed for their structure, surface morphology and elemental composition using X-ray diffraction (XRD), Raman spectroscopy, scanning electron microscopy (SEM), and energy dispersive X-ray spectroscopy (EDX). Amorphous carbon films with sp^3 content as high as 66 percent have been successfully deposited.

INTRODUCTION

The diamond like carbon (DLC) materials are well known for the unique combination of useful properties such as excellent hardness, high thermal conductivity, high electrical resistivity, good wear resistance, chemical inertness and optical transmission over a large range of wavelength. The DLC consists of an amorphous form of the carbon coating containing both graphite type sp^2 bonding and diamond type tetragonal sp^3 bonding. The large range of their useful properties have made the thin films of this material useful in countless applications such as wear resistant coatings in such areas as engine piston rings, biomaterials, optical coatings, tool coatings, and so forth.

The dense plasma focus device [1], historically, has been developed as a fusion device with most of its studies being done in hydrogen and its isotopes in relation to the neutron emission with the aim of finding neutron production mechanisms. The focus device, however, not only is a source of neutrons but also emits highly energetic ions, relativistic electrons and abundant amount of x-rays. More recently, the energetic ions of plasma focus have been used for inducing change of phase in thin films [2] and for deposition of thin films [3]. The electrons from this device have been used for electron microlithography [4]. Soft x-ray lithography, with line width down to below 0.2 µm, has been successfully demonstrated using high repetition rate, high efficiency compact plasma focus [5].

In this paper, we report the systematic study of the variation in structure and morphology of DLC thin films deposited under different conditions on silicon substrates at room temperature.

EXPERIMENTAL SETUP AND METHODOLOGY

The Plasma Focus Device

A simple, 3.3 kJ, Mather-type dense plasma focus device, designated as United Nation University/International Center for Theoretical Physics Plasma Focus Facility (UNU/ICTP PFF), is used for deposition of diamond like carbon thin films. The experimental setup along with the focus subsystems is sketched in Fig. 1. For deposition of diamond like carbon thin films, the conventional central hollow copper anode is fitted with the solid graphite top. The 3.0 cm long solid graphite top is screwed on the copper anode of reduced length so that the total length of the electrode remains 16.0 cm. The DPF device was operated at a charging voltage of 14 kV and argon filling gas pressure of 2.0 mbar. The focusing efficiency was monitored using voltage probe.

The Deposition Process

The qualitative understanding of DLC film deposition process is as follows. A hot dense focused plasma column is formed at the top of the anode during the radial collapse phase of plasma focus device. In low energy plasma focus devices, like ours, sausage instabilities ($m=0$) are seen to set in the focused plasma column. This enhances induced electric field locally, which, coupled with the magnetic field, breaks the focused plasma column and thereby accelerating ions (of the gaseous species), to very high energies, towards the top of the chamber and electrons towards the positively charged anode. Relativistic electrons cause the ablation of anode material. The ablated plasma, from the solid graphite top of anode, has carbon ions and atoms required for DLC formation on the substrate placed downstream the anode axis.

Argon ions, however, are not only formed and accelerated earlier than the ablated carbon ions but also are much more energetic. They, therefore, reach the substrate first causing the etching and hence cleaning of the substrate surface prior to deposition. The energetic argon ions from the next focus shot, as we are using multiple shots to deposit DLC films, cause the processing of the DLC film deposited in previous shot. Actually, short ion pulses of high energy density can cause very rapid heating of the surface region depending on the nature of the surface layer and the energy density of the incident ions.

Film Deposition

The deposition of the DLC thin films is done on 10mm×10mm highly polished silicon substrates. The substrates are mounted, downstream the anode axis, at four different distances of 5.0, 8.0, 11.0 and 14.0 cm from the top of the anode using a specially shaped substrate holder, shown in inset of Fig. 1. At each of these distances the DLC films were deposited using 5 and 10 plasma focus shots. It is well known that in plasma focus most of the ions are emitted in a small solid angle along the anode axis and their flux, generally, decreases with the increasing angle. The specially shaped substrate holder, allowed us to deposit three films, simultaneously, at different angular positions

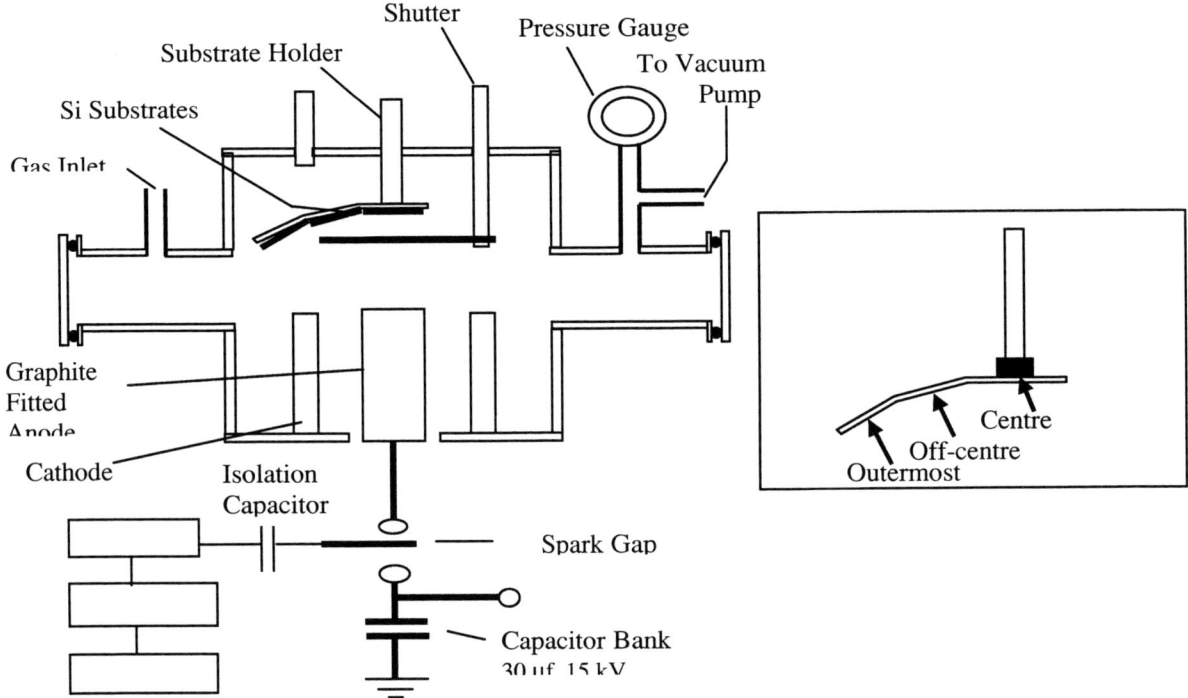

FIGURE 1. Schematic of experimental setup for carbon film deposition. The inset shows specially shaped substrate holder with respect to the anode axis. The three positions, as shown in inset of fig.1, are labeled as (i) center, (ii) off-center, and (iii) outermost.

Characterization Of Thin Films

The characterization of DLC film structure was done using Renishaw Raman Imaging Microscope and Siemen's D 5005 X-ray diffractometer. Jeol JSM 5310-V Scanning Electron Microscope (SEM) was used to study surface morphologies of the films. The recognition of elemental constituents of the film was done using Energy Dispersive X-ray spectroscopy attached to the SEM.

RESULTS AND DISCUSSION

The X-ray diffraction patterns of the deposited films show that the deposited films are amorphous as no diffraction peaks other than that of the silicon substrates were observed. The fluence and the energy of argon ions in plasma focus device is substantially high and so the deposition process is highly energetic one, leading one to could expect that the deposited films may have crystalline diamond phase. The X-ray diffraction, however, shows that the as-deposited carbon film do not contain any diamond phase.

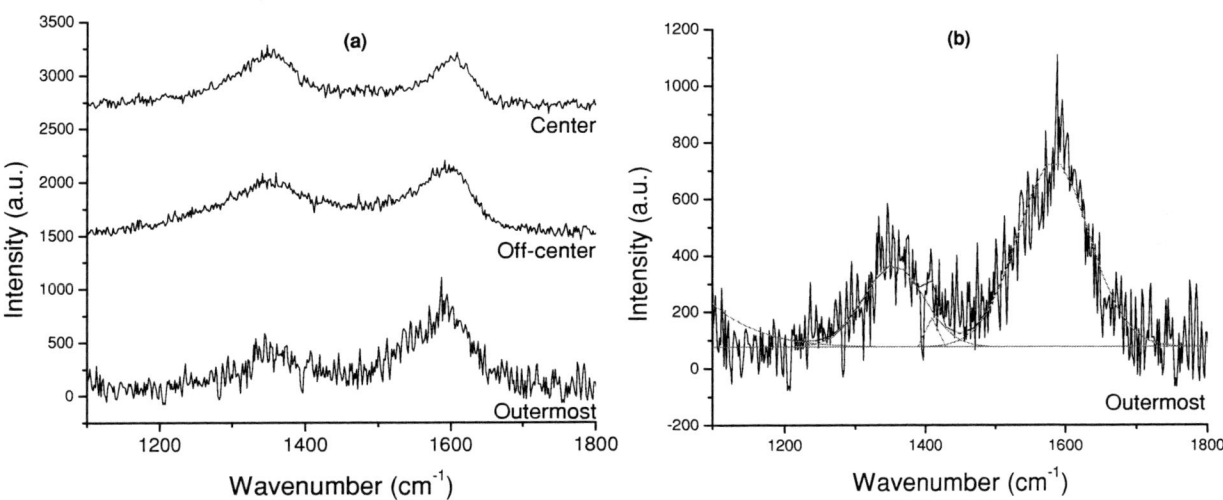

FIGURE 2: (a) Raman spectra of the films deposited at the distance of 5.0 cm from the top of the anode for 5 focus shots. (b) Typical curve fitting to identify G and D peaks.

Typical Raman spectra of the deposited films are shown in fig.2a. Fig.2b shows the typical curve fitting using three Gaussian peaks fit. The Raman spectrum that is typical to our films consists of a broad G peak centered between 1582-1600 cm^{-1} and a D peak between 1350-1420 cm^{-1}. According to Koh and Wong [6], the ratio of D to G peak intensity, I_D/I_G, can be used to estimate the relative amount of diamond sp^3 and graphitic sp^2 bonding of atoms in the deposited thin films. It was found that for the films deposited at center positions at all the distances and for different number of focus deposition shots the intensity of D peak was higher than that of G peak indicating low sp^3 content. Moreover, the SEM pictures, as discussed later, also show patchy surface morphology of for these films. It is for these reasons we have concentrated our discussion of results for films deposited at off-center and outermost positions. The ratio I_D/I_G is found to decrease from 0.76 to 0.59 for the first three distances before increasing to 0.98 for final depositing distance of 14.0cm. Using Koh and Wong's results, on ta-c films, as the benchmark data the best line fit was plotted and extrapolated to estimate the sp^3 fraction of our films. The highest sp^3 content of 66 percent was achieved on the films deposited at the distances 5.0 and 8.0 cm at outermost position with 5 focus shots. Films deposited at the outermost positions have higher sp^3 content as compared to the films deposited at off-center positions. It is important to note that at outermost position the energy and fluence of ions of filling gas species and ablated carbon is least due to highly forward directed emission nature of focus device. This point to the fact that, probably, the energetic argon ions of the filling gas species have detrimental effect on the sp^3 content of the films and that is why the films deposited at center and off-center positions have lower sp^3 content. The poorest sp^3 content of 17 percent was estimated for film deposited at off-center position at 5.0cm distance with 10 focus shots. The results indicate that to tune the focus device to deposit the sp^3 rich DLC films we shall study the effects of other parameters like that of gas pressure, charging voltage of the device, and substrate heating.

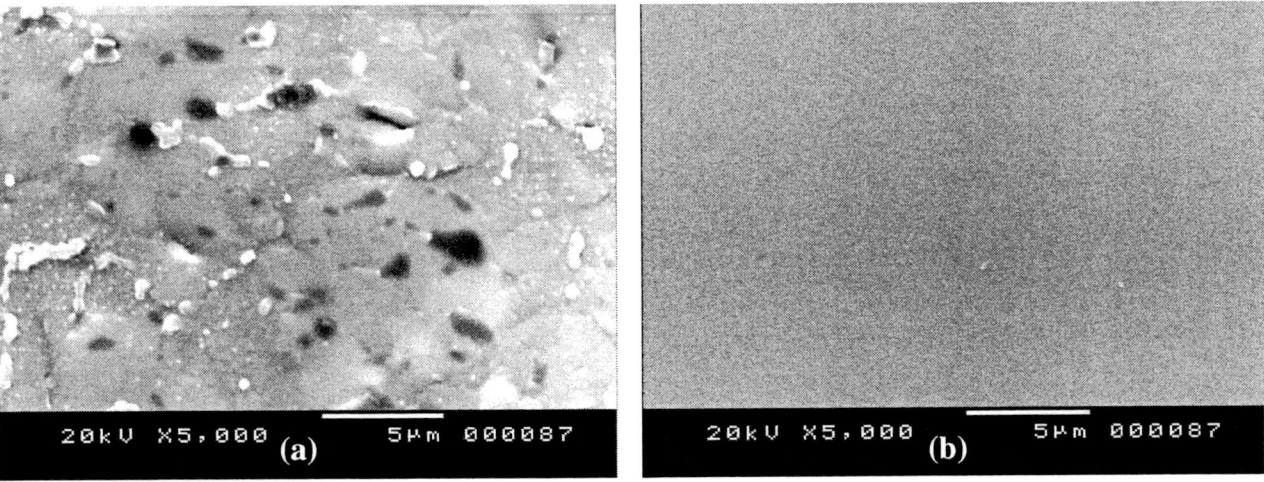

FIGURE 3. SEM pictures of films deposited at (a) center and (b) outermost position at 5.0cm using 5 focus shots.

The surface morphology of the films deposited at center and outermost positions using 5 focus shots at the distance of 5.0 cm from the top of the anode is shown in fig.3. The film deposited at center position is highly non-uniform with lots of holes and cracks on it. These holes and cracks at the film surface are seen at low resolution, the surface of the film shown in fig.3a is actually over relatively smooth area of the films. The EDX background-scan, as well as the point-scan on the whitish spots, on the films deposited at center position shows the presence of copper traces. This copper comes from the copper anode of the focus device. The films deposited at outermost positions are highly uniform with no traces of copper on it. The absence of copper on the outermost position film is because the copper debris from copper is forward directed. The films were also free from particulates, which were there on films deposited on center positions.

CONCLUSIONS

The thin films of amorphous carbon were successfully deposited using dense plasma focus device. The highest diamond sp^3 content of the deposited film is only about 66 percent, which is less than the one reported by others. In order to tune the plasma focus device to deposit the sp^3 content rich amorphous carbon films we need to explore the effects of other parameters like that of gas pressure, charging voltage of the device, and substrate heating.

ACKNOWLEDGMENTS

Authors (RSR and PL) are thankful to National Institute of Education for providing the ARF grant RP17/00/RSR to fund the research project under which this investigation has been performed.

REFERENCES

1. Lee, S., Tou, T.Y., Moo, S.P., Eissa, M.A., Gholap, A.V., Kwek, K.H., Mulyodrono, S., Smith, A.J., Surayadi, Usada, W., and Zakaullah, M., Am. J. Phys. 56, 62-68 (1988).
2. Rawat, R.S., Arun, P., Videshwar, A.G., Lam, Y.L., Lee, P., Liu, M.H., Lee, S., and Huan, A.C.H., Bull. Mat. Res. 35, 477-486 (2000).
3. Rawat, R.S., Lee, P., White, T., Ying, L., and Lee, S., Surf. Coat. Tech. 138, 159-165 (2001).
4. Lee, P., Feng, X., Zhang, G.X., Liu, M.H., and Lee, S., Plasma Sour. Sci. Tech. 6, 343-348 (1997).
5. Lee, S., Lee, P., Zhang, G., Serban, A., Liu, M., Feng, X., Springham, S.V., Selvam, C., Kudryashov, V., and Wong, T.K.S., Sing. J. Phys. 14, 1-10 (1998).
6. Koh, P.E., and Wong, W.O., "Study of thin films using Raman and FTIR spectroscopy", NTU, Singapore (1997).

Charged Particles in the Reactor Chamber of a Remote Plasma Enhanced CVD System

L. Sirghi, G. Popa[*] and Y. Hatanaka

Research Institute of Electronics, Shizuoka University, Hamamatsu 432-8011, Japan
[*]*Faculty of Physics, "Al. I. Cuza" University, Bd. Carol I, no. 11, 6600-Iasi, Romania*

Abstract. A planar Langmuir probe installed at the center of the reactor chamber of a remote plasma enhanced chemical vapor deposition system and the conductive wall of the reactor chamber behave as a double probe. This probe system was used to measure the density and temperature of the electrons reaching the deposition region. In spite of the plasma source remoteness, the measurements revealed relatively large electron density values (10^6 cm^{-3}) for argon plasma.

INTRODUCTION

Depending on the nature of the technologic process, the plasma processing systems use direct plasma, when plasma particles interact with the surface of the solid to be processed, or remote plasma, when only the plasma effluents are of the interest for the process[1]. In reality there is no clear distinction between the two types, because even in the remote plasma systems some of the source plasma particles come into the interaction with the surface of the solid to be processed. To avoid plasma reaching the deposition area, the source plasma is located at certain distance from the deposition chamber[2]. However, due to the diffusion and carrier gas transportation processes charged particles of the source plasma may reach the deposition region.

The goal of this work is to measure the density and temperature of the electrons reaching the deposition region of a remote plasma enhanced chemical vapor deposition (PECVD) system using a microwave discharge. We used double-probe measurements in a special arrangement in which the inner wall of the reactor chamber played the role of one of the probes.

THE EXPERIMENTAL SETUP AND TYPICAL PROBE CHARACTERISTICS

When source plasma is produced far from the reactor chamber, the double-probe may not work properly because of the low density of the charged particles. The probes installed in such low-density plasma adsorb particles and may distort plasma between them, which affects drastically the double-probe current-voltage characteristic[3]. To avoid the plasma distortion phenomenon, we used a probe system consisting of a single-side square gold electrode (9 cm^2) installed at the center of the reactor chamber and, as the reference electrode, the inner wall of the stainless steel chamber of the reactor (see Fig. 1). The cylindrical chamber of the reactor with the inner diameter of 17 cm and length of 47 cm was connected to a quartz tube with the inner diameter of 4 cm and the length of 120 cm. The length of the quartz tube inside the reactor chamber was 14 cm, while the distance between probe and the end of the quartz tube in the reactor chamber was 10 cm. The source plasma was generated by a microwave discharge in the quartz tube at distance d from its end in the reactor chamber (d = 50 and 70 cm in the experiments). The source gas was fed into the system through a mass flow control (MFC) system. Experiments were performed in argon at pressure values ranged from 0.05 to 1 Torr. The $I(V)$ characteristic of the double probe was digitally acquired by a data acquisition system comprised of a digital analog converter (DAC) that generated a rump biasing voltage ranged from -10V to +10V, an differential operational amplifier (OPA) that collected the current signal across a resistor R installed in the probe biasing circuit, an analog digital converter (ADC) and a personal computer (PC).

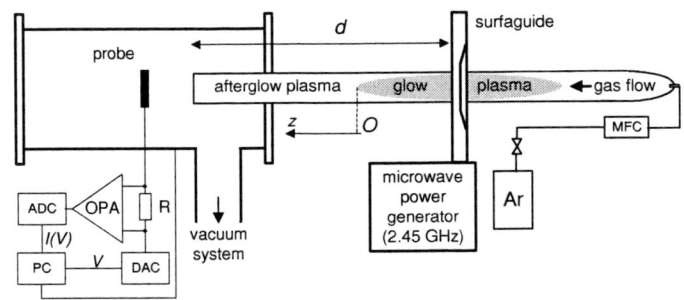

FIGURE 1. Sketch of the experimental setup.

The typical probe current-voltage characteristics $I(V)$, for argon plasma at low (less than 0.2 Torr) and high pressure values are presented in the Figures. 2 (a) and (b), respectively. In spite of the small aspect ratio of the probe and chamber wall areas, because of the big ratio of the plasma densities at the center and near the wall of the reactor chamber, respectively, the values of the ion currents collected by the probe, I_P, and wall, I_W, were comparable. Due to the asymmetry of the double-probe system, the $I(V)$ characteristics did cross the biasing voltage axis at a positive biasing potential ΔV. A finite value of ΔV could arise due to variation of plasma parameters inside the chamber. While the probe characteristics at low pressure shows the typical exponential double-probe current dependence on the probe voltage in collisionless plasmas, the probe characteristic at high pressure shows a linear dependence that is attributed to electron-neutral collisions in the probe sheath. The effect of collisions is easily observed on the first derivative of the probe current, I', which does not have the exponential shape that characterize the noncollisional regime.

THEORETICAL CONSIDERATIONS

The probe characteristics for Ar plasma at low values of the pressure shows the typical double-probe current dependence on the probe voltage in collisionless plasmas[4, 5]. However, because of the low plasma density near the chamber wall, the applicability of classical theory of the double probe to this particular probe system is questionable. For the case of the collisionless probe sheaths, the classical double probe theory yield the following expression for the electron temperature[4, 5]:

$$T_e[eV] = \frac{I_P \cdot I_W}{I_P + I_W}\left(\frac{dV}{dI}\right)_0, \qquad (1)$$

where $(dV/dI)_0$ is the invert of the first derivative of the probe characteristic at $V = \Delta V$. Alternatively, a value of the electron temperature for the plasma nearby the probe can be computed by the slope of the logarithm of the second derivative of $I(V)$, I'', in the negative biasing potential region where it has an exponential dependence on biasing potential [region A-B on the plot of I'' in Fig. 2 (a)]:

$$T_{I''} = -(d\ln I''/dV)^{-1}. \qquad (2)$$

This method is justified by the fact that the second derivative of the probe current in a region where the current

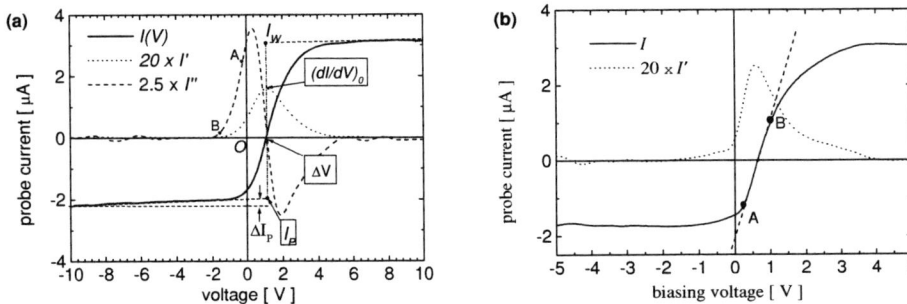

FIGURE 2. The typical current-voltage characteristic $I(V)$ and its derivatives, I' and I'' for the afterglow plasma in argon at low (a) and high (b) pressure values.

does not changed much retains the exponential dependence of the probe electronic current on the biasing potential, and the temperature can be computed on the basis of this dependence. At large gas pressure (more than 0.2 Torr), the probe sheath is collisional also for the electrons and the $I(V)$ loses its exponential shape[6], fact that makes impossible the evaluation of the electron temperature by the methods described above. Computation of the electron mean free path, λ_e, and the ion mean free path, λ_i, shows that at low argon pressure the probe sheath can be considered collisionless for electrons, but collisional for ions. For the collisionless ion sheath, the plasma density, n, can be computed by the Bohm expression of the saturation ion current collected by the probe[7]. For the collisional ion sheath, the existence of a critical Bohm velocity is a controversial issue[8] and an alternative method for computation of plasma density should be used. We assumed that the slope of the linear dependence of the saturation ionic current of the central probe on the biasing potential, dI_P/dV, (see Fig. 2) is determined mainly by the ion mobility in the probe sheath:

$$\frac{dI_P}{dV} = n \cdot e \cdot S \cdot \mu_i(p,T)/L_i, \qquad (3)$$

where $\mu_i(p,T)$ is the ion mobility, S, the probe area, L_i, the probe ionic sheath thickness, of which dependence on V is neglected, and p and T are the gas pressure and temperature, respectively. Neglecting variations of L_i and n, the variation of gas pressure should induce a variation dI_P/dV according to the scaling low $\mu_i \cdot p = const$. Our experimental results fitted well to this dependence, which shows that the model of ohmic conductivity of the ion sheath is reliable and Eq. (3) can be used to determine the plasma density. The ion sheath thickness, L_i, can be approximated by the Debye length, which depends on plasma density. Therefore, the plasma density can be computed by a convergent iteration of Eq. (3) and Debye length formula.

EXPERIMENTAL RESULTS AND DISCUSSION

Relatively close values of the electron temperature were computed by the two methods described above, which proves that ion sheath at the reactive chamber wall is also collisionless for electrons. The electron temperature slightly increased by the increase of the discharge power and decreased by the increase of distance between source plasma and the reactor chamber. Values of temperature varying between 0.4 and 0.7 eV were found. Figures 3 (a) and (b) display the plots showing the dependence of plasma density in the reactor chamber on the discharge power or gas flow rate, respectively. The typical value of electron density in the reactor chamber was situated around 10^6 cm^{-3}. The density increased by the increase of the discharge power, but decreased by the increase of the gas flow rate.

Considering an electron density around 10^{11} cm^{-3} of the microwave discharge plasma[9] and the relative large values of plasma density in the reactor chamber, a plasma density decay factor of 10^{-5} is computed. If the steady-state diffusion equation for the afterglow plasma is solved by assuming loss of the plasma particles by recombination at discharge tube wall and no ionization, a solution $n(z)$ showing exponential decay of plasma density in z direction is found[10]:

$$n(z) = n_0 \cdot \exp(-z/\Lambda_r), \qquad (4)$$

where, n_0 is the plasma density on the column axis at the glow plasma end, z, the distance along the discharge tube in the afterglow plasma (see Fig. 1), $\Lambda_r = R/2.405$, the radial diffusion length and R, the discharge tube radius.

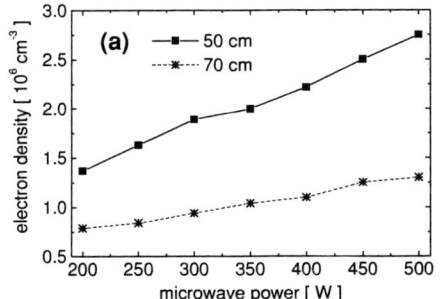
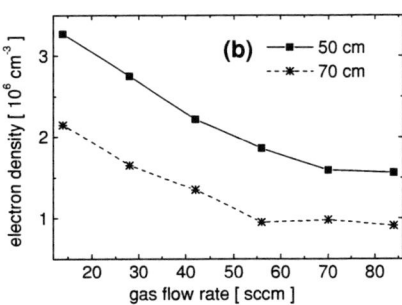

FIGURE 3. Dependence of plasma density in the reactor chamber on microwave power (a) and argon flow rate (b).

Considering a length of 30 cm for the afterglow plasma column and $R = 2$ cm, a plasma decay factor of 10^{-13} is found, which is much lower than the plasma decay factor estimated from the experimental results presented above. If the carrier gas flow plays the major role in plasma particle transportation, the axial decay of the afterglow plasma particles is expressed by the exponential low[2]:

$$n(r,z) = n_0 \cdot \exp(-z/v_g \tau), \qquad (5)$$

where $\tau = (R/2.405)^2/D_a$ is the characteristic time for plasma particle decay by the ambiplor diffusion and recombination at the discharge tube wall, D_a, the ambipolar diffusion coefficient, and v_g, the gas flow velocity. However, according to our experimental results the plasma density decreased by the increase of the gas flow rate, which is contradiction with the predictions based on Eq. (5). For the present experimental conditions, estimation of τ and v_g computes an axial decay characteristic length smaller than 1 cm, which would give also a small plasma density decay factor. Therefore, either the axial diffusion or plasma transportation by the carrier gas models can not describe correctly the decay of the afterglow plasma. To obtain a more realistic solution for the plasma density decay in the afterglow region, ones should take into account the effects of plasma particle trapping into the radial well of potential of the plasma column[11], plasma particle reflection at the discharge tube wall and axial relaxation of the electron temperature. Also, because of the Ramsauer effect[12], a much larger diffusion coefficient for low-energy electrons should be used for the case of argon gas.

CONCLUSION

A planar probe and the conductive wall of the reactor chamber of a remote PECVD system behaved as a double probe. In spite of the small area ratio, the ion saturation currents of the wall and probe had comparable values, which proves a much lower plasma density near the chamber wall. The probe system was used for measurements of the temperature and density of the afterglow argon plasma electrons reaching the deposition region. Taking into account the relatively low density of the plasma in the reactor chamber and the value of the argon gas pressure, it was concluded that the probe sheath is collisional for ions and collisionless for electrons. For the argon gas around 0.1 Torr pressure and microwave discharge power of 300 W, values close to 10^6 cm^{-3} were found for the electron density, while values around 0.6 eV were found for electron temperature. The large values measured for the plasma density at the deposition region are in contradiction with the predictions of the afterglow plasma models that account for either the axial diffusion or the plasma transportation by the carrier gas. The disagreement may come from the unrealistic assumptions of perfectly plasma particle adsorbent wall of the discharge tube and constant axial diffusion rate along the afterglow column.

REFERENCES

1. J. Paraszczak, J. Heidenreich, in Microwave Excited Plasmas edited by Michel Moisan, Jacques Pelletier, Elsiever, Amsterdam, 1992, p. 437.
2. L. Bardos, Vacuum 38, 637 (1988).
3. E. O. Johnson and L. Malter, Phys. Rev. 80, 58 (1950).
4. E. W. Peterson, L. Talbot, AIAA Journal 8, 2215 (1970).
5. B. E. Cherrington, Plasma Chem.Plasma Process 2, 113 (1982).
6. C. H. Su, R. E. Kiel, J. Appl. Phys. 37, 4907 (1966).
7. N. Hershkowitz in Plasma Diagnostics Vol. 1 edited by Orlando Auciello and Daniel L. Flamm, Academic Press, Inc. San Diego, 1989, p. 125.
8. K. U. Riemann, P. Meyer, Phys. Plasmas 3, 4751 (1996).
9. Y. Y. Xu, T. Ogishima, D. Korzec, Y. Nakanishi and Y. Hatanaka, Jpn. J. Appl. Phys. 38, 4538 (1999).
10. Z. Zakrzewski, M. Moisan, J. Margot and G. Sauve, Plasma Sources Sci. Technol. 1, 28 (1992).
11. L. Sirghi, K. Ohe and T. Kimura, Phys. Plasmas 4, 1160 (1997).
12. L. S. Frost, A. V. Phelps, Phys. Rev. A 136, 1538 (1964).

Rotating Cylindrical Magnetrons and Accelerators with Anode Layer for Large – Area Film Deposition Technologies

S.P. Bugaev, A.N. Zakharov, K.V. Oskomov, N.S. Sochugov, and A.A. Solovjev.

High Current Electronics Institute, Siberian Division of the Russian Academy of Sciences, Russia

Abstract. DC magnetrons with a rotating cylindrical cathode and length up to 2.5 m were designed. Results of the experiments for the sputtered film uniformity and target utilization degree increasing are presented. Depending on technological requirements, a magnet system of the magnetron forms either one sputtered particle flow or two flows in diametrically opposite directions. Linear accelerators with an anode layer designed for cleaning the substrate surfaces prior to the coating deposition and for hard carbon films deposition are described as well.

INTRODUCTION

The principal parts of a vacuum setup intended for thin film technologies are the devices creating particle flows from which a coating is formed. Depending on a realized technology, different devices can be used to create flows of this kind. To deposit coatings on large-area substrates, these devices should be extensive at least in one direction. By now, different variants of magnetron sputtering systems (MSS) are widely used to deposit coatings on large-area substrates using PVD methods [1-3]. The main demands made of these MSS are providing with high uniformity of the deposited coating, and achieving high target utilization degree [4].

When developing most of the thin film technologies, we have to solve the problem of increasing the coating adhesion to a substrate. Different variants of devices for preliminary ion-plasma treatment of the substrate surface are used for this purpose. Some of devices are the accelerators with the anode layer [3, 5]. Their merits are the design simplicity and possibility to generate extended (up to 2-3 m) ion-plasma flows. Different CVD technologies can be realized on the basis of this kind of sources.

This paper presents the results of the works carried out at the Institute of High Current Electronics directed on perfection of designs of magnetron sputtering systems with cylindrical cathodes and ion-plasma sources based on the accelerators with anode layer as well as on creation of technologies developed on the basis of these devices.

EXPERIMENTAL

Efficiency increase of MSS with rotating cathode

Among the variants of extended magnetrons the most promising are the systems with cylindrical rotating cathode [4]. A distinctive feature of these systems is using a cathode made as a tube inside which a magnetic system (MS) is placed. The cathode can rotate relative to the immovable MS that provides its uniform wear. The main merit of these magnetrons is high target utilization – up to 80% instead of 20-30% with the planar magnetrons. However, there exist possibilities for further efficiency increase of rotating magnetrons. The shortcoming common for all MSS is the fact that even at high uniformity of magnetic field along the target length, the substrate parts disposed near the magnetron ends are sputtered with less grow rate than the central part of the substrate. The reason is the nonsymmetrical sputtering diagram. Therefore, in order to achieve high uniformity of the coating thickness on the whole substrate area we have to fabricate magnetrons with the target dimensions exceeding the dimensions of the treated substrates by 20-30 cm that results in magnetron cost rise and increase of the vacuum chamber dimensions. The shortcoming of the MPC with a rotating cathode is the accelerated eroding of the end cathode parts owing to high specific power falling to these parts in comparison with the rest part of the target [6]. As a result, in reality the coefficient of the target utilization is far from the maximum possible one.

Target thickness increase at the parts with accelerated eroding or fabrication of these parts from a hard sputtering material [6] solve this problem incompletely as this results in decrease of the magnetic field holding the plasma at the target surface, change of the discharge existing conditions, and appearance of instabilities in plasma. We have carried out a series of experiments on the efficiency increase of a rotating magnetron owing to the expansion of the region of the uniform coating deposition and decrease of the accelerated erosion of the end cathode parts. A cylindrical magnetron with a rotating cathode made of aluminum was used in the experiments. The cathode had an external diameter of 80 mm and the length of the sputtered part of 520 mm. The deposited film thickness was measured with an interference microscope. Erosion uniformity of the target by its length was determined by calculation of the erosion zone cross-sectional area.

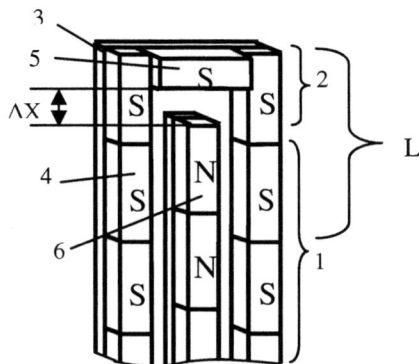

FIGURE 1. Magnetron magnetic system scheme. 1- linear part, 2- turn-around part, 3- core, 4- outer magnets, 5 – end magnets, 6- central magnets.

Figure 1 presents the scheme of the magnetron MS. It consists of a linear *1* and turn-around *2* parts forming a closed sputtering racetrack on the target surface. The linear part consists of three lines of permanent magnets placed at a core *3*. The lines of the outer magnets *4* are connected with each other by the end magnets *5*. Between the end magnets and the line of the central magnets *6* there is a gap ΔX; its value is chosen so that distributions of a longitudinal $B_{||}$ and normal $B_\perp$ components of the magnetic field along the gap ΔX should coincide with the distributions between the outer and central magnets at the linear part of the magnetic system (fig.2a, b) [7].

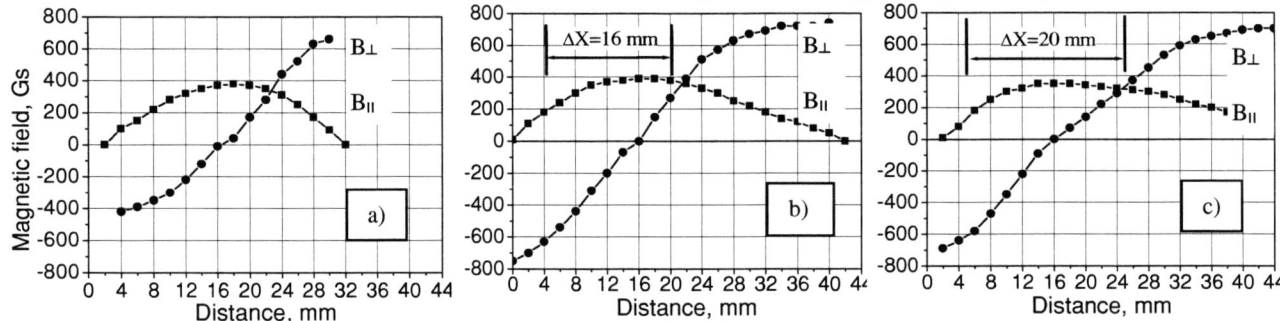

FIGURE 2. Distributions of a longitudinal $B_{||}$ and normal $B_\perp$ components of the magnetic field: a) between the outer and central magnets at the linear part of MS; b) along the gap ΔX =16 mm (initial MS design); c) along the gap ΔX =20 mm (modified MS).

Curves 1 in figs. 3a and 3b present the results of the experiments on measuring the coating thickness uniformity at a substrate and cathode erosion area uniformity. The experiments were carried out with the magnetron having the above-described magnetic system. As it is seen from the figures, the length of the deposition area with the uniformity being no worse than ±1% makes up 22 cm, i.e. at the overall length of the sputtering area equal to 52 cm at the magnetron ends there are areas with the dimensions of 15 cm each that are used ineffectively though they are subjected to sputtering. Obviously, these dimensions are enough large even for magnetrons with the target length of 2-3 meters. From fig. 3b one can see that at the curve 1 on the left there is a region with maximum erosion area that corresponds to the region of accelerated eroding at a turn-around part of the magnetic system. There is the analogous region at the opposite end of the target, but it is not shown in the figure. Sputtering velocity at the target ends is by 20% higher than on average along its length.

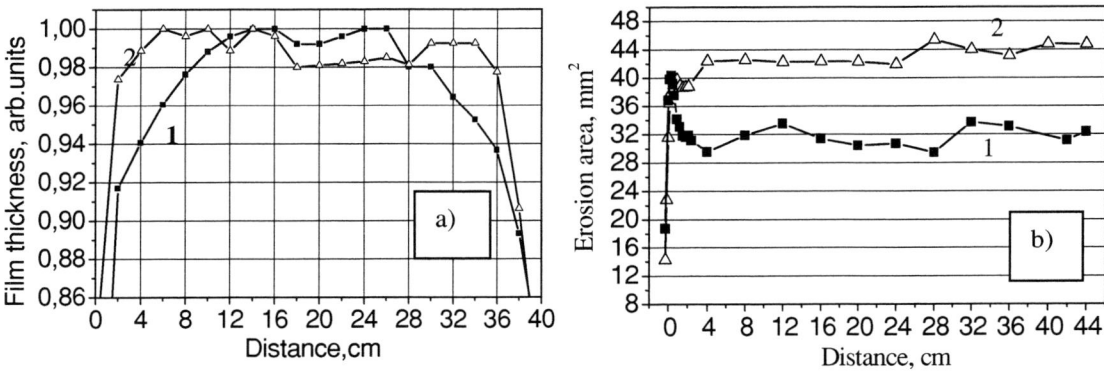

Figure 3. Results of the experiments on measuring the coating thickness uniformity at a substrate (a); and erosion area along the target length (b). 1- initial MS; 2- modified MS.

In order to overcome these shortcomings we've made changes in the MS design. Firstly, magnetic field induction at the parts of the magnetic system adjacent to the turn around parts was increased. This was achieved by means of replacement of the outer magnets (fig. 1) by the magnets having 5-15% higher residual induction of magnetic field. Varying the number of these magnets in the magnetic system we can change the length of the region with the increased magnetic field L (fig.1.). Secondly, by introduction of additional magnets into the gap ΔX the erosion area in the turning region was expanded that allowed decreasing the sputtering power density in these areas. Fig. 2c presents distribution of the magnetic field components corresponding to this case. Curves 2 in figs. 3a and 3b present the results of measurements of the coating thickness uniformity and cathode erosion uniformity. One can see that the length of the deposition area with the uniformity of ±1% expanded by 12 cm and made up 34 cm (fig. 3a). We have also managed to eliminate completely the accelerated wear effect of the rotational cathode part (fig. 3b).

Using the obtained results, we develop and fabricate rotating magnetrons with the target length up to 2.5 m for application in high-production technological setups. Depending on the type and production of a setup, we use either magnetrons with a continuously rotating target or magnetrons with a periodically turned target. A magnetron with a continuously rotating target is more efficient for application at conveyor type technological setups. A magnetron with a periodically turned target can be used at the batch type setups with movable technological sources.

Depending on technological requirements, we use magnetrons with magnetic systems forming either one flow of a sputtered material or two flows in diametrically opposite directions (two-sided magnetron) that allows making coating deposition on two substrates simultaneously. Magnetic system of a two-sided magnetron consists of two extended linear sections forming two sputtering tracks at diametrically opposite parts of the tube and two end sections providing the closure of the racetrack. We use magnetrons having this construction at the technological setups VNUK intended to deposit low- E coatings on architectural glasses having dimensions up to 1.6×2.5 m^2.

Sources of extended ion-plasma flows based on the accelerator with anode layer

We have developed extended ion-plasma sources based on the accelerator with anode layer intended for precleaning of large-area substrates (in particular, architectural glasses) prior to the coating deposition, and realization of CVD technologies. The source operation is characterized by simultaneous existence of the hollow cathode discharge plasma and ion beam in the vacuum chamber. The ion-plasma source has a high uniformity degree of the ion current linear density (current per unit of length of the source) that is achieved by the choice of the electrode system configuration, magnetic field strength, and operating pressure range. Fig. 4 presents measurement results of linear density of the source ion current for different operating parameters. This source has the full length of 350 mm and forms an extended ion beam with the linear part length of 236 mm. Nonuniformity of the ion current linear density doesn't exceed ± 4% at 90% of the source linear part. Ion sources with high uniformity of the ion current with the length up to 2 m have been designed on the basis of the obtained results.

High uniformity of the current linear density at the full extent of the ion source with closed electron drift makes possible to solve the problem of the diamond-like film deposition on large-area substrates. An important feature of the source that we have developed is possibility of simultaneous generation at definite conditions of a directed beam and uniform plasma. This allows realizing combined ion-plasma deposition of a-C:H films differing from ion deposition (ion beam only) by the high grow rate. It is necessary to note that it is not required to supply negative bias potential to the substrate that is important from the technological point of view.

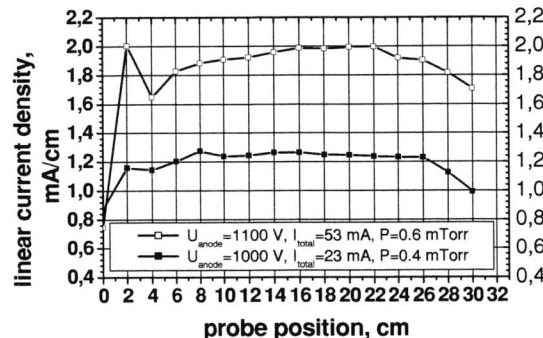

FIGURE 4. Measurement results of linear density of the source ion current for different operating parameters.

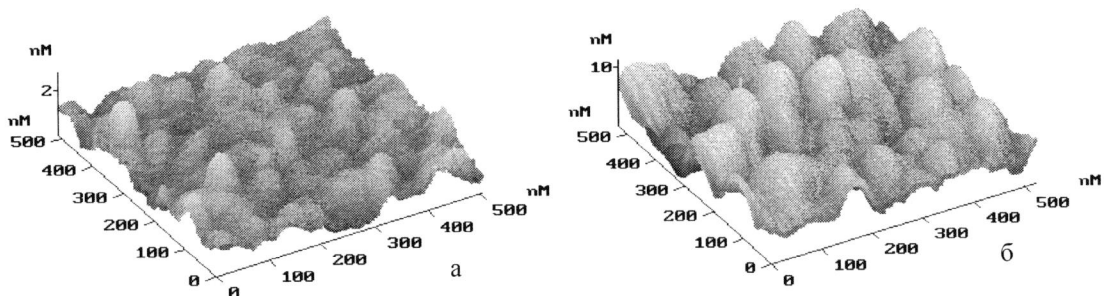

FIGURE 5. A picture of the surface of the diamond-like (a) and polymer-like (b) a-C:H-films obtained by means of an atomic-force microscope.

The experiments that were carried out have shown that the properties of the a-C:H-coatings deposited with the use of the ion-plasma source are determined by the relation of hydrocarbon ions from a beam and hydrocarbon radicals from plasma to the substrate surface in the process of deposition and by the ion energy proper. Optimum deposition parameters (gas pressure, current, discharge voltage) allowing obtaining a diamond-like a-C:H-coating differing by high hardness and adhesion to different substrates (silicon, glass, plastics, and polymers) were determined. Increase of the operating pressure and discharge current as well as voltage decrease at the accelerating gap resulted in deposition of a soft polymer-like film due to insufficient ion bombardment of the coating surface in the process of rising. One of the evidences of transition from the diamond-like to the polymer-like structure of the a-C:H-film can serve increase of the surface roughness from 2-3 to 10-12 nm that was observed by means of an atomic-force microscope (fig. 5).

CONCLUSION

Designs of high-performance cylindrical magnetron sputtering systems and ion – plasma sources allowing realizing the processes of depositing different coatings on large-area substrates have been developed. Use of a modified magnetic system at MSS with a rotating cathode allows widely using possibilities of the cylindrical target in achieving high target utilization degree. It is shown that hard diamond-like films can be obtained at the large area substrates using the chemical vapor deposition by means of an ion source with closed electron drift.

REFERENCES

1. Nadel S. J., and Greene P., Thin Solid Films 392, 174-183 (2001).
2. Safi I., Surf. Surf. Coat. Technol. 127, 203-219 (2000).
3. Bugaev S.P., and Sochugov N.S., Surf. Coat. Technol. 131, 474-480 (2000).
4. Kukla R., Surf. Coat. Technol. 93, 1-6 (1997).
5. Vershinin N., Straumal B, Filonov K., Dimitriou R., Gust W., and Benmalek M., Thin Solid Films 351, 190-193 (1999).
6. Vanderstraeten E., Norgan S., Vanderstraeten J., and Gobin G., patent of Canada, WO98/35070, (1998).
7. Shidoji E., Nemoto M., and Nomura T., J. Vac. Sci. Technol. A 18(6), 2858-2863 (2000).

Development and Application of Vacuum Arc Ion Source at HCEI

A.S. Bugaev, V.I. Gushenets, A.G. Nikolaev, E.M. Oks, K.P. Savkin, and G.Yu. Yushkov

*High Current Electronics Institute, Russian Academy of Sciences,
4 Academichesky ave., Tomsk, 634055, Russia, Ph: +7 (3822) 258776,
Fax: +7 (3822) 259410, E-mail: oks@opee.hcei.tsc.ru*

Abstract. The development of the vacuum arc ion sources is motivated by possibilities of their application in technologies for modification of surface properties and as high-current injectors for heavy ion accelerators. The sources of such a type developed at High Current Electronics Institute provide generation of intensive pulse-periodic ion beams. The present paper deals with peculiarities of vacuum arc ion sources employed to generate beams of gas and metal ions with the controllable ratio of ions of each type in a beam. The design and parameters of the sources are presented.

INTRODUCTION

The development of ion sources based on cold-cathode arc discharges is motivated by possibilities of their wide application in technologies for modification of surface properties of different materials and as high-current injectors for heavy ion accelerators [1, 2]. The sources of such a type developed at High Current Electronics Institute provide generation of intensive pulse-periodic beams of gas and metal ions [3, 4]. These sources are characterized by relative simplicity of their design, convenient operation, reliability, and sufficiently long lifetime. The nonuniformity of the ion current density distribution along the cross section of the beam is no greater than 20 % [5].

Further development of sources of such a type is to improve the design of the "Titan" ion source, to enhance the capabilities of ion sources of the Mevva type [6, 7], and to design new sources of gas and metal ion beams based on cold-cathode arc discharges, where the required beam constitution is determined by establishment of specific conditions, and thus to provide effective gas ionization in a discharge gap.

GAS AND METAL "TITAN" ION SOURCES

The "Titan" ion source was developed at the Institute of High-Current Electronics about ten years ago [3, 5] and found its application in studies on the action of accelerated ions on surface properties of materials that improve the tribological characteristics of the latter [8]. A peculiar feature of the "Titan" source is generation of wide-aperture beams of both gas and metal ions. Metal ions were generated by a vacuum arc discharge, and gas ions were generated by a constricted discharge. The parameters, principles of operation, and the design of the "Titan" sources are described elsewhere [3-5, 9, 10].

The mass-spectrometry studies have shown that in the mode of gas ion beam generation the percentage of metal ions does not exceed 1 % and is determined by sputtering of elements of the discharge chamber. The "Titan" source employs a system of initiation based on an auxiliary gas discharge. This has made possible a substantial increase in the time of continuous operation of the ion source, but, however, required a buildup of pressure of working gas up to $5 \cdot 10^{-5}$ Torr. At the same time, the increased pressure of operating gas brought about the emergence of gas ions in a metal ion beam, though the percentage of these ions did not exceed 10-15 %. Both discharges provide simultaneous generation of gas and metal ions, and the discharge currents or pulse duration can control the ratio of ions of each type in the beam.

CP669, *Plasma Physics: 11th International Congress on Plasma Physics: ICPP 2002*
edited by I. S. Falconer, R. L. Dewar, and J. Khachan
© 2003 American Institute of Physics 0-7354-0133-0/03/$20.00

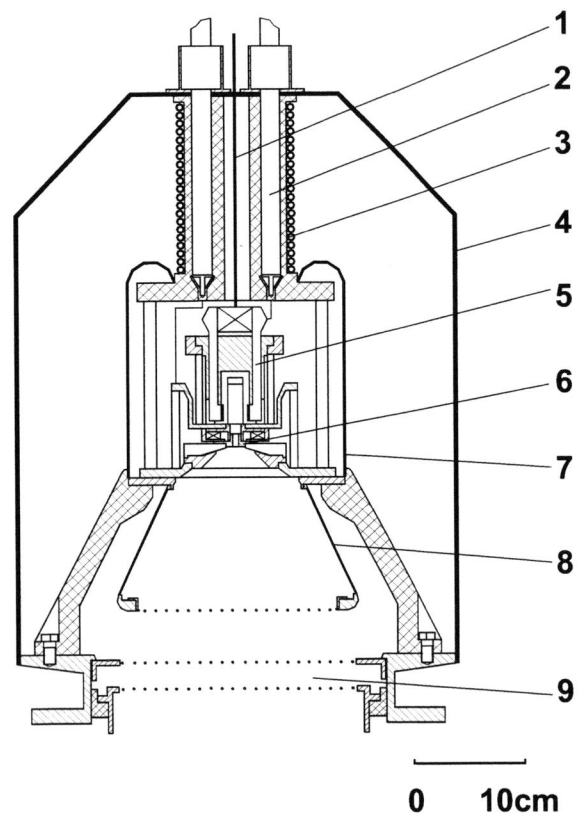

FIGURE. 1. The "Titan-3" ion source. 1- gas tube, 2- cable, 3- cooling water tube, 4- body of source, 5- cathodes of Penning discharge, 6- cathode of vacuum arc discharge, 7- equipotential body, 8- hollow anode, 9- accel-decel system.

Figure 1 shows a schematic of the "Titan-3" source design. The principle of its operation is the same as in the previous versions, but the design features are substantially different.

The water-cooled discharge chamber of the source is located on a base insulator rated at a voltage of up to 80 kV. The chamber is covered with equipotential body 7 made of stainless steel and cooled by distilled water. The latter circulates in the closed circuit due to a water pump. Running water-cools distilled water, in turn. Water is supplied to the discharge chamber through plastic tubes 3 of length more than 10 m to preclude high-voltage water breakdown (discharge) and to decrease leakage currents. The source is covered with hermetically sealing casing 4, which, when needed, can be filled with inert gas at a pressure of several atm to protect elements of the discharge chamber from a corona discharge.

Technological tests of the "Titan-3" source have shown that no carbon and carbide compounds are found in specimens when being treated by an ion beam generated in such a source. Moreover, the "Titan-3" source is very convenient in operation and has lately taken the place of the previous modifications of the "Titan" sources.

GENERATION GAS AND METAL ION BEAMS IN MEVVA-TYPE ION SOURCE

Vacuum arc metal ion sources (Mevva type sources) are designed to implant metal ions [1, 2, 11]. A slight modification of the design of the Mevva type source makes it possible to generate hybrid gas-metal ion beams. Therefore, the conditions established for electron oscillation and gas leak in the discharge chamber provide generation of mixed beams of gas and metal ions [12]. Moreover, with leaking-in gas, it is possible to employ trigger

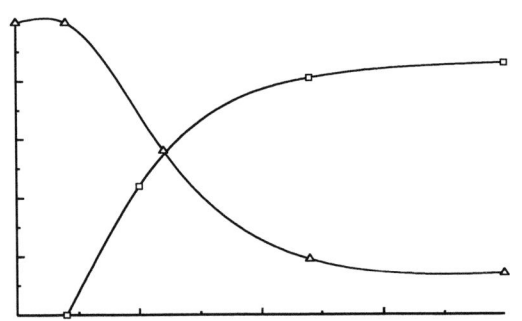

FIGURE 2. Fraction of Al ions (total Al^+, Al^{2+} and Al^{3+}) and O_2 ions (total O_2^+ and O^+) in the beam as a function of magnetic field. $p = 2 \cdot 10^{-4}$ torr.

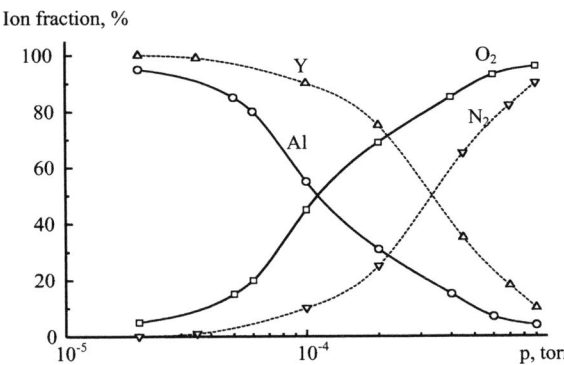

FIGURE 3. Fraction of Al / O_2 ions (solid lines) and Y / N_2 ions (dash lines) in the beam as a function of system pressure. B = 0.1 T.

systems based on ExB discharges [7, 13] in the Mevva type source. The lifetime of such systems, as a rule, is longer than that of conventional systems based on the vacuum arc initiated over the dielectric surface. The given approach was realized in modifications of the Mevva-V sources at Lawrence Berkeley Laboratory, USA [12, 14], the Mevva-IV ones at GSI, Germany [15, 16] and at the Institute of High Current Electronics [7, 13].

To provide more effective generation of gas ions, it is possible to apply a magnetic field of up to 0.2 T to the cathode region. Thus, the change in the value of the magnetic field in the range from 20 mT to 0.2 mT, in the case of oxygen leak in a system with an Al cathode, makes it possible to control the ratio of the atomic percentage of O ions in the beam from nearly 0% to 85% and that of Al ions from 100% to 15% (Fig. 2).

The control of the ratio of the gas ion fractions and those of metal ions in the beam can also be done by varying the value of the gas pressure a constant magnetic field. It can be seen from the dependencies shown in Fig. 3 that the ratio of the O ion percentage for the case of the vacuum arc with the Al cathode ranges from 4 % to 95 %, that of the Ni ion percentage for the case of an Y cathode is from 0 % to 90 % as the pressure is increased from $2 \cdot 10^{-5}$ Torr to 10^{-3} Torr.

REFERENCES

1. I.G. Brown (Ed.) *The physics and technology of ion sources*. - Wiley, New York, 1989.
2. B.H. Wolf (Ed.) *Handbook of ion sources*. – Boca Raton, Fl.: CRS Press, 1995.
3. S.P. Bugaev, A.G. Nikolaev, E.M. Oks, P.M. et al. *Rev. Sci. Instrum.*, **63**, 2422 (1992)
4. A.G. Nikolaev, E.M. Oks, P.M. Schanin, G.Yu. Yushkov. *Rev. Sci. Instrum.*, **67**, 1213 (1996)
5. S.P. Bugaev, A.G. Nikolaev, E.M. Oks, P.M. et al. *Rev. Sci. Instrum.*, **65**, 3119 (1994)
6. A. Bugaev, V. Gushenets, A. Nikolaev, E. Oks, G.Yu.Yushkov, et al. *Proc. 18th Intern. Symp. on Discharges and Electrical Insulation in Vacuum*. Eindhoven, Netherlands, 1998, p. 256.
7. A.G. Nikolaev, G.Yu. Yushkov, E.M. Oks, R.A. MacGill, M.R. Dickinson, I.G. Brown. *Rev. Sci. Instrum.*, **67**, 3095 (1996)
8. A.N. Tyumentsev, Yu.P. Pinzhin, A.D. Korotaev, et al. *Nucl. Instrum. Methods Phys. Res.*, **B 80/81**, 491 (1993)
9. S.P. Bugaev, A.G. Nikolaev, E.M. Oks, et al. *Proc. of 15th Intern. Symp. on Discharges and Electrical Insulation in Vacuum*, Germany, Darmstadt, 1992, p. 686.
10. E.M. Oks and G.Yu. Yushkov. *Proc. Mevva workshop'95*, Berkeley, USA, 1995, p. 24.
11. I.G. Brown. *Rev. Sci. Instrum.*, **65**, 3061 (1994)
12. E.M. Oks, G.Yu. Yushkov, P.J. Evans, A. Oztarhan, I.G. Brown, et al. *Nucl. Instrum. Methods Phys. Res.*, **B 127/128**, 782 (1997)
13. A.G. Nikolaev, G.Yu. Yushkov, E.M. Oks, R.A. MacGill, et al. *Proc. 17th Intern. Symp. on Discharges and Electrical Insulation in Vacuum*, Berkeley, USA, 1996, p. 562.
14. I.G. Brown, A. Anders, S. Anders, M.R. Dickinson, R.A. MacGill, E.M. Oks. *Surface and Coating Technology*, **84**, 550 (1996)
15. P. Spadtke, H. Emig, B.H. Wolf, E.M. Oks. *Rev. Sci. Instrum.*, **65**, 3113 (1994)
16. B.H. Wolf, H. Emig, D.M. Ruck, P. Spadtke, E.M. Oks. *Nucl. Instrum. Methods Phys. Res.*, **B 106**, 651 (1995)

Bulk Plasma Production Using Gaseous Discharge System with External Electron Injection

M.V. Shandrikov, A.V. Vizir, E.M. Oks, and G.Yu. Yushkov

*High Current Electronics Institute, Russian Academy of Sciences,
4 Academichesky ave., Tomsk, 634055, Russia, Ph: +7 (3822) 258776,
Fax: +7 (3822) 259410, E-mail: oks@opee.hcei.tsc.ru*

Abstract. Results of experimental study of low-pressure, high-current gaseous discharge, driven by external electron injection are presented. It is shown that the discharge of this type can be used for efficient generation of dense and uniform gaseous plasma. Physical features of plasma generation are discussed. High efficiency of energy utilization, wide range of operating pressure, absence of thermionic emitter, reliability, and possibility to operate in chemically active environment favor the use of such a system in various surface modification processes. Application areas for this plasma gun are surface cleaning and activation prior and during thin film deposition.

INTRODUCTION

Low-pressure glow discharge is characterized by stability and uniformity of plasma parameters as well as high ion fraction of cathode current. That provides its applications in various ion and plasma sources. Despite of a number of positive properties, drawbacks of a glow discharge are relatively high voltage and operating pressure.

The coefficient of secondary ion-electron emission γ in a low voltage glow dis- charge usually does not exceed 0.1. Hence, the increase of electron fraction of cathode current, even rather small in comparison with the discharge current, can essentially change the discharge parameters and in particular its voltage. Such increase can be carried out "artificially" by external injection of electrons into the cathode region [1]. If the conditions for acceleration of these electrons in cathode potential fall are created, they will be indiscernible from electrons emitted by the cathode as a result of secondary emission. It has been shown [2] that electron injection substantially increases operating pressure range to lower values and reduces discharge voltage.

Based on method of electron injection to support a hollow cathode glow, several ion sources and plasma gun have been made [3]. In these devices injected electrons were generated in additional "keeping" hollow cathode glow. In this paper we present results of investigation of glow discharge with electron injection where to generate electrons a constricted arc is used.

PLASMA SOURCE PRINCIPLE AND PARAMETERS

A schematic view of the electrode system is shown on Fig. 1. The plasma is generated by high-current, two-stage gaseous discharge. First stage (keeping discharge) is a filamentless electron emitter based on constricted arc. Second stage (main discharge) is a low-voltage low-pressure non-selfsustained glow discharge.

Cathodic spots are formed on the internal surface of the hollow cathode of the keeping discharge 1. The electron flow moves to the grid-like keeping discharge anode 2 through the slot of the cathode 1 and slot-like constricting channel in the intermediate electrode 3. Keeping discharge voltage is about 40 V.

The anode of keeping discharge is electrically connected to the cathode of the main discharge. The greater part of electrons is extracted from the keeping discharge plasma by the cathode potential drop adjacent to the cathode 4

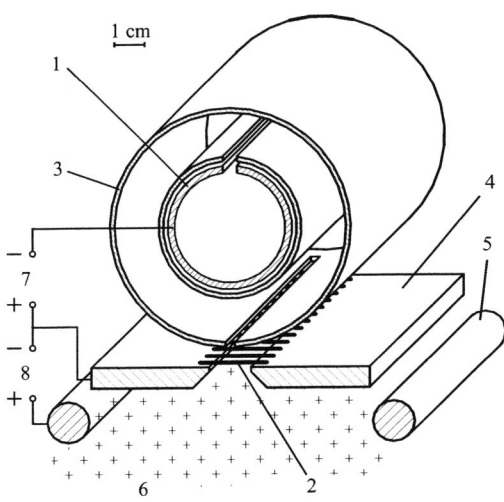

FIGURE 1. Schematic view of the plasma source. 1 – keeping discharge cathode, 2 – keeping discharge anode, 3 – intermediate electrode, 4 – main discharge cathode, 5 – main discharge anode, 6 – main discharge plasma, 7, 8 – power supplies of keeping and main discharges.

through the electrode 2. Injected electrons accelerated by the potential drop effectively ionize the operating gas all over the vacuum chamber.

The potential barrier for the electrons forms near the chamber walls, provided that all electrodes of the system are insulated from the grounded chamber. Measurements show that the plasma potential sets to value about +20 V relatively to the chamber. At the same time, the main discharge voltage, and, correspondingly, the injected electron energy, is about 100 V (eV). Despite of that, the potential barrier is enough to reflect some electrons capable of ionization, because injected electrons loose energy passing through the chamber, and, also, due to elastic collisions, the incidence angle of electron trajectory can differ from normal. The chamber walls are not subjected to intense sputtering that could contaminate the plasma, because of low ion energy (20 eV). Moreover, insulation of the electrode system from ground potential allows to avoid cathode spot formation on the chamber walls.

The configuration and arrangement of the cathode 1 and intermediate electrode 3 (Fig. 1) almost completely excludes contamination of the vacuum chamber volume by products of the keeping discharge cathode erosion.

Hollow shape of the cathode, compared with an open cathode, provides longer cathode lifetime due to re-deposition of cathode material on the opposite cathode wall.

The device operates in DC mode. All electrodes of the source are water-cooled. They are arranged within a vacuum case mounted on the chamber flange. The chamber volume is 0.9 m^3. The operating gas is fed into the keeping discharge cathode cavity. Operating gases were Ar, N$_2$, O$_2$.

Fig. 2 shows current-voltage characteristics of main discharge for constant keeping discharge current and corresponding effect of main discharge voltage on the ion current density measured by Langmuir probe. Between 20 and 60 V, rapid growth of the discharge current and density of its plasma occurs, caused by the increase of electron energy and ionization cross-section. Further voltage increase does not lead to current growth because electrons accelerated by the cathode potential drop of the main discharge start to escape to chamber walls. Nevertheless the ion current density slightly increases. Maximum efficiency of plasma generation is reached with main discharge voltage of 100-120 V.

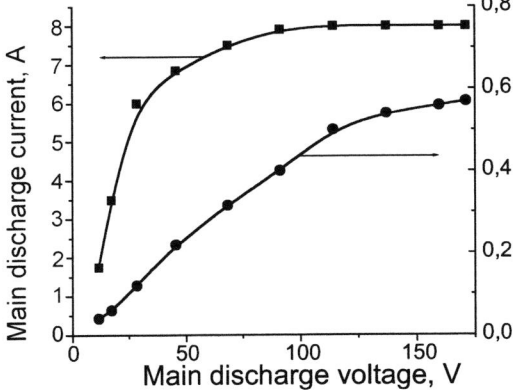

FIGURE 2. Dependence of the main discharge current on its voltage and corresponding variation of the ion current density. Probe bias voltage is –200 V. Keeping discharge current is 8 A. Pressure is 0.4 mTorr of argon.

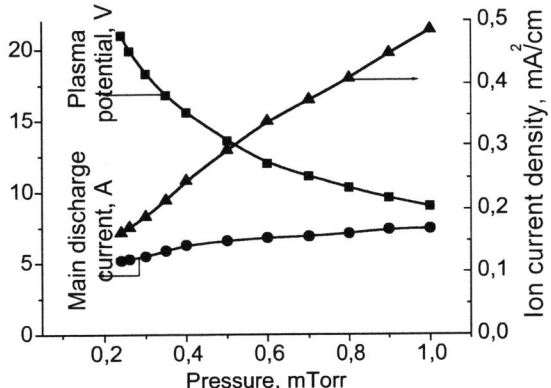

FIGURE 3. Pressure dependencies of main discharge current, density of ion current extracted from plasma, and plasma potential. Keeping discharge current is 7 A. Main discharge voltage is 115 V.

Measurements show that ion current extracted to the probe and, consequently, plasma density is proportional to the discharge current within the current range of 0.1–8 A. This fact shows that the plasma beam discharge does not occur in this system. The plasma is generated exceptionally due to individual particle interaction.

Visually, plasma is distributed sufficiently uniform over the chamber volume. Measurements of ion current density distribution along the chamber axis located 55 cm away from the edge of plasma source show that the current density nonuniformity is only ±10 % on axis length of 50 cm. Because the cross-section of electron scattering on gas substantially exceeds that for inelastic collisions, injected electrons partially loose their original direction at the distance from the source about several tens of centimeters. That leads to a good uniformity of the plasma. Also, "shadow" effect was insignificant. Ion current density at the back side of the sample (plate 10 by 15 cm) was just 16% less than at the sample side facing the source. Reflection of ionizing electrons from the potential barrier near the chamber walls is another factor providing good uniformity.

Plasma density and, consequently, ion current density substantially increases with pressure (Fig. 3), though, main discharge current increases just weakly. This occurs due to reduction of ionization length of injected electrons that, in turn, causes the increase of electron energy fraction that is used for ionization. With low pressure, energy loss becomes considerable because the number of energetic electrons lost at the chamber walls increases. At the same time, plasma potential growth with pressure reduction partially compensates electron loss.

Measured by double probe plasma electron temperature is 8 eV. Calculated using Bohm equation plasma density is $1 \cdot 10^{10}$ cm^{-3}, for discharge current of 8 A.

REFERENCES

1. E.M. Oks, A.V. Vizir, and G.Yu. Yushkov, *Rev. Sci. Instrum.* **69(2)**, 853 (1998)
2. A. V. Vizir, G. Yu. Yushkov, and E. M. Oks, *Rev. Sci. Instrum.* **71(2)**, 728 (2000)
3. A. V. Vizir, G. Yu. Yushkov, and E. M. Oks, "Development of gaseous ion and plasma sources based on hollow cathode discharge with electron injection" in *Proceedings of the 1st International Congress on Radiation Physics, High Current Electronics, and Modification of Materials,* Tomsk, Russia, 24 – 29 September 2000, **3** (Vth Conf. on Modification of Materials with Particle Beams and Plasma Flow), pp. 190-193.

Decomposition Of Benzene By A Low Pressure Glow Discharge

K.Satoh[†], T.Sawada[†], T.Naitoh[†], H.Itoh[†], M.Shimozuma[‡] and H.Tagashira[†]

[†]*Department of Electrical & Electronic Engineering, Muroran Institute of Technology, Muroran 050-8585, Japan*
[‡]*College of Medical Technology, Hokkaido University, Sapporo, 060-0812, Japan*

Abstract. Decomposition characteristics of diluted benzene with nitrogen in a low pressure DC glow discharge plasma are investigated by emission spectroscopy, mass spectrometry, infrared absorption spectroscopy, concentration measurement and gas pressure monitoring. It is likely that benzene is decomposed chiefly by electron collision, and that excite molecules of nitrogen and/or nitrogen ions do not make a large contribution to benzene destruction directly. It is found that H, H_2, H_3, C, CH, CH_4, NH_3, C_2, C_2H, CN, C_2H_2, and C_2H_4 are produced in the glow discharge.

INTRODUCTION

Benzene contained in exhaust fumes of automobiles, coke furnaces, chemical factories, etc. is a toxic substance which causes cancer and other disease, so that it needs to be removed or decomposed before it is released in the air. McCorkle[1] et al. reported that benzene was decomposed by glow discharges in noble gases and that the dissociative attachment of slow electrons to the benzene molecules in high Rydberg state could be dominant decomposition process. Also, Morris[2] suggested that nitrogen ion could contribute decomposition of benzene. Decomposition process of benzene in the glow discharges, however, is not investigated in the papers. In order to improve the efficiency of benzene decomposition in a glow discharge, it is important to clarify the decomposition process.

Recently, Yasui et al.[3] reported that the destruction of benzene ring is one of the dominant reduction processes of dioxins in a pulse corona discharge, so that the information about the decomposition characteristics of benzene in glow discharges would contribute to decomposition of dioxins using discharge plasma.

In the present work, the decomposition process of diluted benzene with nitrogen in a low pressure DC glow discharge is investigated by measuring the emission and mass spectra in the discharges, infrared absorption spectra of the deposited film, which is regarded as some of the fragments and/or the by-products, on silicon wafers, benzene concentration and gas pressure.

EXPERIMENTAL APPARATUS AND PROCEDURE

Fig.1 shows a schematic diagram of experimental apparatus. Parallel plate electrodes of 6.0cm in diameter are set in the discharge chamber of 15.5cm in diameter and 30.0cm in height. The electrodes and the chamber are made of stainless steal, and the lower electrode and the chamber are grounded. Negative DC voltage is applied to the upper electrode to generate glow discharge. Benzene is vaporized in a flask, and then fed into the discharge chamber with nitrogen. The purity of benzene and nitrogen is 99% and 99.999%, respectively.

The temporal variation of emission of the glow discharge is measured by Photonic Multi-Channel Analyzer (Hamamatsu PMA-11), which provides a spectral resolution of <2nm between 200 and 950nm, and mass spectra of neutral species effusing through the orifice with 0.1mm in diameter are measured by Quadrupole Mass Spectrometer (ANELVA M200QA-M), which provides a resolution of $M/\Delta M > 2M$ between 1 and 200 amu. Bonding of molecules of thin films deposited on silicon wafers on the lower electrode is analyzed by Fourier Transform Infrared

Spectrophotometer (SHIMAZU FTIR-8900). The concentration of benzene is measured using gas-detecting tube (GASTEC No.121), and temporal variations of gas pressure and applied voltage are also monitored.

In this work, we start with the emission spectra and benzene concentration measurements in the glow discharge when the mixture gas is confined in the chamber, and then emission and mass spectra measurements are performed when the mixture gas is flowing.

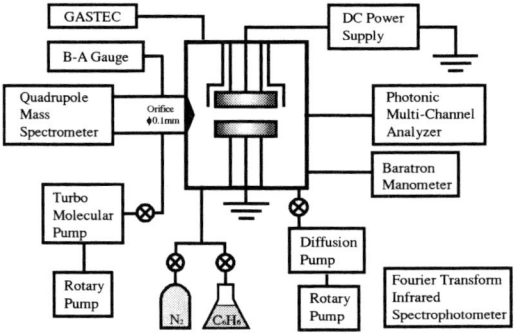

FIGURE 1. Experimental apparatus.

RESULTS AND DISCUSSION

Gas-Confined Experiment

DC glow discharge is generated in benzene(10%)-nitrogen(90%) mixture at p=133Pa. The gap length is 1.0cm and the discharge current is kept 5mA. Figure 2 shows the temporal variations of emission spectra of the discharge. The emissions of CH(431.42nm, not shown clearly in the figure), H_α(656.28nm), C_2(809.51nm) NH_3(791.90nm) and CN(918.85nm, etc.) are observed, and CH, H, C_2, NH_3 and CN are judged fragments of benzene and/or by-products.

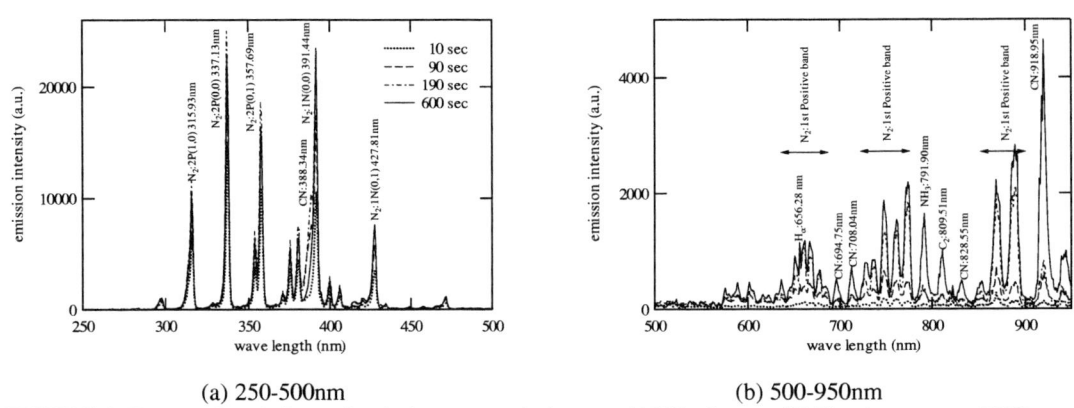

(a) 250-500nm (b) 500-950nm

FIGURE 2. Temporal variations of emission spectra in benzene(10%)-nitrogen(90%) mixture at p=133Pa.

Figure 3 shows the temporal variations of the emission intensities of CH, CN, H_α, second positive (0,0) band (2P00) of nitrogen excited molecule, first negative (0,0) band (1N00) of nitrogen ion, applied voltage, gas pressure and decomposition rate of benzene. It is found that the decomposition rate increases almost linearly against time first, and then it tends to be constant around 200s, and that the temporal profile of the decomposition rate is similar to that of the ratio 1N00/2P00 from which electron mean energy can be inferred[4]. Therefore, it is likely that electrons collide selectively with benzene molecules first because of their large collision cross section[5,6], and that the energy of the electrons is transferred to benzene molecules through inelastic collision. When the benzene concentration decreases (i.e. the decomposition rate increases), the emissions of 1N00 and 2P00 and the ratio 1N00/2P00, namely, electron mean energy increase, so that electrons escaping a collision with benzene molecule can increase their energy and have excitation and ionization collisions with nitrogen molecules. Accordingly, it is estimated that the energy to decompose benzene is lower than the threshold energies of nitrogen electronic excitations.

Since the emission of CH reaches its maximum first, then the emissions of H_α and CN reach their maxima and the gas pressure decreases when the decomposition rate increases, it is likely that benzene is decomposed into CH and other molecules, and that these fragments of benzene immediately deposit on the electrodes or the wall of the discharge chamber.

In the temporal variations of infrared absorption spectra of deposited film on the wafers (figure is not shown in this paper), the absorption spectrum of CH is observed from 50sec, however, those of CN and NH, which are regarded as by-products produced by the interaction between benzene or its fragments and excited nitrogen molecules (N_2^*) and/or nitrogen ions (N_2^+), are observed from 190sec, namely, the time when most of the benzene is decomposed. Accordingly, it seems that benzene is chiefly decomposed by the collision with electrons, and that N_2^* and/or N_2^+ do not make a large contribution to benzene destruction directly.

Since the emission peak of H_α appears behind that of CH, it seems that the destruction of benzene ring is more dominant process than dissociation of hydrogen atom from benzene ring. This is similar to the result obtained by Yasui et al.[3] for decomposition of dioxins, namely, the destruction of benzene ring is more dominant than dissociation of chlorine atom from benzene rings of dioxins.

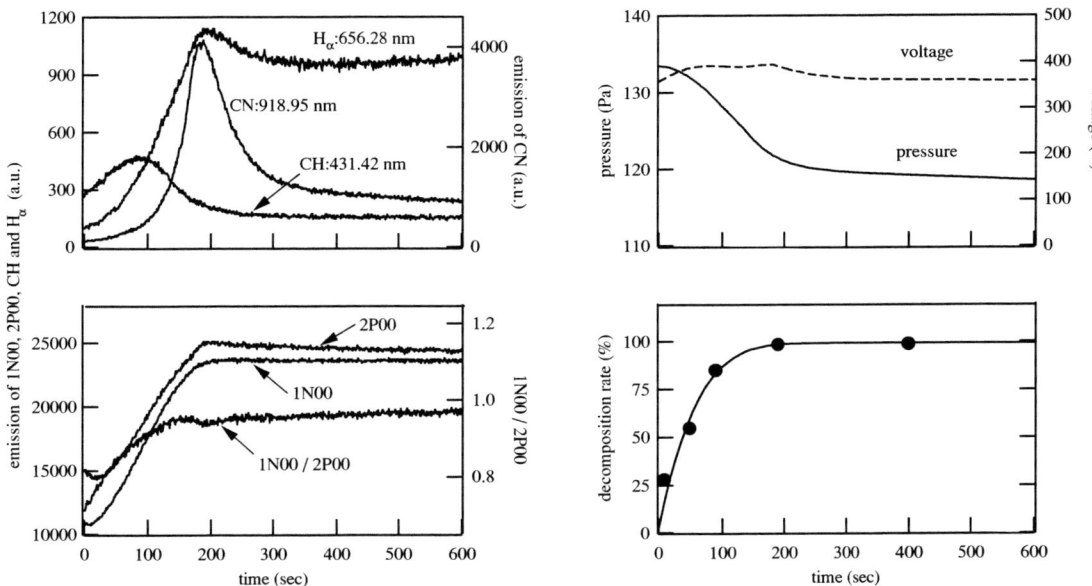

FIGURE 3. Temporal variations of emission intensity of CH, H_α, CN, 1N00 and 2P00, gas pressure, applied voltage and decomposition rate of benzene in benzene (10%)-nitrogen(90%) mixture at p=133Pa.

Gas-Flow Experiment

Benzene (8%)-nitrogen (92%) mixture is fed into the chamber with constant flow rate at p=8.7Pa, and DC voltage is applied to generate a glow discharge. The gap length is 2.5cm and the discharge current is 1mA.

Figure 4 shows the mass spectra when the glow discharge is turned on and off. It is found that many mass spectra are observed in addition to those of nitrogen (28amu) and benzene (78amu) molecules between 1 and 80amu when the discharge is not generated. This means that the benzene and nitrogen molecules effusing through the orifice are decomposed by the electron beam (acceleration voltage:40V) to ionize the neutral molecules in the mass spectrometer. It is, therefore, obvious that electrons play an important role on benzene decomposition.

Figure 5 shows the temporal variations of mass spectra of benzene and molecules the mass of which are less than 29amu, emission spectra, gas pressure and applied voltage. It is clearly seen that the mass spectra of 2, 3, 24, 25 and 26 amu vary in the opposite manner to that of benzene against time, so that it is judged that H_2, H_3, C_2, C_2H and C_2H_2 and/or CN are produced in the glow discharge. Similar tendency is seen in the temporal variations of mass spectra of 12, 13, 16 and 17amu, although the intensities of those spectra are low and fluctuated. It is also judged that C, CH CH_4 and NH_3 are produced in the discharge. Since the spectrum of 28amu increases fractionally when the discharge is generated, it is likely that C_2H_4 is also produced. From the emission spectra, H (1amu) is judged a fragment of benzene, although its mass spectrum does not change temporally.

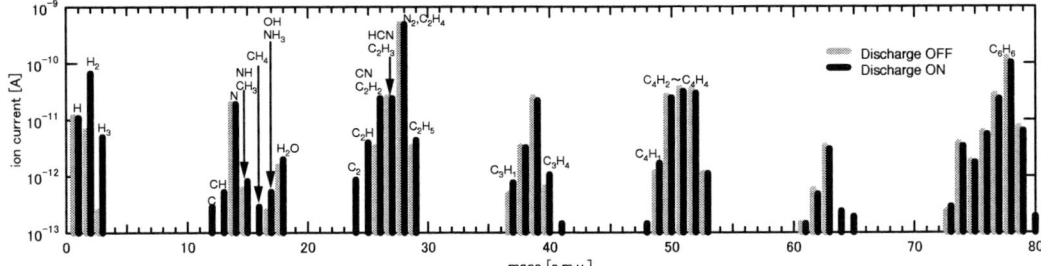

FIGURE 4. Mass spectra in a DC glow discharge in benzene (8%)-nitrogen (92%) mixture at p=8.7Pa.

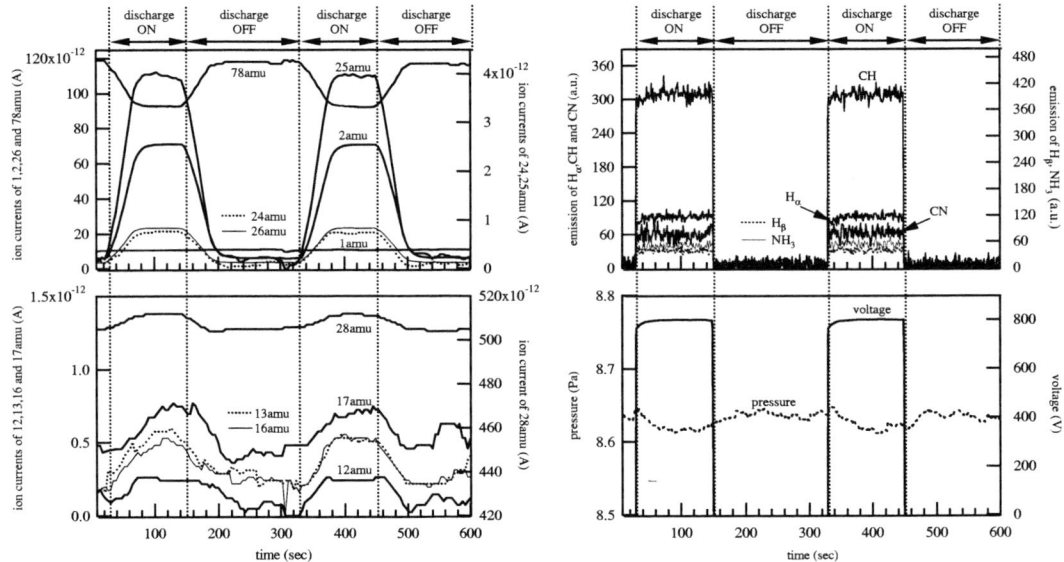

FIGURE 5. Temporal variations of mass spectra, voltage and gas pressure.

CONCLUSIONS

Decomposition process of benzene in the glow discharge is investigated in this work. In the gas-confined experiment, it is likely that benzene is decomposed into CH and other fragments by electrons the energy of which is lower than the threshold energies of nitrogen electronic excitations chiefly, and then some of CH are deposited on the electrodes or wall. It is also likely that N_2^* and/ or N_2^+ do not make a contribution to benzene destruction directly. In the gas-flow experiment, it is found that H, H_2, H_3, C, CH, CH_4, NH_3, C_2, C_2H, CN, C_2H_2, and C_2H_4 are produced in the glow discharge in the benzene-nitrogen mixture gas.

ACKNOWLEDGMENTS

This work was supported by Grant-in-Aid (No.13750236) of Japan Society for the Promotion of Science.

REFERENCES

1. Dennis L McCorkle, Weixing Ding, Cheng-Yu Ma and Lal A Pinnaduwage, J. Phys. D: Appl. Phys., vol.32, 1999, pp.46-54.
2. R.A.Morris, Bulletin of the American Physical Society, Vol.43, No.5, BM1,1998, pp.1412-3.

3. H.Yasui, T.Imai, K.Amemori and M.Yamamoto, Proceedings of the XIII International Conference on Gas Discharges and their Applications, vol.2, 2000, pp.692-5.
4. K.Nishijima and I.Tuneyasu, T. IEE Japan, vol.111-A, No.3, 1991, pp.221-7.
5. P.Mozejko, G.Kasperski, C.Szmytkowski, G.P.Karwasz, R.S.Brusa, A.Zecca, Chem., Phys., Let., vol.257, 1996, pp.309-13
6. R.J.Gulley, S.L.Lunt, J-P Ziesel and D.Field, J.Phys. B: At. Opt. Phys., vol.31, 1998, pp.2735-51

Coupling Power and Information to a Plasma Antenna

M Hargreave[*], J P Rayner, A D Cheetham, G N French and A P Whichello

Plasma Instrumentation Laboratory, University of Canberra, ACT 2601 Australia
[*]CEA Technologies, Canberra, Australia

Abstract. This paper presents a study of several power coupling structures for a plasma antenna and identifies the most effective for plasma generation in coupling to the $m=0$ mode surface wave. Also presented is a study that was undertaken with the aim of identifying the most efficient way of coupling an information signal for transmission using an already existing plasma column. In performing a comparison of various capacitive and inductive coupling structures the most effective structure is identified along with its advantages and disadvantages.

INTRODUCTION

The concept of using plasma as the conductor in a radio frequency (RF) antenna is not new. However the idea was not pursued, as such a system was thought to be too noisy. Recent studies at the Australian National University and the University of Canberra have led to the conclusion that a plasma column may not be as noisy as first suspected. This has regenerated interest in using a plasma column in data communication and radar [1] since there are many possible applications for using plasma elements in antennas including rapidly re-configurable antenna arrays where elements can be rapidly switched; rapidly tunable antennas and stealth antennas.

There are two signals that require coupling into the plasma column if it is to be used for communication purposes. These are the plasma excitation signal used to generate and maintain the plasma column, and the information signal that is to be transmitted by the antenna. The focus of this paper is to present the results of a study into different methods of coupling these signals into a plasma antenna element. The paper also investigates the efficiency of the proposed coupling structures, and problems associated with their implementation. From the information provided in this paper, more informed decisions can be made about which method of coupling would best suit a specific application. Other papers presented at this conference report on the noise aspects [2] and radiation patterns produced by such antennas [3].

The plasma antenna experimental apparatus used in these studies is shown diagrammatically in Figure 1. It comprises a standard 1" fluorescent tube with one end located inside a metal mounting box. The box encloses the coupling structures for both the plasma generation signal (the pump) and the information signal (for communication). For this study the pump was a 500MHz signal at up to 100W. The communications signal was in the VHF band at around 70-100MHz. For most applications it is highly desirable that the plasma is excited from only one end as the presence of wires and feeds to both ends would be deleterious to the radiated signal and also provide a radar target in stealth applications.

In this study the plasma is maintained by exciting the $m=0$ surface wave. Although there are many methods of exciting this wave [4], a simple capacitive collar located 3mm below a hole cut in the top of the earthed mounting box was employed [3]. One objective of this study was therefore to optimize the geometry of the box to maximize the ease with which the plasma could be established and to minimize plasma excitation power requirements.

For the information signal there can be no direct electrical contact with the plasma which means that either capacitive or inductive coupling must be employed. The capacitive option has significant capacitance in the circuit due to the insulating glass tube between the coupler and the plasma, whereas using the magnetic option, it is possible that the antenna may not feed effectively off the ground plane. In deciding which coupling structure is the most efficient, it is important to include a matching network so that all the available power is transferred to the plasma column. Once the coupling structure with the greatest efficiency has been established, it is possible to complete a detailed study into the noise characteristics of the plasma antenna and to determine its effectiveness as a viable alternative to a conventional metal antenna.

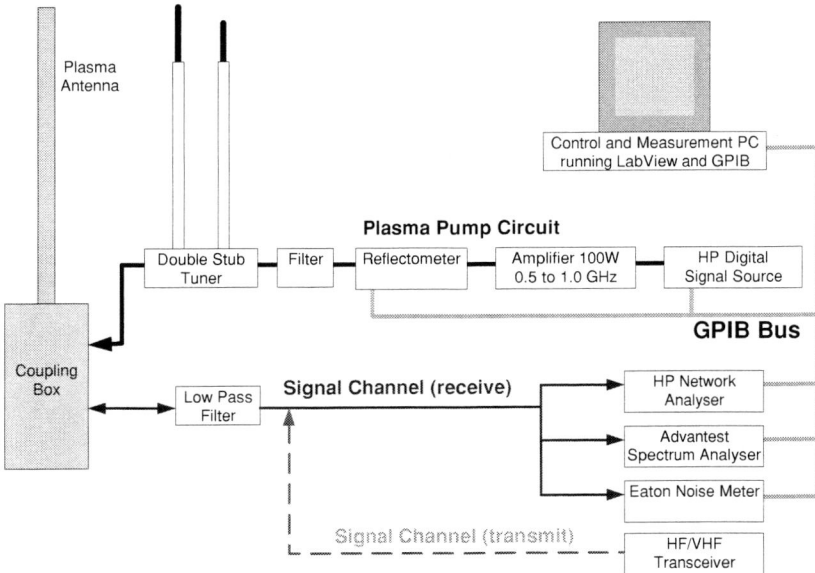

FIGURE 1. Block diagram of the plasma antenna circuit and measurement systems.

COUPLING THE COMMUNICATIONS SIGNAL

An initial simulation study was carried out by replacing the plasma antenna by a similar sized copper tube with a 1 mm mylar insulating layer wrapped around it in order to determine whether an inductive or capacitive coupling structure was more effective. The different coupling structures employed are shown in figure 2. These structures were varied by changing the distance to the ground plane up to a maximum of 10cm. The effectiveness of the various structures was compared from transmission tests made using a network analyzer. Of these structures the double inductive structure was found to be the least effective in coupling RF power into the plasma antenna. It was also found that the separation of the capacitive structure from the ground plane had little effect. The longer inductive coupler was found to be more effective than the short one. From the measured two-port S_{21} transmission data plotted in figure 3 it can be seen that the longer magnetic structure and the capacitive structure were found to be equally as effective in coupling the RF information signal into the copper column.

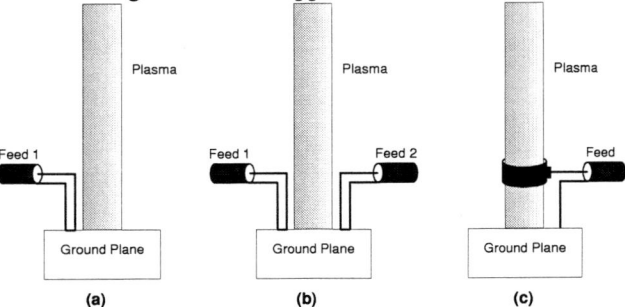

FIGURE 2. Signal Coupling Systems: (a) inductive (b) double inductive (c) capacitive.

When these coupling structures were tested on the plasma column it was found, particularly when the information branch was properly matched, that a significant amount of the RF pump power was absorbed out of the plasma column into the information signal channel. A diplex method of coupling into the plasma antenna was developed to overcome this problem. In this case a single capacitive coupling collar was used for both the pump and the information signal. This system (shown in figure 4) requires two branches: one for the excitation signal and the other for the information signal. Each branch must act independently which implies that each signal when it reaches the T-junction "sees" the other branch as an open circuit. For example, for the plasma antenna to be successful, all the pump power at a frequency f_p = 500 MHz must arrive at the plasma column with none leaking into the information channel. This can be effected by ensuring that the information branch has infinite impedance at the frequency f_p.

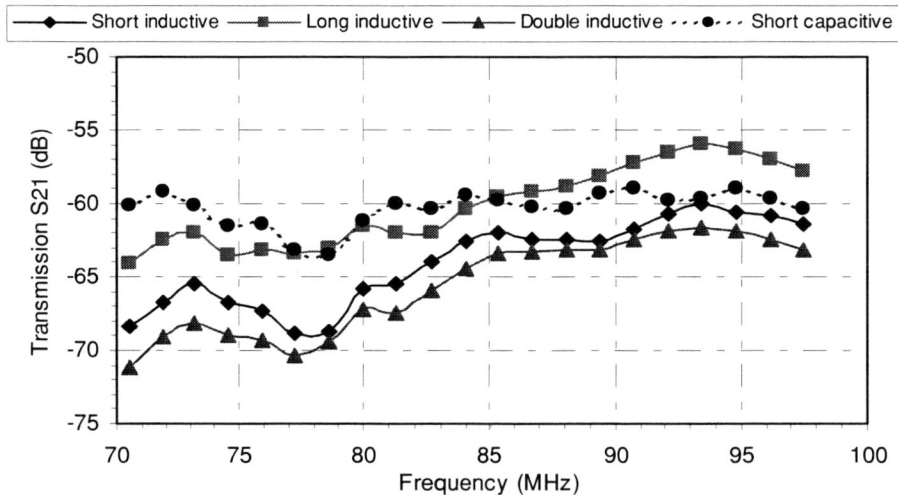

FIGURE 3. Transmission characteristics (S_{21}) of metal rod substituted as plasma column

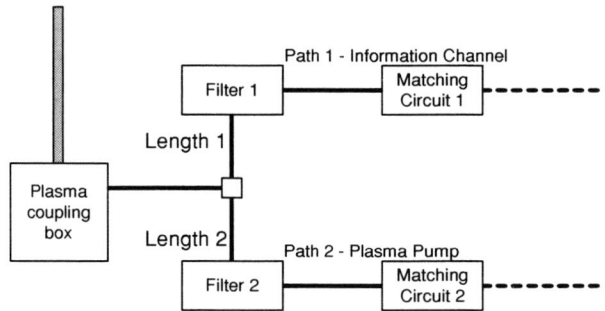

FIGURE 4. Diplex method of separating pump and signal channel.

In this system filter 1 must not affect the band of frequencies f_i carrying the information signal, where, for this test f_i = 78MHz and filter 1 had unity gain over the range DC-250MHz. This filter will either have a short, or open circuit input at the frequency of the pump signal. If the filter has an open circuit input, then length 1 must be $\lambda_P/2$, where λ_P is the wavelength of the pump signal. If the filter has a short circuit input, length (1) will need to be $\lambda_P/4$. These lengths ensure that path (1) has infinite impedance at the frequency of the pump signal.

A network analyzer was used to adjust the lengths (1&2) between the filters and the junction, until the impedance of the lines at the junction was infinite at f_i for path 2 and at f_p for path 1. Connection to the plasma antenna, via the T-Section, created a further mismatch which was tuned out by adjusting length 1 while observing the plasma. When the plasma was at maximum length, the input into the transmission cable was also at maximum impedance for that pump frequency (as the pump power leaking out via this path was minimum).

Tuning the pump path to block out the information signal was more difficult and it was not possible to obtain an accurate match. This was because changing length 1 affected the length required for path 2. Conversely changing length 2 affected the value of length 1.

POWER COUPLING TO THE PLASMA

One of the objectives of this study was to try to maximize the efficiency of the plasma generation in order to minimize additional power requirements for such antennas. While testing the efficiency of the information signal couplers, it was noted that the size of the coupling box affected the power required to ignite and maintain the plasma column. A further study showed that it was the depth of the coupling box that affected the required power. Figure 5 shows that the most efficient size of the coupling box was when the distance between the pump collar and the earthed metal base of the box was $\lambda_P/4$, or 15 cm for a 500 MHz pump signal. Since the coupling structure is at the top of the box, this represents a quarter wavelength standing wave within the box. Subsequent work using an all-glass antenna has confirmed this result [2].

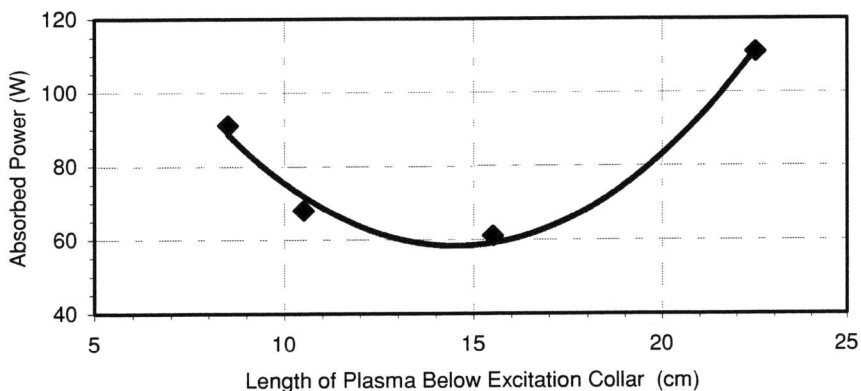

FIGURE 5. Power required to fill plasma column with varying sized excitation box.

SUMMARY

The results of this study into aspects of coupling both the excitation power and information signal into a simple plasma antenna are as follows:
◊ In a simulation using a copper antenna of similar dimensions to that of the plasma antenna, both inductive and capacitive coupling systems performed equally well.
◊ When installed on an actual plasma antenna the well-matched information channel absorbed power from the plasma excitation circuit and significantly reduced the efficiency of plasma generation.
◊ A diplex system was constructed that solved the power leakage problem and allowed effective signal coupling. The structure is advantageous as it looses minimal excitation power from the coupling structure. However, a significant disadvantage is the iterative adjustments needed for the two lines required to maximize both signals.
◊ It was found that the length of the antenna below the coupling point is a very important factor in determining the efficiency of coupling the plasma excitation power. When the length is a quarter wavelength of the pump surface wave then the resulting standing wave leads to high coupling efficiency.

ACKNOWLEDGMENTS

The authors would like to thank Professor J H Harris and Dr G G Borg of the Australian National University for useful discussions. This work was supported by University of Canberra Research Grants, Australian Research Council (ARC) Small Grants and the Australian Institute for Nuclear Science and Engineering (AINSE).

REFERENCES

1. G.G. Borg, J.H. Harris, D.G.Miljak and N.M. Martin *The application of plasma columns to radio frequency antennas.* Applied Physics Letters **74** (1999) 3272-3274.
2. J.P Rayner, A.D Cheetham and A.P Whichello *Physical Characteristics of a Plasma Antenna* Proc International Congress on Plasma Physics 2002, Sydney.
3. A.P Whichello, J.P Rayner and A.D Cheetham *Plasma Antenna Radiation patterns* Proc International Congress on Plasma Physics 2002, Sydney.
4. M. Moisan and Z Zakrzewski, *Plasma sources based on the propagation of electromagnetic surface waves*, J Phys D Appl Phys **24** (1991) 1025-1048.

Physical Characteristics of a Plasma Antenna

JP Rayner, AP Whichello and AD Cheetham

Plasma Instrumentation Laboratory, University of Canberra, ACT, 2601 Australia

Abstract. This experimental and theoretical study examines the excitation of a plasma antenna using an Argon surface wave discharge operating at 500 MHz with RF power levels up to 120 W and pressures between 0.03 and 0.5 mb. The results show that the length of the plasma column increases as the square root of the applied power, that the plasma density decreases linearly from the wave launcher to the end of the column and that these results are consistent with a simple global model of the antenna. Since noise is critical to communication systems, the noise generated by the plasma was measured from 10 MHz to 250 MHz. Between 50 MHz and 250 MHz the excess noise temperature was found to be 17.2 ± 1.0 dB above 290 K. This corresponds to a thermal noise source at 1.4 ± 0.3 eV compared with an electron temperature of 1.65 eV predicted by the model.

INTRODUCTION

A plasma antenna is a radio antenna based on plasma with a number of potential advantages over conventional metal elements for antenna design as electrical, rather than mechanical control of its characteristics becomes a possibility [1]. For example, the effective length of the antenna can be changed by controlling the applied RF power thus allowing rapid reconfiguration of the resonant length of the antenna for different transmitting frequencies. This work therefore aims to determine the RF power required to excite a surface wave discharge to produce a plasma column of a specified length, h, and to establish the form of the electrical conductivity profile along the column.

Noise introduced by the plasma itself is a crucial issue when considering the use of plasma antennas for reception and transmission. Plasmas are well-known sources of noise through to microwave frequencies particularly for DC, or mains driven AC fluorescent tubes. Thus the second objective is to characterize the noise generated by a plasma column excited by a surface wave.

Two different plasma antennas excited by surface waves were employed in this investigation. Antenna One consisted of a conventional fluorescent tube, filled with mercury vapour and argon at a nominal pressure ~ 0.4 mb. The surface wave launcher, to which RF power up to 120 W at 500 MHz was applied, consisted of a copper "pump" collar of length 25 mm mounted 3 mm below a circular hole cut in the top of a grounded diecast box [1,2]. A second "transmission" (or receiving) collar was mounted 10-20 mm above the tube's end cap and ~150 mm below the pump collar.

Experiments were undertaken over the range from 10MHz to 250 MHz for different RF power levels to determine:
- The length of the plasma column as a function of applied RF power.
- The antenna impedance as seen at the transmission collar using a vector network analyser
- The noise recorded at the transmission collar using an RF noise analyser

Line-averaged plasma densities were determined at heights of 15 cm and 36 cm above the top of the launcher using a 10 GHz microwave interferometer.

Antenna Two consisted of a Pyrex glass tube with a length of 1.5 m and diameter of 20 mm connected to a gas handling system that admitted Argon at pressures between 0.01 and 1.0 mb to the tube. A simple surface wave launcher consisting of a square ground plane of side 6 cm and a collar of length 20 mm mounted 2.0 mm above the ground plane could be slid along the tube to determine an optimum position for launching surface waves. This was found to be ~15 cm from the glass end cap and corresponded to a standing wave distance of $\lambda/4$ for the 500 MHz drive signal employed.

THEORY

A global model for a plasma column of length h sustained by a surface wave discharge [3] has been adapted to predict the length, temperature and plasma density of the antenna.

The electron temperature is found from a number density balance where the rate of production of electron-ion pairs by electron-neutral ionising collisions is balanced by the radial loss of electron-ion pairs to the wall. Typical model temperatures range from 2.2 eV at 0.1 mb, down to 1.4 eV at 1.4 mb.

The plasma density, n, is found from a power balance where the power absorbed per unit length by the plasma from the surface wave at a position z along the plasma column, is balanced by the power per unit length lost to the walls by the migration of electron-ion pairs [3].

The attenuation coefficient α for surface waves may be determined from the dispersion relation with allowance being made for losses via collisions and is of the form [3]

$$\alpha(n) = \frac{Cv_m}{(n - n_{res})} \quad (1)$$

where $C \approx 5 \times 10^9 \, \mathrm{m^{-4} s}$ is a constant for a 20 mm diameter plasma and $v_m = v_m(p)$ is the electron-neutral collision frequency for momentum transfer. n_{res} is the density for a plasma frequency corresponding to the RF frequency of the source modified by the dielectric constant of the surrounding insulator.

At the base of the column for an input power of P_0, the density n_0 for $n_0 \gg n_{res}$ can be shown to be

$$n_0 = A(p) P_0^{1/2} \quad (2)$$

where

$$A(p) = \left(2C v_m(p) / K(p)\right)^{1/2} \quad (3)$$

is a constant for a given pressure and $K(p)$ is the power lost to the wall per unit length per electron-ion pair. Similarly the density as a function of distance along the column for $n \gg n_{res}$ is be given by

$$n \approx n_{res} + C v_m(p)(h - z) \quad (4)$$

where for $z = h$, $n = n_{res}$ and from (2), (3) and (4) it follows that:

$$h \approx B(p) P_0^{1/2} \quad (5)$$

where

$$B(p) = \left(CK(p) v_m(p) / 2\right)^{-1/2} \quad (6)$$

In a communication system, identification and control of the various sources of noise is of critical importance in determining the overall link budget. Early attempts to use a DC plasma glow discharge as an antenna were unsuccessful due to the amount of noise introduced by the plasma. In addition to thermal noise, other noise sources might include shot noise, cathode processes, striations in the positive column and ion plasma oscillations. Many of these processes will be absent in a wave-heated discharge, although thermal and ion oscillation noise are still likely to be major contributors.

A real antenna has a radiation resistance R_r and "sees" an effective temperature T_r due to its environment weighted by its radiation pattern. If the plasma column has a resistance R_A at the electron temperature T_e and an efficiency $\eta = R_r / R_T$ where $R_T = R_r + R_A$ is the total resistance, then the total noise temperature is given by [4]

$$T_T = T_r \eta + T_e (1 - \eta) \quad (7)$$

RESULTS AND DISCUSSION

The length of the plasma column for Antenna One as a function of the applied RF power was determined by observing the output of visible light from the plasma. The "end" of the antenna was defined to be where the intensity decreased rapidly over a distance ~5 cm. Based on equation (5), the height, h, was plotted as a function of the square root of the applied power. The resulting straight line confirmed (5) with a slope in this case of 0.095±0.004 m/W$^{1/2}$.

The measurements were repeated for Antenna Two for pressures between 0.03 and 1 mb. Figure 1 shows the slope of the graphs of h verses $P_0^{1/2}$ as a function of pressure compared with the theoretical slope $B(p)$ derived from the global model in equation (6). The comparison shows that Antenna Two at higher pressures agrees with the global model to within the error bars ~15 %, while at lower pressures the low results may be due to some leakage of gas into the system during testing. The result for the fluorescent tube is ~50% higher than predicted, due possibly to the presence of the easily ionisable mercury vapour.

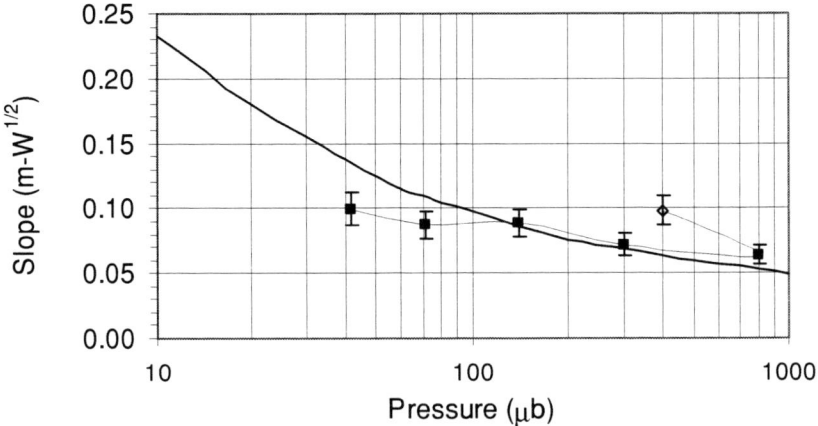

FIGURE 1. Slope of Height v's (Power)$^{1/2}$ plots as a function of Pressure. Experimental slopes compared with the Global Model. Open diamond: Fluorescent tube, Filled squares: Antenna Two.

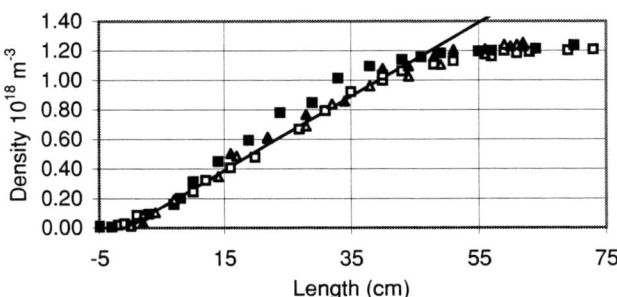

FIGURE 2. Plasma Density monitored at two locations as a Function of the Length of the Column measured from the End of the Column. Squares: $z = 15$ cm, Triangles: $z = 36$ cm. Open symbols: Power decreasing. Filled symbols: Power increasing

Figure 2 shows the plasma density for the fluorescent tube measured at heights of 15 cm and 36 cm above the launcher. The results are plotted as a function of y where $y = h - z$ is the distance measured from the top end of the column. Above $y = 45$ cm the density exceeds the cut-off value for the interferometer. Figure 2 also shows that equation (4) plotted for $p = 0.4$ mb, for which $v_m = 5\times10^8$ Hz. gives good agreement with the data. Based on this result and using the standard expression for the electrical conductivity of a slightly ionised plasma leads to a conductivity profile for $y \leq h$ given by

$$\sigma = 0.84 + 1.4\times10^2\, y(m) \text{ Sm}^{-1} \qquad (8)$$

This investigation has shown that the length of the column for a given RF power can be predicted from a simple global model with a precision ~15%, and that the conductivity profile along the column is essentially linear. From these results it should be possible to design an antenna with an electrically controllable length that would allow it to be rapidly reconfigured for different transmission frequencies. Further work, however is needed to determine the effective electrical length of the antenna due to the finite, tapered ohmic resistance and the consequent effects on the radiation pattern and antenna impedance [2].

Figure 3 plots the excess noise generated by the plasma in dB above a noise temperature of 290 K as a function of frequency for three different values of the RF pump power. The plot shows that above 50 MHz the noise is essentially flat with an average equivalent excess noise of 17.2 ± 1.0 db or 1.4 ± 0.3 eV.

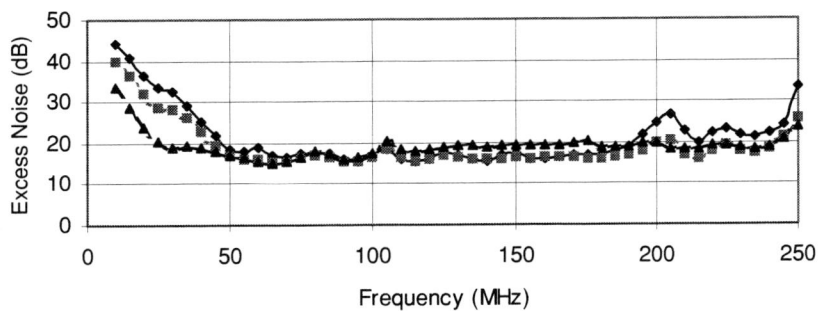

FIGURE 3. Excess Noise Power in dB as a Function of Frequency for different pump Powers
Triangles: 6 W, Squares: 30 W, Diamonds: 100 W

The global model gives an electron temperature of 1.65 eV (18.1 dB) at 0.4 mb which is ~17% higher than the measured noise temperatures but is sufficiently close to infer that the plasma is acting essentially as a thermal noise source operating at the electron temperature. This result is consistent with the long established practice of using discharge tubes as noise sources up to microwave frequencies.

The increase in noise below 50 MHz may be due to oscillations around the ion plasma frequency. Since n ranges from $\sim 10^{17}$ to $\sim 2 \times 10^{18}$ m^{-3} along the column with a corresponding range of ion plasma frequencies from ~3 to ~50 MHz a continuous range of ion plasma frequencies should exist and contribute to the noise up to a maximum ~50 MHz as observed in figure 3.

Equation (7) indicates that the total antenna noise temperature is a function of the antenna efficiency, the weighted temperature of the radiation field and the plasma noise temperature. Whether the plasma noise is a significant factor depends on the circumstances. For example, model calculations based on the conductivity profile given by (8) lead to an efficiency $\eta \approx 66\%$ for a 0.85 m plasma antenna at 100 MHz [2]. At this frequency the sky temperature is ~ 1 eV, which, together with a plasma contribution at 1.4 eV, leads to a total noise temperature of 1.13 eV compared with 1 eV for an ideal antenna.

ACKNOWLEDGEMENTS

The authors would like to thank Professor J H Harris and Dr G G Borg of the Australian National University for useful discussions. This work was supported by University of Canberra Research Grants, Australian Research Council (ARC) Small Grants and the Australian Institute for Nuclear Science and Engineering (AINSE).

REFERENCES

1. Hargreave, M., Rayner, J., Cheetham, A., French, G. and Whichello, A., "Coupling Power and Information to a Plasma Antenna" in *11th International Conference on Plasma Physics*, Sydney, Australia, 2002.
2. Whichello, A., Rayner, J., and Cheetham, A., "Plasma Antenna Radiation Patterns" in *11th International Conference on Plasma Physics*, Sydney, Australia, 2002.
3. Moisan, M. and Zakrzewski, Z., J. Phys. D: Appl. Phys. 24 1025-1048 (1991)
4. Jordan, E. C. and Balmain, K.G., Electromagnetic Waves and Radiating Systems 2nd ed. Prentice Hall, Englewood Cliffs, 1968, pp 414-416.

Plasma Antenna Radiation Patterns

A.P. Whichello[*,†], J.P. Rayner[*] and A.D. Cheetham[*]

[*]*Plasma Instrumentation Laboratory, University of Canberra*
[†]*email:* adrianw@ise.canberra.edu.au

Abstract. Elevation angle pattern measurements of our Plasma Antenna are presented and compared with results from computer models. Good agreement is found for the measurements with patterns predicted by our linearly decreasing conductivity model for the plasma antenna, derived from computer models using readily available packages.

INTRODUCTION

A plasma antenna is a radio frequency (RF) antenna based on plasma elements instead of metal conductors. Such antennas are constructed from insulating tubes filled with low pressure gases. When it is off, plasma is non-conducting and invisible to electromagnetic radiation. When it is on, plasma is an electrical conductor and therefore can provide the conducting medium for the radio signal.

Research undertaken at the University of Canberra [1], [2] has been using straight commercial fluorescent lighting tubes of various dimensions ranging from 200mm by 10mm to 1.2m long by 25mm diameter, using our test rig [3]. The plasma has been excited using 500 MHz at a peak power of 120W, fed through an RF matching circuit to a short copper collar near the base of the tube, as shown in Figure 1. Plasma antennas are normally excited by two signals at different frequencies; one to form the plasma (the "pump" frequency) and one for communications (the signal frequency).

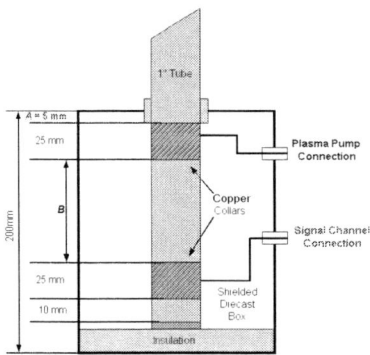

FIGURE 1. Details of the pump and signal coupling to the fluorescent tube forming the plasma antenna column.

ANTENNA RADIATION PATTERNS

The elevation radiation pattern of a simple vertical dipole of length L in free space may be found from the radiation integral[4] where the phase constant β is that of free space:

$$f(\theta) = \int_{-L/2}^{L/2} I(z) e^{j\beta z \cos\theta} \, dz \tag{1}$$

For the case of a good metal conductor, the current distribution $I(z)$ is usually taken to be sinusoidal over the length of the dipole:

$$I(z) = I_0 \sin\left[\beta\left(\frac{L}{2} - |z|\right)\right], \qquad |z| < \frac{L}{2} \tag{2}$$

Substituting this current distribution into (1) gives the (unnormalised) radiation pattern. The θ variation of this expression gives the far-field pattern:

$$f(\theta) = \frac{2I_0}{\beta} \frac{\cos[(\beta L/2)\cos\theta] - \cos(\beta L/2)}{\sin^2\theta} \tag{3}$$

Equation (1) may be interpreted as a Fourier transform of the current distribution to yield a far-field pattern (in terms of the Fourier variables z and the direction cosine $\cos\theta$ plane wave expansion) [5]:

$$I(z) \xleftrightarrow{FT} f(\cos\theta)$$

Our investigations have led us to model the plasma antenna with a linearly decreasing current distribution [6], which multiplies the usual current distribution of equation (2) with a triangular tapering function, $\Lambda(z)$, also known as a Bartlett window. In the Fourier (far-field) domain, the effect of this is to convolve the far-field pattern with a sinc squared filter, which will broaden the far-field pattern lobes and smooth out the pattern nulls:

$$\Lambda(2z/L)I(z) \xleftrightarrow{FT} \text{sinc}^2(L\cos\theta/2) \otimes f(\cos\theta)$$

COMPUTER MODELLING

A simple way to numerically evaluate the integral in equation (1) is to use one of the many variations of the *Numerical Electromagnetics Code* (NEC) Method of Moments computer packages (several are available from http://www.qsl.net/wb6tpu/swindex.html). NEC is widely used for modelling antennas and their environment. The antenna structure is broken down into wires and small surface areas, from which the current distribution and the radiation pattern may be found.

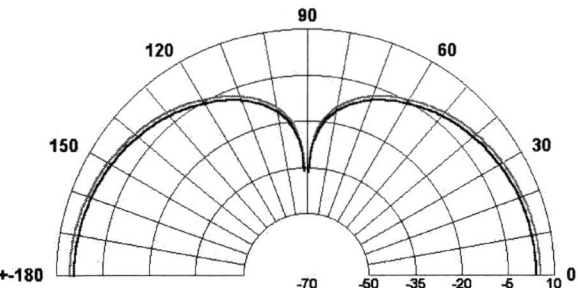

FIGURE 2. Plasma (blue) *vs.* Aluminium (gray) radiation efficiency. 1m columns radiating at 100MHz.

Because of the much lower conductivity of the plasma compared with Copper or Aluminium (≈ 100 S/m *vs.* more than 10^7 S/m), plasma antennas have lower efficiencies; around 50%–60% for the equivalent length of metal [7]. Therefore more RF power is lost in Joule heating of the plasma conducting elements. However, the loss is not serious, as the pattern plot in Figure 2 shows where the Aluminium column conductivity $\sigma = 3.5 \times 10^7$ S/m and the plasma has a linearly decreasing conductivity $100 \geq \sigma \geq 10$ S/m [6]. Only a small drop in the radiated power is observed and the shape of the radiation pattern is practically unchanged.

Over the VHF range (30–300MHz), a plasma column plus its image of length 1m only forms 0.1λ to about λ in length. Therefore the antenna is "electrically short" over most of the VHF range and the elevation angle pattern will be a slice of a torus, as shown in Figure 3. Only at the highest frequencies is there any deviation from the usual (sideways) "figure-8" pattern for a half-wavelength dipole.

Our plasma antenna is excited (or "pumped") using 500MHz to ignite and maintain the plasma. At this frequency, the entire 1m length of the column (plus its image) when fully lit is about 1.67λ, so the pattern structure has some distinct lobes, as shown in the left panel of Figure 4 (blue and green lines). For lower powers when only a short portion of the column length is lit, the pattern becomes more like the that of a half-wavelength dipole (red line).

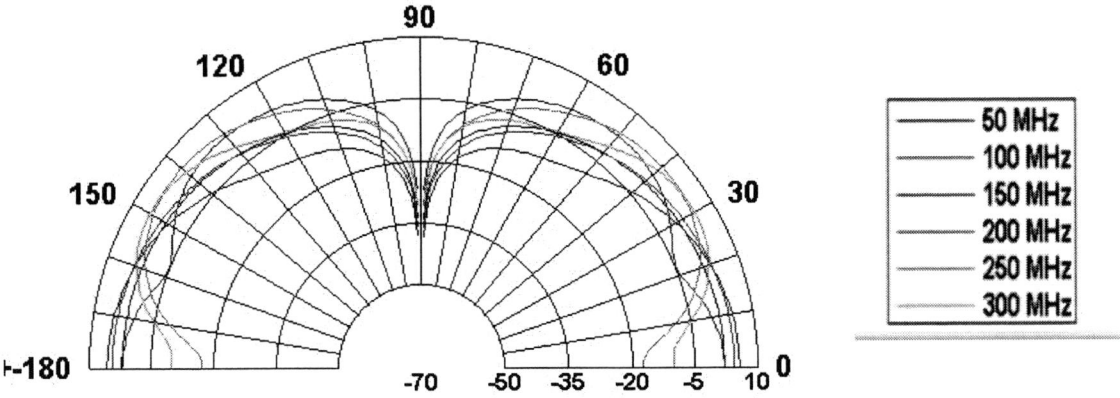

FIGURE 3. Computed elevation radiation patterns for a plasma antenna with linearly varying conductivity $100 \geq \sigma \geq 10$ S/m in the VHF range. Length of plasma column: 1m.

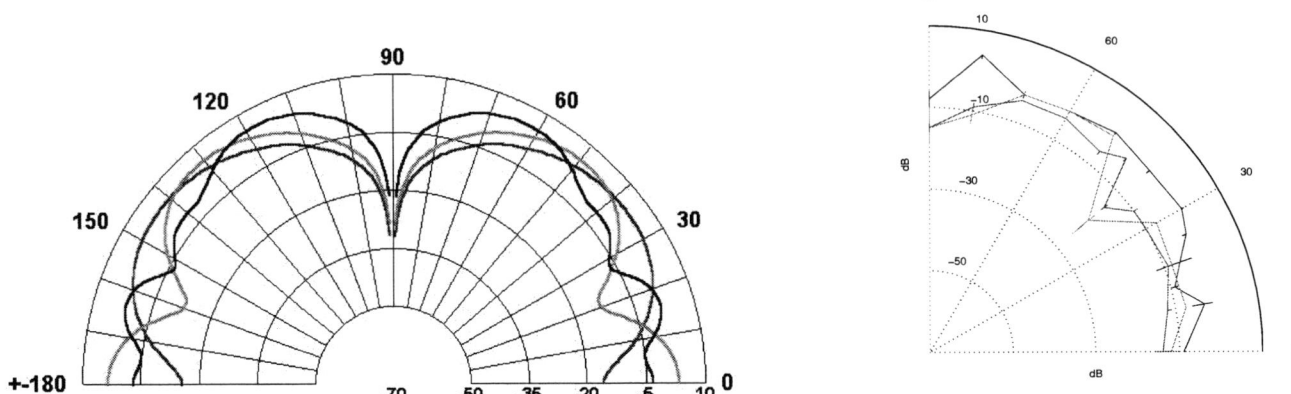

FIGURE 4. Left: Computed elevation pattern for plasma antenna at 500MHz pump frequency. Fully lit column length 1m (blue), 0.85m (green) and 0.5m (red). Right: Measured elevation pattern (and associated measurement errors) for the plasma antenna at 500MHz pump frequency.

PATTERN MEASUREMENT

We measured the elevation patterns of our plasma antenna by moving a probe antenna in an arc over the plasma Antenna Under Test (AUT) in 10° increments from $0° \leq \theta \leq 90°$. The probe antenna was attached to a 5.4m rigid wooden rod, pivoted at the base where the AUT was located. Our plasma antenna is really a monopole, but this may be treated as a dipole [5] when mounted over a perfect ground plane. This produces an image of the monopole below the ground plane; when taken together with the real monopole, a dipole results. We created a suitable ground plane using the large metal mesh area of our ground reflection antenna range.

The main effect of the decreasing conductivity is to smear the pattern, broadening it and reducing the depth of the nulls. This may be seen in the measured results presented in Figures 5 and 4 (right panel), where the pattern has the appearance of that expected for a half-wavelength dipole (compared with Figure 4, left panel). It is only at the highest frequency measured and at the greatest power used that other lobe structure becomes apparent.

The measurements we were constrained to perform were too close to the AUT at maximum plasma column length. The Rayleigh criterion (the distance to the far field) is usually taken to be greater than $2L^2/\lambda$. For our longest effective antenna length (2m), this is about 13m at 500MHz, so we are really measuring the pattern in the radiative near field, leading to filling in of the pattern nulls [8], especially the deep narrow null at zenith.

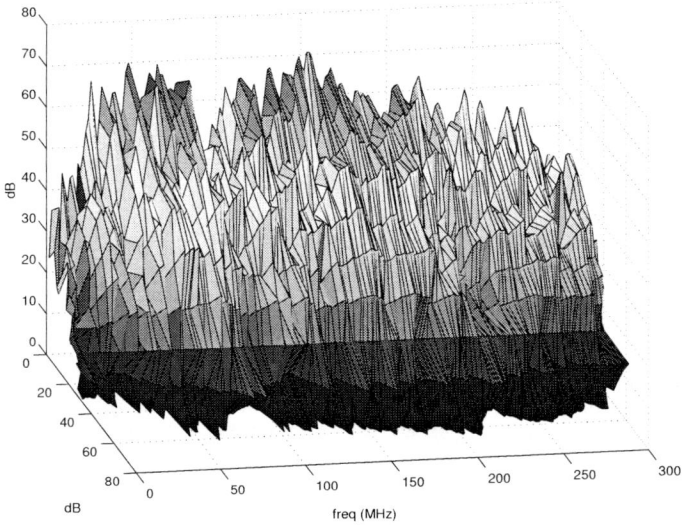
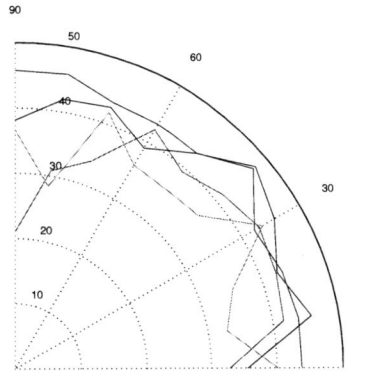

FIGURE 5. Left: A cylindrical hidden line plot showing the measured elevation pattern for the plasma antenna over the VHF frequency band. Plasma column length: 0.85m. Right: Measured elevation patterns for the plasma antenna at various VHF frequencies: 50 MHz (blue), 100 MHz (red), 150 MHz (green) & 200 MHz (black). These are slices of the data shown to the left.

CONCLUSIONS

We have presented both computer modelled and measured results for the radiation pattern of our plasma antenna, at both signal and pump frequencies. These results have shown that readily available computer codes may be used to predict the radiation patterns of plasma antennas. Our investigations have shown that although there is some loss of radiation efficiency due to the lower conductivity of the plasma column, this loss is not serious and may be easily made up by slightly boosting the power used in transmission. We also found that the tapered conductivity profile has little effect on the resulting radiation pattern.

ACKNOWLEDGMENTS

This work was supported by University of Canberra Research Grants, Australian Research Council (ARC) Small Grants and the Australian Institute for Nuclear Science and Engineering (AINSE). We would also like to thank our honours student, Ian Leung for his assistance with the antenna pattern measurements.

REFERENCES

1. Cheetham, A., Rayner, J., Gilbert, B., and French, G., "Surface wave excitation for plasma antenna applications," in *Proc 23rd AINSE Conference on Plasma Science and Technology*, Adelaide, Australia, 2000, pp. 17–19.
2. Whichello, A., Rayner, J., and Cheetham, A., "Plasma Antenna Noise Measurements," in *7th Australian Symposium on Antennas*, Sydney, Australia, 2001.
3. Hargreave, M., Rayner, J., Cheetham, A., French, G., and Whichello, A., "Coupling power and information to a plasma antenna," in *11th International Conference on Plasma Physics*, Sydney, Australia, 2002.
4. Stutzman, W., and Thiele, G., *Antenna Theory and Design, 2nd ed.*, Wiley, New York, USA, 1998, pp. 164–173.
5. Balanis, C., *Antenna Theory*, Wiley, New York, USA, 1997, pp. 165–173.
6. Rayner, J., Whichello, A., and Cheetham, A., "Physical characteristics of a plasma antenna," in *11th International Conference on Plasma Physics*, Sydney, Australia, 2002.
7. Borg, G., Harris, J., Miljak, D., and Martin, N., *Applied Physics Letters*, **74**, 3272–3274 (1999).
8. Evans, G., *Antenna Measurement Techniques*, Artech House, Norwood, USA, 1990.

Study on Laser Induced Plasma Produced in Liquid

N. Tsuda* and J.Yamada*

1247 Yakusa-cho, Yachigusa, Toyota, 4700392, JAPAN

Abstract. When an intense laser light is focused in liquid, a hot plasma is produced at the focal spot. The breakdown threshold and the transmittance of sodium choroids solution are observed using excimer laser or YAG laser. The breakdown threshold decreases with increasing NaCl concentration. Threshold intensity of plasma produced by YAG laser is lower than excimer laser. The behavior of plasma development is observed by a streak camera. The plasma produced by a YAG laser develops only backward. However, the plasma produced by excimer laser develops not only backward but also forward same as the plasma development in high-pressure gases.

INTRODUCTION

When the laser light is focused at the solid or gas, a hot and dense plasma is produced. The studies on laser induced plasma of the solid target have been carried out with aim of the inertial confinement fusion. At the same time, various studies have been carried out on the mechanism of the gas breakdown,[1] the expansion of the plasma[2] and the interaction between the laser light and the plasma.[3] The laser light focused in argon gas up to 150 atm and the development mechanism of the laser induced plasma was studied. The high-pressure plasma produced by a Q-switched ruby laser was studied.[4] The plasma developed only backward towarding the focusing lens, after the breakdown at the focal spot.[5] Recently, a high-power ultraviolet laser light can be easily obtained with development of excimer laser. When the XeCl excimer laser light is focused in the high-pressure argon gas, the plasma develops not only backward but also forward.[6] The forward development mechanism proposed and calculated.[7] However, the laser induced plasma in liquid hasn't been studied enough. When a laser light is focused in liquid, a hot plasma is produced at the focal spot. The mechanisms of the breakdown and the plasma development have not almost been investigated. In liquid, the laser induced plasma may be able to resolve the hazardous material calling the environment material, or the plasma produced in the physiological saline may become to be the basic data when the human body is irradiated by the laser light. Then, plasma produced in liquid by the laser light is studied and the basic data about breakdown threshold of liquid and the development behavior of plasma are examined.

EXPERIMENTAL ARRANGEMENT

Excimer or YAG laser light is focused in liquid. The experiment arrangement is shown in Fig. 1. The maximum laser power of excimer laser is 500 mJ with a wavelength of 308 nm and a pulse half width of 30 ns. While the maximum laser power of YAG laser is 340 mJ with a wavelength of 1064 nm and a pulse half width of 15 ns. Moreover, the YAG laser is able to drive the second harmonic wave oscillation with a power of 180 mJ, a wavelength of 532 nm and a pulse half width of 15 ns. The chamber is made by acrylic and has height of 75 mm, width of 45 mm and length of 70 mm. It has three quarts glass windows of height of 25 mm, width of 30 mm and thickness 2 mm. The ultra pure water or the ultra pure water with a melted NaCl is used as a test liquid. The excimer laser is focused using the concave mirror to avoid influence of refraction at liquid surface. The backside of acrylic chamber is able to attach the concave mirror. The diameter of focal spot is 96 μm when the laser light is focused using the concave mirror of focal length 25.4 mm. On the other hand, the YAG laser light is focused from the out side of the chamber using the lens of focal length 69.5 mm, because the intensity of laser light is so high that concave mirror can't be used. The diameter of focal spot is 130 μm. The laser power is controlled using the optical filter.

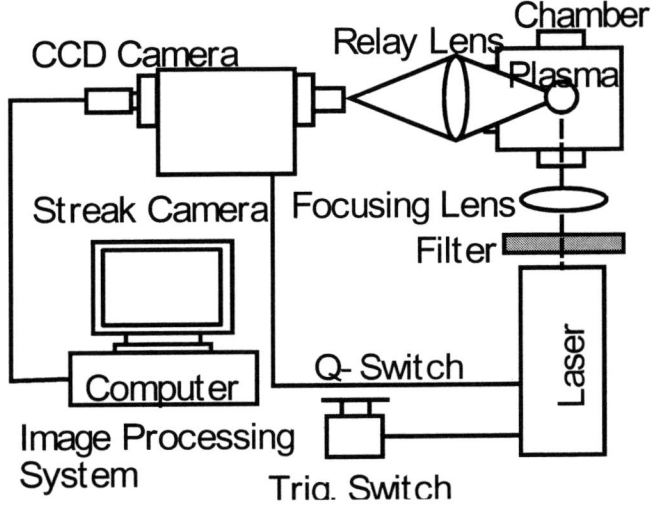

FIGURE 1. Experiment arrangement.

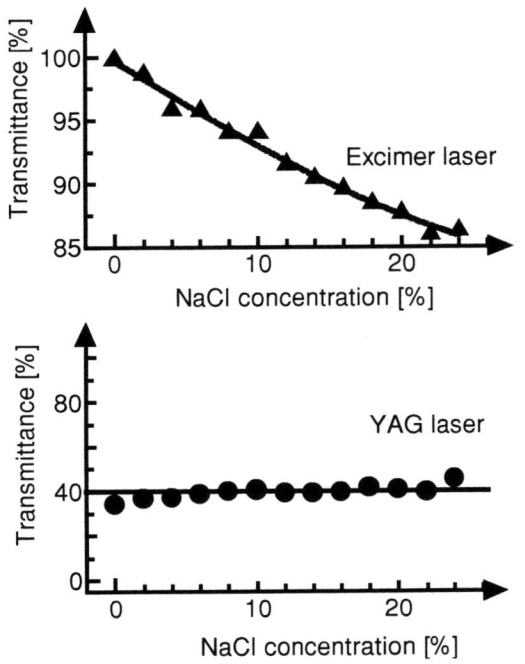

FIGURE 2. Transmittance.

To examine the attenuation of laser light in liquid, the photodiodes are set up at back and front of chamber. The transmittance is observed from the ratio of transmitted laser power to the incident one.

When NaCl concentration is varied, the threshold value of light intensity is measured taking into account the absorption of laser light in liquid. The threshold laser intensity is defined as an intensity at which the plasma production probability is 50 %.

The plasma development is observed using the streak camera. The plasma luminosity is focused on the incident slit of the streak camera by relay lens with a focal length of 100 mm. The streak image is displayed on a monitor by a dummy color of light intensity. The plasma boundary is determined by a threshold intensity because the plasma boundary of the streak image is not so clear. The plasma boundary is drawn by a plotter.

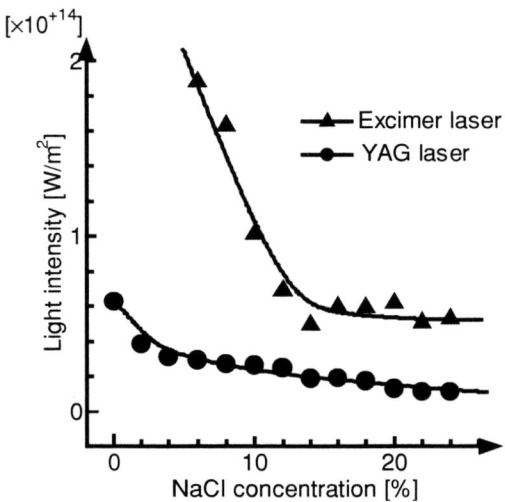

FIGURE 3. Threshold characteristic.

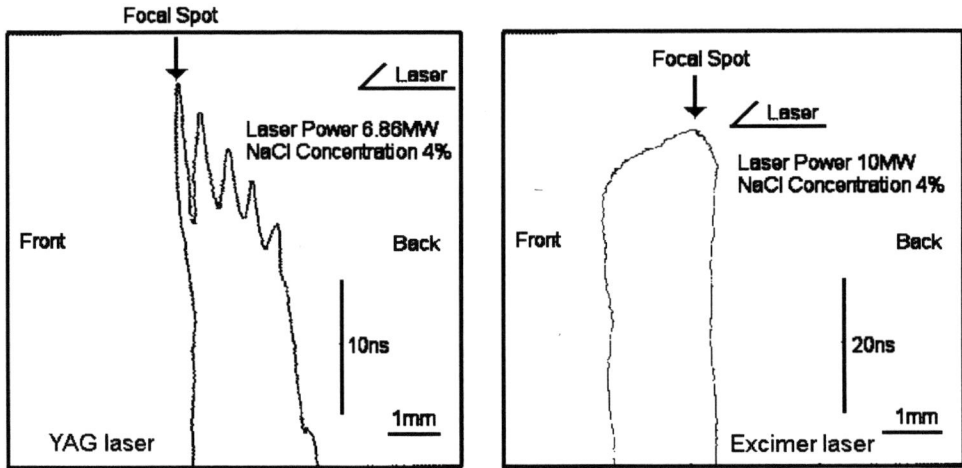

FIGURE 4. Streak image.

EXPERIMENTAL RESULTS

Transmittance

The transmittance of the laser light through NaCl solution is measured. The transmittance as a function of NaCl concentration is shown in Fig. 2. The transmittance decreases with increasing the concentration when the excimer laser is used. The attenuation of laser light in NaCl solution isn't negligible because NaCl molecular absorbs the laser light. However, the transmittance of YAG laser is 40 %, which is almost a constant independent on NaCl concentration. The scattering loss of YAG laser light may be so large that the transmittance turns out large because the temporal coherent of YAG laser is better than one of excimer laser.

Threshold characteristic

When NaCl solution is irradiated by the laser light with the intensity near threshold, we watch whether the plasma is produced or not. From 20 observations, a plasma production probability is examined. The threshold characteristic for NaCl concentration using excimer laser or YAG laser is shown in Fig. 3. The threshold intensity of plasma produced by YAG laser is lower than that by the excimer laser, because the spot size and absorption coefficient of YAG laser are bigger than those of the excimer laser. The threshold values decrease with increasing NaCl concentration because the ionization coefficient decrease with increasing the NaCl concentration. If the NaCl concentration increases moreover, the threshold value may increase again because the collision occurs more frequently with increasing NaCl concentration.

Streak image

The typical streak image of the plasma, which is produced by YAG laser, is shown in left figure of Fig. 4. The laser light is irradiated from the right hand side, the time is scanned from top to bottom, and the inside of the boundary shows the plasma. The YAG laser is operated at 1064 nm. The plasma produced in liquid develops only backward because the plasma frequency is higher than the laser frequency and laser light is absorbed in only backward plasma surface. The plasma consists of a group of plasmas produced from many seed because the electrolytes in liquid as seed may initiate the plasma production.

On the other hand, the plasma produced by the excimer laser is produced at the focal spot and develops backward and forward asymmetrically as shown in right figure of Fig. 4. The development mechanism of forward plasma is different from that of backward plasma. When the concentration of sodium chloride is higher, the plasma develops forward widely. However, when the concentration is too high, the plasma develops hardly because the laser light is absorbed in liquid.

CONCLUSION

When the liquid is irradiated by the ultraviolet laser light or visible laser light, the plasma is produced at the focal spot. The characteristic of transmittance is different when the YAG laser or excimer laser is used. The transmittance of excimer laser light decreases with increasing the NaCl density. However, the transmittance of YAG laser light is constant. Threshold intensity is decreases with increasing NaCl concentration. Threshold intensity of plasma produced by YAG laser is lower than excimer laser, because the focal spot size of YAG laser is bigger than that of the excimer laser. The dynamic behavior of the laser produced plasma is observed by the streak camera. The plasma produced by excimer laser develops not only backward but also forward. However, The plasma produced by YAG laser develops only backward.

ACKNOWLEDGMENTS

This work was partially supported by a grant from Nitto Foundation.

REFERENCES

1. R.G. Meyerand, and A.F.Haught, *Phys.Rev.Lett.* **11**, 401 (1963).
2. Yu.P.Raizer, *Sov.Phys.JETP*, **21**, 1009 (1965).
3. G.V.Ostrovskaya, and A.N.Zaidel', *Sov.Phys.Usp*, **16**, 834 (1974).
4. J.Yamada, and T.Okuda, *Jpn.J.Phys*, **18**, 139 (1979).
5. J.Yamada, and T.Okuda, *Laser Particle Beams*, **7**, 531 (1989).
6. J.Yamada, N.Tsuda, Y.Uchida, H.Huruhashi, and T.Sahashi, *Trans. IEE Jpn*, **114**, 303 (1994).
7. N.Tsuda, and J.Yamada, *J.App.Phy*, **81**, 582 (1997).

Ion and Electron Whistler Wave Dispersion Experiments.

Christian M. Franck[*†], Thomas Klinger[*†] and Olaf Grulke[*†]

[*]*Max-Planck-Institut für Plasmaphysik Greifswald, EURATOM Association, Germany*
[†]*Ernst-Moritz-Arndt Universität Greifswald, Germany*

Abstract. Although whistler waves are studied for almost a century, they are still subject of intense research. Laboratory experiments are of particular value for the interpretation of satellite data, which is often ambiguous or at least difficult to understand. The linear plasma experiment VINETA, which is designed to form a large (4.5 m column length, up to 40 cm diameter) and dense ($n_e \leq 10^{19}\,\mathrm{m}^{-3}$) plasma with great flexibility in the magnetic field configurations (max. field $B_0 = 100\,\mathrm{mT}$). In the present contribution we discuss investigations on electron whistler waves, right-hand polarised electromagnetic waves in the frequency range $\omega_{ci} \ll \omega \leq \omega_{ce}$, and ion whistler waves, left-hand polarised electromagnetic waves below the ion cyclotron frequency $\omega < \omega_{ci}$. Firstly, the dispersion behaviour of electron whistler waves is studied at wavelength up to the plasma dimensions. For smaller wavelengths it turns out that the dispersion relation derived for unbounded whistler waves describes the experimental observations well. For wavelengths greater than the vacuum vessel diameter it is necessary to take the boundary into account. The transition between the two regimes is studied for constant plasma conditions. Secondly, ion whistler waves are considered. In a single-component plasma, the wave dispersion has a real solution only for frequencies below the ion cyclotron frequency ω_{ci}. In a multicomponent plasma, the dispersion relation of ion whistler waves exhibit a different behaviour. In an intermediate frequency regime located between the two neighbouring ion cyclotron frequencies, the wave has a maximum group velocity. This feature is used as a diagnostic tool in satellite measurements to determine the ion composition in atmospheric plasmas, but the results have been subject of a scientific debate. Our laboratory study shows first results on ion whistler wave measurements and aims to contribute to clarify the discussed discrepancies.

THEORETICAL BACKGROUND

The cold plasma dispersion relation for right-hand (R) and left-hand (L) polarised waves of frequency ω propagating parallel to the magnetic field in a large plasma (plasma radius $R_c \gg$ parallel wavelength $\lambda_\parallel$) is given by [1]

$$k_\parallel^2 c^2 = \omega^2 - \frac{\omega \omega_{pe}^2}{\omega \mp \omega_{ce}} - \frac{\omega \omega_{pi}^2}{\omega \pm \omega_{ci}} . \tag{1}$$

Here, the upper sign applies to the R-wave and the lower to the L-wave. $\omega_{pe,i}$ are the electron and ion plasma frequencies and $\omega_{ce,i}$ the electron and ion cyclotron frequencies, respectively. If finite size plasma dimensions is considered, the dispersion is modified due to the specific boundary conditions. A systematic approach treating electromagnetic waves in a plasma-filled cylindrical waveguide was published by Uhm and co-workers [2]. One obtains an analytic expression for the limiting case of a completely plasma-filled waveguide with homogeneous density distribution in the limit of low frequencies ($\omega \ll \omega_{ce}$):

$$\frac{\omega}{\omega_{ce}} = \frac{k_\parallel c}{\omega_{pe}\sqrt{K_c^2 - 1}} \sqrt{K_c^2 \frac{k_\parallel^2 c^2}{\omega_{pe}^2} + 1} \tag{2}$$

with

$$K_c = \frac{\omega_{pe} R_c}{\alpha_{0n} c} . \tag{3}$$

Here, R_c is the plasma radius and α_{0n} denotes the n-th root of the Bessel function $J_0'(x) = 0$.

EXPERIMENTAL SETUP

Experiments were conducted in the VINETA plasma device. Fig. 1 shows a schematic diagram of the experimental setup. The vacuum vessel of 4.5 m length and 40 cm diameter is immersed in a linear magnetic field of induction up to $B_0 = 100$ mT. A plasma is produced in Argon and Helium by an external helical antenna placed around a vacuum glass extension with 10 cm diameter. The antenna is driven at a frequency of $f = 13.56$ MHz at powers of $P = 300\ldots 2500$ W. Depending on the rf power, three qualitatively different discharge modes can be established, capacitive, inductive, and helicon wave sustained plasma. Plasma densities vary between $n_e = 10^{16}\ldots 10^{19}$ m^{-3}. Details of the VINETA device can be found in [3].

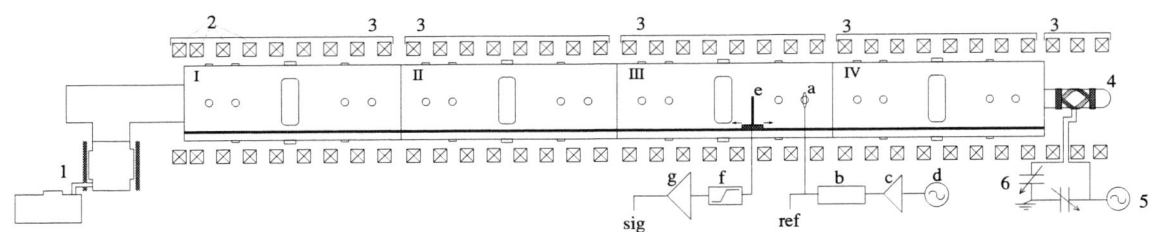

FIGURE 1. Schematic drawing of the VINETA experiment. Vacuum chambers I-IV, vacuum pumps (1), field coils (2) with their dc power supplies (3), glass vacuum extension with helical antenna (4), rf power supply (5), and matching unit (6). Waves are launched with an exciter loop (a), which is fed by a hf source (d) via an amplifier (c) and a matching unit (b). The waves are measured by axially positionable magnetic fluctuation probes (e), which are connected to the digitiser via a filter (f) and an amplifier (g).

High-frequency whistler waves are excited with an additional loop antenna of 30 mm diameter, driven in the frequency range of $f_{ex} = 150\ldots 700$ MHz. Here, the power level is kept below 1 W to ensure minimum plasma disturbance. Three miniature magnetic fluctuation ($\dot{B}$) probes are placed centrally into the chamber mounted on an axially movable positioning system. The probes are calibrated in a test field with Helmholtz coil arrangement and are compensated for electrostatic pickup with a centre-tapped transformer [4]. Low-frequency whistler waves are excited with the same exciter antenna is operated at frequencies between $f_{ex} = 50\ldots 500$ kHz and power levels up to 50 W to enable a reliable measurement to obtain a good signal to noise ratio. The relative phase of the measured signal with respect to the exciter signal and the relative wave amplitude at each axial position are determined from the cross spectral density of the two recorded time series.

EXPERIMENTAL RESULTS AND DISCUSSION

Fig. 2 shows measurements of the magnetic fluctuation at different axial positions relative to the exciter loop antenna. The time series are averaged over 50-500 measurements to improve the signal-to-noise ratio. The slopes indicate waves propagating away from the antenna. A linear best fit to the phase shifts at each axial position yields the parallel wavelength. The wave damping can also be seen in the plots. On the left-hand-side, a wave is excited at $f_{ex} = 250$ MHz. Its wavelength is estimated to be $\lambda_\parallel = (480 \pm 20)$ mm. On the right-hand-side a wave is excited at a much lower frequency of $f_{ex} = 0.25$ MHz. Although such a low-frequency wave is much more difficult to excite and to detect, a wavelength of $\lambda_\parallel = (1530 \pm 110)$ mm can be derived from the averaged signal.

R-wave dispersion measurements

Numerous wavelength measurements for different excitation frequencies ($f_{ex} = 150\ldots 700$ MHz) have been carried out to obtain a dispersion relation $\lambda_\parallel(\omega)$. The left-hand side of Fig. 3 shows the dispersion relation for four different sets of plasma parameters. Shown are the measurements (markers) together with the theoretical dispersion for unbounded R-waves, Eq. (1) (solid lines). Density and magnetic field are determined independently from Langmuir probe measurements [5] and with a Hall sensor, respectively and are directly inserted into the dispersion relation. There is no free fit parameter left. For the first two sets of plasma parameters ($n_e = 3.3 \times 10^{18}$ m^{-3} and 4.5×10^{17} m^{-3}), the unbounded theory explains the wave dispersion behaviour fairly well. For the other two sets of parameters, at low densities of $n_e = 3 \times 10^{16}$ m^{-3}, only the measurement at high frequency (corresponding to small wavelengths) can be explained by unbounded plasma wave dispersion theory. At lower frequencies, the measured wavelengths are larger than expected. This is reasonable as the perpendicular experimental dimensions are smaller than the measured parallel wavelengths. Plotted with dashed lines in these two dispersion diagrams is the curve for bounded plasma theory, Eq. [2], assuming a plasma radius equal to the device radius of $R_c = 0.2\,\text{m}$. The low-frequency part of the measurements shows a good agreement with the bounded plasma theory, but the theoretical curve cannot fully recover the

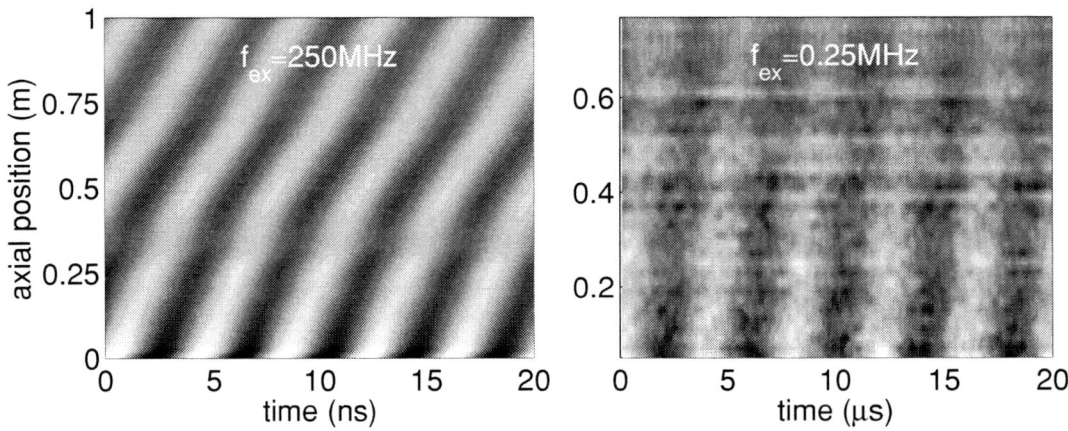

FIGURE 2. Gray-coded diagram of the magnetic fluctuation measurements for many axial positions relative to the exciter loop antenna. Shown are measurements for the R-wave (left hand side, $f_{ex} = 250\,\text{MHz}$) and the L-wave (right hand side, $f_{ex} = 0.25\,\text{MHz}$).

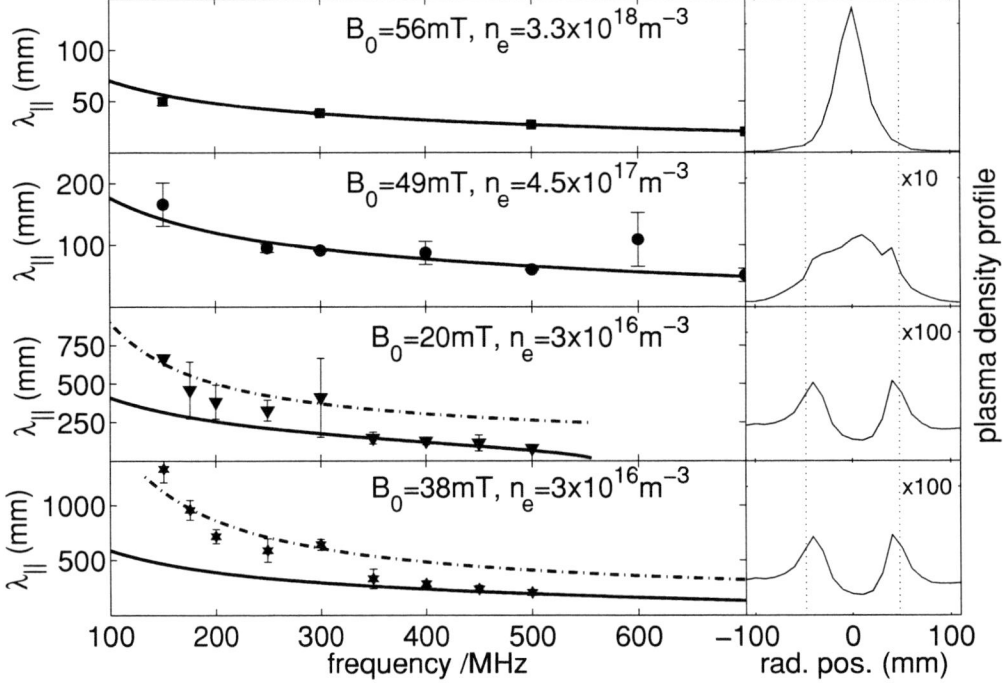

FIGURE 3. Dispersion relations at four different plasma parameters. Shown are the experimental measurements (markers), the theoretical curve for unbounded R-wave dispersion (solid lines), and the dispersion curve for whistler waves bounded in a waveguide of $R_c = 0.2\,\text{m}$ (dashed lines). On the right hand side, the corresponding relative plasma profiles (multiplied with 10, resp. 100 for comparison) are shown. The plasma parameters are determined independently from the wave measurements with Langmuir probes and Hall sensors.

dispersion behaviour. This is due to the somewhat unrealistic assumption of a completely plasma filled waveguide with homogeneous plasma density distribution. Furthermore, Eq. [2] is valid only in the low frequency limit. To underline the discrepancies between assumptions made in the theory and the actual situation, the plasma density profiles are shown on the right-hand side of Fig. 3. The plasma density profiles are given by the different discharge modes of the plasma source and cannot be controlled independently [6, 7]. In a next step, work is in progress to solve the full dispersion relation of a plasma-filled waveguide [2] to take the realistic plasma profile into account and to cover the complete frequency range [8].

L-wave dispersion measurements

Wave dispersion measurements have been done for excitation frequencies $f_{ex} = 100\ldots 500\,\text{kHz}$ as well (markers in Fig. 4). For frequencies above the ion cyclotron frequency $f_{c,He}(75\,\text{mT}) = 284\,\text{kHz}$ again the trend of decreasing wavelength with increasing frequency is observed. This tendency is reversed in the measurements for frequencies below $f_{c,He}$, where the wavelengths decrease for lower frequencies. Such a behaviour is qualitatively unexpected from

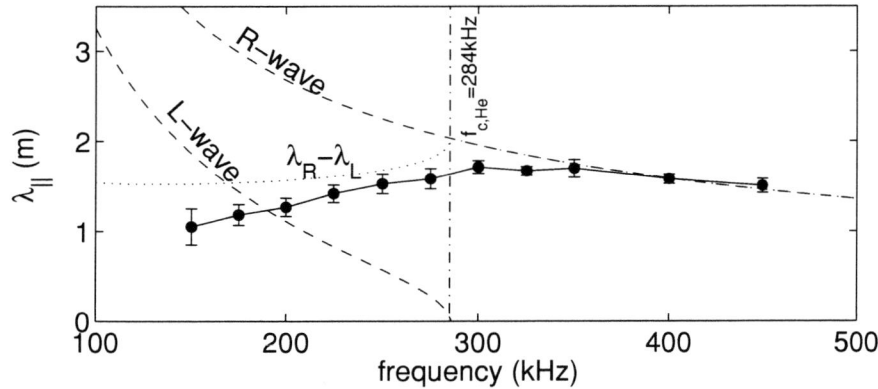

FIGURE 4. Dispersion relation including the ion dynamics at frequencies around the ion cyclotron frequency. Shown are the measurements (markers), the R-wave and L-wave dispersions (dashed lines) and the difference of the latter ones (dotted line). The ion (Helium) plasma frequency is $f_{c,He} = 284\,\text{kHz}$ ($B = 75\,\text{mT}$).

simple R-wave and L-wave dispersion. Both dispersion curves, as obtained for the unbounded Eq. [1], are plotted in Fig. 4 with dashed lines. It is suggested that the exciter antenna launches simultaneously R-waves and L-waves at frequencies below $f_{c,He}$. The measured wave is then a superposition of both. The resulting wavelength is the difference of the two launched waves and can explain the measured dispersion at least qualitatively (dotted line). To prove this assumption, the experimental setup will be changed to measure the polarisation of the wave and the amplitude modulation for more than one wavelength.

ACKNOWLEDGMENTS

This project is partially funded by the "Deutsche Forschungsgemeinschaft": grant number: SFB 198 project A15.

REFERENCES

1. Baumjohann, W., and Treumann, R. A., *Basic Space Plasma Physics*, Imperial College Press, London, UK, 1996.
2. Uhm, H., Hguyen, K., Schneider, R., and Smith, J., *J. Appl. Phys.*, **64**, 1108 (1988).
3. Franck, C., Grulke, O., and Klinger, T., *Phys. of Plasmas*, **9** (2002).
4. Franck, C., Grulke, O., and Klinger, T., *submitted to Rev. Sci. Instrum.* (2002).
5. Demidov, V., Ratynskaia, S., Armstrong, R., and Rypdal, K., *Phys. of Plasmas*, **6**, 350 (1999).
6. Ellingboe, A., and Boswell, R., *Phys. of Plasmas*, **3**, 2797 (1996).
7. Franck, C., Grulke, O., and Klinger, T., *to be published in Plasma Sources Sci. Technol.* (2002).
8. Bonhomme, G., private communication (2002).

Experiments and Observations on Intense Alfvén Waves in the Laboratory and in Space

W. Gekelman, M. VanZeeland, S. Vincena and P. Pribyl

Department of Physics and Astronomy, University of California, Los Angeles, 90095, USA.

Abstract. There are many situations, which occur in space (coronal mass ejections, supernovas), or are man-made (upper atmospheric detonations) in which a dense plasma expands into a background magnetized plasma that can support Alfvén waves. The **LA**rge **P**lasma **D**evice (LAPD) is a machine, at UCLA, in which Alfvén wave propagation in homogeneous and inhomogeneous plasmas has been studied. These will be briefly reviewed. A new class of experiments which involve the expansion of a dense (initially, $\delta n/n_o \gg 1$) laser-produced plasma into an ambient highly magnetized background plasma capable of supporting Alfvén waves will be presented. Measurements are used to estimate the coupling efficiency of the laser energy and kinetic energy of the dense plasma into wave energy. The wave generation mechanism is due to field aligned return currents, coupled to the initial electron current, which replace fast electrons escaping the initial blast.

INTRODUCTION

There are many naturally occurring instances, as well as man-made, in which a dense plasma expands into a much less dense background plasma. On astrophysical scales there are jets which have extreme conditions and may be modeled in the laboratory using intense lasers. In the past decade Coronal Mass Ejections[1] (CME) from the sun have been detected and are now the subject of intense study. These may evolve into structures such as flux ropes and magnetic clouds, which are observed by satellites near the Earth. There have been chemical releases in he Earth's ionosphere which have triggered[2] wave activity and unexpected plasma motion [3]. In the 1962 Starfish experiment a 1.4 megaton bomb was detonated 400 km above the South Pacific[4]. It is believed that these waves coupled to Alfvén waves.

Needless to say a great many phenomena are associated with each of these processes. In this paper we concentrate on waves produced by these rapidly expanding plasmas, specifically Alfvén waves.

The experiment is performed in the Large Plasma Device at UCLA. The plasma is 19 m long, highly magnetized ($133 \leq Dia/Rci \leq 450$), produced by a DC discharge (t_{rep} = 1Hz) and can support Alfvén waves ($n \approx 2 \times 10^{12}$ cm^{-3}, B= 1-2 kG). In these experiments an Al target is placed in the center of the machine and after the background plasma (He,Ne,Ar) is formed, it is struck with a 150 MW (t_{pulse}=7 ns, λ = 1 micron) laser. The laser-produced plasma (lpp) has an initial density, which is several orders of magnitude greater than the background plasma.

Comparisons with space phenomena must be made with scaled parameters such as those given in table I. . For example median CME speeds are 4.5×10^7 cm/s, mean Alfvén speed in corona 10^8 cm/sec, sound speed 10^7 cm sec. The laboratory numbers are similar Valfvén=2.4×10^7-1.2×10^8 and the sound speed is 1×10^5-1×10^6 cm/sec, but the spatial scales are vastly different.

Table. I Parameters and Dimensionless Ratios:

	Laboratory	CME	Starfish
$V_{lpp}/V_{Alfvén}$	1.63-0.125	0.1-2	5
V_{lpp}/V_{sound}	15-150	5	2500
R_{ci}/R_{bubble}	0.13-1.26	2.5×10^{-8}	1.2×10^{-3}
$N_{bubble}/N_{background}$	1-10	1×10^4	250
$V_{te}/V_{Alfvén}$	0.6-1.4	0.9	0.5

Experimental Setup and Results

What also matters is the geometry of the plasma and fields. For the lpp and the "Starfish" they are not very different, but the conjecture for the magnetic field topology in a CME is different as illustrated in figure 1.

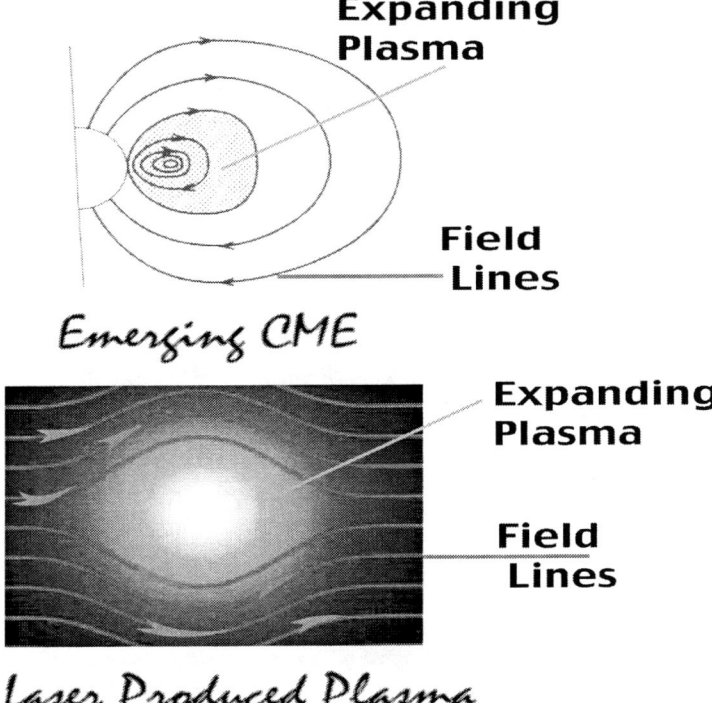

The magnetic field in the CME is dipolar as its initial source is embedded in the sun. The magnetic field lines in the experiment (and blast plasma) are initially straight and uniform. The ejected plasma is dense enough to create a diamagnetic bubble in which the background magnetic field is totally expelled. This has long been observed in lpp experiments in which the lpp expands into vacuum[5]. In this case the expanding plasma generates intense waves. The case in which the lpp expansion is initially parallel to the background field has already been discussed[6]. In this short paper we will concentrate on recent observations in which the lpp initially expands across the background field. Common to both cases is the generation of Alfvén waves both below and above the ion cyclotron frequency. It is now believed that these were generated in the Starfish event[7]. In the laboratory the waves propagate away from the target and are observed to become plasma column resonances[8]. Large density perturbations, ion acoustic and whistler waves have also been observed and will be the subject of future work.

FIGURE 1. Magnetic field lines and plasma density for CME and lpp. The plasma forms a diamagnetic cavity of order several cm in diameter. ($c/\omega_{pe} \approx 4$ mm, $.1$mm $\leq R_{ci} \leq 5$ mm)

Of great importance in plasma expansions is developing an understanding of the mechanism for Alfvén wave generation. The characteristic frequency of the laser burst is $f > 100$ Mhz, the lifetime of the diamagnetic cavity is roughly 1 μsec ($f \approx 1$Mhz) and the observed shear waves have frequencies, $f \approx 100$ kHz. In order to arrive at a picture of this processes detailed measurements of changing magnetic fields, densities and plasma flows were obtained.

The magnetic fields associated with the waves and density perturbations are measured at over 10^4 locations in planes parallel to the background magnetic field ($\mathbf{B}=B_0\mathbf{z}$). Plasma currents along the magnetic field were derived from $\mathbf{j}=\nabla\times\mathbf{H}$. The magnetic probe consisted of three orthogonal, differential 6 turn pickup loops (2.2 X 4.3 mm). The Mach, or double sided Langmuir probe consisted of two 1mm^2 faces oriented parallel/antiparallel to the background magnetic field. The probes were biased at –66 Volts ($T_e \cong 5$ e.V). Data was acquired at temporal intervals as short as 4 ns.

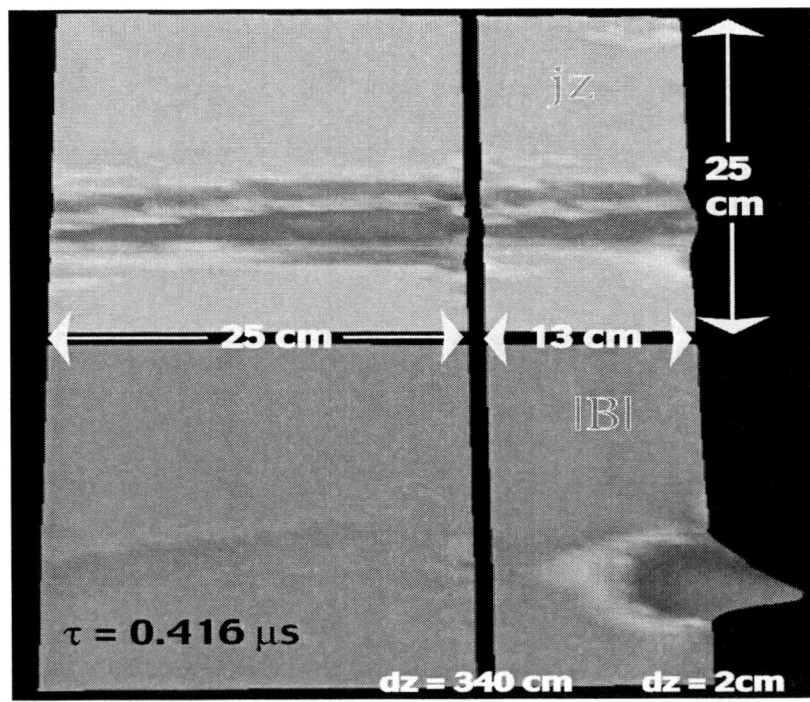

FIGURE 2. (a) The axial current and (b) density (sum of currents on the mach probe faces) 0.416 msec after the target is struck. The lower panel shows the plasma density associated with the expanding bubble. The plane begins 2 cm to the left of the target. The background magnetic field is in the plane of this picture and is parallel to the abscissa. The upper panel indicates that there is considerable plasma current at this time. The "red" part of the surface is associated with fast electrons streaming away from the lpp, and the blue surface is a nearly coaxial return current provided by the background plasma to neutralize the space charge of the expanding plasma. Magnetic fields and currents are shown in figure 3.

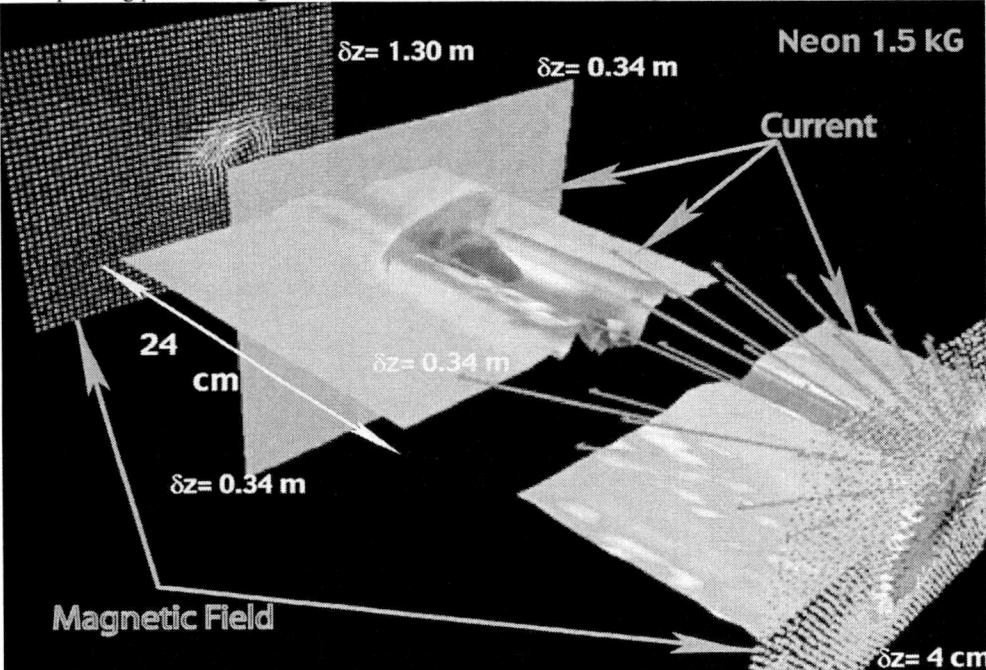

FIGURE 3. Magnetic field shown on two planes which are 4 and 130 cm from the target, and the plasma current on three other planes. The data is acquired at $\tau = 0.496$ μsec after the laser impacts the target. The large arrows emanating from the vertical (x-y) plane on the right is a direct measurement of the magnetic field expelled from the lpp plasma bubble. The current system is shown as well as the magnetic field on a further (dz=1.3m) plane. It is the long lived plasma currents responding to the lpp space charge that produce the Alfvén waves. The spectrum of the waves in a Neon plasma is shown in figure 4.

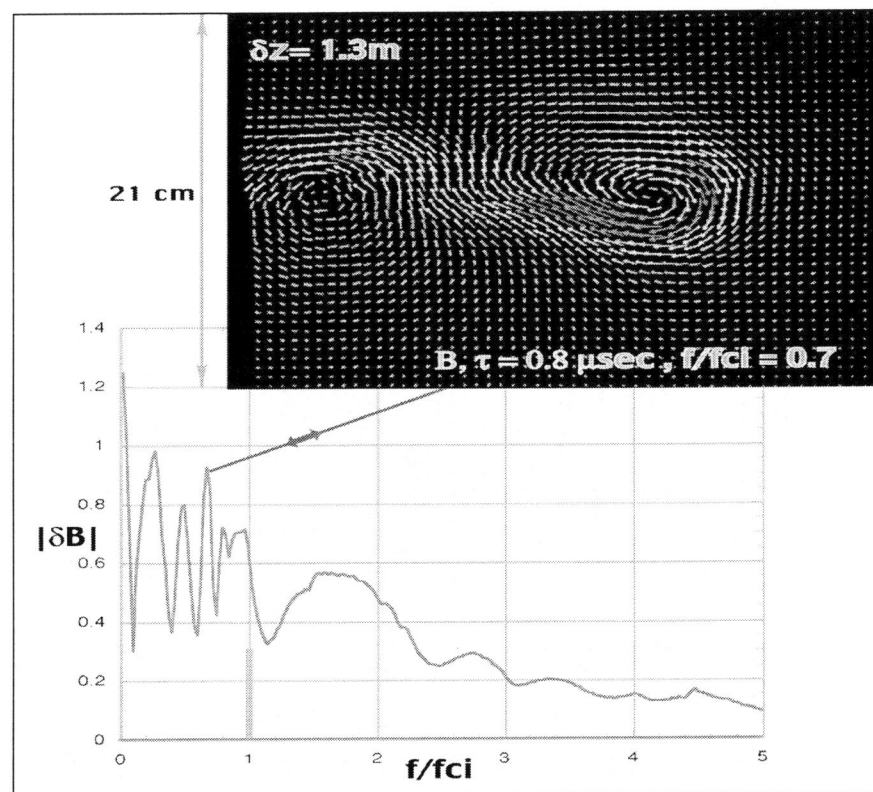

From the spectrum it is apparent that most of the wave energy (70%) is in shear waves. By integrating over the volume data was acquired in, we estimate that the Alfvén wave energy is of order of 1% of the kinetic (expansion) energy of the lpp. For plasma expansion across the background field the current system is dragged across the magnetic field. After several microseconds the coaxial system shown in figure 3 becomes complex and pancake shaped across the field as it tracks the expanding plasma. This is reflected in the wave pattern shown in the upper panel of figure 3. After several Alfvén transit times, and partial reflections from the end of the column, the shear waves become field resonances and the magnetic field pattern that of an m=0 column mode.

FIGURE 4. Lower left panel. Spectrum of Alfvén waves associated wit the lpp. Upper panel: wave magnetic field at early times at f/fci=0.7. In several Alfvén transit times this wave becomes an m=0 column resonance.

CONCLUSIONS

These experiments are the first in which the expansion of a dense lpp into a background plasma, which supports Alfvén waves, has been studied in detail. Wave production is affected by the generation of plasma currents, which flow to neutralize the charge imbalance due to fast particles generated in the plasma burst.

ACKNOWLEDGEMENTS

The authors wish to thank George Morales and Jim Maggs for many informative discussions. We also wish to thank the ONR, DOE and NSF for their support in this research.

REFERENCES

[1] B.C. Low, Jour, Geophys. Res., 106, 25,141-25,163 (2001)
[2] P. Bernhardt et al., J. Geophys. Res, 92, 5777 (1987)
[3] G. Haerendel et. al., IEEE Trans. Geosci. Remote Sensing, 23, 253 (1985), A.T. Lui et al, J. Geophys. Res., 91, 1333 (1986), K. Papadopoulos, A.T. Lui, Geophys. Res. Lett, 13, 925 (1986), A.F. Cheng, J. Geophys. Res., 92, 55 (1987), S.C. Chapman, J. Geophys. Res, 94, 227 (1989)
[4] S. Colgate, J. Geophys. Res., 70, 3161 (1965)
[5] B. Ripin et al, Phys. Rev letts, 59, 2299 (1987), G. Dimonte, L.G. Wiley, Phys. Rev Lett, 67, 1755 (1991)
[6] M. VanZeeland, W. Gekelman, S. Vincena, G Dimonte, *Phys. Rev. Lett.* 87, 105001 (2001).
[7] G. DiPeso, G. Dimonte, D. Hewett, M. VanZeeland, W. Gekelman, Bull APS, 46, 274, (2001)
[8] C. Mitchell, S. Vincena, J. Maggs, W. Gekelman, Geophys. Res. Lett, 28, 923 (2001)

Excitation and Propagation of Alfvén Waves in a Helicon Discharge

Olaf Grulke[*,†], Christian M. Franck[*] and Thomas Klinger[*,†]

[*]Max-Planck-Institute for Plasma Physics, Greifswald Branch, EURATOM Association, Greifswald, Germany
[†]Ernst-Moritz-Arndt University, Greifswald, Germany

Abstract. An experimental study of shear Alfvén waves in a linearly magnetized plasma is presented. Shear Alfvén waves are electromagnetic waves propagating parallel to the background magnetic field without compression of the plasma at a frequency well below the ion cyclotron frequency and a wavelength inversely proportional to the square root of the plasma density. A basic condition on laboratory investigations is that the Alfvén wavelength must be significantly smaller than the device dimension. This makes Alfvén waves difficult to investigate in laboratory experiments and most studies are performed in space, where typical Alfvén wavelengths of several kilometers are observed. The results of these studies are often ambiguous due to difficulties concerning the measurements of plasma parameters and the magnetic field geometry. The primary motivation for the present paper is the investigation of Alfvén wave propagation in a well defined laboratory situation. The experiments are conducted in the linear VINETA device. The necessary operational regime is achieved by the large axial device length of 4.5m and the use of a helicon plasma source providing high density plasmas with ionization degrees of up to 100%. The Argon plasma is magnetized by a set of 36 magnetic field coils, which produce a maximum magnetic field of 0.1T on the device axis. With this configuration a plasma-β of $\geq 10^{-4}$ is achieved, which exceeds the electron to ion mass ration, and the ion cyclotron frequency is $\approx$ 250kHz. Langmuir probes provide detailed informations on the time-averaged plasma profiles. Magnetic field perturbations for the excitation of Alfvén waves are generated by a current loop, which is introduced into the plasma. The surface normal of the current loop is directed perpendicular to the magnetic field. The waves's dispersion relation in dependence of plasma parameters is determined by spatially resolved $\dot{B}$ probe measurements.

INTRODUCTION AND EXPERIMENTAL SETUP

Alfvén waves are low-frequency electromagnetic waves, which are very important for the dynamics of magnetized plasmas in laboratory and space [1]. Experimentally Alfvén waves are difficult to investigate because of their large wavelengths, which are of the order of kilometers in space plasmas and of the order of several meters in typical laboratory plasma experiments. The crucial parameter determining the wavelength is the Alfvén velocity v_A, which scales inversely with the plasma density $v_A = B_0/\sqrt{\mu_0 n m_i}$, where B_0 indicates the ambient magnetic field, n is the plasma density, and m_i is the ion mass. Thus, for laboratory investigations long plasma devices are needed, which provide sufficiently high plasma densities. Basically Alfvén waves can be distinguished between compressional waves, in which magnetic field fluctuations are combined with plasma compression, and shear waves, in which only the magnetic field fluctuates. This paper reports on initial experiments about the excitation and detection of shear Alfvén waves in the well controllable VINETA helicon plasma laboratory experiment [2]. A schematic drawing of the device is shown in Fig. 1. It consists of a 5m length and 0.4m diameter modular vacuum chamber immersed into a set of

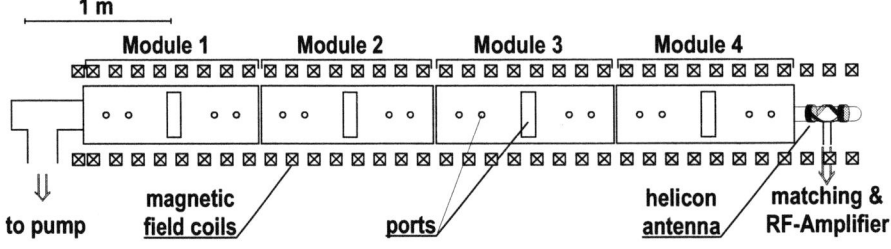

FIGURE 1. Schematic drawing of the VINETA device.

TABLE 1. Typical parameters for the VINETA device.

Parameter	Value
neutral gas pressure (He)	2 Pa
rf frequency	13.56 MHz
rf power	5 kW
plasma density	$5 \cdot 10^{18} \mathrm{m}^{-3}$
ion temperature	<0.1 eV
electron temperature	4 eV
plasma-β	1%
Alfvén velocity	$4.54 \cdot 10^5$ m/s
electron thermal velocity	$1.19 \cdot 10^6$ m/s
ion sound speed	$1 \cdot 10^4$ m/s
collisionless skin depth	$2 \cdot 10^{-3}$ m
effective ion Lamor radius	$6 \cdot 10^{-3}$

36 magnetic field coils, which produce a linear homogeneous magnetic field of max. ≈0.1T with a spatial ripple less than 1%. To achieve high density plasma a helicon plasma source is used for plasma production [3, 4]. Although the detailed discharge mechanism of helicon wave sustained plasmas is not fully understood, the helicon plasma source provides sufficiently high plasma densities at moderate input power, so that the typical Alfvén wavelength is smaller than the device length. The plasma profiles are strongly peaked in the plasma core with strong radial plasma pressure gradients, as it is characteristic for helicon discharges. Typical plasma parameters for VINETA are compiled in Tab. 1 To produce a localized magnetic field perturbation for the excitation of the shear Alfvén wave a current loop (diameter 35mm) is introduced into the plasma with its surface normal perpendicular to the ambient magnetic field to produce perpendicular magnetic field fluctuations $\tilde{B}_\perp$. The loop is fed with an sinusoidal rf signal in the frequency range $\omega = 1 - 2 \cdot 10^6 \mathrm{rad/s} < \omega_{ci}$, where ω_{ci} denotes the ion cyclotron frequency, the cut-off frequency for shear Alfvén waves. The magnetic fluctuations are detected by a triaxial arrangement of magnetic fluctuation probes, which are positionable computer controlled along the magnetic field lines [5]. The probes have been tested and calibrated in a magnetic test field (two magnetic field coils in Helmholtz arrangement) to carefully ensure operation of the probes in the required frequency regime. At each axial position the relative phase shift between the excitation signal and the detected signal are determined via calculation of the cross spectral power density. A linear fit to the axial phase run yields the parallel wavelength. As argued in Hanna *et al.* [6] the magnetic field perturbation can be considered as a localized disturbance within an axial distance of $<$ 1m from the exciter. For larger distances the perpendicular wavelengths lead to a significant spreading of the magnetic field perturbation across the ambient magnetic field. Thus, the measurements are limited to a maximum axial distance of 600mm downstream from the exciter.

EXPERIMENTAL RESULTS

Before the measurements of the phase relation between the excitation signal and the detected signal the amplitudes of the magnetic field fluctuations in radial as well as in axial direction has been determined. Fig. 2 shows the amplitudes

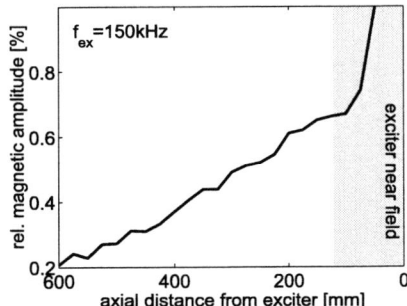
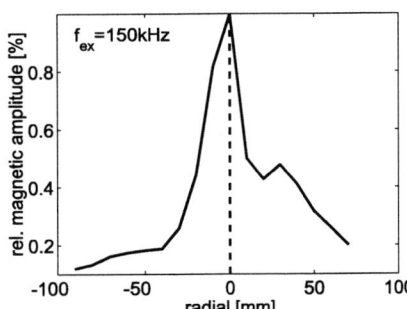

FIGURE 2. Axial and radial run of the relative amplitudes of the detected magnetic fluctuations.

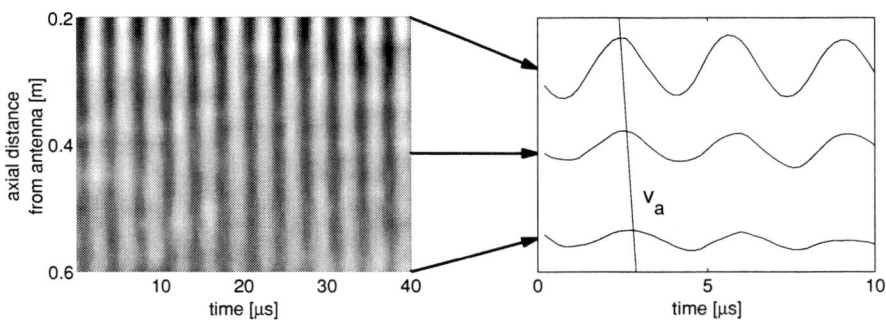

FIGURE 3. Gray-scaled plot of the detected magnetic fluctuations over an axial distance away from the exciter loop located at $z = 0$mm. White indicated large amplitudes, black refers to small amplitudes. For visualization purposes three time series of detected magnetic fluctuations for different axial positions are exemplary shown.

of the detected magnetic field fluctuations normalized to their maximum amplitude for an excitation frequency of 150kHz. In axial direction the wave is almost linearly attenuated to a level of 20% over a distance of $z = 600$mm. Close to the exciter loop the amplitude run deviates significantly from a linear decrease. This is attributed to the fact that in this regions $z < 100$mm the near field of the exciter is also detected and not only the propagating magnetic field perturbation. In radial direction the amplitude profile is strongly peaked at the position of the exciter loop, indicated by the dashed line at $r = 0$. Towards the plasma edges the amplitude strongly decreases to below 20% for $r \geq 50$mm. This radial amplitude profile of the magnetic fluctuations remains essentially unchanged for different axial positions within the measurements distance. For the wavelength measurements the magnetic fluctuation probe is positioned directly in the plasma center and moved axially with a spatial resolution of $\Delta z = 25$mm. A typical result of the axial measurement for a single excitation frequency is shown in Fig. 3. On the l.h.s a space-time diagram of magnetic field fluctuations for the different axial positions is shown gray-scaled. The position of the exciter loop is at $z = 0$mm. A clear but small phase shift along the ambient magnetic field is found. The phase shift is seen very clear if single timeseries at different axial positions are compared, as indicated on the r.h.s. of Fig. 3. Here, it turns out very clear that a propagating wave is observed. The phase velocity of the wave can be determined as the Alfvén velocity. These measurement have been repeated for various excitation frequencies. Calculation of the phase spectrum via the cross power spectral density between the excitation signal and the detected signal yields the phase run along the magnetic field line. Results for three different frequencies are shown in Fig. 4. Except for the near field region around the exciter loop the phases show an almost perfect linear dependence on the axial distance from the exciter loop for all excitation frequencies.

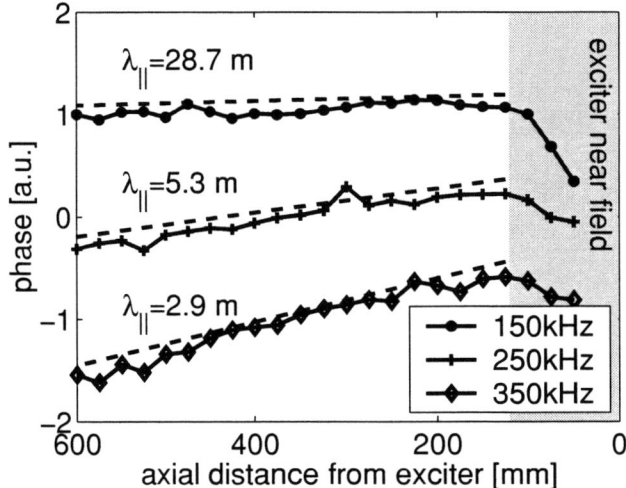

FIGURE 4. Axial run of the relative phase between the exciter signal and the detected signal for three different excitation frequencies.

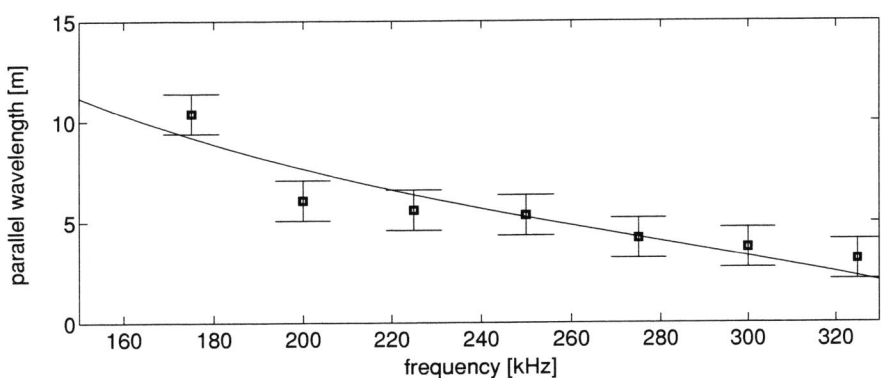

FIGURE 5. Comparison between the measured dispersion relation (squares) and the theoretical dispersion curve (solid line) for Alfvén waves.

The respective wavelengths are obtained by a linear fit to the phase run and differ by an order of magnitude when going from 150kHz to 350kHz. By measurement of the axial wavelengths for different frequencies the dispersion relation is obtained. In Fig. 5 the measurements are compared to the theoretical dispersion relation given by [7]

$$\frac{\omega^2}{k_\parallel^2} = v_A^2 \left(1 - \frac{\omega}{\omega_{ci}}\right). \quad (1)$$

Within the errorbars the agreement of the measured dispersion with theory for Alfvén waves is reasonably well and strongly suggests that a shear Alfvén wave is excited in the VINETA experiment. These results are a promising starting point for the study of different phenomena combined with Alfvén wave dynamics.

ACKNOWLEDGMENTS

This project and the travel expenses are partially funded from the DFG within the SFB 198 project A15.

REFERENCES

1. Cross, R., *An Introduction to Alfvén Waves*, The Adam Hilger Series in Plasma Physics, Adam Hilger, Bristol and Philadelphia, 1988.
2. Franck, C. M., Grulke, O., and Klinger, T., *Phys. Plasmas*, **9** (2002).
3. Boswell, R. W., *Phys. Lett.*, **33A**, 457–458 (1970).
4. Chen, F. F., *Plasma Phys. Controlled Fusion*, **33**, 339–364 (1991).
5. Franck, C. M., Grulke, O., and Klinger, T., *submitted to Rev. Sci. Instrum* (2002).
6. Hanna, J., and Watts, C., *Phys. Plasmas*, **8**, 4251–4254 (2001).
7. Leneman, D., Gekelman, W., and Maggs, J., *Phys. Rev. Lett.*, **82**, 2673–2676 (1999).

Control of Parallel and Perpendicular Flow Shears and Related Low-Frequency Plasma Instabilities

T. Kaneko*, E. Tada*, H. Tsunoyama* and R. Hatakeyama*

Department of Electronic Engineering, Tohoku University, Sendai 980-8579, Japan

Abstract. The two plasma sources using a concentrically three-segmented plasma (ion) emitter are developed, with which the parallel and perpendicular flow shears can be controlled, respectively. In the case of parallel flow-shear experiments, the drift-like, ion-acoustic-like, Kelvin-Helmholtz, and their coupled instabilities are found to be excited by the parallel flow shear. In the case of perpendicular flow-shear experiments, on the other hand, the drift-like instability which exists in the edge (density-gradient) region is suppressed by the perpendicular flow shear in the central region. These results show the importance of combined effects of radial density gradient and parallel/perpendicular flow shears on the low-frequency instabilities in magnetized plasmas.

INTRODUCTION

In recent years a number of studies have been made on magnetic field-aligned (parallel) and transverse (perpendicular) sheared plasma flows in space, laboratory, and fusion-oriented plasmas, which are considered to play important roles in the generation or the suppression of plasma fluctuations and turbulences [1–6]. However, since both of these flow shears often coexist with each other in the space environment and in most of the confinement devices, it is difficult to understand the effects of each flow shear on the turbulences in the real situation. In this sense, we claim that the external control of parallel and perpendicular ion flow shears is a key of experimentally clarifying general features of the topic associated with the origin of induced plasma-turbulence and transport. The aim of the present work is to develop the two plasma sources [7,8] using a concentrically three-segmented plasma (ion) emitter and to carry out basic laboratory experiments on low-frequency instabilities excited and suppressed by the parallel flow shears and the perpendicular flow shears, which are independently controlled using the above newly-developed two plasma sources.

PARALLEL FLOW SHEAR

In the case of parallel flow-shear experiments [7], a plasma is produced by a modified plasma-synthesis method as shown in Fig. 1, where ion and electron emitters are oppositely set at cylindrical machine ends under a strong magnetic field of $B = 2$ kG. The ion emitter is made of a 10.0-cm-diameter tungsten (W) plate and the ions are generated by surface ionization of potassium atoms on the tungsten plate. Here, the tungsten plate is heated to a temperature of 1000 K for the potassium atoms not to contaminate the plate surface under the condition that the thermionic electron is not emitted. The electron emitter using a 10.8-cm-diameter barium oxide (BaO) cathode is mounted at a distance of 170 cm from the ion emitter. Since this cathode is heated to a temperature of 1100 K enough to generate thermionic electrons, the collisionless plasma is synthesized by these ions and electrons. Under our conditions, the plasma density $n_p = 10^8 \sim 10^9$ cm^{-3}, the electron temperature $T_e = 0.2$ eV, and the ion temperature $T_i \leq T_e$. A background gas pressure is less than 1×10^{-6} Torr. A negatively biased stainless (SUS) grid ($V_g = -60$ V) is installed at a distance of 0.5 cm from the ion emitter. An electron velocity distribution function parallel to the magnetic field is considered to become

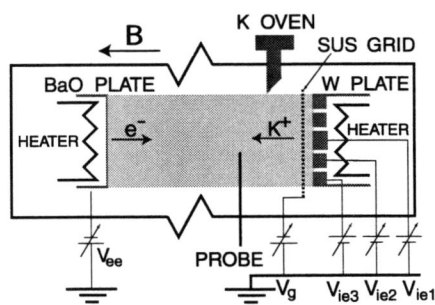

FIGURE 1. Schematic of experimental apparatus of parallel flow shear.

Maxwellian because the grid reflects the electrons flowing from the electron emitter.

Since a voltage applied to the electron emitter ($V_{ee} \simeq -3$ V) determines the plasma potential ϕ in this synthesized plasma, a voltage applied to the ion emitter can control the potential difference between the plasma and the ion emitter. This potential difference can accelerate the ions and generate the field-aligned ion flow. Furthermore, the ion emitter is concentrically segmented into three sections, each of which is electrically isolated and is individually biased. Thus, the field-aligned ion flows with radially-different energies, or parallel flow shears, are generated in the radially-uniform plasma potential. Hereinafter, the electrodes set in order from the center to the outside are called as the first, second, third electrodes and the voltages applied to them are defined as $V_{ie1}, V_{ie2}, V_{ie3}$, respectively.

In the present experiment, V_{ie3} is always kept at 0 V. A small movable Langmuir probe is used to measure plasma parameters and their radial profiles. An ion energy distribution function parallel to the magnetic field is measured by a directional electrostatic energy analyzer. The axial position z is defined as the distance from the SUS grid ($z = 0$ cm).

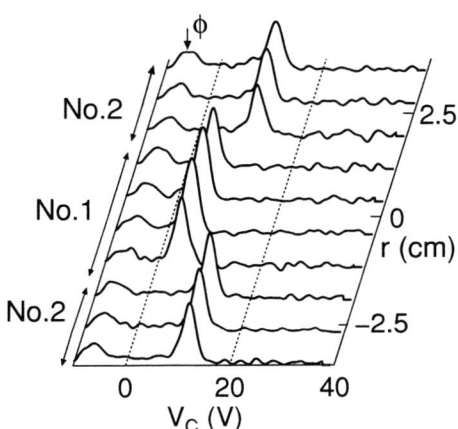

FIGURE 2. Energy distribution functions of ions parallel to the magnetic field $F_{i\parallel}$ with radial position r as a parameter at $z = 100$ cm for $V_{ie1} = 8$ V and $V_{ie2} = 16$ V.

In Fig. 2, ion energy distribution functions parallel to the magnetic field $F_{i\parallel} (\equiv -dI_c/dV_c)$ are presented with the radial position r as a parameter for $V_{ie1} = 8$ V and $V_{ie2} = 16$ V, where I_c is the current flowing to a collector of the energy analyzer and V_c is the collector voltage applied with respect to the ground. Long arrows numbered in this figure indicate the location ranges of the first (No. 1) and the second (No. 2) ion-emitter electrodes. In the third electrode region, however, the plasma density is too small to obtain $F_{i\parallel}$. Large peaks of $F_{i\parallel}$ at $V_c \simeq 5$ V and 13 V are observed in the regions of the first and the second electrodes, respectively, while a small peak of $F_{i\parallel}$ at $V_c \simeq -4$ V, which indicates ϕ, is independent of the radial position.

Since the V_c difference between the small and the large peaks of $F_{i\parallel}$ is considered to denote the ion flow energy ε parallel to the magnetic field, the values of the ion flow energy are $\varepsilon = 9$ eV and 17 eV, which are almost equivalent to V_{ie1} and V_{ie2}, respectively.

Thus, the ion drift difference between adjacent layers, or the field-aligned ion flow velocity shear in the boundary region of these electrodes is found to be easily formed and controlled by means of biasing the ion-emitter electrodes independently. Here, the direction of the radial gradient in the flow velocity, that is the sign of the flow shear, can also be easily reversed by setting V_{ie1} larger or smaller than V_{ie2}. This parallel flow shear is found to give rise to several types of low-frequency instabilities.

One is observed around the radial center where the density profile is relatively uniform. Figure 3 shows the fluctuation amplitude $\tilde{I}_{es}/I_{es}$ of the electron saturation current as functions of V_{ie1} and V_{ie2} at $r = 1.0$ cm (the central boundary region). When V_{ie1} is nearly equal to V_{ie2}, which is given as the dotted line in this figure, the fluctuation is not excited. However, with an increase or a decrease in V_{ie1} at a fixed value of V_{ie2}, namely the parallel flow-velocity difference between the first and the second electrodes increases, the fluctuation amplitude gradually becomes large. Thus, this instability is considered to be related to the Kelvin-Helmholtz instability (KHI) which is excited by the parallel flow shear.

The fluctuation amplitude has the maximum value for $\Delta V (\equiv |V_{ie1} - V_{ie2}|) \simeq 1$ V (solid lines in Fig. 3), which almost corresponds to the theoretical prediction of the threshold for KHI [1]. On the other hand, the fluctuation is also observed at the constant value of $V_{ie1} \simeq 1.5$ V, being independent of V_{ie2}, which is indicated as the dash-dotted line in Fig. 3. This fluctuation is excited depending on the potential difference between ϕ and V_{ie1}, or ion flow energy in the first-electrode region.

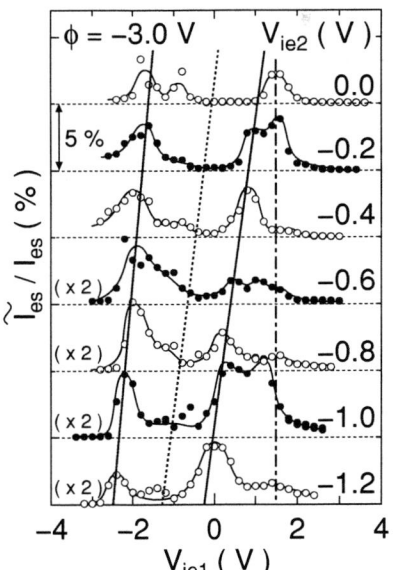

FIGURE 3. Fluctuation amplitude $\tilde{I}_{es}/I_{es}$ of the electron saturation current as functions of V_{ie1} and V_{ie2} at $r = 1.0$ cm.

According to the kinetic analysis [2], the threshold for the ion-acoustic instability becomes very low when the parallel flow shear exists. Thus, this instability is considered to be shear-modified ion-acoustic-like instability driven by the ion current which is determined by the potential difference between ϕ and V_{ie1}.

Another is observed in the peripheral boundary region where the density gradient is relatively large, which is considered to be related to the drift-like instability. As ΔV is increased, however, the fluctuation amplitude is found to increase in the same way as in the central boundary region. Thus, this instability is defined as the coupled drift and Kelvin-Helmholtz instability (CDKHI). The threshold of ΔV for exciting this instability is larger than that in the central boundary region, which means the density gradient tends to stabilize this instability. Furthermore, this CDKHI in the peripheral region is affected by the parallel flow shear in the different region, namely the central boundary region.

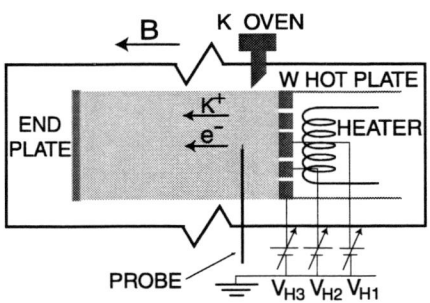

FIGURE 4. Schematic of experimental apparatus of perpendicular flow shear.

PERPENDICULAR FLOW SHEAR

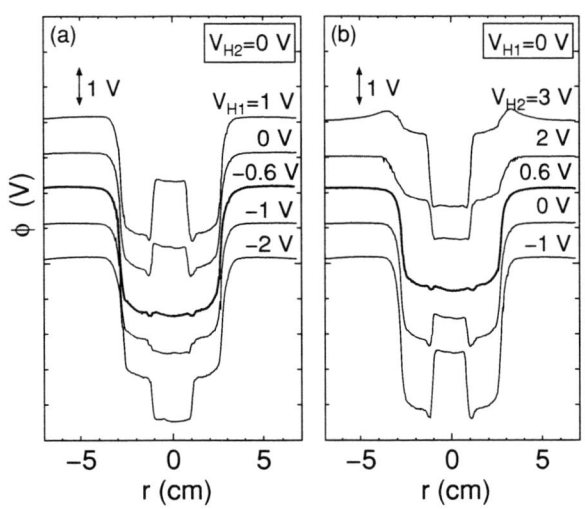

FIGURE 5. Radial profiles of plasma potential ϕ (a) with V_{H1} as a parameter for $V_{H2} = 0$ V and (b) with V_{H2} as a parameter for $V_{H1} = 0$ V.

In the case of perpendicular flow-shear experiments [8], a plasma is produced by surface ionization of potassium atoms on a 10.0-cm-diameter tungsten (W) hot plate under a magnetic field of 1.6 kG in a single-ended Q machine as shown in Fig. 4. The hot plate is concentrically segmented into three section, each of which is electrically isolated and is individually biased.

Thus, the radially-different plasma potential, or radial electric field is generated even in the fully-ionized collisionless plasma. The voltages applied to the electrodes set in order from the center to the outside are defined as V_{H1}, V_{H2}, V_{H3}, respectively. Here, V_{H3} is always kept at 0 V. The experimental conditions of n_p, T_e, T_i are almost the same as in the case of parallel flow-shear experiments.

Figure 5 gives radial profiles of plasma potential ϕ with V_{H1} and V_{H2} as parameters. When V_{H1} is changed from -2 V to 1 V for $V_{H2} = 0$ V [Fig. 5(a)], the plasma potential profile in the central region is changed from the well shape to the hill shape, while that in the peripheral region is almost constant. When V_{H2} is changed with V_{H1} kept constant [Fig. 5(b)], on the other hand, the plasma potential in the region of only the second electrode is changed. From this result, the plasma potentials at the different radial position are found to be independently controlled by the voltage applied to each segmented electrode of the hot plate.

This radially-different plasma potential, or the radial electric field, causes the $E \times B$ drift flow and its shear perpendicular to the magnetic-field lines. The potential difference can be controlled in a range of $-5 \sim 5$ V with the accuracy of 0.05 V. According to measurements with a directional probe, the perpendicular flow velocity shear is confirmed to actually increase when the radial electric field becomes large. This result indicates that the perpendicular flow velocity shear can be controlled by changing the radial electric field, or the applied voltage to the hot plate.

Figure 6 shows a peak amplitude of frequency spectrum $\tilde{I}_{es}/I_{es}$ of the electron saturation current of the probe as a function of V_{H1} for $V_{H2} = 0$ V, which are observed at $r = 2.5$ cm. As V_{H1} is increased, the fluctuation amplitude is found to increase, gradually decreasing for $V_{H1} > -0.6$ V. The frequency of the fluctuation is about 20 kHz, which corresponds to the frequency of the drift instability Doppler-shifted due to the $E \times B$ drift. When the fluctuation amplitude has the maximum value for $V_{H1} \simeq -0.6$ V, the radial profile of plasma potential is almost flat in the central region as shown in Fig. 5(a). This result indicates that the drift instability excited in the edge region is suppressed by the radial electric field, or the perpendicular flow shear in the central region, generated by biasing the segmented hot plate in the Q-machine plasma.

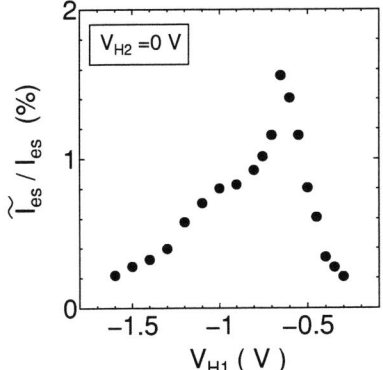

FIGURE 6. Fluctuation amplitude $\tilde{I}_{es}/I_{es}$ of the electron saturation current as a function of V_{H1} for $V_{H2} = 0$ V at $r = 2.5$ cm.

SUMMARY

We have carried out basic laboratory experiments on low-frequency instabilities modified by the parallel flow shears and the perpendicular flow shears, which are independently controlled using the newly-developed two plasma sources. In the case of parallel flow-shear experiments, the drift-like, ion-acoustic-like, Kelvin-Helmholtz, and their coupled instabilities are found to be excited by the parallel flow shear. In the case of perpendicular flow-shear experiments, on the other hand, the drift-like instability which exists in the edge (density-gradient) region is suppressed by the perpendicular flow shear in the central region. It is found that the combined effects of radial density gradient and parallel/perpendicular flow shears are important for controlling the low-frequency instabilities in magnetized plasmas.

ACKNOWLEDGMENTS

We thank Professor N. Sato and Professor M. Inutake for their useful discussion and comments. The authors are indebted to H. Ishida for his technical assistance. This work was supported partly by a Grant-in-Aid for Scientific Research from the Ministry of Education, Culture, Sports, Science and Technology, Japan, and partly by the LHD Joint Planning Research program at National Institute for Fusion Science.

REFERENCES

1. N. D'Angelo, Phys. Fluids **8**, 1748 (1965).
2. V. V. Gavrishchaka, S. B. Ganguli, and G. I. Ganguli, Phys. Rev. Lett. **80**, 728 (1998).
3. E. Agrimson, N. D'Angelo, and R. L. Merlino, Phys. Rev. Lett. **86**, 5282 (2001).
4. C. Teodorescu, E. W. Reynolds, and M. E. Koepke, Phys. Rev. Lett. **88**, 185003 (2002).
5. K. Ida, Plasma Phys. Control. Fusion **40**, 1429 (1998).
6. M. Yoshinuma, M. Inutake, R. Hatakeyama, T. Kaneko, K. Hattori, A. Ando, and N. Sato, Phys. Lett. A **255**, 301 (1999).
7. E. Tada, T. Kaneko, R. Hatakeyama, and N. Sato, J. Plasma Fusion Res. SERIES **4**, 524 (2001).
8. T. Kaneko, R. Hatakeyama, M. Yoshinuma, K. Hattori, A. Ando, M. Inutake, and N. Sato, *Annual Report of National Institute for Fusion Science April 2000 – March 2001*, p.225 (2001).

Periodical Motion of Ionized Front in a Closed Divertor Simulator

A. Matsubara[1], T. Sugimoto[2], T. Shibuya[3],
K. Kawamura[4], S. Sudo[1] and K. Sato[1]

[1] National Institute for Fusion Science, Toki, 509-5292, Japan
[2] Dept. of Fusion Science, The Grad. Univ for Advanced Studies, Toki, 509-5292, Japan
[3] Dept. of Applied Science, Tokai University, Hiratsuka, 259-1292, Japan
[4] Research Institute of Science and Technology, Tokai University, Hiratsuka, 259-1292, Japan

Abstract. The dynamics of plasma contacting with the enormous neutral particles has been investigated experimentally by using a closed-divertor simulator. It has been observed that the ionized front periodically moves in and out of the closed divertor region, as the neutral gas is injected sufficient amount for achieving plasma detachment in the divertor region. Such movement can be related to the reversal flow of the neutral particles injected. Especially friction between ions and neutral particles and recombination are important for the movement.

INTRODUCTION

The study of the plasma-gas interaction is topical subject for gas divertor in confinement device. In gas divertor, plasma-gas interaction results in the recombining plasma, and then the system goes in a state so called "plasma detachment". Several experiments for the plasma detachment have been performed by using linear divertor simulator [1-6]. Here, we report the discovery of the periodical macroscopic motion of an ionized front in a closed divertor regime [7].

EXPERIMENTAL APPARATUS

The experiment was carried out in the linear divertor simulator TPD-II at National Institute for Fusion Science (see Figure 1) [6,7]. The helium plasma was continuously generated by dc discharge between the anode and the LaB_6 cathode. Typical plasma parameters for presented experiment are following: electron density is $10^{19} m^{-3}$, electron temperature is ~20 eV in the axial magnetic field of 0.2 T for the discharge current I_d of 100 A. The plasma goes into the simulated edge plasma region (E-region), and then into the closed divertor region (D-region). The orifice of 20 mm in diameter that is somewhat larger than the plasma diameter is located at 0.7 m distant from the target. This orifice plays a role

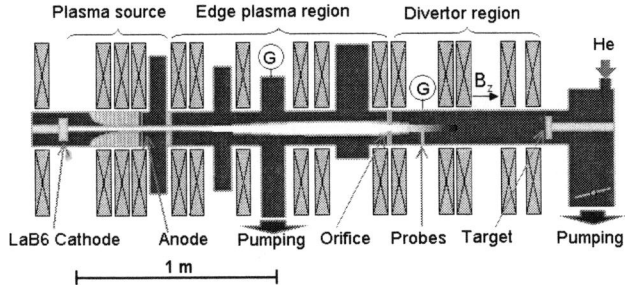

Figure 1. Schematic diagram of TPD-II

of a baffle for the closed divertor in confinement devices. Plasma detachment appears in the D-region where the helium neutral gas is injected with a flow rate of $Q \sim 0.02$ Pa m^{3}s^{-1}, and a neutral gas pressure at the D-region of P_D, is ~1 Pa.

PERIODICAL MOTION OF IONIZED FRONT

The periodical movement of the ionized front was observed for the condition of $0.03 \leq Q \leq 0.1$ Pa m^{3}s^{-1}. If the value of Q was set in such range, the equilibrium was lost; the increase of P_D and the movement of the ionized front toward the upstream began together. Then, periodical movement occurred. Almost no change of the discharge condition was observed (the change of the discharge voltage was within 1%). If the value of Q was set more than 0.2 Pa m^{3}s^{-1}, the system returned to equilibrium state, and the ionized front was settled at the E-region.

Figure 2 shows sequential photographs of the periodical movement for the case of $Q = 0.06$ Pa m^{3}s^{-1}. One can see that the ionized front moves in and out of the D-region. The extent of the movement is ~0.5 m. The period is ~9 s for the present case. We have observed that the period decreases if the value of Q is increased within the range for the appearance of the movement.

During the ionized front moved in that way, the significant change in the neutral pressure was observed. The neutral pressures measured at the D- and E- regions (P_D and P_E) are shown in Figures 3(a) and 3(b), respectively. Vertical lines of t1~t4 in Figure 3 denote each phase which corresponds to the phase indicated respectively as t1~t4 in Figure 2.

Let us look at the relation between the change of the neutral gas pressure and the movement of ionized front. Following the variation of P_D from t1 as a beginning, we can see that P_D increases until t3. In this stage the ionized front moves toward the upstream of the plasma flow (see also Figure 2). At t3 the value of P_D reaches its maximum value, P_D^{max}, of ~13 Pa, and the ionized front moves into the E-region. The limitation of the P_D^{max} will be mentioned in the next section. After t3, P_D decreases immediately while P_E increases drastically. When P_D is reduced to half of its maximum value (at t4), the ionized front comes into the D-region. Then, as P_D increases gradually, the ionized front begins to move toward the upstream again.

The change of the difference between P_D and P_E can be due to the change of the neutral gas flow. For example, the decrease in P_D and the increase in P_E for the moment after t3 imply that the neutral gas accumulated

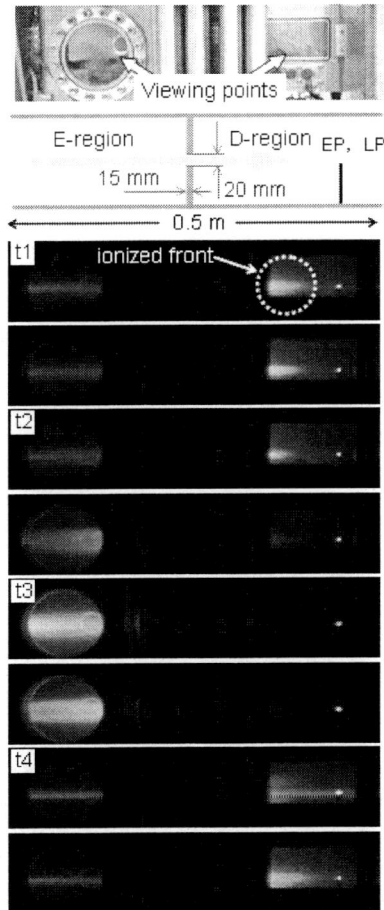

Figure 2. Sequential photographs of the periodical ionized front movement (taken every 1 second). EP and LP mean an emissive probe and a Langmuir probe.

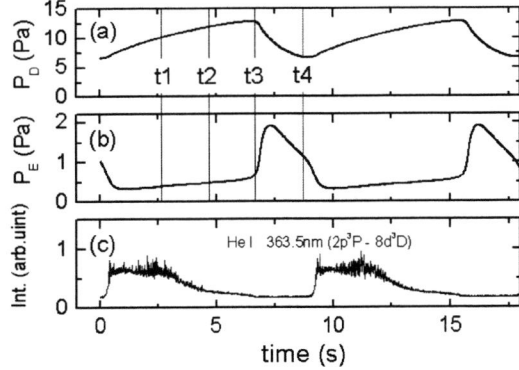

Figure 3. Periodical variations of the neutral pressures at D-region P_D [in (a)], at E- region P_E [in (b)], and the spectral intensity of He I 363.5nm detected from the D-region [in (c)].

in the D-region flows into the E-region. The periodical movement of the ionized front seems to be closely related to the reversal flow of the neutral gas.

INTERPRETATION

A possible interpretation of the periodical movement is given as follows: It is noted that the neutral gas is pumped at the E-region (1 m^{3}s^{-1}), so the neutral particle tends to flow into the E-region through the orifice. During the stage in which the plasma flows through the orifice, the friction between ions and neutral particles prevents the reversal flow. The friction plays a role of suppressing in the effective conductance of the orifice for the reversal flow. In fact, it has been observed that the amount of the reversal flow decreases with increasing the discharge current I_d for the plasma production [6].

As P_D comes to increase, the recombination becomes to be significant. For the present experiment, the strong emission of highly excited level has been observed, indicating that three-body recombination arises, which is shown in Figure 3(c). We can see that the spectral intensity of He I 363.5 nm increases after t4. As a result, the momentum loss of the plasma increases, and the ionized front begins to move toward the upstream.

At the moment when the ionized front goes into the E-region, P_D reaches its maximum value, P_D^{max}, of 13 Pa as mentioned above. It can be considered that P_D^{max} is related to the upstream plasma pressure. Figure 4 shows the discharge current I_d dependence of the electron density n_e, the electron temperature T_e and the plasma pressure

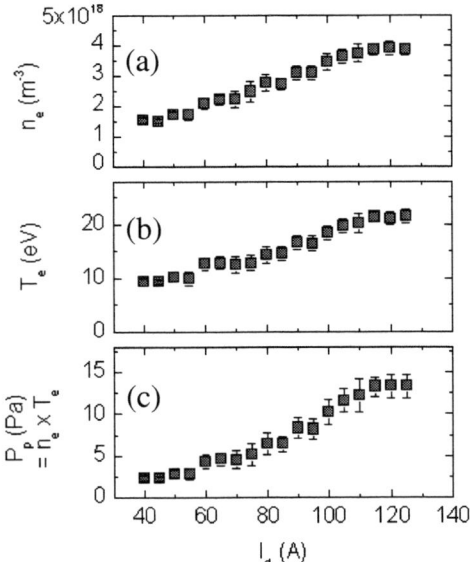

Figure 4. Discharge current I_d dependence of the plasma parameters obtained by single Langmuir probe for the low neutral pressure condition. (a) electron density n_e, (b) electron temperature T_e, and (c) Plasma pressure P_p defined by $n_e \times T_e$.

defined as $n_e \times T_e$, which were obtained by the Langmuir probe located at 12cm from the orifice (see also Figure 2). It is noted that this measurement was performed under the condition of much low pressure of neutral gas, so that those parameters obtained in the D-region can be fairly substituted for those in the upstream. From Figure 4 we can see that P_p at I_d=100A is comparable to P_D^{max} of 13 Pa. This gives us a simple view that P_D^{max} is limited by P_p, i.e., if P_D increases up to P_p, ionized front is extruded from the D-region by the force due to P_D-P_E.

After the ionized front goes into the E-region, the effective conductance of the orifice becomes rapidly weak. The neutral gas accumulated in the D-region flows into the E-region, and then P_D decreases. As a consequence of decrease in P_D, the mean-free-path for ion-neutral collision becomes long. The ionized front moves into the D-region again. The situation of the long mean-free-path can be seen in Figure 2 (at t4); the plasma in the D-region appears to be as energetic as the plasma in the E-region.

SUMMARY

The periodical motion of the ionized front in the closed divertor simulator was investigated. A cause of the observed periodical motion is the friction by the plasma at the orifice. The friction leads to the enhancement of the neutral gas pressure at the D-region, and then the recombination becomes significant. Due to the momentum loss by the recombination, the ionized front shifts to the upstream. The periodicity is sustained by the fact that the friction depends on the position of the ionized front.

ACKNOWLEDGMENTS

This work was supported by the NIFS collaborative research program.

REFERENCES

1. Hsu, W. L., Yamada, M., and Barrett, P. J., Phys. Rev. Lett. **49**, 1001 (1982).
2. Schmitz, L., Blush, L., Chevalier, G., Lehmer, R., Hirooka, Y., Chia, P., Tynan, G., and Conn, R., J. Nucl. Mater. **196-198**, 841 (1992).
3. Ohno, N., Ezumi, N., Takamura, S., Krasheninnikov, S. I., and Pigarov, A. Yu., Phys. Rev. Lett. **81**, 818 (1998).
4. Ezumi, N., Nishijima, D., Kojima, H., Ohno, N., Takamura, S., Krasheninnikov, S. I., and Pigarov, A. Yu., J. Nucl. Mater. **266-269**, 337 (1999).
5. Ohno, N., Nishijima, D., Takamura, S., Uesugi, Y., Motoyama, M., Hattori, N., Arakawa, H., Ezumi, N., Krasheninnikov, S., Pigarov, A., and Wenzel, U., Nucl. Fusion **41**, 1055 (2001).
6. Matsubara, A., Murata, R., Nagata, K., Shibuya, T., Tonegawa, A., Sakamoto, S., Suzuki, H., Ohyabu, N., Watanabe, T., Sato, K., Takayama, K., Kawamura, K., J. Nucl. Sci. Technol. **37**, 555 (2000).
7. A. Matsubara, T. Sugimoto, T. Shibuya, K. Kawamura, S. Sudo and K. Sato, J. Plasma Fusion Res. **78**, 196 (2002).

Particle Acceleration Due to Electric Field Bursts Close to the Lower Hybrid Frequency in a High-Voltage Linear Plasma Discharge

Y. Takeda[1], H. Inuzuka[2]

[1]*Department of Physics, College of Science and Technology*
Nihon University, Kanda Surugadai 1-8, Tokyo 101-8308, Japan
[2]*Department of Electrical and Electronic Engineering*
Faculty of Engineering, Shizuoka University, Johoku 3-5-1, Hamamatsu 432-8561, Japan

Abstract: This paper presents our new findings that strong electric-field bursts, which were generated in a high-voltage linear plasma discharge, give rise to impulsive hard x ray fluxes with energies higher than 20 keV detected dominantly from the direction perpendicular to the magnetic field and sometimes from the direction almost parallel to the magnetic field with much lower intensities. Dependence of the angular distribution of electron acceleration on the spectral profile of an originating electric-field burst whose peak frequency is close to the lower hybrid frequency is explained according to the theory of transit time interaction.

INTRODUCTION

Recently the interests in the study of electron heating and acceleration due to electrostatic waves with a frequency much lower than the electron cyclotron frequency are motivated by satellite observations of space plasmas and nonlinear plasma theories [1,2,3,4].

As reported earlier [5], short electric field bursts observed in a high-voltage linear plasma discharge cause nonthermal electron acceleration evidenced by hard x-ray emissions. They are coincident with the growth of the lower hybrid mode with a peak frequency close to the lower hybrid frequency of a hydrogen plasma, $f_{lh} = (f_{ce} f_{ci})^{1/2}$, where f_{ce} and f_{ci} denote the electron cyclotron- and ion cyclotron frequency, respectively.

In this paper we discuss the key role of the spectral profile of an electric field burst played on the angular distribution of impulsive hard x-ray fluxes observed in a high-voltage linear plasma discharge.

EXPERIMENTAL APPARATUS AND THE METHODS OF MEASUREMENT

We have drawn a high discharge current along the magnetic field with a pre-existing hydrogen (or deuterium) plasma produced by a titanium washer gun. The configuration of the magnetic field is a magnetic mirror with mirror ratio 1.2 and the field intensity at the mirror point is typically 1.2 kG. The discharge is ignited by applying a high voltage V_c =15-20 kV from a capacitor with C=2.2 μF between the cathode (aluminum disk 50 mm in diameter) and the cylindrical muzzle of the plasma gun after a suitable delay time, typically 24 μs from firing the gun.

The parallel electric fields were measured at two axial positions, one at 10 cm in front of the cathode, and the other at the centre of the apparatus (midplane), by using a pair of electric double probes and optically isolated transmission systems. The distance of two measuring points is L=20 cm. The optically isolated transmission system of the E-field fluctuations has frequency characteristics whose upper 3dB cut-off frequency is nearly 80 MHz.

We have detected hard x-ray fluxes in the direction perpendicular to the magnetic field line with a plastic scintillator (NE 102A) and photomultiplier combinations. The x-ray detector attaches an aluminium absorber 1-mm in thickness on the detector head. Thus the energy range of hard x-rays is roughly determined to be larger than 20 keV.

We also detected hard x-ray emissions in the forward (cathode side)- and backward (anode side) directions almost parallel to the magnetic field in order to obtain information on the anisotropy of velocity distribution of energetic electrons produced in the localised wave structure, such as a lower hybrid wave packet.

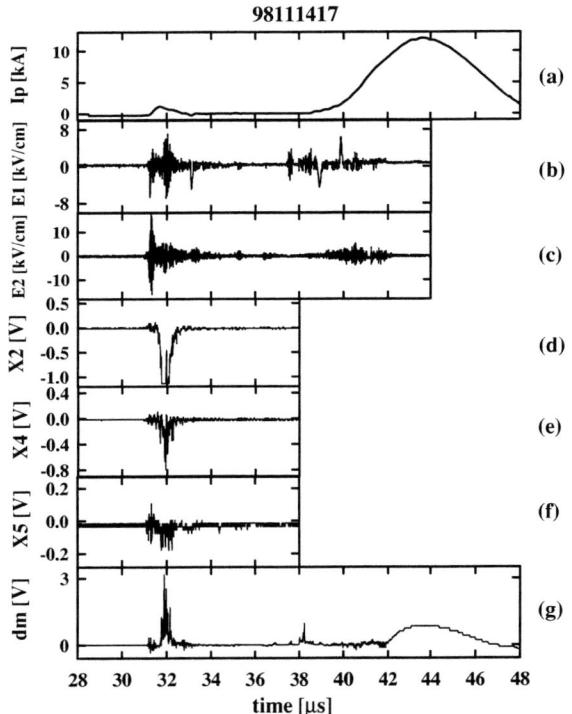

FIGURE 1. Typical data sets of the high-voltage linear plasma discharge. The time trace (a) shows discharge current, (b) the parallel electric field E1 measured 10 cm in front of the cathode, (c) the parallel electric field E2 at the center of the apparatus, (d) x ray emission in the direction perpendicular to the magnetic field, (e) x ray emission almost parallel to the magnetic field detected from the cathode side, and (f) x ray emission almost parallel to the magnetic field detected from the anode side, and (g) magnetic flux of the paramagnetic sense picked up by the diamagnetic loop.

RELATIONSHIP BETWEEN ELECTRIC FIELD BURSTS AND HARD X RAY-EMISSION.

Fig. 1 shows a typical time profile of the discharge current, fluctuations in the longitudinal electric field (abbreviated as E-field henceforth), hard x ray emissions, and magnetic flux which was picked up by a diamagnetic loop and whose polarity is paramagnetic. This data set was simultaneously obtained in the same discharge shot. The start of a time sequence, $t=0$ is taken at the instant when the titanium washer gun is fired to produce and inject a deuterium plasma into the magnetic mirror. As Fig.1 (b) and (c) show, fluctuations in the electric field are highly enhanced at the onset of the plasma discharge and cause large anomalous resistivity.

The hard x-ray emission is coincident with electric field bursts shown in Fig. 1 (b) and (c) and is much stronger in the direction perpendicular to the magnetic field than that in the direction almost parallel to the magnetic field.

Moreover an impulsive magnetic flux of the paramagnetic sense appeared coincident with hard x-ray emissions. Generation of the magnetic flux could be understood as an evidence of excitation of a certain electromagnetic mode, although not identified yet. This phenomenon will be discussed elsewhere in the near future.

For comparison, we show similar data set in Fig.2, which was obtained, in another discharge shot. It should be noted that parallel hard x-ray emissions are correlated with electric field bursts shown in Fig. 2(a) and (b), but hard x-ray emission was not detected in the perpendicular

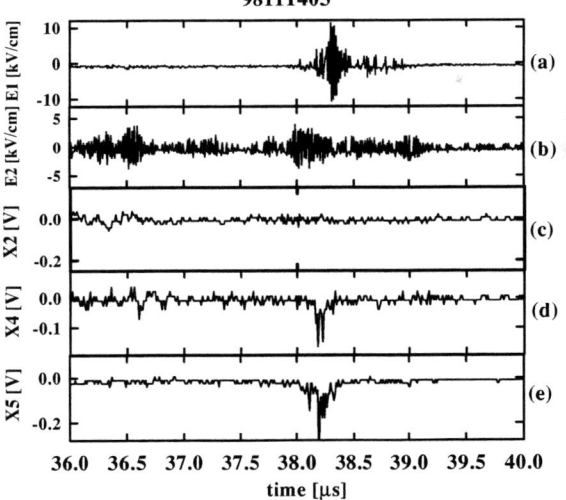

FIGURE 2. Similar data sets of the high-voltage linear plasma discharge. The time trace (a) shows the parallel electric field E1 measured 10 cm in front of the cathode, (b) the parallel electric field E2 at the center of the apparatus, (c) x ray emission in the direction perpendicular to the magnetic field, (d) x ray emission almost parallel to the magnetic field detected from the cathode side, and (e) x ray emission almost parallel to the magnetic field detected from the anode side.

direction.

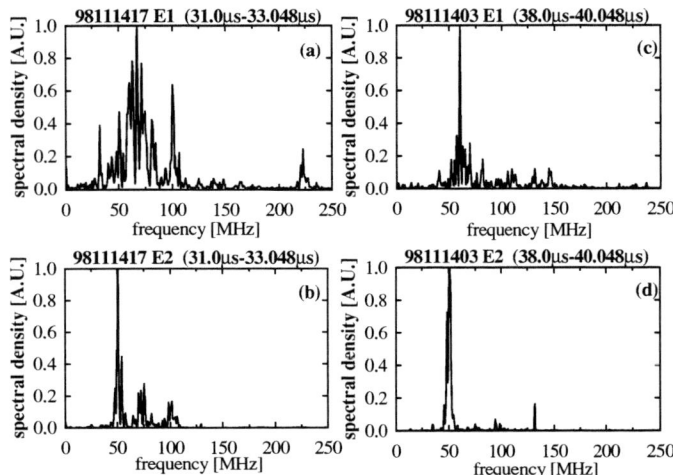

FIGURE 3. Power spectral density functions of the electric field fluctuations calculated with FFT procedures. (a) and (b) show the power spectra using the electric field data shown in Fig.1, and (c) and (d) are calculated from the electric field data shown in Fig. 2.

In order to study this remarkable difference of the angular distribution of hard x-ray emissions, we calculated power spectral density functions of the electric field bursts with FFT (Fast Fourier Transform) procedures. Sampled electric field data are taken from the relevant time interval, t=31.0 μs -33.048 μs for Fig. 1(b) and (c), and t=38.0 μs -40.048 μs for Fig. 2(a) and (b). Fig.3 (a) shows notable broadening of the spectral profile around the lower hybrid frequency in the case of intense hard x-ray emission perpendicular to the magnetic field, while in the case of impulsive hard x-ray emissions preferentially in the direction almost parallel to the magnetic field, the power spectrum of the correlated elec-tric field burst is relatively narrow as shown in Fig. 3(c) and (d).

DISCUSSION AND CONCLUDING REMARKS.

In the present experiment, f_{lh}=75 MHz for a hydrogen plasma, hence f_{ce} is approximately $60 f_{lh}$. Since the longitudinal scale length of a lower hybrid caviton l is estimate to be 5. 2cm, then the transit time is $t_{tr}= l/v_e$= 2.8 ns for electrons with kinetic energy of 1 keV that comprise high-energy tail of the Maxwellian distribution. Then the necessary condition for the transit time acceleration T_{lh} = 13.3ns >> t_{tr} is well satisfied, where v_e denotes the electron velocity and T_{lh} is the period of a lower hybrid wave.

Therefore we can assert that the preferential acceleration of energetic electrons parallel to the magnetic field in the case of fairly monochromatic spectral profile as shown in Fig. 3(c) and (d) could be explained according to the theory of transit time interaction of electrons with a localised lower hybrid wave packet or an elongated lower hybrid caviton [6].

The physics of directive electron acceleration perpendicular to the magnetic field in the case of broadened power spectrum of electric field bursts is not fully understood at the present stage, and will be discussed in another paper referring to the theory of stochastic pitch angle diffusion due to electron-electromagnetic mode (such as whistler wave) interaction [4].

REFERENCES

[1] Tom Chang, Geoffery. B. Crew, John M. Retterer and John R. Jasperse,IEEE Trans. Plasma Sci., **17**, 186(1989).
[2] N. Dubouloz, R. A. Treumann, R. Pottelette, and K. A. Lynch, Geophys. Res. Lett. **22**, 2969(1995).
[3] Liu Chen, Zhihong Lin, and Roscoe, White, Phys. Plasmas **8**, 4713(2001).
[4] W. J. Wykes, S. C. Chapman, and G. Rowlands, Phys. Plasmas **8**, 2953(2001).
[5] Y. Takeda, H. Inuzuka and K. Yamagiwa, *Double Layers* (Proc. of the 5th Symposium on Double Layers, Tohoku University, Sendai, Japan, 1996) pp.193-198, World Scientific, 1997.
[6] P. A. Robinson, A. Melatos, and W. Rozmus, Phys. Plasmas **3**, 133 (1996).

Effect of Neutrals on Alfvén Wave Propagation

Christopher Watts, Jeremy Hanna[†]

Physics Department, New Mexico Tech, Socorro, NM 87801
[†]*Physics Department, Auburn University, Auburn, AL 36849*

Abstract. Alfvén waves are ubiquitous in space plasmas, influencing the dynamics of, for example, solar flares and magnetospheric reconnection. To better understand the role of the waves, in particular in relation to reconnection, we have undertaken a program to study these waves in the laboratory. We present results from the Auburn Linear Experiment for Space Plasma Studies (ALESPI), a two meter device which uses a single helicon source to create the plasma. The high density, highly ionized, quiescent plasma allows detailed study of both shear and compressional waves. Depending on the neutral fraction, the measured dispersion curve changes dramatically. These measurements agree well with theoretical predictions. We present these results and a discussion of the applicability to ionospheric plasmas.

INTRODUCTION

Alfvén waves, low frequency hydromagnetic plasma oscillations, are of fundamental importance in the behavior of many laboratory and space plasmas. These waves communicate information about changes in magnetic field topologies, and are especially important in the dynamics of magnetic reconnection. Low frequency fluctuations observed in the magnetosphere and upper ionosphere by satellites and sounding rockets have been interpreted as Alfvén waves. Of interest is how the propagation characteristics of these waves change as they travel lower in the ionosphere where the neutral population becomes significant.

The Alfvén wave speed is defined as $v_A = B_0/\sqrt{(\mu_0 n_i m_i)}$, where B_0 is the background magnetic field, and n_i and m_i are the ion density and mass. We can define a nominal wavelength, $\lambda_A = v_A/f_{ci} = \sqrt{m_i/\mu_0 n_0 e^2}$ where $2\pi f_{ci} = \omega_{ci} = eZB/m_i$ is the ion cyclotron frequency, with Z as the charge state. In space, Alfvén waves have wavelengths on the order of kilometers, while in the laboratory wavelengths are typically tens of meters, making them difficult to study. However, since this wavelength is inversely proportional to the density, the wavelength can be shortened by operating in a dense plasma. A number of experiments have looked at the basic properties of Alfvén wave propagation[1-5], and several have document neutral effects[6, 7]. In contrast to previous work with pulsed, current-carrying discharges, we present studies using a helicon discharge to create a high-density, steady state plasma. The neutral fraction can be specified by choosing one of several helicon "modes", and the propagation characteristics of Alfvén waves under these steady-state conditions studied.

ALFVÉN WAVES IN PARTIALLY IONIZED PLASMAS

Neglecting resistive effects, the Alfvén wave dispersion relation is given by

$$\frac{\omega^2}{k_\parallel^2} = v_A^2[1-(\omega/\omega_{ci})^2] \tag{1}$$

where $k_\parallel$ is the parallel wavenumber.

A theory by Woods[8] predicts that the dispersion relation for Alfvén waves is modified by ion-neutral collisions, and subsequent studies verified this prediction in a partially ionized plasma column[7]. Following Müller, the MHD equations can be cast in the following form (including the Hall effect) for a perturbation $\tilde{B}$ and unit vector along B_0, b_0.

$$\omega^2 \tilde{B} + \nabla \times [c_A^{*2} b_0 \times (b_0 \times \nabla \times \tilde{B})] - \nabla \times [i \Omega_i^* c_A^{*2} (b_0 \times \nabla \times \tilde{B})] + \nabla \times [i c_\eta^2 \nabla \times \tilde{B}] = 0 \quad (2)$$

where the symbols used are:

$c_A^* = B_0 / \sqrt{\mu_0 m_i n_i S}$;   modified Alfvén speed

$c_\eta = \sqrt{\omega \eta / \mu_0}$;   magnetic diffusion phase speed

$\omega_{ci}^* = eB / m_i^*$;   effective ion cyclotron frequency (incorporating hall effect)

$S = 1 + \dfrac{n_n / n_i}{1 - i\omega / \omega_{ni}}$;   frequency dependent mass loading by neutrals

$\Omega_i^* = \omega / \omega_{ci}^*$; $m_i^* = m_i S$

with η the resistivity and ω_{ni} the ion-neutral collision frequency. For a homogeneous plasma in cylindrical coordinates, the axisymmetric wave solutions of $\tilde{B}$ are Bessel functions, and lead to the dispersion relation

$$[(c_A^{*2} - i c_\eta^2)^2 - \Omega_i^{*2} c_A^{*4}] k^4 + [(c_A^{*2} - i c_\eta^2)(c_A^{*2} x_n^2 - 2(\omega^2 + i c_\eta^2 x_n^2)) - \Omega_i^{*2} c_A^{*4} x_n^2] k^2$$
$$+ (\omega^2 + i c_\eta^2 x_n^2)((\omega^2 + i c_\eta^2 x_n^2) - c_A^{*2} x_n^2) = 0 \quad (3)$$

where $x_n = a_n / r$, a_{nn} is the nth root of the Bessel function $J_1(a)$ and r is the plasma radius. The important adjustable parameters are the electron and ion temperatures, T_e, T_i; the electron and neutral densities, n_e, n_n; and the background magnetic field, B_0.

For the plasma parameters typical of our experiments, $B_0 = 0.085$ T, $T_e = 5$ eV, $T_i = 1$ eV, $n_e = 5 \times 10^{18}$ m^{-3} the neutral fraction has a substantial effect on the Alfvén wave dispersion relation, as seen in Figure 1. The ion-cyclotron resonance is downshifted, and the parallel wavenumber is reduced.

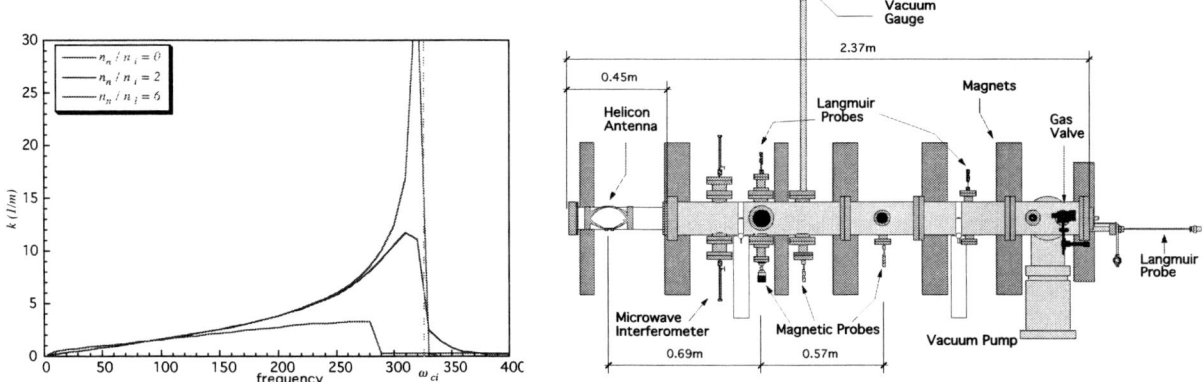

FIGURE 1. Left: Alfvén wave dispersion relation modified by neutral effects. Right: Sketch of the Auburn Linear Experiment for Space Plasma Investigations (ALESPI).

EXPERIMENTAL STUDIES ON ALESPI

The Auburn Linear Experiment for Space Plasma Investigations (ALESPI) is a cylindrical stainless steel vacuum vessel 15 cm in diameter and 2.5 m long, surrounded by seven electromagnets. See Figure 1. These magnets produce an average axial background magnetic field of about 0.085 T, with a spatial magnetic field ripple of ±12% due to the spacing of the magnets. A steady-state plasma discharge is created using a helicon source. There are a number of different antenna configurations; we have made use of a helical m=1 design. The antenna is 10 cm in diameter and 22 cm long surrounding a pyrex tube positioned at one end of the cylindrical vessel, with the turbo pump and gas feed at the other end. For excitation frequencies near the lower hybrid frequency the antenna sets up an oscillating electromagnetic field that inductively couples to the plasma and travels down the plasma column as a circularly polarized wave. The means of power absorption by the plasma are not yet fully understood[9, 10].

Depending on the tuning of the matching circuit several different "modes" of the helicon discharge can be obtained. These modes can be identified by a visual change in the pink core of the discharge. Depicted in Figure 2 are the density and pressure profiles for two of the modes used in this study. In previous work we reported Alfvén wave measurements in what appears to be the most efficient mode, where the gas is nearly completely ionized[11]. In this mode we use He at a base initial fill pressure of 9 mTorr, with 900W of RF power at 10MHz. The neutral edge pressure drops to ~ 1 mTorr during the discharge, and a chord-averaged density of 5.5×10^{18} m^{-3} is found using a 65 GHz microwave interferometer. Density and temperature measurements using Langmuir probes give a core density of about 1.2×10^{19} m^{-3}, with an electron temperature ranging from 3-10 eV. The cord-averaged density of the probe data agrees with the interferometer data to within a few percent.

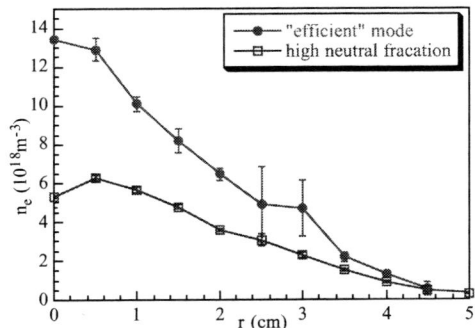 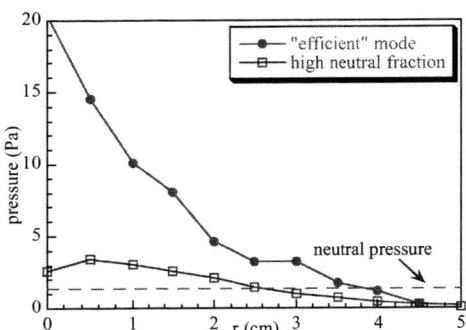

FIGURE 2. Plasma density and pressure profiles for the two helicon modes.

A high neutral fraction mode can also be obtained under similar conditions. In this case the electron density and temperature are significantly lower. As seen in the pressure profile, the edge neutral pressure is a sizable fraction of the central plasma pressure. Thus, the neutral penetration depth and hence the neutral fraction, though not directly measurable, is inferred to be significantly greater for this mode.

Non-axisymmetric Alfvén waves are excited and detected using commercial electronic inductors. The emitter is a 1×10^{-4} H inductor with 0.7 Ω and the detector is a 1-mH inductor with 30.5 Ω. The inductors are inserted into the plasma column by means of protective quartz tubes and water cooled, and the coils can be moved radially. The tubes are positioned 57 cm apart along the z axis in the region where the axial density remains constant.

To excite the waves, a function generator is used to supply a sine wave to the emitter coil with a 20 V peak-to-peak amplitude. This induces an oscillating magnetic field perpendicular to the coil. Calculating the magnetic perturbation from the Alfvén wave, $\tilde{B}$, from the driving voltage and the inductance of the detector, we find the ratio of perturbation to background field to be $\tilde{B}/B_0 \sim 10^{-2}$. The detected signal is a factor of 10^6 smaller, due to poor coupling and attenuation of the wave. The emitted and detected signals are digitized and cross-correlated to obtain phase information. The probes and detection method are calibrated in vacuum to account for any phase shift due to the electronics.

For this series of studies the emitter was placed at fixed radial location oriented so that the perturbed magnetic field is perpendicular to the background axial field, thus exciting a non-axisymmetric shear wave. The detector is positioned downstream where it is stepped in radial position in 5 cm increments. At each radial location the emitted frequency is scanned in 10 kHz increments from 10 kHz to 400 kHz. The emitter is then moved radially and the process repeated. For ALESPI parameters the ion cyclotron frequency is f_{ci} = 320 kHz.

Data are plotted on the left in Figure 3 for the measured wavenumber vs. frequency with both the emitter and detector at the same radius (2.5 cm) for the "efficient" low neutral fraction mode. The approach to the ion cyclotron resonance can be clearly identified, and the data are well fit by the ideal Alfvén wave dispersion relation, Eq (1). The plot on the right shows similar data at two different radii, with both the emitter and detector at the same radius, for a plasma discharge with a high neutral fraction. Near the ion cyclotron frequency in both cases the dispersion relation changes abruptly, as expected. However, the data are poorly fitted by Eq. (1) due to neutral effects.

As a first attempt at modeling the data we ignore the density gradient and treat the plasma as homogeneous with the local plasma parameters at the given radius. As discussed in Cross[12], the density gradient has a negligible effect on Alfvén wave propagation, and for emitter and detector at the same radius we can treat this as a ducted waveguide mode. For detector and emitter at different radii a more complete theory is required. We also ignore the non-axisymmetric nature of the wave.

The data at r = 2.0 cm are fit using the measured local plasma parameters: T_e = 5 eV, T_i = 1 eV, $n_e = 3.5 \times 10^{18}$ m^{-3} and B_0 = 0.080 T. The neutral fraction is adjusted to provide the best fit. In this case, the neutral fraction is 15 $\times 10^{18}$ m^{-3} or n_n/n_e = 4.3, a reasonable value based on Figure 2. For data at r = 2.0 cm a good fit can only be obtained with $n_e = 9.0 \times 10^{18}$, significantly higher than the measured electron density. We suspect that during this measurement scan the helicon discharge changed to a different mode from that used when measuring the density profile. In this case we found $n_n/n_e = 3.1$.

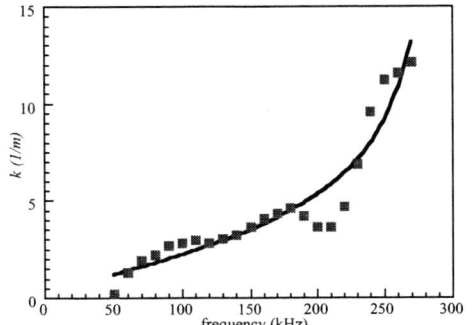
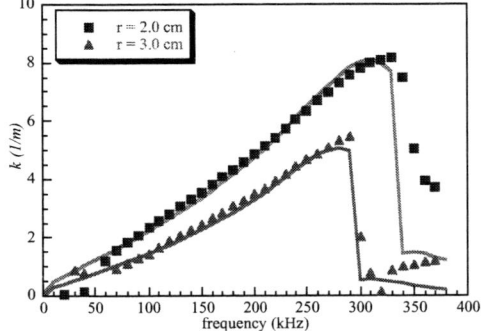

FIGURE 3. Measured Alfvén wave dispersion curve and fit to theory for (left) "efficient" low neutral fraction helicon mode and (right) a high neutral fraction mode. Error bars are of the order of the symbol size.

DISCUSSION

Collisional neutral-ion coupling can play a significant role in the propagation characteristics of Alfvén waves as demonstrated in these laboratory studies. The theoretical predictions of Woods accurately model the experimental data. Beyond the laboratory, using ionospheric parameters, this model predicts a substantial departure from the ideal dispersion relation for $n_n/n_e \sim 100$. Such conditions are present in the F region of the upper ionosphere. At present space-based measurement techniques are unable to map the dispersion relation and confirm this prediction. Conversely, such models, verified by laboratory experiment, may help interpret the morass of data satellites provide.

REFERENCES

1. Y. Amagishi, IEEE Transactions on Plasma Science **20**, 622 (1992).
2. W. Bostick and M. Levine, Physical Review **94**, 815 (1952).
3. D. F. Jephcott and P. M. Stocker, Journal of Fluid Mechanics **13**, 587 (1962).
4. R. C. Cross, Plasma Physics **25**, 1377 (1983).
5. W. Gekelman, D. Leneman, J. Maggs, et al., Physics of Plasmas **1**, 3775 (1994).
6. Y. Amagishi, K. Saeki, and I. J. Donnelly, Plasma Physics and Controlled Fusion **31**, 675 (1989).
7. G. Müller, Plasma Physics **16**, 813 (1974).
8. L. C. Woods, Journal of Fluid Mechanics **13**, 570 (1962).
9. P. A. Keiter, E. E. Scime, and M. M. Balkey, Physics of Plasmas **4**, 2741 (1997).
10. K. P. Shamrai and V. B. Taranov, Plasma Physics and Controlled Fusion **36**, 1719 (1994).
11. Jeremy Hanna and Christopher Watts, Physics of Plasmas **8**, 4251 (2001).
12. R. Cross, *An introduction to Alfvén waves* (IOP Publishing, Bristol, 1988).

Ion Confinement due to Radial Electric Field in a Magnetized Plasma

Yoshinobu Kawamoto*, Satoshi Yasuda*, Masashi Kondo*, Masuo Kondo* and Koichi Saeki*

Department of Physics, Faculty of Science, Shizuoka University, Ohya 836, Shizuoka 422-8529, Japan

Abstract. The decay process of a magnetized plasma having a flow-shear layer after plasma discharge is observed by using two-dimensional multi-probes arranged on a plane perpendicular to magnetic field. The plasma of appropriate potential structure has long lifetime.

INTRODUCTION

Plasma confinement is especially important to achieve nuclear fusion in magnetically confined plasmas. Especially, rotation and flow shear of fusion plasmas are key phenomena leading to both H-mode confinement and formation of edge and core transport barriers. In a narrow shear layer of plasma, the flow velocity shear primarily determines temporal evolution and spatial structure of transport barriers and resultant plasma confinement. These have been investigated in tokamaks and other fusion-oriented machines. Here, we report an investigation of decaying magnetized plasma having a flow shear layer by using a linear machine.

EXPERIMENTAL APPARATUS

The ion confinement time of a magnetized rotating plasma is studied in a linear plasma device. The applied magnetic field is 0.5-2.0 kG. The plasma is created by using a cathode assembly located at the end of a stainless-steel cylindrical chamber. The cathode assembly is shown in Fig. 1. It consists of concentric circle plates of a cathode (radius 6 mm $^\phi$) and an electron emitter (inner-outer radii 10 - 200 mm $^\phi$). Both the electron emitter and the cathode are coated with barium oxide and heated by tungsten filaments from the rear side. A mesh anode is available to be mounted in front of the cathode. In case that the mesh anode is mounted, the large negative pulse voltage V_{K-A} of the order of -100 V between the cathode and the mesh anode is applied for temporal plasma discharge. In case of no anode, a discharged plasma is created by applying the large negative pulse voltage V_{K-E} between the cathode and the emitter. A large negative voltage of 0.1-0.2 ms between the cathode and the anode (or the emitter) creates a discharged argon plasma. The behavior of a decaying plasma after the discharge is investigated in detail. The neutral argon gas pressure P is $(0.4 - 3.0) \times 10^{-3}$ Torr.

The behavior of a decaying plasma after the discharge is detected by multi-probes, which are located at the opposite end against the cathode assembly. The emitter is biased at 10 V in order to detect ion currents by the multi-probes, which are grounded through resisters. The multi-probes are arranged to detect ion-density distributions on a two dimensional plane x,y perpendicular to the direction z of the magnetic field. The multi-probes are consist of 140 probes and arranged in an area of 1.0×2.0 cm^2 on $x-y$ plane.

ION CONFINEMENT TIME

At first, we show a temporal behavior of decaying plasma after the discharge in case that a mesh anode is mounted and $V_{K-A} = -1$ V, in Fig. 1 and Fig. 2. The plasma discharge is performed from time $t = 0.2$ ms to $t = 0.4$ ms by applying

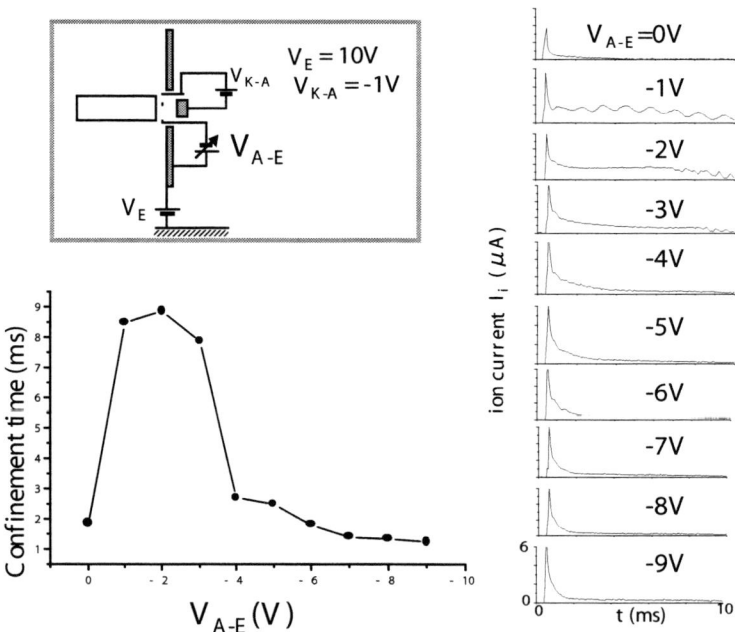

FIGURE 1. Confinement time of a decaying plasma after a discharge in case that a mesh anode is mounted and V_{K-A} = -1 V. The upper left figure show a cathode assembly. V_{K-A} and V_{A-E} means voltages between cathode and anode, and between anode and electron emitter, respectively. V_E is a bias voltage of emitter. The plasma is discharged in front of the mesh anode by applying a large negative voltage V_{K-A}. The right figure shows temporal behaviors with V_{A-E} as a parameter. V_{A-E} dependence of confinement time is summarized in the lower left figure. $P = 1.8 \times 10^{-3}$ Torr, $B = 1.6$ kG.

a negative pulse voltage V_{K-A} of -100 V. V_{K-A} showed in Fig. 1 and Fig. 2 means the voltage between the cathode and the anode after the discharge. V_{K-A} after the discharge and the voltage V_{A-E} between the anode and the emitter are key parameters for ion confinement.

Fig. 2 shows data of $V_{A-E} = -2$ V. The ion current I_i flowed into a probe at the plasma center ($x = 0.3$ cm and $y = -0.3$ cm) is proportional to the ion density at the plasma center. In Fig. 2, the decaying plasma is very stable and ion confinement time is about 9 ms. Here, we define confinement time as the time until which I_i reach $0.1I_{i0}$ from initial value I_{i0}. I_i is almost flat in time region of 2.0-6.0 ms. The density contours is displayed on a $y-t$ plane and $x-y$ planes. After $t = 7$ ms, an instability of $m = 1$ mode takes place and the plasma decays out. In Fig. 1, the V_{A-E} dependence of plasma confinement time is summarized. The maximum ion confinement is achieved when $V_{A-E} = -2$ V.

On the other hand, V_{A-K} dependence of ion confinement time is also investigated. When V_{A-K} is about -1 V, The ion confinement time is maximum. This means that week electron supply from the cathode is necessary to keep plasma stable for long time. The experiment in case of no anode shows strong fluctuation of I_i and results in short confinement time. Thus, steady electron supply from the cathode is a necessary condition for stable plasma. The confinement time increases linearly with gas pressure P. The confinement time depends on magnetic field B in complicate way, but increases simply with B in the week B region.

In a region of stronger magnetic field B compared with the conditions of the case of Fig. 2, we find electrostatic plasma disruption as shown in Fig. 3. Because of stronger magnetic field, the $m = 1$ fluctuation of I_i grows abruptly. The density contours on $y-t$ plane shows that the plasma at the center is disrupted. The density contours on $x-y$ planes reveal clearly that temporally excited spiral rotation brings the central plasma outside.

The development of $m = 1$ mode has two cases. In the case of week magnetic field and low-velocity rotation caused by low V_{A-E}, the plasma continues a decentered rotation and turns to a stable plasma of bell shape. On the other hand, in case of strong magnetic field and rather high-velocity rotation, the plasma develops from a $m = 1$ decentered rotation to a $m = 1$ spiral rotation and disrupts to a stable plasma of ring shape.

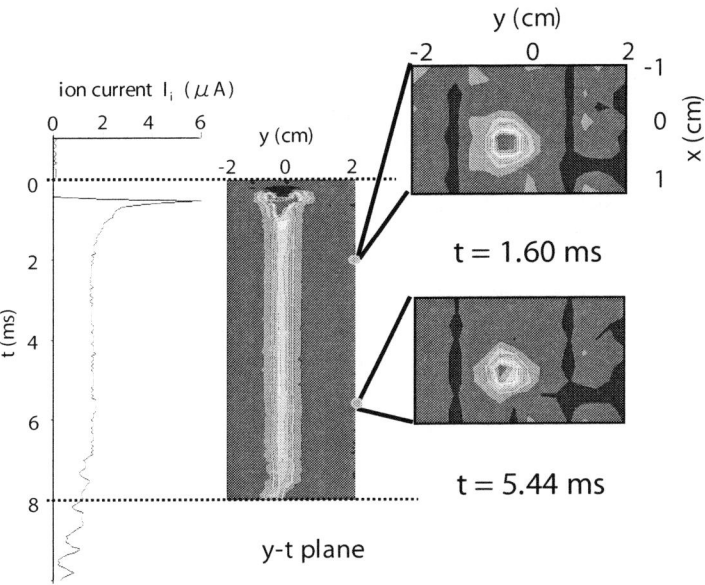

FIGURE 2. Behavior of a decaying plasma in case with a mesh anode. Here, $V_{K-A} = -1$ V and $V_{A-E} = -2$ V. Left figure shows temporal behavior of an ion current at the plasma center ($x = 0.3$ cm, $y = -0.3$ cm). Plasma-density behavior on a $y-t$ plane is shown as density contours in center figure. The right two figures show density distributions on a $x-y$ plane at fixed times.

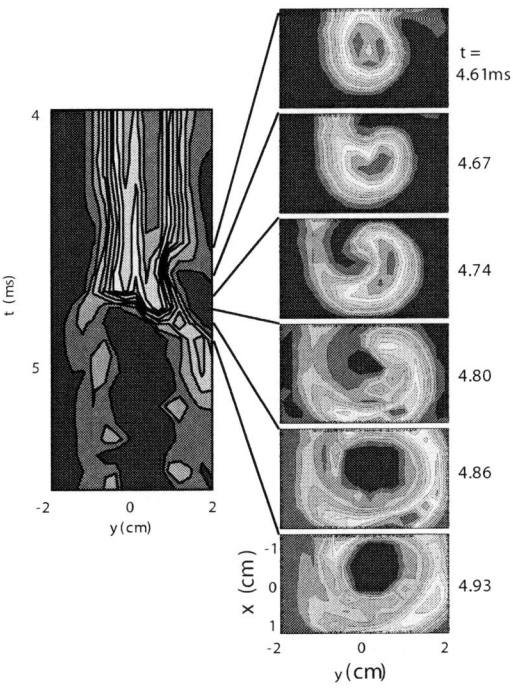

FIGURE 3. Electrostatic plasma disruption. Figures shows ion-density contours on $y-t$ plane and $x-y$ planes. The disruption starts around time $t = 4.7$ ms. the disrupted plasma has a ring shape.

CONCLUSIONS

A decaying magnetized plasma after discharge is investigated in detail. 1) Putting an anode in front of a cathode creates a stable plasma. 2) In order to maintain a stable plasma, weak electron supply is necessary. 3) Weak negative potential of the central plasma confines ions effectively. 4) Stronger negative radial electric field decreases the confinement time. This may be caused by detailed radial potential structure. 5) Stronger magnetic field results in abrupt plasma disruption because the temporally excited spiral rotation brings the central plasma outside.

ACKNOWLEDGMENTS

This work was supported by the Ministry of Education, Science, Sports and Culture in Japan.

Laboratory and Computer Simulations of Non-MHD Flute Instability Structuring the Plasma Clouds During their Artificial Releases at near-Earth Space[*]

Yu. P. Zakharov, H. Nakashima[1], V.M. Antonov, E.L. Boyarintsev, A.V. Melekhov, V.G. Posukh, I.F. Shaikhislamov, D. Mourenas[2], F. Simonet[2]

Institute of Laser Physics (ILP), Russian Academy of Sciences, Novosibirsk, Russia, zakharov@plasma.nsk.ru
[1]*Department of Advanced Energy Eng. Science, Kyushu University(KU), Japan, nakasima@aees.kyushu-u.ac.jp*
[2]*CEA-DAM/Ile-de-France, Departement de Physique Theorique et Appliquee, France*

Abstract. The fast non-MHD development of large-scale flutes at the boundary of plasma clouds expanding into magnetic field is a very common but still unclear phenomenon in a lot of geophysical (AMPTE Barium releases) and laboratory (laser plasma) experiments. To study its specific physics under conditions of the finite value of ion Larmor radius we had compared the data of AMPTE simulation at *KI-1* facility with computer runs by electron Hall/MHD and Hybrid codes with taking into account magnetic field's diffusion relevant to the effect of anomalous electron collision frequency. It was found that the main experimentally observed features of flutes (their increment and non-linear stage) could be correctly described by these models. However, it requires this collision frequency of electrons to be proportional to their gyrofrequency, with values close to estimates derived earlier from *KI-1* experiments through the phenomenology analysis of diamagnetic properties of exploding plasmas.

THE PROBLEM AND SIMULATIVE EXPERIMENTS AT *KI-1* LASER FACILITY

The problem of non-MHD flute instability of exploding *Barium* plasma clouds in space (like AMPTE *Ba*-release [1] of 21.03.85) is unsolved today in spite a lot of related laboratory experiments [2-10] and Hall/MHD [5] or PIC [8,11,12] computer simulations. Its essentially new physics appears due to specific condition of weak magnetization of ions expanding with front velocity V_0 and with directed Larmor radius $R_L=m_icV_0/zeB_0$ comparable with the radius $R_b=(3E_0/B_0^2)^{1/3}$ where such spherical cloud of energy E_0 should be stopped by uniform magnetic field B_0 in ideal case (Raizer, 1963). For the real case of plasma clouds undergoing various micro- and large-scale instabilities the questions would occur about efficiency of plasma deceleration and forms of its propagation across *B*-field when due to finite value of the main criterion $\varepsilon_b=R_L/R_b\sim 1$ of problem [2,11] the radius R_c of diamagnetic cavity of the cloud could be appreciably smaller than its theoretical limit R_b. The development of flutes at the cloud boundary with increment $\gamma >> \gamma_{MHD}$ and still unclear physical scale of its non-linear wavelength λ could be the key process that determines plasma dynamics under such conditions rather well reproducible in Laser-Produced Plasma (LPP) experiments including the influence of background plasma that could stabilize [4,9,10] flute's growth for expansion of clouds at Alfven-Mach number $M_A\sim 1$. The comparison of natural data with both simulative experimental and computational ones allows to understand the basic processes of such non-MHD flute instability which we will name as Large Larmor Radius (LLR) following to [5].

To simulate *Ba*-release by LPP we had used in the first *ILP*-experiments [2-4] a filament plastic target ($||\boldsymbol{B_0}$) irradiated from 2 opposite sides ($\perp \boldsymbol{B_0}$) by single or double CO_2-laser impulses with total output energy Q up to 1kJ. They allowed to study the general diamagnetic and instability's properties of exploding plasmas in a wide range of $0,3 \leq \varepsilon_b \leq 2$ under conditions of large-scale *KI-1* chamber (∅120cm) and high-energy LPP clouds (E_0 up to 350J). Later a new method to produce more quasi-spherically expanding clouds of LPP was applied [7-9] by means of two-sided CO_2-laser irradiation ($Q\sim 50$-100 J in 100ns-impulse) of the spherical pellets ∅4mm from the same Nylon6. Such LPP consisted of H^+ and C^{+4} ions mainly with average $m_i/z \approx 2,8$ a.e.m. that expands with $V_0 \sim 130$

km/s. Recently for the given LPP conditions and in a specially studied range of $1 \leq \varepsilon_b \leq 2$ (typical for the case of AMPTE) we have conducted a new class of simulative experiments [10] with more controlable parameters and advanced diagnostics the main goal of which was to investigate the role of ε_b, M_A criteria in detail and to develop HAWAI hybrid code [12]. For this purpose they were done at their two substantially different values of $\varepsilon_b \sim 0.8$ (at $E_0 \approx 8$ J and $B_0 = 550$ G with $R_b \approx 9$cm) and $\varepsilon_b \sim 2$ (at $E_0 \approx 2$ J and $B_0 = 100$ G with $R_b \approx 18$cm), where all listed parameters of LPP refer namely to its expansion into directions transverse to laser beams (directed nearly from a top and bottom at Fig.1b-d). In these experiments we had used the Image Converter Camera (ICC, gated with 10-30ns duration and view's axis along to B_0), double Langmuir and shielded magnetic (B-dot) probes, ion collectors of their radial flux ($j_i = enV$) and m_i/z-analyzer. To obtain a frame photos of the main energetic C^{+4}-component of LPP we had registered their charge-exchange luminosity in specially added H_2-gas [7] of $P \leq 0.4$ mTorr.

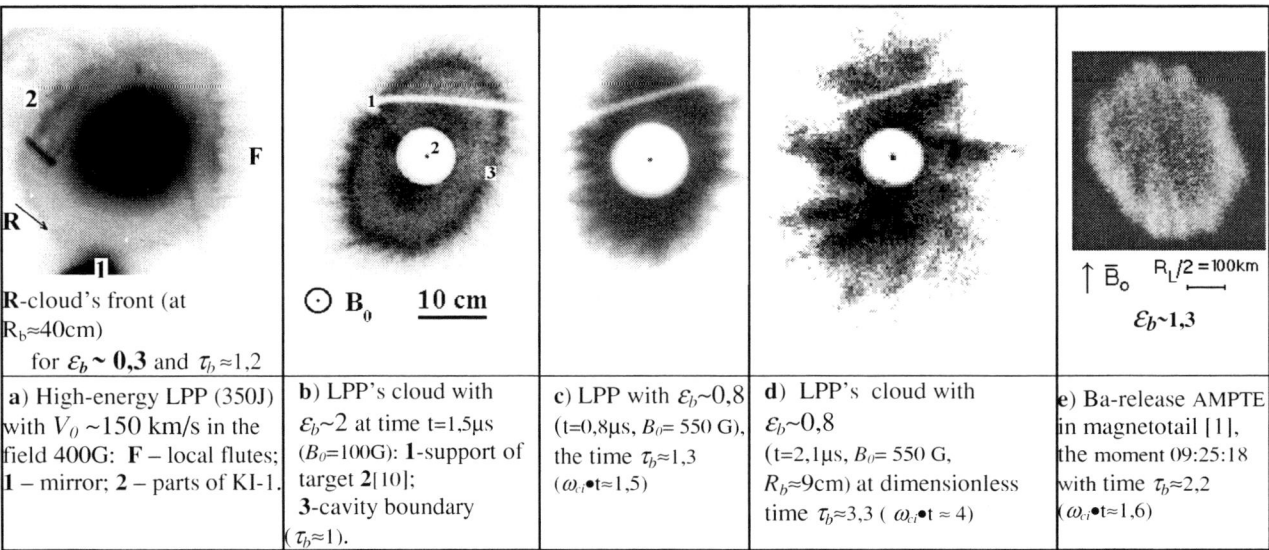

FIGURE 1. Structure of non-MHD flute instability according to data of Laser-Produced Plasma (**a-d**) and AMPTE (**e**) experiments.

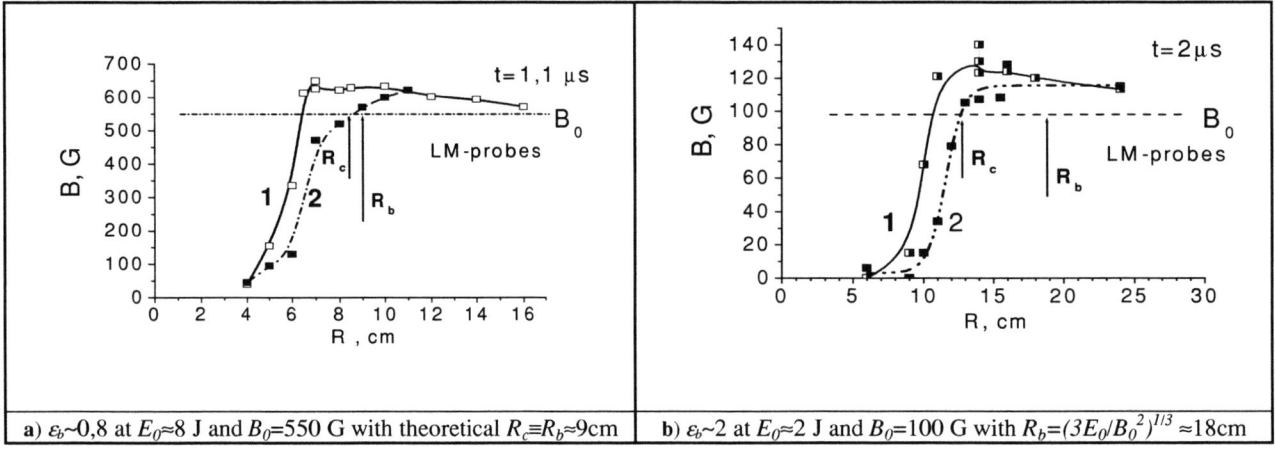

FIGURE 2. The data of **HALL** – experiment [10] on the structure of LPP maximal diamagnetic cavity (of size R_c for curve 2)

THE MAIN EXPERIMENTAL RESULTS IN COMPARISON WITH HALL-MODELS

Under conditions of of AMPTE case ($0.8 \leq \varepsilon_b \leq 1.3$ and $M_A = 0$, in vacuum) we had observed in a previous [3,7,9] and last [10] experiments the structure and dynamics of flutes (and deceleration plasma as a whole) rather close to their geophysical data [1] in dimensionless units $S_b = R/R_b$ and $\tau_b = tV_0/R_b$ of the problem. While laboratory ICC and probe's data allowed to investigate in opposite case ($\varepsilon_b \geq 2$, see Fig. 1b) the "freely" expanding LPP (with max radius $R_f \approx tV_0$ of flute's tips) and in detail for both cases the initial and final stage of LLR-instability as well as the diamagnetizm of exploding plasma. According to them the diamagnetic cavity of LPP could approach to its theoretical limit $R_c=R_b$ only for strong magnetized ions $\varepsilon_b \leq 1,3-1,5$ (see Fig. 2) that is why such plasma only could be decelerated and earlier these relationships were explained [2] by the action of very high effective collision frequency of electrons $v_e \approx \zeta \omega_{ce}$ (with $\zeta \sim 0,3$ as result of the Lower Hybrid Drift Wave – LHDW turbulence). But from the comparison of Fig. 1b and Fig. 2b one can, transverse to laser beams for various regimes of ion magnetization ε_b, in vacuum magnetic fields (**1**) and magnetized background (**2**), conclude that the cavity is bounded by some region where the flutes are developed (which themselves have only local diamagnetic effect $\leq 10\%$) so indeed in the given regime of $\varepsilon_b > 1,5$ the observed enhanced field penetration could be caused by Hall-effect (but not a usual field's diffusion) appearing here in a form of non-MHD LLR-flutes. In a general sense it follows from the field's equation for $b=B/B_0$, $v=V/V_0$: $\partial b /\partial \tau_b = \text{rot}[v \times b] - 0,3\varepsilon_b S_b^3 \text{rot}(\zeta b \bullet \text{rot } b - [\text{rot } b \times b])$, where a second (diffusive) term could achieve the third one (Hall) only when $\zeta \sim 1$ and both of them could be important at scale $R \sim R_b$ only if $\varepsilon_b \geq 1$. Therefore a specific goal of the recent experiment [10] in vacuum (named here as a **HALL**) was to study the correlation between them in detail from the point of view LLR-instability and main channals of LPP-energy losses ΔE in **B**-field since the Hall-term itself acts non-dissipatively.

At Fig. 3a the evolution of flute's size (their radial-r extent Λ_f from cavity's boundary) is presented for both **HALL** regimes for comparison with most suitable LLR-model [5]. We can see that for 100G-case it could describe the onset time $T_I = R_I /V_0$ (of its critical radius $R_I \approx L_n (N_i \bullet r_i / L_n)^{1/2} \approx 7$cm for the r-gradient length $L_n \approx 5$cm and total number of ions $N_i \approx 1,5 \bullet 10^{17}$ with classical radius $r_i = e^2 z_i^2/m_i c^2$) and exponential B-stage with $\gamma_{LLR}=kL_n\gamma_{MHD} \equiv kL_n(g/L_n)^{1/2} \approx 2 \bullet 10^6 s^{-1}$ for the typical flute's wave-length $\lambda \sim 3-5$cm. The larger γ-value at 550G also ~corresponds to LLR-model while according to it in this case the T_I-time could be later ($\geq 1\mu s$). But none of those Hall [5] or extended LHDW [11] linear models could't predict at all such $\lambda \sim L_n$ observed in experiments or PIC-simulations [11] and beside it there are a basic uncertainties relevant to $\lambda(B_0)$ relation. In particular, during **HALL** – experiments with comparable initial L_n and λ (Fig. 1b,c) we had observed an obvious non-linear evolution of λ in strong magnetized case (Fig.1c,d for $\varepsilon_b \approx 0,8$) toward to gross mode structure of AMPTE–type (Fig.1e, $\varepsilon_b \approx 1,3$) while nothing at $\varepsilon_b \sim 2$ (ICC photo at Fig. 1b is similar to itself up to $R_f \geq 30$cm).

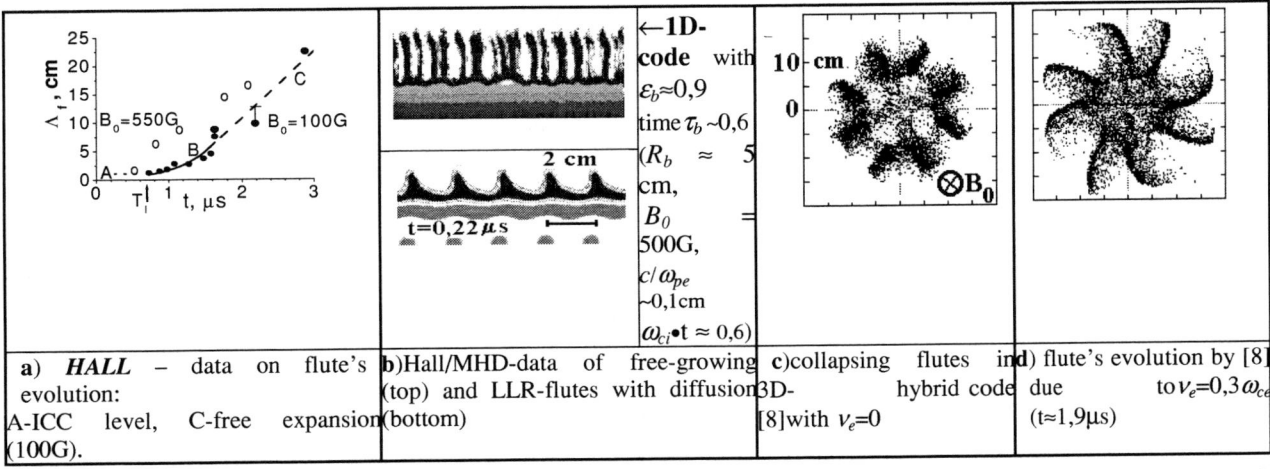

FIGURE 3. Dynamics of LLR-flute instability in simulative LPP-experiments (**a**), 1D/Hall-MHD (**b**) and 3D/PIC (**c,d**) calculations.

To clear this λ-problem of LLR-instability a novel Hall/MHD-model [13] of flutes was developed for 1D-expansion at negligible $M_A=0,1$ with taking into account electron mass $m_e=m_p/1800$ that allowed to study the role of LHDW. Without temperature effect and any dissipation the usual results [11] on the fast growth ($\gamma \leq \omega_{LH}$) of small-scale flutes ($\lambda_e \sim c/\omega_{pe}$) were obtained, the non-linear evolution of which to more large-scale structure at small $\varepsilon_b \leq 0,5$ was revealed. To verify the suggestion that under real plasma conditions the LHDW itself could suppress these λ_e-flutes by its own turbulence namely with specific $v_e = \tilde{\zeta} \omega_{ce}$ and supply the development of experimentally observed gross structures (with $\gamma \sim \gamma_{LLR}$ and $\lambda \geq 10\lambda_e$), we according to [14] add such anomalous diffusive term *ad hoc* into n,V,B - equations of [13] and find out that it could be possible at the near expected value of $\tilde{\zeta} \sim 10^{-3}$. As result a more slower ($\approx \gamma_{LLR} << \omega_{LH}$, depending upon ε_b) development of only a seeded large-scale and 5%-perturbation (of initial density n_0) would be observed (Fig. 3b, bottom). Both small- and large-scale flutes indeed supply an enhanced field's penetration at $\varepsilon_b \geq 1$ so only at at $\varepsilon_b \leq 0,2$ the plasma cavity locally approaches the R_b under conditions of MHD-like value of γ and losses $\Delta E \sim E_0$ could be achieved.

COMPARISON WITH 3D-HYBRID MODEL FOR VACUUM B-FIELD, CONCLUSIONS

Numerical calculations by KU 3D/PIC-code [8] of the Harned type treats ions as individual particles and electrons as a fluid and allow to plasma surface can form very complex shape. The electric field $\vec{E}$ is computed from the momentum equation for an electron fluid with bulk velocity $\vec{V}_e$, temperature T_e (uniform for simplicity) and resistivity η:

$$n_e m_e (d\vec{V}/dt) = -en_e(\vec{E} + \vec{V}_e \times \vec{B}/c - \eta \vec{J}) - \nabla P_e, \text{ where } P_e = n_e T_e \qquad (1)$$

Since electrons are approximated as a massless fluid, so the left-hand side of eq. (1) becomes zero and the fields will be:

$$\vec{E} = -\vec{V}_e \times \vec{B}/c + \eta \vec{J} - (1/en_e)\nabla P_e \quad \text{and} \quad \nabla \times \vec{B} = \frac{4\pi}{c}(\vec{J}_e + \vec{J}_i) \qquad (2a,b)$$

in Darwin approximation. Using $-en_e \vec{V}_e$ for $\vec{J}_e$ in the (2b), solving for $\vec{V}_e$, and substituting it into eq. (2a), we obtain:

$$\vec{E} = (1/4\pi Zen_i)(\nabla \times \vec{B}) \times \vec{B} - (1/Zen_i)\vec{J}_i \times \vec{B}/c + \eta \vec{J} - (T_e/en_i)\nabla n_i \qquad (3)$$

if we assume quasi-neutrality, i.e., $Zn_i = n_e$ in eq. (3) and since eq. (3) includes the terms inversely proportional to the ion's number density n_i, in a vacuum region, these terms must cause numerical infinity. So we solve Laplace equation to obtain the electric field in the vacuum region $\nabla^2 \vec{E} = 0$ while the ion density n_i and the current density $\vec{J}_i$ are calculated by the PIC method from ion's position $\vec{X}_i$ and velocity $\vec{V}_i$, by integrating the equations of its motion:

$$d\vec{V}_i/dt = (Ze/m_i)(\vec{E} + \vec{V}_i \times \vec{B}/c - \eta \vec{J}) \quad \text{and} \quad d\vec{X}_i/dt = \vec{V}_i \qquad (4a,b)$$

Here for total momentum conservation the diffusion term is added to the right-hand of eg. (4a) with $\eta = (m_e/n_e e^2)\nu_e$ hopefully independent from m_e for $\nu_e = 0,3\omega_{ce}$ [2] that allows to study its influence on LLR-flutes. Magnetic field is advanced by Faraday's law $\partial \vec{B}/\partial t = -c\nabla \times \vec{E}$ and Cartesian coordinates with the corresponding full dimensions of calculation grid 61*61*61 and the mesh size $(2cm)^3$. The time levels of each ion position and field quantities are defined at integer time step, and ion velocity and the current density at half time step. Field quantities, number density and current density are spatially defined at the same grid points. The boundary condition adopted here for the field quantities is that the spatial differences of the normal components are set to be zero. These calculations are performed with 100000 particles at SX-4 computer of the Japanese National Institute for Fusion Science (Toki).

The results obtained by this 3D Hybrid model at the regime close to experimental one at $\varepsilon_b \approx 0,8$ (and with 20%-modulation of n_0) clearly show that experimentally observed later stage features of LLR-flutes (their propagation beyond R_b with turn into ion Larmor rotation, Fig. 1b,d) could be described by taking into account the enhanced field's diffusion into them (Fig. 3c,d). On the other hand the problem of their initial evolution into observed large-scale structures also could be related with such kind of anomalous electron collisions ($\nu_e \approx \zeta \omega_{ce}$) but characterized by smaller coefficient ($\tilde{\zeta} \sim 10^{-3}$). Therefore we can conclude as a whole that the diffusion effects of Lower Hybrid turbulence, being a typical feature of LLR-flutes, should play a crucial role in their dynamics (γ) and structure (λ) at all stages of their evolution and consequently in general energetics of plasma cloud's interaction with magnetic field, since a finite part of initial cloud energy E_0 could be converted into transverse (both to V_0 and B_0) ion's motion inside of such flutes [11]. This specifically Hall effect of LLR-flutes (like their turning) could explain the experimentally observed discrepancy between essential radial plasma deceleration at regime $\varepsilon_b \approx 0,8$ and the maximal size $R_c \approx 6-6,5$ cm of its diamagnetic cavity in vacuum (Fig. 2a, curve – 1) rather close but appreciably smaller that its theoretical limit $R_b \approx 9$ cm which could be achieved only at high-density n_* of background (curve–2 at $n_* \approx 3-5 \cdot 10^{13}$ 1/cc of H^+) under conditions of flute's suppression [4,10] at $M_A \sim 1$.

ACKNOWLEDGMENT

This work was supported in part by Grant #02-02-17868 of Russian Fund of Basic Research and by Contract of ILP with CEA.

REFERENCES

1. Bernhardt, P.A., Roussel-Dupre, R.A., Pongratz, M.B., et al., *J. Geoph. Res.*, 92A, 5777-5794 (1987).
2. Zakharov, Yu.P., Orishich, A.M., Ponomarenko, A.G., Posukh, V.G., *Sov. J. Plasma Phys.*, 12, 674-678 (1986).
3. Zakharov, Yu.P., Orishich, A.M., Ponomarenko, A.G., Posukh, V.G., Snytnikov, V.N., in Plasma Astrophysics, ESA SP-251, 37-40 (1986).
4. Antonov, V.M., Zakharov, Yu.P., Orishich, A.M., Ponomarenko, A.G., Posukh, V.G., et al., in Plasma Astrophysics, ESA SP-311, 189-190 (1990).
5. Ripin, B.H., Huba, J.D., McLean, E.A., et al., *Phys. Rev. Let.,* 59, 2299-2302 (1987) and *Phys. Fluids*, B5, 3491-3505 (1993).
6. Dimonte, G., and Wiley, L.G., *Phys. Rev. Let.,* 67, 1755-1758 (1991).
7. Zakharov, Yu.P., Orishich, A.M., Snytnikov, V.N., Shaikhislamov, I.F., *J. Appl. Mech. & Tech. Phys.*, 35, 481-486 (1994).
8. Muranaka, T., Usui, H., Nakashima, H., Zakharov, Yu.P., Ponomarenko, A.G., "Instability Analysis of Exploding Plasma in Magnetic Fields Using a 3D Hybrid Code" in Inertial Fusion Sciences and Application – State of Art 1999, edited by C. Labaune et al., Elsevier, Paris, 2000, pp. 1194-1197.
9. Zakharov, Yu.P., *Adv. in Space Res.,* 29, 1335-1345 (2002).
10. Zakharov, Yu.P., Ponomarenko, A.G., Antonov, V.M., Boyarintsev, E.L., Melekhov, A.V., Posukh, V.G., Shaikhislamov, I.F., Mourenas, D., Simonet, F., in 4th High Energy Density Laboratory Astrophysics (Ann Arbor, 2002), Poster N24, 2p.
11. Winske D., *J. Geophys. Res.*, 93A, 2539-2552 (1988) and *Phys. Fluids*, B1, 1900-1910 (1989).
12. Simonet, F., Lembege, B., *APS Bull.*, 39, 1735 (1990) and Proc. 4th Int. Sch. for Space Simulat. (Kyoto, 1991), pp. 226-229.
13. Shaikhislamov, I.F., in 29th EPS Conf. on Plasma Physics and Contr. Fusion (Montreux, 2002), Poster P-1.007, 4p.
14. Hassam, A.B., and Huba, J.D., *Phys. Fluids*, B2, 2001-2006 (1990).

PLASMA THEORY AND SIMULATION

1. Kinetic Theory and Transport Modeling
2. Statistical Mechanics, Atomic, and Molecular Processes
3. Macroscopic Equilibrium and Stability
4. Waves and Nonlinear Dynamics
5. Flowing and Rapidly Changing Plasmas
6. Particle Acceleration and Transport

Non-Local Kinetic Fluctuations in Plasma

V.V. Belyi

IZMIRAN, Troitsk, Moscow region, 142092, Russia

Abstract. Using the Langevin approach and the multiscale technique, a kinetic theory of the time and space non-local fluctuations in the collisional plasma is constructed. In local equilibrium a generalized version of the Callen-Welton theorem is derived. It is shown that not only the dissipation but also the time and space derivatives of the dispersion determine the amplitude and the width of the spectrum lines of the electrostatic field fluctuations, as well as the form factor. There appear significant differences with respect to the non-uniform plasma. In the kinetic regime the form factor is more sensible to space gradient than the spectral function of the electrostatic field fluctuations. As a result of the inhomogeneity, these proprieties became asymmetric with respect to the inversion of the frequency sign. The differences in amplitude of peaks could become a new tool to diagnose slow time and space variations in the plasma.

Fluctuations attract a great deal of attention. Besides being of interest from the fundamental point of view, there are situations where non-equilibrium fluctuations play an important role, namely in the neighborhood of bifurcations where the system has to choose a branch [1], [2]. Moreover, fluctuations find an application in diagnostic procedures. Indeed, plasma parameters such as temperature, mean velocity, density and their respective profiles can be determined by incoherent (Thomson) scattering diagnostics [3], i.e. by the proper interpretation of data obtained from the scattering of a given electromagnetic field interacting with the system. The key point of interpreting them is the knowledge of the intensity of the dielectric function fluctuations or equally of the electron form factor $(\delta n_e \delta n_e)_{\omega, \mathbf{k}}$. Here ω and $\mathbf{k}$ are, respectively, the frequency and wavevector of the autocorrelations. Due to the Poisson equation the electron form factor in the spatially homogeneous system is directly linked to the electrostatic field fluctuations, which have been the object of active study since the early 1960 [4]. In the thermodynamic equilibrium, the electrostatic field fluctuations satisfy the famous **Callen-Welton** fluctuation-dissipation theorem [6]

$$(\delta \mathbf{E} \delta \mathbf{E})_{\omega \mathbf{k}} = \Theta \frac{8\pi Im \varepsilon(\omega, \mathbf{k})}{\omega |\varepsilon(\omega, \mathbf{k})|^2} \qquad (1)$$

linking their intensity to the imaginary part of the dielectric function $\varepsilon(\omega, \mathbf{k})$, and the temperature Θ in energy units. The spectral function (1) has peaks, corresponding to proper plasma frequencies.

The matter becomes more tricky in the non-equilibrium case. When the state of the plasma is given by Maxwellian distributions characterized by *different constant* temperatures and velocities per species $(\Theta_a, \mathbf{V}_a; a = e, i)$, it is generally admitted that the spectral function of the electron density for a two-component system takes the form [4]:

$$(\delta n_e \delta n_e)_{\omega \mathbf{k}} = \frac{|1 + \chi_i|^2 \, 2n_e k^2 Im\chi_e}{(\omega - \mathbf{k} \cdot \mathbf{V}_e) k_D^2 |\varepsilon(\omega, \mathbf{k})|^2} + \frac{\Theta_i |\chi_e|^2 \, 2n_e k^2 Im\chi_i}{(\omega - \mathbf{k} \cdot \mathbf{V}_i) \Theta_e k_D^2 |\varepsilon(\omega, \mathbf{k})|^2}. \qquad (2)$$

In Eq. (2) k_D is the Debye number and χ_a ($a = e, i$) is the complex dielectric susceptibility of the ath component. This formula has been extensively used to interpret the scattering data mentioned above.

We have indeed shown [5] that in the *collisional regime* Eqs. (1) and (2) should be revisited. We stressed the fact that a *kinetic approach* should be taken. Introducing fluctuations by the Langevin method, we have elaborated a "revisited" Callen-Welton formula containing, beside the terms appearing in (1), new terms explicitly displaying dissipative non-equilibrium contributions:

$$(\delta \mathbf{E} \delta \mathbf{E})_{\omega \mathbf{k}} = \sum_a \frac{8\pi \Theta_a}{(\omega - \mathbf{k} \cdot \mathbf{V}_a) |\varepsilon(\omega, \mathbf{k})|^2}$$

$$\times [\chi_a'' + \nu_{ab}(\Theta_a - \Theta_b)\Phi_1 + \nu_{ab}(\mathbf{k} \cdot \mathbf{V}_a - \mathbf{k} \cdot \mathbf{V}_a)\Phi_2] \qquad (3)$$

It is important that these new terms contain the interparticle collision frequency, the differences in temperatures and velocities, and the functions Φ_1 and Φ_2 of **real** parts of the dielectric susceptibilities.

It is, however, not evident whether the plasma parameters – temperature, velocities and densities can be kept *constant*. Inhomogeneities in space and time of these quantities will certainly also contribute to the fluctuations.

Obviously, to treat the problem, a kinetic approach is *required*, especially when the wavelength of the fluctuations is larger than the Debye wavelength. Using the same formalism as in Reference [5], we have derived a general expression for the electrostatic field fluctuations to the first order in a parameter μ which is taken to be of the order v_{ei}/ω, and measures the strength of the gradients in space and time.

The details of the calculations will be presented elsewhere. The result is that, for local equilibrium with equal temperatures and velocities, the Callen-Welton formula (1) takes the form

$$(\delta \mathbf{E} \delta \mathbf{E})_{\omega \mathbf{k}} = \frac{8\pi\Theta}{\omega} \frac{Im\varepsilon + (\partial^2 Re\varepsilon/\partial\omega\partial\mu t) - (\partial^2 Re\varepsilon/\partial\mu \mathbf{r}\partial \mathbf{k})}{Re\varepsilon^2 + (Im\varepsilon + (\partial^2 Re\varepsilon/\partial\omega\partial\mu t) - (\partial^2 Re\varepsilon/\partial\mu \mathbf{r}\partial \mathbf{k}))^2}. \qquad (4)$$

For the spatially homogeneous system the spectral properties of the electrostatic field fluctuations are the same as for the form factor (the spectral function of the density particles fluctuations): they are linked by the Poisson equation. But for the spatial inhomogeneous plasma this is not evident: the Poisson equation is a non-local relation. In reality, the non-local correction to the spectral function of the electrostatic field fluctuations is different from the correction to the form factor. For example, for the plasma mode we get

$$(\delta n_e \delta n_e)_{\omega,\mathbf{k}}$$
$$= \frac{2n_e \mathbf{k}^2}{\omega k_D^2} \frac{[Im\varepsilon + (\partial^2/\partial\omega\partial\mu t)Re\varepsilon - (1/\mathbf{k}^2)(\partial/\partial\mu r_i)k_j(\partial/\partial k_i)k_j Re\varepsilon]}{Re\varepsilon^2 + (Im\varepsilon + (\partial^2/\partial\omega\partial\mu t)Re\varepsilon - (1/\mathbf{k}^2)(\partial/\partial\mu r_i)k_j(\partial/\partial k_i)k_j Re\varepsilon)^2}. \qquad (5)$$

The inhomogeneous correction in Eq. (5) is greater than in Eq. (4) by a factor $1 + k_D^2/6k^2$.

The structure of the formulae (4) and (5) is very interesting, because it contains both dissipative and dispersive contributions to the fluctuations. These can be of the same order of magnitude. Moreover, the non-homogeneous terms break the symmetry with respect to ω. This leads to an asymmetric spectrum of the fluctuations: the position of the resonances remain the same with respect to the homogeneous case, but their intensity and broadening are different.

We have shown that the amplitude and the width of the spectral lines of the electrostatic field fluctuations and form factor are affected by new non-local dispersive terms. They are not related to Joule dissipation and appear because of an additional phase shift between the vectors of induction and electric field. This phase shift results from the finite time needed to set the polarization in the plasma with dispersion. Such a phase shift in the plasma with space dispersion appears due to the medium inhomogeneity. These results are important for the understanding and the classification of the various phenomena that may be observed in applications; in particular, the asymmetry of lines can be used as a diagnostic tool to measure local gradients in the plasma.

ACKNOWLEDGEMENTS

We acknowledge support from Russian Foundation for Basic Research (grant 00-02-17139).

REFERENCES

1. Glansdorff, P., and Prigogine, I., *Thermodynamic Theory of Structure, Stability and Fluctuations*, Wiley, New York, 1971.
2. Nicolis, G., and Prigogine, I., *Nonequilibrium Systems. From Dissipative Structure to Order Through Fluctuations*, Wiley, New York, 1979.
3. Dougherty, J.P., and Farley, D.T., *Proc. Roy. Soc.* **A259**, 79 (1960).
4. Akhiezer, A., Akhiezer, I., Polovin, R., Sitenko, A., Stepanov, K., *Plasma Electrodynamics, Vol.1, Linear Theory*, Oxford, Pergamon, 1975.
5. Belyi, V.V., and Paiva-Veretennicoff, I., *J. of Plasma Physics*, **43**, 1 (1990).
6. Callen, H.B., and Welton, T.A., *Phys. Rev.* **83**, 34 (1951).

A Space- and Time- Nonlocal Kinetic Equation for a Polarizable Plasma

V.V. Belyi*, Yu.A. Kukharenko† and J. Wallenborn**

*IZMIRAN, Troitsk, Moscow region, 142092, Russia
†IPE RAN, Moscow, Russia
**Physique statistique et Plasmas, CP 231, Université Libre de Bruxelles, 1050 Bruxelles, Belgium

Abstract. A nonmarkovian and nonlocal kinetic equation, which is a generalization of the (nonlinear) Balescu-Lenard equation, is derived for a weakly nonuniform multicomponent polarizable plasma. Balance equations for the momentum and energy densities are calculated in the first order in nonlocality, and potential contributions to the fluxes due to polarization are obtained.

INTRODUCTION

The well known Balescu-Lenard (BL) equation [1][2] - as well as Boltzmann and Landau equations - describes the collisions as instantaneous. That means that the potential energy doesn't play any role on the evolution of the system nor on the resulting stationary state ; the system thus behaves like a perfect gas at equilibrium. On the other hand, in the case of finite collision duration, kinetic and potential energy are exchanged during the collision. Only the total energy is conserved and therefore the solution of the kinetic equation takes into account potential as well as kinetic energy. In other words, the equilibrium properties of the system are those of a nonideal one. The way to account for finite duration of the collisions is to consider non-Markovian (or memory) effects in the kinetic equation.

In order to describe non-Markovian processes in a polarizable plasma Klimontovich [3] derived a set of equations for the particle distribution function and the electric field spectral function. But the equation for the spectral function was not solved, and the collision integral was obtained only in the averaged potential approximation.

More complete results have been obtained by the authors starting from the BBGKY hierarchy [4][5]: a non-Markovian expression for the pair correlation function and the non-Markovian collision integral for a spatially uniform polarizable plasma have been derived.

The non-Markovian results were generalized by Klimontovich [6] to the spatially inhomogeneous case for gases and plasmas. But in the plasma case the static Debye-Hückel potential was used. In the present work [7] the full generalization of BL equation the to spatially inhomogeneous, non-Markovian case is treated. We give the expressions for the non-local pair correlation function and for the nonlinear non-local collision integral in the case of the spatially nonuniform weakly nonideal multicomponent polarizable classical plasma. Explicit expressions for the pressure tensor and heat flux for the same system are also obtained from the kinetic equation. Let us mention that the other attempts to generalize BL equation for nonuniform plasmas [8] stay on a more formal ground and are not easily tractable in practical cases : they don't give explicit expressions ready for calculations.

We consider that the one-particle distribution function varies slowly, both in time and in space, in comparison with the characteristic scales of the pair correlation function, at a stage where all initial correlations have been damped. Our aim is indeed to study the hydrodynamic behaviour of the system and the non-Markovian effects are taken into account for completeness and in view to retain all contributions of same order of magnitude.

PAIR CORRELATION FUNCTION AND KINETIC EQUATION

Out of equilibrium, the evolution of any system can be exactly described by the BBGKY-hierarchy for the reduced distribution functions (df) or, equivalently, for the correlation functions. In the so-called plasma approximation [9] the triple correlation function as well as the direct interaction between correlated particles are neglected. In this way, the hierarchy reduces to a closed set of two equations for $f_a(\mathbf{p}_1,\mathbf{r}_1,t)$, the one-particle df for species a and $g_{ab}(\mathbf{p}_1,\mathbf{p}_2,\mathbf{r}_1,\mathbf{r}_2,t)$, the pair correlation function ($g_{ab} = f_{ab} - f_a f_b$).

The time dependence of the evolution operator has already been treated in refs.[4][5], where we showed how to separate the evolution on the scale of t_c, the correlation time, from the evolution on the scale of t_R, the relaxation time of the one-particle df.

The spatial nonuniformity introduces a new time scale t_H, the "hydrodynamic" time, which is the time necessary for a particle to travel through the characteristic length of an inhomogeneity l_H.

In the present work we consider that the correlation length is less than the mean free path, so that initial correlations are damped at the initial stage. We thus don't consider the formation of correlations at initial stage but we take into account the finite duration of the collisions by retaining the non-Markovian contributions to the kinetic equation.

Generally, in the bulk of a weakly coupled plasma $t_R, t_H \gg t_c$ and sometimes one considers $t_R \approx t_H$, even though local equilibrium needs $t_R < t_H$. In what follows, we shall assume that hydrodynamic and relaxation scales are well separated from the correlation time and, in order to find the balance equation of the conserved densities, in section 5 we shall consider $t_R \approx t_H$. The separation between t_H and t_c scales proceeds in the same way as the separation between t_R and t_c [7].

We have formally solved the equation for the pair correlation function by the method of Green's functions [10].

By the resolvent method we obtained the next expression for pair correlation function:

$$g_{ab}(\mathbf{p}_1,\mathbf{p}_2,\mathbf{k},\mathbf{R},t) = Exp\{i\frac{\partial}{\partial z}\frac{\partial}{\partial t} + i(1/2\frac{\partial}{\partial \widehat{\mathbf{k}}} - \frac{\partial}{\partial \mathbf{K}})\cdot\frac{\partial}{\partial \mathbf{R}} - i(\frac{\partial}{\partial \mathbf{K}} - 1/2\frac{\partial}{\partial \widehat{\mathbf{k}}})\cdot\frac{\partial}{\partial \widehat{\mathbf{R}}}\}$$

$$G_{ab}(z,\mathbf{k},\mathbf{K},\widehat{\mathbf{k}},\mathbf{p}_1,\mathbf{p}_2,\mathbf{R},\widehat{\mathbf{R}},t)\big|_{\widehat{\mathbf{R}}=\mathbf{R},\widehat{\mathbf{k}}=\mathbf{k},\mathbf{K}=\mathbf{0},z=+i0} \quad (1)$$

where

$$G_{ab}(\omega,\mathbf{K},\mathbf{k},\widehat{\mathbf{k}};\mathbf{p}_1,\mathbf{p}_2;\mathbf{R},\widehat{\mathbf{R}},t) = -\frac{1}{\omega - \mathbf{k}_2\cdot\mathbf{v}_2 - \mathbf{k}_1\cdot\mathbf{v}_1 + i0}\sum_c\int d\mathbf{p}_3\{\delta_{bc}\delta(\mathbf{p}_2-\mathbf{p}_3)$$

$$-[\frac{1}{\omega - \mathbf{k}_2\cdot\mathbf{v}_3 - \mathbf{k}_1\cdot\mathbf{v}_1 + i0} + \frac{1}{\mathbf{k}_2\cdot\mathbf{v}_3 - \mathbf{k}_2\cdot\mathbf{v}_2 + i0}]\frac{\Phi_{cc}(\mathbf{k}_2)}{\varepsilon(\mathbf{k}_2\cdot\mathbf{v}_3,\mathbf{k}_2,\widehat{\mathbf{R}})}\mathbf{k}_2\cdot\frac{\partial}{\partial \mathbf{p}_2}f_b(\mathbf{p}_2,\widehat{\mathbf{R}})\}$$

$$\times\{[f_c(\mathbf{p}_3,\widehat{\mathbf{R}}) - \frac{Q^*(\mathbf{k}_2\cdot\mathbf{v}_3,\mathbf{k}_2,\widehat{\mathbf{R}})}{\varepsilon^*(\mathbf{k}_2\cdot\mathbf{v}_3,\mathbf{k}_2,\widehat{\mathbf{R}})}\mathbf{k}_2\cdot\frac{\partial}{\partial \mathbf{p}_3}f_c(\mathbf{p}_3,\widehat{\mathbf{R}})][\Phi_{ab}(\widehat{\mathbf{k}})\widehat{\mathbf{k}}\cdot\frac{\partial}{\partial \mathbf{p}_1}f_a(\mathbf{p}_1,\mathbf{R})$$

$$-\frac{P(\omega - \mathbf{k}_2\cdot\mathbf{v}_3,\mathbf{k}_1,\widehat{\mathbf{k}},\mathbf{R})}{\varepsilon(\omega - \mathbf{k}_2\cdot\mathbf{v}_3,\mathbf{k}_1,\mathbf{R})}\Phi_{ab}(\mathbf{k}_1)\mathbf{k}_1\cdot\frac{\partial}{\partial \mathbf{p}_1}f_a(\mathbf{p}_1,\mathbf{R})]$$

$$-[f_a(\mathbf{p}_1,\mathbf{R}) - \frac{Q(\omega - \mathbf{k}_2\cdot\mathbf{v}_3,\mathbf{k}_1,\mathbf{R})}{\varepsilon(\omega - \mathbf{k}_2\cdot\mathbf{v}_3,\mathbf{k}_1,\mathbf{R})}\mathbf{k}_1\cdot\frac{\partial}{\partial \mathbf{p}_1}f_a(\mathbf{p}_1,\mathbf{R})][\Phi_{ab}(\widehat{\mathbf{k}})\widehat{\mathbf{k}}\cdot\frac{\partial}{\partial \mathbf{p}_3}f_c(\mathbf{p}_3,\widehat{\mathbf{R}})$$

$$-\frac{P^*(\mathbf{k}_2\cdot\mathbf{v}_3,\mathbf{k}_2,\widehat{\mathbf{k}},\widehat{\mathbf{R}})}{\varepsilon^*(\mathbf{k}_2\cdot\mathbf{v}_3,\mathbf{k}_2,\widehat{\mathbf{R}})}\Phi_{ab}(\mathbf{k}_2)\mathbf{k}_2\cdot\frac{\partial}{\partial \mathbf{p}_3}f_c(\mathbf{p}_3,\widehat{\mathbf{R}})]\}. \quad (2)$$

Here we used the following definitions:

$$P(\omega,\mathbf{k}_1,\widehat{\mathbf{k}},\mathbf{R}) = \sum_a \Phi_{aa}(\widehat{\mathbf{k}})\int d\mathbf{p}_1 \frac{1}{\omega - \mathbf{k}_1\cdot\mathbf{v}_1 + i0}\widehat{\mathbf{k}}\cdot\frac{\partial}{\partial \mathbf{p}_1}f_a(\mathbf{p}_1,\mathbf{R},t), \quad (3)$$

$$Q(\mathbf{k}_1, \mathbf{R}, \omega) = \sum_a \Phi_{aa}(\mathbf{k}_1) \int d\mathbf{p}_1 \frac{1}{\omega - \mathbf{k}_1 \cdot \mathbf{v}_1 + i0} f_a(\mathbf{R}, \mathbf{p}_1, t), \quad (4)$$

Using the expression (1) we obtain the collision integral:

$$J_a(\mathbf{p}_1, \mathbf{r}_1, t) = -\frac{i}{8\pi^3} \sum_b \int d\mathbf{p}_2 d\mathbf{k} \Phi_{ab}(\mathbf{k}) \mathbf{k} \cdot \frac{\partial}{\partial \mathbf{p}_1} Exp\{-\frac{i}{2}(\frac{\partial}{\partial \mathbf{k}} + \frac{\partial}{\partial \widehat{\mathbf{k}}}) \cdot (\frac{\partial}{\partial \mathbf{R}} + \frac{\partial}{\partial \widehat{\mathbf{R}}})$$

$$+i\frac{\partial}{\partial z}\frac{\partial}{\partial t} + (i/2\frac{\partial}{\partial \widehat{\mathbf{k}}} - i\frac{\partial}{\partial \mathbf{K}}) \cdot \frac{\partial}{\partial \mathbf{R}} - (i\frac{\partial}{\partial \mathbf{K}} + i/2\frac{\partial}{\partial \widehat{\mathbf{k}}}) \cdot \frac{\partial}{\partial \widehat{\mathbf{R}}}\}$$

$$G_{ab}(z, \mathbf{K}, \mathbf{k}, \widehat{\mathbf{k}}, \mathbf{p}_1, \mathbf{p}_2, \mathbf{R}, \widehat{\mathbf{R}}, t)\big|_{\widehat{\mathbf{R}} = \mathbf{R} = \mathbf{r}_1, \widehat{\mathbf{k}} = \mathbf{k}, \mathbf{K} = 0, z = +i0} \quad (5)$$

Eq. (5) is a generalization of the Balescu-Lenard collision integral to the spatially non-local and non-Markovian case. In the homogeneous case it reduces to the non-Markovian collision integral [4] [5] found earlier by the authors.

BALANCE EQUATIONS

As an application, we have considered the balance equations for the moment density:

$$\rho(\mathbf{r}_1, t)\mathbf{u}(\mathbf{r}_1, t) = \sum_a \int d\mathbf{p}_1 \mathbf{p}_1 f_a(\mathbf{p}_1, \mathbf{r}_1, t), \quad (6)$$

and for the internal kinetic energy density:

$$E^k(\mathbf{r}_1, t) = \sum_a \int d\mathbf{p}_1 \frac{(\mathbf{p}_1 - m_a \mathbf{u})^2}{2m_a} f_a(\mathbf{p}_1, \mathbf{r}_1, t), \quad (7)$$

where $\mathbf{u}(\mathbf{r}_1, t)$ is the average velocity and $\rho(\mathbf{r}_1, t)$ is the mass density.

We have shown that the momentum balance equation can then be written as [7] :

$$\partial_t \rho u_i + \frac{\partial}{\partial r_1^j}(\rho u_i u_j + P_{ij}^{id} + P_{ij}^{cor}) = 0, \quad (8)$$

with

$$P_{ij}^{id} = \sum_a m_a \int d\mathbf{p}_1 (v_1^i - u^i)(v_2^j - u^j) f_a(\mathbf{p}_1, \mathbf{r}_1, t) \quad (9)$$

the kinetic pressure tensor and

$$P_{ij}^{cor} = \frac{1}{16\pi^3} \sum_{ab} \int \Phi_{ab}^2(\mathbf{k}) d\mathbf{k} d\mathbf{p}_1 d\mathbf{p}_2 (\delta_{ij} - 2\frac{k_i k_j}{k^2}) \frac{1}{\mathbf{k} \cdot \mathbf{v}_1 - \mathbf{k} \cdot \mathbf{v}_2} \frac{1}{|\varepsilon(\mathbf{k}, \mathbf{k} \cdot \mathbf{v}_1, \mathbf{r}_1, t)|^2}$$

$$\times [f_b(\mathbf{p}_2, \mathbf{r}_1, t) \mathbf{k} \cdot \frac{\partial}{\partial \mathbf{p}_1} f_a(\mathbf{p}_1, \mathbf{r}_1, t) - f_a(\mathbf{p}_1, \mathbf{r}_1, t) \mathbf{k} \cdot \frac{\partial}{\partial \mathbf{p}_2} f_b(\mathbf{p}_2, \mathbf{r}_1, t)] \quad (10)$$

the potential part of the pressure tensor.

In the same way, the energy balance equation is :

$$\partial_t (\frac{\rho u^2}{2} + E^k + E^p) + \frac{\partial}{\partial r_1^i}[u_i(\frac{\rho u^2}{2} + E^k + E^p) + (P_{ij}^{id} + P_{ij}^{cor})u_j + q_i] = 0, \quad (11)$$

where the heat flux $\mathbf{q}$ is given by

$$\mathbf{q} = \mathbf{q}^{id} + \mathbf{q}^{cor}$$

$$\mathbf{q}^{id} = \sum_a \int d\mathbf{p}_1 (\mathbf{v}_1 - \mathbf{u}) \frac{(\mathbf{p}_1 - m_a \mathbf{u})^2}{2m_a} f_a(\mathbf{p}_1, \mathbf{r}_1, t) \quad (12)$$

$$\mathbf{q}^{cor} = -\frac{1}{8\pi^3}\sum_a \int \frac{\mathbf{k}}{k^2} d\mathbf{p}_1 d\mathbf{k} \Phi_{aa}(\mathbf{k})\mathbf{k}\cdot(\mathbf{v}_1-\mathbf{u})f_a(\mathbf{p}_1)\left[1-\frac{1}{|\varepsilon(\mathbf{k}\cdot\mathbf{v}_1,\mathbf{k})|^2}\right] \quad (13)$$

It is easy to verify that, at equilibrium, P_{ij}^{cor} gives the correct value of the Debye-Hückel pressure while the heat flux is null.

This last result shows the consistency of our theory : out of equilibrium the kinetic equation takes into account the dynamical polarizability of the plasma, while at equilibrium the results of the Debye-Hückel theory are recovered.

CONCLUSION

Starting from the BBGKY hierarchy in the plasma approximation, we have shown that, in the resolvent for the pair correlation function, it is possible to separate time and space scales of the one-particle df from the ones of the pair correlation function. We have then derived a kinetic equation for a spatially nonuniform multicomponent polarizable plasma. We have given the expressions of the nonequilibrium pair correlation function and of the collision integral which are formally valid to any order in nonlocality a long as the system can be considered as locally neutral. We have calculated the balance equations for the conserved densities at first order in the nonlocality and have obtained the explicit expressions of the potential parts of the fluxes. The existence of these potential parts indicates that kinetic and potential energies are not conserved separately. We have shown that at equilibrium the heat flux vanishes while the pressure tensor gives rise to the Debye-Hückel pressure.

Acknowledgements. We acknowledge support from NATO (grant PST.CLG 975001). Two of us (VB and YuK) acknowledge support from Russian Foundation for Basic Research (grant 00-02-17139).

REFERENCES

1. Balescu, R., *Phys. of Fluids*, **3**, 52 (1960).
2. Lenard, A., *Ann. Phys. (N.Y.)*, **10**, 390 (1960).
3. Klimontovich, Yu.L., *Soviet Phys. JETP*, **60**, 1352 (1971).
4. Belyi, V.V., Kukharenko, Yu.A., Wallenborn, J., *Phys. Rev. Lett.*, **76**, 3554 (1996).
5. Belyi, V.V., Kukharenko, Yu.A., Wallenborn, J., *J. Plasma Phys.*, **59**, 657 (1998).
6. Klimontovich, Yu.L., *Kinetic theory of nonideal gases and nonideal plasmas*, Academic Press, New York, 1975.
7. Belyi, V.V., Kukharenko, Yu.A., Wallenborn, J., *Contrib. Plasma Phys.* **42**, 3 (2002).
8. Dorfman, J.R., Cohen, E.G.D., Int. J. *Quantum Chem.* **16**, 63 (1982)
9. Bogoliubov, N.N., in *Studies in statistical mechanics*, edited by J. de Boer and G. E. Uhlenbeck, North-Holland, Amsterdam, 1962.
10. Balescu, R., *Statistical mechanics of charged particles*, Wiley, New York, 1963.

Electroweak Interactions in Dense Plasmas

L. O. Silva*, R. Bingham† and W. B. Mori**

*Centro de Electrodinamica, Instituto Superior Technico, Lisboa, 1096 Cadex, Portugal
†Rutherford Appleton Laboratory, Space Science and Technology Dept., Chilton, Didcot, U.K.
**Department of Physics and Astronomy, University of California Los Angeles, Los Angeles, California 90024-1547

Abstract. Employing the relativistic kinetic equations for neutrinos interacting with dense plasmas via the weak interaction force we explore collective plasma instabilities driven by neutrinos. We examine the anomalous energy transfer between the neutrinos and the background plasma via excitation of electron plasma waves. We present the relativistic equations including the inclusion of external magnetic fields. Solutions of the dispersion equation describing the coupling between a neutrino beam and a plasma wave demonstrates the existence of two regimes a)the kinetic regime and b) the hydrodynamical regime. We demonstrate that the hydrodynamical regime has a growth rate many orders of magnitude larger than the kinetic regime.

INTRODUCTION

There is considerable interest in the propagation dynamics of neutrinos in a background dispersive medium, particularly in the search for a mechanism to explain the dynamics of type II supernovae [1] and solving the solar neutrino problem [2]. Neutrino interactions with matter are usually considered as non self-consistent single particle processes. All the single particle mechanisms describing the dynamical properties of neutrinos in matter have a straight forward analogy with the processes involving single electron interactions with a medium such as Compton scattering, and Cerenkov radiation etc. However, it is well known that the self consistent description of a stream of charged particles moving through a medium such as a plasma give rise to a new class of processes known as collective interactions examples being Landau damping and streaming instabilities which result in either the absorption or generation of plasma waves. Photon beams can also interact strongly with a medium through similar collective interactions commonly known as parametric instabilities [3]. It is well known that collective interactions can lead to stronger coupling between the medium and the photon or charged particle beams. For example the strength of coupling describing collective electromagnetic interactions is of $0(\alpha)$ while (collisional) single particle coupling is of $0(\alpha^2)$, where α is the fine structure constant. Another important point is that collective interactions require the medium be dispersive.

For supernovae the fundamental problem is to understand how gravitational energy released during the collapse or infall of the core is transferred to ejecting the outer mantle. During collapse the central core gets over compressed and releases energy by bouncing back forming a shock wave which rapidly moves outwards, this hydrodynamic shock wave is thought to be responsible for driving the explosion. However, in numerical simulations the shock wave tended to stall because of energy dissipation associated with dissociation of the iron shell it had to move through. Bethe and Wilson [1] have suggested the shock wave can be revived by energy deposition of the neutrino flow behind the shock. Estimates suggest that about 1% of the neutrino energy is required to be deposited behind the shock to revitalise it. Previous estimates of neutrino heating such as direct neutrino scattering [4] and the pair annihilation process $\nu\tilde{\nu} \to e^+e^-$ [5] based on single particle interactions are not sufficient to provide the right amount of heating. An alternative mechanism is the collective neutrino beam instability [6] driving Langmuir or electron plasma waves which heat the material behind the shock, this collective instability is analogous to photon beam or electron beam streaming instabilities in plasmas, and like them has two regimes of interest namely kinetic or hydrodynamic which have quite different dependencies on the coupling parameter.

It was first demonstrated by Bingham et al., [6] that the many body interaction between a neutrino beam and electrons in a plasma results in a ponderomotive force allowing the possibility of collective interactions. The collective neutrino beam plasma interaction resulted in the generation of plasma waves if the flux of neutrinos is sufficiently large [6,7]. The neutrino flux produced during a supernovae explosion was found to be sufficient to induce the generation of plasma waves [7].

Recently Silva et al., [7] developed a self-consistent relativistic kinetic description for neutrinos and electrons in a plasma and obtained a general nonlinear dispersion relation for a neutrino fluid coupled with the plasma. It was demonstrated that a neutrino streaming instability could develop within a supernova leading to the excitation of plasma waves with growth rates which scale with the Fermi constant G_F in agreement with previous results. It was also demonstrated by Bingham et al., [6] and Silva et al., [7] that the coupling of the neutrinos with the plasma giving rise to collective phenomena could be described by the neutrino ponderomotive force and that this coupling gives rise to hydrodynamic instabilities which have growths proportional to G_F, much larger than the kinetic instabilities which have growths that scale as G_F^2, and also larger than collisional coupling, which is also of order G_F^2 whereas the collisional coupling is of order G_F^2. The work of Bingham et al., [6] and Silva et al., [7] agree with earlier work by Semikoz [8] and Oraevskii et al., [9] who also used a self-consistent relativistic kinetic description for neutrino propagation in a dispersive plasma medium. Oraevskii et al., [9] also concludes that the electromagnetic contribution to neutrino dynamics in a dispersive medium can exceed the contribution due to ordinary weak interactions and is larger than the radiative corrections in vacuum.

Most of the papers on neutrino decay into photons or plasma waves consider Cherenkov emission sometimes called radiative decay [10]. Both massless and massive neutrinos have been considered. In these papers single neutrino interactions with matter are considered in a non-selfconsistent manner. By considering forward elastic scattering Wolfenstein [11] showed that neutrinos acquire a refractive index in a dispersive medium like a plasma, similar to photon refraction. This means that neutrinos propagating in an inhomogeneous plasma have an effective potential such that they are repulsed from regions of higher electron density exchanging momentum with the plasma density structures. In a single particle description the plasma recoil can be neglected but for a sufficiently large number of neutrinos such as you have in a supernova explosions the momentum transfer is non-negligible. Therefore the interaction of a neutrino beam with a plasma electron density wave requires a fully self-consistent description since both the neutrino beam and plasma density wave charge simultaneously to be fully self consistent both the radiative decay interactions together with the forward elastic scattering interactions to be included. In most papers the forward elastic scattering interactions are omitted when considering neutrino decay. The omission of the elastic forward scattering processes results in non-dispersive neutrinos and since the essence of any collective process in a medium requires the dispersive terms it is impossible for these calculations to be compared with those of Bingham [6], Silva [7] and Semikoz [8]. For a very intense neutrino flux, such as those in supernovae, propagating in a dense plasma the electromagnetic contributions due to the plasma medium cannot be regarded as a small contribution like the case of radiative corrections in vacuum.

The collisionless relativistic kinetic equations describing the coupling of the neutrinos and electrons developed by Silva et al., [7] are obtained from the effective Hamiltonian and written as

$$\frac{\partial f_\nu}{\partial t} + \mathbf{v}_\nu \frac{\partial f_\nu}{\partial r} - \sqrt{2} G_F \left(\nabla n_e + \frac{1}{c^2} \frac{\partial \mathbf{J}_e}{\partial t} - \frac{\mathbf{v}_\nu}{c^2} \times \nabla \times \mathbf{J}_e \right) \frac{\partial f_\nu}{\partial p_\nu} = 0 \qquad (1)$$

$$\frac{\partial f_e}{\partial t} + \mathbf{v}_e \frac{\partial f_e}{\partial \mathbf{r}} - \sqrt{2} G_F \left(\nabla n_\nu + \frac{1}{c^2} \frac{\partial \mathbf{J}_\nu}{\partial t} - \frac{\mathbf{v}_e}{c^2} \times \nabla \times \mathbf{J}_\nu \right) \cdot \frac{\partial f_e}{\partial p_e} -$$

$$e \left(\mathbf{E} + \mathbf{v}_e \times \mathbf{B} \right) \cdot \frac{\partial f_e}{\partial p_e} = 0 \qquad (2)$$

where f_e (f_ν) is the quasi-classical distribution function for electrons (neutrinos), where $p_{e(\nu)}$ is the electron (neutrino) momentum, G_F is the Fermi coupling constant, $v_{e(\nu)}$ is the electron (neutrino) velocity), $n_{e(\nu)}$ is the electron (neutrino) density, $\mathbf{J}_{e(\nu)}$ is the electron (neutrino) current $n_{e(\nu)} \mathbf{v}_{e(\nu)}$, $\mathbf{E}$ and $\mathbf{B}$ are the electric and magnetic fields set up as a result of space charge and currents. The third term in equation (1) is the force acting on the neutrinos due to electrons and the third term in equation (2) is the force acting on electrons due to neutrinos and the fourth term in equation (2) is the Lorentz force on electrons. These equations describe the collective interactions between neutrinos and plasmas. The main difference between the kinetic approach and the previous calculations by Bingham et al., [6] is that we now can consider a distribution of neutrinos in momentum space and hence also in energy space whereas in the original description the neutrinos were considered to have the same energy and momentum ie a "monochromatic" source which is unrealistic for most cases of interest but is useful for analytical calculations. Another important point to note is that the quasi-classical description holds when the separation distance between particles is greater than the De Broglie wavelength.

A nonlinear dispersion telation for electron plasma waves driven by an intense neutrino flux can easily be obtained from a perturbation treatment of equations (1) and (2) yielding [7]

$$1 + \chi_e(\omega_\ell, \mathbf{k}_\ell) + \chi_v(\omega_\ell, \mathbf{k}_\ell) = 0 \qquad (3)$$

where $\chi_e(\omega_\ell, \mathbf{k}_\ell)$ is the relativistic longitudinal electron susceptibility

$$\chi_e(\omega_\ell, \mathbf{k}_\ell) = \frac{\omega_{po}^2}{k_\ell^2} m_e \int \frac{\mathbf{k}_\ell \frac{\partial \hat{f}_{\ell o}}{\partial \mathbf{p}}}{(\omega_\ell - \mathbf{k}_\ell \cdot \mathbf{v})} d\mathbf{p}$$

and $\chi_v(\omega_\ell, \mathbf{k}_\ell)$ is the relativisitic longitudinal neutrino susceptibility

$$\chi_v(\omega_\ell, \mathbf{k}_\ell) = -2G_F^2 \frac{k_\ell^2 n_{eo} N_{vo}}{m_e \omega_{po}^2} \left(1 - \frac{\omega_\ell^2}{c^2 k_\ell^2}\right)^2 \chi_e(\omega_\ell, \mathbf{k}_\ell) \int \frac{\mathbf{k}_\ell \cdot \frac{\partial \hat{f}_{vo}}{\partial \mathbf{p}}}{(\omega_\ell - \mathbf{k}_\ell \cdot \mathbf{v}_v)} d\mathbf{p}$$

where $\omega_\ell, (\mathbf{k}_\ell)$ is the electron plasma wave frequency (wavenumber), ω_{po} is the electron plasma frequency. Solutions of equation (3) has two important regimes of instability namely the kinetic regime and the hydrodynamic regime sometimes known as a reactive instability.

Equation (3) has been solved by Silva et al., [7] under a number of different assumptions demonstrating that equation (3) has a number of different instability regimes namely a kinetic instability if the phase velocity of the electron plasma waves overlaps the neutrino distribution function and a hydrodynamic regime, if no overlap exists. The kinetic regime because it depends on the shape of the distribution function notably the slope of the neutrino distribution in momentum space. For $\partial f_{ov}/\partial \mathbf{p} > 0$ a kinetic beam instability results with a growth rate proportional to G_F^2 which results in an extremely small growth. In contrast to the kinetic regime the hydrodynamical regime has a much larger growth rate proportional to $G_F^{2/3}$ for the weak beam case and proportional to $G_F^{\frac{1}{2}}$ for the strong beam case which are much larger than the kinetic or single neutrino electron scattering both of which are proportional to G_F^2. The hydrodynamical regime which was the one pursued by Bingham et al., [6] and Silva et al., [7] and shown to be much more efficient. Another mis-representation found in some papers is the role collisions play in the development of the instability. It is well known [7] that the inclusion of collisions only slows down the rate of growth of the instability and provides a threshold condition. The threshold for the instability is set by the product of the electron plasma wave damping and the neutrino electon collision frequency which is extremely small of order G_F^2, this result was derived in the paper by Bingham et al., [6] and shows that the threshold is extremely small for the supernovae parameters. The calculation of the growth rate in the presence of damping takes the damping of both the plasma wave and neutrino beam into account even although it is smaller than one of the damping rates , the instability still occurs, this is also the case for electron beam driven instabilities and photon driven instabilities. The damping only reduces the growth rate and sets a threshold value and does not switch the instability off.

The initial calculation by Silva et al. [7] considered a monochromatic neutrino beam which is unrealistic for supernovae neutrinos. Using two different types of neutrino distributions, in equation 3, namely a waterbag distribution in perpendicular momentum and a waterbag distribution in the polar direction, which are more realistic, Silva [12] categorically demonstrated the viability of the hydrodynamic regime. For broad neutrino distributions the hydrodynamic growth rates scale as [12] G_F^κ, with $2/3 < \kappa < 4/3$, although weaker than the monochromatic beam case it is still much stronger than the kinetic regime $\kappa = 2$.

Finally for the hydrodynamical regime Silva et al., [7] have calculated the growth rates of the instability for supernovae parameters including electron-ion collisions and shown that the growth distance for 20e-foldings is $6km$. The growth distance for the kinetic regime, which scales as G_F^2, is approximately 11 orders of magnitude larger than the hydrodynamical instability. The mean free path for neutrino electron callisions is of the order of 10^{16} cm similar to the growth distance for the kinetic regime which can therefore be ignored in neutrino absorption processes in dense supenovae plasmas. The hydrodynamical regime of the streaming instabilities studied by Bingham [6] and Silva [7] is the only regime of interest and given their much faster growth rates clearly play a significant role in supernovae dynamics.

REFERENCES

1. H A Bethe and J R Wilson, *Ap.J*, **295**, 14, 1985.
2. J Bahcall, Neutrino Astrophysics, Cambridge University Press, Cambridge.
3. P K Kaw, W L Kruer, C S Lui and K Nishikawa in Advances in Plasma Physics, Vol 6, part 1, (Wiley, New York) 1986.
4. J R Wilson, in *Numerical Astrophysics* edited by J Centrella, J Leblanc and R L Bowers, publisher Jones and Bartlett, 1983.
5. J Goodman, A Dar and S Nussinov, *Ap.J.*, **314**, 47, 1987.
6. R Bingham J M Dawson, J J Su and H A Bethe, *Phys. Lett. A*, **220**, 107, 1994.
7. L O Silva, R Bingham, J M Dawson, J T Mendonça and P K Shukla, *Phys. Rev. Lett*, **83**, 2703, 1999. *APJ Supp Series*, **127**, 481, 2000.
8. V B Semikoz, *Physica*, **A142**, 157, 1987.
9. V N Oraevsky and V B Semikoz, *Physica*, **A142**, 135, 1987.
10. G G Raffelt, *Stars as Laboratories*, University Chicago Press, Chicago, 1996.
11. L Wolfenstein, *Phys. Rev.*, **D17**, 2369, 1978.
12. L O Silva, *Phys. Rev.*, **D**, submitted 2002.

Simulation of Hysteresis in Glow Discharge

N. Mizuno[1], U. Tomioka[2], T. Hayashi[2], T. Kawabe[2]

[1] *Department of Physics, College of Humanities & Sciences, Nihon University*
Sakurajosui 3-25-40, Setagayaku, Tokyo, 156-8550 Japan

[2] *Institute of Physics, University of Tsukuba*
Tsukuba, Ibaraki, 305-8571 Japan

Abstract The hysteresis of the glow discharge in VI space has been investigated by the numerical simulation. The simulation code has been formed by combination of the equation of the discharge current and the equation of the circuit. We have shown that the hysteresis is produced by the bistability of the relaxation oscillation and the equilibrium state.

INTRODUCTION

The hysteresis with respect to VI characteristics is observed in the experiments of the glow discharge. The hysteresis appears in around the boundary between the subnormal glow and the normal glow. In experiments, it is known that the relaxation oscillation is observed on the upper branch of the hysteresis curve, while the discharge is stationary on the lower branch [1]. Thus, although the hysteresis is caused by the existence of the above two states, the mechanism of the coexistence has not yet been clear.

In the present paper, we study the mechanism of the generation of the hysteresis in the glow discharge using by the numerical simulation. In order to examine the temporal evolution of the discharge, we use the equation by B. M. Jelenkovic et al.[2] for the discharge current. For the discharge voltage, we use a circuit equation. It is assumed that the electric field of the cathode sheath decreases linearly from the cathode and it is zero at the end of the cathode sheath. The simulation code has been formed by combination of the two equations.

SIMULATION MODEL

The discharge circuit is shown in Fig.1. The discharge tube is filled with argon, which is chosen as the model gas. In order to see the relaxation oscillation, we use the equation of the discharge current by B. M. Jelenkovic et al.[2]. The time evolution of the discharge current is given by

$$\frac{dI}{dt} = \frac{1}{T}\{\gamma[\exp\{\int_0^d \alpha(x)dx\} - 1] - 1\} \qquad (1)$$

where α is the Townsend's first ionization coefficient, γ is the secondary electron emission coefficient, T is the transit time of ions and d is the distance from the cathode to anode. The following formula is used as α:

$$\alpha(x) = pA\exp(-pB/E(x)) \qquad (2)$$

where p is the pressure, E(x) is the electric field. A and B are empirically determined constants [3]. In the discharge experiment, The voltage drop occurs in the cathode dark space. Then, we assume that the value of d in Eq.(1) is the thickness of the cathode dark space. To consider VI characteristics of the discharge tube, we use the following equation of the relation of the cathode fall voltage and the discharge current by Engel-Steenbeck,

$$S\left[(\overline{V}\cdot\overline{I})^{1/3}\right] = \frac{\overline{I}^{2/3}}{\overline{V}^{1/3}} \tag{3}$$

$$\overline{V} = C_1 V, \overline{I} = \left(\frac{C_2 \cdot j}{p^2}\right)$$

where C_1, C_2 are constants, j is the discharge current density and

$$S[a] = \int_0^a \exp\left(-\frac{1}{x}\right) dx$$

Then, we obtain the thickness of the cathode dark space as follows:

$$d = \left\{ \frac{4\mu_i \varepsilon_0 (1+\gamma)}{j} V_c^2 \right\}^{1/3} \tag{4}$$

where V_c is the cathode fall voltage and μ_i is the ion mobility. Potential at the anode is expressed by the equation of the circuit as follows:

$$\frac{dV}{dt} = \frac{1}{RC}(E - V - IR) \tag{5}$$

where R is the load resistance, C is the capacitance and E is the applied voltage (Fig.1).

In the simulation, the values of the parameters are taken as

Degree of vacuum	1.5 [Torr]
Distance from the cathode to the anode	1.0 [cm]
Load resistance	3.0 [MΩ]
Capacitance	180 [pF]

RESULTS AND CONCLUSIONS

We have investigated the hysteresis in the glow discharge using by the numerical simulation. The simulation code has been formed by combination the equation of the discharge current and the equation of the circuit. Fig.2 show the stable periodic orbit and the unstable orbit in the VI space. The trajectory outside the unstable periodic orbit converge the stable periodic orbit. On the other hand, the trajectory inside the unstable periodic orbit converge the equilibrium point. The former is the relaxation oscillation; the latter is the equilibrium state. The hysteresis curve in VI space from the simulation is shown in Fig.3.

It is shown that an unstable periodic orbit is generated by the subcritical Hopf bifurcation of the equilibrium state, which separates phase space into two domains of attraction for two stable states. In one of the domains, all trajectories are attracted by the relaxation oscillation, while in the other domain, all trajectories are attracted by the equilibrium state. The hysteresis is produced by the bistability of the relaxation oscillation and the equilibrium state. The simulation results are qualitative agreement with the experimental results.

REFERENCES

[1] Yu. G. Zakharenko and V. E. Privalov : Sov. Phys. Tech. Phys. 16. 429 (1971)
[2] B. M. Jelenkovic, K. Rozsa and A. V. Phelps : Phys. Rev. E47, 2816 (1993)
[3] Y. P. Raizer : Gas Discharge Physics (Springer-Verlag. Berlin, 1988)

Theory of RF-Induced Current Profile in the Presence of Fluctuations

Masayoshi TAGUCHI

College of Industrial Technology, Nihon University
2-11-1 Shin-ei, Narashino, Chiba 275-0005, Japan

Abstract. The effect of low-frequency fluctuations on the rf-induced current has been studied by using a renormalization technique of statistical dynamics. With the application of the functional integral method to a Fokker-Planck equation in the presence of fluctuations, a closed set of equations is derived for an ensemble-averaged distribution function and a response function to an infinitesimal external perturbation. Explicitly solving this closed set of equations, the radial width of rf-induced current density is discussed for a steady-state current drive.

The radial profile control of toroidal current by rf waves has attracted much attention for stabilizing various instabilities and obtaining attractive MHD equilibrium. In using rf waves for the current profile control, the understanding of radial transport for resonant particles is indispensable. The effect of radial anomalous transport on the rf-induced current has been studied by adding phenomenological radial diffusion terms to a Fokker-Planck equation. In this paper, we derive this radial anomalous term by the use of renormalization technique of statistical dynamics [1, 2] and present governing kinetic equations for the current drive problem in a presence of fluctuations.

The evolution of an electron distribution function f in the presence of rf waves and low-frequency fluctuations is described by a relativistic Fokker-Planck equation

$$\partial_t f + v_\parallel \mathbf{b} \cdot \nabla f + \delta \mathbf{v}_\perp \cdot \nabla_\perp f - C(f) = S_{\text{rf}},$$

where the subscripts $\parallel$ and $\perp$ refer to the parallel and the perpendicular to an averaged magnetic field $\mathbf{B}$, $\mathbf{b} = \mathbf{B}/B$, C is the Fokker-Planck collision operator, and $S_{\text{rf}} = \partial/\partial \mathbf{p} \cdot \mathbf{S}_W$ represents the momentum-space diffusion due to rf waves. The electric and magnetic fluctuations lead to a radial velocity fluctuation of particle. In low-frequency fluctuations, this velocity fluctuation can be written as $\delta \mathbf{v}_\perp = \delta \mathbf{E} \times \mathbf{b}/B$ for electric fluctuations and $\delta \mathbf{v}_\perp = v_\parallel \delta \mathbf{B}_\perp/B$ for magnetic fluctuations, where $\delta \mathbf{E}_\perp$ and $\delta \mathbf{B}_\perp$ are the fluctuating electric and magnetic fields. In the current drive problem, we are interested in the determination of the mean distribution function ensemble-averaged over fluctuations. Thus, applying the functional integral method [2], which is analogous to the Feynman's path integral, to the Fokker-Planck equation, we derive the following coupled equations for an ensemble-averaged distribution function $g(1) = <f(1)>/f_0(p)$ and an averaged response function $G(1,1')$ to an infinitesimal external perturbation:

$$(\partial_t + v_\parallel \mathbf{b} \cdot \nabla - \hat{C})g(1) - \nabla_\perp \cdot \mathbf{F}(1,1')G(1,1') \cdot \nabla'_\perp g(1') = \hat{S}_{\text{rf}}, \qquad (1)$$

$$(\partial_t + v_\parallel \mathbf{b} \cdot \nabla - \hat{C})G(1,1'') - \nabla_\perp \cdot \mathbf{F}(1,1')G(1,1') \cdot \nabla'_\perp G(1',1'') = \delta(1-1''), \qquad (2)$$

where the correlation function $\mathbf{F}(1,1') = <\delta \mathbf{v}(\mathbf{r},\mathbf{p})\delta \mathbf{v}(\mathbf{r}',\mathbf{p}')>$, $\hat{C}(\psi') = C(f_0 \psi')/f_0$, $\hat{S}_{\text{rf}} = S_{\text{rf}}/f_0$, the index 1 denotes $1 = (\mathbf{r}, \mathbf{p}, t)$, $f_0(p)$ is the Maxwellian for bulk particles, and the angular bracket $< \cdot >$ means the ensemble-average over fluctuations.

The coupled equations (1) and (2) are rather complicated nonlinear integro-differential equations. The most serious difficulty in solving these equations comes from the nonlocal feature of the nonlinear terms. First, we use the local approximation to these nonlinear terms. Then, the rf-induced current density in a steady-state current drive is expressed by using adjoint method [3] as

$$J(\mathbf{r}) = -e \int d\mathbf{p} v_\parallel g(\mathbf{r},\mathbf{p}) f_0(p) = -e \int d\mathbf{r}' \int d\mathbf{p}' S_{\text{rf}}(\mathbf{r}',\mathbf{p}') h_1(\mathbf{r}',\mathbf{p}';\mathbf{r}),$$

where the function h_1 is determined by the differential equation

$$(-v_\parallel \mathbf{b} \cdot \nabla - \hat{C} - \nabla_\perp \cdot \mathbf{D} \cdot \nabla_\perp) h_1(\mathbf{r}, \mathbf{p}; \mathbf{r}'') = \frac{p_\parallel}{m\gamma} \delta(\mathbf{r} - \mathbf{r}''). \tag{3}$$

Let define the following function:

$$h_2(\mathbf{r}, \mathbf{p}; \mathbf{r}') = \int d\mathbf{p}' V(\mathbf{p}') \int_{-\infty}^0 d\tau G^\dagger(\mathbf{r}, \mathbf{p}, \tau; \mathbf{r}', \mathbf{p}'),$$

where $V(\mathbf{p}) = 1$ for electric fluctuations and $V(\mathbf{p}) = p_\parallel/m\gamma$ for magnetic fluctuations. Then we find from (2) that the function h_2 satisfies the equation

$$(-v_\parallel \mathbf{b} \cdot \nabla - \hat{C} - \nabla_\perp \cdot \mathbf{D} \cdot \nabla_\perp) h_2(\mathbf{r}, \mathbf{p}; \mathbf{r}') = V(\mathbf{p}) \delta(\mathbf{r} - \mathbf{r}'). \tag{4}$$

Using the function h_2, we define the diffusion tensor $\mathbf{D}$ in (3) and (4) by

$$\mathbf{D}(\mathbf{r}, \mathbf{p}) = V(\mathbf{p}) \int d\mathbf{r}' \hat{\mathbf{F}}(\mathbf{r} - \mathbf{r}') h_2(\mathbf{r}, \mathbf{p}; \mathbf{r}') \tag{5}$$

with $\mathbf{F}(\mathbf{r}, \mathbf{p}; \mathbf{r}', \mathbf{p}') = V(\mathbf{p})V(\mathbf{p}')\hat{\mathbf{F}}(\mathbf{r} - \mathbf{r}')$. We now concentrate on the steady-state current drive in the presence of electrostatic fluctuations. Let assume the spatial homogeneity along the magnetic field, and further assume the correlation function and diffusion tensor can be written as $\hat{\mathbf{F}}(\mathbf{r} - \mathbf{r}') = F(x - x')\mathbf{e}_x \mathbf{e}_x$ and $\mathbf{D} = D(p)\mathbf{e}_x \mathbf{e}_x$ with $\delta \mathbf{v}_\perp = (\delta E_\perp/B)\mathbf{e}_x$. Then, solving the equation (4) by using the high-energy form of collision operator [3] and substituting its solution into (5), we approximately obtain the nonlinear equation for the diffusion coefficient $D(p)$

$$D(p)^2 = \int \frac{dk}{2\pi} \frac{F(k)}{k^2} \left(1 - e^{-\frac{1}{v_c} P(p) D(p) k^2}\right), \tag{6}$$

where $P(p) = p/mc - \tan^{-1}(p/mc)$ and $F(k)$ is the Fourier transform of $F(x - x')$.

The limiting form for the diffusion coefficient is easily obtained from (6). In the weak turbulent limit, we expand the exponential function in the integrand of (6) to obtain

$$D(p) = \frac{1}{v_c} P(p) \int \frac{dk}{2\pi} F(k). \tag{7}$$

The diffusion coefficient in the strong turbulent limit is calculated by neglecting the exponential function in (6) as

$$D(p) = \left[\int \frac{dk}{2\pi} \frac{F(k)}{k^2}\right]^{1/2}, \tag{8}$$

where the integral $\int dk F(k)/k^2$ is assumed to exist. For the correlation function

$$F(x - x') = \beta^2 \exp[-(x - x')^2/2\lambda_\perp^2] \tag{9}$$

with the correlation length $\lambda_\perp$ and the amplitude β of fluctuating electric field, the nonlinear equation (6) can be solved analytically to obtain

$$D(p) = \lambda_\perp \beta C(\xi)$$

with

$$C(\xi) = \left[\frac{1}{2}\sqrt{\xi} + \sqrt{\frac{8}{27} + \frac{1}{4}\xi}\right]^{1/3} - \left[\sqrt{\frac{8}{27} + \frac{1}{4}\xi} - \frac{1}{2}\sqrt{\xi}\right]^{1/3},$$

where $\xi = (2\beta P(p)/v_c \lambda_\perp)^2$. In this correlation function model, the diffusion coefficient for the strong turbulence is proportional to $\beta^{4/3}$ whereas the diffusion coefficient (8) is proportional to β.

We next solve the equation (3) for h_1 to obtain

$$h_1(x, \mathbf{p}; x') = \frac{c}{v_c} \frac{p_\parallel}{p} \left(\frac{\gamma + 1}{\gamma - 1}\right)^{(Z+1)/2} \int_1^\gamma d\gamma' R(\gamma') \frac{1}{\sqrt{2\pi}\sigma_e} e^{-(x - x')^2/(2\sigma_e^2)}, \tag{10}$$

where Z is the charge number,

$$R(\gamma) = \left(\frac{\gamma-1}{\gamma+1}\right)^{(Z+1)/2} \frac{\gamma^2-1}{\gamma^2}, \quad \sigma_e^2 = \frac{2}{v_c} \int_\gamma^\gamma \frac{\bar{p}}{mc\bar{\gamma}} D(\bar{p}) d\bar{\gamma}.$$

Using the expression (10), we have the mean squared width of current density associated with a localized power input at $x = x_0$

$$\sigma_G^2 \equiv \frac{\int J(x)(x-x_0)^2 dx}{\int J(x) dx} = \frac{2}{v_c} \frac{\mathbf{s} \cdot \frac{\partial}{\partial \mathbf{p}} \left(\frac{p_\parallel}{p} G_2(p)\right)}{\mathbf{s} \cdot \frac{\partial}{\partial \mathbf{p}} \left(\frac{p_\parallel}{p} G_1(p)\right)}, \tag{11}$$

where $\mathbf{s} = \mathbf{S_W}/|\mathbf{S_W}|$,

$$G_1(p) = \left(\frac{\gamma+1}{\gamma-1}\right)^{(Z+1)/2} \int_1^\gamma d\gamma' R(\gamma')$$

and

$$G_2(p) = \left(\frac{\gamma+1}{\gamma-1}\right)^{(Z+1)/2} \int_1^\gamma d\gamma' R(\gamma') \int_{\gamma'}^\gamma d\bar{\gamma} \frac{\bar{p}}{mc\bar{\gamma}} D(\bar{p}).$$

In the weak turbulent limit, the mean squared width reduces to the form

$$\sigma_G^2 = \frac{1}{v_c^2} \int \frac{dk}{2\pi} F(k) \frac{\mathbf{s} \cdot \frac{\partial}{\partial \mathbf{p}} \left(\frac{p_\parallel}{p} G_2^{(q)}(p)\right)}{\mathbf{s} \cdot \frac{\partial}{\partial \mathbf{p}} \left(\frac{p_\parallel}{p} G_1(p)\right)}, \tag{12}$$

where

$$G_2^{(q)} = \left(\frac{\gamma+1}{\gamma-1}\right)^{(Z+1)/2} \int_1^\gamma d\gamma' R(\gamma') \left(P(p)^2 - P(p')^2\right).$$

According to (8), the mean squared width in the strong turbulence becomes

$$\sigma_G^2 = \frac{2}{v_c} \left(\int \frac{dk}{2\pi} \frac{F(k)}{k^2}\right)^{1/2} \frac{\mathbf{s} \cdot \frac{\partial}{\partial \mathbf{p}} \left(\frac{p_\parallel}{p} G_2^{(s1)}(p)\right)}{\mathbf{s} \cdot \frac{\partial}{\partial \mathbf{p}} \left(\frac{p_\parallel}{p} G_1(p)\right)},$$

where

$$G_2^{(s1)} = \left(\frac{\gamma+1}{\gamma-1}\right)^{(Z+1)/2} \int_1^\gamma d\gamma' R(\gamma') \left(P(p) - P(p')\right).$$

This mean squared width agrees with that of Rax *et al.* [4] obtained by solving Langevin equations. The mean squared width in the strong turbulence for the model (9) is given by

$$\sigma_G^2 = \frac{\beta^2}{v_c^2} \frac{3}{2} \left(\frac{2}{\bar{\xi}^2}\right)^{1/3} \frac{\mathbf{s} \cdot \frac{\partial}{\partial \mathbf{p}} \left(\frac{p_\parallel}{p} G_2^{(s2)}(p)\right)}{\mathbf{s} \cdot \frac{\partial}{\partial \mathbf{p}} \left(\frac{p_\parallel}{p} G_1(p)\right)},$$

where $\bar{\xi} = \beta/(v_c \lambda_\perp)$ and

$$G_2^{(s2)} = \left(\frac{\gamma+1}{\gamma-1}\right)^{(Z+1)/2} \int_1^\gamma d\gamma' R(\gamma') \left(P(p)^{4/3} - P(p')^{4/3}\right).$$

In Figs. 1 (a) and (b), the mean squared width normalized by $(\beta/v_c)^2$ for the correlation function model (9) is plotted as a function of p/mc for $\beta/(v_c \lambda_\perp) = 10$ and 100. The normalized widths for the weak and strong limits are also shown in these figures by dotted (1) and dashed curves.

Finally we consider the nonlocal effect on the current profile broadening. The spatial local approximation is justified when the ensemble-averaged distribution function and the response function change little over the characteristic

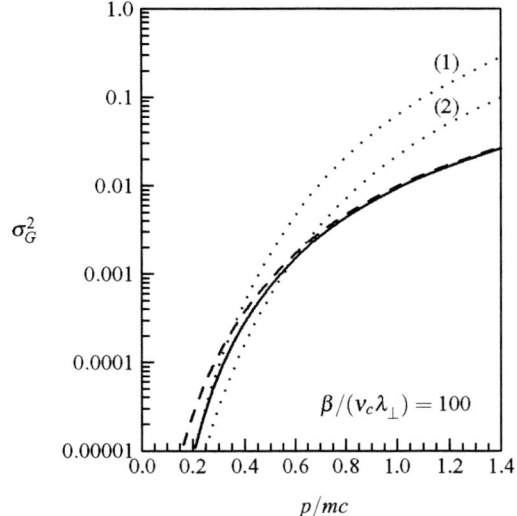
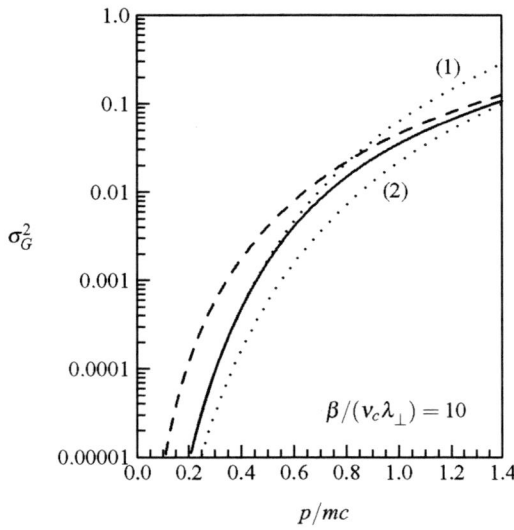

FIGURE 1(a). Plot of the mean squared width σ_G^2 normalized by $(\beta/v_c)^2$ versus p/mc for $\beta/(v_c\lambda_\perp) = 100$.

FIGURE 1(b). Plot of the mean squared width σ_G^2 normalized by $(\beta/v_c)^2$ versus p/mc for $\beta/(v_c\lambda_\perp) = 10$.

length of the correlation function. We here only consider the nonlocal effect in momentum space. Then, retaining this nonlocality, we have the nonlocal equations for $h_1(\mathbf{r},\mathbf{p};\mathbf{r}'')$ and $G^\dagger(\mathbf{r},\mathbf{p};\mathbf{r}'',\mathbf{p}'') = \int_{-\infty}^{0} d\tau G^\dagger(\mathbf{r},\mathbf{p},\tau;\mathbf{r}'',\mathbf{p}'')$:

$$(-v_\parallel \mathbf{b}\cdot\nabla - \hat{C})h_1(\mathbf{r},\mathbf{p};\mathbf{r}'') - \nabla_\perp \cdot \int d\mathbf{p}' \mathbf{D}_N(\mathbf{r},\mathbf{p};\mathbf{p}') \cdot \nabla_\perp h_1(\mathbf{r},\mathbf{p}';\mathbf{r}'') = \frac{p_\parallel}{m\gamma}\delta(\mathbf{r}-\mathbf{r}''), \quad (13)$$

$$(-v_\parallel \mathbf{b}\cdot\nabla - \hat{C})G^\dagger(\mathbf{r},\mathbf{p};\mathbf{r}'',\mathbf{p}'') - \nabla_\perp \cdot \int d\mathbf{p}' \mathbf{D}_N(\mathbf{r},\mathbf{p};\mathbf{p}') \cdot \nabla_\perp G^\dagger(\mathbf{r},\mathbf{p}';\mathbf{r}'',\mathbf{p}'') = \delta(\mathbf{r}-\mathbf{r}'')\delta(\mathbf{p}-\mathbf{p}''), \quad (14)$$

where $\mathbf{D}_N(\mathbf{r},\mathbf{p};\mathbf{p}') = V(\mathbf{p})V(\mathbf{p}') \int d\mathbf{r}' \hat{\mathbf{F}}(\mathbf{r}-\mathbf{r}') G^\dagger(\mathbf{r},\mathbf{p};\mathbf{r}',\mathbf{p}')$. The nonlocal effect in momentum space comes from Coulomb collisions. Therefore, its effect can be neglected in the strong turbulent limit although it becomes important as approaching to the weak turbulent limit. Solving the equations (13) and (14) in the weak turbulent limit, we find that the mean squared width of current density in the presence of electric fluctuations is written by the expression (12) with the following $G_{2N}^{(q)}(p)$ instead of $G_2^{(q)}(p)$:

$$G_{2N}^{(q)}(p) = \left(\frac{\gamma+1}{\gamma-1}\right)^{(Z+1)/2} \int_1^\gamma d\gamma' R(\gamma')[P(p') - P(p)]^2. \quad (15)$$

The dotted (2) curves in Figs. 1 (a) and (b) show the mean squared width obtained by using (15). The ratio $\sigma_G^2(\text{nonlocal})/\sigma_G^2(\text{local})$ is approximately given by $3/(Z+8)$.

This work was partially supported by a Grant-in-Aid for Scientific Research from Japan Society for the Promotion of Science.

REFERENCES

1. P. C. Martin, E. D. Siggia, and H. A. Rose: Phys. Rev. A **8** 423 (1973).
2. R. V. Jensen: J. Stat. Phys. **25** 183 (1981).
3. N. J. Fisch : Rev. Mod. Phys. **59** 175 (1987).
4. J. M. Rax, J. Robiche, and I. Kostyukov: Phys. Plasmas **6** 3233 (1999).

Stabilization of Burn Conditions in a Two-Temperature Fusion Reactor: Preliminary Results

Javier E. Vitela and Julio J. Martinell

Instituto de Ciencias Nucleares
Universidad Nacional Autónoma de México
04510 México D.F., México

Abstract. In this report we show results on the stabilization of the burn conditions of a two-temperature fusion reactor using artificial neural networks. Stabilization at the nominal operating state is obtained for energy confinement times ranging from 4.5 sec to 6.5 sec, in a model with parameters taken from the EDA ITER design but with $T_e = 13$ Kev and $T_i = 12.7$ Kev, as the nominal electron and ion temperatures, respectively. Control actions are implemented through the concurrent modulation of a DT refueling rate, a neutral ^{4}He beam and independent auxiliary heating power to ions and to electrons. Examples of typical transients are used to illustrate the results.

INTRODUCTION

Particle and energy transport in magnetically confined devices are generally estimated using transport losses modelled through a global energy confinement time extrapolated using a comprehensive data base gathered from previous and current tokamak experiments. The reason is that these phenomena depend strongly on turbulent processes and in spite of major advances in theoretical and numerical calculations of plasma transport properties, there still exist significant discrepancies with experimental results even in the absence of significant α heating. In fusion reactors for power production the plasma should be in the so called burning regime, where the heating will rest mainly on the α-particles produced by fusion. In this regime many new phenomena will influence the transport properties making their theoretical and numerical predictions even harder. Hence, if long pulse operation is to be attained, it is important to have appropriate means for the suppression of thermal inestabilities at the nominal burning operating state [1], which can withold transients due to fluctuations as well as to abrupt changes in the scaling law associated to different confinement regimes.

Some studies concerning the stabilization of burn conditions of a single temperature 0-D model of a thermonuclear fusion reactor in the presence of scaling law uncertainties have been reported earlier [2]. Succesful stabilization was achieved, regardless of any particular scaling law for the confinement times within the range $5.0s < \tau_E < 6.5s$, by the concurrent modulation of D-T refueling rate, an auxiliary heating power, and the injection of a neutral He-4 beam. The parameters used were taken from the ITER-EDA design [3] and thus the nominal operating point was: $n_0 = 1.0 \times 10^{20}$ m^{-3}, for the electron density, $T_0 = 12$ keV, for the plasma temperature and $f_0 = 0.09$ for the helium ash fraction; the high-Z impurity density was taken as $n_I = 7.0 \times 10^{17}$ m^{-3} with an effective charge $Z_I = 14.7$. Under this conditions ignition is expected to be reached if $\tau_E = 7.65s$

However, calculations using a two-temperature model show that with the above parameters near ignition conditions can be achieved for an electron temperature of $T_e = 12$ keV, only if the ion temperature is $T_i = 11.8$ keV, and if the energy confinement times is around 7.9 sec; for these conditions we have a Q_G gain factor of around 50. For smaller confinement times , e.g. in the range between 5.5 sec and 6.5 sec, and the same operating conditions the gain factor drops down into the range between 3.5 and 6.5. Nevertheless, increasing the nominal electron temperature to 13 keV while keeping the other plasma parameters the same, near ignition conditions can be achieved, i.e. the total auxiliary power heating is nearly zero, when the ion temperature is approximately 12.64 Kev, at an energy confinement time of 6.9 sec. Resulting in a gain factor Q_G of the order of 141. With these plasma parameters, when the confinement time drops to values between 5.5 sec and 6.5 sec, then the gain factor lies now within the interval value between 6 and 22, a substantial gain compared to the previous values.

FUSION REACTOR

Similarly to previous works, the fusion reactor model considered here is a zero-dimensional plasma system with the ITER-EDA tokamak design parameters [3] describing the time evolution of a quasineutral plasma composed of 50:50 D-T fuel, helium ash, a small amount of high-Z impurities, and electrons, whose densities are n_{DT}, n_α, n_I, and n_e, respectively; quasineutrality condition $n_e = n_{DT} + 2n_\alpha + Z_I n_I$ is assumed. The total thermal energy is determined assuming Maxwellian distribution of the particles: the electrons at a temperature T_e, and the deuterium, tritium, helium ash and impurities ions all at the same temperature T_i. Plasma heating takes place mainly by means of the thermalization of the alpha particles produced by the fusion reactions which are assumed to be instantaneously thermalized: with a fraction $f_e \approx 85\%$, of the original energy $Q_\alpha = 3.5$ Mev carried by the alpha fusion particles deposited to the electrons during the slow down process; and the remaining fraction $f_i \approx 15\%$, deposited directly to the ions. The particle balance equations are the same as in previous reports [2],

$$\frac{d}{dt} n_{DT} = S_f - 2\left(\frac{n_{DT}}{2}\right)^2 <\sigma v> - \frac{n_{DT}}{\tau_P},$$

and

$$\frac{d}{dt} n_\alpha = S_\alpha + \left(\frac{n_{DT}}{2}\right)^2 <\sigma v> - \frac{n_\alpha}{\tau_\alpha};$$

however, in this work the energy balance equation is separated in two parts, one corresponding to the electrons and one to the ions,

$$\frac{d}{dt}\left[\frac{3}{2} n_e T_e\right] = P_{\text{aux},e} + f_e Q_\alpha \left(n_{DT}/2\right)^2 <\sigma v> + \eta j^2 - A_b Z_{eff} n_e^2 T_e^{1/2}$$
$$- \frac{3}{2} n_e (T_e - T_i)/\tau_{ei} - \frac{3}{2} n_e T_e / \tau_{E,e},$$

and

$$\frac{d}{dt}\left[\frac{3}{2}(n_{DT} + n_\alpha + n_I) T_i\right] = P_{\text{aux},i} + f_i Q_\alpha \left(n_{DT}/2\right)^2 <\sigma v> + \frac{3}{2} n_e (T_e - T_i)/\tau_{ei}$$
$$- \frac{3}{2}(n_{DT} + n_\alpha + n_I) T_i / \tau_{E,i};$$

Energy transport losses for the electrons and the ions are accounted for throught the energy confinement times $\tau_{E,e}$ and $\tau_{E,i}$, respectively; although some authors [4] have considered the possibility of these quantities being different, in this work we will assume both are equal at all times. D-T and alpha particle losses are represented by the confinement times, τ_p and τ_α respectively. The model also assumes that the density n_I and effective charge Z_I of the impurities remain constant at all times. Bremsstrahlung is the only radiation loss mechanism considered; and the electron-ion thermal equipartition time is represented by τ_{ei}. The control actions are represented by, S_f the refueling rate, S_α the neutral He-4 injection rate; while $P_{aux,i}$ and $P_{aux,e}$ are the injection rate of the auxiliary heating power density to the ions and to the electrons, respectively. These control variables are constrained to take values exclusively from within the following ranges: $0 \leq S_f \leq 4 \times S_0$, $0 \leq S_\alpha \leq 0.1 \times f_0 n_0$, $0 \leq P_{aux,e} \leq 0.2 \times 1.5 n_0 T_e$, and $0 \leq P_{aux,i} \leq 0.2 \times 1.5 n_0 T_i$.

The dynamical system equations used in the simulations are the set of coupled nonlinear differential equations for the electron density n_e, and the relative fraction of helium ash $f_\alpha = n_\alpha/n_e$, and the electrons and ions temperatures, T_e and T_i, respectively.

The nominal operating state of this report will be assumed: $n_0 = 1.0 \times 10^{20}$ m^{-3}, $T_e = 13.0$ Kev, $T_i = 12.7$ Kev and $f_0 = 0.09$; the high-Z impurity density is $n_I = 7.0 \times 10^{17}$ m^{-3} with an effective charge $Z_I = 14.7$. With $\tau_\alpha = 5\tau_E$ and $\tau_p = 3\tau_E$. With these values of the plasma temperatures, near ignition condition, i.e. with $P_{\text{aux},e} \approx 0$ is achieved with $\tau_E = 6.64$ sec yielding a gain $Q_G \approx 37$. No steady state can be achieved for longer confinement times, since this would require $P_{aux,e} < 0$; while for $5.5s < \tau_E < 6.5s$, the gain lies between $6 < Q_G < 24$. Figure 1 illustrates this behaviour for steady state conditions.

An artificial neural network was trained to stabilize the system, suppressing perturbations in the plasma parameters around $\pm 10\%$ off their nominal operating values for a range of energy confinement times τ_E, which was chosen here to lie between 4.5 sec and 6.5 sec.

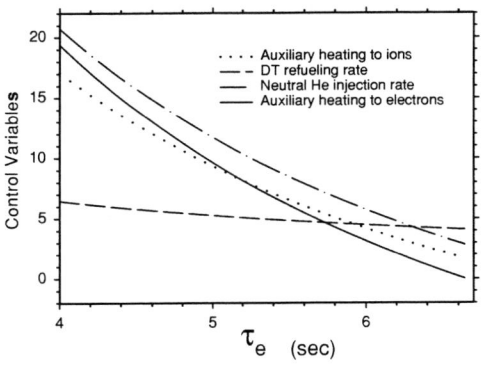
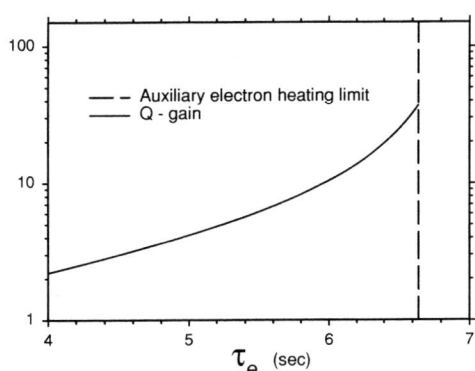

FIGURE 1. DT refueling rate $S_f \times 10^{-18}$ m^{-3} s^{-1}, injection of neutral ^{4}He $S_\alpha \times 10^{-16}$ m^{-3} s^{-1}, and the auxiliary heating powers to electrons and ions $\times 10^{-19}$ Kev m^{-3} s^{-1} at steady state (left) and the gain factor Q_G (right), as function of the energy confinement time τ_E for the same nominal operating state specified in the text.

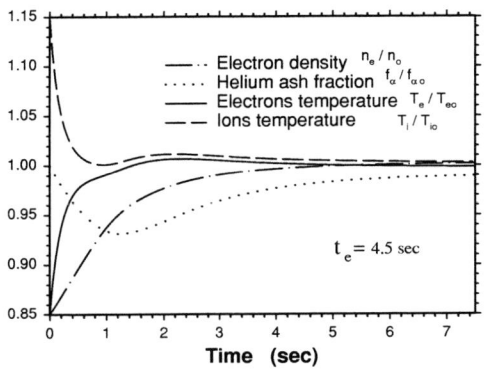
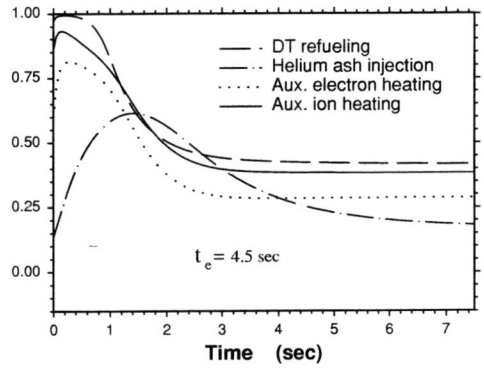

FIGURE 2. Transient behavior of the normalized state variables (left) and control variables (right) as function of time, when the energy confinement time is τ_E=4.5 s. The initial condition is specified in the text.

RESULTS

The results shown here were obtained after 400 iterations using an efficient parallel training algorithm[5] with a radial basis neural network, consisting of 5 input nodes: for the electron density, the fraction of helium ash, the electron and ion temperatures and current value of the energy confinement time; 1024 nodes in the hidden layer and 4 output nodes, corresponding to the values of the DT refueling rate, the neutral helium beam, and the auxiliary heating powers to the electrons and to the ions.

We present two illustrative examples of typical transient behaviour obtained after the neural network training, where the initial state is $n_e = 0.85 \times n_0$, $f_\alpha = f_0$, $T_e = 0.85 \times T_{e0}$ and $T_i = 1.15 \times T_{i0}$. Figures 2 and 3 show the time behavior of the state and control variables for the first 7.5 sec into the transient when the energy confinement time was taken to be $\tau_E = 4.5$ sec and $\tau_E = 6.5$ sec, respectively. In these graphs the state variables are shown normalized to their nominal operating values, while the control variables are normalized to their maxima allowable values. As we see in both cases the neural network is able to succesfully suppress the plasma parameters perturbations within only few seconds into the transient.

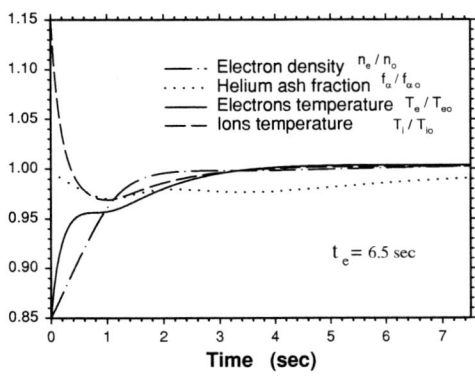 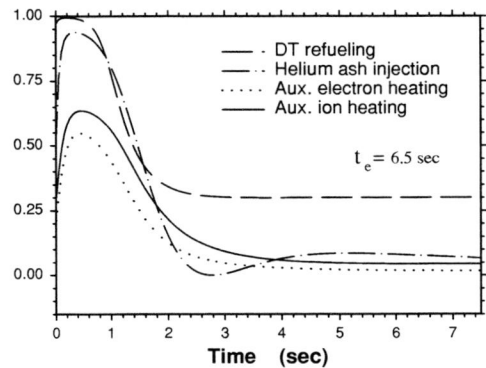

FIGURE 3. Transient behavior of the normalized state variables (left) and control variables (right) as function of time, when the energy confinement time is τ_E=6.5s, for the same initial conditions as the transient shown in Fig.2.

CONCLUSIONS

Simulation results were used to illustrate the stabilization capabilities of a RBNN in a two-temperature fusion reactor of the ITER-EDA type, for two transients starting with the same initial conditions but with quite different energy confinement times. Further testing of the robustneess of the controller using specific scaling laws with embedded uncertainties is underway.

ACKNOWLEDGMENTS

Partial financial support from CONACYT 27974-E and DGAPA-UNAM IN116200 projects is gratefully acknowledged.

REFERENCES

1. F.W. Perkins *et al.*, *1998 ICPP & 25th EPS Conf. on Contr. Fusion and Plasma Physics*, ECA **22C**, 1979 (1998).
2. J.E. Vitela and J.J. Martinell, *Plasma Phys. and Control. Fusion*, **43**, 99-119 (2001). J.E. Vitela and J.J. Martinell, *Proc. 28-th EPS Conf. on Contr. Fusion and Plasma Physics*, ECA **25A** 645 (2001)
3. ITER Physics Basis *Nuclear Fusion* **39**, 2137-2174 (1999)
4. N.J. Fisch and M.C. Herrmann *Nuclear Fusion*, **34**, 1541 (1994).
5. J.E. Vitela, U.R. Hanebutte, J.L. Gordillo and L.M. Cortin *Int. J. Modern Physics C*, **13**, 429-452 (2002).

Direct Kinetic Simulations of Ion Temperature Gradient Driven Turbulence

T.-H. Watanabe and H. Sugama

National Institute for Fusion Science / The Graduate University for Advanced Studies, Toki, 509-5292 Japan

Abstract. Collision frequency dependence of the slab ion temperature gradient (ITG) driven turbulence is investigated by means of direct numerical simulations of a reduced gyrokinetic equation. A statistically steady state of a perturbed ion velocity distribution function δf is obtained in presence of the finite collisionality, while fine-scale fluctuations of δf in the velocity space are continuously generated by the phase mixing in the collisionless ITG turbulence. Power spectrum of δf has a tail which extends longer into a higher velocity-space wave number side for smaller collision frequency ν. A large-scale structure of δf is, nevertheless, independent of ν. It is also shown that the ion heat transport approaches a finite constant value for sufficiently small ν.

INTRODUCTION

Turbulent transport in high-temperature plasmas has been a key issue in the magnetic fusion research [1]. Extensive simulation studies [2] have revealed several important aspects of the ion temperature gradient (ITG) driven turbulence, such as the transport suppression by the self-generated zonal flow. Nevertheless, saturation mechanism of the collisionless ITG turbulence has long been an open question. Since the collisionless gyrokinetic equation has the time-reversal symmetry, in order to consider an irreversible transport processes in a collisionless turbulence, one needs to coarse-grain the one-body velocity distribution function f with small-scale fluctuations caused by the phase mixing. When a steady transport flux is observed, a *quasi*-steady state of collisionless turbulence should be realized [3, 4, 5], where high-order moments of the perturbed distribution function δf continue to grow but the low-order ones reach steady states. Here, δf is defined by $\delta f \equiv f - F_M$ with the Maxwellian F_M.

Existence of the quasi-steady state in the collisionless ITG turbulence has recently been confirmed by means of a direct (so-called Vlasov) numerical simulation (DNS) of the gyrokinetic equation averaged in the perpendicular velocity space [6]. The phase mixing generates fine-scale fluctuations of δf and leads to continuous growth of high-order moments which balances with the transport flux. The phase relation between the temperature and the parallel heat flux has also been examined and compared with the kinetic-fluid closure models [5, 7]. In this paper, we report simulation results of steady ITG turbulence with finite collisionality along with the collisionless one. Introduction of the finite collisionality allows the system to approach a *real* steady state. Even if the collision frequency is much smaller than characteristic ones of the ITG modes, it definitely affects evolution of the system by damping the fine-scale fluctuations of δf in the velocity space. A direct comparison of the kinetic and fluid simulations is also briefly discussed.

MODEL DESCRIPTION

We consider a periodic two-dimensional slab configuration with translational symmetry in the z-direction. The uniform magnetic field is given by $\mathbf{B} = B(\hat{z} + \theta \hat{y})$ where $\theta \ll 1$. The governing equations considered here are derived from the $v_\perp$-integral of gyrokinetic equations [8] by neglecting the parallel nonlinear term and by assuming $\delta f_\mathbf{k}(v_\parallel, v_\perp) = \tilde{f}_\mathbf{k}(v_\parallel) F_M(v_\perp)$. They are written in the wave number space $\mathbf{k} = (k_x, k_y)$ as

$$\partial_t \tilde{f}_\mathbf{k} + i\Theta v_\parallel k_y \tilde{f}_\mathbf{k} + \sum_{\mathbf{k}=\mathbf{k}'+\mathbf{k}''} (k'_y k''_x - k'_x k''_y) \Psi_{\mathbf{k}'} \tilde{f}_{\mathbf{k}''} = -i k_y \Psi_\mathbf{k} \left[1 + (v_\parallel^2 - 1 - k^2)\eta_i/2 + \Theta v_\parallel \right] F_M(v_\parallel) + C_i(\tilde{f}_\mathbf{k}) \quad (1)$$

and
$$[1-\Gamma_0(k^2)]\phi_{\mathbf{k}} = e^{-k^2/2}\int \tilde{f}_{\mathbf{k}}(v_\parallel)dv_\parallel - \tilde{n}_{e,\mathbf{k}}, \qquad (2)$$

where the electric potential $\phi_{\mathbf{k}}$ is related to $\Psi_{\mathbf{k}}$ by $\Psi_{\mathbf{k}} = e^{-k^2/2}\phi_{\mathbf{k}}$ with $k^2 = k_x^2 + k_y^2$. We employ the Lenard-Bernstein model collision operator, $C_i(\tilde{f}_{\mathbf{k}}) = \nu \partial_{v_\parallel}[\partial_{v_\parallel} + v_\parallel]\tilde{f}_{\mathbf{k}}(v_\parallel)$, with the collision frequency ν. The above equations are normalized as follows; $x = x'/\rho_i$, $y = y'/\rho_i$, $v = v'/v_{ti}$, $t = t'v_{ti}/L_n$, $\tilde{f} = \tilde{f}'L_n v_{ti}/\rho_i n_0$, and $\phi = e\phi'L_n/T_i\rho_i$, where v_{ti}, ρ_i, n_0, e, and T_i are the ion thermal velocity, the ion thermal gyro-radius, the background plasma density, the elementary charge, and the background ion temperature ($T_i = m_i v_{ti}^2$; m_i means the ion mass), respectively. Prime means a dimensional quantity. Θ is defined as $\Theta = \theta L_n/\rho_i$. $\Gamma_0(k^2)$ is given by $\Gamma_0(k^2) = \exp(-k^2)I_0(k^2)$. $I_0(z)$ means the 0-th modified Bessel function of z. We also neglect $k_x = 0$ modes of $\tilde{f}_{\mathbf{k}}$, since they are included in the background part with constant density and temperature gradients in the x-direction. The background electron temperature $T_e = T_i$ and the adiabatic electron response are also assumed, such that $\tilde{n}_{e,\mathbf{k}} = \phi_{\mathbf{k}}$ for $k_y \neq 0$.

It is well known that the zonal flow excited by the turbulent stress suppresses the transport flux down to a quite small level [9]. Here, it should be recalled that, in a toroidal geometry, the zonal flow is severely damped by the collisionless transit time magnetic pumping effect [10, 11]. Thus, χ_i observed in the L-mode of a toroidal system is on a much higher level than that obtained in the slab geometry. In order to simulate the large transport observed in the L-mode discharge, we consider a limiting case without the zonal flow components by fixing $\tilde{f}_{k_y=0} = \phi_{k_y=0} = 0$.

In the system described above, defining a functional (increase of the entropy) $\delta S = \sum_{\mathbf{k}} \int dv_\parallel |\tilde{f}_{\mathbf{k}}|^2/2F_M$, the perpendicular ion heat flux $Q_i = \sum_{\mathbf{k}} \int dv_\parallel (-ik_y e^{-k^2/2}\phi_{\mathbf{k}})v_\parallel^2 \tilde{f}_{-\mathbf{k}}/2$, the potential energy $W = \sum_{\mathbf{k}}[1 - \delta_{k_y,0} + (T_e/T_i)(1 - \Gamma_0)]|\phi_{\mathbf{k}}|^2/2$, and the collisional dissipation $D = \sum_{\mathbf{k}} \int dv_\parallel \tilde{f}_{-\mathbf{k}} C_i(\tilde{f}_{\mathbf{k}})/F_M$, one obtains the following relation from Eq.(1),

$$\frac{d}{dt}(\delta S + W) = \eta_i Q_i + D. \qquad (3)$$

We *directly* solve Eq.(1) as a partial differential equation in the phase space without use of an *ad hoc* model such as the finite-sized particles. In order to correctly simulate the phase mixing process, when $\nu = 0$, we have employed 8,193 grid points for discretization of the velocity space ($-5 \leq v_\parallel \leq 5$). Time-integration of Eq.(1) has been carefully calculated by means of the non-dissipative integrator conserving δS [12]. For the case with finite collisionality, we can reduce the grid numbers in accordance to the magnitude of ν, where the fourth-order Runge-Kutta method is used for the time-integration with a sufficiently small time step Δt. Here, $\Delta t = 0.0125$ for $\nu \geq 5 \times 10^{-4}$, but $\Delta t = 6.25 \times 10^{-3}$ for $\nu < 5 \times 10^{-4}$ (and also for the collisionless case) so that the numerical error in Eq.(3) should be much smaller than $\eta_i Q_i$ and D. The minimum and maximum values of the wave number are set to be $k_{\min} = 0.1$ and $k_{\max} = 3.2$ for both of the k_x- and k_y-directions with the 3/2-rule for de-aliasing in the spectral method.

SIMULATION RESULTS

In the collisionless system with $\nu = 0$, according to Eq.(1), the quasi-steady state with monotonical increase of δS should be realized in turbulence for statistical averages of W and Q_i to be constant. On the other hand, average of δS should be constant in the steady state of the collisional system. In order to examine effects of the finite collisionality, we have performed several simulations for different ν, where we set $\eta_i = 10$ and $\Theta = 2.5$. For these parameters, the angular frequency and the linear growth rate of the most unstable mode with $k_x = 0.1$ and $k_y = 0.3$ are -0.957 and 7.73×10^{-2}, respectively.

Time-evolutions of δS are plotted in Fig.1 for the collisionless and weakly-collisional cases. The monotonical increase of δS is observed in the collisionless simulation where the potential energy W and the heat flux Q_i fluctuate around constant levels (not shown here) [6]. The smallest scale-length of δf fluctuating in the velocity space is continuously shortened by the phase mixing. When it reaches to the grid size at $t = 800$, the simulation is stopped in the collisionless case. It is confirmed that the time-derivative of δS balances with $\eta_i Q_i$, that is, $\overline{d(\delta S)/dt} \approx \overline{\eta_i Q_i} \neq 0$, which means the collisionless ITG turbulence is in the *quasi*-steady state. Here, $\overline{\cdots}$ means time averaging on a certain period, longer than characteristic times of the turbulence. In presence of the finite collisionality, the growth of δS is saturated in the turbulence as well as W and Q_i. Thus, the collisional dissipation D balances with the mean transport, that is, $\overline{\eta_i Q_i} \approx -\overline{D}$, which means the *real* statistically steady state where not only the low-order moments of δf but also δf itself are statistically steady.

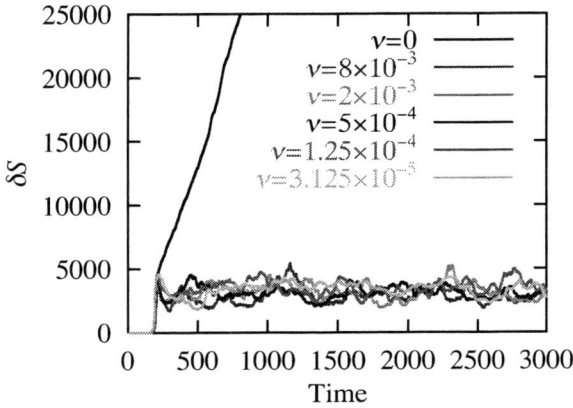

FIGURE 1. Time-evolutions of δS for collisionless and collisional cases.

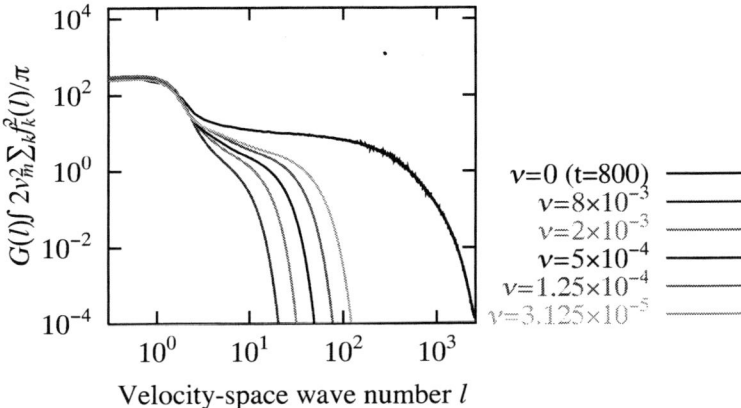

FIGURE 2. Velocity-space spectra of δf for collisionless and collisional cases.

Velocity-space spectra $G(l)$ of δf defined by $G(l) \equiv 2v_m^2 \sum_{\mathbf{k}} |f_{\mathbf{k}}(l)|^2/\pi$ are shown in Fig.2, where v_m and l denote the maximum velocity treated in the simulations and the velocity space wave number, respectively. $f_{\mathbf{k}}(l)$ is the Fourier component of $f_{\mathbf{k}}(v_{\parallel})$. For the collisionless case, a snapshot of the spectrum at $t = 800$ is plotted, while $G(l)$ is time-averaged from $t = 1000$ to 3000 for the collisional ones. In the collisionless case, $G(l)$ has a peaked profile in a low-l side, which is related to the linear eigenfunctions of the unstable modes [6], followed by a long tail in a high-l region. The low-l profile of $G(l)$ is preserved throughout the nonlinear stage, while the tail of $G(l)$ continues to expand as the phase mixing proceeds after the linear growth. The expansion of $G(l)$ into the high-l side, namely, the continuous generation of the fine-scale fluctuations of δf in the velocity space, is responsible for the monotonical increase of δS. In the collisional cases, one obtains steady profiles of $G(l)$ which coincide to the collisionless one in the low-l region, while shorter tails are found in the high-l side for larger values of ν.

Collision frequency dependence of the transport coefficient, $\chi_i \equiv Q_i/\eta_i$, is summarized in Fig.3 where the time-average is taken from $t = 1000$ to 3000. The error bars are estimated from the root-mean-square of differences between running-averaged χ_i for a time period $\tau = 10$ and the total average. In the range of $5 \times 10^{-4} < \nu < 8 \times 10^{-3}$, χ_i has a logarithmic dependence on ν. When χ_i is decreased to a level of the collisionless one, that is, $\chi_i \approx 0.36\rho_i^2 v_{ti}/L_n$, ν-dependence of χ_i becomes quite weak (for $\nu < 5 \times 10^{-4}$ in the present case). This agrees with a concept that the quasi-steady state is regarded as an idealization of the real steady state where growth of high-order moments of δf are saturated as well due to collisional dissipation [5]. It is seen from Figs.2 and 3 that, for small collisionality, microscopic (high-l) structures, which are responsible for dissipation, adjust themselves to the steady state while keeping macroscopic (low-l) structures and heat transport unchanged. This is also consistent with the conjecture of "*flux determines dissipation*" by Krommes and Hu [3].

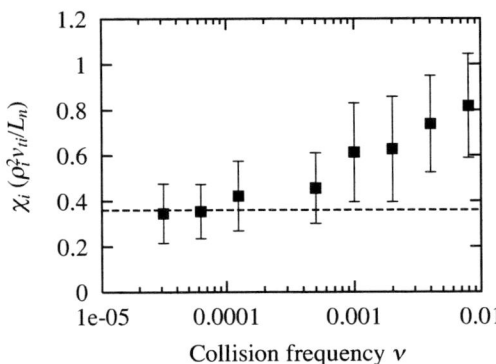

FIGURE 3. Perpendicular ion heat transport coefficient (χ_i) for different collision frequency ν. Dashed line represents χ_i for $\nu = 0$.

We have also compared the collisionless kinetic simulation with kinetic-fluid ones for the slab ITG turbulence [13]. The transport coefficient obtained by the kinetic simulation agrees well with that of the kinetic-fluid simulation using the non-dissipative closure model (NCM) [5], while the Hammett-Perkins (H-P) closure model [7] results in two or three times larger value. The difference is attributed to the constant and dissipative phase relation of the temperature $T_\mathbf{k}$ and the parallel heat flux $q_\mathbf{k}$ in the H-P model, while an oscillatory phase relation of $T_\mathbf{k}$ and $q_\mathbf{k}$ found in the kinetic simulation [6] can be reproduced by NCM.

CONCLUDING REMARKS

We have carried out collisionless and weakly-collisional simulations of the slab ITG turbulence, where the *quasi-* and *real* steady states are observed, respectively. The velocity-space power spectra of δf have a common profile in the low velocity-space wave number (l) side, while a tail of the spectra extends longer into the higher-l region for smaller collision frequency (ν). It is also found that the transport coefficient is insensitive to small ν but has a logarithmic dependence for larger ν. The transport coefficient obtained by the direct kinetic simulation shows a better agreement with that of the NCM fluid simulation than with that of the fluid simulation using the H-P model.

ACKNOWLEDGMENTS

This work is partially supported by grants-in-aid of Ministry of Education, Culture, Sports, Science and Technology (No.12680497 and 14780387).

REFERENCES

1. W. Horton, Rev. Mod. Phys. **71**, 735 (1999).
2. A. M. Dimits, G.Bateman, M. A. Beer, *et al.*, Phys. Plasmas **7**, 969 (2000).
3. J. A. Krommes and G. Hu, Phys. Plasmas **1**, 3211 (1994).
4. H. Sugama, M. Okamoto, W. Horton, and M. Wakatani, Phys. Plasmas **3**, 2379 (1996).
5. H. Sugama, T.-H. Watanabe, and W. Horton, Phys. Plasmas **8**, 2617 (2001).
6. T.-H. Watanabe and H. Sugama, Phys. Plasmas (in press).
7. G. W. Hammett and F. W. Perkins, Phys. Rev. Lett. **64**, 3019 (1990).
8. D. H. E. Dubin, J. A. Krommes, C. Oberman, and W. W. Lee, Phys. Fluids **26**, 3524 (1983).
9. P. W. Terry, Rev. Mod. Phys. **72**, 109 (2000).
10. T. H. Stix, Phys. Fluids **16**, 1260 (1973).
11. Z. Lin, T. S. Hahm, W. W. Lee, W. M. Tang, and R. B. White, Phys. Plasmas **7**, 1857 (2000).
12. T.-H. Watanabe, H. Sugama, and T. Sato, J. Phys. Soc. Jpn. **70**, 3565 (2001).
13. H. Sugama, T.-H. Watanabe, and W. Horton, in proceedings of "*29th EPS Conference on Controlled Fusion and Plasma Physics*", P-5.059 (2002).

Radio Frequency Dielectric Characteristics of a Collisionless Laboratory Dipole Plasma

N.I. Grishanov[*], A.F.D. Loula[*], C.A. de Azevedo[♦], J.P. Neto[♦]

[*]*Laboratório Nacional de Computação Científica, 25651-070 Petropolis, Brazil*
[♦]*Universidade do Estado do Rio de Janeiro, 20550-013 Rio de Janeiro, Brazil*

Abstract. Transverse and longitudinal dielectric permittivity elements have been derived for radio-frequency waves in a laboratory dipole plasma model by solving the Vlasov equation for trapped and untrapped particles accounting for the cyclotron and bounce resonances.

PLASMA MODEL AND VLASOV EQUATION

One of the interesting axisymmetric plasma configurations is a plasma confined in a dipole magnetic field created by a ring current [1, 2]. The point dipole magnetic field is a good approximation for the Earth's magnetosphere, whereas the laboratory dipole magnetic field configuration (with the finite ring current radius) should be considered for LDX (Levitated Dipole eXperiment) plasma [2]. As is well known, the two-dimensional (2D) kinetic wave theory in such plasmas should be based on the solution of Maxwell's equations using the correct kinetic dielectric tensor accounting for the cyclotron and bounce resonances. In this paper, we evaluate the contributions of both the trapped and untrapped (circulating or passing) particles to the dielectric permittivity elements for radio-frequency waves in the 2D laboratory dipole magnetic field plasma (LDMFP) using an approach developed in Refs. [3, 4] for the Earth's magnetosphere.

To describe a 2D axisymmetric LDMFP, we use the quasi-toroidal coordinates (r, θ, ϕ) connected with cylindrical ones (ρ, ϕ, z) as $\rho = a + r\cos\theta$, $z = -r\sin\theta$, $\phi = \phi$. In this case, the cylindrical components of an equilibrium magnetic field, $\mathbf{H}_0$, are

$$H_\rho = \frac{2Ir\sin\theta}{c(a+r\cos\theta)\sqrt{r^2+4a^2+4ar\cos\theta}} \left[K(\kappa) - \frac{r^2+2a^2+2ar\cos\theta}{r^2} E(\kappa) \right] \quad (1)$$

$$H_\phi = 0, \qquad H_z = \frac{2I}{c\sqrt{r^2+4a^2+4ar\cos\theta}} \left[K(\kappa) - \frac{r+2a\cos\theta}{r} E(\kappa) \right] \quad (2)$$

where a is the ring radius, I is the current, c is the speed of light, $K(\kappa)$ and $E(\kappa)$ are the complete elliptic integrals of the first and second kind, respectively,

$$K(\kappa) = \int_0^{\frac{\pi}{2}} \frac{dx}{\sqrt{1-\kappa\sin^2 x}}, \qquad E(\kappa) = \int_0^{\frac{\pi}{2}} \sqrt{1-\kappa\sin^2 x}\, dx, \qquad \kappa = \frac{4a(a+r\cos\theta)}{r^2+4a^2+4ar\cos\theta}. \quad (3)$$

To solve the Vlasov equation we use the standard method of switching to new variables associated with the conservation integrals of the particles energy $v_\parallel^2 + v_\perp^2 = const$, the magnetic moment $v_\perp^2/H_0 = const$, and the equation of an equilibrium magnetic field line $\sqrt{r^2+4a^2+4ar\cos\theta}\,[(2-\kappa)K(\kappa)-2E(\kappa)] = const$, where $H_0 = \sqrt{H_\rho^2 + H_z^2}$ is the module of $\mathbf{H}_0$. Introducing the new variables (v, μ, L) instead of $(v_\parallel, v_\perp, r)$ as

$$v = \sqrt{v_\parallel^2 + v_\perp^2}, \qquad \mu = \frac{v_\perp^2 H_0(r,0)}{v^2 H_0(r,\theta)}, \qquad L = \frac{\pi a}{\sqrt{r^2 + 4a^2 + 4ar\cos\theta}\left[(2-\kappa)K(\kappa) - 2E(\kappa)\right]} \quad (4)$$

the Vlasov equation for the first harmonics of the perturbed distribution function,

$$f(t, \mathbf{r}, \mathbf{v}) = \sum_s^{\pm 1} \sum_l^{\pm\infty} f_l^s(L, \theta, v, \mu) \exp(-i\omega t + in\phi + il\sigma) \quad (5)$$

in the zeroth order of the magnetization parameter, can be reduced to the first order differential equation with respect to one θ variable

$$\sqrt{1 - \frac{\mu}{b(L,\theta)}} \frac{1}{\lambda(L,\theta)} \frac{\partial f_l^s}{\partial \theta} - is\frac{La}{v}(\omega + l\Omega_c)f_l^s = Q_l^s \quad (6)$$

Here

$$\lambda(L,\theta) = \frac{cH_0(r,\theta)(a + r\cos\theta)\sqrt{r^2 + 4a^2 + 4ar\cos\theta}}{2Ila\left[(r^2 + 2a^2 + 3ar\cos\theta)E(\kappa) - r(a + r\cos\theta)K(\kappa)\right]} \quad (7)$$

$$b(L,\theta) = \frac{H_0(r(L,0),0)}{H_0(r(L,\theta),\theta)}, \quad \Omega_c = \frac{eH_0}{Mc}, \quad F = \frac{N}{\pi^{1.5} v^3}\exp\left(-\frac{v^2}{v_T^2}\right), \quad v_T = \sqrt{\frac{2T}{M}} \quad (8)$$

$$Q_0^s = \frac{e}{T}LaF\sqrt{1 - \frac{\mu}{b(L,\theta)}} E_\parallel, \qquad Q_{\pm 1}^s = \frac{se}{2T}LaF\sqrt{\frac{\mu}{b(L,\theta)}}(E_n \mp iE_b) \quad (9)$$

where $l = 0, \pm 1$ is the number of the cyclotron harmonics, σ is the gyrophase angle in velocity space; F is the equilibrium distribution function of particles with the density N, mass M, charge e, temperature T; E_n, E_b and $E_\parallel$ are, respectively, the normal, binormal and parallel perturbed electric field components relative to $\mathbf{H}_0$; further $E_l = E_n \mp ilE_b$, The index of particles kind is omitted in Eqs. (5,6,8,9). By $s = \pm 1$ we differ the particles with positive and negative parallel velocity, $v_\parallel$, relative to $\mathbf{H}_0$: $v_\parallel = sv\sqrt{1 - \mu/b(L,\theta)}$. Note, the "old" r variable in Eqs. (7) and (8) should be determined by $r(L,\theta)$ satisfying Eq. (4), $L=L(r,\theta)$.

Since LDMFP is a configuration with one minimum of $\mathbf{H}_0$, the plasma particles should be split in the two populations of the so-called trapped and untrapped particles. In the phase volume such separation can be done by the μ variable. In dependence on μ and θ we have 1) $0 \leq \mu \leq \mu_0$, $-\pi \leq \theta \leq \pi$ for the untrapped particles, where $\mu_0 = b(L,\pi)$, and 2) $\mu_0 \leq \mu \leq 1$, $-\theta_t \leq \theta \leq \theta_t$ for the trapped particles, where $\pm\theta_t(\mu,L)$ are the reflection points (or stop points, or mirror points) of the trapped particles, i.e., $v_\parallel(v,\mu,L,\pm\theta_t) = 0$.

DIELECTRIC TENSOR ELEMENTS

After solving Eq. (6), the 2D transverse and longitudinal (relative to $\mathbf{H}_0$) current density components, respectively $j_{\pm 1}$ and $j_\parallel$, can be found as

$$j_\parallel(L,\theta) = \frac{\pi e}{b(L,\theta)}\sum_s^{\pm 1} s\int_0^\infty v^3 dv\left[\int_0^{\mu_0} f_{0,u}^s(L,\theta,v,\mu)d\mu + \int_{\mu_0}^{b(L,\theta)} f_{0,t}^s(L,\theta,v,\mu)d\mu\right] \quad (10)$$

$$j_l(L,\theta) = \frac{\pi e}{2b(L,\theta)}\sum_s^{\pm 1} \int_0^\infty v^3 dv\left[\int_0^{\mu_0} \frac{f_{l,u}^s \sqrt{\mu}d\mu}{\sqrt{b(L,\theta) - \mu}} + \int_{\mu_0}^{b(L,\theta)} \frac{f_{l,t}^s \sqrt{\mu}d\mu}{\sqrt{b(L,\theta) - \mu}}\right], \quad l = \pm 1 \quad (11)$$

where the subscribed indexes u and t correspond to the untrapped and trapped particles, respectively. Of course, the population of u-particles is very small at the external magnetic surfaces since $\mu_0 \to 0$ if L increases. Note, the normal and binormal to $\mathbf{H}_0$ current density components, j_n and j_b in our notation, are equal to $j_n = j_{+1} + j_{-1}$ and $j_b = i(j_{+1} - j_{-1})$. The expressions for $j_l|_{l=\pm 1}$ are convenient to analyze the cyclotron resonance effects on the fundamental cyclotron frequency of both the ions (if $l=-1$) and electrons (if $l=+1$) in the explicit form.

To describe the bounce-periodic motion of the trapped and untrapped particles along $\mathbf{H}_0$-field line, it is convenient to introduce the new time-like variable τ instead of θ,

$$\tau(\theta) = \int_0^\theta \frac{\lambda(L,\theta')}{\sqrt{1-\mu/b(L,\theta')}} d\theta', \tag{12}$$

accounting for that the bounce-periods of the u- and t-particles are proportional to $T_{b,u}=2\tau(\pi)$ and $T_{b,t}=4\tau(\theta_\tau)$. After this, the distribution functions of u- and t-particles can be defined by the corresponding Fourier series ($\alpha=u,t$)

$$f_{l,\alpha}^s = \sum_p^{\pm\infty} f_{l,\alpha}^{s,p} \exp\left[ip\frac{2\pi}{T_{b,\alpha}}\tau(\theta) + isl\frac{La}{v}\int_0^\theta \frac{\tilde{\Omega}_{c,\alpha}\lambda(L,\theta')}{\sqrt{1-\mu/b(L,\theta')}}d\theta'\right] \tag{13}$$

where $\tilde{\Omega}_{c,u} = \Omega_{c,u} - \overline{\Omega}_{c,u}$ and $\tilde{\Omega}_{c,t} = \Omega_{c,t} - \overline{\Omega}_{c,t}$ are the oscillating parts of the cyclotron frequencies of the u- and t-particles; and $\overline{\Omega}_{c,u}$ and $\overline{\Omega}_{c,t}$ are the corresponding bounce-averaged cyclotron frequencies (secular part of Ω_c) of untrapped and trapped particles:

$$\overline{\Omega}_{c,u} = \frac{2}{T_{b,u}}\int_0^\pi \frac{\Omega_c \lambda(L,\theta)}{\sqrt{1-\mu/b(L,\theta)}}d\theta, \qquad \overline{\Omega}_{c,t} = \frac{4}{T_{b,t}}\int_0^{\theta_t} \frac{\Omega_c \lambda(L,\theta)}{\sqrt{1-\mu/b(L,\theta)}}d\theta. \tag{14}$$

The Fourier amplitudes $f_{l,u}^{s,p}$ and $f_{l,t}^{s,p}$ (for $l=0,\pm 1$) should be found, as usual, by the corresponding bounce-averagings ($\alpha=u,t$):

$$f_{l,\alpha}^{s,p} = \frac{-iv}{2\pi p v - s T_{b,\alpha} La(\omega + l\overline{\Omega}_{c,\alpha})} \int_{-T_{b,\alpha}}^{T_{b,\alpha}} Q_l^s \exp\left[-ip\frac{\tau}{T_{b,\alpha}} - isl\frac{La}{v}\int_0^\tau \tilde{\Omega}_{c,\alpha} d\tau'\right] d\tau. \tag{15}$$

Note, using Eq. (13) we satisfy automatically the corresponding boundary conditions for the perturbed distribution functions: a namely, 1) the periodicity of the untrapped particles circulating along $\mathbf{H}_0$-field line

$$f_{l,u}^s(L,\theta) = f_{l,u}^s(L,\theta+2\pi) \qquad \text{or} \qquad f_{l,u}^s(L,\tau) = f_{l,u}^s(L,\tau+T_{b,u}) \tag{16}$$

and 2) the continuity of the perturbed distribution functions of the trapped particles at the reflection points

$$f_{l,t}^s(L,\pm\theta_t) = f_{l,t}^{-s}(L,\pm\theta_t) \qquad \text{or} \qquad f_{l,t}^s(L,\tau) = f_{l,t}^s(L,\tau+T_{b,t}). \tag{17}$$

To evaluate the dielectric tensor elements we use the Fourier expansions of the perturbed electric field and current density components over θ

$$b(L,\theta)\mathbf{j}(L,\theta) = \sum_m^{\pm\infty} \mathbf{j}^{(m)}(L)\exp(im\theta), \qquad \lambda(L,\theta)\mathbf{E}(L,\theta) = \sum_{m'}^{\pm\infty} \mathbf{E}^{(m')}(L)\exp(im'\theta). \tag{18}$$

As a result,

$$\frac{4\pi i}{\omega} j_l^{(m)} = \frac{2i}{\omega}\int_{-\pi}^{\pi} b(L,\theta) j_l(L,\theta)\exp(-im\theta)d\theta = \sum_{m'}^{\pm\infty}\left(\varepsilon_{l,u}^{m,m'} + \varepsilon_{l,t}^{m,m'}\right)E_l^{(m')}, \quad l=\pm 1 \tag{19}$$

$$\frac{4\pi i}{\omega} j_\parallel^{(m)} = \frac{2i}{\omega}\int_{-\pi}^{\pi} b(L,\theta) j_\parallel(L,\theta)\exp(-im\theta)d\theta = \sum_{m'}^{\pm\infty}\left(\varepsilon_{\parallel,u}^{m,m'} + \varepsilon_{\parallel,t}^{m,m'}\right)E_\parallel^{(m')} \tag{20}$$

where $\varepsilon_{l,u}^{m,m'}$, $\varepsilon_{l,t}^{m,m'}$ and $\varepsilon_{\parallel,u}^{m,m'}$, $\varepsilon_{\parallel,t}^{m,m'}$ are the contribution of untrapped (u) and trapped (t) particles to the transverse and longitudinal permittivity elements, respectively. After the s-summation, the dielectric permittivity elements can be expressed as

$$\varepsilon_{l,u}^{m,m'} = \frac{\omega_p^2 L\,a}{8\omega\pi^{2.5} v_T} \sum_{p=-\infty}^{\infty} \int_0^{\mu_0} \mu\, d\mu \int_{-\infty}^{\infty} \frac{u^4 \exp(-u^2)}{pu - Z_{l,u}} A_{l,p}^m(u,\mu) A_{l,p}^{m'}(u,\mu) du \tag{21}$$

$$\varepsilon_{l,t}^{m,m'} = \frac{\omega_p^2 L\,a}{8\omega\pi^{2.5} v_T} \sum_{p=-\infty}^{\infty} \int_{\mu_0}^1 \mu\, d\mu \int_{-\infty}^{\infty} \frac{u^4 \exp(-u^2)}{pu - Z_{l,t}} \hat{B}_{l,p}^m(u,\mu) B_{l,p}^{m'}(u,\mu) du \tag{22}$$

$$\varepsilon_{\parallel,u}^{m,m'} = \frac{\omega_p^2 L^2 a^2}{8\pi^3 v_T^2} \sum_{p=1}^{\infty} \frac{1}{p^2} \int_0^{\mu_0} T_{b,u} C_p^m C_p^{m'}\left[1 + 2u_p^2 + 2i\sqrt{\pi}u_p^3 W(u_p)\right] d\mu \tag{23}$$

$$\varepsilon_{\parallel,t}^{m,m'} = \frac{\omega_p^2 L^2 a^2}{8\pi^3 v_T^2} \sum_{p=1}^{\infty} \frac{1}{p^2} \int_{\mu_0}^1 T_{b,t} D_p^m D_p^{m'}\left[1 + 2v_p^2 + 2i\sqrt{\pi}v_p^3 W(v_p)\right] d\mu. \tag{24}$$

Here we have introduced the following definitions

$$A_{l,p}^m(u,\mu) = \int_{-\pi}^{\pi} \cos\left[m\theta - \left(\frac{2\pi p}{T_{b,u}} - \frac{lLa}{uv_T}\overline{\Omega}_{c,u}\right)\tau(\theta) - \frac{lLa}{uv_T}\int_0^\theta \frac{\Omega_c \lambda(L,\eta)d\eta}{\sqrt{1-\mu/b(L,\eta)}}\right] \frac{d\theta}{\sqrt{b(L,\theta)-\mu}} \quad (25)$$

$$\hat{B}_{l,p}^m(u,\mu) = \int_{-\theta_t}^{\theta_t} \cos\left[m\theta - \left(\frac{2\pi p}{T_{b,t}} - \frac{lLa}{uv_T}\overline{\Omega}_{c,t}\right)\tau(\theta) - \frac{lLa}{uv_T}\int_0^\theta \frac{\Omega_c \lambda(L,\eta)d\eta}{\sqrt{1-\mu/b(L,\eta)}}\right] \frac{d\theta}{\sqrt{b(L,\theta)-\mu}} \quad (26)$$

$$B_{l,p}^m(u,\mu) = \hat{B}_{l,p}^m(u,\mu) + (-1)^p \hat{B}_{l,-p}^m(-u,\mu), \qquad u_p = \frac{v}{v_T}, \qquad \omega_p^2 = \frac{4\pi N e^2}{M} \quad (27)$$

$$Z_{l,u} = \frac{LaT_{b,u}}{2\pi v_T}(\omega + l\overline{\Omega}_{c,u}), \quad Z_{l,t} = \frac{LaT_{b,t}}{2\pi v_T}(\omega + l\overline{\Omega}_{c,t}), \quad u_p = \frac{\omega LaT_{b,u}}{2\pi p v_T}, \quad v_p = \frac{\omega LaT_{b,t}}{2\pi p v_T} \quad (28)$$

$$C_p^m(\mu) = \int_{-\pi}^{\pi} \cos\left[m\theta - \frac{2\pi p}{T_{b,u}}\tau(\theta)\right]d\theta, \qquad W(z) = \exp(-z^2)\left[1 + \frac{2i}{\sqrt{\pi}}\int_0^z \exp(t^2)dt\right] \quad (29)$$

$$D_p^m(\mu) = \int_{-\theta_t}^{\theta_t} \cos\left[m\theta - \frac{2\pi p}{T_{b,t}}\tau(\theta)\right]d\theta + (-1)^{p-1}\int_{-\theta_t}^{\theta_t} \cos\left[m\theta + \frac{2\pi p}{T_{b,t}}\tau(\theta)\right]d\theta. \quad (30)$$

As was mentioned above, Eqs. (21-24) describe the contribution of any kind of untrapped and trapped particles to the dielectric tensor elements. The corresponding expressions for plasma electrons and ions can be obtained from (21)-(24) by replacing T (temperature), N (density), M (mass), e (charge) by the electron T_e, N_e, m_e, e_e and ion T_i, N_i, M_i, e_i parameters, respectively. To obtain the total expressions of the permittivity elements, as usual, it is necessary to carry out the summation over all species of plasma particles. Note, since the phase coefficients $C_p^m(\mu)$ and $D_p^m(\mu)$ for the longitudinal permittivity elements, are independent of the wave frequency, ω, and the particle energy, v, it is possible the analytical Landau integration of the perturbed distribution functions of both the trapped and untrapped particles in velocity space. As a result, $\varepsilon_{\parallel,u}^{m,m'}$ and $\varepsilon_{\parallel,t}^{m,m'}$ are written by the summation of bounce-resonant terms including the well known plasma dispersion function $W(z)$, i.e, by the probability integral of the complex argument, Eq. (29). After this, the numerical estimations of both the real and imaginary parts of the longitudinal permittivity elements become simpler, and their dependence on the wave frequency ω is defined only by the arguments $u_p(\mu,\omega,...)$ and $v_p(\mu,\omega,...)$ of the plasma dispersion functions, $W(u_p)$ and $W(v_p)$. Using the periodicity of the perturbed distribution function of the trapped and untrapped particles over the τ variable, Eqs. (16) and (17), the wave-particle resonance conditions in the LDMFP are defined as usually by the zeros of the corresponding denominators of the $f_{l,u}^s$ and $f_{l,t}^s$ harmonics of the perturbed distribution functions of the untrapped and trapped particles. These resonant denominators are presented in Eqs. (21) and (22) in the explicit form.

ACKNOWLEDGMENTS

This research was supported by CNPq of Brazil (Conselho Nacional de Desenvolvimento Científico e Tecnológico, project 300637/01-2).

REFERENCES

1. Hasegawa, A., Comments on Plasma Phys. Controlled Fusion **1**, 147-163 (1987).
2. Kesner, J., Bromberg, L., Mauel, M.E., and Garnier, D.T., 17th Int. Conf. Plasma Phys. Control. Fusion Res., Yokahama, Japan, 19-23 Oct, 1998, paper IAEA-F1-CN-69/ICP/09.
3. Grishanov, N.I., de Azevedo, C.A., and de Assis, A.S., Phys. Plasmas **5**, 4384-4394 (1998).
4. Grishanov, N.I., de Azevedo, C.A., and de Assis, A.S., 1998 Int. Congress Plasma Phys. & 25th EPS Conf. Control. Fusion Plasma Phys., Praha, Czech Republic, 29 June - 3 July, 1998, pp. 1146-1149.

Wave Dissipation by Electron Landau Damping in Axisymmetric D-shaped Tokamaks

N.I. Grishanov[*], A.F.D. Loula[*], C.A. de Azevedo[♦], J.P. Neto[♦]

[*]*Laboratório Nacional de Computação Científica, 25651-070 Petropolis, Brazil*
[♦]*Universidade do Estado do Rio de Janeiro, 20550-013 Rio de Janeiro, Brazil*

Abstract. Longitudinal permittivity elements have been derived for radio-frequency waves in a toroidal plasma with D-shaped magnetic surfaces. These dielectric characteristics are suitable to estimate the wave dissipation by electron Landau damping (e.g., during the plasma heating and current drive generation) in the frequency range of Alfvén, fast magnetosonic, and lower hybrid waves, for both the large and low aspect ratio tokamaks.

PLASMA MODEL AND DRIFT-KINETIC EQUATION

As is well known, the wave electron Landau damping in any magnetized plasma should be described in the scope of kinetic theory. However, for axisymmetric tokamaks, this problem is not simple since to solve the two-dimensional Maxwell's equations it is necessary to use the correct kinetic dielectric tensor valid in a given frequency range for realistic plasma models. Many nowadays tokamaks, essentially the spherical ones, have the elongated cross-sections of the magnetic surfaces. In this paper, longitudinal permittivity elements are derived for radio-frequency waves in a toroidal plasma with D-shaped magnetic surfaces, under arbitrary aspect ratio and elongation, but small triangularity. A collisionless plasma model is considered. Drift-kinetic equation is solved separately for untrapped and usual t-trapped particles as a boundary-value problem in the case when the so-called d-trapped particles are absent in the plasma, using an approach developed for low aspect ratio tokamaks with circular [1] and elliptic [2] magnetic surfaces.

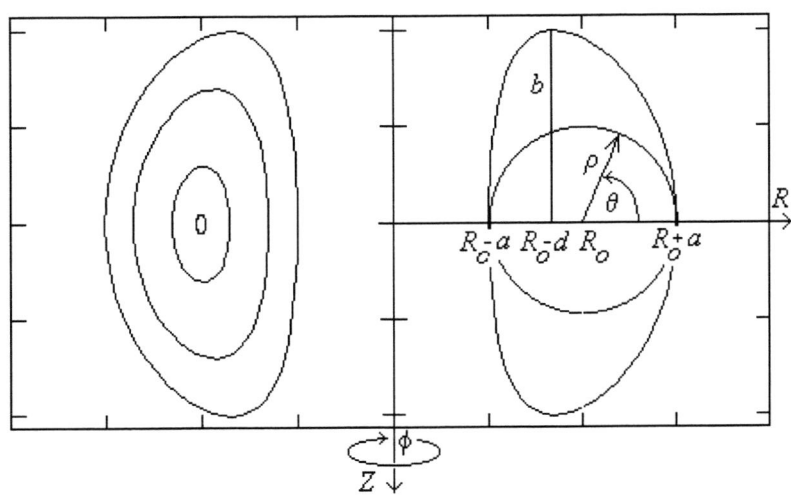

FIGURE 1. Cylindrical (R, ϕ, Z) and quasi-toroidal (ρ, θ, ϕ) coordinates to describe an axisymmetric tokamak with D-shaped magnetic surfaces.

To describe an axisymmetric D-shaped tokamak we use the quasi-toroidal coordinates (ρ,θ,ϕ) connected with the cylindrical ones (R, ϕ, Z) as

$$R = R_0 + \rho\cos\theta - \frac{d\rho^2}{a^2}\sin^2\theta, \qquad \phi = \phi, \qquad Z = -\frac{b}{a}\rho\sin\theta, \qquad (1)$$

where R_0 is the radius of the magnetic axis; a and b are, respectively, the minor and major semiaxes of the cross-section of the external magnetic surface, see Fig. 1. In this model, all magnetic surfaces are similar each other with the same elongation equal to b/a; the triangularity is small $d/a \ll 1$; Shafranov shift is not accounted. In the (ρ,θ)-coordinates, the initial D-shaped cross-sections are transformed to the circles with the corresponding radius ρ in $0 < \rho < a$; and the cylindrical components of an equilibrium magnetic field $\mathbf{H}$ are

$$H_\phi = H_{\phi 0}\frac{R_0}{R}, \qquad H_R = H_{\theta 0}\frac{R_0}{R}\sin\theta\left(1+\frac{d\rho}{a^2}\cos\theta\right), \qquad H_Z = H_{\theta 0}\frac{b}{a}\frac{R_0}{R}\cos\theta. \qquad (2)$$

Here $H_{\phi 0}$ and $H_{\theta 0}$ are, respectively, the toroidal and poloidal magnetic field amplitudes at a given (by ρ) magnetic surface. Thus, the module $H=|\mathbf{H}|$ of an equilibrium magnetic field is

$$H(\rho,\theta) = \frac{\sqrt{H_{\phi 0}^2 + H_{\theta 0}^2}}{g(\rho,\theta)}, \qquad g(\rho,\theta) = \frac{1+\varepsilon\cos\theta - \delta\varepsilon\sin^2\theta}{\sqrt{1+\lambda\cos^2\theta + \kappa\cos\theta\sin^2\theta}} \qquad (3)$$

where

$$\varepsilon = \frac{\rho}{R_0}, \qquad \delta = \frac{d\rho}{a^2}, \qquad \lambda = h_\theta^2\left(\frac{b^2}{a^2}-1\right), \qquad \kappa = 4\delta h_\theta^2, \qquad h_\theta = \frac{H_{\theta 0}}{\sqrt{H_{\phi 0}^2 + H_{\theta 0}^2}}. \qquad (4)$$

To solve the linearized drift-kinetic equation for the perturbed distribution functions, $f(t,\mathbf{r},\mathbf{v}) = f(\rho,\theta,v_\parallel,v_\perp)\exp(-i\omega t + in\phi)$, we use the standard method of switching to new variables associated with conservation integrals of energy, $v_\parallel^2 + v_\perp^2 = const$, and magnetic moment, $v_\perp^2/2H = const$. Introducing the variables v and μ in velocity space (instead of $v_\parallel$ and $v_\perp$),

$$v^2 = v_\parallel^2 + v_\perp^2, \qquad \mu = \frac{v_\perp^2}{v_\parallel^2 + v_\perp^2}g(\rho,\theta), \qquad (5)$$

the drift-kinetic equation for harmonics f_s, $f(\rho,\theta,v_\parallel,v_\perp) = \sum_{s=\pm 1} f_s(\rho,\theta,v,\mu)$, in the zeroth order over the magnetization parameter, after averaging over the gyrophase angle in velocity space, can be reduced to the first order differential equation with respect to poloidal angle θ:

$$\sqrt{\frac{1-\mu/g(\rho,\theta)}{1+\lambda\cos^2\theta + \kappa\cos\theta\sin^2\theta}}\left(\frac{\partial f_s}{\partial\theta} + \frac{inqf_s}{1+\varepsilon\cos\theta - \varepsilon\delta\sin^2\theta}\right) - i\frac{s\rho\omega}{vh_\theta}f_s = \frac{2e\rho E_\parallel N}{Mh_\theta\pi^{1.5}v_T^5}\exp\left(-\frac{v^2}{v_T^2}\right)\sqrt{1-\frac{\mu}{g(\rho,\theta)}} \qquad (6)$$

where the variables ρ, v, μ appear as the parameters; $q = \varepsilon h_\phi/h_\theta$, $v_T^2 = 2T/M$, and N, M, e, T are, respectively, the particle density, mass, charge, temperature. By $s = \pm 1$ we distinguish the perturbed distribution functions, f_s, with positive and negative values of the parallel velocity $v_\parallel = sv\sqrt{1-\mu/g(\rho,\theta)}$ relative to $\mathbf{H}$. After solving Eq. (6), the longitudinal (parallel to $\mathbf{h}=\mathbf{H}/H$) component of the current density, $j_\parallel = \mathbf{j}\cdot\mathbf{h}$, can be expressed as

$$j_\parallel(\rho,\theta) = \frac{\pi e}{g(\rho,\theta)}\sum_s^{\pm 1} s\int_0^\infty v^3 \int_0^{g(\rho,\theta)} f_s(\rho,\theta,v,\mu)d\mu dv. \qquad (7)$$

Depending on μ and θ the phase volume of plasma particles should be split in the phase volumes of untrapped, t-trapped and d-trapped particles by the following inequalities:

$$0 \leq \mu \leq \mu_u \qquad -\pi \leq \theta \leq \pi \qquad \text{- for untrapped particles} \qquad (8)$$

$$\mu_u \leq \mu \leq \mu_t \qquad -\theta_t \leq \theta \leq \theta_t \qquad \text{- for } t\text{-trapped particles} \qquad (9)$$

$$\mu_t \leq \mu \leq \mu_d \qquad -\theta_t \leq \theta \leq -\theta_d \qquad \text{- for } d\text{-trapped particles} \qquad (10)$$

$$\mu_t \leq \mu \leq \mu_d \qquad \theta_d \leq \theta \leq \theta_t \qquad \text{- for } d\text{-trapped particles} \qquad (11)$$

where the reflection points $\pm\theta_t$ and $\pm\theta_d$ for the t- and d-trapped particles, respectively, can be defined analysing the condition $v_\parallel(\mu,\theta)=0$. In particular, $\mu_u = \mu(\rho,\pm\pi) = (1-\varepsilon)/\sqrt{1+\lambda}$ and $\mu_t = \mu(\rho,0) = (1+\varepsilon)/\sqrt{1+\lambda}$. Further, we solve Eq. (6) only for untrapped and usual t-tapped particles in the case when the d-particles are absent, i.e., considering the tokamak magnetic field configuration as a system with one minimum of $\mathbf{H}(\rho,\theta)$; i.e., when the criterion of the d-particle existence, $\varepsilon < \lambda$ (or $b/a > \sqrt{1+\varepsilon+q^2/\varepsilon}$), cannot be satisfied.

The solution of Eq. (6) should be found by the specific boundary conditions of the trapped and untrapped particles. For untrapped particles, we use the periodicity of f_s over θ. Whereas, the boundary condition for the t-trapped particles is the continuity of f_s at the corresponding stop-points $\pm\theta_t$. As a result, we seek the perturbed distribution functions of untrapped, f_s^u, and t-trapped, f_s^t, particles as

$$f_s^u = \sum_p^{\pm\infty} f_{s,p}^u \exp\left[i2\pi(p+nq_t)\frac{\tau(\theta)}{T_{b,u}} - inq_t\bar{\theta}(\theta)\right], \qquad f_s^t = \sum_p^{\mp\infty} f_{s,p}^t \exp\left[i2\pi p\frac{\tau(\theta)}{T_{b,t}} - inq_t\bar{\theta}(\theta)\right] \quad (12)$$

where p is the number of the bounce resonances;

$$\tau(\theta) = \frac{\varepsilon + \delta\left(1 - \sqrt{1-\varepsilon^2}\right)}{\varepsilon\sqrt{1-\varepsilon^2}} \int_0^\theta \frac{1+\varepsilon\cos\eta - \varepsilon\delta\sin^2\eta}{g(\rho,\eta)\sqrt{1-\mu/g(\rho,\eta)}} d\eta \quad (13)$$

is the new time-like variable (instead of θ) to describe the bounce-periodic motion of untrapped and t-trapped particles along the magnetic field line with the corresponding bounce periods proportional to $T_{b,u} = 2\tau(\pi)$ and $T_{b,t} = 4\tau(\theta_t)$, respectively;

$$\bar{\theta}(\theta) = \frac{2(\varepsilon+\delta)}{\varepsilon+\delta\left(1-\sqrt{1-\varepsilon^2}\right)}\arctan\left(\sqrt{\frac{1-\varepsilon}{1+\varepsilon}}\tan\frac{\theta}{2}\right) - \frac{\delta\sqrt{1-\varepsilon^2}}{\varepsilon+\delta\left(1-\sqrt{1-\varepsilon^2}\right)}\left(\theta - \frac{\varepsilon\sin\theta}{1+\varepsilon\cos\theta}\right) \quad (14)$$

is the new poloidal angle where the magnetic field lines are straight; and

$$q_t = \frac{q}{\sqrt{1-\varepsilon^2}}\left(1 + \frac{\delta}{\varepsilon}\left(1-\sqrt{1-\varepsilon^2}\right)\right) \quad (15)$$

is the tokamak safety factor. The Fourier harmonics $f_{s,p}^u$ and $f_{s,p}^t$ for untrapped and t-trapped particles can be readily derived after the corresponding bounce-averaging.

LONGITUDINAL PERMITTIVITY ELEMENTS AND WAVE DISSIPATION

To evaluate the dielectric tensor elements we use the Fourier expansions of the current density and electric field over the poloidal angle $\bar{\theta}$:

$$j_\parallel(\theta)g(\rho,\theta) = \sum_m^{\pm\infty} j_\parallel^m \exp(im\bar{\theta}), \qquad E_\parallel(\theta)\frac{(1+\varepsilon\cos\theta - \varepsilon\delta\sin^2\theta)^2}{g(\rho,\theta)} = \sum_{m'}^{\pm\infty} E_\parallel^{m'} \exp(im'\bar{\theta}). \quad (16)$$

As a result, the whole spectrum of electric field, $E_\parallel^{m'}$, is present in the given m-th harmonic $j_\parallel^m$ of the current density:

$$\frac{4\pi i}{\omega}j_\parallel^m = \sum_{m'}^{\pm\infty}\varepsilon_\parallel^{m,m'}E_\parallel^{m'} = \sum_{m'}^{\pm\infty}(\varepsilon_{\parallel,u}^{m,m'} + \varepsilon_{\parallel,t}^{m,m'})E_\parallel^{m'} \quad (17)$$

where $\varepsilon_{\parallel,u}^{m,m'}$ and $\varepsilon_{\parallel,t}^{m,m'}$ are the separate contributions of untrapped and t-trapped particles, respectively, to the longitudinal (parallel) permittivity elements:

$$\varepsilon_{\parallel,u}^{m,m'} = \frac{\omega_p^2 \rho^2}{h_\theta^2 v_T^2 \pi^3}\sum_{p=-\infty}^{\infty}\int_0^{\mu_u}\frac{\tau(\pi)C_p^m C_p^{m'}}{(p+nq_t)^2}\left[1 + 2u_p^2 + 2i\sqrt{\pi}u_p^3 W(u_p)\right]d\mu, \quad (18)$$

$$\varepsilon_{\parallel.t}^{m,m'} = \frac{2\omega_p^2 \rho^2}{h_\theta^2 v_T^2 \pi^3} \sum_{p=1}^{\infty} \int_{\mu_u}^{\mu_t} \frac{\tau(\theta_t)}{p^2} D_p^m D_p^{m'} \left[1 + 2v_p^2 + 2i\sqrt{\pi} v_p^3 W(v_p)\right] d\mu. \tag{19}$$

Here we have used the following definitions:

$$u_p = \frac{\rho\omega\sqrt{1-\varepsilon^2}\tau(\pi)}{h_\theta |p+nq_t| v_T \pi}\left(1 - \frac{\delta}{\varepsilon}\left(1-\sqrt{1-\varepsilon^2}\right)\right), \qquad v_p = \frac{2\rho\omega\sqrt{1-\varepsilon^2}\tau(\theta_t)}{h_\theta p v_T \pi \left(1 + \frac{\delta}{\varepsilon}\left(1-\sqrt{1-\varepsilon^2}\right)\right)}$$

$$C_p^m = \frac{\varepsilon\sqrt{1-\varepsilon^2}}{\varepsilon + \delta\left(1-\sqrt{1-\varepsilon^2}\right)} \int_0^\pi \frac{\cos\left[(m+nq_t)\bar{\theta}(\eta) - (p+nq_t)\pi \frac{\tau(\eta)}{\tau(\pi)}\right]}{1 + \varepsilon\cos\eta - \varepsilon\delta\sin^2\theta} d\eta$$

$$\hat{D}_p^m = \frac{\varepsilon\sqrt{1-\varepsilon^2}}{\varepsilon + \delta\left(1-\sqrt{1-\varepsilon^2}\right)} \int_0^{\theta_t} \frac{\cos\left[(m+nq_t)\bar{\theta}(\eta) - p\frac{\pi\tau(\eta)}{2\tau(\theta_t)}\right]}{1 + \varepsilon\cos\eta - \varepsilon\delta\sin^2\theta} d\eta, \qquad \omega_p^2 = \frac{4\pi N e^2}{M}$$

$$D_p^m = \hat{D}_p^m + (-1)^{p-1}\hat{D}_{-p}^m, \qquad W(z) = \exp(-z^2)\left(1 + \frac{2i}{\sqrt{\pi}}\int_0^z \exp(t^2) dt\right).$$

It should be noted, Eqs. (18, 19) describe the contribution of any kind of untrapped and trapped particles to the dielectric elements. The corresponding expressions for plasma electrons and ions can be obtained from (18) and (19) replacing T, N, M, e by the electron T_e, N_e, m_e, e_e and ion T_i, N_i, M_i, e_i parameters, respectively. To obtain the total expressions of the permittivity elements, as usual, it is necessary to carry out the summation over all species of plasma particles.

One of the main mechanisms of the radio frequency plasma heating is the electron Landau damping of waves due to the Cherenkov resonance interaction of $E_\parallel$ with the trapped and untrapped electrons. Cherenkov resonance conditions are different for trapped and untrapped particles in the tokamak plasmas and have nothing in common with the wave-particle resonance condition in the cylindrical magnetized plasmas. Another important feature of tokamak plasmas is the contributions of all $E_\parallel^{m'}$-harmonics to the given $j_\parallel^m$-harmonic, Eq. (17). As a result, after averaging in time and poloidal angle, the wave power absorbed due to the trapped and untrapped electrons, $P = \text{Re}(E_\parallel \cdot j_\parallel^*)$, can be estimated by the expression

$$P = \frac{\omega}{8\pi} \sum_m^{\pm\infty} \sum_{m'}^{\pm\infty} \left(\text{Im}\,\varepsilon_{\parallel,u}^{m,m'} + \text{Im}\,\varepsilon_{\parallel,t}^{m,m'}\right)\left(\text{Re}\,E_\parallel^m \text{Re}\,E_\parallel^{m'} + \text{Im}\,E_\parallel^m \text{Im}\,E_\parallel^{m'}\right), \tag{20}$$

where $\text{Im}\,\varepsilon_{\parallel,u}^{m,m'}$ and $\text{Im}\,\varepsilon_{\parallel,t}^{m,m'}$ are the contributions of untrapped and t-trapped electrons to the imaginary part of the longitudinal permittivity elements. The expressions (18, 19) have a natural limit to the corresponding results for tokamak plasmas with elliptic magnetic surfaces [2] if $\delta=0$, and circular magnetic surfaces [1] if $b=a$ and $\lambda \to 0$. Our longitudinal permittivity elements can be applied for both the large and low aspect ratio tokamaks with D-shaped magnetic surfaces to study the wave processes with a regular frequency such as the wave propagation and wave dissipation during the plasma heating and current drive generation; when the wave frequency has been done, e.g., by the antenna-generator system.

REFERENCES

1. Grishanov, N.I., de Azevedo, C.A., and Neto, J.P., Plasma Phys. Controlled Fusion **43**, 1003-1021 (2001).
2. Grishanov, N.I., Ludwig, G.O., de Azevedo, C.A., and Neto, J.P., Phys. Plasmas **9**, 4089-4092 (2002).

Some Remarks on the Theory of the Plasma-Wall Transition (PWT) Layer

D.D. Tskhakaya[a,b] and S. Kuhn[b]

[a]*Institute of Physics, Georgian Academy of Sciences, 380077 Tbilisi, Georgia*
[b]*Department of Theoretical Physics, University of Innsbruck, A-6020 Innsbruck, Austria*

Abstract: Starting from the Poisson equation and the kinetic equations for the plasma plasma particles, the equation of the balance between the pressure forces and the particle momentum fluxes is constructed. We demonstrate that the well-known Bohm criterion [1, 2] and Harrison-Thompson solution [3] (for the Tonks-Langmuir model of the PWT layer [4]) are consequences of this balance equation. For the latter model it is shown that far from the wall the dependence of the potential on position is parabolic even if the assumption of quasineutrality is not used and Poisson's equation is solved instead. A condition on the ion fluid velocity (necessary for a monotonic potential profile) is formulated. This condition, having a form similar to the Bohm criterion, must be fulfilled at any spatial point of a plasma.

GENERAL THEORY

The plasma constituents are electrons (e), ions (i), dust particles (d) and neutrals (n). In the stationary state, the basic eqs. are Poisson's equation

$$-\frac{\partial^2}{\partial z^2}\varphi(z) = 4\pi \sum_\alpha e_\alpha n_\alpha, \qquad (1)$$

and the kinetic equations for the distribution functions of plasma particles, f_α,

$$v_z \frac{\partial f_\alpha}{\partial z} - \frac{e_\alpha}{m_\alpha} \frac{\partial \varphi}{\partial z} \frac{\partial f_\alpha}{\partial v_z} = \sum_\beta C_{\alpha\beta}\{f_\alpha; f_\beta\}, \qquad (2)$$

where $\alpha = e, i, d$ and $\beta = e, i, d, n$; m_α is the particle mass, and $C_{\alpha\beta}\{f_\alpha; f_\beta\}$ is the collision integral for collisions between the particles. In what follows we consider a semi-bounded plasma (except in the third section, where we consider a plasma bounded from both sides). We assume the boundary wall to be plane and the z axis to be directed perpendicularly to the wall. The boundary wall is absorbing and negatively charged, and the electric potential in the plasma is negative, $\varphi(z) \leq 0$. The plasma occupies the half space with $z < 0$. For the electron charge we have $e_e = -e$, $(e > 0)$, for the ions $e_i = e$, and for the dust particles $e_d = -Z_d e$. Multiplying Eq. (1) by $\partial \varphi(z)/\partial z$ we obtain

$$-\frac{1}{2}\frac{\partial}{\partial z}\left(\frac{\partial \varphi(z)}{\partial z}\right)^2 = 4\pi \sum_\alpha e_\alpha n_\alpha \frac{\partial \varphi(z)}{\partial z}. \qquad (3)$$

The expression in the right-hand-side of (3) can be found from the kinetic Eq. (2) by multiplying the latter by v_z and integrating over velocity space. By standard definition we have [5]:

$$\int d\mathbf{v}\, m_\alpha v_z^2 f_\alpha(z,\mathbf{v}) = m_\alpha n_\alpha(z) u_\alpha^2(z) + n_\alpha(z) T_\alpha(z), \qquad (4)$$

where $u_\alpha(z)$ and $T_\alpha(z)$ are the fluid velocity and temperature of species α. From Eq. (2) we then obtain

$$\frac{\partial}{\partial z}\left\{-\frac{1}{8\pi}\left(\frac{\partial\varphi(z)}{\partial z}\right)^2 + \sum_\alpha \left(m_\alpha n_\alpha(z)u_\alpha^2(z) + n_\alpha(z)T_\alpha(z)\right)\right\} = \sum_{\alpha\beta} m_\alpha \int d\mathbf{v}\, v_z C_{\alpha\beta}\{f_\alpha; f_\beta\}. \quad (5)$$

According to momentum conservation [5], in the summation by β there will remain only the term with $\beta = n$, which describes the momentum loss of the charged particles due to collisions with neutrals. The assumptions made in the following are as follows:

1. The electron and ion densities are less than the neutral density, $n_e, n_i \ll n_n$.
2. The electron thermal velocity V_{Te} is much higher than the ion and neutral thermal velocities, $V_{Te} \gg V_{Ti}, V_{Tn}$
3. The ionization processes do not contribute in the electron and ion momentum transfer – the velocities of the newly generated ions will be distributed chaotically as the velocity of the parent particles (neutrals) are assumed to be chaotic. We consider the case when $V_i \approx \nu_{ion}L$, (V_i is the characteristic velocity of ions, ν_{ion} is the frequency of ionization due to electron-neutral collision, L is the characteristic spatial scale of electric potential distribution, which can be of the order of the ion mean-free path or the Debye length λ_D [2]). For the electrons we have $V_{Te} \gg \nu_{ion}L$ and in the electron kinetic equation the ionization collision integral can be neglected. The physical meaning of such an assumption is as follows: The inequality $V_{Te} \gg \nu_{ion}L$ assumes that the number of electrons newly born during the time interval required for electrons to pass the distance L is negligibly small. By contrast, the time interval for ions to pass the distance L is much larger and the number of newly generated ions can be considerable due to their accumulation during this time.
4. The cross section of ion-neutral charge-exchange collisions is larger than the cross section of ion-neutral elastic scattering [2, 6].

According to these assumptions Eq. (5) yields

$$\frac{\partial}{\partial z}\left\{-\frac{1}{8\pi}\left(\frac{\partial\varphi(z)}{\partial z}\right)^2 + \sum_\alpha \left(m_\alpha n_\alpha(z)u_\alpha^2(z) + n_\alpha(z)T_\alpha(z)\right)\right\} = -|R_{in}^{cx}|, \quad (6)$$

where $R_{in}^{cx} < 0$ is the friction force acting on the ions due to their charge-exchange collisions with the neutrals. According to [2, 7], for cold neutrals this force equals

$$R_{in}^{cx} = \int d\mathbf{v}\, m_i v_z C_{in}^{exc}\{f_i; f_n\} = -\int d\mathbf{v}\, m_i v_z \frac{|\mathbf{v}|}{\lambda_{cx}(|\mathbf{v}|)} f_i(z, \mathbf{v}), \quad (7)$$

where $\lambda_{cx}(|\mathbf{v}|) = \{n_n \sigma_{in}^{exc}(|\mathbf{v}|)\}^{-1}$ is the ion mean free path for charge-exchange collisions.

The relation (6) represents the balance of the following physical quantities: the electric-field pressure force, the total momentum flux associated with the particles, the gas-kinetic pressure force, and the ion friction force. Note that the friction force connected with the ionization does not appear explicitly in (6), but in calculations of hydrodynamic quantities (densities, velocities) we have to take into account the ionization collisions.

BOHM CRITERION

Let us assume that the fluid velocity of the electrons is smaller than their thermal velocity, $m_e u_e^2(z) \ll T_e$. We will follow the procedure described in [8] and [2], where different spatial scales for the collisional "presheath" and the collisionless "sheath" are introduced. On the "sheath" scale, the potential and the electric field are zero at the sheath edge. In the sheath $R_{in}^{cx} = 0$. Integrating Eq. (6) over the interval (z_s, z), where z_s denotes the position of the sheath edge, we get

$$\frac{1}{8\pi}\left(\frac{\partial\varphi(z)}{\partial z}\right)^2 = m_i n_{is} u_{is}\left(u_i(z) - u_{is}\right) + n_e(z)T_e(z) - n_{es}T_{es}, \quad (8)$$

where $n_{\alpha s}$, $u_{\alpha s}$ and $T_{\alpha s}$ are the density, the fluid velocity and the temperature, respectively, for particle species α at the sheath edge. In (8), the relations $n_i(z)u_i(z) = n_{is}u_{is}$ and $u_i(z) = \{u_{is}^2 + 2e|\varphi(z)|/m_i\}^{1/2}$ are used for the ion density and fluid velocity. Due to their large mass, the dust grains can be considered as immobile. Then the influence of the dust grains will be reduced to the mere fulfilment of the quasineutrality condition at the sheath edge, $n_{is} = n_{es} + Z_d n_{ds}$. The electrons are Boltzmann-distributed, $n_e \cong n_{es}(-e|\varphi(z)|/T_e)$. The electron temperature is assumed to be constant throughout the plasma, $T_e(z) = T_{es}$. Then, expanding the right-hand side for small $|\varphi(z)|$, we obtain the condition for the existence of the nonoscillatory solution for $\varphi(z)$ in the form

$$u_{is}^2 > c_s^2 \left(1 + \frac{Z_d n_{ds}}{n_{es}}\right), \qquad (9)$$

where $c_s = (T_{es}/m_i)^{1/2}$ is the ion-sound velocity. So the presence of the dust particles makes the Bohm criterion more restrictive – the quantity in parentheses in (9) is larger than one.

PROFILE OF POTENTIAL FAR FROM THE WALL

In the Tonks-Langmuir model [3, 4], there are no dust grains, $n_d(z)=0$ and the ion kinetics are governed only by the ionization of cold neutrals by electrons. Then in (6) the friction force R_{in}^{exc} can be neglected. We consider the plane symmetric case, when a one-dimensional plasma is placed between two plane walls at the points $z = \pm l$ ($l \gg \lambda_D$) [3, 4]. For the potential $\varphi(z)$ we choose the boundary conditions $\varphi(z) = 0$ and $\partial \varphi(z)/\partial z = 0$ at $z = 0$. The electrons exhibit a Boltzmann distribution. The kinetic equation for the ions reads

$$v_z \frac{\partial f_i}{\partial z} + \frac{e}{m_i}\frac{\partial |\varphi(z)|}{\partial z}\frac{\partial f_i}{\partial v_z} = v_{ion} n_e(z)\delta(v_z), \qquad (10)$$

where $\delta(v_z)$ appears due to the distribution of the cold neutrals. Solving Eq. (10) for ions moving in the positive direction, $v_z > 0$, we can construct the expression for the ion momentum flux (4). And for the potential far from the wall, $\chi = e|\varphi|/T_e \ll 1$, we obtain from (6)

$$\frac{1}{2}\left(\frac{\partial \chi}{\partial z}\right)^2 + U(\chi) = 0, \qquad (11)$$

$$U(\chi) = (1/\lambda_D^2)\left\{\chi(z) - \sqrt{2}(v_{ion}/c_s)\int_0^z dz' \sqrt{\chi(z) - \chi(z')}\right\}. \qquad (12)$$

The solution of this equation for $(v_{ion}\lambda_D/c_s) \ll 1$ is

$$\chi(z) = \frac{\pi^2}{8}\frac{v_{ion}^2}{c_s^2} z^2 \left\{1 - \frac{\pi^2}{2}\frac{v_{ion}^2 \lambda_D^2}{c_s^2}\right\} \qquad (13)$$

The parabolic dependence of the potential on position, $|\varphi(z)| \approx z^2$, far from the wall was also found in [3] considering only the "presheath" region, $n_i(z) = n_e(z)$, and neglecting charge separation, the first term in (11). From (11)–(13) it follows that the parabolic dependence far from the wall is valid even in the case of charge separation. The second small term in the braces in (13) appears due to charge separation and is necessary to keep the effective potential energy (12) negative.

GENERALIZED BOHM CRITERION

For a semi-bounded plasma, we choose the "natural" boundary conditions $\varphi(z) \to 0$, $\partial \varphi(z)/\partial z \to 0$, $n_{e,i}(z) \to n_0$ and $u_i(z) \to 0$, at $z \to -\infty$. The plasma constituents are only electrons, ions and neutrals, whereas $n_d(z) = 0$. We assume the electron gas to be isothermal, $T_e(z) = T_{e0} = const$, and neglect the ion thermal motion. After integration of (6) the positiveness of $(\partial \varphi(z)/\partial z)^2$ requires the fulfilment of the inequality

$$u_i^2(z) \geq c_s^2 \left\{ \frac{n_0 - n_e(z)}{n_i(z)} - \frac{1}{n_i(z) T_{e0}} \int_{-\infty}^{z} dz' \, | R_{in}^{cx}(z') | \right\}, \tag{14}$$

where R_{in}^{cx} is defined by (7). Similar to (6), this inequality must be fulfilled at any point of the plasma. It has the form of a generalized Bohm criterion. The ion and electron densities both decrease monotonically when approaching the wall, $n_\alpha(z) = n_0 - \Delta n_\alpha(z)$, $\Delta n_\alpha(z) > 0$, ($\alpha = e, i$), and due to the shielding of the wall's negative charge we have $\Delta n_e(z) \geq \Delta n_i(z)$. Therefore $n_i(z) \geq n_e(z)$, where the equality sign applies only for $z \to -\infty$. Depending on the wall potential, the first term in the braces in (14) can be larger or smaller than one. From (14) it follows that the friction force decreases the threshold value of the ion fluid velocity.

ACKNOWLEDGMENTS

This work was supported by Austrian Science Fund (FWF) Project P15013 and has been carried out within the Association Euratom-ÖAW. Its content is the sole responsibility of the authors and does not necessarily represent the views of the Commission or its Services.

REFERENCES

1. Riemann, K.-U., General Invited Lecture XXIVthe ICPIG, Warsaw 1999, J.Tech. Phys.,**41**, 89 (2000).
2. Riemann, K.-U., Phys. Fluids, 25, 2163 (1981).
3. E. R. Harrison, E. R., and W.B. Thompson, W. B., Proc. Phys. Soc., (London) **74**, 145 (1959).
4. Tonks L., and Langmuir, I., Phys. Rev., **34**, 876 (1929).
5. Braginskii, S. I., in: "Reviews of Plasma Physics", vol.1,Consultants Bureau, New York, 1964.
6. McDaniel, E. W., and Mason, E. A., "The Mobility and Diffusion of Ions in Gases", Wiley, New York, 1973, Chaps. 5, 6 and 7.
7. Holstein, T., J. Phys. Chem., **56**, 832 (1952).
8. Chen, F.F., "Plasma Physics and Controlled Fusion" Plenum, New York, 1984, vol.1, chap.8.

New Physics of the Positive Column

Graeme G. Lister

OSRAM SYLVASNIA INC., 71 Cherry Hill Drive, Beverly, MA 01915, USA

Abstract. Recent advances in experiment and modelling of the positive column have brought new insight into the fundamental physics of low pressure discharges. This is particularly valuable for understanding important processes in the new range of "electrodeless" fluorescent lamps which operate at higher power loading than standard fluorescent lamps.

INTRODUCTION

The physics of the positive column has been studied, both experimentally and theoretically, for more than a century. Its length and uniformity make it readily accessible to virtually all types of experimental diagnostics and one dimensional theoretical treatments are appropriate because variations in plasma parameters occur principally in the radial direction.

In the last decade, two new developments in theory and modelling have led to renewed interest in the physics of the positive column. The first is the rediscovery of "non-local" kinetics, first formulated by Bernstein and Holstein [1] and the second has been the rapid increase in computer speed, which has allowed Monte Carlo calculations to investigate regions of parameter space which were previously inaccessible [2,3]. These developments are important to other low pressure discharges, such as those used for plasma processing, because it enables the fundamental physics incorporated in numerical models to be validated before extending them to more complex geometry.

Prior to the early 1990's, most numerical models of the electron distribution function (EEDF) in the positive column assumed the "local" approximation, in which the isotropic distribution function is characterised by the kinetic energy of the electrons ε. This approximation is valid provided the electron energy relaxation length $\lambda_\varepsilon \ll \Lambda$ where Λ is the typical inhomogeneity scale of the plasma. Bernstein and Holstein showed that there was another limit $\lambda_\varepsilon \gg \Lambda$, in which the EEDF is characterised by the *total* energy of the electrons $\varepsilon^* = \varepsilon + e\Phi$, where Φ is the local electrostatic potential. Applications of this theory [4] have proved valuable in understanding the physics of the positive column and other low pressure discharges.

An important parameter in modeling the positive column is the product pd of gas pressure and internal diameter (i.d) of the discharge. Recent Monte-Carlo calculations in rare gases [2,3] have extended the range of application to $pd>4$ torr cm. This is close to the operating regime of fluorescent lamps, in which the positive column is the principal source of light output.

FLUORESCENT LAMPS

The standard fluorescent lamp is a long cylindrical discharge filled with a mixture of a few mtorr of mercury and a rare gas, typically a few torr of argon. Mercury atoms are excited to produce ultra violet radiation, principally at 254 nm. The UV radiation is then converted to visible light by a set of phosphor coatings. This process is one of the most efficient ways of producing light, since about 25% of the electrical power applied to the lamps is converted to visible radiation.

Theoretical and numerical models of the positive column in fluorescent lamps have been developed for more than 40 years, since the early work of Waymouth and Bitter [5] and Cayless [6]. Since then, all major lighting companies and many university groups have developed numerical programs to describe the positive (cf. [7])

In the early 1960's, researchers at Philips conducted an extensive series of experiments in Hg-Ar discharges, for discharge parameters corresponding to the standard fluorescent lamp at that time. They measured electric fields, electron densities and temperatures using Langmuir probes [8] and mercury excited state densities [9] and radiation balance using spectroscopic analysis [10]. Despite the lack of fundamental data, numerical models reproduced the experimental results fairly well over a wide parameter range.

In the last decade, many lighting companies have developed inductively coupled "electrodeless" fluorescent lamps, many of which operate at power densities up to a factor 10 higher than standard lamps. Models have done less well in reproducing experimental results in these highly loaded lamps, in particular overestimating the maintenance electric field and hence the electrical power in the lamp [11].

POSITIVE COLUMN MODELS

Fluorescent lamp positive column models [7] are used to estimate the efficiency of conversion of electrical power to useful radiation as a function of discharge parameters by solving a set of coupled equations:

Ohm's Law relates the discharge current density to the axial field.

Particle diffusion equations determine the densities of the ground state and excited atoms, electrons and ions. This requires a description of the electron energy distribution function (EEDF).

A *thermal conduction equation* calculates the gas temperature.

A power *balance equation* partitions the electrical power supplied to the positive column into radiation, wall and volume losses.

The cross sections for many important atomic processes are not well known, particularly those for electron impact excitation to the higher states of mercury. The ability to compute electron impact cross sections for a number of atomic systems is showing encouraging progress in this regard. Other processes for which better data are required include the interaction between excited mercury atoms, particularly "chemi-ionization" producing an atomic or molecular ion. Accurate modeling of mercury resonance radiation transport under fluorescent lamp operating conditions is also vital if the radiation output is to be correctly calculated.

Many of these issues were recently addressed by a consortium under the joint auspices of the Electric Power Research Institute (EPRI) and OSRAM SYLVANIA, in the framework of the first ALITE project. Participants were from University of Wisconsin (UW), Polytechnic University, New York (PU), Los Alamos National Laboratory (LANL) and the National Institute of Standards and Technology (NIST). The program produced a considerable amount of new data (see [11]) and the principal results have been published or will be published in the near future. The new data is being incorporated into numerical models and some important discrepancies between theoretical predictions and experimental measurements remain. In particular, it has been difficult to reconcile the electric field characteristics of the discharges with Langmuir probe measurements.

DISCHARGE MODELING: THE ELECTRIC FIELD

The first step in modelling the behaviour of the positive column is a correct representation of the electric field as a function of the discharge parameters. This is represented by Ohm's law

$$I = 2\pi E \int_0^R \sigma_e(r) r dr \tag{1}$$

where I and E are the discharge current and electric field and σ_e is the electrical conductivity of the plasma

$$\sigma_e = -\frac{n_e}{3N}\left(\frac{2e}{m_e}\right)^{1/2} \int_0^\infty \frac{\varepsilon}{q_t(\varepsilon)} \frac{\partial f_0}{\partial \varepsilon} d\varepsilon \tag{2}$$

n_e is the electron density, N is the total gas density, $f_0(\varepsilon)$ is the isotropic component of the electron energy *probability* function for electron energy ε and $q_t(\varepsilon)$ is the total electron momentum transfer cross section, including argon, mercury and Coulomb collisions. Fig. 1 shows the result of comparing the experimentally measured discharge current with that calculated from equation (1), using the Langmuir probe data for n_e and $f_0(\varepsilon)$ for two sets of experimental data, corresponding to the Philips measurements [8] in T12 lamps and recent measurements at OSI in highly loaded ICETRON lamps [12], using the latest available cross sections for electron momentum transfer with Hg and Ar (see 11]). Coulomb collisions are included using the formalism of Spitzer and Härm [13] for fully ionized plasmas. Results are compared with those obtained without Coulomb collisions.

The results show that even for the previously reported results for T12, there is considerable disagreement between the measured and calculated results. Earlier models ignored Coulomb collisions, so the previous good agreement between theory and experiment may have been fortuitous. The discrepancy is of such a fundamental nature that we are led to one of two conclusions. Either the theoretical derivation of the electron momentum cross section is in error, or the validity of applying Langmuir probes for these discharge parameters is questionable.

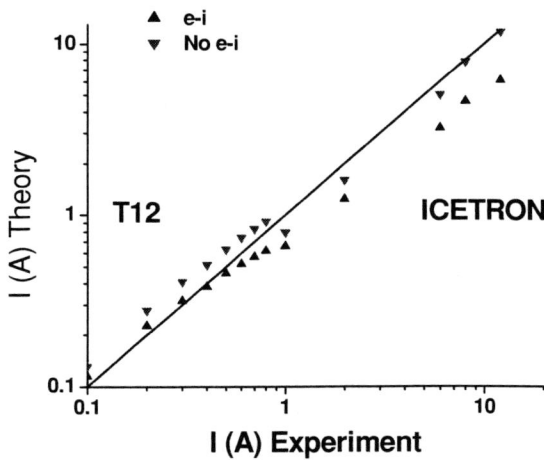

FIGURE 1. Comparison of discharge currents in T12 and ICETRON calculated from equation (1) with the measured values. Calculations used the experimental measurements of electron density and temperature for T12 (i.d=36mm, 3 torr Ar, 7 mtorr Hg) from Verweij [8] and for ICETRON (i.d.=50mm, 300 mtorr Ar, 6 mtorr Hg) from Godyak *et al* [12].

CONCLUSIONS

Significant advances have been made recently in the theory and modelling of the positive column of gas discharges. The ALITE project produced a considerable amount of new data, which are being applied to models of fluorescent lamps. A number of discrepancies between experiment and theory remain. In particular, the electrical conductivities calculated from Langmuir probe measurements are inconsistent with the measured electrical characteristics in these discharges.

REFERENCES

1. Bernstein I B and Holstein T, 1954, *Phys. Rev.*, **94**, 1475
2. Lawler J. E., and Kortshagen U., *J. Phys. D: Appl. Phys.*, **32**, 3188 (1999)
3. Kawamura, E. , and Ingold, J.H., *J. Phys. D: Appl. Phys.*, **34**, 3150 (2001)
4. Kortshagen U., Busch C., and Tsendin L. D., *Plasma Sources Sci. Technol.*,**5**,1(1996)
5. Waymouth, J.F., and Bitter, F., *J. Appl. Phys.* **27**, 122 (1956)
6. Cayless, M.A., in *Proc. 5th Int. Conf. On Ionization Phenomena in Gases*, Munich (1961)
7. Lister G G, 2000, in *Low Temperature Plasma Physics* (eds. R. Hippler, S. Pfau, M. Schmidt, K H Schoenbach; Wiley-VCH)
8. Verweij, W., *Philips Res. Rep. Suppl.* **2**, 1 (1961)
9. Koedam, M., and Kruithof, A.A., *Physica* **28**, 80 (1962);
10. Koedam, M., Kruithof, A.A., and Riemens, J., *Physica* **29**, 565 (1963)
11. Lister, G.G, in , in *Proc. 9th Int. Symp. On Science and Technology of Light Sources*, Ithaca, USA (Cornell Print and Digital Copy Service) pp 459-468 (2001)
12. Godyak, V., Piejak, R., and Alexandrovich B., *ibid*, pp 157-158 (2001)
13. Spitzer, L., and Härm, R., *Phys. Rev.* **89**, 977 (1953)

Kinetic Modeling Of Axially Non-Uniform Cylindrical Magnetron Discharge

I. A. Porokhova[1,2], Yu. B. Golubovskii[2], P. Kudrna[3], M. Tichý[3], J. F. Behnke[1]

[1]EMA-University, Institute for Physics, Domstrasse 10a, D-17491, Greifswald, Germany
[2]St. Petersburg State University, Department of Optics, Ulianovskaia 1, St.Petersburg, 195904, Russia
[3]Charles University in Prague, Faculty of Math. and Physics, V Holesovickach 2, 180 00 Prague, Czech Republic

Abstract. Cylindrical magnetron discharges (CMD) with two co-axial electrodes and a uniform magnetic field being directed axially, represent relatively simple systems for theoretical studies that can contribute to better understanding of discharge operation. The dc discharge in Ar at pressure 3 Pa, magnetic field strength 100 G and current 150 mA is considered where the pronounced non-local regime of the electron distribution function (EDF) formation is realized. The electron component is analyzed on the basis of radially and axially inhomogeneous Boltzmann kinetic equation. Unmagnetized electrons, which move in axial direction are trapped in the axial potential well, their energy relaxation length is large, and the kinetic equation can be averaged over discharge length. Using a model two-dimensional potential profile, the EDFs at different axial positions are obtained and spatial distributions of the electron density and ionization rate are represented and discussed. The results of the modeling can be used for explanation of recent experiments that have detected axial inhomogeneity of the plasma in magnetron discharge.

INTRODUCTION

The growing importance of magnetron discharges as devices used for deposition of thin films requires comprehensive experimental and theoretical studies directed to better understanding of both the fundamental physical processes and technological aspects of magnetron discharge operation control. For this purpose such relatively simple system as cylindrical magnetron discharge where electrodes are the co-axial cylinders and homogeneous magnetic field is directed axially represents an attractive object for experimental studies and modeling.

In left part of Fig. 1 the schematic view of magnetron discharge is shown. Both ends of the discharge tube are closed with shields that are at the potential of the cathode. The dc discharge in Ar at pressure 3 Pa, magnetic field strength 100 G and current 150 mA will be considered here. The radial behavior of the plasma parameters from cathode to anode has been studied in [1] by PIC simulation and in [2] by self-consistent numerical solving the spatially inhomogeneous Boltzmann kinetic equation and the equations of ion motion, current balance and Poisson equation. The experimental investigations performed in [3] have detected a noticeable axial inhomogeneity of the discharge (right part of Fig. 1), which becomes more pronounced at higher pressures and larger magnetic fields.

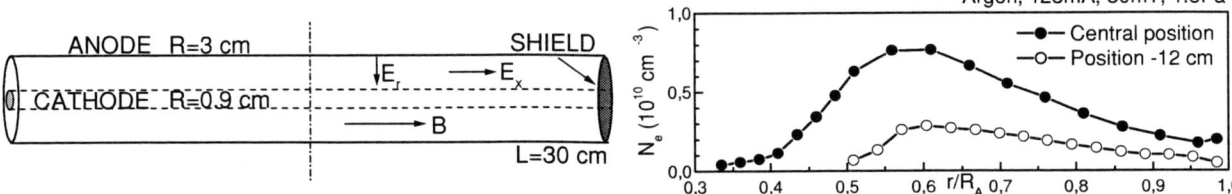

FIGURE 1. Schematic view of the cylindrical magnetron discharge (left). Electron densities measured in [3] (right) for two axial positions.

The kinetics of electrons in inhomogeneous axial and radial fields that are formed in magnetron discharge will be studied on the basis of the Boltzmann kinetic equation. The so-called non-local approach will be applied to explain the axial inhomogeneity of plasma parameters obtained in experiment.

ELECTRON KINETICS IN NON-UNIFORM FIELDS

The electron motion in CMD will be analyzed basing on the spatially inhomogeneous Boltzmann equation

$$\vec{v}\cdot\nabla_{\vec{r}}F-(e/m)(\vec{E}+[\vec{v}\vec{B}])\nabla_{\vec{v}}F=C(F), \quad (1)$$

where m and $-e$ are the mass and charge of electron. Electron distribution function (EDF) $F(\vec{v},\vec{r})$ is formed under the influence of spatial gradients, the actions of axial magnetic field $\vec{B}$, electric field $\vec{E}$ which has radial and axial components and collision processes $C(F)$. Employing the two-term expansion of the EDF in spherical harmonics and integrating the kinetic equation over solid angles we come to the following expressions for the radial, axial and azimuth components of the EDF anisotropy

$$f_{1r}(U,r,x)=\frac{\lambda_e}{1+(\lambda_e/r_{ec})^2}\left[-\frac{\partial f_0}{\partial r}+eE_r\frac{\partial f_0}{\partial U}\right]; \quad f_{1x}(U,r,x)=\lambda_e\left[-\frac{\partial f_0}{\partial x}+eE_x\frac{\partial f_0}{\partial U}\right]; \quad f_{1\theta}=\frac{\lambda_e}{r_{ec}}f_{1r} \quad (2)$$

Here λ_e is the electron free path and r_{ec} is the radius of electron cyclotron motion, $U=mv^2/2$ is the kinetic energy, functions F and f differ by the normalization factor $2\pi(2/m)^{3/2}$. The kinetic equation for the isotropic distribution function can be represented as follows

$$\frac{U}{3}\text{div}\,\vec{f}_1-\frac{\partial}{\partial U}\left(\frac{U}{3}(e\vec{E}\cdot\vec{f}_1)\right)=S_0^{el}(f_0)+S_0^{in}(f_0)+S_0^{di}(f_0),$$

where the collision operator includes elastic and inelastic collisions and direct ionization.

It is seen from Eq. (2) that the magnetic field does not influence the axial anisotropy, while it reduces the radial anisotropic distribution by the factor $(r_{ec}/\lambda_e)^2$.

It is more convenient to solve the kinetic equation in the variables of total energy and coordinates since in the new variables the projections of electric field and mixed derivatives disappear. By introducing the total energy $\varepsilon=U+e\Phi(r,x)$, where $\Phi(r,x)$ is the two-dimensional spatial potential, and substituting anisotropic components we obtain the final partial differential equation for the isotropic distribution.

$$\frac{1}{r}\frac{\partial}{\partial r}r\frac{U^2 NQ_\Sigma(U)}{U(NQ_\Sigma)^2+m\omega_{eB}^2/2}\frac{\partial f_0(r,x,\varepsilon)}{\partial r}+\frac{\partial}{\partial x}\frac{U}{NQ_\Sigma(U)}\frac{\partial f_0(r,x,\varepsilon)}{\partial x}+2\frac{m}{M}\frac{\partial}{\partial \varepsilon}U^2 NQf_0(r,x,\varepsilon)$$
$$=UN[Q_{ex}(U)+Q_{ion}(U)]f_0(r,x,\varepsilon)-(U+U_{ex})NQ_{ex}(U+U_{ex})f_0(r,x,\varepsilon+U_{ex}) \quad (3)$$
$$-(U/\beta+U_{ion})NQ_{ion}(U/\beta+U_{ion})f_0(r,x,\varepsilon+U(1-\beta)/\beta+U_{ion})/\beta$$
$$-(U/(1-\beta)+U_{ion})NQ_{ion}(U/(1-\beta)+U_{ion})f_0(r,x,\varepsilon+U\beta/(1-\beta)+U_{ion})/(1-\beta)$$

Here Q_Σ is the total collision cross-section, i.e. the sum of the cross sections for momentum transfer Q, excitation Q_{ex} with the threshold U_{ex} and ionization Q_{ion} with ionization potential U_{ion}, M is the atom mass, N is the gas density, β and $(1-\beta)$ are the fractions in which the remaining energy of the colliding electron is shared in ionization event between the two outgoing electrons.

Characteristic Scales of the EDF Formation

When describing electron component of magnetron discharge plasma, it is necessary to take into account the characteristic spatial scales of the EDF formation. These characteristic scales that determine electron kinetics are the free paths, energy relaxation lengths with respect to various collision processes, and plasma inhomogeneity scales.

The presence of the axial magnetic field, as it is seen from Eq. (2), does not influence plasma in axial direction but reduces electron drift velocity in radial direction by the factor of $1+(\lambda_e/r_{ec})^2$. This is equivalent to reduction of the effective electron free path in radial direction λ_{eB} which can be defined as $\lambda_{eB}=\lambda_e/(1+(\lambda_e/r_{ec})^2)$.

Electron energy relaxation length may be understood as a distance on which electrons diffuse at a time τ of energy exchange in collisions, $\lambda_\varepsilon \sim (D_{eB}\tau)^{1/2}$, where $D_{eB} \sim v\lambda_{eB}/3$ is the electron diffusion coefficient in direction perpendicular to magnetic field. For electrons with energies smaller than the excitation threshold $U<U_{ex}$

$$\lambda_\varepsilon \sim \left(\frac{M}{m}\right)^{1/2}\frac{\lambda_e}{(1+(\lambda_e/r_{ec})^2)^{1/2}} \approx \begin{cases}(M/m)^{1/2}\lambda_e & r_{ec} \gg \lambda_e \\ (M/m)^{1/2}r_{ec} & r_{ec} \ll \lambda_e\end{cases} \quad (4)$$

For unmagnetized electrons the energy relaxation length is proportional to the free path, and in strong magnetic fields ($\lambda_e \gg r_{ec}$, magnetized electrons) it is determined by the Larmor radius. For the electrons whose energies exceed the excitation potential $U>U_{ex}$, under condition that $(M/m)\lambda_e >> \lambda^*_e$ (λ^*_e is the free path for inelastic impact) the energy relaxation length with respect to inelastic processes can be estimated as follows

$$\lambda^*_\varepsilon \sim \left(\frac{\lambda_e \lambda^*_e}{1+(\lambda_e/r_{eL})^2}\right)^{1/2} \approx \begin{cases} (\lambda_e \lambda^*_e)^{1/2} & r_{eL} >> \lambda_e \\ (\lambda^*_e/\lambda_e)^{1/2} r_{eL} & r_{eL} << \lambda_e \end{cases} \quad (5)$$

The characteristic scales are represented in Fig. 2 for the specified discharge conditions.

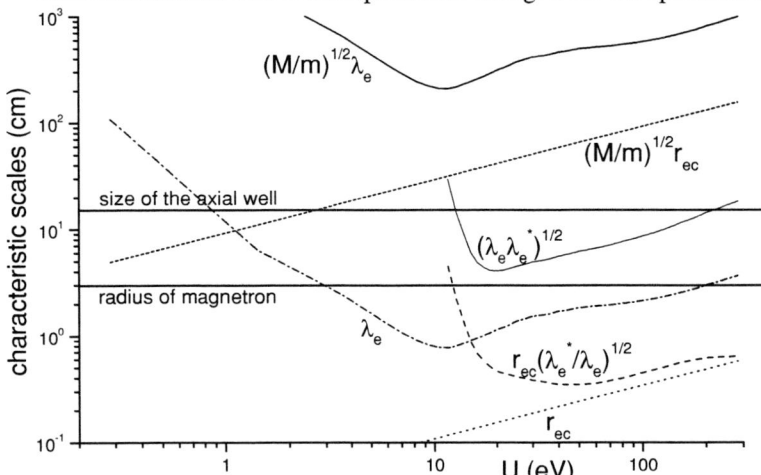

FIGURE 2. Characteristic scales of the electron motion. Ar, p=3 Pa, B=100 G. For unmagnetized electrons who move in axial direction the energy relaxation length with respect to elastic collisions $(M/m)^{1/2}\lambda_e$ much exceeds the discharge dimensions; the energy relaxation length with respect to inelastic collisions $(\lambda_e\lambda^*_e)^{1/2}$ is comparable with the size of the axial potential well. For magnetized electrons who move in radial direction the energy relaxation length with respect to elastic collisions $(M/m)^{1/2}r_{ec}$ exceeds the radius of magnetron; the energy relaxation length with respect to inelastic collisions $r_{ec}(\lambda^*_e/\lambda_e)^{1/2}$ is short. Both in axial and radial directions the EDF is formed non-locally.

Solution of the Kinetic Equation and Results

The model potential profile is shown in Fig. 3 for half of the discharge length. The cathode and shields are at potential of the cathode that is taken here equal to zero. Anode is the equipotential surface. The radial distribution of the potential at the center of the magnetron is typical for that obtained self-consistently from the solutions of one-dimensional problems by PIC and non-local kinetic techniques. The axial distribution of the potential is taken in the way that boundary conditions are satisfied and at the central part of magnetron the potential is rather flat. Thus, the radial profiles of the potential are almost identical at the central region of magnetron and the potential well is formed at the periphery of the discharge, which traps the electron motion in axial direction.

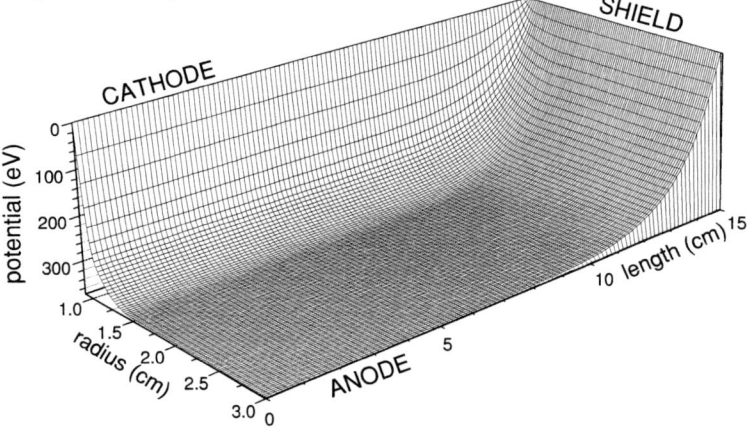

FIGURE 3. Profile of the model potential in axially symmetric cylindrical magnetron discharge.

As soon as the energy relaxation length in the axial direction considerably exceeds the length of the magnetron, the EDF is formed by whole axial potential, and electrons travel across the axial well many times before they acquire energy in radial field. Therefore, the kinetic equation (3) can be averaged over axial coordinate to yield the EDF at the center as a function of the total energy and radial coordinate. The electron distributions in the axial points distanced by x_0 from the center can be obtained by rearrangement of the variables of the axial EDF according to the potential $\Phi(x_0,r)$. This procedure is analogous to calculations of the non-local distribution functions in the positive column of a discharge [4]. The resultant EDFs are shown in Fig. 4 for the axial point and that shifted by 10 cm.

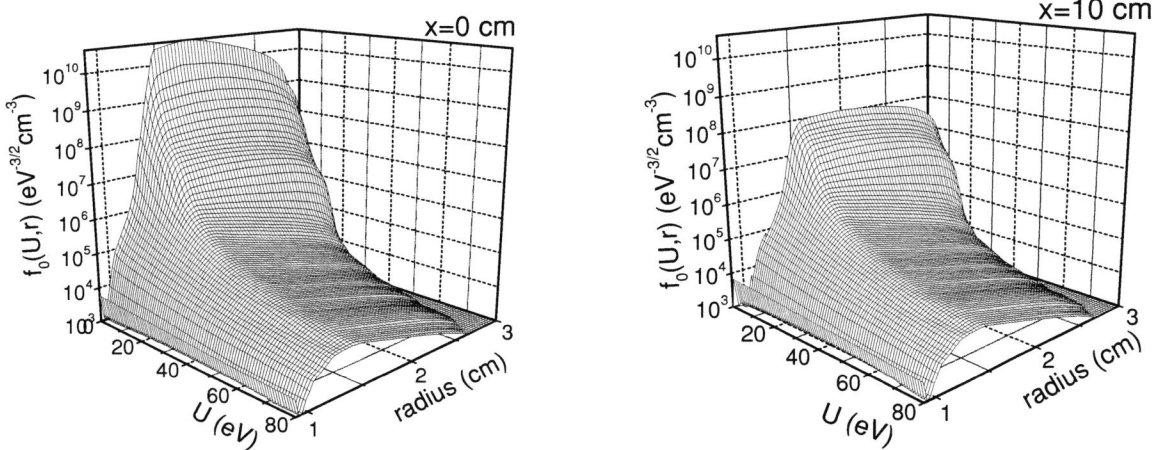

FIGURE 4. EDFs as the functions of the kinetic energy and radial position at the discharge center (left) and 10 cm aside (right).

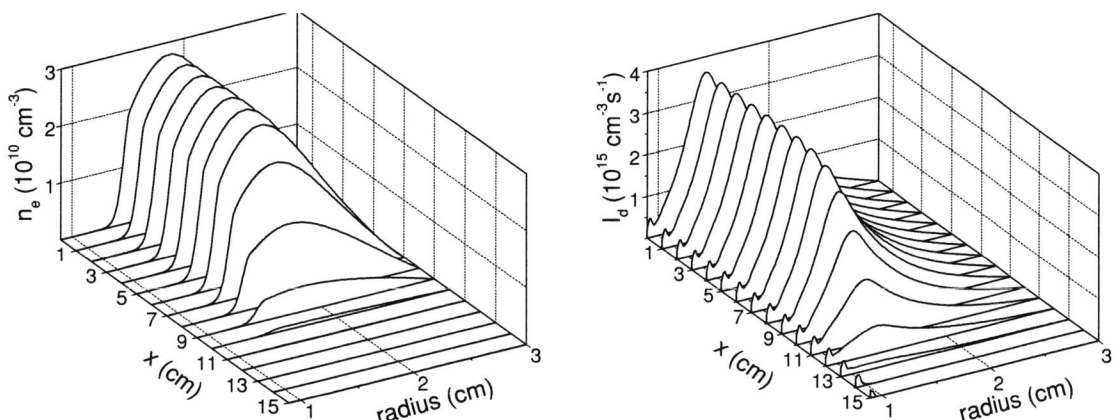

FIGURE 5. Electron density (left) and direct ionization rate (right) as the functions of axial and radial coordinates.

Using the EDFs at different axial positions the radial profiles, for instance, of the electron density (n_e) and direct ionization rate (I_d) can be calculated as the functions of axial position (Fig. 5). The depletion of the macroscopic properties caused by the disturbing action of the shields is seen in the discharge periphery that correlates with the experimentally observed effects (Fig. 1).

ACKNOWLEDGMENTS

Work supported by the DFG SFB 198 Greifswald "Kinetik partiell ionisierter Plasmen", DAAD trilateral project of Universities of Greifswald, St.Petersburg and Paris-Sud, the Grant Agency of Czech Republic, Grant No. 202/00/1689, 202/00/1217, 202/01/D095 by the Grant Agency of Charles University, Grant No. 171/2000/B FYZ MFF, COST 527.70 and by the Ministry of Education, Youth and Sports, Research plan No. MSM 113200002.

REFERENCES

1. Kudrna P., Holik M., and Tichy M., *Czech J. Phys.* **52**, Suppl. D, D666-D672 (2002).
2. Porokhova I. A., Golubovskii Yu. B., Bretagne J., Tichy M., and Behnke J. F., *Phys. Rev. E* **63**, 056408 (2001).
3. Kudrna P., Tichy M., Behnke J. F., Csambal C., and Rusz J., *Proc. 15th ISPC*, Orleans (France) **4**, 2119-2124 (2001).
4. Golubovskii Yu. B., Porokhova I. A., Behnke J. F., Behnke J., *J. Phys. D: Appl. Phys.* **32** 456-470 (1999).

PIC-MCC Modeling of the Cylindrical Magnetron Discharge

P. Kudrna[1,2], M. Holík[1], I. A. Porokhova[1,3], Yu. B. Golubovskii[3], M. Tichý[1,2], J. F. Behnke[1]

[1] Charles University in Prague, Faculty of Mathematics and Physics, Department of Electronics and Vacuum Physics, V Holesovickach 2, 180 00 Praha 8, Czech Republic
[2] University of Greifswald, Institute of Physics, Domstrasse 10a, 17487 Greifswald, FRG
[3] St. Petersburg State University, Department of Optics, Ulianovskaia 1, St.Petersburg, 195904, Russia

Abstract. Plasma of the DC discharge in cylindrical magnetron is studied on the basis of 1-D particle-in-cell simulation. In 1-D model the radial dependence of plasma parameters is calculated. The main features of the model are described and the obtained results discussed. The electron particle densities, electric field, and the radial course of plasma potential obtained by model are compared with experimental results.
Measurements and calculations have been performed in dc discharge in Ar at the pressures in the range p=1÷6 Pa, magnetic fields B=10÷40 mT and discharge currents I_D=100÷400 mA. Presented are results at p=3 Pa and B=10 mT.
This model study concerns the novel experimental system constructed in Greifswald University and described e.g. in [1]. In this new system the length of the discharge vessel is 300 mm, radii of inner cathode and outer anode are 9 mm and 29 mm respectively. The axially segmented cathode and axially movable Langmuir probe were implemented.
We present comparison of experimental results of radial profiles of electron density, plasma potential, and from it derived electric field, in the discharge plasma with the results obtained from the numerical model. While for the electron density a reasonable agreement is obtained the calculated electric field in the positive column is higher than that obtained from experiment. On the contrary the computed discharge voltage is lower that that observed experimentally.

INTRODUCTION

Cylindrical magnetrons are nowadays used for deposition of thin films with special optical properties and for high-temperature superconductive thin films [2,3,4,5]. Moreover, as a relatively simple system the DC discharge in cylindrical magnetron offers a possibility to create a numerical model and compare its results with experiment. Our experimental system of cylindrical magnetron has been described in detail e.g. in [1] so that only brief details will be given here. Six evenly distributed coils create the axial magnetic field with the homogeneity of ±0.2 % over the whole discharge vessel length 300 mm. The outer electrode is grounded and serves as anode. Its diameter is 58 mm. The cathode has the diameter of 18 mm and is positioned co-axially with the anode. One half of the cathode length is divided into 14 segments, i.e. one segment has a length of about 10 mm. Segments are isolated between each other so the axial distribution of discharge current can be measured. The discharge volume is axially limited by means of two stainless steel discs that are connected to the cathode potential. The view of the discharge vessel is in Figure 1.

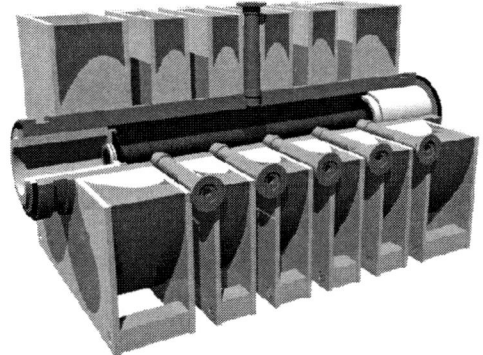

FIGURE 1. Cross-sectional view of the experimental apparatus. On the left-hand side of the cathode it is possible to observe the cathode segments. Also evident are the vacuum ports. The three ports on the left-hand side are used for Langmuir probes. The cross-sections of the coil windings are also seen. The windings profiles are calculated in the way to keep the longitudinal magnetic field as constant as possible. In reality the homogeneity of the magnetic field at the vessel axis of ±0.2 % over the whole discharge vessel length has been achieved.

EXPERIMENTAL

Presented experimental results have been obtained by probe measurements. The cylindrical Langmuir probe was movable in radial direction approximately in the middle of the reactor length. The probe construction is described in detail in [6]. The dimensions of the tungsten cylindrical Langmuir probe were 47 μm in diameter and 2.5 mm in length. In this report the probe axis has been positioned perpendicular to the magnetic field lines. In accord with the results of the work [7,8] there is no substantial influence of the presence of the magnetic field to the probe data in such arrangement at conditions of our experiment. The apparatus enables the measurements of radial profiles of the floating and plasma potential, radial electric field, plasma density, electron mean energy and the electron energy distribution function (EEDF) in the discharge plasma. The radial profiles of the mentioned parameters are measurable only at such radial distances where plasma can be supposed to be quasi-neutral. The region of the cathode fall is therefore excluded. In the close vicinity of the anode the probe measurements may be affected by the presence of the orifice necessary for the insertion of the probe into the reactor vessel. Hence, the plasma parameters in the close vicinity of the anode wall are burden with larger error than that at intermediate positions between the anode and cathode. In addition, the discharge (cathode) voltage, total discharge current and the current to each segment of the cathode have been measured at every discharge condition. The distribution of the discharge current gives the information about the homogeneity of the discharge along the axis of the system.

NUMERICAL SIMULATION

We have modeled the CMD in argon with a one-dimensional cylindrical PIC model, using a modified XPDC1-code (Berkeley) [9,10]. This code includes binary collisions using Monte Carlo methods. Electron-neutral, electron-metastable and ion-neutral collisions are taken into account. Electron-electron interactions are realized via binary collisions, treating them as being equivalent to collisions of electrons with neutral species [11]. In doing so, individual pseudo-electron particles collide with an energy-resolved electron fluid. Energy is exchanged with this fluid using the Coulomb cross section. The electron energy distribution function (EEDF) used for the Coulomb collisions is calculated via the following procedure: (1) At each simulation step the radial position and the energy of every electron are calculated. (2) A radially and energetically resolved counter $F_0(E,r)$ is incremented for every electron (energy resolution 0.1 eV, radial resolution $(r_A-r_K)/100$) (3) The EEDF $f_0(E,r)$ is calculated from $F_0(E,r)$. (4) The mean energy $E_{mean}(r)$ is calculated from the EEDF $f_0(E,r)$. The collision processes that have been taken into account in the PIC model of the cylindrical magnetron discharge presented in this work are: (i) the process of excitation of Ar atoms in collision with electron to the 1s and 2p levels, (ii) the ionization of Ar atoms from the excited as well as from the ground state levels, (iii) the processes of charge exchange between the Ar atoms and the Ar ions, (iv) the elastic collisions of electrons and ions with the neutral particles and (v) the electron-electron interactions. The cross sections of 3 most important collision types included in our model are depicted in Figure 2.

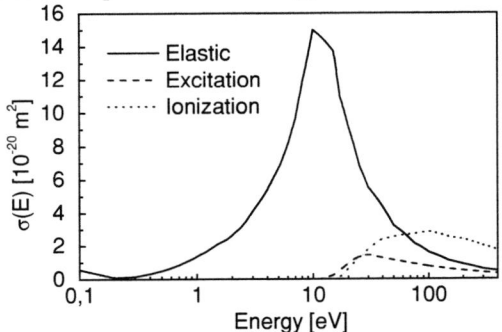

FIGURE 2. Cross sections for elastic, ionization and excitation collisions used in the PIC-MCC model.

Owing to the comparatively low pressure used in experiments and modeling the formation of molecular ions and the relevant processes that include the molecular ions have not been considered in the present model. The PIC-MCC model yields the radial dependence of the electron and the ion energy distribution function and consequently the radial distribution of the mean electron energy, of the electric potential (field) and of the plasma density. Also, cathode potential is calculated and can be compared with experiment. It was possible to make the calculation at the conditions similar to that in experiment and to obtain the calculated radial profiles of the above mentioned plasma characteristic parameters. The calculations have been made on the single-processor PC system Pentium II/450MHz with 512 MB of memory. The anode-cathode distance was divided into 100 cells; the energy range of the EEDF was 0-100 eV and was divided into 500 energy intervals. The time steps were chosen in the range 3-10 ps while the sizes of macroparticles were 6×10^7-3×10^8. The computational time to reach the steady state corresponded to simulated discharge time of about few tens of microseconds and was approximately 3-5 days.

RESULTS

The calculated radial profile of the plasma density shows good agreement with Langmuir probe measurements at experimental conditions where the discharge is axially homogeneous. This good homogeneity was checked by measurement of the current to the segmented cathode. An example of such situation at magnetic field B=10 mT, pressure p=3 Pa and the discharge currents 150, 200 and 300 mA depicted in Figure 3.

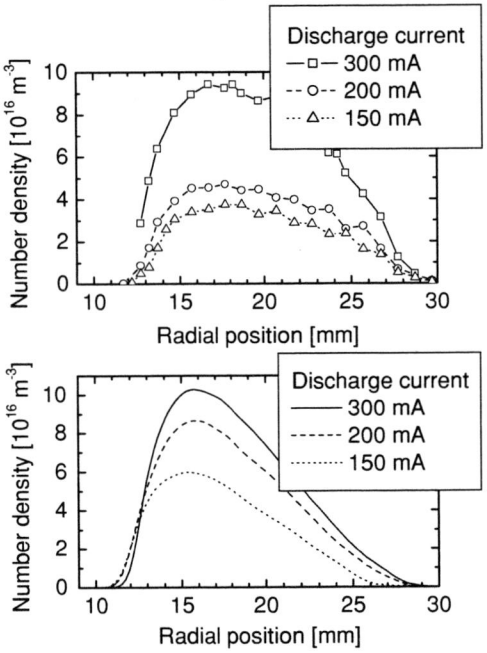

FIGURE 3. Comparison of the radial profile of the density measured by the probe positioned in the middle of the vessel length (top graph) with the profile calculated using PIC model (bottom graph). Experimental conditions: Ar, B=10 mT, p=3 Pa.

The measured current to the cathode segments was found constant (±10%) over the whole length of the vessel save for the distance of about 3-6 cm near both shields (limiters). In Figure 3 both the radial positions of the maximums and the absolute values of plasma density measured by the probe correspond well to the calculated ones.

The radial course of plasma potential measured by probe and calculated using PIC model is shown in Figure 4. The plasma potential cannot be measured up to anode surface because of the orifice, which is needed to insert the probe. The measured curves show almost no radial electric field (~0.3 Vcm^{-1}) and most of the potential profile is very close to anode potential (0 V). The calculated curves show the anode fall of about 18 V and the radial electric field is ~2 Vcm^{-1} in the positive column of the discharge. The measured discharge voltage reached at discharge current 150 mA the value -600 V, while the computed one at the same discharge current was only -351 V.

DISCUSSION AND CONCLUSION

The comparison between the experiment and model shows almost quantitative agreement in the radial profile of the plasma density. The discharge conditions were chosen such that the real discharge was stable and homogeneous along the axis of the cylindrical magnetron. The computed electric field in the positive column, even at such conditions, was higher than that obtained experimentally. On the contrary, the computed discharge voltage was lower than the measured value. This difference cannot be attributed to end effects, since the measuring probe was positioned in the middle of the vessel length and the experimental investigation [13] as well as the theoretical calculations [14] showed that the discharge is influenced by the limiters only to the depth of 3-6 cm. The central region of the discharge therefore remains free of the effects at the vessel ends. At present the work on the application of the 2-D PIC-MCC model, which could describe also the axial variations of the plasma parameters is in progress.

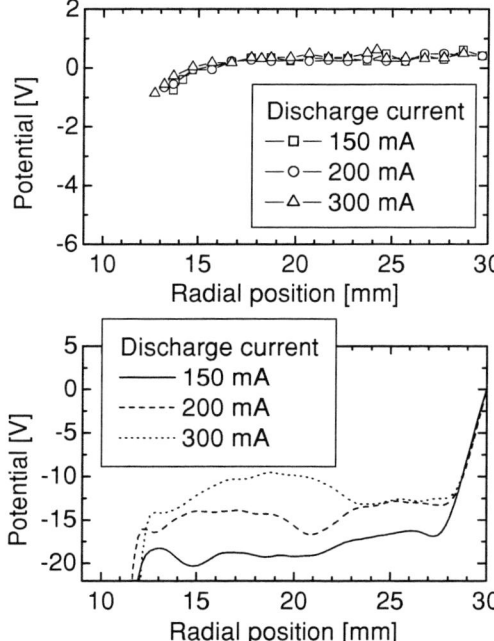

FIGURE 4. Comparison of the plasma potential profile measured by the Langmuir probe (top graph) with the profile calculated using PIC model (bottom graph). Experimental conditions are the same as in Figure 3.

ACKNOWLEDGMENTS

The work in Greifswald was financially supported by the Deutsche Forschungs-Gemeinschaft (DFG) in frame of the project SFB 198 Greifswald "Kinetik partiell ionisierter Plasmen" and by DAAD trilateral project of the University in Greifswald, Paris-Sud, St. Petersburg. The work in Prague was partially financially supported by the Grant Agency of Czech Republic, Grant No. 202/00/1689, 202/00/1217, 202/01/D095, by the Grant Agency of Charles University, Grant No. 171/2000/B FYZ /MFF, by project COST action 527.70 and by the Ministry of Education, Youth and Sports, Research plan MSM 113200002.

REFERENCES

1. Kudrna, P., Tichy, M., Behnke, J. F., Csambal, C., Rusz, J., *Proc. 15th International Symposium on Plasma Chemistry*, edited by A. Bouchoule et al., Orleans, (France), July, 9-13, 2001, Vol. VI, pp. 2119-2124.
2. Zhang, Qi-Chu, Zhao, K., Li, B. F., *Journal of vacuum science & technology* **17**, 2885 (1999).
3. Zhang, Qi-Chu, Zhao, K., Li, B. F., *Journal of vacuum science & technology* **16**, 628 (1998).
4. Di Chiara, A., Lombardi, F., Russo, M., *IEEE transactions on applied superconductivity* **5**, 2782 (1995).
5. Yu, Zengqi Collins, Harbison, Barry George, *Japanese journal of applied physics* **31**, 3969 (1992).
6. Passoth, E., Kudrna, P., Csambal, C., Behnke, J. F., Tichy, M., Helbig, V., *J. Phys. D: Appl. Phys.*, **30**, 1763-1777 (1997).
7. Kudrna, P., Passoth, E., *Contributions to Plasma Physics*, **37**, 417-429 (1997).
8. Passoth, E., Behnke, J. F., Csambal, C., Tichy, M., Kudrna, P., Golubovskii, Yu. B., Porokhova, I. A., *J. Phys. D: Appl. Phys.*, **32**, 2655-2665 (1999).
9. http://langmuir.eecs.berkeley.edu.
10. Birdsall, C. K. and Langdon, A. B., *Plasma Physics Via Computer Simulation*, McGraw-Hill 1995, Adam-Hilger 1991.
11. van der Straaten, T. A., Cramer, N. F., *Physics of Plasmas*, **7**, 391-402 (2000).
12. Weng, Y., Kushner, M. J., *Phys. Rev. A* **42**, 6192 (1990).
13. Kudrna, P., Holik, M., Bilyk, O., Porokhova, I. A., Golubovskii, Yu. B., Tichy, M., Behnke, J. F., "Experimental Study Of Axial Plasma Parameter Variations In The Cylindrical Magnetron Discharge" in *Proc. International Congress on Plasma Physics 2002 (this volume)*, accepted.
14. Porokhova, I. A., Golubovskii, Yu. B., Kudrna, P., Tichy, M., Behnke, J. F., "Kinetic Modelling of Axially Non-Uniform Cylindrical Magnetron Discharge" in *Proc. International Congress on Plasma Physics 2002 (this volume)*, accepted.

Model Kinetic Description in Plasma

V.V. Belyi

IZMIRAN, Troitsk, Moscow region, 142092, Russia

Abstract. A consistent derivation of the model linearized collision operator for a multicomponent plasmas in the approximation of local (not full) equilibrium is presented. In these results an ambiguity in the choice of coefficients is eliminated, in contrast to the BGK type models. Furthermore, the linearized collision operator constructed by us describes correctly the Spitzer values of viscosity and thermal conductivity. A technique to reconstruct the model linearized operator is proposed. It is shown that the model collision integral in the local equilibrium approximation does not contain a complicated exponential which is common for the BGK type integral.

The most widely used kinetic model equation is Bhatnagar, Gross & Krook (BGK) model [1]. Recall that BGK model collision term for the one-component system is the departure of distribution function (df) from Maxwellian the parameters of which are the momenta of df.

$$I\{f\} = -\nu\{f - \Phi(f)\} \tag{1}$$

This model term vanishes at equilibrium and satisfies the conservation laws and H-theorem. In the problems of linear transport and fluctuations one usually uses its linearized form:

$$\delta \widehat{I} |h\rangle = -\nu \left(|h\rangle - \sum_{\alpha=1}^{5} |\Psi_\alpha\rangle \langle \Psi_\alpha|h\rangle \right), \tag{2}$$

where $|h\rangle$ defined by

$$f = f^0 + \delta f = f^0(1+h), \tag{3}$$

and $|\Psi_\alpha\rangle$ are the first 5 Hermit polynomials.

The corresponding expression for Langevin's source in kinetic equation [2].

$$(\frac{\partial}{\partial t} + v\frac{\partial}{\partial \mathbf{r}} + F\frac{\partial}{\partial \mathbf{p}} + \delta \widehat{I}_\mathbf{p})\delta f(x,t) + \delta F \frac{\partial}{\partial \mathbf{p}} f(x,t) = y(x,t) \tag{4}$$

is of the form:

$$(yy)_{\omega,\mathbf{k},\mathbf{p}_1,\mathbf{p}_2} = -(\delta \widehat{I}_{\mathbf{p}_1} + \delta \widehat{I}_{\mathbf{p}_2}) f^0(p_1)\delta(p_1-p_2)$$

$$= 2\nu f^0(p_1) \left(\delta(\mathbf{p}_1 - \mathbf{p}_2) - f^0(\mathbf{p}_2) \sum_{i=1}^{5} \Psi_i(\mathbf{p}_1)\Psi_i(\mathbf{p}_2) \right). \tag{5}$$

The Langevin's source satisfies laws of conservations too.

The advantage of BGK model is that the solution of kinetic equation reduces to a system of algebraic equations. A weak point is that the model implies Prandtl number equals one and does not lead to the correct formula of Landau-Lifshitz [3] for Langevin's sources of hydrodynamic equations.

A graver situation arises in many component system. According to the Gross & Krook (GK) model [4] the collision operator has the form of deviation of the df from a "mythical" exponent.

$$I_a\{f_a\} = -\sum_b \nu_{ab}[f_a - \frac{n_a}{(2\pi m_a T_{ab})^{3/2}} \exp -\frac{m_a(\mathbf{v} - \mathbf{V}_{ab})^2}{2T_{ab}}], \tag{6}$$

where the parameters V_{ab} and T_{ab} are connected with the momenta of df via linear relations:

$$V_{ab} = \alpha_{aa}V_a + \alpha_{ab}V_b; \quad T_{ab} = \beta_{aa}T_a + \beta_{ab}T_b. \tag{7}$$

Coefficients $\alpha_{aa}; \alpha_{ab}; \beta_{aa}; \beta_{ab}$ are chosen in such manner that both conservation laws and balance equations for the momenta and energy for each component hold. Since the number of equations that the parameters of the model should satisfy (for five-moment description of two-component system these equations are 4: 2 for balance for momenta and 2 for balance for temperature) is less than the number of unknown parameters (in this approximation they are 5: $v_{ab}; \alpha_{aa}; \alpha_{ab}; \beta_{aa}; \beta_{ab}$) there is an arbitrariness in the choice of parameters. Therefore there exist various modifications (see, for example, [5]) of the collision model which correctly describe relaxation of five momenta. Distinctions of these models reveal when the higher momenta are taken into account: such as the pressure tensor and heat flux vector (although none of these models gives the right Prandtl number). Use of the GK model gives the same estimate for viscosity and heat conductivity both for light and heavy components, that is the viscosity of the heavy component is in the square root of the masses ratio underestimated.

But probably the most dubious point of GK model is the complicated exponential dependence on df.

Using the technique developed in Ref. [6], for a one-component system one may get the following expression for the linearized model collision operator of many-component system. The only difference is that the reference state one takes local equilibrium state for each components.

$$\delta\widehat{I}|h\rangle_a = -v_a|h\rangle_a + v_a \sum_{i=1}^{13} |\Psi_i^a\rangle\langle\Psi_i^a|h\rangle + \sum_b \sum_{i,j=1}^{13} |\Psi_i^a\rangle\langle\Psi_i^a|\,\delta\widehat{I}\,|\Psi_j^b\rangle\langle\Psi_j^b|h\rangle \tag{8}$$

In the state of local equilibrium with the use of the identity

$$\delta\widehat{I}\{f_a^0\} = 2\widehat{I}\{f_a^0\} \tag{9}$$

the collision integral takes a simple form:

$$\widehat{I}\{f_a^0\} = -\frac{1}{2}\sum_b \sum_{i=1}^{5} f_a^0 \Psi_i^a \langle\Psi_i^a|\,\delta\widehat{I}|1\rangle \tag{10}$$

Non-diagonal matrix elements are essentially non-equilibrium values and the state of thermodynamical equilibrium vanishes. Calculating the matrix elements for many-component plasma we get:

$$\widehat{I}\{f_a^0\} = -\sum_b v_{ab} f_a^0 [\delta\mathbf{v}_a \frac{T_b\mathbf{V}_a - T_a\mathbf{V}_b}{T_aT_b} \frac{m_a}{T_a} \frac{T_am_b + T_bm_a}{m_b + m_a}$$
$$+ (\frac{m_a}{T_a}\delta v_a^2 - 3)\frac{T_a - T_b}{T_aT_b}\frac{m_a}{m_b + m_a}] \tag{11}$$

where v_{ab} is the momentum relaxation frequency. Thus the complicated exponential dependency typical for G-K model appears to be unfounded and does not hold for the states remote from the full equilibrium. Acknowledgements. I acknowledge support from Russian Foundation for Basic Research (grant 00-02-17139).

REFERENCES

1. Bhatnagar, P.L., Gross, E.P., and Krook, M., *Phys. Rev.*, **94**, 511-525 (1954).
2. Lifshitz, E.M., and Pitaevskii, L.P., *Physical Kinetics*, Nauka, Moscow, 1979.
3. Landau, L.D., and Lifshitz, E.M., *Statistical Physics*, Nauka, Moscow, 1978.
4. Gross, E.P., and Krook, M. *Phys. Rev.*, **103**, 593-604 (1956).
5. Green, J.M., *Phys. Fluids.*, **16**, 2022-2023 (1973).
6. Belyi, V.V., and Paiva-Veretennicoff, I., *Teor.i Mat. Fizika*, **62**, 291-303 (1985).

Evaluation of the Bootstrap Current in Stellarators

W. Kernbichler*, S. V. Kasilov†, V. V. Nemov†, G. Leitold* and M. F. Heyn*

*Institut für Theoretische Physik, Technische Universität Graz, Petersgasse 16, A–8010 Graz, Austria
†Institute of Plasma Physics, National Science Center "Kharkov Institute of Physics and Technology",
Ul. Akademicheskaya 1, 61108 Kharkov, Ukraine

Abstract. It is well-known that a small bootstrap current is desirable for stellarators since in contrast to a tokamak this current can aggravate the stellarator confinement properties. In the present work, the computation of the bootstrap current in the long mean-free-path regime is performed for various stellarator configurations. For this purpose, the technique of integration along magnetic field lines is used. In particular, stellarator configurations with various relations between toroidal and helical inhomogeneities of the magnetic field and with quasi-helical symmetry are considered. The obtained results are discussed and compared to the results from the W7-X configuration where one of the design goals was the minimization of the bootstrap current. It is also well-known that expressions for the bootstrap current in stellarators possess resonances at rational magnetic surfaces in the low collision frequency regime. A method to smooth out these singularities is implemented in the present work.

INTRODUCTION

An investigation of the bootstrap current in stellarators is very important since it can negatively affect the plasma equilibrium and the magnetic configuration. In the present work, the bootstrap current is numerically analyzed for configurations with quasi-helical and quasi-axial symmetry as well as for the standard stellarator with different ratio of the toroidal and the helical inhomogeneity of the magnetic field. For this purpose, the technique of integration along magnetic field lines [1, 2, 3] is used. The results are benchmarked to the W7-X configuration which has been optimized in order to reduce the bootstrap current. In addition, results are given for two configurations in Boozer coordinates, where the influence of Coulomb collisions on the resonant behavior is discussed.

COMPUTATIONAL TECHNIQUE AND BASIC PARAMETERS

The method to calculate the bootstrap current is based on an integration procedure along the magnetic field lines [1, 2, 3] and is similar to the one proposed in [4]. However, the contribution of trapped particles which has been neglected in [4] is recovered. Also, in contrast to [4] this method allows one to do the computation in real-space coordinates and to analyze the bootstrap current for the magnetic fields given in real-space as well as in Boozer coordinates. The evaluation of the bootstrap current is performed using the geometrical factor λ_b which corresponds to λ_{b1} in [1, 2, 3] and which is similar to the corresponding factor in [4]. The dimensionless factor λ_b is obtained from the drift kinetic equation in the long mean-free-path-regime with a Lorentz collision operator which describes pitch-angle scattering but does not conserve momentum. However, it has been shown in [4] that if the fraction of trapped particles is small, the geometrical factor is also valid in case of momentum conservation. In this case, the general expression for the bootstrap current is represented by λ_b multiplied by some combination of the density and the temperature gradients with factors that do not depend directly on the magnetic field geometry. As an example, the following equation for the total bootstrap current can be used, which represents Eq. (4) of [4]:

$$\frac{j_b}{B} = -c\,\lambda_b \frac{1}{B_0^2}\left[1.67(T_e+T_i)\frac{dn}{dr} + 0.47n\frac{dT_e}{dr} - 0.29n\frac{dT_i}{dr}\right]. \tag{1}$$

In the present work, λ_b is calculated for the quasi-helically symmetric (QHS) configuration [5], for the quasi-axially symmetric CHS-qa configuration [6], and for the W7-X configuration [7], all in real space with $\beta=0$. To complement this, results are also given for a finite-β W7-X configuration [9] and a slightly different CHS-qa vacuum configuration, both in Boozer coordinates.

Additionally, two cases of a simplified $l=3$ configuration with different ratios of the toroidal and the helical magnetic field inhomogeneity are considered in order to analyze the influence of the helical field inhomogeneity, ε_h, on the bootstrap current. The first case corresponds to the idealized Uragan-3M torsatron ($\beta=0$) with only one toroidal harmonic function. The second case corresponds to a strongly decreased toroidal inhomogeneity, ε_t, of the magnetic field with approximately the same ε_h.

BOOTSTRAP RESONANCES

It had been shown in [1] that bootstrap current resonances arise from a considerable increase of particle displacement of passing particles from the magnetic surface. They are caused by the fact that on almost rational surfaces particles frequently come close to the global maximum of B where for particles at the trapped-passing boundary $v_\parallel$ approaches zero. In the presence of collisions, this situation changes due to collisional particle displacements. This fact can be simulated by introducing the parameter Δ into the expression for $v_\parallel$

$$v_\parallel = v\sqrt{1 - y\hat{B} + \Delta}, \qquad (2)$$

where $\hat{B}=B/B_0$ is the normalized magnetic field module, $y=J_\perp B_0/v^2$ with $J_\perp$ the perpendicular adiabatic invariant, and $B_0=B_{max}^{abs}$ is the reference magnetic field. The parameter Δ can be calculated from

$$\Delta = y\hat{B}\sqrt{-\gamma \ln[-y^2\gamma \ln(y^2\gamma)]}, \qquad (3)$$

with $\gamma=vL/(2v_{th})$, $L=(-\hat{B}_0'')^{-1/2}$, and $\hat{B}_0''=d^2\hat{B}/ds^2$ at the global maximum of $\hat{B}$, where s is the distance measured along the field line, v is the collision frequency, and v_{th} is the thermal velocity.

COMPUTATIONAL RESULTS FOR THE BOOTSTRAP CURRENT

Computational results for λ_b are shown in Fig. 1 and 2 as function of the mean magnetic surface radius r in units of the mean boundary surface radius a. To simplify the comparison for the CHS-qa and for the $l=3$ configuration, the normalized quantity $\lambda_{bn}=\lambda_b \iota \sqrt{r/R}$ is introduced, where ι is the rotational transform and R is the big radius of the torus. This quantity is equal to 1.46 for a tokamak with a large aspect ratio (see, e.g., [4]).

The calculations are performed for non-resonant magnetic surfaces. The behavior of λ_b in the vicinity of some resonant magnetic surfaces is also shown in Fig. 1. For the QHS configuration ι reaches its minimum near $r/a \approx 0.5$ and is somewhat smaller than 24/17. The quantity λ_{bn}, changes from approximately 0.6 to 1.36 near the boundary. For the CHS-qa configuration $|\lambda_b|$ is bigger than for QHS but smaller than for an equivalent tokamak. For non-resonant magnetic surfaces $|\lambda_{bn}|$ changes from 0.47 to 0.83 near the boundary. The results related to W7-X were already partly considered in [1, 2]. Here, in more detail the tendency of the λ_b behavior in the vicinity of the island surfaces corresponding to $\iota=10/11$ and $\iota=1$ is studied. Rather close to the boundary $|\lambda_b|$ is very small. Over the whole r region $|\lambda_{bn}|$ does not exceed 0.25. Note that for all three configurations in Fig. 1 the rotational transform ι has an anti-clockwise direction (with increasing toroidal angle). In this case, a negative λ_b corresponds to a tokamak-like direction of bootstrap current since it produces the poloidal magnetic field in an anti-clockwise direction (CHS-qa, W7-X).

For comparison, Fig. 2 gives λ_b for W7-X [9]. The results are rather close to the results in Fig. 1 (line 3) with a slightly different position of the 10/11 rational surface. It can easily be seen that resonances resulting from resonances at high order rational surfaces are smoothed by collisional effects. In contrast to this, it becomes evident that the main resonance at 10/11 (2/11 per field period) survives even in the presence of collisions.

For CHS-qa the normalized quantity λ_{bn} is given in Fig. 3. Again, results are comparable with line 4 in Fig. 1. The 4/11 resonance is less pronounced in the Boozer equilibrium from VMEC. Since ι has a non-monotonic behavior with s, the 4/11 resonance appears twice and it can be seen that collisions smooth the effect of this resonance but don't remove it completely. The rather steep change at $s^{1/2}=0.8$ corresponds roughly to the minimum value of $|\iota|=0.35$.

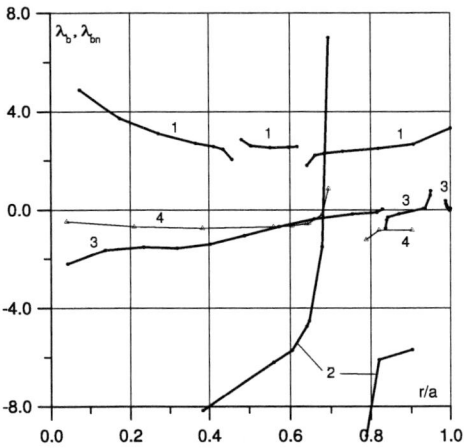

FIGURE 1. Parameter λ_b for QHS (curve 1), for CHS-qa (curve 2) and for W7-X (curve 3); curve 4 shows the λ_{bn} parameter for CHS-qa; the discontinuities in the curves correspond to the resonant magnetic surfaces for ι close to 24/17 for QHS, 4/11 for CHS-qa and 10/11 and unity for W7-X.

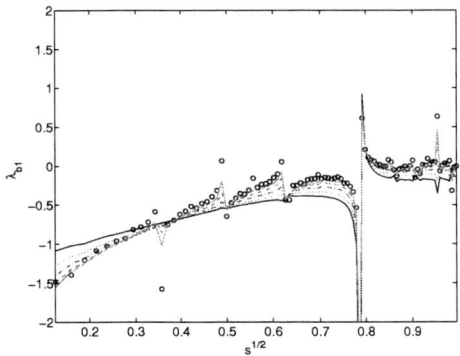

FIGURE 2. Parameter λ_b for W7-X. Black circles show results on given flux surfaces in the collisionless limit. The curves correspond to the following values of the collision parameter $\nu/(2\nu_{th})$: 10^{-6} (blue), 10^{-7} (red), 10^{-8} (magenta), 10^{-9} (green), and 10^{-10} (cyan), respectively. The dominant resonance is at the $\iota=10/11$ rational surface.

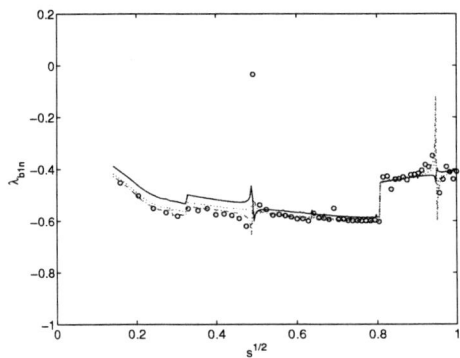

FIGURE 3. Parameter λ_{bn} for CHS-qa. Black circles mark results in collisionless limit. The curves correspond to the following values of $\nu/(2\nu_{th})$: 10^{-6} (blue), 10^{-7} (red), and 10^{-8} (magenta).

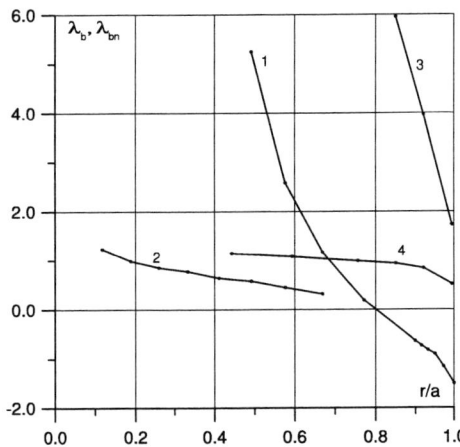

FIGURE 4. Parameters λ_b and λ_{bn} for the $l=3$ magnetic field with a strongly decreased (ten times) value of the toroidal ripple ε_t (curves 1 and 2, respectively) and with the original relation between the toroidal and the helical magnetic field ripples ε_t and ε_h (curves 3 and 4, respectively). Here the rotational transform ι has a clockwise direction. Therefore, the positive values of λ_b correspond to a tokamak-like direction of the bootstrap current.

The results related to the $l=3$ configurations are presented in Fig. 4. From this it follows that in case of decreased ε_t the quantity λ_b changes its sign for magnetic surfaces rather close to the boundary (curve 1). The reason is that for these values of r, ε_h becomes essentially bigger than ε_t and the bootstrap current is produced mainly due to the helical field inhomogeneity. In this case, λ_{bn} reaches the value of -1.93 near the boundary. The role of the helical field inhomogeneity is also seen from curves 3 and 4 for the ordinary $l=3$ configuration. In this case λ_{bn} becomes smaller than unity near the boundary.

SUMMARY

Employing a newly developed technique [1, 2, 3] which is based on an integration along magnetic field lines, the geometric factor in the expression for the bootstrap current is studied numerically for various stellarator configurations. The method is valid in the long mean-free path limit. In addition, the influence of collisions on bootstrap resonances which are observed in all configurations (both, given in real-space variables and in Boozer coordinates) is discussed in two cases. For that purpose an approximate analysis of this effect is presented, clearly showing that low order resonances are not easily destroyed by collisional effects.

ACKNOWLEDGMENTS

This work has been carried out within the Association EURATOM-ÖAW and with funding from the Austrian Academy of Sciences.

REFERENCES

1. S. V. Kasilov et al., in *27th EPS Conf. on Contr. Fusion and Plasma Physics*, 12-16 June 2000, Budapest, Hungary.
2. W. Kernbichler et al., *Problems of Atomic Science and Technology. Series: Plasma Physics (6)*, 8 (2000).
3. W. Kernbichler et al., *Proceedings of the 13th Int. Stellarator Workshop*, 25 February - 1 March 2002, Canberra, Australia.
4. A. H. Boozer and H. J. Gardner, *Physics of Fluids B* **2**, 2408 (1990).
5. J. Nührenberg and R. Zille, *Phys. Lett. A* **129**, 113 (1988).
6. S. Okamura et al., in *27th EPS Conf. on Contr. Fusion and Plasma Physics*, 12-16 June 2000, Budapest, Hungary.
7. C. Nührenberg, *Phys. Plasmas* **3**, 2401 (1996).
8. M. F. Heyn, M. Isobe, S. V. Kasilov et al., *Plasmas Phys. Control. Fusion* **43**, 1311 (2001).
9. C. D. Beidler et al., *Proceedings of the 13th Int. Stellarator Workshop*, 25 February - 1 March 2002, Canberra, Australia.

Influence of Nongyrotropy in the Electron Beam-Plasma Interaction

M. A. E. de Moraes[*], Y. Omura[†], M. Virgínia Alves[*]

[*]*Instituto Nacional de Pesquisas Espaciais, INPE-LAP, PO Box 515, 12245-970, S. J. Campos, SP, Brazil*
[†]*Radio Science Center for Space & Atmosphere, Kyoto University, Japan,*

Abstract. Nongyrotropic particle species have been detected in most regions of geoplasma, the distant solar wind, and cometary environments. In this work we performed particle simulations of beam-plasma interaction in a one-dimensional system taken along the magnetic field. We introduced a nongyrotropy in the particle population of an electron beam drifting against the background plasma. We study possible electromagnetic emissions. In the nongyrotropic case, we found that the magnetic field energy became much larger than in the gyrotropic case, indicating a strong electromagnetic wave emission.

INTRODUCTION

Distribution functions in magnetoplasmas of the type $F(v_{//}, v_\perp)$, where velocities occur both parallel ($v_{//}$) and perpendicular ($v_\perp$) to the background magnetic field ($\vec{B}_0$) are symmetric with respect to the magnetic field and are termed gyrotropic. When this symmetry is broken, the distribution becomes gyrophase dependent or nongyrotropic [1].

Nongyrotropic magnetoplasmas with a background field $\vec{B}_0 = B_0 \hat{x}$ have at least one particle population whose unperturbed distribution function depends on the gyrophase angle $\varphi = \tan^{-1}(v_z/v_y)$ [2]. The effects of nongyrotropy on linear wave dispersion were first studied in the context of fusion plasmas [3,4]. Several studies followed these pioneering researches. They showed that the introduction of gyrophase organization (bunching) can bring about coupling among the parallel eigenmodes, with the associated free energy enhancing previously existing (gyrotropic) instabilities or, in otherwise stable media, generating wave growth [2,5-8].

Nongyrotropic particle populations are frequently encountered in space plasmas. Nongyrotropy has been observed in ion populations in the region at and just upstream of Earth's bow schok [9], several Earth radii upstream [10] in the ion foreshock and downstream in the magnetosheath [11]. Measurements by the ISEE1 and 2 indicates the existence of nongyrotropic electrons in these same regions [12].

In this work we performed particle simulations of electron beam-plasma interaction in a one-dimensional system taken along the magnetic field. We introduced a nongyrotropy in the particle population of an electron beam drifting against the background plasma. We study possible electromagnetic emissions. In the nongyrotropic case, we found that the magnetic field energy became much larger than in the gyrotropic case, indicating a strong electromagnetic wave emission.

SIMULATION MODEL

We use a particle-in-cell code, KEMPO1 [13], that allows for spatial variations along *x*-direction. Since we are interested about parallel propagation, the wave-vector of the modes is aligned with the *x*-direction, $\vec{k} = k\hat{x}$, with the ambient magnetic field defined by $\vec{B}_0 = B_0\hat{x}$.

For the proposed study the simulation code incorporates three species of charged particles: background electrons and ions, and an electron beam with a given drift velocity. We assume the ion species to be of infinite mass, providing a neutralizing background. Both beam and plasma electrons have Maxwellian population. For gyrotropic and nongyrotropic cases the electrons of the beam are distributed with a pitch angle $\theta = 45^o$. For the nongyrotropic case the electron beam presents also a drift velocity at the z-direction, v_{z0}, introduced at $t=0$, which gives a gyrophase angle $\varphi = 0^o$. Velocity distribution of the moving particles, gyrotropic and nongyrotropic cases, are shown in Figure 1, v_y and v_z as a function of position x, at $t=0$. Boundary conditions are periodic and preexisting wave packets are not assumed, and all the waves grow self-consistently out of noise. Electrostatic modes are investigated by observing the longitudinal wave electric fields ($\vec{E} // \vec{k}$) whereas the electromagnetic modes by observing the wave field components (E_y, E_z), and (B_y, B_z).

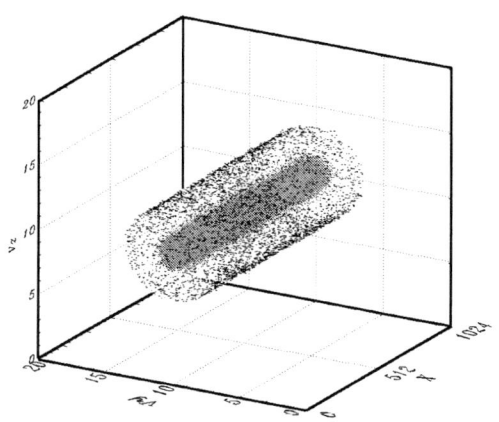
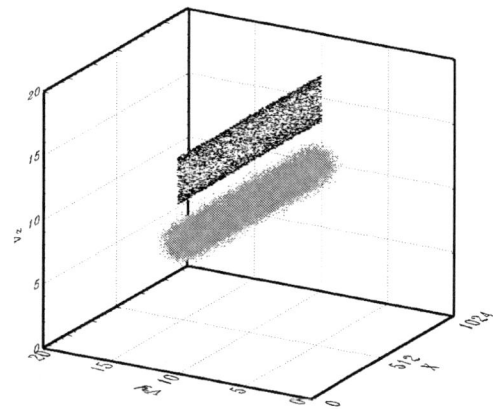

FIGURE 1. Particle velocity distribution, v_y and v_z as a function of propagation position x, at $t=0$, for the gyrotropic case (rigth) and nongyrotropic case (left).

RESULTS AND DISCUSSION

Simulation results presented in this section were obtained using the following computational parameters: electron plasma frequency, $\omega_p^2 = \omega_{pe}^2 + \omega_{pb}^2 = 1$; electron cyclotron frequency $\Omega_e = 0.5$; electron thermal velocity, $v_{th} = 0.02c$; electron beam thermal velocity, $v_{bth} = 0.02c$; electron beam drift velocity $v_{drift} = 0.1c$; grid spacing, $\Delta x = 0.1c/\omega_p$; number of grid points, 1024; number of superparticles, 4096000; time step, $\Delta t = 0.005\omega_p^{-1}$; beam to plasma density ratio, $n_b/n_0 = 0.04$.

Figure 2 presents the time evolution of electrostatic and kinetic energy for the gyrotropic (left) and nongyrotropic case (right). At $t=0$, the nongyrotropic case presents a higher kinetic energy due to the introduction of $v_{z0} \approx 20v_{th}$. We can observe that until $t \approx 50\omega_p^{-1}$ both cases present similar behaviour, the corresponding decreasing of kinetic energy appearing as an increasing of electrostatic energy. For $t \approx 100\omega_p^{-1}$, the decreasing of kinetic energy does not show up as an increasing of electrostatic energy, for the nongyrotropic case (Figure 2 (right)).

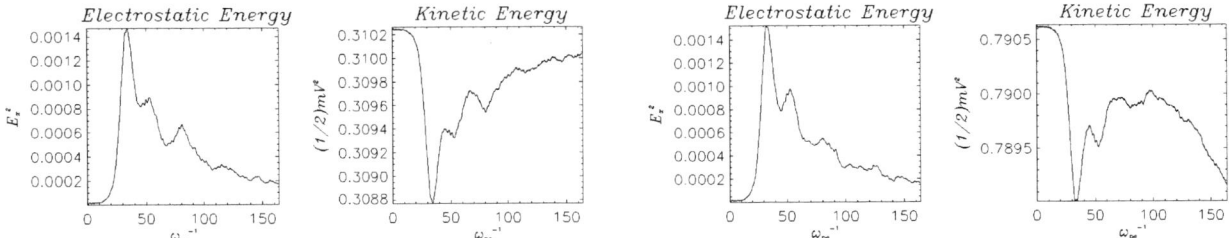

FIGURE 2. Time history of the electrostatic energy, $\propto E_x^2$, and the kinetic energy, $1/2mv^2$ for the gyrotropic case (left), and for the nongyrotropic case (right).

Concerning the electromagnetic energy, we see that for the gyrotropic case there is no variation along the time, as shown in Figure 3 (left). For the nongyrotropic case, we see an increasing of the electromagnetic energy as shown in Figure 3 (right). The growing of electromagnetic energy starts at $t \approx 80\omega_p^{-1}$ reaching the first maximum at $t \approx 100\omega_p^{-1}$, coincident with the point where the kinetic energy starts to decrease (see Figure 2).

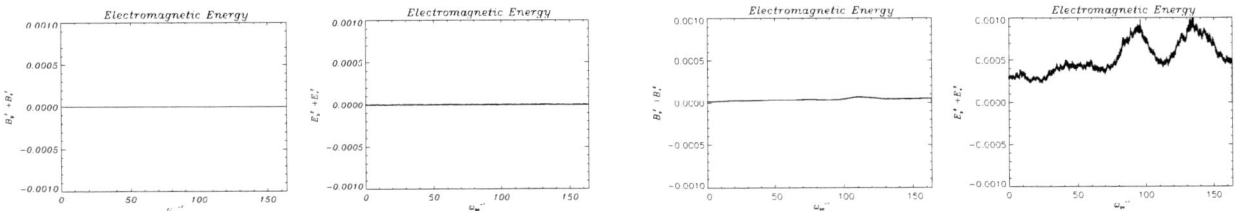

FIGURE 3. Time history of the electromagnetic energy, $\propto B_y^2 + B_z^2$, and $\propto E_y^2 + E_z^2$, for the gyrotropic case (left) and the nongyrotropic case (right).

The diagram $\omega \times k$ tells us the modes that are present in the system. We constructed the $\omega \times k$ diagram for the electromagnetic fields components (E_y^2, E_z^2, B_y^2, and B_z^2). We will show the $\omega \times k$ diagram only for the electromagnetic component E_z, since the behaviour for the electrostatic component is very similar for gyrotropic and nongyrotropic cases. For both cases we observe Langmuir waves, forward and backward propagating and also the beam mode forward propagating. Figure 4 shows the $\omega \times k$ diagram for the E_z component (electromagnetic mode) for the gyrotropic (left) and the nongyrotropic case (right). For both cases we observe the RCP and LCP high frequency modes, forward and backward propagating. We also observe the whistler mode (RCP low frequency) in both cases, but for the nongyrotropic case this mode emission is intensified. Grey scale is related to the intensity of the field component (in dB). We also observe in the nongyrotropic case a localized emission (nonpropagating) and a second branch in the whistler mode backward propagating, yet to be explained.

The behaviour of the system can also be illustrated by the phase space of the particles ($v_x \times x$), not shown here. At early times, up to $t \approx 90\omega_p^{-1}$, the behaviour of the gyrotropic and nongyrotropic systems is similar to each other and also similar to an electrostatic electron beam-plasma interaction. For later times, the nongyrotropic case presents a cell formation in the phase space, indicating that electrons are trapped in their own potential.

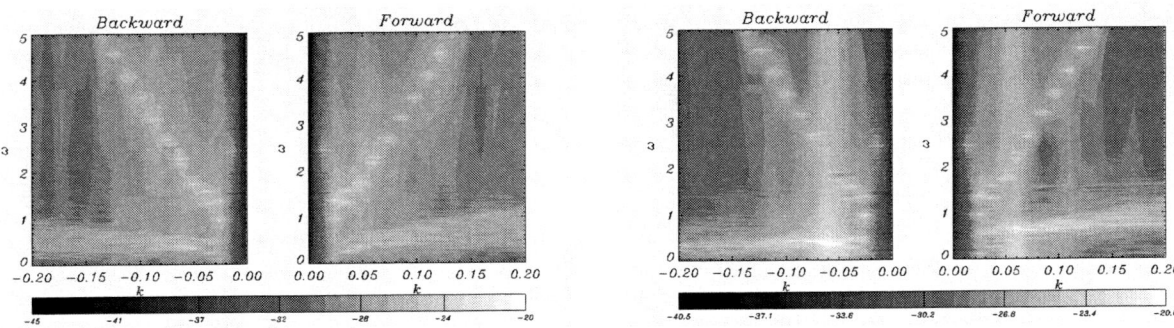

FIGURE 4. $\omega \times k$ diagram for the electric field component, E_z, electromagnetic mode, for the gyrotropic case (right) and for the nongyrotropic case (left).

CONCLUSIONS

In this work we performed particle simulations of electron beam-plasma interaction in a one-dimensional system taken along the magnetic field. We introduced a nongyrotropy in the particle population of an electron beam drifting against the background plasma. We compare the behaviour of two systems, gyrotropic and nongyrotropic. We observe that at early times, up $t \approx 90\omega_p^{-1}$, both systems have similar behaviour. For times larger than $90\omega_p^{-1}$, there is an enhancement of the electromagnetic energy for the nongyrotropic case. An intensification of the emission of the whistler mode can be observed in the $\omega \times k$ diagram for the E_z component (see Figure 4). Phase space of the beam and the plasma electrons also shows the different behaviour for the late times for the gyrotropic and nongyrotropic cases. Different pitch-angle and gyrophase angle, as well beam to plasma density ratios should be investigated in the near future.

ACKNOWLEDGMENTS

This work is supported by FAPESP- Fundação de Amparo à Pesquisa do Estado de São Paulo.

REFERENCES

1. Motschamann, U, Kafemann, H. and Scholer, M. *Ann. Geophysicae* **15**, 603-613 (1997)
2. Romeiras, F. J. and Brinca, A. L., *J. Geophys. Res.* **104**, 12407-12413 (1999)
3. Sudan, R. N. *Phys. of Plasmas* **8**, 1915-1918 (1965)
4. Eldridge, O, *Phys. of Plasmas* **13**, 1791-1794 (1970)
5. Brinca, A. L., Borda de Água, L. and Winske, D., *Geophys. Res. Let.* **12**, 2445-2448 (1992)
6. Brinca, A. L., Borda de Água, L, and Winske, D., *J. Geophys. Research*, **98**, 7549-7560 (1993)
7. Brinca, A. L., Omura, Y., Matsumoto, H., *J. Geophys. Research*, **98**, 21071-21076 (1993)
8. Brinca, A. L., J. Atmospheric and Solar-Terrestrial Physics, **62**, 701-709 (2000)
9. Thomsen, M. F., Gosling, J. T., Bames, S. J., and Russel, C. T., *J. Geophys. Res.*, **90**, 270-273, (1985)
10. Gurgiolo, C., Parks, G. K., Mauk, B. H., Lin, C. S., Anderson, A., Lin, R. P., and Réme, H., *J. Geophys. Res.*, **86**, 4415-4424 (1981)
11. Sckope, N., Paschmann, G., Brinca, A. L., Carlson, C. W., Lür, H., *J. Geophy. Res.*, **95**, 6337-6352 (1990)
12. Anderson, A. K., Lin, R. P., Gurgiolo, C., Parks, G. H., Potter, D. W., Werden, S., Rème, H., *J. Geophys. Res.*, **90**, 10809-10814 (1985)
13. Omura, Y., Matsumoto, H., in: *Computer Space Plasma Physics*, ed. by H. Matsumoto and Y. Omura, Chap. 2, 21-84 (1993)

PIC Simulation of Collisionless Negative Ion Plasma Expansion into a Vacuum

M. Cercek[1], T. Gyergyek[1,2] and V. Ignatescu[1,3]

[1] *J. Stefan Institute, Jamova 39, 1000 Ljubljana, Slovenia*
[2] *Faculty of Electrical Engineering, University of Ljubljana, 1000 Ljubljana, Slovenia*
[3] *Faculty of Physics, A.I.Cuza University, 6600 Iaşi, România*

Abstract. In the contribution the preliminary results of the simulation experiment on expansion of a collisionless plasma with negative ions into a vacuum are presented. The electric field profile development in time and phase spaces for all three species are presented and discussed. Dependencies on negative ion mass are also studied. Double layer-like potential structure is observed to form in the expanding plasma under certain conditions.

INTRODUCTION

Plasmas consisting of positive ions, negative ions and electrons attracted in the past years a considerable interest. This is understandable since such negative ion plasmas are found in many plasma assisted technological processes, they are used also as sources for negative ion beam production in neutral beam heating schemes in controlled fusion research. Recently, the expansion of negative ion plasma into a vacuum received considerable attention. Several reports on the results of theoretical studies were published. El-Zein *et al.* [1] modeled the expansion of a negative ion plasma into a vacuum by using nonlinear fluid equations for positive and negative ions and the Boltzmann assumption for the electron density distribution, together with Poisson's equation to determine the electric potential in the expanding plasma. In addition to the self-similar expansion, which is usually found in a quasi-neutral plasma expansion model, they identified also a burst of positive ions accelerated into the vacuum and a burst of negative ions accelerated back into the plasma. These accelerated ions appeared since quasi-neutrality was not imposed in the model. On the other hand, Garcia et al. [2] used in their study a one-dimensional Vlasov-Poisson model without imposing quasi neutrality or Boltzmann densities for any of the species. They came to the same result that in the early stages some electrons run ahead the bulk of the plasma generating an electric field that accelerates a group of positive ions into the vacuum. The less massive electrons and positive ions were later followed by the more massive negative ions. The expansion process was clearly presented by particle distribution functions in real phase space. In a very recent investigation Medvedev [3] used a PIC simulation to study preferentially the phenomena in the initially unperturbed region of the expanding plasma. The most important result of his investigation is the conclusion that at some special values of negative ion mass and initial density ratio a collision-less shock arises and moves backward into the unperturbed region. The shock was classified as a rarefactive shock. It was particularly emphasized that the rarefaction shock represents not only a rarefaction of positive ions, but also a compression of negative ions. The positive ion velocity and the potential profile in the expanding region are found to be independent of negative ion species. The density profile of positive ions was decreasing but the ion velocity was increasing almost linearly. Positive ions were accelerated by the electric field of the rarefaction wave. In the present work we study, by using a one-dimensional particle-in-cell simulation code, negative ion plasma expansion from a planar source region towards a floating wall. It is positioned at the other end of an initially empty experimental system. The code was developed by Birdsall and co-workers at Berkeley [4] and, among many other applications, used also to study the potential formation in a plasma with warm ions flowing towards a floating collector [5]. It was used by the present authors to study the potential formation in a two-electron temperature plasma [6] and also applied in an investigation of the two-electron temperature plasma expansion into a vacuum [7].

SIMULATION

The simulation region is bounded at x = 0 by a planar plasma source and at the opposite side, $x = L$, by a floating wall. The system is L = 0.5 m wide and initially empty (vacuum). We assume that fully ionized plasma consisting of Maxwellian positive ions, negative ions and electrons fills the source. Plasmas with various ion compositions, $\mu = M^-/M^+$, are studied. The temperature of the ion species is set to $T_+ = T_- = 0.1$ eV and the electron temperature to $T_e = 2$ eV. At $t = 0$ equal and temporally constant fluxes of negative and positive particles with half-Maxwellian velocity distributions are injected into the system. The ratio of negative ion to positive ion flux is chosen in the presentation to be $\alpha = j_-/j_+ = 0.1$ and 0.9, which gives low and high values of the negative ion fraction at the source. A decreasing (negative) potential profile is expected to form in the early times of expansion in the system since the almost massless electrons are the fastest particles in this process. As in the case of two-electron temperature plasma expansion, special double layer-like potential structures are also expected to form in the system. The reflected negative particles are refluxed to the system, no charge is accumulated at the source boundary and the electric field is always zero there. The positive ions are expected to be accelerated toward the floating wall. The simulation run is stopped when the first ions reach the wall. At the final phases of expansion systems with 10^5 particles of majority species are investigated. The evolution of various plasma parameters is investigated through diagnostic windows built into the code. In the present contribution the temporal evolution of the electric field profiles and the particle phase spaces for various specially chosen values of ion mass ratio μ and of injected ion flux ratio α are displayed. All profiles are not time averaged but are snapshots at the last time step before the "measurement". Some oscillations of the measured quantities are in this way also captured and shown on the plots. There is no axial magnetic field in these one-dimensional simulations. The system is also assumed to be collision-less.

RESULTS

We present the simulation results for three typical values of ion mass ratio: μ = 0.025, 1 and 3.65, which correspond to mixtures of H^-/Ar^+, H^-/H^+ and SF_6^-/Ar^+, respectively. These mixtures are frequently found in experimental systems. In all cases, without exception, we observe that the plasma expansion is initiated by electrons, which expand into the vacuum region due to their high velocity and without forming a front. The transition time in the 0.5 long system is for the fastest electrons from a 2 eV Maxwellian distribution $t_{tre} = 3 \times 10^{-7}$ s ($v_{max} = 1.7 \times 10^{-6}$ m/s ~ $3 \times v_{the}$). The charge separation induces a potential gradient, which retards the electrons and accelerates positive ions. In their expansion they form a front which can be observed as a corresponding peak in electric field profiles at far right side on all plots in Fig 1(a) and Fig 2(a). It must be mentioned, that in the case of heavy ion mixture no such electric field peak is observed (Fig. 3(a), positive ion front is located e.g. at $x/L = 0.32$ after $t = 10$ μs of expansion!). In Figs 1(b), 2(b) and 3(b) we show on the lowest plot the positive ion phase space v_i-x. It can be observed that the ions are in average steadily accelerated in the system. The electrons form (plots on the top) almost a Maxwellian distribution due to retardation and reflection during the expansion. The behavior of the negative ions during the expansion process is very much dependent on the ion mass ratio. In the case of very light negative ions (H^- vs Ar^+) they scarcely expand from the source region during the first 10 μs of the process, as it can be seen on the middle plot in Fig 1(b). A steep potential drop is formed at their front and positive ions are strongly accelerated there. The negative ions are reflected at the potential drop and have an almost full-Maxwellian velocity distribution. Similar results were obtained by Medvedev in his investigations [3]. The situation is somehow similar to the so-called "structured" negative ion discharge. In the $\mu = 1$ mixture the negative ions are accelerated (!) from the source region, but they still form a well-defined front bounded by a negative double-layer like potential structure. It can be observed as a second very strong electric field peak at the left end of the profiles in Fig 2(a). Again, the positive ions are strongly accelerated at this points. Such double-layer potential structure was also observed to form during a two-electron temperature plasma expansion into vacuum, theoretically, in simulations and also experimentally [7, 8]. In this case, the situation is somehow similar to the "double-layer structured" negative ion discharge. During the expansion process of SF_6^-/Ar^+ plasma both ion species are accelerated into the system but negative ions still lag behind and are confined by a potential drop which is represented by a single electric field peak in Fig 3(a). No additional acceleration of positive ions is observed at these points. Similar results were obtained by Garcia et al. [2]. In fact, they observed also similar acceleration of electrons, a behavior, which was specially pointed out. We have not yet investigated the dependence of the expansion on the negative ion to electron temperature ratio and on the negative ion density ratio which are very important parameters for DL formation in two-negative particle species plasmas.

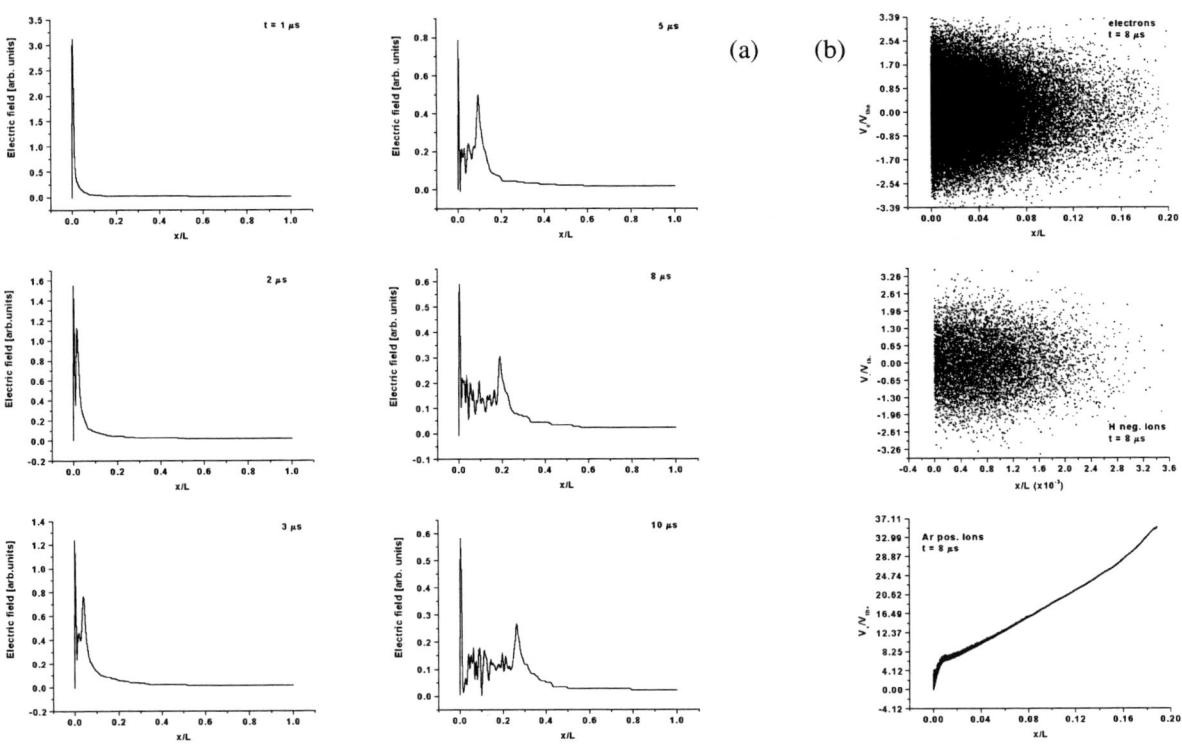

FIGURE 1. (a) Temporal evolution of the axial electric field profiles for $\mu = 0.025$ (H^-/Ar^+) and $\alpha = 0.1$. (b) Spatial profiles of electron, negative ion and positive ion velocity scatter (phase space) plots for $\alpha = 0.1$ at $t = 8$ μs.

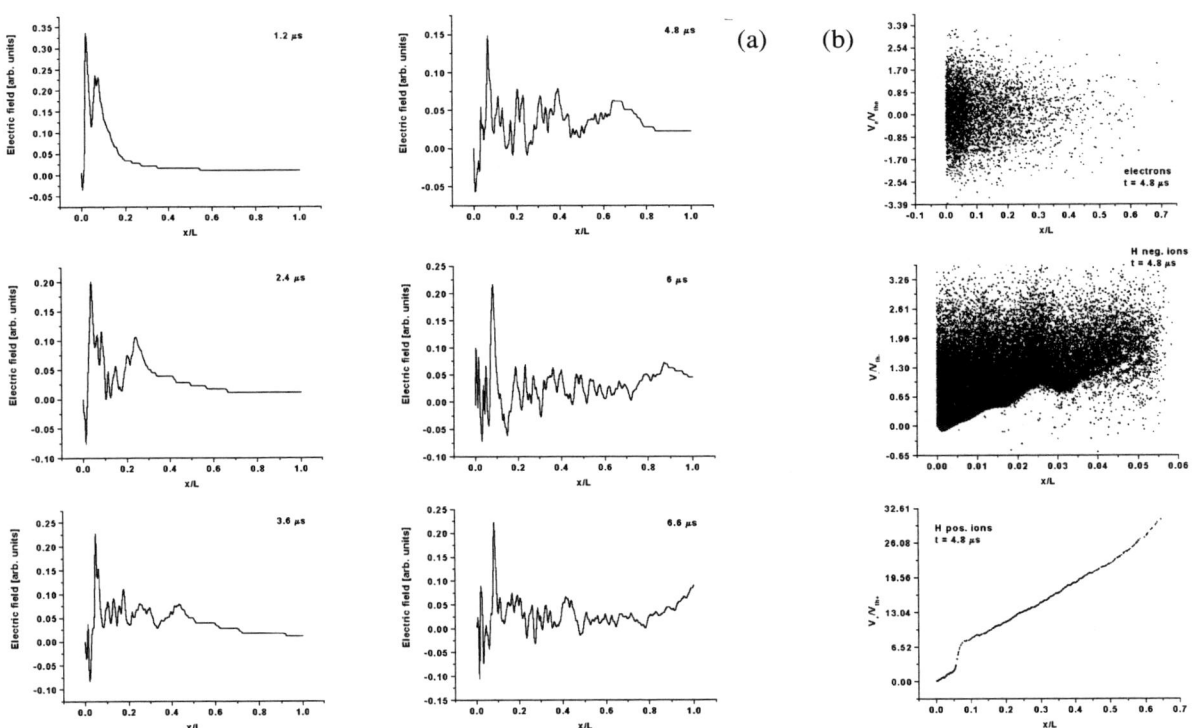

FIGURE 2. (a) Temporal evolution of the axial electric field profiles for $\mu = 1$ (H^-/H^+) and $\alpha = 0.9$. (b) Spatial profiles of electron, negative ion and positive ion velocity scatter (phase space) plots for $\alpha = 0.9$ at $t = 4.8$ μs.

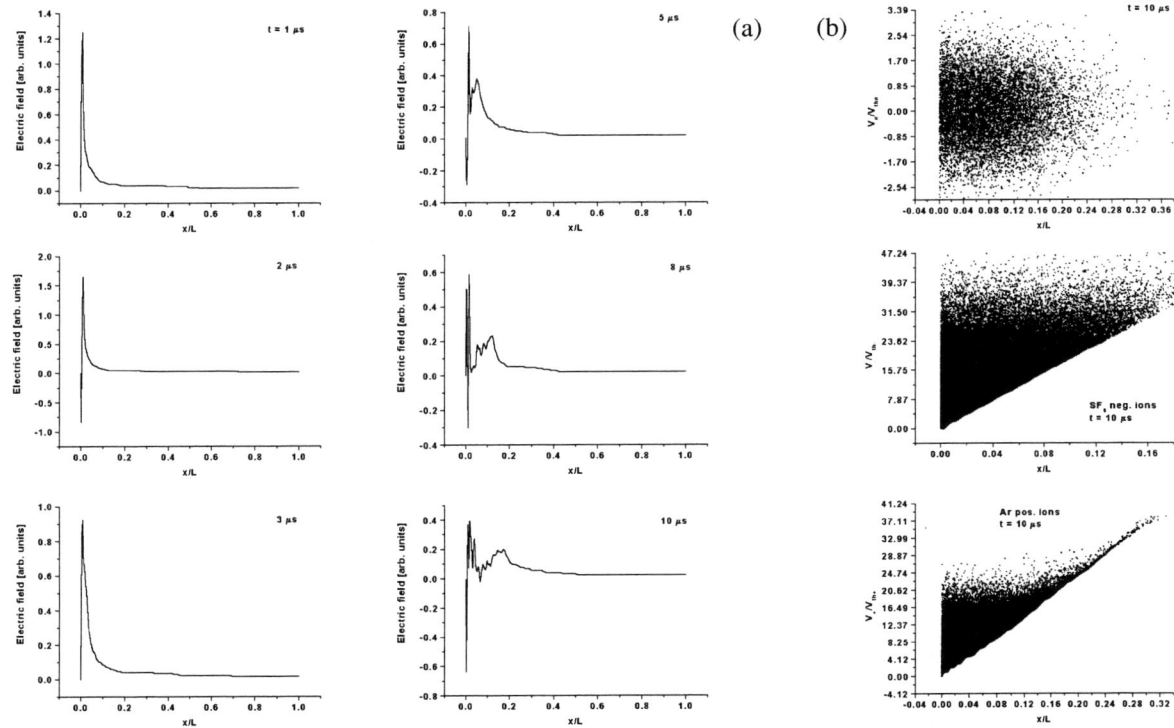

FIGURE 3. (a) Temporal evolution of the axial electric field profiles for $\mu = 3.65$ (SF_6^-/Ar^+) and $\alpha = 0.9$. (b) Spatial profiles of electron, negative ion and positive ion velocity scatter (phase space) plots for $\alpha = 0.9$ at $t = 10$ μs.

CONCLUSIONS

The PIC simulations presented in this contribution verify the main features of plasma expansion into a vacuum. A charge separation between electrons and positive ions induces a potential gradient that accelerates ions to very high velocities. In their expansion they form a well-defined front with a corresponding peak in the electric field profile. For the first time, to our knowledge, a second double layer-like potential structure was observed to form in certain cases and which confines the expanding negative ions. The phenomenon is similar to that observed in a two-electron temperature plasma expansion into a vacuum.

ACKNOWLEDGMENTS

Part of the work was done during the stay of one of the authors (V. I.) at the J. Stefan Institute within the frame of the CEEPUS (project A-103) exchange program.

REFERENCES

1. El-Zein, Y., Amin, A., Kim, H-S., Yi, S., and Lonngren K. E., Phys. Plasmas **2** 1073-1076 (1995).
2. Garcia, L. G., Goedert, J., Figua, H., Fijalkow, E., and Feix, M. R., Phys. Plasmas **4** 4240-4253 (1997).
3. Medvedev, Y. V., Plasma Phys. Control. Fusion **41** 303-313 (1999).
4. Vahedi, V., Verboncoeur, J., Birdsall, C.K., 1966 XPDP1-4.1 Plasma Device 1 Dim Bounded Electrostatic Code (PTSG, UC Berkeley).
5. Schwager, L. A., Birdsall, C. K., Phys. Fluids **B 2** 1057-1068 (1990).
6. Cercek, M., Gyergyek, T., and Stanojevic, M., Contrib. Plasma Phys. **39** 541-556 (1999).
7. Cercek, M., Gyergyek, T., and Stanojevic, M., Proc. XXIV ICPIG **4** 207 (1999).
8. Hairapetian, G., Stenzel, R., Phys. Rev. Letters 61, 1607-1610 (1988)

Calculation of Self-consistent Radial Electric Field in Presence of Convective Electron Transport in a Stellarator

W. Kernbichler*, S. V. Kasilov† and M. F. Heyn*

*Institut für Theoretische Physik, Technische Universität Graz, Petersgasse 16, A–8010 Graz, Austria
†Institute of Plasma Physics, National Science Center "Kharkov Institute of Physics and Technology",
Ul. Akademicheskaya 1, 61108 Kharkov, Ukraine

Abstract. Convective transport of supra-thermal electrons can play a significant role in the energy balance of stellarators in case of high power electron cyclotron heating. Here, together with neoclassical thermal particle fluxes also the supra-thermal electron flux should be taken into account in the flux ambipolarity condition, which defines the self-consistent radial electric field. Since neoclassical particle fluxes are non-linear functions of the radial electric field, one needs an iterative procedure to solve the ambipolarity condition, where the supra-thermal electron flux has to be calculated for each iteration. A conventional Monte-Carlo method used earlier for evaluation of supra-thermal electron fluxes [1] is rather slow for performing the iterations in reasonable computer time. In the present report, the Stochastic Mapping Technique [2, 3] (SMT), which is more effective than the conventional Monte Carlo method, is used instead. Here, the problem with a local monoenergetic supra-thermal particle source is considered and the effect of supra-thermal electron fluxes on both, the self-consistent radial electric field and the formation of different roots of the ambipolarity condition are studied.

FLUX BALANCE AND NEOCLASSICAL PARTICLE FLUXES

In a stellarator the constraint that the ion and electron fluxes be equal determines the radial electric field. Thus, the equation for the flux balance, $\Gamma_e^{nc} + \Gamma_e^s = Z_i \Gamma_i^{nc}$, has do be fulfilled on each flux surface. Here, Γ_α^{nc} with $\alpha = e, i$ are the neoclassical particle fluxes [1],

$$\Gamma_\alpha^{nc} = -n_\alpha \left\{ D_{11}^\alpha \left(\frac{n_\alpha'}{n_\alpha} - \frac{q_\alpha E_r}{T_\alpha} \right) + D_{12}^\alpha \frac{T_\alpha'}{T_\alpha} \right\},$$

with $q_\alpha, n_\alpha, T_\alpha, E_r$ being the particle charge, density, temperature and the radial electric field, respectively, and prime denotes a derivative with respect to a formal radius.

The neoclassical diffusion coefficients D_{11}^α and D_{12}^α are computed according to the Shaing-Houlberg-model [4], where instead of ε_h the effective ripple ε_{eff} [5] is used [6]. The balance equation is a non-linear equation in the radial electric field which might have multiple roots.

SUPRA-THERMAL PARTICLE FLUXES

The supra-thermal particle flux, Γ_e^s, is of particular importance for the confinement since it can influence the radial electric field through the ambipolarity condition [1]. Following the SMT approach [2, 3], the usual expression for particle flux through the magnetic surface $\hat{\psi} = \hat{\psi}_0$ defined in guiding center variables and flux coordinates $(\hat{\psi}, \theta, \varphi)$ can be written as an average over Poincaré cuts of the phase space flux density,

$$\Gamma_e^s = 2\pi \sum_{\mathbf{m}} \int d^5 u \Gamma_{\mathbf{m}}(\mathbf{u}) \delta(t - u^5) \left[\Theta(\hat{\psi}(\mathbf{Z}(\mathbf{z_m}, \tau_{b\mathbf{m}})) - \hat{\psi}_0) - \Theta(\hat{\psi}(\mathbf{Z}(\mathbf{z_m}, 0))) - \hat{\psi}_0) \right].$$

Here, $\Gamma_{\mathbf{m}}(\mathbf{u})$ is the pseudo-scalar particle flux density through those Poincaré cuts, $\mathbf{u}$ denotes the five variables x^1, x^2, p, λ, t, where x^1, x^2 are contravariant coordinates in a local magnetic coordinate system, and p, λ, t are the

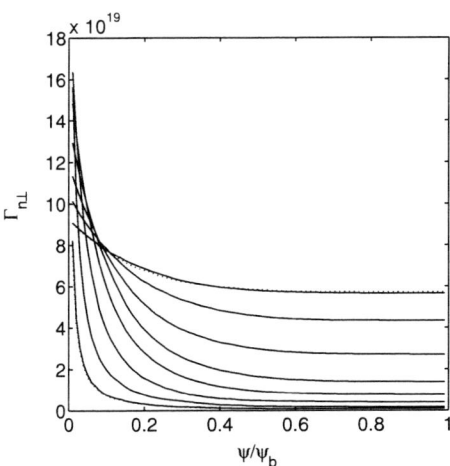
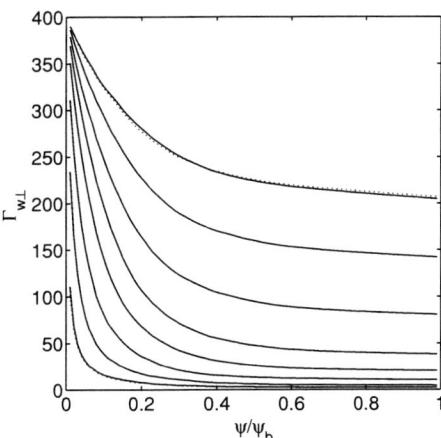

FIGURE 1. Particle flux (left) and energy flux (right) versus normalized flux label $\hat{\psi}$, respectively. The particle energy ranges from $w_0 = 2T_0$ to $w_0 = 9T_0$ from bottom to top.

momentum modulus, the particle pitch and the time, respectively. The summation over **m** is a summation over contributions from different Poincaré cuts and Θ is the Heaviside step function. In SMT, $\mathbf{Z}(\mathbf{z_m}, \tau)$ is the solution to equations of particle drift motion with $\mathbf{z_m}$ being the initial value of phase space variables on the Poincaré cut with index **m**, and with τ_{bm} being a transition time between cuts. All details of SMT can be found in Ref. [2, 3]. The energy flux is obtained in the same way and it differs from Γ_e^s by the factor $w_k(\mathbf{u})$ in the sub-integrand, where w_k is the value of the kinetic energy of the particle at the phase space point **u** on the cut.

COMPUTATIONAL RESULTS

For numerical computations, the magnetic field from the W7-AS stellarator [7] was used in its real space representation. In Figure 1, particle and energy fluxes of supra-thermal particles are shown, respectively. The particle source is on the magnetic axis in the magnetic field minimum located at the elliptic cross section of W7-AS. Trapped particles with a pitch value $\lambda_0 = 0.1$ and fixed energies w_0 ranging from $w_0 = 2T_0$ to $w_0 = 9T_0$ are generated there. The source rate in these computation was $v_{\text{stat}} = P_{source}/w_0$ where $P_{source} = 400$ kW is the source power. The profiles of the equilibrium parameters were the following, $T_\alpha(\hat{\psi}) = T_0(1.2 - \hat{\psi})$, $n_\alpha(\hat{\psi}) = n_0(1.2 - \hat{\psi})^2$, $\Phi(\hat{\psi}) = T_0\hat{\psi}/e$, where $T_0 = 3$ keV, $n_0 = 3 \cdot 10^{13}$ cm^{-3}, and $a = 17.4$ cm, respectively. The quantity $\psi/\psi_b = \hat{\psi} = (r/a)^2$ was chosen as a formal flux label, where the radius $r = R - R_0$ is computed in the mid plane of the symmetric cross section and R_0 is the radius of the magnetic axis. It can be seen that the energy of the source particles has a significant influence on the profiles of supra-thermal fluxes.

The results of self-consistent modeling are presented in Figures 2, 3, 4 and 5, where a modified density profile $n_0(1.2 - \hat{\psi}^2)$ is used. Figure 2 shows the self-consistent E_r-profile with neoclassical fluxes only. In Figure 3 supra-thermal electron fluxes are given for $\lambda_0 = 0.1$ and two energies, in each case with its respective E_r-profile. One can clearly see that particles with lower energy ($4T_0$) have time to slow down, whereas particles with higher energy ($9T_0$) quickly drift out of the plasma.

Figure 4 shows the formation of the "electron root" in a rather narrow region near the magnetic axis. In addition, Figure 5 shows the dependence of fluxes on E_r in two radial positions. One can see that at $r = 5.5$cm two stable solutions exist, which finally result in the formation of the "electron root". The decision which root has to be chosen is based on the minimization of a generalized heat production rate [1]. Following that approach, the position of the poloidal shear layer can be determined from

$$P(r) = \int_{E_r^i}^{E_r^e} (Z_i \Gamma_i^{nc} - \Gamma_e^{nc} - \Gamma_i^s) dE_r = 0,$$

where E_r^i and E_r^e are the stable solutions for E_r in the "ion" or "electron root", respectively. The "ion root" is then realized for $P > 0$ and the "electron root" for $P < 0$ [1]. Basically, the ion root is realized almost everywhere. When

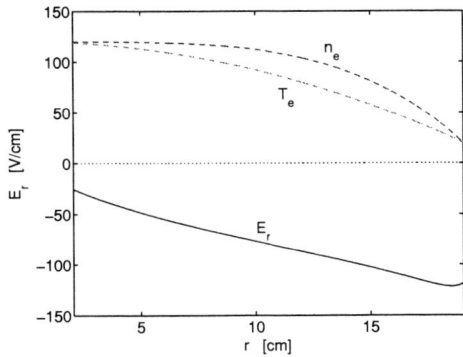

FIGURE 2. Radial electric field E_r resulting from neoclassical fluxes only versus radius r. In addition, the profiles of n_e and T_e in dimensionless units are given.

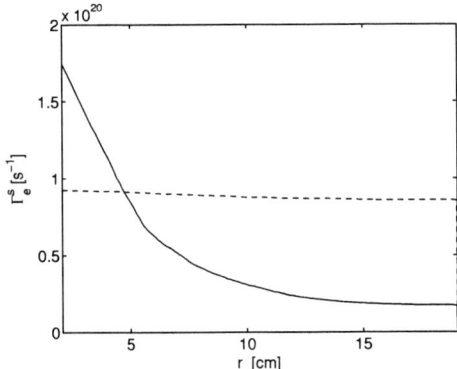

FIGURE 3. Supra-thermal particle flux Γ_e^s versus radius r. The supra-thermal fluxes are computed for $w_0 = 4T_0$ (full) and $w_0 = 9T_0$ (dashed) in each case with a selfconsistent E_r.

approaching the axis, the neoclassical fluxes are decreasing together with the magnetic surface area, but at the same time the supra-thermal flux is increasing. Finally, the neoclassical bifurcation occurs and the root is changed from "ion" to "electron". Further inward, the "electron root" disappears which can be seen in Figure 4 where the pertinent root vanishes. This event is an artifact of the neoclassical transport model used in the present computation where the ion flux is decreasing with increasing E_r and cannot balance the supra-thermal flux anymore. As discussed in Ref. [1], the validity of the neoclassical theory may be violated in such a case of a very strong radial electric field.

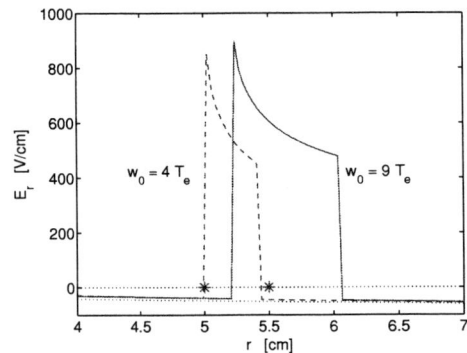

FIGURE 4. Radial electric field E_r versus radius r for two energies of source particles and for the neoclassical equilibrium.

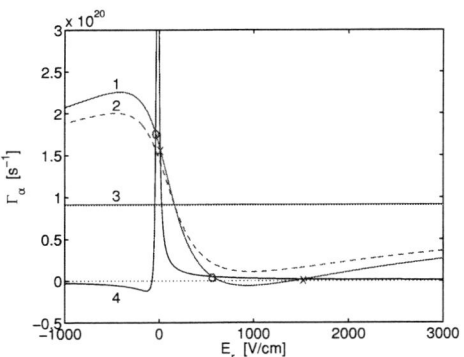

FIGURE 5. Total electron flux at $r = 5.5$cm (1) and at $r = 5.0$cm (2), supra-thermal electron flux (3) and ion flux (4) versus radial electric field E_r, respectively, for the $w_0 = 9T_0$-case. Circles mark stable roots, whereas crosses mark unstable ones.

SUMMARY

The application of SMT to a "global" computation of supra-thermal particle fluxes in a stellarator shows that this method is fast enough to allow for iterations of the radial electric field using the ambipolarity condition taking into account fluxes from supra-thermal particles. For this purpose, SMT is the ideal tool, because the computation of one self-consistent profile requires only tens of minutes on a DEC Alphastation 500 depending on accuracy. Therefore, SMT combined with a neoclassical balance code permits the self-consistent modeling of particle and energy balance in a stellarator with strong electron or ion cyclotron heating where the convective transport of supra-thermal particles plays a significant role. It is also shown that convective fluxes are very sensitive to the detailed structure of the supra-thermal particle source. In the case of ECRH, non-linear effects of wave-particle interaction are dominant in the formation of such a source [8]. The method for modeling this effects has been recently developed and will be included in future models based on SMT.

ACKNOWLEDGMENTS

This work has been carried out within the Association EURATOM-ÖAW and with funding from the Austrian Academy of Sciences.

REFERENCES

1. Maassberg, H., Beidler, C. D., Gasparino, U., Rome, M., Team, W.-A., Dyabilin, K. S., Marushchenko, N. B., and Murakami, S., *Phys. Plasmas*, **7**, 295–311 (2000).
2. Kasilov, S. V., Kernbichler, W., Nemov, V., and Heyn, M., *Phys. Plasmas*, **August** (2002).
3. Kasilov, S. V., Kernbichler, W., and Heyn, M. F., "Mapping Technique for Stellarators with Realistic Magentic Field," in *28th EPS Conf. on Contr. Fusion and Plasma Physics, Funchal, Portugal, 18–22 June 2001*, edited by C. Silva, C. Varandas, and D. Campbell, European Physical Society, 2001, vol. 25A, pp. 1981–1984.
4. Hastings, D. E., Houlberg, W. A., and Shaing, K. C., *Nuclear Fusion*, **25**, 445–454 (1985).
5. Nemov, V. V., Kasilov, S. V., Kernbichler, W., and Heyn, M. F., *Phys. Plasmas*, **6**, 4622–4632 (1999).
6. Reiman, A., Ku, L., Monticello, D., Hirshman, S., Hudson, S., Kessel, C., Lazarus, E., Mikkelsen, D., Zarnstorff, M., Berry, L. A., Boozer, A., Brooks, A., Cooper, W. A., Drevlak, M., Fredrickson, E., Fu, G., Goldston, R., Hatcher, R., Isaev, M., Jun, C., Knowlton, S., Lewandowski, J., Lin, Z., Lyon, J. F., Merkel, P., Mikhailov, M., Miner, W., Mynick, H., Neilson, G., Nelson, B. E., Nührenberg, C., Pomphrey, N., Redi, M., Reiersen, W., Rutherford, P., Sanchez, R., Schmidt, J., Spong, D., Strickler, D., Subbotin, A., Valanju, P., and White, R., *Phys. Plasmas*, **8**, 2083–2094 (2001).
7. Dommaschk, W., Lotz, W., and Nührenberg, J., *Nuclear Fusion*, **24**, 794 (1984).
8. Kamendje, R., Kasilov, S. V., Kernbichler, W., and Heyn, M. F., "Effects of Nonlinear Wave-Particle Interaction on the Electron Distribution Function During ECRH," in *28th EPS Conf. on Contr. Fusion and Plasma Physics, Funchal, Portugal, 18–22 June 2001*, edited by C. Silva, C. Varandas, and D. Campbell, European Physical Society, 2001, vol. 25A, pp. 817–820.

Particle-In-Cell Simulations of Laser-Produced Plasma Experiments to Study Thrust Conversion Processes in a Laser Fusion Rocket

Hideki Nakashima, Konstantin V. Vchivkov*, Tomonori Esaki, Yuri P. Zakharov[1], Toshihiko Kawano and Takanobu Muranaka[2]

Department of Advanced Energy Engineering Science, Kyushu University, Kasuga-Kouen, Kasuga, Fukuoka 816-8580, Japan
[1] *Institute of Laser Physics (ILP), Novosibirsk 630090, Russia*
[2] *Institute of Laser Engineering, Osaka University, Suita, Osaka 565-0871, Japan*

Abstract. Here we present the comparison analysis of Laser-produced Plasma Cloud (LPC) expansion in a dipole field between experimental data and simulation results obtained by a three-dimensional (3D) hybrid code. The thrust conversion process in a laser fusion rocket (LFR) and the plasma behaviour are examined. We found a real thrust efficiency as high as 60% can be achieved in a future LFR. It was also found that the experiments with a "usual" LPC will be very useful to develop LFR design and test the basic features of large - scale simulative LFR experiment proposed for National Ignition Facility (NIF).

INTRODUCTION

A propulsion system driven by a laser-induced fusion called Laser Fusion Rocket (LFR) is an attractive candidate for future interplanetary missions [1]. A fusion reaction can produce easily the plasma of a high temperature and density. The resulting plasma flow can be controlled by a magnetic thrust chamber. In the laser fusion rocket, the chamber composes of the solenoidal superconducting coil.

In this paper we discuss the experimental and the simulation work to examine the thrust conversion process in LFR. The expansion of LPC in the dipole field is studied under the conditions close to the LFR magnetic thrust chamber [2].

NUMERICAL MODEL

The calculation model considered here is illustrated in Figure 1 and it is based on the experiment performed by Zakharov, et.al. [2]. The equations controlling our system can be derived from Maxwell's equations and the equations of motion of the particles. The details are given in Refs. [3] and [4].

The plasma behaviour was calculated by the 3D hybrid Particle-In-Cell (PIC) code. The general calculation parameters used in the simulation are shown in Table 1.

Initial distributions of the particle positions and velocities were assumed to be uniform. Here the simulation starts from $t = 0.15\mu s$ to take account of the time elapsed for the plasma expansion to 2cm in the radius. The initial plasma location is the same as the experiment.

---

* On leave from ILP.

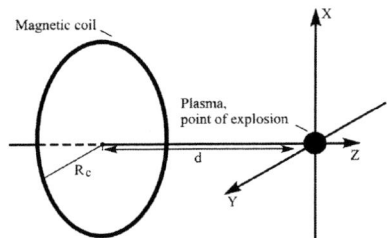

FIGURE 1. Scheme of the calculation model.

TABLE 1. The calculation parameters.

Coil radius (m)	0.05	Plasma mass (g)	0.6×10^{-6}	Electron temperature (eV)	0
Coil current (A)	2.52×10^5	Atomic mass (AMU)	12.0	Initial max. velocity (km/s)	140
Coil position on Z (m)	-0.15	Atomic number	6.5	Time step Δt (μs)	0.00025
Plasma position (m)	(0,0,0)	Magnetic moment (Gcm^3)	2×10^6	Calculation region (m)	$0.6\times0.6\times0.45$
Plasma radius (m)	0.021	Initial magnetic field		Mesh size	$40\times40\times30$
Plasma energy (J)	3.5	strength at the plasma (T)	0.1	Number of particles	100000

For the comparative analysis we used the data of an "Impulse" experiment at KI-1 facility [2]. The target was placed in the vacuum chamber near a current coil. The experimental parameters are as follows: the initial expansion velocity $V_0 \sim 150\text{-}200$ km/s and total kinetic energy $E_0 \sim 3\text{-}8$ J at the axis of quasi-stationary (~1 ms) dipole with a moment $\mu_d = (1\text{-}2)\times 10^6$ G·cm^3 and stainless-steel spherical shell of radius 8 cm. The shell was used to measure the impulse transfer due to the plasma expansion. The plasma was generated at the background pressure ~0,001mTorr by means of CO_2-laser beams with the total energy ~50-100 J irradiating the Nylon 6 pellet target of diameter 3-4mm (suspended at a thin metallic wire). The laser pulse of 70ns-duration was short enough to fulfill the condition of instantaneous exploding - like energy release of LFR.

RESULTS AND DISCUSSION

The numerical results for the time evolutions of particle position are shown in Figures 2(a) and 2(b), where they are projected onto the XZ and XY planes, respectively. As we can see in these figures the plasma shape is spherical at the initial stage and the plasma expands almost isotropically, and then the ions in the direction of the coil are reflected back by the magnetic field, and its shape changes to follow the dipole magnetic field line. On the other hand, as shown in Figure 2(b) the shape of plasma is symmetric in the all stages at the XY plane.

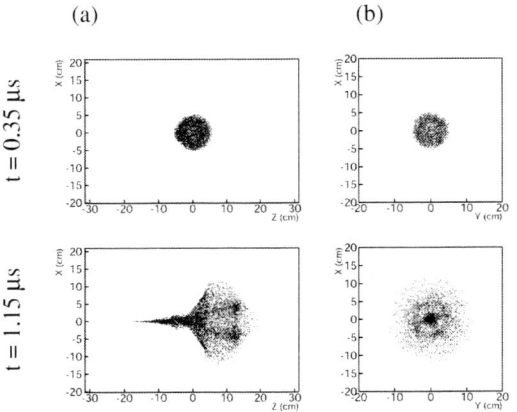

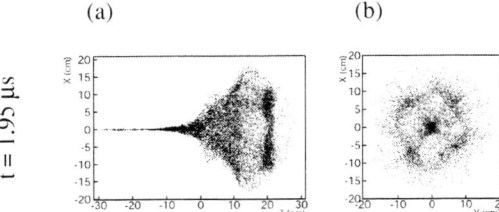

FIGURE 2 Simulation results of particle positions projected onto the (a) XZ plane and (b) XY plane.

Figure 3 shows the experimental results of the particles position projected onto the XZ plane at the time 0.75 μs. As seen from the picture and Figure 2(a) the plasma front expands almost identically between the simulation and experiment. Figure 4 shows time-integrated picture of LPC: (a) experimental results and (b) simulation results. The angle of LPC expansion is about 60^0 for the both pictures.

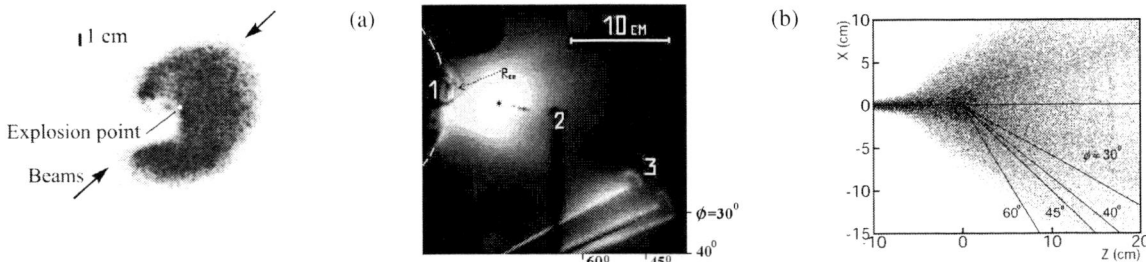

FIGURE 3 The experimental results of the particles position projected onto the XZ plane at 0.75 μs.

FIGURE 4 Time-integrated picture of plasma cloud; (a) the experimental results, 1 shell surface (with magnetic probe), 2 target supports (with a pellet on wire), 3 Langmuir probes, R_{ce} maximum effective radius of LPC diamagnetic cavity, ϕ angle of LPC expansion; (b) the simulation results.

Figures 5 (a) and (b) show the time evolution of velocity distributions projected onto the XZ and XY planes at early stages. The velocity vectors are directed and displaced according to the magnetic strength lines. In Figure 5(b), we can see the ion Larmor rotation of the plasma particles related with the diamagnetic exclusion of magnetic field in the plasma.

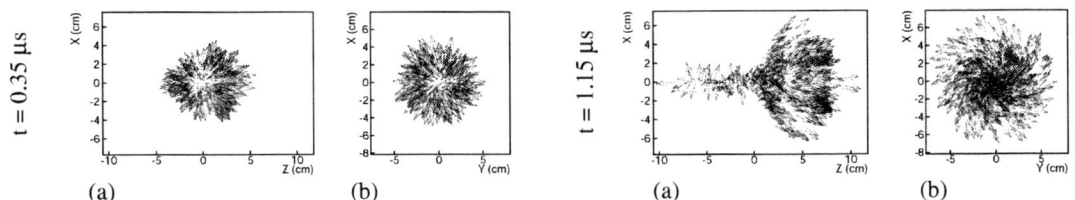

FIGURE 5 Time evolution of velocity distributions at early stages projected onto the (a) XZ plane, and (b) XY plane at Z≈5cm (simulation results).

Figure 6 shows the angular distribution of the plasma flow obtained by the experiment and Figure 7 obtained by the simulation. Here for the comparison we consider an arbitrary unit in Figure 7, because we use the discrete particles in the simulation. We can see from these figures that the plasma flow is collimated by the magnetic field within an angle of $\sim 60^0$.

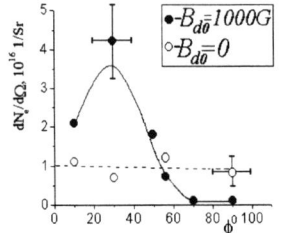 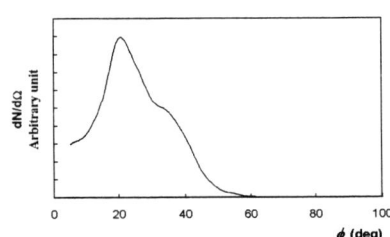

FIGURE 6 Angular distribution of time-integrated plasma flow under conditions of LPC with $E_0 \sim$3-4 J, R≥20cm (experiment).

FIGURE 7 Angular distribution of plasma flow, spatially integrated at t≈3μs (simulation results).

The experimental data on magnetic field disturbances ΔB caused by the diamagnetic cavity of plasma obtained at various position of magnetic probes are illustrated in Figures 8 (a) and (b). Figure 9 shows the simulation data of magnetic field disturbances ΔB.

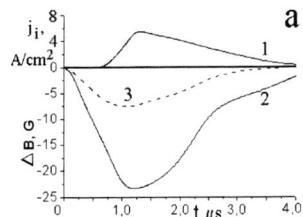 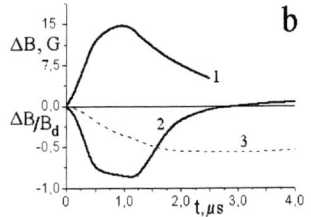 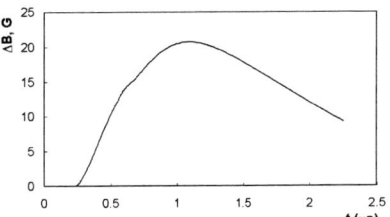

FIGURE 8 In FIGURE (a): 1 plasma flow (Rp≈25cm., $\phi \approx 10^0$), 2 ΔB (Rp≈16cm., $\phi \approx 0^0$), 3 ΔB (Rp≈21cm., $\phi \approx 0^0$); in FIGURE (b): 1 ΔB (Rp≈14cm., $\phi \approx 90^0$ – outside of LPC), 2 ΔB (Rp≈4.5cm., $\phi \approx 45^0$), 3 ΔB (Rp≈21cm., $\phi \approx 0^0$), E$_0$=8J, B$_0$≈300G.

FIGURE 9 The simulation data of magnetic field disturbances ΔB, (Rp=14cm, $\phi \approx 90^0$), (cf. the curve 1 in Fig. 8b).

The thrust efficiency η in terms of the momentum is calculated as follows:

$$\eta = \sum M v_z \Big/ \sum M |v_0|,$$

where M is the ion mass, v_z the z-component of its velocity and $|v_0|$ the absolute value of the initial velocity. The sum is carried over the all plasma particles. The results are shown in Figure 10 as a function of time.

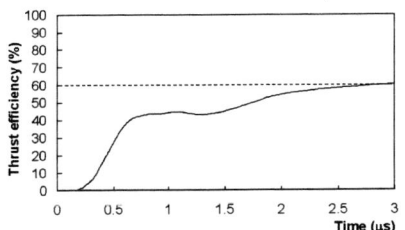

FIGURE 10 Time evolution of the thrust efficiency (simulation result).

FIGURE 11 Time evolution of the thrust efficiency (experimental result).

The value of η increases rapidly between 0.1 μs and 0.5 μs and then its value saturates around 60% at 3.0 μs. The value of about 60% was obtained from the experiment as shown in Figure 11 by measuring the impulse to the shell, obtained by new electrodynamic method [2].

CONCLUSION

We carried out analysis on plasma behaviours in the dipole filed using the 3D hybrid code, and the comparison was made between the experimental data and the results from the 3D hybrid code. An overall good agreement in the expansion behaviour of LPC between these results was found. It was then concluded that the thrust efficiency as high as 60% is achievable in LFR.

REFERENCES

1. Orth, C.D., et.al., AIAA-87-1904, 1987.
2. Zakharov, Yu.P., et.al., Current Trends in International Fusion Research – Book of abstracts of 4th Symposium, pp. 31-34, 2001.
3. Harned, D.S., J. Comput. Phys. Vol 47, pp. 452-462, 1982.
4. Horowitz, J.E., et al., J. Comput. Phys., Vol. 84, pp. 279-310, 1989.

Integration of ALE Hydro and Collective PIC Codes for Fast Ignition Simulations

Hitoshi Sakagami* and Kunioki Mima†

Computer Engineering, Himeji Institute of Technology, 2167 Shosha, Himeji, Hyogo 671-2201, Japan
† Institute of Laser Engineering, Osaka University, 2-6 Yamada-oka, Suita, Osaka 565-0871, Japan

Abstract. An overall implosion process, laser-plasma interaction, hot electron transport and hot electron energy deposition are key subjects for the Fast Ignition. All these phenomena couple with each other, and more studies by simulations are essential. We have a plan to simulate the whole Fast Ignition self-consistently with four individual codes. Four codes are integrated into one big system in a Fast Ignition Integrated Interconnecting code project. In first stage of this project, we integrate the ALE hydro code with the collective PIC code. The PIC code obtains density profile at maximum compression from the ALE hydro code to introduce plasma corresponding to the profile, and we can simulate interaction between ignition laser and realistic plasma. We have evaluated absorption rate and reflected laser spectrum and found mush differences between the realistic plasma profile and the conventional plasma profile in PIC simulations.

INTRODUCTION

Relativistic laser-plasma interaction near the critical density and subsequent energy transport of super hot electrons, which are generated through the interaction, to an overdense core are mandatory subjects for the Fast Ignition. In the interaction region where an ultrahigh intense ignition laser pulse is propagating, relativistic self-focusing and parametric instability take place [1] and these phenomena make the issue more complicated. On the other hand, current that is carried by super hot electrons should be accompanied by return current by cold background electrons to keep charge neutrality. A magnetic field can be generated by these currents and feed back to motions of both electrons, introducing small scale Weibel instability [2]. The interaction between magnetic field structure and electron motions will affect conversion efficiency from driver energy to ignition energy. In addition, energy deposition of super hot electrons at the target core must be solved in details to prove fuel burning in the Fast Ignition scheme. And furthermore, each phenomenon is strongly affected by density and temperature profiles that are determined by an overall implosion process. Thus we must compute the implosion dynamics simultaneously, coupling with above phenomena. All these physics are intertwined and more studies by simulations are essential.

We have planned to simulate the Fast Ignition with self consistent fields in a full range of parameters and just started challenging Fast Ignition Integrated Interconnecting code (FI^3) project.

FI^3 PROJECT

To prove fuel burning in the Fast Ignition scheme, we must consider 1) overall fluid dynamics of the implosion, 2) laser-plasma interaction and super hot electron generation, 3) super hot electron transport to the target core and 4) super hot electron energy deposition within the core. These phenomena are coupling with each other and both time and space scales of them are much different. Each range of physical quantities that should be considered in respective phenomena also varies in a large way. Thus it is impossible to simulate all phenomena with one code, and we must simulate each phenomenon with individual codes and integrate them.

First, we introduce ALE (Arbitrary Lagrangian Eulerian) hydro code [3] to calculate the overall fluid dynamics because a space scale varies in a wide range and the target is continuously deformed during the implosion process. The ALE hydro code should also include absorption of implosion lasers that determines the implosion manner. Second, collective PIC code must be used to simulate the interaction between ultrahigh intense ignition lasers and plasmas [4]. The relativistic PIC code can compute generation of super hot electrons and get their distribution functions without any model assumptions. As the target core becomes 1000 times solid density at a maximum compression, it is impossible to compute up to this range with the PIC code. We should, therefore, introduce a hybrid code in which super hot electrons are treated as a particle and background electrons as a fluid [5]. The hybrid code receives the distribution function of super hot electrons from the PIC code and calculates their transport into the target core. As the super hot current is in the order of 100-1000 MA, return current electrons becomes too warm and should not be treated as same temperature as cold background electrons. Thus we should employ two temperature electron fluids for return current and background. This hybrid approach, however, has uncertainty about return current temperature and density of both fluids. In our code integration, the hybrid code can get not only the distribution function of super hot electrons but also the temperature and density for both return currents and background fluids from the PIC code to solve this ambiguous problem. As surface connection between the hybrid and PIC codes usually causes numerical errors, we would install volume connection scheme between them. Finally, Fokker-Planck code [6] is used to compute their energy deposition and to estimate the fuel burning. A code integration diagram and physical quantities that should be communicated between codes are shown in Fig. 1 (a). Physical quantities with solid arrows are primary data that should be transferred between codes as the law of causality, and those with shaded arrows govern a feedback effect that could be neglected according to simulation circumstances.

If we integrate four individual codes into big one, it is very hard to maintain this code because each code has been developed by separate teams at different sites and this approach is never realistic. Additionally, we can not run this code efficiently on single parallel computer system, because computation grains of each code are very different and parallel performance is strongly degraded. Thus we will run the code individually at each site, collaborating each other with data transfer via the computer network. As each code can be independently executed, we can select appropriate architecture of computers, such as a vector supercomputer, a symmetrical multiprocessor type parallel computer or a massive parallel scalar computer, to run the code. Since communication in our project is very straightforward and not complex, we design a lightweight protocol, Distributed Computing Collaboration Protocol, to transfer data between codes and implement two kinds of daemon programs. One of them is called Communicator that actually transfers data instead of code itself, and the other is called Arbitrator that manages communication between Communicators. The code only asks the Communicator, which runs in the background at his site, to send data to another code, and then the Communicator passes data to another Communicator corresponding to the code that should be received data via the Internet. Finally that code receives the data from the domestic Communicator. If a broadband dedicated line is available between Communicators, the Arbitrator tells both Communicators to use that connected line and high speed communication can be done even both codes do not know about details of network connections. This concept is shown in Fig. 1 (b) as the code communication chart.

INTEGRATION OF HYDRO CODE AND PIC CODE

In first stage of the project, we integrated the ALE hydro code with the collective PIC code. The PIC code obtains density profile at maximum compression from the ALE hydro code and introduces plasma corresponding to the profile. Then an ultrahigh intense ignition laser is launched into the plasma, and we can simulate laser-plasma interactions in a realistic situation with the PIC code. The typical density profile at the maximum compression that is computed by the ALE hydro code is shown in Fig. 2. As a low density corona plasma is too long to simulate

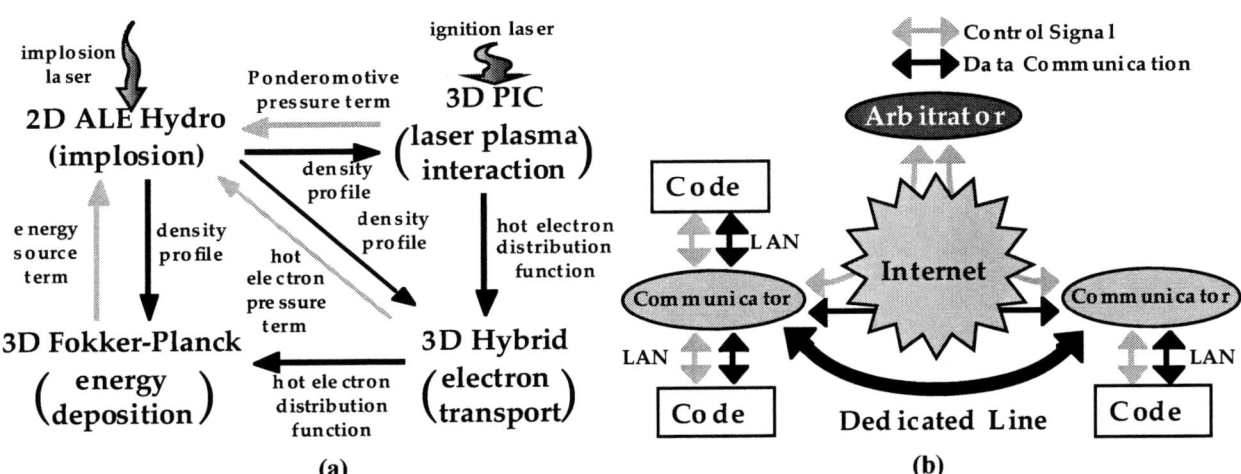

FIGURE 1. The concept of Fast Ignition Integrated Interconnecting code (FI^3) project, (a) the code integration diagram and (b) the code communication chart.

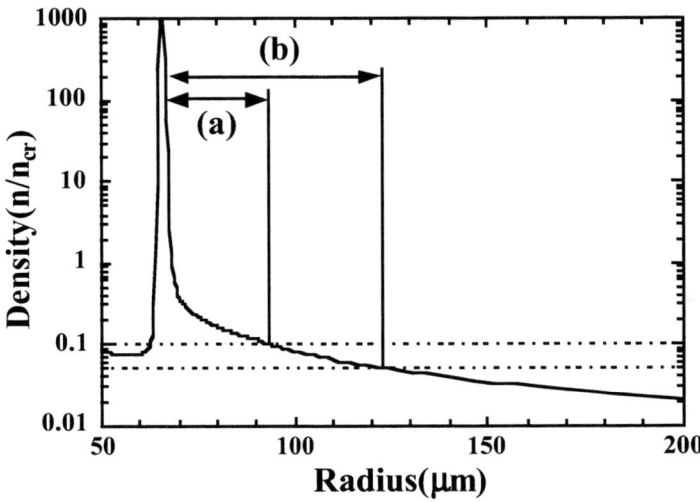

FIGURE 2. The typical density profile that is transferred from the ALE hydro code to the PIC code, (a) cut the plasma at 0.1 n_{cr} with 25 [μm] underdense plasma and (b) at 0.05 n_{cr} with 55 [μm].

with the PIC code, we must cut the plasma at an appropriate point. Parametric instabilities, however, produced by laser plasma couplings, such as stimulated Ramman scattering, predominantly occur within this underdense plasma, and an ignition laser may lose its energy before reaching a dense plasma. Thus a thickness of the underdense plasma can affect an energy conversion ratio from the ignition laser to super hot electrons and could be an important factor for simulations of the Fast Ignition. We introduced two cases, namely (a) cut the plasma at 0.1 n_{cr} with 25 [μm] underdense plasma and (b) at 0.05 n_{cr} with 55 [μm]. On the other hand, we cut the profile at 100 n_{cr} to avoid difficulty to compute ultra dense plasma with the PIC algorithm. We gradually increase the number of electrons per mesh to track the obtained density profile up to 200 electrons per mesh at twice critical density. In the overdense region, we preserve the number of electrons to 200 per mesh but increase a collective factor to fit the profile. We also introduced another typical profile that is mostly used in PIC simulations, namely (c) linear profile and (d) flattop profile to evaluate effects due to existence of realistic underdense plasmas. Plasma density increases up to 100 n_{cr} within 25 [μm] long in the linear profile using the same fitting technique, and 20 [μm] long plasma of 100 n_{cr} with 200 electrons per mesh is employed in the flattop profile.

It is noted that ions are immobile and the profile is not communicated between the codes but is passed with a file at this moment.

IGNITION LASER ABSORPTION

As a pilot test problem, we have injected the ignition laser with Gaussian pulse of wavelength $\lambda_L = 1.06$ [μm], intensity $I_L\lambda_L^2 = 1\times10^{20}$ [W/cm^2-μm^2] and HWHM = $4\lambda_L$ (50 [fsec]).

Absorption rates of the ignition laser for different profiles are summarized in Table 1. Only a few percent of the ignition laser energy is absorbed with the flattop plasma because electrons have very little chance to be pulled out from the plasma into vacuum and to get energy in the laser field due to large plasma density. The linear profile plasma shows better absorption rate than that of the flattop plasma because there is room for laser plasma interaction near the critical density. As the ignition laser can interact with long underdense plasma, more than 40 % of the energy of the ignition laser can be converted to super hot electrons.

TABLE 1. Absorption rate of the ignition laser for different profiles.

ALE profile cut at 0.1 n_{cr}	ALE profile cut at 0.05 n_{cr}	linear profile	flattop profile
40%	47%	16%	2%

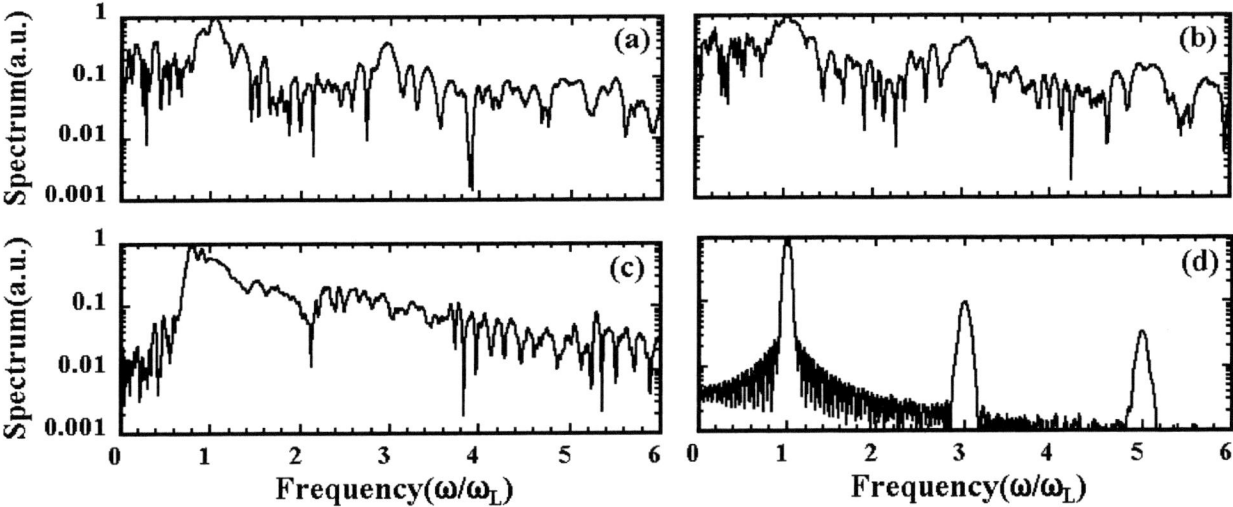

FIGURE 3. Spectrums of the reflected laser for (a) the ALE profile cut at 0.1 n_{cr}, (b) the ALE profile cut at 0.05 n_{cr}, (c) the linear profile and (d) the flattop profile. ω_L is the ignition laser frequency.

Spectrums of the reflected laser are shown in Fig. 3 for (a) the ALE profile cut at 0.1 n_{cr}, (b) the ALE profile cut at 0.05 n_{cr}, (c) the linear profile and (d) the flattop profile. Odd numbers of higher harmonics are generated by coupling between the laser frequency and $2\omega_L$ oscillating Ponderomotive force in Fig. 3 (d). The parametric instabilities excited by the laser plasma couplings in the underdense region can produce spectrum below the laser frequency with ALE profiles and it is clearly seen in Fig. 3 (a) and (b). The oscillating Ponderomotive force also generates third harmonics due to steep profile in overdense regions. The ignition laser can penetrate into the overdense plasma due to the relativistic effect, and the anomalously penetrating pulse is reflected at the recession front [7]. Thus the reflected laser is Doppler shifted as shown in Fig. 3 (c). Since there is neither steep profile nor long underdense region, no odd higher harmonics are found and lower frequency modes than ω_L are not remarkable in the spectrum.

SUMMARY

We have just started the Fast Ignition Integrated Interconnecting code project to simulate an entire extent of the Fast Ignition, including four individual different codes. These codes, which exchange appropriate physical quantities each other during execution, will be integrated into one big system.

In first stage of this project, the collective PIC code obtains density profile at maximum compression from the ALE hydro code and introduces plasma corresponding to the profile. Then an ultrahigh intense ignition laser is launched into the plasma, and we can simulate laser-plasma interactions in a realistic situation with the PIC code.

We have evaluated absorption rate and reflected laser spectrum and found mush differences between the realistic plasma profile and the conventional plasma profile in PIC simulations.

REFERENCES

1. Pukhov, A. M., and Meyer-ter-Vehn, J., *Phys. Rev. Lett.* **79**, 2686-2689 (1997).
2. Honda, M., Meyer-ter-Vehn, J. and Pukhov, A. M., *Phys. Plasmas* **7**, 1302-1307 (2000).
3. Nagatomo, H. et. al., "Analysis of Hydrodynamic Instabilities in Implosion using High-accuracy Integrated Implosion Code," in *2nd Int. Conf. on Inertial Fusion Sciences and Applications,* Kyoto, Japan, 2001, IFSA1165.
4. Sakagami, H. and Mima, K., "Fast Ignition Simulations with Collective PIC Code," in *2nd Int. Conf. on Inertial Fusion Sciences and Applications,* Kyoto, Japan, 2001, IFSA242.
5. Taguchi, T. et. al., *Phys. Rev. Lett.* **86**, 5055-5058 (2001).
6. Nakao, Y. et. al., "Two-Dimensional Transport Code for Alpha-Particles in ICF Plasmas," in *2nd Int. Conf. on Inertial Fusion Sciences and Applications,* Kyoto, Japan, 2001, IFSA531.
7. Sakagami, H. and Mima, K., *Phys. Rev. E* **54**, 1870-1875 (1996).

A Kinetic Model of Solar Wind

Y.M. Vasenin[1], N.R. Minkova[2], A. Shamin[3]

[1]*Department of Applied Aeromechanics, Tomsk State University, 634050 Russia*
[2]*Department of Mathematical Physics, Tomsk State University, 634050 Russia*
[3]*Department of Applied Aeromechanics, Tomsk State University, 634050 Russia*
(now at Encom Technology, 118 Alfred St Milsons Pt NSW 2061 Australia)

Abstract. A kinetic model of quasi-neutral collisionless plasma flow for description of solar wind is presented. The model is based on the stationary equation for two-particles (electron-proton) velocity distribution. The influence of magnetic field of the Sun has been neglected. The obtained exact analytic solution for the distribution function allows to deduce the dependencies of the solar wind density and speed on heliocentric distance that agree with observational data.

Most models of solar wind that have been developed during the last forty years were based on the hydrodynamic approach. They faced difficulties such as lack of plasma flow energy to provide the results that would match the observed values of solar wind speed at large heliocentric distances. The kinetic concept was also used mostly to obtain the equations for the average parameters of plasma flow as a continuum [1,2]. The first attempt to estimate the solar wind speed and density from the kinetic point of view was made probably in [3]. This paper presents a mathematical model based on the kinetic approach.

TWO-PARTICLES VELOCITY DISTRIBUTION

A kinetic model of quasi-neutral plasma flow with spherical symmetry for description of solar wind presented in this paper is based on the stationary equation for two-particles (electron-proton) velocity distribution. The collisions between particles beyond the exobase and the influence of magnetic field have been neglected. Then the kinetic equations for distribution functions of electrons (f_e) and protons (f_p) can be written in the following form:

$$u_{rj}\frac{\partial f_j}{\partial r} + \frac{F_{rj}}{m_j}\cdot\frac{\partial f_j}{\partial u_{rj}} + \frac{u_{\perp j}^2}{r}\cdot\frac{\partial f_j}{\partial u_{rj}} - \frac{u_{rj}u_{\perp j}}{r}\cdot\frac{\partial f_j}{\partial u_{\perp j}} = 0, \quad j=e,p, \qquad (1)$$

where u_r, $u_\perp$ are radial and tangential components of particle velocity, r is a heliocentric distance, F_r is a radial component of force applied to a particle, m is particle mass. The indices e and p refer to electrons and protons respectively. The equations (1), written in variables r, $\varepsilon_e = 0.5 \cdot m_e \cdot (u_{re}^2 + u_{\perp e}^2)$, $u_{\perp e}$, $\varepsilon_p = 0.5 \cdot m_p \cdot (u_{rp}^2 + u_{\perp p}^2)$, $u_{\perp p}$ yield the following equation for a two-particle distribution function f provided the electron and proton velocity distributions are statistically independent ($f = f_e \cdot f_p$):

$$\frac{\partial f}{\partial r} + F_{re}\cdot\frac{\partial f}{\partial \varepsilon_e} + F_{rp}\cdot\frac{\partial f}{\partial \varepsilon_p} - \frac{u_{\perp e}}{r}\cdot\frac{\partial f}{\partial u_{\perp e}} - \frac{u_{\perp p}}{r}\cdot\frac{\partial f}{\partial u_{\perp p}} = 0 \qquad (2)$$

The forces applied to a particle in gravitational and electrical fields are expressed through potentials of these fields ψ and φ: $F_{rp,e} = \pm e \cdot \psi_r' - m_{p,e}\varphi_r'$, where $\pm e$ are the charges of a proton and electron respectively, $\varphi = -\gamma M/r$, γ is the gravitational constant, M is the mass of the Sun. The electrical potential $\psi(r) < 0$ is produced by plasma polarization.

The general solution to (2) is the arbitrary differentiable function of the first integrals E_e, E_p, M_e, M_p of the characteristic equations that express the laws of energy and moments of momentum conservation for particles:

$$f = f(E_e, E_p, M_e, M_p),$$
$$E_j = \varepsilon_j + m_j\varphi \pm e\psi = \varepsilon_{j0} + m_j\varphi_0 \pm e\psi_0, \quad M_j = ru_{\perp j} = ru_{\perp j0}, \quad j = e, p \quad (3)$$

where $m = m_e + m_p$, r_0 is the heliocentric distance, at which the initial values of parameters (marked by a zero index) are specified. Let us consider a case of a Maxwell-Boltzmann distribution for the velocity at the exobase ($r=r_0$):

$$f_0 = f(r_0, u_{rp0}, u_{\perp p0}) = 4N_0^2 \left(\frac{\sqrt{m_e m_p}}{2\pi k T_0}\right)^3 \exp\left(-\frac{\varepsilon_0}{kT_0}\right) \quad (4)$$

where $\varepsilon_0 = m_e(u_{re0}^2 + u_{\perp e0}^2) + m_p(u_{rp0}^2 + u_{\perp p0}^2)$, $m = m_e + m_p$, k - Boltzmann's constant, T_0 - coronal electron and proton temperature at the exobase. The solution of a problem (3), (4) is easy to deduce by substituting the expression $\varepsilon_0 = \varepsilon - m\cdot(\varphi_0 - \varphi)$, resulted from (3), in the function (4):

$$f = f(r, u_{re}, u_{\perp e}, u_{rp}, u_{\perp p}) = N_0^2 \cdot \left(\frac{\sqrt{m_e m_p}}{2\pi k T_0}\right)^3 \exp\left(-\frac{\varepsilon - m(\varphi_0 - \varphi)}{kT_0}\right), \quad (5)$$

where $\varepsilon = m_e(u_{re}^2 + u_{\perp e}^2) + m_p(u_{rp}^2 + u_{\perp p}^2)$. It is assumed that a half of total number of particles ($0.5\cdot N_0$) at the exobase are moving in the direction away from the sun. The particles whose kinetic energy is less than the sum of potential energy of gravitational attraction to the Sun and plasma polarization electrostatic field:

$$\varepsilon_j < -\Phi_j = -(m_j\varphi \pm e\psi), \quad j = e, p \quad (6)$$

form the atmosphere. The high-energy electrons and protons overcome the potential well and escape to infinity. The particles with radial component of velocity $u_r > 0$ correspond to characteristics (3) coming from the first quadrant of the initial velocity planes $u_{re0}, u_{\perp e0}$ and $u_{rp0}, u_{\perp p0}$. The atmosphere particles falling back to the Sun relate to the characteristics (3) reflected from the potential well. These two sets of characteristics form the domain of definition for the two-particle distribution function (5) in velocity space at any heliocentric distance. This domain is restricted by the inequalities that result from the energy and moment of momentum conservation laws (3) and the expression for a potential well (6):

$$\begin{cases} (0 \leq u_{\perp j}) \wedge (\varepsilon_{\perp j} \leq \frac{\varepsilon_{rj} - \Delta\Phi_j}{\bar{r}^2 - 1}), & 0 \leq u_{rj} \leq \infty, \\ (0 \leq u_{\perp j}) \wedge (\varepsilon_{\perp j} \leq \frac{\varepsilon_{rj} - \Delta\Phi_j}{\bar{r}^2 - 1}) \wedge (\varepsilon_{\perp j} \leq -\Phi_j - \varepsilon_{rj}), & u_{rj} \leq 0. \end{cases} \quad (j = e, p) \quad (7)$$

Where $\Phi_j = m_j\varphi \pm e\psi$, $\Delta\Phi_j = \Phi_{j0} - \Phi_j$, $j = e, p$.

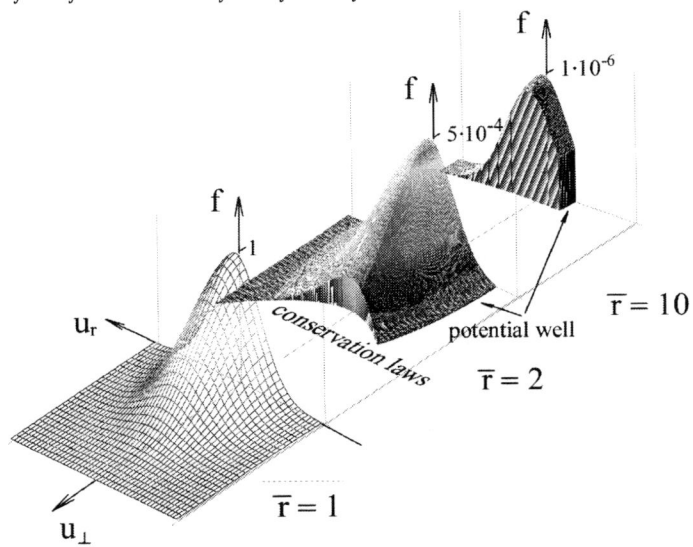

FIGURE 1. Distribution of electron (proton) velocities at different heliocentric distances $\bar{r}$.

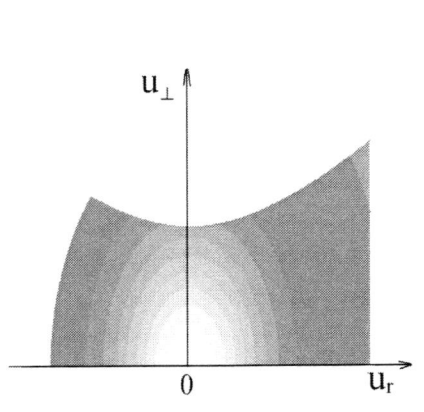

FIGURE 2. Domain of definition for velocity distribution f at $\bar{r} = 2$ (shade is modulated by f values)

The shape of the initial distribution f_0 (4) and distributions f (5), (7) at two heliocentric distances $\bar{r}=2$ and $\bar{r}=10$ in the space u_{rj}, $u_{\perp j}$ ($j=e,p$) are shown in the Fig.1. The scales of f marked on the axis are relative. Fig. 2 presents the domain of definition (7) for velocity distribution (5) at $\bar{r}=2$ in the same coordinates. Indices j on the figures are omitted.

The distribution (5), (7) allows to obtain the average solar wind's parameters such as density and speed.

SOLAR WIND DENSITY AND SPEED

For evaluation of **plasma flow density** N the two-particle distribution function (5) should be integrated in velocity space over the domain (7). This yields:

$$N = 2N_+ - N_\to, \tag{8}$$

$$N_+(\bar{r}) = \frac{N_0}{2} e^{\Delta\bar{\varphi}} \sqrt{\left(1 - \sqrt{1-\bar{r}^{-2}}\, e^{\frac{\Delta\bar{\Phi}_e}{\bar{r}^2-1}}\right)\left(1 - \sqrt{1-\bar{r}^{-2}}\, e^{\frac{\Delta\bar{\Phi}_p}{\bar{r}^2-1}}\right)}, \tag{9}$$

$$N_\to(\bar{r}) = \frac{N_0}{2} e^{\Delta\bar{\varphi}} \prod_{j=e,p} \left[Erf\left(\sqrt{-\bar{\Phi}_j}\right) - e^{\frac{\Delta\bar{\Phi}_j}{\bar{r}^2-1}} \sqrt{1-\bar{r}^{-2}}\, Erf\left(\sqrt{\frac{\bar{\varepsilon}_{rj}^\times}{1-\bar{r}^{-2}}}\right) + \frac{2}{\sqrt{\pi}} e^{\bar{\Phi}_j}\left(\sqrt{-\bar{\Phi}_j} - \sqrt{\bar{\varepsilon}_{rj}^\times}\right) \right], \tag{10}$$

where $\bar{r}=r/r_0$, $\Delta\bar{\varphi}=m(\varphi_0-\varphi)/2kT_0$, $\Delta\bar{\Phi}_j=\Delta\Phi_j/kT_0$, $\bar{\varepsilon}_{rj}^\times = \left(\Phi_{j0}\bar{r}^{-2} - \Phi_j\right)/kT_0$, $j=e,p$. Here N_+ and $N_\to$ mean the density of particles moving away from the sun and escaping particles density respectively. The formulae (8)-(10) are obtained assuming $N_e=N_p=N$ (quasi-neutral plasma).

The main term of asymptotic decomposition of the relations (8-10) at $\bar{r}\to\infty$ is inversely proportional to square of heliocentric distance:

$$N(\bar{r}) \sim \frac{N_0}{2} e^{\Delta\bar{\varphi}} \sqrt{\left(\frac{1}{2}-\Delta\bar{\Phi}_e\right)\cdot\left(\frac{1}{2}-\Delta\bar{\Phi}_p\right)}\, \bar{r}^{-2} \tag{11}$$

The expression for the Sun's atmosphere density is possible to deduce from (9) and (10) by using the following relation: $N_{atm}(\bar{r}) = 2\cdot(N_+(\bar{r}) - N_\to(\bar{r}))$. Hence $N_{atm} \propto \bar{r}^{-2.5}$ at $\bar{r}\gg 1$. It means that atmospheric density reduces in $\sqrt{\bar{r}}$ times faster, than density of escaping particles.

The solar wind speed, as the statistical average velocity of the particles in the radial direction, is equal to the ratio of a specific rate of escaping particles' flux $(Nu_r)_\to$ to solar wind density N:

$$u(\bar{r}) = \frac{(Nu_r)_\to(\bar{r})}{N(\bar{r})}, \qquad (Nu)_\to = \frac{N_0}{2} e^{\bar{\varphi}_0} \sqrt{\bar{u}_{e0}\bar{u}_{p0}} \sqrt{(1-\bar{\Phi}_{e0})(1-\bar{\Phi}_{p0})}\, \bar{r}^{-2}, \tag{12}$$

where $\bar{\varphi}_0 = \varphi_0/2kT_0$; $\bar{u}_{j0} = \sqrt{2kT_0/\pi m_j}$ ($j=e,p$) are electron and proton thermal velocities divided by $\sqrt{\pi}$. The formulae (12) are deduced by assuming the equality of electrons and protons fluxes (quasi-neutral plasma). The speed $u(\bar{r})$ is an increasing function of heliocentric distance due to the fact that the relative part of atmosphere particles in the total density $N = N_{atm} + N_\to$ is decreasing with the distance what was shown above. These particles play the role of ballast because the total stationary flux of atmosphere particles is equal to zero and does not contribute to the rate of the total flux $(Nu_r)_\to$. Therefore $u(\bar{r})$ increases with the distance and reaches the terminal value which is an average velocity of escaping particles. According to (12) the solar wind speed at $\bar{r}\to\infty$ has the limit u_∞:

$$u_\infty^2 = \bar{u}_{e0}\bar{u}_{p0} \frac{(1-\bar{\Phi}_{e0})(1-\bar{\Phi}_{p0})}{(0.5-\bar{\Phi}_{e0})(0.5-\bar{\Phi}_{p0})} \tag{13}$$

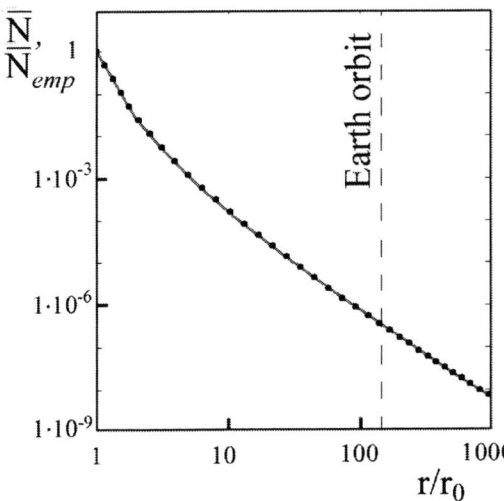
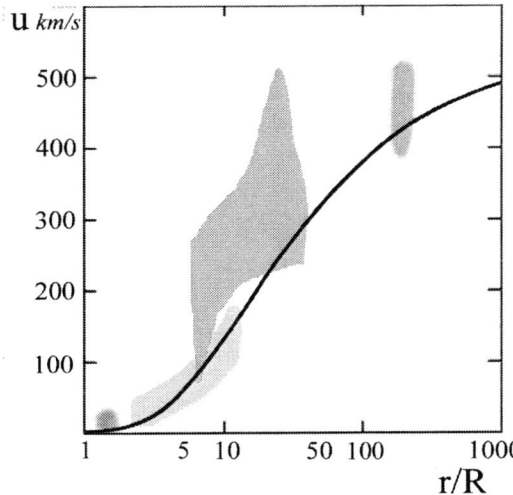

FIGURE 3. Empirical data [4] (squares) and theoretical (8)-(10), (15) (solid line) results for solar wind density

FIGURE 4. Observational data [5] (filled area) and theoretical results (12),(15) (solid line) for solar wind speed

By applying assumption (13) and using parameter values proper to the Sun (see below) the limit value u_∞ can be evaluated: $u_\infty \approx 1.031\sqrt{\overline{u}_{e0}\overline{u}_{p0}} \cong 540\,km/s$. At the Earth's orbit, $\overline{r} \cong 143$, (12,15) gives about 460 km/s.

To close the presented model and obtain the numeric values of the solar wind density and speed some model of the polarization potential of plasma is required. Taking advantage of the relation between gravitational and electric potential for quasi-neutral plasma thermodynamic equilibrium in a gravitational field ([1], p. 28):

$$\psi = \varphi/2 \quad \forall \overline{r} \geq 1 \qquad (14)$$

the necessary closing relations can be expressed as

$$\overline{\Phi}_e = \overline{\Phi}_p = \overline{\varphi}, \quad \Delta\overline{\Phi}_e = \Delta\overline{\Phi}_p = \Delta\overline{\varphi} \qquad (15)$$

The obtained dependencies for solar wind density and speed (8)-(10), (12), (15) agree with available observational data [4,5] (Fig. 3 and Fig. 4). The values of parameters are as follows: $r_0 = 1.5R$ (R - radius of the Sun), $T_0 = 1.2 \cdot 10^6$ K (chosen so that theoretical $\overline{N} = N(\overline{r})/N(1)$ and empirical $\overline{N}_{emp} = N_{emp}(\overline{r})/N_{emp}(1)$ [4] values of density at $r = 1$ AU differed no more than by 3 %). This yields the initial density $N_0 = 2.3027 \cdot 10^{13}$ m^{-3}. The difference between theoretical $N(r)$ (a solid line on the Fig.3) and corresponding empirical dependence [4] (squares on the Fig.3.) is about 10% ($r_0 = 1.5R_\epsilon$, $T_0 = 1200K$). The Fig.4 shows $u(r/R)$ (a solid line) compared with the observational data [5] taken from different sources and rendered as filled areas (T_0, r_0 are the same).

The presented results show efficiency of the suggested kinetic approach and also allow to conclude that the solar wind at large heliocentric distances, including the Earth's orbit ($r/R \cong 215$) is formed mostly by escaping particles (~64% at 1 AU) which constitute a flow that could be considered collisionless.

REFERENCES

1. Hundhausen, A.J., *Coronal Expansion and Solar Wind*, Mir, Moscow, 1976.
2. Baranov, V.B., Krasnobaev, K.V., *Hydrodynamic Theory of Space plasma*, Nauka, Moscow, 1977.
3. Meyer-Vernet, N., *Eur.J.Phys.*, **20**, 167-176 (1999).
4. Rubtsov, S.N., etc., *Space Research*, **25**, 4, 620-625 (1987).
5. Yakubov, V.P., *Doppler superlargebase interferometery*, Vodoley, Tomsk, 1997, p.136.

Evolution of Electron-Acoustic Wave in Auroral Region

Paulo H. Sakanaka[*] and Roberto da Trindade Faria Jr.[†]

Instituto de Física "Gleb Wataghin", C.P.6165
Universidade Estadual de Campinas
13083-970, Campinas, SP, Brazil
e-mail: sakanaka@ifi.unicamp.br
[†]*LCFIS-Universidade Estadual do Norte Fluminense*
28015-620, Campos dos Goytacazes, RJ, Brazil

Abstract. Broadband noise in auroral region is related to an interesting coupled mode, namely, electron-acoustic and kinetic Alfvén coupled mode, which are derived from a set of nonlinear equations for low-frequency short wavelength electromagnetic waves in a nonuniform magnetized plasma with sheared plasma flows. In the linear limit it is found that sheared equilibrium flows can be the cause of instability of Alfvén-like electromagnetic waves and electron-acoustic waves in magnetospheric measurements. It is also shown that possible stationary solutions of the nonlinear equations without dissipation can be represented in the form of novel electron-acoustic street vortices.

INTRODUCTION

The presence of modern satellites and similar measurement devices has been playing an important role in understanding better various phenomena in space-plasmas.

We focus our attention on linear and nonlinear electron-acoustic waves close to plasma sheet boundaries in the Space, which could be the source of the well-known broadband noise that has been observed in the geomagnetic tail, in the Earth's bow shock, in the heliospheric termination shock and, mainly, in the magnetosphere and ionosphere, specially in the auroral acceleration region [1, 2]. Normally, for this region, newly theoretical models have pointed out to the presence of two species of electrons (hot and cold). But, recently, the FREJA and FAST satellites also detected hot ion fluxes in ionospheric and magnetospheric regions [3]. Thus, it is possible to occur interesting coupled modes between magnetospheric electromagnetic waves and ionospheric ion thermal flows. Indeed, in this region of magnetized plasma, it was observed the presence of sheared flows and that the ion temperature was close to ten times the electron temperature [4, 5, 6]. This interaction could be the source of nonlinear electron-acoustic structures like vortices [7]. These vortices can be responsible for the appearance of holes and cavities in the auroral zone.

In face of these observations, we consider a model consisting of inhomogeneous magnetized plasma with sheared plasma flows where the ions are hotter than the electrons in a linear and nonlinear short wavelength regime taking the two-fluid equations. Various drift-velocities are considered so that couplings among electron-acoustic waves and Alfvén waves are found to occur in a well-localized way.

We want to demonstrate the possibility of the existence of linear and nonlinear electron acoustic waves connected with kinetic Alfvén waves in the Earth's auroral region. They are analytically and numerically analyzed and the results are plotted, showing clearly the mechanism of saturation of the instabilities for the above-cited region.

SYSTEM OF NONLINEAR EQUATIONS

We implement the study of nonlinear propagation of low-frequency (in comparison with the electron gyrofrequency $\omega_{ce} = eB_0/m_e c$, where, B_0 is the external magnetic field, m_e is the electron mass, and c is the light speed) electromagnetic waves in a nonuniform magnetized plasma. Our model contains an equilibrium density and velocity gradients.

For low-frequency, the electron fluid velocity perturbation is given by [8]

$$\mathbf{v}_e \approx \mathbf{v}_{EB} + \mathbf{v}_{De} + \mathbf{v}_{pe} + (v_{e0} + v_{ez})\mathbf{B}_\perp/B_0 + \hat{\mathbf{z}} v_{ez}, \qquad (1)$$

where $\mathbf{v}_{EB} = (c/B_0)\hat{\mathbf{z}} \times \nabla\phi$, $\mathbf{v}_{De} = -(cT_e/eB_0 n_e)\hat{\mathbf{z}} \times \nabla n_e$, and $\mathbf{v}_{pe} = (c/B_0 \omega_{ce})[\partial_t + v_{e0}\partial_z - \mu_e \nabla_\perp^2 + (\mathbf{v}_{EB} + \mathbf{v}_{De})\cdot\nabla + v_{ez}\partial_z]\nabla_\perp\phi$ are the $\mathbf{E} \times \mathbf{B}_0$, the diamagnetic, and the polarization drift velocities, respectively, $\mathbf{E} = -\nabla\phi - c^{-1}\partial_t A_z \hat{\mathbf{z}}$ is the electric field vector, ϕ is the electrostatic potential, and A_z is the component of the vector potential along the z axis. The electron number density is n_e, T_e the constant electron temperature, $\mu_e = 0.51 v_e \rho_e^2$ is the electron gyroviscosity, v_e is the electron collision frequency, and $\mathbf{B}_\perp = \nabla A_z \times \hat{\mathbf{z}}$ is the two-dimensional magnetic field perturbation. The z-component of the electron fluid velocity perturbation is obtained from the Ampère's law $v_{ez} \approx (c/4\pi n_e e)\nabla_\perp^2 A_z$.

In order to obtain the nonlinear electron and ion dynamics in the presence of the electromagnetic fields, we substitute (1) into the continuity and momentum equation and taking Boltzmann distribution for both electrons and ions, we have from Poisson equation, $n_e = (1/4\pi e)[\nabla^2\phi + (T_i \lambda_{Di}^{-2}/e)\exp(-e\phi/T_i)]$, where $\lambda_{Di} = (T_i/4\pi n_0 e^2)^{1/2}$ is the ion Debye length, we get in the limit of $\rho_e^2 \nabla^2 \ll 1$

$$\mathcal{M}_t - D_c(1 - 0.51\rho_a^2 \nabla_\perp^2)\nabla_\perp^2 \phi + \rho_a^2 \frac{\omega_{ce}}{n_0}(\hat{\mathbf{z}} \times \nabla n_0)\cdot\nabla\phi$$
$$+ \frac{4\pi\lambda_{Di}^2}{B_0}(\hat{\mathbf{z}} \times \nabla j_{e0})\cdot\nabla A_z - c\lambda_{Di}^2 \mathcal{M}_z \nabla_\perp^2 A_z = 0, \qquad (2)$$

and

$$(d_t + \mathbf{v}_{D0}\cdot\nabla)A_z - \lambda_e^2(\mathcal{M}_t + v_e)\nabla_\perp^2 A_z + c(\partial_z + \mathbf{S}_{v0}\cdot\nabla)\phi + c\sigma \mathcal{M}_z \phi = 0. \qquad (3)$$

where $\rho_a = c_a/\omega_{ce}$, $c_a = (T_i/m_e)^{1/2}$ is the electron-acoustic velocity, and $\sigma = T_e/T_i$. In (7), we have assumed that $\nabla^2 \lambda_{Di}^2 \ll 1$, $\partial_t \phi^2 \ll \omega_{ce}\rho_a^4 |\hat{\mathbf{z}} \times \nabla\phi \cdot \nabla \nabla_\perp^2 \phi|$, $\mathcal{M}_t \equiv \partial_t + v_{e0}\partial_z + \mathbf{v}_{EB}\cdot\nabla + v_{ez}\partial_z$, $\mathcal{M}_z = \partial_z + B_0^{-1}\nabla A_z \times \hat{\mathbf{z}}\cdot\nabla$, $j_{e0} = -n_0 e v_{e0}$, $D_c = v_e \rho_e^2$ is the coefficient of the electron diffusion, $n_{e1}(= n_e - n_0 \ll n_0)$ is the perturbed electron number density, $d_t = \partial_t + \mathbf{v}_{EB}\cdot\nabla$, $\mathbf{v}_{D0} = -(cT_e/eB_0 n_0)\hat{\mathbf{z}} \times \nabla n_0$ is the equilibrium electron diamagnetic drift velocity, $\lambda_e = c/\omega_{pe}$ is the collisionless electron skin depth, $\omega_{pe} = (4\pi n_0 e^2/m_e)^{1/2}$ is the electron plasma frequency, and $\mathbf{S}_{v0} = \hat{\mathbf{z}} \times \nabla v_{e0}/\omega_{ce}$.

DISPERSION RELATION

We reach the local linear dispersion relation, neglecting the nonlinear terms in the governing equations which are Fourier transformed. Thus,

$$\omega^2 - (\omega_{i*} + \omega_{m*})\omega + \omega_{i*}\omega_{m*} - S_s = 0, \qquad (4)$$

where $\omega_{i*} = \rho_a^2 \omega_{ce}\mathbf{k}\cdot\mathbf{k}_n/(1+b_a)$, $b_a = k_\perp^2 \rho_a^2$, $\omega_{m*} = \omega_{e*}/(1+b_e)$, $\omega_{e*} = \mathbf{k}\cdot\mathbf{v}_{D0}$, $b_e = k_\perp^2 \lambda_e^2$, $k^2 \lambda_{De}^2 \ll 1$, $\mathbf{k}_n = \hat{\mathbf{z}} \times \nabla n_0$, and $S_s = [c\lambda_{Di}^2/(1+b_a)(1+b_e)][(4\pi k_y(\partial j_{e0}/\partial x)/B_0) + ck_z k_\perp^2][(1+\sigma)k_z + k_y(\partial v_{e0}/\partial x)/\omega_{ce}]$.

Next, we numerically analyze equation (4). Figure 1 shows the short wavelength coupling-mode, ω versus k_z where $k_y = -0.2$ cm^{-1}, which is the maximal value of the growth rate ω_i in the plasma. We note that the coupling begins in $k_z \approx 0$ and finish in $k_z \approx 0.00015$. The full line represents the real part ω_r which is an electron-acoustic wave and the imaginary part ω_i, dashed line, represents the growth rate. For this case, $T_e = 1$ eV, $T_i = 10$ eV, $n_{e0} = n_{i0} = 2 \times 10^6$ cm^{-3}, and $B_0 = 0.4 \times 10^{-4}$ G.

NONLINEAR SOLUTIONS

The nonlinear interaction between finite amplitude modes can be responsible for the formation of vortices [7]. We present here an interesting electron-acoustic street vortex.

Now we discuss some appropriate and approximate solutions. We assume that $c\omega_{ce}|\nabla_\perp^2 A_z \partial_z| \ll \omega_{pe}^2 |\hat{\mathbf{z}} \times \nabla\phi\cdot\nabla|$ and $\partial_z^2 \ll \nabla_\perp^2$. We transform the inertial frame so that $\xi = y + \alpha z - ut$, where α and u are constants, and assume that ϕ and A_z are functions of x and ξ only.

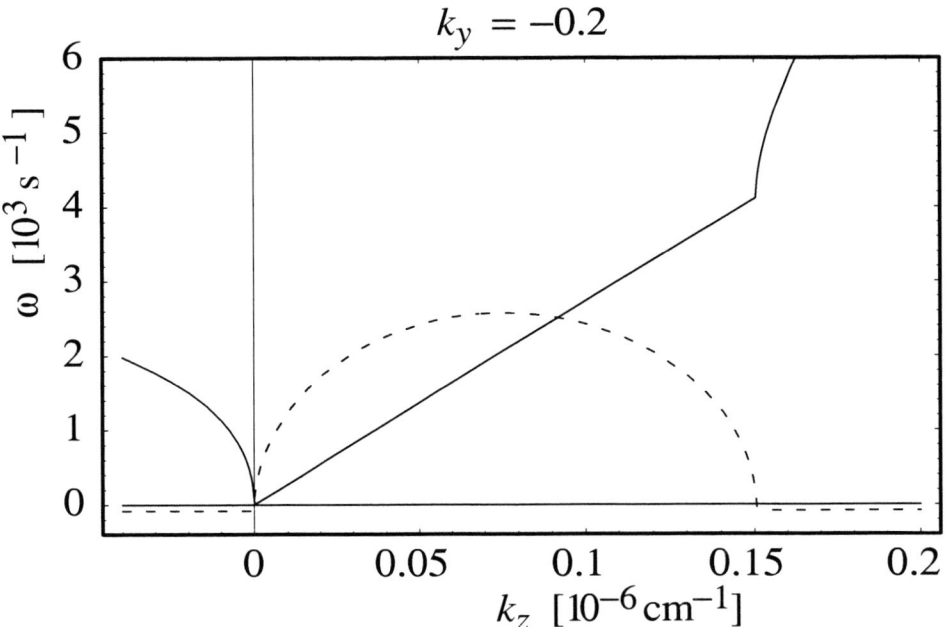

FIGURE 1. Dispersion relation for the electron-acoustic mode. The real part of the frequency (full line) and the growth rate (dashed line) for $k_y = -0.2\,\text{cm}^{-1}$, which is the maximal value of the growth rate are presented.

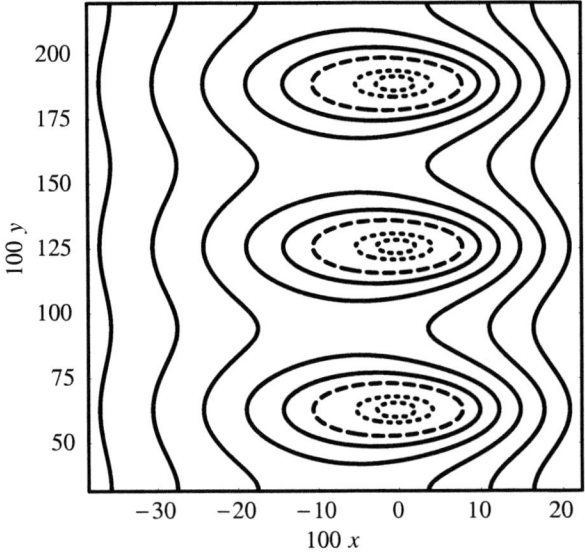

FIGURE 2. Street vortex localized in the auroral zone. The dashed lines represent negative values and the full ones positive values. The rotation is counter-clockwise for a magnetic field going out from the graphics plane. The same parameters of figure 1 are used here.

We rewrite (2) and (3), neglecting dissipations, so that the final nonlinear equation takes the form of the Euler equation

$$\partial_\xi \nabla_\perp^2 \phi - \frac{\mu^* c}{u_\alpha B_0} J(\phi, \nabla_\perp^2 \phi) = 0, \tag{5}$$

where $\mu^* = (\rho_a^2 - \alpha_0^{*2} c^2 \lambda_{Di}^2/u_*^2)/(\rho_a^2 - \alpha \alpha_0^* c^2 \lambda_{Di}^2/u_\alpha u_*) > 0$, $u_\alpha = u - \alpha v_{e0}$, $\alpha_0^* = \alpha_0 + \sigma\alpha$, $\alpha_0 = \alpha + (\partial v_{e0}/\partial x)/\omega_{ce}$, $u_* = u = V_*$, $V_* = cT_e/eB_0 L_n$, and $L_n = n_0/(\partial n_0/\partial x)$. Equation (4) is satisfied by

$$\nabla_\perp^2 \phi = \frac{4\phi_s^* K^{*2}}{a^{*2}} \exp\left[-\frac{2}{\phi_s^*}\left(\phi - \frac{uB_0}{\mu^* c}x\right)\right], \tag{6}$$

where ϕ_s^*, K^* and a^* are arbitrary constants. Equation (5) represents a chain of vortices, called street vortex, which can be found in cavities in the auroral region and should be one of the probable sources of instabilities in our study, where nonlinear structures like solitons and vortices (which are shown in figure 2) can be observed.

CONCLUSIONS

We have investigated the formation of linear and nonlinear electron-acoustic and kinetic Alfvén coupling modes, which could be the cause of holes and/or cavities in the auroral region.

We have derived a set of nonlinear mode coupling equations, considering a nonuniform magnetoplasmas with sheared flows. In the linear regime, our analysis has suggested that electromagnetic disturbances could be driven on account of the free energy stored in the sheared equilibrium plasma flows. In the nonlinear aspect we have shown that electromagnetic waves and sheared flows could lead to the formation of street vortex. Since the vortex solution exists locally, our theory requires that the vortex sizes are much smaller than the scalelengths of the equilibrium density and velocity gradients.

We stress that the electron-acoustic and the kinetic Alfvén waves could form a coupling mode that, probably, explain the linear and nonlinear dynamics of the region.

ACKNOWLEDGMENTS

This work was supported by a Brazilian research funding agency FAPESP-Fundação de Amparo à Pesquisa do Estado de São Paulo.

REFERENCES

1. Mamun A. A., Shukla P. K., and Stenflo L., *Phys. Plasmas* **9**, 1474 (2002).
2. Singh S. V. and Lakhina G. S., *Plan. Space Sci.* **49**, 107 (200).
3. Pottelette R., and Treumann R. A., *J. Geophys. Res.* **103**, 9299 (1998).
4. Jovanović D., Shukla P. K., and Schamel H., *Phys. Plasmas* **7**, 3247 (2000).
5. Pottelette R., Ergun R. E., Treumann R. A., Berthomier M., Carlson C. W., McFadden J. P., and Roth I., *Geophys. Res. Lett.* **26**, 2629 (1999).
6. Seyler, C. E. and Wahlund, J. E., *J. Geophys. Res.* **101**, 21795 (1996).
7. Mikhailovskii A. B., *Electromagnetic Instabilities in an Inhomogeneous Plasma* (Adam Hilger, Institute of Physics Publishing, Bristol, UK), (1992).
8. Shukla P. K., Mirza A. M., and Faria Jr. R. T., *Phys. Plasmas* **5**, 616 (1998).

Effect of Electron Heat Transport on Spatial Distribution of Surface Wave Plasma Parameters

I.B. Denysenko[1,2], K.N. Ostrikov[2,3], N.A. Azarenkov[1], M.Y. Yu[4], and S. Xu[2]

[1]*Kharkiv National University, 4 Svobody sq., Kharkiv 61077, Ukraine*
[2]*Plasma Sources and Applications Center, NIE, Nanyang Technological University, 637616 Singapore*
[3]*SOCPES, Flinders University, GPO Box 2100, Adelaide 5001, Australia*
[4]*Theoretical Physics I, Ruhr University, 44780 Bochum, Germany*

Abstract. The structure of steady-state surface-wave sustained long (L >> R) and short (L ~ R) cylindrical argon plasma columns with the radius R and length L is investigated using a fluid model with 3-moment equations for electrons and 2-moment equations for ions, and accounting for the electron heat transport. The radial profiles of the plasma density, electron temperature, as well as the number density of argon atoms in the $3p^56d$ excited state are computed. Different approaches towards the electron heat flux with the appropriate boundary conditions are analyzed. The best agreement with the existing experimental data is achieved when the electron heat transport is calculated assuming the dependence of the electron-neutral collision rates on the electron temperature. The electron heat transport appears to affect the spatial distribution of the plasma parameters (especially that of the excited-state argon atoms) in the discharge bulk. A 2D fluid code has been applied for modeling the spatial structure of a short, metal-shielded argon discharge sustained by the edge-localized surface waves propagating azimuthally along the plasma edge. It is shown that self-consistent accounting for axial plasma diffusion and radial non-uniformity of the electron temperature can explain the frequently reported deviations of the measured radial density profiles from that of the conventional linear diffusion models. The simulation results are in a good agreement with the experiments on the surface-wave sustained large-diameter plasmas.

RADIAL STRUCTURE OF LONG CYLINDRICAL ARGON PLASMA COLUMN

We are interested in plasmas produced and sustained by axial-symmetrical surface waves (SW) of frequency ω propagating in axial (z) direction along a cylindrical dielectric vessel of inner diameter 2a, outer diameter 2b, and relative permittivity ε_d. The vessel is surrounded by air (vacuum). Fig.1a illustrates the configuration considered here. Plasma permittivity is $\varepsilon_p = 1 - \omega_{pe}^2/(\omega \cdot (\omega + i\nu_e))$. Here ν_e is rate of electron-neutral collision, ω_{pe} plasma frequency of electrons. The latter depends on r. High frequency waves are considered ($\nu_e << \omega$). It is assumed that SW electromagnetic field components depend on time t as $G(r,z,t) = R(r)\exp(i(k_3 z - \omega t))$, where k_3 is the axial SW number. The equations for the axial (E_z) and the radial (E_r) components of the wave electric field inside the plasma are: $\partial^2 E_z/\partial r^2 + (1/r + k_3^2/(\kappa_1^2 \varepsilon_p)\partial \varepsilon_p/\partial r)\partial E_z/\partial r - \kappa_1^2 E_z = 0$; $E_r = -(ik_3/\kappa_1^2)\partial E_z/\partial r$, where $k = \omega/c$, c is the speed of light in vacuum, $\kappa_1 = \sqrt{k_3^2 - k^2 \varepsilon_p}$. The expressions for field components in dielectric and vacuum regions can be easy obtained from Maxwell equations and can be found elsewhere [1]. k_3 was obtained from the dispersion equation for SW considered [1]. To describe the electron density and temperature distributions in plasma column a hydrodynamic model has been applied. The electrons and ions were assumed to be Maxwellian. The continuity and momentum equations for the electrons or the ions are:

$$\nabla \cdot (n\vec{v}) = n\nu^i, \tag{1}$$

$$m_i(\mathbf{v}\cdot\nabla)\mathbf{v} \approx -\nabla(nT_e)/n - (V_i + v^i)\mathbf{v} m_i, \qquad (2)$$

where v^i is ionization rate that depends on electron temperature, n and $\mathbf{v}$ are electron (or ion) density and velocity (In the long tubular case it is equal to the radial component of the velocity), $V_i = (v_{ti} + v)/\lambda_i$, m_i and λ_i are the ion mass and mean free path. The electron energy equation is:

$$\nabla \cdot \mathbf{q}_e \approx -n I_e + S_{ext}, \qquad (3)$$

Here $\mathbf{q}_e \approx -K_e \nabla T_e$ is the heat flux density, K_e is the electron thermal conductivity, I_e is the collision integral for electron energy loss, and $S_{ext} \approx n v_e e^2 (E_r^2(r) + E_z^2(r))/2m_e\omega^2$. It was assumed that the power Q absorbed in the discharge per unit length is fixed. Q is connected with S_{ext} by the following mean $Q = \int_0^a S_{ext} 2\pi r dr$. At the column boundary it was assumed that the ion velocity is equal to Bohm velocity v_B. The boundary electron heat flux density is:

$$T_{eb}(2 + \ln(\sqrt{m_i/m_e}))n_{eb}v_B. \qquad (4)$$

Here n_{eb}, T_{eb} are boundary electron density and temperature, respectively. For the numerical calculations we consider a SW discharge with argon as the working gas and $3p^56d$ the excite atoms under typical experimental conditions [2]. The 600MHz SW sustained plasma is in a dielectric vessel with inner diameter $2a=26$ mm, outer diameter $2b=30$ mm, and $\varepsilon_d = 3.78$ (Fig.1a). In Figs.1b,c typical density profiles of the $3p^56d$ argon atoms are presented. The electron density and temperature distributions are shown in Figs.2a,b. In order to investigate the effect of the thermal flux on the radial distribution of the excited atoms we have invoked several different models for the electron thermal conductivity: (i) $K_e = K_{en} = (5/2)nT_e/m_e v_e$, which is most commonly used in the literature and is valid for Maxwellian electrons when the electron-neutral collision frequency v_e is independent of the electron temperature [3]. (ii) When v_e depends on T_e, the expression for the electron heat flux is rather complicated and is given by, for example, Eqs. (7.24) and (7.48) of al. [3], and one finds $K_e < K_{en}$. For the parameters considered here we can use, say, $K_e = 0.1 K_{en}$. (iii) For the commonly used adiabatic approximation [4] we have $K_e \sim 0.01 K_{en}$.

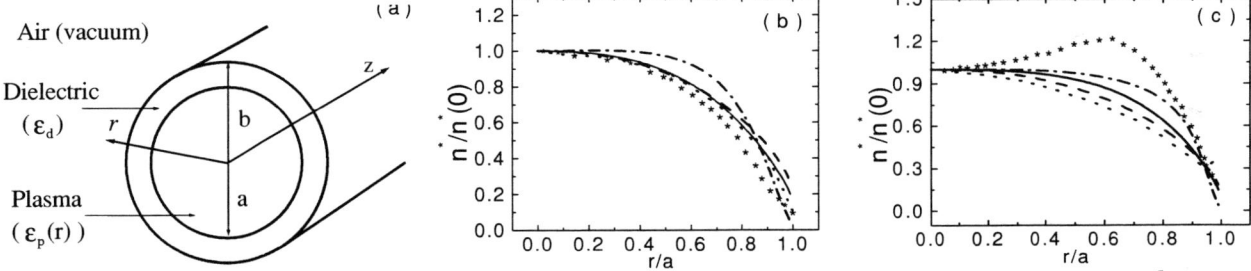

FIGURE 1. Schematic diagram of the SW discharge system (a); Theoretical and experimental radial profiles of the $3p^56d$ excited argon atom density in 600 MHz SW sustained plasma with cross-section-averaged electron density 10^{11} cm^{-3} for 50 mTorr (b) and with the density 4.5×10^{10} cm^{-3} for 150 mTorr (c): present model with $K_e = K_{en}$ (dash line); $K_e = 0.1K_{en}$ (solid line); and $K_e = 0.01K_{en}$ (dash-dotted line); model of Peres *et al* (dotted line); and experimental data (stars).

We can see from Fig.1b that except very close to the plasma edge the density profile (solid curve) of the $3p^56d$ argon atoms corresponding to Model (ii) practically coincides with that (dot curve) of Peres *et al.* [4] and with the experiment [2] for the averaged density 10^{11} cm^{-3}. The results for the small density of the excited argon atoms are some different from experiment. When compared with that from the existing experiment [2] and theory [5], one can see that with $\lambda_i = const$ the results fit the experimental values better than that from Peres et al. [4] (Fig.1c). In general the agreement is better for the models with K_e smaller than K_{en}. The discrepancy at low plasma densities could be due to the fact that the low density region is located at the end of the plasma column. There the axial motion of the charged particles may be essential when the density is low since the axial plasma density gradient is usually large in this region. Accounting for the T_e dependence of v_e and the corresponding heat flux can thus improve the theoretical density profile of the excited argon atoms over that from the other thermal conductivity models. With decreasing thermal conductivity the density profiles of the excited atoms became flatter in the center of the discharge and steeper near the plasma boundary, as can be expected because of the dependence of T_e (and thus n_e) on K_e (Fig.2).

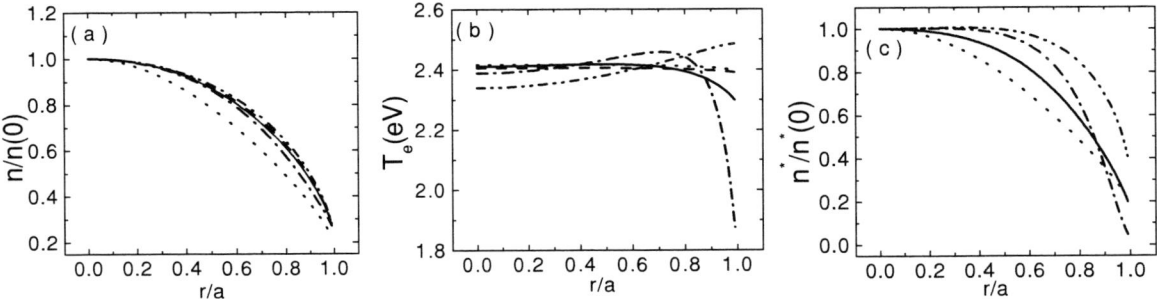

FIGURE 2. Same conditions as in Fig.1b. Radial profiles of the plasma parameters for $K_e = K_{en}$ (dashed line), $K_e = 0.1K_{en}$ (solid line), and $K_e = 0.01K_{en}$ (dash-dotted line), $K_e = 0.01K_{en}$ with the boundary condition $\nabla T_e = 0$ (dash-dot-dotted curve), in the all cases the ion mean free path is constant; ion-neutral-collision rate is constant with $K_e = K_{en}$ (dotted line).

With a decrease of K_e, T_e becomes more nonuniform. In fact, we see from Fig.2b that for $r/a > 0.7$, the T_e curve of Model (ii) starts to deviate from that of Model (i), where $K_e = K_{en}$. With Model (iii), where $K_e \sim 0.01K_{en}$, T_e is smaller near the plasma axis, and increases slowly in the region *0.4 <r/a <0.7*, and decreases rapidly near the plasma boundary. The decrease of T_e with a decrease of K_e is due to the boundary condition for the electron heat flux (4). At $r=a$, one has $|\nabla T_e| \sim 1/K_e$. Thus a decrease in K_e leads to a decrease of T_e near plasma boundary. The density profile also depends on the choice of K_e. With increase of the latter the electron temperature decreases near the plasma boundary as a result of a decrease of ionization in that part of the discharge. Therefore, smaller density gradients are needed to evacuate charged particles from that part of the discharge and the gradient of $n(r)/n(0)$ at $r=a$ decreases with increasing K_e for this case. In Fig. 2 we have also shown the profiles (dotted curves) for the case $v_i = const$, corresponding to $K_e = K_{en}$. The density profile for $\lambda_i = const$ is flatter in the center and steeper at the boundary of the plasma as compared to that for constant ion mean free time (see Fig.2a). We now consider the effect of the boundary conditions of the electron heat flux on the profiles of the plasma parameters. The alternative boundary condition $\nabla T_e = 0$ is often used [3] for evaluating the spatial profile of the plasma parameters. In Fig.2 the profiles given by the dash-dot-dotted curves are for the adiabatic case (iii) with $\nabla T_e = 0$ at the discharge boundary. In Fig.2a one can see that for this case the density profile is only slightly flatter for most of the plasma and steeper near $r=a$ compared to that (given by the dash-dot curves) for the same (adiabatic) case when the normal boundary condition for the heat flux is used. However, the corresponding electron temperature (Fig.2b) at the center of the plasma is now lower and it increases all the way up to the plasma boundary. In Fig.2c the radial profiles of excited argon atoms in the $3p^56d$ state are presented for different models of the thermal conductivity. One can see that in comparison with that for $v_i = const$, the density profile n^* of the excited atoms for $\lambda_i = const$ is flatter in the center and steeper near the plasma boundary. That is, the value of K_e and the boundary condition for q_e can affect the density profile of the excited atoms. In general, with increasing K_e the n^* profile becomes flatter at center and steeper near $r=0$. Furthermore, the boundary condition $\nabla T_e = 0$ leads to the largest number of excited atoms.

RADIAL STRUCTURE OF SHORT CYLINDRICAL ARGON PLASMA COLUMN

The spatial structure of a short, metal-shielded large-diameter argon discharge sustained by the edge-localized surface waves propagating azimuthally along the plasma edge has been studied too. The schematic diagram of the SW sustained discharge is shown in Fig. 3a. The SW propagates in azimuthal direction. The radial electric field profiles of pure SWs assuming that the chamber is filled by uniform plasma [5]: $E_\varphi \approx A\exp[-\kappa(a-r)]$ and $E_r \approx i(k_\varphi/\kappa)E_\varphi$, where $\kappa^2 \approx k_\varphi^2 - k^2\varepsilon_p$. The propagation constant k_φ is obtained from the dispersion relation: $\kappa\varepsilon_d + \varepsilon_p\kappa_d \tanh[\kappa_d(b-a)] = 0$, where $\kappa_d^2 \approx k_\varphi^2 - k^2\varepsilon_d$. The particle and energy balance equations were the same as in long tubular discharge case (1,3). But now $S_{ext} \approx n v_e e^2 (E_r^2(r,z) + E_\varphi^2(r,z))/2m_e\omega^2$ and $Q = \int_0^L dz \int_0^a S_{ext} 2\pi r dr$, where L is discharge length. . The equation for the charged particle velocity was simplified and was $\vec{v} \approx -\nabla(nT_e)/nm_i v_{in}$. The expression for electron heat flux was chosen to take into account the dependence of electron-neutral collision rate on the temperature: $q_e \approx -(5/2 - \Delta)nT_e/(m_e v_e)\nabla T_e$ with $\Delta = (T_e/v_e)\partial v_e/\partial T_e$. The case $\Delta = 0$ was considered too. The plasma parameters have been calculated at the experimental conditions of [6]. The external parameters were: $a = 6.3$ cm, $b = 7.5$ cm, $L = 14$ cm, $P = 0.76$ Torr. The discharge was sustained with a RF power of $Q = 300$ W at 2.45 GHz. The results of the calculation are shown in Figs. 3b-3d. At self-consistent accounting for axial plasma diffusion and radial non-uniformity of the electron temperature the plasma density profile is essentially differ from one obtained for constant electric field in the discharge volume. When the dependence of v_e on the electron temperature is taken into account in the discharge model the electron temperature is more nonuniform in comparison with case when $\Delta = 0$. The discrepancy in the density profiles was small (about 3%). So, the electron heat transport affects the spatial profile of the electron/ion number density and temperature, as well as the spatially averaged plasma density. At the experimental conditions the volume recombination may be important. Accordingly, we added a term $\beta_r n^2$ ($\beta_r \sim 10^{-8} cm^3/s$ is an effective coefficient of volume recombination, to the right-hand side of Eq.1). The agreement between our model and the experiment is now significantly better (Fig.3d).

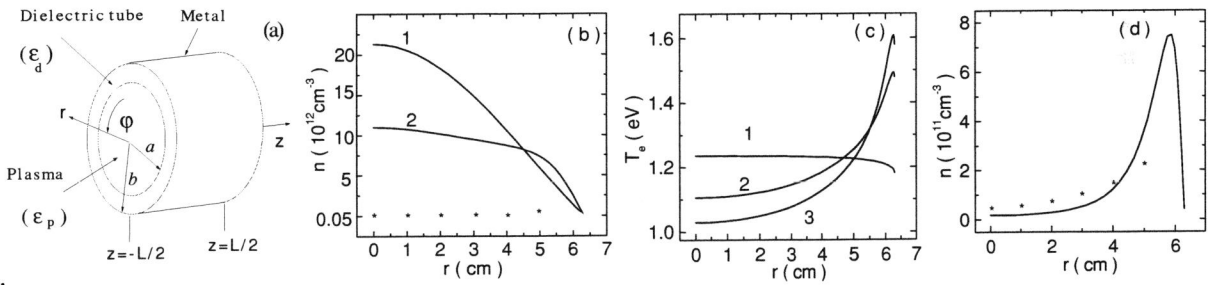

FIGURE 3. Schematic diagram of a short SW sustained discharge (a). The radial plasma parameters at the experimental conditions of [6]. Curve 1 is obtained for case when HF electric field components are constant in whole discharge volume. Curves 2 and 3 are obtained for SW sustained discharge with $\Delta = 0$ and $\Delta \neq 0$ in the expression for electron heat flux density, respectively. Stars are experimental points taking from [6]. Solid line in Fig.3d was calculated accounting for volume recombination.

ACKNOWLEDGMENTS

This work was supported by the Science and Technology Center in Ukraine (Project No. 1112), Agency for Science, Technology, and Research of Singapore (Project No. 012 101 00247), the NATO (Grant PST.CLG.978083), and the Sonderforschungsbereich 191/A8, Germany.

REFERENCES

1. Denysenko, I.B., Azarenkov, N.A., and Yu, M.Y., Phys. Scripta **65**, 76 (2002).
2. Margot, J., Moisan, M., and Ricard, A., Appl. Spectrosc. **45**, 260 (1991).
3. Golant, V. E., Zhilinskii, A. P., and Sakharov I. E., Fundamentals of plasma physics, Wiley, New York, 1980.
4. Peres, I., Fortin, M., and Margot, J., Phys. Plasmas, **3**, 1754 (1996).
5. Denysenko, I.B., Gapon, A.V., Azarenkov, N.A., Ostrikov, K.N., and Yu, M.Y., Phys. Review E **65**, 046419 (2002).
6. Korzec, D., Werner, F., Winter, R., and Engemann, J. Plasma Sources Sci. Technol. **5**, 216 (1996).

Dust Growth and Gravitation-like Instabilities in Astrophysical Plasmas

R. Bingham*† and V.N.Tsytovich**‡

Rutherford Appleton Laboratory, Space Science and Technology Dept., Chilton, Didcot, U.K.
†*University of St Andrews, Fife, Scotland KY16 9SS*
**General Physics Institute, Russian Academy of Science, Vavilova str.38, Moscow, Russia*
‡*Moscow Physical Technical Institute, Moscow, Russia*

Abstract. Dust growth by agglomeration due to a new mechanism of dust attraction is shown to compete in space plasmas with the standard accretion mechanism by neutral particles. We demonstrate that dust growth by accretion of neutrals and plasma particles is aided by a pressure force due to a shadowing effect first proposed by Bingham & Tsytovich (*IEEE Transactions on Plasma Science* **29**, 158, (2001)) to explain rapid growth of dust in laboratory discharges. Consequences of this new force promoting dust growth by agglomeration in astrophysics are discussed including the role of a new gravitational-like instability which produces clumping enhancing the rate of gravitational collapse.

INTRODUCTION

The subject of dust formation in space and astrophysical plasmas is important in diverse environments such as dust molecular clouds, proto-planetary nebulae, stellar outbursts, and supernova explosions. The formation of dust proceeds the formation of stellar objects and planets. And the ultimate fate of most of the dust is to be consumed in the process of making stars, planets and smaller bodies such as comets and asteroids.

Dust agglomeration has been shown to be important in laboratory etching experiments [4] where growth of dust is extremely rapid and is due in the final stages to dust-dust attraction by a plasma or neutral bombardment force known as the shadow force ([9][10]. Since small grains of dust are attaching themselves together the rate of growth to large sized grains of order $10\mu m$ or larger is extremely rapid of order of hours in laboratory experiments and as we will demonstrate it is of order 10 years in *C*-stars.

This rapid formation of large dust grains through agglomeration can lead to a new type of instability formed as a result of the net attractive force existing between the dust grains. By replacing the force of gravity by this force we derive a linear dispersion relation describing the growth rate of a gravitation like instability leading to clumping of the dust structures with a typical length for collapse smaller than the Jeans length for gravitational collapse. We find that the formation of clump like structures proceeds faster than gravitational collapse. The consequence is that this type of collapse is perhaps necessary before the onset of gravitational collapse resulting in a self similar or scale invariant process describing self-organization of dust structures.

DUST GROWTH

Our understanding of dust growth has benefited greatly from laboratory experiments [4], [3] carried out within the last 10 years. These experiments have been successfully explained by a combination of a series of stages of dust growth. In the first stage nucleation occurs with the chemical formation of atom clusters this stage is well documented for both laboratory and space scenarios. It is understood that the Van der Waals and chemical binding forces play a key role in forming these nm sized molecular clusters.

The second stage observed in laboratory experiments carried out in the presence of a plasma is rapid dust agglomeration up to grain sizes of order $0.1 - 1.0\mu m$. One of the main differences introduced by the plasma is the presence of charge on the grains which normally acquire a negative charge in the absence of UV radiation due to the greater electron mobility. Estimates show that for grains of 10nm in size charges of $10-50$ electron charges are possible on isolated grains.

The third stage of growth is the plasma deposition or plasma accretion stage where ions are attracted to the charged dust, increasing the size of dust to about 100μm or larger. During this stage irregular shaped dust soon becomes spherical due to a spherical converging flow of ions. The main difference between neutral accretion and ion accretion or deposition is that the ions are accelerated by the field produced by the charged dust to energies of the order of the electron temperature. The potential of the charged dust grain is of the order of the electron thermal energy in some cases this can be as high as 3×10^4K. This energy which is imparted to the ions causes ion implantation.

The shadow force is the main force which acts in the fourth stage of dust growth. The shadow force due to neutrals and ions caused by dust is given by a simple expression which can easily be obtained qualitatively (for a more exact theory see [1], [5]).

$$F = \frac{a^2}{r^2} nT \pi a^2 A \tag{1}$$

where a^2/r^2 is the solid angle, a is the dust size and r is the distance separating the dust particles, nT is the pressure of either neutrals or ions we can neglect electrons because of their smaller mass and A is a numerical coefficient of order of unity. In the calculation we have assumed that the dust particles are of equal size.

The physics of the shadow force is very simple. For a single dust particle the flux of absorbed charged and neutral particles on the dust surface is symmetric resulting in no net momentum transfer. Another dust particle at distance r shadows the flux to the first dust particle with a solid angle $\approx a^2/r^2$. The net momentum transfer will be proportional to the solid angle, to the surface area of the dust particle πa^2 and to the pressure nT as written in Eq.(1). The numerical coefficient A depends on whether the distance between the dust particle is less or larger than the mean free path of dust/neutral or dust/ion collisons being in both cases of the order of 1 [5].

The fourth stage of dust growth observed under laboratory conditions but not all situations ([4]) is the advanced stage or second stage of agglomeration which leads to dust grains of 0.1mm in size which are observed optically. The physics of this stage is dominated by dust-dust attraction, which is proportional to the fourth power of the dust size, over-coming Coulomb repulsion between the charged dust, which is proportional to the square of the dust size [9][10].

The four stages of dust growth observed in laboratory experiments in the presence of plasma will also be important in space plasmas. For example in carbon rich star outbursts [12] dust formation takes about 10 years in regions where there is a mixture of neutrals and plasma at relatively high temperatures. The gas density in these stellar outbursts is as high as 10^{10}cm^{-3} with temperatures large enough that ionization is also present.

The force given by Eq.(1) operates in the presence or absence of plasmas. In the presence of plasma the force due to the ion and neutral flux should be compared to the Coulomb force F_c between two dust grains given by

$$F_c = \frac{Z_d^2 e^2}{r^2} = \left(\frac{Z_d e^2}{aT_e}\right)^2 \frac{n_i T_e^2 a^2}{e^2 n_i r^2} = z^2 \frac{a^2}{r^2} n_i T_e 4\pi d_e^2 \tag{2}$$

where $z = \frac{Z_d e^2}{aT_e}$ is the dimensionless dust charge of order 1-2 and $d_e = \sqrt{\frac{T_e}{4\pi n_e e^2}}$ is the electron Debye radius.

Comparison of Eqs.(1)-(2) demonstrates that the bombardment force due to ions is $\frac{4d_e^2 z^2}{a^2}$ less than the Coulomb repulsive force for $r \ll d_e$. For $r \gg d_e$ the Coulomb force is screened while the attractive shadow force is not, this can lead to a contraction of the dust cloud until the inter-dust distances are comparable to 10 times the Debye radius. For distances larger than the Debye radius the ion accretion force [9][10] can dominate Coulomb repulsion and should be added to the attraction due to gas particle bombardment but this contribution is usually small. For outbursts of carbon rich stars where $n_n \simeq 10^{10}$cm^{-3}, n_i is of order n_n and $T_e \sim 1eV$, resulting in a Debye radius of between 100 and 1000μm, attraction can occur between two equally charged grains. Using $r \sim 4d_e$ where the repulsive force can be overcome by the attractive force forming an attractive potential well, with the potential given by

$$V \simeq F_c r \approx z^2 n_i d_e a^2 T_e \tag{3}$$

We find that a "dust molecule" is formed if

$$T_d < T_e n_i d_e a^2 z^2 \tag{4}$$

where T_d is the dust kinetic temperature not the surface temperature. In environments where dust is cooled by emission of radiation condition (4) can easily be satisfied. For $n_i \sim 10^{10}$cm^{-3}, $d_e \sim 10^{-2}$cm i.e. $r \sim 10^3 \mu$m and $a \sim 1\mu$m Eq.(4) results in $T_d < T_e z^2 \sim 10 T_e$ which is easily satisfied in carbon rich star outbursts [12].

Laboratory experiments [8] demonstrate the possibility of producing regular dust structures known as dust crystals which occur for $Z_d e^2 n_d^{\frac{1}{3}} > 170 T_d$. In space this condition is more difficult to satisfy namely

$$T_d > T_e n_i a d_e^2 z^2 \left(n_d^{\frac{1}{3}} a\right)/170 \tag{5}$$

but it may be possible which has lead to suggestions that dust crystals could exist in space environments [7]. For $T_i \ll T_e$ numerical results [1] show that the attraction force Eq.(1) can dominate the Coulomb force for $a > (8-60)d_i$ thus leading to dust agglomeration, d_i is the ion Debye radius.

In dust molecular clouds d_e is very large and the degree of ionization is very low. An estimate of the agglomeration time is found by taking into account the relative number of particles in phase space with energies less than the attractive potential well

$$\frac{dm_d}{dt} = 2 m_d v_d n_d^{\frac{1}{3}} \left(\frac{V}{T_d}\right)^{\frac{3}{2}} ; \quad v_d = \sqrt{\frac{T_d}{m_d}} \tag{6}$$

Since the force acting is in the direction separating the two dust particles we can use for the time scale the time for the dust to travel the inter-dust distance which is a direct consequence of (6) given by

$$\frac{1}{\tau_{\text{aggl.}}} \simeq \sqrt{\frac{T_d}{m_d}} n_d^{\frac{1}{3}} \left(\frac{T_e n_i a^2 d_e z^2}{T_d}\right)^{3/2} \tag{7}$$

for $a \sim 1 \mu m$, $T_d \sim 100 K$, $n_d \sim 10^{-3} cm^{-3}$ we estimate $\tau_{\text{aggl.}}$ to be about 10^9 seconds $\simeq 30$ years.

GRAVITATIONAL LIKE INSTABILITY

The shadow force also creates a gravitational like instability. The dispersion relation for the agglomeration instability for the dust component can easily be written using the dispersion relation for the Jeans instability by replacing the force of gravity by the shadow force resulting in

$$\omega^2 = k^2 v_s^2 - G_{\text{eff}} n_d m_d \tag{8}$$

where according to (1)

$$G_{\text{eff}} = \frac{n T_n \pi a^4}{m_d^2} \tag{9}$$

and

$$v_s^2 = \frac{n_d T_d + n_n T_n}{n_d m_d + n_n m_n} \sim \frac{n_n T_n}{n_d m_d + n_n m_n} \tag{10}$$

if $n_n T_n \gg n_d T_d$. Although the attraction forces are acting only on dust particles the expression for v_s will correspond to the sound speed for frequencies much less than the dust-neutral collision frequency [11].

The corresponding Jeans type length is determined by the relation

$$L = \frac{2\pi v_s}{\sqrt{G_{\text{eff}} n_d m_d}} \simeq \frac{2\sqrt{\pi}}{n_d a^2 \sqrt{(1 + n_n m_n / n_d m_d)}} \tag{11}$$

For $a \sim 1 \mu m$, $n_d \sim 10^{-3} cm^{-3}$ we have $L \sim 10^{11} cm$, resulting in dust clumps of order $10^{11} - 10^{12} cm$ in size.

These clumps are much smaller than the Jeans length for gravitational collapse and will proceed stellar and planetary formation. It is envisaged that clumping could proceed through this mechanism forming larger and larger scales all the way to the Jeans length scales, at each scale the structures are similar and scale invariant. These structures are a result of self-organization found in complex physical systems, each scale could collapse forming structures from asteroids in size all the way up to star formation which could be enhanced by additional compression produced by shock waves. The characteristic time for the Jeans like instability to develop is given by $\tau_{\text{ag.inst}}$ i.e. the agglomeration instability time scale

$$\tau_{\text{ag.inst}} \simeq \sqrt{\frac{m_d}{n_n T_n \pi a^4 n_d}} \tag{12}$$

which for $n_d \sim 10^{-3} \text{cm}^{-3}, n_n \sim 10^3, T_n \sim 50\text{K}, a \sim 1-3\mu\text{m}$ is

$$\tau_{\text{ag.inst}} \sim 10 - 100 \text{ years.} \tag{13}$$

This is 3 orders of magnitude faster than the Jeans gravitational collapse time and the Jeans length is 10^3 orders of magnitude larger. The result is a clumpy structure which is often observed in dust-molecular clouds.

CONCLUSIONS

In this article we have described a new force of attraction giving rise to rapid dust agglomeration in astrophysical plasmas, the rate of dust formation is much more rapid than that due to the accretion of neutral particles. This new force arises due to a shadowing of the dust particles in either a neutral or plasma atmosphere. The force of attraction that results from the shadow effect gives rise to the formation of smaller scale structurs on faster time scales than that due to the standard Jeans instability. The structures that are then formed could easily be responsible for the formation of planetesimals.

ACKNOWLEDGMENTS

Authors acknowledge the support of the EU INTAS grant 97-2149, and EU grant .

REFERENCES

1. Bingham, R., & Tsytovich, V.N., *IEEE Transactions on Plasma Science* **29**, 158, (2001).
2. Blum, J., et al. *Phys. Rev. Lett.* **85**, 2426, (2000).
3. Boufendi, L., & Bouchoule, A., *Plasma Sources Sci. Technology* **3**, 262, (1994).
4. Garscadden, A., Ganguly, B.N., Healand, P.D., & Williams, J., *Plasma Sources Sci. Technology* **3**, 239, (1994).
5. Khodataev, Ya.K., Morfill, G.E., Tsytovich, V.N., *J. Plasma Phys.*, in press.
6. Meaking, P., *Rev. Geophys.* **29**, 317, (1991).
7. Morfill, G.E., 2000, Private communication
8. Thomas, H.M., & Morfill, G.E., *Nature* **379**, 806, (1996).
9. Tsytovich, V.N., Khodataev, Y., & Bingham, R., 1996a, *Comments on Plasma Physics and Controlled Fusion* **17**, 249, (1996).
10. Tsytovich, V.N., Khodataev, Y., Bingham, R., & Tarakanov, V., 1996b, *Advances in Dusty Plasma*, ed. P.K.Shukla, D.A. Mendis and T. Desai, World Scientific, Singapore, 1996, p. 212.
11. Tsytovich, V.N., *Physics Uspekhy* **40**, 53, (1997).
12. Winters, J.M., Keady, J.J., Gauger, A. & Sada, P.V., *Astron. Astrophys.*, **359**, 651, (2000).

Thermal instability of an optically thin dusty plasma

Madhurjya P. Bora[1]

Physics Department, Gauhati University, Guwahati 781014, India.

Abstract. We investigate the role of thermal instability, arising from radiative cooling of an optically thin, dusty plasma, by linear stability analysis. The corresponding isobaric stability condition for condensation mode is found to be modified significantly by the concentration of finite sized, relatively heavy, and negatively charged dust particles. It has been shown that the radiation condensation mode is severely affected by the presence of dust particles.

In this work, we consider the possible effect of dusts in a radiating plasma. It is now well known that the dust particles constitute an ubiquitous and important component of many astrophyical plasmas including interstellar clouds, stellar and planetary atmospheres, and planetary nebulae (PNE) etc. The presence of dust in an ionized astrophysical structure such as PNE and giant H-II regions (GHR) can greatly modify its thermal structure in many ways. One of the important ways the dust can modify the stability properties is by capturing energetic electrons and cooling through radiation [1, 2] in the infrared region and we expect that the presence of dust particles significantly modifies the corresponding growth rates. In this regard, the phenomena is also similar to what is known as the multifaceted asymmetric radiation from the edge (MARFE) in tokamaks [3], which is due to the impurity ions.

It has however been proposed that majority of the dust grains could be destroyed as a hot ionized regions of plasma such as a PNE, as it evolves, either because of spallation by hard UV photons or shocks. Subsequently, a PNE or a GHR should be free of dust grains [4, 5]. But based on recent observations of the infrared spectra of the evolved PNE nebula NGC 6445, it has been shown that only a little destruction of the dust grains inside the ionized region of the nebula could have occurred and dust-grain separation inside the ionized region is not plausible [6]. Therefore it is of particular interest to examine the effect of dust on the thermal instability inside a hot, ionized plasma.

We begin by writing the linearized equations for a unbounded, collisional dusty plasma with a heat-loss function [8] $L(n_j, T_j)$, which represents the rate of net loss of energy per unit mass, through radiation and the subscripts refer to different species of the plasma i.e. electron, ions, and dust particles. Thus we have a set of Braginskii-type [7] equations

$$\frac{\partial n_j}{\partial t} + n_{j0}(\nabla \cdot \mathbf{v}_{j1}) = 0, \tag{1}$$

$$\nabla p_{e1} + en_{e0}\mathbf{E}_1 + m_e n_{e0} v_{ei}(\mathbf{v}_{e1} - \mathbf{v}_{i1}) + \alpha n_{e0}\nabla T_{e1} = 0, \tag{2}$$

$$m_i n_{i0}\frac{\partial \mathbf{v}_{i1}}{\partial t} = -\nabla p_{i1} + en_{i0}\mathbf{E}_1 + v_i \nabla^2 \mathbf{v}_{i1}, \tag{3}$$

$$\frac{3}{2}n_{e0}\frac{\partial T_{e1}}{\partial t} + p_{e0}(\nabla \cdot \mathbf{v}_{e1}) = -\alpha \nabla \cdot [n_{e0}T_{e0}(\mathbf{v}_{e1} - \mathbf{v}_{i1})] + \chi_\parallel^e \nabla^2 T_{e1} - m_e n_{e0} L_{e1}, \tag{4}$$

$$\frac{3}{2}n_{i0}\frac{\partial T_{i1}}{\partial t} + p_{i0}(\nabla \cdot \mathbf{v}_{i1}) = \chi_\parallel^i \nabla^2 T_{i1} - m_i n_{i0} L_{i1}, \tag{5}$$

$$e(n_{e1} - n_{i1}) + n_{d0} Q_{d1} = 0, \tag{6}$$

$$\left(\frac{\partial}{\partial t} + \eta\right) Q_{d1} = -|I_e|\left(\frac{n_{i1}}{n_{i0}} - \frac{n_{e1}}{n_{e0}}\right), \tag{7}$$

where the subscripts 0 and 1 refer to equilibrium and perturbed quantities and $j = i, e$ refers to ions and electrons,

---

[1] E-mail: mpb@guphysics.ac.in

respectively. In writing the above equations, we have assumed that the dust particles are negatively charged. We further neglect the dust dynamics as the dust acoustic velocity is negligibly smaller ($\sim 10^{-6}$ cm) than the ion-acoustic velocity ($\sim 10^5$ cm) for a typical range of parameters of PNEs and GHRs. However the effect of the presence of dust particles are self-consistently included through the electron-depletion (quasineutrality condition) and grain-charging equations [9] (6) and (7). and In the equations, $Q_{d0} = eZ_{d0}$ and Z_{d0} is the equilibrium charge number on the surface of a dust grain, $\eta = e|I_e|(T_{e0}^{-1} - W_0^{-1})/C$, with $W_0 = T_{i0} - e\phi_{f0}$, ϕ_{f0} is the potential at the dust surface, $|I_e| \sim I_i \sim en_{i0}\pi a^2 c_s$ is the equilibrium electron (or ion) current at the dust surface, and $C \sim a$ is the grain capacitance. All other symbols have their usual meanings. The quantity α is a constant equal to 0.71. The equilibrium pressure is given as $p_{j0} = n_{j0}T_{j0}$ and the perturbed pressure has the form

$$p_{j1} = \frac{n_{j1}}{n_{j0}} + \frac{T_{j1}}{T_{j0}}. \tag{8}$$

We now consider a small perturbation of the form $f(\mathbf{r},t) \sim f_1 e^{-i(\omega t - \mathbf{k}\cdot\mathbf{r})}$. From Eqs.(7) and (8) we get the relation between the perturbed densities

$$\tilde{n}_e \left[1 + i\frac{\hat{I}_0}{\delta(\omega + i\eta)}\right] = \tilde{n}_i \left[\delta_e^{-1} + i\frac{\hat{I}_0}{\delta(\omega + i\eta)}\right] \tag{9}$$

where the '~' quantities are the normalized by their respective equilibrium values, and δ_j represents (n_{j0}/n_{i0}) for the electrons and dust particles, $\delta = \delta_e/\delta_d$ is the electron to dust ratio, and $\hat{I}_0 = |I_e|/(eZ_{d0})$. Using Eqs.(1), (3), and (5) along with the quasineutrality condition we get another relation between the perturbed ion-density and the electrostatic potential

$$\tilde{n}_i \left[1 - \frac{\omega_i^2}{\omega(\omega + i\hat{\mu})} + \frac{\omega_i^2(n_{i0}L_{n_i}/c_i^2 + i\omega)}{\omega(\omega + i\hat{\mu})(T_{i0}L_{T_i}/c_i^2 + \chi_\parallel^i k^2/n_{i0} - 3i\omega/2)}\right] = \frac{\omega_s^2}{\omega(\omega + i\hat{\mu})}\tilde{\phi}, \tag{10}$$

where $\tilde{\phi} = e\phi_1/T_{e0}$ is normalized electrostatic potential. Note that the temperatures are expressed in energy units throughout this analysis. The perturbed radiative loss functions are given by

$$L_{j1} = L_{n_j}n_{j1} + L_{T_j}T_{j1}, \tag{11}$$

where the subscripts n_j and T_j denotes the derivative of the equilibrium heat-loss function with respect to the corresponding equilibrium densities and temperatures. The normalized ion viscosity is expressed as $\hat{\mu} = k^2\mu_i/(m_i n_{i0})$. The various ωs are respective sound frequencies i.e.

$$\omega_j = kc_j, \quad \omega_s = kc_s, \tag{12}$$

with $c_j^2 = T_{j0}/m_{j0}$ are electron and ion thermal velocities and $c_s^2 = T_{e0}/m_i$ is the acoustic velocity. To close the set of equations, we use the electron equations to get a relation between the perturbed densities and the perturbed electrostatic potential

$$\tilde{n}_e \left[1 - i\frac{\nu_{ei}\omega}{\omega_e^2} - \frac{\beta(n_{e0}L_{n_e}/c_e^2 + i\beta\omega)}{(T_{e0}L_{T_e}/c_e^2 + \chi_\parallel^e k^2/n_{e0} - 3i\omega/2)}\right] - \tilde{\phi} =$$
$$\tilde{n}_i \left[-i\frac{\nu_{ei}\omega}{\omega_e^2} - \frac{i\beta\alpha\omega}{(T_{e0}L_{T_e}/c_e^2 + \chi_\parallel^e k^2/n_{e0} - 3i\omega/2)}\right], \tag{13}$$

where the quantity $\beta = (1 + \alpha)$ and is a constant equal to 1.71.

We finally derive the linear dispersion relation with the help of equations (9), (10), and (13) which can be written as

$$\left[1 - \frac{\omega_i^2}{\omega(\omega + i\hat{\mu})} + \frac{\omega_i^2(i\omega + \omega_{n_i})}{\omega(\omega + i\hat{\mu})(\omega_{T_i} + \hat{\chi}_\parallel^i - 3i\omega/2)}\right]\left[1 + i\frac{\hat{I}_0}{\delta(\omega + i\eta)}\right] =$$
$$\frac{\omega_s^2}{\omega(\omega + i\hat{\mu})}\left[\left\{\delta_e^{-1} + i\frac{\hat{I}_0}{\delta(\omega + i\eta)}\right\}\left\{1 - 3.2i\frac{\omega}{\hat{\chi}_\parallel^e} - \frac{\beta(i\beta\omega + \omega_{n_e})}{(\omega_{T_e} + \hat{\chi}_\parallel^e - 3i\omega/2)}\right\} + \left\{1 + i\frac{\hat{I}_0}{\delta(\omega + i\eta)}\right\}\left\{3.2i\frac{\omega}{\hat{\chi}_\parallel^e} + \frac{\beta(i\beta\omega + \omega_{n_e})}{(\omega_{T_e} + \hat{\chi}_\parallel^e - 3i\omega/2)}\right\}\right]. \tag{14}$$

We have expressed the thermal conductivities in their normalized form given by

$$\hat{\chi}_\parallel^j = \frac{k^2 \chi_\parallel^j}{n_{j0}}. \tag{15}$$

We have further introduced the frequencies of isothermal and isochoric perturbations as $\omega_{n_j} = k_{n_j} c_j$ and $\omega_{T_j} = k_{T_j} c_j$ with the corresponding wave numbers defined as $k_{n_j} = n_{j0} L_{n_j}/c_e^3$ and $k_{T_j} = T_{j0} L_{T_j}/c_e^3$.

In what follows, we ignore the radiation loss terms (ω_{n_j, T_j}) in the ion-energy equation, Eq.(5). At temperatures above $10^4\,°K$, which is typical in ionized hot regions of PNEs and GHRs, it is effectively the electron energy which gets radiated away through the ions and the dust particles. To have an idea on the complex growth rate of the instability, we assume that the charge fluctuation on the dust surface is smaller ($\hat{I}_0 \ll 1$) and $\omega \gg \eta$. We can then simplify the expressions containing $\hat{I}_0$ as

$$\left[\delta_e^{-1} + i\frac{\hat{I}_0}{\delta(\omega + i\eta)}\right]^{-1} \approx \delta_e \left(1 - i\frac{\hat{I}_0}{\omega}\frac{\delta_e}{\delta}\right). \tag{16}$$

For a prototype heat-loss function [10] given by $L \sim 10^{-3} n_H T^{1/2}$ and critical wave number [8] given by $k_c = \rho_0(\rho_0 L_\rho/T_0 - L_T)/\chi$, the ion thermal conductivity ($\hat{\chi}_\parallel^i$) and ion viscosity ($\hat{\mu}$) are $\sim 10^{-6} - 10^{-7}$ times smaller than the corresponding acoustic frequency at $k = k_c$ for a typical set of parameters for PNEs ($n_i \sim n_H \sim 10^4\,cm^{-3}$, $T_e \sim T_i \sim 10^4\,°K$) and can be neglected. With these simplifications, the dispersion relation (14) can be written as

$$\left(1 - \frac{\omega_i^2}{\omega^2}\right)\left[1 + i\frac{\hat{I}_0}{\delta\omega}(1 - \delta_e)\right] = \frac{\omega_s^2}{\delta_e \omega^2}\left[1 - 3.2i\frac{\omega}{\hat{\chi}_\parallel^e} - \frac{\beta(i\beta\omega + \omega_{n_e})}{\omega_{T_e} + \hat{\chi}_\parallel^e} + \delta_e\left\{1 + i\frac{\hat{I}_0}{\delta\omega}(1 - \delta_e)\right\}\left\{3.2i\frac{\omega}{\hat{\chi}_\parallel^e} + i\frac{\beta\alpha\omega}{\omega_{T_e} + \hat{\chi}_\parallel^e}\right\}\right], \tag{17}$$

where, we have further assumed that $\omega_{T_e} + \hat{\chi}_\parallel^e \gg \omega$. Assuming that $\omega = \omega_r + i\gamma$, and $\omega_r \gg \gamma$, with an ordering of $\omega_{n_e} \sim O(\gamma)$, and using $\gamma/\omega_r^{(0)}$ as an expansion parameter, the real part of ω, to the first order can be written as

$$\omega_r^2 \approx (\omega_r^{(0)})^2 - \omega_s^2 \frac{\beta}{\delta_e}\frac{\omega_{n_e}}{\omega_{T_e} + \hat{\chi}_\parallel^e} - \hat{I}_0 \omega_s^2 (\delta_e^{-1} - 1)\left(\frac{3.2}{\hat{\chi}_\parallel^e} + \frac{\beta\alpha}{\omega_{T_e} + \hat{\chi}_\parallel^e}\right), \tag{18}$$

with the growth rate given by

$$\gamma \approx -1.6\frac{\omega_s^2}{\hat{\chi}_\parallel^e}(\delta_e^{-1} - 1) - \frac{1}{2}\frac{\omega_s^2}{\omega_{T_e} + \hat{\chi}_\parallel^e}[\beta^2(\delta_e^{-1} - 1) + \beta], \tag{19}$$

where

$$(\omega_r^{(0)})^2 = k^2 c_s^2 \left(\frac{5}{3}\frac{T_{i0}}{T_{e0}} + \frac{n_{i0}}{n_{e0}}\right). \tag{20}$$

From Eq.(19), we see that for a growing wave, we should have

$$\omega_{T_e} < 0, \quad |\omega_{T_e}| < \hat{\chi}_\parallel^e \left[1 + 0.3\beta\left\{\beta + \frac{\delta_e}{1 - \delta_e}\right\}\right]. \tag{21}$$

For a purely condensation mode ($\omega = i\gamma$), with an ordering of $\omega_{T_e} \sim \omega_{n_e} \sim \hat{\chi}_\parallel^e \gg \gamma$, we have

$$\hat{k} < \frac{1}{\zeta}\left[\frac{\beta}{(5/3\delta_e + 1)} - \hat{k}_T\right], \tag{22}$$

where, $\hat{k} = k/k_{n_e}$, $\hat{k}_T = k_{T_e}/k_{n_e}$, and $\zeta = k_{n_e}\hat{\chi}_\parallel^e/(c_e n_{e0})$. We distinctly see the effect of dust on the cut-off wave number beyond which the system becomes stable to condensation mode.

In Fig.1, we have shown the growth rate of the condensation mode with a very negligible heat conduction. However, with only $\delta_e = 0.9$, the growth rate drops significantly. With more heat-conduction (Fig.2), we get the usual behavior

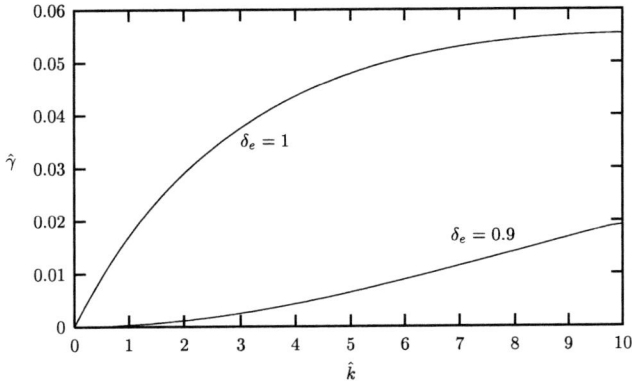

FIGURE 1. The normalized growth vs. normalized wave number, when the conductivity parameter $\zeta = 0.0001$.

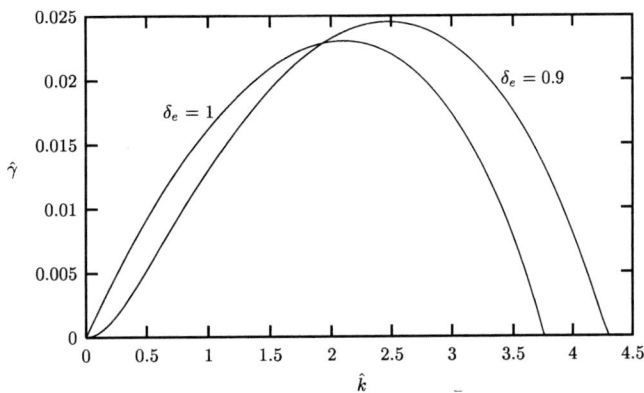

FIGURE 2. The normalized growth vs. normalized wave number, when the conductivity parameter $\zeta = 0.01$.

with a critical wave number beyond which the mode does not grow. However, note the lengthening of the cut-off point in the second case

To conclude, we have shown that the presence of negatively charged, massive dust particles can significantly alter the regime of stability of a radiating plasma. A more detailed analysis is needed which should also take into account the dynamics of the neutral particles.

REFERENCES

1. Oliveria, S., and Maciel, W. J., *Astro. Sp. Sc.*, **126**, 211 (1986).
2. Baldwin, J. A. et. al., *Ap. J.*, **374**, 580 (1991).
3. Jensen, R., *Nucl. Fusion*, **17**, 1187 (1977).
4. Natta, A., and Panagia, N., *Ap. J.*, **248**, 189 (1981).
5. Pottasch, S. R., in *Late Stages of Stellar Evolution*, edited by S. Kwok and S. R. Pottasch (D. Reidel, Dordrecht, 1987), pp. 355.
6. van Hoof, P. A. M., Van de Steene, G. C., Beintema, D. A., Martin, P. G., Pottash, S. R., and Ferland, G. J., *Ap. J.*, **532**, 384 (2000).
7. Barginskii, S. I., in *Revies of Plasma Physics*, edited by M. A. Leontovich (Consulyants Bureau, New York, 1965), Vol. 1, pp. 205.
8. Field, G. B., *Ap. J.*, **142**, 531 (1965).
9. Jana, M. R., Sen, A., and Kaw, P. K., *Phys. Rev. E*, **48**, 3930 (1993).
10. Gomez-Pelaez, A. J., and Moreno-Insertis, F., *Ap. J.*, **569**, 766 (2002).

Dynamics of Dust Particulates in Magnetic Field

O. Ishihara,[1] T. Kamimura,[2] K. Hirose,[2] and N. Sato[3]

[1]*Faculty of Engineering, Yokohama National University, Yokohama 240-8501 Japan*
[2]*National Institute for Fusion Science, Toki, 509-5292 Japan*
[3]*Graduate School of Engineering, Tohoku University, Sendai 980-8579 Japan*

Abstract. Dynamic behavior of a cluster of charged dust particulates in a plasma in the presence of magnetic field is studied. Dust particulates confined in a harmonic potential are found to move in a circular motion, driven by the external magnetic field, with harmonic oscillations around the equilibrium orbit. The collisional effects of ions and neutral particles on dust particulates are examined.

INTRODUCTION

As a complex plasma, dust particulates placed in a plasma show unique features not expected in an ordinary plasma. Recent study of laboratory complex plasmas reveals fundamental features of dust particulates in a plasma such as a variable charge Q on the order of 10^3 to 10^4 negative electric charges on the surface of a dust particulate of 1 to 10 μm in radius as well as a plasma crystal formation. A dust particulate with a large mass m_d with a large negative charge may be considered as a macro-electron. However, the specific charge of a dust particulate is quite different from the one for an electron, i.e.,

$$\frac{|Q|}{m_d} \sim 10^{-12}\left(\frac{e}{m_e}\right) \qquad (1)$$

for a typical spherical dust particulate of 1 μm in radius with 10^4 electric charge, where Q/m_d is the specific charge of a dust particulate and e/m_e is the specific charge of an electron. The charge neutrality condition

$$Z_i e n_i + Z_d e n_d - e n_e = 0 \qquad (2)$$

may cause a drastic change in collective nature known to exist in an ordinary plasma. Here, Z_i and Z_d are the charge state of an ion and a dust particulate, respectively, n_i, n_e, n_d are densities of ions, electrons, and dust particulates, respectively. Recent extensive study of a complex plasma involving dust particulates is summarized in some textbooks,[1] but limited in a plasma in the absence of magnetic field. Recent laboratory experiments show the rotational motion of a cloud of dust particulates in a confined plasma in the presence of magnetic field, while an individual dust particulate shows a spinning motion.[2,3,4] A dust particulate is reasonably expected, from the analogy of a macro-electron, to behave like an electron in the presence of magnetic field. However, the nature of a large mass makes a dust particulate difficult to be magnetized, resulting in non-existence of circular behavior by cyclotron motion alone. A close observation of a rotating dust trajectory revealed the superimposed harmonic oscillation on the orbital circular motion.[5] We review the recent theoretical attempts to understand the rotational motion of dust particulates in the presence of magnetic field, and clarify the role of coupling of harmonic oscillations and the orbital motion of a dust particulate. The roles of collisions of plasma particles are also discussed.

FORMULATION

We consider a motion of a charged body (a dust particulate) with mass m_d and a charge Q placed in a potential which is given by a combination of a harmonic potential ϕ_h and an electrostatic potential ϕ_e and a vector potential **A**. A cylindrical coordinate system (r, θ, z) is chosen to describe the motion of the dust particulate. The kinetic energy of the charged body in the absence of the magnetic field is given by $p^2/2m_d$, which should be modified as $|\mathbf{p} - Q\mathbf{A}/c|^2/2m_d$ in the presence of a vector potential. The kinetic energy is given by

$$\frac{(p_r - QA_r/c)^2}{2m_d} + \frac{(p_\theta - QrA_\theta/c)^2}{2m_d r^2} + \frac{(p_z - QA_z/c)^2}{2m_d}. \tag{3}$$

We now consider the vector potential which gives only a z-directional magnetic field. In other words, our canonical momenta are given by

$$p_r = m_d \dot{r} \tag{4}$$

$$p_\theta = m_d r \left(r\dot\theta + \frac{QA_\theta}{m_d c} \right) \tag{5}$$

$$p_z = m_d \dot{z}. \tag{6}$$

The dust particulate is situated in the influence of potentials and collisions from plasma particles. The potentials include the harmonic potential and the electrostatic Coulomb potential resulted from neighboring charges. The harmonic potential will confine the dust particulate toward the center of the system. The Coulomb potential prohibits neighboring charges to come too close to the charge. The particle simulation carried out recently revealed that the collisional effect due to ions and neutrals in a plasma would slow down the seemingly random motion of charges and settle the charged bodies near the equilibrium circular position.[6] The equilibrium condition for a Coulomb cluster making a circle together with a dynamic motion of such a cluster was studied recently.[7] The equilibrium condition determines the radius of the circular cluster r_{eq} as a function of the charge $Q!$, a constant K and a number N, or

$$r_{eq} = r_{eq}(Q, K, N), \tag{7}$$

where K is a constant to define the strength of the harmonic potential and N is the number of the charged particulates in the circular orbit. The equilibrium radius becomes larger when the number of the charges in the circle increases due to the Coulomb repulsion. The dynamic motion of the dust particulate can be described by

$$m_d \frac{dv_r}{dt} = -Q\frac{\partial \phi}{\partial r} - K_1 r + m_d(\omega + \omega_c)v_\theta + F_r, \tag{8}$$

$$m_d \frac{dv_\theta}{dt} = -Q\frac{1}{r}\frac{\partial \phi}{\partial \theta} - m_d(\omega + \omega_c)v_r + F_\theta \tag{9}$$

$$m_d \frac{dv_z}{dt} = -Q\frac{\partial \phi}{\partial z} - K_2 z - m_d g + F_z, \tag{10}$$

where ω is the angular frequency defined by $d\theta/dt$, ω_c is the cyclotron frequency defined by QB/mc, K_1 and K_2 are constants to define the strength of the radial and vertical harmonic motion, ϕ is the Coulomb potential, mg is the gravitational force, and F is the force resulting from the collisions from plasma particles. The three equations describe the dynamic motion of the dust particulates with three different time scales, i.e., ω, ω_c and ω_h (the frequency associated with the harmonic oscillation). The harmonic oscillation in the (r, θ) plane is rather fast in time and its oscillation is proportional to the square root of the ratio of the harmonic constant and the mass, or

$$\omega_h \sim \sqrt{\frac{K_1}{m_d}} \tag{11}$$

The circular cluster will rotate along the circle as a result of the coupling of rapid harmonic motion with the slow rotational motion. We neglect for the moment the collisional effect for the computational purpose and the effects will be discussed later. The coupled equations (8) and (9) with $\mathbf{F} = 0$ were solved as an initial value problem and the trajectory of a dust particulate in the (r, θ) plane is shown in Fig. 1, in which a rotational period is about $50/\sqrt{3K_1/m_d}$ corresponding to a second for $m_d = 2\times 10^{-16}$ kg and $K_1 = 10$ keV/cm². The rotational motion is accompanied by the oscillatory harmonic nature as shown in the Figure. Such a unique feature has been observed in the laboratory experiment.[5]

EFFECT OF COLLISIONS

Now we consider the effect of collisions on the dynamic motion of the dust particulate. The collisional force may be expressed as

$$\mathbf{F} = -m_d [\nu_i (\mathbf{v} - \mathbf{v}_i) + \nu_n (\mathbf{v} - \mathbf{v}_n)], \quad (12)$$

where ν_i is the ion collision frequency, ν_n is the neutral collision frequency, and $\mathbf{v}_i, \mathbf{v}_n$ are ion and neutral velocity, respectively. The ion velocity is determined by the equation of motion in which the Lorenz force drives ions in the azimuthal direction. Equation (12), together with the equation of motion for ions, suggests that ions are responsible for driving the dust particulate in the azimuthal direction, while neutrals are responsible for relaxing the dynamic motion. In a laboratory plasma, the neutral flow velocity is controlled by the gas pumping. As was observed in particle simulations,[6] the dynamic movement of dust particulates will come to an end rather quickly if the Epstein neutral drag is sufficiently large. If the ion friction dominates the process, the azimuthal velocity of a dust particulate is determined by the azimuthal velocity of magnetized ions as was studied by Kaw et al[8] and Shukla.[9] The nature of the azimuthal velocity depends on the ion-neutral collision frequency compared with the ion cyclotron frequency. However, the observed rotational movement with harmonic oscillations[5] may only be explained with the dynamic model expressed by Eqs. (8) and (9). The ion and neutral drag forces together with other forces acting on a dust particulate in a typical laboratory conditions have been estimated in a typical laboratory parameter range as[10]

$$F_{ion} = 6.5 \times 10^{-8} \text{ dyne}$$
$$F_{neutral} = 2.8 \times 10^{-15} \sim 10^{-9} \text{ dyne}$$
$$F_{Coulomb} = 7 \times 10^{-9} \text{ dyne}$$

Here a dust particulate of 2μm in radius with Q=-4000e is placed in a helium plasma of ion density $n_i = 10^9 cm^{-3}$, neutral density $n_n = 3.3 \times 10^{15} cm^{-3}$ and $T_e = 1eV > T_i$. The ion flow speed of 1.5 times of the ion acoustic speed and the neutral speed of 1 to 10^3 cm/s are assumed. The Coulomb force is estimated at a distance of Debye length. The neutral collision time and ion collision time associated with dust particulates may be evaluated by $\tau_n = 3m/4\pi a^2 m_n n_n \langle v_n \rangle$ and $\tau_{ion} = m/\sigma_i m_i n_i \langle v_i \rangle$, where a is a radius of the dust particulate, σ_i is the cross section of ions, $m_n, n_n, \langle v_n \rangle$ and $m_i, n_i, \langle v_i \rangle$ are mass, number density, and average velocity of neutrals and ions, respectively. They are conveniently expressed as

$$\tau_n = 600 \left(\frac{a}{1\mu m}\right)\left(\frac{\rho_d}{10^3 kg \cdot m^{-3}}\right)\left(\frac{10^{21} m^{-3}}{n_n}\right)\left(\frac{10 m \cdot s^{-1}}{\langle v_n \rangle}\right)\left(\frac{m_p}{m_n}\right) \sec \quad (13)$$

$$\tau_i = 500 \left(\frac{a}{1\mu m}\right)^3\left(\frac{\rho_d}{10^3 kg \cdot m^{-3}}\right)\left(\frac{10^{15} m^{-3}}{n_i}\right)\left(\frac{500 m \cdot s^{-1}}{\langle v_i \rangle}\right)\left(\frac{10^{-8} m^2}{\sigma_i}\right)\left(\frac{m_p}{m_n}\right) \sec, \quad (14)$$

where ρ_d is the mass density of a dust particulate and m_p is the proton mass. As shown in Equations (13) and (14), the time scale may meet the observed dynamics depending on the parameters and the collisional effect could play a role in the dynamics of dust particulates.

CONCLUSIONS

Dynamics of dust particulates placed in a magnetic field is studied by a model equation in which confined potential, Coulomb potential and the magnetic field are considered. A cluster of dust particulates is formed in a ring structure and rotates in a circle with oscillatory behavior in the magnetic field as a result of coupling between harmonic oscillation and the circular motion.

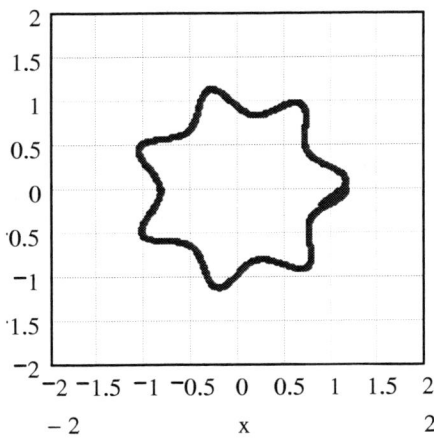

FIGURE. 1. Trajectory of a dust particulate. Initial location is at $r = 1.16r_0$ (r_0 is the equilibrium radius), $\theta = 0$ with the initial angular frequency of $\omega = -0.1$. The cyclotron frequency is set $\omega_c = -0.01$. The frequency is in the unit of $(3K_1/m_d)^{1/2}$.

ACKNOWLEDGMENTS

This work was supported by the Grants-in-Aid for Scientific Research (C) of the Japan Society for the Promotion of Science.

REFERENCES

1. Verheest, F., *Waves in Dusty Plasmas* (Kluwer Academic, Dordrecht, 2000) ; Shukla, P.K. and Mamun, A.A.: *Introduction to Dusty Plasma Physics* (Institute of Physics, Bristol, UK, 2002); Kikuchi, H., *Electrohydrodynamics in Dusty and Dirty Plasmas* (Kluwer Academic, Dordrecht, 2001).
2. Sato, N., Uchida, G., Ozaki, R., Iizuka, S., and Kamimura T., in *Frontiers in Dusty Plasmas,* ed. by Nakamura Y., Yokota, T., and Shukla, P.K. (Elsevier, Amsterdam, 2000), pp. 329-336.,
3. Konopka, U., Samsonov, D., Ivlev, A.V., Goree, J., Steinberg, V., and Morfill, G.E., Phys. Rev. E **61**, 1890 (2000).
4. Ishihara, O., and Sato, O., IEEE Trans. Plasma Sci.. **29**, 179 (2001)
5. Sato, N., Uchida, G., Kaneko, T., Shimizu, S., and Iizuka, S., Phys. Plasmas **8**, 1786 (2001).
6. Kamimura, T., Hirose. K., Uchida, G., Iizuka, S., and Sato, N. in *Potential and Structure in Plasmas*, J. Plasma and Fusion Research Series **4**, 480-482 (2001).
7. Ishihara, O., Kamimura, T., Hirose, K I., and Sato, N. Phys. Rev. E **66**, 046406-1 (2002).
8. Kaw, P. K., Nishikawa, K., and Sato, N., Phys. Plasmas **9**, 387 (2002).
9. Shukla, P. K., Phys. Lett. A **299**, 258 (2002).
10. Ishihara, O. Phys. Plasmas **5**, 357 (1998).

Algebraic Improvement For The Numerical Treatment Of The Ion Acoustic Modes In Ion Beam Dusty-Plasma Considering Dust Charging Effects

J. Puerta*, J. Silva* and C. Cereceda*

Departamento de Fisica, Universidad Simón Bolivar, Apdo. 89000, Caracas, Venezuela.

Abstract. Dust charging effects to analyze the ion-beam plasma system is of vital importance due to the fact that the dust particles could accept very high charge states. In this paper we propose a suitable dispersion relation for a one dimensional dust ion-beam plasma system to calculate the growing rates and phase velocities of ion acoustic waves propagating in the system. We use an hybrid model to describe electrons as a fluid while ions are described by kinetic theory. The charge state of dust particles has a strong effect on the threshold on beam velocity of the onset of the instability. We apply also an algebraic method described in previous works to simplify the calculations.

INTRODUCTION

Nowadays there has been a positive growing of interest in the physics of dusty plasma, which means solid particles with diameters between 1 to 100 μm. Under usual circumstances the dust particulates could achieve high states of negative charge, 10^3 - 10^4 electron charges. The dusty plasma system can support a vast variety of physical phenomena, in our case we are interested to calculate numerically the ion-beam instability using a simple hybrid model, that means that we treat the electrons as a fluid and the ions kinetically in order to find a suitable dispersion relation.

MODEL

Weak ion acoustic perturbation of a dusty plasma in the presence of an ion beam with the electrons as neutralizing background can be treated in the following description of our model [1].

For the electrons:

$$m_e n_e (\frac{\partial \vec{v}_e}{\partial t} + (\vec{v}_e . \nabla) \vec{v}_e) = -e n_e \vec{E} \tag{1}$$

$$\frac{\partial n_e}{\partial t} + \nabla . (n_e \vec{v}_e) = \frac{dn_e}{dt}|_{charging} \tag{2}$$

where the charging process is defined by [3]

$$\frac{dn_e}{dt}|_{charging} = -\frac{1}{q_e}(\frac{d\rho_{de}}{dt}) \; ; \; \rho_{de} = Q_{de} n_e \tag{3}$$

with ρ_{de} is the charge density due to the electron charging current.

For dust:

$$\frac{\partial n_d}{\partial t} + \nabla . (n_d \vec{v}_d) = 0 \tag{4}$$

$$m_d n_d (\frac{\partial \vec{v}_d}{\partial t} + (\vec{v}_d . \nabla) \vec{v}_d) = -Z_d e n_d \vec{E} \tag{5}$$

Ambipolar field (without collisions):

$$e(\vec{E}_i + \vec{E}_i) = \frac{\nabla(n_i T_i)}{n_i} + \frac{\nabla(n_b T_b)}{n_b} \tag{6}$$

Quasineutrality:

$$n_e = n_i - Z_d n_d + n_b \tag{7}$$

Dust charging equation [2, 3]:

$$e\frac{\partial Z_d}{\partial t} = I_e + I_d + I_b \tag{8}$$

From (1) - (8) we obtain the perturbation relations:

$$eE^1 = \frac{ik(n_e^1 + Z_d^1 n_d^0 + Z_d^0 n_d^1 - n_b^1)T_i^0}{n_e^0 + Z_d^0 n_d^0 - n_b^0} + \frac{ikn_b^1 T_b^0}{n_b^0} \tag{9}$$

$$n_e^1 = \frac{en_e^0(ik + n_e^0 \pi a^2)E^1}{\omega^2 m_e(Z_d^0 - 1)} \tag{10}$$

$$n_d^1 = \frac{kn_d^0 Z_d^0 E^1}{i\omega^2 m_d} \tag{11}$$

Now from equation (8) we get

$$ei\omega Z_d^1 = I_e^1 + I_i^1 + I_b^1 \tag{12}$$

with

$$I_e^0 + I_i^0 + I_b^0 = 0 \tag{13}$$

From equation (13) we can obtain Z_d^0. The basic equations for the ions and for the beam are given by the linearized Vlasov equation

$$\frac{\partial f_\alpha^1}{\partial t} + \vec{v} \cdot \frac{\partial f_\alpha^1}{\partial \vec{r}} + \frac{\vec{F}}{m} \cdot \frac{\partial f_\alpha^0}{\partial \vec{v}} = 0 \tag{14}$$

with α = ions, ion-beam.

DERIVATION OF THE DISPERSION RELATION

Now, using the equations (8) - (12), we obtain an expression for $f_\alpha^1(\vec{v})$. Introducing these results into equation (14) we obtain the dispersion relation. This can be transformed using the multipolar technique for the Z dispersion function in order to write down this dispersion relation in a polynomial form $\frac{\sum_i A_i \Omega_i}{\sum_j B_j \Omega_j} = 0$, where $\sum_j B_j \Omega_j \neq 0$ and $\Omega = \frac{\omega}{\omega_{pi}}$, and $K = \frac{k}{k_{De}}$. The A_i coefficients depends on the following dimensionless parameters:

$$\zeta_i = \left(\frac{\Omega}{k}\right)\sqrt{\frac{T_e}{2T_i}} \quad ; \quad \zeta_p = \left(\frac{\Omega}{K} - U_b\right)\sqrt{\frac{T_e}{2T_i}} \quad ; \quad k_{De} = \sqrt{\frac{n_e^0 e^2}{\varepsilon_0 T_e}} \quad ;$$

$$\Gamma = \frac{e^2 Z_d^0}{4\pi \varepsilon_0 a T_e} \quad ; \quad f_1(u_0) = \left(\frac{\sqrt{\pi}}{4 u_0}\right)\left(1 + 2u_0^2\right) erf(u_0) \quad ; \quad f_2(u_0) = \left(\frac{\sqrt{\pi}}{2 u_0}\right) erf(u_0)$$

$$M_d = \frac{m_d}{m_i} \quad ; \quad M_i = \frac{m_i}{m_e} \quad ; \quad \theta_b = \frac{T_i^0}{T_b^0} \quad ; \quad \theta_i = \frac{T_e^0}{T_i^0} \ .$$

where Γ is the coupling parameter. Solving the numerator we get the roots, and a careful analysis of these roots gives us the real and imaginary part of Ω.

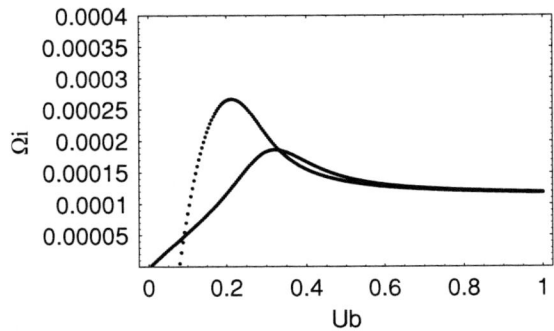

FIGURE 1. Ω_i vs. U_b, $a = 0.001 \mu m$.

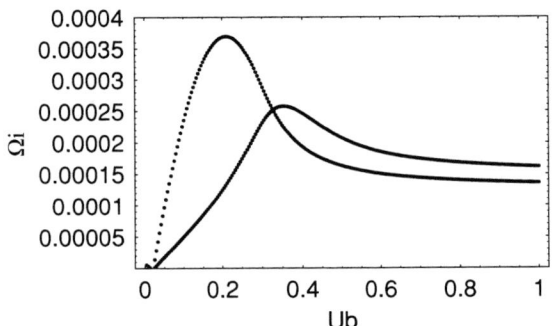

FIGURE 2. Ω_i vs. U_b, $a = 0.1 \mu m$.

CONCLUSION

Figures 1 and 2 show the growth rate ($\Omega_i = Im(\Omega)$) versus the ion streaming velocity U_b. In all figures the dimensionless wave number used is $K = 0.01$, the beam temperature to ion temperature ratio is $\theta_b = 10$. The equilibrium value of dust charge is calculated to be $Z_d^0 = 4 \times 10^4$. In Figs. 1 and 2 the electron temperature to ion temperature ratio is set equal to 10.

We observe two unstable modes for beam velocities in the range between 0 to 1, in good agreement with the results found by the numerical simulation by Gyoo-Soo Chae et al. [3]. It is believed that this ion acoustic instability is due to dust charging fluctuations. For smaller grain radius the maximum instability goes lower, this may be explained by the decreasing of the current fluctuations with the reduction of the cross section for smaller grain radius (see Figs. 1 and 2). This hybrid method appears to be a well suited technique to analyze such complex systems.

REFERENCES

1. J. Silva, J. Puerta and C. Cereceda, "New numerical model to analyse ion acoustic modes in a low collisional ion-beam dusty plasma." *Proceedings of the ICPP 2000*, Quebec, 2000.
2. K. Ostrikov, M. Y. Yu, L. Stenflo, *IEEE Transactions on Plasma Science*, **29**, 175 (2001).
3. G. Chae, W. A. Scales, G. Ganguli, P. A. Bernhardt and M. Lampe, *EE Transactions on Plasma*, **29**, 186 (2001).

Theory of Dust Ion-acoustic Solitary and Shock Waves

P.K.Shukla* and A.A.Mamun*†

*Fakultät für Physik und Astronomie, Ruhr-Universität Bochum, D-44780 Bochum, Germany
†Department of Physics, Jahangirnagar University, Savar, Dhaka, Bangladesh

Abstract. We present theories for dust ion-acoustic (DIA) solitary and shock waves in an unmagnetized dusty plasma including the effects of dust charge fluctuations. We derive the Kortweg-de Vries equation as well as the Kortweg-de Vries-Burgers equation by employing the reductive perturbation method. It is shown that the dust charge fluctuation does not only modify the properties of DIA solitary waves, but it also introduces a dissipation which is responsible for DIA shock waves.

INTRODUCTION

Shukla and Silin [1] have theoretically shown that due to the conservation of the equilibrium charge density, $n_{e0}e + n_{d0}Z_{d0}e - n_{i0}e = 0$, and the strong inequality $n_{e0} \ll n_{i0}$ [where n_{s0} is the particle number density of the species s with $s = e$ (i) d for the electrons (ions) and dust, Z_{d0} is the number of the electrons residing onto the dust grain surface, and e is the magnitude of the electron charge] a dusty plasma (with negatively charged static dust grains) supports low-frequency (in comparison with the ion plasma frequency) dust ion-acoustic (DIA) waves whose phase speed is much smaller (larger) than the electron (ion) thermal speed.

The DIA wave spectrum is similar to the usual ion-acoustic wave spectrum for a plasma with $n_{i0} = n_{e0}$ and $T_i \ll T_e$, where T_i (T_e) is the ion (electron) temperature. However, in dusty plasmas we usually have $n_{i0} \gg n_{e0}$ and $T_i \simeq T_e$. Obviously, a dusty plasma cannot support the usual ion-acoustic waves, but it does support the DIA waves [1]due to reduced electron Landau damping. The DIA waves have been observed in laboratory experiments [2, 3]. The linear properties of the DIA waves in dusty plasmas are now fully understood from both theoretical and experimental points of view [1, 2, 3].

The DIA solitary and shock waves have been investigated theoretically by several authors [4, 5, 6, 7, 8]. However, all these investigations [4, 5, 6, 7, 8] are limited to constant dust grain charge which may not be a realistic situation in space and laboratory devices, since in space as well in laboratory devices the charge on a dust grain is not constant but varies with space and time [9, 10, 11]. In the present paper, we consider an unmagnetized dusty plasma model consisting of the ion fluid, Boltzmann electrons, and stationary charge fluctuating dust grains, and analyze the nonlinear propagation of the DIA waves.

GOVERNING EQUATIONS

The nonlinear dynamics of one-dimensional DIA waves, whose phase speed is much smaller (larger) than the electron (ion) thermal speed, is governed by

$$\frac{\partial n_i}{\partial t} + \frac{\partial}{\partial x}(n_i u_i) = 0, \qquad (1)$$

$$\frac{\partial u_i}{\partial t} + u_i \frac{\partial u_i}{\partial x} = -\frac{\partial \phi}{\partial x}, \qquad (2)$$

$$\frac{\partial^2 \phi}{\partial x^2} = \mu \exp(\phi) - n_i + (1-\mu)Z_d, \qquad (3)$$

where n_i is the ion number density normalized by its equilibrium value n_{i0}, u_i is the ion fluid velocity normalized by $C_i = (k_B T_e/m_i)^{1/2}$ (k_B is the Boltzmann constant and m_i is the ion mass), ϕ is the electrostatic wave potential normalized by $k_B T_e/e$, and Z_d is the number of electrons residing onto the dust grain surface normalized by its equilibrium value Z_{d0}. The time and space variables are in units of the ion plasma period $\omega_{pi}^{-1} = (m_i/4\pi n_{i0} e^2)^{1/2}$ and the Debye radius $\lambda_{Dm} = (k_B T_e/4\pi n_{i0} e^2)^{1/2}$, respectively. We have denoted $\mu = n_{e0}/n_{i0}$. We note that Z_d is not constant but varies with space and time. Thus, Eq. (3) is completed by the normalized dust grain charging equation

$$\eta \frac{\partial Z_d}{\partial t} = \mu \beta \exp(\phi - \alpha Z_d) - \beta_i n_i u_i \left(1 + \frac{2\alpha Z_d}{u_i^2}\right), \tag{4}$$

where $\eta = \sqrt{\alpha m_e(1-\mu)/2m_i}$, $\beta = (r_d/a)^{3/2}$, $\beta_i = \beta \sqrt{\pi m_e/8m_i}$, $\alpha = Z_{d0} e^2/k_B T_e r_d$, and $a = n_{d0}^{-1/3}$. We note that at equilibrium $\mu \beta \exp(-\alpha) = \beta_i u_0(1 + 2\alpha/u_0^2)$, where u_0 is the ion streaming speed normalized by C_i.

DIA SOLITARY WAVES

To study small but finite amplitude DIA solitary waves, we introduce the stretched coordinates $\xi = \varepsilon^{1/2}(x - v_0 t)$, $\tau = \varepsilon^{3/2} t$, and $\eta = \varepsilon \eta_0$, where ε is the expansion parameter. Expanding n_i, u_i, ϕ, and Z_d in a power series of ε, and developing equations in various powers of ε we finally obtain a Kortweg-de Vries (K-dV) equation

$$\frac{\partial \phi^{(1)}}{\partial \tau} + A \phi^{(1)} \frac{\partial \phi^{(1)}}{\partial \xi} + B \frac{\partial^3 \phi^{(1)}}{\partial \xi^3} = 0, \tag{5}$$

where the coefficients A and B are given in [12].

Now, transforming the independent variables ξ to $\zeta = \xi - U_0 \tau$, where U_0 is a constant speed normalized by C_i, and imposing the appropriate boundary conditions for localized perturbations, the stationary solitary wave solution of Eq. (5) is

$$\phi^{(1)} = \psi \operatorname{sech}^2\left[(\xi - U_0 \tau)/\Delta\right], \tag{6}$$

where $\psi = 3U_0/A$ and $\Delta = \sqrt{4B/U_0}$ represent the amplitude and the width of the solitary waves, respectively. It is obvious from Eq. (6) that there exists compressive (rarefactive) solitary waves if $A > 0$ ($A < 0$). We have numerically analyzed A for the parameters corresponding to space dusty plasma situations (e.g. Saturn's E-ring [9, 10, 11]: $n_{d0} = 10^{-7}$ cm^{-3}, $r_d = 1$ μm, $k_B T_e = 50$ eV, and $Z_d = 10^3$ as well as for the parameters corresponding to laboratory dusty plasma devices [2, 3] ($n_{d0} = 10^5$ cm^{-3}, $r_d = 5$ μm, $k_B T_e = 0.2$ eV, and $Z_d = 10^3$ and found that A is always positive. This means that our present dusty plasma model can support only compressive solitary waves (solitary waves with $\phi > 0$).

DIA SHOCK WAVES

We have seen in the preceding section that the parameter η does not play any role in our analysis of the DIA solitary waves. This is because of the scaling that we have used. We now consider a situation in which we can scale the parameter η as $\eta = \varepsilon^{1/2} \eta_0$. Using this additional scaling and retaining stretching/scaling of other variables as before, we obtain a K-dV-Burgers equation

$$\frac{\partial \phi^{(1)}}{\partial \tau} + A \phi^{(1)} \frac{\partial \phi^{(1)}}{\partial \xi} + B \frac{\partial^3 \phi^{(1)}}{\partial \xi^3} = C \frac{\partial^2 \phi^{(1)}}{\partial \xi^2}, \tag{7}$$

where the coefficient C is given in [12].

An exact analytical solution of Eq. (7) is not possible. However, we can deduce some approximate analytical shock solutions. The nature of these shock structures depends on the relative values between the dispersive and dissipative coefficients B and C. If the value of C is very small, the energy of the particle decreases very slowly and the first few oscillations at the wave front will be close to solitary waves. However, if the value of C is larger than a critical value, the motion of the particle will be aperiodic, and we obtain a shock wave with a monotonic structure. A monotonic shock wave solution is

$$\phi^{(1)} \simeq \psi_{sh} - \psi_{sh} \tanh\left[(\xi - U_0 \tau)/\Delta_{sh}\right], \tag{8}$$

where $\psi_{sh} = U_0/A$ and $\Delta_{sh} = 2C/U_0$ represent the amplitude and the width of the shock wave, respectively.

The DIA shock waves were experimentally excited in a dusty double plasma (DP) device by Nakamura *et al.* [13]. The experimental results of Nakamura *et al.* [13] reveal that the oscillatory wave structure behind the shock becomes less in number with increasing dust particle number density and finally disappears completely at a sufficiently high dust particle number density, leaving only the laminar shock front. The effect of the dust particle number density on the ion acoustic compressional pulses has also been experimentally studied by Luo *et al.* [14] who observed a steepening of the ion-acoustic pulses. Our results are capable of explaining the observations [13, 14].

DISCUSSION

We have studied the properties of DIA solitary and shock waves in an unmagnetized dusty plasma including the dust charge fluctuation dynamics. Our results show that the effects of the dust grain charge fluctuations modify the properties of the DIA solitary waves. It has been found that the effects of dust grain charge fluctuations reduce the speed of compressive DIA solitary waves. The characteristics of these DIA solitary waves in the space dusty plasma condition are found to be different from those in laboratory dusty plasma condition. Furthermore, we have shown that dust grain charge fluctuations are responsible for the formation of DIA shock waves in both space and laboratory dusty plasmas. The shock wave amplitude is one third of the solitary wave amplitude.

REFERENCES

1. Shukla, P. K. and Silin, V. P., *Physica Scripta* **45**, 508 (1992).
2. Barkan, A., D'Angelo, N. and Merlino, R. L., *Planet. Space Sci.* **44**, 239–242 (1996).
3. Merlino, R. L., Barkan, A., Thompson, C. and D'Angelo, N., *Physics of Plasmas* **5**, 1607-1614 (1998).
4. Bharuthram, R. and Shukla, P. K., *Planet. Space Sci.* **40**, 973–977 (1992).
5. Shukla, P. K., *Phys. Plasmas* **7**, 1044–1046 (2000).
6. Nakamura, Y. and Sharma, A., *Phys. Plasmas* **8**, 3921–3926 (2001).
7. Popel, S. I. and Yu, M. Y., *Contrib. Plasma Phys.* **35**, 103–108 (1995).
8. Mamun, A. A. and Shukla, P. K., *Phys. Plasmas* **9**, 1468–1470 (2002).
9. Mendis, D. A. and Rosenberg, M., *Anu. Rev. Astron. Astrophys.* **32**, 418–463 (1994).
10. Verheest, F., *Waves in Dusty Space Plasmas* Kluwer Academic Publishers, Dordrecht, 2000.
11. Shukla, P. K. and Mamun, A. A., *Introduction to Dusty Plasma Physics* Institute of Physics Publishing Ltd., Bristol, 2002.
12. Mamun, A. A. and Shukla, P. K., *IEEE Trans. Plasma Sci.* **30**, in press (2002).
13. Nakamura, Y., Bailung, H. and Shukla, P. K., *Phys. Rev. Lett.* **83**, 1602–1605 (1999).
14. Luo, Q. Z., D'Angelo, N. and Merlino, R. L., *Phys. Plasmas* **6**, 3455–3458 (1999).

Finite Amplitude Double-Layer Solutions In Two-Component-Charged Dusty Plasmas With Trace Electrons And Ions

I. Spassovska[1], P.H. Sakanaka[1], P.K. Shukla[2]

[1]Instituto de Física "Gleb Wataghin", Universidade Estadual de Campinas, Campinas, SP, Brazil.
[2]Institute für Theoretische Physik IV, Ruhr-Universität Bochum, Bochum, Germany

Abstract. Finite amplitude double-layer solution in unmagnetized two-component dusty plasma is shown to exist when a very small amount of electrons and even smaller amount of light ion component are present. Assuming that the constituents of dusty plasmas are warm electrons, warm positive light ions, and an admixture of negatively and positively charged cold dust grains which are simultaneously present, it is shown that stationary solutions of the fluid equations combined with Poisson's equation can be expressed in terms of the energy integral of a classical particle with a modified Sagdeev potential. Conditions under which double-layers arise are given. In particular, we have applied the theory in the laboratory plasma and we can predict that a double-layer might be possible to be launched if a trace ions component is added.

INTRODUCTION

Since the discovery of the dust acoustic wave (DAW) [1], there been a great interest in investigating numerous collective processes in dusty plasmas. In their paper, Rao, Shukla, and Yu [1], discovered the dust-acoustic wave (DAW), and introduced a theory for dust-acoustic solitons in a three-component dusty plasma with negatively charged dust grains. Recently, it has been suggested that positively and negatively charged dust grains can co-exist in space [2]-[4] and laboratory [5] plasmas. Therefore, it is desirable to investigate the linear and nonlinear properties of dust-acoustic waves in four-component plasma that consists electrons, ions and positively and negatively charged dust grains.

Here we present the governing equations for the DAW when both the negative and positive dust components are simultaneously present (see [6]). We discuss the properties of the DAW in the presence of positive and negative dust components, and define parameters that are relevant for the analysis of the nonlinear DAW. Stationary solutions of the governing nonlinear equations for arbitrary large amplitudes are discussed. Here, we derive the energy integral with a modified Sagdeev potential. The latter is analyzed both analytically and numerically to obtain the parameter regimes where DA double-layers are possible. It turns out that the presence of a positive dust component in a multi-component dusty plasma gives rise to such novel features of the nonlinear structures as the compressional DA potential distribution and the monotonic double-layers, which otherwise are absent. Finally, a possible application of our investigation in laboratory plasmas is given.

GOVERNING EQUATIONS

We consider an unmagnetized dusty plasma consisting of the electrons, the ions, negatively and positively charged massive dust particles, with similar masses.
The quasi-neutrality at equilibrium is written

$$N_{e0} + Z_n N_{n0} = N_{i0} + Z_p N_{p0}, \qquad (1)$$

where, N_{e0} and N_{i0} are the average electron and average ion number densities, Z_n and Z_p are the negative and positive dust particle charge, N_{n0} and N_{p0} are the average dust particles number density, respectively.

The dust particles are assumed to be point charges and their sizes are much smaller than the effective Debye length. For low phase velocity (compared to the electron and ion thermal velocities) dust-acoustic waves, both the electrons and ions can be considered inertialess fluid and their number densities can be given by the Boltzmann distribution, respectively,

$$N_e = N_{e0} e^{e\Phi/T_e} \quad \text{and} \quad N_i = N_{i0} e^{-e\Phi/T_i}, \tag{2}$$

where, Φ is the electrostatic potential and e is the magnitude of the electron charge.

The dynamics of charged dust grains are governed by the equations of the continuity and the momentum, which are, respectively,

$$\frac{\partial N_p}{\partial t} + \frac{\partial (N_p V_p)}{\partial x} = 0 \quad \text{and} \quad \frac{\partial N_n}{\partial t} + \frac{\partial (N_n V_n)}{\partial x} = 0 \tag{3}$$

and

$$\frac{\partial V_p}{\partial t} + V_p \frac{\partial V_p}{\partial x} = -\frac{Z_p e}{M_p} \frac{\partial \Phi}{\partial x} \quad \text{and} \quad \frac{\partial V_n}{\partial t} + V_n \frac{\partial V_n}{\partial x} = +\frac{Z_n e}{M_n} \frac{\partial \Phi}{\partial x} \tag{4}$$

for positively and negatively charged dust grain. Here V_p, V_n, M_p, M_n are the fluid velocities and mass of the positively and negatively charged dust grains, respectively. We are assuming cold dust particles, so no pressure term is present. The system of equations is closed with the Poisson's equation

$$\frac{\partial^2 \Phi}{\partial x^2} = 4\pi e \left(N_e - N_i + Z_n N_n - Z_p N_p \right) \tag{5}$$

FINITE AMPLITUDE NONLINEAR DUST ACOUSTIC WAVES

We can get a plane wave solution for the set of linearized equations of (1) to (5) for small amplitude disturbances with angular frequency ω and wave number k, with the dispersion relation:

$$\frac{\omega}{k} = \frac{C_{da}}{\sqrt{1 + \lambda_{Dd}^2 k^2}}, \quad \text{where} \quad C_{da}^2 = \frac{T_0}{M_0} \quad \text{and} \quad \lambda_{Da} = \frac{C_{da}}{\omega_{pd}}. \tag{6}$$

We have introduced the symbols ω_{pd}, N_0, T_0 and M_0 as

$$\omega_{pd}^2 = \frac{4\pi e^2 N_0}{M_0}, \quad N_0 = N_{e0} + N_{i0}, \quad \frac{N_0}{T_0} = \frac{N_{e0}}{T_e} + \frac{N_{i0}}{T_i}, \quad \frac{N_0}{M_0} = \frac{Z_n^2 N_{n0}}{M_n} + \frac{Z_p^2 N_{p0}}{M_p}. \tag{7}$$

Here, C_{da} is the dust-acoustic velocity, λ_{Dd} the effective Debye length, ω_{pd} the dust plasma frequency, N_0, M_0, and T_0 are the effective number density, the mass and the temperature, respectively.

With the purpose of understanding the parametric space, which limits the existence of double-layers, we are normalizing all the parameters. The natural quantities for the normalization are T_0, N_0 and M_0, the effective temperature (in unit of energy), the plasma particle number density and the mass, respectively. From these we get the normalizing quantities for the time, $t \to t\omega_{pd}$, the space, $x \to x/\lambda_{Da}$, and the mass, $M_j \to M_j/M_0$. The normalized velocities are expressed as the Mach number, $M = V_0/C_{da}$.

We define, then

$$\phi = \frac{e\Phi}{T_0} \quad \text{and} \quad u(\phi) = \frac{U(\Phi)}{4\pi N_0 T_0}, \tag{8}$$

where $U(\Phi)$ is the potential energy density, and which will appear in the later context,

$$n_{e0} = \frac{N_{e0}}{N_0}, \quad n_{i0} = \frac{N_{i0}}{N_0}, \quad a_e = \frac{T_0}{T_e}, \quad a_i = \frac{T_0}{T_i} \tag{9}$$

for the electron and ion number densities and the temperatures,

$$n_n = \frac{Z_n N_{n0}}{N_0}, \quad n_p = \frac{Z_p N_{p0}}{N_0}, \quad a_n = \frac{Z_n T_0}{M_n V_0^2}, \quad a_p = \frac{Z_p T_0}{M_p V_0^2} \tag{10}$$

for the dust particle number density.

From equation (1), (7) and (8)-(10) for the nonlinear dust acoustic wave parameters we have

$$n_{e0} + n_n = n_{i0} + n_p \tag{11}$$

$$n_{e0} + n_{i0} = 1 \tag{12}$$

$$n_{e0}a_e + n_{i0}a_i = 1 \tag{13}$$

$$n_n a_n + n_p a_p = \frac{1}{M^2} \tag{14}$$

These last four equations are very important because they show the relation between the four component dusty plasma parameters. We will use them for our calculations to define the regions of existence of the double-layer.

For weakly nonlinear perturbations we can use a higher order perturbation method and get a soliton solution or a double-layer. Taking up to the second order perturbation and using the reductive perturbation technique we get Korteweg - De Vries (KdV) equation with soliton solutions. For the third order perturbation we get the modified KdV equation with a double-layer solution (see [6]).

Our main thrust for this work is to find the parametric limits for the existence of double-layers using equations (7) and (8) of reference [6], which we reproduce here

$$\frac{1}{2}\left(\frac{\partial \phi}{\partial \zeta}\right)^2 + u(\phi) = 0, \tag{15}$$

where the modified Sagdeev potential [6] for our purposes is

$$u(\phi) = -\left\{\frac{n_{e0}}{a_e}\left(e^{a_e\phi} - 1\right) + \frac{n_{i0}}{a_i}\left(e^{-a_i\phi} - 1\right) + \frac{n_n}{a_n}\left(\sqrt{1 + 2a_n\phi} - 1\right) + \frac{n_p}{a_p}\left(\sqrt{1 - 2a_p\phi} - 1\right)\right\}, \tag{16}$$

with the conditions for the existence of double-layers:

$$\begin{aligned}(i) \quad & u(\phi) = u'(\phi) = 0 \text{ at } \phi = 0, \\ (ii) \quad & u(\phi) = u'(\phi) = 0 \text{ at } \phi = \phi_1 \neq 0, \\ (iii) \quad & u(\phi) < 0 \text{ for } 0 < |\phi| < |\phi_1|.\end{aligned} \tag{17}$$

Now the conditions in the item (ii) provide two relations:

$$n_{e0}a_e + n_{i0}a_i - n_n a_n - n_p a_p = 0 \text{ and} \tag{18}$$

$$n_{e0}a_e^2 - n_{i0}a_i^2 + 3n_n a_n^2 - 3n_p a_p^2 = 0, \tag{19}$$

which are conditions under which double-layers exist.

NUMERICAL RESULTS

We proceed to obtain the parametric regions where conditions (17) are satisfied. Starting with 9 parameters defined in (9)-(10) with the inclusion of 4 equations (11)-(14), we have a 5-parameter region. We introduce parameters α and β in substitution of a_i and a_n, $\alpha = a_e/a_i = T_i/T_e$ and $\beta = a_p/a_n = Z_p M_n/Z_n M_p$. So, we have to deal which a function $f(n_{e0}, n_p, M, \alpha, \beta)$ which satisfies relations given in (17).

Furthermore we reduce the 5 parameters to a even smaller number by taking a reasonable physical values for α, β and M, resulting in a two parametric space: $g(n_{e0}, n_p)$. In Figure 1a, we show the curves where the double-layer solutions are found. We have chosen $\alpha = 0.09$ and $\beta = 0.10$. For each given value of M, from 1.01 to 3.0, a curve is drawn on $n_{e0} \times n_p$ space where DL exists.

The same treatment was applied for the particular case of laboratory plasma reported by Oohara et al [7], where a fullerene-ion plasma of the same mass (C_{60}) was produced in the process of a hollow electron-beam impact ionization. Authors observed two low-frequency electrostatic waves. For calculations we used main characteristics of the dusty plasma, i.e. $n_e / n_p \sim 10^{-6}$, $n_e = 1.0$, $n_p = n_n \sim 10^{-6}$ and $M_p = M_n$. Moreover, we introduce, on their experimental conditions, a small quantity of ions to fulfill conditions of the four component dusty plasma. Thus we have parameter $\alpha = 0.09$ and $\beta = 1.0$ that is different from the case discussed above.

In the Figure1b the result of the double-layer conditions for different M values is shown. As we can see, the authors [7] have possibility to obtain a double-layer in laboratory plasma. It is interesting to observe the different comportment of the curves for small values of the n_p. In contrast to the case of low β, for the DL exist it is necessary increasing both n_e and n_p for some constant M. Furthermore, the limit of the double-layer existence decrease with increasing β.

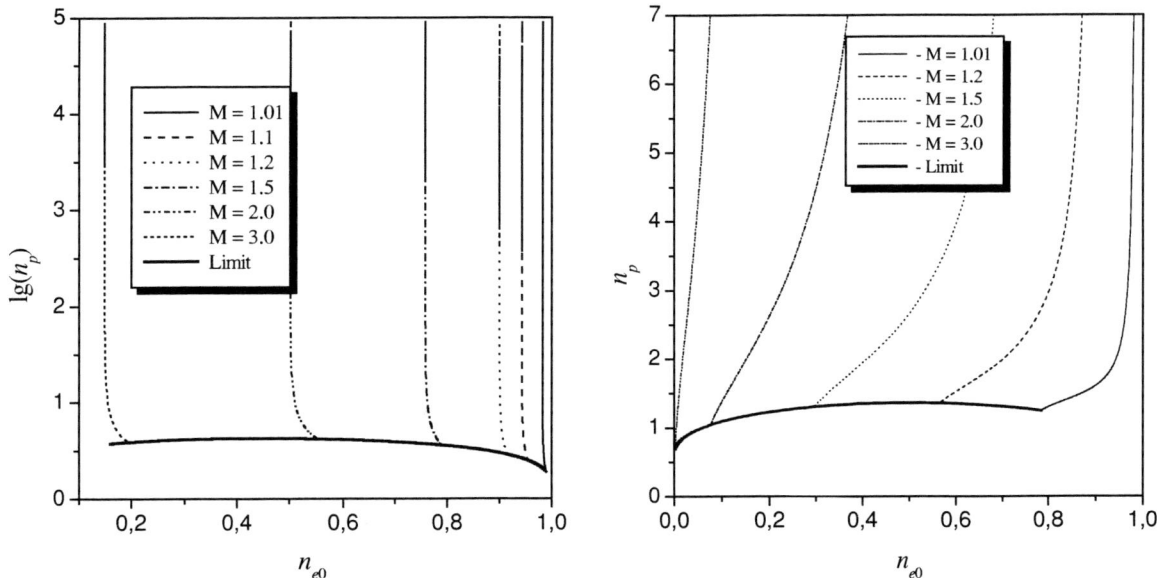

FIGURE 1. Relations between the parameters n_p and n_e for different values of Mach number M and parameter $\alpha = 0.09$: a) $\beta = 0.10$; b) $\beta = 1.0$.

SUMMARY

The linear and nonlinear properties of dust-acoustic waves (DAW) were studied. We used the model of multi-component dusty plasma with inertialess electrons and ions as well as positively and negatively charged inertial dust grains. We found that in four component dusty plasma there are remarkable changes in the nonlinear properties of the DAW. The presence of positively charged dust grains produces double-layers in those parameter regimes. The theory was applied to the laboratory plasma reported by Oohara et al. We predict that a double-layer might be possible to be launched in their experiment if a trace ions component is added. The results of the investigation can be useful for designing laboratory experiments dealing with the demonstration of DAW in multi-component dusty plasma with the positive and negative dust grains. Our parametric studies and double-layers should be useful in identifying coherent nonlinear structures in the Earth's mesosphere. Furthermore, non-stationary double-layers could be potential accelerators for dust particulates in space plasmas..

ACKNOWLEDGMENTS

We acknowledge for the financial support of CAPES (Fundação Coordenação de Aperfeiçoamento de Pessoal de Nível Superior) and FAPESP (Fundação de Amparo à Pesquisa do Estado de São Paulo) under Proc. 98/14711-4.

REFERENCES

1. Rao, N. N., Shukla, P. K., and Yu, M. Y., Planet. Space Sci. 38, 543 (1990).
2. Nakamura, Y., Odagiri, T., and Tsukabayashi, I., Plasma Phys. Control. Fusion 39, 105 (1997), Nakamura, Y., and Tsukabayashi, I., Phys. Rev. Lett. 52, 2356 (1984).
3. Watanabe, S., J. Phys. Soc. Japan 53, 950 (1984), Tajiri, M., and Tilda, M., ibid. 54,19 (1985).
4. Sheridan, T. E., J. Plasma Phys. 60,17 (1998).
5. Shukla, P. K., Phys. Plasmas 1, 1362 (1994).
6. Sakanaka, P.H., and Shukla, P.K., Phys. Scripta, 84, 181 (2000)
7. W. Oohara, N. Tomioka, T. Hirata, R. Hatakeyama, and N. Sato, Proceedings of the 2000 International Congress on Plasma Physics, Quebec, October, 2000., Vol 1, pag. 116-119 (2000).

Charge Variations in Planar RF Discharge

O.S. Vaulina, A.A. Samarian

School of Physics, University of Sydney, NSW 2006, Australia
Institute of High Energy Densities, Russian Academy of Sciences, Moscow, Russia

Abstract. In a complex plasma the dust particles achieve electrostatic equilibrium with respect to the plasma by acquiring negative charge. This charge is extremely large compared to the ionic charge. In addition, the particle charge is not fixed but is coupled self-consistently to the surrounding plasma parameters. There are two mechanisms that can lead to random fluctuations of particle charge relative to its equilibrium (time averaged) value [1]. The first is related to the random nature of ionic and electronic currents which charge the particles. The second is a result of random motion of the particles (for example due to their Brownian motion) in the spatially inhomogeneous plasma. Here we consider the influence of charge variations on the dynamics of a mono-layered structure levitated above the powered electrode in a planar rf discharge. Vertical oscillations in such a structure were first reported in [2].

THEORY

For explanation of the observed self-excited oscillations we have taken into account the random fluctuations of the particle charges caused by the stochastic variations of charging current, and also fluctuations of charges due to their random motion in the presence of gradient of particle charge in the vertical direction. In this 2-D case the dynamics of dust particles levitated in the plasma sheath are described by the following equations

$$m_p y'' = -m_p v_{fr} y' - (\alpha_y e\langle Z \rangle + \beta_y E_y) y + e\tilde{Z}_f E_y + F_y^{br}, \tag{1a}$$

$$m_p r'' = -m_p v_{fr} r' - \alpha_r e\langle Z \rangle r + e(\tilde{Z}_f + \tilde{Z}_s) E_r + F_r^{br}, \tag{1б}$$

where r'', y'', r', y' are derivatives with respect to time, and $\alpha_y = dE_y/dy$, $\alpha_r = dE_r/dr$ are spatial gradients of the electric field $\mathbf{E} = (E_y, E_r)$. The charge variations $\tilde{Z}_f$, $\tilde{Z}_s$ and their mean square displacements $\langle \tilde{Z}_f \rangle = \sqrt{\langle \tilde{Z}_f^2 \rangle_t}$, $\langle \tilde{Z}_s \rangle = \sqrt{\langle \tilde{Z}_s^2 \rangle_t}$ ($\langle \tilde{Z}^2 \rangle_t$) are given by

$$\frac{d\tilde{Z}_f}{dt} = -\eta \tilde{Z}_f + \tilde{F}_f, \quad \langle \tilde{Z}_f \rangle = \xi\sqrt{\langle Z \rangle}, \tag{2a}$$

$$\tilde{Z}_s = \beta_y y, \quad \langle \tilde{Z}_s \rangle = \beta_y \sqrt{\langle y^2 \rangle}, \tag{2b}$$

where η and ξ are parameters determined by the plasma condition. Assuming no correlation between $\tilde{F}_f, F_y^{br}, F_r^{br}$ the total kinetic energy of a dust particle can be defined as $2\langle K \rangle = 3T_o + \Delta^f T + \Delta^s T$, where T_o is gas temperature, and $\Delta^{f(s)}T$ depends on the nature of the charge fluctuation. By solving the equations (1)-(2a) we found an expression for the value of oscillation kinetic energy acquired due to fluctuations of the charging currents:

$$\Delta^f T_{y(r)} = \frac{e^2 Z \xi^2 E_{y(r)}^2}{m_p v_{fr}(v_{fr} + \eta)}, \tag{3}$$

where the value of E_r is $\sim (eZ/l)^2$. Estimating $\Delta^f T$ for typical experimental conditions yields $\Delta^f T = 2\Delta^f T_r + \Delta^f T_y < 10 T_o$. This is less than experimentally observed values which are about 100eV.

The expression for the kinetic energy acquired due to random motion in the presence of a charge gradient can be obtain from equations (1)-(2b):

$$\Delta^s T_r = (T_o + \Delta^f T_y + \Delta^s T_y)\theta_1, \quad (4)$$

where

$$\theta_1 \approx \left(\frac{\beta_y}{e\langle Z\rangle}\right)^2 \frac{2\alpha\omega_y^2 e^4 \langle Z\rangle^4}{m_p^2 g(\alpha_y/E_y + \beta_y/\langle Z\rangle)((\omega_r^2 - \omega_y^2)^2 + 2(\omega_r^2 - \omega_y^2)\nu_{fr}^2 + 4\omega_y^2\nu_{fr}^2)l_p^4} \quad (4a)$$

And $\omega_r^2 = e\langle Z\rangle\alpha_r/m_p \sim (e\langle Z\rangle)^2 n_p$, $\omega_y^2 = (\beta_y E_y + e\langle Z\rangle\alpha_y)/m_p$. For $(\omega_r^2 - \omega_y^2) \ll \nu_{fr}^2$:

$$\theta_1 \approx \left(\frac{\beta_y}{e\langle Z\rangle}\right)^2 \frac{\alpha e^4 \langle Z\rangle^4}{2m_p^2 g(\alpha_y/E_y + \beta_y/\langle Z\rangle)\nu_{fr}^2 l_p^4}. \quad (4b)$$

Taking into account that $\Delta^f T_y \approx \Delta^f T/3$ $l_p^2 \cong l^2 + \langle y^2\rangle$, where

$$\langle y\rangle^2 = \frac{(T_o + \Delta^f T_y + \Delta^s T_y)}{m_p \omega_y^2},$$

$\Delta^f T_y \approx \Delta^f T/3$, $\Delta^s T_y = \gamma \Delta^s T_r$, and γ is the parameter of kinetic energy redistribution, for $l^2 \gg \langle y^2\rangle$ and $\gamma = 1$ ($\Delta^s T = \Delta^s T_y \equiv \Delta^s T_r$) we finally obtain:

$$\Delta^s T = \frac{(T_o + \Delta^f T/3)}{1 - \theta_1} \quad (5)$$

Estimates of the maximum value of kinetic energy and oscillation amplitude using a linear approximation in the case $l^2 \ll \langle y^2\rangle$ give us

$$\Delta^s T^{max} = \theta_1 l^2 \omega_y^2 m_p, \qquad A_y^{max} \approx l\sqrt{2\sqrt{\theta_1}} \quad (6)$$

where $A_y^{max} = \sqrt{2\langle y^2\rangle}$.

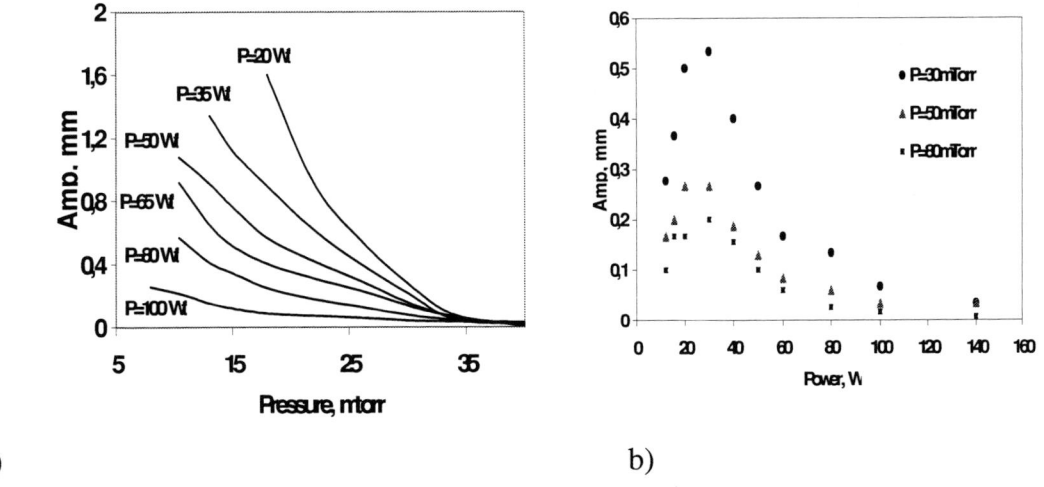

a) b)

FIGURE 1. Experimental chamber and image of test gains levitated above the electrode.

EXPERIMENT

The one layer dust structure in Ar plasma of planar rf discharge was investigated (the experiment details can be find in [2]). It was found that when the pressure was decreased bellow a critical value the dust particles began to oscillate spontaneously in the vertical direction. The amplitude and frequency of the oscillations are several millimetres and about ten Hz, respectively. When the rf input power was decreased the oscillation amplitude was found to increase. Fig. 1 shows the dependence of oscillation amplitude on gas pressure and rf power. For pressures below 35 mTorr the oscillation amplitude increased dramatically. This increase is greater for lower rf powers. For the 6.28 mm particles, saturation and decrease of the oscillation amplitude is observed.

Comparison of theoretical and experimental dates better provide using the relative parameters. In our case we normalised the oscillation amplitude to the definite value $A_y(P_o)$ and look on the dependence on pressure.

The dependencies of the ratio $A_y^{max}(P_i)/A_y(P_o)$ on pressure is shown in Fig.2. Taking into account $\theta_1(P_i) = \theta_1(P_o)(P_o/P_i)^2$, the value of $\theta_1(P_o)$ can be obtained using equation (Fig.3) by fitting the simulation data to the experimental results. Using this routine $\theta_1 \approx 0.5$ for $P_o = 50$ mTorr ($a=1\mu m$) and $\theta_1 \approx 0.28$ for $P_o = 100$ mTorr, ($a=3.07$ μm) were obtained.

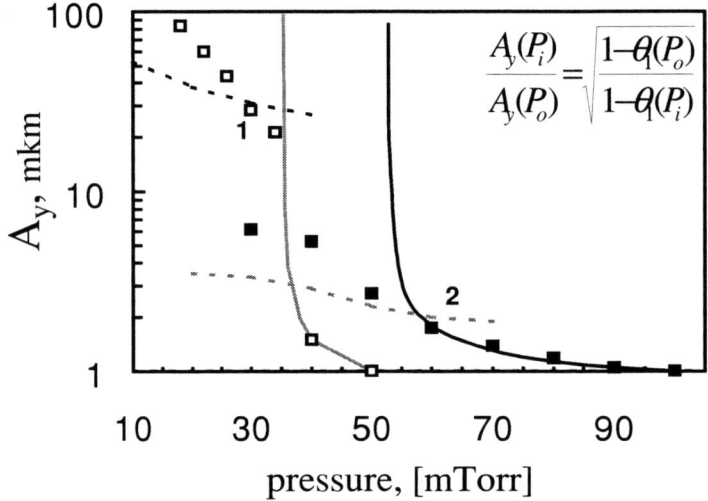

FIGURE 2. Normalised amplitude versus pressure. 1-for a=1.05 μm, 2- for a=3.07 μm.

Solid lines are simulation results. Dotted lines $A_y \approx n_p^{-1/3}$. $A_y(P_i)$ is oscillation measured for different pressures P_i.

ACKNOWLEDGMENTS

This work was supported by Australian Research Council, the Science Foundation for Physics within the University of Sydney. AAS was supported by University of Sydney U2000 Fellowship.

REFERENCES

1. Khrapak S.A., Nefedov A.P., Petrov O.F., Vaulina O.S., *Phys.Rev. E* **59**, 6017 (1999).
2. Samarian, A., James, B., Vladimirov, S., and Cramer, N., *Phys. Rev.* **64**, 025402(R) (2001).

Control of Equilibrium Structure of a Toroidal Non-neutral Plasma in Proto-RT

H. Saitoh*, Z. Yoshida*, H. Himura*, C. Nakashima*, J. Morikawa† and M. Fukao†

*Graduate School of Frontier Sciences, The University of Tokyo, Bunkyo-ku, Tokyo 113-0033, Japan
†The High Temperature Plasma Center, The University of Tokyo, Bunkyo-ku, Tokyo 113-0033, Japan

Abstract. The structure and its control method of toroidal non-neutral plasma equilibria have been studied in an internal conductor geometry. It was demonstrated that the potential profiles of toroidal electron plasmas can be altered by use of control electrodes. With a ring electrode negatively biased, the hollow potential of the plasma was eliminated, and consequently, the improvement in containment time was observed. Electrostatic fluctuations shows that the electrons were confined in dipole fields for up to 100 $msec$.

INTRODUCTION

Non-neutral plasmas have attracted a wide range of interests in its diverse applications as well as the fundamental physics. The self-electric field and the resultant strong flow induce various types of structure formations such as vortex dynamics. A high β equilibrium state with strong flow has been predicted by a two-fluid model [1]. Traps of charged particles including anti-matters [2] are also one of the main topics in atomic physics experiments.

Toroidal trap systems have no open ends of magnetic field lines and consequently, no electrostatic potentials are required along the toroidal direction. This advantage allows toroidal systems to trap high energy or multiple species of different charges. For several decades, confinement of electron plasmas in toroidal magnetic fields have been studied [3, 4]. The toroidal field and the electric fields in a plasma produce an $\mathbf{E} \times \mathbf{B}$ drift, which overcome curvature drifts, and the motion of a single particle takes a closed orbit. However, the toroidal effects cause cross field transport of the particles [5], which prevent the long time containment of non-neutral plasmas in pure toroidal fields.

The Proto-RT (Prototype Ring Trap) device is a toroidal trap constructed for exploring various phenomena in non-neutral plasma physics [6]. Besides conventional toroidal field coils, Proto-RT has two kinds of poloidal field coils (a dipole field coil and vertical field coils) and can generate a variety of field configurations. Recently, the possibility of stable and long time confinement of non-neutral plasmas on magnetic surfaces has been predicted in a stellarator configuration [7]. In contrast to pure toroidal field devices, the poloidal coils of Proto-RT have made possible a containment of non-neutral plasmas on similar kinds of magnetic surfaces. However, Proto-RT has an internal conductor (IC) in the confinement region of the torus and the outer shell of IC is electrically grounded. Thus the confined non-neutral plasmas have hollow distribution, which might cause instability.

In this work, we have studied the structure of electron plasma equilibria and its control method in a toroidal device of the Proto-RT type (i.e., toroidal-internal conductor system). Externally applied electric fields can control potential and flow profiles in the plasma and thereby might contribute to the improvement in the containment properties. Experimentally, potential structures and the effects of biased electrodes upon the equilibria and containment time were evaluated. It was demonstrated that the potential profiles were successfully controlled by external electric fields. The negatively biased electrode cancelled the hollow potential profile in the plasma, and a long time containment of a torus electron plasma was observed.

EXPERIMENTAL SETUP

Experiments were carried out on the Proto-RT device (Fig. 1). The torus has a rectangular poloidal cross section of $0.9 \, m \times 0.472 \, m$ and is pumped to $3 \times 10^{-7} \, Torr$. In the chamber, toroidal fild (TF) coils are penetrating through its

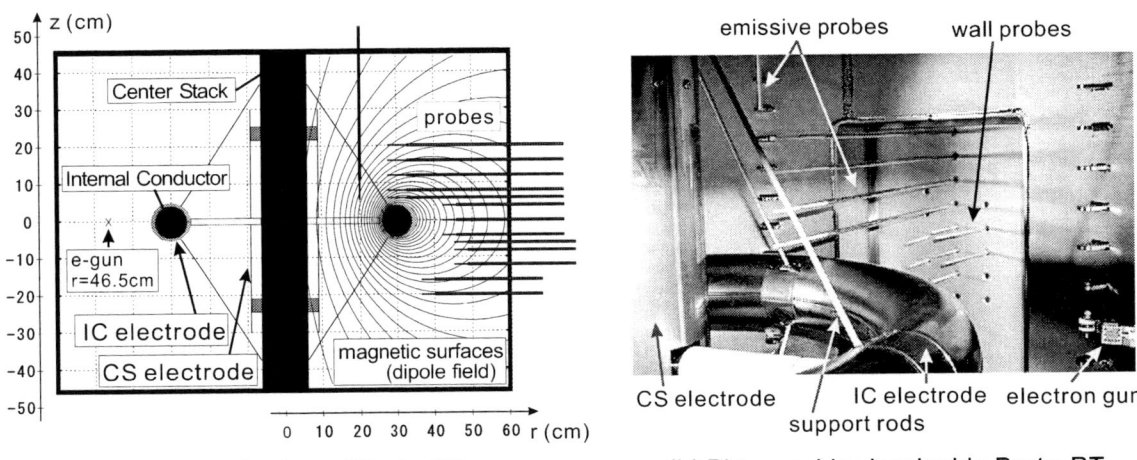

FIGURE 1. (a) The poloidal cross section and (b) the inside view of the Proto-RT device. Support rods, electric feeders, and a coolant nozzle are connected to the internal conductor, covered by ceramic insulators. To minimize the disturbance to the plasma, each probe was independently inserted into the chamber. Vertical field coils are located at outside of the chamber, $r = 90\ cm$.

hollow center stack (CS) of $0.114\ m$ diameter. Proto-RT also has two kinds of poloidal field coils; an internal conductor (IC) coil for dipole fields and a pair of vertical field (VF) coils. Electrons are injected into the torus of Proto-RT from an electron gun of LaB_6 cathode located at the equatorial ($z = 0$) plane. In this study, the gun is fixed at $r = 0.465\ m$ and fired at an acceleration voltage of $300\ V$. In the typical magnetic field strength of Proto-RT, $\sim 0.01\ T$ near the electron gun, the beam current obtained was $2\ mA$. The details of the device is given in literatures [6].

Equilibria of a torus non-neutral plasma are formed by a help of external electric fields from outside of the space charges. As far as an equilibrium is found in a device, the required external fields have automatically been generated by the induced image charges on the chamber. However, more actively, the equilibrium structure of the plasma can be externally controlled by applied electric fields. Proto-RT has a pair of plasma control electrodes on IC coil and CS in the vessel. These electrodes are electrically insulated from the chamber or each other, and can produce external fields around $1\ kV m^{-3}$ in a vacuum chamber.

Two kinds of probes have been used in the experiments. One was a Langmuir probe for the measurements of potential profiles. The probe configuration in the device is shown in Fig. 1. To measure space potentials in non-neutral plasmas, where the strong flow is induced, we have employed emissive Langmuir probes [8]. The probe consists of a tip of $0.1\ mm$ diameter thoria-tungsten spiral filament and molybdenum wire of $1.0\ mm$ diameter covered with an insulating ceramic tube. It was terminated across a resistance of $100\ M\Omega$ and used as a floating probe, where the disturbance to the plasma is relatively small. In comparison with probe characteristics (I-V curves), the potential ϕ_H measured by these high-impedance emissive probes showed a good agreement with the space potential ϕ_p in a electron plasma. The probes were configured to form an array and each probe is movable along the radial direction (or z direction, for a z probe), so that the spatial distributions of the plasma and the effects of external fields are obtained.

The other probe was a wall probe for the measurements of image charges of the electron clouds. As a wall foil, a copper sheet was enclosed in an insulating glass tube and located inside the chamber. The foil is connected to the chamber through a resistance of $470\ \Omega$ or a current amplifier. Electrostatic fluctuation of the plasma and, in the current integrating mode, the variation of trapped charge are measured. The wall probe is also used for the evaluation of the confinement time of the plasma.

RESULTS AND DISCUSSION

Potential structures of electron plasmas in Proto-RT were measured in a variation of applied voltage on the IC electrode (Fig. 2). In dipole field configurations, the confinement region of the plasma agrees with the outer shell of the closed magnetic surfaces, but the distribution of the plasma is inwardly shifted, surrounding the IC electrode [9]. This is supposed to be due to the diffusion of electrons from the original beam orbit and the formation of the magnetized bulk component of the plasma. When the potential is not externally controlled (i.e., the IC electrode is grounded) or the

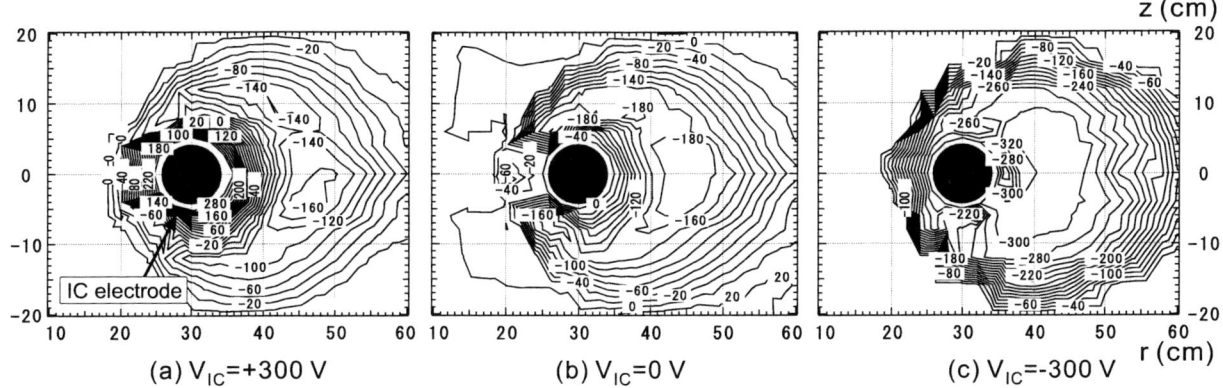

FIGURE 2. Potential profiles and its control in the poloidal cross section of Proto-RT measured by emissive probes. V_{IC} is an electrostatic potential on the internal conductor. Electrons were injected by an acceleration voltage of 300 V in a dipole magnetic field (coil current $I_{dipole} = 5.25\ kAt$). The CS electrode is grounded. The potential is not equal to 0 at $r = 59\ cm$, the inside diameter of the chamber, because the inside surface of the diagnostic flange is located at $r = 65\ cm$. The corresponding number density of the electron $n \sim 2 \times 10^{13}\ m^{-3}$, and the total trapped charge $Q \sim 2 \times 10^{-7}\ C$.

electrode is positively biased, the potential profiles also take hollow distributions around the IC electrode (Fig. 2 (a) and (b)). In these cases, the ridge of the potential peak is surrounding the IC electrode, and the resultant $\mathbf{E} \times \mathbf{B}$ drift flow takes a sheer distribution, which may cause a diocotron instability. In contrast, negatively biased IC electrode eliminates the hollow potential profile (Fig. 2 (c)).

Figure 3 shows the electrostatic fluctuations from the plasma. When the potential structure is not externally controlled (i.e., the IC electrode is grounded), the electrostatic fluctuation from the plasma decays with a 40 μsec exponential curve after the stop of the electron supply at $t = 0$. However, when the IC electrode is negatively biased so that the hollow potential profile is eliminated, the fluctuation signals were observed for much longer period. In Fig. 3 (b), the amplitude of the oscillation rapidly grow again during the first damping oscillation ($t = 300\ \mu sec$). This signal shows an unstable change in the amplitude until $t = 1.4\ msec$ with decreasing its frequency. After $t = 1.4\ msec$, the amplitude of the signal decreases with an approximately $3msec$ exponential decay, keeping an almost constant frequency. The fundamental frequency of the oscillation of this period is proportional to $1/B$ and approximately proportional to V_{IC}, the externally applied electrostatic potential on the IC electrode. These properties agree with a diocotron oscillation in an electron plasma, and the corresponding density for the $l = 1$ mode is $n \sim 3 \times 10^{11}\ m^{-3}$.

When V_{IC} was more negatively biased than $-250\ V$, the electrostatic fluctuation was detected with a longer delay from $t = 0$. Figure 4 (a) is the typical oscillation signal of the plasma in a strong external field. A large amplitude

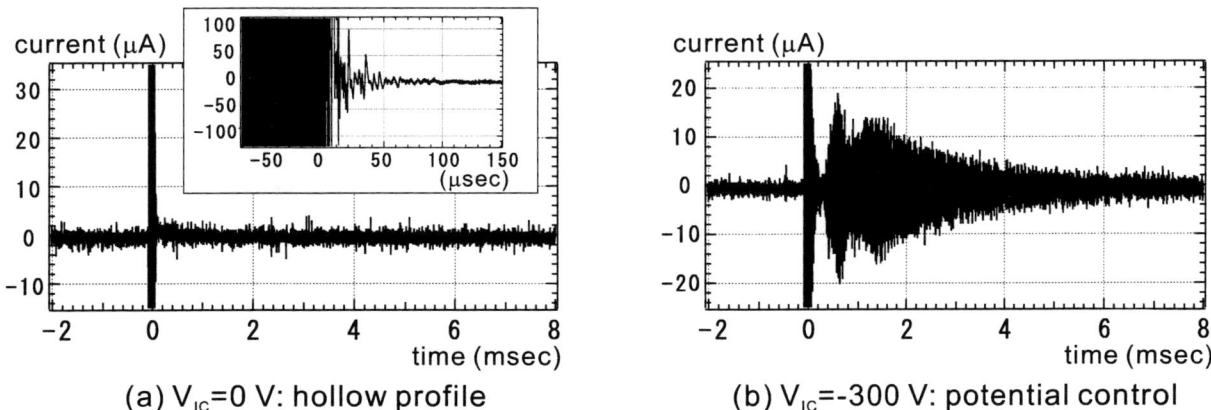

FIGURE 3. Electrostatic fluctuations measured by wall probes. The IC electrode is (a) grounded or (b) negatively ($-300\ V$ against the chamber) biased. Electrons were injected by a gun with an acceleration voltage of 300 V from $t = -75$ to 0 μsec. When the potential profile were adjusted to eliminate the hollow profile, the oscillating signal were continuously observed.

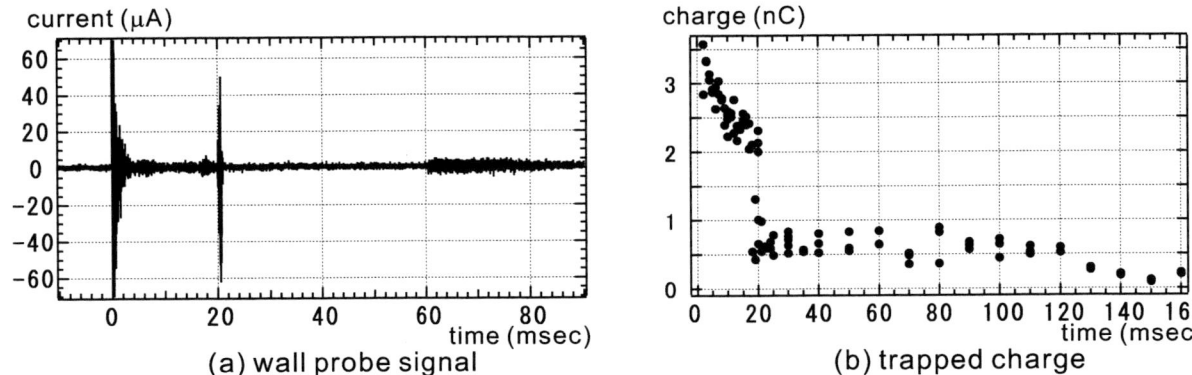

FIGURE 4. (a) The electrostatic fluctuation of the plasma and (b) the integrated current of image charges flow to the wall probe, when a strong external electric field and magnetic field ($V_{IC} = -350\,V$ and $I_{dipole} = 10.5\,kA$) is applied. The electron gun is stopped at $t = 0$.

oscillation and small long signal appear at $t = 20$ *msec* and after 60 *msec*. The fundamental frequency of each oscillation is 42.5 *kHz* and 4.4 *kHz*. We have estimated the confinement time of the plasma by using the image charge measurements. The potential on the IC electrode was changed from $V_{IC} = -350\,V$ to $0\,V$ at t, and the image charges flow to the wall probe was measured. The damp of V_{IC} causes a destabilization and the plasma decays at t (cf. 3 (a)). By the comparison of the image charges of trapped electrons with potential profiles obtained by the Langmuir probe array, the contained charge and its decay was determined. Before the large amplitude oscillation appears at $t = 20$ *msec*, about 2 % of the total electrons during the injection from the gun shows a good containment property and slowly decreases with an approximately 50 *msec* exponential decay. The confined charge rapidly drops at this point, but about 0.3 % of the electrons are still trapped for another 100 *msec*.

In summary, the optimization of the potential profiles has made possible a long time containment of toroidal electron plasmas in dipole field configurations. Plasmas with density $n \sim 3 \times 10^{11}\,m^{-3}$ were confined for 20 *msec* and those with $n \sim 4 \times 10^{10}\,m^{-3}$ for 120 *msec* after the stop of the electron supply, although the oberbed disruptions caused sudden decay of the trapped charges. These confinement time has a sensitive dependence on the back pressure of the device and the strength of the magnetic field. It is probable that the electron-neutral collision set this limit, and the improvement in these parameters might contribute to the longer confinement of torus electron plasmas.

ACKNOWLEDGMENTS

This work was supported by a Grant-in-Aid for Scientific Research from the Japanese Ministry of Education, Science, Sports and Culture No. 09308011. The work of HS was partly supported by JSPS research fellowship.

REFERENCES

1. S. M. Mahajan and Z. Yoshida, Phys. Rev. Lett. **81**, 4863 (1998); Z. Yoshida et al., Phys. Plasmas **8**, 2125(2001)
2. C. M. Surko et al., Phys. Rev. Lett. **62**, 901 (1989); S. J. Gilbert et al., Phys. Plasmas **8**, 4982 (2001)
3. J. D. Daugherty and R. H. Levy, Phys. Fluids **10**, 155 (1967); K. Avinash, Phys. Fluids B **3**, 3226 (1991); Leaf Turner and D. C. Barns, Phys. Rev. Lett. **70**, 798 (1993); S. Kondoh et al., Phys.Plasmas **8**, 2635 (2001)
4. G. S. Janes, Phys. Rev. Lett. **15**, 135 (1965); J. D. Daugherty, et al., Phys. Fluids, **12**, 2677 (1969); W. Clark et al., Phys. Rev. Lett. **37**, 592 (1976); Puravi Zaveri et al., Phys. Rev. Lett. **68**, 3265 (1992); S. S. Khirwadkar et al., Phys. Rev. Lett. **71**, 4334 (1993); M. R. Stoneking et al., Phys. Plasmas, **9**, 766 (2002); C. Nakashima, et al., Phys. Rev. E **65**, 036409 (2002)
5. S. M. Crocks and T. M. O'Neil, Phys. Plasmas **3**, 2533 (1996)
6. Z. Yoshida et al., in *Nonneutral Plasma Physics III*, AIP, 397-416 (1999); S. Kondoh and Z. Yoshida, Nucl. Inst. and Meth. in Phys. Res. A **382**, 561 (1996); C. Nakashima and Z. Yoshida, Nucl. Inst. and Meth. in Phys. Res. A **428**, 284 (1998)
7. Thomas Sunn Pedersen et al., Phys. Rev. Lett. **88**, 205002 (2002)
8. H. Himura, C. Nakashima, H. Saito, Z. Yoshida, Phys. Plasmas **8**, 4651 (2001)
9. H. Saitoh, Z. Yoshida, C. Nakashima, Rev. Sci. Instrum. **73**, 87 (2002)

Simply Connected High-β Magnetic Configurations

F. Rogier[1], G. Bracco, A. Mancuso, P. Micozzi and F. Alladio[2]

[1]*ONERA, Toulouse, France*
[2]*Associazione Euratom-ENEA sulla Fusione, Frascati (Roma), Italy*

Abstract. Simply connected Chandrasekhar-Kendall-Furth plasma configurations (CKF) contain a magnetic separatrix, dividing a spherical torus (ST), two secondary tori and a surrounding discharge. Axisymmetric CKF equilibria are calculated imposing a constant relaxation parameter $\mu = \mu_0 \mathbf{j} \cdot \mathbf{B}/B^2$ at the edge of the plasma and a surface averaged $<\mu>$ decreasing toward the axis of the ST. A poloidal current flows in the surrounding discharge and injects magnetic helicity into the ST. ∇p is supposed to be concentrated in the same region where $\nabla <\mu>$ has the largest variation. CKF equilibria are stable to all ideal MHD perturbations with low toroidal mode numbers (n=1,2,3), up to $\beta = 2\mu_0 <p>_{Vol}/<B^2>_{Vol} \approx 1$, even in absence of a nearby conducting shell. The poloidal current driven in the surrounding discharge by different orbits of co/counter-circulating promptly lost fusion products could help in sustaining a CKF reactor.

INTRODUCTION

The most investigated magnetic fusion configurations (tokamaks) are not simply connected: a central post, containing the inner part of the toroidal magnet and the ohmic transformer, links the plasma torus. The development of simply connected magnetic configuration would strongly simplify the design of a fusion reactor. The engineering advantages include the simplification of the confining magnetic field (solenoid), the absence of damage and maintenance of the critical central post and the ease of access to a cylindrical reactor chamber. The physics advantages are mainly connected with the "open" structure of the magnetic field: the presence of a minimum of the field (magnetic well) inside the confinement region allows for plasma beta values approaching unity; a confinement system with two "ends" eases the refueling/exhausting of the plasma; the emerging field lines ease the control of the electric potential within the plasma. A simply connected configuration would be particularly suitable for magnetic fusion space propulsion: the channeling of the charged fusion products into particle jets could transform the fusion power into propulsive power, with a high efficiency available for thrust. Furthermore D-^{3}He fusion reactions are much more attractive than D-T for space thrusters [1], as they would yield mainly charged fusion products (maximum thrust) and reduce the shield mass (minimum weight). However the requirements upon the plasma temperature T and upon the product of confinement time and plasma density $n\tau_E$, for maintaining ignition, are both 5 times larger for D-^{3}He than for D-T. As the fusion power density scales as $P_{fusion} \sim \beta^2 B^4$ and the Lawson parameter as $nT\tau_E \sim \beta/\chi \{L^2 B^2\}$, the achievement of a high plasma beta, $\beta \approx O(1)$, becomes mandatory, in order to minimize the magnetic field B, at given plasma thermal diffusivity χ and system size L. Obvious considerations about weight of coils and cyclotron radiation forbid the maximization of B for space thrusters. Compact tori such as Spheromaks and Field Reversed Configurations (FRC) are attractive candidates for magnetic fusion propulsion. However Spheromaks are limited by ideal MHD stability to $\beta \approx 0.15$, whereas FRC have $\beta \approx 1$, but the theoretical understanding of their macroscopic stability remains elusive. This paper introduces a new simply connected high beta magnetic confinement scheme, which could be worthwhile of experimental study.

EQUILIBRIUM AND STABILITY OF CKF CONFIGURATIONS

The problem of finding a stable force-free magnetic field, by minimizing (over all domains of given volume) the magnetic energy, given nonzero magnetic helicity, has received [2] as a tentative answer a singular domain: an extreme "apple", in which north and south pole are pressed together. So the Spheromak has the right topology, but needs to be embedded in a surrounding stabilizing plasma. In line with this result, a simply connected magnetic confinement scheme is obtained superposing two axisymmetric homogeneous force-free fields, each with $\nabla \wedge \mathbf{B} = \mu \mathbf{B}$, both having the same value of the relaxation parameter $\mu = \mu_0 \mathbf{j} \cdot \mathbf{B}/B^2$. The first is the Chandrasekhar-Kendall force-free field [3] $\psi_{\mu,1}^{CK}$; fixing $\mu = 14.066$, four zeroes of $\psi_{\mu,1}^{CK}$ are present within a unity circle. The second is the Furth square-toroid force-free field [4] $\psi_{\mu,\lambda}^{F}$; λ is chosen so that the point (R=0, Z=0.775), which lies upon the third zero of $\psi_{\mu,1}^{CK}$, lies also upon the first zero of $\psi_{\mu,\lambda}^{F}$. The superpositions are written: $\psi(r,\theta) = \psi_{\mu,1}^{CK} + \gamma \psi_{\mu,\lambda}^{F}$ (Fig.1); when $0.402 \leq \gamma \leq 0.69$, they contain, in a simply connected region, a toroidal current density j_ϕ with the same sign and will be called Chandrasekhar-Kendall-Furth force-free fields (CKF). A magnetic separatrix with ordinary X-points ($B \neq 0$), divides a main spherical torus (ST), two secondary tori (SC) on top and bottom of the main torus and a surrounding discharge (P), embedding the three tori. Two degenerate magnetic X-points (B=0) are present on the symmetry axis.

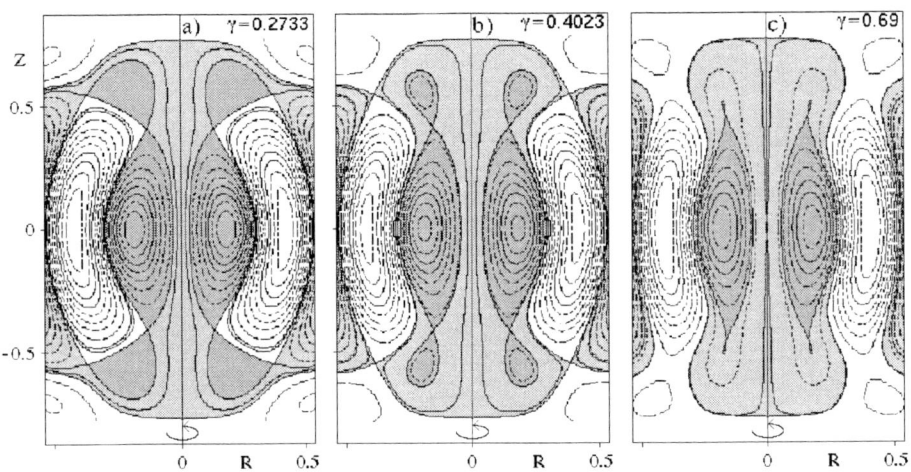

FIGURE 1. Force-free CKF: $\psi = \psi_{\mu,1}^{CK} + \gamma \psi_{\mu,\lambda}^{F}$. Colored regions mean toroidal current density $j_\phi > 0$, white regions mean $j_\phi < 0$.

However CKF force-free fields are pressureless ($\nabla p = 0$) and unable to confine plasmas of fusion interest. Unrelaxed ($\nabla \mu \neq 0$, $\nabla p \neq 0$) MHD equilibria (Fig. 2a), similar to CKF force-free fields, are calculated imposing that μ is constant only at the edge of the plasma and that the surface averaged $<\mu> = \mu_0 <\mathbf{j} \cdot \mathbf{B}/B^2>$ decreases from the edge toward the axis of the ST. If the surrounding discharge is sustained by driving a total poloidal current I_e, magnetic helicity, flowing down the gradient of $<\mu>$, is injected into the ST. Magnetic reconnections at the X-points, produce a total toroidal current I_{ST} flowing inside the main spherical torus, while converting part of the magnetic energy into kinetic plasma energy: in the equilibrium calculations the pressure profile is assumed such that the region of maximum ∇p coincides with that of maximum $\nabla \mu$ (Fig. 2b).

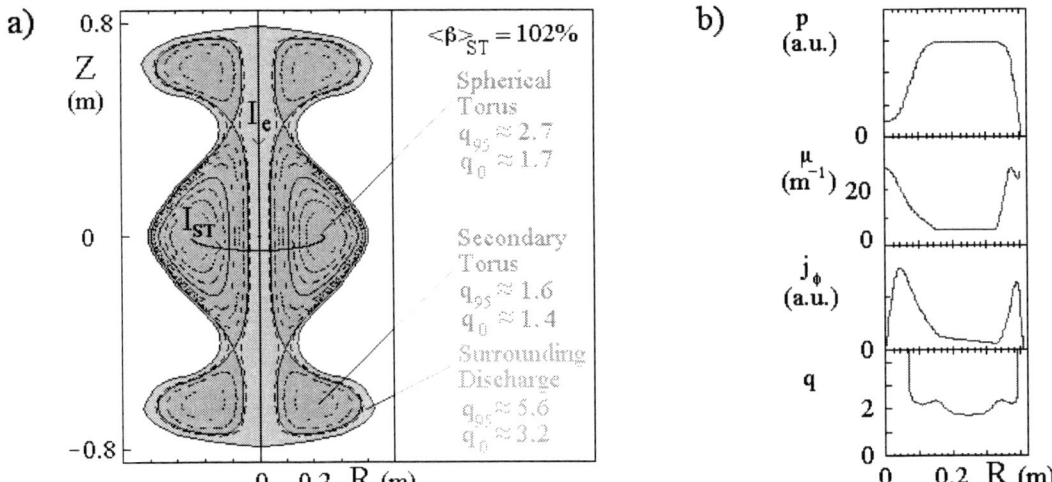

FIGURE 2. Unrelaxed CKF equilibrium and profiles of pressure, relaxation parameter, toroidal current density and safety factor.

A parametric scan of unrelaxed CKF equilibria, at fixed shape of the plasma edge, has been calculated in terms of the currents ratio I_{ST}/I_e and of the safety factor at the ST magnetic axis q_0^{ST}. The vanishing of the toroidal current density j_ϕ on the ST axis, a too large ∇j_ϕ at the edge and $\nabla \mu \neq 0$ extending up to the ST axis set, respectively: the low I_{ST}/I_e, the low and the high q_0^{ST} equilibrium limits in the plot of Fig. 3. The ideal MHD stability has been calculated by solving the normal mode problem for the perturbed displacement ξ: the equilibria are analyzed in magnetic Boozer coordinates (ψ_T radial, θ poloidal, ϕ toroidal, Jacobian $\sqrt{g} \propto 1/B^2$). An innovative numerical MHD stability code has been written; it can analyze plasmas with multiple magnetic axes; it allows for the presence of the plasma on the symmetry axis (R=0) and on the degenerate X-points (B=0), defining the displacement variables as: $\xi=\xi\cdot\nabla\psi_T/R^2$, $\eta=\xi\cdot(\nabla\theta-\nabla\phi/q)/B$, $\nu=-\sqrt{g}\xi\cdot\nabla\phi$; a 2D FEM computes the perturbed energy in the vacuum region around the plasma, accounting for conducting shells of any shape. Only the normal displacement $\xi=\xi\cdot\nabla\psi_T/R^2$ is continuous at the magnetic separatrix, whereas the other components of ξ can make jumps. Global MHD modes extended to the whole plasma must be periodic in the poloidal angle θ over all the three tori: therefore the continuity of ξ forces to multiples of 3 the poloidal mode numbers allowed inside the surrounding discharge. Limited MHD modes, restricted to the surrounding discharge, require instead all integer poloidal mode numbers: their radial displacement ξ goes to zero at the separatrix, avoiding any requirement of periodicity on the three tori. Fig. 3 shows the ideal MHD free-boundary stability to modes with low toroidal numbers (n=1,2,3), without any conducting shell around the plasma.

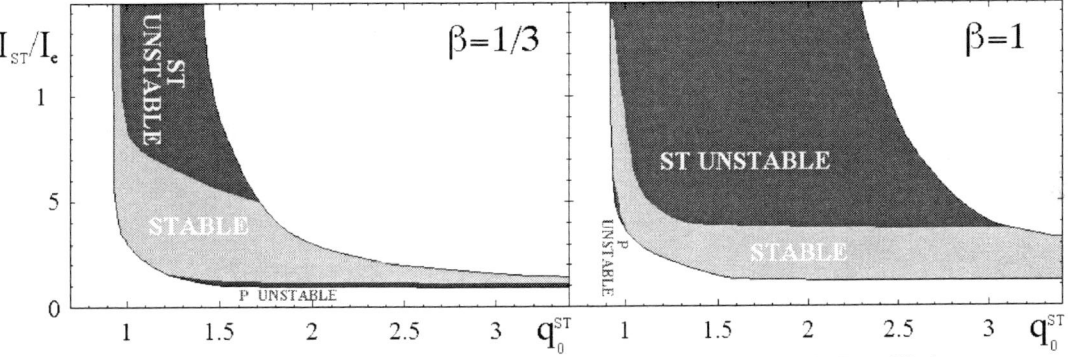

FIGURE 3. Ideal MHD stability plots at $\beta=1/3$ and $\beta=1$ for unrelaxed CKF equilibria.

At $\beta=1/3$ all the equilibria with a current ratio $1.5 < I_{ST}/I_e < 5$ are stable, irrespective of the internal profiles (q_0^{ST}). At $\beta=1$ the stable range is limited to $1.5 < I_{ST}/I_e < 3.5$. Fig. 4a shows the arrow plot of an n=1 global mode at $\beta=1$, $I_{ST}/I_e=5$ low q_0^{ST}, due to a too large ∇p inside the surrounding discharge. Fig. 4b shows the arrow plot of an n=3 global mode, again at $\beta=1$, $I_{ST}/I_e=5$ but at larger q_0^{ST}, due to a too large ∇p inside the ST. At higher values of I_{ST}/I_e the stable window shrinks or disappears, due to global modes with n=2 and n=3, confirming that a classical Spheromak, which is obtained in the limit $I_{ST}/I_e \rightarrow \infty$ with $q_0^{ST} \approx 0.9$, becomes unstable at a rather low β value. At a current ratio $I_{ST}/I_e \approx 1$, an n=1 mode, limited to the surrounding discharge, kinks the central column (Fig. 4c), due to a too large longitudinal (poloidal) current flowing inside the central column. The kink mode at $I_{ST}/I_e \approx 1$ shows up even at $\beta=0$.

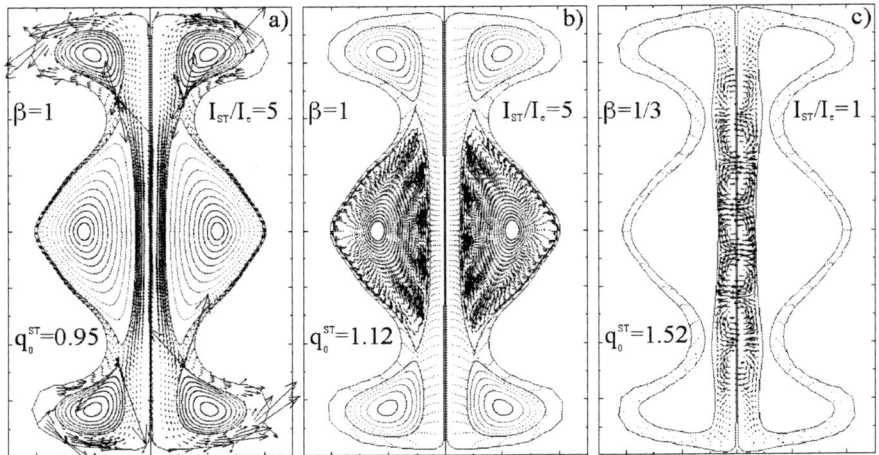

FIGURE 4. Arrow plots for unstable CKF equilibria: a) and b) global modes; c) mode limited to the surrounding discharge.

LOSSES OF CHARGED FUSION PRODUCTS FROM A CKF BURNER

The fusion capabilities of the unrelaxed CKF configurations are investigated. Given a plasma diffusivity, the spherical torus current I_{ST} required for achieving a volume averaged burning temperature $<T_{ST}>$ is derived as a function of the ST major radius R_{ST}. Table 1 assumes a confinement enhanced over L-mode scaling by a factor H=2 and presents the minimum size CKF burners, both for the D-T reaction (where the minimization of the size is limited by the neutron wall loading), as well as for the D-^{3}He reaction (where the minimization of the size is limited by the maximum strength of the magnetic field on the poloidal field coils B_{coils}).

TABLE 1. Minimum size CKF burners.

Reaction	R_{ST} (m)	$<T_{ST}>$(keV)	$<n_{ST}>(10^{20}$ m$^{-3})$	I_{ST}, I_e (MA)	β	B_{coils} (T)	$P_{neutrons}$, $P_{charged}$ (MW)
D-T	1.09	12	4.6	17.5, 2.7	0.99	2.4	468, 117
D-^{3}He	1.54	70	3.9	95, 14.4	0.27	9.2	35, 467

A CKF burner, if endowed with the right helicity injection, β limit and energy confinement, allows for an unimpeded outflow, from the burning ST, of a part of the charged fusion products, easing direct energy conversion and a possible use as a space thruster. The promptly lost charged fusion products drift across the magnetic separatrix, to the (extended) degenerate magnetic X-points (B=0) at the two ends of the configuration (see Fig. 5). Even in axisymmetry the large Larmor radii of the charged fusion products induce remarkable prompt losses, enhanced by the large variation of the magnetic field strength (see Fig 5a). In the case of the D-T burner of Table 1 about 75% of the α-particles born in the outermost 15% of the minor radius of the ST are promptly lost. A few examples of α-particle orbits escaping from this D-T burner are shown in Fig. 5b. The co- and the counter-circulating fusion products have different orbits on the inboard and on the outboard of the surrounding discharge; they escape preferentially at opposite ends of the CKF, producing a mechanical torque and a current density $j_{charged}$ inside the surrounding discharge. The direction of this current density is such as to reinforce the pre-existing equilibrium current density, at least near the divertor-like magnetic structures at the two ends of the configuration. A fully dynamical model is required for evaluating whether the cinematic effect of the promptly lost charged fusion products, which can be increased by the non-axisymmetric resistive instabilities connected with the helicity injection, could be so large as to self-sustain the poloidal current I_e inside the surrounding discharge.

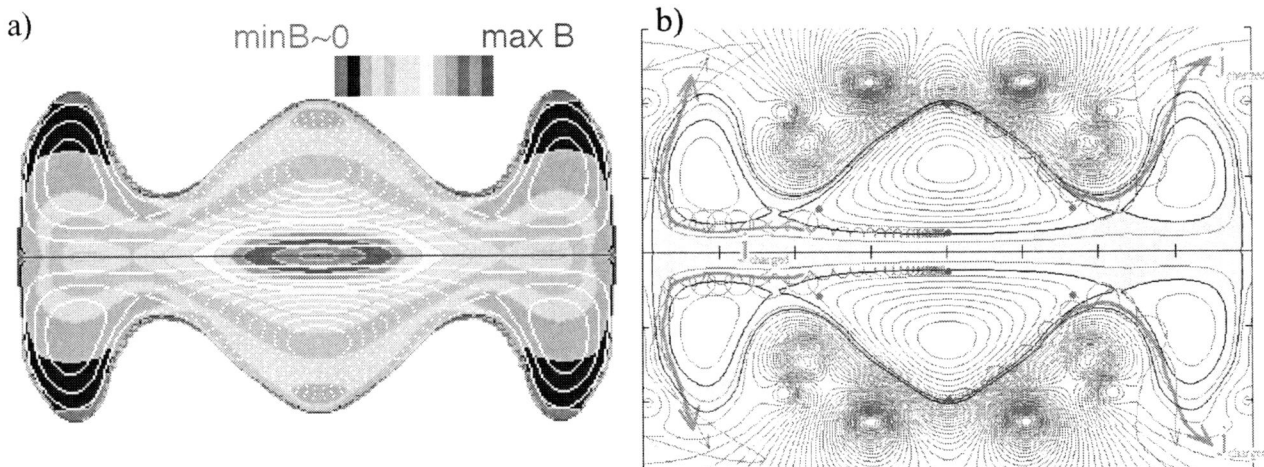

FIGURE 5. CKF (D-T) burner of Table 1: a) contours of constant B; b) orbits of promptly lost α-particles.

REFERENCES

1. Post, R. F., and Santarius, J.F., Fusion Technology **22**, 13-26, (1992).
2. Cantarella, J., De Turck, D., Gluck, H. and Teytel, M., J. Math. Phys. **41**, 5615-5641 (2000).
3. Chandrasekhar, S. and Kendall, P.C., Astrophys. J. **41**, 457, (1957)
4. Furth, H.P., Levine, M.A. and Waniek, R.W., Rev. Sci. Instr. **28**, 949 (1957)

Constructing Integrable Full-pressure Full-current Free-boundary Stellarator Magnetohydrodynamic Equilibria

S.R.Hudson*, D.A.Monticello*, A.H.Reiman*, D.J.Strickler[†] and S.P.Hirshman[†]

*Princeton Plasma Physics Laboratory, P.O. Box 451, Princeton NJ 08543, USA.
[†]Oak Ridge National Laboratory, P.O. Box 2009, Oak Ridge TN 37831, USA.

Abstract. For the (non-axisymmetric) stellarator class of plasma confinement devices to be feasible candidates for fusion power stations it is essential that, to a good approximation, the magnetic field lines lie on nested flux surfaces; however, the inherent lack of a continuous symmetry implies that magnetic islands are guaranteed to exist. Magnetic islands break the smooth topology of nested flux surfaces and chaotic field lines result when magnetic islands overlap. An analogous case occurs with $1\frac{1}{2}$-dimension Hamiltonian systems where resonant perturbations cause singularities in the transformation to action-angle coordinates and destroy integrability. The suppression of magnetic islands is a critical issue for stellarator design, particularly for small aspect ratio devices.

Techniques for 'healing' vacuum fields and fixed-boundary plasma equilibria have been developed, but what is ultimately required is a procedure for designing stellarators such that the self-consistent plasma equilibrium currents and the coil currents combine to produce an integrable magnetic field, and such a procedure is presented here for the first time.

Magnetic islands in free-boundary full-pressure full-current stellarator magnetohydrodynamic equilibria are suppressed using a procedure based on the Princeton Iterative Equilibrium Solver [A.H.Reiman & H.S.Greenside, *Comp. Phys. Comm.*, 43:157, 1986.] which iterates the equilibrium equations to obtain the plasma equilibrium. At each iteration, changes to a Fourier representation of the coil geometry are made to cancel resonant fields produced by the plasma. As the iterations continue, the coil geometry and the plasma simultaneously converge to an equilibrium in which the island content is negligible. The method is applied to a candidate plasma and coil design for the National Compact Stellarator eXperiment [G.H.Neilson et.al., *Phys. Plas.*, 7:1911, 2000.].

The magnetic field lines of toroidal plasma confinement devices, such as stellarators, are $1\frac{1}{2}$ dimensional Hamiltonian systems and magnetic flux-surfaces are the analog of constant action surfaces [1]. This may be seen by noting that in arbitrary toroidal coordinates (r,θ,ζ) any vector, in particular the magnetic vector potential, may be written $\mathbf{A} = \psi\nabla\theta - \chi\nabla\zeta + \nabla g$, where ψ, χ and g are functions of (r,θ,ζ): from which $\mathbf{B} = \nabla\psi\times\nabla\theta + \nabla\zeta\times\nabla\chi$. Using the toroidal angle ζ as the independent (time) coordinate, and considering $\chi = \chi(\psi,\theta,\zeta)$, the magnetic field line flow equations may be recast in a form identical to Hamilton's equations: $d_\zeta\theta = \partial_\psi\chi, d_\zeta\psi = -\partial_\theta\chi$.

Integrable $1\frac{1}{2}$ dimensional Hamiltonians naturally occur only in systems with a continuous symmetry, and stellarators have no continuous symmetry. Integrability can be studied by perturbing an integrable field $\mathbf{B}_0$. Writing $\mathbf{B} = \mathbf{B}_0 + \mathbf{B}_1$ and $p = p_0 + p_1$, the perturbed system is in equilibrium if $\mathbf{B}_0\cdot\nabla p_1 + \mathbf{B}_1\cdot\nabla p_0 = 0$. In magnetic coordinates this becomes $\iota\partial_\theta p_1 + \partial_\zeta p_1 = -p'_0 B_1^\psi/B_0^\zeta$. If this can be non-trivially solved for p_1, new magnetic coordinates exist and the perturbed state preserves integrability; however, the coefficients of $p_1 = \sum p_{1mn}\cos(m\theta - n\zeta)$ are given by the coefficients of (B_1^ψ/B_0^ζ) divided by $(\iota m - n)$. At rational rotational-transform surfaces, $\iota = n/m$, a singularity exists. This is the classical problem of small denominators and magnetic islands are formed with width in ψ given by $[(B_1^\psi/B_0^\zeta)_{mn}/\iota' m]^{1/2}$. Islands and the chaotic field lines caused by island overlap result in poor plasma confinement.

Changes in coil geometry will change $(B_1^\psi/B_0^\zeta)_{mn}$ and can be used to reduce islands and their associated stochastic regions. It is perhaps impossible to completely eliminate all resonant perturbation terms in non-symmetric systems, but this is too stringent a requirement as sufficiently small islands will have little, if any, effect on plasma confinement. All that is required for practical purposes is that the magnetic islands occupy less than a tolerable percentage of the plasma volume. Such a magnetic field is said to have 'good-flux-surfaces'.

The construction of vacuum magnetic fields with good-flux-surfaces is not trivial [2], but is simpler than the case when a plasma is present. The complexity arises as the plasma currents modify $(B_1^\psi/B_0^\zeta)_{mn}$, and the self-consistent solution requires that the plasma equilibrium field and the coil field combine to give zero resonant component at the rational rotational-transform surfaces. Previous studies of stellarator MHD equilibria have computed finite pressure equilibria with islands and even showed that 'self-healing' [3] can occur.

The article gives a procedure for adjusting the coil geometry to remove islands at the operating plasma configuration, without degrading the previous optimization of either the coils or the self-consistent plasma equilibrium. Plasma and coil design optimization relies on equilibrium codes. The fastest equilibrium codes presuppose the existence of perfect magnetic surfaces — the existence or size of magnetic islands cannot be addressed. To do this a more computationally intensive code is used to remove the islands by iterating between the plasma equilibrium and the coil geometry.

A recent article [4] presented a method by which high-pressure full-current *fixed*-boundary solutions may be constructed with good-flux-surfaces. Although stellarator coils must balance the normal field B_n produced by the plasma currents on the plasma surface, balancing B_n at each point on a arbitrary surface generically leads to singular coil currents. Nevertheless, a number of constraints on the normal field distribution may be satisfied.

The fixed-boundary healing work [4] showed that a given (m,n) island is controlled by a single spatial distribution of normal field B_n. One constraint that must be achieved is these distributions be nulled. This is achievable since, in most cases, it is the few low-order islands which are most problematic. Additional constraints in the present context are properties that must be preserved — in particular the optimized plasma properties (ideal stability, quasi-axisymmetry, ...) and the optimized engineering properties of the coil design.

Compared to traditional stellarator designs, the problem of resonances is enhanced for the National Compact Stellarator eXperiment (NCSX) [5], which is the present motivation for this work. NCSX is both compact, thus the lack of symmetry is pronounced, and has large transform per period and large shear, which results in multiple low order resonances. NCSX will have significant plasma current, thus the rotational transform profile will be quite different to that of the vacuum state, and will operate at high plasma pressure, thus the magnetic surfaces will be different to those of the vacuum. The introduction of a technique for 'healing' a given plasma-coil configuration to ensure good-flux-surfaces at the operating plasma configuration is critical for the design. The operating configuration of NCSX has been designed using an optimization algorithm to maximize quasi-axisymmetry (for control of particle orbits), subject to the constraint that the plasma is stable at an averaged plasma energy above 4% of the averaged magnetic energy.

Our method is based on the free-boundary Princeton Iterative Equilibrium Solver (PIES) code [6] which iterates the MHD equilibrium equations to solve for plasma equilibria in stellarator geometry and allows for the existence of islands. Island suppression is achieved by adding to the standard PIES algorithm a procedure that alters the coil geometry at each iteration so that the coil magnetic field cancels the resonant components of the plasma magnetic field. By continuously adjusting the coils as required, the inherent non-linearity of the plasma response is effectively controlled. The changes in coil geometry are constrained to preserve engineering constraints on minimum bend radius and minimum coil-coil separation, as well as the plasma constraint of ideal kink stability. As the iterations continue, the coil geometry and the plasma simultaneously converge to an island-free, stable-plasma with buildable coils.

The total magnetic field is the sum of the plasma field, $\mathbf{B}_P$, and the magnetic field produced by the confining coils, $\mathbf{B}_C$, which is a function of a set of Fourier harmonics, ξ, that describe the coil geometry, at the nth iteration

$$\mathbf{B}^n = \mathbf{B}^n_P + \mathbf{B}_C(\xi^n). \qquad (1)$$

The initial plasma state is provided by the VMEC code [7], which imposes the artificial constraint that the plasma has nested flux-surfaces, and the initial coil geometry is provided by the COILOPT code [8]. The method presented in this article can be considered as removing the constraint of nested surfaces and allowing the VMEC initialization to relax into an equilibrium, potentially with broken flux-surfaces (islands), while making adjustments to the coils to remove islands as they develop. The PIES iterations solve for the plasma current $\mathbf{J}$ given $\mathbf{B}$ and given pressure profile p

$$\nabla p = \mathbf{J}^{n+1} \times \mathbf{B}^n. \qquad (2)$$

A magnetic-differential equation $\mathbf{B} \cdot \nabla(J_\parallel/B) = \nabla \cdot J_\perp$ gives the parallel current which is solved using magnetic coordinates [9]. The PIES code allows the field topology to break up into islands and chaos. In such regions the magnetic-differential equation need not be solved because the current and pressure profiles are flattened, which eliminates the singular parallel currents. The plasma magnetic field is then solved given $\mathbf{J}$, and blended to provide numerical stability :

$$\mathbf{J}^{n+1} = \nabla \times \mathbf{B}_P \quad ; \quad \mathbf{B}^{n+1}_P = \alpha \mathbf{B}^n_P + (1-\alpha)\mathbf{B}_P. \qquad (3)$$

Typically the blending parameter $\alpha = 0.99$ for NCSX style equilibria. The standard PIES algorithm makes no changes to the coil geometry and iterates through equations (2,3) to calculate the free-boundary equilibrium for a given pressure profile and coil set. The additional steps in the implementation of the coil-healing are as follows. The total magnetic field $\bar{\mathbf{B}}$ is written

$$\bar{\mathbf{B}} = \mathbf{B}_P^{n+1} + \mathbf{B}_C(\xi^n). \qquad (4)$$

We may consider $\bar{\mathbf{B}}$ as a *nearly* integrable field and that magnetic islands are caused by fields normal to and resonant with rational rotational-transform flux-surfaces of a *nearby* integrable field.

A set of resonances that are to be suppressed is selected. The selection is determined by the rotational-transform profile. Islands associated with low-order rationals are typically the largest, but where the shear is small higher-order islands can easily overlap and result in chaotic field lines. A set of toroidal surfaces matching the selected resonances is constructed. Each such surface (in fact a quadratic-flux-minimizing surface [10]) may be considered as a rational rotational-transform flux-surface of an underlying integrable field [11], with each surface passing directly through its associated island chain and containing the stable and unstable periodic orbits. The construction of the quadratic-flux-minimizing surfaces provides an optimal magnetic coordinate system, or equivalently an optimal nearby integrable magnetic field, and in these coordinates resonant perturbation harmonics are clearly identified. The method is computationally efficient as the quadratic-flux-minimizing surfaces may be constructed exactly and only where required — at the rational rotational-transform surfaces where islands develop.

The amplitude of each of the N selected resonant field harmonics, denoted $\{\bar{B}_i : i = 1, N\}$, is calculated by Fourier decomposing the magnetic field normal to the quadratic-flux-minimizing surface. The Fourier decomposition is performed using an angle coordinate which corresponds to magnetic coordinate of the underlying integrable field on that surface. The COILOPT [8] code provides a convenient Fourier representation of the coil geometry and a set of M coil harmonics $\{\xi_j : j = 1, M\}$ is systematically varied to set $\bar{B}_i = 0$ using a Newton method. The coupling matrix, ∇B_{Cij}^n, is defined as the partial derivatives of the selected resonant harmonics of the coil magnetic field normal to the quadratic-flux-minimizing surface, which is updated every PIES iteration, with respect to the chosen coil harmonics and is calculated using finite-differences. A multi-dimensional Newton method is applied to find the coil changes $\delta \xi_j$ that set $\bar{B}_i = 0$

$$-\bar{B}_i = \nabla B_{Cij}^n \cdot \delta \xi_j^n. \qquad (5)$$

This equation is solved for the $\delta \xi_j$ in a few iterations by inverting the $N \times M$ matrix ∇B_{Cij}^n using singular-value decomposition [12] and the coil set is adjusted $\xi_j^{n+1} = \xi_j^n + \delta \xi_j^n$, at every PIES iteration, such that resonant components of the combined plasma-coil field are eliminated. The algorithm returns to Eqn(1). As the iterations proceed, the coil geometry and the plasma simultaneously converge to coil geometry-plasma solution with good-flux-surfaces.

To be 'build-able', the minimum coil-curvature and coil-coil separation, for example, of the coils must exceed certain limits. Such constraints are calculated by the COILOPT code and the initial coil set, described by ξ^0, is satisfactory from an engineering perspective. The healing algorithm is modified to preserve the minimum curvature and coil-coil separation by adding to the set of resonant fields to be eliminated the (appropriately weighted) differences in minimum curvature and coil-separation of the nth coil set, described by ξ^n, from the initial coil set. This constrains the island-eliminating coil variations to lie in the nullspace of these measures of engineering acceptability. In a similar manner, the algorithm preserves kink stability. The VMEC initialization is kink-stable, and by calculating kink stability using VMEC and the TERPSICHORE code [13], the coil changes are constrained to preserve kink stability.

The method is routinely applied to NCSX [5] candidate coil and plasma designs. NCSX is a proposed proof-of-principle device with three field periods, aspect ratio A=4.4, major radius R=1.4m and magnetic field B=1.7T. The stellarator symmetric coil design consists of 18 modular coils (3 distinct coil types), 18 toroidal field coils, and six pairs of poloidal field coils and some additional trim coils. The plasma is designed to be quasi-axisymmetric to give good transport, and is stable to kink modes at $\beta \sim 4\%$, but is marginally unstable to infinite-n ballooning modes. The rotational-transform profile has $\iota \sim 0.4$ on axis, maximum $\iota \sim 0.66$ near the edge and $\iota \sim 0.65$ at the edge: including the low order resonances $\iota = 3/7, 3/6$ and $3/5$. Note that the shear vanishes near the $\iota = 6/9$ resonance.

Considering a candidate coil set (named M45) and selecting the $(n,m) = (3,6), (3,5)$ islands to be suppressed, subject to the constraint that the minimum coil curvatures, the coil-coil separation and the kink stability be preserved (9 constraints), and allowing some $m = 3,4,5,6,7,8$ modular coil harmonics to vary (36 independent variables), a healed coil-plasma state is achieved. The engineering measures are preserved and the plasma is stable with respect to kink modes. Also, the plasma retains quasi-axisymmetry and is stable to ballooning modes $n < 45$.

Several hundred iterations are required to approach convergence in both the plasma field and the coil geometry. To confirm convergence several hundred additional standard PIES iterations are performed with the coil set unchanged. A

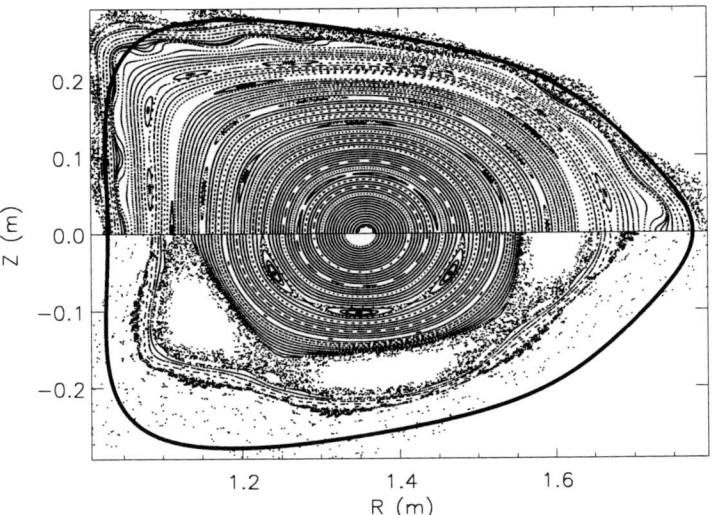

FIGURE 1. Poincaré plot of the converged healed coil-plasma field (upper) and for the original, unhealed coils after 180 standard PIES iterations (lower) for the NCSX candidate coil set M45. The VMEC initialization boundary is shown as the thick solid line. The island content in the healed configuration is negligible, though there is some resonant $m = 18$ deformation near the zero shear location and some high order ($m = 10, 11, 12,$ and 14) island chains. For the unhealed case there is a large $m = 5$ island and the configuration deteriorates into large regions of chaos.

Poincaré plot of the final field is shown on an up-down symmetric toroidal cross-section in the upper half of fig(1). For comparison, a Poincaré plot of the unhealed configuration is shown after 180 standard PIES iterations in the lower half of the figure. The maximum coil alteration is about 2cm, which comfortably exceeds manufacturing tolerances, but is not so large that 'healing' significantly impacts other design concerns, such as diagnostic access. The coil harmonics varied actually describe the toroidal variation of the modular coils on a toroidal winding surface. The calculation shown used 63 radial surfaces, 12 poloidal and 6 toroidal modes. Similar results have been obtained using up to 93 radial surfaces and 20 poloidal modes.

The flux-surface quality of the 'healed' equilibrium shows remarkable improvement compared to the unhealed configuration. The coils have been described with a filamentary model, and a finite thickness model of the healed coils shows further improvement, in particular the $m = 18$ deformation and the high order islands are reduced. In principle, in the limit of suppressing additional islands, this approach can lead to non-axisymmetric coil-plasma configurations with integrable magnetic fields. The procedure amounts to a stellarator design optimization routine that for the first time provides a mechanism for suppressing magnetic islands, while providing ideal stability and satisfying engineering constraints and is an important advance for the design of stellarator experiments.

We thank the NCSX design team, in particular Long-Poe Ku and Guo-Yong Fu for stability analysis. One of us (SRH) is deeply indebted to Allen Boozer for suggesting the coupling matrix approach and advice regarding the text. This work was supported in part by US Department of Energy contract DE-AC02-76CH03073.

REFERENCES

1. J.R. Cary and R.G. Littlejohn. *Annals of Physics* 151:1, 1983; A.H. Boozer. *Phys. Plas.*, 26(4):1288, 1983.
2. J.R. Cary. *Physical Review Letters*, 49(4):276, 1982; S.R. Hudson and R.L. Dewar. *Phys. Lett. A*, 226:85, 1997.
3. A. Bhattacharjee et al. *Phys. Plas.*, 2(3):883, 1995; S.S. Lloyd et al. *J. Plas. and Fus. Res. SERIES*, 1:484, 1997.
4. S.R. Hudson, D.A. Monticello, and A.H. Reiman. *Phys. Plas.*, 8(7):3377, 2001.
5. G.H. Neilson et al. *Phys. Plas.*, 7:1911, 2000.
6. A.H. Reiman and H.S. Greenside. *Comp. Phys. Comm.*, 43:157, 1986.
7. S.P. Hirshman and O. Betancourt. *J.Comp. Phys.*, 96:99, 1991.
8. D.J. Strickler, L.A. Berry, and S.P. Hirshman. *Fusion Science and Technology*, 41, 2001.
9. A.H. Reiman and H.S. Greenside. *J.Comp. Phys.*, 75(2):423, 1988.
10. R.L. Dewar, S.R. Hudson, and P. Price. *Phys. Lett. A*, 194:49, 1994.
11. S.R. Hudson and R.L. Dewar. *Phys. Plas.*, 6(5):1532, 1999.
12. W.H. Press, B.P. Flannery et al. *Numerical Recipes in Fortran 77*. Cambridge University Press, Cambridge, U.K., 1992.
13. D.V.Anderson, W.A.Cooper et al. *Scient. Comp. Supercomputer* II, 159 1990.

Transport and Stability Analysis of Low q_a Discharges

S.Lahiri[*] and S.Mukhopadhyay[†]

[*]*Department of Physics, Techno India Institute of Technology, India*
[†]*EHEP division, Saha Institute of Nuclear Physics, India*

Abstract. An one-dimensional stability transport code (PROSIM) [1] has been used to analyze the accessibility of low q_a discharges obtained in the SINP tokamak. In order to evolve the transport properties, the code solves the electron temperature evolution equation and the poloidal magnetic field diffusion equation. To incorporate the effects of various MHD instabilities which are likely to have substantial effect on the evolution of the discharge, equations governing the MHD mixing and tearing modes have been solved at each time step. Since reverse shear profiles have been observed in the rising phases of low q_a discharges, the stability calculations have been carried out for plasmas having multiple resonant surfaces. Experimental observations [2] indicate the presence of large positive spikes just before the ultra low q_a (ULQ) regime is entered. In this paper we present detailed analysis of plasma evolution near the positive spike in the loop voltage. We have tried to relate our results to the complicated question of accessibility of low q_a discharges. It has been found that during this phase, the $m = 1, n = 1$ plays a very important role despite the fact that the edge safety factor remains considerably above one. In a parametric study, the effect of the rate of rise of the plasma current on the growth of various instabilities and on the accessibility of ULQ discharges have been carried out. Finally, we have compared the numerical results with experimental ones and found the agreement to be satisfactory. Thus, it has been concluded that the stability transport solver, despite its simplicity, incorporates most of the important physics processes occuring in the rising phase of the ULQ discharges of the SINP tokamak.

INTRODUCTION

Computer simulation has been found to be extremely successful in exploring intricate physical phenomena, and thus help in a better understanding of experimental situations. We have attempted to develop a simple one-dimensional numerical model to simulate the evolution of current density (j_ϕ) profile and related parameters for different discharges obtained in the SINP tokamak [3]. In this paper, the principal aim of the present study was to simulate the current density profile evolution in the rising phase of the low q_a ($q_a < 1$) discharges obtained experimentally in the SINP tokamak because the setting up phase has been found to be utmost importance for these discharges. It has been found that consideration of the evolution of particle density improves the numerical prediction by a significant amount.

DESCRIPTION OF THE PRESENT MODEL

The geometry of the model [4] is cylindrical and is applicable to a tokamak provided the aspect ratio is sufficiently large. The model is divided into two parts: 1) Transport and 2) Stability.

In the transport model, we have considered two coupled differential equations (1 and 2) where eq 1 is the poloidal magnetic field (B_θ) diffusion equation and eq 2 is the electron temperature (T_e) evolution equation. The diffusion equation is written as

$$\frac{\partial B_\theta}{\partial t} = \frac{\partial E(r)}{\partial r} \qquad (1)$$

where E(r) is the axial electric field. The temperature evolution is represented by:

$$\frac{\partial T_e}{\partial t} = \frac{1}{3n_e} \times [jE + \frac{1}{r} \times \frac{\partial}{\partial r}(rn_e\chi_e\frac{\partial T_e}{\partial r} + \frac{5}{2}rT_eD_e\frac{\partial n_e}{\partial r}) - R(T)] \qquad (2)$$

where j, χ D and R are the current density, thermal conductivity, diffusion and the radiation loss factors respectively. The boundary conditions for eq (2) are $\frac{\partial T_e}{\partial r} = 0$, at $r = 0$ and $T_e = T_a$ at $r = a_p$, where T_a is the edge value of tempetature, a_p is the minor radius of the plasma.

In order to incorporate the effect of MHD instabilities, resistive stability analysis has been carried out to follow the growth and decay rates of a large number of MHD modes. Different methods like empirical, mhd mixing [5] and tearing mode stability [6] analysis has been incorporated as and when required to incorporate the effects of resistive instabilities which are expected to be of importance for the low q_a discharges.

RESULTS AND DISCUSSIONS

Different models to simulate the diffusion, thermal conductivity and radiation suitable for the SINP tokamak discharges [1] have been obtained through an extensive numerical scan of the available parameters of equation 2. Going through numerical experiments, we concluded that for diffusion Bohm model, for thermal conductivity Ohkawa model are suitable for the low q_a discharges. A satisfactory model of impurity distribution for the SINP tokamak low q_a plasma may be the following: 1) The distribution approximately follows the plasma density distribution 2) Low Z impurities remain below 2% 3) High Z impurities remain below 0.5%

To reconstruct the experimental scenario of a low q_a discharge several parameters were supplied as inputs. In fig.1, we show such a representative discharge with $q_a = 0.8$. In this paper, we have presented numerical results for the same discharge. The various input parameters were as follows: Major Radius: 0.3 m, Minor Radius: 0.055 m, Toroidal magnetic field: 0.308 T (ULQ) and 0.88 T for (NQ), Thermal Conductivity: Ohkawa Model, Diffusion: Bohm Model, Impurity content: Carbon: 2.0%, Oxygen: 2.0%, Iron: 0.5%, Tungsten: 0.5%, Modes Considered: 4/1, 3/1, 2/1, 3/2, 1/1, 4/5, 3/4, 2/3, 1/2 The value of the total plasma current at each instant (obtained experimentally) was taken into account.

Intially, we assumed that the plasma particle density remains an invariant throughout the evolution of the discharge. In this case, plasma particle density at the centre: $2 \times 10^{19} m^{-3}$ and at the edge: $1 \times 10^{18} m^{-3}$. Such an assumption led to an acceptable result [1] but there were some discrepancies particularly near the positive jump of the loop voltage. In particular the loop voltage was found to be considerably higher than the experimentally measured value in the rising phase of the discharge (figure. 3). In order to obtain a more realistic prediction from the simulation, we have upgraded the solver and considered an evolution of the particle density (in both magnitude and shape) in the present work. Since we do not have experimentally measured data of the above parameters for the ULQ discharges of the SINP tokamak, we assumed a certain evolution of the magnitude and shape of the particle density distribution using information available in the references [7], [8].

For the results presented here, the magnitude and shape of the particle density is shown in figure 2. It can be seen that the density increases with time in the rising phase. The shape is assumed to be hollow in the early rising phase, after which it changes into a flat parabola and after the positive jump in the loop voltage, a sharp peaked one.

In figure 3, we compare the experimental loop volatge with the simulated ones. It can be easily seen that the computation which took care of the temporal evolution of particle density yields considerably better match with the experimental result, especially in the early rising phase.

In figure 4, we compare the measured and simulated values of the resistance. It is again noted that the computation which took care of the temporal evolution of particle density yields considerably better match with the experimental result, especially in the early rising phase. The Physics of the behaviour around the beginning of the positive jump is rather complicated and is under serious investigation at present.

CONCLUSIONS

The current penetration problem is known to be a highly nonlinear one, involving a competition between current diffusion, Ohmic heating, thermal transport and stability. Despite the simplicity of the numerical model, it is found that the simulation reproduced the global features of the experimental observations. When the density evolution is considered the match is considerably better compared to the case when the magnitude and shape of the density is fixed throughout the discharge. We plan to improve the code by incorporating anomaly in the values of χ, η and D which may be present during the discharge evolution (especially in the fluctuation dominated ULQ regime), surface wall interaction, interaction among different modes etc.

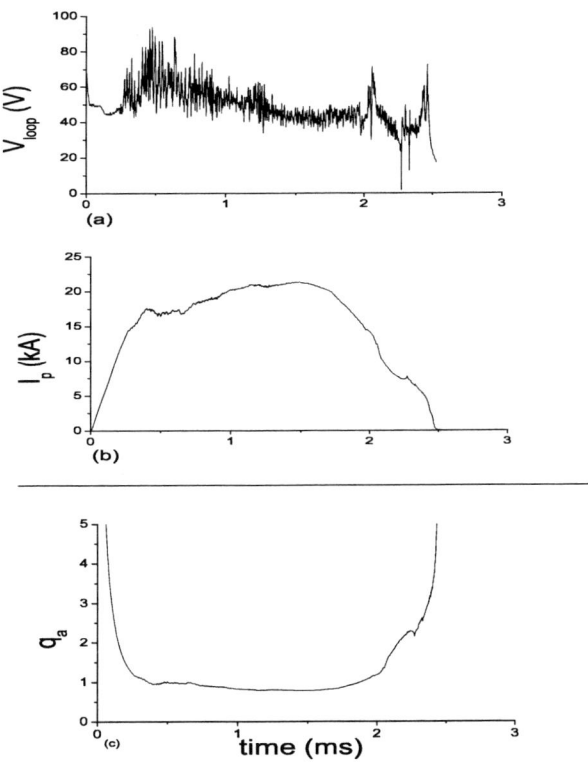

FIGURE 1. (a) Loop voltage, (b) Plasma current, (c) Edge safety factor (q_a)

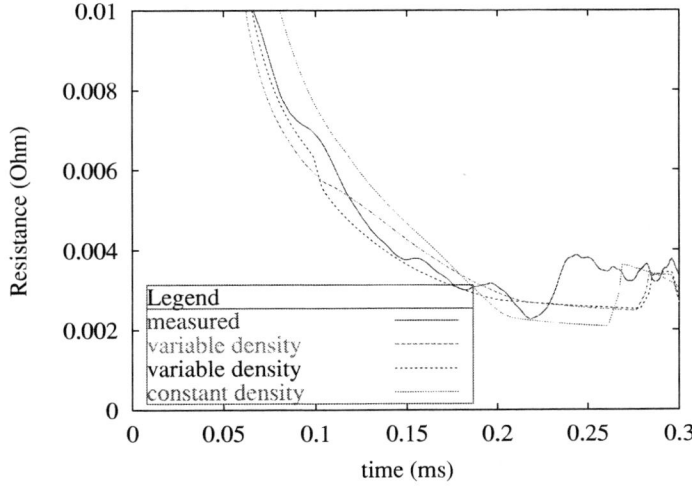

FIGURE 2. Evolution of particle density

ACKNOWLEDGMENTS

We acknowledge the Directors of Techno India and Saha Institute of Nuclear Physics for their constant encouragements and all the members of EHEP division, S.I.N.P for their kind help to carry out this work and to prepare the document.

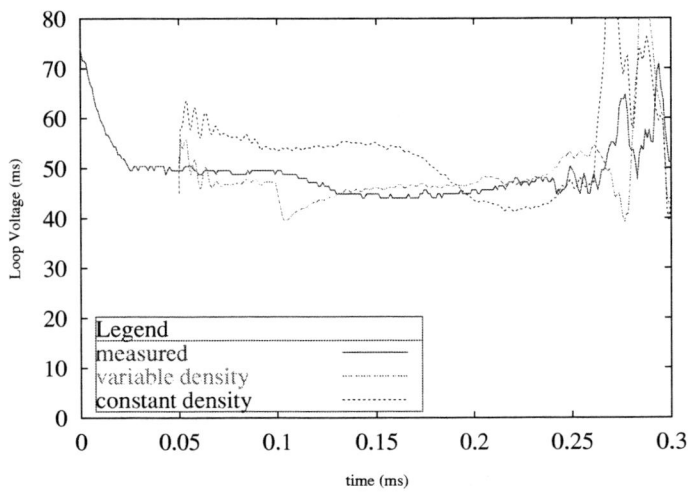

FIGURE 3. Comparison of experimental and simulated loop voltage

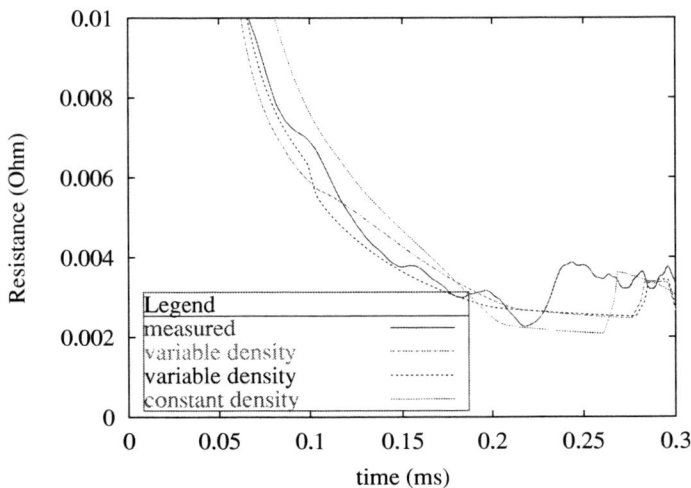

FIGURE 4. Comparison of measured and simulated plasma resistance

REFERENCES

1. S.Lahiri, S.Mukhopadhyay, A.N.S. Iyengar, R. Pal, Pramana -j. Phys, 56, 615 (2001)
2. S.Lahiri, A.N.S. Iyengar, S. Mukhopadhyay, Nuclear Fusion, 36, 254 (1996) 567 (2005)
3. S.Lahiri, A.N.S. Iyengar, S.Mukhopadhyay and R. Pal, Pramana -j. Phys, 58, No. 1, 79 (2002)
4. Turner, M.F., Wesson, J.A., *Nucl. Fusion*, **22**, 1069 (1982).
5. Dnestrovskii, Y.N., Kostomarov, D.P., *Numerical Simulation of Plasmas*, (Springer-Verlag, Berlin, Heidelberg 1986).
6. Wesson, J., *Tokamaks*, (Clarendon Press, Oxford, 1987).
7. H. Morimoto, Y. Kamada, T. Fujita, Y. Murakami, Y. Fuke, K. Saito, H. Nihei, J. Morikawa, Z. Yoshida, N. Inoue., *Nuclear Fusion*, **29**, 1171 (1989).
8. Yu. N. Dnestrovskij, G.V. Pereverzev, *Nuclear Fusion*, **23**, 633 (1983).

Magnetic Island Growth
A comparison of local and global effects

S.S.Lloyd[*] and H.J.Gardner[†]

[*]Department of Theoretical Physics, Australian National University, ACT 0200, Australia
[†]Department of Computer Science, Australian National University

Abstract. In stellarators a hot plasma is confined to a torus by a magnetic field with both toroidal and poloidal components generated by external currents. Plasma currents develop to balance the pressure gradient with a $\mathbf{J} \times \mathbf{B}$ force which in turn change the shape of confining magnetic field.

Self-consistent equilibrium magnetic fields and plasma currents for some H-1NF configurations were calculated using the HINT code. This code relaxes a simplified set of resistive MHD equations on a coordinate grid until an equilibrium is reached [1].

Islands can occur in the equilibrium magnetic field, surrounding field lines with low-order rational rotational transform. The island widths are influenced by four types of currents. External currents determine the vacuum island widths. Global resonant and non-resonant currents increase linearly with plasma pressure and can act in or out of phase to the external currents. Local resonant currents are caused by the presence of an island and reinforce or counteract the island depending on the field strength gradient [2].

We compare the impact of local resonant and global non-resonant currents by comparing the results of HINT for several related configurations of H-1NF. Two configurations with slightly different rotational transforms (but otherwise very similar parameters) will have very different resonant plasma currents but nearly identical non-resonant plasma currents. Comparing the effect of the currents of the two configurations on island width gives an insight into the different contributions of resonant and non-resonant plasma currents to island growth or self-healing.

INTRODUCTION

The current carrying coils of a stellarator are arranged to produce a magnetic field with nested, toroidal magnetic surfaces. Transport of charged particles across these surfaces is slow so a plasma pressure can be contained, with constant pressure on each magnetic surface.

On a magnetic surface the field lines wind around with rotational transform ι. If ι is irrational a single field line will cover the entire magnetic surface. If ι is a low order rational the field lines close on themselves after a small number of periods. Magnetic islands may occur at these rational surfaces. Even if islands are not present in the vacuum field, they are likely to develop in a finite pressure equilibrium from perturbations caused by the plasma current. The pressure is flat within an island and so islands degrade plasma confinement. Islands can also be surrounded by regions of chaotic fieldlines which further degrade confinement. A better understanding of island growth may allow islands to be avoided or minimised in future stellarator designs enabling them to reach higher plasma pressure.

PLASMA CURRENTS AND MAGNETIC ISLANDS

Electric currents in the plasma produce a force which balances the pressure gradient so that at equilibrium $\mathbf{J} \times \mathbf{B} = \nabla p$. This force balance equation determines the component of the plasma current perpendicular to the magnetic field.

The parallel component of the plasma current is determined by charge conservation, giving $\nabla \cdot \mathbf{J}_\parallel = -\nabla \cdot \mathbf{J}_\perp$. It is convenient to work in magnetic coordinates where the contained toroidal flux, ψ is used to label magnetic surfaces and poloidal and toroidal coordinates θ and ϕ are chosen to give straight field lines[3]. The parallel current can then be separated into harmonics of the magnetic coordinates θ and ϕ.

$$\mathbf{J}_\| = \sum_{l,m} Q_{l,m}(\psi) e^{i(l\theta - m\phi)} (\nabla\psi \times \nabla\theta + \iota(\psi)\nabla\phi \times \nabla\psi) \qquad (1)$$

If the plasma pressure is small the plasma equilibrium can be treated as a linear perturbation from the vacuum field. This gives a solution for $Q_{l,m}$ in terms of properties of the vacuum field[4]:

$$Q_{lm} = \frac{-p'(\psi) \mathcal{J}_{lm}(\psi)}{\iota(\psi) - m/l} \qquad (2)$$

Here $\mathcal{J}_{lm}$ are the harmonics of the Jacobian for the transformation to magnetic field coordinates of the vacuum field.

Near a rational surface, where $\iota \approx m/l$, the resonant harmonic of the parallel current will be very large, changing sign as the surface is crossed. This resonant parallel current will influence the size of the magnetic island that develops at the rational surface. If the magnetic surfaces were circular the resonant component of the parallel current would be the only portion of the plasma current that would contribute to the growth of this island. If the three dimensional shape of the plasma is taken into account there can be cross coupling with non-resonant plasma currents. There will also be a local non-linear resonant plasma current due to the changing shape of the magnetic surfaces near an island[4, 5].

The width of a magnetic island at finite pressure is determined by the size of the vacuum island (w_v), a contribution (C) from the resonant component of the linear current, a contribution (E) from non-resonant components of the linear current and a contribution (G) from the resonant component of the local non-linear current. These components combine to give a total island width:

$$w = G/2 + \sqrt{(G/2)^2 + |E + C \pm w_v^2|} \qquad (3)$$

At small pressure G, C and E are all proportional to the pressure. G may be positive or negative depending on vacuum field parameters. Depending on the values of G, C and E for a given vacuum field, the island size may increase or decrease as pressure increases, or it may decrease to zero and then grow again with opposite phase.

In order to better understand, predict and minimise island sizes in stellarators it would be helpful to be able to separate the contributions of the various currents to island growth.

A COMPUTATIONAL EXPERIMENT

Consider a vacuum field $\mathbf{B_v}$ that contains a rational surface ψ_r where $\iota(\psi_r) = m_r/l_r$. This would give $Q_{m_r,l_r}(\psi)$ that was large near ψ_r, as was discussed in the previous section. The equilibrium field $\mathbf{B_v} + \mathbf{B_p}$ will contain an island with width determined by equation 3.

Compare this to a vacuum field $\mathbf{B_v^*}$ that is very similar to B_v except that the rotational transform has been shifted so that there is no rational surface $\iota^*(\psi) = \iota(\psi) + \Delta\iota$. This would give a smooth (and comparitively small) resonant parallel current, $Q^*_{m_r,l_r}(\psi)$, but leave the non-resonant components almost unchanged. A magnetic field could be created from the original vacuum field $\mathbf{B_v}$ with the field $\mathbf{B_p^*}$ due to the plasma currents of the vacuum field $\mathbf{B_v^*}$ with ι shifted. The magnetic field $\mathbf{B_v} + \mathbf{B_p^*}$ would have a magnetic island with width $w = \sqrt{|E + C^* \pm w_v^2|}$.

Comparing the island size in the field $\mathbf{B_v} + \mathbf{B_p^*}$ with that of the equilibrium field, $\mathbf{B_v} + \mathbf{B_p}$, will allow us to determine how much of the island pressure response is due to the linear resonant current and how much is due to local, nonlinear resonant currents or to non-resonant currents.

METHOD

The H-1NF Heliac is a flexible stellarator built at ANU using mostly planar, circular coils[6]. A helical winding wound around the main poloidal field coil allows the rotational transform to be altered. For small helical currents the rotational transform is alteed uniformly across the plasma radius and other characteristics of the magnetic field are not significantly effected. Because the variation of ι across the plasma is small (low shear) only a small change is needed to remove the rational surface and alter the resonant current. Vertical field coils can be used to position and shape the magnetic surfaces, as well as alter other field parameters such as well depth.

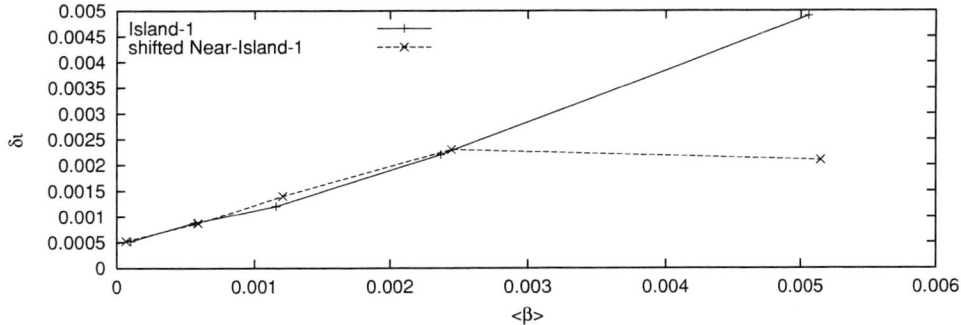

FIGURE 1. A comparison of island width found in the Island-1 configuration and for a shifted Near-Island-1 configuration.

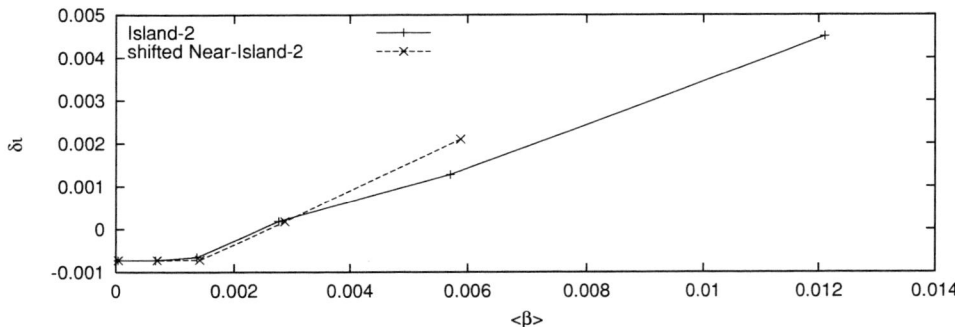

FIGURE 2. A comparison of island width found in the Island-2 configuration and for a shifted Near-Island-2 configuration.

H-1NF provides the closely related magnetic fields needed to carry out the above experiment. This method was used to examine the islands in two configurations. In one of these the island width increased gradually with increasing pressure (Fig. 1), while in the other the islands changed phase before growing (Fig. 2).

The equilibrium currents and their related magnetic fields were calculated using the HINT code [1]. HINT solves for a self consistent equilibrium field and plasma current on a fixedgrid. The initial state of the system is given by the vacuum field and a low pressure specified as a function of the contained flux. This state is relaxed to equilibrium via a simplified set of resistive MHD equations, with artificially high resistivity.

$$\mathbf{J} = \nabla \times \mathbf{B}$$
$$\frac{\partial \mathbf{v}}{\partial t} = \mathbf{J} \times \mathbf{B} - \nabla p - \mu \nabla^2 \mathbf{v}$$
$$\frac{\partial \mathbf{B}}{\partial t} = \nabla \times (\mathbf{v} \times \mathbf{B} - \eta \mathbf{J}) \qquad (4)$$

This field relaxation proceeds with a fixed pressure distribution. The field relaxation is alternated with a pressure relaxation which averages the pressure along field lines. The pressure relaxation algorithm has been improved to allow for the long field line tracing needed to acheive pressure flattening in islands of low shear stellarators[7].

RESULTS AND CONCLUSIONS

For both configurations we looked at the island sizes of the equilibrium magneticfields were almost identical to those in the corresponding shifted fields (Fig. 1 and 2). This indicates that most of the change in island size at low pressure is due to non-resonant current. At higher pressure the island widths of the equilibrium fields deviate significantly from the shifted fields. At this point the island widths have become significant compared to the HINT grid spacing and it is likely that nonlinear local currents are strongly influencing the island size.

ACKNOWLEDGMENTS

The authors would like to acknowledge the support of the Australian Institute for Nuclear Science and Engineering (AINSE).

REFERENCES

1. Harafuji, K., Hayashi, T., and Sato, T., *Journal of Computational Physics*, **81**, 169–192 (1989).
2. Bhattacharjee, A., Hayashi, T., Hegna, C. C., Nakajima, N., and Sato, T., *Physics of Plasmas*, **2**, 883–888 (1995).
3. Boozer, A. H., *Phys Fluids*, **24**, 1999 (1981).
4. Cary, J. R., and Kotschenreuther, M., *Phys. Fluids*, **28**, 1392–1401 (1985).
5. Hegna, C. C., and Bhattacharjee, A., *Phys. Fluids B*, **1**, 392–397 (1989).
6. Hamberger, S. M., Blackwell, B. D., Sharp, L. E., and Shenton, D. B., *Fusion Technology*, **17**, 123–130 (1990).
7. Lloyd, S. S., *Simulating Magnetic Islands in the H-1NF Heliac with the HINT Code*, Ph.D. thesis, Australian National University (2002).

Calculation of Eddy Currents in the ETE Spherical Torus

G.O. Ludwig

Instituto Nacional de Pesquisas Espaciais, 12227-010 S. J. Campos, SP, Brazil

Abstract. The currents induced during startup in the vacuum vessel of the ETE spherical torus (Experimento Tokamak Esférico) are evaluated using a circuit model based on the Green's function method. The distribution of eddy currents is calculated using a thin shell approximation for the vessel and local curvilinear coordinates. The predicted results agree quite well with values of the eddy currents measured in ETE.

INTRODUCTION

This paper presents a magnetostatic model developed to evaluate the currents induced during startup in the continuous vacuum vessel of the ETE spherical torus. The distribution of eddy currents is modeled using a thin shell approximation for the vessel. The equation governing the induction of surface current on the thin shell is derived using the Green's function method. Symmetry considerations, and adoption of both a local curvilinear coordinates system and a spectral representation for the contour of the vacuum vessel reduce this three-dimensional problem in space to one dimension. The resulting one-dimensional integral equation for the surface current can be solved expanding the current in a Fourier series in the poloidal angle. Finally, by the introduction of Laplace transformation in time, the problem for the set of Fourier components of the surface current is reduced to a circuit model that can be solved by matrix procedures. The results are compared with preliminary measurements of the eddy currents in ETE.

FORMULATION OF THE MAGNETOSTATIC PROBLEM

The surface current density in a thin shell of thickness δ is given in terms of the current density by $\vec{K} = \delta \vec{j}$ [1], where the current density is related to the electric field by Ohm's law, $\vec{j} = \sigma \vec{E}$. Application of Faraday's law for a constant conductivity σ leads to

$$\nabla \times \vec{K} = \sigma \delta \nabla \times \vec{E} = -\sigma \delta \left(\partial \vec{B} / \partial t \right),$$

where $\vec{B}$ corresponds to the total induction. The condition of current continuity gives $\nabla \cdot \vec{K} = \sigma \delta \nabla \cdot \vec{E} = 0$.

For an axisymmetric configuration the problem is independent of the toroidal angle ζ. Furthermore, the variation of the toroidal flux in time, $\partial \Phi_T / \partial t$, is neglected during startup, and so no poloidal currents are induced on the vacuum vessel. In this case, axisymmetry and the solenoidal property of the magnetic field, $\nabla \cdot \vec{B} = 0$, imply a single toroidal component of the vector potential. In vector form the potential is given in terms of the poloidal flux Φ_P by $\vec{A} = (2\pi)^{-1} \Phi_P \nabla \zeta$. In the same way, the surface current vector is expressed in terms of the single toroidal component K_T by $\vec{K} = h_\zeta K_T \nabla \zeta$, where the scale factor $h_\zeta = |\partial \vec{r} / \partial \zeta|$ corresponds to the radial distance to the symmetry axis in cylindrical coordinates. Now, the magnetic induction is calculated in terms of the poloidal flux by

$$\vec{B} = \nabla \times \vec{A} = -(2\pi)^{-1} \nabla \zeta \times \nabla \Phi_P.$$

This equation, combined with the previous one and the assumption of an uniform distribution over the small thickness δ, leads to a relation between the toroidal surface current density and the local value of the poloidal flux:

$$K_T = -\sigma \delta (2\pi h_\zeta)^{-1} (\partial \Phi_P / \partial t).$$

In general, the vector potential at any point $\vec{r}$ not lying on the surface S' is given by the extension of the Biot-Savart law

$$\vec{A}(\vec{r}) = \frac{\mu_0}{4\pi} \iint_{S'} \frac{\vec{K}(\vec{r}')}{|\vec{r} - \vec{r}'|} d^2 r' + \vec{A}_{ext}(\vec{r}),$$

where $\vec{A}_{ext}$ stands for the external sources. The differential element of area in the coordinate surface ρ that coincides with the surface layer of current is $d^2r(\rho) = h_\zeta d\ell(\theta) d\zeta$. Using the properties $|\nabla \zeta|^2 = h_\zeta^{-2}$ and $\nabla \zeta \cdot \nabla \zeta' = \cos(\zeta - \zeta')/(h_\zeta h_{\zeta'})$ the equivalent integral relation for the flux function is

$$\Phi_P(\vec{r}) = \mu_0 \oint K_T(\vec{r}') \left\langle \frac{\pi h_\zeta h_{\zeta'} \cos(\zeta - \zeta')}{|\vec{r} - \vec{r}'|} \right\rangle_{\zeta'} d\ell(\theta') + \Phi_{ext}(\vec{r}),$$

where $\langle ... \rangle_\zeta = (2\pi)^{-1} \int (...) d\zeta$. This defines the Green's function for the axisymmetric Ampère's law, $G(\vec{r}, \vec{r}') = \langle \pi h_\zeta h_{\zeta'} \cos(\zeta - \zeta')/|\vec{r} - \vec{r}'|\rangle_{\zeta'}$. The Green's function integral for Φ_P automatically satisfies the boundary condition $\hat{n} \cdot \nabla \Phi_P = -2\pi \mu_0 h_\zeta K_T$, which corresponds to the discontinuity of the magnetic induction across the surface layer of current, $\hat{n} \times [\vec{B}]_S = \mu_0 \vec{K}$ ($\hat{n}$ is the unit normal). Finally, taking the derivative with respect to time and using the relation between K_T and $\partial \Phi_P / \partial t$ provided by Faraday's law, the excitation of Foucault currents in a thin axisymmetric shell is governed by the equation

$$\frac{2\pi h_\zeta}{\sigma \delta} K_T(\vec{r}) = -\mu_0 \oint \frac{\partial K_T(\vec{r}')}{\partial t} G(\vec{r}, \vec{r}') d\ell(\theta') - \frac{\partial \Phi_{ext}(\vec{r})}{\partial t}. \quad (1)$$

This equation has local terms depending on the shell resistivity and non-local terms depending on mutual inductance effects between diverse regions of the current distribution. The total toroidal current induced in the shell is $I_T = \int_0^{2\pi} K_T(\theta) h_\theta d\theta$, where the scale factor $h_\theta = |\partial \vec{r}/\partial \theta|$.

SPECTRAL REPRESENTATION OF THE ETE VACUUM VESSEL

In order to apply effectively the one-dimensional integral equation for the eddy currents obtained in the previous section, it is necessary to use a coordinate system coinciding with the contour of the axisymmetric shell. The centerline of the ETE vacuum vessel has an exact sectionally (piecewise) continuous representation shown on the left side of Figure 1 as a continuous line.

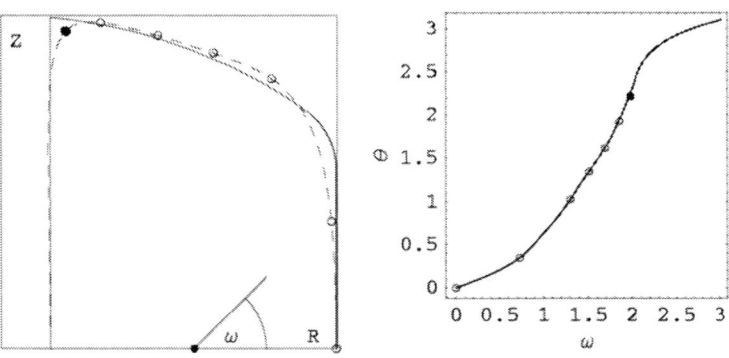

FIGURE 1. The centerline of the ETE vacuum vessel (continuous line) and its spectral fit (dashed line) are shown on the left side. The white circles correspond to the center of flanges used to measure the eddy current distribution, and the black circle on the spectral fit corresponds to the inner edge of the vessel centerline. The adjusted $\theta - \omega$ mapping is shown on the right side.

In accordance with Figure 1 the sectional continuous representation specifies the cylindrical coordinates $R(\omega)$, $Z(\omega)$ as functions of the poloidal angle ω in a pseudo-toroidal coordinate system centered in the cross-section of the vacuum vessel. Now, the centerline of the vacuum vessel can be represented approximately by a truncated spectral expansion in Chebyshev polynomials:

$$\begin{cases} R(\theta) = C_0 + C_1 \cos\theta - a \sum_{n=1}^{N} C_n [1 - T_n(\cos\theta)] \\ Z(\theta) = E_V \sin\theta \left[C_1 - a \sum_{n=1}^{N} C_n U_{n-1}(\cos\theta) \right] \end{cases}$$

The coefficients C_0 and C_1 are determined by the constraints $R(0) = R_0 + a$, $R(\pi) = R_0 - a$, where R_0 and

a are the major and minor radii of the toroidal vessel, respectively. The elongation E_V and the remaining spectral coefficients $C_2, C_3, \ldots C_N$ can be determined by a least-squares fitting procedure. In the case of the ETE vacuum vessel a reasonable spectral representation can be obtained including only elongation, triangularity and quadrangularity (squareness) corrections. The least-squares calculation gives $E_V = 2.164$, $C_0 = 0.320$, $C_1 = 0.287$, $C_2 = 0.0981$ and $C_3 = -0.110$, and the resulting spectral fit is show on the left side of Figure 1 as a dashed line. The least-squares fitting procedure includes also a determination of the best mapping between the pseudo-toroidal angle coordinate ω and the poloidal angle θ in the local curvilinear coordinate system. The adjusted $\theta - \omega$ mapping is shown on the right side of Figure 1.

FOURIER COMPONENTS OF THE SURFACE CURRENT

The integral equation (1) for the Foucault currents in a thin shell that was derived in the second section of this paper can be solved by expansion of $K_T(\theta, t)$ in a Fourier series

$$K_T(\theta, t) = \frac{1}{2\pi h_\theta(\theta)} \left(I_T(t) + \sum_{n=1}^{\infty} I_n(t) \cos n\theta \right). \tag{2}$$

The total toroidal current flowing in the axisymmetric shell is $I_T(t)$ according to the definition in the second section. Substitution of the Fourier series (2) in the integral equation (1) gives

$$\frac{h_\zeta(\theta)}{\sigma \delta h_\theta(\theta)} \left(I_T(t) + \sum_{n=1}^{\infty} I_n(t) \cos n\theta \right) = -\mu_0 \left(\frac{\partial I_T}{\partial t} \langle G(\theta, \theta') \rangle_{\theta'} + \sum_{n=1}^{\infty} \frac{\partial I_n}{\partial t} \langle G(\theta, \theta') \cos n\theta' \rangle_{\theta'} \right) - \frac{\partial \Phi_{ext}}{\partial t},$$

where $\langle \ldots \rangle_\theta = (2\pi)^{-1} \int (\ldots) d\theta$. Limiting the Fourier coefficients to order ℓ, the $\cos m\theta$ harmonics of this equation result in a set of $\ell + 1$ linear equations for $I_T(t)$ and $I_n(t)$ that can be written in the form

$$R_{0m} I_T(t) + L_{0m} \frac{\partial I_T}{\partial t} + \sum_{n=1}^{\ell} \left(R_{nm} I_n(t) + L_{nm} \frac{\partial I_n}{\partial t} \right) = -\frac{\partial}{\partial t} \langle \Phi_{ext}(\theta, t) \cos m\theta \rangle_\theta,$$

where R_{nm} and L_{nm} are resistance and mutual inductance coefficients defined by:

$$\begin{cases} R_{nm} = \frac{1}{\sigma \delta} \left\langle \frac{h_\zeta(\theta)}{h_\theta(\theta)} \cos n\theta \cos m\theta \right\rangle_\theta \\ L_{nm} = \mu_0 \left\langle \langle G(\theta, \theta') \cos n\theta' \rangle_{\theta'}, \cos m\theta \right\rangle_\theta \end{cases}$$

These definitions and the symmetry of the Green's function show that R_{nm} and L_{nm} are symmetric matrices.

In general, the external flux is the sum of the magnetizing flux $\Phi_M(t)$ produced by an ideal transformer and the fluxes $\Phi_k(\vec{r}, t)$ produced by sets of poloidal field coils:

$$\Phi_{ext}(\vec{r}, t) = \Phi_M(t) + \sum_k \Phi_k(\vec{r}, t) = \Phi_M(t) + \mu_0 \sum_k I_k(t) [G_k(\vec{r}, R_k, Z_k) + G_k(\vec{r}, R_k, -Z_k)],$$

where it was assumed that the external coils are formed by pairs of coils placed symmetrically with respect to the equatorial plane, and connected in series. Defining the mutual inductance coefficients, $L_{km} = \mu_0 \langle [G_k(\theta) + G_k(-\theta)] \cos m\theta \rangle_\theta$, the equations for the Fourier coefficients of the surface current density may be written (δ_{nm} is the Kronecker delta)

$$R_{0m} I_T(t) + L_{0m} \frac{\partial I_T}{\partial t} + \sum_{n=1}^{\ell} \left(R_{nm} I_n(t) + L_{nm} \frac{\partial I_n}{\partial t} \right) = -\frac{\partial \Phi_M}{\partial t} \delta_{0m} - \sum_k L_{km} \frac{\partial I_k}{\partial t}. \tag{3}$$

In this way the problem of Foucault currents induced in a thin axisymmetric shell is reduced to the solution of a set of circuit-like coupled linear equations for the Fourier components of the surface current density. The calculation of the mutual coefficients L_{nm} requires some attention due to self-field effects related to the singular character of the Green's function [2].

SOLUTION OF THE CIRCUIT MODEL AND RESULTS

It is now an easy matter to solve the set of circuit equations (3) for the Fourier components of the Foucault current.

Introducing Laplace transformation in time and denoting the complex frequency by s, the equations for $I_T(s)$ and $I_n(s)$ $(n = 1, 2, \ldots, \ell)$ can be written in matrix form, $[[R + sL]][I(s)] = -s[\Phi(s)]$, where $[[R + sL]]$ is a symmetric matrix. The initial values of the magnetizing sources are taken equal to zero at startup. The solution of the circuit model is obtained simply by multiplying the vector of electromotive forces $-s[\Phi(s)]$ by the inverse matrix $[[R + sL]]^{-1}$ and then calculating the inverse Laplace transform. One advantage of the method is that the inverse matrix depends only on the geometry of the problem, which is independent of the detailed excitation.

The resistance and inductance components scale as $A/(E_V \sigma \overline{\delta})$ and $\mu_0 R_0 [\ln(8A) - 2]$, respectively, where $E_V = 2.164$ is the elongation, $A = 1.346$ is the aspect ratio and $R_0 = 0.348\,\text{m}$ is the major radius of the vacuum vessel. The conductivity of Inconel at room temperature is $\sigma \cong 7.8 \times 10^5\,(\Omega \cdot \text{m})^{-1}$. Now, the thickness of the vacuum vessel is $\Delta_V = 6.35\,\text{mm}$ for both the torispherical head and the external cylindrical wall, and $\delta_V = 1.00\,\text{mm}$ for the internal cylindrical wall. In the calculation of R_{nm} the θ integration is split in two sections to account for the change in the wall thickness, and an average thickness $\overline{\delta} \cong 4.59\,\text{mm}$ is defined taking into account the length of the two segments. The average surface current scales as $\overline{K}_T \sim A I_T / (2\pi R_0)$.

Calculations of the eddy current behavior in space and time were performed and compared with measurements taken in the ETE vacuum vessel. Satisfactory results were obtained including only three harmonics, $\ell = 3$, in the calculations. Figure 2 shows the distribution of the surface current at the instant $\tau_0/4$ that corresponds approximately to the maximum negative value of the induced current ($\tau_0 = \mu_0 \sigma \overline{\delta} R_0 / A \cong 1.16\,\text{ms}$ sets the time scale). From the plot in Figure 2 and the mapping $\theta - \omega$ shown in Figure 1 one verifies that the eddy current distribution has a peak at $\omega \sim 113°$, $\theta \sim 127°$, near the inside corner of the vacuum vessel contour and in accordance with rough measurements of the distribution excited by the ohmic heating system in ETE. The bar chart in Figure 2 compares the relative currents measured over several sectors of the vacuum chamber (light gray bars) with the calculated values (dark bars).

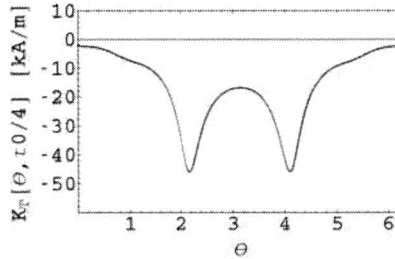 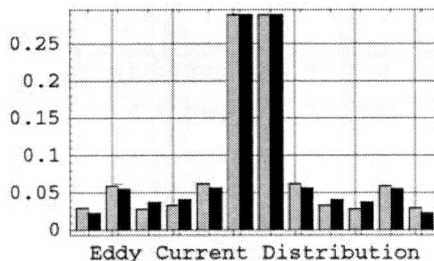

FIGURE 2. The calculated eddy current distribution in the vacuum vessel is shown on the left at the instant of maximum current induced during tests. The measured and calculated distributions are compared on the right.

Based on these results the operation of a pair of compensation coils is being optimized to apply a vertical field bias during plasma breakdown in ETE. In addition, the eddy current distribution is being used to model the vacuum vessel effects in plasma discharge simulations during the early phase. In these zero-dimensional simulations the external inductance of the low aspect ratio ETE plasma and the mutual inductance coefficients between the plasma, the vacuum vessel and the external poloidal field coils are calculated in accordance with a previous work [3].

ACKNOWLEDGMENTS

This work was partially supported by The State of São Paulo Research Foundation (FAPESP) and by the International Atomic Energy Agency (IAEA).

REFERENCES

1. J.A. Stratton, *Electromagnetic Theory*, McGraw-Hill, New York, 1941.
2. S.P. Hirshman and G.H. Neilson, *Phys. Fluids*, **29**, 790 (1986).
3. G.O. Ludwig and M.C.R. Andrade, *Phys. Plasmas*, **5**, 2274 (1998).

A Broken Degeneracy in the Resistive MHD Spectrum

B.F. McMillan*, R.L. Dewar* and R.G. Storer [†]

*Department of Theoretical Physics, Research School of Physical Sciences and Engineering, The Australian National University, Australia.
[†]School of Chemistry, Physics and Earth Sciences, Faculty of Science and Engineering, Flinders University, Australia

Abstract. We present work relating to features in the stable resistive MHD spectrum, which is qualitatively different from the ideal MHD spectrum even for vanishing resistivity. In the cases under consideration stable eigenvalues are spaced along loci in the complex plane. The loci become densely populated with eigenvalues as the resistivity increases. We examine the eigenvalue spectrum of a cylindrical model and relate apparently conflicting results about the shape of the locus. In particular, in a specific model we show how a broken degeneracy produces a pair of precursor loci which converge towards the infinitesimal resistivity locus. We also examine the effects of compressibility on this spectrum, in order to demonstrate the qualitative difference between compressible and incompressible spectra.

INTRODUCTION

Several papers have been published on the stable resistive MHD spectrum, and many of the early papers [1] [2] [8] focused on cylindrical models. These papers have established certain generic features of the resistive spectrum. However, it is not clear how the "fork" structure seen in many model cases is related to the "double locus" structure found in an analytic cylindrical model due to Storer ([8]).

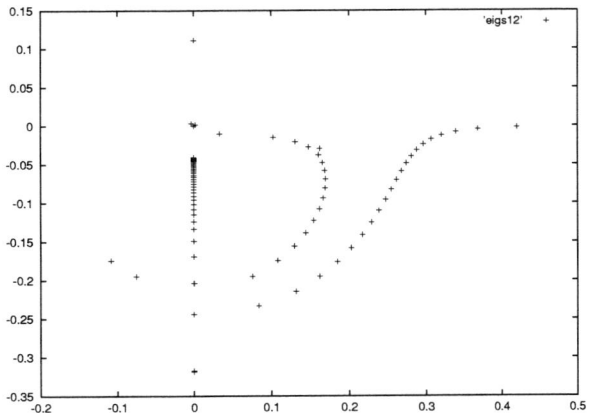

FIGURE 1. The resistive spectrum of the Storer model

The cause of this qualitative difference is the condition of incompressibility present in the Storer model. The decoupling of the sound waves from the Alfvén waves which is introduced by finite compressibility reduces the double-locus structure to a single locus. Varying the compressibility allows one to see the locus structures evolving between the two qualitatively different pictures. We performed this numerically using a code based on [6] which is described later in the paper.

The presence of two loci in the incompressible model is a consequence of the behaviour of the slow continuum as the compressibility is reduced. It can be easily seen from an expression for the position of these continua that the slow spectrum has the same frequency range as the Alfvén continuum, and thus the spectra are 'degenerate' in the compressible case.

The fork structure has been qualitatively explained in terms of WKB analysis by examining turning points within the plasma [2]. The fork has three lines joining at a point below the ideal MHD continuum. Lines run between the intersection point and either end of the Alfvén continuum. Another line runs around approximately in a semicircle to touch the imaginary axis.

Because the Alfvén continuum is a point for the Storer model, and the fork does not have a finite extent, a simple WKB analysis would suggest a line of modes inside the Alfvén annulus- i.e. on a circle.

The full structure can be recovered analytically in this case because the equations can be manipulated into a much simpler differential equation, which can then be solved.

It turns out that a slightly perturbed variant of this model, in which a slight shear is added to the magnetic field, is still amenable to the type of manipulation performed above. By introducing shear we produce a model which has a finite Alfvén continuum, in which we might hope to recover the generic fork structure shown in other analyses. We therefore solved this model using WKB analysis to explain the qualitatively different spectrum.

WKB ANALYSIS OF A SIMPLE EQUILIBRIUM

The model case is derived from [8], which considers cylindrical plasma with a constant axial field and no shear. This model has been studied earlier in [3],[4].

The equations used for this analysis are those of linearised, resistive, incompressible MHD ($\nabla \cdot \mathbf{v} = 0$), with the curl of the equation of motion taken in order to suppress the perturbed pressure:

$$\rho \mu_0 \frac{\partial}{\partial t}(\nabla \times \mathbf{v}) = \nabla \times (\mathbf{B} \cdot \nabla \mathbf{b} + \mathbf{b} \cdot \nabla \mathbf{B}) \tag{1}$$

$$\frac{\partial \mathbf{b}}{\partial t} = \nabla \times (\mathbf{v} \times \mathbf{B}) - \nabla \times (\frac{\eta}{\mu_0} \nabla \times \mathbf{b}) \tag{2}$$

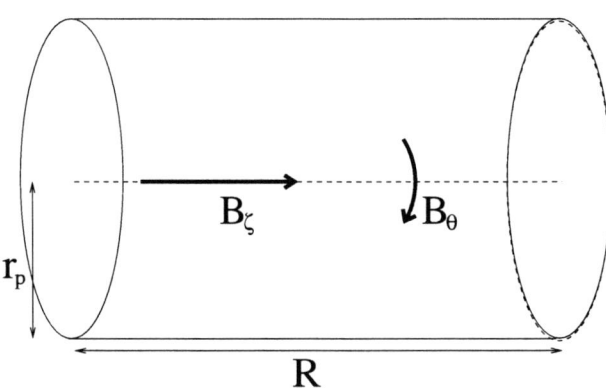

FIGURE 2. One period of the cylindrical model

The idea is to introduce the shear as a small quantity, of the same order as the inverse wave number. The analysis then follows by analogy to the shear free case, up to two orders in the inverse wavenumber. The radial dependence is included in the dispersion relation in the radially varying quantities: $B_0(r)$, $B_p(r)$ and $q(r)$.

The first take the large wavenumber limit by choosing $\nabla \simeq O(1/\varepsilon)$. For significantly dissipative modes, maximal balance of Ampere's law (Eq. 2) occurs for $\varepsilon = O(S^{-\frac{1}{2}})$, so for typical $S \simeq 1000$ we have $\varepsilon = 0.03$.

The magnetic field is expressed as $\mathbf{B} = \hat{z}B_0(r) + r\hat{\theta}B_p(r)$ with $d(\log[B_p(r)])/dr, d(\log[B_0(r)])/dr \simeq O(\varepsilon)$, in order to satisfy the requirement of small shear.

By looking at perturbations of the form $b = \exp(im\theta - in\zeta)b(r)$ we can approximately reduce Eqs 1 and 2 to:

$$\frac{1}{r}\frac{d}{dr}r\frac{d}{dr}v_z = \frac{\alpha^2}{r_p^2}v \qquad (3)$$

With α given by:

$$\alpha(r,\omega) = \frac{2(m - nq(r))\kappa}{\frac{B_0(r)^2}{B_p(r)^2}i\omega(i\omega - \frac{\alpha(r,\omega)^2}{S}) + (m - nq(r))^2} \qquad (4)$$

This is amenable to standard WKB analysis.

Away from singularities, $\alpha(r,\omega)$ is changing on a length scale r_p so the dispersion relation is a slowly varying quantity. We use WKB solution for:

$$\frac{d^2}{dr^2}v_z = Q(r)v_z \qquad (5)$$

Which is:

$$v_z \simeq a_{out}e^{i\phi} + a_{in}e^{-i\phi} \qquad (6)$$

with

$$\phi(r|c) = \int_c^r Q^{\frac{1}{2}}(r')dr' \qquad (7)$$

For our case, we have $\alpha^2/r_p^2 = Q$.

We have a cubic dispersion relation in terms of α. Thus solving the dispersion relation (Eq. 4) for $\alpha(r,\omega)$ gives three roots. These are the radial wavenumbers of three different local dissipative wave modes of the plasma. Several kinds of wave modes may be present in the plasma.

Global modes are found in the usual way: we look for paths C in complex space joining the axis and boundary where $\int_B \alpha(x)dx$ is real for any subpath B of C. These paths will be WKB solutions if the integral $\int_C \alpha(x)dx = \int_{[0,1]} \alpha(x)dx$ which can be guaranteed if there are no singularities of our differential equation coefficients in the region. In particular, this requires that the circular path $C - [0,1]$ does not enclose any Stokes' points.

Localised modes proceed from the axis or outer boundary of the plasma and propagate along ray trajectories (which will in this case be anti-Stokes lines) to a Stokes point. They are then evanescent past this point, so it must be possible to draw a path connecting the relevant Stokes point to the other boundary without crossing a Stokes line.

APPLICATION OF THE WKB METHOD

For explicit studies, we use a 'small shear' test case with a uniform axial magnetic field, and a varying poloidal field of the form $B_\theta(r) = B_{\theta 0}/(1 - r\delta)$.

The WKB trajectories in the complex r plane were determined numerically, permitting a search for the ray trajectories corresponding to global modes.

To find the relevant global modes by the WKB method, we need to look for ray trajectories which start on the magnetic axis and end at the outer boundary of the plasma. Finding localised modes require a determination of the Stokes points and lines. This was examined interactively, and then refined by automated procedures. This system allowed us to determine the spectrum of various model cases. We describe results pertaining to the following test case:

$$\begin{array}{lll} B_\theta(r) = \frac{B_{\theta 0}}{1 - r\delta}, & \delta = 0.1, & B_\zeta(r) = 1, \\ R = \frac{50}{7}, & n = 10, & m = 1, \\ r_p = 1, & & \end{array} \qquad (8)$$

Several loci found using the code. The loci can be characterised by the branch of the dispersion relation which they lie on, and whether the corresponding wave modes are fully global modes, or have a turning point inside the plasma.

The eigenvalues along these loci are determined by the quantisation condition, which we obtain from examining the behaviour at the boundaries.

The eigenvalues are displayed in Fig. 3, together with the numerical result from a code based on [6]. The spectrum is qualitatively similar to a fork structure, but also shares the features of the original simple model. Note that the double loci (running parallel to each other in an arc) are still present in this model.

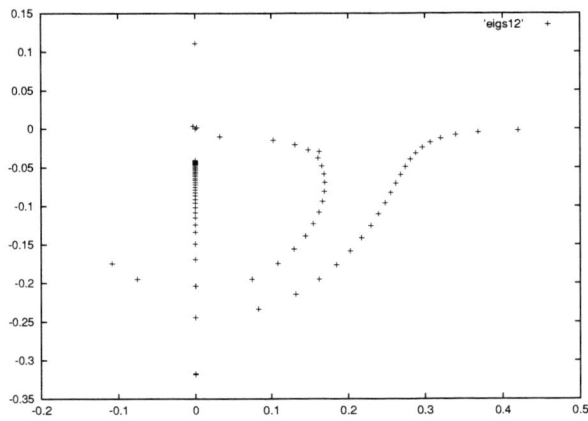

FIGURE 3. The resistive spectrum deduced from numerics together with the WKB result.

CONCLUSIONS

Although incompressible MHD fluids exist in some physical situations, the assumption of incompressibility is often used as an approximation, especially for numerics.

There has been significant discussion of the consequences of the incompressibility assumption on the MHD spectrum. One important result is that configurations with unstable compressible ideal MHD modes will have unstable incompressible MHD modes, and vice versa; incompressibility does not affect marginal stability in the ideal MHD case (the result does not extend to resistive MHD analysis). In general, we would expect that where we have phenomena with significantly differing time scales, like the Alfvén and sound wave timescales, there would be some form of simplification possible.

This has lead to a number of incompressible MHD codes being developed: of particular relevance to resistive analysis, the 3D resistive MHD code Spector3D relies on the incompressibility assumption.

We have shown how the stable spectrum of the incompressible case evolves as compressibility is introduced. This addresses any apparent inconsistency between [8] and papers such as [1] [2].

The shape of the incompressible spectrum for a more general model, with shear present, was determined numerically and by WKB analysis. The resulting spectrum included the "split eigenvalues" of the zero shear model, and also demonstrated the fork structure that is seen generically for stable resistive MHD spectra.

In this case the WKB analysis was able to produce a good description of the spectra. A non-standard evaluation of the dispersion relation by analogy with an analytic case allowed the inclusion of all the relevant details. The dispersion relation for this model leads to an interesting WKB analysis because of its unusual branch structure and behaviour at Stokes points.

It is clear that although the incompressible resistive MHD spectra shares many features in common with the compressible spectrum, there are significant qualitative and quantitative differences.

REFERENCES

1. Davies, B. *Phys. Letters* **100A**, 144-148 (1984).
2. Dewar, R.L., and Davies, B., *J. Plasma Physics* **32**, 443-461 (1984).
3. Tayler, R.J., *Rev. Mod. Phys.* **32**, 907-913 (1960)
4. S.M.Breus, *Zh.tech. Fiz.* **30**, 1030-1034 (1960); *Soviet Phys. Tech. Phys.* **5**, 960-965 (1960)
5. S. Wolfram, *The Mathematica Book*, 3rd ed., Wolfram Media/Cambridge University Press, Champagne, Illinois, 1996.
6. Kerner, W., Lerbinger, K., Gruber, R., Tsunematsu, T., *Computer Phys. Commun.* **36**, 225-240 (1985).
7. Gruber, R., Rappaz, J., *Finite Element Methods in Linear Ideal Magnetohydrodynamics*, Springer-Verlag, New York, 1985
8. Storer, R.G., *Plasma Phys.*, **25** 1279-1282 (1983)

On the Nature of Visco-Resistive MHD Steady States

L.P.J. Kamp* and D.C. Montgomery**

*Department of Applied Physics
Eindhoven University of Technology
P.O. Box 513, NL-5600 MB Eindhoven, The Netherlands
**Department of Physics and Astronomy
Dartmouth College
6127 Wilder Laboratory
Hanover, NH 03755-3528, U.S.A.

Abstract. Static ideal MHD equilibria continue to provide the conceptual framework and vocabulary by which toroidal steady-state plasmas are described, despite the growing recognition that virtually all toroidal plasmas involve non-trivial flows. A more promising mathematical approach to realizable states would appear to be through the inclusion of finite transport coefficients (viscosity as well as resistivity) and non-ideal boundary conditions, where it has been known for some time that flows are a necessary consequence of the demands of axisymmetry and time independence. Heretofore [1,2], we have been able to describe toroidal resistive steady states in perturbation theory, expanding the solutions in powers of the Reynolds number or, more accurately, the Hartmann number. Using new numerical techniques that have become available, we are now able to lift this limitation and calculate voltage-driven toroidal steady states through a range of Hartmann numbers that runs from <<1 to >>1. The flow pattern ranges, as the Hartmann number is raised, from a previously identified pair of counter-rotating toroidal vortices (poloidal convection cells) to a pattern in which the flow is primarily in the toroidal direction. None of the flows identified is a simple rotation, poloidal or toroidal, sheared or otherwise. Detailed "weather maps" can now be drawn.

INTRODUCTION

In fluid mechanics, steady-state solutions of the ideal Euler equations are very plentiful, but Navier-Stokes solutions obeying viscous boundary conditions are difficult to come by. The explicit possibilities are limited essentially to one-dimensional cases: pipe flow, plane and rotating Couette flow, plane Poiseuille flow, etc., all standard textbook examples. The non-ideal solutions dominate the discussion of laboratory situations, and relegate most of the ideal Euler solutions to the status of elegant mathematical curiosities.

Magnetohydrodynamics (MHD) contains fluid mechanics as a special case, and its overall mathematical structure is highly similar. Yet standard textbook presentations of MHD steady states deemed to be relevant for fusion confinement are almost invariably ideal equilibria. In recent years, we have been attempting [1-6] to alter this asymmetry and explore what the possibilities are for driven, dissipative, MHD steady states, believing them to be the most likely candidates for physically relevant solutions.

It has been considered to be essential to treat the equation of motion, Ohm's law, and Faraday's law on an equal footing, along with imposing viscous and resistive boundary conditions. Earlier treatments of "resistive equilibria" typically ignored one or more of these pieces of the story, and they are very restrictive requirements, taken together. Combined with toroidal (as opposed to periodic straight cylinder) geometry, they turn out to imply features of allowed steady states that it would have been difficult to guess.

In Sec. 2, we summarize briefly the difficulties of identifying non-ideal steady states with ideal ones that might be hoped to be similar in character. Sec. 3 states the problem and in Sec. 4, we report some typical recent results [6] for toroidal geometry.

RESISTIVE STEADY STATES

If, in accord with standard practice, we were to assume no flow of the magnetofluid, then any resistive steady state must obey the simple Ohm's law, $\mathbf{J}=\mathbf{E}/\sigma$, where $\mathbf{J}$ is the current density, $\mathbf{E}$ is the electric field, and σ is the conductivity (we assume for simplicitiy that it is a scalar). If $\mathbf{J}$ is proportional to the curl of $\mathbf{B}$, the magnetic field, then the divergence of $\mathbf{E}/\sigma$ must be zero. If $\mathbf{E}$ is to be a time-independent field, then its curl must vanish and it may be expressed as the gradient of a scalar potential, $\mathbf{E} = -\nabla\Phi$. Thus the divergence of $\nabla\Phi/\sigma$ must vanish. If σ is uniform, this is simply Laplace's equation for Φ, and thus Φ is determined by boundary conditions alone. If σ is a function of position, then Φ still obeys a Laplace-like equation, with spatially-dependent coefficients, with every reason to believe that its solutions are still uniquely determined by boundary conditions.

Thus the arbitrariness of ideal MHD (Grad-Shafranov theory [7]) has disappeared, and we have not yet even brought in the equation of motion. With boundary conditions on $\mathbf{B}$, $\mathbf{J}$ will determine $\mathbf{B}$ through the Biot-Savart law or its equivalent. We are led abruptly to the conclusion that the electromagnetic configuration is unique, given only the boundary conditions on $\mathbf{E}$, $\mathbf{J}$, and $\mathbf{B}$, and the electrical conductivity!

When the equation of motion is brought in, finding a zero flow steady state for MHD now becomes a problem of finding a pressure whose gradient will balance $\mathbf{J} \times \mathbf{B}$. In cylindrical geometry, this is not difficult for axisymmetric cases. This is because for any situation in which σ is a function of radius r alone, $\mathbf{J} \times \mathbf{B}$ can be chosen rather generally to be radial and to be a function of r alone. A radially dependent pressure can typically be found to balance the $\mathbf{J} \times \mathbf{B}$ force without the need for mass flow. It is important to notice that the only freedom left in the problem is that which arises through a possible spatial dependence of conductivity on radius. We have already come far beyond the Grad-Shafranov theory, and we have not mentioned toroidal geometry.

It turns out that in toroidal geometry [3-5], geometrical considerations alone mandate that the curl of $\mathbf{J} \times \mathbf{B}$ must be non-vanishing. It is thus impossible to cancel $\mathbf{J} \times \mathbf{B}$ with any gradient, and flow is demanded for force balance. This flow is only calculable analytically in the limit of small Hartmann number or high viscosity. Recently [6], we have found it possible numerically to go beyond this limitation using newly available software [8], and in Sec. 4, we describe some typical results of these numerical investigations of possible toroidal steady states in the non-ideal case.

STATEMENT OF THE PROBLEM

The MHD equation of motion for a uniform-density, incompressible conducting fluid is, in a familiar set of dimensionless units (``Alfvénic'' units):

$$\frac{\partial \mathbf{v}}{\partial t} + \mathbf{v} \cdot \nabla \mathbf{v} = -\nabla p + \mathbf{J} \times \mathbf{B} + \nabla \cdot (\nu \nabla \mathbf{v}).$$

where p is the pressure, and ν is the kinematic viscosity. In these dimensionless units. ν is the reciprocal of the viscous Lundquist number M. The velocity field $\mathbf{v}$ obeys the incompressibility condition,

$$\nabla \cdot \mathbf{v} = 0.$$

Furthermore we need Faraday's law

$$\nabla \times \mathbf{E} = -\frac{\partial \mathbf{B}}{\partial t},$$

Ampère's law

$$\nabla \times \mathbf{B} = \mathbf{J},$$

and Ohm's law

$$\mathbf{E} + \mathbf{v} \times \mathbf{B} = \eta \mathbf{J},$$

where η is the reciprocal of the resistive Lundquist number, S. The Hartmann number is given by $H = \sqrt{MS}$.

In the next section we will present numerically obtained resistive steady state solutions of the equations above for which the field components are all independent of the azimuthal coordinate φ, in a set of cylindrical coordinates (r, φ, z). The z-axis is the axis of symmetry, and $z = 0$ is the mid-plane of a toroid the cross section of which is considered to be a rectangle. The center line of this toroid will lie in the midplane at a major radius $r = r_0$, say.

These steady states are assumed to be sustained by externally applied curl-free electric and magnetic fields, $\mathbf{E_0}$ and $\mathbf{B_0}$ respectively, that each point in the φ-direction. The boundary conditions that have been imposed are that any tangential viscous stress, and the normal components of $\mathbf{v}$, $\mathbf{J}$, and $\mathbf{B}$, should vanish at the wall of the toroid.

RESULTS

In Fig. 1 we show a typical example of a solution obtained with FEMLAB [8] with $|\mathbf{E_0}| = 1 = |\mathbf{B_0}|$, $\eta = 1$, and Hartmann number taken to be 100. We should note that because of the symmetries about the mid-plane $z = 0$, we are showing only the upper half of the toroidal cross section in these and similar figures; the variables in the lower half can be inferred from obvious symmetries. The figures do not illustrate realistic parameter regimes for currently-operating confinement devices.

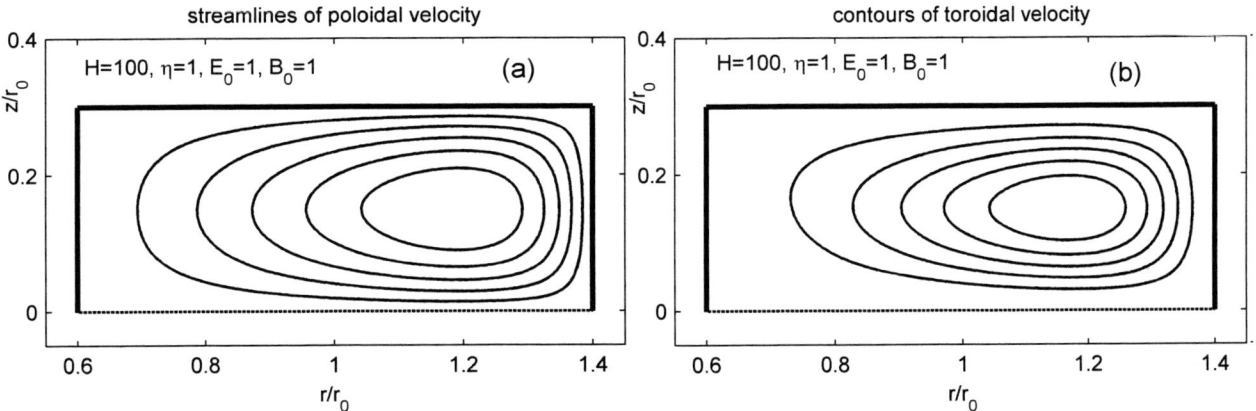

FIGURE 1. (a) Streamlines of the poloidal velocity. (b) Contours of constant toroidal velocity.

Adding the toroidal velocity component (Fig. 1(b)) vectorially to the poloidal flow (Fig.1(b)) results in a steady-state streamline configuration that topologically is equivalent to helices, which circle the toroid in alternate senses, amounting to what are essentially four ``convection cells''.

Figure 2 contains a vector plot of ∇p for the same set of parameters as in Fig. 1 but now the Hartmann number is taken to be unity. At the rectangular toroidal boundary, ∇p has a finite tangential component, indicating that the bounding wall is not an isobaric surface. However, our boundary conditions are such that tangential stress at the wall is absent since we require toroidal vorticity to vanish there.

Figure 3 finally shows the ratio of kinetic energies contained in the toroidal and poloidal components of the flow versus an increasing Hartmann number for $|\mathbf{E_0}| = 1 = |\mathbf{B_0}|$ and also for $|\mathbf{E_0}| = 10 = |\mathbf{B_0}|$. In order to have both plots in the same figure, what is actually plotted is this ratio divided by $|\mathbf{B_0}|^2$. For $|\mathbf{E_0}| = 10 = |\mathbf{B_0}|$ we see that for sufficiently low viscosity toroidal flow overtakes the poloidal one.

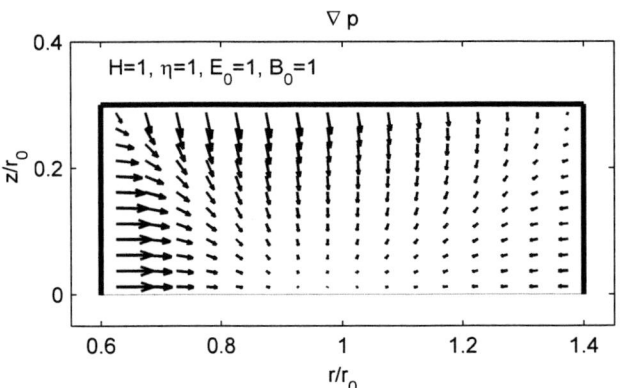

FIGURE 2. Arrow plot of the vector field ∇p for the same set of parameters as in Fig. 1 but with Hartmann number taken to be unity.

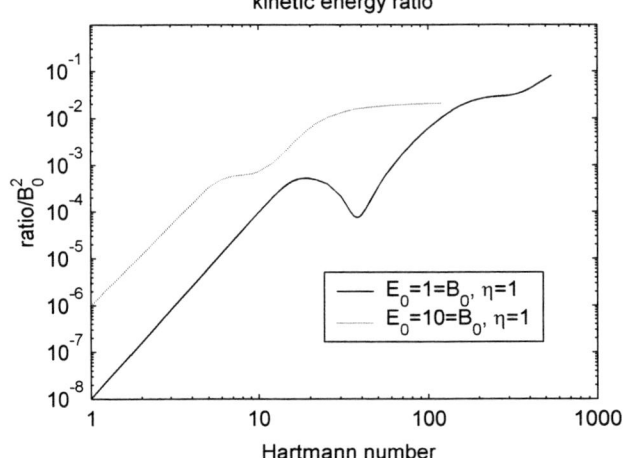

FIGURE 3. Ratio of kinetic energies contained in the toroidal component of the flow to that contained in the poloidal part of it.

ACKNOWLEDGMENTS

The work described here was financially supported by the Netherlands Organization for Scientific Research (NOW) and in part by the U.S. Department of Energy.

REFERENCES

1. Kamp, L.P.J., Montgomery, D.C., and Bates, J.W., Phys. Fluids 9, 1757-1766 (1998).
2. Bates, J.W., and Montgomery, D.C., Phys. Plasmas 5, 2649-2653 (1998).
3. Montgomery, D., and Shan, X., Comm. Plasma Phys. & Contr. Fusion 15, 315-320 (1994).
4. Montgomery, D., Bates, J.W., and Lewis, H.R., Phys. Plasmas 4, 1080-1086 (1997)).
5. Montgomery, D., Bates, J.W., and Kamp, L.P.J., Plasma Phys. Contr. Fusion 41, A507-A517 (1999).
6. Kamp, L.P.J., and Montgomery, D.C., "Toroidal flows in resistive magnetohydrodynamic steady states," submitted to Phys. Plasmas (2002).
7. Wesson, J.A., TOKAMAKS, Oxford University Press, Oxford, U.K., 1987.
8. FEMLAB Reference Manual, version 2.2 COMSOL AB, Stockholm, Sweden, 2001.

Effect of Equilibrium Flow on Plasma Parameters

S. Mukhopadhyay[*], S. Lahiri[†], P.H. Sakanaka[**] and B. Dasgupta[‡]

[*]*Experimental High Energy Physics Division, Saha Institute of Nuclear Physics, India*
[†]*Department of Physics, Techno India, India*
[**]*Projecto Fibras Opticas, Instituto de Fisica, Brazil*
[‡]*NASA Jet Propulsion Laboratory, USA*

Abstract. The transition to high confinement modes have been identified with the occurence of strong shear flow near the plasma boundary. Plasma flow has also been associated with various instabilites, heating and other physical processes. As a result, it has become very important to study the effect of such flows on various plasma parameters. In this paper, we present the numerical solution of plasma equilibrium with incompressible toroidal and poloidal flows in several magnetic confinement configurations including tokamaks. The code, which was reported [1] in the last conference, has been used to solve the problem in both circular and D-shaped devices. A parametric study on the generation of shear flow due to radial electric fields has been carried out. Through this study, it has been possible to generate plasma equilibria having sharp pressure gradients which are remarkably close to those reported in various H-mode experiments. The effects of flow on reverse shear equlibria and on the position of the magnetic axis has been studied. Finally, a detailed study has been carried out to understand the effect of flows on important plasma parameters, such as the poloidal flux function, β, energy confinement time.

INTRODUCTION

Additional heating, application of radial electric fields, transition to H-mode by various approaches normally lead to large amount of flow in fusion device plasmas. These velocities can reach upto and, sometimes, become larger than sonic velocities. The magneto hydrostatic equation (Grad Shafranov Schluter) is too drastic a simplification for the description of the plasma transport in such contemporary tokamak experiments. The problem of plasma equilibrium with flow has been addressed by several workers in the recent past [1], [2], [3]. Following the formulation of [3], a computer code has been developed, validated and applied to solve several scenarios as presented in [4]. These and investigations carried out by other workers ([4], [5]) concentrated mostly on purely toroidal flows and flows parallel to the magnetic field. However, in general, plasma flows are neither in the direction of the toroidal field (sonic flows have been observed in the poloidal directions), nor are they strictly parallel to the total magnetic field. In the present work, we present the upgradation of the solver to take into account of such non-parallel flows which are known to occur due to the application of radial electric fields which play a role in the transitions to improved confinement regimes [ref. 20 of Tasso]. The solver has been validated against analytical solution and applied to carry out a parametric study of the effect of the effect of flow on various plasma parameters.

THEORETICAL BACKGROUND

In an axisymmetric torus, the magnetohydrodynamic equilibrium of an incompressible plasma column with flow can be described by the following equation [3]

$$(1-M_p^2)\Delta^*\psi - \frac{1}{2}(M^2)'|\nabla\psi^2| + \frac{1}{2}(\frac{X^2}{1-M^2})' + R(P_s - \frac{XF'\phi'}{1-M^2}) + \frac{R^4}{2}(\frac{\rho(\Phi')^2}{1-M^2}) = 0 \qquad (1)$$

where $M \equiv \frac{v_p}{v_{AP}} = \frac{(F')^2}{\rho}$ is the Mach number and $v_{AP} = \frac{|\vec{\nabla}\psi|^2}{R^2\rho}$ is the Alfven velocity associated with the magnetic field. In equation (1), $F(\psi), \Phi(\psi), X(\psi), \rho(\psi)$ and $P_s(\psi)$ are surface quantities with their usual meanings which must be obtained from other physical considerations before proceeding to solve the problem of equilibrium. The solution of (1) also requires the specification of ψ at a suitable boundary.

Assuming $\Phi' \propto \psi^{-k/2}$ and $\rho \propto \psi^k$, where k is a paramter, the electric field can be obtained as $\vec{E} \propto -\psi^{-k/2}\nabla\psi$ and the above equation can be written as:

$$\Delta^*\psi + \frac{1}{(1-M_c^2)^2}[XX' + R^2((1-M_c^2)P_s - C_{F'}C_{\Phi'}M_cX)'] = 0 \tag{2}$$

where $C_{F'}$, $C_{\Phi'}$ are constants of proportionality. Equation (2) is the governing equation for the present study.

Numerical Methods

Direct solution of equations (1) and (2) generally yields trivial solutions. Hence, it has been posed as an eigenvalue problem which has then been solved by adopting a finite-difference successive over relaxation (SOR) scheme. The constant flux approach outlined in [4] has been followed. The ψ field has been assumed to have converged when $max|\psi^{n+1} - \psi^n| < 10^{-6}$ where n denotes $n-th$ iteration.

In order to compare with the analytical results, we had to use a computational geometry which represented a very low aspect ratio D-shaped torus. In fact, this geometry was the representation of a sphere in the form of a torus with a very thin center-post. Otherwise, a normal tokamak geometry (R = 1.0, a = 0.25) has been used as shown in figure 1 of [4].

The code has been developed to handle boundary of any shape defined in a cartesian coordinate system and supplied to the code as points of intersection of the boundary with the grid lines. Evenly spaced nodal points have been used throughout the computational domain. Extra nodal points (some of which have been indicated using arrows in figure 1 of [4]) have been used to ensure that the surface boundary points are included in the calculation procedure. So, near the boundary, we have used an uneven mesh and the f-d equivalent of the partial differential equation has lost its simplicity.

RESULTS AND DISCUSSION

Equation (2) can be solved analytically only under certain simplifying assumptions. For example, if $\hat{X} = constant$, the poloidal flux function can be obtained as

$$\psi = \psi_c \frac{R^2}{R_c^4}(2R_c^2 - R^2 - 4d^2Z^2) \tag{3}$$

and the electric field becomes

$$\vec{E} = -\psi^{-k/2}[\frac{\psi_c}{R_c^4}(4R_c^2R - 4R^3 - 8d^2RZ^2)\hat{i} + \frac{\psi_c}{R_c^4}(-8d^2R^2Z)\hat{j}] \tag{4}$$

The numerical solver has been used to reproduce the same result. Results from the two have been compared and have been found to agree quite well.

In the following, we present the parametric study based on Mach number and k. We have started from purely toroidal flows (indicated by Mach number 0.0) and went up to Mach number 0.9. The variation of velocity in the Z-direction with the major axis has been presented in figure 1. As expected, the velocity is found to increase in magnitude as the Mach number increases.

In figure 2, we present the variation of velocity in the Z-direction due to variation of the parameter k. It has been observed that negative values of k lead to physically relaistic results whereas a positive k leads to unrealistic results. Thus, there seems to be a constraint in the parameter space and the forms of Φ, ρ seem to have restrictions. In figure 2, we show the variation of the electric field due to the variation in the above parameter. Relating figure 2 and figure 3, we can find out the effect of applied electric field on the generation of the shear flow.

CONCLUSION

The numerical solver has been found to be working satisfactorily for solving problems related to plasma equilibrium with a variety of flows. It has been possible to handle purely toroidal flows, flows parallel to $\vec{B}$ and non-parallel flows.

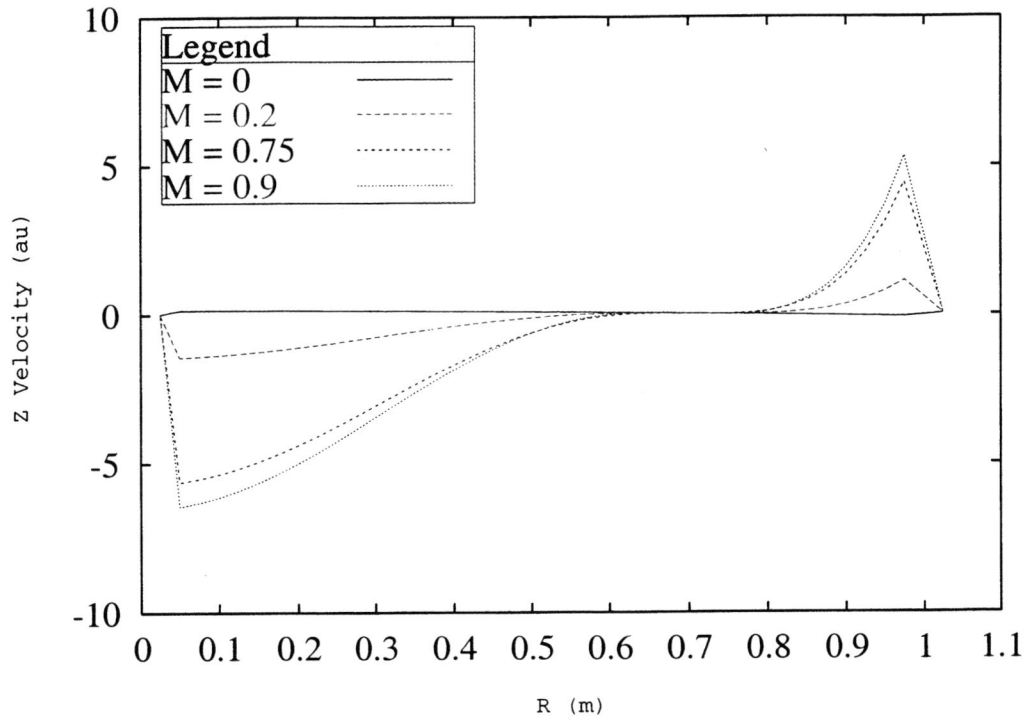

FIGURE 1. Variation of Z velocity with poloidal Mach number

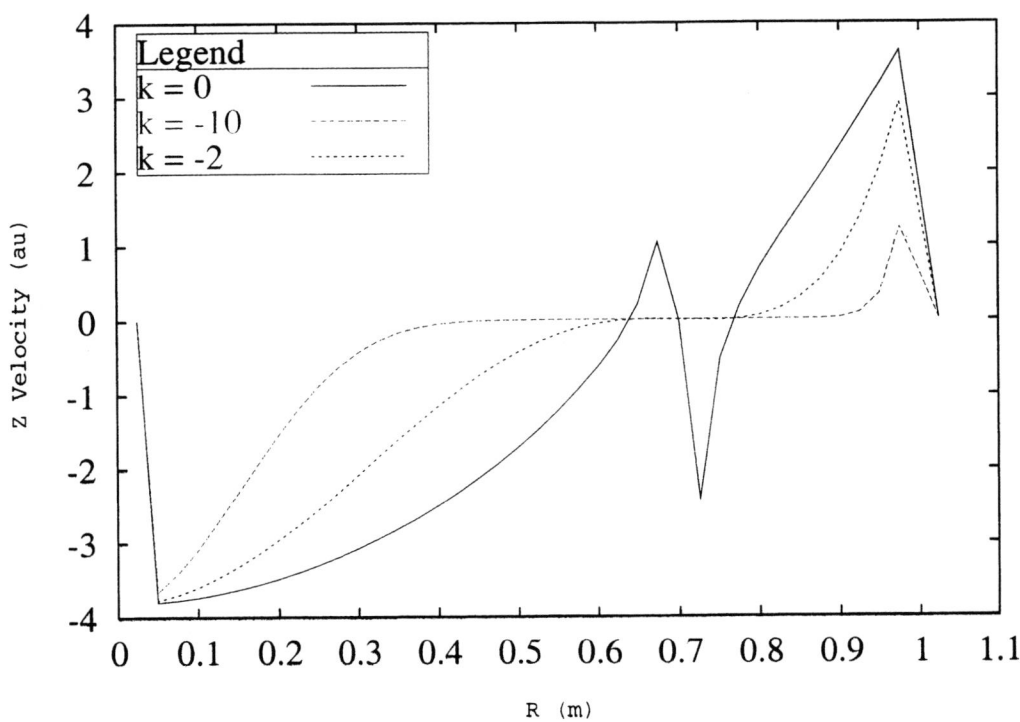

FIGURE 2. Variation of Z velocity with k

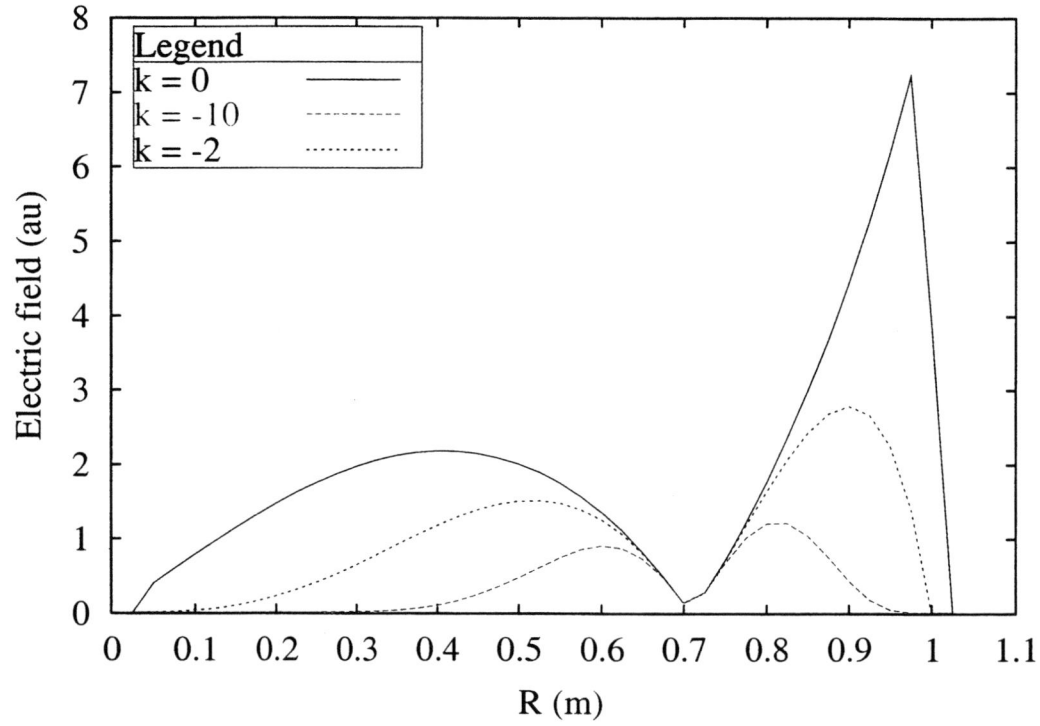

FIGURE 3. Variation of Electric Field with *k*

The possible parameter space, especially in the latter case, is quite large. Although some calculations have been carried out to find out the effects of flow on final plasma parameters such as confinement time, β, further study is being carried out to evolve a consistent understanding of these complicated relationships.

REFERENCES

1. Zelazny, R., Stankiewicz, R., Potempski, S., JET-R (88) 16 (1988)
2. Avinash, K., Bhattacharya, S.N., Green, B.J., Plasma Phys. Controlled Fusion 34, (1992) 465
3. Tasso, H., Throumoulopoulos, G.N., Physics of Plasmas, 5, (1998) 2378.
4. Mukhopadhyay, S., Lahiri, S., Sakanaka, P.H., Dasgupta, B., Proceedings of ICPP-2000, **2**, 672 (2000)
5. Furukawa, M., Nakamura, Y., Hamaguchi, S., and Watakani, M., Proceedings of ICPP-2000, **2**, 628 (2000)
6. Lackner, K., Computer Physics Communications, 12, (1976) 33.

Neoclassical MHD Equilibria in Low Aspect Ratio Reversed Field Pinch

Y. Nagamine[1], Y. Yoshioka[1], M. Harada[1], Y. Osanai[1], Y. Kondoh[2], M. Taguchi[3], E. Uchimoto[4], K.H. Saito[1], K.N. Saito[1], S. Shiina[1]

[1] *Institute of Quantum Science, College of Science and Technology, Nihon Univ., Tokyo, Japan*
[2] *Department of Electronic Engineering, Gunma Univ., Gunma, Japan*
[3] *College of Industrial Technology, Nihon Univ., Chiba, Japan*
[4] *Department of Physics and Astronomy, Univ. of Montana, Missoula, U.S.A.*

Abstract. The magnetohydrodynamic (MHD) equilibrium and local mode stability of a low aspect ratio reversed field pinch (RFP) plasma is studied. The dependences of the equilibrium and stability properties, and bootstrap current (BSC) profile on the plasma pressure profile, β value and cross sectional shape are investigated with a view of obtaining stable, high β and high BSC-driven equilibria against high-m localized Mercier mode. In addition, the neoclassical MHD RFP equilibria including bootstrap current effects are calculated self-consistently and compared with the classical MHD equilibria.

INTRODUCTION

In order to realize a fusion reactor with a low cost of electricity on the base of reversed field pinch (RFP) plasmas, it is considered that one of the major problems to be solved is to maintain the dynamo-free, stable RFP configuration in steady state with the less power requirements of non-inductive radio-frequency wave current drive. We are investigating magnetohydrodynamic (MHD) equilibrium and stability properties of RFP focusing on its low aspect ratio and bootstrap current. In our previous studies on the MHD equilibria with partially relaxed state model (PRSM) with uniform j_θ/B_θ-profile, which is reasonably close to a dynamo-free, stable minimum energy state at finite beta, it was found that i) as for the PRSM-RFP cylindrical plasma being stable to Suydam modes, the stability beta limit is $<\beta>_V$(volume averaged beta)$\cong 25\%$ against ideal m=1 MHD mode at the pinch parameter of $\Theta=2.1$ and the field reversal parameter of $F=-1.4$, ii) as for the PRSM-RFP low aspect ratio plasma with parabolic pressure profile, there exists an optimum combination of equilibrium parameters in a certain region of aspect ratio ($A\leq 3.0$) and beta ($<\beta>_V\geq 25\%$) to keep a high bootstrap current ratio $F_{bs}\equiv I_\phi^{bs}/I_\phi^{eq}$ (I_ϕ^{bs}: total toroidal bootstrap current, I_ϕ^{eq}: plasma current) giving a high magnetic shear relative to that for the cylindrical plasma[1]. As for the dependency on plasma cross sectional shape, increasing elongation κ and decreasing triangularity δ enhance the bootstrap current simultaneously the magnetic shear. Preliminary approach to reactor concept shows the possibility of attaining an economical, compact steady-state reactor with a lower aspect ratio and then a higher power density compared with reverse shear tokamak.

LOCAL MODE STABILITY

In the RFP equilibrium, the safety factor q increases on the magnetic axis and decreases at the edge, while its profile flattens in the center region and sharpens in the edge as the aspect ratio (A) decreases. Therefore, it is considered that the stability of localized modes is an important issue because it may limit the stability beta limit although the global magnetic shear is enhanced. Pressure profile control is necessary so that the stability condition for pressure-driven high m localized modes, which is known as Mercier's criterion, can be satisfied with a high BSC ratio F_{bs}. The Mercier criterion is given by [2]

$$D_M - \frac{1}{4} < 0. \qquad \text{a)}$$

Here, the coefficient D_M is a function only of the equilibrium quantities. This criterion is valid for arbitrary β, A and cross sectional shape and can be tested separately, one flux surface at a time. It must be satisfied on each surface to guarantee interchange stability. D_M is evaluated by use of the numerical solution to the equilibrium Grad-Shafranov equation. Various types of RFP equilibria are constructed by the equilibrium solver, TEAP-MFL code [3] which was improved by us recently for the purpose of the analysis of neoclassical MHD equilibria. The pressure profile is taken to be of the form

$$p(\psi) = p_0 (1 - \hat{\psi}^{b_p})^{a_p}. \qquad \text{b)}$$

Here, ψ with a hat means the normalized poloidal flux and p_0 the pressure at the magnetic axis. The poloidal current flux function $I(\psi)$ is determined by the PRSM condition [4]

$$\mu_0 \, dI(\psi)/d\psi = \lambda(\psi) \qquad \text{c)}$$

with assuming $\lambda=\lambda_0(const.)$. The geometry of plasma boundary is assumed to be fixed and its non-circularity expressed by both ellipticity κ and triangularity δ. The bootstrap current is calculated by use of Hirshman model [5] which is the single ion model in the collisionless limit and valid for arbitrary aspect ratio and flux surface geometry. The results of equilibrium calculations for various pressure profiles are summarized in Fig. 1 and Table I. The q-profiles, Mercier coefficients(D_M-1/4) and BSC profiles of PRSM-RFP configuration with A=2.0, p_0=8.4kPa, κ=1.4 and δ=0.4 are illustrated in Figure 1.

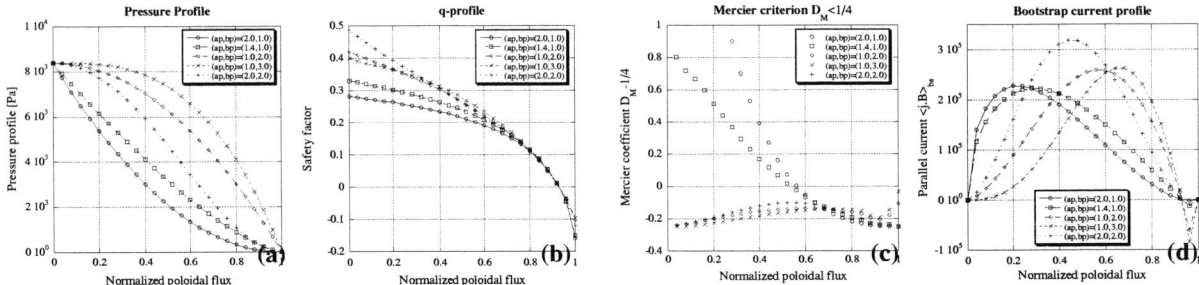

FIGURE 1. PRSM-RFP equilibrium profiles for each pressure profile. (a) Pressure profile control. (b) q-profile. (c) Mercier coefficient. (d) Parallel bootstrap current profile.

TABLE 1. Pressure Profile Dependence of Low Aspect Ratio PRSM-RFP Equilibrium (A=2.0, p_0=8.4kPa, κ=1.4, δ=0.4).

(a_p,b_p)	(a_T,b_T)	$\lambda_0[m^{-1}]$	F	Θ	<β>[%]	β_p[%]	β_t[%]	q_0/q_a	I_φ^{eq}[MA]	F_{bs}	C_{bs}
(2.0,1.0)	(1.75,1.0)	42.2	-0.27	1.90	23.9	40.7	23.1	0.28/-0.16	0.138	0.70	2.45
(1.4,1.0)	(1.15,1.0)	44.9	-0.26	1.90	35.3	41.4	31.5	0.33/-0.15	0.160	0.74	2.51
(1.0,2.0)	(0.75,2.0)	45.6	-0.22	2.16	65.1	55.1	54.1	0.42/-0.10	0.187	0.61	1.56
(1.0,3.0)	(0.75,3.0)	46.3	-0.22	2.26	78.4	58.7	61.5	0.40/-0.10	0.194	0.48	1.16
(2.0,2.0)	(1.75,2.0)	46.9	-0.22	1.93	50.2	44.8	44.6	0.49/-0.12	0.186	0.77	2.44

$<\beta>\equiv 2\mu_0<p>/B^2$, $\beta_t\equiv 2\mu_0<p>/B_{\varphi 0}^2$, $\beta_p\equiv 2\mu_0<p>/B_{pw}^2$, $B_{pw}\equiv \mu_0 I_\varphi^{eq}/L_p$, $L_p\equiv 2\pi a[(1+\kappa^2)/2]^{1/2}$, $F_{bs}\equiv I_\varphi^{bs}/I_\varphi^{eq}$, $C_{bs}\equiv (A^{1/2}/\beta_p)F_{bs}$, $T(\psi)=T_0(1-\Psi^{bt})^{at}$.

Fig. 1 shows that equilibria having peaked pressure profiles, for example (a_p,b_p)=(2.0,1.0) i.e. parabolic profile and (1.4,1.0) in this case, can be unstable in the center region against the Mercier criterion. The parabolic pressure profile is close to Suydam mode stable one in cylindrical geometry and also preferable from the viewpoint of decreasing the average temperature required to generate the maximum fusion thermal power and to attain the ignition condition for a given density profile [6]. Therefore a pressure profile close to parabolic which satisfies the Mercier criterion is needed to be chosen by its profile control. The criterion can be satisfied by a broad profile with a relatively high β value when the p_0, $I(\psi)$, κ and δ are fixed. A broad pressure brings a larger BSC in the edge region(Fig.1(d)), but a smaller F_{bs} as shown in Table 1. It is noted that Mercier criterion can be satisfied even for a relatively peaked pressure profile $(a_p,b_p)=(1.4,1.0)$ close to parabolic one by increasing the ellipticity κ of plasma cross section as shown in Figure 2.

NEOCLASSICAL MHD EQUILIBRIA

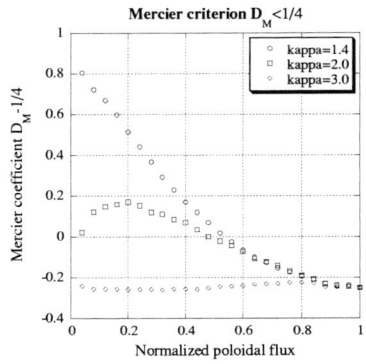

FIGURE 2. Dependence of the Mercier coefficient on ellipticity κ in the case of A=2.0, p_0=8.4kPa and $(a_p,b_p)=(1.4,1.0)$.

It is an important issue to evaluate the bootstrap current effect on RFP equilibria from the viewpoint of sustaining the plasma current non-inductively or generating the steady-state equilibrium configuration. For this purpose we demonstrate self-consistent calculation of equilibria with neoclassical current effects (neoclassical conductivity and bootstrap current), so called, neoclassical MHD equilibria. When the current sources in equilibria are expressed as the parallel current $\langle \vec{j} \cdot \vec{B} \rangle$, the toroidal field function $F(\psi)$ is represented by the pressure gradient and the surface averaged parallel current [7]

$$F' + \left(p'/\langle B^2 \rangle\right)F + \left(\langle \vec{j} \cdot \vec{B} \rangle / \langle B^2 \rangle\right) = 0 \qquad \text{d)}$$

Here, prime denotes the derivative with respect to ψ and $\langle \cdot \rangle$ the flux surface average. Thus the $F(\psi)$ is numerically determined by solving this ordinary differential equation when the parallel current is specified by the generalized Ohm's law. The equilibrium equation is solved by use of this numerical solution, iteratively. As for the RFP equilibrium, auxiliary current sources are needed around the reversal surface of the toroidal field because the Ohmic current does not flow there. Here is included this external term as rf-driven current assumed. The result of numerical analysis for neoclassical, low aspect ratio RFP MHD equilibria is shown in Figure 3 and Table 2, where the classical and neoclassical equilibrium profiles are compared, in the case of A=2.0, p_0=8.4kPa, parabolic pressure profile and circular cross section.

Fig.3(a) and (b) illustrate the profiles of parallel currents in the classical(Ohmic) equilibrium (without bootstrap current) and in the neoclassical equilibrium. Characteristic values of each equilibrium are summarized in Table II. Here I_ϕ^{Ohm}, I_ϕ^{PS}, I_ϕ^{ex}, F and Θ denote the Ohmic current, Pfirsch-Schluter current, external current, reversal and pinch parameters, respectively. The effect of bootstrap current on RFP equilibria is obvious from these figures and it is observed that a hollow current profile, which is one of the typical features of the neoclassical equilibria, is generated even with F_{bs}=0.40. The Ohmic current in the neoclassical equilibrium is reduced because of the large bootstrap current, thus the loop voltage is reduced. Fig.3(c) shows the comparison of the q-profile in each equilibrium. In the neoclassical case, the value of q is increased rapidly near the magnetic axis due to the hollow equilibrium current profile, then it enhances the local and global magnetic shears, which is favorable to the stability. To compare

TABLE 2. Comparison of the classical and neoclassical equilibra (A=2, p_0=8.4kPa, κ=1.0, δ=0.0).

Equil. quantity	classical	neoclassical
I_ϕ [kA]	136.0	136.0
I_ϕ^{Ohm}[kA](V_l[V])	102.7(1.88)	41.7(0.8)
I_ϕ^{PS} [kA]	30.8	31.0
I_ϕ^{BS} [kA] (F_{bs})	-	54.2(0.40)
I_ϕ^{ex} [kA]	2.6	9.1
F/Θ	-0.24/1.89	-0.24/1.86
β_p / β_t	0.45/0.19	0.47/0.20

the MHD stability properties of such neoclassical and classical equilibria we are investigating the stabilities against both Mercier and m=1 MHD kink modes, simultaneously the dependencies of F_{bs} and magnetic shear on β, pressure-temperature profiles, non-circularity, aspect ratio and other device parameters for an optimization of the equilibria.

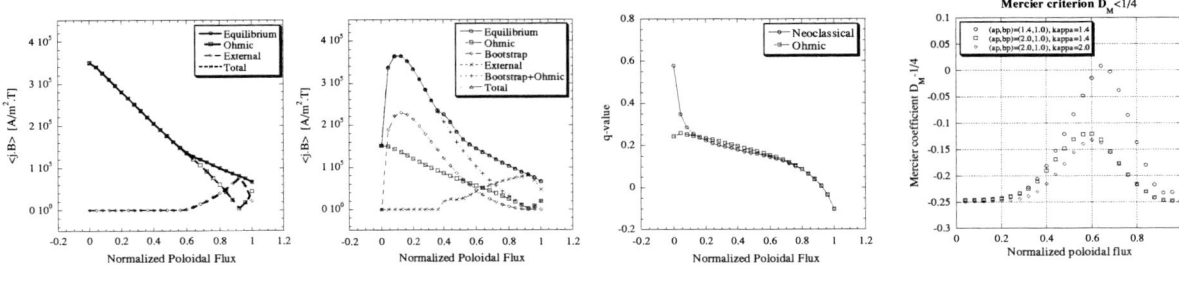

FIGURE 3. (a) The flux surface averaged parallel current profiles $<j\cdot B>_{total}(=<j\cdot B>_{eq})=<j\cdot B>_E+<j\cdot B>_{ex}$ in the classical equilibrium. (b) The parallel current profiles $<j\cdot B>_{total}(=<j\cdot B>_{eq})=<j\cdot B>_E+<j\cdot B>_B+<j\cdot B>_{ex}$ in the neoclassical equilibrium. (c) The comparison of q-profile in each equilibrium.

FIGURE 4. Mercier coefficients in the neoclassical equilibria.

It is possible to obtain neoclassical equilibria with larger F_{bs} by applying more peaked density profile for a fixed pressure profile and/or larger ellipticity κ. For example, F_{bs} value goes to 0.67 when κ=2.0 and other parameters are the same as Fig.3. Also, it could be attainable to obtain equilibria with much larger F_{bs} even for a relatively flat pressure profile in the central region if the local current drive at center of the plasma is possible without Ohmic current, for instance, by taking account of the α-particle contribution around the magnetic axis[8] or applying radio-frequency wave which can travel into high-density regime. The local stability of RFP configuration is improved owing to such neoclassical effect because its flattened q-profile around the axis is sloped. Figure 4 shows that Mercier modes are stabilized in the neoclassical equilibria for the case of a peaked pressure profile, which is shown to be unstable in Section 2. It is noteworthy that the neoclassical equilibrium for any pressure profile mentioned above is stable against Mercier mode by increasing κ. To obtain stable equilibria against the localized modes with high-β and high-F_{bs}, one must search for the optimum combination of the parameters such as pressure profile, ellipticity and so on. Stability for the kink mode is also required to survey because the PRSM condition of uniform j_θ/B_θ-profile is not satisfied in neoclassical equilibria.

SUMMARY

The equilibrium and stability properties in low aspect ratio PRSM-RFP configuration are investigated. The Mercier criterion can be satisfied even for a peaked pressure profile close to parabolic one by increasing the ellipticity of plasma cross section. We demonstrated the equilibrium analysis including neoclassical effects self-consistently and its effects on the RFP configuration were studied. The increase of the magnetic shear suggests the possibility of attaining the neoclassical RFP equilibrium with simultaneous high bootstrap current fraction and high stability beta limit.

REFERENCES

1. Y. Nagamine, S. Shiina, et al., 28th EPS Conf. Contl. Fusion and Plasma Phys. (Madeira, 2001).
2. C. Mercier, Nucl. Fusion 1, 47 (1960).
3. Y. Kondoh, Nucl. Fusion 18, 421 (1978).
4. Y. Kondoh, J. Phys. Soc. Jpn. 58, 489 (1989).
5. S. P. Hirshman, Phys. Fluids 31, 3150 (1988).
6. D. A. Ehst, E. Kenneth, Jr., M. S. Weston, Jr., Nucl. Technology 43, 28(1979).
7. S. Tokuda, T. Takeda, M. Okamoto, J. Phys. Soc. Jpn 58, 871 (1989).
8. M. Taguchi, J. of Plasma and Fusion Research 77, 153 (2001).

An Equilibrium Equation of a Magnetized Rotating Plasma

Koichi Saeki[*], Akira Tsushima[†] and Heiji Sanuki[**]

[*]*Department of Physics, Faculty of Science, Shizuoka University, Ohya 836, Shizuoka 422-8529, Japan*
[†]*Department of Physics, Faculty of Engineering, Yokohama National University, Tokiwadai 156, Hodogaya-ku, Yokohama 240-8501, Japan*
[**]*National Institute for Fusion Sciences, Toki, Gifu 509-5292, Japan*

Abstract. We derive an equilibrium equation of a magnetized plasma including electric fields, which gives the description of both tokamaks and magnetoelectric tori. Thus, this equation includes both the Grad-Shafranov equation and the equilibrium equations derived by T. H. Stix who proposed the plasma confinement by a magnetoelectric torus.

INTRODUCTION

The existence of equilibrium solution is one of the most important keys to construct nuclear fusion devices. The equilibrium of a tokamak plasma is sustained by a toroidal current and described by the Grad-Shafranov equation [1]. On the other hand, Stix [2] proposed the plasma confinement by a magnetoelectric torus where the poloidal plasma rotation reduces the charge accumulation.

DERIVATION OF AN EQUILIBRIUM EQUATION

Here, we derive an equilibrium equation of a magnetized plasma including electric fields, which gives the description of both tokamaks and magnetoelectric tori. In order to get an equilibrium equation, we start from MHD equations. The momentum balance equation of a steady-state plasma is

$$nm(\boldsymbol{v}\cdot\boldsymbol{\nabla})\boldsymbol{v} = -mC_s^2\boldsymbol{\nabla}n - \rho_c\boldsymbol{\nabla}\phi + \boldsymbol{j}\times\boldsymbol{B}. \qquad (1)$$

Here, $\boldsymbol{v}$ is the plasma flow velocity. n and m are the plasma density and the ion mass. $\boldsymbol{j}$ and $\boldsymbol{B}$ are the electric current density and the magnetic field. ρ_c and ϕ are the electric charge density and the plasma potential, respectively. $C_s^2 = (\kappa T_e + \gamma\kappa T_i)/m$. Here, C_s, κ, T_e, T_i and γ are the ion acoustic velocity, Boltzmann constant, the electron and ion temperatures and the specific-heat ratio of ions. Ohm's law in case of zero resistivity is

$$-\boldsymbol{\nabla}\phi + \boldsymbol{v}\times\boldsymbol{B} = 0. \qquad (2)$$

The electric charge density ρ_c is determined by Poisson's equation,

$$\nabla^2\phi = -\frac{\rho_c}{\epsilon_0}. \qquad (3)$$

Here, ϵ_0 is the dielectric constant of vacuum. The equations of continuity for the magnetic field $\boldsymbol{B}$, the electric current density $\boldsymbol{j}$ and the plasma flow density $n\boldsymbol{v}$ are

$$\boldsymbol{\nabla}\cdot\boldsymbol{B} = 0, \qquad (4)$$
$$\boldsymbol{\nabla}\cdot\boldsymbol{j} = 0, \qquad (5)$$
$$\boldsymbol{\nabla}\cdot n\boldsymbol{v} = 0. \qquad (6)$$

We treat an axisymmetric plasma by using the cylindrical coordinates r,θ,z. We use following flux functions. The poloidal magnetic flux function $\psi(r,z)$, the poloidal current flux function $I(r,z)$ and the poloidal flow flux function

$S(r,z)$ are defined by

$$B = \frac{1}{2\pi r}(\boldsymbol{\nabla}\psi \times \hat{\boldsymbol{\theta}} + \mu_0 I \hat{\boldsymbol{\theta}}), \tag{7}$$

$$\boldsymbol{j} = \frac{1}{2\pi r}(\boldsymbol{\nabla}I \times \hat{\boldsymbol{\theta}} - \frac{1}{\mu_0}\mathcal{L}\psi \hat{\boldsymbol{\theta}}), \tag{8}$$

$$n\boldsymbol{v} = \frac{1}{2\pi r}\boldsymbol{\nabla}S \times \hat{\boldsymbol{\theta}} + nv_\theta \hat{\boldsymbol{\theta}}. \tag{9}$$

respectively. Here, $\mathcal{L}\psi = r\partial/\partial r(\partial\psi/r\partial r) + \partial^2\psi/\partial z^2$. $\hat{\boldsymbol{\theta}}$ is the unit vector in the azimuth direction. The plasma density n and the plasma potential ϕ are also functions of r,z. By taking the scalar products of Eq. (2) with $\boldsymbol{B}$, $\hat{\boldsymbol{\theta}}$ and $\boldsymbol{\nabla}\psi/|\boldsymbol{\nabla}\psi|^2$, we get the following equations,

$$\phi = \phi(\psi), \tag{10}$$
$$S = S(\psi), \tag{11}$$
$$\boldsymbol{v} = \frac{1}{2\pi rn}\frac{\partial S}{\partial \phi}\boldsymbol{\nabla}\phi \times \hat{\boldsymbol{\theta}} + v_\theta \hat{\boldsymbol{\theta}}. \tag{12}$$

Similarly, the scalar products of Eqs. (1) with $\boldsymbol{B}$, $\hat{\boldsymbol{\theta}}$ and $\boldsymbol{\nabla}\psi/|\boldsymbol{\nabla}\psi|^2$, we get the following equations,

$$[(\boldsymbol{v}\cdot\boldsymbol{\nabla})\boldsymbol{v}]\cdot\boldsymbol{B} = \frac{C_s^2}{2\pi rn}\frac{\partial(n,\psi)}{\partial(r,z)}, \tag{13}$$

$$[(\boldsymbol{v}\cdot\boldsymbol{\nabla})\boldsymbol{v}]_\theta = -\frac{1}{(2\pi r)^2 nm}\frac{\partial(I,\psi)}{\partial(r,z)}, \tag{14}$$

$$\frac{1}{(2\pi r)^2}\left[\frac{\mathcal{L}\psi}{\mu_0} + \mu_0 I\frac{\boldsymbol{\nabla}I\cdot\boldsymbol{\nabla}\psi}{|\boldsymbol{\nabla}\psi|^2}\right] - \epsilon_0\nabla^2\phi\frac{\partial\phi}{\partial\psi} + \left[mC_s^2\boldsymbol{\nabla}n + nm(\boldsymbol{v}\cdot\boldsymbol{\nabla})\boldsymbol{v}\right]\cdot\frac{\boldsymbol{\nabla}\psi}{|\boldsymbol{\nabla}\psi|^2} = 0. \tag{15}$$

Thus, Eqs. (13) and (14) indicates the deviation of the equidensity surface n from the magnetic surface ψ and the current surface I. The convective derivative term $nm[(\boldsymbol{v}\cdot\boldsymbol{\nabla})\boldsymbol{v}]$ induces the derivations of the current and equidensity surfaces I, n from the magnetic surface ψ. Conversely, if the convective derivative term is negligibly small, all of the surfaces I, n, ϕ, S, ψ are coincident. Now, we get the Grad-Shafranov equation including electric fields in case of neglecting the convective derivative term, as follows.

$$\frac{1}{(2\pi r)^2}\left[\frac{1}{\mu_0}\mathcal{L}\psi + \mu_0 I\frac{\partial I}{\partial \psi}\right] + mC_s^2\frac{\partial n}{\partial \psi} - \epsilon_0\nabla^2\phi\frac{\partial\phi}{\partial\psi} = 0, \tag{16}$$

If we neglect the plasma potential ϕ, Eq. (16) is reduced to Grad-Shafranov equation describing the tokamak equilibrium. Eq. (16) describes both the tokamak and the magnetoelectric torus at the same time.

DERIVATION OF MAGNETIC AND POTENTIAL SURFACES

Here, we derive magnetic and potential surfaces ψ, ϕ of a torus plasma where the minor and major radii are a and R, respectively. Thus, we use the toroidal coordinates ρ, θ, ω; $r = R + \rho\cos\omega$, $z = \rho\sin\omega$, and solve Eq. (16) under simple assumptions as follows. $\psi(a) = 0$, and $\phi(a) = 0$. In the plasma, $\partial p/\partial\psi = $ constant, $I\partial I/\partial\psi = $ constant, and $\partial\phi/\partial\psi = C$ (= constant). The inverse aspect ratio a/R and the toroidal beta is considerably smaller than 1.

It is convenient to use the ratio δ_0 of $(\epsilon_0\nabla^2\phi)\partial\phi/\partial\psi$ with $\mathcal{L}\psi/(2\pi r)^2\mu_0$. Under the above assumption, this is rewritten as $\delta_0 = (\epsilon_0 E_\rho^2(a)/2)/(B_\omega^2(a)/2\mu_0) = (2\pi R)^2\epsilon_0\mu_0 C^2$. We put $\delta = (1-\delta_0)/(1+\delta_0)$. Then, $\delta = 1$ and -1 means the pure tokamak and the pure magnetoelectric torus, respectively. The change of δ leads the plasma from the tokamak to the magnetoelectric torus continuously.

The outer solutions of Eq. (16) under the above assumption are

$$\psi_{out} = -\psi(0)\{2\ln\frac{\rho}{a} + [\ln\frac{\rho}{a} + \lambda_1(1-\frac{\rho^2}{a^2})]\frac{\rho}{R}\cos\omega\}, \qquad (17)$$

$$\phi_{out} = -C\psi(0)\{2\ln\frac{\rho}{a} + [-ln\frac{\rho}{a} + \lambda_2(1-\frac{\rho^2}{a^2})]\frac{\rho}{R}\cos\omega\}, \qquad (18)$$

The inner solution is

$$\psi_{in} = -\psi(0)\{-(1-\frac{\rho^2}{a^2})(1+\lambda_3\frac{\rho}{R}\cos\omega)\}, \qquad (19)$$

and $\phi_{in} = C\psi_{in}$. Here, $\lambda_1 = (\beta+0.75)/\delta - 1$, $\lambda_2 = (\beta+0.75)/\delta$, $\lambda_3 = (\beta+0.75)/\delta - 0.5$. $\beta = <nmC_s^2>/(\epsilon_0 E_\rho^2(a)/2 + B_\omega^2(a)/2\mu_0)$ and indicates the generalized poloidal beta.

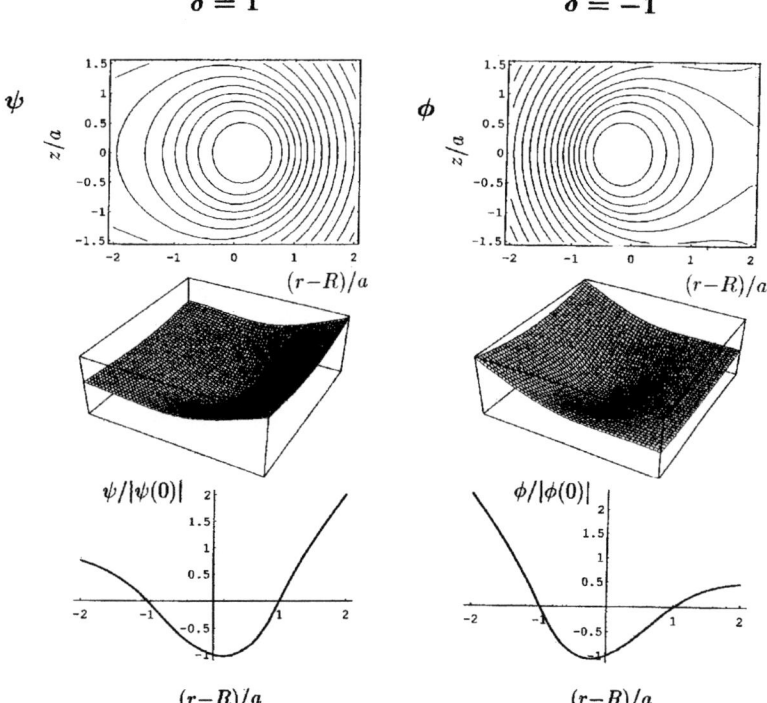

FIGURE 1. Equilibrium magnetic surfaces ψ of a Tokamak ($\delta = 1$) and equilibrium potential surfaces ϕ of a magnetoelectric torus ($\delta = -1$). $\beta = 1.4, a/R = 0.2$ and $\nu \ll 1$.

The parameter of the convective derivative term ν is defined as

$$\begin{aligned}\nu &= \frac{nm(\boldsymbol{v}\cdot\boldsymbol{\nabla})\boldsymbol{v}\cdot\boldsymbol{\nabla}\psi/|\boldsymbol{\nabla}\psi|^2}{\mathcal{L}\psi/(2\pi r)^2\mu_0 - (\epsilon_0\nabla^2\phi)\partial\phi/\partial\psi} \\ &= 0.25(\omega_{pi0}/\omega_{ci0})^2(1-\delta)/\delta.\end{aligned}$$

Thus, the convective derivative term is able to be neglected when $\nu \ll 1$. Figure 1 shows equilibrium magnetic surfaces ψ of a Tokamak ($\delta = 1$), and equilibrium potential surfaces ϕ of a magnetoelectric torus ($\delta = -1$). Here, $\beta = 1.4$, $a/R = 0.2$ and $\nu \ll 1$.

CONCLUSIONS

The derived new equilibrium equation of the magnetized plasma including electric fields describes the equilibrium plasma from the tokamak to the magnetoelectric torus, continuously.

ACKNOWLEDGMENTS

We appreciate Prof. D. Grésillon for his cooperation. A part of this work is supported by the Ministry of Education, Science, Sports and Culture in Japan.

REFERENCES

1. Shafranov V. D., *Soviet Physics JETP* **6**, 545 (1958).
2. Stix T. H., *Phys. Rev. Lett.* **24**, 135 (1970).

High-Mode Configuration Solutions from Generalized Beltrami Equation

Ana Márcia A. Taveira, Paulo H. Sakanaka,
Carlos E. Scussiatto, and Brahmananda Dasgupta

Instituto de Física "Gleb Wataghin", C.P.6165
Universidade Estadual de Campinas
13083-970, Campinas, SP, Brazil
e-mail: taveira@ifi.unicamp.br

Abstract. It was reported by Dasgupta and Sakanaka that starting with a warm, homogeneous, non-relativistic electron-positron plasma, described by the two-fluid model, a *Triple* Beltrami equation is derived, describing its steady-state solution with flow. This is equivalent to the *Simple* Beltrami equation shown by J.B. Taylor to model the force-free relaxed state of plasma, a static solution of plasma, and the *Double* Beltrami equation shown by S.M. Mahajan and Z. Yoshida to model a steady-state solution of electron-ion plasma with flow velocity. We explore the eigenvalue solutions obtained by imposing a consistent set of boundary values for the magnetic and velocity fields.

INTRODUCTION

Electron-positron plasmas are frequently encountered in the early universe, in active galactic nuclei, in the pulsar magnetosphere, as well as in laser irradiated plasmas, [1, 2, 3, 4, 5]. The physics of electron-positron plasmas, as reported by Dasgupta and Sakanaka [6], is different [7, 8, 9, 10, 11] from that of an electron-ion plasma, because of the equality of the masses of the two components. Specifically, some of the temporal and spatial scales (viz. the ion plasma period, the ion gyroperiod, the ion Debye and ion gyro radii etc., as well as the ion acoustic and Alfvén waves), which are ubiquitous in an electron-ion plasma, cease to exist in electron-positron plasmas. Thus, the magnetohydrodynamics (MHD) of an electron-positron plasma is expected to show some new physical results, which could be significantly different from those of an electron-ion plasma. In fact, present day laboratory experiments [12, 13] show that it is possible to obtain nonrelativistic electron-positron plasmas which can stay for sufficiently longer time if the annihilation time for the electrons and positrons is larger than the electron plasma and gyroperiods.

In this work, we show that, the triple Beltrami equation for (e-p) plasma gives configuration of the magnetic field with a wide range of novel structures.

TRIPLE BELTRAMI EQUATION

We start with a warm, homogeneous, non-relativistic (e-p) plasma. The dynamics of low phase velocity (in comparison with the speed of light c) e-p plasma are governed by the electron positron momentum equations, which are, respectively,

$$\frac{\partial \mathbf{v_e}}{\partial t} + \frac{1}{2}\nabla \mathbf{v_e}^2 - \mathbf{v_e} \times \nabla \times \mathbf{v_e} = -\frac{e}{m}\left[\mathbf{E} + \frac{1}{c}\mathbf{v_e} \times \mathbf{B}\right] - \frac{1}{mn}\nabla p_e, \qquad (1)$$

and

$$\frac{\partial \mathbf{v_p}}{\partial t} + \frac{1}{2}\nabla \mathbf{v_p}^2 - \mathbf{v_p} \times \nabla \times \mathbf{v_p} = \frac{e}{m}\left[\mathbf{E} + \frac{1}{c}\mathbf{v_p} \times \mathbf{B}\right] - \frac{1}{mn}\nabla p_p \qquad (2)$$

supplemented by Faraday's law

$$\frac{\partial \mathbf{B}}{\partial t} = -c\nabla \times \mathbf{E} \qquad (3)$$

and Ampère's law

$$\nabla \times \mathbf{B} = \frac{4\pi}{c}\mathbf{J} = \frac{4\pi}{c}ne(\mathbf{v_p} - \mathbf{v_e}) \equiv \frac{8\pi}{c}ne\mathbf{U} \tag{4}$$

where $\mathbf{v_e}, (\mathbf{v_p})$ is the electron (positron) fluid velocity, $\mathbf{U} = \frac{1}{2}(\mathbf{v_p} - \mathbf{v_e})$, and, $\mathbf{E}(\mathbf{B})$ is the electric (magnetic) field, n is the uniform electron (positron) number density (given by the continuity equation), e is the magnitude of the electron charge, m is the electron/positron mass, and $p_e, (p_p)$ is the scalar electron (positron) pressure. The fluid velocity $\mathbf{V}$ for e-p plasma is defined by,

$$\mathbf{V} = \frac{1}{2}(\mathbf{v_p} + \mathbf{v_e}). \tag{5}$$

When equations (1)-(5) are processed one gets,

$$\nabla \times \nabla \times \nabla \times \mathbf{B} + p\nabla \times \nabla \times \mathbf{B} + q\nabla \times \mathbf{B} + \frac{p}{2}\mathbf{B} = 0 \tag{6}$$

and

$$V = g\mathbf{B} + h\nabla \times \mathbf{B} + g\nabla \times \nabla \times \mathbf{B} \tag{7}$$

where $g = \frac{2}{\sqrt{4+p^2-4q}}$ and $h = \frac{pg}{2}$.

It may be mentioned that for an ideal coupled magnetofluid, with flow velocity, where the ion mass is much greater than that of electrons, Mahajan and Yoshida [14] have reported a similar type equation, a double Beltrami equation.

The general solution of equation (6) can be obtained as a linear superposition of the Chandrasekhar-Kendall eigenfunctions [15], which are the eigenfunctions of the "curl" operator, i.e. $\nabla \times \mathbf{B} = \lambda \mathbf{B}$. A formal solution of equation (6) is obtained as,

$$\mathbf{B} = \alpha_1 \mathbf{B_1} + \alpha_2 \mathbf{B_2} + \alpha_3 \mathbf{B_3} \tag{8}$$

where, $\alpha_i (i = 1, 2, 3)$'s are (three) arbitrary constants, and $\mathbf{B_i}$ are the eigenfunctions of

$$\nabla \times \mathbf{B_i} = \lambda_i \mathbf{B_i} \tag{9}$$

where, eigenvalues λ_i's are the roots of the algebraic equation,

$$\lambda^3 + p\lambda^2 + q\lambda + p/2 = 0. \tag{10}$$

We have numerically constructed a particular solution of the equilibrium state of e-p plasma, governed by the equation (6), by the superposition of the solutions of equation (9), in an axisymmetric spherical geometry. It is to be noted that, in axisymmetric case, with $m = 0, n = 0$ (where, k is replaced by n), the general solution of equation (9) given (Chandrasekhar-Kendall representation) by

$$\mathbf{B_i} = -\nabla \times (\mathbf{r}\psi_i) - \frac{1}{\lambda_i}\nabla \times \nabla \times (\mathbf{r}\psi_i) \tag{11}$$

with

$$\nabla^2 \psi + \lambda_i^2 \psi = 0, \quad \text{therefore} \quad \psi_i = j_m P_m^n(\cos\theta)\exp(in\phi) \tag{12}$$

We consider axisymmetric state, with $m = 1, n = 0$ (Rosenbluth and Bussac), for this the components of the magnetic fields are given by, $B_{ir} = -\frac{2}{\lambda_i r}j_1(\lambda_i r)\cos\theta$, $B_{i\theta} = \frac{1}{\lambda_i r}\frac{d}{dr}[rj_1(\lambda_i r)]\sin\theta$, and $B_{i\phi} = -j_1(\lambda_i r)\sin\theta$.
Taking as boundary conditions, for example:

$$\mathbf{B} \cdot \hat{\mathbf{n}}|_{r=a} = 0, \quad \mathbf{J} \cdot \hat{\mathbf{n}}|_{r=a} = 0 \tag{13}$$

which can be rewritten as

$$j_1(a) + \frac{\alpha_2}{\lambda_2}j_1(\lambda_2 a) + \frac{\alpha_3}{\lambda_3}j_1(\lambda_3 a) = 0 \tag{14}$$

and

$$j_1(a) + \alpha_2 j_1(\lambda_2 a) + \alpha_3 j_1(\lambda_3 a) = 0 \tag{15}$$

we can obtain the parameters which satisfy the given boundary conditions.

Figures 1 and 2 show the contour plots for the ψ-function and velocity-functions, respectively. We have depicted a special solution which has the plasma of the central region pushed away towards equatorial region. Its parameters are a=4.5; α_2= -0.5694; α_3=-0.01935; λ_2=1.718; λ_3=1.116. In figure 2, the left plot refers to the toroidal velocity field and the right poloidal one. The sgins "+" and "-" indicate the positive and the negative values, repsectively.

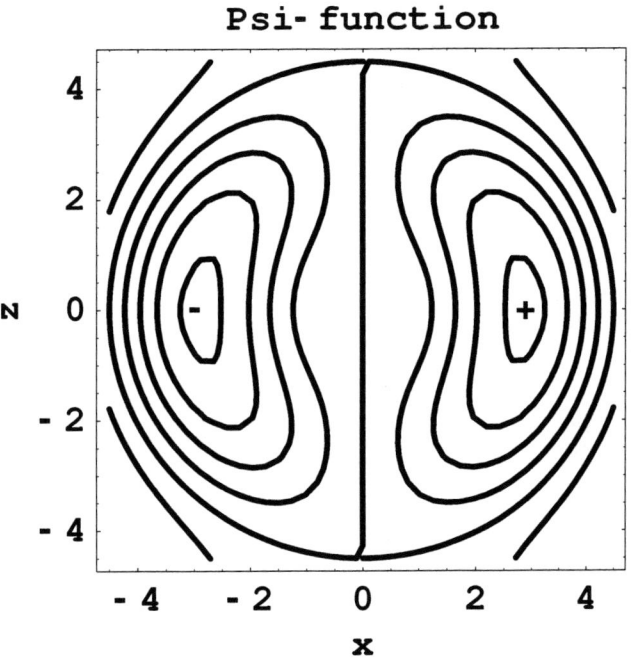

FIGURE 1. Contourplots of the ψ-function. The parameters taken where: a=4.5; α_2= -0.5694; α_3=-0.01935; λ_2=1.718; λ_3=1.116.

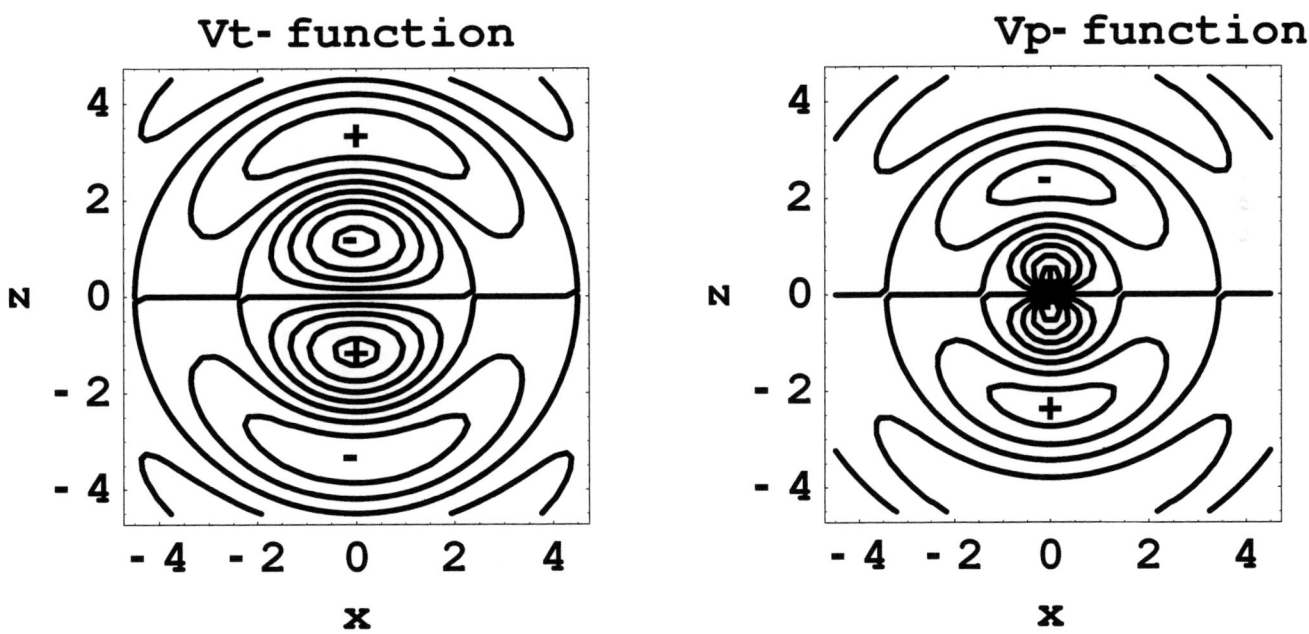

FIGURE 2. Contourplots of the toroidal (left) and poloidal (right) velocity fields, respectively. The parameters taken where: a=4.5; α_2= -0.5694; α_3=-0.01935; λ_2=1.718; λ_3=1.116.

CONCLUSIONS

Starting from two fluid equations for two-species magnetized plasma, with one negatively and one positively charged species, with arbitrary mass and charge, it is possible to derive a general equilibrium solution given as:

$$s\nabla \times \nabla \times \nabla \times \vec{B} + p\nabla \times \nabla \times \vec{B} + q\nabla \times \vec{B} + r\vec{B} = 0$$

where $\vec{B}$ is the magnetic field, s, p, q, and r are constants to be determined through boundary conditions. In the case of electron-ion plasma, $s = 0$ and the solution is that obtained by S. Mahajan and Z. Yoshida. In the case of electron-positron plasma, s is a non-zero quantity and the equation is the same one shown by B. Dasgupta, P. H. Sakanaka, and P. Shukla [7]. For a dusty plasma with electrons entirely attached to dust, it can be shown that the value of s can be either zero or non-zero. In the case of $s = p = 0$, the solution is a well known J. B. Taylor force-free relaxed solution, obtained from a one-fluid equation, where the pressure is zero. In general, for the electron-ion plasma, one requires a tensorial pressure to balance the plasma for equilibrium [10]. Electron-positron plasma is not a far fetched concept. It is presently an well established physical system, found in astrophysical objects as well as in the laboratory. Its equilibrium state is also attained only when a tensorial pressure is present. Presently, it is also possible to produce a dusty plasma with equally charged, with opposite charges, with particles of same mass, for instance, C60 (fullerene) plasma. In this latter case, the coefficient s is again non-zero. We explore various equilibrium states of such a varied two-species plasma. There are cases of equilibrium in which it resembles an H-mode in tokamak plasma.

ACKNOWLEDGMENTS

The authors gratefully acknowledge the financial support from FAPESP - Fundação de Amparo à Pesquisa do Estado de São Paulo, under Proc. 1999/02942-4 and Proc. 2001/0462508, and CNPq - Conselho Nacional de Desenvolvimento Científico e Tecnológico-Bolsa de Iniciação Científica.

REFERENCES

1. B. Dasgupta and P.H. Sakanaka, Proceed. of 2000 International Congress on Plasma Physics, October (2000).
2. P. Goldreich and W. H. Julian, Astrophys. J. **157**, 869 (1969).
3. F. C. Michel, Rev. Mod. Phys. **54**, 1 (1982).
4. M. J. Rees, in *The Very Early Universe*, edited by G. W. Gibbons, S. W. Hawking, and S. Siklas (Cambridge University Press, Cambridge, 1983).
5. H. R. Miller and P. J. Witta, in *Active Galactic Nuclei* (Springer-Verlag, Berlin, 1987), p. 202.
6. V. I. Berezhiani, D. D. Tskhakaya, and P. K. Shukla, Phys. Rev. A **46**, 6608 (1992); E. P. Laing, S. C. Wilks, and M. Tabak, Phys. Rev. Lett. **81**, 4887 (1998).
7. P. K. Shukla, N. N. Rao, M. Y. Yu, and N. L. Tsintsadze, Phys. Rep. **135**, 1 (1986).
8. N. Iwamoto, Phys. Rev. E **47**, 604 (1993); G. P. Zank and R. G. Greaves, *ibid.* **51**, 6079 (1995).
9. V. Sarka, V. I. Berezhiani, and G. Carlini, Physica Scripta **57**, 456 (1998).
10. J. I. Sakai, T. Haruki, and Y. Kazimura, Phys. Rev. E **60**, 899 (1999).
11. P.K. Shukla, B. Dasgupta, and P.H. Sakanaka,Phys. Lett.,A**269**, 144 (2000).
12. C. M. Surko, M. Leventhal, and A. Pasner, Phys. Rev. Lett. **69**, 901 (1989); R. G. Greaves and C. M. Surko, *ibid.* **75**, 3846 (1995).
13. R. G. Greaves, M. D. Tinkle, and C. M. Surko, Phys. Plasmas **1**, 1439 (1994).
14. S.M. Mahajan and Z. Yoshida, Phys. Rev. Lett., **81**, 4863 (1998).
15. S. Chandrasekhar and P.C. Kendall, Astrophys. J., **126**, 457 (1957).

Filamentary Magnetohydrodynamic Simulation Using MDGRAPE-2

Y. Yatsuyanagi*, T. Ebisuzaki*, T. Hatori† and T. Kato**

RIKEN (The Institute of Physical and Chemical Research), Wako 351-0198, Japan
†*Department of Information Science, Kanagawa University, Hiratsuka 259-1293, Japan*
**Department of Applied Physics, Waseda University, Shinjuku 169-8555, Japan*

Abstract. A simulation model of the "magnetohydrodynamic" vortex method, current-vortex method, is developed. The concept is based on the current-vortex filament model. In general, the vortex method needs much calculation time for the Biot-Savart integral. To overcome this problem, we use MDGRAPE-2. The MDGRAPE-2 is a special-purpose computer and is originally developed for the numerical simulations of molecular dynamics. It is found that the MDGRAPE-2 is capable for not only molecular dynamics simulations but also magnetohydrodynamic simulations, because the MDGRAPE-2 accelerates the calculation of the Biot-Savart integral.

INTRODUCTION

We extend the vortex method to two-dimensional MHD one, which we call "current-vortex method" [1]. The concept of our model is based on the current-vortex filament model [2, 3]. In the current-vortex method, electric current density and vorticity are assumed to be proportional to the Dirac's delta function. The electric current and the vorticity always share the same point in the space. The spatial profiles of the electric current and the vorticity are described by the sum of such points. The magnetic and the velocity fields are obtained by the Biot-Savart integral. However, one must notice the calculation cost of the Biot-Savart integral. It is proportional to N^2 where N is the number of points on which the electric current and the vorticity exist. It is likely that simulation time is dominated by the calculation time for the Biot-Savart integral. Thus, fast calculation of the Biot-Savart integral successfully yields short simulation time. To improve the performance of the Biot-Savart integral, we use MDGRAPE-2 [4]. It is a special-purpose computer for classical molecular dynamics simulations. The MDGRAPE-2 accelerates calculations of non-bonding forces, i.e., Coulomb force, van der Waals force and so on. Rewriting the Biot-Savart integral, one finds that the MDGRAPE-2 can calculate the integral.

FILAMENTARY MAGNETOHYDRODYNAMICS

We use the two-dimensional ideal MHD equations:

$$\frac{\partial \omega_z}{\partial t} = -(\mathbf{u} \cdot \nabla)\omega_z + (\mathbf{B} \cdot \nabla)j_z, \tag{1}$$

$$\frac{\partial A_z}{\partial t} = -(\mathbf{u} \cdot \nabla)A_z, \tag{2}$$

$$\nabla \cdot \mathbf{u} = 0, \tag{3}$$

$$\mathbf{E} + \mathbf{u} \times \mathbf{B} = 0, \tag{4}$$

$$\mathbf{B} = -\hat{\mathbf{z}} \times \nabla A_z, \tag{5}$$

$$\omega_z(\mathbf{r},t) = \hat{\mathbf{z}} \cdot \nabla \times \mathbf{u}, \tag{6}$$

$$j_z(\mathbf{r},t) = \frac{\hat{\mathbf{z}}}{\mu_0} \cdot \nabla \times \mathbf{B}, \tag{7}$$

where **B** and **u** are the magnetic and the velocity fields on the x-y plane, A_z, j_z and ω_z are the z components of the magnetic vector potential, electric current density and the vorticity, respectively. The unit vector in z direction is denoted by $\hat{z}$. The mass density is normalized to unity.

We assume that the electric current and the vorticity have discontinuous distributions, and they are confined in the same filament coaxially,

$$j_z(\mathbf{r},t) = \sum_i J_i(t)\delta(\mathbf{r}-\mathbf{r}_i(t)) \quad (=-\nabla^2 A_z(\mathbf{r},t)), \tag{8}$$

$$\omega_z(\mathbf{r},t) = \sum_i \Omega_i(t)\delta(\mathbf{r}-\mathbf{r}_i(t)) \quad (=-\nabla^2 \psi(\mathbf{r},t)), \tag{9}$$

where $\delta(\mathbf{r})$ is the Dirac's two-dimensional delta function, $\mathbf{r}_i(t)$ is the position vector of the i-th filament which has total electric current $J_i(t)$ and circulation $\Omega_i(t)$. Equations (8) and (9) directly shows that there is a current-vortex filament at $\mathbf{r}_i(t)$. Then, the magnetic vector potential $A_z(\mathbf{r},t)$ and the stream function $\psi(\mathbf{r},t)$, the magnetic field $\mathbf{B}(\mathbf{r},t)$ and the velocity field $\mathbf{u}(\mathbf{r},t)$ are defined by

$$A_z(\mathbf{r},t) = \sum_i J_i(t)G(\mathbf{r}-\mathbf{r}_i(t)), \tag{10}$$

$$\psi(\mathbf{r},t) = \sum_i \Omega_i(t)G(\mathbf{r}-\mathbf{r}_i(t)), \tag{11}$$

$$\mathbf{B}(\mathbf{r},t) = -\hat{z}\times\nabla A_z(\mathbf{r},t) = \sum_i J_i(t)\nabla G(\mathbf{r}-\mathbf{r}_i(t))\times\hat{z}, \tag{12}$$

$$\mathbf{u}(\mathbf{r},t) = -\hat{z}\times\nabla\psi(\mathbf{r},t) = \sum_i \Omega_i(t)\nabla G(\mathbf{r}-\mathbf{r}_i(t))\times\hat{z}. \tag{13}$$

Function $G(\mathbf{r})$ is the two-dimensional Green's function for the Poisson equation. The right hand sides of Eqs. (12) and (13) are the discretized form of the Biot-Savart integral. At this point, all the basic quantities are explicitly given by the filamentary representations.

In the next, we rewrite the vorticity equation (1) and the magnetic induction equation (2) using the filamentary representations. Substituting Eqs. (8) - (13) into Eqs. (1) and (2), we obtain the following two equations

$$\sum_i \frac{d\Omega_i(t)}{dt}\delta(\mathbf{r}-\mathbf{r}_i) - \sum_i \Omega_i(t)\nabla\cdot[\mathbf{U}_i\delta(\mathbf{r}-\mathbf{r}_i)] = 0, \tag{14}$$

$$\sum_i \frac{dJ_i(t)}{dt}G(\mathbf{r}-\mathbf{r}_i) - \sum_i J_i(t)\frac{d\mathbf{r}_i}{dt}\cdot\nabla G(\mathbf{r}-\mathbf{r}_i) = -\sum_i J_i(t)(\mathbf{u}\cdot\nabla)G(\mathbf{r}-\mathbf{r}_i), \tag{15}$$

where $\mathbf{U}_i$ is defined by

$$\mathbf{U}_i = \frac{d\mathbf{r}_i}{dt} - \mathbf{u}(\mathbf{r}_i,t) - \frac{J_i(t)}{\Omega_i(t)}\mathbf{B}(\mathbf{r}_i,t), \tag{16}$$

$$\mathbf{u}(\mathbf{r}_i,t) = \sum_{l\neq i}\Omega_l(t)\nabla G(\mathbf{r}_i-\mathbf{r}_l)\times\hat{z}, \tag{17}$$

$$\mathbf{B}(\mathbf{r}_i,t) = \sum_{l\neq i}J_l(t)\nabla G(\mathbf{r}_i-\mathbf{r}_l)\times\hat{z}. \tag{18}$$

To obtain the solution concerning the specific filament, say k-th filament, we integrate Eqs. (14) and (15) over a circle area whose center and radius are $\mathbf{r}_k$ and ε, respectively. We assume that the radius ε is small enough and that the distances between the filaments are much larger than ε. After some calculations, the two equations are reduced to the following forms:

$$\frac{d\Omega_k(t)}{dt} = 0, \tag{19}$$

$$\frac{d\mathbf{r}_k}{dt} = \mathbf{u}(\mathbf{r}_k,t) - \frac{J_k(t)}{\Omega_k(t)}\mathbf{B}(\mathbf{r}_k,t), \tag{20}$$

$$\frac{dJ_k(t)}{dt} = 0. \tag{21}$$

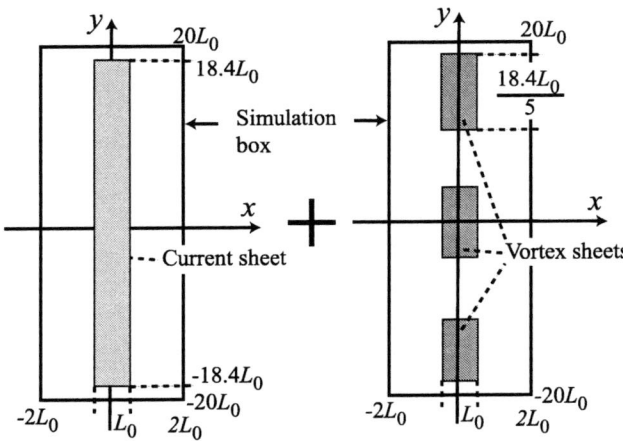

FIGURE 1. Initial condition of the simulation is shown.

Now we have a set of equations that determine the motion of the current-vortex filaments. Equations (19) and (20) are obtained from the vorticity equation (1), and Eq. (21) is obtained from the magnetic induction equation (2). Equation (19) gives the conservation of circulation inside the k-th filament. Equation (20) gives the equation of motion of the k-th filament. Equation (21) gives the conservation of total electric current inside the k-th filament.

SIMULATION RESULTS

The main purpose of the MDGRAPE-2 is to calculate the Coulomb force very fast. After some manipulations of the formulae, one finds that the Biot-Savart integral can be calculated by the MDGRAPE-2. The details are given in [1].

We show a simulation result obtained by the current-vortex method on the MDGRAPE-2 and one obtained by the ordinary MHD code on the general-purpose computer. We have been investigating the evolution of system that contains both the electric current and the vorticity via numerical simulations [5, 6, 7]. Illustrations of the initial condition are shown in Fig. 1. There are an electric current sheet and three vortex sheets in the system. Initial widths of the current and the vortex sheets are denoted by L_0. Size of the simulation box is $4L_0$ in the x direction and $40L_0$ in the y direction. Both the current and the vortex sheets have finite lengths in the y direction. Initial length of the current sheet is $36.8L_0$. Initial lengths of the vortex sheets are reduced to $1/5$ each, compared with that of current sheet. Initial current density and vorticity are given by j_0 and ω_0, respectively, and uniform inside the sheet. We adopt the free boundary condition at all the edges of the simulation box, because the sheets do not deform in x direction near the upper and lower boundaries, if periodic boundary conditions are used. The time scale of simulations is normalized by the Alfvén transit time τ_A. The resistive time scale is given by $\tau_R = 1000\tau_A$.

In the simulations done by the MDGRAPE-2, we use the current-vortex method. Four MDGRAPE-2 boards are used in the simulation. In the simulations done by the general-purpose computer we use the traditional MHD simulation code.

The time evolutions of the current and vortex sheets are shown in Fig. 2. The result shown in Fig. 2 is obtained by the general-purpose computer. In Fig. 2, we can see that the electric current sheet is split into some pieces with time, following the distribution of the vorticity. After $T = 1.5 \times 10^2 \tau_A$, new configurations appear in both the distributions of the electric current and the vorticity. They exhibit the very similar forms. The electric current and the vorticity evolve to create more overlapping regions where they coexist. The structure is rather filament-like than sheet-like. Once the filaments are formed, they survive stably. This is due to the strong correlation between the electric current and the vorticity. The above comments are for the result obtained by the general-purpose computer. One finds that basically the same phenomenon can be demonstrated by the MDGRAPE-2 that is shown in Fig. 3. The time-evolved distributions of the electric current and the vorticity show very similar ones. The calculation time is 110 minutes for the MDGRAPE-2 and 350 minutes for the general-purpose computer. Because the number of the filaments for the MDGRAPE-2 is larger than the mesh number for the general-purpose computer, we cannot compare the calculation time directly. However, it gives evidence that the MHD simulation using 10^5 filaments can be carried out by the MDGRAPE-2 within realistic calculation time.

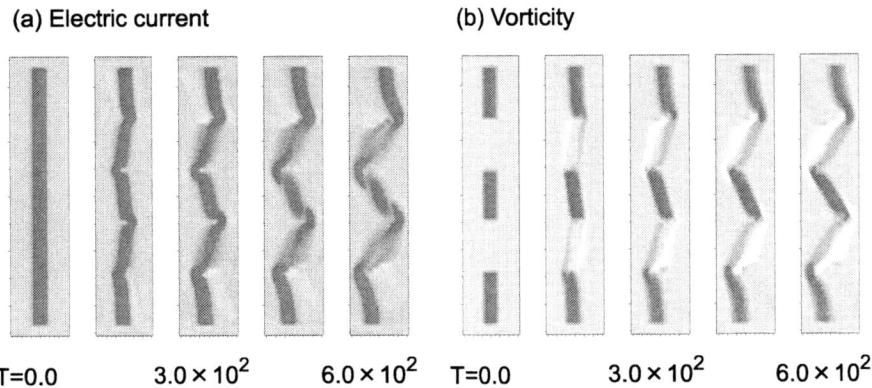

FIGURE 2. A simulation result done by the general-purpose computer is shown. The snap shots of the distribution of (a) electric current and (b) vorticity at T= 0.0, $1.5 \times 10^2 \tau_A$, $3.0 \times 10^2 \tau_A$, $4.5 \times 10^2 \tau_A$ and $6.0 \times 10^2 \tau_A$ are given.

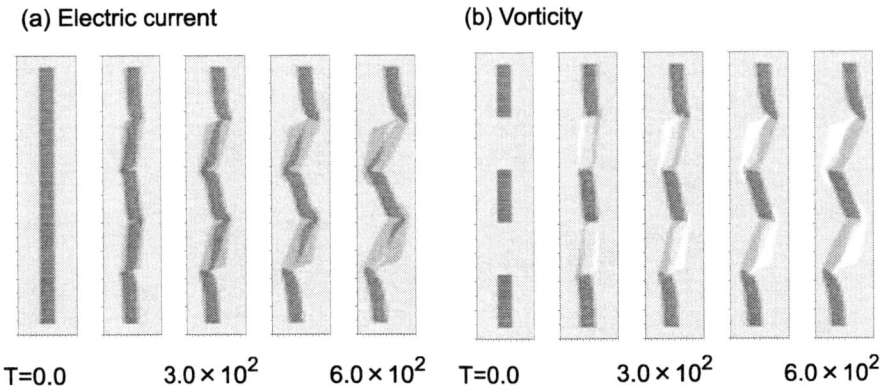

FIGURE 3. A simulation result done by the MDGRAPE-2 is shown The snap shots of the distribution of (a) electric current and (b) vorticity at T= 0.0, $1.5 \times 10^2 \tau_A$, $3.0 \times 10^2 \tau_A$, $4.5 \times 10^2 \tau_A$, $6.0 \times 10^2 \tau_A$ are given.

ACKNOWLEDGMENTS

This work is supported by Special Postdoctoral Research Program at RIKEN.

REFERENCES

1. Y. Yatsuyanagi, T. Ebisuzaki, T. Hatori and T. Kato, submitted to J. Comp. Phys.
2. Y. Yatsuyanagi, T. Hatori and T. Kato, J. Phys. Soc. Jpn. **65**, 745 (1996).
3. Y. Yatsuyanagi, T. Hatori and T. Kato, J. Phys. Soc. Jpn. **67**, 166 (1998).
4. T. Narumi, R. Susukita, T. Ebisuzaki, G. McNiven and B. Elmegreen, Mol. Siml. **21**, 401 (1999).
5. Y. Yatsuyanagi, T. Ebisuzaki, T. Hatori and T. Kato, Phys. Plasmas **9**, 446 (2002).
6. Y. Yatsuyanagi, T. Hatori and T. Kato, Earth, Planets, Space **53**, 615 (2001).
7. Y. Yatsuyanagi, T. Hatori and T. Kato, J. Plasma Phys. **62**, 493 (1999).

Advances in Sun-Earth Connection Modeling

S.B. Ganguli and V.V. Gavrishchaka

Science Applications International Corporation, McLean, Virginia

Abstract. Space weather forecasting is a focus of a multidisciplinary research effort motivated by a sensitive dependence of many modern technologies on geospace conditions. Adequate understanding of the physics of the Sun-Earth connection and associated multi-scale magnetospheric and ionospheric processes is an essential part of this effort. Modern physical simulation models such as multimoment multifluid models with effective coupling from small-scale kinetic processes can provide valuable insight into the role of various physical mechanisms operating during geomagnetic storm/substorm activity. However, due to necessary simplifying assumptions, physical models are still not well suited for accurate real-time forecasting. Complimentary approach includes data-driven models capable of efficient processing of multi-scale spatio-temporal data. However, the majority of advanced nonlinear algorithms, including neural networks (NN), can encounter a set of problems called dimensionality curse when applied to high-dimensional data. Forecasting of rare/extreme events such as large geomagnetic storms/substorms is of the most practical importance but is also very challenging for many existing models. A very promising algorithm that combines the power of the best nonlinear techniques and tolerance to high-dimensional and incomplete data is support vector machine (SVM). We have summarized advantages of the SVM and described a hybrid model based on SVM and extreme value theory (EVT) for rare event forecasting. Results of the SVM application to substorm forecasting and future directions are discussed.

INTRODUCTION

Over the last several years space weather forecasting has been the focus of a multi-disciplinary research effort [12] motivated by a sensitive dependence of many modern technologies on geospace conditions. One of the practical goals of this research is to develop the capability to reliably forecast geomagnetic storms and substorms. Accurate forecasting of the extreme/rare events such as large geomagnetic storms/substorms is of the most practical importance but is also very challenging.

Global physical models of the magnetosphere system [5,15] can provide valuable insight into general physical mechanisms of storm/substorm activity. However, due to necessary simplifying *a priori* assumptions, physical models are still not well suited for accurate real-time forecasting. Data-driven forecasting models derived from the actual data are significantly more accurate than complex physical models. The most successful global nonlinear models currently available are based on artificial neural networks (NN) [6,8,9,18], which have been applied to forecast both geomagnetic storms (*Dst* index) and substorms (*AE* and *AL* indicies).

Despite impressive general performance of the NN-based and other data-driven models models, reliable predictability of the large substorm events remains to be an outstanding problem [6]. Significant improvement can be achieved only by incorporating more detailed multi-scale data that are available or can become available in real time in the near future.

Simplifying feature extraction based on *a priori* assumptions can lead to the loss of important information and can result in less accurate forecasting models. For example, *Gleisner and Lundstedt* [8] have shown that NN substorm models based on raw solar wind (density and velocity) and IMF data are more accurate that those based on various coupling functions. It has been also shown that incorporation of the ionospheric dynamics into low-dimensional analogue models results in more realistic and accurate description of the substorm events [10,11]. The same approach can be used with data-driven substorm models once real-time ionospheric data become available. However, incorporation of the multi-scale ionospheric and other relevant data can dramatically increase input dimensionality of the model. Moreover reliable forecasting of the extreme/rare events would require ability to efficiently extract information from the incomplete data sets.

Thus further advancement in the substorm forecasting will require efficient processing of the high-dimensional and incomplete data. However, for many algorithms (including NN) increase in the input-data dimensionality can cause a set of problems collectively called the "dimensionality curse" [1]. One of the very promising alternative techniques is support vector machine (SVM) [16,17], which is currently under active investigation [3]. SVM combines the training efficiency and simplicity of linear algorithms with the accuracy of the best nonlinear techniques. In many practical application SVMs can tolerate high-dimensional and/or incomplete data and often demonstrate performance superior to analogous NN-based systems [3].

Adequate modeling of the large substorm events may require combination of the SVM or similar technique with extreme value theory (EVT) [4] that can give a reasonable probability estimation of the extremely rare events. Recent studies have demonstrated power of EVT in integrated risk management applications in finance. Risk management goals and framework are quite similar to those of practical space weather forecasting.

FORECASTING SYSTEM BASED ON SVM AND EVT

Despite the initial success of the NN and other techniques applied to substorm (auroral electrojet) modeling, an outstanding issue remains unresolved, i.e., the significant underestimation of large-amplitude events [6,8]. Optimization techniques to improve accuracy of the large substorm prediction have been proposed by *Gavrishchaka and Ganguli* [6]. One of those is symbolic encoding of the *AE* index and usage of the classification framework instead of regression. The simplest approach is to consider two-class or threshold classification problem, i.e., classification of the future *AE* index as supercritical ($AE > AE_c$) or subcritical ($AE < AE_c$) to the application-specific threshold AE_c. This approach is natural for majority of practical needs where the most important issue is to forecast whether the future geomagnetic event will be above or below some application-specific threshold.

Classification approach is well suited when prediction of the exact *AE* value is not necessary and allows significant improvement of the prediction accuracy of large substorm events [6]. However still forecasting accuracy significantly decreases for larger values of AE_c as illustrated in Figure 2. Here prediction accuracy is given by the ratio of the number of the correctly classified supercritical events ($AE > AE_c$) to the total number of these events in the test set. Results obtained from the NN model [6] are shown by dashed line.

Significant improvement in large substorm forecasting can be achieved only by incorporating more detailed multi-scale data and by using an adequate framework for rare event modeing. Thus further advancement in the substorm forecasting will require efficient processing of the high-dimensional and incomplete data where NN and many other algorithms can easily become inadequate.

SVMs developed by *Vapnik* [16,17] have recently been receiving significant interest due to excellent results in various applications [3]. We have provided a short introduction to the main ideas used in SVM in our recent paper [7]. Here we give only a brief description of the SVM and its main advantages. Although we will describe only SVM for classification, SVM for regression is based on similar principles.

SVM is a combination of a kernel-based approach and a structural risk minimization (SRM) principle [16,17]. First step is a nonlinear mapping from the input to a higher-dimensional feature space. This approach is motivated by Cover's theorem stating that any pattern-recognition problem is linearly separable in a sufficiently high dimensional space [14]. The kernel-based approach allows the representation of the discriminant function in high-dimensional feature space without explicit dependence on the feature space dimensionality. Kernel-based machine decouples the number of free parameters (related to the machine capacity) from the size of the input space which can be very large or even infinite. SRM provides solid theoretical grounds for optimization of the SVM generalization ability that is superior to empirical risk minimization approaches used in other machine learning algorithms including NN. In general, the SVM training reduces to the minimization problem with constraints that constitutes a typical quadratic programming problem. That is, in contrast to NN, there is no multiple local minimums. Here we use SVM implementation described by *Chang et al.* [2].

Thus, SVM combines the learning effectiveness of linear machines with the classification/regression power of the best nonlinear algorithms. Unlike typical nonlinear techniques such as NNs, the size of the SVM input space is decoupled from the number of free parameters and allows one to process high-dimensional data without encountering the "high-dimensionality curse". This makes SVM an ideal candidate for processing real-time space data of constantly increasing resolution and complexity. Other advantages of SVM include built-in optimization of the generalization performance based on the SRM principle and fast training (up to orders of magnitude faster compared to analogous NN models).

Another important advantage of the SVM is the ability to efficiently extract information from the incomplete data. For example in many practical cases SVM can be trained with a number of samples that is significantly smaller than the input dimensionality [3]. This feature gives SVM an important advantage over other approaches in modeling rare events.

Although SVM may be able to capture dynamics of the rare events such as large substorms from just a few examples of such events, it is still will be impossible to make a prediction of very rare events that are absent in the available data set. A modern statistical approach in such cases is EVT [4]. EVT consists of several key theorems and results for universal description of the distribution tails and distribution of extreme events.

The two most important EVT results include Fisher-Tippett theorem and theorem by Balkema, de Haan, and Pickands [4]. The latter theorem is of direct importance to our problem. We are interested in the conditional excess distribution function F_u (cedf) defined by $F_u(y) = P(X-u <= y \mid X > u)$, where X is a random variable of interest (e.g., AE or $|dAE/dt|$), u is a given threshold (e.g., AE_c), and $y = (x - u)$ are the excesses. Thus given that $X > u$, F_u gives the tail of the distribution that can be used to estimate probability of extreme events. Theorem by Balkema, de Haan, and Pickands states that for a large class of underlying distribution functions F the cedf $F_u(y)$, for u large, is well approximated by generalized Pareto distribution (GPD) function, given by $G_{\xi,\sigma}(y) = 1-(1+\xi y/\sigma)^{-1/\xi}$ for nonzero ξ and by $G_{\xi,\sigma}(y) = 1- e^{-y/\sigma}$ for $\xi=0$. Given the sample data set (e.g., AE index), the acceptable value of u can be found using standard techniques such as quantile plots (*QQ-plots*). Parameters (ξ, σ) of the GPD can be found by maximum likelihood estimation, i.e., by optimization of the log-likelihood function.

Combining SVM forecasting of the supercritical/subcritical substorm event and EVT for probability estimation of the event with a particular $AE > AE_c$, we can construct a robust risk management system with respect to disruptive substorm events. Similar hybrid frameworks based on combination of standard volatility models and EVT for excess estimations are successfully used in finance for portfolio risk management.

SUBSTORM FORECASTING AND OTHER APPLICATIONS

SVM-based model described in the previous section can be used in many different space weather applications including substorm forecasting from the solar wind (density and velocity) and IMF (B_z and B_y) data. Results of this first space-related application of the SVM have been discussed in details by *Gavrishchaka and Ganguli* [7] and are summarized in Figure 2. Here prediction rates obtained from the linear classifier (solid line), NN (dashed line), and one of the best SVMs (dashed-dotted line) are shown. Linear machine was modeled as an SVM with a linear kernel. Total number of inputs used is 84 (5 min resolution data with up to 100 min of delay). We see that prediction rates obtained with the SVM are superior to those obtained with the simple linear machine, and comparable to or better than those obtained from the analogous NN model. Note also that SVM training in this case was up to 2 orders of magnitude faster compared to the analogous NN model.

Described SVM model for substorm prediction can be combined with GPD function estimated on historical AE data with some reasonable threshold value. Obtained hybrid model could be used not only to predict arrival of the large supercritical substorm but also to estimate conditional probability that the disturbance will reach a particular value of AE ($AE > AE_c$). Potentially this information can be used at the design and operation stage of the ground and spacecraft equipment sensitive to geospace conditions.

There are a number of other space applications for SVM-based models. These include forecasting geo-effective coronal mass ejections (CMEs). Although real-time availability of the solar data constantly increases (e.g., SOHO program), the accurate forecasting of the Earth-directed CMEs, that can potentially cause large non-recurrent geomagnetic storms, remains to be an outstanding problem. The usage of the global average characteristics of the solar magnetic field and/or X-ray flux as low-dimensional inputs may not provide adequate accuracy. Then more detailed solar magnetogram and/or X-ray images should be used. This will lead to a model with a very high-dimensional input where SVM could be an algorithm of choice. For example SVM have been successfully used for image classification where raw pixel data have been used as high-dimensional inputs [13].

SUMMARY AND CONCLUSION

Limitations of the existing space weather forecasting models have been reviewed and new approaches for the next generation models have been proposed. In particular, SVM have been shown to be very promising machine learning technique that can handle high-dimensional and incomplete data. Presented preliminary results confirmed usefulness of the SVM for the substorm forecasting. In the future forecasting models dealing with high-dimensional inputs, that may include real-time ionospheric data, SVM is believed to have performance superior to the best existing algorithms. Hybrid model based on SVM and EVT have been proposed for modeling of rare events such as large substorms. Possibility of the SVM application for geo-effective CME forecasting have been also discussed.

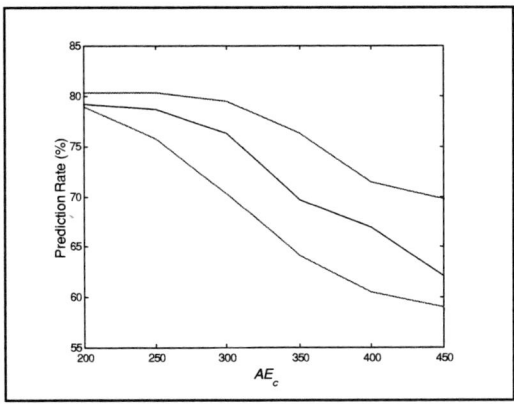

FIGURE 1. Prediction rate obtained from a linear machine (solid line), neural network (dashed line), and support vector machine with hyperbolic tangent kernel (dashed-dotted line).

ACKNOWLEDGMENTS

This work was supported by Science Applications International Corporation.

REFERENCES

1. Bishop, C.M., *Neural Networks for Pattern Recognition*, Clarendon Press, Oxford, 1995.
2. Chang, C.-C., C.-W. Hsu, and C.-J. Lin, The analysis of decomposition methods for support vector machines, *IEEE Trans. Neural Networks*, **11**, 1003, 2000.
3. Cristianini, N., and J. Shawe-Taylor, *Introduction to Support Vector Machines and other Kernel-Based Learning Methods*, Cambridge University Press, 2000.
4. Embrechts, P., et al., Modeling Extremal Events, Springer-Verlag, Berlin-Heidelberg, 2001.
5. Fedder, J.A., J.G. Lyon, S.P. Slinker, and C.M. Mobarry, Topological structure of the magnetotail as a function of interplanetary magnetic field direction, *J. Geophys. Res.*, **100**, 3613, 1995.
6. Gavrishchaka, V.V., and S.B. Ganguli, Optimization of the neutral-network geomagnetic model for forecasting large-amplitude substorm events, *J. Geophys. Res.*, **106**, 6247, 2001a.
7. Gavrishchaka, V.V., and S.B. Ganguli, Support vector machine as an efficient tool for high-dimensional data processing: Application to substorm forecasting, *J. Geophys. Res.*, **106**, 29911, 2001b.
8. Gleisner, H., and H. Lundstedt, Response of the auroral electrojets to the solar wind modeled with neural networks, *J. Geophys. Res.*, **102**, 14269, 1997.
9. Gleisner, H., H. Lundstedt, and P. Wintoft, Predicting geomagnetic storms from solar-wind data using time-delay neural networks, *Ann. Geophys.*, **14**, 679, 1996.
10. Horton, W., and I. Doxas, A low-dimensional dynamical model for the solar wind driven geotail-ionosphere system, *J. Geophys. Res.*, **103**, 4561, 1998.
11. Horton, W., et al., The solar-wind driven magnetosphere-ionosphere as a complex dynamical system, *Phys. Plasmas*, **6**, 4178, 1999.
12. Joselyn, J.A., Geomagnetic activity forecasting: The state of the art, *Rev. Geophys.*, **33**, 383, 1995.
13. Pontil, M., and A. Verri, Object recognition with support vector machines, *IEEE Trans. on PAMI*, 20, 637, 1998.
14. Principe, J.C., N.R. Euliano, and W.C. Lefebvre, *Neural and adaptive systems*, John Wiley, NY, 2000.
15. Schunk, R.W., and J.J. Sojka, The global ionosphere-polar wind system during changing magnetic activity, *J. Geophys. Res.*, **102**, 11625, 1997.
16. Vapnik, V., *The Nature of Statistical Learning Theory*, Springer Verlag, 1995.
17. Vapnik, V., *Statistical Learning Theory*, Wiley, 1998.
18. Weigel, R.S., W. Horton, T. Tajima, and T. Detman, Forecasting auroral electrojet activity from solar wind input with neural networks, *Geophys. Res. Lett.*, **26**, 1353, 1999.

Dynamics of Vortex Type Wave Structures in Plasmas and Fluids

V.Yu. Belashov, R.M. Singatulin[a]

Kazan State Power Engineering University, Russia

Abstract. The results of numerical study of evolution and interaction of the vortex structures in the continuum, and, specifically, in plasmas and fluids in two-dimensional approach, when the Euler-type equations are valid, are presented. The set of the model equations $e_i d_t x_i = \partial_{y_i} H/B$, $e_i d_t y_i = -\partial_{x_i} H/B$, $\partial_t \rho + \mathbf{v} \cdot \nabla \rho = 0$, $\mathbf{v} = -(\hat{\mathbf{z}} \times \nabla \psi)/B$, $\Delta \psi = -\rho$ describing the a continuum or quasi-particles (filaments) with Coulomb interaction models, where ρ is a vorticity or charge density and ψ is a stream function or potential for inviscid fluid and guiding-centre plasma, respectively, and H is a Hamiltonian, was considered. For numerical simulation the CD method specially modified was used. In terms of vortex motion of fluids the results of numerical experiments, specifically, showed that for some conditions the interaction of vortexes in continuum may be nontrivial and, as for the "classic" FAVRs, lead to formation of complex forms of vorticity regions, for example, the vorticity filaments and sheets, and also can ended to formation of the turbulent field. The undertaken approach may be effective in studying of the atmospheric and Alfvén vortex dynamics, and also useful for the interpretation of effects associated with turbulent processes in fluids and plasmas.

BASIC EQUATIONS

In this paper we study numerically the evolution and interaction of the vortex structures (so-called FAVRs [1]) in the continuum, and, specifically, in plasmas and fluids in 2D approach, when the Euler-type equations are valid. In general case the set of the model equations describing the a continuum (inviscid incompressible fluid) or quasi-particles (charged filaments aligned with a uniform field **B**) with Coulomb interaction models is the following:

$$e_i d_t x_i = \partial_{y_i} H/B, \quad e_i d_t y_i = -\partial_{x_i} H/B, \quad \partial_v \equiv \partial/\partial v,$$
$$\partial_t \rho + \mathbf{v} \cdot \nabla \rho = 0, \quad \mathbf{v} = -(\hat{\mathbf{z}} \times \nabla \psi)/B, \quad (1)$$
$$\Delta \psi - f = -\rho$$

where e_i is the charge per unit length of the filaments or the strength (circulation) of discrete vortex, ϕ is a z-component of vorticity ζ or charge density ρ, and ψ is a stream function or potential for 2D flow of inviscid fluid and guiding-centre plasma, respectively, and H is a Hamiltonian. Note, that in the continuum (fluid) model $B=1$ in the Hamiltonian eqs. (1). Function $f=0$ for the continuum or quasi-particles (filaments) with Coulomb interaction models [2], and $f = k^2 \psi$ for a screened Coulomb interaction model [3]. We will consider here only a case $f=0$, and generalization of our approach for $f = k^2 \psi$ is rather trivially.

For numerical simulation the contour dynamics (CD) method [1], to some extent modified, was used. Its idea is that the interaction between the boundaries of the constant ϕ regions is considered, and thanks to that the problem dimension decreases on one. Analytical solution of the Poisson equation (1) for current function _ has form

$$\psi = -\frac{1}{2\pi} \iint d\xi d\eta [\ln r] \phi(\xi, \eta) \quad (2)$$

---

[a] Work supported by the Russian Foundation of Basic Research (grant N 01-02-16116).

where $r = [(x-\xi)^2 + (y-\eta)^2]^{1/2}$. Then a value of velocity can be obtained by differentiation of integral (2), namely:

$$\mathbf{u}(x,y) = \phi_0 \oint_\Gamma [\ln r][\mathbf{e}_x d\xi + \mathbf{e}_y d\eta]. \tag{3}$$

Further, let us define the changing of the contour coordinates with time by solving differential equations $\mathbf{u}(x, y) = \dot{x}\mathbf{e}_x + \dot{y}\mathbf{e}_y$.

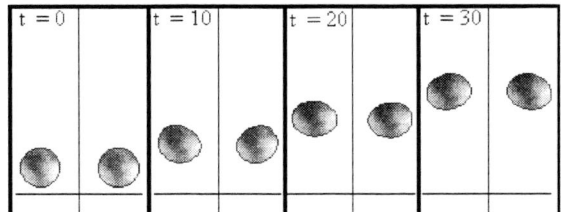

FIGURE 1. Evolution of two vortices having opposite polarities ζ at $\delta = d$.

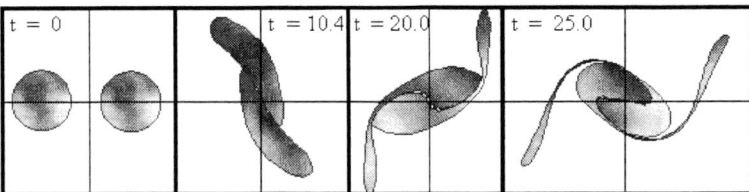

FIGURE 2. Evolution of two vortices having the same polarities ζ at $\delta = d/2$.

For the computer simulation of the vortex structures the contour boundary is divided into N lattice points (moreover the point quantity should be rather great), and the temporal evolution is defined for each point. At this, eq. (3) is written in discrete form and it allows us to define a value of velocity of each point of contour in dependence on influence to it of the points of both the same contour and the contour interacted with it. So, you can observe the temporal evolution of the vortex structure setting its initial form. Besides, the CD method gives a possibility not only to observe evolution of single contour, but also to study the interaction between vortexes having different symmetry order (different modes). Let us demonstrate it in the next paragraph.

NUMERICAL RESULTS AND DISCUSSION

Let us consider some results of numerical simulation in terms of the vortex motion of the inviscid incompressible fluid, as more visual. In general, to study the evolution of vortex structures with different symmetry orders it is necessary to insert a small amplitude disturbance $r = R_0[1 + \varepsilon \cos(m\alpha - \widetilde{\omega}_m t)]$ (where R_0 is a conditional radius, ε is an eccentricity, m is symmetry order (mode), α is an angle and $\widetilde{\omega}_m = \zeta_0(m-1)/2$) to the constant vorticity circle region. But, accounting that the results of evolution for one and two vortices with different m were described in detail in [4], let us stay on results on interaction of vortices and consider the most simple case of circle vortices when $m=1$ and, therefore, $\widetilde{\omega}_m = 0$.

For two vortices the result of the interaction depends on sign of vorticity ζ ("polarity") and the distance δ between boundaries of vortices. Fig.1 shows the evolution of two vortices having opposite polarities for initial distance $\delta = d$ where d is a vortex diameter. One can see that in first case the vortices, rotating in opposite directions, move in the same direction and, practically don't interact independently on value of δ. For the vortices having opposite polarity the result of evolution depends essentially on δ.

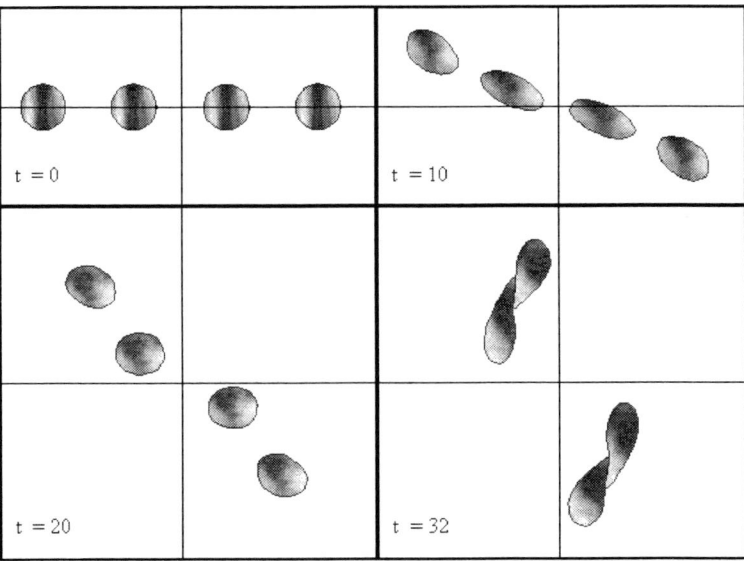

FIGURE 3. Interaction of four linearly disposed vortices with the same polarities and $\delta_i = d$.

So, for rather small δ the vortices, on a level with rotation about their own axes and around of their common center, interact forming a common vortex region which consists of the vorticities of more small scales (see fig. 2). For rather big δ the vortices on a level with rotation about their own axes rotate around of common center, at this, their interaction is reduced to a cyclic change of their shape (so-called "quasi-return" phenomenon [1] is observed). In our numerical experiments we have found that critical initial distance dividing these two types of interaction $\delta_{cr} = 3d/4$.

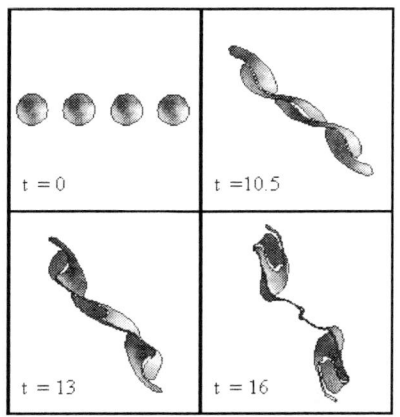
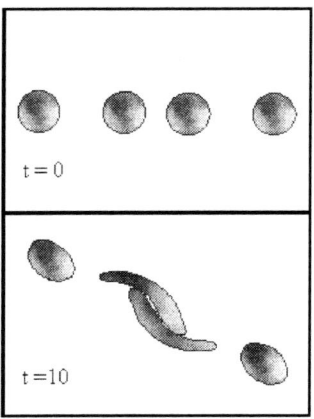

FIGURE 4. Interaction of four linearly disposed vortices with the same polarities: (a) $\delta_i = d/2$, (b) $\delta_{out} = d, \delta_{inn} = d/2$.

To study the interaction of the linearly disposed vortices in more details we have consider the problem for four vortices being at initial time along one line. Figs. 3 and 4 shows the results of the interaction at linear disposition of four vortices. One can see that for rather big and equal initial distance between vortices the evolution leads to formation of two vorticity regions as a result of more strong interaction of the "outer" vortices with the "inner" ones. At this, the interaction of forming pairs is similar to that of two vortices case (fig. 2). In case $\delta_i = d/2$ (fig. 4,a) we can observe the formation of a complex vortex structure which consists of many vorticities of more small scales. The further evolution of this structure leads to formation of complex turbulent field. Let us note that in last case we can also see that the interaction between outer vortices is stronger. This fact can be explained by the fact of more strong "attraction" of outer vortices to the "center of mass" of the vortex system because the outer vortex is attracted to the center by three other vortices, and the inner vortex is attracted to the center by two vortices and, in opposite side, by one outer vortex. To test this statement in the next series of numerical experiments we have arranged outer and inner vortices on different initial distances. As a result, we observed the formation of vortex structure from two inner vortices (see fig. 4,b).

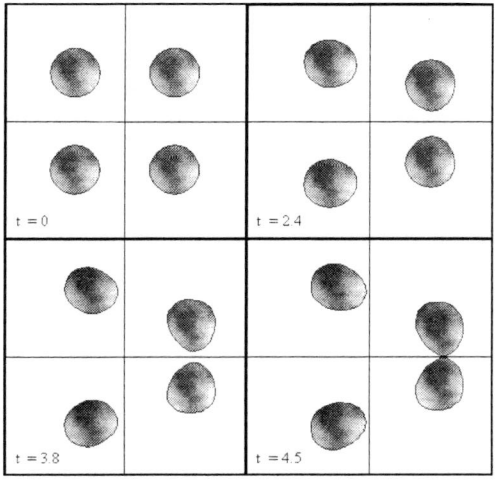

FIGURE 5. Interaction of four vortices with $\zeta_1 = \zeta_2 > 0$, $\zeta_3 = \zeta_4 < 0$ (numbering - clockwise, since the upper left corner) for $\delta_i = d$.

In the next series of experiments we studied the interaction between the vortices disposed at initial time in the corners of appropriate equilateral figures. At this, the following results were obtained.

In case of evolution of three vortices with opposite signs of ζ being at initial time in the corners of triangle, we have obtained that a pair of them, having opposite polarities, behaves as well as pair of vortices with opposite polarities in two vortices case, and third vortex does not participate in interaction almost, practically independently on value of δ_i $(i = 1, 2, 3)$. The similar character of interaction is observed for four vortices with opposite signs of ζ being at $t = 0$ in the corners of square (see figs. 5 and 6).

The character of interaction of three and four vortices having the same polarities depends essentially on the distances between them like in the two vortices case. An example of such interaction for $\delta = d/2 < \delta_{cr}$ is

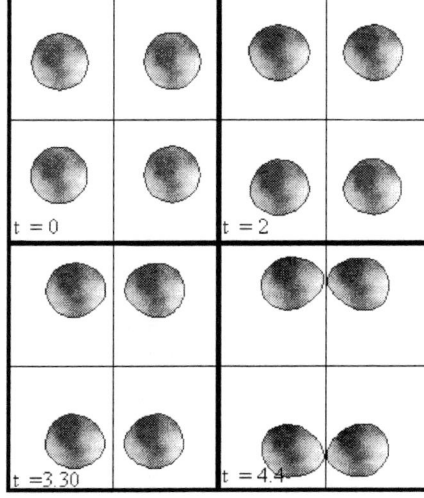

FIGURE 6. Interaction of four vortices with $\zeta_1 = \zeta_3 > 0, \zeta_2 = \zeta_4 < 0$ for $\delta_i = d$.

shown in fig. 7. One can see that in this case three vortices are rotated forming one big vortex which consists of many vorticities of more small scales. Similar picture is observed for four vortices being at $t=0$ in the corners of square on distances smaller critical one from another.

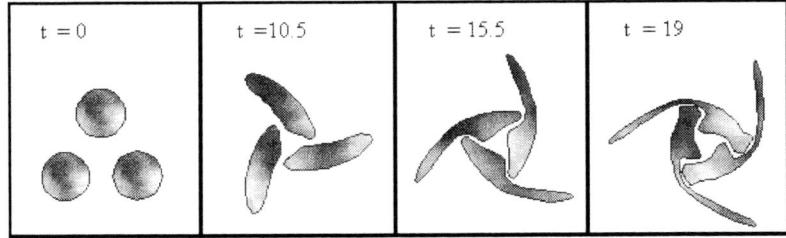

FIGURE 7. Interaction of three vortices with the same polarities and $\delta_i = d/2$.

CONCLUSION

So, we have presented, as more visual, some results of numerical simulation of eqs. (1) in terms of the vortex motion of the inviscid incompressible fluid. But, as we noted in the beginning, the set of eqs. (1) with $f=0$ may describe also the quasi-particles with Coulomb interaction model. At this, the results presented above can be easily extended on the 2D simple system where the plasma is represented by charged filaments, aligned with a uniform field **B**, that move with the guiding-centre velocity $\mathbf{E} \times \mathbf{B} / B^2$. Moreover, the approach undertaken can be useful and also for other 2D continuum models when $f \ne 0$ in the Poisson eq. (1). They can describe the vortices or filaments with the non-Coulomb interaction. In the last case it is assumed that ions move with the guiding-centre velocity but electrons have a Boltzman distribution, at this, the additional term $f = k^2 \psi$ will describe the Debye screening [3]. Another plasma model that can be investigated using the described approach is the Hasegawa-Mima model [5], its equations can also be put in form (1) by way of inclusion of ion polarization drift through the ion equation of motion [3] $d_t \mathbf{v} = (e/M)(-\nabla \varphi + \mathbf{v} \times \mathbf{B})$. As to a hydrodynamic fluid model corresponding to a screened interaction that such model having form (1) has been proposed in [6] to describe the Earth's atmosphere with the equation of motion for the horizontal atmospheric flow $d_t \mathbf{v} = -g \nabla h + R \mathbf{v} \times \hat{\mathbf{z}})$ where h is the atmospheric depth and R is the Coriolis force. Another possible application can take a place in the study of the problems associated with the dynamics of the Alfven vortices in the ionospheric and magnetospheric plasma [7].

In conclusion, let us note that the applications of the dynamics of different types' vortex structures will not rest the problems of study of the processes discussed above. in the power units. Such investigations are now one of the most perspective trends in some technical fields, for example in the problems of study of the vortex motions in working chambers of the varied types' power units, where the working substance may be water, vapor-gas mixtures and plasma [8]. At this, the most actual problem is the investigation of the vortex motions in the spatial regions where there are the flows with velocity shear (for example, near the walls of the boiler of thermal power units and the chambers of the gas turbines), and also the study of the vortex motions in magnetized plasma without dissipation (controlled thermonuclear fusion systems) where the electric currents can lead to formation of very complicated forms of the vortex regions.

REFERENCES

1. N.J. Zabusky, M.H. Hughes, K.V. Roberts, J. Comput. Phys., **135**, 220 (1997)
2. J.B.Taylor, B.McNamara, Phys. Fluids **14**, 1492 (1971)
3. J.B. Taylor, Plasma Phys. and Contr. Fusion, **39**, A1 (1997)
4. V.Yu. Belashov, R.M.Singatulin, Proc. RNSPE-3, Kazan, Russia, **2**, 61 (2001)
5. A. Hasegawa and K. Mima, Phys. Fluids, **21**, 87 (1978)
6. J.G. Charney, Geophs. Public. Kosjones Nors. Videnshap-Acad. Oslo, **17**, 3 (1948)
7. O.A. Pokhotelov, L.Stenflo, and P.K.Shukla, Plasma Physics Reports, **22**, 941 (1996)
8. V.Yu. Belashov, R.M. Singatulin, and Izv. Vuzov, Pow. Eng. Probl. **9**, 103 (2001)

Electrostatic Turbulence and Transport in Edge Plasmas: Bursts and Zonal Flows, Stochastic Field Lines, and Transport Barriers

P. Beyer*, S. Benkadda*, X. Garbet†, P. Ghendrih† and Y. Sarazin†

*LPIIM, CNRS – Université de Provence, St. Jérôme, Case 321, 13397 Marseille Cedex 20, France
†Association Euratom – CEA sur la Fusion, CEA Cadarache, 13108 St-Paul-lez-Durance, France

Abstract. Turbulent transport at the edge of a tokamak plasma is characterized by the formation of two types of structures. The first are related to large scale radial transport events (bursts), the second are fluctuations of the poloidal velocity (zonal flows) that regulate transport events. These structures and their dynamics and interplay are studied by 3D numerical simulations of resistive ballooning turbulence. Additionally, a layer of stochastic magnetic field lines is considered where sheared poloidal flows are found to be strongly reduced and long lived eddies appear. As a consequence, the level of convective flux associated with fluctuations is not quenched by the magnetic field perturbation. Finally, the dynamics of transport barriers, generated by externally imposed shear flows is studied. They are found to be intermittently eroded by successions of large bursts, leading to relaxation oscillations of the barrier.

INTRODUCTION

It has been observed now in many different turbulence simulations [1, 2, 3, 4, 5] dealing with a variety of instabilities in different radial locations of a tokamak plasma, that electrostatic turbulence and associated transport are governed by the interplay between two types of structures. The first concern large scale radial transport events or "bursts" that are related to radially elongated convection cells called "streamers". The latter can appear due to a non-linear process when smaller eddies, moving arbitrarily in the turbulent velocity field, line up to form an elongated cell. Streamers give rise to ballistic transport events corresponding to radially propagating deformations of the density, temperature, and/or pressure profile. Such events are also interpreted in terms of "avalanches" [6] and studied in models including self organized criticality [7] or front dynamics [8]. The second structure playing an important role in self-organized turbulence are fluctuations of the poloidal velocity related to poloidally elongated convection cells called "zonal flows" [9, 10]. These flows are known to be stabilizing and are indeed found to regulate transport events. There are also some experimental evidences for the appearance of bursts [11, 12, 13, 14] and zonal flows [15, 16]. In this paper, we present results from 3D fluid simulations of resistive ballooning modes at the plasma edge.

Additionally, the impact of static magnetic field perturbations on electrostatic turbulence and transport is studied. The interest on this topic is twofold: First, the behavior of turbulent cross field transport in regions of stochastic magnetic field lines is highly important for the performance of ergodic divertors [17] as well as many other experimental situations such as the stochastic boundary of stellarators and transport in the vicinity of the separatrix of standard tokamak divertors. Second, the model of turbulence in the presence of magnetic field perturbations allows to study transport and poloidal flow modification due to the so called magnetic flutter [18, 19]. Experimental observations of density fluctuations on TEXT [20] and Tore Supra [21, 22] with ergodic divertor show a decrease of the fluctuation level and a stabilization of large scale structures in the stochastic region. Surprisingly, there is no evidence of a change of the turbulent cross field diffusivity [17, 23]. Furthermore, large scale transport events are expected to be affected not only directly by the shearing due to the magnetic field perturbation but also indirectly due to modifications of poloidal mean and zonal flows in the stochastic layer. These are in fact found to be competing effects.

Finally, transport barriers and their interplay with bursts are studied. These barriers play an important role in improved confinement regimes. They are associated with localized $E \times B$ velocity shear and/or zero magnetic shear. Here, a transport barrier is generated by an externally imposed strong velocity shear. The behavior of bursts in the presence of the barrier and the dynamics of the latter due to an intermittent erosion by bursts are then studied.

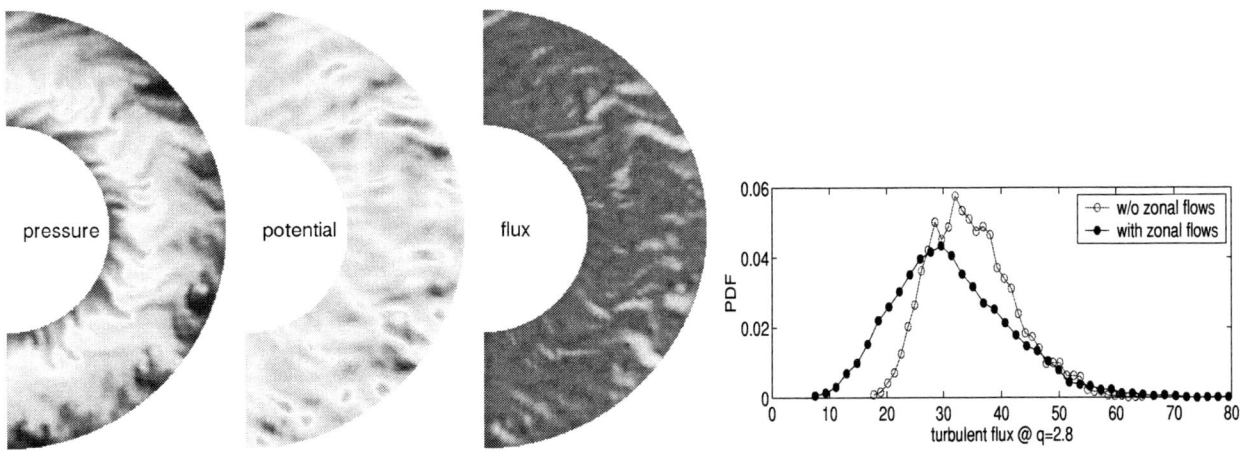

FIGURE 1. (left) Snapshot of pressure, potential fluctuations and radial flux at the low field side of a poloidal cross section. The annulus has been stretched radially by a factor of 4. (up) Probability distribution functions of the turbulent flux at the radial position $r = r_{q=2.8}$.

MODEL FOR RESISTIVE BALLOONING TURBULENCE

The model consists of reduced resistive MHD equations for the complete fields of electrostatic potential ϕ and pressure p,

$$(\partial_t + \vec{v}_E \cdot \nabla) \nabla_\perp^2 \phi = -\nabla_\parallel^2 \phi - Gp + \nu \nabla_\perp^4 \phi , \tag{1}$$

$$(\partial_t + \vec{v}_E \cdot \nabla) p = \delta_c G\phi + \chi_\parallel \nabla_\parallel^2 p + \chi_\perp \nabla_\perp^2 p + S(r) , \tag{2}$$

where $\vec{v}_E$ is the $E \times B$ drift, turbulence is driven by a constant incoming flux $\Gamma_{tot} = \int S\,dr$, and the operator $G = \sin\theta\,\partial_r + \cos\theta\,\frac{1}{r}\partial_\theta$ accounts for the compressibility of diamagnetic current and $E \times B$ drift due to toroidal magnetic curvature. Diamagnetic drift is neglected with respect to $\vec{v}_E$ and no (self-consistent) magnetic fluctuations are included. The (unperturbed) magnetic field is written as $\vec{B}_0 = B_\varphi(\hat{e}_\varphi + \frac{r}{Rq}\hat{e}_\theta)$ in toroidal coordinates and the inverse of the safety factor q is linearly approximated at the vicinity of a reference surface $r = r_0$ at the plasma edge. Our computational domain covers the region between $q = 2$ and $q = 3$ and a slab geometry is introduced at the vicinity of $r_0 = r_{q=2.5}$. Finally, all quantities are suitably normalized [24]. The parameters $\nu, \chi_\parallel, \chi_\perp, \delta_c$ correspond to the collisional viscosity, parallel and perpendicular heat conductivity, and the ratio between a reference pressure gradient length and the large radius R, respectively.

BURSTS & ZONAL FLOWS

In a statistically stationary state, series of snapshots of pressure show localized regions of high pressure propagating radially outwards and regions of low pressure extending to the inner boundary of the computational domain. These intermittently appearing "fingers" corresponding both to events of positive radial flux, connect the inner and outer regions of the plasma over large radial distances (Fig. 1, left). The propagation can be visualized in plots of poloidally and toroidally averaged pressure or flux versus radius and time [4, 25]. When such an event appears, a large, radially elongated convection cell can be observed in the potential field (Fig. 1) which has been created by percolation of smaller, symmetric eddies.

The "fingers" in the pressure field are found to be distorted by zonal flows. In fact, when artificially suppressing mean and zonal flows, less small events and more large burst are observed. More precisely, a comparison of the probability distribution functions (PDFs) of the turbulent flux at a given radial position (Fig. 1, up) shows that the maximum of the curve is shifted to lower values and the PDF gets broader when zonal flows are included self consistently. This corresponds to an increase of small events and a reduction of a significant part of large bursts (for a turbulent flux between 30 and 45 in Fig. 1) due to zonal flows. This observation is qualitatively the same for all radial positions [25].

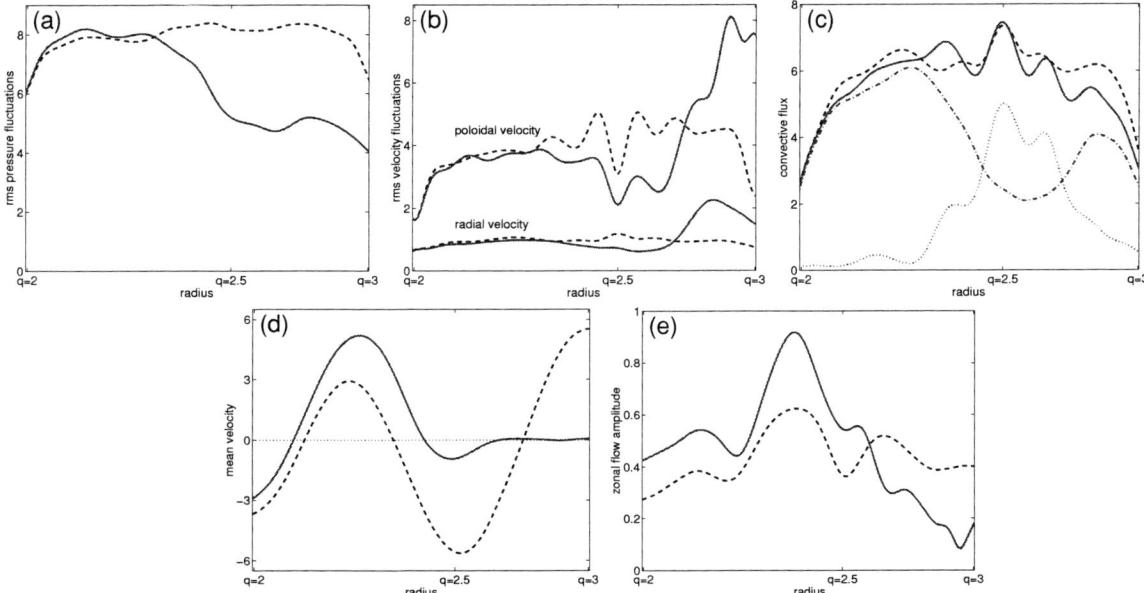

FIGURE 2. Radial profiles of pressure (a) & velocity fluctuations (b), radial flux due to fluctuations (c), mean poloidal velocity (d), and zonal flows (e). Full red lines correspond to a simulation with magnetic field perturbation, dashed blue lines indicate a reference case without perturbation. In (c), the dashed dotted and dotted lines indicate the components due to long-lived stationary eddies and turbulent fluctuations, respectively.

STOCHASTIC MAGNETIC FIELD LINES

A perturbation is added to the magnetic field $\vec{B} = \vec{B}_0 + \nabla \delta \psi \times \hat{e}_\varphi$ where $\delta \psi$ represents a sum of different poloidal harmonics corresponding to overlapping island chains. The amplitude of $\delta \psi$ increases exponentially with radius and field lines become ergodic in the radial domain $q > 2.7$. The effect of such a perturbation on electrostatic fluctuations and associated transport is found to be threefold: First, non vanishing time averages of the fields of pressure and potential are observed that vary poloidally and toroidally and give rise to a positive convective flux. This is due to the appearance of long-lived stationary eddies. Second, time varying pressure fluctuations decrease (Fig. 2a) but velocity fluctuations tend to increase (Fig. 2b) especially in the region of strong magnetic perturbation (close to $q = 3$) where the stationary eddies are intermittently destroyed by a secondary instability. In total, this leads to a roughly unchanged convective flux due to fluctuations (Fig. 2c). Third, the mean plasma flow is suppressed (Fig. 2d) and zonal flows are strongly reduced (Fig. 2e) by an anomalous friction in the stochastic region. This effect, as well as the existence of long-lived eddies providing natural "channels" for the propagation of bursts, is beneficial for the appearance of large bursts that can be observed even in the stochastic layer [26].

DYNAMICS OF TRANSPORT BARRIERS

Transport barriers are generated in our model by externally imposing a strong localized shear flow v_0. This is done by adding a driving term to the poloidally and toroidally averaged equation (1) which in fact is the equation governing the dynamics of the mean poloidal flow $\bar{v}_\theta = \partial_r \bar{\phi}$:

$$\partial_t \bar{v}_\theta + \partial_r \langle \tilde{v}_r \tilde{v}_\theta \rangle = \nu \partial_r^2 \bar{v}_\theta - \mu (\bar{v}_\theta - v_0). \tag{3}$$

The characteristic time scale μ^{-1} of the additional friction is chosen small compared to the viscous time scale $(\nu \sigma^2)^{-1}$ (where σ is the width of the shear layer) and to a characteristic non linear time of the fluctuations $\tilde{v}_r, \tilde{v}_\theta$. In this case, $\bar{v}_\theta$ relaxes rapidly to the imposed flow v_0 and time variations due to the Reynolds stress are negligible. It has been checked that for lower values of μ, where these variations of $\bar{v}_\theta$ become significant, the results do not change qualitatively. In the shear layer, the mean turbulent flux is found to be significantly reduced and the pressure profile steepens, i.e. a barrier

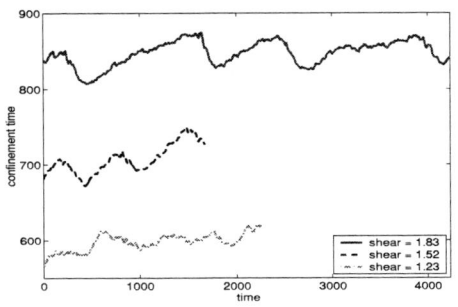 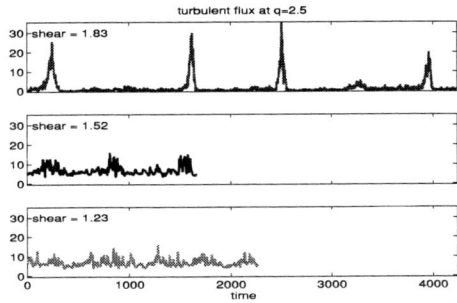

FIGURE 3. Time evolution of the confinement time (left) and the turbulent flux at the center of the barrier (right) for three different values of the maximal velocity shear.

forms. For strong velocity shear, relaxation oscillations of the barrier can be observed (Fig. 3): The pressure on the inner side of the barrier grows slowly during quiet phases that alternate with intermittent fast relaxation events. During quiet phases (for times typically of the order of the confinement time) low frequency components of the turbulent flux are strongly reduced and bursts are suppressed in the center of the barrier. During a relaxation event, the barrier is found to be eroded by a succession of large bursts.

CONCLUSIONS

Turbulent transport is characterized by the interplay between large scale transport events and zonal flows. In a layer of stochastic magnetic field lines, transport associated to fluctuations is not quenched despite a lower level of turbulent pressure. Mean flows are suppressed and zonal flows are strongly reduced due to an anomalous friction. Transport barriers generated by a strong velocity shear exhibit relaxation oscillations where bursts do not cross the barrier during quiet phases but the latter is intermittently eroded by successions of large bursts.

REFERENCES

1. J. F. Drake, P. N. Guzdar, A. B. Hassam, *Phys. Rev. Lett.* **61**, 2205 (1988).
2. Y. Sarazin, Ph. Ghendrih, *Phys. Plasmas* **5**, 4214 (1998).
3. X. Garbet, Y. Sarazin, P. Beyer, et al., *Nucl. Fusion* **39**, 2063 (1999).
4. P. Beyer, S. Benkadda, X. Garbet, et al., *Phys. Rev. Lett.* **85** 4892 (2000).
5. F. Jenko, W. Dorland, M. Kotschenreuther, *Phys. Plasmas* **7**, 1904 (2000).
6. B. A. Carreras, D. Newman, V. E. Lynch, et al., *Phys. Plasmas* **3**, 2903 (1996).
7. P. H. Diamond and T. S. Hahm, *Phys. Plasmas* **2**, 3640 (1995).
8. D. E. Newman, B. A. Carreras, D. Lopez-Bruna, et al.,*Phys. Plasmas* **5**, 938 (1998).
9. Z. Lin, T. S. Hahm, W. W. Lee et al., *Phys. Rev. Lett.* **83**, 3645 (1999).
10. P. H. Diamond et al., *Plasma Phys. Control. Nucl. Fusion Res.* (IAEA, Vienna, 1998).
11. B. A. Carreras, B. van Milligen, M. A. Pedrosa et al., *Phys. Rev. Lett.* **80**, 4438 (1998).
12. I. Garcia-Cortez, R. Balbin, A. Loarte, et al., *Plasma Phys. Control. Fusion* **42**, 389 (2000).
13. P. A. Politzer, *Phys. Rev. Lett.* **84**, 1192 (2000).
14. P. A. Politzer, M. E. Austin, M. Gilmore, et al., *Phys. Plasmas* **9**, 1962 (2002).
15. P. H. Diamond, M. N. Rosenbluth, E. Sanchez, et al., *Phys. Rev. Lett.* **84**, 4842 (2000).
16. S. Coda, M. Porkolab, and K. H. Burell, *Phys. Rev. Lett.* **86**, 4835 (2001).
17. P. Ghendrih, A. Grosman, and H. Capes, *Plasma Phys. Control. Fusion* **38**, 1653 (1996)
18. J. D. Callen, *Phys. Rev. Lett.* **39**, 1540 (1977).
19. B. Scott, *Plasma Phys. Control. Fusion* **39**, 1635 (1997).
20. S. C. MacCool, A. J. Wootton, Aydemir, et al., *Nucl. Fusion* **29**, 547 (1989).
21. J. Payan, X. Garbet, J. H. Chatenet, et al., *Nucl. Fusion* **35**, 1357 (1995).
22. P. Devynck, H. Capes, J. Gunn, et al., to appear in *Nucl. Fusion* (2002)
23. A. Grosman, P. Ghendrih, H. Capes, et al., *Contrib. Plasma Physics* (1997).
24. P. Beyer, X. Garbet, and P. Ghendrih, *Physics of Plasmas* **5**, 4271 (1998).
25. P. Beyer, S. Benkadda, N. Bian, et al., *28th EPS Conf., Funchal, ECA Vol.* **25A**, 2137 (2001).
26. P. Beyer, X. Garbet, S. Benkadda, et al., to appear in *Plasma Phys. Control. Fusion*.

Turbulent Evolution of a Plasma Described Through Classical Mechanics Only

D.F.Escande* and Y. Elskens*

*UMR 6633 CNRS–Université de Provence,
case 321, Centre de Saint-Jérôme, F-13397 Marseille cedex 20*

Abstract. For the first time an old dream of the XIXth century comes true: the non trivial evolution of a macroscopic many-body system is described through classical mechanics only. This is done for the relaxation of a warm electron beam in a plasma, which results in the generation of Langmuir turbulence and in the formation of a plateau in the velocity distribution function of the electrons. Our derivation starts from the hamiltonian describing the one-dimensional N-body system corresponding to the beam and plasma bulk electrons in electrostatic interaction. For such a system, the dynamics can be reduced to the resonant interaction of M Langmuir waves with $N'(\ll N)$ beam particles. The rigorous analytical calculation of a quasilinear diffusion coefficient is performed for the chaotic motion of a particle in a quite general set of prescribed longitudinal waves with random phases and large amplitudes. This result proves to be extendable within controllable approximations to the self-consistent evolution of $M \gg 1$ Langmuir waves with $N' \gg 1$ beam particles. This yields the proof of the classical quasilinear equations describing the coupled evolution of the wave spectrum and of the beam velocity distribution function in the strongly nonlinear regime where their validity is the matter of a longstanding controversy.

This paper deals with the Langmuir turbulence due to the weak warm beam-plasma instability. Quasilinear (QL) theory was initially developed to describe the saturation of this instability [1, 2]. It correctly predicted the development of a Langmuir turbulence and the formation of a plateau in the electron velocity distribution function. However the original derivations of the QL equations assume that particle orbits are weakly perturbed (quasi linear description), though the plateau formation be the result of a strong chaotic diffusion of the beam particles. Over two decades a controversy has developed over the validity of the QL equations in the chaotic saturation regime (see [3, 4] and references therein) within the Vlasovian description of the problem, and is not yet settled. We propose here a proof of these equations in both the weakly and the strongly chaotic regimes without resorting to this description. Instead we describe the Langmuir wave-beam system as a finite degree-of-freedom hamiltonian system, following the path initially elaborated for the cold beam-plasma instability [5, 6, 7], and later on extended to more general cases [8, 9, 10]. The derivation of the corresponding hamiltonian is done in [9, 10] by starting from that describing all the particles of the beam-plasma system coupled by the sum of their two-body electrostatic potentials. Therefore the QL equations are now derived through classical mechanics only. Since going from the QL equations to the description of the saturation of the weak warm beam-plasma instability is a straightforward step, we may say that the turbulent evolution of a plasma is now described through classical mechanics only.

This paper sketches the main ideas of the full derivation given in [10]. For brevity it focuses on the calculation of the diffusion coefficient. In the present approach this calculation for the beam-plasma system reduces after a few controllable steps to the similar one for the chaotic motion of a particle in a given spectrum of Langmuir waves. Therefore we first consider this simpler problem before considering the self-consistent evolution of many Langmuir waves with many particles.

We now consider a particle being acted upon by a set of Langmuir waves with random phases and large amplitudes, as defined by the Hamiltonian

$$H(p,q,t) = \frac{p^2}{2} + \sum_{m=1}^{M} A_m \cos(k_m q - \omega_m t + \varphi_m), \tag{1}$$

where the φ_m's are independent random variables uniformly distributed on $[0, 2\pi]$, where the (A_m, k_m)'s are prescribed pairs of positive parameters, and where $\omega_m = \omega(k_m)$, with $\omega = \omega(k)$ defining the dispersion relation of Langmuir waves. Let A be the typical value of the A_m's; the large A limit (dynamically speaking, the limit of strong resonance overlap parameter) corresponds to the limit of continuous spectrum often encountered in physics. In this case the diffusion coefficient has been found numerically to take on the quasilinear value [11, 12, 13, 14, 15] we define now.

Let $v_m = \omega_m/k_m$, $\Delta v_m = v_{m+1} - v_m$. Let p_0 be a parameter, $\Omega_m = k_m p_0 - \omega_m$, and $\Delta\Omega_m = \Omega_{m+1} - \Omega_m$. We assume that k_m is an increasing function of m. Let

$$D_m \equiv \frac{\pi A_m^2 k_m}{2|\Delta v_m|} = \lim_{p_0 \to v_m} \frac{\pi (A_m k_m)^2}{2|\Delta\Omega_m|}. \tag{2}$$

D_m may fluctuate with m (this will be important for describing the chaotic self-consistent regime), but we assume that, for some $L \geq 0$, $\bar{D}_m = \sum_{j=-L}^{L} D_{m+j}|\Delta v_{m+j}|/|v_{m+L+1} - v_{m-L}|$ is varying with m over a scale much larger than L. We define $D_{QL}(p)$ as a slowly varying function of p interpolating the values $D_{QL}(v_m) = \bar{D}_m$; $D_{QL}(p)$ is called the quasilinear diffusion coefficient at velocity p. We introduce the local resonance broadening frequency [17] $\gamma_{Dn} \equiv (k_n^2 D_{QL}(v_n))^{1/3}$ and take $\gamma_D \equiv \max_n \gamma_{Dn}$.

Let $\Delta\Omega_{LM} = \max|\Omega_{m+L+1} - \Omega_{m-L}|$, $\tau_{discr} = \Delta\Omega_{LM}^{-1}$ and $\tau_c = (\Omega_{max} - \Omega_{min})^{-1}$; τ_{discr} and τ_c are respectively the discretization time and the correlation time of the wave spectrum as seen by the particle.

We evaluate $\Delta p(t) = p(t) - p(0)$ by integrating formally the equation of motion for p. We compute the average of its square over all phases (in the following $\langle...\rangle$ means averaging over all φ_m's). This yields $\langle \Delta p^2(t) \rangle = \Delta_0 + \Delta_+ + \Delta_-$, with

$$\Delta_j = -\eta_j \int_0^t \int_0^t \sum_{m_1=1}^{M} \sum_{m_2=1}^{M} \frac{A_{m_1} k_{m_1} A_{m_2} k_{m_2}}{2} \langle \cos[\Phi_{m_1}(t_1) + \eta_j \Phi_{m_2}(t_2)] \rangle dt_1 dt_2, \tag{3}$$

where $\Phi_m(t) = k_m \Delta q(t) + \Omega_m t + \varphi_m$, with $\eta_\pm = \pm 1$ and $\eta_0 = -1$, and under condition $m_1 \neq m_2$ for $j = -$, and condition $m_1 = m_2$ for $j = 0$. As was pointed out in Ref. [14], $\Delta_\pm$ vanishes provided that the dependence of Δq over any two phases with all other phases fixed is weak. Under such a condition $\langle \Delta p(t)^2 \rangle = \Delta_0 = 2D_{QL}t$. Therefore the quasilinear estimate for the diffusion coefficient holds over the time τ_{QL} where the dependence of Δq over any two phases is negligible.

To estimate τ_{QL} we study how the orbit $\mathcal{O}$, which is at (q_0, p_0) at $t = 0$, is modified when phases φ_n and $\varphi_m, m \neq n$, change from 0 to a finite value. Let $(q_\eta(t), p_\eta(t))$ be the orbit for $\varphi_n = \varphi_m = 0$, let $\delta q_n(t) = q(t) - q_\eta(t)$. Since $\delta q_n(0) = 0$, for t small enough $\delta q_n(t)$ is small, and the evolution of $(q_\eta(t), p_\eta(t))$ may be computed by linearizing the motion [10, 16]. Then

$$\langle \delta q_n(t)^2 \rangle \leq 0.14 \mathscr{B} k_n^{-2} (e^{t'} - 1 + 2g(t')), \tag{4}$$

with $\mathscr{B} = \min k_m |\Delta v_m|/\gamma_{Dm}$, $t' \equiv 4^{1/3} \gamma_{Dn} t$ and $g(t') = e^{-t'/2} \cos(t'\sqrt{3}/2) - 1$. This rigorous estimate for the variance of $\delta q_n(t)$ starts from zero at $t = 0$ and diverges exponentially for $t \to \infty$. Its exponentiation time scale $\tau_{Lyap} \sim \gamma_{Dn}^{-1}$ is the reciprocal of the Lyapunov characteristic instability rate (this is reminiscent of Ref. [11]). $\mathscr{B}$ scales as $5s^{-4/3}$ where s is the typical size of the Chirikov resonance overlap parameters $s(v_n) = 2[A_n^{1/2} + A_{n+1}^{1/2}]/|\Delta v_n|$. For a strong overlap it is small, and the time needed for $k_n^2 \langle \delta q_n(t)^2 \rangle$ to reach unity is of the order of

$$\tau_{QL} = \gamma_D^{-1} |\ln \mathscr{B}| \tag{5}$$

This time is $O(\ln \mathscr{B}^{-1}) \gg 1$ times larger than the time $\tau_{spread} = 4\gamma_D^{-1}$ over which the QL approximation is traditionally justified. As the latter time is the timescale for memory loss by the dynamics, our result suggests that the QL approximation may hold over longer times.

Indeed we now show that the quasilinear estimate holds for asymptotic times for s large. We define $\delta q(\tau|p,q,t) = q(t+\tau) - q - p\tau$, where $q(t')$ is the position at time t' of an orbit which is at (p,q) at time t: $\delta q(\tau|p,q,t)$ tells the departure of this orbit from the free motion during the time interval τ.

Integrating formally the equation of motion for p yields

$$\langle \Delta p^2(t) \rangle = -\sum_{m,n=1}^{M} \sum_{\varepsilon=\pm 1} \varepsilon \frac{A_m k_m A_n k_n}{2} \int_0^t \int_0^t \langle \cos[(k_m + \varepsilon k_n)q(t'') + k_m p(t'')\tau - \omega_m t' - \varepsilon \omega_n t''$$
$$+ k_m \delta q(\tau|X(t'')) + \varphi_m + \varepsilon \varphi_n] \rangle dt' dt'', \tag{6}$$

where $\tau = t' - t''$, and where $X(t'')$ stands for $[p(t''), q(t''), t'']$. Note that $\langle \cos[k_m \delta q(\tau|X(t''))]\rangle$ is independent from $q(t'')$, as all phases φ_m are drawn independently from a uniform distribution on the circle.

We introduce the probability distribution $P(\delta p, t|p_0)$ of $\delta p = p(t) - p_0$ for the orbit started at p_0 at $t_0 = 0$, and we write the integrand of (6) for diagonal ($m = n$, $\varepsilon = -1$) terms

$$\langle \cos[k_m \delta q(\tau|X(t'')) + (k_m p(t'') - \omega_m)\tau]\rangle = \Re \int \langle e^{i[k_m \delta q(\tau|p_0 + \delta p, q, t'') + (\Omega_m + k_m \delta p)\tau]}\rangle_* P(\delta p, t''|p_0) d\delta p, \qquad (7)$$

where the star subscript means the average conditioned by the constraint $p(t'') = p_0 + \delta p$. The quantity $\delta q(\tau|p_0 + \delta p, q, t'')$ may be Taylor expanded [10, 16].

The contribution to $\langle \Delta p^2(t)\rangle/(2t)$ of the first term in the Taylor expansion is

$$B = \lim_{t \to \infty} \sum_{m=1}^{M} \frac{(A_m k_m)^2}{8t} \Re \int_0^t \int_{-t''}^{t-t''} \langle e^{ik_m \delta q(\tau|p_0, q, t'')}\rangle_* \tilde{P}(k_m \tau, t''|p_0) \exp[i\Omega_m \tau] d\tau dt'', \qquad (8)$$

with $\tilde{P}(\alpha, t''|p_0) = \int_{-\infty}^{\infty} P(\delta p, t''|p_0) \exp(i\alpha \delta p) d\delta p$. We assume t to be large enough for the width of P to be large over most of the time-integration interval, so that $\tilde{P}$ is narrow enough in $k\tau$. Then the width with respect to τ is also small enough for the spread of δq to be negligible. Then $\langle \exp[ik_m \delta q(\tau|p_0, q, t'')]\rangle_* \simeq 1$ in the part of the integration domain over τ where $\tilde{P}$ takes appreciable values in (8), and $B = \lim_{t \to \infty} \sum_{m=1}^{M} \frac{\pi A_m^2 k_m}{2t} \int_0^t P(v_m - p_0, t''|p_0) dt'' = \int_0^t \sum_{m=1}^{M} \frac{D_m \Delta v_m}{2t} P(v_m - p_0, t''|p_0) dt''$ where the inverse Fourier transform was provided by the integral over τ. Now, if t is large enough for P to be almost constant over the range $[v_{m-L}, v_{m+L+1}]$ for all m's, we use the definition of $D_{QL}(p)$ and substitute the sum over v_m by an integral: $B = \frac{1}{2t} \int_0^t \int D_{QL}(v) P(v - p_0, t''|p_0) dv dt'' = D_{QL}(p_0)$.

The contribution of the next terms in the Taylor expansion of equation (7) and of the general term in (6) can be shown to be small [10, 16]. This ends our proof of the quasilinear estimate for asymptotic times [19].

We now turn to the Langmuir wave-beam system described as a finite degree-of-freedom system with the self-consistent hamiltonian [8, 9, 10]

$$H_{sc} = \sum_{r=1}^{N} \frac{p_r^2}{2} + \sum_{j=1}^{M} \omega_{j0} \frac{X_j^2 + Y_j^2}{2} + \sum_{r=1}^{N} \sum_{j=1}^{M} \lambda_j (Y_j \sin k_j x_r - X_j \cos k_j x_r), \qquad (9)$$

where $\lambda_j = \frac{1}{k_j}\left(\frac{\partial \varepsilon}{\partial \omega}(k_j, \omega_j)\right)^{-1/2} \omega_p \sqrt{\frac{2\eta}{N}}$, where ω_p is the plasma frequency, where ε is the plasma dielectric function, and where η is the ratio of the beam density to the background plasma density. This hamiltonian describes N beam particles coupled to M Langmuir waves which show up as harmonic oscillators. When going to the action-angle variables (I_j, θ_j) of these oscillators, the hamiltonian becomes $H_{sc} = \sum_{r=1}^{N} \frac{p_r^2}{2} + \sum_{j=1}^{M} \omega_{j0} I_j - \sum_{r=1}^{N} \sum_{j=1}^{M} \lambda_j \sqrt{2I_j} \cos(k_j x_r - \theta_j)$ where the coupling term takes on the natural structure of a sinusoidal wave potential with an amplitude scaling like the square root of the wave energy. Such an expression makes obvious that the total momentum $P_{sc} = \sum_{r=1}^{N} p_r + \sum_{j=1}^{M} k_j I_j$ is a constant of the motion. It is convenient to define the complex wave envelope $z_j = (X_j + iY_j) \exp(i\omega_{j0} t)$. The evolution equation for z_j is $\dot{z}_j = i\lambda_j \sum_{l=1}^{N} e^{i(\omega_{j0} t - k_j x_l)}$.

As for the non self-consistent case, we compute how the variance of the velocity of particles with given initial velocity p_0 grows with time. An initial non intrinsically chaotic quasilinear diffusion can be recovered over a time of the order of τ_{QL} estimated as previously by substituting $\lambda_j |z_j|$ for A_m. The argument is quite similar to the non self-consistent case, because the averages may be split into averages involving the z_j's and averages involving the x_l's thanks to the weak correlation between any wave and any particle. Knowledge of the initial quasilinear diffusion regime is sufficient to deal with the case $\tau_w \ll \tau_{spread}$, where τ_w is the time of growth or decay of Langmuir waves. Indeed in that case particle motions are nearly ballistic on the time scale over which the wave changes, and the perturbation to these ballistic motions was found in [20] to obey the quasilinear equations in agreement with the original vlasovian construction of quasilinear theory [1, 2]. Thus we now focus on the case $\tau_w \gg \tau_{spread}$ where the chaotic motion of particles has an overwhelming impact on the global features of the beam-plasma system.

Formal integration of the equation of motion of p_l in terms of the z_j's yields its variance as

$$\langle \Delta p_l^2(t) \rangle = -\frac{1}{2}\Re \sum_{j=1}^{M} \sum_{j'=1}^{M} \int_0^t \int_0^t \lambda_j k_j \lambda_{j'} k_{j'} \Big(\langle z_j(t') z_{j'}(t'') e^{i[\Psi_{lj}(t') + \Psi_{lj'}(t'')]} \rangle$$
$$- \langle z_j(t') z_{j'}^*(t'') e^{i[\Psi_{lj}(t') - \Psi_{lj'}(t'')]} \rangle \Big) dt'' dt' \qquad (10)$$

where $\Psi_{lj}(t) = k_j x_l(t) - \omega_{j0} t$. As for the non self-consistent case, the evolutions of $x_l(t')$ and $x_l(t'')$ are correlated only for $|t'' - t'| < \Delta t \sim \tau_{\text{spread}} \ll \tau_{\text{QL}}$. This bounds the relevant integration domain.

We first consider the second term in (10) with $j = j'$ for $t \gg \Delta t$, and call it B. We compute the integrand of this term by expressing $z_j(t'')$ in terms of $z_j(t')$ by formal integration of the evolution equation for z_j

$$F \equiv \langle z_j(t') z_j^*(t'') e^{i[\Psi_{lj}(t') - \Psi_{lj}(t'')]} \rangle$$
$$= \langle |z_j(t')|^2 e^{i[\Psi_{lj}(t') - \Psi_{lj}(t'')]} \rangle - i\lambda_j \sum_{l'=1}^{N} \int_{t'}^{t''} \langle z_j(t') e^{i[\Psi_{lj}(t') - \Psi_{lj}(t'') + \Psi_{l'j}(t_1)]} \rangle dt_1. \qquad (11)$$

We first discuss the first term in (11). For $|t'' - t'| > \Delta t$, $\Psi_{lj}(t'')$ is uncorrelated with $\Psi_{lj}(t')$. Since many waves are acting on a particle, this is also true for the conditional distribution of Ψ given a fixed value of $|z_j(t')|^2$. Therefore in the first term of (11), $\langle e^{i\Psi_{lj}(t'')} \rangle$ may be factorized out, and vanishes over most of the integration interval of t''. For $|t'' - t'| \leq \Delta t$, the effect of $z_j(t')$ upon $\Psi_{lj}(t') - \Psi_{lj}(t'')$ may be computed perturbatively. It is small in the limit of a continuous spectrum where the wave energy is shared among an infinite number of waves. As a result we may factorize the average $\langle |z_j(t')|^2 e^{i[\Psi_{lj}(t') - \Psi_{lj}(t'')]} \rangle$ into $\langle |z_j(t')|^2 \rangle \langle e^{i[\Psi_{lj}(t') - \Psi_{lj}(t'')]} \rangle$. Anticipating that the second term of (11) is negligible, we are then left with

$$B \simeq \frac{1}{2} \sum_{j=1}^{M} (\lambda_j k_j)^2 \int_0^t \int_0^t \langle |z_j(t')|^2 \rangle \langle \cos[\Psi_{lj}(t') - \Psi_{lj}(t'')] \rangle dt'' dt'. \qquad (12)$$

As $\tau_w \gg \tau_{\text{spread}}$, we consider a time t such that $\tau_w \gg t \gg \tau_{\text{spread}}$. Then $\langle |z_j(t')|^2 \rangle$ may be considered as a constant, and this equation has exactly the same structure as equation (6) obtained for the non self-consistent case. The contribution of the first term in (11) to $B/(2t)$ then yields the quasilinear diffusion coefficient.

Now we consider the second term in (11). If $l' = l$, since $t \gg \tau_{\text{spread}}$, in most of the integration interval for t' and t'' the three phases in the exponential cannot cancel, and at fixed $z_j(t')$ the argument of the exponential has large fluctuations. Therefore the expectation is small, and this contribution is negligible.

Finally we consider the last term of (11) for $l' \neq l$. Then

$$\langle z_j(t') e^{i[\Psi_{lj}(t') - \Psi_{lj}(t'') + \Psi_{l'j}(t_1)]} \rangle = \langle z_j(t') e^{i[\Psi_{lj}(t') - \Psi_{lj}(t'')]} \langle e^{i\Psi_{l'j}(t_1)} \rangle_* \rangle \qquad (13)$$

where the star denotes conditional averaging with fixed $z_j(t')$ and fixed increment $\Psi_{lj}(t') - \Psi_{lj}(t'')$, which sets three conditions over $N + M$ random phases. The value of $\Psi_{l'j}(t_1)$ follows from an integration over a long time interval so that it is uniformly distributed (modulo 2π) even under the two conditions over $N + M$ random phases, and $\langle e^{i\Psi_{l'j}(t_1)} \rangle_* = 0$. Hence (13) vanishes and in (11) only the first term must be taken into account.

The other terms in (10) are negligible, as shown by a reasoning similar to that in the non self-consistent case for non-diagonal terms. As a result the diffusion coefficient is again given by the QL estimate. At this point it is natural to assume that, for $\tau_w \gg \tau_{\text{spread}}$, τ_{spread} is the decorrelation time of the dynamics, i.e. that the decorrelation time for a single particle is also the decorrelation time for the waves. This enables the derivation of the QL (Fokker-Planck) equation ruling the evolution of the particle distribution function. Then the Landau growth rate for the waves can be shown to be related to the QL diffusion coefficient due to a local conservation of the average wave-particle momentum. This ends the derivation of the QL equations in the chaotic regime. Their full and more rigorous derivation is provided in [10].

The reason for obtaining the same quasilinear equations in the two time limits of interest can be traced back to several reasons : the weak correlation of any wave with any particle, the even weaker correlation between any two waves or any two particles, the local conservation of the average wave-particle momentum, and the fact that the diffusion is quasilinear in the non self-consistent case for both the initial non-chaotic regime, and for the chaotic regime in the limit of a continuous spectrum (strong resonance overlap).

To summarize, starting from the N-body description of the beam-plasma system, new tools of chaotic classical mechanics enable the derivation of the quasilinear equations describing the turbulent evolution of the system. For the first time an old dream of the XIXth century comes true: the non trivial evolution of a macroscopic many-body system is described through classical mechanics only.

ACKNOWLEDGMENTS

Thomas O'Neil, Patrick Diamond, and Marshall Rosenbluth made useful comments to one of us (DE). Comments on this work by Didier Bénisti are gratefully acknowledged.

REFERENCES

1. A.A. Vedenov, E.D. Velikhov, R.Z. Sagdeev, Nucl. Fusion Suppl. 2 (1962) 465.
2. W.E. Drummond, D. Pines, Nucl. Fusion Suppl. 3 (1962) 1049.
3. I. Doxas, J.R. Cary, Phys. Plasmas 4 (1997) 2508.
4. G. Laval, D. Pesme, Plasma Phys. Control. Fusion 41 (1999) A239.
5. I.N. Onishchenko, A.R. Linetskiĭ, N.G. Matsiborko, V.D. Shapiro, V.I. Shevchenko, ZhETF Pis. Red. 12 (1970) 407; transl. JETP Lett. 12 (1970) 281.
6. T.M. O'Neil, J.H. Winfrey, J.H. Malmberg, Phys. Fluids 14 (1971) 1204.
7. H.E. Mynick, A.N. Kaufman, Phys. Fluids 21 (1978) 653.
8. J.L. Tennyson, J.D. Meiss, P.J. Morrison, Physica D 71 (1994) 1.
9. M. Antoni, Y. Elskens, D.F. Escande, Phys. Plasmas 5 (1998) 841.
10. Y. Elskens, D.F. Escande, *Microscopic dynamics of plasmas and chaos* (IOP, Bristol, to be published in 2002).
11. A.B. Rechester, M.N. Rosenbluth, R.B. White, Phys. Rev. Lett. 42 (1979) 1247.
12. J.R. Cary, D.F. Escande, A.D. Verga, Phys. Rev. Lett. 65 (1990) 3132.
13. O. Ishihara, H. Xia, S. Watanabe, Phys. Fluids B 5 (1993) 2786.
14. D. Bénisti, D.F. Escande, Phys. Plasmas 4 (1997) 1576.
15. B.R. Ragot, J. Plasma Phys. 60 (1998) 299.
16. D.F. Escande and Y. Elskens, Acta Phys. Pol. B33 (2002) 1073.
17. T.H. Dupree, Phys. Fluids 9 (1966) 1773.
18. D. Bénisti, D.F. Escande, J. Stat. Phys. 92 (1998) 909.
19. D. Pesme and A. Brisset assert to have proved a quasilinear diffusion equation for a spatially averaged velocity distribution function using a diagrammatic technique (private communication, 1981; D. Pesme, D.F. DuBois, Los Alamos report LA-UR-81-2234 *Nonlinear problems : present and future* Bishop A ed. (North Holland, Amsterdam, 1982); D. Pesme, Physica Scripta T 50 (1994) 7).
20. D.F. Escande, S. Zekri, Y. Elskens, Phys. Plasmas 3 (1996) 3534.

A Theory of Longitudinal Plasma Waves With the Motion of Ions Taken into Account

G.N.Kichigin

Abstract. This paper is concerned with the problem of propagation of steady longitudinal plasma waves of a large amplitude in collisionless cold plasma. The conditions are determined, for which there exist periodic potential waves. Taking into account the motion of ions in the wave led to a nonmonotonic dependence of the frequency on the wave velocity: with the increasing velocity, the frequency that is equal at small velocities to the plasma electron frequency for linear waves, then decreases slightly to a certain minimum value; after that, it begins to increase and, subsequently, already for relativistic waves, it again becomes equal to the plasma frequency; when the wave velocity tends to the velocity of light, it increases infinitely.

INTRODUCTION

Results of theoretical studies of stationary waves in collisionless plasma are covered adequately in a fundamental paper [1] (see also a monograph [2]). Akhiezer et al. [1] addressed plane waves in unbounded plasma consisting of electrons which are considered cold, and of ions which are assumed to be infinitely heavy and immobile. However, the validity of the assumption about the immobility of ions when describing nonlinear waves with large amplitudes is questionable. Indeed, from recent publications devoted to the study of relativistic waves in plasma [3-7], as well as from investigations related to the interaction of laser radiation with plasma [8-10], it became clear that in the case of sufficiently large amplitudes of the electric field in relativistic waves it is necessary to take into account the motion of the ion component of plasma. The influence of the ion dynamics on the wave structure was considered by Khachatryan [7], who investigated - for longitudinal plasma waves - the dependence of an ultimate electric field and the wavelength on the parameter μ for different values of the relativistic factor $\gamma = (1-u^2/c^2)^{-1/2}$ (here μ is the ion to electron mass ratio, u is the phase velocity of the wave, and c is the velocity of light in a vacuum). Bulanov et al. [8] made estimates of the magnitudes of the field of the laser impulse acting on plasma where it is necessary to take into account the motion of plasma ions. Furthermore, taking into account the mass of ions they determined the dependencies of the wavelength, the potential amplitude and the electric field of the wake waves on the magnitude of a maximum field in the laser impulse.

SETTING OF THE PROBLEM

In this paper we investigate the wave motions of unbounded cold plasma consisting of ions with the rest mass M and of electrons with the rest mass m ($\mu = M/m$), and ions are represented by protons; therefore, the charge of an ion q_i is equal to the charge of an electron: $q_i = e$, where e is the absolute value of the electron charge. Unlike [1,2] where in the nonlinear plasma wave the ions are considered fixed, we take into account the motion of ions in a very general treatment, including the relativistic case. Assume that the external magnetic field is absent. We limit ourselves to considering the longitudinal plane wave propagating along the axis x which we shall consider steady-state.

In the steady wave it is convenient to carry out all our treatment in the wave's frame of reference in which the problem solved here is a stationary one, and all desired variables are a function of the coordinate x only. The equations that are needed for the solution of the formulated problem are the Maxwell equations, the relativistic equations of motion, and the electron and ion equations of continuity.

We seek the solutions to these equations in the form of a periodic sign-changing potential wave. In this case, on the scale equal to the wavelength λ, at the points lying between a maximum and minimum of the potential, the electric field will have extreme values. From the Maxwell equation for the electric field $E(x)$

$$dE/dx = 4\pi e [(n_i(x) - n_e(x)] \tag{1}$$

it then follows that at these points the ion concentration $n_i(x)$ equals the electron concentration $n_e(x)$. Let the coordinate of one of the extreme points $x = 0$, then at this point $n_i(0) = n_e(0) = n$, $E(0) = E_0$ and here we designated the extreme value of the electric field by E_0. Without any loss of generality, at the extreme points we set the wave potential $\varphi(x)$ equal to zero, then we have $\varphi(0) = 0$ for $x = 0$.

From the electron and ion equations of continuity it follows that $n_i(x)v_i(x) = C_1$ and $n_e(x)v_e(x) = C_2$, where $v_e(x)$, $v_i(x)$ is the velocity of electrons and ions, respectively, and C_1, C_2 are x-independent constants. We find these constants by assuming $x = 0$. Since $n_i(0) = n_e(0) = n$, we obtain $C_1 = nv_i(0)$, $C_2 = nv_e(0)$, where $v_e(0)$ and $v_i(0)$ are constant velocities. We make the following assumption: $v_e(0) = v_i(0) = u$, where u is some constant velocity. Then a total current at all points on the wave profile will be zero (as well in [1]): $e[n_i(x)v_i(x) - n_e(x)v_e(x)] = 0$, and this means that the wave under consideration does not have any disturbed magnetic field.

For a cold plasma in the absence of a magnetic field, the dynamics of the motion of electrons and ions in the wave's electric field can be considered using relativistic equations of motion which in the wave's frame of reference have the form:

$$v_e(x)\frac{dp_e(x)}{dx} = mc^2 \frac{d\gamma_e(x)}{dx} = -eE(x), \quad v_i(x)\frac{dp_i(x)}{dx} = Mc^2 \frac{d\gamma_i(x)}{dx} = eE(x), \tag{2}$$

where $\gamma_e(x) = [1 - v_e(x)/c]^{-1/2}$, $\gamma_i(x) = [1 - v_i(x)/c]^{-1/2}$.

It is an easy matter to obtain from (2) the equations for total energy of electrons and ions:

$$Mc^2 \gamma_e(x) + e\varphi(x) = Mc^2 \gamma, \quad mc^2 \gamma_i(x) - e\varphi(x) = mc^2 \gamma. \tag{3}$$

Constants in (3) we have found, determining values of energy and potential in a point $x = 0$, in which we have accepted, that $\varphi(0)=0$, $v_e(0) = v_i(0) = u$. Here we also have entered the denotation $\gamma = 1/\sqrt{1-\beta^2}$, in which $\beta = u/c$. It is obvious that with the resulting parameter γ, the problem under consideration has a physical meaning only for the values of the velocity u not exceeding the velocity of light c.

If in equation (1) the ion and electron concentrations are expressed in terms of the velocity: $n_e(x) = nu/v_e(x)$, $n_i(x) = nu/v_i(x)$, the right-hand and left-hand sides of equation (1) are multiplied by $E(x)$, and if, further, the resulting combinations $eE(x)/v_e(x)$ and $eE(x)/v_i(x)$ from equations (2) are expressed in terms of the impulses of electrons and ions, respectively, then we can obtain the relation $\frac{d}{dx}\{E^2(x) - 8\pi n u [p_e(x) + p_i(x)]\} = 0$, from which we obtain one further law of conservation:

$$E^2(x) - 8\pi n u [p_e(x) + p_i(x)] = E_0^2 - 8\pi n \gamma (M + m) u^2. \tag{4}$$

Here $E_0 = E(0)$ and a constant we have finded at $x = 0$.

With introduction of speed u in the task_one more frame of reference has appeared which goes with this speed concerning the system of counting of a wave. We call it the laboratory frame of reference (LFR). The quantities in the LFR will be denoted by the superscript "L". In this frame of reference the wave moves opposite to the axis x with the velocity u, and is characterized by the wavelength λ^L and by the period of potential and electric field oscillations $T = u/\lambda^L$. In LFR for point $x=0$ we have $n_i^L(0) = n_e^L(0) = n_0$. Our obtained system of equations (1)-(4) is sufficient for solving the formulated problem, the parameters of which are the quantities n_0, μ and γ.

DETERMINING THE WAVE PROFILE

We introduce the dimensionless variables for the coordinate $\xi = x\omega_p\sqrt{\beta}/c$ and the potential $\psi(\xi) = \varphi(x)/(mc^2)$, where $\omega_p = \sqrt{4\pi e^2 n/m}$ is the electron plasma frequency. if the electron and ion momenta are expressed in terms of the potential, which is possible to do using equations (3), then in dimensionless variables equation (4) may be written as:

$$V(\psi) = \epsilon - (d\psi(\xi)/d\xi)^2/2 =$$
$$= \mu\beta\gamma - \sqrt{\mu^2\beta^2\gamma^2 - \psi(2\mu\gamma - \psi)} + \beta\gamma - \sqrt{\beta^2\gamma^2 + \psi(2\gamma + \psi)} \tag{5}$$

where the variable ψ is a function of ξ, the constant parameters μ, β and γ are defined above, a constant parameter $\epsilon = (1/2)(d\psi/d\xi)_0^2$ is the value of the dimensionless energy density of the electric field at the point $\xi = 0$, at which $\psi = 0$, and the electric field is maximal.

From (5) we can find the desired dependencies of the potential and electric field on the coordinate. In the present case the function $V(\psi)$, which in its meaning is the dimensionless energy density of the electric field, obviously satisfies the equation: $d^2\psi/d\xi^2 = -dV/d\psi$, from the form of which it follows that $V(\psi)$ has the role of the potential energy of the system for the problem of the motion of a particle with unit mass, where the coordinate is represented by the variable ψ, and the time is represented by the variable ξ. The parameter ϵ has the role of the total energy of a particle moving in the potential well under consideration.

An analysis of the function $V(\psi)$ readily reveals that the value of the function at the extreme point of the segment where $\psi = \psi_+^*$ ($\psi_+^* = \mu(\gamma - 1)$ is a maximum value of the positive range of the potential) is always larger than at the other extreme point where $\psi = \psi_-^* = -(\gamma - 1)$, which, on the other hand, is a maximum value of the negative range of the potential. It is obvious that with the specified value of the parameter ϵ, a total range of potential oscillations is determined from (5) for $V(\psi) = \epsilon$. Thus assuming $V(\psi) = \epsilon$ when $\psi = \psi_-^*$, from (5) it is possible to find the limiting value of the parameter ϵ and, hence, the limiting value of the electric field amplitude in the wave: $\epsilon_m = \mu\beta\gamma + \beta\gamma - \sqrt{\mu^2\beta^2\gamma^2 + (\gamma-1)(2\mu\gamma + \gamma - 1)}$. From this relation for nonrelativistic waves ($\beta \ll 1$, $\gamma \approx 1 + \beta^2/2$): $\epsilon_m \approx \beta/2$, which in a dimensional form is represented by the formula: $E_0^2/(8\pi) \approx n_0 m u^2/2$. For relativistic waves ($\beta \approx 1$, $\gamma > 1$) we have: $\epsilon_m \approx (\gamma - 1)/(\beta\gamma)$. From here we obtain in the LFR for the limiting value of the electric field in the wave in a dimensional form: $(E_0)_m \approx (mc/e)\omega_{po}\sqrt{2(\gamma-1)}$, where $\omega_{po} = \sqrt{4\pi n_0 e^2/m}$ is the electron plasma frequency of linear oscillations of plasma in the LFR.

With given values of the parameters γ and ϵ, as has already been pointed out, the range of potential oscillations can be deduced from the equation $V(\psi) = \epsilon$. By putting $\mu \gg 1$ and assuming that $\epsilon = \epsilon_m$, from this relation we can obtain formulas for ultimate potential amplitudes of the wave for the case $\gamma \gg 1$. we know that with a given value of the parameter γ, a maximum value of the amplitude of the negative range of potential oscillations equals the boundary value: $\psi_-^m = \psi_-^*$. We determine the amplitude of the positive range by taking into consideration that when $\gamma \gg 1$ the value of $\epsilon_m \approx 1$. Finally, we obtain

$$\psi_-^m = \psi_-^* \approx -\gamma, \quad \psi_+^m \approx 2\beta\mu^2\gamma^2/(\mu^2 + 2\beta\mu\gamma). \tag{6}$$

DETERMINING THE WAVE FREQUENCY

To determine the wave frequency in the LFR we avail ourselves of the formula $\omega = 2\pi/T = 2\pi u\gamma/\lambda$. Here λ is the spatial period of potential oscillations in the wave's system: $\lambda = c\sqrt{\beta}\omega_p \int_{\psi_-}^{\psi_+} \frac{d\psi}{\sqrt{\epsilon - V(\psi)}}$, where $V(\psi)$ is determined from (5). From this we obtain for the value of ω the relation

$$\omega = \omega_{p0}\frac{\pi\sqrt{2}(\beta\gamma)^{3/2}}{\int_{\psi_-}^{\psi_+}\frac{d\psi}{\sqrt{\epsilon - V(\psi)}}}. \tag{7}$$

The exact value of the frequency can only be obtained by a numerical evaluation of the integral involved in formula

(7). However, the value of ω can be approximately estimated in two limiting cases: 1) $\gamma \approx 1$ and 2) $\gamma \gg 1$. We carry out all calculations and estimations for plasma waves having a maximum amplitude, i.e. it is assumed that $\epsilon = \epsilon_m$. In the case of small oscillations ($\gamma \approx 1$, $\beta \ll 1$), when $\epsilon_m \approx \beta/2$ and $\psi_+ \approx \psi_- \approx \beta^2$, the "integral" is taken, and we find that the frequency of small oscillations equals the Langmuir frequency of plasma oscillations which is written, in view of the motion of ions [2], as: $\omega = \omega_{p0}\sqrt{1+1/\mu} \approx \omega_{p0}$. In the case of the relativistic plasma wave ($\beta \approx 1$, $\gamma \gg 1$) the integral can be evaluated by noting that the value of $\epsilon_m - V(\psi)$ at large values of the parameter γ is virtually constant at every ψ, and in the case of ultimate amplitudes of the electric field ($\epsilon \approx \epsilon_m$) we have $\epsilon_m - V(\psi) \approx 1$ (see formula (5)). Taking this into consideration, from (7) we can obtain an estimate of $\omega \approx \omega_{p0}\pi\sqrt{2}\,(\gamma)^{3/2}/(\psi_+^m - \psi_-^m)$. Assuming that $\beta \approx 1$ and $\epsilon_m \approx 1$, we substitute here the values of ψ_-^m and ψ_+^m involved in (6) to give:

$$\omega \approx \omega_{p0} \frac{\pi(\mu^2 + 2\mu\gamma)}{\mu^2 \sqrt{2\gamma}}. \tag{8}$$

It is evident from formula (8) that in the range of the values $1 \ll \gamma \ll \mu$ the wave frequency decreases as $\gamma^{-1/2}$ (as well in [1]), and when $\gamma > \mu$ it increases with an increase of γ as $\gamma^{1/2}$. At a certain value of γ the frequency has a minimum value: $\omega_{min} \approx 2\pi\,\omega_{p0}/\sqrt{\mu}$ when $\gamma_{min} \approx \mu/2$. Obviously, at a certain $\gamma = \gamma_0 \gg 1$, that is, for relativistic waves, the wave frequency again, as in the case of linear oscillations ($\gamma \approx 1$), equals the plasma frequency. We find from (8) the value of γ_0 at a certain $\omega = \omega_{p0}$: $\gamma_0 \approx \mu^2/(2\pi^2)$.

For the purposes of illustration we now examine the values of the resulting quantities for longitudinal waves propagating in plasma consisting of electrons and protons. In this case a minimum frequency $\omega_{min} \approx \omega_{p0}/7$ and $\gamma_{min} \approx 10^3$, and for ultrarelativistic waves the frequency equals the plasma frequency for $\gamma_0 \approx 2\cdot 10^5$. The approximate dependence of ω on γ and the estimates of the values of γ_0, γ_{min} and ω_{min} are confirmed by accurate calculations obtained by a numerical integration of the expression involved in formula (7).

REFERENCES

1. *Akhiezer A.I.* and *Polovin R.V.* ZhETF, 1956, v.30, p.915.
2. *Akhiezer A.I., Akhiezer I.A., Polovin R.V. et al.* Plasma Electrodynamics. Edited by A.I.Akhiezer. Moscow: Nauka, 1974.
3. *Lünow W.* Plasma Physics, 1968, v.10, p.879.
4. *Max C.* Phys. Fluids, 1973, v.16, p.1277.
5. *Kozlov V.A., Litvak A.G.* and *Suvorov E.V.* ZhETF, 1979, v.76, p.148.
6. *Farina D.* and *Bulanov S.V.* Fizika plazmy, 2001, v.27, p.680.
7. *Khachatryan A.G.* Physical Review, v.58, p.7799.
8. *Bulanov S.V., Vshivkov V.A., Dudnikova G.I. et al.* Fizika plazmy, 1999, v.25, p.701.
9. *Gorbunov L.M., Mora P., Ramazashvili R.R.,* and *Solodov A.A.* Physics of Plasmas, 2000, v.7, p.375.
10. *Gorbunov L.M., Mora P.* and *Solodov A.A.* Physical Review Letters, 2001, v.86, p.3332.

Electromagnetic Transport Components and Sheared Flows in Plasma Edge Turbulence

Volker Naulin

Association
EURATOM-Risø National Laboratory
OFD-128, Risø,
DK-4000 Roskilde, Denmark

Abstract. The paradigm for transport barriers is based on the simple argument that sheared flows suppress turbulence and transport via a decorrelation mechanism. Here we present results from 3D simulations of drift-Alfvén turbulence in a sheared geometry. An intrinsic relationship between the shear flows and magnetic field perturbations leads to a more complex behavior, where for larger values of beta the transport might rise in the presence of shear flows.

INTRODUCTION

In hot magnetized plasmas the main cross-field transport is anomalous and often ascribed to low frequency electrostatic fluctuations of the driftwave type [1]. The transition to H-mode is generally believed to be due to shear flows setting up transport barriers, reducing transport via a decorrelation mechanism (see f.x. [2, 3] and references therein). For drift wave dynamics electromagnetic effects become important as they compete with parallel electron resistivity, which is responsible for the phase relation between density and potential fluctuations. This in turn determines transport and growth rates of the instabilities. While the net transport due to magnetic flutter in electromagnetic drift wave turbulence is close to zero in the closed field line domain – as opposed to the Scrape Off Layer (SOL) where the field lines end on material surfaces – it still can influence the level of transport. For a steady state situation the transport caused by fluctuations in the $E \times B$ drift velocity and the flutter associated transport add up to a function with radially vanishing divergence. We here investigate the relationship between electromagnetic and electrostatic transport components with a focus on their behavior in the vicinity of self-consistently occuring shear flows.

MODEL EQUATIONS

The fluid equations for drift-Alfvén micro-turbulence in three dimensional flux tube geometry result from standard ordering based upon the slowness of the dynamics compared to the ion gyrofrequency $\Omega_i = eB/M_ic$. Normalisation of quanteties is with $e\phi/T_e$, n/n_0, u/c_s, J/n_0ec_s and the small parameter $\delta = \rho_s/L_\perp$. The scale perpendicular to the magnetic field is ρ_s; the parallel scale is the closed flux surface connection length $L_\parallel = 2\pi qR$, with R the toroidal major radius and q the field line pitch parameter. The time scale is $L_\perp/c_s$. Details can be found in [4]. The parallel coordinate is s with x and y defining the "drift plane" perpendicular to **B**. The model dynamical system is described by the following set of equations:

$$\frac{\partial \Omega}{\partial t} + \mathbf{v}_E \cdot \nabla \Omega = \mathcal{K}(n) + \nabla_\parallel J + \mu_w \nabla_\perp^2 \Omega \tag{1}$$

$$\frac{\partial n}{\partial t} + \mathbf{v}_E \cdot \nabla(n_0 + n) = \mathcal{K}(n - \phi) + \nabla_\parallel (J - u) + \mu_n \nabla_\perp^2 n \tag{2}$$

$$\frac{\partial}{\partial t}\left(\hat{\beta} A_\parallel + \hat{\mu} J\right) + \hat{\mu} \mathbf{v}_E \cdot \nabla J = \nabla_\parallel (n_0 + n - \phi) - CJ \tag{3}$$

$$\hat{\varepsilon}\left(\frac{\partial u}{\partial t}+\mathbf{v}_E\cdot\nabla u\right)=-\nabla_\|(n_0+n) \tag{4}$$

with the vorticity, current (Ampere's law), and background profile described by

$$\Omega=\nabla_\perp^2\phi \qquad J=-\nabla_\perp^2 A_\| \qquad n_0=-x \tag{5}$$

The advective $\mathbf{v}_E\cdot\nabla=\{\phi,\}$ and parallel $\nabla_\|=\frac{\partial}{\partial s}-\{\hat{\beta}A_\|,\}$ derivatives carry nonlinearities entering through ϕ and $A_\|$, which can be expressed in terms of Poisson brackets in the xy-plane: $\{f,g\}=\frac{\partial f}{\partial x}\frac{\partial g}{\partial y}-\frac{\partial f}{\partial y}\frac{\partial g}{\partial x}$. The curvature operator $\mathcal{K}$ is written as $\mathcal{K}=\omega_B\left(\sin s\frac{\partial}{\partial x}+\cos s\frac{\partial}{\partial y}\right)$. The viscous terms in Eqs. (1) and (2) are introduced artificially to provide sub-grid dissipation. The parameters reflect the competition between parallel and perpendicular dynamics, via the scale ratio $\hat{\varepsilon}=(qR/L_\perp)^2$. The electron parallel dynamics is controlled by

$$\hat{\beta}=\frac{4\pi p_e}{B^2}\hat{\varepsilon} \qquad \hat{\mu}=\frac{m_e}{M_i}\hat{\varepsilon} \qquad C=0.51\frac{L_\perp}{\tau_e c_s}\hat{\mu} \tag{6}$$

where τ_e is the electron collision time. The competition between the three effects representing magnetic induction, electron inertia, and resistive relaxation determines the response of J to the static force imbalance in Eq. (3). The only other physical parameter is ω_B signifying the effects of magnetic curvature (equivalently magnetic gradient, in a toroidal model). All magnetic induction ($\partial A_\|/\partial t$) and flutter ($\{A_\|,\}$) effects enter through the finite beta ($\beta_e=4\pi p_e/B^2$ or c_s^2/v_A^2, where v_A is the Alfvén velocity, and $\hat{\beta}=\beta_e\hat{\varepsilon}$).

The geometry is a three dimensional flux tube with local slab like coordinates $\{x,y,s\}$ set up generally as described above, and then further specialized to a local domain in the drift plane while following the structure of the metric in the s-direction along $\mathbf{B}$. Herein, we use the simplified forms

$$x=r-a; \qquad y_k=\frac{a}{q_a}[q(\theta-\theta_k)-\zeta]; \qquad s=q_a R\theta; \tag{7}$$

where R and a are constants describing the toroidal major and minor radii, r is the minor radial coordinate (flux surface label), and θ and ζ are the poloidal and toroidal angles, $q=q_a(1+\hat{s}x/a)$ is the field line pitch parameter (contravariant component ratio B^ζ/B^θ for $\mathbf{B}$) with q_a and $\hat{s}$ constant, and θ_k is a constant describing the poloidal location where the k-th coordinate system is referenced. The coordinate system is therefore defined differently on each drift plane referenced at $s=s_k=q_a R\theta_k$ to avoid the effects of coordinate deformation. Partial derivatives with respect to s expressed as finite differences incur shifts in the y-coordinate. The coordinates are normalized in terms of ρ_s for x and y, and $q_a R$ for s. The domain is bounded in x, cyclic in y and field-line connected in s.

Fluctuation energetics

Multiplying Eq. (1) by $-\phi$, Eq. (2) by n, Eq. (3) by J and finally Eq. (4) by u and integrating over the whole domain, we find the evolution equation for the positive definite integral giving the total free energy, $E=E_n+E_f+E_B+E_J+E_u$ where the free energy pieces are defined as

$$E_n=\frac{1}{2}\int n^2 dV \qquad E_f=\frac{1}{2}\int(\nabla_\perp\phi)^2 dV$$

$$E_B=\frac{1}{2}\int\hat{\beta}(\nabla_\perp A_\|)^2 dV \qquad E_u=\frac{1}{2}\int\hat{\varepsilon}u^2 dV \qquad E_J=\frac{1}{2}\int\hat{\mu}J^2 dV.$$

The time dependence of the total free energy is given by

$$\frac{d}{dt}E=\Gamma_n+\Gamma_J-\Gamma_c \tag{8}$$

The driving terms are due to the background gradient and depict the $E\times B$ transport $\Gamma_n=-\int n\frac{\partial\phi}{\partial y}dV$ and the the flutter-flux along perturbed magnetic field lines $\Gamma_J=\int(u-J)\hat{\beta}\frac{\partial A_\|}{\partial y}dV$. On the dissipative side we only have the finite parallel resistivity: $\Gamma_c=\int CJ^2 dV$ It is worthwhile to note that curvature represents an energy exchange term between density and potential fluctuations but not a source of energy [4]. Thus curvature can act destabilizing by coupling the density background gradient to the potential fluctuations, but it acts in no way as "effective gravity" does in simpler paradigmatic models [5].

GENERAL DYNAMICS

The simulations were performed on a $64 \times 256 \times 16$ grid, with dimensions $64 \times 256 \times 2\pi$ in x, y, s. Standard parameters for the runs were $\hat{\mu} = 5$, $\hat{s} = 1$, $\omega_B = 0.05$ with the artificial viscosities set to $\mu_w = 0.025$ and $\mu_n = 0.025$. Scans were performed in the collisionality C, with C taking the values $\{2.55, 5.10, 7.65, 12.75, 25.5\}$ and $\hat{\beta}$ taken from $\{0.3, 1.0, 2.0, 3.0, 4.0.5.0, 7.0, 10.0\}$. The turbulence is characterized by a high correlation (>0.9) between density and potential fluctuations, as appropriate for drift wave turbulence. The scaling of the flux with the essential parameters is depicted in Fig. 1, which to a large degree reproduces the results of the DALF Family of codes. For a more detailed comparison see [Naulin and Scott].

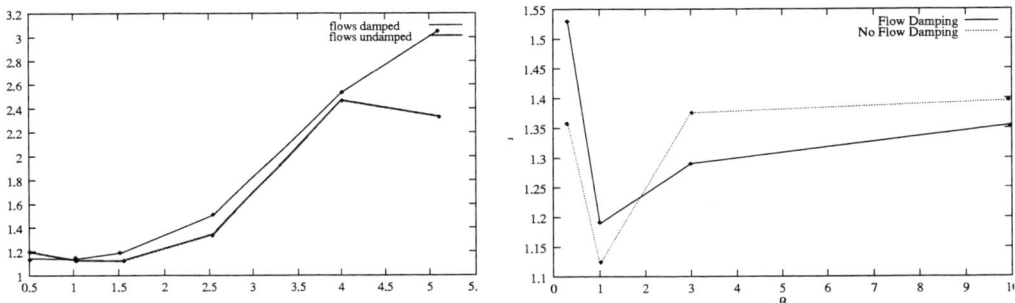

FIGURE 1. Flux Γ versus collisionality ν (left) and $\hat{\beta}$, for runs with viscous damping of the flow modes and without.

CORRELATION BETWEEN $E \times B$ AND ELECTROMAGNETIC FLUXES

We now investigate the flux scaling with collisionality and plasma beta, changing the level of self-consistently developing flows. Due to viscosity flows are damped damped at a rate proportional to $\mu_w \partial_{xx} \phi_{k_y=0}$. We now switch off flow damping by explicitly setting $\mu_w = 0$ for $k_y = 0$.

A comparison of net transport with flow damping and without shows for low values of $\hat{\beta}$ that indeed higher flow levels hinder the transport. In the density profile a local steepening up of the gradient is see where the flow shear is strongest. Fig.1 shows that this behavior prevails for all values of collisionality scanned. It should be noted, that the flow shear is however, weaker than the fluctuating vorticity. Thus no real quenching of the turbulence takes place.

For the transport scaling with $\hat{\beta}$ and removed flow damping the picture changes. For large values of $\hat{\beta}$ we now observe an actual increase in the overall transport as depicted in Fig.1.

We additionally check if the absence of flow damping indeed increases flow level. Fig. 2 shows transport and flow level as measured by the variation in plasma potential versus collisionality for $\hat{\beta} = 6$. Indeed flow level and transport scale proportional to each other for large $\hat{\beta}$, so that opposite to the low $\hat{\beta}$ situation and increase in flow activity increases transport as well. The reason for this can be the following: The parallel current J is convected with the $E \times B$ velocity as well, so that a shear region poses a barrier to the current as well see Fig. 3. Thus, on average, current fluctuations will be strongest in the barrier region. Large parallel currents indicate large magnetic field perturbations, which make

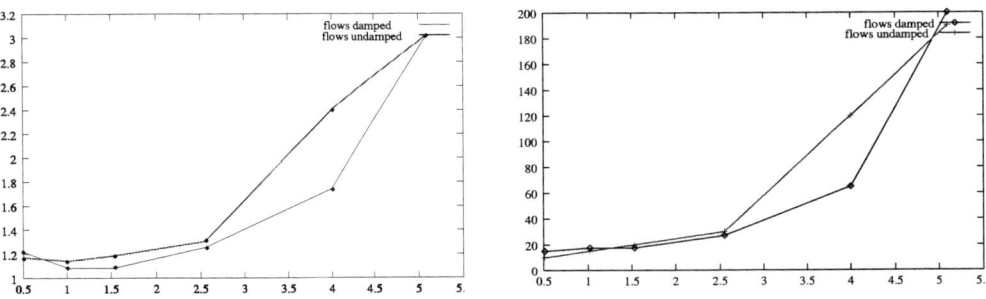

FIGURE 2. Flux Γ (left) and maximum of $<\phi>_{av}$ (right) versus collisionality ν for $\beta = 6$

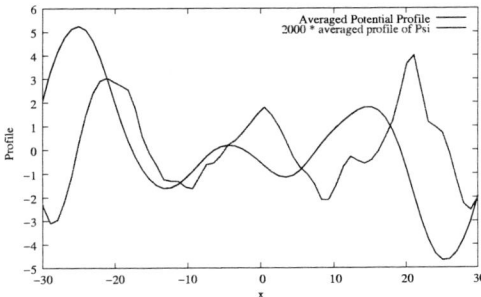

FIGURE 3. The profiles of the magnetic vector potential and the potential show that shear regions pose a barrier for the parallel current convection.

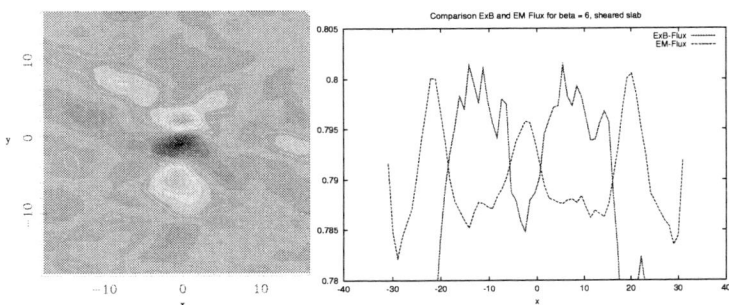

FIGURE 4. $E \times B$ and flutter flux are anti-correlated (left) and the variation in the $E \times B$ flux is compensated by the flutter flux

a locally large flutter-transport possible. The averaged flux is determined, in steady state, by the minimum value the flux takes radial. In the low $\hat{\beta}$ case this is the minimum value at the shear flow. However, in the large $\hat{\beta}$ regime, just in that region the flutter-flux can locally rise the total flux, thus tunneling the transport barrier and rising the overall value of the flux.

This is seen in Fig. 4, showing that flutter flux and $E \times B$ flux weakly anti-correlated and that the flutter flux profile adds up with the $E \times B$ flux profile to a radially more or less constant value, compensating the effects of the shear flows partially.

CONCLUSION

EM flutter transport adds to the dynamics in the presence of shear flows, barriers for transport also build up current sheaths. Thus, the magnetic field perturbations increase, which results in an enhanced magnetic transport and a local tunnelling through the shear flow induced transport barrier. This raises the overall transport, which is determined by the location, where the transport is minimum in a the quasi-steady situation.

ACKNOWLEDGMENTS

Discussions with Jens Juul Rasmussen and Bruce Scott are gratefully acknowledged

REFERENCES

1. Wagner, F., and Stroth, U., *Plasma Phys. Controlled Fusion*, **35**, 1321–1371 (1993).
2. Terry, P. W., *Rev. Mod. Phys.*, **72**, 109–165 (2000).
3. Burell, K. H., *Phys. Plasmas*, **6**, 4418–4435 (1999).
4. Scott, B. D., *Plasma Phys. Control. Fusion*, **39**, 1635–1668 (1997).
5. Carreras, B. A., Hidalgo, C., Sánchez, E., Pedrosa, M., Balbín, R., García-Cortés, van Milligen, B., Newman, D. E., and Lynch, V. E., *Phys. Plasmas*, **3**, 2664–2672 (1996).

Naulin and Scott. Naulin, V., and Scott, B., *to be published*.

Self-Organization and Coupling of Waves in a Plasma

Lj. Nikolić[*], S. Ishiguro[†] and T. Sato[**]

[*]*The Graduate University for Advanced Studies, National Institute for Fusion Science, 322-6 Oroshi-cho, Toki-shi 509-5292, Japan*
[†]*Theory and Computer Simulation Center*
National Institute for Fusion Science, 322-6 Oroshi-cho, Toki-shi 509-5292, Japan
[**]*Earth-Simulator Center, JAMSTEC, 3173-25 Showa-cho, Yokohama-shi 236-0001, Japan*

Abstract. Numerous systems in nature during their evolution exhibit self-organization. The ability of a system to evolve in such a way is a consequence of the fact that the real systems are open to their environment. In such systems among natural modes, as a result of nonlinear interactions, a significant number of non-fundamental modes exists. We investigate the excitation of these modes and their contribution to the development of quasi-stationary nonequilibrium states. As a model of an open system we consider the coupling of an intense laser light to an underdense plasma placed in vacuum. It is shown that the response of the plasma system can be very high due to the generation of waves with frequency below the plasma frequency which can interact with a large number of bulk electrons.

INTRODUCTION

The emergence of ordered states in nature is one of the most interesting phenomena and a central theme of modern interdisciplinary science - the Science of Complexity. It is often very difficult to describe dynamics and predict behavior of real systems. The systems are open to the environment, governed by nonlinear dynamics with the common property that the response to an external excitation is not a simple superposition of responses of their parts. Therefore, for most of problems analytical solutions in the form of known functions cannot be obtained. However, the lack of analytical description, i.e. mathematical intractability, does not stop the progress. In contrast, from traditional studies of steady equilibrium structures in isolated systems, the frontiers are now expanding to consider dynamics of open systems. One of the advances is the development of statistical physics of open systems [1] and description of (quasi)stable states far-from equilibrium [2]. These efforts provide an extension of Boltzmann-Gibbs statistics and redefinition of traditional thermodynamics. Yet, the tracing of the evolution of complex systems and finding of universal rules which can be applied to the class of equivalent systems are great challenges of computer simulations. Indeed, computer simulations are established as a powerful research methodology and the present-day supercomputer status offers a good base for handling complexity from first principles [3].

Although the view on self-organization suffers from even conceptual difficulties, it has been demonstrated that for dissipative systems nonequilibrium dynamics and instabilities play a key role in the process of self-organization [4]. In general, a complex system supports a wide range of natural modes and the coupling of external excitations to these modes can lead to the generation of instabilities. It is generally believed that the dominant responses of systems are governed by such a type of connection. However, the manifestation of nonlinear interactions can be a significant presence of modes that do not belong to the set of fundamental ones. A typical example of the multi-wave nature of the interaction in plasma physics is the parametric coupling of waves. Theory and particle-in-cell simulations are used to investigate the coupling and stability of a large amplitude laser light in an underdense electron plasma. It is well known that in a plasma below quarter-critical density stimulated Raman scattering (coupling of fundamental modes) plays an important role. Much work has been devoted to this subject and it is still an active research topic. The purpose of our study is, however, to see how much of the plasma response can be due to the non-Langmuir electrostatic (ES) beatings. In particular, we consider excitation of ES waves below the plasma frequency. The phase velocity of these waves can be relatively low so that the waves can interact with a large number of bulk electrons. In order to explain this instability, a 3-wave model of the backscattering is discussed.

MODEL AND RESULTS

Parametric instabilities in an underdense plasma have long been an issue for inertial fusion oriented research. One of these instabilities is stimulated Raman scattering (SRS), which is the resonant decay of an electromagnetic (EM) wave (ω_0, k_0) into a scattered light (ω_s, k_s) and an electron plasma (Langmuir) wave (ω_{EPW}, k_{EPW}), where $\omega_0 = \omega_s + \omega_{EPW}$ and $\mathbf{k}_0 = \mathbf{k}_s + \mathbf{k}_{EPW}$ (ω and k are the frequency and wave number). The dispersion equations of these waves are $\omega_{0/s}^2 \approx \omega_p^2 + c^2 k_{0/s}^2$ and $\omega_{EPW}^2 \approx \omega_p^2 + 3v_{th}^2 k_{EPW}^2$, where v_{th} is the thermal velocity. As can be concluded from the frequency matching condition, SRS can occur for plasma densities $n \leq 0.25 n_{cr}$ ($n_{cr} = n\omega_p^2/\omega_0^2$ is the critical density). SRS is of particular interest since Langmuir waves significantly scatter laser light and generate high energy electrons which preheat the core of plasma. Although SRS itself is well understood its implications are difficult to determine and predictions are often in a large disagreement with experimental observations. Major complications arise from the fact that the created waves can couple with other waves in the plasma, they can interact with each other and transfer part of the energy of the laser light to the plasma [5], [6].

To investigate generation of ES waves in an underdense electron plasma a relativistic 1d3v particle-in cell code with minimum 50 particles per cell was used. A series of runs was performed with plasma layers placed in vacuum and excited by an intense linearly polarized laser light. The ions were kept immobile and the length of layers, densities, laser strength and the temperature were varied.

As an example of typical responses of the considered plasma system, in Fig. 1 simulation results for two connected plasma layers L_1 and L_2 are shown. The densities and lengths of the layers were $n_1 = 0.15 n_{cr}$, $n_2 = 0.8 n_{cr}$, $L_1 = 0.30 c/\omega_0$ and $L_2 = 0.50 c/\omega_0$, the temperature was $T = 1$keV and the laser strength was $\beta = 0.37$ ($\beta = eE_0/(mc\omega_0)$, e and m are the electron charge and mass, E_0 is the amplitude of the electric field). In fact, in Fig. 1a the time history of the reflectivity ($R = \langle S_r \rangle / \langle S_0 \rangle$, S_r and S_0 are Poynting vectors for reflected and incident wave, respectively, and $\langle \rangle$ denotes time averaged values) averaged over four periods of the incident light and in Fig. 1b the longitudinal distribution functions of the system for $t\omega_0 = 1120, 2500$ and 4000 are shown. As can be seen, the respose of the system exhibits three regimes of evolution. In the first phase ($t\omega_0 < 1500$) characterized by reflectivity pulsations, backward SRS in the layer L_1 is the dominant instability. A rapid growth of Langmuir waves and suppresion of the instability through the generation of high energy particles (see Fig. 1b, for $t\omega_0 = 1120$) are known signatures of SRS. Thus, the behavior of the system in this phase is almost "standard-like". However, for later times two large pulses in Fig. 1a, with reflectivities over 100% of the incident laser light indicate unexpected features. Detailed investigations were carried out to determine the source of these signals. It was concluded that this instability is not an onset SRS instability, but the scattering from an acoustic ES wave ($\omega < \omega_p$) in the layer L_2. Moreover, this ES wave has a large amplitude and relatively low phase velocity and therefore, it interacts with a large number of bulk electrons. This is confirmed by Fig. 1b which displays the distributions for the states when instability is saturated ($t\omega_0 = 2500$ and 4000). We note that the hierarchy of instabilities in time scale causes successive transitions from one quasi-steady nonequilibrium state into another. There are at least two reasons for such behavior: the nature of instabilities (different growth rates, saturation levels etc.) and the status of resonant conditions (matching/mismatching) given by the current dynamics of the system. In fact, it is known that a laser plasma possesses a very rich interplay between instabilities.

To clarify the nature of the instability observed in a plasma with over-critical density for SRS ($n > 0.25 n_{cr}$), in Fig. 2 spectra of EM (Fig. 2a) and ES (Fig. 2b) waves are shown. The simulation parameters were $n = 0.6 n_{cr}$, $L = 40 c/\omega_0$, $T = 500$eV and the laser strength was $\beta = 0.4$. Fig. 2 clearly shows that the instability is the decay of the laser wave ($\omega/\omega_0 = 1$) into a scattered wave with frequency near the plasma frequency ($\omega_p/\omega_0 \approx 0.84$), and an acoustic ES wave at $\omega/\omega_0 \approx 0.15$. Note that apart from ES noise around $\omega/\omega_0 \approx 0.15$, the plasma wave ($\omega_p/\omega_0 \approx 0.84$), a driven wave at $\omega/\omega_0 = 2$ and its product of the coupling ($\omega/\omega_0 = 1.85$) with the low-frequency ES wave are visible in Fig. 2b. Note also that, the nonlinear theory predicts ES wave at ($2\omega_0, 2k_0$) [7]. However, this wave is non-trapped since its phase velocity exceeds the velocity of light.

As is apparent from the results, the transfer of energy to an underdense plasma and high reflectivities can be caused by the growth of non-Langmuir ES waves. However, such a wave does not appear in the linear plasma theory as a fundamental mode. In order to examine the nonlinear properties of coupling of waves, we consider a general dispersion equation for light wave scattering in a cold plasma,

$$D_- D_+ = G(D_- D_{1+} + D_+ D_{1-}) + G^2 (D_2^2 - D_{1+} D_{1-}), \qquad (1)$$

where $G = \frac{\omega_p^2 \beta}{4}$, $D_{+/-} = (\omega \pm \omega_0)^2 - \omega_p^2 - c^2 (k \pm k_0)^2$, $D_{1+/1-} = \frac{k^2 c^2}{\omega^2 - \omega_p^2} + \frac{(k \pm 2k_0)^2 c^2}{(\omega \pm 2\omega_0)^2 - \omega_p^2} - 3$, $D_2 = \frac{k^2 c^2}{\omega^2 - \omega_p^2} + \frac{1}{2}\left(\frac{4k_0^2 c^2}{4\omega_0^2 - \omega_p^2} - 3\right)$. This dispersion equation includes the relativistic shift of the plasma frequency along with the den-

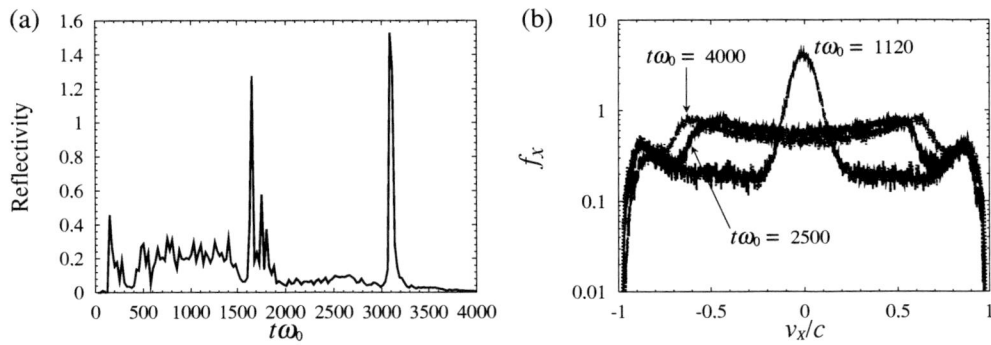

FIGURE 1. (a) Reflectivity from two connected plasma layers ($n_1 = 0.15 n_{cr}$, $L_1 = 40 c/\omega_0$, $n_2 = 0.8 n_{cr}$, $L_2 = 50 c \omega_0$, $T = 1$keV, $\beta = 0.37$) and (b) longitudinal distribution functions of the system for $t\omega_0 = 1120$, 2500 and 4000.

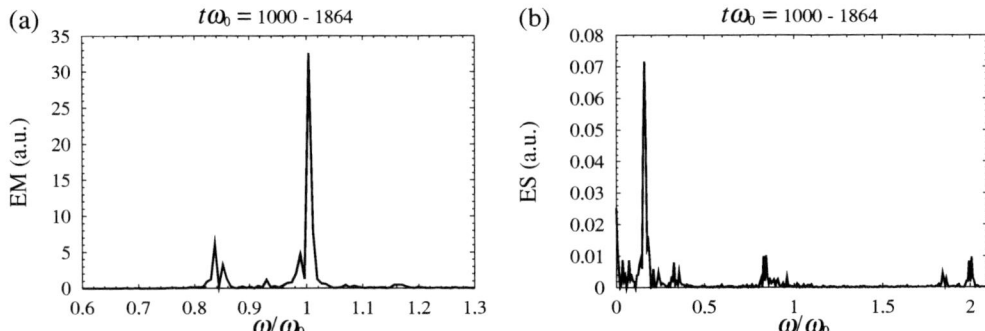

FIGURE 2. Spectrum of (a) electromagnetic (EM) and (b) electrostatic (ES) waves in the plasma layer ($n = 0.9 n_{cr}$, $L = 50 c/\omega_0$, $T = 500$eV, $\beta = 0.4$) for time interval $t\omega_0 = 1000 - 1864$.

sity perturbations at $2\omega_0$, and can be obtained by following the procedure outlined in [7]. As an illustration, in Fig. 3 we show unstable solutions of Eq. 1 for $n = 0.4 n_{cr}$ and laser strengths $\beta = 0.4$ (Fig. 3a) and $\beta = 0.5$ (Fig. 3b) as a function of the wave number. It is well known that, for sufficiently high laser intensity, the relativistic correction to the mass of electrons oscillating in the incident electric field causes the relativistic modulation instability (RMI). In the parts of plasma with densities $> 0.25 n_{cr}$, when SRS is halted, only RMI is found with a low growth rate (see Fig 3a). When the laser intensity is increased the RMI branch brodens, and relativistically shifted to the high densities Raman branch appears (Fig. 3b). As we can see, the dispersion equation predicts a low frequency ES wave but does not explain the growth of instability shown in Fig. 2. Namely, RMI is a long-wavelength low-frequency perturbation of the laser wave envelope, causing a growing instability in the frame moving with the group velocity of the laser wave, while the observed acoustic instability is backward scattering. Moreover, from our simulations we found that the frequency of the scattered light is always $\approx \omega_p$, thus $k_s \approx 0$, and the frequency and wave number of the unstable ES wave are $\omega_a \approx \omega_0 - \omega_p$ and $k_a \approx k_0$.

Although much of our understanding of processes camnes from cold plasma and especially from linear theory, higher levels of description of collisionless plasmas show new modes and features [8], [9]. Therefore, in order to interpret the instability let us consider a general 3-wave ($a_i(x,t) \exp[i(k_i x - \omega_i t)]$) interaction described by the coupled wave equations,

$$\frac{\partial a_0}{\partial t} + V_0 \frac{\partial a_0}{\partial x} = -M_0 a_s a_a, \qquad (2)$$

$$\frac{\partial a_s}{\partial t} - V_s \frac{\partial a_s}{\partial x} = M_s a_0^* a_a, \qquad (3)$$

$$\frac{\partial a_a}{\partial t} + V_a \frac{\partial a_a}{\partial x} + \Gamma_a a_a = M_a a_0^* a_s, \qquad (4)$$

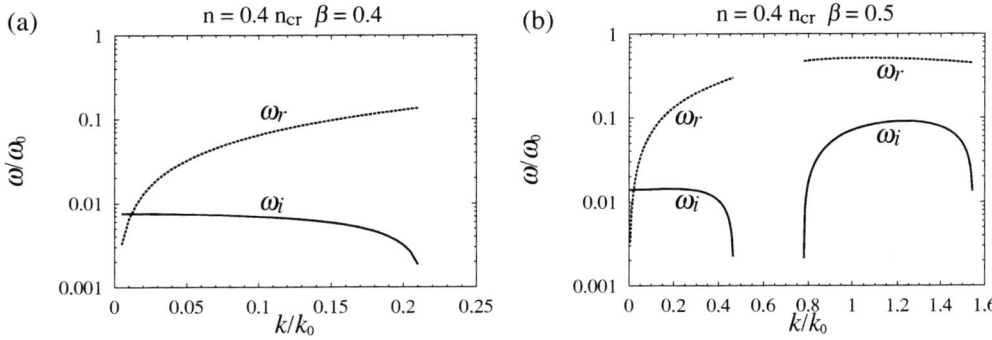

FIGURE 3. Solutions of the dispersion equation (Eq. 1) for $n = 0.4n_{cr}$ and (a) $\beta = 0.4$ and (b) $\beta = 0.5$.

where $V_i > 0$ is the wave group velocity, Γ_a is the damping rate ($\Gamma_0 = \Gamma_s = 0$ for EM waves is used), $M_i > 0$ are the coupling coefficients and a_i are the wave amplitudes, where $i = 0, s, a$, denote the pump, backscattered and ES wave, respectively. The plasma slab thickness L with the boundary conditions for backscattering, $a_0(0,t) = E_0$, $a_s(L,t) = a_a(0,t) = 0$, leads to the condition for an absolute instability $L/L_0 > \pi/2$, [10], where $L_0 = (V_s V_a)^{1/2}/\gamma_0$ is the basic gain length and $\gamma_0 = E_0 (M_s M_a)^{1/2}$ is the uniform growth rate. Since $\Gamma_a \neq 0$, ES wave is characterized by the longitudinal absorption length $L_a = V_a/\Gamma_a$ and the condition for instability has the following additional requirement, $L_0/L_a < 2$ [10]. In general, a laser wave interacts with a whole spectrum of ES waves, but most of them are strongly damped $L_a \approx 0$, and thus the condition for instability is not satisfied. However, as can be seen, the key factor for the growth of the observed instability is a slowly propagating backscattered EM wave, so that $V_s \approx 0$ ($\omega_s \approx \omega_p$) minimizes L_0 and the threshold for instabilities $\gamma_0 > 0.5\Gamma_a(V_s/V_a)^{1/2}$ [10]. Note that for observed acoustic ES waves, described by Eq. 4, no dispersion equation in analytical form exists and additional questions arise concerning possible connection with electron acoustic resonance [9].

CONCLUSIONS

Simulation results and theoretical considerations presented in this paper suggest that a significant fraction of incident laser energy can be scattered from a trapped ES wave with frequency below the plasma frequency. The possibility of generation of such ES waves in a plasma deserves future attention. From the point of view of general complex systems, the results demonstrate that through a multi-mode interaction, a mode that does not appear as a fundamental mode of a system, can efficiently lead the system to the new nonequilibrium state.

ACKNOWLEDGMENTS

One of us (Lj.N.) is grateful to M.M. Škorić for helpful discussions and H. Takamaru for providing his PIC code.

REFERENCES

1. Yu. L. Klimontovich, *Statistical Theory of Open Systems* Vol.1, (Kluwer Academic Publishing, Netherland 1995).
2. R. A. Treumann, Astrophys. and Space Sci. **277**, 81 (2001).
3. T. Sato and the Complexity Simulation Group, Phys. Plasmas **3**, 2135 (1996).
4. G. Nicolis and I. Prigogine, *Exploring Complexity* (Freeman, New York, 1989).
5. D. S. Montgomery et al., Phys. Plasmas **9**, 2311 (2002).
6. J. C. Fernández et al., Phys. Plasmas **7**, 3743 (2000).
7. C. J. McKinstrie and R. Bingham, Phys Fluids **B4**, 2626 (1992).
8. H. Schamel, Phys. Plasmas **7**, 4831 (2000).
9. H. A. Rose and D. A. Russell, Phys. Plasmas **8**, 4784 (2001).
10. V. Fuchs, Phys. Fluids **20**, 1535 (1977).

Modeling of Small Dense-Plasma Helicon Source

S. Shinohara[1] and K.P. Shamrai[2]

[1]*Interdisciplinary Graduate School of Engineering Sciences, Kyushu University, Fukuoka, Japan*
[2]*Institute for Nuclear Research, National Academy of Sciences, Kiev, Ukraine*

Abstract. The power absorption in a small helicon source is examined theoretically over a broad range of driving frequencies, magnetic fields, and gas pressures. The operation window is evaluated where the large plasma resistance, i.e. the high power efficiency of the source, can be attained at low gas pressure, i.e. at reduced gas consumption. Effects of the radial and axial plasma nonuniformity on the power absorption and mode structure are considered, and a possible scenario for attaining the high plasma density is discussed.

INTRODUCTION

Helicon sources of dense plasma are advanced tools for various applications. Due to the high ionization efficiency and the ability to operate at low gas pressures the helicon source is attractive for the space propulsion. The potentiality of a small-scale helicon source to serve as an ion thruster, e.g. for a satellite orbital control, was recently demonstrated by exploring the devices with accelerating grid [1]. A large-scale source is promising as an element of the plasma rocket for deep space mission [2]. In these schemes the thrust is produced directly, by emanating the plasma jet. There is also an indirect scheme, which is being developed within the concept of mini-magnetospheric plasma propulsion [3]; it assumes the utilization of the solar wind power by means of the "plasma sail" inflated by the helicon source.

To provide a high efficiency of the power utilization in the source, one should sustain the discharge regimes with the power absorption in plasma considerably exceeding losses in the feeding circuitry. For that, the plasma load resistance should amount the value at least on the order of a few ohms. The gas pressure in these regimes should be not very high to provide sufficient ionization efficiency. In this report, we examine the power absorption for a small source excited by a double-loop $m = 0$ antenna. A broad range of magnetic fields, plasma densities, driving frequencies, and gas pressure is considered to reveal the operation regimes in which both the power and ionization efficiencies are expected to be simultaneously high. The effects of radial and axial plasma nonuniformity on the mode structure and power absorption are discussed.

RF POWER ABSORPTION

The rf power absorption in plasma was examined on the assumption of the cold-plasma model of the helicon source [4]. The length of the source and the plasma radius were assumed to be $L_z = 4$ cm and $r_0 = 2$ cm, respectively. The rf antenna consists of two loops, of radius 2.5 cm and width 0.5 cm, placed symmetrically relative to the midplane with a distance 2 cm between them. The analysis was carried out for the broad ranges of the plasma density $n_0 \leq 10^{13}$ cm^{-3}, driving frequency $f \leq 100$ MHz, magnetic field $B_0 = 0$–200 G, and Ar gas pressure $p_{Ar} = 5$–100 mTorr. The lower boundary of the pressure range was chosen considering the empiric fact that the ignition of the helicon discharge is possible provided that the electron mean free path is less than the chamber size.

We first considered the rf power absorption in the ICP mode with unmagnetized electrons, i.e. provided that the external magnetic field is below 30 G, so that the electron cyclotron frequency, ω_{ce}, is less than the electron collision frequency, $v_e = v_{en} + v_{ei}$. Under this condition, the resistance was found to increase with pressure and to amount the eligible values above 1 ohm at very high Ar pressures only, above 70 mTorr. For this reason, the ICP regime seems to be improper for thruster needs as long as the power efficiency arrives at acceptable level at such high pressure that the ionization is expected to be low.

A. Absorption in Uniform Helicon Plasma

The absorptance is found to be considerably larger in plasma with magnetized electrons, i.e. when $\omega_e > \nu_e$. The plasma resistance, R_p, rapidly increases with magnetic field at $B_0 > 30$ G and can amount the values well above 1 ohm in the range of magnetic fields about 100G. However, the resistance turns out to be large in a narrow band on the plasma density and driving frequency as a result of a single-mode excitation of the helicon waves. In a short device under consideration the helicon waves appear as axially standing modes with wavelengths quantized by the device length, $\lambda_z = 2L_z/s$ ($s = 1,2,...$). Figure 1 demonstrates the resistance computed for the parallel (anti-parallel) currents in the antenna loops when the first (second) axial mode is excited. Position of the resistance peak scales with parameters according to the helicon dispersion

$$n_0 f \lambda_z / B_0 = \text{const.} \tag{1}$$

The band of high absorption apparently can be moved to higher densities in several ways. The first, by lowering of the driving frequency, is fraught with a considerable decrease of the resistance as seen from Fig. 1. The second, by decreasing the wavelength, can be operated by the antenna spectrum; however, it also results in the resistance reduction (see Fig. 1). The most appropriate way is the increase of magnetic field which gives rise to the increase of R_p approximately linearly with B_0. Contrary to the ICP regime, the plasma resistance in the helicon regime increases with decreasing pressure, so that the power utilization and ionization efficiency can be simultaneously high.

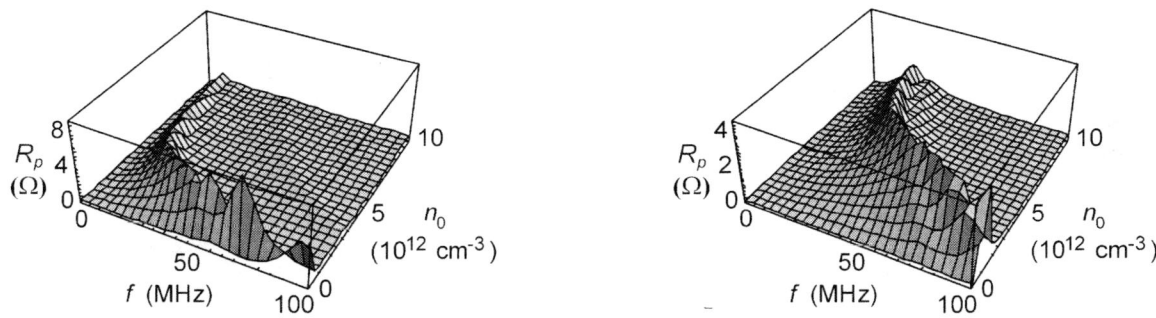

FIGURE 1. Plasma load resistance as a function of plasma density and driving frequency for the same (left) and opposite (right) current directions in the antenna loops. $B_0 = 100$ G and $p_{Ar} = 10$ mTorr.

B. Influence of Radial Plasma Nonuniformity

Computations of the resistance for the radially nonuniform plasma were performed with the parabolic density profile for various values of density at the radial plasma edge. As seen form Fig. 2, the nonuniformity has no crucial effect on the absorptance, even if the edge density is very low. The absorption bands are found to move to higher densities and the magnitudes of the absorption peaks slightly change.

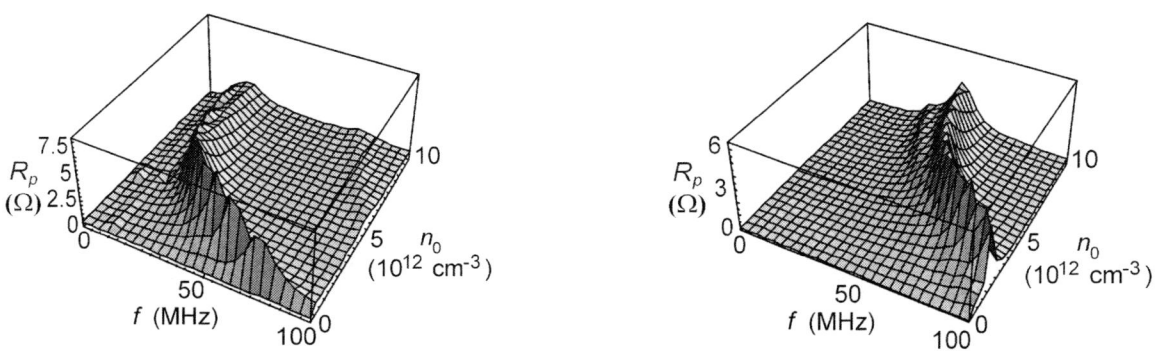

FIGURE 2. The same as in Fig. 1 but for the radially nonuniform plasma with zero edge density.

EFFECT OF AXIAL PLASMA NONUNIFORMITY

If the plasma column is axially uniform and limited by conducting end plates, the fields are superpositions of helicon and Trivelpiece-Gould (TG) waves (see, e.g., [4]). They appear as axial eigenmodes with z-dependences as $\cos k_z z$ or $\sin k_z z$, where the wave number k_z is quantized by the source length, as mentioned above. However, axial plasma profile is really nonuniform, especially in front of the end plates and/or near the accelerating grid or near the open source outlet where the density is considerably reduced. This effect can substantially alter the eigenmodes [5].

We model the edge density reduction by introducing the vacuum gaps between the plasma and the source ends (see Fig. 3) and suppose a uniform density inside. The presence of gap sets non-conducting boundary conditions at the plasma ends; this results in a necessity to introduce, in addition to the helicon and TG waves, a new, surface wave that is located at the plasma edge. Plasma eigenmodes are then found as superpositions of the bulk (helicon or TG) and surface waves. The dispersion of all three waves in collisionless plasma is shown in Fig. 4.

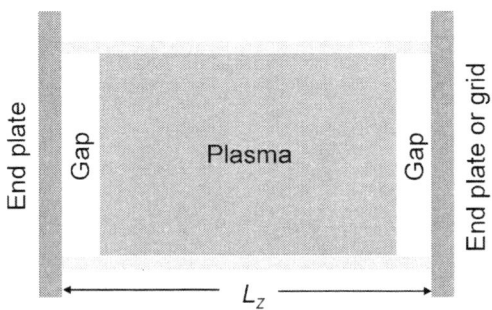

FIGURE 3. Model of the plasma source with vacuum gaps substituted for the edge density drops.

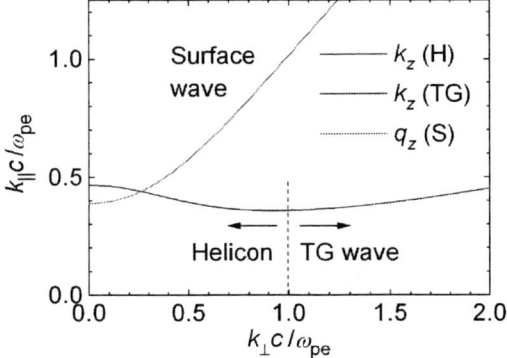

FIGURE 4. Dispersion of oscillations in plasma with non-conducting axial boundaries, at $B_0 = 100$ G.

At zero vacuum gaps, the dispersion curves for helicon eigenmodes are shown in Fig. 5, for $L_z = 4$ cm. Noticeable are the wide bands on density (gray) where the eigenmodes are forbidden. The presence of these bands was the reason for the strong separation of absorption peaks in above calculations. With finite gap widths, the quantization on k_z is no more the case, but the eigenmodes still exist. If the gaps are equal in width, the modes are found to be either symmetric or asymmetric relative to the system midplane. With z-axis origin placed at the midplane, the plasma E_θ-field of axially symmetric $m = 0$ eigenmode takes the following form

$$E_\theta = (A_B \cos k_z z + A_S \cosh q_z z) J_1(k_\perp r), \quad (2)$$

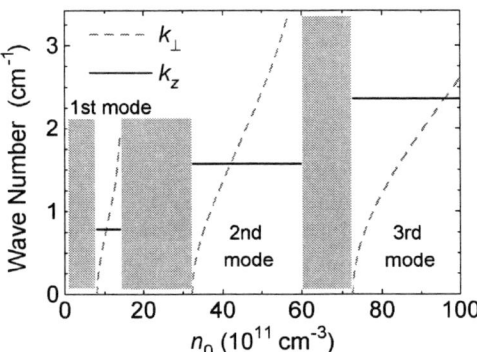

FIGURE 5. Dispersion of helicon eigenmodes at zero gap width. $B_0 = 100$ G and $f = 50$ MHz.

where J_1 is the Bessel function; k_z and q_z are some functions of $k_\perp$ and plasma parameters. The asymmetric mode takes the form similar to Eq. (2) but with $\sin k_z z$ and $\sinh k_z z$ in brackets. Constants A_B and A_S for the bulk and surface waves in Eq. (2) are defined, together with the eigenvalue of $k_\perp$, from joining conditions at the axial boundaries.

The eigenvalues as functions of density are plotted in Fig. 6, for the gap widths of 0.5 cm and the same parameters as in Fig. 5. One can see that k_z is no more constant and there are no substantial forbidden gaps where the eigenfunctions do not exist. Noticeable is that over wide ranges on density the eigenmodes has very small $k_\perp$; these oscillations are characterized by small perpendicular fields while by strong E_z field in the gaps. In the regions where $k_\perp$ is increased and thus the perpendicular fields acquire substantial values, the dispersion of symmetric mode is similar to the third (symmetric) mode in Fig. 5. The asymmetric mode with finite $k_\perp$ is quite relative to the second (asymmetric) mode in Fig. 5 whereas the first mode has no the analogue.

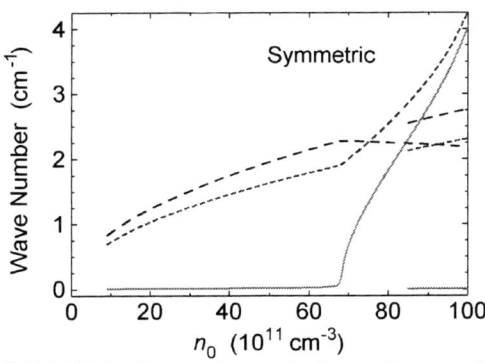 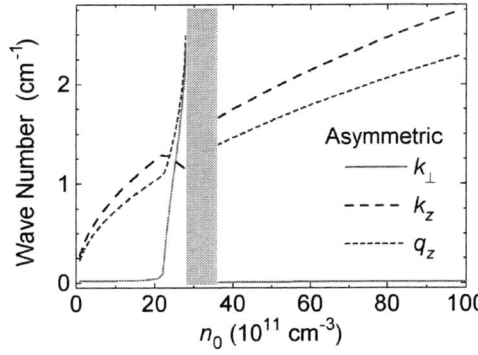

FIGURE 6. Dispersion of helicon eigenmodes at the gap width of 0.5 cm. $B_0 = 100$ G and $f = 50$ MHz.

Axial mode profiles are shown in Fig. 7 for $k_\perp = 0.92$ cm^{-1}, i.e. when $E_\theta \sim J_1(k_\perp r)$ is maximum at the boundary, $r = r_0$, and thus the maximum absorption is expected due to the mode conversion of the helicon wave into TG wave [4].

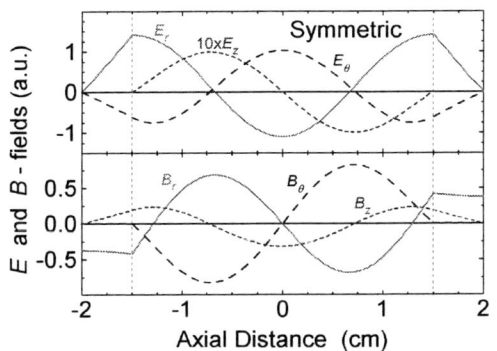 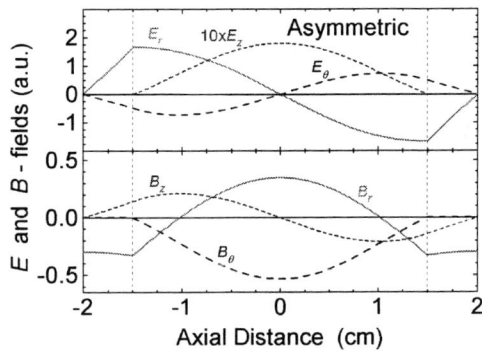

FIGURE 7. Axial profiles of eigenmodes at $n_0 = 7.2 \times 10^{12}$ cm^{-3} (left) and 2.4×10^{12} cm^{-3} (right).

CONCLUSIONS

In conclusion, we evaluated the operation window for a small helicon source where both the power efficiency and gas ionization can be high. A scenario for attaining a dense plasma mode seems to be as follows. It is advisable to ignite the discharge at higher frequency and lower magnetic field when the absorption peaks at lower densities. The further density rise can be provided by increasing the magnetic field and/or by decreasing the driving frequency.

ACKNOWLEDGMENTS

We would like to thank Prof. Y. Kawai for his continuous encouragement.

REFERENCES

1. Shamrai, K.P., Aleksandrov, A.F., Bougrov, G.E., Virko, V.F., Katiukha, V.P., Koh, S.K., Kralkina, E.A., Kirichenko, G.S., and Rukhadze, A.A., J. Physique (Coll. IV) 7, 365-381 (1997).
2. Chang Díaz, F.R., Trans. Fusion Technology 35, 87-93 (1999).
3. Ziemba, T., Winglee, R., Euripides, P., and Slough J., Bull. Am. Phys. Soc. 46, No. 8, 193-194 (2001).
4. Shinohara, S., and Shamrai, K.P., Plasma Phys. Control. Fusion 42, 865-880 (2000); Shamrai, K.P., and Shinohara, S., Phys. Plasmas 8, 4659-4674 (2001).
5. Chen, F.F., Phys. Fluids 22, 2346-2358 (1979).

Gyrokinetic Theory and Simulations of Alfvénic Instabilities in Dipole Plasmas

Sean Dettrick* and Liu Chen*

Department of Physics and Astronomy, University of California, Irvine, Irvine, California 92697-4575

Abstract. A one-dimensional linear hybrid gyrokinetic-magnetohydrodynamic δf PIC simulation code is developed to study the detailed mechanisms of energetic particle drift-bounce resonant destabilization of Alfvén modes in the ring-current region of the magnetosphere. The code is based on a reduced initial-value formulation of eigenmode equations which were solved perturbatively in a previous paper. The model plasma is composed of a cold (~ 100eV) component which provides inertia plus a tenuous energetic (~ 10keV) "ring-current" component which provides drift-bounce resonant destabilization of odd and even parity modes and gyrokinetic compressional stabilization of the fundamental even modes. Full kinetic effects such as finite Larmor radii and particle magnetic bounce and precessional drift motions are retained nonperturbatively. A simple finite β dipolar equilibrium model is assumed. Simulations show excellent agreement with earlier perturbative analyses. Results show that, when the energetic ion thermal velocity is super-Alfvénic, the ions destabilize shear Alfvén MHD modes via the drift-bounce resonances. The growth rates of the resulting modes scale linearly with plasma β. The most unstable of these modes are found to be drift-bounce resonance destabilized modes with odd parity, with wavenumbers such that the equatorial thermal value of $k_\perp \rho \approx 0.5$. The destabilization first occurs at a critical wavenumber which is usually $k_\phi \rho_0 \approx 0.3$. When the wavenumber is close to this critical value and the plasma β is close to the ideal MHD critical value, the mode frequency is determined by the energetic particle dynamics, similar to the Energetic Particle Modes (EPM's) that have been observed in tokamaks. When the energetic particles have Alfvénic or sub-Alfvénic thermal velocity, they contribute to damping of the MHD modes via the bounce resonance.

MODEL

Satellite observations [1,2] have found that compressional Pc 5 and radial Pc 4–5 (Pc 4: $\tau = 45 - 150s$, Pc 5: $\tau = 150 - 600s$) pulsations have: (1) large azimuthal wave numbers, $m \sim 100$ (2) westward phase velocity similar to the ion magnetic drift velocity, (3) frequencies similar to those of standing shear Alfvén waves, and (4) antisymmetric structures along the field lines (i.e. $\delta B_\parallel$, $\delta E_\perp$ are antismmetric, $\delta B_\perp$ is symmetric) [3,4]. Satellite observations also show [5] that there are two plasma components, consisting of a core plasma, with temperature $T_c \sim 100$ eV, and energetic ring-current protons with temperature $T_E \sim 10$ keV.

Assuming a separate core and energetic plasma component, as observed above, perturbative gyrokinetic analysis by Chen and Hasegawa [6,7] makes predictions in good agreement with these observations. In particular, the most unstable Alfvén instabilities excited by wave-particle interactions are predicted to: (1) have $k_\phi \rho \approx 0.5$, implying that $m \sim O(100)$ for 30 keV ring current protons, (2) have westward phase velocity, (comparable to the drift velocity of the energetic ions) (3) be destabilized by ring-current pressure inhomogeneities via the drift-bounce resonances, $\omega - \omega_d = K \omega_b$, of hot ($\sim 10^2$ keV) ions (ω_d, ω_b are the drift and bounce frequencies, the K are integers), and (4) have antisymmetric structure along the field line (i.e. $K = -1$). The purpose of the present work is to verify and extend the *perturbative* analysis by developing a *non-perturbative* initial value code. Related simulations have been independently performed by Belova [8,9].

Converting the eigenmode equations of reference 7 to an initial value formulation, the MHD stability of shear Alfvén waves is determined by a generalized vorticity equation:

$$B\frac{\partial}{\partial l}\left(\frac{k_\perp^2}{B}\frac{\partial}{\partial l}\right)\delta\psi - \frac{k_\perp^2}{V_A^2}\frac{\partial^2}{\partial t^2}\delta\psi + \frac{e}{m}\langle J_0^2 \omega_d \hat{\omega}_* F_0 \delta\psi\rangle_E = \frac{4\pi e}{c^2}\left\langle \omega_d J_{0i}\frac{\partial}{\partial t}\delta K\right\rangle_E, \qquad (1)$$

where l is the distance along the field line; $\delta\psi$ represents the shear Alfvén perturbation via $\delta A_\parallel \equiv -i(c/\omega)\mathbf{b}\cdot\nabla\delta\psi$; $\langle\cdots\rangle_E = \int d^3v$ over the energetic particle component; $k_\perp$ is the perpendicular wavenumber; V_A is the equatorial Alfvén speed; $\hat{\omega}_*$ and ω_d are the diamagnetic drift and precessional drift; F_0 is the energetic particle equilibrium distribution function; $J_0 = J_0(k_\perp \rho)$ is due to the gyroaveraging; and ρ is the energetic particle gyroradius. Terms in equation 1 represent the effects of field-line bending, inertia, interchange-ballooning drive, and energetic particle compression, respectively.

The nonadiabatic gyrokinetic response δK to the fluctuation $\delta \psi$ is described by the gyrokinetic equation:

$$\left(v_\parallel \frac{\partial}{\partial l} + \frac{\partial}{\partial t} + i\omega_d \right) i \frac{\partial \delta K}{\partial t} = i \frac{e}{m} J_0 \omega_d Q F_0 \delta \psi \qquad (2)$$

where

$$Q F_0 \delta \psi = i \frac{\partial \delta \psi}{\partial t} \frac{\partial F_0}{\partial E} + \delta \psi \mathbf{k}_\perp \times \mathbf{b} \cdot \nabla F_0$$

describes the spatial and velocity space inhomogeneities in F_0. We assume that $k_\perp \approx k_\phi$, where ϕ is the azimuthal coordinate. The validity of this assumption is supported by global theory [10]. We also assume the plasma is isotropic, i.e. $\partial F_0 / \partial \mu = 0$.

Coupling between these two equations is incorporated in the model using the Particle in Cell (PIC) method. The equations are linear and are presented in a δf formulation [11] for efficient statistics. A self-consistent finite-β model equilibrium is assumed [12] where the energetic particles have a Maxwellian distribution.

We employ the resulting simulation code to study instabilities of field lines at $L \sim 6.6$ (L is the dipole L-shell parameter), which is in the outer edge of the ring current. Here the spatial gradient is negative and therefore tends to be destabilizing. Since we choose a Maxwellian energetic particle distribution, the energy gradient is also negative, which tends to be stabilizing.

SIMULATION RESULTS

Assuming the energetic particles have a thermal velocity V_{thE} which is super-Alfvénic, i.e. $V_A/V_{thE} \ll 1$, bounce averaging shows that the approximate solution of equation 2 is [7]

$$\delta K \approx \frac{e}{m} Q F_0 \left[\overline{\omega_d \delta \psi / \omega} P(\overline{\omega}_d - \omega)^{-1} + i\pi \sum_{K=-\infty}^{\infty} \delta(\omega - \overline{\omega}_d - K\omega_b) c_{l_1}^l \overline{\delta S_1 c_{l_1}^l} \right], \qquad (3)$$

where ω_b is the bounce frequency; the overbar denotes bounce averaging; P is the principal value; the K are integers; $\delta S = \omega_d J_0 \delta \psi / \omega$; l_1 is the south end of the field line; $c_a^b = \cos I_a^b$; and $I_a^b = \int_a^b \frac{dl}{|v_\parallel|}(\omega - \omega_d)$. Thus, when $\omega - \overline{\omega}_d - K\omega_b = 0$, δK can be expected to be dominated by the delta functions in equation 3. This feature is confirmed in the simulations as shown in figure 1, where the kinetic responses of typical marker particle populations are shown. The simulation parameters are $\alpha = 7$, $\beta_{0v} = 0.6$, $\eta = 1$, and $k_\phi \rho_0 = 0.5$ and $k_\phi \rho_0 = 1$ for the left and right plots, respectively. The peaks in the plots correspond to the delta functions with $K = -1$ and $K = -2$ in equation 3 for, respectively, $k_\phi \rho_0 = 0.5$ and $k_\phi \rho_0 = 1$. The parities of the unstable wavefunctions are odd and even respectively for the $K = -1$ and $K = -2$ resonances, and have the same shape as the ideal MHD eigenfunctions.

The destabilization of such modes by finite $k_\phi \rho_0$ is shown by a typical set of results in figure 2. Here, $\beta_{0v} = 0.6$, $\alpha = 7$, $\eta = 1$, $L = 6.6$, $V_A/V_{thE} = 0.1$, which is stable in ideal MHD. The first panel shows the growth rate, the second panel the real frequency, and the third panel shows the value of $(\omega_r - \overline{\omega}_d)/\omega_b$ at which the gyrokinetic response is a maximum. When kinetic effects are included (finite $k_\phi \rho_0$) an odd-parity mode is driven unstable at $k_\phi \rho_0 > 0.3$. The gyrokinetic response peaks at $(\omega_r - \overline{\omega}_d)/\omega_b = -1$ for the $0.3 < k_\phi \rho_0 < 1$ modes, indicating a $K = -1$ drift-bounce resonance. The most unstable mode is at $k_\phi \rho_0 = 0.5$ and is odd, in agreement with the theoretical predictions of CH88. At $k_\phi \rho_0 > 1$ the dominant unstable mode has even parity, and is destabilized by the $K = -2$ drift-bounce resonance. The real frequency is a strong function of $k_\phi \rho_0$, transitioning between three ideal MHD limits shown in the figure: (i) the eigenfrequency for the first odd mode eigenmode, (ii) the eigenfrequency for the same mode, calculated assuming that large $k_\phi \rho_0$ makes the ballooning term negligible due to the J_0^2 factor in equation 1, and (iii) the eigenfrequency corresponding to the second even eigenmode, assuming the same large $k_\phi \rho_0$ effect.

The results also indicate (figure 2, right) that, close to the ideal MHD stability boundary, the real frequency is predominantly determined by the resonant energetic ions (as is the growth rate), suggesting the possibility of Energetic Particle Modes (EPM's) similar to those predicted [13] and observed [14] in tokamaks.

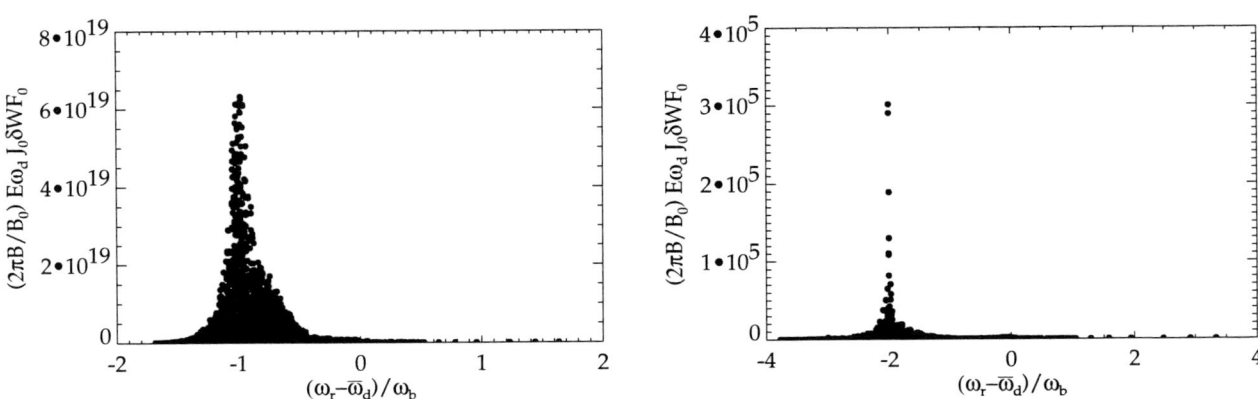

FIGURE 1. The contribution of each marker particle to $\langle \omega_d J_0 \delta G \rangle_h$. $K = -1$ and $K = -2$ wave-particle resonances are evident.

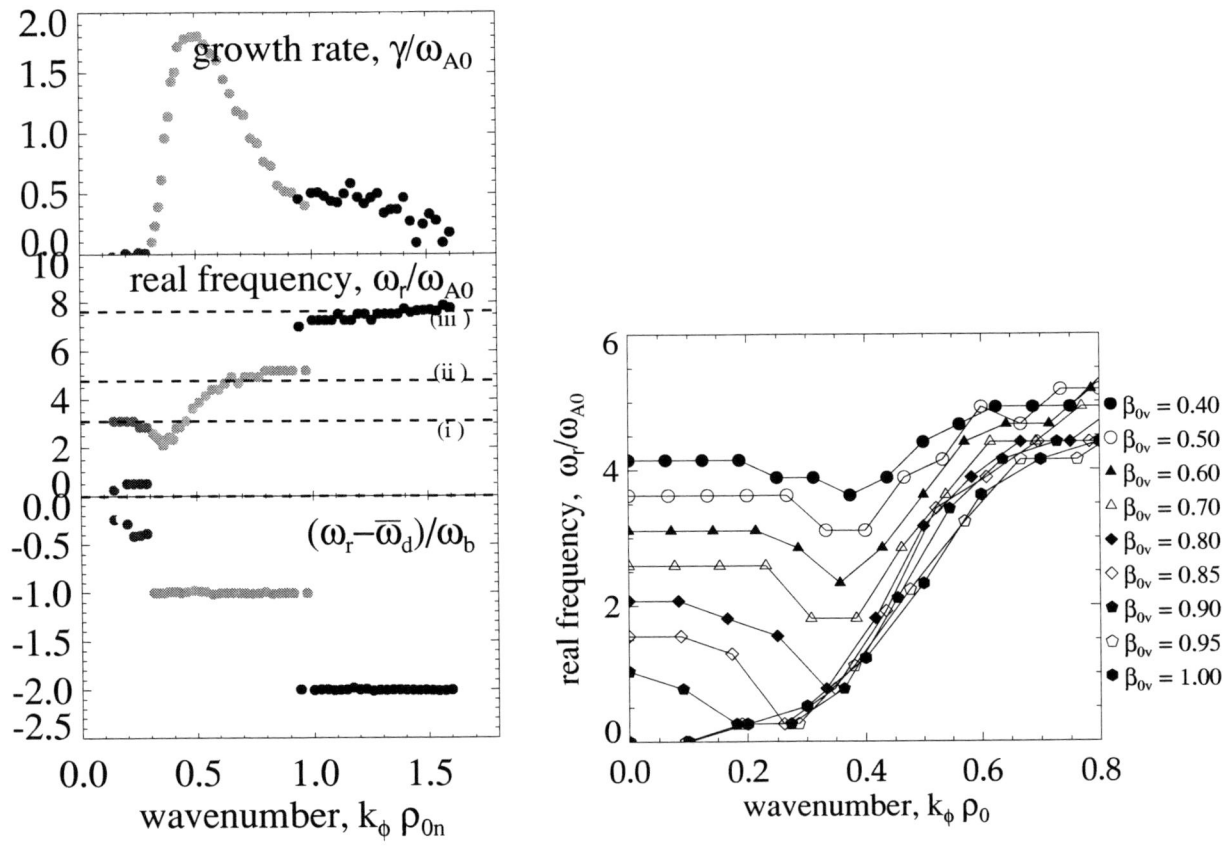

FIGURE 2. **Left:** Dependences of wave stability properties on the azimuthal wavenumber $k_\phi \rho_0$. **Right:** Frequency of Unstable Modes is sometimes determined by particle dynamics rather than by MHD.

The results up to this point have been obtained using an energetic particle distribution with super-Alfvénic thermal velocity ($V_A/V_{thE} = 0.1$), to which perturbative and bounce-averaged analysis of reference 7 is applicable. With the PIC code, we are able to relax this assumption and vary V_A/V_{thE}. The result is shown in figure 3. For $V_A/V_{thE} \ll 1$, the critical $k_\phi \rho_0 \approx 0.3$ and the most unstable $k_\phi \rho_0 \approx 0.5$ as before. However, as V_A/V_{thE} becomes $O(1)$, the critical and most unstable $k_\phi \rho_0$ increase until, at $V_A/V_{thE} \approx 1$, mode destabilization is no longer observed. This is due to the fact that, as V_A/V_{thE} increases to 1, fewer particles contribute to the destabilizing $K = -1$ resonance, and more particles

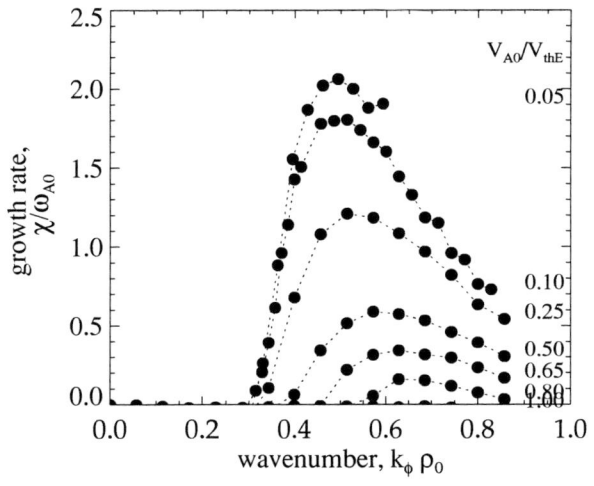 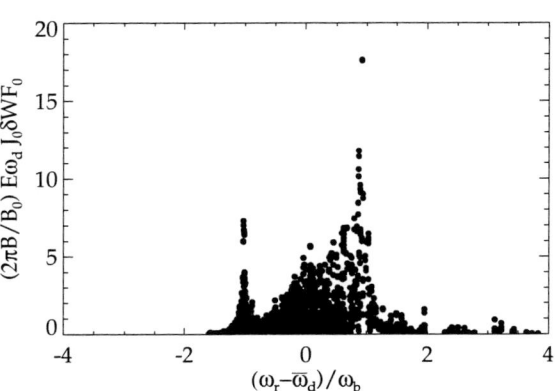

FIGURE 3. **Left:** Effect of V_A/V_{thE} ratio on instability. **Right:** When $V_A/V_{thE} = 0.8$ the $K = +1$ resonance can be observed.

contribute to the $K = +1$ *stabilizing* resonance, as indicated in figure 3 (right) (for $V_A/V_{thE} = 0.8$, $k_\phi \rho_0 = 0.57$). The diminishing drift-bounce destabilization is eventually dominated by this bounce-resonant stabilization, leading to a complete stabilization at around $V_A/V_{thE} = 1$. This suggests that if a warm (ie sub-Alfvénic and finite $k_\phi \rho_0$) ion population were to coexist with the energetic (super-Alfvénic) ion population, a competition between stabilizing $K > 0$ warm-particle resonances and destabilizing $K < 0$ energetic-particle resonances would ensue. This would likely produce a threshold value in the hot particle β for destabilization.

Finally, we observe that in all of our simulations, the fundamental even eigenmode (which would correspond to $K = 0$) has been stabilized by the trapped-particle gyrokinetic compression.

ACKNOWLEDGEMENTS

This work was supported by NSF and DoE grants. Computations were performed on the UC Irvine NACS Beowulf cluster and on Seaborg at NERSC.

REFERENCES

1. Takahashi, K., P.R. Higbie, and D.N. Baker, *J. Geophys. Res.*, **90**, 1473, 1985
2. Allan, W., E.M. Pulter, and E. Nielsen, *J. Geophys. Res.*, **87** 6163, 1982
3. Takahashi, K., and P.K. Higbie, *J. Geophys. Res.*, **91**, 11163, 1986
4. Takahashi, K., J.F. Fennell, E. Amata, and P.R. Higbie, *J. Geophys. Res.*, **92** 5857, 1987
5. Baumjohann, W., et. al., *J. Geophys. Res.*, **92**, 12,203, 1987
6. Chen, L. and A. Hasegawa, *J. Geophys. Res.*, **93** (A8) 8763-8767, 1988
7. Chen, L. and A. Hasegawa, *J. Geophys. Res.*, **96** (A2) 1503-1512, 1991
8. Belova, E.V., PhD Thesis, Dartmouth College, 1997
9. Belova, E.V., M.K. Hudson and R.E. Denton, *J. Comput. Phys.*, **136**, 324, 1997
10. Vetoulis, G. and L. Chen, *J. Geophys. Res.*, **101** (A7) 15441-15456, 1996
11. Parker, S.E. and W.W. Lee, *Phys. Fluids*, **B5**, 77, 1993
12. Chan, A.A., M. Xia and L. Chen, *J. Geophys. Res.*, **99** (A9) 17351-17366, 1994
13. Chen, L., *Phys. Plasmas*, **1**, 1519, 1994
14. Briguglio, S., et. al., 28th EPS Conference on Controlled Fusion and Plasma Physics, Portugal, 18 - 22 June 2001

Analogies of Rapidly Rotating Tokamaks and Accretion Disks

J.P. Goedbloed*, A.J.C. Beliën[†] and B. van der Holst**

FOM-Institute for Plasma Physics, Nieuwegein, the Netherlands
[†]*Shell International Exploration and Production, Rijswijk, the Netherlands*
**Centre for Plasma-Astrophysics, KU Leuven, Belgium*

Abstract. Equilibrium, waves, and instabilities of tokamaks and accretion disks that are rotating with arbitrary transonic velocities have been solved by means of advanced numerical and analytical techniques. The different transonic flow regimes yield a surprisingly large number of new MHD waves and instabilities that (1) are relevant for turbulent processes in accretion disks, (2) provide a clear correspondence between tokamaks and accretion disk dynamics, with different influence of rotation profiles, gravity, and magnetic pressure, (3) provide a new angle on rapid transition phenomena in transonic MHD flows of rotating astrophysical plasmas.

The new angle entails a complete revision of all previously obtained spectral results. The reason is that transonic flows upset the standard theoretical approach to plasma dynamics, consisting of a separate study of the equilibrium state and of the perturbations of this background. We will discuss a new approach to this dichotomy consisting of a study of the similarities of the nonlinear stationary flow patterns and the different linear wave structures that occur when the background speed traverses the full range of critical speeds (from 'slow magnetosonic' to 'Alfvén' to 'fast magnetosonic'). This has required the development of new computational tools that yield the mentioned plethora of new waves and instabilities.

INTRODUCTION

The study of waves and instabilities of magnetically confined plasmas in tokamaks is traditionally based on the assumption of *static* equilibrium. However, divertors, neutral beams, and other heating techniques create sizeable poloidal and toroidal plasma flows which completely upset the standard paradigm of a split in background equilibrium and linear perturbations. The reason is that poloidal rotation may cause transonic transitions from elliptic to hyperbolic flows. For hyperbolic stationary states a suitable stability formulation does not exist. Hence, for spectral studies, the best thing we can do at present is to stay in the elliptic flow regimes and to study the perturbations in the limit of approaching the hyperbolic regimes. This is what we do in the present paper. The nonlinear counterpart where the hyperbolic regions may be crossed will be studied later.

Once tokamaks with rapid rotations are considered, the study may be extended to include accretion disks rotating and gravitating about a compact object as well since the background stationary state is of equal complexity, requiring advanced analytical [1, 2] and numerical solution techniques [3, 4]. Thus, we may contribute to one of the outstanding problems in accretion disk dynamics, which is breaking the ideal MHD constraint of parallel poloidal flow and magnetic field needed to get magnetized jets emanating from the accretion flows. The necessary dissipation process is unknown but generally assumed to involve small-scale MHD waves and turbulence. The frequency spectrum and eigenfunction distributions of these waves and their stability properties are largely determined by the magnetic fields and the flows in the axisymmetric geometry of the disk. So far, no calculations have been performed where all these different effects (i.e. both toroidal and poloidal magnetic fields and flows in a genuine two-dimensional disk geometry in the presence of a strong gravitational field) are incorporated simultaneously. Yet, from tokamak experience with MHD waves and instabilities, it is crucial to represent the two-dimensional equilibrium and perturbations in a self-consistent manner.

We have solved the equations for stationary flows in tokamaks and accretion disks in the different transonic flow regimes by means of a new numerical program (called FINESSE) which solves the coupled system of flux and Bernoulli equations by means of an efficient and accurate iterative procedure [3]. This program produces equilibria for quite different physical systems (like stellar winds, tokamaks, accretion disks, solar loops), all with the requisite

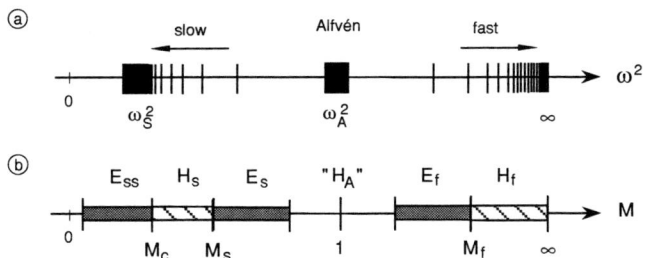

FIGURE 1. (a) Schematic spectrum of the three MHD waves for a *static* background equilibrium. For large wave numbers, the discrete eigenvalues accumulate at the continua $\{\omega_S^2\}$, $\{\omega_A^2\}$, and $\{\omega_F^2\} \equiv \infty$; (b) Flow regimes characterized by the value of the poloidal Alfvén Mach Number $M \equiv v_p/v_{A,p}$ of a *stationary* equilibrium flow. The flow turns from elliptic to hyperbolic at the boundaries of the hatched regions H_s and H_f, whereas the Alfvén region "H_A" collapses into the point $M \equiv 1$.

accuracy needed for wave and stability studies. The flexibility of this approach guarantees that experience obtained in one field (e.g. tokamak stability) can be transferred to another (e.g. accretion disk dynamics), and vice versa.

We have also derived the equations for the waves in this system of nested magnetic surfaces produced by the electric currents and mass flows and the central gravitating object. These equations are solved by means of a new spectral program (called PHOENIX) which yields the complete complex spectrum of ideal and resistive waves and overstable modes of both tokamaks and accretion disks with transonic flows [4]. We here present the method of solution and a glimpse of the large number of new modes and instabilities that occur.

TRANSONIC FLOWS

The difficulty of transonic plasma dynamics is that, somewhere in the middle of the region of interest, the speed of the plasma crosses a critical speed where the character of the flow suddenly changes dramatically. The implications of these transitions are manifold, but the theory on where and when this happens and on what precisely goes on, i.e. *the spatial resolution of the flows*, is poorly known. We expect significant progress in the understanding of these phenomena from the new mentioned spectral and nonlinear methods that we recently developed [5, 6, 7].

We have previously [8, 9] demonstrated that linear waves and non-linear stationary equilibria are not separate issues in magnetohydrodynamics. This is illustrated once more in Fig. 1, which shows the deep analogy between the waves and the stationary states. The short-wavelength limit determines the spectral structure (in ω^2) with three singular continuous spectra. For equilibria with flow, these continua are Doppler shifted and split into forward and backward ones so that the spectrum (in ω now) becomes asymmetric with respect to the value $\omega = 0$. Moreover, the problem is no longer self-adjoint so that *overstable modes* may occur. We have already investigated this new structure for toroidal flows, where we found new gaps in the continua and new toroidal flow-induced Alfvén eigenmodes (TFAE's) inside those gaps [7]. Here, we consider the much more involved spectral issues for poloidal flows.

Three elliptic flow regimes are available: (1) a *sub-slow* regime where the squared poloidal Alfvén Mach number $M^2 \equiv \rho v_p^2 / B_p^2 < M_c^2$, (2) a *slow* regime (sub-Alfvénic, but faster than the slow critical speed M_c) for $M_s^2 < M^2 < 1$, and a (3) *fast* regime (super-Alfvénic) where $1 < M^2 < M_f^2$. Here,

$$M_c^2 \equiv \frac{\gamma p}{\gamma p + B^2}, \qquad M_{f,s}^2 \equiv \frac{\gamma p + B^2}{2 B_p^2} \left[1 \pm \sqrt{1 - \frac{4\gamma p B_p^2}{(\gamma p + B^2)^2}} \right]. \tag{1}$$

One of the problems addressed in the numerics is the fact that these transition values are not known beforehand, but are to be determined together with the solutions. Hence, staying in the elliptic flow regimes is a delicate numerical problem. Moreover, even outside the hyperbolic flow regimes there is no guarantee that solutions will be found: As schematically indicated in Fig. 1, forbidden regimes exist, bordering the elliptic ones, where no solutions are found. E.g., the Alfvén value $M = 1$ may or may not be embedded in an allowed flow regime. All this depends on the specific flux functions chosen leading to implicit restrictions on the Bernoulli function by the physical values of the poloidal magnetic field.

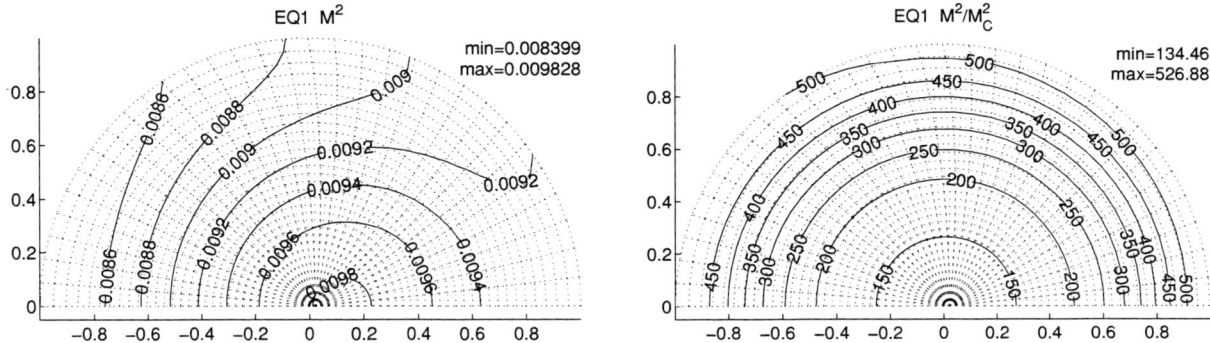

FIGURE 2. Top half of the poloidal cross-section of a rapidly rotating tokamak, in the slow elliptic flow regime, with the distribution of the squared poloidal Alfvén Mach number, M^2, and the ratio M^2/M_c^2; the inverse aspect ratio $\varepsilon = 0.05$.

NUMERICAL TOOLS

For the study of waves in plasmas with a background poloidal rotation, one first needs to compute stationary equilibrium states that are accurate enough to justify a spectral analysis. Typically, this involves accurate computation of currents, vorticities, and field line curvatures. The essential variables are the poloidal magnetic field $\mathbf{B}_p = R^{-1}\mathbf{e}_\varphi \times \nabla \psi$ and velocity $\mathbf{v}_p = (\rho R)^{-1}\mathbf{e}_\varphi \times \nabla \chi$, derived from streamfunctions $\psi(R,Z)$ and $\chi(\psi)$, and the square of the poloidal Alfvén Mach number, $M^2(R,Z)$. The poloidal flux ψ satisfies a nonlinear partial differential equation that may change from elliptic to hyperbolic, depending on the value of M^2, and an intricate algebraic equation for the determination of M^2. The remaining physics is described by five arbitrary flux functions $\Lambda_i(\psi)$. The basic problem is to determine, for a given choice of these flux functions, the poloidal distribution of $\psi(R,Z)$ and $M^2(R,Z)$. This problem has been solved and incorporated in the new finite element equilibrium solver FINESSE [3]. The only restriction of this code is that the poloidal flows should be in the elliptic regimes (i.e. either one of the three shaded areas of Fig. 1(b)).

A typical result for a tokamak equilibrium in the slow flow regime is shown in Fig. 2. The plots for M^2 and M^2/M_c^2 are superposed on the computational (ψ, ϑ) grid, where ϑ is constructed such that the equilibrium magnetic field lines are straight on each magnetic surface. Notice that M^2 is a strongly varying function of ϑ and that $M_c^2 \ll M^2 \ll 1$. Similar distributions are found for accretion disks when the gravitational field of the compact object is "switched on".

Next, one needs to solve for the ideal and resistive waves and instabilities of this system, described by ψ, M^2, and the five flux functions. The formidable analysis has been completed and implemented in a Galerkin scheme with finite elements for the normal and Fourier harmonics for the poloidal direction. This leads to a large non-symmetric eigenvalue problem. Presently, the parallel Jacobi-Davidson algorithm developed by Sleijpen and van der Vorst [10], permits us to zoom in onto a target eigenvalue and then produce all eigenvalues in a wide neighborhood of it with unprecedented accuracy! This has been incorporated in our new spectral code PHOENIX [4], which has produced the first spectra ever of poloidally rotating plasmas.

TRANSONIC INSTABILITIES

The first spectral results for localized continuum modes of a gravitating disk with both poloidal and toroidal magnetic fields and flows are shown in Fig. 3. The eigenvalues of the stable waves are located along the Im λ-axis, the two curves in the complex λ-plane correspond to forward and backward propagating instabilities driven by the poloidal flow and gravity. The spectrum is quite characteristic for flows in the second elliptic flow regime and instability will generally occur when the value of the poloidal Alfvén Mach number for the accretion flow has surpassed the critical value M_c. The instabilities are localized on magnetic ($\equiv$ flow) surfaces and occupy a large fraction of the outer part of the disk so that they may be considered as suitable candidates for anomalous dissipation by MHD turbulence in accretion disks. A supplementary investigation of a 1D model of accretion flows also exhibits a large number of non-local instabilities when the usual restrictions of high β and axisymmetry of the modes are dropped [11].

In conclusion: We have analyzed the waves and instabilities of tokamaks and thick accretion disks in the second elliptic flow regime and found that significant instabilities operate there that are absent in the first elliptic flow regime.

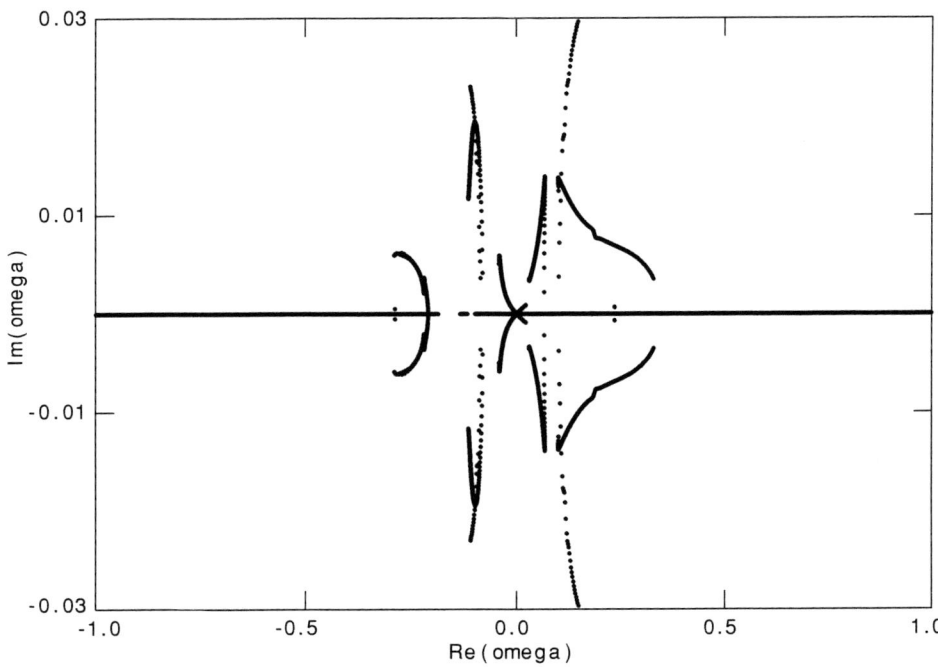

FIGURE 3. Continuous spectrum of waves (purely real ω) and overstable modes (with an additional imaginary part of ω) of a thick accretion disk; the continuous distribution of the eigenvalue parameter ω is shown in the complex ω-plane with the radial location $s \equiv \sqrt{\psi}$ as a parameter.

These instabilities should be ascribed to the transonic transition at $M = M_c^2$. Hence, there appears to be a strong correlation between singularities and discontinuities that occur in the background nonlinear stationary states when the critical values of the poloidal Alfvén Mach number are surpassed (i.e. in the hyperbolic flow regimes, which are inaccessible for spectral studies) and the instabilities that are found in the next elliptic flow regime.

ACKNOWLEDGEMENTS

The authors wish to thank Rony Keppens and Fabien Casse for stimulating discussions, and Mike Botchev for numerical support. This work was performed as part of the research program of the association agreement of Euratom and FOM with financial support from NWO, Euratom, and the EC Human Potential program under contract HPRN-CT-2000-00153. NCF is acknowledged for providing computer facilities.

REFERENCES

1. J.P. Goedbloed and A. Lifschitz, *Phys. Plasmas* **4**, 3544 (1997).
2. J.P. Goedbloed and A.E. Lifschitz, to be published (2002).
3. A.J.C. Beliën, M.A. Botchev, J.P. Goedbloed, B. van der Holst, R. Keppens, *J. Comp. Phys.*, submitted (2002).
4. B. van der Holst, A.J.C. Beliën, J.P. Goedbloed, to be published (2002).
5. G. Tóth, R. Keppens, and M.A. Botchev, *Astron. & Astrophys.* **332**, 1159 (1998).
6. R. Keppens and J.P. Goedbloed, *Ap. J.* **530**, 1036 (2000).
7. B. van der Holst, A.J.C. Beliën, and J.P. Goedbloed, *Phys. Rev. Lett.* **84**, 2865 (2000); *Phys. Plasmas* **7**, 4208 (2000).
8. J.P. Goedbloed, A.J.C. Beliën, B. van der Holst, and R. Keppens., Proc. 27th Eur. Conf. on *Controlled Fusion and Plasma Physics*, Budapest, OR.011 (2000).
9. J.P. Goedbloed, in *New Plasma Horizons*, International Topical Conference on Plasma Physics, 3–7 September 2001, Faro, Ed. Lennart Stenflo, *Physica Scripta* **T98**, 43–47 (2002).
10. G.L.G. Sleijpen and H.A. van der Vorst, *SIAM J. Matrix Anal. Appl.* **17**, 401 (1996).
11. R. Keppens, F. Casse, J.P. Goedbloed, *Astrophys. J.* **569**, L121–L126 (2002).

Kinetic Effects on the Parametric Decay of Circularly Polarized Electromagnetic Waves in a Relativistic Pair Plasma

V. Muñoz[*] and L. Gomberoff[*]

[*]*Departamento de Física, Facultad de Ciencias, Universidad de Chile*

Abstract. Parametric decays of a circularly polarized electromagnetic wave in an electron-positron plasma are studied. Relativistic effects on the particle motion in the wave field are included. The analysis is based on the Vlasov equation in order to account for kinetic effects. There are two types of decays: one in which the pump wave decays into two electromagnetic waves, and the other where the pump wave decays into electromagnetic waves and electroacoustic pseudomodes. These decays have been found in the weakly relativistic regime for both the fluid and the kinetic theory, even though the electroacoustic pseudomodes satisfy $\omega/kv_{\text{th}} \sim 1$, and therefore they are strongly Landau damped. In this work we calculate the dispersion relation for the parametric decays in the relativistic regime, and it is studied numerically assuming the equilibrium distribution function is a one-dimensional Maxwell-Jüttner distribution. Results are compared with the weakly relativistic case.

INTRODUCTION

Electron-positron plasmas are different from electron-ion plasmas, because in the absence of a mass difference, there are no high or low natural frequency scales. Understanding interactions between waves and relativistic electron-positron plasmas is relevant to proposed pulsar emission mechanisms [1, 2], and may give insight into structure formation in the early Universe [3].

Parametric instabilities of a large amplitude electromagnetic wave in electron-positron plasmas have been studied based on fluid theory [4, 5, 6, 7, 8]. Some of these treatments include weakly relativistic effects on the particle motion in the wave field. There are two types of decays: one in which the pump wave decays into two electromagnetic waves, and the other where the pump wave decays into electromagnetic waves and electroacoustic pseudomodes. Since the electroacoustic pseudomodes satisfy $\omega/k \simeq v_s$ (where ω is the frequency, k the wavenumber, and v_s is the electroacoustic velocity), these modes are expected to be strongly damped. Therefore, the study of parametric decays taking into account possible damping effects on the electro-acoustic pseudomodes, is an important problem. In this paper we consider the dispersion relation for parametric decays in an unmagnetized relativistic electron-positron plasma using the Vlasov equation [9]. We study it numerically assuming the equilibrium distribution function is a one-dimensional Maxwell-Jüttner distribution. The dispersion relation is solved for given ultra-relativistic temperature and velocity, and a decay instability is found. Results are compared with the weakly relativistic case.

KINETIC DISPERSION RELATION FOR A RELATIVISTIC PLASMA

The basic equations are Maxwell equations and the collisionless Vlasov equation,

$$0 = \frac{\partial f_j}{\partial t} + \vec{v} \cdot \vec{\nabla} f_j + q_j \left(\vec{E} + \frac{1}{c} \vec{v} \times \vec{B} \right) \cdot \vec{\nabla}_{\vec{p}} f_j , \qquad (1)$$

where $\vec{E} = -\vec{\nabla}\varphi$, $\vec{B} = \vec{\nabla} \times \vec{A}$ are the electric and magnetic fields, respectively, φ, $\vec{A}$ are the electrostatic and vector potentials, and $f_j(\vec{r},\vec{p},t)$ is the distribution function for species j ($j = e$ for electrons, and $j = p$ for positrons).

We assume that kinetic effects are only important along the longitudinal direction, so that we may approximate the distribution function in the transverse direction by the cold plasma expression,

$$f_j(z,\vec{p},t) = n_0 g_j(z,p_z,t)\delta\left(p_x + \frac{q_j A_x}{c}\right)\delta\left(p_y + \frac{q_j A_y}{c}\right),\qquad(2)$$

where g_j is the one-dimensional distribution function along the z direction, normalized to unity. n_0 represents the constant density of each species at equilibrium. This approximation is justified as long as energy transfer along the perpendicular direction can be neglected.

In Ref. [9], a perturbative analysis of the system is performed, assuming an equilibrium state consisting of a plasma composed by electrons and positrons, characterized by a common distribution function $g_{e0}(p_z) = g_{p0}(p_z) \equiv g_0(p_z)$, and a circularly polarized electromagnetic wave of frequency ω_0 and wavenumber k_0. This yields the pump wave dispersion relation

$$\omega_0^2 = c^2 k_0^2 + 2\omega_p^2 \int_{-\infty}^{\infty} dp_z \frac{g_0}{\gamma_0},\qquad(3)$$

where $\gamma_0^2 = 1 + (p_z/mc)^2 + \alpha^2$, $\alpha = eA/mc^2$, and the dispersion relation for the parametric decays:

$$(D_+ - \omega_p^2 \alpha^2 I_4)(D_- - \omega_p^2 \alpha^2 I_4) = -\omega_p^2 mc^2 \alpha^2 k I_3 (D_+ + D_- - 2\omega_p^2 \alpha^2 I_4),\qquad(4)$$

where

$$D_\pm = -(\omega_0 \pm \omega)^2 + c^2(k_0 \pm k)^2 + 2\omega_p^2 \int_{-\infty}^{\infty} dp_z \frac{g_0}{\gamma_0},\qquad(5)$$

$$I_3 = \int_{-\infty}^{\infty} dp_z \frac{1}{\gamma_0^2} \frac{\partial g_0/\partial p_z}{v_z k - \omega},\qquad(6)$$

$$I_4 = \int_{-\infty}^{\infty} dp_z \frac{g_0}{\gamma_0^3}.\qquad(7)$$

In the nonrelativistic limit, Eq. (4) reduces to the dispersion relation in [10].

NUMERICAL RESULTS

In Ref. [9], Eq. (4) was solved numerically in the weakly relativistic regime. In this case, g_0 can be taken to be a Maxwellian, and the dispersion relation can be written in terms of the plasma dispersion function Z.

In the relativistic case, the equilibrium distribution is the Maxwell-Boltzmann-Jüttner distribution,

$$g_0(p_z) = \frac{1}{2mcK_1(\mu)} e^{-\mu[1+(p_z/mc)^2]^{1/2}}, \quad \mu = \frac{mc^2}{k_B T},\qquad(8)$$

where $K_1(\mu)$ is the modified Bessel function of the second kind. Integrals in (5) and (7) can be calculated via standard numerical algorithms (we used Gaussian quadratures), but Eq. (6) is more subtle since its integrand has both branching points and poles, and I_3 is not an analytical function of ω.

Since the pole in I_3 is evident as a function of v_z, but the integral is done in momentum p_z, we first write I_3 as an integral in velocity space:

$$I_3 = \frac{1}{2mcK_1(\mu)} \frac{1}{ck} \frac{1}{1+\alpha^2} \int_{-1}^{1} du \frac{1}{u - \omega/ck} \left(2u + \frac{1-u^2}{u-\omega/ck}\right) e^{-\mu\alpha\left(\frac{1/\alpha^2+u^2}{1-u^2}\right)^{1/2}}.\qquad(9)$$

First, we note that the integrand has four branching points, at $\pm 1, \pm i/\alpha$, and we choose four branch cuts, parallel to the imaginary axis, as shown in Fig. 1. In this way, the integrand in (9) is single valued along the path of integration.

With this choice of branch cuts, I_3 is a discontinuous function of ω/ck at the real axis. In order to continue it analytically, we deform the integration path in velocity space as is done for the plasma dispersion function (see, e.g., [11]), so that the continuous version of I_3 is:

$$I_3^c = I_3 + \delta i\pi \mathrm{Res}_{u=\omega/ck},\qquad(10)$$

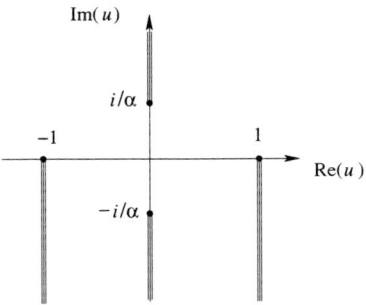

FIGURE 1. Branch cuts chosen in order to define a Riemann sheet for the integrand in (9).

where $\delta = 1$ if $\text{Im}\,\omega = 0$, $\delta = 2$ if $\text{Im}\,\omega < 0$, and

$$\text{Res}_{u=\omega/ck} = -\frac{1}{ck}\frac{1}{2mcK_1(\mu)}\frac{1}{1+\alpha^2}\mu\alpha\frac{1}{(1/\alpha^2+u^2)^{1/2}}\frac{u(1+1/\alpha^2)}{(1-u^2)^{1/2}}e^{-\mu\alpha\left(\frac{1/\alpha^2+u^2}{1-u^2}\right)^{1/2}}\bigg|_{u=\omega/ck} \quad (11)$$

is the residue of the integrand in I_3. To evaluate this residue we use the same branch cuts chosen in Fig. 1.

In order to study the dispersion relation we set $\mu = 0.1$, $\alpha = 10$ (ultra-relativistic temperatures and transverse velocities), and choose a value of $x_0 = \omega_0/\omega_p = 1.5$.

Integration of I_3^c is performed in the interval $[-1+\varepsilon, 1-\varepsilon]$, with $\varepsilon = 10^{-4}$, to avoid the branch points, and is evaluated using 39-point Gaussian quadrature. For the parameters considered, this algorithm fails to give an accurate answer if $|\text{Im}\,x| = |\text{Im}\,\omega/\omega_p| < 0.1$ due to the highly oscillatory nature of the integrand in this case. By inspection we found this could be fixed by subdividing the interval in three, $[-1+\varepsilon, \text{Re}\,x - 10\cdot\text{Im}\,x]$, $[\text{Re}\,x - 10\cdot\text{Im}\,x, \text{Re}\,x + 10\cdot\text{Im}\,x]$, $[\text{Re}\,x + 10\cdot\text{Im}\,x, 1-\varepsilon]$, and performing a 39-point Gaussian quadrature in each subinterval. This yields satisfactory results for a preliminary analysis of Eq. (4), and numerical instability near the real axis is significantly reduced.

For $ck/\omega_p = 1$, Eq. (4) yields Fig. 2. Here, the roots of the real part (labeled "1") and imaginary part (labeled "2") of the dispersion relation are plotted in the same graph, in a certain zone of the $\text{Re}\,\omega - \text{Im}\,\omega$ plane. Roots of the dispersion relation correspond to the crossings between both lines. A root is found at $\omega/\omega_p \simeq 0.87 + 0.12i$. Since $\text{Im}\,\omega > 0$, this corresponds to an unstable solution.

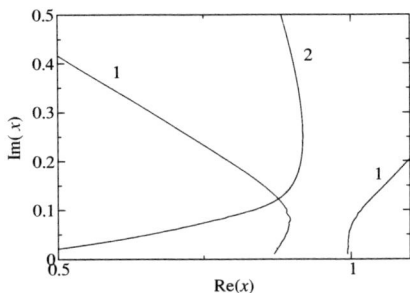

FIGURE 2. Roots of the dispersion relation Eq. (4), for $x_0 = \omega_0/\omega_p = 1.5$, $\mu = 0.1$, $\alpha = 10$, $y = ck/\omega_p = 1$. Curves 1: Roots of the real part of the dispersion relation. Curve 2: Roots of the imaginary part.

Following this root for higher values of ck/ω_p we obtain the curves in Fig. 3. In Fig. 3(a), $\text{Im}\,\omega$ vs. k is plotted. An instability is evident. In Fig. 3(b) k vs. $\text{Re}\,\omega$ is plotted, so that the unstable branch of the dispersion relation can be identified. For reference, the electromagnetic branches of the dispersion relation, corresponding to $D_\pm = 0$ for $\alpha = 0$, have also been plotted. It is found that the unstable branch corresponds approximately to the $D_- = 0$ branch. No exact matching is expected, since branches are strongly deformed for $\alpha > 1$. These results are consistent with previous findings for the non-relativistic and weakly relativistic regimes [10, 9], that a resonant decay instability is present in an electron-positron plasma, involving the coupling of a circularly polarized pump wave of frequency ω_0 with an electro-acoustic pseudomode of frequency ω and an electromagnetic wave of frequency $\omega_0 - \omega$. In this work we have found evidence that this decay instability is also present for ultra-relativistic velocities and temperatures.

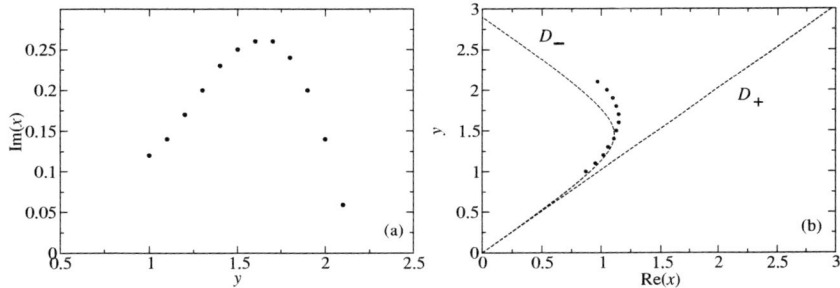

FIGURE 3. (a) Growth rate and (b) unstable branch for the decay instability. Parameters are the same as in Fig. 2.

SUMMARY

We have studied the parametric decays of a circularly polarized electromagnetic wave in an electron-positron plasma using kinetic theory. Fully relativistic effects have been considered, and kinetic effects have been assumed to be relevant only along the propagation direction of the pump wave.

For ultra-relativistic temperature and velocity, an instability is found. The unstable branch is found to correspond approximately to a forward propagating normal mode of frequency $\omega_0 - \omega$, given by $D_- = 0$ in the zero amplitude pump wave limit. Exact matching is not expected, because branches of the dispersion relation are strongly deformed for $\alpha > 1$, but these results are consistent with previous ones obtained both for a fluid and a kinetic theory in the weakly relativistic limits, namely, that a resonant decay instability develops for $\alpha \neq 0$, involving an electroacoustic pseudomode of frequency ω and an electromagnetic wave of frequency $\omega_0 - \omega$. In this work we have found that the decay instability is also present in the relativistic regime.

In the weakly relativistic regime, two modulational instabilities are also found in the system. Future work will involve study of these instabilities, and the analysis of the behaviour of all instabilities under various values of temperature and transverse particle velocity. In particular, considering $\mu \gg 1$, $\alpha \ll 1$ would allow a closer comparison with the fluid and kinetic results previously obtained. In this case, our numeric algorithms must me modified because the integrand in Eq. (9) becomes highly oscillatory in all the interval of integration. The algorithms must also be improved in order to better study the region $\text{Im}\,\omega/\omega_p \ll 1$.

ACKNOWLEDGMENTS

This work has been partially supported by Fondecyt, grant Nos. 1020558 and 1020152.

REFERENCES

1. Luo, Q., and Melrose, D. B., *Mon. Not. R. Astron. Soc.*, **258**, 616–620 (1992).
2. Mahajan, S. M., *Astrophys. J. Lett.*, **479**, L129–L132 (1997).
3. Berezhiani, V. I., and Mahajan, S. M., *Phys. Rev. Lett.*, **73**, 1110–1113 (1994).
4. Gangadhara, R. T., Krishan, V., and Shukla, P. K., *Mon. Not. R. Astron. Soc.*, **262**, 151–163 (1993).
5. Shukla, P. K., Rao, N. N., Yu, M. Y., and Tsintsadze, N. L., *Phys. Rep.*, **138**, 1–149 (1986).
6. Gomberoff, L., Muñoz, V., and Galvão, R. M. O., *Phys. Rev. E*, **56**, 4581–4590 (1997).
7. Muñoz, V., and Gomberoff, L., *Phys. Plasmas*, **5**, 3171–3179 (1998).
8. Muñoz, V., and Gomberoff, L., *Phys. Rev. E*, **57**, 994–1004 (1998).
9. Muñoz, V., and Gomberoff, L., *Phys. Plasmas*, **9**, 2534–2540 (2002).
10. Shukla, P. K., and Stenflo, L., *Phys. Plasmas*, **7**, 2726–2730 (2000).
11. Goldston, R. J., and Rutherford, P. H., *Introduction to Plasma Physics*, Institute of Physics, Bristol, 1995.

Generation of Zonal Flows by Interchange Mode Turbulences

Zh.N.Andrushchenko[1], V.P.Pavlenko[2], K.Schoepf[3], S.Kuhn[3]

[1]*Institute for Nuclear Research, Kiev, Ukraine*
[2]*Department of Astronomy and Space Physics, Uppsala University, SE – 751 20, Uppsala, Sweden (EURATOM-VR Fusion Association)*
[3]*Institute for Theoretical Physics, University of Innsbruck, Innsbruck, Austria (Association EURATOM-OEAW)*

Abstract. The theory of zonal flow generation is extended by the interchange mode turbulence. A multiple scale expansion is employed to describe the dynamics of a large-scale plasma flow that varies on a longer time scale compared to the small-scale fluctuations. Averaging the model equations over small scales yields the evolution equations for mean flow generation with sources consisting of the standard and diamagnetic Reynolds stresses. Analysis of these equations shows that diamagnetic effects may significantly increase the total Reynolds force. The rate at which a zonal flow is generated by an interchange mode turbulence is calculated and the physics of the acceleration mechanism is elucidated.

INTRODUCTION

Zonal flows are poloidally and toroidally symmetric ($q_\theta=q_z=0$), zero-frequency vortex modes with a finite radial scale significantly larger than the scale of the underlying small-scale turbulence, i.e. $q_r<<k_r$, where **q** represents the wave vector for the large-scale perturbations and **k** is the wave vector of the background small-scale turbulence [1]. Lacking a radial component, zonal flows do not cause cross-field transport. Their importance lies in their ability to limit the radial size of eddies through a shear de-correlation mechanism [2] and thus to effectively regulate turbulent transport. Zonal flow structures have been experimentally observed in tokamaks, pinches and stellarartors, e.g. in TEXT [3], ISTTOK, RFX [4], TJ-IU [5]. Indirect experimental observations in DIII-D [6] were interpreted as a clear signature of zonal flows.

When zonal flows are excited, they form an active environment for the parent waves. Therefore the coupled dynamics of the large-scale flow and the small-scale turbulence should be considered. Coupling of small-scale fluctuations to the mean flow is described by the kinetic equation for wave packets, and corresponds to the conservation of an action-like invariant of the wave turbulence with slowly varying parameters due to the mean sheared flow. In our study here we extend the theory of zonal flow generation by turbulent Reynolds stress to the interchange mode turbulence.

MODEL AND BASIC EQUATIONS

The interchange instability arises from an unfavorable curvature of the magnetic field with an antiparallel density gradient. The mechanism of this instability is the interchange of flux tubes to tap free energy. We consider low frequency interchange perturbations ($\omega/\Omega_i<<1$ with $\Omega_i=eB_0/(m_ic)$ and ω being the frequency of the mode) in a weakly inhomogeneous magnetized plasma, which are uniform in the direction of the external magnetic field. The nonlinear evolution of the interchange modes is described by the continuity equation for electrons and ions, assuming quasi-neutrality. To simplify the problem, we use a local slab coordinate system with the x-direction corresponding to the radial and the y-direction being analogous to the poloidal direction in toroidal geometry. Suppose that such a dense plasma ($4\pi n_0 m_i c^2/B_0^2>1$), having the finite temperature $T_i=T_e=T$, is immersed in a constant magnetic field $\mathbf{B}_0=(0,0,B_0)$. The curvature of magnetic field lines is imitated by the fictional gravity $\mathbf{g}=(g,0,0)$. Further, since we treat a low-β plasma, we refer to the electrostatic approximation, $\mathbf{E}=-\nabla\Phi$. The plasma density is expressed as $N(\mathbf{r},t)=n_0(x)+\delta n(\mathbf{r},t)$ with $\mathbf{r}\equiv(x,y)$, where the equilibrium density $n_0(x)$ is assumed to vary slowly with x over a characteristic scale length. Then, to the first order in the parameter ω/Ω_i, we can reduce the basic equations in this study to a pair of coupled nonlinear equations for the dimensionless density $n=\delta n/n_0$ and the potential Φ [7],

$$\frac{\partial n}{\partial t}+\frac{\kappa c}{B_0}\frac{\partial \Phi}{\partial y}=-\frac{c}{B_0}\{\Phi,n\}, \qquad (1)$$

$$\left(\frac{\partial}{\partial t}-v_*\frac{\partial}{\partial y}\right)\nabla_\perp^2\Phi+\frac{gB_0}{c}\frac{\partial n}{\partial y}=-\frac{c}{B_0}\left[\{\Phi,\nabla_\perp^2\Phi\}+\frac{T}{e}div\{n,\nabla_\perp\Phi\}\right]. \qquad (2)$$

Here is $v_*=\kappa cT/eB_0$ the diamagnetic ion drift velocity, $\kappa=-d\ln n_0/dx>0$, $g=T/(m_iR)$ and R the characteristic scale length of magnetic field inhomogeneities. The { }-brackets denote Poisson brackets, $\{a,b\}=\mathbf{z}\cdot[\nabla a\times\nabla b]$, and $\nabla=\nabla_\perp$. The finite Larmor radius effect is incorporated to the lowest order by the term proportional to T. Further we neglected the inertia of electrons, the effect of a temperature gradient and the collisional viscosity of the stress tensor.

Linearizing Eqs.(1)-(2) for small perturbations $(n,\Phi)\sim\exp(-i\omega t+i\mathbf{kr})$, we obtain the dispersion relation with the wave eigenfrequency

$$\omega_k=-\frac{k_y v_*}{2}\pm\frac{k_y v_*}{2}\left(1-\frac{4L_n}{R}\frac{1}{k^2\rho_i^2}\right)^{1/2}, \qquad (3)$$

where $k^2=k_x^2+k_y^2$, $k>>\kappa$, $\rho_i=V_{Ti}/\Omega_i$. For $T=0$ Eq.(3) describes interchange modes with a growth rate $\gamma_k=k_y/k(\kappa g)^{1/2}$. Accounting for the finite Larmor radius effect leads to the stabilization of interchange instabilities for $k^2\rho_i^2\geq 4L_n/R$, where $L_n=\kappa^{-1}$ is the characteristic scale length of the plasma.

COUPLED DYNAMICS OF INTERCHANGE MODES AND ZONAL FLOWS

We start with a scale separation in order to distinguish between the small-scale fluctuations of interchange modes and the long-scale shear flow pattern generated in the final state. The electrostatic potential Φ is expressed as the sum of a large-scale flow quantity $\overline{\Phi}$ and a small-scale turbulent part $\tilde{\Phi}$. The mean potential $\overline{\Phi}$ is the average of the total potential over the fast small-scale variables. A similar representation has been chosen for the plasma density: $n=\overline{n}+\tilde{n}$. Note that it is the total electrostatic potential Φ and the total density n that enter Eqs.(1) and (2).

Averaging Eqs.(1) and (2) over small scales yields the evolution equations for the large-scale structures:

$$\frac{\partial \overline{n}}{\partial t}+\frac{\kappa c}{B_0}\frac{\partial \overline{\Phi}}{\partial y}=-\frac{c}{B_0}\overline{\{\tilde{\Phi},\tilde{n}\}}, \qquad (4)$$

$$\left(\frac{\partial}{\partial t}-v_*\frac{\partial}{\partial y}\right)\nabla_\perp^2\overline{\Phi}+\frac{gB_0}{c}\frac{\partial \overline{n}}{\partial y}=-\frac{c}{B_0}\overline{\{\tilde{\Phi},\nabla_\perp^2\tilde{\Phi}\}}-\frac{cT}{eB_0}\left(\overline{\{\tilde{n},\nabla_\perp^2\tilde{\Phi}\}}+\overline{\{\nabla_\perp\tilde{n},\nabla_\perp\tilde{\Phi}\}}\right). \qquad (5)$$

The first term on the RHS of Eq.(5) contains the standard Reynolds force and the second one the diamagnetic Reynolds force. Thus, a small-scale turbulence can drive, via Reynolds stresses, large-scale structures. These flow structures, in turn, modulate the turbulence dynamics. When large-scale flows are excited, they form an environment for the parent waves. Assuming that the like-scale interactions are small compared to the interaction between disparate scales, we derive from Eqs.(1),(2) the following set of equations for the small-scale fluctuations:

$$\frac{\partial \tilde{n}}{\partial t}+\frac{\kappa c}{B_0}\frac{\partial \tilde{\Phi}}{\partial y}=-\frac{c}{B_0}\{\overline{\Phi},\tilde{n}\}-\frac{c}{B_0}\{\tilde{\Phi},\overline{n}\}, \qquad (6)$$

$$\left(\frac{\partial}{\partial t}-v_*\frac{\partial}{\partial y}\right)\nabla_\perp^2\tilde{\Phi}+\frac{gB_0}{c}\frac{\partial \tilde{n}}{\partial y}=-\frac{c}{B_0}[\{\overline{\Phi},\nabla_\perp^2\tilde{\Phi}\}+\{\tilde{\Phi},\nabla_\perp^2\overline{\Phi}\}]-\frac{cT}{eB_0}[div\{\overline{n},\nabla_\perp\tilde{\Phi}\}+div\{\tilde{n},\nabla_\perp\overline{\Phi}\}] \qquad (7)$$

For zonal flows ($q_x\neq 0$), the evolution equations, Eqs. (4) and (5), are reduced to

$$\frac{\partial}{\partial t}\overline{n}=0, \qquad (8)$$

$$\frac{\partial}{\partial t}\overline{V}_y = \frac{c^2}{B_0^2}\frac{\partial}{\partial x}\sum_{\mathbf{k}} k_x k_y \left(1 - \frac{R}{2L_n}k^2\rho_i^2\right)\left|\widetilde{\Phi}_{\mathbf{k}}\right|^2. \tag{9}$$

As apparent from Eq.(8), the mean plasma density associated with zonal flows does not evolve with time, i.e. it can be considered as a constant. The first term in the curly bracket on the RHS of Eq.(9) corresponds to the standard Reynolds stress and the second one to the diamagnetic Reynolds stress. Analysis of Eq.(9) shows that the diamagnetic effects significantly modify the total Reynolds stress. Indeed, if the maximum of this distribution is located in the long-wave-length limit ($k^2\rho_i^2 < 2L_n/R$), the diamagnetic Reynolds stress is not essential for flow generation. However, if this maximum is shifted towards shorter wave lengths, then the diamagnetic contribution significantly modifies the large-scale flow dynamics. It is seen that the diamagnetic effects can also suppress the generation of large-scale structures, when the standard and diamagnetic Reynolds stresses balance one another.

The propagation of interchange modes in such weakly inhomogeneous media with slowly varying parameters can be conveniently described with the help of a wave kinetic equation for the wave action density. The sources of these slow spatial and temporal variations are flow induced velocity and density perturbations. To find appropriate canonical variables, we introduce a generalized wave action N_k via a useful combination of field perturbations [8]: instead of the potential $\widetilde{\Phi}_k$ and of the plasma density $\widetilde{n}_k$, we consider the normal variables $\Psi_k = \widetilde{n}_k + \alpha_k \widetilde{\Phi}_k$, where α_k is a coefficient to be determined. After straightforward calculations from Eqs.(6) and (7) we find

$$N_k = \left|\Psi_k\right|^2 = k^2\rho_i^2 \frac{4R}{L_n}\left(1 - k^2\rho_i^2 \frac{R}{4L_n}\right)\left|\frac{e\widetilde{\Phi}_k}{T}\right|^2. \tag{10}$$

Coupling of small-scale fluctuations to the large-scale flow is described by the kinetic equation for wave packets,

$$\frac{\partial N_k}{\partial t} + \frac{\partial \omega_k^{NL}}{\partial \mathbf{k}}\frac{\partial N_k}{\partial \mathbf{r}} - \frac{\partial \omega_k^{NL}}{\partial \mathbf{r}}\frac{\partial N_k}{\partial \mathbf{k}} = S. \tag{11}$$

The "collisional" term S accounts for the wave growth and damping due to linear and nonlinear mechanisms, as well as local (i.e. non-action-conserving) wave interactions. Symbolically, it can be represented as

$$S(N_k) = 2\gamma_k N_k - St\{N_k\}, \tag{12}$$

where γ_k is the linear growth rate and $St\{N_k\}$ is the damping term due to nonlinear broadening effects. Below we assume that the equilibrium small-scale turbulence is close to a stationary state, so that $S \to 0$. The local frequency of interchange modes entering this equation is modified in the presence of flows because of the Doppler shift induced by $\mathbf{E} \times \mathbf{B}$ flows and the lowest-order finite Larmor radius corrections due to finite ion temperature,

$$\omega_k^{NL} = \omega_k^{Re} + \mathbf{k}\mathbf{V}_0, \quad \mathbf{V}_0 = -\frac{c}{2B_0}\left[\nabla\overline{\Phi} \times \mathbf{z}\right] - \frac{cT}{2eB_0}\left[\nabla\overline{n} \times \mathbf{z}\right]. \tag{13}$$

Only the real part of the frequency is presumed for ω_k here. It is seen that the small-scale turbulence will be sheared by both the flow shear and the diamagnetic effect. Equation (11) generalizes the wave kinetic equation for the case of unstable interchange modes in the presence of a mean plasma flow. Equations (8), (9) and (11) are the final set of equations for the coupled zonal flow–interchange turbulence problem.

Now we assume the wave spectrum to consist of an equilibrium part N_k^0, averaged over the short spatial scales but evolving in time and $\mathbf{k}$, and of a perturbed part $\widetilde{N}_k$, i.e. $N_k = N_k^0 + \widetilde{N}_k$. Considering small perturbations, $(\widetilde{N}_k, \overline{\Phi}) \sim \exp(-i\Omega T + i\mathbf{q}\mathbf{r})$, and linearizing Eqs.(8),(9) and (11), we obtain the set of equations

$$\frac{\partial N_k^0}{\partial t} - \frac{\partial}{\partial \mathbf{k}}D_{\mathbf{k}}\frac{\partial N_k^0}{\partial \mathbf{k}} = 0, \tag{14}$$

$$\frac{\partial \widetilde{N}_k}{\partial t} + \mathbf{V}_g \frac{\partial \widetilde{N}_k}{\partial \mathbf{r}} - \frac{\partial}{\partial \mathbf{r}}(\mathbf{k}\mathbf{V}_0)\frac{\partial N_k^0}{\partial \mathbf{k}} = 0, \tag{15}$$

where the $\mathbf{k}$-space diffusion coefficient for the zonal flows is calculated as a function of their intensity,

$$D_{k_x} = \frac{c^2}{4B_0^2}k_y^2 q_x^4 \left(\overline{\Phi}^2 + \frac{T^2}{4e^2}\overline{n}^2\right)R(\Omega, q_x). \tag{16}$$

The quantity $R(\Omega, q_x)$, entering above, is the response function,

$$R(\Omega, q_x) = \frac{i}{\Omega - q_x V_{gx} + i\Delta\omega_k}, \qquad (17)$$

with V_{gx} being the x-component of the interchange mode group velocity and $\Delta\omega_k$ denoting the total de-correlation frequency which may also include the linear growth rate and a nonlinear frequency shift. In the weakly nonlinear regime $R(\Omega,q_x) \to \pi\delta(\Omega - q_x V_{gx})$, one obtains $R(\Omega,q_x) \to 1/\Delta\omega_k$ for a wide spectrum of fluctuations. Note that the validity of the assumption of random shearing and, therefore, of the applicability of quasilinear theory requires the overlap of resonances between the wave group velocity (V_{gx}) and the flow phase velocity (Ω/q_x).

The growth rate for zonal flows may be straightforwardly calculated via linearization of Eq.(15) to yield

$$\gamma = -\frac{c^2 T^2}{8e^2 B_0^2} \int d\mathbf{k} \frac{k_x k_y^2 q_x^2}{k^2 \rho_i^2} \frac{\partial N_k^0}{\partial k_x} R(\Omega, q_x). \qquad (18)$$

Note that the condition for growth, $\gamma > 0$, equivalent to

$$k_x \frac{\partial N_k^0}{\partial k_x} < 0, \qquad (19)$$

is typically satisfied in interchange mode turbulences and therefore, population inversion is not required. Physically, this instability is a manifestation of an inverse cascade and shows that energy is pumped into longer scales.

CONCLUSIONS

The mechanism for the generation of a mean flow by the interchange mode turbulence could be elucidated. It was shown that a disbalance between standard and diamagnetic Reynolds stresses is required for flow generation. Diamagnetic effects (finite density fluctuations) may significantly modify the total Reynolds force when the maximum of the energy spectrum of the turbulence is shifted towards short wavelengths, and can even suppress the formation of large-scale structures.

On the other hand, a large-scale flow, in turn, modulates and regulates the turbulence dynamics. The wave kinetic equation has been formulated and an appropriate adiabatic invariant for the small-scale turbulence in the presence of a mean flow has been determined. Further, the presence of large-scale structures was found to shear the turbulence. Both the **k**-space diffusion coefficients for zonal flows and the rates at which zonal flows are generated by the interchange mode turbulence have been calculated.

ACKNOWLEDGMENTS

This work was supported by the European Communities under an association contract between EURATOM and the Swedish Research Council (VR), Grant No F 5102-8092001. Further it has been partially carried out within the Association EURATOM-OEAW projects P4 and P1. The content of the publication is the sole responsibility of its authors and does not necessarily represent the views of the European Commission or its services.

REFERENCES

1. Hasegawa, A. and Wakatani, M., Phys. Rev. Lett. 59, 1581-1584 (1987).
2. Biglari, H., Diamond, P.H., and Terry, P.W., Phys. Fluids B2, 1-4 (1990).
3. Ritz, Ch.P., Lin, H., Rhodes, T.L. and Wootton, A.J., Phys. Rev. Lett. 65, 2543-2546 (1990).
4. Antoni, V., Desideri, D., Martines, E., Serianni, G. and Tramontin, L., Phys. Rev. Lett. 79, 4814-4817 (1997).
5. Hidalgo, C., Silva, C., Pedrosa, M.A., Sánchez, E., Fernandes, H. and Varandas, C.A.F., Phys. Rev. Lett. 83, 2203-2205 (1999).
6. Coda, S., Porkolab, M. and Burrell, K.H., Phys. Rev. Lett. 86, 4835-4838 (2001).
7. Kodama, Y., Pavlenko, V.P., Phys. Rev. Lett. 60, 1506-1509 (1988).
8. Smolyakov, A.I., Diamond, P.H., Medvedev, M.V., Phys. Plasmas 7, 3987-3992 (2000).

A Model of Self-Consistent L-H Transition Based on Finite-β Drift Waves

Julio J. Martinell[*] and Parvez N. Guzdar[†]

[*]*ICN - Universidad Nacional Autónoma de México, A. Postal 70-543, 04510 México D. F., MEXICO*
[†]*IREAP, University of Maryland, College Park, MD 20742, U.S.A.*

Abstract. A set of equations is derived that describes the slow and fast evolution of a tokamak plasma, for finite β drift waves. This provides a self-consistent model for the L-H transition showing the interrelation of fast fluctuations and slow equilibrium quantities, that explains the zonal flows and fields in terms of drift wave turbulence. The relevant equations are separated in axisymmetric and fluctuating parts and transformed to a helical mode-aligned coordinate, thus reducing the problem to 2D. Only two toroidal modes are included. The slow equations, which describe the transport, are coupled to the fast equations, which have been shown to predict the L-H transition threshold [2] for several experiments, given in terms of the mode dimensionless parameters α_{MHD} and α_D that measure ideal stability and diamagnetic effects, respectively. This approach allows to relate this threshold, to the experimentally measured injected power. The full system of equations is numerically solved to get a power threshold for the L-H transition.

INTRODUCTION

It is thought that the underlying cause for the transition to the H-mode in a tokamak is the turbulence suppression due to a sheared radial electric field near the edge region of the plasma. However, the specific behavior of the transition from the L-mode to the H-mode, as observed in experiments, depend on the specific turbulence model assumed. It is belived that in this cooler region, magnetic effects are relevant in the fluctuations and resistive ballooning modes have been considered. A transition threshold has been determined in 3D numerical simulations of edge turbulence that consider electron and ion dynamics, by Rogers, Drake and Zeiler [3]. It is given in terms of $\alpha_{MHD} = \beta q^2 R/L_p$ and $\alpha_D = \rho_s c_s t_0/L_n L_0$ (with $t_0 = (RL_n/2c_s^2)^{1/2}$ for ideal ballooning growth time and L_0 the characteristic length) as, $\alpha_{MHD}\alpha_D^2 = const$. This threshold has been checked to agree with experimental data from Alcator C-mod, DIII-D and ASDEX.

The same threshold can be obtained from the stability properties of finite-β drift wave turbulence as shown by Guzdar et al. [2]. With a non-linear analysis, they can interpret the L-H transition as a finite β effect turbulence quenching. However, the threshold obtained is not in terms of parameters that are experimentally controlled, like injected power, so it would be convenient to check if it agrees with experimental fits for the L-H transition power threshold, which scales as, $P_{th} = knBS$ where S is the plasma surface area ($k = const$). To this end, plasma transport should be added to the turbulence simulations.

It is possible to get an estimate of the scaling for the power theshold in terms of the critical temperature T_c, obtained for turbulence models. For instance, for drift-resistive ballooning modes, characteristic time and space scales entering the α's are $t_0 = (RL_n/2c_s^2)^{1/2}$ and $L_0 = 2\pi q(v_{ei}R\rho_s/2\omega_{ce})^{1/2}(2R/L_n)^{1/4}$. Then the theshold condition gives a critical temperature: $T_c^3 = \text{const} \left(\frac{L_n^3}{R}\right)^{1/2} B^2 \left(\frac{m_e}{m_i}\right)^{1/2} \frac{2\sqrt{2\pi}\lambda e^4}{3}$. This is in turn related to the power through an energy diffusion equation in steady state, i.e. $P = \frac{D_{an}}{L_n^2}nT = \frac{L_0^2}{t_0}\frac{nT}{L_n^2}$. Then, a simplistic power scaling would be,

$$P_{th} = \text{const} \left(\frac{m_e}{m_i}\right)^{1/12} \frac{q^2 n^2 R^{11/12}}{L_n^{11/4} B^{5/3}}.$$

For drift wave turbulence: $t_0 \sim \omega_*^{-1} \sim (c_s^2/\omega_{ci}L_nL_0)^{-1}$ and $L_0 \sim R/t_0\omega_{ci} \sim (R\rho_s^2/L_nL_0)$, and so,

$$P_{th} = \text{const } B^{-1/3}L_n^{-9/4}R^{1/12}n.$$

These do not quite agree with the experimental one, but in order to get a more realistic scaling it is necessary to include transport.

TRANSPORT AND FLUCTUATIONS EQUATIONS

We separate fluid equations in fast and slow components, the former describing turbulence and the later describing transport. As in most turbulence simulations, computations are made in a flux tube. According to what was shown in [1], the equations are first separated in fast and slow components and then transformed to flux-tube geometry. The analysis is an extension of a model by Guzdar and Hassam [4], where a self consistent description of L-H transition was developed, based on low β resistive ballooning modes. The model is 2D. Slow quantities are axisymmetrical, and fluctuating quantities are assumed to be dominated by a narrow range in toroidal mode numbers, as suggested by simulations: Low toroidal mode numbers are stabilized by coupling to acoustic waves, while high toroidal mode numbers are stabilized by finite Larmor radius effects. In [4] a single mode number was considered, since non-linear couplings were neglected. This reduces the problem to 2D. Here, in order to consider non-linear coupling, more than a single mode number has to be included. We take two toroidal mode numbers as the simplest extension, assuming the others are damped. Slow quantities enter in fast equations and vice versa, creating a coupling that takes into account the effect of turbulence on transport and the effect of transport on shearing (hence turbulence suppression).

Braginskii equations are used under assumption that $t_0^{-1} < \omega_{ci}$ and $L_0 > \rho_i$. This was considered by Zeiler et al. [5], who performed a partial linearization to obtain turbulence equations. Equations for ions and electrons are added to get a single fluid, and magnetic field variations are included through the magnetic potential ψ, determined by Faraday's law. The simplifying assumptions assumed are: (1)Equal temperatures $T = T_e = T_i$, (2)total power $P = P_i + P_e$ is used, (3)only $E \times B$ drift is included in polarization drift v_{pol}, (4)$v_d = v_{de} = v_{di}$. The relevant variables are then $n, \phi, v_{\|i}, T$ and ψ.

Each variable is separated these in slow and fast components: $\xi(x,y,z,t) = \overline{\xi}(x,y,t) + \tilde{\xi}(x,y,z,t)$, with $x = r, y = a\theta, z = R\zeta$. The equations are then averaged over z to get slow equations and subtract from full equations to get fast evolution. After separation, partial linearization of fast equations is performed, keeping non-linear terms related to v_E advection and parallel variations along poloidal field ($\nabla\psi$). We assume average B-field gives zonal field associated to zonal flow, so equation for $\overline{\psi}$ is kept. However, $\tilde{v}_{\|}$ is ignored since acoustic term does not affect stability of the modes, for the mode numbers considered, and, for simplicity, temperature fluctuations are not considered, $\tilde{T} = 0$.

The fast equations are equivalent to reduced Zeiler's equations [5] used by Guzdar et al.[2], but including curvature and effects of variation of averaged quantities. Fast variables are changed to helical coordinates $\varphi = q\theta - \zeta$, and separate in two modes with toroidal mode numbers n and 2n; define $\varphi' = n\varphi$:

$$\tilde{\xi}(x,y,z,t) = \xi_{1s}(x,y,t)\sin\varphi' + \xi_{1c}(x,y,t)\cos\varphi' + \xi_{2s}(x,y,t)\sin 2\varphi' + \xi_{2c}(x,y,t)\cos 2\varphi'$$

explicit variations in y and x are slow. The dimensionless variables used are: $\hat{n} = n/n_0$, $\hat{\phi} = e\phi/mT_0$, $\hat{v}_{\|} = v_{\|i}/c_s$, $\hat{T} = T/T_0$, $\hat{j}_{\|} = j_{\|}/en_0V_A$, $\hat{\psi} = \Omega v_A\psi/c_s^2B_0$, and the time and radial length scales are $t_0 = \sqrt{RL_n/2}/c_s$ and $L_0 = c_s^2 t_0/a\Omega$. In the following the overbar has been dropped from average quantities, and dissipative terms are thrown away: $\eta_{\|} \to 0$. For the mode number, the poloidal number will be used, $m = nq$, and also a normalized number, $\hat{m} = mL_0/a$.

Computations are for circular flux surfaces with, $\mathbf{B} = (\hat{\zeta} + \Theta\hat{\theta})B_0/R_0(1 + \varepsilon\cos\theta)$. Define the vorticities and currents for averages and modes with $i = 1,2$ and $k = c,s$,

$$\omega = \frac{\partial^2\phi}{\partial x^2}, \quad \omega_{ik} = \left(\frac{\partial^2}{\partial x^2} - (im)^2\right)\phi_{ik}, \quad j = \frac{c_s v_A}{(\Omega L_0)^2}\frac{\partial^2\psi}{\partial x^2}, \quad j_{ik} = \frac{c_s v_A}{(\Omega L_0)^2}\left(\frac{\partial^2}{\partial x^2} - (im)^2\right)\psi_{ik}$$

The final equations, keeping only dominant terms, can be written explicitly for the 17 variables, $n, \phi, v_{\|}, \psi, T, n_{1s}, n_{1c}, n_{2s}, n_{2c}, \phi_{1s}, \phi_{1c}, \phi_{2s}, \phi_{2c}, \psi_{1s}, \psi_{1c}, \psi_{2s}, \psi_{2c}$, in the following form.

EQUATIONS FOR AXISYMMETRIC QUANTITIES (TRANSPORT)

$$\frac{\partial n}{\partial t} + \frac{1}{m}(\phi_{,x}n_{,y} - \phi_{,y}n_{,x}) + 2\varepsilon\sin\theta\left[\frac{n}{m}\phi_{,x} - (nT)_{,x}\right] - \frac{c_s}{v_A}(\psi_{,x}(nv-j)_{,y} - \psi_{,y}(nv-j)_{,x}) + \langle\tilde{\phi}\tilde{n}\rangle = 0 \quad (1)$$

$$\frac{\partial\omega}{\partial t} + \frac{1}{m}(\phi_{,x}\omega_{,y} - \phi_{,y}\omega_{,x}) - \frac{\hat{m}L_n c_s}{2n\varepsilon L_0 v_A}(\psi_{,x}j_{,y} - \psi_{,y}j_{,x}) + \frac{2\hat{m}L_n}{nL_0}\sin\theta(nT)_{,x} + \langle\tilde{\phi}\tilde{\omega}\rangle = 0 \quad (2)$$

$$\frac{\partial v}{\partial t} + \frac{1}{m}(\phi_{,x}v_{,y} - \phi_{,y}v_{,x}) + \frac{2c_s}{nv_A}\left(\psi_{,x}[(nT)_{,y} + \frac{n}{4}v^2_{,y}] - \psi_{,y}[(nT)_{,x} + \frac{n}{4}v^2_{,x}]\right) = 0 \quad (3)$$

$$\frac{\partial\psi}{\partial t} + [\psi_{,y}(\frac{\phi}{m} - n - 1.71T)_{,x} - \psi_{,x}(\frac{\phi}{m} - n - 1.71T)_{,y}] - m\langle\tilde{\psi}(\frac{\tilde{\phi}}{m} - T\tilde{n})\rangle = 0 \quad (4)$$

$$\frac{\partial T}{\partial t} + \frac{4T}{3n}\varepsilon\sin\theta[(nT)_{,x} - \frac{n}{m}\phi_{,x}] + \frac{2Tc_s}{3nv_A}\{\psi_{,x}[(nv)_{,y} - 1.35j_{,y}] - \psi_{,y}[(nv)_{,x} - 1.35j_{,x}]\} + \frac{\phi_{,x}}{m}T_{,y} - \frac{\phi_{,y}}{m}T_{,x}$$
$$- \frac{2T}{3n}\langle\tilde{\phi}\tilde{n}\rangle - \frac{\hat{\kappa}c_s}{3nv_A^2}\left(\frac{\partial\psi}{\partial x}\frac{\partial}{\partial y}\right)^2 T - \frac{\hat{P}}{3n} = 0 \quad (5)$$

with

$$2\langle\tilde{f}\tilde{g}\rangle = (f_{1c}g_{1s} - f_{1s}g_{1c})_{,x} + 2(f_{2c}g_{2s} - f_{2s}g_{2c})_{,x}, \quad \hat{\kappa} = \frac{\kappa_\parallel}{n_0}\sqrt{\frac{2}{RL_n}}, \quad \hat{P} = \frac{P}{n_0 T_0}$$

EQUATIONS FOR THE TWO TOROIDAL MODES (TURBULENCE)

These equations involve more coupling terms. Not all of them are given here, but only a few representative ones.

$$\frac{\partial n_{1s}}{\partial t} + \left(-\phi_{1s,x}n_{2s} - \phi_{1c,x}n_{2c} + n_{1s,x}\phi_{2s} + n_{1c,x}\phi_{2c} + \frac{n_{1s}\phi_{2s,x} + n_{1c}\phi_{2c,x} - \phi_{1s}n_{2s,x} - \phi_{1c}n_{2c,x}}{2}\right)$$
$$+ \phi_{1c}n_{,x} - \phi_{,x}n_{1c} - \frac{c_s}{v_A}[\psi_{,x}j_{1s,y} - \psi_{,y}j_{1s,x}] + \frac{c_s m}{v_A}\left(\psi_{1s,x}j_{2s} + \psi_{1c,x}j_{2c} - j_{1s,x}\psi_{2s}\right.$$
$$\left. - j_{1c,x}\psi_{2c} + \frac{\psi_{1s}j_{2s,x} + \psi_{1c}j_{2c,x} - \psi_{2s,x}j_{1s} - \psi_{2c,x}j_{1c}}{2}\right) = 0 \quad (6)$$

$$\frac{\partial n_{2c}}{\partial t} + \left(\frac{\phi_{1c,x}n_{1s} + \phi_{1s,x}n_{1c} - n_{1c,x}\phi_{1s} - n_{1s,x}\phi_{1c}}{2}\right) - 2(\phi_{2s}n_{,x} - n_{2s}\phi_{,x})$$
$$- \frac{c_s}{v_A}\left(\psi_{,x}j_{2c,y} - \psi_{,y}j_{2c,x} + \frac{m}{2}[\psi_{1c,x}j_{1s} - \psi_{1c}j_{1s,x} + \psi_{1s,x}j_{1c} - \psi_{1s}j_{1c,x}]\right) \quad (7)$$

$$\frac{\partial\omega_{1c}}{\partial t} + \frac{1}{2}\left[\frac{\partial\phi_{1c}}{\partial x}\omega_{2s} - \frac{\partial\phi_{1s}}{\partial x}\omega_{2c} - \frac{\partial\omega_{1c}}{\partial x}\phi_{2s} + \frac{\partial\omega_{1s}}{\partial x}\phi_{2c} + \frac{\partial}{\partial x}(\phi_{1c}\omega_{2s} - \phi_{1s}\omega_{2c} - \phi_{2s}\omega_{1c} + \phi_{2c}\omega_{1s})\right]$$
$$- \phi_{1s}\omega_{,x} + \omega_{1s}\phi_{,x} + \frac{2L_n T}{L_0 n}\left[\hat{m}^2(\cos\theta - \varepsilon)n_{1s} + \hat{m}\sin\theta n_{1c,x}\right] - \frac{\hat{m}L_n c_s}{2nL_0 v_A}\left[\psi_{,x}j_{1c,y} - \psi_{,y}j_{1c,x}\right] \quad (8)$$
$$+ m[\frac{\partial j_{1s}}{\partial x}\psi_{2c} - \frac{\partial j_{1c}}{\partial x}\psi_{2s} - \frac{\partial\psi_{1s}}{\partial x}j_{2c} + \frac{\partial\psi_{1c}}{\partial x}j_{2s} + \frac{\partial}{\partial x}(\psi_{1c}j_{2s} - \psi_{1s}j_{2c} - \psi_{2s}j_{1c} + \psi_{2c}j_{1s})] = 0$$

$$\frac{\partial\omega_{2s}}{\partial t} + \frac{\phi_{1s,x}\omega_{1s} - \phi_{1c,x}\omega_{1c} - \phi_{1s}\omega_{1s,x} + \phi_{1c}\omega_{1c,x}}{2} + 2\phi_{2c}\omega_{,x} - 2\omega_{2c}\phi_{,x} - \frac{4L_n T}{L_0 n}\left[\hat{m}^2(\cos\theta - \varepsilon)n_{2c}\right. \quad (9)$$
$$\left. - \frac{\hat{m}}{2}\sin\theta n_{2s,x}\right] - \frac{\hat{m}L_n c_s}{2nL_0 v_A}\left[\psi_{,x}j_{2s,y} - \psi_{,y}j_{2s,x} + m(\psi_{1s,x}j_{1s} - \psi_{1c,x}j_{1c} - \psi_{1s}j_{1s,x} + \psi_{1c}j_{1c,x})\right] = 0$$

$$\frac{\partial \psi_{1s}}{\partial t} + m\psi_{1c}\left[\frac{\phi_{,x}}{m} - \frac{T}{n}n_{,x} - 1.71T_{,x}\right] + \psi_{,x}\left[\frac{\phi_{1s,y}}{m} - Tn_{1s,y}\right] - \psi_{,y}\left[\frac{\phi_{1s,x}}{m} - Tn_{1s,x}\right] + m\left[\left(\frac{\phi_{1s,x}}{m}\right.\right.$$
$$- Tn_{1s,x})\psi_{2s} + \psi_{2c}(\frac{\phi_{1c,x}}{m} - Tn_{1c,x}) - \psi_{1s,x}(\frac{\phi_{2s}}{m} - Tn_{2s}) - \psi_{1c,x}(\frac{\phi_{2c}}{m} - Tn_{2c})\Big] \quad (10)$$
$$+ \frac{m}{2}\left(\psi_{2s,x}(\frac{\phi_{1s}}{m} - Tn_{1s}) + \psi_{2c,x}(\frac{\phi_{1c}}{m} - Tn_{1c}) - \psi_{1s}(\frac{\phi_{2s,x}}{m} - Tn_{2s,x}) - \psi_{1c}(\frac{\phi_{2c,x}}{m} - Tn_{2c,x})\right)\Big] = 0$$

$$\frac{\partial \psi_{2c}}{\partial t} - 2m\psi_{2s}\left[\frac{\partial \phi}{m\partial x} - \frac{T}{n}\frac{\partial n}{\partial x} - 1.71\frac{\partial T}{\partial x}\right] + \psi_{,x}\left[\frac{\phi_{2c,y}}{m} - Tn_{2c,y}\right] - \psi_{,y}\left[\frac{\phi_{2c,x}}{m} - Tn_{2c,x}\right] \quad (11)$$
$$+ \frac{m}{2}\left(\psi_{1s,x}(\frac{\phi_{1c}}{m} - Tn_{1c}) + \psi_{1c,x}(\frac{\phi_{1s}}{m} - Tn_{1s}) - \psi_{1s}(\frac{\phi_{1c,x}}{m} - Tn_{1c,x}) - \psi_{1c}(\frac{\phi_{1s,x}}{m} - Tn_{1s,x})\right) = 0$$

The missing equations have a similar structure. Here, there is the presence of coupling of zonal flows and fields (ϕ, ψ) with a pump wave, which gives rise to sideband modes (ϕ_{ik}, ψ_{ik}), as in the approach of [2].

SIMULATIONS AND CONCLUSIONS

A numerical code based on these equations has been developed and describes the coupling of transport and turbulence properties (shearing stabilization, saturation). The region considered is a narrow ring of width L in the outer plasma region. The boundary conditions used are: periodicity in the variable $y = a\theta$; for all fluctuating quantities $\xi_{ik}(0,y) = \xi_{ik}(L,y) = 0$. This code is capable of simulating the L-H transition and predict a power threshold. In contrast to the previous work of [4], where the L-mode was forced by setting the $m = 0$ component of the potential to zero in order to prevent shear stabilization, this model has a trigger for the H-mode in the power P. This 2D model gives a simple way to understand the underlying mechanism for the transition, but it relies on results of full 3D simulations.

ACKNOWLEDGMENTS

This work was partially supported by Conacyt project 27974-E and DGAPA-UNAM project IN116200.

REFERENCES

1. Martinell, J.J., Guzdar, P.N. and Hassam, A.B., *Phys. Plasmas* **5**, 1273 (1998)
2. Guzdar, P.N., Kleva, R.G., Das, A. and Kaw, P.K., *Phys. Rev. Lett.* **87**, 15001 (2001)
3. Rogers, B.N., Drake, J.F. and Zeiler, A., *Phys. Rev. Lett.* **81**, 4396 (1998)
4. Guzdar, P.N., and Hassam, A.B., *Phys. Plasmas* **3**, 3701 (1996)
5. Zeiler, A., Drake, J.F. and Rogers, B.N., *Phys. Plasmas* **4**, 2134 (1997)

Dynamics of Transport Barriers and ELM-Like Behaviour in Electrostatic Turbulence

V. Naulin[*], J. Nycander[†] and J. Juul Rasmussen[*]

[*]Association EURATOM – Risø National Laboratory
Optics and Fluid Dynamics Department, OFD-128, Risø
4000 Roskilde, Denmark
[†]Department of Meteorology
University of Stockholm
106 91 Stockholm, Sweden

Abstract. The self-consistent development of transport barriers is investigated analytically and numerically in flux driven interchange turbulence. Numerical simulations show the turbulence leading to a homogenization of Lagrangian invariants by mixing, resulting in quasi-steady pressure profiles predicted by turbulent equipartition. Below a critical aspect ratio $\alpha = L_y/L_x \approx 3.8$, which is related to the rotational transform, large scale poloidal flows develop. They quench the turbulence, constitute transport barriers for the turbulent fluxes, but are intermittently disrupted by strong bursts in the transport, which may be related to the strong Edge Localized Modes (ELM) observed in toroidal devices.

INTRODUCTION

In hot magnetized plasmas the cross-field transport is anomalous and ascribed to low frequency electrostatic fluctuations [1]. It is generally recognized that self-consistently developing large scale poloidal flows strongly reduce the radial turbulent transport by "quenching" the turbulence (see [2, 3] and references therein). This mechanism may be responsible for the transition to H-mode which [4] is often found to be accompanied by bursts in the transport, related to edge localized modes (ELM) [5, 6].
We consider the evolution and dynamics of transport barriers and their interplay with the turbulent transport, which reveals an intermittent behavior with very strong burst events. A preliminary account of the results was presented previously [7].

MODEL EQUATIONS

Our model describes the plasma motion in the presence of an inhomogeneous, curved magnetic field including adiabatic compression and heating of fluid parcels displaced into regions of larger B. All profiles are allowed to evolve self-consistently under the influence of eventual external heating:

$$\frac{\partial n}{\partial t} + \{\phi, n\} + \mathcal{K}(n + T - \phi) = \nu \nabla^2 n, \tag{1}$$

$$\frac{\partial T}{\partial t} + \{\phi, T\} + \frac{2}{3}\mathcal{K}(n + \frac{7}{2}T - \phi) = \kappa \nabla^2 T, \tag{2}$$

$$\frac{\partial \nabla^2 \phi}{\partial t} + \{\phi, \nabla^2 \phi\} + \mathcal{K}(n + T) = \mu \nabla^4 \phi. \tag{3}$$

$\{f,g\} = (\partial f/\partial x)(\partial g/\partial y) - (\partial f/\partial x)(\partial g/\partial y)$ denotes the Poisson bracket. The potential ϕ is normalized by $\mathcal{T}/e$, the time by $\omega_{ci}^{-1} = m_i/(eB_0)$, and the length by $\rho = (\mathcal{T}/m_i)^{1/2}/\omega_{ci}$. The curvature operator is $\mathcal{K} = -\nabla \cdot \frac{\mathbf{B} \times \nabla}{B^2}$. The coefficients ν, κ and μ model diffusion. It is assumed that the fields deviate only slightly from constant reference levels $\mathcal{N}$, $\mathcal{T}$, with, e.g., $\tilde{n} = \mathcal{N}(1 + n(x,y,t))$. We model the outboard mid-plane of a Tokamak with the curvature operator $\mathcal{K}(f) = \omega_B \partial f/\partial y$, where $\omega_B = 2\rho/R_0$. This corresponds to the large aspect ratio approximation for the magnetic field $\mathbf{B} = (B_0 R_0/R)\hat{\mathbf{b}}$, where $\hat{\mathbf{b}}$ is the unit vector in the toroidal direction, locally along the z-coordinate. R is the distance from the torus axis and B_0 is the magnetic field at $R = R_0$. In the coordinate system of the considered slab x corresponds to the radial direction and y to the poloidal direction.

In the in-viscid limit the equations possess the Lagrangian invariants:

$$l_{\pm} = \pm\sqrt{5/2}(n + \omega_B x) + 3T/2 - n, \tag{4}$$

advected by the pseudo velocities $\mathbf{v}_{\pm} = \hat{z} \times \nabla[\phi - n - (1 \pm (5/2)^{1/2})T]$. This leads to TEP profiles

$$n + \omega_B x \approx const \text{ and } T + \frac{2}{3}\omega_B x \approx const. \tag{5}$$

In contrast to models, where curvature is modeled by an effective gravity, the curvature here only couples the equations: it allows the potential energy in the pressure gradient to be converted to turbulent motion of the plasma. The field inhomogeneity does not act as a source of free energy itself.

We linearize Eqs. (1)-(3) around the background profiles $n_0(x)$, $T_0(x)$ assuming a waveform for the potential perturbation $\psi_k(x)exp(iky - i\omega t)$ respecting the boundary conditions in the x-direction, and similarly for n and T, where k is the wavenumber in the y-direction. The equation for the wave amplitude ψ_k reads:

$$\frac{d^2\psi_k}{dx^2} + \left[-k^2 + D(c,x)\right]\psi_k = 0, \tag{6}$$

where

$$D = \frac{cN - \frac{5}{3}(\omega_B)^2(n_0' + \omega_B)}{c(c^2 - \frac{10}{3}c\omega_B + \frac{5}{3}\omega_B^2)}. \tag{7}$$

We have introduced $c = \omega/k$, and the "buoyancy" frequency $N \equiv \omega_B(n_0' + T_0' + \frac{5}{3}\omega_B)$. The "prime" denotes derivative with respect to x.

Equation (6) may be solved for a given profile of N and with boundary conditions $\psi_k|_{(x=0)} = \psi_k|_{(x=L_x)} = 0$ resulting in an expression for the complex phase velocity c - the dispersion relation. L_x is the width of the slab in the radial direction. In order to illustrate the features of the instability we consider within a local approximation and the long wave limit, i.e., $D \approx N/(c^2 - \frac{10}{3}c\omega_B + \frac{5}{3}\omega_B^2)$, we obtain:

$$c = \frac{5}{3}\omega_B \pm \sqrt{\frac{10}{9}\omega_B^2 + \frac{N}{K^2}}, \tag{8}$$

where $K^2 = k_x^2 + k^2$. Thus, we have instability for $N < -K^2 \frac{10}{9}\omega_B^2$ and for a given negative N we have stability for sufficiently high K-values. This is the "standard" Rayleigh-Taylor instability with growth rate

$$\gamma = k\sqrt{\frac{|N|}{K^2} - \frac{10}{9}\omega_B^2}, \tag{9}$$

With $Re(c) = \frac{5}{3}\omega_B$ we have propagation in the positive y direction.

SIMULATION RESULTS

We consider a two-dimensional domain bounded in x (length L_x) and periodic in y (length L_y). The poloidal periodicity length is interpreted as the recurrence length of a magnetic field line: Assuming an infinite correlation along magnetic field lines for the $q = 3$ rational surface we would f.x. find $L_y = 2\pi r/3$. L_x is related to the gradient scale length or the width of the Scrape Off Layer (SOL). The aspect ratio $\alpha \equiv L_y/L_x$ is then be directly related to the safety factor q, i.e., $\alpha \propto 1/q$.

We performed numerical runs for numerous values of the different parameters of the system and observe the following general scenarios depending on aspect ratio (Fig. 1)).

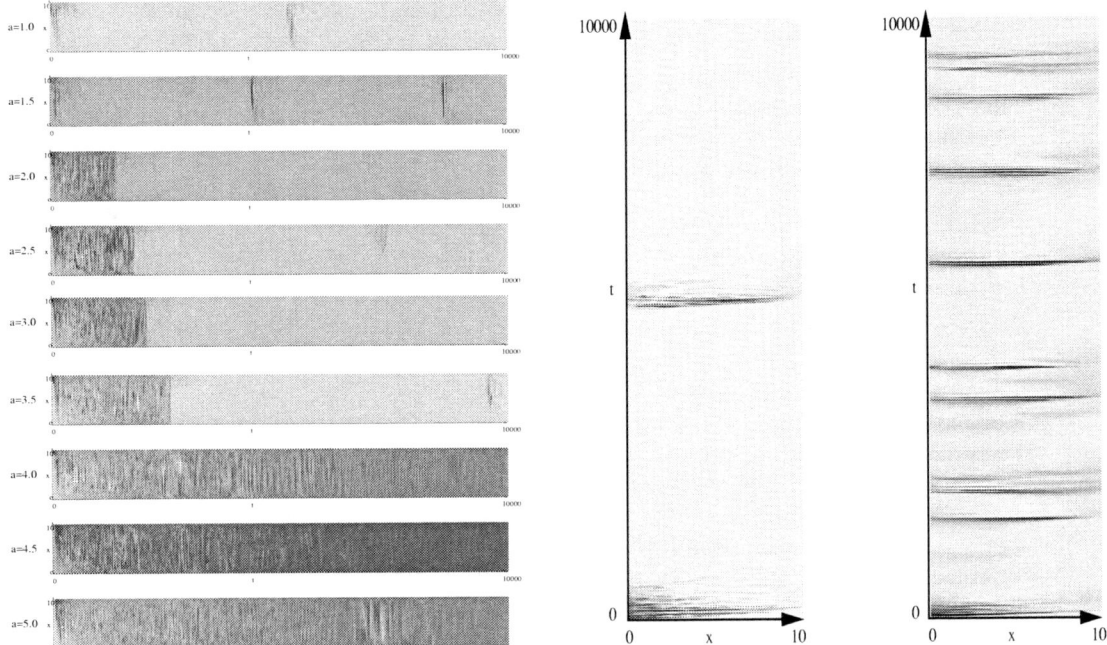

FIGURE 1. Poloidally averaged heat flux versus time for different aspect ratio. For an aspect ratio larger than 3.8 no transport suppression is observed.

FIGURE 2. Poloidally averaged heat flux $\Gamma_T = <uT>$ versus (x,t) for aspect ratio $\alpha = 1$ and dissipation coefficients $= 10^{-3}$ (left frame) and 10^{-2} (right frame).

a) For sufficiently large $\alpha \geq \alpha_c \approx 3.8$ the system develops into the TEP state described in Eq. (5) regardless of the value of T_0, demonstrating profile consistency and resilience [8]. There is a radial, turbulent heat flux, which is persistent but intermittent.

b) For smaller α in an initial phase the turbulence establishes the TEP profiles with a high flux-level. Then flux is interrupted – an H-mode like state with steeper averaged gradients and lower diffusion develops. For long times the system is very quiescent, as seen in Fig. 1) and Figs. 2) for $\alpha = 1$. Sporadic flux bursts of high amplitude – ELMs – are observed at somewhat random intervals. The time scale of the quiescent periods decreases almost proportionally to μ^{-1}, as illustrated in Fig. 2b).

The quiet periods are associated with the establishment of a strong poloidal mean flow which characterizes the

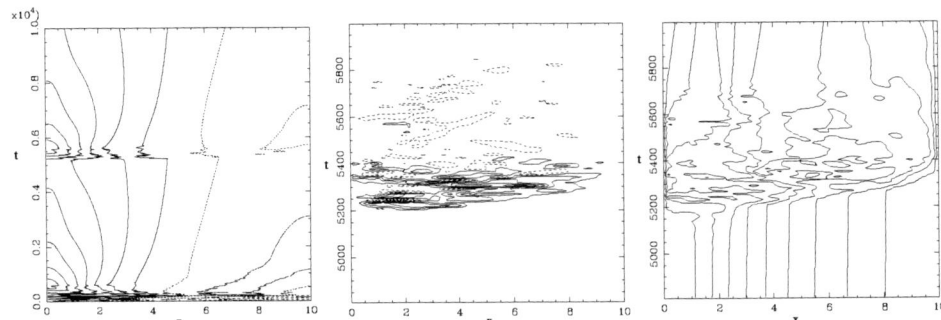

FIGURE 3. Contours of the zonal velocity versus time and radial coordinate with $dz = 0.2$ (left). Enlarged with respect to time for the flux burst: (center) $\Gamma_T(t)$ plotted versus (x,t) with $dz = 0.5$ and (right) the temperature profile versus time with $dz = 1$. Same parameters as in Fig. 1 for $\mu = 10^{-3}$

transport barrier (see Fig. 3)). There is no longer sufficient mixing by the turbulence to maintain the TEP profiles.

The profiles start to steepen via the diffusive inflow of heat from the hot boundary. This behavior is detailed in Fig. 3) where we depict the evolution of the zonal flow showing the contours of V in the x-t-plane in Fig. 3a). After the initial phase where the turbulence develops and establishes the TEP profiles a zonal flow is seen to form at $t \approx 100$. The flow is strongest at the heated boundary where it is in the positive y-direction. Near the other wall it is reversed, resulting in a strong shear. This flow dominates the evolution for a long period, during which the turbulence is suppressed and the heat flux is negligible, compare Fig. 1a). The zonal flow is slowly decaying and at around $t = 5200$ it has become weak enough for allowing the onset of the instability. If the viscosity is larger, the zonal flow decays faster, and turbulent bursts are more frequent, as in Fig. 2b). The burst in the heat flux results in a flattening of the temperature profile as observed in Figs. 2b) and c). The flux is first established near the heated wall propagates outwards as a sharp front accompanied by the change in the temperature profile. Again a zonal flow builds up at around $t = 5600$. This quenches the turbulence and the flux and the previous scenario repeats.

It remains to be explained why the zonal flow is only dominating the evolution for aspect ratios below a critical value α_c. It is observed that the flow stabilizes the steep super-critical (with respect to the RTI) pressure profile that builds up due to viscous diffusion in the quite periods. This may be explained by the recent analysis of Benilov et al [9] (see also [10]), who demonstrates that an induced shear flow, irrespectively of its actual profile, will tend to stabilize the larger wave numbers of the RTI instability of an inversely stratified fluid. This implies that for a given geometry of the domain, i.e., a given aspect ratio, the shear flow may stabilize all the modes allowed by the geometry $k > 2\pi/L_y$. Thus, even if the shear flow is not stabilizing in the global sense [11] it may be stabilizing in a finite system.

We recall the interpretation of the aspect ratio being inversely proportional to the safety factor in a toroidal configuration with a sheared magnetic field. Then the stabilization of the shear flow instability, the Kelvin-Helmholtz instability, for low aspect ratio appears to be equivalent with the well-known stabilization by a sheared magnetic field [2].

CONCLUSION

We have shown that the nonlinear evolution of pressure driven turbulence depends strongly on the aspect ratio. For low aspect ratio, which corresponds to either high irrational q or q values close to rational ones, the evolution is characterized by long lasting quiescent H-mode periods. The turbulent transport is suppressed by a zonal shear flow. These periods are separated by short violent flux bursts (ELMs) during which the zonal flow breaks down. A similar behavior was observed in toroidal gyro-kinetic ITG turbulence simulations [12, 13]. For large aspect ratio we find a continuous strong turbulent flux, which is temporally intermittent and spatially localized in narrow channels between the dominating convective rolls.

The model contributes to the fundamental understanding of the role of zonal flows in controlling turbulence and confinement and shows it's fundamental relationship to the ELM type of behavior observed in the H-mode. It reproduces the spatial and temporal intermittent evolution of the turbulent fluxes generally observed and offers a consistent non-linear picture of ELM behavior, allowing some prediction of transport barrier formation in relation to the q profile.

REFERENCES

1. F. Wagner and U. Stroth, Plasmas Phys. Control. Fusion **35**, 1321, (1993).
2. P.W. Terry, Rev. Mod. Phys. **72**, 109 (2000).
3. K.H. Burrell, Phys. Plasmas **6**, 4418 (1999).
4. The Asdex Team, Nucl Fusion **29**, 1959 (1989).
5. H. Zohm, Plasma Phys. Contr. Fusion **38**, 105 (1996).
6. W. Suttrop, Plasma Phys. Contr. Fusion **42**, A1 (2000).
7. V. Naulin, J. Nycander and J. Juul Rasmussen; Proc. ICPP-2000 (October 23 - 27, 2000, Quebec City, Canada) Vol. **1**, 208 (2000).
8. V. Naulin, J. Nycander and J. Juul Rasmussen Phys. Rev. Lett. **81**, 4148 (1998).
9. E.S. Benilov, V. Naulin and J. Juul Rasmussen, Phys. Fluids **14**, 1674 (2002).
10. H.L. Kuo, Phys. Fluids **6**, 195 (1963); A.B. Hassam, Phys. Fluids B **4**, 485 (1992)
11. J.W. Miles, J. Fluid Mech. **10**, 496 (1961); L.N. Howard, ibid. p. 509.
12. Z. Lin, T.S. Hahm, W.W. Lee, W.M. Tang and P.H. Diamond, Phys. Rev. Lett. **83**, 3645 (1999).
13. M.A. Malkov, P.H. Diamond and M.N. Rosenbluth, Phys. Plasmas **8**, 5073 (2001).

Modelling the Formation of Large Scale Zonal Flows in Drift Wave Turbulence in a Rotating Fluid Experiment

V. Naulin, J.Juul Rasmussen, B. Stenum, L.J.A. van Bokhoven, and J. van de Konijnenberg

Association EURATOM – Risø National Laboratory Optics and Fluid Dynamics Department, OFD-128 Risø, 4000 Roskilde, Denmark

Abstract. The formation of large-scale flows - zonal flows - is demonstrated in a laboratory experiment with a fluid in a rotating container with bottom topography. The fluid is efficiently stirred by external forcing. The flow generation is explained by mixing and homogenization of the potential vorticity. By employing the analogy between drift wave dynamics in a magnetized plasma and flow dynamics in the rotating tank the results can be considered as a model for zonal flow generation in drift wave turbulence.

INTRODUCTION

The self-consistent generation of large scale flows - zonal flows - by the rectification of small scale turbulent fluctuations is of great importance importance both in geophysical flows and in magnetically confined plasmas [1, 2]. These flows will regulate the turbulence suppressing the small scale structures and set up transport barriers. A relatively simple description of the generation of zonal flows is provided by the idea of homogenization of the so-called potential vorticity which is a Lagrangian invariant of the flow. This idea was put forward by Rhines [2] in the context of geophysical flows. In a rotating fluid layer the dynamics is quasi two-dimensional. In the inviscid limit it can be shown that the flow is governed by the conservation of the potential vorticity (PV) (see, e.g., [3, 2]):

$$\frac{DPV}{Dt} = \frac{D}{Dt}\frac{\omega + f}{h} = 0, \qquad (1)$$

where $D/Dt = \partial/\partial t + \mathbf{v}\cdot\nabla$ is the Lagrangian derivative, $\omega = (\nabla \times \mathbf{v})\cdot\hat{z}$ is the (relative) vorticity, h is the depth of the fluid, and f is the Coriolis parameter, $f = 2\Omega\sin\varphi$ (planetary vorticity), for a fluid layer on a rotating planet, with Ω the planetary angular velocity and φ the geographic latitude. For the case of a rotating fluid container in the laboratory $f = 2\Omega$ is a constant (Ω being the angular velocity of the container), and a spatially varying depth $h(x,y)$ - a bottom topography - can mimic the variation in the planetary vorticity $f(\varphi)$ [4]. Under the assumption of small Rossby number ($Ro = \omega/f$) and weak perturbations of the depth ($\Delta h/h \ll 1$) Eq. (1) reduces to the Charney-Obukov equation or the the equivalent barotropic vorticity equation [3]:

$$\frac{\partial}{\partial t}(\nabla^2\phi - \frac{\phi}{\rho^2}) + [\phi, \nabla^2\phi] + \beta\frac{\partial\phi}{\partial x} = 0, \qquad (2)$$

with $[a,b] = a_x b_y - a_y b_x$. For the planetary case positive x is eastward and positive y is northward (the direction of the gradient in the Coriolis force). For the laboratory case positive y is towards the most shallow part (the direction of the gradient of the bottom topography). ϕ is the geostrophic stream-function ($\mathbf{v} = -\nabla\phi \times \hat{z}$, where $\mathbf{v}$ is the velocity), $\omega = \nabla^2\phi$, β is proportional to the gradient in the Coriolis force and/or the gradient in the bottom topography, and $\rho = \sqrt{gH_0/f^2}$ (H_0 is the averaged fluid depth) is the Rossby radius. The term ϕ/ρ^2 accounts for the perturbation of the fluid depth via perturbations of a free surface. When this is suppressed, as in a rotating container with a rigid lid, the term vanishes, formally by $\rho \to \infty$. Then the conserved PV reads:

$$PV = \omega + \beta y. \qquad (3)$$

We have performed laboratory experiments in a rotating fluid to investigating the formation of large-scale flows by mixing and homogenization of the potential vorticity. The experiments are performed in a container with radial symmetric bottom topography and a rigid lid. The bottom has a constant negative slope, b, in the radial direction. For this system $PV = \omega + \beta r$ with β negative (Eq. 3 in cylindrical coordinates). It is readily seen that an effective mixing that homogenizes PV will lead to replacing the high PV near the center with low PV from the outside. This will appear as an anticyclonic vortex over the central region and a large scale flow has been generated by a random (small scale) mixing. This is the essence of Rhines's theory [2] for zonal flow generation. We note that the flux of vorticity is given by the gradient of the Reynolds stress, $Re = \langle uv \rangle$ (u, v are the x and y component of the fluctuating velocity). This implies that the homogenization of PV is mitigated by Re, which makes links to the "standard" description of zonal flow being driven by the gradient of Re, see, e.g., [5, 1] in the plasma context.

It is well-known that the Rossby wave dynamics in the atmosphere of a rotating planet is similar to the drift wave dynamics in a magnetized plasma. Indeed a homomorphic equation to Eq. (2) appears in plasma physics and is here referred to as Hasegawa-Mima (HM) equation [6]. It governs the evolution of non-linear drift-waves in a magnetized plasma with a density gradient and β is here related to the density gradient scale length and ρ is the ion Lamor radius at the electron temperature. The magnetic field is in the z-direction, drift waves propagate in the negative x direction and the density gradient is pointing in the negative y-direction. Thus, in the rotating fluid experiment the deepest part corresponds to the highest density for the plasma analogue. This is readily seen by comparing the ion vorticity equation, governing the dynamics of drift waves, with Eq. 1. The ion vorticity equation is directly obtained from the ion momentum equation in the limit of cold ions and expresses the conservation of "ion potential vorticity" PV_i:

$$\frac{DPV_i}{Dt} = \frac{D}{Dt}\frac{\omega + \omega_c}{n} = 0, \qquad (4)$$

where n is the ion density and ω_c is the ion cyclotron frequency. Applying the drift scaling, the convection velocity is the $E \times B$-velocity, and assuming adiabatic electron response we obtain the "HM"-equation 2.

Although the zonal flow generation in the plasma case cannot be treated directly by the HM-equation, because the zonal flow component will not obey the assumption of adiabatic electron response, we believe it can be described by a similar mechanism as the one discussed above. Using the drift scaling the PV_i expands like:

$$PV_i = \omega + \tilde{n} + \beta y. \qquad (5)$$

where $\tilde{n}$ is the density perturbation. Homogenization of PV_i will lead to a redistribution of the density - e.g., a flattening of the gradient by $\langle \tilde{n} \rangle_x$ - and the generation of a zonal flow through $\langle \omega \rangle_x$.

EXPERIMENT

The experiments are performed in a circular tank with radius $R = 19.4\,cm$, maximum depth $D = 20\,cm$ and a rigid lid, see Fig. 1. A topographic "β-plane" was created with a false bottom, consisting of a cone with the hight difference

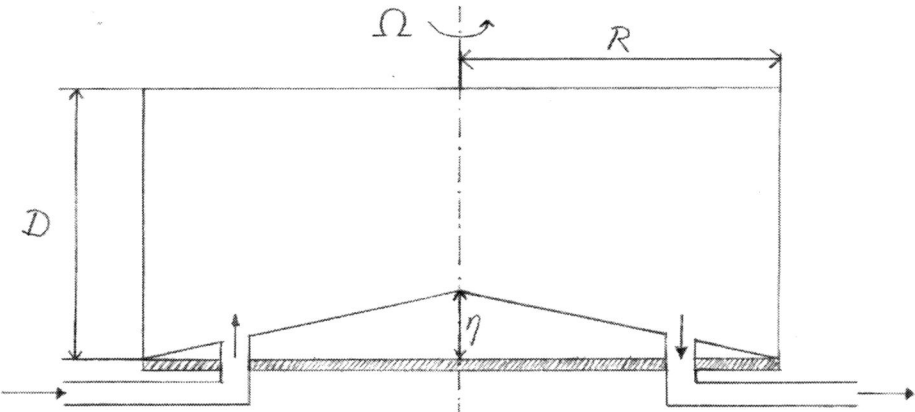

FIGURE 1. Experimental setup.

of $\eta = 5\,cm$. The tank was placed on a rotating table, and filled with demineralized water of room temperature.

The table is rotating counter-clockwise and in the experiments we report here, the angular velocity of the table is $\Omega = 1.26\,rad/s$ (12rpm). The angular velocity is controlled electronically, and is accurate to 0.01%. With these parameters the topographic $\beta = 2\Omega s/(D - \eta/2) = -0.037\,s^{-1}cm^{-1}$, where $s = -0.26$ is the slope of the bottom, with respect to the cylindrical coordinates (r, θ, z). The full velocity field of the flow in a horizontal plane is measured by particle tracking. For this purpose the water is seeded with small ($\sim 50\,\mu m$) neutrally buoyant particles and the tank is intersected by a horizontal light sheet. First, a video recording of the flow was made with a video camera co-rotating with the tank. Then, after the experiment, the recording was processed by a PC equipped with a frame grabber employing the image processing system, DigImage [7], that allows the tracking of particles and calculating their velocities. The velocity field is then mapped to a regular grid. The vorticity is obtained by matching the data with spline functions and manipulating the coefficients of this expansion.

EXPERIMENTAL RESULTS

The mixing is forced by periodically pumping water in and out of two holes (diameter $2\,cm$) in the bottom of the tank (see Fig. 1). They are placed azimuthally symmetric near the outer boundary of the tank with their center at a radius of $14.2\,cm$. Pumping water out of the tank leads to the formation of cyclonic vorticity - i.e., rotating in the same direction as the background rotation - over the hole, while pumping water into the tank leads to the formation of anticyclonic vorticity. With this setup the azimuthally averaged forced vorticity will be zero. The period of the forcing T_F is larger than rotation period of the tank and much smaller than the period of Rossby waves ($\approx 60\,s$) and a typical dissipation time, the Ekman spin down time ($\approx 90\,s$). Typically we used $6s < T_F < 15s$ and obtained similar results. In the experiments discussed here we used $T_F = 6.6\,s$. The pumping is initiated when the fluid has spun up to solid body rotations, around $30\,min$ after the start of the rotation.

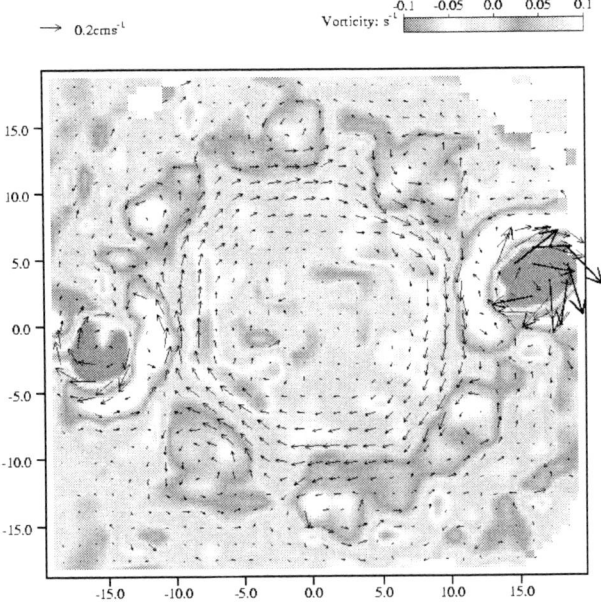

FIGURE 2. Velocity field shown by arrows and vorticity contours averaged over 10 forcing periods. The scales for the fields are indicated above the figure.

In Fig. 2 we show the velocity and vorticity field obtained by particle tracking several tens of periods after the forcing was started. The fields are averaged over 10 forcing periods. We clearly observe the formation of an anticyclonic circulation around the center of the tank associated with a region of negative vorticity as expected from the homogenization of *PV* discussed in the introduction. Relative strong perturbations are visible around the holes. The averaged vorticity maximum $\langle \omega \rangle$ is found to be 10 - 20% of the rms value of the fluctuating vorticity.

The temporally averaged velocity field in Fig. 2 is averaged over the azimuthal direction to obtain the zonal velocity. The result is shown in Fig. 3 (left panel). It is observed that the negative (anticyclonic) zonal velocity peaks in a region away from the forcing regime. While there is a band with a positive zonal velocity in- and outside the

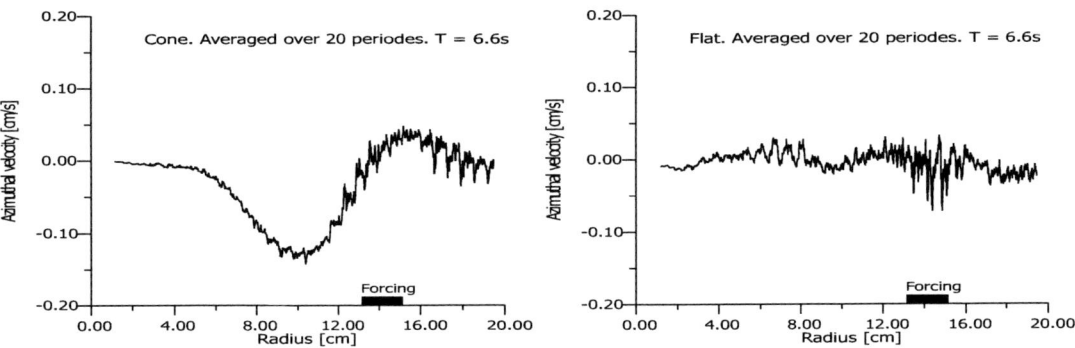

FIGURE 3. Averaged azimuthal velocity versus radius. The left frame shows the velocity for the case of the cone shaped bottom, while the right frame shows the reference case with a flat bottom. The black bar shows the position of the forcing region.

forcing regime. Thus, keeping the total circulation zero. By calculating the azimutally averaged vorticity we have verified that $\langle PV \rangle_\theta = \langle \omega \rangle_\theta + \beta r$ is roughly constant over the region of the flow generation. This supports the idea of homogenization of PV as being the cause for the flow generation.

In the experiment dissipative effects cannot be neglected. The flow field decays mainly due to the Ekman spin down caused by the friction both at the bottom and at the lid. The Ekman number is given by $E = \nu/D^2\Omega$ (μ is the kinematic viscosity), corresponding to a spin down time of $\tau_E \approx 90\,s$. The dissipation may explain the fact that the mixing appear to be ineffective inside a certain radius ($\approx 6\,cm$).

As a control case, we performed experiments with the same forcing parameters for the case of a flat bottom. For this case *no* zonal flow was generated as seen in the right panel in Fig. 3. Furthermore, initial investigations where the cone bottom topography is replaced by a "bowl" topography with the deepest part at the center have revealed the formation of a cyclonic zonal flow.

The experimental results are supported by direct numerical solutions of the quasi-geostrophic vorticity equation (Eq. 2) equipped with forcing and dissipation terms on a disk with no-slip boundary conditions at the walls.

CONCLUSION

We have demonstrated that large scale zonal flows may be generated by the mixing and homogenization of potential vorticity. Results from a laboratory experiment with a rotating fluid compare well with numerical solutions of the barotropic vorticity equation. Due to the similarities of the dynamics in a rotating fluid with varying depth profile and the low frequency dynamics in magnetized plasma with density gradient, we believe that similar mechanisms for zonal flow generation are operable in drift wave turbulence. Thus, our results contribute to the important issue of understanding and describing zonal flow generation by turbulence in general.

REFERENCES

1. Terry, P.W., *Rev. Mod. Phys.* **72**, 109 (2000).
2. Rhines, P.B, *Ann. Rev. Fluid Mech.* **11**, 401 (1979).
3. Pedlosky, J. *Geophysical Fluid Dynamics*, Springer Verlag, New York 1987.
4. van Heijst, G.J.F., *Meccanica* **29**, 431 (1994).
5. Diamond, P.H. and Kim, Y.-B., *Phys. Fluids B* **3**, 1626 (1991).
6. Hasegawa, A. and Mima, K., *Phys. Fluids* **21**, 87 (1978).
7. Dalziel, S.B., *Applied Scientific Research* **49**, 217 (1992).

Destabilizing effect of shear flow
- beyond Kelvin-Helmholtz instability

T. Tatsuno*, M. Hirota*, A. Ito*, S. Kondoh*, Z. Yoshida* and S. M. Mahajan[†]

*Graduate School of Frontier Sciences, The University of Tokyo, Tokyo 113-0033, Japan
[†]Institute for Fusion Studies, The University of Texas at Austin, Austin, Texas 78712 USA

Abstract. A few examples of shear flow destabilization are presented. The destabilization is driven by a different mechanism from Kelvin-Helmholtz instability. One is from the non-Hermiticity of the operator, which shows the secular growth of the field. The other is from the Alfvén wave. The Alfvén wave may act on the mode to escape from the shear flow stretching, and standing mode is allowed.

INTRODUCTION

Shear flow is recently considered one of the most expected candidate of stabilizing utilities for otherwise existing instabilities due to its stretching effect [1]. Up to now, the famous destabilizing effect of the shear flow is, perhaps, only the Kelvin-Helmholtz (KH) instability. The KH instability is an old, well-known instability first found in the neutral fluid [2]. However, only to draw the old established understandings is evidently insufficient for a variety of plasmas in both mathematical and physical sense.

After the first discovery of KH instability, the functional analysis was considerably developed in the context of quantum mechanics [3]. Therefore, the linear spectral theory has to be developed also for KH instabilities focusing on the continuous spectra, non-Hermiticity, etc. On the other hand, the physical effect of shear flow is not yet investigated deeply in plasmas, either. Being different from the classical fluid, the plasma contains a lot of varieties because of its inherent waves and instabilities.

Recently, we have developed a spectral theory for the surface wave model of KH instability [4]. Rayleigh [5] had first considered an infinitesimal displacement of the surface which separates two constant vorticity regions, and found an exponentially growing instability. However, according to complete spectral theory on a proper Hilbert space, we pointed out that the system contains frequency overlapping between continuous and point spectra. The frequency overlapping is shown to give rise to the infinite dimensional analogue of Jordan's canonical representation.

By considering the electron plasma, we can introduce the coupling of the vorticity with parallel electron motions [6] (see also the abstract of this presentation). In addition to the above 'resonance' between continuum and point, another continuous spectrum inhabits the system. Due to the higher order 'resonance' and the complexity of the couplings among those modes, localized asymptotic secularity is observed.

Moreover, with the careful look on the Rayleigh's original paper [5], we find that the surface wave can have an effect to escape from the stretching effect of the shear flow. Based on the knowledge that the wave can have an effect to prevent the mode from shear-flow stretching, we have investigated the effect of linear shear flow on the interchange instability in the incompressible, ideal plasma. It is found that even the linear shear flow — which is free from KH instability — can destabilize the mode when the system contains the Alfvén wave.

SECULAR BEHAVIOR OF ELECTRON PLASMAS

In this section, we show a localized secular growth of the perturbed density (vorticity) by considering the coupling of parallel electron motion with the Rayleigh equation [5]. For a slab electron plasma with a sharp density profile, the

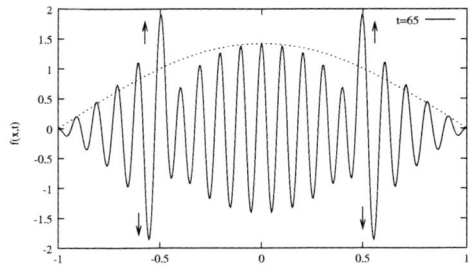 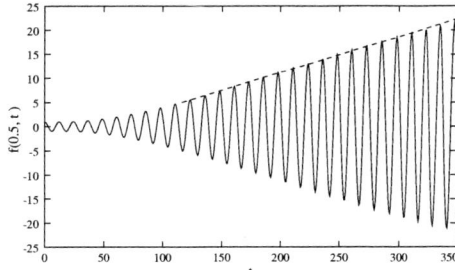

FIGURE 1. Left figure shows the profile of the inner perturbed density $f(x,t)$ at time $t = 65$ (solid) from a sinusoidal initial condition (dashed). Right figure shows the time evolution of f at $x = 0.5$. The linear amplification in time is asymptotically observed.

governing equations look

$$i\partial_t n_1 = (k \cdot v_0)n_1 + \frac{k_y n_0'}{B}\text{K}\ n_1 + n_0 k_z v_z, \tag{1}$$

$$i\partial_t v_z = (k \cdot v_0)v_z - \frac{1}{s^2} k_z \text{K}\ n_1, \tag{2}$$

where n_1 is the perturbed density, v_z is the perturbed velocity parallel to the ambient magnetic field, the prime denotes the x-derivative, and $\text{K} = \Delta^{-1}$ with the integral kernel $K(x,\xi)$. Here, the length and the time are normalized by the half of the width of plasma slab, and diocotron frequency (= flow shear), respectively. The magnetic field is supposed on the z direction.

When we assume a sharp boundary of electron plasma, n_0' gives a combination of delta functions which corresponds to that in v_0'' in Rayleigh equation. Equation (1) additionally contains the coupling with parallel dynamics at the last term on the right hand side. We decompose $n_1(x,t) = a(t)\delta(x+1) + b(t)\delta(x-1) + f(x,t)$. As for v_z, we do not need decomposition since the operator K on the right hand side of (2) will regularize it. Then, (1)-(2) are combined to give

$$i\partial_t \begin{pmatrix} a \\ b \\ f \\ v_z \end{pmatrix} = \begin{pmatrix} -\lambda_1 & \frac{k_y}{B}K(-1,1) & \frac{k_y}{B}\int K(-1,x)\cdot dx & 0 \\ -\frac{k_y}{B}K(1,-1) & \lambda_1 & -\frac{k_y}{B}\int K(1,x)\cdot dx & 0 \\ 0 & 0 & \lambda_x & k_z \\ -\frac{k_z}{s^2}K(x,-1) & -\frac{k_z}{s^2}K(x,1) & -\frac{k_z}{s^2}\int K(x,\xi)\cdot d\xi & \lambda_x \end{pmatrix} \begin{pmatrix} a \\ b \\ d \\ v_z \end{pmatrix}, \tag{3}$$

where $\lambda_1 = k_y(2k-1)/(2kB)$ and $\lambda_x = k_y v_y(x)$.

By means of the numerical integration of the system (3) as an initial value problem, we have observed a secular growth of the amplitude. We have chosen the parameter as $s = 10^{-2}$, $k_y = 1$ and $k_z = 10^{-3}$, and taken the initial condition by $f(x,0) = \cos(\pi x/2)$, $v_z(x,0) = 0$, and $a = b = 0$. It is noted that for $k_y < k_c \simeq 0.639$, the system exhibits exponential instability corresponding to KH instability. We are investigating the mode which is KH stable. In Fig. 1, the inner density fluctuation at $t = 65$ is plotted by the solid line (left figure), which shows the local amplification around $x \sim 0.5$. The time evolution at $x = 0.5$ is also shown in Fig. 1 (right figure). The algebraic amplification is clearly observed in $f(x)$ after $t \sim 100$ $[f(x,t) \propto t]$.

It is noted that the localized secularity is appearing at the places where $\lambda_x = \pm\lambda_1$, which are written as $x = \mu_\pm$. Let us pick up the point $x = \mu_+$ and neglect the mixing (integration) term of the generator on the right hand side of (3) (we denote it by A). Then, the obtained matrix may be written as

$$\text{A}\,\mu_+ = \text{T} \begin{pmatrix} -\lambda_1 & 0 & 0 & 0 \\ 0 & \lambda_1 & 1 & 0 \\ 0 & 0 & \lambda_1 & 1 \\ 0 & 0 & 0 & \lambda_1 \end{pmatrix} \text{T}^{-1}, \tag{4}$$

where we have introduced a transform matrix T for the canonization. As we see in (4) the system contains the third order Jordan block, which should lead to the divergence parabolic in time. The reason why $f(\mu_+, t)$ only shows the linear growth is explained by the damping of the surface wave. The damping of the surface wave is originated from the integral terms in the generating matrix which is neglected in (4). It is also noted that such damping can be estimated by means of the renormalization technique when we take $k_z \ll 1$ in (3) and solve it perturbatively. The detailed analysis will be presented elsewhere.

DESTABILIZING EFFECT OF LINEAR SHEAR FLOW

Let us now move on to the problem in the charge-neutral plasma. We consider an incompressible, ideal magnetized plasma in a finite domain $[x \in (-a, a)]$ with a homogeneous interchange (Rayleigh-Taylor) drive. Then, the governing equations for linear dynamics are:

$$(\partial_t + v_y \partial_y)\Delta\phi - v_y'' \partial_y \phi = \frac{B_0 \cdot \nabla}{\mu_0 \rho_0}\Delta\psi + \frac{g}{\rho_0}\partial_y \rho_1, \tag{5}$$

$$(\partial_t + v_y \partial_y)\rho_1 = -\rho_0' \partial_y \phi, \tag{6}$$

$$(\partial_t + v_y \partial_y)\psi = B_0 \cdot \nabla \phi, \tag{7}$$

where ρ, ϕ and ψ denote the mass density, stream function and flux function, respectively. The subscript 0 and 1 denote the ambient (equilibrium) and perturbed quantities. Here we assumed that the velocity field $v_0 = (0, v_y(x), 0)$, and the constant magnetic field $B_0 = (0, B_y, B_z)$ in the Cartesian coordinate. The gravity is supposed to be in the x direction with the gravitational constant g.

If we suppose there is no magnetic field and $\rho_0' > 0$, then we obtain the same set of equations with (1)-(2). Therefore, we conclude that, with the same velocity profile, this system will also exhibit the same secularity with the previous section. On the other hand, if we neglect the density gradient and take $B_0 \neq 0$, then the obtained equation is again quite similar to (1)-(2). The only difference is the existence of the Laplacian operator on the right hand side of (5). The physical situation is quite similar, i.e. the Rayleigh equation coupled with a wave (Alfvén wave or plasma oscillation). It is interesting how the difference appears in the secular behavior, but we do not treat it here.

Hereafter, we assume that the flow profile is linear $[v_y(x) = \sigma x]$. If we replace ∂_t by $-i\omega$ for considering exponential instability, we obtain the spectral ODE:

$$\frac{d^2\phi}{dx^2} - k_y^2 \phi - \frac{2k_y \sigma k_\parallel^2}{\Omega^2(\Omega^2 - k_\parallel^2)}\left(k_y \sigma \phi + \Omega \frac{d\phi}{dx}\right) - \frac{k_y^2 G}{\Omega^2 - k_\parallel^2}\phi = 0, \tag{8}$$

where $G = -\rho_0' g / \rho_0$ and $\Omega = \omega - k_y v_y$ denote the instability drive (when positive) and the locally Doppler shifted frequency of the mode, respectively. The quantities are normalized in space and time by a and a/v_A, where $v_A = B_0/\sqrt{\mu_0 \rho_0}$ is the Alfvén velocity. The $\parallel$ denotes the parallel direction to the ambient magnetic field, and $k_\parallel$ denotes the effect of the Alfvén wave. For simplicity, we assume here that the coefficients of all terms are assumed constant except for the shear flow term.

Let us see how the flow shear acts on interchange instability. First, the eigenvalue problem (8) without shear flow is readily solved to give; the eigenvalues

$$\omega^2 = k_\parallel^2 - \frac{k_y^2 G}{k_y^2 + n^2 \pi^2 / 4}, \tag{9}$$

and the corresponding eigenfunctions

$$\phi^{(n)} = \begin{cases} \cos(n\pi x/2) & \text{for } n: \text{odd} \\ \sin(n\pi x/2) & \text{for } n: \text{even} \end{cases}, \tag{10}$$

for $n \in \mathbf{N}$. Since the zero-th order eigenfunctions span the function space completely, we can invoke the perturbative analysis.

For the most unstable mode $n = 1$, we carry out the perturbative analysis by expanding eigenfunctions and eigenvalues in terms of $\sigma \ll 1$: $\phi = \phi_0^{(1)} + \phi_1 + \phi_2 + \cdots$, and $\omega = \omega_0 + \omega_1 + \omega_2 + \cdots$, where $|\omega_1|/|\omega_0| \sim |\phi_1|/|\phi_0| \sim O(\sigma)$ and so on. The zero-th order eigenvalue and the eigenfunction are (9) and (10) with $n = 1$, respectively.

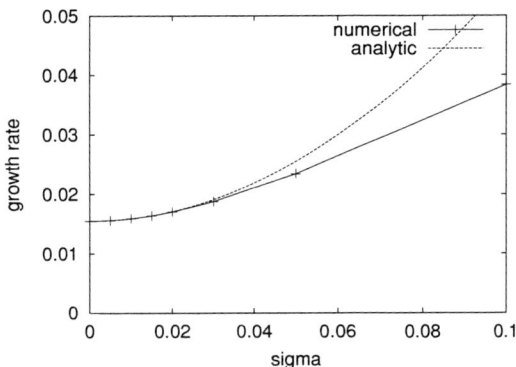

FIGURE 2. Comparison of the growth rate between analytic and numerical solutions for $G = 2.72$ and $k_y = k_\| = 0.5$.

In the first order of σ, we have $\omega_1 = 0$ from the solubility condition. The corresponding eigenfunction ϕ_1 is obtained as

$$\phi_1 = \sum_{m=1}^{\infty} q_m \sin(m\pi x), \qquad (11)$$

where

$$q_m = \frac{2\sigma}{\pi^2 k_y G \omega_0}\left(k_y^2 + \frac{\pi^2}{4}\right)\frac{m(-1)^m}{(m^2-1/4)^2}\left[k_\|^2 + \left(k_y^2 + \frac{\pi^2}{4}\right)\frac{2\omega_0^2}{(m^2-1/4)\pi^2}\right]. \qquad (12)$$

By substituting (11) into the second order equation and invoking the solubility condition again, we finally obtain the following dispersion relation for ω_2:

$$2\omega_0 \omega_2 = -\frac{k_\|^2 \sigma^2}{G}\left(1 + 4P_{4+}C + \frac{16}{\pi^2}P_{4+}^2 B\right) + k_y^2 \sigma^2 \left(\frac{16}{\pi^6} P_{4+} B + 3C\right), \qquad (13)$$

where $B = \sum[m^2/(m^2-1/4)^5] \simeq 4.219581$ and $C = 1/3 - (2/\pi^2) \simeq 0.130691$ are the constants, and $P_{4+} = k_y^2 + (\pi^2/4)$. The first term of (13) denotes the destabilizing effect due to the shear flow and Alfvén wave. It is explicitly shown that the mode with high enough $k_\|$ will be destabilized for given G and k_y.

The result of the perturbative analysis is compared with the numerical shooting in Fig. 2. It is clearly seen that the perturbative analysis agrees perfectly with numerical one when $\sigma \lesssim 0.03$.

SUMMARY

We have presented a few examples of new shear flow destabilization. Firstly, the non-Hermiticity of the generator brings about the unremovable couplings between modes, which is expressed in the form of Jordan matrix. It is also shown that the shear flow is not only a stabilizing tool even if the flow shear is linear. The Alfvén wave is considered to have an effect on the mode to escape from the shear flow stretching.

REFERENCES

1. Z. Lin, T. S. Hahm, W. W. Lee, W. M. Tang, and R. B. White, Science **281**, 1835 (1998); P. W. Terry, Rev. Mod. Phys. **72**, 109 (2000); T. Tatsuno, F. Volponi, and Z. Yoshida, Phys. Plasmas **8**, 399 (2001).
2. H. Helmholtz, Phil. Mag., Ser. 4, **36**, 337 (1868); W. Thomson (Lord Kelvin), Phil. Mag., Ser. 4, **42**, 362 (1871).
3. J. von Neumann, *Mathematical foundations of quantum mechanics* (Princeton Univ., Princeton, 1955).
4. Z. Yoshida and T. Tatsuno, submitted to J. Math. Phys.
5. J. W. S. Rayleigh, Proc. London Math. Soc. **9**, 57 (1880).
6. M. Hirota, T. Tatsuno, S. Kondoh, and Z. Yoshida, Phys. Plasmas **9**, 1177 (2002).

Dynamics and Stability of Multidimensional Solitons in a Plasma

V.Yu. Belashov[1a], S.V. Vladimirov[2b]

[1]*Kazan State Power Engineering University, Russia*
[2]*School of Physics, University of Sydney, NSW 2006, Australia*

Abstract. The formation, structure, stability and dynamics of multidimensional nonlinear waves in a plasma with $\beta \equiv 4\pi nT/B^2 \ll 1$ and $\beta > 1$ are studied. The problem of soliton evolution and collision dynamics is investigated. To study the stability of multidimensional solitons, the variation problem of the Hamiltonian bounding with respect to deformations conserving momentum is used. To study evolution of solitons and their collision dynamics the equations are integrated numerically. It was obtained that in both cases the formation of multidimensional solitons can be observed. These results may be also interpreted in terms of self-focusing phenomenon for the wave beam as a stationary beam formation, scattering and self-focusing. It is found that the soliton elastic collisions can lead to formation of complex structures including the multisoliton bound states.

BASIC EQUATIONS

In this paper, we study formation, structure, stability and dynamics of multidimensional solitons formed on the low-frequency branch of oscillations in a plasma for $\beta \equiv 4\pi nT/B^2 \ll 1$ and $\beta > 1$. These oscillations are described by the equation

$$\partial_t u + A(t,u)u = f, \qquad f = \kappa \int_{-\infty}^{x} \Delta_\perp u \, dx, \qquad \Delta_\perp = \partial_y^2 + \partial_z^2 \;. \tag{1}$$

For

$$A(t,u) = \alpha u \partial_x - \partial_x^2 (\nu - \beta \partial_x - \gamma \partial_x^3) \;, \tag{2}$$

Eq. (1) falls into the GKP (Generalized Kadomtsev-Petviashvili) class of equations, and in the case when $\beta \equiv 4\pi nT/B^2 \ll 1$ for $\omega < \omega_B = eB/Mc$, $k\lambda_D \ll 1$, describe propagation of the fast magnetosonic (FMS) wave in a magnetized plasma with $k_x^2 \gg k_\perp^2$, $v_x \ll c_A$ near the cone of $\theta = \arctan(M/m)^{1/2}$ [1]. In this case, the function u is the dimensionless amplitude of the magnetic field of the wave, $h = B_\sim / B$, the factors at the terms describing nonlinearity, dissipation and dispersion effects, respectively, are defined by plasma parameters and angle $\theta = (\mathbf{B}, \mathbf{k})$. In the opposite case, when

$$A(t,u) = 3s|p|^2 u^2 \partial_x - \partial_x^2 (i\lambda + \nu) \;, \tag{3}$$

Eq. (1) converts into 3-dimensional derivative nonlinear Schrödinger (3-DNLS) equation class and in the case when $\beta > 1$ describes dynamics of the finite-amplitude Alfvén waves propagating nearly parallel to B for $u = h = (B_y + iB_z)/2B|1-\beta|$, $\mathbf{h} = \mathbf{B}_\perp / B_0$ where $p = (1 + ie)$, and e is the "eccentricity" of the polarization ellipse of the Alfvén wave [2]. The upper and lower signs of $\lambda = \pm 1$ correspond to the right and left circularly polarized wave, respectively; the sign of nonlinearity is accounted by the factor $s = \text{sgn}(1-p) = \pm 1$ in the nonlinear term; and $\kappa = -r_A/2$, $r_A = v_A/\omega_{0i}$.

Eqs. (1,2) and/or (1,3) are not completely integrable. Therefore, in the analytical study of these sets we use qualitative analysis of stability and asymptotics of their multidimensional solutions. To study evolution of solitons and

---

[a] Work supported by the Russian Foundation of Basic Research (grant N 01-02-16116).
[b] Work supported by the Australian Research Council.

their collision dynamics the equations are integrated numerically using the special simulation codes. Below, we consider these problems separately.

STABILITY OF 2D AND 3D SOLUTIONS (*GKP EQUATION*)

To study the solutions stability, we perform coordinate transformation and rewrite Eqs. (1,2) into the Hamiltonian form

$$\partial_t u = \partial_x (\delta H/\delta u), \qquad (4)$$

where

$$H = \int \left[-\frac{\varepsilon}{2}(\partial_x u)^2 + \frac{\lambda}{2}(\partial_x^2 u)^2 + \frac{1}{2}(\nabla_\perp \partial_x v)^2 - u^3 \right] d\mathbf{r}, \qquad (5)$$

$\partial_x^2 v = u$, $\varepsilon = \beta|\gamma|^{-1/2}$, $\lambda = \text{sgn}\,\gamma$. The stationary solutions of Eq. (4) are defined from the variation problem, δ (H + vP_x) = 0, where $P_x = \frac{1}{2}\int u^2 d\mathbf{r}$ is the momentum projection onto the x axis, v is the Lagrange's factor, illustrating the fact that all finite solutions of Eq. (4) are the stationary points of the Hamiltonian for fixed P_x. Conforming with Lyapunov's theorem, the stationary points of a dynamical system realizing maximum or minimum of H are absolutely stable; if the extremum is local then the locally stable solutions are possible. The unstable states correspond to monotonous dependence of H on its variables, i.e. to the case when the stationary point is a saddle point. Thus, it is needed to prove the Hamiltonian's boundedness (from below) for fixed P_x.

Let us consider in real vector space R the scale transformations $u(x,\mathbf{r}_\perp) \to \zeta^{-1/2}\eta^{(1-d)/2} u(x/\zeta, \mathbf{r}_\perp/\eta)$ (where d is the problem dimension, and $\zeta,\eta \in$ R) conserving the momentum projection P_x. The Hamiltonian as a function of parameters ζ, η takes a form

$$H(\zeta,\eta) = a\zeta^{-2} + b\zeta^2\eta^{-2} - c\zeta^{-1/2}\eta^{(1-d)/2} + e\zeta^{-4}, \qquad (6)$$

where $a = -(\varepsilon/2)\int(\partial_x u)^2 d\mathbf{r}$, $b=(1/2)\int(\nabla_\perp \partial_x v)^2 d\mathbf{r}$, $c=\int u^3 d\mathbf{r}$, $e = (\lambda/2)\int(\partial_x^2 u)^2 d\mathbf{r}$. In 2D case [d=2 in expression (6)] one can obtain that for $\lambda = 1, \varepsilon \leq 0$ the Hamiltonian at fixed P_x is bounded from below, and, hence, the 2D solitons are absolutely stable in this case. In the cases $\lambda = 1, \varepsilon > 0$ and $\lambda = -1, \varepsilon < 0$ the Hamiltonian H has local minima, and Eq. (4) may have the locally stable solutions for some parameters (see [3] for details). All other cases correspond to unstable 2D solutions.

In 3D case we obtain that the absolutely stable 3D solutions take place for $\lambda = 1, \varepsilon > 0$, and the locally stable solutions can be observed for $\lambda = 1, \varepsilon \leq 0$ if the condition $ab^2 e/c^4 < 9/512$ is satisfied.

It is interesting to note that GKP-equation taking into account, unlike the usual KP-equation, the next order dispersive corrections, has stable 3D solutions. The analysis to the problem of the FMS waves beam's propagation in magnetized plasma enables us to prove [1], for example, that the 3D beam propagating at θ angle to the magnetic field is not focusing and therefore becomes stationary and stable within the cone $\theta < \arctan(M/m)^{1/2}$ when the inequality $(m/M - \cot^2\theta)^2[\cot^4\theta(1+\cot^2\theta)]^{-1} > 4/3$ is satisfied. We also note that obtained results give us the possibility to interpret correctly some numerical and theoretical results on the dynamics of the internal gravity wave solitons induced by the pulse-type sources in the F-region of the ionosphere [4].

STABILITY OF 2D AND 3D SOLUTIONS (*3-DNLS EQUATION*)

We rewrite 3-DNLS equation (4) by performing the formal change $u \to h$ with the Hamiltonian [5]

$$H = \int_{-\infty}^{\infty} \left[\tfrac{1}{2}|h|^4 + \lambda s h h^* \partial_x \varphi + \tfrac{1}{2}\kappa(\nabla_\perp \partial_x w)^2 \right] d\mathbf{r}, \quad \partial_x^2 w = h, \; \varphi = \arg(h), \qquad (7)$$

and (see also [6]) investigate the boundedness of H under its deformations conserving. Consider the scale transformation $h(x,\mathbf{r}_\perp) \to \zeta^{-1/2}\eta^{-1}h(x/\zeta, \mathbf{r}_\perp/\eta)$ ($\zeta, \eta \in \mathbf{C}$) conserving P_x, in complex vector space $\mathbf{C}$. The Hamiltonian as a function of ζ, η is given by

$$H(\zeta,\eta) = a\zeta^{-1}\eta^{-2} + b\zeta^{-1} + c\zeta^2\eta^{-2}, \tag{8}$$

where $a = (1/2)\int |h|^4 d\mathbf{r}$, $b = \lambda s \int hh^* \partial_x \varphi \, d\mathbf{r}$, $c = (\sigma/2)\int (\nabla_\perp \partial_x w)^2 d\mathbf{r}$. Solving the extremum problem for functional (8) we obtain that Hamiltonian (7) is bounded from below, i.e.,

$$H > -3bd/(1+2d^2), \quad b < 0, \tag{9}$$

if $ac^{-1} < d = \left(2\sqrt{2}\right)^{-1}\sqrt{13+\sqrt{185}}$, and in this case 3D solutions of 3-DNLS equation are stable. The solutions are unstable in the opposite case, $ac^{-1} \geq d$, $b < 0$. Condition $b < 0$ corresponds to the right circularly polarized wave with $p = 4\pi nT/B^2 > 1$, i.e. when $\lambda = 1, s = -1$ in Eqs. (1,3), and to the left circularly polarized wave when $\lambda = -1$, $s = 1$. It is necessary to note that the sign change $\lambda = 1 \to -1$, $s = -1 \to 1$ is equivalent to the change $t \to -t, \kappa \to -\kappa$ and for negative κ the Hamiltonian becomes negative in the area "occupied" by the 3D wave weakly limited in the $\mathbf{k}_\perp$-direction; in this case condition (9) is not satisfied. Change of sign of b to positive [when $\lambda = 1$, $s = 1$ or $\lambda = -1$, $s = -1$ in Eqs. (1,3)] is equivalent to the analytical extension of solution from real y, z to pure imaginary values: $y \to -iy$, $z \to -iz$ and, therefore, equivalent to the change of sign of κ in the basic equations. In this case instead of inequality (9) the opposite inequality will take place. From the physical point of view this means that if the opposite inequality is satisfied, the right polarized wave with the positive nonlinearity and the left polarized wave with the negative nonlinearity are stable. Note that in the particular case $\kappa = 0$ in Eqs. (1,3) (1D approximation), instead of inequality (9) and the opposite one, it is easy to obtain the conditions $H > 0$ and $H < 0$, respectively, that is completely in agreement with the results obtained [7] for the 1-DNLS equation.

STRUCTURE AND DYNAMICS OF 2D AND 3D SOLUTIONS (*GKP EQUATION*)

Let us first consider 2D case when $\partial_z = 0$ in Eq. (1). Initial conditions are taken in the form of an exact 2D soliton solution of the usual 2D KP equation [9]. It is obtained that for $\lambda = 1$, $\varepsilon \leq 0$ the formation of stable hump solutions is observed with the asymptotics close to that of algebraic KP soliton (Fig. 1,a). In the case $\varepsilon > 0$ the formation of solitons oscillating in the direction of their propagation and monotonic in the transverse direction was observed (Fig. 1,b).

FIGURE 1. General view of 2D soliton of eqs. (1), (2) with at $\nu=0$: (a) $\lambda=1$, $\varepsilon=-0.6$ ($t=0.2$); (b) $\lambda=1$, $\varepsilon=3.16$ ($t=0.5$).

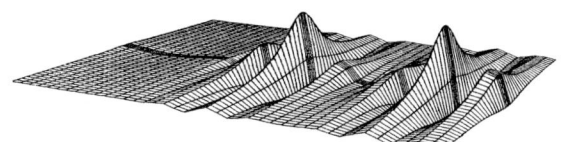

FIGURE 2. 2D bisoliton solution formed from initial pulses with $u_1(0)=1.35$, $u_2(0)=1.3$, $\Delta x(0)=6$ for $\lambda=1$, $\varepsilon=1.9$, $v=0$ ($t=1.3$).

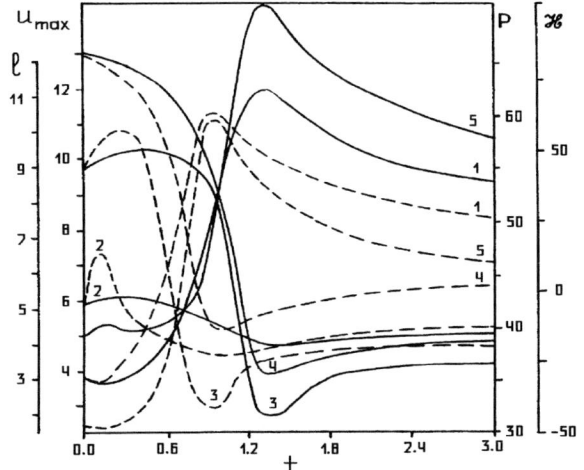

FIGURE 3. Changing with time of the solution parameters, momentum and H of the system for $\lambda=1$ (solid lines - $\varepsilon=-0.45$, dashed lines - $\varepsilon=1.34$): (1) u_{max}; (2) l_x; (3) l_p; (4) P; (5) H.

For $\lambda=-1$ and $\varepsilon\geq 0$, and small $\varepsilon<0$ the evolution of the initial condition leads to the spreading of wave packet which is formed at the first stage. For large absolute values of $\varepsilon<0$, however, we observed the formation of a stable soliton with oscillating asymptotics, that corresponds to above mentioned analytical results. It is interesting to note that the 2D soliton interaction dynamics is not trivial for GKP equation unlike usual KP equation [9]. So, for example, for $\lambda=1$, $\varepsilon>0$ the formation of a stable two-soliton structure (so-called "bisoliton") can be observed as a final result of interaction of two initial pulses (fig. 2). Let us note that for all cases the analysis of the Hamiltonian H deformations on the numerical solutions confirmed the stability of solutions considered above.

In 3D case we obtain three stages of formation and evolution of soliton-like structures [10]. On the first stage, the self-focusing instability is developed (fig. 3), the pulse "wings" fall behind its center, and the amplitude increases sharply. Then, in the instability saturation stage, the equation term being proportional to the fifth derivative begins to play the dominant role owing to the decrease of the pulse characteristic dimensions. On the next stage, the defocusing of the wave field is observed. For $\varepsilon\leq 0$ it leads to the pulse spreading, and for $\varepsilon>0$ the evolution ends with the formation of the 3D soliton (Fig. 3). For $\lambda=-1$ and arbitrary ε, the solutions are in the form of 3D wave packets which spread in time. These results are confirmed by the analysis of the Hamiltonian H bounding with its deformation on numerical solutions. So, unlike 3D KP equation, there are no collapsing solutions for FMS waves propagating at $\theta<\arctan(M/m)^{1/2}$, but stable 3D solutions may be observed.

STRUCTURE AND DYNAMICS OF 2D AND 3D SOLUTIONS (*3-DNLS EQUATION*)

The initial conditions are chosen in two different forms: soliton-like solution, and modulated "plane" wave (see [2]). It is obtained that for $\lambda=1$ and $s=-1$ with large $\kappa>0$ and the initial pulse weakly bounded in the transverse direction [when inequality (9) is satisfied] we observe formation of stable 3D soliton-like solution (Fig. 4). For the opposite signs of λ and s [i.e. equivalent to $t\to -t$, $\kappa\to-\kappa$ in Eq. (1)], the Hamiltonian becomes negative, and the 3D wave is spread. For $\lambda=1$ and $s=-1$ with small $\kappa>0$ and the initial pulse rather strongly bounded in the transverse direction, we observe formation of the 3D collapsing solutions. This effect is typical for all nonlinear systems where there are both unlimited H for fixed "junior" integrals and positive-defined quadratic terms in (7). The series of numerical experiments done for $b>0$ when $\lambda=1$, $s=1$ and $\lambda=-1$, $s=-1$ demonstrates that the initial 3D pulse is unstable and spreads with time. These results are well confirmed by the analysis of H bounding on the numerical solutions.

CONCLUSION

To conclude, we have considered two types of the low-frequency oscillations in a plasma with $\beta \ll 1$ and $\beta > 1$ which correspond to two types of waves and can lead to the formation of the multidimensional solitary wave structures. As a result, we have obtained that for FMS waves the 2D and 3D soliton formation can be observed. In particular, in the 3D case for the FMS wave beam having the small angular distribution, the stationary propagation may be observed as a result of the nonlinear beam stabilization. In the case of Alfvén waves propagating along the magnetic field lines, we have obtained that 3D stable solutions may be observed, with 3D spreading and collapsing ones. These results can be also interpreted in terms of the self-focusing phenomenon for the Alfvén waves' beam as the stationary beam formation, scattering, and self-focusing. Let us note that we observed the dynamics of the Alfvén waves' beam propagating in a plasma with $\beta > 1$ at angles near $0°$ with respect to the magnetic field, and the dynamics of the FMS wave beam propagating in plasma with $\beta \ll 1$ at angles near $\pi/2$ with respect to the magnetic field.

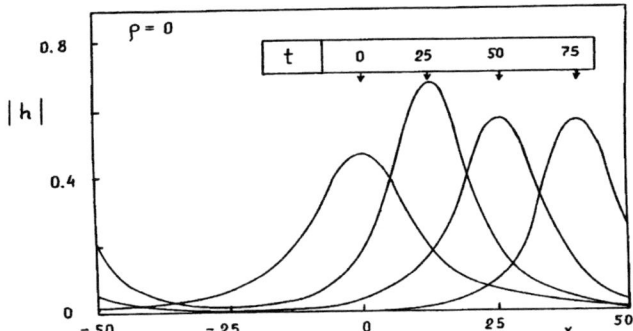

FIGURE 4. Evolution of a 3D right circularly polarized nonlinear pulse for $\lambda=1$, $s=-1$, $\kappa=1$; $H>-3bd/(1+2d^2)>0$.

REFERENCES

1. V.Yu.Belashov, Plasma Phys. and Contr. Fusion, **36** (1994) 1661.
2. V.Yu.Belashov, URSI and STEP/GAPS Workshop, Warsaw, Poland (1995) 9.3.
3. V.Yu.Belashov, Dokl. Acad. Nauk SSSR, **320** (1991) 85.
4. V.Yu.Belashov, Proc. 1989 Intern. Symp. on EMC, Nagoya, Japan, **1** (1989) 228.
5. V.Yu.Belashov, Double layers - Potential Formation and Related Nonlinear Phenomena in Plasmas, World Sci. (1996) 337.
6. V.Yu.Belashov, Proc. ICPP 1996, Contributed Papers, Nagoya, Japan, **1** (1997) 954.
7. S.P.Dawson, C.F.Fontán, Phys. Fluids, **31**, 1 (1988) 83.
8. V.Yu.Belashov, Proc. ISSS-5, Kyoto, Japan (1997) 118.
9. V.I.Karpman, V.Yu.Belashov, Phys. Lett. **154A** (1991) 131.
10. V.I.Karpman, V.Yu.Belashov, Phys. Lett. **154A** (1991) 140.

Laser Envelope Solitons in Plasmas

S. Poornakala*, A. Das*, A. Sen*, P. K. Kaw*, Z. M. Sheng†, Y. Sentoku†, K. Mima† and K. Nishikawa†

Institute for Plasma Research, Bhat, Gandhinagar 382428, India
†*Institute for Laser Engineering, Osaka Univ., Osaka, Japan*

Abstract. Modulated light pulses in which the modulation envelope propagates as an isolated solitary plasma wave are an interesting class of exact one dimensional nonlinear solutions of the relativistic cold plasma model. They have been investigated in great detail in recent years due to their potential applications in various intense laser plasma interaction scenarios including plasma based particle and photon acceleration schemes, fast ignition method of laser fusion and radiation dynamics around a pulsar. We review some of the interesting properties of these solitons and discuss a few fundamental issues related to their existence, spectral properties and the influence of ion dynamics and finite temperature effects. We also present a new class of solitary wave solutions that exhibit an oscillatory structure in the amplitude of the electrostatic potential.

INTRODUCTION

The nature of intense light pulse propagation in a plasma is a subject of great current interest because of its potential application to a variety of physical situations including modern plasma based particle acceleration schemes, table top terawatt (T^3) laser experiments, fast ignition method of heating pre-compressed pellets in laser fusion and astrophysical scenarios like the propagation of radiation in the vicinity of a pulsar. When the intensity of the light wave is such that the electron quiver velocity approaches the speed of light, two dominant nonlinearities affect the light propagation characteristics, namely, the relativistic mass variation of the electron and the ponderomotive effects arising from the **v** × **B** forces, where **B** is the magnetic field associated with the light wave and **v** is the electron fluid velocity. Relativistic mass variation has interesting physical consequences like the lowering of the effective plasma frequency and thereby permitting propagation into overdense plasmas. Ponderomotive forces introduce coupling to longitudinal waves and the excitation of large space charge fields. A good physical model that captures the essential nonlinear features of intense light wave propagation consists of the relativistic cold electron fluid equations and the full set of Maxwell equations. Such a model, introduced and studied in detail long ago by Akhiezer and Polovin [1], permits a wide variety of nonlinear solutions. One particular class of exact one dimensional solutions, that has attracted particular attention in recent times, consists of modulated light pulses coupled to electron plasma waves. The solutions are in the form of isolated solitary pulses (solitons) and in general appear to fall into two distinct classes [2, 3]. The first kind is characterized by a single peak of the laser field inside the soliton and arises as a result of a balance of electrostatic and ponderomotive forces on the electron fluid. The typical physical extent of these solitons are of the order of a few skin depths. In the second kind where the soliton size can be considerably larger, the laser field can have multiple peaks in the near evacuated central region of the soliton and the electon fluid is pushed out to the edges to have a sharp rise there. The spectra of these two kinds of solitons also appear to be different - the multi-peak solitons have a discrete spectrum while the single pulse solitons appear to have a continuum spectrum. Several authors in the past have studied these solutions in detail particularly in the context of the cold plasma model[2, 3, 6, 7, 8, 9, 10, 11]. For example, Esirkepov, et al.,[6] have established the existence of single hump solitons in the limit of zero group velocity and obtained exact analytic expressions of these solutions. The corresponding question of the existence of such solitons for finite group velocity and the nature of transition from the standing single humped solitons to moving solitons of single hump or multihumped-type has not been fully settled or completely understood yet. We address this issue here on the basis of recent numerical investigations and supporting analytic work [8]. A composite picture of the entire spectrum of single peak and multi-peak solitons (in the space of eigenvalues and group velocity) is presented. We next discuss the important modifications in the nature of this spectrum as well as in the properties of the solitons due to ion dynamics[7, 8] and plasma temperature[11] effects. Finally we present a new class of solitons that we have found with our numerical studies that are characterized by multiple peaks in the space charge potential.

MODEL EQUATIONS

The basic equations we adopt for describing the propagation of a circularly polarized intense laser pulse in a warm plasma are the one dimensional relativistic fluid equations and the Maxwell's equations. For nonlinear traveling wave solutions these can be conveniently reduced to the following normalized set of coupled equations [3, 8]

$$\phi'' = n_e - n_i, \tag{1}$$

$$R'' + \frac{R}{1-\beta^2}\left[\frac{\lambda^2}{1-\beta^2} - \frac{n_e}{\gamma_e} - a\frac{n_i}{\gamma_i}\right] = 0, \tag{2}$$

where $n_j = \beta/(\beta - u_j)$ is the density of the species $j = i, e$. The longitudinal velocity u_j can be obtained from the conservation equation for the perpendicular canonical momentum for the electron and ion fluids viz $\gamma_e(1-\beta u_e) - \phi + \alpha_e \Gamma_e \log n_e = \gamma_{e0}$, $\gamma_i(1-\beta u_i) + a(\phi + \alpha_i \Gamma_i \log n_i) = \gamma_{i0}$. Here, $a = m_e/m_i$ is the electron to ion mass ratio, α_j is the temperature (nonrelativistic) of the species j and Γ_j is the adiabatic factor. The other definitions and normalizations are the same as used in Ref.[3, 8] R is the modulation envelope of the light pulse and is nonlinearly coupled to the electrostatic potential ϕ of the plasma wave. Our interest is to seek coupled solitary pulse solutions of the above equations that are propagating with speed β. The boundary conditions for such pulses are $R = R_0, R' = 0$ for $\xi = \pm\infty$, $\gamma_{e0} = \gamma_{i0} = 1$ for bright solitons and $R = R_0, R' = 0$ at $\xi = \pm\infty$ for dark solitons.

COLD PLASMA

We first discuss the topic of cold plasma bright solitons [2, 3, 7, 8] of Eq.(1-2) (i.e. $\alpha_e = \alpha_i = 0$) with an immobile ion background. The interesting limit is the overdense region with slowly propagating solitons (i.e. $\beta << 1$). Such solitons would be of interest for example to transport energy into a pre-compressed pellet in the fast ignition scheme of laser fusion. In past work, Esirkepov, et al.,[6] have obtained an exact analytical solution of a soliton in the limit of zero group velocity. It has a single hump in the light wave field and a continuous spectrum over a limited range of the normalized frequency $\lambda = \omega(1-\beta^2)$. The upper limit to the amplitude (lower limit on λ) arises from the physical constraint that the electron density remains non-negative. In a subsequent work Farina, et al.,[7] have stated that the model equations do not permit any moving single hump soliton and hence no connection exists between the standing ($\beta = 0$) and moving pulses ($\beta \to 1$). Our numerical investigations [8] however reveal that moving single pulse solitons do exist and can have fairly large amplitudes with significant charge separation fields. They appear to have a smooth spectrum over a range of λ and a smooth transition to the $\beta = 0$ solutions of Esirkepov et al[6]. In Fig. 1a. the thickened portion on the vertical axis corresponds to the existence region of the Esirkepov, et al solitons and the shaded region shows the existence space of the single hump solitons obtained by our numerical studies. The upper curve given by $\lambda = 1 - \beta^2$ corresponds to the vanishing amplitude limit of the solitons. Close to that curve the soliton solutions are akin to exact solutions of the nonlinear Schrodinger equation and this analytic limit has been discussed earlier in the work of Kaw et al[3]. Note that near $\beta = 0$, the shaded region has a broad width ($\Delta\lambda/\lambda \approx 20\%$) and connects smoothly to standing solutions. As one decreases λ for a given β, single hump solutions with significant density perturbations exist up to a certain critical λ. In Fig.1b we show an example of a single hump solution with significant density perturbation. Within our numerical accuracy, the eigenvalue spectrum appears to be continuous in the overdense plasma. However the question of the true nature of the spectrum (in the strictest mathematical sense) still remains an open issue and a perturbative analysis of the type carried out by Dimant, et al.,[5] for the underdense case remains to be done in this overdense region. For finite ion mass case, i.e ($a = 5 \times 10^{-4}$), the smooth connection to the $\beta = 0$ solutions is broken. Litvak et al.,[2] and Farina, et al.,[7] have earlier investigated the ion mass effect on the envelope solitons. However, their analysis was restricted to quasineutral solitons. Our numerical analysis reveals the existence of non-quasineutral single hump solutions with finite amplitudes[8].

The solutions corresponding to the curves below the single hump region in Fig.1a (for immobile ion case) have multiple peaks in R with a large single peak in ϕ. These solutions occur for discrete eigen values. One such solution is shown in Fig 1c. Farina, et al.,[7] have shown that when ion dynamics is included, these multipeak solutions undergo wave breaking near a critical value of β and lead to ion acceleration.

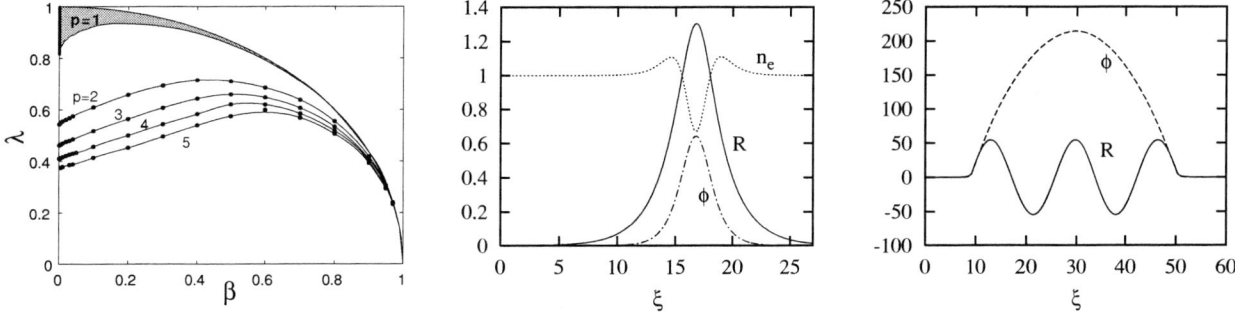

FIGURE 1. *a.* $\lambda - \beta$ diagram showing the region of existence of single hump and multi-peak solitons for immobile ion case. *b.* Finite amplitude solution for $\beta = 0.01$ and $\lambda = 0.87$. *c.* Multi-peak solution for $\beta = 0.01$ and $\lambda = 0.37719$.

WARM PLASMA

In the limit when the group velocity β of the pulse tends to zero, the cold plasma approximation is not strictly valid. Infact there is now an interesting interplay of effects arising from ion dynamics and temperature effects as the group velocity crosses various limits associated with electron and ion thermal velocities and the ion acoustic speed. To delineate the nature of the solutions in these regions we have carried out extensive numerical investigation of a somewhat reduced model in which a weak relativistic limit has been taken and the temperature terms are modeled in a nonrelativistic fashion. Our results are summarized in the schematic diagram of Fig. 2a. where we show the existence regions of bright and dark solitons in various ranges of β. As is known, in a cold plasma with ion motion included, bright solitons cease to exist for $\beta < \sqrt{a}$. However, when thermal effects are included, we find another regime of bright solitons for group speeds below the ion acoustic speed. In the intermediate region, dark solitons exist[11]. It is possible to understand the basic characteristics of these solutions from simple analytic arguments. For example, in the case of bright solitons propagating with subsonic speeds, when the amplitude is small i.e $R^2 \ll 1$, the ion inertia becomes negligible, and also plasma becomes quasineutral as the density perturbations becomes small. Hence the light wave equation i.e Eq.(2) can be reduced to a nonlinear Schrodinger equation whose solution is $R(\xi) = \sqrt{\alpha} R_0(\lambda,\beta) \text{sech}(k(\lambda,\beta)\xi)$ This solution can be contrasted with the cold plasma solutions. In the small amplitude limit, the solutions have identical form except for the fact that the warm plasma solutions have an explicit factor of $\sqrt{\alpha}$. Also, the space charge field is negative. For these speeds, the dominant nonlinearity comes from striction nonlinearity rather than the relativistic nonlinearity. In Fig. 2b we compare the warm and cold plasma solutions of the same amplitude. Clearly for solutions having same amplitude, the warm plasma solutions have smaller width than the cold plasma. In Fig .2c we present the numerical solution of Eqs.(1- 2) for cold ions ($\alpha_i = 0, a = 5 \times 10^{-4}$) and warm isothermal electrons ($\alpha_e = 0.1$). The warm plasma supports existence of multiple peak structures in R with a negative ϕ. This solution is different from that of the cold plasma where ϕ is always positive.

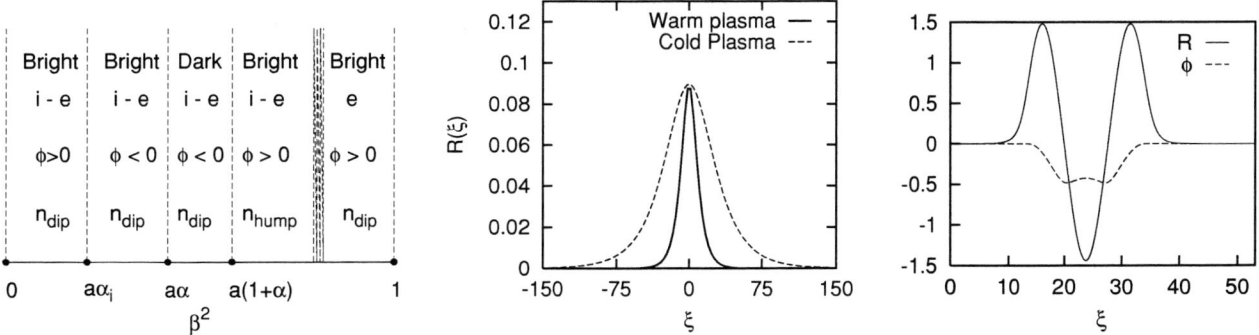

FIGURE 2. *a.* Schematic showing different regimes of existence of bright and dark solitons. *b.* Comparison of warm and cold plasma quasi-neutral solutions with same amplitude. *c.* Two peak structure in R for $\beta = 0.0004$ and $\lambda = 0.4279439$, the scale potential ϕ is negative in contrast to cold plasma.

NEW CLASS OF SOLITONS IN COLD ELECTRON PLASMA

Our numerical investigations have also revealed a whole new class of envelope soliton solutions which are characterized by multiple oscillations in ϕ. For these solutions the electron density structure resembles a grating-like structure. Fig.3a shows one such class with oscillations in ϕ and R has one positive and one negative peak. In Fig.3b we show

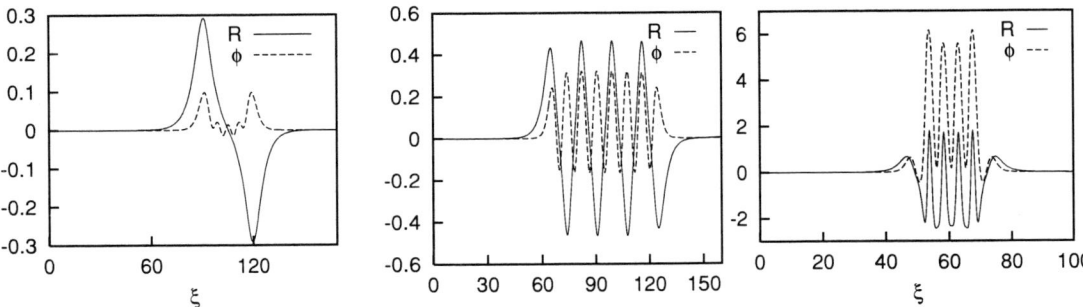

FIGURE 3. *a.* Oscillating ϕ solutions with one positive and one negative peak in R for $\Lambda = 0.041525$. *b.* Multiple peaks in R ϕ for $\Lambda = 0.084409$, ϕ oscillates about zero. *c.* Multiple peaks in R ϕ for $\Lambda = 0.16987$, ϕ oscillates about a mean value and R can have multiple peaks

another class of solutions with R and ϕ having multiple peak structures where ϕ oscillates about zero. In Fig.3c, we show yet another class with ϕ oscillating about a mean value and enclosing the R solutions. These solutions which display a rich variety of structures offer a new class of nonlinear stationary solutions whose properties and potential applications need to be explored.

CONCLUSION

In conclusion we have reviewed a class of laser envelope solitons and discussed various aspects of their properties and existence regimes. We have also pointed out some interesting open issues concerning the nature of their spectra and discussed warm plasma and ion dynamical effects. The model equations permit a large variety of solitonic solutions and we have presented one such new class with characteristic potential oscillations. The nature and applications of these solutions need further detailed studies and such work is in progress.

REFERENCES

1. A. I. Akhiezer, and R. V. Polovin, *Sov. Phys. JETP* **3**, 696 (1956).
2. V. A. Kozlov, A. G. Litvak, and E. V. Suvorov, *Sov. Phys. JETP* **49**, 75 (1979).
3. P. K. Kaw, A. Sen and T. Katsouleas, *Phys. Rev. Lett.* **68**, 3172 (1992).
4. H. H. Kuehl and C. Y. Zhang, *Phys. Rev.* **E48**, 1316 (1993).
5. Y. S. Dimant, R. N. Sudan, and O. B. Shiryaev, *Phys. Plasmas* **4**, 1489 (1997).
6. T. Zh. Esirkepov, F. F.Kamenets, S. V. Bulanov, *et al*, *JETP. Lett.* **68**, 36 (1999).
7. D. Farina and S. V. Bulanov, *Phys. Rev. Lett.* **86**, 5289 (2001); *Plasma Phys. Rep.* **27**, 641 (2001).
8. S. Poornakala, A. Das, A. Sen, and P. K. Kaw, *Phys. Plasmas* **9**, 1820 (2002)
9. M. Y. Yu, P. K. Shukla, and K. H. Spatschek, *Phys.Rev. A* **18**, 1591 (1978).
10. N. Nagesha Rao, R. K. Verma, and P. K. Shukla *et al.*, *Phys.Fluids* **26**, 2488 (1983).
11. S. Poornakala and A. Das, A. Sen, *et al* To be published in *Phys. Plasmas* **9**, (2002);

Ion-Acoustic Cnoidal Waves In A Plasma With Negative Ions

Lakhan Lal Yadav

Department of Physics, Faculty of Science, Kigali Institute of Education, P.O. Box 5039, Kigali, Rwanda

Abstract. Using the reductive perturbation method, we present a theory of different nonlinear periodic waves, viz. the Korteweg-de Vries and modified KdV (mKdV) cnoidal waves, in a plasma with negative ions, which in the limiting case reduce to localized structures, namely KdV compressive or rarefactive solitons, and mKdV compressive and rarefactive solitons, respectively. It is found that the amplitude dependence of frequency is different for KdV and mKdV cnoidal waves.

BASIC EQUATIONS

There has been a great deal of interest in the theoretical and experimental study of Korteweg-de Vries (KdV) and modified KdV (mKdV) ion-acoustic solitons in a plasma with negative ions [1-5 and references therein]. The aim of this paper is to present a theory of nonlinear periodic waves, namely KdV and mKdV cnoidal waves, in a plasma with negative ions which in the limiting case reduce to KdV compressive or rarefactive solitons, and compressive and rarefactive mKdV solitons, respectively.

The dynamics of the one-dimensional ion-acoustic waves in a plasma with cold positive and negative ion species, and hot isothermal Maxwellian electrons is described by the following normalized fluid equations:

$$\frac{\partial n_i}{\partial t} + \frac{\partial}{\partial x}(n_i v_i) = 0, \tag{1}$$

$$\frac{\partial v_i}{\partial t} + v_i \frac{\partial v_i}{\partial x} = \beta_i \frac{\partial \phi}{\partial x}, \tag{2}$$

$$\frac{\partial^2 \phi}{\partial x^2} = n_e + z_2 n_2 - z_1 n_1, \tag{3}$$

where $n_e = \exp(\phi)$ is the density of electrons. Subscripts $i=1, 2$ in equations (1, 2) refer to positive and negative ion species. $\beta_1 = -1/\beta$, $\beta_2 = \varepsilon_z/(\beta\mu)$, $\beta = \alpha_{10}\gamma z_1$, $\gamma = 1 + \varepsilon_z^2 \alpha/\mu$, $\varepsilon_z = z_2/z_1$, $\alpha = \alpha_{20}/\alpha_{10}$, $\alpha_{10} = n_{10}/n_{e0}$, $\alpha_{20} = n_{20}/n_{e0}$, and $\mu = M_2/M_1$. In equations (1-3) densities, fluid velocities, space, time and potential are normalized by the equilibrium electron density, ion-acoustic velocity, Debye length, inverse of effective ion-plasma frequency and electron thermal potential, respectively.

THE KDV EQUATION AND ITS PERIODIC WAVE SOLUTION

To solve equations (1-3), we introduce the stretched coordinates $\xi = \varepsilon^{1/2}(x-st)$, $\tau = \varepsilon^{3/2}t$, where ε is a small parameter and s is the phase velocity of the wave, and expand the variable quantities about their equilibrium values in powers of ε. Using the reductive perturbation method and appropriate boundary conditions [6], we obtain the KdV equation

$$\frac{\partial \phi}{\partial \tau} + a\phi\frac{\partial \phi}{\partial \xi} + C\frac{\partial \phi}{\partial \xi} + \frac{1}{2}\frac{\partial^3 \phi}{\partial \xi^3} = 0, \tag{4}$$

where

$$a = \frac{1}{2}\left[3\left(\frac{z_1\alpha_{10}}{\beta^2} - \frac{z_2\alpha_{20}\varepsilon_z^2}{\beta^2\mu^2}\right) - 1\right], \tag{5}$$

$C = \frac{z_1\alpha_{10}}{\beta}C_1^{(1)} + \frac{z_2\alpha_{20}\varepsilon_z}{\beta\mu}C_2^{(1)}$, $\alpha_{10} = [z_1(1-\alpha\varepsilon_z)]^{-1}$, $\alpha_{20} = \alpha\varepsilon_z[z_2(1-\alpha\varepsilon_z)]^{-1}$, $\phi = \phi^{(1)}$. $C_1^{(1)}$ and $C_2^{(1)}$ are integration constants, the perturbed fluid velocities of positive and negative ions where ϕ vanishes.

For the steady state solution of the KdV equation, we consider $\eta = \xi - u_1\tau$, where u_1 is a constant velocity. Integrating twice with respect to η, we obtain

$$\frac{1}{2}\left(\frac{d\phi}{d\eta}\right)^2 + V(\phi) = 0 \tag{6}$$

where the Sagdeev potential $V(\phi)$ is given by

$$V(\phi) = \frac{a}{3}\phi^3 - u\phi^2 + \rho_0\phi - \frac{1}{2}E_0^2, \tag{7}$$

where ρ_0 and E_0 are, respectively, the charge density and electric field where ϕ vanishes, and $u = u_1 - C$. A cnoidal wave solution of equation (6) is given by

$$\phi = \alpha_2 + (\alpha_1 - \alpha_2)cn^2\{D\eta, m\}, \tag{8}$$

where cn is the Jacobi elliptic function. The parameter m, called modulus, and D can be expressed in terms of the three real zeros of Sagdeev potential, α_1, α_2 and α_3, as:

$$m^2 = \frac{\alpha_2 - \alpha_1}{\alpha_3 - \alpha_1}, \quad D = \sqrt{\frac{(\alpha_1 - \alpha_3)a}{6}}. \tag{9}$$

The α_1, α_2 and α_3 are such that $\alpha_1 > \alpha_2 \geq \alpha_3$ for $a > 0$ and $\alpha_1 < \alpha_2 \leq \alpha_3$ for $a < 0$.

$$u_1 = C \pm \frac{a}{3}(\alpha_1 + \alpha_2 + \alpha_3). \tag{10}$$

The upper (lower) sign in equation (10) corresponds to $a > 0$ (< 0). The wavelength λ of the cnoidal wave is defined as

$$D\lambda = 2K(m), \tag{11}$$

$K(m)$ is the first kind of complete elliptic integral. Using the conservation condition of number density

$$\int_0^\lambda (z_1 n_1 - z_2 n_2 - 1) d\eta = 0, \quad (12)$$

we can find the three real zeros of Sagdeev potential in terms of modulus m. To evaluate (12), we assume that $n_i^{(2)}$ and higher order terms are very much less than $n_i^{(1)}$.

Using first order solutions and relations (8), (10), and (12), we obtain

$$u_1 = C \pm \frac{Aa}{3} \left\{ \frac{1}{m^2}(2 - 3H(m)) - 1 \right\} \quad (13)$$

The frequency $f = V/\lambda$, of the cnoidal wave can be given by (8), (9), (11) and (12), as

$$f = \frac{V}{2K(m)m} \left[\pm \frac{aA}{6} \right]^{1/2}. \quad (14)$$

The velocity V of the cnoidal wave in the lab frame is given by, $V = s + u_1 = 1 + u_1$, using equation (13)

$$V = 1 + C \pm \frac{Aa}{3} \left\{ \frac{1}{m^2}(2 - 3H(m)) - 1 \right\}. \quad (15)$$

THE MKDV EQUATION AND ITS PERIODIC WAVE SOLUTION

The KdV equation (5) is no longer valid when 'a' becomes zero. For example, for an (Ar^+, F^-) plasma, at a critical value of $\alpha = \alpha_c = 0.1024$, '$a$' vanishes (eq. (5)). To consider the ion-acoustic waves in this situation, higher order nonlinearity must be taken into account. Hence, we use the modified stretching and appropriate boundary conditions [6] to derive the mKdV equation

$$\frac{\partial \phi}{\partial \tau} + b\phi^2 \frac{\partial \phi}{\partial \xi} + C \frac{\partial \phi}{\partial \xi} + \frac{1}{2} \frac{\partial^3 \phi}{\partial \xi^3} = 0, \quad (16)$$

where

$$b = \frac{1}{4} \left[15 \left(\frac{z_1 \alpha_{10}}{\beta^3} + \frac{z_2 \alpha_{20} \varepsilon_z^3}{\beta^3 \mu^3} \right) - 1 \right]. \quad (17)$$

For the steady state solution of the mKdV equation (16), we consider $\eta = \xi - u_1 \tau$, where u_1 is a constant velocity. Integrating twice with respect to η, we obtain

$$\frac{1}{2} \left(\frac{d\phi}{d\eta} \right)^2 + V(\phi) = 0, \quad (18)$$

where the Sagdeev potential V(φ) is given by

$$V(\phi) = \frac{b}{6} \phi^4 - u\phi^2 - \frac{1}{2} E_0^2 = \frac{b}{6} \left[(\phi^2 - \alpha_5)(\alpha_4 - \phi^2) \right], \quad (19)$$

$u = u_1 - C$. For the plasma considered here, b is positive for any possible values of μ, z_1 and z_2, therefore, equation (18) gives the mKdV cnoidal wave solution [6]

$$\phi = \sqrt{\alpha_4} cn\{D\eta, m\}, \quad (20)$$

where

$$m^2 = \frac{\alpha_4}{\alpha_4 - \alpha_5}, \quad D = \left[\frac{b(\alpha_4 - \alpha_5)}{3}\right]^{1/2}, \tag{21}$$

with $\alpha_4 > 0$ and $\alpha_5 \leq 0$. The coefficients α_4 and α_5 can also be expressed in terms of modulus m and the amplitude of the cnoidal wave

$$\alpha_4 = A^2/4 \text{ and } \alpha_5 = A^2(m^2 - 1)/(4m^2). \tag{22}$$

The frequency f of the mKdV cnoidal wave is given by equations (11), (21) and (22), as

$$f = \frac{VA}{4mK(m)}\left[\frac{b}{3}\right]^{1/2}, \tag{23}$$

where the velocity of the mKdV cnoidal wave $V = s + u_1 = 1 + u_1$ is given by

$$V = 1 + C + b\{2 - (1/m^2)\}A^2/24. \tag{24}$$

DISCUSSION

As the coefficient b in mKdV equation (16) is positive for any possible values of μ, z_1 and z_2 for negative-ion plasma, therefore, unlike the two-electron-temperature plasma [6], the plasma with negative ions does not support the snoidal waves. From equations (14) and (23), it is clear that the amplitude dependence of frequency is different for KdV and mKdV cnoidal waves. From equation (15), we find that for values of modulus m above (below) a critical value $m_c \approx 0.9486$, the velocity of the KdV cnoidal wave increases (decreases) linearly with amplitude. It is clear from equation (25) that the velocity of mKdV cnoidal wave increases (decreases) with its amplitude above (below) a critical value of m, $m_c = 1/\sqrt{2}$.

In the limiting case, $m \to 1$, the KdV cnoidal wave solution (8) is reduced to the KdV soliton solution [2 and references therein]

$$\phi = \phi_m \sec h^2(\eta/\delta), \tag{25}$$

where the amplitude of the soliton $\phi_m = 3u_1/a$ and width of the soliton $\delta = (2/u_1)^{1/2}$. Here, positive (negative) 'a' corresponds to the compressive (rarefactive) soliton. In the limit, $m \to 1$, the mKdV cnoidal wave solution (20) is reduced to the mKdV soliton solution [1, 2]

$$\phi = \phi_m \sec h(\eta/\delta), \tag{26}$$

where the amplitude of the soliton $\phi_m = \pm(3u_1/b)^{1/2}$ and width of the soliton $\delta = (1/2u_1)^{1/2}$. The plus and minus signs in the equation of amplitude represent the compressive and rarefactive solitons, respectively. It shows that compressive and rarefactive mKdV solitons can exist in the plasma with same parameters. It should be noted that the coexistence of compressive and rarefactive mKdV solitons has been experimentally verified in a plasma with negative ions [3-5]. The periodic signals appear frequently in auroral and magnetospheric plasmas [7], which may be due to nonlinear periodic waves. The periodic solitons, which are similar to nonlinear periodic waves, have been excited in laboratory plasmas [8]. We infer that the present theory may be useful to explain the periodic signals as well as localized solitary structures in laboratory and space plasmas where positive and negative ion species exist.

ACKNOWLEDGEMENTS

The author thanks the Local Organizing Committee of ICPP-2002 for providing the financial support to present this work at ICPP. He also thanks the Management of KIE for the encouragement and support.

REFERENCES

1. Watanabe, S., J. Phys. Soc. Jpn. **53**, 950-956 (1984).
2. Mishra, M.K. and Chhabra, R.S., Phys. Plasmas **3**, 4446-4454 (1996).
3. Nakamura, Y. and Tsukabayashi, I., Phys. Rev. Lett. **52**, 2356-2359 (1984).
4. Cooney, J.L., Aossey, D.W., Williams, J.E. and Lonngren, K.E., Phys. Rev. E **47**, 564-569 (1993).
5. Nakamura, Y., Bailung, H. and Lonngren, K.E., Phys. Plasmas **6**, 3466-3470 (1999).
Yadav, L.L., Tiwari, R.S., Maheshwari K.P. and Sharma, S.R., Phys. Rev. E **52**, 3045-3052(1995).
6. Boström, R., Gustafsson, G., Holback, B., Holmgren, G., Koskinen, H. and Kintner, P., Phys. Rev. Lett. **61**, 82-85 (1988).
7. Nagasawa, T., Phys. Plasmas **6**, 3471-3476 (1999).

Turbulence-Double-Layer Synergetic Auroral Electron Acceleration

Altair Souza de Assis

UFF – GMA, 24001-970 Niterói, RJ - Brasil

Abstract. We discuss afresh the problem of the auroral electron acceleration based on the controversy reports of D. A. Bryant, et al., Phys. Rev. Lett. 68, **37** (1992), and J. Borovsky, Phys. Rev. Lett. **69**, 1054 (1992), related to which mechanism is more tenable to accelerate auroral electrons: dc electric field generated by double–layers or wave-particle interaction due to wave turbulence? Here, we show that both mechanisms are important, and what is most likely to happen in aurora is that the turbulence and the double-layer will assist each other so as to synergetically accelerate those electrons.

INTRODUCTION

In earlier papers, several space plasma scientists have shown the importance of static electric fields to the auroral acceleration process and reported that intense auroral electron fluxes are generated by quasistatic potential structures [1,2,3]. On the other hand, Bryant and collaborators [4], have shown that auroral electron fluxes can also be formed by pure wave turbulence activities. These two theories have successfully explained a majority of ground and spacecraft measurements showing that the acceleration pattern is related to the structure of the observed background electric fields or wave turbulence. Extending this discussion further, Bryant et al. [5], criticized the former acceleration model saying that double-layer could not accelerate auroral electrons to create auroras, and cited several references and physical conjectures to support their ideas. Then Borovsky [6] answered this criticism, and showed that double–layers indeed do accelerate auroral electrons and can then form auroras. However, after all the discussions the former still maintains his position that auroral electron acceleration by double-layer is fundamentally untenable [8,9]. Later new papers by the same authors, on the same subject, were published, leaving untouched the possibility of the synergy in the turubulence-double-layer auroral electron acceleration. However, examining observations of rockets and satellites, it is clear that the two structures coexist, and the flux enhancement is clearly seen [10 and references therein]. Though the reports on such events are few in literature, their existence cannot readily be explained by the two current theories cited above, and therefore a further explanation is necessary. In this paper, we present a theoretical discussion that supports the conjecture of wave-particle interaction assisting the auroral electron acceleration due to a dc electric field working so as to enhance the electron flux. We access the field aligned electron acceleration using the Fokker-Planck equation, where the turbulence is modeled by the weak turbulence theory. The result is valid for the turbulence induced by lower hybrid waves - LHW, kinetic Alfvén waves - KAW, and electromagnetic ion cyclotron waves - EMICW.

AURORAL ELECTRON FLUX

The electron flux is accessed by the equation:

$$\frac{\partial}{\partial t} f(v_\parallel, t) = \frac{\partial}{\partial v_\parallel}\left[D_{wave}(v_\parallel, t)\frac{\partial}{\partial v_\parallel} f(v_\parallel, t)\right] + \frac{eE_{DL}}{m_e}\frac{\partial}{\partial v_\parallel} f(v_\parallel, t)$$
$$+ \frac{\partial}{\partial v_\parallel}\left\{v_A\left[v_\parallel f(v_\parallel, t) + (v_{the})^2 \frac{\partial}{\partial v_\parallel} f(v_\parallel, t)\right]\right\}$$
(1)

where $f(v_\parallel, t)$ is the time averaged field aligned velocity electron distribution function, which evolves in the slow time scale due to turbulence, anomalous collisions induced by turbulence and the double-layer dc electric field. The terms: $D_{wave}(v_{//}, t)$, E_{DL}, $v_A v_{//}$, $v_A v_{the}^2 = D_C^A$, v_{the}, e, and m_e are the wave diffusion coefficient, the field aligned dc electric field, the anomalous friction term, the anomalous collisional diffusion, the electrons thermal speed, the electron's charge, the electron's mass, respectively. The last term on the right hand side of equation (1) is used to permit the electron distribution function to return to the original starting Maxwellian state, it simulates, via turbulence, the effect of the Fokker-Planck's Coulomb collision operator. The Maxwellization of the electrons by turbulence is called the "Langmuir paradox" [11].

Considering now the steady-state regime equation (1) can be written as follows:

$$\frac{d}{dv_\parallel}\left\{v_A\left[v_\parallel f(v_\parallel) + v_{the}^2 \frac{d}{dv_\parallel} f(v_\parallel)\right]\right\} \equiv$$
$$\frac{d}{dv_\parallel}\left[D_{wave}(\frac{d}{dv_\parallel} f(v_\parallel))\right] + \frac{eE_{DL}}{m_e}\frac{d}{dv_\parallel} f(v_\parallel) + A\delta(v_\parallel) = 0 \quad , \tag{2}$$

where A is a source term and $\delta(v_\parallel)$ is the Dirac's delta distribution. In the above equation, A should be interpreted as the electron flux induced by the dc electric field [E_{DL}] assisted by the wave turbulence [$D_{wave}(v_{//})$]. The auroral emission is related to A, and this is the important quantity to be known. After introducing the step function $H(x)$ [$\frac{d}{dx}H(x) = \delta(x)$] in the source term [$A\delta(x) \Rightarrow A\frac{d}{dx}H(x)$], in equation (2), all terms are total derivatives with respect to $v_\parallel$, and can then be immediately integrated to yield,

$$\Pi(v_\parallel) f(v_\parallel) + \Theta(v_\parallel)\frac{df}{dv_\parallel} = C - AH(v_\parallel) \quad , \tag{3}$$

where a convenient regrouping of terms has been carried out. In equation (3): $\Pi(v_\parallel) = v_A v_z + eE_{DL}/m_e$ and $\Theta(v_\parallel) = D_{wave}(v_\parallel) + D_C^A$. In order to determine the constant of integration C, we consider the limit $v_\parallel \rightarrow +\infty$, where $H(v_\parallel)$ is equal to unity, $df(v_\parallel)/dv_\parallel$ as well as v_A are equal to zero. Hence, (3) becomes

$$f(+\infty)\frac{eE_{DL}}{m_e} = C - A \tag{4}$$

The term on the left-hand side of (4) is the flux of the runaway electrons, which in the steady-state must be equal to the source constant A. Consequently, consistency requires that the integration constant C be zero. As a consequence, we can write the auroral electron flux as:

$$A = \frac{eE_D^A f(v_\parallel = 0)}{\sqrt{2\pi} m_e}\left(\frac{E_{DL}}{E_D^A}\right)^{\frac{3}{2}} \exp\left(-\frac{1 + \frac{v_{\parallel}}{v_C^A}\frac{D_{wave}}{D_C^A}}{2\frac{E_{DL}}{E_C^A}(1 + \frac{D_{wave}}{D_C^A})}\right). \tag{5}$$

This equation shows that the electron flux induced by the dc electric field structure is enhanced by the presence of the turbulence since the exponential goes to zero smoother if $D_{wave} \neq 0$.

Solving equation (5), for the general case, numerically, we can show: (1) the electron fluxes can be enhanced by orders of magnitude over that without turbulence, (2) for any wave phase velocity condition and turbulence spectral width, the enhancement is weaker the weaker the double-layer electric field, (3) the enhancement is caused by the Cherenkov damping and it saturates when the plateau in the distribution function is reached, the spectral width of the turbulence plays a more important role on the enhancement than the spectral shape does, and finally (5) the following synergetic effects can be present in this kind of acceleration process:

(a) Electric field-turbulence: $[A_{E_{DL} \neq 0}^{D \neq 0} > A_{E_{DL}=0}^{D \neq 0} + A_{E_{DL} \neq 0}^{D=0}]$,

(b) Turbulence-turbulence $[A_{E_{DL} \neq 0}^{D_{wave1}+D_{wave2} \neq 0} > A_{E_{DL} \neq 0}^{D_{wave1} \neq 0} + A_{E_{DL} \neq 0}^{D_{wave2} \neq 0}]$.

CONCLUSION

In conclusion, we have shown that auroral magnetic field aligned dc electric field structures and auroral Alfvénic and/or lower hybrid turbulence can work synergetically so as to enhance the observed auroral electron fluxes induced by the former. Therefore, it is not possible to explain all the auroral electron energy observations, by the satellites and rockets [10 and references therein; 12], if one considers the acceleration by these two mechanisms separately, the synergetic effects must be considered as well, being the main case the enhancement the one observed at the edges of auroral arcs. Note that $f(v_\parallel = 0)$ depend on the local plasma density, so showers of electrons will be generated at different auroral regions.

ACKNOWLEDGMENTS

The trip to attend the ICPP-2002 was partially funded by FAPERJ – The Rio de Janeiro State Research Foundation. This work was done within the framework of the Associateship Scheme of the International Centre for Theoretical Physics -ICTP, Trieste, Italy. The author would like to thank ICTP and the Royal Institute of Technology – The Alfvén Laboratory, Stockholm, Sweden for the research support and also to thank G. Marklund, T. J. Karlsson, and C-G Falthammar, for fundamental discussions on auroral plasmas.

REFERENCES

[1] T. Sato, and H. Okuda, Numerical Simulation on Ion Acoustic Double-Layer, Princeton Plasma Physics Laboratory, Report PPPL – 1681, UC 20 (1980)
[2] F. S. Mozer, C. A. Cattell, M. K. Hudson, R. L. Lysak, M. Temerin, and R. B. Torbert, Space Science Reviews **27**, 155 (1980).
[3] G. T. Marklund, I. Sandahl, and H. Opgenoorte, P. Spac. Sci., **30**, 179 (1982).
[4] D. A. Bryant,, Contemp. Phys., **35**, 165 (1994).
[5] D. A. Bryant, R. Birgham, and U. de Angelis, Phys. Rev. Lett., **68**, 37 (1992).
[6] J. E. Borovsky, Phys. Rev. Lett. **69**, 1054 (1992).
[7] D. A. Bryant, Electron Acceleration and Beyond (IPP, Bristol, 1998a).
[8] K. G. Mc Clements, Book Review, Nuclear Fusion **39**, 1071 (1999).
[9] J. E. Borovsky, J. Geophys. Res. **98**, 6101 (1993)
[10] N. Ivchenko, G. Marklund, K. Lynch, D. Pietrowski,, R. Torbert, Prinmddahl, and A. Ranta, Geophys. Res. Lett. **26**, 3365(1999).
[11] B. B. Kadomtsev, Plasma Turbulence, Section 5b, Page 129, Academic Press, 1965
[12] K. A. Lynch, R. B. Torbet, N. Ivenko, G. Marklund, and F. Primdahl, Geophys. Res. Lett., **26**, 3365(1999)

Sub-Grid-Scale Parameterisations for Large-Scale Eddy Simulations

Jorgen S. Frederiksen

CSIRO Division of Atmospheric Research

Abstract. The accuracy of simulations of turbulent flows is strongly dependent on the appropriate treatment of subgrid-scale processes that cannot be explicitly resolved at the finite resolutions of the simulations. Subgrid-scale parameterizations of eddy viscosity, stochastic backscatter and net eddy viscosity are formulated for two-dimensional turbulence based on eddy damped quasi-normal Markovian (EDQNM) and direct interaction approximation closures. The focus of the study is geophysical flows described by the barotropic vorticity equation but the results should also be relevant to the low frequency drift wave dynamics of magnetized plasmas described by the Hasegawa-Mima equation. The subgrid scale parameterizations are found to have a cusp behaviour at the cutoff wavenumber where they have their largest magnitudes. The conventional net eddy viscosity is shown to be the relatively small difference between the eddy drain viscosity and the backscatter viscosity. Large-eddy simulations (LES) with the barotropic vorticity equation have been performed incorporating these dynamic sub-grid scale parameterizations and compared with higher-resolution direct numerical simulations (DNS), which are regarded as the benchmark or "truth" for comparisons. Good comparisons are found between kinetic energy spectra for the LES and DNS at the scales of the LES for both nonrotating two-dimensional turbulence and differentially rotating Rossby wave (drift wave) turbulence. This is contrasted with much poorer comparisons when using a number of ad hoc eddy viscosity parameterizations is LES. The applications of the results to more complex circulations models is discussed.

INTRODUCTION

The accuracy of simulations of turbulent flows is strongly dependent on the appropriate treatment of subgrid-scale processes that cannot be explicitly resolved at the finite resolutions of the simulations. For example, eddy viscosity parameterisations strongly affect both the mean climate and kinetic energy spectra of atmospheric circulation models as reviewed by Frederiksen et al.[1]

Subgrid-scale parameterisations of eddy viscosity, stochastic backscatter and net eddy viscosity have been formulated for two-dimensional turbulence based on eddy damped quasi-normal Markovian (EDQNM) and direct interaction approximation (DIA) closures. The results should also be applicable to low frequency drift-wave dynamics of magnetized plasmas described by the Hasegawa-Mima equation.

Large eddy simulations (LES) with the barotropic vorticity equation have been performed incorporating these dynamic subgrid scale parameterizations and compared with higher resolution direct numerical simulations (DNS) which are regarded as the benchmark or 'truth' for comparisons. Good comparisons are found between kinetic energy spectra for the LES and the DNS at the scales of the LES for both non-rotating two-dimensional turbulence and differentially rotating Rossby-wave (drift-wave) turbulence. This is in contrast with much poorer comparisons when using a number of ad hoc eddy viscosity parameterizations.

EDQNM BASED SUBGRID SCALE PARAMETERIZATIONS

We base our derivation of subgrid parameterizations of turbulent eddies on the EDQNM statistical closure equations describing the statistics of large ensembles of flows satisfying the barotropic vorticity equation on the sphere. These equations for isotropic two-dimensional turbulence describe the evolution of

$$C_{mn}(t,t) = <\zeta_{mn}(t)\zeta_{mn}^*(t)>, \tag{1}$$

twice the enstrophy in the nondimensional vorticity spectral component ζ_{mn} where m is zonal wave number and n total wave number. For isotropic turbulence C_{mn} in fact does not depend on m.

If the resolved scales have triangular truncation T_R and the total truncation of the system under consideration is $T > T_R$, then the prognostic equations for C_{mn} for the resolved scales can be written as

$$\frac{\partial C_{mn}(t,t)}{\partial t} + (2\nu_o(n)n(n+1) + 2\eta_{mn}^S(t))C_{mn}(t,t) - (F_o(n) + 2S_{mn}^S(t))$$
$$= -2\eta_{mn}^R(t)C_{mn}(t,t) + 2S_{mn}^R(t). \tag{2}$$

Here $\nu_o(n)$ is the 'bare' viscosity, $F_o(n)$ is the white noise 'bare' random forcing variance and

$$\eta_{mn}(t) = \sum_p \sum_q \sum_r \sum_s B_{nqs}^{mpr} \theta_{nqs}^{mpr}(t) C_{rs}(t,t), \tag{3}$$

$$S_{mn}(t) = \sum_p \sum_q \sum_r \sum_s B_{nqs}^{mpr} \theta_{nqs}^{mpr}(t) C_{pq}(t,t) C_{rs}(t,t), \tag{4}$$

$$B_{nqs}^{mpr} = -4 I_{nqs}^{mpr} I_{qns}^{pmr}, \tag{5}$$

$$I_{nqs}^{mpr} = \frac{1}{2}\left(\frac{q(q+1)-s(s+1)}{q(q+1)s(s+1)}\right)\int_{-1}^{1} d\mu P_n^m \left(pP_q^p \frac{dP_s^r}{d\mu} - rP_q^p \frac{dP_q^p}{d\mu}\right), \tag{6}$$

where $P_n^m(\mu)$ are orthonormalized Legendre function and μ is sin (latitude). Also θ_{nqs}^{mpr} is the triad relaxation time; we use the asymptotic form given by

$$\theta_{nqs}^{mpr}(\infty) = (\mu_{mn} + \mu_{pq} + \mu_{rs})^{-1}. \tag{7}$$

We follow Leith[2] and write the damping in the form

$$\mu_{mn} = \gamma[n(n+1)C_{mn}(t,t)]^{1/2} + \nu_o(n)n(n+1), \tag{8}$$

where $\gamma = 0.6$ is a dimensionless coefficient.

In Eqn. 2, the terms with superscript R refer to resolved scales which lie within the triangular truncation region TR in the sums in Eqns. 3 and 4 and the superscript S refers to the subgrid scales. From Eqn. 2, we see that η_{mn}^S modifies the viscous damping and S_{mn}^S modifies the random forcing variance. In fact, S_{mn}^S is positive semidefinite and thus provides a stochastic backscatter source term from the subgrid scales to the resolved scales. We therefore define the stochastic backscatter by $F_b(n) = 2S_{mn}^S$, the renormalized backscatter by $F_r(n) = F_b(n) + F_o(n)$, the eddy drain viscosity by $\nu_d(n) = [n(n+1)]^{-1}\eta_{mn}^S$, the renormalized drain viscosity by $\nu_r(n) = \nu_d(n) + \nu_o(n)$, the eddy backscatter viscosity by $\nu_b(n) = -[n(n+1)C_{mn}(t,t)]^{-1}S_{mn}^S$, the net eddy viscosity by $\nu_n(n) = \nu_d(n) + \nu_b(n)$ and the renormalized net viscosity by $\nu_{rn}(n) = \nu_n(n) + \nu_o(n)$.

COMPARISON OF DNS AND LES FOR BAROTROPIC VORTICITY EQUATION

In our comparisons of DNS and LES, we use T63 as the triangular truncation in the DNS of the barotropic vorticity equation and T31 and T15 in the LES. The bare viscosity in the DNS is given by

$$v_o(n) = \begin{cases} \dfrac{1.014 \times 10^{-2}}{n(n+1)} & \text{for } 2 \le n \le 15 \\ \dfrac{1.014 \times 10^{-2}}{n(n+1)} + 4.223 \times 10^{-5} & \text{for } 16 \le n \le 63 \end{cases} \qquad (9)$$

The drag contribution (1.014×10^{-2}) corresponds to 7.4×10^{-7} s^{-1}, or an e-folding time of 15.6 days, and the Laplacian contribution (4.223×10^{-5}) corresponds to 1.25×10^{5} m^2 s^{-1} in dimensional units. We shall attempt to simulate the isotropized January 1979 monthly average kinetic energy spectrum in the DNS. To do this we calculate within the EDQNM, the bare random forcing $F_o(n)$ which will maintain the January 1979 kinetic energy spectrum as a stationary solution. Using this $F_o(n)$ to force the barotropic vorticity equation then yields a DNS spectrum very close to the January 1979 spectrum as shown in Figure 1. Here the kinetic energy $\bar{E}(m)$ as a function of zonal wave number m is averaged over the last 100 days of a 150 day integration.

For the LES we use renormalized backscatter and renormalized drain viscosity derived from the EDQNM theory. Figure 2, shows the related dissipation functions for the LES at T31. They have a cusp behaviour at the smallest retained scales and the net eddy dissipation is the small difference of the eddy drain dissipation and the eddy backscatter dissipation. Figure 1 also shows the kinetic energy spectra $\bar{E}(m)$ for LES at T31 and T15 and the DNS and January 1979 spectra truncated back to these resolutions. The comparisons between DNS and LES are very good. Similar results are found using a renormalized net viscosity, if DIA based parameterisations are employed, and also when differential rotation is included as in Rossby-wave turbulence[3].

CONCLUSIONS AND ACKNOWLEDGMENTS

The EDQNM based subgrid scale parameterizations described here are able to maintain approximately the same large scale kinetic energy spectra in LES with differing horizontal resolutions.

Thanks to Tony Davies and Steve Kepert for assistance.

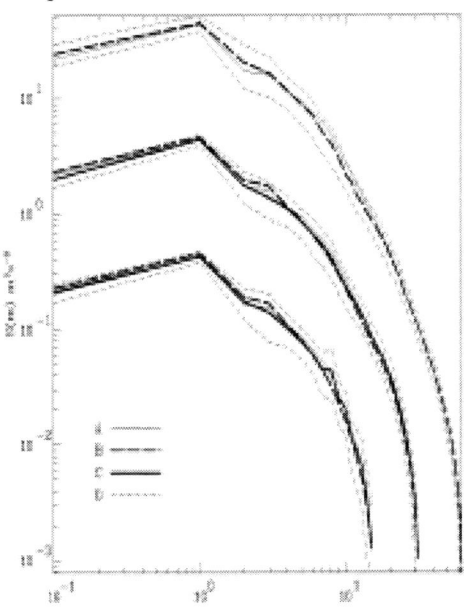

FIGURE 1. Kinetic energy spectra $\bar{E}(m)$ in m^2 s^{-2} for (A) January 1979, (B) DNS at T63, (C) LES at T31 or T15 and (E) $\bar{E}(m) \pm \sum(m)$ where $\sum(m)$ is the standard deviation in the simulations. The T31 results are scaled by 10^{-1} and T15 by 10^{-2}.

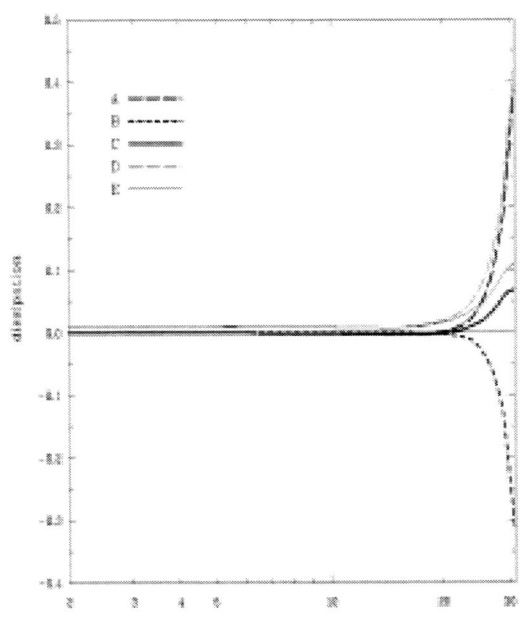

FIGURE 2. Dissipation functions (A) $v_d(n)n(n+1)$, (B) $v_b(n)n(n+1)$, (C) $v_n(n)n(n+1)$, (D) $v_r(n)n(n+1)$ and (E) $v_{rn}(n)n(n+1)$.

REFERENCES

[1] Frederiksen, J.S., Dix, M.R. and Kepert, S.M., J. Atmos. Sci., **53**, 887-904 (1996)
[2] Leith, C.E., J. Atmos. Sci., **28**, 145-161 (1971)
[3] Frederiksen, J.S. and Davies, A.G., J. Atmos. Sci., **56**, 2475-2492 (1997)

Understanding the Simple Magnetized Torus

J.-V. Paulsen*, O.E. Garcia[†] and K. Rypdal[†]

*PIIM UMR case 321, CNRS, Université de Provence, Centre de St Jerome, F-13397 Marseille Cedex 20, France
[†]Department of Physics, University of Tromsø, N-9037 Tromsø, Norway

Abstract. The role that small tokamaks and non-tokamaks can play in the understanding of tokamak transport should not be overlooked because it is often easier to understand the simpler case before moving on to the more complex one. Indeed, conversely to large tokamaks experiments where the comparison between theory and experiment is often hard to achieve, a small and simple magnetized torus such as the Blåmann device is well suited for detailed checks of theoretical models. In this paper results from numerical simulation of transport in the Blåmann device are presented, and some comparisons with experimental data are performed. Both the experiments and simulations demonstrate the appearers of a large rotating dipole vortex, which has some similarity to the convective cells observed in Rayleigh-Bénard convection.

INTRODUCTION

A simple magnetized torus (SMT) is a experimental device operated with a simple magnetic field configuration. The torus is a vacuum vessel described by two geometrical quantities, that is the major R_0 and minor r_0 radius. The magnetic field $\mathbf{B}_0$, produced by external field coils, is purely toroidal and have only a spatial dependence on the distance from the major axis R as $\mathbf{B}_0 = (B_0 R_0/R)\mathbf{b}$, where $\mathbf{B}_0$ is the magnetic field strength in the center of the poloidal cross-section and $\mathbf{b}$ is a unit vector in the toroidal direction. The plasma is in a low-β state, so that any induced magnetic field $\tilde{\mathbf{B}}$ can be ignored, hence making the device a simple magnetized torus (SMT).

The most common method of producing plasma in many small laboratory experiments, is to apply a discharge current to ionize the neutral gas present in the vacuum vessel. The discharge current represents a constant source of charge as well as plasma in these experiments, hence there must be some losses in order to achieve a turbulent equilibria [1, 2]. The turbulent equilibria is defined, in this article, as a time averaging over some turbulent time interval. In the turbulent equilibria, the production of the discharge current must be equal to the time averaged losses as far as the averaging is taken over a long time. The typical state in a discharge plasma is therefore a situation of continuous production and losses, which on averaged is equal. Normally the production and loss areas are separated, since the discharge current is applied in the interior of the vacuum vessel by a hot cathode filament whereas the losses is mostly close to the boundary (the vacuum vessel walls).

The second section of this article is devoted to the modeling of a SMT, and in the third section some experimental results obtained from such a device (the Blåmann plasma device at the university of Tromsø, Norway) is reviewed. A comparison between theoretical concepts and experimental data is essential in order to obtain an understanding of a physical system, thus in the forth section such an comparison is attempted in order to achieve a better understanding of a SMT.

MODELING A DISCHARGE IN A SMT

To gain any understanding of the behavior in a plasma discharge three ingredients must be modeled; the production, the transport and the losses. In this work the focus is on modeling the transport, hence the production and losses are represented in a simple manner. The most fundamental equations for a discharge plasma is

$$\nabla \cdot \mathbf{j} = \nabla \cdot \mathbf{j}_{\text{dis}} - L_q = S_q - L_q \qquad (1)$$

and
$$\frac{\partial \rho_m}{\partial t} + \nabla \cdot (\rho_m \mathbf{u}) = S_m - L_m = \frac{\alpha m_i}{e} S_q - L_m, \tag{2}$$

which is the charge density conservation equation, under the quasineutral approximation, and the mass density conservation equation. In these two equations S_q and S_m is the sources due to the discharge current density $\mathbf{j}_{\text{dis}}$, L_q and L_m is the losses, α is the number of neutral particles that the electron from the hot cathode is ionizing, m_i is the ion mass and e is the unit charge.

To apply the two conservation equations (1) and (2), the evolution for the current density $\mathbf{j}$ and flux density $\rho_m \mathbf{u}$, which represents the transport, must be found. In order to be at a simple level of plasma description, the one-fluid theory will be employed. Invoking the following momentum equation

$$\rho_m \frac{D\rho_m}{Dt} = \mathbf{j} \times \mathbf{B}_0 - \nu_{in}\rho_m \mathbf{u} - S_m \mathbf{u} - \nabla p + \rho_m \mu \nabla^2 \mathbf{u} \tag{3}$$

and Ohm's law

$$\mathbf{E} + \mathbf{u} \times \mathbf{B}_0 = \frac{m_i}{Ze\rho_m}(\mathbf{j} \times \mathbf{B}_0 - \nabla p) + \eta \mathbf{j}, \tag{4}$$

with the Hall term, where μ is the kinematic viscosity, ν_{in} is the collision frequency between ions and the neutral gas particles, Z is the charge number for the ions, $\mathbf{B}_0$ is the external magnetic field, p is the pressure and $D/Dt = \partial_t + \mathbf{u} \cdot \nabla$, the expressions needed for the evolution of the transport can be computed.

An expression for five different current densities can be established from the momentum equation (3), under the assumption of $\mathbf{u} = \mathbf{E} \times \mathbf{B}_0/B^2$, that is

$$\mathbf{j}_m = -\frac{\rho_m}{B^2}\frac{d}{dt}\nabla_\perp \phi, \; \mathbf{j}_p = \frac{1}{B^2}\mathbf{B}_0 \times \nabla_\perp p, \; \mathbf{j}_v = -\frac{1}{B^2}\nu_{in}\rho_m \nabla_\perp \phi, \; \mathbf{j}_s = -\frac{1}{B^2}S_m \nabla_\perp \phi$$

$$\text{and } \mathbf{j}_\mu = \frac{1}{B^2}\mu \nabla^2 \nabla_\perp \phi \; \text{ with } \; \frac{d}{dt} = \frac{\partial}{\partial t} + \frac{\mathbf{B}_0 \times \nabla_\perp \phi}{B^2} \cdot \nabla_\perp$$

and $\nabla_\perp = \nabla - \mathbf{b} \cdot \nabla$. A closed toroidal surface at a given radius r and spanned by the poloidal and toroidal angel (the toroidal surface at $r = r_0$ is the vacuum vessel wall) can be applied to integrate the current densities above according to Gauss' law, hence the currents $I_m(r)$, $I_p(r)$, $I_v(r)$, $I_s(r)$ and $I_\mu(r)$ can be found. Averaging these currents in time, where $\langle \rangle$ denotes time averaging, and ignoring losses, the following $\langle I_m \rangle + \langle I_p \rangle + \langle I_v \rangle + \langle I_s \rangle + \langle I_\mu \rangle = I_{\text{dis}}$ should hold in a turbulent equilibria. Separating the mass density, or the density since $\rho_m = m_i n$, and the potential into perturbations ($\delta\phi$ and δn) and turbulent equilibria ($\langle \phi \rangle$ and $\langle n \rangle$) as $\delta\phi = e(\phi - \langle \phi \rangle)/T_e$ and $\delta n = (n - \langle n \rangle)/\langle n \rangle$, thus $\langle \delta\phi \rangle = 0$ and $\langle \delta n \rangle = 0$, it is easily seen that only $\langle I_v \rangle$ and $\langle I_m \rangle$ can have any contribution from the perturbations. The notation $\langle \tilde{I}_v \rangle$ and $\langle \tilde{I} \rangle_m$ is going to be employed to these contributions, hence $\langle I_{v0} \rangle = \langle I_v \rangle - \langle \tilde{I}_v \rangle$ and $\langle I_{m0} \rangle = \langle I_m \rangle - \langle \tilde{I}_m \rangle$ introduces the notation for the contribution from the equilibria.

The drift approximation makes it possible to solve Ohm's law (4) as an iteration with the $\mathbf{E} \times \mathbf{B}$-drift as the zero order drift, hence the zero order flux density is

$$\Gamma_0 = \rho_m \mathbf{u}_E = \rho_m \frac{\mathbf{B}_0 \times \nabla \phi}{B^2} \equiv \Gamma_E.$$

The next iteration provides the first order flux density Γ_1, and it can be shown that

$$\Gamma_1 = \frac{m_i}{Ze}(\mathbf{j}_m + \mathbf{j}_s + \mathbf{j}_v + \mathbf{j}_\mu) = \frac{m_i}{Ze}(\mathbf{j}_\perp - \mathbf{j}_p).$$

Notice that $\mathbf{j}_p$ do not contribute to Γ_1. As for the current densities, the flux density can be integrated to yield the flux $F_E(r)$ and $F_1(r)$. In the numerical simulations a diffusion $\Gamma_d = -D\nabla_\perp \rho_m = -\delta D_B \nabla \rho_m$ is included, where $D_B = T_e/eB$ and δ is a free parameter, hence in the turbulent equilibria the following relation $\langle F_E \rangle + \langle F_1 \rangle + \langle F_d \rangle = S_m$ must be satisfied if losses are ignored. Both $\langle F_E \rangle$ and $\langle F_1 \rangle$ will have contribution from the perturbations, and the notation $\langle \tilde{F}_E \rangle$ is applied.

Now that some expressions for the charge densities and flux densities is established, the charge density conservation equation (1) and mass density conservation equation (2) can be invoked to perform numerical simulation of a discharge plasma in a SMT. Taking the divergence introduces many terms and the calculation will not be presented here, instead only the model equations

$$\frac{d}{dt}\omega = \frac{1}{\rho_m}\left(\frac{1}{\alpha\delta R_a^2}\left(\frac{r_0}{\rho_s}\right)^2 S_q + \rho_b \mathcal{R}\mathcal{P}\frac{2a}{R_a^2}\mathbf{e}_z\cdot\nabla_\perp\rho_m + \nabla_\perp\rho_m\cdot\frac{d}{dt}\nabla_\perp\phi + \frac{2a}{R_a}S_m\mathbf{e}_R\cdot\nabla_\perp\phi + \right.$$

$$\left. \nabla_\perp S_m\cdot\nabla_\perp\phi - S_m\omega + \frac{v_{in}}{\delta\omega_{ci}}\left(\frac{r_0}{\rho_s}\right)^2\nabla_\perp\rho_m\cdot\nabla_\perp\phi\right) + \frac{v_{in}}{\delta\omega_{ci}}\left(\frac{r_0}{\rho_s}\right)^2\frac{2a}{R_a}\mathbf{e}_R\cdot\nabla_\perp\phi - \frac{v_{in}}{\delta\omega_{ci}}\left(\frac{r_0}{\rho_s}\right)^2\omega + $$

$$\frac{2a}{R_a}\mathbf{e}_R\cdot\frac{\partial\nabla_\perp\phi}{\partial t} + 3a\,\mathbf{e}_R\cdot(\{\mathbf{b}\times\nabla_\perp\phi\}\cdot\nabla_\perp)\nabla_\perp\phi + \mathcal{P}\nabla_\perp^2\omega$$

and

$$\frac{d}{dt}\rho_m = -\frac{2a}{\delta}\mathbf{e}_z\cdot\nabla_\perp\rho_m + \nabla_\perp^2\rho_m + 2a\rho_m\mathbf{e}_z\cdot\nabla_\perp\phi + \frac{1+\alpha}{\alpha}S_m,$$

with $\omega = \nabla_\perp^2\phi$ and the unit vectors are $\mathbf{e}_r$, $\mathbf{e}_\theta$ and $\mathbf{e}_z$ in a cylindrical coordinate system ($\mathbf{e}_\theta$ in the toroidal direction), after they have been normalized is given in this work. In the equations above $(d/dt) = (\partial/\partial t) + R_a(\mathbf{b}\times\nabla_\perp)\cdot\nabla_\perp$.

Notice that the isothermal approximation is applied, which relates the pressure p to the mass density as $p = (T/m_i)\rho_m$. In the equations above time is normalized by the dissipative time scale r_0^2/D, the potential is normalized by $B_0 D$, the vorticity (or charge density) is normalized with $\varepsilon_0 B_0 D/r_0^2$ and the mass density is normalized with $\alpha m_i S_{q0} r_0^2/(eD)$, where $\alpha m_i S_{q0}/e = S_{m0}$. As mentioned before $D = \delta D_B$, and $\rho_s = (m_i/T_e)\omega_{ci}$ is the Larmor radius and $\omega_{ci} = eB_0/m_i$ is the ion gyration frequency. The sources are normalized with $S_{q0} = I_{dis}/V$ such that

$$\int S_q dV = I_{dis},$$

hence V depend on the spatial shape of the source S_q. The other dimensionless parameters are $a = r_0/R_0$, $R_a = R/R_0$, which is taking into account the curvature of the magnetic field, $\mathcal{P}$ is the Prandtl number $\mathcal{P} = \mu/D$ and

$$\mathcal{R}\mathcal{P} = \frac{\alpha}{\delta^3 \omega_{ci}}\frac{I_{dis}}{n_b Ve}\left(\frac{r_0}{\rho_s}\right)^4,$$

where $\mathcal{R}$ is the Rayleigh number, $\rho_b = m_i n_b$ is the mass density at the boundary and the discharge frequency I_f defined as $I_f \equiv I_{dis}/en_b V$ will be applied later.

If the assumption that the mass density is constant in the charge conservation equation (1) except where the pressure has been related to the mass density, otherwise there would not be any diamagnetic current density, and that the friction force $S_m\mathbf{u}$ along with higher order curvature terms can be neglected, the following equation

$$\frac{d}{dt}\omega = \frac{I_f}{\omega_{ci}\delta^2 R_a^2}\left(\frac{r_0}{\rho_s}\right)^4 S_q + \mathcal{R}\mathcal{P}\frac{2a}{R_a^2}\mathbf{e}_z\cdot\nabla_\perp\rho_m - \frac{v_{in}}{\delta\omega_{ci}}\left(\frac{r_0}{\rho_s}\right)^2\omega + \mathcal{P}\nabla_\perp^2\omega$$

is obtained. The mass density conservation equation is as before. These two equations have some similarity to that of Rayleigh-Bénard (RB) convection, although some differences exists. There are three distinctions that ought to be mention; the driving of the system is now the source S_m and a cold boundary (and not a cold and warm boundary normally applied in RB convection), the source S_q generates a flow that give rise to a centrifugal force, and there is a damping in the system due to collisions with the neutrals. Nevertheless the phenomena observed in RB convection should also be found in a numerical simulation of a discharge plasma in a SMT, even if the model presented here differ from that of RB convection.

EXPERIMENTS IN THE SIMPLE MAGNETIZED TORUS BLÅMANN

In order to perform numerical simulations of a discharge plasma the quantities of R_0, r_0, B_0, I_{dis}, α, ρ_b, m_i, v_{in} and T_e must be chosen, thus in this work the typical discharge plasma observed in the Blåmann device at the university of Tromsø, Norway, will be simulated. The Blåmann device is a SMT device with a major and minor radius of $R_0 = 65.1$ cm and $r_0 = 13.3$ cm, and a more detailed description of the device is given by Rypdal et al.[3]. In the simulation presented in the next section the parameters are chosen as follows; $R_0 = 65.1$ cm, $r_0 = 13.3$ cm, $B_0 = 0.13$ T, $I_{dis} = 0.5$

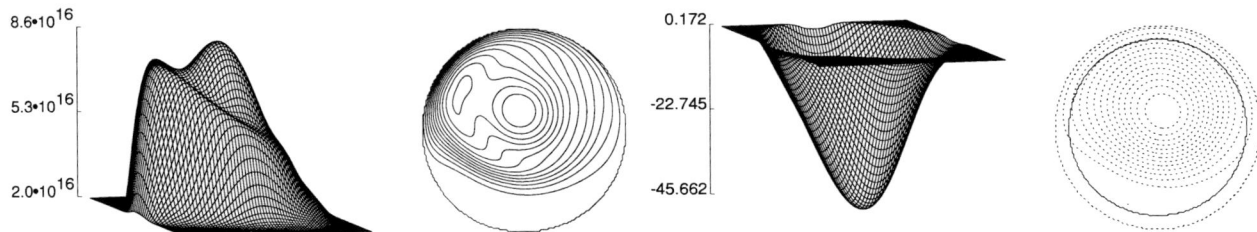

FIGURE 1. The turbulent equilibria of the density (left) and the potential (right). The unit for the density is m^{-3}, whereas the unit for the potential is Volts.

A, $\alpha = 1$, $n_b = 2 \cdot 10^{16}$ m^{-3}, $mi = 2$u, $v_{in} = 5 \cdot 10^4$ Hz and $T_e = 2$eV. The choice of $m_i = 2$u is applied in order to simulate a H_2^+-plasma and the value of v_{in} and α is an estimate corresponding to a pressure of $p = 10^{-3}$ mbar and $V_{\text{dis}} = 140$ V, respectively. Choosing $\delta = 0.12391$ and $\mathscr{P} = 1$, the Rayleigh number $\mathscr{R}$ becomes approximately 10^8. These parameters were chosen to reproduce the experimental data obtained by O. M. Olsen in his Phd thesis [6], but it also gives basically the same plasma condition, with a potential well of the order -40 V, as in the work by Øynes et al.[4, 5].

Øynes et al. [4, 5] observe large scale coherent structures, employing conditional sampling, with a size determined by the minor radius, which propagate with the mean $\mathbf{E} \times \mathbf{B}$-drift. The coherent structures take the from of a dipole with a life-time of approximately one rotation, and the flutelike nature of these structures is explicitly demonstrated. The plasma condition in the Blåmann experiment have both the magnetic curvature and the centrifugal force that can act as a effective gravitation, thus the phenomena of RB convection ought to be observed. In fact, the large scale coherent structures reported by Øynes et al. [4, 5] have many of the features that is found in RB convection.

In the density profile Øynes et al. [4, 5] identify three regions, region A, B and C, and they conclude that the large coherent structures have a more dominant role in region B giving rise to a smaller density gradient in this region. This phenomena is typical for RB convection, where region *B* can be interpreted as the flat turbulent profile and A and C as the boundary layers.

SIMULATION OF A DISCHARGE PLASMA

In the numerical simulation the parameter mentioned in the last section are applied and the shape of the source is Gaussian with a standard deviation $\sigma = 1.9$ cm at the center of the poloidal cross-section. There is only a limiter for the vorticity, or the charge in a 2D simulation, which is modeled employing a term $-v_L \omega$ with v_L zero outside the limiter area. In our modeling it has been assumed that the plasma behave as a fluid, which is only the case of a large number of particles in the device. Hence a initial mass density $\rho_m(t=0) = \rho_b = m_i \cdot 2 \cdot 10^{16}$m^{-3} is applied, so the boundary condition becomes $\rho_m = \rho_b$ and $\omega_b = 0$ in accordance with the initial mass density and the vorticity limiter (ρ_b and ω_b is the value at the boundary).

When the discharge current is turned on a density peak and a potential well is formed. The turbulent equilibria, defined as an averaging in time, are presented in figure 1. The potential forms a potential well as the equilibria, whereas the density take the shape of a density peak superimposed on a density slope or gradient. The density n slope is approximately given by the relation $R_a(\mathbf{b} \times \nabla_\perp \phi) \cdot \nabla_\perp n = 2an\, \mathbf{e}_z \cdot \nabla_\perp \phi$. The simulations shows that the perturbations are contributing significantly to the total radial current, through $\tilde{I}_m$ and $\tilde{I}_v$, thus the turbulence is important in order to provide a current channel in a discharge plasma. The standard deviation of $\langle I_p \rangle$, $\langle I_m \rangle$ and $\langle I_v \rangle$ is between 0 and 20 % of the discharge current. Thus even though the perturbations contribute significantly to the current channel, the currents them self do not fluctuate much in time.

The plasma transport is dominated by F_E in the interior of the plasma, whereas F_d provide must of the transport close to the boundary. This is also typical for RB convection, where the turbulent contribution close to the boundary is prohibited by the boundary condition (which is also the case here). There is a large difference between that plasma transport and the currents when it comes to the standard deviation, thus the bursty behavior of the plasma is primarily associated with the plasma transport.

Employing conditional sampling, in the same way as Øynes et al. [4, 5], a spatial and temporal evolution of the large events, taking the form of a dipole structure rotating with the mean $\mathbf{E} \times \mathbf{B}$-drift, is observed in the simulation. The plume shape in the density and the dipole in the potential, seen in the simulations, are typical for RB convection.

The simulations has demonstrated that the bursty behavior, of a discharge plasma in a SMT, is associated with large avalanches in the plasma transport, and that these avalanches are connected with the plumes observed in the density perturbations. The turbulence forms both a plasma transport channel, which is strongly varying in time, and a current channel, which is almost stationary and therefore disallowing any large charge accumulation. Hence the plasma transport displays a bursty behavior, whereas the currents have a fluctuating behavior.

REFERENCES

1. K. Rypdal, H. Fredriksen, J.-V. Paulsen, and O. M. Olsen. *Phys. Scripta*, **T63** 167-173, (1996).
2. K. Rypdal, O. E. Garcia, and J.-V. Paulsen. *Phys. Rev. Lett.*, **79** 1857-1860, (1997).
3. K. Rypdal, E. Grønvoll, F. Øynes, Å Fredriksen, R. J. Armstrong, J. Trulsen, and H. L. Pécseli. *Plasma Phys. Control. Fusion*, **36** 1099-1114, (1994).
4. F. J. Øynes, O. M. Olsen, H. L. Pécseli, Å Fredriksen, and K. Rypdal. *Phys. Rev. E*, **57** 2242-2255(1998).
5. F. J. Øynes, H. L. Pécsel, and K. Rypdal. *Phys. Rev. Lett.*, **75** 81-84, (1995).
6. O. M. Olsen. *Studies of the plasma state in a magnetized torus*. PhD thesis, University of Tromsø, Norway, 1998.

Impact of large scale flows on turbulent transport

Y.Sarazin[*], Ph.Ghendrih[*], S.Benkadda[†], P.Beyer[†], G.Falchetto[*], C.Figarella[†], X.Garbet[**] and V.Grandgirard[*]

[*]Association EURATOM-CEA, CE de Cadarache, 13108 St-Paul-Lez-Durance, France
[†]Equipe Dynamique Systèmes Complexes, PIIM CNRS-Univ. de Provence, 13397 Marseille, France
[**]EFDA JET, Abingdon OX14 3EA, United Kingdom

Abstract. The respective impact on turbulent transport of zonal flows (ZF) and poloidally localized flows is investigated. A 2D model for interchange instability in the Scrape-off Layer of tokamaks is used. The turbulent transport features intermittent large scale transport events, called avalanches. ZF are driven by Reynolds stress. Our results suggest that ZF alone participate only weakly in the self-regulation of turbulence, as compared to non-zonal components of the flow.

1. Introduction. Instabilities in tokamaks naturally arise from the magnetic field line curvature, and the departure from thermodynamical equilibrium, namely gradients of quantities such as density, temperature and pressure. They are interchange-like in essence. A significant change in the analysis of plasma turbulence has taken place as the drive of the system was considered to be a flux rather than a gradient. Indeed, the standard analysis of systems out of thermodynamical equilibrium aims at determining the flux resulting from an imposed gradient. Let us consider the density field and hence the relationship between the particle flux and density. In all fusion plasmas, the system is driven away from equilibrium by the particle source. As such, our systems are flux driven rather than gradient driven. This change in driving force leads to significant modifications in the dynamics of the turbulent response, with continuous reorganization of the turbulent eddies and locally, repeated evolution from the linear stage to the nonlinear saturation of the turbulence.

Recent numerical simulations show that, when driven by an external heat or particle flux, such systems exhibit long range transport events, called avalanches [1]. This behavior is observed in numerous fluid models for edge and core turbulence [2, 3, 4]. It is also reported for Scrape-off Layer (SOL) turbulence [5]. Experimental evidence of large scale and/or bursty transport is also reported [6]. Intermittent bursts propagate almost ballisticaly, and can be understood either in terms of radially elongated convective cells, namely streamers [7], or in terms of front propagation [8]. A simple analogy with the latter description is provided by sand pile automaton models [9].

In addition to this change in approach, the interest in the poloidal flows self-generated by the turbulence has increased [10]. Indeed, any radially localized poloidal flow v_θ of "sufficient" lifetime and poloidal extent k_θ^{-1} is expected to actively control the turbulent transport by shearing apart convective cells, leading to the reduction of the radial correlation length of turbulence. Present studies focus on zonal flows (ZF), characterized by $k_\theta = 0$. These mean flows $\langle v_\theta \rangle$ are constant on a magnetic surface. ZF are specific in that they are always stable linearly. As a consequence, they can only be excited non-linearly, either via turbulence through the Reynolds stress (RS) [11], or via the so-called Stringer-Winsor term [12]. Such flows appear to play a role in the non-linear saturation of turbulence [13, 14]. The impact of the lifetime of ZF on the transport has been investigated theoretically [15]. It has been proposed that only those flows of frequency smaller than the one of turbulence would lead to an efficient shearing, and to a further reduction of the transport. Ultimately, a sheared $\langle v_\theta \rangle$, constant-in-time, imposed externally to the system, is able to trigger a transport barrier.

In the present study, we aim at discriminating the impact of ZF from that of poloidally localized flows on the dynamics and the level of avalanche-like turbulent transport. The 2D system used as a paradigm to investigate this question is presented in section 2. It models the interchange instability in the SOL of tokamaks, where magnetic field lines intercept the wall. Section 3 exemplifies how two systems with similar linear properties but different viscosities exhibit strong differences in the transport, although ZF exhibit essentially the same behavior in both runs. Section 4 shows that the artificial suppression of ZF modifies only slightly the transport level and its dynamics, the $k_\theta \neq 0$ modes remaining unaffected.

2. 2D model for avalanche-like transport. Interchange instability is supposed to play a key role in turbulent transport in the SOL. Indeed, the low sheath conductivity destabilizes interchange electrostatic modes. These modes are driven by the pressure gradient in the bad curvature – Low Field Side – region [16]. A field line average reduces the problem to 2D. Assuming a constant electron temperature T_e and cold ions reduces the number of parameters to the normalized sheath conductivity $\sigma_{\parallel}$ and the average field line curvature g. The normalized system is then derived from particle balance and charge conservation:

$$d_t n = \sigma_{\parallel} n \exp(\Lambda - \phi) + D\nabla_{\perp}^2 n + S \qquad (1)$$

$$d_t \nabla_{\perp}^2 \phi = \sigma_{\parallel}\{1 - \exp(\Lambda - \phi)\} - g\, \partial_y Log(n) + \nu \nabla_{\perp}^4 \phi \qquad (2)$$

where the total time derivative stands for the Lagrangian derivative $d_t \equiv \partial_t + [\phi,.]$. n and ϕ are the *total* density and electric potential, and $[f,g] \equiv \partial_x f \partial_y g - \partial_y f \partial_x g$ the Poisson brackets. The particle source is constant in time and poloidally, and has a Gaussian shape $S(x) = S_0 \exp[-(x/\lambda_S)^2]$. Finally, radial and poloidal lengths (x and y, respectively) are normalized to the Larmor radius ρ_s, and time to the Larmor frequency $\omega_s = c_s/\rho_s$. The sound speed is equal to $c_s^2 = T_e/m$. Typical parameters are $\sigma_{\parallel} = 2.27 \times 10^{-4}$, $g = 5.72 \times 10^{-4}$ and $S_0 = 10^{-2}$. The box grid is 256×256, with $1\rho_s$ per grid point.

Most of the dynamical properties of the system are discussed in reference [5]. The source builds up the density profile. Above a given threshold, the gradient combines with the curvature to excite turbulence. Two saturation mechanisms then compete: (*i*) non-linear coupling leads to energy transfer to large and small scales where the energy is dissipated; (*ii*) turbulent transport leads to the relaxation of the density profile, leading to a local decrease of the linear drive. Point (*i*) refers to energy cascades, involving the generation of poloidal flows and the associated shearing process. Point (*ii*) can lead to large scale transport, as a kind of domino effect. As a result of these mechanisms, turbulent transport is dominated by long range transport events that propagate almost ballisticaly. These bursts are called avalanches. Their effective velocity is a fraction of the acoustic speed in this model. The dynamical properties of the transport model are reminiscent to those observed in 3D codes modeling of the ion temperature gradient instability in the core [3] and to the resistive ballooning modes at the plasma edge [4].

In 2D, ZF correspond to $k_y = 0$ modes, and $\langle v_y \rangle$ satisfies the following equation:

$$\partial_t \langle v_y \rangle = -\partial_x \langle \tilde{v}_x \tilde{v}_y \rangle + \sigma_{\parallel} \int_0^x \langle 1 - \exp(\Lambda - \phi) \rangle dx + \nu \partial_{xx} \langle v_y \rangle \qquad (3)$$

In the system Eq. 2, they are essentially driven by the RS, which dominates by about one order of magnitude over the sheath term. Linear damping of the mean poloidal velocity $\langle v_y \rangle$ is then ensured by the viscosity ν.

3. Impact of viscosity on Zonal Flows and transport. Two simulations with low (L) and high (H) values of the viscosity ν are compared: $\nu_L = 5.10^{-3}$ and $\nu_H = 2.10^{-2}$. To keep the linear properties constant, the diffusion coefficients are also different. More precisely, the product $D\nu$ is equal in both runs.

An efficient way to quantify the difference between both simulations, in the non linear regime, is to estimate the characteristic radial extent of convective cells at each time, $\lambda_c(t)$. Details of the computation can be found in reference [5]. The larger ν, the larger λ_c, as exemplified by the probability distribution functions (PDF) of λ_c in Fig. 1a. In the ν_H case, λ_c exhibits very large transient excursions, almost covering the entire size of the system along the radial direction, reminiscent of streamers. Accordingly, the largest avalanches are observed in this simulation, where density fronts extend up to the very far SOL, namely $x > 100$. On average, λ_c is 1.4 times larger for ν_H than for ν_L. Consistently, turbulent transport is also increased. Defining an effective diffusivity $D_{eff} = -\langle \Gamma \rangle / \langle \partial_x n \rangle$, one finds that D_{eff} is larger by a factor of about 2.2 in the far SOL in the largest ν case, Fig. 1b. In the present model, the increase of the turbulent transport at large viscosity is however not due to the viscous damping of ZF. Indeed, Fig. 1c shows that the magnitude of the velocity shear generated by ZF is roughly the same in both simulations. The reason is the following: the large ν case is also the low D case. The poloidal diffusive transport is then reduced, leading to large poloidal gradients of the density. As a consequence, the curvature term increases the linear drive of turbulence, Eq. 1. This further leads to the increase of the drive of ZF (i.e. of the RS) in Eq. 3, which balances the increase of the viscous damping. This result suggests that ZF do not play a dominant role in the shearing process of turbulent eddies.

4. Respective impact of ZF and $k_y \neq 0$ modes on the transport. The present study investigates the effect of suppressing the main drive of ZF, namely the RS term in Eq. 3. The remainder of the velocity shear in the no-RS case,

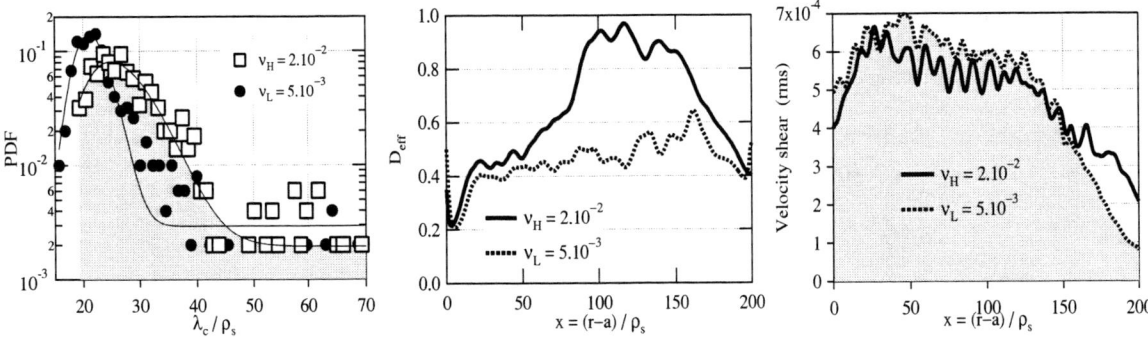

FIGURE 1. Characteristics of 2 simulations with different viscosities and the same γ_{lin} ($D\nu = Cst$): (a) PDF of λ_c (see text), (b) effective turbulent diffusivities, (c) magnitude of $\partial_x \langle v_y \rangle$ (rms).

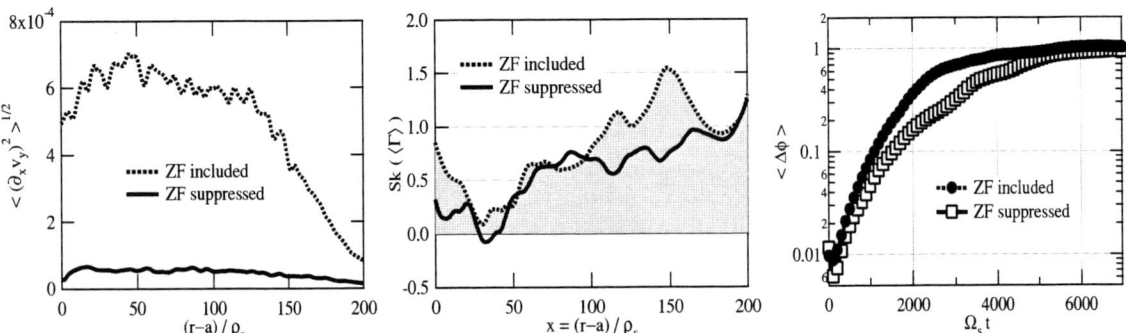

FIGURE 2. Characteristics of 2 simulations with and without RS: (a) magnitude of $\partial_x \langle v_y \rangle$ (rms), (b) Skewness of the flux, (c) time evolution of $\langle \Delta \phi \rangle$ after the addition of white noise to ϕ at $t = 0$ (see text).

Fig. 2a, is driven by the sheath non linearity (second term in the RHS of Eq. 3). Clearly, the root mean square (rms) of the ZF shearing $\partial_x \langle v_y \rangle$ differs in the two simulations by one order of magnitude.

Let us first investigate the impact of such a difference on the dynamics of the turbulent transport. The radial profile of the skewness Sk of the radial turbulent flux (defined as $Sk(f) \equiv \langle f^3 \rangle / \langle f^2 \rangle^{3/2}$) is plotted in Fig. 2b. A Gaussian PDF has a vanishing skewness. In both cases, Sk is positive and increases in the far SOL, indicating that large bursts dominate the transport in this region. Also, the presence of strong ZF tends to increase Sk in the far SOL. This modification remains however moderate. Adding white noise (WN) to the potential ϕ at a given time then allows one to investigate the sensitivity to initial conditions. The quantity $\langle \Delta \phi \rangle \equiv \langle (\tilde{\phi}_{WN} - \tilde{\phi})^2 \rangle^{1/2} / \langle \tilde{\phi}_{WN}^2 + \tilde{\phi}^2 \rangle^{1/2}$ (here, $\langle ... \rangle$ refers to $\int\int ... dx\, dy$) then quantifies this effect. The result is plotted on Fig. 2c. The departure looks roughly exponential in both cases in the initial phase, before reaching a saturation level where the dynamics are totally uncorrelated. In this final state, fluctuations remain of the same order of magnitude, as emphasized by the saturation value equal to unity. The exponents at which simulations diverge are analogous to Lyapunov exponents. The faster dynamics of turbulence is observed when ZF are present. This result agrees with the larger shearing expected with ZF. Again, Lyapunov exponents depart from each other by less than a factor of 2, and only in the late time evolution of $\langle \Delta \phi \rangle$. Let us then investigate how spatial scales are affected by the suppression of RS. Fig. 3a shows the Fourier spectrum of ϕ at a given time, for the modes $k_y = 0$ and the first harmonic $k_y \approx 0.025$. The first harmonic is hardly affected by the absence of ZF. Only the small radial scales with negligible magnitude are different. Hence, the two simulations only differ by their spectrum of ZF: poloidally localized flows (i.e. of finite k_y) have the same magnitude in each case. As far as the mode $k_y = 0$ is concerned, it appears that the RS term generates intermediate and small radial scales. The spectra look similar for $k_x \to 0$, suggesting that the large radial scales of the fluctuations are governed by the parallel boundary conditions, namely the sheath physics.

Finally, the impact on the turbulence correlation length and on the transport magnitude is presented in Fig.3b and 3c. The case without ZF leads to minor changes in the PDF of λ_c. The time average correlation length is only increased by less than 10% (from $\sim 30 \rho_s$ to $\sim 33 \rho_s$) while the *rms* velocity shear due to ZF is reduced by a factor of 10. As a

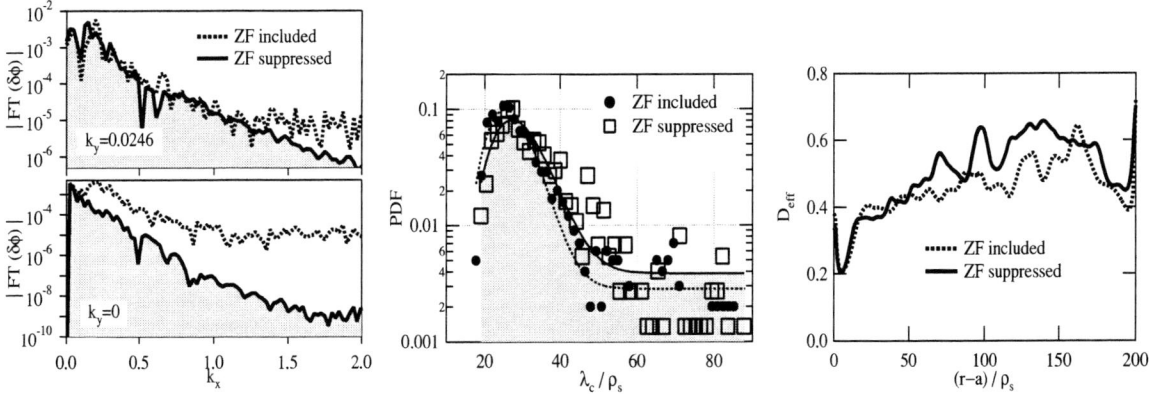

FIGURE 3. Characteristics of 2 simulations with and without RS term: (a) Fourier spectrum of ϕ at a given time, for the modes $k_y = 0$ and the first harmonic $k_y \approx 0.025$, (b) PDF of λ_c (see text), (c) effective turbulent diffusivities.

result, the effective turbulent diffusivity D_{eff} remains almost unaffected.

5. Conclusions. The respective impact on avalanche-like turbulent transport of ZF and poloidally localized poloidal flows is investigated, using a 2D model for interchange instability in the SOL of tokamaks. It is found that the turbulence correlation length and the transport level increase with the viscosity at constant linear growth rate. ZF are not responsible for this increase since both their dynamics and magnitude remain unaffected. In the present model, suppressing artificially the Reynolds stress leads to a reduction of the *rms* of the velocity shear due to ZF by one order of magnitude. Only ZF ($k_y = 0$ modes) are affected by this suppression, the poloidally localized flows (of finite k_y) having the same magnitude. The transport dynamic is slightly less intermittent when ZF vanish. However, the turbulence correlation length and the transport level remain essentially unaffected. These results strongly suggest that, if ZF participate in the self-regulation of turbulence, they do not play a prominent role. In that respect, the wake turbulence generated by avalanche events is suspected to have a strong impact.

REFERENCES

1. P.H. Diamond, T.S. Hahm, *Phys. Plasmas* **2**, 3640 (1995)
2. B.A. Carreras et al., *Phys. Plasmas* **3**, 2903 (1996); A. Thyagaraja, *Plasma Phys. Control. Fusion* **42**, B255 (2000)
3. X. Garbet, R. Waltz, *Phys. Plasmas* **5**, 2836 (1998)
4. P. Beyer et al., *Phys. Rev. Let.* **85**, 4892 (2000)
5. Y. Sarazin, Ph. Ghendrih, *Phys. Plasmas* **5**, 4214 (1998); Y. Sarazin et al., to appear in *J. Nucl. Materials* (2002)
6. C. Hidalgo et al., *Fusion Energy 1996 (Proc. 16th Int. Conf. Montreal, 1996)*, Vol. 1, IAEA, Vienna 617 (1997); R.A. Moyer et al., *J. Nucl. Matter.* **241-243**, 633 (1997); J. Boedo et al., *Rev. Sci. Instrum.* **70**, 2997 (1999); P.A. Politzer, *Phys. Rev. Let.* **84**, 1192 (2000) G.Y. Antar et al., *Phys. Plasmas* **8**, 1612 (2001) G.F. Counsell et al., *Plasma Phys. Control. Fusion* **44**, 827 (2002)
7. S. Champeaux, P.H. Diamond, *Phys. Let. A* **288**, 214 (2001)
8. Y. Sarazin et al., *Phys. Plasmas* **7**, 1085 (2000)
9. D.L. Newman et al., *Phys. Plasmas* **3**, 1858 (1996)
10. X. Garbet et al., *Proc. 12th IAEA Fusion Energy Conf., Nice, 1988* (Vienna, Austria, 1989), vol. 2, p.163.
11. P.H. Diamond et al., *Proc. 17th IAEA Fusion Energy Conf., Yokohama, 1998* (Vienna, Austria, 1999), vol. 4, p. 1421
12. K. Hallatschek, D. Biskamp, *Phys. Rev. Let.* **86**, 1223 (2001)
13. A.M. Dimits et al., *Phys. Rev. Let.* **77**, 71 (1996); Z. Lin et al., *Science* **281**, 1835 (1998); L. Chen et al., *Phys. Plasmas* **7**, 3129 (2000); M.A. Malkov, P.H. Diamond, *Phys. Plasmas* **8**, 3996 (2001).
14. S. Coda et al., *Phys. Rev. Let.* **86**, 4835 (2001); R.A. Moyer et al., *Phys. Rev. Let.* **87**, 135001 (2001)
15. T.S. Hahm et al., *Phys. Plasmas* **6**, 922 (1999)
16. A.V. Nedospasov, *Sov. J. Plasma Phys.* **15**, 659 (1989); X. Garbet et al., *Nucl. Fusion* **31**, 967 (1991)

Electron Beam Driven Cyclotron Maser Radiation

R. Bingham*†, R. A. Cairns** and B. J. Kellett*

Rutherford Appleton Laboratory, Space Science and Technology Dept., Chilton, Didcot, U.K.
†*Department of Physics and Applied Physics, Univ. of Strathclyde, Glasgow, U.K.*
**University of St Andrews, Fife, Scotland KY16 9SS*

Abstract. We present results of a new cyclotron maser radiation mechanism driven by a crescent or horseshoe electron distribution function. Such distribution functions are easily created by an electron beam moving into a stronger magnetic field region, where conservation of the first adiabatic invariant causes an increase in their pitch angle. This produces a broad region on the distribution function where there is a +ve slope in the perpendicular component. of the velocity space distribution function. Planetary dipole magnetic fields are examples where these types of distribution can be found, giving rise to, for example, the auroral kilometric radiation. We examine the stability of these electron horseshoe distribution functions for radiation close to the electron cyclotron frequency propagating perpendicular to the magnetic field for both non-relativistic and relativistic beams.

CYCLOTRON MASER RADIATION

The role of coherent emission mechanisms in explaining planetary and solar observations is firmly established, for example the emission from planetary magnetospheres is considered to be electron cyclotron maser radiation. It was first introduced by Twiss [1] to explain radio astronomical sources. The electron cyclotron maser [2] is a collective electromagnetic emission producing intense extraordinary mode radiation close to the electron cyclotron frequency Ω_e, with small bandwidth.

The electron cyclotron maser instability is a powerful mechanism for producing non-thermal stimulated radiation in a plasma. The free energy source for the electron cyclotron instability giving rise to stimulated emission of radiation is an anisotropic electron distribution function such that $\partial f_e/\partial v_\perp > 0$, which constitutes a population inversion. The loss-cone distribution function with $\partial f_e/\partial v_\perp > 0$, is commonly considered as the free energy source [3]. However, in many situations where intense radio emission is observed energetic particle beams are also observed or invoked. Electron beams propagating in a varying magnetic field give rise to a characteristically shaped horseshoe distribution as a result of the first adiabatic invariant. Horseshoe distributions also have $\partial f_e/\partial v_\perp > 0$ a necessary structure for generating radiation [4]. The horseshoe distribution is extremely efficient [4] in generating cyclotron maser radiation. The electron cyclotron maser produces stimulated emission with a narrow bandwidth polarized in the R-X mode and occurs in regions where the cold background plasma component is depleted in comparison to the hot component. The maser instability has maximum growth for low background plasma density such that the beam density is greater than the ambient density. This is supported by satellite measurements of depleted density regions known as the auroral cavity, where a horseshoe electron distribution function is also observed [5]. Another type of electron distribution function invoked to explain non-thermal radiation from space and astrophysical objects is a ring type distribution in perpendicular velocity space. The ring distribution is similar to those used in laboratory gyrotron devices where intense maser radiation is generated from relativistic electrons [2]. In astrophysics ring type distributions can also be found for example at quasi-perpendicular shocks [6].

We first examine the stability of a horseshoe type of electron distribution function for $R-X$ mode waves propagating perpendicular to the steady magnetic field. To obtain the distribution function we start with a drifting Maxwellian, with a drift velocity well above the thermal speed, which is typical of electrons accelerated by lower-hybrid turbulence. This is then considered to move into an increasing magnetic field where the distribution function is readily calculated using invariance of total energy and magnetic moment. The distribution function is used in the dispersion relation for the $R-X$ mode which is easily obtainable from the susceptibility tensor given by Stix [7].

We shall assume that the frequency is close to the electron cyclotron frequency, and also assume that the Larmor radius is much less than the wavelength for typical electron velocities. This latter condition means that we need only consider the susceptibility to lowest order in $\frac{k_\perp v_\perp}{\Omega_e}$. If we neglect all but the zero order terms we get the cold plasma result. To a first approximation we need only take account of the velocity distribution of the electrons in the resonant integral which involves $1/(\omega - \Omega_e)$ where Ω_e is the relativistic electron cyclotron frequency $\frac{eB}{\gamma m_e}$ with e the electron charge, B the magnetic field γ the Lorenz factor and m_e the electron rest mass. In this resonant term we must take account of the relativistic shift of the cyclotron frequency, since this picks out a particular group of resonant electrons and produces damping or growth of the wave.

In terms of momentum p we have

$$\Omega_e = \Omega_{e0}(1 + \frac{p^2}{m^2 c^2})^{-1/2}$$

where Ω_{e0} is the non-relativistic electron cyclotron frequency. For the real part of the resonant integral we can simply take the cold plasma value. Although this goes as $1/(\omega - \Omega_{e0})$ and appears to be near singular at the resonance, the $1/(\omega - \Omega_{e0})$ factors in the real part of the dispersion relation cancel out, as we shall see, and it behaves quite smoothly in the vicinity of the cyclotron frequency. It is not crucial to include small corrections to the cyclotron frequency in the real part of the dispersion relation. The refractive index n for the R-X mode, which propagate perpendicular to the magnetic field, is given by Stix [7].

$$n^2 = \left(\varepsilon_\perp^2 - \varepsilon_{xy}^2\right)/\varepsilon_\perp \quad (1)$$

and the dielectric tensor elements are given by

$$\varepsilon_\perp = 1 - \frac{1}{2}\frac{\omega_p^2}{\omega(\omega + \Omega_{e0})} + A \quad (2)$$

$$\varepsilon_{xy} = \frac{1}{2}\frac{\omega_p^2}{\omega(\omega + \Omega_{e0})} + A \quad (3)$$

with

$$A = \frac{1}{4}\frac{\omega_p^2}{\omega} \int_0^\infty 2\pi p_\perp dp_\perp \int_{-\infty}^\infty dp_\parallel \frac{1}{\omega - \Omega_e} p_\perp \frac{\partial f_0}{\partial p_\perp} \quad (4)$$

To obtain this we have included only the ± 1 terms in the sum over harmonics which appears in the dielectric tensor elements, and used the small argument expansion $J_1(x) \approx \frac{x}{2}$ where $x = \frac{k_\perp v_\perp}{\Omega_e}$ changing the variables to (p, μ, ϕ), spherical polars with the usual angle θ replaced with $\mu = \cos\theta = p_\parallel/p$ then

$$A = -\frac{1}{2}\frac{\omega_p^2}{\omega(\omega - \Omega_o)} - \frac{i}{2}\frac{\omega_p^2}{\Omega_o} \int_1^{-1} (1-\mu^2) p_o^2 \gamma_o^2 \left\{\frac{\partial f_0}{\partial p} - \frac{\mu}{p_o}\frac{\partial f_0}{\partial \mu}\right\}\Big|_{p=p_o} d\mu \quad (5)$$

with $\gamma = (1+p^2)^{\frac{1}{2}}$, p_o is the resonant momentum given by $mc(2(\Omega_{eo} - \omega)\Omega_e)^{\frac{1}{2}}$. A full discussion of the derivation of the dispersion relation for perpendicularly propagating waves can be found in Stix [7].

Using equation (4) in equations (2) and (3) we can analyse the stability of equation (1) with respect to induced emission of right hand polarized radiation.

The initial beam is considered to be a drifting Maxwellian. Figure 1 illustrates the characteristic crescent or horseshoe distribution formed when the beam moves into a stronger field region. Using the evolved distribution in equation (1) we obtain the spatial growth rate shown in figure 2 for two initial beam energies. Figure 2a,2b represents the imaginary part of the refractive index as a function of frequency for a mean beam energy of a)100KeV, b)500KeV both with a 1% energy spread and a magnetic field ratio of 20. The maximum growth rate for the 500KeV beam is more tan 4 times greater than for the 100KeV beam. The analysis presented above considers strictly perpendicular propagation. However we have calculated the growth rates for modes which also contain a parallel wavenumber component and find that modes which propagate more than 5° to the perpendicular do not grow. The fastest growing component is for purely perpendicular propagation. The region of instability in frequency space is extremely narrow with a bandwidth $\Delta\omega/\omega$ of order 0.5%.

We now consider the case of a velocity ring distribution shown in figure 3a. The ratio of the ring density to background density is 0.2 and a ring energy much greater than the Maxwellian background. Using the electron ring distribution function defined by

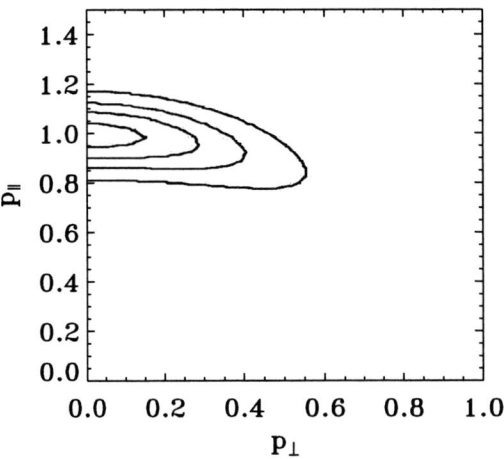

FIGURE 1. Contour plot in momentum space of the perpendicular and parallel electron momentum components of the evolved horseshoe distribution function. The contours represent constant phase space density.

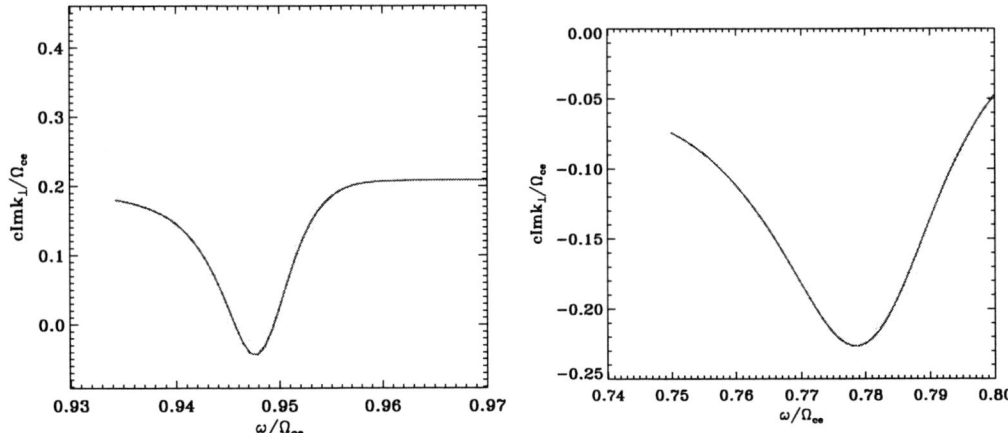

FIGURE 2. (a) The imaginary part of the refractive index as a function of frequency for a mean beam energy of 100 keV and a thermal spread of 1 keV. The magnetic field ratio is taken to be 20. (b) As (a), except the mean beam energy is 500 keV and the spread is 5 keV.

$$f_{\text{ring}}(p_\perp, p_\parallel) = \text{const.} \exp\left[-\frac{(p_\perp - p_o)^2}{m_e^2 v_{T_\perp}^2} - \frac{p_\parallel^2}{m_e^2 v_{T_\parallel}^2}\right] \qquad (6)$$

and using equation (1) we obtain the spatial growth rate for a ring distribution shown in figure 3b. It should be noted that the emission from a horseshoe or ring distribution is primarily in the plane perpendicular to the magnetic field and the emission is close to the cyclotron frequency. For higher energy beams or rings the frequency decreases due to the relativistic mass increase.

A parallel component of wavenumber introduces a Doppler shift into the resonance condition, so that the resonant particles no longer lie on a sphere centred on the origin in momentum space, these obliquely propagating waves are confined to within 5° of the perpendicular. This means, in turn, that for any significant Doppler shift, the resonant particles will no longer all lie in the part of the distribution function where there is a positive slope towards increasing energy. For this reason the growth rate will fall off as we go away from perpendicular propagation and the maximum emission is expected to be in the perpendicular direction. The instability is found to be sensitive to the ratio of the cyclotron frequency to the plasma frequency Ω_{ce}/ω_{pe}. By changing the ratio of Ω_{ce}/ω_{pe} we find that growth is suppressed when Ω_{ce}/ω_{pe} is close to or less than 1. The instability would be expected to occur in regions of low

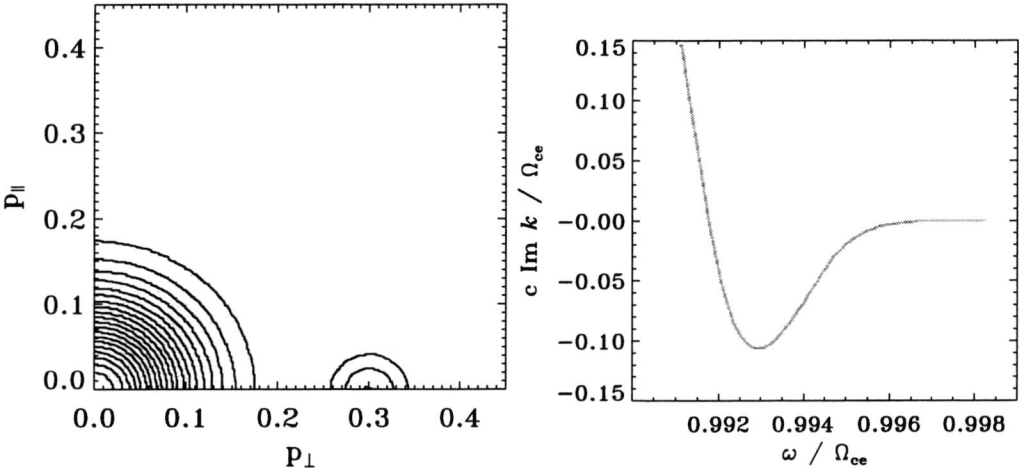

FIGURE 3. (a) Contour plot in momentum space of the electron momentum components depicting a background Maxwellian and a perpendicular ring distribution. The ratio of ring electron density to background Maxwellian density is 0.2. (b) Spatial growth rate $Imk_\perp$ of the R-X mode for the ring distribution shown in (a).

background plasma density where two-stream and other instabilities are not strong enough to disrupt the beam. Two stream instabilities have growth rates that are proportional to ω_{pe} (Krall & Trivelpiece [8]) whereas the instability described here is proportional to ω_{ce}.

SUMMARY & CONCLUSIONS

In this article we have developed an analytical approach describing the cyclotron maser instability convective spatial growth rate. Various types of distribution functions similar to those found at shocks lead to rapid growth of R-X mode radiation. We suggest that the spreading in perpendicular velocity as a beam moves into a region of increasing magnetic field or the formation of a velocity ring distribution, both produced at collisionless shocks, may trigger the instability and radio emission. Our model seems to be capable of explaining, at least in outline, strong radio emission, the importance of low density regions and the fact that the emission is perpendicular to the magnetic field. However it is clear that a lot more detailed work is needed to bring the various aspects together into a comprehensive model. Future particle in cell simulations will investigate self consistently the instability starting with an isotropic beam moving into a stronger magnetic field which will reveal a more realistic comparison with the observations.

ACKNOWLEDGMENTS

We would like to pay tribute to the late Professor John Dawson who played such an important role in this research.

REFERENCES

1. Twiss, R.Q., *Aust. J. Phys.*, **11**, 564, 1958.
2. Sprangle. P., & Drobot. A.T., *Proc. IEEE Trans. Microwave Theory Tech*, **MTT-25**, 528, 1977.
3. Wu, C.S., & Lee, L.C., *Ap.J.*, **230**, 621, 1979.
4. Bingham, R., & Cairns, R.A., *Phys. of Plasma*, **7**, 3089, 2000.
5. Ergun, R.E., Carlion, C.W., McFadden, J.P., Delory, G.T., Strangway, R.J., & Pritchett, P.L., *Ap.J.*, **538**, 456, 2000.
6. Tokar, R.L., Aldrich, C.H., Forslund. D.W., & Quast. K.B., *Phys. Rev. Lett.*, **56**, 1059, 1986.
7. Stix, S.H., *Waves in Plasmas*, American Institute of Physics, New York, 1992.
8. Krall. N., & Trivelpiece, A.W., *Principles of Plasma Physics*, McGraw Hill, New York, 1973.

A Detailed Model of the X-ray Emission from Comets

B. J. Kellett*, R. Bingham*†, C. M. Lisse**, M. Torney†, H. P. Summers† and V. D. Shapiro‡

*Rutherford Appleton Laboratory, Space Science and Technology Dept., Chilton, Didcot, U.K.
†Department of Physics and Applied Physics, Univ. of Strathclyde, Glasgow, U.K.
**Univ. of Maryland, Dept. of Astronomy, College Park, MD, USA 20792.
‡Univ. of California, San Diego, La Jolla, Dept. of Physics & Electrical & Computer Science, USA.

Abstract. Comet C/LINEAR 1999 S4 was observed by the CHANDRA ACIS-S CCD spectrometer in mid July, 2000. The resultant spectrum is currently the best and most detailed example available to help elucidate the underlying physical processes that enable the comet (or the cometary interaction with the solar wind) to generate X-ray emission. We have utilized the ADAS (Atomic Data and Analysis Structure) atomic physics code to accurately simulate the X-ray spectrum of a cometary atmosphere of assumed composition (hydrogen, carbon, nitrogen and oxygen) based on the plasma interaction model of Bingham et al. (1997) and Shapiro et al. (1999). We find an excellent agreement between theory and data provided we assume an unusually low carbon abundance for the comet. This low carbon abundance was also found in the infrared spectral data of the comet — thus confirming the X-ray result.

AIM OF THIS WORK

This work started with the straightforward target of developing a model to account for the CHANDRA ACIS-S X-ray spectrum of C/LINEAR 1999 S4 with the minimum number of assumptions and model parameters. The basis of this work is our previous theoretical treatment of the interaction of a comet with the solar wind [1, 2]. This model uses the plasma interaction between the incoming solar wind and out flowing newly ionized cometary ions to generate plasma waves and then keV-energy electrons. It is these electrons that can then interact with the cometary atmosphere to generate X-ray emission — both continuum (bremsstrahlung) emission and line radiation. It is this fact that our model predicts that the emission is a combination of bremsstrahlung and K-shell line emission that is crucial to understanding and explaining the CHANDRA X-ray spectrum.

Since this work is in its early development phase, we will not attempt to fit the entire observed spectrum from first principles. Rather, we will use the observed CHANDRA spectrum to guide and constrain the likely range of parameters. We can then check these parameters for internal consistency.

MODEL ASSUMPTIONS

As with any modelling procedure, we do need to make some initial assumptions to get the fitting started. However, we will endeavor to keep these assumptions to a minimum. We will assume:-

1. The X-ray spectrum contains a continuum emission component that can be fitted with a thermal bremsstrahlung emission model.
2. The line radiation comes from the three most abundant cometary elements — oxygen, carbon and nitrogen.

It should be noted that the most abundant atomic species in the cometary atmosphere is, of course, hydrogen. However, in the X-ray emission band, hydrogen has no line emission – it only contributes to the bremsstrahlung component. Also, it should be pointed out that it is the abundances of oxygen, carbon and nitrogen of the comet that are important — rather than the solar wind abundances.

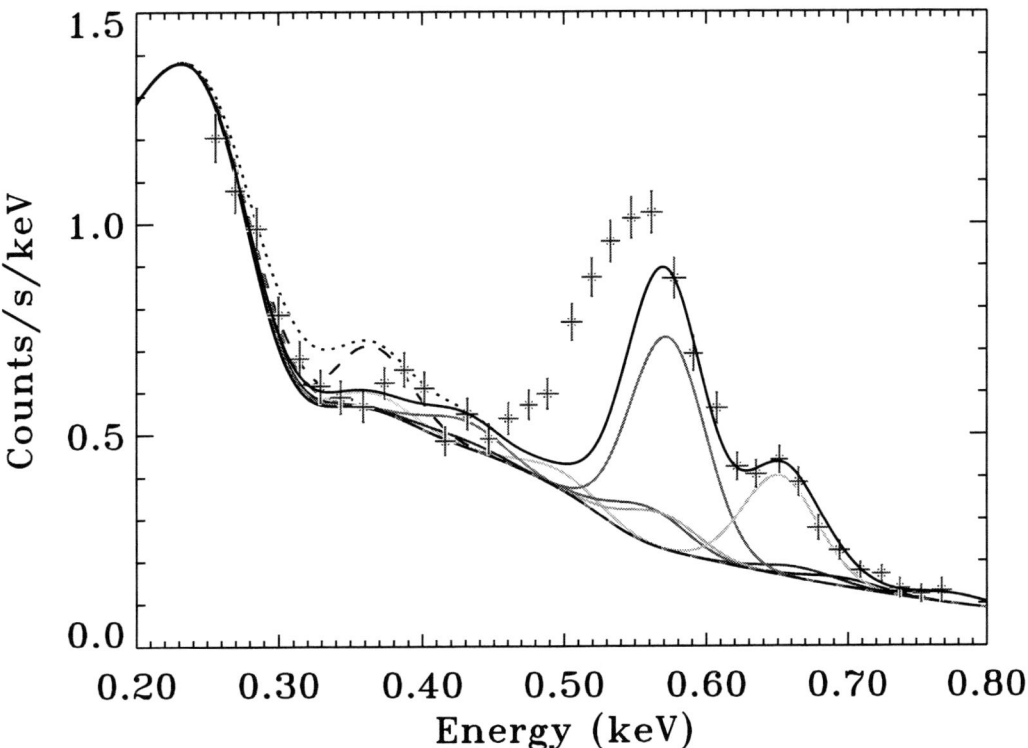

FIGURE 1. Intermediate spectral fit with no fluorescence lines included. The dotted line at low energies shows the result of assuming "normal" carbon abundances.

The starting point of the fitting process is obviously the observed spectrum of comet C/LINEAR 1999 S4 as presented by Lisse et al. [3]. The first model component to fit is the underlying bremsstrahlung continuum emission. We started with the temperature derived by Lisse et al. [3], but since we were using the very latest CHANDRA ACIS-S energy responses, we obtained a slightly hotter temperature of 4.5–5 million K (rather than 4 million K of Lisse et al. [3] [actually, 435 eV versus 390 eV]). However, in either case it is well worth pointing out that these results actually suggest that comets are the hottest objects in the solar system — given that the solar corona is only 1–2 million K. It also presents a serious problem for the only other cometary X-ray emission model – the charge exchange model, since this model is driven by the input solar coronal temperature. It seems rather difficult to explain a derived temperature that is some 2–3 times hotter than the input/driving spectrum.

Having derived the continuum emission spectrum, it is now time to add the K-shell line emission component. This is where our method starts to diverge from Lisse et al. [3]. Our model assumes that it is the comet itself that provides the "impurities" to produce the impurity line radiation. Therefore, we needed to assume a composition for comet LINEAR 1999 S4. We decided to initially adopt the ratios of O:C:N (oxygen:carbon:nitrogen) = 10:2:1 [4] measured for comet Halley. We used the Atomic Data and Analysis Structure, (ADAS) [5], to then calculate the equilibrium ionization balance of each element and return the emission functions for all required spectral lines from the various ions of each element. ADAS uses generalized collisional-radiative modelling for these calculations and so the fractional abundances of the ions of an element and their emission functions depend on both the electron temperature and electron density in the plasma.

ADAS provided data for around 40-50 emission lines in the range 0.1–1 keV. Of these, some were prominent, but most were likely to be negligibly weak. From this list of emission lines we started selecting individual lines to add to our bremsstrahlung continuum. This started with the strongest lines from oxygen (the most abundant cometary species). It became immediately apparent that the strongest line around 0.57 keV was due to highly ionised oxygen, as was the somewhat weaker line around 0.66 keV. However, when we came to add the lines from carbon and nitrogen, we

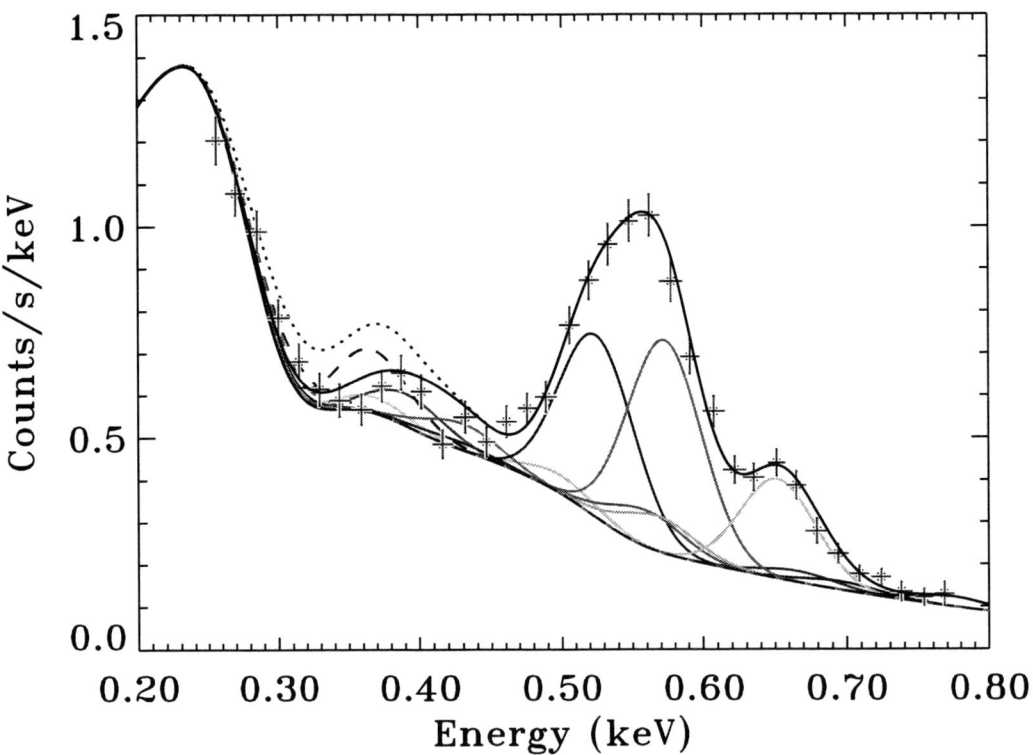

FIGURE 2. Final spectral fit with fluorescence lines included.

came across something of a dilemma. The strongest lines from these ions fell in parts of the CHANDRA spectrum that required little or no additional flux — *i.e.* in parts of the spectrum that were already well fitted by the bremsstrahlung continuum and oxygen lines. This was despite the fact that there was still clearly "missing" flux in the feature around 0.4 keV and also on the low energy side of the strong oxygen line, *i.e.* around 0.5 keV. This was particularly striking for the 368 eV line from C V which was expected to be the strongest line from carbon and given the expected carbon abundance of comet LINEAR 1999 S4, should have been clearly present in the CHANDRA spectrum. Since our modelling effort was directed at fitting the observed spectrum, the only option at this stage of the procedure was to reduce the relative strength of the carbon and (to a lesser extent) nitrogen lines to ensure that our model did not exceed the observed CHANDRA spectrum. The result at the end of this stage is shown in Figure 1. Our model is clearly doing very well to explain the low and high energy ends of the spectrum, but is clearly deficient in the 0.35–0.55 keV energy band.

MISSING PHYSICS: FLUORESCENCE FROM DUST

Our previous model only included the three main atomic species in the gas phase of the comet. However, we omitted another very important component of the cometary atmosphere — dust! We knew dust was important in comets with respect to their total X-ray luminosity. That is, very dust comets tend to be relatively weaker X-ray objects and also very dust poor comets seem to emit relatively strongly. This we explain as due to electron/dust interactions — the dust effectively "soaks up" the energetic electrons. However, in our model here we have ignored this point.

Now, cometary dust particles are almost certainly going to be coated in frozen ices – containing water ice, ammonia, carbon dioxide, etc. These will clearly then be a reservoir of neutral atoms. Therefore, when one of our energetic electrons hits such a dust particle, there will be a finite probability of the electron actually hitting one of these neutral ice atoms and possibly even displacing one of the inner K-shell electrons. This would then lead to the generation of fluorescent X-ray line emission from the same three elements that are present in the gas phase. The oxygen and nitrogen neutral fluorescent lines are at 0.53 and 0.40 keV, respectively — exactly where our model is deficient in line emission!

Figure 2 shows the result of adding the oxygen and nitrogen fluorescence lines due to dust to our model. Clearly, we now have a very good representation of the observed comet X-ray spectrum. The question of whether there is any carbon fluorescence is rather more difficult to determine because of the presence of the carbon edge in the window/filter of the detector.

This final fit still had the same "problem" of why the carbon abundance required to fit the spectrum is significantly reduced from the assumed comet Halley abundance. In fact the abundances used here is ×4–5 less. However, comet C/LINEAR had already been found to be unusual in the initial infrared spectrometry. Mumma et al. [6] report carbon depletion of 5–10 when compared to comet Halley. This is in excellent agreement with the X-ray spectral results reported here.

CONCLUSIONS

The plasma interaction model that we previously developed [1, 2] has now been extended with the addition of ADAS to produce a more complete description of the expected observable spectrum. Using this model and the CHANDRA spectrum we have been able to model the full spectrum of the comet and even detect the same low carbon abundance has seen in the infrared data. However, probably the most important result is the detection of neutral fluorescence lines of oxygen and nitrogen (and possibly also carbon). This means that X-ray spectroscopy is capable of simultaneous detecting and quantifying both the gas and dust components of the cometary atmosphere from a distance of 1 AU (150 million km).

ACKNOWLEDGMENTS

We would like to pay tribute to the late Professor John Dawson who played such an important role in this research.

REFERENCES

1. Bingham, R., Dawson, J. M., Shapiro, V. D., Mandis, A., Kellett, B. J. *Science*, **275**, 49, (1997).
2. Shapiro, V. D., Bingham, R., Dawson, J. M., Dobe, Z., Kellett, B. J., & Mendis, D. A. *JGR*, **104**, 2537, (1999).
3. Lisse, C. M., Christian, D. J., Dennerl, K., Meech, K. J., Petre, R., Weaver, H. A., & Wolk, S. J. *Science*, **292**, 1343, (2001).
4. Landolt and Börnstein. *Astronomy and Astrophysics*, **3A**, ed H. H. Voigt Berlin, Springer, (1993).
5. Summers, H.P. *The ADAS User Manual* Second edition, release V2.3, 2002, http://adas.phys.strath.ac.uk/
6. Mumma, M. J. et al. , *Science*, **292**, 1334 (2001).

Radiation Resonance Emission from Steep Overcritical Plasma Profiles Illuminated by Femtosecond Laser Pulses

R. Ondarza-Rovira* and T.J.M. Boyd[†]

*ININ, A.P. 18-1027, México 11801, D.F., Mexico
[†]Centre for Physics, University of Essex, Wivenhoe Park, Colchester CO4 3SQ, U.K.

Abstract. A radiation resonance effect observed in the reflection spectra from overdense plasmas illuminated by femtosecond laser pulses at normal incidence is reported from particle-in-cell simulations. Harmonic emission at multiple orders of the fundamental is found to exhibit resonance phenomena, with the number of resonances and power emitted depending on the electron plasma density. For relatively low laser intensities the reflected light at the laser frequency shows prominent resonant emission at specific values of the plasma density, mainly at 4, 16 and 36 times critical. For increasing laser intensities, strong harmonic emission at 4 and 16 times critical dominates the reflection spectra. In the case of the third laser harmonic, the emission is found to be resonant at 4 and 36 times critical and presents, additionally, a distinctive resonant region around 9 times critical. For higher harmonic numbers, weak radiation resonances persist in the emission spectra, with their number increasing with order. The resonance effect reported in this paper is found to occur at densities that satisfy $n_e/n_c = 4n^2$, where n_c is the critical density and n an integer. For the third harmonic, the second resonance is shown to take place at $n_e/n_c = 9$, which instead corresponds to $n = 1.5$.

INTRODUCTION

The laser-dense plasma interaction physics involves a number of processes that may occur when a target medium is irradiated by a light source. The main aspect of our interest is harmonic generation reflected from the vacuum-plasma interface. The motivation behind the study of harmonic emission has been the search for the development of coherent short x-ray sources, that may offer potential applications in many fields.

Strong emission detected up to the 46th harmonic was first reported by Carman *et al.* [1] from nanosecond CO_2 laser pulses at intensities above 5×10^{14} W/cm^2 incident on carbon targets. The highest harmonic emitted was interpreted as corresponding to the upper shelf density of the highly steepened profile with emission attributed to nonlinear resonant absorption.

Recent numerical simulations, with more accurate resolution, have shown no evidence of a cutoff at greater irradiances, with spectra extending to higher harmonic numbers [2]. Experiments have confirmed emission to high orders, up to the 75th harmonic [3]. The vacuum heating effect, consisting of energetic electrons that are dragged out of the plasma by the electric field of the driver and reinjected again to the plasma, is thought to be responsible for part of the emission [2].

A mechanism for harmonic emission, based on phase modulation by the laser light upon reflection from an oscillating plasma-vacuum interface, was proposed by Bulanov *et al.* [4], who took into account the oscillatory motion of a reflecting charge sheet of plasma electrons. Lichters *et al.* [5] obtained that the model spectra from plasma oscillations at the boundary agreed with particle-in-cell (PIC) simulations.

In this paper we report on a radiation effect observed from PIC simulations of laser interactions with overcritical plasmas characterized by very steep density profiles. It is found that laser harmonic emission reflected from the plasma and generated at the critical layer exhibits a resonance phenomena, with enhancement of the emission around specific multiples of the critical density. It is shown that the power of the reflected emission depends on the plasma density and strong resonant enhancements become apparent for densities 4, 16 and 36 times critical, for the lower harmonic multiples of the laser frequency. In the case of the third harmonic the second resonance appeared instead at 9 times critical. For the fifth and higher harmonics, weak resonances persist in the radiation spectra, with their number increasing with density.

RESONANT ENHANCEMENT

The nonlinear response from the plasma under the influence of an external radiation field was studied by means of PIC kinetic simulations. We used a 1 1/2-D, fully relativistic and electromagnetic code with immobile ions as a neutralizing background. For all the simulations, the density scale length was chosen as a small fraction of a laser wavelength, allowing to prepare the simulation box with a very steep density gradient at the front boundary. In this paper we show particle simulations for laser light of femtosecond pulse duration. The laser pulses were launched to the front of the plasma interface at normal incidence. The plasma had an extension of 4-6 laser wavelengths and 2000 grid cells per wavelength were used, containing 2×10^6 particles. This allowed a spatial grid refined enough to resolve a Debye length with acceptable accuracy. Two vacuum gaps to the left and to the right boundaries of the simulation box of the order of half a wavelength each were used to allow for particle and field propagation. The electron temperature was taken as 1 keV.

Considering the input intensities, time scales, and the sharpness of the vacuum-plasma interface, collisions between plasma particles were neglected.

Ref. 6 suggested a nonlinear resonance effect for the third laser harmonic at a density 4 times critical. This resonance phenomena is captured by our simulations at that precise density, exhibiting a prominent feature in the emission spectra.

Even harmonics can be generated for oblique incident interactions since the component E_x of the electric field of the laser pulse in the direction of wave propagation and along the density gradient excites electron plasma waves that couple the radiation field generating both odd and even harmonics.

For our simulations we have used laser light of wavelength 0.248 μm and pulse duration of 40 fs normally incident on an overdense plasma characterized by a steep density profile at the interface. We have Fourier-analised the reflected signal to obtain the emission spectra. Figs. 1-2 show the reflected power of the fundamental and its first 3 odd harmonics, as a function of the plasma density. The power emission calculated at different points are shown in the plots, where interpolation was applied over the entire range of plasma densities. Laser pulses of different strength were used in the simulations. In this paper we show results for $a_0 = 0.5$, which corresponds to a laser intensity of 5.57×10^{18} W/cm^2.

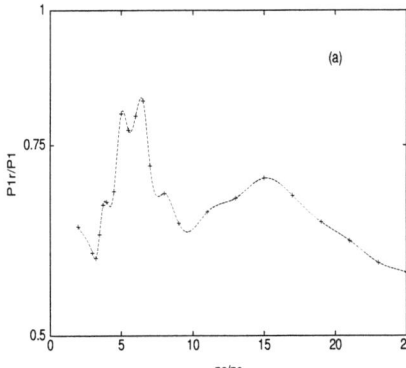 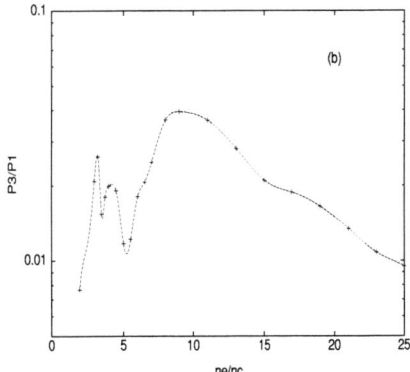

FIGURE 1. Reflected power: (a) at the laser frequency and (b) of the 3rd laser harmonic as a function of the plasma density, for $a_0 = 0.5$ and $\lambda_L = 0.248\,\mu$m.

Fig. 1a shows the radiated power at the fundamental frequency. Resonance effects characterize the emission around multiples of the critical density. It is found a resonance at densities around 4 and 16 times critical. The conversion efficiency for the first resonance is nearly 80% of the input energy, and decreases for higher densities, with a resonant gain of 70% at $m = n_e/n_c = 16$. For the third harmonic, two resonances at 4 times critical and at $m = 9$ are found. This is shown in Fig. 1b. As expected, the conversion efficiency is lower than the emission at the fundamental and decreases again with density, although the reflected power of the resonance at $m = 9$ result to be greater than the one at $m = 4$. This is a characteristic feature that distinguishes the third harmonic from higher orders for any laser intensity.

For the fifth laser harmonic, resonances are found at 4 and at 36 times critical, with similar strength in the power emitted. The strength of the emission is found to decay at higher densities. This is shown in Fig. 2a.

For the seventh harmonic, resonant emission occurs at 4, 16, and 36 times critical, with strength of the emission decaying with order, as shown in Fig. 2b. One remarkable feature observed in the emission plots for the third and fifth laser harmonics is the resonance structure around the density 4 times critical. A number of two and three emission

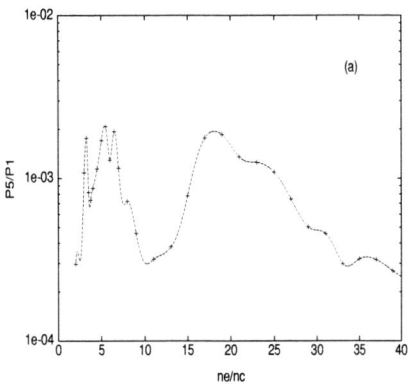 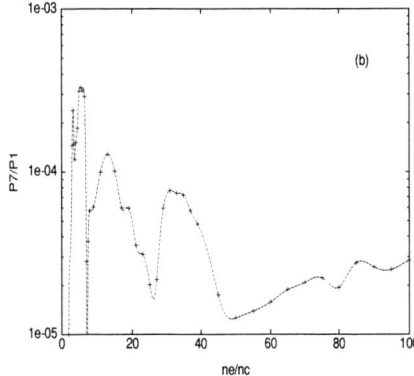

FIGURE 2. Reflected power: (a) of the 5th and (b) of the 7th laser harmonic as a function of the plasma density, parameters as in Fig. 1.

peaks can be discerned in this region, indicating a strong coupling effect, predominantly among oscillation modes of the plasma at densities equal to critical and 4 times critical.

CONCLUSION

We have reported resonance emission effects from the interaction of short laser pulses with plasmas of overcritical density. The resonances occur at specific values of the plasma density and satisfy $m = n_e/n_c = 4n^2$, where n_c is the critical plasma density and n is an integer. The reflected light at the fundamental frequency showed two prominent resonances at 4 and at 16 times critical with the power of the emission decaying with increasing density. For the third laser harmonic, the second resonance is shown to be produced at $n_e/n_c = 9$, which instead corresponds to $n = 1.5$. For this harmonic number the resonance at $m = 9$ is stronger than the resonance at $m = 4$. The opposite is true for the higher harmonics, where more resonances are observed with conversion efficiency decreasing with density. One possible mechanism that may explain the phenomenon reported here is resonance absorption. Nevertheless, for very steep density profiles the mode-coupling mechanism breaks down since the electron oscillation amplitude becomes comparable to the density scalelength. On the other hand, the $v \times B$ mechanism can explain the efficiency of converting the incident energy into energetic electrons that arise from wavebreaking of the density oscillations which generate the harmonics. Thus, we can assume that the harmonic enhancement effect involves a mechanism resulting from mode-coupling of laser harmonics and density oscillations.

ACKNOWLEDGMENTS

One of us (ROR) acknowledges support from CONACyT under Contract No. 33251-E.

REFERENCES

1. R. L. Carman, C. K. Rhodes, and R. F. Benjamin, *Phys. Rev. A* **24**, 2649 (1981).
2. P. Gibbon, *Phys. Rev. Lett.* **76**, 50 (1996).
3. P. A. Norreys et al., *Phys. Rev. Lett.* **76**, 1832 (1996).
4. S. V. Bulanov and N. M. Naumova, *Phys. Plasmas* **1**, 745 (1994).
5. R. Lichters, J. Meyer-ter-Vehn, and A. Pukhov, *Phys. Plasmas* **3**, 3425 (1996).
6. S. C. Wilks, W. L. Kruer, and W. B. Mori, *IEEE Trans. Plasma Sci.* **21**, 120 (1993).

Bifurcation and Metamorphosis of Plasma Turbulence–Shear Flow Dynamics: the Path to the Top of the Hill

R. Ball*, R. L. Dewar* and H. Sugama[†]

Department of Theoretical Physics, The Australian National University, Canberra ACT 0200 Australia
[†]*National Institute For Fusion Science, Oroshi-cho, Toki GIFU 509-5292 Japan*

Abstract. The structural properties of an economical model for a confined plasma turbulence governor are investigated through bifurcation and stability analyses. Two types of discontinuous low to high confinement transition are found. One involves classical hysteresis, governed by viscous dissipation. The other is intrinsically oscillatory and non-hysteretic, and thus provides a model for observed "dithering" transitions. This metamorphosis of the system dynamics is an important late side-effect of symmetry-breaking, which manifests as an unusual non-symmetric transcritical bifurcation induced by a significant shear flow drive.

INTRODUCTION

Fusion plasmas in magnetic containers may undergo a more-or-less dramatic transition from a low to a high confinement state — the L–H transition — as the power input is increased, with the desirable outcome confinement is greatly improved due to localized transport reduction [9]. Since L–H transitions were first observed [12] research to understand this change has been copious and diverse, because control over access to high confinement régimes is one of the keys to economical fusion power [11].

Efforts to model plasma mass dynamics around L–H transitions and concomitant oscillatory phenomena have concentrated on the development of low-dimensional descriptions, or systems of coupled ODE in a few dynamical variables and parameters that represent physical properties or external controls. These are reviewed in [1]. What drives this approach is the predictive power that such a unified description would have in designing and controlling confinement states. For example, a model that characterises hysteresis in the L–H transition would help engineers to manage access to H-modes.

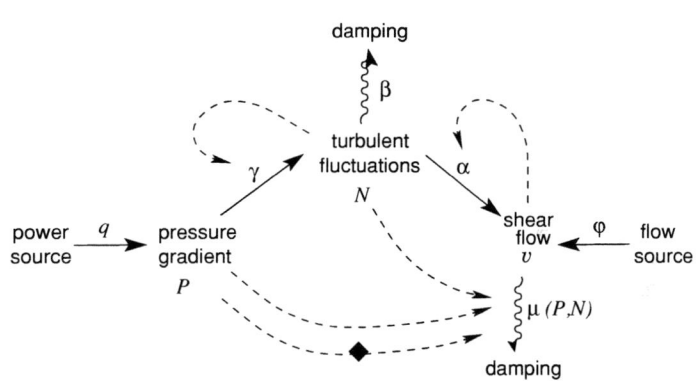

FIGURE 1. A power input q creates a pressure gradient P from which the turbulent density fluctuation intensity N grows at a rate γ and is damped at a rate β. The turbulence feeds energy into the poloidal shear flow v via the Reynolds stress α. The shear flow is generated externally at rate φ and damped by the ion viscosity μ.

However, there have been some rocky shores in the quest for a low-dimensional sub-space that captures the dynamics of L–H transitions [2, 3, 4]. The heart of the matter is the relationship between the bifurcation structure of a dynamical model and the physics of the process it is supposed to represent. If we probe this relationship we find that degenerate bifurcations ought to reflect some matching physics — the point of onset of hysteresis, for example, or a fragile symmetry — or they are pathological.

In this work we describe a consensus model that captures the association between discontinuous and oscillatory behavior and supplies intelligence on the big issues such as turbulence suppression by sheared flows. We present graphically a bifurcation and stability analysis that reveals a a radical change, or *metamorphosis*, in the dynamics.

THEORY AND RESULTS

The primitive of a plasma turbulence governor schematised in Fig. 1 is modeled by the dynamical system of Eqs 1–4. The first and second terms in Eq. 4 model the neoclassical (b) and anomalous (a) viscosity damping respectively. A high pressure gradient tends to inhibit the b contribution but enhance the a contribution.

The formal bifurcation and stability analysis of Eqs 1–3 was reported in [4]. Here we present a graphical interpretation.

$$\varepsilon \frac{dP}{dt} = q - \gamma PN \qquad (1)$$

$$\frac{dN}{dt} = \gamma PN - \alpha v^2 N - \beta N^2 \qquad (2)$$

$$2\frac{dv}{dt} = \alpha vN - \mu(P,N)v + \varphi \qquad (3)$$

$$\mu(P,N) = bP^{-3/2} + aPN. \qquad (4)$$

In Fig. 2 bifurcation diagrams for non-critical (main figure) and critical (inset) values of b and φ are plotted from equilibrium solutions of Eqs 1–3. We see that the symmetry of the inset figure is broken, by selection of a non-zero value of φ, which determines the direction of the shear flow.

The stable solutions along the $+v$ curve to the limit point s_L at $(v,q) \approx (0.2, 0.495)$ comprise the L-mode branch, where the pressure gradient feeds the turbulence but does not inhibit the neoclassical viscosity damping enough for the shear flow to increase much.

At s_L the solutions become unstable and the system must jump to a stable attractor, normally the stable solution at $(v,q) \approx (1.5, 0.495)$. This section of the curve with high positive shear flow represents the $+v$ H-mode branch.

If the system is given an opposing kick near s_L it may move onto the stable $-v$ curve. This is the $-v$ H-mode branch. Shear flow reversal may also be forced from the $-v$ curve, but is *spontaneous* if the power input slips below the $(-v,q)$ limit point, s_{H-}.

Continuing from left to right we arrive at points on the $+v$ and $-v$ H-mode branches where the solutions again become unstable. These points are Hopf bifurcations. In Fig. 2 the first Hopf bifurcations h_A^+ and h_A^- occur at $(+v,q) \approx (1.74, 2.06)$ and $(-v,q) \approx (1.72, 3.06)$. Here the system develops stable limit cycles, which grow in amplitude then shrink as the anomalous viscosity takes over, until they are extinguished at the second Hopf bifurcations h_B^+ and h_B^-, at $(+v,q) \approx (1.66, 9.83)$ and $(-v,q) \approx (1.67, 8.69)$. This passage through an oscillatory régime is typical of type III ELMs [8]. The relative phases of v, N, and P in this region are shown in Fig. 3.

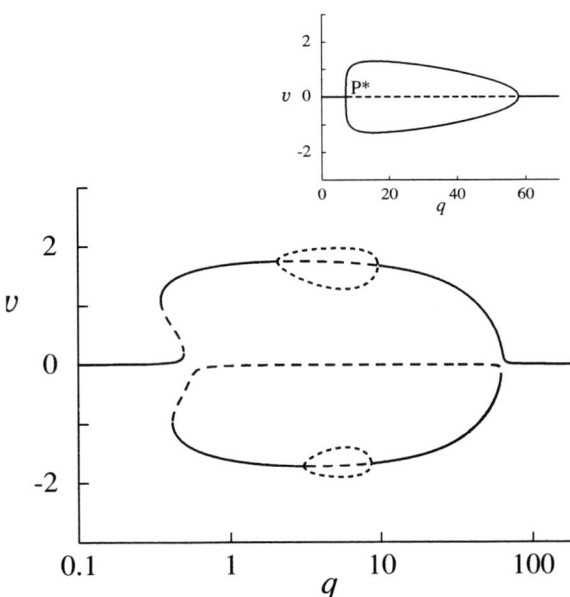

FIGURE 2. Bifurcation diagram for $b=1$, $\varphi = 0.02$ (main figure), and for the critical set (P*): $\varphi = 0$, $b = 18.58$ (inset). $\alpha = 2.4$, $\beta = 1$, $\gamma = 1$, $a = 0.3$, $\varepsilon = 1.5$. Solid lines: stable solutions; dashed lines: unstable solutions; dotted lines: amplitude envelope of limit cycle branches.

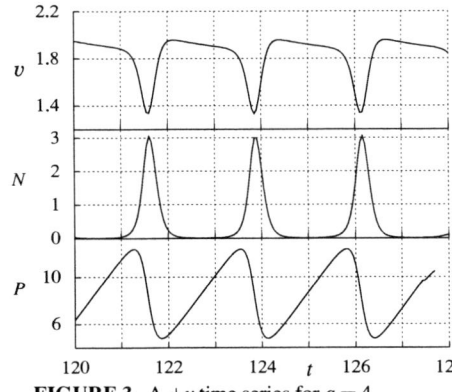

FIGURE 3. A $+v$ time series for $q = 4$.

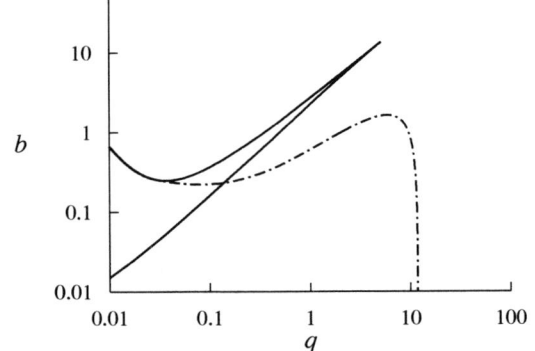

FIGURE 4. Curves of the singular points in Fig. 2. Solid lines: loci of limit points, dot-dash lines: Hopf bifurcation loci.

Figure 2 also shows hysteresis, which can be shown rigorously to be structurally endemic to Eqs 1–4 [4].

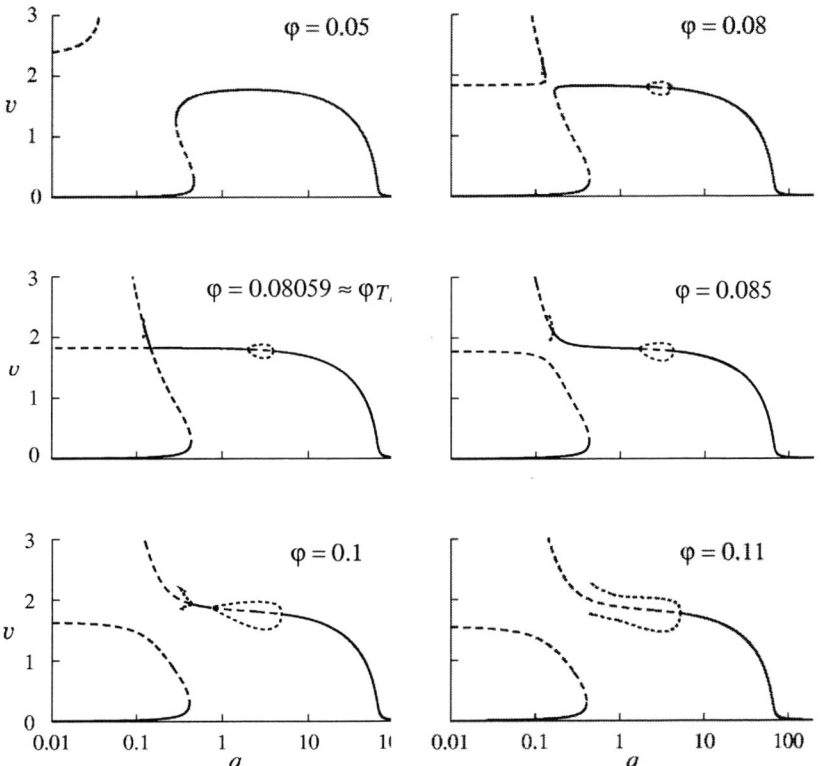

FIGURE 5. A series of bifurcation diagram snapshots illustrates the exchange at φ_{T_m} and its aftermath. Here $\varepsilon = 1.0$ and other parameters are the same as in Fig. 2.

Transitions with hysteresis have been observed in many experiments and simulations [10, 7, 6, 5], although it is not a universal feature.

The extent of hysteresis over b is shown in the *two-parameter bifurcation diagram* for the $+v$ curve in Fig. 4, in which computed curves of the singular points in Fig. 2 are shown. The minimum in the limit point curve is a *transcritical bifurcation* in the equilibrium solutions of Eqs 1–3. It may have important issues concerning access to and control of confinement states.

Consider the series of snapshots in Fig. 5.

In the first frame, for $\varphi = 0.05$, a separate branch of solutions, which is trapped at $(q, v) = (0, \infty)$ for $\varphi = 0$, has appeared. It does not remain unstable. At a singular point $0.05 < \varphi < 0.08$ a degenerate Hopf bifurcation appears, then in the frame for $\varphi = 0.08$ the new limit point s_C and Hopf bifurcation h_C have separated, with stable solutions between them. Note that h_A and h_B have appeared on the hysteretic curve.

At a metamorphic value of φ that we designate φ_{Tm} the arms of the two separate steady-state branches are exchanged at a transcritical bifurcation. (In numbers $(v, q, \varphi)_{Tm} = (1.8247.., 0.1468.., 0.08059..).$)

In the frames for $\varphi = 0.085$ and $\varphi = 0.1$ h_C and h_A are moving together, and in the last frame, for $\varphi = 0.11$, they have collided and annihilated each other (at the minimum in the dot-dash curve).

On the L-mode branch in the last frame we see that the L–H transition is *very* different from the hysteretic transition in Fig 2. Here the stable attractor on the H-mode branch is a limit cycle rather than a fixed point. *Furthermore, it can be shown rigorously that hysteresis is (locally) forbidden.* Thus it reflects the dithering or L–H–L transitions, followed by a quiescent H-mode, that are often reported.

In terms of the governor in Fig. 1 interesting non-linear dynamics are expected, because more kinetic energy in the shear flow leads to more turbulence suppression but also a larger damping effect, which alters the competitive distribution of energy from the pressure gradient.

CONCLUSIONS

Bifurcation and stability analysis of Eqs 1–3 reveals two qualitatively different transitions. The hysteretic transition is controlled by the damping rate coefficients. The non-hysteretic transition occurs when a relatively strong shear flow drive interacts with the internal dynamics to cause the metamorphosis shown in Fig. 5. The top of the hill, or local organizing center, is the point φ_{T_m}, the view from which encompasses all of the qualitative behavior that the system is capable of.

Finally, we see in the results of this analysis strong evidence that *remarkably low-dimensional models can capture and help explain essential aspects of turbulent flows that elude understanding from numerical simulations that include resolved spatial scales, and that physical deductions can be made from observations of bifurcations.*

ACKNOWLEDGMENTS

R.B. would like to thank the Australian Research Council for support.

REFERENCES

1. Ball, R., 2002, Low-dimensional models for plasma dynamics, preprint.
2. Ball, R., and R. L. Dewar, 2000, Phys. Rev. Lett. **84**(14), 3077.
3. Ball, R., and R. L. Dewar, 2001, J. Plasma Fus. Res. **4**, 266.
4. Ball, R., R. L. Dewar, and H. Sugama, 2002, Phys. Rev. E (to appear, preprint at http://arXiv.org/abs/physics/0206078).
5. Hubbard, A. E. *et al.*, 1998, Plasma Phys. Control. Fusion **40**, 689.
6. Igitkhanov, Y. *et al.*, 1998, Plasma Phys. Control. Fusion **40**, 837.
7. Ryter, F. *et al.*, 1998, Plasma Phys. Control. Fusion **40**, 725.
8. Suttrop, W., 2000, Plasma Phys. Control. Fusion **42**, A1.
9. Terry, P. W., 2000, Reviews of Modern Physics **72**(1), 109.
10. Thomas, D. *et al.*, 1998, Plasma Phys. Control. Fusion **40**, 707.
11. U.S. DoE, 2000, Report of the Integrated Program Planning Activity for the DoE's Fusion Energy Sciences Program, available at www.pppl.gov/common_pages/fusion_policy_docs.html.
12. Wagner, F. *et al.*, 1982, Phys. Rev. Lett. **49**(19), 1408.

Onset of Alfvén Turbulence via Boundary Crisis

A. C.-L. Chian*†, F. A. Borotto** and E. L. Rempel*†

*World Institute for Space Environment Research-WISER, NITP, University of Adelaide, SA 5005, Australia
†National Institute for Space Research-INPE, P.O. Box 515, 12227-010 São José dos Campos, SP, Brazil
**Universidad de Concepción, Departamento de Física, Concepción, Chile

Abstract. Alfvén waves are important for heating and particle acceleration in space, astrophysical and laboratory plasmas. Chaos and intermittent turbulence of Alfvén waves have been observed in the solar wind. We show that the onset/destruction of Alfvén chaos can occur via a transition mechanism known as boundary crisis. A crisis is a global bifurcation whereby a chaotic (strange) attractor abruptly expands in size or vanishes along with its basin of attraction. Recent theoretical and experimental studies have confirmed that interior and boundary crises can occur in plasmas. We identify a boundary crisis in a complex plasma region in the presence of a large number of coexisting attractors. An example of double boundary crises is characterised using the unstable periodic orbit determined numerically from the numerical solution of the driven-dissipative nonlinear Schrödinger equation, which describes the nonlinear dynamics of a large-amplitude, left-hand circularly polarised Alfvén wave propagating along an ambient magnetic field. We demonstrate how the same unstable periodic orbit causes the appearance/disappearance of two chaotic attractors due to two successive homoclinic tangencies. The theoretical tools developed in this paper can be applied to detect Alfvén crises and other nonlinear dynamical Alfvénic phenomena in laboratory experiments as well as in space observations of interplanetary Alfvén turbulence.

INTRODUCTION

Alfvén waves are ubiquitous in cosmic plasmas and play a key role in the heating and confinement of fusion laboratory plasmas [1]. Recent theoretical studies of Alfvén waves have improved our understanding of their nonlinear behaviors such as the rich variety of solitons and traveling waves [2, 3] and parametric instabilities [4]. In particular, the theory of deterministic chaos has helped to elucidate the relation between nonlinear Alfvén waves and Alfvén turbulence. In Chian et al. [5], it was shown that nonlinear Alfvén waves can evolve to chaos via period-doubling bifurcation, saddle-node bifurcation or interior crisis.

The aim of this paper is to show that the onset/destruction of Alfvén chaos can occur via a transition mechanism known as *boundary crisis*. A crisis is a global bifurcation whereby a chaotic (strange) attractor abruptly expands in size or vanishes along with its basin of attraction [6, 7]. The former is called an interior crisis and the latter a boundary crisis. Both types of crises involve the tangency (or collision) of a chaotic attractor with an unstable fixed point or unstable periodic orbit (or its invariant stable manifolds) as some control parameter of the system is varied. Two kinds of tangency can occur: homoclinic and heteroclinic [6]. For the interior crisis the tangency takes place in the interior of the basin of attraction of the chaotic attractor, whereas for the boundary crisis the tangency takes places on the boundary of the basin of attraction of the chaotic attractor. Hence, interior crisis can occur in the presence of one or more attractors whereas the boundary crisis requires the coexistence of at least two attractors. Recently, the Alfvén interior crisis [5] and the interior crisis-induced transition to spatiotemporal chaos of drift plasma waves [8] were discussed. In this paper, we report the existence of boundary crisis in plasmas.

THE DERIVATIVE NONLINEAR SCHRÖDINGER EQUATION

The nonlinear dynamics of a large-amplitude Alfvén wave traveling along an ambient magnetic field in the x-direction is described by the following driven-dissipative derivative nonlinear Schrödinger equation

$$\partial_t b + \alpha \partial_x (|b|^2 b) - i(\mu + i\eta)\partial_x^2 b = S(b,x,t) , \qquad (1)$$

where η is the dissipative scale length, $b = b_y + ib_z$ is the complex transverse wave magnetic field normalized to the constant ambient magnetic field B_0, time t is normalized to the inverse of the ion cyclotron frequency $\omega_{ci} = eB_0/m_i$, space x is normalized to c_A/ω_{ci}, $c_A = B_0/(\mu_0\rho_0)^{1/2}$ is the Alfvén velocity, $\alpha = 1/[4(1-\beta)]$, $\beta = c_S^2/c_A^2$, $c_S = (P_0/\gamma\rho_0)^{1/2}$ is the acoustic velocity and μ is the dispersive parameter. We take the external driving force $S(b,x,t) = A\exp(ik\phi)$ to be a monochromatic left-hand circularly polarized wave with a wave phase $\phi = x - Vt$, where V is a constant wave velocity, A and k are arbitrary constants.

The first integral of (1) can be reduced to a system of ordinary differential equations by seeking stationary wave solutions with $b = b(\phi)$, yielding

$$\dot{b}_y - v\dot{b}_z = \partial H/\partial b_z + a\cos\theta, \quad (2)$$

$$\dot{b}_z + v\dot{b}_y = -\partial H/\partial b_y + a\sin\theta, \quad (3)$$

$$\dot{\theta} = \Omega, \quad (4)$$

where $H = (\mathbf{b}^2 - 1)^2/4 - (\lambda/2)(\mathbf{b} - \hat{\mathbf{y}})^2$, the overdot denotes derivative with respect to the phase variable $\tau = \alpha b_0^2 \phi/\mu$, the normalized dissipation parameter $v = \eta/\mu$, $b \to b/b_0$ (where b_0 is an integration constant), $\mathbf{b} = (b_y, b_z)$, $\theta = \Omega\phi$, $\Omega = \mu k/(\alpha b_0^2)$, $a = A/(\alpha b_0^2 k)$, $\lambda = -1 + V/(\alpha b_0^2)$. We assume $\beta < 1$, hence $\alpha > 0$.

NONLINEAR DYNAMICAL ANALYSIS

A bifurcation diagram can be numerically computed from (2)-(4) by varying the dissipation control parameter v while keeping other control parameters fixed ($a = 0.1, \Omega = -1, \lambda = 1/4, \mu = 1/2$). We identified a complex plasma region, where five different attractors are present and a wealth of dynamical features are found. The bifurcation diagram of this region is given in Fig. 1. In this region the attractor A_1 remains a period-one limit cycle throughout, whereas the other four attractors appear only during certain intervals of v. Only three out of five attractors (A_1, A_2 and A_3) are plotted in Fig. 1. Attractors A_4 and A_5 are confined within the region indicated by the bar in Fig. 1.

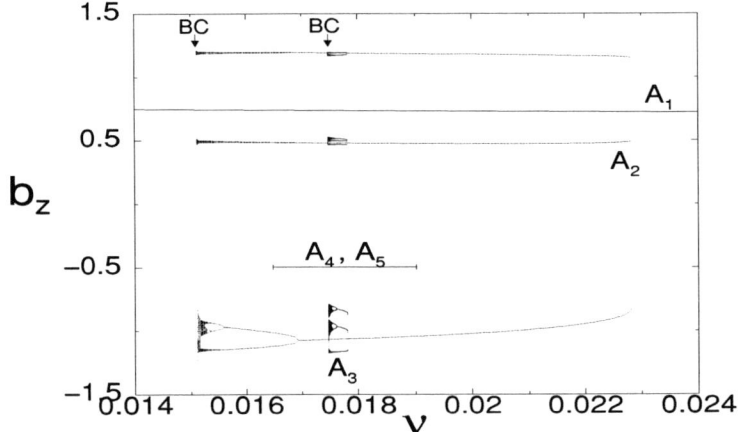

FIGURE 1. Bifurcation diagram $b_z(v)$ for $a = 0.1$, $\Omega = -1$, $\mu = 1/2$, $\lambda = 1/4$ and $\beta < 1$, showing attractors (A_1, A_2, A_3); BC denotes boundary crisis.

The bifurcation diagram Fig. 1 shows that apart from a small range of v wherein a single attractor (A_1) exists by itself, in most regions of the bifurcation diagram there is coexistence of two or more attractors.

Unstable periodic orbit (UPO) is the key for the characterization of nonlinear dynamical phenomena such as the boundary crisis. We determine the UPO from the numerical solution of (2)-(4) and analyze in detail the role played by the UPO in the onset of Alfvén boundary crisis. The complex plasma region we identified exhibits a large number of crises. In this paper, we shall focus only on the characterization of the double boundary crises indicated in Fig. 1.

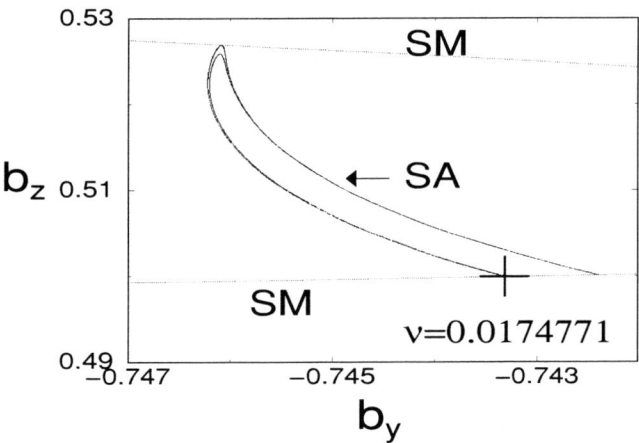

FIGURE 2. Homoclinic tangency at boundary crisis BC of Fig. 1. SA denotes the Poincaré map of the middle branch of strange attractor A_3. The cross denotes one of the Poincaré points of the unstable periodic orbit of period-9 and the light lines represent its stable manifolds (SM).

An examination of Fig. 1 shows that a saddle-node bifurcation occurs for the attractor A_3 at $v \sim 0.0178162$ (attractor A_2 at $v \sim 0.02279$), giving rise to a pair of period-9 (period-3) stable and unstable periodic orbits, respectively. For each attractor, the resulting stable periodic orbit undergoes a cascade of period-doubling bifurcations which leads to the formation of a chaotic attractor. The chaotic attractor disappears abruptly at $v \sim 0.0174771$ for A_3 and at $v \sim 0.01514$ for A_2 due to double boundary crises. In both cases, the period-9 UPO of A_3, arising from the saddle-node bifurcation at $v \sim 0.0178162$, is involved.

The characterization of Alfvén double boundary crises can be performed using the Poincaré method. The Poincaré map of the middle branch of the strange attractor (SA) A_3 near the first crisis point $v \sim 0.0174771$ is shown in Fig. 2. The cross denotes one of the Poincaré points associated with the middle branch of the period-9 UPO. The invariant stable manifolds (SM) associated with the period-9 UPO (saddle) are represented by the light lines. Evidently, Fig. 2 shows that at the crisis point the strange attractor collides head-on with the saddle and its stable manifolds via a homoclinic tangency. Before the first boundary crisis, the system has four coexisting attractors (A_1, A_2, A_3, A_4). After the crisis, both the attractor A_3 as well as its basin of attraction vanish, leaving the system with only three coexisting attractors (A_1, A_2, A_4).

The period-9 UPO survives the first boundary crisis that causes the destruction of A_3 and continues to participate in the second boundary crisis that causes the destruction of A_2. A Poincaré map of the middle branch of the strange attractor A_2 at $v \sim 0.01514$ (marked BC in Fig. 1) is plotted in Fig. 3. The three crosses denote the three Poincaré points associated with the middle branch of the period-9 UPO, and the light lines denote its stable manifolds. This second boundary crisis occurs via homoclinic tangency, as shown in Fig. 3. Before the second crisis, the system has two coexisting attractors (A_1, A_2). After the crisis, both the attractor A_2 as well as its basin of attraction vanish, leaving the system with only the attractor A_1. Although we only demonstrated the double homoclinic tangencies of the middle branch of the attractors A_2 and A_3, the same dynamics applies to the upper and lower branches. In a homoclinic tangency such as shown in Fig. 2 and Fig. 3, the strange attractor coincides with the invariant unstable manifolds of the saddle (period-9 UPO) as confirmed by numerical solutions.

CONCLUSIONS

Boundary crisis with its associated sudden appearance/disappearance of a chaotic attractor was seen in the numerical simulations of a theoretical heart model described by the Van der Pol oscillator [9]. Recently, this phenomenon was observed in a dripping faucet laboratory experiment [10]. These observations suggest that Alfvén boundary crisis is likely to take place in cosmic and laboratory plasmas. In fact, in a recent tokamak plasma experiment, a period-one unstable periodic orbit was identified in a weakly turbulent edge-localized-mode high-performance regime (ELM H mode) [11]. As shown in the present paper, the identification of UPO is essential for the characterization of chaotic

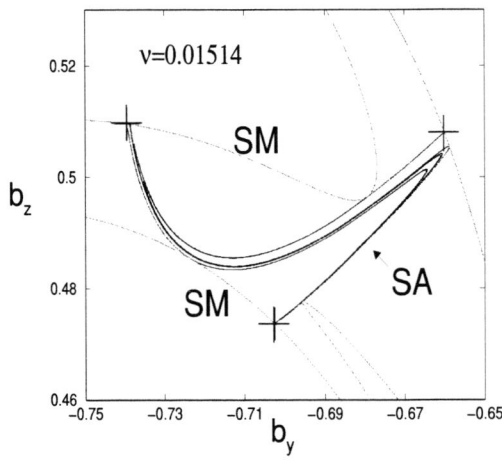

FIGURE 3. Homoclinic tangency of attractor A_2 with the stable manifolds of the saddle at the boundary crisis for $v = 0.01514$. SA denotes the strange attractor, the crosses denote the Poincaré points of the middle branch of the unstable periodic orbit of period-9, the light lines denote the stable manifolds (SM) of the period-9 saddle.

system dynamics. It is worth mentioning that Hada et al. [12] noticed sudden termination of Alfvén chaos in certain "sporadic" attractors. Our analysis indicated that the termination of those "sporadic" chaotic attractors are actually due to the occurrence of boundary crisis discussed in this paper.

In conclusion, the theoretical tools developed in this paper can be applied to the detection of Alfvén boundary crisis and other nonlinear dynamical Alfvén phenomena in laboratory experiments [13] as well as in space observations of interplanetary Alfvén turbulence [5, 14, 15].

ACKNOWLEDGMENTS

The authors would like to acknowledge the support of the Australian Institute for Nuclear Science and Engineering (AINSE), CNPq (Brazil), FAPESP (Brazil) and AFOSR. A. C.-L. Chian and E. L. Rempel wish to thank Professors Tony Thomas and Tony Williams of the University of Adelaide for their kind hospitality.

REFERENCES

1. Chian, A. C.-L., de Assis, A. S., de Azevedo, C. A., Shukla, P. K., and Stenflo, C., *Alfvén Waves in Cosmic and Laboratory Plasmas*, *Physica Scripta*, **T60** (Royal Swedish Academy of Sciences, Stockholm),1995.
2. Kennel, C. F., Buti, B., Hada, T., and Pellat, R., *Phys. Fluids*, **31**, 1949-1961 (1988).
3. Hada, T., Kennel, C. F., and Buti, B., *J. Geophys. Res.*, **94**, 65-77 (1989).
4. Oliveira, L. P. L., and Chian, A. C.-L., *J. Plasma Phys.* **56**, 251-264 (1996).
5. Chian, A. C.-L., Borotto, F. A., and Gonzalez, W. D., *Astrophys. J.*, **505**, 993-998 (1998).
6. Grebogi, C., Ott, E., and Yorke, J. A., *Phys. Rev. Lett*, **48**, 1507-1510 (1982).
7. Chian, A. C.-L., Borotto, F. A., and Rempel, E. L., *Int. J. Bifurcation Chaos*, in press.
8. He, K., *Phys. Rev. Lett.*, **80**, 696-699 (1998).
9. Abraham, R. H., and Stewart, H. B., *Physica*, **D21**, 394-400 (1986).
10. Pinto, R. D., and Sartorelli, J. C., *Phys. Rev. E*, **61**, 342-347 (2000).
11. Bak, P. E., Yoshino, R., Akasura, N., and Nakano, T., *Phys. Rev. Lett.*, **83**, 1339-1342 (1999).
12. Hada, T., Kennel, C.F., Buti, B., and Mjølhus, E., *Phys. Fluids*, **B2**, 2581-2590 (1990).
13. Gekelman, W., *J. Geophys. Res.*, **104**, 14417-14435 (1999).
14. Tu, C.-Y., and Marsch, E., *Space Sci. Rev.*, **73**, 1-210 (1995).
15. Baumgärtel, K., *J. Geophys. Res.*, **104**, 28295-28308 (1999).

Nonlinear Dynamical Analysis of Two Current-Driven Low-Frequency Instabilities in a Magnetised Plasma Column

D. Dimitriu,[1] V. Ignatescu,[1] C. Ionita, E. Lozneanu,[1] M. Sanduloviciu,[1] R. Schrittwieser

Department of Ion Physics, University of Innsbruck, Innsbruck, Austria,
[1]*Faculty of Physics, University "Al. I. Cuza", Iaşi, Romania*

Abstract. By drawing an electron current along the field lines of a magnetised plasma column, several low-frequency instabilities can be excited. Many such experiments have been performed in Q-machines which produce an alkaline plasma of about 10^9 cm^{-3} density, confined by a *B*-field of about 0.2 T. If the collector, which draws the current, has a radius smaller than that of the plasma column but larger than a few ion gyroradii, the electrostatic ion-cyclotron instability (EICI) is predominant, but also the potential relaxation instability (PRI) can appear and there can be a strong interaction between the two. The frequency of the EICI f_{EICI} is slightly above the ion gyrofrequency $\Omega_i = eB/m_i$, whereas that of the PRI f_{PRI} is given by the axial ion transit time through the system. We have investigated this interaction in the Innsbruck Q-machine in a potassium plasma in the case when the amplitudes of the two instabilities have similar values. In our case, for $B = 0.13$ T the interaction leads not only to a strong modulation of the amplitude but also of the frequency of the EICI by the PRI. Our results have indicated that the EICI frequency as such decreases with increasing strength of the exciting electron current, which effect can be explained by electron-neutral impact ionisation processes in front of the collector. In order to clarify the strong interaction between the two instabilities, the experimental results have been subject to a nonlinear dynamical analysis, which delivered the autocorrelation function and enabled us to reconstruct the state space of the phenomenon and to simulate the signal in a special case. These results are presented in this contribution.

INTRODUCTION

Q-machines were developed as laboratory plasma sources in order to produce a quiescent plasma [1,2], i.e., a plasma that is free from low frequency instabilities, unless they are excited on purpose. Under well defined external experimental conditions, the plasma of a Q-machine becomes unstable and a variety of mainly electrostatic instabilities can be excited. These instabilities appear as coherent oscillations of the plasma parameters, sometimes showing a strongly nonlinear behaviour [3,4].

The **p**otential **r**elaxation **i**nstability (PRI) and the **e**lectrostatic **i**on-**c**yclotron **i**nstability (EICI) are excited by drawing an electron current parallel to the magnetic field to a circular collector, which is inserted into the plasma column perpendicular to the axis. For exciting the PRI, the radius of the collector has to be sufficiently larger than the ion gyroradius so that the ion trajectories can be considered as one-dimensional. For exciting the EICI, the radius of the collector must be considerably smaller than that of the plasma column, but still in the range of a few ion gyroradii. A transition from the PRI into the EICI by increasing the diameter of the collector was reported earlier [5]. A certain range of collector radii was found, where both instabilities could be excited simultaneously. This phenomenon led to an amplitude modulation of the EICI by the PRI, with the amplitude of the latter being much larger than that of the former. Both instabilities lead to strong modulations of the electron current through the plasma, however, on different frequency scales.

We have reported [6] on an investigation in the magnetised alkaline plasma of a single-ended Q-machine where both instabilities, PRI and EICI, are excited simultaneously with comparable amplitudes by gradually increasing the bias of the collector, with the EICI appearing at first, and later the PRI. This led to a strong modulation of the EICI by PRI, which affected not only the amplitude but also the frequency of the EICI. Because of this, sidebands appear in the spectrum around f_{EICI} with a frequency difference equal to $\pm f_{PRI}$. We have found a linear decrease of the EICI frequency with increasing collector current.

Now, we have made a comprehensive nonlinear dynamical analysis of the recorded signals, which gives us a broad picture of the state space dynamics of our system. For this purpose, we have recorded the time series of the collector current with a sampling rate of 500 kHz. We have plotted the histograms of the signal values and the 3D Poincaré maps through the reconstructed space and we have simulated the modulated signal in a special case.

EXPERIMENTAL RESULTS

The experiments have been performed in the Innsbruck single-ended Q machine in a potassium plasma, produced on a 6 cm diameter tungsten hot plate, heated to about 2200 K. A circular tantalum limiter, inserted 3.6 cm in front of the hot plate, reduces the diameter of the plasma column to 3.5 cm, thus providing sharper edges and a flatter radial density profile. The distance between the plasma source (the hot plate) and the collector is 27.5 cm. The plasma density was in the range $10^8 < n_{pl} < 10^9$ cm^{-3}, and the ion and electron temperatures where $T_i \cong T_e \cong 0.2$ eV. The confining magnetic field was $0.05 < B < 0.2$ T.

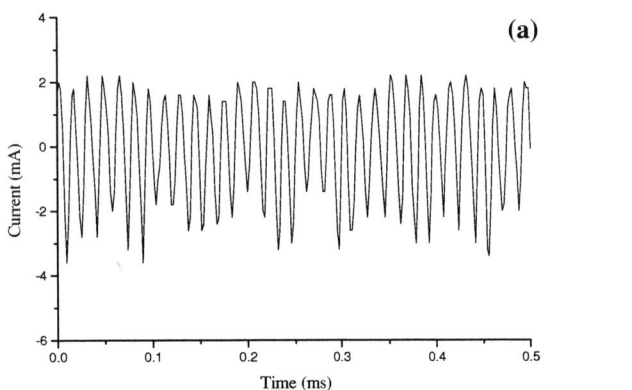
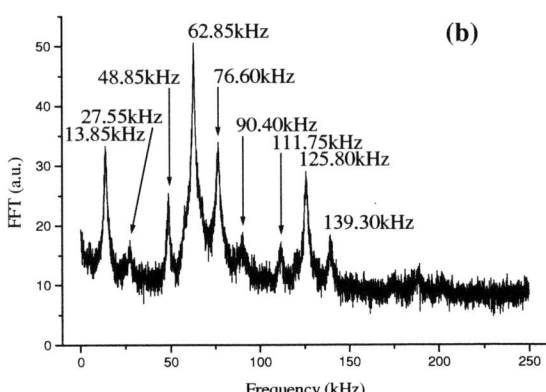

FIGURE 1. Time series (a) and FFT (b) of the ac component of the current through collector in the case where both instabilities are excited simultaneously

Both instabilities were excited simultaneously by drawing an electron current to a 10 mm diameter heated tantalum collector [7]. By slowly increasing the voltage on the collector, for $B = 0.13$ T, first the EICI appears with 67 kHz approximately, later also PRI appears with about 12 kHz. In a certain range there is a strong interaction between the two instabilities, which leads to an amplitude and frequency modulation of the EICI by PRI (see Fig. 1). The mechanism at the origin of this modulation was already described earlier [8].

We observe that the amplitude modulation of the time series with the PRI frequency affects more strongly the negative amplitudes. The upward excursions of the oscillating current are always much less pronounced, a fact which appears in all the time series, sometimes even as an absolute limitation of the collector current, especially for higher values of the potential on the collector, so that the positive amplitude seemed to be clipped off. A current limitation due to the formation of thermal barriers in front of a periodically travelling double layer, is a well-known feature of both instabilities [9,10,11].

NONLINEAR DYNAMICAL ANALYSIS

The nonlinear dynamical analysis provides us with powerful tools for analysing the evolution of a nonlinear system such as time histories, histograms, Fourier spectra, state space plots, Poincaré maps, autocorrelation functions, Lyapunov exponents, dimension calculations, etc. Since the phenomena studied here are clearly strongly nonlinear, we presume that a nonlinear dynamical analysis of the ac components of the collector current can offer an excellent insight into the state space dynamics of our system. For this purpose, we have recorded the time series of the collector current with a sampling rate of 500 kHz delivering 15000 points in 0.03 s, i.e., the sampling time was $\tau_s = 2$ μs.

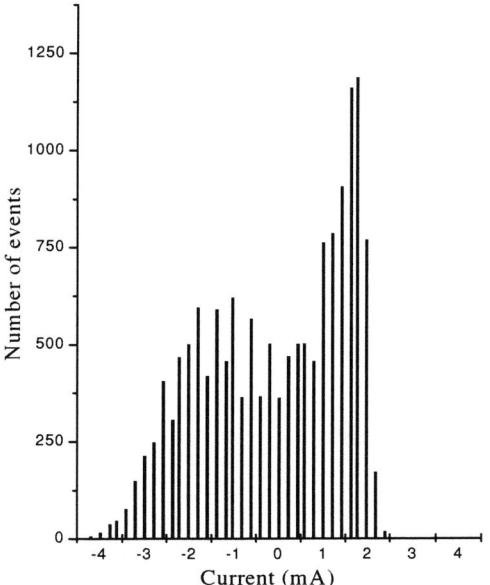

FIGURE 2. Histogram of the time series values from Fig. 1

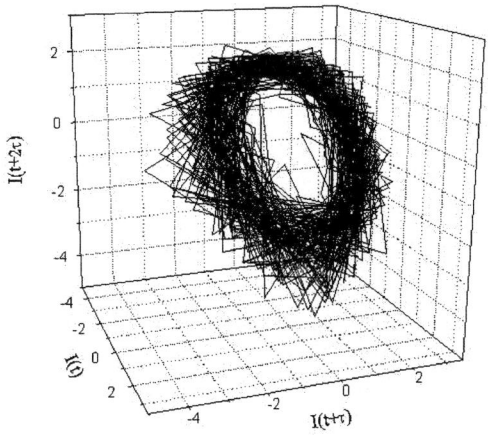

FIGURE 3. 3D Poincaré map of the reconstructed space for the case of Fig. 1

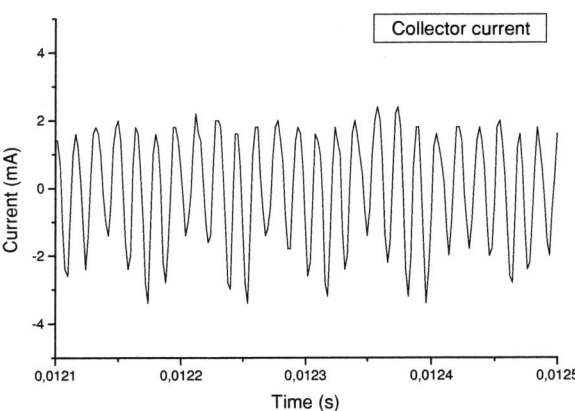

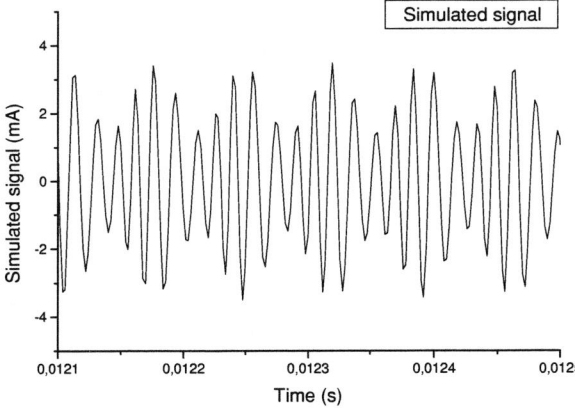

FIGURE 4. (a) Time series of the collector current. (b) Computer simulation of an amplitude and frequency modulated signal given by: $x(t) = 2.5(1 + 0.4 \cos \omega_{PRI} t) \cos(\omega_{EICl} t + 0.7 \cos \omega_{PRI} t)$.

Fig. 2 shows the histogram of the signal value corresponding to Fig. 1a, i.e. the number of events in the specified intervals of the collector current. From the strong asymmetry of the histogram we can discern the limitation of the collector current. Since this asymmetry appears simultaneously with the PRI, we conclude that this is mainly due to the PRI.

Fig. 3 shows the 3D Poincaré map by means of the reconstructed space corresponding to the signal from Fig. 1a. We obtained this map using the method of delays proposed by Packard et al. [12], Ruelle [13] and Takens [14]. We point out that the trajectories in the phase space, constructed with the derivatives, have almost the same shape as in Fig. 3. We remark the torus shape which is the usual form in the case of two coupled oscillators as they are present in our case [6].

To describe an amplitude and frequency modulated signal, Nayfeh and Balachandran [15] have proposed an analytical expression given by:

$$x(t) = a(t) \cos[\omega t + \beta(t)] \qquad (1)$$

where

$$a(t) = a_o [1 + \alpha \cos(\omega_m t + \theta)] \quad \text{and}$$

$$\beta(t) = \beta_o \cos(\omega_m t) \qquad (2)$$

We used this expression to fit our modulated signal of Fig. 1a, for $\omega = \omega_{EICl}$ and $\omega_m = \omega_{PRI}$. As can be seen in Fig. 4, we obtain a very good simulation of the time series for the values: $a_0 = 2.5$ mA, $\alpha = 0.4$, $\theta = 0$

and $\beta_0 = 0.7$. So, the function describing this reads:

$$x(t) = 2.5(1 + 0.4\cos\omega_{PRI} t)\cos(\omega_{EICI} t + 0.7\cos\omega_{PRI} t) \tag{3}$$

Naturally, we obtain a good fit only for the downward part of the signal (negative amplitudes), since the upward part of the signal is affected by the phenomenon of current limitation, which cannot be described by the simulation above.

CONCLUSIONS

We performed a nonlinear dynamical analysis of the experimentally determined times series in the case of the simultaneously excitation of two low-frequency ion-instabilities in magnetized plasma. The construction of the signal value histogram clearly shows the limitation of the current phenomenon and allows us to conclude that this is due, in our case, to the PRI. The nonlinear analysis allows us to reconstruct the state space dynamics of our plasma system. We found a very good analytical fit for one of our time series, thereby obtaining a simulation model of the two coupled oscillators EICI and PRI.

ACKNOWLEDGMENTS

This work was supported by the Fonds zur Förderung der Wissenschaftlichen Forschung (Austria) under grant No. P-14545-PHY and by the University of Innsbruck.

REFERENCES

[1] N. Rynn and N. D'Angelo, *Rev. Sci. Instrum.* **31** (1960), 1326.
[2] R. C. Knechtli and J. Y. Wada, *Phys. Rev. Lett.* **6** 1961), 215.
[3] C. Avram, R. Schrittwieser and M. Sanduloviciu, *J. Phys. D: Appl. Phys.* **32** (1999), 2750 and 2758.
[4] R. Schrittwieser, C. Avram, P.C. Balan, V. Pohoaţă, M. Sanduloviciu, C. Stan, *Physica Scripta* **T84** (2000), 122.
[5] R. Schrittwieser, *Phys. Fluids* **26** (1983), 2250.
[6] D.-G. Dimitriu, V. Ignatescu, E. Lozneanu, C. Ioniţă, M. Sanduloviciu, R. Schrittwieser, *Int. J. Mass Spectrom.* (2002), in print.
[7] D.G. Dimitriu, V. Ignatescu, C. Ioniţă, R. Schrittwieser, M. Sanduloviciu, *Proc. 10th Intern. Cong. Plasma Phys./42nd Ann. Meeting Div. Plasma Phys., American Phy. Soc.*, Québec City, Québec, Canada, 2000, p. 272, *Bull. American Phys. Soc.* **45** (2000), p. 71.
[8] D.G. Dimitriu, V. Ignatescu, E. Lozneanu, M. Sanduloviciu, C. Ioniţă, R. Schrittwieser, *Proc. 28th EPS Conf. Contr. Fusion Plasma Phys.*, (Madeira, Portugal, 2001), *Europhys. Conf. Abst.* **25A** (2001), p. 1733.
[9] S. Iizuka, P. Michelsen, J.J. Rasmussen, R. Schrittwieser, R. Hatakeyama, K. Saeki, N. Sato, *Phys. Rev. Lett.* **48** (1982), 145.
[10] J. J. Rasmussen and R. Schrittwieser, *IEEE Trans. Plasma Sci.* **19** (1991), 457.
[11] N. Sato and R. Hatakeyama, *J. Phys. Soc. Japan* **54** (1985), 1661; R. Hatakeyama, F. Muto and N. Sato, *Japan J. Appl. Phys.* **24** (1985), L285.
[12] N.H. Packard, J.P. Crutchfield, J.D. Farmer and R.S. Shaw, *Phys. Rev. Lett.* **45** (1980), 712.
[13] D. Ruelle, *Chaotic Evolution and Strange Attractors*, Cambridge University Press, 1989.
[14] F. Takens, in *Dynamical Systems and Turbulence, Lecture Notes in Mathematics* vol. 898, D.A. Rand and L.S. Young editors, Springer-Verlag, 1981.
[15] A. H. Nayfeh and B. Balachandran, *Applied Nonlinear Dynamics – Analytical, Computational and Experimental Methods*, John Wiley & Sons, 1995.

High-Dimensional Interior Crisis in Plasmas

E. L. Rempel*†, A. C.-L. Chian*†, E. E. Macau†, R. R. Rosa† and F. Christiansen**

*World Institute for Space Environment Research-WISER, NITP, University of Adelaide, SA 5005, Australia
†National Institute for Space Research-INPE, P.O. Box 515, 12227-010 São José dos Campos, SP, Brazil
**Solar-Terrestrial Physics Division, Danish Meteorological Institute, Lyngbyvej 100, DK-2100 Copenhagen Ø, Denmark

Abstract. The study of periodic orbits and their invariant manifolds can be essential for a proper understanding of the onset of chaos and intermittent turbulence in space plasmas, as well as for monitoring and controlling instabilities in tokamak experiments. Plasma turbulence can be generated by an event known as crisis, characterized by the sudden expansion of a chaotic attractor due to the collision of the chaotic attractor with an unstable periodic orbit. Most previous analysis of crises are restricted to low-dimensional dynamical systems described by maps or systems with a few ordinary differential equations. In this work we identify an interior crisis in a spatiotemporal model for plasma turbulence with a high-dimensional representation in the Fourier space. We numerically solve the Kuramoto-Sivashinsky partial differential equation, which was first derived to describe the nonlinear saturation of the collisional trapped-ion mode in plasmas confined in toroidal devices. We numerically find an unstable periodic orbit and show that, after its collision with the coexisting chaotic attractor, the attractor is abruptly enlarged, with a respective jump in the value of the maximum Lyapunov exponent. The methodology presented in this work may be followed in further studies of high-dimensional dynamical systems, and can be used to characterize crises in other strongly dissipative spatiotemporal systems.

INTRODUCTION

The Kuramoto-Sivashinsky (K-S) equation is a widely studied nonlinear reaction-diffusion equation that exhibits a wealth of nonlinear and turbulent states found in spatially extended systems. It was first derived to describe the nonlinear saturation of the collisional trapped-ion mode, a drift wave associated with the oscillation of plasma particles trapped in magnetic wells created by the inhomogeneous magnetic field of a tokamak [1]. This equation is also relevant for other nonlinear plasma phenomena such as the edge-localized-mode in tokamaks [2], and nonlinear coupling of Langmuir and ion-acoustic waves [3], all of which can be modeled by the Ginzburg-Landau type equation. It has been proved that the K-S equation is closely related to the Ginzburg-Landau equation since under certain approximations it governs the phase evolution of the complex amplitude of the Ginzburg-Landau equation [4].

Crises are sudden changes in chaotic attractors caused by the collision of the chaotic attractor with an unstable periodic orbit (UPO), as some control parameter of the system is changed [5]. Recent theoretical studies have indicated that crises can appear in plasmas [6]-[10]. Intermittency of Alfvén waves in the solar wind plasma can be induced by an interior crisis [6]. Double boundary crises of Alfvén waves are seen in a complex plasma region in the presence of a large number of coexisting attractors [7]. Other types of global bifurcations that lead to crisis and torus breakdown in plasmas have been identified theoretically [8] and experimentally [9].

Most previous analysis of crises are restricted to low-dimensional dynamical systems described by maps or ordinary differential equations [5]-[7]. In this paper, we report interior crisis in an extended, spatiotemporal system described by the K-S equation. An interior crisis is a type of crisis characterized by the sudden expansion of the chaotic attractor. We show in this paper that unstable periodic orbits and their associated invariant manifolds in the Poincaré plane can be an effective tool for characterizing high-dimensional interior crisis in the K-S equation, in the same way as has been done in the deterministic dynamical systems of low-dimension.

THE KURAMOTO-SIVASHINSKY EQUATION

The one-dimensional Kuramoto-Sivashinsky equation can be written as

$$\partial_t u = -\partial_x^2 u - \nu \partial_x^4 u - \partial_x u^2, \tag{1}$$

where ν is a 'viscosity' damping parameter. We assume that $u(x,t)$ is subject to periodic boundary conditions $u(x,t) = u(x+2\pi,t)$ and expand the solutions in a discrete spatial Fourier series

$$u(x,t) = \sum_{k=-\infty}^{\infty} b_k(t) e^{ikx}. \tag{2}$$

Substituting Eq. (2) into Eq. (1) yields an infinite set of ordinary differential equations for the complex Fourier coefficients $b_k(t)$

$$\dot{b}_k(t) = (k^2 - \nu k^4) b_k(t) - ik \sum_{m=-\infty}^{\infty} b_m(t) b_{k-m}(t), \tag{3}$$

where the dot denotes derivative with respect to t. Reality of $u(x,t)$ implies that $b_{-k} = b_k^*$. We restrict our investigation to the subspace of odd solutions $u(x,t) = -u(-x,t)$ assuming that $b_k(t)$ are purely imaginary by setting $b_k(t) = -ia_k(t)/2$, where $a_k(t)$ are real. Equation (3) then becomes

$$\dot{a}_k(t) = (k^2 - \nu k^4) a_k(t) + \frac{k}{2} \sum_{m=-\infty}^{\infty} a_m(t) a_{k-m}(t), \tag{4}$$

where $a_0 = 0$, $1 \leq k \leq N$, N is the truncation order. We integrate the high-dimensional dynamical system given by Eq. (4) using a fourth-order variable step Runge-Kutta integration routine. We choose $N = 16$, since numerical tests indicate that for the range of the control parameter ν used in this paper the solution dynamics remains essentially unaltered for $N > 16$. We adopt a Poincaré map with the $(N-1)$ dimensional hyperplane defined by $a_1 = 0$, with $\dot{a}_1 > 0$, as in [11].

NONLINEAR DYNAMICAL ANALYSIS

A bifurcation diagram can be obtained from the numerical solutions of the 16-mode truncation of Eq. (4) by varying the control parameter ν and ploting the Poincaré points of one Fourier mode after discarding the initial transient. Figure 1a shows a period-3 (p-3) window where we plot the Poincaré points of the Fourier mode a_6 as a function of ν. The dotted lines in Fig. 1a denote the Poincaré points of the p-3 unstable periodic orbit which emerges via a saddle-node bifurcation at $\nu = 0.02992498$, marked SN in Fig. 1a. In this paper, we will analyze the role played by this p-3 UPO in the onset of interior crisis at $\nu_{IC} = 0.02992021$, marked IC in Fig. 1. The interior crisis at ν_{IC} occurs when the p-3 UPO collides head on with the 3-band weak strange attractor evolved from the cascade of period-doubling bifurcations, as seen in Fig. 1a.

The interior crisis leads to a sudden expansion of the strange attractor, turning the weak strange attractor (WSA) into a strong strange attractor (SSA), as seen in Fig. 2. Figure 2 is a 3-dimensional projection (a_1, a_{10}, a_{16}) of the strong strange attractor (light line) defined in the 15-dimensional Poincaré hyperplane right after crisis ($\nu = 0.02992020$), superimposed by the 3-band weak strange attractor (dark line) at crisis ($\nu = 0.02992021$). The abrupt increase in the system's chaoticity after the interior crisis can be characterized by the value of the maximum Lyapunov exponent (λ_{max}), plotted in Fig. 1(b). At crisis ($\nu_{IC} = 0.02992021$), $\lambda_{max} = 0.35$, and after the crisis at $\nu = 0.02992006$, $\lambda_{max} = 0.62$.

We proceed next with the characterization of the high-dimensional crisis at ν_{IC} by showing in Fig. 3 the collision of the weak strange attractor with the p-3 UPO in the reduced 2-dimensional Poincaré plane (a_5 vs. a_6), in the vicinity of the upper branch of the attractor in Fig. 2. The dark line denotes the strange attractor, and the light line denotes the numerically computed invariant unstable manifold of the saddle point. Figures 3a,b,c display the dynamics before, at, and after the crisis, respectively. Note that the strange attractor always "overlaps" the invariant unstable manifold. Figure 3b shows the "head-on" collision of the weak strange attractor with the p-3 UPO at ν_{IC}, which proves the occurrence of an interior crisis [5, 7, 12]. Although we are showing only two of the 16 modes, the collision can be seen in any choice of modes. This collision leads to an abrupt expansion of the strange attractor, as seen in Fig. 3c.

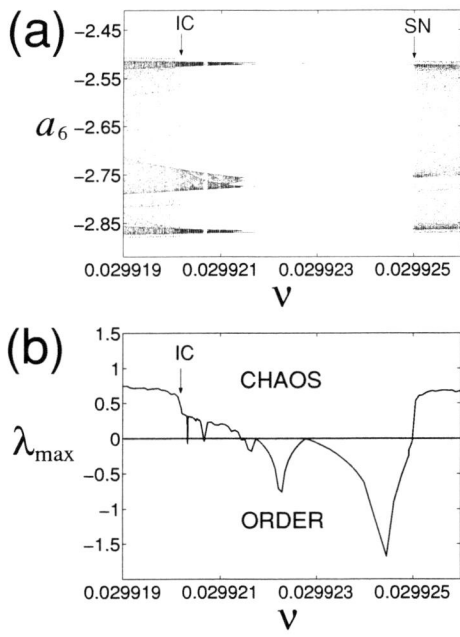

FIGURE 1. (a) Bifurcation diagram of a_6 as a function of v. IC denotes interior crisis and SN denotes saddle-node bifurcation. The dotted lines represent a period-3 unstable periodic orbit. (b) Variation of the maximum Lyapunov exponent λ_{max} with v.

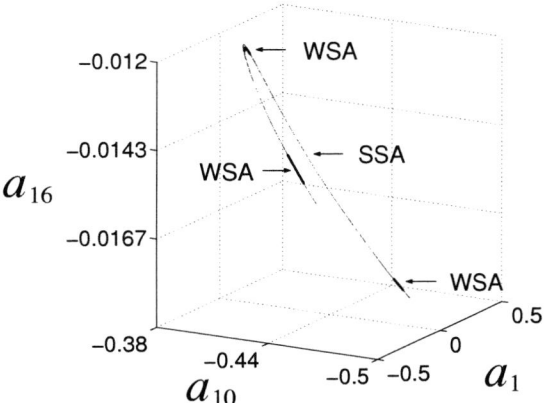

FIGURE 2. Three-dimensional projection (a_1, a_{10}, a_{16}) of the strong strange attractor SSA (light line) defined in the 15-dimensional Poincaré hyperplane right after crisis at $v = 0.02992020$, superimposed by the 3-band weak strange attractor WSA (dark line) at crisis ($v = 0.02992021$).

CONCLUSIONS

In conclusion, we have shown that high-dimensional interior crisis can be found in spatially extended systems exemplified by the Kuramoto-Sivashinsky equation. Although we have adopted a 16-mode truncated system in our analysis, all the calculations performed can be extended to an arbitrary high number ($N < \infty$) of modes for an appropriate choice of v and L. The identification of the unstable periodic orbits and their invariant manifolds is fundamental for monitoring and controlling the instabilities, chaos and turbulence in tokamak experiments [13]. Further theoretical and experimental studies of high-dimensional dynamical systems, following the methodology developed in this paper, may improve confinement in tokamaks and the understanding of other complex systems.

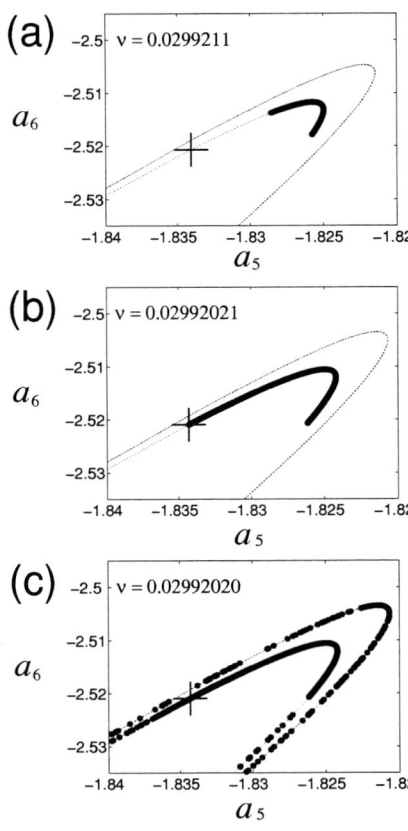

FIGURE 3. The plots of the strange attractor (dark line) and invariant unstable manifolds (light lines) of the saddle before (a), at (b) and after (c) crisis. The cross denotes one of the saddle points.

ACKNOWLEDGMENTS

The authors would like to acknowledge the support of the Australian Institute for Nuclear Science and Engineering (AINSE), CNPq (Brazil), FAPESP (Brazil) and AFOSR. A. C.-L. Chian and E. L. Rempel wish to thank Professors Tony Thomas and Tony Williams of the University of Adelaide for their kind hospitality.

REFERENCES

1. LaQuey, R. E. *et al.*, *Phys. Rev. Lett.*, **34**, 391-394 (1975); Cohen, B. I. *et al.*, *Nuclear Fusion*, **16**, 971-992 (1976).
2. Itoh, S. I. *et al.*, *Phys. Rev. Lett.*, **67**, 2485-2488 (1991).
3. Erichsen, R., Brunnet, L. G., and Rizzato, F. B., *Phys. Rev. E*, **60**, 6566-6570 (1999).
4. Kuramoto, Y., and Tsuzuki, T., *Prog. Theor. Phys.*, **55**, 356-369 (1976).
5. Grebogi, C., Ott, E., and Yorke, J. A. *Phys. Rev. Lett.*, **48**, 1507-1510 (1982); Grebogi, C., Ott, E., and Yorke, J. A., *Physica D* **7**, 181-200 (1983); Szabó, K. G. *et al.*, *Phys. Rev. E*, **61**, 5019-5032 (2000).
6. Chian, A. C.-L., Borotto, F. A., and Gonzalez, W. D., *Astrophys. J.*, **505**, 993-998 (1998).
7. A. C.-L. Chian, F. A. Borotto, and E. L. Rempel, *Int. J. Bifurcation Chaos*, in press (2002).
8. Kaifen He, *Phys. Rev. Lett.*, **80**, 696-699 (1998); Kaifen He, *Phys. Rev. Lett.*, **84**, 3290-3293 (2000); Kaifen He, *Phys. Rev. E*, **63**, 016218 (2001).
9. Letellier, C. *et al.*, *Phys. Rev. E*, **63**, 042702 (2001).
10. Chian, A. C.-L., Rempel, E. L., Macau, E. E., Rosa, R. R., and Christiansen, F., *Phys. Rev. E*, **65**, 035203(R) (2002).
11. Christiansen, F., Cvitanović, P., and Putkaradze, V., *Nonlinearity*, **10**, 55-70 (1997).
12. Alligood, K. T., Sauer, T. D., and Yorke, J.A. *Chaos: An Introduction to Dynamical Systems*, Springer-Verlag, New York, 1996.
13. Bak, P. E. *et al.*, *Phys. Rev. Lett.*, **83**, 1339-1342 (1999).

Reconnection Bifurcation in Tokamaks

M. Roberto[*], E.C. Silva[*] and I.L. Caldas[†]

[*]*Technological Aeronautic Institute, S. José dos Campos, S. Paulo, 12228-900, Brazil.*
[†]*Institute of Physics, University of São Paulo, S.Paulo, Brazil.*
e-mail: marisar@fis.ita.br

Abstract. An ergodic magnetic limiter has been proposed to control the plasma-wall interactions in tokamaks. This apparatus generates resonant magnetic fields that interact with the equilibrium field, causing a selective destruction of magnetic surfaces. In this work, a numerical study of magnetic islands induced by a magnetic limiter in a large-aspect ratio tokamak with non-monotonic safety factor profiles is performed using a symplectic non-twist mapping. The dominant resonance leads to dimerized island chains and reconnection. The main bifurcation associated with this transition and the onset of global chaos is identified.

INTRODUCTION

In order to confine plasmas in tokamaks it is necessary to control the plasma contamination by impurities released from the inner wall by surface processes [1]. A way to reduce these effects is to use resonant helicoidal conductors which modify the magnetic surface topology creating magnetic islands and chaotic field lines [2,3]. Thus, an ergodic magnetic limiter [4] has been used to generate external magnetic fields that interact with the equilibrium field causing a selective destruction of the magnetic surfaces.

In this work, a numerical study of magnetic islands induced by a magnetic limiter in a large aspect-ratio tokamak, with non-monotonic safety factor radial profiles, is described. The equilibrium model field is analytically obtained by solving a Grad-Shafranov equation in toroidal polar coordinates. A symplectic nontwist mapping is introduced to analyze the magnetic islands associated to the perturbing resonances induced by the magnetic limiter. A hamiltonian theory for the magnetic field structure is used to obtain the symplectic mapping. The dominant resonance leads to dimerized island chains and reconnection phenomenon. The main bifurcation associated to this transition and the onset of global chaos is identified.

THE EQUILIBRIUM PROFILES

We consider in this paper MHD equilibrium plasmas confined in tokamaks in the large aspect-ratio limit, $R \gg a$, where R is the major radius and a is the plasma column radius. In order to obtain the equilibrium magnetic field the Grad-Shafranov equation is used in the non-orthogonal polar toroidal coordinate system $(r_t, \theta_t, \varphi_t)$ [5], which reduces to the local system (r, θ, φ) in the large aspect-ratio limit. An approximate solution can be written as $\psi_p(r_t, \theta_t) = \psi_{p0}(r_t) + \delta \psi_p(r_t, \theta_t)$ where $|\delta \psi_p| \ll |\psi_{p0}|$. In the large aspect ratio limit ψ_{p0} does not depend on θ_t. The intersections of magnetic surfaces $\psi_{p0}(r_t)$ = constant with a toroidal plane $\varphi = 0$ are not concentrical circles but have a Shafranov shift toward the exterior equatorial region.

In order to solve the Grad-Shafranov equation we choose a toroidal current density profile which is given by

$$j = j_0 \left(1 + 2\frac{r_t^2}{a^2}\right)\left(1 - \frac{r_t^2}{a^2}\right) \quad (1)$$

Here, a is the plasma column radius and $j_0 = I_p R_0'/(2\pi a^2)$, where I_p is the total plasma current and R_0' is the radius of the magnetic axis, which is shifted with respect to the geometric major (R_0) due to a toroidicity effect (Shafranov shift). This current profile leads to a non-monotonic satefy factor radial profile. For some given q values, there are

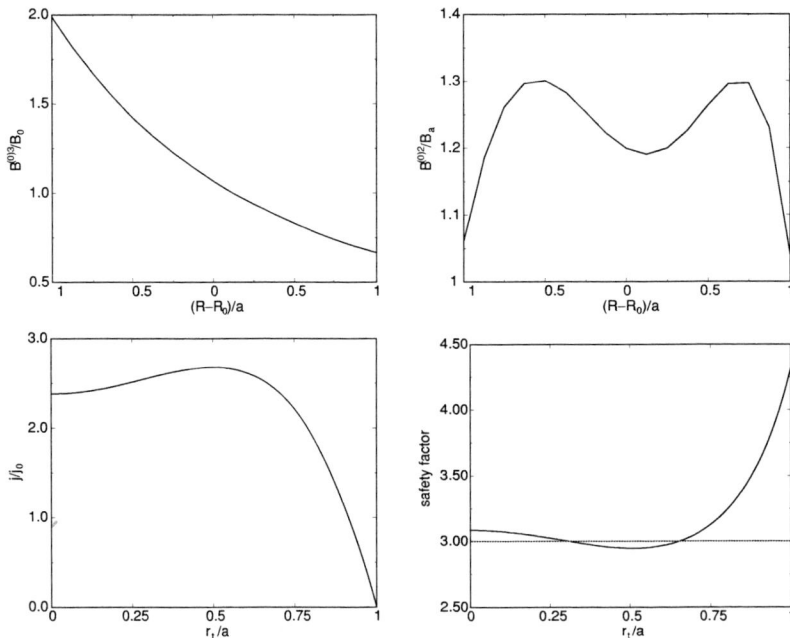

FIGURE 1. Toroidal and poloidal equilibrium field components normalized to B_0 and B_a respectively, current density and safety factor profiles.

two rational surfaces with different radii in the internal region. These non monotonic profiles have been observed in tokamaks, at the initial stages of electrical discharges [6]. Recently, they have been observed in transitions from L (low) to H (high) discharge regimes [7].

The equilibrium magnetic contravariant components can be written in terms of the surfaces functions ψ_p and the poloidal current function according to Ref. [8].

The poloidally-averaged safety factor is given by

$$q(r_t) = q_c(r_t) \left(1 - \frac{4r_t^2}{R_0^{'2}}\right)^{-1/2} \qquad (2)$$

where

$$q_c(r_t) = \frac{I_e}{I_p} \frac{a^2}{R_0^{'2}} 0.42 \frac{1}{\left(\frac{1}{2} + \frac{r_t^2}{4a^2} - \frac{r_t^4}{3a^4}\right)} \qquad (3)$$

where I_e is the external current that generates the equilibrium toroidal field, which is related to the poloidal current function I through the expression $I = -I_e/(2\pi)$ [8]. We normalize the minor radius b_t and plasma radius a to the major radius R_0', so that $a/R_0' = 0.25$ and $b_t/R_0' = 0.33$, which are consistent with typical tokamak discharges [9].

Figure 1 shows the toroidal and poloidal equilibrium field components, the current density and safety factor profiles. We choose $q(a) = 4.32$ and $q(0) = 3.10$. The magnetic field profiles are normalized, so that $\bar{B}_0^3 = B_0^3/B_0$ and $\bar{B}_0^2 = B_0^2/B_a$, where $B_0 = \mu_0 I_e/(2\pi R_0^{'2})$ and $B_a = \mu_0 I_p/(2\pi a^2)$. The effect of toroidal geometry is noticeable at the plasma edge increasing the magnetic shear at this region.

PERTURBING MAGNETIC FIELDS DUE TO ERGODIC MAGNETIC LIMITER

The ergodic magnetic limiter is a device which has N_a current rings of length l located symmetrically along the toroidal circunference of the tokamak. These current rings may be regarded as slices of a pair of external helical windings located at the tokamak minor radius $r_t = b_t$, and conducting a current I_h in opposite senses for adjacent conductors. To induce a resonant perturbation we choose a helical winding with the same pitch as the field lines in the rational surface

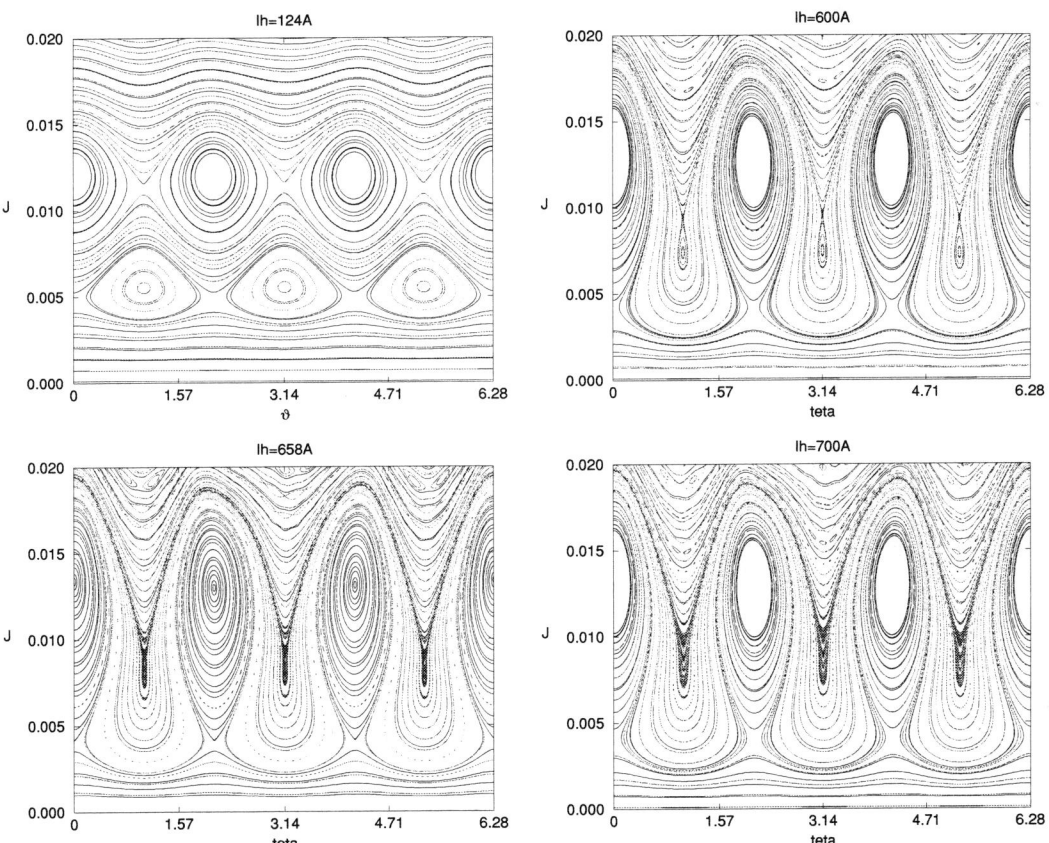

FIGURE 2. Poincaré maps of the equilibrium plasma perturbed by ergodic magnetic limiter. (a) $I_h/I_p = 0.86\%$; (b) $I_h/I_p = 4.11\%$; (c) $I_h/I_p = 4.54\%$ and (d) $I_h/I_p = 4.80\%$.

we want to perturb. This surface has a safety factor $q = m_0/n_0$, where m_0 and n_0 are positive integers. The winding law can be written as $m_0\theta_t - n_0\varphi_t =$ constant. In our case, we choose the resonant effect to occur at the equilibrium rational magnetic surface with q = 3/1, once for this value there are two rational surfaces with different radii in the internal region according to Figure 1d.

The derivation of the symplectic mapping used in this work can be found in Ref. [10]. The field line equations, $\vec{B} \times \vec{dl} = 0$, can be put in a Hamiltonian form [10]. It depends on the variables $\mathscr{J}$ and ϑ which are angle-action variables. They are related to the toroidal polar coordinates [10].

The canonical area-preserving mapping for this near-integrable system can be written as

$$\mathscr{J}_{n+1} = \mathscr{J}_n + \varepsilon f(\mathscr{J}_{n+1}, \vartheta_n, t_n), \tag{4}$$

$$\vartheta_{n+1} = \vartheta_n + \left(\frac{2\pi}{N_a q(\mathscr{J}_{n+1})}\right) + \varepsilon g(\mathscr{J}_{n+1}, \vartheta_n, t_n), \tag{5}$$

$$t_{n+1} = t_n + \frac{2\pi}{N_a} \tag{6}$$

where $f = -\partial H_1/\partial \vartheta$, $g = \partial H_1/\partial \mathscr{J}$, and $\varepsilon = 2I_h \ell/I_e R'_0$. Here t is the canonical time. The perturbing Hamiltonian which describes the action of the ergodic magnetic limiter is periodic in ϑ and in t.

A resonant effect at the surfaces with $q = m_0/n_0$, due to the considered perturbative field, give rise to chains of m islands. In this work we consider $m_0 = 3$ and $n_0 = 1$. Figure 2a shows twin Poincaré chains with 3 islands at the two surfaces with $q = 3/1$. This case corresponds to $I_h/I_p = 0.86\%$. Increasing values of I_h leads to the enlargement of the external islands (higher J values) and decrease of internal island, according to Figure 2b, which corresponds to $I_h/I_p = 4.11\%$. New islands and chaotic regions can be recognized in this figure. There is a I_h value for which

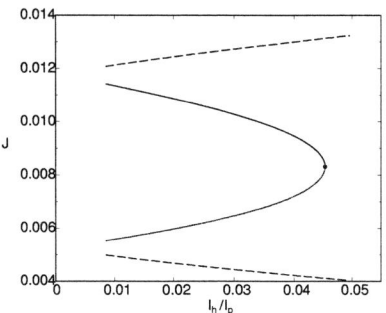

FIGURE 3. Action variables $\mathcal{J}$ of the elliptic and hyperbolic points as a function of I_h/I_p.

the island chains join together. From this critical value, starts the reconnection of field lines, which gives rise to two dimerized island chains, as shown in Figure 2b. Furthermore, a bifurcation occurs for $I_h/I_p = 4.54\%$, according to Figure 2c, where the external hyperbolic and internal elliptic points disappear. After that only the external dimerized chain remains, as can be seen in Figure 2d, with $I_h/I_p = 4.80\%$.

In order to determine the critical I_h^*/I_p value for which the bifurcation occurs, we determine the $\mathcal{J}$ variables of the hyperbolic and elliptic points of the two island chains as a function of I_h/I_p. In Figure 3 the solid line represents the action of the hyperbolic external points of the external island chain and the dotted line represents the action of the elliptic internal point of the internal island chain. The upper dashed line refers to elliptic external points and the lower dashed line refers to hyperbolic internal points. As one can see, as I_h/I_p increases the distance between the external hyperbolic points and internal elliptic points decreases. For $I_h/I_p = 4.54\%$, the bifurcation takes place, remaining the elliptic external points and hyperbolic internal points. Thus, the increase of I_h values causes the superposition and destruction of these islands, giving rise to chaotic regions.

CONCLUSION

Magnetic surfaces were studied on MHD equilibrium plasmas with a non-monotonic safety factor radial profiles, confined on large aspect-ratio tokamaks. An ergodic magnetic limiter was considered in order to perturbe the magnetic surfaces. A symplectic nontwist mapping was applied showing the magnetic islands associated to the perturbing resonances induced by the magnetic limiter. It is shown that the dominant resonance leads to dimerized island chains and reconnection phenomenon. For high I_h values chaotic regions appear.

ACKNOWLEDGMENTS

This work has been supported by grants of FAPESP, Proc. 00/11777-6R, Fundação de Amparo à Pesquisa do Estado de São Paulo and CNPq, Conselho Nacional de Pesquisas, Proc. 98/15125-1.

REFERENCES

1. McCool, S., et al. Nuclear Fusion **29**, 547 (1989)
2. Vannucci, A., Nascimento, I.C., Caldas, I.L. Plasma Phys. Controlled Fusion **31**, 147 (1989
3. Lichtenberg, A.J., Lieberman, M.A. Regular and Chaotic Motion, 2nd. ed., Spring-Verlag, NY, 1992.
4. Viana, R.L., Caldas, I.L. Zeitschrift für Naturforschung **47**, 941 (1992)
5. Corso, G., Oda, G., Caldas, I.L. Chaos, Solitons and Fractals **8**, 1891 (1997)
6. Dimock, D., Johnson, H. Nuclear Fusion **25**, 1101 (1985)
7. Mazzucato, E., et al. Phys. Rev. Letters **77**, 3145 (1996)
8. Silva, E.C., Caldas, I.L., Viana, R.L. IEEE Trans. Plasma Science **29**, 617 (2001)
9. Silva, E.C., Caldas, I.L., Viana, R.L. Phys. of Plasmas **8**, 2855 (2001)
10. Silva, E.C., Caldas, I.L., Viana, R.L. Phys. Plasmas **29**, 617 (2001)

Nonlinear Shear Flow Structures in Magnetic Curvature Driven Rayleigh Taylor Instability

A. Das*, A. Sen* and P. K. Kaw*

Institute for Plasma Research, Bhat, Gandhinagar 382428, India

Abstract. The evolution of nonlinear shear flow structures in the form of streamers or zonal flows is studied in the context of the magnetic curvature driven Rayleigh Taylor instability. For a two dimensional electrostatic model it has been shown earlier [1] that the path to the final state is a sensitive function of the initial evolution of spectral power in the short scale fluctuations. A key parameter influencing this evolution is the amount of dissipation in the system. Low dissipation was found to favor the formation of zonal flow patterns while high dissipation led to streamer formation. In this work we demonstrate that the unconstrained growth of streamer structures can get limited by the onset of a secondary Kelvin Helmholtz instability and eventually steer the system to zonal flow structures. We also extend the earlier electrostatic model to include electromagnetic effects and find preliminary evidence of streamer stabilization due to coupling to stable Alfven waves.

INTRODUCTION

Shear flow structures are known to play an important role in governing the transport properties of a magnetically confined plasma. Depending on their symmetry pattern they can either enhance or diminish transport in a particular direction. In a tokamak for example, the presence of elongated radial flow patterns with a narrow poloidal extent, commonly known as streamers, can give rise to rapid radial transport. Zonal flows, on the other hand, which are poloidal flow patterns with narrow radial extent, help in inhibiting radial transport by reducing the radial correlation length [2, 3, 4, 5, 6, 7]. The onset of such shear flow structures is a common phenomena for many low frequency plasma instabilities such as drift waves and its several variants which have been studied in the context of tokamaks [8, 9, 10, 11, 12, 13]. As was shown in [1] similar nonlinear behaviour can also occur for the magnetic curvature driven Rayleigh Taylor instability. The RT model is particularly suited for the tokamak edge and is also a useful paradigm for understanding nonlinear behaviour in a number of other physical situations like the ionospheric F layer, magnetospheric plasmas, ablating laser fusion targets etc. In this work we continue our investigation of the RT model and explore further aspects of its nonlinear behaviour by examining secondary instabilities like the Kelvin Helmholtz instability and the effect of including finite $k_\parallel$ and electromagnetic corrections. Our principal new findings are that the onset of a secondary Kelvin Helmholtz instability can limit the growth of streamers and ultimately lead to a stable zonal pattern. Numerical simulations carried out for the extended electromagnetic model also suggest that streamers can also experience stabilizing effects from coupling to the stable Alfven branch.

MODEL EQUATIONS AND SUMMARY OF EARLIER WORK

A generalized set of model equations for the magnetic curvature driven Rayleigh Taylor instability which incorporates electromagnetic effects can be obtained as follows. The total magnetic field can be expressed as $\vec{B} = B_0(x)\hat{z} + \hat{z} \times \nabla \psi$, where $B_0\hat{z}$ is the equilibrium magnetic field along the $\hat{z}$ direction and has finite curvature due to its functional dependence on x. A local slab geometry approximation is made so that x represents the radial coordinate, y is the poloidal coordinate and z is the toroidal direction. We further restrict ourselves to incompressible perturbations, thereby ignoring coupling to sound waves and retain only coupling to Alfven waves represented by the magnetic perturbations of the vector potential $-\psi\hat{z}$. Now by using the ion continuity equation, quasi-neutrality condition and the parallel momentum equation for electrons, it is straightforward to obtain the following coupled set of equations:

$$\frac{\partial n}{\partial t} + V_g \frac{\partial n}{\partial y} + (V_n - V_g)\frac{\partial \varphi}{\partial y} - V_A^2 \frac{\partial}{\partial z}\nabla_\perp^2 \psi \hat{z} \times \vec{\nabla}\varphi \cdot \vec{\nabla} n - V_A^2 \hat{z} \times \vec{\nabla}\psi \cdot \vec{\nabla}\nabla_\perp^2 \psi = D\nabla^2 n \qquad (1)$$

$$\frac{\partial}{\partial t}\nabla^2 \varphi + V_g \frac{\partial n}{\partial y} - V_A^2 \frac{\partial}{\partial z}\nabla_\perp^2 \psi \hat{z} \times \vec{\nabla}\varphi \cdot \vec{\nabla}\nabla^2 \varphi - V_A^2 \hat{z} \times \vec{\nabla}\psi \cdot \vec{\nabla}\nabla^2 \psi = \mu \nabla^4 \varphi \qquad (2)$$

$$\frac{\partial \psi}{\partial t} + \frac{\partial}{\partial z}(n - \varphi) + \hat{z} \times \vec{\nabla}\psi \cdot \vec{\nabla}(n - \varphi) = \eta \nabla_\perp^2 \psi \qquad (3)$$

Here $V_g = c_s/(R\Omega_i)$ is the normalized gravitational drift arising through the magnetic curvature terms, $V_n = c_s/(L_n\Omega_i)$ is the diamagnetic drift speed arising due to the equilibrium density gradient, c_s is the ion acoustic speed, R is the major radius of curvature, Ω_i is the ion cyclotron frequency, L_n is the equilibrium density scale-length and μ and D are the dynamical viscosity and diffusion coefficients respectively. The equilibrium density profile is chosen as $N_0 = N_{00}\exp(-x/L_n)$, φ is the perturbed potential and n is given by $n = \ln(N/N_0)$. We have chosen to normalize N_0 the equilibrium density by N_{00}, ϕ by T_e/e, time by Ω_i, length by c_s/Ω_i and the magnetic potential ψ with $\Omega_i/(B_0 c_s)$. Other notations are identical to those adopted in some of the earlier works [1, 14].

The linearized dispersion relation obtained from Eqs.(1,2,3) can be written for $D = \mu = \eta = 0$ as

$$\gamma^3 + ik_y V_g \gamma^2 + \left\{ k_z^2 V_A^2 (1 + k_\perp^2) - \frac{k_y^2}{k_\perp^2} V_g (V_n - V_g) \right\} \gamma + ik_y (V_n - V_g) k_z^2 V_A^2 = 0 \qquad (4)$$

It is readily seen that for the limit of $k_z = 0$ the above equation reduces to the electrostatic dispersion relation discussed in Das et al. [14] and for $V_n = V_g = 0$ it reduces to the Alfven wave dispersion relation. The nonlinear equations thus constitute a coupling between the Rayleigh Taylor mode and the stable Alfven mode.

We now briefly recapitulate the earlier fluid simulation studies of the electrostatic model as reported in [1] where two dimensional perturbations with $\psi = 0$ and $\partial/\partial z = 0$ were considered. For this case the model reduces to a coupled set of two equations in the variables n (density) and φ (potential). The linear evolution of this set of equations is essentially governed by the Rayleigh Taylor growth rate. In the nonlinear regime, i.e. when the fluctuation amplitude is sufficiently large, the evolution is influenced by the two nonlinear terms viz. the convective nonlinearity in the density equation and the polarization drift nonlinearity in the vorticity equation. These two terms cascade power in opposite directions with the polarization drift nonlinearity pumping power into long scales and the convective term favoring power transfer into shorter scales. From the numerical simulations it was seen that the polarization drift nonlinearity assumed the dominant role, resulting in a cascade of power towards the longest permissible scale in the simulation box size for both the variables n and φ. It was further found that for symmetric initial and symmetric boundary conditions the amount of dissipation in the system played a crucial role in determining the final symmetry of the shear flow patterns. Low dissipation promoted nonlinear interaction amongst a large spectrum of short scale excitations to give rise to the formation of zonal flow patterns. However, when the dissipation was large such that only a few long scale modes were unstable, nonlinear interactions were inhibited and the evolution was characterized by the development of the linearly unstable streamer mode. Within the limitations of the model the streamers grew unrestrained with no mechanism available for their containment. As we will discuss in the next section one of the primary limitations of the model used in [1] was the choice of a square box size which did not permit the existence of scale lengths that were long enough in the x direction to permit the onset of any secondary instability like the Kelvin-Helmholtz instability. In our present simulations we remove this constraint and further explore the effects of retaining electromagnetic effects as well.

SECONDARY INSTABILITY AND ELECTROMAGNETIC EFFECTS

Eqs.(1-3) are three coupled nonlinear partial differential equations in three space dimensions. In a preliminary study we have numerically solved these equations for different fixed values of k_z, thereby reducing them to two dimensional equations in each case. The three coupled two dimensional equations are then solved with the help of a pseudospectral code in which individual Fourier modes are evolved in time. The nonlinear terms are, however, evaluated in real space.

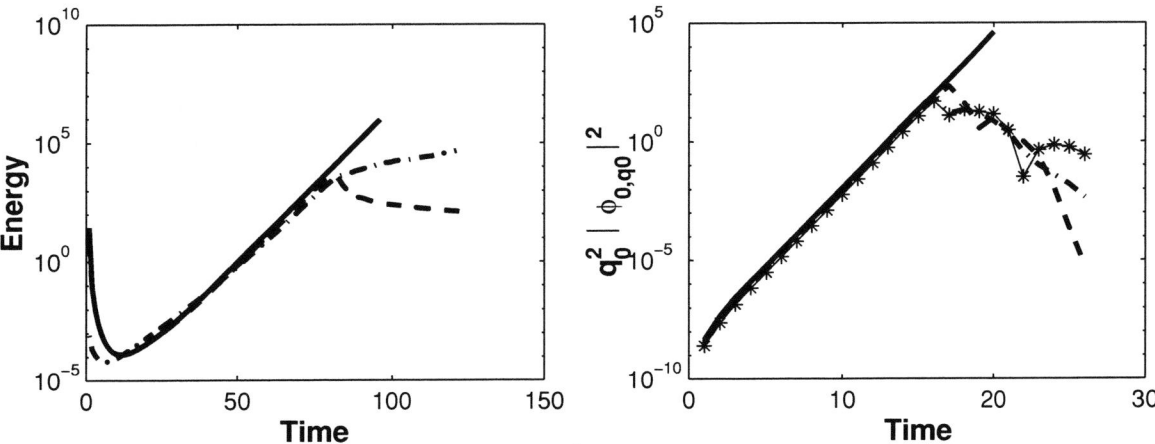

FIGURE 1. The left plot shows the evolution of energy for three different cases (i) solid line electrostatic case with $L_x = L_y$, (ii) dotted lines electrostatic case with $L_x = 2L_y$ (iii) dotted dash lines again has $L_x = L_y$ but electromagnetic effects due to field line bending have been incorporated in the simulation. The right plot shows the evolution of power in streamer mode for various different choices of the ratio of L_x/L_y. Solid line, dashed line, dot - dash line and solid lines with starts correspond to $L_x/L_y = 1, 1.25, 1.5$ and 2 respectively

Fast Fourier transform schemes are employed to go back and forth in the real and Fourier spaces.

We report on two sets of studies. In the first set we reexamined the previously studied electrostatic case [1] but removed the restriction of a square box. In particular we selected the region of parametric space for which a square simulation box had yielded growing streamer solutions and reran these cases in a simulation box where L_x, the simulation box size along the x, was longer than L_y, thereby permitting longer scale lengths along the x direction. This choice immediately led to a stabilization of the streamers as can be seen in the energy evolution diagrams of Fig.1. This behaviour can be easily understood from physical arguments. The Kelvin Helmholtz instability associated with shear can only be excited provided the excitation scales are longer than the equilibrium shear scale length. By choosing the box size to be identical in both the directions one was putting a restriction on the permitted secondary scales of excitations. Relaxing this contraint led the shear flow in streamers to get ultimately stabilized by the excitation of a Kelvin Helmholtz instability. We show in Fig.2 the growth of power in the streamer mode for various choices of the ratio of L_x/L_y. From the figure it is clear that as soon as the ratio exceeds unity the growth in streamer mode gets contained and reversed.

In the second set of studies we carried out simulations of the entire set of three equations (for fixed values of k_z) thereby retaining electromagnetic perturbations. In this case too we observed (see Fig.1) that there was a considerable slowing down of the original growth of the streamers. However the containment of instability in this case is for a different reason. We see that the additional nonlinear terms introduced in the model equations in the presence of electromagnetic perturbations are of such a nature that they lead to an accumulation of power at short scales in both density and potential perturbations. The resulting nonlinear interaction amidst this short scales then again lead to the nonlinear growth of zonal shear patterns. The finite amplitude zonal modes then tend to stabilize the growing streamer structures. Fig. 2 showing a comparison between the contour plots of the potential of the electrostatic and electromagnetic cases illustrates this point clearly.

DISCUSSION

To summarize, in this work we have extended our earlier work [14, 1] on magnetic curvature driven Rayleigh Taylor instability to demonstrate two different stabilizing effects that can influence nonlinear shear flow structures. In the electrostatic limit we have shown that by permitting long scale perturbations in the x direction (with a choice of a suitable simulation box geometry) it is possible to stabilize the streamer structure by the onset of a secondary Kelvin

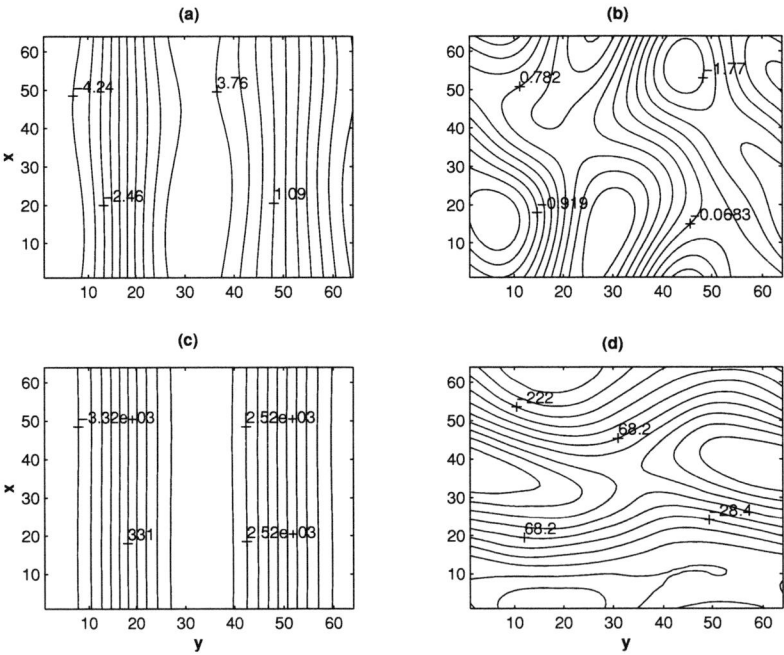

FIGURE 2. Comparison of potential contours at $t = 50$(subplots(a,b)) and $t = 100$(subplots(c,d)) for electrostatic (subplots(a,c)) and electromagnetic cases (subplots(b,d)).

Helmholtz instability. Further we have generalized the earlier electrostatic model to include electromagnetic effects arising due to magnetic field line bending. The model equations thus couple the Rayleigh Taylor mode to the stable Alfven waves. Initial simulations runs in the regime where streamers were nonlinearly unstable in the electrostatic case are found to be stabilized by the electromagnetic terms. However more detailed studies are needed to further validate this observation and to arrive at a better understanding of the physical mechanism of the stabilizing scheme. Further work in this direction as well as towards a fully three dimensional solution of the generalized equations are in progress.

REFERENCES

1. A. Das, A. Sen, S. Mahajan and P. Kaw, Phys. Plasmas **8** 5104 (2001).
2. F. H. Busse, CHAOS **4** 123 (1994).
3. R. Z. Sagdeev, V. D. Shapiro and V. I. Shevchenko, Sov. J. Plasma Physics **4** 306 (1978).
4. Z. Lin, T. S. Hahm, W. W Lee, W. M. Tang and R. B. White, Science **281** 1835 (1998).
5. G. Hammet, M. Beer, W. Dorland, S. C. Cowley and S. A. Smith; Plasma Phys. Contr. Fusion **35** 973 (1993).
6. P. H. Diamond, M. N. Rosenbluth, F. L. Hilton et al., in Plasma Physics and Controlled Nuclear Fusion Research (International Atomic Energy Agency, Vienna, 1998), IAEA-CN-69/TH3/1.
7. P. H. Diamond, S. Champeaux, M. Malkov, et al.,Nuclear Fusion, **41** 1067 (2001).
8. S. Cowley et al. Phys. Fluids, **3**, 2767 (1991).
9. P. N. Guzdar et al. Phys. Rev. Lett. **57**, 2818 (1986).
10. B. N. Rogers, W. Dorland and M. Kotschenreuther, Phys. Rev. Lett., **85**, 5336 (2000).
11. W. Dorland, F. Jenko, M. Kotschenreuther and B. N. Rogers, Phys. Rev. Lett **85**, 5579 (2000).
12. R. Singh, P. K. Kaw and J. Weilend, Nucl. Fusion **41**, 1219 (2001).
13. C. Holland and P. H. Diamond, 'Electromagnetic Secondary instabilities in electron temperature gradient turbulence' (to be published).
14. A. Das, S. Mahajan, P.Kaw, A. Sen, S. Benkadda and A. Verga, Phys. Plasmas **4**, 1018 (1997).

Two-Dimensional Particle Simulation of Electrostatic Solitary Waves with an Open Boundary Condition

T. Umeda, Y. Omura, H. Matsumoto and H. Usui

Radio Science Center for Space and Atmosphere, Kyoto University
Gokasho, Uji Kyoto 611-0011, JAPAN

Abstract. We study formation process of two-dimensional electrostatic solitary waves observed by the GEOTAIL spacecraft. Previous simulation studies have confirmed that the electrostatic solitary waves correspond to BGK electron holes formed through nonlinear evolution of electron beam instabilities. In the present study, we performed two-dimensional electrostatic particle simulations of an electron bump-on-tail instability with an open boundary condition. We inject a weak electron beam from the open boundary into the background plasma. In the two-dimensional open system, spatial and temporal development of the electron bump-on-tail instability is different from that in the uniform periodic system. A lower hybrid mode is excited locally in the region close to the source of the electron beam through coupling with electron holes at the same parallel phase velocity, because the drift velocity of the electron holes is much faster than the parallel group velocities of the lower hybrid mode. As a result, spatial structures of electron holes vary depending on the distance from the source of the electron beam.

INTRODUCTION

Electrostatic solitary waves (ESW) were first observed by the GEOTAIL spacecraft in the magnetotail [1]. ESW are bipolar electric pulses longitudinal to the geomagnetic field. ESW are modeled as electron phase-space density holes which are Bernstein-Greene-Kruskal (BGK) modes [2]. Electron holes are formed through nonlinear evolution of electron beam instabilities. In a long time nonlinear evolution of a instability, electron holes coalesce with adjacent holes and merge into larger, more intense and isolated holes. The previous one- and two-dimensional simulation studies have confirmed that ESW in the magnetotail are one-dimensional potentials generated through nonlinear evolution of an electron bump-on-tail instability [3,4].

The recent spacecraft observed bipolar parallel electric fields accompanied by strong perpendicular electric fields [5,6]. Formation processes of such multi-dimensional electron holes were studied by recent computer simulations [7-10]. These simulations have demonstrated that electron holes drifting along the external magnetic field can radiate obliquely propagating electrostatic waves such as electrostatic whistler mode (oblique Langmuir mode [7,8] and lower hybrid mode [10]. The electron holes become multi-dimensional through modulation by the oblique modes.

Simulations of electron beam instabilities were conventionally performed in uniform periodic systems. In simulation runs with periodic boundary conditions, unstable velocity distribution functions are assumed to exist uniformly in space as initial conditions. In the real space plasma, however, wave-particle interaction of electron beam instabilities does not necessarily take place uniformly. Sources of electron beams are localized, when the electron beams result from acceleration by localized electric fields that appear in a magnetic reconnection process or a shock transition.

Recently a number of simulations of electron holes were also performed in non-periodic systems [11-13]. In the present study, we injected a weak electron beam from an open boundary into homogeneous background plasma. We studied spatial and temporal development of electron beam instability from a localized source of an electron beam. Note that a similar simulation was performed by Mandrake et al. [12] with an injection of a narrow electron beam into non-uniform background plasma. In the present study, we assume a wide electron beam with perpendicular dimension much longer than both electron and ion gyro-radii. We focus on the interaction of electron holes with lower hybrid mode that occurs under weak magnetic fields with the electron plasma frequency equal to the electron cyclotron frequency.

SIMULATION MODEL

In the present study, we assumed that a weak electron beam drifts against the background majority electrons. As the initial condition, we assume that the background electrons and ions with mass ratio $m_i/m_e = 100$ exist uniformly in the simulation box without the electron beam. When a computer simulation is started, the electron beam is continuously injected from the left boundary of the simulation box into the background homogeneous plasma. The injected particles (beam electrons, background electrons and ions) have flux velocity distribution functions given by $vf(v)$ [15]. Particles are injected with the constant flux, while exiting particles are removed at the open boundaries as if there is no boundary. In the present study, we assume that the beam electrons are distributed uniformly in the direction perpendicular to the external magnetic field.

We developed a two-dimensional electrostatic particle code modified from Kyoto university ElectroMagnetic Particle cOde (KEMPO) [14]. The external magnetic field is taken in the x direction. We used open boundaries in the x direction and periodic boundaries in the y direction. The two-dimensional system consists of $N_x \times N_y = 1024 \times 64$ cells. We used 1,600 superparticles per cell respectively for electrons and ions. The density ratio of the electron beam is defined as $R = n_b/(n_b + n_e) = 0.06$. The simulation parameters are listed in TABLE 1. The bump-on-tail velocity distribution function of the present study is based on the previous one- and two-dimensional studies [3,4]. Note that electron bump-on-tail velocity distribution functions are consistent with the recent statistical analyses of the GEOTAIL data. In the particle measurement of the GEOTAIL spacecraft, electron beams are observed as enhanced non-thermal fluxes with flat diffused velocity distributions that result in after saturation of electron beam instabilities [16].

TABLE 1. Simulation Parameters.

Electron cyclotron frequency	Ω_e/Π_e	1.0
Drift velocity of electron beam	V_d/V_{te}	2.0
Thermal velocity of electron beam	V_{tb}/V_{te}	0.1
Thermal energy ratio	T_e/T_i	64

SIMULATION RESULT

Spatial profiles of the electrostatic potentials at different times are shown in FIGURE 1. In the very early phase of the simulation run ($\Pi_e t = 0 \sim 100$), potential structures of electron holes formed by the bump-on-tail instability are uniform in the direction perpendicular to the external magnetic field. As time elapses, amplitude of the electron oblique beam modes becomes larger and larger in the generation region of electron holes. The excited oblique beam modes modulate the background electron densities in the y direction. Potentials structures initially uniform in the y direction become twisted through modulation by the oblique beam mode. According to the linear dispersion relation of the electron beam modes, the parallel beam mode has the maximum growth rate, and oblique beam modes with larger wave normal angle have smaller growth rates. The present simulation result is consistent with the linear dispersion relation.

As the twisted electron holes and two-dimensional electron holes propagate farther from the generation region, the electron holes are more and more isolated in the y direction through coalescence. The perpendicular dimensions of two-dimensional electron holes also become longer and longer through coalescence of the electron holes. Since the electrons trapped therein are mixed in the y direction at the coalescence because of their cyclotron motions, the twisted potentials and two-dimensional potentials are aligned in the y direction as schematically illustrated in Figure 13b. The perpendicular electric fields of the electron holes decay and potential structures become one-dimensional. The decay process of the perpendicular electric fields through the coalescence is described by Miyake et al. [4,10]. One-dimensional potentials in the leading edge of the electron beam can propagate without changing their characteristics. In the leading edge of the electron beam, potential structures are free from the electron oblique beam modes propagating slower than the parallel beam mode in the x direction.

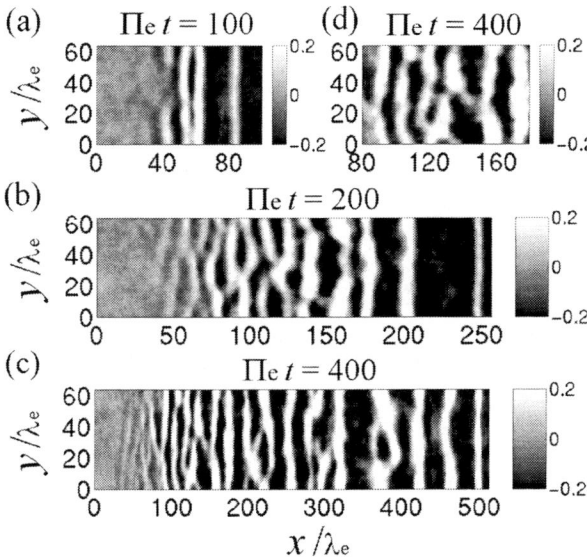

FIGURE 1. Spatial profiles of the electrostatic potentials at different times.

In a long time evolution of the bump-on-tail instability up to $\omega_{LHR} t > 50$, strong oblique electrostatic waves are excited in a region close to the source of the electron beam ($x/\lambda_e = 100 \sim 200$). As shown in FIGURE 2a, E_y components with larger k_y are enhanced only in this region. We analyzed Fourier spectrum of the oblique mode, and found that this mode is a lower hybrid mode propagating $87°$ relative to the external magnetic field (FIGURE 2b). Excitation of the lower hybrid mode is localized, because the parallel group velocity of the lower hybrid mode is much slower than the drift velocity of the electron holes [17]. The electron holes excited in the later phases are more modulated by the lower hybrid mode in the y direction. As a result, the electron holes are accompanied by the perpendicular electric fields to form "modulated" one-dimensional potentials. The perpendicular electric fields of electron holes are carried by the electron holes at the drift velocity of the electron holes.

We performed additional simulation runs with different mass ratios. We found that the mass ratio is not essential in the emission process of the lower hybrid mode, but it changes the time scale of growth of the lower hybrid mode and its propagation angle. The propagation angle of the lower hybrid mode is approximately given by

$$\theta = \tan^{-1}\left(\sqrt{\frac{2m_i}{m_e}} \frac{V_d}{V_{te}}\right) \quad (1)$$

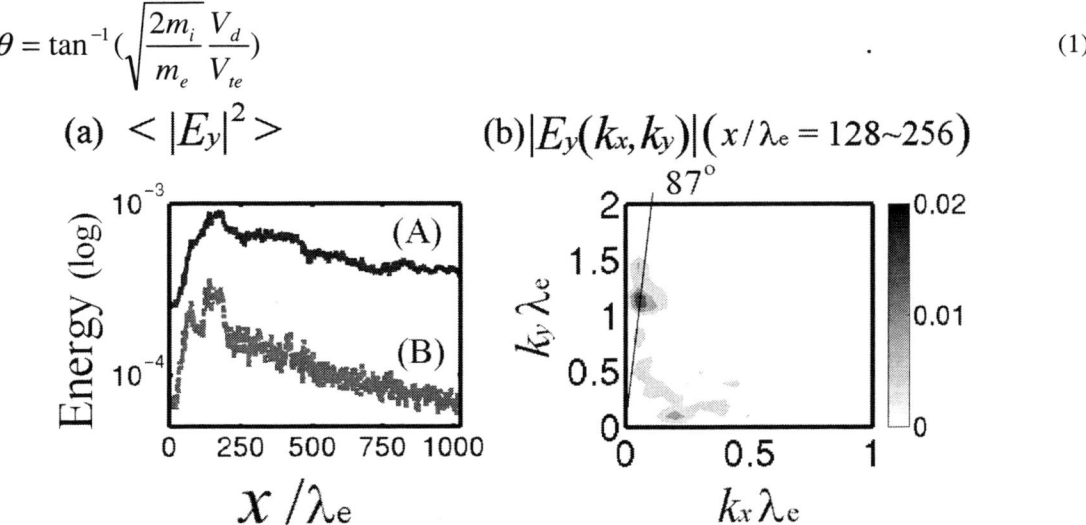

FIGURE 2. (a) Spatial profiles of $|E_y|^2$ along the magnetic field: (A) Energy density of $|E_y|^2$ for $k_x \lambda_e = 0 \sim \pi$, and (B) energy density of $|E_y|^2$ for $k_x \lambda_e = 1.0 \sim 1.6$. We took time average of $|E_y|^2$ for $\Pi_e t = 819.2 \sim 1024.0$. (b) $k_x - k_y$ spectrum of $|E_y|$ at $\Pi_e t = 1024.0$.

CONCLUSION

In the open system, spatial structures of electron holes vary depending on the distance from the source of the electron beam. Since both oblique electron beam mode and lower hybrid mode propagate much slower than electron holes in the parallel direction, electron holes in the leading edge of the electron beam are not affected by these oblique modes. On the other hand, electron holes excited at the later times are more modulated by these oblique modes. In the early phase of the two-dimensional simulation runs, spatial structures of electron holes are determined by the oblique electron beam modes. In the generation region of the electron holes, electron holes initially uniform in the y direction are twisted thorough modulation by the oblique beam modes. In regions far from the source of the electron beam, electron holes become one-dimensional through the coalescence/decay process. In the later phase of the two-dimensional simulation run, the wave-wave coupling process takes place in the localized region, because the parallel group velocity of the lower hybrid mode is much smaller than the drift velocity of the electron holes. Electron holes are accompanied by the perpendicular electric fields through modulation by the lower hybrid mode, resulting in formation of modulated one-dimensional potentials. The perpendicular electric fields of electron holes can be observed even at a distant place from the source, because they are carried by the electron holes at the drift velocity of the electron holes. In the present simulation study, we found that both perpendicular electric fields of electron holes and strong lower hybrid modes are observed in the generation region of the electron holes. We can make use of these characteristics to identify source regions of electron holes from waveform data of plasma wave observations.

ACKNOWLEDGMENTS

This work was supported by the research fellowship of Japan Society for the Proportion of Science. Computer simulations were performed on the KDK computer system at Radio Science Center for Space and Atmosphere, Kyoto University.

REFERENCES

1. Matsumoto, H., H. Kojima, T. Miyatake, Y. Omura, M. Okada, I. Nagano, and M. Tsutsui, *Geophys. Res. Lett.*, **21**, 2915-2918 (1994).
2. Krasovsky, V. L., H. Matsumoto, and Y. Omura, *J. Geophys. Res.*, **102**, 22131-22139 (1997).
3. Omura, Y., H. Matsumoto, T. Miyake, and H. Kojima, *J. Geophys. Res.*, **101**, 2685-2697 (1996).
4. Miyake, T., Y. Omura, H. Matsumoto, and H. Kojima, *J. Geophys. Res.*, **103**, 11841-11850 (1998).
5. Omura, Y., H. Kojima, N. Miki, and H. Matsumoto, *Adv. Space Res.*, **24**, 55-58 (1999).
6. Ergun, R. E., C. W. Carlson, J. P. McFadden, F. S. Mozer, L. Muschietti, I. Roth, and R. Strangeway, *Phys. Rev. Lett.*, **81**, 826-829 (1998).
7. Goldman, M. V., M. M. Oppenheim, and D. L. Newmann, *Geophys. Res. Lett.*, **26**, 1821-1924 (1999).
8. Oppenheim, M. M., D. L. Newmann, and M. V. Goldman, *Phys. Rev. Lett.*, **83**, 2344-2347 (1999).
9. Muschietti, L., I. Roth, C. W. Carlson, and R. E. Ergun, *Phys. Rev. Lett.*, **85**, 94-97 (2000).
10. Miyake, T., Y. Omura, and H. Matsumoto, *J. Geophys. Res.*, **105**, 23239-23249 (2000).
11. Muschietti, L., R. E. Ergun, I. Roth, and C. W. Carlson, *Geophys. Res. Lett.*, **26**, 1093-1097 (1999).
12. Mandrake. L., P. L. Pritchett, and F. V. Coroniti, *Geophys. Res. Lett.*, **27**, 2869-2872 (2000).
13. Omura, Y., H. Kojima, T. Umeda, and H. Matsumoto, *Astrophys. Space Sci.*, **277**, 45-57 (2001).
14. Omura, Y., and H. Matsumoto, "KEMPO1: Technical guide to one-dimensional electromagnetic particle code", in *Computer Space Plasma Physics* edited by H. Matsumoto and Y. Omura, Terra Scientific, Tokyo, 1993, pp.21-65.
15. Birsall, C. K., and A. B. Langdon, *Plasma Physics via Computer Simulation*, McGraw-Hill, New York, 1985.
16. Omura, Y., H. Kojima, N. Miki, T. Mukai, H. Matsumoto, and R. Anderson, *J. Geophys. Res.*, **104**, 14627- 14637 (1999).
17. Umeda, T., Y. Omura, H. Matsumoto, and H. Usui, *J. Geophys. Res.*, in press (2002).

Laser Giant Ion Source and the Prepulse Effects for Picosecond Interaction for High Gain Laser Fusion

Heinrich Hora[1,2], J. Badziak[3], F.P. Boody[4], R. Höpfl[1], K. Jungwirth[5],
B. Králikova[6], J. Krása[6], L. Láska[6], P. Parys[3], V. Perina[7], M. Pfeifer[6], K. Rohlena[6],
J. Skála[6], J. Ullschmied[5], J. Wolowski[3], and E. Woryna[3]

[1]*University of Applied Science, 94453 Deggendorf, Germany*
[2]*Dept. Theoretical Physics, University of New South Wales, Sydney 2052, Australia*
[3]*Institute of Plasma Physics and Laser Microfusion, 00-908 Warsaw, Poland*
[4]*Ion Light Technologies GmbH, 93077 Bad Abbach, Germany*
[5]*Institute of Plasma Physics ASCR, 18221 Prague 8, Czech Republic;* [6]*Institute of Physics ASCR, 18221 Prague 8, Czech Republic*
[7] *Nuclear Physics Institute, ASCR, 25068 Rez, Czech Republic*

Abstract. By studying laser driven ion sources which produce giant ion emission current densities exceeding the few mA/cm^2 of classical ion sources (MEVVA or ECR) by more than six orders of magnitude, we unexpectedly measured an anomalous low ion energy with ps laser pulses. The emission is basically different from that with the fastest ion energies in the MeV to GeV range due to relativistic self focusing and from the second fastest ion group due to quiver-thermalization processes. We report on specifically designed experiments with gold targets where 0.5 ns laser pulses produce MeV Au-ions in accordance with relativistic self focusing in strong contrast to ps pulses where a 400 times higher intensity from TW pulses is needed to arrive at the same ion energies. These can be explained by a basically new model without self-focusing as a skin layer effect where the absence of a prepulse is essential. This has consequences for the application of laser driven ion sources and may improve the hitherto highest published laser fusion gains with 50 TW-ps laser pulses without the usual spherical precompression.

INTRODUCTION

Lasers irradiating solid targets produce very energetic, highly charged ions with ion current densities more than million times above the Langmuir-Child space charge limit [1-3]. This is of interest for ion sources in accelerators as the LHC [1] or for ion implantation for surface hardening and reduction of dry friction [4]. Beyond these settled facts it turns out that the interaction regime with picosecond laser pulses shows strong differences [5] to that of the nanosecond pulses which we are analyzing under the aspects of relativistic self focusing [2,6,7] of the laser beam in the generated plasma.

The drastic contrast to the classically limited ion emission can be seen from the following facts. The most advanced and highly improved metal vapor vacuum arc (MEVVA) technology [8] arrived finally at ion emission current densities of 13 mA/cm^2 at the emitter surface and even considerably less from ECR sources [9] where even most refined immersion techniques [10] or the sheath drift [11] cannot overcome the classical limits. Laser driven ion sources produced mA/cm^2 in 1 meter distance from the source at 1.2 ps and irradiation by one mJ on a 0.03 mm focus diameter [9,10]. Since the emission in one meter distance has a cross section of much more than 100 cm^2, the emission current of 0.1 A based on a diameter at the emission surface of 3.14x10^{-6} cm^2 corresponds to 31 kA/cm^2 where a temporal spreading of the ion pulse and a possible smaller emission area due to self focusing will correspond even to very much higher current densities. While the ion number for ps laser interaction is in the same way by many orders of magnitudes higher than the classical limits similar to the ns pulses, we report here on a new anomaly [5] about the energy of the emitted ions for which a basically new skin layer interaction model [12,13] is necessary, and where conclusions can be drawn for laser fusion experiments [14].

FAST IONS BY RELATIVISTIC SELF FOSCUSING AND QUIVER-COLLISION

Since the sixties it was observed that laser produced plasma emitted fast ions and a slow thermal ion group. The fast ions are in the keV, MeV and up to GeV range, show a separation of the ion energy on the charge number Z and are of numbers beyond 10^{13} such that is impossible to explain these by ambipolar acceleration in surface sheaths which would result in less than 10^9 ions [1,3]. When looking into other acceleration mechanisms than sheaths to result in the more than 10^{13} ions, self-focusing was an explanation. With the first derived dielectric modification of the ponderomotive force generalized to the nonlinear force [1,3] the ponderomotive self focusing threshold was derived arriving at a minimum power limit of about 1 MW [15]. This process including the beam diameter and the measurements of the plasma depletion in the beam axis was measured in all details in agreement [16] with the theory derived later by a number of authors in a different way and all resulting in the same numbers [15].

This showed how the MW threshold led to a laser intensity in the filament such that the nonlinear force accelerated the whole volume of the electron cloud with the attached electrons, but separated by Z, to the 10 keV energies. This process to the electron cloud with the strong gradient of the electromagnetic laser field energy [18] was like gravitation but with differentiating according to Z.

Apart from this plasma dynamics (non-thermal) driving by the laser field above 1 MW, another instantaneous self-focusing appeared purely dielectric when the quiver energy of the electrons in the laser field were at least about one thousandth of the relativistic electron energy mc^2 [6]. The laser beam was shrinking down to a diameter of multiples δ of the laser wave length λ. The extremely high laser intensities in the filament resulted then in a nonlinear force acceleration to the MeV energies. This agreed with all the highest ion energies measured all different times at different laboratories [1] (Fig. 10) before the ps laser pulses were available. The maximum ion energy ε_{ion} is then independent of the wave length λ and is determined only by the laser power P in watts as [6,7]

$$\varepsilon_{trans} = Z\, mc^2\, P\, e^2/(\pi^2 \delta^2 m^2 c^5) = Z\, mc^2\, (P/\delta^2)\, 5.7 \times 10^{-12}\, W \qquad (1)$$

where δ is a self diffraction factor given in multiples of the wave length and may vary around one.

One example from recent experiments with the PALS iodine laser is the following. A third harmonics iodine laser pulse of 0.4 ns duration and 6×10^{11} W power irradiating tantalum produced fastest 81 MeV Ta^{+50} ions. The theory of relativistic self-focusing, Eq. (1) with $\delta = 1$ arrives at 89 MeV. This is a rather sufficient confirmation that the very fast ions were produced by relativistic self focusing and subsequent nonlinear force acceleration.

Apart from this fastest ion group following relativistic self-focusing, a second fastest nonthermal ion group was discovered by Ehler [17]. Their origin was explained before from the observed properties [1] that this is due to the partial thermalization of the high electron quiver energy where a subsequent acceleration process of this hot plasma occurs [18]. What is convincing (Chap. 5 of Ref. 1) is that the quantum modified collision frequency fully agrees with the dependence on the laser intensity I and laser pulse length τ_L of the hot plasma temperature T_x as measured from the hot x-ray component as

$$T_x = const\, \tau_L\, I^{1/2} \qquad (2)$$

while the use of the classical collision frequency arrives at

$$T_x \propto I^{-1/2} \qquad (3)$$

This result can be extended to the ps laser plasma interaction experiments by Clark et al [19]. 50 TW 0.9-2.3 ps neodymium glass laser pulses were focused to 10 wave length beam diameter at an obliquely arranged target where strong suppression of any prepulse produced a plasma in front of the target of less then 10 wave length thickness. The highest energies of Pb^{+48} was 430 MeV while the fastest C^{+6} ions had 60 MeV energy. The fact that the lead ions should have 7.1 GeV and the carbon ions about 880 MeV energy if relativistic self-focusing had been involved [Eq. (1)], indicates that the ions were not due to relativistic self-focusing. The conditions of the 10 wave length focus and the thin preplasma may indicate qualitatively that relativistic self focusing could not have been established. It is very important to note that the measurement of the intensity dependence (Fig. 4 of Ref. [19]) at the unchanged wavelength showed an increase of the maximum ion energy ε_{imax} at constant laser pulse duration of the fastest ions close to

$$\varepsilon_{imax} \propto I^{0.5} \qquad (4)$$

in agreement with Eq. (3).

DRASTIC ANOMALY AT PS INTERACTION WITHOUT PREPULSE

After this clarification of the fastest and the second fastest ion groups including our PALS-experiments and that of Clark et al. [19] we are discussing the results with very low maximum ion energies [5] and report on new experiments which specifically were performed to support these very unexpected results.

Focusing a neodymium glass laser pulse of 1.5 ps duration at power P up to 2.1×10^{11} W produced Cu^{+13} ion or up to 450 keV energy. Using the theory of relativistic self focusing, Eq. (1), the ion energy should be 22.6 MeV. Again we see that relativistic self focusing could not have occurred in this experiment. It is important to note that any prepulse (aspect ratio) had been reduced between 10^{-8} and 10^{-4}. Another very strange result is that not any change of the ion number was observed on the variation of the laser power (energy) over a factor 25 (Fig. 5a of Ref. 5) while the ion velocity increased on a square root law (Fig. 5b of Ref. 5) such that the energy of the fastest ions showed the following dependence on the laser intensity I

$$\varepsilon_{imax} = I \qquad (5)$$

Therefore, the measured fast ions were different from the quiver-collision property, Eq. (4) but were different also from the relativistic-self-focusing property due to the 50-times too low maximum ion energy. Furthermore the number of the fast ions was independent on the laser power (Fig. 5a of Ref. 5).

In order to clarify this discrepancy, measurements were specifically designed for these conditions using the laser system described before [5]. This neodymium glass laser system produced 0.5 ns laser pulses of 1.4×10^9 W/cm^2 in the focus for irradiation on gold targets at an intensity of 2×10^{14} W/cm^2. In other measurements the same pulse has been compressed in time to 1.2 ps to a power of $P = 6 \times 10^{11}$ W for an intensity of 8×10^{16} W/cm^2 on the gold target. For clarifying the discrepancy between ps and ns irradiation, the whole optical system was identical in both cases of experiments with a short or long pulses where the same laser energy of 0.7 J was incident in both cases. The energy of the emitted ions was measured by the time of flight probes and by ion electrostatic analyzers. A most significant result is that of a maximum ion energy of 0.95 MeV for Au^{+34} produced by the 0.5 ns pulses, while laser pulses of 1.2 ps duration produced fastest 1.03 MeV Au^{+26} ions. How is it possible that the 30-times ionized gold ions had nearly the same maximum energy of MeV though the irradiation power and the intensity differed by more than 400 between the cases of 0.5 ns and 1.2 ps laser pulse duration?

SKIN LAYER INTERACTION WITH NONLINEAR FORCE ACCELERATION

We are now explaining the fifty times lower energy of the fastest gold ions in the 1.2 ps cases compared with relativistic self focusing. In our experiment we had an aspect ratio of less than 10^{-8} for any prepulse until less than 1 ns before the main pulse: no plasma could then have been produced. For the following 100 ps, the prepulse aspect ratio was less than 10^{-4}. It can be estimated from plasma hydrodynamics in view of the 30 μm diameter of the beam in the focus that the thin plasma layer in front of the target will not permit relativistic self focusing. The laser beam within its whole 30 μm cross section penetrates then into the superdense target plasma only one skin depth, i.e. less than or about one wave length deep. This is an optical property and therefore independent on the laser intensity. The thin prepulse layer may permit a dielectric swelling to a factor 2 to 4 as can be estimated from similar plane geometry computations [2]. At these conditions the maximum quiver energy of 19.5 eV in vacuum is increased by the factor of swelling. These conditions provided then the ideal plane wave geometry (without any filamentation) for a plane geometry acceleration of the electrons against the laser beam by the multiple of the swelling to produce the MeV maximum energy of the Au^{+26} ions. After the clarification that the plane skin layer model explains the maximum gold and oxygen ions for the ps interaction without relativistic self focusing, we can explain also the result of Fig. 5a of Ref. [5] that under these ps irradiation without prepulse, the number of the accelerated fast ions form the intensity independent skin layer volume of Fig. 1b is constant too. The deposited energy is proportional to the laser energy and the quiver energy of the electrons resulting then in the linear increase of the maximum ion energy on the intensity, Eq. (5).

The transition of our skin layer model towards the relativistic self focusing conditions has been seen before [20] in experiments when ps laser pulses irradiated solid targets at a systematic variation of a prepulse of an intensity where plasma is generated. If the prepulse is at a too short time τ_p before the main pulse, the x-ray emission is very low in agreement with the skin layer model. As soon as the τ_p was 70 ps before the main pulse, the x-ray signals were strong and ion energies appear as expected from the relativistic self focusing in the high density plasma. This is due to the fact that the prepulse has produced a high density plasma plume about two times the beam diameter above the target. This agrees with the here presented experiments confirming the skin layer mechanisms presented as a basically new scheme.

CONSEQUENCES FOR PICOSECOND-PETAWATT LASER FUSION

A very interesting consequence of the skin layer interaction to intentionally apply or to suppress prepulses in experiments [14] where 6 to 15 TW - 1.3 ps neodymium glass laser pulses produced very high fusion gains at irradiation on D_2 or deuterated polyethylene. Converting to DT and assuming isotropy, the 8 J ps pulses produced gains of up to 3% without the usual spherical plasma compression. This is an essential alternative for controlled generation of fusion energy compared with the fast ignitor [21]. Going back to the scheme of Norreys et al. [14] and giving special attention to provide the conditions of skin-depth geometry of the interaction by fully controlling or suppressing a prepulse and avoiding self focusing, may lead to laser fusion with ps-PW pulses in plane geometry without very strong compression of the fusion fuel.

ACKNOWLEDGMENTS

Supported by the European Community (contract No. HPRI-CT-1999-00053, project PALS/005), the International Atomic Energy Agency in Vienna (Contract No. 11535/RO), by the Bavarian Ministry of Science, and in parts by the State Committee for Scientific Research (KBN) of Poland (grant No. 2 PO3B 08219), and by the Australian Institute of Nuclear Science (AINSE) for one of the authors (H.H.)

REFERENCES

[1] H. Haseroth and H. Hora, Laser and Particle Beams, **14**, 393 (1996)
[2] H. Hora, Plasmas at High Temperature and Density (S. Roderer, Regensburg-Germany 2000)
[3] E. Woryna, J. Wolowski, B. Kralikowa, J. Kraska, L. Laska, M. Pfeifer, K. Rohlena, J. Skala, V. Perina, R. Höpfl, and H. Hora, Rev. Scient. Instrum. **71**, 949 (2000)
[4] F. Boody, R. Höpfl, H. Hora and J.C. Kelly, Laser and Particle Beams **14**, 443 (1996)
[5] J. Badziak, A.A. Kozlov, J. Makowski, P. Parys, L. Ryc, J. Wolowski, E. Woryhna, and A.B. Vankov, Laser and Particle Beams, **17**, 323 (1999)
[6] H. Hora, J. Opt. Soc. Am. **65**, 882 (1975)
[7] T. Häuser, W. Scheid and H. Hora, Phys. Rev. **A45**, 1278 (1992)
[8] I.G. Brown et al Rev. Sci. Inst. 57, 1069 (1986); Shuangbao Wang et al., Phys. Stat. Sol. **A179**, 95 (2000)
[9] J. Geller, Rev. Sci. Instrum. 63, 2759 (1992)
[10] A. Anders, Surface Coatings Technology **93**, 158 (1997)
[11] I.G. Brown, O.R. Monteiro and M.M.M. Bilek, Appl.Phys.Lett. **74**, 2426 (1999)
[12] H. Hora, F. Osman, R. Höpfl, J. Badziak, P. Parys, J. Wolowski, E. Woryna, F. Boody, K. Jungwirth, B. Kralikowa, J. Kraska, L. Laska, M. Pfeifer, K. Rohlena, J. Skala, and J. Ullschmied, Czech.. J. Phys. **52**, D349 (2002)
[13] H. Hora, J. Badziak, F.P. Boody, R. Höpfl, K. Jungwirrth, B. Kralikova, J. Kraska, L. Laska, P. Parys, V. Perina, M. Pfeifer, K. Rohlena, J. Skala, J.Ullschmied, J.Wolowski, R. Woryna, Optics Commun. **207**, 333 (2002)
[14] P.A. Norreys, A.P. Fews, F.N. Beg, A.R. Bell, A.E. Cangor, P. Lee, M.B. Nelson, H. Schmidt, M. Tatarakis, and M.D. Cable, Plasma Phys. Control. Fusion **40**, 175 (1998)
[15] H. Hora, Z. Physik **226**, 15 (1969)
[16] M.C. Richardson and A.J. Alcock, Applied Physics Letters **18**, 357 (1971)
[17] A.W. Ehler, J. Appl. Phys. **46**, 2464 (1975)

[18] R.L. Morse, and C.W. Nielson, Physics of Fluids 16, 909 (1973)
[19] E.L. Clark, K.Krushelnik, et al Phys. Rev. Letters, 85, 1654 (2001)
[20] P. Zhang, J.T. He, D.B. Chen, et al. Phys. Rev. E57, R3746 (1998)
[21] M. Tabak et al., Physics of Plasmas 1, 1626 (1994)

Laser ICF with Single Event Solution

H.Hora[1,2], P. Toups[2], P. Evans[2], F. Osman[2], R. Castillo[2], K. Mima[3], M. Murakami[3], S. Nakai[3], K. Nishihara[3], C. Yamanaka[3], T. Yamanaka[3], and G.H.Miley[4]

[1] University of New South Wales, Sydney 2052, Australia
[2] University of Western Sydney, Kingsford 2747, Australia,
[3] Insitute for Laser Engineering, Osaka University, Suita 565, Japan
[4] Fusion Studies Lab., University of Illinois, Urban IL. 61801, USA

Abstract. In contrast to the *double event* scheme of the fast ignitor, we follow the results of measured high fusion gains to derive *single event* schemes for laser fusion. Following the results of the volume ignition and the results of low temperature ignition, we report about conditions where a single-event classical ns laser pulse may be used. Several details were elaborated where the laser compression to more than 3000 times the solid state and with lasers pulses above 5 MJ produce very high gain fusion efficiencies though the initial optimum temperatures are in the range of 500 eV or even less. - Another one-step laser fusion using the new PW-ps laser pulses combines the Norreys-Fews scheme where the effect of a new discovered skin layer interaction is applied under control of prepulses as used for explaining the effect of anomalous low energy of ions emitted from ps irradiated targets measured by Badziak et al.. A drastic increase of the fusion gains for the experiment by Norreys et al. can be expected by application of the interpenetration model discussed earlier for the ANTARES laser.

INTRODUCTION

Recent measurements of the fast ignitor laser fusion experiments were compared by Key [1] with "single step" or "one step" [2] mechanisms in contrast to the two steps, the first compression of DT plasma to more than 3000 times the solid state density by conventional ns laser pulses and the following ps laser pulse for depositing laser energy into the center of the compressed plasma, as needed for the fast ignitor [3]. This second heating pulse seemed to be necessary after compression of polyethylene to 2000 times the solid state results in energies of 300 eV or less [4] which is too low for fusion reactions. The deposition of the ps laser pulse energy into the center the for a fusion detonation wave for spark ignition appears to be extremely difficult as shown by experiments [5] and the known effects [6] which in the best case for highly selected conditions may lead to a uniform distribution of the ps pulse energy to ions spread over the whole compressed plasma to be then used for volume ignition [7]. Experiments with 30TW-04ps laser pulses demonstrated that the laser light accelerates the electron cloud like in vacuum [8,9] and the ions follow then the electron beam jet as an ion jet. Furthermore, the ps laser pulses above TW power produce ions of energies up to and beyond 100 MeV and gammas of comparable energy causing all kinds of secondary nuclear reactions.

While fast ignitor research is strongly in the focus of interest [1] we are going back here to the alternative question whether a one step (single event) process for laser interaction may lead to the generation of fusion energy. We follow here two lines: a) use of ns laser pulses b) use ps laser pulses.

There seem to be both options open for the development of laser fusion. For the case a) we are following the lines of volume ignition [7,10] where for very large plasmas and laser pulses, the ignition at low temperatures is possible such that the studies may well continue from the results of measured high densities and low temperatures [4]. In the case b), the Mourou technique provides clearly the conditions for relativistic effects compared with the carbon dioxide laser ANTARES 20 years ago. For this earlier case, the generation of a reaction wave similar to that at spark ignition [10] or for light ion fusion was elaborated as an ion interpenetration process [11]. Our theory of the skin depth interaction [12,13] for explaining the effect of an anomaly of ion emission by lasers measured by Badziak et.al

[14] permits an extrapolation of conditions where the necessary 10^7 J/cm^2 and space charge neutral ion beams of 10^{10} A/cm^2 are possible for a reaction front as calculated before [11]. This is related to the very high gain laser fusion experiments with ps pulses without plasma pre-compression by Norreys, Fews et al [15] for which the skin depth theory predicts drastic improvements by prepulse control.

LASER FUSION WITH NANOSECOND PULSES

Laser fusion inertial confinement reached 1.8% fusion energy per laser energy which corresponds to a gain of 30% [16]. The simple burn of a DT plasma of an initial density n_o (in relation to the solid state density n_s) where an energy E_o was deposited as scaled by a break-even energy E_{BE} arrived at an optimum (core) gain G

$$G = \begin{cases} (E_o/E_{BE})^{1/3}(n_o/n_s)^{2/3} & \text{(Hora 1964, Hora and Pfirsch 1970)} \\ \text{const } n_o R & \text{(Kidder 1974, Fraley et al 1974)} \end{cases} \quad (1)$$

where both expressions are algebraically identical [10,17] if the optimum temperature of 17 keV is provided. The break-even energy for DT is 6 MJ for adiabatic compression and expansion using the self-similarity model for the hydrodynamics [10,17] showing the advantage of high compression.

After yields of 10^{13} DT fusion neutrons were detected [18] compression of CD$_2$ targets containing some tritium were compressed to 2000 times the solid state density [4]. The measured low temperature of <300 eV motivated E.M. Campbell to introduce the fast ignitor [3]. However the following described volume ignition will also permit a very low ignition temperature such that this may be an alternative to the fast ignitor.

RADIATION RE-ABSORPTION FOR LOW TEMPERATURE VOLUME IGNITION

Fusion reactions in a stationary low density plasma with continuous emission of bremsstrahlung needs a minimum temperature of about 4.5 keV for DT to compensate the radiation loss by fusion energy generation. This is no problem at the optimum reaction temperature of 17 keV for the (non-stationary) expanding and adiabatic cooling uniform reaction sphere from where Eq. (1) was derived.

A first detailed treatment of this low temperature inertial confinement fusion was automatically possible when the main omissions of the simple fusion burn covered by Eq. (1) was overcome by including

(a) the fuel depletion, (b) the reheat by fusion reaction products within the spherical plasma, and (c) the partial re-absorption of the bremsstrahlung in the detailed computation.

For the reheat by alphas the collective model was used [19] which is basically the same as that of Gabor [20]. For the adiabatic expansion of the plasma sphere, the self-similarity model [17,21] was used as before with a box-like density instead of a Gaussian profile. Since the re-absorption of the bremsstrahlung is essential for the low temperature ignition, this is based on the fulfillment of transparency as checked before such that the Kramers spectrum for each temperature in each time step was used to calculate the spectral maximum. For this frequency the collision absorption in the plasma given by the temperature and density of each of the steps was calculated and for the then present radius the integrated absorption length determined the percentage of the re-absorbed bremsstrahlung during this time step causing a further increase of the temperature. These were the main ingredients in the computations (Hora et al 1978 [7]) which led to the discovery of volume ignition characterized by the fact that the time dependence of the temperature can increase drastically over the initial temperature of the compressed DT sphere.

The fusion gains (core gains related to the energy E_o which was deposited uniformly in the whole sphere for the subsequent adiabatic expansion) were rather similar to that of Eq. (1) if the gain G was below 8 for DT. The ignition leads to plasma temperatures above 100 keV and the fusion reaction profits then mostly from the alpha-reheat as a kind of very much extra driver energy added to that of the laser. If the initial temperature is increased again a little bit, the adiabatic expansion is so fast that the fusion gain drops remarkably. It was confirmed [10] that the measurements with the highest fusion gains [16] ideally followed the self-similarity model [10,17,21] for volume burn.

RESULTS OF LOW TEMPERATURE VOLUME IGNITION

The following evaluation of cases of low temperature volume ignition is based on the diagrams of Ref.[10]. For higher densities, the question of the Fermi energy is to be taken into account. The Fermi energy is given by

$$E_F = (h^2/2m)(3n_e/8\pi)^{2/3} \qquad (2)$$

The computation results were performed all for non-degenerate plasma. Therefore the Fermi energy of the electrons had to be less than the electron temperature of the plasma. Since the compression and the expansion of the plasma occurs adiabatically, the once determined property of non-degeneracy is preserved over the whole compression and expansion process.

Taking under the these conditions the case of a laser pulse energy of 10 MJ with a 10% hydrodynamic efficiency (E_o = 1 MJ) for the compression to $5000n_s$ for the 50:50 DT fuel, the optimum temperature for volume ignition is 560 eV arriving at a core gain of G = 1300. This corresponds to a total fusion gain per incident laser energy of G_{tot} = 130. Lightly higher total gains of 130 are achieved with 20 MJ laser pulse energy where the optimum temperature decreases to 425 eV for the same compression to ten thousand times the solid state.

The just mentioned result of very low ignition temperature in continuation of Johzaki et al. [22] - mostly due to the self absorption of the bremsstrahlung but highly favored by the very high extra self-generated driver energy from the re-heat - may well be an orientation for the conditions how the rather "robust" (Lackner et al 1994 [7]) volume ignition may work within the here mentioned extrapolated parameters. Since the measurements of high compression and low temperature [4] were performed with laser pulses in the 10 kJ range, the experiments with 10MJ pulses may well differ and may arrive at higher temperatures than that required for the optimum ignition temperature concluded form the computations. Any higher temperature will then relax the requirements of the high compression and the high laser pulse energy, and not the contrary. Even more optimistic conditions may be expected due to the here not included re-heating by neutrons and if the thermal non-equilibrium decoupling of the electrons ions and the blackbody radiation are included as e.g. in the specific computations by Martinez-Val et al. (1994), see Ref. [7].

LASER FUSION WITH PICOSECOND PULSES

Discussing [23] the measurements of x-ray emission from sub-ps laser irradiated targets using the technique of Jie Zhang for a very careful control of the prepulse [24], resulted in very low x-rays if the prepulse was arriving few ps before the main pulse, but was growing to very strong emission for a prepulse from >70 ps producing a plasma up to 70 μm in front of the target. Since the focused beam had a diameter of 30 μm, relativistic self focusing has then reduced the beam diameter to about one wave length at the target producing high energetic ions and the very intense x-rays. Without this plasma in front, the beam had no self-focusing and the whole laser power was uniformly spread over the whole focus cross section with 1000 times lower intensity producing much less x-rays.

This result was most helpful for explaining the effect observed by Badziak et al [14] where (a) 50 times lower energies of the emitted Cu^{+13} were observed than the expected 22 MeV energy due to relativistic self focusing, and (b) the most strange observation was reported that changing of the laser intensity by a factor 30 did not at all change the number of the fast ions. The experience with the observations of Zhang et al [24] led to the explanation: In the case of Badziak et al [14] - confirmed by subsequent specific measurements [13] - arrived at similar conditions. The low maximum ion energies are a proof that there is no relativistic self focusing due to careful elimination of a prepulse until 50 ps before the main pulse arrived, and the interaction of the laser happened only in the skin layer of the target. Since this skin volume is determined by optical constants only and not by the laser intensity, the number of the emitted ions is unchanged whatever the laser intensity I is. The detailed analysis of the experiments of Badziak et al fully explained [12] the energy of the emitted Cu^{+13} [14] and the MeV Au^{+27} [13] ions if a swelling of the quiver energy by factor 2 to 3 could be assumed as expected from the higher prepulse during the 50 ps before the main pulse generating an about 10 μm plasma in front. This plasma layer is too thin for relativistic self focusing but just sufficient for the swelling process as numerically confirmed [13,25]. The measured increase of the maximum ion energy $\varepsilon_{imax} \propto I$, fully confirms this skin layer interaction model [12]. The analysis also clarified the two different fast ion groups. The fastest based on the relativistic self-focusing [26] and agrees with the measured 80 MeV Ta^{+50} ions from the 0.7 TW-05 ns PALS-laser experiments [13] as in very numerous earlier experiments. The 50 TW-ps experiments with 430 MeV Pb^{+48} ions [27] did not agree with relativistic self focusing where the theory arrived at 7.1 GeV ion energy. The result [27], however, ideally confirmed the quiver-collision model (Section 5 of [28]) with the number of the ion energy and with the measured dependence $\varepsilon_{imax} \propto I^{1/2}$. The geometry of this experiment did not at all permit relativistic self-focusing.

The connection to the single event laser fusion experiment of Norreys et al [15] with 6TW–1.3ps neodymium glass laser pulses on a DD targets producing a fusion gain per laser energy of 3% when converted to DT, is given by the fact that very probably prepulse and laser geometry conditions as in the following experiment [27] may have been the case. If we may project these results to the theory of the skin depth interaction [12,13,25] it may be concluded that a preparation of a pure skin layer experiment will arrive at very much higher fusion gains far above break even. If a skin layer experiment with a 6PW-ps pulse focused to a cross section of 1 mm^2 of a DT target with such suppression of a prepulses is performed that only a few μm thick preplasma would be produced for a 2-3 times swelling in the plane geometry (see Sect. 10 of Ref. [17]), relativistic particles with >100 MeV could be avoided but two ideal blocks of ions of about 100 keV energy would be produced, one moving against the laser but the other moving into the target with a (space charge neutral) ion beam current density of 10^{10} Amp/cm^2 and an energy density of nearly 10^5 J/cm^2. This comes close to the ignition conditions for the ion interpenetration fusion front [11] as it was calculated for the similar conditions of the ANTARES laser. Continuation of the reaction into pB(11) was discussed before and should lead to a new type of laser fusion reactor for direct conversion of the gained nuclear energy into electricity with a minimum of heat generation [29].

ACKNOWLEDGEMENTS

The authors from the Australian Universities would like to acknowledge the support of AINSE.

REFERENCES

[1] M.H. Key, Nature 412, 775 (2001); R. Kodama, P.A. Norreys, K. Mima, A.E. Dangor, R.G. Evans et al, 2001, Nature **421**, 798
[2] Heinrich Hora, Peter Evans, F. Osman and R. Castillo, K. Mima, M. Murakami, S. Nakai, K. Nishihara, C. Yamanaka, and T. Yamanaka, Inertial Fusion Science and Applications Conf. Kyoto 2001, (Elsevier) in print
[3] M. Tabak et al. Physics of Plasmas **1**, 2010 (1993)
[4] H. Azechi, et al. Laser and Particle Beams **9**, 167 (1991)
[5] A. P. Fews, P.A. Norreys et al Phys. Rev. Letters **73**, 1801 (1994)
[6] H. Hora, H. Azechi, S. Eliezer, Y. Kitagawa, J.-M. Martinez-Val, K. Mima, M. Murakami, K. Nishihara, M. Piera, H. Takabe, M. Yamanaka, and T. Yamanaka, *Laser Interaction and Related Plasma Phenomena*, AIP Conf. Proceedings No. 406, G.H. Miley and E. Michael Campbell eds., (Am. Inst. Physics, New York 1997) p. 165
[7] H. Hora and P.S. Ray, Zeitschr. f. Naturforschung **A33**, 890 (1978); R.C. Kirkpatrick and J.A. Wheeler, Nucl. Fusion **21**, 389 (1981); R.J. Stening, R. Khoda-Bakhsh, P.Pieruschka, G. Kasotakis, E. Kuhn, G.H. Miley and H. Hora, Laser Interaction and Related Plasma Phenomena, G.H. Miley et al ed., (Plenum New York 1991) Vol. 10, p. 347; X.T. He and Y.S. Li, Laser Interaction and Related Plasma Phenomena. G.H. Miley ed. AIP Conf Proceedings No. 318 (Am. Inst. Physics, New York 1994) p. 334; K.S. Lackner, S.A. Colgate, N.I. Johnson, R. Kirkpatrick, and A.G. Petschek, Laser Interaction and Related Plasma Phenomena, G.H. Miley ed., AIP Conf. Proceed. No. 318 (Am. Inst. Phys., New York 1994) p. 356; J.-M- Martinez-Val, S. Eliezer and M. Piera, Laser and Particle Beams **12**, 681 (1994); N.A. Tahir and D.H.H. Hoffmann, Fusion Engin. and Design **24**, 418 (1994); S. Atzeni, Jap. J. Physics **34**, 1986 (1995)
[8] H. Hora, Nature **333**, 337 (1988)
[9] H. Hora, M. Hoels, W. Scheid, J.W. Wang, Y.K. Ho, F.Osman and R.Castillo, Laser & Part. Beams **18**, 135 (2000)
[10] H. Hora, H. Azechi, Y. Kitagawa, K. Mima, M. Murakami, S. Nakai, K. Nishihara, H. Takabe, C. Yamanaka, M. Yamanaka, and T. Yamanaka, J. Plasma Physics **60**, 743 (1998)
[11] H. Hora, Atomkernenergie-Kerntechnik **42**, 7 (1983)
[12] H. Hora, F. Osman, R. Höpfl, J. Badziak, P. Parys, J. Wolowski, E. Woryna, F. Boody, K. Jungwirth, B. Kralikowa,J.Kraska,L.Laska,M.Pfeifer,K.Rohlena,J.Skala,and J.Ullschmied, Czechoslov. J. Phys. **52**, D349 (2002)

[13] H. Hora, J. Badziak, F.P. Boody, R. Höpfl, K. Jungwirth, B. Kralikova, J. Kraska, L. Laska, P. Parys, V. Perina, M. Pfeifer, K. Rohlena, J. Skala, J.Ullschmied, J. Wolowski, R. Woryna, Optics Communic. **207**, 333 (2002)

[14] J. Badziak, et al., Laser and Particle Beams **17**, 323 (1999)

[15] P.A. Norreys, A.P. Fews, et al., Plasma Physics and Controlled Fusion **40**, 175 (1998)

[16] J.M. Soures, R.L. McCrory et al, Phys. Plasmas **3**, 2108 (1996)

[17] H. Hora, Plasmas at High Temperature and Density (Springer Berlin 1991; S. Roderer, Regensburg 2000)

[18] C. Yamanaka and S. Nakai, 1986 Nature **319**, 757; H. Takabe et al 1988 Phys. Fluids **31**, 2884

[19] P.S. Ray and H. Hora, Nuclear Fusion **16**, 535 (1976)

[20] D. Gabor, Proc. Roy. Soc. (London) **A213**, 73 (1953)

[21] R.F. Schmalz, Physics of Fluids **29**, 1389 (1986)

[22] T. Johzaki, Y. Nakao, M. Murakami, and K. Nishihara, 1998 Nuclear Fusion **38**, 467

[23] H. Hora and Long Wang, Contributed paper at the Regional Plasma Physics College, Islamabad, Febr. 2001

[24] P. Zhang, J.T. He, D.B. Chen, Z.H. Li, Y. Zhang, J.G. Wang, Z.L. Li, B.H. Feng, X.L. Zhang, X.W. Tang, J. Zhang, Phys. Rev. E **57**, 3746 (1998)

[25] H. Hora, H. Peng, W. Zhang, and F. Osman, Bull. Am. Phys. Soc. **47**, 879 (2002); SPIE Photonics Asia Conference "High Power Lasers and Applications" 14-18 October 2002

[26] H. Hora, J. Opt. Soc. Am. **65**, 882 (1975); D.A. Jones, E.L. Kane et al. Phys. Fluids, **25**, 2295;T. Häuser, W. Scheid and H. Hora, Phys. Rev. **A45**, 1278 (1992);

[27] E.L. Clark, K.Krushelnik, M. Zepf, M. Tatasakis, M. Machacek, M.I.K. Santala, I. Watts, P.A. Norreys, and A.E. Dangor, Phys. Rev. Letters, **85**, 1654 (2001)

[28] H. Haseroth and H. Hora, Laser and Particle Beams **14**, 393 (1996)

[29] H. Hora, German Patent Application 1033 08 515.3 (28.2.2002)

Dynamics of a Supersonic Plume Moving Along a Magnetized Plasma

G.J. Morales, F.S.Tsung and J.N. Leboeuf

*Physics and Astronomy Department, University of California, Los Angeles,
Los Angeles, CA 90095 USA*

Abstract. A recently developed particle-in-cell-code is used to investigate the dynamics and evolution of a microscopic density plume moving through a background plasma with supersonic speed directed along the confinement magnetic field.

OVERVIEW

The motion of a dense plasma plume (or filament) along an ambient magnetized plasma is a phenomenon that can arise in a wide range of natural and controlled environments. Events of this type occur in astrophysical plasmas, are a by-product of mass ejections in the solar corona, give rise to structures in the solar wind, and can play a role in the formation of prominent features sampled by spacecraft in near-earth plasmas. Density plumes can also be generated in the laboratory by laser ablation of solid targets. Recently, the controlled expansion of a microscopic plume in a large magnetized plasma has been observed [1] to generate large amplitude Alfvén waves. In general, the localized release of substantial energy can cause the generation of swiftly moving plasma plumes.

Although there are many interesting and challenging problems related to plasma plumes, we have been primarily motivated by issues related to microscopic interactions in which kinetic processes mediate non-local modifications in the ambient plasma. We are particularly interested in the role played by plume dynamics in energy and mass transport as well as in the formation of long-lived structures. This perspective is also of contemporary interest to magnetic confinement studies.

Motivated by these various issues we have chosen at this stage to investigate the behavior of plumes whose extent across the magnetic field is comparable to the electron skin-depth. This is the fundamental screening length that regulates parallel electron currents and recent observations by spacecraft [2] with enhanced resolution suggest that structures at this scale are ubiquitous to plasmas that are far from thermal equilibrium.

In the present study we ignore the issue of how the initial plume is generated. Although this is an important topic, it detracts from the immediate goal, i.e., to identify the major nonlinear processes triggered by its motion. In here we legislate the initial density, temperature and speed of the plume and follow its space-time evolution as it propagates through a uniform background plasma. The choice of parameters is dictated by practical considerations set by the available computing power.

COMPUTATIONAL TOOL

This study uses an electromagnetic, particle-in-cell code that is based on an object-oriented code (OSIRIS) originally developed at UCLA to study relativistic laser-plasma interactions. The present configuration follows the relativistic, fully magnetized trajectories of ions and electrons and has the capability to run with periodic conditions or finite boundaries. It has been verified that the code correctly simulates the linear properties of shear and compressional Alfvén waves of relevance to the electrodynamics of plume expansion. Depending on the nature of the issue to be isolated, the code is run in extended mode in a supercomputer or in smaller versions in a Mac cluster.

For the results reported here the code is run in 2-1/2 dimensions, i.e, two spatial coordinates and three velocity coordinates. One of the spatial dimensions across the confinement field is taken as ignorable.

The ion to electron mass ratio is M/m=20 which implies that the ratio of the ion acoustic speed to the electron thermal velocity is c_s/v_e=.22. This allows for the practical separation of the relevant time scales within a run spanning several ion gyroperiods while retaining full electron dynamics. The ratio of the electron plasma frequency to the electron gyrofrequency is ω_{pe}/Ω_e=.83, the ratio of the electron thermal velocity to the speed of light is v_e/c=.05, and the ratio of the Alfvén speed to the speed of light is v_A/c=.29. Accordingly, the ordering of the relevant speeds is $c > v_A \gg v_e \gg c_s$ which implies that the Alfvénic phenomena is in the inertial regime, i.e., the electron plasma beta parameter is $\beta_e = 1.7 \times 10^{-3}$ which is smaller than the mass ratio m/M.

The initial density profile of the plume is $n_p(x,z,t=0) = n_p(0)[1-(z/L_\parallel)^2][1-(x/L_\perp)^2]$. The cases described at this conference correspond to $L_\parallel$ =5 c/ω_{pe} and $L_\perp$ =.5 and 2 c/ω_{pe} with plume peak densities equal or twice as large as the density of the background plasma n_0. The coordinate system used is (x,y,z) with the confinement magnetic field B_0 pointing in the -z direction and x is transverse to z with y the ignorable spatial coordinate. The plume is initialized with its center of mass velocity v_p directed along the magnetic field, i.e., towards the left in the figures presented. We consider here the supersonic regime with $v_p = v_e$. The plume and background ions are initially cold while the plume and background electrons have equal and isotropic temperatures in their respective rest frames. In the figures presented the scaled time coordinate is $T = t\omega_{pe}$, the axial and transverse coordinates are scaled to c/ω_{pe} and density to the stationary plasma density.

MAJOR RESULTS

A key feature in the expansion of a plasma density enhancement is that the electrons move very rapidly out of the peak density and leave behind the more massive ions. If the expansion occurs in vacuum, an ambipolar electric field develops that holds the electrons back and results in a self-similar expansion, as has been documented in an early simulation study [3]. However, in the presence of a background plasma of sufficient density, shielding for the outgoing electrons can be provided and a significant ballistic current can thus develop. Since the original density enhancement does not carry a current, collective processes involving the induction electric field set up a current system that closes on itself. In the presence of a strong magnetic field this requires that the cross-field currents be carried by the ions and thus a configuration of the type associated with whistler and/or Alfvén waves develops. In the present study it is found that this process plays a central role. An unexpected feature that has been uncovered is that the plasma develops a finite positive charge in the region of shock compression. This results in a macroscopic dipolar current that generates a quasi-static perturbed magnetic field which is convected with the supersonic plume.

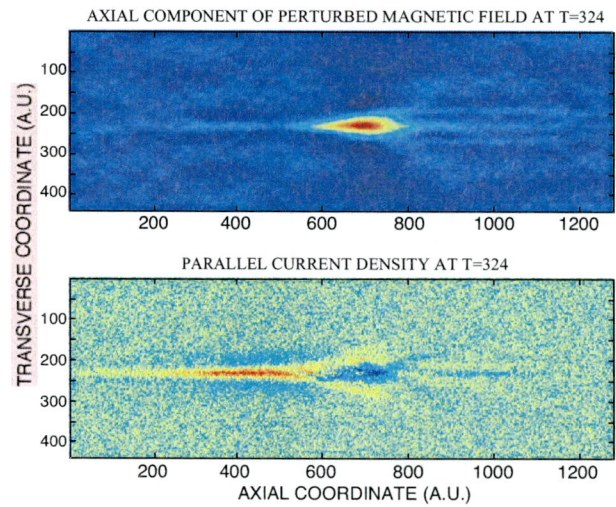

FIGURE 1. Top: diamagnetic field. Bottom: parallel current density.

The early stage in the plume evolution consists of the ballistic outflow of electrons and formation of a region of large positive potential. The background ions are ejected rapidly across the magnetic field while the background electrons are trapped in the deep potential well anchored to the center of mass of the plume. If the plume remained stationary an equilibrium would be reached leading to equalization of the pressure gradients at the sound speed. However, for a supersonic plume an equilibrium is not reached because the number of electrons reaching the pressure enhancement from the right is different from that from the left and within a time scale comparable to the sound transit time across the axial gradient, a shock develops. Fine scale structures develop within the leading edge of the moving plume and an oscillatory wake is formed. Late in time an extended region of depleted density is formed behind a small number of plume ions that lose almost all of their initial energy. These processes are accompanied by the generation of perturbed magnetic fields along the three spatial axes.

The top panel in Fig. 1 displays the two-dimensional color contour of the perturbed magnetic field component along the axial direction at a scaled time T=324 for a case with $L_\perp = .5\ c/\omega_{pe}$ and $n_p(0)=2n_0$. Red corresponds to the largest positive value (pointing to the right) and dark blue to the largest negative value (pointing to the left). This component of the field is generated by diamagnetic currents along the y-direction associated with the transverse pressure gradient. The total field is depressed inside the plume and it is compressed (not visible in the figure) above and below the plume. The faint light trace ahead of the plume and along the center line arises from the additional plasma pressure provided by the ballistic electrons. In essence they generate a dilute density filament shielded by cross-field currents due to background ions much earlier than the arrival of the plume ions. Similar faint traces to the right of the figure indicate that multiple filaments also appear in the wake region.

The bottom panel in Fig. 1 exhibits the two-dimensional contour of the parallel current density at the same time as the diamagnetic field previously discussed. In here again, red corresponds to right-going currents and dark blue to left-going currents. The extended positive current ahead of the plume center is generated by the ballistic electrons going left. Surrounding the bright red channel there is a faint blue region associated with the return currents driven by the induction field , i.e., a coaxial current system develops ahead of the plume. The dark blue region around location 700 in the horizontal axis is not due to ballistic electrons because it is collocated with the peak pressure region seen in the top panel. This corresponds to a positive charge moving to the left and experiencing its own coaxial shielding by background electrons at radial distances larger than $L_\perp$. The faint blue line well behind the plume corresponds to the right-going ballistic electrons.

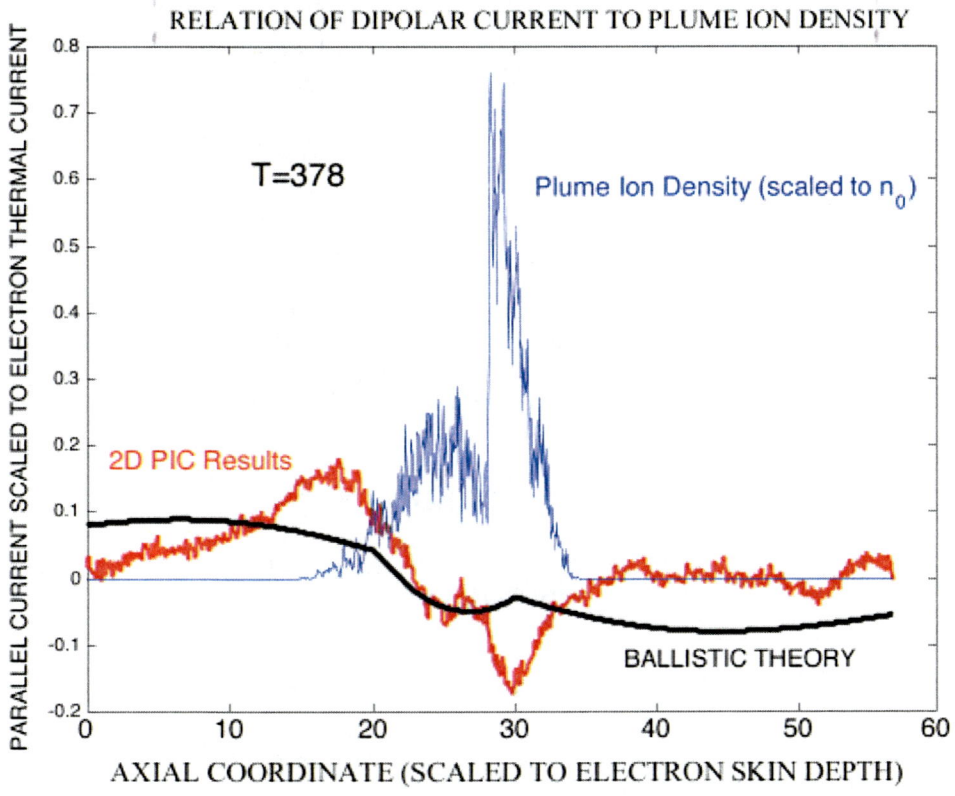

FIGURE 2. Dipolar current involves a positive charge at shock position

Figure 2 clearly identifies the relationship between the dipolar current and the plume ion density for a case corresponding to $L_\perp = 4\ c/\omega_{pe}$ and $n_p(0)=n_0$. The noise-free line corresponds to the linear ballistic theory prediction of the current density at the time T=378. The oscillatory curve close to it is the simulation result and the top trace is the instantaneous density of the plume ions. It is seen that the negative part of the dipolar current is associated with the shock compression of the plume ions and results in a contribution well in excess of the ballistic prediction. The forward ballistic current is held back by the charge separation while the backward current is severely curtailed and upon further scrutiny is found to break into spatial oscillations.

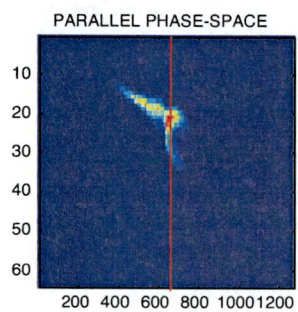

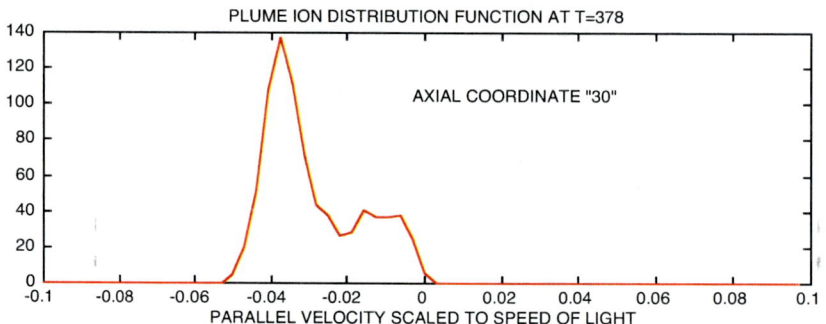

FIGURE 3. Top: phase-space of plume ions. Bottom: velocity distribution along cut.

The top panel of Fig. 3 shows the parallel phase-space contour of the plume ions at the same time as the spatial density trace shown in Fig. 2. It is evident that the original beam-like distribution has developed multiple components, including some particles with small velocity. The corresponding distribution function along $z=30\ c/\omega_{pe}$ is shown in the bottom panel. The original distribution is a delta function centered at a scaled parallel velocity of -.05. It is seen that the peak of the distribution has recoiled by a factor of .01 corresponding to c_s and that a slowing down tail has developed. From the energy lost by the slowest ions it is inferred that potential layers on the order of ten times the electron temperature are developed in narrow layers within the plume.

ACKNOWLEDGEMENTS

This work is sponsored by the Office of Naval Research and by the National Science Foundation. The original OSIRIS code was developed by Prof. W. Mori and his research group at UCLA.

REFERENCES

1. VanZeeland, M., Gekelman, W., Vincena, S., and Dimonte, G., *Phys. Rev. Lett.* **87**, 105001 (2001).
2. Stasiewiccz, K., et al., *Space Sci. Rev.* **92**, 423 (2000)
3. Denavit, J., *Phys. Fluids* **22**, 1384 (1979).

MHD Simulation of Reflection Dynamics of Field-Reversed Configuration Plasma

T. Kanki[*], S. Okada[†], and S. Goto[†]

[*] *Japan Coast Guard Academy, 5-1 Wakaba, Kure, Hiroshima 737-8512, Japan*
[†] *Plasma Physics Laboratory, Graduate School of Engineering, Osaka University,
2-1 Yamada-oka, Suita, Osaka 565-0871, Japan*

Abstract. A two-dimensional magnetohydrodynamic simulation of a reflection dynamics on a field-reversed configuration (FRC) plasma is carried out by numerically modeling the confinement region of the FRC Injection Experiment (FIX) machine. It is shown from the simulation results that the FRC plasma is reflected by the downstream magnetic mirror field without severe destruction of the closed magnetic flux surfaces even when injected with supersonic velocity into the magnetic mirror region, showing the robustness of the FRC against external perturbations. The effects of the magnetic mirror field and the translation velocity on the FRC plasma are discussed.

INTRODUCTION

Numerous translation experiments [1-4] of field-reversed configuration (FRC) plasma have been performed intensively to establish the convenient way of separating formation, heating, and burn regions in a fusion reactor design. The translation technology is an attractive and essential ingredient in present experiments as well as D-^{3}He fuelled reactor studies. In the translation experiment on the FRC Injection Experiment (FIX) machine [3-4] at Osaka, FRC plasma is created in a source region by a field-reversed theta-pinch method, and is injected into an adjacent confinement region with magnetic mirror fields at either end. The injected FRC moves along the magnetic field of the confinement region (confining magnetic field) at supersonic velocity. Subsequently, the FRC reflects at the downstream magnetic mirror. The reflected FRC moves back toward the source region and reflects at the upstream magnetic mirror. It eventually settles down the center of the confinement region without severe degradation of the confinement. As the result, the FRC translation has been successfully achieved by empirically adjusting the confining field. However, it is observed in the first reflection from the downstream mirror that the separatrix radius of the FRC expands excessively when the confining field is reduced, and that more energy of the translated FRC is lost in the case where the translation velocity is larger and the magnetic mirror ratio is smaller. Details concerning how such a reflection process of a FRC plasma is performed by interaction with the magnetic mirror field remain unclear. Under this experimental background, the purpose of this study is to investigate the fundamental physics of reflection dynamics of a FRC plasma by means of an axisymmetric numerical simulation. In particular, we will focus our attention to examine the time evolution of the poloidal flux contours and velocity fields.

NUMERICAL MODEL

In Fig. 1 we show the numerical model of the confinement region of the FIX machine that will be used throughout this paper. The confinement region consists of a central part with mirror field regions at both ends. The central part has 0.8 m inner diameter and 3.0 m long metal chamber. The confining magnetic field in this part applied in axial is produced by solenoid coils. The confining vacuum field B_0 is uniform over a 3.0 m length and can be varied from discharge-to-discharge to values of up to 0.08 T. In our simulation, we fix B_0 at 0.04 T as a typical experimental parameter. The both ends of the confinement region are tapered to mirror field region with an inner diameter of 0.5 m. In the experiment, the typical strength of the upstream and downstream mirror fields B_m are 0.13

and 0.17 T, respectively. In the simulation, we set $B_m = 0.1$ T as the strength of upstream mirror field, and change that of downstream mirror field. Mirror ratio R_m (defined as the ratio of the maximum vacuum mirror field B_m to B_0) at the downstream region varies in accordance with the strength of downstream mirror field.

Next, we explain the governing equation and the boundary condition used in the simulation. Let us use a cylindrical coordinate system (z, r, ϕ), take the symmetry axis of the confinement chamber as z-axis, and choose the center of the confinement region as the origin. We assume the axisymmetry, and carry out numerical simulations in a two dimensional (z, r) plane. As shown in Fig. 1, we divide the confinement region where the simulation is performed into two regions. One is a vacuum subregion where the magnetic field is calculated only, and the other is a plasma subregion where the full set of magnetohydrodynamic (MHD) equations are solved. A perfect conducting boundary at the chamber wall is used. Further, in order to reduce the reflection of MHD flow, free boundary conditions are imposed on the open end of the confinement region. Under the boundary conditions, plasma is allowed to go out freely through the boundary. We use a rezoned Lagrangian mesh which employs an adaptive algorithm [5] to concentrate the grid because a FRC has the sharp plasma pressure gradients in the vicinity of the separatrix. Furthermore, a two temperature MHD model [6], which calculates electron and ion separately, is used, since the electron-ion energy transfer time in the confinement region is much longer than the time scale of the reflection. Resistive diffusion of the magnetic field is calculated using classical resistivity. The effects of unequal parallel and perpendicular thermal conduction, which follow Braginskii [7], are included for the both the electrons and ions. Details of the numerical method have been described elsewhere [8-9].

Finally, as the initial value for the simulation, we use a numerically computed MHD equilibrium with separatrix radius (normalized by the radius of the chamber wall at $z = 0$), $x_s \cong 0.4$ such as is typically observed in the FIX experiment [10]. By applying the thermodynamic equation an ideal fluid, the plasma density profile can be estimated, assuming that the temperature is spatially uniform in the plasma region. Also, we provide the initial axial velocity, v_z that is spatially constant in the plasma region because of assuming that the FRC plasma is in equilibrium, even when it is translating. The parameters of the initial condition for the simulation are listed in Table 1.

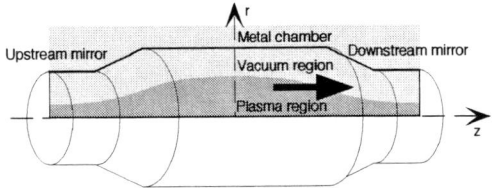

FIGURE 1. Schematic diagram of the computation region in cylindrical coordinates (z, r, ϕ), corresponding to the confinement region in the FIX machine. The shaded poloidal plane shows the two dimensional simulation region.

TABLE 1. Parameters for Initial Plasma Condition.

Parameter	Initial Value
Separatrix Radius r_s	0.17 m
Separatrix Length l_s	3.15 m
Average Mass Density ρ	2.2×10^{-7} kg/m^3
Axial Velocity v_z	1.5×10^5 m/sec
Radial Velocity v_r	0.0 m/sec
Ion Temperature T_i	50.0 eV
Electron Temperature T_e	50.0 eV
Confining Magnetic Field B_0	0.04 T

NUMERICAL RESULTS

We present the numerical results for the reflection dynamics of a FRC plasma at the downstream mirror region. Poloidal flux contours and velocity fields of the FRC are shown in Fig. 2 for the mirror ratio, $R_m = 3.0$ at the downstream region, and the initial axial velocity, $v_z = 1.5 \times 10^5$ m/sec. Here, the plasma region is plotted only in the poloidal plane. Also, this initial axial velocity is faster than the sound velocity of 9.0×10^4 m/sec at the confinement region. This translation velocity, $v_z \leq 2.0 \times 10^5$ m/sec is observed in the FIX experiment for $B_0 = 0.04$ T. The top panel in Fig. 2 represents the initial poloidal flux contours and velocity fields before the reflection process, which are an equilibrium with the initial axial velocity of the FRC. The FRC plasma travels towards the downstream region along the confining field and the front part of this arrives there at about $t = 10$ μsec. Then as the result of strong interaction with the magnetic mirror, this front part is compressed, not only axially but also radially until about $t = 20$ μsec. Also, the whole FRC shrinks axially by about 80 %. This axial contraction lasts until about $t = 25$ μsec, and then the separatrix radius expands in the vicinity of the central part of the FRC. At this time, excessive expansion of the separatrix radius does not occur regardless of significant volume compression since the FRC deeply creeps into the mirror region. Subsequently, a front part of the FRC has passed through the downstream region. After $t = 20$ μsec, a precursor plasma inside the separatrix ($z > 2.2$ m) passes through the downstream region, while the following plasma ($z < 2.2$ m) reflects from the mirror and then goes back towards the upstream region. The opposite direction of plasma flow can stretch out the magnetic field lines, and these field lines at axial position $z = 2.2$ m reconnect at about $t = 30$ μsec. Due to this magnetic reconnection, axial plasma flow is found in the end of the downstream region. This suggests that the magnetic mirror may play a role in triggering the magnetic reconnection in the case where the translation velocity is larger. Actually, in the experiment, considerable diamagnetic signals measured by magnetic probes have been observed behind mirror field when the FRC is reflected. After about $t = 35$ μsec, the FRC expands again towards the upstream region and leaves the downstream mirror. It is observed at about $t = 40$ μsec that a transverse wave propagates along an axial magnetic field in the FRC periphery. This observation implies that Alfven wave is induced within the FRC when the plasma is reflected by the magnetic mirror field. This reflection is inelastic so that the reflected FRC moves slower than the injected FRC. Consequently, the velocity of the reflected FRC is reduced to subsonic. The rebound coefficient e predicted from the average velocity inside the separatrix is calculated to be 0.56. This prediction form the MHD simulation is in agreement with the magnitude of the translation velocity observed in the FIX experiment which is reduced by 40 % to 70 % after the first reflection [3]. In this reflection process, the FRC collides with the magnetic mirror at supersonic velocity. The severe destruction of the FRC is not observed regardless of this fact, and the FRC still maintains most of its closed magnetic field configuration. Therefore, this suggests that the FRC plasma is robust against external perturbations.

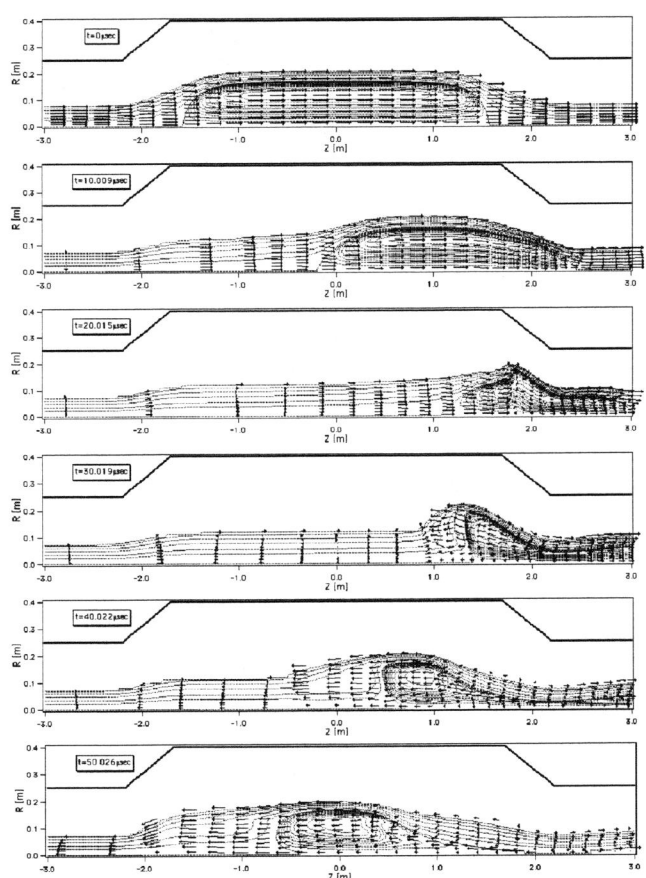

FIGURE 2. Time evolution of the poloidal flux contours and the velocity fields of the FRC for the parameter with $R_m = 3.0$ and $v_z = 1.5 \times 10^5$ m/sec. The dotted lines and the arrows represent flux contours and flow vectors, respectively.

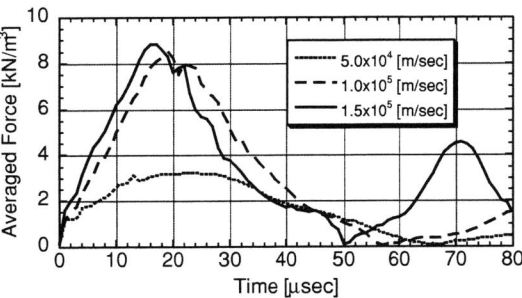

FIGURE 3. Time evolution of the spatially averaged force $<| j \times B - \nabla p |>$ for the translation velocities: 5.0×10^4, 1.0×10^5 and 1.5×10^5 m/sec. Here j, B and p denote the current density, magnetic field and plasma pressure, respectively.

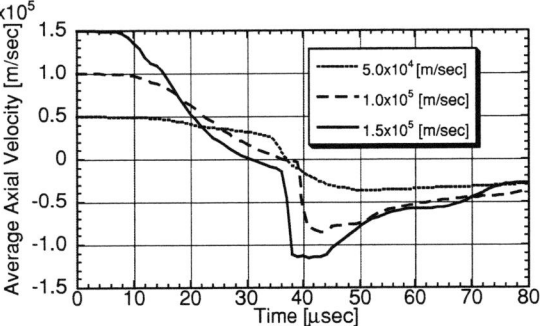

FIGURE 4. Time evolution of the average axial velocity on the midplane ($z = 0.0$ m) for the same parameters as Fig. 3.

Next, in order to examine the dependence of the motion of the FRC during reflection on the translation velocity, we plot the temporal evolution of the spatially averaged force $\triangleleft |\boldsymbol{j} \times \boldsymbol{B} - \nabla p| \triangleright$ in Fig. 3. It is illustrated for any translation velocities that the configuration is far out of MHD equilibrium condition of $\boldsymbol{j} \times \boldsymbol{B} = \nabla p$ in the reflection process. By about $t = 20$ μsec, there is a strong deceleration of the FRC by the magnetic mirror. After about $t = 20$ μsec the average force essentially approaches zero. In the translation case of supersonic velocity, $v_z = 1.5 \times 10^5$ m/sec, the average force has peaked value at about $t = 70$ μsec due to the effect of the second reflection at the upstream magnetic mirror. It is predicted for any translation velocities that the FRC eventually settles down in the confinement region and approaches MHD equilibrium condition.

Figure 4 shows the time evolution of the average axial velocity, $<v_z>$ on the midplane ($z = 0.0$ m) for the same parameters as Fig. 3. The value of $<v_z>$ is obtained by taking an average of axial flows on the midplane. As the translation velocity is increased, $<v_z>$ decreases rapidly at about $t = 40$ μsec. The dip of $<v_z>$ at about $t = 40$ μsec is attributed to the reflected plasma flows coming back to the midplane. This discontinuity of $<v_z>$ is generated in the case that the translation velocity exceeds the sound velocity at the confinement region. Therefore, this suggests that the formation of the discontinuous front is caused by a shock wave when the plasma is reflected by the magnetic mirror.

SUMMARY

In summary, we have investigated the dynamics of a reflection process on a FRC plasma by using an axisymmetric MHD simulation. As an initial state we use a MHD equilibrium configuration such as is typically observed in FIX experiment. We give an appropriate initial velocity for the equilibrium configuration. It is found from the simulation results that remarkably, in this reflection process, the FRC still maintains most of its closed magnetic field configuration even when injected with supersonic velocity into the magnetic mirror region, showing the robustness of the FRC against external perturbations. This is a significant feature of the FRC.

From the temporal evolution of the spatially averaged force $\triangleleft |\boldsymbol{j} \times \boldsymbol{B} - \nabla p| \triangleright$, it is predicted for any translation velocities that the FRC eventually settles down in the confinement region and approaches MHD equilibrium condition. Also, it is found from time evolution of the average axial velocity, $<v_z>$ on the midplane that the formation of the discontinuous front may be caused by a shock wave when the FRC with supersonic velocity is reflected by the magnetic mirror.

REFERENCES

1. Armstrong, W. T., Linford R. K., Lipson, J., Platts, D. A., and Sherwood, E. G., *Phys. Fluids* **24**, 2068-2088 (1981).
2. Rej, D. J., Armstrong, W. T., Chrien, R. E., Klingner, P. L., Linford R. K., McKenna, K. F., Sherwood, E. G., Siemon, R. E., and Tuszewski, M., *Phys. Fluids* **29**, 852-862 (1986).
3. Himura, H., Okada, S., Sugimoto, S., and Goto, S., *Phys. Plasmas* **2**, 191-197 (1995).
4. Himura, H., Ueoka, S., Hase, M., Yoshida, R., Okada, S., and Goto, S., *Phys. Plasmas* **5**, 4262-4270 (1998).
5. Brackbill, J. U., and Pracht, W. E., *J. Comput. Phys.* **13**, 455-482 (1973).
6. Milroy, R. D., and Brackbill, J. U., *Phys. Fluids* **25**, 775-783 (1982).
7. Braginskii, S. I., "Transport Processes in a Plasma," in *Review of Plasma Physics* **1**, edited by M. A. Leontovich, Consultants Bureau, New York, 1965, pp. 205-311.
8. Kanki, T., Suzuki, Y., Okada, S., and Goto, S., *Phys. Plasmas* **6**, 4672-4678 (1999).
9. Kanki, T., *IEEE Trans. Magn.* **38**, 1205-1208 (2002).
10. Suzuki, Y., Okada, S., and Goto, S., *Phys. Plasmas* **7**, 4062-4069 (2000).

Electron Heating In Thomson Scattering Measurements Of Plasma Temperature: Are Thermal Plasmas Thermal?

A. B. Murphy[a], J. Aubreton[b], and M. F. Elchinger[b]

[a] CSIRO *Telecommunications and Industrial Physics, P.O. Box 218, Lindfield NSW 2070, Australia*
[b] *SPCTS, University of Limoges, 123 av. Albert Thomas, 87060 Limoges cedex, France*

Abstract. Thomson scattering measurements of electron temperature have indicated that the electron temperature in thermal plasmas at 1 bar is up to 7000 K greater than the ion temperature. In these experiments, the heating of the electrons by the laser beam was taken into account by measuring their temperature as a function of laser pulse energy, and linearly extrapolating the results to zero pulse energy to obtain an unperturbed electron temperature. It is shown here that the absorption of laser energy by the electrons, and the specific heat and thermal conductivity of the plasma, are strongly dependent on the electron temperature, and therefore on the pulse energy. We use a single-fluid model of the plasma to show that the linear extrapolation is invalid. We calculate unperturbed electron temperatures in agreement with ion and spectroscopic temperatures.

INTRODUCTION

The question of whether thermal plasmas, such as welding arcs and plasma jets, at atmospheric pressure are in local thermodynamic equilibrium (LTE) has long been a subject of debate. While it has been demonstrated that deviations from LTE occur close to the electrodes and in the fringes of the plasma [1–3], the bulk of the theoretical and experimental evidence is that the remainder of the plasma is in LTE for electron densities above about 10^{23} m^{-3}, owing to the high particle collision rates. Thomson scattering measurements of electron and ion temperatures in thermal plasmas have however indicated that all regions of the plasma are far from LTE; in particular that the electron temperature is many thousands of kelvin higher than the ion temperature. Snyder *et al.* [4] measured an electron temperature of 20 900 ± 1700 K and an ion temperature of 14 200 ± 700 K at a position 2 mm below the cathode of a 100 A free-burning arc in argon. Bentley [5] repeated the electron temperature measurements, obtaining a temperature of 20 400 ± 500 K. Spectroscopic measurements of a similar arc yielded an excitation temperature of 16 600 K, in agreement with the ion temperature given by a laser-scattering technique [6]. Tanaka and Ushio [7] compared Thomson scattering measurements of electron and ion temperatures and spectroscopic measurements of excitation temperature of 50 and 150 A arcs in argon. They found that the electron temperature was about 5000 K higher than the ion and excitation temperatures. Measurements performed in an atmospheric-pressure plasma jet gave similar results [8].

Gregori *et al.* [9] and Snyder, Crawford and Fincke [10] have investigated the influence of the scattering angle and laser wavelength on the measured electron temperature. Gregori *et al.* [11] suggested that the standard method of deriving the electron temperature from the spectrum of the scattered signal should be replaced by a memory function method. This gave significantly lower electron temperatures, and may be an important factor to be considered in resolving the controversy.

LASER HEATING OF ELECTRONS

The measurement of electron temperature from the spectral profile of the Thomson scattered signal requires the use of a high-power pulsed laser such as a frequency-doubled Nd-YAG laser (wavelength = 532 nm). The interaction of the laser beam with the plasma heats the electrons by linear inverse bremsstrahlung [12]. To take this effect into account, electron temperature has been measured as a function of laser pulse energy. A straight line has been fitted to the results, and extrapolated to zero pulse energy to obtain the electron temperature free of influence from laser heating. The use of a straight-line fit has been justified [4,7,9] by reference to Hughes [12]. Hughes, however, gives an expression for the coefficient of absorption of laser light by a thermal plasma that depends strongly on the electron temperature T_e, the electron density n_e and the ion density n_i, both directly, and through the average Gaunt factor $\bar{g}$:

$$\alpha = \frac{n_e n_i Z^2 e^6 [1-\exp(-\hbar\omega/k_B T_e)]}{6\pi\varepsilon_0^3 c \hbar \omega^3 m_e^3} \left(\frac{m_e}{2\pi k_B T_e}\right)^{1/2} \frac{\pi}{3} \bar{g} . \tag{1}$$

The absorption will therefore vary with time during a laser pulse, and the absorbed energy will have a nonlinear dependence on the laser pulse energy. In Eq. (1), Z is the average ionisation level of the plasma, ω is the laser frequency, and the constants e, m_e, $\hbar$, k_B, ε_0, and c are respectively the electron charge and mass, Planck's constant, Boltzmann's constant, the permeability of free space, and the speed of light in vacuum.

FLUID DYNAMIC MODELLING

During the laser pulse, the electrons in a narrow cylindrical region (defined by the laser beam) are heated rapidly. These electrons are cooled by thermal conduction to the surrounding plasma. We can write a one-dimensional energy conservation equation in polar geometry to describe the heating and cooling of electrons due to a laser pulse of energy E_p and duration τ_p. In this equation, the plasma is treated as a single fluid, and the material properties of the plasma are calculated as a function of the electron and heavy particle temperatures.

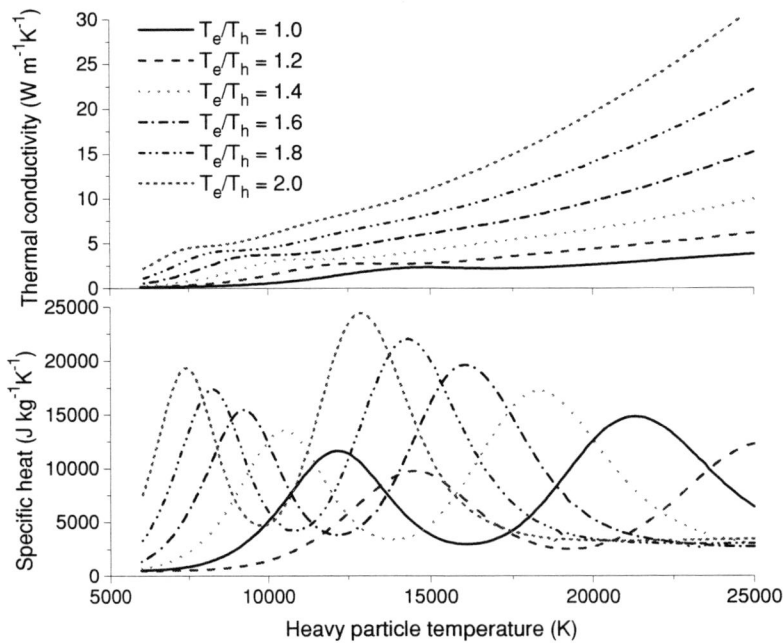

FIGURE 1. Dependence of thermal conductivity and specific heat on heavy particle temperature T_h, for different ratios of electron temperature and heavy particle temperature.

$$\frac{\partial \rho h}{\partial t} = \frac{1}{r}\left(\frac{rk}{c_p}\frac{\partial h}{\partial r}\right) + \frac{\alpha E_p}{At_p} - U , \tag{2}$$

where ρ, c_p, h, k and U are respectively the density, specific heat, enthalpy, thermal conductivity and radiative emission coefficient of the plasma. These are calculated as a function of T_e and the heavy particle temperature using standard methods [13] (we assume that all heavy species have the same temperature). Equation (2) is valid if the plasma composition can be described by a two-temperature Saha equilibrium. It can be shown, in fact, that the level of ionisation is lower, owing to the short duration of the laser pulse [14]. Nevertheless, results derived from Eq. (2) provide results very similar to those obtained taking this incomplete ionisation into account [14].

Figure 1 shows the dependence of the thermal conductivity and specific heat on the heavy particle temperature, for different ratios of the electron and heavy particle temperatures. It can be seen that, for a given heavy particle temperature, the specific heat and the thermal conductivity are strong functions of the electron temperature. The increase in electron temperature per unit absorbed energy and the rate of cooling by thermal conduction are hence both strongly dependent on electron temperature. Radiative cooling also increases with electron temperature.

We have solved Eq. (2) numerically, using the finite-difference method of Patankar [15]. We assumed the laser beam profile to be Gaussian, that the pulse shape was square in time, and that initially the ion and electron temperatures were equal. We also assumed that the ion temperature was constant during the pulse. This is consistent with Thomson scattering measurements, which show that the ion temperature is independent of laser pulse energy [4]. The laser beam diameter (FWHM) and pulse duration were chosen to be 200 μm and 7 ns, respectively, in accordance with the experimental parameters [4,5]. The calculation region extended over a radius of 350 μm, with an evenly-spaced 1 μm grid. The time step is 0.1 ns. Doubling the time and grid resolution, and the calculation region resulted in a less than 0.1% change in the calculated electron temperatures.

RESULTS

Figure 2 shows typical results for the evolution of the electron temperature during the laser pulse, calculated using Eq. (2). It is clear that the electron temperature increases more in the early stages of the laser pulse.

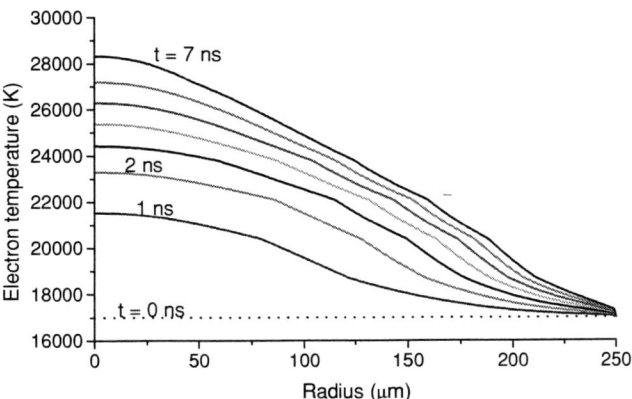

FIGURE 2. Evolution of electron temperature distribution during a 7 ns laser pulse. The radius of the laser beam is 100 mm, the laser pulse energy is 100 mJ, and the heavy particle temperature and initial electron temperature are 17 000 K

.As noted earlier, the electron temperature has a nonlinear dependence in the pulse energy as a result of the decrease in the absorption of laser energy, and the increased rate of cooling, as the electron temperature increases. The extent of the deviation from the linear relationship used to derive unperturbed electron temperatures in previous works is shown in Fig. 3. The figure compares least-squares linear fits to the measurements of Snyder *et al.* [4] and Bentley [5] with least-square fits to solutions of Eq. (2). These latter least-square fits were obtained using two free parameters; the initial electron temperature T_{e0}, which was set equal to the heavy particle temperature, and a constant C by which the absorption coefficient α was multiplied. We allow deviations from $C = 1$ to take into account uncertainties in the laser beam's spatial profile and diameter, in the time dependence of the laser pulse energy, and in the spatial and time averaging of the electron temperature

The least-square best fit to the measurements of Snyder *et al.* shown in Fig. 4(b) is obtained for $T_{e0} = 15\,900$ K. The ion temperature measured by Thomson scattering of $14\,200 \pm 700$ K [4]. The least-square best fit to the measurements of Bentley shown in Fig. 4(a) is obtained with $T_{e0} = 16\,800$ K. The excitation temperature, and the ion temperature measured by laser scattering, were around 16 600 K for the same conditions [5].

The values of T_{e0} calculated using best fits to the solution of Eq. (2) are between 3000 K and 5000 K lower than those obtained using a linear fit, and are within 1000 K of the ion temperature and excitation temperature. Since there are significant uncertainties in the transport and thermodynamic data used in the calculations, and since the results of calculations using lower values of T_{e0} fit the measured data almost equally well, we conclude that the results are consistent with values of electron temperature that are in fair agreement with ion and spectroscopic temperatures.

This problem has also been analysed using a two-fluid approach in which the electrons are treated separately to the heavy particles. The results obtained are very similar [14].

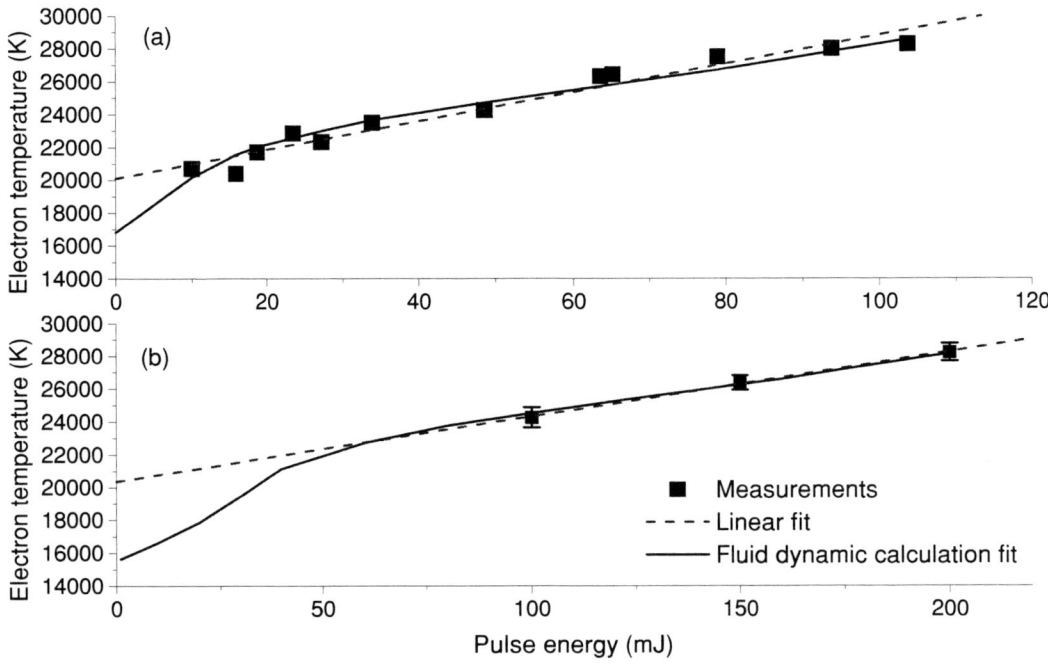

FIGURE 3. Comparison between a least-squares fit to the fluid dynamic calculation and a linear fit to the measurements of (a) Bentley [5] and (b) Snyder et al. [4].

CONCLUSIONS

We have shown that the method of deriving an unperturbed electron temperature by linearly extrapolating measurements of electron temperature as a function of laser pulse energy is physically invalid. The electron-temperature dependence of the absorption of laser energy, of the cooling of the heated electrons by thermal conduction and radiative emission, and of the specific heat of the plasma, must all be considered. We have used a one-dimensional single-fluid model of the electron heating and cooling to show that these effects are substantial, and that electron temperatures are 3000 to 5000 K lower than given by the linear extrapolation, and are comparable with ion and excitation temperatures. A two-fluid model, in which electrons are treated separately from the heavy particles, gives similar results. We conclude that Thomson-scattering measurements of electron temperature in thermal plasmas do not provide reliable evidence for significant deviations from LTE.

REFERENCES

1. Haidar, J., *J. Phys. D* **32**, 263-272 (1999).
2. Cram, L. E., Poladian, L., and Roumeliotis G., *J. Phys. D* **21**, 418-425 (1988).
3. Snyder, S. C., Murphy, A. B., Hofeldt, D. L., and Reynolds, L. D., *Phys. Rev. E* **52**, 2999-3009 (1995).
4. Snyder, S. C., Lassahn, G. D., and Reynolds, L. D., *Phys. Rev. E* **48**, 4124-4127 (1993).
5. Bentley, R. E., *J. Phys. D* **30**, 2880-2886 (1997).
6. Murphy, A. B., Farmer, A. J. D., and Haidar, J., *Appl. Phys. Lett.* **60**, 1304-1306 (1992).
7. Tanaka, M., and Ushio, M., *J. Phys. D* **32**, 1153-1162 (1999).
8. Snyder, S. C., Reynolds, L. D., Fincke, J. R., Lassahn, G. D., Grandy, J. D., and Repetti, T. E., *Phys. Rev. E* **50**, 519-525 (1994).

9. Gregori, G., Schein, J., Schwendinger, P., Kortshagen, U., Heberlein, J., and Pfender, E., *Phys. Rev. E* **59**, 2268-2291 (1999).
10. Snyder, S. C., Crawford, D. M., and Fincke, J. R., *Phys. Rev. E* **61**, 1920-1924 (2000).
11. Gregori, G., Kortshagen, U., Heberlein, J., and Pfender, E., *Phys. Rev. E* **65**, 046411 (2002).
12. Hughes, T. P., *Plasmas and Laser Light*, Adam Hilger, Bristol, 1975.
13. Andre, P., Aubreton, J., Elchinger, M. F., Fauchais, P., and Lefort, A., *Plasma Phys. Plasma Process.* **21**, 83-105 (2001).
14. Murphy, A. B., *Phys. Rev. Lett.* **89**, 025002 (2002).
15. Patankar, S. V., *Numerical Heat Transfer and Fluid Flow*, Hemisphere, Washington DC, 1980.

Magnetic field generation and electrostatic shock wave formation driven by counter-streaming pair plasmas

T. Haruki* and J. I. Sakai*

Laboratory for Plasma Astrophysics, Faculty of Engineering, Toyama University, Toyama 930-8555, Japan

Abstract. By using two-Dimensional (2-D) and 3-D fully electromagnetic and relativistic Particle-In-Cell (PIC) codes, we investigated a collision process of the counter-streaming electron-positron pair plasmas. The collisionless counter-streaming plasmas become unstable against the counter-streaming instability similar to Weibel instability. This instability plays important role in conversion process from particle flow energy to magnetic field energy. We found two kinds of instability in counter-streaming plasmas. The magnetic fields are generated by the electromagnetic counter-streaming instability and the electric fields are formed by the electrostatic two-stream instability. High-energy particles are also produced due to the generated electric fields. In astrophysical plasmas, Gamma-Ray Burst (GRB) can release huge energy over a few seconds in a small volume. It is widely believed that GRB afterglows are observed by synchrotron emission. However, this emission mechanism must require strong magnetic fields in the background. Our simulation results are useful for understanding synchrotron GRB model.

INTRODUCTION

Gamma-Ray Burst (GRB) is the most luminous object in the universe which releases $10^{51} - 10^{53}$ ergs over a few seconds in a small volume for isotropic emission (See reviews by Piran [1, 2]). Recently, Frail et al. [3] suggested that the release energy of GRB is about 5×10^{50} ergs with the conical jet model. The mechanism of GRB is thought to be the relativistic internal-external shocks fireball model. A compact source may produce a variable relativistic wind, which forms relativistic shocks. The relativistic internal shock produces the GRB. The two relativistic external shocks are formed interacting with the surrounding circumstellar matter like Inter-Stellar Medium (ISM). The forward shock produces a gamma-ray or X-ray signal and continues an afterglow at the late. The reverse shock produces a prompt optical flash and early radio emission. The GRB afterglows are believed to be observed in the synchrotron emission of relativistic plasmas accelerated by shock waves. The synchrotron emission must always require the background magnetic field.

The generation mechanism of magnetic field has been studied under the motivation of GRB synchrotron emission [4, 5] and the origin of magnetic field in the universe. The magnetic field generation is explained by the collisionless electromagnetic counter-streaming instability similar to Weibel instability [6]. Plasma flows in the one direction show effective high temperature. Such an anisotropic particle velocity distribution tends to be Maxwellian involving generation of the electromagnetic fluctuations. Califano et al. [7, 8] framed a theory of the counter-streaming instability by describing the dynamics of the two electron counter-streaming populations in the fluid approximation. On the other hand, Kazimura et al. [9, 10] got the growth rate of this instability by numerical calculation and investigated the magnetic field generation by using 2-D PIC simulation.

In this paper we investigate the counter-streaming instability by both 2-D and 3-D PIC simulations. Most results are from a 2-D simulation, because both simulations show almost the same results. The magnetic fields with the scale of skin depth size are formed through the counter-streaming instability, coalescence each other and eventually change into larger units. For a 3-D simulation results, there appears a lattice structure of magnetic field associated with two counter current filaments and two X-points due to the limitation of small system size perpendicular to plasma flow. Magnetic fields are generated once, but they are dissipated due to particle thermalization. We also found the electrostatic waves through the two-stream instability. These waves produce high-energy particles faster than the initial plasma flow. We concluded that the magnetic fields are generated behind the electrostatic shock waves driven by counter-streaming pair plasmas.

SIMULATION MODEL

The code used here is a 2-D relativistic fully electromagnetic PIC code modified from 3-D code [11]. The lengths of system in two dimensions are $L_x = 4000\Delta$ and $L_y = 64\Delta$, respectively, where $\Delta(=1)$ is the grid size. The periodic boundary conditions are imposed on fields and particles in both the x- and y-directions. We consider electron-positron plasmas with equal mass and opposite charge. The total particle number is 15,360,000 electron-positron pairs, where the number density is 60 per cell. There are no magnetic and electric fields at the initial condition. Therefore, charge neutrality is also satisfied. In order to set counter-streaming plasma configuration, we divided all particles into two components and imposed them drift velocity. Both electrons and positrons are given by drift velocity, $v_d/c = +0.5$, in the left side region ($x \leq L_x/2$) and $v_d/c = -0.5$, in the right side region ($x \geq L_x/2$), where c is light speed. The other parameters as follows: the simulation time step, $\omega_{pe}\Delta t = 0.05$, where ω_{pe} is electron plasma frequency; Debye length, $\lambda_D = v_{te}/\omega_{pe} = 1\Delta$; skin depth, $c/\omega_{pe} = 10\Delta$; the ratio of electron thermal velocity to light speed, $v_{te}/c = 0.1$; the ratio of positron temperature to electron temperature, $T_p/T_e = 1$.

SIMULATION RESULTS

The collisionless counter-streaming plasmas become unstable against the counter-streaming instability. The magnetic fields with the scale of skin depth are generated dominantly perpendicular to plasma streaming. The small-scale magnetic fields merge, evolve by repeating coalescence and change into larger units. Figure 1 shows the time development of magnetic field component, B_z, in the X-Y plane at (a) $\omega_{pe}t=50$, (b) 100, (c) 150 and (d) 200, where the maximum intensity, 0.4, in this figure corresponds to $\omega_{ce}/\omega_{pe}=0.41$. Fig. 1 (a) shows that there appear magnetic fields through counter-streaming instability near the contact surface, $X = 2000\Delta$. As seen in Figs. 1 (a)-(d), magnetic fields propagate along the x-direction whose speed is about $0.4v_d$ by repeating coalescence. We found that the magnetic fields are generated once, but they are dissipated later near the center of system. Here, we could find two regions: the magnetic field generation and dissipation regions. In the magnetic field generation region, we estimated that the Larmor radius is $\rho_e \sim 2.5\Delta$ by using the initial thermal velocity and the magnetic field maximum intensity. We concluded that plasmas in the generation region are magnetized because Larmor radius is smaller than magnetic field scale. The magnetic field required in GRB synchrotron emission is confirmed. On the other hand, in the magnetic field dissipation region at the center of system, the mixing of counter-streaming plasma can lead to thermalization behind the generation region. Plasmas are heated to approximately 4 times the initial thermal velocity, which corresponds to 16 times the initial temperature. We estimated that Larmor radius is much larger than skin depth size. Therefore, plasmas are unmagnetized in this region. This thermalization is also expected to be behind shock in GRB.

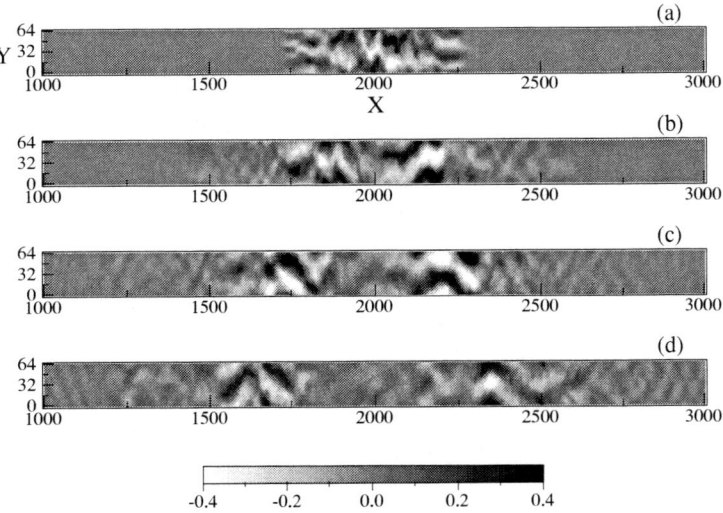

FIGURE 1. The time development of magnetic field component, B_z, in the X-Y plane at (a) $\omega_{pe}t=50$, (b) 100, (c) 150 and (d) 200.

Figure 2 shows the time development of electric field component, E_x, in the X-Y plane at (a) $\omega_{pe}t=50$, (b) 100, (c) 150 and (d) 200. The electric fields are generated by the electrostatic two-stream instability. The generated electric fields are known as Langmuir waves. In fact, the electron phase snapshot (v_x vs. x) displays several electron holes. These fields propagate oblique to the plasma flow, whose angle is 45 degree and speed is about v_d. We found also that they can accelerate a part of particles to about $0.8c$.

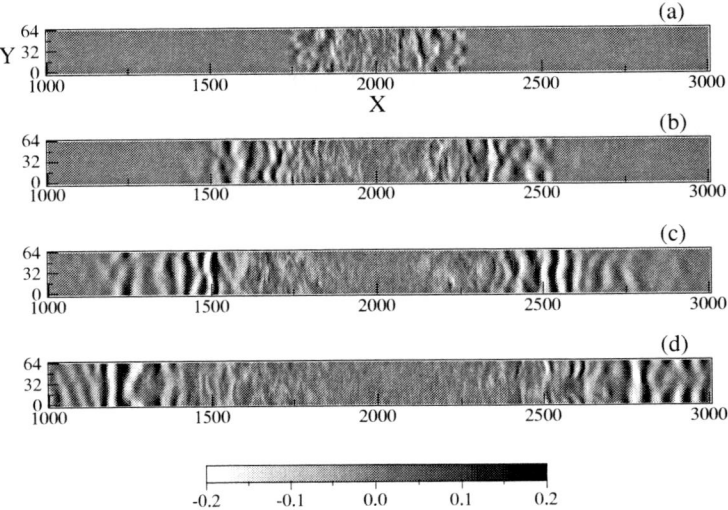

FIGURE 2. The time development of electric field component, E_x, in the X-Y plane at (a) $\omega_{pe}t=50$, (b) 100, (c) 150 and (d) 200.

Figure 3 shows the time histories of (a) magnetic field energy, B_z^2, and (b) electric field energy, E_x^2, in the system, where these energies are normalized by the initial electron relativistic kinetic energy. The other components represent noise level, except of the electric field, E_y, associated with oblique propagation. First, Fig. 3 (a) shows that magnetic field energy increases gradually in spite of no magnetic field in the initial state. Maximum conversion rate is 2.5% of the electron flow energy at $\omega_{pe}t = 150$. We found that the growth rate of counter-streaming instability obtained by simulation result is $\gamma/\omega_{pe} = 0.58$. Theoretical growth rate is $\gamma/\omega_{pe} \sim 0.46$ with propagation angle, 76 degree. We concluded that almost same orders are obtained by both simulation and theoretical approach. Next, from Fig. 3 (b), electric field energy also increases gradually, whose maximum conversion rate is just below 1.0%. The growth rate of two-stream instability represents $\gamma/\omega_{pe} = 0.42$ in our simulation. Theoretical maximum growth rate is $\gamma/\omega_{pe} = 0.50$ for complete parallel propagation. Therefore, we can get almost the same order values. For our simulation, we found both electromagnetic and electrostatic instabilities occur at the same time and their growth rates obtained by our simulation agree well with the theoretical values.

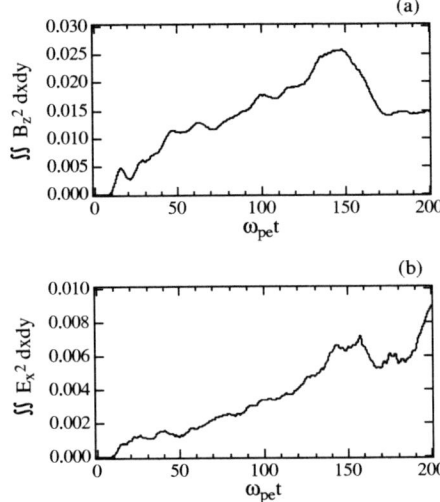

FIGURE 3. The time histories of (a) magnetic field energy, B_z^2, and (b) electric field energy, E_x^2, in the system, where these energies are normalized by the initial electron relativistic kinetic energy

CONCLUSIONS

By using both 2-D and 3-D fully electromagnetic and relativistic PIC (Particle-In-Cell) codes, we investigated a collision process of the counter-streaming pair plasmas. Both cases show almost the same results. We found both of the electromagnetic counter-streaming instability and the electrostatic two-stream instability at the same time in our simulations. The counter-streaming instability leads to generation of magnetic field with the scale of skin depth size, which merge by repeating coalescence and eventually change into larger units. The magnetic fields are generated once, but they are dissipated due to particle thermalization backward. On the other hand, the electric fields are formed through two-stream instability. The generated electric fields are Langmuir waves propagating oblique to plasma flow, which can accelerate a part of particles with drift velocity $0.5c$ to about $0.8c$. We concluded that the growth rate of the counter-streaming instability is just faster than that of two-stream instability. It is clear that the magnetic fields are generated behind the electrostatic shock waves driven by counter-streaming pair plasmas. The particles are also heated behind the magnetic field dissipation region.

ACKNOWLEDGMENTS

T. Haruki is supported by JSPS (Japan Society for the promotion of Science). Sakai is also supported by a Grand-in-Aid for Scientific Research from Japan Ministry of Education (11695028).

REFERENCES

1. Piran, T. *Phys. Rep.*, **314**, 57 (1999)
2. Piran, T. *Phys. Rep.*, **333**, 529 (2000)
3. Frail, D. A., *et al. Ap. J. Letters*, **562**, L55 (2000)
4. Medvedev, M. V., and Loeb, A. *Ap. J. Letters*, **26**, 697 (1999)
5. Gruzinov, A. *Ap. J. Letters*, **563**, L15 (2001)
6. Weibel, E. W. *Phys. Rev. Lett.*, **2**, 83 (1959)
7. Califano, F., Pegoraro, F., Bulanov, S. V., and Mangeney, A. *Phys. Rev. E*, **57**, 7048 (1998)
8. Califano, F., Prandi, R., Pegoraro, F., and Bulanov, S. V. *Phys. Rev. E*, **58**, 7837 (1998)
9. Kazimura, Y., Califano, F., Sakai, J. I., Neubert, T., Pegoraro, F., and Bulanov, S. *J. Phys. Soc. Jpn.*, **67**, 1079 (1998)
10. Kazimura, Y., Sakai, J. I., Neubert, T., and Bulanov, S. V. *Ap. J. Letters*, **498**, L183 (1998)
11. Buneman, O., *Computer Space Plasma Physics, Simulation Techniques and Software*, edited by Matsumoto, H., and Omura, Y., Terra Scientific, Tokyo, 1993, p. 67.

Extreme Plasmas Near Pulsars and Strange Stars

Vladimir V. Usov

Department of Condensed Matter Physics, Weizmann Institute, Rehovot 76100, Israel

Abstract. Pulsars are fast rotating neutron stars with extremely strong magnetic fields ($\sim 10^9 - 10^{13}$ G) at their surface. Current models for the generation of coherent radio emission of pulsars require formation of electron-positron pairs in pulsar magnetospheres. Such pair plasma may be created via conversion of γ-rays into electron-positron pairs in a strong magnetic field. Generation of γ-rays and creation of pair plasma in the magnetospheres of pulsars are discussed. Plasma instabilities in the magnetized pair plasma are reviewed.

Bare strange stars, which are entirely made of deconfined quarks, may be extremely powerful sources of pair plasma. In this case, electron-positron pairs are created in a thin layer with a very strong electric field at the quark surface. The luminosity of a bare strange star in pairs is a function of the surface temperature, and it may be as high as $\sim 10^{51} - 10^{52}$ ergs s^{-1} at the moment of the strange star formation when the surface temperature may be $\sim$ a few $\times 10^{10}$ K or even higher. Creation of electron-positron pairs at the surface of bare strange stars and physical processes in the outflowing pair plasma are discussed.

INTRODUCTION

There is now compelling evidence that plasma composed of electrons and positrons is ejected from many astronomical objects and flows away at relativistic speeds. The Lorentz factor of such plasma wind is ~ 10 for the jets associated with active galactic nuclei [1], $\sim 10^2 - 10^3$ for gamma-ray bursters [2], and $\sim 10^6 - 10^7$ or even more for pulsars [3]. Below, we discuss creation of electron-positron plasma and its properties in the vicinities of pulsars and bare strange-quark-matter stars.

PLASMA PROCESSES IN PULSAR MAGNETOSPHERES

A common point of all acceptable models of pulsars is that the spin-down power of strongly magnetized neutron stars is the energy source of non-thermal emission of pulsars. The rotational energy of the neutron star may be transformed into the pulsar emission by the following sequence of processes [3-6].

Strong electric fields are generated in the magnetospheres of rotating magnetized neutron stars. The component of the electric field $\mathbf{E}_\parallel$ along the magnetic field $\mathbf{B}$ is non-zero. Primary particles (electrons or positrons, depending on the direction of the electric field $\mathbf{E}_\parallel$) are accelerated by this electric field to ultrarelativistic energies (up to $\sim 10^7 m_e c^2$). In the pulsar frame the density of primary particles is about

$$n_{\rm cr} = |\mathbf{\Omega} \cdot \mathbf{B}|/(2\pi c e) \simeq 10^{11}(R/r)^3 \text{ cm}^{-3}, \tag{1}$$

where Ω is the angular velocity of the pulsar rotation and $R \simeq 10^6$ cm is the radius of the neutron star. We have adopted here the following characteristic values for the pulsar parameters: $\Omega \simeq 10$ s^{-1} and $B \simeq 10^{12}(R/r)^3$ G is the magnetic field in the magnetosphere.

Ultrarelativistic primary particles in the process of their outflow generate γ-rays that are absorbed in a strong magnetic field and create electron-positron pairs, $\gamma + B \to e^+ + e^- + B$. The created pairs screen the electric field $\mathbf{E}_\parallel$ in the pulsar magnetosphere everywhere except for compact regions. The compact regions where $\mathbf{E}_\parallel$ is unscreened are called gaps. These gaps are "engines" that are responsible for the non-thermal radiation of pulsars.

For the outflowing pair plasma the typical values of the Lorentz factor and the density are, respectively,

$$\Gamma_\pm \simeq 10 - 10^3, \qquad n_\pm \simeq 10 - 10^5 n_{\rm cr}. \tag{2}$$

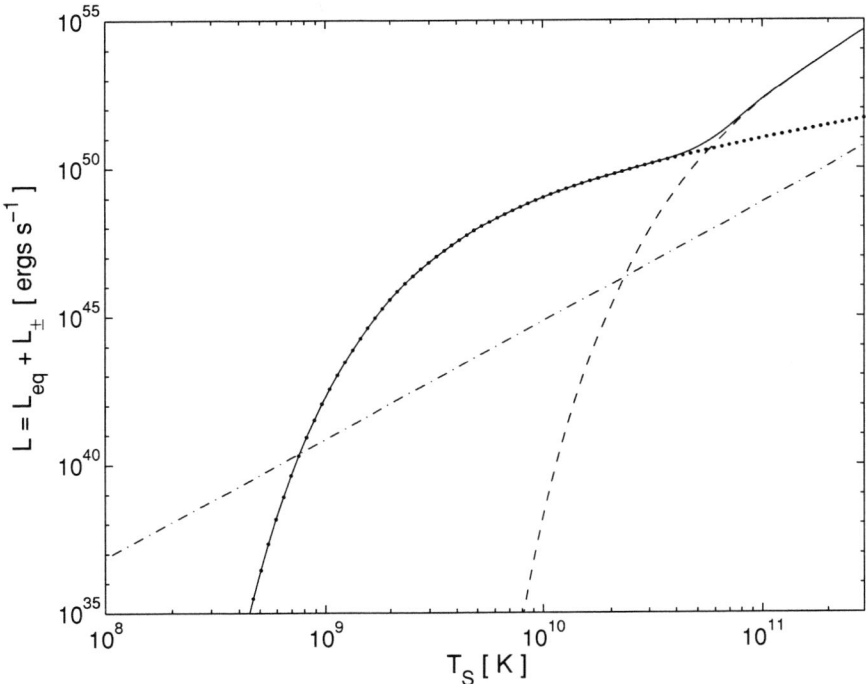

FIGURE 1. The total luminosity of a bare strange star $L = L_{\text{eq}} + L_{\pm}$ (solid line), where L_{eq} and $L_{\pm}$ are the luminosities in thermal equilibrium photons (dashed line) and electron-positron pairs (dotted line), respectively. The upper limit on the luminosity in non-equilibrium photons, $L_{\text{neq}} < 10^{-4} 4\pi R^2 F_{\text{BB}}$, is shown by the dot-dashed line.

In the vicinities of pulsars electrons and positrons very quickly lose the momentum component transverse to the magnetic field due to synchrotron losses, and their distribution becomes one-dimensional. Plasma instabilities developed in this one-dimensional, relativistic, multi-component plasma of electrons and positrons are responsible for the generation of coherent radio emission of pulsars.

Development of instabilities in a magnetized plasma when transverse electromagnetic waves (t-waves) are generated depends on the ratio of the Langmuir and cyclotron frequencies. Near the pulsar surface ($r \sim R$) this ratio is very small ($\omega_p/\omega_B \sim 10^{-7}$), and therefore, these instabilities cannot be developed. The ratio ω_p/ω_B increases with increase of the distance r from the pulsar, $\omega_p/\omega_B \propto r^{3/2}$. Near the light cylinders of pulsars ($r \sim c/\Omega$) the ratio ω_p/ω_B may be as high as $10^{-2} - 10^{-1}$, and the outflowing plasma may be unstable with respect to excitation of cyclotron t-waves [7-9]. For typical pulsars the frequency of these waves is in the radio range, from $\sim 10^2$ MHz to a few $\times 10^3$ MHz [7]. The model based on the cyclotron instability also can explain many data on both coherent radio emission of pulsars [8,9] and their high-frequency emission [10].

However, observations of pulsars suggest that the radio emission regions are far inside the light cylinder [12],

$$r_{\text{radio}} \simeq 50R \simeq (0.02 - 0.04) c/\Omega \ll c/\Omega, \tag{3}$$

In this case, the model of coherent radio emission of pulsars that is based on the development of the two-stream instability in their magnetospheres has no an alternative. The two-stream instability that develops due to strong nonhomogeneity of the outflowing plasma is the most plausible reason for the generation of extremely intense radio emission of pulsars [13-15]. Longitudinal (nonescaping) Langmuir waves (l-waves) that are generated in the process of development of this instability may be converted by means of different non-linear effects into t-waves that can escape from the magnetospheres of pulsars [16-19]. The conversion of l-waves into t-waves is poorly known. It is a "bottle-neck" that impedes the meeting of the model of radio emission of pulsars and observational data.

PAIR WINDS FROM HOT BARE STRANGE STARS

Strange stars that are entirely made of deconfined quarks have been long ago proposed as an alternative to neutron stars. The bulk properties (size, moment of inertia, etc.) of strange and neutron stars in the observed mass range ($1 < M/M_\odot < 2$) are rather similar, and it is very difficult to discriminate between strange and neutron stars.

The possible existence of strange stars is a direct consequence of the conjecture by Witten [20] that strange quark matter (SQM) composed of roughly equal numbers of up, down, and strange quarks plus a smaller numbers of electrons (to neutralize the electric charge of the quarks) may be the absolute ground state of the strong interaction, i.e., absolutely stable with respect to ^{56}Fe. If Witten's idea is true, SQM with the density of $\sim 5 \times 10^{14}$ g cm^{-3} might exist up to the surface of strange stars [21]. This would open a possibility to observe super-dense quark matter.

Since the plasma frequency of SQM is very high ($\hbar\omega_p \simeq 20$ MeV) a bare strange star is a very poor emitter of thermal photons with energies $\epsilon_\gamma < 20$ MeV [21,22]. Recently, it was shown [22,23] that the bare surface of a hot strange star may be a very powerful source of electron-positron pairs which are created in an extremely strong electric field at the quark surface. Figure 1 shows the thermal luminosity of a bare strange star in both electron-positron pairs and photons as a function of the surface temperature T_s.

Recently, the kinetics of electron-positron winds and the terminal emission radiated from the wind photospheres were studied [24]. Annihilation of pairs into two and three photons, creation of pairs by photons, bremsstrahlung radiation of pairs and scattering of particles and photons were taken into account. We have shown that at very high luminosities, $L > 10^{41}$ ergs s^{-1}, the photon spectrum of the terminal emission is nearly blackbody. At $L < 10^{41}$ ergs s^{-1}, the photon spectrum differs considerably from the blackbody spectrum, and its hardness increases when L decreases. This is because photons that form in annihilation of electron-positron pairs do not have enough time for thermalization before they escape from the star's vicinity. When the total luminosity decreases from $\sim 10^{41}$ ergs s^{-1} to $\sim 10^{36}$ ergs s^{-1}, the mean energy of photons increases from ~ 30 keV to ~ 500 keV while the spectrum of photons changes eventually into a very wide ($\Delta E/E \sim 0.3$) annihilation line of energy $E \sim 500$ keV [24,25].

ACKNOWLEDGMENTS

This research was supported by the MINERVA Foundation, Munich, Germany and by the Israel Science Foundation of the Israel Academy of Sciences and Humanities.

REFERENCES

1. Levinson, A., and Blandford, R. Astrophys. J. Letters, **456**, L29, 1996.
2. Baring, M.G., and Harding, A.K. Astrophys. J., **491**, 663, 1997.
3. Michel, F.C. Theory of Neutron Star Magnetospheres, Univ. of Chicago Press, 1991.
4. Ruderman, M.A., and Sutherland P.G. Astrophys. J, **196**, 51, 1975.
5. Arons, J. In Proc. IAU Symp. 95, Pulsars, eds. W. Sieber and R. Wielebinski, Reidel, Dordrecht, 69, 1981.
6. Usov, V.V. In Proc. Pulsars: Problems & Progress, ASP Conf. Ser. 105, eds. S. Johnston, M.A. Walker & M. Bailes, 323, 1996.
7. Machabeli, G.Z., and Usov, V.V. Soviet Astron. Letters, **5**, 238, 1979.
8. Machabeli, G.Z., and Usov, V.V. Soviet Astron. Letters, **15**, 393, 1989.
9. Lyutikov, M., Machabeli, G.Z., and Blandford, R.D., Mon. Not. R. Astron. Soc., **305**, 338, 1999.
10. Lominadze, J.G., Machabeli, G.Z., and Usov, V.V. Astrophys. Sp. Sci., **90**, 19, 1983.
11. Malov, I.F., and Machabeli, G.Z. Astrophys. J., **554**, 587, 2001.
12. Kijak, J., and Gil, J.A. Mon. Not. R. Astron. Soc., **288**, 631, 1997.
13. Usov, V.V. Astrophys. J., **320**, 333, 1987.
14. Ursov, V.N., and Usov, V.V. Astrophys. Sp. Sci., **140**, 325, 1988.
15. Asseo, E., and Melikidze, G.I. Mon. Not. R. Astron. Soc., 301, 59, 1998.
16. Melikidze, G.I., and Pataraya, A.D. Astrophizika, **20**, 157, 1984.
17. Lesch, H., Gil J.A., and Shukla, P.K. Space Sci. Rev., **68**, 349, 1994.
18. Lyubarsky, Yu.E. Astron. Astrophys., **308**, 809, 1996.
19. Melikidze, G.I., Gil, J.A., and Pataraya, A.D. Astrophys. J., **544**, 1081, 2000.
20. Witten, E. Phys. Rev. D, **30**, 272, 1984.
21. Alcock, C., Farhi, E., and Olinto, A., Astrophys. J., **310**, 261, 1986.
22. Usov, V.V. Astrophys. J. Letters, **550**, L179, 2001.
23. Usov, V.V. Phys. Rev. Letters, **80**, 230, 1998.
24. Aksenov, A., Milgrom, M., and Usov, V.V. Astrophys. J., in preparation.
25. Page, D., and Usov, V.V. Phys. Rev. Letters, submitted.

Driven Reconnection Controlled By Particle Dynamics In A Collisionless Open System

R. Horiuchi[1], W. B. Pei[2], and T. Sato[3]

[1]*National Institute for Fusion Science, Toki, Gifu 509-5292, Japan*
[2]*Institute of Applied Physics and Computational Mathematics, Beijing 100088, China*
[3]*Earth Simulator Center, Kanazawa-ku, Yokohama 236-0001, Japan*

Abstract. Long time scale evolution of collisionless driven reconnection in an open system is investigated by using newly developed electromagnetic particle simulation codes. It is found that there are two evolving regimes in the temporal behavior of collisionless reconnection, dependently on the spatial size of plasma inflow through the upstream boundary (window size), i.e., a steady regime and an intermittent regime. The system evolves toward a steady regime for a narrow input window, in which collisionless reconnection is controlled by ion dynamics. On the other hand, an intermittent behavior appears due to the frequent formation of magnetic islands for a wide input window, in which an electron dynamics plays an important role.

OPEN BOUNDARY MODEL

Collisionless magnetic reconnection plays a crucial role in a number of interesting phenomena with fast magnetic energy release, plasma acceleration and heating both in space plasmas [1] and in laboratory plasmas [2]. Magnetic reconnection is often discussed with steady models. On the other hand, long time scale MHD simulations have demonstrated that magnetic reconnection takes place intermittently in the presence of a constant energy supply from the exterior region[3]. Which situation is realized in collisionless reconnection? When and under what condition intermittent reconnection is triggered? It is believed that steady reconnection is realized when the flux input rate into the system is balanced with the reconnection rate. However, this balance condition is not always assured because magnetic reconnection is controlled by two different processes with different time scales, i.e., an external global process and an internal microscopic process.

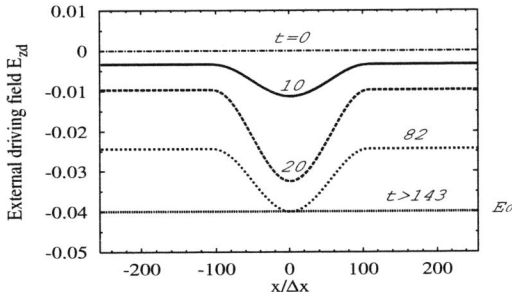

FIGURE 1. Spatial profiles of the driving electric field imposed at the upstream boundary at five different time periods for the case of $x_d = 0.84\, x_b$ and $E_0 = -0.04\, B_0$, where the time unit is ω_{ce}^{-1}.

In order to study long time scale behavior of collisionless driven reconnection we developed a new open boundary model [4,5,6] in which a free condition is used at the downstream boundary ($x = \pm x_b$) and an input condition is used at the upstream boundary ($y = \pm y_b$). The plasma inflows are symmetrically driven from two upstream boundaries by the external electric field imposed in the z direction. Figure 1 shows the spatial profiles of the driving electric field imposed at the upstream boundary at five different time periods. The amplitude of driving field $E_{zd}(x,t)$ is set for zero at $t=0$, and increases with time while keeping a bell-shaped profile near the center and a flat profile in the periphery for an initial short moment. After then a constant profile is kept with maximum flux input rate E_0. We call the spatial size of initial bell-shaped profile x_d input window size, because the inflow velocity is locally enhanced within this region.

As an initial condition we adopt a one-dimensional equilibrium with the Harris-type anti-parallel magnetic configuration. The two-dimensional electromagnetic particle simulation is carried out on a 512×128 point grid by making use of 6.4 million particles. The main parameters are in the followings: $m_i/m_e=25$, time step width $\omega_{ce}\,\Delta t=$ 0.02, initial scale height $L= 0.8\, y_b \sim 3\, \rho_i$, where ρ_i is the ion Larmor radius.

SIMULATION RESULTS

We have examined the dependence of collisionless driven reconnection on the input window size x_d and the flux input rate E_0, and found that the long time scale behavior is strongly dependent on the value of x_d, but insensitive to the value of E_0. In the following subsections we discuss the simulation results in two cases, i.e., the narrow input case ($E_0=-0.04\, B_0$, $x_d = 18\, \rho_i$) and the wide input case ($E_0=-0.04\, B_0$, $x_d = 36\, \rho_i$).

Steady Reconnection

First, let us consider the case of narrow input window ($x_d = 18\, \rho_i$) [4,5,6]. Figure 2 shows the perspective view of spatiotemporal structure of off-plane electric field E_z in the t-y plane where the reconnection point is located in the mid y-axis. After experiencing the initial transient phase, the system relaxes into a steady state in which the off-plane electric field becomes uniform in space and constant in time, and thus must be equal to the external driving field E_0 at the upstream boundary. In other words, the reconnection rate is balanced with the flux input rate at the upstream boundary in the steady state.

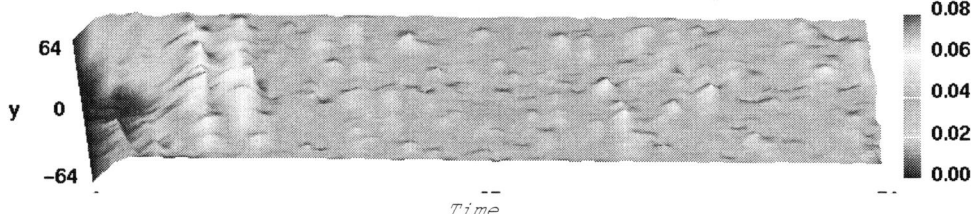

FIGURE 2. Spatiotemporal structure of off-plane electric field.

The dissipation region has a two-scale structure underlying the quite different characteristic scale lengths of electron and ion dynamics. The ion motion decouples from the magnetic field due to the inertia effect within a region of $|y|<c/\omega_{pi}$, while the electrons remain frozen in the magnetic field until they enter a region of c/ω_{pe} which is slight larger than an electron thermal scale [6]. As is shown in Fig. 3(a), however, the current density profile has the same scale as the mass density profile, which is always kept equal to the ion thermal scale l_{mi}. Here, the thermal scale of a charged particle in the vicinity of neutral sheet is given by the average orbit amplitude of meandering motion [7]. That is, the spatial structure of the current sheet in the steady state is exclusively controlled by the dynamics of meandering ions.

The meandering ions bounce back and forth across the neutral sheet in y direction with the average amplitude l_{mi}. An electrostatic field is generated in the inflow direction through the finite Larmor radius effect in the region of $|y| < l_{mi}$, i.e., $\boldsymbol{E}_{st}\cdot \boldsymbol{u}>0$ in the upstream region of the current sheet. Because the meandering ions are unmagnetized and the electrostatic field dominates over the restoring force of ion bounce motion ($|\boldsymbol{E}_{st}| > |\boldsymbol{u}\times\boldsymbol{B}|$), it greatly intensifies the bounce frequency and energy of meandering ions. Thus, anomalous ion heating takes place in the meandering ion scale at the neutral sheet under the action of an electrostatic field, as is shown in Fig. 3(b)[6].

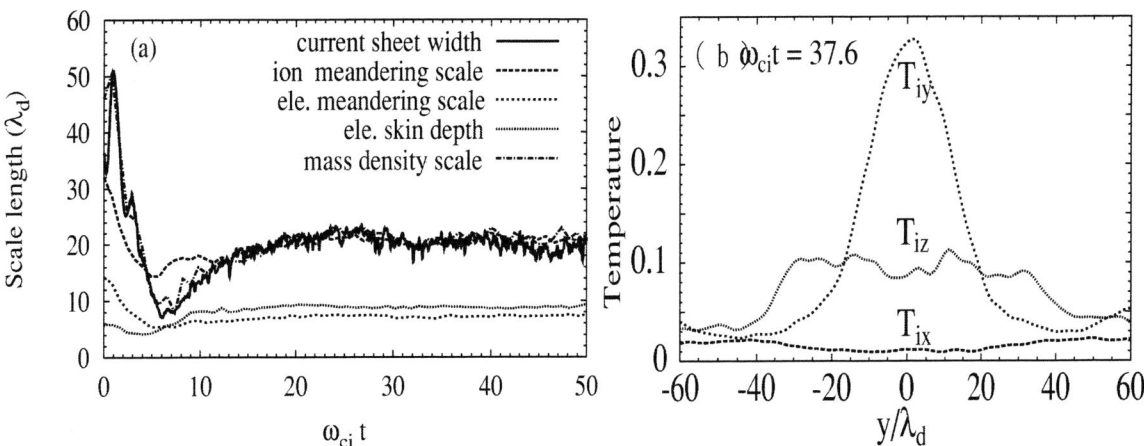

FIGURE 3. (a) Temporal evolution of five spatial scales and (b) spatial profile of three components of ion temperature in the steady state for the narrow input window.

Intermittent Reconnection

When the input window size x_d increases twofold ($x_d=36\,\rho_i$) [4,5], the system reveals a quite different behavior from the narrow window case ($x_d=18\,\rho_i$). An elongated current sheet along the x-axis is created as a result of the plasma compression over a relatively long range in the wide window case. The length of the current sheet is roughly estimated as $L_{cs}\sim 10\,\rho_i$, which means that the current sheet becomes unstable against a collisionless tearing instability. Magnetic islands are frequently created in the central region of the current sheet through the collisionless tearing instability, as is shown in Fig. 4. Consequently, the system never reaches the steady state with a single reconnection point and a constant reconnection rate in the wide window case. Furthermore, it is also found that the growth of magnetic islands is caused by the increase of electron current density through the electron trapping inside magnetic islands.

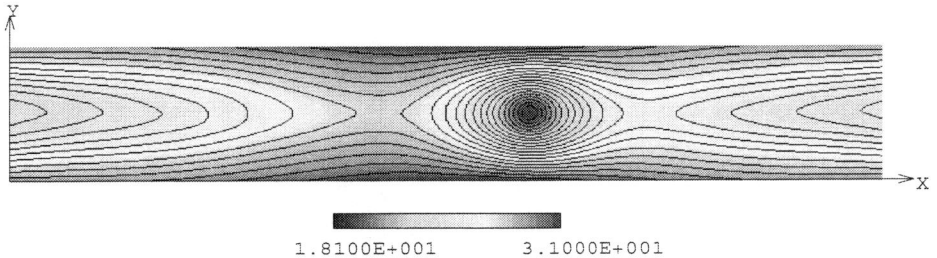

FIGURE 4. Contour plots of vector potential $A_z(x,y)$ at $\omega_{ce}t=755$.

Figure 5 illustrates the temporal evolutions of (a) three spatial scales in the current sheet and (b) the reconnection electric field and effective resistivity where the effective resistivity is defined as the ratio of electric field to current density at the reconnection point. The width of the current layer d_{jz} changes between the ion meandering scale l_{mi} and the electron meandering scale l_{me} in the wide window case. When a magnetic island starts to grow in the neutral sheet ($\omega_{ce}t=570, 960$), the reconnection field exceeds the driving field ($|E_{rec}|>|E_0|=0.04$) and the current sheet width decreases below the ion scale toward the electron scale. The width returns to the ion scale from the electron scale after an island moves away from the central region. The effective resistivity increases in accordance with the growth of magnetic islands and returns to a small value when an island disappears in the central region. In this way, the intermittent behavior in collisionless driven reconnection is deeply related to the electron dynamics.

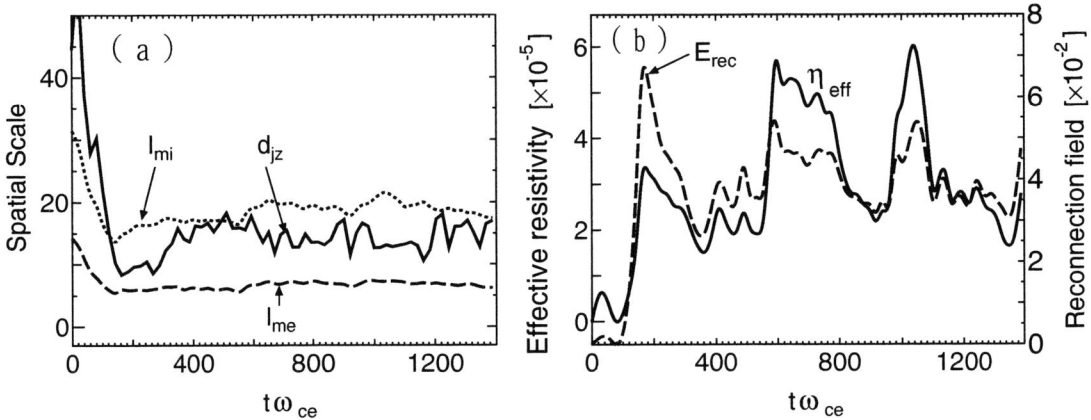

FIGURE 5. Temporal evolutions of (a) three spatial scales in the current sheet, and (b) reconnection electric field (dashed line) and effective resistivity (solid line) at reconnection point for the wide input window.

SUMMARY

Long time scale evolution of collisionless driven reconnection in an open system is investigated by using newly developed electromagnetic particle simulation codes [4,5,6,]. It is found that there are two evolving regimes in the temporal behavior of collisionless reconnection, dependently on the window size of plasma inflow, i.e., a steady regime and an intermittent regime.

The steady collisionless reconnection is realized in the case of small input window size, in which the reconnection rate is balanced with the flux input rate at the upstream boundary. The spatial scale of current density profile is the same as that of the mass density profile, which is always kept equal to the ion thermal scale. It is also found that anomalous ion heating takes place in the meandering ion scale at the neutral sheet under the action of an electrostatic field. Thus, the global dynamic process of steady magnetic reconnection is dominantly controlled by ion dynamics.

As the window size increases, the current sheet becomes longer, which is favorable to the excitation of a collisionless electron tearing instability. The system evolves toward the intermittent regime, in which magnetic islands are frequently generated in the current sheet. When an island grows up, the current scale decreases below the ion thermal scale and the reconnection rate becomes larger compared with that in no island case. The current density needed for the growth of a magnetic island is supplied by trapping electrons in it. Thus, the intermittent behavior in collisionless driven reconnection is deeply related to the electron dynamics.

REFERENCES

1. A. Nishida, "Geomagnetic Diagnostics of the Magnetoshpere" (Springer-Verlang, New York, 1978), p. 38.
2. Y. Ono, M. Yamada, T. Akao, T. Tajima, and R. Matsumoto, Phys. Rev. Lett. 76, 3328 (1996).
3. H. Amo, T. Sato, A. Kageyama, K. Watanabe, R. Horiuchi, T. Hayashi, Y. Todo, T. H. Watanabe, and H. Takamaru, Phys. Rev., E51, 3838-3841 (1995).
4. R. Horiuchi, W. Pei and T. Sato, Earth Planets Space, **53**, 439-445 (2001).
5. W. Pei, R. Horiuchi and T. Sato, Phys. Plasmas, 8, 3251-3257 (2001).
6. W. Pei, R. Horiuchi and T. Sato, Phys. Rev.Lett., 87, 235003-1-235003-4 (2001).
7. R. Horiuchi and T. Sato, Phys. Plasmas, 1, 3587-3597 (1994).

Magnetic Reconnection: MHD and Beyond

Giovanni Lapenta* and J.U.Brackbill*

*Los Alamos National Laboraory, Los Alamos, NM 87545, USA

Abstract. A new paradigm is considered for 3D magnetic reconnection where the interaction of reconnection processes with current aligned instabilities plays an important role. According to the new paradigm, the initial equilibrium is rendered unstable by current aligned instabilities (lower-hybrid drift instability first, drift-kink instability later) and the non-uniform development of kinking modes leads to compression of magnetic field lines in certain locations and rarefaction in others. The areas where the flow is compressional undergo driven reconnection on the time scale of the driving mechanism (the kink mode). In the present paper we illustrate this series of events with a selection of simulation results.

INTRODUCTION

We consider a new paradigm for 3D magnetic reconnection where the interaction of reconnection processes with current aligned instabilities plays an important role. The new paradigm is suggested by kinetic models where the lower-hybrid drift instability (LHDI) and kink modes (KM) [15, 8] drive field lines together and promote the onset of reconnection [14]. The new paradigm is suggested by MHD models where the presence of velocity shear induces a Kelvin-Helmholtz instability (KHI) that drives reconnection by locally compressing field lines [5, 6, 11, 12, 16].

According to the new paradigm, reconnection in 3D is eminently a naturally driven process. The driving force is determined by the instabilities developing in the current aligned direction. For the typical magnetotail configuration with the current in the dawn dusk direction (y) and the field mostly in the tailward direction (x) and the gradients mostly in the north-south direction (z), the tearing instability develops in the (x,z) plane. The collisionless tearing mode has long been considered the best hope to explain reconnection onset. However, accurate studies have not yet resolved the issue of the instability of the tearing mode in actual realistic magnetotail configurations. Most of the results seem to conclude that the tearing mode is stable in realistic configurations and that the cause of reconnection onset must be searched elsewhere [17].

The new paradigm for 3D reconnection cited above deals with the issue of reconnection onset. It is a new explanation of how reconnection can start, of what mechanism breaks the frozen in condition in the first place. In a spatially varying magnetic field configuration (such as the Earth magnetotail [15] or a corona arcade [16]) or in presence of spatially varying flow shear (such as at different latitudes along the magnetopause [6, 11]) the non-uniform development of kinking modes leads to compression of magnetic field lines in certain locations and rarefaction in others. The areas where the flow is compressional are subjected to a driven reconnection process on the time scale of the driving mechanism (the kink mode). In the new paradigm, two main physics processes must be considered.

First, the kink modes, including both the KHI and the drift kink instability (DKI) [13, 7], can drive field lines together, causing a localized compression that drives field lines to reconnect [6].

Second, new oblique modes are excited and contribute to the process of reconnection [14].

This new paradigm is distinct but complementary to the other recent remarkable progress in understanding fast reconnection through the role played by the Hall physics [2]. The Hall fast reconnection that has attracted so much recent attention is relevant to the fully developed reconnection process in the non-linear phase. The progress in that area has brought about understanding of the fast reconnection rates observed naturally. But still leaves us at a loss in trying to understand reconnection onset. It is the new paradigm for the role of current aligned instabilities that gives us the tools to understand onset of reconnection.

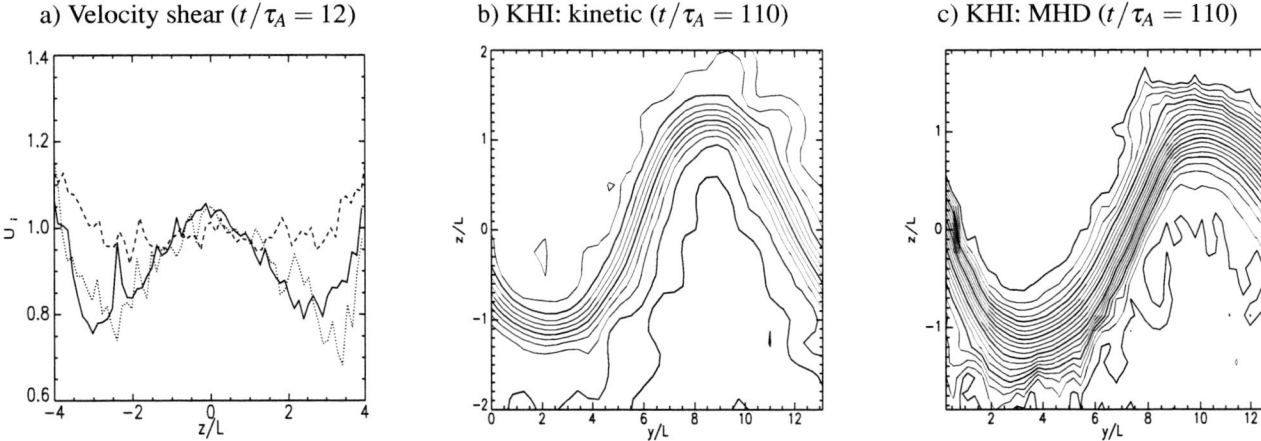

FIGURE 1. Harris equilibrium with $u_i/v_i = 1$, $m_i/m_e = 180$. Velocity shear at $t/\tau_A = 12$ (a) produced by the LHDI in a kinetic simulation. Three different temperature ratios are considered: $T_i/T_e = 10$ (solid); $T_i/T_e = 4$ (dotted); $T_i/T_e = 2$ (dashed). Contour plot of $B_x(y,z)$ at the end of two simulations ($t/\tau_A = 110$): b) a fully kinetic simulation with no initial shear; b) a MHD simulation with an initial shear equal to the shear observed in the kinetic simulation after saturation of the LHDI. Details in (Lapenta and Brackbill, 2002).

PHYSICAL SYSTEM

The Earth's magnetotail is described with the Birn's empirical profile chosen to best fit the actual magnetotail [1]. The system is simulated using the Vlasov-Maxwell model. The Vlasov-Maxwell system is solved using the CELESTE 3D implicit particle in cell code. A detailed description of the implicit moment method used in CELESTE 3D can be found in the review paper [3] and the details of the implementation can be found in [18]. For comparison, we will also consider simulations conducted with the 3D MHD code FLIP–MHD [4].

VELOCITY SHEAR

The initial equilibrium considered above is unstable to a number of current aligned instabilities, propagating along y. Two are of particular importance for understanding reconnection onset: the lower hybrid drift instability (LHDI) and kinking modes (KM).

Recent simulation work [15, 8] has shown that the early dynamic of current sheet is dominated by the LHDI. It has been observed [15, 8] that the nonlinear evolution of the LHDI changes the initial density and current profile and modifies the initial flat velocity profile by creating a velocity shear. Figure 1-a shows the velocity profile after saturation of the LHDI. The creation of a robust velocity shear is evident. It sould also be noted that the LHDI growth rate and saturation level is directly proportional to the $\sqrt{T_i/T_e}$ just like the velocity shear observed in Fig. ??, an indication that the velocity shear is caused by the LHDI.

It is interesting to note that the process of creation of velocity shear by the LHDI is reminiscent of the creation of zonal flows [10] observed in tokamaks in relation with the L-H transition [9].

Once a velocity shear is created in the short time scales of the LHDI a much slower fluid instability arises, the Kelvin-Helmholtz instability (KHI). The process can be modeled very simply using MHD. Figure 1-b shows a fully kinetic simulation where the velocity shear is created initially by the LHDI and is destabilized later by the KHI. At the end of the simulation the distinctive kinking caused by the KHI is observed. Figure 1-c shows a simple MHD simulation with the same initial equilibrium used for the kinetic simulation but with the addition of an initial velocity shear equal to the velocity shear formed naturally by the LHDI in the kinetic simulation. Clearly, the comparison of Fig. 1-b and Fig. 1-c proves that the evolution following the creation of shear is purely a fluid instability. However, note that in the kinetic simulation the velocity shear arises naturally while in the MHD simulation the shear is artificially introduced as an initial condition.

RECONNECTION ONSET

The evolution of the KHI and the kinking of the initial current sheet has an important consequence on reconnection onset. The presence of kinking causes the compression of the field lines in some regions and the rarefaction in others. Compression of field lines can drive reconnection. Such mechanism has already been observed in MHD simulations of the evolution of the KHI and tearing instability [5, 12] for application to the Earth magnetopause [6, 11] and to the solar corona [16]. The present paper is the first instance where the same mechanism is observed in the magnetotail.

In the references just cited, the mechanism was proposed in presence of externally driven shears (driven by the solar wind in the case of the magnetopause [6] or driven by the photosphere in the case of the corona [16]). In the magnetotail we suggest that, instead, the shear is created naturally by the LHDI without requiring any external action. Once the current sheet is thin enough for the LHDI to cause the velocity shear the chain of events sets in place and causes the onset of reconnection. The only required external action is the transfer of flux from the dayside to cause the thinning of the magnetotail.

An illustration of how the KHI can drive reconnection onset is presented in Fig. 2. A typical magnetotail equilibrium (Fig. 2-a) is rendered unstable to the KHI by the velocity shear induced by the LHDI. The KHI moves the current sheet up and down, compressing field lines in some regions and rarefying in others. A region of field line compression is evident in the northern side of the near Earth tail in Fig. 2-b. Eventually, field line compression drives reconnection (Fig. 2-c). Note that a 2D simulation of the same initial configuration cannot capture the presence of the KHI and the mechanism just proposed would not be observed. Indeed, in a 2D simulation of the equilibrium shown in Fig. 2-a, the tearing mode is stable due to the electron compressibility [17] and reconnection never sets in. Only 3D simulations can capture the chain of events proposed in the new paradigm for naturally driven reconnection onset.

CONCLUSIONS

A new scenario for magnetotail reconnection has been presented. The scenario is based on a sequence of three events.

First, the LHDI grows and saturates changing the initial equilibrium. A consequence of this change is the creation of a velocity shear that renders the equilibrium unstable to the KHI.

Second, the KHI grows and kinks the current sheet causing regions of compression of field lines.

Third, local compression drives field lines together and causes the onset of reconnection. Once reconnection is started by the driving force of the KHI it can progress via the action of the Hall term mediated fast reconnection process of recent discovery [2].

ACKNOWLEDGMENTS

The author is very grateful to Jerry Brackbill and Dana Knoll for the useful discussions on the physics of reconnection.

This research is supported by the United States Department of Energy, under Contract No. W-7405-ENG-36 and by NASA, under the "Sun Earth Connection Theory Program". The supercomputer used in this investigation was provided by funding from JPL Institutional Computing and Information Services and the NASA Offices of Space Science and Earth Science.

REFERENCES

1. Birn, J., *J. Geophys. Res.*, **92**, 11101 (1987).
2. Biskamp, D., *Magnetic Reconnection in Plasmas* (Cambridge University Press, 2000).
3. Brackbill, J. U., D. W. Forslund, Simulation of Low-Frequency, Electromagnetic Phenomena in Plasmas, *in Multiple Time Scales*, J. U. Brackbill and B. I. Cohen Eds. (Academic Press, Orlando, 1985), pp. 271–310.
4. Brackbill, J.U., *J. Computat. Phys.*, **96**, 163 (1991).
5. Brackbill, J.U., *J. Computat. Phys.*, **108**, 38 (1993).
6. Brackbill, J.U., Knoll, D.A., *Phys. Rev. Lett.*, **86**, 2329 (2001)
7. Daughton, W.S., *Phys. Plasmas*, **6**, 1329-1343 (1999).
8. Daughton, W.S., *Phys. Plasmas*, submitted (2002).
9. Diamond P.H., M.N. Rosenbluth, E. Sanchez, C. Hidalgo, B. VanMilligen, T. Estrada, B. Branas, M. Hirsch, H.J. Hartfuss, B.A. Carreras, *Phys. Rev. Lett.*, **84**, 4842 (2000).

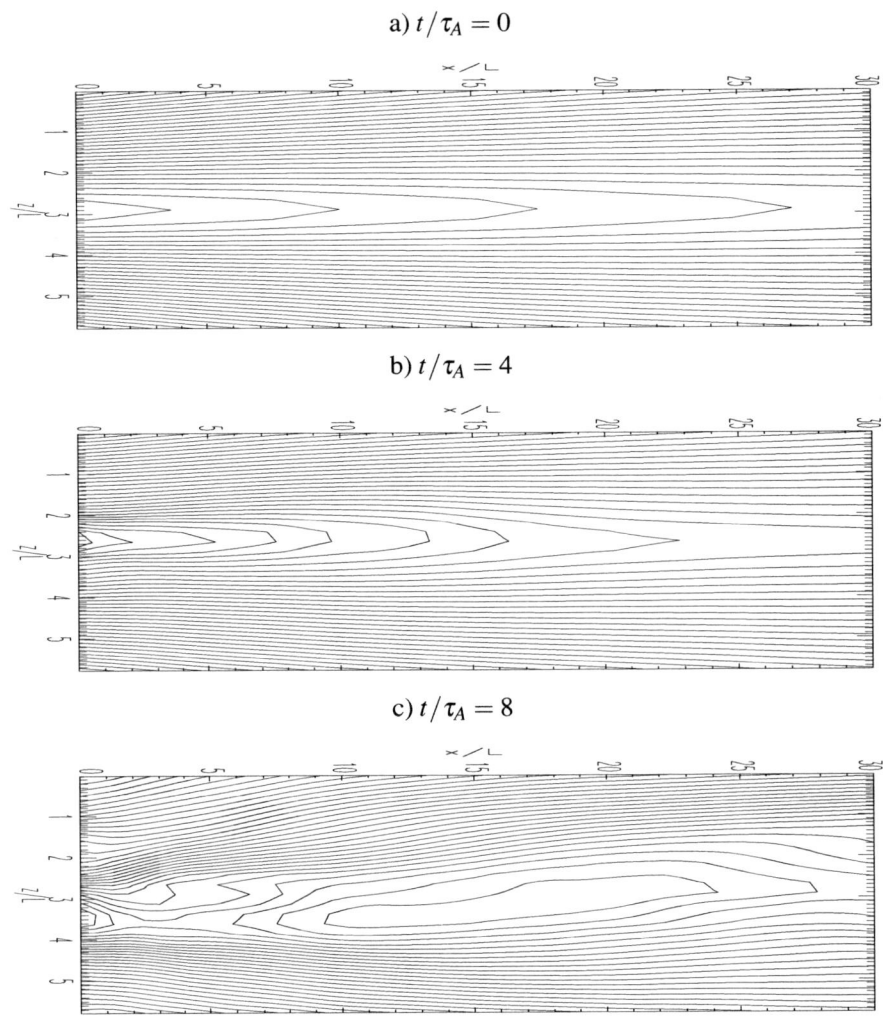

FIGURE 2. Cross section of the flux surfaces for a 3D MHD simulation at $y = L_y/2$ and three different times: initial ($t/\tau_A = 0$), $t/\tau_A = 4$, $t/\tau_A = 8$.

10. Drake, J.F., J.M. Finn, P. Guzdar, V. Shapiro, V. Shevchenko, *Phys. Fluids B*, **4**, 488 (1992).
11. Knoll, D.A., Brackbill, J.U., *Phys. Plasmas*, submitted (2002).
12. Knoll, D.A., Chacon L., *Phys. Rev. Lett.*, **8821**, 5003 (2002).
13. Lapenta, G., J.U. Brackbill, *J. Geophys. Res.*, *102*, 27099 (1997).
14. Lapenta, G., J.U. Brackbill, *Nonlinear Processes Geophys.*, **7**, 151 (2000).
15. Lapenta, G., J.U. Brackbill, *Physics Plasmas*, **9**, 1544 (2002).
16. Lapenta, G., D.A. Knoll, *Solar Phys.*, submitted (2002).
17. Quest, K. B., H. Karimabadi, and M. Brittnacher, *J. Geophys. Res.*, **101**, 179, (1996).
18. Ricci, P., G. Lapenta, J.U. Brackbill, *J. Computat. Phys.*, submitted (2002).

Alfvén Resonances, Forced Magnetic Reconnection and Model of Solar Flares

C. Uberoi

Department of Mathematics, Indian Institute of Science, Bangalore 560 012, India

Abstract. This paper briefly reviews the surface wave induced magnetic reconnection (SWIMR) model based on the Alfvén resonance theory near the neutral point. Extending this model to magnetosonic waves it is seen that it not only has features of formation of current sheets and plasmoids as in the incompressible case but in addition energy considerations show the triggering of solar flares by the catastrophic loss of equilibrium. The similarity of SWIMR model to the emerging flux and plasmoid induced magnetic reconnection model for solar flares is discussed.

INTRODUCTION

In the early days of understanding of magnetic reconnection in solar flares it became clear that to explain the impulsive phase of solar flares or the sequential triggering phases as observed by Vorphol [1] we need driven or forced magnetic reconnection. There can be two ways either by flow velocity or by wave dispersion. The observations [1] hints a mechanism of triggering of solar flares by magnetosonic waves. Following this suggestion theoretical model to explain the role of the nonlinearity of these waves in triggering of solar flares was given by Sakai and Washami [2].

In recent years the wealth of high resolution observations from various space missions as well as ground based observations have emphasised the structural magnetic fields and discontinuities in the form of magnetic flux tubes and coronal loops to understand the solar plasma processes. The wave propagation in such a structured medium can become an important mechanism for the reconnection.

Further, recent numerical simulations with high spatial resolutions have revealed that common feature of reconnection in solar flares are formation of current sheets, time-dependence of the reconnection process and the formation and ejection of plasmoids. These features are taken into account in giving a unified model of flares, the plasmon-induced-reconnection model [3]. The ejection of plasmoids is also a feature of models in which the flare is triggered by catastrophic loss of equilibrium of the current sheet.

Considering the MHD waves in structural and inhomogeneous magnetic fields Uberoi [4] proposed long wavelength Alfvén surface waves as possible source for producing conditions suitable for inducing reconnection, the SWIMR model. This model is based on the Alfvén resonance theory near the neutral point. We first briefly review this model and then show that when extended to compressional or magnetosonic waves not only has features of formation of current sheets and plasmoids but in addition energy considerations show the triggering of solar flares by the catastrophic loss of equilibrium.

BASIC CONCEPT OF SWIMR MODEL

The equation governing the dynamic of the hydromagnetic waves in the incompressible media propagating in inhomogeneous magnetic fields with variation in the direction perpendicular to the plane of the magnetic field $\vec{B}_0[0,0,B_0(x)]$ is given as

$$\frac{d}{dx}\left(\varepsilon\frac{dv_x}{dx}\right) - k_{\|}^2 \varepsilon v_x = 0 \quad (1)$$

where $\varepsilon = [\omega^2 4\pi\rho_0(x) - k_{\|}^2 B_0^2(x)]$, $\vec{k} = (k_{\|}, k_{\perp})$ is the wave vector in the (y,z) plane. Here v_x is the perturbed velocity component.

From the extensive studies of eqn. (1) in the literature [5,6], it can be emphasised that the time evolution of Alfvén waves shows that the current density increases secularly with time t, thus showing the development of current sheets as the Alfvén wave propagates in the inhomogeneous system. The thickness of the current sheet decreases as $1/t$. These current sheets arise not due to any instability but due to the accumulation of energy around the resonance point, where Alfvén wave energy is absorbed resonantly. This situation continues till time $t = t_h$. For $t > t_h$, where t_h is the intrinsic time scale, the resistivity effects become important and the regularisation of the singularity takes place by resonant mode conversion of surface waves along a sharp discontinuity to the resistivity modified Alfvén wave at the non-zero singular point at the centre of the resistivity layer. However, as pointed out by Uberoi [4] the mathematical structure of the hydromagnetic wave eqn. (1) remains same at zero and non-zero singular points, but the role played by surface waves in the resonant absorption mechanism at these two points is different.

The theories of resonant absorption of Alfvén waves consider surface waves propagating along a sharp discontinuity separating two infinitely extended plasma regions. The structural discontinuities are not taken into account. Near a neutral point the structure of the discontinuity becomes important as waves are now of long wavelengths. In this case wave propagation is to be considered for a plasma layer, which can support two types of surface modes of oscillations, symmetric and asymmetric.

Consider a plasma layer with thickness 2a, with the magnetic field profile $B_0(x)$

$$B_1, |x| < -a; \quad B_0 x/a, -a < x < a; \quad B_2, |x| > a \tag{2}$$

and the density being, $\rho_1, \rho(x)$ and ρ_2 in the three regions respectively. The linear variation of magnetic fields is taken only for simplicity, the results given in [4] are for a general variation of the magnetic field. Writing the dispersion equation for the surface waves within the plasma layer and taking the long wavelength limit $ka \ll 1$, it is noted that symmetric surface modes couple to the low frequency end of the Alfvén continuum [7,8].

When the resistivity is switched on, the long wave length surface waves couples with the tearing mode of the layer, thus inducing magnetic reconnection on the tearing mode time scale. This mode is unstable for certain wavelengths greater than a critical value, $\lambda = 2\pi a/0.64$, and begins to grow until magnetic islands are formed.

The important time scales in this model are:

$$t_h = \tau_A^{2/3} \tau_R^{1/3} \quad \text{or} \quad t_h = \tau_A S^{1/3},$$

where $\tau_R = 4\pi a^2/\eta$, $\tau_A = a/V_A$ and $S = \tau_R/\tau_A$, the Lundquist number. Here η is the finite conductivity. The resistivity effects begin to play role for $t \geq t_h$.

The other important time scale is the reconnection time:

$$t_r = \tau_A^{2/5} \quad \text{or} \quad t_r = \tau_A S^{3/5}.$$

The width of the resistive layer scales as $\Delta x \propto S^{-1/3}$.

SWIMR AND TAYLOR MODEL

It is interesting to note that the basic concept of SWIMR model is similar to the Taylor model [9] of forced reconnection, in which the reconnection is induced by perturbing the boundary of a simple slab equilibrium of an incompressible plasma with a resonant surface. In fact the SWIMR model is a general formulation of Taylor's model and complete correspondence was pointed out by Uberoi and Zweibel [10].

The Taylor model concept has been used to understand coronal heating [11,12]. Specifically Wang and Bhattacharjee [11] show that the Taylor model and Parker's model [13] for coronal heating have similar qualitative features. SWIMR model therefore, unifies the coronal heating models at low frequencies due to magnetic reconnection.

THERMAL CATASTROPHE IN THE SWIMR MODEL

For the magnetosonic waves eqn. (1) takes the form, near the resonant point, as

$$\frac{d}{dx}\left(\varepsilon \frac{dv_x}{dx}\right) - k_\perp^2 \varepsilon v_x = 0 \tag{3}$$

for low β plasma and for wave numbers such that $k_\perp^2 \gg k_\parallel^2$. As eqn. (8) is similar to eqn. (1) the basic features of the development of current sheets and magnetic reconnection induced by the resonant coupling of compressional surface waves leading to formation of plasmoids will be same as in the incompressible case. We note here that though the dispersion relation for the surface waves along the density discontinuities are different from that of incompressible case [14], the long wavelength compressional surface waves for the layered structure have the similar dispersion equation and so without any modification the SWIMR model can be used both for the shear Alfvén and magnetosonic waves.

For developing the energy equation consider the plasma layer as given by the eqn. (2). Following the earlier works [15] the heating rate Q_A per unit area in the (y,z) plane due to resonant absorption of compressional waves is:

$$Q_A = \frac{2\pi \omega a b_{1x}^2(a)/4\pi}{\left[\frac{k_\parallel^2 B_2^2}{4\pi \omega^2 \rho_1}\frac{k}{k_\perp} - 1\right]^2 + \pi^2 k^2 a^2} \qquad (4)$$

where $k^2 = k_\perp^2 + k_\parallel^2$ and b_{1x} is the amplitude of perturbation of surface waves at $x = a$. In writing Q_A the following assumptions (only for mathematical simplicity, but can be easily generalized) are made: $B_2 \gg B_1$, in medium (2) $4\pi \rho_2 \omega^2 \ll k_\parallel^2 B_2^2$ and in medium (1) $V_{A_1}^2 \ll V_{Ph}^2$, which is in accordance with the fact that resonance occurs in the region $-a < x < a$.

Balancing the heating solely against convective transport towards the reconnection layer around $x = 0$ with convection velocity V_x, the energy equation is

$$\frac{d}{dx}\left[\frac{5}{2}V_x P\right] = q_0 + \frac{1}{2}Re J.E^*, \qquad (5)$$

the last term on the right is the contribution from resonant absorption. q_0 is the combined heating rates of other heating mechanisms and $P(x) = n(x)T(x)$.

Integrating eqn. (5) from 0 to a and defining.

$$T_w = \frac{1}{2}m_i\frac{\omega^2}{k_\parallel^2}\frac{k_\perp}{k}, \quad X^2 = \pi^2 k^2 a^2, \quad T' = T/T_\omega, T_0' = T_0/T_\omega, \quad W = \frac{\omega a b_{1x}^2}{5n_0 V_x T_w}$$

and noting from the equilibrium conditions $B_2^2/4\pi = 2n_0 T$, eqn. (5) becomes

$$W = (T' - T_0')[(T' - 1)^2 + X^2], \qquad (6)$$

which has a family of solution in the form of mathematical catastrophe for $0 < T_0' < 1 - \sqrt{3}X$ as shown in Fig. 1. For a critical value $W = W^*$ the equilibrium temperature T' jumps discontinuously to a higher value. The thermal catastrophe of the current sheet corresponds to the triggering of the solar flare.

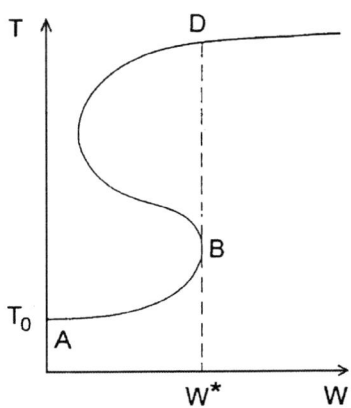

FIGURE 1. Schematic of solutions $T(T_0, X, Y) = 0$ Eqn. (6) showing the unphysical catastrophic jumps in temperature from B to D for the critical energy W^*.

For constant T_0, the initial temperature, and constant value of the convective flow V_x, W increases slowly through increased coupling of the photospheric anchored magnetic field lines foot-point motion, inducing the surface waves along the coronal loops, to the compressional waves in the current sheets in the coronal magnetic field, till it reaches the critical value W^*.

The maximum heating rate when $T = T_w$ over an area A on both sides of the current sheet $2AQ_A$ is

$$E_{\max} = \frac{2AV_{sw}b_{1x}^2(a)}{2\pi X} \tag{7}$$

with $\omega/k = V_{sw}$, the surface phase velocity. Considering the current sheet with $a = 10^4$ kms and length as 10^5 km, in the coronal plasma, we have $V_{sw} = 10^8$ cm/s, $b_{1x} = 0.1B_2$, with $B_2 = 100$ gauss and $X = 0.03$ (the $X_{\max} = 1/\sqrt{3}$), $E_{\max} = 10^{29}$ ergs/s which compares with the observed values for solar flares [3].

We note that there are two processes arising due to resonant absorption of MHD waves in the neutral sheet in a structured plasma medium. One is the resonant excitation of the tearing mode after time $t \geq t_h$, giving the magnetic reconnection. It is seen that when $t \sim t_h$, the reconnection energy, in the incompressible care, equals to resonant absorption energy [16]. In the other process the Alfvén wave heating energy when balanced with the convective transport on energy towards the current sheet gives the thermal catastrophic loss of equilibrium. Assuming that time scale t_c for the system to reach another state of equilibrium being such that $t_c < t_h$, we can say that resonant energy leads to catastrophic jump in the temperature and as the resistivity effects become important for $t \geq t_h$, tearing of th sheet and formation of magnetic islands takes place. The spatial scales of the plasmoids calculated from λ. For a ranging from 10^3 to 10^4 kms, are 1×10^4 to 2.7×10^4 kms. These numbers agree well with those calculated from the plasmon-induced-reconnection model [3].

REFERENCES

1. T.A. Vorphol, Ap. J. **205**, 868 (1976).
2. J. Sakai and H. Washimi, Ap. J. **258**, 823 (1982).
3. K. Shibata, Observational Plasma Astrophysics: Five years of Yokoh and Beyond, eds. T. Watanabe et al. (Kluwer Academic Publishers, Netherlands, 1998) p. 187.
4. C. Uberoi, J. Plasma Physics **52**, 215 (1994)
5. A. Hasegawa and C. Uberoi, The Alfvén Wave (Technical Information Center, U.S. Department of Energy, 1982).
6. C. Uberoi and Z. Sedlacek, Phys. Fluids **B4**, 6 (1992)
7. C. Uberoi, L.J. Lanzerotti and A. Wolfe, J. Geophys. Res. **101**, 24, 979 (1996)
8. C. Uberoi, L.J. Lanzerotti and C.G. Maclennan, J. Geophys. Res. **104**, 25, 153 (1999).
9. T.S. Hahm and R.M. Kulsrud, Phys. Fluids **28**, 2412 (1985)
10. C. Uberoi and E.G. Zweibel, J. Plasma Phys. **62**, 345 (1999).
11. X. Wang and A. Bhattacharjee, Phys. Fluids **B4**, 1745 (1992).
12. G.E. Vekstein and R. Jain, Phys. Plasmas **5**, 1506 (1998).
13. E.N. Parker, Ap. J. **174**, 499 (1972).
14. C. Uberoi, J. Geophys. Res. **94**, 6941 (1989).
15. R.A. Smith, C.K. Goertz and W. Grossman, Geophys. Res. Lett. **13**, 1380 (1986).
16. G. Vekstein, Phys. Plasmas, **7**, 3808 (2000).

Dynamo Activity In Imposed DC Magnetic Fields

D.C. Montgomery [1], W.H. Matthaeus [2], L.J. Milano [2] and P. Dmitruk [2]

[1] *Physics & Astronomy Dept., Dartmouth College, Hanover, NH 03755, U.S.A.*
[2] *Bartol Research Institute, Univ. of Delaware, Newark, DE 19716, U.S.A.*

Abstract. A widely accepted mechanism for the spontaneous appearance of large-scale magnetic fields in magnetofluids is the turbulent inverse cascade of magnetic helicity. The evidence for the effect is largely computational, and has been acquired by using spectral-method codes and imposing three-dimensional (3D) rectangular periodic boundary conditions. We report here similar investigations of the phenomenon that result when a uniform external dc magnetic field is present. With no imposed dc magnetic field, the 1981 results of Meneguzzi et al are recovered, but by adding an externally-imposed dc magnetic field, we find that it is possible to suppress entirely the inverse magnetic helicity cascade phenomenon. This is a somewhat puzzling result, and we attribute it not to any fundamental change in the physical processes involved as much as to the inconsistency of 3D rectangular periodic boundary conditions (which permit no net current through the basic computational box). These boundary conditions seem inadequate as a representation of the nonlinear dynamics of helically-driven MHD in the presence of a dc magnetic field (which encourage such a dc net current to flow). We believe that if the turbulent computations were to be repeated with more realistic geometry and boundary conditions (e.g., toroids, disks, or periodic cylinders), the inverse magnetic helicity cascade phenomenon would reappear. But at the moment, this must be regarded as an open question.

INTRODUCTION

The spontaneous appearance of large-scale magnetic fields is common in the Universe, and the "dynamo problem," seeking mechanisms for their generation, goes back at least to Gauss in 1838 [1]. Large-scale magnetic fields seem likely to appear any time there is sufficient mechanical turbulence in an electrically-conducting fluid. Magnetic flux-tube stretching by turbulent velocity fields in electrically-conducting media is a straightforward candidate for magnetic field amplification, but by itself can only account for magnetic field generation at length scales as small as or smaller than the characteristic length scales of the mechanical turbulence. How the magnetic excitations get transferred back to scale sizes as large as or larger than the size of the system has been a puzzle. Laboratory attempts at producing spontaneous dynamo action have proved notoriously hard to complete, though several groups are presently active in the area. Most of the detailed information we have to go on currently comes from numerical solutions of the magnetohydrodynamic (MHD) equations, and qualitative insights from such calculations as those of the "alpha model" [2,3].

In 1975, Frisch et al [4] suggested a promising mechanism: the inverse cascade of magnetic helicity, formulated in analogous terms to those of the inverse cascade of kinetic energy in two-dimensional Navier-Stokes turbulence. Laboratory experiments on inverse helicity cascades have proved inconclusive so far, but the effect was well documented numerically by Meneguzzi et al [5] in 1981, using a pseudospectral three-dimensional code for solving the incompressible one-fluid magnetohydrodynamic (MHD) equations. A mechanical band-limited, random forcing function in the equation of motion which injected mechanical helicity was employed in the MHD description, and spontaneous long-wavelength magnetic excitations appeared. Further elaboration and development has occurred since [e.g., 6,7,8].

Dynamo Suppression By DC Magnetic Fields

Recently, we have repeated the computations of Meneguzzi et al in the presence of a uniform dc magnetic field [9], we believe for the first time. (Most of the results we discuss here appear in detail, with supporting graphics, in

Ref. [9]). We otherwise retained the rectangular triply periodic boundary conditions that have been in essentially universal usage for decades, both for fluid and for MHD turbulence computations. Such triply-periodic spatial boundary conditions are almost a universal feature of modern turbulence computations, both for fluids and magnetofluids. A sharply discontinuous and novel behavior was observed, in that the imposed dc magnetic field was seen to suppress the inverse magnetic cascade at a rather low level (with the dc magnetic field at a small fraction of the rms turbulent magnetic field strength). While not totally unexpected, this computational result requires clarification and an assessment of whether it is physically to be expected or is in some way only an artifact of the way the computations were performed.

We incline to the latter view, if for no other reason than that a generation of experience with laboratory reversed-field pinches provides us with numerous examples of dynamo phenomena in the presence of externally-supported, imposed, dc magnetic fields [10]. Earlier, various puzzling features and inconsistencies of triply periodic rectangular boundary conditions in MHD had been noted [11,12,13]. Here, we wish to offer a more detailed physical explanation of what we think is important and limiting about triply periodic boundary conditions in conjunction with an imposed dc magnetic field for this particular problem. In the process, we will explain why we think that removing the artificiality of the triply periodic boundary conditions in future computations will likely allow a dynamo effect to reappear.

Prohibition of Net DC Currents By Periodicity

The presence of a dc magnetic field (**Bo**, say) in MHD locks together the dynamics of the variable magnetic field **B** and the velocity field **v** in a much more intimately detailed way than they are coupled in the absence of the finite dc magnetic field. In the limit of small-amplitude excitations, every three-dimensional excitation in incompressible MHD becomes in fact an Alfven wave, with a group velocity directed parallel or anti-parallel to **Bo**. For a plane-polarized Alfven waves, there is no net <**v** x **B**> associated with the fluctuating wave fields. (Here, the angle brackets < > imply a spatial average.) But for a circularly-polarized Alfven wave (one with helical magnetic field lines, of the kind that helical mechanical excitations are bound to excite), there is a definite <**v** x **B**> emf directed parallel to, or opposite to, **Bo** [14]. If one sign of helicity predominates, there is a net global emf across the entire system. Without the artificiality of the periodic boundary conditions, in real life the effect of this emf would be to drive a net current through the basic box inside which the computation is taking place, parallel or anti-parallel to **Bo**. However, triply periodic rectangular boundary conditions explicitly forbid any net current to flow, since we still are required to have <curl **B**> = 0, by Ampere's law--so that no net dc current is permitted through the system. In effect, "open-circuit" boundary conditions have been imposed. In real life, what would happen under such circumstances is that the charges would separate along **Bo** and collect on opposite faces of the box until an electrostatic field capable of balancing <**v** x **B**> would develop and be sufficient to oppose the dynamo emf. However, rectangular periodic boundary conditions (together with the overall charge neutrality implied in MHD) prohibit this as well. Essentially what seems to have happened is that the dynamo effect, which would generate magnetic flux perpendicular to **Bo** and the driven current, has been shut down by an inappropriate boundary condition.

What we think most likely is that if mechanical helicity injection can be used to excite MHD turbulence in the presence of a dc magnetic field without requiring spatially periodic boundary conditions in all three directions, the inverse magnetic cascade phenomenon, no longer homogeneous and isotropic, but recognizable as an inverse magnetic helicity cascade, will reappear. This can be done in periodic cylinders with rigid walls, with spheres, and perhaps in restricted rectangular regions using overall periodic 3D codes where symmetries are invoked to permit net currents to flow through isolated subregions--ones which do not communicate with other subregions in which equal and opposite net currents are flowing.

CONCLUSIONS

The situation seems ripe for a new generation of nonlinear MHD codes for studying turbulence in more realistic geometries, ones which are not restricted to the now limited-seeming restrictions of rectangular periodicity in all three directions. The most promising geometries seem to be cylinders (of either square or circular cross sections and material walls), spheres, or disks. The numerical task is far from easy, but seems necessary if the subject is to advance further.

This work was supported in part by National Science Foundation Grant No. ATM-9713595.

REFERENCES

1. See, e.g., G.E. Backus, Ann. Phys. (NY) 4, 372 (1958).
2. F. Krause and K.-H. Raedler, *Mean-Field Magnetohydrodynamics and Dynamo Theory* (Oxford, U.K.: Pergamon Press, 1980).
3. H.K. Moffatt *Magnetic Field Generation in Electrically Conducting Fluids*(Cambridge, U.K.: Cambridge University Press, 1978).
4. U. Frisch, A. Pouquet, J. Leorat, and A. Mazure, J. Fluid Mech. **68**, 769 (1975).
5. M. Meneguzzi, U. Frisch, and A. Pouquet, Phys. Rev. Lett. **47**, 1060 (1981).
6. G.B. Field, E.G. Blackman, and H. Chou, Astrophys. J. **513**, 638 (1999).
7. F. Cattaneo and D.W. Hughes, Phys. Rev. E**54**, 4532 (1996).
(In this paper, the effect of a dc magnetic field on depressing the spatial average <**v x B**> was noted , but the necessary conclusion was not reached that the "alpha" coefficient of magnetic amplification mean-field theory must itself be a function of wavenumber, small but non-zero, and distinct from <**v x B**>. See Ref. [13] for a more thorough explanation of the distinction, and a discussion of long but finite wavelength magnetic field amplification and how it is affected by the presence of a spatially uniform **Bo**.
8. A. Brandenburg and K.J. Donner, Mon. Not. R. Astron. Soc.**288**, L29 (1997).
9. D. C. Montgomery, W.H. Matthaeus, L.J. Milano, and P. Dmitruk, Phys. Plasmas **9**, 1221 (2002).
10. e.g., J.P. Dahlburg, D. Montgomery, G.D. Doolen, and L. Turner, J. Plasma Phys. **40**, 39 (1988).
11. D.C. Montgomery and J.W. Bates, Phys. Plasmas **6**, 2727 (1999).
12. T. Stribling, W.H. Matthaeus, and S. Ghosh, J. Geophys. Res. **99**, 2567 (1994).
13. D. Montgomery and H. Chen, Plasma Phys. Controlled Fusion **26**, 1189 (1984).
14. All that is required to see this is a straightforward generalization of the calculation given by H.K. Moffatt in *Magnetic Field Generation in Electrically Conducting Fluids* (Cambridge, U.K., Cambridge University Press, 1978, pp. 162 ff.), using the framework of full MHD rather than the "kinematic dynamo" formulation. (Moffatt, however, does not draw the inference of the inappropriateness of periodic boundary conditions from the result.)

Explosion of Plasma Foils in the Petawatt Regime: Generation of MeV Particle Beams

M. Eloy[*,†], A. Guerreiro[**,‡], J. T. Mendonça[†] and R. Bingham[**,§]

[*]*FEUCP-Faculdade de Engenharia da Universidade Católica Portuguesa, Estrada de Talaide, 2635-631 Rio de Mouro, Portugal.*
[†]*GoLP/Centro de Física dos Plasmas, Instituto Superior Técnico, Av. Rovisco Pais 1096 Lisboa Codex, Portugal.*
[**]*Rutherford Appleton Laboratory, Space Science and Technology Dept., Chilton, Didcot, U.K.*
[‡]*CLOQ/Centro de Lasers e Óptica Quântica, Departamento de Física, Faculdade de Ciências, Universidade do Porto, R. Campo Alegre, 687,4169-007 Porto, Portugal.*
[§]*Department of Physics and Applied Physics, Univ. of Strathclyde, Glasgow, U.K.*

Abstract. We report on simulation studies of thin plasma foils explosions upon interacting with high-intensity, ultra-short laser pulses. By using a fully relativistic Particle-in-cell code we describe the time-resolved position, momentum and energy of electrons and ions, for laser pulses with durations of tens of fs and intensities ranging from 10^{20} W/cm^2 to 10^{23} W/cm^2. Results show the generation of a Mev electron beam as well as supra-Mev ions. Ponderomotive Boost and Colombic Explosions are mechanisms used to explain the results.

INTRODUCTION

When an ultra-fast ultra-high intensity laser pulse is focused on a thin foil target with width that is comparable to the wavelength of the incident laser pulse, there is little attenuation of the laser pulse as it propagates through the whole width of the material. As a result all of the foil will take part in the interaction. A direct consequence of the interaction is the generation of fast electron beam, in conjunction with the production of extremely energetic ion beams as well X-rays and γ-rays.

In contrast, when a laser pulse is focused onto a thick target, with a width much larger when compared to the incident laser wavelength, the physical processes involved are much more complex, since most of the pulse will be absorbed at the entrance surface, giving rise to a very dense plasma, and will therefore not directly interact with the rear surface.

Thin foil targets are closely related with atomic clusters which has became in recent years an important area of theoretical and experimental research. Atomic clusters exhibit near solid densities and intermediate sizes between solid and gaseous targets. Recent experiments have shown that the interaction of short (<ps), intense (about 10^{17}W/cm^2) laser pulses with rare gas clusters is responsible for the production of highly energetic electrons and ions, X-ray emission in the Kev range and coherent harmonic generation.

For intensities larger than 10^{17}W/cm^2 and pulse durations shorter than 150fs the cluster expansion mechanism can be described using Coulomb repulsion model. In this regime, a sufficient number of electrons exit the cluster core leaving behind a positively charged cluster which explodes due to electrostatic repulsion between the ions. This is an extremely fast process which takes place at time scales much shorter to allow normal transport processes such as plasma heating and thermalization.

Numerical simulations using particle dynamics to investigate the interaction of the laser pulse with dense targets require substantial computational resources and cannot account for a large number of atoms, whereas a fluid description of the process cannot provide as detailed an insight of the process.

To circumvent these limitations, a novel approach to describe the dynamics of foil explosion under irradiation with intense laser beams is presented, through the application of a Particle-in-Cell (PIC) code which has already been used to calculate the Coulombic expansion of atomic clusters [1], [2], [3], [4], [5], [6]. This powerful tool can provide an accurate description of the physics involved by determining for each time step the position, momentum and energy of

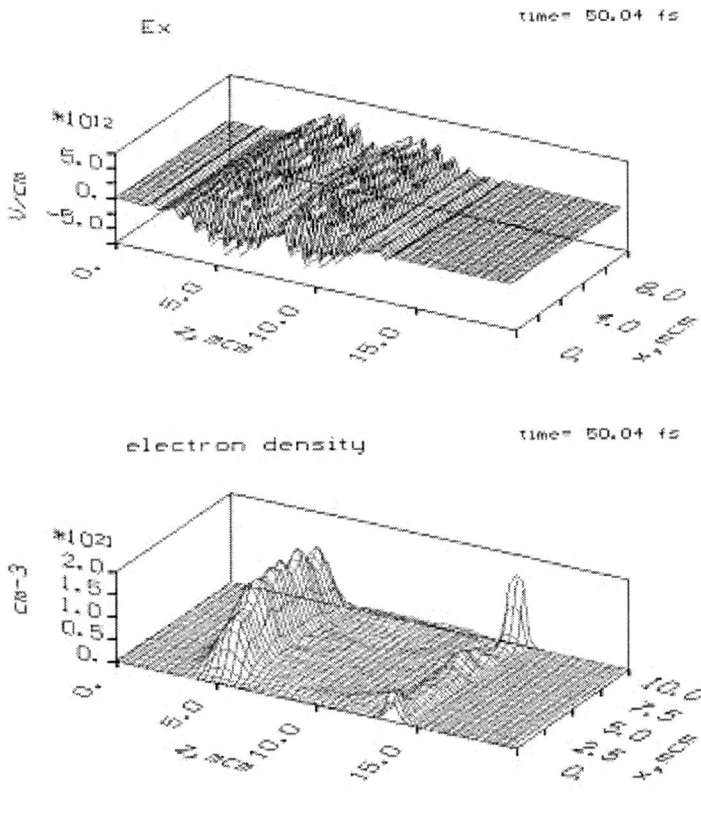

FIGURE 1. Generation of a Mev electron beam. Above: Snapshot of the laser e-field propagating through the simulation region, perturbed at $z=5\mu m$ by the foil ions left behind. Below: Snapshot of the electron density profile, evidencing an electron bunch in the front of the laser pulse for $I = 10^{22} W/cm^2$

the particles under the influence of both the external electric field of the laser and the electric fields generated by the particles themselves, for a very large number of particles and within reasonable computational times.

Like in the case of atomic clusters, the thin target, upon interaction with an ultra-short, high-intensity laser pulse, will become fully ionized through its whole width within the first laser cycles. Hence it is possible to treat the problem in a similar fashion as a cluster explosion, and a PIC simulation of the process will produce an accurate and detailed description of the physical processes involved.

THE SIMULATION PARAMETERS

A thin foil of H, modelled as a rectangular pre-formed plasma with 1 μm width, height equal to that of the simulation box and ion densities varying from 10^{19} cm^{-3} to 10^{21} cm^{-3} and with ions exhibiting charge states of +1, is placed at coordinate z = 5 μm.

The laser beam is the same used in 2D simulations of cluster explosions, a 800 nm, plane polarized in the X-direction pulse, with a gaussian temporal profile with duration of a few tens of fs (FWHM) and a 10 μm radius gaussian spatial profile, propagating in the Z-direction, from left to right, injected at the leftmost (z = 0 μm) boundary of the simulation box at coordinates (x, z) = (5, 0) μm or (x, z) = (10, 0) μm, depending on the problem specifications.

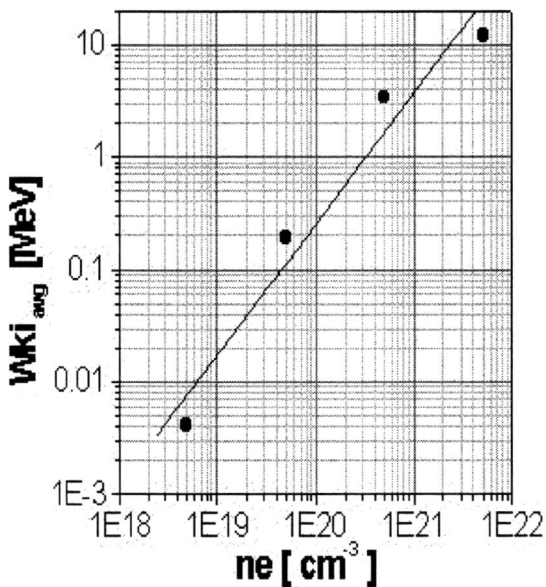

FIGURE 2. Linear scaling of ion energy obtained in simulations with electron (and ion) density confirms the explosion to be Coulombic type.

The laser intensities used ranged from 10^{20} W/cm^2 up to 10^{23} W/cm^2.

The choice of these laser intensities as well as the simulation box parameters was decided upon verifying, in early simulations, that the data obtained for the thin foil explosion was, in all aspects, identical to those generated by the 2D single-cluster explosion, and therefore, different regimes of interest needed to be studied.

SIMULATION RESULTS

As the laser pulse reaches the foil, ponderomotive longitudinal force, now higher due to the increase in laser intensity, boosts electrons in the forward direction and laser trapping accelerates these particles to relativistic velocities. The instant when the trapping occurs will determine the electron dynamics: electrons trapped at the head of the pulse have the highest velocities, approaching the speed of light, c, and will move with the front of the pulse, giving rise to the formation of a cold, extremely energetic electron beam, which will only occur for the higher laser intensities This effect, referred to as **snowplow** is shown in Figures 1. Eventually, these electrons will lapse into the center of the laser pulse.

The electrons initially trapped at the center of the pulse feel an extremely intense **quasi-static** e-m field, and are expelled transversely with vx close to c, moving rapidly out of the region where the laser field is present.

Finally, the electrons trapped at the trailing edge of the pulse move relatively slower and soon go out of phase with the laser e-m field, undergoing scattering.

These three regimes are, however, not so visible at the lowest intensity, 10^{20}W/cm^2. Snowplow will only occur at the highest intensities, since only then the initial longitudinal ponderomotive push will be sufficiently large to instantly accelerate the electrons to longitudinal velocities close to c, enabling them to move with the front of the pulse for long distances, generating the cold electron bunch.

The ion energy yields are similar to those expected from a Coulombic explosion of atomic clusters, varying linearly with the cluster ion density, as predicted by the model.

Scaling of the maximum ion energy with the charge state of the target ions is a valid procedure when handling Coulombic explosion of foils, as verified previously in the cluster PIC simulations already described. Changing the foil species to Ar^{9+} will then result in the production of GeV ion beams (Figure 2).

CONCLUSIONS

The interaction of a ultra-short high-intensity laser pulses with a thin foil were modelled by a PIC-code, the problem being similar to that of an exploding 2D cluster. For higher laser intensities, ranging from 10^{20} - 10^{23} W/cm^2, the H plasma foil electrons are trapped and accelerated to relativistic velocities, close to c, by longitudinal snowplow effect and via transverse quasistatic e-m acceleration. Snowplow is responsible for the formation of a cold, MeV electron beam, moving with the front of the laser pulse. Space-charge separation of electrons causes Coulombic explosion of ions, mainly in the forward and backward directions, exhibiting energies up to 13 MeV. Moreover, since in the tail of the ionic distribution 100 MeV ions are observed, it is valid to assume, based on the linear scaling of ionic final energy with the ion charge state, already observed in the cluster expansion simulations, that GeV ions can be produced, as long as the target ions have appropriate charge states, such as, for instance, the case of Ar^{13+}.

REFERENCES

1. M. Eloy, R. Azambuja, J. T. Mendonça and R. Bingham, *RAL Annual Report* (1999).
2. M. Eloy, R. Azambuja, J.T. Mendonca and R. Bingham, *Inertial Fusion Sciences and Applications 99*, Elsevier 1038 (1999).
3. M. Eloy, R. Azambuja, J. T. Mendonça and R. Bingham, *Phys. Plasmas* **8,** 1084 (2001).
4. M. Eloy, R. Azambuja, J. T. Mendonça and R. Bingham, *Phys. Scripta* **60,** T89 (2001).
5. M. Eloy, R. Azambuja, J. T. Mendonça and R. Bingham, *Proceedings of SPIE* **4424,** 418 (2001).
6. J. T. Mendonça, J. R. Davies and M. Eloy, *Meas. Sci. Technol.* **12,** 1801 (2001).

Theory of Anomalous Bohm Diffusion and the related concept of Photo-Field Fusion Phenomena

V.S. Belyaev[1], V.N. Mikhaylov[2]

[1] *The Central Research Institute of Machine Building of Russian Aviation and Space Agency*
[2] *The Institute of Strategic Stability of Ministry for Atomic Energy of the Russian Federation*

Abstract. Anomalous (or Bohm) diffusion is plasma diffusion across the magnetic field lines. The expression for diffusion coefficient was only suggested due to analyzed experimental data but not derived theoretically. Our analysis of this process was performed using three independent approaches — method of adiabatic invariant, magnetic-hydrodynamic approach, quantum continuity equation. The quantum nature of anomalous diffusion in plasmas in magnetic field as a tunneling transition of potential barrier was demonstrated. The coefficient of anomalous diffusion was determined. It is demonstrated that this coefficient controls cyclotron emission, dynamic pinch, field-induced (tunneling) ionization of atoms, plasma particle transport in magnetic. Super strong magnetic fields generated in laser-produced plasma, super-high pressure connected with these fields, important and determining role of magnetic interaction, quantum character of transport mechanism of particles as well high efficiency of electromagnetic fields transformation through atomic structures to scale close to nuclear are the basis of the principal new concept put forward by us — Photo-Field Fusion.

INTRODUCTION

Anomalous diffusion [1] is known since 1948 when American physicist D. Bohm having analyzed experimental data assumed that plasma diffusion across the magnetic field lines occurs much faster than it is predicted by the classical theory. Bohm assumed that collective processes in plasma (instabilities) may give rise to an electron diffusion with a coefficient equal to $D_B \approx c\kappa T/16eB$, where c - velocity of light, κ - Boltzmann constant, T - temperature, e - electron charge, B - magnetic field induction. This expression was only suggested but not derived theoretically.

Criteria of classical and quantum plasma treatment have been analyzed in works [2,3] indicate that laser-produced plasma is a quantum object and demand for its consideration the quantum theory methods for intensities higher than 10^{15} W/cm^2.

In work [2] our analysis was performed with the use of three independent approaches. We demonstrated that the coefficient of anomalous diffusion is an adiabatic invariant of the same nature as the adiabatic invariant $W_\perp / B$, where $W_\perp$ is the energy of transverse rotation of an electron in a magnetic field B. We have shown that the quantum continuity equation is reduced, under certain conditions, to the diffusion equation with the determined coefficient of anomalous diffusion. The same value of the diffusion coefficient has been found with the use of the magnetic-hydrodynamic approach.

Our analysis has allowed us to understand the nature of anomalous diffusion. Due to the high numerical value of the coefficient of such a diffusion, this process can be considered as a field-induced (tunneling) instability. In other words it is tunnel effect.

THEORY OF ANOMALOUS BOHM DIFFUSION

Anomalous Diffusion Coefficient as an Adiabatic Invariant

Generally, the diffusion coefficient is determined using mobility μ by the relationship [4] $D = \mu \kappa T$.

Mobility μ is defined via τ - mean time between collisions and m – particle mass $\mu = \tau/m$.

Consider the diffusion of electrons involved in larmor rotation in a magnetic field. The electron motion on larmor orbits in the plane normal to the magnetic field can be related to the temperature $T_\perp$ associated with the energy of larmor rotation as a limited and hence quantized process as $\kappa T_\perp = \hbar \omega_L$, where $\omega_L = eB/mc$ - Larmor frequency.

Suppose that the share of this energy in the total energy κT is determined via the fine structure constant $\alpha = e^2/\hbar c$ by $\kappa T_\perp = \alpha^2 \kappa T$. Thus in hydrogen-like atoms the spin-orbit interaction is proportional to $\alpha^2 R_y$, where R_y – the Rydberg constant.

Take into account these relation and assume $\tau = 1/\omega_L$, obtain

$$D = \alpha^{-2} \hbar/m, \qquad (1)$$

The relation between the diffusion coefficient obtained and the empirical expression for the Bohm diffusion can be illustrated by substituting $\hbar \omega_L = \alpha^2 \kappa T$ and $\omega_L = eB/mc$ into $D = \alpha^{-2} \hbar \omega_L \cdot 1/m\omega_L$, we have the relationship $D = c\kappa T/eB$, which agrees with that obtained by Bohm with an accuracy of the coefficient.

Note the linkage between the obtained diffusion coefficient independent of magnetic induction since $D = c\kappa T/eB = \alpha^{-2} \hbar/m$ and the adiabatic invariant $W_\perp/B$, where $W_\perp$ - the energy of electron cross motion in a magnetic field. These are two adiabatic invariants of the same nature. Their essence is in the quantum nature of electron rotation energy within larmor radius in a magnetic field.

Quantum-Mechanical Consideration

Below we obtain an expression for the Bohm diffusion coefficient by quantum mechanics methods.

The continuity equation for probability density $\psi^*\psi$ and its associated electric current looks like

$$\frac{\partial(\psi^*\psi)}{\partial t} + div\left\{\frac{i\hbar}{2m}[\psi\nabla\psi^* - \psi^*\nabla\psi] - \frac{e}{mc}\mathbf{A}\psi^*\psi\right\} = 0, \qquad (2)$$

where $\mathbf{A}$ - vector-potential.

Take into account gauge invariance $\mathbf{A} \to \mathbf{A} + \nabla\psi^*\psi$ the equation (2) reduces as showed in [5] to diffusion equation $\frac{\partial(\psi^*\psi)}{\partial t} = \alpha^{-2} \frac{\hbar}{m} \nabla^2 \psi^*\psi$ with diffusion coefficient (1).

Magnetohydrodynamic Consideration

Below the Bohm diffusion process is treated under the framework of classical magnetic hydrodynamics.

Following [1], the magnetic hydrodynamic equation which stems from the combined Ohm's law for plasma and associates $\mathbf{B}$ and $\mathbf{V}$ can be written:

$$\frac{\partial \mathbf{B}}{\partial t} + \mathbf{B}(\nabla \cdot \mathbf{V}) - (\mathbf{B} \cdot \nabla)\mathbf{V} + (\mathbf{V} \cdot \nabla)\mathbf{B} = \frac{c^2}{4\pi\sigma} \nabla^2 \mathbf{B},, \qquad (3)$$

where electroconductivity σ is expressed via density n and mean (effective) collision rate v_{eff}

$$\sigma = \frac{ne^2}{mv_{эфф}} = \frac{1}{4\pi} \frac{\omega_{pe}^2}{v_{эфф}}.$$

Consider case of quick rise of magnetic field. Then we can ignore inertial terms in (3). Considering $\omega_{pe} \approx \omega_a$, $v_{eff} \approx \omega_a = \alpha^2 mc^2/\hbar$, obtain

$$\frac{\partial B}{\partial t} = \alpha^{-2} \frac{\hbar}{m} \nabla^2 B. \qquad (4)$$

This is diffusion equation with same diffusion coefficient (1).

Magnetic hydrodynamics is applicable when a sufficiently strong magnetic field is present in a liquid. This requirement is met with the Lundquist criterion $L = \dfrac{\sigma B_c l_c}{n^{1/2} c^2} \gg 1$.

Note that the condition is met quite well for the space plasma. Therefore, the superdense plasma produced by action of high-level ultrashort laser radiation on a target adequately reproduces the space plasma conditions and can model the physical processes going on space scales [5].

CONCERNED WITH BOHM DIFFUSION PHENOMENA

Induced Cyclotron Radiation and Dynamic Pinch

The process of electron diffusion in a magnetic field – the Bohm diffusion – by nature represents transitions between Landau energy levels induced by the magnetic field. This is a pure quantum effect. In this case the population of a level changes whereupon an electromagnetic quantum is either emitted or absorbed. Then the diffusion equation with the coefficient $\alpha^{-2} \hbar/m$, when combined with the kinetics equation which governs changes in population of the energy level involved in the process of photon emission/absorption, has the wave equation descriptive of electromagnetic wave as a joint resolution.

Indeed from two above-mentioned equations:

$$\begin{cases} \dfrac{dn}{dt} = D \dfrac{d^2 n}{dx^2} \\ \dfrac{dn}{dt} = \omega_{эфф} n \end{cases} \quad (5)$$

where n – energy level population, $D = \alpha^{-2} \hbar/m = c^2/\omega_{eff}$ - diffusion coefficient, ω_{eff} – effective frequency.

Assuming the vector-potential to be defined by the current which in turn is defined by the electron density, i.e. $j = nev \sim n$, $j = ne^2/mc\, A \sim A$, we obtain the wave equation for the vector-potential

$$\dfrac{d^2 A}{dt^2} - c^2 \dfrac{d^2 A}{dx^2} = 0. \quad (6)$$

Thus the diffusion process in a magnetic field produces electromagnetic radiation. This cyclotron radiation arises from the induced transitions between the magnetic field energy levels – Landau levels. The Einstein coefficients for induced transitions are known to be $B_{nm} = B_{mn}$, i.e. the probability of induced downward transition is equal to that of induced upward one. Here it gives diffusion paths both away from the center of larmor circles, that is electron trajectories in a magnetic field, and towards the center. The latter case relates to the above-mentioned phenomenon of fall to center which is addressed in [6,7] for an atom.

On macro-scales the fall to center is referred to as pinch-effect. The case under consideration, that is diffusion of electrons rotating in the larmor circles and "falling" to center, corresponds to θ-pinch such that a large axial field arises near the plasma spout surface in a short space of time. However, the magnetic hydrodynamic equation (4) which stems from the combined Ohm's law for plasma (3) is also true for Z-pinch wherein the current flows along the plasma spout axis.

The present considerations brings us to a conclusion that the dynamic pinch-effect represents the diffusion process with the large diffusion coefficient $D = \alpha^{-2} \hbar / m = c^2 / \omega_a$.

Note that the atomic-scale quantum phenomenon – the fall to center – can be qualified as an atomic θ-pinch.

Anomalous Diffusion – Tunnel Effect. Plasma Bleaching in Magnetic Field

The anomalous diffusion is considered in work [2] as a tunnel transition of a potential barrier which is formed by an oscillator potential and has a periodic structure of Landau levels. The periodic structure provides condition for overcoming the potential barrier when the barrier transmittance is equal to unity for the wide range or particle energy. Analogy is drawn between this plasma "bleaching" by a magnetic field and the optical phenomenon of electron transit through a solid periodic lattice. As examples of periodic potential barrier overcoming, α-decay probabilities in a magnetic field and field (tunnel) ionization probability are considered.

The plasma bleaching effect associated with the periodic structure of charged particles rotation energy levels in a magnetic field must be taken into account in calculations of the systems of plasma confinement by a magnetic field.

PHOTO-FIELD FUSION CONCEPT

The developed Bohm diffusion consideration has general character and it is correct in skin-layer and in atom as well right up to Compton wavelength.

The vortical electrons structures arised in laser plasma created superstrong magnetic fields can effectually transform macroscale own magnetic flux into Bohr atom radius [3]. A unit magnetic flux can be trapped by a single atom under these conditions during the formation of an electron vortical structure with a magnetic field inside this structure. This process may play an important role since the "fall" on the center may occur in an atom at later stages due to the predominance of the magnetic potential with a threshold value $U = -\hbar^2/8mr^2$ under conditions when the magnetic flux is conserved. As a result, magnetic fields are induced around a nucleus, and these fields are sufficient for the excitation of the nucleus and subsequent processes of decay and fusion. Fusion under these conditions differs in its nature from thermonuclear fusion, as it has a character of field-induced nuclear fusion. Note that the fall on the center in the regime when condition is satisfied was first described by Landau [8]. The possibility of excitation of nuclei under conditions of the fall on the center was considered in [7].

The fall on the center is determined by an anomalous diffusion also. It lead to an electron condensation in nearest vicinity of nucleus.

The vortical electronic structures in the laser-produced plasmas provide a fertile ground for nuclear decay and/or nuclear fusion within these structures. These structures exhibit many parameters sought for in the design of large-scale plants of TokamaK type. Thus, even the megagauss magnetic fields, surrounding an atom in these structure, provide the pressure of order of one million atm (bar). Then the pressure grows with the square of magnetic field intensity H^2, and there is an evidence of pressures of 10^6 Mbar present in the laser-produced plasmas. Moreover, the process of electron near-nucleus condensation noticeably suppresses Coulomb repulsion of nuclei – the phenomenon much similar to the well-known μ-catalysis. These nuclei, similar to μ-nuclei, feature a high probability of coalescence/fusion under superhigh pressures due to the tunnel effect. It is favorable energetically for light nuclei. For heavy nuclei, the probability increases for decay of any kind, including cluster-fission decay. The peculiarities and probability of α-decay for nuclei in strong magnetic fields, as we see them, is considered in [7]. Thus, the above-discussed process of fusion in the laser-produced plasmas brings us to a fundamentally new concept which can be called "photo-fusion". More precisely, considering the collective nature of the processes and the prevalence of the tunnel (field) processes at all stages of progress, the concept, put forward here by us, can be given the name "photo-field fusion".

The photo-field fusion is based on the efficiency of electromagnetic fields transformation in the laser-produced plasmas through atomic structures and atoms to scales close to nuclear.

ACKNOWLEDGMENTS

Work was supported by International Science and Technology Center, Project 2155.

REFERENCES

1. Krall, N.A. and Trivelpiece, A.W., Principles of Plasma Physics, Academic, New York, 1973.
2. Belyaev, V.S., and Mikhaylov, V.N., Laser Physics 11, No.8, p.957 (2001).
3. Belyaev, V.S., and Arefyev, V.I., "Nature of atomic-nuclear processes in intense short laser-matter interaction" in Proseeding of International Forum on Advanced High-Power Lasers and Applications AHPLA'99, Osaka, Japan, 1999.
4. Feymann, R., Leyton, R., Sendz, M., Feymann Physic Lectures, Mir, Moscow, 1967.
5. Peratt, L.P., Physics of the Plasma Universe, Springer-Verlag, New York, 1992.
6. Landau, L.D., and Lifshits, E.M., Quantum Mechanics, Nauka, Moscow, 1974.
7. Belyaev, V.S., Arefyev, V.I., and Mikhaylov, V.N., Laser Physics 8, No.6, p.157 (1998).
8. Landau, L.D. and Lifshits, E.M., Course of Theoretical Physics, vol. 3: Quantum Mechanics: Non-Relativistic Theory, 3rd ed., Pergamon, New York, 1977.

High Energy Electrons Formation in Laser-Produced Plasma: Theory and Experiment

V.S. Belyaev[1], V.N. Mikhaylov[2]

[1] *The Central Research Institute of Machine Building of Russian Aviation and Space Agency*
[2] *The Institute of Strategic Stability of Ministry for Atomic Energy of the Russian Federation*

Abstract. The mechanism of high-energy electrons formation in ultra-high intensity laser pulse interaction with solid targets has been suggested and investigated. The relationship between kinetic energy of hot electrons generated from laser-produced plasma and intensity of laser radiation in wide range of frequencies has been established. The phenomenon has been investigated using methods of relativistic mechanics, classical electrodynamics and plasma instabilities. The theoretical results are well verified by numerous experiments including original ones by the authors.

INTRODUCTION

This investigation is an attempt to reveal and describe, based on the model suggested, the high-energy electron formation mechanism in laser plasmas so as to derive theoretical dependences which would represent specific relations between the parameters of fast electrons, laser radiation and target substance.

Any theory can be accepted only after reliable experimental verification. The degree of reliability is determined not only by the sufficient diversity of independent experimental data, but also by the ability to choose out of these data those best representative of the overall pattern. Analysis of numerous experiments to measure energy of fast electrons formed in laser plasmas shows that with a particular laser facility, given its available radiation intensity, fast electron maximum energy can be determined most closely. Generally, it is electron maximum energy values that are most widely presented in experimental investigations. This is motivated not only by experimenters' striving to get extreme record-breaking output parameters, but also by the possibility to most closely determine the electron maximum energy around their spectrum extrapolation at specified intensity of laser radiation incident on a target. On this basis we will establish our theoretical model of the maximum-energy electron formation process for a given laser radiation intensity.

THEORETICAL MODEL

Without going into details of magnetic field generation mechanisms, it can be noted that a vortical electron structure develops eventually in plasma. Given the applied electric field (constituent of the incident laser radiation) and the dominance of tunnel ionization, a great number of electrons (practically determined by solid density) are accelerated. This current of electrons generates a magnetic field which bends their trajectory. Under certain conditions these trajectories can close at skin-layer depth within Larmor-radius circle. The high electron density and, correspondingly, the circular current strength cause super-strong magnetic fields generation.

Condition for such fields generation can be written as a condition for electron movement around such a circle in the form of a balance between the centrifugal force and the Lorentz force:

$$\frac{mV^2}{r} = \frac{eVB}{c}, \qquad (1)$$

where $r = \delta/2$, δ - skin-layer thickness, e, m, V – charge, mass, electron velocity, c – velocity of light, B – magnetic induction in the electron orbit.

Taking electromagnetic field penetration depth δ to be equal to incident radiation wave length λ, we have $r = \lambda/2$.

Given the relationship between mass and velocity, the kinetic energy change due to the action of the forces applied is always equal to

$$E_{KIN} = (m_V - m_0)c^2 = m_0 c^2 (\gamma - 1), \qquad (2)$$

where m_V - relativistic mass, m_0 – electron rest mass, $\gamma = 1/\sqrt{1 - V^2/c^2}$ - relativistic factor.

Considering:
- relativistic expression of electron momentum

$$P = m_0 V \gamma, \qquad (3)$$

where V – electron velocity;
- use of generalized momentum

$$\mathbf{P} = \mathbf{p} + \frac{e}{c}\mathbf{A}, \qquad (4)$$

where $\mathbf{p}$ – ordinary momentum (3), $\mathbf{A}$ – vector-potential;

- magnetic field B cylindrical symmetry: $B_X = 0$; $B_Y = 0$; $B_Z = -\frac{\partial A_x}{\partial y} = B$, which allows to put $A = Br$ and

$$P = m_0 V = \frac{e}{c} Br, \qquad (5)$$

So from (2) – (5) we have

$$E_{KIN} = m_0 c^2 \left[1 + \left(\frac{eBr}{m_0 c^2}\right)^2\right]^{\frac{1}{2}} - m_0 c^2, \qquad (6)$$

To find the electron maximum kinetic energy at specified intensity of laser radiation incident on the target we need the maximum value of B – magnetic field induced in laser plasma. This value can be estimated using the energy conservation law.

Omitting calculations we can use the following formula easy to keep in mind:

$$B_{MAX}\,[\text{Gs}] = 10^{-1}\sqrt{J[\text{W/cm}^2]}, \qquad (7)$$

Substitution this formulae in expression (6) for kinetic electron energy gives:

$$E_{KIN} = 0.5\left[1 + \frac{9J\lambda^2}{10^{18}}\right]^{\frac{1}{2}} - 0.5, \qquad (8)$$

where intensity J expressed in W/cm^2, λ – in micrometer, kinetic energy – MeV.

Graph of this dependence show Fig. 1 by curve 1.

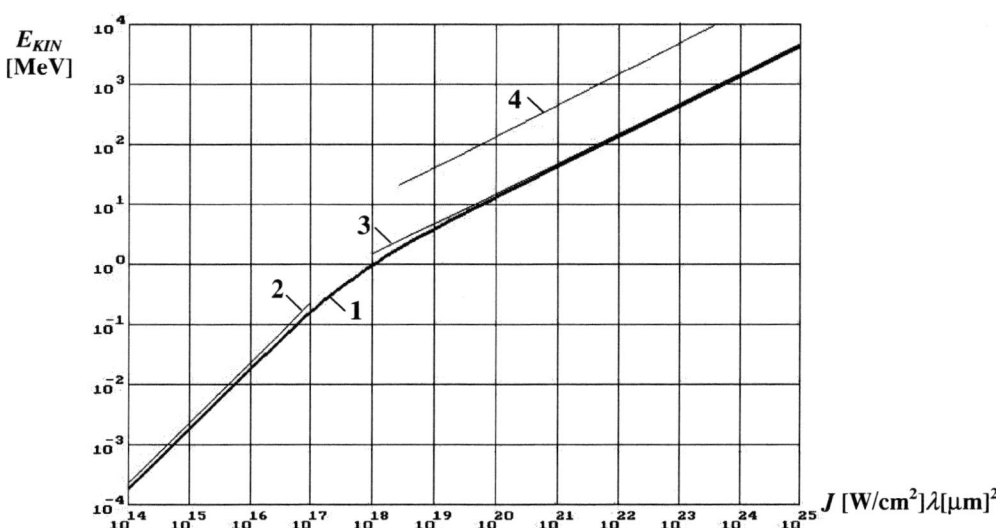

FIGURE 1. Dependence of electron kinetic energy on laser radiation intensity.

Consider limiting cases.

1. $E_{KIN} = m_0 c^2 = 0.5$ MeV.

Expression (8) gives this value at intensity

$$J_R = \frac{1}{3} 10^{18} \left[\frac{1}{\lambda\,[\mu m]}\right] \frac{W}{cm^2}, \qquad (9)$$

The intensity J_R can be called as relativistic intensity.

2. $E_{KIN} \ll m_0 c^2$; $J < J_R$.

In this case
$$E_{KIN} = 2.25 \frac{J\lambda^2}{10^{18}}, \qquad (10)$$

Graph of this dependence show at Fig. 1 by curve 2.

3. $E_{KIN} \gg m_0 c^2$; $J > J_R$.

For this case
$$E_{KIN} = 1.5 \left(\frac{J\lambda^2}{10^{18}}\right)^{\frac{1}{2}}, \qquad (11)$$

and graph of this dependence show on Fig. 1 by curve 3.

Equations obtained for small ($<m_0c^2$) and large ($>m_0c^2$) values of kinetic energy agree with those in use for calculations of particle energy in a cyclotron and in a betatron, correspondingly. In both cases electrons are accelerated under the action of an electric field. In a cyclotron, this is a periodically changing electric field applied externally. In a betatron, this is a vortex electric field occurring with axisymmetric magnetic field rise in time. In laser plasmas a magnetic field is generated giving rise to a vortex electric field accelerating electrons. Thus the laser-plasma electron acceleration mechanism resembles the betatron case.

Equation (6) for electron kinetic energy was derived on the assumption that the electron acceleration is governed only by the laser radiation incident on the target without considering the processes going within the target substance, specifically, ionization process. Formally, it is reflected in the fact that the skin-layer size is determined by the laser radiation frequency

$$\delta = c/\nu = 2\pi c/\omega = \lambda, \qquad (12)$$

meaning that the laser radiation frequency ω is an effective frequency. This assumption is true only at the first stage of interaction with the substance when a vortical electron structure develops on skin-layer scales, its characteristic size being in accordance with (12). This structure is unstable and there is a possibility of its transformation to smaller-scale structures. This process is known as a dynamic pinch.

It is demonstrated in [1] that in case of laser plasmas produced by the action of high-intensity ($J > 10^{16}$ W/cm^2) laser radiation of ultrashort duration ($\tau < 10^{-12}$ sec) on a solid target this process is of quantum nature and can be described by the diffusion equation. Without going into the process nature, note that under tunnel ionization the vortical electron structure generated on skin-layer scales (12) transforms to another one, its characteristic size now being determined by the ionization frequency as an effective frequency at the next stage of laser radiation interaction with the substance, i.e. at the stage of tunnel ionization development:

$$l_i = 2\pi c/\omega_i, \qquad (13)$$

Assuming that the vortical electron structure transformation process goes with the magnetic flow kept unchanged, we have

$$B_0 \lambda^2 = B_i l_i^2, \qquad (14)$$

where B_i – magnetic field within the vortical structure, its characteristic size l_i, being determined by (13). Such a vortical structure provides the following kinetic energy to the electrons:

$$E_{кин} = eB_i l_i = eB_0 \frac{\lambda^2}{l_i}, \qquad (16)$$

Equation (16) determines the maximum energy of the small group (tail) of high-energy electrons. This dependence can be represented via the energy or ionization potential of the target substance atoms:

$$E_{KIN} = eB_0 \lambda \frac{\omega_i}{\omega} = 1.5 \left(\frac{J\lambda^2}{10^{18}}\right)^{\frac{1}{2}} \frac{\omega_i}{\omega} = 1.5 \left(\frac{J\lambda^2}{10^{18}}\right)^{\frac{1}{2}} \frac{I}{\hbar\omega} \text{ [MeV]}, \qquad (17)$$

Here J is in W/cm^2, λ - in μm, I and $\hbar\omega$ - in eV. This dependence is plotted in Fig. 1 (curve 4).

The equation obtained demonstrates the proportionality between the electron energy and ionization frequency, hich determines physical nature of the electron acceleration process. The physics of the electron acceleration processes as a result of high-intensity laser radiation action on a substance is closely related to the physics of the ionization processes in superatomic intensity fields.

The ionization frequency is generally one or two orders higher than the laser one. This results in the high acceleration rate and electron energy.

The process of dynamic pinch development gives rise to formation of high-energy tail (17) and has threshold nature. Our estimations give value of threshold $0.31 \cdot 10^{18} - 3.2 \cdot 10^{19}$ W/cm^2. Threshold smearing evidences for stochastic character of the process.

Table 1 presents experimental results of electron formation investigations.

TABLE 1.

N	Country, Facility	Laser Radiation Intensity, W/cm^2	Wave length, μm	Pulse duration	Target material, Ionization potential	Experimentally obtainer E_{KIN}^{MAX}, MeV	Theoretically predicted E_{KIN}^{MAX}, MeV	Formulae for E_{KIN}^{MAX}	Reference
1	Russia, "Neodym"	~10^{17}	1.06	1 ps	Be, Al, Cu, Ta	0.24	0.25	(10)	[2]
2	France, CEA/LV	$2 \cdot 10^{18}$	1.056	330 – 500 fs	Ct, 10.4	2.4	2.12	(11)	[3]
3	Russia, "Progress"	$5 \cdot 10^{18}$	1.053	1.4 – 1.5 ps	Sn, 7.34	< 22.0	20.8	(17)	[4]
4	Germany, France	10^{18} - 10^{19}	0.78	30 fs	W, 7.97	~ 1.0	0.77 – 3.23	(8)	[5]
5	England, VULCAN	< 10^{19}	1.053	~ 1 ps	Ta, 7.8	30.0	31.3	(17)	[6]
6	USA	$8 \cdot 10^{19}$	1.06	0.5 ps	Au, 9.2	100.0	104.6	(17)	[7]

The good agreement between theory and experiment suggests the realizability of the proposed high-energy electron formation mechanism in laser plasmas.

ACKNOWLEDGMENTS

Work was supported by International Science and Technology Center, Project 2155.

REFERENCES

1. Belyaev, V.S., and Mikhaylov, V.N., Laser Physics 11, No.8, p.957 (2001).
2. Matafonov, A., and Belyaev, V., "Experimental Investigation of Gamma and Neutron Radiation from Laser Produced Plasma" in 28th European Physical Society Conference on Controlled Fusion and Plasma Physics-2001, edited by C. Silva et al., Madeira Tecnopolo, Funchal, Portugal, 2001, p. 730.
3. Malka, G., and Miquel, J.L., Phys. Rev. Letters, 77, No.1, p.75 (1996).
4. Borodin, V.G. et al., JETPh Letters, 71, No.6, p.354 (2000).
5. Nickles, P.V. et al., Quantum Electronics, 27, No.2, p.165 (1999).
6. Ledingham, K.W.D., and Norreys, P.A., Contemporary Physics, 40, No.6, p.367 (1999).
7. Cowan, T.E. et al., Laser and Particle Beams, 17, No.4, p.773 (1999).

Millimeter Wave Generation by a Relativistic Electron Beam in a Plasma Filled Sheath Helix Loaded Waveguide

N.K. Jaiman* and V.K. Tripathi

Physics Department Indian Institute of Technology
New Delhi, India
*Permanent address : Physics Department, Raj Rishi college, Alwar – 301001, India.
*Email: nkjaiman@rediffmail.com

Abstract. A relativistic electron beam propagating through a plasma filled sheath helix, a slow wave device, excites a Trivelpiece-Gould (TG) mode around 10 GHz with high efficiency. A large amplitude TG mode undergoes stimulated Compton backscattering off the electron beam parametrically driving an electromagnetic wave at millimeter wave length. The growth rate of the instability scales as one third power of the beam density and decreases with the operating frequency of the free electron laser (FEL). The FEL frequency increases with γ_0^2, γ_0 is the relativistic gamma factor. As the beam mode looses energy, the phase matching gets detuned, resulting in the saturation of the instability.

INTRODUCTION

Travelling wave tubes are a versatile source of microwave generation upto 30 GHz. frequency. It employs a slow wave structure, e.g. a sheath helix to slow down the transverse magnetic (TM) or Trivelpiece-Gould (TG) modes which exchange energy with a co-propagating electron beam via Cerenkov resonance. The device can not operate at higher frequencies as the slow modes are physically detached from the beam and are localized near the slow wave structure. Walsh et.al[1]. have proposed an elegant scheme of employing relativistic electron beam for the generation of shorter wavelengths. In this case TM mode needs to be slowed down just a little from its velocity in free space and its mode structure has large radial width. Frequency up-conversion via. free electron laser (FEL) mechanism using a background propagating wave at 8.4 GHz (excited linearly in a backward wave oscillator (BWO) has been experimetally observed by Kehs et al[2]. Balakirev et al[3]. and Joshi at al[4]. have examined the feasibility of a plasma wave wiggler for FEL operating in the Compton regime. Jaiman and Tripathi[5] have studied a cylindrical plasma configuration as a slow wave structure

In this paper we explore the possibility of millimeter wave generation by a relativistic electron beam as two stage process as plasma filled sheath helix loaded wave guide. The electron beam excites a TG mode via Cerenkov interaction. The TG mode has both the forward propagating and backward propagating components. As the mode acquires large amplitude, it undergoes stimulated Compton backscattering from the electron beam producing short wavelength radiation. In section II we derive a dispersion relation for a slow electromagnetic mode in plasma filled sheath helix. In section III we carry out the fluid treatment of the nonlinear three mode coupling. The coupled mode equations are solved using first order perturbation theory to obtain growth rate. A discussion of the results is presented in section IV.

PUMP WAVE DISPERSION RELATION

Consider a sheath helix loaded waveguide. It is filled with a uniform plasma of density N_P and is immersed in a strong d.c. magnetic field $B_s \hat{z}$. The radius of sheath helix is r_0. We treat helix as an anisotropically conducting cylinder with infinite conductivity along the windings of the helix which is at an angle ψ to the plane perpendicular to the z-axis. This is reasonable as long as the phase change over each element of length d of the periodic structure is less than π. A solid electron beam of radius r_0 and density $N_b = N_{0b}\delta(r - r_b)$, propagates through the sheath helix with velocity $v_b \hat{z}$. The magnetized plasma acts as an anisotropic medium for the modes supported by the structure with dielectric constant given by the permittivity tensor $\underline{\underline{\epsilon}}$. The electric field vector of the modes inside the plasma region obey the wave equation

$$\nabla^2 \vec{E} - \nabla(\nabla \cdot \vec{E}) + \frac{\omega^2}{c^2} \underline{\underline{\epsilon}} \cdot \vec{E} = 0, \qquad \epsilon = 1 - \frac{\omega_p^2}{\omega^2} - \frac{\omega_{pb}^2}{\gamma_0^3(\omega - kv_{0b})^2}, \tag{1}$$

where $\underline{\underline{\epsilon}}$ is the plasma dielectric tensor and ω and $\vec{k}$ are the eigen frequency and wave vector respectively of the slow electromagnetic wave and ω_{pb} is the beam-plasma frequency, ω_p is the plasma frequency and γ_0 is the relativistic gamma factor. The z-comp. of (1) is,

$$\frac{\partial^2 E_z}{\partial r^2} + \frac{1}{r}\frac{\partial E_z}{\partial r} - \xi^2 E_z = 0, \text{ where } \xi^2 = \left(k_z^2 - \frac{\omega^2}{c^2}\right)\left(1 - \frac{\omega_p^2}{\omega^2} - \frac{\omega_{pb}^2}{\gamma_0^3(\omega - kv_{0b})^2}\right). \tag{2}$$

Solution of Eq. (2) is written as

$$E_z^I = A_1 J_0(\xi_1 r), \tag{3}$$

as in region I, $\omega_p > \omega$ and $k > \frac{\omega}{c}$;

$$E_z^{II} = B_1 I_0(\xi_1 r) + B_2 k_0(\xi_1 r)$$

as $E_z^{II} = 0$ at r==b, $E_z^{II} = B_1 \left[I_0(\xi_1 r) - \frac{I_0(\xi_1 b)}{k_0(\xi_1 r)} k_0(\xi_1 r) \right]. \tag{4}$

Similarly the B_z wave equation is

$$\frac{\partial^2 B_z}{\partial r^2} + \frac{1}{r}\frac{\partial B_z}{\partial r} - \left(k^2 - \frac{\omega^2}{c^2}\right) B_z. \tag{5}$$

The solution of Eq. (5) is

$$B_z^I = C_1 I_0(\alpha r), \quad B_z^{II} = D_1 I_0(\alpha r) + D_2 K_0(\alpha r), \tag{6}$$

where $\alpha = (k_z^2 - \omega^2/c^2)^{1/2}(1 - \omega_p^2/w^2)$. \tag{7}

The other field components are written as

$$E_r, B_\theta = \frac{i(k, \frac{\omega}{c})}{(\frac{\omega^2}{c^2} - k_z^2)} \frac{\partial E_z}{\partial r}, \text{ and } E_\theta, B_r = \frac{i(-\frac{\omega}{c}, k)}{(\frac{\omega^2}{c^2} - k_z^2)} \frac{\partial B_z}{\partial r}. \tag{8}$$

Using the boundary condition, $E_\theta^{II} = 0$ we obtain

$$B_z^{II} = D_1[I_0(\alpha r) - \frac{I_0'(\alpha b)}{K_0'(\alpha b)} - k_0(\alpha r)].$$

Using the boundary condition that the electric field should disappear in the direction of current flow (infinite conductivity), i.e.

$$E_z^I \sin\psi + E_\theta^I \cos\psi = 0, \qquad E_z^{II} \sin\psi + E_\theta^{II} \cos\psi = 0 \text{ at } r = r_0$$

and also that E_z must be continuous across the boundary, i.e.

$$E_z^{II} \cos\psi - E_\theta^{II} \sin\psi = E_z^I \cos\psi - E_\theta^I \sin\psi \text{ at r=b and } E_\theta^I = E_\theta^{II} \text{ at } r = r_0$$

and then using the boundary condition that the component of magnetic field in the direction of conduction must be continuous across the boundary since no current flows in the perpendicular direction to it, we obtain a dispersion relation of the slow electromagnetic mode.

$$\tan^2 \psi = \frac{-(\omega^2/c^2)\xi}{(\frac{\omega^2}{c^2} - k_z^2)^{3/2}} \left[\frac{J_0'(\xi r_0) K_0(\xi r_0) - J_0(\xi r_0) K_0'(\xi r_0)}{J_0'(\alpha r_0) K_0(\alpha r_0) - J_0(\alpha r_0) K_0'(\alpha r_0)} \right] \frac{I_0(\xi b)}{I_0'(\xi b)}. \qquad (9)$$

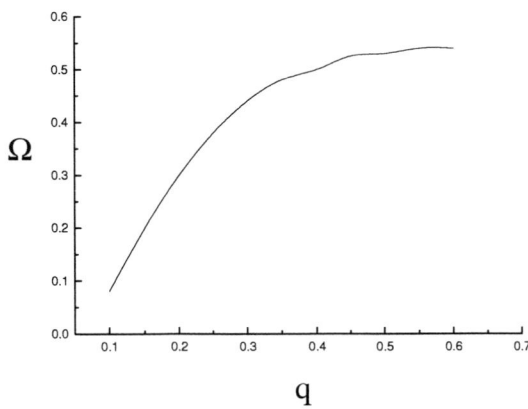

FIGURE 1. Dispersion curve of the pump (TG) mode for $\psi = 10^0$.

INSTABILITY ANALYSIS

Now we examine the stimulated Compton scattering of a large amplitude TG mode. The electric field of the pump is

$$\vec{E}_0 = \vec{A}_0(\vec{r}) \, e^{-i(\omega_0 t - k_o z)}. \qquad (10)$$

The beam acquires an oscillatory velocity and density perturbation due to the pump,

$$v_{0bz} = \frac{eE_{0z}}{mi\gamma_0^3(\omega_0 - k_{0z}V_b)},$$

$$n_{0b} = \frac{-iek_{oz}N_b E_{oz}}{m\gamma_0^3(\omega_0 - k_{0z}V_b)}. \qquad (11)$$

N_b is the beam density, e and m are electron charge and mass and $\omega_0 \ll \omega_c$ has been assumed. The pump couples with a high frequency electromagnetic wave, $\vec{E}_1 = \vec{A}_1 \, e^{-i(\omega_1 t - k_1 z)}$, and a space charge perturbation. The pump and the high frequency em wave exert a ponderomotive force on electrons whose z-component is given as $F_{pz} = -iek_z \phi_p$, where ϕ_p is the ponderomotive potential written as

$$\phi_p = \frac{-eA_0 A_1}{2m\gamma_0^3(\omega_1 - k_{1z}V_b)(\omega_0 - k_{0z}V_b)}, \qquad (12)$$

where $E_{1r}k_{1r} + E_{1z}k_{1z} = 0$ has been used.

In the presence of of F_{pz} and self consistent space charge field $\vec{E} = -\nabla \phi$ the electrons acquire an oscillatory velocity and density

$$v = \frac{F_{pz} + iek_z \phi}{mi\gamma_0^3(\omega - k_z V_b)} \quad , \quad n = \frac{N_b k_z v_z}{(\omega - k_z V_b)} \qquad (13)$$

The phase matching conditions $\omega_1 = \omega + \omega_0$ and $|k_1| = |k| - |k_0|$ yield the operating frequency of the FEL, $\omega_1 = 2\gamma_0^2(\omega_0 + k_0 c)$, where γ_0 is the relativistic gamma factor. The wave equation for the FEL wave is written as

$$\frac{\partial^2 \vec{E}_1}{\partial r^2} + \frac{1}{r}\frac{\partial \vec{E}_1}{\partial r} + \alpha^2 \vec{E}_1 = -4\pi i \omega_1 \vec{J}_{1b}^{NL}, \qquad (14)$$

where, $\vec{J}_{1b}^{NL} = -(1/2) \, nev_{0z}^* \, \hat{z}$. $\qquad (15)$

The complementary solution of the above equation is
$$E_1 = A_1 J_1(\alpha r). \qquad (16)$$

The pump wave couples to a beam space charge mode ($\omega, \vec{k}$) and a FEL radiation mode ($\omega_1, \vec{k}_1$) satisfying the dispersion relation,

$$\alpha^2 a^2 = X_{0p}^2, \tag{17}$$

where a is the outer radius of the plasma, or, $\frac{\omega_{1r}^2}{c^2} - k_{1z}^2 = \frac{X_{0p}^2}{a^2}$.

Also, $\quad \omega_1^2 - \omega_{1r}^2 = \{\int_0^a (-4i\pi\omega J_{1b}^{NL}) J_1(\alpha r) r \, dr\} / \int_0^a J_1^2(\alpha r) r \, dr \tag{18}$

Substituting the values of n and v_{obz}^* from Eqns. (13) and (11) respectively in Eq. (15) and then putting the resulting value of J_{1b}^{NL} in equation (18) we obtain,

$$(\omega_1^2 - \omega_{1r}^2)(\omega - k_z V_b)^2 = P, \tag{19}$$

where $P \equiv -\frac{e}{2} a_1 |A_{1z}|^2 v_{oz}^*$,

$$a_1 \equiv \frac{N_b e^2 k_z^2}{2m^2 \gamma_0^8 (\omega_1 - k_{1z} V_b)(\omega_0 - k_{0z} V_b)} + \frac{N_b e^2 k_{0z} k_{1z} \left(\frac{E_{1r}}{E_{1z}} + \frac{v_b k_{1z}}{(\omega_1 - k_{1z} V_b)}\right)}{2m^2 \gamma_0^6 \omega_1 (\omega_0 - k_{0z} V_b)}. \tag{20}$$

The growth rate (γ) of FEL instability turns out to be

$$\gamma \sim \frac{\sqrt{3}}{2} \left(\frac{P}{2\omega_1}\right)^{1/3}. \tag{21}$$

DISCUSSION

We solve the nonlinear dispersion relation for the following set of typical parameters: equilibrium plasma density, $N_{op} \approx 10^{11}$ cm.$^{-3}$, beam current $I_b \approx 1.5$ KA, beam radius b = 0.2 cm. and helix radius $r_0 \approx 0.4$ cm, pump frequency $\omega_0 / 2\pi \approx 8.4$ GHz, $\gamma_0 \approx 2$, pump power density ≈ 17 MW/cm.^{2} and $N_{0b} \approx 10^{10}$ cm.$^{-3}$. We have observed that the FEL frequency ω_1 increases more than linearly with γ_0. Also the growth rate of the parametric instability decreases with ω_1. It may be mentioned that the choice of the pump power density ≈ 17 MW/cm.^{2} is reasonable as powers of the order 400 MW at 8.6 GHz have been observed in BWO. The growth rate of the FEL instability scales as 1/3rd power of beam density. As the beam mode loses energy, the phase matching gets detuned, resulting in the saturation of the instability. In figure (1), we have plotted the variation of dimensionless frequency ($\Omega = \omega_0 / \omega_p$) with the dimensionless pump wave vector $q = k_{0z} c / \omega_p$.

REFERENCES

1. J.E. Walsh, T.C. Marshal and S.P. Schlesinger, Phys. Fluids, **20**, 709 (1977).
2. R.A. Kehs, Y. Carmel, V.L. Granatstein and W-W. Dester, Phys. Rev. Lett.; **60**, 279 (1988)
3. U.A. Balakirev, V.I. Miroshniechenka and Ya. B. Iainberg, Sov. J. Plasma Phys; **12**, 563 (1986).
4. C. Josji, T. Katsouleas, J.M. Dawson, Y.T. Yan, and J.M. Slater, IEEE J. Quantum Electronics, QE **24**, 1571 (1987).
5. N.K. Jaiman and V.K. Tripathi, Physics of Plasmas **4**(7), 2687 (1997).

Non-thermal particle populations in space plasmas

M. P. Leubner

Institute for Theoretical Physics, University of Innsbruck
A-6020 Innsbruck, Austria
and
Space Research Institute, Austrian Academy of Sciences
A-8042 Graz, Austria

Abstract. Numerous in situ observations indicate clearly the presence of non-thermal electron and ion structures as ubiquitous and persistent feature of most space plasma environments. The three detected dominant deviations from multi-temperature Maxwellians are suprathermal particle populations loss-cone structures and, provided by recent high resolution data analysis, gyrophase-bunched electron and ion distributions at plasma boundary layers. After clarifying the occurrence of specific non-thermal features in space plasmas different generation mechanisms of energetic particles are discussed in view of coronal/solar wind and magnetospheric conditions. We demonstrate from a Fokker-Planck approach how Landau interaction ultimately leads to kappa-distributions, favored in astrophysical plasma modeling and show that these distributions turn out theoretically as consequence of a nonextensive entropy approach. In addition, magnetic field gradients can act as catalyst for the generation of energetic populations up to relativistic energy due to synergetic effects in a multi-stage acceleration process, relevant for solar flare conditions. With regard to wave-particle interaction processes and instability analysis we present a class of highly general analytical representations of velocity space distributions, shown to model accurately observed complicated situations ranging from multi-component, two-temperature high energy tail formations and loss-cone features to non-gyrotropic distributions, where Maxwellians are recovered as special case.

INTRODUCTION

Depending on their symmetry properties velocity space distributions can be classified into velocity space dependent, gyrotropic functions ranging from isotropic Maxwellians to two-temperature distributions in uniform magnetic fields where symmetry is retained also in presence of suprathermal tails, beam or loss-cone features. Dealing with velocity and space dependence any symmetry is lost and we talk about anisotropic, non-gyrotropic distributions that may be time dependent as well, if e.g. rotating. Space plasma observations indicate three major characteristics of non-thermal features: (a) suprathermal tails [1, 2], (b) loss-cone structures [3, 4] and (c) non-gyrotropic conditions [5, 6], or a combination of all. Examples of experimental verifications for (a) are auroral particle detections by DE2 observations, Freja electron spectra, ring current particles, observations of the solar wind near 1AU by WIND and beyond by Ulysses. Loss cone structures (b) are inherent in any magnetospheric plasma or in the solar atmosphere and recently also Jupiter's moon Ganymed was observed to create a large loss cone in all particle distributions. Non-gyrotropicity (c) for ions was detected predominantly in the plasma sheath boundary, by GEOTAIL observations in reconnection regions [7, 8] and at comets. Since the electron and ion gyro-radii are related as $r_{ge} << r_{gi}$ only a few observation for non-gyrotropic electron distributions are available, as WIND detections at the Earth's bow shock [6].

THEORETICAL OUTLINE

The general equation for the collective, time dependent development of the velocity distribution $f(v,t)$ in response to a wave spectrum by resonant Landau interaction, to particle collisions and external electric fields reads

$$\frac{\partial f}{\partial t} = \left[\frac{\partial f}{\partial t}\right]_{waves} + \left[\frac{\partial f}{\partial t}\right]_{collisions} + \left[\frac{\partial f}{\partial t}\right]_{E-field} \quad (1)$$

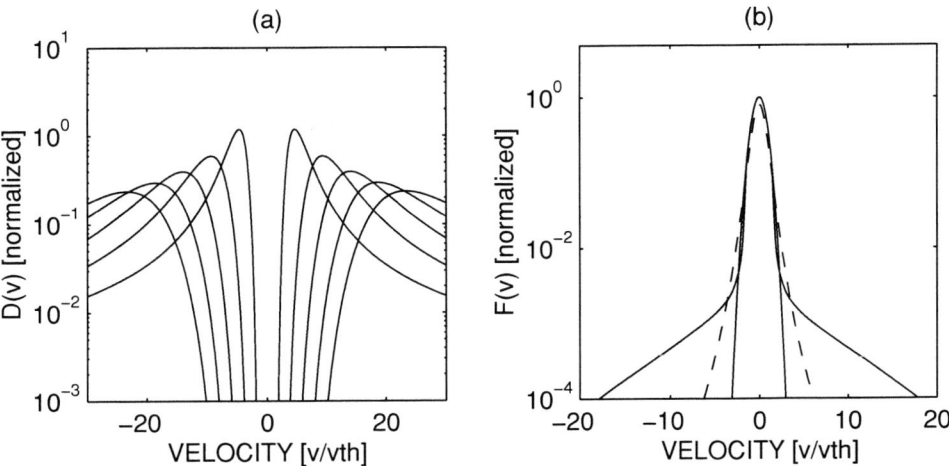

FIGURE 1. (a) Diffusion properties in increasing magnetic fields from 500G (innermost curve) to 2500G (outermost curve). (b) A Maxwellian, a κ-distribution and a saturated structure from synergetic acceleration subject to highly energetic electron populations.

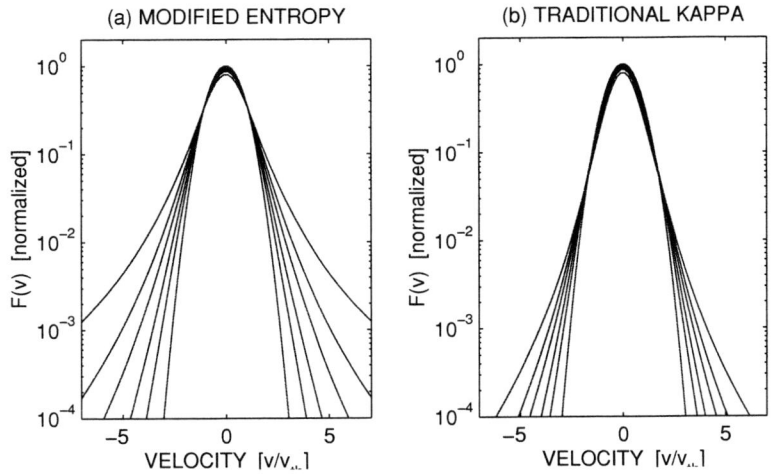

FIGURE 2. Schematic plot of the family of kappa-distributions. The subplots (a) and (b) demonstrate the difference between the nonextensive entropy solution and the traditional family of kappa-distributions. For both, $\kappa = 2$ corresponds to the outermost curve followed by values $\kappa = 3, 4, 6$ and 10. The innermost curve represents with $\kappa = \infty$ an isotropic Maxwellian.

Here $f(v, t = 0)$ shall be represented by a starting Maxwellian equilibrium distribution function. Modeling the wave-particle interaction by a diffusive process [2, 9] we consider the time evolution of the parallel velocity distribution function with respect to the magnetic field $\mathbf{B}_0$, being regulated by the Fokker-Planck equation as

$$\frac{\partial f}{\partial t} = \frac{\partial}{\partial v} \nu(v) \left[vf + v_{th}^2 \frac{\partial f}{\partial v} \right] + \frac{\partial}{\partial v} D(v) \frac{\partial f}{\partial v} \qquad (2)$$

Here $\nu(v)$ and $D(v)$ denote the velocity dependent collision and diffusion operators, respectively and v_{th} is the thermal speed. The first term on the right hand side of equation (2) restores the distribution function to a Maxwellian. For appropriate approximations we note that for low collisionality changes in the pitch angle are negligible and that the main dynamics of Alfvén wave-particle energy exchange due to Landau interactions is regulated in parallel direction. Introducing the quasi-linear, one dimensional diffusion operator and adopting a broadband spectrum of Alfvén waves trapped inside an envelop of specific extension, Equ. (2) provides the formalism for numerical simulations of wave-particle Landau energy exchange [2, 9]. As result, in uniform space plasmas suprathermal particle populations can be generated by resonant energy transfer from a broadband Alfvén wave spectrum yielding power law (kappa-like)

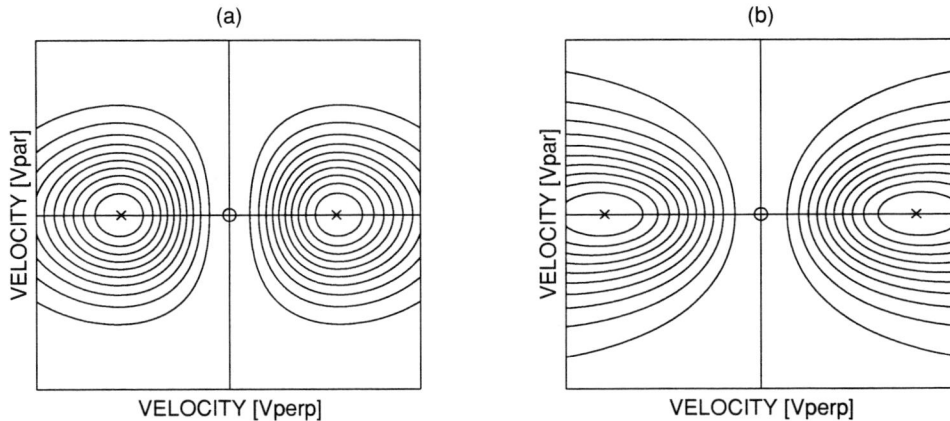

FIGURE 3. Schematic plot of a mixed suprathermal loss-cone distribution: (a) the transition from a Maxwellian-type $\kappa = \infty$, $j = 1$ loss-cone to (b) a mixed $\kappa = 2$, $j = 1$ suprathermal loss-cone distribution. The circle indicates the zero velocity position and the x the peaks of the distribution function.

velocity distributions [2]. In case of magnetic field or density gradients energetic particle populations are generated by synergetic effects in a multi-stage acceleration process. Multiple wave packets of increasing phase velocity provide highly efficient particle acceleration up to relativistic energies since the gradients act as catalyst [9], see Fig. 1.

Astrophysical environments are subject to long range interactions requiring a generalization of the Boltzmann-Gibbs-Shannon entropy. A non-extensive entropy approach [10] yields an undisturbed distribution $f(v) = B_\kappa (1 + v^2/v_{th}^2)^{-\kappa}$ (B_κ is a normalization constant) that differs only in the exponent by a constant in comparison to the traditional form of κ-distributions and results in structures exhibiting more pronounced tails for the same κ values, see Fig. 2. This provides naturally a justification of the basic analytical form of κ-distributions hitherto used from fundamental physics [10], but suggests to use the nonextensive entropy solution for future data fitting.

Magnetospheric observations require a generalization of Maxwellians into mixed suprathermal loss-cone distributions, provided e.g. by merging κ- distributions with a Dory-Guest-Harris (DGH) type loss-cone as

$$f_{\kappa j}(v_\parallel, v_\perp) = A_{\kappa j} \left(\frac{v_\perp}{v_{th\perp}}\right)^{2j} \left[1 + \frac{v_\parallel^2}{\kappa v_{th\parallel}^2} + \frac{v_\perp^2}{\kappa v_{th\perp}^2}\right]^{-(\kappa+1)} \tag{3}$$

see Fig. 3. Here the parameter κ shapes predominantly the suprathermal tails of this distribution and j is a measure of the loss-cone strength. $v_{th\parallel,\perp}$ are the thermal speeds parallel and perpendicular to $\mathbf{B}_0$ and the normalization is performed with respect to the particle density N, where $A_{\kappa j}$ denotes the normalization constant. It was shown that the particular structure of the distribution modeled by the spectral index κ and loss-cone index j dominates as regulating mechanism for instability thresholds over changes in the plasma parameters β and temperature anisotropy [11].

Finally, if plasma inhomogeneity scales are smaller than the charged particles gyroradius gyrophase-bunched distributions are generated. This occurs predominantly at boundary layers and at reconnection structures [5-6] and requires an appropriate model distribution in view of wave-particle interaction analysis. Introducing a ring distribution at a shift $v_{0\perp}$ and superimposing the dependence on the gyrophase angle Φ in a $x - y$ plane perpendicular to $\mathbf{B}_0$ where particle concentration maximizes at some gyroradius r_g yields in suprathermal tail generalization

$$f_{g\kappa} = A_{g\kappa}[1 + \frac{v_\parallel^2}{\kappa v_{th\parallel}^2} + \frac{(v_\perp - v_{0\perp})^2}{\kappa v_{th\perp}^2}]^{-(\kappa+1)}[1 + \frac{(v_x \cos\Phi + v_y \sin\Phi - v_{0\perp})^2}{\kappa v_{th\perp}^2}]^{-(\kappa+1)} \tag{4}$$

modeling accurately structures observed at boundary layers, see Fig. 4. Non-thermal features of the general velocity space representation $f_{g\kappa}$ ($A_{g\kappa}$ is a normalization constant) include temperature anisotropy, ring structures, stationary and, in case of time dependency, non-stationary non-gyrotropy as well as suprathermal populations. This enables us to handle any combination with a minimum of free parameters providing excellent fitting of observed structures [12].

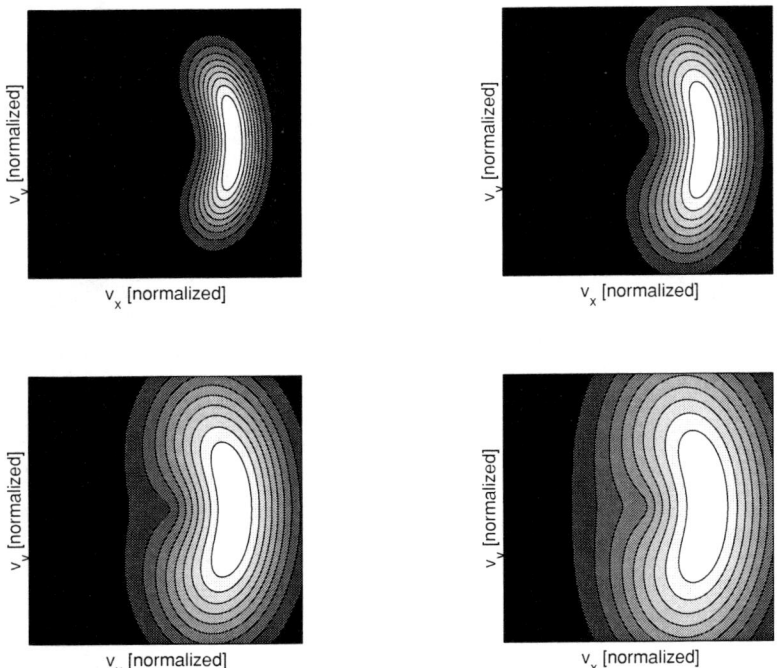

FIGURE 4. Contour plots of a non-gyrotropic distribution. The four panels illuminate changes in the distribution structure due to a variation of the thermal speed.

SUMMARY

Kappa-like distributions fitting observed suprathermal particle populations can be generated by resonant Alfvén wave - particle interaction. In non-uniform plasmas relativistic energies can be generated via synergetic effects, i.e. resonant interaction with wave packets of increasing phase velocity. Due to inclusion of long range interactions a nonextensive entropy approach justifies naturally the analytical form of kappa-distributions from fundamental physics. With regard to magnetospheric applications a mixed suprathermal loss-cone distribution model is proposed, which can serve as appropriate tool for wave-particle interaction analysis in a variety of mirror type environments. Finally, consistent with in situ observations a highly general analytical representation of suprathermal, gyrophase-bunched distributions is introduced, providing the basis for instability analysis at space plasma boundary layers.

REFERENCES

1. Mendis, D. A., and Rosenberg, M., *Ann. Rev. Astron. Astrophys.*, **32**, 419 (1994).
2. Leubner, M. P., *Planet. Space Sci.*, **48**, 133 (2000).
3. Summers, D., and Thorne, R. M., *J. Plasma Phys.*, **53**, 293 (1995).
4. Williams, D. J., and Mauk, B., *J. Geophys. Res.*, **102**, 24283 (1997).
5. Motschmann, U., Glassmeier, K. H., and Brinca, A. L., *Ann. Geophys.*, **17**, 613 (1999).
6. Gurgiolo, C., *Geophys. Res. Lett.*, **27**, 3153 (2000).
7. Frank, L. A., Paterson, W. R., and Kivelson, M. G., *J. Geophys. Res.*, **99**, 14887 (1994).
8. Tu, J.-N., Mukai, T., Hoshino, M., Saito, T., Matusno, Y., Yamamoto, T., and Kokubun, S., *Geophys. Res. Lett.*, **24**, 2247 (1997).
9. Leubner, M. P., in *Highly Energetic Physical Processes and Mechanisms for Emission from Astrophysical Plasmas*, edited by P. C. H. Martens, S. Tsuruta, and M. A. Weber, International Astronomical Union, IAU 195, 2000, 315.
10. Leubner, M. P., *Astrophys. Space Sci.*, in print (2002).
11. Leubner, M. P., and Schupfer, N., *J. Geophys. Res.*, **106**, 12993 (2001).
12. Leubner, M. P., to appear in *Planet. Space Sci.*, Special Issue (2002).

Collimation of Hot Electrons by Spontaneous Magnetic Fields in the Interaction of a Short-Pulse and High-Intensity Laser with a Relativistic Plasma

Hong Liu[1] and Xian Tu He[2]

[1]*Graduate School, China Academy of Engineering Physics, Beijing P.O.Box2101, 100088, China (Email: Liuhong_cc@yahool.com.cn)*
[2]*Institute of Applied Physics and Computational Mathematics, BeijingP.O.Box8009, 100088, China (Email: XTHE@mail.iapcm.ac.cn)*

Abstract. The emittance of hot electrons produced from the interaction of a short-pulse and high intense laser with a relativistic plasma is an obstacle for igniting hot spot formation in the concept of fast ignition scheme for inertial confinement fusion. Our numerical results show that the tens of megagauss spontaneous quasistatic magnetic field generated by the short-pulse and high-intensity laser with a relativistic plasma may collimate hot electron motion. The spontaneous axial magnetic field driven by a helical current, parallel to laser propagation, behaves as a guiding center with a submicron Larmor radius. And the spontaneous toroidal magnetic field driven by ponderomotive current causes a pinch effect on hot electrons.

INTRODUCTION

To show this complex question we use a laser fast igniting model. A test particle, such as an electron, moves in a channel produced by a precursor hole-boring pulse within a pre-compressed fusion fuel. With a circularly polarized short-pulse laser pulse (Intensity I = 2×10^{19}W/cm^2, 100fs). We can get 40MG $\mathbf{B}_z(x,y,z,t)$ and 80MG $\mathbf{B}_\theta(x,y,z,t)$ from numerical results[13]. Figure 1 (over page) shows the trajectories of a test electron with a relativistic motion in a laser field $\mathbf{a}(x,y,z,t)$, where $\mathbf{a}$ is the normalized vector potential ($a_0 = 4$), and self-magnetic fields: $\mathbf{B}_z(x,y,z,t)$, $\mathbf{B}_\theta(x,y,z,t)$. The first line only has $\mathbf{a}$, the second line has $\mathbf{a}$ and $\mathbf{B}_z$, the third line has $\mathbf{a}$ and $\mathbf{B}_\theta$, and the fourth has $\mathbf{a}$, $\mathbf{B}_z$ and $\mathbf{B}_\theta$. We use a Gaussian profile to describe the laser propagating (z direction) in the channel.

$$\vec{a}_1 = a_0 e^{-\frac{x^2+y^2}{R_0^2}} e^{-\frac{(k_1 z - \omega_1 t)^2}{L^2}} e^{-(k_1 z - \omega_1 t)t} (\hat{x} + i\hat{y}) \quad (1)$$

$$\vec{a}_2 = a_0 e^{-\frac{x^2+y^2}{R_0^2}} e^{-\frac{(k_2 z - \omega_2 t)^2}{L^2}} e^{-(k_2 z - \omega_2 t)t} (\hat{x} + i\hat{y}) \quad (2)$$

$$\vec{B}_z = B_{10} \times \vec{a}_1(k_1, \varpi_1) \times \vec{a}_2^{*}(k_2, \varpi_2) \quad (3)$$

$$\vec{B}_\theta = B_{20} \times \sqrt{x^2 + y^2} \cdot a^2$$

L is the pulse width $L = 10\lambda$. R_0 is the minimum spot size. $R_0 = 5\lambda$. λ is the laser wave length. $\lambda = 1.06\mu m$. Electrons initially having 0.001c velocity escape the pulse axis. From the last line we can see good collimation exists when $\mathbf{a}$, $\mathbf{B}_z$ and $\mathbf{B}_\theta$ coexist. When $n_0 = 5 \times 10^{20}$/cm^3 we can get $\mathbf{B}_z = 1.4 \times 10^8$G from Ref. [13] and $\mathbf{B}_\theta = 1.7 \times 10^8$G from $\nabla \times \vec{B} = \frac{4\pi}{c}\vec{j}$ and $\vec{J} = -env_z$. In Figure 1 we choose $B_{zmax} = 14 \times 10^8$G and $B_{\theta max} = 17 \times 10^8$G.

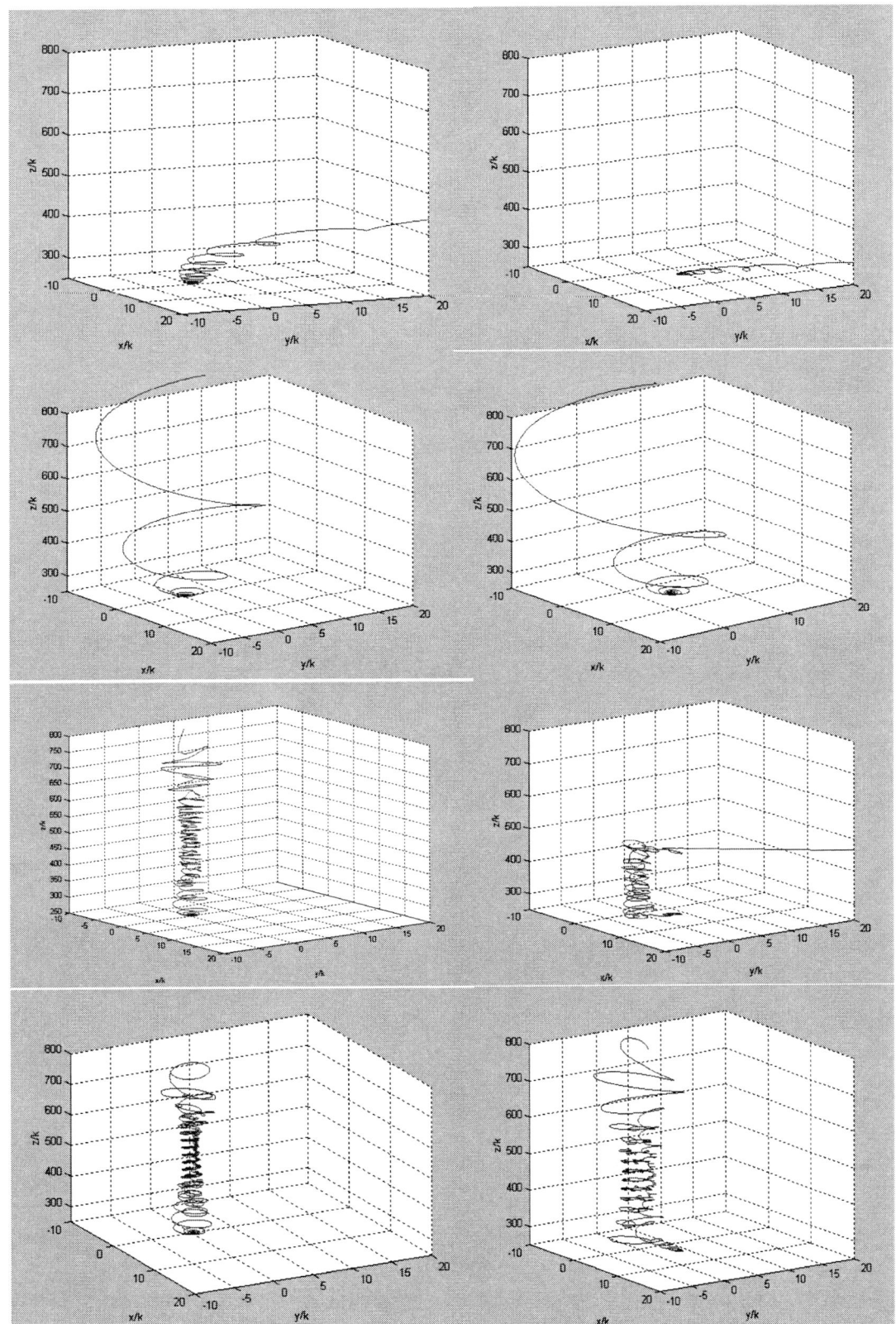

FIGURE1. Trajectories of a test electron in a laser field. In the left column the electron is initially on the axis of the laser beam, $x_0 = y_0 = 0$. In the right column the electron is initially one wavelength from the axis, $x_0 = y_0 = 4/k$.

MODEL AND FORMULAE

The basic equations are:

$$\frac{d\vec{p}}{dt} = -e\left(\vec{E} + \frac{\vec{v}}{c} \times \vec{B}\right) \quad (5)$$

$$\frac{dE}{dt} = -e\vec{v} \cdot \vec{E} \quad (6)$$

Then we have:

$$\frac{d\vec{p}_t}{dt} = \frac{\partial \vec{a}}{\partial t} - \vec{v} \times (\nabla \times \vec{a}) \quad (7)$$

$$\frac{d}{dt}\gamma = \vec{v} \cdot \frac{\partial \vec{a}}{\partial t} \quad (8)$$

Let

$$\vec{u} = \frac{\vec{v}}{c}, \quad \vec{p}' = \frac{\vec{p}}{m_0 c} = \frac{\gamma m_o v}{m_o c} = \gamma \vec{u}, \quad v_g \approx c \text{ (in underdense plasma)}$$

$$\vec{a} = \frac{e\vec{A}}{m_0 c^2}, \quad \phi = \frac{e\phi}{m_0 c^2}, \quad \vec{x}' = k\vec{x}, \quad t' = t\omega, \quad c = \frac{x'}{t'} = \frac{x/k}{t/\omega} = \frac{x}{t} \cdot \frac{\omega}{k} \quad (9)$$

We have

$$\begin{cases} \frac{dp_x}{dt} = (u_z - 1)\frac{\partial a_1}{\partial y} + u_y\left(\frac{\partial a_1}{\partial y} - \frac{\partial a_2}{\partial x}\right) - u_y B_1 + u_z B_2 \cos\theta & \frac{dx}{dt} = \frac{p_x}{\gamma} \\ \frac{dp_y}{dt} = (u_z - 1)\frac{\partial a_2}{\partial y} + u_x\left(\frac{\partial a_2}{\partial y} - \frac{\partial a_1}{\partial y}\right) + u_x B_1 + u_z B_2 \sin\theta & \text{and} \quad \frac{dy}{dt} = \frac{p_y}{\gamma} \\ \frac{dp_z}{dt} = -u_x \frac{\partial a_1}{\partial z} - u_y \frac{\partial a_2}{\partial z} + [-u_x(\cos\theta) \cdot B_2 - u_y(\sin\theta) \cdot B_2] & \frac{dz}{dt} = \frac{p_z}{\gamma} \end{cases} \quad (10)$$

where $\vec{B}_z$ comes from[13]

$$\frac{e\vec{B}_z}{m_e c \omega_p} = -i\frac{\omega_p^2 \beta_3}{\omega_0^2} \vec{E}_1 \times \vec{E}_2^* + i\frac{\omega_p^2 \beta_3}{\omega_0^2} \vec{E}_1^* \times \vec{E}_2$$

$$= -i\frac{\omega_p^2}{\omega_0^2} \beta_3 \left[\frac{4\eta^2}{L^4} + 1\right](\vec{a}_1 \times \vec{a}_2^*) + i\frac{\omega_p^2 \beta_3}{\omega_0^2}\left[\frac{4\eta^2}{L^4} + 1\right](\vec{a}_1^* \times \vec{a}_2) \quad (11)$$

$$\nabla \times \vec{B} = \frac{4\pi}{c}\vec{j} \quad (12)$$

$B_{zmax} = 1.4 \times 10^8 G$ ($n_c = 5 \times 10^{20}/cm^3$, $I = 2 \times 10^{19} W/cm^2$)
$T = 50 keV$, $\beta_3 = 0.33$[13]
and $B_\theta = B_2$ comes from Eq. (12).

$$\vec{j}_z = -en_0(z)v_z$$
$$= -en_0(z)\frac{a^2}{2\gamma} \quad (13)$$
$$(v_z = \frac{\gamma - 1}{\gamma}) \quad (14)$$

from

$$\gamma = \sqrt{1 + p_\perp^2 + p_z^2}$$

$$= \sqrt{1 + a^2 + (\gamma v_z)^2} \tag{15}$$

We can have

$$\gamma = 1 + \frac{a^2}{2} \tag{16}$$

Then $\quad \gamma v_z = \frac{a^2}{2}, \ B_\theta = (-e \cdot n \cdot 4\pi \cdot \frac{1}{2\gamma} \cdot \frac{1}{2}) \cdot r \cdot a^2 \tag{17}$

so $B_\theta = 1.7 \times 10^8 G$ when ($n_c = 5 \times 10^{20}/cm^3$, $I = 2 \times 10^{19} W/cm^2$).

RESULTS AND DISCUSSION

In summary, we use single particle theory first given the typical trajectories which can avoid the emittance of hot electrons in laser fusion plasmas. We found a circularly polarized short pulse and high intensity laser can generate self magnetic field both in axial or in toroidal case. These two kinds of magnetic fields can cause collimation of hot electrons when laser gives them an emittance. These results of collimation of electrons with special trajectories are of benefit for fast ignition and will have an implications in astrophysics.

ACKNOWLEDGMENTS

We thank Heinrich Hora for useful discussions. This work was supported by National Hi-Tech Inertial Confinement Fusion Committee of China and National Natural Science Foundation of China. Grant No.19735002 and National Basic Research Project "Nonlinear Science" in China.

REFERENCES

[1] M. Tabak, J. Hammer, M. E. Glinsky, et al., Ignition and high gain with ultrapowerful lasers. Phys. Plasmas, **1**, 1626 (1994).
[2] P. Sprangle, E. Esarey, and J. Krall, Laser driven electron acceleration in vacuum, gases, and plasmas. Phys. Plasmas, **3**, 2183 (1996).
[3] D. Umstader, S.-Y. Chen, A. Maksimchuk, G. Mourou, and R. Wagner, Nonlinear optics in relativistic plasmas and laser wakefield acceleration of electrons. Science, **273**, 472 (1996).
[4] G. Malka, E. Lefebvre, and J. L. Miquel, Experimental Observation of Electrons accelerated in Vacuum to Relativistic Energies by a High-Intensity Laser. Phys. Rev. Lett. **78**, 3314 (1997)
[5] M. H. Key, M. D. Cable, T.E. Cowan, et al Hot electron production and heating by hot electrons in fast ignitor research. Phys. Plasmas, **5**, 1966 (1998).
[6] M. G. Haines, Magnetic-field generation in laser fusion and hot-electron transport. Can. J. Phys. **64**, 912 (1986)
[7] L. Gorbunov, P. Mora, and T. M. Antonsen, Magnetic field of a plasma wake driven by a laser pulse. Phys. Rev. Lett. **76**, 2495 (1996).
[8] S. V. Bulanov and m. Lontano, Electron Vortices produced by Ultraintense Laser Pulses, Phys. Rev. Lett. **76**, 3562 (1996).
[9] H. Suk, N. Barov, and J. B. Rosenzweig, Plasma Electron Trapping and Acceleration in a Plasma Wake Field Using a Density Transition. Phys. Rev. Lett. **86**, 1011 (2001).
[10] A. J. W. Reitsma, V. V. Golovizmin, L.P.J. Kamp, and T. J. Schep, Bunch Self-Focusing Regime of laser Wakefield Acceleration with Reduced Emittance Growth. Phys. Rev. Lett. **88**, 014802 (2002).
[11] Z. Donko, G. J. Kalman, and K.I. Golden, Caging of Particles in One-Component plasmas, Phys. Rev. Lett. 88, 225001 (2002).
[12] M. Roth, T. E. Cowan, M. H. Key, et al, Fast ignition by intense laser-accelerated proton beams. Phys. Rev. Lett. **86**, 436 (2001).
[13] Shao-ping Zhu, X. T. He, and C. Y. Zheng, Slow-time-scale magnetic fields driven by fast-time-scale waves in an underdense relativistic Vlasov plasma. Phys. Plasmas **8**, 321 (2001).

Cross Field Diffusion of Cosmic Rays: Dependence on 2-D Field Turbulence Models

F. Otsuka and T. Hada

*Department of Earth System Science Technology, Kyushu University,
Japan*

Abstract. Cross field diffusion is important in efficient diffusive shock acceleration of cosmic rays for quasi-perpendicular geometry. Cross field diffusion of cosmic rays have been discussed theoretically by quasi-linear approximation for 'slab' geometry of magnetic turbulence (for example, reference [1]). In the solar wind, the presence of quasi-two dimensional, nearly incompressional turbulence was reported by Matthaeus et. al [2]. We discuss cross field diffusion of energetic particles numerically using compressional (*C*) and non-compressional (*NC*) two dimensional turbulence models. For both cases of *C* and *NC* models, the diffusion is found to be composed of several regimes with different statistics. The diffusion coefficient defined in the usual way in each regime exhibits dependence to the time scale, suggesting that the underlining physical process should be described by Levy statistics. When $\rho/L_b \ll 1$, where ρ is particle Larmor radius and L_b is field turbulence correlation length, particles tend to follow equi-contour lines of field magnitude (b_z) and those of field vector potential magnitude (a_z) for *C* and *NC* model, respectively. For small field turbulence level, the diffusion coefficients for *C* and *NC* models are characterized by b_z and a_z islands statistics. For the same turbulence level, Levy type orbits are observed due to switching of the contour lines particles are attached.

INTRODUCTION

Cross field transport of particles are essentially important for diffusive shock acceleration to efficiently operate in quasi-perpendicular shock geometries. We discuss transport of energetic charged particles perpendicular to average magnetic field, using test particle simulations with two different turbulence models. The transport of cosmic rays are aften discussed using three dimensional turbulence models, which include the slab and the quasi-two dimensional fluctuations[1]. In three dimensional model, particles can cross the field lines essentially in two ways. One is due to gradient-B drift, and in this case the particle motion is restricted within a plane perpendicular to the field. The other is the crossing via the particle motion alone the field. In this paper, we perform test particle simulations using only two dimensional turbulence models so that we can focus on the former process. Jokipii et al[2] indicated that no cross field diffusion should take place in a system with less than two spatial dimensions. However, the two dimensional system we consider here is an exception since the orientation of ignorable spatial coordinate consider with that of the magnetic field. In the present paper we use two different methods of the turbulence field: Compressional $\boldsymbol{B}=(0,0,b_z(x,y)+B_0)$, and Non-Compressional $\boldsymbol{B}=(b_x(x,y),b_y(x,y), B_0)$.

FORMULATION

Since we are dealing with energetic particles whose velocities are much lager than MHD velocities, we assume the field turbulence to be the stationary (fossil turbulence). Then, the particle energy is conserved, and the spatial position $\boldsymbol{r}=(x,y,z)$ and the velocity $\boldsymbol{v}=(v_x,v_y,v_z)$ obey equation of motion.

$$\dot{\boldsymbol{v}} = \boldsymbol{v} \times \boldsymbol{B}, \quad \dot{\boldsymbol{r}} = \boldsymbol{v}, \quad \boldsymbol{B} = \boldsymbol{b} + \boldsymbol{z} \tag{1}$$

where $\boldsymbol{z}$ is a unit vector in the z direction, $\boldsymbol{b}$ is the fluctuation part of the normalized field, and time is normalized to the reciprocal of $\Omega_0 = qB_0/mc$.

The two models for $\boldsymbol{b}$ are summarized in **TABLE 1**.

TABLE 1. Magnetic field turbulence models.

Compressible (C model)	Non-compressible (NC model)
$b=(0,0,b_z(x,y))$	$b=(b_x(x,y),b_y(x,y),0)$, $a=(0,0,a_z(x,y))$
$b_z = \sum_{k_x}\sum_{k_y} P_b(k)\cos(k_x x + k_y y + \phi_k)$	$b = \nabla \times a$, $a_z = \sum_{k_x}\sum_{k_y} P_a(k)\cos(k_x x + k_y y + \phi_k)$
$P_b(k) = \begin{cases} k^{-\gamma} & (k_{min} \le k \le k_{max}) \\ k_{min}^{-\gamma} & (k_{sys} \le k < k_{min}) \end{cases}$	$P_a(k) = \begin{cases} k^{-\gamma-1} & (k_{min} \le k \le k_{max}) \\ k_{min}^{-\gamma}/k & (k_{sys} \le k < k_{min}) \end{cases}$
$L_b^2 = \left\langle \dfrac{P_b(k)}{(k/2\pi)^2} \right\rangle / \langle P_b(k) \rangle$	$L_b^2 = \left\langle \dfrac{kP_a(k)}{(k/2\pi)^2} \right\rangle / \langle P_a(k) \rangle$, $L_a^2 = \left\langle \dfrac{P_a(k)}{(k/2\pi)^2} \right\rangle / \langle P_a(k) \rangle$
$\delta B / B_0 = \sqrt{\langle b_z^2 \rangle}$	$\delta B / B_0 = \sqrt{\langle b_x^2 + b_y^2 \rangle}$

Where $k = \sqrt{k_x^2 + k_y^2}$, $k_{sys} = 2\pi/\max(L_x, L_y)$, $k_{max} = 16\pi/\max(L_x, L_y)$, $k_{min} = 4\sqrt{2}\pi/\max(L_x, L_y)$, and simulation box size L_x, L_y=512. We let choose $\gamma=1.5$, and other field parameters in such a way that the correlation length of magnetic turbulence $L_b \sim 164$ for both models (in the NC model, $L_a \sim 251$).

For the *NC* model, canonical moment alone ignorable coordinate p_z is conserved, namely, $p_z = a_z + v_z = const$.

$$D(\tau) = \frac{\langle \Delta r(\tau)^2 \rangle}{\tau} \quad (2)$$

We numerically time integrate (1) for N particles using the forth order Runge-Kutta method, and evaluate the spatial diffusion coefficient, keeping the time scale dependence, where $\langle \, \rangle$ denote an ensemble average of N particles, and $\Delta r(\tau)^2$ is the squared deviation of particle position within elapsed time, τ. For computing the ensemble average, we typically used $N = 256$ particles.

The guiding center trajectory is approximately defined $r_g = r + v \times z$.

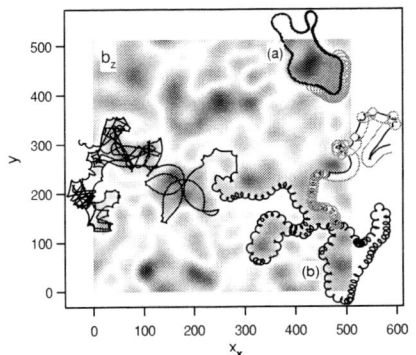
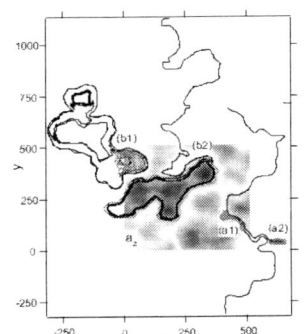

Exact orbit ———
Guiding Center Orbit ———

FIGURE 1. Typical particle orbit in the *C* model (left) and in the *NC* model (right). Background colors represent variations of the magnetic field strength (b_z) in the left panel, and variations of the vector potential (a_z) in the right panel, respectively. In each panel, several particle orbits are superposed. For all the orbits, v/L_b=0.1. Orbits denoted as for (a) and (b) respectively correspond to the runs in which the turbulence level is given the value of $\delta B/B_0$=0.1 and 1.

Let us define the guiding center of the particle,

$$r_g = r + \frac{v \times \Omega}{\Omega^2} \quad (3)$$

where $\Omega = z + b$. By differentiating in time, we have

$$v_{g\perp} = \left(\frac{v \cdot \Omega \Omega}{\Omega^2}\right)_\perp + \left(v \times \left(v \cdot \nabla\left(\frac{\Omega}{\Omega^2}\right)\right)\right) \quad (4)$$

The first term in the r.h.s. represents the deviation of the field line from a straight line parallel to z as experienced by a particle, and is related to the field line random walk. The second term, on the other hand, reduces to the gradient-B drift velocity,

$$v_{\nabla B} = \frac{v_\perp^2}{2}(z \times \nabla_\perp |b|) \quad (5)$$

In the limit of $\rho/L_b \ll 1$, by averaging over the gyroperiod and by taking only the leading order. As seen in **Figure1.** particles tend to follow contour lines of b_z, due to the grad-B drift and a_z due to the first term in the r.h.s of eq(4) in the C and the NC models, respectively. However, detrapped of the particles from these contour lines do occur sometimes also. Trapping in closed orbits and occasional switching of orbits constitute the main elements of the cross-field diffusion we consider in the present study.

RESULTS

In this paper, we only discussed particles in the regime, $\rho/L_b(=1)<1$. At first, diffusion coefficients for the NC model (D_{NC}) are estimated, given constant pitch angle of velocity defined by d.c. field direction. Second we calculated diffusion coefficients in two turbulence models, given randomly isotropic distribution of velocity space. We compare diffusion coefficient for the NC model (D_{NC}) and that for the C model (D_C).

Dependence on the pitch angle

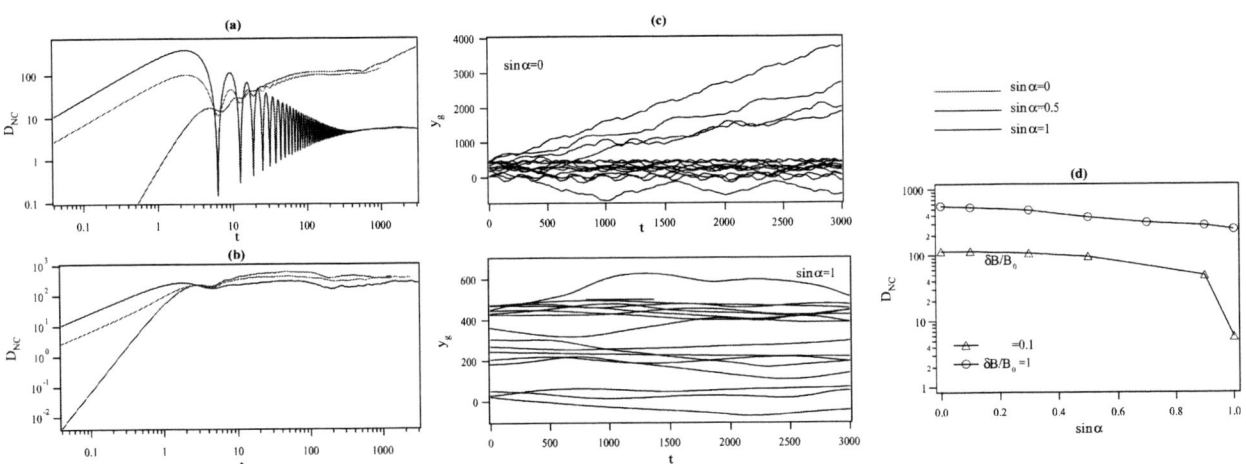

FIGURE 2. Panels **(a) (b)** show diffusion coefficient for NC model (D_{NC}) vs. time. Different colars correspond to $\sin\alpha = 0, 0.5, 1$. The turbulence level is $\delta B/B_0=0.1$ (upper panel) and 1 (lower panel). Panel **(c)** shows the y component of the guiding center position of vs. time for $\delta B/B_0=0.1$. Panel **(d)** shows the diffusion coefficients vs. $\sin\alpha$.

In **Figure 2(a)**, we plot D_{NC} for several different pitch angles. When $\alpha=\pi/2$, the regime t<400 is still under influence of Larmor rotations. At large time scale, the diffusion coefficient is almost constant. When $\sin\alpha<1$, D_{NC} now appears to have dependence on time, this is due to the fact that some particles can travel long perpendicular distances via field-aligned motion, as shown in **Figure2(c)**. **Figure 2(d)** shows D_{NC} is about 10 times smaller for $\sin\alpha=1$ than $\sin\alpha<1$ for $\delta B/B_0=0.1$. When $\delta B/B_0=1$, D_{NC} become only weakly dependent on pitch angle.

Dependence on the turbulence model

Next, diffusion coefficients are compared in different turbulence models for both $\delta B/B_0=0.1$. 1. From **Figure 3(a)** we found that for both of the two models, D represents Larmor rotation in short time scale (regime i). At longer time scale (regime ii), D is approximately constant. From **Figure 3(b), (c)**, guiding center positions are spread due to following the a_z and b_z contour lines, respectively. The value of D_{NC}/D_C is approximately the same as $L_a^2/L_b^2 \sim 2.3$. At longer time scale (regime iii), D become proportional to time, because the free streaming particles which follow open field contour lines contributes dominantly to the value of D. In **Figure 4(a)**, the Larmor motion regime is not apparently seen, implying that the process basically is a classical random walk, but the time series of particle positions in **Figure 4(b).** are clearly different from classical random walk. From right panel of **Figure 2(b)** we find particles follow contour lines of a_z, but some time their orbits that switch to other contour lines. For the C model, diffusion coefficient is decrease as the time scale is increased, implying that the particles tend to be trapped. At longer time scale, D_C is almost constant. But also from **Figure 4(c).**, various orbits, such as trapped and Levy like orbits, are observed. From left panel of **Figure 3.** we see guiding center orbit meanders, producing the jagged time series in **Figure 4(c)**.

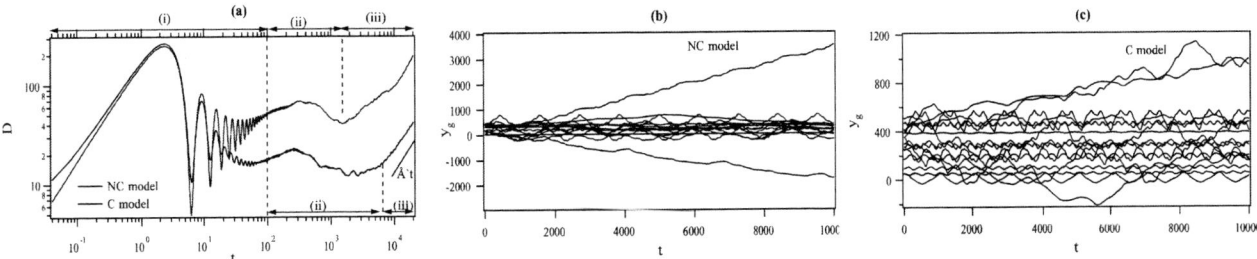

FIGURE 3. (a) is same as **Figure 2(a)** but initial velocities are randomly given isotropic for two turbulence models. (b), (c) are the same as **Figure 2(c)** but corresponding to diffusion coefficient in (a).

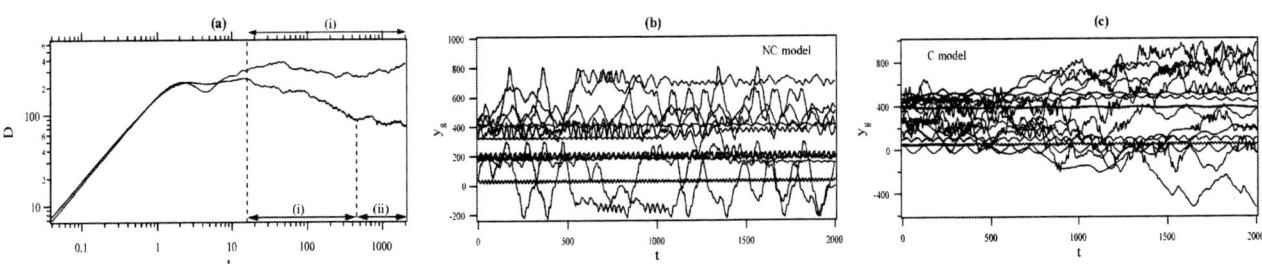

FIGURE 4. Same as **Figure3.** but for $\delta B/B_0 = 1$.

DISCUSSION

Cross field diffusion of energetic particles is discussed for both compressional (C) and non-compressional (NC) two dimensional turbulence magnetic field models, for the case $\rho/L_b \ll 1$, where ρ is the particle Larmor radius (~ $v\sin\alpha$ since $\Omega_0 = 1$) and L_b is the correlation length of the turbulence. Particles tend to follow contour lines of field turbulence b_z and that of the vector potential field a_z for C and NC models, respectively. Various orbits are observed due to switching of the field contour lines particle follow for both turbulence models. Although we restrict our discussion to the regime $\rho/L_b \ll 1$ in the present study, regime $\rho/L_b \gg 1$, and in particular, $\rho/L_b \sim 1$ are of much interest. When $\rho/L_b \gg 1$, the quasi-linear formulation[4] presumably provides a valid description. However, in general, observed characteristics (type of orbit, time scale dependence of the diffusion coefficient) of the present test particle simulations suggest that a qualitatively different approach from the classical Brownian-type diffusion theory is necessary for a proper description of the cross-field diffusion process. The key to the understanding of the issue may in the analysis of the time series consisting of the trapped (sticking) and free streaming orbits, as analyzed by Eric et. al[5]

REFERENCES

1. J. Giacalone and J.R. Jokipii, Astrophys. J., 1999, pp204, **520**.
2. W.H. Matthaeus, M.L. Golstein, and D.A. Roberts, J. Geophys. Res., 1990, pp.20673, 95, **12A**.
3. J.R. Jokipii, J.Kota and J. Giacalone, Geophys. Res. Lett., 1993, pp1759, **20**.
4. J.R. Jokipii, Rev. Geophys.Sp.Phys., 1971, pp27, **9**.
5. Eric R. Weeks, and Harry L. Swinney, 1998, pp4915, 57, **5**.

Resonant Solar Acceleration and its Astrophysical Implications

Ilan Roth

Space Sciences, University of California, Berkeley, CA 94720, USA

Abstract. Valuable information on active solar sites can be obtained from particle abundances as measured *in situ* by satellites. Energization processes in the strongly magnetized and dense solar corona or in the weakly magnetized and dilute interplanetary medium form a convenient prototype of laboratories for stellar and galactic plasma processes. Coronal energization is due to resonant interaction of a subset of coronal elements and isotopes with electromagnetic waves, which are excited during a major reconfiguration of the flaring magnetic field. These resonant processes together with the intense, less frequent gradual acceleration by interplanetary shocks forms a new heliospheric baseline abundances of minority elements during active periods. Similar process may enrich the planetary nebulae with ^{3}He, solving partly the "astrophysical ^{3}He problem", with important implications for the stellar and galactic evolution.

INTRODUCTION

Our Sun, as well as most stars with convective envelopes which support strong magnetic activity, fill their nebulae with material which is ejected from the star through (a) stellar winds, (b) as a result of hydromagnetic shock waves or (c) via plasma processes requiring magnetic field reconfiguration. Solar energetic ions with energies of 0.01-10.0 MeV/nucleon, which are frequently observed in the interplanetary space, carry important information about physical processes in the solar corona and in the heliosphere, and by implication in other astrophysical sites. Large, solar gradual SEP events are believed to involve propagating shock waves with inhomogeneously self-generated waves and scattering of particles back to the propagating shock [Lee, 1983], with a transport to the observing location [Reames and Ng, 1998] and abundances similar to the coronal values [Luhn et al., 1985]. Smaller, impulsive SEP events involve selective enrichment of a subset of coronal ions satisfying resonance conditions with electromagnetic waves which propagate along the coronal field lines due to flare reconfiguration of a strongly stressed magnetic field; since plasma resonant processes on magnetized field lines are selective with respect to the characteristics (charge, mass and energy) of the ionized elements which participate in the interaction, the abundances, charge and isotopic states of some elements are often distinctly different from the coronal or solar wind values. As a result one observes a dramatic abundance enhancement by factor of 10^3-10^5 in the isotopic ratio of ^{3}He/^{4}He [Hsieh and Simpson, 1970; Mason et al., 2000] and smaller enrichments by factor ~ 10 (with respect to O) in heavier ions up through Fe [Mason et al., 1986; Reames et al., 1994].

ENRICHMENT OF RARE ELEMENTS IN THE HELIOSPHERE

Recent Advanced Composition Explorer (ACE) observations point out to an existence of a subset of ^{3}He impulsive events in which both ^{3}He and Fe exhibit similar energy distribution form, peaking in rigidity around 40-50 MV (200-400keV/n and 100keV/n, respectively). This similar shape, which differs from all other observed ions spectra, indicates a commonality in their acceleration mechanism. Additionally, the velocity spectrograms of the observed ions show that ^{3}He and Fe arrive along the same energy-time curves, confirming a simultaneous injection at the Sun [Mason et al., 2000]. Recent study of 45 ^{3}He-rich events concluded that ^{3}He /^{4}He and Fe/O enhancements are clearly correlated [Ho et al., 2000].

The acceleration model is based on the resonant interaction between specific ions which satisfy the resonance condition with the propagating electromagnetic cyclotron waves at the frequency range below the hydrogen gyrofrequency

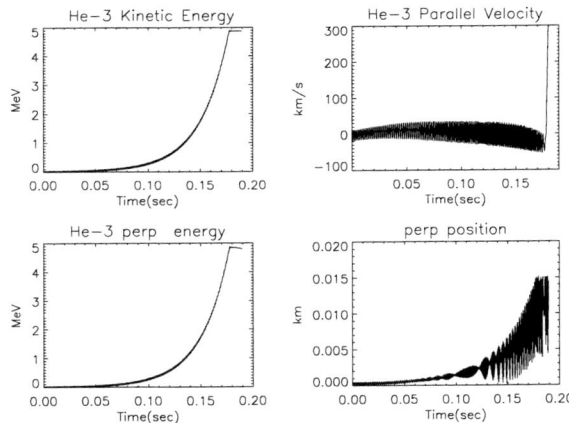

FIGURE 1. Energization of ^{3}He ion on coronal field lines: B = 500G, n=$10^9 cm^{-3}$.

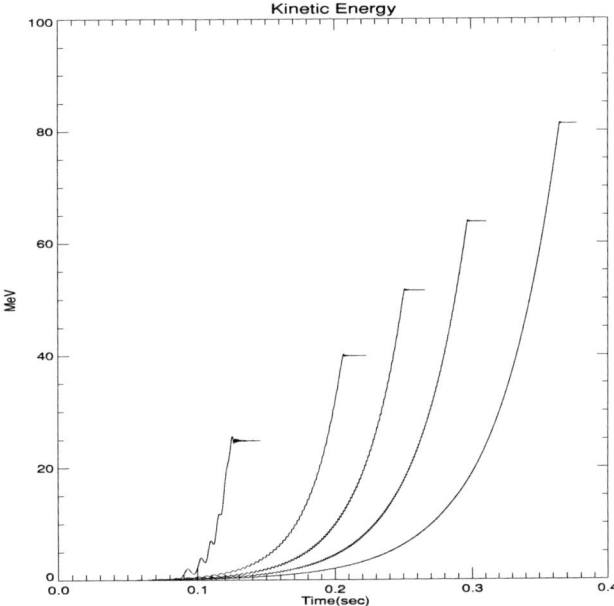

FIGURE 2. Energization of Fe ions with varying coronal charge states: Z= 16, 18, 20, 22, 24 (from left to right), A=56; same parameters as in Fig 1.

Ω_H and above the ion-hybrid frequency [Roth and Temerin, 1997]. ^{3}He satisfies the lowest, while Fe satisfies higher harmonic resonance conditions with their cyclotron frequencies.

When the propagating electromagnetic perturbation can be approximated by a monochromatic wave, as the ion approaches the resonance region the wave slows it down while its perpendicular energy increases exponentially (Roth and Temerin, 1997). The value of maximum energization is given with the help of the roots of consecutive Bessel functions for ^{3}He and Fe, respectively. For a broader spectrum the ion may jump between several resonant waves but on the average its final acceleration is not modified.

Figure 1 depicts the total and the perpendicular energy per nucleon of ^{3}He ion as it traverses the resonance region, as well as its parallel velocity and perpendicular coordinates (gyroradius) with respect to the magnetic field. In the example shown the thermal ion with an energy of 300 eV and a small initial pitch angle with velocities $v_\parallel$ =250km /s and $v_\perp$ = 20 km/s starts at $z = -5$km, position distant from the resonance region. As it enters the resonance region its perpendicular gyroradius increases while it slows down along the field line, and finally is ejected when its adiabatic invariant reaches a threshold value.

Figure 2 shows the time history of the energization process of five Fe ions interacting along the coronal field lines with monochromatic waves of given perpendicular wavenumbers. For $k_\perp \rho$ too small, ($\rho = v_\perp/\Omega$ denotes the gyroradius) the interaction is weak due to the small value of the Bessel function $J(k_\perp \rho)$ before the ion is ejected from the interaction region by the mirror force. For $k_\perp$ too large the maximum energization decreases since it satisfies approximately $\rho \sim v/k_\perp$, where v is the root of the Bessel function. The time history of Fe ions with different q/m ratios is taken from coronal distribution with varying charge states: Z= 16-24, and A=56. The frequencies of the waves are adjusted accordingly to allow resonant interaction. We observe that for a given wavenumber the final energy increases with q/m, or equivalently with the gyrofrequency. Hence, higher charge states are more likely to be energized to higher energies, provided the waves can satisfy the resonant condition. Figures 1 and 2 show that both ions are accelerated by the same mechanism adding correlation to the experimental observations.

Although the resonant processes which accelerate ^{3}He and a subset of heavy ions (mainly Fe) are similar, there exist several reasons for the much less effective energization of Fe isotopes. While ^{3}He resonates with the lowest harmonic of the waves (n=1), Fe and other heavy ions require higher harmonics, n= 2, 3 ... [Roth and Temerin, 1997]; therefore the initial heavy-ion gyroradius must be a non-negligible fraction of the perpendicular wavelength and only the tail of the distribution can be affected by the interaction.

ENRICHMENT OF RARE ELEMENTS IN PLANETARY NEBULAE (PN)

^{3}He is one of the light elements which was produced in the big bang nucleosynthesis (BBN) and therefore can be used as a cosmological "baryometer". Lack of consistency between abundance observations and galactic evolutionary models poses a major problem for the cosmological theory (Astrophysical ^{3}He problem) [Galli et al., 1995].

^{3}He measurements in the protosolar material, the local interstellar medium and galactic H II regions indicate that the ^{3}He/H abundance ratio is similar to the non-processsed galactic value of $\sim 2 \times 10^{-5}$. However, in a series of observations [Balser et al., 1997] several PN sources were detected with ^{3}He/H = $10^{-4} - 10^{-3}$, i.e. more than an order of magnitude larger than those found in any H II region, local interstellar medium or protosolar system. The modification of the stellar evolution relies on the observed anomaly in the isotopic ^{12}C/^{13}C ratio in a number of evolved stars, which was shown to be below the standard models (5 vs 30) for masses $< 2M_s$, and was suggested as an indicator for a non-standard mixing after the first dredge-up before the end of RGB phase (Hogan, 1995), with a similar destruction of ^{3}He (Charbonnel, 1995). This scenario would require that the PN mass progenitors have masses $> 2M_s$, however new analysis of the relevant PN showed that their masses satisfy $1.2\, M_s < M < 2.2 M_s$ [Galli et al., 1997]. Therefore the consistency with presolar and local ISM values is possible if most of the low-mass stars have undergone enhanced ^{3}He depletion and then the solution of the observed enhanced ^{3}He abundances in PN must be related to a different physical process.

It is suggested, that in addition to the specific requirements from the stellar evolution to produce ^{3}He in low-mass stars in PN and deplete it in H II regions, one may consider a planetary nebula with its very hot central star and very intense magnetic activity as a source of steady enhancement of ^{3}He over a broad range of energies. This results in an average higher value of ^{3}He/^{4}He and ^{3}He/H in the surrounding environment. The large H II regions which consist of many stars with on the average much less intense magnetic activity will not reach such high baseline values.

We acknowledge the support of NASA grant NAG5-10866.

REFERENCES

1. Balser, D. S., Bania, T. M., Rood, R. T., and Wilson, T. L., *Astrophys. J.*, **483**, 320, 1997.
2. Charbonnel, C., *Astrophys. J. Lett.*, **453**, L41, 1995.
3. Galli, D., Stanghellini L., Tossi M., and Palla, F., *Astrophys. J.*, **443**, 536, 1995.
4. Galli, D., Stanghellini L., Tossi M., and Palla, F., *Astrophys. J.*, **477**, 218, 1997.
5. Hogan, C. J., *Astrophys. J. Lett.*, **441**, L17, 1995.
6. Hsieh, K. C. and Simpson, J. A. *Astrophys. J. Lett.*, **162**, L191, 1970.
7. Lee, K. A., *J. Geophys. Res.*, **88**, 6109, 1983.
8. Luhn, A., Klecker, B., Hovestadt, D., and Möbius, E., *Astrophys. J.*, **317**, 951, 1987.
9. Mason, G. M., Reames, D. V., et al., *Astrophys. J.*, **303**, 849, 1986.
10. Mason, G. M., Dwyer, J. R. and Mazur, J. E., *Astrophys. J. Lett.*, **545**, L157, 2000.
11. Reames, D. V., Meyer, J. P., and von Rosevinge, T. T., *Astrophys. J. Supp.*, **90**, 649, 1994.

12. Reames, D. V., and Ng, C. K., *Astrophys. J.*, **504**, 1002, 1998.
13. Roth, I. and Temerin, M., *Astrophys. J.*, **477**, 940, 1997

Cross-Field Plasma Acceleration and Potential Formation Induced by Electromagnetic Waves in a Relativistic Magnetized Plasma

R. Sugaya

*Department of Physics, Faculty of Science, Ehime University,
2-5 Bunkyo-cho, Matsuyama 790-8577, Japan*

Abstract. It has been proved theoretically that particle acceleration along and across a magnetic field and electric field across a magnetic field can be induced by almost perpendicularly propagating electromagnetic waves in a relativistic magnetized plasma.

INTRODUCTION

Relativistic and non-relativistic particle acceleration along and across a magnetic field, and the generation of an electric field across a magnetic field, both induced by almost perpendicularly propagating electromagnetic waves in a relativistic magnetized plasma, are investigated theoretically based on relativistic quasilinear transport equations derived from relativistic Vlasov-Maxwell equations[1,3-5]. The electromagnetic waves accelerate plasma particles in the k direction. As a result, the strong plasma acceleration or transport across a magnetic field appears. Simultaneously the intense cross-field electric field $E_0 = B_0 \times v_d/c$ is generated via the dynamo effect of perpendicular particle drift to satisfy the generalized Ohm's law, that is, the electromagnetic waves can produce the cross-field particle drift that is identical to $E \times B$ drift. Namely, the same result as that for the electromagnetic waves in a non-relativistic magnetized plasma[5] and as that for the electrostatic waves in non-relativistic and relativistic magnetized plasma[2-4] was obtained. Moreover, the relativistic quasilinear transport equations for the relativistic cross-field particle drift were derived by the Lorentz transformation of the relativistic quasilinear transport equations in the coordinate without the relativistic cross-field particle drift[3,4]. They can be applied to the theoretical investigation of the relativistic cross-field particle acceleration that may occur in space plasmas.

NON-RELATIVISTIC CROSS-FIELD ACCELERATION

Relativistic quasilinear momentum-space diffusion equation

We consider the non-relativistic cross-field particle acceleration in the uniform magnetic and electric fields $B_0=(0, 0, B_0)$ and $E_0=(0, E_0, 0)$. The dielectric tensor for electromagnetic waves propagating obliquely to the magnetic field can be obtained as

$$\varepsilon_k = I + \sum_s \varepsilon_k^{(s)} \quad , \tag{1}$$

$$\varepsilon_k^{(s)jl} = -\frac{\omega_{ps}^2}{\omega_k^2} \sum_{r=-\infty}^{\infty} \int d\mathbf{p} \frac{1}{\gamma_s} [u_k^{*j}(\mathbf{p}) J_r(\mu_k)] Z_{k,r}^l(\mathbf{p}) g_{s0} \quad , \tag{2}$$

$$Z^l_{k,r}(p) = \frac{[w^l_k(p)J_r(\mu_k)]Y^l_{k,r}(p) + s_l v_d J_r(\mu_k) U_r(k)}{k_\parallel v_\parallel + k_\perp v_d - \omega_k + r\omega_{cs}} \quad , \tag{3a}$$

$$Y^x_{k,r}(p) = Y^y_{k,r}(p) = k_\parallel v_\perp \frac{\partial}{\partial p_\parallel} + (\omega_k - k_\perp v_d - k_\parallel v_\parallel)\frac{\partial}{\partial p_\perp} \quad , \tag{3b}$$

$$Y^z_{k,r}(p) = (\omega_k - k_\perp v_d - r\omega_{cs})\frac{\partial}{\partial p_\parallel} + r\omega_{cs}\frac{p_\parallel}{p_\perp}\frac{\partial}{\partial p_\perp} \quad , \tag{3c}$$

$$U_r(k) = k_\parallel \frac{\partial}{\partial p_\parallel} + \frac{r\omega_{cs}}{v_\perp}\frac{\partial}{\partial p_\perp} \quad , \tag{3d}$$

$$u^x_k(p)J_r(\mu_k) = (\frac{rp_\perp}{\mu_k} + p_{ds})J_r(\mu_k) \quad , \quad u^y_k(p) = i\sigma_s p_\perp \frac{\partial}{\partial \mu_k} \quad , \quad u^z_k(p) = p_\parallel \quad , \tag{3e}$$

$$w^x_k(p)J_r(\mu_k) = \frac{r}{\mu_k}J_r(\mu_k) \quad , \quad w^y_k(p) = i\sigma_s \frac{\partial}{\partial \mu_k} \quad , \quad w^z_k(p) = 1 \quad , \tag{3f}$$

$$g_s = g_{s0}(((p_x - p_{ds})^2 + p_y^2)^{1/2}, p_\parallel, t) \tag{4}$$

where $p=p'+p_{ds}$, $p'=\gamma_s' m_s v'$, $p_{ds}=\gamma_s' m_s v_d$, $\gamma_s'=(1-v'^2/c^2)^{1/2}=(1+p'^2/m_s^2 c^2)^{1/2}$, J_r is the Bessel function of rth order, $\mu_k=k_\perp v_\perp/\omega_{cs}$, $\omega_{ps}=(4\pi n_s e_s^2/m_s)^{1/2}$, $\omega_{cs}=|e_s|B_0/\gamma_s' m_s c$, $p_x=p_\perp \cos\theta + p_{ds}$, $p_y=p_\perp \sin\theta$, $p_z=p_\parallel$, $p_\perp=\gamma_s' m_s v_\perp$, $p_\parallel=\gamma_s' m_s v_\parallel$, $I_{jl}=\delta_{jl}$, $s_x=1$, $s_y=s_z=0$, $\sigma_s=e_s/|e_s|$. The momentum-space integration in Eq. (2) is performed in the cylindrical coordinates ($d\mathbf{p}= p_\perp dp_\perp d\theta dp_\parallel$) which are displaced by p_{ds} in the x direction, $\mathbf{v}_d=(v_d, 0, 0)=c\mathbf{E}_0 \times \mathbf{B}_0/B_0^2$ equals the $\mathbf{E} \times \mathbf{B}$ drift velocity, $v_d/c<<1$, $\mathbf{k}=(k_\perp, 0, k_\parallel)$, g_s is the background momentum distribution function containing the fluctuation-induced cross-field drift velocity $v_d=\int d\mathbf{p} v_x g_s$. When $v_d=0$, g_s is reduced to g_{s0} being symmetric with respect to the magnetic field. Equation (4) is the solution of the unperturbed relativistic Vlasov equation which leads to the generalized Ohm's law

$$\mathbf{E}_0 + \mathbf{v}_d \times \mathbf{B}_0/c = 0 \quad . \tag{5}$$

This means that $\mathbf{E}_0=\mathbf{B}_0 \times \mathbf{v}_d/c$ is produced via the dynamo effect of the cross-field particle drift arising from the acceleration due to electromagnetic waves[2-5].

Next the relativistic quasilinear momentum-space diffusion equation governing the temporal evolution of g_s can be derived from relativistic Vlasov-Maxwell equations as follows:

$$\frac{\partial g_s}{\partial t} = \sum_{k \neq 0}\sum_{j,l} E^{*j}_k E^l_k Q^{jl}_k g_{s0} \tag{6}$$

with the θ–dependent momentum-space diffusion coefficient

$$Q^{jl}_k = P_{AH}[\frac{e_s^2}{\omega_k^2}\sum_{n,r=-\infty}^{\infty} e^{j(n-r)\theta}(H^j_{k,n}(p)[w^{*j}_k(p)J_n(\mu_k)] + K^j_{n,r}(p))Z^l_{k,r}(p)] \quad , \tag{7}$$

$$H^x_{k,n}(p) = H^y_{k,n}(p) = k_\parallel v_\perp \frac{\partial}{\partial p_\parallel} + \frac{1}{p_\perp}(\omega_k - k_\perp v_d - k_\parallel v_\parallel)\frac{\partial}{\partial p_\perp}p_\perp \quad , \tag{8a}$$

$$H^z_{k,n}(p) = Y^z_n(p) \quad , \tag{8b}$$

$$K^x_{n,r}(p) = v_d U_n(k)J_n(\mu_k) - \frac{1}{p_\perp}(n-r)(\omega_k - k_\parallel v_\parallel)J'_n(\mu_k) \quad , \tag{8c}$$

$$K^y_{n,r}(p) = \frac{i}{p_\perp}(n-r)[\frac{n}{\mu_k}(\omega_k - k_\perp v_d - k_\parallel v_\parallel) - k_\perp v_\perp]J_n(\mu_k) \quad , \tag{8c}$$

$$K^z_{n,r}(p) = -\frac{k_\perp v_\parallel}{p_\perp}(n-r)J'_n(\mu_k) \quad , \tag{8d}$$

where P_{AH} indicates the anti-Hermitian part of the tensor and J_n' is the first derivative of J_n with respect to the argument. The diffusion coefficient Q_k^{jl} is expressed in displaced cylindrical coordinates in momentum space, thereby g_{s0} appears on the right-hand side of Eq. (6). The azimuthal dependence of Q_k^{jl} represents the anisotropy of momentum-space diffusion around the magnetic field and the resulting cross-field particle acceleration or transport in the x direction[3-5].

Relativistic quasilinear transport equations

Carrying out the momentum-space integration of the relativistic quasilinear momentum-space diffusion equation multiplied by $w_s = n_s \gamma_s'(1 + v_d v_x'/c^2) m_s c^2$ or $p_s = n_s p$, we can obtain the relativistic quasilinear transport equations showing the temporal development of the energy and momentum densities of magnetized particles of species s as follows[3-5]:

$$\frac{\partial U_s}{\partial t} = -\sum_{k \neq 0} 2\gamma_k^{(s)} U_k \quad , \tag{9}$$

$$\frac{\partial \mathbf{P}_s}{\partial t} = -\sum_{k \neq 0} \frac{2\gamma_k^{(s)} \mathbf{k}}{\omega_k} U_k \quad , \tag{10}$$

where $U_k = \frac{1}{8\pi} \mathbf{E}_k^* \cdot \{\frac{\partial}{\partial \omega_k}[(\varepsilon'_k - N_k)\omega_k]\} \cdot \mathbf{E}_k$ is the wave energy density, $\mathbf{k} U_k/\omega_k$ is the wave momentum density, $\varepsilon_k = \varepsilon_k' + i\varepsilon_k''$, $\varepsilon_k^{(s)} = \varepsilon_k'^{(s)} + i\varepsilon_k''^{(s)}$, $N_k^{jl} = (c^2/\omega_k^2)(k^2 \delta_{jl} - k_j k_l)$, $k_x = k_\perp$, $k_y = 0$, $k_z = k_\parallel$, $U_s = \int d\mathbf{p} w_s g_s$, $\mathbf{P}_s = \int d\mathbf{p} \mathbf{p}_s g_s$ are the energy and momentum densities of particles of species s,

$$\gamma_k^{(s)} = -\frac{\omega_k}{8\pi U_k}(\mathbf{E}_k^* \cdot \varepsilon_k''^{(s)} \cdot \mathbf{E}_k) \quad , \quad (\gamma_k = \sum_s \gamma_k^{(s)})$$

is the linear damping rate ascribed to particles of species s, and $\mathbf{P}_s = (P_{s\perp}, 0, P_{s\parallel})$, $P_{s\perp} = n_s p_{ds}$, $P_{s\parallel} = n_s \gamma_s' m_s v_{s\parallel}$. The transport equations (9) and (10) predict obviously that the electromagnetic waves can generate strong particle acceleration or transport along and across the magnetic field via Landau or cyclotron damping due to quasilinear wave-particle interaction of electromagnetic waves. Thus the electromagnetic waves propagating almost perpendicularly or obliquely can accelerate particles ($k_\parallel, \gamma_k^{(s)}$ 0). The relation $P_{s\parallel}/P_{s\perp} = k_\parallel/k_\perp$ ($\mathbf{P}_s \parallel \mathbf{k}$) can be proved from Eq. (10) with $\mathbf{P}_s(0) = 0$, thereby the particle acceleration occurs in the $\mathbf{k}$ direction and the small parallel and large perpendicular particle acceleration appears simultaneously.

When the electromagnetic waves are governed by the linear kinetic wave equation, the following conservation laws for the total energy and momentum densities of waves and particles are satisfied as has been shown previously[2-5]:

$$\frac{\partial}{\partial t}(\sum_k U_k + \sum_s U_s) = 0 \quad , \tag{11}$$

$$\frac{\partial}{\partial t}(\sum_k \frac{\mathbf{k}}{\omega_k} U_k + \sum_s \mathbf{P}_s) = 0 \quad . \tag{12}$$

RELATIVISTIC CROSS-FIELD ACCELERATION

Finally we investigate relativistic cross-field particle acceleration in a relativistic magnetized plasma. The relativistic quasilinear transport equations in the laboratory (stationary) frame of reference C with an electric field and $\mathbf{E} \times \mathbf{B}$ drift can be obtained by means of Lorentz transformation for the momentum-space integration of the relativistic quasilinear momentum-space diffusion equation in frame of reference C' moving with the $\mathbf{E} \times \mathbf{B}$ drift velocity v_d. The stationary magnetic and electric fields in C' are found by means of Lorentz transformation[3,4]:

$$\mathbf{E}_0' = 0 \quad , \quad \mathbf{B}_0' = \mathbf{B}_0/\gamma_d \quad , \tag{13}$$

where $\gamma_d = (1 - \beta^2)^{-1/2}$ and $\beta = v_d/c$. In C', the stationary electric field and $\mathbf{E} \times \mathbf{B}$ drift vanish. Further the relation between the frequencies and wave numbers in C and C' is represented as

$$\tilde{k}_\perp = \gamma_d(k_\perp - \frac{\beta \omega_k}{c}) \quad , \quad \tilde{k}_\parallel = k_\parallel \quad , \quad \tilde{\omega}_k = \gamma_d(\omega_k - k_\perp v_d) \quad , \tag{14}$$

$$k_\perp = \gamma_d(\tilde{k}_\perp + \frac{\beta \tilde{\omega}_k}{c}) \quad , \quad k_\parallel = \tilde{k}_\parallel \quad , \quad \omega_k = \gamma_d(\tilde{\omega}_k + \tilde{k}_\perp v_d) \quad , \tag{15}$$

where ω_k and $\mathbf{k} = (k_\perp, 0, k_\parallel)$ are defined in C, $\tilde{\omega}_{\tilde{k}}$ and $\tilde{\mathbf{k}} = (\tilde{k}_\perp, 0, \tilde{k}_\parallel)$ in C'. Thus we derive the relativistic quasilinear transport equations in C as follows:

$$\frac{\partial U_{Rs}}{\partial t} = -\sum_{k\neq 0} 2\gamma_{Rk}^{(s)} U_{Rs} \quad , \qquad (16)$$

$$\frac{\partial \boldsymbol{P}_{Rs}}{\partial t} = -\sum_{k\neq 0} \frac{2\gamma_{Rk}^{(s)} \boldsymbol{k}}{\omega_k} U_{Rk} \quad . \qquad (17)$$

$U_{Rs} = \int d\boldsymbol{p} w_s g_s$ and $\boldsymbol{P}_{Rs} = \int d\boldsymbol{p} \boldsymbol{p}_s g_s$ are the energy and momentum densities of particles of species s in C, and $U_{Rk} = \frac{1}{8\pi} \tilde{\boldsymbol{E}}_{\tilde{\boldsymbol{k}}}^* \cdot \{ \frac{\partial}{\partial \omega_k}[(\varepsilon_{Rk}^l - N_{Rk})\omega_k]\} \cdot \tilde{\boldsymbol{E}}_{\tilde{\boldsymbol{k}}}$ and $\boldsymbol{k} U_{Rk}/\omega_k$ are the effective wave energy and momentum densities in C. $\varepsilon_{Rk} = \tilde{\varepsilon}_{\tilde{\boldsymbol{k}}}$, $\varepsilon_{Rk}^{(s)} = \tilde{\varepsilon}_{\tilde{\boldsymbol{k}}}^{(s)}$ and $N_{Rk} = \tilde{N}_{\tilde{\boldsymbol{k}}}$ are given by $\tilde{\varepsilon}_{\tilde{\boldsymbol{k}}}$, $\tilde{\varepsilon}_{\tilde{\boldsymbol{k}}}^{(s)}$ and $\tilde{N}_{\tilde{\boldsymbol{k}}}$ substituted with $\tilde{\omega}_{\tilde{\boldsymbol{k}}} = \gamma_d(\omega_k - k_\perp v_d)$, $\tilde{k}_\perp = \gamma_d(k_\perp - \beta\omega_k/c)$ and $\tilde{k}_\parallel = k_\parallel$. The linear damping rate in C is provided similarly by $\gamma_{Rk}^{(s)} = -\frac{\omega_k}{8\pi U_{Rk}}(\tilde{\boldsymbol{E}}_{\boldsymbol{k}}^* \cdot \varepsilon_{Rk}^{\prime\prime(s)} \cdot \tilde{\boldsymbol{E}}_{\boldsymbol{k}})$.

When the linear kinetic wave equation holds, we also find the same conservation laws as Eqs. (11) and (12).

REFERENCES

[1] R. Sugaya, J. Plasma Phys. **56**, 193 (1996); Phys. Plasmas **1**, 2768 (1996); Phys. Plasmas **3**, 3485 (1996).
[2] R. Sugaya, *in Proceedings of the 26th EPS Conference on Controlled Fusion and Plasma Physics, Maastricht*, ECA Vol. **23J**, p. 497-500 (1999).
[3] R. Sugaya, J. Plasma Phys. **64**, 109 (2000).
[4] R. Sugaya, *in Proceedings of the 27th EPS Conference on Controlled Fusion and Plasma Physics, Budapest*, ECA Vol. **24B**, p. 37-40 (2000).
[5] R. Sugaya, J. Plasma Phys. **66**, 143 (2001).

Nonlinear Development of Current-Driven Instabilities and Energy Transport to Heavy Ions

Mieko Toida and Hayato Okumura

*Department of Physics, Nagoya University,
Nagoya 464-8602, Japan*

Abstract. Nonlinear evolution of strong current-driven instabilities and associated energy transport among different particles species are studied by means of a two and a half dimensional electrostatic particle code with full ion and electron dynamics. The plasma consists of electrons, H, ^{4}He and ^{3}He ions. The abundance of ^{3}He ions is small. Initially, the electrons are assumed to drift along a uniform magnetic field with a speed larger than the electron thermal speed. Simulations show that Buneman waves grow at first and the electron temperature rapidly rises owing to the trapping by the Buneman waves. Then, H cyclotron waves are destabilized and grow to large amplitudes. Their frequencies are higher than those predicted by the linear theory based on the initial conditions. The waves with the frequencies around the second times as high as the ^{3}He cyclotron frequency eventually become dominant and transfer their energies preferentially to ^{3}He ions. Thus, the selective energy transfer to ^{3}He ions is caused by the nonlinear development of strong current-driven instabilities.

INTRODUCTION

Selective heating and acceleration of heavy ions have been one of the most important issues in space plasma physics. For example, it is well known that, in some solar flares, the abundance of high-energy ^{3}He ions is extremely increased.[1-3] Especially, the abundance ratio of ^{3}He/^{4}He becomes very high. As a mechanism for the ^{3}He rich events, current-driven instabilities are believed to be important and several theoretical models have been proposed.[4-6] Those theories are mainly based on the linear theories for weak-current driven instabilities. For strong current-driven instabilities, it was expected that many kinds of waves with wide frequencies would be destabilized and the selective acceleration of ^{3}He ions would be difficult. However, some observations show that ^{3}He rich events are accompanied by energetic electrons with several tens of keV.[7,8] Those electrons will cause the strong current-driven instabilities. In order to solve this discrepancy, nonlinear development of instabilities and energy transport among different particle species should be investigated.

Recently, the nonlinear development of strong current-driven instabilities in a plasma consisting of electrons, H and ^{4}He ions has been studied by means of two-dimensional, electrostatic, particle simulation.[9-11] Electrons are assumed to drift along a uniform magnetic field with an initial speed larger than the electron thermal speed. It was demonstrated that because of the electron trapping by the Buneman waves which grow at first, the shape of the electron velocity distribution function $f_e(v_\parallel)$ is drastically changed and is significantly broadened. Then, H cyclotron waves, which are marginal in the initial state, are destabilized and grow to large amplitudes. Although ^{4}He ions do not gain energies from the Buneman waves, they are slightly heated by H cyclotron waves.

In this paper, we consider a plasma containing H, ^{4}He and ^{3}He ions. It is demonstrated that ^{3}He ions are heated much more than ^{4}He ions, as a result of the nonlinear development of instabilities. The heating of ^{3}He ions is caused by the H cyclotron waves with $\omega \sim 2\Omega_{3He}$. The frequencies of these waves are significantly increased compared to the theoretical value based on the initial conditions. The present result would be an important clue to the problem of preferential energy transfer to ^{3}He ions in solar flares.

SIMULATION METHOD AND PARAMETERS

We perform simulation using a two and a half dimensional (two space and three velocity components), electrostatic particle code with full ion and electron dynamics. The system size is $(L_x, L_y)=(256\Delta_g, 1024\Delta_g)$, where Δ_g is the grid spacing. We use periodic boundary conditions. The code has electrons and three species ions, H, ^{4}He and ^{3}He. Their total particle numbers are N(e)=16,777,216, N(H)= 12,783,616, N(4He)=1,597,440, and N(3He)=399,360. The mass ratios are $m_H/m_e=100$, $m_{4He}/m_H=4$, $m_{3He}/m_H=3$; the charge ratios are $q_H/|q_e|=1$, $q_{4He}/q_H=2$, $q_{3He}/q_H=2$, the temperatures are equal among all the particle species, $T_e=T_H=T_{4He}=T_{3He}$. The electron cyclotron frequency is $|\Omega_e|/\omega_{pe}=2.0$. The Debye length is $\lambda_{De}=\Delta_g$. The uniform external magnetic field is in the y direction. Initially, the electrons have a shifted Maxwell distribution with the drift speed parallel to the magnetic field, $v_d = 4\ v_{Te}$. The ions have isotropic Maxwellian velocity distribution functions. According to the linear theory based on these initial conditions and parameters, Buneman waves are unstable, while H cyclotron waves are almost stable.

SIMULATION RESULTS

Wave evolution

Figure 1 shows time variations of the Buneman wave with $(k_{\parallel}\rho_H, k_{\perp}\rho_H)=(0.64, 0.0)$, the H cyclotron wave with $(k_{\parallel}\rho_H, k_{\perp}\rho_H)=(0.031, 0.43)$, and electron temperatures. Here, the amplitude of the waves $|E_k|^2$ is normalized to $L_x L_y m_e v_{Te}^2$. In the bottom panel of Fig. 1, the solid and dashed lines represent temperatures parallel and perpendicular to the magnetic field, respectively. They are normalized to their initial values.

As predicted by the linear theory based on the initial conditions, the Buneman wave rapidly grows. However, it soon saturates owing to the electron trapping. Then, as shown in the previous paper,[11] the electron velocity distribution function $f_e(v_{\parallel})$ is drastically changed and is significantly broadened. Thus, the electron parallel temperature T_{ey} rapidly rises. The H cyclotron waves, which are marginal in the initial state, are destabilized because of the change of the shape of $f_e(v_{\parallel})$ and continue to grow during the period from $\omega_{pe}t=200$ to 600. (The amplitude oscillation is caused by the beating between the H cyclotron wave and the oblique Buneman wave with much smaller amplitude.) Since the H cyclotron waves are destabilized after the rise of T_{ey}, we can expect that the wave properties are determined in a plasma with high electron temperatures.

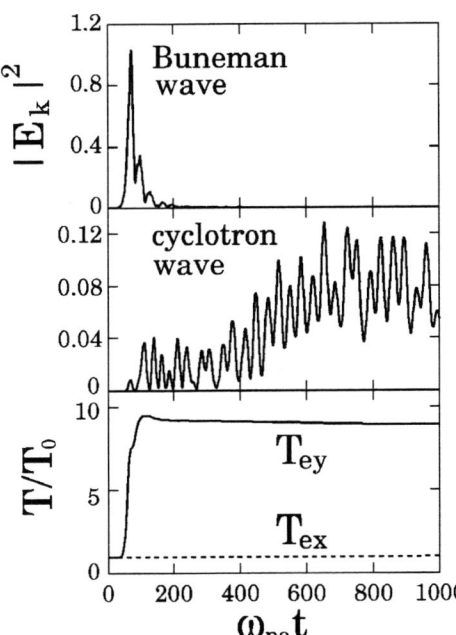

FIGURE 1. Time variations of amplitudes of two typical modes and electron temperatures.

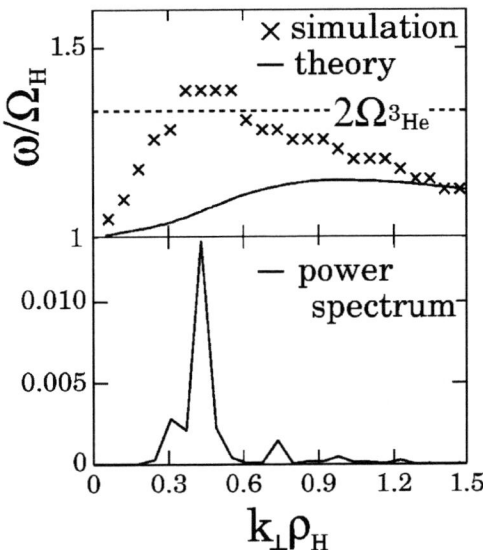

FIGURE 2. Dispersion relation and power spectrum of H cyclotron waves.

Figure 2 shows the dispersion relation and the power spectrum of H cyclotron waves with $k_\parallel \rho_H$=0.031. Here, the solid line in the upper panel denotes the dispersion relation based on the initial conditions. The crosses denotes the simulation results obtained from the data from $\omega_{pe}t$ =0 to 2000. In the long wavelength region, $k_\perp \rho_H <1$, the frequencies observed in the simulation are much higher than the theoretical values based on the initial conditions. The frequency increase is due to the electron temperature rise. The lower panel in Fig. 2 shows the power spectrum of these modes. The wave with $\omega \sim 2\Omega_{3He}$ have the largest amplitude. Although the waves with $\omega \sim 2\Omega_{3He}$ do not exist in the initial state, they are excited by the nonlinear effect and eventually become dominant.

Energy transport to Ions

As demonstrated in the previous paper, the Buneman waves reflect some H ions and accelerate them in the direction parallel to the magnetic field. However, the heavy ions are not reflected by the Buneman waves, and their parallel energies do not increase. On the other hand, the perpendicular energies of the heavy ions can increase through the cyclotron resonance with the H cyclotron waves.

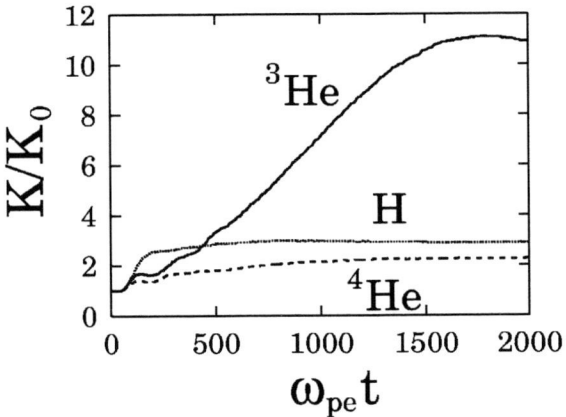

FIGURE 3. Time variations of total perpendicular energies of ions.

Figure 3 shows time variations of the total perpendicular energies of H, ^{3}He, and ^{4}He ions. The energies are normalized to their initial values. The energy gain of ^{3}He ions is the most noticeable. This is because the waves with $\omega \sim 2\Omega_{3He}$ have large amplitudes as shown in Fig.2, and their energies are preferentially transferred to ^{3}He ions. Furthermore, the energy of ^{3}He ions continues to increase during the long period from $\omega_{pe}t$ =200 to 1600, while those of H and ^{4}He ions are rapidly saturated at $\omega_{pe}t$ ~200. Since the abundance of ^{3}He ions is quite small, the wave damping due to the resonance of ^{3}He ions is also small. Thus, the waves with $\omega \sim 2\Omega_{3He}$ can keep their amplitudes for a long time and heat ^{3}He ions efficiently.

Figure 4 shows the energy distribution functions of H, ^{4}He, and ^{3}He ions. Here, v_{Ti} is the initial ion thermal velocity and dashed lines are the initial distributions. In ^{3}He ions, many particles are accelerated to high energies, $v^2/v_{T3He}^2>100$ at $\omega_{pe}t$ =2000. The fraction of energetic particles in ^{3}He ions is much higher than that in H or ^{4}He ions.

SUMMARY

By means of a two and a half dimensional, electrostatic, particle simulation code, we have studied nonlinear development of strong current-driven instabilities and associated energy transport in a plasma containing H, ^{4}He, and ^{3}He ions. After the development of the Buneman waves, H cyclotron waves are destabilized, although they are almost stable in the initial state. We have investigated the dispersion relation of the H cyclotron waves. It is shown that, in the long wavelength region, their frequencies are higher than the theoretical values based on the linear theory. Because of this frequency change, the waves with $\omega \sim 2\Omega_{3He}$ eventually grow to large amplitudes and transfer their energies preferentially to ^{3}He ions. Thus, it has been demonstrated that, even by a strong current-driven instabilities, the ^{3}He ions are selectively accelerated as a result of the nonlinear development of the instabilities.

Our simulation result indicates that, in solar flares, the selective acceleration of ^{3}He ions would be possible much more than the conventional theory predicts. In order to explain the ^{3}He rich events, we will need a quantitative discussion. In the future, we will study the energies of accelerated ions.

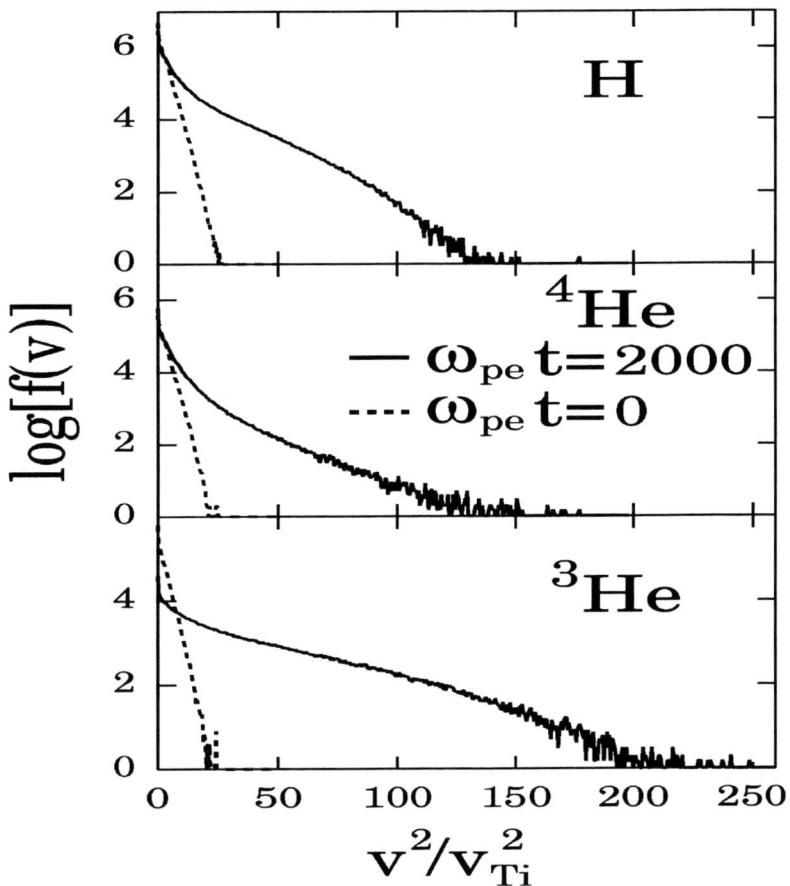

FIGURE 4. Energy distribution of H, ^{4}He, and ^{3}He ions. Here, v_{Ti} is the initial thermal velocity.

REFERENCES

1. Anglin, J. D., Astrophys. J. 198, 733 (1975)
2. Reames, D. V., Astrophys. J. 73, 235 (1990)
3. Reames, D. V., Meyer, J. P., and von Rosenvinge, T. T., Astrophys. J., Suppl. Ser., 90, 649 (1994)
4. Fisk, L. A., Astrophys. J. 224, 1048 (1978)
5. Zhang, T. X., Toida, M., and Ohsawa, Y., J. Phys. Soc. Jpn. 62, 2545 (1993)
6. Nakazawa, S., Zhang, T. X., and Ohsawa, Y., Solar Phys., 166, 159 (1996)
7. Reames, D. V., von Rosenvinge, T. T., & Lin, R. P., Astrophys. J. 292, 716 (1985)
8. Ho, G. C., Roelof, E. C., Hawkins III, S. E., Gold, R. E., Mason, G. M., Dwyer, and J. R., Mazur, J. E., Astrophys. J. 552, 863-870 (2001)
9. Toida, M., Maeda, T., Shiiba, I., and Y. Ohsawa, Phys. Plasmas 7, 4882-4888 (2000)
10. Toida, M., and Sugishima, A., J. Phys. Soc. Jpn. 70, 3285-3290 (2001)
11. Toida, M., and Sugishima, A., and Ohsawa, Y., Phys. Plasmas, 9, 2541-2548 (2002)

Stochastic Motion of Relativistic Particles in the Field of a Wide Wave Packet

E. Nagornykh* and A. Tel'nikhin*

Department of Physics, Altai State University, Dimitrova 66, Barnaul, 656061, Russia

Abstract. Stochastic motion of relativistic particles in the field of a wave packet propagating under an angle to the external magnetic field are investigated. The interplay of the dynamical and statistical aspects of the behavior of the relativistic particle-potential wave packet system is considered. Dynamics of this system are described by nonlinear mapping and corresponding Fokker-Planck-Kolmogorov equation in phase space possesses canonical Hamiltonian structure. The following general problems of stochastic motion are disscussed: local instability and the Lyapunov exponents and the Kolmogorov entropy; a fractal structures and its dimension; bifurcations of a vector fields and the boundaries of the region of dynamical chaos. The results of numerical simulation are presented. A possible astrophysical application of the results obtained is discussed.

INTRODUCTION

Numerous observations have shown that Jupiter is a source of synchrotron emission from energetic electrons gyrating in Jupiter's magnetic field. These energetic electrons might become repeatedly accelerated through an interaction with plasma waves, which can transfer energy into the electrons. The observation data [1] suggest that electrons with high energies ($\sim 100\ MeV$) are more numerous than previously expected. The inner radiation belts of Jupiter and Earth contain extremely high-energy electrons that require substantial acceleration by processes other than wave-particle resonant diffusion [1]. A possible mechanism can be local stochastic acceleration.

Stochasticity of deterministic dynamical systems is nowadays recognized to be universal. In the field of plasma physics, there are a number of aspects in which coherent systems can produce stochastic motion. One have looked at fundamental properties of the wave particle system, and showed that the stochastisity occurs when the amplitude of the wave is large enough [2]. For example the method of Poincare mappings used to solve problems involving the motion of charged particles in the field a time-like wave packet [3]. The stochastic dynamics of relativistic particles in the space-like packet and a possible astrophysical application of the results have been discussed by [4].

In the present work we describe the particle stochastic acceleration in the electromagnetic modes under conditions of strong nonlinearity using a model of resonance interaction between particle and wave. In Section II we describe stochastic heating of relativistic particles in the field of wide packet of electrostatic waves. We present numerical experiments for stochastic motion, described in terms of the Poincare maps. In the stochasticity region, the dynamics of a particle can be studied by a kinetic equation which describes the energy distribution of the particles. Section III is devoted to application to particle acceleration in the magnetosphere. The results demonstrate the existence of ultra-relativistic electrons in Jupiter's and Earth's radiation belts and extend the nonthermal energy spectra. In Section V we give the conclusions of our studies of stochastic heating pronounced in radiation belts.

STOCHASTIC ACCELERATION OF RELATIVISTIC PARTICLES IN THE FIELD OF A WIDE WAVE PACKET

Let us consider a relativistic particle of charge $|e|$ and mass m in homogeneous static magnetic field of strength B and potential field $U = \sum \Phi_k \sin(zk_z + k_\perp x - \omega_k t)$, of the wave packet of the Langmuir wave (LWP) propagating under an arbitrary small angle to the external magnetic field ($k_\perp/k_z \ll 1$), where k is the wave number and ω_k is the wave frequency. In order to write down equations of the particle motion we must specify a coordinate system.

We have chosen a Cartesian spatial coordinates system whose z to be a distance parameter, measured in the direction of the magnetic field at the guiding center position. The plane perpendicular to this direction is spanned by the local orthogonal coordinates x and y. We shall assume that the L characteristic length scales of inhomogeneity are sufficiently great that the wave length. Taken into account of axial symmetry of system we can introduce new variables the action I, the angle θ by canonical transformation. In this representation the Hamiltonian becomes

$$H(p_z, z, I, \theta, t) = H_0(p_z, I) + \sum_{k \in R, s \in Z} U_k \sin(zk_z + s\theta - \omega_k t), \tag{1}$$

$$U_{ks} = \Phi_k J_s(k_\perp r), \qquad H_0(p_z, I) = (m^2 + p_z^2 + 2m\omega_B I)^{1/2}$$

where Φ_k is amplitude of the potentials of the modes in the wave packet, $k_\perp$ is perpendicular wave number, $r = \sqrt{2m\omega_B I}/m\omega_B$, the gyroradius, $J_s(\cdot)$ is Bessel function, ω_B is cyclotron frequency, p_z is momentum of the particle.

In fact, the nonlinear resonance wave-particle interaction play the dominant role in determination of the motion characters. Hence one must consider dynamics determines by the condition

$$k_z p_z H_0^{-1} + s\omega_B m H_0^{-1} - \omega_k = 0. \tag{2}$$

If the inequality $V \gg V_g$ is valid also, V is the particle velocity, V_g is group velocity of the wave, then wave field (2) has space-like packet form [5]. In such a case, the nonlinear wave-particle interaction may be described by the following set of nonlinear equations

$$\begin{array}{ll} \dot{p}_z = -k_0 U_0 J_s(k_\perp r) L \sum \delta(z-nL) \cos(\psi), & \dot{I} = -s U_0 J_s(k_\perp r) L \sum \delta(z-nL) \cos(\psi), \\ \dot{z} = p_z H_0^{-1}, & \dot{\theta} = \omega_B m H_0^{-1}, \quad n \in Z, \end{array} \tag{3}$$

where $\psi = k_0 z + s\theta - \omega_0 t$ is the phase of fundamental mode, $s = 0, \pm 1$, and $\delta(\cdot)$ is the Dirac delta function. From equations (3) we obtain the additional integral of motion

$$I = \left|\frac{sp_z}{k_0}\right| + I_0, \tag{4}$$

here I_0 the constant value determining by initial conditions. Using (4) and integrating equations (3) we arrive at the dynamical system

$$G^n: \begin{array}{ll} u_{n+1} = & u_n + Q\sin 2\pi q_n \\ q_{n+1} = & q_n + (1/4\pi)u_{n+1}^{-2} \mod 1, \end{array} \tag{5}$$

where u is the normalized energy, q is the phase of fundamental mode, Q characterizes the intensity of the wave packet. The mapping was obtained by [4].

Then Jacobian of the system is $\det J = \det \frac{\partial(u_{j+1}, \psi_{j+1})}{\partial(u_j, \psi_j)} = 1$, denote by λ_1, λ_2 the eigenvalues and tr the trace of the J-matrix. Taken into account the conditions

$$\det J = \lambda_1 \lambda_2 = 1, \qquad tr J = \lambda_1 + \lambda_2 = 3 \tag{6}$$

it is easy to see that (5) is area-preserving mapping with eigenvalues $\lambda_1 = \frac{3+\sqrt{5}}{2}$, $\lambda_1 \cdot \lambda_2 = 1$, and that conditions (6) lead to the following range of values of u for the region of dynamics chaos [6]

$$u_m \leq sup\{u\} = (Q)^{1/3}. \tag{7}$$

We have numerically integrated equations (5) for fixed values of Q. Our results are shown in Fig. 1 for $Q = 2\pi 10^{-3}$. The figure shows that the stochastisity region is bounded in phase space. The $sup\{u\}$ calculated from formula (7) agrees with numerical results.

To discuss possible situations one assumed the inequalities $k_\perp r \ll 1$, $k_\perp/k \ll 1$ are fulfilled.

In the Cherenkov's resonance (CR) events (2) at $s = 0$, $J_0(k_\perp r) \simeq 1$ is valid, the variable u and the parameter Q can be written as

$$Q = \sqrt{kL}\phi/m, \qquad u = \varepsilon \frac{1}{\sqrt{kL}}. \tag{8}$$

In view of (7), (8) we arrive at

$$\varepsilon_m = sup\{\varepsilon\} = \sqrt{kL}(\sqrt{kL}\frac{\phi}{m})^{1/3}, \qquad \varepsilon = w/m, \tag{9}$$

here ε is the dimensionless energy of particle along the magnetic field lines.

Further using FPK-equations

$$\frac{\partial f}{\partial t} = \frac{1}{2}\frac{\partial}{\partial u}D\frac{\partial f}{\partial u}, \qquad (10)$$

along with (5), (7) we find diffusion coefficient $D = Q^2 c/2L$ and the characteristic time of stochastic heating

$$\tau_d \sim 2Q^{-4/3}L/c. \qquad (11)$$

In the case of the anomalous cyclotron resonance (ACR) at $s = -1$ in (2), the variable u and parameter Q have form

$$Q = \frac{k_\perp}{k}\frac{\omega}{\omega_B}\sqrt{kL}\frac{\phi}{m}, \qquad u = \sqrt{\frac{\omega}{2\omega_B}\frac{1}{kL}}\varepsilon, \qquad (12)$$

where the auxiliary conditions $\varepsilon \gg \omega/2\omega_B$, $J_1(k_\perp r) \simeq (1/2)k_\perp r$ are used.
According to (7) and (12) along with (4) we get

$$\varepsilon_m = sup\{\varepsilon\} = Q^{2/3}kL\frac{2\omega_B}{\omega}, \qquad \varepsilon_\perp = \sqrt{\frac{2\omega_B}{\omega}}\varepsilon, \qquad (13)$$

where $\varepsilon_\perp$ is the perpendicular energy of particle.

The kinetic equation with the diffusion coefficient is solved together with the boundary condition and expression

$$D\frac{\partial f}{\partial u}\bigg|_{u=u_m} = 0, \qquad \|f\| = \int_{-\infty}^{\infty} f(u,t)du = 1. \qquad (14)$$

With this condition the time-independent distribution over energy spectrum

$$f(\varepsilon) = (2\varepsilon_m)^{-1}, \qquad \varepsilon \in (1, \varepsilon_m) \qquad (15)$$

and characteristic time for the energy redistribution over the spectrum (11).

Let us examine the fractal structure of stochastic set. Introducing the fractal dimension $d_F = 1 - \ln\lambda_1/\ln\lambda_2$ and the information dimension $d_I = -\int f \ln f du/\sqrt{2u_m}$ [7] and substitute the results (7), (15) into this expressions to obtain $d_F = d_I = 2$.

It is important to note that dynamical chaos leads to a "pancake" angle distribution during resonant interaction, τ_d. Denoting $\tan\alpha = \varepsilon/\varepsilon_\perp$, where ε and $\varepsilon_\perp$ obey the expressions (13) in ultra relativistic limit we find following angle distribution function

$$\alpha(\varepsilon) = arctan\sqrt{\frac{\omega}{2\omega_B}\varepsilon}. \qquad (16)$$

The analytical solution for diffusion curves have following form

$$\frac{\varepsilon_\perp}{\varepsilon} = \sqrt{\frac{2\omega_B}{\omega}}\varepsilon^{-1}. \qquad (17)$$

The solution given by (17), (16) are significantly different from the relativistic solution discussed by [8]. From expressions (9) and (13) supervene following dependencies

$$sup\{w\} \propto m^{2/3}, \quad s = 0; \qquad sup\{w\} \propto m^{1/6}, \quad s = -1.$$

This expressions imply that stochastic acceleration of protons is possible.

APPLICATION TO PARTICLE ACCELERATION IN THE MAGNETOSPHERE

It is well known that high energy particles exist in Jupiter's and Earth's radiation belts. Wave-particle interaction produce relativistic electron population with soft broad energy distribution. The stochastic heating of relativistic charged particles can be regarded as a possible mechanism for the formation of the energy spectrum. To obtain quantitative estimates, we specify the parameters of the problem.

By way of illustration, we apply the results of our investigation to the Earth's magnetosphere. Accordingly to [9], we take the following typical mean parameters of this problem: let the electron plasma frequency is $\omega_p \simeq 10^6 s^{-1}$, $\omega/\omega_B \simeq 10$, the wavelength is $\lambda \simeq 10^5 cm$. Assuming that the length L of the mapping is equals to the characteristic scale length of the Earth's radiation belt, we can set $L \simeq 2 \cdot 10^8 cm$.

Now we discuss stochastic particle heating in CR case. By formulas (9),(11) we evaluate diffusion time over spectrum $\sim 10s$ and $sup\{w\} \simeq 20 MeV$. From expressions (13) and (16) at $k_\perp/k = 0.1$ in just same way in ACR case we estimate $\tau_d \simeq 10s$ and $sup\{w\} \simeq 25 MeV$.

Now we treat the feathers of stochastic heating electrons, in the background magnetic field $B = 1.2G$ spiralling with frequency $\omega_B = 2 \cdot 10^7 s^{-1}$ in Jupiter's radiation belt with the length $L \simeq 2 \cdot 10^9 cm$ [1]. By the Poisson formula we estimate the magnitude of the plasma field $\phi/m \simeq (\omega^2/k^2)(\delta n/n)|(\omega/k=c=1)$, where $\delta n/n$ is the relative level of fluctuations in the electron density. Let us $\delta n/n \simeq 10^{-6}$.

Under existing conditions we have following results for heating in the wave packet field in CR case, by formulas (9) and (11) we calculate diffusion time $\tau_d \simeq 4 \cdot 10^3 s$, electron's energy $sup\{w\} \simeq 500 MeV$. And in case ACR by formulas (13) and (16) we obtain $\tau_d \simeq 10^4 s$ and electron's energy up to $sup\{w\} \simeq 100 MeV$ at $k_\perp/k = 0.01$.

CONCLUSION

Mapping equations which describe relativistic particle motion in the high frequency wave fields have been used. The particle motion can be stochastic when the wave intensity is strong enough. From FPK description have found energy distribution function and characteristic time for the energy redistribution over the spectrum. We are shown that significant electron energy diffusion can leading to a "pancake" angle distribution.

The study pointed out the importance of the stochastic acceleration mechanism to explain the appearance of relativistic particles in Earth's Van Allen belts and in Jupiter's radiation belts. Our results are in reasonable agreement with the observations data.

REFERENCES

1. Bolton,S.J., Janssen,M., Thorne, R. M., et. al., *Nature*, **415**, 987-991, (2002).
2. Sagdeev,R.Z., Zaslavsky,G.M., *Non-linear Phenomena in Plasma Physics and Hydrodynamics*, Moscow, Mir, 1986.
3. Zaslavsky,G.M., Sagdeev, R.Z., Usikov,D.A., Chernikov, A.A., *Weak Chaos and Quasi-Regular Patterns*, Moscow, Nauka, 1991.
4. Krotov,A., Tel'nikhin, A., *Plasma Phys. Rep.*, **4**,767-771, (1998).
5. Chirikov,B.V., *Phys. Rep.*, **2**, 463, (1979).
6. Arnold V. I., Avez, A., *Ergodic problems of classical mechanics*, Benjamin, New York, 1968.
7. Schuster, H. *Deterministic chaos*, Weinheim, Physik Verlag, (1984).
8. Summers, D., R. Thorne, R.M., and Xiao,F., *J.Geophys. Res.*, **103**, 20487-20500, (1998).
9. Walker, A.D.M., *Plasma Waves in the Magnetosphere*, Springer-Verlag, New York, 1993.

OBSERVATIONS OF NATURAL PLASMAS

1. Planetary Atmospheres, Ionospheres, and Auroral Regions

2. Planetary Magnetospheres and Space Weather

3. Solar Wind and Outer Heliosphere

4. Solar and Coronal Physics

On Property of Scattering Structure in Random Continua

Jian-shan Guo, Manlian Zhang, Jiankui Shi, Sheping Shang, Xigui Luo, Hong Zheng

Laboratory of Space Weather Study, Center for Space Science and Applied Research, Chinese Academy of Sciences, Email:guojs@center.cssar.ac.cn

Abstract. This paper attempt to study scattering property of turbulently advected medium. A new concept, scattering structure, is defined. A cross section is given and discussed. The results revealed that the scattering structure results from Bragg and time sampling actions. Properties of the scattering structure, e.g. scattering efficiency, aspect sensitivity, frequency dependence and discreteness, are analyzed. Some of them are demonstrated by Digital ionosonde measurement. The particle-behaved scattering structures within resolution volume are the scattering sources of pixels of the radar interferometry image.

INTRODUCTION

The turbulence is ubiquitous at the Solar-terrestrial space, and could occur at anytime. However, the turbulent refractive index field is usually either difficult to be formulated to be able to have a solution for EM parabolic equation or remains an unambiguously reasoning. This difficulty or ambiguity come from solving of EM parabolic equation calls for instantaneous refractive index field but the turbulent refractive field is usually described by statistics. For example, classical ionospheric drift analyses are based on the correlation analysis of ground diffraction pattern that is a result of interference of echoes from scatters in ionosphere. The diffraction pattern used to be considered as statistic field, However, the echoes, which forms the pattern, is usually to be supposed as a deterministic point scatter or point reflector[Briggs, 1980]. Though some theory starting with a random medium field obtains rigorously correlation function of the echoes, the so-called Bragg scatters was supposed to be important and distributed randomly in the medium in the theory. Doppler image interferometer (DII) adopts a more straightforward point scatter (or reflector) concept, has observed discrete scatter distribution in view of antenna (called Skymap in Digital ionosonde [Bible and Reinisch, 1978]). A sequence of Skymap could describe evolution of spatial structure of the medium process [Kudeki, et al., 1991, Woodman, 1997, Reinisch, et al., 1998]. These stimulated our original attempt to gain insight into the pixel echoes and its scattering structure in the refractive index field. It looks as if the point (or discrete) scattering model has to be preserved, because all the techniques mentioned above have been quite successful for decades, however, should be improved because it theoretically remains confusion.

In the section2 the formula for cross section of resolution volume will be given, upon which we discuss Bragg and time (range) sampling actions and define the scattering structure. The scattering efficiency, aspect sensitivity, frequency dependence and discreteness of scattering structure are discussed in section 3. The simulation results and comparison with observation are given and discuss in section 4. At last we will conclude our research and discuss some issues about atmosphere turbulence.

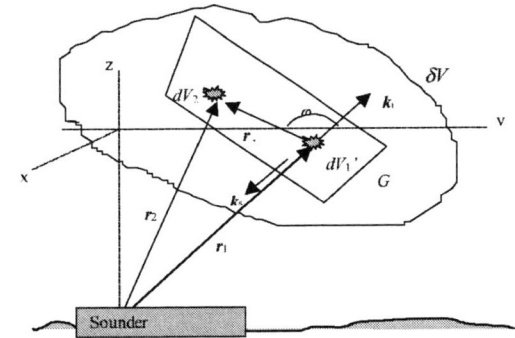

FIGURE 1. Schematically showing the scattering volume δ back scatters a plane wave at dV' in the direction k within the sampling volume G

CROSS SECTION IN A TURBULENTLY ADVECTED RANDOM CONTINUA AND THE SAMPLING ACTIONS

We now consider a piece of random medium in volume δV illuminated by an incident wave at direction $\mathbf{k}_i$ from a sounder Fig.1. We wish to find the cross section of resolution volume G. We start with some general formulae of differential cross section per unit volume $\sigma(\mathbf{k}_s)$ given by previous authors [Ishimaru, 1978, Yeh and Liu. 1972]. The assumptions and conditions made in the paper are as follows. As a first step the discussion below is limited to the case of back scattering. The resolution volume G can be small enough by taking advantage of modern data processing technique. Therefore, the far field condition is met. The medium meets the requirement of locally frozen-in condition i.e. The shape of fluctuation does not change significantly during the sampling interval. The medium fluctuation is weak comparing with its average value, so the Born approximation could be used. Turbulently advected refractivity field has Gaussian velocity fluctuation

The differential scattering cross section per unit volume of medium is given by

$$\sigma(k_s) = A\phi_n(k_s), \qquad A = 2k^4 \pi \sin^2 \chi \tag{1}$$

$$\phi_n(k_s) = \frac{1}{(2\pi)^3} \int_G B_n(r_d) \exp(ik \cdot r_d) dV_d \tag{2}$$

Here $\phi_n(k)$ is spatial spectral density of the medium fluctuation $n_1(r)$. χ the angle between polarization and direction of the observation. $B(r_d)$ the correlation function of $n_1(r)$ at r_1', r_2', and $r_d = r_1' - r_2'$. $k_s = k_i - k_o$, the subscripts s, i, o stand for mirror, incident and scattering wave number respectively. $\phi_n(k_s)$ is the $\phi_n(k)$ evaluated at k_s, this is equivalent to $\phi_n(k)\delta(k-k_s)$ integrated over k space. Therefore (1) can be written as

$$\sigma(k_s) = A \int_\infty \phi_n(k)\delta(k - k_s) dk. \tag{3}$$

Convert the integral for $\phi(k)$ from wave number k space into spatial vector r_d space. According to Parseval theorem, we have cross section of medium in G

$$\sigma_G(k_s) = A \int_G B_n(r_d) \exp(ik_s \cdot r_d) \, dr_d. \tag{4}$$

We should noticed from (1) or (3) that the differential scattering cross section per unit volume is proportional to the value of spatial spectrum $\phi n(k)$ at the mirror spatial spectral component ks in both direction and scalar value. However the limited resolution volume G cuts off the medium (4) so that the component ks should not be infinitive in space so as the confined medium has more than one Fourier component, in turn, has a certain structure. This does not show explicitly on (1), however after sampling by δ function in (4), and by resolution volume G in (4), becomes visible. The certain structure of medium that resulted in by the sampling actions, effectively contributes to the scattering field is called scattering structure in this paper. The property of the scattering structure obviously could not remain statistic but deterministic property. For purpose of simulation we put the digitized and real form of (4) as

$$\sigma(k_s) = A \sum_m B_n(m\Delta r_d) \sin(k_s \cdot m\Delta r_d) \Delta r_d \tag{5}$$

For the turbulently advected refractivity field with Gaussian velocity fluctuation, average velocity V_0, and variance σ_v^2, the differential scattering cross section per unit volume [Ishimaru, 1978, Liu, et al., 1990, Doviak, et al., 1996]

$$\sigma(k_s, \tau) = A\phi_n(k_s, 0) \exp\left(-\frac{1}{2}k_s^2 \sigma_v^2 \tau^2\right) \exp(i k_s \cdot V_0 \tau). \tag{6}$$

Here $\varphi_n(k_s, 0)$ is the spectral density of medium $\varphi_n(k_s, \tau)$ at $\tau = 0$ and evaluated at k_s. The temporal frequency spectrum

$$\sigma(k_s, \omega) = A\phi_n(k_s) K_D = A\phi_n(k_s) \frac{2\sqrt{2\pi}}{k_s^2 \sigma_v^2} \exp\left(-\frac{(\omega + k_s \cdot V_0)^2}{2k_s^2 \sigma_v^2}\right).$$

We can see that V_0 results in Doppler shift frequency and σ_v^2 in Doppler frequency spread. When $\sigma_v \to 0$ and note (2), the formula (6) becomes

$$\sigma_{G,\omega}(k_s, \omega) = A \int_G \int_{\omega_\tau} \phi_n(k_s) \delta_k(k - k_s) \delta_\omega(\omega + k_s \cdot V_0) dk d\omega_\tau. \tag{7}$$

The $\varphi_n(k)$ undergoes two sampling actions δ_k (we call it Bragg sampling for convenience) and Doppler sampling δ_ω In the back scattering case at Doppler shift spectral line $\omega_D = -\mathbf{ks} \cdot V_0 = -2\mathbf{k} \cdot V_0$, the formula (7) behaves just as (3) or (4) in position space. The expression (4) could be considered as zero-order approximation to more complicate case. However it covers the basic properties of scattering action. Therefore we shall concentrate on the analysis of expression (4).

SINUOUS STRUCTURE

Suppose $B_n(z)$ is a vector $k_b(z)=k_b(z)\mathbf{z}$ that directs along z-axis with an arbitrarily scalar value, i.e. the medium is vertically stratified. The volume of G is a cuboid with length $\pm L/2$ along z-axis and unit area at x, y plane. In lag coordinate system, $k_d = k_s - k_b$, $B_n(z)$ can be expanded as Fourier series over all k_b

$$B_n(z) = \sum_b B_m \exp(-ik_b \cdot r_d). \qquad (8)$$

Substitution and integration in (4) give

$$\sigma_b(k_s) = AB_m \sum_b \frac{2\sin(k_d \cos(\vartheta)L/2)}{k_d \cos(\vartheta)}. \qquad (10)$$

θ is the angle between k_s and r_d. In case of $k_d = k_s - k_b = 0$, or $k_b = k_s = 2k$,

$$\lim_{k_d \to 0} \sigma_b(k_s) = AB_m L, \qquad (11)$$

or
$$\sigma_b(k_s) = AB_m L. \qquad (12)$$

When k_b k_s, $\theta=(\pi-\theta_g)/2$, $k=k_s\sin(\theta_g/2)$. θ_g is zenith angle of k. Substitution in (10) gives $\sigma_b(k_s)$ as a function of θ_g

$$\sigma_b(k_s) = AB_m L \sum_b \frac{\sin(kL\sin^2(\theta_g/2))}{kL\sin^2(\theta_g/2)}$$

An aspect sensitive factor $\sin(\theta_g)$ appears in the expression of $\sigma_b(k_d)$. When $\theta_g =0$, i.e. $\theta=\pi/2$, or vertically incident, the $\sigma(k_d)$ reaches maximum value $\sigma_{bm}(k_d)$. Any value of k_d makes $\sigma_{bm}(k_d)$ getting smaller, the larger k_d the faster the $\sigma_{bm}(k_d)$ vanishes.

That the correlation function $B_n(z)$ sinuously varies with z does not certainly means refractivity field $n_1(z)$ does so. Suppose medium fluctuation $n_1(z) = N \sin(z')$, convolution is needed only within a period $2\pi/k_s$, so we have correlation function $B_n(z)=(N^2/2)\cos(k_n z)$. It shows that fluctuation $n_1(z)$ has $\pi/2$ phase shift to it's correlation function $B_n(z)$. Because the peaks of $B_n(z)$ corresponds to the peaks of scattering wave $E(k_s z)$, so at points of $n_1(z) = 0$, $E(k_s z)$ reaches its maximum value.

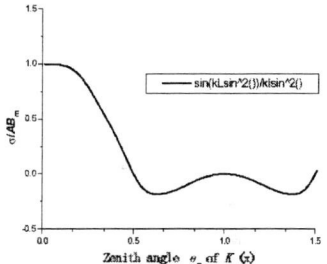

FIGURE 2. Cross section variation with zenith angle θ_g

SIMULATION AND MEASURED RESULTS

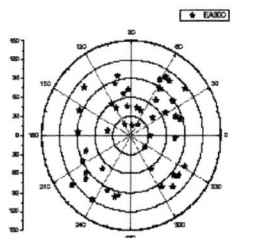

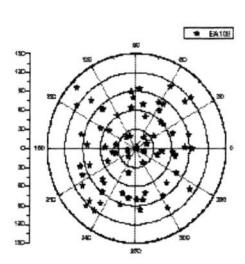

Fig.3a Scattering structure distribution(for f=1.5MHz)

Fig.3b Scattering structure distribution(for f=0.5MHz)

We now construct a 3-D statistically homogenous and isotropic $B_n(r_d)$ field with Kolmogorov spectrum density, $L_0=1\ km$ and $l_0=0.001 km$ based on the digitized formula (5). λ is taken from $10 m$ to $1000 m$ and $m=1,2,\ldots 100$. We study $\sigma_G(k_s)$ in 2-D plane. The value of $\sigma_G(k_s)$ on the x,y plane less than threshold ws ascut off, so where the value exists at are where the scattering structure locates at. Fig.3 shows a distribution pattern of scattering structures at $z=200k$ at x,y plane. Fig.3a for $\lambda=100m$ and Fig.3b for $\lambda=300m$. The pattern shows frequency dependence and discreteness. The measured Skymaps on Feb. 4, 1988 at 01:17:26 LT in Qaanaaq (77.5 N, 290.8 E) is given in fig.4. Four Skymaps correspond to two frequencies 2.0025MHz, 2.4025 MHz and four heights 280km, 290km, 315km and 325km and also shows frequency dependency and discreteness. There are no scattering structures existing at height 290km and frequency 2.0025MHz, though this is an often encountered case in practice.

CONCLUSION AND DISCUSSION

The scattering structure is a medium structure that holds a space interval larger than about $\lambda/2$ and less than L, see (12), along k direction, transverse area principally does not limited on the sphere with radiu $|r|$, ssampling range or confined by Fresnel radius on the plane perpendicular to k. It results in Bragg sampling action so that effectively scatters wave power. However it is not a 'Bragg scatter'. The medium in resolution volume can be one scattering structure (in case of high resolution) or the more arranged upon Bragg equation. Spatial variation of refractivity in the structure can be considered as deterministic but distribution of the structure within the large volume, say, δV in Fig.1 is randomly. However up to now we don't yet have agreed-upon terms for this turbulence description, and tentatively call it hybrid approximation. We also did not yet prove the sinuous structure has highest scattering efficiency in this paper. In measured Skymap scatters appears to be clustered (Fig.4), but the simulated one tends to be uniform. This is because that natural turbulence is frequently not homogeneous but the homogeneous was supposed in the simulation. Both Fig.3 and Fig.4 show that different frequency corresponds to different distribution Spattern.

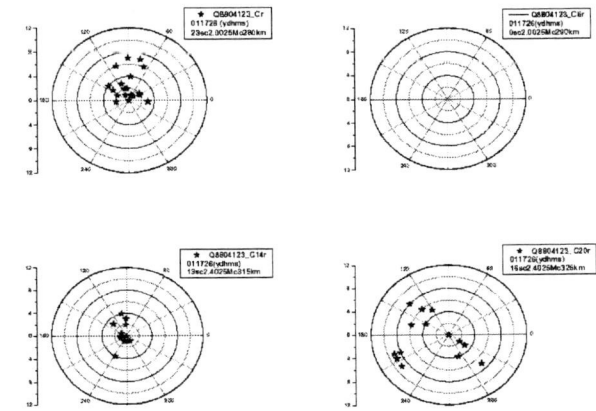

Fig.4 Measured Skymap at Qaanaaq with Digisonde

In summary we conclude that the scattering structure refers to the refractivity fluctuation structure effectively scattering wave power. The scattering structure possesses itself of the efficiency, aspect sensitivity, frequency dependence and discreteness that described well by its cross section. Scattering structure results from Bragg and time (range) sampling actions. Radar Imaging Interferometry (RII) technique observes spatial structure evolution of plasma flow, the particle-behaved scattering structures are the sources of pixels of the image. For description of refractivity field turbulence theory calls for statistic, yet EM wave equation calls for instantaneous field. This stems confusions. It seems the theory needs to be improved. This paper attempts to do so.

ACKNOWLEDGEMENTS

The work was supported by NSFC at No's 44944008_49990454_49574240. Thanks Dr. B.W. Reinisch of UMLCAR for providing Digisonde data of Qaanaaq.

REFERENCES

1. Bible, K.., and B.W.Reinisch, The universal Digital Ionosonde, Radio Sci.., 13(3), 519-530, 1978
2. Briggs,B.H., Radar observation of atmospheric wind and turbulence; a comparison of techniques, J. Atmos. Terr. Phys., 42, 823-833, 1980
3. Doviak, R.J., R.J.Lataitis, C.L.Holloway, Cross correlation and cross spectra for spaced antenna wind profilers, Radio Sci., 31(1), 157-180, 1996
4. Ishimaru, A., Wave propagation and scattering in random media, Academic Press, 1978
5. Kudeki, E. and S_s_c_, F., Radar interferometric image of field aligned plasma irregularities in the equatorial electrojet, Geophys. Res. Lett., 18,41-44, 1991
6. Liu, C.H., J. Rouger, C.J.Pan, and S.J.Franke,A model for spaced antenna observational mode for MST radars, Radio Sci., 25(4), 551-563, 1990
7. Reinisch, B.W., J.L., Scali and D.M. Haines, Ionospherric drift measurements with ionosondes, Annali di geofisica, 41, 695-702, 1998
8. Woodman, R.F., Coherent radar imaging: signal processing and statistical property, Radio Sci., 6, 2373-2391, 997
9, Yeh, K.C. and C.H. Liu, Ionospheric wave theory, Academic, San Diago, Calf., 1972

Ionospheric Response to Flux Transfer Events at the Earth's Magnetopause

F. Pitout[a,b] and P.L. Blelly[b]

[a]*Swedish Institute of Space Physics, Box 537, 751 21 Uppsala, Sweden*
[b]*CESR, 6, avenue du Colonel Roche, 31028 Toulouse, France*

Abstract. Flux transfer events (FTE) are thought to be the manifestation of the pulsed reconnection process by which interplanetary magnetic field lines and Earth's magnetic field lines merge. They are observed at the magnetopause as bipolar signatures in the magnetic field component perpendicular to the magnetopause (normal component) as the distorted newly reconnected field lines get dragged by the magnetic tension and the solar wind. FTEs are also observed in the ionosphere by ground-based instrumentations. Optical instruments commonly record poleward moving auroral forms in the red line (630nm) corresponding to low-energy electron precipitation. SuperDARN radars observe the convection associated to FTEs as enhanced flow channels. The EISCAT radars on the Svalbard archipelago have revealed the transient and impulsive nature of the ionosphere under the polar cusp for southward IMF. We have performed a simulation of the ionospheric footprint of a FTE by including the Southwood model of FTE in the TRANSCAR model of ionosphere. We can trace the evolution of an open flux tube and the behaviour and dynamics of the surrounding ionospheric plasma. We particularly focus on the signature of the two areas of return flow on both side of the FTE footprint. Those return flow may actually be responsible of some of the structures observed by Incoherent Scatter radars such as the Eiscat Svalbard Radar (ESR). We then compare our simulations with observations and finally discuss the pros and cons of the Southwood model of FTE.

INTRODUCTION

The ionospheric response to flux transfer events (FTE) is still unclear. Several models have been proposed among which the most credible ones are Southwood [1] and Cowley and Lockwood [2] models. They are actually very closed to each other. While the Southwood model describes rather the local effect due to a FTE, the Cowley and Lockwood model tries also to explain the large-scale convection enhancement due to an extra region of open flux in the polar cap. The Cowley and Lockwood model may be more realistic, but the Southwood model remains much easier to put into equations. Nevertheless, both models predict two region of return flow on both sides of the FTE footprint. It has been quite controversial whether these regions of return flow actually exist or not. Recent observations seem to indicate that they do exist but that they are not as localised as the Southwood model suggests it [3,4]. A first attempt to model the local ionospheric plasma flow due to an FTE has been already performed [5] by using the Wei and Lee model [6], which is nothing but a simple formalism of the Southwood model. Those simulations were designed to be compared to SuperDARN HF radar observations. In this work, we propose the same approach but we use an ionospheric model to compute all the main parameters of the ionospheric plasma, electron density and temperature, ion temperature and velocity.

MODEL

TRANSCAR ionospheric model

The TRANSCAR ionospheric model [7] consists of a 13-moment fluid code able to resolve the transport equations for electrons and 6 ion species in 1-D along a flux tube within the 100-3000km altitude range, coupled to a kinetic code handling suprathermal populations at low altitudes within the same flux tube. The code makes this flux tube convect horizontally (in 2-D) thanks to electric field and potential inputs. TRANSCAR is therefore virtually a 3-D code. The other inputs are the solar insulation, the neutral atmosphere, and the precipitating particles characteristics. The outputs are, among others, the plasma parameters measured by incoherent scatter radars: electron density and temperature, ion temperature and velocity.

Southwood model of FTE and Wei and Lee formalism

Southwood [1] proposed a model of ionospheric signature of a FTE describing the flow pattern inside and outside the FTE footprint as well as the field-aligned current system associated to the FTE. We have chosen to use this model mainly for its simplicity. Wei and Lee [6] have proposed a quantitative model describing the ionospheric plasma flow, electric field and potential patterns associated to the Southwood model. This quantitative model considers two regions: inside and outside the elliptic core of the FTE footprint. It assumes that the plasma flow is: uniform inside, uncompressible, vanishing at infinity, non-rotational outside. Figure 1 shows the FTE structure with electric field and potential projected on a 160x160 km area. Inside the ellipse, the plasma is uniform in direction and velocity (V_0). Particles precipitations inside the ellipse are assumed to be cusp-like, that is, low-energy electrons of the order of 100eV (Maxwellian distribution assumed). Outside the ellipse, the plasma flows around the ellipse to form the so-called return flow. The precipitation there is taken according to observations in the dayside polar ionosphere for quiet conditions.

SIMULATIONS

Following a flux tube

We consider a fixed FTE structure centred on ESR latitude, that is 75° MLAT. The aim is to make a flux tube convect along an isopotential inside and outside the structure and to study the evolution of the parameters directly linked to the convection, namely, the electric field and the ion temperature. It is clear that this does not correspond to the reality but this appears to be a good way to emphasize the effects of electric field in each. Figure 2 shows in its upper panel the electric field as the flux tube experiences it. The eastward component of the electric field (green) reaches its maximum, corresponding to a strong poleward flow, inside the ellipse whereas the total electric field (red) reaches its maximum in the return flow region (on both sides of the ellipse). The bottom panel of figure 2 displays the ion temperature inside the flux tube between 100 and 600 km. Interestingly enough, it is not inside the core the ion temperature is the highest but outside, in the region of return flow, where as we saw above, the total electric field is the strongest.

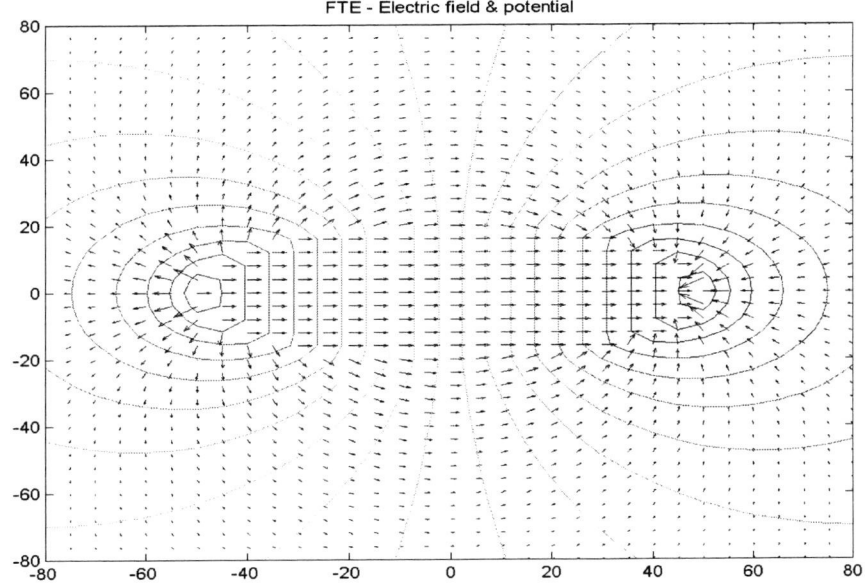

FIGURE 1. Electric potential iso-contours and electric field vectors inside and outside a FTE.

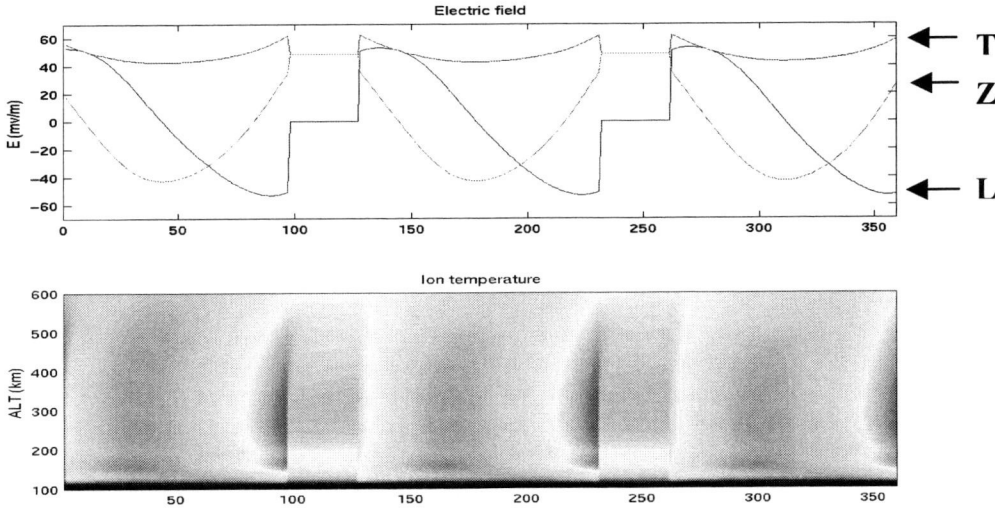

FIGURE 2. Upper panel. Electric field experienced by the flux tube: total (curve indicated by T), zonal (curve indicated by Z) and latitudinal (curve indicated by L) versus time (min). Lower panel. Ion temperature (normalised unit) as a function of altitude and time (min)

Signature at ESR

In this section, we model a radar observation. Therefore, we model the ionosphere at a fixed geographic location, that of the ESCAT Svalbard Radar in our case. Once again, we consider a fixed FTE structure in terms of magnetic local time. Only the Earth's rotation will make the radar pass under the structure. We are able to sound the ionospheric response to each part of the FTE: first the distant region of return flow, then progressively, a region of more and more intense return flow and finally the core of the FTE in which the solar wind particles precipitate.

In figure 3, one can see 30 min of simulation. The figure shows, from top to bottom, the electron concentration Ne, the electron temperature, the field-parallel ion temperature Ti and the line-of-sight ion velocity Vi as a function of time and altitude. The radar enters the FTE core at t=13 min after the beginning. Then both electron density and temperature increase due to soft electron precipitation. Interestingly enough, the ion temperature start to increase well before. As we saw in the previous section, this is due to the return flow regions. The flow velocity is strongly sunward there. This makes the plasma moves significantly downward as the line-of-sight velocity shows it.

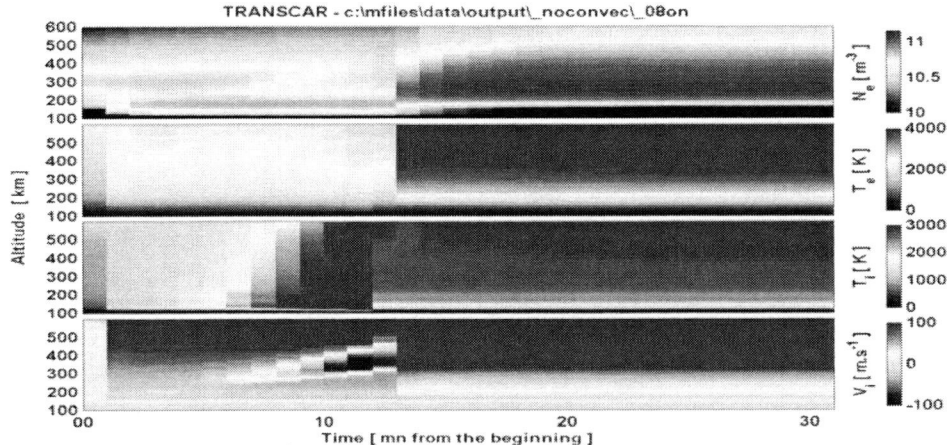

FIGURE 3. Altitude profiles of, from top to bottom, electron density, electron temperature, ion temperature, and ion velocity.

INTERPRETATION AND DISCUSSION

Expected consequences of FTEs in the ionosphere

The consequences on the ionospheric plasma inside the FTE core are obvious. As expected, the electron temperature is high reaching 5000K., the ion frictional heating due to strong electric field is effective (Ti~3000K). The ion upflow is due to the combined effects of the strong poleward convection and high electron temperature.
On the sides of the FTE, the localised high electric field (60 mV/m) induces a strong return flow (1200km/s), which in turn, drives locally a strong ion heating. It is interesting to note that with the Wei and Lee models, the electric field just outside the FTE core is higher than inside (figure 3).

Validity of the model and future improvements

There is observational evidence that the return flow (outside the central channel of the vortex) is usually spread over a very large area of the ionosphere [3]. This is one explanation for why it is so very hard to detect. So the enhanced convection channel may well only occupy about 100-150 km, but the return flow must occupy many 100's of km. Furthermore, the original Southwood model suggested that the enhanced convection channel, that is the main jet of plasma, perhaps only 100 to 200 km in length. Observations suggest that it actually spans at least 1000 km [4]. The model used here doesn't take into account those observations. The return flow region is in fact quite local in our model and therefore its effects over-estimated.

First of all, the Southwood model of FTE needs obviously to be slightly modified in order to take into account the remarks made above. The ion heating found at the edge of the FTE structure is very likely over-estimated. By enlarging the regions of return flow; we expect a less dramatic effect of this region. Secondly, we need to put this FTE structure into the right context. In other words, we have to make it move across the dayside polar ionosphere and reconstruct a more realistic expected radar signature.

From a more technical point of view, the Wei and Lee model, although quite simple, has a couple of disadvantages. The most obvious are the two singularities located in the centre of the two return flow regions. The second is the non-continuity of the electric field at the ellipse boundary. That causes difficulties for our ionospheric model to handle the convection.

CONCLUSION

We have inserted the Southwood model of FTE with help of the Wei and Lee formalism in the TRANSCAR ionospheric model. The results of the modelling are not so surprising inside the FTE structure. As expected, low-energy electrons together with high electric field make both ion and electron temperatures high. We were in fact more interested in the surroundings of the structure and more particularly in the two regions of return flow.

The simulations reveal that these regions experience a strong electric field responsible for fast flow and strong ion heating. However, it appears that this feature reflects more the limitations of the models than the reality. After this first modelling attempt, the model of FTE and definitely the formalism need to be refined.

REFERENCES

1. Southwood, D.J., The ionospheric signature of flux transfer events, *J. Geophys. Res.*, **92**, 3207-3213, (1987).
2. Cowley, S.W.H., and M. Lockwood, Excitation and decay of solar wind-driven flows in the magnetosphere-ionosphere system, , *Annales Geophys.*, **1**, 103-115, (1992).
3. Pinnock, M., A.S. Rodger, J.R. Dudeney, K.B. Baker, P.T. Newell, R.A. Greenwald, and M.E. Greespan, Observations of an enhanced convection channel in the cusp ionosphere, *J. Geophys. Res.*, **98**, 3767-3776, (1993).
4. Milan, S.E., M. Lester, R.A. Greenwald, G. Sofko, The ionospheric signature of transient reconnection and the associated pulsed convection return flow, *Annales Geophys..*, **17**, 1166-1171, (1999).
5. Thorolfsson, A., J.-C. Cerisier, M. Lockwood, P.E. Sandholt, C. Senior, M. Lester, Simulataneous optical aurora and radar signatures of polar-moving auroral forms, *Annales Geophys.*, **18**, 1054-1066, (2000).
6. Wei, C.Q., and L.C. Lee, Ground magnetic signatures of moving elongated plasma clouds, *J. Geophys. Res.*, **95**, 2405-2418, (1990).
7. Diloy, P.-Y., A. Robineau, J. Lilensten, P.-L. Blelly, J. Fontanari, A numerical model of the ionosphere, including the E-region above EISCAT, *Annales Geophys.*, **14**, 191-200, (1996).

Phase Coherence of Large Amplitude MHD Waves in the Earth's Foreshock: Geotail Observations

Tohru Hada*, Eiko Yamamoto† and Daiki Koga*

*E.S.S.T., Kyushu University, Japan
†NASDA, Japan

Abstract. Large amplitude MHD turbulence is commonly found in the earth's foreshock region. It can be represented as a superposition of Fourier modes with characteristic frequency, amplitude, and phase. Nonlinear interactions between the Fourier modes are likely to produce finite correlation among the wave phases. For discussions of various transport processes of energetic particles, it is fundamentally important to determine whether the wave phases are randomly distributed (as assumed in quasi-linear theories) or they have a finite coherence. However, naive inspection of wave phases does not reveal anything, as the wave phase is sensitively related to the choice of origin of the coordinate, which should be arbitrary. Using a method based on a surrogate data technique and a fractal analysis, we analyzed Geotail magnetic field data to evaluate the phase coherence among the MHD waves in the earth's foreshock region. We show that the correlation of wave phases does exist, indicating that the nonlinear interactions between the waves is in progress. Furthermore, by introducing an index to represent the degree of the phase coherence, we discuss that the wave phases become more coherent as the turbulence amplitude increases, and also as the propagation angle of the most dominant wave mode becomes oblique. Details of the analysis as well as implications of the present results to transport processes of energetic particles will be discussed.

INTRODUCTION

Magnetohydrodynamic (MHD) waves are ubiquitously present in space. In particular, those found in the foreshock region of the earth's bowshock typically have order of unity normalized magnetic field amplitude ($\delta B/B_0 \sim 1$), and thus they have been served as an excellet subject for the research of nonlinear wave theories. In fact, various unique waveforms observed by spacecraft experiments (such as the so-called shocklets, SLAM's, etc) have motivated a number of theoretical as well as numerical simulation studies.

Since the MHD is a nonlinear system, finite amplitude MHD waves nonlinearly interact each other, leading to various interesting nonlinear wave phenomena. We should note here that, strictly speaking, these finite amplitude MHD waves are not at all true eigenmodes of the MHD system: they are simply the finite amplitude version of the eigenmodes of the linearized MHD. In nonlinear systems, in general it is a formidable task to seek for eigenmodes. In integrable systems, the eigenmodes are solitons (plus the 'radiation'), and since they do not interact each other, the description of the system evolution becomes extremely simple. Nevertheless, it is conventional and also handy to deal with the eigenmodes of the linearized system, i.e., the linear MHD waves with finite wave amplitude, and so we use this approach also.

In this paper we focus our attention to the distribution of wave phases of the observed MHD turbulence in space. The phase distribution contains the same amount of information as the power spectrum, but it has been often neglected in past studies of turbulence. This may at least partially be due to a simple fact that, when one plots the phase distribution (phase versus the wavenumber), it will almost certainly appear to be a random distribution. This is because the wave phase depends on the origin of the coordinate: if we write the fluctuation field as $\delta B = \sum_k B_k \sin(kx + \phi_k)$, then a shift of the x coordinate by δx shuffle the distribution of ϕ_k to $\phi_k + kx_0$. Instead of examining the information in the Fourier space, we characterize the coherence among the wave phases by inspection of the characteristics of the waveform, the information in the real space, which off course is independent of the choice of the coordinate origin.

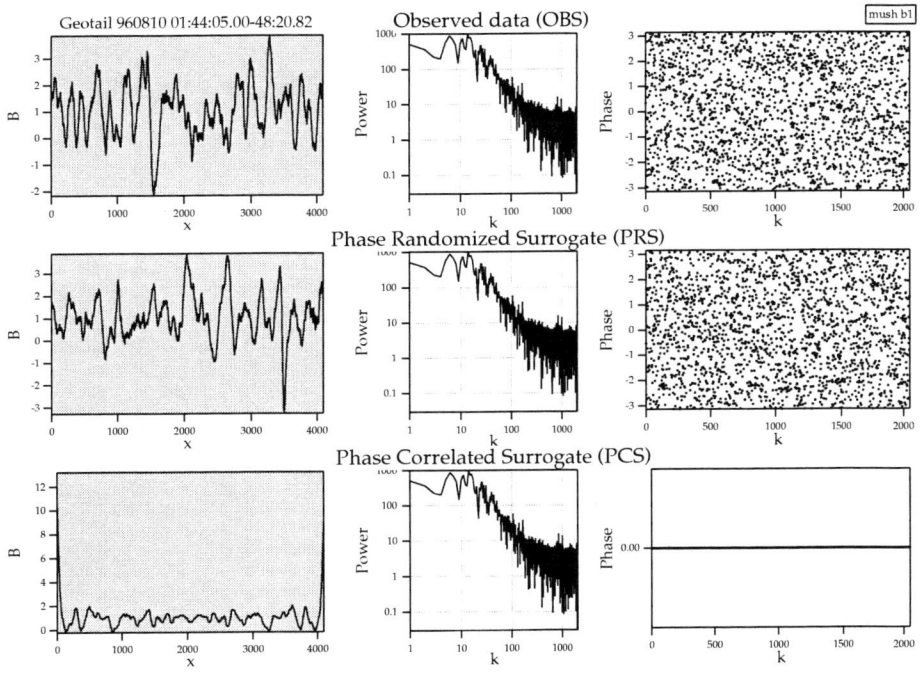

FIGURE 1.

METHOD OF ANALYSIS

In this section we define the phase coherence index, C_ϕ, for a given time series. Suppose we have a sequence of the magnetic field, $\mathbf{B(t)}$. Since it is usually defined in a coordinate not much relevant to the nature of the turbulence, we should rotate the coordinate axes by, for example, using the minimum variance method. Then, suppose we pick one transverse component of $\mathbf{B(t)}$ in the new coordinate for our analysis. Let us write it $B(t)$.

We first make two different surrogate datasets (Figure 1). The original data (top, left) is Fourier decomposed into the power spectrum (top, middle) and the phase (top, right). Keeping the power spectrum unchanged, the phase is shuffled (middle column) or made completely equal (bottom). These are then inverse Fourier transformed to make the Phase Randomized Surrogate (PRS) and the Phase Correlated Surrogate (PCS).

As mentioned earliear, the distribution of phases of the original time series (ORG) look almost the same as that of the PRS. We characterize the differences in the phase distribution by the differences in the curves in real space: when the phases are correlated, there appears a sharp peak at a certain location in space (in case of Fig. 1, it is at $x = 0$), and thus the rest part of the wave curve becomes smoother compared with curves of the ORG and the PRS. These differences can be most naturally captured by the structure function,

$$S(m,d) = \sum_x |b(x+d) - b(x)|^m$$

where the unit norm, d, is a measure characterizing the magnification level of the curve.

Left panel of Figure 2 shows $S(1,d)$ for the ORG, PRS, and the PCS datasets. Since $m = 1$, the structure function gives the path length measured with a unit length, d. The lengths are slightly different for the three datasets. We define the phase coherence index, C_ϕ, as

$$C_\phi = \frac{L_{PRS} - L_{OBS}}{L_{PRS} - L_{PCS}}$$

where L_* denotes the value of $S(1,d)$ for the dataset $*$. When the original data is random, C_ϕ takes a value close to 0, while C_ϕ equals unity when the data is completely phase correlated. Right panel of Figure 2 shows that the phase coherence does exist up to d 100, which roughly corresponds to the time scale of the ion cyclotron period in the plasma rest frame.

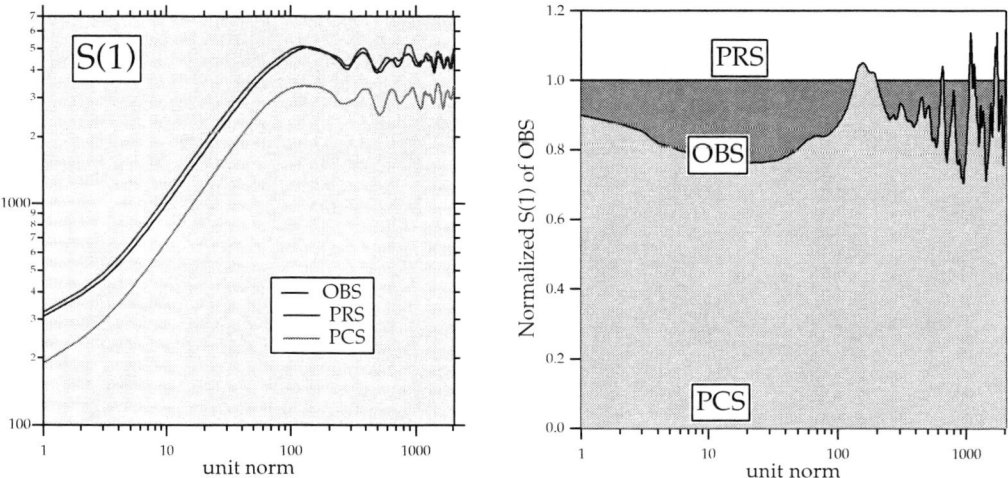

FIGURE 2.

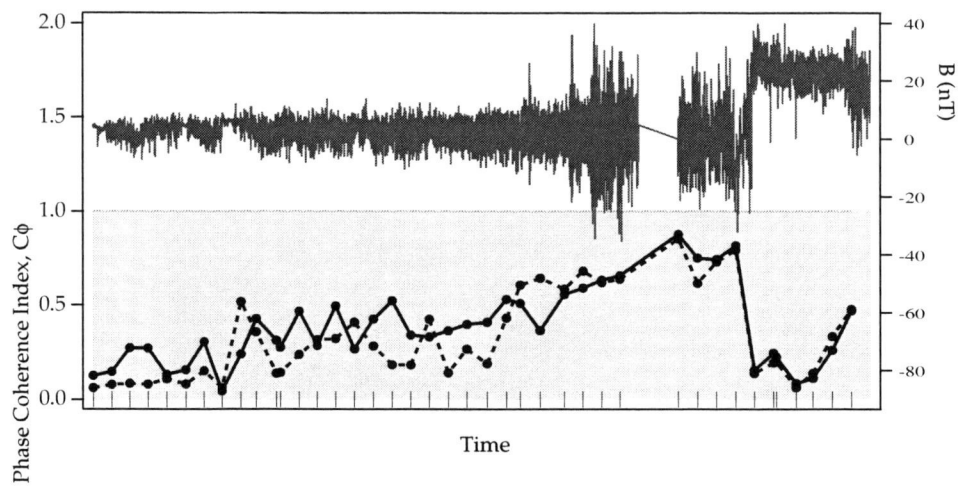

FIGURE 3.

In Figure 3, evolution of C_ϕ (bottom part of the figure) is evaluated for a long sequence of data, as the Geotail spacecraft approaches the bow shock from far upstream. The solid and dashed lines represent the phase coherence index of the magnetic field component transverse and parallel to the minimum variance axis, respectively. The magnetic field turbulence (top panel) increases as the spacecraft comes closer to the bowshock, becomes extremely large within the magnetosheath, and decreases again as the spacecraft enters the magnetosphere (decrease of the normalized turbulence level is apparent). Apparently, the evolution of C_ϕ approximately follow the evolution of the magnetic field turbulence level. This is a natural consequence if the phase coherence is generated via nonlinear interaction between the finite amplitude MHD waves.

We have examined also the dependence of C_ϕ to an angle θ between orientations of the average magnetic field and the minimum variance. This angle approximately represents the propagation direction of the main MHD wave (containing the largest amount of energy) relative to the background magnetic field. It was found that C_ϕ and θ are positively correlated. This result also can be understood since the nonlinear coupling between the MHD waves takes place at a lower order of the wave amplitude for obliquely propagating waves: i.e., if we write the background field $\mathbf{B_0} = (\mathbf{B_0}\cos\theta, \mathbf{B_0}\sin\theta, \mathbf{0})$ and the fluctuatin field $\delta\mathbf{B} = (\mathbf{0}, \delta\mathbf{B_y}, \delta\mathbf{B_z})$, fluctuation of the ponderomotive term, $|B|^2$ is the second order of $\delta\mathbf{B}$ for parallel, and the first order for obliquely propagation waves.

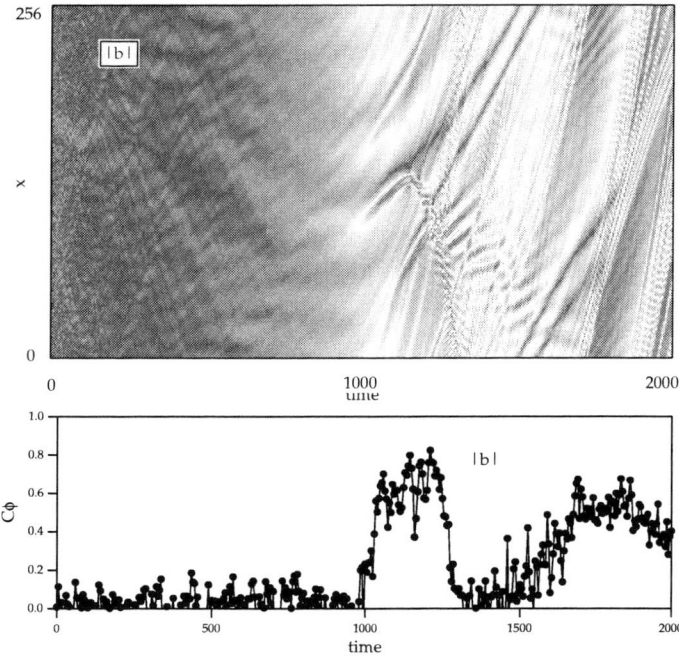

FIGURE 4.

DISCUSSION

By employing the surrogate technique and the fractal analysis method, we defined the phase coherence index, C_ϕ, to characterize the correlation of phases among waves present in a given time series. We found that C_ϕ is almost never equal to zero (or very small) when MHD waves of moderate amplitude ($\delta B/B_0$ 0.05 exist. This bears important implications in discussions of various transport processes of charged particles in space: in most of the conventional arguments of particle transport, it is customary to assume that the wave phases are completely random (random phase approximation). Without this assumption, the diffusion process may become a qualitatively different one[1].

We note that the coherence of phase is tightly related to the localization in space (cf. the sharp peak in the PCS dataset in Figure 1). In Figure 4, we show numerical integration of the driven DNLS (derivative nonlinear Schroedinger equation), a subset of the MHD system including both right and left hand polarized quasi-parallel Alfven waves propagating in the same direction. The top panel is the evolution of $|B|$, and the bottom panel shows the associated evolution of C_ϕ. Initially the system contains a very small white noise, and at first ($t < 300$) a single wave mode is continuously amplified by the convolution type driving term added to the DNLS. As the wave amplitude grows, it becomes modulationally unstable, and a series of solitary waves are generated (t 1000). After this time stage is a turbulence represented by interacting multiple solitons. The phase coherence index C_ϕ is eminently enhanced at the birth of new solitary waves, i.e., generation of locality in the real space. In this sense, the enhancement of the phase coherence can be regarded as one example of self-organization. More detailed discussions of the analysis of the solar wind field data as well as the numerical simulation results will be reported elsewhere.

REFERENCE

1. Kuramitsu and Hada, Geophys. Res. Lett., 629-32, vol.27, no.5, 2000.

Relationship Between Horizontal Flow Velocity and Cell Size for Supergranulation using SOHO Dopplergrams

U. Paniveni, V. Krishan, Jagdev Singh and R. Srikanth

Indian Institute of Astrophysics, Koramangala, Bangalore-34, India

Abstract. A study of 90 supergranular cells obtained from SOHO Dopplergrams was undertaken in order to investigate a possible relation between the sizes and peak horizontal velocities of the cells. For, the sample obtained the two parameters are found to be correlated with a relation horizontal velocity is proportional to size$^{1/3}$. This is in agreement with the Kolmogorov theory of turbulence as applied to large scale solar convection.

INTRODUCTION

The solar photosphere shows an irregular network pattern of typical scale 30000 km and a typical lifetime of 24 hours. Supergranules are convective cells whose horizontal flows sweep magnetic fields to the boundaries. Excess heating in the overlying atmosphere produces the chromospheric network. A dependence of network size on magnetic activity, solar cycle and cell lifetime has been reported.

In this paper we study the possible interrelationship between the supergranular size and convective flow field of the cell. This can shed light on how convection scales are related to the convective energy throughput and on the possible turbulent origin of the velocity field.

The source of data was 20 hours of full disk Dopplergrams obtained on 28th June 1996 by the Michelson-Doppler Interferometer (MDI) experiment on board the solar and Heliospheric Observatory (SOHO) (Scherrer et al. 1995).

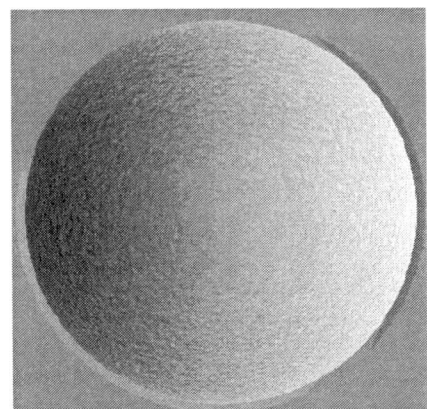

FIGURE 1. Processed SoHO dopplergrams.

DATA ANALYSIS

Dopplergrams were time-averaged for 10 minutes to remove signal due to p-modes using IRAF (Image reduction and analysis facility). Dopplershifts caused by the solar rotation were corrected for and 90 supergranules with high contrast boundaries and lying between 15 degree and 60 degree angular distance from the disk center were selected. Cell size and maximum horizontal velocity were obtained by scanning each cell using IDL (Interactive data language).

RESULTS

The horizontal velocity and size of the supergranular cell are fairly well correlated (correlation coefficient = 0.55). Maximum, minimum, mean and standard deviation for the horizontal velocity computed using $r = 0.6$.

TABLE 1. Cell size and maximum horizontal velocity.

	Max	Min	Mean	Σ
L (Mm)	57.8	15.5	33.7	8.96
v_h (m/sec)	757.4	285.5	491.1	74.10

A power law of the form

$$v_h = f L^\alpha \qquad (1)$$

was fitted to the data using least squares method. For $r = 0.6$, we find that $f = 0.014 \pm 0.007$ and $\alpha = 0.34 \pm 0.046$.

THEORETICAL DISCUSSION

The ideas of the direct [1] and the inverse cascade [2] of energy in a turbulent medium predict Kolmogorov's law of $L^{1/3}$ for supergranulation. The $K^{-5/3}$ law for granulation and $K^{-0.7}$ for mesogranulation have been inferred from observations [1,3,4]. The hypothesis for inverse cascade predicts a K^{-1} spectrum for mesogranulation and $K^{-5/3}$ for granulation and supergranulation [2]. Thus the observed entire spectrum for granulation–mesogranulation–supergranulation of $K^{-5/3}$–$K^{-0.7}$–$K^{-5/3}$ seems to favour the inverse cascade hypothesis. The somewhat large average cell size we obtain in comparison with the traditional cell size of 23Mm obtained by visual inspection [5] suggests that our method for choosing cells might be biased towards bigger cells. More data on supergranulation, with particular emphasis to remove possible size-oriented biasing effects, is required before the velocity-size relation could be confirmed however.

According to Kolmogorov's hypothesis for a turbulent medium,

$$v = \varepsilon^{1/3} L^{1/3} \qquad (2)$$

where ε is the energy injection rate. Equation (2) matches our fit for Eq. (1) provided we set $f = \varepsilon^{1/3}$ and $\alpha = 1/3$.

According to the turbulent convection theory, $\varepsilon = v^2 / \tau$ where τ is the eddy turnover time, and can be identified with the supergranular lifetime, and v is a typical velocity.

Thus for $v = 0.5$ km/s, $\varepsilon = (0.5 \times 0.5)/(24 \times 60 \times 60) \approx 2.89 \times 10^{-6}$ km^2/s^3. Hence $\varepsilon^{1/3}$ is of the order of 10^{-2} km^2/sec^3 and $\alpha = 0.34 \approx 1/3$ and theoretical and observed values are in close agreement. Both direct [1] and inverse cascade [2] of energy in turbulent media predict Kolmogorov's law of $v(L) \propto L^{1/3}$ for supergranulation. The observations are found to be in agreement with the theoretical predictions.

ACKNOWLEDGMENTS

We thank Dr. P.H. Scherrer and the SOHO consortium for providing us with MDI/SOI data.

REFERENCES

1. Zahn, J. P., 1987, Solar and Stellar Physics eds. E. H. Schroter and M. Schussler (Berlin: Springer-Verlag), p55.
2. Krishan, V., 1991, MNRAS, **250**, 50.
3. Malherbe, J. M., Mein, P., Muller, R., Roudier, R., Coutard, C., Hellier, R, 1987, eds. Z. Mouradian and P. Lemaire, Colloque Themis, College de France, 5-6 March, 1987 p.53.
4. Keil, S et al. 1994, eds. D. M. Rabin et al. Infrared Solar Physics p. 251.
5. Singh, J. and Bappu, M. K. V.,1981, Solar Physics, **71**, 16

AUTHOR INDEX

A

Adámek, J., 207, 211
Ahmed Rudwan, I. F. M., 294
Alejaldre, C., 162
Ali, H., 244
Alimov, S. S., 350
Alladio, F., 557
Allen, M., 289
Almoguera, L., 162
Alonso, J., 162
Alves, M. Virgínia, 496
Amagishi, Y., 18
Anderson, J. K., 219
Ando, A., 137, 298, 306, 314
Ando, M., 48
Andreiev, V. V., 350
Andrenucci, M., 302
Andrushchenko, Z. N., 650
Antoni, V., 191, 302
Antonov, V. M., 435
Aramaki, E. A., 339
Arun, P., 343
Ascasíbar, E., 162
Askinazi, A., L. G., 175
Aubreton, J., 757
Audebert, P., 289
Avinash, K., 93
Azarenkov, N. A., 71, 524
Azechi, H., 269
Azis, A., 361

B

Baciero, A., 162
Badziak, J., 739
Bagatin, M., 302
Bak, P. E., 223
Balan, P., 207, 211
Balbín, R., 162
Baldzuhn, J., 166
Ball, R., 711
Bateman, G., 232
Behnke, J. F., 79, 482
Belashov, V. Y., 609, 670
Belien, A. J. C., 642
Belyaev, V. S., 788, 792
Belyi, V. V., 443, 445, 490
Benkadda, S., 613, 696
Benstetter, G., 327
Beyer, P., 613, 696
Bhattacharjee, A., 93

Biewer, T. M., 219
Bilek, M. M. M., 319
Bilyk, O., 79
Bingham, R., 449, 528, 700, 704, 784
Blackwell, B. D., 158
Blanco, E., 162
Blaumoser, M., 162
Blazevic, A., 289
Blelly, P. L., 835
Block, D., 64, 105, 109
Bobkov, V. V., 350
Bogomaz, A. A., 34
Bohmeyer, W., 149
Bolkhovitinov, E. A., 265
Boody, F. P., 739
Boozer, A., 244
Bora, M. P., 532
Borg, G. G., 158
Borotto, F. A., 715
Borovitskaya, I. V., 335
Botija, J., 162
Bourgoin, D., 248
Boyarintsev, E. L., 435
Boyd, T. J. M., 708
Bracco, G., 557
Brackbill, J. U., 773
Brakel, R., 166
Brañas, B., 162
Budaev, V., 240
Budin, A. V., 34
Bugaev, A. S., 377
Bugaev, S. P., 373
Burachevsky, Y., 358
Burdovitsin, V., 358
Burhenn, R., 166

C

Cabral, J. A., 211
Cairns, R. A., 700
Caldas, I. L., 727
Cappa, A., 162
Carrasco, R., 162
Castejón, F., 162
Castillo, R., 744
Cavazzana, R., 191
Cepero, J. R., 162
Cercek, M., 273, 500
Cereceda, C., 540
Chapman, B. E., 219
Charles, C., 158
Cheetham, A. D., 388, 392, 396

Chen, J. Y., 330
Chen, L., 638
Cheung, F. M. H., 87
Chian, A. C.-L., 715, 723
Chmyga, A. A., 162, 175
Christiansen, F., 723
Chu, P. K., 330
Collis, S. M., 158
Cowan, T., 289
Craig, D., 219

D

Dallaqua, R. S., 323
Das, A., 261, 675, 731
Dasgupta, B., 585, 597
da Silva, R. P., 195
da Trindade Faria Jr., R., 520
de Azevedo, C. A., 467, 471
Dedurin, A. I., 335
Degeling, A. W., 223
de la Luna, E., 162
de la Peña, A., 162
Del Bosco, E., 323
de Moraes, M. A. E., 496
de Moraes, M. B., 339
Den-Hartog, D. J., 219
Denniss, P., 319
Denysenko, I. B., 71, 524
Dettrick, S., 638
Dewar, R. L., 577, 711
Dimitriu, D., 719
Dmitruk, P., 781
Doncel, J., 162
Doskach, I. Y., 265
Dreval, N. B., 162, 175
Ďuran, I., 207

E

Ebisuzaki, T., 601
Eguilior, S., 162
Ehmler, H., 166
Elchinger, M. F., 757
Elfimov, A. G., 195
Eliseev, L., 162
Eloy, M., 784
Elskens, Y., 617
Erokhin, A. A., 265
Erzen, D., 273
Esaki, T., 508
Escande, D. F., 617
Espino, F. P., 281, 285
Essiptchouk, A. M., 22

Estrada, T., 162
Evans, P., 744

F

Fahrbach, H.-U., 199
Falchetto, G., 696
Fedorov, M., 358
Fedotov, A. S., 335
Fedotov, S. I., 265
Fedyanin, O., 162
Feng, Y., 166
Feoktistov, L. P., 265
Fernandes, H., 211
Fernández, A., 162
Figarella, C., 696
Figueiredo, H. F. C., 211
Fonseca, A. M. M., 195
Franck, C. M., 26, 404, 412
Frederiksen, J. S., 687
Freire Jr., F. L., 354
French, G. N., 388
Fuchs, J., 289
Fuentes, C., 162
Fujimura, S., 298
Fujioka, S., 269
Fujita, H., 52
Fujita, H., 257
Fukao, M., 154, 553
Fussmann, G., 149

G

Gabriel, S. B., 294
Gadelmeier, F., 166
Galvão, R. M. O., 195
Ganguli, S. B., 605
Gao, H., 215
Garbet, X., 613, 696
García, A., 162
Garcia, O. E., 691
García-Cortés, I., 162
Gardner, H. J., 158, 569
Gauthier, J. C., 289
Gavrishchaka, V. V., 605
Geissel, M., 289
Gekelman, W., 408
Ghendrih, P., 613, 696
Ghoranneviss, M., 327, 347
Giannone, L., 166
Glass, F. J., 158
Glosik, J., 60
Goedbloed, J. P., 642
Gohda, T., 14

Golant, V. E., 175
Golubovskii, Y. B., 79, 482, 486
Gomberoff, L., 646
Gonçalves, B., 162
Goree, J., 93
Goto, S., 753
Grandgirard, V., 696
Greiner, F., 64
Grigull, P., 166
Grishanov, N. I., 467, 471
Grulke, O., 26, 404, 412
Guasp, J., 162
Guenther, W., 289
Guerreiro, A., 784
Guo, J.-S., 831
Gushenets, V. I., 377
Gus'kov, S. Y., 265
Guzdar, P. N., 654
Gyergyek, T., 273, 500

H

Habs, D., 289
Hada, T., 808, 840
Hanna, J., 427
Hantehzadeh, M. R., 327, 347
Harada, M., 589
Hargreave, M., 388
Harris, J. H., 158
Hartmann, D., 166
Haruki, T., 762
Hatakeyama, R., 3, 117, 416
Hatanaka, Y., 369
Hatori, T., 601
Hattori, K., 137, 298, 306, 314
Hayashi, T., 449
He, X.-T., 804
Hegelich, M., 289
Helblom, G., 64
Herranz, J., 162
Heyn, M. F., 492, 504
Hidalgo, A., 162
Hidalgo, C., 162, 211
Higaki, H., 170
Hildebrandt, D., 166
Himura, H., 154, 553
Hirose, K., 536
Hirota, M., 666
Hirsch, M., 166
Hirshman, S. P., 561
Hishida, T., 18
Hojo, H., 170
Hole, M. J., 323
Holik, M., 79, 486
Höpfl, R., 327, 739

Hora, H., 327, 347, 739, 744
Horinouchi, K., 170
Horiuchi, R., 769
Hosokawa, Y., 137
Howard, J., 158
Hron, M., 207, 211
Hu, S., 93
Hua, X., 158
Huang, N., 330
Hudson, S. R., 561

I

Ichimura, M., 170
Ida, K., 203
Idehara, T., 68, 145, 179
Idei, H., 187
Ignatescu, V., 500, 719
Iizuka, S., 14
Ikeya, N., 141
Inagaki, S., 187
Inutake, M., 137, 298, 306, 314
Inuzuka, H., 424
Ioniţă, C., 207, 211, 719
Ishiguro, S., 14, 630
Ishihara, O., 125, 536
Ishii, N., 48
Isobe, M., 154
Ito, A., 666
Itoh, H., 383
Itoh, K., 228
Itoh, S.-I., 228
Ivanov, L. I., 335

J

Jaenicke, R., 166
Jaiman, N. K., 796
Jain, N., 261
James, B. W., 87, 90, 97
Jiménez, J. A., 162
Jitsuno, T., 257
Jungwirth., K., 739

K

Kakimoto, S., 170
Kamimura, T., 536
Kamp, L. P. J., 581
Kanai, Y., 269
Kaneko, T., 416
Kanki, T., 753
Kano, H., 170

Karsch, S., 289
Kasilov, S. V., 492, 504
Kato, T., 601
Kaw, P. K., 261, 675, 731
Kawabe, T., 449
Kawahata, K., 68, 179
Kawamoto, Y., 431
Kawamura, K., 420
Kawano, T., 508
Kawasaki, M., 228
Kellett, B. J., 700, 704
Kemp, A., 289
Kernbichler, W., 492, 504
Khrebtov, S. M., 162, 175
Kichigin, G. N., 622
Kirov, K. K., 199
Kirpitchev, I., 162
Kisslinger, J., 166
Kiss'ovski, Z., 149
Kitagawa, Y., 257
Klinger, T., 26, 166, 404, 412
Knauer, J., 166
Kobayashi, M., 18
Koch, B., 149
Kodama, R., 257
Koga, D., 840
Kolikov, V. A., 34
Komarov, A. S., 162, 175
Kondo, Masashi, 431
Kondo, Masuo, 431
Kondoh, S., 666
Kondoh, Y., 589
König, R., 166
Kornev, V. A., 175
Kozachok, A. S., 162
Králiková, B., 739
Krása, J., 739
Krishan, V., 844
Kritz, A. H., 232
Krokhin, O. N., 335
Kruglov, B. V., 265
Krupnik, L. I., 162, 175
Krushelnick, K., 257
Kubo, S., 187
Kudrna, P., 60, 79, 482, 486
Kuhn, S., 475, 650
Kukharenko, Y. A., 445
Kumagai, R., 137, 298

L

Lahiri, S., 565, 585
Lapayese, F., 162
Lapenta, G., 773
Láska, L., 739

Lebedev, S. V., 175
Leboeuf, J. N., 749
Lee, P., 343, 365
Lee, S., 343, 365
Leitold, G., 492
Leng, Y., 330
Leng, Y. X., 330
Lerche, E. A., 195
Leubner, M. P., 800
Leuterer, F., 199
Likin, K., 162
Lim, C. K., 361
Liniers, M., 162
Lisse, C. M., 704
Lister, G. G., 479
Lister, J. B., 223
Liu, H., 804
Llobet, X., 223
Lloyd, S. S., 569
López-Bruna, D., 162
López-Fraguas, A., 162
López-Rázola, J., 162
López-Sánchez, A., 162
Loula, A. F. D., 467, 471
Lozneanu, E., 719
Lu, R., 215
Ludwig, G. O., 573
Lui, W., 215
Luo, X.-G., 831

M

Macau, E. E., 723
Macharaga, G., 365
Machida, M., 339
Mahajan, S. M., 666
Malaquias, A., 162
Mamun, A. A., 543
Mancuso, A., 557
Mardanian, M., 327, 347
Marotta, A., 22
Martín, R., 162
Martin, Y. R., 223
Martinell, J. J., 459, 654
Martines, E., 191, 207, 211
Matsubara, A., 420
Matsuda, K., 30
Matsumoto, H., 735
Matsumoto, N., 310
Matsuoka, K., 154
Matsuyama, S., 236
Matthaeus, W. H., 781
McCarthy, K. J., 162
McCormick, K., 166
Mckenzie, D. R., 319

McMillan, B. F., 577
Medina, F., 162
Medrano, M., 162
Melekhov, A. V., 435
Melnikov, A. V., 162
Méndez, P., 162
Mendonça, J. T., 784
Michael, C. A., 158
Micozzi, P., 557
Mieno, T., 310
Mikhaylov, V. N., 788, 792
Milanese, M., 277
Milano, L. J., 781
Miley, G. H., 744
Mima, K., 253, 257, 512, 675, 744
Minkova, N. R., 516
Misawa, T., 101
Mitsudo, S., 68
Miyakoshi, T., 257
Miyanaga, N., 257, 269
Miyazaki, H., 306
Mizuno, N., 449
Mizuuchi, T., 30
Möller, I., 56
Montgomery, D. C., 581, 781
Monticello, D. A., 561
Morales, G. J., 749
Moreno, J., 83
Mori, W. B., 449
Morikawa, J., 553
Morimoto, S., 30
Morita, S., 187
Moroso, R., 277
Mourenas, D., 435
Mukhopadhyay, S., 565, 585
Muñoz, V., 646
Murakami, M., 269, 744
Murakami, T., 101
Muranaka, T., 269, 508
Murphy, A. B., 757
Muto, S., 187

N

Nagamine, Y., 589
Nagatomo, H., 253, 269
Nagayama, Y., 187
Nagornykh, E., 824
Naitoh, T., 383
Nakai, M., 269
Nakai, S., 744
Nakamura, M., 170
Nakamura, Y., 7, 10, 121
Nakashima, C., 553
Nakashima, H., 435, 508

Nakashima, Y., 170
Narihara, K., 187, 203
Narushima, Y., 203
Naujoks, D., 166
Naulin, V., 626, 658, 662
Nedzelskiy, I. S., 162
Nemov, V. V., 492
Neto, J. P., 467, 471
Niedermeyer, H., 166
Nieto, J., 285
Nikolaev, A. G., 377
Nikolić, L., 630
Nikulin, V. Y., 335
Nishihara, K., 253, 269, 744
Nishikawa, K., 675
Nishikino, M., 269
Nishimura, H., 269
Norimatsu, T., 257
Norreys, P. A., 257
Notake, T., 187
Novotny, O., 60
Nunomura, S., 93
Nycander, J., 658

O

Oates, T. W. H., 319
Obiki, T., 30
Ochando, M., 162
Ogawa, I., 68, 145, 179
Ohdachi, S., 203
Ohkubo, K., 187
Ohnishi, N., 253
Ohno, N., 101, 113, 240
Okada, H., 30
Okada, S., 753
Oks, E. M., 358, 377, 380
Okumura, H., 820
Omura, Y., 496, 735
Ondarza-Rovira, R., 708
Onjun, T., 232
Oohara, W., 3, 117
Osanai, Y., 589
Osipov, M. V., 265
Oskomov, K. V., 373
Osman, F., 744
Ostrikov, K. N., 44, 71, 524
Otani, F., 257
Otsuka, F., 808
Ozono, E. M., 195

P

Pacios, L., 162
Paganucci, F., 302
Pan, G., 215

Paniveni, U., 844
Parail, V., 232
Parys, P., 739
Pasch, E., 166
Pastor, I., 162
Paulsen, J.-V., 691
Pavlenko, V. P., 650
Pavlichenko, R., 68, 179
Pedrosa, M. A., 162, 211
Peeters, A. G., 199
Pei, W. B., 769
Pereverzev, G., 199
Perina, V., 739
Petrov, A., 162
Pfeifer, M., 739
Piel, A., 64, 105, 109
Pigott, J., 319
Pitout, F., 835
Plasil, R., 60
Poornakala, S., 675
Popa, G., 369
Porokhova, I. A., 79, 482, 486
Portas, A., 162
Posukh, V. G., 435
Poterya, V., 60
Pouzo, J., 277
Pretty, D. G., 158
Pretzler, G., 289
Pribyl, P., 408
Puerta, J., 540
Punjabi, A., 244
Punzmann, H., 158
Puzirev, V. N., 265
Pysanenko, A., 60

R

Rahman, S. A., 361
Ramasubramanian, N., 166
Rasmussen, J. J., 658, 662
Ratynskaja, S., 64
Rawat, R. S., 343, 365
Rayner, J. P., 388, 392, 396
Regnoli, G., 191
Reiman, A. H., 561
Rempel, E. L., 715, 723
Roberto, M., 727
Robledo-Martinez, A., 281, 285
Rodríguez-Rodrigo, L., 162
Rogier, F., 557
Rohlena, K., 739
Romero, J., 162
Rosa, R. R., 723
Rosales, L., 83
Ross, G. G., 248

Rossetti, P., 302
Roth, I., 812
Roth, M., 289
Rozanov, V. B., 265
Ruchko, L. F., 195
Rupasov, A. A., 265
Rutberg, P. G., 34
Rypdal, K., 64, 691
Ryter, F., 199

S

Saeki, K., 431, 593
Saito, K. H., 589
Saito, K. N., 589
Saitoh, H., 553
Saitou, Y., 10, 75
Sakagami, H., 512
Sakai, J. I., 762
Sakaiya, T., 269
Sakakibara, S., 203
Sakakita, H., 52, 219
Sakanaka, P. H., 520, 546, 585, 597
Sakawa, Y., 40
Salas, A., 162
Samarian, A. A., 87, 90, 97, 550
Sánchez, E., 162
Sánchez, J., 162
Sanduloviciu, M., 719
Sano, F., 30
Sanuki, H., 593
Saosaki, S., 170
Sarazin, Y., 613, 696
Sardei, F., 166
Sari, A. H., 327, 347
Sarksian, K., 162
Sato, G., 117
Sato, K., 420
Sato, K. N., 52
Sato, M., 3, 179
Sato, N., 14, 536
Sato, S., 121
Sato, T., 257, 630, 769
Satoh, K., 383
Savkin, K. P., 377
Savvateev, A. F., 34
Sawada, K., 253
Sawada, T., 383
Sawai, M., 101
Schchepetov, S., 162
Schoepf, K., 650
Schrittwieser, R., 207, 211, 719
Scussiatto, C. E., 597
Sen, A., 675, 731
Sengupta, S., 261

Sentoku, Y., 257, 675
Serianni, G., 191, 302
Shaikhislamov, I. F., 435
Shamin, A., 516
Shamrai, K. P., 634
Shandrikov, M. V., 380
Shang, S.-P., 831
Shapiro, V. D., 704
Sharakhovsky, L. I., 22
Shats, M. G., 133, 158
Sheng, Z. M., 675
Shevkin, E. A., 175
Shi, J.-K., 831
Shibuya, T., 420
Shigemori, K., 257, 269
Shiina, S., 589
Shikanov, A. S., 265
Shimozuma, M., 383
Shimozuma, T., 187
Shinohara, S., 236, 634
Shiraga, H., 269
Shoji, T., 40
Shukla, P. K., 543, 546
Silva, E. C., 727
Silva, J., 540
Silva, L. O., 449
Simonet, F., 435
Simpson, S. W., 323
Singatulin, R. M., 609
Singh, J., 844
Sirghi, L., 369
Skála, J., 739
Skvortsova, N., 162
Slyusarenko, Y. V., 350
Sochugov, N. S., 373
Solomon, W. M., 133, 158
Solovjev, A. A., 373
Soltwisch, H., 56
Soto, L., 83
Souza de Assis, A., 684
Spada, E., 191
Spassovska, I., 546
Spolaore, M., 191
Srinkanth, R., 844
Starovoitov, R. I., 350
Stenum, B., 662
Stöckel, J., 207, 211
Storer, R. G., 71, 577
Strickler, D. J., 561
Studenov, V. B., 265
Sudo, S., 420
Sugama, H., 463, 711
Sugaya, R., 816
Sugimori, K., 236
Sugimoto, T., 420
Summers, H. P., 704

Sun, H., 330
Sunahara, A., 257, 269
Suttrop, W., 199

T

Tabarés, F., 162
Tada, E., 416
Tafalla, D., 162
Tagashira, H., 383
Taguchi, M., 455, 589
Takabe, H., 253, 269
Takado, T., 3
Takahashi, M., 48
Takai, T., 141
Takamura, S., 101, 113, 240
Takayama, T., 269
Takeda, T., 141
Takeda, Y., 424
Tamari, Y., 269
Tanaka, K., 203
Tanaka, K. A., 257
Tanaka, M., 269
Tanaka, N., 240
Tanpo, M., 257
Tardini, G., 199
Tatsuno, T., 666
Taveira, A. M. A., 597
Tel'nikhin, A., 824
Tendler, M., 175
Terry, S. D., 219
Tichý, M., 60, 79, 207, 211, 482, 486
Tikhomirov, A. A., 335
Tobari, H., 306, 314
Tohyama, Y., 257
Toi, K., 203
Toida, M., 820
Tomioka, U., 449
Tonogaki, T., 113
Torney, M., 704
Toups, P., 744
Tribaldos, V., 162
Tripathi, V. K., 796
Tsakadze, E. L., 44, 71
Tsakadze, Z. L., 44
Tsang, W., 90
Tskhakaya, D. D., 475
Tsuda, N., 400
Tsukabayashi, I., 121
Tsung, F. S., 749
Tsunoyama, H., 416
Tsushima, A., 18, 75, 113, 125, 145, 593
Tsytovich, V. N., 7, 528
Tukachinsky, A. S., 175

U

Uberoi, C., 777
Uchimoto, E., 589
Ullschmied, J., 739
Umeda, T., 735
Usov, V. V., 766
Usui, H., 735

V

van Bokhoven, L. J. A., 662
van de Konijnenberg, J., 662
van der Holst, B., 642
van Milligen, B., 162
Van Oost, G., 175, 207, 211
VanZeeland, M., 408
Varandas, C. F. A., 162, 211
Vasenin, Y. M., 516
Vasin, B. L., 265
Vaulina, O. S., 550
Vchivkov, K. V., 508
Vega, J., 162
Vianello, N., 191, 302
Videshwar, A. G., 343
Vincena, S., 408
Vitela, J. E., 459
Vizir, A. V., 380
Vladimirov, S. V., 97, 670

W

Wagner, F., 166
Wakabayashi, H., 154
Wallenborn, J., 445
Wan, G. J., 330
Wan, S., 215
Wang, C., 215
Wang, J., 215, 330
Wang, X., 93
Wang, Z., 215
Watanabe, K. Y., 203
Watanabe, T., 170
Watanabe, T.-H., 463
Watts, C., 427
Weller, A., 166
Wen, Y., 215
Whichello, A. P., 388, 392, 396
Witte, K. J., 289
Wolowski, J., 739
Wong, A. Y., 129
Woryna, E., 739

X

Xia, H., 133
Xu, S., 44, 71, 365, 524

Y

Yadav, L. L., 679
Yagai, T., 137, 298
Yagi, M., 228
Yakushev, O. F., 265
Yamada, H., 154, 203
Yamada, I., 187
Yamada, J., 400
Yamada, K., 145
Yamada, S., 253
Yamagiwa, K., 141
Yamaguchi, S., 145
Yamaguchi, Y., 170
Yamamoto, E., 840
Yamamoto, S., 203
Yamamoto, T., 48
Yamanaka, C., 744
Yamanaka, T., 257, 269, 744
Yang, P., 330
Yasaka, Y., 48
Yasuda, S., 431
Yatsu, K., 170
Yatsuyanagi, Y., 601
Yoshida, Z., 154, 553, 666
Yoshimura, Y., 187
Yoshino, K., 306, 314
Yoshioka, Y., 589
Yu, C., 215
Yu, M. Y., 524
Yushkov, G. Y., 377, 380

Z

Zakharov, A. N., 373
Zakharov, Y. P., 435, 508
Zakouril, P., 60
Zambra, M., 83
Zepf, M., 257
Zhang, G., 183
Zhang, M.-L., 831
Zheng, H., 831
Zheng, W., 183
Zhubr, N. A., 175
Zuin, M., 302
Zurro, B., 162